SOLUTIONS MANUAL

Prepared by

Neil Wigley
University of Windsor
Albert Herr

To accompany

CALCULUS

Seventh Edition

Howard Anton
Drexel University
Irl C. Bivens
Davidson College
Stephen L. Davis
Davidson College

John Wiley & Sons, Inc.

Cover Design: Norm Christensen

To order books or for customer service call 1-800-CALL-WILEY (225-5945).

ISBN 0-471-43495-7

Printed in the United States of America

10 9 8 7 6 5 4 3 2 1

Printed and bound by Bradford & Bigelow, Inc.

CONTENTS

CHAPTER 1
Functions

EXERCISE SET 1.1

1. **(a)** around 1943 **(b)** 1960; 4200
(c) no; you need the year's population **(d)** war; marketing techniques
(e) news of health risk; social pressure, antismoking campaigns, increased taxation

2. **(a)** 1989; \$35,600 **(b)** 1975, 1983; \$32,000
(c) the first two years; the curve is steeper (downhill)

3. **(a)** $-2.9, -2.0, 2.35, 2.9$ **(b)** none **(c)** $y = 0$
(d) $-1.75 \le x \le 2.15$ **(e)** $y_{\text{max}} = 2.8$ at $x = -2.6$; $y_{\text{min}} = -2.2$ at $x = 1.2$

4. **(a)** $x = -1, 4$ **(b)** none **(c)** $y = -1$
(d) $x = 0, 3, 5$ **(e)** $y_{\text{max}} = 9$ at $x = 6$; $y_{\text{min}} = -2$ at $x = 0$

5. **(a)** $x = 2, 4$ **(b)** none **(c)** $x \le 2$; $4 \le x$ **(d)** $y_{\text{min}} = -1$; no maximum value

6. **(a)** $x = 9$ **(b)** none **(c)** $x \ge 25$ **(d)** $y_{\text{min}} = 1$; no maximum value

7. **(a)** Breaks could be caused by war, pestilence, flood, earthquakes, for example.
(b) C decreases for eight hours, takes a jump upwards, and then repeats.

8. **(a)** Yes, if the thermometer is not near a window or door or other source of sudden temperature change.
(b) No; the number is always an integer, so the changes are in movements (jumps) of at least one unit.

9. **(a)** The side adjacent to the building has length x, so $L = x + 2y$. Since $A = xy = 1000$, $L = x + 2000/x$.
(b) $x > 0$ and x must be smaller than the width of the building, which was not given.
(c)

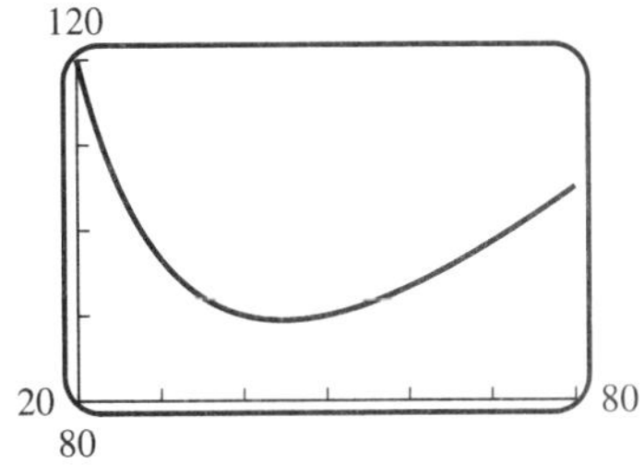

(d) $L_{\text{min}} \approx 89.44$ ft

10. **(a)** $V = lwh = (6 - 2x)(6 - 2x)x$ **(b)** From the figure it is clear that $0 < x < 3$.
(c)

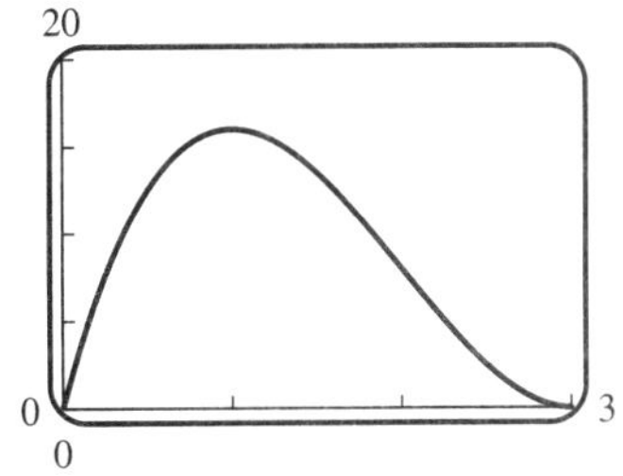

(d) $V_{\text{max}} \approx 16$ in^3

11. (a) $V = 500 = \pi r^2 h$ so $h = \dfrac{500}{\pi r^2}$. Then

$$C = (0.02)(2)\pi r^2 + (0.01)2\pi rh = 0.04\pi r^2 + 0.02\pi r\frac{500}{\pi r^2}$$
$$= 0.04\pi r^2 + \frac{10}{r};\ C_{\min} \approx 4.39 \text{ at } r \approx 3.4,\ h \approx 13.8.$$

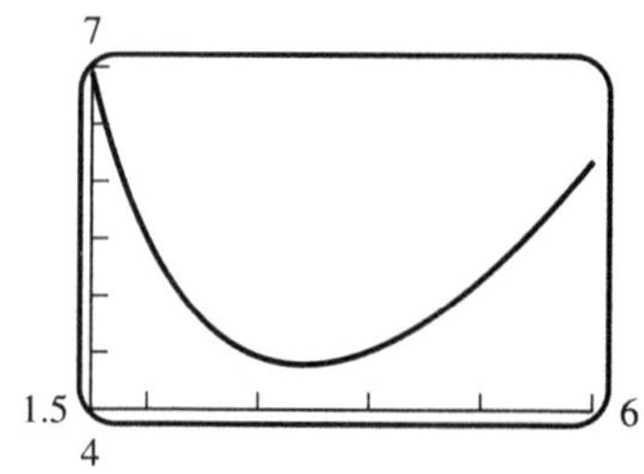

(b) $C = (0.02)(2)(2r)^2 + (0.01)2\pi rh = 0.16r^2 + \dfrac{10}{r}$. Since $0.04\pi < 0.16$, the top and bottom now get more weight. Since they cost more, we diminish their sizes in the solution, and the cans become taller.

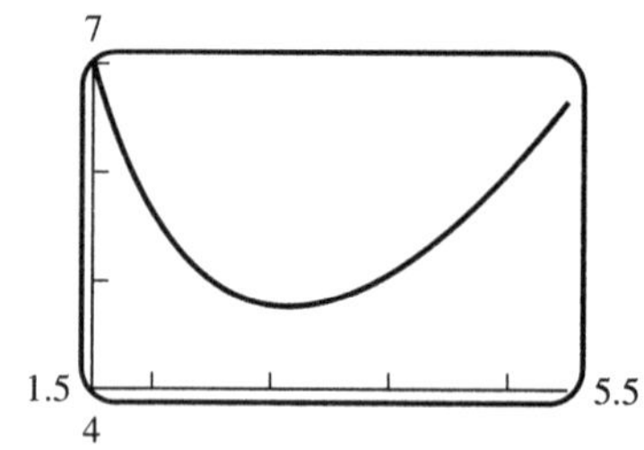

(c) $r \approx 3.1$ cm, $h \approx 16.0$ cm, $C \approx 4.76$ cents

12. (a) The length of a track with straightaways of length L and semicircles of radius r is $P = (2)L + (2)(\pi r)$ ft. Let $L = 360$ and $r = 80$ to get $P = 720 + 160\pi = 1222.65$ ft. Since this is less than 1320 ft (a quarter-mile), a solution is possible.

(b) $P = 2L + 2\pi r = 1320$ and $2r = 2x + 160$, so
$L = \frac{1}{2}(1320 - 2\pi r) = \frac{1}{2}(1320 - 2\pi(80 + x))$
$= 660 - 80\pi - \pi x.$

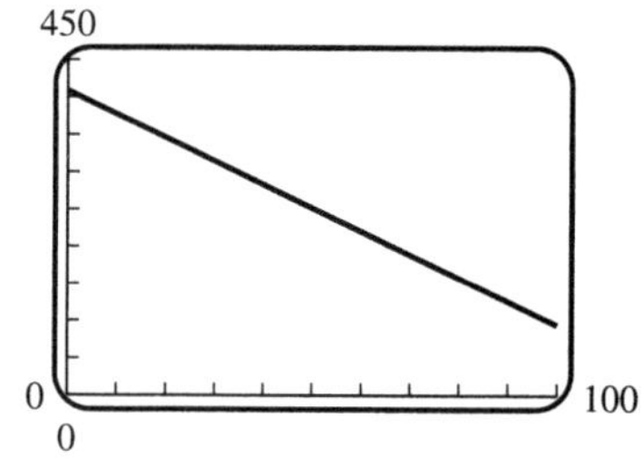

(c) The shortest straightaway is $L = 360$, so $x = 15.49$ ft.

(d) The longest straightaway occurs when $x = 0$, so $L = 660 - 80\pi = 408.67$ ft.

EXERCISE SET 1.2

1. (a) $f(0) = 3(0)^2 - 2 = -2$; $f(2) = 3(2)^2 - 2 = 10$; $f(-2) = 3(-2)^2 - 2 = 10$; $f(3) = 3(3)^2 - 2 = 25$; $f(\sqrt{2}) = 3(\sqrt{2})^2 - 2 = 4$; $f(3t) = 3(3t)^2 - 2 = 27t^2 - 2$

(b) $f(0) = 2(0) = 0$; $f(2) = 2(2) = 4$; $f(-2) = 2(-2) = -4$; $f(3) = 2(3) = 6$; $f(\sqrt{2}) = 2\sqrt{2}$; $f(3t) = 1/3t$ for $t > 1$ and $f(3t) = 6t$ for $t \le 1$.

2. (a) $g(3) = \dfrac{3+1}{3-1} = 2$; $g(-1) = \dfrac{-1+1}{-1-1} = 0$; $g(\pi) = \dfrac{\pi+1}{\pi-1}$; $g(-1.1) = \dfrac{-1.1+1}{-1.1-1} = \dfrac{-0.1}{-2.1} = \dfrac{1}{21}$; $g(t^2-1) = \dfrac{t^2-1+1}{t^2-1-1} = \dfrac{t^2}{t^2-2}$

(b) $g(3) = \sqrt{3+1} = 2$; $g(-1) = 3$; $g(\pi) = \sqrt{\pi+1}$; $g(-1.1) = 3$; $g(t^2-1) = 3$ if $t^2 < 2$ and $g(t^2-1) = \sqrt{t^2-1+1} = |t|$ if $t^2 \ge 2$.

3. (a) $x \ne 3$ (b) $x \le -\sqrt{3}$ or $x \ge \sqrt{3}$

(c) $x^2 - 2x + 5 = 0$ has no real solutions so $x^2 - 2x + 5$ is always positive or always negative. If $x = 0$, then $x^2 - 2x + 5 = 5 > 0$; domain: $(-\infty, +\infty)$.

(d) $x \ne 0$ (e) $\sin x \ne 1$, so $x \ne (2n + \frac{1}{2})\pi$, $n = 0, \pm1, \pm2, \ldots$

4. (a) $x \neq -\frac{7}{5}$

(b) $x - 3x^2$ must be nonnegative; $y = x - 3x^2$ is a parabola that crosses the x-axis at $x = 0, \frac{1}{3}$ and opens downward, thus $0 \le x \le \frac{1}{3}$

(c) $\dfrac{x^2 - 4}{x - 4} > 0$, so $x^2 - 4 > 0$ and $x - 4 > 0$, thus $x > 4$; or $x^2 - 4 < 0$ and $x - 4 < 0$, thus $-2 < x < 2$

(d) $x \neq -1$

(e) $\cos x \le 1 < 2$, $2 - \cos x > 0$, all x

5. (a) $x \le 3$ (b) $-2 \le x \le 2$ (c) $x \ge 0$ (d) all x (e) all x

6. (a) $x \ge \frac{2}{3}$ (b) $-\frac{3}{2} \le x \le \frac{3}{2}$ (c) $x \ge 0$ (d) $x \neq 0$ (e) $x \ge 0$

7. (a) yes (b) yes
(c) no (vertical line test fails) (d) no (vertical line test fails)

8. The sine of $\theta/2$ is $(L/2)/10$ (side opposite over hypotenuse), so that $L = 20\sin(\theta/2)$.

9. The cosine of θ is $(L - h)/L$ (side adjacent over hypotenuse), so $h = L(1 - \cos\theta)$.

10.

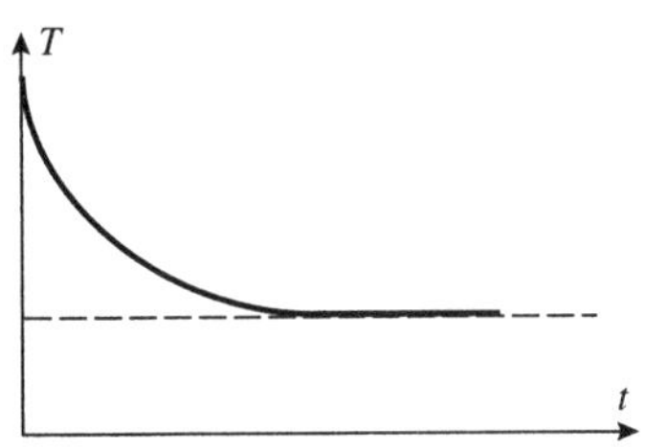

11.

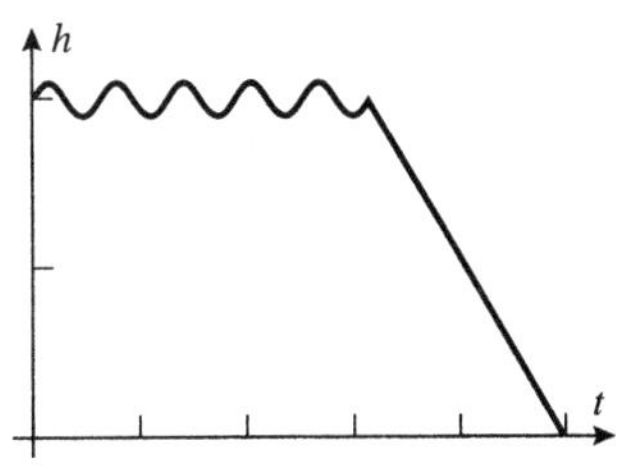

12.

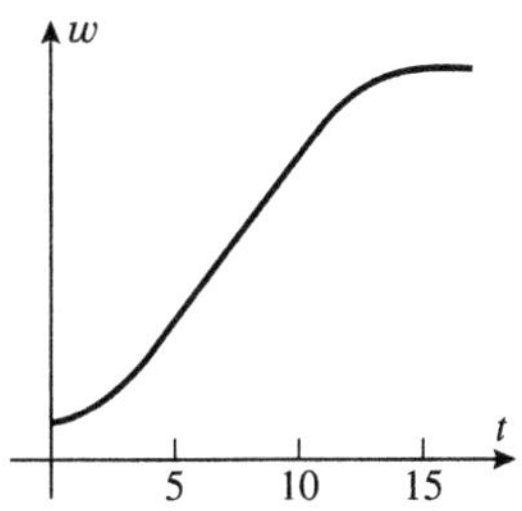

13. (a) If $x < 0$, then $|x| = -x$ so $f(x) = -x + 3x + 1 = 2x + 1$. If $x \ge 0$, then $|x| = x$ so $f(x) = x + 3x + 1 = 4x + 1$;

$$f(x) = \begin{cases} 2x + 1, & x < 0 \\ 4x + 1, & x \ge 0 \end{cases}$$

(b) If $x < 0$, then $|x| = -x$ and $|x - 1| = 1 - x$ so $g(x) = -x + 1 - x = 1 - 2x$. If $0 \le x < 1$, then $|x| = x$ and $|x - 1| = 1 - x$ so $g(x) = x + 1 - x = 1$. If $x \ge 1$, then $|x| = x$ and $|x - 1| = x - 1$ so $g(x) = x + x - 1 = 2x - 1$;

$$g(x) = \begin{cases} 1 - 2x, & x < 0 \\ 1, & 0 \le x < 1 \\ 2x - 1, & x \ge 1 \end{cases}$$

14. (a) If $x < 5/2$, then $|2x - 5| = 5 - 2x$ so $f(x) = 3 + (5 - 2x) = 8 - 2x$. If $x \ge 5/2$, then $|2x - 5| = 2x - 5$ so $f(x) = 3 + (2x - 5) = 2x - 2$;

$$f(x) = \begin{cases} 8 - 2x, & x < 5/2 \\ 2x - 2, & x \ge 5/2 \end{cases}$$

(b) If $x < -1$, then $|x - 2| = 2 - x$ and $|x + 1| = -x - 1$ so $g(x) = 3(2 - x) - (-x - 1) = 7 - 2x$. If $-1 \le x < 2$, then $|x - 2| = 2 - x$ and $|x + 1| = x + 1$ so $g(x) = 3(2 - x) - (x + 1) = 5 - 4x$. If $x \ge 2$, then $|x - 2| = x - 2$ and $|x + 1| = x + 1$ so $g(x) = 3(x - 2) - (x + 1) = 2x - 7$;

$$g(x) = \begin{cases} 7 - 2x, & x < -1 \\ 5 - 4x, & -1 \le x < 2 \\ 2x - 7, & x \ge 2 \end{cases}$$

15. **(a)** $V = (8 - 2x)(15 - 2x)x$ **(b)** $-\infty < x < +\infty, -\infty < V < +\infty$ **(c)** $0 < x < 4$
(d) minimum value at $x = 0$ or at $x = 4$; maximum value somewhere in between (can be approximated by zooming with graphing calculator)

16. **(a)** $x = 3000 \tan\theta$ **(b)** $\theta \neq n\pi + \pi/2$ for n an integer, $-\infty < n < \infty$
(c) $0 \leq \theta < \pi/2, 0 \leq x < +\infty$ **(d)** 3000 ft

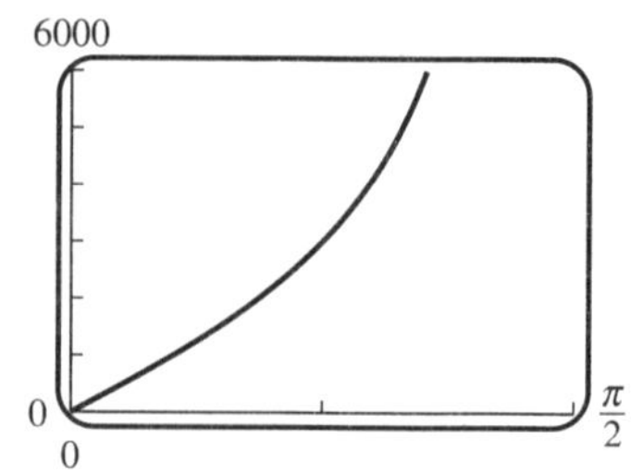

17. **(i)** $x = 1, -2$ causes division by zero **(ii)** $g(x) = x + 1$, all x

18. **(i)** $x = 0$ causes division by zero **(ii)** $g(x) = \sqrt{x} + 1$ for $x \geq 0$

19. **(a)** 25°F **(b)** 2°F **(c)** −15°F

20. If $v = 48$ then $-60 = \text{WCI} = 1.6T - 55$; thus $T = (-60 + 55)/1.6 \approx -3$°F.

21. If $v = 8$ then $-10 = \text{WCI} = 91.4 + (91.4 - T)(0.0203(8) - 0.304\sqrt{8} - 0.474)$; thus $T = 91.4 + (10 + 91.4)/(0.0203(8) - 0.304\sqrt{8} - 0.474)$ and $T = 5$°F

22. The WCI is given by three formulae, but the first and third don't work with the data. Hence $-15 = \text{WCI} = 91.4 + (91.4 - 20)(0.0203v - 0.304\sqrt{v} - 0.474)$; set $x = \sqrt{v}$ so that $v = x^2$ and obtain $0.0203x^2 - 0.304x - 0.474 + (15 + 91.4)/(91.4 - 20) = 0$. Use the quadratic formula to find the two roots. Square them to get v and discard the spurious solution, leaving $v \approx 25$.

23. Let t denote time in minutes after 9:23 AM. Then $D(t) = 1000 - 20t$ ft.

EXERCISE SET 1.3

1. (e) seems best, though only (a) is bad.

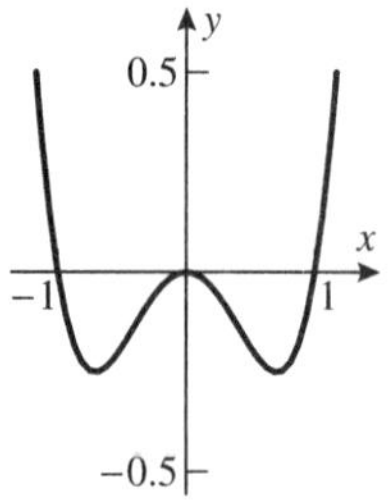

2. (e) seems best, though only (a) is bad and (b) is not good.

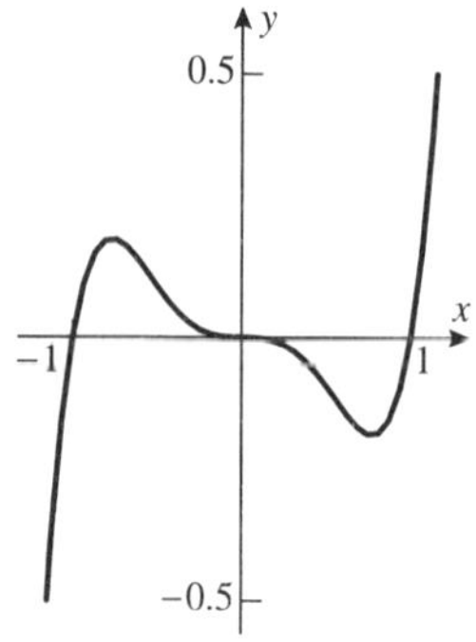

3. (b) and (c) are good; (a) is very bad.

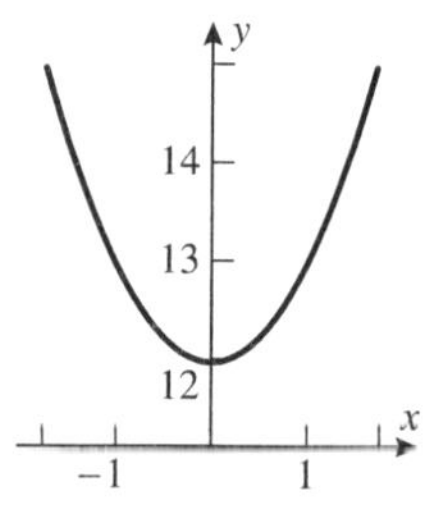

4. (b) and (c) are good; (a) is very bad.

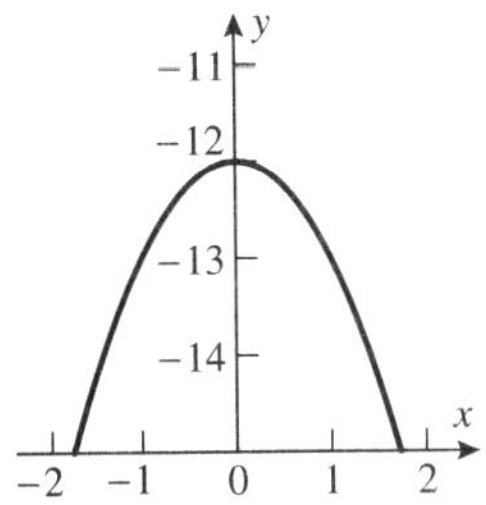

5. $[-3, 3] \times [0, 5]$

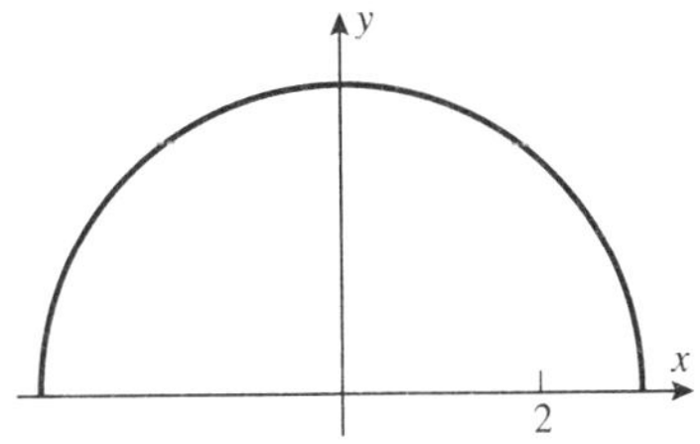

6. $[-4, 2] \times [0, 3]$

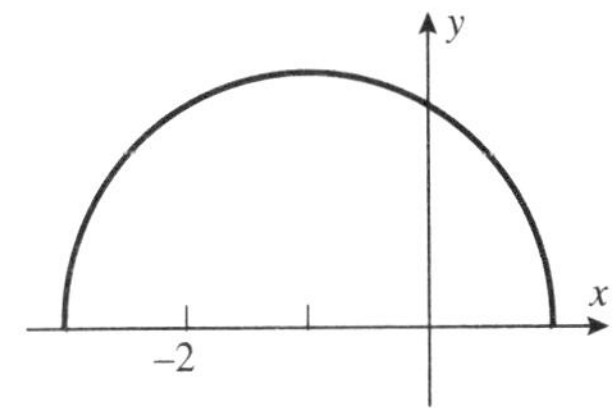

7. **(a)** window too narrow, too short
(b) window wide enough, but too short
(c) good window, good spacing
(d) window too narrow, too short

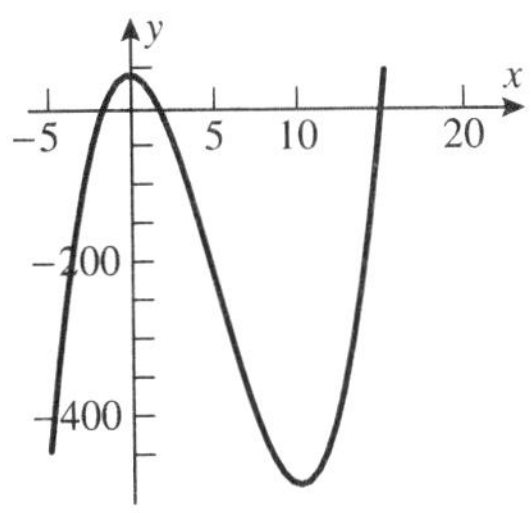

(e) window too narrow, too short

8. **(a)** window too narrow
(b) window too short
(c) good window, good tick spacing
(d) window too narrow, too short

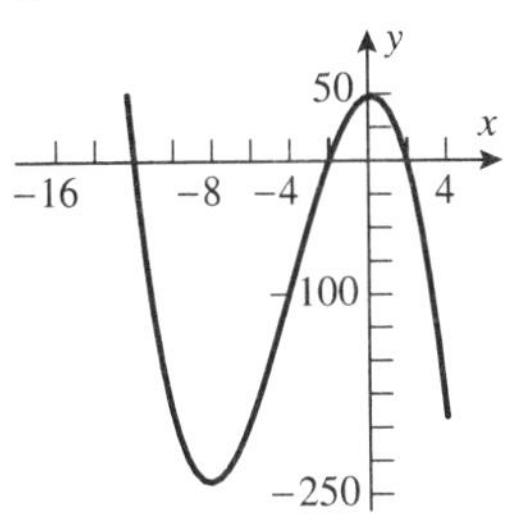

(e) shows one local minimum only, window too narrow, too short

9. $[-5, 14] \times [-60, 40]$

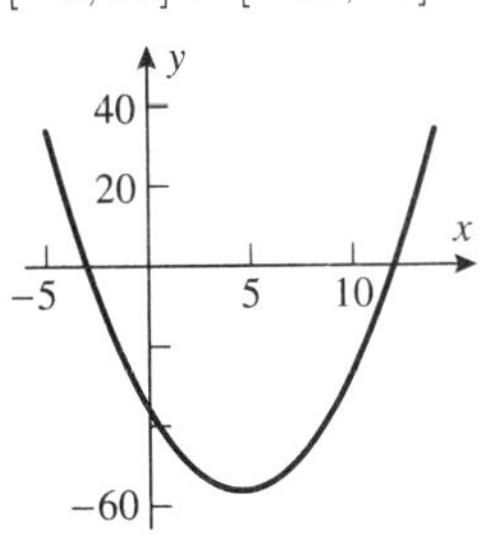

10. $[6, 12] \times [-100, 100]$

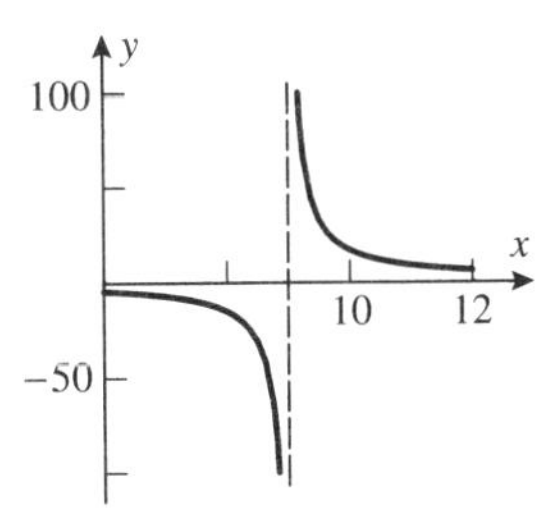

11. $[-0.1, 0.1] \times [-3, 3]$

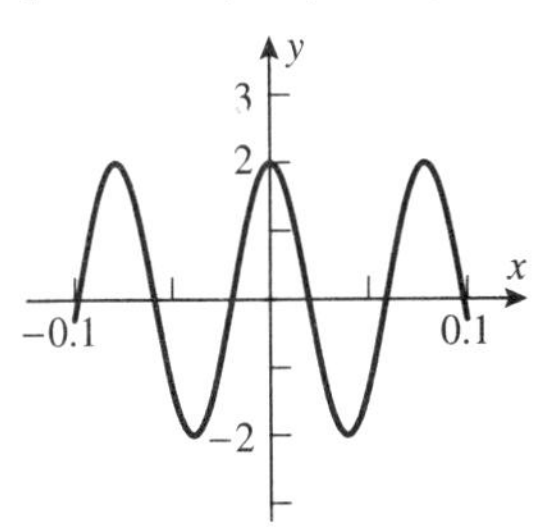

12. $[-1000, 1000] \times [-13, 13]$

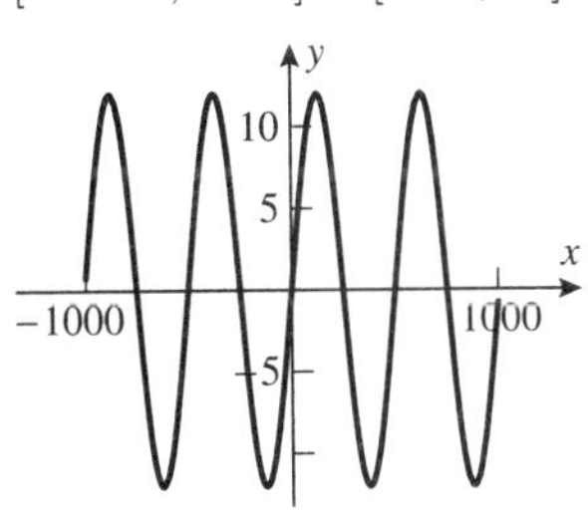

13. $[-250, 1050] \times [-1500000, 600000]$

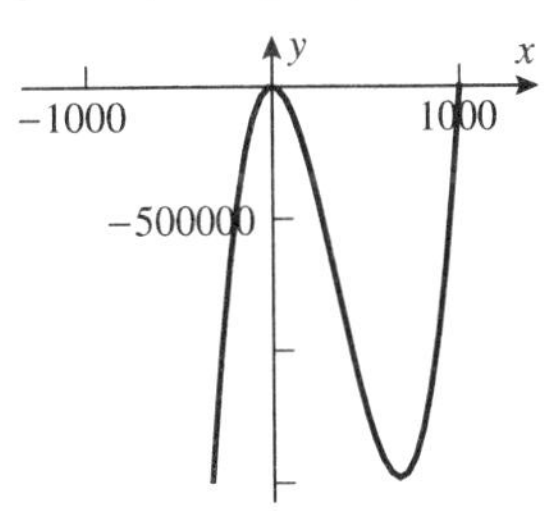

14. $[-3, 20] \times [-3500, 3000]$

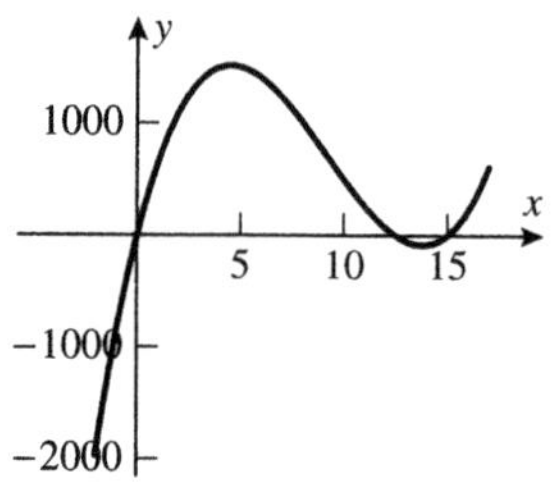

15. $[-2, 2] \times [-20, 20]$

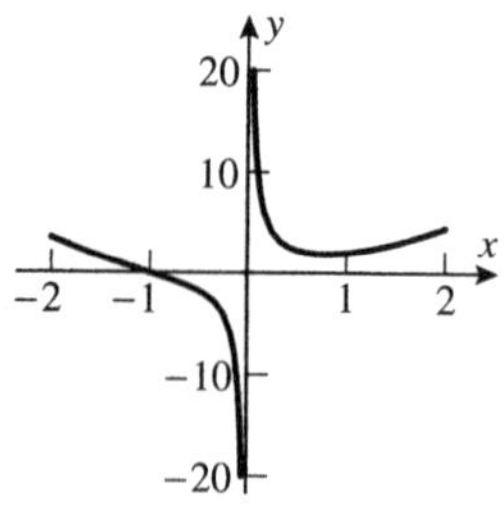

16. $[1.6, 2] \times [0, 2]$

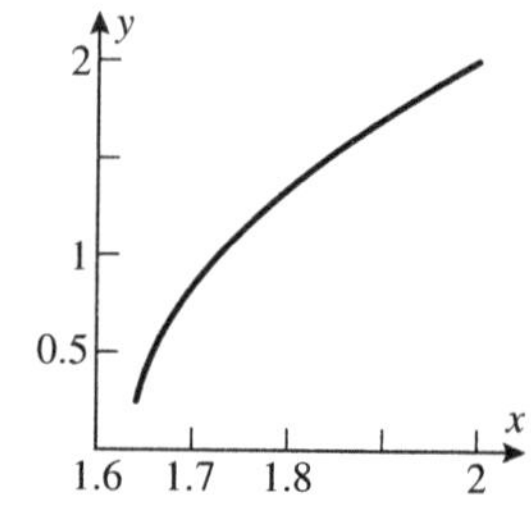

17. depends on graphing utility

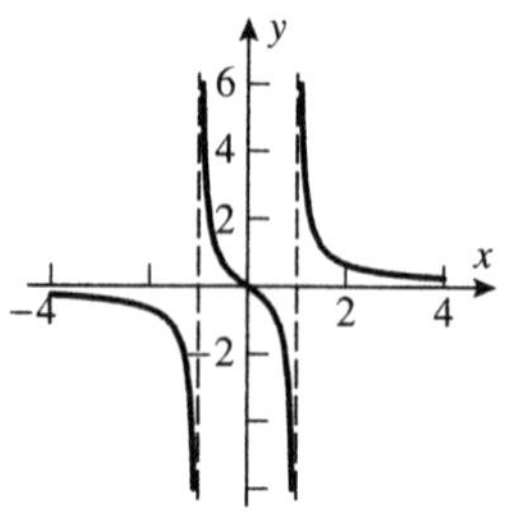

18. depends on graphing utility

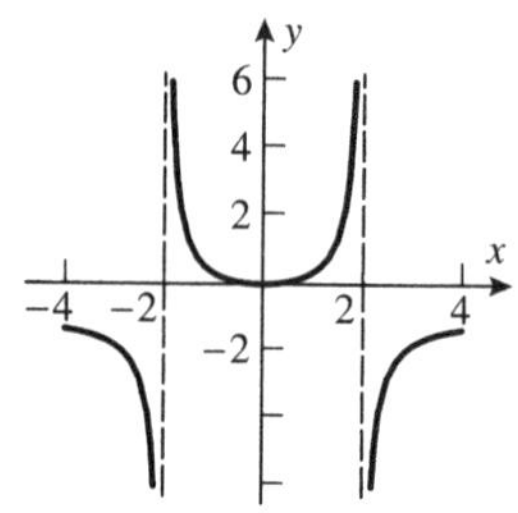

19. **(a)** $f(x) = \sqrt{16 - x^2}$

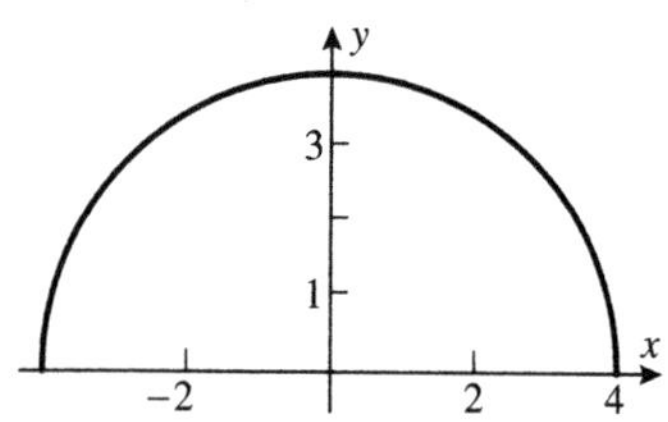

(b) $f(x) = -\sqrt{16 - x^2}$

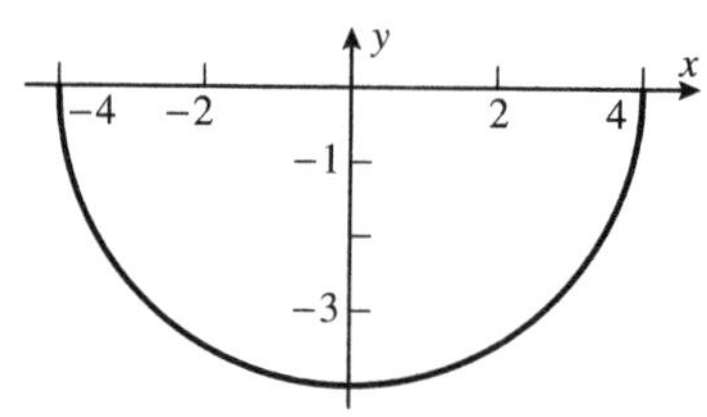

(c)

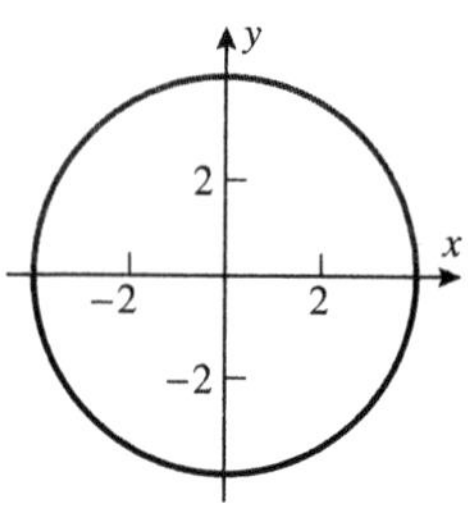

(d)

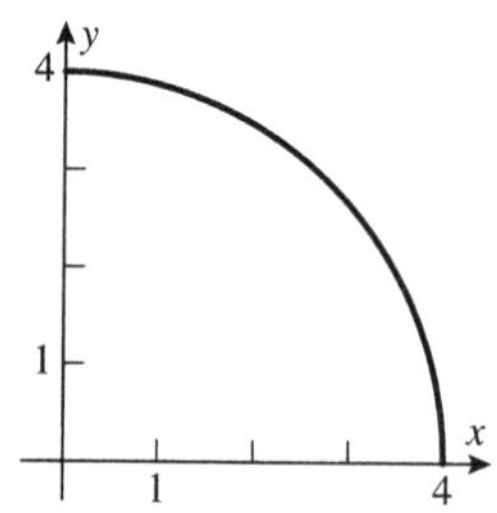

(e) No; the vertical line test fails.

20. **(a)** $y = \pm 3\sqrt{1 - x^2/4}$

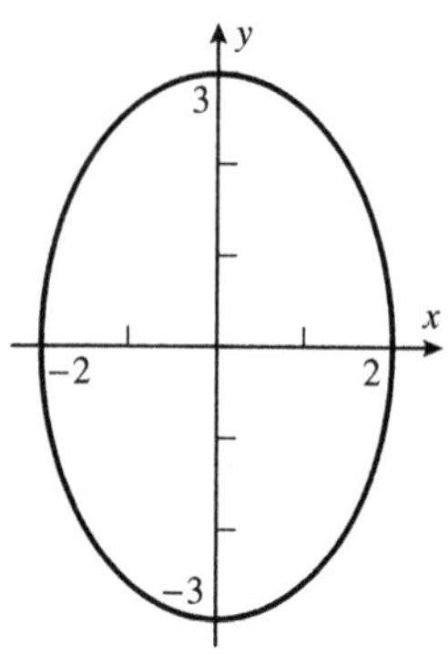

(b) $y = \pm\sqrt{x^2 + 1}$

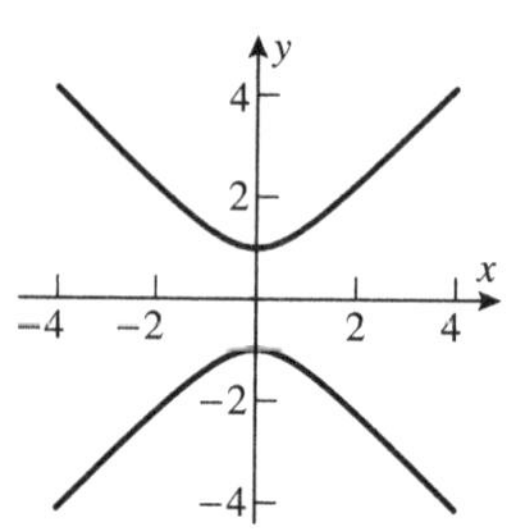

21. **(a)**

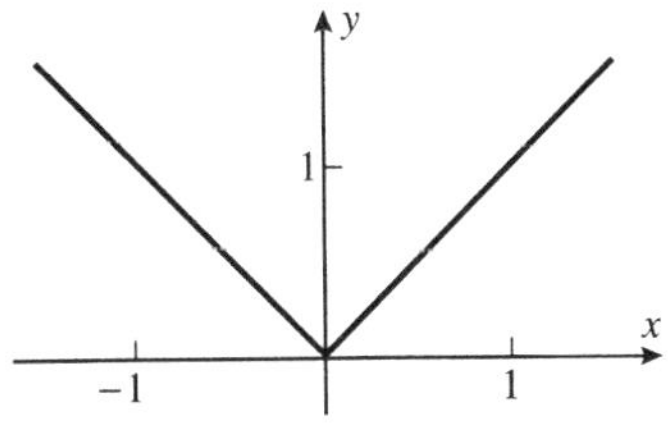

(b)

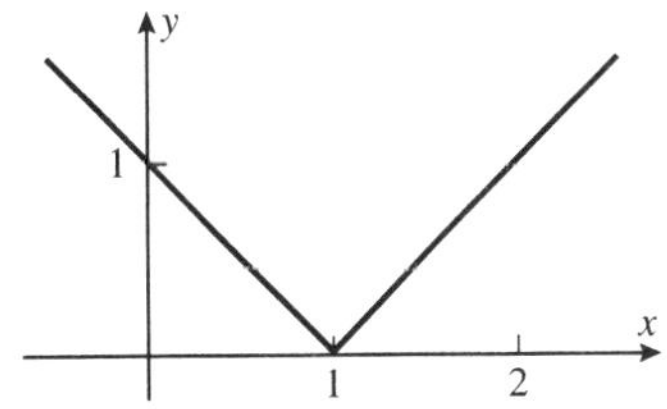

(c)

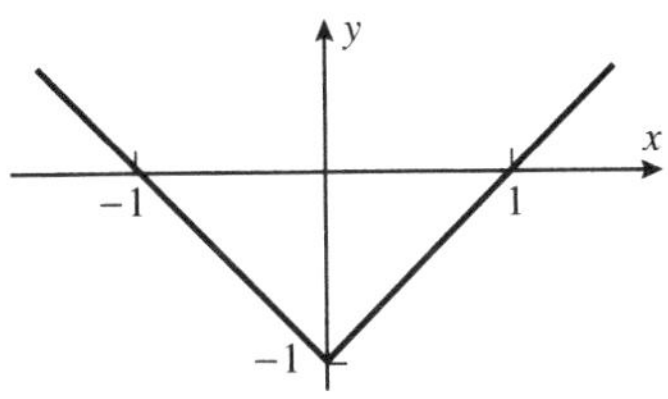

(d)

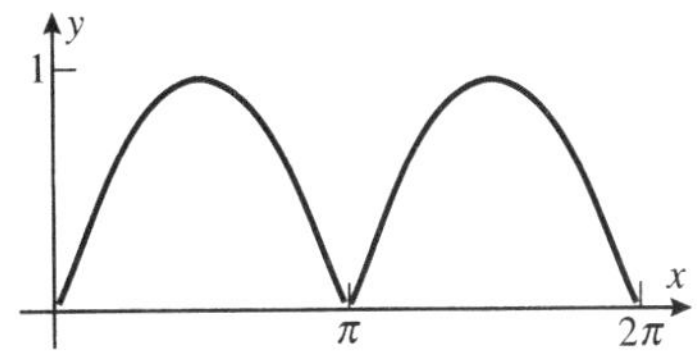

(e)

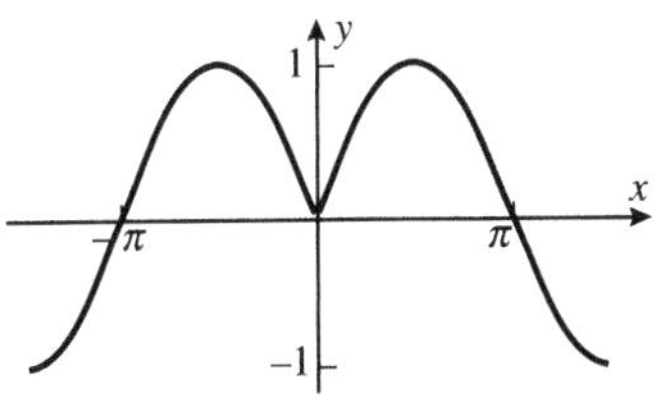

(f)

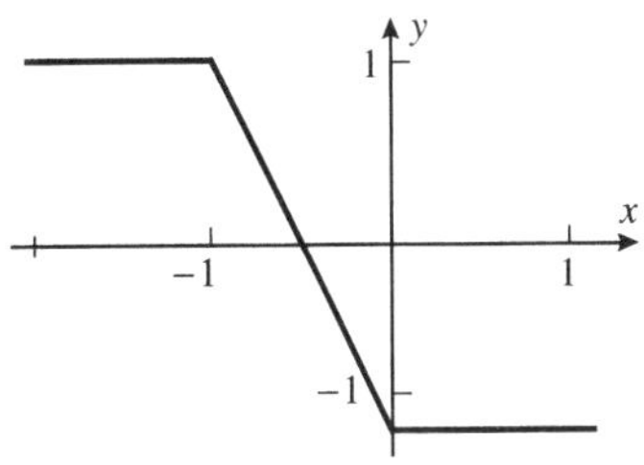

22.

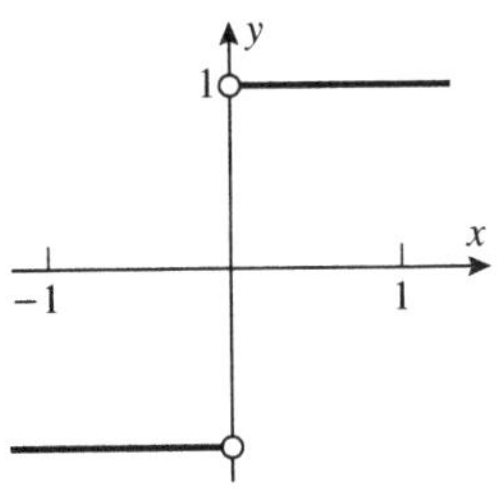

23. The portions of the graph of $y = f(x)$ which lie below the x-axis are reflected over the x-axis to give the graph of $y = |f(x)|$.

24. Erase the portion of the graph of $y = f(x)$ which lies in the left-half plane and replace it with the reflection over the y-axis of the portion in the right-half plane (symmetry over the y-axis) and you obtain the graph of $y = f(|x|)$.

25. **(a)** for example, let $a = 1.1$

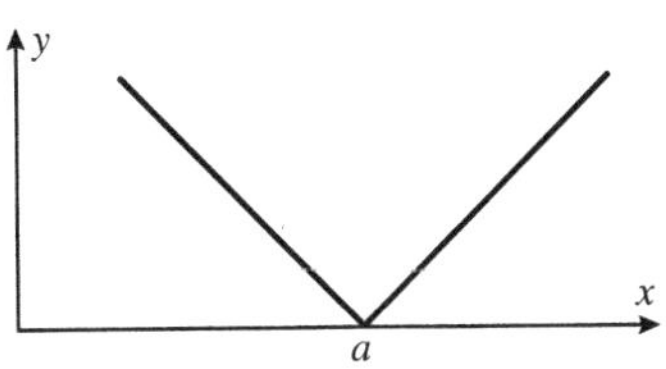

(b)

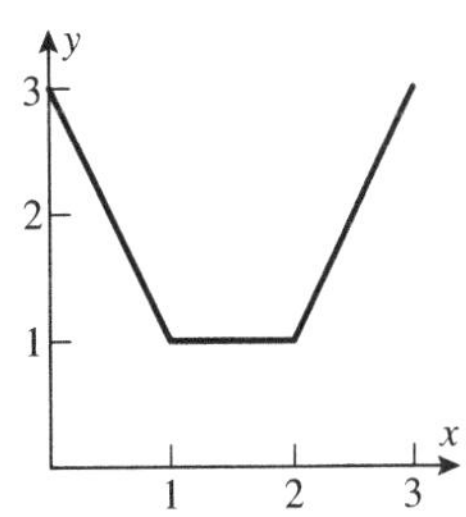

26. They are identical.

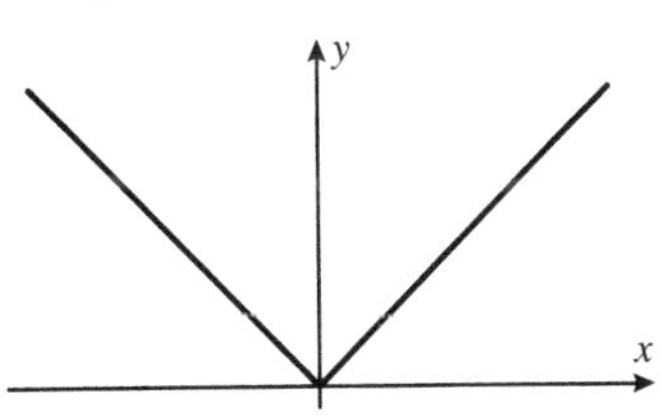

27.

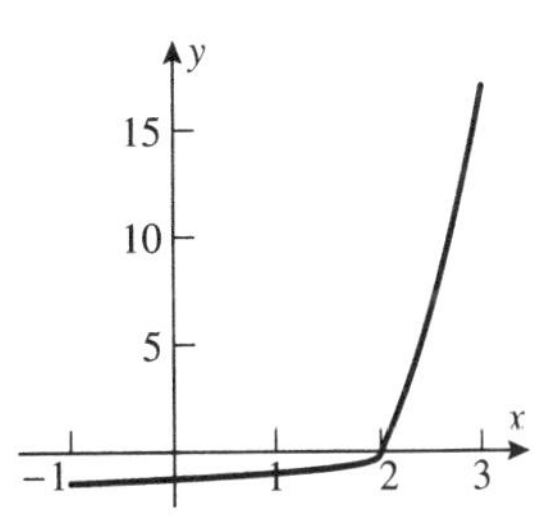

28. This graph is very complex. We show three views, small (near the origin), medium and large:

(a)

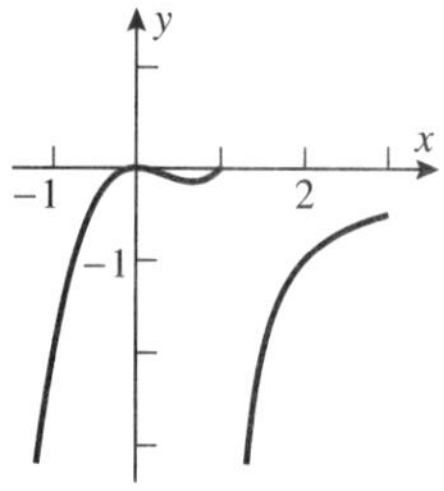

(b)

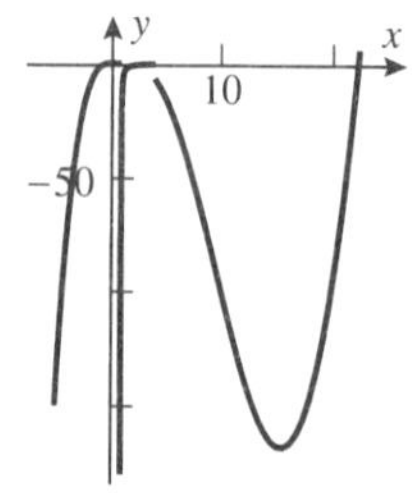

(c)

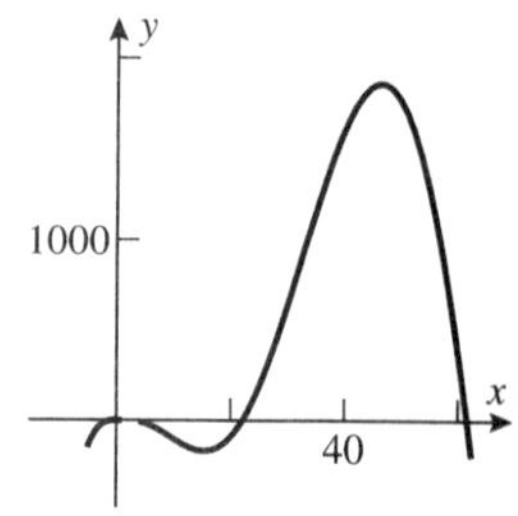

29. (a)

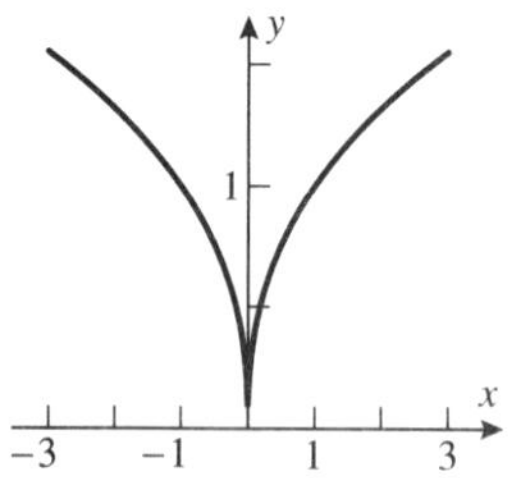

(b)

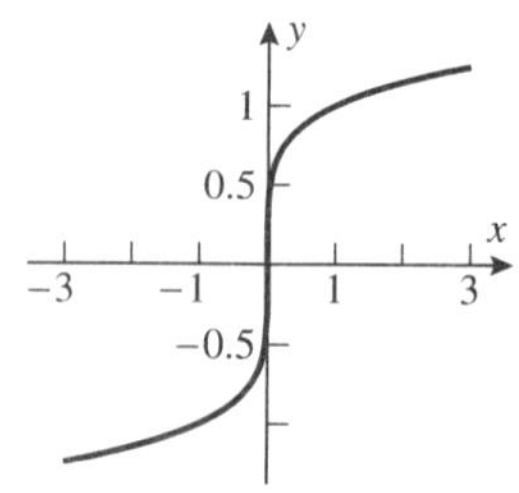

(c)

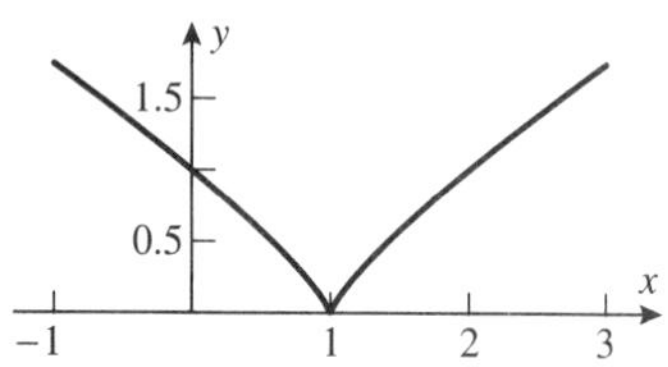

(d)

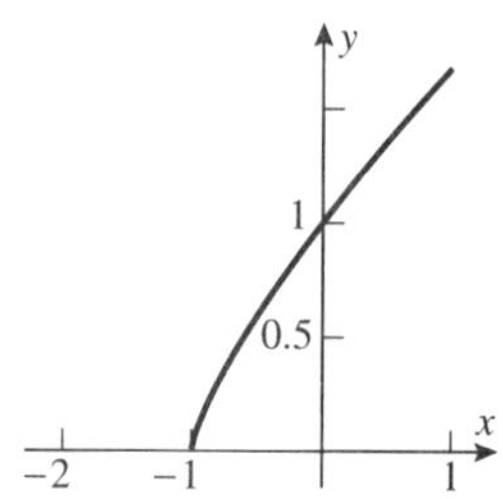

30.

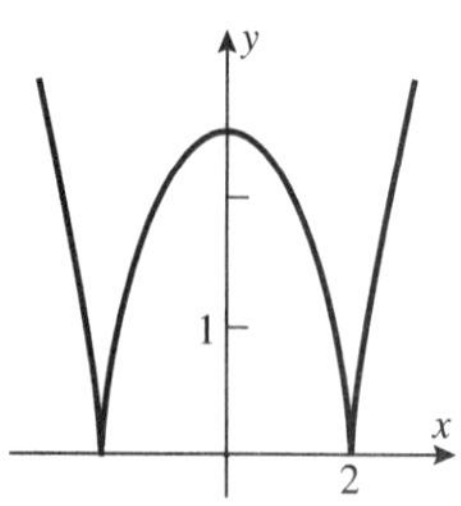

31. (a) stretches or shrinks the graph in the y-direction; reflects it over the x-axis if c changes sign

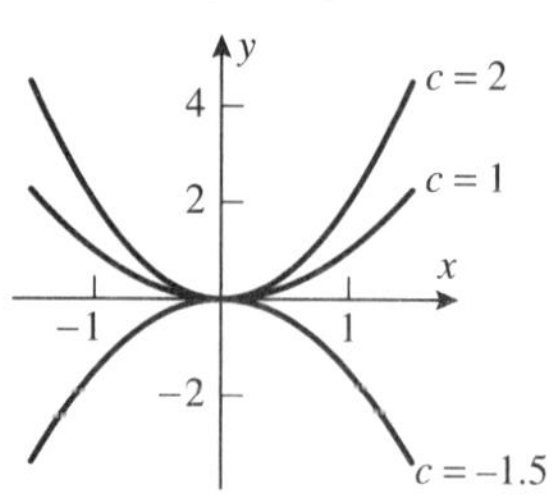

(b) As c increases, the parabola moves down and to the left. If c increases, up and right.

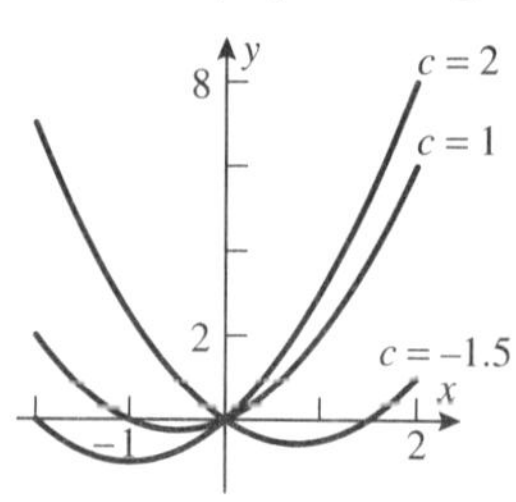

(c) The graph rises or falls in the y-direction with changes in c.

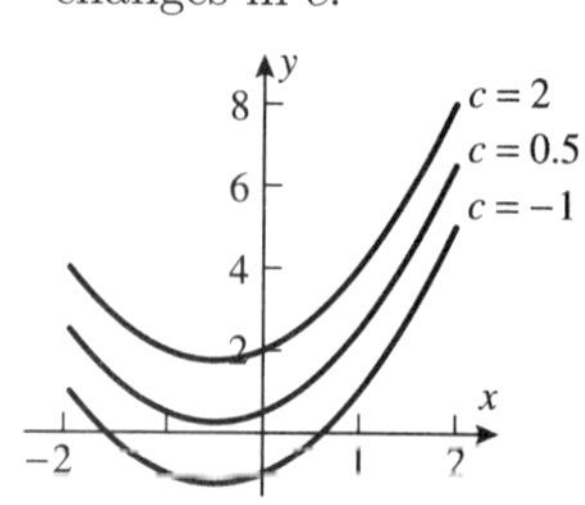

32. (a)

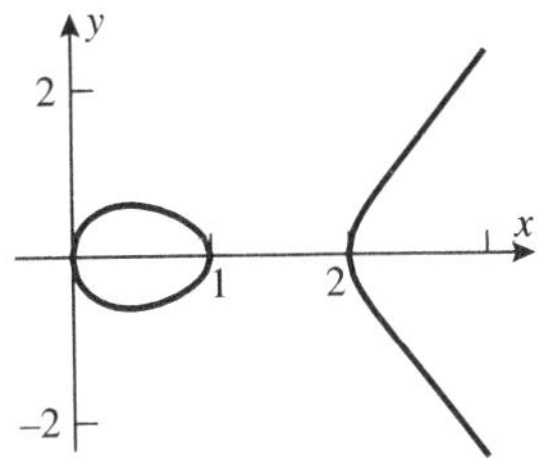

(b) x-intercepts at $x = 0, a, b$. Assume $a < b$ and let a approach b. The two branches of the curve come together. If a moves past b then a and b switch roles.

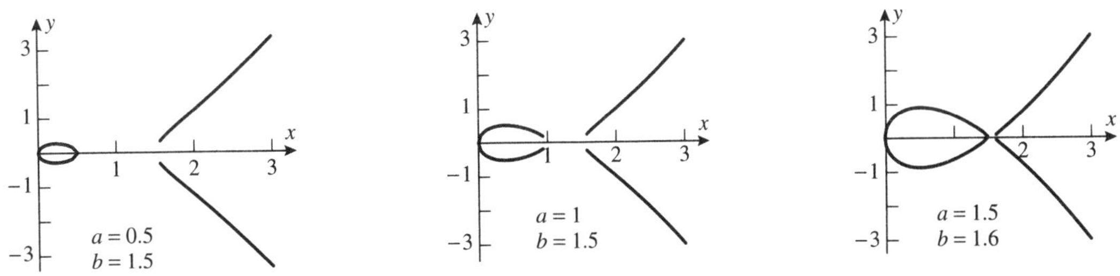

33. The curve oscillates between the lines $y = x$ and $y = -x$ with increasing rapidity as $|x|$ increases.

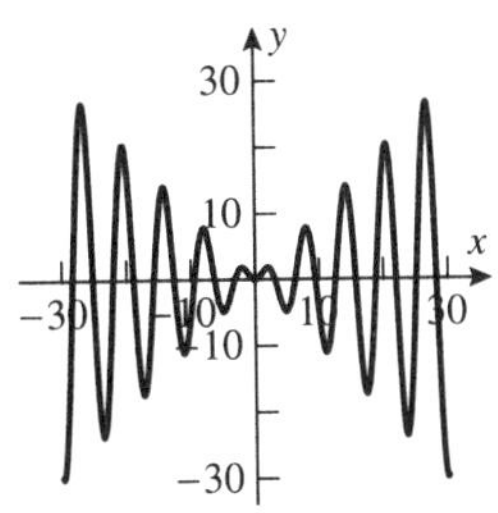

34. The curve oscillates between the lines $y = +1$ and $y = -1$, infinitely many times in any neighborhood of $x = 0$.

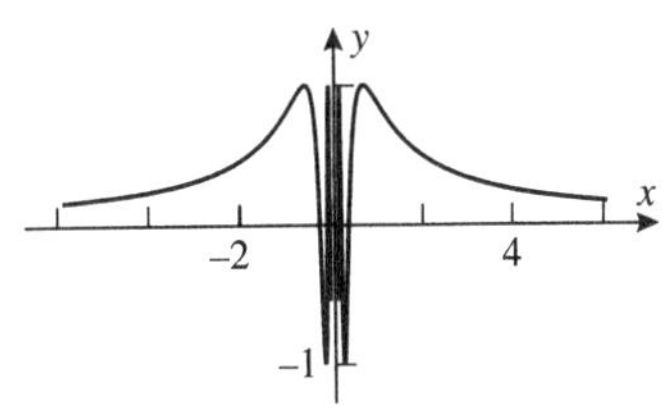

EXERCISE SET 1.4

1. (a)

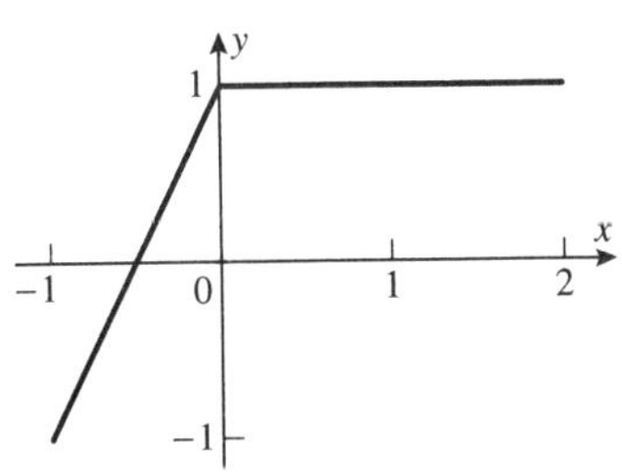

(b)

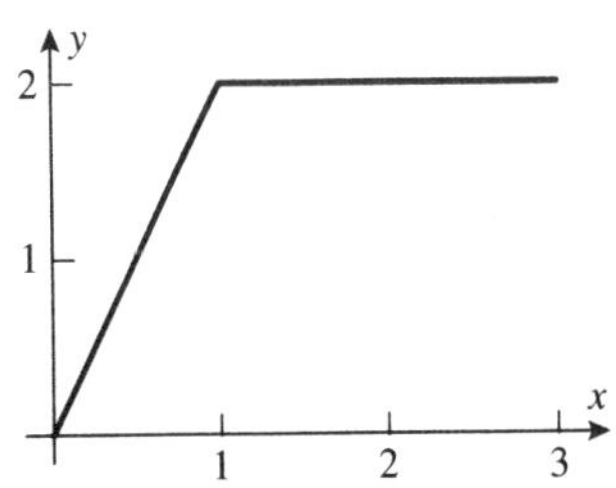

(c)

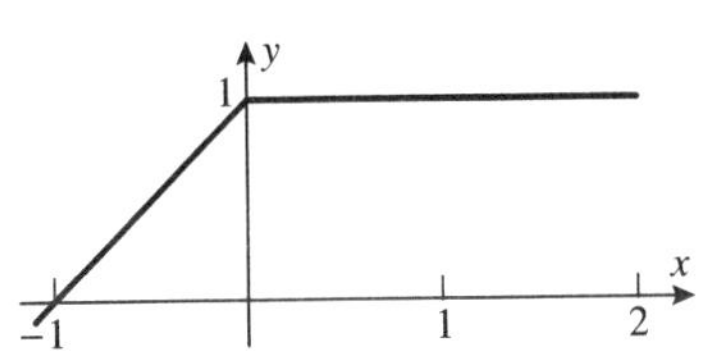

(d)

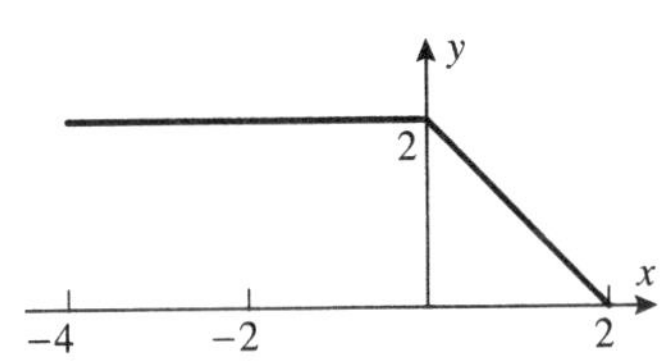

2. **(a)**

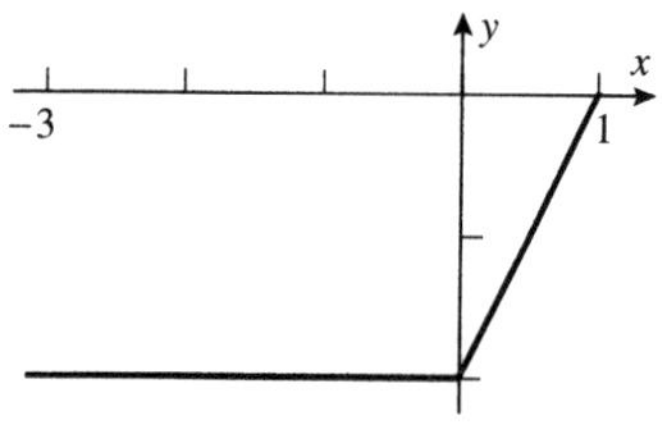

(b)

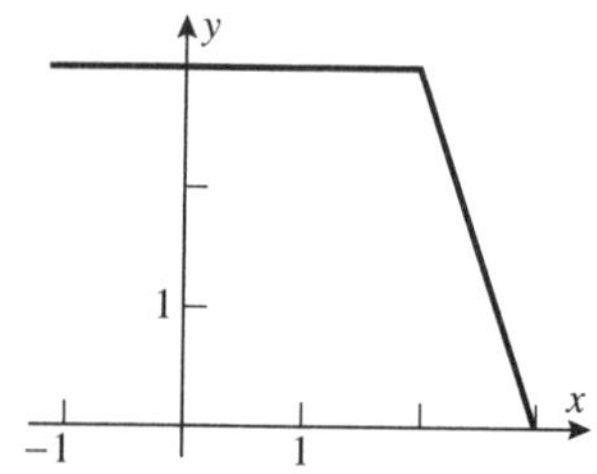

(c)

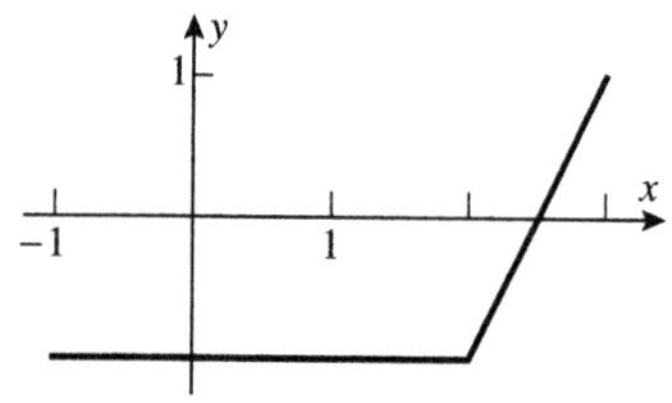

(d)

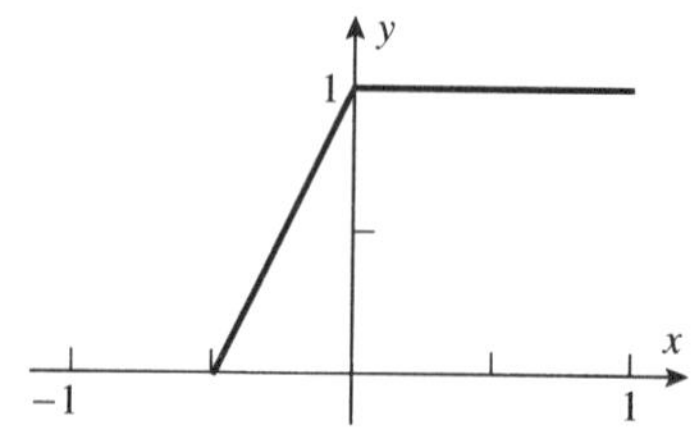

3. **(a)**

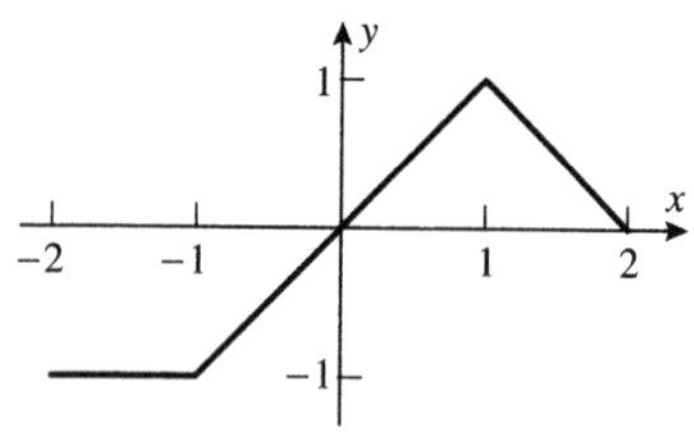

(b)

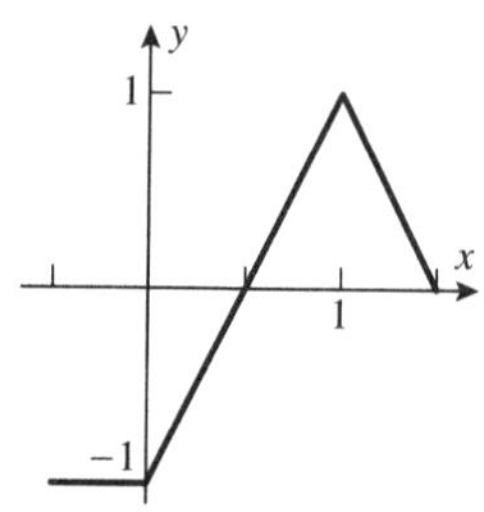

(c)

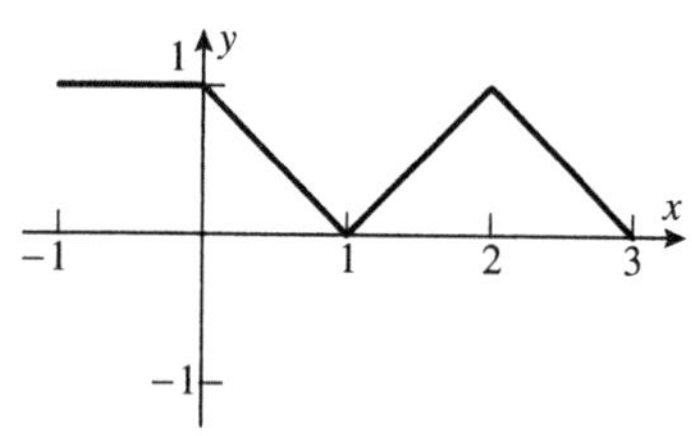

(d)

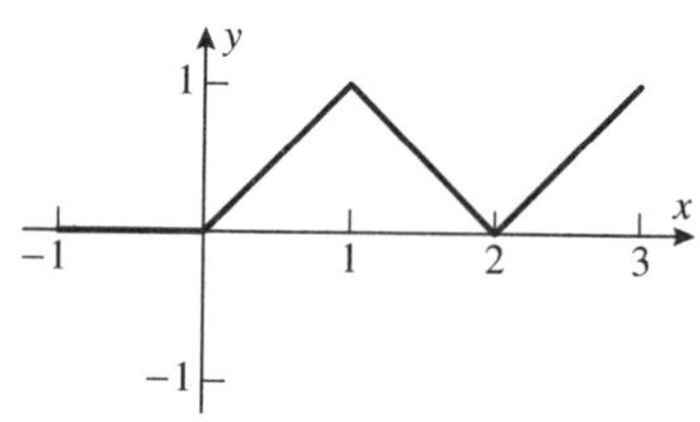

4.

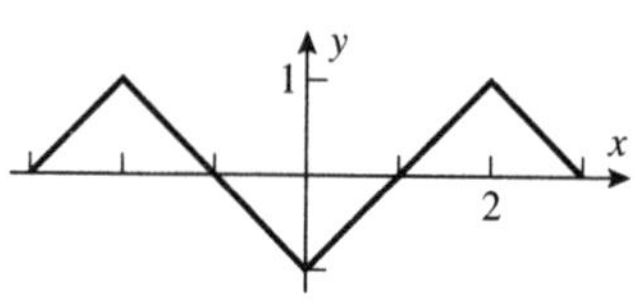

5. Translate right 2 units, and up one unit.

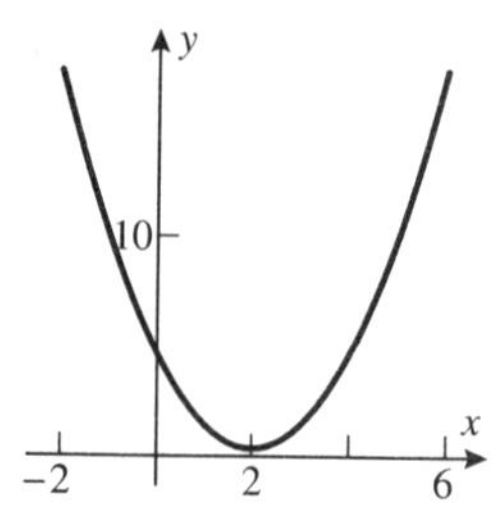

6. Translate left 1 unit, reflect over x-axis, and translate up 2 units.

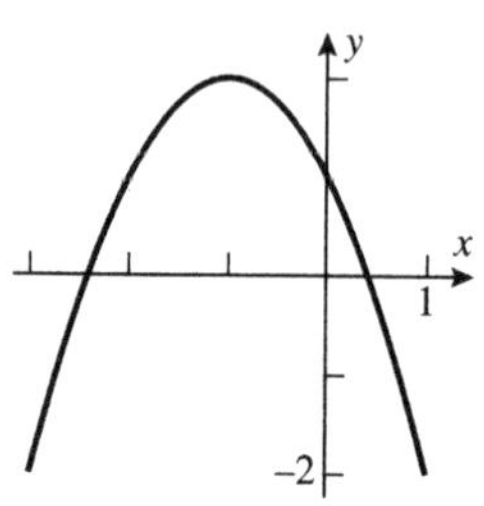

7. Translate left 1 unit, stretch vertically by a factor of 2, reflect over x-axis, translate down 3 units.

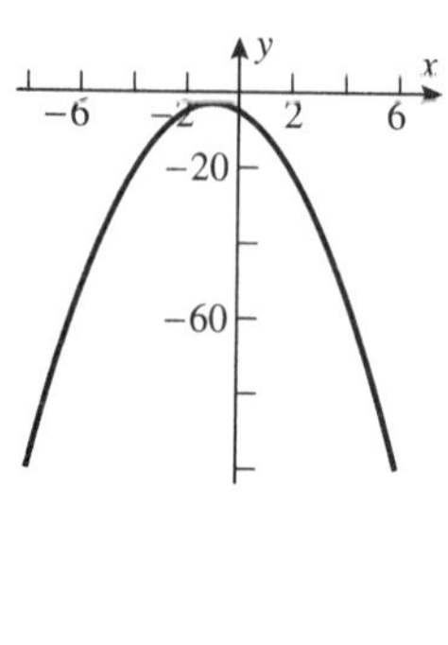

8. Translate right 3 units, compress vertically by a factor of $\frac{1}{2}$, and translate up 2 units.

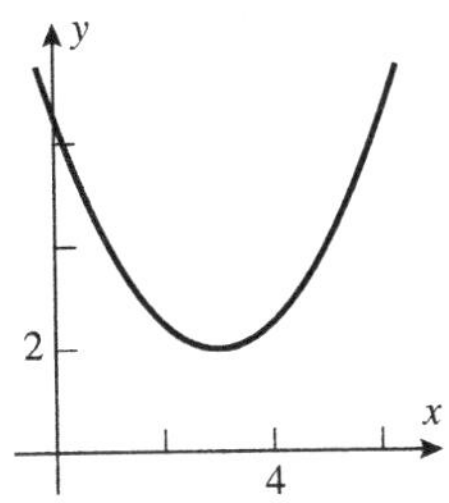

9. $y = (x+3)^2 - 9$; translate left 3 units and down 9 units.

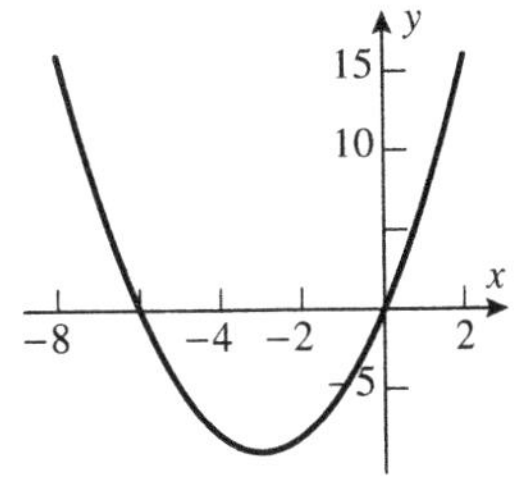

10. $y = (x+3)^2 - 19$; translate left 3 units and down 19 units.

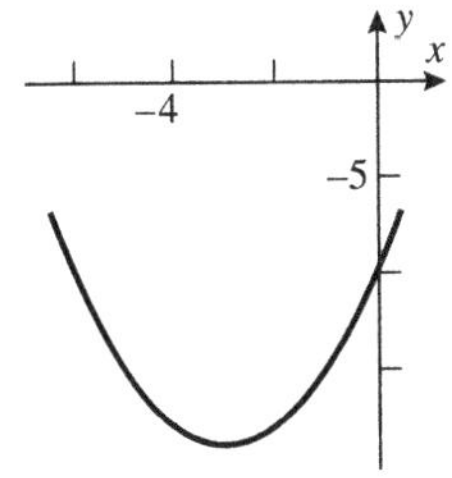

11. $y = -(x-1)^2 + 2$; translate right 1 unit, reflect over x-axis, translate up 2 units.

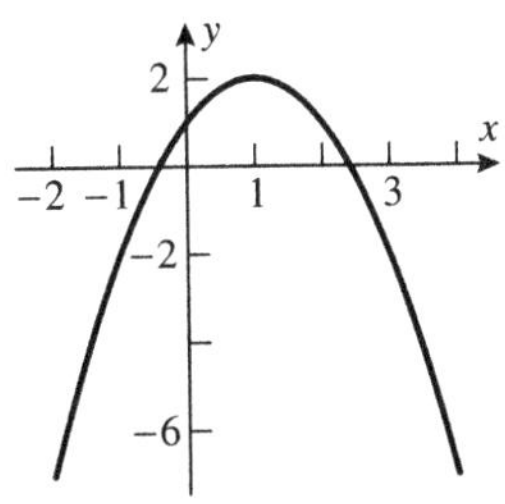

12. $y = \frac{1}{2}[(x-1)^2 + 2]$; translate left 1 unit and up 2 units, compress vertically by a factor of $\frac{1}{2}$.

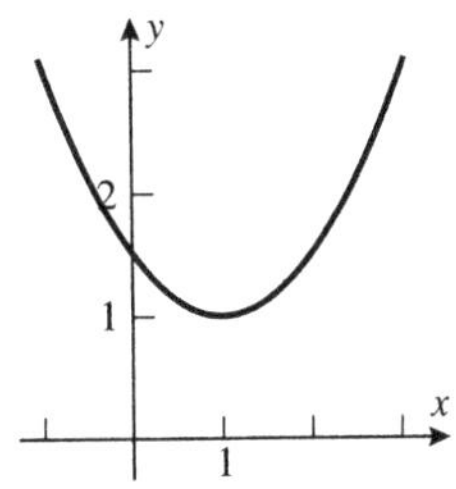

13. Translate left 1 unit, reflect over x-axis, translate up 3 units.

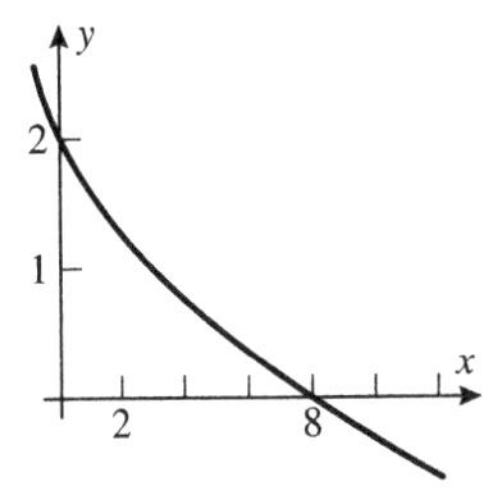

14. Translate right 4 units and up 1 unit.

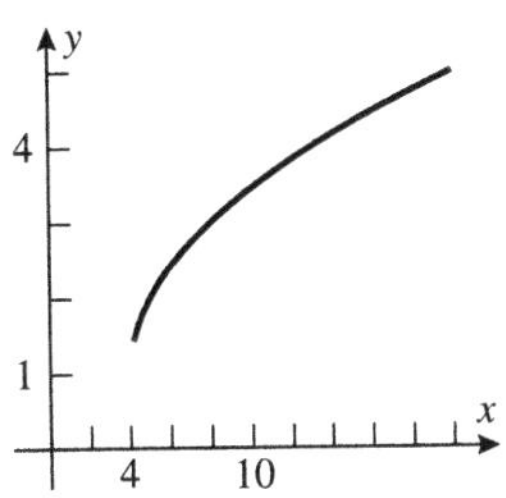

15. Compress vertically by a factor of $\frac{1}{2}$, translate up 1 unit.

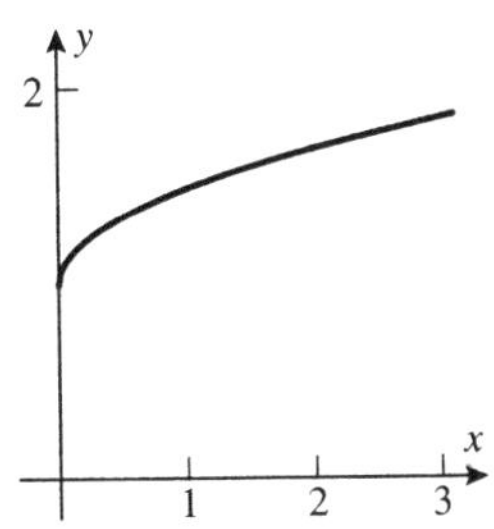

16. Stretch vertically by a factor of $\sqrt{3}$ and reflect over x-axis.

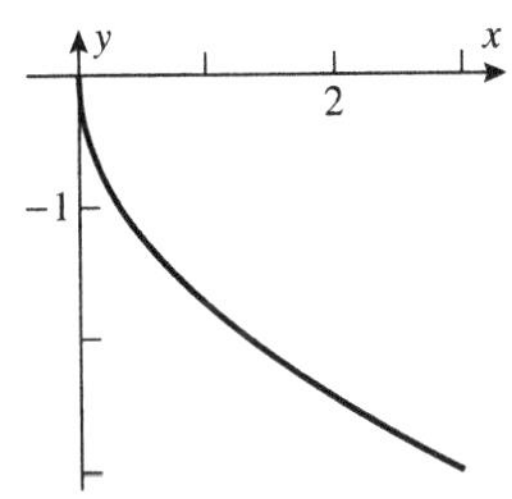

17. Translate right 3 units.

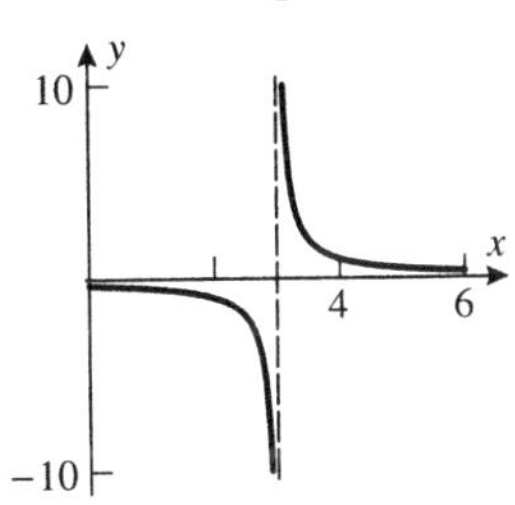

18. Translate right 1 unit and reflect over x-axis.

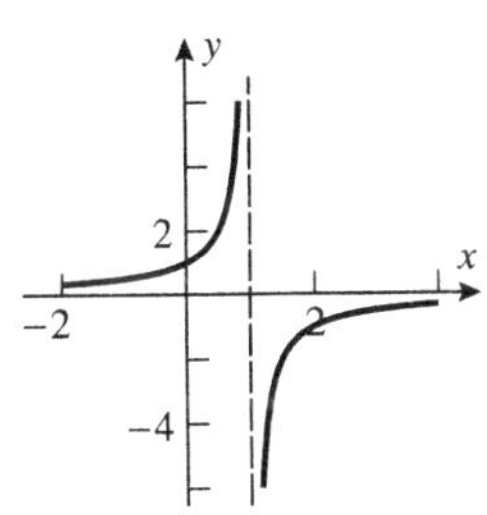

19. Translate left 1 unit, reflect over x-axis, translate up 2 units.

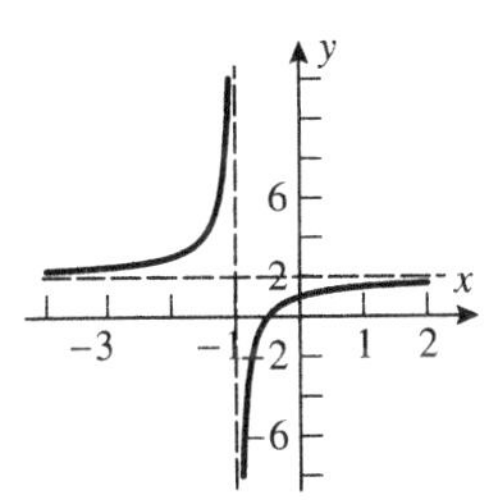

20. $y = 1 - 1/x$;
reflect over x-axis,
translate up 1 unit.

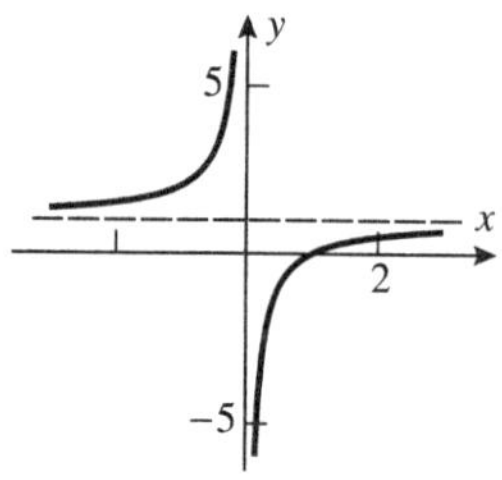

21. Translate left 2 units
and down 2 units.

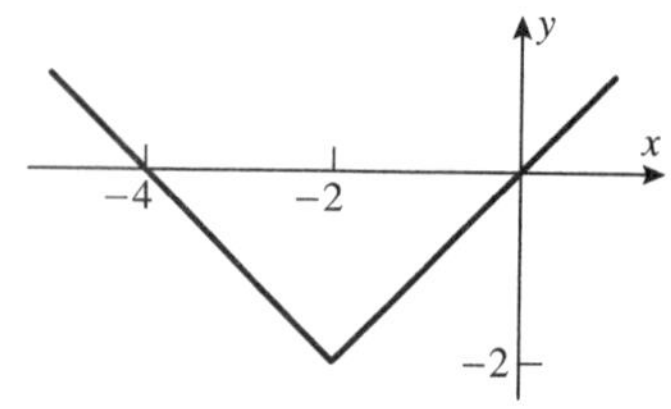

22. Translate right 3 units, reflect
over x-axis, translate up 1 unit.

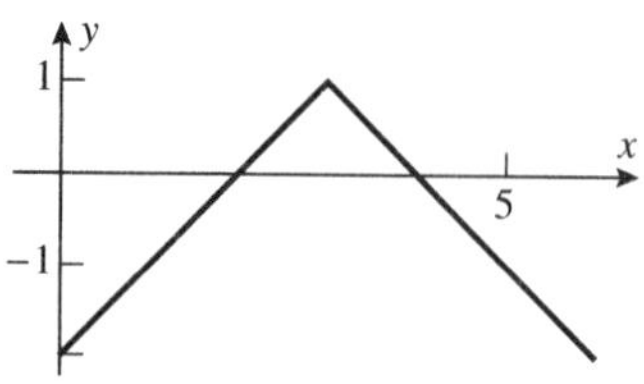

23. Stretch vertically by a factor of 2,
translate right 1 unit and up 1 unit.

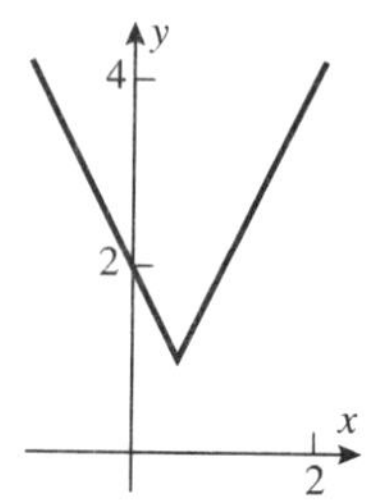

24. $y = |x - 2|$; translate right 2 units.

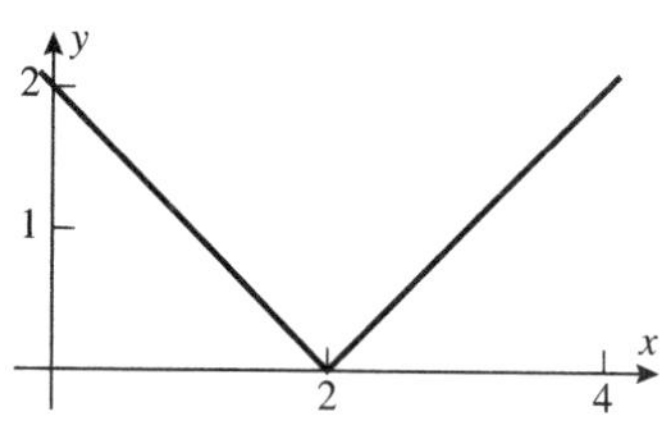

25. Stretch vertically by a factor of 2,
reflect over x-axis, translate up 1 unit.

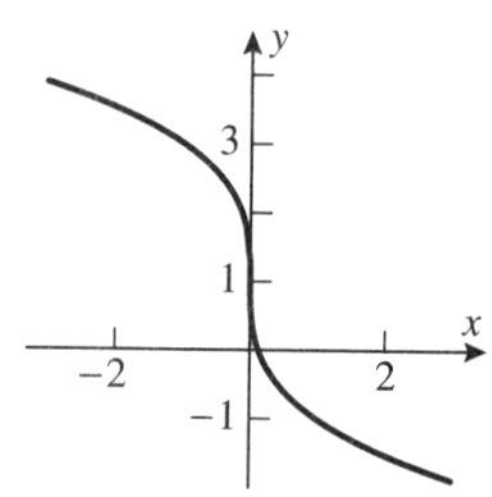

26. Translate right 2 units
and down 3 units.

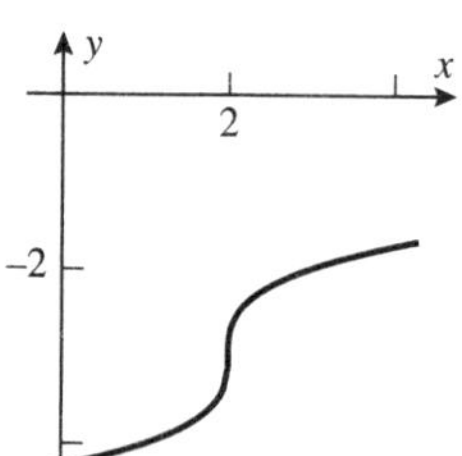

27. Translate left 1 unit
and up 2 units.

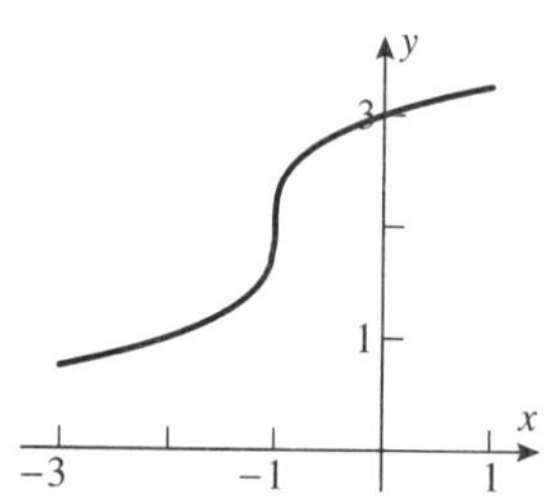

28. Translate right 2 units,
reflect over x-axis.

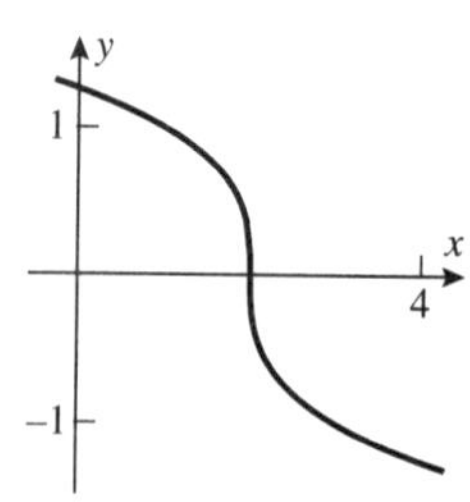

29. (a)

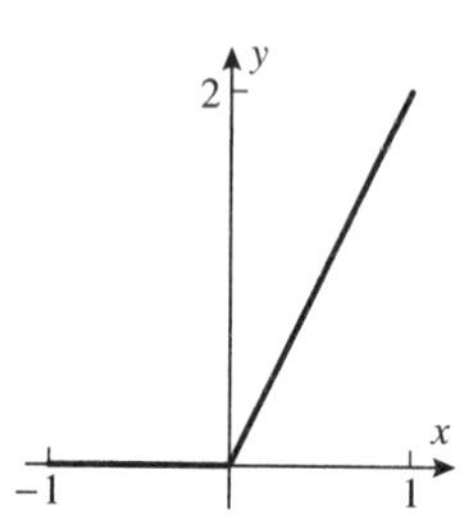

(b) $y = \begin{cases} 0 \text{ if } x \le 0 \\ 2x \text{ if } 0 < x \end{cases}$

30. 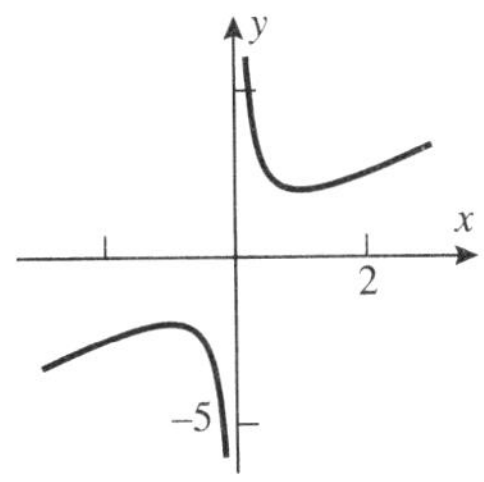

31. $(f+g)(x) = x^2+2x+1$, all x; $(f-g)(x) = 2x - x^2 - 1$, all x; $(fg)(x) = 2x^3+2x$, all x; $(f/g)(x) = 2x/(x^2+1)$, all x

32. $(f+g)(x) = 3x-2+|x|$, all x; $(f-g)(x) = 3x-2-|x|$, all x; $(fg)(x) = 3x|x| - 2|x|$, all x; $(f/g)(x) = (3x-2)/|x|$, all $x \neq 0$

33. $(f+g)(x) = 3\sqrt{x-1}$, $x \geq 1$; $(f-g)(x) = \sqrt{x-1}$, $x \geq 1$; $(fg)(x) = 2x-2$, $x \geq 1$; $(f/g)(x) = 2, x > 1$

34. $(f+g)(x) = (2x^2+1)/[x(x^2+1)]$, all $x \neq 0$; $(f-g)(x) = -1/[x(x^2+1)]$, all $x \neq 0$; $(fg)(x) = 1/(x^2+1)$, all $x \neq 0$; $(f/g)(x) = x^2/(x^2+1)$, all $x \neq 0$

35. **(a)** 3 **(b)** 9 **(c)** 2 **(d)** 2

36. **(a)** $\pi - 1$ **(b)** 0 **(c)** $-\pi^2 + 3\pi - 1$ **(d)** 1

37. **(a)** t^4+1 **(b)** t^2+4t+5 **(c)** x^2+4x+5 **(d)** $\dfrac{1}{x^2}+1$
(e) $x^2+2xh+h^2+1$ **(f)** x^2+1 **(g)** $x+1$ **(h)** $9x^2+1$

38. **(a)** $\sqrt{5s+2}$ **(b)** $\sqrt{\sqrt{x}+2}$ **(c)** $3\sqrt{5x}$ **(d)** $1/\sqrt{x}$
(e) $\sqrt[4]{x}$ **(f)** 0 **(g)** $1/\sqrt[4]{x}$ **(h)** $|x-1|$

39. $(f \circ g)(x) = 2x^2 - 2x + 1$, all x; $(g \circ f)(x) = 4x^2 + 2x$, all x

40. $(f \circ g)(x) = 2 - x^6$, all x; $(g \circ f)(x) = -x^6 + 6x^4 - 12x^2 + 8$, all x

41. $(f \circ g)(x) = 1 - x$, $x \leq 1$; $(g \circ f)(x) = \sqrt{1-x^2}$, $|x| \leq 1$

42. $(f \circ g)(x) = \sqrt{\sqrt{x^2+3}-3}$, $|x| \geq \sqrt{6}$; $(g \circ f)(x) = \sqrt{x}$, $x \geq 3$

43. $(f \circ g)(x) = \dfrac{1}{1-2x}$, $x \neq \dfrac{1}{2}, 1$; $(g \circ f)(x) = -\dfrac{1}{2x} - \dfrac{1}{2}$, $x \neq 0, 1$

44. $(f \circ g)(x) = \dfrac{x}{x^2+1}$, $x \neq 0$; $(g \circ f)(x) = \dfrac{1}{x} + x$, $x \neq 0$

45. $x^{-6}+1$

46. $\dfrac{x}{x+1}$

47. **(a)** $g(x) = \sqrt{x}$, $h(x) = x+2$ **(b)** $g(x) = |x|$, $h(x) = x^2 - 3x + 5$

48. **(a)** $g(x) = x+1$, $h(x) = x^2$ **(b)** $g(x) = 1/x$, $h(x) = x - 3$

49. **(a)** $g(x) = x^2$, $h(x) = \sin x$ **(b)** $g(x) = 3/x$, $h(x) = 5 + \cos x$

50. **(a)** $g(x) = 3\sin x$, $h(x) = x^2$ **(b)** $g(x) = 3x^2 + 4x$, $h(x) = \sin x$

51. **(a)** $f(x) = x^3$, $g(x) = 1 + \sin x$, $h(x) = x^2$ **(b)** $f(x) = \sqrt{x}$, $g(x) = 1 - x$, $h(x) = \sqrt[3]{x}$

52. **(a)** $f(x) = 1/x$, $g(x) = 1 - x$, $h(x) = x^2$ **(b)** $f(x) = |x|$, $g(x) = 5 + x$, $h(x) = 2x$

53.

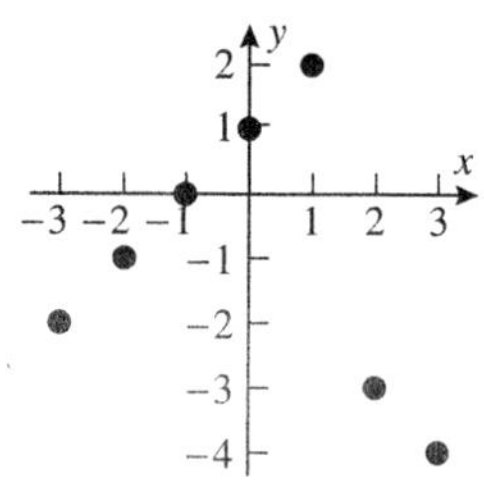

54. $\{-2, -1, 0, 1, 2, 3\}$

55. Note that $f(g(-x)) = f(-g(x)) = f(g(x))$, so $f(g(x))$ is even.

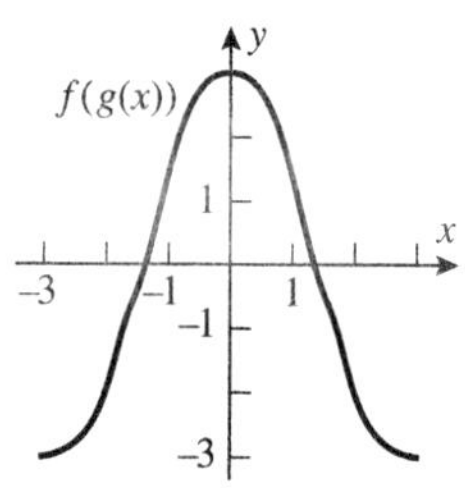

56. Note that $g(f(-x)) = g(f(x))$, so $g(f(x))$ is even.

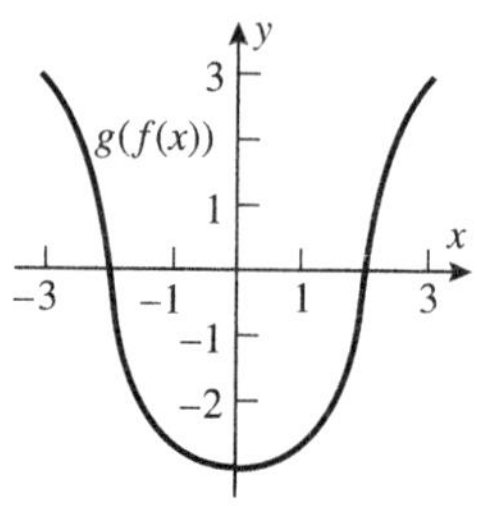

57. $f(g(x)) = 0$ when $g(x) = \pm 2$, so $x = \pm 1.4$; $g(f(x)) = 0$ when $f(x) = 0$, so $x = \pm 2$.

58. $f(g(x)) = 0$ at $x = -1$ and $g(f(x)) = 0$ at $x = -1$

59. $$\frac{3(x+h)^2 - 5 - (3x^2 - 5)}{h} = \frac{6xh + 3h^2}{h} = 6x + 3h;$$

$$\frac{3w^2 - 5 - (3x^2 - 5)}{w - x} = \frac{3(w-x)(w+x)}{w - x} = 3w + 3x$$

60. $$\frac{(x+h)^2 + 6(x+h) - (x^2 + 6x)}{h} = \frac{2xh + h^2 + 6h}{h} = 2x + h + 6;$$

$$\frac{w^2 + 6w - (x^2 + 6x)}{w - x} = w + x + 6$$

61. $$\frac{1/(x+h) - 1/x}{h} = \frac{x - (x+h)}{xh(x+h)} = \frac{-1}{x(x+h)}; \quad \frac{1/w - 1/x}{w - x} = \frac{x - w}{wx(w - x)} = -\frac{1}{xw}$$

62. $$\frac{1/(x+h)^2 - 1/x^2}{h} = \frac{x^2 - (x+h)^2}{x^2h(x+h)^2} = -\frac{2x+h}{x^2(x+h)^2}; \quad \frac{1/w^2 - 1/x^2}{w - x} = \frac{x^2 - w^2}{x^2w^2(w - x)} = -\frac{x + w}{x^2w^2}$$

63. **(a)** the origin **(b)** the x-axis **(c)** the y-axis **(d)** none

64. (a)

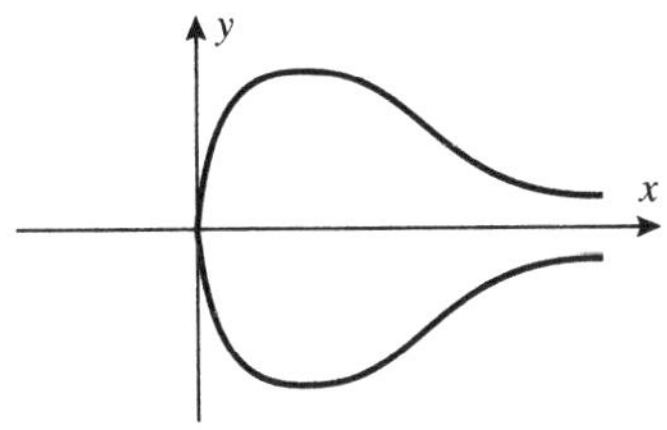

(b)

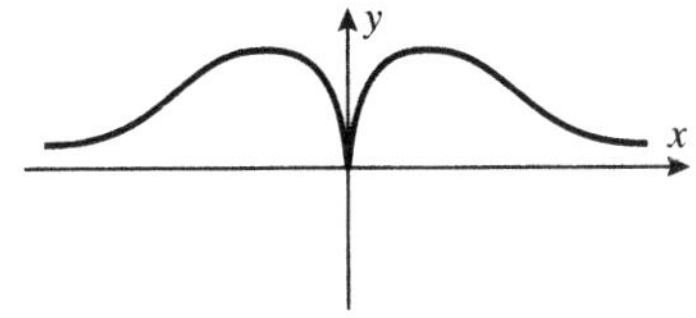

(c)

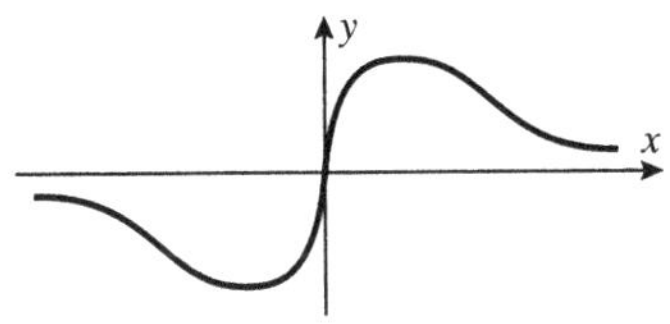

65. (a)

x	-3	-2	-1	0	1	2	3
$f(x)$	1	-5	-1	0	-1	-5	1

(b)

x	-3	-2	-1	0	1	2	3
$f(x)$	1	5	-1	0	1	-5	-1

66. (a)

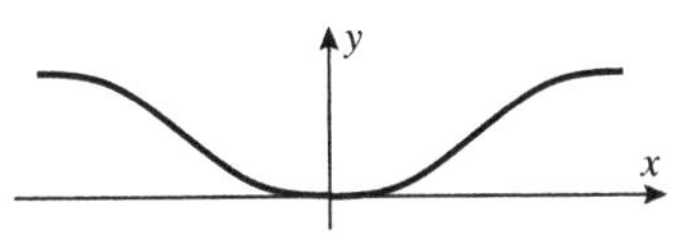

(b)

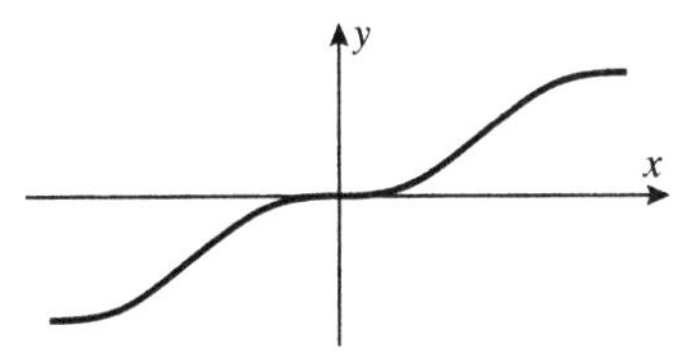

67. (a) even (b) odd (c) odd (d) neither

68. neither; odd; even

69. (a) $f(-x) = (-x)^2 = x^2 = f(x)$, even (b) $f(-x) = (-x)^3 = -x^3 = -f(x)$, odd

(c) $f(-x) = |-x| = |x| = f(x)$, even (d) $f(-x) = -x + 1$, neither

(e) $f(-x) = \dfrac{(-x)^3 - (-x)}{1 + (-x)^2} = -\dfrac{x^3 + x}{1 + x^2} = -f(x)$, odd

(f) $f(-x) = 2 = f(x)$, even

70. (a) x-axis, because $x = 5(-y)^2 + 9$ gives $x = 5y^2 + 9$

(b) x-axis, y-axis, and origin, because $x^2 - 2(-y)^2 = 3$, $(-x)^2 - 2y^2 = 3$, and $(-x)^2 - 2(-y)^2 = 3$ all give $x^2 - 2y^2 = 3$

(c) origin, because $(-x)(-y) = 5$ gives $xy = 5$

71. (a) y-axis, because $(-x)^4 = 2y^3 + y$ gives $x^4 = 2y^3 + y$

(b) origin, because $(-y) = \dfrac{(-x)}{3 + (-x)^2}$ gives $y = \dfrac{x}{3 + x^2}$

(c) x-axis, y-axis, and origin because $(-y)^2 = |x| - 5$, $y^2 = |-x| - 5$, and $(-y)^2 = |-x| - 5$ all give $y^2 = |x| - 5$

72.

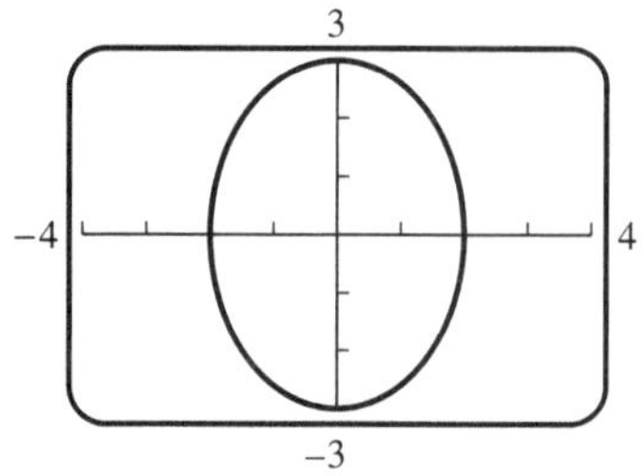

73.

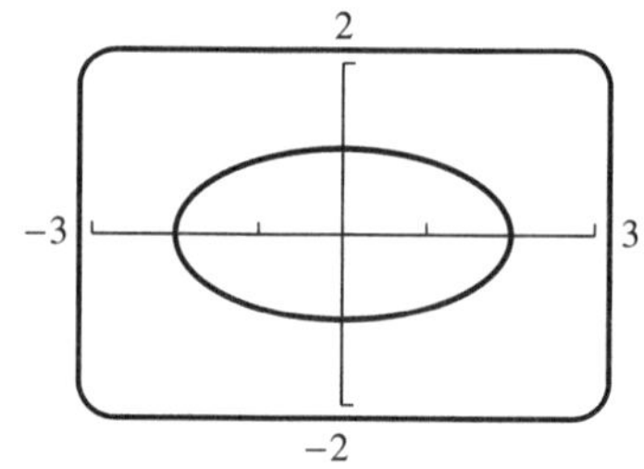

74. **(a)** Whether we replace x with $-x$, y with $-y$, or both, we obtain the same equation, so by Theorem 1.4.3 the graph is symmetric about the x-axis, the y-axis and the origin.

(b) $y = (1 - x^{2/3})^{3/2}$

(c) For quadrant II, the same; for III and IV use $y = -(1 - x^{2/3})^{3/2}$. (For graphing it may be helpful to use the tricks that precede Exercise 29 in Section 1.3.)

75.

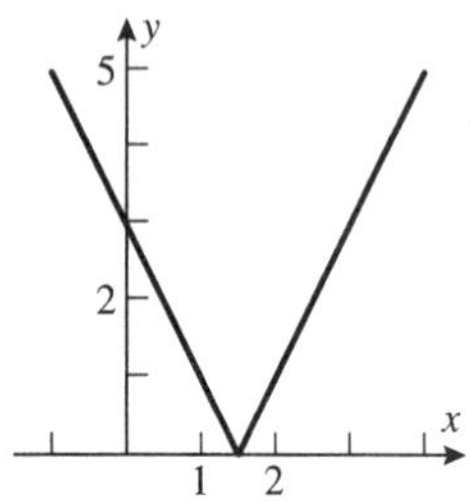

76.

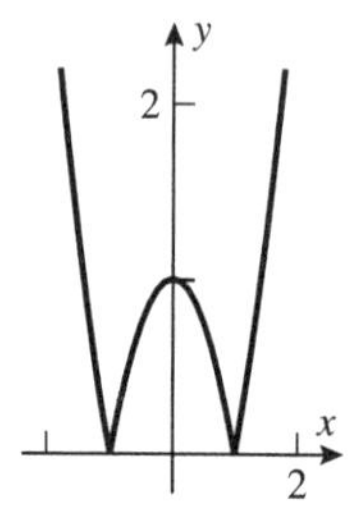

77. **(a)**

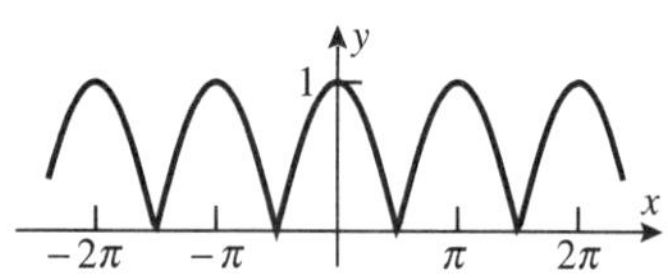

(b)

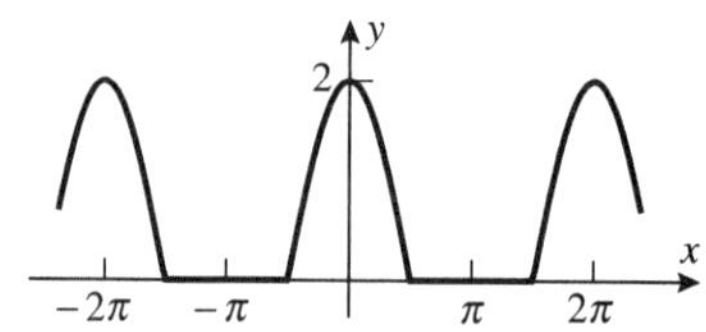

78. **(a)**

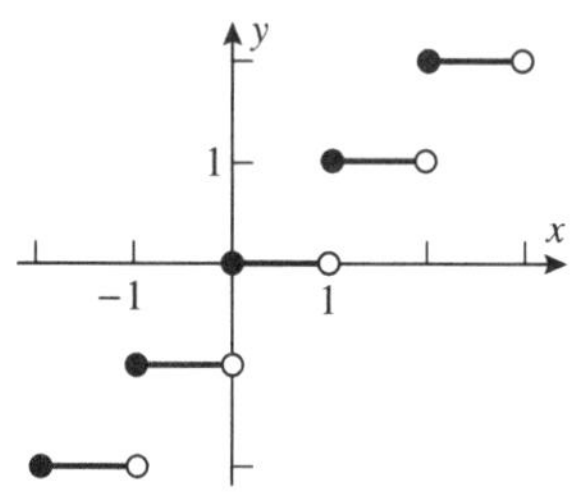

(b)

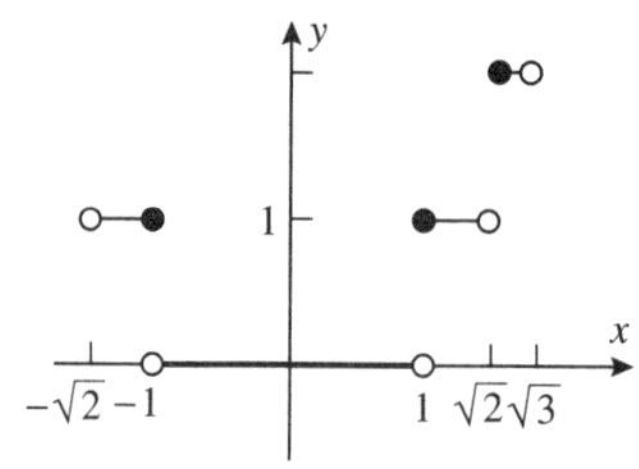

(c)

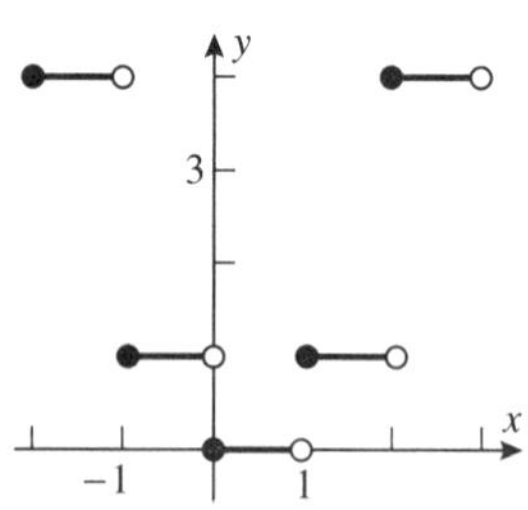

(d)

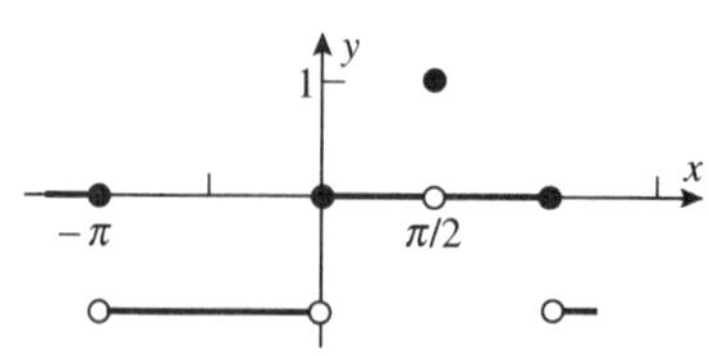

79. Yes, e.g. $f(x) = x^k$ and $g(x) = x^n$ where k and n are integers.

80. If $x \geq 0$ then $|x| = x$ and $f(x) = g(x)$. If $x < 0$ then $f(x) = |x|^{p/q}$ if p is even and $f(x) = -|x|^{p/q}$ if p is odd; in both cases $f(x)$ agrees with $g(x)$.

EXERCISE SET 1.5

1. **(a)** $\dfrac{3-0}{0-2}=-\dfrac{3}{2}, \dfrac{3-(8/3)}{0-6}=-\dfrac{1}{18}, \dfrac{0-(8/3)}{2-6}=\dfrac{2}{3}$

(b) Yes; the first and third slopes above are negative reciprocals of each other.

2. **(a)** $\dfrac{-1-(-1)}{-3-5}=0, \dfrac{-1-3}{5-7}=2, \dfrac{3-3}{7-(-1)}=0, \dfrac{-1-3}{-3-(-1)}=2$

(b) Yes; there are two pairs of equal slopes, so two pairs of parallel lines.

3. III $<$ II $<$ IV $<$ I

4. III $<$ IV $<$ I $<$ II

5. **(a)** $\dfrac{1-(-5)}{1-(-2)}=2, \dfrac{-5-(-1)}{-2-0}=2, \dfrac{1-(-1)}{1-0}=2$. Since the slopes connecting all pairs of points are equal, they lie on a line.

(b) $\dfrac{4-2}{-2-0}=-1, \dfrac{2-5}{0-1}=3, \dfrac{4-5}{-2-1}=\dfrac{1}{3}$. Since the slopes connecting the pairs of points are not equal, the points do not lie on a line.

6. The slope, $m=-2$, is obtained from $\dfrac{y-5}{x-7}$, and thus $y-5=-2(x-7)$.

(a) If $x=9$ then $y=1$.

(b) If $y=12$ then $x=7/2$.

7. The slope, $m=3$, is equal to $\dfrac{y-2}{x-1}$, and thus $y-2=3(x-1)$.

(a) If $x=5$ then $y=14$.

(b) If $y=-2$ then $x=-1/3$.

8. **(a)** Compute the slopes: $\dfrac{y-0}{x-0}=\dfrac{1}{2}$ or $y=x/2$. Also $\dfrac{y-5}{x-7}=2$ or $y=2x-9$. Solve simultaneously to obtain $x=6, y=3$.

9. **(a)** The first slope is $\dfrac{2-0}{1-x}$ and the second is $\dfrac{5-0}{4-x}$. Since they are negatives of each other we get $2(4-x)=-5(1-x)$ or $7x=13$, $x=13/7$.

10. **(a)** $27°$ **(b)** $135°$ **(c)** $63°$ **(d)** $91°$

11. **(a)** $153°$ **(b)** $45°$ **(c)** $117°$ **(d)** $89°$

12. **(a)** $m=\tan\phi=-\sqrt{3}/3$, so $\phi=150°$

(b) $m=\tan\phi=4$, so $\phi=76°$

13. **(a)** $m=\tan\phi=\sqrt{3}$, so $\phi=60°$

(b) $m=\tan\phi=-2$, so $\phi=117°$

14. $y=0$ and $x=0$ respectively

15. $y=-2x+4$

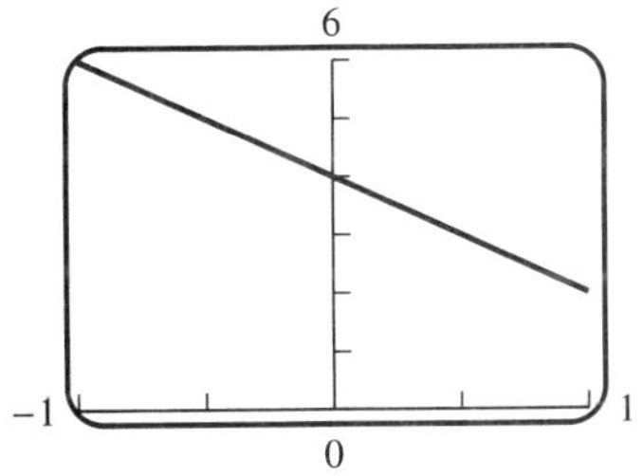

16. $y=5x-3$

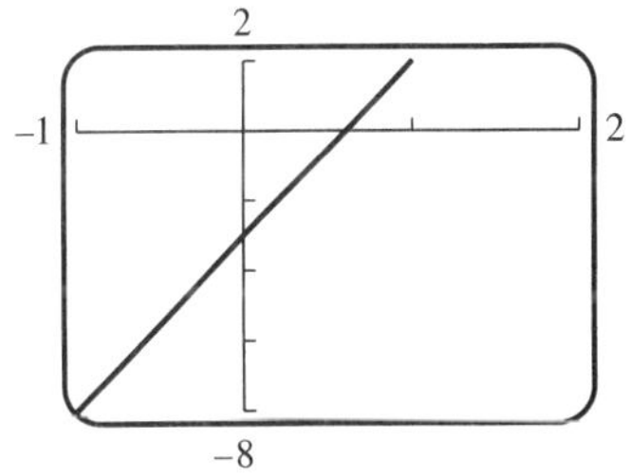

17. Parallel means the lines have equal slopes, so $y = 4x + 7$.

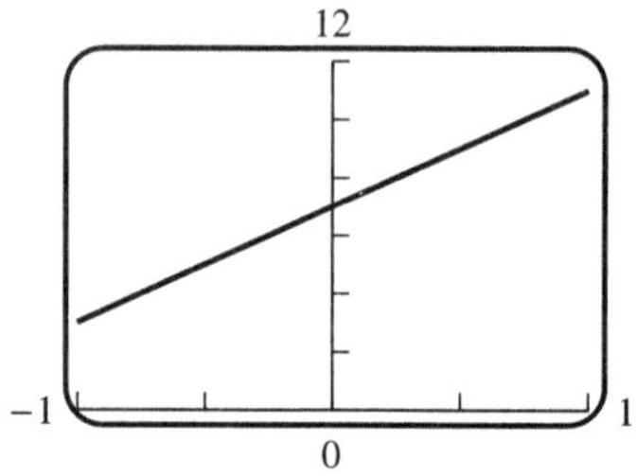

18. The slope of both lines is $-3/2$, so $y - 2 = (-3/2)(x - (-1))$, or $y = -\frac{3}{2}x + \frac{1}{2}$

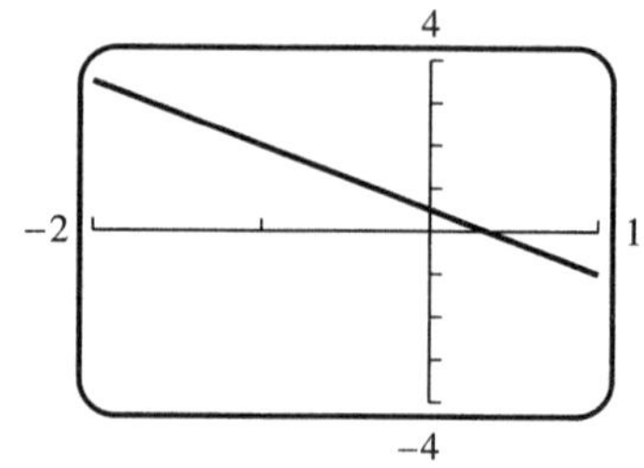

19. The negative reciprocal of 5 is $-1/5$, so $y = -\frac{1}{5}x + 6$.

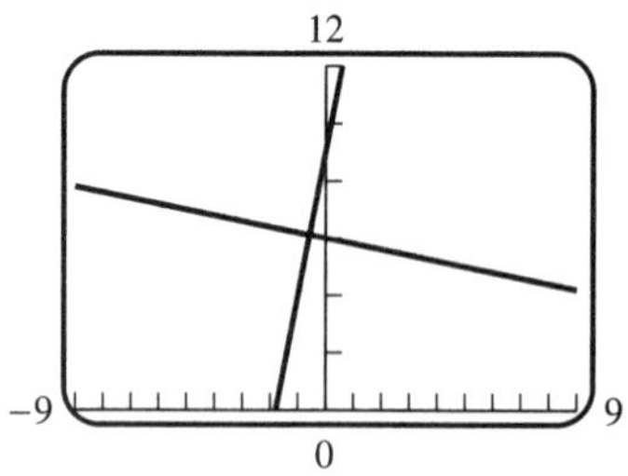

20. The slope of $x - 4y = 7$ is $1/4$ whose negative reciprocal is -4, so $y - (-4) = -4(x - 3)$ or $y = -4x + 8$.

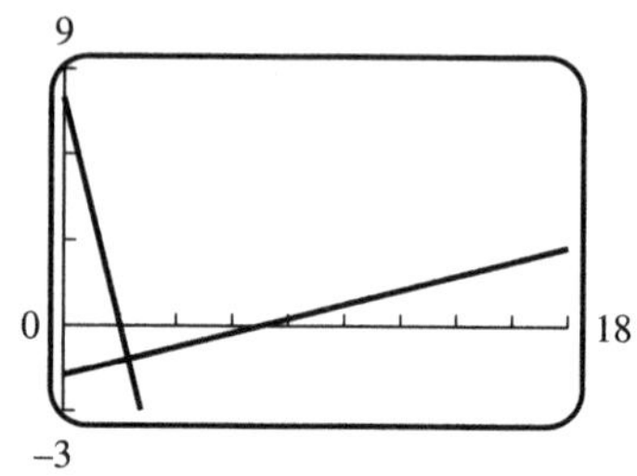

21. $m = \dfrac{4 - (4 - 7)}{2 - 1} = 11$, so $y - (-7) = 11(x - 1)$, or $y = 11x - 18$

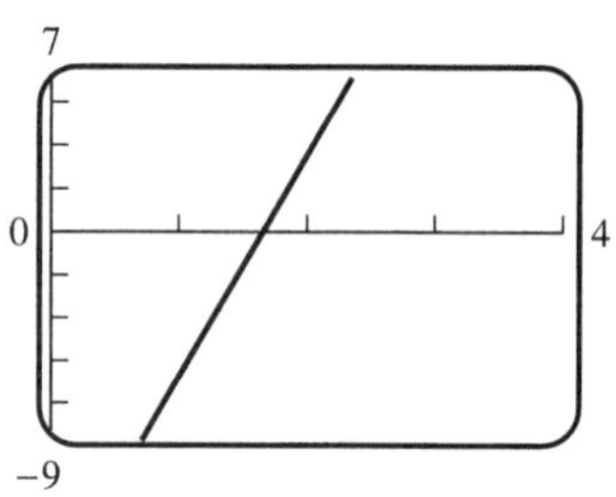

22. $m = \dfrac{6 - 1}{-3 - (-2)} = -5$, so $y - 6 = -5(x - (-3))$, or $y = -5x - 9$

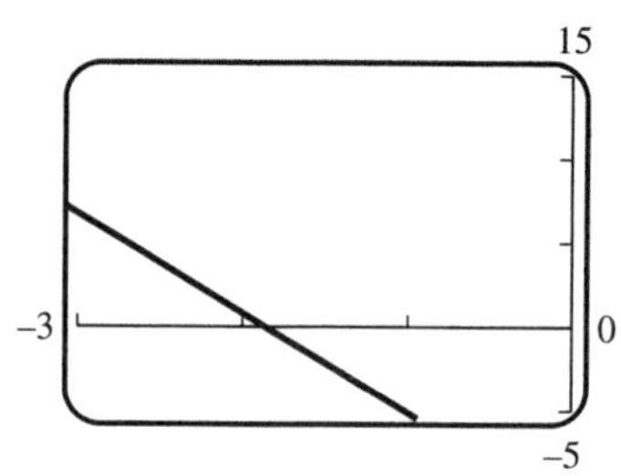

23. **(a)** $m_1 = m_2 = 4$, parallel **(b)** $m_1 = 2 = -1/m_2$, perpendicular
(c) $m_1 = m_2 = 5/3$, parallel
(d) If $A \neq 0$ and $B \neq 0$ then $m_1 = -A/B = -1/m_2$, perpendicular; if $A = 0$ or $B = 0$ (not both) then one line is horizontal, the other vertical, so perpendicular.
(e) neither

24. **(a)** $m_1 = m_2 = -5$, parallel **(b)** $m_1 = 2 = -1/m_2$, perpendicular
(c) $m_1 = -4/5 = -1/m_2$, perpendicular **(d)** If $B \neq 0$, $m_1 = m_2 = -A/B$; if $B = 0$ both are vertical, so parallel
(e) neither

25. **(a)** $m = (0 - (-3))/(2 - 0)) = 3/2$ so $y = 3x/2 - 3$
(b) $m = (-3 - 0)/(4 - 0) = -3/4$ so $y = -3x/4$

26. **(a)** $m = (0 - 2)/(2 - 0)) = -1$ so $y = -x + 2$
(b) $m = (2 - 0)/(3 - 0) = 2/3$ so $y = 2x/3$

27. **(a)** The velocity is the slope, which is $\frac{5-(-4)}{10-0} = 9/10$ ft/s.

(b) $x = -4$

(c) The line has slope 9/10 and passes through $(0, -4)$, so has equation $x = 9t/10 - 4$; at $t = 2$, $x = -2.2$.

(d) $t = 80/9$

28. **(a)** $v = \frac{5-1}{4-2} = 2$ m/s **(b)** $x - 1 = 2(t-2)$ or $x = 2t - 3$ **(c)** $x = -3$

29. **(a)** The acceleration is the slope of the velocity, so $a = \frac{3-(-1)}{1-4} = -\frac{4}{3}$ ft/s^2.

(b) $v - 3 = -\frac{4}{3}(t-1)$, or $v = -\frac{4}{3}t + \frac{13}{3}$ **(c)** $v = \frac{13}{3}$ ft/s

30. **(a)** The acceleration is the slope of the velocity, so $a = \frac{0-5}{10-0} = -\frac{5}{10} = -\frac{1}{2}$ ft/s^2.

(b) $v = 5$ ft/s **(c)** $v = 4$ ft/s **(d)** $t = 4$ s

31. **(a)** It moves (to the left) 6 units with velocity $v = -3$ cm/s, then remains motionless for 5 s, then moves 3 units to the left with velocity $v = -1$ cm/s.

(b) $v_{\text{ave}} = \frac{0-9}{10-0} = -\frac{9}{10}$ cm/s

(c) Since the motion is in one direction only, the speed is the negative of the velocity, so $s_{\text{ave}} = \frac{9}{10}$ cm/s.

32. It moves right with constant velocity $v = 5$ km/h; then accelerates; then moves with constant, though increased, velocity again; then slows down.

33. **(a)** If x_1 denotes the final position and x_0 the initial position, then $v = (x_1 - x_0)/(t_1 - t_0) = 0$ mi/h, since $x_1 = x_0$.

(b) If the distance traveled in one direction is d, then the outward journey took $t = d/40$ h. Thus $s_{\text{ave}} = \frac{\text{total dist}}{\text{total time}} = \frac{2d}{t + (2/3)t} = \frac{80t}{t + (2/3)t} = 48$ mi/h.

(c) $t + (2/3)t = 5$, so $t = 3$ and $2d = 80t = 240$ mi round trip

34. **(a)** down, since $v < 0$ **(b)** $v = 0$ at $t = 2$ **(c)** It's constant at 32 ft/s^2.

35. **(a)**

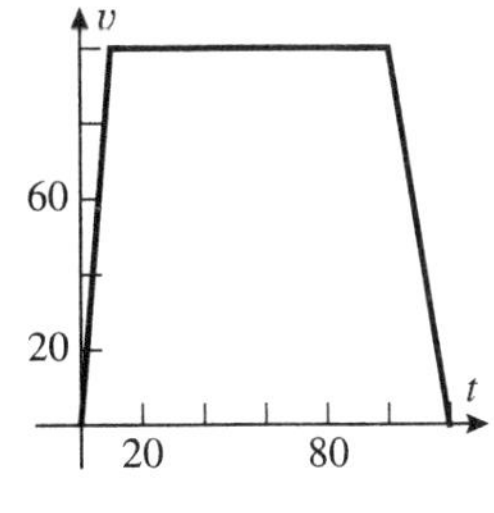

(b) $v = \begin{cases} 10t & \text{if } 0 \le t \le 10 \\ 100 & \text{if } 10 \le t \le 100 \\ 600 - 5t & \text{if } 100 \le t \le 120 \end{cases}$

36.

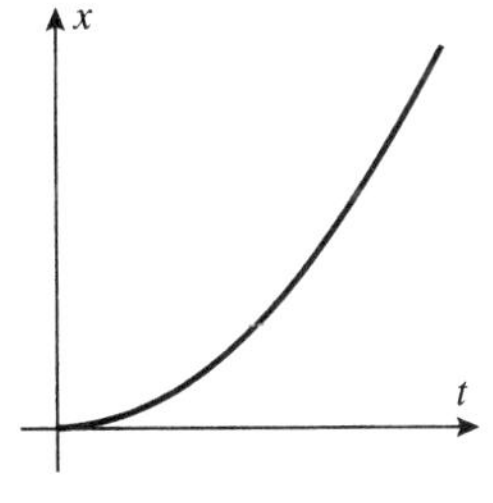

37. **(a)** $y = 20 - 15 = 5$ when $x = 45$, so $5 = 45k$, $k = 1/9$, $y = x/9$

(b)

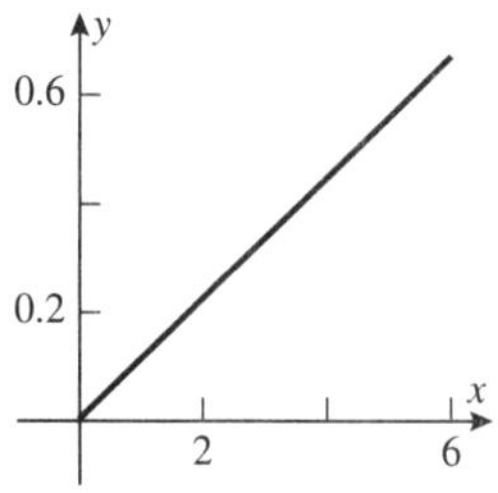

(c) $l = 15 + y = 15 + 100(1/9) = 26.11$ in.

(d) If $y_{\text{max}} = 15$ then solve $15 = kx = x/9$ for $x = 135$ lb.

38. **(a)** Since $y = 0.2 = (1)k$, $k = 1/5$ and $y = x/5$

(b)

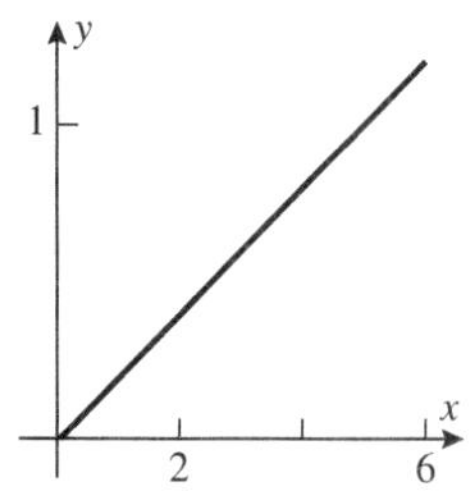

(c) $y = 3k = 3/5$ so 0.6 ft.

(d) $y_{\text{max}} = (1/2)3 = 1.5$ so solve $1.5 = x/5$ for $x = 7.5$ tons

39. Each increment of 1 in the value of x yields the increment of 1.2 for y, so the relationship is linear. If $y = mx + b$ then $m = 1.2$; from $x = 0$, $y = 2$, follows $b = 2$, so $y = 1.2x + 2$

40. Each increment of 1 in the value of x yields the increment of -2.1 for y, so the relationship is linear. If $y = mx + b$ then $m = -2.1$; from $x = 0$, $y = 10.5$ follows $b = 10.5$, so $y = -2.1x + 10.5$

41. **(a)** With T_F as independent variable, we have $\dfrac{T_C - 100}{T_F - 212} = \dfrac{0 - 100}{32 - 212}$, so $T_C = \dfrac{5}{9}(T_F - 32)$.

(b) 5/9

(c) Set $T_F = T_C = \frac{5}{9}(T_F - 32)$ and solve for T_F: $T_F = T_C = -40°$ (F or C).

(d) 37° C

42. **(a)** One degree Celsius is one degree Kelvin, so the slope is the ratio $1/1 = 1$. Thus $T_C = T_K - 273.15$.

(b) $T_C = 0 - 273.15 = -273.15°$ C

43. **(a)** $\dfrac{p - 1}{h - 0} = \dfrac{5.9 - 1}{50 - 0}$, or $p = 0.098h + 1$ **(b)** when $p = 2$, or $h = 1/0.098 \approx 10.20$ m

44. **(a)** $\dfrac{R - 123.4}{T - 20} = \dfrac{133.9 - 123.4}{45 - 20}$, so $R = 0.42T + 115$. **(b)** $T = 32.38°$C

45. **(a)** $\dfrac{r - 0.80}{t - 0} = \dfrac{0.75 - 0.80}{4 - 0}$, so $r = -0.0125t + 0.8$ **(b)** 64 days

46. **(a)** Let the position at rest be y_0. Then $y_0 + y = y_0 + kx$; with $x = 11$ we get $y_0 + kx = y_0 + 11k = 40$, and with $x = 24$ we get $y_0 + kx = y_0 + 24k = 60$. Solve to get $k = 20/13$ and $y_0 = 300/13$.

(b) $300/13 + (20/13)W = 30$, so $W = (390 - 300)/20 = 9/2$ g.

47. **(a)** For x trips we have $C_1 = 2x$ and $C_2 = 25 + x/4$

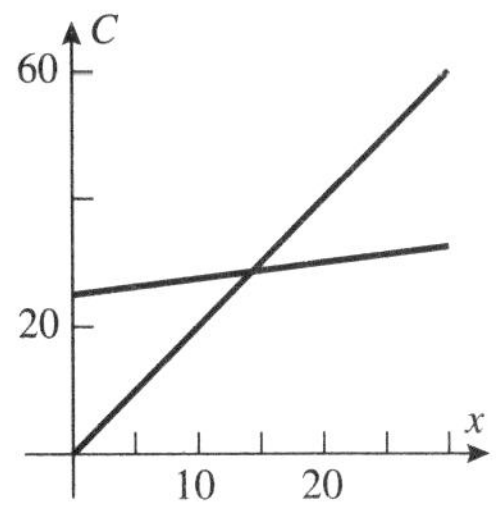

(b) $2x = 25 + x/4$, or $x = 100/7$, so the commuter pass becomes worthwhile at $x = 15$.

48. If the student drives x miles, then the total costs would be $C_A = 4000 + (1.25/20)x$ and $C_B = 5500 + (1.25/30)x$. Set $4000 + 5x/80 = 5500 + 5x/120$ and solve for $x = 72,000$ mi.

EXERCISE SET 1.6

1. **(a)** $y = 3x + b$ **(b)** $y = 3x + 6$ **(c)**

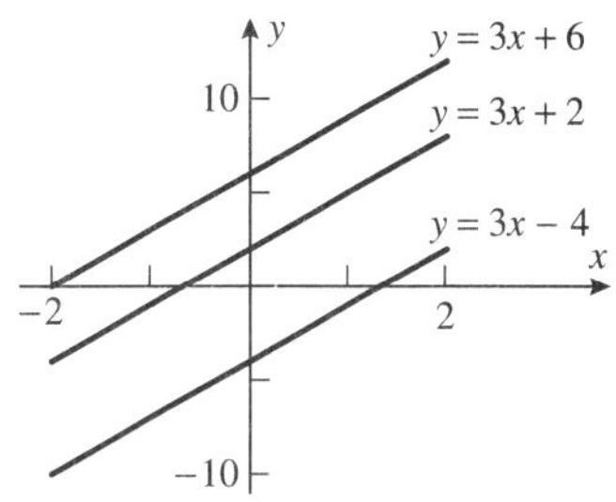

2. Since the slopes are negative reciprocals, $y = -\frac{1}{3}x + b$.

3. **(a)** $y = mx + 2$

(b) $m = \tan\phi = \tan 135° = -1$, so $y = -x + 2$

(c)

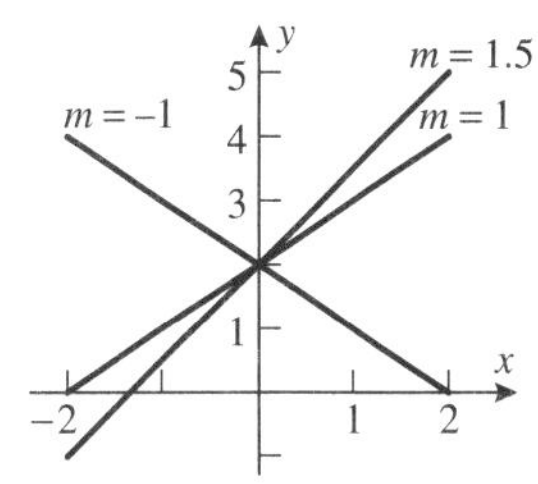

4. **(a)** $y = mx$ **(b)** $y = m(x - 1)$

(c) $y = -2 + m(x - 1)$ **(d)** $2x + 4y = C$

5. **(a)** The slope is -1.

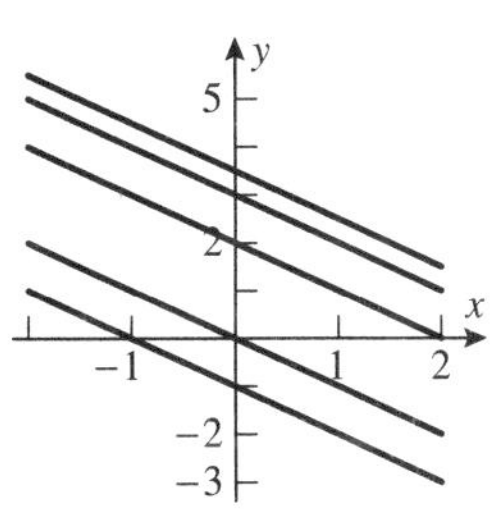

(b) The y-intercept is $y = -1$.

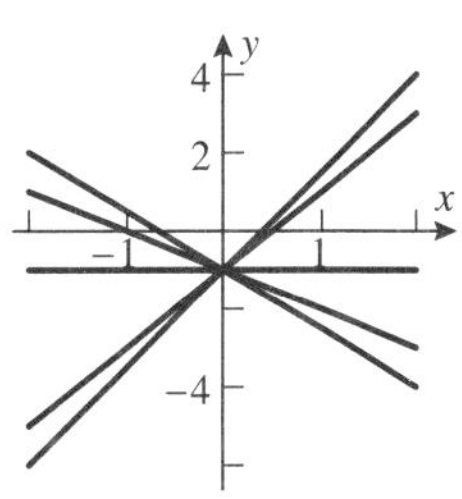

(c) They pass through the point $(-4, 2)$.

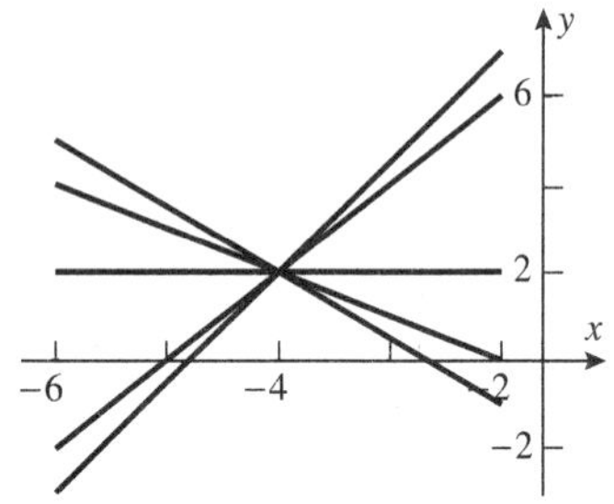

(d) The x-intercept is $x = 1$.

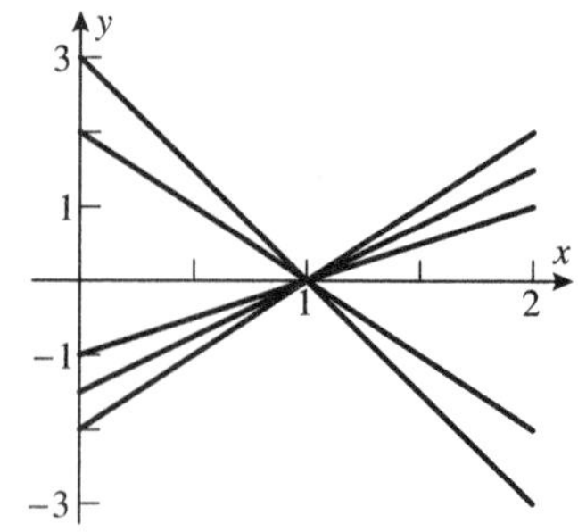

6. **(a)** horizontal lines

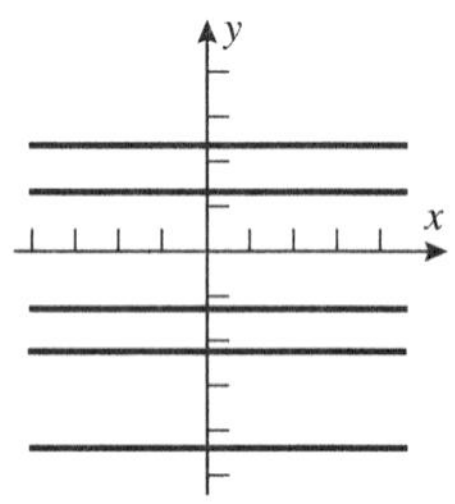

(b) The y-intercept is $y = -1/2$.

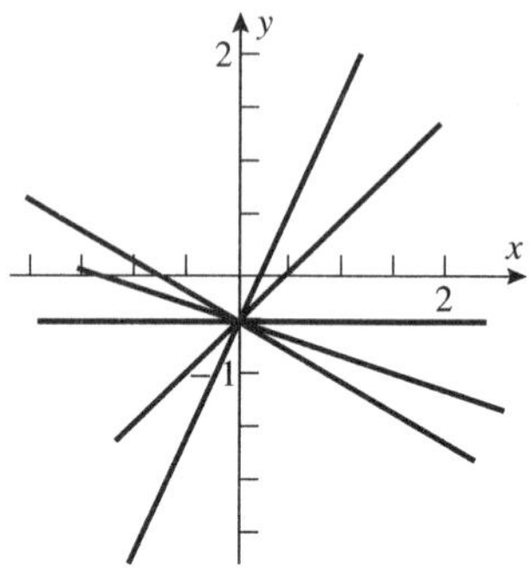

(c) The x-intercept is $x = -1/2$.

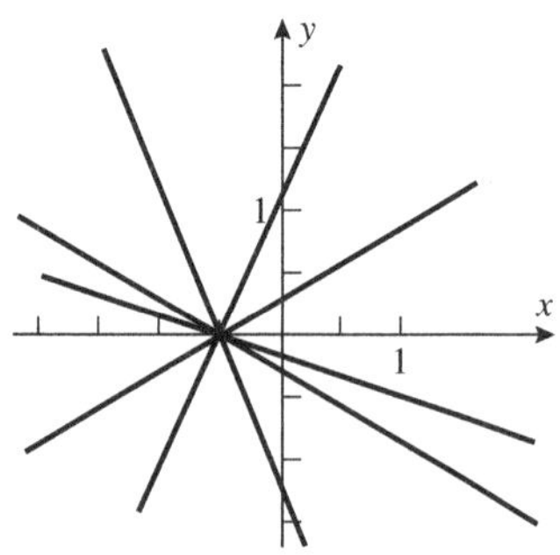

(d) They pass through $(-1, 1)$.

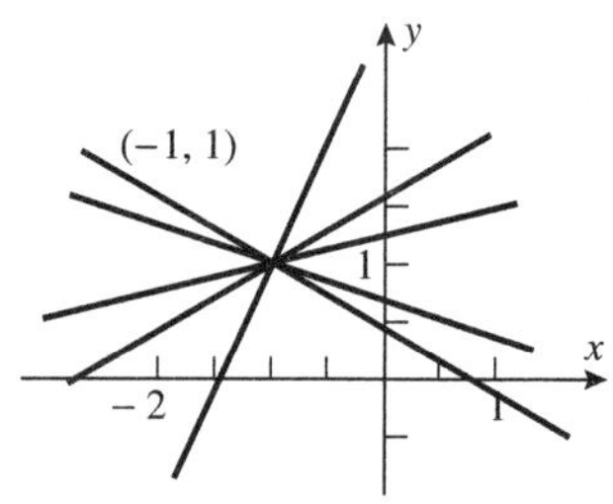

7. Let the line be tangent to the circle at the point (x_0, y_0) where $x_0^2 + y_0^2 = 9$. The slope of the tangent line is the negative reciprocal of y_0/x_0 (why?), so $m = -x_0/y_0$ and $y = -(x_0/y_0)x + b$. Substituting the point (x_0, y_0) as well as $y_0 = \pm\sqrt{9 - x_0^2}$ we get $y = \pm\dfrac{9 - x_0 x}{\sqrt{9 - x_0^2}}$.

8. Solve the simultaneous equations to get the point $(-2, 1/3)$ of intersection. Then $y = \frac{1}{3} + m(x + 2)$.

9. The x-intercept is $x = 10$ so that with depreciation at 10% per year the final value is always zero, and hence $y = m(x - 10)$. The y-intercept is the original value.

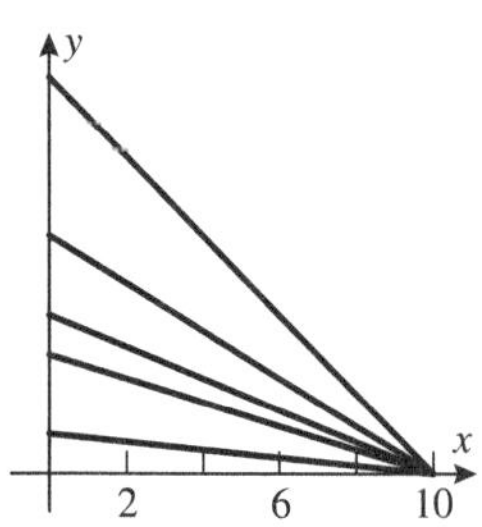

10. A line through $(6, -1)$ has the form $y + 1 = m(x - 6)$. The intercepts are $x = 6 + 1/m$ and $y = -6m - 1$. Set $-(6 + 1/m)(6m + 1) = 3$, or $36m^2 + 15m + 1 = (12m + 1)(3m + 1) = 0$ with roots $m = -1/12, -1/3$; thus $y + 1 = -(1/3)(x - 6)$ and $y + 1 = -(1/12)(x - 6)$.

11. **(a)** VI **(b)** IV **(c)** III **(d)** V **(e)** I **(f)** II

12. In all cases k must be positive, or negative values would appear in the chart. Only kx^{-3} decreases, so that must be $f(x)$. Next, kx^2 grows faster than $kx^{3/2}$, so that would be $g(x)$, which grows faster than $h(x)$ (to see this, consider ratios of successive values of the functions). Finally, experimentation (a spreadsheet is handy) for values of k yields (approximately) $f(x) = 10x^{-3}$, $g(x) = x^2/2$, $h(x) = 2x^{1.5}$.

13. **(a)**

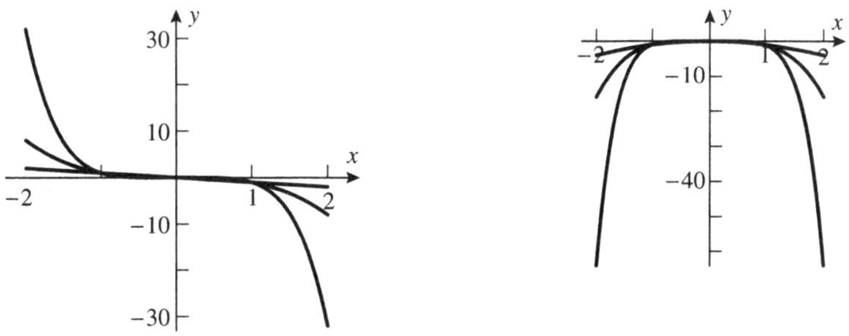

(b)

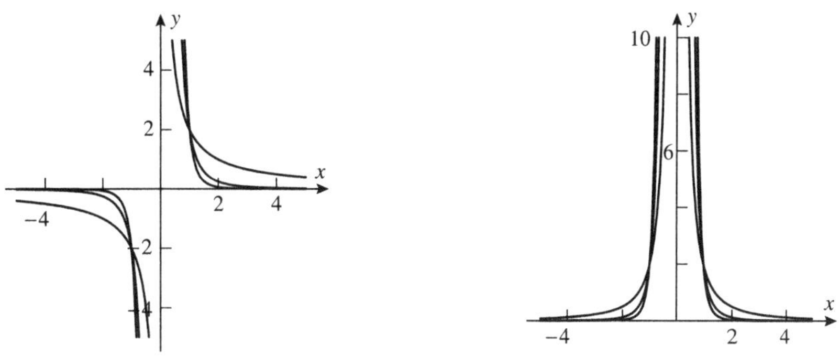

(c)

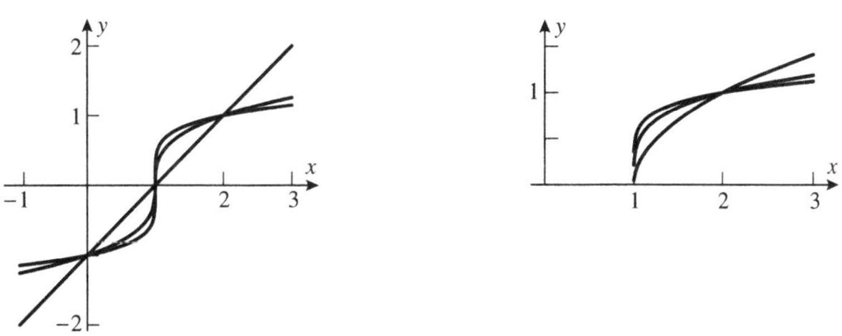

14. **(a)**

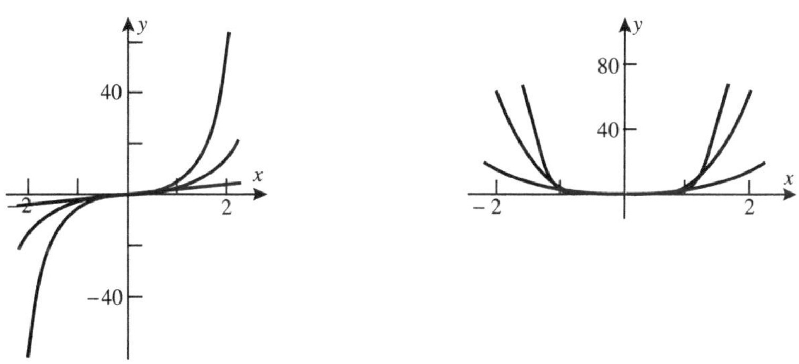

(b)

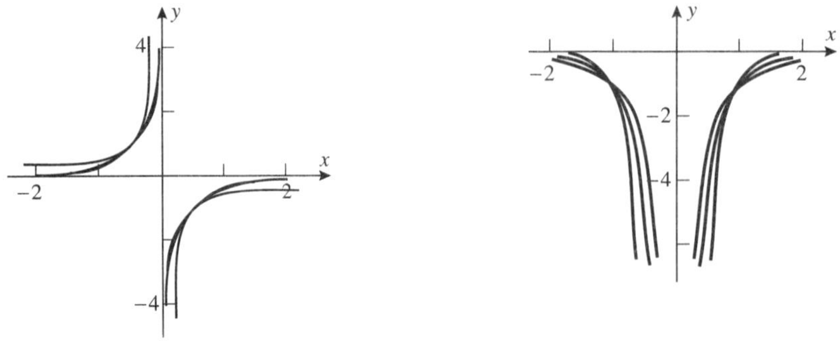

(c)

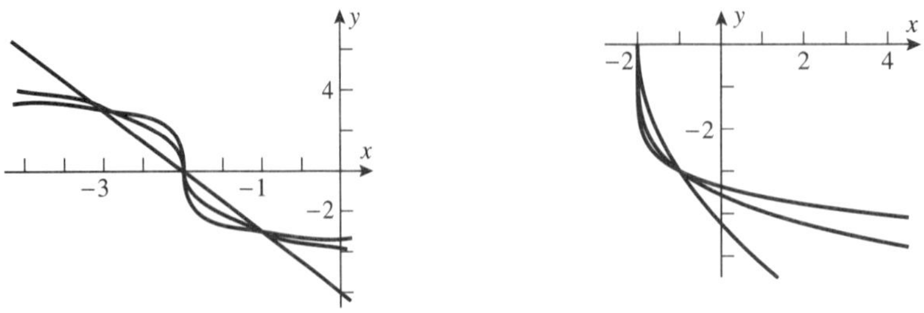

15. (a)

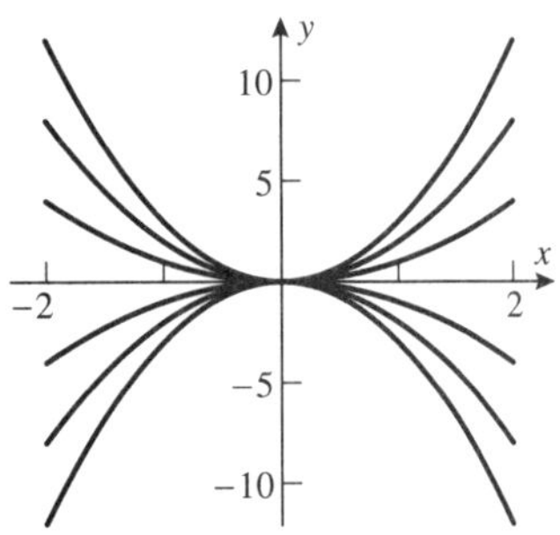

(b)

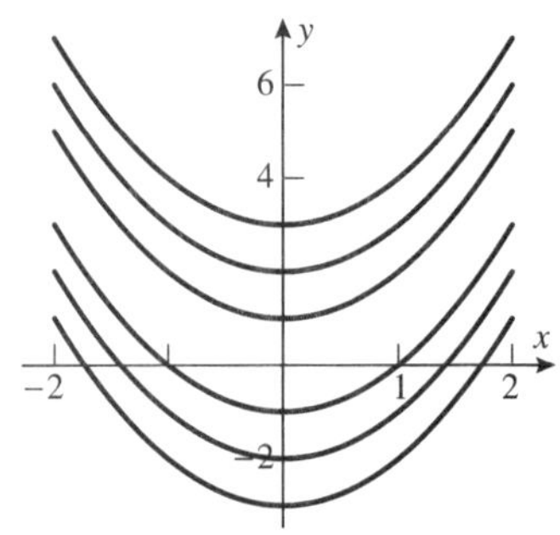

(c)

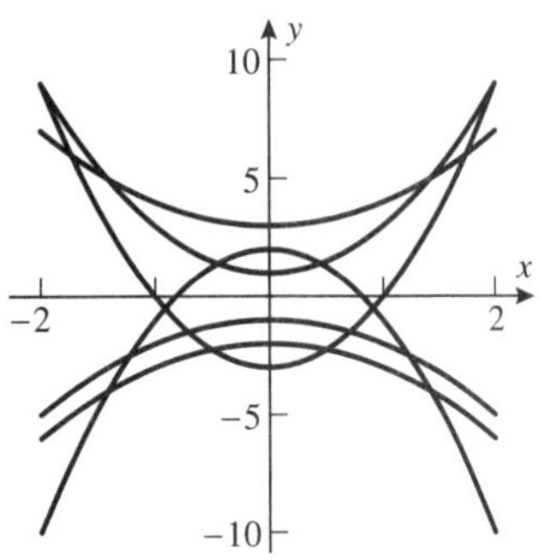

16. (a)

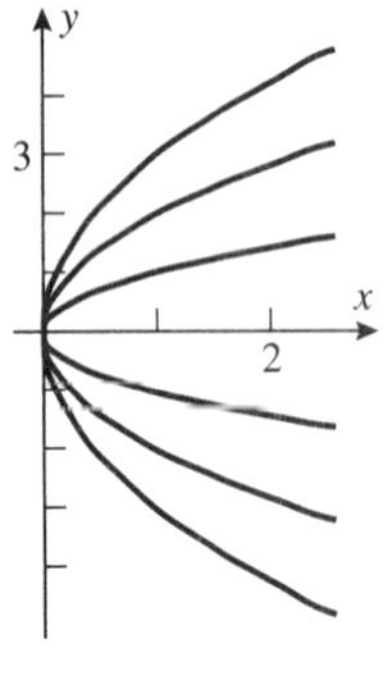

(b)

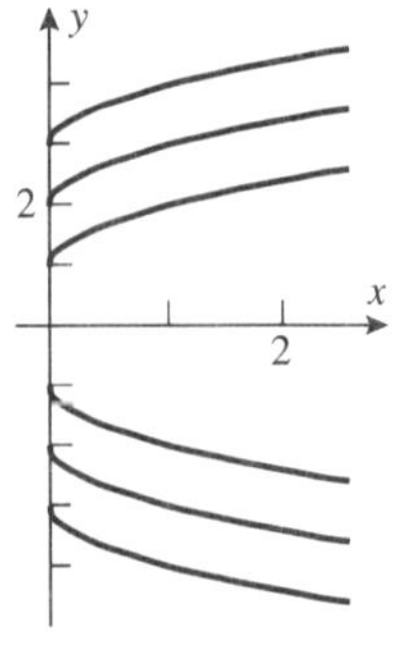

(c)

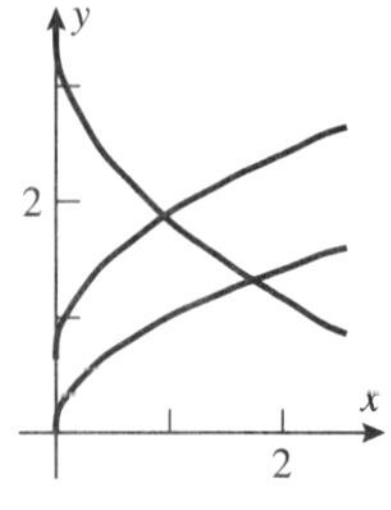

17. (a)

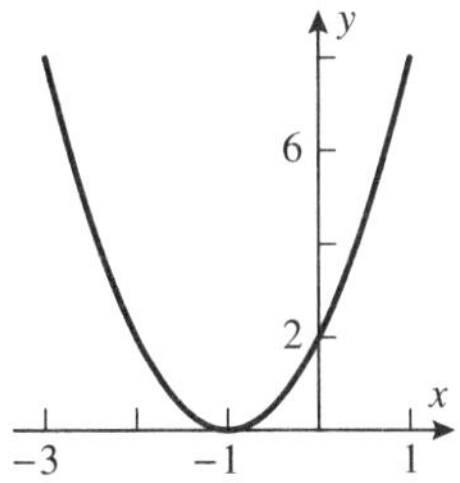

(b)

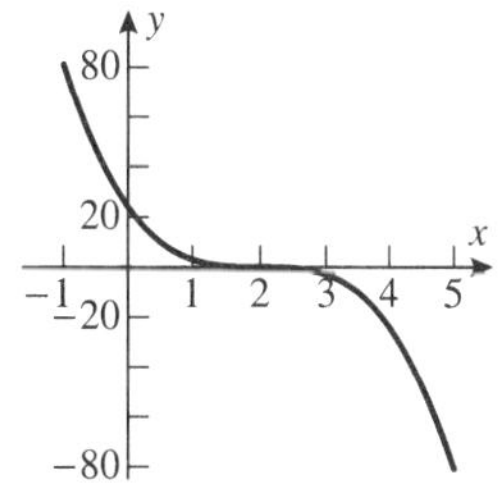

(c)

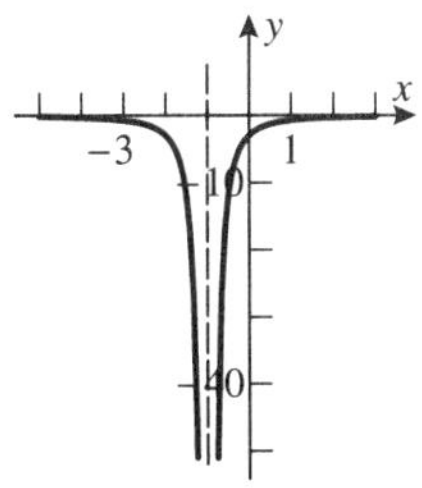

(d)

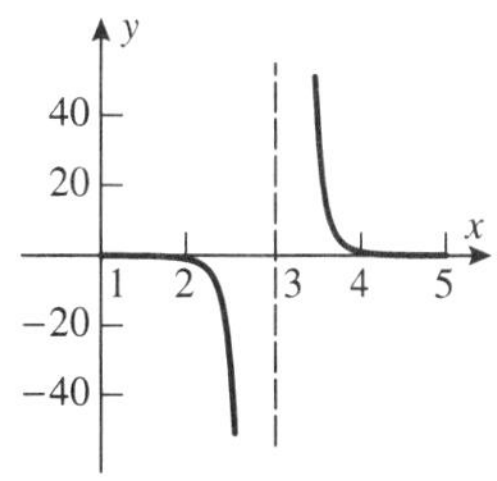

18. (a)

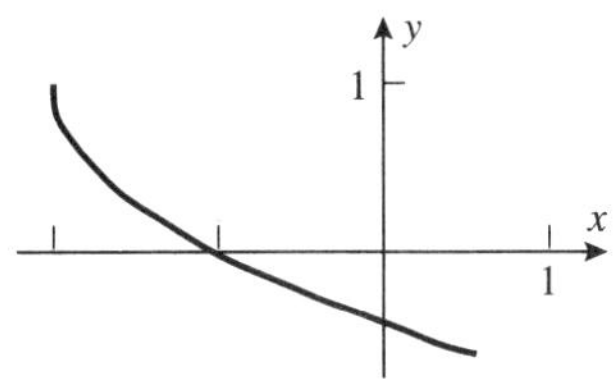

(b)

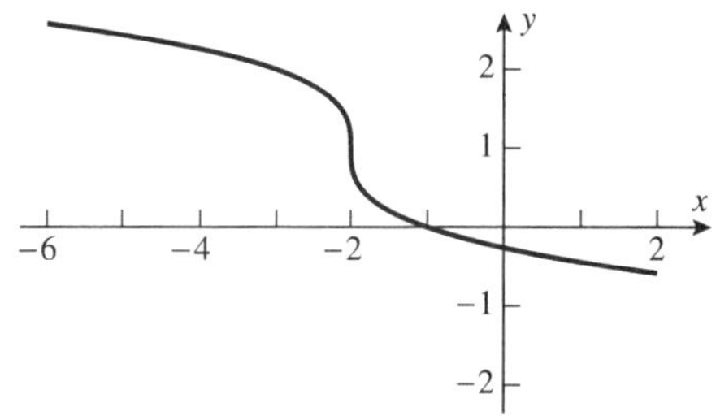

(c)

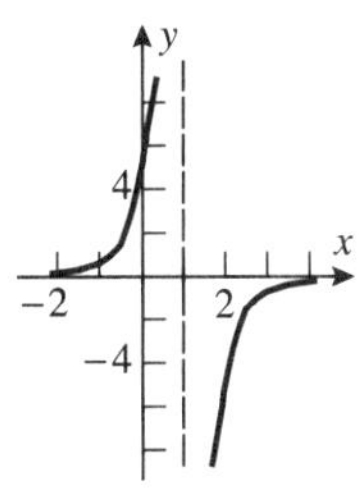

(d)

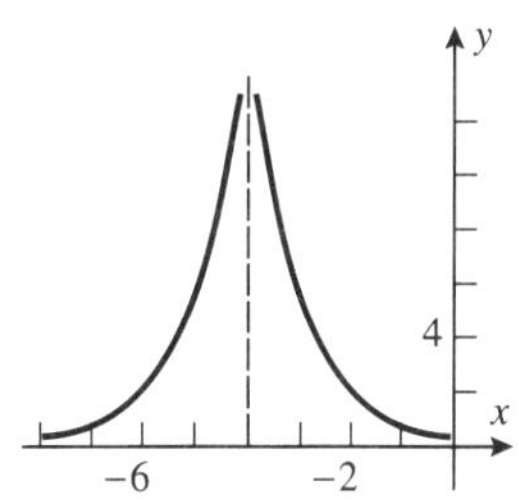

19. (a)

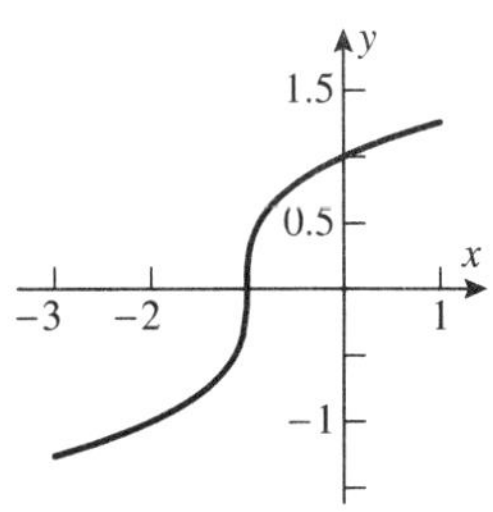

(b)

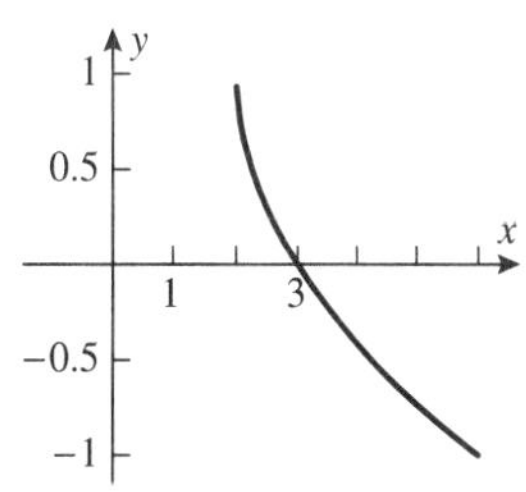

(c)

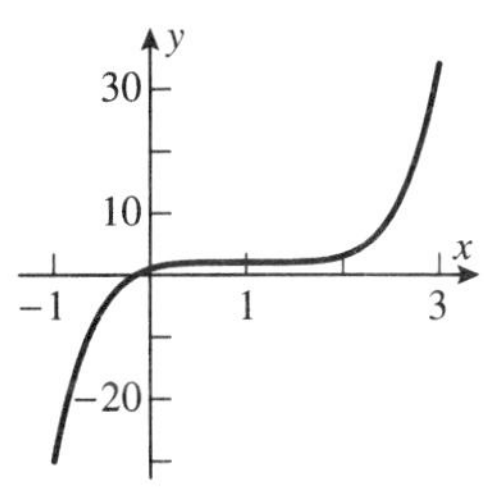

(d)

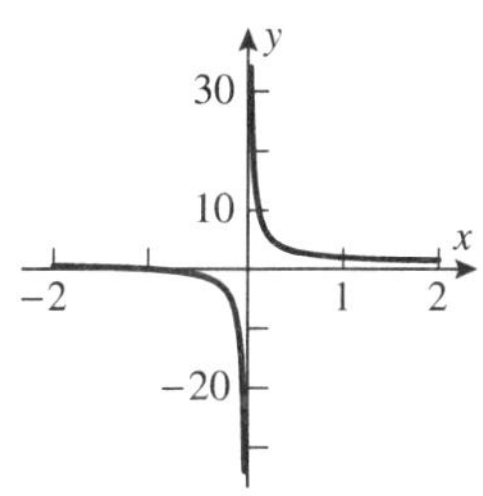

20. (a)

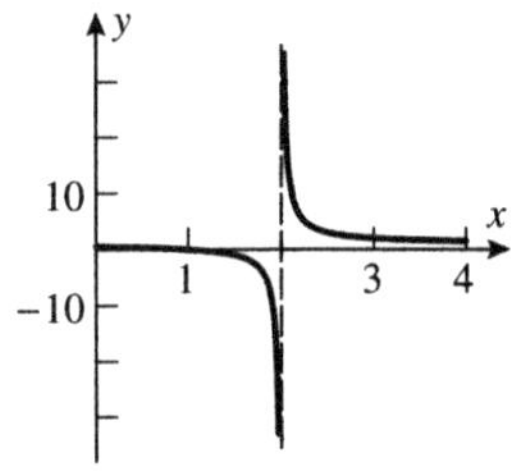

(b)

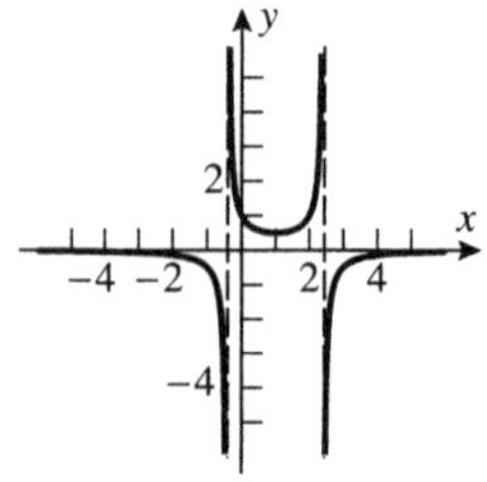

(c)

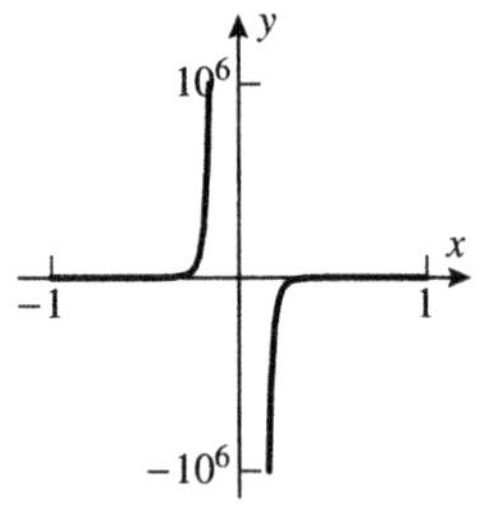

(d)

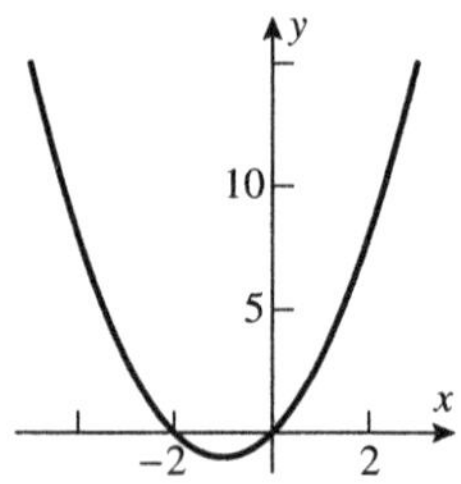

21. $y = x^2 + 2x = (x+1)^2 - 1$

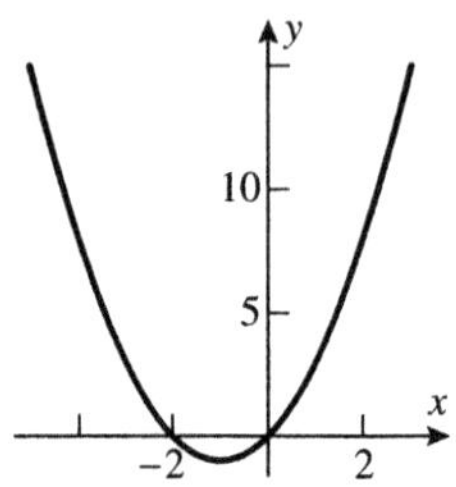

22. (a)

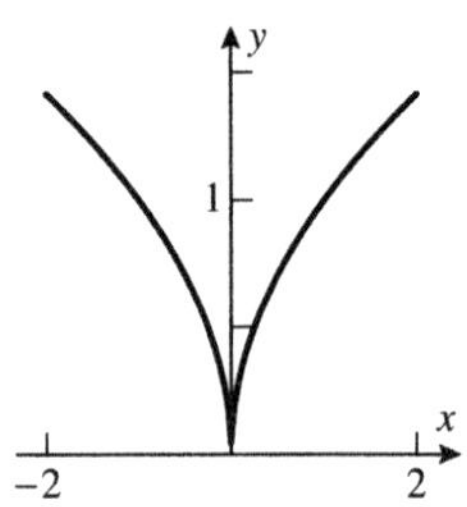

(b)

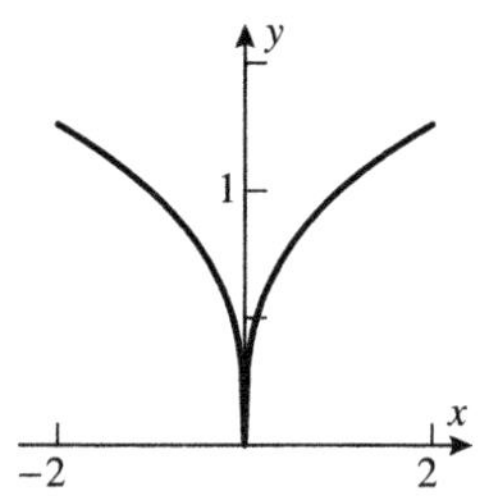

23. (a) N·m

(b) k = 20 N·m

(c)

V(L)	0.25	0.5	1.0	1.5	2.0
P (N/m^2)	80×10^3	40×10^3	20×10^3	13.3×10^3	10×10^3

(d)

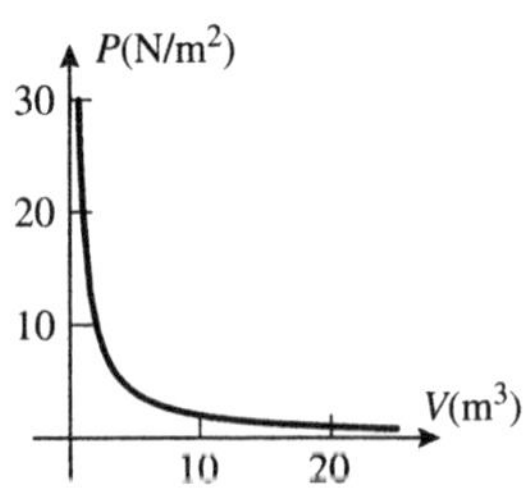

24. If the side of the square base is x and the height of the container is y then $V = x^2y = 100$; minimize $A = 2x^2 + 4xy = 2x^2 + 400/x$. A graphing utility with a zoom feature suggests that the solution is a cube of side $100^{\frac{1}{3}}$ cm.

25. (a) $F = k/x^2$ so $0.0005 = k/(0.3)^2$ and $k = 0.000045$ N·m^2.

(b) $F = 0.000005$ N

(c)

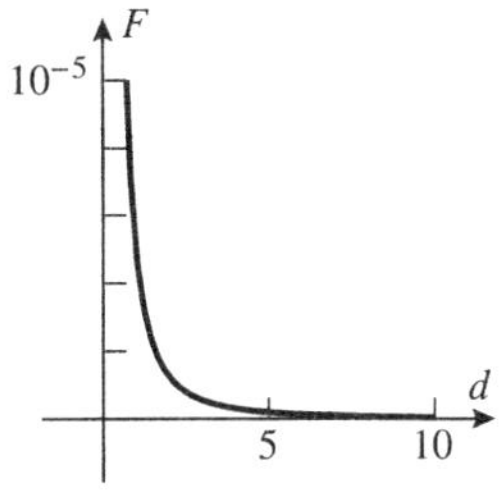

(d) When they approach one another, the force becomes infinite; when they get far apart it tends to zero.

26. (a) $2000 = C/(4000)^2$, so $C = 3.2 \times 10^{10}$ lb·mi^2

(b) $W = C/5000^2 = (3.2 \times 10^{10})/(25 \times 10^6) = 1280$ lb.

(c)

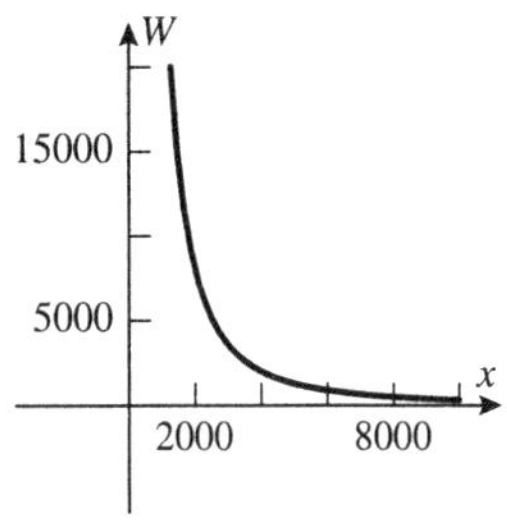

(d) No, but W is very small when x is large.

27. (a) II; $y = 1$, $x = -1, 2$

(b) I; $y = 0$, $x = -2, 3$

(c) IV; $y = 2$

(d) III; $y = 0$, $x = -2$

28. The denominator has roots $x = \pm 1$, so $x^2 - 1$ is the denominator. To determine k use the point $(0, -1)$ to get $k = 1$, $y = 1/(x^2 - 1)$.

29. Order the six trigonometric functions as sin, cos, tan, cot, sec, csc:

(a) pos, pos, pos, pos, pos, pos

(b) neg, zero, undef, zero, undef, neg

(c) pos, neg, neg, neg, neg, pos

(d) neg, pos, neg, neg, pos, neg

(e) neg, neg, pos, pos, neg, neg

(f) neg, pos, neg, neg, pos, neg

30. (a) neg, zero, undef, zero, undef, neg

(b) pos, neg, neg, neg, neg, pos

(c) zero, neg, zero, undef, neg, undef

(d) pos, zero, undef, zero, undef, pos

(e) neg, neg, pos, pos, neg, neg

(f) neg, neg, pos, pos, neg, neg

31. (a) $\sin(\pi - x) = \sin x$; 0.588

(b) $\cos(-x) = \cos x$; 0.924

(c) $\sin(2\pi + x) = \sin x$; 0.588

(d) $\cos(\pi - x) = -\cos x$; -0.924

(e) $\cos^2 x = 1 - \sin^2 x; 0.655$

(f) $\sin^2 2x = 4\sin^2 x \cos^2 x = 4\sin^2 x(1 - \sin^2 x); 0.905$

32. (a) $\sin(3\pi + x) = -\sin x$; -0.588

(b) $\cos(-x - 2\pi) = \cos x$; 0.924

(c) $\sin(8\pi + x) = \sin x$; 0.588

(d) $\sin(x/2) = \pm\sqrt{(1 - \cos x)/2}$; use the negative sign for x small and negative; -0.195

(e) $\cos(3\pi + 3x) = -4\cos^3 x + 3\cos x$; -0.384

(f) $\tan^2 x = \dfrac{1 - \cos^2 x}{\cos^2 x}$; 0.172

33. (a) $-a$ (b) b (c) $-c$ (d) $\pm\sqrt{1-a^2}$
(e) $-b$ (f) $-a$ (g) $\pm 2b\sqrt{1-b^2}$ (h) $2b^2-1$
(i) $1/b$ (j) $-1/a$ (k) $1/c$ (l) $(1-b)/2$

34. (a) The distance is $36/360 = 1/10$th of a great circle, so it is $(1/10)2\pi r = 2,513.27$ mi.
(b) $36/360 = 1/10$

35. If the arc length is x, then solve the ratio $\dfrac{x}{1} = \dfrac{2\pi r}{27.3}$ to get $x \approx 87,458$ km.

36. The distance travelled is equal to the length of that portion of the circumference of the wheel which touches the road, and that is the fraction $225/360$ of a circumference, so a distance of $(225/360)(2\pi)3 = 11.78$ ft

37. The second quarter revolves twice (720°) about its own center.

38. Add r to itself until you exceed $2\pi r$; since $6r < 2\pi r < 7r$, you can cut off 6 pieces of pie, but there's not enough for a full seventh piece. Remaining pie is $\dfrac{2\pi r - 6r}{2\pi r} = 1 - \dfrac{3}{\pi}$ times the original pie.

39. (a) $y = 3\sin(x/2)$ (b) $y = 4\cos 2x$ (c) $y = -5\sin 4x$

40. (a) $y = 1 + \cos \pi x$ (b) $y = 1 + 2\sin x$ (c) $y = -5\cos 4x$

41. (a) $y = \sin(x + \pi/2)$ (b) $y = 3 + 3\sin(2x/9)$ (c) $y = 1 + 2\sin(2(x - \pi/4))$

42. $V = 120\sqrt{2}\sin(120\pi t)$

43. (a) $3, \pi/2, 0$

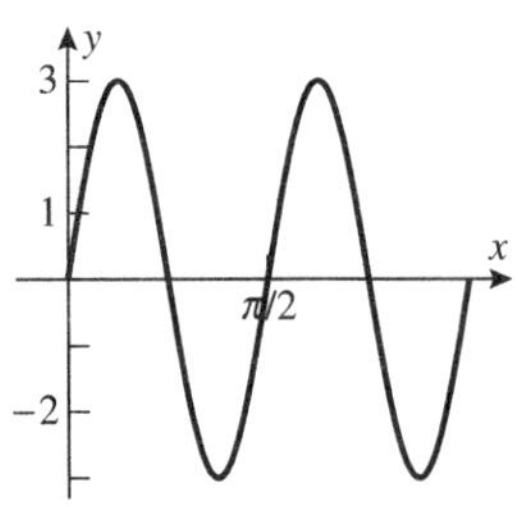

(b) $2, 2, 0$

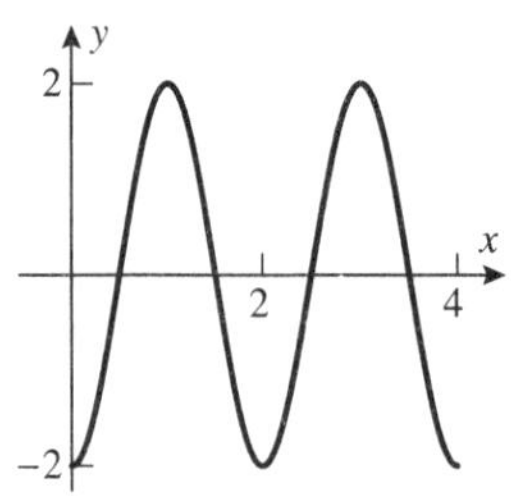

(c) $1, 4\pi, 0$

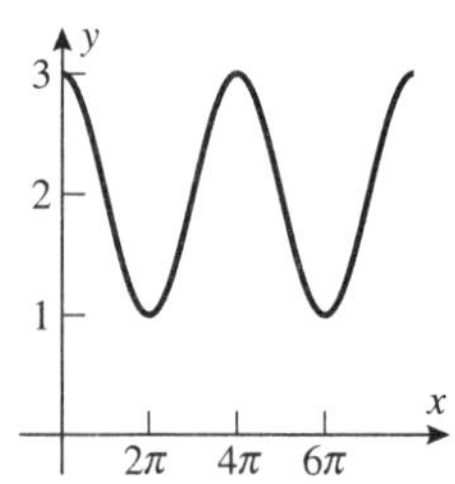

44. (a) $4, \pi, 0$

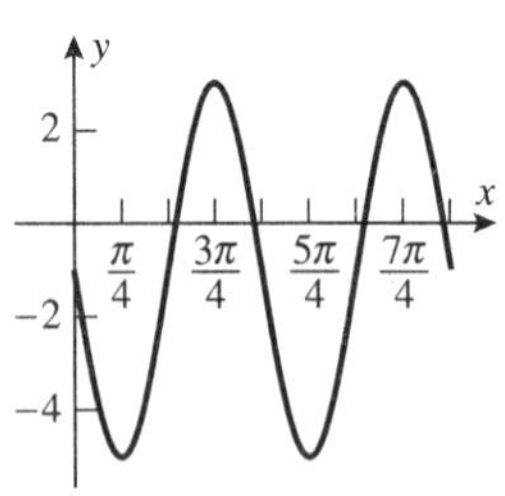

(b) $1/2, 2\pi/3, \pi/3$

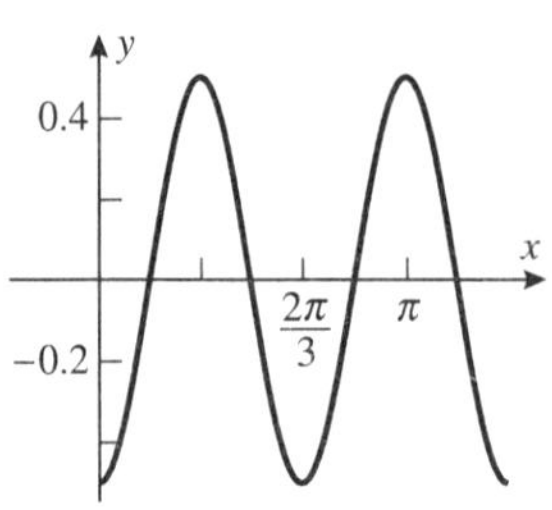

(c) $4, 6\pi, -6\pi$

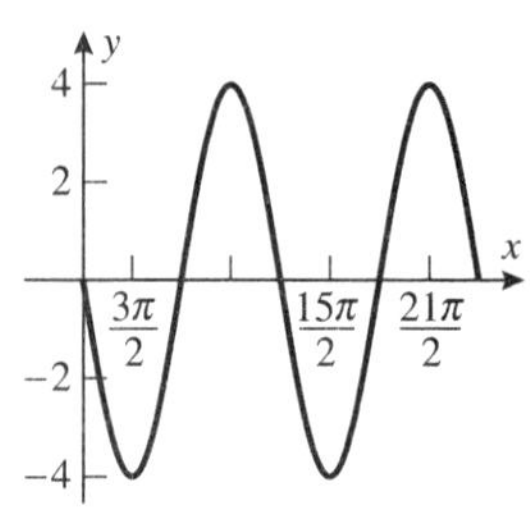

45. (a) $A\sin(\omega t + \theta) = A\sin(\omega t)\cos\theta + A\cos(\omega t)\sin\theta = A_1\sin(\omega t) + A_2\cos(\omega t)$
(b) $A_1 = A\cos\theta$, $A_2 = A\sin\theta$, so $A = \sqrt{A_1^2 + A_2^2}$ and $\theta = \tan^{-1}(A_2/A_1)$.

(c) $A = 5\sqrt{13}/2$, $\theta = \tan^{-1}\dfrac{1}{2\sqrt{3}}$;

$$x = \frac{5\sqrt{13}}{2}\sin\left(2\pi t + \tan^{-1}\frac{1}{2\sqrt{3}}\right)$$

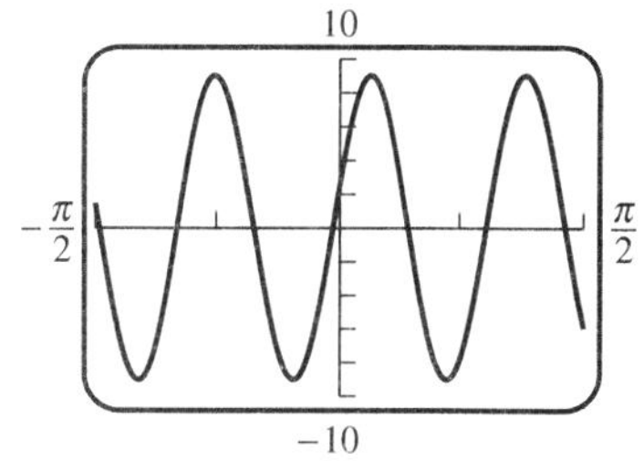

46. three; $x = 0$, $x = \pm 1.8955$

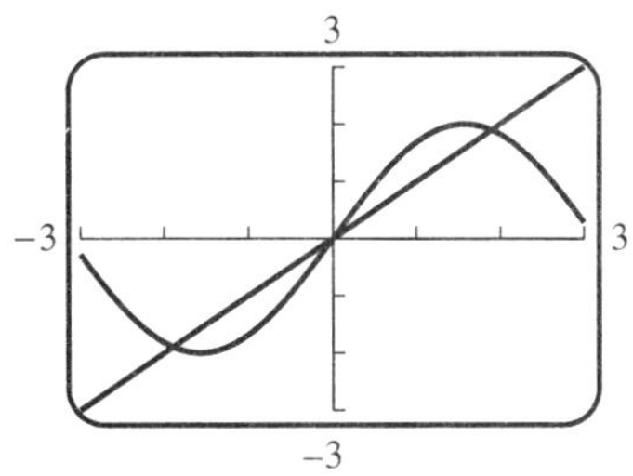

EXERCISE SET 1.7

1. The sum of the squares for the residuals for line I is approximately $1^2 + 1^2 + 1^2 + 0^2 + 2^2 + 1^2 + 1^2 + 1^2 = 10$, and the same for line II is approximately $0^2 + (0.4)^2 + (1.2)^2 + 0^2 + (2.2)^2 + (0.6)^2 + (0.2)^2 + 0^2 = 6.84$; line II is the regression line.

2. (a) The data appear to be periodic, so a trigonometric model may be appropriate.
(b) The data appear to lie near a parabola, so a quadratic model may be appropriate.
(c) The data appear to lie near a line, so a linear model may be appropriate.
(d) The data appear to be randomly distributed, so no model seems to be appropriate.

3. Least squares line $S = 1.5388t - 2842.9$, correlation coefficient 0.83409

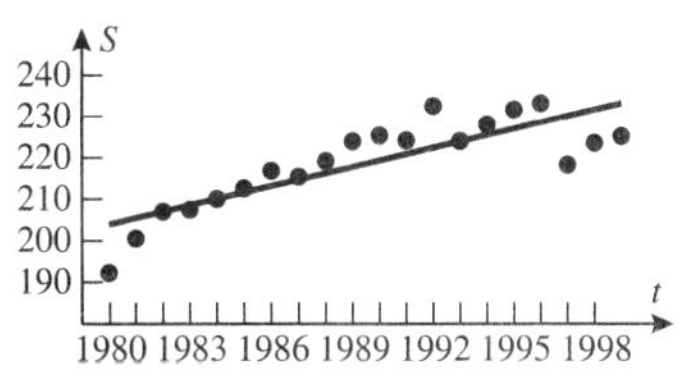

4. (a) $p = 0.0154T + 4.19$ (b) 3.42 atm

5. (a) Least squares line $p = 0.0146T + 3.98$, correlation coefficient 0.9999
(b) $p = 3.25$ atm (c) $T = -272°\text{C}$

6. (a) $p = 0.0203\,T + 5.54, r = 0.9999$ (b) $T = -273$
(c) $1.05\,p = 0.0203(T)(1.1) + 5.54$ and $p = 0.0203T + 5.54$, subtract to get $.05p = 0.1(0.0203T)$, $p = 2(0.0203T)$. But $p = 0.0203T + 5.54$, equate the right hand sides, get $T = 5.54/0.0203 \approx 273°\text{C}$

7. (a) $R = 0.00723T + 1.55$ (b) $T = -214°\text{C}$

8. **(a)** $y = 0.0236x + 4.79$ **(b)** $T = -203°\text{C}$

9. **(a)** $S = 0.50179w - 0.00643$ **(b)** $S = 8$, $w = 16$ lb

10. **(a)** $S = 0.756w - 0.0133$
(b) $2S = 0.756(w + 5) - 0.0133$, subtract to get $S = 5(0.756) = 3.78$

11. **(a)** Let h denote the height in inches and y the number of rebounds per minute. Then $y = 0.00630h - 0.266$, $r = 0.313$

(b)

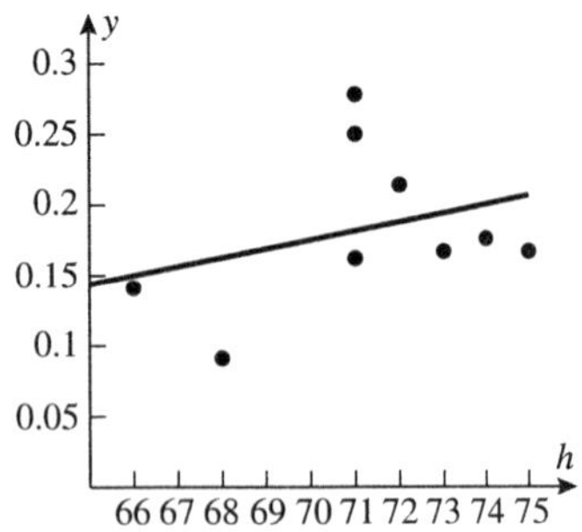

(c) No, the data points are too widely scattered.

12. **(a)** Let h denote the height in inches and y the weight in pounds. Then $y = 7.73h - 391$

(b)

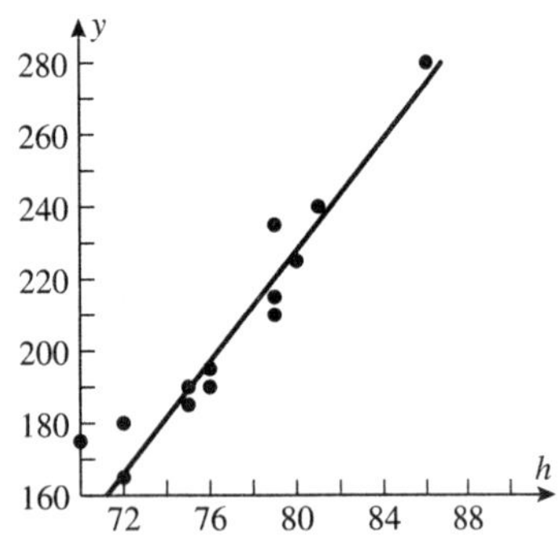

(c) 259 lb

13. **(a)** $H \approx 20000/110 \approx 181$ km/s/Mly
(b) One light year is 9.408×10^{12} km and
$$t = \frac{d}{v} = \frac{1}{H} = \frac{1}{20\text{km/s/Mly}} = \frac{9.408 \times 10^{18}\text{km}}{20\text{km/s}} = 4.704 \times 10^{17}\text{ s} = 1.492 \times 10^{10}\text{ years.}$$
(c) The Universe would be even older.

14. **(a)** $f = 0.906m + 5.93$ **(b)** $f = 85.7$

15. **(a)** $P = 0.322t^2 + 0.0671t + 0.00837$ **(b)** $P = 1.43$ cm

16. The population is modeled by $P = 0.0045833t^2 - 16.378t + 14635$; in the year 2000 the population would be $P = 212,200,000$. This is far short; in 1990 the population of the US was approximately 250,000,000.

17. As in Example 4, a possible model is of the form $T = D + A\sin\left[B\left(t - \frac{C}{B}\right)\right]$. Since the longest day has 993 minutes and the shortest has 706, take $2A = 993 - 706 = 287$ or $A = 143.5$. The midpoint between the longest and shortest days is 849.5 minutes, so there is a vertical shift of $D = 849.5$. The period is about 365.25 days, so $2\pi/B = 365.25$ or $B = \pi/183$. Note that the sine function takes the value -1 when $t - \frac{C}{B} = -91.8125$, and T is a minimum at about $t = 0$. Thus the phase shift $\frac{C}{B} \approx 91.5$. Hence $T = 849.5 + 143.5\sin\left[\frac{\pi}{183}t - \frac{\pi}{2}\right]$ is a model for the temperature.

18. As in Example 4, a possible model for the fraction f of illumination is of the form $f = D + A\sin\left[B\left(t - \dfrac{C}{B}\right)\right]$. Since the greatest fraction of illumination is 1 and the least is 0, $2A = 1$, $A = 1/2$. The midpoint of the fraction of illumination is $1/2$, so there is a vertical shift of $D = 1/2$. The period is approximately 30 days, so $2\pi/B = 30$ or $B = \pi/15$. The phase shift is approximately $49/2$, so $C/B = 49/2$, and $f = 1/2 + 1/2\sin\left[\dfrac{\pi}{15}\left(t - \dfrac{49}{2}\right)\right]$

19. $t = 0.445\sqrt{d}$

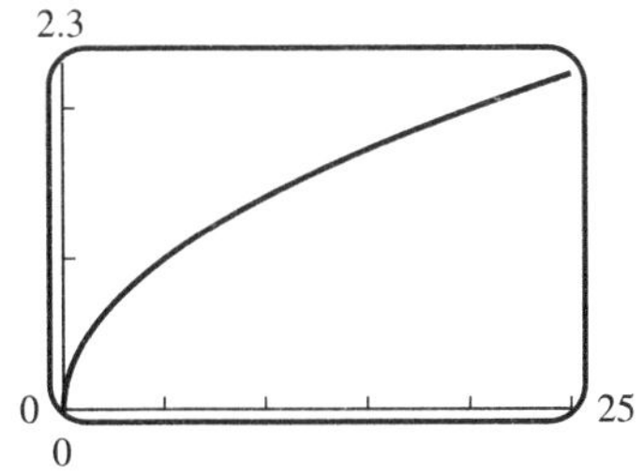

20. (a) $t = 0.373\ r^{1.5}$ (b) 238,000 km (c) 1.89 days

EXERCISE SET 1.8

1. (a) $x + 1 = t = y - 1$, $y = x + 2$

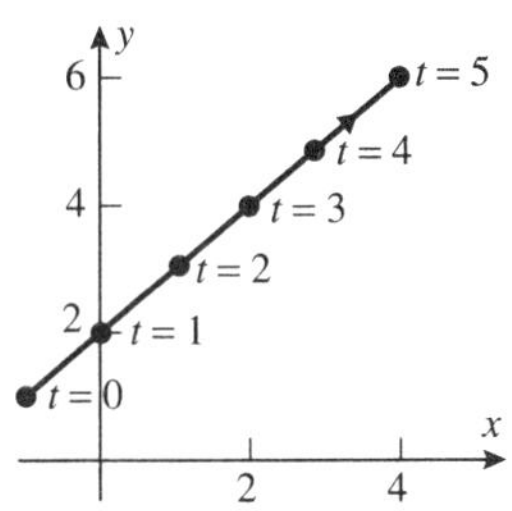

(c)

t	0	1	2	3	4	5
x	−1	0	1	2	3	4
y	1	2	3	4	5	6

2. (a) $x^2 + y^2 = 1$

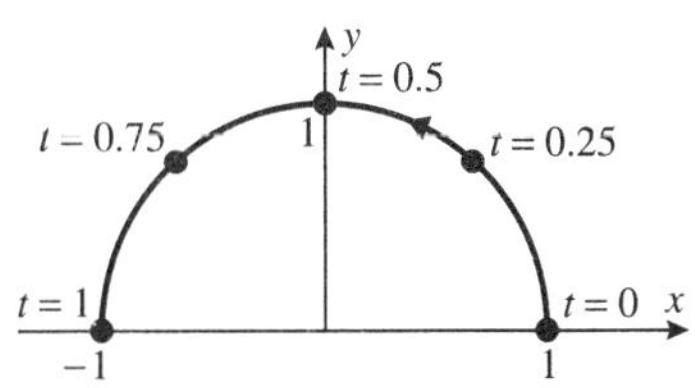

(c)

t	0	0.2500	0.50	0.7500	1
x	1	0.7071	0.00	−0.7071	−1
y	0	0.7071	1.00	0.7071	0

3. $t = (x + 4)/3; y = 2x + 10$

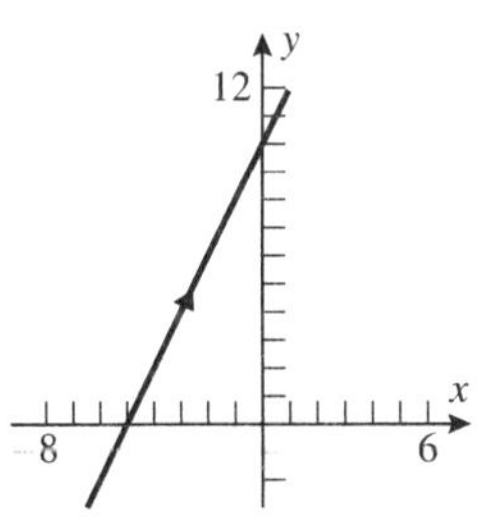

4. $t = x + 3; y = 3x + 2, -3 \le x \le 0$

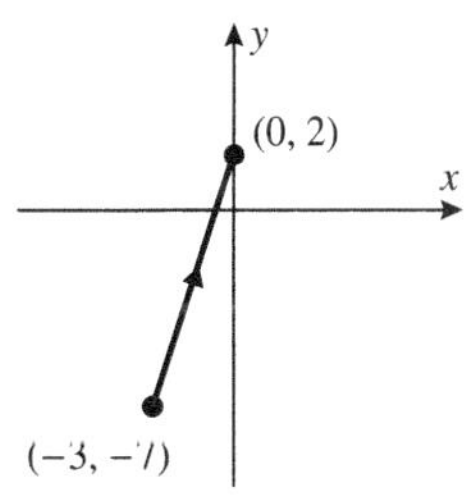

5. $\cos t = x/2$, $\sin t = y/5$;
$x^2/4 + y^2/25 = 1$

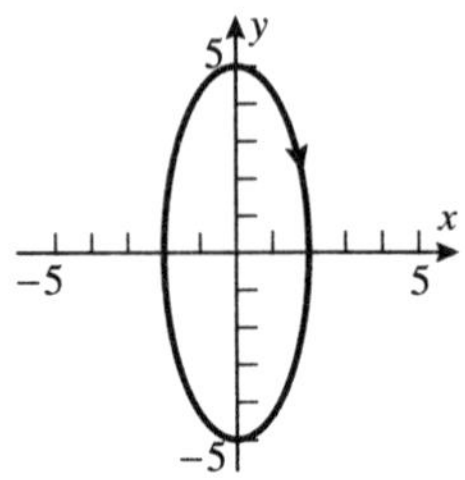

6. $t = x^2$; $y = 2x^2 + 4$,
$x \geq 0$

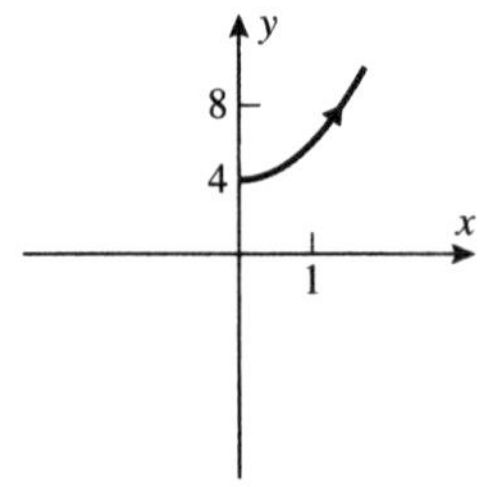

7. $\cos t = (x-3)/2$,
$\sin t = (y-2)/4$;
$(x-3)^2/4 + (y-2)^2/16 = 1$

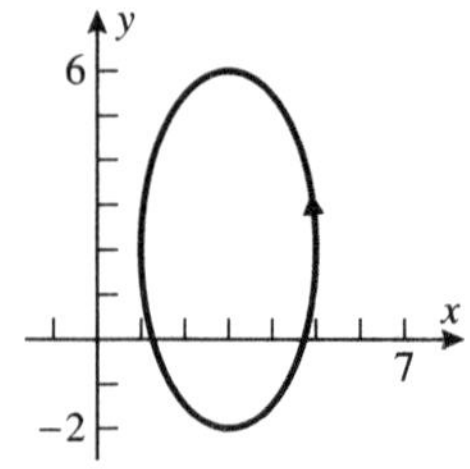

8. $\sec^2 t - \tan^2 t = 1$;
$x^2 - y^2 = 1$, $x \leq -1$
and $y \geq 0$

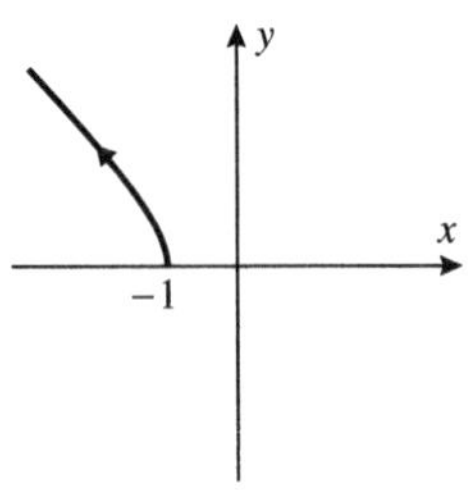

9. $\cos 2t = 1 - 2\sin^2 t$;
$x = 1 - 2y^2$,
$-1 \leq y \leq 1$

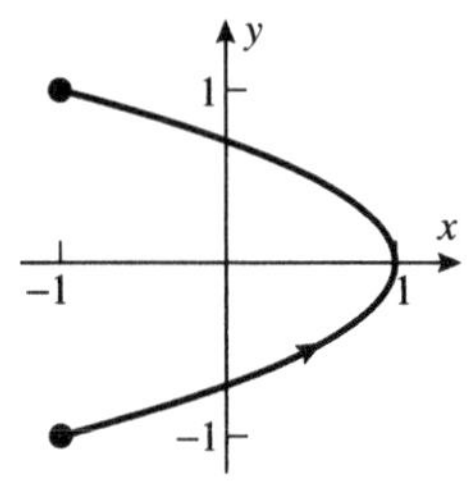

10. $t = (x-3)/4$;
$y = (x-3)^2 - 9$

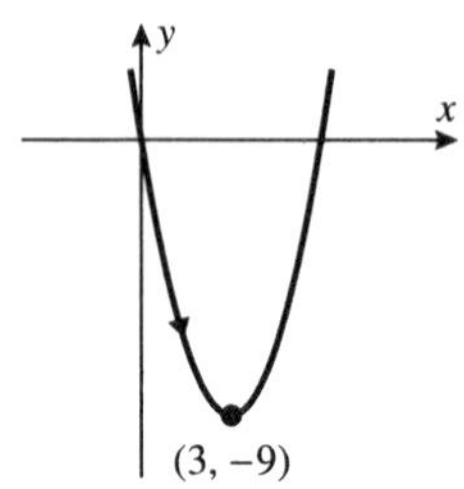

11. $x/2 + y/3 = 1, 0 \leq x \leq 2, 0 \leq y \leq 3$

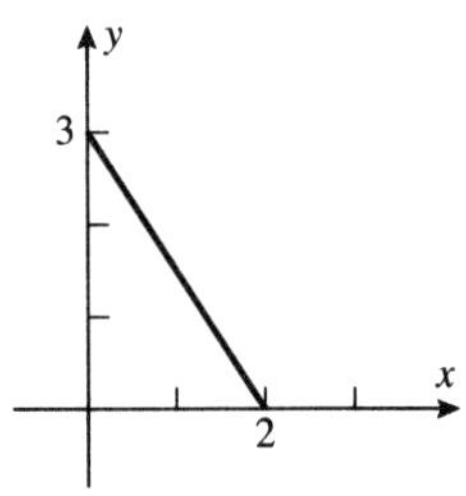

12. $y = x - 1, x \geq 1, y \geq 0$

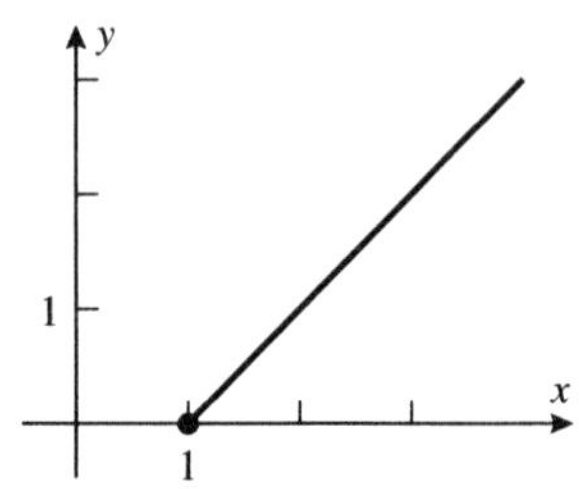

13. $x = 5\cos t$, $y = -5\sin t$, $0 \leq t \leq 2\pi$

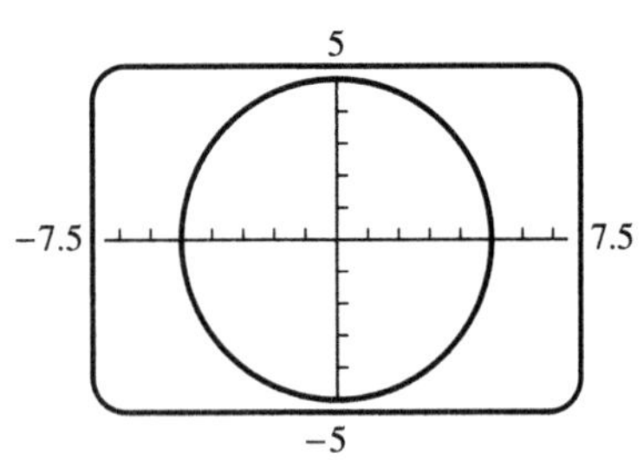

14. $x = \cos t$, $y = \sin t$, $\pi \leq t \leq 3\pi/2$

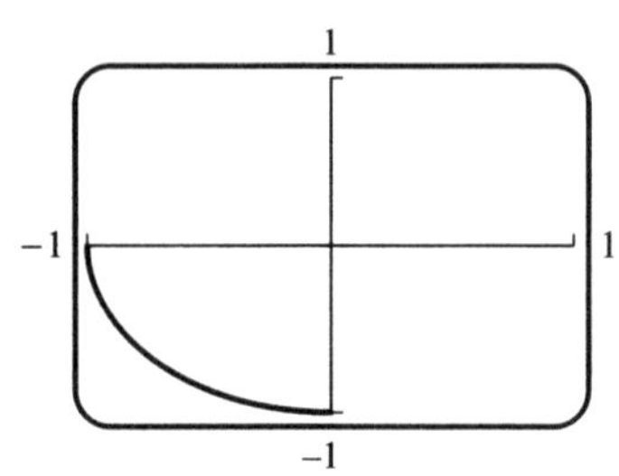

15. $x = 2$, $y = t$

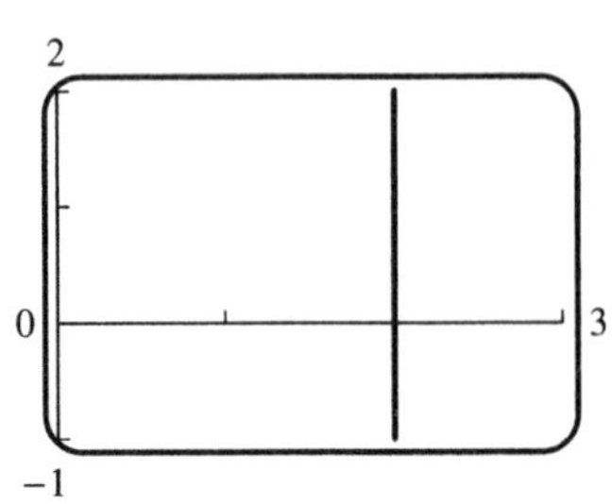

16. $x = 2\cos t$, $y = 3\sin t$, $0 \leq t \leq 2\pi$

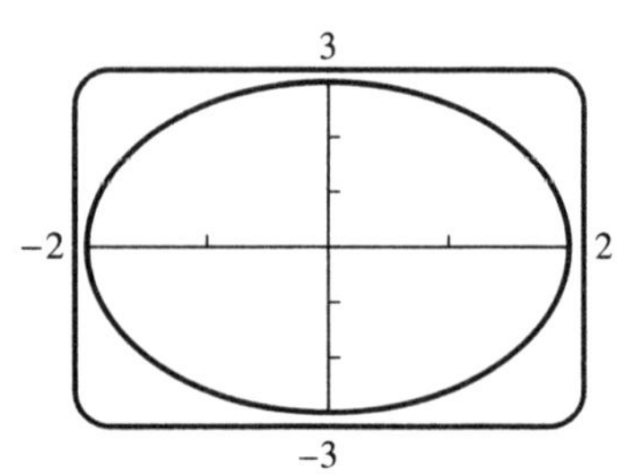

17. $x = t^2,\ y = t,\ -1 \le t \le 1$

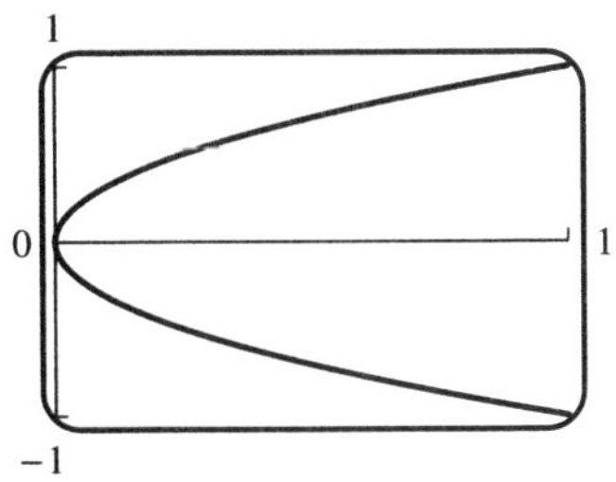

18. $x = 1 + 4\cos t,\ y = -3 + 4\sin t,\ 0 \le t \le 2\pi$

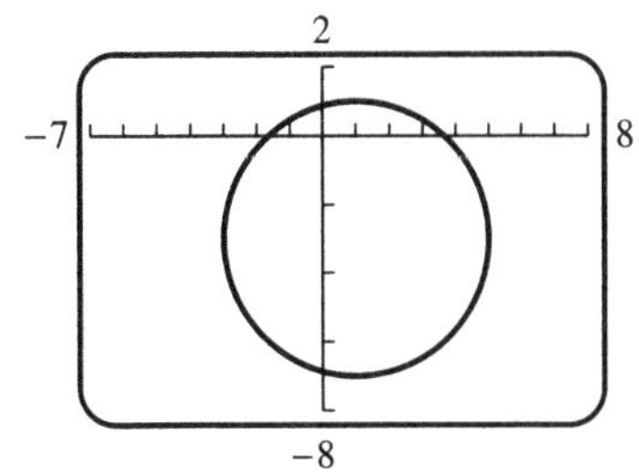

19. **(a)** IV, because x always increases whereas y oscillates.

(b) II, because $(x/2)^2 + (y/3)^2 = 1$, an ellipse.

(c) V, because $x^2 + y^2 = t^2$ increases in magnitude while x and y keep changing sign.

(d) VI; examine the cases $t < -1$ and $t > -1$ and you see the curve lies in the first, second and fourth quadrants only.

(e) III because $y > 0$.

(f) I; since x and y are bounded, the answer must be I or II; but as t runs, say, from 0 to π, x goes directly from 2 to -2, but y goes from 0 to 1 to 0 to -1 and back to 0, which describes I but not II.

20. **(a)** from left to right **(b)** counterclockwise **(c)** counterclockwise

(d) As t travels from $-\infty$ to -1, the curve goes from (near) the origin in the third quadrant and travels up and left. As t travels from -1 to $+\infty$ the curve comes from way down in the second quadrant, hits the origin at $t = 0$, and then makes the loop clockwise and finally approaches the origin again as $t \to +\infty$.

(e) from left to right

(f) Starting, say, at $(1, 0)$, the curve goes up into the first quadrant, loops back through the origin and into the third quadrant, and then continues the figure-eight.

21. **(a)**

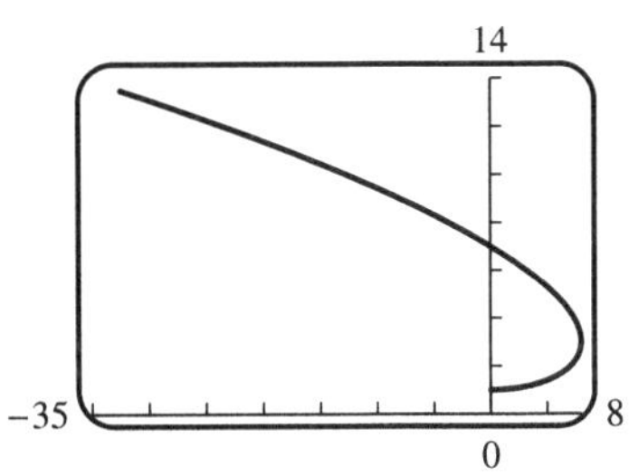

(b)

t	0	1	2	3	4	5
x	0	5.5	8	4.5	-8	-32.5
y	1	1.5	3	5.5	9	13.5

(c) $x = 0$ when $t = 0, 2\sqrt{3}$.

(d) for $0 < t < 2\sqrt{2}$

(e) at $t = 2$

22. **(a)**

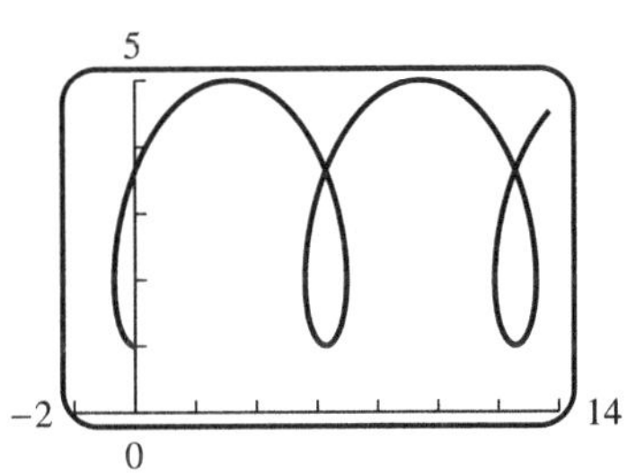

(b) y is always ≥ 1 since $\cos t \le 1$

(c) greater than 5, since $\cos t \ge -1$

23. **(a)** 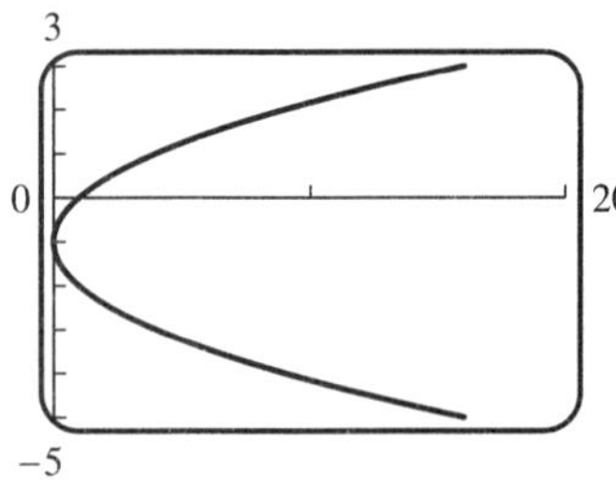

(b) 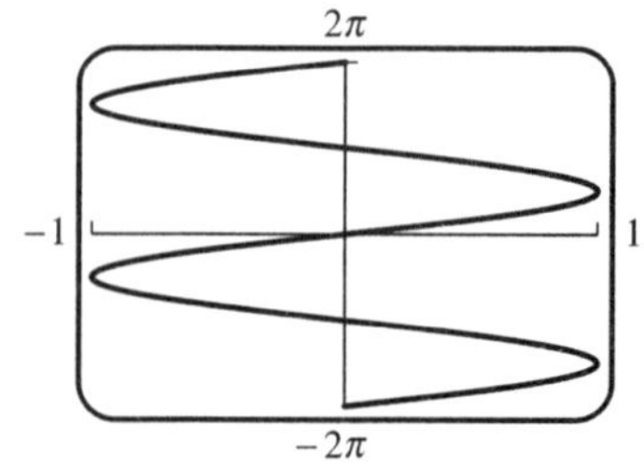

24. **(a)** 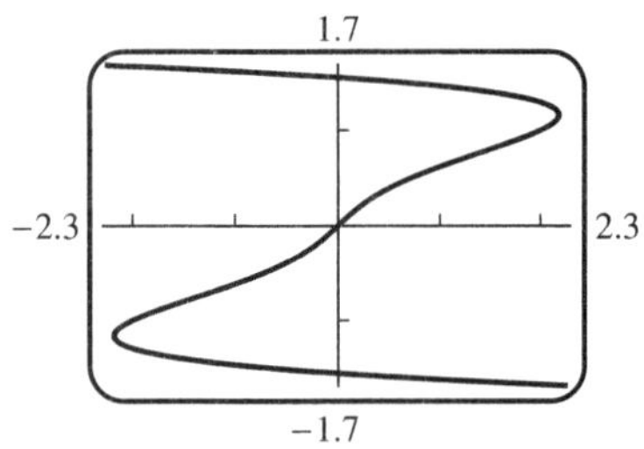

(b) 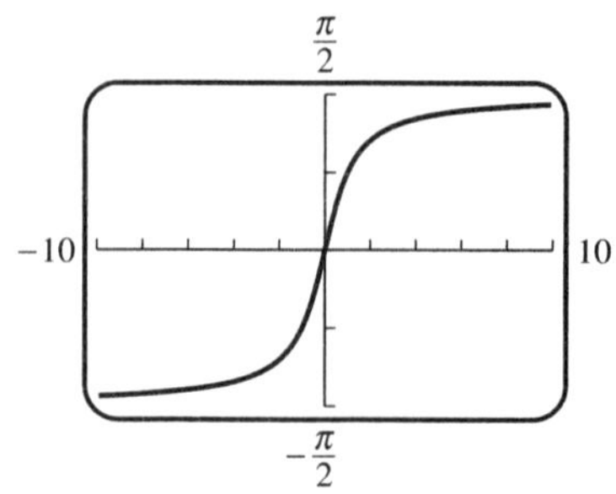

25. **(a)** Eliminate t to get $\dfrac{x-x_0}{x_1-x_0}=\dfrac{y-y_0}{y_1-y_0}$

(b) Set $t=0$ to get (x_0,y_0); $t=1$ for (x_1,y_1).

(c) $x=1+t$, $y=-2+6t$

(d) $x=2-t$, $y=4-6t$

26. **(a)** $x=-3-2t$, $y=-4+5t, 0\le t\le 1$ **(b)** $x=at$, $y=b(1-t), 0\le t\le 1$

27. **(a)** $|R-P|^2=(x-x_0)^2+(y-y_0)^2=t^2[(x_1-x_0)^2+(y_1-y_0)^2]$ and $|Q-P|^2=(x_1-x_0)^2+(y_1-y_0)^2$, so $r=|R-P|=|Q-P|t=qt$.

(b) $t=1/2$ **(c)** $t=3/4$

28. $x=2+t$, $y=-1+2t$

(a) $(5/2,0)$ **(b)** $(9/4,-1/2)$ **(c)** $(11/4,1/2)$

29. The two branches corresponding to $-1\le t\le 0$ and $0\le t\le 1$ coincide.

30. **(a)** Eliminate $\dfrac{t-t_0}{t_1-t_0}$ to obtain $\dfrac{y-y_0}{x-x_0}=\dfrac{y_1-y_0}{x_1-x_0}$.

(b) from (x_0,y_0) to (x_1,y_1)

(c) $x=3-2(t-1)$, $y=-1+5(t-1)$

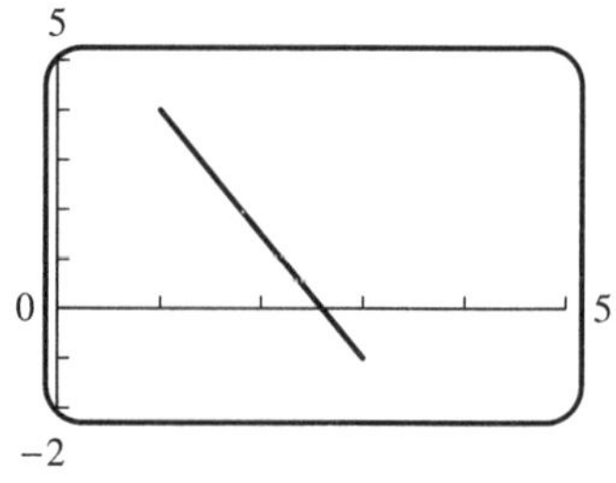

31. **(a)** $\dfrac{x-b}{a} = \dfrac{y-d}{c}$

(b)
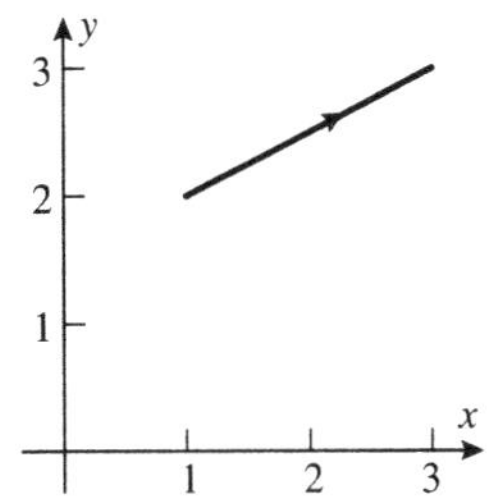

32. **(a)** If $a = 0$ the line segment is vertical; if $c = 0$ it is horizontal.

(b) The curve degenerates to the point (b, d).

33.
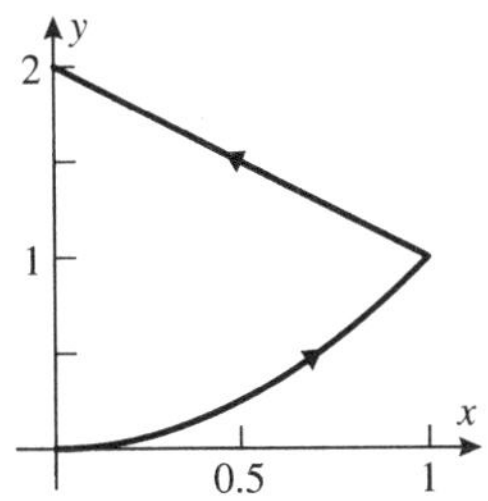

34.

$x = 1/2 - 4t,\quad y = 1/2 \quad \text{for} \quad 0 \le t \le 1/4$
$x = -1/2,\quad y = 1/2 - 4(t - 1/4) \quad \text{for} \quad 1/4 \le t \le 1/2$
$x = -1/2 + 4(t - 1/2),\quad y = -1/2 \quad \text{for} \quad 1/2 \le t \le 3/4$
$x = 1/2,\quad y = -1/2 + 4(t - 3/4) \quad \text{for} \quad 3/4 \le t \le 1$

35. **(a)** $x = 4\cos t,\ y = 3\sin t$

(b) $x = -1 + 4\cos t,\ y = 2 + 3\sin t$

(c)
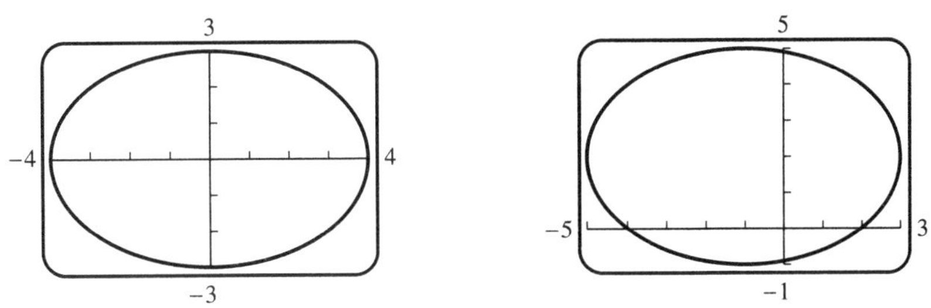

36. **(a)** $t = x/(v_0 \cos\alpha)$, so $y - x\tan\alpha - gx^2/(2v_0^2\cos^2\alpha)$.

(b)
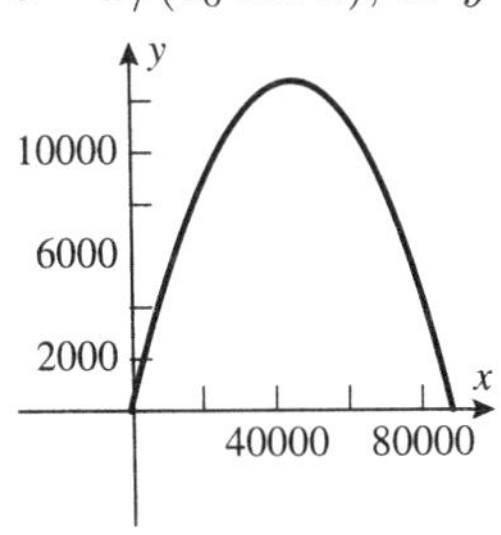

37. **(a)** From Exercise 36, $x = 400\sqrt{2}t$, $y = 400\sqrt{2}t - 4.9t^2$.

(b) 16,326.53 m

(c) 65,306.12 m

38. (a)

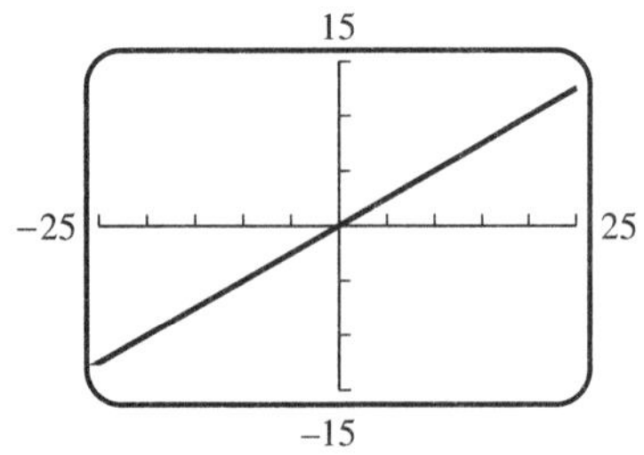

(b)

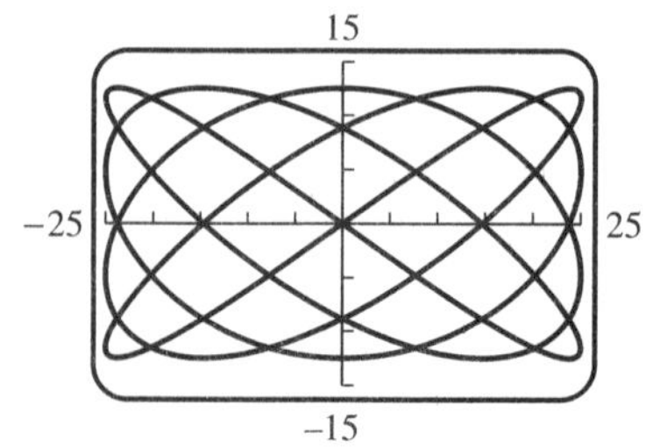

(c)

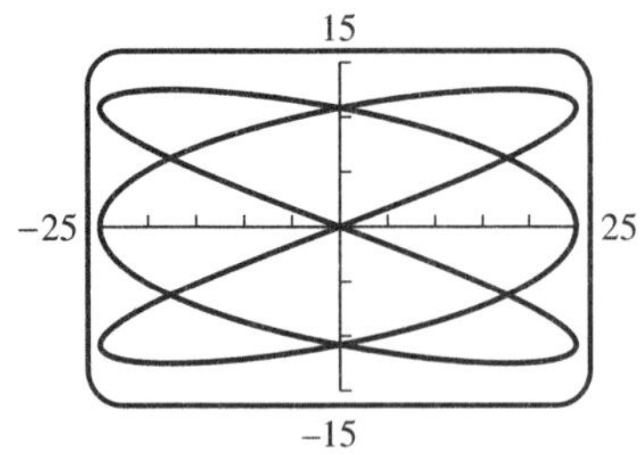

$a = 3, b = 2$

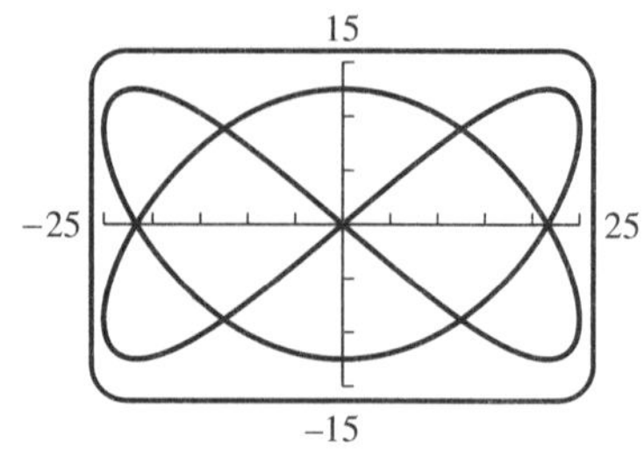

$a = 2, b = 3$

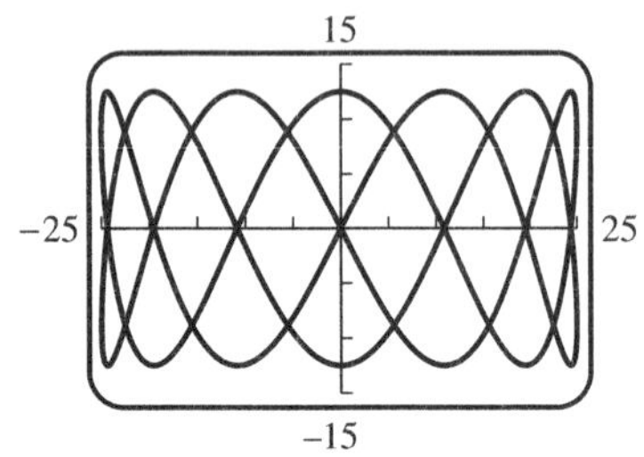

$a = 2, b = 7$

39. Assume that $a \neq 0$ and $b \neq 0$; eliminate the parameter to get $(x-h)^2/a^2 + (y-k)^2/b^2 = 1$. If $|a| = |b|$ the curve is a circle with center (h, k) and radius $|a|$; if $|a| \neq |b|$ the curve is an ellipse with center (h, k) and major axis parallel to the x-axis when $|a| > |b|$, or major axis parallel to the y-axis when $|a| < |b|$.

(a) ellipses with a fixed center and varying axes of symmetry

(b) (assume $a \neq 0$ and $b \neq 0$) ellipses with varying center and fixed axes of symmetry

(c) circles of radius 1 with centers on the line $y = x - 1$

40. Refer to the diagram to get $b\theta = a\phi$, $\theta = a\phi/b$ but

$\theta - \alpha = \phi + \pi/2$ so $\alpha = \theta - \phi - \pi/2 = (a/b - 1)\phi - \pi/2$

$$x = (a-b)\cos\phi - b\sin\alpha$$
$$= (a-b)\cos\phi + b\cos\left(\frac{a-b}{b}\right)\phi,$$
$$y = (a-b)\sin\phi - b\cos\alpha$$
$$= (a-b)\sin\phi - b\sin\left(\frac{a-b}{b}\right)\phi.$$

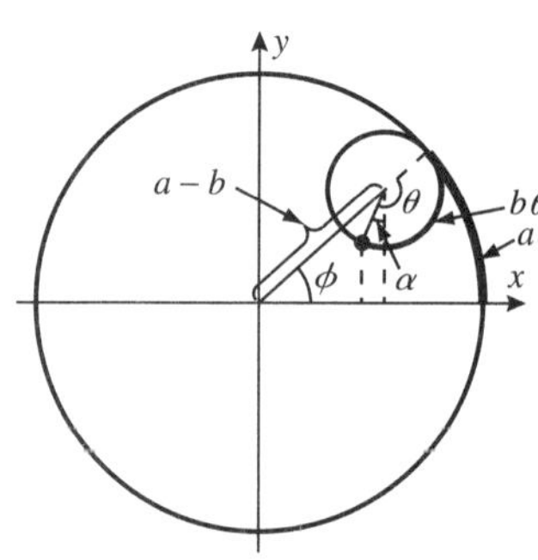

41. (a) 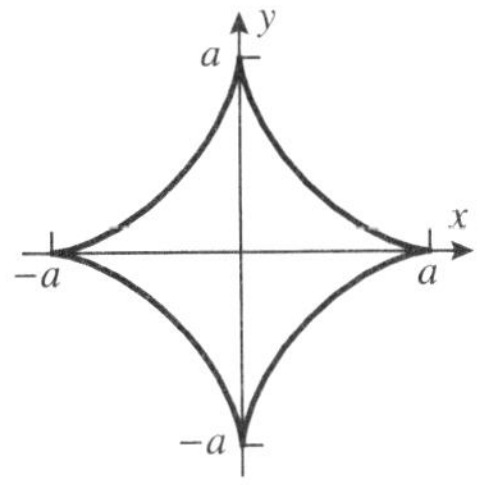

(b) Use $b = a/4$ in the equations of Exercise 40 to get
$x = \frac{3}{4}a\cos\phi + \frac{1}{4}a\cos 3\phi,\ y = \frac{3}{4}a\sin\phi - \frac{1}{4}a\sin 3\phi;$
but trigonometric identities yield $\cos 3\phi = 4\cos^3\phi - 3\cos\phi,\ \sin 3\phi = 3\sin\phi - 4\sin^3\phi$,
so $x = a\cos^3\phi,\ y = a\sin^3\phi$.

(c) $x^{2/3} + y^{2/3} = a^{2/3}(\cos^2\phi + \sin^2\phi) = a^{2/3}$

42. (a) 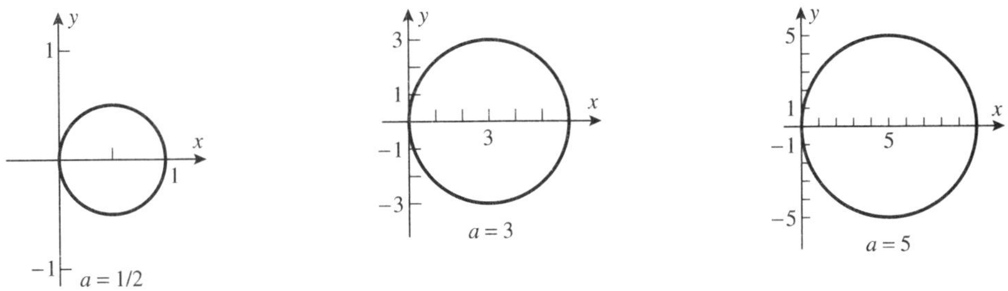

(b) $$\begin{aligned}(x-a)^2 + y^2 &= (2a\cos^2 t - a)^2 + (2a\cos t\sin t)^2\\ &= 4a^2\cos^4 t - 4a^2\cos^2 t + a^2 + 4a^2\cos^2 t\sin^2 t\\ &= 4a^2\cos^4 t - 4a^2\cos^2 t + a^2 + 4a^2\cos^2 t(1-\cos^2 t) = a^2,\end{aligned}$$
a circle about $(a, 0)$ of radius a

CHAPTER 1 SUPPLEMENTARY EXERCISES

1. 1940-45; the greatest five-year slope

2. (a) $f(-1) = 3.3,\ g(3) = 2$
(b) $x = -3, 3$
(c) $x < -2, x > 3$
(d) the domain is $-5 \le x \le 5$ and the range is $-5 \le y \le 4$
(e) the domain is $-4 \le x \le 4.1$, the range is $-3 \le y \le 5$
(f) $f(x) = 0$ at $x - -3, 5$; $g(x) = 0$ at $x = -3, 2$

3. 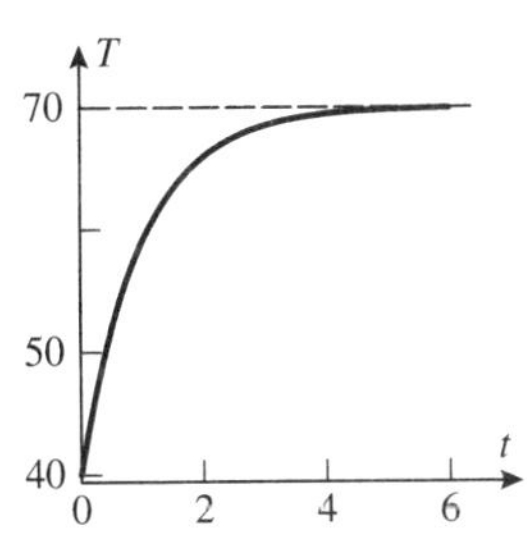

4. 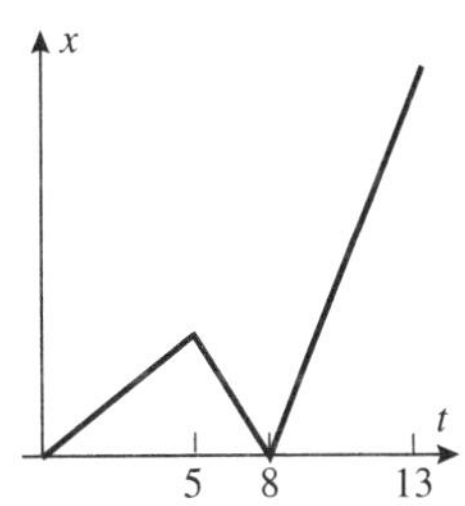

5. If the side has length x and height h, then $V = 8 = x^2h$, so $h = 8/x^2$. Then the cost $C = 5x^2 + 2(4)(xh) = 5x^2 + 64/x$.

6. Assume that the paint is applied in a thin veneer of uniform thickness, so that the quantity of paint to be used is proportional to the area covered. If P is the amount of paint to be used, $P = k\pi r^2$. The constant k depends on physical factors, such as the thickness of the paint, absorption of the wood, etc.

7.

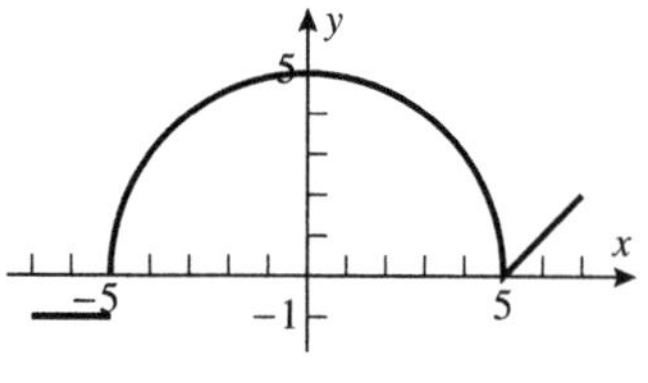

8. Suppose the radius of the uncoated ball is r and that of the coated ball is $r+h$. Then the plastic has volume equal to the difference of the volumes, i.e.
$V = \frac{4}{3}\pi(r+h)^3 - \frac{4}{3}\pi r^3 = \frac{4}{3}\pi h[3r^2 + 3rh + h^2]$ in^3.

9. (a) The base has sides $(10-2x)/2$ and $6-2x$, and the height is x, so $V = (6-2x)(5-x)x$ ft^3.
(b) From the picture we see that $x < 5$ and $2x < 6$, so $0 < x < 3$.
(c) 3.57 ft $\times$3.79 ft $\times$1.21 ft

10. $\{x \neq 0\}$ and $\emptyset$ (the empty set)

11. $f(g(x)) = (3x+2)^2+1, g(f(x)) = 3(x^2+1)+2$, so $9x^2+12x+5 = 3x^2+5, 6x^2+12x = 0, x = 0, -2$

12. (a) $(3-x)/x$
(b) no; $f(g(x))$ can be defined at $x = 1$, whereas g, and therefore $f \circ g$, requires $x \neq 1$

13. $1/(2-x^2)$

14. $g(x) = x^2 + 2x$

15.

x	−4	−3	−2	−1	0	1	2	3	4
$f(x)$	0	−1	2	1	3	−2	−3	4	−4
$g(x)$	3	2	1	−3	−1	−4	4	−2	0
$(f \circ g)(x)$	4	−3	−2	−1	1	0	−4	2	3
$(g \circ f)(x)$	−1	−3	4	−4	−2	1	2	0	3

16. (a) $y = |x-1|$, $y = |(-x)-1| = |x+1|$,
$y = 2|x+1|$, $y = 2|x+1| - 3$,
$y = -2|x+1| + 3$

(b)

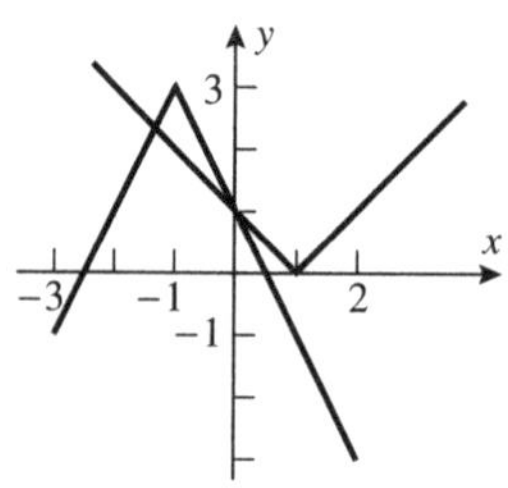

17. (a) even $\times$ odd = odd
(b) a square is even
(c) even + odd is neither
(d) odd $\times$ odd = even

18. **(a)** $y = \cos x - 2\sin x\cos x = (1 - 2\sin x)\cos x$, so $x = \pm\frac{\pi}{2}, \pm\frac{3\pi}{2}, \frac{\pi}{6}, \frac{5\pi}{6}, -\frac{7\pi}{6}, -\frac{11\pi}{6}$

(b) $\left(\pm\frac{\pi}{2},0\right), \left(\pm\frac{3\pi}{2},0\right), \left(\frac{\pi}{6},\sqrt{3}/2\right), \left(\frac{5\pi}{6},-\sqrt{3}/2\right), \left(-\frac{7\pi}{6},-\sqrt{3}/2\right), \left(-\frac{11\pi}{6},\sqrt{3}/2\right)$

19. **(a)** If x denotes the distance from A to the base of the tower, and y the distance from B to the base, then $x^2+d^2 = y^2$. Moreover $h = x\tan\alpha = y\tan\beta$, so $d^2 = y^2-x^2 = h^2(\cot^2\beta - \cot^2\alpha)$, $h^2 = \dfrac{d^2}{\cot^2\beta - \cot^2\alpha} = \dfrac{d^2}{1/\tan^2\beta - 1/\tan^2\alpha} = \dfrac{d^2\tan^2\alpha\tan^2\beta}{\tan^2\alpha - \tan^2\beta}$, which yields the result.

(b) 295.72 ft.

20. **(a)**

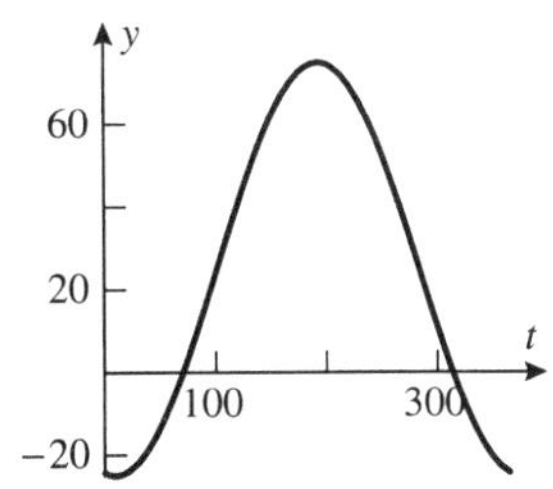

(b) when $\frac{2\pi}{365}(t - 101) = \frac{3\pi}{2}$, or $t = 374.75$, which is the same date as $t = 9.75$, so during the night of January 10th-11th

(c) from $t = 0$ to $t = 70.58$ and from $t = 313.92$ to $t = 365$ (the same date as $t = 0$) , for a total of about 122 days

21. When $x = 0$ the value of the green curve is higher than that of the blue curve, therefore the blue curve is given by $y = 1 + 2\sin x$.
The points A, B, C, D are the points of intersection of the two curves, i.e. where $1+2\sin x = 2\sin(x/2)+2\cos(x/2)$. Let $\sin(x/2) = p, \cos(x/2) = q$. Then $2\sin x = 4\sin(x/2)\cos(x/2)$, so the equation which yields the points of intersection becomes $1 + 4pq = 2p + 2q$, $4pq - 2p - 2q + 1 = 0, (2p-1)(2q-1) = 0$; thus whenever either $\sin(x/2) = 1/2$ or $\cos(x/2) = 1/2$, i.e. when $x/2 = \pi/6, 5\pi/6, \pm\pi/3$. Thus A has coordinates $(-2\pi/3, 1-\sqrt{3})$, B has coordinates $(\pi/3, 1+\sqrt{3})$, C has coordinates $(2\pi/3, 1+\sqrt{3})$, and D has coordinates $(5\pi/3, 1-\sqrt{3})$.

22. Let $y = A + B\sin(at + b)$. Since the maximum and minimum values of y are 35 and 5, $A + B = 35$ and $A - B = 5$, $A = 20$, $B = 15$. The period is 12 hours, so $12a = 2\pi$ and $a = \pi/6$. The maximum occurs at $t = 2$, so $1 = \sin(2a + b) = \sin(\pi/3 + b)$, $\pi/3 + b = \pi/2$, $b = \pi/2 - \pi/3 = \pi/6$ and $y = 20 + 15\sin(\pi t/6 + \pi/6)$.

23. **(a)** The circle of radius 1 centered at (a, a^2); therefore, the family of all circles of radius 1 with centers on the parabola $y = x^2$.

(b) All parabolas which open up, have latus rectum equal to 1 and vertex on the line $y = x/2$.

24. **(a)** $x = f(1 - t), y = g(1 - t)$

25.

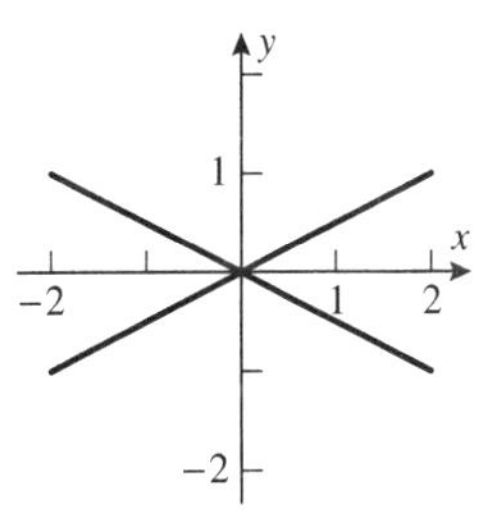

26. Let $y = ax^2 + bx + c$. Then $4a + 2b + c = 0, 64a + 8b + c = 18, 64a - 8b + c = 18$, from which $b = 0$ and $60a = 18$, or finally $y = \frac{3}{10}x^2 - \frac{6}{5}$.

27.

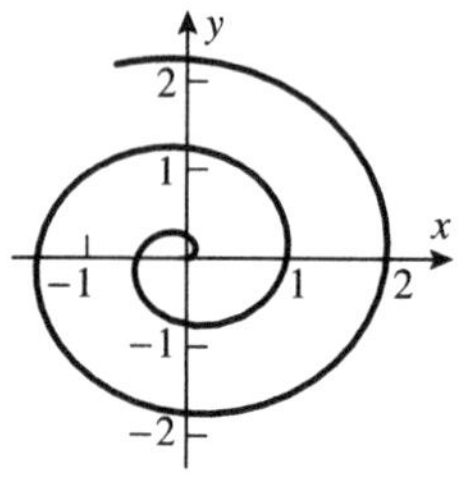

28. **(a)** $R = R_0$ is the R-intercept, $R_0 k$ is the slope, and $T = -1/k$ is the T-intercept

(b) $-1/k = -273$, or $k = 1/273$

(c) $1.1 = R_0(1 + 20/273)$, or $R_0 = 1.025$

(d) $T = 126.55°\text{C}$

29. $d = \sqrt{(x-1)^2 + (\sqrt{x} - 2)^2}$;

$d = 9.1$ at $x = 1.358094$

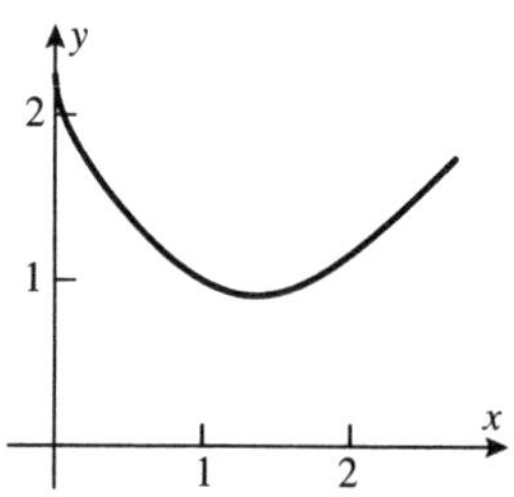

30. $d = \sqrt{(x-1)^2 + 1/x^2}$;

$d = 0.82$ at $x = 1.380278$

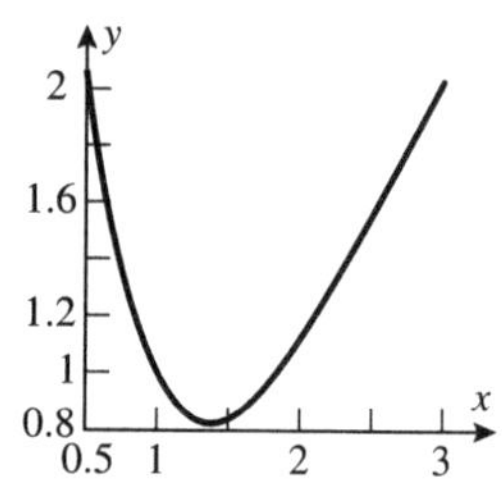

31. $w = 63.9V$, $w = 63.9\pi h^2(5/2 - h/3)$; $h = 0.48$ ft when $w = 108$ lb

32. **(a)**

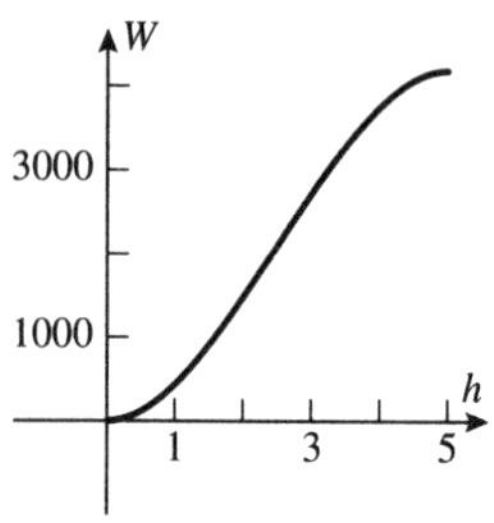

(b) $w = 63.9\pi h^2(5/2 - h/3)$; at $h = 5/2$, $w = 2091.12$ lb

33. **(a)**

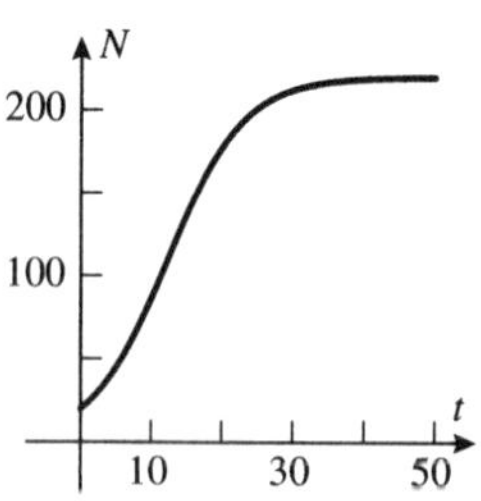

(b) $N = 80$ when $t = 9.35$ yrs

(c) 220 sheep

34. **(a)**

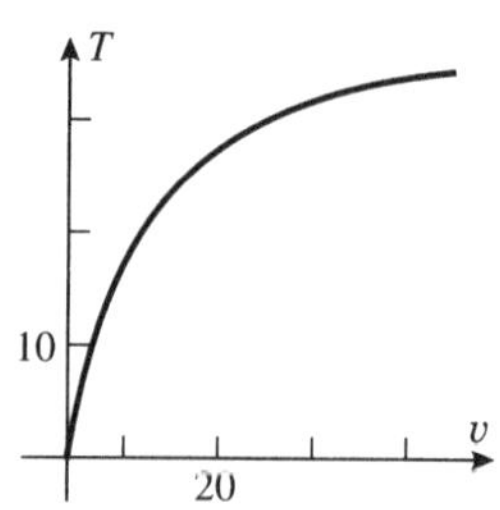

(b) $T = 17°\text{F}, 27°\text{F}, 32°\text{F}$

35. **(a)**

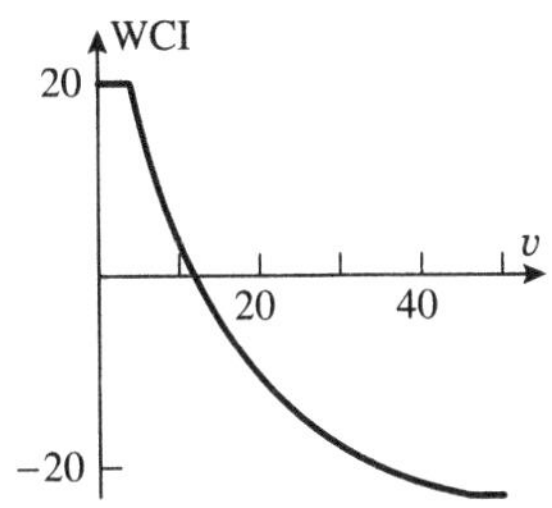

(b) $T = 3°\text{F}, -11°\text{F}, -18°\text{F}, -22°\text{F}$

(c) $v = 35, 19, 12, 7$ mi/h

36. The domain is the set of all x, the range is $-0.1746 \le y \le 0.1227$.

37. The domain is the set $-0.7245 \le x \le 1.2207$, the range is $-1.0551 \le y \le 1.4902$.

38. **(a)** The potato is done in the interval $27.65 < t < 32.71$.
(b) 91.54 min.

39. **(a)**

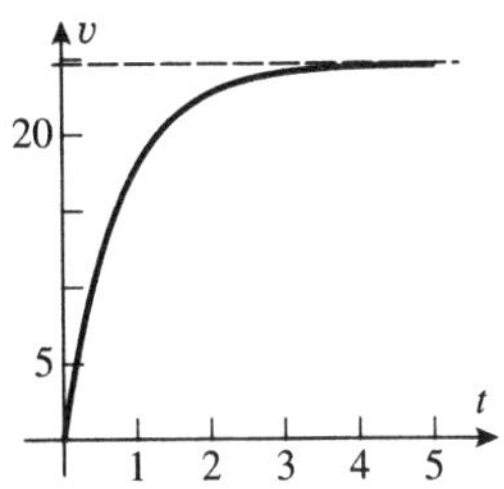

(b) As $t \to \infty$, $(0.273)^t \to 0$, and thus $v \to 24.61$ ft/s.
(c) For large t the velocity approaches c.
(d) No; but it comes very close (arbitrarily close).
(e) 3.013 s

40. **(a)** $y = -0.01716428571x + 1.433827619$

41. **(a)**

1.90	1.92	1.94	1.96	1.98	2.00	2.02	2.04	2.06	2.08	2.10
3.4161	3.4639	3.5100	3.5543	3.5967	3.6372	3.6756	3.7119	3.7459	3.7775	3.8068

(b) $y = 1.9589x - 0.2910$
(c) $y - 3.6372 = 1.9589(x - 2)$, or $y = 1.9589x - 0.2806$
(d) As one zooms in on the point $(2, f(2))$ the two curves seem to converge to one line.

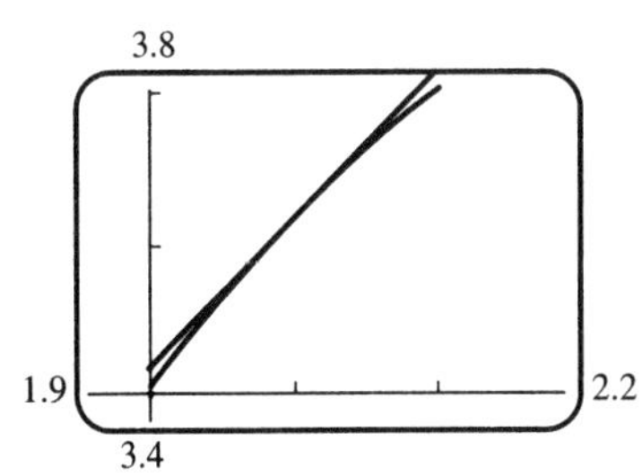

42. **(a)**

−0.10	−0.08	−0.06	−0.04	−0.02	0.00	0.02	0.04	0.06	0.08	0.10
0.9950	0.9968	0.9982	0.9992	0.9998	1.0000	0.9998	0.9992	0.9982	0.9968	0.9950

(b) $y = -\frac{1}{2}x^2 + 1$

(c) $y = -\frac{1}{2}x^2 + 1$

(d) As one zooms in on the point $(0, f(0))$ the two curves seem to converge to one curve.

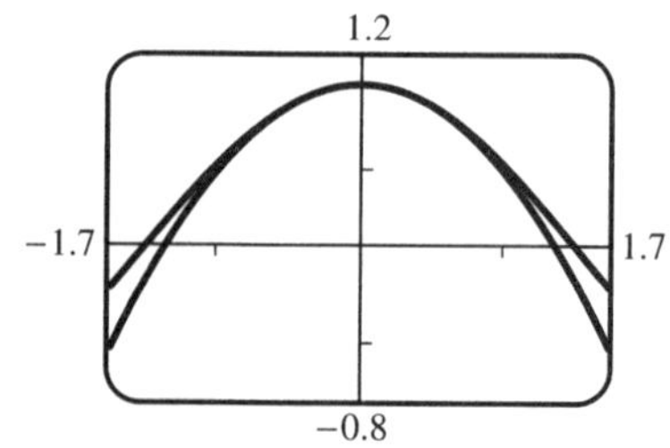

43. The data are periodic, so it is reasonable that a trigonometric function might approximate them. A possible model is of the form $T = D + A\sin\left[B\left(t - \frac{C}{B}\right)\right]$. Since the highest level is 1.032 meters and the lowest is 0.045, take $2A = 1.032 - 0.042 = 0.990$ or $A = 0.495$. The midpoint between the lowest and highest levels is 0.537 meters, so there is a vertical shift of $D = 0.537$. The period is about 12 hours, so $2\pi/B = 12$ or $B = \pi/6$. The phase shift $\frac{C}{B} \approx 6.5$. Hence $T = 0.537 + 0.495\sin\left[\frac{\pi}{6}(t - 6.5)\right]$ is a model for the temperature.

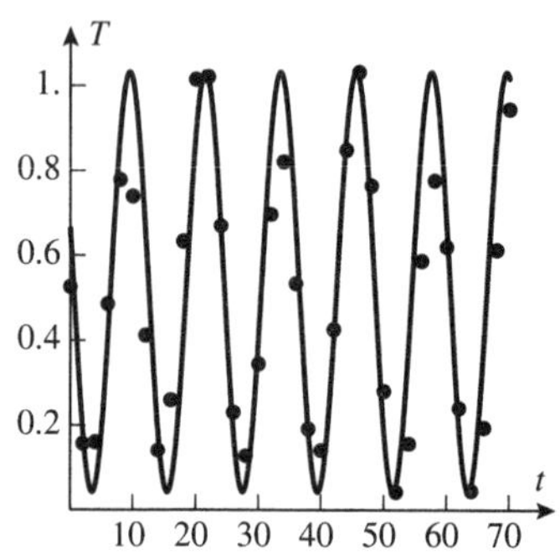

CHAPTER 1 HORIZON MODULE

1. **(a)** $0.25, 6.25\times10^{-2}, 3.91\times10^{-3}, 1.53\times10^{-5}, 2.32\times10^{-10}, 5.42\times10^{-20}, 2.94\times10^{-39}, 8.64\times10^{-78}, 7.46\times10^{-155}, 5.56\times10^{-309}$;
$1, 1, 1, 1, 1, 1, 1, 1, 1, 1$;
$4, 16, 256, 65536, 4.29\times10^{9}, 1.84\times10^{19}, 3.40\times10^{38}, 1.16\times10^{77}, 1.34\times10^{154}, 1.80\times10^{308}$

2. $1, 3, 2.3333333, 2.23809524, 2.23606890, 2.23606798, \ldots$

3. **(a)** $\frac{1}{2}, \frac{1}{4}, \frac{1}{8}, \frac{1}{16}, \frac{1}{32}, \frac{1}{64}$ **(b)** $y_n = \frac{1}{2^n}$

4. **(a)** $y_{n+1} = 1.05y_n$

(b) $y_0 =\$1000$, $y_1 =\$1050$, $y_2 =\$1102.50$, $y_3 =\$1157.62$, $y_4 =\$1215.51$, $y_5 =\$1276.28$

(c) $y_{n+1} = 1.05y_n$ for $n \geq 1$

(d) $y_n = (1.05)^n 1000$; $y_{15} =\$2078.93$

5. (a) $x^{1/2}, x^{1/4}, x^{1/8}, x^{1/16}, x^{1/32}$

(b) They tend to the horizontal line $y = 1$, with a hole at $x = 0$.

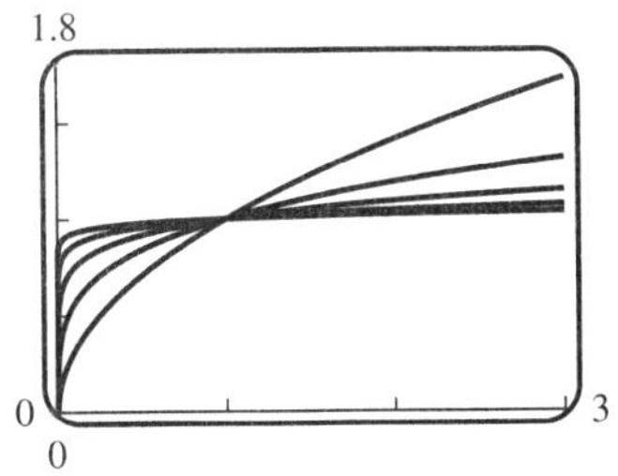

6. (a) $\frac{1}{2}\ \frac{2}{3}\ \frac{3}{5}\ \frac{5}{8}\ \frac{8}{13}\ \frac{13}{21}\ \frac{21}{34}\ \frac{34}{55}\ \frac{55}{89}\ \frac{89}{144}$

(b) 1, 2, 3, 5, 8, 13, 21, 34, 55, 89;
each new numerator is the sum of the previous two numerators.

(c) $\frac{144}{233}, \frac{233}{377}, \frac{377}{610}, \frac{610}{987}, \frac{987}{1597}, \frac{1597}{2584}, \frac{2584}{4181}, \frac{4181}{6765}, \frac{6765}{10946}, \frac{10946}{17711}$

(d) $F_0 = 1$, $F_1 = 1$, $F_n = F_{n-1} + F_{n-2}$ for $n \geq 2$.

(e) the positive solution

7. (a) $y_1 = cr$, $y_2 = cy_1 = cr^2$, $y_3 = cr^3$, $y_4 = cr^4$ (b) $y_n = cr^n$

(c) If $r = 1$ then $y_n = c$ for all n; if $r < 1$ then y_n tends to zero; if $r > 1$, then y_n gets ever larger (tends to $+\infty$).

8. The first point on the curve is $(c, kc(1-c))$, so $y_1 = kc(1-c)$ and hence y_1 is the first iterate. The point on the line to the right of this point has equal coordinates (y_1, y_1), and so the point above it on the curve has coordinates $(y_1, ky_1(1-y_1))$; thus $y_2 = ky_1(1-y_1)$, and y_2 is the second iterate, etc.

9. (a) $0.261, 0.559, 0.715, 0.591, 0.701$

(b) It appears to approach a point somewhere near 0.65.

CHAPTER 2
Limits and Continuity

EXERCISE SET 2.1

1. (a) -1 (b) 3 (c) does not exist
(d) 1 (e) -1 (f) 3

2. (a) 2 (b) 0 (c) does not exist
(d) 2 (e) 0 (f) 2

3. (a) 1 (b) 1 (c) 1 (d) 1 (e) $-\infty$ (f) $+\infty$

4. (a) 3 (b) 3 (c) 3 (d) 3 (e) $+\infty$ (f) $+\infty$

5. (a) 0 (b) 0 (c) 0 (d) 3 (e) $+\infty$ (f) $+\infty$

6. (a) 2 (b) 2 (c) 2 (d) 3 (e) $-\infty$ (f) $+\infty$

7. (a) $-\infty$ (b) $+\infty$ (c) does not exist
(d) undef (e) 2 (f) 0

8. (a) $+\infty$ (b) $+\infty$ (c) $+\infty$ (d) undef (e) 0 (f) -1

9. (a) $-\infty$ (b) $-\infty$ (c) $-\infty$ (d) 1 (e) 1 (f) 2

10. (a) 1 (b) $-\infty$ (c) does not exist
(d) -2 (e) $+\infty$ (f) $+\infty$

11. (a) 0 (b) 0 (c) 0
(d) 0 (e) does not exist (f) does not exist

12. (a) 3 (b) 3 (c) 3
(d) 3 (e) does not exist (f) 0

13. for all $x_0 \neq -4$

14. for all $x_0 \neq -6, 3$

19. (a)

2	1.5	1.1	1.01	1.001	0	0.5	0.9	0.99	0.999
0.1429	0.2105	0.3021	0.3300	0.3330	1.0000	0.5714	0.3690	0.3367	0.3337

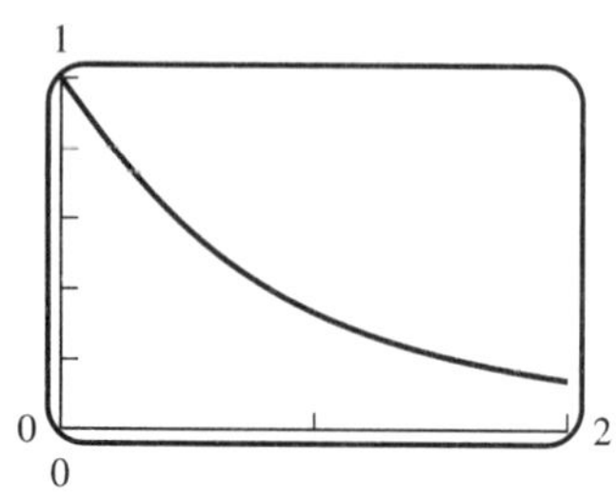

The limit is 1/3.

(b)

2	1.5	1.1	1.01	1.001	1.0001
0.4286	1.0526	6.344	66.33	666.3	6666.3

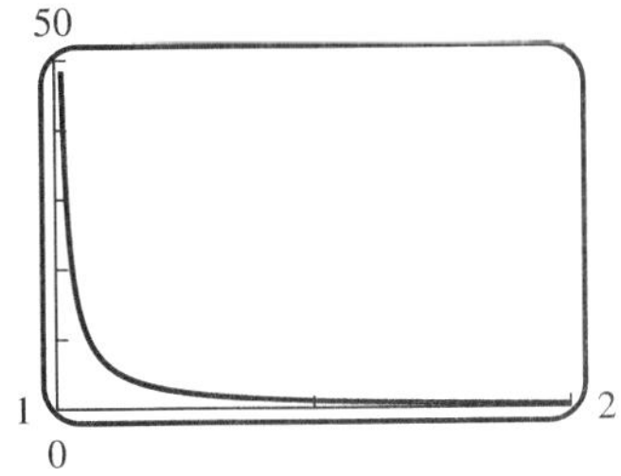

The limit is $+\infty$.

(c)

0	0.5	0.9	0.99	0.999	0.9999
−1	−1.7143	−7.0111	−67.001	−667.0	−6667.0

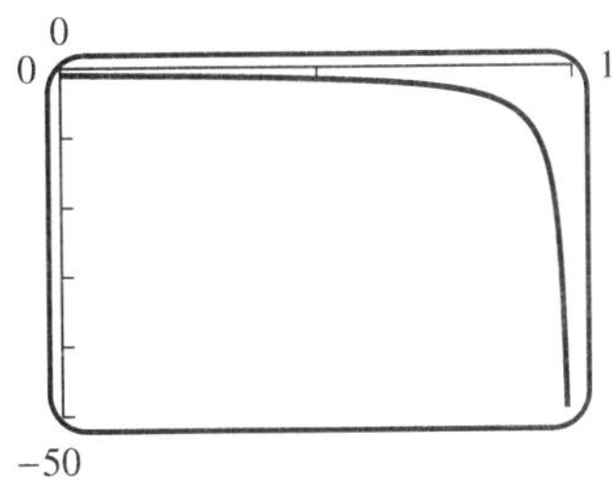

The limit is $-\infty$.

20. (a)

−0.25	−0.1	−0.001	−0.0001	0.0001	0.001	0.1	0.25
0.5359	0.5132	0.5001	0.5000	0.5000	0.4999	0.4881	0.4721

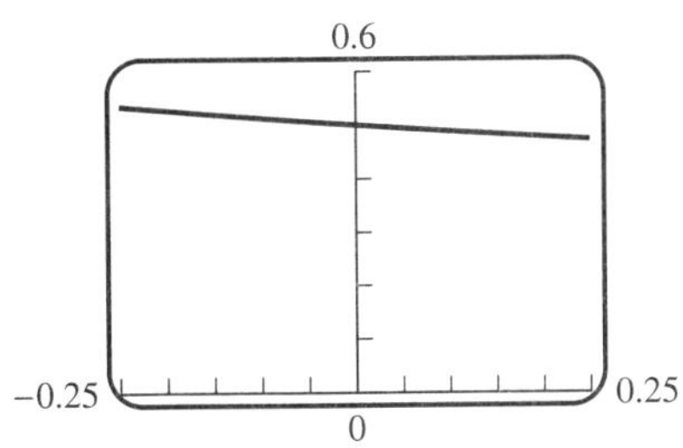

The limit is $1/2$.

(b)

0.25	0.1	0.001	0.0001
8.4721	20.488	2000.5	20001

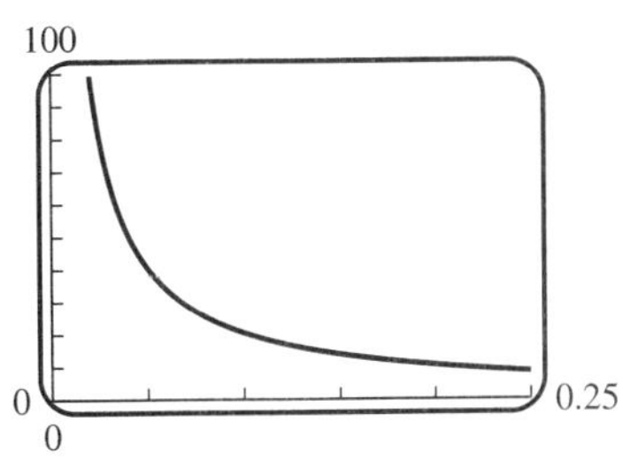

The limit is $+\infty$.

(c)

−0.25	−0.1	−0.001	−0.0001
−7.4641	−19.487	−1999.5	−20000

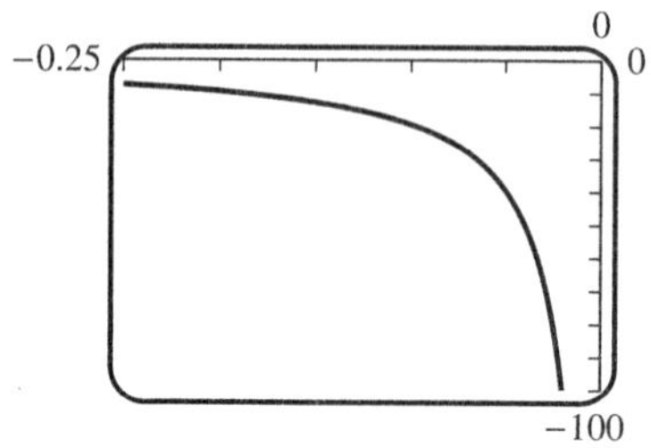

The limit is $-\infty$.

21. (a)

−0.25	−0.1	−0.001	−0.0001	0.0001	0.001	0.1	0.25
2.7266	2.9552	3.0000	3.0000	3.0000	3.0000	2.9552	2.7266

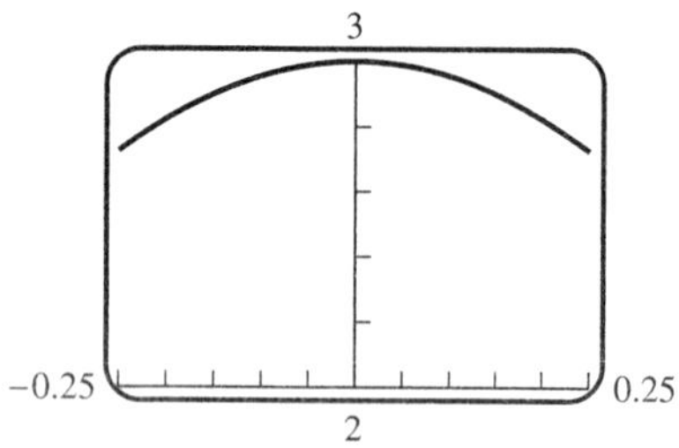

The limit is 3.

(b)

0	−0.5	−0.9	−0.99	−0.999	−1.5	−1.1	−1.01	−1.001
1	1.7552	6.2161	54.87	541.1	−0.1415	−4.536	−53.19	−539.5

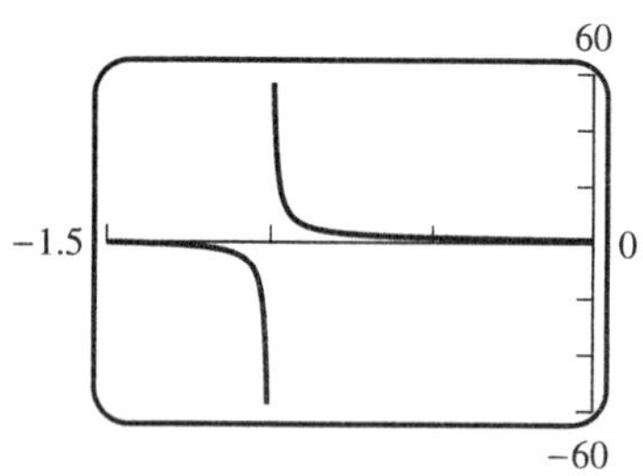

The limit does not exist.

22. (a)

0	−0.5	−0.9	−0.99	−0.999	−1.5	−1.1	−1.01	−1.001
1.5574	1.0926	1.0033	1.0000	1.0000	1.0926	1.0033	1.0000	1.0000

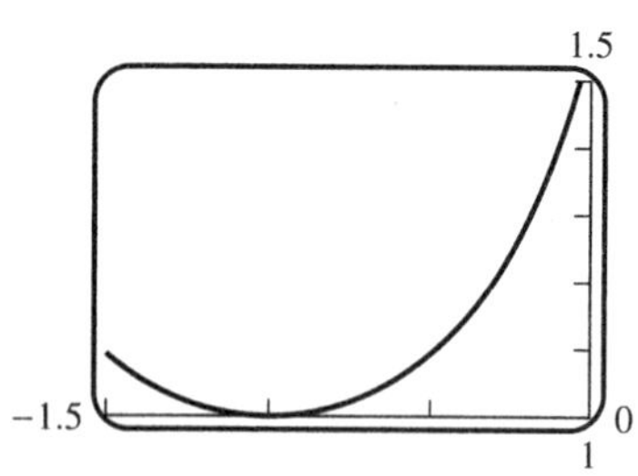

The limit is 1.

(b)

−0.25	−0.1	−0.001	−0.0001	0.0001	0.001	0.1	0.25
1.9794	2.4132	2.5000	2.5000	2.5000	2.5000	2.4132	1.9794

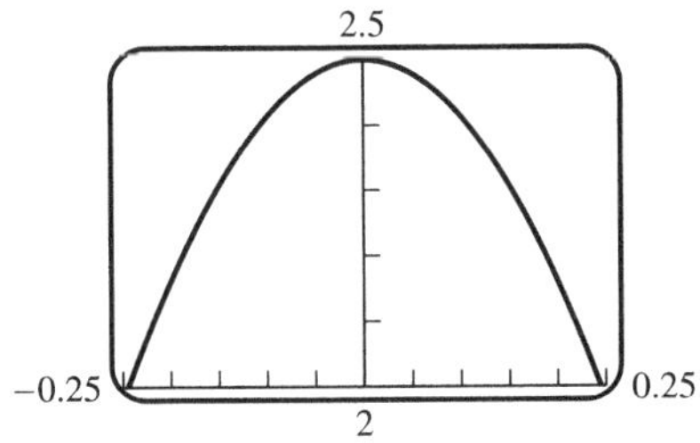

The limit is 5/2.

23. The height of the ball at time $t = 0.25 + \Delta t$ is $s(0.25 + \Delta t) = -16(0.25 + \Delta t)^2 + 29(0.25 + \Delta t) + 6$, so the distance traveled over the interval from $t = 0.25 - \Delta t$ to $t = 0.25 + \Delta t$ is $s(0.25 + \Delta t) - s(0.25 - \Delta t) = -64(0.25)\Delta t + 58\Delta t$.
Thus the average velocity over the same interval is given by
$v_{\text{ave}} = [s(0.25 + \Delta t) - s(0.25 - \Delta t)]/2\Delta t = (-64(0.25)\Delta t + 58\Delta t)/2\Delta t = 21$ ft/s,
and this will also be the instantaneous velocity, since it happens to be independent of Δt.

24. The height of the ball at time $t = 0.75 + \Delta t$ is $s(0.75 + \Delta t) = -16(0.75 + \Delta t)^2 + 29(0.75 + \Delta t) + 6$, so the distance traveled over the interval from $t = 0.75 - \Delta t$ to $t = 0.75 + \Delta t$ is $s(0.75 + \Delta t) - s(0.75 - \Delta t) = -64(0.75)\Delta t + 58\Delta t$.
Thus the average velocity over the same interval is given by
$v_{\text{ave}} = [s(0.75 + \Delta t) - s(0.75 - \Delta t)]/2\Delta t = (-64(0.75)\Delta t + 58\Delta t)/2\Delta t = 5$ ft/s,
and this will also be the instantaneous velocity, since it happens to be independent of Δt.

25. (a)

−100,000,000	−100,000	−1000	−100	−10	10	100	1000
2.0000	2.0001	2.0050	2.0521	2.8333	1.6429	1.9519	1.9950

100,000	100,000,000
2.0000	2.0000

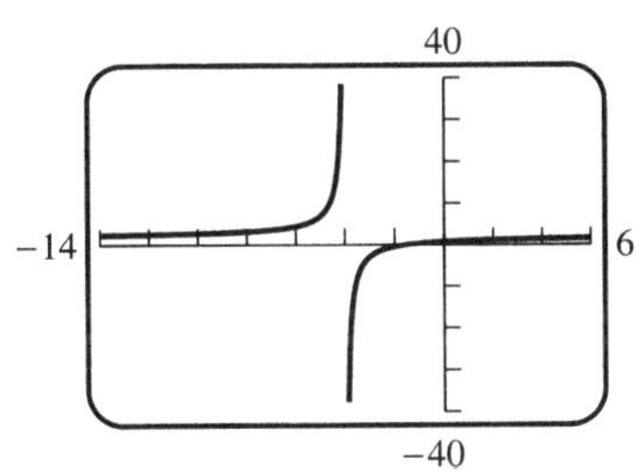

asymptote $y = 2$ as $x \to \pm\infty$

(b)

−100,000,000	−100,000	−1000	−100	−10	10	100	1000
20.0855	20.0864	20.1763	21.0294	35.4013	13.7858	19.2186	19.9955

100,000	100,000,000
20.0846	20.0855

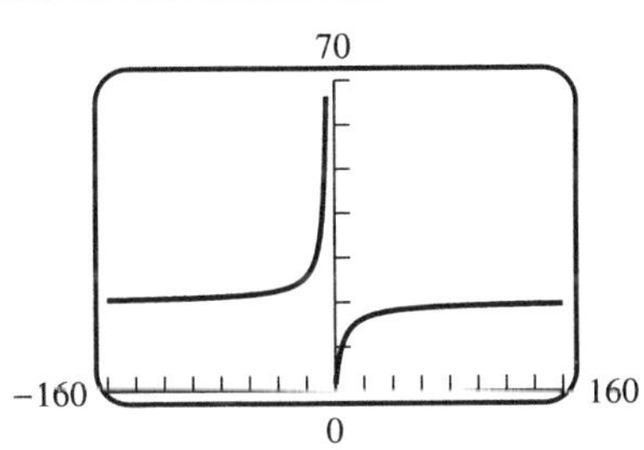

asymptote $y = 20.086$.

(c)

−100,000,000	−100,000	−1000	−100	−10	10	100	1000	100,000	100,000,000
−100,000,001	−100,000	−1001	−101.0	−11.2	9.2	99.0	999.0	99,999	99,999,999

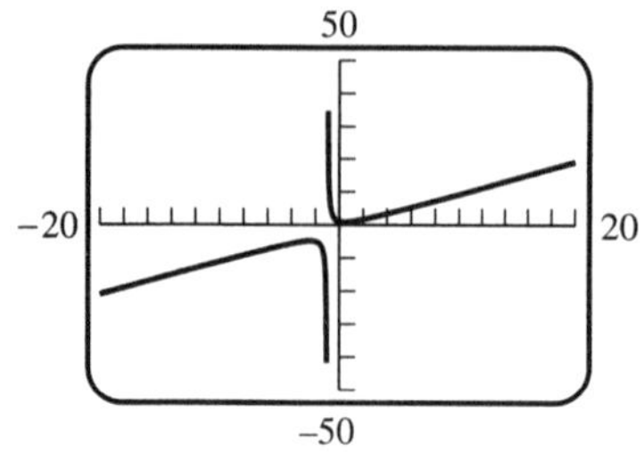

no horizontal asymptote

26. (a)

−100,000,000	−100,000	−1000	−100	−10	10	100	1000	100,000	100,000,000
0.2000	0.2000	0.2000	0.2000	0.1976	0.1976	0.2000	0.2000	0.2000	0.2000

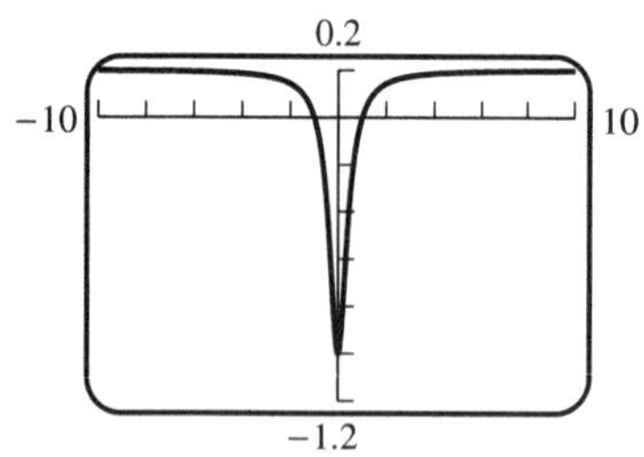

asymptote $y = 1/5$ as $x \to \pm\infty$

(b)

−100,000,000	−100,000	−1000	−100	−10	10	100
0.0000	0.0000	0.0000	0.0000	0.0016	1668.0	2.09×10^{18}

1000	100,000	100,000,000
1.77×10^{301}	?	?

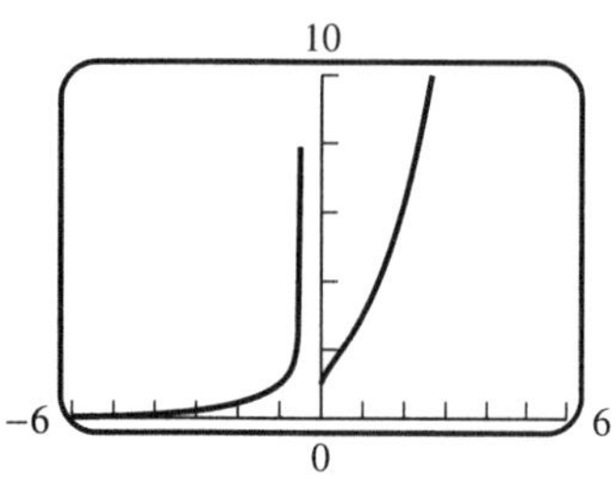

asymptote $y = 0$ as $x \to -\infty$, none as $x \to +\infty$

(c)

−100,000,000	−100,000	−1000	−100	−10	10	100
0.0000	0.0000	0.0008	−0.0051	−0.0544	−0.0544	−0.0051

1000	100,000	100,000,000
0.0008	0.0000	0.0000

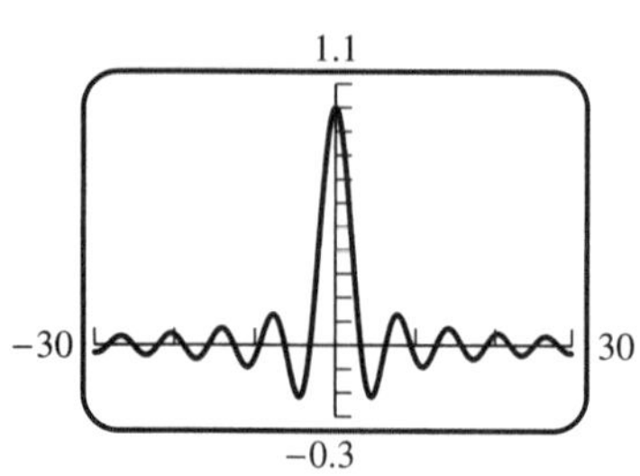

asymptote $y = 0$ as $x \to \pm\infty$

27. It appears that $\lim_{t\to+\infty} n(t) = +\infty$, and $\lim_{t\to+\infty} e(t) = c$.

28. **(a)** It is the initial temperature of the oven.
(b) It is the ambient temperature, i.e. the temperature of the room.

29. **(a)** $\lim_{t\to 0^+} \dfrac{\sin t}{t}$ **(b)** $\lim_{t\to 0^+} \dfrac{t-1}{t+1}$ **(c)** $\lim_{t\to 0^-} (1+2t)^{1/t}$

30. **(a)** $\lim_{t\to 0^+} \dfrac{\cos \pi t}{\pi t}$ **(b)** $\lim_{t\to 0^+} \dfrac{1}{t+1}$ **(c)** $\lim_{t\to 0^-} \left(1+\dfrac{2}{t}\right)^t$

31. $\lim_{x\to-\infty} f(x) = L$ and $\lim_{x\to+\infty} = L$

32. **(a)** no
(b) yes; $\tan x$ and $\sec x$ at $x = n\pi + \pi/2$, and $\cot x$ and $\csc x$ at $x = n\pi$, $n = 0, \pm 1, \pm 2, \ldots$

33. **(a)** The limit appears to be 3.

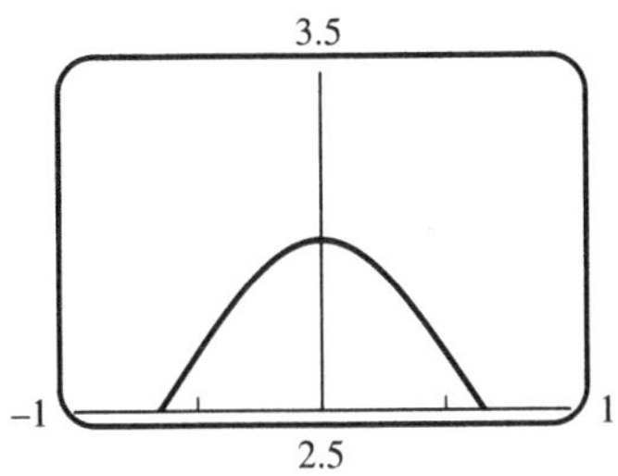

(b) The limit appears to be 3.

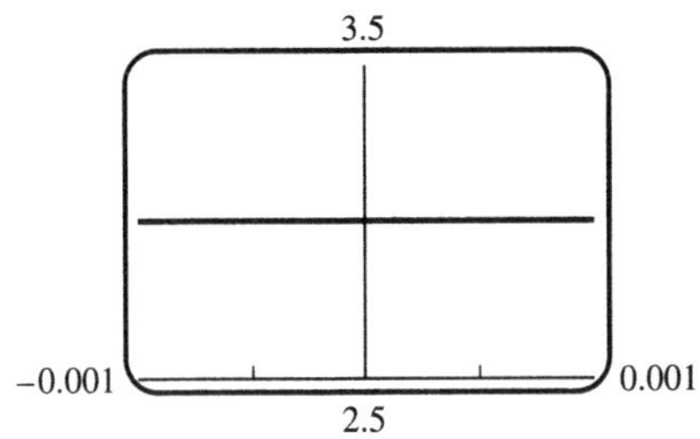

(c) The limit does not exist.

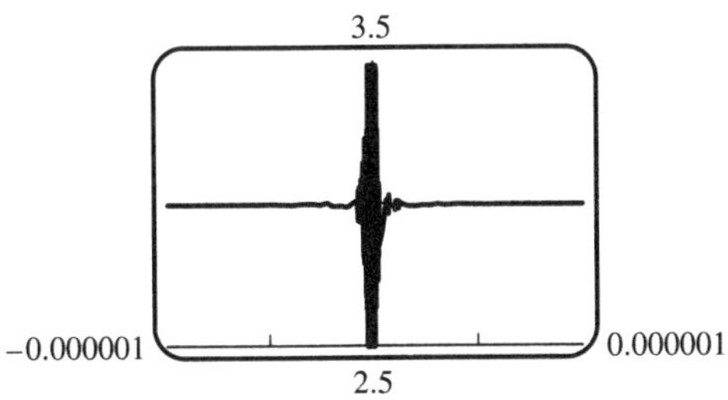

35. **(a)** The plot over the interval $[-a, a]$ becomes subject to catastrophic subtraction if a is small enough (the size depending on the machine).
(c) It does not.

EXERCISE SET 2.2

1. **(a)** 7 **(b)** π **(c)** -6 **(d)** 36

2. **(a)** 1 **(b)** -1 **(c)** 1 **(d)** -1

3. **(a)** -6 **(b)** 13 **(c)** -8 **(d)** 16 **(e)** 2 **(f)** $-1/2$
(g) The limit doesn't exist because the denominator tends to zero but the numerator doesn't.
(h) The limit doesn't exist because the denominator tends to zero but the numerator doesn't.

4. (a) 0
(b) The limit doesn't exist because $\lim f$ doesn't exist and $\lim g$ does.
(c) 0 (d) 3 (e) 0
(f) The limit doesn't exist because the denominator tends to zero but the numerator doesn't.
(g) The limit doesn't exist because $\sqrt{f(x)}$ is not defined for $0 \le x < 2$.
(h) 1

5. 0 6. 3/4 7. 8 8. −3

9. 4 10. 12 11. −4/5 12. 0

13. 3/2 14. 4/3 15. $+\infty$ 16. $-\infty$

17. does not exist 18. $+\infty$ 19. $-\infty$ 20. does not exist

21. $+\infty$ 22. $-\infty$ 23. does not exist 24. $-\infty$

25. $+\infty$ 26. does not exist 27. $+\infty$ 28. $+\infty$

29. 6 30. 4 32. −19

33. (a) 2 (b) 2 (c) 2

34. (a) −2 (b) 0 (c) does not exist

35. (a) 3 (b)

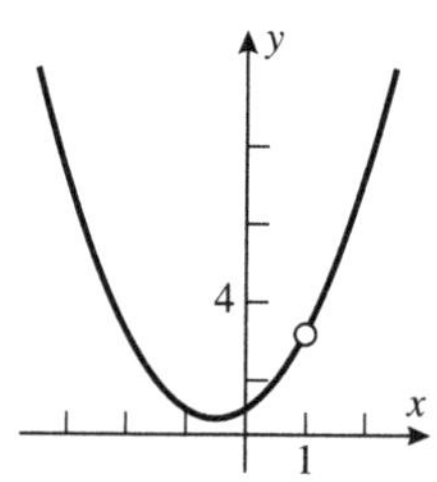

36. (a) −6 (b) $F(x) = x - 3$

37. (a) Theorem 2.2.2(a) doesn't apply; moreover one cannot add/subtract infinities.
(b) $\lim_{x\to 0^+}\left(\frac{1}{x} - \frac{1}{x^2}\right) = \lim_{x\to 0^+}\left(\frac{x-1}{x^2}\right) = -\infty$

38. $\lim_{x\to 0^-}\left(\frac{1}{x} + \frac{1}{x^2}\right) = \lim_{x\to 0^-}\frac{x+1}{x^2} = +\infty$

39. $\lim_{x\to 0}\frac{x}{x\left(\sqrt{x+4}+2\right)} = \frac{1}{4}$

40. $\lim_{x\to 0}\frac{x^2}{x\left(\sqrt{x+4}+2\right)} = 0$

41. The left and/or right limits could be plus or minus infinity; or the limit could exist, or equal any preassigned real number. For example, let $q(x) = x - x_0$ and let $p(x) = a(x - x_0)^n$ where n takes on the values $0, 1, 2$.

EXERCISE SET 2.3

1. (a) -3 (b) $-\infty$

2. (a) 1 (b) -1

3. (a) -12 (b) 21 (c) -15 (d) 25
(e) 2 (f) $-3/5$ (g) 0
(h) The limit doesn't exist because the denominator tends to zero but the numerator doesn't.

4. (a) 20 (b) 0 (c) $+\infty$ (d) $-\infty$
(e) $-42^{1/3}$ (f) $-6/7$ (g) 7 (h) $-7/12$

5. $+\infty$

6. 5

7. $-\infty$

8. $+\infty$

9. $+\infty$

10. $+\infty$

11. $3/2$

12. $5/2$

13. 0

14. 0

15. 0

16. $5/3$

17. $-5^{1/3}/2$

18. $\sqrt[3]{3/2}$

19. $-\sqrt{5}$

20. $\sqrt{5}$

21. $1/\sqrt{6}$

22. $-1/\sqrt{6}$

23. $\sqrt{3}$

24. $\sqrt{3}$

25. $-\infty$

26. $+\infty$

27. $-1/7$

28. $4/7$

29. (a) $+\infty$ (b) -5

30. (a) 0 (b) -6

31. $\lim_{x\to+\infty}(\sqrt{x^2+3}-x)\dfrac{\sqrt{x^2+3}+x}{\sqrt{x^2+3}+x} = \lim_{x\to+\infty}\dfrac{3}{\sqrt{x^2+3}+x} = 0$

32. $\lim_{x\to+\infty}(\sqrt{x^2-3x}-x)\dfrac{\sqrt{x^2-3x}+x}{\sqrt{x^2-3x}+x} = \lim_{x\to+\infty}\dfrac{-3x}{\sqrt{x^2-3x}+x} = -3/2$

33. $\lim_{x\to+\infty}\left(\sqrt{x^2+ax}-x\right)\dfrac{\sqrt{x^2+ax}+x}{\sqrt{x^2+ax}+x} = \lim_{x\to+\infty}\dfrac{ax}{\sqrt{x^2+ax}+x} = a/2$

34. $\lim_{x\to+\infty}\left(\sqrt{x^2+ax}-\sqrt{x^2+bx}\right)\dfrac{\sqrt{x^2+ax}+\sqrt{x^2+bx}}{\sqrt{x^2+ax}+\sqrt{x^2+bx}} = \lim_{x\to+\infty}\dfrac{(a-b)x}{\sqrt{x^2+ax}+\sqrt{x^2+bx}} = \dfrac{a-b}{2}$

35. $\lim_{x\to+\infty}p(x)=(-1)^n\infty$ and $\lim_{x\to-\infty}p(x)=+\infty$

36. If $m>n$ the limits are both zero. If $m=n$ the limits are both 1. If $n>m$ the limits are $(-1)^{n+m}\infty$ and $+\infty$, respectively.

37. If $m>n$ the limits are both zero. If $m=n$ the limits are both equal to a_m, the leading coefficient of p. If $n>m$ the limits are $\pm\infty$ where the sign depends on the sign of a_m and whether n is even or odd.

38. (a) $p(x)=q(x)=x$ (b) $p(x)=x$, $q(x)=x^2$
(c) $p(x)=x^2$, $q(x)=x$ (d) $p(x)=x+3$, $q(x)=x$

39. If $m > n$ the limit is 0. If $m = n$ the limit is -3. If $m < n$ and $n - m$ is odd, then the limit is $+\infty$; if $m < n$ and $n - m$ is even, then the limit is $-\infty$.

40. If $m > n$ the limit is zero. If $m = n$ the limit is c_m/d_m. If $n > m$ the limit is $\pm\infty$, where the sign depends on the signs of c_n and d_m.

41. $f(x) = x + 2 + \dfrac{2}{x-2}$, so $\lim\limits_{x\to\pm\infty}(f(x) - (x+2)) = 0$ and $f(x)$ is asymptotic to $y = x + 2$.

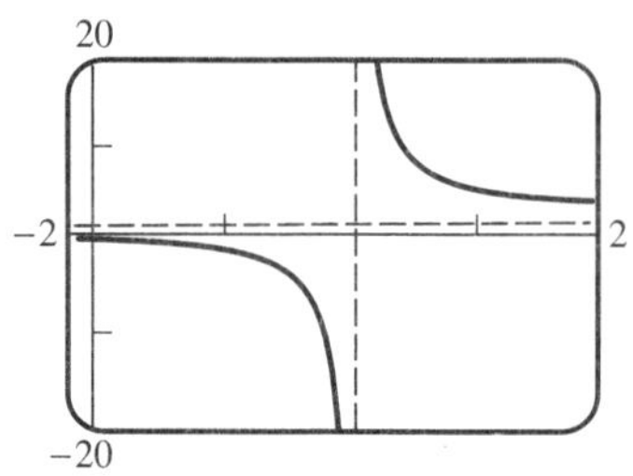

42. $f(x) = x^2 - 1 + 3/x$, so $\lim\limits_{x\to\pm\infty}[f(x) - (x^2 - 1) = 0]$ and $f(x)$ is asymptotic to $y = x^2 - 1$.

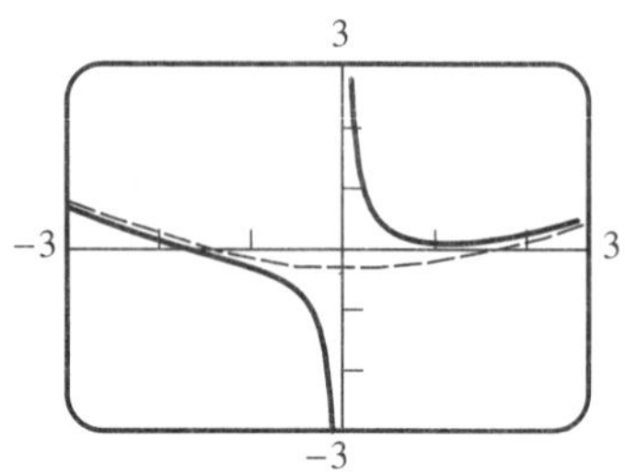

43. $f(x) = -x^2 + 1 + 2/(x-3)$ so $\lim\limits_{x\to\pm\infty}[f(x) - (-x^2+1)] = 0$ and $f(x)$ is asymptotic to $y = -x^2 + 1$.

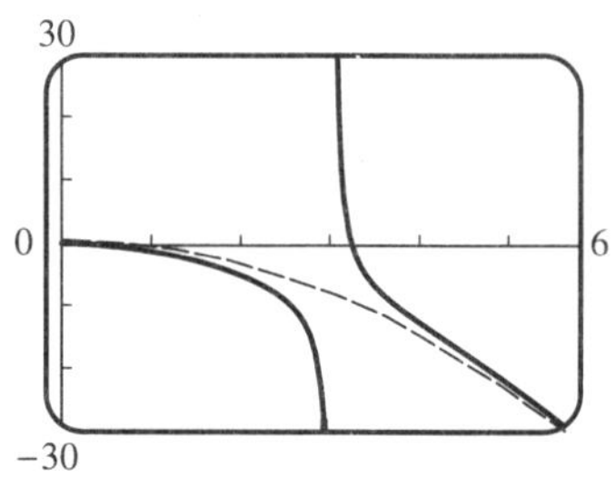

44. $f(x) = x^3 + \dfrac{3}{2(x-1)} - \dfrac{3}{2(x+1)}$ so $\lim\limits_{x\to\pm\infty}[f(x) - x^3] = 0$ and $f(x)$ is asymptotic to $y = x^3$.

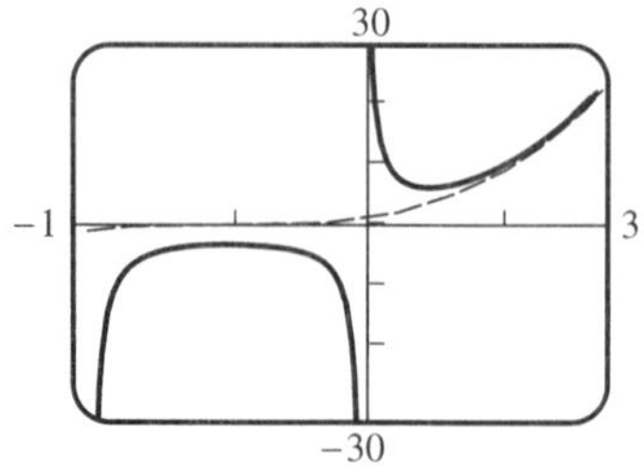

45. $f(x) - \sin x = 0$ and $f(x)$ is asymptotic to $y = \sin x$.

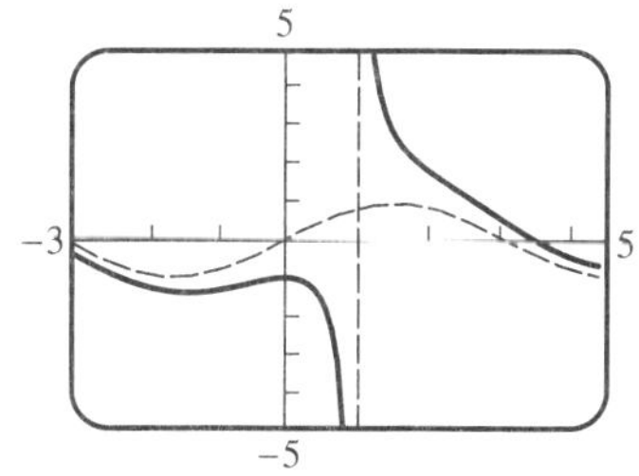

46. Note that the function is not defined for $-1 < x <= 1$. For x outside this interval we have $f(x) = \sqrt{x^2 + \dfrac{2}{x-1}}$ which suggests that $\lim_{x\to\pm\infty}[f(x) - |x|] = 0$ (this can be checked with a CAS) and hence $f(x)$ is asymptotic to $y = |x|$.

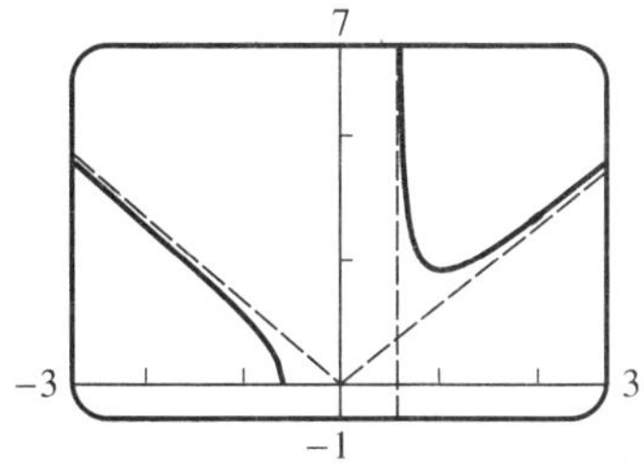

EXERCISE SET 2.4

1. **(a)** $|f(x) - f(0)| = |x + 2 - 2| = |x| < 0.1$ if and only if $|x| < 0.1$
(b) $|f(x) - f(3)| = |(4x - 5) - 7| = 4|x - 3| < 0.1$ if and only if $|x - 3| < (0.1)/4 = 0.0025$
(c) $|f(x) - f(4)| = |x^2 - 16| < \epsilon$ if $|x - 4| < \delta$. We get $f(x) = 16 + \epsilon = 16.001$ at $x = 4.000124998$, which corresponds to $\delta = 0.000124998$; and $f(x) = 16 - \epsilon = 15.999$ at $x = 3.999874998$, for which $\delta = 0.000125002$. Use the smaller δ: thus $|f(x) - 16| < \epsilon$ provided $|x - 4| < 0.000125$ (to six decimals).

2. **(a)** $|f(x) - f(0)| = |2x + 3 - 3| = 2|x| < 0.1$ if and only if $|x| < 0.05$
(b) $|f(x) - f(0)| = |2x + 3 - 3| = 2|x| < 0.01$ if and only if $|x| < 0.005$
(c) $|f(x) - f(0)| = |2x + 3 - 3| = 2|x| < 0.0012$ if and only if $|x| < 0.0006$

3. **(a)** $x_1 = (1.95)^2 = 3.8025, x_2 = (2.05)^2 = 4.2025$
(b) $\delta = \min(|4 - 3.8025|, |4 - 4.2025|) = 0.1975$

4. **(a)** $x_1 = 1/(1.1) = 0.909090\ldots, x_2 = 1/(0.9) = 1.111111\ldots$
(b) $\delta = \min(|1 - 0.909090|, |1 - 1.111111|) = 0.0909090\ldots$

5. $|(x^3-4x+5)-2| < 0.05$, $-0.05 < (x^3-4x+5)-2 < 0.05$, $1.95 < x^3-4x+5 < 2.05$; $x^3-4x+5 = 1.95$ at $x = 1.0616$, $x^3 - 4x + 5 = 2.05$ at $x = 0.9558$; $\delta = \min(1.0616 - 1, 1 - 0.9558) = 0.0442$

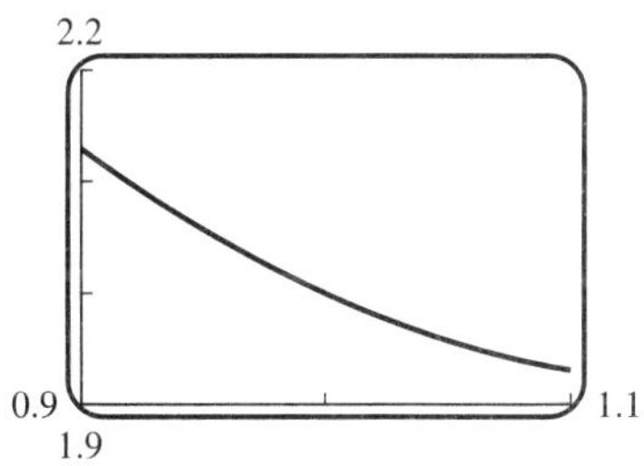

6. $\sqrt{5x+1} = 3.5$ at $x = 2.25$, $\sqrt{5x+1} = 4.5$ at $x = 3.85$, so $\delta = \min(3 - 2.25, 3.85 - 3) = 0.75$

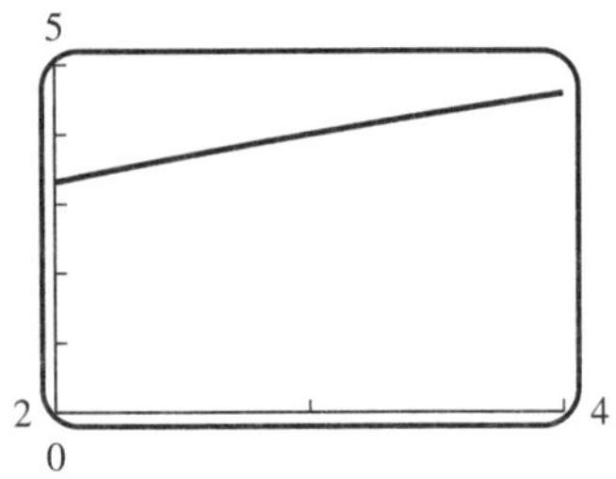

7. With the TRACE feature of a calculator we discover that (to five decimal places) (0.87000, 1.80274) and (1.13000, 2.19301) belong to the graph. Set $x_0 = 0.87$ and $x_1 = 1.13$. Since the graph of $f(x)$ rises from left to right, we see that if $x_0 < x < x_1$ then $1.80274 < f(x) < 2.19301$, and therefore $1.8 < f(x) < 2.2$. So we can take $\delta = 0.13$.

8. From a calculator plot we conjecture that $\lim_{x\to 1} f(x) = 2$. Using the TRACE feature we see that the points $(\pm 0.2, 1.94709)$ belong to the graph. Thus if $-0.2 < x < 0.2$ then $1.95 < f(x) \le 2$ and hence $|f(x) - L| < 0.05 < 0.1 = \epsilon$.

9. $|2x - 8| = 2|x - 4| < 0.1$ if $|x - 4| < 0.05$, $\delta = 0.05$

10. $|x/2 + 1| = (1/2)|x - (-2)| < 0.1$ if $|x + 2| < 0.2$, $\delta = 0.2$

11. $|7x + 5 - (-2)| = 7|x - (-1)| < 0.01$ if $|x + 1| < \frac{1}{700}$, $\delta = \frac{1}{700}$

12. $|5x - 2 - 13| = 5|x - 3| < 0.01$ if $|x - 3| < \frac{1}{500}$, $\delta = \frac{1}{500}$

13. $\left|\dfrac{x^2-4}{x-2} - 4\right| = \left|\dfrac{x^2 - 4 - 4x + 8}{x-2}\right| = |x - 2| < 0.05$ if $|x - 2| < 0.05$, $\delta = 0.05$

14. $\left|\dfrac{x^2-1}{x+1} - (-2)\right| = \left|\dfrac{x^2 - 1 + 2x + 2}{x+1}\right| = |x + 1| < 0.05$ if $|x + 1| < 0.05$, $\delta = 0.05$

15. if $\delta < 1$ then $|x^2 - 16| = |x - 4||x + 4| < 9|x - 4| < 0.001$ if $|x - 4| < \frac{1}{9000}$, $\delta = \frac{1}{9000}$

16. if $\delta < 1$ then $|\sqrt{x} - 3|\left|\dfrac{\sqrt{x}+3}{\sqrt{x}+3}\right| = \dfrac{|x-9|}{|\sqrt{x}+3|} < \dfrac{|x-9|}{\sqrt{8}+3} < \dfrac{1}{4}|x - 9| < 0.001$ if $|x - 9| < 0.004$, $\delta = 0.004$

17. if $\delta \le 1$ then $\left|\dfrac{1}{x} - \dfrac{1}{5}\right| = \dfrac{|x-5|}{5|x|} \le \dfrac{|x-5|}{20} < 0.05$ if $|x - 5| < 1$, $\delta = 1$

18. $|x - 0| = |x| < 0.05$ if $|x| < 0.05$, $\delta = 0.05$

19. $|3x - 15| = 3|x - 5| < \epsilon$ if $|x - 5| < \frac{1}{3}\epsilon$, $\delta = \frac{1}{3}\epsilon$

20. $|(4x - 5) - 7| = |4x - 12| = 4|x - 3| < \epsilon$ if $|x - 3| < \frac{1}{4}\epsilon$, $\delta = \frac{1}{4}\epsilon$

21. $|2x - 7 - (-3)| = 2|x - 2| < \epsilon$ if $|x - 2| < \frac{1}{2}\epsilon$, $\delta = \frac{1}{2}\epsilon$

22. $|2 - 3x - 5| = 3|x + 1| < \epsilon$ if $|x + 1| < \frac{1}{3}\epsilon$, $\delta = \frac{1}{3}\epsilon$

23. $\left|\dfrac{x^2+x}{x} - 1\right| = |x| < \epsilon$ if $|x| < \epsilon$, $\delta = \epsilon$

24. $\left|\dfrac{x^2-9}{x+3}-(-6)\right| = |x+3| < \epsilon$ if $|x+3| < \epsilon$, $\delta = \epsilon$

25. if $\delta < 1$ then $|2x^2-2| = 2|x-1||x+1| < 6|x-1| < \epsilon$ if $|x-1| < \frac{1}{6}\epsilon$, $\delta = \min(1, \frac{1}{6}\epsilon)$

26. if $\delta < 1$ then $|x^2-5-4| = |x-3||x+3| < 7|x-3| < \epsilon$ if $|x-3| < \frac{1}{7}\epsilon$, $\delta = \min(1, \frac{1}{7}\epsilon)$

27. if $\delta < \dfrac{1}{6}$ then $\left|\dfrac{1}{x}-3\right| = \dfrac{3\left|x-\frac{1}{3}\right|}{|x|} < 18\left|x-\dfrac{1}{3}\right| < \epsilon$ if $\left|x-\dfrac{1}{3}\right| < \frac{1}{18}\epsilon$, $\delta = \min\left(\dfrac{1}{6}, \dfrac{1}{18}\epsilon\right)$

28. If $\delta < \frac{1}{2}$ and $|x-(-2)| < \delta$ then $-\frac{5}{2} < x < -\frac{3}{2}$, $x+1 < -\frac{1}{2}$, $|x+1| > \frac{1}{2}$; then $\left|\dfrac{1}{x+1}-(-1)\right| = \dfrac{|x+2|}{|x+1|} < 2|x+2| < \epsilon$ if $|x+2| < \dfrac{1}{2}\epsilon$, $\delta = \min\left(\dfrac{1}{2}, \dfrac{1}{2}\epsilon\right)$

29. $|\sqrt{x}-2| = \left|(\sqrt{x}-2)\dfrac{\sqrt{x}+2}{\sqrt{x}+2}\right| = \left|\dfrac{x-4}{\sqrt{x}+2}\right| < \dfrac{1}{2}|x-4| < \epsilon$ if $|x-4| < 2\epsilon$, $\delta = 2\epsilon$

30. $|\sqrt{x+3}-3|\left|\dfrac{\sqrt{x+3}+3}{\sqrt{x+3}+3}\right| = \dfrac{|x-6|}{\sqrt{x+3}+3} \le \dfrac{1}{3}|x-6| < \epsilon$ if $|x-6| < 3\epsilon$, $\delta = 3\epsilon$

31. $|f(x)-3| = |x+2-3| = |x-1| < \epsilon$ if $0 < |x-1| < \epsilon$, $\delta = \epsilon$

32. If $\delta < 1$ then $|(x^2+3x-1)-9| = |(x-2)(x+5)| < 8|x-2| < \epsilon$ if $|x-2| < \frac{1}{8}\epsilon$, $\delta = \min(1, \frac{1}{8}\epsilon)$

33. **(a)** $|f(x)-L| = \dfrac{1}{x^2} < 0.1$ if $x > \sqrt{10}$, $N = \sqrt{10}$

(b) $|f(x)-L| = \left|\dfrac{x}{x+1}-1\right| = \left|\dfrac{1}{x+1}\right| < 0.01$ if $x+1 > 100$, $N = 99$

(c) $|f(x)-L| = \left|\dfrac{1}{x^3}\right| < \dfrac{1}{1000}$ if $|x| > 10$, $x < -10$, $N = -10$

(d) $|f(x)-L| = \left|\dfrac{x}{x+1}-1\right| = \left|\dfrac{1}{x+1}\right| < 0.01$ if $|x+1| > 100$, $-x-1 > 100$, $x < -101$, $N = -101$

34. **(a)** $\left|\dfrac{1}{x^3}\right| < 0.1$, $x > 10^{1/3}$, $N = 10^{1/3}$ **(b)** $\left|\dfrac{1}{x^3}\right| < 0.01$, $x > 100^{1/3}$, $N = 100^{1/3}$

(c) $\left|\dfrac{1}{x^3}\right| < 0.001$, $x > 10$, $N = 10$

35. **(a)** $\dfrac{x_1^2}{1+x_1^2} = 1-\epsilon$, $x_1 = -\sqrt{\dfrac{1-\epsilon}{\epsilon}}$; $\dfrac{x_2^2}{1+x_2^2} = 1-\epsilon$, $x_2 = \sqrt{\dfrac{1-\epsilon}{\epsilon}}$

(b) $N = \sqrt{\dfrac{1-\epsilon}{\epsilon}}$ **(c)** $N = -\sqrt{\dfrac{1-\epsilon}{\epsilon}}$

36. **(a)** $x_1 = -1/\epsilon^3$; $x_2 = 1/\epsilon^3$ **(b)** $N = 1/\epsilon^3$ **(c)** $N = -1/\epsilon^3$

37. $\dfrac{1}{x^2} < 0.01$ if $|x| > 10$, $N = 10$

38. $\dfrac{1}{x+2} < 0.005$ if $|x+2| > 200$, $x > 198$, $N = 198$

39. $\left|\dfrac{x}{x+1}-1\right| = \left|\dfrac{1}{x+1}\right| < 0.001$ if $|x+1| > 1000$, $x > 999$, $N = 999$

40. $\left|\frac{4x-1}{2x+5}-2\right| = \left|\frac{11}{2x+5}\right| < 0.1$ if $|2x+5| > 110$, $2x > 105$, $N = 52.5$

41. $\left|\frac{1}{x+2}-0\right| < 0.005$ if $|x+2| > 200$, $-x-2 > 200$, $x < -202$, $N = -202$

42. $\left|\frac{1}{x^2}\right| < 0.01$ if $|x| > 10$, $-x > 10$, $x < -10$, $N = -10$

43. $\left|\frac{4x-1}{2x+5}-2\right| = \left|\frac{11}{2x+5}\right| < 0.1$ if $|2x+5| > 110$, $-2x-5 > 110$, $2x < -115$, $x < -57.5$, $N = -57.5$

44. $\left|\frac{x}{x+1}-1\right| = \left|\frac{1}{x+1}\right| < 0.001$ if $|x+1| > 1000$, $-x-1 > 1000$, $x < -1001$, $N = -1001$

45. $\left|\frac{1}{x^2}\right| < \epsilon$ if $|x| > \frac{1}{\sqrt{\epsilon}}$, $N = \frac{1}{\sqrt{\epsilon}}$

46. $\left|\frac{1}{x}\right| < \epsilon$ if $|x| > \frac{1}{\epsilon}$, $-x > \frac{1}{\epsilon}$, $x < -\frac{1}{\epsilon}$, $N = -\frac{1}{\epsilon}$

47. $\left|\frac{1}{x+2}\right| < \epsilon$ if $|x+2| > \frac{1}{\epsilon}$, $-x-2 < \frac{1}{\epsilon}$, $x > -2-\frac{1}{\epsilon}$, $N = -2-\frac{1}{\epsilon}$

48. $\left|\frac{1}{x+2}\right| < \epsilon$ if $|x+2| > \frac{1}{\epsilon}$, $x+2 > \frac{1}{\epsilon}$, $x > \frac{1}{\epsilon}-2$, $N = \frac{1}{\epsilon}-2$

49. $\left|\frac{x}{x+1}-1\right| = \left|\frac{1}{x+1}\right| < \epsilon$ if $|x+1| > \frac{1}{\epsilon}$, $x > \frac{1}{\epsilon}-1$, $N = \frac{1}{\epsilon}-1$

50. $\left|\frac{x}{x+1}-1\right| = \left|\frac{1}{x+1}\right| < \epsilon$ if $|x+1| > \frac{1}{\epsilon}$, $-x-1 > \frac{1}{\epsilon}$, $x < -1-\frac{1}{\epsilon}$, $N = -1-\frac{1}{\epsilon}$

51. $\left|\frac{4x-1}{2x+5}-2\right| = \left|\frac{11}{2x+5}\right| < \epsilon$ if $|2x+5| > \frac{11}{\epsilon}$, $-2x-5 > \frac{11}{\epsilon}$, $2x < -\frac{11}{\epsilon}-5$, $x < -\frac{11}{2\epsilon}-\frac{5}{2}$,
$N = -\frac{5}{2}-\frac{11}{2\epsilon}$

52. $\left|\frac{4x-1}{2x+5}-2\right| = \left|\frac{11}{2x+5}\right| < \epsilon$ if $|2x+5| > \frac{11}{\epsilon}$, $2x > \frac{11}{\epsilon}-5$, $x > \frac{11}{2\epsilon}-\frac{5}{2}$, $N = \frac{11}{2\epsilon}-\frac{5}{2}$

53. **(a)** $\frac{1}{x^2} > 100$ if $|x| < \frac{1}{10}$ **(b)** $\frac{1}{|x-1|} > 1000$ if $|x-1| < \frac{1}{1000}$

(c) $\frac{-1}{(x-3)^2} < -1000$ if $|x-3| < \frac{1}{10\sqrt{10}}$ **(d)** $-\frac{1}{x^4} < -10000$ if $x^4 < \frac{1}{10000}$, $|x| < \frac{1}{10}$

54. **(a)** $\frac{1}{(x-1)^2} > 10$ if and only if $|x-1| < \frac{1}{\sqrt{10}}$

(b) $\frac{1}{(x-1)^2} > 1000$ if and only if $|x-1| < \frac{1}{10\sqrt{10}}$

(c) $\frac{1}{(x-1)^2} > 100000$ if and only if $|x-1| < \frac{1}{100\sqrt{10}}$

55. if $M > 0$ then $\frac{1}{(x-3)^2} > M$, $0 < (x-3)^2 < \frac{1}{M}$, $0 < |x-3| < \frac{1}{\sqrt{M}}$, $\delta = \frac{1}{\sqrt{M}}$

56. if $M < 0$ then $\dfrac{-1}{(x-3)^2} < M$, $0 < (x-3)^2 < -\dfrac{1}{M}$, $0 < |x-3| < \dfrac{1}{\sqrt{-M}}$, $\delta = \dfrac{1}{\sqrt{-M}}$

57. if $M > 0$ then $\dfrac{1}{|x|} > M$, $0 < |x| < \dfrac{1}{M}$, $\delta = \dfrac{1}{M}$

58. if $M > 0$ then $\dfrac{1}{|x-1|} > M$, $0 < |x-1| < \dfrac{1}{M}$, $\delta = \dfrac{1}{M}$

59. if $M < 0$ then $-\dfrac{1}{x^4} < M$, $0 < x^4 < -\dfrac{1}{M}$, $|x| < \dfrac{1}{(-M)^{1/4}}$, $\delta = \dfrac{1}{(-M)^{1/4}}$

60. if $M > 0$ then $\dfrac{1}{x^4} > M$, $0 < x^4 < \dfrac{1}{M}$, $x < \dfrac{1}{M^{1/4}}$, $\delta = \dfrac{1}{M^{1/4}}$

61. if $x > 2$ then $|x+1-3| = |x-2| = x-2 < \epsilon$ if $2 < x < 2+\epsilon$, $\delta = \epsilon$

62. if $x < 1$ then $|3x+2-5| = |3x-3| = 3|x-1| = 3(1-x) < \epsilon$ if $1-x < \frac{1}{3}\epsilon$, $1-\frac{1}{3}\epsilon < x < 1$, $\delta = \frac{1}{3}\epsilon$

63. if $x > 4$ then $\sqrt{x-4} < \epsilon$ if $x-4 < \epsilon^2$, $4 < x < 4+\epsilon^2$, $\delta = \epsilon^2$

64. if $x < 0$ then $\sqrt{-x} < \epsilon$ if $-x < \epsilon^2$, $-\epsilon^2 < x < 0$, $\delta = \epsilon^2$

65. if $x > 2$ then $|f(x)-2| = |x-2| = x-2 < \epsilon$ if $2 < x < 2+\epsilon$, $\delta = \epsilon$

66. if $x < 2$ then $|f(x)-6| = |3x-6| = 3|x-2| = 3(2-x) < \epsilon$ if $2-x < \frac{1}{3}\epsilon$, $2-\frac{1}{3}\epsilon < x < 2$, $\delta = \frac{1}{3}\epsilon$

67. **(a)** if $M < 0$ and $x > 1$ then $\dfrac{1}{1-x} < M$, $x-1 < -\dfrac{1}{M}$, $1 < x < 1-\dfrac{1}{M}$, $\delta = -\dfrac{1}{M}$

(b) if $M > 0$ and $x < 1$ then $\dfrac{1}{1-x} > M$, $1-x < \dfrac{1}{M}$, $1-\dfrac{1}{M} < x < 1$, $\delta = \dfrac{1}{M}$

68. **(a)** if $M > 0$ and $x > 0$ then $\dfrac{1}{x} > M$, $x < \dfrac{1}{M}$, $0 < x < \dfrac{1}{M}$, $\delta = \dfrac{1}{M}$

(b) if $M < 0$ and $x < 0$ then $\dfrac{1}{x} < M$, $-x < -\dfrac{1}{M}$, $\dfrac{1}{M} < x < 0$, $\delta = -\dfrac{1}{M}$

69. **(a)** Given any $M > 0$ there corresponds $N > 0$ such that if $x > N$ then $f(x) > M$, $x+1 > M$, $x > M-1$, $N = M-1$.

(b) Given any $M < 0$ there corresponds $N < 0$ such that if $x < N$ then $f(x) < M$, $x+1 < M$, $x < M-1$, $N = M-1$.

70. **(a)** Given any $M > 0$ there corresponds $N > 0$ such that if $x > N$ then $f(x) > M$, $x^2-3 > M$, $x > \sqrt{M+3}$, $N = \sqrt{M+3}$.

(b) Given any $M < 0$ there corresponds $N < 0$ such that if $x < N$ then $f(x) < M$, $x^3+5 < M$, $x < (M-5)^{1/3}$, $N = (M-5)^{1/3}$.

71. if $\delta \le 2$ then $|x-3| < 2$, $-2 < x-3 < 2$, $1 < x < 5$, and $|x^2-9| = |x+3||x-3| < 8|x-3| < \epsilon$ if $|x-3| < \frac{1}{8}\epsilon$, $\delta = \min\left(2, \frac{1}{8}\epsilon\right)$

72. **(a)** We don't care about the value of f at $x = a$, because the limit is only concerned with values of x near a. The condition that f be defined for all x (except possibly $x = a$) is necessary, because if some points were excluded then the limit may not exist; for example, let $f(x) = x$ if $1/x$ is not an integer and $f(1/n) = 6$. Then $\lim_{x \to 0} f(x)$ does not exist but it would if the points $1/n$ were excluded.

(b) when $x < 0$ then $\sqrt{x}$ is not defined

(c) yes; if $\delta \le 0.01$ then $x > 0$, so $\sqrt{x}$ is defined

EXERCISE SET 2.5

1. (a) no, $x = 2$ (b) no, $x = 2$ (c) no, $x = 2$ (d) yes
(e) yes (f) yes

2. (a) no, $x = 2$ (b) no, $x = 2$ (c) no, $x = 2$ (d) yes
(e) no, $x = 2$ (f) yes

3. (a) no, $x = 1, 3$ (b) yes (c) no, $x = 1$ (d) yes
(e) no, $x = 3$ (f) yes

4. (a) no, $x = 3$ (b) yes (c) yes (d) yes
(e) no, $x = 3$ (f) yes

5. (a) At $x = 3$ the one-sided limits fail to exist.
(b) At $x = -2$ the two-sided limit exists but is not equal to $F(-2)$.
(c) At $x = 3$ the limit fails to exist.

6. (a) At $x = 2$ the two-sided limit fails to exist.
(b) At $x = 3$ the two-sided limit exists but is not equal to $F(3)$.
(c) At $x = 0$ the two-sided limit fails to exist.

7. (a) 3 (b) 3

8. $-2/5$

9. (a)

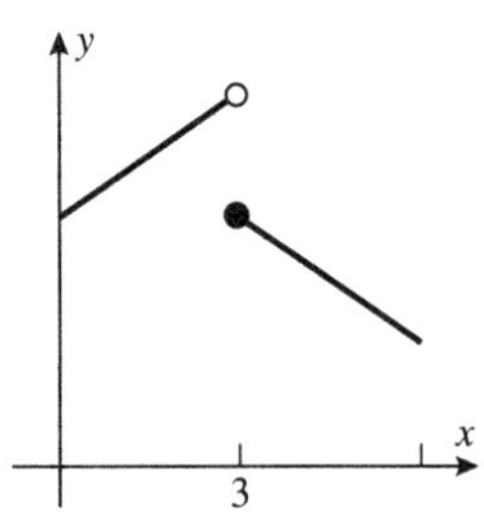

(b)

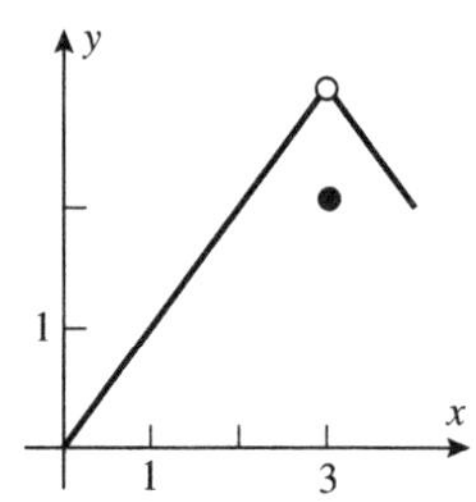

(c)

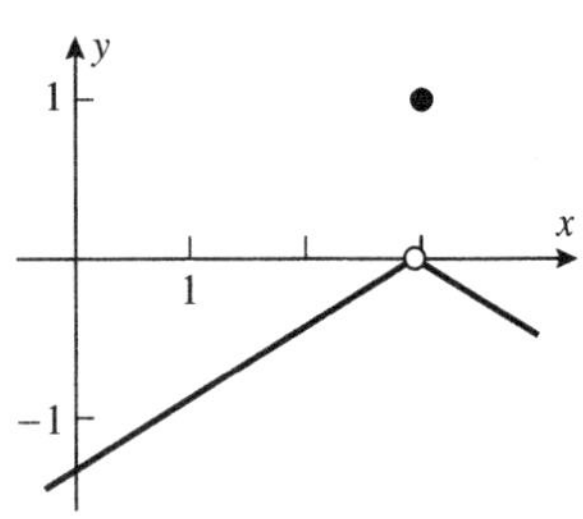

(d)

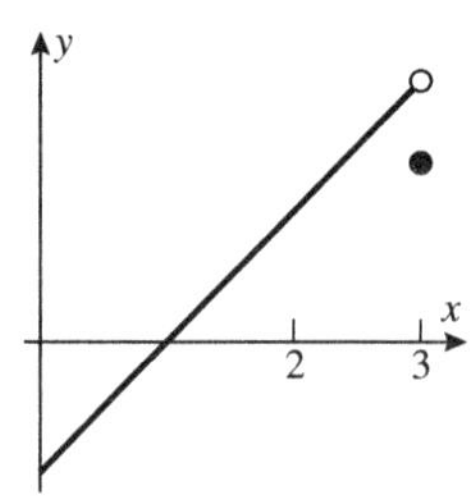

10. $f(x) = 1/x,\ g(x) = \begin{cases} 0 & \text{if } x = 0 \\ \sin\dfrac{1}{x} & \text{if } x \neq 0 \end{cases}$

11. (a)

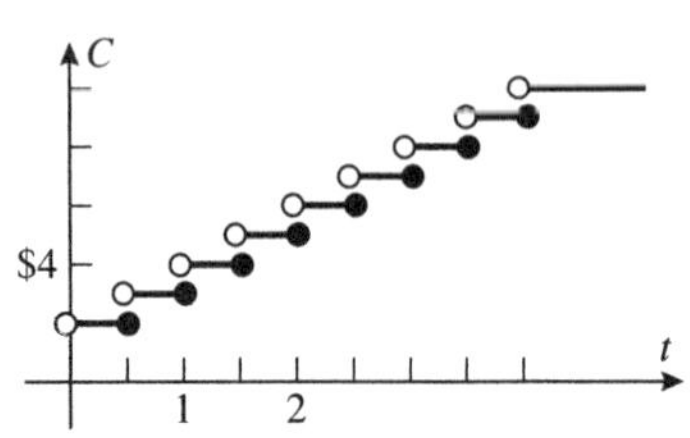

(b) One second could cost you one dollar.

12. **(a)** no; disasters (war, flood, famine, pestilence, for example) can cause discontinuities
(b) continuous
(c) not usually continuous; see Exercise 11
(d) continuous

13. none

14. none

15. none

16. f is not defined at $x = \pm 1$

17. f is not defined at $x = \pm 4$

18. f is not defined at $x = \dfrac{-7 \pm \sqrt{57}}{2}$

19. f is not defined at $x = \pm 3$

20. f is not defined at $x = 0, -4$

21. none

22. f is not defined at $x = 0, -3$

23. none; $f(x) = 2x + 3$ is continuous on $x < 4$ and $f(x) = 7 + \dfrac{16}{x}$ is continuous on $4 < x$; $\lim\limits_{x \to 4^-} f(x) = \lim\limits_{x \to 4^+} f(x) = f(4) = 11$ so f is continuous at $x = 4$

24. $\lim\limits_{x \to 1} f(x)$ does not exist so f is discontinuous at $x = 1$

25. **(a)** f is continuous for $x < 1$, and for $x > 1$; $\lim\limits_{x \to 1^-} f(x) = 5$, $\lim\limits_{x \to 1^+} f(x) = k$, so if $k = 5$ then f is continuous for all x

(b) f is continuous for $x < 2$, and for $x > 2$; $\lim\limits_{x \to 2^-} f(x) = 4k$, $\lim\limits_{x \to 2^+} f(x) = 4 + k$, so if $4k = 4 + k$, $k = 4/3$ then f is continuous for all x

26. **(a)** no, f is not defined at $x = 2$
(b) no, f is not defined for $x \le 2$
(c) yes
(d) no, f is not defined for $x \le 2$

27. **(a)**

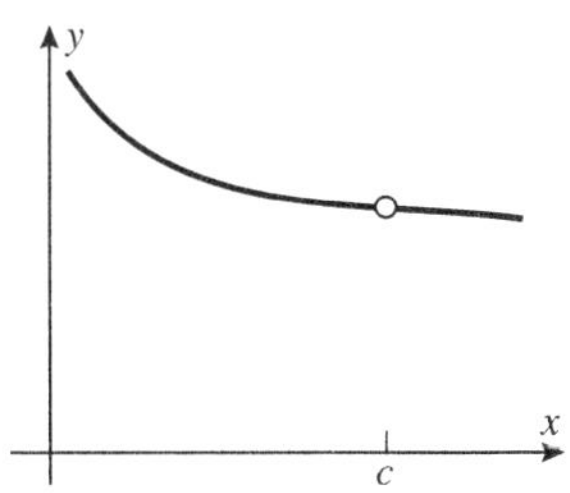

(b)

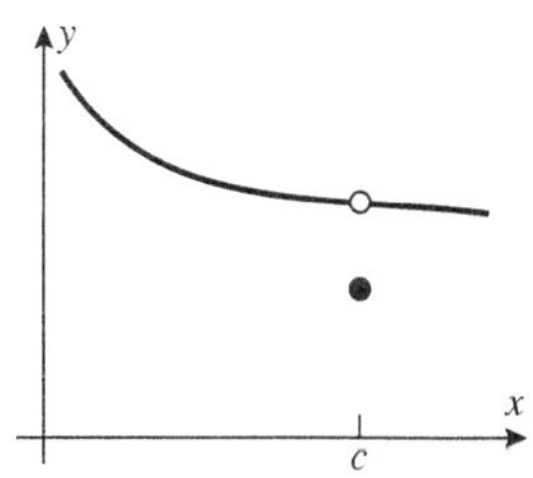

28. **(a)** $f(c) = \lim\limits_{x \to c} f(x)$

(b) $\lim\limits_{x \to 1} f(x) = 2$

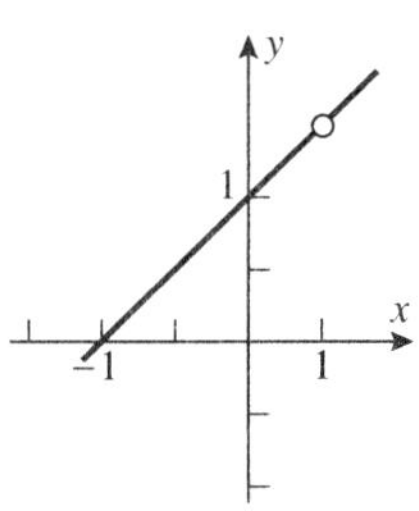

$\lim\limits_{x \to 1} g(x) = 1$

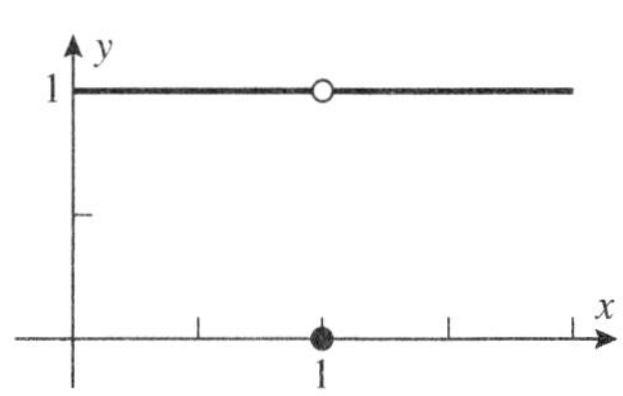

(c) Define $f(1) = 2$ and redefine $g(1) = 1$.

29. **(a)** $x = 0$, $\lim_{x\to 0^-} f(x) = -1 \neq +1 = \lim_{x\to 0^+} f(x)$ so the discontinuity is not removable

(b) $x = -3$; define $f(-3) = -3 = \lim_{x\to -3} f(x)$, then the discontinuity is removable

(c) f is undefined at $x = \pm 2$; at $x = 2$, $\lim_{x\to 2} f(x) = 1$, so define $f(2) = 1$ and f becomes continuous there; at $x = -2$, $\lim_{x\to -2}$ does not exist, so the discontinuity is not removable

30. **(a)** f is not defined at $x = 2$; $\lim_{x\to 2} f(x) = \lim_{x\to 2} \dfrac{x+2}{x^2+2x+4} = \dfrac{1}{3}$, so define $f(2) = \dfrac{1}{3}$ and f becomes continuous there

(b) $\lim_{x\to 2^-} f(x) = 1 \neq 4 = \lim_{x\to 2^+} f(x)$, so f has a nonremovable discontinuity at $x = 2$

(c) $\lim_{x\to 1} f(x) = 8 \neq f(1)$, so f has a removable discontinuity at $x = 1$

31. **(a)** discontinuity at $x = 1/2$, not removable; at $x = -3$, removable

(b) $2x^2 + 5x - 3 = (2x-1)(x+3)$

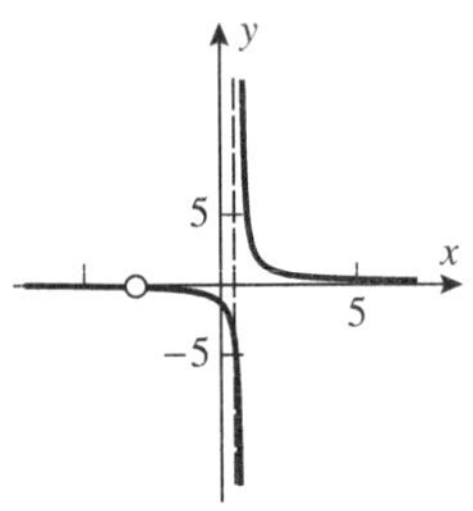

32. **(a)** there appears to be one discontinuity near $x = -1.52$

(b) one discontinuity at $x = -1.52$

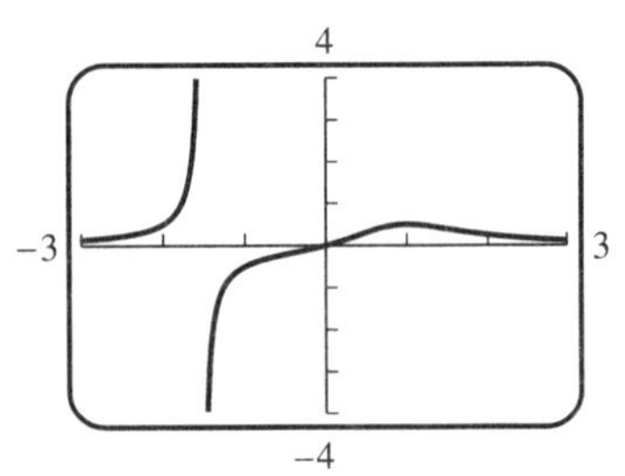

33. For $x > 0$, $f(x) = x^{3/5} = (x^3)^{1/5}$ is the composition (Theorem 2.4.6) of the two continuous functions $g(x) = x^3$ and $h(x) = x^{1/5}$ and is thus continuous. For $x < 0$, $f(x) = f(-x)$ which is the composition of the continuous functions $f(x)$ (for positive x) and the continuous function $y = -x$. Hence $f(-x)$ is continuous for all $x > 0$. At $x = 0$, $f(0) = \lim_{x\to 0} f(x) = 0$.

34. $x^4 + 7x^2 + 1 \geq 1 > 0$, thus $f(x)$ is the composition of the polynomial $x^4 + 7x^2 + 1$, the square root $\sqrt{x}$, and the function $1/x$ and is therefore continuous by Theorem 2.5.6.

35. **(a)** Let $f(x) = k$ for $x \neq c$ and $f(c) = 0$; $g(x) = l$ for $x \neq c$ and $g(c) = 0$. If $k = -l$ then $f + g$ is continuous; otherwise it's not.

(b) $f(x) = k$ for $x \neq c$, $f(c) = 1$; $g(x) = l \neq 0$ for $x \neq c$, $g(c) = 1$. If $kl = 1$, then fg is continuous; otherwise it's not.

36. A rational function is the quotient $f(x)/g(x)$ of two polynomials $f(x)$ and $g(x)$. By Theorem 2.5.2 f and g are continuous everywhere; by Theorem 2.5.3 f/g is continuous except when $g(x) = 0$.

37. Since f and g are continuous at $x = c$ we know that $\lim_{x\to c} f(x) = f(c)$ and $\lim_{x\to c} g(x) = g(c)$. In the following we use Theorem 2.2.2.

(a) $f(c) + g(c) = \lim_{x\to c} f(x) + \lim_{x\to c} g(x) = \lim_{x\to c}(f(x) + g(x))$ so $f + g$ is continuous at $x = c$.

(b) same as (a) except the $+$ sign becomes a $-$ sign

(c) $\dfrac{f(c)}{g(c)} = \dfrac{\lim_{x\to c} f(x)}{\lim_{x\to c} g(x)} = \lim_{x\to c} \dfrac{f(x)}{g(x)}$ so $\dfrac{f}{g}$ is continuous at $x = c$

38. $h(x) = f(x) - g(x)$ satisfies $h(a) > 0$, $h(b) < 0$. Use the Intermediate Value Theorem or Theorem 2.5.9.

39. Of course such a function must be discontinuous. Let $f(x) = 1$ on $0 \le x < 1$, and $f(x) = -1$ on $1 \le x \le 2$.

40. A square whose diagonal has length r has area $f(r) = r^2/2$. Note that $f(r) = r^2/2 < \pi r^2/2 < 2r^2 = f(2r)$. By the Intermediate Value Theorem there must be a value c between r and $2r$ such that $f(c) = \pi r^2/2$, i.e. a square of diagonal c whose area is $\pi r^2/2$.

41. The cone has volume $\pi r^2 h/3$. The function $V(r) = \pi r^2 h$ (for variable r and fixed h) gives the volume of a right circular cylinder of height h and radius r, and satisfies $V(0) < \pi r^2 h/3 < V(r)$. By the Intermediate Value Theorem there is a value c between 0 and r such that $V(c) = \pi r^2 h/3$, so the cylinder of radius c (and height h) has volume equal to that of the cone.

42. If $f(x) = x^3 - 4x + 1$ then $f(0) = 1$, $f(1) = -2$. Use Theorem 2.5.9.

43. If $f(x) = x^3 + x^2 - 2x$ then $f(-1) = 2$, $f(1) = 0$. Use the Intermediate Value Theorem.

44. Since $\lim_{x\to-\infty} p(x) = -\infty$ and $\lim_{x\to+\infty} p(x) = +\infty$ (or vice versa, if the leading coefficient of p is negative), it follows that for $M = -1$ there corresponds $N_1 < 0$, and for $M = 1$ there is $N_2 > 0$, such that $p(x) < -1$ for $x < N_1$ and $p(x) > 1$ for $x > N_2$. Choose $x_1 < N_1$ and $x_2 > N_2$ and use Theorem 2.5.9 on the interval $[x_1, x_2]$ to find a solution of $p(x) = 0$.

45. For the negative root, use intervals on the x-axis as follows: $[-2, -1]$; since $f(-1.3) < 0$ and $f(-1.2) > 0$, the midpoint $x = -1.25$ of $[-1.3, -1.2]$ is the required approximation of the root. For the positive root use the interval $[0, 1]$; since $f(0.7) < 0$ and $f(0.8) > 0$, the midpoint $x = 0.75$ of $[0.7, 0.8]$ is the required approximation.

46. $x = -1.25$ and $x = 0.75$.

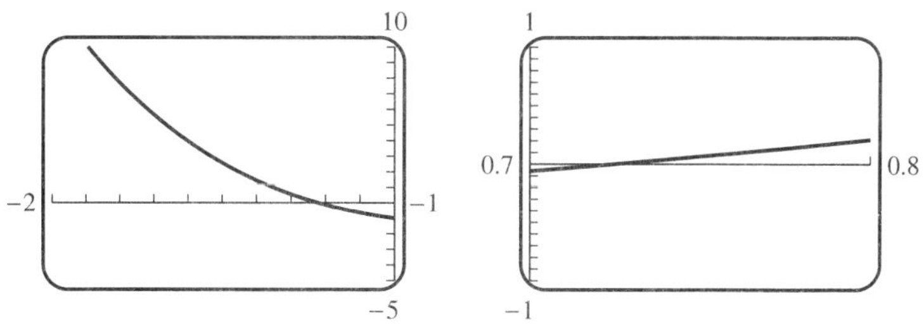

47. For the negative root, use intervals on the x-axis as follows: $[-2, -1]$; since $f(-1.7) < 0$ and $f(-1.6) > 0$, use the interval $[-1.7, -1.6]$. Since $f(-1.61) < 0$ and $f(-1.60) > 0$ the midpoint $x = -1.605$ of $[-1.61, -1.60]$ is the required approximation of the root. For the positive root use the interval $[1, 2]$; since $f(1.3) > 0$ and $f(1.4) < 0$, use the interval $[1.3, 1.4]$. Since $f(1.37) > 0$ and $f(1.38) < 0$, the midpoint $x = 1.375$ of $[1.37, 1.38]$ is the required approximation.

48. $x = -1.605$ and $x = 1.375$.

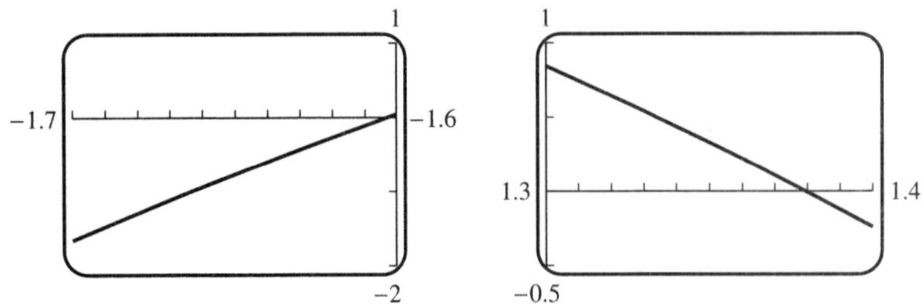

49. $x = 2.24$

50. Set $f(x) = \dfrac{a}{x-1} + \dfrac{b}{x-3}$. Since $\lim\limits_{x\to 1^+} f(x) = +\infty$ and $\lim\limits_{x\to 3^-} f(x) = -\infty$ there exist $x_1 > 1$ and $x_2 < 3$ (with $x_2 > x_1$) such that $f(x) > 1$ for $1 < x < x_1$ and $f(x) < -1$ for $x_2 < x < 3$. Choose x_3 in $(1, x_1)$ and x_4 in $(x_2, 3)$ and apply Theorem 2.5.9 on $[x_3, x_4]$.

51. The uncoated sphere has volume $4\pi(x-1)^3/3$ and the coated sphere has volume $4\pi x^3/3$. If the volume of the uncoated sphere and of the coating itself are the same, then the coated sphere has twice the volume of the uncoated sphere. Thus $2(4\pi(x-1)^3/3) = 4\pi x^3/3$, or $x^3 - 6x^2 + 6x - 2 = 0$, with the solution $x = 4.847$ cm.

52. Let $g(t)$ denote the altitude of the monk at time t measured in hours from noon of day one, and let $f(t)$ denote the altitude of the monk at time t measured in hours from noon of day two. Then $g(0) < f(0)$ and $g(12) > f(12)$. Use Exercise 38.

53. We must show $\lim\limits_{x\to c} f(x) = f(c)$. Let $\epsilon > 0$; then there exists $\delta > 0$ such that if $|x - c| < \delta$ then $|f(x) - f(c)| < \epsilon$. But this certainly satisfies Definition 2.4.1.

EXERCISE SET 2.6

1. none

2. $x = \pi$

3. $x = n\pi$, $n = 0, \pm 1, \pm 2, \ldots$

4. $x = n\pi + \pi/2$, $n = 0, \pm 1, \pm 2, \ldots$

5. $x = n\pi$, $n = 0, \pm 1, \pm 2, \ldots$

6. none

7. none

8. $x = n\pi + \pi/2$, $n = 0, \pm 1, \pm 2, \ldots$

9. $2n\pi + \pi/6, 2n\pi + 5\pi/6$, $n = 0, \pm 1, \pm 2, \ldots$

10. none

11. **(a)** $\sin x, x^3 + 7x + 1$ **(b)** $|x|, \sin x$ **(c)** $x^3, \cos x, x + 1$
(d) $\sqrt{x}, 3 + x, \sin x, 2x$ **(e)** $\sin x, \sin x$ **(f)** $x^5 - 2x^3 + 1, \cos x$

12. **(a)** Use Theorem 2.5.6. **(b)** $g(x) = \cos x$, $g(x) = \dfrac{1}{x^2+1}$, $g(x) = x^2 + 1$

13. $\cos\left(\lim\limits_{x\to+\infty} \dfrac{1}{x}\right) = \cos 0 = 1$

14. $\sin\left(\lim\limits_{x\to+\infty} \dfrac{2}{x}\right) = \sin 0 = 0$

15. $\sin\left(\lim\limits_{x\to+\infty} \dfrac{\pi x}{2 - 3x}\right) = \sin\left(-\dfrac{\pi}{3}\right) = -\dfrac{\sqrt{3}}{2}$

16. $\dfrac{1}{2}\lim\limits_{h\to 0} \dfrac{\sin h}{h} = \dfrac{1}{2}$

17. $3 \lim_{\theta \to 0} \frac{\sin 3\theta}{3\theta} = 3$

18. $\left(\lim_{\theta \to 0^+} \frac{1}{\theta}\right) \lim_{\theta \to 0^+} \frac{\sin \theta}{\theta} = +\infty$

19. $-\lim_{x \to 0^-} \frac{\sin x}{x} = -1$

20. $\frac{1}{3}\left(\lim_{x \to 0} \frac{\sin x}{x}\right)^2 = \frac{1}{3}$

21. $\frac{1}{5} \lim_{x \to 0^+} \sqrt{x} \lim_{x \to 0^+} \frac{\sin x}{x} = 0$

22. $\frac{\sin 6x}{\sin 8x} = \frac{6}{8} \frac{\sin 6x}{6x} \frac{8x}{\sin 8x}$, so $\lim_{x \to 0} \frac{\sin 6x}{\sin 8x} = \frac{6}{8} = \frac{3}{4}$

23. $\frac{\tan 7x}{\sin 3x} = \frac{7}{3 \cos 7x} \frac{\sin 7x}{7x} \frac{3x}{\sin 3x}$ so $\lim_{x \to 0} \frac{\tan 7x}{\sin 3x} = \frac{7}{3(1)}(1)(1) = \frac{7}{3}$

24. $\left(\lim_{\theta \to 0} \sin \theta\right) \lim_{\theta \to 0} \frac{\sin \theta}{\theta} = 0$

25. $\left(\lim_{h \to 0} \cos h\right) \lim_{h \to 0} \frac{h}{\sin h} = 1$

26. $\frac{\sin h}{1 - \cos h} = \frac{\sin h}{1 - \cos h} \frac{1 + \cos h}{1 + \cos h} = \frac{\sin h(1 + \cos h)}{1 - \cos^2 h} = \frac{1 + \cos h}{\sin h}$; no limit

27. $\frac{\theta^2}{1 - \cos \theta} \frac{1 + \cos \theta}{1 + \cos \theta} = \frac{\theta^2(1 + \cos \theta)}{1 - \cos^2 \theta} = \left(\frac{\theta}{\sin \theta}\right)^2 (1 + \cos \theta)$ so $\lim_{\theta \to 0} \frac{\theta^2}{1 - \cos \theta} = (1)^2 2 = 2$

28. $\cos(\frac{1}{2}\pi - x) = \sin(\frac{1}{2}\pi) \sin x = \sin x$, so $\lim_{x \to 0} \frac{x}{\cos\left(\frac{1}{2}\pi - x\right)} = 1$

29. 0

30. $\frac{t^2}{1 - \cos^2 t} = \left(\frac{t}{\sin t}\right)^2$, so $\lim_{t \to 0} \frac{t^2}{1 - \cos^2 t} = 1$

31. $\frac{1 - \cos 5h}{\cos 7h - 1} = \frac{(1 - \cos 5h)(1 + \cos 5h)(1 + \cos 7h)}{(\cos 7h - 1)(1 + \cos 5h)(1 + \cos 7h)} = -\frac{25}{49}\left(\frac{\sin 5h}{5h}\right)^2 \left(\frac{7h}{\sin 7h}\right)^2 \frac{1 + \cos 7h}{1 + \cos 5h}$ so $\lim_{h \to 0} \frac{1 - \cos 5h}{\cos 7h - 1} = -\frac{25}{49}$

32. $\lim_{x \to 0^+} \sin\left(\frac{1}{x}\right) = \lim_{t \to +\infty} \sin t$; limit does not exist

33. $\lim_{x \to 0^+} \cos\left(\frac{1}{x}\right) = \lim_{t \to +\infty} \cos t$; limit does not exist

34. $\lim_{x \to 0} x - 3 \lim_{x \to 0} \frac{\sin x}{x} = -3$

35. $2 + \lim_{x \to 0} \frac{\sin x}{x} = 3$

36.

5.1	5.01	5.001	5.0001	5.00001	4.9	4.99	4.999	4.9999	4.99999
0.098845	0.099898	0.99990	0.099999	0.100000	0.10084	0.10010	0.10001	0.10000	0.10000

The limit is 0.1.

37.

2.1	2.01	2.001	2.0001	2.00001	1.9	1.99	1.999	1.9999	1.99999
0.484559	0.498720	0.499875	0.499987	0.499999	0.509409	0.501220	0.500125	0.500012	0.500001

The limit is 0.5.

38.

−1.9	−1.99	−1.999	−1.9999	−1.99999	−2.1	−2.01	−2.001	−2.0001	−2.00001
−0.898785	−0.989984	−0.999000	−0.999900	−0.999990	−1.097783	−1.009983	−1.001000	−1.000100	−1.000010

The limit is −1.

39.

-0.9	-0.99	-0.999	-0.9999	-0.99999	-1.1	-1.01	-1.001	-1.0001	-1.00001
0.405086	0.340050	0.334001	0.333400	0.333340	0.271536	0.326717	0.332667	0.333267	0.333327

The limit is 1/3.

40. $k = f(0) = \lim\limits_{x\to 0} \dfrac{\sin 3x}{x} = 3 \lim\limits_{x\to 0} \dfrac{\sin 3x}{3x} = 3$, so $k = 3$

41. $\lim\limits_{x\to 0^-} f(x) = k \lim\limits_{x\to 0} \dfrac{\sin kx}{kx \cos kx} = k$, $\lim\limits_{x\to 0^+} f(x) = 2k^2$, so $k = 2k^2$, $k = \dfrac{1}{2}$

42. No; $\sin x/|x|$ has unequal one-sided limits.

43. **(a)** $\lim\limits_{t\to 0^+} \dfrac{\sin t}{t} = 1$ **(b)** $\lim\limits_{t\to 0^-} \dfrac{1-\cos t}{t} = 0$ (Theorem 2.6.3)

(c) $\sin(\pi - t) = \sin t$, so $\lim\limits_{x\to\pi} \dfrac{\pi - x}{\sin x} = \lim\limits_{t\to 0} \dfrac{t}{\sin t} = 1$

44. $\cos\left(\dfrac{\pi}{2} - t\right) = \sin t$, so $\lim\limits_{x\to 2} \dfrac{\cos(\pi/x)}{x-2} = \lim\limits_{t\to 0} \dfrac{(\pi - 2t)\sin t}{4t} = \lim\limits_{t\to 0} \dfrac{\pi - 2t}{4} \lim\limits_{t\to 0} \dfrac{\sin t}{t} = \dfrac{\pi}{4}$

45. $t = x - 1$; $\sin(\pi x) = \sin(\pi t + \pi) = -\sin \pi t$; and $\lim\limits_{x\to 1} \dfrac{\sin(\pi x)}{x-1} = -\lim\limits_{t\to 0} \dfrac{\sin \pi t}{t} = -\pi$

46. $t = x - \pi/4$; $\tan x - 1 = \dfrac{2\sin t}{\cos t - \sin t}$; $\lim\limits_{x\to\pi/4} \dfrac{\tan x - 1}{x - \pi/4} = \lim\limits_{t\to 0} \dfrac{2\sin t}{t(\cos t - \sin t)} = 2$

47. $-|x| \le x \cos\left(\dfrac{50\pi}{x}\right) \le |x|$

48. $-x^2 \le x^2 \sin\left(\dfrac{50\pi}{\sqrt[3]{x}}\right) \le x^2$

49. $\lim\limits_{x\to 0} f(x) = 1$ by the Squeezing Theorem

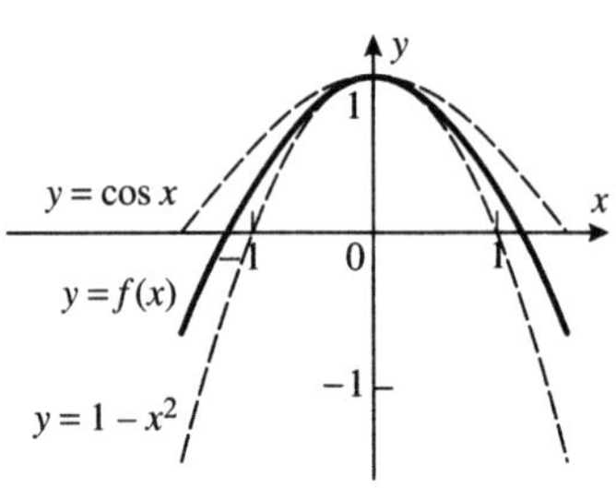

50. $\lim\limits_{x\to +\infty} f(x) = 0$ by the Squeezing Theorem

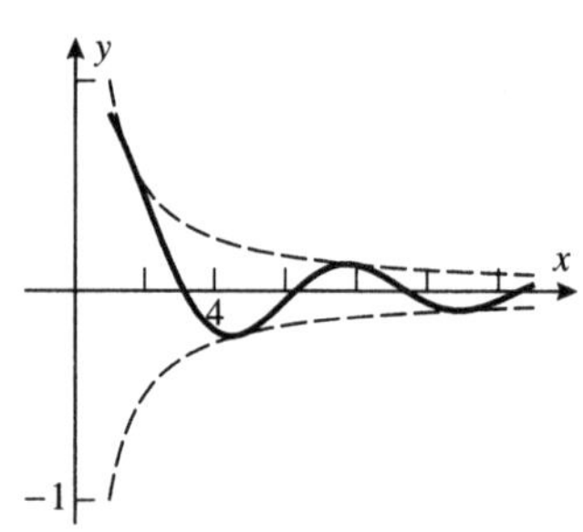

51. Let $g(x) = -\dfrac{1}{x}$ and $h(x) = \dfrac{1}{x}$; thus $\lim\limits_{x\to +\infty} \dfrac{\sin x}{x} = 0$ by the Squeezing Theorem.

52.

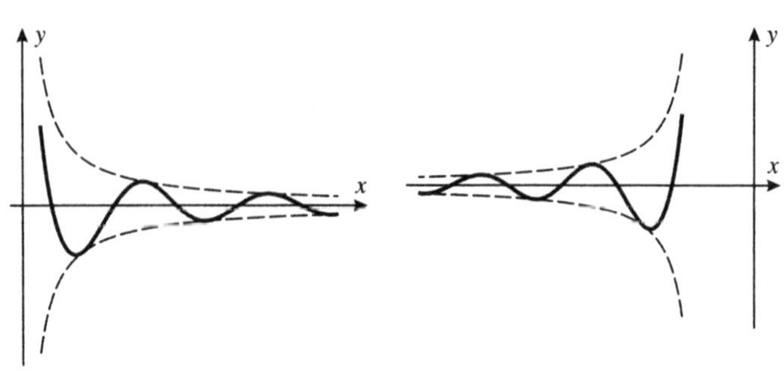

53. **(a)** $\sin x = \sin t$ where x is measured in degrees, t is measured in radians and $t = \frac{\pi x}{180}$. Thus $\lim_{x\to 0} \frac{\sin x}{x} = \lim_{t\to 0} \frac{\sin t}{(180t/\pi)} = \frac{\pi}{180}$.

54. $\cos x = \cos t$ where x is measured in degrees, t in radians, and $t = \frac{\pi x}{180}$. Thus $\lim_{x\to 0} \frac{1-\cos x}{x} = \lim_{t\to 0} \frac{1-\cos t}{(180t/\pi)} = 0$.

55. **(a)** $\sin 10° = 0.17365$ **(b)** $\sin 10° = \sin \frac{\pi}{18} \approx \frac{\pi}{18} = 0.17453$

56. **(a)** $\cos\theta = \cos 2\alpha = 1 - 2\sin^2(\theta/2)$
$\approx 1 - 2(\theta/2)^2 = 1 - \frac{1}{2}\theta^2$
(b) $\cos 10° = 0.98481$
(c) $\cos 10° = 1 - \frac{1}{2}\left(\frac{\pi}{18}\right)^2 \approx 0.98477$

57. **(a)** 0.08749 **(b)** $\tan 5° \approx \frac{\pi}{36} = 0.08727$

58. **(a)** $h = 52.55$ ft
(b) Since α is small, $\tan\alpha° \approx \frac{\pi\alpha}{180}$ is a good approximation.
(c) $h \approx 52.36$ ft

59. **(a)** Let $f(x) = x - \cos x$; $f(0) = -1$, $f(\pi/2) = \pi/2$. By the IVT there must be a solution of $f(x) = 0$.
(b)

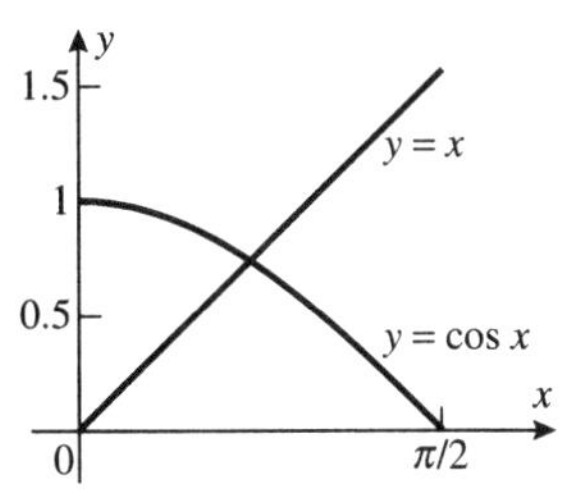

(c) 0.739

60. **(a)** $f(x) = x + \sin x - 1$; $f(0) = -1$, $f(\pi/6) = \pi/6 - 1/2 > 0$. By the IVT there must be a solution of $f(x) = 0$.
(b)

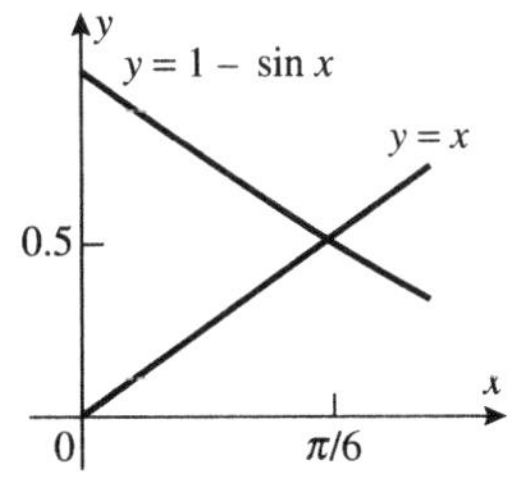

(c) $x = 0.511$

61. **(a)** There is symmetry about the equatorial plane.
(b) Let $g(\phi)$ be the given function. Then $g(38) < 9.8$ and $g(39) > 9.8$, so by the Intermediate Value Theorem there is a value c between 38 and 39 for which $g(c) = 9.8$ exactly.

62. (a) does not exist (b) the limit is zero

(c) For part (a) consider the fact that given any $\delta > 0$ there are infinitely many rational numbers x satisfying $|x| < \delta$ and there are infinitely many irrational numbers satisfying the same condition. Thus if the limit were to exist, it could not be zero because of the rational numbers, and it could not be 1 because of the irrational numbers, and it could not be anything else because of *all* the numbers. Hence the limit cannot exist. For part (b) use the Squeezing Theorem with $+x$ and $-x$ as the 'squeezers'.

CHAPTER 2 SUPPLEMENTARY EXERCISES

1. (a) 1 (b) no limit (c) no limit
(d) 1 (e) 3 (f) 0
(g) 0 (h) 2 (i) 1/2

2. (a) $f(x) = 2x/(x-1)$ (b)

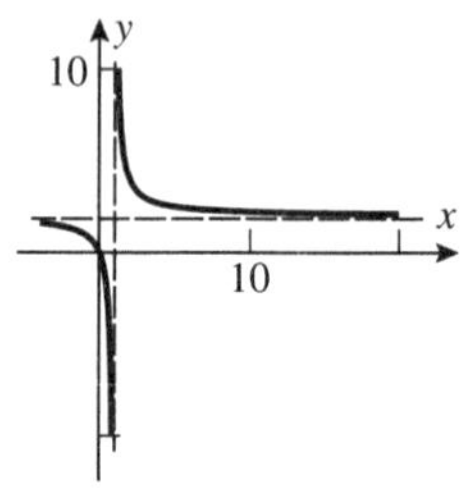

4. $f(x) = -1$ for $a \le x < \dfrac{a+b}{2}$ and $f(x) = 1$ for $\dfrac{a+b}{2} \le x \le b$

5. (a) $0.222\ldots, 0.24390, 0.24938, 0.24994, 0.24999, 0.25000$; for $x \ne 2$, $f(x) = \dfrac{1}{x+2}$, so the limit is 1/4.

(b) $1.15782, 4.22793, 4.00213, 4.00002, 4.00000, 4.00000$; to prove, use $\dfrac{\tan 4x}{x} = \dfrac{\sin 4x}{x \cos 4x} = \dfrac{4}{\cos 4x}\dfrac{\sin 4x}{4x}$, the limit is 4.

6. (a) $y = 0$ (b) none (c) $y = 2$

7. (a)

x	1	0.1	0.01	0.001	0.0001	0.00001	0.000001
$f(x)$	1.000	0.443	0.409	0.406	0.406	0.405	0.405

(b)

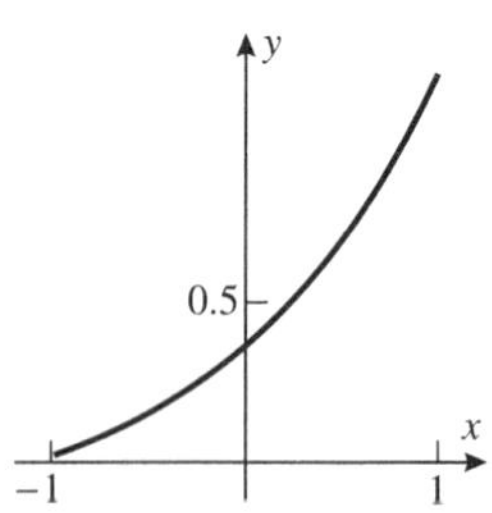

8. (a) 0.4 amperes (b) $[0.3947, 0.4054]$ (c) $\left[\dfrac{3}{7.5+\delta}, \dfrac{3}{7.5-\delta}\right]$

(d) 0.0187 (e) It becomes infinite.

9. **(a)**

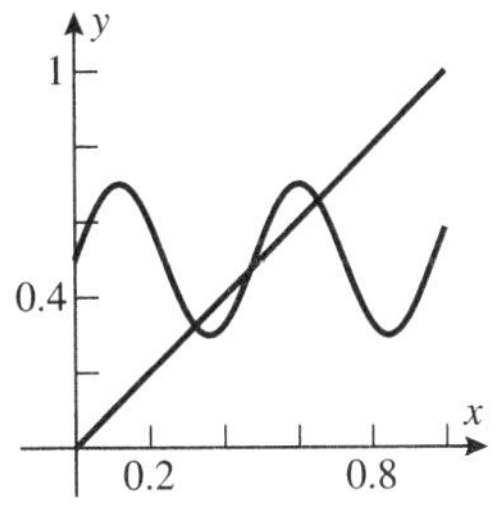

(b) Let $g(x) = x - f(x)$. Then $g(1) \geq 0$ and $g(0) \leq 0$; by the Intermediate Value Theorem there is a solution c in $[0, 1]$ of $g(c) = 0$.

10. **(a)** $\lim_{\theta\to 0} \tan\left(\frac{1-\cos\theta}{\theta}\right) = \tan\left(\lim_{\theta\to 0}\frac{1-\cos\theta}{\theta}\right) = \tan\left(\lim_{\theta\to 0}\frac{1-\cos^2\theta}{\theta(1+\cos\theta)}\right) = \tan 0 = 0$

(b) $\frac{t-1}{\sqrt{t}-1} = \frac{t-1}{\sqrt{t}-1}\frac{\sqrt{t}+1}{\sqrt{t}+1} = \frac{(t-1)(\sqrt{t}+1)}{t-1} = \sqrt{t}+1$; $\lim_{t\to 1}\frac{t-1}{\sqrt{t}-1} = \lim_{t\to 1}(\sqrt{t}+1) = 2$

(c) $\frac{(2x-1)^5}{(3x^2+2x-7)(x^3-9x)} = \frac{(2-1/x)^5}{(3+2/x-7/x^2)(1-9/x^2)} \to 2^5/3 = 32/3$ as $x \to +\infty$

(d) $\sin(\theta+\pi) = \sin\theta\cos\pi - \cos\theta\sin\pi = -\sin\theta$, so $\lim_{\theta\to 0}\cos\left(\frac{\sin(\theta+\pi)}{2\theta}\right) = \lim_{\theta\to 0}\cos\left(\frac{-\sin\theta}{2\theta}\right)$
$= \cos\left(\lim_{\theta\to 0}\frac{-\sin\theta}{2\theta}\right) = \cos\left(-\frac{1}{2}\right)$

11. If, on the contrary, $f(x_0) < 0$ for some x_0 in $[0, 1]$, then by the Intermediate Value Theorem we would have a solution of $f(x) = 0$ in $[0, x_0]$, contrary to the hypothesis.

12. For $x < 2$ f is a polynomial and is continuous; for $x > 2$ f is a polynomial and is continuous. At $x = 2$, $f(2) = -13 \neq 13 = \lim_{x\to 2^+} f(x)$ so f is not continuous there.

13. $f(-6) = 185$, $f(0) = -1$, $f(2) = 65$; apply Theorem 2.4.9 twice, once on $[-6, 0]$ and once on $[0, 2]$

14. 3.317

15. Let $\epsilon = f(x_0)/2 > 0$; then there corresponds $\delta > 0$ such that if $|x - x_0| < \delta$ then $|f(x) - f(x_0)| < \epsilon$, $-\epsilon < f(x) - f(x_0) < \epsilon$, $f(x) > f(x_0) - \epsilon = f(x_0)/2 > 0$ for $x_0 - \delta < x < x_0 + \delta$.

16.

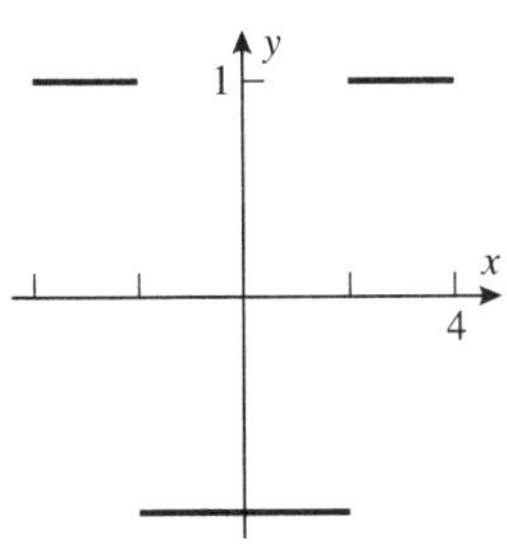

17. **(a)** $-3.449, 1.449$ **(b)** $x = 0, \pm 1.896$

18. Since $\lim_{x\to 0}\sin(1/x)$ does not exist, no conclusions can be drawn.

19. **(a)** $\sqrt{5}$, no limit, $\sqrt{10}$, $\sqrt{10}$, no limit, $+\infty$, no limit

(b) $5, 10, 0, 0, 10, -\infty, +\infty$

20. **(a)** $-1/5, +\infty, -1/10, -1/10$, no limit, 0, 0 **(b)** $-1, +1, -1, -1$, no limit, $-1, +1$

21. a/b **22.** 1 **23.** does not exist

24. 2 **25.** 0 **26.** k^2 **27.** $3 - k$

28. The numerator satisfies: $|2x + x\sin 3x| \le |2x| + |x| = 3|x|$. Since the denominator grows like x^2, the limit is 0.

30. **(a)** $\dfrac{\sqrt{x^2+4}-2}{x^2}\dfrac{\sqrt{x^2+4}+2}{\sqrt{x^2+4}+2} = \dfrac{x^2}{x^2(\sqrt{x^2+4}+2)} = \dfrac{1}{\sqrt{x^2+4}+2}$, so

$$\lim_{x\to 0}\frac{\sqrt{x^2+4}-2}{x^2} = \lim_{x\to 0}\frac{1}{\sqrt{x^2+4}+2} = \frac{1}{4}$$

(b)

x	1	0.1	0.01	0.001	0.0001	0.00001
$f(x)$	0.236	0.2498	0.2500	0.2500	0.25000	0.00000

The division may entail division by zero (e.g. on an HP 42S), or the numerator may be inaccurate (catastrophic subtraction, e.g.).

(c) in the 3d picture, catastrophic subtraction

31.

x	0.1	0.01	0.001	0.0001	0.00001	0.000001
$f(x)$	2.59	2.70	2.717	2.718	2.7183	2.71828

32.

x	3.1	3.01	3.001	3.0001	3.00001	3.000001
$f(x)$	5.74	5.56	5.547	5.545	5.5452	5.54518

33.

x	1.1	1.01	1.001	1.0001	1.00001	1.000001
$f(x)$	0.49	0.54	0.540	0.5403	0.54030	0.54030

34.

x	0.1	0.01	0.001	0.0001	0.00001	0.000001
$f(x)$	99.0	9048.8	368063.3	4562.7	3.9×10^{-34}	0

35.

x	100	1000	10^4	10^5	10^6	10^7
$f(x)$	0.48809	0.49611	0.49876	0.49961	0.49988	0.49996

36. For large values of x (not much more than 100) the computer can't handle 5^x or 3^x, yet the limit is 5.

37. $\delta \approx 0.07747$ (use a graphing utility) **38.** \$2,001.60, \$2,009.66, \$2,013.62, \$2013.75

39. **(a)** $x^3 - x - 1 = 0$, $x^3 = x + 1$, $x = \sqrt[3]{x+1}$.

(b)

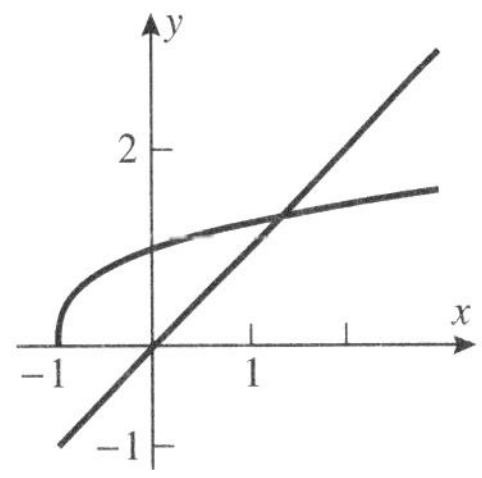

(c)

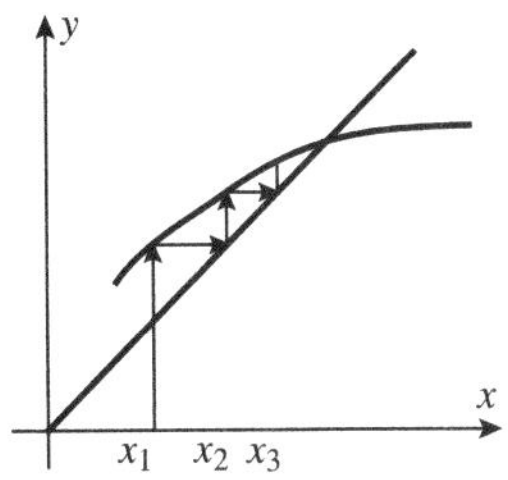

(d) $1, 1.26, 1.31, 1.322, 1.324, 1.3246, 1.3247$

40. **(a)**

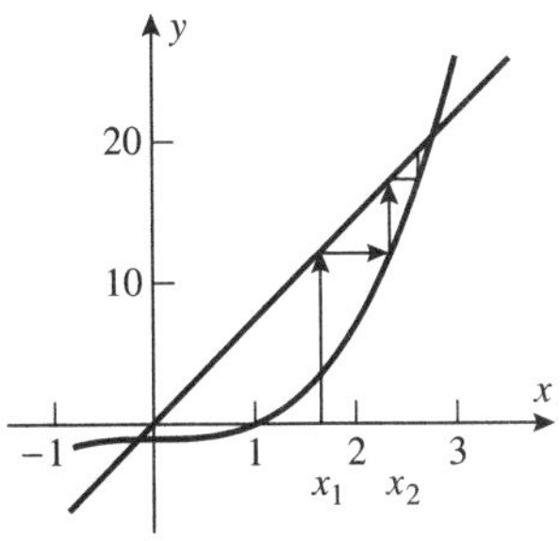

(b) $0, -1, -2, -9, -730$

41. $x = \sqrt[5]{x+2}$; 1.267168

42. $x = \cos x$; 0.739085 (after 33 iterations!).

CHAPTER 3
The Derivative

EXERCISE SET 3.1

1. **(a)** $m_{\tan} = (50 - 10)/(15 - 5)$
$= 40/10$
$= 4$ m/s

(b)

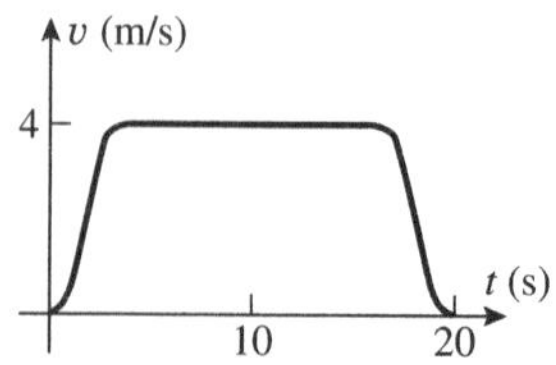

2. **(a)** $(10 - 10)/(3 - 0) = 0$ cm/s

(b) $t = 0$, $t = 2$, and $t = 4.2$ (horizontal tangent line)

(c) maximum: $t = 1$ (slope > 0) minimum: $t = 3$ (slope < 0)

(d) $(3 - 18)/(4 - 2) = -7.5$ cm/s (slope of estimated tangent line to curve at $t = 3$)

3. From the figure:

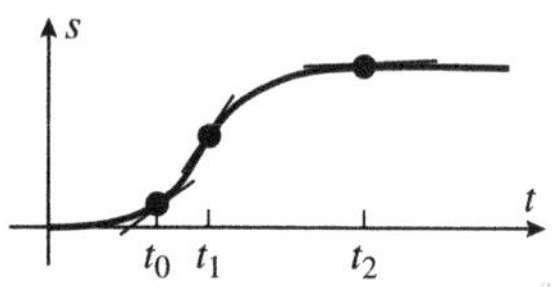

(a) The particle is moving faster at time t_0 because the slope of the tangent to the curve at t_0 is greater than that at t_2.

(b) The initial velocity is 0 because the slope of a horizontal line is 0.

(c) The particle is speeding up because the slope increases as t increases from t_0 to t_1.

(d) The particle is slowing down because the slope decreases as t increases from t_1 to t_2.

4.

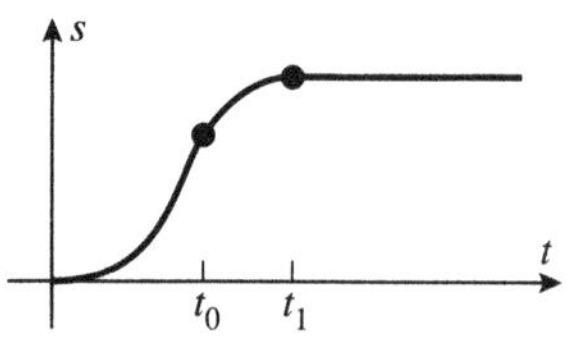

5. It is a straight line with slope equal to the velocity.

6. **(a)** decreasing (slope of tangent line decreases with increasing time)

(b) increasing (slope of tangent line increases with increasing time)

(c) increasing (slope of tangent line increases with increasing time)

(d) decreasing (slope of tangent line decreases with increasing time)

7. **(a)** $m_{\sec} = \dfrac{f(4) - f(3)}{4 - 3} = \dfrac{(4)^2/2 - (3)^2/2}{1} = \dfrac{7}{2}$

(b) $m_{\tan} = \lim\limits_{x_1 \to 3} \dfrac{f(x_1) - f(3)}{x_1 - 3} = \lim\limits_{x_1 \to 3} \dfrac{x_1^2/2 - 9/2}{x_1 - 3}$

$$= \lim_{x_1 \to 3} \frac{x_1^2 - 9}{2(x_1 - 3)} = \lim_{x_1 \to 3} \frac{(x_1 + 3)(x_1 - 3)}{2(x_1 - 3)} = \lim_{x_1 \to 3} \frac{x_1 + 3}{2} = 3$$

(c) $$m_{\tan} = \lim_{x_1 \to x_0} \frac{f(x_1) - f(x_0)}{x_1 - x_0} = \lim_{x_1 \to x_0} \frac{x_1^2/2 - x_0^2/2}{x_1 - x_0} = \lim_{x_1 \to x_0} \frac{x_1^2 - x_0^2}{2(x_1 - x_0)} = \lim_{x_1 \to x_0} \frac{x_1 + x_0}{2} = x_0$$

(d)

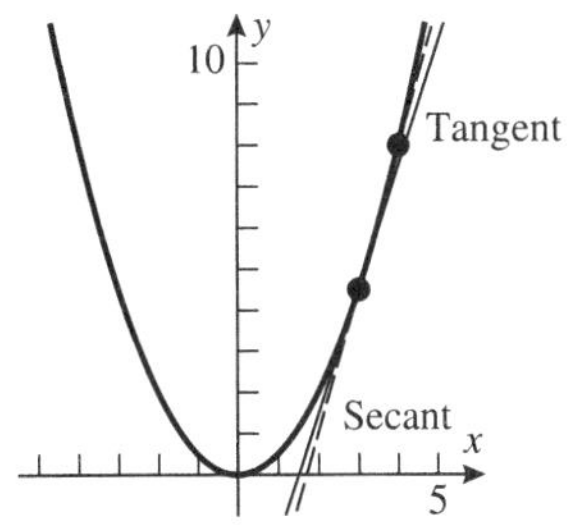

8. (a) $$m_{\sec} = \frac{f(2) - f(1)}{2 - 1} = \frac{2^3 - 1^3}{1} = 7$$

(b) $$m_{\tan} = \lim_{x_1 \to 1} \frac{f(x_1) - f(1)}{x_1 - 1} = \lim_{x_1 \to 1} \frac{x_1^3 - 1}{x_1 - 1} = \lim_{x_1 \to 1} \frac{(x_1 - 1)(x_1^2 + x_1 + 1)}{x_1 - 1} = \lim_{x_1 \to 1} (x_1^2 + x_1 + 1) = 3$$

(c) $$m_{\tan} = \lim_{x_1 \to x_0} \frac{f(x_1) - f(x_0)}{x_1 - x_0} = \lim_{x_1 \to x_0} \frac{x_1^3 - x_0^3}{x_1 - x_0} = \lim_{x_1 \to x_0} (x_1^2 + x_1 x_0 + x_0^2) = 3x_0^2$$

(d)

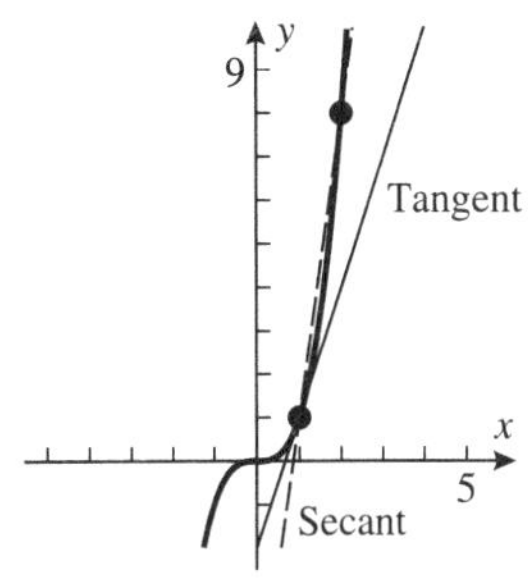

9. (a) $$m_{\sec} = \frac{f(3) - f(2)}{3 - 2} = \frac{1/3 - 1/2}{1} = -\frac{1}{6}$$

(b) $$m_{\tan} = \lim_{x_1 \to 2} \frac{f(x_1) - f(2)}{x_1 - 2} = \lim_{x_1 \to 2} \frac{1/x_1 - 1/2}{x_1 - 2} = \lim_{x_1 \to 2} \frac{2 - x_1}{2x_1(x_1 - 2)} = \lim_{x_1 \to 2} \frac{-1}{2x_1} = -\frac{1}{4}$$

(c) $$m_{\tan} = \lim_{x_1 \to x_0} \frac{f(x_1) - f(x_0)}{x_1 - x_0} = \lim_{x_1 \to x_0} \frac{1/x_1 - 1/x_0}{x_1 - x_0} = \lim_{x_1 \to x_0} \frac{x_0 - x_1}{x_0 x_1 (x_1 - x_0)} = \lim_{x_1 \to x_0} \frac{-1}{x_0 x_1} = -\frac{1}{x_0^2}$$

(d)

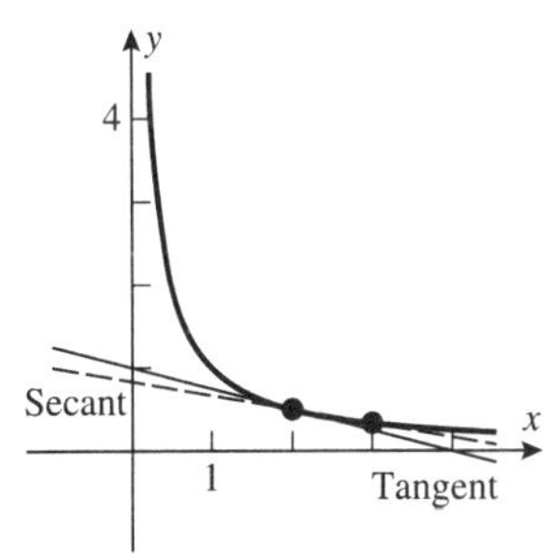

10. (a) $$m_{\sec} = \frac{f(2) - f(1)}{2 - 1} = \frac{1/4 - 1}{1} = -\frac{3}{4}$$

(b) $$m_{\tan} = \lim_{x_1 \to 1} \frac{f(x_1) - f(1)}{x_1 - 1} = \lim_{x_1 \to 1} \frac{1/x_1^2 - 1}{x_1 - 1} = \lim_{x_1 \to 1} \frac{1 - x_1^2}{x_1^2(x_1 - 1)} = \lim_{x_1 \to 1} \frac{-(x_1 + 1)}{x_1^2} = -2$$

(c)
$$m_{\tan} = \lim_{x_1 \to x_0} \frac{f(x_1) - f(x_0)}{x_1 - x_0} = \lim_{x_1 \to x_0} \frac{1/x_1^2 - 1/x_0^2}{x_1 - x_0} = \lim_{x_1 \to x_0} \frac{x_0^2 - x_1^2}{x_0^2 x_1^2 (x_1 - x_0)} = \lim_{x_1 \to x_0} \frac{-(x_1 + x_0)}{x_0^2 x_1^2} = -\frac{2}{x_0^3}$$

(d)

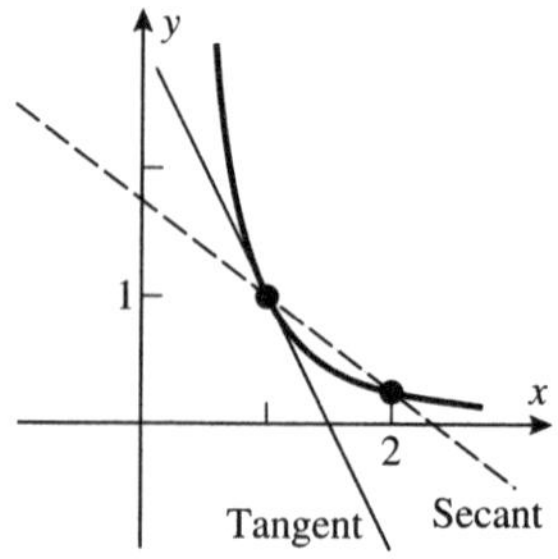

11. (a)
$$m_{\tan} = \lim_{x_1 \to x_0} \frac{f(x_1) - f(x_0)}{x_1 - x_0} = \lim_{x_1 \to x_0} \frac{(x_1^2 + 1) - (x_0^2 + 1)}{x_1 - x_0} = \lim_{x_1 \to x_0} \frac{x_1^2 - x_0^2}{x_1 - x_0} = \lim_{x_1 \to x_0} (x_1 + x_0) = 2x_0$$

(b) $m_{\tan} = 2(2) = 4$

12. (a)
$$m_{\tan} = \lim_{x_1 \to x_0} \frac{f(x_1) - f(x_0)}{x_1 - x_0} = \lim_{x_1 \to x_0} \frac{(x_1^2 + 3x_1 + 2) - (x_0^2 + 3x_0 + 2)}{x_1 - x_0} = \lim_{x_1 \to x_0} \frac{(x_1^2 - x_0^2) + 3(x_1 - x_0)}{x_1 - x_0} = \lim_{x_1 \to x_0} (x_1 + x_0 + 3) = 2x_0 + 3$$

(b) $m_{\tan} = 2(2) + 3 = 7$

13. (a)
$$m_{\tan} = \lim_{x_1 \to x_0} \frac{f(x_1) - f(x_0)}{x_1 - x_0} = \lim_{x_1 \to x_0} \frac{\sqrt{x_1} - \sqrt{x_0}}{x_1 - x_0} = \lim_{x_1 \to x_0} \frac{1}{\sqrt{x_1} + \sqrt{x_0}} = \frac{1}{2\sqrt{x_0}}$$

(b) $m_{\tan} = \dfrac{1}{2\sqrt{1}} = \dfrac{1}{2}$

14. (a)
$$m_{\tan} = \lim_{x_1 \to x_0} \frac{f(x_1) - f(x_0)}{x_1 - x_0} = \lim_{x_1 \to x_0} \frac{1/\sqrt{x_1} - 1/\sqrt{x_0}}{x_1 - x_0} = \lim_{x_1 \to x_0} \frac{\sqrt{x_0} - \sqrt{x_1}}{\sqrt{x_0}\,\sqrt{x_1}\,(x_1 - x_0)} = \lim_{x_1 \to x_0} \frac{-1}{\sqrt{x_0}\,\sqrt{x_1}\,(\sqrt{x_1} + \sqrt{x_0})} = -\frac{1}{2x_0^{3/2}}$$

(b) $m_{\tan} = -\dfrac{1}{2(4)^{3/2}} = -\dfrac{1}{16}$

15. (a) 72°F at about 4:30 P.M. **(b)** about $(67 - 43)/6 = 4$°F/h
(c) decreasing most rapidly at about 9 P.M.; rate of change of temperature is about -7°F/h (slope of estimated tangent line to curve at 9 P.M.)

16. For $V = 10$ the slope of the tangent line is about -0.25 atm/L, for $V = 25$ the slope is about -0.04 atm/L.

17. (a) during the first year after birth
(b) about 6 cm/year (slope of estimated tangent line at age 5)
(c) the growth rate is greatest at about age 14; about 10 cm/year

(d) 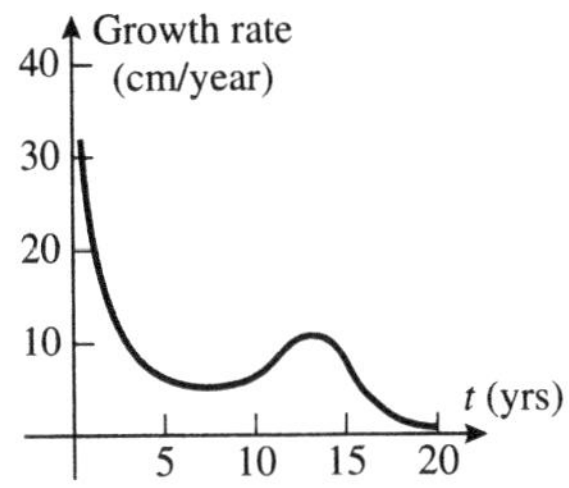

18. **(a)** The rock will hit the ground when $16t^2 = 576$, $t^2 = 36$, $t = 6$ s (only $t \geq 0$ is meaningful)

(b) $v_{\text{ave}} = \dfrac{16(6)^2 - 16(0)^2}{6 - 0} = 96$ ft/s

(c) $v_{\text{ave}} = \dfrac{16(3)^2 - 16(0)^2}{3 - 0} = 48$ ft/s

(d) $$v_{\text{inst}} = \lim_{t_1 \to 6} \frac{16t_1^2 - 16(6)^2}{t_1 - 6} = \lim_{t_1 \to 6} \frac{16(t_1^2 - 36)}{t_1 - 6} = \lim_{t_1 \to 6} 16(t_1 + 6) = 192 \text{ ft/s}$$

19. **(a)** $5(40)^3 = 320{,}000$ ft

(b) $v_{\text{ave}} = 320{,}000/40 = 8{,}000$ ft/s

(c) $5t^3 = 135$ when the rocket has gone 135 ft, so $t^3 = 27$, $t = 3$ s; $v_{\text{ave}} = 135/3 = 45$ ft/s.

(d) $$v_{\text{inst}} = \lim_{t_1 \to 40} \frac{5t_1^3 - 5(40)^3}{t_1 - 40} = \lim_{t_1 \to 40} \frac{5(t_1^3 - 40^3)}{t_1 - 40} = \lim_{t_1 \to 40} 5(t_1^2 + 40t_1 + 1600) = 24{,}000 \text{ ft/s}$$

20. **(a)** $v_{\text{ave}} = \dfrac{[3(3)^2 + 3] - [3(1)^2 + 1]}{3 - 1} = 13$ mi/h

(b) $$v_{\text{inst}} = \lim_{t_1 \to 1} \frac{(3t_1^2 + t_1) - 4}{t_1 - 1} = \lim_{t_1 \to 1} \frac{(3t_1 + 4)(t_1 - 1)}{t_1 - 1} = \lim_{t_1 \to 1} (3t_1 + 4) = 7 \text{ mi/h}$$

21. **(a)** $v_{\text{ave}} = \dfrac{6(4)^4 - 6(2)^4}{4 - 2} = 720$ ft/min

(b) $$v_{\text{inst}} = \lim_{t_1 \to 2} \frac{6t_1^4 - 6(2)^4}{t_1 - 2} = \lim_{t_1 \to 2} \frac{6(t_1^4 - 16)}{t_1 - 2} = \lim_{t_1 \to 2} \frac{6(t_1^2 + 4)(t_1^2 - 4)}{t_1 - 2} = \lim_{t_1 \to 2} 6(t_1^2 + 4)(t_1 + 2) = 192 \text{ ft/min}$$

EXERCISE SET 3.2

1. $f'(1) = 2$, $f'(3) = 0$, $f'(5) = -2$, $f'(6) = -1/2$

2. $f'(4) < f'(0) < f'(2) < 0 < f'(-3)$

3. **(b)** $m = f'(2) = 3$

(c) the same, $f'(2) = 3$

4. $f'(-1) = m = \dfrac{4 - 3}{0 - (-1)} = 1$

5.

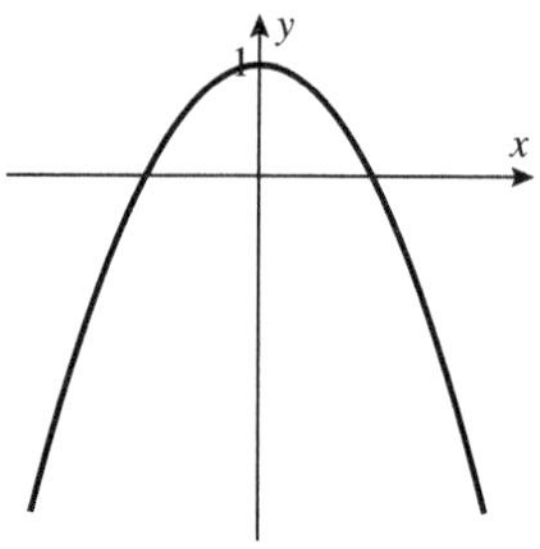

6.

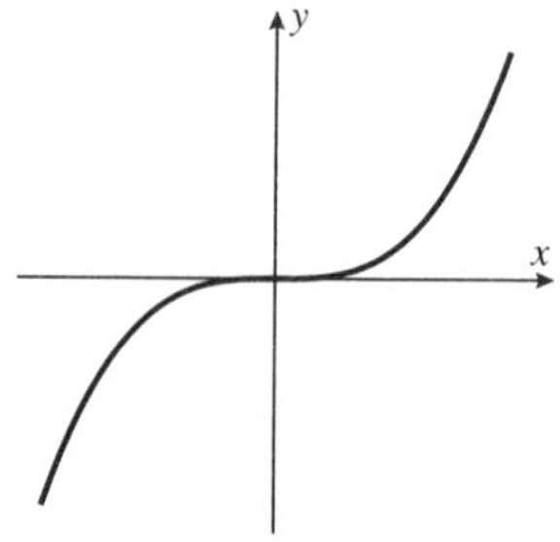

7. $y-(-1)=5(x-3),\ y=5x-16$

8. $y-3=-4(x+2),\ y=-4x-5$

9. $f'(x)=\lim_{w\to x}\frac{f(w)-f(x)}{w-x}=\lim_{w\to x}\frac{3w^2-3x^2}{w-x}=\lim_{w\to x}3(w+x)=6x;\ f(3)=3(3)^2=27,\ f'(3)=18$ so $y-27=18(x-3),\ y=18x-27$

10. $f'(x)=\lim_{w\to x}\frac{f(w)-f(x)}{w-x}=\lim_{w\to x}\frac{w^4-x^4}{w-x}=\lim_{w\to x}(w^3+w^2x+wx^2+x^3)=4x^3;$ $f(-2)=(-2)^4=16,\ f'(-2)=-32$ so $y-16=-32(x+2),\ y=-32x-48$

11. $f'(x)=\lim_{w\to x}\frac{f(w)-f(x)}{w-x}=\lim_{w\to x}\frac{w^3-x^3}{w-x}=\lim_{w\to x}(w^2+wx+x^2)=3x^2;\ f(0)=0^3=0,$ $f'(0)=0$ so $y-0=(0)(x-0),\ y=0$

12. $f'(x)=\lim_{w\to x}\frac{f(w)-f(x)}{w-x}=\lim_{w\to x}\frac{2w^3+1-(2x^3+1)}{w-x}=\lim_{w\to x}2(w^2+wx+x^2)=6x^2;$ $f(-1)=2(-1)^3+1=-1,\ f'(-1)=6$ so $y+1=6(x+1),\ y=6x+5$

13. $f'(x)=\lim_{w\to x}\frac{f(w)-f(x)}{w-x}=\lim_{w\to x}\frac{\sqrt{w+1}-\sqrt{x+1}}{w-x}$

$$=\lim_{w\to x}\frac{\sqrt{w+1}-\sqrt{x+1}}{w-x}\,\frac{\sqrt{w+1}+\sqrt{x+1}}{\sqrt{w+1}+\sqrt{x+1}}=\lim_{w\to x}\frac{1}{(\sqrt{w+1}+\sqrt{x+1})}=\frac{1}{2\sqrt{x+1}};$$

$f(8)=\sqrt{8+1}=3,\ f'(8)=\frac{1}{6}$ so $y-3=\frac{1}{6}(x-8),\ y=\frac{1}{6}x+\frac{5}{3}$

14. $f'(x)=\lim_{w\to x}\frac{f(w)-f(x)}{w-x}=\lim_{w\to x}\frac{\sqrt{2w+1}-\sqrt{2x+1}}{w-x}$

$$=\lim_{w\to x}\frac{2}{\sqrt{2w+1}+\sqrt{2x+1}}=\lim_{w\to x}\frac{2}{\sqrt{9+2h}+3}=\frac{1}{\sqrt{2x+1}}$$

$f(4)=\sqrt{2(4)+1}=\sqrt{9}=3, f'(4)=1/3$ so $y-3=\frac{1}{3}(x-4), y=\frac{1}{3}x+\frac{5}{3}$

15. $f'(x)=\lim_{\Delta x\to 0}\frac{\frac{1}{x+\Delta x}-\frac{1}{x}}{\Delta x}=\lim_{\Delta x\to 0}\frac{\frac{x-(x+\Delta x)}{x(x+\Delta x)}}{\Delta x}$

$$=\lim_{\Delta x\to 0}\frac{-\Delta x}{x\Delta x(x+\Delta x)}=\lim_{\Delta x\to 0}-\frac{1}{x(x+\Delta x)}=-\frac{1}{x^2}$$

16. $f'(x) = \lim_{\Delta x \to 0} \dfrac{\dfrac{1}{(x+\Delta x)+1} - \dfrac{1}{x+1}}{\Delta x} = \lim_{\Delta x \to 0} \dfrac{\dfrac{(x+1)-(x+\Delta x+1)}{(x+1)(x+\Delta x+1)}}{\Delta x}$

$= \lim_{\Delta x \to 0} \dfrac{x+1-x-\Delta x-1}{\Delta x(x+1)(x+\Delta x+1)} = \lim_{\Delta x \to 0} \dfrac{-\Delta x}{\Delta x(x+1)(x+\Delta x+1)}$

$= \lim_{\Delta x \to 0} \dfrac{-1}{(x+1)(x+\Delta x+1)} = -\dfrac{1}{(x+1)^2}$

17. $f'(x) = \lim_{\Delta x \to 0} \dfrac{[a(x+\Delta x)^2+b]-[ax^2+b]}{\Delta x} = \lim_{\Delta x \to 0} \dfrac{ax^2+2ax\Delta x+a(\Delta x)^2+b-ax^2-b}{\Delta x}$

$= \lim_{\Delta x \to 0} \dfrac{2ax\Delta x+a(\Delta x)^2}{\Delta x} = \lim_{\Delta x \to 0}(2ax+a\Delta x) = 2ax$

18. $f'(x) = \lim_{\Delta x \to 0} \dfrac{(x+\Delta x)^2-(x+\Delta x)-(x^2-x)}{\Delta x} = \lim_{\Delta x \to 0} \dfrac{2x\Delta x+\Delta x^2-\Delta x}{\Delta x}$

$= \lim_{\Delta x \to 0}(2x-1+\Delta x) = 2x-1$

19. $f'(x) = \lim_{\Delta x \to 0} \dfrac{\dfrac{1}{\sqrt{x+\Delta x}} - \dfrac{1}{\sqrt{x}}}{\Delta x} = \lim_{\Delta x \to 0} \dfrac{\sqrt{x}-\sqrt{x+\Delta x}}{\Delta x\sqrt{x}\sqrt{x+\Delta x}}$

$= \lim_{\Delta x \to 0} \dfrac{x-(x+\Delta x)}{\Delta x\sqrt{x}\sqrt{x+\Delta x}(\sqrt{x}+\sqrt{x+\Delta x})} = \lim_{\Delta x \to 0} \dfrac{-1}{\sqrt{x}\sqrt{x+\Delta x}(\sqrt{x}+\sqrt{x+\Delta x})} = -\dfrac{1}{2x^{3/2}}$

20. $f'(x) = \lim_{\Delta x \to 0} \dfrac{\dfrac{1}{(x+\Delta x)^2} - \dfrac{1}{x^2}}{\Delta x} = \lim_{\Delta x \to 0} \dfrac{\dfrac{x^2-(x+\Delta x)^2}{x^2(x+\Delta x)^2}}{\Delta x}$

$= \lim_{\Delta x \to 0} \dfrac{x^2-x^2-2x\Delta x-\Delta x^2}{x^2\Delta x(x+\Delta x)^2} = \lim_{\Delta x \to 0} \dfrac{-2x\Delta x-\Delta x^2}{x^2\Delta x(x+\Delta x)^2} = \lim_{\Delta x \to 0} \dfrac{-2x-\Delta x}{x^2(x+\Delta x)^2} = -\dfrac{2}{x^3}$

21. $f'(t) = \lim_{h \to 0} \dfrac{f(t+h)-f(t)}{h} = \lim_{h \to 0} \dfrac{[4(t+h)^2+(t+h)]-[4t^2+t]}{h}$

$= \lim_{h \to 0} \dfrac{4t^2+8th+4h^2+t+h-4t^2-t}{h}$

$= \lim_{h \to 0} \dfrac{8th+4h^2+h}{h} = \lim_{h \to 0}(8t+4h+1) = 8t+1$

22. $\dfrac{dV}{dr} = \lim_{h \to 0} \dfrac{\dfrac{4}{3}\pi(r+h)^3 - \dfrac{4}{3}\pi r^3}{h} = \lim_{h \to 0} \dfrac{\dfrac{4}{3}\pi(r^3+3r^2h+3rh^2+h^3-r^3)}{h}$

$= \lim_{h \to 0} \dfrac{4}{3}\pi(3r^2+3rh+h^2) = 4\pi r^2$

23. **(a)** D **(b)** F **(c)** B **(d)** C **(e)** A **(f)** E

24. Any function of the form $f(x) = x + k$ has slope 1, and thus the derivative must be equal to 1 everywhere.

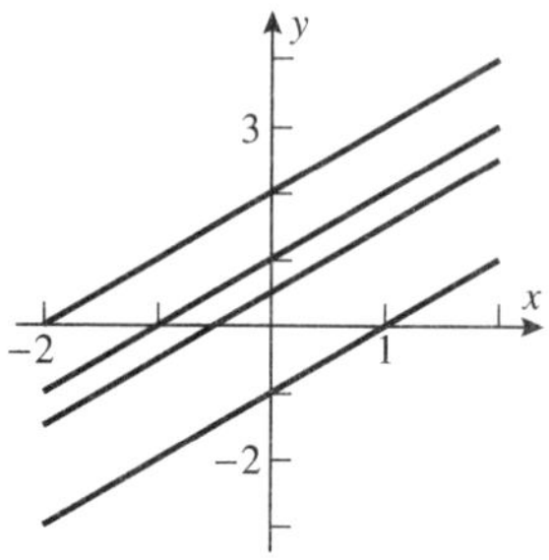

25. (a)

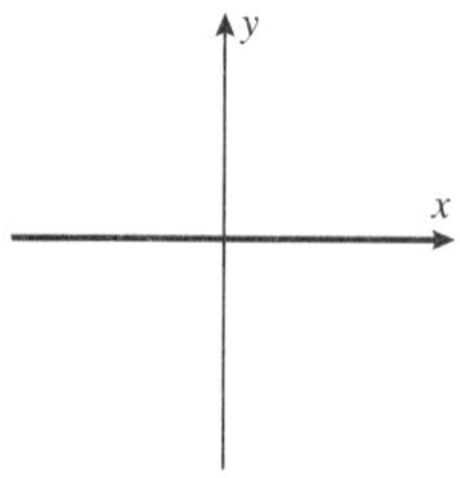

(b)

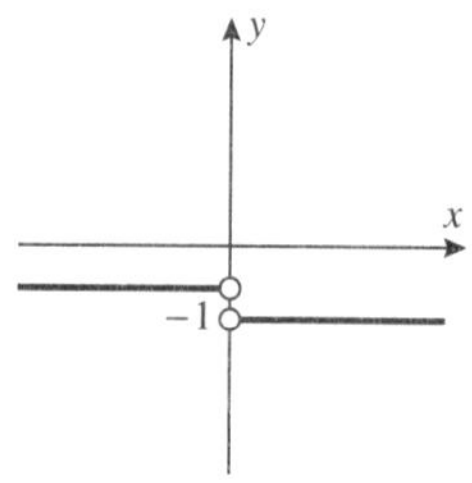

(c)

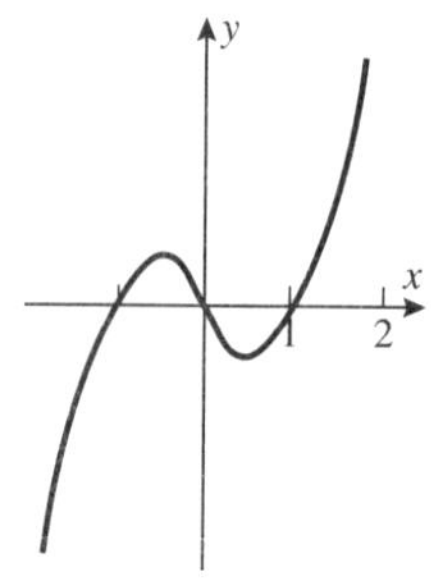

26. (a)

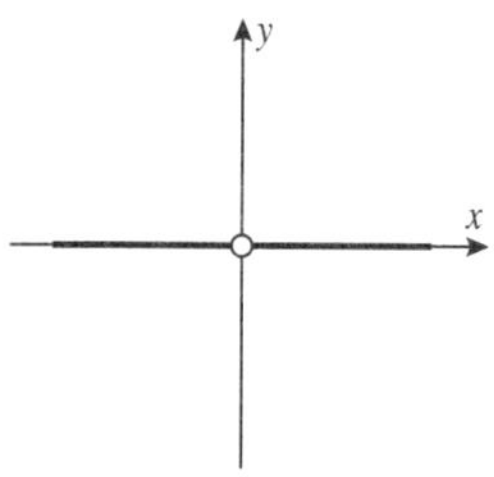

(b)

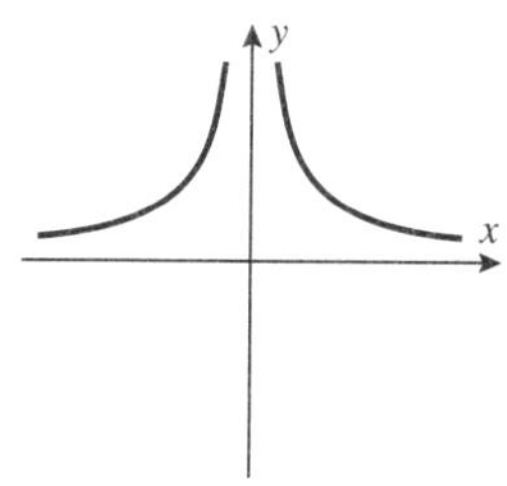

(c)

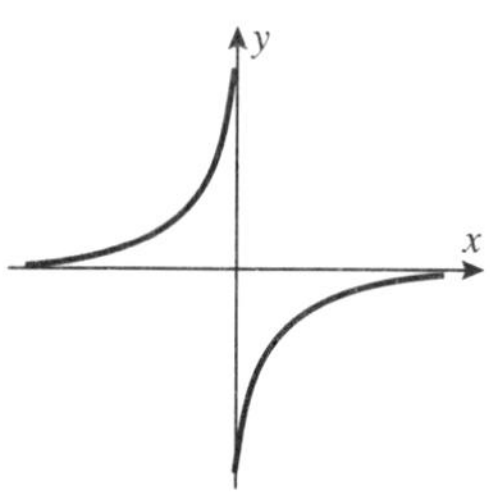

27. (a) $f(x) = x^2$ and $a = 3$ (b) $f(x) = \sqrt{x}$ and $a = 1$

28. (a) $f(x) = x^7$ and $a = 1$ (b) $f(x) = \cos x$ and $a = \pi$

29. $\dfrac{dy}{dx} = \lim_{h \to 0} \dfrac{[4(x+h)^2 + 1] - [4x^2 + 1]}{h} = \lim_{h \to 0} \dfrac{4x^2 + 8xh + 4h^2 + 1 - 4x^2 - 1}{h} = \lim_{h \to 0} (8x + 4h) = 8x$

$\left.\dfrac{dy}{dx}\right|_{x=1} = 8(1) = 8$

30. $\dfrac{dy}{dx} = \lim_{h\to 0} \dfrac{\left(\dfrac{5}{x+h}+1\right) - \left(\dfrac{5}{x}+1\right)}{h} = \lim_{h\to 0} \dfrac{\dfrac{5}{x+h} - \dfrac{5}{x}}{h} = \lim_{h\to 0} \dfrac{\dfrac{5x-5(x+h)}{x(x+h)}}{h}$

$= \lim_{h\to 0} \dfrac{5x-5x-5h}{hx(x+h)} = \lim_{h\to 0} \dfrac{-5}{x(x+h)} = -\dfrac{5}{x^2}$

$\left.\dfrac{dy}{dx}\right|_{x=-2} = -\dfrac{5}{(-2)^2} = -\dfrac{5}{4}$

31. $y = -2x + 1$

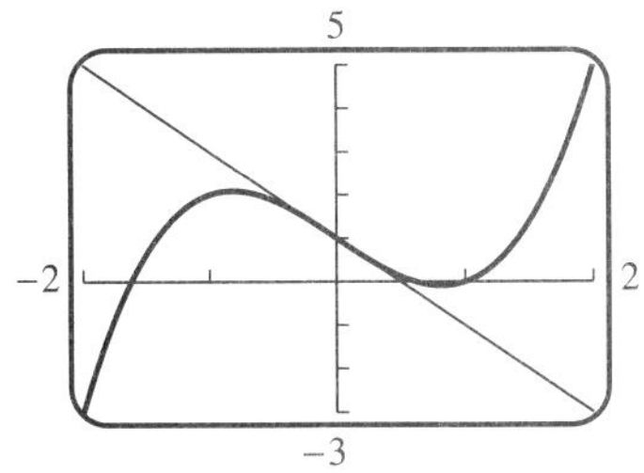

32.

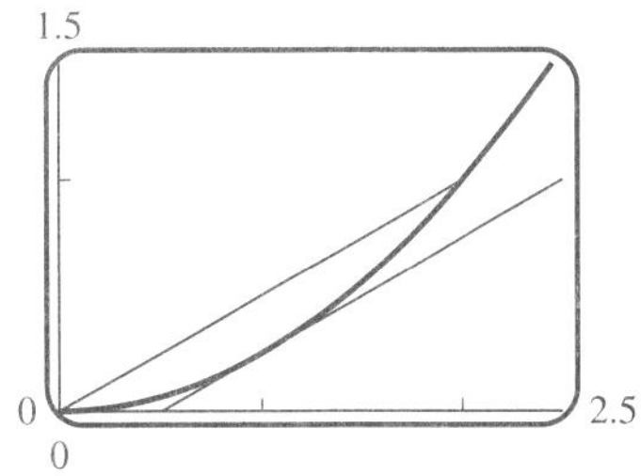

33. **(b)**

h	0.5	0.1	0.01	0.001	0.0001	0.00001
$(f(1+h)-f(1))/h$	1.6569	1.4355	1.3911	1.3868	1.3863	1.3863

34. **(b)**

h	0.5	0.1	0.01	0.001	0.0001	0.00001
$(f(1+h)-f(1))/h$	0.50489	0.67060	0.70356	0.70675	0.70707	0.70710

35. **(a)** dollars/ft

(b) As you go deeper the price per foot may increase dramatically, so $f'(x)$ is roughly the price per additional foot.

(c) If each additional foot costs extra money (this is to be expected) then $f'(x)$ remains positive.

(d) From the approximation $1000 = f'(300) \approx \dfrac{f(301) - f(300)}{301 - 300}$ we see that $f(301) \approx f(300) + 1000$, so the extra foot will cost around \$1000.

36. **(a)** gallons/dollar

(b) The increase in the amount of paint that would be sold for one extra dollar.

(c) It should be negative since an increase in the price of paint would decrease the amount of paint sold.

(d) From $-100 = f'(10) \approx \dfrac{f(11) - f(10)}{11 - 10}$ we see that $f(11) \approx f(10) - 100$, so an increase of one dollar would decrease the amount of paint sold by around 100 gallons.

37. **(a)** $F \approx 200$ lb, $dF/d\theta \approx 50$ lb/rad **(b)** $\mu = (dF/d\theta)/F \approx 50/200 = 0.25$

38. **(a)** The slope of the tangent line $\approx \dfrac{10 - 2.2}{2050 - 1950} = 0.078$ billion, or in 2050 the world population was increasing at the rate of about 78 million per year.

(b) $\dfrac{dN}{dt} =\approx \dfrac{0.078}{6} = 0.013 = 1.3$ %/year

39. **(a)** $T \approx 115°\text{F}$, $dT/dt \approx -3.35°\text{F/min}$

(b) $k = (dT/dt)/(T - T_0) \approx (-3.35)/(115 - 75) = -0.084$

41. $\lim_{x \to 0} f(x) = \lim_{x \to 0} \sqrt[3]{x} = 0 = f(0)$, so f is continuous at $x = 0$.

$\lim_{h \to 0} \dfrac{f(0+h) - f(0)}{h} = \lim_{h \to 0} \dfrac{\sqrt[3]{h} - 0}{h} = \lim_{h \to 0} \dfrac{1}{h^{2/3}} = +\infty$, so $f'(0)$ does not exist.

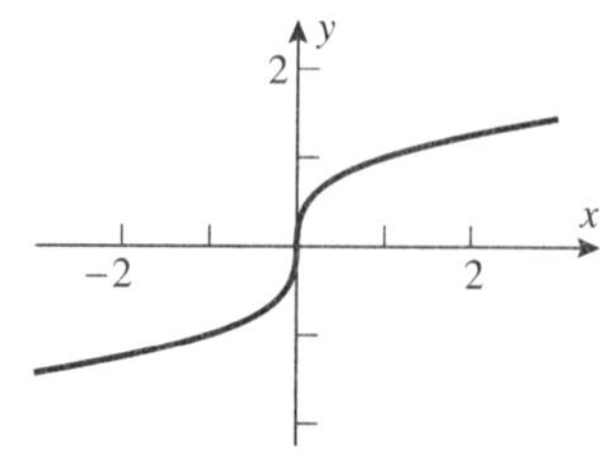

42. $\lim_{x \to 2} f(x) = \lim_{x \to 2} (x-2)^{2/3} = 0 = f(2)$ so f is continuous at $x = 2$. $\lim_{h \to 0} \dfrac{f(2+h) - f(2)}{h} = \lim_{h \to 0} \dfrac{h^{2/3} - 0}{h} = \lim_{h \to 0} \dfrac{1}{h^{1/3}}$ which does not exist so $f'(2)$ does not exist.

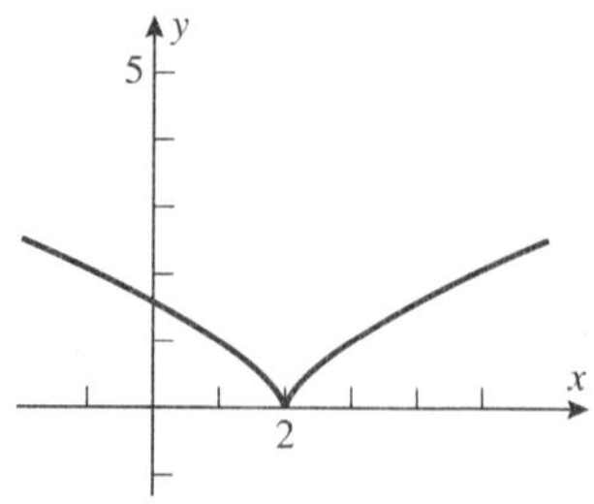

43. $\lim_{x \to 1^-} f(x) = \lim_{x \to 1^+} f(x) = f(1)$, so f is continuous at $x = 1$.

$\lim_{h \to 0^-} \dfrac{f(1+h) - f(1)}{h} = \lim_{h \to 0^-} \dfrac{[(1+h)^2 + 1] - 2}{h} = \lim_{h \to 0^-} (2+h) = 2;$

$\lim_{h \to 0^+} \dfrac{f(1+h) - f(1)}{h} = \lim_{h \to 0^+} \dfrac{2(1+h) - 2}{h} = \lim_{h \to 0^+} 2 = 2$, so $f'(1) = 2$.

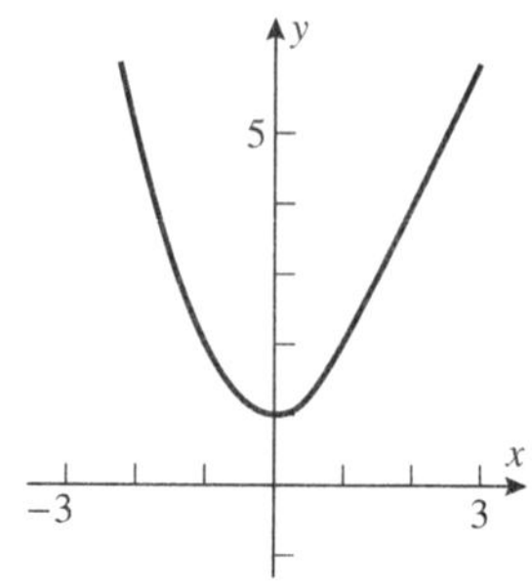

44. $\lim_{x \to 1^-} f(x) = \lim_{x \to 1^+} f(x) = f(1)$ so f is continuous at $x = 1$.

$\lim_{h \to 0^-} \dfrac{f(1+h) - f(1)}{h} = \lim_{h \to 0^-} \dfrac{[(1+h)^2 + 2] - 3}{h} = \lim_{h \to 0^-} (2+h) = 2;$

$\lim_{h \to 0^+} \dfrac{f(1+h) - f(1)}{h} = \lim_{h \to 0^+} \dfrac{[(1+h) + 2] - 3}{h} = \lim_{h \to 0^+} 1 = 1,$

so $f'(1)$ does not exist.

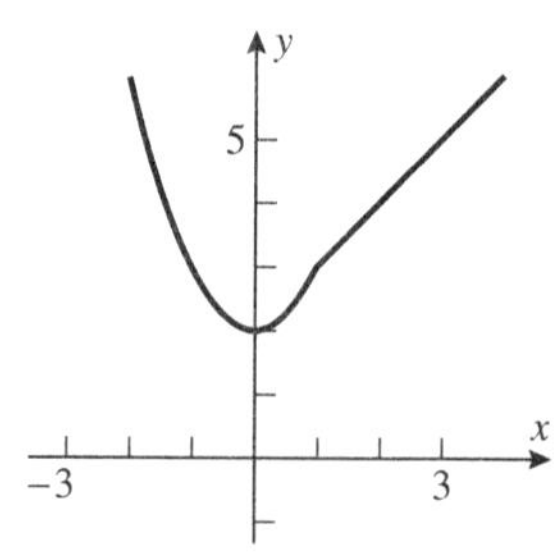

45. Since $-|x| \le x \sin(1/x) \le |x|$ it follows by the Squeezing Theorem (Theorem 2.6.2) that $\lim_{x \to 0} x \sin(1/x) = 0$. The derivative cannot exist: consider $\dfrac{f(x) - f(0)}{x} = \sin(1/x)$. This function oscillates between -1 and $+1$ and does not tend to zero as x tends to zero.

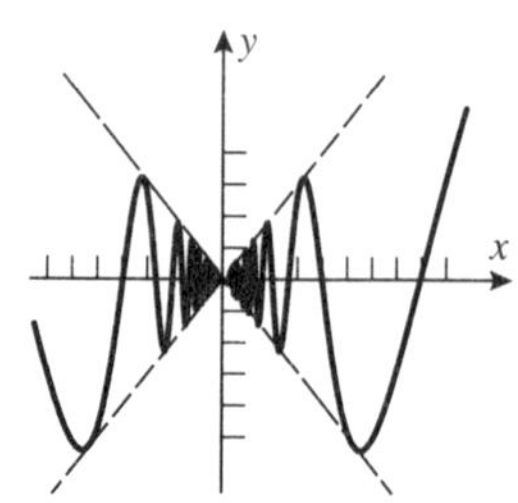

46. For continuity, compare with $\pm x^2$ to establish that the limit is zero. The differential quotient is $x \sin(1/x)$ and (see Exercise 45) this has a limit of zero at the origin.

47. f is continuous at $x = 1$ because it is differentiable there, thus $\lim_{h \to 0} f(1+h) = f(1)$ and so $f(1) = 0$ because $\lim_{h \to 0} \dfrac{f(1+h)}{h}$ exists; $f'(1) = \lim_{h \to 0} \dfrac{f(1+h) - f(1)}{h} = \lim_{h \to 0} \dfrac{f(1+h)}{h} = 5$.

48. Let $x = y = 0$ to get $f(0) = f(0) + f(0) + 0$ so $f(0) = 0$. $f'(x) = \lim_{h\to 0} \dfrac{f(x+h) - f(x)}{h}$, but $f(x+h) = f(x) + f(h) + 5xh$ so $f(x+h) - f(x) = f(h) + 5xh$ and

$$f'(x) = \lim_{h\to 0} \frac{f(h) + 5xh}{h} = \lim_{h\to 0}\left(\frac{f(h)}{h} + 5x\right) = 3 + 5x.$$

49.
$$f'(x) = \lim_{h\to 0} \frac{f(x+h) - f(x)}{h} = \lim_{h\to 0} \frac{f(x)f(h) - f(x)}{h} = \lim_{h\to 0} \frac{f(x)[f(h) - 1]}{h}$$
$$= f(x) \lim_{h\to 0} \frac{f(h) - f(0)}{h} = f(x)f'(0) = f(x)$$

EXERCISE SET 3.3

1. $28x^6$ **2.** $-36x^{11}$ **3.** $24x^7 + 2$ **4.** $2x^3$

5. 0 **6.** $\sqrt{2}$ **7.** $-\frac{1}{3}(7x^6 + 2)$ **8.** $\frac{2}{5}x$

9. $3ax^2 + 2bx + c$ **10.** $\frac{1}{a}\left(2x + \frac{1}{b}\right)$ **11.** $24x^{-9} + 1/\sqrt{x}$ **12.** $-42x^{-7} - \frac{5}{2\sqrt{x}}$

13. $-3x^{-4} - 7x^{-8}$ **14.** $\frac{1}{2\sqrt{x}} - \frac{1}{x^2}$

15.
$$f'(x) = (3x^2 + 6)\frac{d}{dx}\left(2x - \frac{1}{4}\right) + \left(2x - \frac{1}{4}\right)\frac{d}{dx}(3x^2 + 6) = (3x^2 + 6)(2) + \left(2x - \frac{1}{4}\right)(6x)$$
$$= 18x^2 - \frac{3}{2}x + 12$$

16.
$$f'(x) = (2 - x - 3x^3)\frac{d}{dx}(7 + x^5) + (7 + x^5)\frac{d}{dx}(2 - x - 3x^3)$$
$$= (2 - x - 3x^3)(5x^4) + (7 + x^5)(-1 - 9x^2)$$
$$= -24x^7 - 6x^5 + 10x^4 - 63x^2 - 7$$

17.
$$f'(x) = (x^3 + 7x^2 - 8)\frac{d}{dx}(2x^{-3} + x^{-4}) + (2x^{-3} + x^{-4})\frac{d}{dx}(x^3 + 7x^2 - 8)$$
$$= (x^3 + 7x^2 - 8)(-6x^{-4} - 4x^{-5}) + (2x^{-3} + x^{-4})(3x^2 + 14x)$$
$$= -15x^{-2} - 14x^{-3} + 48x^{-4} + 32x^{-5}$$

18.
$$f'(x) = (x^{-1} + x^{-2})\frac{d}{dx}(3x^3 + 27) + (3x^3 + 27)\frac{d}{dx}(x^{-1} + x^{-2})$$
$$= (x^{-1} + x^{-2})(9x^2) + (3x^3 + 27)(-x^{-2} - 2x^{-3}) = 3 + 6x - 27x^{-2} - 54x^{-3}$$

19. $12x(3x^2 + 1)$ **20.** $f(x) = x^{10} + 4x^6 + 4x^2$, $f'(x) = 10x^9 + 24x^5 + 8x$

21.
$$\frac{dy}{dx} = \frac{(5x - 3)\frac{d}{dx}(1) - (1)\frac{d}{dx}(5x - 3)}{(5x - 3)^2} = -\frac{5}{(5x - 3)^2};\quad y'(1) = -5/4$$

22. $\dfrac{dy}{dx} = \dfrac{(\sqrt{x}+2)\dfrac{d}{dx}(3) - 3\dfrac{d}{dx}(\sqrt{x}+2)}{(\sqrt{x}+2)^2} = -3/(2\sqrt{x}(\sqrt{x}+2)^2);\ y'(1) = -3/18 = -1/6$

23. $\dfrac{dx}{dt} = \dfrac{(2t+1)\dfrac{d}{dt}(3t) - (3t)\dfrac{d}{dt}(2t+1)}{(2t+1)^2} = \dfrac{(2t+1)(3)-(3t)(2)}{(2t+1)^2} = \dfrac{3}{(2t+1)^2}$

24. $\dfrac{dx}{dt} = \dfrac{(3t)\dfrac{d}{dt}(t^2+1) - (t^2+1)\dfrac{d}{dt}(3t)}{(3t)^2} = \dfrac{(3t)(2t)-(t^2+1)(3)}{9t^2} = \dfrac{t^2-1}{3t^2}$

25. $\dfrac{dy}{dx} = \dfrac{(x+3)\dfrac{d}{dx}(2x-1) - (2x-1)\dfrac{d}{dx}(x+3)}{(x+3)^2}$

$= \dfrac{(x+3)(2)-(2x-1)(1)}{(x+3)^2} = \dfrac{7}{(x+3)^2};\ \left.\dfrac{dy}{dx}\right|_{x=1} = \dfrac{7}{16}$

26. $\dfrac{dy}{dx} = \dfrac{(x^2-5)\dfrac{d}{dx}(4x+1) - (4x+1)\dfrac{d}{dx}(x^2-5)}{(x^2-5)^2}$

$= \dfrac{(x^2-5)(4)-(4x+1)(2x)}{(x^2-5)^2} = -\dfrac{4x^2+2x+20}{(x^2-5)^2};\ \left.\dfrac{dy}{dx}\right|_{x=1} = \dfrac{13}{8}$

27. $\dfrac{dy}{dx} = \left(\dfrac{3x+2}{x}\right)\dfrac{d}{dx}\left(x^{-5}+1\right) + \left(x^{-5}+1\right)\dfrac{d}{dx}\left(\dfrac{3x+2}{x}\right)$

$= \left(\dfrac{3x+2}{x}\right)\left(-5x^{-6}\right) + \left(x^{-5}+1\right)\left[\dfrac{x(3)-(3x+2)(1)}{x^2}\right]$

$= \left(\dfrac{3x+2}{x}\right)\left(-5x^{-6}\right) + \left(x^{-5}+1\right)\left(-\dfrac{2}{x^2}\right);$

$\left.\dfrac{dy}{dx}\right|_{x=1} = 5(-5)+2(-2) = -29$

28. $\dfrac{dy}{dx} = (2x^7-x^2)\dfrac{d}{dx}\left(\dfrac{x-1}{x+1}\right) + \left(\dfrac{x-1}{x+1}\right)\dfrac{d}{dx}(2x^7-x^2)$

$= (2x^7-x^2)\left[\dfrac{(x+1)(1)-(x-1)(1)}{(x+1)^2}\right] + \left(\dfrac{x-1}{x+1}\right)(14x^6-2x)$

$= (2x^7-x^2)\cdot\dfrac{2}{(x+1)^2} + \left(\dfrac{x-1}{x+1}\right)(14x^6-2x);$

$\left.\dfrac{dy}{dx}\right|_{x=1} = (2-1)\dfrac{2}{4} + 0(14-2) = \dfrac{1}{2}$

29. $f'(1) \approx \dfrac{f(1.01)-f(1)}{0.01} = \dfrac{0.999699-(-1)}{0.01} = 0.0301$, and by differentiation, $f'(1) = 3(1)^2 - 3 = 0$

30. $f'(1) \approx \dfrac{f(1.01)-f(1)}{0.01} = \dfrac{1.01504-1}{0.01} = 1.504$, and by differentiation,

$f'(1) = \left.\left(\sqrt{x}+\dfrac{x}{2\sqrt{x}}\right)\right|_{x=1} = 1.5$

31. $f'(1) = 0$ **32.** $f'(1) = 1$ **33.** $32t$

34. 2π **35.** $3\pi r^2$ **36.** $-2\alpha^{-2} + 1$

37. **(a)** $\dfrac{dV}{dr} = 4\pi r^2$ **(b)** $\left.\dfrac{dV}{dr}\right|_{r=5} = 4\pi(5)^2 = 100\pi$

38. $\dfrac{d}{d\lambda}\left[\dfrac{\lambda\lambda_0 + \lambda^6}{2 - \lambda_0}\right] = \dfrac{1}{2 - \lambda_0}\dfrac{d}{d\lambda}(\lambda\lambda_0 + \lambda^6) = \dfrac{1}{2 - \lambda_0}(\lambda_0 + 6\lambda^5) = \dfrac{\lambda_0 + 6\lambda^5}{2 - \lambda_0}$

39. **(a)** $g'(x) = \sqrt{x}f'(x) + \dfrac{1}{2\sqrt{x}}f(x)$, $g'(4) = (2)(-5) + \dfrac{1}{4}(3) = -37/4$

(b) $g'(x) = \dfrac{xf'(x) - f(x)}{x^2}$, $g'(4) = \dfrac{(4)(-5) - 3}{16} = -23/16$

40. **(a)** $g'(x) = 6x - 5f'(x)$, $g'(3) = 6(3) - 5(4) = -2$

(b) $g'(x) = \dfrac{2f(x) - (2x+1)f'(x)}{f^2(x)}$, $g'(3) = \dfrac{2(-2) - 7(4)}{(-2)^2} = -8$

41. **(a)** $F'(x) = 5f'(x) + 2g'(x), F'(2) = 5(4) + 2(-5) = 10$
(b) $F'(x) = f'(x) - 3g'(x), F'(2) = 4 - 3(-5) = 19$
(c) $F'(x) = f(x)g'(x) + g(x)f'(x), F'(2) = (-1)(-5) + (1)(4) = 9$
(d) $F'(x) = [g(x)f'(x) - f(x)g'(x)]/g^2(x), F'(2) = [(1)(4) - (-1)(-5)]/(1)^2 = -1$

42. **(a)** $F'(x) = 6f'(x) - 5g'(x), F'(\pi) = 6(-1) - 5(2) = -16$
(b) $F'(x) = f(x) + g(x) + x(f'(x) + g'(x)), F'(\pi) = 10 - 3 + \pi(-1 + 2) = 7 + \pi$
(c) $F'(x) = 2f(x)g'(x) + 2f'(x)g(x) = 2(20) + 2(3) = 46$
(d) $F'(x) = \dfrac{(4 + g(x))f'(x) - f(x)g'(x)}{(4 + g(x))^2} = \dfrac{(4-3)(-1) - 10(2)}{(4-3)^2} = -21$

43. $y - 2 = 5(x + 3)$, $y = 5x + 17$

44. $\dfrac{dy}{dx} = \dfrac{(1+x)(-1) - (1-x)(1)}{(1+x)^2} = -\dfrac{2}{(1+x)^2}$, $\left.\dfrac{dy}{dx}\right|_{x=2} = -\dfrac{2}{9}$ and $y = -\dfrac{1}{3}$ for $x = 2$ so an equation of the tangent line is $y - \left(-\dfrac{1}{3}\right) = -\dfrac{2}{9}(x - 2)$, or $y = -\dfrac{2}{9}x + \dfrac{1}{9}$.

45. **(a)** $dy/dx = 21x^2 - 10x + 1$, $d^2y/dx^2 = 42x - 10$
(b) $dy/dx = 24x - 2$, $d^2y/dx^2 = 24$
(c) $dy/dx = -1/x^2$, $d^2y/dx^2 = 2/x^3$
(d) $y = 35x^5 - 16x^3 - 3x$, $dy/dx = 175x^4 - 48x^2 - 3$, $d^2y/dx^2 = 700x^3 - 96x$

46. **(a)** $y' = 28x^6 - 15x^2 + 2$, $y'' = 168x^5 - 30x$
(b) $y' = 3$, $y'' = 0$
(c) $y' = \dfrac{2}{5x^2}$, $y'' = -\dfrac{4}{5x^3}$
(d) $y = 2x^4 + 3x^3 - 10x - 15$, $y' = 8x^3 + 9x^2 - 10$, $y'' = 24x^2 + 18x$

47. **(a)** $y' = -5x^{-6} + 5x^4$, $y'' = 30x^{-7} + 20x^3$, $y''' = -210x^{-8} + 60x^2$
(b) $y = x^{-1}$, $y' = -x^{-2}$, $y'' = 2x^{-3}$, $y''' = -6x^{-4}$
(c) $y' = 3ax^2 + b$, $y'' = 6ax$, $y''' = 6a$

48. (a) $dy/dx = 10x - 4$, $d^2y/dx^2 = 10$, $d^3y/dx^3 = 0$

(b) $dy/dx = -6x^{-3} - 4x^{-2} + 1$, $d^2y/dx^2 = 18x^{-4} + 8x^{-3}$, $d^3y/dx^3 = -72x^{-5} - 24x^{-4}$

(c) $dy/dx = 4ax^3 + 2bx$, $d^2y/dx^2 = 12ax^2 + 2b$, $d^3y/dx^3 = 24ax$

49. (a) $f'(x) = 6x$, $f''(x) = 6$, $f'''(x) = 0$, $f'''(2) = 0$

(b) $\dfrac{dy}{dx} = 30x^4 - 8x$, $\dfrac{d^2y}{dx^2} = 120x^3 - 8$, $\left.\dfrac{d^2y}{dx^2}\right|_{x=1} = 112$

(c) $\dfrac{d}{dx}\left[x^{-3}\right] = -3x^{-4}$, $\dfrac{d^2}{dx^2}\left[x^{-3}\right] = 12x^{-5}$, $\dfrac{d^3}{dx^3}\left[x^{-3}\right] = -60x^{-6}$, $\dfrac{d^4}{dx^4}\left[x^{-3}\right] = 360x^{-7}$,
$\left.\dfrac{d^4}{dx^4}\left[x^{-3}\right]\right|_{x=1} = 360$

50. (a) $y' = 16x^3 + 6x^2$, $y'' = 48x^2 + 12x$, $y''' = 96x + 12$, $y'''(0) = 12$

(b) $y = 6x^{-4}$, $\dfrac{dy}{dx} = -24x^{-5}$, $\dfrac{d^2y}{dx^2} = 120x^{-6}$, $\dfrac{d^3y}{dx^3} = -720x^{-7}$, $\dfrac{d^4y}{dx^4} = 5040x^{-8}$,
$\left.\dfrac{d^4y}{dx^4}\right|_{x=1} = 5040$

51. $y' = 3x^2 + 3$, $y'' = 6x$, and $y''' = 6$ so
$y''' + xy'' - 2y' = 6 + x(6x) - 2(3x^2 + 3) = 6 + 6x^2 - 6x^2 - 6 = 0$

52. $y = x^{-1}$, $y' = -x^{-2}$, $y'' = 2x^{-3}$ so
$x^3y'' + x^2y' - xy = x^3(2x^{-3}) + x^2(-x^{-2}) - x(x^{-1}) = 2 - 1 - 1 = 0$

53. $F'(x) = xf'(x) + f(x)$, $F''(x) = xf''(x) + f'(x) + f'(x) = xf''(x) + 2f'(x)$

54. (a) $F'''(x) = xf'''(x) + 3f''(x)$

(b) Assume that $F^{n)}(x) = xf^{(n)}(x) + nf^{(n-1)}(x)$ for some n (for instance $n = 3$, as in part (a)). Then $F^{(n+1)}(x) = xf^{(n+1)}(x) + (1 + n)f^{(n)}(x) = xf^{(n+1)}(x) + (n + 1)f^{(n)}(x)$, which is an inductive proof.

55. The graph has a horizontal tangent at points where $\dfrac{dy}{dx} = 0$, but $\dfrac{dy}{dx} = x^2 - 3x + 2 = (x - 1)(x - 2) = 0$ if $x = 1, 2$. The corresponding values of y are $5/6$ and $2/3$ so the tangent line is horizontal at $(1, 5/6)$ and $(2, 2/3)$.

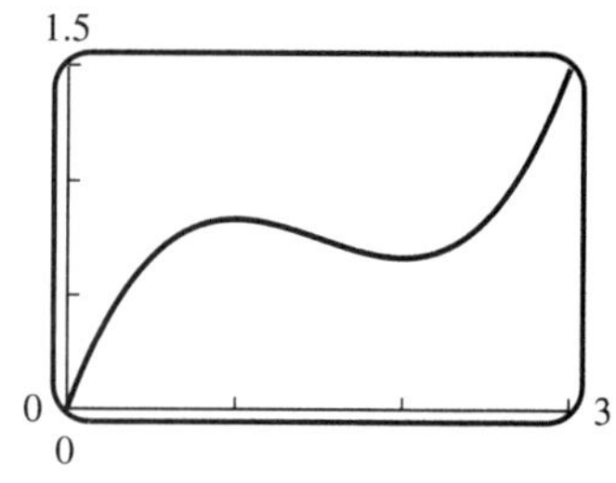

56. $\dfrac{dy}{dx} = \dfrac{9 - x^2}{(x^2 + 9)^2}$; $\dfrac{dy}{dx} = 0$ when $x^2 = 9$ so $x = \pm 3$.
The points are $(3, 1/6)$ and $(-3, -1/6)$.

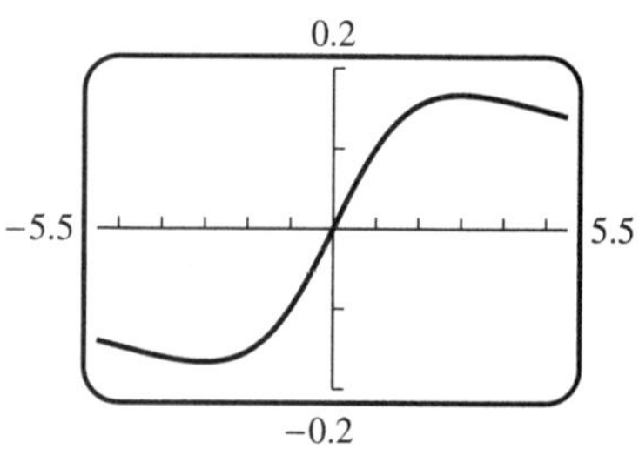

57. The y-intercept is -2 so the point $(0,-2)$ is on the graph; $-2 = a(0)^2 + b(0) + c$, $c = -2$. The x-intercept is 1 so the point (1,0) is on the graph; $0 = a + b - 2$. The slope is $dy/dx = 2ax + b$; at $x = 0$ the slope is b so $b = -1$, thus $a = 3$. The function is $y = 3x^2 - x - 2$.

58. Let $P(x_0, y_0)$ be the point where $y = x^2 + k$ is tangent to $y = 2x$. The slope of the curve is $\dfrac{dy}{dx} = 2x$ and the slope of the line is 2 thus at P, $2x_0 = 2$ so $x_0 = 1$. But P is on the line, so $y_0 = 2x_0 = 2$. Because P is also on the curve we get $y_0 = x_0^2 + k$ so $k = y_0 - x_0^2 = 2 - (1)^2 = 1$.

59. The points $(-1, 1)$ and $(2, 4)$ are on the secant line so its slope is $(4-1)/(2+1) = 1$. The slope of the tangent line to $y = x^2$ is $y' = 2x$ so $2x = 1$, $x = 1/2$.

60. The points $(1, 1)$ and $(4, 2)$ are on the secant line so its slope is $1/3$. The slope of the tangent line to $y = \sqrt{x}$ is $y' = 1/(2\sqrt{x})$ so $1/(2\sqrt{x}) = 1/3$, $2\sqrt{x} = 3$, $x = 9/4$.

61. $y' = -2x$, so at any point (x_0, y_0) on $y = 1 - x^2$ the tangent line is $y - y_0 = -2x_0(x - x_0)$, or $y = -2x_0x + x_0^2 + 1$. The point $(2, 0)$ is to be on the line, so $0 = -4x_0 + x_0^2 + 1$, $x_0^2 - 4x_0 + 1 = 0$. Use the quadratic formula to get $x_0 = \dfrac{4 \pm \sqrt{16-4}}{2} = 2 \pm \sqrt{3}$.

62. Let $P_1(x_1, ax_1^2)$ and $P_2(x_2, ax_2^2)$ be the points of tangency. $y' = 2ax$ so the tangent lines at P_1 and P_2 are $y - ax_1^2 = 2ax_1(x - x_1)$ and $y - ax_2^2 = 2ax_2(x - x_2)$. Solve for x to get $x = \frac{1}{2}(x_1 + x_2)$ which is the x-coordinate of a point on the vertical line halfway between P_1 and P_2.

63. $y' = 3ax^2 + b$; the tangent line at $x = x_0$ is $y - y_0 = (3ax_0^2 + b)(x - x_0)$ where $y_0 = ax_0^3 + bx_0$. Solve with $y = ax^3 + bx$ to get

$$\begin{aligned}
(ax^3 + bx) - (ax_0^3 + bx_0) &= (3ax_0^2 + b)(x - x_0) \\
ax^3 + bx - ax_0^3 - bx_0 &= 3ax_0^2x - 3ax_0^3 + bx - bx_0 \\
x^3 - 3x_0^2x + 2x_0^3 &= 0 \\
(x - x_0)(x^2 + xx_0 - 2x_0^2) &= 0 \\
(x - x_0)^2(x + 2x_0) &= 0, \text{ so } x = -2x_0.
\end{aligned}$$

64. Let (x_0, y_0) be the point of tangency. Refer to the solution to Exercise 65 to see that the endpoints of the line segment are at $(2x_0, 0)$ and $(0, 2y_0)$, so (x_0, y_0) is the midpoint of the segment.

65. $y' = -\dfrac{1}{x^2}$; the tangent line at $x = x_0$ is $y - y_0 = -\dfrac{1}{x_0^2}(x - x_0)$, or $y = -\dfrac{x}{x_0^2} + \dfrac{2}{x_0}$. The tangent line crosses the x-axis at $2x_0$, the y-axis at $2/x_0$, so that the area of the triangle is $\dfrac{1}{2}(2/x_0)(2x_0) = 2$.

66. $f'(x) = 3ax^2 + 2bx + c$; there is a horizontal tangent where $f'(x) = 0$. Use the quadratic formula on $3ax^2 + 2bx + c = 0$ to get $x = (-b \pm \sqrt{b^2 - 3ac})/(3a)$ which gives two real solutions, one real solution, or none if

(a) $b^2 - 3ac > 0$ **(b)** $b^2 - 3ac = 0$ **(c)** $b^2 - 3ac < 0$

67. $F = GmMr^{-2}$, $\dfrac{dF}{dr} = -2GmMr^{-3} = -\dfrac{2GmM}{r^3}$

68. $dR/dT = 0.04124 - 3.558 \times 10^{-5}T$ which decreases as T increases from 0 to 700. When $T = 0$, $dR/dT = 0.04124\,\Omega/°\text{C}$; when $T = 700$, $dR/dT = 0.01633\,\Omega/°\text{C}$. The resistance is most sensitive to temperature changes at $T = 0°\text{C}$, least sensitive at $T = 700°\text{C}$.

69. $f'(x) = 1 + 1/x^2 > 0$ for all $x \neq 0$

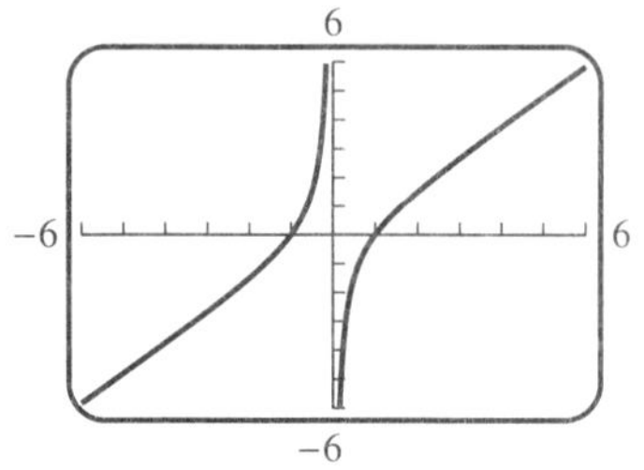

70. $f'(x) = -5\dfrac{x^2-4}{(x^2+4)^2}$;
$f'(x) > 0$ when $x^2 < 4$, i.e. on $-2 < x < 2$

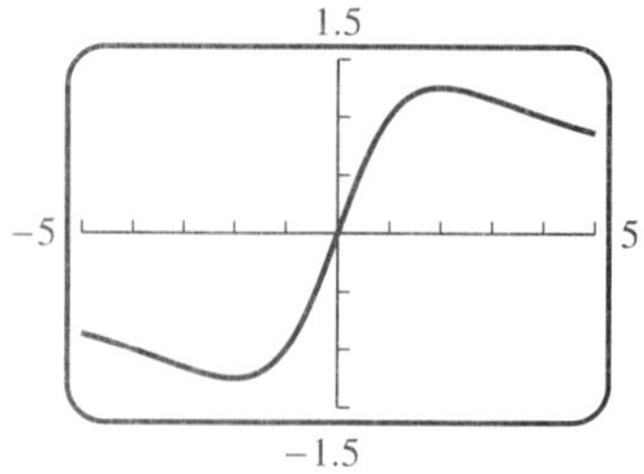

71. $(f \cdot g \cdot h)' = [(f \cdot g) \cdot h]' = (f \cdot g)h' + h(f \cdot g)' = (f \cdot g)h' + h[fg' + f'g] = fgh' + fg'h + f'gh$

72. $(f_1 f_2 \cdots f_n)' = (f_1' f_2 \cdots f_n) + (f_1 f_2' \cdots f_n) + \cdots + (f_1 f_2 \cdots f_n')$

73. **(a)** $2(1 + x^{-1})(x^{-3} + 7) + (2x + 1)(-x^{-2})(x^{-3} + 7) + (2x + 1)(1 + x^{-1})(-3x^{-4})$

(b) $(x^7 + 2x - 3)^3 = (x^7 + 2x - 3)(x^7 + 2x - 3)(x^7 + 2x - 3)$ so

$$\begin{aligned}\frac{d}{dx}(x^7 + 2x - 3)^3 = &(7x^6 + 2)(x^7 + 2x - 3)(x^7 + 2x - 3)\\ &+(x^7 + 2x - 3)(7x^6 + 2)(x^7 + 2x - 3)\\ &+(x^7 + 2x - 3)(x^7 + 2x - 3)(7x^6 + 2)\\ = &\, 3(7x^6 + 2)(x^7 + 2x - 3)^2\end{aligned}$$

74. **(a)** $-5x^{-6}(x^2 + 2x)(4 - 3x)(2x^9 + 1) + x^{-5}(2x + 2)(4 - 3x)(2x^9 + 1)$
$+x^{-5}(x^2 + 2x)(-3)(2x^9 + 1) + x^{-5}(x^2 + 2x)(4 - 3x)(18x^8)$

(b) $(x^2 + 1)^{50} = (x^2 + 1)(x^2 + 1)\cdots(x^2 + 1)$, where $(x^2 + 1)$ occurs 50 times so

$$\begin{aligned}\frac{d}{dx}(x^2 + 1)^{50} &= [(2x)(x^2 + 1)\cdots(x^2 + 1)] + [(x^2 + 1)(2x)\cdots(x^2 + 1)]\\ &\quad + \cdots + [(x^2 + 1)(x^2 + 1)\cdots(2x)]\\ &= 2x(x^2 + 1)^{49} + 2x(x^2 + 1)^{49} + \cdots + 2x(x^2 + 1)^{49}\\ &= 100x(x^2 + 1)^{49} \text{ because } 2x(x^2 + 1)^{49} \text{ occurs 50 times.}\end{aligned}$$

75. f is continuous at 1 because $\lim\limits_{x\to 1^-} f(x) = \lim\limits_{x\to 1^+} f(x) = f(1)$; also $\lim\limits_{x\to 1^-} f'(x) = \lim\limits_{x\to 1^-} (2x + 1) = 3$ and $\lim\limits_{x\to 1^+} f'(x) = \lim\limits_{x\to 1^+} 3 = 3$ so f is differentiable at 1.

76. f is not continuous at $x = 9$ because $\lim\limits_{x\to 9^-} f(x) = -63$ and $\lim\limits_{x\to 9^+} f(x) = 36$.
f cannot be differentiable at $x = 9$, for if it were, then f would also be continuous, which it is not.

77. f is continuous at 1 because $\lim\limits_{x\to 1^-} f(x) = \lim\limits_{x\to 1^+} f(x) = f(1)$, also $\lim\limits_{x\to 1^-} f'(x) = \lim\limits_{x\to 1^-} 2x = 2$ and $\lim\limits_{x\to 1^+} f'(x) = \lim\limits_{x\to 1^+} \dfrac{1}{2\sqrt{x}} = \dfrac{1}{2}$ so f is not differentiable at 1.

78. f is continuous at $1/2$ because $\lim\limits_{x\to 1/2^-} f(x) = \lim\limits_{x\to 1/2^+} f(x) = f(1/2)$, also
$\lim\limits_{x\to 1/2^-} f'(x) = \lim\limits_{x\to 1/2^-} 3x^2 = 3/4$ and $\lim\limits_{x\to 1/2^+} f'(x) = \lim\limits_{x\to 1/2^+} 3x/2 = 3/4$ so $f'(1/2) = 3/4$, and f is differentiable at $x = 1/2$.

79. **(a)** $f(x) = 3x - 2$ if $x \geq 2/3$, $f(x) = -3x + 2$ if $x < 2/3$ so f is differentiable everywhere except perhaps at $2/3$. f is continuous at $2/3$, also $\lim_{x \to 2/3^-} f'(x) = \lim_{x \to 2/3^-}(-3) = -3$ and $\lim_{x \to 2/3^+} f'(x) = \lim_{x \to 2/3^+}(3) = 3$ so f is not differentiable at $x = 2/3$.

(b) $f(x) = x^2 - 4$ if $|x| \geq 2$, $f(x) = -x^2 + 4$ if $|x| < 2$ so f is differentiable everywhere except perhaps at ± 2. f is continuous at -2 and 2, also $\lim_{x \to 2^-} f'(x) = \lim_{x \to 2^-}(-2x) = -4$ and $\lim_{x \to 2^+} f'(x) = \lim_{x \to 2^+}(2x) = 4$ so f is not differentiable at $x = 2$. Similarly, f is not differentiable at $x = -2$.

80. **(a)** $f'(x) = -(1)x^{-2}$, $f''(x) = (2 \cdot 1)x^{-3}$, $f'''(x) = -(3 \cdot 2 \cdot 1)x^{-4}$

$$f^{(n)}(x) = (-1)^n \frac{n(n-1)(n-2)\cdots 1}{x^{n+1}}$$

(b) $f'(x) = -2x^{-3}$, $f''(x) = (3 \cdot 2)x^{-4}$, $f'''(x) = -(4 \cdot 3 \cdot 2)x^{-5}$

$$f^{(n)}(x) = (-1)^n \frac{(n+1)(n)(n-1)\cdots 2}{x^{n+2}}$$

81. **(a)** $$\frac{d^2}{dx^2}[cf(x)] = \frac{d}{dx}\left[\frac{d}{dx}[cf(x)]\right] = \frac{d}{dx}\left[c\frac{d}{dx}[f(x)]\right] = c\frac{d}{dx}\left[\frac{d}{dx}[f(x)]\right] = c\frac{d^2}{dx^2}[f(x)]$$

$$\frac{d^2}{dx^2}[f(x) + g(x)] = \frac{d}{dx}\left[\frac{d}{dx}[f(x) + g(x)]\right] = \frac{d}{dx}\left[\frac{d}{dx}[f(x)] + \frac{d}{dx}[g(x)]\right]$$
$$= \frac{d^2}{dx^2}[f(x)] + \frac{d^2}{dx^2}[g(x)]$$

(b) yes, by repeated application of the procedure illustrated in Part (a)

82. $(f \cdot g)' = fg' + gf'$, $(f \cdot g)'' = fg'' + g'f' + gf'' + f'g' = f''g + 2f'g' + fg''$

83. **(a)** $f'(x) = nx^{n-1}$, $f''(x) = n(n-1)x^{n-2}$, $f'''(x) = n(n-1)(n-2)x^{n-3}, \ldots,$ $f^{(n)}(x) = n(n-1)(n-2)\cdots 1$

(b) from Part (a), $f^{(k)}(x) = k(k-1)(k-2)\cdots 1$ so $f^{(k+1)}(x) = 0$ thus $f^{(n)}(x) = 0$ if $n > k$

(c) from Parts (a) and (b), $f^{(n)}(x) = a_n n(n-1)(n-2)\cdots 1$

84. $\lim_{h \to 0} \dfrac{f'(2+h) - f'(2)}{h} = f''(2)$; $f'(x) = 8x^7 - 2$, $f''(x) = 56x^6$, so $f''(2) = 56(2^6) = 3584$.

85. **(a)** If a function is differentiable at a point then it is continuous at that point, thus f' is continuous on (a, b) and consequently so is f.

(b) f and all its derivatives up to $f^{(n-1)}(x)$ are continuous on (a, b)

EXERCISE SET 3.4

1. $f'(x) = -2\sin x - 3\cos x$

2. $f'(x) = \sin x(-\sin x) + \cos x(\cos x) = \cos^2 x - \sin^2 x = \cos 2x$

3. $f'(x) = \dfrac{x(\cos x) - (\sin x)(1)}{x^2} = \dfrac{x\cos x - \sin x}{x^2}$

4. $f'(x) = x^2(-\sin x) + (\cos x)(2x) = -x^2 \sin x + 2x \cos x$

5. $f'(x) = x^3(\cos x) + (\sin x)(3x^2) - 5(-\sin x) = x^3 \cos x + (3x^2 + 5)\sin x$

6. $f(x) = \dfrac{\cot x}{x}$ (because $\dfrac{\cos x}{\sin x} = \cot x$), $f'(x) = \dfrac{x(-\csc^2 x) - (\cot x)(1)}{x^2} = -\dfrac{x \csc^2 x + \cot x}{x^2}$

7. $f'(x) = \sec x \tan x - \sqrt{2} \sec^2 x$

8. $f'(x) = (x^2 + 1)\sec x \tan x + (\sec x)(2x) = (x^2 + 1)\sec x \tan x + 2x \sec x$

9. $f'(x) = \sec x(\sec^2 x) + (\tan x)(\sec x \tan x) = \sec^3 x + \sec x \tan^2 x$

10. $$f'(x) = \frac{(1 + \tan x)(\sec x \tan x) - (\sec x)(\sec^2 x)}{(1 + \tan x)^2} = \frac{\sec x \tan x + \sec x \tan^2 x - \sec^3 x}{(1 + \tan x)^2}$$
$$= \frac{\sec x(\tan x + \tan^2 x - \sec^2 x)}{(1 + \tan x)^2} = \frac{\sec x(\tan x - 1)}{(1 + \tan x)^2}$$

11. $f'(x) = (\csc x)(-\csc^2 x) + (\cot x)(-\csc x \cot x) = -\csc^3 x - \csc x \cot^2 x$

12. $f'(x) = 1 + 4\csc x \cot x - 2\csc^2 x$

13. $f'(x) = \dfrac{(1 + \csc x)(-\csc^2 x) - \cot x(0 - \csc x \cot x)}{(1 + \csc x)^2} = \dfrac{\csc x(-\csc x - \csc^2 x + \cot^2 x)}{(1 + \csc x)^2}$ but

$1 + \cot^2 x = \csc^2 x$ (identity) thus $\cot^2 x - \csc^2 x = -1$ so

$$f'(x) = \frac{\csc x(-\csc x - 1)}{(1 + \csc x)^2} = -\frac{\csc x}{1 + \csc x}$$

14. $$f'(x) = \frac{\tan x(-\csc x \cot x) - \csc x(\sec^2 x)}{\tan^2 x} = -\frac{\csc x(1 + \sec^2 x)}{\tan^2 x}$$

15. $f(x) = \sin^2 x + \cos^2 x = 1$ (identity) so $f'(x) = 0$

16. $f(x) = \dfrac{1}{\cot x} = \tan x$, so $f'(x) = \sec^2 x$

17. $f(x) = \dfrac{\tan x}{1 + x \tan x}$ (because $\sin x \sec x = (\sin x)(1/\cos x) = \tan x$),

$$f'(x) = \frac{(1 + x \tan x)(\sec^2 x) - \tan x[x(\sec^2 x) + (\tan x)(1)]}{(1 + x \tan x)^2}$$
$$= \frac{\sec^2 x - \tan^2 x}{(1 + x \tan x)^2} = \frac{1}{(1 + x \tan x)^2} \text{ (because } \sec^2 x - \tan^2 x = 1)$$

18. $f(x) = \dfrac{(x^2 + 1)\cot x}{3 - \cot x}$ (because $\cos x \csc x = (\cos x)(1/\sin x) = \cot x$),

$$f'(x) = \frac{(3 - \cot x)[2x \cot x - (x^2 + 1)\csc^2 x] - (x^2 + 1)\cot x \csc^2 x}{(3 - \cot x)^2}$$
$$= \frac{6x \cot x - 2x \cot^2 x - 3(x^2 + 1)\csc^2 x}{(3 - \cot x)^2}$$

19. $dy/dx = -x\sin x + \cos x$, $d^2y/dx^2 = -x\cos x - \sin x - \sin x = -x\cos x - 2\sin x$

20. $dy/dx = -\csc x\cot x$, $d^2y/dx^2 = -[(\csc x)(-\csc^2 x) + (\cot x)(-\csc x\cot x)] = \csc^3 x + \csc x\cot^2 x$

21. $dy/dx = x(\cos x) + (\sin x)(1) - 3(-\sin x) = x\cos x + 4\sin x$,
$d^2y/dx^2 = x(-\sin x) + (\cos x)(1) + 4\cos x = -x\sin x + 5\cos x$

22. $dy/dx = x^2(-\sin x) + (\cos x)(2x) + 4\cos x = -x^2\sin x + 2x\cos x + 4\cos x$,
$d^2y/dx^2 = -[x^2(\cos x) + (\sin x)(2x)] + 2[x(-\sin x) + \cos x] - 4\sin x = (2 - x^2)\cos x - 4(x+1)\sin x$

23. $dy/dx = (\sin x)(-\sin x) + (\cos x)(\cos x) = \cos^2 x - \sin^2 x$,
$d^2y/dx^2 = (\cos x)(-\sin x) + (\cos x)(-\sin x) - [(\sin x)(\cos x) + (\sin x)(\cos x)] = -4\sin x\cos x$

24. $dy/dx = \sec^2 x$; $d^2y/dx^2 = 2\sec^2 x\tan x$

25. Let $f(x) = \tan x$, then $f'(x) = \sec^2 x$.
(a) $f(0) = 0$ and $f'(0) = 1$ so $y - 0 = (1)(x - 0)$, $y = x$.
(b) $f\left(\frac{\pi}{4}\right) = 1$ and $f'\left(\frac{\pi}{4}\right) = 2$ so $y - 1 = 2\left(x - \frac{\pi}{4}\right)$, $y = 2x - \frac{\pi}{2} + 1$.
(c) $f\left(-\frac{\pi}{4}\right) = -1$ and $f'\left(-\frac{\pi}{4}\right) = 2$ so $y + 1 = 2\left(x + \frac{\pi}{4}\right)$, $y = 2x + \frac{\pi}{2} - 1$.

26. Let $f(x) = \sin x$, then $f'(x) = \cos x$.
(a) $f(0) = 0$ and $f'(0) = 1$ so $y - 0 = (1)(x - 0)$, $y = x$
(b) $f(\pi) = 0$ and $f'(\pi) = -1$ so $y - 0 = (-1)(x - \pi)$, $y = -x + \pi$
(c) $f\left(\frac{\pi}{4}\right) = \frac{1}{\sqrt{2}}$ and $f'\left(\frac{\pi}{4}\right) = \frac{1}{\sqrt{2}}$ so $y - \frac{1}{\sqrt{2}} = \frac{1}{\sqrt{2}}\left(x - \frac{\pi}{4}\right)$, $y = \frac{1}{\sqrt{2}}x - \frac{\pi}{4\sqrt{2}} + \frac{1}{\sqrt{2}}$

27. **(a)** If $y = x\sin x$ then $y' = \sin x + x\cos x$ and $y'' = 2\cos x - x\sin x$ so $y'' + y = 2\cos x$.
(b) If $y = x\sin x$ then $y' = \sin x + x\cos x$ and $y'' = 2\cos x - x\sin x$ so $y'' + y = 2\cos x$; differentiate twice more to get $y^{(4)} + y'' = -2\cos x$.

28. **(a)** If $y = \cos x$ then $y' = -\sin x$ and $y'' = -\cos x$ so $y'' + y = (-\cos x) + (\cos x) = 0$;
if $y = \sin x$ then $y' = \cos x$ and $y'' = -\sin x$ so $y'' + y = (-\sin x) + (\sin x) = 0$.
(b) $y' = A\cos x - B\sin x$, $y'' = -A\sin x - B\cos x$ so
$y'' + y = (-A\sin x - B\cos x) + (A\sin x + B\cos x) = 0$.

29. **(a)** $f'(x) = \cos x = 0$ at $x = \pm\pi/2, \pm 3\pi/2$.
(b) $f'(x) = 1 - \sin x = 0$ at $x = -3\pi/2, \pi/2$.
(c) $f'(x) = \sec^2 x \geq 1$ always, so no horizontal tangent line.
(d) $f'(x) = \sec x\tan x = 0$ when $\sin x = 0$, $x = \pm 2\pi, \pm\pi, 0$

30. **(a)**

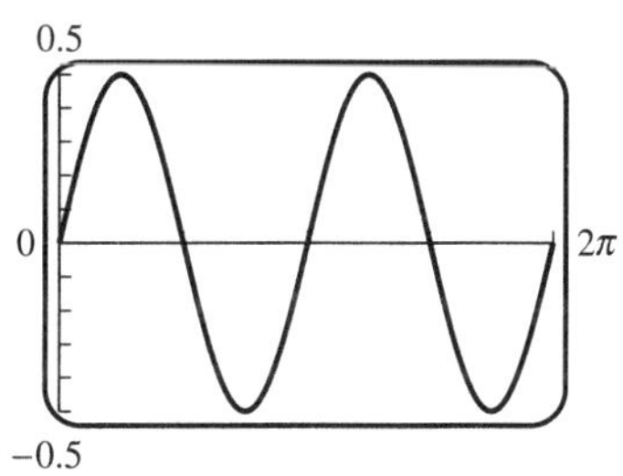

(b) $y = \sin x\cos x = (1/2)\sin 2x$ and $y' = \cos 2x$. So $y' = 0$ when $2x = (2n+1)\pi/2$ for $n = 0, 1, 2, 3$ or $x = \pi/4, 3\pi/4, 5\pi/4, 7\pi/4$

31. $x = 10\sin\theta$, $dx/d\theta = 10\cos\theta$; if $\theta = 60°$, then
$dx/d\theta = 10(1/2) = 5$ ft/rad $= \pi/36$ ft/deg ≈ 0.087 ft/deg

32. $s = 3800\csc\theta, ds/d\theta = -3800\csc\theta\cot\theta$; if $\theta = 30°$, then
$ds/d\theta = -3800(2)(\sqrt{3}) = -7600\sqrt{3}$ ft/rad $= -380\sqrt{3}\pi/9$ ft/deg ≈ -230 ft/deg

33. $D = 50\tan\theta, dD/d\theta = 50\sec^2\theta$; if $\theta = 45°$, then
$dD/d\theta = 50(\sqrt{2})^2 = 100$ m/rad $= 5\pi/9$ m/deg ≈ 1.75 m/deg

34. **(a)** From the right triangle shown, $\sin\theta = r/(r+h)$ so $r + h = r\csc\theta$, $h = r(\csc\theta - 1)$.
(b) $dh/d\theta = -r\csc\theta\cot\theta$; if $\theta = 30°$, then
$dh/d\theta = -6378(2)(\sqrt{3}) \approx -22,094$ km/rad ≈ -386 km/deg

35. **(a)** $\dfrac{d^4}{dx^4}\sin x = \sin x$, so $\dfrac{d^{4k}}{dx^{4k}}\sin x = \sin x$; $\dfrac{d^{87}}{dx^{87}}\sin x = \dfrac{d^3}{dx^3}\dfrac{d^{4\cdot 21}}{dx^{4\cdot 21}}\sin x = \dfrac{d^3}{dx^3}\sin x = -\cos x$

(b) $\dfrac{d^{100}}{dx^{100}}\cos x = \dfrac{d^{4k}}{dx^{4k}}\cos x = \cos x$

36. $\dfrac{d}{dx}[x\sin x] = x\cos x + \sin x$ $\qquad \dfrac{d^2}{dx^2}[x\sin x] = -x\sin x + 2\cos x$

$\dfrac{d^3}{dx^3}[x\sin x] = -x\cos x - 3\sin x$ $\qquad \dfrac{d^4}{dx^4}[x\sin x] = x\sin x - 4\cos x$

By mathematical induction one can show

$\dfrac{d^{4k}}{dx^{4k}}[x\sin x] = x\sin x - (4k)\cos x$; $\qquad \dfrac{d^{4k+1}}{dx^{4k+1}}[x\sin x] = x\cos x + (4k+1)\sin x$;

$\dfrac{d^{4k+2}}{dx^{4k+2}}[x\sin x] = -x\sin x + (4k+2)\cos x$; $\qquad \dfrac{d^{4k+3}}{dx^{4k+3}}[x\sin x] = -x\cos x - (4k+3)\sin x$;

Since $17 = 4\cdot 4 + 1$, $\dfrac{d^{17}}{dx^{17}}[x\sin x] = x\cos x + 17\sin x$

37. **(a)** all x **(b)** all x
(c) $x \neq \pi/2 + n\pi$, $n = 0, \pm 1, \pm 2, \ldots$ **(d)** $x \neq n\pi$, $n = 0, \pm 1, \pm 2, \ldots$
(e) $x \neq \pi/2 + n\pi$, $n = 0, \pm 1, \pm 2, \ldots$ **(f)** $x \neq n\pi$, $n = 0, \pm 1, \pm 2, \ldots$
(g) $x \neq (2n+1)\pi$, $n = 0, \pm 1, \pm 2, \ldots$ **(h)** $x \neq n\pi/2$, $n = 0, \pm 1, \pm 2, \ldots$
(i) all x

38. **(a)** $\dfrac{d}{dx}[\cos x] = \lim\limits_{h\to 0}\dfrac{\cos(x+h) - \cos x}{h} = \lim\limits_{h\to 0}\dfrac{\cos x\cos h - \sin x\sin h - \cos x}{h}$

$$= \lim_{h\to 0}\left[\cos x\left(\frac{\cos h - 1}{h}\right) - \sin x\left(\frac{\sin h}{h}\right)\right] = (\cos x)(0) - (\sin x)(1) = -\sin x$$

(b) $\dfrac{d}{dx}[\sec x] = \dfrac{d}{dx}\left[\dfrac{1}{\cos x}\right] = \dfrac{\cos x(0) - (1)(-\sin x)}{\cos^2 x} = \dfrac{\sin x}{\cos^2 x} = \sec x\tan x$

(c) $\dfrac{d}{dx}[\cot x] = \dfrac{d}{dx}\left[\dfrac{\cos x}{\sin x}\right] = \dfrac{\sin x(-\sin x) - \cos x(\cos x)}{\sin^2 x}$

$$= \frac{-\sin^2 x - \cos^2 x}{\sin^2 x} = \frac{-1}{\sin^2 x} = -\csc^2 x$$

(d) $\dfrac{d}{dx}[\csc x] = \dfrac{d}{dx}\left[\dfrac{1}{\sin x}\right] = \dfrac{(\sin x)(0) - (1)(\cos x)}{\sin^2 x} = -\dfrac{\cos x}{\sin^2 x} = -\csc x\cot x$

39. $f'(x) = -\sin x$, $f''(x) = -\cos x$, $f'''(x) = \sin x$, and $f^{(4)}(x) = \cos x$ with higher order derivatives repeating this pattern, so $f^{(n)}(x) = \sin x$ for $n = 3, 7, 11, \ldots$

40. (a) $\lim\limits_{h\to 0} \dfrac{\tan h}{h} = \lim\limits_{h\to 0} \dfrac{\left(\dfrac{\sin h}{\cos h}\right)}{h} = \lim\limits_{h\to 0} \dfrac{\left(\dfrac{\sin h}{h}\right)}{\cos h} = \dfrac{1}{1} = 1$

(b) $$\frac{d}{dx}[\tan x] = \lim_{h\to 0} \frac{\tan(x+h) - \tan x}{h} = \lim_{h\to 0} \frac{\dfrac{\tan x + \tan h}{1 - \tan x \tan h} - \tan x}{h}$$

$$= \lim_{h\to 0} \frac{\tan x + \tan h - \tan x + \tan^2 x \tan h}{h(1 - \tan x \tan h)} = \lim_{h\to 0} \frac{\tan h(1 + \tan^2 x)}{h(1 - \tan x \tan h)}$$

$$= \lim_{h\to 0} \frac{\tan h \sec^2 x}{h(1 - \tan x \tan h)} = \sec^2 x \lim_{h\to 0} \frac{\dfrac{\tan h}{h}}{1 - \tan x \tan h}$$

$$= \sec^2 x \frac{\lim\limits_{h\to 0} \dfrac{\tan h}{h}}{\lim\limits_{h\to 0}(1 - \tan x \tan h)} = \sec^2 x$$

41. $\lim\limits_{x\to 0} \dfrac{\tan(x+y) - \tan y}{x} = \lim\limits_{h\to 0} \dfrac{\tan(y+h) - \tan y}{h} = \dfrac{d}{dy}(\tan y) = \sec^2 y$

43. Let t be the radian measure, then $h = \dfrac{180}{\pi} t$ and $\cos h = \cos t$, $\sin h = \sin t$.

(a) $\lim\limits_{h\to 0} \dfrac{\cos h - 1}{h} = \lim\limits_{t\to 0} \dfrac{\cos t - 1}{180t/\pi} = \dfrac{\pi}{180} \lim\limits_{t\to 0} \dfrac{\cos t - 1}{t} = 0$

(b) $\lim\limits_{h\to 0} \dfrac{\sin h}{h} = \lim\limits_{t\to 0} \dfrac{\sin t}{180t/\pi} = \dfrac{\pi}{180} \lim\limits_{t\to 0} \dfrac{\sin t}{t} = \dfrac{\pi}{180}$

(c) $\dfrac{d}{dx}[\sin x] = \sin x \lim\limits_{h\to 0} \dfrac{\cos h - 1}{h} + \cos x \lim\limits_{h\to 0} \dfrac{\sin h}{h} = (\sin x)(0) + (\cos x)(\pi/180) = \dfrac{\pi}{180} \cos x$

EXERCISE SET 3.5

1. $(f\circ g)'(x) = f'(g(x))g'(x)$ so $(f\circ g)'(0) = f'(g(0))g'(0) = f'(0)(3) = (2)(3) = 6$

2. $(f\circ g)'(2) = f'(g(2))g'(2) = 5(-3) = -15$

3. (a) $(f\circ g)(x) = f(g(x)) = (2x-3)^5$ and $(f\circ g)'(x) = f'(g(x)g'(x) = 5(2x-3)^4(2) = 10(2x-3)^4$
(b) $(g\circ f)(x) = g(f(x)) = 2x^5 - 3$ and $(g\circ f)'(x) = g'(f(x))f'(x) = 2(5x^4) = 10x^4$

4. (a) $(f\circ g)(x) = 5\sqrt{4+\cos x}$ and $(f\circ g)'(x) = f'(g(x))g'(x) = \dfrac{5}{2\sqrt{4+\cos x}}(-\sin x)$
(b) $(g\circ f)(x) = 4 + \cos(5\sqrt{x})$ and $(g\circ f)'(x) = g'(f(x))f'(x) = -\sin(5\sqrt{x})\dfrac{5}{2\sqrt{x}}$

5. (a) $F'(x) = f'(g(x))g'(x) = f'(g(3))g'(3) = -1(7) = -7$
(b) $G'(x) = g'(f(x))f'(x) = g'(f(3))f'(3) = 4(-2) = -8$

6. (a) $F'(x) = f'(g(x))g'(x)$, $F'(-1) = f'(g(-1))g'(-1) = f'(2)(-3) = (4)(-3) = -12$
(b) $G'(x) = g'(f(x))f'(x), G'(-1) = g'(f(-1))f'(-1) = -5(3) = -15$

7. $f'(x) = 37(x^3+2x)^{36}\dfrac{d}{dx}(x^3+2x) = 37(x^3+2x)^{36}(3x^2+2)$

8. $f'(x) = 6(3x^2+2x-1)^5\dfrac{d}{dx}(3x^2+2x-1) = 6(3x^2+2x-1)^5(6x+2) = 12(3x^2+2x-1)^5(3x+1)$

9. $f'(x) = -2\left(x^3-\dfrac{7}{x}\right)^{-3}\dfrac{d}{dx}\left(x^3-\dfrac{7}{x}\right) = -2\left(x^3-\dfrac{7}{x}\right)^{-3}\left(3x^2+\dfrac{7}{x^2}\right)$

10. $f(x) = (x^5-x+1)^{-9}$,
$f'(x) = -9(x^5-x+1)^{-10}\dfrac{d}{dx}(x^5-x+1) = -9(x^5-x+1)^{-10}(5x^4-1) = -\dfrac{9(5x^4-1)}{(x^5-x+1)^{10}}$

11. $f(x) = 4(3x^2-2x+1)^{-3}$,
$f'(x) = -12(3x^2-2x+1)^{-4}\dfrac{d}{dx}(3x^2-2x+1) = -12(3x^2-2x+1)^{-4}(6x-2) = \dfrac{24(1-3x)}{(3x^2-2x+1)^4}$

12. $f'(x) = \dfrac{1}{2\sqrt{x^3-2x+5}}\dfrac{d}{dx}(x^3-2x+5) = \dfrac{3x^2-2}{2\sqrt{x^3-2x+5}}$

13. $f'(x) = \dfrac{1}{2\sqrt{4+3\sqrt{x}}}\dfrac{d}{dx}(4+3\sqrt{x}) = \dfrac{3}{4\sqrt{x}\sqrt{4+3\sqrt{x}}}$

14. $f'(x) = 3\sin^2 x\dfrac{d}{dx}(\sin x) = 3\sin^2 x\cos x$

15. $f'(x) = \cos(x^3)\dfrac{d}{dx}(x^3) = 3x^2\cos(x^3)$

16. $f'(x) = 2\cos(3\sqrt{x})\dfrac{d}{dx}[\cos(3\sqrt{x})] = -2\cos(3\sqrt{x})\sin(3\sqrt{x})\dfrac{d}{dx}(3\sqrt{x}) = -\dfrac{3\cos(3\sqrt{x})\sin(3\sqrt{x})}{\sqrt{x}}$

17. $f'(x) = 20\cos^4 x\dfrac{d}{dx}(\cos x) = 20\cos^4 x(-\sin x) = -20\cos^4 x\sin x$

18. $f'(x) = -\csc(x^3)\cot(x^3)\dfrac{d}{dx}(x^3) = -3x^2\csc(x^3)\cot(x^3)$

19. $f'(x) = \cos(1/x^2)\dfrac{d}{dx}(1/x^2) = -\dfrac{2}{x^3}\cos(1/x^2)$

20. $f'(x) = 4\tan^3(x^3)\dfrac{d}{dx}[\tan(x^3)] = 4\tan^3(x^3)\sec^2(x^3)\dfrac{d}{dx}(x^3) = 12x^2\tan^3(x^3)\sec^2(x^3)$

21. $f'(x) = 4\sec(x^7)\dfrac{d}{dx}[\sec(x^7)] = 4\sec(x^7)\sec(x^7)\tan(x^7)\dfrac{d}{dx}(x^7) = 28x^6\sec^2(x^7)\tan(x^7)$

22. $f'(x) = 3\cos^2\left(\dfrac{x}{x+1}\right)\dfrac{d}{dx}\cos\left(\dfrac{x}{x+1}\right) = 3\cos^2\left(\dfrac{x}{x+1}\right)\left[-\sin\left(\dfrac{x}{x+1}\right)\right]\dfrac{(x+1)(1)-x(1)}{(x+1)^2}$
$= -\dfrac{3}{(x+1)^2}\cos^2\left(\dfrac{x}{x+1}\right)\sin\left(\dfrac{x}{x+1}\right)$

23. $f'(x) = \dfrac{1}{2\sqrt{\cos(5x)}}\dfrac{d}{dx}[\cos(5x)] = -\dfrac{5\sin(5x)}{2\sqrt{\cos(5x)}}$

24. $f'(x) = \dfrac{1}{2\sqrt{3x-\sin^2(4x)}}\dfrac{d}{dx}[3x-\sin^2(4x)] = \dfrac{3-8\sin(4x)\cos(4x)}{2\sqrt{3x-\sin^2(4x)}}$

25. $f'(x) = -3\left[x+\csc(x^3+3)\right]^{-4}\dfrac{d}{dx}\left[x+\csc(x^3+3)\right]$

$$= -3\left[x+\csc(x^3+3)\right]^{-4}\left[1-\csc(x^3+3)\cot(x^3+3)\frac{d}{dx}(x^3+3)\right]$$

$$= -3\left[x+\csc(x^3+3)\right]^{-4}\left[1-3x^2\csc(x^3+3)\cot(x^3+3)\right]$$

26. $f'(x) = -4\left[x^4-\sec(4x^2-2)\right]^{-5}\dfrac{d}{dx}\left[x^4-\sec(4x^2-2)\right]$

$$= -4\left[x^4-\sec(4x^2-2)\right]^{-5}\left[4x^3-\sec(4x^2-2)\tan(4x^2-2)\frac{d}{dx}(4x^2-2)\right]$$

$$= -16x\left[x^4-\sec(4x^2-2)\right]^{-5}\left[x^2-2\ \sec(4x^2-2)\tan(4x^2-2)\right]$$

27. $\dfrac{dy}{dx} = x^3(2\sin 5x)\dfrac{d}{dx}(\sin 5x)+3x^2\sin^2 5x = 10x^3\sin 5x\cos 5x + 3x^2\sin^2 5x$

28. $\dfrac{dy}{dx} = \sqrt{x}\left[3\tan^2(\sqrt{x})\sec^2(\sqrt{x})\dfrac{1}{2\sqrt{x}}\right]+\dfrac{1}{2\sqrt{x}}\tan^3(\sqrt{x}) = \dfrac{3}{2}\tan^2(\sqrt{x})\sec^2(\sqrt{x})+\dfrac{1}{2\sqrt{x}}\tan^3(\sqrt{x})$

29. $\dfrac{dy}{dx} = x^5\sec\left(\dfrac{1}{x}\right)\tan\left(\dfrac{1}{x}\right)\dfrac{d}{dx}\left(\dfrac{1}{x}\right)+\sec\left(\dfrac{1}{x}\right)(5x^4)$

$$= x^5\sec\left(\frac{1}{x}\right)\tan\left(\frac{1}{x}\right)\left(-\frac{1}{x^2}\right)+5x^4\sec\left(\frac{1}{x}\right)$$

$$= -x^3\sec\left(\frac{1}{x}\right)\tan\left(\frac{1}{x}\right)+5x^4\sec\left(\frac{1}{x}\right)$$

30. $\dfrac{dy}{dx} = \dfrac{\sec(3x+1)\cos x - 3\sin x\sec(3x+1)\tan(3x+1)}{\sec^2(3x+1)} = \cos x\cos(3x+1)-3\sin x\sin(3x+1)$

31. $\dfrac{dy}{dx} = -\sin(\cos x)\dfrac{d}{dx}(\cos x) = -\sin(\cos x)(-\sin x) = \sin(\cos x)\sin x$

32. $\dfrac{dy}{dx} = \cos(\tan 3x)\dfrac{d}{dx}(\tan 3x) = 3\sec^2 3x\cos(\tan 3x)$

33. $\dfrac{dy}{dx} = 3\cos^2(\sin 2x)\dfrac{d}{dx}[\cos(\sin 2x)] = 3\cos^2(\sin 2x)[-\sin(\sin 2x)]\dfrac{d}{dx}(\sin 2x)$

$$= -6\cos^2(\sin 2x)\sin(\sin 2x)\cos 2x$$

34. $\dfrac{dy}{dx} = \dfrac{(1-\cot x^2)(-2x\csc x^2\cot x^2)-(1+\csc x^2)(2x\csc^2 x^2)}{(1-\cot x^2)^2} = -2x\csc x^2\,\dfrac{1+\cot x^2+\csc x^2}{(1-\cot x^2)^2}$

35. $\dfrac{dy}{dx} = (5x+8)^{13}12(x^3+7x)^{11}\dfrac{d}{dx}(x^3+7x)+(x^3+7x)^{12}13(5x+8)^{12}\dfrac{d}{dx}(5x+8)$

$$= 12(5x+8)^{13}(x^3+7x)^{11}(3x^2+7)+65(x^3+7x)^{12}(5x+8)^{12}$$

36. $\dfrac{dy}{dx} = (2x-5)^2 3(x^2+4)^2(2x)+(x^2+4)^3 2(2x-5)(2)$

$$= 6x(2x-5)^2(x^2+4)^2+4(2x-5)(x^2+4)^3$$

$$= 2(2x-5)(x^2+4)^2(8x^2-15x+8)$$

37. $\dfrac{dy}{dx} = 3\left[\dfrac{x-5}{2x+1}\right]^2 \dfrac{d}{dx}\left[\dfrac{x-5}{2x+1}\right] = 3\left[\dfrac{x-5}{2x+1}\right]^2 \cdot \dfrac{11}{(2x+1)^2} = \dfrac{33(x-5)^2}{(2x+1)^4}$

38. $\dfrac{dy}{dx} = 17\left(\dfrac{1+x^2}{1-x^2}\right)^{16} \dfrac{d}{dx}\left(\dfrac{1+x^2}{1-x^2}\right) = 17\left(\dfrac{1+x^2}{1-x^2}\right)^{16} \dfrac{(1-x^2)(2x)-(1+x^2)(-2x)}{(1-x^2)^2}$

$= 17\left(\dfrac{1+x^2}{1-x^2}\right)^{16} \dfrac{4x}{(1-x^2)^2} = \dfrac{68x(1+x^2)^{16}}{(1-x^2)^{18}}$

39. $\dfrac{dy}{dx} = \dfrac{(4x^2-1)^8(3)(2x+3)^2(2)-(2x+3)^3(8)(4x^2-1)^7(8x)}{(4x^2-1)^{16}}$

$= \dfrac{2(2x+3)^2(4x^2-1)^7[3(4x^2-1)-32x(2x+3)]}{(4x^2-1)^{16}} = -\dfrac{2(2x+3)^2(52x^2+96x+3)}{(4x^2-1)^9}$

40. $\dfrac{dy}{dx} = 12[1+\sin^3(x^5)]^{11}\dfrac{d}{dx}[1+\sin^3(x^5)]$

$= 12[1+\sin^3(x^5)]^{11} 3\sin^2(x^5)\dfrac{d}{dx}\sin(x^5) = 180x^4[1+\sin^3(x^5)]^{11}\sin^2(x^5)\cos(x^5)$

41. $\dfrac{dy}{dx} = 5\left[x\sin 2x+\tan^4(x^7)\right]^4 \dfrac{d}{dx}\left[x\sin 2x \tan^4(x^7)\right]$

$= 5\left[x\sin 2x+\tan^4(x^7)\right]^4\left[x\cos 2x\dfrac{d}{dx}(2x)+\sin 2x+4\tan^3(x^7)\dfrac{d}{dx}\tan(x^7)\right]$

$= 5\left[x\sin 2x+\tan^4(x^7)\right]^4\left[2x\cos 2x+\sin 2x+28x^6\tan^3(x^7)\sec^2(x^7)\right]$

42. $\dfrac{dy}{dx} = 4\tan^3\left(2+\dfrac{(7-x)\sqrt{3x^2+5}}{x^3+\sin x}\right)\sec^2\left(2+\dfrac{(7-x)\sqrt{3x^2+5}}{x^3+\sin x}\right)$

$\times\left(-\dfrac{\sqrt{3x^2+5}}{x^3+\sin x}+3\dfrac{(7-x)x}{\sqrt{3x^2+5}\,(x^3+\sin x)}-\dfrac{(7-x)\sqrt{3x^2+5}\,(3x^2+\cos x)}{(x^3+\sin x)^2}\right)$

43. $\dfrac{dy}{dx} = \cos 3x - 3x\sin 3x$; if $x=\pi$ then $\dfrac{dy}{dx} = -1$ and $y=-\pi$, so the equation of the tangent line is $y+\pi = -(x-\pi), y = x$

44. $\dfrac{dy}{dx} = 3x^2\cos(1+x^3)$; if $x=-3$ then $y = -\sin 26, \dfrac{dy}{dx} = -27\cos 26$, so the equation of the tangent line is $y+\sin 26 = -27(\cos 26)(x+3), y = -27(\cos 26)x - 81\cos 26 - \sin 26$

45. $\dfrac{dy}{dx} = -3\sec^3(\pi/2-x)\tan(\pi/2-x)$; if $x=-\pi/2$ then $\dfrac{dy}{dx} = 0, y=-1$ so the equation of the tangent line is $y+1=0, y=-1$

46. $\dfrac{dy}{dx} = 3\left(x-\dfrac{1}{x}\right)^2\left(1+\dfrac{1}{x^2}\right)$; if $x=2$ then $y=\dfrac{27}{8}, \dfrac{dy}{dx} = 3\dfrac{9}{4}\dfrac{5}{4} = \dfrac{135}{16}$ so the equation of the tangent line is $y-27/8 = (135/16)(x-2), y = \dfrac{135}{16}x - \dfrac{108}{8}$

47. $\frac{dy}{dx} = \sec^2(4x^2)\frac{d}{dx}(4x^2) = 8x\sec^2(4x^2)$, $\left.\frac{dy}{dx}\right|_{x=\sqrt{\pi}} = 8\sqrt{\pi}\sec^2(4\pi) = 8\sqrt{\pi}$. When $x = \sqrt{\pi}$, $y = \tan(4\pi) = 0$, so the equation of the tangent line is $y = 8\sqrt{\pi}(x - \sqrt{\pi}) = 8\sqrt{\pi}x - 8\pi$.

48. $\frac{dy}{dx} = 12\cot^3 x \frac{d}{dx}\cot x = -12\cot^3 x\csc^2 x$, $\left.\frac{dy}{dx}\right|_{x=\pi/4} = -24$. When $x = \pi/4, y = 3$, so the equation of the tangent line is $y - 3 = -24(x - \pi/4)$, or $y = -24x + 3 + 6\pi$.

49. $\frac{dy}{dx} = 2x\sqrt{5 - x^2} + \frac{x^2}{2\sqrt{5 - x^2}}(-2x)$, $\left.\frac{dy}{dx}\right|_{x=1} = 4 - 1/2 = 7/2$. When $x = 1, y = 2$, so the equation of the tangent line is $y - 2 = (7/2)(x - 1)$, or $y = \frac{7}{2}x - \frac{3}{2}$.

50. $\frac{dy}{dx} = \frac{1}{\sqrt{1 - x^2}} - \frac{x}{2}(1 - x^2)^{3/2}(-2x)$, $\left.\frac{dy}{dx}\right|_{x=0} = 1$. When $x = 0, y = 0$, so the equation of the tangent line is $y = x$.

51. $\frac{dy}{dx} = x(-\sin(5x))\frac{d}{dx}(5x) + \cos(5x) - 2\sin x\frac{d}{dx}(\sin x)$

$$= -5x\sin(5x) + \cos(5x) - 2\sin x\cos x = -5x\sin(5x) + \cos(5x) - \sin(2x),$$

$$\frac{d^2y}{dx^2} = -5x\cos(5x)\frac{d}{dx}(5x) - 5\sin(5x) - \sin(5x)\frac{d}{dx}(5x) - \cos(2x)\frac{d}{dx}(2x)$$

$$= -25x\cos(5x) - 10\sin(5x) - 2\cos(2x)$$

52. $\frac{dy}{dx} = \cos(3x^2)\frac{d}{dx}(3x^2) = 6x\cos(3x^2)$,

$$\frac{d^2y}{dx^2} = 6x(-\sin(3x^2))\frac{d}{dx}(3x^2) + 6\cos(3x^2) = -36x^2\sin(3x^2) + 6\cos(3x^2)$$

53. $\frac{dy}{dx} = \frac{(1 - x) + (1 + x)}{(1 - x)^2} = \frac{2}{(1 - x)^2} = 2(1 - x)^{-2}$ and $\frac{d^2y}{dx^2} = -2(2)(-1)(1 - x)^{-3} = 4(1 - x)^{-3}$

54. $\frac{dy}{dx} = x\sec^2\left(\frac{1}{x}\right)\frac{d}{dx}\left(\frac{1}{x}\right) + \tan\left(\frac{1}{x}\right) = -\frac{1}{x}\sec^2\left(\frac{1}{x}\right) + \tan\left(\frac{1}{x}\right)$,

$$\frac{d^2y}{dx^2} = -\frac{2}{x}\sec\left(\frac{1}{x}\right)\frac{d}{dx}\sec\left(\frac{1}{x}\right) + \frac{1}{x^2}\sec^2\left(\frac{1}{x}\right) + \sec^2\left(\frac{1}{x}\right)\frac{d}{dx}\left(\frac{1}{x}\right) = \frac{2}{x^3}\sec^2\left(\frac{1}{x}\right)\tan\left(\frac{1}{x}\right)$$

so $y - \frac{27}{8} = \frac{135}{16}(x - 2)$, $y = \frac{135}{16}x - \frac{27}{2}$

55. $y = \cot^3(\pi - \theta) = -\cot^3\theta$ so $dy/dx = 3\cot^2\theta\csc^2\theta$

56. $6\left(\frac{au + b}{cu + d}\right)^5 \frac{ad - bc}{(cu + d)^2}$

57. $\frac{d}{d\omega}[a\cos^2\pi\omega + b\sin^2\pi\omega] = -2\pi a\cos\pi\omega\sin\pi\omega + 2\pi b\sin\pi\omega\cos\pi\omega$

$$= \pi(b - a)(2\sin\pi\omega\cos\pi\omega) = \pi(b - a)\sin 2\pi\omega$$

58. $2\csc^2(\pi/3 - y)\cot(\pi/3 - y)$

59. (a)

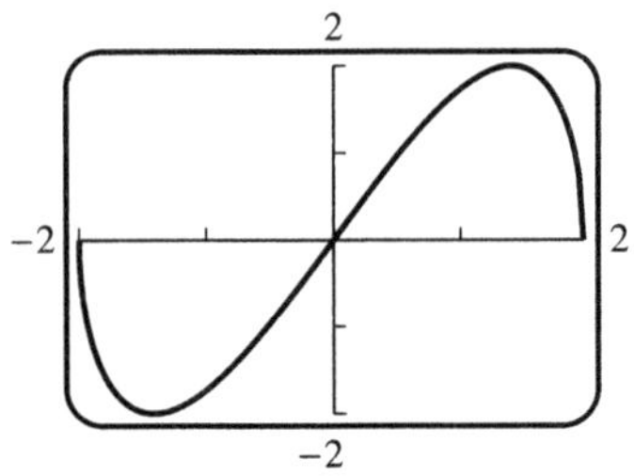

(c) $f'(x) = x\dfrac{-x}{\sqrt{4-x^2}} + \sqrt{4-x^2} = \dfrac{4-2x^2}{\sqrt{4-x^2}}$

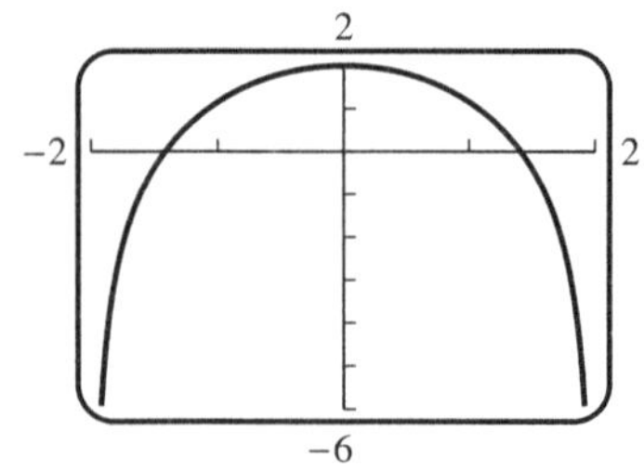

(d) $f(1) = \sqrt{3}$ and $f'(1) = \dfrac{2}{\sqrt{3}}$ so the tangent line has the equation $y - \sqrt{3} = \dfrac{2}{\sqrt{3}}(x-1)$.

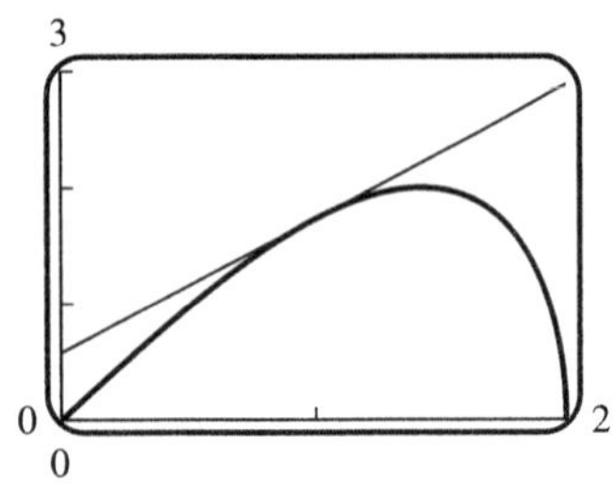

60. (a)

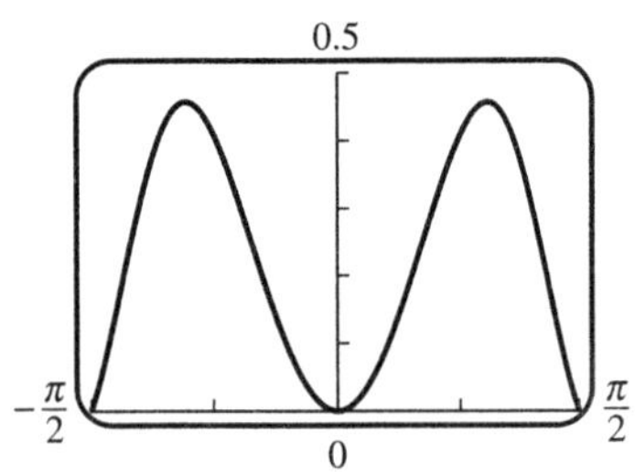

(c) $f'(x) = 2x\cos(x^2)\cos x - \sin x \sin(x^2)$

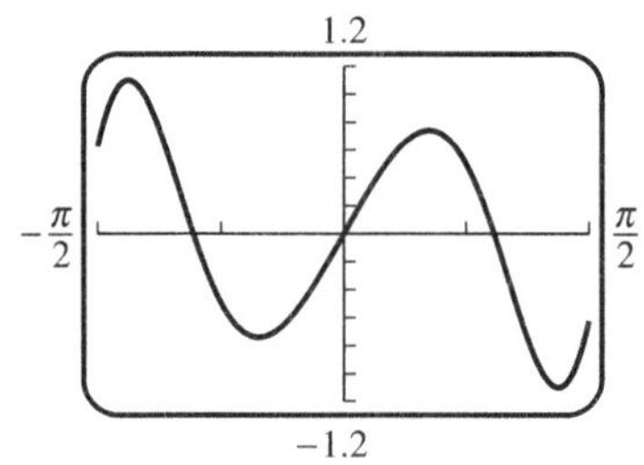

(d) $f(1) = \sin 1 \cos 1$ and $f'(1) = 2\cos^2 1 - \sin^2 1$, so the tangent line has the equation $y - \sin 1 \cos 1 = (2\cos^2 1 - \sin^2 1)(x-1)$.

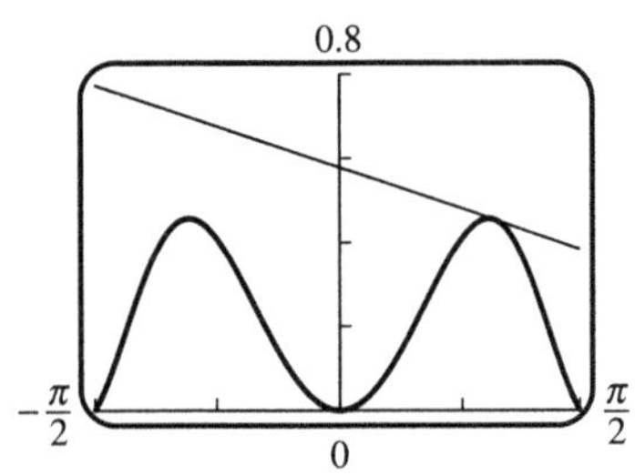

61. (a) $dy/dt = -A\omega \sin \omega t$, $d^2y/dt^2 = -A\omega^2 \cos \omega t = -\omega^2 y$

(b) one complete oscillation occurs when ωt increases over an interval of length 2π, or if t increases over an interval of length $2\pi/\omega$

(c) $f = 1/T$

(d) amplitude = 0.6 cm, $T = 2\pi/15$ s/oscillation, $f = 15/(2\pi)$ oscillations/s

62. $dy/dt = 3A\cos 3t$, $d^2y/dt^2 = -9A\sin 3t$, so $-9A\sin 3t + 2A\sin 3t = 4\sin 3t$, $-7A\sin 3t = 4\sin 3t$, $-7A = 4$, $A = -4/7$

63. **(a)** $p \approx 10$ lb/in^2, $dp/dh \approx -2$ lb/in^2/mi

(b) $\dfrac{dp}{dt} = \dfrac{dp}{dh}\dfrac{dh}{dt} \approx (-2)(0.3) = -0.6$ lb/in^2/s

64. **(a)** $F = \dfrac{45}{\cos\theta + 0.3\sin\theta}, \dfrac{dF}{d\theta} = -\dfrac{45(-\sin\theta + 0.3\cos\theta)}{(\cos\theta + 0.3\sin\theta)^2}$;
if $\theta = 30°$, then $dF/d\theta \approx 10.5$ lb/rad ≈ 0.18 lb/deg

(b) $\dfrac{dF}{dt} = \dfrac{dF}{d\theta}\dfrac{d\theta}{dt} \approx (0.18)(-0.5) = -0.09$ lb/s

65. With $u = \sin x$, $\dfrac{d}{dx}(|\sin x|) = \dfrac{d}{dx}(|u|) = \dfrac{d}{du}(|u|)\dfrac{du}{dx} = \dfrac{d}{du}(|u|)\cos x = \begin{cases} \cos x, & u > 0 \\ -\cos x, & u < 0 \end{cases}$

$$= \begin{cases} \cos x, & \sin x > 0 \\ -\cos x, & \sin x < 0 \end{cases} = \begin{cases} \cos x, & 0 < x < \pi \\ -\cos x, & -\pi < x < 0 \end{cases}$$

66. $\dfrac{d}{dx}(\cos x) = \dfrac{d}{dx}[\sin(\pi/2 - x)] = -\cos(\pi/2 - x) = -\sin x$

67. **(a)** For $x \neq 0$, $|f(x)| \leq |x|$, and $\lim\limits_{x\to 0}|x| = 0$, so by the Squeezing Theorem, $\lim\limits_{x\to 0} f(x) = 0$.

(b) If $f'(0)$ were to exist, then the limit $\dfrac{f(x) - f(0)}{x - 0} = \sin(1/x)$ would have to exist, but it doesn't.

(c) for $x \neq 0$, $f'(x) = x\left(\cos\dfrac{1}{x}\right)\left(-\dfrac{1}{x^2}\right) + \sin\dfrac{1}{x} = -\dfrac{1}{x}\cos\dfrac{1}{x} + \sin\dfrac{1}{x}$

(d) $\lim\limits_{x\to 0}\dfrac{f(x) - f(0)}{x - 0} = \lim\limits_{x\to 0}\sin(1/x)$, which does not exist, thus $f'(0)$ does not exist.

68. **(a)** $-x^2 \leq x^2\sin(1/x) \leq x^2$, so by the Squeezing Theorem $\lim\limits_{x\to 0} f(x) = 0$.

(b) $f'(0) = \lim\limits_{x\to 0}\dfrac{f(x) - f(0)}{x - 0} = \lim\limits_{x\to 0} x\sin(1/x) = 0$ by Exercise 67, Part a.

(c) For $x \neq 0$, $f'(x) = 2x\sin(1/x) + x^2\cos(1/x)(-1/x^2) = 2x\sin(1/x) - \cos(1/x)$

(d) If $f'(x)$ were continuous at $x = 0$ then so would $\cos(1/x) = f'(x) - 2x\sin(1/x)$ be, since $2x\sin(1/x)$ is continuous there. But $\cos(1/x)$ oscillates at $x = 0$.

69. **(a)** $g'(x) = 3[f(x)]^2 f'(x)$, $g'(2) = 3[f(2)]^2 f'(2) = 3(1)^2(7) = 21$
(b) $h'(x) = f'(x^3)(3x^2)$, $h'(2) = f'(8)(12) = (-3)(12) = -36$

70. $F'(x) = f'(g(x))g'(x) = \sqrt{3(x^2 - 1) + 4}(2x) = 2x\sqrt{3x^2 + 1}$

71. $F'(x) = f'(g(x))g'(x) = f'(\sqrt{3x - 1})\dfrac{3}{2\sqrt{3x - 1}} = \dfrac{\sqrt{3x - 1}}{(3x - 1) + 1}\dfrac{3}{2\sqrt{3x - 1}} = \dfrac{1}{2x}$

72. $\dfrac{d}{dx}[f(x^2)] = f'(x^2)(2x)$, thus $f'(x^2)(2x) = x^2$ so $f'(x^2) = x/2$ if $x \neq 0$

73. $\dfrac{d}{dx}[f(3x)] = f'(3x)\dfrac{d}{dx}(3x) = 3f'(3x) = 6x$, so $f'(3x) = 2x$. Let $u = 3x$ to get $f'(u) = \dfrac{2}{3}u$;
$\dfrac{d}{dx}[f(x)] = f'(x) = \dfrac{2}{3}x$.

74. **(a)** If $f(-x) = f(x)$, then $\dfrac{d}{dx}[f(-x)] = \dfrac{d}{dx}[f(x)]$, $f'(-x)(-1) = f'(x)$, $f'(-x) = -f'(x)$ so f' is odd.

(b) If $f(-x) = -f(x)$, then $\dfrac{d}{dx}[f(-x)] = -\dfrac{d}{dx}[f(x)]$, $f'(-x)(-1) = -f'(x)$, $f'(-x) = f'(x)$ so f' is even.

75. For an even function, the graph is symmetric about the y-axis; the slope of the tangent line at $(a, f(a))$ is the negative of the slope of the tangent line at $(-a, f(-a))$. For an odd function, the graph is symmetric about the origin; the slope of the tangent line at $(a, f(a))$ is the same as the slope of the tangent line at $(-a, f(-a))$.

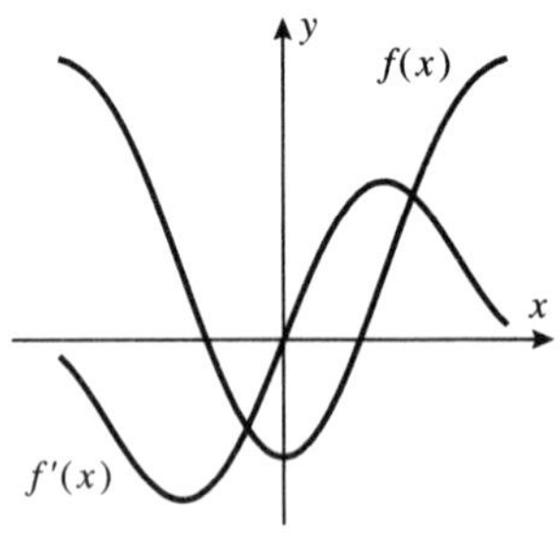

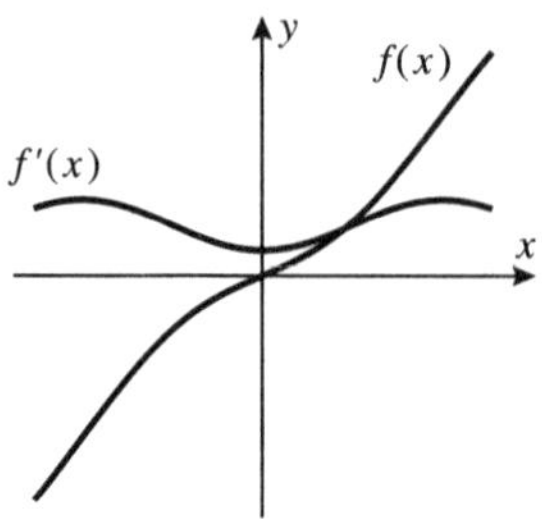

76. $\dfrac{dy}{dx} = \dfrac{dy}{du}\dfrac{du}{dv}\dfrac{dv}{dw}\dfrac{dw}{dx}$

77. $$\frac{d}{dx}[f(g(h(x)))] = \frac{d}{dx}[f(g(u))], \quad u = h(x)$$
$$= \frac{d}{du}[f(g(u))]\frac{du}{dx} = f'(g(u))g'(u)\frac{du}{dx} = f'(g(h(x)))g'(h(x))h'(x)$$

EXERCISE SET 3.6

1. $y = (2x-5)^{1/3}$; $dy/dx = \dfrac{2}{3}(2x-5)^{-2/3}$

2. $dy/dx = \dfrac{1}{3}\left[2+\tan(x^2)\right]^{-2/3}\sec^2(x^2)(2x) = \dfrac{2}{3}x\sec^2(x^2)\left[2+\tan(x^2)\right]^{-2/3}$

3. $dy/dx = \dfrac{3}{2}\left[\dfrac{x-1}{x+2}\right]^{1/2}\dfrac{d}{dx}\left[\dfrac{x-1}{x+2}\right] = \dfrac{9}{2(x+2)^2}\left[\dfrac{x-1}{x+2}\right]^{1/2}$

4. $dy/dx = \dfrac{1}{2}\left[\dfrac{x^2+1}{x^2-5}\right]^{-1/2}\dfrac{d}{dx}\left[\dfrac{x^2+1}{x^2-5}\right] = \dfrac{1}{2}\left[\dfrac{x^2+1}{x^2-5}\right]^{-1/2}\dfrac{-12x}{(x^2-5)^2} = -\dfrac{6x}{(x^2-5)^2}\left[\dfrac{x^2+1}{x^2-5}\right]^{-1/2}$

5. $dy/dx = x^3\left(-\dfrac{2}{3}\right)(5x^2+1)^{-5/3}(10x) + 3x^2(5x^2+1)^{-2/3} = \dfrac{1}{3}x^2(5x^2+1)^{-5/3}(25x^2+9)$

6. $dy/dx = \dfrac{x^2\frac{4}{3}(3-2x)^{1/3}(-2) - (3-2x)^{4/3}(2x)}{x^4} = \dfrac{2(3-2x)^{1/3}(2x-9)}{3x^3}$

7. $dy/dx = \dfrac{5}{2}[\sin(3/x)]^{3/2}[\cos(3/x)](-3/x^2) = -\dfrac{15[\sin(3/x)]^{3/2}\cos(3/x)}{2x^2}$

8. $dy/dx = -\frac{1}{2}\left[\cos(x^3)\right]^{-3/2}\left[-\sin(x^3)\right](3x^2) = \frac{3}{2}x^2\sin(x^3)\left[\cos(x^3)\right]^{-3/2}$

9. **(a)** $3x^2 + x\frac{dy}{dx} + y - 2 = 0,\ \frac{dy}{dx} = \frac{2-3x^2-y}{x}$

(b) $y = \frac{1+2x-x^3}{x} = \frac{1}{x} + 2 - x^2$ so $\frac{dy}{dx} = -\frac{1}{x^2} - 2x$

(c) from Part (a), $\frac{dy}{dx} = \frac{2-3x^2-y}{x} = \frac{2-3x^2-(1/x+2-x^2)}{x} = -2x - \frac{1}{x^2}$

10. **(a)** $\frac{1}{2}y^{-1/2}\frac{dy}{dx} - e^x = 0$ or $\frac{dy}{dx} = 2e^x\sqrt{y}$

(b) $y = (2+e^x)^2 = 2 + 4e^x + e^{2x}$ so $\frac{dy}{dx} = 4e^x + 2e^{2x}$

(c) from Part (a), $\frac{dy}{dx} = 2e^x\sqrt{y} = 2e^x(2+e^x) = 4e^x + 2e^{2x}$

11. $2x + 2y\frac{dy}{dx} = 0$ so $\frac{dy}{dx} = -\frac{x}{y}$

12. $3x^2 - 3y^2\frac{dy}{dx} = 6(x\frac{dy}{dx} + y),\ -(3y^2+6x)\frac{dy}{dx} = 6y - 3x^2$ so $\frac{dy}{dx} = \frac{x^2-2y}{y^2+2x}$

13. $x^2\frac{dy}{dx} + 2xy + 3x(3y^2)\frac{dy}{dx} + 3y^3 - 1 = 0$

$(x^2+9xy^2)\frac{dy}{dx} = 1 - 2xy - 3y^3$ so $\frac{dy}{dx} = \frac{1-2xy-3y^3}{x^2+9xy^2}$

14. $x^3(2y)\frac{dy}{dx} + 3x^2y^2 - 5x^2\frac{dy}{dx} - 10xy + 1 = 0$

$(2x^3y - 5x^2)\frac{dy}{dx} = 10xy - 3x^2y^2 - 1$ so $\frac{dy}{dx} = \frac{10xy-3x^2y^2-1}{2x^3y-5x^2}$

15. $-\frac{1}{y^2}\frac{dy}{dx} - \frac{1}{x^2} = 0$ so $\frac{dy}{dx} = -\frac{y^2}{x^2}$

16. $2x = \frac{(x-y)(1+dy/dx) - (x+y)(1-dy/dx)}{(x-y)^2}$,

$2x(x-y)^2 = -2y + 2x\frac{dy}{dx}$ so $\frac{dy}{dx} = \frac{x(x-y)^2+y}{x}$

17. $\cos(x^2y^2)\left[x^2(2y)\frac{dy}{dx} + 2xy^2\right] = 1,\ \frac{dy}{dx} = \frac{1-2xy^2\cos(x^2y^2)}{2x^2y\cos(x^2y^2)}$

18. $2x = \frac{(1+\csc y)(-\csc^2 y)(dy/dx) - (\cot y)(-\csc y\cot y)(dy/dx)}{(1+\csc y)^2}$,

$2x(1+\csc y)^2 = -\csc y(\csc y + \csc^2 y - \cot^2 y)\frac{dy}{dx}$,

but $\csc^2 y - \cot^2 y = 1$, so $\frac{dy}{dx} = -\frac{2x(1+\csc y)}{\csc y}$

19. $3\tan^2(xy^2+y)\sec^2(xy^2+y)\left(2xy\frac{dy}{dx} + y^2 + \frac{dy}{dx}\right) = 1$

so $\frac{dy}{dx} = \frac{1-3y^2\tan^2(xy^2+y)\sec^2(xy^2+y)}{3(2xy+1)\tan^2(xy^2+y)\sec^2(xy^2+y)}$

20. $\dfrac{(1+\sec y)[3xy^2(dy/dx)+y^3]-xy^3(\sec y\tan y)(dy/dx)}{(1+\sec y)^2}=4y^3\dfrac{dy}{dx}$,

multiply through by $(1+\sec y)^2$ and solve for $\dfrac{dy}{dx}$ to get

$$\frac{dy}{dx}=\frac{y(1+\sec y)}{4y(1+\sec y)^2-3x(1+\sec y)+xy\sec y\tan y}$$

21. $\dfrac{dy}{dx}=\dfrac{3x}{4y}$, $\dfrac{d^2y}{dx^2}=\dfrac{(4y)(3)-(3x)(4dy/dx)}{16y^2}=\dfrac{12y-12x(3x/(4y))}{16y^2}=\dfrac{12y^2-9x^2}{16y^3}=\dfrac{-3(3x^2-4y^2)}{16y^3}$,

but $3x^2-4y^2=7$ so $\dfrac{d^2y}{dx^2}=\dfrac{-3(7)}{16y^3}=-\dfrac{21}{16y^3}$

22. $\dfrac{dy}{dx}=-\dfrac{x^2}{y^2}$, $\dfrac{d^2y}{dx^2}=-\dfrac{y^2(2x)-x^2(2ydy/dx)}{y^4}=-\dfrac{2xy^2-2x^2y(-x^2/y^2)}{y^4}=-\dfrac{2x(y^3+x^3)}{y^5}$,

but $x^3+y^3=1$ so $\dfrac{d^2y}{dx^2}=-\dfrac{2x}{y^5}$

23. $\dfrac{dy}{dx}=-\dfrac{y}{x}$, $\dfrac{d^2y}{dx^2}=-\dfrac{x(dy/dx)-y(1)}{x^2}=-\dfrac{x(-y/x)-y}{x^2}=\dfrac{2y}{x^2}$

24. $\dfrac{dy}{dx}=\dfrac{y}{y-x}$,

$$\frac{d^2y}{dx^2}=\frac{(y-x)(dy/dx)-y(dy/dx-1)}{(y-x)^2}=\frac{(y-x)\left(\frac{y}{y-x}\right)-y\left(\frac{y}{y-x}-1\right)}{(y-x)^2}$$

$=\dfrac{y^2-2xy}{(y-x)^3}$ but $y^2-2xy=-3$, so $\dfrac{d^2y}{dx^2}=-\dfrac{3}{(y-x)^3}$

25. $\dfrac{dy}{dx}=(1+\cos y)^{-1}$, $\dfrac{d^2y}{dx^2}=-(1+\cos y)^{-2}(-\sin y)\dfrac{dy}{dx}=\dfrac{\sin y}{(1+\cos y)^3}$

26. $\dfrac{dy}{dx}=\dfrac{\cos y}{1+x\sin y}$,

$$\frac{d^2y}{dx^2}=\frac{(1+x\sin y)(-\sin y)(dy/dx)-(\cos y)[(x\cos y)(dy/dx)+\sin y]}{(1+x\sin y)^2}$$

$$=-\frac{2\sin y\cos y+(x\cos y)(2\sin^2 y+\cos^2 y)}{(1+x\sin y)^3},$$

but $x\cos y=y$, $2\sin y\cos y=\sin 2y$, and $\sin^2 y+\cos^2 y=1$ so

$$\frac{d^2y}{dx^2}=-\frac{\sin 2y+y(\sin^2 y+1)}{(1+x\sin y)^3}$$

27. By implicit differentiation, $2x+2y(dy/dx)=0$, $\dfrac{dy}{dx}=-\dfrac{x}{y}$; at $(1/\sqrt{2},1/\sqrt{2})$, $\dfrac{dy}{dx}=-1$; at $(1/\sqrt{2},-1/\sqrt{2})$, $\dfrac{dy}{dx}=+1$. Directly, at the upper point $y=\sqrt{1-x^2}$, $\dfrac{dy}{dx}=\dfrac{-x}{\sqrt{1-x^2}}=-1$ and at the lower point $y=-\sqrt{1-x^2}$, $\dfrac{dy}{dx}=\dfrac{x}{\sqrt{1-x^2}}=+1$.

28. If $y^2-x+1=0$, then $y=\sqrt{x-1}$ goes through the point $(10,3)$ so $dy/dx=1/(2\sqrt{x-1})$. By implicit differentiation $dy/dx=1/(2y)$. In both cases, $dy/dx|_{(10,3)}=1/6$. Similarly $y=-\sqrt{x-1}$ goes through $(10,-3)$ so $dy/dx=-1/(2\sqrt{x-1})=-1/6$ which yields $dy/dx=1/(2y)=-1/6$.

29. $4x^3 + 4y^3\frac{dy}{dx} = 0$, so $\frac{dy}{dx} = -\frac{x^3}{y^3} = -\frac{1}{15^{3/4}} \approx -0.1312$.

30. $3y^2\frac{dy}{dx} + x^2\frac{dy}{dx} + 2xy + 2x - 6y\frac{dy}{dx} = 0$, so $\frac{dy}{dx} = -2x\frac{y+1}{3y^2 + x^2 - 6y} = 0$ at $x = 0$

31. $4(x^2 + y^2)\left(2x + 2y\frac{dy}{dx}\right) = 25\left(2x - 2y\frac{dy}{dx}\right)$,

$\frac{dy}{dx} = \frac{x[25 - 4(x^2 + y^2)]}{y[25 + 4(x^2 + y^2)]}$; at $(3, 1)$ $\frac{dy}{dx} = -9/13$

32. $\frac{2}{3}\left(x^{-1/3} + y^{-1/3}\frac{dy}{dx}\right) = 0$, $\frac{dy}{dx} = -\frac{y^{1/3}}{x^{1/3}} = \sqrt{3}$ at $(-1, 3\sqrt{3})$

34. **(a)**

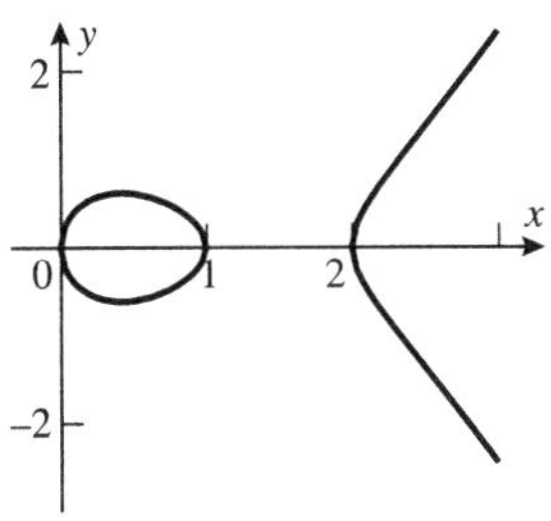

(b) $2y\frac{dy}{dx} = (x - a)(x - b) + x(x - b) + x(x - a) = 3x^2 - 2(a + b)x + ab$. If $\frac{dy}{dx} = 0$ then $3x^2 - 2(a + b)x + ab = 0$. By the Quadratic Formula

$$x = \frac{2(a + b) \pm \sqrt{4(a + b)^2 - 4 \cdot 3ab}}{6} = \frac{1}{3}\left[a + b \pm (a^2 + b^2 - ab)^{1/2}\right].$$

(c) $y = \pm\sqrt{x(x - a)(x - b)}$. The square root is only defined for nonnegative arguments, so it is necessary that all three of the factors x, $x - a$, $x - b$ be nonnegative, or that two of them be nonpositive. If, for example, $0 < a < b$ then the function is defined on the disjoint intervals $0 < x < a$ and $b < x < +\infty$, so there are two parts.

35. **(a)**

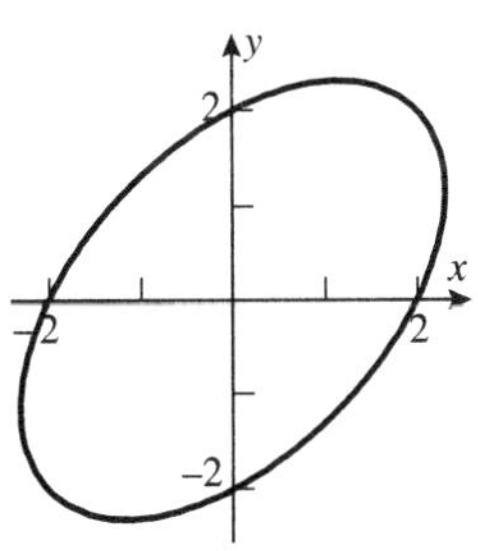

(b) ± 1.1547

(c) Implicit differentiation yields $2x - x\frac{dy}{dx} - y + 2y\frac{dy}{dx} = 0$. Solve for $\frac{dy}{dx} = \frac{y - 2x}{2y - x}$. If $\frac{dy}{dx} = 0$ then $y - 2x = 0$ or $y = 2x$. Thus $4 = x^2 - xy + y^2 = x^2 - 2x^2 + 4x^2 = 3x^2$, $x = \pm\frac{2}{\sqrt{3}}$.

36. $\frac{1}{2}u^{-1/2}\frac{du}{dv} + \frac{1}{2}v^{-1/2} = 0$ so $\frac{du}{dv} = -\frac{\sqrt{u}}{\sqrt{v}}$

37. $4a^3\frac{da}{dt} - 4t^3 = 6\left(a^2 + 2at\frac{da}{dt}\right)$, solve for $\frac{da}{dt}$ to get $\frac{da}{dt} = \frac{2t^3 + 3a^2}{2a^3 - 6at}$

38. $1 = (\cos x)\dfrac{dx}{dy}$ so $\dfrac{dx}{dy} = \dfrac{1}{\cos x} = \sec x$

39. $2a^2\omega\dfrac{d\omega}{d\lambda} + 2b^2\lambda = 0$ so $\dfrac{d\omega}{d\lambda} = -\dfrac{b^2\lambda}{a^2\omega}$

40. Let $P(x_0, y_0)$ be the required point. The slope of the line $4x - 3y + 1 = 0$ is $4/3$ so the slope of the tangent to $y^2 = 2x^3$ at P must be $-3/4$. By implicit differentiation $dy/dx = 3x^2/y$, so at P, $3x_0^2/y_0 = -3/4$, or $y_0 = -4x_0^2$. But $y_0^2 = 2x_0^3$ because P is on the curve $y^2 = 2x^3$. Elimination of y_0 gives $16x_0^4 = 2x_0^3$, $x_0^3(8x_0 - 1) = 0$, so $x_0 = 0$ or $1/8$. From $y_0 = -4x_0^2$ it follows that $y_0 = 0$ when $x_0 = 0$, and $y_0 = -1/16$ when $x_0 = 1/8$. It does not follow, however, that $(0,0)$ is a solution because $dy/dx = 3x^2/y$ (the slope of the curve as determined by implicit differentiation) is valid only if $y \neq 0$. Further analysis shows that the curve is tangent to the x-axis at $(0,0)$, so the point $(1/8, -1/16)$ is the only solution.

41. The point (1,1) is on the graph, so $1 + a = b$. The slope of the tangent line at (1,1) is $-4/3$; use implicit differentiation to get $\dfrac{dy}{dx} = -\dfrac{2xy}{x^2 + 2ay}$ so at (1,1), $-\dfrac{2}{1+2a} = -\dfrac{4}{3}$, $1 + 2a = 3/2$, $a = 1/4$ and hence $b = 1 + 1/4 = 5/4$.

42. Use implicit differentiation to get $dy/dx = (y - 3x^2)/(3y^2 - x)$, so $dy/dx = 0$ if $y = 3x^2$. Substitute this into $x^3 - xy + y^3 = 0$ to obtain $27x^6 - 2x^3 = 0, x^3 = 2/27$, $x = \sqrt[3]{2}/3$ and hence $y = \sqrt[3]{4}/3$.

43. Let $P(x_0, y_0)$ be a point where a line through the origin is tangent to the curve $x^2 - 4x + y^2 + 3 = 0$. Implicit differentiation applied to the equation of the curve gives $dy/dx = (2 - x)/y$. At P the slope of the curve must equal the slope of the line so $(2 - x_0)/y_0 = y_0/x_0$, or $y_0^2 = 2x_0 - x_0^2$. But $x_0^2 - 4x_0 + y_0^2 + 3 = 0$ because (x_0, y_0) is on the curve, and elimination of y_0^2 in the latter two equations gives $x_0^2 - 4x_0 + (2x_0 - x_0^2) + 3 = 0$, $x_0 = 3/2$ which when substituted into $y_0^2 = 2x_0 - x_0^2$ yields $y_0^2 = 3/4$, so $y_0 = \pm\sqrt{3}/2$. The slopes of the lines are $(\pm\sqrt{3}/2)/(3/2) = \pm\sqrt{3}/3$ and their equations are $y = (\sqrt{3}/3)x$ and $y = -(\sqrt{3}/3)x$.

44. By implicit differentiation, $dy/dx = k/(2y)$ so the slope of the tangent to $y^2 = kx$ at (x_0, y_0) is $k/(2y_0)$ if $y_0 \neq 0$. The tangent line in this case is $y - y_0 = \dfrac{k}{2y_0}(x - x_0)$, or $2y_0y - 2y_0^2 = kx - kx_0$. But $y_0^2 = kx_0$ because (x_0, y_0) is on the curve $y^2 = kx$, so the equation of the tangent line becomes $2y_0y - 2kx_0 = kx - kx_0$ which gives $y_0y = k(x + x_0)/2$. If $y_0 = 0$, then $x_0 = 0$; the graph of $y^2 = kx$ has a vertical tangent at $(0,0)$ so its equation is $x = 0$, but $y_0y = k(x + x_0)/2$ gives the same result when $x_0 = y_0 = 0$.

45. By the chain rule, $\dfrac{dy}{dx} = \dfrac{dy}{dt}\dfrac{dt}{dx}$. Use implicit differentiation on $2y^3t + t^3y = 1$ to get

$\dfrac{dy}{dt} = -\dfrac{2y^3 + 3t^2y}{6ty^2 + t^3}$, but $\dfrac{dt}{dx} = \dfrac{1}{\cos t}$ so $\dfrac{dy}{dx} = -\dfrac{2y^3 + 3t^2y}{(6ty^2 + t^3)\cos t}$.

46. $2x^3y\dfrac{dy}{dt} + 3x^2y^2\dfrac{dx}{dt} + \dfrac{dy}{dt} = 0$, $\dfrac{dy}{dt} = -\dfrac{3x^2y^2}{2x^3y + 1}\dfrac{dx}{dt}$

47. $2xy\dfrac{dy}{dt} = y^2\dfrac{dx}{dt} = 3(\cos 3x)\dfrac{dx}{dt}$, $\dfrac{dy}{dt} = \dfrac{3\cos 3x - y^2}{2xy}\dfrac{dx}{dt}$

48. **(a)** $f'(x) = \dfrac{4}{3}x^{1/3}$, $f''(x) = \dfrac{4}{9}x^{-2/3}$

(b) $f'(x) = \dfrac{7}{3}x^{4/3}$, $f''(x) = \dfrac{28}{9}x^{1/3}$, $f'''(x) = \dfrac{28}{27}x^{-2/3}$

(c) generalize parts (a) and (b) with $k = (n - 1) + 1/3 = n - 2/3$

49. $y' = rx^{r-1}, y'' = r(r-1)x^{r-2}$ so $3x^2\left[r(r-1)x^{r-2}\right] + 4x\left(rx^{r-1}\right) - 2x^r = 0$,
$3r(r-1)x^r + 4rx^r - 2x^r = 0, (3r^2 + r - 2)x^r = 0$,
$3r^2 + r - 2 = 0$, $(3r-2)(r+1) = 0$; $r = -1, 2/3$

50. $y' = rx^{r-1}, y'' = r(r-1)x^{r-2}$ so $16x^2\left[r(r-1)x^{r-2}\right] + 24x\left(rx^{r-1}\right) + x^r = 0$,
$16r(r-1)x^r + 24rx^r + x^r = 0, (16r^2 + 8r + 1)x^r = 0$,
$16r^2 + 8r + 1 = 0, (4r+1)^2 = 0$; $r = -1/4$

51. We shall find when the curves intersect and check that the slopes are negative reciprocals. For the intersection solve the simultaneous equations $x^2 + (y-c)^2 = c^2$ and $(x-k)^2 + y^2 = k^2$ to obtain $cy = kx = \frac{1}{2}(x^2 + y^2)$. Thus $x^2 + y^2 = cy + kx$, or $y^2 - cy = -x^2 + kx$, and $\frac{y-c}{x} = -\frac{x-k}{y}$. Differentiating the two families yields (black) $\frac{dy}{dx} = -\frac{x}{y-c}$, and (gray) $\frac{dy}{dx} = -\frac{x-k}{y}$. But it was proven that these quantities are negative reciprocals of each other.

52. Differentiating, we get the equations (black) $x\frac{dy}{dx} + y = 0$ and (gray) $2x - 2y\frac{dy}{dx} = 0$. The first says the (black) slope is $= -\frac{y}{x}$ and the second says the (gray) slope is $\frac{x}{y}$, and these are negative reciprocals of each other.

EXERCISE SET 3.7

1. $\frac{dy}{dt} = 3\frac{dx}{dt}$

(a) $\frac{dy}{dt} = 3(2) = 6$

(b) $-1 = 3\frac{dx}{dt}, \frac{dx}{dt} = -\frac{1}{3}$

2. $\frac{dx}{dt} + 4\frac{dy}{dt} = 0$

(a) $1 + 4\frac{dy}{dt} = 0$ so $\frac{dy}{dt} = -\frac{1}{4}$ when $x = 2$.

(b) $\frac{dx}{dt} + 4(4) = 0$ so $\frac{dx}{dt} = -16$ when $x = 3$.

3. $2x\frac{dx}{dt} + 2y\frac{dy}{dt} = 0$

(a) $2\frac{1}{2}(1) + 2\frac{\sqrt{3}}{2}\frac{dy}{dt} = 0$, so $\frac{dy}{dt} = -\frac{1}{\sqrt{3}}$.

(b) $2\frac{\sqrt{2}}{2}\frac{dx}{dt} + 2\frac{\sqrt{2}}{2}(-2) = 0$, so $\frac{dx}{dt} = 2$.

4. $2x\frac{dx}{dt} + 2y\frac{dy}{dt} = 2\frac{dx}{dt}$

(a) $-2 + 2\frac{dy}{dt} = -2$, $\frac{dy}{dt} = 0$

(b) $\frac{2+\sqrt{2}}{2}\frac{dx}{dt} + \frac{\sqrt{2}}{2}(3) = \frac{dx}{dt}$, $(2 + \sqrt{2} - 2)\frac{dx}{dt} + 3\sqrt{2} = 0$, $\frac{dx}{dt} = -3$

5. (b) $A = x^2$

(c) $\frac{dA}{dt} = 2x\frac{dx}{dt}$

Find $\left.\frac{dA}{dt}\right|_{x=3}$ given that $\left.\frac{dx}{dt}\right|_{x=3} = 2$. From Part (c), $\left.\frac{dA}{dt}\right|_{x=3} = 2(3)(2) = 12$ ft^2/min.

6. **(b)** $A = \pi r^2$ **(c)** $\dfrac{dA}{dt} = 2\pi r \dfrac{dr}{dt}$

(d) Find $\left.\dfrac{dA}{dt}\right|_{r=5}$ given that $\left.\dfrac{dr}{dt}\right|_{r=5} = 2$. From Part (c), $\left.\dfrac{dA}{dt}\right|_{r=5} = 2\pi(5)(2) = 20\pi$ cm^2/s.

7. **(a)** $V = \pi r^2 h$, so $\dfrac{dV}{dt} = \pi\left(r^2\dfrac{dh}{dt} + 2rh\dfrac{dr}{dt}\right)$.

(b) Find $\left.\dfrac{dV}{dt}\right|_{\substack{h=6,\\ r=10}}$ given that $\left.\dfrac{dh}{dt}\right|_{\substack{h=6,\\ r=10}} = 1$ and $\left.\dfrac{dr}{dt}\right|_{\substack{h=6,\\ r=10}} = -1$. From Part (a),

$\left.\dfrac{dV}{dt}\right|_{\substack{h=6,\\ r=10}} = \pi[10^2(1) + 2(10)(6)(-1)] = -20\pi$ in^3/s; the volume is decreasing.

8. **(a)** $\ell^2 = x^2 + y^2$, so $\dfrac{d\ell}{dt} = \dfrac{1}{\ell}\left(x\dfrac{dx}{dt} + y\dfrac{dy}{dt}\right)$.

(b) Find $\left.\dfrac{d\ell}{dt}\right|_{\substack{x=3,\\ y=4}}$ given that $\dfrac{dx}{dt} = \dfrac{1}{2}$ and $\dfrac{dy}{dt} = -\dfrac{1}{4}$.

From Part (a) and the fact that $\ell = 5$ when $x = 3$ and $y = 4$,

$\left.\dfrac{d\ell}{dt}\right|_{\substack{x=3,\\ y=4}} = \dfrac{1}{5}\left[3\left(\dfrac{1}{2}\right) + 4\left(-\dfrac{1}{4}\right)\right] = \dfrac{1}{10}$ ft/s; the diagonal is increasing.

9. **(a)** $\tan\theta = \dfrac{y}{x}$, so $\sec^2\theta\dfrac{d\theta}{dt} = \dfrac{x\dfrac{dy}{dt} - y\dfrac{dx}{dt}}{x^2}$, $\dfrac{d\theta}{dt} = \dfrac{\cos^2\theta}{x^2}\left(x\dfrac{dy}{dt} - y\dfrac{dx}{dt}\right)$

(b) Find $\left.\dfrac{d\theta}{dt}\right|_{\substack{x=2,\\ y=2}}$ given that $\left.\dfrac{dx}{dt}\right|_{\substack{x=2,\\ y=2}} = 1$ and $\left.\dfrac{dy}{dt}\right|_{\substack{x=2,\\ y=2}} = -\dfrac{1}{4}$.

When $x = 2$ and $y = 2$, $\tan\theta = 2/2 = 1$ so $\theta = \dfrac{\pi}{4}$ and $\cos\theta = \cos\dfrac{\pi}{4} = \dfrac{1}{\sqrt{2}}$. Thus from Part (a), $\left.\dfrac{d\theta}{dt}\right|_{\substack{x=2,\\ y=2}} = \dfrac{(1/\sqrt{2})^2}{2^2}\left[2\left(-\dfrac{1}{4}\right) - 2(1)\right] = -\dfrac{5}{16}$ rad/s; θ is decreasing.

10. Find $\left.\dfrac{dz}{dt}\right|_{\substack{x=1,\\ y=2}}$ given that $\left.\dfrac{dx}{dt}\right|_{\substack{x=1,\\ y=2}} = -2$ and $\left.\dfrac{dy}{dt}\right|_{\substack{x=1,\\ y=2}} = 3$.

$\dfrac{dz}{dt} = 2x^3y\dfrac{dy}{dt} + 3x^2y^2\dfrac{dx}{dt}$, $\left.\dfrac{dz}{dt}\right|_{\substack{x=1,\\ y=2}} = (4)(3) + (12)(-2) = -12$ units/s; z is decreasing

11. Let A be the area swept out, and θ the angle through which the minute hand has rotated. Find $\dfrac{dA}{dt}$ given that $\dfrac{d\theta}{dt} = \dfrac{\pi}{30}$ rad/min; $A = \dfrac{1}{2}r^2\theta = 8\theta$, so $\dfrac{dA}{dt} = 8\dfrac{d\theta}{dt} = \dfrac{4\pi}{15}$ in^2/min.

12. Let r be the radius and A the area enclosed by the ripple. We want $\left.\dfrac{dA}{dt}\right|_{t=10}$ given that $\dfrac{dr}{dt} = 3$. We know that $A = \pi r^2$, so $\dfrac{dA}{dt} = 2\pi r\dfrac{dr}{dt}$. Because r is increasing at the constant rate of 3 ft/s, it follows that $r = 30$ ft after 10 seconds so $\left.\dfrac{dA}{dt}\right|_{t=10} = 2\pi(30)(3) = 180\pi$ ft^2/s.

13. Find $\left.\dfrac{dr}{dt}\right|_{A=9}$ given that $\dfrac{dA}{dt}=6$. From $A=\pi r^2$ we get $\dfrac{dA}{dt}=2\pi r\dfrac{dr}{dt}$ so $\dfrac{dr}{dt}=\dfrac{1}{2\pi r}\dfrac{dA}{dt}$. If $A=9$ then $\pi r^2=9$, $r=3/\sqrt{\pi}$ so $\left.\dfrac{dr}{dt}\right|_{A=9}=\dfrac{1}{2\pi(3/\sqrt{\pi})}(6)=1/\sqrt{\pi}$ mi/h.

14. The volume V of a sphere of radius r is given by $V=\dfrac{4}{3}\pi r^3$ or, because $r=\dfrac{D}{2}$ where D is the diameter, $V=\dfrac{4}{3}\pi\left(\dfrac{D}{2}\right)^3=\dfrac{1}{6}\pi D^3$. We want $\left.\dfrac{dD}{dt}\right|_{r=1}$ given that $\dfrac{dV}{dt}=3$. From $V=\dfrac{1}{6}\pi D^3$ we get $\dfrac{dV}{dt}=\dfrac{1}{2}\pi D^2\dfrac{dD}{dt}$, $\dfrac{dD}{dt}=\dfrac{2}{\pi D^2}\dfrac{dV}{dt}$, so $\left.\dfrac{dD}{dt}\right|_{r=1}=\dfrac{2}{\pi(2)^2}(3)=\dfrac{3}{2\pi}$ ft/min.

15. Find $\left.\dfrac{dV}{dt}\right|_{r=9}$ given that $\dfrac{dr}{dt}=-15$. From $V=\dfrac{4}{3}\pi r^3$ we get $\dfrac{dV}{dt}=4\pi r^2\dfrac{dr}{dt}$ so $\left.\dfrac{dV}{dt}\right|_{r=9}=4\pi(9)^2(-15)=-4860\pi$. Air must be removed at the rate of 4860π cm^3/min.

16. Let x and y be the distances shown in the diagram. We want to find $\left.\dfrac{dy}{dt}\right|_{y=8}$ given that $\dfrac{dx}{dt}=5$. From $x^2+y^2=17^2$ we get $2x\dfrac{dx}{dt}+2y\dfrac{dy}{dt}=0$, so $\dfrac{dy}{dt}=-\dfrac{x}{y}\dfrac{dx}{dt}$. When $y=8$, $x^2+8^2=17^2$, $x^2=289-64=225$, $x=15$ so $\left.\dfrac{dy}{dt}\right|_{y=8}=-\dfrac{15}{8}(5)=-\dfrac{75}{8}$ ft/s; the top of the ladder is moving down the wall at a rate of 75/8 ft/s.

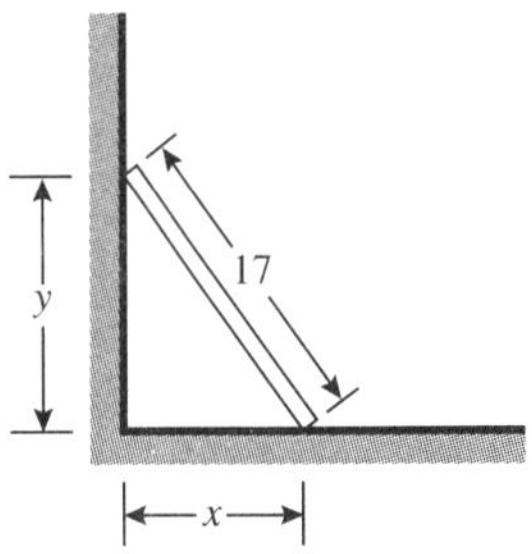

17. Find $\left.\dfrac{dx}{dt}\right|_{y=5}$ given that $\dfrac{dy}{dt}=-2$. From $x^2+y^2=13^2$ we get $2x\dfrac{dx}{dt}+2y\dfrac{dy}{dt}=0$ so $\dfrac{dx}{dt}=-\dfrac{y}{x}\dfrac{dy}{dt}$. Use $x^2+y^2=169$ to find that $x=12$ when $y=5$ so $\left.\dfrac{dx}{dt}\right|_{y=5}=-\dfrac{5}{12}(-2)=\dfrac{5}{6}$ ft/s.

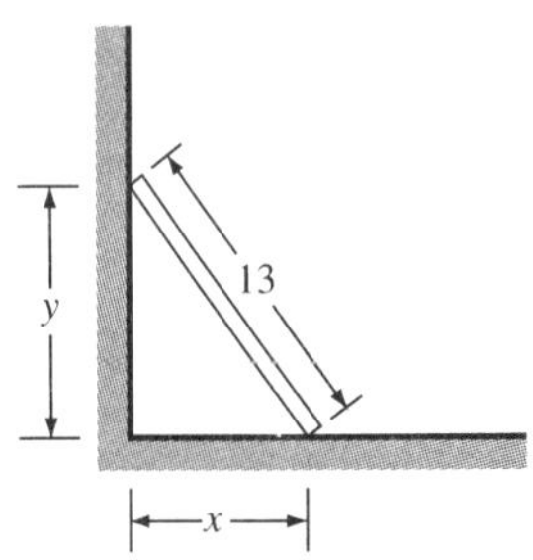

18. Let θ be the acute angle, and x the distance of the bottom of the plank from the wall. Find $\left.\dfrac{d\theta}{dt}\right|_{x=2}$ given that $\left.\dfrac{dx}{dt}\right|_{x=2}=-\dfrac{1}{2}$ ft/s. The variables θ and x are related by the equation $\cos\theta=\dfrac{x}{10}$ so $-\sin\theta\dfrac{d\theta}{dt}=\dfrac{1}{10}\dfrac{dx}{dt}$, $\dfrac{d\theta}{dt}=-\dfrac{1}{10\sin\theta}\dfrac{dx}{dt}$. When $x=2$, the top of the plank is $\sqrt{10^2-2^2}=\sqrt{96}$ ft above the ground so $\sin\theta=\sqrt{96}/10$ and $\left.\dfrac{d\theta}{dt}\right|_{x=2}=-\dfrac{1}{\sqrt{96}}\left(-\dfrac{1}{2}\right)=\dfrac{1}{2\sqrt{96}}\approx 0.051$ rad/s.

19. Let x denote the distance from first base and y the distance from home plate. Then $x^2 + 60^2 = y^2$ and $2x\dfrac{dx}{dt} = 2y\dfrac{dy}{dt}$. When $x = 50$ then $y = 10\sqrt{61}$ so $\dfrac{dy}{dt} = \dfrac{x}{y}\dfrac{dx}{dt} = \dfrac{50}{10\sqrt{61}}(25) = \dfrac{125}{\sqrt{61}}$ ft/s.

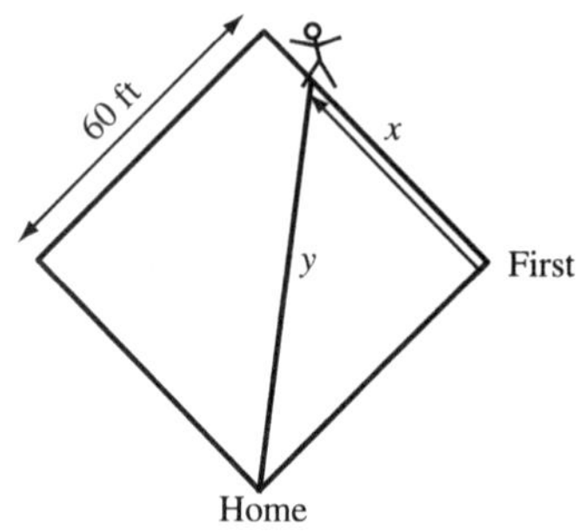

20. Find $\left.\dfrac{dx}{dt}\right|_{x=4}$ given that $\left.\dfrac{dy}{dt}\right|_{x=4} = 2000$. From $x^2 + 5^2 = y^2$ we get $2x\dfrac{dx}{dt} = 2y\dfrac{dy}{dt}$ so $\dfrac{dx}{dt} = \dfrac{y}{x}\dfrac{dy}{dt}$. Use $x^2 + 25 = y^2$ to find that $y = \sqrt{41}$ when $x = 4$ so $\left.\dfrac{dx}{dt}\right|_{x=4} = \dfrac{\sqrt{41}}{4}(2000) = 500\sqrt{41}$ mi/h.

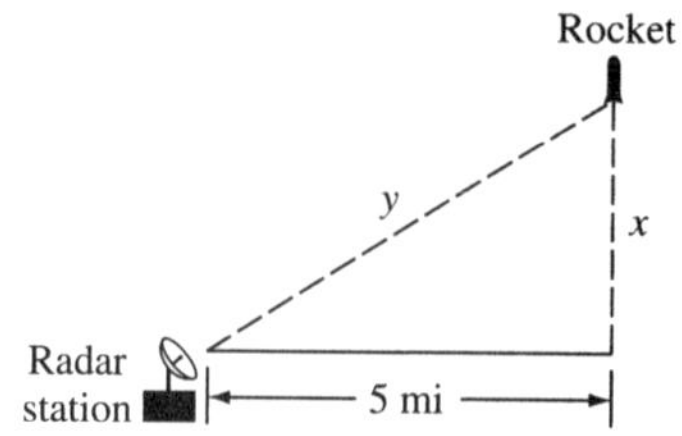

21. Find $\left.\dfrac{dy}{dt}\right|_{x=4000}$ given that $\left.\dfrac{dx}{dt}\right|_{x=4000} = 880$. From $y^2 = x^2 + 3000^2$ we get $2y\dfrac{dy}{dt} = 2x\dfrac{dx}{dt}$ so $\dfrac{dy}{dt} = \dfrac{x}{y}\dfrac{dx}{dt}$. If $x = 4000$, then $y = 5000$ so

$$\left.\frac{dy}{dt}\right|_{x=4000} = \frac{4000}{5000}(880) = 704 \text{ ft/s}.$$

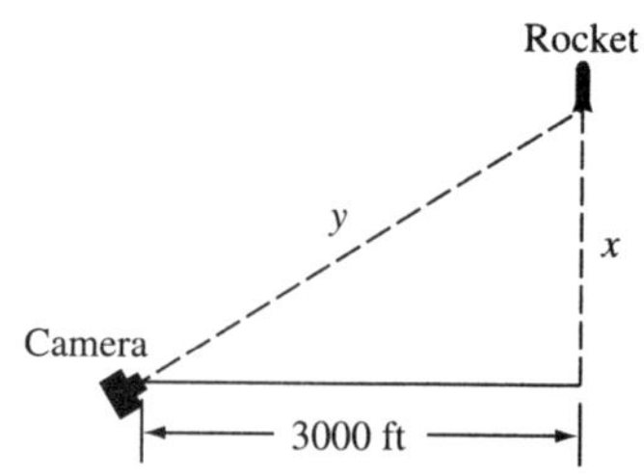

22. Find $\left.\dfrac{dx}{dt}\right|_{\phi=\pi/4}$ given that $\left.\dfrac{d\phi}{dt}\right|_{\phi=\pi/4} = 0.2$. But $x = 3000\tan\phi$ so

$$\frac{dx}{dt} = 3000(\sec^2\phi)\frac{d\phi}{dt},\ \left.\frac{dx}{dt}\right|_{\phi=\pi/4} = 3000\left(\sec^2\frac{\pi}{4}\right)(0.2) = 1200 \text{ ft/s}.$$

23. (a) If x denotes the altitude, then $r - x = 3960$, the radius of the Earth. $\theta = 0$ at perigee, so $r = 4995/1.12 \approx 4460$; the altitude is $x = 4460 - 3960 = 500$ miles. $\theta = \pi$ at apogee, so $r = 4995/0.88 \approx 5676$; the altitude is $x = 5676 - 3960 = 1716$ miles.

(b) If $\theta = 120°$, then $r = 4995/0.94 \approx 5314$; the altitude is $5314 - 3960 = 1354$ miles. The rate of change of the altitude is given by

$$\frac{dx}{dt} = \frac{dr}{dt} = \frac{dr}{d\theta}\frac{d\theta}{dt} = \frac{4995(0.12\sin\theta)}{(1 + 0.12\cos\theta)^2}\frac{d\theta}{dt}.$$

Use $\theta = 120°$ and $d\theta/dt = 2.7°/\text{min} = (2.7)(\pi/180)$ rad/min to get $dr/dt \approx 27.7$ mi/min.

24. (a) Let x be the horizontal distance shown in the figure. Then $x = 4000\cot\theta$ and

$\dfrac{dx}{dt} = -4000\csc^2\theta\dfrac{d\theta}{dt}$, so $\dfrac{d\theta}{dt} = -\dfrac{\sin^2\theta}{4000}\dfrac{dx}{dt}$. Use $\theta = 30°$ and $dx/dt = 300$ mi/h $= 300(5280/3600)$ ft/s $= 440$ ft/s to get

$d\theta/dt = -0.0275$ rad/s $\approx -1.6°$/s; θ is decreasing at the rate of $1.6°$/s.

(b) Let y be the distance between the observation point and the aircraft. Then $y = 4000\csc\theta$ so $dy/dt = -4000(\csc\theta\cot\theta)(d\theta/dt)$. Use $\theta = 30°$ and $d\theta/dt = -0.0275$ rad/s to get $dy/dt \approx 381$ ft/s.

25. Find $\left.\frac{dh}{dt}\right|_{h=16}$ given that $\frac{dV}{dt} = 20$. The volume of water in the tank at a depth h is $V = \frac{1}{3}\pi r^2 h$. Use similar triangles (see figure) to get $\frac{r}{h} = \frac{10}{24}$ so $r = \frac{5}{12}h$ thus $V = \frac{1}{3}\pi\left(\frac{5}{12}h\right)^2 h = \frac{25}{432}\pi h^3$, $\frac{dV}{dt} = \frac{25}{144}\pi h^2\frac{dh}{dt}$;

$\frac{dh}{dt} = \frac{144}{25\pi h^2}\frac{dV}{dt}$, $\left.\frac{dh}{dt}\right|_{h=16} = \frac{144}{25\pi(16)^2}(20) = \frac{9}{20\pi}$ ft/min.

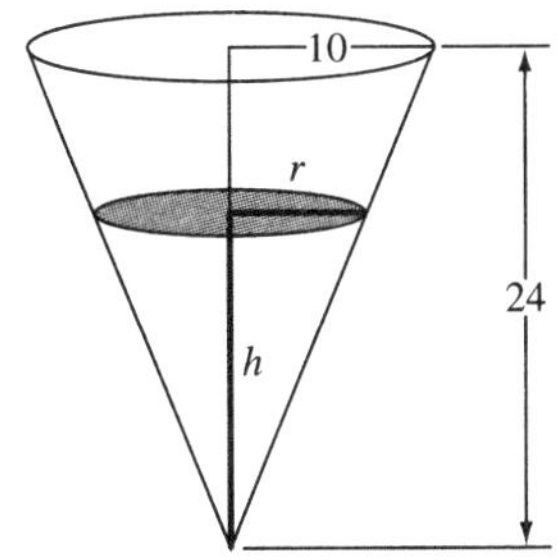

26. Find $\left.\frac{dh}{dt}\right|_{h=6}$ given that $\frac{dV}{dt} = 8$. $V = \frac{1}{3}\pi r^2 h$, but $r = \frac{1}{2}h$ so $V = \frac{1}{3}\pi\left(\frac{h}{2}\right)^2 h = \frac{1}{12}\pi h^3$, $\frac{dV}{dt} = \frac{1}{4}\pi h^2\frac{dh}{dt}$,

$\frac{dh}{dt} = \frac{4}{\pi h^2}\frac{dV}{dt}$, $\left.\frac{dh}{dt}\right|_{h=6} = \frac{4}{\pi(6)^2}(8) = \frac{8}{9\pi}$ ft/min.

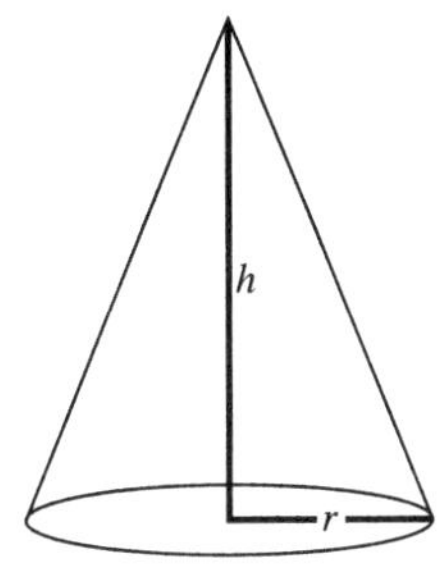

27. Find $\left.\frac{dV}{dt}\right|_{h=10}$ given that $\frac{dh}{dt} = 5$. $V = \frac{1}{3}\pi r^2 h$, but $r = \frac{1}{2}h$ so $V = \frac{1}{3}\pi\left(\frac{h}{2}\right)^2 h = \frac{1}{12}\pi h^3$, $\frac{dV}{dt} = \frac{1}{4}\pi h^2\frac{dh}{dt}$,

$\left.\frac{dV}{dt}\right|_{h=10} = \frac{1}{4}\pi(10)^2(5) = 125\pi$ ft^3/min.

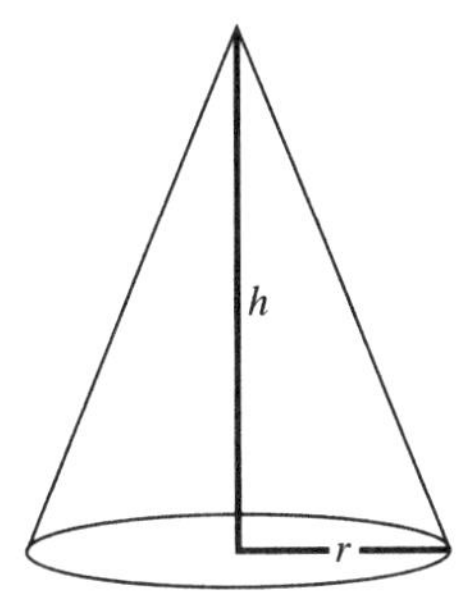

28. Let r and h be as shown in the figure. If C is the circumference of the base, then we want to find $\left.\frac{dC}{dt}\right|_{h=8}$ given that $\frac{dV}{dt} = 10$.

It is given that $r = \frac{1}{2}h$, thus $C = 2\pi r = \pi h$ so

$$\frac{dC}{dt} = \pi\frac{dh}{dt} \qquad (1)$$

Use $V = \frac{1}{3}\pi r^2 h = \frac{1}{12}\pi h^3$ to get $\frac{dV}{dt} = \frac{1}{4}\pi h^2\frac{dh}{dt}$, so

$$\frac{dh}{dt} = \frac{4}{\pi h^2}\frac{dV}{dt} \qquad (2)$$

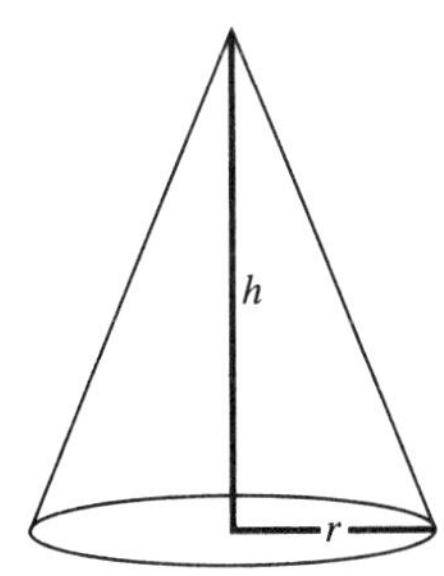

Substitution of (2) into (1) gives $\frac{dC}{dt} = \frac{4}{h^2}\frac{dV}{dt}$ so $\left.\frac{dC}{dt}\right|_{h=8} = \frac{4}{64}(10) = \frac{5}{8}$ ft/min.

29. With s and h as shown in the figure, we want to find $\frac{dh}{dt}$ given that $\frac{ds}{dt} = 500$. From the figure, $h = s \sin 30° = \frac{1}{2}s$ so $\frac{dh}{dt} = \frac{1}{2}\frac{ds}{dt} = \frac{1}{2}(500) = 250$ mi/h.

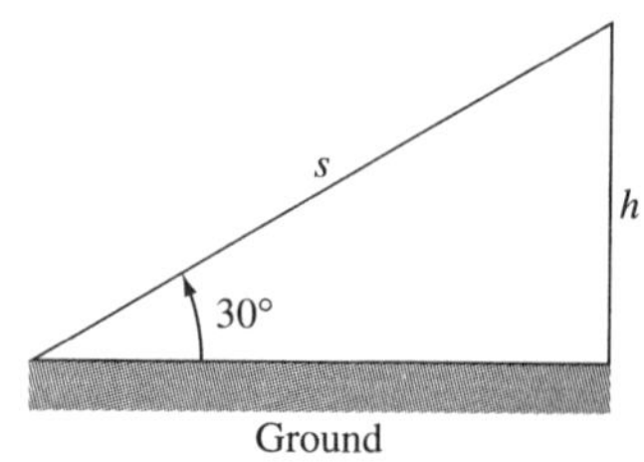

30. Find $\left.\frac{dx}{dt}\right|_{y=125}$ given that $\frac{dy}{dt} = -20$. From $x^2 + 10^2 = y^2$ we get $2x\frac{dx}{dt} = 2y\frac{dy}{dt}$ so $\frac{dx}{dt} = \frac{y}{x}\frac{dy}{dt}$. Use $x^2 + 100 = y^2$ to find that $x = \sqrt{15,525} = 15\sqrt{69}$ when $y = 125$ so $\left.\frac{dx}{dt}\right|_{y=125} = \frac{125}{15\sqrt{69}}(-20) = -\frac{500}{3\sqrt{69}}$.

The boat is approaching the dock at the rate of $\frac{500}{3\sqrt{69}}$ ft/min.

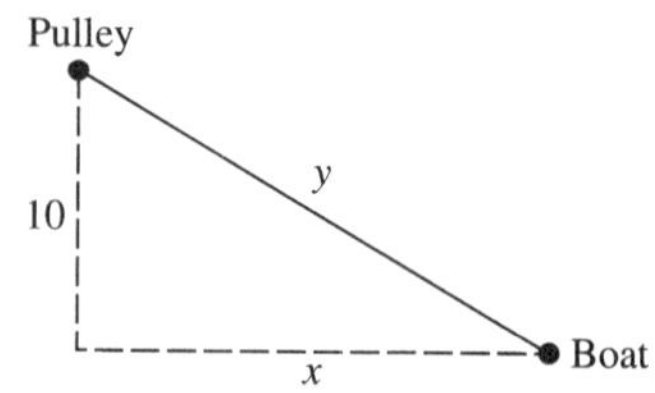

31. Find $\frac{dy}{dt}$ given that $\left.\frac{dx}{dt}\right|_{y=125} = -12$. From $x^2+10^2 = y^2$ we get $2x\frac{dx}{dt} = 2y\frac{dy}{dt}$ so $\frac{dy}{dt} = \frac{x}{y}\frac{dx}{dt}$. Use $x^2 + 100 = y^2$ to find that $x = \sqrt{15,525} = 15\sqrt{69}$ when $y = 125$ so $\frac{dy}{dt} = \frac{15\sqrt{69}}{125}(-12) = -\frac{36\sqrt{69}}{25}$. The rope must be pulled at the rate of $\frac{36\sqrt{69}}{25}$ ft/min.

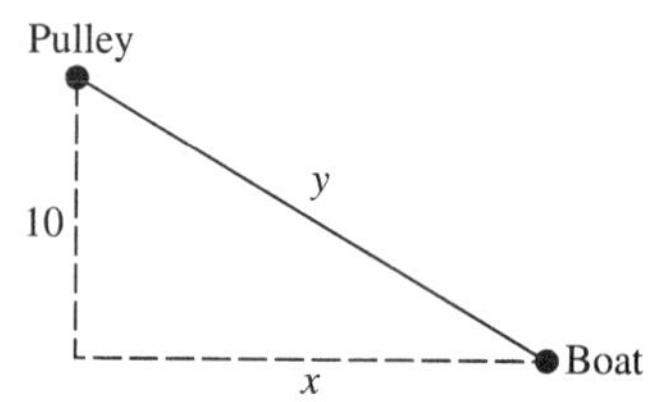

32. (a) Let x and y be as shown in the figure. It is required to find $\frac{dx}{dt}$, given that $\frac{dy}{dt} = -3$. By similar triangles, $\frac{x}{6} = \frac{x+y}{18}$, $18x = 6x + 6y$, $12x = 6y$, $x = \frac{1}{2}y$, so $\frac{dx}{dt} = \frac{1}{2}\frac{dy}{dt} = \frac{1}{2}(-3) = -\frac{3}{2}$ ft/s.

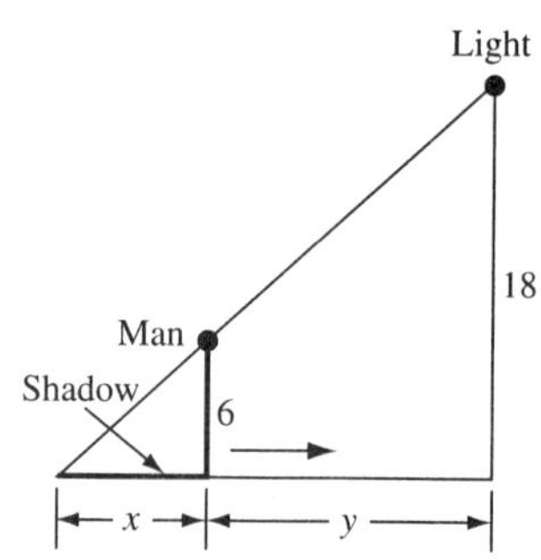

(b) The tip of the shadow is $z = x + y$ feet from the street light, thus the rate at which it is moving is given by $\frac{dz}{dt} = \frac{dx}{dt} + \frac{dy}{dt}$. In part (a) we found that $\frac{dx}{dt} = -\frac{3}{2}$ when $\frac{dy}{dt} = -3$ so $\frac{dz}{dt} = (-3/2) + (-3) = -9/2$ ft/s; the tip of the shadow is moving at the rate of 9/2 ft/s toward the street light.

33. Find $\left.\dfrac{dx}{dt}\right|_{\theta=\pi/4}$ given that $\dfrac{d\theta}{dt} = \dfrac{2\pi}{10} = \dfrac{\pi}{5}$ rad/s.

Then $x = 4\tan\theta$ (see figure) so $\dfrac{dx}{dt} = 4\sec^2\theta\dfrac{d\theta}{dt}$,

$\left.\dfrac{dx}{dt}\right|_{\theta=\pi/4} = 4\left(\sec^2\dfrac{\pi}{4}\right)\left(\dfrac{\pi}{5}\right) = 8\pi/5$ km/s.

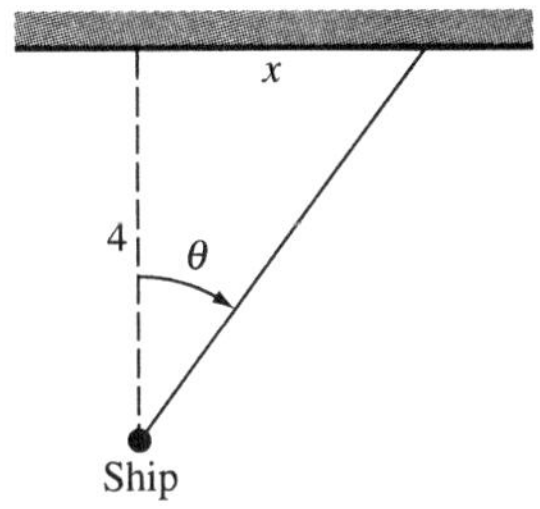

34. If x, y, and z are as shown in the figure, then we want $\left.\dfrac{dz}{dt}\right|_{\substack{x=2,\\y=4}}$ given that $\dfrac{dx}{dt} = -600$ and $\left.\dfrac{dy}{dt}\right|_{\substack{x=2,\\y=4}} = -1200$. But

$z^2 = x^2 + y^2$ so $2z\dfrac{dz}{dt} = 2x\dfrac{dx}{dt} + 2y\dfrac{dy}{dt}$, $\dfrac{dz}{dt} = \dfrac{1}{z}\left(x\dfrac{dx}{dt} + y\dfrac{dy}{dt}\right)$.

When $x = 2$ and $y = 4$, $z^2 = 2^2 + 4^2 = 20$, $z = \sqrt{20} = 2\sqrt{5}$ so

$\left.\dfrac{dz}{dt}\right|_{\substack{x=2,\\y=4}} = \dfrac{1}{2\sqrt{5}}[2(-600) + 4(-1200)] = -\dfrac{3000}{\sqrt{5}} = -600\sqrt{5}$ mi/h;

the distance between missile and aircraft is decreasing at the rate of $600\sqrt{5}$ mi/h.

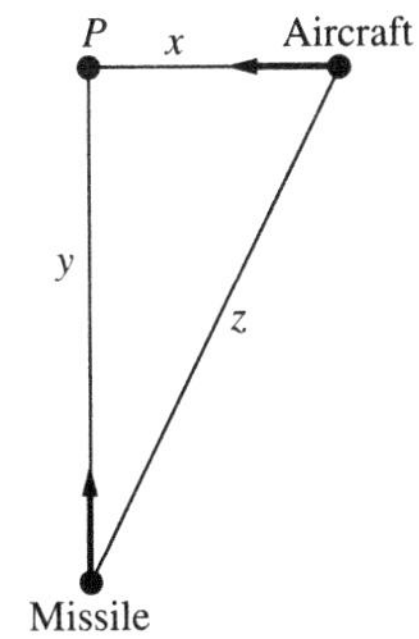

35. We wish to find $\left.\dfrac{dz}{dt}\right|_{\substack{x=2,\\y=4}}$ given $\dfrac{dx}{dt} = -600$ and $\left.\dfrac{dy}{dt}\right|_{\substack{x=2,\\y=4}} = -1200$ (see figure). From the law of cosines,

$z^2 = x^2 + y^2 - 2xy\cos 120° = x^2 + y^2 - 2xy(-1/2)$

$= x^2+y^2+xy$, so $2z\dfrac{dz}{dt} = 2x\dfrac{dx}{dt}+2y\dfrac{dy}{dt}+x\dfrac{dy}{dt}+y\dfrac{dx}{dt}$,

$\dfrac{dz}{dt} = \dfrac{1}{2z}\left[(2x+y)\dfrac{dx}{dt} + (2y+x)\dfrac{dy}{dt}\right]$.

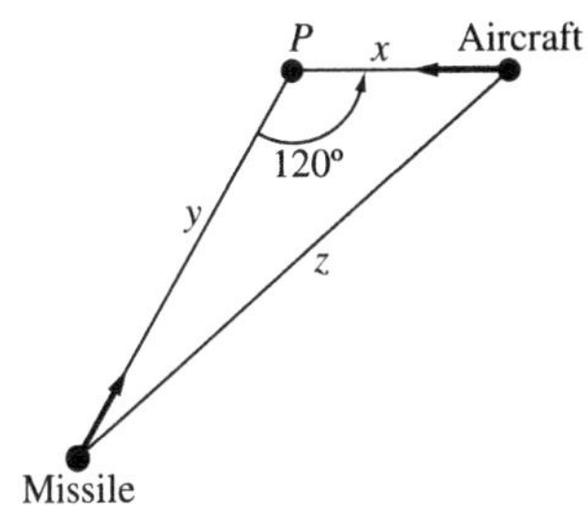

When $x = 2$ and $y = 4$, $z^2 = 2^2 + 4^2 + (2)(4) = 28$, so $z = \sqrt{28} = 2\sqrt{7}$, thus

$\left.\dfrac{dz}{dt}\right|_{\substack{x=2,\\y=4}} = \dfrac{1}{2(2\sqrt{7})}[(2(2)+4)(-600) + (2(4)+2)(-1200)] = -\dfrac{4200}{\sqrt{7}} = -600\sqrt{7}$ mi/h;

the distance between missile and aircraft is decreasing at the rate of $600\sqrt{7}$ mi/h.

36. (a) Let P be the point on the helicopter's path that lies directly above the car's path. Let x, y, and z be the distances shown in the first figure. Find $\left.\dfrac{dz}{dt}\right|_{\substack{x=2,\\y=0}}$ given that $\dfrac{dx}{dt} = -75$ and $\dfrac{dy}{dt} = 100$. In order to find an equation relating x, y, and z, first draw the line segment that joins the point P to the car, as shown in the second figure. Because triangle OPC is a right triangle, it follows that PC has length $\sqrt{x^2 + (1/2)^2}$; but triangle HPC is also a right triangle so $z^2 = \left(\sqrt{x^2 + (1/2)^2}\right)^2 + y^2 = x^2 + y^2 + 1/4$ and $2z\dfrac{dz}{dt} = 2x\dfrac{dx}{dt} + 2y\dfrac{dy}{dt} + 0$, $\dfrac{dz}{dt} = \dfrac{1}{z}\left(x\dfrac{dx}{dt} + y\dfrac{dy}{dt}\right)$. Now, when $x = 2$ and $y = 0$, $z^2 = (2)^2 + (0)^2 + 1/4 = 17/4$,

$$z = \sqrt{17}/2 \text{ so } \left.\frac{dz}{dt}\right|_{\substack{x=2,\\ y=0}} = \frac{1}{(\sqrt{17}/2)}[2(-75) + 0(100)] = -300/\sqrt{17} \text{ mi/h}$$

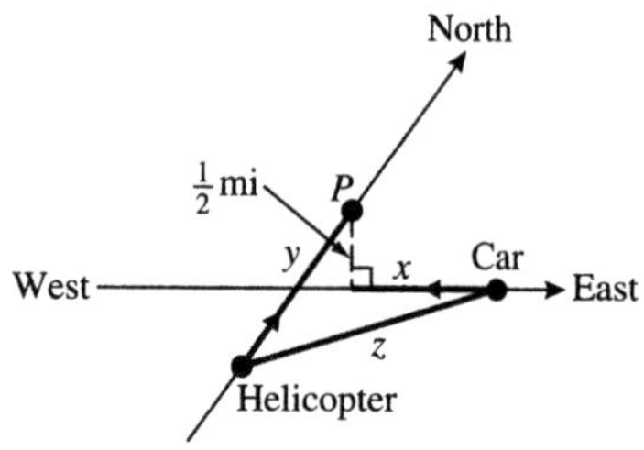

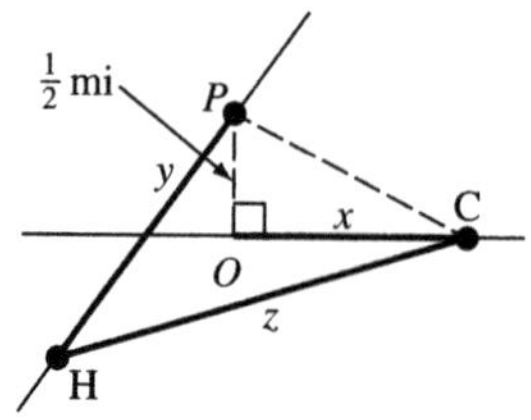

(b) decreasing, because $\frac{dz}{dt} < 0$.

37. **(a)** We want $\left.\frac{dy}{dt}\right|_{\substack{x=1,\\ y=2}}$ given that $\left.\frac{dx}{dt}\right|_{\substack{x=1,\\ y=2}} = 6$. For convenience, first rewrite the equation as

$xy^3 = \frac{8}{5} + \frac{8}{5}y^2$ then $3xy^2\frac{dy}{dt} + y^3\frac{dx}{dt} = \frac{16}{5}y\frac{dy}{dt}$, $\frac{dy}{dt} = \frac{y^3}{\frac{16}{5}y - 3xy^2}\frac{dx}{dt}$ so

$$\left.\frac{dy}{dt}\right|_{\substack{x=1,\\ y=2}} = \frac{2^3}{\frac{16}{5}(2) - 3(1)2^2}(6) = -60/7 \text{ units/s.}$$

(b) falling, because $\frac{dy}{dt} < 0$

38. Find $\left.\frac{dx}{dt}\right|_{(2,5)}$ given that $\left.\frac{dy}{dt}\right|_{(2,5)} = 2$. Square and rearrange to get $x^3 = y^2 - 17$

so $3x^2\frac{dx}{dt} = 2y\frac{dy}{dt}$, $\frac{dx}{dt} = \frac{2y}{3x^2}\frac{dy}{dt}$, $\left.\frac{dx}{dt}\right|_{(2,5)} = \left(\frac{5}{6}\right)(2) = \frac{5}{3}$ units/s.

39. The coordinates of P are $(x, 2x)$, so the distance between P and the point $(3, 0)$ is

$D = \sqrt{(x-3)^2 + (2x-0)^2} = \sqrt{5x^2 - 6x + 9}$. Find $\left.\frac{dD}{dt}\right|_{x=3}$ given that $\left.\frac{dx}{dt}\right|_{x=3} = -2$.

$\frac{dD}{dt} = \frac{5x-3}{\sqrt{5x^2 - 6x + 9}}\frac{dx}{dt}$, so $\left.\frac{dD}{dt}\right|_{x=3} = \frac{12}{\sqrt{36}}(-2) = -4$ units/s.

40. **(a)** Let D be the distance between P and $(2, 0)$. Find $\left.\frac{dD}{dt}\right|_{x=3}$ given that $\left.\frac{dx}{dt}\right|_{x=3} = 4$.

$D = \sqrt{(x-2)^2 + y^2} = \sqrt{(x-2)^2 + x} = \sqrt{x^2 - 3x + 4}$ so $\frac{dD}{dt} = \frac{2x-3}{2\sqrt{x^2 - 3x + 4}}$;

$$\left.\frac{dD}{dt}\right|_{x=3} = \frac{3}{2\sqrt{4}} = \frac{3}{4} \text{ units/s.}$$

(b) Let θ be the angle of inclination. Find $\left.\frac{d\theta}{dt}\right|_{x=3}$ given that $\left.\frac{dx}{dt}\right|_{x=3} = 4$.

$\tan\theta = \frac{y}{x-2} = \frac{\sqrt{x}}{x-2}$ so $\sec^2\theta\frac{d\theta}{dt} = -\frac{x+2}{2\sqrt{x}(x-2)^2}\frac{dx}{dt}$, $\frac{d\theta}{dt} = -\cos^2\theta\frac{x+2}{2\sqrt{x}(x-2)^2}\frac{dx}{dt}$.

When $x = 3$, $D = 2$ so $\cos\theta = \frac{1}{2}$ and $\left.\frac{d\theta}{dt}\right|_{x=3} = -\frac{1}{4}\frac{5}{2\sqrt{3}}(4) = -\frac{5}{2\sqrt{3}}$ rad/s.

41. Solve $\frac{dx}{dt} = 3\frac{dy}{dt}$ given $y = x/(x^2+1)$. Then $y(x^2+1) = x$. Differentiating with respect to x, $(x^2+1)\frac{dy}{dx} + y(2x) = 1$. But $\frac{dy}{dx} = \frac{dy/dt}{dx/dt} = \frac{1}{3}$ so $(x^2+1)\frac{1}{3} + 2xy = 1$, $x^2+1+6xy = 3$, $x^2+1+6x^2/(x^2+1) = 3$, $(x^2+1)^2+6x^2-3x^2-3 = 0$, $x^4+5x^2-3 = 0$. By the binomial theorem applied to x^2 we obtain $x^2 = (-5 \pm \sqrt{25+12})/2$. The minus sign is spurious since x^2 cannot be negative, so $x^2 = (\sqrt{33}-5)/2$, $x \approx \pm 0.6101486081, y = \pm 0.4446235604$.

42. $32x\frac{dx}{dt} + 18y\frac{dy}{dt} = 0$; if $\frac{dy}{dt} = \frac{dx}{dt} \neq 0$, then $(32x+18y)\frac{dx}{dt} = 0$, $32x + 18y = 0$, $y = -\frac{16}{9}x$ so $16x^2 + 9\frac{256}{81}x^2 = 144$, $\frac{400}{9}x^2 = 144$, $x^2 = \frac{81}{25}$, $x = \pm\frac{9}{5}$. If $x = \frac{9}{5}$, then $y = -\frac{16}{9}\frac{9}{5} = -\frac{16}{5}$. Similarly, if $x = -\frac{9}{5}$, then $y = \frac{16}{5}$. The points are $\left(\frac{9}{5}, -\frac{16}{5}\right)$ and $\left(-\frac{9}{5}, \frac{16}{5}\right)$.

43. Find $\left.\frac{dS}{dt}\right|_{s=10}$ given that $\left.\frac{ds}{dt}\right|_{s=10} = -2$. From $\frac{1}{s} + \frac{1}{S} = \frac{1}{6}$ we get $-\frac{1}{s^2}\frac{ds}{dt} - \frac{1}{S^2}\frac{dS}{dt} = 0$, so $\frac{dS}{dt} = -\frac{S^2}{s^2}\frac{ds}{dt}$. If $s = 10$, then $\frac{1}{10} + \frac{1}{S} = \frac{1}{6}$ which gives $S = 15$. So $\left.\frac{dS}{dt}\right|_{s=10} = -\frac{225}{100}(-2) = 4.5$ cm/s. The image is moving away from the lens.

44. Suppose that the reservoir has height H and that the radius at the top is R. At any instant of time let h and r be the corresponding dimensions of the cone of water (see figure). We want to show that $\frac{dh}{dt}$ is constant and independent of H and R, given that $\frac{dV}{dt} = -kA$ where V is the volume of water, A is the area of a circle of radius r, and k is a positive constant. The volume of a cone of radius r and height h is $V = \frac{1}{3}\pi r^2 h$. By similar triangles $\frac{r}{h} = \frac{R}{H}$, $r = \frac{R}{H}h$ thus $V = \frac{1}{3}\pi\left(\frac{R}{H}\right)^2 h^3$ so

$$\frac{dV}{dt} = \pi\left(\frac{R}{H}\right)^2 h^2\frac{dh}{dt} \qquad (1)$$

But it is given that $\frac{dV}{dt} = -kA$ or, because

$A = \pi r^2 = \pi\left(\frac{R}{H}\right)^2 h^2$, $\frac{dV}{dt} = -k\pi\left(\frac{R}{H}\right)^2 h^2$,

which when substituted into equation (1) gives

$-k\pi\left(\frac{R}{H}\right)^2 h^2 = \pi\left(\frac{R}{H}\right)^2 h^2\frac{dh}{dt}$, $\frac{dh}{dt} = -k$.

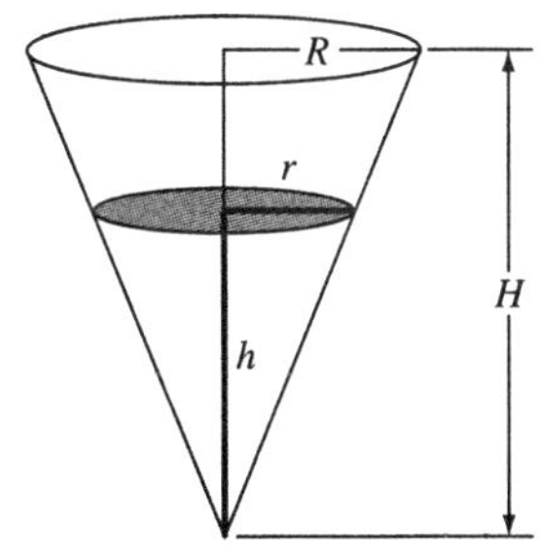

45. Let r be the radius, V the volume, and A the surface area of a sphere. Show that $\frac{dr}{dt}$ is a constant given that $\frac{dV}{dt} = -kA$, where k is a positive constant. Because $V = \frac{4}{3}\pi r^3$,

$$\frac{dV}{dt} = 4\pi r^2\frac{dr}{dt} \qquad (1)$$

But it is given that $\frac{dV}{dt} = -kA$ or, because $A = 4\pi r^2$, $\frac{dV}{dt} = -4\pi r^2 k$ which when substituted into equation (1) gives $-4\pi r^2 k = 4\pi r^2\frac{dr}{dt}$, $\frac{dr}{dt} = -k$.

46. Let x be the distance between the tips of the minute and hour hands, and α and β the angles shown in the figure. Because the minute hand makes one revolution in 60 minutes, $\dfrac{d\alpha}{dt} = \dfrac{2\pi}{60} = \pi/30$ rad/min; the hour hand makes one revolution in 12 hours (720 minutes), thus $\dfrac{d\beta}{dt} = \dfrac{2\pi}{720} = \pi/360$ rad/min. We want to find $\left.\dfrac{dx}{dt}\right|_{\substack{\alpha=2\pi,\\ \beta=3\pi/2}}$ given that $\dfrac{d\alpha}{dt} = \pi/30$ and $\dfrac{d\beta}{dt} = \pi/360$.

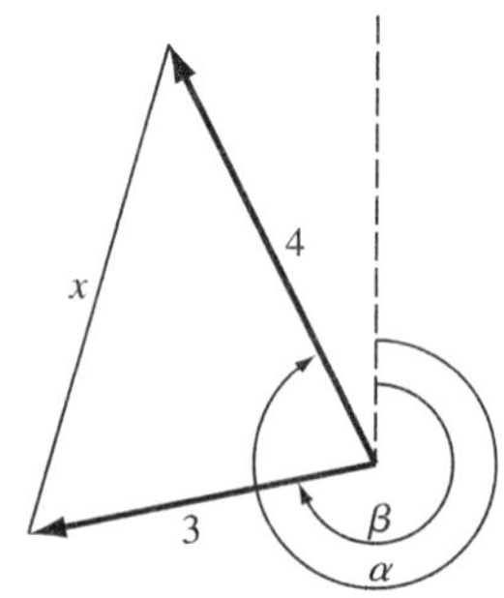

Using the law of cosines on the triangle shown in the figure,
$x^2 = 3^2 + 4^2 - 2(3)(4)\cos(\alpha - \beta) = 25 - 24\cos(\alpha - \beta)$, so
$2x\dfrac{dx}{dt} = 0 + 24\sin(\alpha - \beta)\left(\dfrac{d\alpha}{dt} - \dfrac{d\beta}{dt}\right)$,
$\dfrac{dx}{dt} = \dfrac{12}{x}\left(\dfrac{d\alpha}{dt} - \dfrac{d\beta}{dt}\right)\sin(\alpha - \beta)$. When $\alpha = 2\pi$ and $\beta = 3\pi/2$,
$x^2 = 25 - 24\cos(2\pi - 3\pi/2) = 25$, $x = 5$; so
$\left.\dfrac{dx}{dt}\right|_{\substack{\alpha=2\pi,\\ \beta=3\pi/2}} = \dfrac{12}{5}(\pi/30 - \pi/360)\sin(2\pi - 3\pi/2) = \dfrac{11\pi}{150}$ in/min.

47. Extend sides of cup to complete the cone and let V_0 be the volume of the portion added, then (see figure)
$V = \dfrac{1}{3}\pi r^2 h - V_0$ where $\dfrac{r}{h} = \dfrac{4}{12} = \dfrac{1}{3}$ so $r = \dfrac{1}{3}h$ and
$V = \dfrac{1}{3}\pi\left(\dfrac{h}{3}\right)^2 h - V_0 = \dfrac{1}{27}\pi h^3 - V_0$, $\dfrac{dV}{dt} = \dfrac{1}{9}\pi h^2\dfrac{dh}{dt}$,
$\dfrac{dh}{dt} = \dfrac{9}{\pi h^2}\dfrac{dV}{dt}$, $\left.\dfrac{dh}{dt}\right|_{h=9} = \dfrac{9}{\pi(9)^2}(20) = \dfrac{20}{9\pi}$ cm/s.

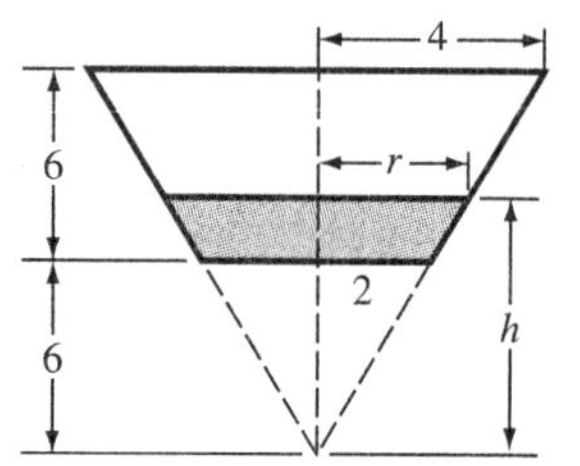

EXERCISE SET 3.8

1. **(a)** $f(x) \approx f(1) + f'(1)(x-1) = 1 + 3(x-1)$
(b) $f(1 + \Delta x) \approx f(1) + f'(1)\Delta x = 1 + 3\Delta x$
(c) From Part (a), $(1.02)^3 \approx 1 + 3(0.02) = 1.06$. From Part (b), $(1.02)^3 \approx 1 + 3(0.02) = 1.06$.

2. **(a)** $f(x) \approx f(2) + f'(2)(x-2) = 1/2 + (-1/2^2)(x-2) = (1/2) - (1/4)(x-2)$
(b) $f(2 + \Delta x) \approx f(2) + f'(2)\Delta x = 1/2 - (1/4)\Delta x$
(c) From Part (a), $1/2.05 \approx 0.5 - 0.25(0.05) = 0.4875$, and from Part (b), $1/2.05 \approx 0.5 - 0.25(0.05) = 0.4875$.

3. **(a)** $f(x) \approx f(x_0) + f'(x_0)(x - x_0) = 1 + (1/(2\sqrt{1})(x-0) = 1 + (1/2)x$, so with $x_0 = 0$ and $x = -0.1$, we have $\sqrt{0.9} = f(-0.1) \approx 1 + (1/2)(-0.1) = 1 - 0.05 = 0.95$. With $x = 0.1$ we have $\sqrt{1.1} = f(0.1) \approx 1 + (1/2)(0.1) = 1.05$.

(b)

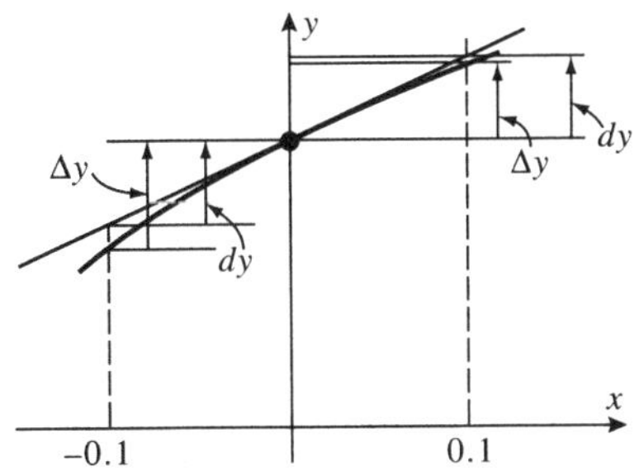

4. **(a)** $f(x) \approx f(x_0)+f'(x_0)(x-x_0) = 1/2-\left[1/(2\cdot 4^{3/2})\right](x-4) = 1/2-(x-4)/16$, so with $x_0 = 4$ and $x = 3.9$ we have $1/\sqrt{3.9} = f(3.9) \approx 0.5-(-0.1)/16 = 0.50625$. If $x_0 = 4$ and $x = 4.1$ then $1/\sqrt{4.1} = f(4.1) \approx 0.5-(0.1)/16 = 0.49375$

(b)

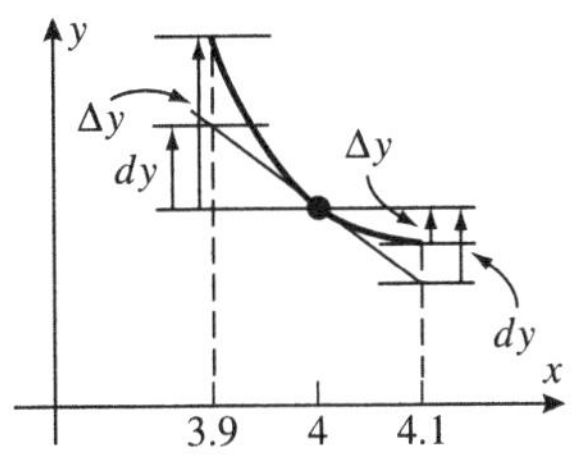

5. $f(x) = (1+x)^{15}$ and $x_0 = 0$. Thus $(1+x)^{15} \approx f(x_0)+f'(x_0)(x-x_0) = 1+15(1)^{14}(x-0) = 1+15x$.

6. $f(x) = \dfrac{1}{\sqrt{1-x}}$ and $x_0 = 0$, so $\dfrac{1}{\sqrt{1-x}} \approx f(x_0)+f'(x_0)(x-x_0) = 1+\dfrac{1}{2(1-0)^{3/2}}(x-0) = 1+x/2$

7. $\tan x \approx \tan(0)+\sec^2(0)(x-0) = x$

8. $\dfrac{1}{1+x} \approx 1+\dfrac{-1}{(1+0)^2}(x-0) = 1-x$

9. $x^4 \approx (1)^4+4(1)^3(x-1)$. Set $\Delta x = x-1$; then $x = \Delta x+1$ and $(1+\Delta x)^4 = 1+4\Delta x$.

10. $\sqrt{x} \approx \sqrt{1}+\dfrac{1}{2\sqrt{1}}(x-1)$, and $x = 1+\Delta x$, so $\sqrt{1+\Delta x} \approx 1+\Delta x/2$

11. $\dfrac{1}{2+x} \approx \dfrac{1}{2+1}-\dfrac{1}{(2+1)^2}(x-1)$, and $2+x = 3+\Delta x$, so $\dfrac{1}{3+\Delta x} \approx \dfrac{1}{3}-\dfrac{1}{9}\Delta x$

12. $(4+x)^3 \approx (4+1)^3+3(4+1)^2(x-1)$ so, with $4+x = 5+\Delta x$ we get $(5+\Delta x)^3 \approx 125+75\Delta x$

13. $f(x) = \sqrt{x+3}$ and $x_0 = 0$, so
$\sqrt{x+3} \approx \sqrt{3}+\dfrac{1}{2\sqrt{3}}(x-0) = \sqrt{3}+\dfrac{1}{2\sqrt{3}}x$, and
$\left|f(x)-\left(\sqrt{3}+\dfrac{1}{2\sqrt{3}}x\right)\right| < 0.1$ if $|x| < 1.692$.

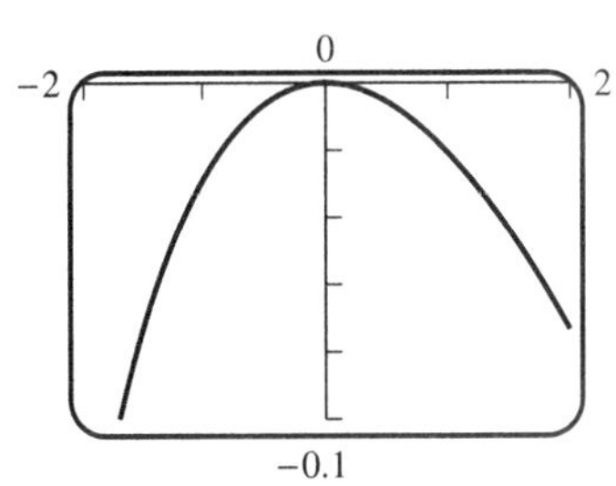

$\left|f(x)-\left(\sqrt{3}+\frac{1}{2\sqrt{3}}x\right)\right|$

14. $f(x) = \dfrac{1}{\sqrt{9-x}}$ so

$$\frac{1}{\sqrt{9-x}} \approx \frac{1}{\sqrt{9}} + \frac{1}{2(9-0)^{3/2}}(x-0) = \frac{1}{3} + \frac{1}{54}x,$$

and $\left|f(x) - \left(\dfrac{1}{3} + \dfrac{1}{54}x\right)\right| < 0.1$ if $|x| < 5.5114$

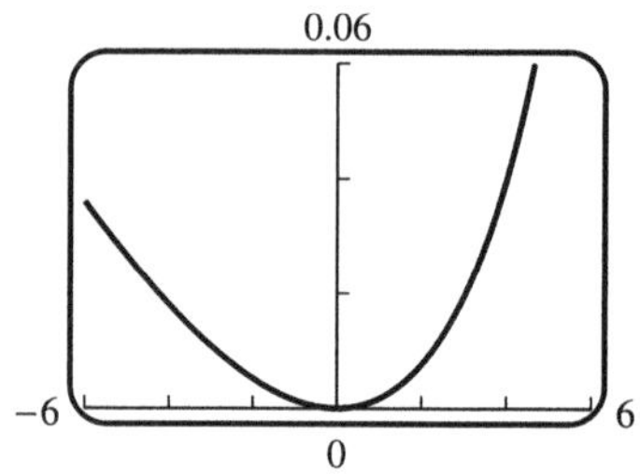

$\left|f(x) - \left(\frac{1}{3} + \frac{1}{54}x\right)\right|$

15. $\tan x \approx \tan 0 + (\sec^2 0)(x - 0) = x$,
and $|\tan x - x| < 0.1$ if $|x| < 0.6316$

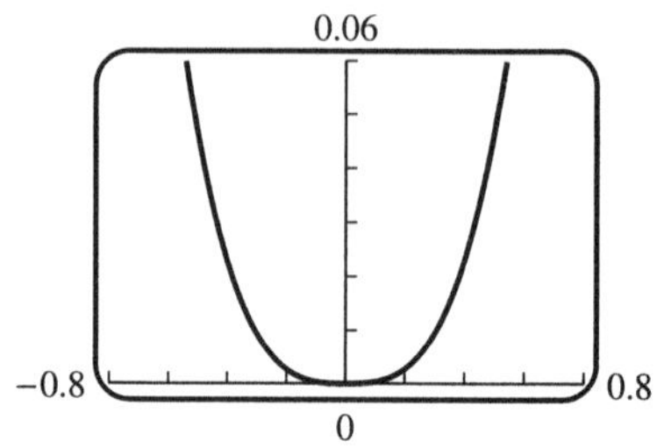

$|f(x) - x|$

16. $\dfrac{1}{(1+2x)^5} \approx \dfrac{1}{(1+2\cdot 0)^5} + \dfrac{-5(2)}{(1+2\cdot 0)^6}(x-0) = 1 - 10x,$
and $|f(x) - (1 - 10x)| < 0.1$ if $|x| < 0.0372$

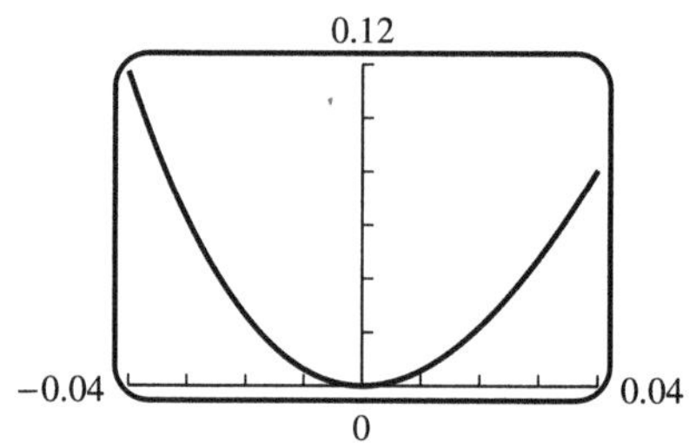

$|f(x) - (1 - 10x)|$

17. **(a)** The local linear approximation $\sin x \approx x$ gives $\sin 1° = \sin(\pi/180) \approx \pi/180 = 0.0174533$ and a calculator gives $\sin 1° = 0.0174524$. The relative error $|\sin(\pi/180) - (\pi/180)|/(\sin \pi/180) = 0.000051$ is very small, so for such a small value of x the approximation is very good.

(b) Use $x_0 = 45°$ (this assumes you know, or can approximate, $\sqrt{2}/2$).

(c) $44° = \dfrac{44\pi}{180}$ radians, and $45° = \dfrac{45\pi}{180} = \dfrac{\pi}{4}$ radians. With $x = \dfrac{44\pi}{180}$ and $x_0 = \dfrac{\pi}{4}$ we obtain $\sin 44° = \sin\dfrac{44\pi}{180} \approx \sin\dfrac{\pi}{4} + \left(\cos\dfrac{\pi}{4}\right)\left(\dfrac{44\pi}{180} - \dfrac{\pi}{4}\right) = \dfrac{\sqrt{2}}{2} + \dfrac{\sqrt{2}}{2}\left(\dfrac{-\pi}{180}\right) = 0.694765$. With a calculator, $\sin 44° = 0.694658$.

18. **(a)** $\tan x \approx \tan 0 + \sec^2 0(x - 0) = x$, so $\tan 2° = \tan(2\pi/180) \approx 2\pi/180 = 0.034907$, and with a calculator $\tan 2° = 0.034921$

(b) use $x_0 = \pi/3$ because we know $\tan 60° = \tan(\pi/3) = \sqrt{3}$

(c) with $x_0 = \dfrac{\pi}{3} = \dfrac{60\pi}{180}$ and $x = \dfrac{61\pi}{180}$ we have

$$\tan 61° = \tan\frac{61\pi}{180} \approx \tan\frac{\pi}{3} + \left(\sec^2\frac{\pi}{3}\right)\left(\frac{61\pi}{180} - \frac{\pi}{3}\right) = \sqrt{3} + 4\frac{\pi}{180} = 1.8019,$$

and with a calculator $\tan 61° = 1.8040$

19. $f(x) = x^4$, $f'(x) = 4x^3$, $x_0 = 3$, $\Delta x = 0.02$; $(3.02)^4 \approx 3^4 + (108)(0.02) = 81 + 2.16 = 83.16$

20. $f(x) = x^3$, $f'(x) = 3x^2$, $x_0 = 2$, $\Delta x = -0.03$; $(1.97)^3 \approx 2^3 + (12)(-0.03) = 8 - 0.36 = 7.64$

21. $f(x) = \sqrt{x}$, $f'(x) = \dfrac{1}{2\sqrt{x}}$, $x_0 = 64$, $\Delta x = 1$; $\sqrt{65} \approx \sqrt{64} + \dfrac{1}{16}(1) = 8 + \dfrac{1}{16} = 8.0625$

22. $f(x) = \sqrt{x}$, $f'(x) = \dfrac{1}{2\sqrt{x}}$, $x_0 = 25$, $\Delta x = -1$; $\sqrt{24} \approx \sqrt{25} + \dfrac{1}{10}(-1) = 5 - 0.1 = 4.9$

23. $f(x) = \sqrt{x}$, $f'(x) = \dfrac{1}{2\sqrt{x}}$, $x_0 = 81$, $\Delta x = -0.1$; $\sqrt{80.9} \approx \sqrt{81} + \dfrac{1}{18}(-0.1) \approx 8.9944$

24. $f(x) = \sqrt{x}$, $f'(x) = \dfrac{1}{2\sqrt{x}}$, $x_0 = 36$, $\Delta x = 0.03$; $\sqrt{36.03} \approx \sqrt{36} + \dfrac{1}{12}(0.03) = 6 + 0.0025 = 6.0025$

25. $f(x) = \sin x$, $f'(x) = \cos x$, $x_0 = 0$, $\Delta x = 0.1$; $\sin 0.1 \approx \sin 0 + (\cos 0)(0.1) = 0.1$

26. $f(x) = \tan x$, $f'(x) = \sec^2 x$, $x_0 = 0$, $\Delta x = 0.2$; $\tan 0.2 \approx \tan 0 + (\sec^2 0)(0.2) = 0.2$

27. $f(x) = \cos x$, $f'(x) = -\sin x$, $x_0 = \pi/6$, $\Delta x = \pi/180$;
$\cos 31^\circ \approx \cos 30^\circ + \left(-\dfrac{1}{2}\right)\left(\dfrac{\pi}{180}\right) = \dfrac{\sqrt{3}}{2} - \dfrac{\pi}{360} \approx 0.8573$

28. **(a)** Let $f(x) = (1+x)^k$ and $x_0 = 0$. Then $(1+x)^k \approx 1^k + k(1)^{k-1}(x-0) = 1 + kx$. Set $k = 37$ and $x = 0.001$ to obtain $(1.001)^{37} \approx 1.037$.

(b) With a calculator $(1.001)^{37} = 1.03767$.

(c) The approximation is $(1.1)^{37} \approx 1 + 37(0.1) = 4.7$, and the calculator value is 34.004. The error is due to the relative largeness of $f'(1)\Delta x = 37(0.1) = 3.7$.

29. **(a)** $dy = f'(x)dx = 2xdx = 4(1) = 4$ and
$\Delta y = (x + \Delta x)^2 - x^2 = (2+1)^2 - 2^2 = 5$

(b)

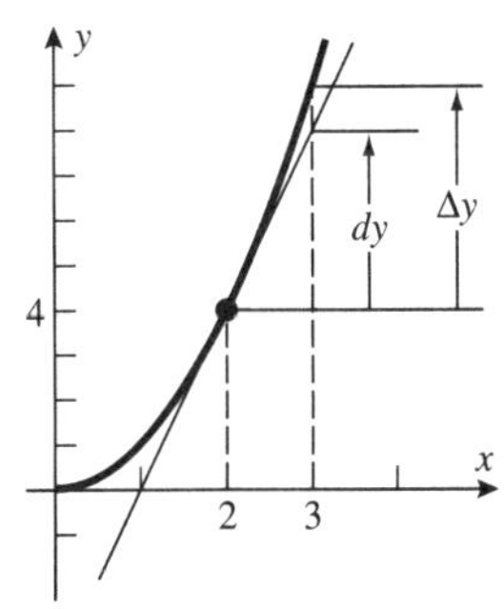

30. **(a)** $dy = 3x^2 dx = 3(1)^2(1) = 3$ and
$\Delta y = (x + \Delta x)^3 - x^3 = (1+1)^3 - 1^3 = 7$

(b)

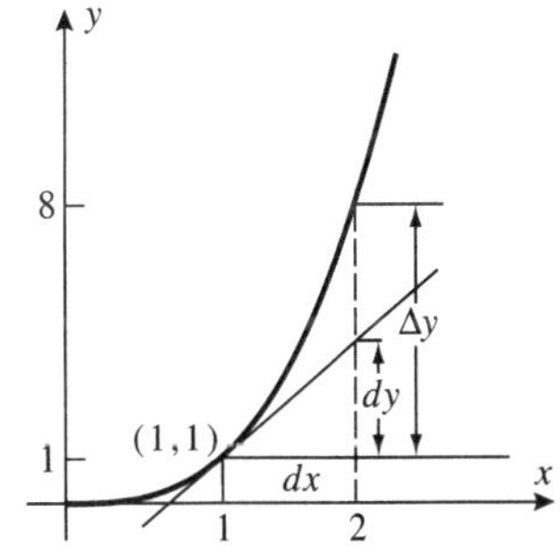

31. (a) $dy = (-1/x^2)dx = (-1)(-0.5) = 0.5$ and
$\Delta y = 1/(x + \Delta x) - 1/x$
$= 1/(1 - 0.5) - 1/1 = 2 - 1 = 1$

(b) 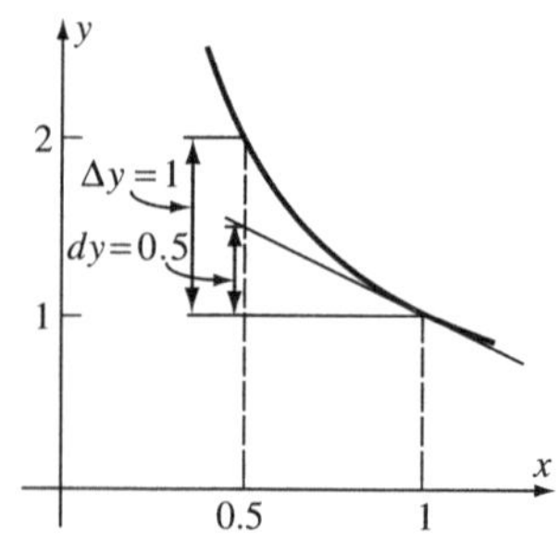

32. (a) $dy = (1/2\sqrt{x})dx = (1/(2 \cdot 3))(-1) = -1/6 \approx -0.167$ and
$\Delta y = \sqrt{x + \Delta x} - \sqrt{x} = \sqrt{9 + (-1)} - \sqrt{9} = \sqrt{8} - 3 \approx -0.172$

(b) 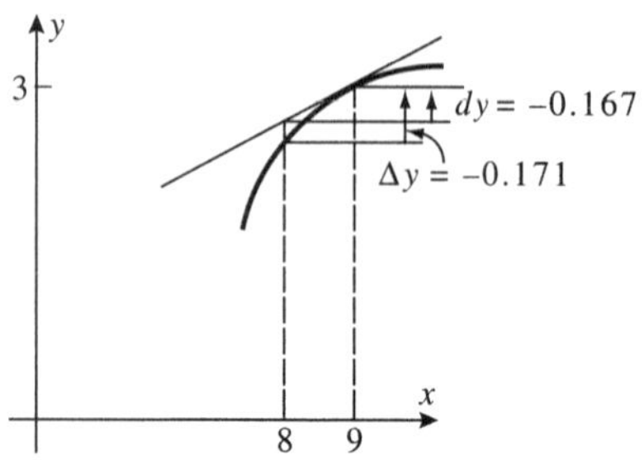

33. $dy = 3x^2dx$;
$\Delta y = (x + \Delta x)^3 - x^3 = x^3 + 3x^2\Delta x + 3x(\Delta x)^2 + (\Delta x)^3 - x^3 = 3x^2\Delta x + 3x(\Delta x)^2 + (\Delta x)^3$

34. $dy = 8dx$; $\Delta y = [8(x + \Delta x) - 4] - [8x - 4] = 8\Delta x$

35. $dy = (2x - 2)dx$;
$\Delta y = [(x + \Delta x)^2 - 2(x + \Delta x) + 1] - [x^2 - 2x + 1]$
$= x^2 + 2x\ \Delta x + (\Delta x)^2 - 2x - 2\Delta x + 1 - x^2 + 2x - 1 = 2x\ \Delta x + (\Delta x)^2 - 2\Delta x$

36. $dy = \cos x\, dx$; $\Delta y = \sin(x + \Delta x) - \sin x$

37. (a) $dy = (12x^2 - 14x)dx$
(b) $dy = x\, d(\cos x) + \cos x\, dx = x(-\sin x)dx + \cos x dx = (-x \sin x + \cos x)dx$

38. (a) $dy = (-1/x^2)dx$ (b) $dy = 5 \sec^2 x\, dx$

39. (a) $dy = \left(\sqrt{1-x} - \dfrac{x}{2\sqrt{1-x}}\right) dx = \dfrac{2 - 3x}{2\sqrt{1-x}}dx$

(b) $dy = -17(1 + x)^{-18}dx$

40. (a) $dy = \dfrac{(x^3 - 1)d(1) - (1)d(x^3 - 1)}{(x^3 - 1)^2} = \dfrac{(x^3 - 1)(0) - (1)3x^2dx}{(x^3 - 1)^2} = -\dfrac{3x^2}{(x^3 - 1)^2}dx$

(b) $dy = \dfrac{(2 - x)(-3x^2)dx - (1 - x^3)(-1)dx}{(2 - x)^2} = \dfrac{2x^3 - 6x^2 + 1}{(2 - x)^2}dx$

41. $dy = \dfrac{3}{2\sqrt{3x - 2}}dx$, $x = 2$, $dx = 0.03$; $\Delta y \approx dy = \dfrac{3}{4}(0.03) = 0.0225$

42. $dy = \dfrac{x}{\sqrt{x^2 + 8}}dx$, $x = 1$, $dx = -0.03$; $\Delta y \approx dy = (1/3)(-0.03) = -0.01$

43. $dy = \dfrac{1-x^2}{(x^2+1)^2}dx$, $x = 2$, $dx = -0.04$; $\Delta y \approx dy = \left(-\dfrac{3}{25}\right)(-0.04) = 0.0048$

44. $dy = \left(\dfrac{4x}{\sqrt{8x+1}} + \sqrt{8x+1}\right)dx$, $x = 3$, $dx = 0.05$; $\Delta y \approx dy = (37/5)(0.05) = 0.37$

45. **(a)** $A = x^2$ where x is the length of a side; $dA = 2x\,dx = 2(10)(\pm 0.1) = \pm 2$ ft^2.

(b) relative error in x is $\approx \dfrac{dx}{x} = \dfrac{\pm 0.1}{10} = \pm 0.01$ so percentage error in x is $\approx \pm 1\%$; relative error in A is $\approx \dfrac{dA}{A} = \dfrac{2x\,dx}{x^2} = 2\dfrac{dx}{x} = 2(\pm 0.01) = \pm 0.02$ so percentage error in A is $\approx \pm 2\%$

46. **(a)** $V = x^3$ where x is the length of a side; $dV = 3x^2dx = 3(25)^2(\pm 1) = \pm 1875$ cm^3.

(b) relative error in x is $\approx \dfrac{dx}{x} = \dfrac{\pm 1}{25} = \pm 0.04$ so percentage error in x is $\approx \pm 4\%$; relative error in V is $\approx \dfrac{dV}{V} = \dfrac{3x^2dx}{x^3} = 3\dfrac{dx}{x} = 3(\pm 0.04) = \pm 0.12$ so percentage error in V is $\approx \pm 12\%$

47. **(a)** $x = 10\sin\theta$, $y = 10\cos\theta$ (see figure),

$$dx = 10\cos\theta d\theta = 10\left(\cos\frac{\pi}{6}\right)\left(\pm\frac{\pi}{180}\right) = 10\left(\frac{\sqrt{3}}{2}\right)\left(\pm\frac{\pi}{180}\right)$$
$$\approx \pm 0.151 \text{ in},$$
$$dy = -10(\sin\theta)d\theta = -10\left(\sin\frac{\pi}{6}\right)\left(\pm\frac{\pi}{180}\right) = -10\left(\frac{1}{2}\right)\left(\pm\frac{\pi}{180}\right)$$
$$\approx \pm 0.087 \text{ in}$$

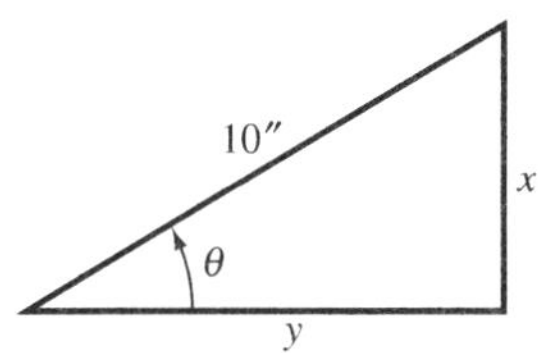

(b) relative error in x is $\approx \dfrac{dx}{x} = (\cot\theta)d\theta = \left(\cot\dfrac{\pi}{6}\right)\left(\pm\dfrac{\pi}{180}\right) = \sqrt{3}\left(\pm\dfrac{\pi}{180}\right) \approx \pm 0.030$
so percentage error in x is $\approx \pm 3.0\%$;
relative error in y is $\approx \dfrac{dy}{y} = -\tan\theta d\theta = -\left(\tan\dfrac{\pi}{6}\right)\left(\pm\dfrac{\pi}{180}\right) = -\dfrac{1}{\sqrt{3}}\left(\pm\dfrac{\pi}{180}\right) \approx \pm 0.010$
so percentage error in y is $\approx \pm 1.0\%$

48. **(a)** $x = 25\cot\theta$, $y = 25\csc\theta$ (see figure);

$$dx = -25\csc^2\theta d\theta = -25\left(\csc^2\frac{\pi}{3}\right)\left(\pm\frac{\pi}{360}\right)$$
$$= -25\left(\frac{4}{3}\right)\left(\pm\frac{\pi}{360}\right) \approx \pm 0.291 \text{ cm},$$
$$dy = -25\csc\theta\cot\theta d\theta = -25\left(\csc\frac{\pi}{3}\right)\left(\cot\frac{\pi}{3}\right)\left(\pm\frac{\pi}{360}\right)$$
$$= -25\left(\frac{2}{\sqrt{3}}\right)\left(\frac{1}{\sqrt{3}}\right)\left(\pm\frac{\pi}{360}\right) \approx \pm 0.145 \text{ cm}$$

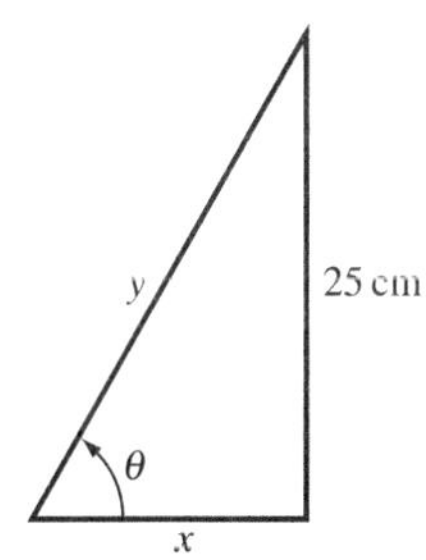

(b) relative error in x is $\approx \dfrac{dx}{x} = -\dfrac{\csc^2\theta}{\cot\theta}d\theta = -\dfrac{4/3}{1/\sqrt{3}}\left(\pm\dfrac{\pi}{360}\right) \approx \pm 0.020$ so percentage error in x is $\approx \pm 2.0\%$; relative error in y is $\approx \dfrac{dy}{y} = -\cot\theta d\theta = -\dfrac{1}{\sqrt{3}}\left(\pm\dfrac{\pi}{360}\right) \approx \pm 0.005$
so percentage error in y is $\approx \pm 0.5\%$

49. $\dfrac{dR}{R} = \dfrac{(-2k/r^3)dr}{(k/r^2)} = -2\dfrac{dr}{r}$, but $\dfrac{dr}{r} \approx \pm 0.05$ so $\dfrac{dR}{R} \approx -2(\pm 0.05) = \pm 0.10$; percentage error in R is $\approx \pm 10\%$

50. $h = 12\sin\theta$ thus $dh = 12\cos\theta d\theta$ so, with $\theta = 60° = \pi/3$ radians and $d\theta = -1° = -\pi/180$ radians, $dh = 12\cos(\pi/3)(-\pi/180) = -\pi/30 \approx -0.105$ ft

51. $A = \frac{1}{4}(4)^2\sin 2\theta = 4\sin 2\theta$ thus $dA = 8\cos 2\theta d\theta$ so, with $\theta = 30° = \pi/6$ radians and $d\theta = \pm 15' = \pm 1/4° = \pm\pi/720$ radians, $dA = 8\cos(\pi/3)(\pm\pi/720) = \pm\pi/180 \approx \pm 0.017$ cm^2

52. $A = x^2$ where x is the length of a side; $\frac{dA}{A} = \frac{2x\,dx}{x^2} = 2\frac{dx}{x}$, but $\frac{dx}{x} \approx \pm 0.01$ so $\frac{dA}{A} \approx 2(\pm 0.01) = \pm 0.02$; percentage error in A is $\approx \pm 2\%$

53. $V = x^3$ where x is the length of a side; $\frac{dV}{V} = \frac{3x^2dx}{x^3} = 3\frac{dx}{x}$, but $\frac{dx}{x} \approx \pm 0.02$ so $\frac{dV}{V} \approx 3(\pm 0.02) = \pm 0.06$; percentage error in V is $\approx \pm 6\%$.

54. $\frac{dV}{V} = \frac{4\pi r^2 dr}{4\pi r^3/3} = 3\frac{dr}{r}$, but $\frac{dV}{V} \approx \pm 0.03$ so $3\frac{dr}{r} \approx \pm 0.03$, $\frac{dr}{r} \approx \pm 0.01$; maximum permissible percentage error in r is $\approx \pm 1\%$.

55. $A = \frac{1}{4}\pi D^2$ where D is the diameter of the circle; $\frac{dA}{A} = \frac{(\pi D/2)dD}{\pi D^2/4} = 2\frac{dD}{D}$, but $\frac{dA}{A} \approx \pm 0.01$ so $2\frac{dD}{D} \approx \pm 0.01$, $\frac{dD}{D} \approx \pm 0.005$; maximum permissible percentage error in D is $\approx \pm 0.5\%$.

56. $V = x^3$ where x is the length of a side; approximate ΔV by dV if $x = 1$ and $dx = \Delta x = 0.02$, $dV = 3x^2dx = 3(1)^2(0.02) = 0.06$ in^3.

57. V = volume of cylindrical rod $= \pi r^2 h = \pi r^2(15) = 15\pi r^2$; approximate ΔV by dV if $r = 2.5$ and $dr = \Delta r = 0.001$. $dV = 30\pi r\,dr = 30\pi(2.5)(0.001) \approx 0.236$ cm^3.

58. $P = \frac{2\pi}{\sqrt{g}}\sqrt{L}$, $dP = \frac{2\pi}{\sqrt{g}}\frac{1}{2\sqrt{L}}dL = \frac{\pi}{\sqrt{g}\sqrt{L}}dL$, $\frac{dP}{P} = \frac{1}{2}\frac{dL}{L}$ so the relative error in $P \approx \frac{1}{2}$ the relative error in L. Thus the percentage error in P is $\approx \frac{1}{2}$ the percentage error in L.

59. **(a)** $\alpha = \Delta L/(L\Delta T) = 0.006/(40 \times 10) = 1.5 \times 10^{-5}/°\text{C}$
(b) $\Delta L = 2.3 \times 10^{-5}(180)(25) \approx 0.1$ cm, so the pole is about 180.1 cm long.

60. $\Delta V = 7.5 \times 10^{-4}(4000)(-20) = -60$ gallons; the truck delivers $4000 - 60 = 3940$ gallons.

CHAPTER 3 SUPPLEMENTARY EXERCISES

4. **(a)**
$$\frac{dy}{dx} = \lim_{h\to 0}\frac{\sqrt{9-4(x+h)} - \sqrt{9-4x}}{h} = \lim_{h\to 0}\frac{(9-4(x+h) - (9-4x)}{h(\sqrt{9-4(x+h)} + \sqrt{9-4x})}$$
$$= \lim_{h\to 0}\frac{-4h}{h(\sqrt{9-4(x+h)} + \sqrt{9-4x})} = \frac{-4}{2\sqrt{9-4x}} = \frac{-2}{\sqrt{9-4x}}$$

(b)
$$\frac{dy}{dx} = \lim_{h\to 0}\frac{\dfrac{x+h}{x+h+1} - \dfrac{x}{x+1}}{h} = \lim_{h\to 0}\frac{(x+h)(x+1) - x(x+h+1)}{h(x+h+1)(x+1)}$$
$$= \lim_{h\to 0}\frac{h}{h(x+h+1)(x+1)} = \frac{1}{(x+1)^2}$$

5. Set $f'(x)=0$: $f'(x)=6(2)(2x+7)^5(x-2)^5+5(2x+7)^6(x-2)^4=0$, so $2x+7=0$ or $x-2=0$ or, factoring out $(2x+7)^5(x-2)^4$, $12(x-2)+5(2x+7)=0$. This reduces to $x=-7/2$, $x=2$, or $22x+11=0$, so the tangent line is horizontal at $x=-7/2,2,-1/2$.

6. Set $f'(x)=0$: $f'(x)=\dfrac{4(x^2+2x)(x-3)^3-(2x+2)(x-3)^4}{(x^2+2x)^2}$, and a fraction can equal zero only if its numerator equals zero. So either $x-3=0$ or, after factoring out $(x-3)^3$, $4(x^2+2x)-(2x+2)(x-3)=0$, $2x^2+12x+6=0$, whose roots are (by the quadratic formula) $x=\dfrac{-6\pm\sqrt{36-4\cdot 3}}{2}=-3\pm\sqrt{6}$. So the tangent line is horizontal at $x=3,-3\pm\sqrt{6}$.

7. Set $f'(x)=\dfrac{3}{2\sqrt{3x+1}}(x-1)^2+2\sqrt{3x+1}(x-1)=0$. If $x=1$ then $y'=0$. If $x\neq 1$ then divide out $x-1$ and multiply through by $2\sqrt{3x+1}$ (at points where f is differentiable we must have $\sqrt{3x+1}\neq 0$) to obtain $3(x-1)+4(3x+1)=0$, or $15x+1=0$. So the tangent line is horizontal at $x=1,-1/15$.

8. $f'(x)=3\left(\dfrac{3x+1}{x^2}\right)^2\dfrac{d}{dx}\dfrac{3x+1}{x^2}=3\left(\dfrac{3x+1}{x^2}\right)^2\dfrac{x^2(3)-(3x+1)(2x)}{x^4}$

$=-3\left(\dfrac{3x+1}{x^2}\right)^2\dfrac{3x^2+2x}{x^4}=0$. If $f'(x)=0$

then $(3x+1)^2(3x^2+2x)=0$. The tangent line is horizontal at $x=-1/3,-2/3$ ($x=0$ is ruled out from the definition of f).

9. **(a)** $x=-2,-1,1,3$
(b) $(-\infty,-2)$, $(-1,1)$, $(3,+\infty)$
(c) $(-2,-1)$, $(1,3)$
(d) $g''(x)=f''(x)\sin x+2f'(x)\cos x-f(x)\sin x$; $g''(0)=2f'(0)\cos 0=2(2)(1)=4$

10. **(a)** $f'(1)g(1)+f(1)g'(1)=3(-2)+1(-1)=-7$
(b) $\dfrac{g(1)f'(1)-f(1)g'(1)}{g(1)^2}=\dfrac{-2(3)-1(-1)}{(-2)^2}=-\dfrac{5}{4}$
(c) $\dfrac{1}{2\sqrt{f(1)}}f'(1)=\dfrac{1}{2\sqrt{1}}3=\dfrac{3}{2}$
(d) 0 (because $f(1)g'(1)$ is constant)

11. The equation of such a line has the form $y=mx$. The points (x_0,y_0) which lie on both the line and the parabola and for which the slopes of both curves are equal satisfy $y_0=mx_0=x_0^3-9x_0^2-16x_0$, so that $m=x_0^2-9x_0-16$. By differentiating, the slope is also given by $m=3x_0^2-18x_0-16$. Equating, we have $x_0^2-9x_0-16=3x_0^2-18x_0-16$, or $2x_0^2-9x_0=0$. The root $x_0=0$ corresponds to $m=-16, y_0=0$ and the root $x_0=9/2$ corresponds to $m=-145/4, y_0=-1305/8$. So the line $y=-16x$ is tangent to the curve at the point $(0,0)$, and the line $y=-145x/4$ is tangent to the curve at the point $(9/2,-1305/8)$.

12. The slope of the line $x+4y=10$ is $m_1=-1/4$, so we set the negative reciprocal

$4=m_2=\dfrac{d}{dx}(2x^3-x^2)=6x^2-2x$ and obtain $6x^2-2x-4=0$ with roots

$x=\dfrac{1\pm\sqrt{1+24}}{6}=1,-2/3$.

13. The line $y - x = 2$ has slope $m_1 = 1$ so we set $m_2 = \dfrac{d}{dx}(3x - \tan x) = 3 - \sec^2 x = 1$, or $\sec^2 x = 2$, $\sec x = \pm\sqrt{2}$ so $x = n\pi \pm \pi/4$ where $n = 0, \pm 1, \pm 2, \ldots$.

14. $f(x)$ is continuous and differentiable at any $x \neq 1$, so we consider $x = 1$.

(a) $\lim_{x\to 1^-}(x^2 - 1) = \lim_{x\to 1^+} k(x-1) = 0 = f(1)$, so any value of k gives continuity at $x = 1$.

(b) $\lim_{x\to 1^-} f'(x) = \lim_{x\to 1^-} 2x = 2$, and $\lim_{x\to 1^+} f'(x) = \lim_{x\to 1^+} k = k$, so only if $k = 2$ is $f(x)$ differentiable at $x = 1$.

15. The slope of the tangent line is the derivative

$y' = 2x\Big|_{x=\frac{1}{2}(a+b)} = a + b$. The slope of the secant is $\dfrac{a^2 - b^2}{a - b} = a + b$, so they are equal.

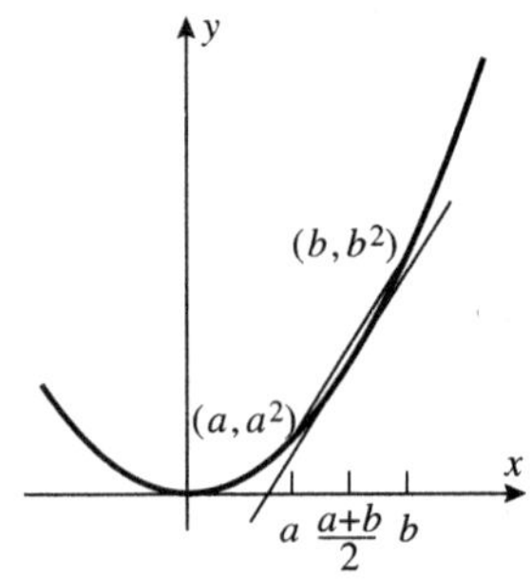

16. To average 60 mi/h one would have to complete the trip in two hours. At 50 mi/h, 100 miles are completed after two hours. Thus time is up, and the speed for the remaining 20 miles would have to be infinite.

17. **(a)** $\Delta x = 1.5 - 2 = -0.5$; $dy = \dfrac{-1}{(x-1)^2}\Delta x = \dfrac{-1}{(2-1)^2}(-0.5) = 0.5$; and

$$\Delta y = \frac{1}{(1.5-1)} - \frac{1}{(2-1)} = 2 - 1 = 1.$$

(b) $\Delta x = 0 - (-\pi/4) = \pi/4$; $dy = \left(\sec^2(-\pi/4)\right)(\pi/4) = \pi/2$; and $\Delta y = \tan 0 - \tan(-\pi/4) = 1$.

(c) $\Delta x = 3 - 0 = 3$; $dy = \dfrac{-x}{\sqrt{25 - x^2}} = \dfrac{-0}{\sqrt{25 - (0)^2}}(3) = 0$; and

$$\Delta y = \sqrt{25 - 3^2} - \sqrt{25 - 0^2} = 4 - 5 = -1.$$

18. **(a)** $\dfrac{4^3 - 2^3}{4 - 2} = \dfrac{56}{2} = 28$ **(b)** $(dV/d\ell)|_{\ell=5} = 3\ell^2\big|_{\ell=5} = 3(5)^2 = 75$

19. **(a)** $\dfrac{dW}{dt} = 200(t - 15)$; at $t = 5$, $\dfrac{dW}{dt} = -2000$; the water is running out at the rate of 2000 gal/min.

(b) $\dfrac{W(5) - W(0)}{5 - 0} = \dfrac{10000 - 22500}{5} = -2500$; the average rate of flow out is 2500 gal/min.

20. $\cot 46^\circ = \cot\dfrac{46\pi}{180}$; let $x_0 = \dfrac{\pi}{4}$ and $x = \dfrac{46\pi}{180}$. Then

$\cot 46^\circ = \cot x \approx \cot\dfrac{\pi}{4} - \left(\csc^2\dfrac{\pi}{4}\right)\left(x - \dfrac{\pi}{4}\right) = 1 - 2\left(\dfrac{46\pi}{180} - \dfrac{\pi}{4}\right) = 0.9651$;

with a calculator, $\cot 46^\circ = 0.9657$.

21. **(a)** $h = 115\tan\phi$, $dh = 115\sec^2\phi\, d\phi$; with $\phi = 51^\circ = \dfrac{51}{180}\pi$ radians and

$d\phi = \pm 0.5^\circ = \pm 0.5\left(\dfrac{\pi}{180}\right)$ radians, $h \pm dh = 115(1.2349) \pm 2.5340 = 142.0135 \pm 2.5340$, so the height lies between 139.48 m and 144.55 m.

(b) If $|dh| \le 5$ then $|d\phi| \le \dfrac{5}{115}\cos^2\dfrac{51}{180}\pi \approx 0.017$ radians, or $|d\phi| \le 0.98°$.

22. **(a)** $\dfrac{dT}{dL} = \dfrac{2}{\sqrt{g}}\dfrac{1}{2\sqrt{L}} = \dfrac{1}{\sqrt{gL}}$ **(b)** s/m

(c) Since $\dfrac{dT}{dL} > 0$ an increase in L gives an increase in T, which is the period. To speed up a clock, decrease the period; to decrease T, decrease L.

(d) $\dfrac{dT}{dg} = -\dfrac{\sqrt{L}}{g^{3/2}} < 0$; a decrease in g will increase T and the clock runs slower

(e) $\dfrac{dT}{dg} = 2\sqrt{L}\left(\dfrac{-1}{2}\right)g^{-3/2} = -\dfrac{\sqrt{L}}{g^{3/2}}$ **(f)** s^3/m

23. **(a)** $f'(x) = 2x, f'(1.8) = 3.6$
(b) $f'(x) = (x^2 - 4x)/(x-2)^2, f'(3.5) \approx -0.777778$

24. **(a)** $f'(x) = 3x^2 - 2x, f'(2.3) = 11.27$
(b) $f'(x) = (1-x^2)/(x^2+1)^2, f'(-0.5) = 0.48$

25. $f'(2) \approx 2.772589; f'(2) = 4\ln 2$

26. $f'(2) = 0.312141; f'(2) = 2^{\sin 2}(\cos 2\ln 2 + \sin 2/2)$

27. $v_{\text{inst}} = \lim\limits_{h\to 0}\dfrac{3(h+1)^{2.5} + 580h - 3}{10h} = 58 + \dfrac{1}{10}\dfrac{d}{dx}3x^{2.5}\Big|_{x=1} = 58 + \dfrac{1}{10}(2.5)(3)(1)^{1.5} = 58.75$ ft/s

28. 164 ft/s

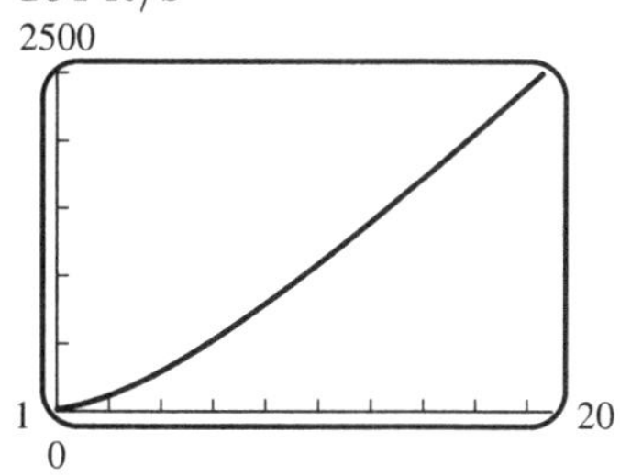

29. Solve $3x^2 - \cos x = 0$ to get $x = \pm 0.535428$.

30. When $x^4 - x - 1 > 0$, $f(x) = x^4 - 2x - 1$; when $x^4 - x - 1 < 0$, $f(x) = -x^4 + 1$, and f is differentiable in both cases. The roots of $x^4 - x - 1 = 0$ are $x_1 = -0.724492$, $x_2 = 1.220744$. So $x^4 - x - 1 > 0$ on $(-\infty, x_1)$ and $(x_2, +\infty)$, and $x^4 - x - 1 < 0$ on (x_1, x_2). Then $\lim\limits_{x\to x_1^-} f'(x) = \lim\limits_{x\to x_1^-}(4x^3 - 2) = 4x_1^3 - 2$ and $\lim\limits_{x\to x_1^+} f'(x) = \lim\limits_{x\to x_1^+} -4x^3 = -4x_1^3$ which is not equal to $4x_1^3 - 2$, so f is not differentiable at $x = x_1$; similarly f is not differentiable at $x = x_2$.

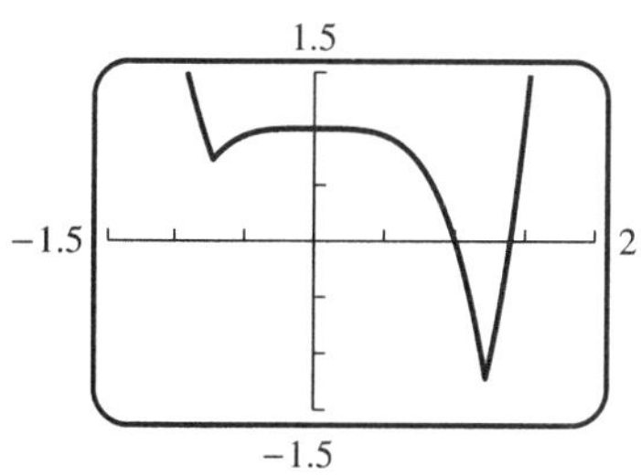

31. **(a)** $f'(x) = 5x^4$ **(b)** $f'(x) = -1/x^2$ **(c)** $f'(x) = -1/2x^{3/2}$
(d) $f'(x) = -3/(x-1)^2$ **(e)** $f'(x) = 3x/\sqrt{3x^2+5}$ **(f)** $f'(x) = 3\cos 3x$

32. $f'(x) = 2x\sin x + x^2\cos x$

33. $f'(x) = \dfrac{1 - 2\sqrt{x}\sin 2x}{2\sqrt{x}}$

34. $f'(x) = \dfrac{6x^2 + 8x - 17}{(3x+2)^2}$

35. $f'(x) = \dfrac{(1+x^2)\sec^2 x - 2x\tan x}{(1+x^2)^2}$

36. $f'(x) = \dfrac{x^2\cos\sqrt{x} - 2x^{3/2}\sin\sqrt{x}}{2x^{7/2}}$

37. $f'(x) = \dfrac{-2x^5\sin x - 2x^4\cos x + 4x^4 + 6x^2\sin x + 6x - 3x\cos x - 4x\sin x + 4\cos x - 8}{2x^2\sqrt{x^4 - 3} + 2(2-\cos x)^2}$

38. Differentiating, $\dfrac{2}{3}x^{-1/3} - \dfrac{2}{3}y^{-1/3}y' - y' = 0$. At $x = 1$ and $y = -1$, $y' = 2$. The tangent line is $y + 1 = 2(x - 1)$.

39. Differentiating, $(xy' + y)\cos xy = y'$. With $x = \pi/2$ and $y = 1$ this becomes $y' = 0$, so the equation of the tangent line is $y - 1 = 0(x - \pi/2)$ or $y = 1$.

40. Find $\left.\dfrac{d\theta}{dt}\right|_{\substack{x=1\\y=1}}$ given $\dfrac{dz}{dt} = a$ and $\dfrac{dy}{dt} = -b$. From the figure $\sin\theta = y/z$; when $x = y = 1$, $z = \sqrt{2}$. So $\theta = \sin^{-1}(y/z)$ and

$$\frac{d\theta}{dt} = \frac{1}{\sqrt{1 - y^2/z^2}}\left(\frac{1}{z}\frac{dy}{dt} - \frac{y}{z^2}\frac{dz}{dt}\right) = -b - \frac{a}{\sqrt{2}}$$

when $x = y = 1$.

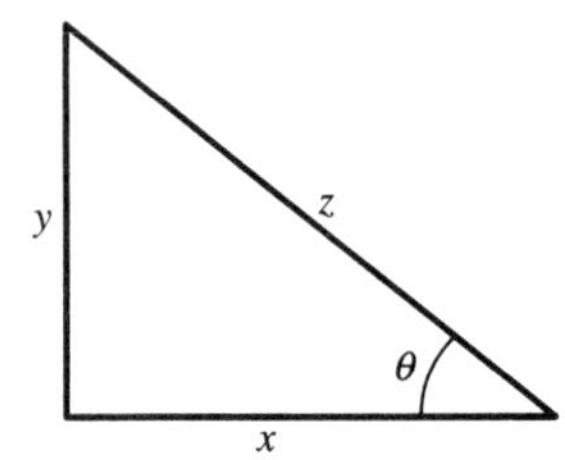

CHAPTER 3 HORIZON MODULE

1. $x_1 = l_1\cos\theta_1, x_2 = l_2\cos(\theta_1+\theta_2)$, so $x = x_1 + x_2 = l_1\cos\theta_1 + l_2\cos(\theta_1+\theta_2)$ (see Figure 3 in text); similarly $y_1 = l_1\sin\theta_1 + l_2\sin(\theta_1+\theta_2)$.

2. Fix θ_1 for the moment and let θ_2 vary; then the distance r from (x, y) to the origin (see Figure 3 in text) is at most $l_1 + l_2$ and at least $l_1 - l_2$ if $l_1 \ge l_2$ and $l_2 - l_1$ otherwise. For any fixed θ_2 let θ_1 vary and the point traces out a circle of radius r.

(a) $\{(x, y) : 0 \le x^2 + y^2 \le 2l_1\}$
(b) $\{(x, y) : l_1 - l_2 \le x^2 + y^2 \le l_1 + l_2\}$
(c) $\{(x, y) : l_2 - l_1 \le x^2 + y^2 \le l_1 + l_2\}$

3.
$$\begin{aligned}(x, y) &= (l_1\cos\theta + l_2\cos(\theta_1+\theta_2), l_1\sin\theta_1 + l_2\sin(\theta_1+\theta_2))\\ &= (\cos(\pi/4) + 3\cos(5\pi/12), \sin(\pi/4) + 3\sin(5\pi/12)) = \left(\frac{\sqrt{2}+3\sqrt{6}}{4}, \frac{7\sqrt{2}+3\sqrt{6}}{4}\right)\end{aligned}$$

4. $x = (1)\cos 2t + (1)\cos(2t + 3t) = \cos 2t + \cos 5t$,
$y = (1)\sin 2t + (1)\sin(2t + 3t) = \sin 2t + \sin 5t$

5.

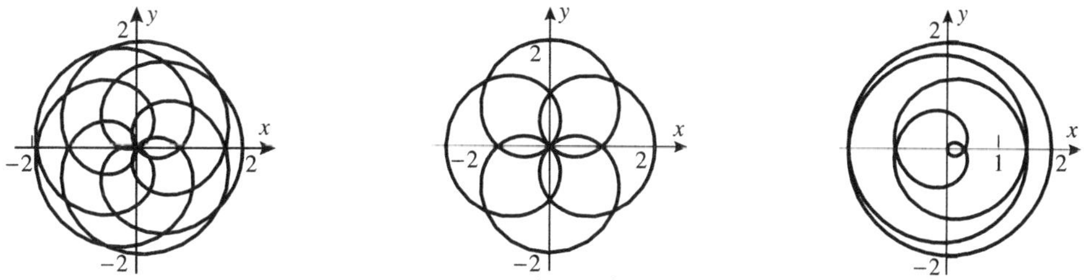

$\omega_1 = 3, \omega_2 = 5$ $\quad$ $\omega_1 = 1, \omega_2 = 4$ $\quad$ $\omega_1 = 4, \omega_2 = 1$

6. $x = 2\cos t$, $y = 2\sin t$, a circle of radius 2

7. **(a)** $9 = [3\sin(\theta_1+\theta_2)]^2 + [3\cos(\theta_1+\theta_2)]^2 = [5 - 3\sin\theta_1]^2 + [3 - 3\cos\theta_1]^2$
$= 25 - 30\sin\theta_1 + 9\sin^2\theta_1 + 9 - 18\cos\theta_1 + 9\cos^2\theta_1 = 43 - 30\sin\theta_1 - 18\cos\theta_1$,
so $15\sin\theta_1 + 9\cos\theta_1 = 17$

(b) $1 = \sin^2\theta_1 + \cos^2\theta_2 = \left(\dfrac{17 - 9\cos\theta_1}{15}\right)^2 + \cos\theta_1$, or $306\cos^2\theta_1 - 306\cos\theta_1 = -64$

(c) $\cos\theta_1 = \left(153 \pm \sqrt{(153)^2 - 4(153)(32)}\right)/306 = \dfrac{1}{2} \pm \dfrac{5\sqrt{17}}{102}$

(e) If $\theta_1 = 0.792436$ rad, then $\theta_2 = 0.475882$ rad $\approx 27.2660°$;
if $\theta_1 = 1.26832$ rad, then $\theta_2 = -0.475882$ rad $\approx -27.2660°$.

8. $$\frac{dx}{dt} = -3\sin\theta_1\frac{d\theta_1}{dt} - (3\sin(\theta_1+\theta_2))\left(\frac{d\theta_1}{dt} + \frac{d\theta_2}{dt}\right)$$
$$= -3\frac{d\theta_1}{dt}(\sin\theta_1 + \sin(\theta_1+\theta_2)) - 3(\sin(\theta_1+\theta_2))\frac{d\theta_2}{dt}$$
$$= -y\frac{d\theta_1}{dt} - 3(\sin(\theta_1+\theta_2))\frac{d\theta_2}{dt};$$

similarly $\dfrac{dy}{dt} = x\dfrac{d\theta_1}{dt} + 3(\cos(\theta_1+\theta_2))\dfrac{d\theta_2}{dt}$. Now set $\dfrac{dx}{dt} = 0$, $\dfrac{dy}{dt} = 1$.

9. **(a)** $x = 3\cos(\pi/3) + 3\cos(-\pi/3) = 6\dfrac{1}{2} = 3$ and $y = 3\sin(\pi/3) - 3\sin(\pi/3) = 0$; equations (4) become $3\sin(\pi/3)\dfrac{d\theta_2}{dt} = 0$, $3\dfrac{d\theta_1}{dt} + 3\cos(\pi/3)\dfrac{d\theta_2}{dt} = 1$ with solution $d\theta_2/dt = 0$, $d\theta_1/dt = 1/3$.

(b) $x = -3$, $y = 3$, so $-3\dfrac{d\theta_1}{dt} = 0$ and $-3\dfrac{d\theta_1}{dt} - 3\dfrac{d\theta_2}{dt} = 1$, with solution $d\theta_1/dt = 0$, $d\theta_2/dt = -1/3$.

CHAPTER 4

The Derivative in Graphing and Applications

EXERCISE SET 4.1

1. (a) $f' > 0$ and $f'' > 0$

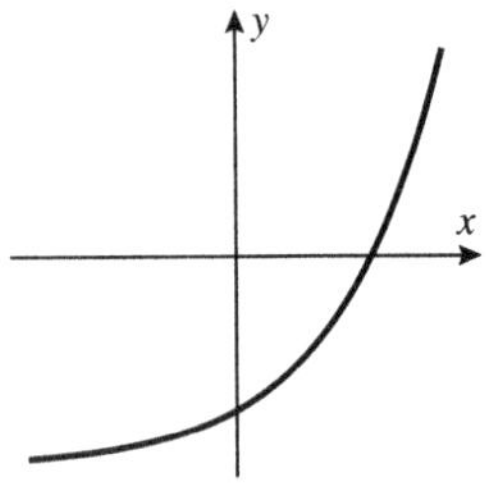

(b) $f' > 0$ and $f'' < 0$

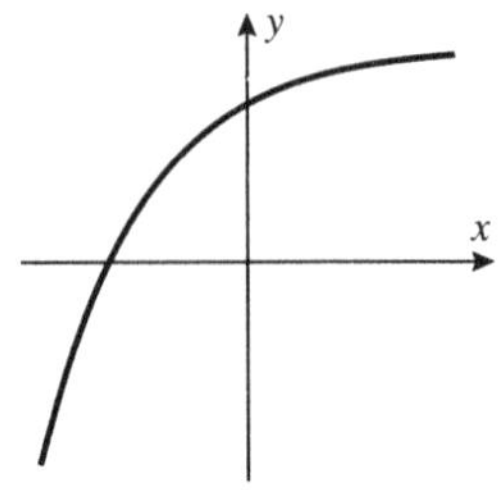

(c) $f' < 0$ and $f'' > 0$

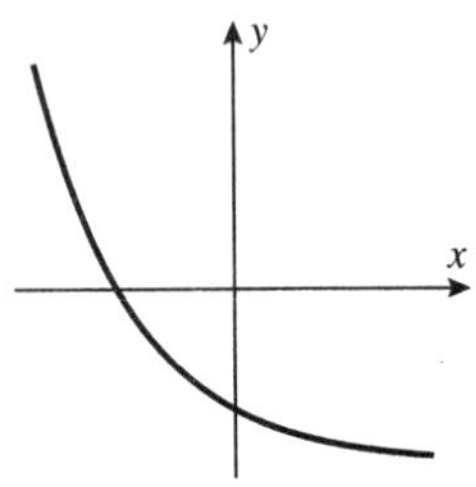

(d) $f' < 0$ and $f'' < 0$

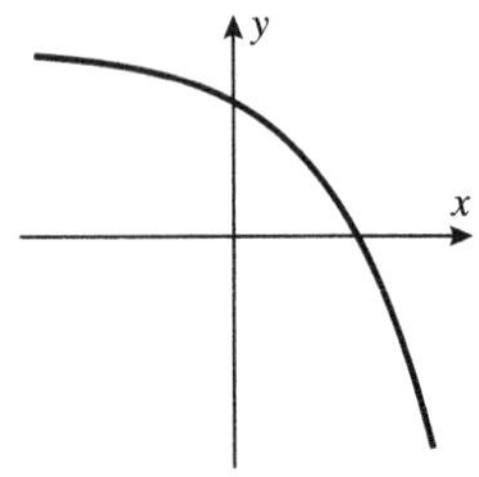

2. (a)

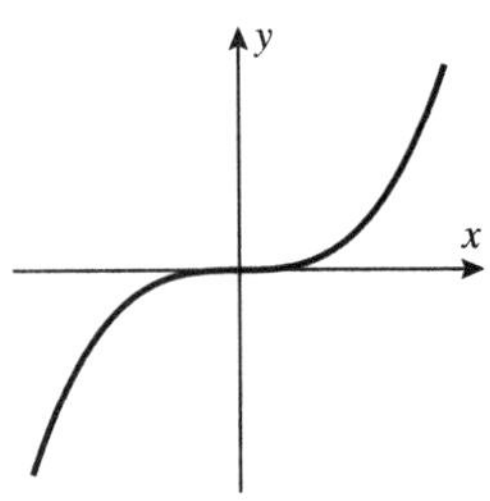

(b)

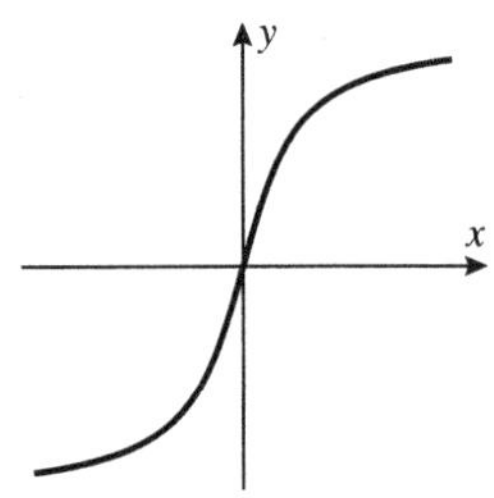

(c)

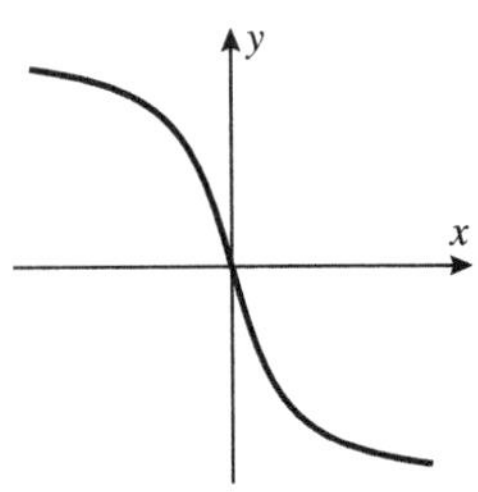

(d)

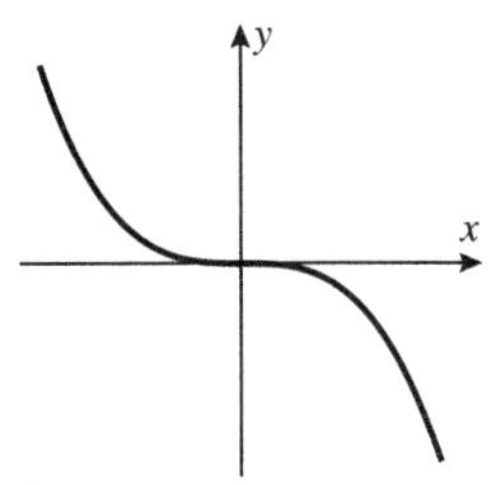

3. A: $dy/dx < 0$, $d^2y/dx^2 > 0$
B: $dy/dx > 0$, $d^2y/dx^2 < 0$
C: $dy/dx < 0$, $d^2y/dx^2 < 0$

4. A: $dy/dx < 0$, $d^2y/dx^2 < 0$
B: $dy/dx < 0$, $d^2y/dx^2 > 0$
C: $dy/dx > 0$, $d^2y/dx^2 < 0$

5. An inflection point occurs when f'' changes sign: at $x = -1, 0, 1$ and 2.

6. (a) $f(0) < f(1)$ since $f' > 0$ on $(0, 1)$.
(b) $f(1) > f(2)$ since $f' < 0$ on $(1, 2)$.
(c) $f'(0) > 0$ by inspection.
(d) $f'(1) = 0$ by inspection.
(e) $f''(0) < 0$ since f' is decreasing there.
(f) $f''(2) = 0$ since f' has a minimum there.

7. (a) $[4, 6]$
(b) $[1, 4]$ and $[6, 7]$
(c) $(1, 2)$ and $(3, 5)$
(d) $(2, 3)$ and $(5, 7)$
(e) $x = 2, 3, 5$

8.

	$(1,2)$	$(2,3)$	$(3,4)$	$(4,5)$	$(5,6)$	$(6,7)$
f'	$-$	$-$	$-$	$+$	$+$	$-$
f''	$+$	$-$	$+$	$+$	$-$	$-$

9. **(a)** f is increasing on $[1,3]$ **(b)** f is decreasing on $(-\infty,1], [3,+\infty]$
(c) f is concave up on $(-\infty,2), (4,+\infty)$ **(d)** f is concave down on $(2,4)$
(e) points of inflection at $x=2,4$

10. **(a)** f is increasing on $(-\infty,+\infty)$ **(b)** f is nowhere decreasing
(c) f is concave up on $(-\infty,1), (3,+\infty)$ **(d)** f is concave down on $(1,3)$
(e) f has points of inflection at $x=1,3$

11. $f'(x)=2x-5$
$f''(x)=2$
(a) $[5/2,+\infty)$ **(b)** $(-\infty,5/2]$
(c) $(-\infty,+\infty)$ **(d)** none
(e) none

12. $f'(x)=-2(x+3/2)$
$f''(x)=-2$
(a) $(-\infty,-3/2]$ **(b)** $[-3/2,+\infty)$
(c) none **(d)** $(-\infty,+\infty)$
(e) none

13. $f'(x)=3(x+2)^2$
$f''(x)=6(x+2)$
(a) $(-\infty,+\infty)$ **(b)** none
(c) $(-2,+\infty)$ **(d)** $(-\infty,-2)$
(e) -2

14. $f'(x)=3(4-x^2)$
$f''(x)=-6x$
(a) $[-2,2]$ **(b)** $(-\infty,-2], [2,+\infty)$
(c) $(-\infty,0)$ **(d)** $(0,+\infty)$
(e) 0

15. $f'(x)=12x^2(x-1)$
$f''(x)=36x(x-2/3)$
(a) $[1,+\infty)$ **(b)** $(-\infty,1]$
(c) $(-\infty,0)$, $(2/3,+\infty)$ **(d)** $(0,2/3)$
(e) $0, 2/3$

16. $f'(x)=4x(x^2-4)$
$f''(x)=12(x^2-4/3)$
(a) $[-2,0]$, $[2,+\infty)$ **(b)** $(-\infty,-2]$, $[0,2]$
(c) $(-\infty,-2/\sqrt{3})$, $(2/\sqrt{3},+\infty)$ **(d)** $(-2/\sqrt{3},2/\sqrt{3})$
(e) $-2/\sqrt{3}$, $2/\sqrt{3}$

17. $f'(x)=\dfrac{4x}{(x^2+2)^2}$ $\quad f''(x)=-4\dfrac{3x^2-2}{(x^2+2)^3}$

(a) $[0,+\infty)$ **(b)** $(-\infty,0]$ **(c)** $(-\sqrt{2/3},+\sqrt{2/3})$
(d) $(-\infty,-\sqrt{2/3})$, $(+\sqrt{2/3},+\infty)$ **(e)** $-\sqrt{2/3}$, $\sqrt{2/3}$

18. $f'(x)=\dfrac{2-x^2}{(x^2+2)^2}$ $\quad f''(x)=\dfrac{2x(x^2-6)}{(x^2+2)^3}$

(a) $[-\sqrt{2},\sqrt{2}]$ **(b)** $(-\infty,-\sqrt{2}]$, $[\sqrt{2},+\infty)$ **(c)** $(-\sqrt{6},0)$, $(\sqrt{6},+\infty)$
(d) $(-\infty,-\sqrt{6})$, $(0,\sqrt{6})$ **(e)** $-\sqrt{6},0,\sqrt{6}$

19. $f'(x)=\frac{1}{3}(x+2)^{-2/3}$
$f''(x)=-\frac{2}{9}(x+2)^{-5/3}$
(a) $(-\infty,+\infty)$ **(b)** none
(c) $(-\infty,-2)$ **(d)** $(-2,+\infty)$
(e) -2

20. $f'(x) = \frac{2}{3}x^{-1/3}$
$f''(x) = -\frac{2}{9}x^{-4/3}$

(a) $[0, +\infty)$ **(b)** $(-\infty, 0]$
(c) none **(d)** $(-\infty, 0)$, $(0, +\infty)$
(e) none

21. $f'(x) = \dfrac{4(x+1)}{3x^{2/3}}$
$f''(x) = \dfrac{4(x-2)}{9x^{5/3}}$

(a) $[-1, +\infty)$ **(b)** $(-\infty, -1]$
(c) $(-\infty, 0)$, $(2, +\infty)$ **(d)** $(0, 2)$
(e) $0, 2$

22. $f'(x) = \dfrac{4(x-1/4)}{3x^{2/3}}$
$f''(x) = \dfrac{4(x+1/2)}{9x^{5/3}}$

(a) $[1/4, +\infty)$ **(b)** $(-\infty, 1/4]$
(c) $(-\infty, -1/2)$, $(0, +\infty)$ **(d)** $(-1/2, 0)$
(e) $-1/2, 0$

23. $f'(x) = -\sin x$
$f''(x) = -\cos x$
(a) $[\pi, 2\pi]$ **(b)** $[0, \pi]$
(c) $(\pi/2, 3\pi/2)$ **(d)** $(0, \pi/2)$, $(3\pi/2, 2\pi)$
(e) $\pi/2$, $3\pi/2$

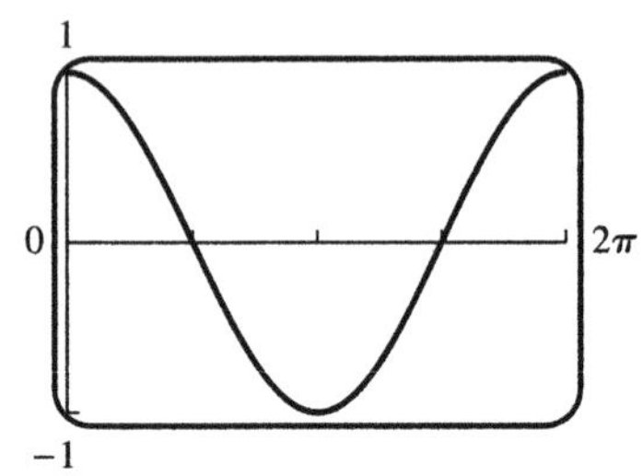

24. $f'(x) = 2\sin 4x$
$f''(x) = 8\cos 4x$
(a) $(0, \pi/4]$, $[\pi/2, 3\pi/4]$
(b) $[\pi/4, \pi/2]$, $[3\pi/4, \pi]$
(c) $(0, \pi/8)$, $(3\pi/8, 5\pi/8)$, $(7\pi/8, \pi)$
(d) $(\pi/8, 3\pi/8)$, $(5\pi/8, 7\pi/8)$
(e) $\pi/8$, $3\pi/8$, $5\pi/8$, $7\pi/8$

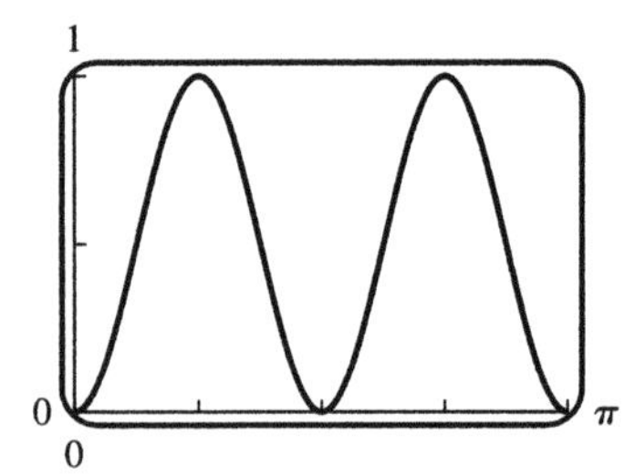

25. $f'(x) = \sec^2 x$
$f''(x) = 2\sec^2 x \tan x$
(a) $(-\pi/2, \pi/2)$ **(b)** none
(c) $(0, \pi/2)$ **(d)** $(-\pi/2, 0)$
(e) 0

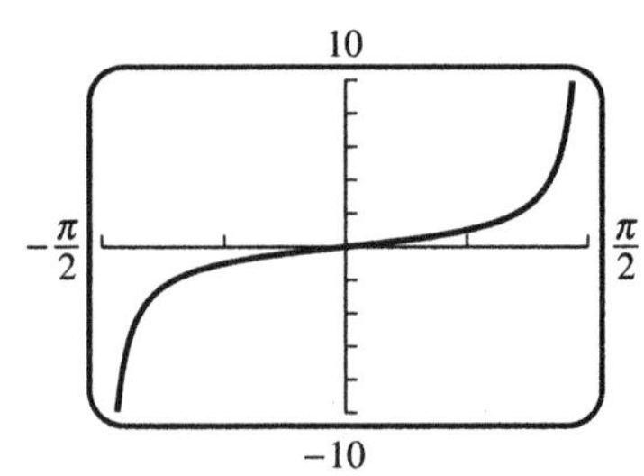

26. $f'(x) = 2 - \csc^2 x$
$f''(x) = 2\csc^2 x \cot x = 2\dfrac{\cos x}{\sin^3 x}$
(a) $[\pi/4, 3\pi/4]$ **(b)** $(0, \pi/4], [3\pi/4, \pi)$
(c) $(0, \pi/2)$ **(d)** $(\pi/2, \pi)$
(e) $\pi/2$

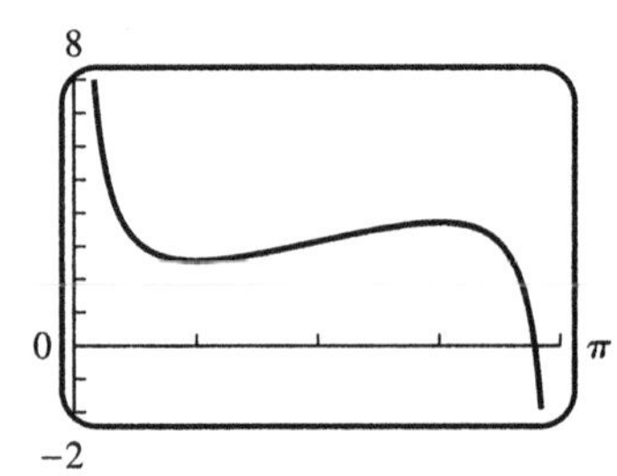

27. $f'(x) = \cos 2x$
$f''(x) = -2\sin 2x$

(a) $[0, \pi/4]$, $[3\pi/4, \pi]$ **(b)** $[\pi/4, 3\pi/4]$
(c) $(\pi/2, \pi)$ **(d)** $(0, \pi/2)$
(e) $\pi/2$

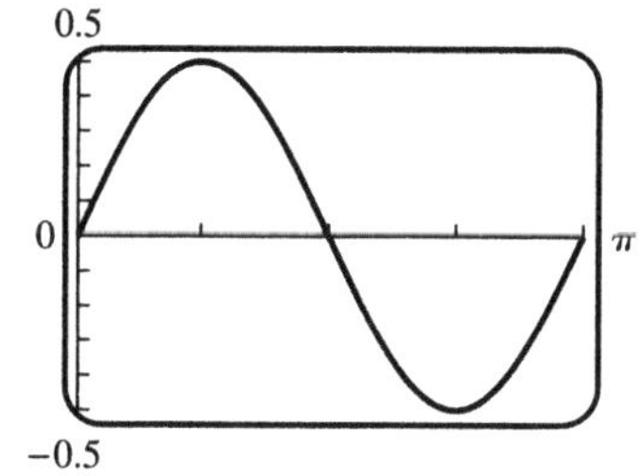

28. $f'(x) = -2\cos x \sin x - 2\cos x = -2\cos x(1 + \sin x)$
$f''(x) = 2\sin x\,(\sin x + 1) - 2\cos^2 x = 2\sin x(\sin x + 1) - 2 + 2\sin^2 x = 4(1 + \sin x)(\sin x - 1/2)$
Note: $1 + \sin x \geq 0$

(a) $[\pi/2, 3\pi/2]$ **(b)** $[0, \pi/2]$, $[3\pi/2, 2\pi]$
(c) $(\pi/6, 5\pi/6)$ **(d)** $(0, \pi/6)$, $(5\pi/6, 2\pi)$
(e) $\pi/6$, $5\pi/6$

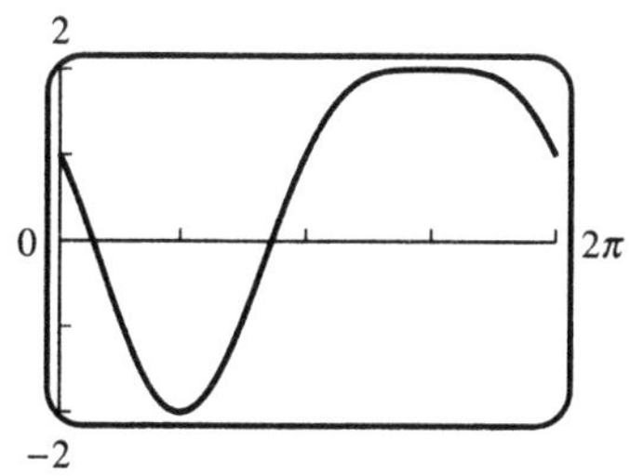

29. **(a)**

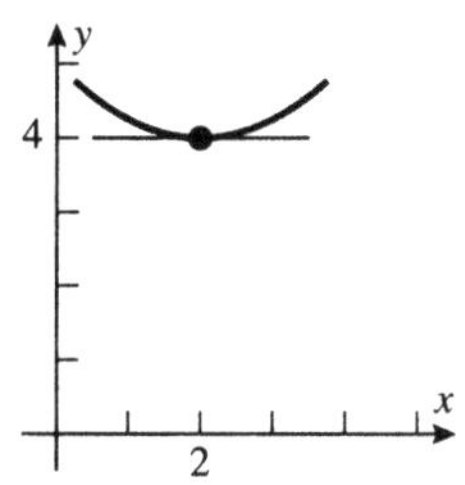

(b)

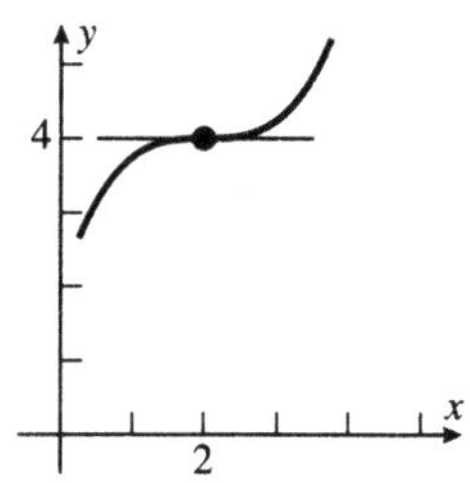

(c)

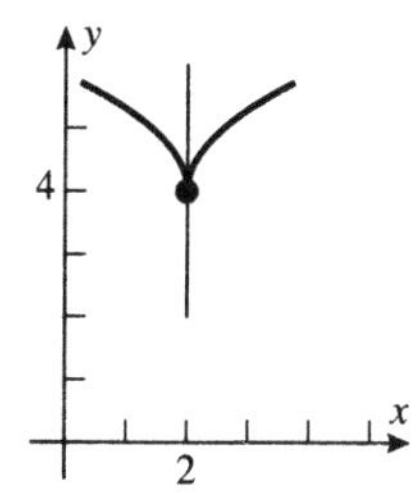

30. **(a)**

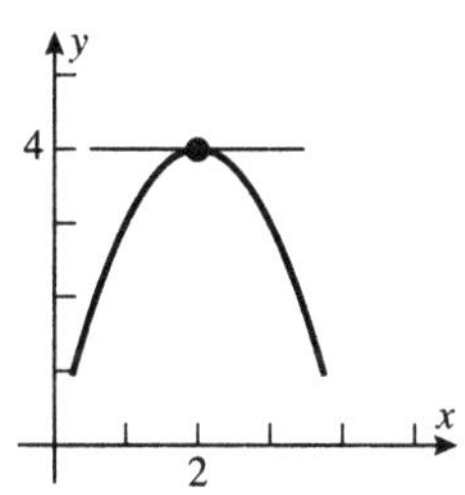

(b)

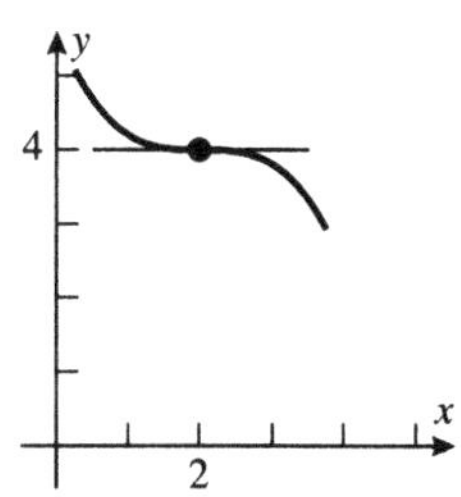

(c)

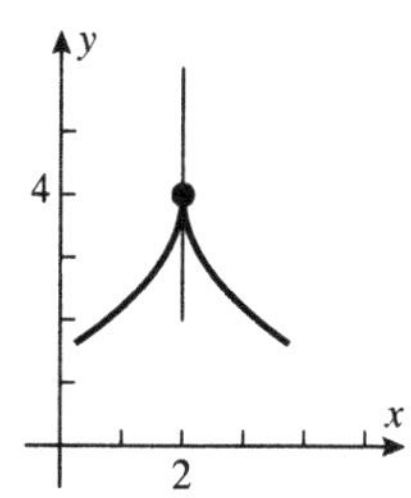

31. **(a)** $f'(x) = 3(x - a)^2$, $f''(x) = 6(x - a)$; inflection point is $(a, 0)$
(b) $f'(x) = 4(x - a)^3$, $f''(x) = 12(x - a)^2$; no inflection points

32. For $n \geq 2$, $f''(x) = n(n-1)(x - a)^{n-2}$; there is a sign change of f'' (point of inflection) at $(a, 0)$ if and only if n is odd. For $n = 1$, $y = x - a$, so there is no point of inflection.

33. $f'(x) = 1/3 - 1/[3(1 + x)^{2/3}]$ so f is increasing on $[0, +\infty)$ thus if $x > 0$, then $f(x) > f(0) = 0$, $1 + x/3 - \sqrt[3]{1 + x} > 0$, $\sqrt[3]{1 + x} < 1 + x/3$.

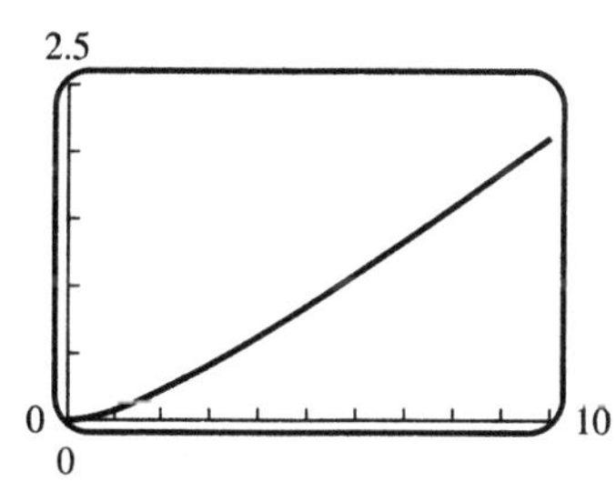

34. $f'(x) = \sec^2 x - 1$ so f is increasing on $[0, \pi/2)$ thus if $0 < x < \pi/2$, then $f(x) > f(0) = 0$, $\tan x - x > 0$, $x < \tan x$.

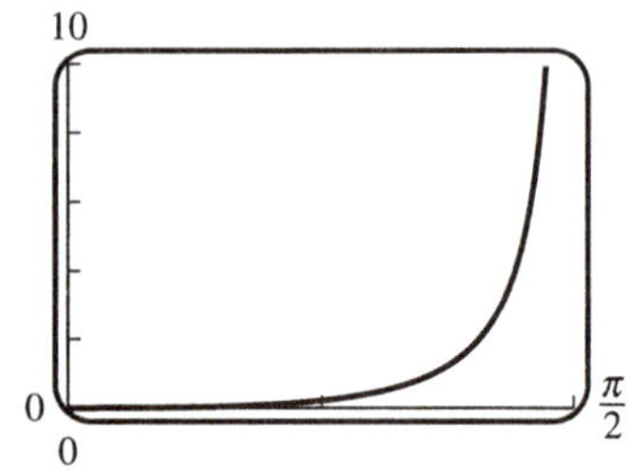

35. $x \geq \sin x$ on $[0, +\infty)$: let $f(x) = x - \sin x$. Then $f(0) = 0$ and $f'(x) = 1 - \cos x \geq 0$, so $f(x)$ is increasing on $[0, +\infty)$.

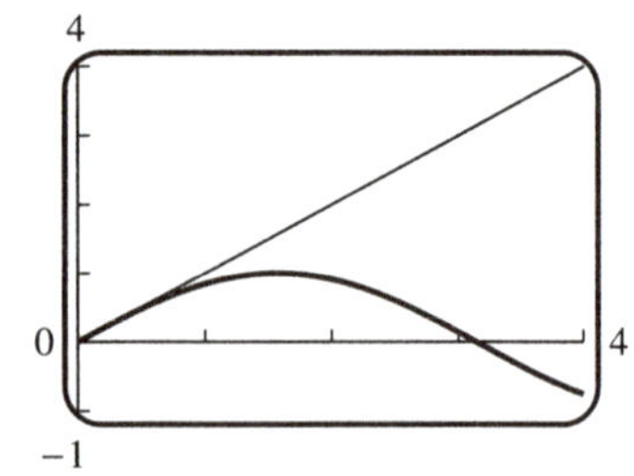

36. Let $f(x) = 1 - x^2/2 - \cos x$ for $x \geq 0$. Then $f(0) = 0$ and $f'(x) = -x + \sin x$. By Exercise 35, $f'(x) \leq 0$ for $x \geq 0$, so $f(x) \leq 0$ for all $x \geq 0$, that is, $\cos x \geq 1 - x^2/2$.

37. Points of inflection at $x = -2, +2$. Concave up on $(-5, -2)$ and $(2, 5)$; concave down on $(-2, 2)$. Increasing on $[-3.5829, 0.2513]$ and $[3.3316, 5]$, and decreasing on $[-5, -3.5829]$ and $[0.2513, 3.3316]$.

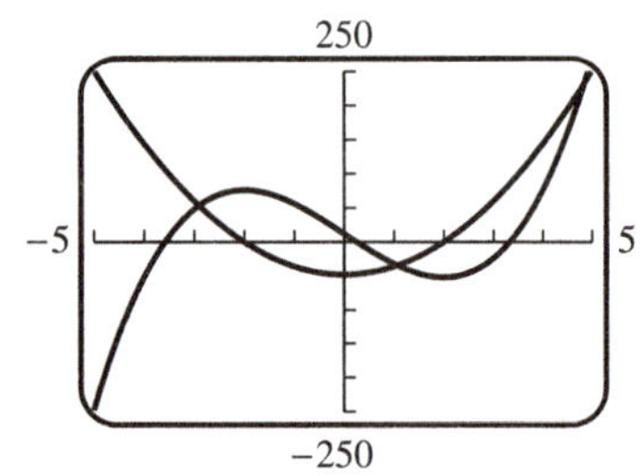

38. Points of inflection at $x = \pm 1/\sqrt{3}$. Concave up on $[-5, -1/\sqrt{3}]$ and $[1/\sqrt{3}, 5]$, and concave down on $[-1/\sqrt{3}, 1/\sqrt{3}]$. Increasing on $[-5, 0]$ and decreasing on $[0, 5]$.

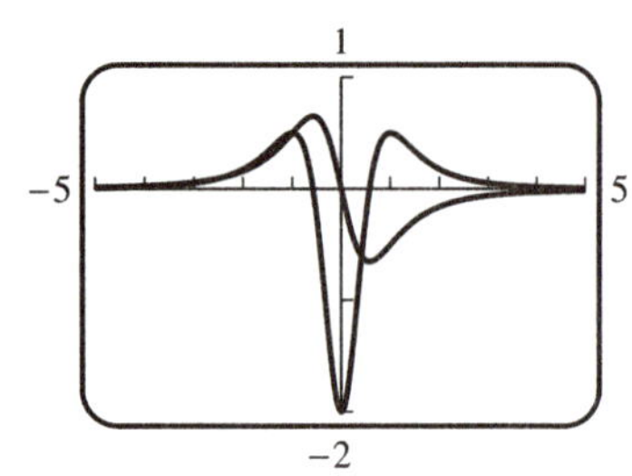

39. $f''(x) = 2\dfrac{90x^3 - 81x^2 - 585x + 397}{(3x^2 - 5x + 8)^3}$. The denominator has complex roots, so is always positive; hence the x-coordinates of the points of inflection of $f(x)$ are the roots of the numerator (if it changes sign). A plot of the numerator over $[-5, 5]$ shows roots lying in $[-3, -2]$, $[0, 1]$, and $[2, 3]$. To six decimal places the roots are $x = -2.464202, 0.662597, 2.701605$.

40. $f''(x) = \dfrac{2x^5 + 5x^3 + 14x^2 + 30x - 7}{(x^2 + 1)^{5/2}}$. Points of inflection will occur when the numerator changes sign, since the denominator is always positive. A plot of $y = 2x^5 + 5x^3 + 14x^2 + 30x - 7$ shows that there is only one root and it lies in $[0, 1]$. To six decimal place the point of inflection is located at $x = 0.210970$.

41. $f(x_1) - f(x_2) = x_1^2 - x_2^2 = (x_1 + x_2)(x_1 - x_2) < 0$ if $x_1 < x_2$ for x_1, x_2 in $[0, +\infty)$, so $f(x_1) < f(x_2)$ and f is thus increasing.

42. $f(x_1) - f(x_2) = \dfrac{1}{x_1} - \dfrac{1}{x_2} = \dfrac{x_2 - x_1}{x_1 x_2} > 0$ if $x_1 < x_2$ for x_1, x_2 in $(0, +\infty)$, so $f(x_1) > f(x_2)$ and thus f is decreasing.

43. **(a)** If $x_1 < x_2$ where x_1 and x_2 are in I, then $f(x_1) < f(x_2)$ and $g(x_1) < g(x_2)$, so $f(x_1) + g(x_1) < f(x_2) + g(x_2)$, $(f+g)(x_1) < (f+g)(x_2)$. Thus $f+g$ is increasing on I.

(b) Case I: If f and g are ≥ 0 on I, and if $x_1 < x_2$ where x_1 and x_2 are in I, then $0 < f(x_1) < f(x_2)$ and $0 < g(x_1) < g(x_2)$, so $f(x_1)g(x_1) < f(x_2)g(x_2)$, $(f \cdot g)(x_1) < (f \cdot g)(x_2)$. Thus $f \cdot g$ is increasing on I.

Case II: If f and g are not necessarily positive on I then no conclusion can be drawn: for example, $f(x) = g(x) = x$ are both increasing on $(-\infty, 0)$, but $(f \cdot g)(x) = x^2$ is decreasing there.

44. **(a)** $f(x) = x$, $g(x) = 2x$ **(b)** $f(x) = x$, $g(x) = x + 6$ **(c)** $f(x) = 2x$, $g(x) = x$

45. **(a)** $f''(x) = 6ax + 2b = 6a\left(x + \dfrac{b}{3a}\right)$, $f''(x) = 0$ when $x = -\dfrac{b}{3a}$. f changes its direction of concavity at $x = -\dfrac{b}{3a}$ so $-\dfrac{b}{3a}$ is an inflection point.

(b) If $f(x) = ax^3 + bx^2 + cx + d$ has three x-intercepts, then it has three roots, say x_1, x_2 and x_3, so we can write $f(x) = a(x - x_1)(x - x_2)(x - x_3) = ax^3 + bx^2 + cx + d$, from which it follows that $b = -a(x_1 + x_2 + x_3)$. Thus $-\dfrac{b}{3a} = \dfrac{1}{3}(x_1 + x_2 + x_3)$, which is the average.

(c) $f(x) = x(x^2 - 3x^2 + 2) = x(x-1)(x-2)$ so the intercepts are 0, 1, and 2 and the average is 1. $f''(x) = 6x - 6 = 6(x-1)$ changes sign at $x = 1$.

46. $f''(x) = 6x + 2b$, so the point of inflection is at $x = -\dfrac{b}{3}$. Thus an increase in b moves the point of inflection to the left.

47. **(a)** Let $x_1 < x_2$ belong to (a, b). If both belong to $(a, c]$ or both belong to $[c, b)$ then we have $f(x_1) < f(x_2)$ by hypothesis. So assume $x_1 < c < x_2$. We know by hypothesis that $f(x_1) < f(c)$, and $f(c) < f(x_2)$. We conclude that $f(x_1) < f(x_2)$.

(b) Use the same argument as in Part (a), but with inequalities reversed.

48. By Theorem 4.1.2, f is increasing on any interval $[(2n-1)\pi, 2(n+1)\pi]$ $(n = 0, \pm 1, \pm 2, \cdots)$, because $f'(x) = 1 + \cos x > 0$ on $((2n-1)\pi, (2n+1)\pi)$. By Exercise 47 (a) we can piece these intervals together to show that $f(x)$ is increasing on $(-\infty, +\infty)$.

49. By Theorem 4.1.2, $f$is decreasing on any interval $[(2n\pi + \pi/2, 2(n+1)\pi + \pi/2]$ $(n = 0, \pm 1, \pm 2, \cdots)$, because $f'(x) = -\sin x + 1 < 0$ on $((2n\pi + \pi/2, 2(n+1)\pi + \pi/2)$. By Exercise 47 (b) we can piece these intervals together to show that $f(x)$ is decreasing on $(-\infty, +\infty)$.

50. By zooming on the graph of $y'(t)$, maximum increase is at $x = -0.577$ and maximum decrease is at $x = 0.577$.

51.

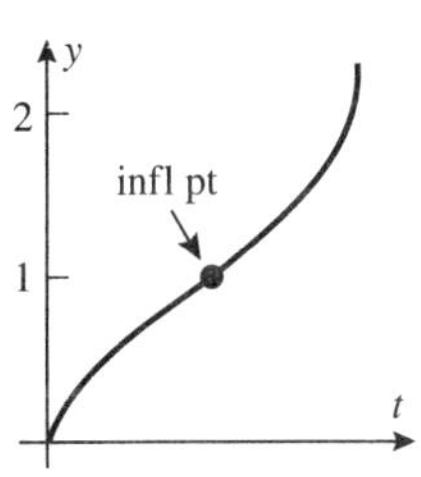

52.

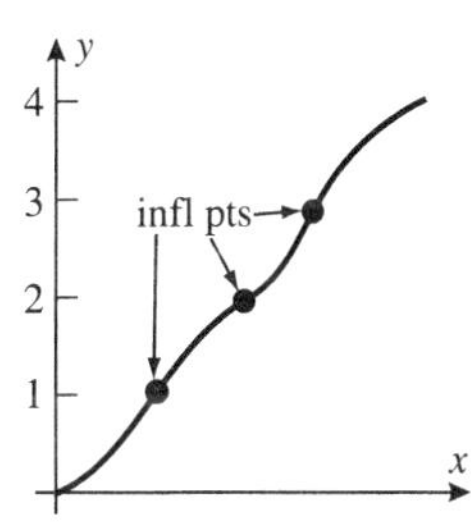

53. **(a)** $g(x)$ has no zeros:
There can be no zero of $g(x)$ on the interval $-\infty < x < 0$ because if there were, say $g(x_0) = 0$ where $x_0 < 0$, then $g'(x)$ would have to be positive between $x = x_0$ and $x = 0$, say $g'(x_1) > 0$ where $x_0 < x_1 < 0$. But then $g'(x)$ cannot be concave up on the interval $(x_1, 0)$, a contradiction.

There can be no zero of $g(x)$ on $0 < x < 4$ because $g(x)$ is concave up for $0 < x < 4$ and thus the graph of $g(x)$, for $0 < x < 4$, must lie above the line $y = -\dfrac{2}{3}x + 2$, which is the tangent line to the curve at $(0, 2)$, and above the line $y = 3(x - 4) + 3 = 3x - 9$ also for $0 < x < 4$ (see figure). The first condition says that $g(x)$ could only be zero for $x > 3$ and the second condition says that $g(x)$ could only be zero for $x < 3$, thus $g(x)$ has no zeros for $0 < x < 4$.

Finally, if $4 < x < +\infty$, $g(x)$ could only have a zero if $g'(x)$ were negative somewhere for $x > 4$, and since $g'(x)$ is decreasing there we would ultimately have $g(x) < -10$, a contradiction.

(b) one, between 0 and 4

(c) We must have $\lim\limits_{x\to+\infty} g'(x) = 0$; if the limit were -5 then $g(x)$ would at some time cross the line $x = -10$; if the limit were 5 then, since g is concave down for $x > 4$ and $g'(4) = 3$, g' must decrease for $x > 4$ and thus the limit would be < 4.

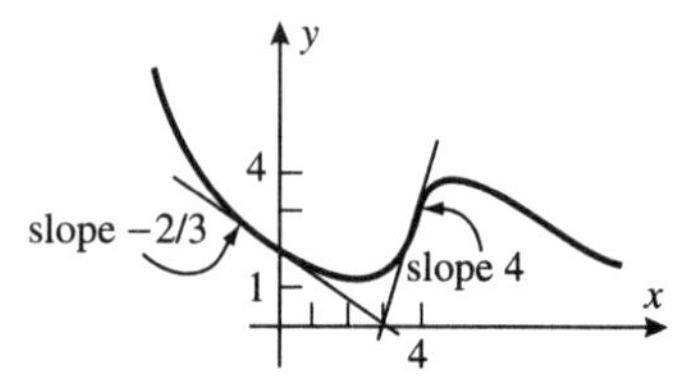

EXERCISE SET 4.2

1. **(a)**

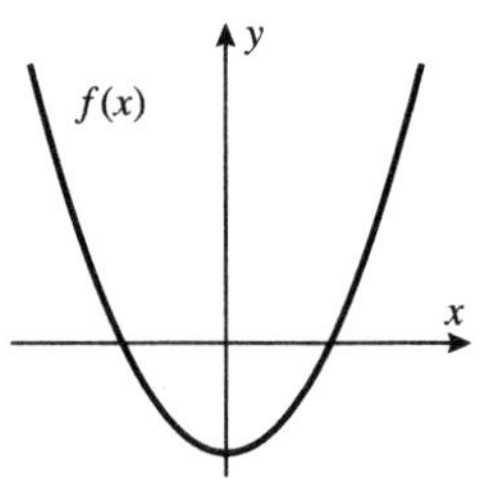

(b)

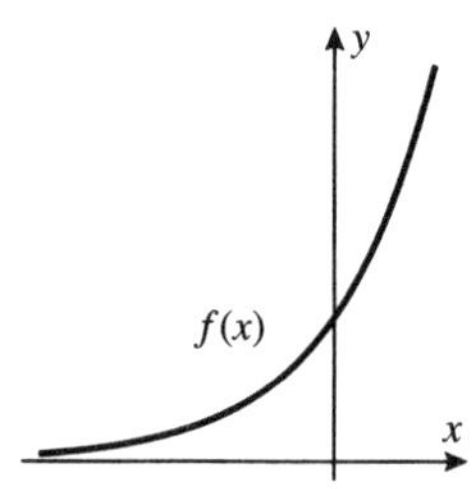

(c)

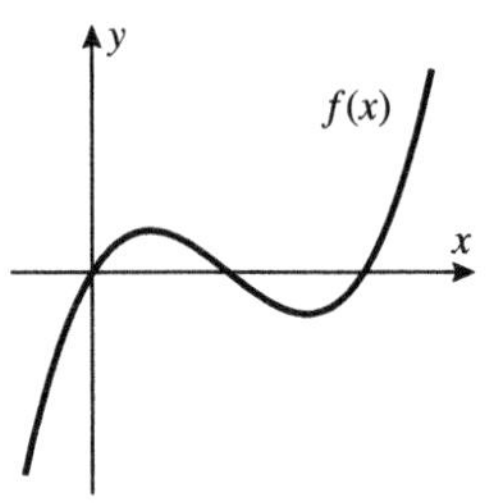

(d)

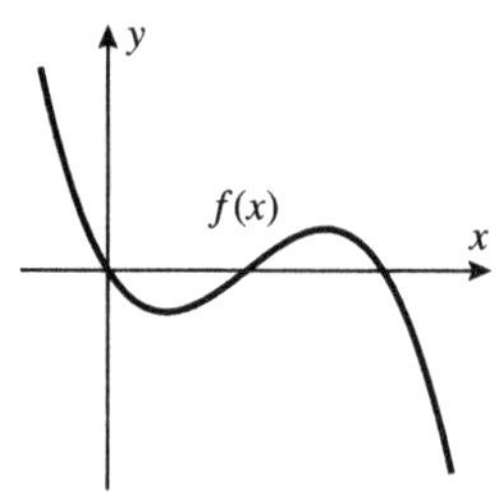

2. **(a)**

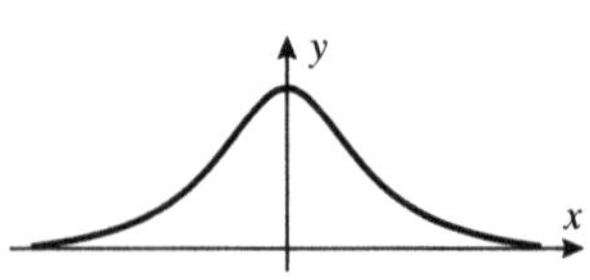

(b)

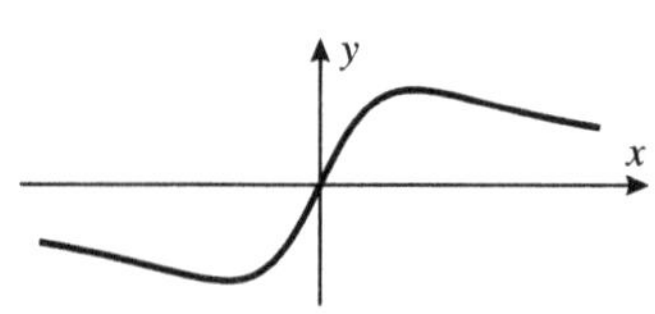

(c)

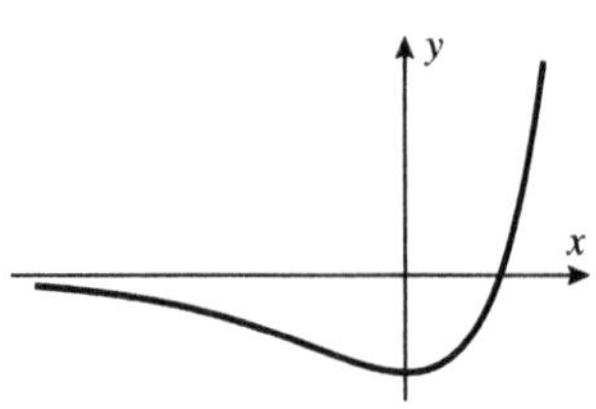

(d)

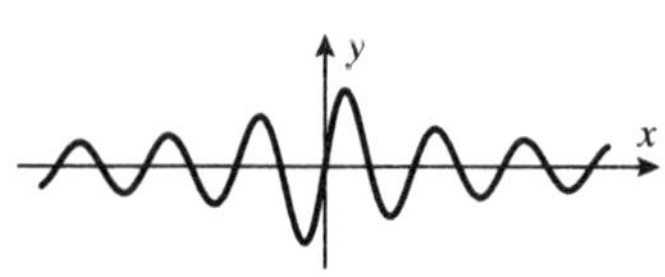

3. **(a)** $f'(x) = 6x - 6$ and $f''(x) = 6$, with $f'(1) = 0$. For the first derivative test, $f' < 0$ for $x < 1$ and $f' > 0$ for $x > 1$. For the second derivative test, $f''(1) > 0$.

(b) $f'(x) = 3x^2 - 3$ and $f''(x) = 6x$. $f'(x) = 0$ at $x = \pm 1$. First derivative test: $f' > 0$ for $x < -1$ and $x > 1$, and $f' < 0$ for $-1 < x < 1$, so there is a relative maximum at $x = -1$, and a relative minimum at $x = 1$. Second derivative test: $f'' < 0$ at $x = -1$, a relative maximum; and $f'' > 0$ at $x = 1$, a relative minimum.

4. **(a)** $f'(x) = 2\sin x \cos x = \sin 2x$ (so $f'(0) = 0$) and $f''(x) = 2\cos 2x$. First derivative test: if x is near 0 then $f' < 0$ for $x < 0$ and $f' > 0$ for $x > 0$, so a relative minimum at $x = 0$. Second derivative test: $f''(0) = 2 > 0$, so relative minimum at $x = 0$.

(b) $g'(x) = 2\tan x \sec^2 x$ (so $g'(0) = 0$) and $g''(x) = 2\sec^2 x(\sec^2 x + 2\tan^2 x)$. First derivative test: $g' < 0$ for $x < 0$ and $g' > 0$ for $x > 0$, so a relative minimum at $x = 0$. Second derivative test: $g''(0) = 2 > 0$, relative minimum at $x = 0$.

(c) Both functions are squares, and so are positive for values of x near zero; both functions are zero at $x = 0$, so that must be a relative minimum.

5. **(a)** $f'(x) = 4(x-1)^3$, $g'(x) = 3x^2 - 6x + 3$ so $f'(1) = g'(1) = 0$.

(b) $f''(x) = 12(x-1)^2$, $g''(x) = 6x - 6$, so $f''(1) = g''(1) = 0$, which yields no information.

(c) $f' < 0$ for $x < 1$ and $f' > 0$ for $x > 1$, so there is a relative minimum at $x = 1$; $g'(x) = 3(x-1)^2 > 0$ on both sides of $x = 1$, so there is no relative extremum at $x = 1$.

6. **(a)** $f'(x) = -5x^4$, $g'(x) = 12x^3 - 24x^2$ so $f'(0) = g'(0) = 0$.

(b) $f''(x) = -20x^3$, $g''(x) = 36x^2 - 48x$, so $f''(0) = g''(0) = 0$, which yields no information.

(c) $f' < 0$ on both sides of $x = 0$, so there is no relative extremum there; $g'(x) = 12x^2(x-2) < 0$ on both sides of $x = 0$ (for x near 0), so again there is no relative extremum there.

7. **(a)** $f'(x) = 3x^2 + 6x - 9 = 3(x+3)(x-1)$, $f'(x) = 0$ when $x = -3, 1$ (stationary points).

(b) $f'(x) = 4x(x^2 - 3)$, $f'(x) = 0$ when $x = 0, \pm\sqrt{3}$ (stationary points).

8. **(a)** $f'(x) = 6(x^2 - 1)$, $f'(x) = 0$ when $x = \pm 1$ (stationary points).

(b) $f'(x) = 12x^3 - 12x^2 = 12x^2(x-1)$, $f'(x) = 0$ when $x = 0, 1$ (stationary points).

9. **(a)** $f'(x) = (2 - x^2)/(x^2+2)^2$, $f'(x) = 0$ when $x = \pm\sqrt{2}$ (stationary points).

(b) $f'(x) = \frac{2}{3}x^{-1/3} = 2/(3x^{1/3})$, $f'(x)$ does not exist when $x = 0$.

10. **(a)** $f'(x) = 8x/(x^2+1)^2$, $f'(x) = 0$ when $x = 0$ (stationary point).

(b) $f'(x) = \frac{1}{3}(x+2)^{-2/3}$, $f'(x)$ does not exist when $x = -2$.

11. **(a)** $f'(x) = \dfrac{4(x+1)}{3x^{2/3}}$, $f'(x) = 0$ when $x = -1$ (stationary point), $f'(x)$ does not exist when $x = 0$.

(b) $f'(x) = -3\sin 3x$, $f'(x) = 0$ when $\sin 3x = 0, 3x = n\pi, n = 0, \pm 1, \pm 2, \ldots$
$x = n\pi/3, n = 0, \pm 1, \pm 2, \ldots$ (stationary points)

12. **(a)** $f'(x) = \dfrac{4(x - 3/2)}{3x^{2/3}}$, $f'(x) = 0$ when $x = 3/2$ (stationary point), $f'(x)$ does not exist when $x = 0$.

(b) $f(x) = |\sin x| = \begin{cases} \sin x, & \sin x \geq 0 \\ -\sin x, & \sin x < 0 \end{cases}$ so $f'(x) = \begin{cases} \cos x, & \sin x > 0 \\ -\cos x, & \sin x < 0 \end{cases}$ and $f'(x)$ does not exist when $x = n\pi$, $n = 0, \pm 1, \pm 2, \ldots$ (the points where $\sin x = 0$) because $\lim_{x \to n\pi^-} f'(x) \neq \lim_{x \to n\pi^+} f'(x)$ (see Theorem preceding Exercise 75, Section 3.3). Now $f'(x) = 0$ when $\pm\cos x = 0$ provided $\sin x \neq 0$ so $x = \pi/2 + n\pi$, $n = 0, \pm 1, \pm 2, \ldots$ are stationary points.

13. (a) none
(b) $x = 1$ because f' changes sign from $+$ to $-$ there
(c) none because $f'' = 0$ (never changes sign)

14. (a) $x = 1$ because $f'(x)$ changes sign from $-$ to $+$ there
(b) $x = 3$ because $f'(x)$ changes sign from $+$ to $-$ there
(c) $x = 2$ because $f''(x)$ changes sign there

15. (a) $x = 2$ because $f'(x)$ changes sign from $-$ to $+$ there.
(b) $x = 0$ because $f'(x)$ changes sign from $+$ to $-$ there.
(c) $x = 1, 3$ because $f''(x)$ changes sign at these points.

16. (a) $x = 1$ (b) $x = 5$ (c) $x = -1, 0, 3$

17. (a) critical numbers $x = 0, \pm\sqrt{5}$; f':

− − 0 + + 0 − − 0 + +
at $-\sqrt{5}$, 0, $\sqrt{5}$

$x = 0$: relative maximum; $x = \pm\sqrt{5}$: relative minimum

(b) critical number $x = 1, -1$; f':

+ + + 0 − − − 0 + + +
at -1, 1

$x = -1$: relative maximum; $x = 1$: relative minimum

18. (a) critical numbers $x = 0, -1/2, 1$; f':

+ + + 0 − 0 − − − 0 +
at $-\frac{1}{2}$, 0, 1

$x = 0$: neither; $x = -1/2$: relative maximum; $x = 1$: relative minimum

(b) critical numbers: $x = \pm 3/2, -1$; f':

+ + 0 − − ? + + 0 − −
at $-\frac{3}{2}$, -1, $\frac{3}{2}$

$x = \pm 3/2$: relative maximum; $x = -1$: relative minimum

19. $f'(x) = -2(x + 2)$; critical number $x = -2$; $f'(x)$:

+ + + 0 − − −
at -2

$f''(x) = -2$; $f''(-2) < 0$, $f(-2) = 5$; relative maximum of 5 at $x = -2$

20. $f'(x) = 6(x - 2)(x - 1)$; critical numbers $x = 1, 2$; $f'(x)$:

+ + + 0 − − − 0 + + +
at 1, 2

$f''(x) = 12x - 18$; $f''(1) < 0$, $f''(2) > 0$, $f(1) = 5$, $f(2) = 4$; relative minimum of 4 at $x = 2$, relative maximum of 5 at $x = 1$

21. $f'(x) = 2\sin x \cos x = \sin 2x$;
critical numbers $x = \pi/2,\ \pi,\ 3\pi/2$; $f'(x)$:

+ + 0 − − 0 + + 0 − −
at $\frac{\pi}{2}$, π, $\frac{3\pi}{2}$

$f''(x) = 2\cos 2x$; $f''(\pi/2) < 0$, $f''(\pi) > 0$, $f''(3\pi/2) < 0$, $f(\pi/2) = f(3\pi/2) = 1$,
$f(\pi) = 0$; relative minimum of 0 at $x = \pi$, relative maximum of 1 at $x = \pi/2,\ 3\pi/2$

22. $f'(x) = 1/2 - \cos x$; critical numbers $x = \pi/3, 5\pi/3$; $f'(x)$:

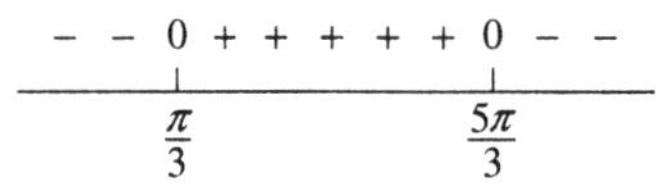

$f''(x) = -\sin x$; $f''(\pi/3) < 0$, $f''(5\pi/3) > 0$
$f(\pi/3) = \pi/6 - \sqrt{3}/2$, $f(5\pi/3) = 5\pi/6 + \sqrt{3}/2$;
relative minimum of $\pi/6 - \sqrt{3}/2$ at $x = \pi/3$, relative maximum of $5\pi/6 + \sqrt{3}/2$ at $x = 5\pi/3$

23. $f'(x) = 3x^2 + 5$; no relative extrema because there are no critical numbers.

24. $f'(x) = 4x(x^2 - 1)$; critical numbers $x = 0, 1, -1$
$f''(x) = 12x^2 - 4$; $f''(0) < 0$, $f''(1) > 0$, $f''(-1) > 0$
relative minimum of 6 at $x = 1, -1$, relative maximum of 7 at $x = 0$

25. $f'(x) = (x - 1)(3x - 1)$; critical numbers $x = 1, 1/3$
$f''(x) = 6x - 4$; $f''(1) > 0$, $f''(1/3) < 0$
relative minimum of 0 at $x = 1$, relative maximum of 4/27 at $x = 1/3$

26. $f'(x) = 2x^2(2x + 3)$; critical numbers $x = 0, -3/2$
relative minimum of $-27/16$ at $x = -3/2$ (first derivative test)

27. $f'(x) = 4x(1 - x^2)$; critical numbers $x = 0, 1, -1$
$f''(x) = 4 - 12x^2$; $f''(0) > 0$, $f''(1) < 0$, $f''(-1) < 0$
relative minimum of 0 at $x = 0$, relative maximum of 1 at $x = 1, -1$

28. $f'(x) = 10(2x - 1)^4$; critical number $x = 1/2$; no relative extrema (first derivative test)

29. $f'(x) = \frac{4}{5}x^{-1/5}$; critical number $x = 0$; relative minimum of 0 at $x = 0$ (first derivative test)

30. $f'(x) = 2 + \frac{2}{3}x^{-1/3}$; critical numbers $x = 0, -1/27$
relative minimum of 0 at $x = 0$, relative maximum of 1/27 at $x = -1/27$

31. $f'(x) = 2x/(x^2 + 1)^2$; critical number $x = 0$; relative minimum of 0 at $x = 0$

32. $f'(x) = 2/(x + 2)^2$; no critical numbers ($x = -2$ is not in the domain of f) no relative extrema

33. $f'(x) = 2x$ if $|x| > 2$, $f'(x) = -2x$ if $|x| < 2$,
$f'(x)$ does not exist when $x = \pm 2$; critical numbers $x = 0, 2, -2$
relative minimum of 0 at $x = 2, -2$, relative maximum of 4 at $x = 0$

34. $f'(x) = -1$ if $x < 3$, $f'(x) = 2x$ if $x > 3$, $f'(3)$ does not exist;
critical number $x = 3$, relative minimum of 6 at $x = 3$

35. $f'(x) = 2\cos 2x$ if $\sin 2x > 0$,
$f'(x) = -2\cos 2x$ if $\sin 2x < 0$,
$f'(x)$ does not exist when $x = \pi/2, \pi, 3\pi/2$;
critical numbers $x = \pi/4, 3\pi/4, 5\pi/4, 7\pi/4, \pi/2, \pi, 3\pi/2$
relative minimum of 0 at $x = \pi/2, \pi, 3\pi/2$;
relative maximum of 1 at $x = \pi/4, 3\pi/4, 5\pi/4, 7\pi/4$

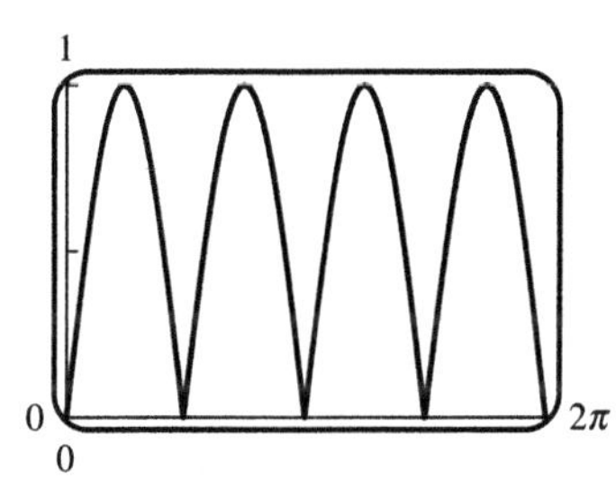

36. $f'(x) = \sqrt{3} + 2\cos x$; critical numbers $x = 5\pi/6, 7\pi/6$
relative minimum of $7\sqrt{3}\pi/6 - 1$ at $x = 7\pi/6$;
relative maximum of $5\sqrt{3}\pi/6 + 1$ at $x = 5\pi/6$

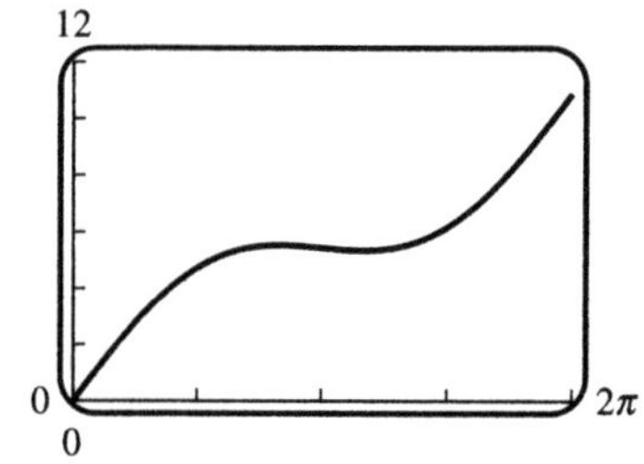

37. $f'(x) = -\sin 2x$; critical numbers $x = \pi/2, \pi, 3\pi/2$
relative minimum of 0 at $x = \pi/2, 3\pi/2$;
relative maximum of 1 at $x = \pi$

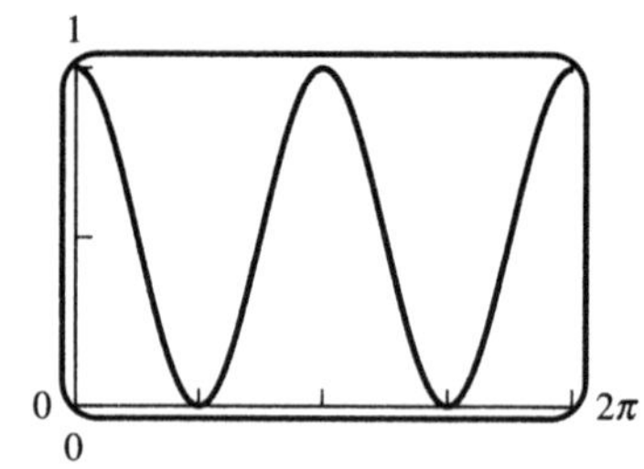

38. $f'(x) = (2\cos x - 1)/(2 - \cos x)^2$;
critical numbers $x = \pi/3, 5\pi/3$
relative maximum of $\sqrt{3}/3$ at $x = \pi/3$,
relative minimum of $-\sqrt{3}/3$ at $x = 5\pi/3$

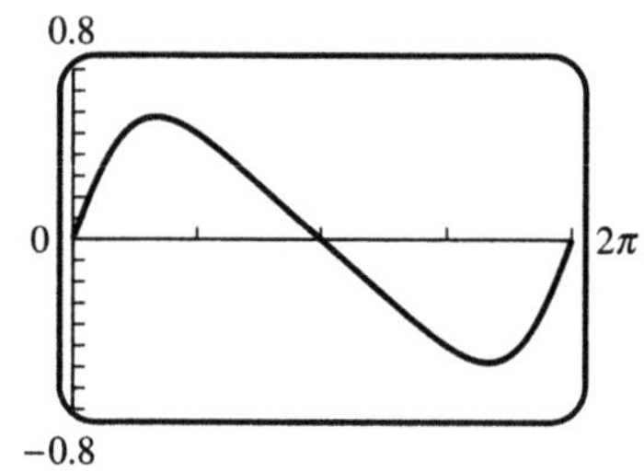

39. Relative minima at $x = -3.58, 3.33$;
relative maximum at $x = 0.25$

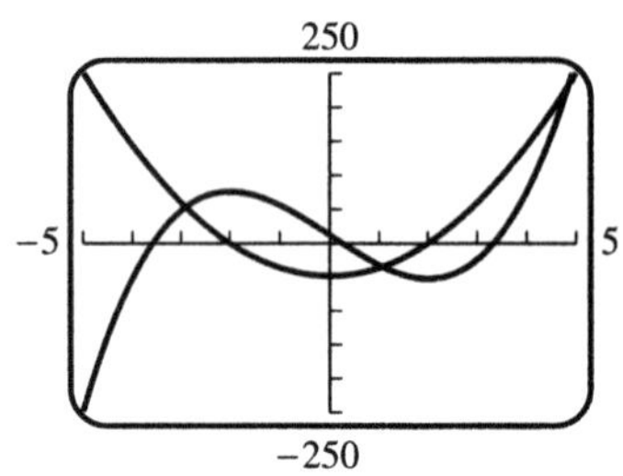

40. Relative minimum at $x = -0.84$;
relative maximum at $x = 0.84$

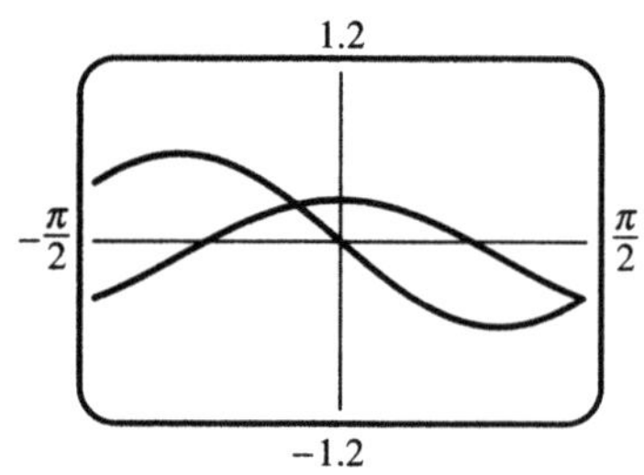

41. Relative minimum at $x = -1.20$ and a relative maximum at $x = 1.80$

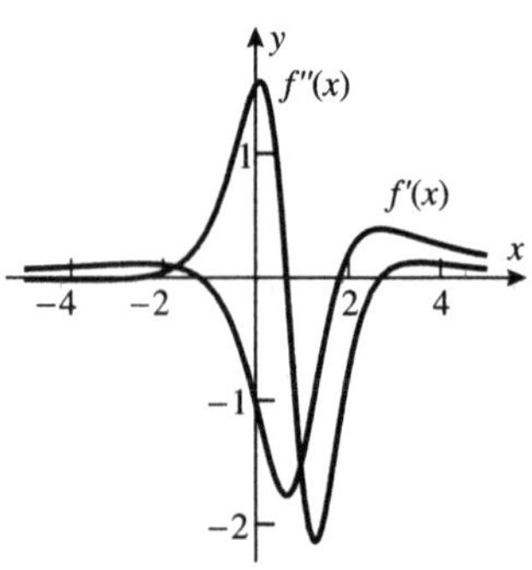

42. Relative maximum at $x = -0.78$ and a relative minimum at $x = 1.55$

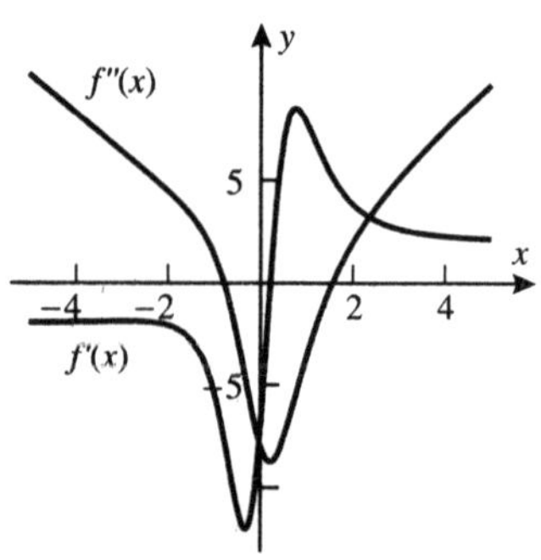

43. $f'(x) = \dfrac{x^4 + 3x^2 - 2x}{(x^2+1)^2}$

$f''(x) = -2\dfrac{x^3 - 3x^2 - 3x + 1}{(x^2+1)^3}$

Relative maximum at $x = 0$, relative minimum at $x \approx 0.59$

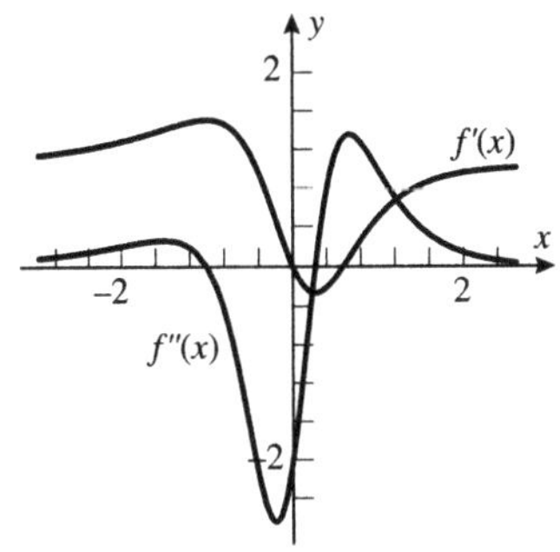

44. $f'(x) = \dfrac{4x^3 - \sin 2x}{2\sqrt{x^4 - \cos^2 x}}$,

$f''(x) = \dfrac{6x^2 - \cos 2x}{\sqrt{x^4 + \cos^2 x}} - \dfrac{(4x^3 - \sin 2x)(4x^3 - \sin 2x)}{4(x^4 + \cos^2 x)^{3/2}}$

Relative minima at $x \approx \pm 0.62$, relative maximum at $x = 0$

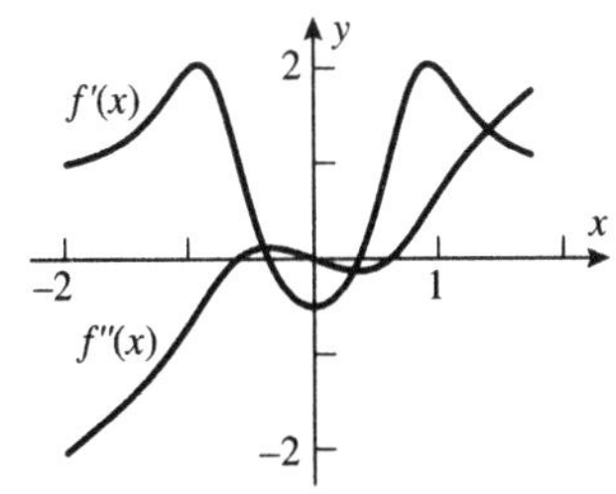

45. (a) Let $f(x) = x^2 + \dfrac{k}{x}$, then $f'(x) = 2x - \dfrac{k}{x^2} = \dfrac{2x^3 - k}{x^2}$. f has a relative extremum when $2x^3 - k = 0$, so $k = 2x^3 = 2(3)^3 = 54$.

(b) Let $f(x) = \dfrac{x}{x^2 + k}$, then $f'(x) = \dfrac{k - x^2}{(x^2 + k)^2}$. f has a relative extremum when $k - x^2 = 0$, so $k = x^2 = 3^2 = 9$.

46. (a) relative minima at $x \approx \pm 0.6436$, relative maximum at $x = 0$

(b) $x \approx \pm 0.6436, 0$

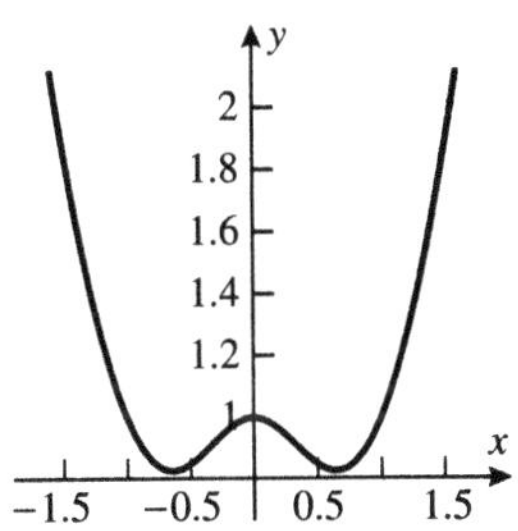

47. (a) $(-2.2, 4), (2, 2.2), (4.2, 3)$

(b) f' exists everywhere, so the critical numbers are when $f' = 0$, i.e. when $x = \pm 2$ or $r(x) = 0$, so $x \approx -5.1, -2, 0.2, 2$. At $x = -5.1$ f' changes sign from $-$ to $+$, so minimum; at $x = -2$ f' changes sign from $+$ to $-$, so maximum; at $x = 0.2$ f' doesn't change sign, so neither; at $x = 2$ f' changes sign from $-$ to $+$, so minimum.
Finally, $f''(1) = (1^2 - 4)r'(1) + 2r(1) \approx -3(0.6) + 2(0.3) = -1.2$.

48. $g'(x)$ exists everywhere, so the critical points are the points when $g'(x) = 0$, or $r(x) = x$, so $r(x)$ crosses the line $y = x$. From the graph it appears that this happens precisely when $x = 0$.

49. $f'(x) = 3ax^2 + 2bx + c$ and $f'(x)$ has roots at $x = 0, 1$, so $f'(x)$ must be of the form $f'(x) = 3ax(x-1)$; thus $c = 0$ and $2b = -3a$, $b = -3a/2$. $f''(x) = 6ax + 2b = 6ax - 3a$, so $f''(0) > 0$ and $f''(1) < 0$ provided $a < 0$. Finally $f(0) = d$, so $d = 0$; and $f(1) = a + b + c + d = a + b = -a/2$ so $a = -2$. Thus $f(x) = -2x^3 + 3x^2$.

50. **(a)** Because h and g have relative maxima at x_0, $h(x) \le h(x_0)$ for all x in I_1 and $g(x) \le g(x_0)$ for all x in I_2, where I_1 and I_2 are open intervals containing x_0. If x is in both I_1 and I_2 then both inequalities are true and by addition so is $h(x)+g(x) \le h(x_0)+g(x_0)$ which shows that $h+g$ has a relative maximum at x_0.

(b) By counterexample; both $h(x) = -x^2$ and $g(x) = -2x^2$ have relative maxima at $x = 0$ but $h(x) - g(x) = x^2$ has a relative minimum at $x = 0$ so in general $h - g$ does not necessarily have a relative maximum at x_0.

51. **(a)**

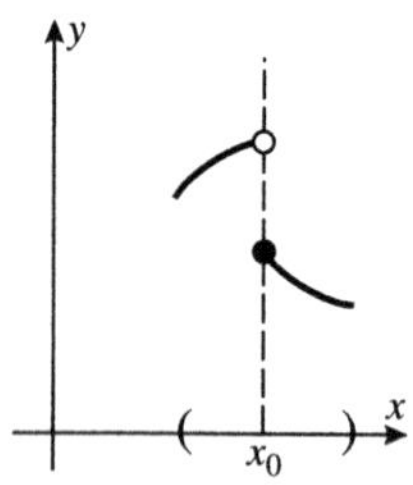

$f(x_0)$ is not an extreme value.

(b)

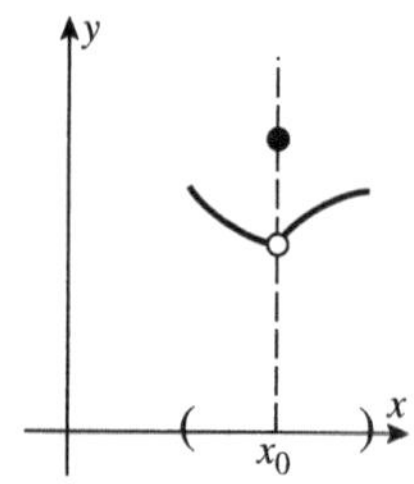

$f(x_0)$ is a relative maximum.

(c)

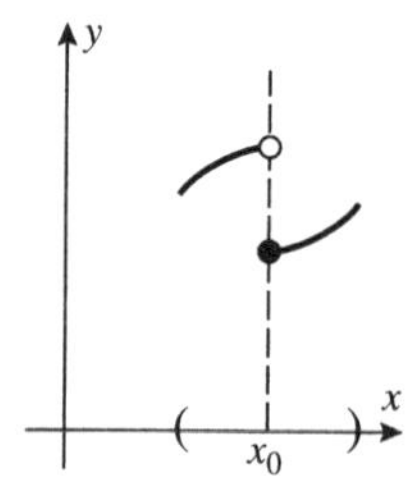

$f(x_0)$ is a relative minimum.

EXERCISE SET 4.3

1. $y = x^2 - 2x - 3;$
$y' = 2(x - 1);$
$y'' = 2$

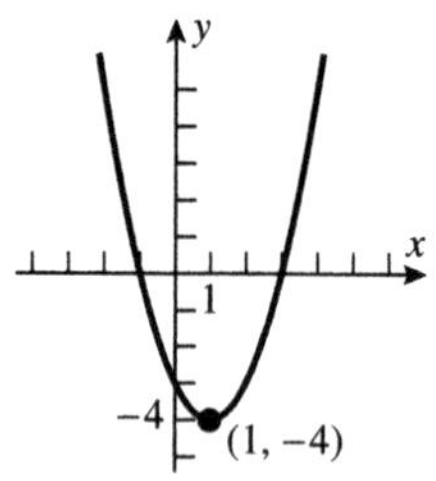

2. $y = 1 + x - x^2;$
$y' = -2(x - 1/2);$
$y'' = -2$

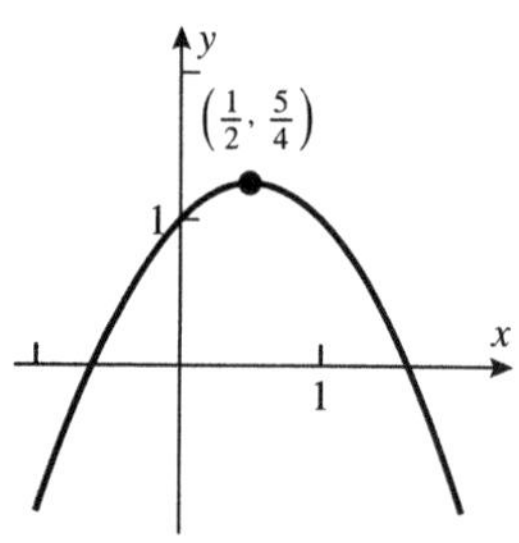

3. $y = x^3 - 3x + 1;$
$y' = 3(x^2 - 1);$
$y'' = 6x$

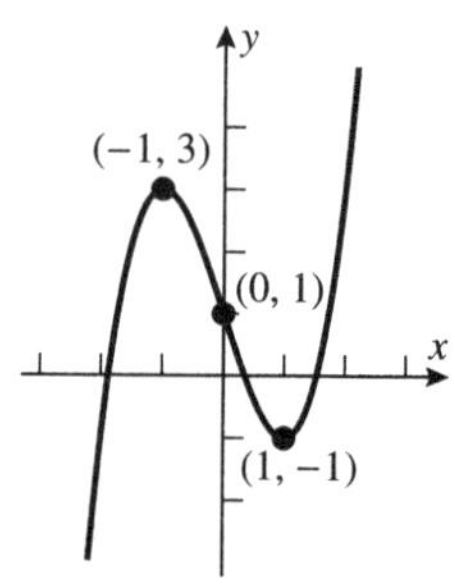

4. $y = 2x^3 - 3x^2 + 12x + 9;$
$y' = 6(x^2 - x + 2);$
$y'' = 12(x - 1/2)$

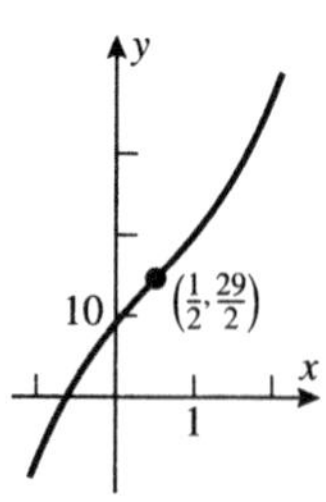

5. $y = x^4 + 2x^3 - 1;$
$y' = 4x^2(x + 3/2);$
$y'' = 12x(x + 1)$

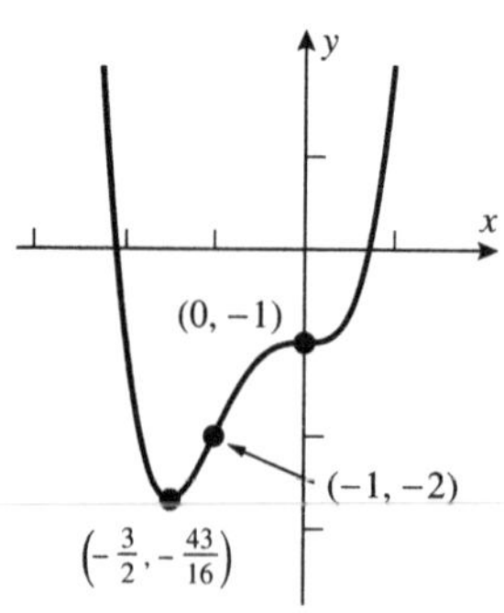

6. $y = x^4 - 2x^2 - 12;$
$y' = 4x(x^2 - 1);$
$y'' = 12(x^2 - 1/3)$

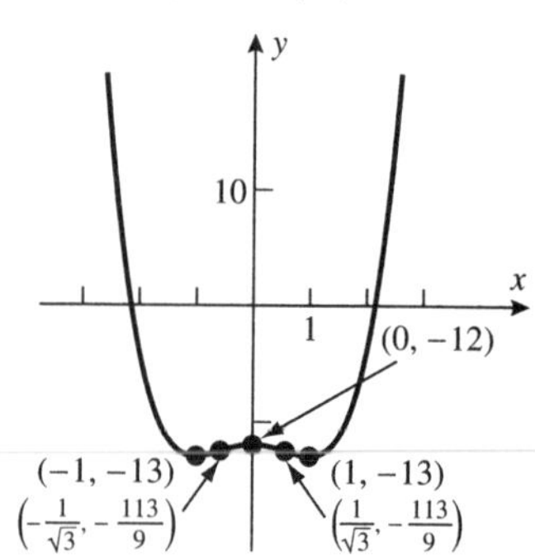

7. $y = x^3(3x^2 - 5)$;
$y' = 15x^2(x^2 - 1)$;
$y'' = 30x(2x^2 - 1)$

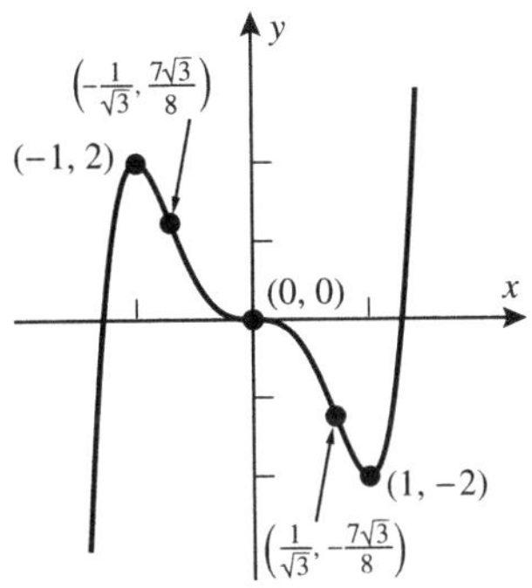

8. $y = 3x^3(x + 4/3)$;
$y' = 12x^2(x + 1)$;
$y'' = 36x(x + 2/3)$

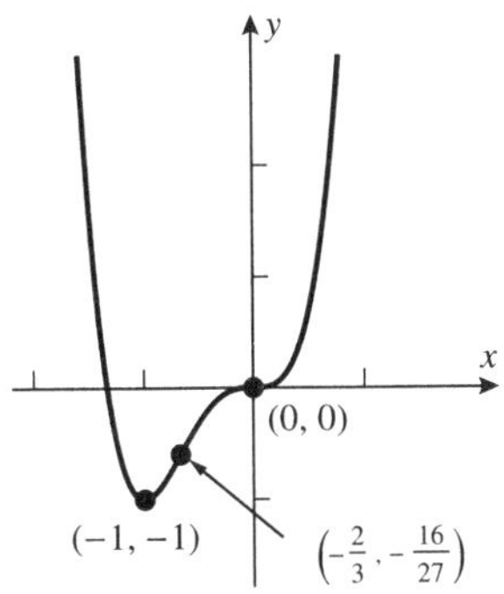

9. $y = x(x - 1)^3$;
$y' = (4x - 1)(x - 1)^2$;
$y'' = 6(2x - 1)(x - 1)$

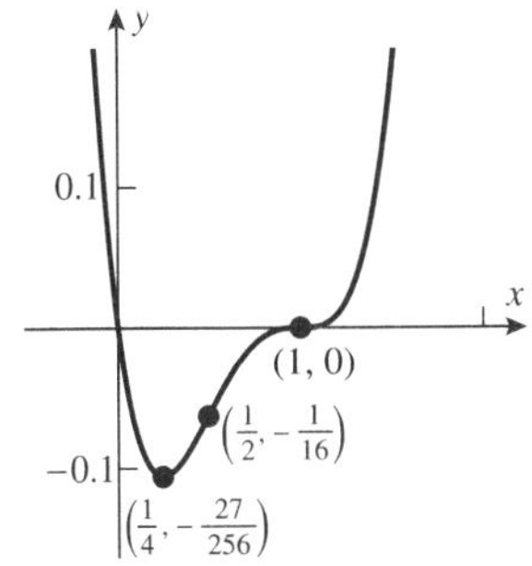

10. $y = x^4(x + 5)$;
$y' = 5x^3(x + 4)$;
$y'' = 20x^2(x + 3)$

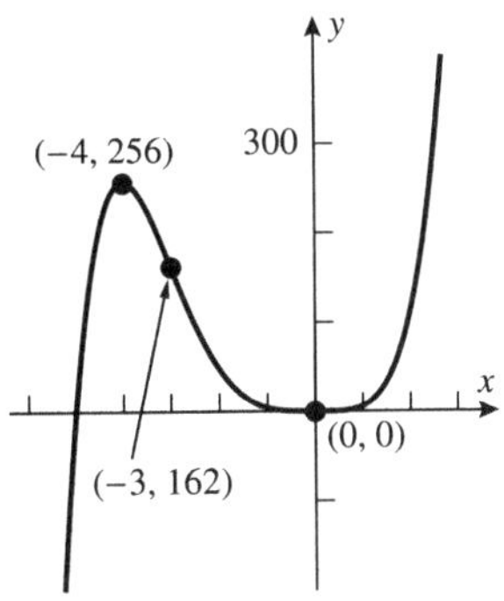

11. $y = 2x/(x - 3)$;
$y' = -6/(x - 3)^2$;
$y'' = 12/(x - 3)^3$

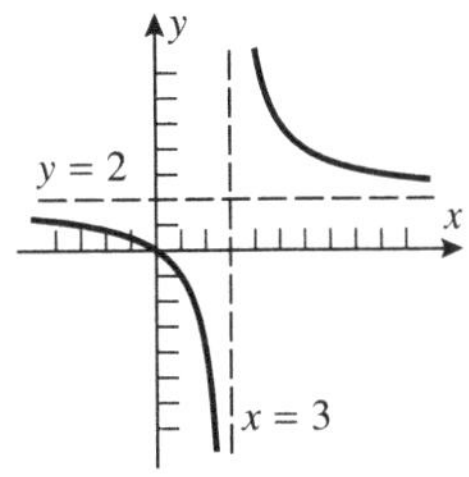

12. $y = \dfrac{x}{x^2 - 1}$;

$y' = -\dfrac{x^2 + 1}{(x^2 - 1)^2}$;

$y'' = \dfrac{2x(x^2 + 3)}{(x^2 - 1)^3}$

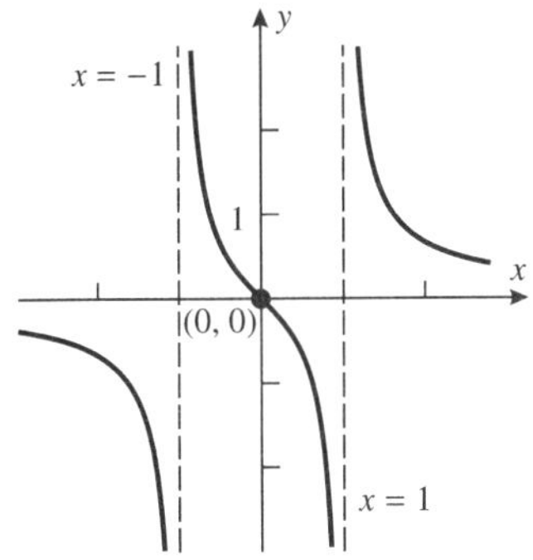

13. $y = \dfrac{x^2}{x^2 - 1}$;

$y' = -\dfrac{2x}{(x^2 - 1)^2}$;

$y'' = \dfrac{2(3x^2 + 1)}{(x^2 - 1)^3}$

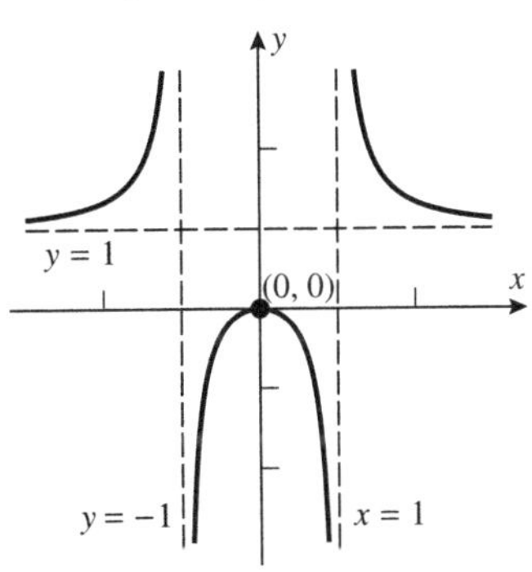

14. $y = \dfrac{x^2 - 1}{x^2 + 1}$;

$y' = \dfrac{4x}{(x^2 + 1)^2}$;

$y'' = \dfrac{4(1 - 3x^2)}{(x^2 + 1)^3}$

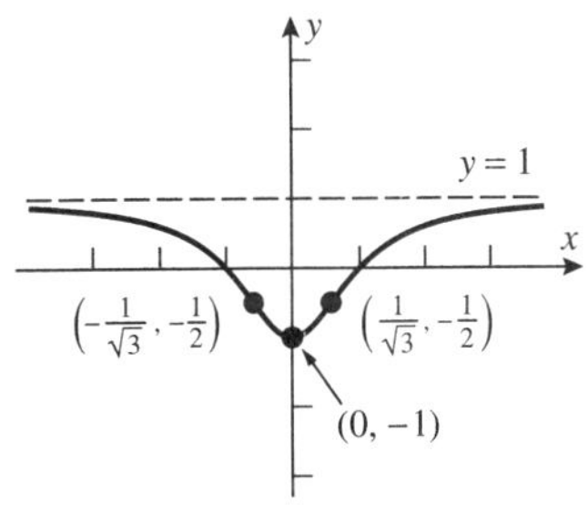

15. $y = x^2 - \dfrac{1}{x} = \dfrac{x^3 - 1}{x}$;

$y' = \dfrac{2x^3 + 1}{x^2}$,

$y' = 0$ when $x = -\sqrt[3]{\dfrac{1}{2}} \approx -0.8$;

$y'' = \dfrac{2(x^3 - 1)}{x^3}$

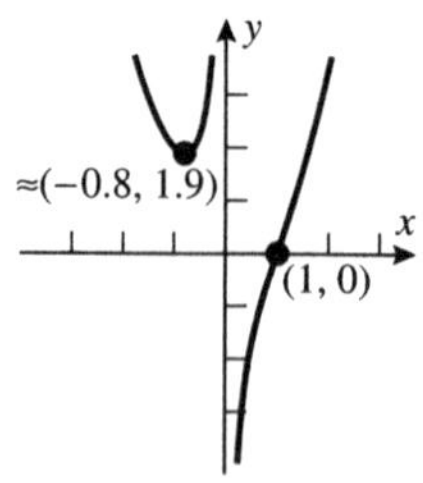

16. $y = \dfrac{2x^2 - 1}{x^2}$;

$y' = \dfrac{2}{x^3}$;

$y'' = -\dfrac{6}{x^4}$

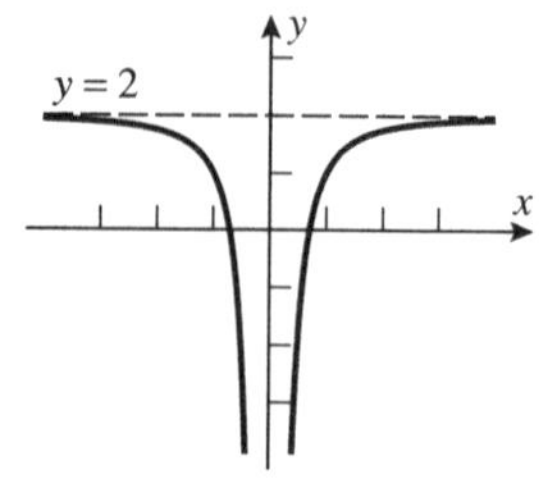

17. $y = \dfrac{x^3 - 1}{x^3 + 1}$;

$y' = \dfrac{6x^2}{(x^3 + 1)^2}$;

$y'' = \dfrac{12x(1 - 2x^3)}{(x^3 + 1)^3}$

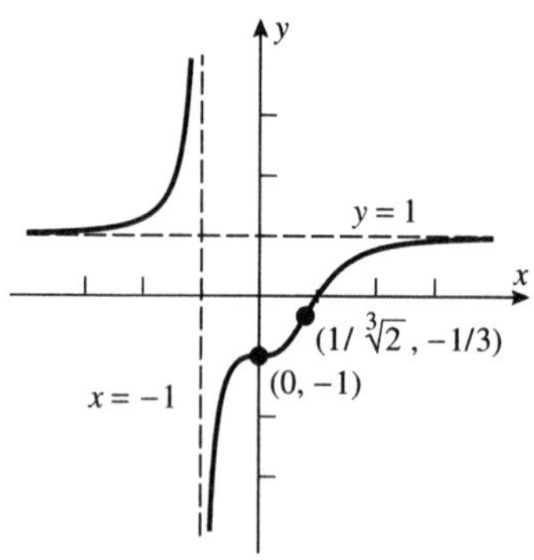

18. $y = \dfrac{8}{4 - x^2}$;

$y' = \dfrac{16x}{(4 - x^2)^2}$;

$y'' = \dfrac{16(3x^2 + 4)}{(4 - x^2)^3}$

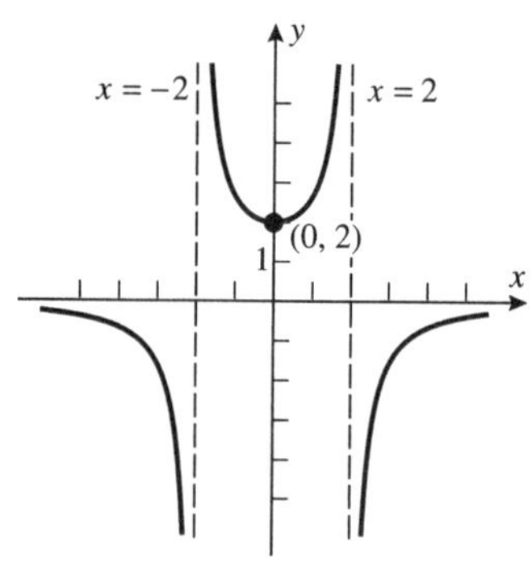

19. $y = \dfrac{x - 1}{x^2 - 4}$;

$y' = -\dfrac{x^2 - 2x + 4}{(x^2 - 4)^2}$

$y'' = 2\dfrac{x^3 - 3x^2 + 12x - 4}{(x^2 - 4)^3}$

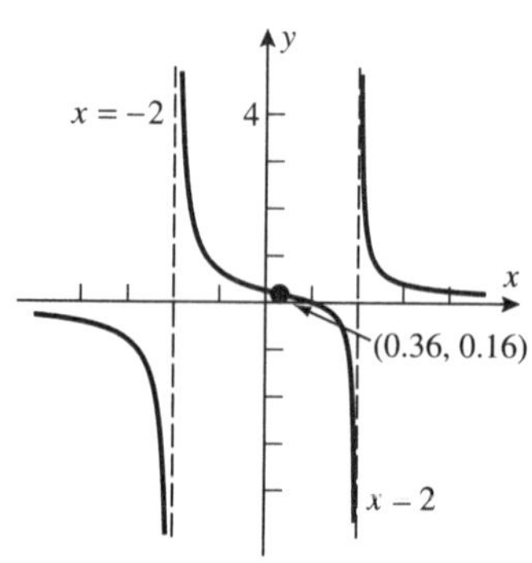

20. $y = \dfrac{x+3}{x^2-4}$;
$y' = -\dfrac{(x^2+6x+4)}{(x^2-4)^2}$
$y'' = 2\dfrac{x^3+9x^2+12x+12}{(x^2-4)^3}$

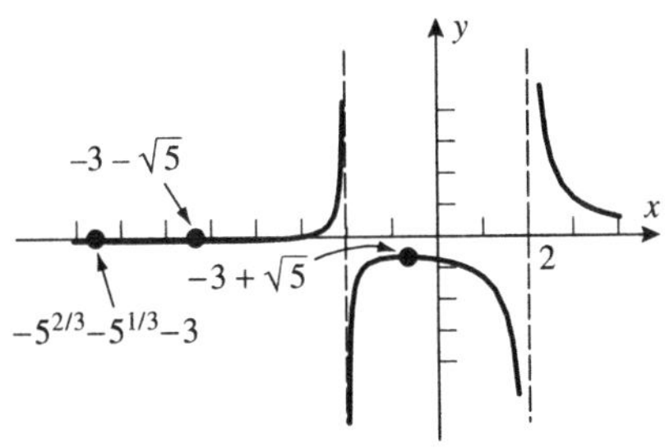

21. $y = \dfrac{1}{x-2}$;
$y' = \dfrac{-1}{(x-2)^2}$

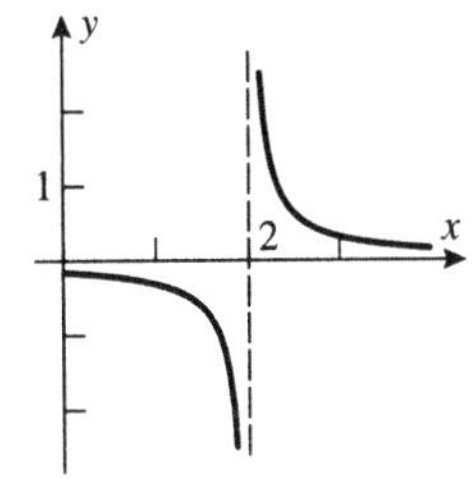

22. $y = \dfrac{x^2-1}{x^3-1} = \dfrac{x+1}{x^2+x+1}$;
$y' = -\dfrac{x(x+2)}{(x^2+x+1)^2}$
$y'' = 2\dfrac{x^3+3x^2-1}{(x^2+x+1)^3}$

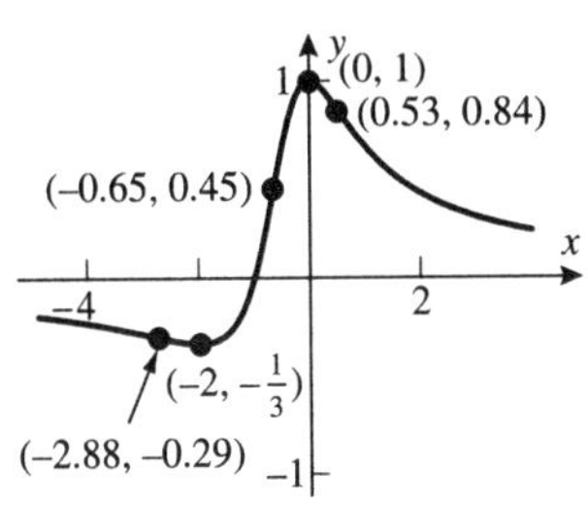

23. $y = \dfrac{(x-1)^2}{x^2}$;
$y' = \dfrac{2(x-1)}{x^3}$;
$y'' = \dfrac{2(3-2x)}{x^4}$

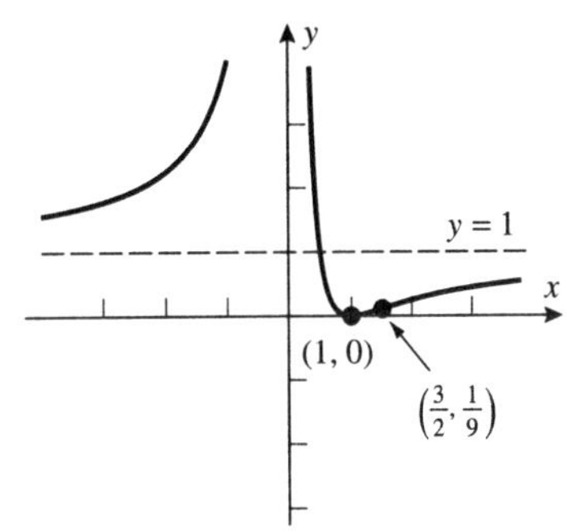

24. $y = 3 - \dfrac{4}{x} - \dfrac{4}{x^2}$;
$y' = \dfrac{4(x+2)}{x^3}$;
$y'' = -\dfrac{8(x+3)}{x^4}$

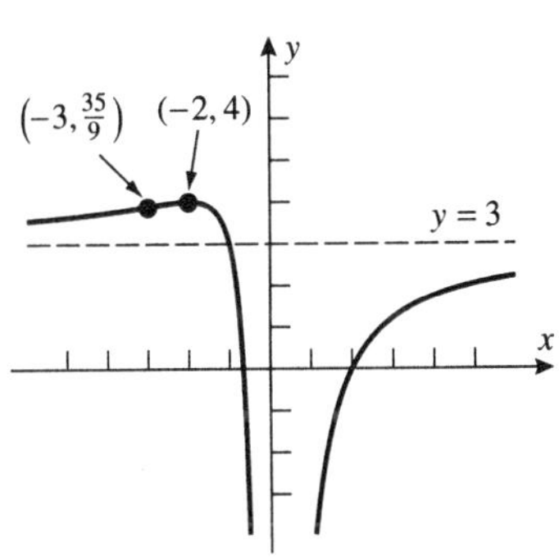

25. $y = 4 + \dfrac{x-1}{x^4}$;
$y' = -\dfrac{3x-4}{x^5}$;
$y'' = 4\dfrac{3x-5}{x^6}$

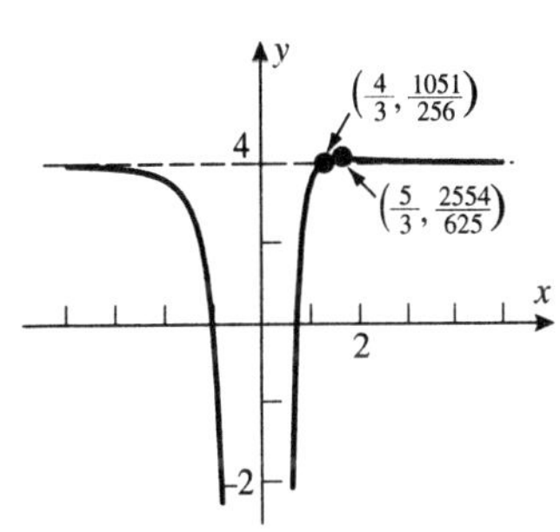

26. $y = 2 + \frac{3}{x} - \frac{1}{x^3}$;

$y' = \frac{3(1-x^2)}{x^4}$;

$y'' = \frac{6(x^2-2)}{x^5}$

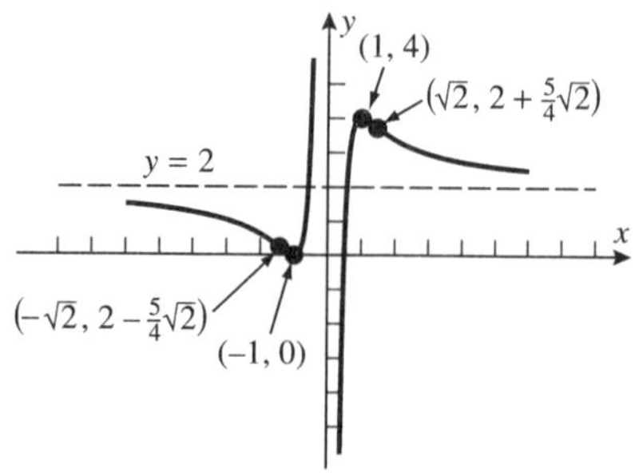

27. **(a)** VI **(b)** I **(c)** III **(d)** V **(e)** IV **(f)** II

28. **(a)** When n is even the function is defined only for $x \geq 0$; as n increases the graph approaches the line $y = 1$ for $x > 0$.

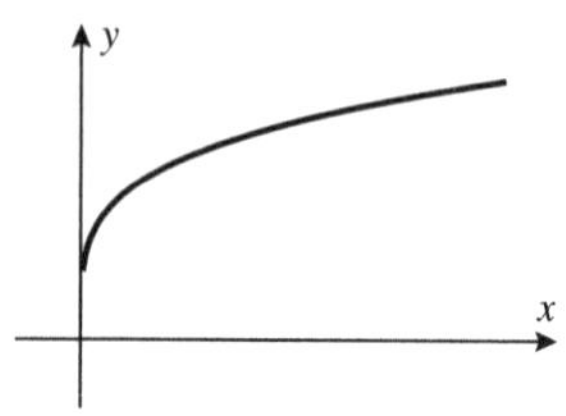

(b) When n is odd the graph is symmetric with respect to the origin; as n increases the graph approaches the line $y = 1$ for $x > 0$ and the line $y = -1$ for $x < 0$.

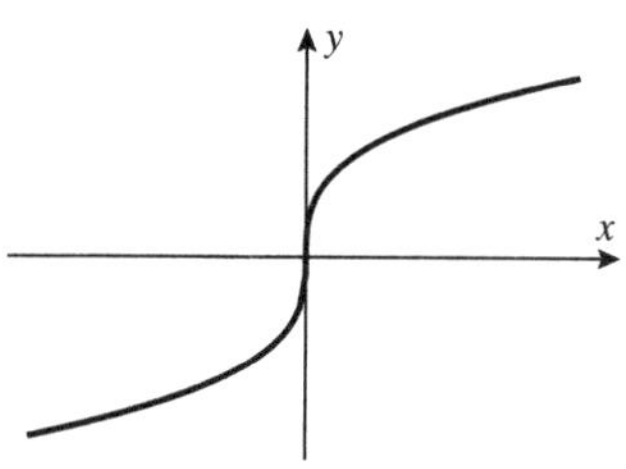

29. $y = \sqrt{x^2 - 1}$;

$y' = \frac{x}{\sqrt{x^2-1}}$;

$y'' = -\frac{1}{(x^2-1)^{3/2}}$

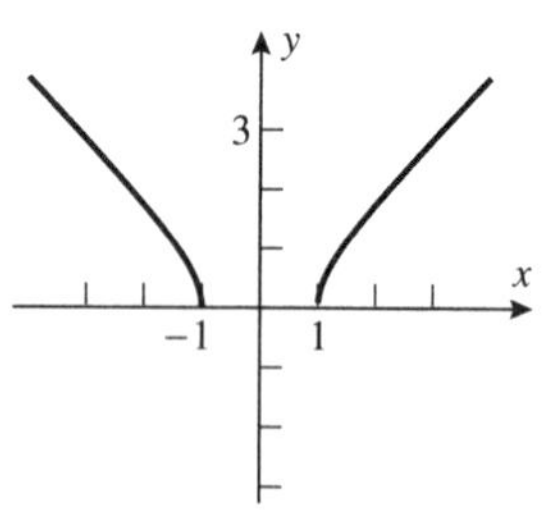

30. $y = \sqrt[3]{x^2 - 4}$;

$y' = \frac{2x}{3(x^2-4)^{2/3}}$;

$y'' = -\frac{2(3x^2+4)}{9(x^2-4)^{5/3}}$

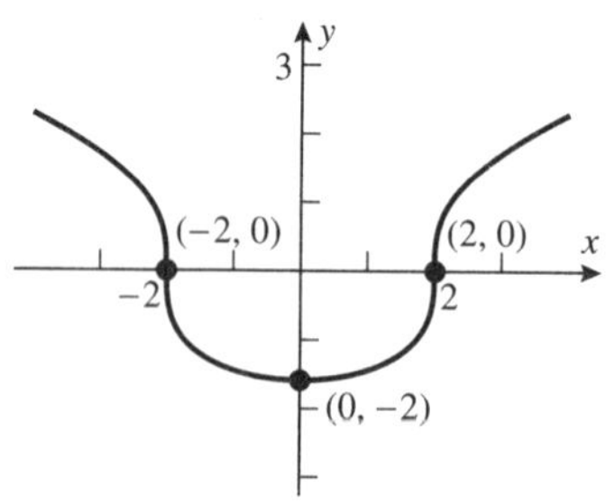

31. $y = 2x + 3x^{2/3}$;
$y' = 2 + 2x^{-1/3}$;
$y'' = -\frac{2}{3}x^{-4/3}$

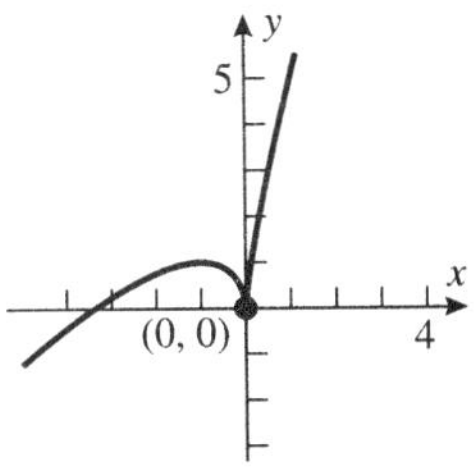

32. $y = 4x - 3x^{4/3}$;
$y' = 4 - 4x^{1/3}$;
$y'' = -\frac{4}{3}x^{-2/3}$

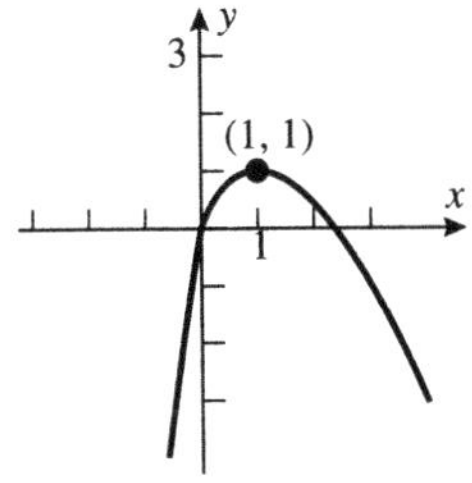

33. $y = x(3 - x)^{1/2}$;
$y' = \frac{3(2 - x)}{2\sqrt{3 - x}}$;
$y'' = \frac{3(x - 4)}{4(3 - x)^{3/2}}$

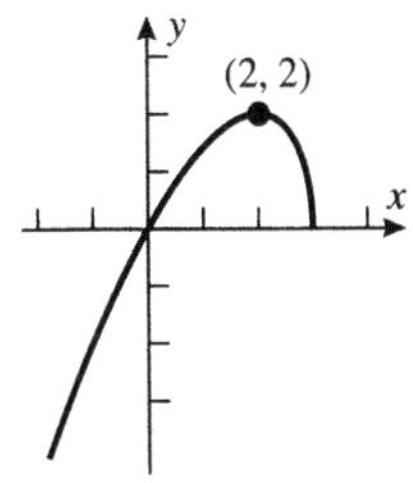

34. $y = x^{1/3}(4 - x)$;
$y' = \frac{4(1 - x)}{3x^{2/3}}$;
$y'' = -\frac{4(x + 2)}{9x^{5/3}}$

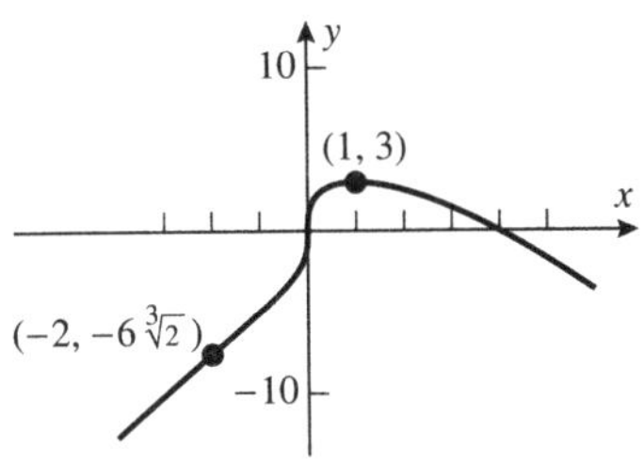

35. $y = \frac{8(\sqrt{x} - 1)}{x}$;
$y' = \frac{4(2 - \sqrt{x})}{x^2}$;
$y'' = \frac{2(3\sqrt{x} - 8)}{x^3}$

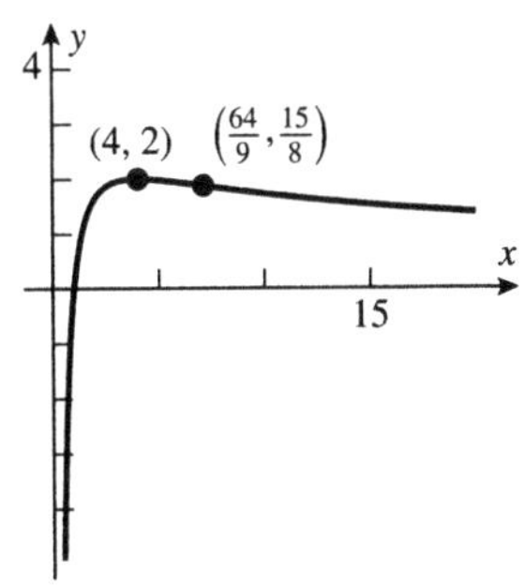

36. $y = \frac{1 + \sqrt{x}}{1 - \sqrt{x}}$;
$y' = \frac{1}{2\sqrt{x}(1-\sqrt{x})^2}$;
$y'' = \frac{3\sqrt{x} - 1}{2x^{3/2}(1 - \sqrt{x})^3}$

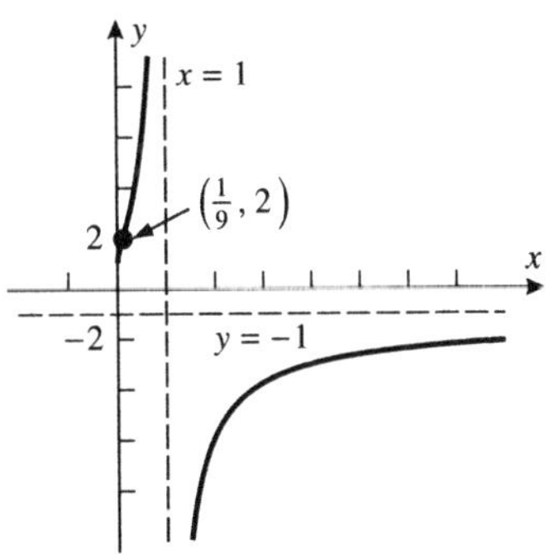

37. $y = x + \sin x$;
$y' = 1 + \cos x$, $y' = 0$ when $x = \pi + 2n\pi$;
$y'' = -\sin x$; $y'' = 0$ when $x = n\pi$
$n = 0, \pm 1, \pm 2, \ldots$

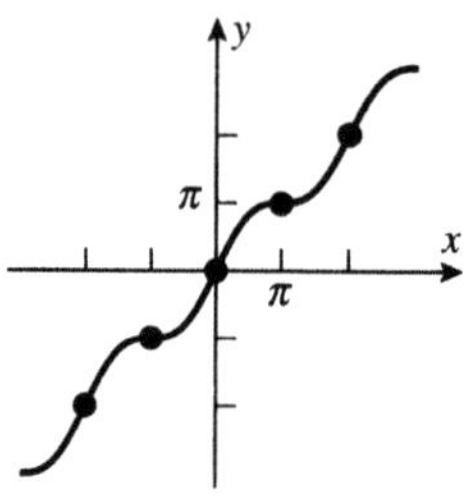

38. $y = x - \cos x$;
$y' = 1 + \sin x$;
$y' = 0$ when $x = -\pi/2 + 2n\pi$;
$y'' = \cos x$;
$y'' = 0$ when $x = \pi/2 + n\pi$
$n = 0, \pm 1, \pm 2, \ldots$

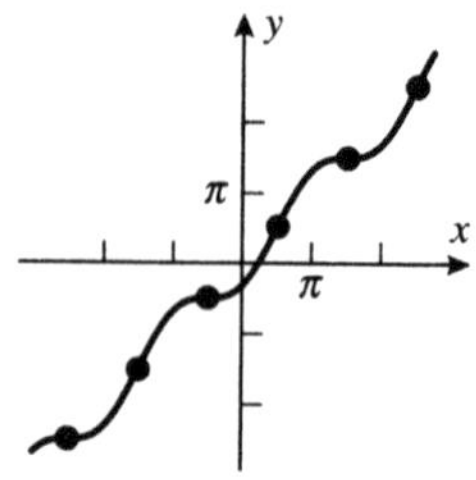

39. $y = \sin x + \cos x$;
$y' = \cos x - \sin x$;
$y' = 0$ when $x = \pi/4 + n\pi$;
$y'' = -\sin x - \cos x$;
$y'' = 0$ when $x = 3\pi/4 + n\pi$

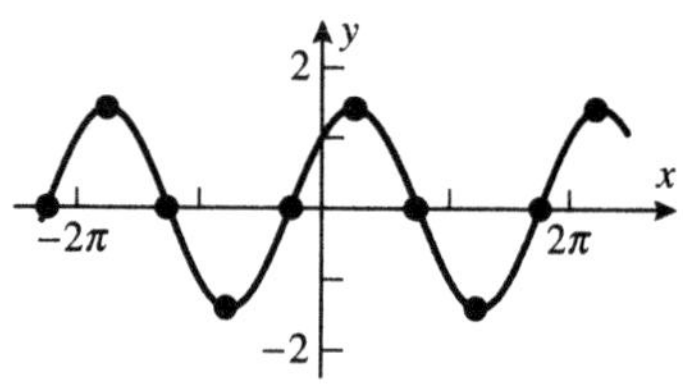

40. $y = \sqrt{3}\cos x + \sin x$;
$y' = -\sqrt{3}\sin x + \cos x$;
$y' = 0$ when $x = \pi/6 + n\pi$;
$y'' = -\sqrt{3}\cos x - \sin x$;
$y'' = 0$ when $x = 2\pi/3 + n\pi$

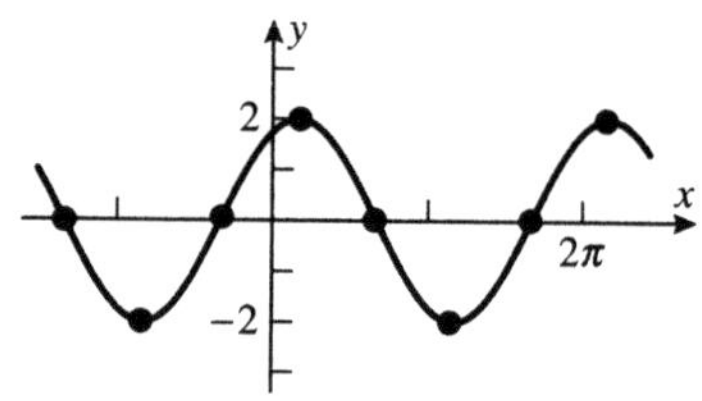

41. $y = \sin^2 x$, $0 \le x \le 2\pi$;
$y' = 2\sin x \cos x = \sin 2x$;
$y'' = 2\cos 2x$

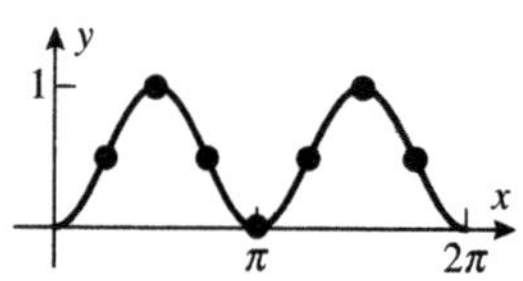

42. $y = x\tan x$, $-\pi/2 < x < \pi/2$;
$y' = x\sec^2 x + \tan x$;
$y' = 0$ when $x = 0$;
$y'' = 2\sec^2 x(x\tan x + 1)$, which is always positive for $-\pi/2 < x < \pi/2$

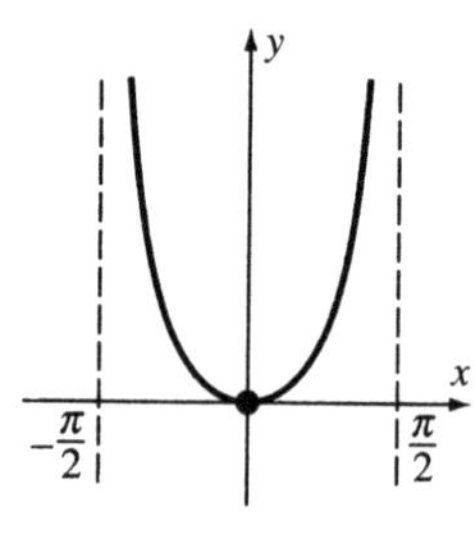

43. (a) $\lim\limits_{x\to-\infty} y = -\infty$, $\lim\limits_{x\to+\infty} y = +\infty$;
curve crosses x-axis at $x = 0, 1, -1$

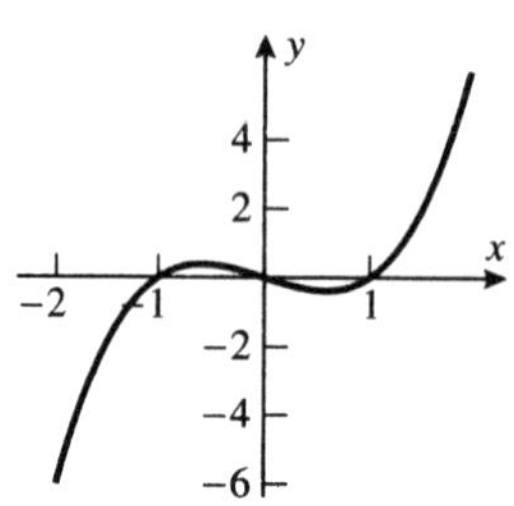

(b) $\lim\limits_{x\to\pm\infty} y = +\infty$;
curve never crosses x-axis

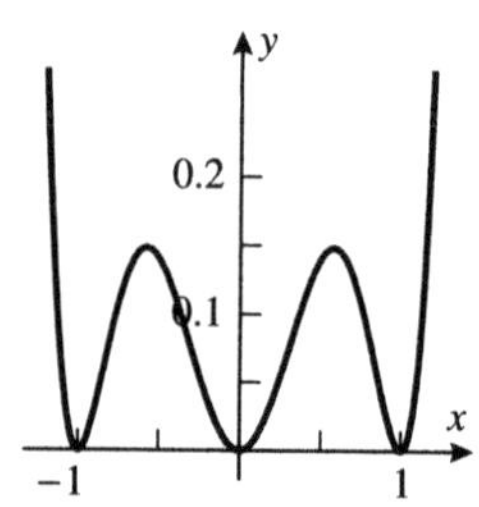

(c) $\lim_{x\to-\infty} y = -\infty$, $\lim_{x\to+\infty} y = +\infty$;
curve crosses x-axis at $x = -1$

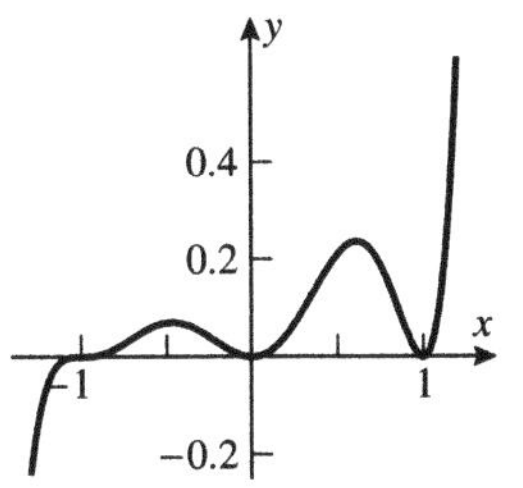

(d) $\lim_{x\to\pm\infty} y = +\infty$;
curve crosses x-axis at $x = 0, 1$

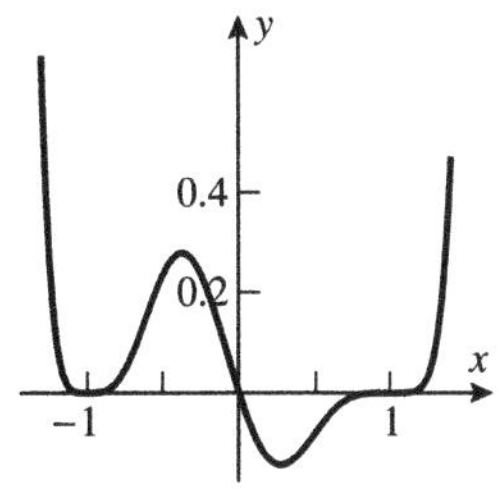

44. **(a)**

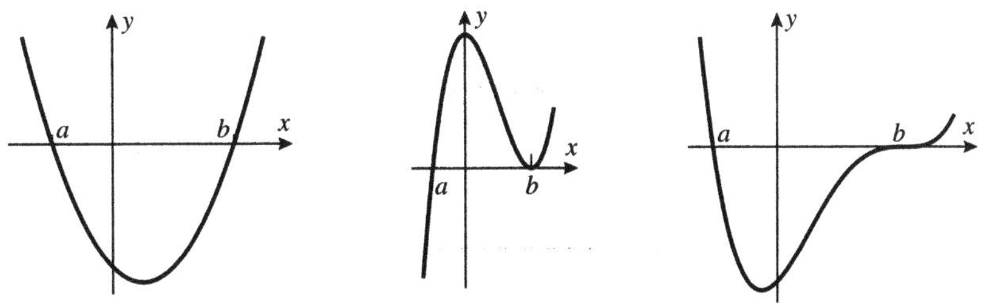

(b)

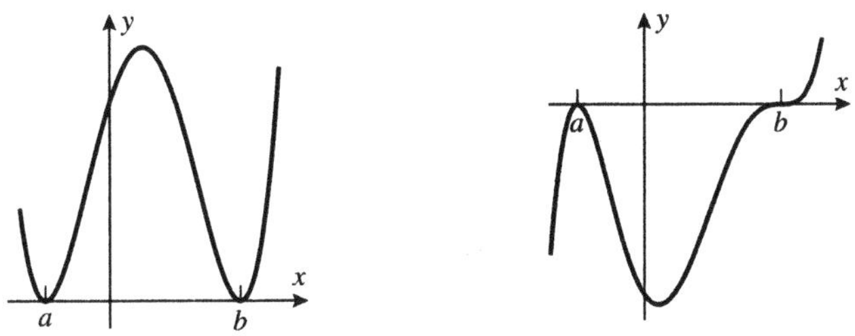

(c)

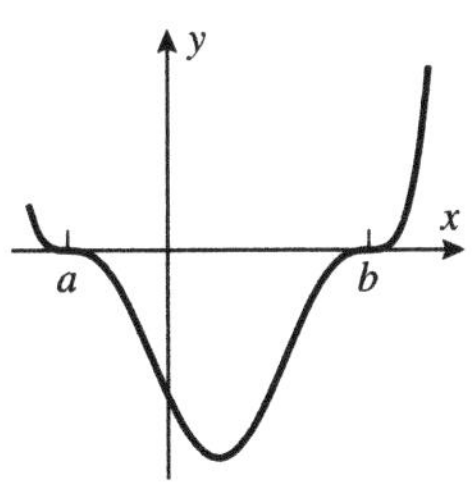

45. **(a)** horizontal asymptote $y = 3$
as $x \to \pm\infty$, vertical asymptotes
at $x = \pm 2$

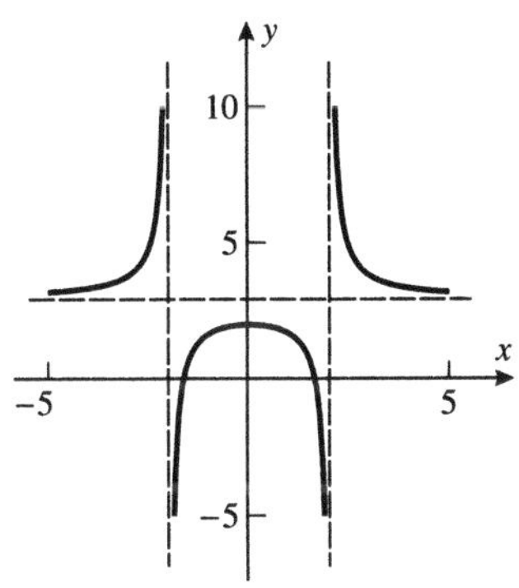

(b) horizontal asymptote $y = 1$
as $x \to \pm\infty$, vertical asymptotes
at $x = \pm 1$

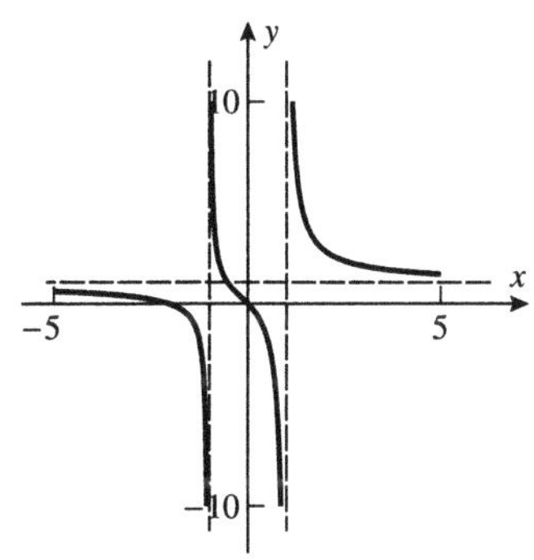

(c) horizontal asymptote $y = -1$ as $x \to \pm\infty$, vertical asymptotes at $x = -2, 1$

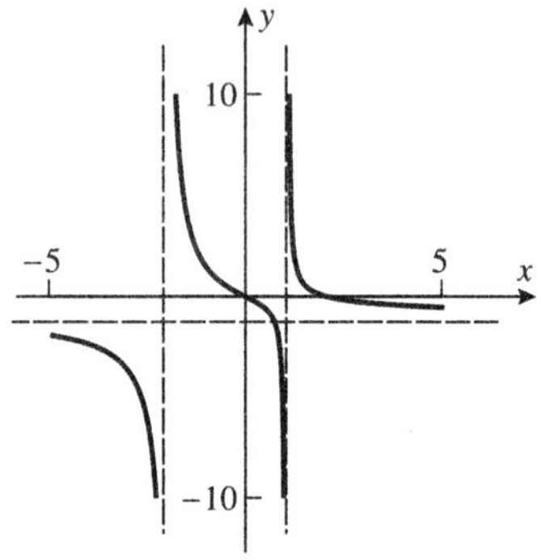

(d) horizontal asymptote $y = 1$ as $x \to \pm\infty$, vertical asymptote at $x = -1, 2$

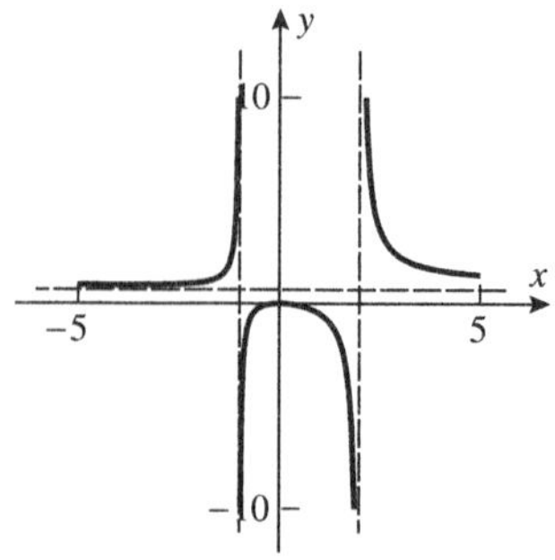

46.

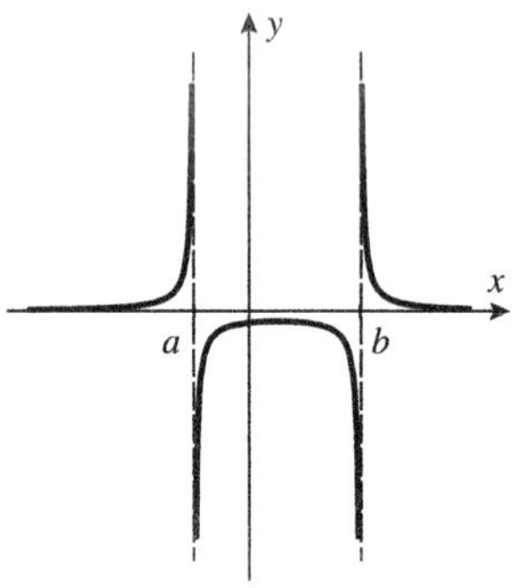

47. Symmetric about the line $x = \dfrac{a+b}{2}$ means $f\left(\dfrac{a+b}{2} + x\right) = f\left(\dfrac{a+b}{2} - x\right)$ for any x.

Note that $\dfrac{a+b}{2} + x - a = x - \dfrac{a-b}{2}$ and $\dfrac{a+b}{2} + x - b = x + \dfrac{a-b}{2}$, and the same equations are true with x replaced by $-x$. Hence

$$\left(\frac{a+b}{2} + x - a\right)\left(\frac{a+b}{2} + x - b\right) = \left(x - \frac{a-b}{2}\right)\left(x + \frac{a-b}{2}\right) = x^2 - \left(\frac{a-b}{2}\right)^2$$

The right hand side remains the same if we replace x with $-x$, and so the same is true of the left hand side, and the same is therefore true of the reciprocal of the left hand side. But the reciprocal of the left hand side is equal to $f\left(\dfrac{a+b}{2} + x\right)$. Since this quantity remains unchanged if we replace x with $-x$, the condition of symmetry about the line $x = \dfrac{a+b}{2}$ has been met.

48. $\displaystyle\lim_{x\to\pm\infty}\left[\frac{P(x)}{Q(x)} - (ax+b)\right] = \lim_{x\to\pm\infty}\frac{R(x)}{Q(x)} = 0$ because the degree of $R(x)$ is less than the degree of $Q(x)$.

49. $y = \dfrac{x^2 - 2}{x} = x - \dfrac{2}{x}$ so

$y = x$ is an oblique asymptote;

$y' = \dfrac{x^2 + 2}{x^2}$,

$y'' = -\dfrac{4}{x^3}$

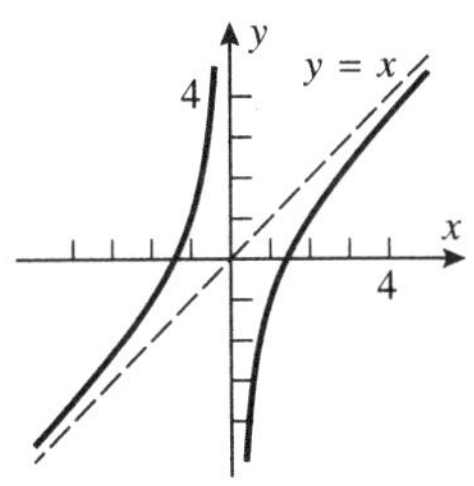

50. $y = \dfrac{x^2 - 2x - 3}{x + 2} = x - 4 + \dfrac{5}{x + 2}$ so

$y = x - 4$ is an oblique asymptote;

$y' = \dfrac{x^2 + 4x - 1}{(x + 2)^2}$, $y'' = \dfrac{10}{(x + 2)^3}$

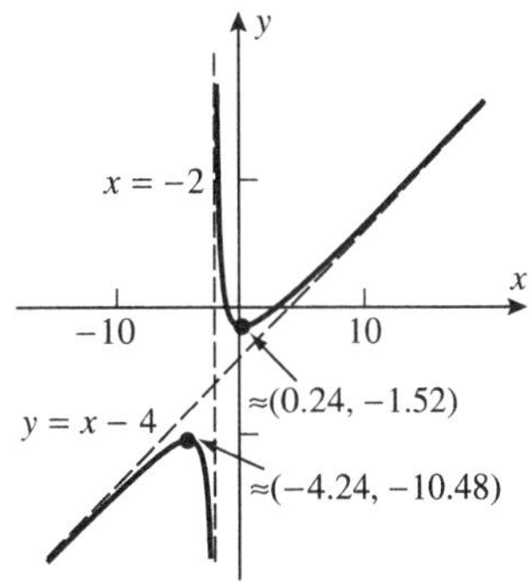

51. $y = \dfrac{(x - 2)^3}{x^2} = x - 6 + \dfrac{12x - 8}{x^2}$ so

$y = x - 6$ is an oblique asymptote;

$y' = \dfrac{(x - 2)^2(x + 4)}{x^3}$,

$y'' = \dfrac{24(x - 2)}{x^4}$

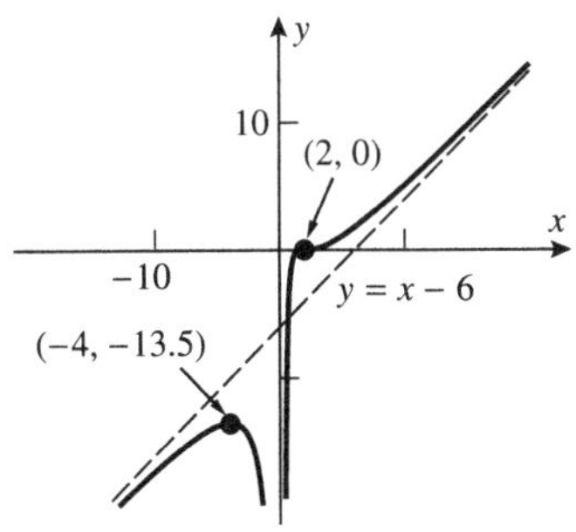

52. $y = \dfrac{4 - x^3}{x^2}$,

$y' = -\dfrac{x^3 + 8}{x^3}$,

$y'' = \dfrac{24}{x^4}$

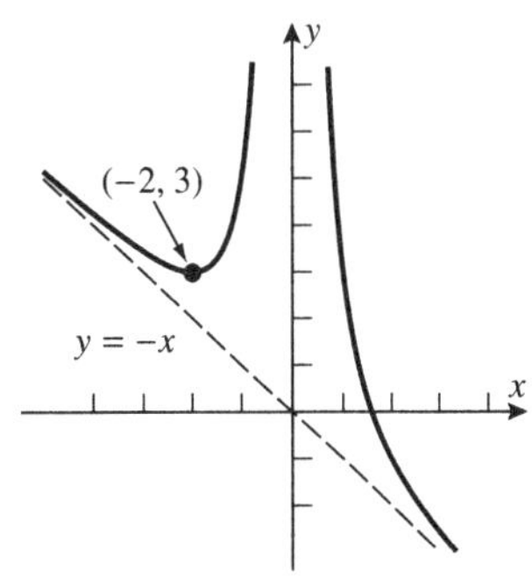

53. $y = x + 1 - \dfrac{1}{x} - \dfrac{1}{x^2} = \dfrac{(x - 1)(x + 1)^2}{x^2}$,

$y = x + 1$ is an oblique asymptote;

$y' = \dfrac{(x + 1)(x^2 - x + 2)}{x^3}$,

$y'' = -\dfrac{2(x + 3)}{x^4}$

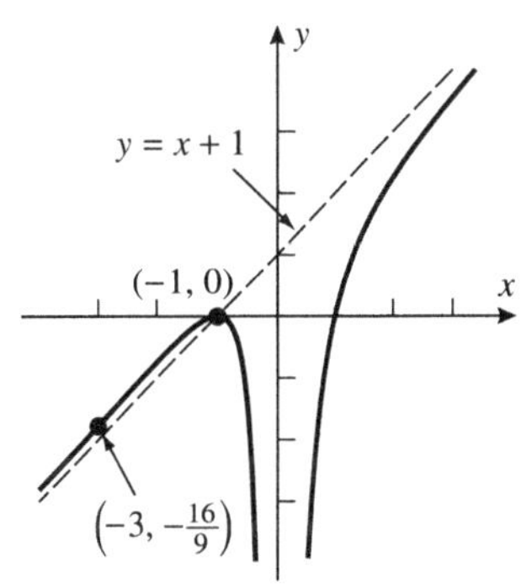

54. The oblique asymptote is $y = 2x$ so $(2x^3 - 3x + 4)/x^2 = 2x$, $-3x + 4 = 0$, $x = 4/3$.

55. $\lim_{x\to\pm\infty}[f(x)-x^2] = \lim_{x\to\pm\infty}(1/x) = 0$

$y = x^2 + \dfrac{1}{x} = \dfrac{x^3+1}{x}$, $y' = 2x - \dfrac{1}{x^2} = \dfrac{2x^3-1}{x^2}$,

$y'' = 2 + \dfrac{2}{x^3} = \dfrac{2(x^3+1)}{x^3}$, $y' = 0$ when $x = 1/\sqrt[3]{2} \approx 0.8$,

$y = 3\sqrt[3]{2}/2 \approx 1.9$; $y'' = 0$ when $x = -1, y = 0$

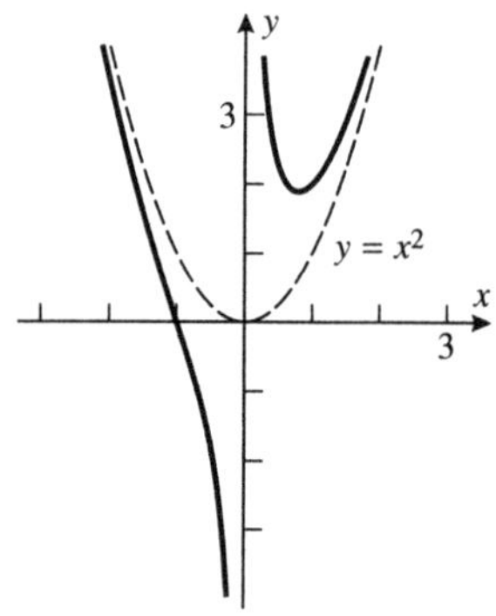

56. $\lim_{x\to\pm\infty}[f(x)-(3-x^2)] = \lim_{x\to\pm\infty}(2/x) = 0$

$y = 3 - x^2 + \dfrac{2}{x} = \dfrac{2+3x-x^3}{x}$, $y' = -2x - \dfrac{2}{x^2} = -\dfrac{2(x^3+1)}{x^2}$,

$y'' = -2 + \dfrac{4}{x^3} = -\dfrac{2(x^3-2)}{x^3}$, $y' = 0$ when $x = -1, y = 0$;

$y'' = 0$ when $x = \sqrt[3]{2} \approx 1.3$, $y = 3$

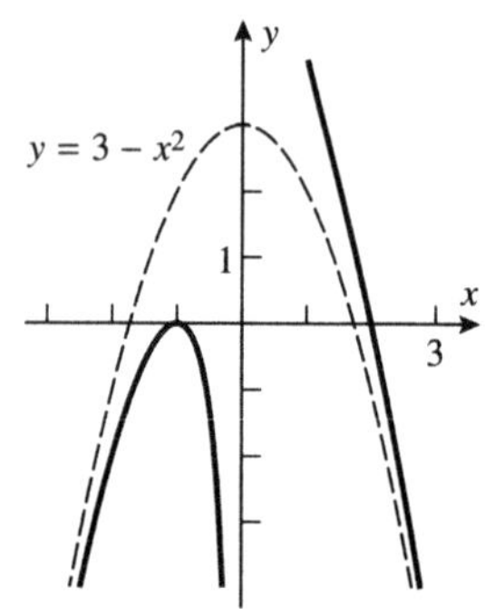

57. Let y be the length of the other side of the rectangle, then $L = 2x+2y$ and $xy = 400$ so $y = 400/x$ and hence $L = 2x + 800/x$. $L = 2x$ is an oblique asymptote (see Exercise 48)

$L = 2x + \dfrac{800}{x} = \dfrac{2(x^2+400)}{x}$,

$L' = 2 - \dfrac{800}{x^2} = \dfrac{2(x^2-400)}{x^2}$,

$L'' = \dfrac{1600}{x^3}$,

$L' = 0$ when $x = 20, L = 80$

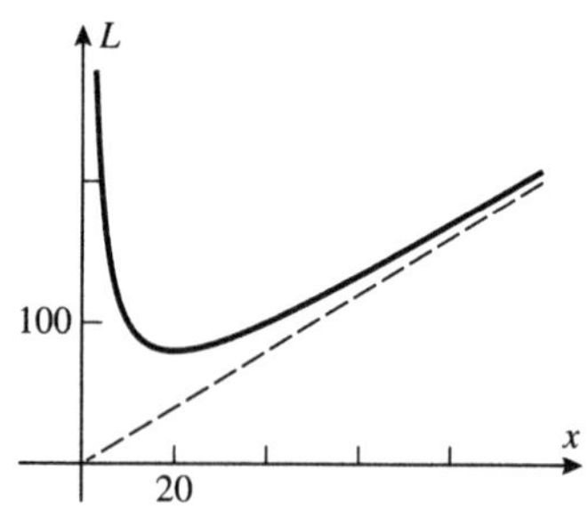

58. Let y be the height of the box, then $S = x^2 + 4xy$ and $x^2y = 500$ so $y = 500/x^2$ and hence $S = x^2 + 2000/x$. The graph approaches the curve $S = x^2$ asymptotically (see Exercise 63)

$S = x^2 + \dfrac{2000}{x} = \dfrac{x^3+2000}{x}$,

$S' = 2x - \dfrac{2000}{x^2} = \dfrac{2(x^3-1000)}{x^2}$,

$S'' = 2 + \dfrac{4000}{x^3} = \dfrac{2(x^3+2000)}{x^3}$,

$S'' = 0$ when $x = 10, S = 300$

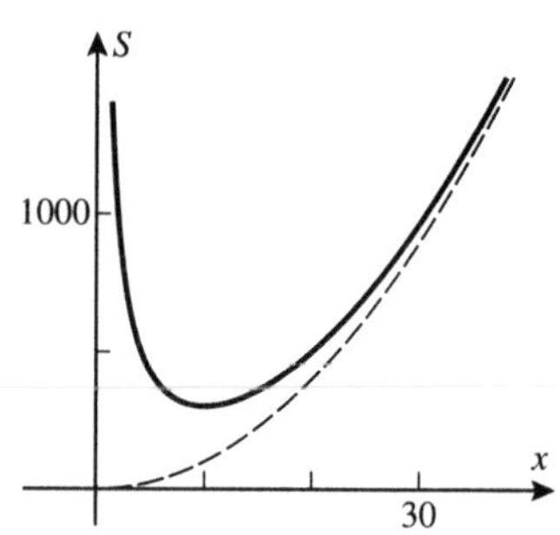

59. $y' = 0.1x^4(6x - 5)$;
critical numbers: $x = 0$, $x = 5/6$;
relative minimum at $x = 5/6$,
$y \approx -6.7 \times 10^{-3}$

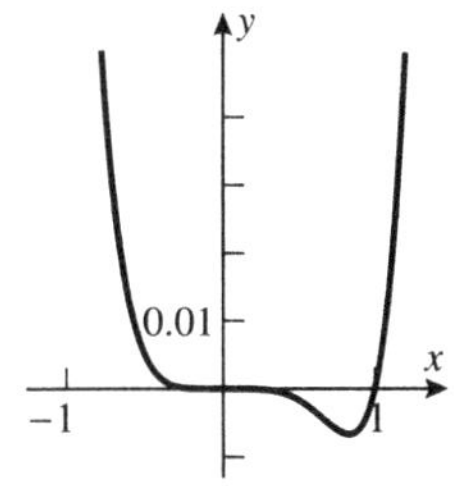

60. $y' = 0.1x^4(x+1)(7x+5)$;
critical numbers: $x = 0$, $x = -1$, $x = -5/7$,
relative maximum at $x = -1$, $y = 0$;
relative minimum at $x = -5/7$, $y \approx -1.5 \times 10^{-3}$

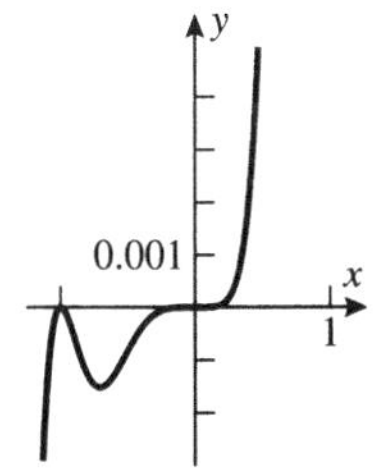

EXERCISE SET 4.4

1. **(a)** positive, negative, slowing down
(b) positive, positive, speeding up
(c) negative, positive, slowing down

2. **(a)** positive, slowing down
(b) negative, slowing down
(c) positive, speeding up

3. **(a)** left because $v = ds/dt < 0$ at t_0
(b) negative because $a = d^2s/dt^2$ and the curve is concave down at $t_0 (d^2s/dt^2 < 0)$
(c) speeding up because v and a have the same sign
(d) $v < 0$ and $a > 0$ at t_1 so the particle is slowing down because v and a have opposite signs.

4. **(a)** III **(b)** I **(c)** II

5.

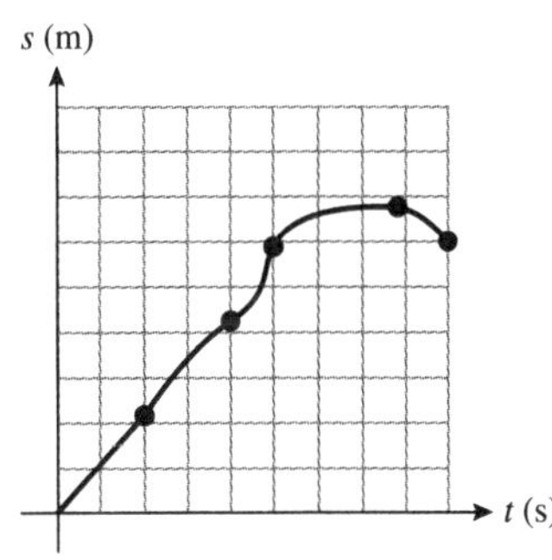

6. **(a)** when $s \geq 0$, so $0 < t < 2$ and $4 < t \leq 8$ **(b)** when the slope is zero, at $t = 3$
(c) when s is decreasing, so $0 \leq t < 3$

7.

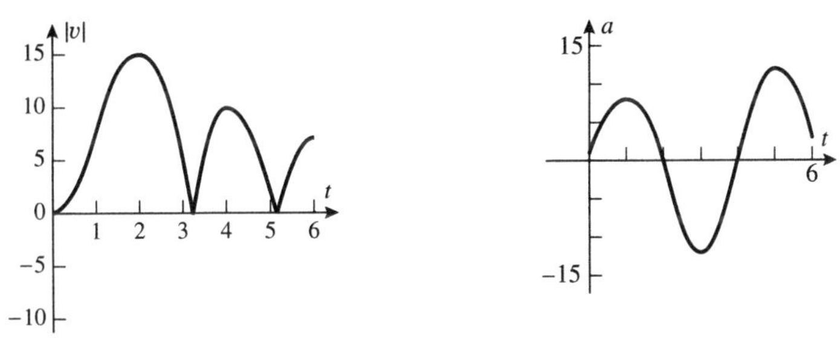

8. (a) $v \approx (30 - 10)/(15 - 10) = 20/5 = 4$ m/s

(b)

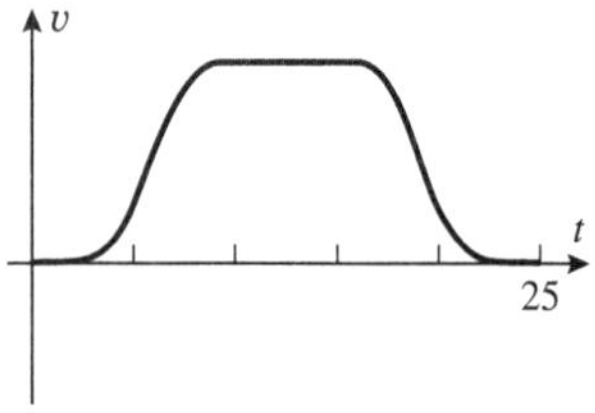

(1)

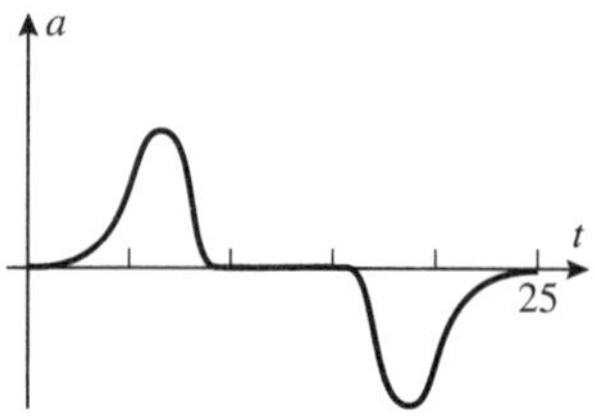

(2)

9. (a) At 60 mi/h the slope of the estimated tangent line is about 4.6 mi/h/s. Use 1 mi $= 5,280$ ft and 1 h $= 3600$ s to get $a = dv/dt \approx 4.6(5{,}280)/(3600) \approx 6.7$ ft/s^2.

(b) The slope of the tangent to the curve is maximum at $t = 0$ s.

10. (a)

t	1	2	3	4	5
s	0.71	1.00	0.71	0.00	−0.71
v	0.56	0.00	−0.56	−0.79	−0.56
a	−0.44	−0.62	−0.44	0.00	0.44

(b) to the right at $t = 1$, stopped at $t = 2$, otherwise to the left

(c) speeding up at $t = 3$; slowing down at $t = 1, 5$; neither at $t = 2, 4$

11. (a) $v(t) = 3t^2 - 12t$, $a(t) = 6t - 12$

(b) $s(1) = -5$ ft, $v(1) = -9$ ft/s, speed $= 9$ ft/s, $a(1) = -6$ ft/s^2

(c) $v = 0$ at $t = 0, 4$

(d) for $t \geq 0$, $v(t)$ changes sign at $t = 4$, and $a(t)$ changes sign at $t = 2$; so the particle is speeding up for $0 < t < 2$ and $4 < t$ and is slowing down for $2 < t < 4$

(e) total distance $= |s(4) - s(0)| + |s(5) - s(4)| = |-32 - 0| + |-25 - (-32)| = 39$ ft

12. (a) $v(t) = 4t^3 - 4$, $a(t) = 12t^2$

(b) $s(1) = -1$ ft, $v(1) = 0$ ft/s, speed $= 0$ ft/s, $a(1) = 12$ ft/s^2

(c) $v = 0$ at $t = 1$

(d) speeding up for $t > 1$, slowing down for $0 < t < 1$

(e) total distance $= |s(1) - s(0)| + |s(5) - s(1)| = |-1 - 2| + |607 - (-1)| = 611$ ft

13. (a) $v(t) = -(3\pi/2)\sin(\pi t/2)$, $a(t) = -(3\pi^2/4)\cos(\pi t/2)$

(b) $s(1) = 0$ ft, $v(1) = -3\pi/2$ ft/s, speed $= 3\pi/2$ ft/s, $a(1) = 0$ ft/s^2

(c) $v = 0$ at $t = 0, 2, 4$

(d) v changes sign at $t = 0, 2, 4$ and a changes sign at $t = 1, 3, 5$, so the particle is speeding up for $0 < t < 1$, $2 < t < 3$ and $4 < t < 5$, and it is slowing down for $1 < t < 2$ and $3 < t < 4$

(e)
$$\begin{aligned}\text{total distance } &= |s(2) - s(0)| + |s(4) - s(2)| + |s(5) - s(4)| \\ &= |-3 - 3| + |3 - (-3)| + |0 - 3| = 15 \text{ ft}\end{aligned}$$

14. (a) $v(t) = \dfrac{4 - t^2}{(t^2 + 4)^2}$, $a(t) = \dfrac{2t(t^2 - 12)}{(t^2 + 4)^3}$

(b) $s(1) = 1/5$ ft, $v(1) = 3/25$ ft/s, speed $= 3/25$ ft/s, $a(1) = -22/125$ ft/s^2

(c) $v = 0$ at $t = 2$

(d) a changes sign at $t = 2\sqrt{3}$, so the particle is speeding up for $2 < t < 2\sqrt{3}$ and it is slowing down for $0 < t < 2$ and for $2\sqrt{3} < t$

(e) total distance $= |s(2) - s(0)| + |s(5) - s(2)| = \left|\dfrac{1}{4} - 0\right| + \left|\dfrac{5}{29} - \dfrac{1}{4}\right| = \dfrac{19}{58}$ ft

15. $v(t) = \dfrac{5-t^2}{(t^2+5)^2}$, $a(t) = \dfrac{2t(t^2-15)}{(t^2+5)^3}$

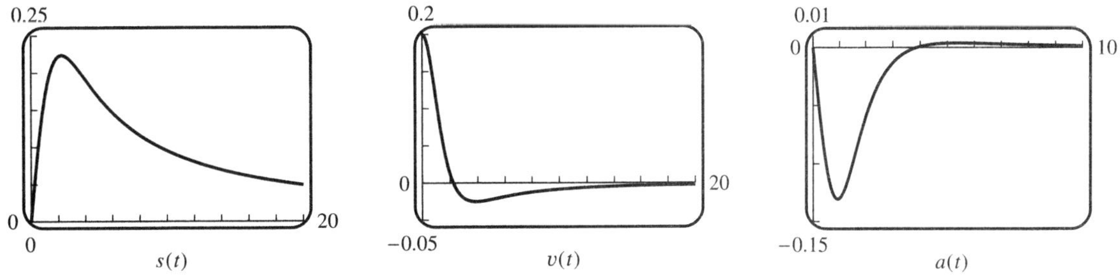

(a) $v = 0$ at $t = \sqrt{5}$

(b) $s = \sqrt{5}/10$ at $t = \sqrt{5}$

(c) a changes sign at $t = \sqrt{15}$, so the particle is speeding up for $\sqrt{5} < t < \sqrt{15}$ and slowing down for $0 < t < \sqrt{5}$ and $\sqrt{15} < t$

16. $v(t) = \dfrac{-t^2+1}{(t^2+1)^2}$, $a(t) = 2\dfrac{t(t^3-3)}{(t^2+1)^3}$

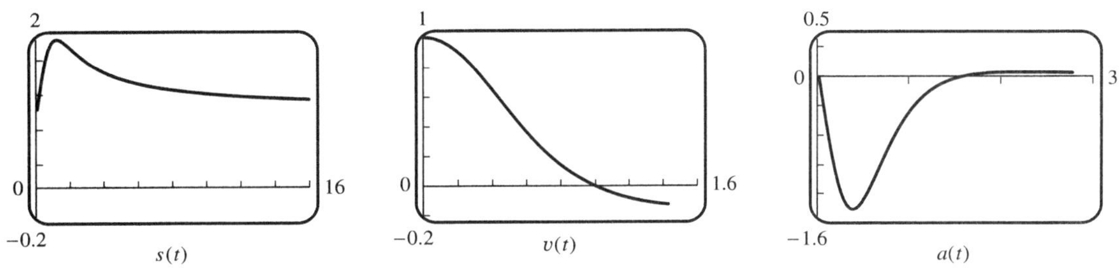

(a) $v = 0$ at $t = 1$

(b) $s = 3/2$ at $t = 1$

(c) a changes sign at $t = 3^{1/3}$, so the particle is slowing down for $t < 3^{1/3}$ and speeding up for $3^{1/3} < t$.

17. $s = -3t + 2$
$v = -3$
$a = 0$

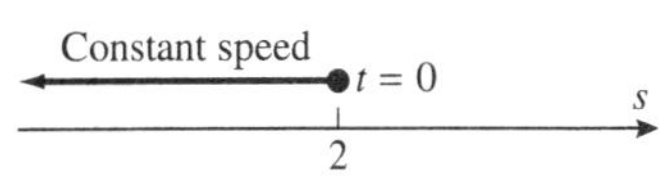

18. $s = t^3 - 6t^2 + 9t + 1$
$v = 3(t-1)(t-3)$
$a = 6(t-2)$

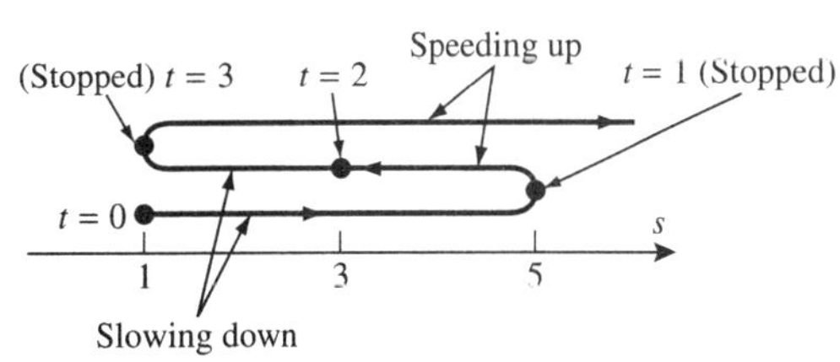

19. $s = t^3 - 9t^2 + 24t$
$v = 3(t-2)(t-4)$
$a = 6(t-3)$

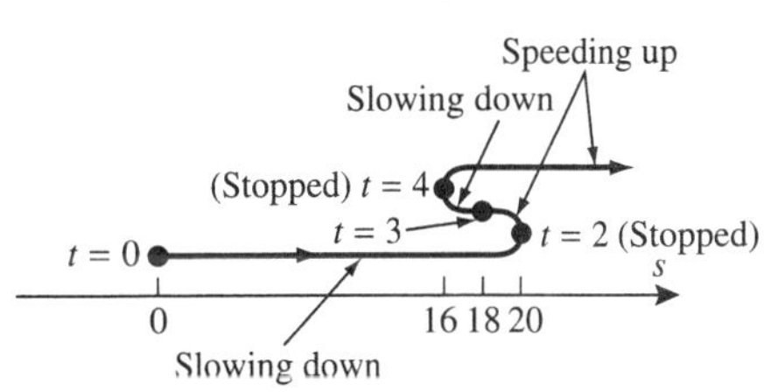

20. $s = t + \dfrac{9}{t+1}$

$v = \dfrac{(t+4)(t-2)}{(t+1)^2}$

$a = \dfrac{18}{(t+1)^3}$

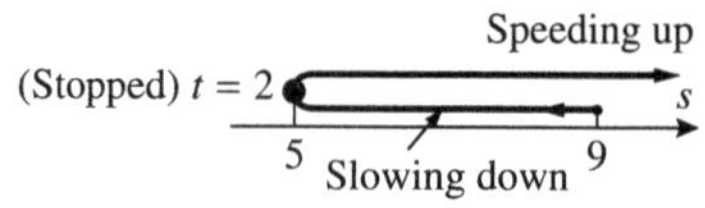

21. $s = \begin{cases} \cos t, & 0 \le t \le 2\pi \\ 1, & t > 2\pi \end{cases}$

$v = \begin{cases} -\sin t, & 0 \le t \le 2\pi \\ 0, & t > 2\pi \end{cases}$

$a = \begin{cases} -\cos t, & 0 \le t < 2\pi \\ 0, & t > 2\pi \end{cases}$

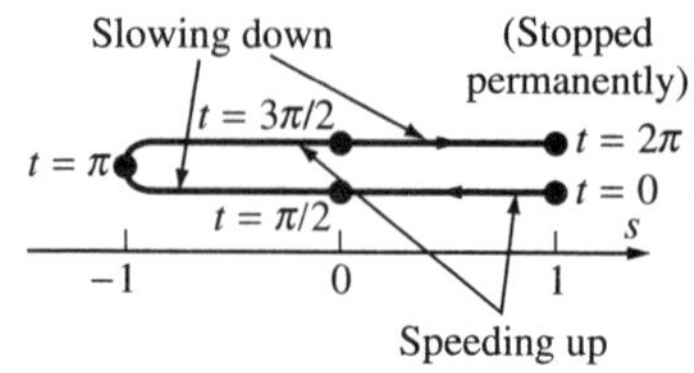

22. $v(t) = \dfrac{5t^2 - 6t + 2}{\sqrt{t}}$ is always positive,

$a(t) = \dfrac{15t^2 - 6t - 2}{2t^{3/2}}$ has a positive root at $t = \dfrac{3 + \sqrt{39}}{15}$

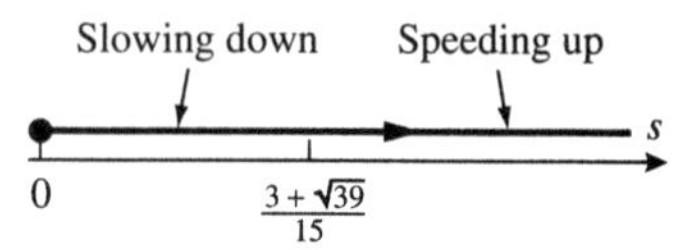

23. **(a)** $v = 10t - 22$, speed $= |v| = |10t - 22|$. $d|v|/dt$ does not exist at $t = 2.2$ which is the only critical point. If $t = 1, 2.2, 3$ then $|v| = 12, 0, 8$. The maximum speed is 12 ft/s.

(b) the distance from the origin is $|s| = |5t^2 - 22t| = |t(5t - 22)|$, but $t(5t - 22) < 0$ for $1 \le t \le 3$ so $|s| = -(5t^2 - 22t) = 22t - 5t^2$, $d|s|/dt = 22 - 10t$, thus the only critical point is $t = 2.2$. $d^2|s|/dt^2 < 0$ so the particle is farthest from the origin when $t = 2.2$. Its position is $s = 5(2.2)^2 - 22(2.2) = -24.2$.

24. $v = -\dfrac{200t}{(t^2+12)^2}$, speed $= |v| = \dfrac{200t}{(t^2+12)^2}$ for $t \ge 0$. $\dfrac{d|v|}{dt} = \dfrac{600(4 - t^2)}{(t^2+12)^3} = 0$ when $t = 2$, which is the only critical point in $(0, +\infty)$. By the first derivative test there is a relative maximum, and hence an absolute maximum, at $t = 2$. The maximum speed is 25/16 ft/s to the left.

25. $s(t) = s_0 - \frac{1}{2}gt^2 = s_0 - 4.9t^2$ m, $v = -9.8t$ m/s, $a = -9.8$ m/s^2

(a) $|s(1.5) - s(0)| = 11.025$ m

(b) $v(1.5) = -14.7$ m/s

(c) $|v(t)| = 12$ when $t = 12/9.8 = 1.2245$ s

(d) $s(t) - s_0 = -100$ when $4.9t^2 = 100$, $t = 4.5175$ s

26. **(a)** $s(t) = s_0 - \frac{1}{2}gt^2 = 800 - 16t^2$ ft, $s(t) = 0$ when $t = \sqrt{\dfrac{800}{16}} = 5\sqrt{2}$

(b) $v(t) = -32t$ and $v(5\sqrt{2}) = -160\sqrt{2} \approx 226.27$ ft/s $= 154.28$ mi/h

27. $s(t) = s_0 + v_0 t - \frac{1}{2}gt^2 = 60t - 4.9t^2$ m and $v(t) = v_0 - gt = 60 - 9.8t$ m/s

(a) $v(t) = 0$ when $t = 60/9.8 \approx 6.12$ s

(b) $s(60/9.8) \approx 183.67$ m

(c) another 6.12 s; solve for t in $s(t) = 0$ to get this result, or use the symmetry of the parabola $s = 60t - 4.9t^2$ about the line $t = 6.12$ in the t-s plane

(d) also 60 m/s, as seen from the symmetry of the parabola (or compute $v(6.12)$)

28. **(a)** they are the same

(b) $s(t) = v_0t - \frac{1}{2}gt^2$ and $v(t) = v_0 - gt$; $s(t) = 0$ when $t = 0, 2v_0/g$;
$v(0) = v_0$ and $v(2v_0/g) = v_0 - g(2v_0/g) = -v_0$ so the speed is the same at launch $(t = 0)$ and at return $(t = 2v_0/g)$.

29. If g $= 32$ ft/s^2, $s_0 = 7$ and v_0 is unknown, then $s(t) = 7 + v_0t - 16t^2$ and $v(t) = v_0 - 32t$; $s = s_{\max}$ when $v = 0$, or $t = v_0/32$; and $s_{\max} = 208$ yields
$208 = s(v_0/32) = 7 + v_0(v_0/32) - 16(v_0/32)^2 = 7 + v_0^2/64$, so $v_0 = 8\sqrt{201} \approx 113.42$ ft/s.

30. **(a)** Use (6) and then (5) to get $v^2 = v_0^2 - 2v_0gt + g^2t^2 = v_0^2 - 2g(v_0t - \frac{1}{2}gt^2) = v_0^2 - 2g(s - s_0)$.

(b) Add v_0 to both sides of (6): $2v_0 - gt = v_0 + v$, $v_0 - \frac{1}{2}gt = \frac{1}{2}(v_0 + v)$;
from (5) $s = s_0 + t(v_0 - \frac{1}{2}gt) = s_0 + \frac{1}{2}(v_0 + v)t$

(c) Add v to both sides of (6): $2v + gt = v_0 + v$, $v + \frac{1}{2}gt = \frac{1}{2}(v_0 + v)$; from Part (b),
$s = s_0 + \frac{1}{2}(v_0 + v)t = s_0 + vt + \frac{1}{2}gt^2$

31. $v_0 = 0$ and $g = 9.8$, so $v^2 = -19.6(s - s_0)$; since $v = 24$ when $s = 0$ it follows that $19.6s_0 = 24^2$ or $s_0 = 29.39$ m.

32. $s = 1000 + vt + \frac{1}{2}(32)t^2 = 1000 + vt + 16t^2$; $s = 0$ when $t = 5$, so $v = -(1000 + 16 \cdot 5^2)/5 = -280$ ft/s.

33. **(a)** $s = s_{\max}$ when $v = 0$, so $0 = v_0^2 - 2g(s_{\max} - s_0)$, $s_{\max} = v_0^2/2g + s_0$.

(b) $s_0 = 7$, $s_{\max} = 208$, $g = 32$ and v_0 is unknown, so from Part (a) $v_0^2 = 2g(208 - 7) = 64 \cdot 201$, $v_0 = 8\sqrt{201} \approx 113.42$ ft/s.

34. $s = t^3 - 6t^2 + 1$, $v = 3t^2 - 12t$, $a = 6t - 12$.

(a) $a = 0$ when $t = 2$; $s = -15$, $v = -12$.

(b) $v = 0$ when $3t^2 - 12t = 3t(t - 4) = 0$, $t = 0$ or $t = 4$. If $t = 0$, then $s = 1$ and $a = -12$; if $t = 4$, then $s = -31$ and $a = 12$.

35. **(a)**

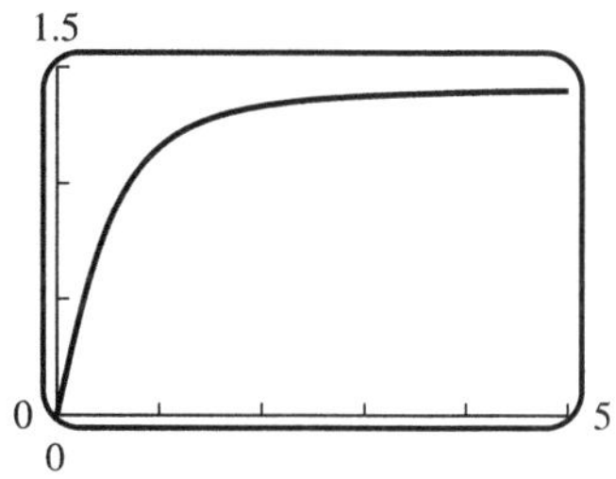

(b) $v = \dfrac{2t}{\sqrt{2t^2 + 1}}$, $\lim_{t \to +\infty} v = \dfrac{2}{\sqrt{2}} = \sqrt{2}$

36. **(a)** $a = \dfrac{dv}{dt} = \dfrac{dv}{ds}\dfrac{ds}{dt} = v\dfrac{dv}{ds}$ because $v = \dfrac{ds}{dt}$

(b) $v = \dfrac{3}{2\sqrt{3t + 7}} = \dfrac{3}{2s}$; $\dfrac{dv}{ds} = -\dfrac{3}{2s^2}$; $a = -\dfrac{9}{4s^3} = -9/500$

37. **(a)** $s_1 = s_2$ if they collide, so $\frac{1}{2}t^2 - t + 3 = -\frac{1}{4}t^2 + t + 1$, $\frac{3}{4}t^2 - 2t + 2 = 0$ which has no real solution.

(b) Find the minimum value of $D = |s_1 - s_2| = \left|\frac{3}{4}t^2 - 2t + 2\right|$. From Part (a), $\frac{3}{4}t^2 - 2t + 2$ is never zero, and for $t = 0$ it is positive, hence it is always positive, so $D = \frac{3}{4}t^2 - 2t + 2$.

$\dfrac{dD}{dt} = \frac{3}{2}t - 2 = 0$ when $t = \frac{4}{3}$. $\dfrac{d^2D}{dt^2} > 0$ so D is minimum when $t = \frac{4}{3}$, $D = \frac{2}{3}$.

(c) $v_1 = t - 1$, $v_2 = -\frac{1}{2}t + 1$. $v_1 < 0$ if $0 \le t < 1$, $v_1 > 0$ if $t > 1$; $v_2 < 0$ if $t > 2$, $v_2 > 0$ if $0 \le t < 2$. They are moving in opposite directions during the intervals $0 \le t < 1$ and $t > 2$.

38. (a) $s_A - s_B = 20 - 0 = 20$ ft
(b) $s_A = s_B$, $15t^2 + 10t + 20 = 5t^2 + 40t$, $10t^2 - 30t + 20 = 0$, $(t-2)(t-1) = 0$, $t = 1$ or $t = 2$ s.
(c) $v_A = v_B$, $30t + 10 = 10t + 40$, $20t = 30$, $t = 3/2$ s. When $t = 3/2$, $s_A = 275/4$ and $s_B = 285/4$ so car B is ahead of car A.

39. $r(t) = \sqrt{v^2(t)}$, $r'(t) = 2v(t)v'(t)/[2\sqrt{v(t)}] = v(t)a(t)/|v(t)|$ so $r'(t) > 0$ (speed is increasing) if v and a have the same sign, and $r'(t) < 0$ (speed is decreasing) if v and a have opposite signs.
If $v(t) > 0$ then $r(t) = v(t)$ and $r'(t) = a(t)$, so if $a(t) > 0$ then the particle is speeding up and a and v have the same sign; if $a(t) < 0$, then the particle is slowing down, and a and v have opposite signs.
If $v(t) < 0$ then $r(t) = -v(t)$, $r'(t) = -a(t)$, and if $a(t) > 0$ then the particle is speeding up and a and v have opposite signs; if $a(t) < 0$ then the particle is slowing down and a and v have the same sign.

EXERCISE SET 4.5

1. relative maxima at $x = 2, 6$; absolute maximum at $x = 6$; relative and absolute minimum at $x = 4$

2. relative maximum at $x = 3$; absolute maximum at $x = 7$; relative minima at $x = 1, 5$; absolute minima at $x = 1, 5$

3. (a)

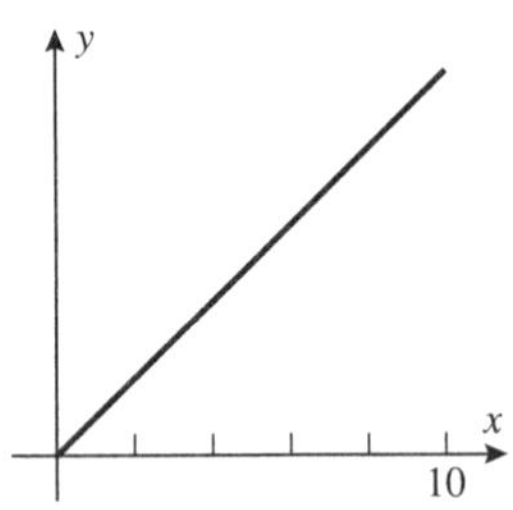

(b)

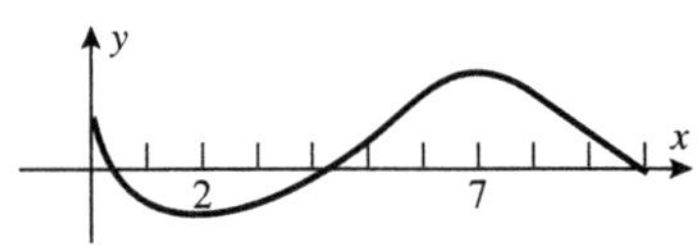

(c)

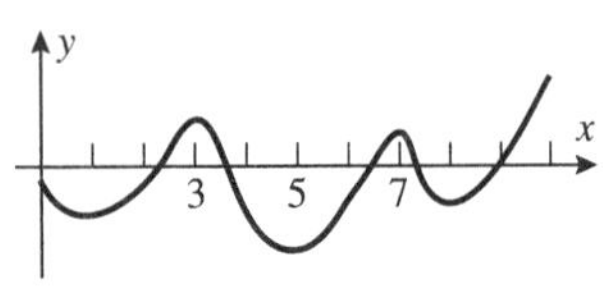

4. (a)

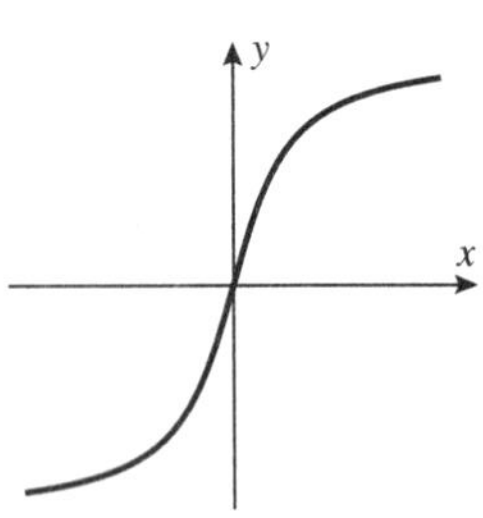

(b)

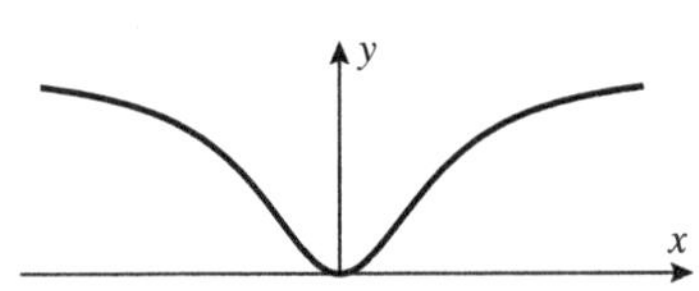

(c)

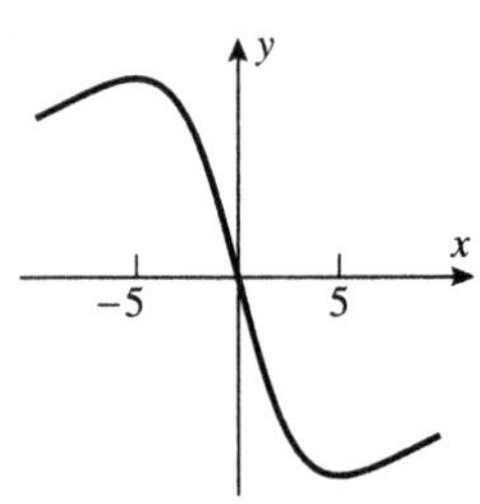

5. $f'(x) = 8x - 4$, $f'(x) = 0$ when $x = 1/2$; $f(0) = 1$, $f(1/2) = 0$, $f(1) = 1$ so the maximum value is 1 at $x = 0, 1$ and the minimum value is 0 at $x = 1/2$.

6. $f'(x) = 8 - 2x$, $f'(x) = 0$ when $x = 4$; $f(0) = 0$, $f(4) = 16$, $f(6) = 12$ so the maximum value is 16 at $x = 4$ and the minimum value is 0 at $x = 0$.

7. $f'(x) = 3(x-1)^2$, $f'(x) = 0$ when $x = 1$; $f(0) = -1$, $f(1) = 0$, $f(4) = 27$ so the maximum value is 27 at $x = 4$ and the minimum value is -1 at $x = 0$.

8. $f'(x) = 6x^2 - 6x - 12 = 6(x+1)(x-2)$, $f'(x) = 0$ when $x = -1, 2$; $f(-2) = -4$, $f(-1) = 7$, $f(2) = -20$, $f(3) = -9$ so the maximum value is 7 at $x = -1$ and the minimum value is -20 at $x = 2$.

9. $f'(x) = 3/(4x^2+1)^{3/2}$, no critical points; $f(-1) = -3/\sqrt{5}$, $f(1) = 3/\sqrt{5}$ so the maximum value is $3/\sqrt{5}$ at $x = 1$ and the minimum value is $-3/\sqrt{5}$ at $x = -1$.

10. $f'(x) = \dfrac{2(2x+1)}{3(x^2+x)^{1/3}}$, $f'(x) = 0$ when $x = -1/2$ and $f'(x)$ does not exist when $x = -1, 0$; $f(-2) = 2^{2/3}$, $f(-1) = 0$, $f(-1/2) = 4^{-2/3}$, $f(0) = 0$, $f(3) = 12^{2/3}$ so the maximum value is $12^{2/3}$ at $x = 3$ and the minimum value is 0 at $x = -1, 0$.

11. $f'(x) = 1 - \sec^2 x$, $f'(x) = 0$ for x in $(-\pi/4, \pi/4)$ when $x = 0$; $f(-\pi/4) = 1 - \pi/4$, $f(0) = 0$, $f(\pi/4) = \pi/4 - 1$ so the maximum value is $1 - \pi/4$ at $x = -\pi/4$ and the minimum value is $\pi/4 - 1$ at $x = \pi/4$.

12. $f'(x) = \cos x + \sin x$, $f'(x) = 0$ for x in $(0, \pi)$ when $x = 3\pi/4$; $f(0) = -1$, $f(3\pi/4) = \sqrt{2}$, $f(\pi) = 1$ so the maximum value is $\sqrt{2}$ at $x = 3\pi/4$ and the minimum value is -1 at $x = 0$.

13. $f(x) = 1 + |9 - x^2| = \begin{cases} 10 - x^2, & |x| \le 3 \\ -8 + x^2, & |x| > 3 \end{cases}$, $f'(x) = \begin{cases} -2x, & |x| < 3 \\ 2x, & |x| > 3 \end{cases}$ thus $f'(x) = 0$ when $x = 0$, $f'(x)$ does not exist for x in $(-5, 1)$ when $x = -3$ because $\lim_{x \to -3^-} f'(x) \ne \lim_{x \to -3^+} f'(x)$ (see Theorem preceding Exercise 75, Section 3.3); $f(-5) = 17$, $f(-3) = 1$, $f(0) = 10$, $f(1) = 9$ so the maximum value is 17 at $x = -5$ and the minimum value is 1 at $x = -3$.

14. $f(x) = |6 - 4x| = \begin{cases} 6 - 4x, & x \le 3/2 \\ -6 + 4x, & x > 3/2 \end{cases}$, $f'(x) = \begin{cases} -4, & x < 3/2 \\ 4, & x > 3/2 \end{cases}$, $f'(x)$ does not exist when $x = 3/2$ thus $3/2$ is the only critical point in $(-3, 3)$; $f(-3) = 18$, $f(3/2) = 0$, $f(3) = 6$ so the maximum value is 18 at $x = -3$ and the minimum value is 0 at $x = 3/2$.

15. $f'(x) = 2x - 3$; critical point $x = 3/2$. Minimum value $f(3/2) = -13/4$, no maximum.

16. $f'(x) = -4(x+1)$; critical point $x = -1$. Maximum value $f(-1) = 5$, no minimum.

17. $f'(x) = 12x^2(1-x)$; critical points $x = 0, 1$. Maximum value $f(1) = 1$, no minimum because $\lim_{x \to +\infty} f(x) = -\infty$.

18. $f'(x) = 4(x^3+1)$; critical point $x = -1$. Minimum value $f(-1) = -3$, no maximum.

19. No maximum or minimum because $\lim_{x \to +\infty} f(x) = +\infty$ and $\lim_{x \to -\infty} f(x) = -\infty$.

20. No maximum or minimum because $\lim_{x \to +\infty} f(x) = +\infty$ and $\lim_{x \to -\infty} f(x) = -\infty$.

21. $f'(x) = x(x+2)/(x+1)^2$; critical point $x = -2$ in $(-5, -1)$. Maximum value $f(-2) = -4$, no minimum.

22. $f'(x) = -6/(x-3)^2$; no critical points in $[-5, 5]$ ($x = 3$ is not in the domain of f). No maximum or minimum because $\lim_{x \to 3^+} f(x) = +\infty$ and $\lim_{x \to 3^-} f(x) = -\infty$.

23. $(x^2-1)^2$ can never be less than zero because it is the square of x^2-1; the minimum value is 0 for $x=\pm 1$, no maximum because $\lim_{x\to+\infty} f(x)=+\infty$.

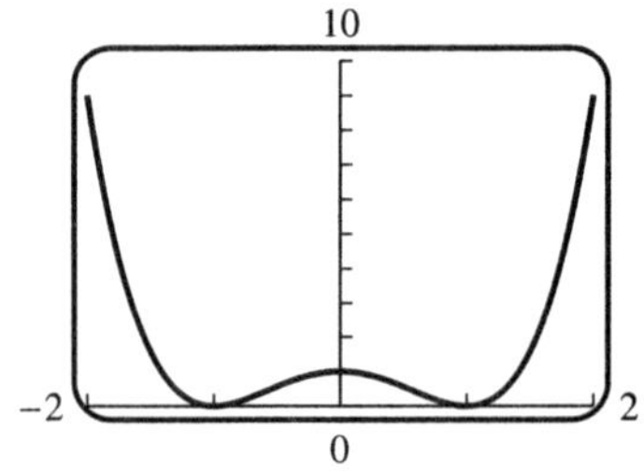

24. $(x-1)^2(x+2)^2$ can never be less than zero because it is the product of two squares; the minimum value is 0 for $x=1$ or -2, no maximum because $\lim_{x\to+\infty} f(x)=+\infty$.

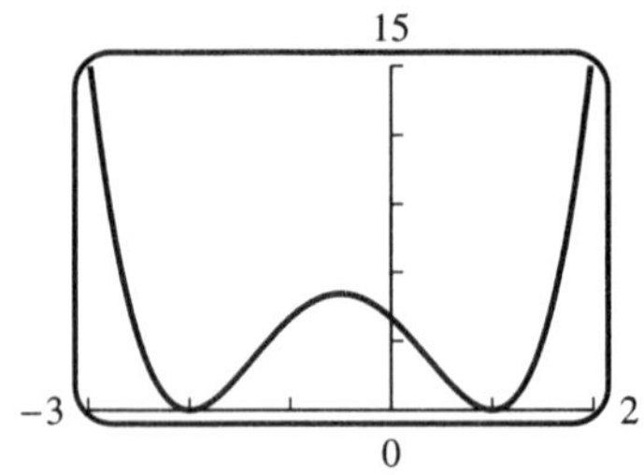

25. $f'(x)=\dfrac{5(8-x)}{3x^{1/3}}$, $f'(x)=0$ when $x=8$ and $f'(x)$ does not exist when $x=0$; $f(-1)=21$, $f(0)=0$, $f(8)=48$, $f(20)=0$ so the maximum value is 48 at $x=8$ and the minimum value is 0 at $x=0,20$.

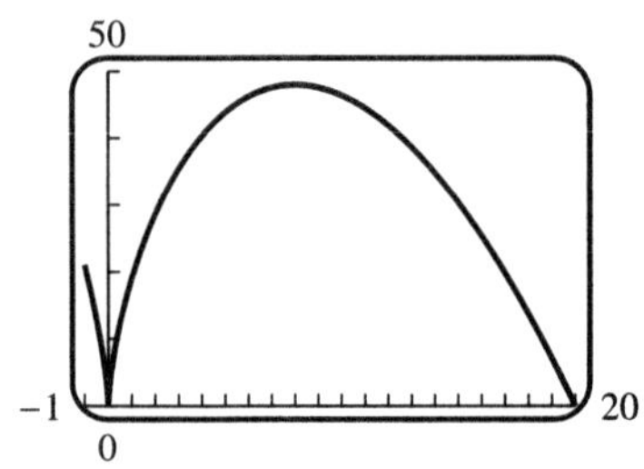

26. $f'(x)=(2-x^2)/(x^2+2)^2$, $f'(x)=0$ for x in the interval $(-1,4)$ when $x=\sqrt{2}$; $f(-1)=-1/3$, $f(\sqrt{2})=\sqrt{2}/4$, $f(4)=2/9$ so the maximum value is $\sqrt{2}/4$ at $x=\sqrt{2}$ and the minimum value is $-1/3$ at $x=-1$.

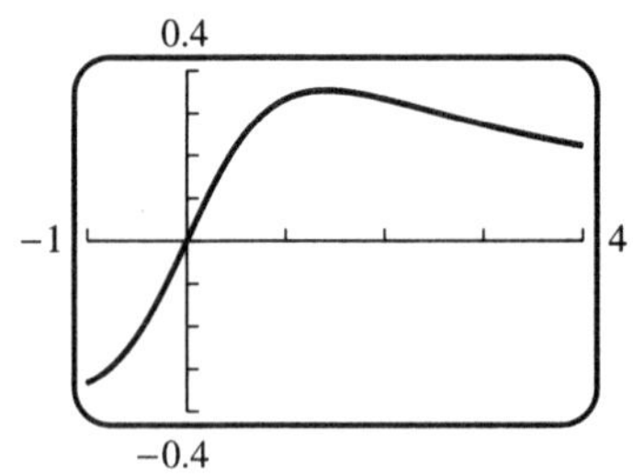

27. $f'(x)=-1/x^2$; no maximum or minimum because there are no critical points in $(0,+\infty)$.

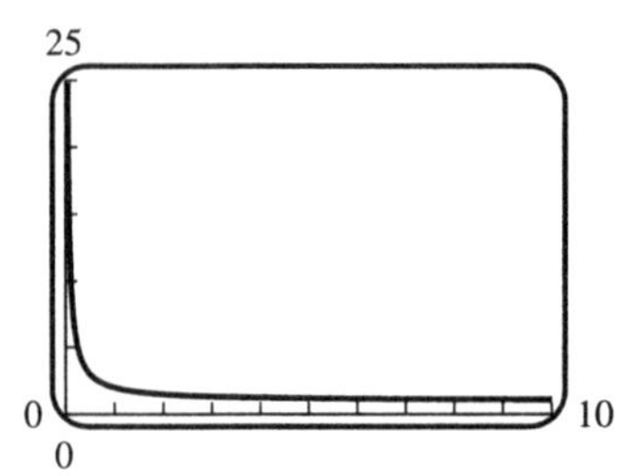

28. $f'(x) = (1 - x^2)/(x^2 + 1)^2$; critical point $x = 1$. Maximum value $f(1) = 1/2$, minimum value 0 because $f(x)$ is never less than zero on $[0, +\infty)$ and $f(0) = 0$.

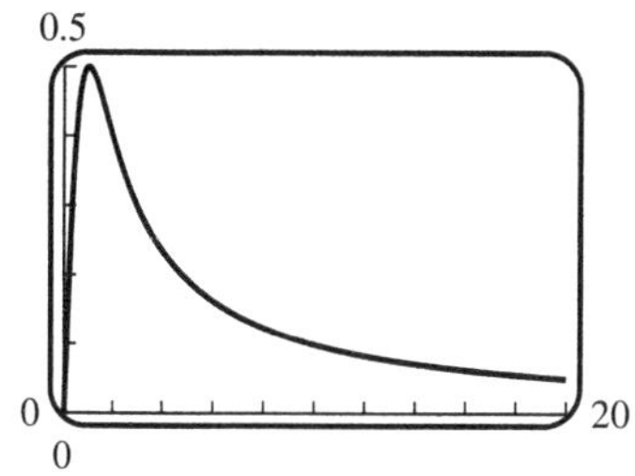

29. $f'(x) = 2\sec x \tan x - \sec^2 x = (2\sin x - 1)/\cos^2 x$, $f'(x) = 0$ for x in $(0, \pi/4)$ when $x = \pi/6$; $f(0) = 2$, $f(\pi/6) = \sqrt{3}$, $f(\pi/4) = 2\sqrt{2} - 1$ so the maximum value is 2 at $x = 0$ and the minimum value is $\sqrt{3}$ at $x = \pi/6$.

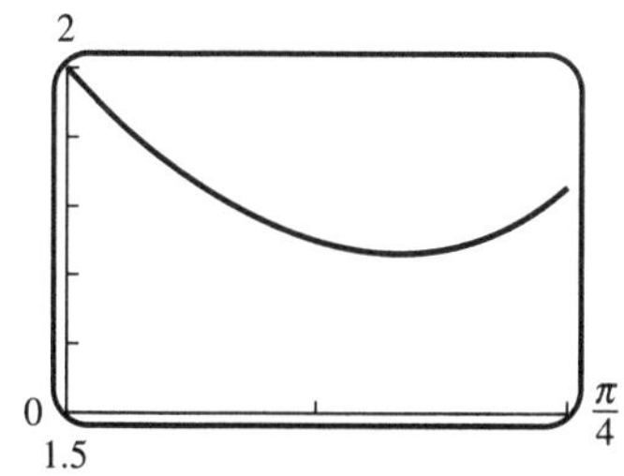

30. $f'(x) = 2\sin x \cos x - \sin x = \sin x(2\cos x - 1)$, $f'(x) = 0$ for x in $(-\pi, \pi)$ when $x = 0, \pm\pi/3$; $f(-\pi) = -1$, $f(-\pi/3) = 5/4$, $f(0) = 1$, $f(\pi/3) = 5/4$, $f(\pi) = -1$ so the maximum value is $5/4$ at $x = \pm\pi/3$ and the minimum value is -1 at $x = \pm\pi$.

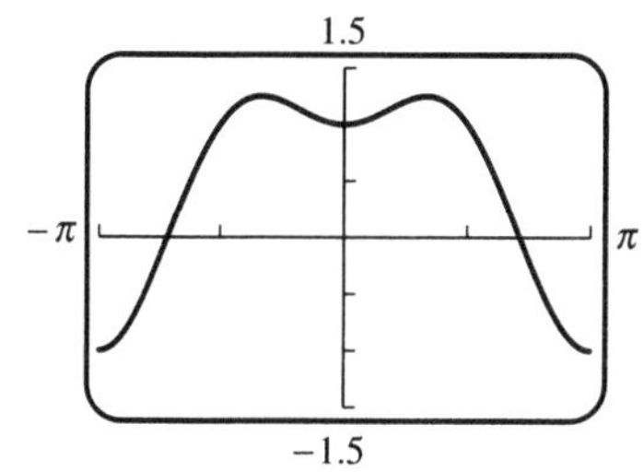

31. $f'(x) = -[\cos(\cos x)]\sin x$; $f'(x) = 0$ if $\sin x = 0$ or if $\cos(\cos x) = 0$. If $\sin x = 0$, then $x = \pi$ is the critical point in $(0, 2\pi)$; $\cos(\cos x) = 0$ has no solutions because $-1 \le \cos x \le 1$. Thus $f(0) = \sin(1)$, $f(\pi) = \sin(-1) = -\sin(1)$, and $f(2\pi) = \sin(1)$ so the maximum value is $\sin(1) \approx 0.84147$ and the minimum value is $-\sin(1) \approx -0.84147$.

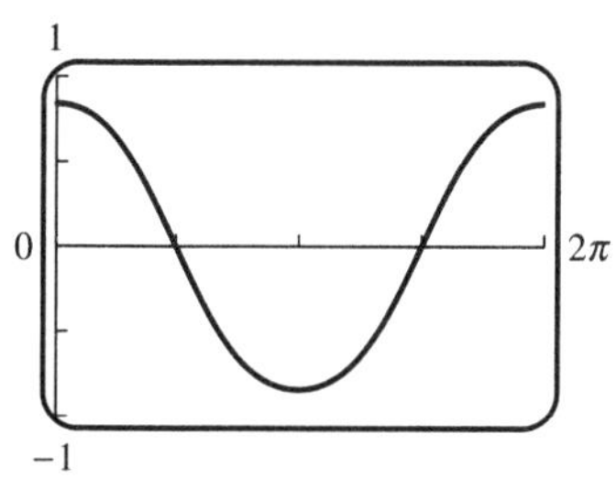

32. $f'(x) = -[\sin(\sin x)]\cos x$; $f'(x) = 0$ if $\cos x = 0$ or if $\sin(\sin x) = 0$. If $\cos x = 0$, then $x = \pi/2$ is the critical point in $(0, \pi)$; $\sin(\sin x) = 0$ if $\sin x = 0$, which gives no critical points in $(0, \pi)$. Thus $f(0) = 1$, $f(\pi/2) = \cos(1)$, and $f(\pi) = 1$ so the maximum value is 1 and the minimum value is $\cos(1) \approx 0.54030$.

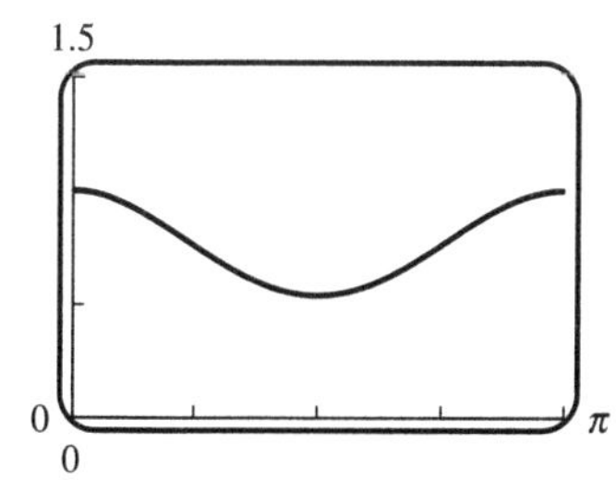

33. $f'(x) = \begin{cases} 4, & x < 1 \\ 2x - 5, & x > 1 \end{cases}$ so $f'(x) = 0$ when $x = 5/2$, and $f'(x)$ does not exist when $x = 1$ because $\lim_{x \to 1^-} f'(x) \ne \lim_{x \to 1^+} f'(x)$ (see Theorem preceding Exercise 75, Section 3.3); $f(1/2) = 0$, $f(1) = 2$, $f(5/2) = -1/4$, $f(7/2) = 3/4$ so the maximum value is 2 and the minimum value is $-1/4$.

34. $f'(x) = 2x + p$ which exists throughout the interval $(0, 2)$ for all values of p so $f'(1) = 0$ because $f(1)$ is an extreme value, thus $2 + p = 0$, $p = -2$. $f(1) = 3$ so $1^2 + (-2)(1) + q = 3$, $q = 4$ thus $f(x) = x^2 - 2x + 4$ and $f(0) = 4$, $f(2) = 4$ so $f(1)$ is the minimum value.

35. $\sin 2x$ has a period of π, and $\sin 4x$ a period of $\pi/2$ so $f(x)$ is periodic with period π. Consider the interval $[0, \pi]$. $f'(x) = 4\cos 2x + 4\cos 4x$, $f'(x) = 0$ when $\cos 2x + \cos 4x = 0$, but $\cos 4x = 2\cos^2 2x - 1$ (trig identity) so

$$\begin{aligned} 2\cos^2 2x + \cos 2x - 1 &= 0 \\ (2\cos 2x - 1)(\cos 2x + 1) &= 0 \\ \cos 2x = 1/2 \quad \text{or} \quad \cos 2x &= -1. \end{aligned}$$

From $\cos 2x = 1/2$, $2x = \pi/3$ or $5\pi/3$ so $x = \pi/6$ or $5\pi/6$. From $\cos 2x = -1$, $2x = \pi$ so $x = \pi/2$. $f(0) = 0$, $f(\pi/6) = 3\sqrt{3}/2$, $f(\pi/2) = 0$, $f(5\pi/6) = -3\sqrt{3}/2$, $f(\pi) = 0$. The maximum value is $3\sqrt{3}/2$ at $x = \pi/6 + n\pi$ and the minimum value is $-3\sqrt{3}/2$ at $x = 5\pi/6 + n\pi$, $n = 0, \pm 1, \pm 2, \ldots$.

36. $\cos\dfrac{x}{3}$ has a period of 6π, and $\cos\dfrac{x}{2}$ a period of 4π, so $f(x)$ has a period of 12π. Consider the interval $[0, 12\pi]$. $f'(x) = -\sin\dfrac{x}{3} - \sin\dfrac{x}{2}$, $f'(x) = 0$ when $\sin\dfrac{x}{3} + \sin\dfrac{x}{2} = 0$ thus, by use of the trig identity $\sin a + \sin b = 2\sin\dfrac{a+b}{2}\cos\dfrac{a-b}{2}$, $2\sin\left(\dfrac{5x}{12}\right)\cos\left(-\dfrac{x}{12}\right) = 0$ so $\sin\dfrac{5x}{12} = 0$ or $\cos\dfrac{x}{12} = 0$. Solve $\sin\dfrac{5x}{12} = 0$ to get $x = 12\pi/5$, $24\pi/5$, $36\pi/5$, $48\pi/5$ and then solve $\cos\dfrac{x}{12} = 0$ to get $x = 6\pi$. The corresponding values of $f(x)$ are $-4.0450, 1.5450, 1.5450, -4.0450, 1, 5, 5$ so the maximum value is 5 and the minimum value is -4.0450 (approximately).

37. Let $f(x) = x - \sin x$, then $f'(x) = 1 - \cos x$ and so $f'(x) = 0$ when $\cos x = 1$ which has no solution for $0 < x < 2\pi$ thus the minimum value of f must occur at 0 or 2π. $f(0) = 0$, $f(2\pi) = 2\pi$ so 0 is the minimum value on $[0, 2\pi]$ thus $x - \sin x \geq 0$, $\sin x \leq x$ for all x in $[0, 2\pi]$.

38. Let $h(x) = \cos x - 1 + x^2/2$. Then $h(0) = 0$, and it is sufficient to show that $h'(x) \geq 0$ for $0 < x < 2\pi$. But $h'(x) = -\sin x + x \geq 0$ by Exercise 37.

39. Let $m =$ slope at x, then $m = f'(x) = 3x^2 - 6x + 5$, $dm/dx = 6x - 6$; critical point for m is $x = 1$, minimum value of m is $f'(1) = 2$

40. (a) $f'(x) = -\dfrac{64\cos x}{\sin^2 x} + \dfrac{27\sin x}{\cos^2 x} = \dfrac{-64\cos^3 x + 27\sin^3 x}{\sin^2 x\cos^2 x}$, $f'(x) = 0$ when

$27\sin^3 x = 64\cos^3 x$, $\tan^3 x = 64/27$, $\tan x = 4/3$ so the critical point is $x = x_0$ where $\tan x_0 = 4/3$ and $0 < x_0 < \pi/2$. To test x_0 first rewrite $f'(x)$ as

$$f'(x) = \frac{27\cos^3 x(\tan^3 x - 64/27)}{\sin^2 x\cos^2 x} = \frac{27\cos x(\tan^3 x - 64/27)}{\sin^2 x};$$

if $x < x_0$ then $\tan x < 4/3$ and $f'(x) < 0$, if $x > x_0$ then $\tan x > 4/3$ and $f'(x) > 0$ so $f(x_0)$ is the minimum value. f has no maximum because $\lim_{x\to 0^+} f(x) = +\infty$.

(b) If $\tan x_0 = 4/3$ then (see figure)
$\sin x_0 = 4/5$ and $\cos x_0 = 3/5$
so $f(x_0) = 64/\sin x_0 + 27/\cos x_0$
$= 64/(4/5) + 27/(3/5)$
$= 80 + 45 = 125$

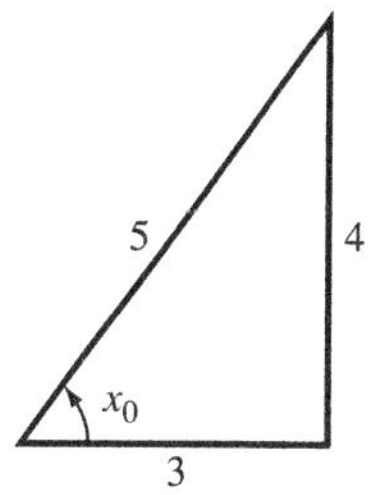

41. $f'(x) = \dfrac{2x(x^3 - 24x^2 + 192x - 640)}{(x-8)^3}$; real root of $x^3 - 24x^2 + 192x - 640$ at $x = 4(2 + \sqrt[3]{2})$. Since $\lim_{x\to 8^+} f(x) = \lim_{x\to +\infty} f(x) = +\infty$ and there is only one relative extremum, it must be a minimum.

42. Assume A and B nonzero, since if A or $B = 0$ the solution is obvious. Otherwise
$f'(x) = -A\sin x + B\cos x = 0$ when $x = x_0 = \tan^{-1}(B/A)$.
From the figure we get
$f(x_0) = A^2/\sqrt{A^2+B^2} + B^2/\sqrt{A^2+B^2} = \sqrt{A^2+B^2}$.
This must be a maximum since $f''(x_0) = -f(x_0) < 0$ there.

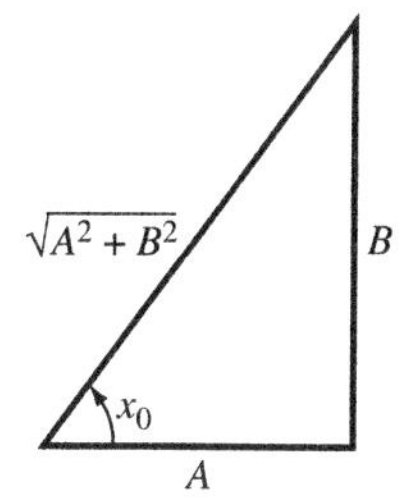

43. The slope of the line is -1, and the slope of the tangent to $y = -x^2$ is $-2x$ so $-2x = -1$, $x = 1/2$. The line lies above the curve so the vertical distance is given by $F(x) = 2 - x + x^2$; $F(-1) = 4$, $F(1/2) = 7/4$, $F(3/2) = 11/4$. The point $(1/2, -1/4)$ is closest, the point $(-1, -1)$ farthest.

44. The slope of the line is $4/3$; and the slope of the tangent to $y = x^3$ is $3x^2$ so $3x^2 = 4/3$, $x^2 = 4/9$, $x = \pm 2/3$. The line lies below the curve so the vertical distance is given by $F(x) = x^3 - 4x/3 + 1$; $F(-1) = 4/3$, $F(-2/3) = 43/27$, $F(2/3) = 11/27$, $F(1) = 2/3$. The closest point is $(2/3, 8/27)$, the farthest is $(-2/3, -8/27)$.

45. The absolute extrema of $y(t)$ can occur at the endpoints $t = 0, 12$ or when $dy/dt = 2\sin t = 0$, i.e. $t = 0, 12, k\pi$, $k = 1, 2, 3$; the absolute maximum is $y = 4$ at $t = \pi, 3\pi$; the absolute minimum is $y = 0$ at $t = 0, 2\pi$.

46. **(a)** The absolute extrema of $y(t)$ can occur at the endpoints $t = 0, 2\pi$ or when
$dy/dt = 2\cos 2t - 4\sin t\cos t = 2\cos 2t - 2\sin 2t = 0$, $t = 0, 2\pi, \pi/8, 5\pi/8, 9\pi/8, 13\pi/8$;
the absolute maximum is $y = 3.4142$ at $t = \pi/8, 9\pi/8$; the absolute minimum is $y = 0.5859$ at $t = 5\pi/8, 13\pi/8$.

(b) The absolute extrema of $x(t)$ occur at the endpoints $t = 0, 2\pi$ or when

$\dfrac{dx}{dt} = -\dfrac{2\sin t + 1}{(2+\sin t)^2} = 0$, $t = 7\pi/6, 11\pi/6$. The absolute maximum is $x = 0.5774$ at $t = 11\pi/6$

and the absolute minimum is $x = -0.5774$ at $t = 7\pi/6$.

47. $f'(x) = 2ax + b$; critical point is $x = -\dfrac{b}{2a}$

$f''(x) = 2a > 0$ so $f\left(-\dfrac{b}{2a}\right)$ is the minimum value of f, but

$f\left(-\dfrac{b}{2a}\right) = a\left(-\dfrac{b}{2a}\right)^2 + b\left(-\dfrac{b}{2a}\right) + c = \dfrac{-b^2+4ac}{4a}$ thus $f(x) \geq 0$ if and only if

$f\left(-\dfrac{b}{2a}\right) \geq 0$, $\dfrac{-b^2+4ac}{4a} \geq 0$, $-b^2+4ac \geq 0$, $b^2 - 4ac \leq 0$

48. Use the proof given in the text, replacing "maximum" by "minimum" and "largest" by "smallest" and reversing the order of all inequality symbols.

EXERCISE SET 4.6

1. Let x = one number, y = the other number, and $P = xy$ where $x + y = 10$. Thus $y = 10 - x$ so $P = x(10-x) = 10x - x^2$ for x in $[0, 10]$. $dP/dx = 10 - 2x$, $dP/dx = 0$ when $x = 5$. If $x = 0, 5, 10$ then $P = 0, 25, 0$ so P is maximum when $x = 5$ and, from $y = 10 - x$, when $y = 5$.

2. Let x and y be nonnegative numbers and z the sum of their squares, then $z = x^2 + y^2$. But $x + y = 1$, $y = 1 - x$ so $z = x^2 + (1-x)^2 = 2x^2 - 2x + 1$ for $0 \leq x \leq 1$. $dz/dx = 4x - 2$, $dz/dx = 0$ when $x = 1/2$. If $x = 0, 1/2, 1$ then $z = 1, 1/2, 1$ so

(a) z is as large as possible when one number is 0 and the other is 1.

(b) z is as small as possible when both numbers are 1/2.

3. If $y = x + 1/x$ for $1/2 \leq x \leq 3/2$ then $dy/dx = 1 - 1/x^2 = (x^2-1)/x^2$, $dy/dx = 0$ when $x = 1$. If $x = 1/2, 1, 3/2$ then $y = 5/2, 2, 13/6$ so

(a) y is as small as possible when $x = 1$.

(b) y is as large as possible when $x = 1/2$.

4. $A = xy$ where $x + 2y = 1000$ so $y = 500 - x/2$ and $A = 500x - x^2/2$ for x in $[0, 1000]$; $dA/dx = 500 - x$, $dA/dx = 0$ when $x = 500$. If $x = 0$ or 1000 then $A = 0$, if $x = 500$ then $A = 125{,}000$ so the area is maximum when $x = 500$ ft and $y = 500 - 500/2 = 250$ ft.

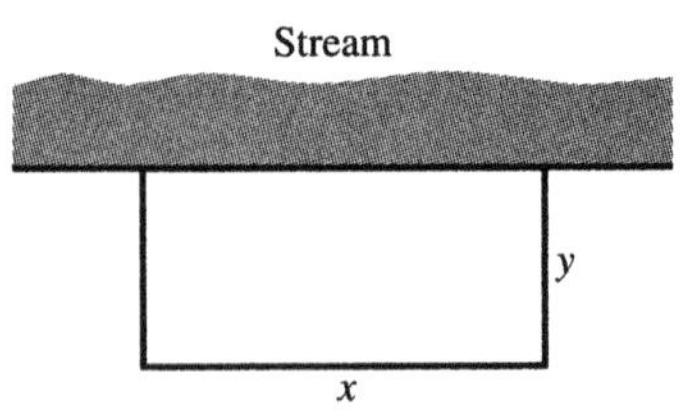

5. Let x and y be the dimensions shown in the figure and A the area, then $A = xy$ subject to the cost condition $3(2x) + 2(2y) = 6000$, or $y = 1500 - 3x/2$. Thus $A = x(1500 - 3x/2) = 1500x - 3x^2/2$ for x in $[0, 1000]$. $dA/dx = 1500 - 3x$, $dA/dx = 0$ when $x = 500$. If $x = 0$ or 1000 then $A = 0$, if $x = 500$ then $A = 375{,}000$ so the area is greatest when $x = 500$ ft and (from $y = 1500 - 3x/2$) when $y = 750$ ft.

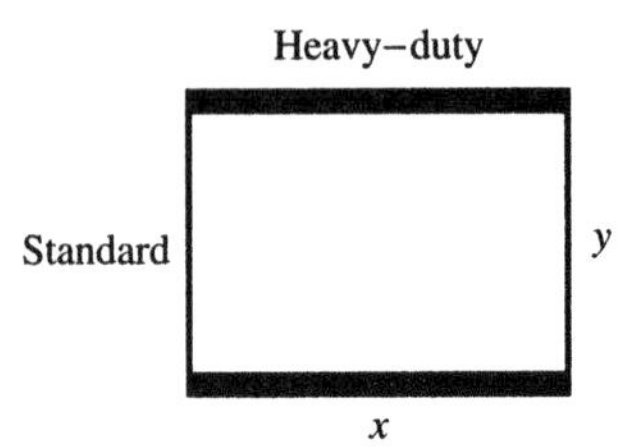

6. Let x and y be the dimensions shown in the figure and A the area of the rectangle, then $A = xy$ and, by similar triangles, $x/6 = (8-y)/8$, $y = 8 - 4x/3$ so $A = x(8 - 4x/3) = 8x - 4x^2/3$ for x in $[0, 6]$. $dA/dx = 8 - 8x/3$, $dA/dx = 0$ when $x = 3$. If $x = 0, 3, 6$ then $A = 0, 12, 0$ so the area is greatest when $x = 3$ in and (from $y = 8 - 4x/3$) $y = 4$ in.

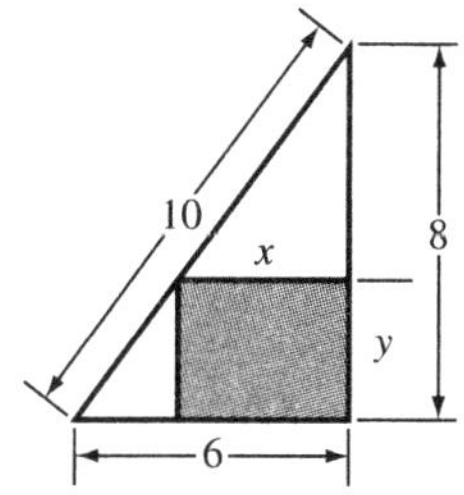

7. Let x, y, and z be as shown in the figure and A the area of the rectangle, then $A = xy$ and, by similar triangles, $z/10 = y/6$, $z = 5y/3$; also $x/10 = (8-z)/8 = (8 - 5y/3)/8$ thus $y = 24/5 - 12x/25$ so $A = x(24/5 - 12x/25) = 24x/5 - 12x^2/25$ for x in $[0, 10]$. $dA/dx = 24/5 - 24x/25$, $dA/dx = 0$ when $x = 5$. If $x = 0, 5, 10$ then $A = 0, 12, 0$ so the area is greatest when $x = 5$ in. and $y = 12/5$ in.

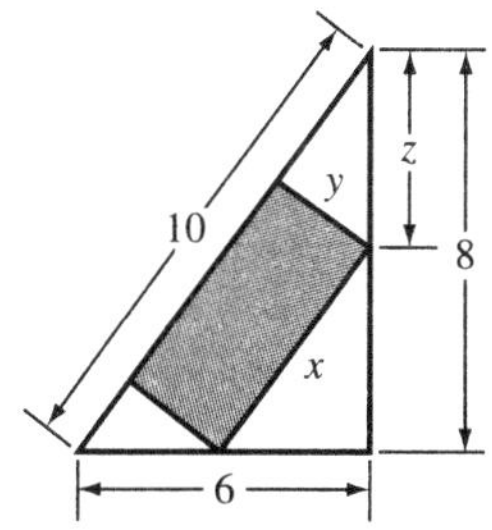

8. $A = (2x)y = 2xy$ where $y = 16 - x^2$ so $A = 32x - 2x^3$ for $0 \le x \le 4$; $dA/dx = 32 - 6x^2$, $dA/dx = 0$ when $x = 4/\sqrt{3}$. If $x = 0, 4/\sqrt{3}, 4$ then $A = 0, 256/(3\sqrt{3}), 0$ so the area is largest when $x = 4/\sqrt{3}$ and $y = 32/3$. The dimensions of the rectangle with largest area are $8/\sqrt{3}$ by $32/3$.

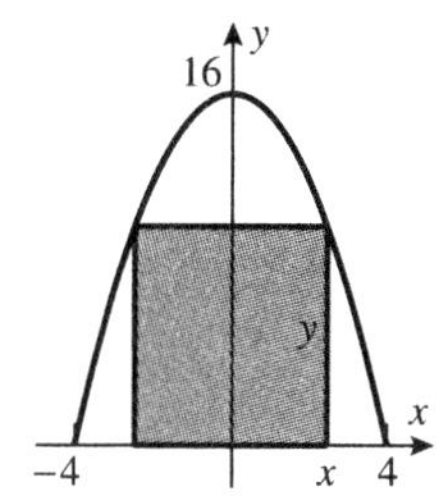

9. $A = xy$ where $x^2 + y^2 = 20^2 = 400$ so $y = \sqrt{400 - x^2}$ and $A = x\sqrt{400 - x^2}$ for $0 \le x \le 20$;
$dA/dx = 2(200 - x^2)/\sqrt{400 - x^2}$, $dA/dx = 0$ when $x = \sqrt{200} = 10\sqrt{2}$. If $x = 0, 10\sqrt{2}, 20$ then $A = 0, 200, 0$ so the area is maximum when $x = 10\sqrt{2}$ and $y = \sqrt{400 - 200} = 10\sqrt{2}$.

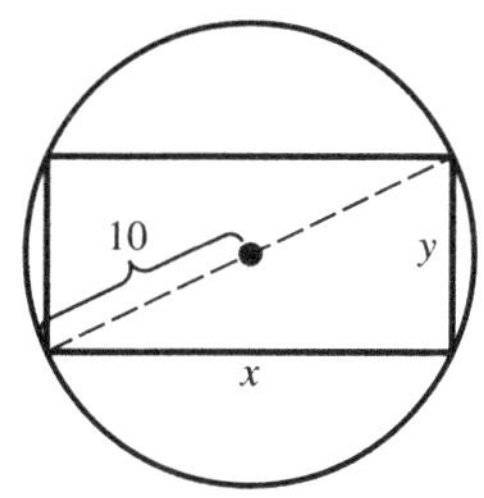

10. Let x and y be the dimensions shown in the figure, then the area of the rectangle is $A = xy$.

But $\left(\frac{x}{2}\right)^2 + y^2 = R^2$, thus

$y = \sqrt{R^2 - x^2/4} = \frac{1}{2}\sqrt{4R^2 - x^2}$ so

$A = \frac{1}{2}x\sqrt{4R^2 - x^2}$ for $0 \le x \le 2R$.

$dA/dx = (2R^2 - x^2)/\sqrt{4R^2 - x^2}$, $dA/dx = 0$ when $x = \sqrt{2}R$. If $x = 0, \sqrt{2}R, 2R$ then $A = 0, R^2, 0$ so the greatest area occurs when $x = \sqrt{2}R$ and $y = \sqrt{2}R/2$.

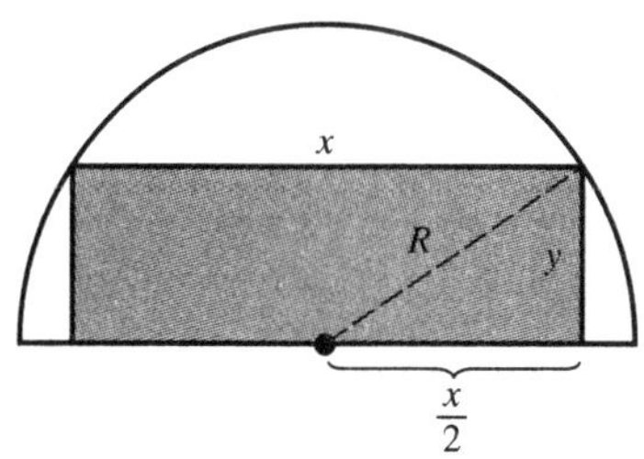

11. Let $x =$ length of each side that uses the \$1 per foot fencing,
$y =$ length of each side that uses the \$2 per foot fencing.
The cost is $C = (1)(2x) + (2)(2y) = 2x + 4y$, but $A = xy = 3200$ thus $y = 3200/x$ so

$$C = 2x + 12800/x \text{ for } x > 0,$$
$$dC/dx = 2 - 12800/x^2,\ dC/dx = 0 \text{ when } x = 80,\ d^2C/dx^2 > 0 \text{ so}$$

C is least when $x = 80$, $y = 40$.

12. $A = xy$ where $2x + 2y = p$ so $y = p/2 - x$ and $A = px/2 - x^2$ for x in $[0, p/2]$; $dA/dx = p/2 - 2x$, $dA/dx = 0$ when $x = p/4$. If $x = 0$ or $p/2$ then $A = 0$, if $x = p/4$ then $A = p^2/16$ so the area is maximum when $x = p/4$ and $y = p/2 - p/4 = p/4$, which is a square.

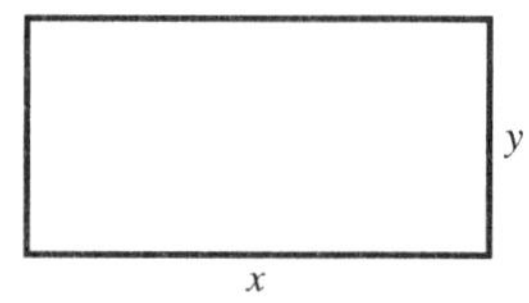

13. Let x and y be the dimensions of a rectangle; the perimeter is $p = 2x + 2y$. But $A = xy$ thus $y = A/x$ so $p = 2x + 2A/x$ for $x > 0$, $dp/dx = 2 - 2A/x^2 = 2(x^2 - A)/x^2$, $dp/dx = 0$ when $x = \sqrt{A}$, $d^2p/dx^2 = 4A/x^3 > 0$ if $x > 0$ so p is a minimum when $x = \sqrt{A}$ and $y = \sqrt{A}$ and thus the rectangle is a square.

14. With x, y, r, and s as shown in the figure, the sum of the enclosed areas is $A = \pi r^2 + s^2$ where $r = \dfrac{x}{2\pi}$ and $s = \dfrac{y}{4}$ because x is the circumference of the circle and y is the perimeter of the square, thus $A = \dfrac{x^2}{4\pi} + \dfrac{y^2}{16}$. But $x + y = 12$, so $y = 12 - x$ and

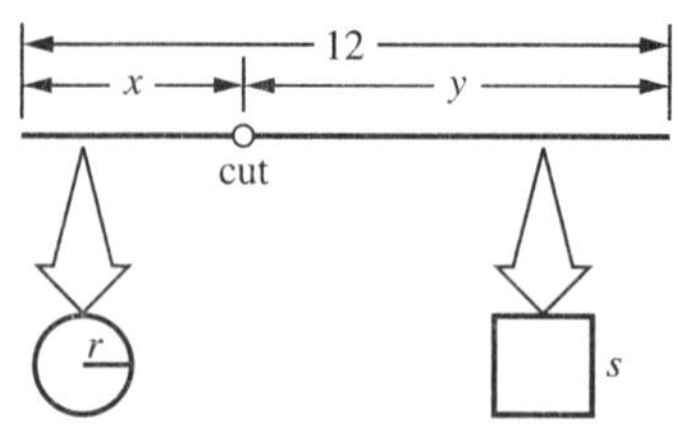

$A = \dfrac{x^2}{4\pi} + \dfrac{(12 - x)^2}{16} = \dfrac{\pi + 4}{16\pi}x^2 - \dfrac{3}{2}x + 9$ for $0 \le x \le 12$.

$\dfrac{dA}{dx} = \dfrac{\pi + 4}{8\pi}x - \dfrac{3}{2}$, $\dfrac{dA}{dx} = 0$ when $x = \dfrac{12\pi}{\pi + 4}$. If $x = 0, \dfrac{12\pi}{\pi + 4}, 12$

then $A = 9, \dfrac{36}{\pi + 4}, \dfrac{36}{\pi}$ so the sum of the enclosed areas is

(a) a maximum when $x = 12$ in. (when all of the wire is used for the circle)

(b) a minimum when $x = 12\pi/(\pi + 4)$ in.

15. The area of the field is $A = 2bh$, where $b = y/2$ is the base of the half-triangle and h is the height. The length of fencing is $2x = 300$. Thus $x = 150$ and

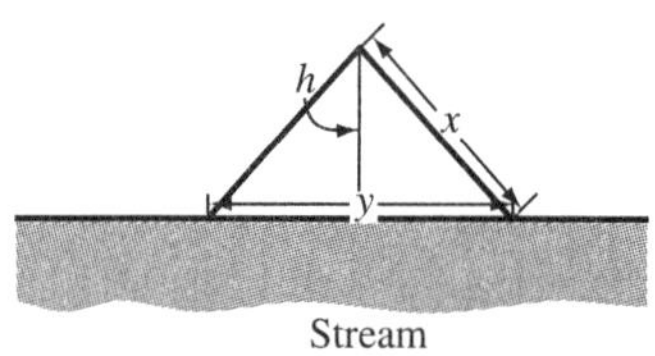

$$A = yh = y\sqrt{x^2 - y^2/4} = y\sqrt{(150)^2 - y^2/4}$$

$$dA/dy = \frac{y(-y/2)}{2\sqrt{(150)^2 - y^2/4)}} + \sqrt{(150)^2 - y^2/4} = 0$$

when $-y^2/4 + (150)^2 - y^2/4 = 0$, or $y^2 = 2(150)^2$. Thus $y = 150\sqrt{2}$ yd, $x = 150$ yd; $b = 75\sqrt{2}$ yd, $h = 75\sqrt{2}$ yd, and $A = yh = (150)^2 = 22{,}500$ yd^2.

16. The area of the window is $A = 2rh + \pi r^2/2$, the perimeter is $p = 2r + 2h + \pi r$ thus

$h = \dfrac{1}{2}[p - (2 + \pi)r]$ so

$A = r[p - (2 + \pi)r] + \pi r^2/2$

$\quad = pr - (2 + \pi/2)r^2$ for $0 \le r \le p/(2 + \pi)$,

$dA/dr = p - (4 + \pi)r$, $dA/dr = 0$ when $r = p/(4 + \pi)$ and $d^2A/dr^2 < 0$, so A is maximum when $r = p/(4 + \pi)$.

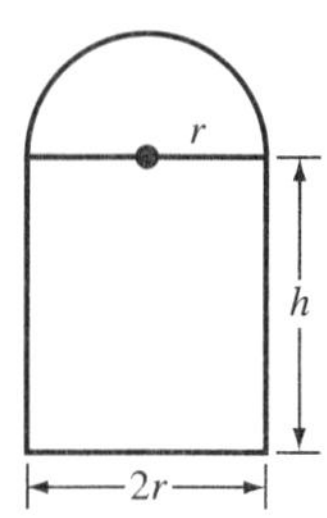

17. $V = x(12-2x)^2$ for $0 \le x \le 6$;
$dV/dx = 12(x-2)(x-6)$, $dV/dx = 0$ when $x = 2$ for $0 < x < 6$. If $x = 0, 2, 6$ then $V = 0, 128, 0$ so the volume is largest when $x = 2$ in.

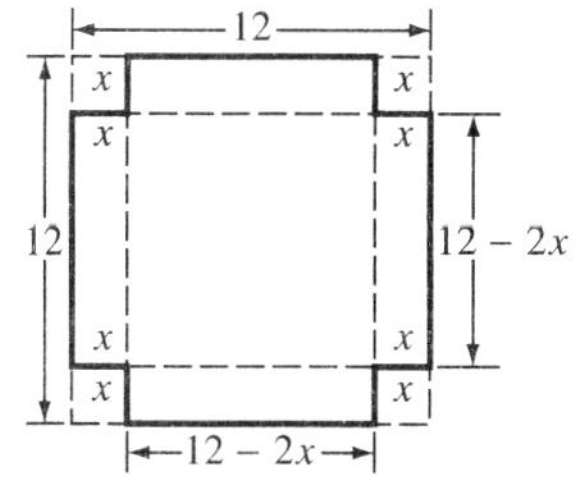

18. The dimensions of the box will be $(k-2x)$ by $(k-2x)$ by x so $V = (k-2x)^2 x = 4x^3 - 4kx^2 + k^2 x$ for x in $[0, k/2]$. $dV/dx = 12x^2 - 8kx + k^2 = (6x-k)(2x-k)$, $dV/dx = 0$ for x in $(0, k/2)$ when $x = k/6$. If $x = 0, k/6, k/2$ then $V = 0, 2k^3/27, 0$ so V is maximum when $x = k/6$. The squares should have dimensions $k/6$ by $k/6$.

19. Let x be the length of each side of a square, then $V = x(3-2x)(8-2x) = 4x^3 - 22x^2 + 24x$ for $0 \le x \le 3/2$; $dV/dx = 12x^2 - 44x + 24 = 4(3x-2)(x-3)$, $dV/dx = 0$ when $x = 2/3$ for $0 < x < 3/2$. If $x = 0, 2/3, 3/2$ then $V = 0, 200/27, 0$ so the maximum volume is $200/27$ ft^3.

20. Let x = length of each edge of base, y = height. The cost is
$C =$ (cost of top and bottom) + (cost of sides) $= (2)(2x^2) + (3)(4xy) = 4x^2 + 12xy$, but
$V = x^2 y = 2250$ thus $y = 2250/x^2$ so $C = 4x^2 + 27000/x$ for $x > 0$, $dC/dx = 8x - 27000/x^2$,
$dC/dx = 0$ when $x = \sqrt[3]{3375} = 15$, $d^2C/dx^2 > 0$ so C is least when $x = 15$, $y = 10$.

21. Let x = length of each edge of base, y = height, $k = \$/\text{cm}^2$ for the sides. The cost is
$C = (2k)(2x^2) + (k)(4xy) = 4k(x^2 + xy)$, but $V = x^2 y = 2000$ thus $y = 2000/x^2$ so
$C = 4k(x^2 + 2000/x)$ for $x > 0$; $dC/dx = 4k(2x - 2000/x^2)$, $dC/dx = 0$ when
$x = \sqrt[3]{1000} = 10$, $d^2C/dx^2 > 0$ so C is least when $x = 10$, $y = 20$.

22. Let x and y be the dimensions shown in the figure and V the volume, then $V = x^2 y$. The amount of material is to be 1000 ft^2, thus (area of base) + (area of sides) $= 1000$, $x^2 + 4xy = 1000$, $y = \dfrac{1000 - x^2}{4x}$ so

$$V = x^2 \frac{1000 - x^2}{4x} = \frac{1}{4}(1000x - x^3) \text{ for } 0 < x \le 10\sqrt{10}.$$

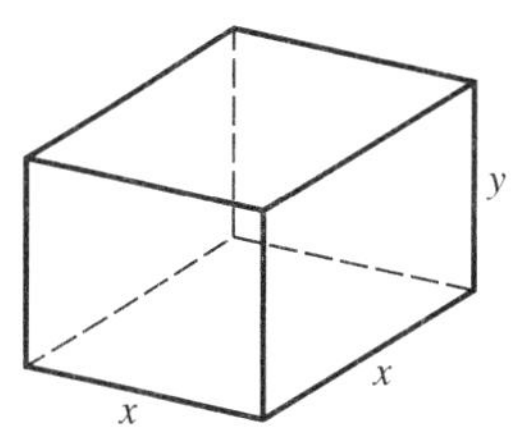

$$\frac{dV}{dx} = \frac{1}{4}(1000 - 3x^2), \quad \frac{dV}{dx} = 0$$

when $x = \sqrt{1000/3} = 10\sqrt{10/3}$.

If $x = 0, 10\sqrt{10/3}, 10\sqrt{10}$ then $V = 0, \dfrac{5000}{3}\sqrt{10/3}, 0$;

the volume is greatest for $x = 10\sqrt{10/3}$ ft and $y = 5\sqrt{10/3}$ ft.

23. Let x = height and width, y = length. The surface area is $S = 2x^2 + 3xy$ where $x^2 y = V$, so $y = V/x^2$ and $S = 2x^2 + 3V/x$ for $x > 0$; $dS/dx = 4x - 3V/x^2$, $dS/dx = 0$ when $x = \sqrt[3]{3V/4}$,

$d^2S/dx^2 > 0$ so S is minimum when $x = \sqrt[3]{\dfrac{3V}{4}}$, $y = \dfrac{4}{3}\sqrt[3]{\dfrac{3V}{4}}$.

24. Let r and h be the dimensions shown in the figure, then the volume of the inscribed cylinder is $V = \pi r^2 h$. But $r^2 + \left(\frac{h}{2}\right)^2 = R^2$ thus $r^2 = R^2 - \frac{h^2}{4}$ so $V = \pi\left(R^2 - \frac{h^2}{4}\right)h = \pi\left(R^2 h - \frac{h^3}{4}\right)$ for $0 \le h \le 2R$. $\frac{dV}{dh} = \pi\left(R^2 - \frac{3}{4}h^2\right)$, $\frac{dV}{dh} = 0$ when $h = 2R/\sqrt{3}$. If $h = 0, 2R/\sqrt{3}, 2R$ then $V = 0, \frac{4\pi}{3\sqrt{3}}R^3, 0$ so the volume is largest when $h = 2R/\sqrt{3}$ and $r = \sqrt{2/3}R$.

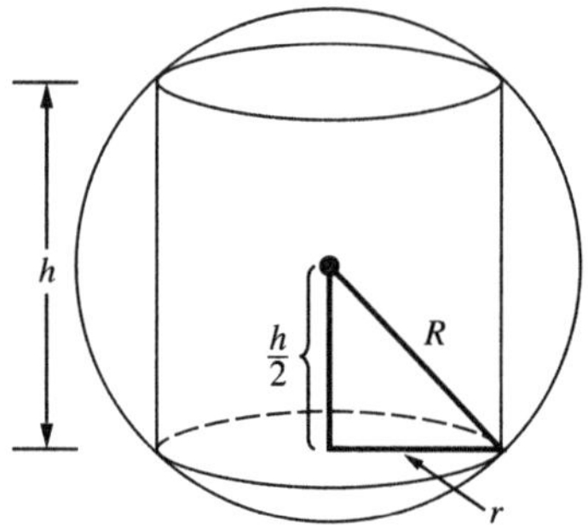

25. Let r and h be the dimensions shown in the figure, then the surface area is $S = 2\pi rh + 2\pi r^2$.
But $r^2 + \left(\frac{h}{2}\right)^2 = R^2$ thus $h = 2\sqrt{R^2 - r^2}$ so

$S = 4\pi r\sqrt{R^2 - r^2} + 2\pi r^2$ for $0 \le r \le R$,

$\frac{dS}{dr} = \frac{4\pi(R^2 - 2r^2)}{\sqrt{R^2 - r^2}} + 4\pi r$; $\frac{dS}{dr} = 0$ when

$$\frac{R^2 - 2r^2}{\sqrt{R^2 - r^2}} = -r \qquad (1)$$

$$R^2 - 2r^2 = -r\sqrt{R^2 - r^2}$$

$$R^4 - 4R^2r^2 + 4r^4 = r^2(R^2 - r^2)$$
$$5r^4 - 5R^2r^2 + R^4 = 0$$

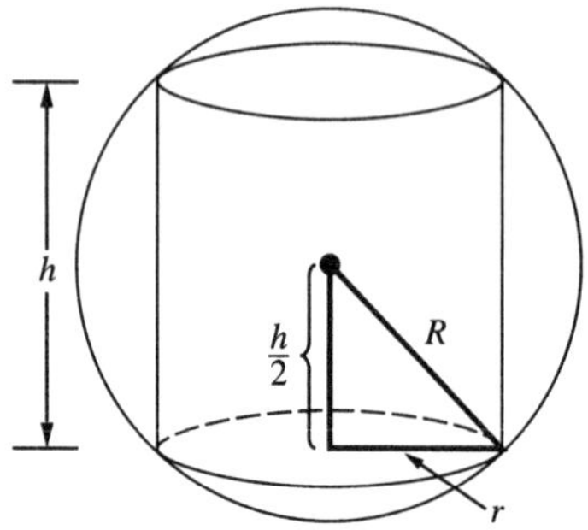

and using the quadratic formula $r^2 = \frac{5R^2 \pm \sqrt{25R^4 - 20R^4}}{10} = \frac{5 \pm \sqrt{5}}{10}R^2$, $r = \sqrt{\frac{5 \pm \sqrt{5}}{10}}R$, of which only $r = \sqrt{\frac{5 + \sqrt{5}}{10}}R$ satisfies (1). If $r = 0, \sqrt{\frac{5 + \sqrt{5}}{10}}R, 0$ then $S = 0, (5 + \sqrt{5})\pi R^2, 2\pi R^2$ so the surface area is greatest when $r = \sqrt{\frac{5 + \sqrt{5}}{10}}R$ and, from $h = 2\sqrt{R^2 - r^2}$, $h = 2\sqrt{\frac{5 - \sqrt{5}}{10}}R$.

26. Let R and H be the radius and height of the cone, and r and h the radius and height of the cylinder (see figure), then the volume of the cylinder is $V = \pi r^2 h$. By similar triangles (see figure) $\frac{H - h}{H} = \frac{r}{R}$ thus $h = \frac{H}{R}(R - r)$ so $V = \pi\frac{H}{R}(R - r)r^2 = \pi\frac{H}{R}(Rr^2 - r^3)$ for $0 \le r \le R$. $\frac{dV}{dr} = \pi\frac{H}{R}(2Rr - 3r^2) = \pi\frac{H}{R}r(2R - 3r)$, $\frac{dV}{dr} = 0$ for $0 < r < R$ when $r = 2R/3$. If $r = 0, 2R/3, R$ then $V = 0, 4\pi R^2 H/27, 0$ so the maximum volume is $\frac{4\pi R^2 H}{27} = \frac{4}{9}\frac{1}{3}\pi R^2 H = \frac{4}{9}\cdot$ (volume of cone).

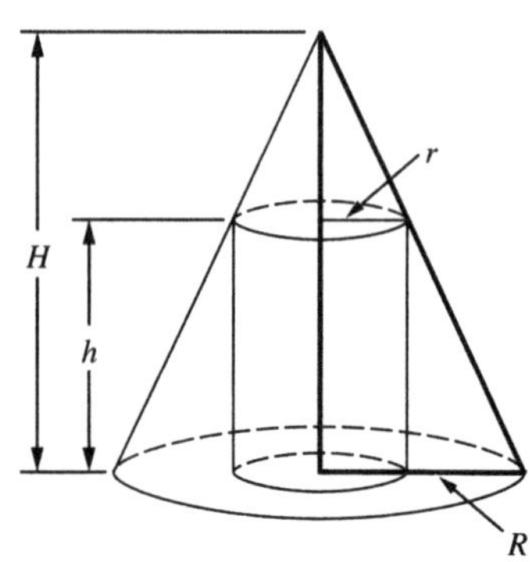

27. From (13), $S = 2\pi r^2 + 2\pi rh$. But $V = \pi r^2 h$ thus $h = V/(\pi r^2)$ and so $S = 2\pi r^2 + 2V/r$ for $r > 0$. $dS/dr = 4\pi r - 2V/r^2, dS/dr = 0$ if $r = \sqrt[3]{V/(2\pi)}$. Since $d^2S/dr^2 = 4\pi + 4V/r^3 > 0$, the minimum surface area is achieved when $r = \sqrt[3]{V/2\pi}$ and so $h = V/(\pi r^2) = [V/(\pi r^3)]r = 2r$.

28. $V = \pi r^2 h$ where $S = 2\pi r^2 + 2\pi rh$ so $h = \dfrac{S - 2\pi r^2}{2\pi r}$, $V = \dfrac{1}{2}(Sr - 2\pi r^3)$ for $r > 0$.
$\dfrac{dV}{dr} = \dfrac{1}{2}(S - 6\pi r^2) = 0$ if $r = \sqrt{S/(6\pi)}$, $\dfrac{d^2V}{dr^2} = -6\pi r < 0$ so V is maximum when $r = \sqrt{S/(6\pi)}$ and $h = \dfrac{S - 2\pi r^2}{2\pi r} = \dfrac{S - 2\pi r^2}{2\pi r^2}r = \dfrac{S - S/3}{S/3}r = 2r$, thus the height is equal to the diameter of the base.

29. The surface area is $S = \pi r^2 + 2\pi rh$ where $V = \pi r^2 h = 500$ so $h = 500/(\pi r^2)$ and $S = \pi r^2 + 1000/r$ for $r > 0$;
$dS/dr = 2\pi r - 1000/r^2 = (2\pi r^3 - 1000)/r^2$, $dS/dr = 0$ when $r = \sqrt[3]{500/\pi}$, $d^2S/dr^2 > 0$ for $r > 0$ so S is minimum when $r = \sqrt[3]{500/\pi}$ cm, and

$$h = \frac{500}{\pi r^2} = \frac{500}{\pi}\left(\frac{\pi}{500}\right)^{2/3} = \sqrt[3]{500/\pi} \text{ cm.}$$

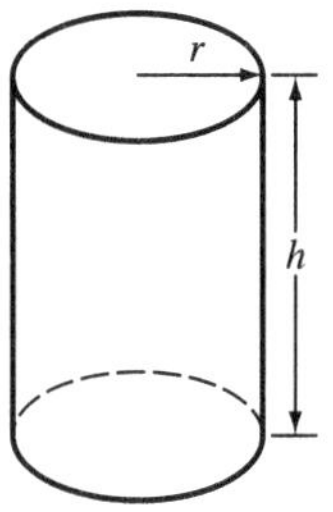

30. The total area of material used is
$A = A_{\text{top}} + A_{\text{bottom}} + A_{\text{side}} = (2r)^2 + (2r)^2 + 2\pi rh = 8r^2 + 2\pi rh$.
The volume is $V = \pi r^2 h$ thus $h = V/(\pi r^2)$ so $A = 8r^2 + 2V/r$ for $r > 0$,
$dA/dr = 16r - 2V/r^2 = 2(8r^3 - V)/r^2$, $dA/dr = 0$ when $r = \sqrt[3]{V}/2$. This is the only critical point, $d^2A/dr^2 > 0$ there so the least material is used when $r = \sqrt[3]{V}/2$, $\dfrac{r}{h} = \dfrac{r}{V/(\pi r^2)} = \dfrac{\pi}{V}r^3$ and, for $r = \sqrt[3]{V}/2$, $\dfrac{r}{h} = \dfrac{\pi}{V}\dfrac{V}{8} = \dfrac{\pi}{8}$.

31. Let x be the length of each side of the squares and y the height of the frame, then the volume is $V = x^2y$. The total length of the wire is L thus $8x + 4y = L$, $y = (L - 8x)/4$ so $V = x^2(L - 8x)/4 = (Lx^2 - 8x^3)/4$ for $0 \le x \le L/8$. $dV/dx = (2Lx - 24x^2)/4$, $dV/dx = 0$ for $0 < x < L/8$ when $x = L/12$. If $x = 0, L/12, L/8$ then $V = 0, L^3/1728, 0$ so the volume is greatest when $x = L/12$ and $y = L/12$.

32. **(a)** Let x = diameter of the sphere, y = length of an edge of the cube. The combined volume is $V = \dfrac{1}{6}\pi x^3 + y^3$ and the surface area is $S = \pi x^2 + 6y^2$ = constant. Thus $y = \dfrac{(S - \pi x^2)^{1/2}}{6^{1/2}}$ and $V = \dfrac{\pi}{6}x^3 + \dfrac{(S - \pi x^2)^{3/2}}{6^{3/2}}$ for $0 \le x \le \sqrt{\dfrac{S}{\pi}}$;
$\dfrac{dV}{dx} = \dfrac{\pi}{2}x^2 - \dfrac{3\pi}{6^{3/2}}x(S - \pi x^2)^{1/2} = \dfrac{\pi}{2\sqrt{6}}x(\sqrt{6}x - \sqrt{S - \pi x^2})$. $\dfrac{dV}{dx} = 0$ when $x = 0$, or when $\sqrt{6}x = \sqrt{S - \pi x^2}$, $6x^2 = S - \pi x^2$, $x^2 = \dfrac{S}{6 + \pi}$, $x = \sqrt{\dfrac{S}{6 + \pi}}$. If $x = 0$, $\sqrt{\dfrac{S}{6 + \pi}}$, $\sqrt{\dfrac{S}{\pi}}$, then $V = \dfrac{S^{3/2}}{6^{3/2}}$, $\dfrac{S^{3/2}}{6\sqrt{6 + \pi}}$, $\dfrac{S^{3/2}}{6\sqrt{\pi}}$ so that V is smallest when $x = \sqrt{\dfrac{S}{6 + \pi}}$, and hence when $y = \sqrt{\dfrac{S}{6 + \pi}}$, thus $x = y$.

(b) From Part (a), the sum of the volumes is greatest when there is no cube.

33. Let h and r be the dimensions shown in the figure, then the volume is $V = \frac{1}{3}\pi r^2 h$. But $r^2 + h^2 = L^2$ thus $r^2 = L^2 - h^2$ so $V = \frac{1}{3}\pi(L^2 - h^2)h = \frac{1}{3}\pi(L^2 h - h^3)$ for $0 \le h \le L$. $\frac{dV}{dh} = \frac{1}{3}\pi(L^2 - 3h^2)$. $\frac{dV}{dh} = 0$ when $h = L/\sqrt{3}$. If $h = 0, L/\sqrt{3}, 0$ then $V = 0, \frac{2\pi}{9\sqrt{3}}L^3, 0$ so the volume is as large as possible when $h = L/\sqrt{3}$ and $r = \sqrt{2/3}L$.

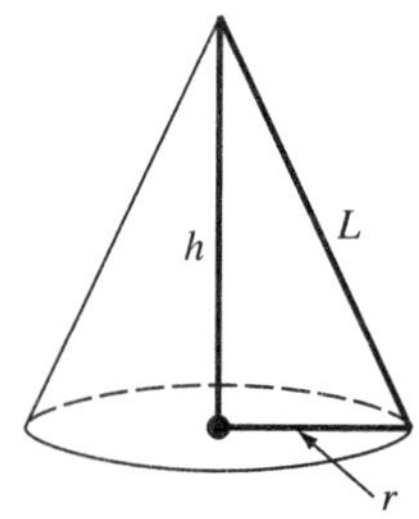

34. Let r and h be the radius and height of the cone (see figure). The slant height of any such cone will be R, the radius of the circular sheet. Refer to the solution of Exercise 33 to find that the largest volume is $\frac{2\pi}{9\sqrt{3}}R^3$.

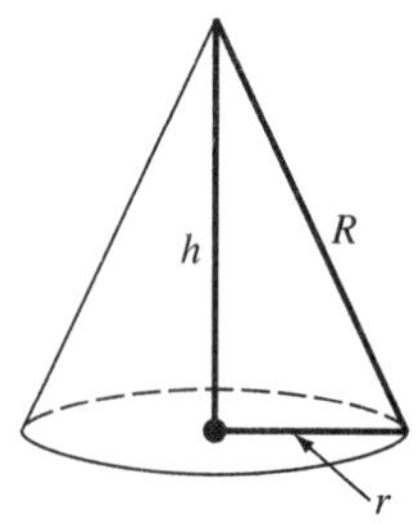

35. The area of the paper is $A = \pi r L = \pi r\sqrt{r^2 + h^2}$, but $V = \frac{1}{3}\pi r^2 h = 10$ thus $h = 30/(\pi r^2)$
so $A = \pi r\sqrt{r^2 + 900/(\pi^2 r^4)}$.
To simplify the computations let $S = A^2$,
$S = \pi^2 r^2\left(r^2 + \frac{900}{\pi^2 r^4}\right) = \pi^2 r^4 + \frac{900}{r^2}$ for $r > 0$,
$\frac{dS}{dr} = 4\pi^2 r^3 - \frac{1800}{r^3} = \frac{4(\pi^2 r^6 - 450)}{r^3}$, $dS/dr = 0$ when $r = \sqrt[6]{450/\pi^2}$, $d^2S/dr^2 > 0$, so S and hence A is least when $r = \sqrt[6]{450/\pi^2}$ cm, $h = \frac{30}{\pi}\sqrt[3]{\pi^2/450}$ cm.

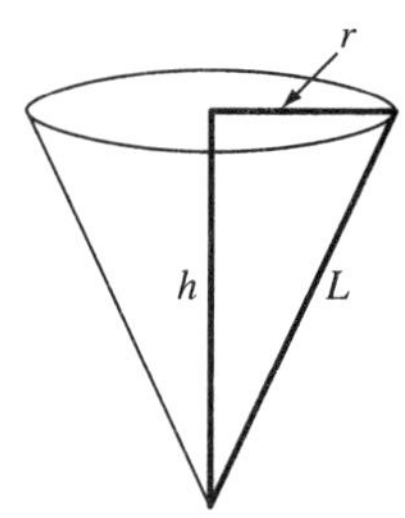

36. The area of the triangle is $A = \frac{1}{2}hb$. By similar triangles (see figure) $\frac{b/2}{h} = \frac{R}{\sqrt{h^2 - 2Rh}}$, $b = \frac{2Rh}{\sqrt{h^2 - 2Rh}}$
so $A = \frac{Rh^2}{\sqrt{h^2 - 2Rh}}$ for $h > 2R$, $\frac{dA}{dh} = \frac{Rh^2(h - 3R)}{(h^2 - 2Rh)^{3/2}}$,
$\frac{dA}{dh} = 0$ for $h > 2R$ when $h = 3R$, by the first derivative test A is minimum when $h = 3R$. If $h = 3R$ then $b = 2\sqrt{3}R$ (the triangle is equilateral).

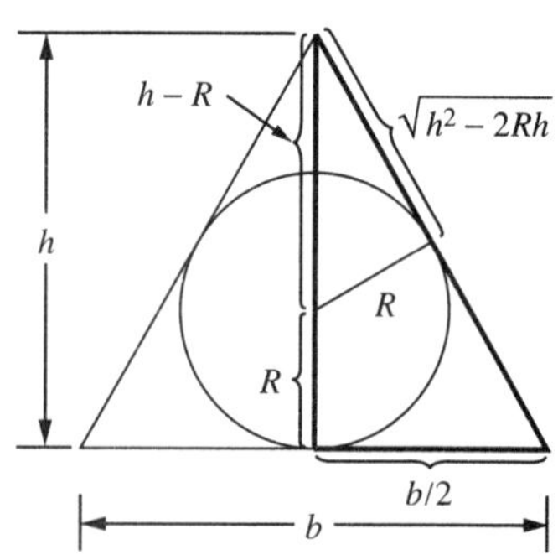

37. The volume of the cone is $V = \dfrac{1}{3}\pi r^2 h$. By similar triangles (see figure) $\dfrac{r}{h} = \dfrac{R}{\sqrt{h^2 - 2Rh}}$, $r = \dfrac{Rh}{\sqrt{h^2 - 2Rh}}$ so $V = \dfrac{1}{3}\pi R^2 \dfrac{h^3}{h^2 - 2Rh} = \dfrac{1}{3}\pi R^2 \dfrac{h^2}{h - 2R}$ for $h > 2R$, $\dfrac{dV}{dh} = \dfrac{1}{3}\pi R^2 \dfrac{h(h - 4R)}{(h - 2R)^2}$, $\dfrac{dV}{dh} = 0$ for $h > 2R$ when $h = 4R$, by the first derivative test V is minimum when $h = 4R$. If $h = 4R$ then $r = \sqrt{2}R$.

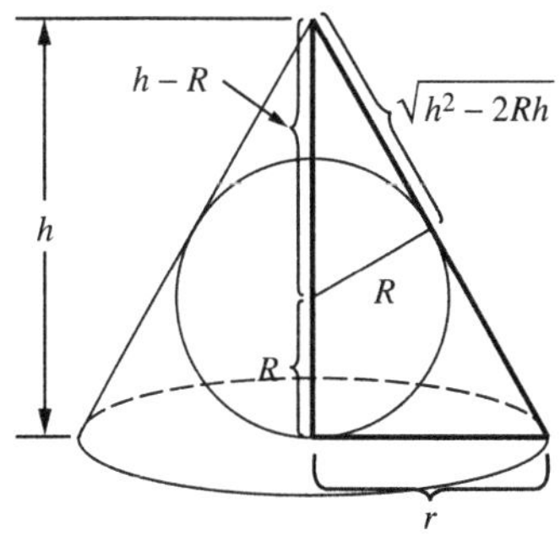

38. The area is (see figure)

$$A = \frac{1}{2}(2\sin\theta)(4 + 4\cos\theta)$$
$$= 4(\sin\theta + \sin\theta\cos\theta)$$

for $0 \le \theta \le \pi/2$;

$$dA/d\theta = 4(\cos\theta - \sin^2\theta + \cos^2\theta)$$
$$= 4(\cos\theta - [1 - \cos^2\theta] + \cos^2\theta)$$
$$= 4(2\cos^2\theta + \cos\theta - 1)$$
$$= 4(2\cos\theta - 1)(\cos\theta + 1)$$

$dA/d\theta = 0$ when $\theta = \pi/3$ for $0 < \theta < \pi/2$. If $\theta = 0, \pi/3, \pi/2$ then $A = 0, 3\sqrt{3}, 4$ so the maximum area is $3\sqrt{3}$.

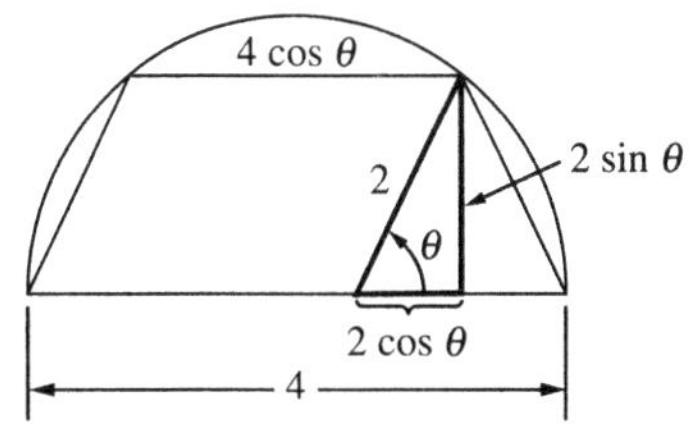

39. Let b and h be the dimensions shown in the figure, then the cross-sectional area is $A = \dfrac{1}{2}h(5 + b)$. But $h = 5\sin\theta$ and $b = 5 + 2(5\cos\theta) = 5 + 10\cos\theta$ so $A = \dfrac{5}{2}\sin\theta(10 + 10\cos\theta) = 25\sin\theta(1 + \cos\theta)$ for $0 \le \theta \le \pi/2$.

$$dA/d\theta = -25\sin^2\theta + 25\cos\theta(1 + \cos\theta)$$
$$= 25(-\sin^2\theta + \cos\theta + \cos^2\theta)$$
$$= 25(-1 + \cos^2\theta + \cos\theta + \cos^2\theta)$$
$$= 25(2\cos^2\theta + \cos\theta - 1) = 25(2\cos\theta - 1)(\cos\theta + 1).$$

$dA/d\theta = 0$ for $0 < \theta < \pi/2$ when $\cos\theta = 1/2$, $\theta = \pi/3$. If $\theta = 0, \pi/3, \pi/2$ then $A = 0, 75\sqrt{3}/4, 25$ so the cross-sectional area is greatest when $\theta = \pi/3$.

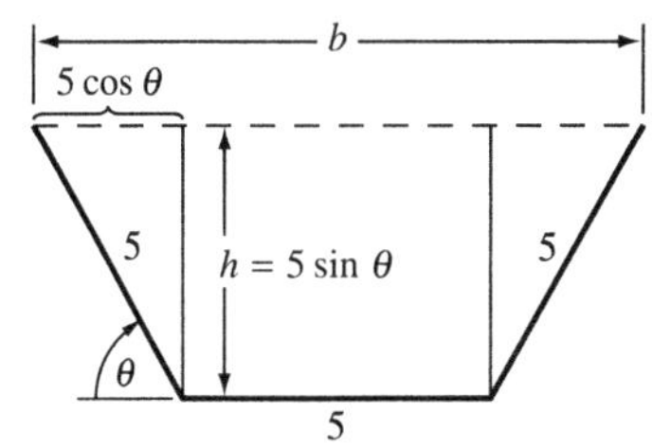

40. $I = k\dfrac{\cos\phi}{\ell^2}$, k the constant of proportionality. If h is the height of the lamp above the table then $\cos\phi = h/\ell$ and $\ell = \sqrt{h^2 + r^2}$ so $I = k\dfrac{h}{\ell^3} = k\dfrac{h}{(h^2 + r^2)^{3/2}}$ for $h > 0$, $\dfrac{dI}{dh} = k\dfrac{r^2 - 2h^2}{(h^2 + r^2)^{5/2}}$, $\dfrac{dI}{dh} = 0$ when $h = r/\sqrt{2}$, by the first derivative test I is maximum when $h = r/\sqrt{2}$.

41. Let L, L_1, and L_2 be as shown in the figure, then $L = L_1 + L_2 = 8\csc\theta + \sec\theta$,

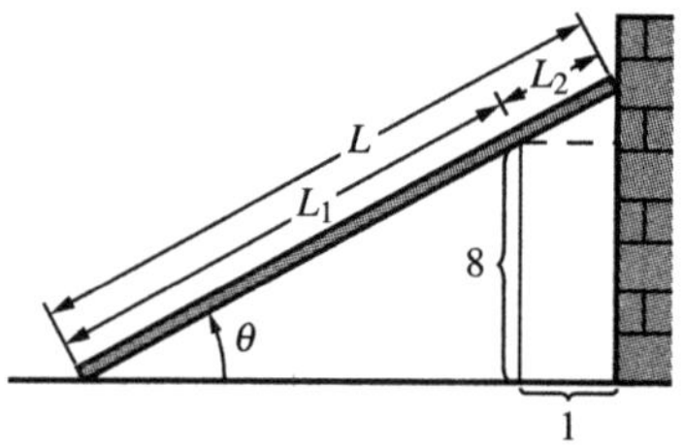

$$\frac{dL}{d\theta} = -8\csc\theta\cot\theta + \sec\theta\tan\theta,\ 0 < \theta < \pi/2$$
$$= -\frac{8\cos\theta}{\sin^2\theta} + \frac{\sin\theta}{\cos^2\theta} = \frac{-8\cos^3\theta + \sin^3\theta}{\sin^2\theta\cos^2\theta};$$

$\dfrac{dL}{d\theta} = 0$ if $\sin^3\theta = 8\cos^3\theta$, $\tan^3\theta = 8$, $\tan\theta = 2$ which gives the absolute minimum for L because $\lim_{\theta\to 0^+} L = \lim_{\theta\to\pi/2^-} L = +\infty$.
If $\tan\theta = 2$, then $\csc\theta = \sqrt{5}/2$ and $\sec\theta = \sqrt{5}$ so $L = 8(\sqrt{5}/2) + \sqrt{5} = 5\sqrt{5}$ ft.

42. Let $x =$ number of steers per acre
$w =$ average market weight per steer
$T =$ total market weight per acre
then $T = xw$ where $w = 2000 - 50(x - 20) = 3000 - 50x$
so $T = x(3000 - 50x) = 3000x - 50x^2$ for $0 \le x \le 60$,
$dT/dx = 3000 - 100x$ and $dT/dx = 0$ when $x = 30$. If $x = 0, 30, 60$ then $T = 0, 45000, 0$ so the total market weight per acre is largest when 30 steers per acre are allowed.

43. **(a)** The daily profit is
$P =$ (revenue) $-$ (production cost) $= 100x - (100,000 + 50x + 0.0025x^2)$
$= -100,000 + 50x - 0.0025x^2$
for $0 \le x \le 7000$, so $dP/dx = 50 - 0.005x$ and $dP/dx = 0$ when $x = 10,000$. Because 10,000 is not in the interval $[0, 7000]$, the maximum profit must occur at an endpoint. When $x = 0$, $P = -100,000$; when $x = 7000$, $P = 127,500$ so 7000 units should be manufactured and sold daily.

(b) Yes, because $dP/dx > 0$ when $x = 7000$ so profit is increasing at this production level.

(c) $dP/dx = 15$ when $x = 7000$, so $P(7001) - P(7000) \approx 15$, and the marginal profit is \$15.

44. **(a)** $R(x) = px$ but $p = 1000 - x$ so $R(x) = (1000 - x)x$

(b) $P(x) = R(x) - C(x) = (1000 - x)x - (3000 + 20x) = -3000 + 980x - x^2$

(c) $P'(x) = 980 - 2x$, $P'(x) = 0$ for $0 < x < 500$ when $x = 490$; test the points $0, 490, 500$ to find that the profit is a maximum when $x = 490$.

(d) $P(490) = 237{,}100$

(e) $p = 1000 - x = 1000 - 490 = 510$.

45. The profit is

$$P = \text{(profit on nondefective)} - \text{(loss on defective)} = 100(x - y) - 20y = 100x - 120y$$

but $y = 0.01x + 0.00003x^2$ so $P = 100x - 120(0.01x + 0.00003x^2) = 98.8x - 0.0036x^2$ for $x > 0$, $dP/dx = 98.8 - 0.0072x$, $dP/dx = 0$ when $x = 98.8/0.0072 \approx 13,722$, $d^2P/dx^2 < 0$ so the profit is maximum at a production level of about 13,722 pounds.

46. The total cost C is
$C = c\cdot$ (hours to travel 3000 mi at a speed of v mi/h)
$= c\cdot\dfrac{3000}{v} = (a+bv^n)\dfrac{3000}{v} = 3000(av^{-1}+bv^{n-1})$ for $v>0$,
$dC/dv = 3000[-av^{-2}+b(n-1)v^{n-2}] = 3000[-a+b(n-1)v^n]/v^2$,
$dC/dv = 0$ when $v = \left[\dfrac{a}{b(n-1)}\right]^{1/n}$. This is the only critical point and dC/dv changes sign from $-$ to $+$ at this point so the total cost is least when $v = \left[\dfrac{a}{b(n-1)}\right]^{1/n}$ mi/h.

47. The distance between the particles is $D = \sqrt{(1-t-t)^2+(t-2t)^2} = \sqrt{5t^2-4t+1}$ for $t\ge 0$. For convenience, we minimize D^2 instead, so $D^2 = 5t^2-4t+1$, $dD^2/dt = 10t-4$, which is 0 when $t = 2/5$. $d^2D^2/dt^2 > 0$ so D^2 and hence D is minimum when $t = 2/5$. The minimum distance is $D = 1/\sqrt{5}$.

48. The distance between the particles is $D = \sqrt{(2t-t)^2+(2-t^2)^2} = \sqrt{t^4-3t^2+4}$ for $t\ge 0$. For convenience we minimize D^2 instead so $D^2 = t^4-3t^2+4$, $dD^2/dt = 4t^3-6t = 4t(t^2-3/2)$, which is 0 for $t>0$ when $t = \sqrt{3/2}$. $d^2D^2/dt^2 = 12t^2-6>0$ when $t=\sqrt{3/2}$ so D^2 and hence D is minimum there. The minimum distance is $D = \sqrt{7}/2$.

49. Let $P(x,y)$ be a point on the curve $x^2+y^2=1$. The distance between $P(x,y)$ and $P_0(2,0)$ is $D = \sqrt{(x-2)^2+y^2}$, but $y^2 = 1-x^2$ so $D = \sqrt{(x-2)^2+1-x^2} = \sqrt{5-4x}$ for $-1\le x\le 1$, $\dfrac{dD}{dx} = -\dfrac{2}{\sqrt{5-4x}}$ which has no critical points for $-1<x<1$. If $x=-1,1$ then $D=3,1$ so the closest point occurs when $x=1$ and $y=0$.

50. Let $P(x,y)$ be a point on $y=\sqrt{x}$, then the distance D between P and $(2,0)$ is $D = \sqrt{(x-2)^2+y^2} = \sqrt{(x-2)^2+x} = \sqrt{x^2-3x+4}$, for $0\le x\le 3$. For convenience we find the extrema for D^2 instead, so $D^2 = x^2-3x+4$, $dD^2/dx = 2x-3 = 0$ when $x=3/2$. If $x = 0,3/2,3$ then $D^2 = 4,7/4,4$ so $D = 2,\sqrt{7}/2,2$. The points $(0,0)$ and $(3,\sqrt{3})$ are at the greatest distance, and $(3/2,\sqrt{3/2})$ the shortest distance from $(2,0)$.

51. Let (x,y) be a point on the curve, then the square of the distance between (x,y) and $(0,2)$ is $S = x^2+(y-2)^2$ where $x^2-y^2=1$, $x^2=y^2+1$ so
$S = (y^2+1)+(y-2)^2 = 2y^2-4y+5$ for any y, $dS/dy = 4y-4$, $dS/dy = 0$ when $y=1$,
$d^2S/dy^2>0$ so S is least when $y=1$ and $x=\pm\sqrt{2}$.

52. The square of the distance between a point (x,y) on the curve and the point $(0,9)$ is
$S = x^2+(y-9)^2$ where $x=2y^2$ so $S = 4y^4+(y-9)^2$ for any y,
$dS/dy = 16y^3+2(y-9) = 2(8y^3+y-9)$, $dS/dy=0$ when $y=1$ (which is the only real solution),
$d^2S/dy^2>0$ so S is least when $y=1$, $x=2$.

53. If $P(x_0,y_0)$ is on the curve $y=1/x^2$, then $y_0 = 1/x_0^2$. At P the slope of the tangent line is $-2/x_0^3$ so its equation is $y-\dfrac{1}{x_0^2} = -\dfrac{2}{x_0^3}(x-x_0)$, or $y = -\dfrac{2}{x_0^3}x+\dfrac{3}{x_0^2}$. The tangent line crosses the y-axis at $\dfrac{3}{x_0^2}$, and the x-axis at $\dfrac{3}{2}x_0$. The length of the segment then is $L = \sqrt{\dfrac{9}{x_0^4}+\dfrac{9}{4}x_0^2}$ for $x_0>0$. For convenience, we minimize L^2 instead, so $L^2 = \dfrac{9}{x_0^4}+\dfrac{9}{4}x_0^2$, $\dfrac{dL^2}{dx_0} = -\dfrac{36}{x_0^5}+\dfrac{9}{2}x_0 = \dfrac{9(x_0^6-8)}{2x_0^5}$, which is 0 when $x_0^6=8$, $x_0=\sqrt{2}$. $\dfrac{d^2L^2}{dx_0^2}>0$ so L^2 and hence L is minimum when $x_0=\sqrt{2}$, $y_0=1/2$.

54. If $P(x_0, y_0)$ is on the curve $y = 1 - x^2$, then $y_0 = 1 - x_0^2$. At P the slope of the tangent line is $-2x_0$ so its equation is $y - (1 - x_0^2) = -2x_0(x - x_0)$, or $y = -2x_0 x + x_0^2 + 1$. The y-intercept is $x_0^2 + 1$ and the x-intercept is $\frac{1}{2}(x_0 + 1/x_0)$ so the area A of the triangle is

$A = \frac{1}{4}(x_0^2 + 1)(x_0 + 1/x_0) = \frac{1}{4}(x_0^3 + 2x_0 + 1/x_0)$ for $0 \le x_0 \le 1$. $dA/dx_0 = \frac{1}{4}(3x_0^2 + 2 - 1/x_0^2) = \frac{1}{4}(3x_0^4 + 2x_0^2 - 1)/x_0^2$ which is 0 when $x_0^2 = -1$ (reject), or when $x_0^2 = 1/3$ so $x_0 = 1/\sqrt{3}$. $d^2A/dx_0^2 = \frac{1}{4}(6x_0 + 2/x_0^3) > 0$ at $x_0 = 1/\sqrt{3}$ so a relative minimum and hence the absolute minimum occurs there.

55. At each point (x, y) on the curve the slope of the tangent line is $m = \frac{dy}{dx} = -\frac{2x}{(1 + x^2)^2}$ for any x, $\frac{dm}{dx} = \frac{2(3x^2 - 1)}{(1 + x^2)^3}$, $\frac{dm}{dx} = 0$ when $x = \pm 1/\sqrt{3}$, by the first derivative test the only relative maximum occurs at $x = -1/\sqrt{3}$, which is the absolute maximum because $\lim\limits_{x \to \pm\infty} m = 0$. The tangent line has greatest slope at the point $(-1/\sqrt{3}, 3/4)$.

56. Let x be how far P is upstream from where the man starts (see figure), then the total time to reach T is

$t =$ (time from M to P) + (time from P to T)

$= \frac{\sqrt{x^2 + 1}}{r_R} + \frac{1 - x}{r_W}$ for $0 \le x \le 1$,

where r_R and r_W are the rates at which he can row and walk, respectively.

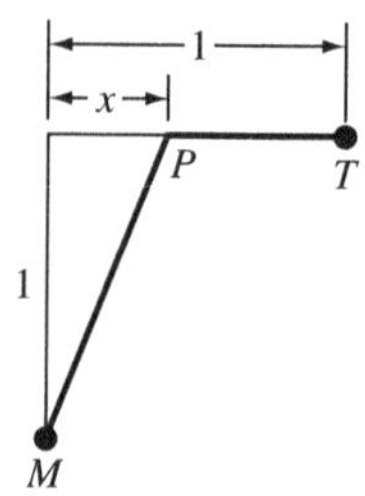

(a) $t = \frac{\sqrt{x^2 + 1}}{3} + \frac{1 - x}{5}$, $\frac{dt}{dx} = \frac{x}{3\sqrt{x^2 + 1}} - \frac{1}{5}$ so $\frac{dt}{dx} = 0$ when $5x = 3\sqrt{x^2 + 1}$, $25x^2 = 9(x^2 + 1)$, $x^2 = 9/16$, $x = 3/4$. If $x = 0, 3/4, 1$ then $t = 8/15, 7/15, \sqrt{2}/3$ so the time is a minimum when $x = 3/4$ mile.

(b) $t = \frac{\sqrt{x^2 + 1}}{4} + \frac{1 - x}{5}$, $\frac{dt}{dx} = \frac{x}{4\sqrt{x^2 + 1}} - \frac{1}{5}$ so $\frac{dt}{dx} = 0$ when $x = 4/3$ which is not in the interval $[0, 1]$. Check the endpoints to find that the time is a minimum when $x = 1$ (he should row directly to the town).

57. With x and y as shown in the figure, the maximum length of pipe will be the smallest value of $L = x + y$. By similar triangles

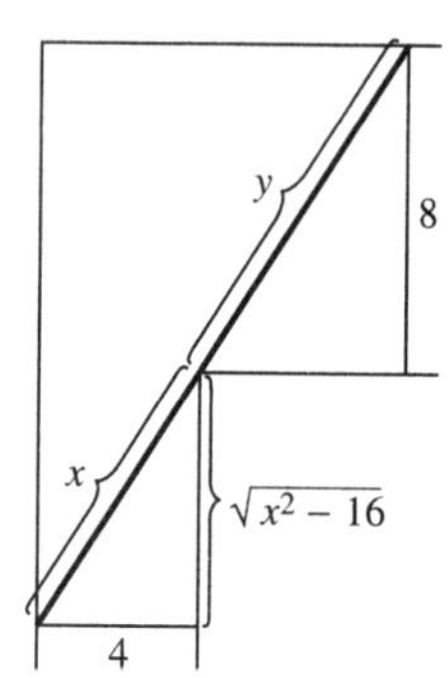

$\frac{y}{8} = \frac{x}{\sqrt{x^2 - 16}}$, $y = \frac{8x}{\sqrt{x^2 - 16}}$ so

$L = x + \frac{8x}{\sqrt{x^2 - 16}}$ for $x > 4$, $\frac{dL}{dx} = 1 - \frac{128}{(x^2 - 16)^{3/2}}$,

$\frac{dL}{dx} = 0$ when

$$
\begin{aligned}
(x^2 - 16)^{3/2} &= 128 \\
x^2 - 16 &= 128^{2/3} = 16(2^{2/3}) \\
x^2 &= 16(1 + 2^{2/3}) \\
x &= 4(1 + 2^{2/3})^{1/2},
\end{aligned}
$$

$d^2L/dx^2 = 384x/(x^2 - 16)^{5/2} > 0$ if $x > 4$ so L is smallest when $x = 4(1 + 2^{2/3})^{1/2}$. For this value of x, $L = 4(1 + 2^{2/3})^{3/2}$ ft.

58. $s = (x_1 - \bar{x})^2 + (x_2 - \bar{x})^2 + \cdots + (x_n - \bar{x})^2$,
$ds/d\bar{x} = -2(x_1 - \bar{x}) - 2(x_2 - \bar{x}) - \cdots - 2(x_n - \bar{x})$,
$ds/d\bar{x} = 0$ when

$$(x_1 - \bar{x}) + (x_2 - \bar{x}) + \cdots + (x_n - \bar{x}) = 0$$
$$(x_1 + x_2 + \cdots x_n) - (\bar{x} + \bar{x} + \cdots + \bar{x}) = 0$$
$$(x_1 + x_2 + \cdots + x_n) - n\bar{x} = 0$$
$$\bar{x} = \frac{1}{n}(x_1 + x_2 + \cdots + x_n),$$

$d^2s/d\bar{x}^2 = 2 + 2 + \cdots + 2 = 2n > 0$, so s is minimum when $\bar{x} = \frac{1}{n}(x_1 + x_2 + \cdots + x_n)$.

59. Let x = distance from the weaker light source, I = the intensity at that point, and k the constant of proportionality. Then

$I = \dfrac{kS}{x^2} + \dfrac{8kS}{(90 - x)^2}$ if $0 < x < 90$;

$$\frac{dI}{dx} = -\frac{2kS}{x^3} + \frac{16kS}{(90 - x)^3} = \frac{2kS[8x^3 - (90 - x)^3]}{x^3(90 - x)^3} = 18\frac{kS(x - 30)(x^2 + 2700)}{x^3(x - 90)^3},$$

which is 0 when $x = 30$; $\dfrac{dI}{dx} < 0$ if $x < 30$, and $\dfrac{dI}{dx} > 0$ if $x > 30$, so the intensity is minimum at a distance of 30 cm from the weaker source.

60. If $f(x_0)$ is a maximum then $f(x) \le f(x_0)$ for all x in some open interval containing x_0 thus $\sqrt{f(x)} \le \sqrt{f(x_0)}$ because $\sqrt{x}$ is an increasing function, so $\sqrt{f(x_0)}$ is a maximum of $\sqrt{f(x)}$ at x_0. The proof is similar for a minimum value, simply replace $\le$ by $\ge$.

61. Let v = speed of light in the medium. The total time required for the light to travel from A to P to B is

$t =$ (total distance from A to P to B)$/v = \dfrac{1}{v}(\sqrt{(c - x)^2 + a^2} + \sqrt{x^2 + b^2})$,

$$\frac{dt}{dx} = \frac{1}{v}\left[-\frac{c - x}{\sqrt{(c - x)^2 + a^2}} + \frac{x}{\sqrt{x^2 + b^2}}\right]$$

and $\dfrac{dt}{dx} = 0$ when $\dfrac{x}{\sqrt{x^2 + b^2}} = \dfrac{c - x}{\sqrt{(c - x)^2 + a^2}}$. But $x/\sqrt{x^2 + b^2} = \sin\theta_2$ and $(c - x)/\sqrt{(c - x)^2 + a^2} = \sin\theta_1$ thus $dt/dx = 0$ when $\sin\theta_2 = \sin\theta_1$ so $\theta_2 = \theta_1$.

62. The total time required for the light to travel from A to P to B is

$t =$ (time from A to P) $+$ (time from P to B) $= \dfrac{\sqrt{x^2 + a^2}}{v_1} + \dfrac{\sqrt{(c - x)^2 + b^2}}{v_2}$,

$\dfrac{dt}{dx} = \dfrac{x}{v_1\sqrt{x^2 + a^2}} - \dfrac{c - x}{v_2\sqrt{(c - x)^2 + b^2}}$ but $x/\sqrt{x^2 + a^2} = \sin\theta_1$ and $(c - x)/\sqrt{(c - x)^2 + b^2} = \sin\theta_2$ thus $\dfrac{dt}{dx} = \dfrac{\sin\theta_1}{v_1} - \dfrac{\sin\theta_2}{v_2}$ so $\dfrac{dt}{dx} = 0$ when $\dfrac{\sin\theta_1}{v_1} = \dfrac{\sin\theta_2}{v_2}$.

63. **(a)** The rate at which the farmer walks is analogous to the speed of light in Fermat's principle.

(b) the best path occurs when $\theta_1 = \theta_2$ (see figure).

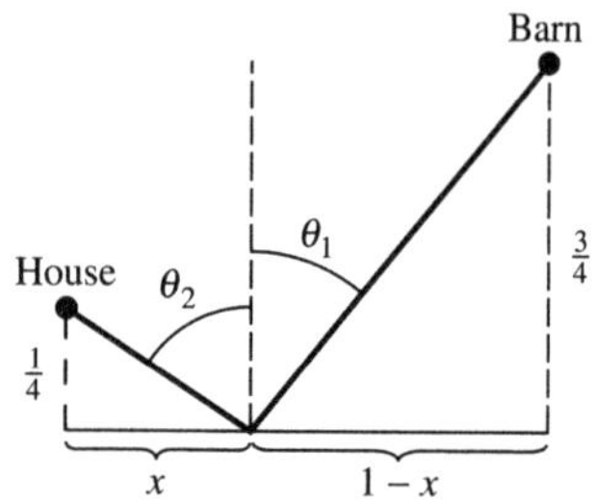

(c) by similar triangles,

$$x/(1/4) = (1-x)/(3/4)$$
$$3x = 1 - x$$
$$4x = 1$$
$$x = 1/4 \text{ mi.}$$

EXERCISE SET 4.7

1. $f(x) = x^2 - 2,\ f'(x) = 2x,\ x_{n+1} = x_n - \dfrac{x_n^2 - 2}{2x_n}$

$x_1 = 1,\ x_2 = 1.5,\ x_3 = 1.416666667, \ldots,\ x_5 = x_6 = 1.414213562$

2. $f(x) = x^2 - 7,\ f'(x) = 2x,\ x_{n+1} = x_n - \dfrac{x_n^2 - 7}{2x_n}$

$x_1 = 3,\ x_2 = 2.666666667,\ x_3 = 2.645833333, \ldots,\ x_5 = x_6 = 2.645751311$

3. $f(x) = x^3 - 6,\ f'(x) = 3x^2,\ x_{n+1} = x_n - \dfrac{x_n^3 - 6}{3x_n^2}$

$x_1 = 2,\ x_2 = 1.833333333,\ x_3 = 1.817263545, \ldots,\ x_5 = x_6 = 1.817120593$

4. $x^n - a = 0$

5. $f(x) = x^3 - x + 3,\ f'(x) = 3x^2 - 1,\ x_{n+1} = x_n - \dfrac{x_n^3 - x_n + 3}{3x_n^2 - 1}$

$x_1 = -2,\ x_2 = -1.727272727,\ x_3 = -1.673691174, \ldots, x_5 = x_6 = -1.671699882$

6. $f(x) = x^3 + x - 1,\ f'(x) = 3x^2 + 1,\ x_{n+1} = x_n - \dfrac{x_n^3 + x_n - 1}{3x_n^2 + 1}$

$x_1 = 1,\ x_2 = 0.75,\ x_3 = 0.686046512, \ldots,\ x_5 = x_6 = 0.682327804$

7. $f(x) = x^5 + x^4 - 5,\ f'(x) = 5x^4 + 4x^3,\ x_{n+1} = x_n - \dfrac{x_n^5 + x_n^4 - 5}{5x_n^4 + 4x_n^3}$

$x_1 = 1,\ x_2 = 1.333333333,\ x_3 = 1.239420573, \ldots, x_6 = x_7 = 1.224439550$

8. $f(x) = x^5 - x + 1,\ f'(x) = 5x^4 - 1,\ x_{n+1} = x_n - \dfrac{x_n^5 - x_n + 1}{5x_n^4 - 1}$

$x_1 = -1,\ x_2 = -1.25,\ x_3 = -1.178459394, \ldots, x_6 = x_7 = -1.167303978$

9. $f(x) = x^4 + x - 3$, $f'(x) = 4x^3 + 1$,

$$x_{n+1} = x_n - \frac{x_n^4 + x_n - 3}{4x_n^3 + 1}$$

$x_1 = -2$, $x_2 = -1.645161290$,
$x_3 = -1.485723955, \ldots, x_6 = x_7 = -1.452626879$

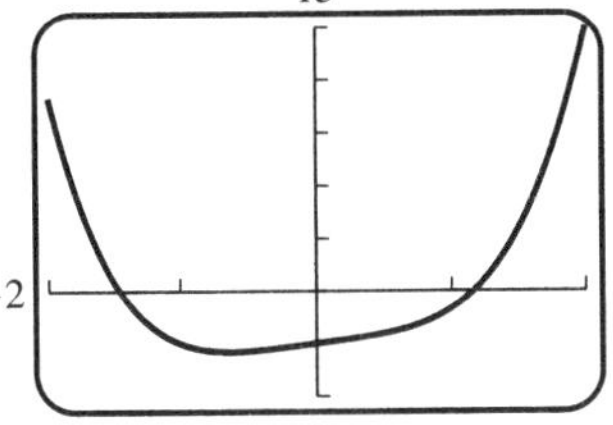

10. $f(x) = x^5 - 5x^3 - 2$, $f'(x) = 5x^4 - 15x^2$,

$$x_{n+1} = x_n - \frac{x_n^5 - 5x_n^3 - 2}{5x_n^4 - 15x_n^2}$$

$x_1 = 2$, $x_2 = 2.5$,
$x_3 = 2.327384615, \ldots, x_7 = x_8 = 2.273791732$

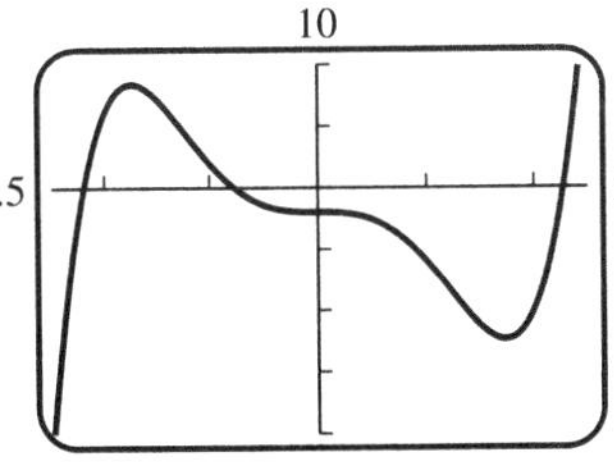

11. $f(x) = 2\sin x - x$, $f'(x) = 2\cos x - 1$,

$$x_{n+1} = x_n - \frac{2\sin x_n - x_n}{2\cos x_n - 1}$$

$x_1 = 2$, $x_2 = 1.900995594$,
$x_3 = 1.895511645$, $x_4 = x_5 = 1.895494267$

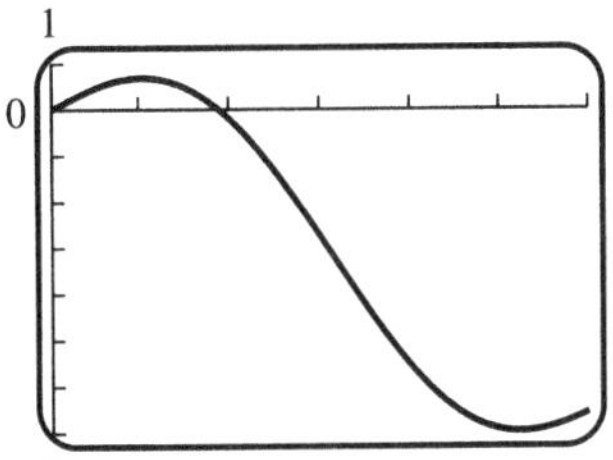

12. $f(x) = \sin x - x^2$,
$f'(x) = \cos x - 2x$,

$$x_{n+1} = x_n - \frac{\sin x_n - x_n^2}{\cos x_n - 2x_n}$$

$x_1 = 1$, $x_2 = 0.891395995$,
$x_3 = 0.876984845, \ldots, x_5 = x_6 = 0.876726215$

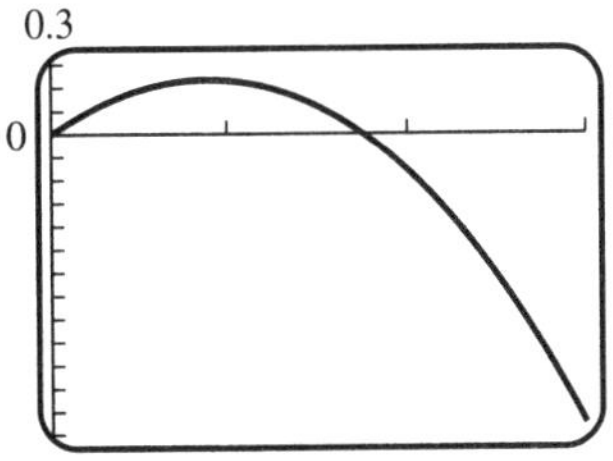

13. $f(x) = x - \tan x$,

$f'(x) = 1 - \sec^2 x = -\tan^2 x$, $x_{n+1} = x_n + \dfrac{x_n - \tan x_n}{\tan^2 x_n}$

$x_1 = 4.5$, $x_2 = 4.493613903$, $x_3 = 4.493409655$,
$x_4 = x_5 = 4.493409458$

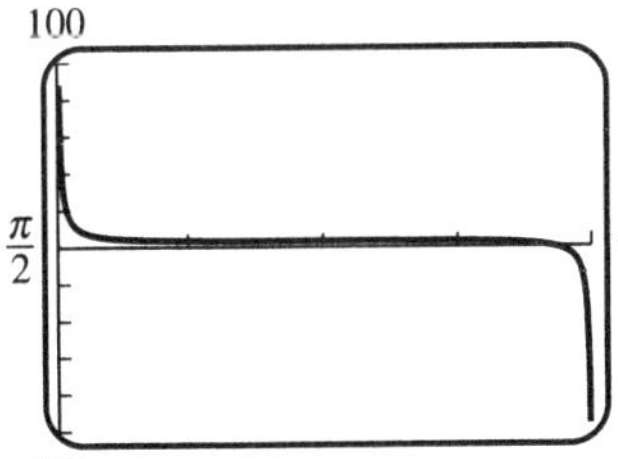

14. $f(x) = 1 + x^2 \cos x$, $f'(x) = 2x \cos x - x^2 \sin x$

$$x_{n+1} = x_n - \frac{1 + x_n^2 \cos x_n}{2x_n \cos x_n - x_n^2 \sin x_n}$$

$x_1 = 2$, $x_2 = 1.8746$, $x_3 = 1.8631$,
$x_4 = 1.863045316, x_5 = x_6 = 1.863045308$

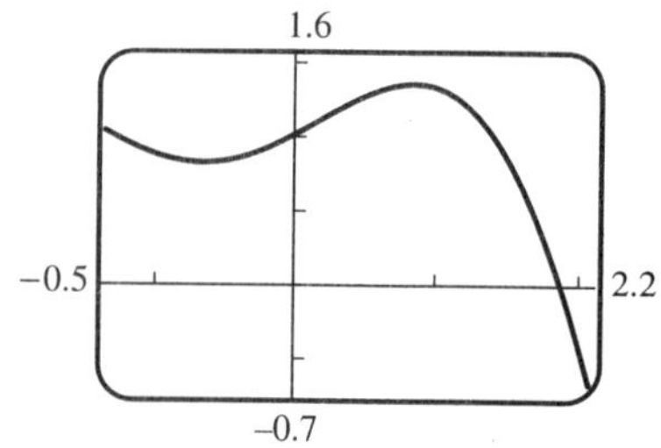

15. At the point of intersection, $x^3 = 0.5x - 1$, $x^3 - 0.5x + 1 = 0$. Let $f(x) = x^3 - 0.5x + 1$. By graphing $y = x^3$ and $y = 0.5x - 1$ it is evident that there is only one point of intersection and it occurs in the interval $[-2, -1]$; note that $f(-2) < 0$ and $f(-1) > 0$. $f'(x) = 3x^2 - 0.5$ so

$$x_{n+1} = x_n - \frac{x_n^3 - 0.5x + 1}{3x_n^2 - 0.5};$$

$x_1 = -1$, $x_2 = -1.2$,

$x_3 = -1.166492147, \ldots,$
$x_5 = x_6 = -1.165373043$

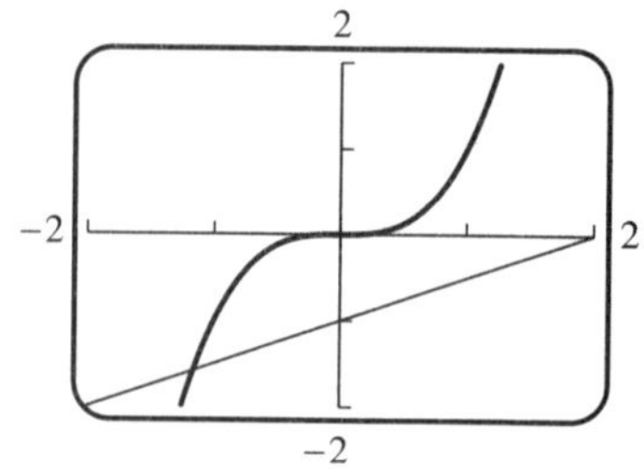

16. The graphs of $y = \sin x$ and $y = x^3 - 2x^2 + 1$ intersect at points near $x = -0.8$ and $x = 0.6$ and $x = 2$. Let $f(x) = \sin x - x^3 + 2x^2 - 1$, then $f'(x) = \cos x - 3x^2 + 4x$, so

$$x_{n+1} = x_n - \frac{\cos x - 3x^2 + 4x}{\sin x - x^3 + 2x^2 + 1}.$$

If $x_1 = -0.8$, then $x_2 = -0.783124811$,
$x_3 = -0.782808234$,
$x_4 = x_5 = -0.782808123$; ; if $x_1 = 0.6$, then
$x_2 = 0.568003853, x_3 = x_4 = 0.568025739$; if $x_1 = 2$, then
$x_2 = 1.979461151$, $x_3 = 1.979019264, x_4 = x_5 = 1.979019061$

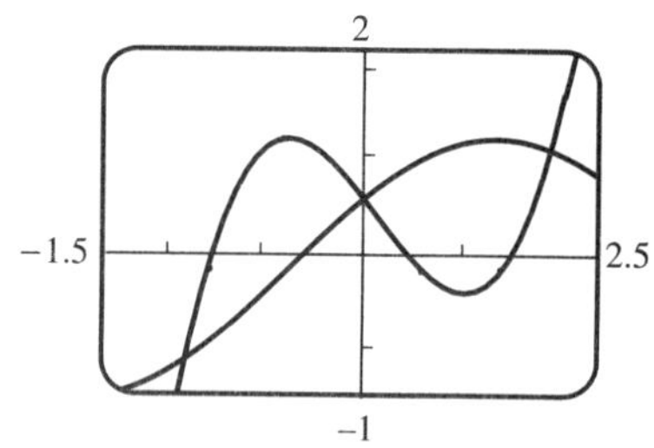

17. The graphs of $y = x^2$ and $y = \sqrt{2x + 1}$ intersect at points near $x = -0.5$ and $x = 1$; $x^2 = \sqrt{2x + 1}$, $x^4 - 2x - 1 = 0$. Let $f(x) = x^4 - 2x - 1$, then $f'(x) = 4x^3 - 2$ so

$$x_{n+1} = x_n - \frac{x_n^4 - 2x_n - 1}{4x_n^3 - 2}.$$

If $x_1 = -0.5$, then $x_2 = -0.475$,
$x_3 = -0.474626695$,
$x_4 = x_5 = -0.474626618$; if
$x_1 = 1$, then $x_2 = 2$,
$x_3 = 1.633333333, \ldots, x_8 = x_9 = 1.395336994$.

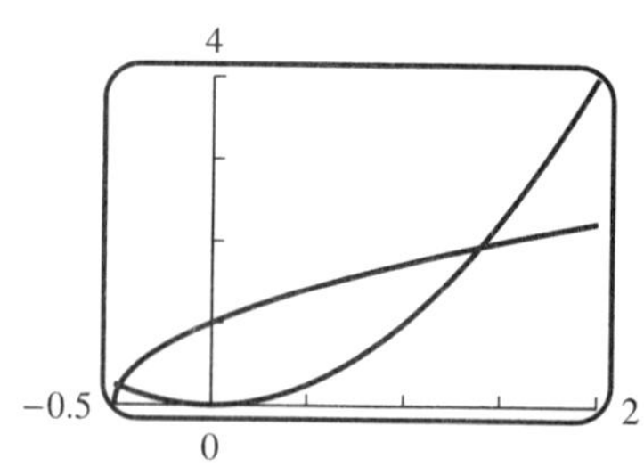

18. The graphs of $y = x^3/8 + 1$ and $y = \cos 2x$ intersect at $x = 0$ and at a point near $x = -2$; $x^3/8 + 1 = \cos 2x$, $x^3 - 8\cos 2x + 8 = 0$. Let $f(x) = x^3 - 8\cos 2x + 8$, then $f'(x) = 3x^2 + 16\sin 2x$ so

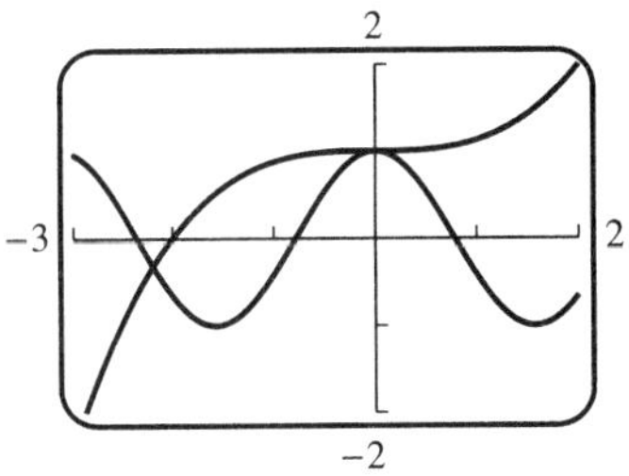

$$x_{n+1} = x_n - \frac{x_n^3 - 8\cos 2x_n + 8}{3x_n^2 + 16\sin 2x_n}.$$

$x_1 = -2$, $x_2 = -2.216897577$,
$x_3 = -2.193821581, \ldots, x_5 = x_6 = -2.193618950$.

19. (a) $f(x) = x^2 - a$, $f'(x) = 2x$, $x_{n+1} = x_n - \dfrac{x_n^2 - a}{2x_n} = \dfrac{1}{2}\left(x_n + \dfrac{a}{x_n}\right)$

(b) $a = 10$; $x_1 = 3$, $x_2 = 3.166666667$, $x_3 = 3.162280702$, $x_4 = x_5 = 3.162277660$

20. (a) $f(x) = \dfrac{1}{x} - a$, $f'(x) = -\dfrac{1}{x^2}$, $x_{n+1} = x_n(2 - ax_n)$

(b) $a = 17$; $x_1 = 0.05$, $x_2 = 0.0575$, $x_3 = 0.058793750$, $x_5 = x_6 = 0.058823529$

21. $f'(x) = x^3 + 2x + 5$; solve $f'(x) = 0$ to find the critical points. Graph $y = x^3$ and $y = -2x - 5$ to see that they intersect at a point near $x = -1$; $f''(x) = 3x^2 + 2$ so $x_{n+1} = x_n - \dfrac{x_n^3 + 2x_n + 5}{3x_n^2 + 2}$.

$x_1 = -1$, $x_2 = -1.4$, $x_3 = -1.330964467, \ldots, x_5 = x_6 = -1.328268856$ so the minimum value of $f(x)$ occurs at $x \approx -1.328268856$ because $f''(x) > 0$; its value is approximately -4.098859132.

22. From a rough sketch of $y = x\sin x$ we see that the maximum occurs at a point near $x = 2$, which will be a point where $f'(x) = x\cos x + \sin x = 0$. $f''(x) = 2\cos x - x\sin x$ so

$$x_{n+1} = x_n - \frac{x_n\cos x_n + \sin x_n}{2\cos x_n - x_n\sin x_n} = x_n - \frac{x_n + \tan x_n}{2 - x_n\tan x_n}.$$

$x_1 = 2$, $x_2 = 2.029048281$, $x_3 = 2.028757866$, $x_4 = x_5 = 2.028757838$; the maximum value is approximately 1.819705741.

23. Let $f(x)$ be the square of the distance between $(1, 0)$ and any point (x, x^2) on the parabola, then $f(x) = (x-1)^2 + (x^2 - 0)^2 = x^4 + x^2 - 2x + 1$ and $f'(x) = 4x^3 + 2x - 2$. Solve $f'(x) = 0$ to find the critical points; $f''(x) = 12x^2 + 2$ so $x_{n+1} = x_n - \dfrac{4x_n^3 + 2x_n - 2}{12x_n^2 + 2} = x_n - \dfrac{2x_n^3 + x_n - 1}{6x_n^2 + 1}$.

$x_1 = 1$, $x_2 = 0.714285714$, $x_3 = 0.605168701, \ldots, x_6 = x_7 = 0.589754512$; the coordinates are approximately $(0.589754512, 0.347810385)$.

24. The area is $A = xy = x\cos x$ so $dA/dx = \cos x - x\sin x$. Find x so that $dA/dx = 0$;

$$d^2A/dx^2 = -2\sin x - x\cos x \text{ so } x_{n+1} = x_n + \frac{\cos x_n - x_n\sin x_n}{2\sin x_n + x_n\cos x_n} = x_n + \frac{1 - x_n\tan x_n}{2\tan x_n + x_n}.$$

$x_1 = 1$, $x_2 = 0.864536397$, $x_3 = 0.860339078$, $x_4 = x_5 = 0.860333589$; $y \approx 0.652184624$.

25. (a) Let s be the arc length, and L the length of the chord, then $s = 1.5L$. But $s = r\theta$ and $L = 2r\sin(\theta/2)$ so $r\theta = 3r\sin(\theta/2)$, $\theta - 3\sin(\theta/2) = 0$.

(b) Let $f(\theta) = \theta - 3\sin(\theta/2)$, then $f'(\theta) = 1 - 1.5\cos(\theta/2)$ so $\theta_{n+1} = \theta_n - \dfrac{\theta_n - 3\sin(\theta_n/2)}{1 - 1.5\cos(\theta_n/2)}$.

$\theta_1 = 3$, $\theta_2 = 2.991592920$, $\theta_3 = 2.991563137$, $\theta_4 = \theta_5 = 2.991563136$ rad so $\theta \approx 171°$.

26. $r^2(\theta - \sin\theta)/2 = \pi r^2/4$ so $\theta - \sin\theta - \pi/2 = 0$. Let $f(\theta) = \theta - \sin\theta - \pi/2$, then $f'(\theta) = 1 - \cos\theta$ so $\theta_{n+1} = \dfrac{\theta_n - \sin\theta_n - \pi/2}{1 - \cos\theta_n}$.
$\theta_1 = 2$, $\theta_2 = 2.339014106$, $\theta_3 = 2.310063197, \ldots, \theta_5 = \theta_6 = 2.309881460$ rad; $\theta \approx 132°$.

27. If $x = 1$, then $y^4 + y = 1$, $y^4 + y - 1 = 0$. Graph $z = y^4$ and $z = 1 - y$ to see that they intersect near $y = -1$ and $y = 1$. Let $f(y) = y^4 + y - 1$, then $f'(y) = 4y^3 + 1$ so $y_{n+1} = y_n - \dfrac{y_n^4 + y_n - 1}{4y_n^3 + 1}$.
If $y_1 = -1$, then $y_2 = -1.333333333$, $y_3 = -1.235807860, \ldots, y_6 = y_7 = -1.220744085$;
if $y_1 = 1$, then $y_2 = 0.8$, $y_3 = 0.731233596, \ldots, y_6 = y_7 = 0.724491959$.

28. If $x = 1$, then $2y - \cos y = 0$. Graph $z = 2y$ and $z = \cos y$ to see that they intersect near $y = 0.5$. Let $f(y) = 2y - \cos y$, then $f'(y) = 2 + \sin y$ so $y_{n+1} = y_n - \dfrac{2y_n - \cos y_n}{2 + \sin y_n}$.
$y_1 = 0.5$, $y_2 = 0.450626693$, $y_3 = 0.450183648$, $y_4 = y_5 = 0.450183611$.

29. $S(25) = 250{,}000 = \dfrac{5000}{i}\left[(1+i)^{25} - 1\right]$; set $f(i) = 50i - (1+i)^{25} + 1$, $f'(i) = 50 - 25(1+i)^{24}$; solve $f(i) = 0$. Set $i_0 = .06$ and $i_{k+1} = i_k - \left[50i - (1+i)^{25} + 1\right] / \left[50 - 25(1+i)^{24}\right]$. Then $i_1 = 0.05430$, $i_2 = 0.05338$, $i_3 = 0.05336$, ..., $i = 0.053362$.

30. (a) $x_1 = 2$, $x_2 = 5.3333$,
$x_3 = 11.055$, $x_4 = 22.293$,
$x_5 = 44.676$

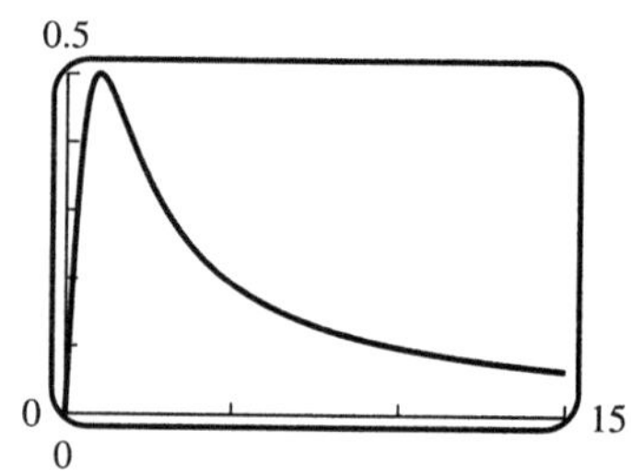

(b) $x_1 = 0.5$, $x_2 = -0.3333$, $x_3 = 0.0833$, $x_4 = -0.0012$, $x_5 = 0.0000$ (and $x_n = 0$ for $n \geq 6$)

31. (a)

x_1	x_2	x_3	x_4	x_5	x_6	x_7	x_8	x_9	x_{10}
0.5000	−0.7500	0.2917	−1.5685	−0.4654	0.8415	−0.1734	2.7970	1.2197	0.1999

(b) The sequence x_n must diverge, since if it did converge then $f(x) = x^2 + 1 = 0$ would have a solution. It seems the x_n are oscillating back and forth in a quasi-cyclical fashion.

EXERCISE SET 4.8

1. $f(0) = f(4) = 0$; $f'(3) = 0$; $[0, 4]$, $c = 3$

2. $f(-3) = f(3) = 0$; $f'(0) = 0$

3. $f(2) = f(4) = 0$, $f'(x) = 2x - 6$, $2c - 6 = 0$, $c = 3$

4. $f(0) = f(2) = 0$, $f'(x) = 3x^2 - 6x + 2$, $3c^2 - 6c + 2 = 0$; $c = \dfrac{6 \pm \sqrt{36 - 24}}{6} = 1 \pm \sqrt{3}/3$

5. $f(\pi/2) = f(3\pi/2) = 0$, $f'(x) = -\sin x$, $-\sin c = 0$, $c = \pi$

6. $f(-1) = f(1) = 0,\ f'(x) = \dfrac{x^2 - 4x + 1}{(x-2)^2},\ \dfrac{c^2 - 4c + 1}{(c-2)^2} = 0,\ c^2 - 4c + 1 = 0$

$c = \dfrac{4 \pm \sqrt{16-4}}{2} = 2 \pm \sqrt{3}$, of which only $c = 2 - \sqrt{3}$ is in $(-1, 1)$

7. $f(0) = f(4) = 0,\ f'(x) = \dfrac{1}{2} - \dfrac{1}{2\sqrt{x}},\ \dfrac{1}{2} - \dfrac{1}{2\sqrt{c}} = 0,\ c = 1$

8. $f(1) = f(3) = 0,\ f'(x) = -\dfrac{2}{x^3} + \dfrac{4}{3x^2},\ -\dfrac{2}{c^3} + \dfrac{4}{3c^2} = 0,\ -6 + 4c = 0,\ c = 3/2$

9. $\dfrac{f(8) - f(0)}{8 - 0} = \dfrac{6}{8} = \dfrac{3}{4} = f'(1.54);\ c = 1.54$

10. $\dfrac{f(4) - f(0)}{4 - 0} = 1.19 = f'(0.77)$

11. $f(-4) = 12,\ f(6) = 42,\ f'(x) = 2x + 1,\ 2c + 1 = \dfrac{42 - 12}{6 - (-4)} = 3,\ c = 1$

12. $f(-1) = -6,\ f(2) = 6,\ f'(x) = 3x^2 + 1,\ 3c^2 + 1 = \dfrac{6 - (-6)}{2 - (-1)} = 4,\ c^2 = 1,\ c = \pm 1$ of which only $c = 1$ is in $(-1, 2)$

13. $f(0) = 1,\ f(3) = 2,\ f'(x) = \dfrac{1}{2\sqrt{x+1}},\ \dfrac{1}{2\sqrt{c+1}} = \dfrac{2-1}{3-0} = \dfrac{1}{3},\ \sqrt{c+1} = 3/2,\ c + 1 = 9/4,\ c = 5/4$

14. $f(3) = 10/3,\ f(4) = 17/4,\ f'(x) = 1 - 1/x^2,\ 1 - 1/c^2 = \dfrac{17/4 - 10/3}{4 - 3} = 11/12,\ c^2 = 12,\ c = \pm 2\sqrt{3}$ of which only $c = 2\sqrt{3}$ is in $(3, 4)$

15. $f(-5) = 0,\ f(3) = 4,\ f'(x) = -\dfrac{x}{\sqrt{25 - x^2}},\ -\dfrac{c}{\sqrt{25 - c^2}} = \dfrac{4 - 0}{3 - (-5)} = \dfrac{1}{2},\ -2c = \sqrt{25 - c^2}$,
$4c^2 = 25 - c^2,\ c^2 = 5,\ c = -\sqrt{5}$
(we reject $c = \sqrt{5}$ because it does not satisfy the equation $-2c = \sqrt{25 - c^2}$)

16. $f(2) = 1,\ f(5) = 1/4,\ f'(x) = -1/(x-1)^2,\ -\dfrac{1}{(c-1)^2} = \dfrac{1/4 - 1}{5 - 2} = -\dfrac{1}{4},\ (c-1)^2 = 4,\ c - 1 = \pm 2$, $c = -1$ (reject), or $c = 3$

17. (a) $f(-2) = f(1) = 0$
The interval is $[-2, 1]$

(b) $c = -1.29$

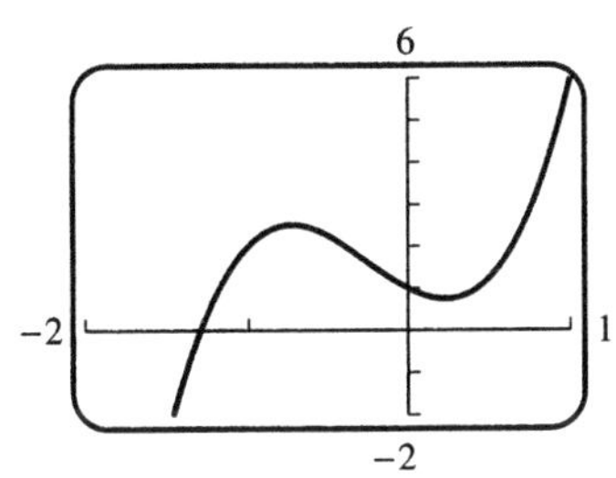

(c) $x_0 = -1,\ x_1 = -1.5,\ x_2 = -1.32,\ x_3 = -1.290,\ x_4 = -1.2885843$

18. **(a)** $m = \dfrac{f(-2) - f(1)}{-2 - 1} = \dfrac{-16 - 5}{-3} = 7$ so $y - 5 = 7(x - 1)$, $y = 7x - 2$

(b) $f'(x) = 3x^2 + 4 = 7$ has solutions $x = \pm 1$; discard $x = 1$, so $c = -1$

(c) $y - (-5) = 7(x - (-1))$ or $y = 7x + 2$ **(d)**

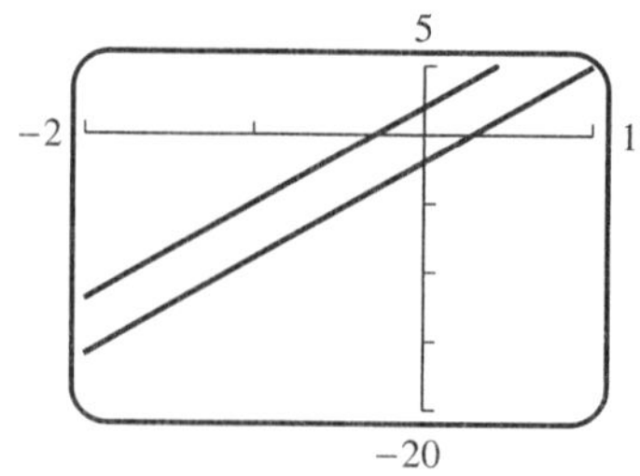

19. **(a)** $f'(x) = \sec^2 x$, $\sec^2 c = 0$ has no solution **(b)** $\tan x$ is not continuous on $[0, \pi]$

20. **(a)** $f(-1) = 1$, $f(8) = 4$, $f'(x) = \dfrac{2}{3}x^{-1/3}$

$\dfrac{2}{3}c^{-1/3} = \dfrac{4 - 1}{8 - (-1)} = \dfrac{1}{3}$, $c^{1/3} = 2$, $c = 8$ which is not in $(-1, 8)$.

(b) $x^{2/3}$ is not differentiable at $x = 0$, which is in $(-1, 8)$.

21. **(a)** Two x-intercepts of f determine two solutions a and b of $f(x) = 0$; by Rolle's Theorem there exists a point c between a and b such that $f'(c) = 0$, i.e. c is an x-intercept for f'.

(b) $f(x) = \sin x = 0$ at $x = n\pi$, and $f'(x) = \cos x = 0$ at $x = n\pi + \pi/2$, which lies between $n\pi$ and $(n + 1)\pi$, $(n = 0, \pm 1, \pm 2, \ldots)$

22. $\dfrac{f(x_1) - f(x_0)}{x_1 - x_0}$ is the average rate of change of y with respect to x on the interval $[x_0, x_1]$. By the Mean-Value Theorem there is a value c in (x_0, x_1) such that the instantaneous rate of change $f'(c) = \dfrac{f(x_1) - f(x_0)}{x_1 - x_0}$.

23. Let $s(t)$ be the position function of the automobile for $0 \le t \le 5$, then by the Mean-Value Theorem there is at least one point c in $(0, 5)$ where
$s'(c) = v(c) = [s(5) - s(0)]/(5 - 0) = 4/5 = 0.8$ mi/min $= 48$ mi/h.

24. Let $T(t)$ denote the temperature at time with $t = 0$ denoting 11 AM, then $T(0) = 76$ and $T(12) = 52$.

(a) By the Mean-Value Theorem there is a value c between 0 and 12 such that
$T'(c) = [T(12) - T(0)]/(12 - 0) = (52 - 76)/(12) = -2°$ F/h.

(b) Assume that $T(t_1) = 88°$F where $0 < t_1 < 12$, then there is at least one point c in $(t_1, 12)$ where $T'(c) = [T(12) - T(t_1)]/(12 - t_1) = (52 - 88)/(12 - t_1) = -36/(12 - t_1)$. But $12 - t_1 < 12$ so $T'(c) < -36/12 = -3°$F/h.

25. Let $f(t)$ and $g(t)$ denote the distances from the first and second runners to the starting point, and let $h(t) = f(t) - g(t)$. Since they start (at $t = 0$) and finish (at $t = t_1$) at the same time, $h(0) = h(t_1) = 0$, so by Rolle's Theorem there is a time t_2 for which $h'(t_2) = 0$, i.e. $f'(t_2) = g'(t_2)$; so they have the same velocity at time t_2.

26. $f(x) = x^6 - 2x^2 + x$ satisfies $f(0) = f(1) = 0$, so by Rolle's Theorem $f'(c) = 0$ for some c in $(0, 1)$.

27. **(a)** By the Constant Difference Theorem $f(x) - g(x) = k$ for some k; since $f(x_0) = g(x_0)$, $k = 0$, so $f(x) = g(x)$ for all x.

(b) Set $f(x) = \sin^2 x + \cos^2 x$, $g(x) = 1$; then $f'(x) = 2 \sin x \cos x - 2 \cos x \sin x = 0 = g'(x)$. Since $f(0) = 1 = g(0)$, $f(x) = g(x)$ for all x.

28. **(a)** By the Constant Difference Theorem $f(x) - g(x) = k$ for some k; since $f(x_0) - g(x_0) = c$, $k = c$, so $f(x) - g(x) = c$ for all x.

(b) Set $f(x) = (x-1)^3$, $g(x) = (x^2+3)(x-3)$. Then
$f'(x) = 3(x-1)^2$, $g'(x) = (x^2+3) + 2x(x-3) = 3x^2 - 6x + 3 = 3(x^2 - 2x + 1) = 3(x-1)^2$, so $f'(x) = g'(x)$ and hence $f(x) - g(x) = k$. Expand $f(x)$ and $g(x)$ to get
$h(x) = f(x) - g(x) = (x^3 - 3x^2 + 3x - 1) - (x^3 - 3x^2 + 3x - 9) = 8$.

(c) $h(x) = x^3 - 3x^2 + 3x - 1 - (x^3 - 3x^2 + 3x - 9) = 8$

29. **(a)** If x, y belong to I and $x < y$ then for some c in I, $\dfrac{f(y) - f(x)}{y - x} = f'(c)$,
so $|f(x) - f(y)| = |f'(c)||x - y| \le M|x - y|$; if $x > y$ exchange x and y; if $x = y$ the inequality also holds.

(b) $f(x) = \sin x$, $f'(x) = \cos x$, $|f'(x)| \le 1 = M$, so $|f(x) - f(y)| \le |x - y|$ or
$|\sin x - \sin y| \le |x - y|$.

30. **(a)** If x, y belong to I and $x < y$ then for some c in I, $\dfrac{f(y) - f(x)}{y - x} = f'(c)$,
so $|f(x) - f(y)| = |f'(c)||x - y| \ge M|x - y|$; if $x > y$ exchange x and y; if $x = y$ the inequality also holds.

(b) If x and y belong to $(-\pi/2, \pi/2)$ and $f(x) = \tan x$, then $|f'(x)| = \sec^2 x \ge 1$ and
$|\tan x - \tan y| \ge |x - y|$

(c) y lies in $(-\pi/2, \pi/2)$ if and only if $-y$ does; use Part (b) and replace y with $-y$

31. **(a)** Let $f(x) = \sqrt{x}$. By the Mean-Value Theorem there is a number c between x and y such that
$\dfrac{\sqrt{y} - \sqrt{x}}{y - x} = \dfrac{1}{2\sqrt{c}} < \dfrac{1}{2\sqrt{x}}$ for c in (x, y), thus $\sqrt{y} - \sqrt{x} < \dfrac{y - x}{2\sqrt{x}}$

(b) multiply through and rearrange to get $\sqrt{xy} < \dfrac{1}{2}(x + y)$.

32. Suppose that $f(x)$ has at least two distinct real solutions r_1 and r_2 in I. Then $f(r_1) = f(r_2) = 0$ so by Rolle's Theorem there is at least one number between r_1 and r_2 where $f'(x) = 0$, but this contradicts the assumption that $f'(x) \ne 0$, so $f(x) = 0$ must have fewer than two distinct solutions in I.

33. **(a)** If $f(x) = x^3 + 4x - 1$ then $f'(x) = 3x^2 + 4$ is never zero, so by Exercise 32 f has at most one real root; since f is a cubic polynomial it has at least one real root, so it has exactly one real root.

(b) Let $f(x) = ax^3 + bx^2 + cx + d$. If $f(x) = 0$ has at least two distinct real solutions r_1 and r_2, then $f(r_1) = f(r_2) = 0$ and by Rolle's Theorem there is at least one number between r_1 and r_2 where $f'(x) = 0$. But $f'(x) = 3ax^2 + 2bx + c = 0$ for
$x = (-2b \pm \sqrt{4b^2 - 12ac})/(6a) = (-b \pm \sqrt{b^2 - 3ac})/(3a)$, which are not real if $b^2 - 3ac < 0$ so $f(x) = 0$ must have fewer than two distinct real solutions.

34. $f'(x) = \dfrac{1}{2\sqrt{x}}$, $\dfrac{1}{2\sqrt{c}} = \dfrac{\sqrt{4} - \sqrt{3}}{4 - 3} = 2 - \sqrt{3}$. But $\dfrac{1}{4} < \dfrac{1}{2\sqrt{c}} < \dfrac{1}{2\sqrt{3}}$ for c in $(3, 4)$ so
$\dfrac{1}{4} < 2 - \sqrt{3} < \dfrac{1}{2\sqrt{3}}$, $0.25 < 2 - \sqrt{3} < 0.29$, $-1.75 < -\sqrt{3} < -1.71$, $1.71 < \sqrt{3} < 1.75$.

35. **(a)** $\dfrac{d}{dx}[f^2(x) + g^2(x)] = 2f(x)f'(x) + 2g(x)g'(x) = 2f(x)g(x) + 2g(x)[-f(x)] = 0$,
so $f^2(x) + g^2(x)$ is constant.

(b) $f(x) = \sin x$ and $g(x) = \cos x$

36. $\dfrac{d}{dx}[f^2(x) - g^2(x)] = 2f(x)f'(x) - 2g(x)g'(x) = 2f(x)g(x) - 2g(x)f(x) = 0$ so $f^2(x) - g^2(x)$ is constant.

37. If $f'(x) = g'(x)$, then $f(x) = g(x) + k$. Let $x = 1$,
$f(1) = g(1) + k = (1)^3 - 4(1) + 6 + k = 3 + k = 2$, so $k = -1$. $f(x) = x^3 - 4x + 5$.

38. Let $h = f - g$, then h is continuous on $[a, b]$, differentiable on (a, b), and $h(a) = f(a) - g(a) = 0$, $h(b) = f(b) - g(b) = 0$. By Rolle's Theorem there is some c in (a, b) where $h'(c) = 0$. But $h'(c) = f'(c) - g'(c)$ so $f'(c) - g'(c) = 0$, $f'(c) = g'(c)$.

39.

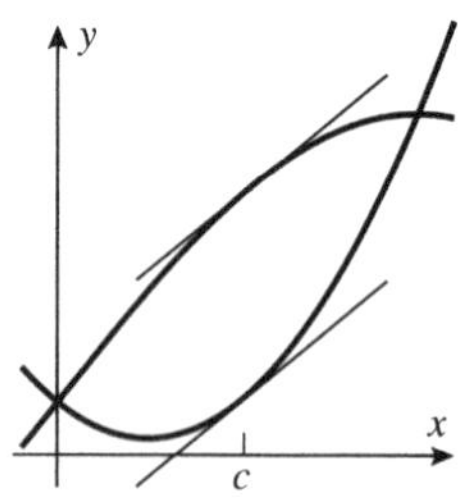

40. **(a)** Suppose $f'(x) = 0$ more than once in (a, b), say at c_1 and c_2. Then $f'(c_1) = f'(c_2) = 0$ and by using Rolle's Theorem on f', there is some c between c_1 and c_2 where $f''(c) = 0$, which contradicts the fact that $f''(x) > 0$ so $f'(x) = 0$ at most once in (a, b).

(b) If $f''(x) > 0$ for all x in (a, b), then f is concave up on (a, b) and has at most one relative extremum, which would be a relative minimum, on (a, b).

41. **(a)** similar to the proof of Part (a) with $f'(c) < 0$

(b) similar to the proof of Part (a) with $f'(c) = 0$

42. Let $x \neq x_0$ be sufficiently near x_0 so that there exists (by the Mean-Value Theorem) a number c (which depends on x) between x and x_0, such that

$$\frac{f(x) - f(x_0}{x - x_0} = f'(c).$$

Since c is between x and x_0 it follows that

$$\begin{aligned} f'(x_0) &= \lim_{x \to x_0} \frac{f(x) - f(x_0)}{x - x_0} && \text{(by definition of derivative)} \\ &= \lim_{x \to x_0} f'(c) && \text{(by the Mean-Value Theorem)} \\ &= \lim_{x \to x_0} f'(x) && \text{(since } \lim f'(x) \text{ exists and } c \text{ is between } x \text{ and } x_0\text{).} \end{aligned}$$

43. If f is differentiable at $x = 1$, then f is continuous there;
$\lim\limits_{x \to 1^+} f(x) = \lim\limits_{x \to 1^-} f(x) = f(1) = 3$, $a + b = 3$; $\lim\limits_{x \to 1^+} f'(x) = a$ and $\lim\limits_{x \to 1^-} f'(x) = 6$ so $a = 6$ and $b = 3 - 6 = -3$.

44. **(a)** $\lim\limits_{x \to 0^-} f'(x) = \lim\limits_{x \to 0^-} 2x = 0$ and $\lim\limits_{x \to 0^+} f'(x) = \lim\limits_{x \to 0^+} 2x = 0$; $f'(0)$ does not exist because f is not continuous at $x = 0$.

(b) $\lim\limits_{x \to 0^-} f'(x) = \lim\limits_{x \to 0^+} f'(x) = 0$ and f is continuous at $x = 0$, so $f'(0) = 0$;
$\lim\limits_{x \to 0^-} f''(x) = \lim\limits_{x \to 0^-} (2) = 2$ and $\lim\limits_{x \to 0^+} f''(x) = \lim\limits_{x \to 0^+} 6x = 0$, so $f''(0)$ does not exist.

45. From Section 3.2 a function has a vertical tangent line at a point of its graph if the slopes of secant lines through the point approach $+\infty$ or $-\infty$. Suppose f is continuous at $x = x_0$ and $\lim_{x \to x_0^+} f(x) = +\infty$. Then a secant line through $(x_1, f(x_1))$ and $(x_0, f(x_0))$, assuming $x_1 > x_0$, will have slope $\dfrac{f(x_1) - f(x_0)}{x_1 - x_0}$. By the Mean Value Theorem, this quotient is equal to $f'(c)$ for some c between x_0 and x_1. But as x_1 approaches x_0, c must also approach x_0, and it is given that $\lim_{c \to x_0^+} f'(c) = +\infty$, so the slope of the secant line approaches $+\infty$. The argument can be altered appropriately for $x_1 < x_0$, and/or for $f'(c)$ approaching $-\infty$.

SUPPLEMENTARY EXERCISES FOR CHAPTER 4

4. **(a)** False; an example is $y = \dfrac{x^3}{3} - \dfrac{x^2}{2}$ on $[-2, 2]$; $x = 0$ is a relative maximum and $x = 1$ is a relative minimum, but y = 0 is not the largest value of y on the interval, nor is $y = -\dfrac{1}{6}$ the smallest.

(b) true

(c) False; for example $y = x^3$ on $(-1, 1)$ which has a critical number but no relative extrema

6. **(a)**

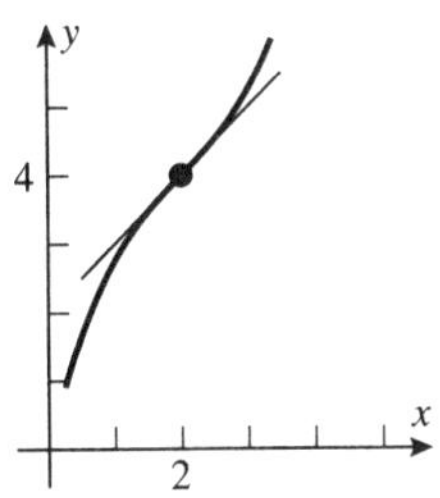

(b)

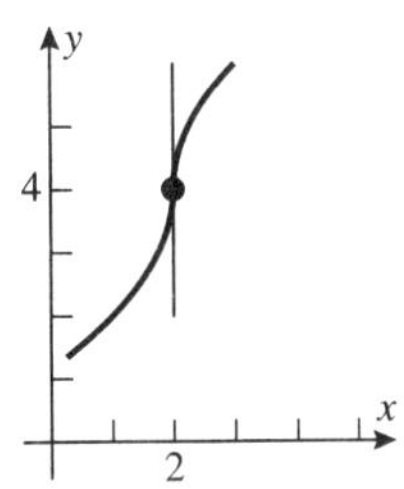

(c)

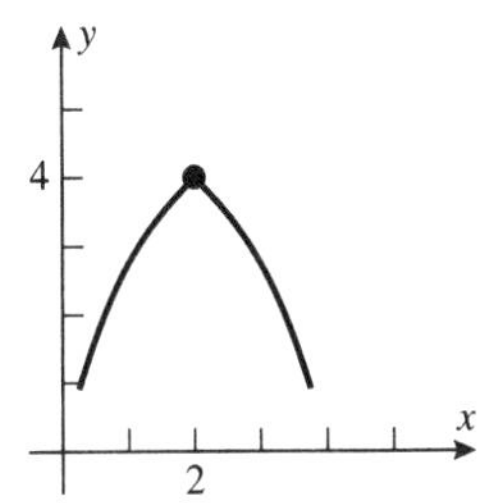

7. **(a)** $f'(x) = \dfrac{7(x-7)(x-1)}{3x^{2/3}}$; critical numbers at $x = 0, 1, 7$;
neither at $x = 0$, relative maximum at $x = 1$, relative minimum at $x = 7$ (First Derivative Test)

(b) $f'(x) = 2\cos x(1 + 2\sin x)$; critical numbers at $x = \pi/2, 3\pi/2, 7\pi/6, 11\pi/6$;
relative maximum at $x = \pi/2, 3\pi/2$, relative minimum at $x = 7\pi/6, 11\pi/6$

(c) $f'(x) = 3 - \dfrac{3\sqrt{x-1}}{2}$; critical numbers at $x = 5$; relative maximum at $x = 5$

8. **(a)** $f'(x) = \dfrac{x-9}{18x^{3/2}}$, $f''(x) = \dfrac{27-x}{36x^{5/2}}$; critical number at $x = 9$;
$f''(9) > 0$, relative minimum at $x = 9$

(b) $f'(x) = 2\dfrac{x^3-4}{x^2}$, $f''(x) = 2\dfrac{x^3+8}{x^3}$;
critical number at $x = 4^{1/3}$, $f''(4^{1/3}) > 0$, relative minimum at $x = 4^{1/3}$

(c) $f'(x) = \sin x(2\cos x + 1)$, $f''(x) = 2\cos^2 x - 2\sin^2 x + \cos x$; critical numbers at $x = 2\pi/3, \pi, 4\pi/3$; $f''(2\pi/3) < 0$, relative maximum at $x = 2\pi/3$; $f''(\pi) > 0$, relative minimum at $x = \pi$; $f''(4\pi/3) < 0$, relative maximum at $x = 4\pi/3$

9. $\lim_{x\to-\infty} f(x) = +\infty$, $\lim_{x\to+\infty} f(x) = +\infty$
$f'(x) = x(4x^2 - 9x + 6)$, $f''(x) = 6(2x-1)(x-1)$
relative minimum at $x = 0$,
points of inflection when $x = 1/2, 1$,
no asymptotes

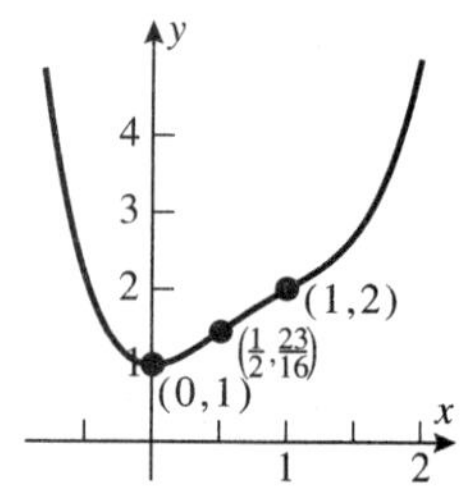

10. $\lim_{x\to-\infty} f(x) = -\infty$, $\lim_{x\to+\infty} f(x) = +\infty$
$f(x) = x^3(x-2)^2$, $f'(x) = x^2(5x-6)(x-2)$,
$f''(x) = 4x(5x^2 - 12x + 6)$

critical numbers at $x = 0, \dfrac{8 \pm 2\sqrt{31}}{5}$

relative maximum at $x = \dfrac{8 - 2\sqrt{31}}{5} \approx -0.63$

relative minimum at $x = \dfrac{8 + 2\sqrt{31}}{5} \approx 3.83$

points of inflection at $x = 0, \dfrac{6 \pm \sqrt{66}}{5} \approx 0, -0.42, 2.82$
no asymptotes

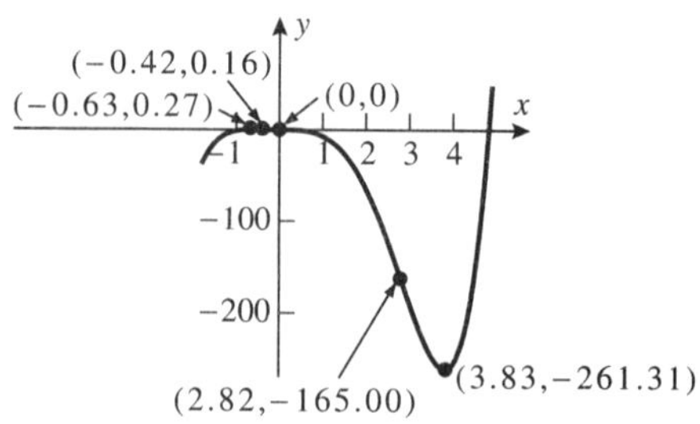

11. $\lim_{x\to\pm\infty} f(x)$ doesn't exist
$f'(x) = 2x\sec^2(x^2+1)$,
$f''(x) = 2\sec^2(x^2+1)\left[1 + 4x^2\tan(x^2+1)\right]$
critical number at $x = 0$; relative minimum at $x = 0$
point of inflection when $1 + 4x^2\tan(x^2+1) = 0$
vertical asymptotes at $x = \pm\sqrt{\pi(n + \frac{1}{2}) - 1}$, $n = 0, 1, 2, \ldots$

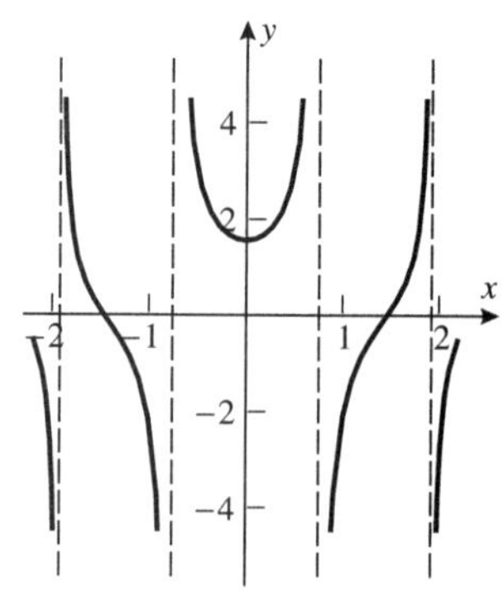

12. $\lim_{x\to-\infty} f(x) = -\infty$, $\lim_{x\to+\infty} f(x) = +\infty$
$f'(x) = 1 + \sin x$, $f''(x) = \cos x$
critical numbers at $x = 2n\pi + \pi/2$, $n = 0, \pm1, \pm2, \ldots$,
no extrema because $f' \geq 0$ and by Exercise 51 of Section 5.1,
f is increasing on $(-\infty, +\infty)$
inflections points at $x = n\pi + \pi/2$, $n = 0, \pm1, \pm2, \ldots$
no asymptotes

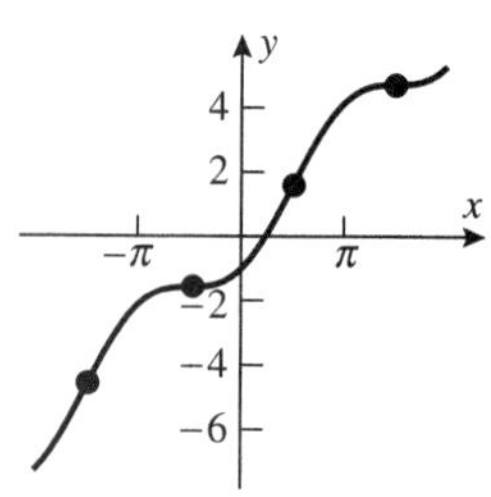

13. $f'(x) = 2\dfrac{x(x+5)}{(x^2+2x+5)^2}$, $f''(x) = -2\dfrac{2x^3+15x^2-25}{(x^2+2x+5)^3}$
critical numbers at $x = -5, 0$;
relative maximum at $x = -5$,
relative minimum at $x = 0$
points of inflection at $x = -7.26, -1.44, 1.20$
horizontal asymptote $y = 1$ as $x \to \pm\infty$

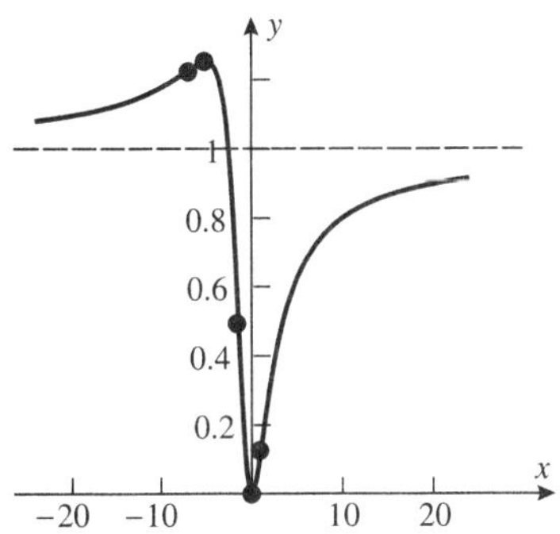

14. $f'(x) = 3\dfrac{3x^2-25}{x^4}$, $f''(x) = -6\dfrac{3x^2-50}{x^5}$
critical numbers at $x = \pm 5\sqrt{3}/3$;
relative maximum at $x = -5\sqrt{3}/3$,
relative minimum at $x = +5\sqrt{3}/3$
inflection points at $x = \pm 5\sqrt{2/3}$
horizontal asymptote of $y = 0$ as $x \to \pm\infty$,
vertical asymptote $x = 0$

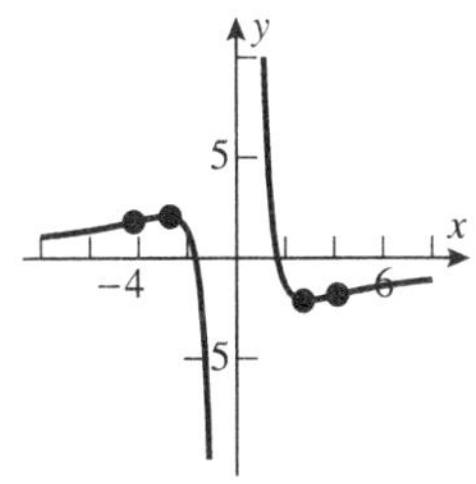

15. $\lim_{x\to-\infty} f(x) = +\infty$, $\lim_{x\to+\infty} f(x) = -\infty$

$$f'(x) = \begin{cases} x \\ -2x \end{cases} \text{if } \begin{cases} x \le 0 \\ x > 0 \end{cases}$$

critical number at $x = 0$, no extrema
inflection point at $x = 0$ (f changes concavity)
no asymptotes

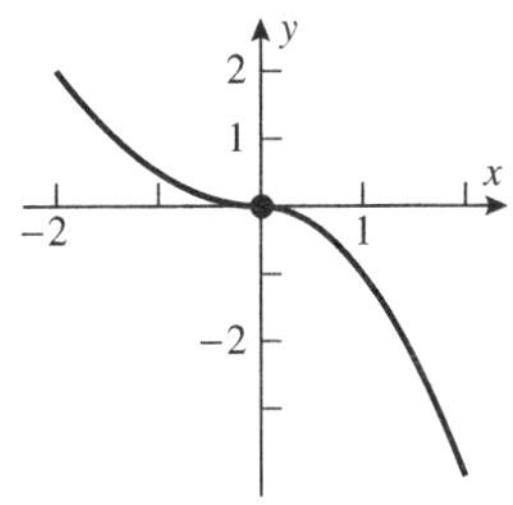

16. $f'(x) = \dfrac{5-3x}{3(1+x)^{1/3}(3-x)^{2/3}}$,

$f''(x) = \dfrac{-32}{9(1+x)^{4/3}(3-x)^{5/3}}$

critical number at $x = 5/3$;
relative maximum at $x = 5/3$
cusp at $x = -1$;
point of inflection at $x = 3$
oblique asymptote $y = -x$ as $x \to \pm\infty$

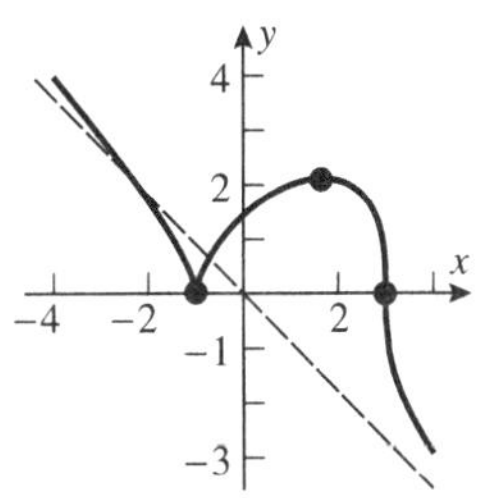

17. (a)

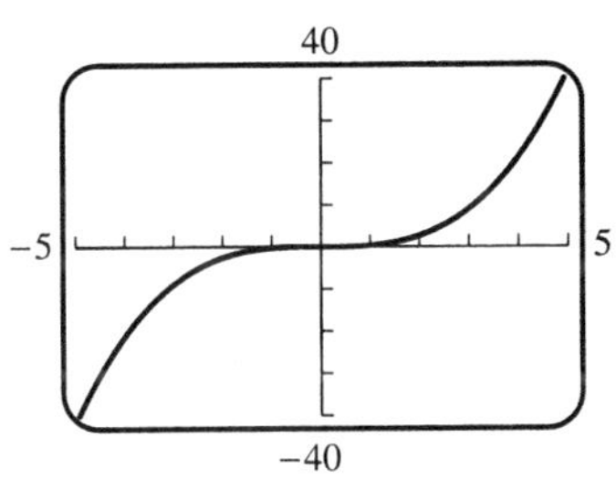

(b) $f'(x) = x^2 - \dfrac{1}{400}$, $f''(x) = 2x$

critical points at $x = \pm\dfrac{1}{20}$;

relative maximum at $x = -\dfrac{1}{20}$,

relative minimum at $x = \dfrac{1}{20}$

(c) The finer details can be seen when graphing over a much smaller x-window.

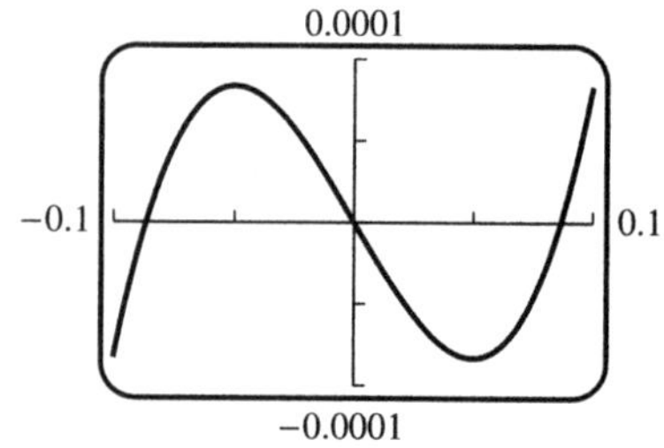

18. (a)

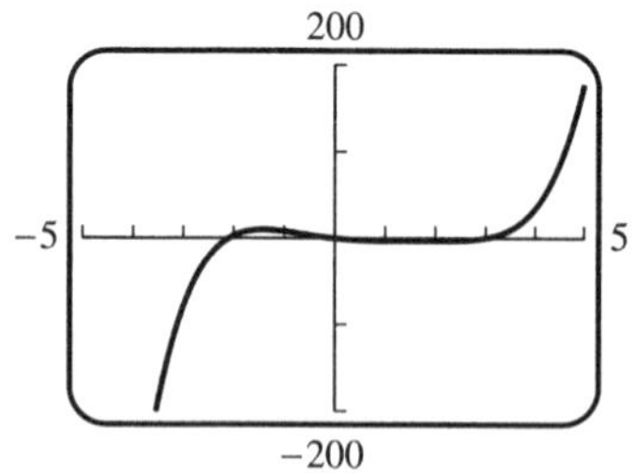

(b) critical points at $x = \pm\sqrt{2}, \frac{3}{2}, 2$;
relative maximum at $x = -\sqrt{2}$,
relative minimum at $x = \sqrt{2}$,
relative maximum at $x = \frac{3}{2}$,
relative minimum at $x = 2$

(c)

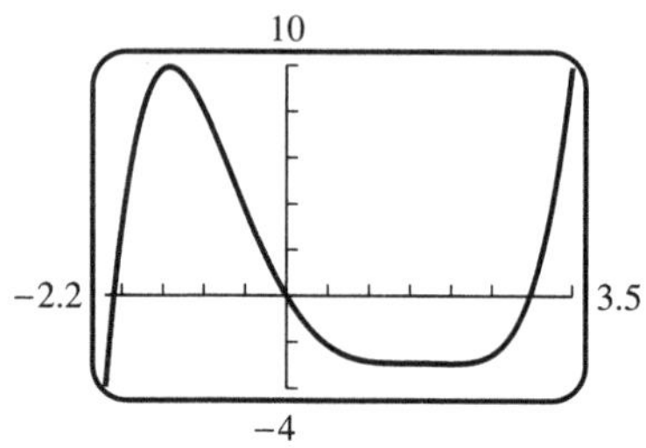

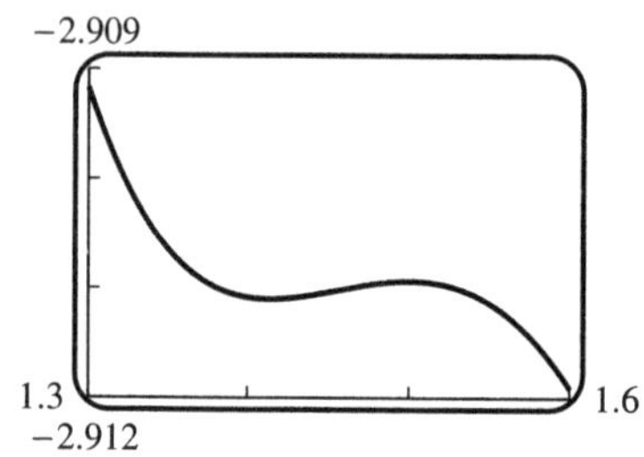

19. (a)

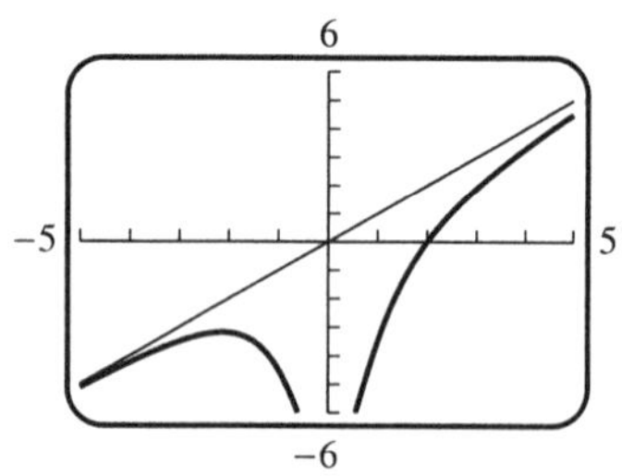

(b) Divide $y = x^2 + 1$ into $y = x^3 - 8$ to get the asymptote $ax + b = x$

20. (a) $p(x) = x^3 - x$ (b) $p(x) = x^4 - x^2$
(c) $p(x) = x^5 - x^4 - x^3 + x^2$ (d) $p(x) = x^5 - x^3$

21. $f'(x) = 4x^3 - 18x^2 + 24x - 8$, $f''(x) = 12(x-1)(x-2)$
$f''(1) = 0$, $f'(1) = 2$, $f(1) = 2$; $f''(2) = 0$, $f'(2) = 0$, $f(2) = 3$,
so the tangent lines at the inflection points are $y = 2x$ and $y = 3$.

22. $\cos x - (\sin y)\dfrac{dy}{dx} = 2\dfrac{dy}{dx}$; $\dfrac{dy}{dx} = 0$ when $\cos x = 0$. Use the first derivative test: $\dfrac{dy}{dx} = \dfrac{\cos x}{2 + \sin y}$ and $2 + \sin y > 0$, so critical points when $\cos x = 0$, relative maxima when $x = 2n\pi + \pi/2$, relative minima when $x = 2n\pi - \pi/2$, $n = 0, \pm 1, \pm 2, \ldots$

23. $f(x) = \dfrac{(2x-1)(x^2+x-7)}{(2x-1)(3x^2+x-1)} = \dfrac{x^2+x-7}{3x^2+x-1}, \qquad x \neq 1/2$
horizontal asymptote: $y = 1/3$,
vertical asymptotes: $x = (-1 \pm \sqrt{13})/6$

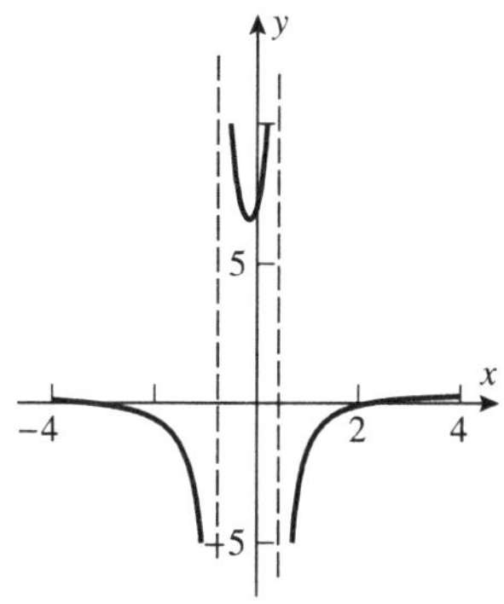

24. **(a)** $f(x) = \dfrac{(x-2)(x^2+x+1)(x^2-2)}{(x-2)(x^2-2)^2(x^2+1)} = \dfrac{x^2+x+1}{(x^2-2)(x^2+1)}$

(b)

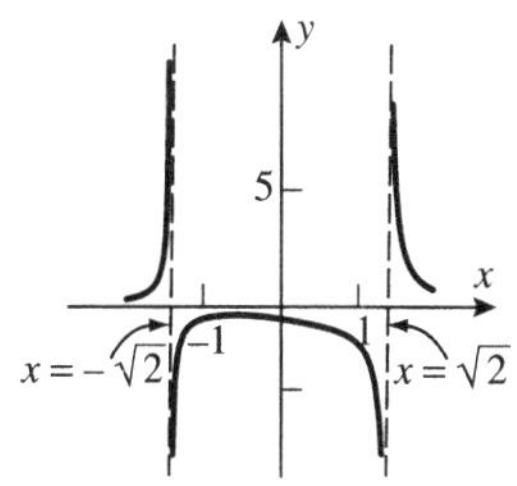

25. $f'(x) = 2ax + b$; $f'(x) > 0$ or $f'(x) < 0$ on $[0, +\infty)$ if $f'(x) = 0$ has no positive solution, so the polynomial is always increasing or always decreasing on $[0, +\infty)$ provided $-b/2a \leq 0$.

26. $f'(x) = 3ax^2 + 2bx + c$; $f'(x) > 0$ or $f'(x) < 0$ on $(-\infty, +\infty)$ if $f'(x) = 0$ has no real solutions so from the quadratic formula $(2b)^2 - 4(3a)c < 0$, $4b^2 - 12ac < 0$, $b^2 - 3ac < 0$. If $b^2 - 3ac = 0$, then $f'(x) = 0$ has only one real solution at, say, $x = c$ so f is always increasing or always decreasing on both $(-\infty, c]$ and $[c, +\infty)$, and hence on $(-\infty, +\infty)$ because f is continuous everywhere. Thus f is always increasing or decreasing if $b^2 - 3ac \leq 0$.

27. **(a)** relative minimum -0.232466 at $x = 0.450184$

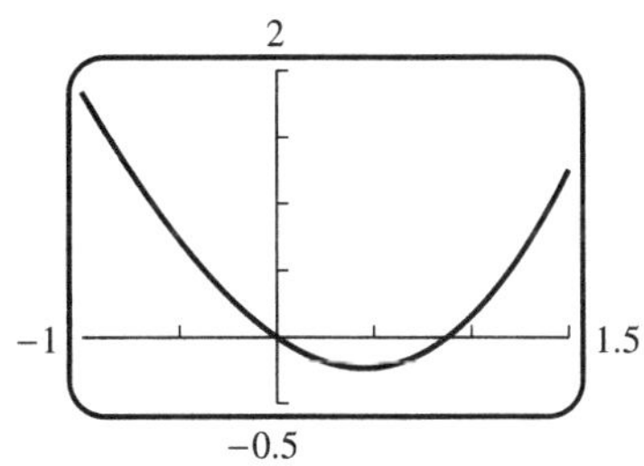

(b) relative maximum 0 at $x = 0$; relative minimum -0.107587 at $x = \pm 0.674841$

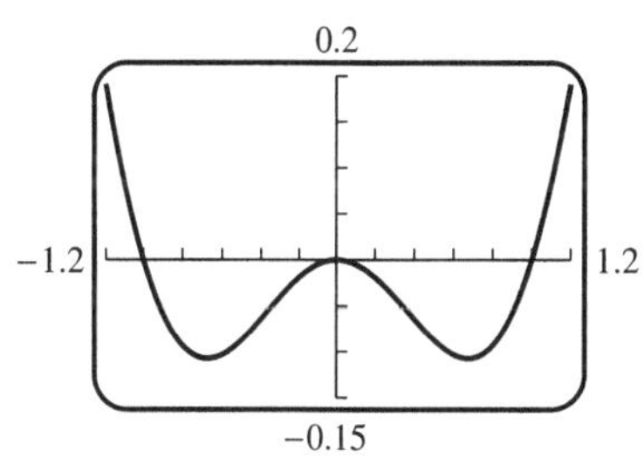

(c) relative maximum 0.876839; at $x = 0.886352$; relative minimum -0.355977 at $x = -1.244155$

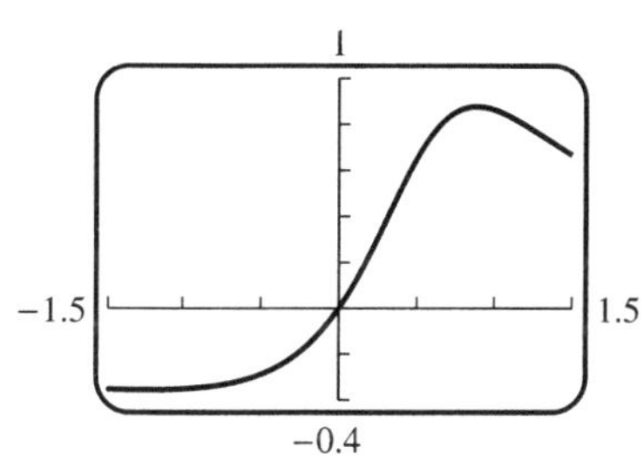

28. $f'(x) = 2 + 3x^2 - 4x^3$ has one real root at $x_0 \approx 1.136861168$, so $f(x)$ is increasing on $(-\infty, k))$ at least for $k = x_0$. But $f''(x_0) < 0$, so $f(x)$ has a relative maximum at $x = x_0$, and is thus decreasing to the right of $x = x_0$. So f is increasing on $(-\infty, x_0]$, where $x_0 \approx 1.136861$.

29. (a) If $a = k$, a constant, then $v = kt + b$ where b is constant; so the velocity changes sign at $t = -b/k$.

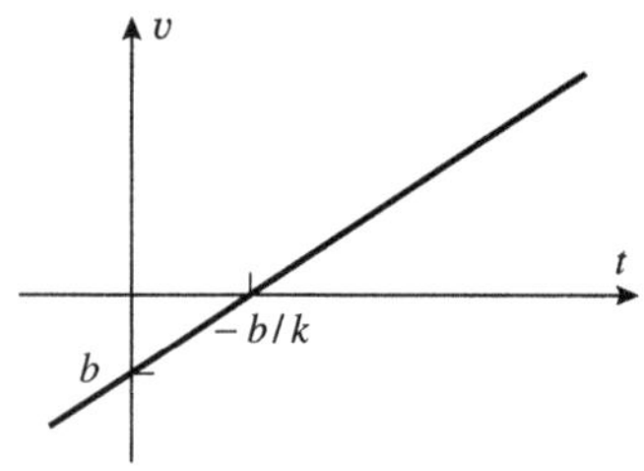

(b) Consider the equation $s = 5 - t^3/6$, $v = -t^2/2$, $a = -t$. Then for $t > 0$, a is decreasing and $av > 0$, so the particle is speeding up.

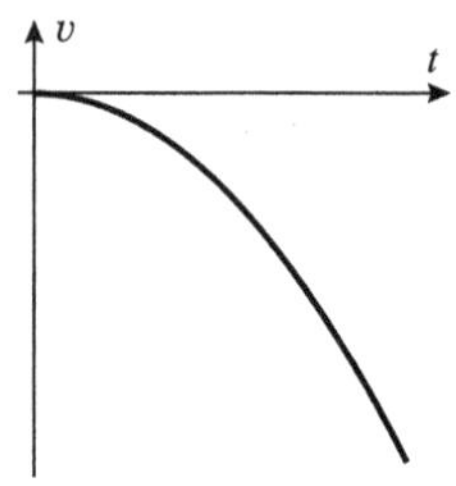

30. $s(t) = t/(t^2 + 5)$, $v(t) = (5 - t^2)/(t^2 + 5)^2$, $a(t) = 2t(t^2 - 15)/(t^2 + 5)^3$

(a)

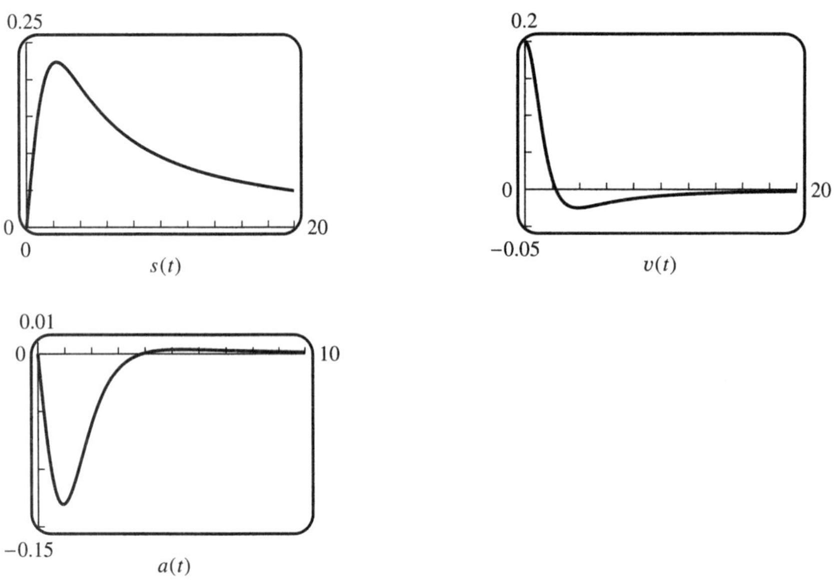

(b) v changes sign at $t = \sqrt{5}$

(c) $s = \sqrt{5}/10$, $v = 0$, $a = -\sqrt{5}/50$

(d) a changes sign at $t = \sqrt{15}$, so the particle is speeding up for $\sqrt{5} < t < \sqrt{15}$, and it is slowing down for $0 < t < \sqrt{5}$ and $\sqrt{15} < t$

(e) $v(0) = 1/5$, $\lim_{t \to +\infty} v(t) = 0$, $v(t)$ has one t-intercept at $t = \sqrt{5}$ and $v(t)$ has one critical point at $t = \sqrt{15}$. Consequently the maximum velocity occurs when $t = 0$ and the minimum velocity occurs when $t = \sqrt{15}$.

31. **(a)** $s(t) = s_0 + v_0 t - \frac{1}{2}gt^2 = v_0 t - 4.9t^2$, $v(t) = v_0 - 9.8t$; $s_{\max}$ occurs when $v = 0$, i.e. $t = v_0/9.8$, and then $0.76 = s_{\max} = v_0(v_0/9.8) - 4.9(v_0/9.8)^2 = v_0^2/19.6$, so $v_0 = \sqrt{0.76 \cdot 19.6} = 3.86$ m/s and $s(t) = 3.86t - 4.9t^2$. Then $s(t) = 0$ when $t = 0, 0.7878$, $s(t) = 0.15$ when $t = 0.0410, 0.7468$, and $s(t) = 0.76 - 0.15 = 0.61$ when $t = 0.2188, 0.5689$, so the player spends $0.5689 - 0.2188 = 0.3501$ s in the top 15.0 cm of the jump and $0.0410 + (0.7878 - 0.7468) = 0.0820$ s in the bottom 15.0 cm.

(b) The height *vs* time plot is a parabola that opens down, and the slope is smallest near the top of the parabola, so a given change Δh in height corresponds to a large time change Δt near the top of the parabola and a narrower time change at points farther away from the top.

32. **(a)** $s(t) = s_0 + v_0 - 4.9t^2$; assume $s_0 = v_0 = 0$, so $s(t) = -4.9t^2$, $v(t) = -9.8t$

t	0	1	2	3	4
s	0	−4.9	−19.6	−44.1	−78.4
v	0	−9.8	−19.6	−29.4	−39.2

(b) The formula for v is linear (with no constant term).

(c) The formula for s is quadratic (with no linear or constant term).

33. **(a)** $v = -2\dfrac{t(t^4 + 2t^2 - 1)}{(t^4+1)^2}$, $a = 2\dfrac{3t^8 + 10t^6 - 12t^4 - 6t^2 + 1}{(t^4+1)^3}$

(b)

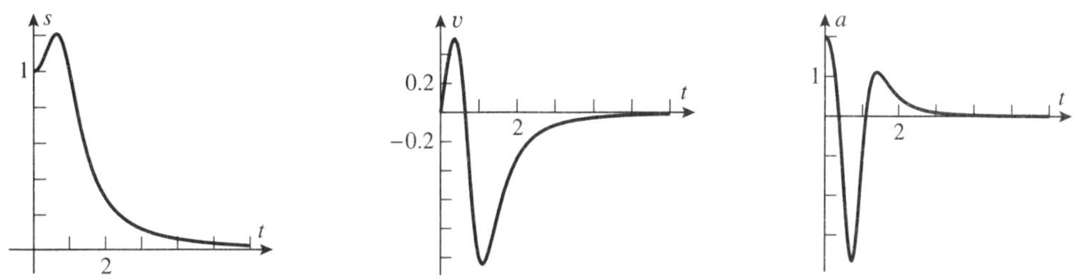

(c) It is farthest from the origin at approximately $t = 0.64$ (when $v = 0$) and $s = 1.2$

(d) Find t so that the velocity $v = ds/dt > 0$. The particle is moving in the positive direction for $0 \le t \le 0.64$ s.

(e) It is speeding up when $a, v > 0$ or $a, v < 0$, so for $0 \le t < 0.36$ and $0.64 < t < 1.1$, otherwise it is slowing down.

(f) Find the maximum value of $|v|$ to obtain: maximum speed $= 1.05$ m/s when $t = 1.10$ s.

34. No; speeding up means the velocity and acceleration have the same sign, i.e. $av > 0$; the velocity is increasing when the acceleration is positive, i.e. $a > 0$. These are not the same thing. An example is $s = t - t^2$ at $t = 1$, where $v = -1$ and $a = -2$, so $av > 0$ but $a < 0$.

37. **(a)** If f has an absolute extremum at a point of (a, b) then it must, by Theorem 6.1.4, be at a critical point of f; since f is differentiable on (a, b) the critical point is a stationary point.

(b) It could occur at a critical point which is not a stationary point: for example, $f(x) = |x|$ on $[-1, 1]$ has an absolute minimum at $x = 0$ but is not differentiable there.

38. Yes; by the Mean-Value Theorem there is a point c in (a, b) such that $f'(c) = \dfrac{f(b) - f(a)}{b - a} = 0$.

39. **(a)** $f'(x) = -1/x^2 \neq 0$, no critical points; by inspection $M = -1/2$ at $x = -2$; $m = -1$ at $x = -1$

(b) $f'(x) = 3x^2 - 4x^3 = 0$ at $x = 0, 3/4$; $f(-1) = -2$, $f(0) = 0$, $f(3/4) = 27/256$, $f(3/2) = -27/16$, so $m = -2$ at $x = -1$, $M = 27/256$ at $x = 3/4$

(c) $f'(x) = \dfrac{x(7x-12)}{3(x-2)^{2/3}}$, critical points at $x = 12/7, 2$; $m = f(12/7) = \dfrac{144}{49}\left(-\dfrac{2}{7}\right)^{1/3} \approx -1.9356$ at $x = 12/7$, $M = 9$ at $x = 3$

(d) $\lim\limits_{x\to 0^+} f(x) = \lim\limits_{x\to+\infty} f(x) = +\infty$ and $f'(x) = \dfrac{e^x(x-2)}{x^3}$, stationary point at $x = 2$; by Theorem 6.1.5 $f(x)$ has an absolute minimum at $x = 2$, and $m = e^2/4$.

40. (a) $f'(x) = 2\dfrac{3-x^2}{(x^2+3)^2}$, critical point at $x = \sqrt{3}$. Since $\lim\limits_{x\to 0^+} f(x) = 0$, $f(x)$ has no minimum, and $M = \sqrt{3}/3$ at $x = \sqrt{3}$.

(b) $f'(x) = 10x^3(x-2)$, critical points at $x = 0, 2$; $\lim\limits_{x\to 3^-} f(x) = 88$, so $f(x)$ has no maximum; $m = -9$ at $x = 2$

(c) critical point at $x = 2$; $m = -3$ at $x = 3$, $M = 0$ at $x = 2$

42. $x = -2.11491, 0.25410, 1.86081$

43. $x = 2.3561945$

44. Let $a < x < x_0$. Then $\dfrac{f(x) - f(x_0)}{x - x_0} > 0$ since f is increasing on $[a, b]$. Similarly if $x_0 < x < b$ then $\dfrac{f(x) - f(x_0)}{x - x_0} > 0$. Thus, since the limit exists, $\lim\limits_{x\to x_0} \dfrac{f(x) - f(x_0)}{x - x_0} \geq 0$.

45. (a) yes; $f'(0) = 0$
(b) no, f is not differentiable on $(-1, 1)$
(c) yes, $f'(\sqrt{\pi/2}) = 0$

46. (a) no, f is not differentiable on $(-2, 2)$

(b) yes, $\dfrac{f(3) - f(2)}{3-2} = -1 = f'(1+\sqrt{2})$

(c) $\lim\limits_{x\to 1^-} f(x) = 2$, $\lim\limits_{x\to 1^+} f(x) = 2$ so f is continuous on $[0, 2]$; $\lim\limits_{x\to 1^-} f'(x) = \lim\limits_{x\to 1^-} -2x = -2$ and $\lim\limits_{x\to 1^+} f'(x) = \lim\limits_{x\to 1^+} (-2/x^2) = -2$, so f is differentiable on $(0, 2)$; and $\dfrac{f(2) - f(0)}{2 - 0} = -1 = f'(\sqrt{2})$.

47. Let k be the amount of light admitted per unit area of clear glass. The total amount of light admitted by the entire window is
$T = k\cdot$ (area of clear glass) $+ \dfrac{1}{2}k\cdot$ (area of blue glass) $= 2krh + \dfrac{1}{4}\pi k r^2$.
But $P = 2h + 2r + \pi r$ which gives $2h = P - 2r - \pi r$ so

$$T = kr(P - 2r - \pi r) + \frac{1}{4}\pi k r^2 = k\left[Pr - \left(2 + \pi - \frac{\pi}{4}\right)r^2\right]$$

$$= k\left[Pr - \frac{8+3\pi}{4}r^2\right] \text{ for } 0 < r < \frac{P}{2+\pi},$$

$$\frac{dT}{dr} = k\left(P - \frac{8+3\pi}{2}r\right), \ \frac{dT}{dr} = 0 \text{ when } r = \frac{2P}{8+3\pi}.$$

This is the only critical point and $d^2T/dr^2 < 0$ there so the most light is admitted when $r = 2P/(8+3\pi)$ ft.

48. If one corner of the rectangle is at (x, y) with $x > 0$, $y > 0$, then $A = 4xy$, $y = 3\sqrt{1-(x/4)^2}$, $A = 12x\sqrt{1-(x/4)^2} = 3x\sqrt{16-x^2}$, $\dfrac{dA}{dx} = 6\dfrac{8-x^2}{\sqrt{16-x^2}}$, critical point at $x = 2\sqrt{2}$. Since $A = 0$ when $x = 0, 4$ and $A > 0$ otherwise, there is an absolute maximum $A = 24$ at $x = 2\sqrt{2}$.

49. (a)

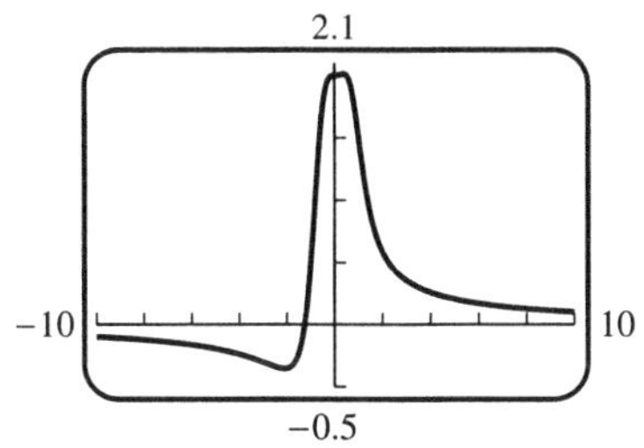

(b) minimum: $(-2.111985, -0.355116)$
maximum: $(0.372591, 2.012931)$

50. (a)

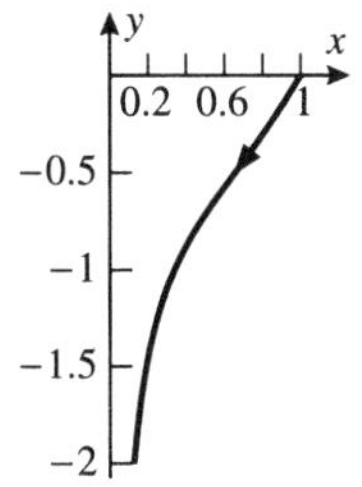

(b) The distance between the boat and the origin is $\sqrt{x^2+y^2}$, where $y = (x^{10/3}-1)/(2x^{2/3})$. The minimum distance is 0.8247 mi when $x = 0.6598$ mi. The boat gets swept downstream.

(c) Use the equation of the path to obtain $dy/dt = (dy/dx)(dx/dt)$, $dx/dt = (dy/dt)/(dy/dx)$. Let $dy/dt = -4$ and find the value of dy/dx for the value of x obtained in Part (b) to get $dx/dt = -3$ mi/h.

51. Solve $\phi - 0.0167\sin\phi = 2\pi(90)/365$ to get $\phi = 1.565978$ so $r = 150 \times 10^6(1 - 0.0167\cos\phi) = 149.988 \times 10^6$ km.

52. Solve $\phi - 0.0934\sin\phi = 2\pi(1)/1.88$ to get $\phi = 3.325078$ so $r = 228 \times 10^6(1 - 0.0934\cos\phi) = 248.938 \times 10^6$ km.

CHAPTER 5
Integration

EXERCISE SET 5.1

1. Endpoints $0, \frac{1}{n}, \frac{2}{n}, \ldots, \frac{n-1}{n}, 1$; using right endpoints

$$A_n = \left[\sqrt{\frac{1}{n}} + \sqrt{\frac{2}{n}} + \cdots + \sqrt{\frac{n-1}{n}} + 1\right]\frac{1}{n}$$

n	2	5	10	50	100
A_n	0.853553	0.749739	0.710509	0.676095	0.671463

2. Endpoints $0, \frac{1}{n}, \frac{2}{n}, \ldots, \frac{n-1}{n}, 1$; using right endpoints

$$A_n = \left[\frac{n}{n+1} + \frac{n}{n+2} + \frac{n}{n+3} + \cdots + \frac{n}{2n-1} + \frac{1}{2}\right]\frac{1}{n}$$

n	2	5	10	50	100
A_n	0.583333	0.645635	0.668771	0.688172	0.690653

3. Endpoints $0, \frac{\pi}{n}, \frac{2\pi}{n}, \ldots, \frac{(n-1)\pi}{n}, \pi$; using right endpoints

$$A_n = [\sin(\pi/n) + \sin(2\pi/n) + \cdots + \sin(\pi(n-1)/n) + \sin\pi]\,\frac{\pi}{n}$$

n	2	5	10	50	100
A_n	1.57080	1.93376	1.98352	1.99935	1.99984

4. Endpoints $0, \frac{\pi}{2n}, \frac{2\pi}{2n}, \ldots, \frac{(n-1)\pi}{2n}, \frac{\pi}{2}$; using right endpoints

$$A_n = [\cos(\pi/2n) + \cos(2\pi/2n) + \cdots + \cos((n-1)\pi/2n) + \cos(\pi/2)]\,\frac{\pi}{2n}$$

n	2	5	10	50	100
A_n	0.555359	0.834683	0.919405	0.984204	0.992120

5. Endpoints $1, \frac{n+1}{n}, \frac{n+2}{n}, \ldots, \frac{2n-1}{n}, 2$; using right endpoints

$$A_n = \left[\frac{n}{n+1} + \frac{n}{n+2} + \cdots + \frac{n}{2n-1} + \frac{1}{2}\right]\frac{1}{n}$$

n	2	5	10	50	100
A_n	0.583333	0.645635	0.668771	0.688172	0.690653

6. Endpoints $-\frac{\pi}{2}, -\frac{\pi}{2} + \frac{\pi}{n}, -\frac{\pi}{2} + \frac{2\pi}{n}, \ldots, -\frac{\pi}{2} + \frac{(n-1)\pi}{n}, \frac{\pi}{2}$; using right endpoints

$$A_n = \left[\cos\left(-\frac{\pi}{2} + \frac{\pi}{n}\right) + \cos\left(-\frac{\pi}{2} + \frac{2\pi}{n}\right) + \cdots + \cos\left(-\frac{\pi}{2} + \frac{(n-1)\pi}{n}\right) + \cos\left(\frac{\pi}{2}\right)\right]\frac{\pi}{n}$$

n	2	5	10	50	100
A_n	1.99985	1.93376	1.98352	1.99936	1.99985

7. Endpoints $0, \frac{1}{n}, \frac{2}{n}, \ldots, \frac{n-1}{n}, 1$; using right endpoints

$$A_n = \left[\sqrt{1-\left(\frac{1}{n}\right)^2} + \sqrt{1-\left(\frac{2}{n}\right)^2} + \cdots + \sqrt{1-\left(\frac{n-1}{n}\right)^2} + 0\right]\frac{1}{n}$$

n	2	5	10	50	100
A_n	0.433013	0.659262	0.726130	0.774567	0.780106

8. Endpoints $-1, -1+\frac{2}{n}, -1+\frac{4}{n}, \ldots, -1+\frac{2(n-1)}{n}, 1$; using right endpoints

$$A_n = \left[\sqrt{1-\left(\frac{n-2}{n}\right)^2} + \sqrt{1-\left(\frac{n-4}{n}\right)^2} + \cdots + \sqrt{1-\left(\frac{n-2}{n}\right)^2} + 0\right]\frac{2}{n}$$

n	2	5	10	50	100
A_n	1	1.423837	1.518524	1.566097	1.569136

9. $3(x-1)$ **10.** $5(x-2)$ **11.** $x(x+2)$ **12.** $\frac{3}{2}(x-1)^2$

13. $(x+3)(x-1)$ **14.** $\frac{3}{2}x(x-2)$

15. The area in Exercise 13 is always 3 less than the area in Exercise 11. The regions are identical except that the area in Exercise 11 has the extra trapezoid with vertices at $(0,0), (1,0), (0,2), (1,4)$ (with area 3).

16. **(a)** The region in question is a trapezoid, and the area of a trapezoid is $\frac{1}{2}(h_1+h_2)w$.

(b) From Part (a), $A'(x) = \frac{1}{2}[f(a)+f(x)] + (x-a)\frac{1}{2}f'(x)$

$$= \frac{1}{2}[f(a)+f(x)] + (x-a)\frac{1}{2}\frac{f(x)-f(a)}{x-a} = f(x)$$

17. B is also the area between the graph of $f(x)=\sqrt{x}$ and the interval $[0,1]$ on the y–axis, so $A+B$ is the area of the square.

18. If the plane is rotated about the line $y=x$ then A becomes B and vice versa.

EXERCISE SET 5.2

1. **(a)** $\int \frac{x}{\sqrt{1+x^2}}dx = \sqrt{1+x^2}+C$ **(b)** $\int x^2\cos(1+x^3)dx = \frac{1}{3}\sin(1+x^3)+C$

2. **(a)** $\frac{d}{dx}(\sin x - x\cos x + C) = \cos x - \cos x + x\sin x = x\sin x$

(b) $\frac{d}{dx}\left(\frac{x}{\sqrt{1-x^2}}+C\right) = \frac{\sqrt{1-x^2}+x^2/\sqrt{1-x^2}}{1-x^2} = \frac{1}{(1-x^2)^{3/2}}$

3. $\frac{d}{dx}\left[\sqrt{x^3+5}\right] = \frac{3x^2}{2\sqrt{x^3+5}}$ so $\int \frac{3x^2}{2\sqrt{x^3+5}}dx = \sqrt{x^3+5}+C$

4. $\frac{d}{dx}\left[\frac{x}{x^2+3}\right] = \frac{3-x^2}{(x^2+3)^2}$ so $\int \frac{3-x^2}{(x^2+3)^2}dx = \frac{x}{x^2+3} + C$

5. $\frac{d}{dx}\left[\sin\left(2\sqrt{x}\right)\right] = \frac{\cos\left(2\sqrt{x}\right)}{\sqrt{x}}$ so $\int \frac{\cos\left(2\sqrt{x}\right)}{\sqrt{x}}dx = \sin\left(2\sqrt{x}\right) + C$

6. $\frac{d}{dx}[\sin x - x\cos x] = x\sin x$ so $\int x\sin x\,dx = \sin x - x\cos x + C$

7. **(a)** $x^9/9 + C$ **(b)** $\frac{7}{12}x^{12/7} + C$ **(c)** $\frac{2}{9}x^{9/2} + C$

8. **(a)** $\frac{3}{5}x^{5/3} + C$ **(b)** $-\frac{1}{5}x^{-5} + C = -\frac{1}{5x^5} + C$ **(c)** $8x^{1/8} + C$

9. **(a)** $\frac{1}{2}\int x^{-3}dx = -\frac{1}{4}x^{-2} + C$ **(b)** $u^4/4 - u^2 + 7u + C$

10. $\frac{3}{5}x^{5/3} - 5x^{4/5} + 4x + C$

11. $\int (x^{-3} + x^{1/2} - 3x^{1/4} + x^2)dx = -\frac{1}{2}x^{-2} + \frac{2}{3}x^{3/2} - \frac{12}{5}x^{5/4} + \frac{1}{3}x^3 + C$

12. $\int (7y^{-3/4} - y^{1/3} + 4y^{1/2})dy = 28y^{1/4} - \frac{3}{4}y^{4/3} + \frac{8}{3}y^{3/2} + C$

13. $\int (x + x^4)dx = x^2/2 + x^5/5 + C$

14. $\int (4 + 4y^2 + y^4)dy = 4y + \frac{4}{3}y^3 + \frac{1}{5}y^5 + C$

15. $\int x^{1/3}(4 - 4x + x^2)dx = \int (4x^{1/3} - 4x^{4/3} + x^{7/3})dx = 3x^{4/3} - \frac{12}{7}x^{7/3} + \frac{3}{10}x^{10/3} + C$

16. $\int (2 - x + 2x^2 - x^3)dx = 2x - \frac{1}{2}x^2 + \frac{2}{3}x^3 - \frac{1}{4}x^4 + C$

17. $\int (x + 2x^{-2} - x^{-4})dx = x^2/2 - 2/x + 1/(3x^3) + C$

18. $\int (t^{-3} - 2)dt = -\frac{1}{2}t^{-2} - 2t + C$

19. $-4\cos x + 2\sin x + C$ **20.** $4\tan x - \csc x + C$

21. $\int (\sec^2 x + \sec x\tan x)dx = \tan x + \sec x + C$

22. $\int (\sec x\tan x + 1)dx = \sec x + x + C$ **23.** $\int \frac{\sec\theta}{\cos\theta}\,d\theta = \int \sec^2\theta\,d\theta = \tan\theta + C$

24. $\int \sin y\,dy = -\cos y + C$ **25.** $\int \sec x\tan x\,dx = \sec x + C$

26. $\int(\phi + 2\csc^2\phi)d\phi = \phi^2/2 - 2\cot\phi + C$

27. $\int(1+\sin\theta)d\theta = \theta - \cos\theta + C$

28. $\int \frac{2\sin x\cos x}{\cos x}dx = 2\int \sin x\,dx = -2\cos x + C$

29. $\int \frac{1-\sin x}{1-\sin^2 x}dx = \int \frac{1-\sin x}{\cos^2 x}dx = \int \left(\sec^2 x - \sec x\tan x\right)dx = \tan x - \sec x + C$

30. $\int \frac{1}{1+\cos 2x}dx = \int \frac{1}{2\cos^2 x}dx = \int \frac{1}{2}\sec^2 x\,dx = \frac{1}{2}\tan x + C$

31. **(a)**

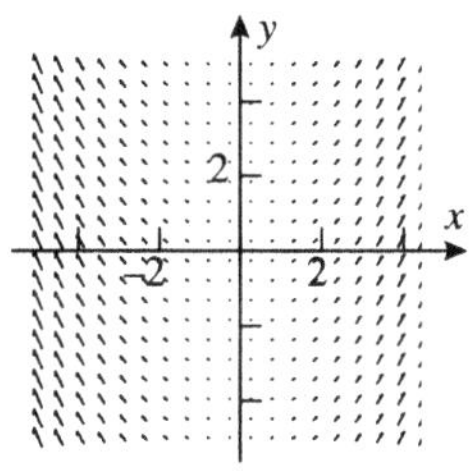

(b)

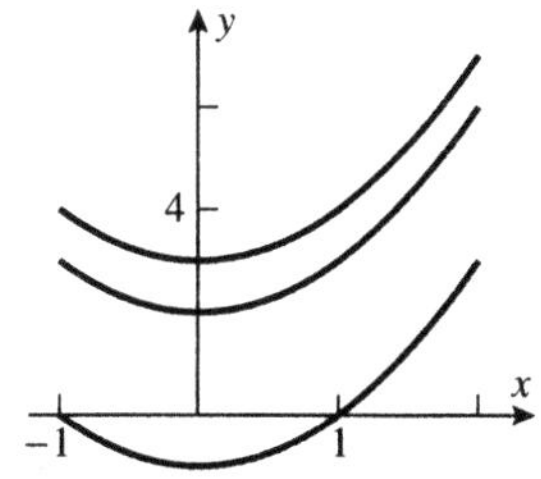

(c) $f(x) = x^2/2 - 1$

32. **(a)**

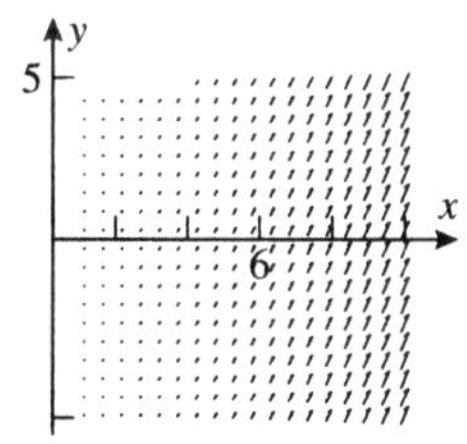

(b)

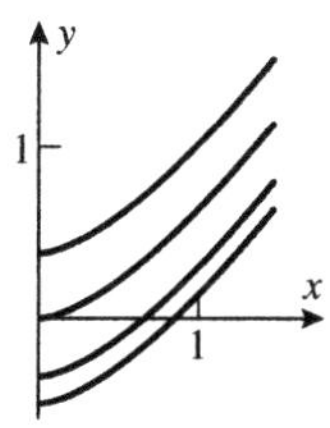

(c) $y = \frac{2}{3}x^{3/2} - 2$

33.

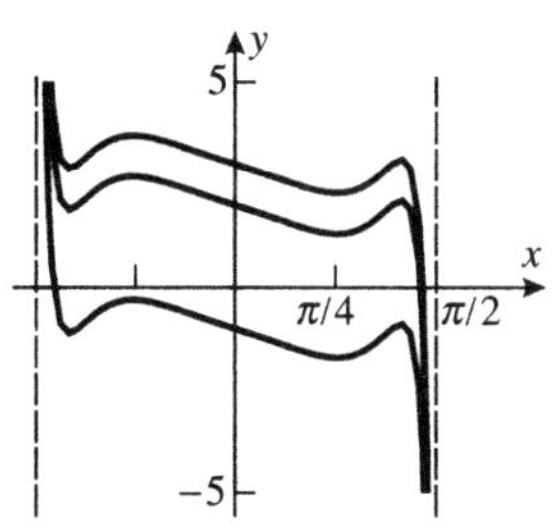

34.

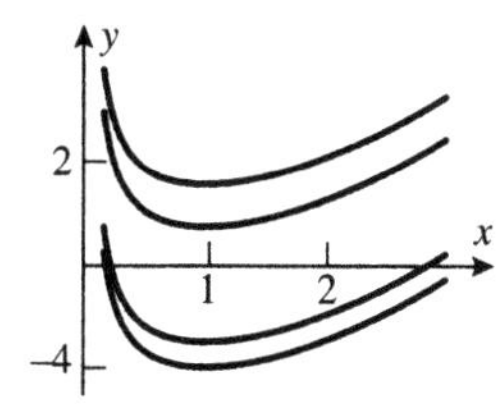

35. $f'(x) = m = -\sin x$ so $f(x) = \int(-\sin x)dx = \cos x + C$; $f(0) = 2 = 1 + C$
so $C = 1$, $f(x) = \cos\ x\ +\ 1$

36. $f'(x) = m = (x+1)^2$, so $f(x) = \int(x+1)^2dx = \frac{1}{3}(x+1)^3 + C$;
$f(-2) = 8 = \frac{1}{3}(-2+1)^3 + C = -\frac{1}{3} + C, = 8 + \frac{1}{3} = \frac{25}{3}$, $f(x) = \frac{1}{3}(x+1)^3 + \frac{25}{3}$

37. **(a)** $y(x) = \int x^{1/3}dx = \frac{3}{4}x^{4/3} + C$, $y(1) = \frac{3}{4} + C = 2, C = \frac{5}{4}$; $y(x) = \frac{3}{4}x^{4/3} + \frac{5}{4}$

(b) $y(t) = \int(\sin t + 1)\,dt = -\cos t + t + C$, $y\left(\frac{\pi}{3}\right) = -\frac{1}{2} + \frac{\pi}{3} + C = \frac{1}{2}, C = 1 - \frac{\pi}{3}$;
$y(t) = -\cos t + t + 1 - \frac{\pi}{3}$

(c) $y(x) = \int (x^{1/2} + x^{-1/2})dx = \frac{2}{3}x^{3/2} + 2x^{1/2} + C$, $y(1) = 0 = \frac{8}{3} + C$, $C = -\frac{8}{3}$,
$y(x) = \frac{2}{3}x^{3/2} + 2x^{1/2} - \frac{8}{3}$

38. **(a)** $y(x) = \int \frac{1}{8}x^{-3}dx = -\frac{1}{16}x^{-2} + C$, $y(1) = 0 = -\frac{1}{16} + C$, $C = \frac{1}{16}$; $y(x) = -\frac{1}{16}x^{-2} + \frac{1}{16}$

(b) $y(t) = \int (\sec^2 t - \sin t)\, dt = \tan t + \cos t + C$, $y(\frac{\pi}{4}) = 1 = 1 + \frac{\sqrt{2}}{2} + C$, $C = -\frac{\sqrt{2}}{2}$;
$y(t) = \tan t + \cos t - \frac{\sqrt{2}}{2}$

(c) $y(x) = \int x^{7/2}dx = \frac{2}{9}x^{9/2} + C$, $y(0) = 0 = C$, $C = 0$; $y(x) = \frac{2}{9}x^{9/2}$

39. $f'(x) = \frac{2}{3}x^{3/2} + C_1$; $f(x) = \frac{4}{15}x^{5/2} + C_1 x + C_2$

40. $f'(x) = x^2/2 + \sin x + C_1$, use $f'(0) = 2$ to get $C_1 = 2$ so $f'(x) = x^2/2 + \sin x + 2$,
$f(x) = x^3/6 - \cos x + 2x + C_2$, use $f(0) = 1$ to get $C_2 = 2$ so $f(x) = x^3/6 - \cos x + 2x + 2$

41. $dy/dx = 2x + 1, y = \int (2x + 1)dx = x^2 + x + C$; $y = 0$ when $x = -3$
so $(-3)^2 + (-3) + C = 0, C = -6$ thus $y = x^2 + x - 6$

42. $dy/dx = x^2, y = \int x^2 dx = x^3/3 + C$; $y = 2$ when $x = -1$ so $(-1)^3/3 + C = 2, C = 7/3$
thus $y = x^3/3 + 7/3$

43. $dy/dx = \int 6x dx = 3x^2 + C_1$. The slope of the tangent line is -3 so $dy/dx = -3$ when $x = 1$.
Thus $3(1)^2 + C_1 = -3$, $C_1 = -6$ so $dy/dx = 3x^2 - 6$, $y = \int (3x^2 - 6)dx = x^3 - 6x + C_2$. If $x = 1$,
then $y = 5 - 3(1) = 2$ so $(1)^2 - 6(1) + C_2 = 2, C_2 = 7$ thus $y = x^3 - 6x + 7$.

44. $dT/dx = C_1$, $T = C_1 x + C_2$; $T = 25$ when $x = 0$ so $C_2 = 25$, $T = C_1 x + 25$. $T = 85$ when $x = 50$
so $50C_1 + 25 = 85$, $C_1 = 1.2$, $T = 1.2x + 25$

45. **(a)** $F'(x) = G'(x) = 3x + 4$
(b) $F(0) = 16/6 = 8/3$, $G(0) = 0$, so $F(0) - G(0) = 8/3$
(c) $F(x) = (9x^2 + 24x + 16)/6 = 3x^2/2 + 4x + 8/3 = G(x) + 8/3$

46. **(a)** $F'(x) = G'(x) = 10x/(x^2 + 5)^2$
(b) $F(0) = 0$, $G(0) = -1$, so $F(0) - G(0) = 1$
(c) $F(x) = \frac{x^2}{x^2 + 5} = \frac{(x^2 + 5) - 5}{x^2 + 5} = 1 - \frac{5}{x^2 + 5} = G(x) + 1$

47. $\int (\sec^2 x - 1)dx = \tan x - x + C$

48. $\int (\csc^2 x - 1)dx = -\cot x - x + C$

49. **(a)** $\frac{1}{2}\int (1 - \cos x)dx = \frac{1}{2}(x - \sin x) + C$

(b) $\frac{1}{2}\int (1 + \cos x)\, dx = \frac{1}{2}(x + \sin x) + C$

50. **(a)** $F'(x) = G'(x) = f(x)$, where $f(x) = \begin{cases} 1, & x > 0 \\ -1, & x < 0 \end{cases}$

(b) $G(x) - F(x) = \begin{cases} 2, & x > 0 \\ 3, & x < 0 \end{cases}$ so $G(x) \neq F(x)$ plus a constant

(c) no, because $(-\infty, 0) \cup (0, +\infty)$ is not an interval

51. $v = \dfrac{1087}{2\sqrt{273}} \displaystyle\int T^{-1/2}\, dT = \dfrac{1087}{\sqrt{273}} T^{1/2} + C$, $v(273) = 1087 = 1087 + C$ so $C = 0$, $v = \dfrac{1087}{\sqrt{273}} T^{1/2}$ ft/s

EXERCISE SET 5.3

1. **(a)** $\displaystyle\int u^{23} du = u^{24}/24 + C = (x^2+1)^{24}/24 + C$

(b) $\displaystyle -\int u^3 du = -u^4/4 + C = -(\cos^4 x)/4 + C$

(c) $\displaystyle 2\int \sin u\, du = -2\cos u + C = -2\cos\sqrt{x} + C$

(d) $\displaystyle \frac{3}{8}\int u^{-1/2} du = \frac{3}{4}u^{1/2} + C = \frac{3}{4}\sqrt{4x^2+5} + C$

2. **(a)** $\displaystyle \frac{1}{4}\int \sec^2 u\, du = \frac{1}{4}\tan u + C = \frac{1}{4}\tan(4x+1) + C$

(b) $\displaystyle \frac{1}{4}\int u^{1/2} du = \frac{1}{6}u^{3/2} + C = \frac{1}{6}(1+2y^2)^{3/2} + C$

(c) $\displaystyle \frac{1}{\pi}\int u^{1/2} du = \frac{2}{3\pi}u^{3/2} + C = \frac{2}{3\pi}\sin^{3/2}(\pi\theta) + C$

(d) $\displaystyle \int u^{4/5} du = \frac{5}{9}u^{9/5} + C = \frac{5}{9}(x^2+7x+3)^{9/5} + C$

3. **(a)** $\displaystyle -\int u\, du = -\frac{1}{2}u^2 + C = -\frac{1}{2}\cot^2 x + C$

(b) $\displaystyle \int u^9 du = \frac{1}{10}u^{10} + C = \frac{1}{10}(1+\sin t)^{10} + C$

(c) $\displaystyle \frac{1}{2}\int \cos u\, du = \frac{1}{2}\sin u + C = \frac{1}{2}\sin 2x + C$

(d) $\displaystyle \frac{1}{2}\int \sec^2 u\, du = \frac{1}{2}\tan u + C = \frac{1}{2}\tan x^2 + C$

4. **(a)** $\displaystyle \int (u-1)^2 u^{1/2} du - \int (u^{5/2} - 2u^{3/2} + u^{1/2}) du = \frac{2}{7}u^{7/2} - \frac{4}{5}u^{5/2} + \frac{2}{3}u^{3/2} + C$

$$= \frac{2}{7}(1+x)^{7/2} - \frac{4}{5}(1+x)^{5/2} + \frac{2}{3}(1+x)^{3/2} + C$$

(b) $\displaystyle \int \csc^2 u\, du = -\cot u + C = -\cot(\sin x) + C$

(c) $\displaystyle \int \sin u\, du = -\cos u + C = -\cos(x-\pi) + C$

(d) $\displaystyle \int \frac{du}{u^2} = -\frac{1}{u} + C = -\frac{1}{x^5+1} + C$

5. $u = 2 - x^2$, $du = -2x\,dx$; $-\frac{1}{2}\int u^3 du = -u^4/8 + C = -(2-x^2)^4/8 + C$

6. $u = 3x - 1$, $du = 3dx$; $\frac{1}{3}\int u^5 du = \frac{1}{18}u^6 + C = \frac{1}{18}(3x-1)^6 + C$

7. $u = 8x$, $du = 8dx$; $\frac{1}{8}\int \cos u\,du = \frac{1}{8}\sin u + C = \frac{1}{8}\sin 8x + C$

8. $u = 3x$, $du = 3dx$; $\frac{1}{3}\int \sin u\,du = -\frac{1}{3}\cos u + C = -\frac{1}{3}\cos 3x + C$

9. $u = 4x$, $du = 4dx$; $\frac{1}{4}\int \sec u \tan u\,du = \frac{1}{4}\sec u + C = \frac{1}{4}\sec 4x + C$

10. $u = 5x$, $du = 5dx$; $\frac{1}{5}\int \sec^2 u\,du = \frac{1}{5}\tan u + C = \frac{1}{5}\tan 5x + C$

11. $u = 7t^2 + 12$, $du = 14t\,dt$; $\frac{1}{14}\int u^{1/2} du = \frac{1}{21}u^{3/2} + C = \frac{1}{21}(7t^2+12)^{3/2} + C$

12. $u = 4 - 5x^2$, $du = -10x\,dx$; $-\frac{1}{10}\int u^{-1/2} du = -\frac{1}{5}u^{1/2} + C = -\frac{1}{5}\sqrt{4-5x^2} + C$

13. $u = x^3 + 1$, $du = 3x^2 dx$; $\frac{1}{3}\int u^{-1/2} du = \frac{2}{3}u^{1/2} + C = \frac{2}{3}\sqrt{x^3+1} + C$

14. $u = 1 - 3x$, $du = -3dx$; $-\frac{1}{3}\int u^{-2} du = \frac{1}{3}u^{-1} + C = \frac{1}{3}(1-3x)^{-1} + C$

15. $u = 4x^2 + 1$, $du = 8x\,dx$; $\frac{1}{8}\int u^{-3} du = -\frac{1}{16}u^{-2} + C = -\frac{1}{16}(4x^2+1)^{-2} + C$

16. $u = 3x^2$, $du = 6x\,dx$; $\frac{1}{6}\int \cos u\,du = \frac{1}{6}\sin u + C = \frac{1}{6}\sin(3x^2) + C$

17. $u = 5/x$, $du = -(5/x^2)dx$; $-\frac{1}{5}\int \sin u\,du = \frac{1}{5}\cos u + C = \frac{1}{5}\cos(5/x) + C$

18. $u = \sqrt{x}$, $du = \frac{1}{2\sqrt{x}}dx$; $2\int \sec^2 u\,du = 2\tan u + C = 2\tan\sqrt{x} + C$

19. $u = x^3$, $du = 3x^2 dx$; $\frac{1}{3}\int \sec^2 u\,du = \frac{1}{3}\tan u + C = \frac{1}{3}\tan(x^3) + C$

20. $u = \cos 2t$, $du = -2\sin 2t\,dt$; $-\frac{1}{2}\int u^3 du = -\frac{1}{8}u^4 + C = -\frac{1}{8}\cos^4 2t + C$

21. $u = \sin 3t$, $du = 3\cos 3t\,dt$; $\frac{1}{3}\int u^5 du = \frac{1}{18}u^6 + C = \frac{1}{18}\sin^6 3t + C$

22. $u = 5 + \cos 2\theta$, $du = -2\sin 2\theta\,d\theta$; $-\frac{1}{2}\int u^{-3} du = \frac{1}{4}u^{-2} + C = \frac{1}{4}(5+\cos 2\theta)^{-2} + C$

23. $u = 2 - \sin 4\theta,\ du = -4\cos 4\theta\, d\theta;\ -\dfrac{1}{4}\displaystyle\int u^{1/2} du = -\dfrac{1}{6}u^{3/2} + C = -\dfrac{1}{6}(2 - \sin 4\theta)^{3/2} + C$

24. $u = \tan 5x,\ du = 5\sec^2 5x\, dx;\ \dfrac{1}{5}\displaystyle\int u^3 du = \dfrac{1}{20}u^4 + C = \dfrac{1}{20}\tan^4 5x + C$

25. $u = \sec 2x,\ du = 2\sec 2x \tan 2x\, dx;\ \dfrac{1}{2}\displaystyle\int u^2 du = \dfrac{1}{6}u^3 + C = \dfrac{1}{6}\sec^3 2x + C$

26. $u = \sin\theta,\ du = \cos\theta\, d\theta;\ \displaystyle\int \sin u\, du = -\cos u + C = -\cos(\sin\theta) + C$

27. $u = x - 3,\ x = u + 3,\ dx = du$

$$\int (u+3)u^{1/2} du = \int (u^{3/2} + 3u^{1/2}) du = \frac{2}{5}u^{5/2} + 2u^{3/2} + C = \frac{2}{5}(x-3)^{5/2} + 2(x-3)^{3/2} + C$$

28. $u = y + 1,\ y = u - 1,\ dy = du$

$$\int \frac{u-1}{u^{1/2}} du = \int (u^{1/2} - u^{-1/2}) du = \frac{2}{3}u^{3/2} - 2u^{1/2} + C = \frac{2}{3}(y+1)^{3/2} - 2(y+1)^{1/2} + C$$

29. $\displaystyle\int \sin^2 2\theta \sin 2\theta\, d\theta = \int (1 - \cos^2 2\theta)\sin 2\theta\, d\theta;\ u = \cos 2\theta,\ du = -2\sin 2\theta\, d\theta,$

$$-\frac{1}{2}\int (1 - u^2) du = -\frac{1}{2}u + \frac{1}{6}u^3 + C = -\frac{1}{2}\cos 2\theta + \frac{1}{6}\cos^3 2\theta + C$$

30. $\sec^2 3\theta = \tan^2 3\theta + 1, u = 3\theta, du = 3d\theta$

$$\int \sec^4 3\theta\, d\theta = \frac{1}{3}\int (\tan^2 u + 1)\sec^2 u\, du = \frac{1}{9}\tan^3 u + \frac{1}{3}\tan u + C = \frac{1}{9}\tan^3 3\theta + \frac{1}{3}\tan 3\theta + C$$

31. $u = a + bx, du = bdx,$

$$\int (a+bx)^n dx = \frac{1}{b}\int u^n du = \frac{(a+bx)^{n+1}}{b(n+1)} + C$$

32. $u = a + bx,\ du = b\, dx,\ dx = \dfrac{1}{b}du$

$$\frac{1}{b}\int u^{1/n} du = \frac{n}{b(n+1)}u^{(n+1)/n} + C = \frac{n}{b(n+1)}(a+bx)^{(n+1)/n} + C$$

33. $u = \sin(a+bx),\ du = b\cos(a+bx)dx$

$$\frac{1}{b}\int u^n du = \frac{1}{b(n+1)}u^{n+1} + C = \frac{1}{b(n+1)}\sin^{n+1}(a+bx) + C$$

35. **(a)** with $u = \sin x,\ du = \cos x\, dx;\ \displaystyle\int u\, du = \dfrac{1}{2}u^2 + C_1 = \dfrac{1}{2}\sin^2 x + C_1;$

with $u = \cos x,\ du = -\sin x\, dx;\ -\displaystyle\int u\, du = -\dfrac{1}{2}u^2 + C_2 = -\dfrac{1}{2}\cos^2 x + C_2$

(b) because they differ by a constant:

$$\left(\frac{1}{2}\sin^2 x + C_1\right) - \left(-\frac{1}{2}\cos^2 x + C_2\right) = \frac{1}{2}(\sin^2 x + \cos^2 x) + C_1 - C_2 = 1/2 + C_1 - C_2$$

36. **(a)** First method: $\int (25x^2 - 10x + 1)dx = \frac{25}{3}x^3 - 5x^2 + x + C_1$;

second method: $\frac{1}{5}\int u^2 du = \frac{1}{15}u^3 + C_2 = \frac{1}{15}(5x-1)^3 + C_2$

(b) $\frac{1}{15}(5x-1)^3 + C_2 = \frac{1}{15}(125x^3 - 75x^2 + 15x - 1) + C_2 = \frac{25}{3}x^3 - 5x^2 + x - \frac{1}{15} + C_2$;

the answers differ by a constant.

37. $y(x) = \int \sqrt{3x+1}dx = \frac{2}{9}(3x+1)^{3/2} + C$,

$y(1) = \frac{16}{9} + C = 5$, $C = \frac{29}{9}$ so $y(x) = \frac{2}{9}(3x+1)^{3/2} + \frac{29}{9}$

38. $y(x) = \int (6 - 5\ \sin 2x)dx = 6x + \frac{5}{2}\cos 2x + C$,

$y(0) = \frac{5}{2} + C = 3$, $C = \frac{1}{2}$ so $y(x) = 6x + \frac{5}{2}\cos 2x + \frac{1}{2}$

39. $f'(x) = m = \sqrt{3x+1}$, $f(x) = \int (3x+1)^{1/2}dx = \frac{2}{9}(3x+1)^{3/2} + C$; $f(0) = 1 = \frac{2}{9} + C$, $C = \frac{7}{9}$,

so $f(x) = \frac{2}{9}(3x+1)^{3/2} + \frac{7}{9}$

40.

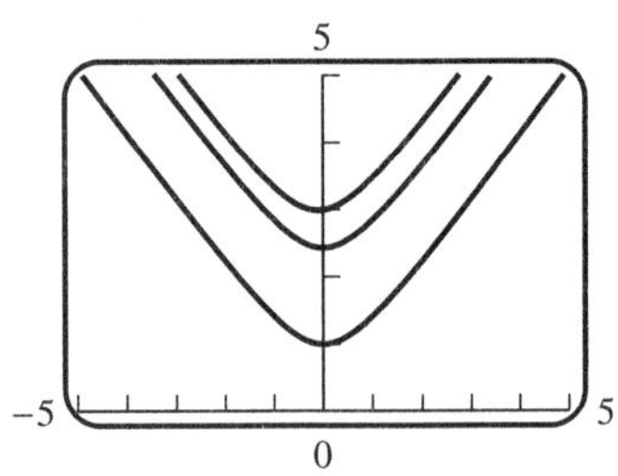

41. $p(t) = \int (4 + 0.15t)^{3/2}dt = \frac{8}{3}(4 + 0.15t)^{5/2} + C$; $p(0) = 100{,}000 = \frac{8}{3}4^{5/2} + C = \frac{256}{3} + C$,

$C = 100{,}000 - \frac{256}{3} \approx 99{,}915$, $p(t) \approx \frac{8}{3}(4+0.15t)^{5/2} + 99{,}915$, $p(5) \approx \frac{8}{3}(4.75)^{5/2} + 99{,}915 \approx 100{,}046$

EXERCISE SET 5.4

1. **(a)** $1 + 8 + 27 = 36$ **(b)** $5 + 8 + 11 + 14 + 17 = 55$

(c) $20 + 12 + 6 + 2 + 0 + 0 = 40$ **(d)** $1 + 1 + 1 + 1 + 1 + 1 = 6$

(e) $1 - 2 + 4 - 8 + 16 = 11$ **(f)** $0 + 0 + 0 + 0 + 0 + 0 = 0$

2. **(a)** $1 + 0 - 3 + 0 = -2$ **(b)** $1 - 1 + 1 - 1 + 1 - 1 = 0$

(c) $\pi^2 + \pi^2 + \cdots + \pi^2 = 14\pi^2$ (14 terms) **(d)** $2^4 + 2^5 + 2^6 = 112$

(e) $\sqrt{1} + \sqrt{2} + \sqrt{3} + \sqrt{4} + \sqrt{5} + \sqrt{6}$

(f) $1 - 1 + 1 - 1 + 1 - 1 + 1 - 1 + 1 - 1 + 1 = 1$

3. $\sum_{k=1}^{10} k$ **4.** $\sum_{k=1}^{20} 3k$ **5.** $\sum_{k=1}^{10} 2k$

6. $\sum_{k=1}^{8}(2k-1)$ **7.** $\sum_{k=1}^{6}(-1)^{k+1}(2k-1)$ **8.** $\sum_{k=1}^{5}(-1)^{k+1}\frac{1}{k}$

9. (a) $\sum_{k=1}^{50} 12k$ (b) $\sum_{k=1}^{50}(2k-1)$

10. (a) $\sum_{k=1}^{5}(-1)^{k+1}a_k$ (b) $\sum_{k=0}^{5}(-1)^{k+1}b_k$ (c) $\sum_{k=0}^{n}a_k x^k$ (d) $\sum_{k=0}^{5}a^{5-k}b^k$

11. $\frac{1}{2}(100)(100+1) = 5050$

12. $7\sum_{k=1}^{100} k + \sum_{k=1}^{100} 1 = \frac{7}{2}(100)(101) + 100 = 35{,}450$

13. $\frac{1}{6}(20)(21)(41) = 2870$

14. $\sum_{k=1}^{20} k^2 - \sum_{k=1}^{3} k^2 = 2870 - 14 = 2856$

15. $\sum_{k=1}^{30} k(k^2-4) = \sum_{k=1}^{30}(k^3-4k) = \sum_{k=1}^{30} k^3 - 4\sum_{k=1}^{30} k = \frac{1}{4}(30)^2(31)^2 - 4\cdot\frac{1}{2}(30)(31) = 214{,}365$

16. $\sum_{k=1}^{6} k - \sum_{k=1}^{6} k^3 = \frac{1}{2}(6)(7) - \frac{1}{4}(6)^2(7)^2 = -420$

17. $\sum_{k=1}^{n}\frac{3k}{n} = \frac{3}{n}\sum_{k=1}^{n} k = \frac{3}{n}\cdot\frac{1}{2}n(n+1) = \frac{3}{2}(n+1)$

18. $\sum_{k=1}^{n-1}\frac{k^2}{n} = \frac{1}{n}\sum_{k=1}^{n-1} k^2 = \frac{1}{n}\cdot\frac{1}{6}(n-1)(n)(2n-1) = \frac{1}{6}(n-1)(2n-1)$

19. $\sum_{k=1}^{n-1}\frac{k^3}{n^2} = \frac{1}{n^2}\sum_{k=1}^{n-1} k^3 = \frac{1}{n^2}\cdot\frac{1}{4}(n-1)^2n^2 = \frac{1}{4}(n-1)^2$

20. $\sum_{k=1}^{n}\left(\frac{5}{n} - \frac{2k}{n}\right) = \frac{5}{n}\sum_{k=1}^{n} 1 - \frac{2}{n}\sum_{k=1}^{n} k = \frac{5}{n}(n) - \frac{2}{n}\cdot\frac{1}{2}n(n+1) = 4 - n$

22. $\frac{n(n+1)}{2} = 465$, $n^2 + n - 930 = 0$, $(n+31)(n-30) = 0$, $n = 30$.

23. $\frac{1+2+3+\cdots+n}{n^2} = \sum_{k=1}^{n}\frac{k}{n^2} = \frac{1}{n^2}\sum_{k=1}^{n} k = \frac{1}{n^2}\cdot\frac{1}{2}n(n+1) = \frac{n+1}{2n}$; $\lim_{n\to+\infty}\frac{n+1}{2n} = \frac{1}{2}$

24. $\dfrac{1^2+2^2+3^2+\cdots+n^2}{n^3}=\displaystyle\sum_{k=1}^{n}\frac{k^2}{n^3}=\frac{1}{n^3}\sum_{k=1}^{n}k^2=\frac{1}{n^3}\cdot\frac{1}{6}n(n+1)(2n+1)=\frac{(n+1)(2n+1)}{6n^2}$;

$\displaystyle\lim_{n\to+\infty}\frac{(n+1)(2n+1)}{6n^2}=\lim_{n\to+\infty}\frac{1}{6}(1+1/n)(2+1/n)=\frac{1}{3}$

25. $\displaystyle\sum_{k=1}^{n}\frac{5k}{n^2}=\frac{5}{n^2}\sum_{k=1}^{n}k=\frac{5}{n^2}\cdot\frac{1}{2}n(n+1)=\frac{5(n+1)}{2n};\ \lim_{n\to+\infty}\frac{5(n+1)}{2n}=\frac{5}{2}$

26. $\displaystyle\sum_{k=1}^{n-1}\frac{2k^2}{n^3}=\frac{2}{n^3}\sum_{k=1}^{n-1}k^2=\frac{2}{n^3}\cdot\frac{1}{6}(n-1)(n)(2n-1)=\frac{(n-1)(2n-1)}{3n^2}$;

$\displaystyle\lim_{n\to+\infty}\frac{(n-1)(2n-1)}{3n^2}=\lim_{n\to+\infty}\frac{1}{3}(1-1/n)(2-1/n)=\frac{2}{3}$

27. (a) $\displaystyle\sum_{j=0}^{5}2^j$ (b) $\displaystyle\sum_{j=1}^{6}2^{j-1}$ (c) $\displaystyle\sum_{j=2}^{7}2^{j-2}$

28. (a) $\displaystyle\sum_{k=1}^{5}(k+4)2^{k+8}$ (b) $\displaystyle\sum_{k=9}^{13}(k-4)2^k$

29. Endpoints $2,3,4,5,6;\Delta x=1$;

(a) Left endpoints: $\displaystyle\sum_{k=1}^{4}f(x_k^*)\Delta x=7+10+13+16=46$

(b) Midpoints: $\displaystyle\sum_{k=1}^{4}f(x_k^*)\Delta x=8.5+11.5+14.5+17.5=52$

(c) Right endpoints: $\displaystyle\sum_{k=1}^{4}f(x_k^*)\Delta x=10+13+16+19=58$

30. Endpoints $1,3,5,7,9,\Delta x=2$;

(a) Left endpoints: $\displaystyle\sum_{k=1}^{4}f(x_k^*)\Delta x=\left(1+\frac{1}{3}+\frac{1}{5}+\frac{1}{7}\right)2=\frac{352}{105}$

(b) Midpoints: $\displaystyle\sum_{k=1}^{4}f(x_k^*)\Delta x=\left(\frac{1}{2}+\frac{1}{4}+\frac{1}{6}+\frac{1}{8}\right)2=\frac{25}{12}$

(c) Right endpoints: $\displaystyle\sum_{k=1}^{4}f(x_k^*)\Delta x=\left(\frac{1}{3}+\frac{1}{5}+\frac{1}{7}+\frac{1}{9}\right)2=\frac{496}{315}$

31. Endpoints: $0,\pi/4,\pi/2,3\pi/4,\pi;\Delta x=\pi/4$

(a) Left endpoints: $\displaystyle\sum_{k=1}^{4}f(x_k^*)\Delta x=\left(1+\sqrt{2}/2+0-\sqrt{2}/2\right)(\pi/4)=\pi/4$

(b) Midpoints: $\displaystyle\sum_{k=1}^{4}f(x_k^*)\Delta x=[\cos(\pi/8)+\cos(3\pi/8)+\cos(5\pi/8)+\cos(7\pi/8)](\pi/4)$

$=[\cos(\pi/8)+\cos(3\pi/8)-\cos(3\pi/8)-\cos(\pi/8)](\pi/4)=0$

(c) Right endpoints: $\displaystyle\sum_{k=1}^{4}f(x_k^*)\Delta x=\left(\sqrt{2}/2+0-\sqrt{2}/2-1\right)(\pi/4)=-\pi/4$

32. Endpoints $-1, 0, 1, 2, 3; \Delta x = 1$

(a) $\sum_{k=1}^{4} f(x_k^*)\Delta x = -3 + 0 + 1 + 0 = -2$

(b) $\sum_{k=1}^{4} f(x_k^*)\Delta x = -\frac{5}{4} + \frac{3}{4} + \frac{3}{4} + \frac{15}{4} = 4$

(c) $\sum_{k=1}^{4} f(x_k^*)\Delta x = 0 + 1 + 0 - 3 = -2$

33. (a) 0.718771403, 0.705803382, 0.698172179
(b) 0.668771403, 0.680803382, 0.688172179
(c) 0.692835360, 0.693069098, 0.693134682

34. (a) 0.761923639, 0.712712753, 0.684701150
(b) 0.584145862, 0.623823864, 0.649145594
(c) 0.663501867, 0.665867079, 0.666538346

35. (a) 4.884074734, 5.115572731, 5.248762738
(b) 5.684074734, 5.515572731, 5.408762738
(c) 5.34707029, 5.338362719, 5.334644416

36. (a) 0.919403170, 0.960215997, 0.984209789
(b) 1.076482803, 1.038755813, 1.015625715
(c) 1.001028824, 1.000257067, 1.000041125

37. $\Delta x = \frac{3}{n}$, $x_k^* = 1 + \frac{3}{n}k$; $f(x_k^*)\Delta x = \frac{1}{2}x_k^*\Delta x = \frac{1}{2}\left(1 + \frac{3}{n}k\right)\frac{3}{n} = \frac{3}{2}\left[\frac{1}{n} + \frac{3}{n^2}k\right]$

$$\sum_{k=1}^{n} f(x_k^*)\Delta x = \frac{3}{2}\left[\sum_{k=1}^{n}\frac{1}{n} + \sum_{k=1}^{n}\frac{3}{n^2}k\right] = \frac{3}{2}\left[1 + \frac{3}{n^2}\cdot\frac{1}{2}n(n+1)\right] = \frac{3}{2}\left[1 + \frac{3}{2}\frac{n+1}{n}\right]$$

$$A = \lim_{n\to+\infty}\frac{3}{2}\left[1 + \frac{3}{2}\left(1 + \frac{1}{n}\right)\right] = \frac{3}{2}\left(1 + \frac{3}{2}\right) = \frac{15}{4}$$

38. $\Delta x = \frac{5}{n}$, $x_k^* = 0 + k\frac{5}{n}$; $f(x_k^*)\Delta x = (5 - x_k^*)\Delta x = \left(5 - \frac{5}{n}k\right)\frac{5}{n} = \frac{25}{n} - \frac{25}{n^2}k$

$$\sum_{k=1}^{n} f(x_k^*)\Delta x = \sum_{k=1}^{n}\frac{25}{n} - \frac{25}{n^2}\sum_{k=1}^{n}k = 25 - \frac{25}{n^2}\cdot\frac{1}{2}n(n+1) = 25 - \frac{25}{2}\left(\frac{n+1}{n}\right)$$

$$A = \lim_{n\to+\infty}\left[25 - \frac{25}{2}\left(1 + \frac{1}{n}\right)\right] = 25 - \frac{25}{2} = \frac{25}{2}$$

39. $\Delta x = \frac{3}{n}$, $x_k^* = 0 + k\frac{3}{n}$; $f(x_k^*)\Delta x = \left(9 - 9\frac{k^2}{n^2}\right)\frac{3}{n}$

$$\sum_{k=1}^{n} f(x_k^*)\Delta x = \sum_{k=1}^{n}(9 - 9\frac{k^2}{n^2})\frac{3}{n} = \frac{27}{n}\sum_{k=1}^{n}\left(1 - \frac{k^2}{n^2}\right) = 27 - \frac{27}{n^3}\sum_{k=1}^{n}k^2$$

$$A = \lim_{n\to+\infty}\left[27 - \frac{27}{n^3}\sum_{k=1}^{n}k^2\right] = 27 - 27\left(\frac{1}{3}\right) = 18$$

40. $\Delta x = \frac{3}{n},\ x_k^* = k\frac{3}{n}$

$$f(x_k^*)\Delta x = \left[4 - \frac{1}{4}(x_k^*)^2\right]\Delta x = \left[4 - \frac{1}{4}\frac{9k^2}{n^2}\right]\frac{3}{n} = \frac{12}{n} - \frac{27k^2}{4n^3}$$

$$\sum_{k=1}^{n} f(x_k^*)\Delta x = \sum_{k=1}^{n}\frac{12}{n} - \frac{27}{4n^3}\sum_{k=1}^{n}k^2$$

$$= 12 - \frac{27}{4n^3}\cdot\frac{1}{6}n(n+1)(2n+1) = 12 - \frac{9}{8}\frac{(n+1)(2n+1)}{n^2}$$

$$A = \lim_{n\to+\infty}\left[12 - \frac{9}{8}\left(1+\frac{1}{n}\right)\left(2+\frac{1}{n}\right)\right] = 12 - \frac{9}{8}(1)(2) = 39/4$$

41. $\Delta x = \frac{4}{n},\ x_k^* = 2 + k\frac{4}{n}$

$$f(x_k^*)\Delta x = (x_k^*)^3\Delta x = \left[2 + \frac{4}{n}k\right]^3\frac{4}{n} = \frac{32}{n}\left[1 + \frac{2}{n}k\right]^3 = \frac{32}{n}\left[1 + \frac{6}{n}k + \frac{12}{n^2}k^2 + \frac{8}{n^3}k^3\right]$$

$$\sum_{k=1}^{n} f(x_k^*)\Delta x = \frac{32}{n}\left[\sum_{k=1}^{n}1 + \frac{6}{n}\sum_{k=1}^{n}k + \frac{12}{n^2}\sum_{k=1}^{n}k^2 + \frac{8}{n^3}\sum_{k=1}^{n}k^3\right]$$

$$= \frac{32}{n}\left[n + \frac{6}{n}\cdot\frac{1}{2}n(n+1) + \frac{12}{n^2}\cdot\frac{1}{6}n(n+1)(2n+1) + \frac{8}{n^3}\cdot\frac{1}{4}n^2(n+1)^2\right]$$

$$= 32\left[1 + 3\frac{n+1}{n} + 2\frac{(n+1)(2n+1)}{n^2} + 2\frac{(n+1)^2}{n^2}\right]$$

$$A = \lim_{n\to+\infty} 32\left[1 + 3\left(1+\frac{1}{n}\right) + 2\left(1+\frac{1}{n}\right)\left(2+\frac{1}{n}\right) + 2\left(1+\frac{1}{n}\right)^2\right]$$

$$= 32[1 + 3(1) + 2(1)(2) + 2(1)^2] = 320$$

42. $\Delta x = \frac{2}{n},\ x_k^* = -3 + k\frac{2}{n};\ f(x_k^*)\Delta x = [1 - (x_k^*)^3]\Delta x = \left[1 - \left(-3 + \frac{2}{n}k\right)^3\right]\frac{2}{n}$

$$= \frac{2}{n}\left[28 - \frac{54}{n}k + \frac{36}{n^2}k^2 - \frac{8}{n^3}k^3\right]$$

$$\sum_{k=1}^{n} f(x_k^*)\Delta x = \frac{2}{n}\left[28n - 27(n+1) + 6\frac{(n+1)(2n+1)}{n} - 2\frac{(n+1)^2}{n}\right]$$

$$A = \lim_{n\to+\infty} 2\left[28 - 27\left(1+\frac{1}{n}\right) + 6\left(1+\frac{1}{n}\right)\left(2+\frac{1}{n}\right) - 2\left(1+\frac{1}{n}\right)^2\right]$$

$$= 2(28 - 27 + 12 - 2) = 22$$

43. $\Delta x = \frac{3}{n},\ x_k^* = 1 + (k-1)\frac{3}{n}$

$$f(x_k^*)\Delta x = \frac{1}{2}x_k^*\Delta x = \frac{1}{2}\left[1 + (k-1)\frac{3}{n}\right]\frac{3}{n} = \frac{1}{2}\left[\frac{3}{n} + (k-1)\frac{9}{n^2}\right]$$

$$\sum_{k=1}^{n} f(x_k^*)\Delta x = \frac{1}{2}\left[\sum_{k=1}^{n}\frac{3}{n} + \frac{9}{n^2}\sum_{k=1}^{n}(k-1)\right] = \frac{1}{2}\left[3 + \frac{9}{n^2}\cdot\frac{1}{2}(n-1)n\right] = \frac{3}{2} + \frac{9}{4}\frac{n-1}{n}$$

$$A = \lim_{n\to+\infty}\left[\frac{3}{2} + \frac{9}{4}\left(1 - \frac{1}{n}\right)\right] = \frac{3}{2} + \frac{9}{4} = \frac{15}{4}$$

44. $\Delta x = \dfrac{5}{n},\ x_k^* = \dfrac{5}{n}(k-1)$

$$f(x_k^*)\Delta x = (5 - x_k^*)\Delta x = \left[5 - \frac{5}{n}(k-1)\right]\frac{5}{n} = \frac{25}{n} - \frac{25}{n^2}(k-1)$$

$$\sum_{k=1}^{n} f(x_k^*)\Delta x = \frac{25}{n}\sum_{k=1}^{n} 1 - \frac{25}{n^2}\sum_{k=1}^{n}(k-1) = 25 - \frac{25}{2}\frac{n-1}{n}$$

$$A = \lim_{n\to+\infty}\left[25 - \frac{25}{2}\left(1 - \frac{1}{n}\right)\right] = 25 - \frac{25}{2} = \frac{25}{2}$$

45. $\Delta x = \dfrac{3}{n}, x_k^* = 0 + (k-1)\dfrac{3}{n}; f(x_k^*)\Delta x = (9 - 9\dfrac{(k-1)^2}{n^2})\dfrac{3}{n}$

$$\sum_{k=1}^{n} f(x_k^*)\Delta x = \sum_{k=1}^{n}\left[9 - 9\frac{(k-1)^2}{n^2}\right]\frac{3}{n} = \frac{27}{n}\sum_{k=1}^{n}\left(1 - \frac{(k-1)^2}{n^2}\right) = 27 - \frac{27}{n^3}\sum_{k=1}^{n}k^2 + \frac{54}{n^3}\sum_{k=1}^{n}k - \frac{27}{n^2}$$

$$A = \lim_{n\to+\infty} = 27 - 27\left(\frac{1}{3}\right) + 0 + 0 = 18$$

46. $\Delta x = \dfrac{3}{n},\ x_k^* = (k-1)\dfrac{3}{n}$

$$f(x_k^*)\Delta x = \left[4 - \frac{1}{4}(x_k^*)^2\right]\Delta x = \left[4 - \frac{1}{4}\frac{9(k-1)^2}{n^2}\right]\frac{3}{n} = \frac{12}{n} - \frac{27k^2}{4n^3} + \frac{27k}{2n^3} - \frac{27}{4n^3}$$

$$\sum_{k=1}^{n} f(x_k^*)\Delta x = \sum_{k=1}^{n}\frac{12}{n} - \frac{27}{4n^3}\sum_{k=1}^{n}k^2 + \frac{27}{2n^3}\sum_{k=1}^{n}k - \frac{27}{4n^3}\sum_{k=1}^{n}1$$

$$= 12 - \frac{27}{4n^3}\cdot\frac{1}{6}n(n+1)(2n+1) + \frac{27}{2n^3}\frac{n(n+1)}{2} - \frac{27}{4n^2}$$

$$= 12 - \frac{9}{8}\frac{(n+1)(2n+1)}{n^2} + \frac{27}{4n} + \frac{27}{4n^2} - \frac{27}{4n^2}$$

$$A = \lim_{n\to+\infty}\left[12 - \frac{9}{8}\left(1 + \frac{1}{n}\right)\left(2 + \frac{1}{n}\right)\right] + 0 + 0 - 0 = 12 - \frac{9}{8}(1)(2) = 39/4$$

47. $\Delta x = \dfrac{1}{n}, x_k^* = \dfrac{2k-1}{2n}$

$$f(x_k^*)\Delta x = \frac{(2k-1)^2}{(2n)^2}\frac{1}{n} = \frac{k^2}{n^3} - \frac{k}{n^3} + \frac{1}{4n^3}$$

$$\sum_{k=1}^{n} f(x_k^*)\Delta x = \frac{1}{n^3}\sum_{k=1}^{n}k^2 - \frac{1}{n^3}\sum_{k=1}^{n}k + \frac{1}{4n^3}\sum_{k=1}^{n}1$$

Using Theorem 5.4.4,

$$A = \lim_{n\to+\infty}\sum_{k=1}^{n} f(x_k^*)\Delta x = \frac{1}{3} + 0 + 0 = \frac{1}{3}$$

48. $\Delta x = \dfrac{2}{n}, x_k^* = -1 + \dfrac{2k-1}{n}$

$$f(x_k^*)\Delta x = \left(-1 + \frac{2k-1}{n}\right)^2\frac{2}{n} = \frac{8k^2}{n^3} - \frac{8k}{n^3} + \frac{2}{n^3} - \frac{2}{n}$$

$$\sum_{k=1}^{n} f(x_k^* k)\Delta x = \frac{8}{n^3}\sum_{k=1}^{n} k^2 - \frac{8}{n^3}\sum_{k=1}^{n} k + \frac{2}{n^2} - 2$$

$$A = \lim_{n\to+\infty}\sum_{k=1}^{n} f(x_k^*)\Delta x = \frac{8}{3} + 0 + 0 - 2 = \frac{2}{3}$$

49. $\Delta x = \frac{2}{n}, x_k^* = -1 + \frac{2k}{n}$

$$f(x_k^*)\Delta x = \left(-1 + \frac{2k}{n}\right)\frac{2}{n} = -\frac{2}{n} + 4\frac{k}{n^2}$$

$$\sum_{k=1}^{n} f(x_k^*)\Delta x = -2 + \frac{4}{n^2}\sum_{k=1}^{n} k = -2 + \frac{4}{n^2}\frac{n(n+1)}{2} = -2 + 2 + \frac{2}{n}$$

$$A = \lim_{n\to+\infty}\sum_{k=1}^{n} f(x_k^*)\Delta x = 0$$

The area below the x-axis cancels the area above the x-axis.

50. $\Delta x = \frac{3}{n}, x_k^* = -1 + \frac{3k}{n}$

$$f(x_k^*)\Delta x = \left(-1 + \frac{3k}{n}\right)\frac{3}{n} = -\frac{3}{n} + \frac{9}{n^2}k$$

$$\sum_{k=1}^{n} f(x_k^*)\Delta x = -3 + \frac{9}{n^2}\frac{n(n+1)}{2}$$

$$A = \lim_{n\to+\infty}\sum_{k=1}^{n} f(x_k^*)\Delta x = -3 + \frac{9}{2} + 0 = \frac{3}{2}$$

The area below the x-axis cancels the area above the x-axis that lies to the right of the line $x = 1$; the remaining area is a trapezoid of width 1 and heights 1, 2, hence its area is $\frac{1+2}{2} = \frac{3}{2}$

51. $\Delta x = \frac{2}{n}, x_k^* = \frac{2k}{n}$

$$f(x_k^*) = \left[\left(\frac{2k}{n}\right)^2 - 1\right]\frac{2}{n} = \frac{8k^2}{n^3} - \frac{2}{n}$$

$$\sum_{k=1}^{n} f(x_k^*)\Delta x = \frac{8}{n^3}\sum_{k=1}^{n} k^2 - \frac{2}{n}\sum_{k=1}^{n} 1 = \frac{8}{n^3}\frac{n(n+1)(2n+1)}{6} - 2$$

$$A = \lim_{n\to+\infty}\sum_{k=1}^{n} f(x_k^*)\Delta x = \frac{16}{6} - 2 = \frac{2}{3}$$

52. $\Delta x = \frac{2}{n}, x_k^* = -1 + \frac{2k}{n}$

$$f(x_k^*)\Delta x = \left(-1 + \frac{2k}{n}\right)^3\frac{2}{n} = -\frac{2}{n} + 12\frac{k}{n^2} - 24\frac{k^2}{n^3} + 16\frac{k^3}{n^4}$$

$$\sum_{k=1}^{n} f(x_k^*)\Delta x = -2 + \frac{12}{n^2}\frac{n(n+1)}{2} - \frac{24}{n^3}\frac{n(n+1)(2n+1)}{6} + \frac{16}{n^4}\left(\frac{n(n+1)}{2}\right)^2$$

$$A = \lim_{n\to+\infty}\sum_{k=1}^{n} f(x_k^*) = -2 + \frac{12}{2} - \frac{48}{6} + \frac{16}{2^2} = 0$$

53. $\Delta x = \dfrac{b-a}{n},\ x_k^* = a + \dfrac{b-a}{n}(k-1)$

$$f(x_k^*)\Delta x = mx_k^*\Delta x = m\left[a + \frac{b-a}{n}(k-1)\right]\frac{b-a}{n} = m(b-a)\left[\frac{a}{n} + \frac{b-a}{n^2}(k-1)\right]$$

$$\sum_{k=1}^{n} f(x_k^*)\Delta x = m(b-a)\left[a + \frac{b-a}{2}\cdot\frac{n-1}{n}\right]$$

$$A = \lim_{n\to+\infty} m(b-a)\left[a + \frac{b-a}{2}\left(1-\frac{1}{n}\right)\right] = m(b-a)\frac{b+a}{2} = \frac{1}{2}m(b^2-a^2)$$

54. $\Delta x = \dfrac{b-a}{n},\ x_k^* = a + \dfrac{k}{n}(b-a)$

$$f(x_k^*)\Delta x = \frac{ma}{n}(b-a) + \frac{mk}{n^2}(b-a)^2$$

$$\sum_{k=1}^{n} f(x_k^*)\Delta x = ma(b-a) + \frac{m}{n^2}(b-a)^2\frac{n(n+1)}{2}$$

$$A = \lim_{n\to+\infty}\sum_{k=1}^{n} f(x_k^*)\Delta x = ma(b-a) + \frac{m}{2}(b-a)^2 = m(b-a)\frac{a+b}{2}$$

55. **(a)** With x_k^* as the right endpoint, $\Delta x = \dfrac{b}{n},\ x_k^* = \dfrac{b}{n}k$

$$f(x_k^*)\Delta x = (x_k^*)^3\Delta x = \frac{b^4}{n^4}k^3,\ \sum_{k=1}^{n} f(x_k^*)\Delta x = \frac{b^4}{n^4}\sum_{k=1}^{n}k^3 = \frac{b^4}{4}\frac{(n+1)^2}{n^2}$$

$$A = \lim_{n\to+\infty}\frac{b^4}{4}\left(1+\frac{1}{n}\right)^2 = b^4/4$$

(b) $\Delta x = \dfrac{b-a}{n},\ x_k^* = a + \dfrac{b-a}{n}k$

$$f(x_k^*)\Delta x = (x_k^*)^3\Delta x = \left[a + \frac{b-a}{n}k\right]^3\frac{b-a}{n}$$

$$= \frac{b-a}{n}\left[a^3 + \frac{3a^2(b-a)}{n}k + \frac{3a(b-a)^2}{n^2}k^2 + \frac{(b-a)^3}{n^3}k^3\right]$$

$$\sum_{k=1}^{n} f(x_k^*)\Delta x = (b-a)\left[a^3 + \frac{3}{2}a^2(b-a)\frac{n+1}{n} + \frac{1}{2}a(b-a)^2\frac{(n+1)(2n+1)}{n^2} + \frac{1}{4}(b-a)^3\frac{(n+1)^2}{n^2}\right]$$

$$A = \lim_{n\to+\infty}\sum_{k=1}^{n} f(x_k^*)\Delta x$$

$$= (b-a)\left[a^3 + \frac{3}{2}a^2(b-a) + a(b-a)^2 + \frac{1}{4}(b-a)^3\right] = \frac{1}{4}(b^4-a^4).$$

56. Let A be the area of the region under the curve and above the interval $0 \le x \le 1$ on the x-axis, and let B be the area of the region between the curve and the interval $0 \le y \le 1$ on the y-axis. Together A and B form the square of side 1, so $A + B = 1$.
But B can also be considered as the area between the curve $x = y^2$ and the interval $0 \le y \le 1$ on the y-axis. By Exercise 47 above, $B = \dfrac{1}{3}$, so $A = 1 - \dfrac{1}{3} = \dfrac{2}{3}$.

57. If $n = 2m$ then $2m + 2(m-1) + \cdots + 2\cdot 2 + 2 = 2\sum_{k=1}^{m} k = 2\cdot\dfrac{m(m+1)}{2} = m(m+1) = \dfrac{n^2+2n}{4}$;

if $n = 2m+1$ then $(2m+1) + (2m-1) + \cdots + 5 + 3 + 1 = \sum_{k=1}^{m+1}(2k-1)$

$$= 2\sum_{k=1}^{m+1} k - \sum_{k=1}^{m+1} 1 = 2\cdot\frac{(m+1)(m+2)}{2} - (m+1) = (m+1)^2 = \frac{n^2+2n+1}{4}$$

58. $50\cdot 30 + 49\cdot 29 + \cdots + 22\cdot 2 + 21\cdot 1 = \sum_{k=1}^{30} k(k+20) = \sum_{k=1}^{30} k^2 + 20\sum_{k=1}^{30} k = \dfrac{30\cdot 31\cdot 61}{6} + 20\dfrac{30\cdot 31}{2} = 18{,}755$

59. both are valid

60. none is valid

61. $\sum_{k=1}^{n}(a_k - b_k) = (a_1 - b_1) + (a_2 - b_2) + \cdots + (a_n - b_n)$

$$= (a_1 + a_2 + \cdots + a_n) - (b_1 + b_2 + \cdots + b_n) = \sum_{k=1}^{n} a_k - \sum_{k=1}^{n} b_k$$

62. $\sum_{k=1}^{n}\left[(k+1)^4 - k^4\right] = (n+1)^4 - 1$ (telescoping sum), expand the

quantity in brackets to get $\sum_{k=1}^{n}(4k^3 + 6k^2 + 4k + 1) = (n+1)^4 - 1$,

$$4\sum_{k=1}^{n} k^3 + 6\sum_{k=1}^{n} k^2 + 4\sum_{k=1}^{n} k + \sum_{k=1}^{n} 1 = (n+1)^4 - 1$$

$$\sum_{k=1}^{n} k^3 = \frac{1}{4}\left[(n+1)^4 - 1 - 6\sum_{k=1}^{n} k^2 - 4\sum_{k=1}^{n} k - \sum_{k=1}^{n} 1\right]$$

$$= \frac{1}{4}[(n+1)^4 - 1 - n(n+1)(2n+1) - 2n(n+1) - n]$$

$$= \frac{1}{4}(n+1)[(n+1)^3 - n(2n+1) - 2n - 1]$$

$$= \frac{1}{4}(n+1)(n^3 + n^2) = \frac{1}{4}n^2(n+1)^2$$

63. **(a)** $\sum_{k=1}^{n} 1$ means add 1 to itself n times, which gives the result.

(b) $\dfrac{1}{n^2}\sum_{k=1}^{n} k = \dfrac{1}{n^2}\dfrac{n(n+1)}{2} = \dfrac{1}{2} + \dfrac{1}{2n}$, so $\lim_{n\to+\infty}\dfrac{1}{n^2}\sum_{k=1}^{n} k = \dfrac{1}{2}$

(c) $\dfrac{1}{n^3}\sum_{k=1}^{n} k^2 = \dfrac{1}{n^3}\dfrac{n(n+1)(2n+1)}{6} = \dfrac{2}{6} + \dfrac{3}{6n} + \dfrac{1}{6n^2}$, so $\lim_{n\to+\infty}\dfrac{1}{n^3}\sum_{k=1}^{n} k^2 = \dfrac{1}{3}$

(d) $\dfrac{1}{n^4}\sum_{k=1}^{n} k^3 = \dfrac{1}{n^4}\left(\dfrac{n(n+1)}{2}\right)^2 = \dfrac{1}{4} + \dfrac{1}{2n} + \dfrac{1}{4n^2}$, so $\lim_{n\to+\infty}\dfrac{1}{n^4}\sum_{k=1}^{n} k^3 = \dfrac{1}{4}$

EXERCISE SET 5.5

1. (a) $(4/3)(1) + (5/2)(1) + (4)(2) = 71/6$ (b) 2

2. (a) $(\sqrt{2}/2)(\pi/2) + (-1)(3\pi/4) + (0)(\pi/2) + (\sqrt{2}/2)(\pi/4) = 3(\sqrt{2} - 2)\pi/8$
   (b) $3\pi/4$

3. (a) $(-9/4)(1) + (3)(2) + (63/16)(1) + (-5)(3) = -117/16$
   (b) 3

4. (a) $(-8)(2) + (0)(1) + (0)(1) + (8)(2) = 0$ (b) 2

5. $\int_{-1}^{2} x^2\,dx$

6. $\int_{1}^{2} x^3 dx$

7. $\int_{-3}^{3} 4x(1-3x)dx$

8. $\int_{0}^{\pi/2} \sin^2 x\,dx$

9. (a) $\lim_{\max \Delta x_k \to 0} \sum_{k=1}^{n} 2x_k^* \Delta x_k;\ a = 1,\ b = 2$ (b) $\lim_{\max \Delta x_k \to 0} \sum_{k=1}^{n} \frac{x_k^*}{x_k^* + 1} \Delta x_k;\ a = 0,\ b = 1$

10. (a) $\lim_{\max \Delta x_k \to 0} \sum_{k=1}^{n} \sqrt{x_k^*}\, \Delta x_k,\ a = 1,\ b = 2$

    (b) $\lim_{\max \Delta x_k \to 0} \sum_{k=1}^{n} (1 + \cos x_k^*)\, \Delta x_k,\ a = -\pi/2,\ b = \pi/2$

11. (a) $A = \frac{1}{2}(3)(3) = 9/2$

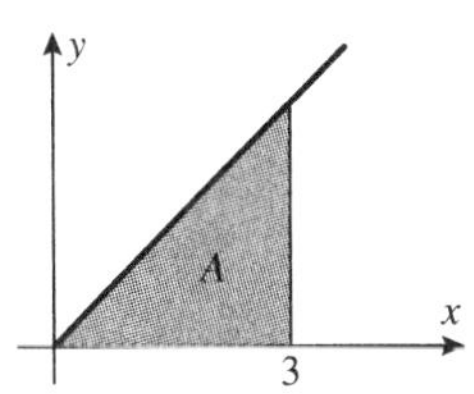

(b) $-A = -\frac{1}{2}(1)(1+2) = -3/2$

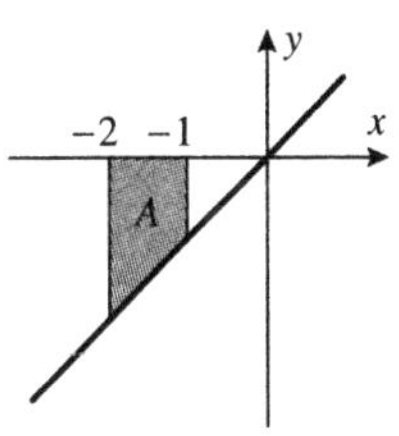

(c) $-A_1 + A_2 = -\frac{1}{2} + 8 = 15/2$

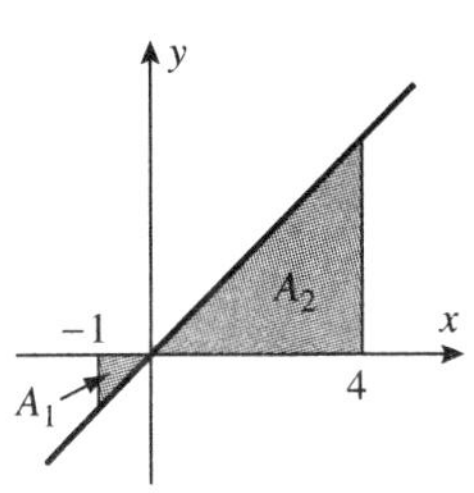

(d) $-A_1 + A_2 = 0$

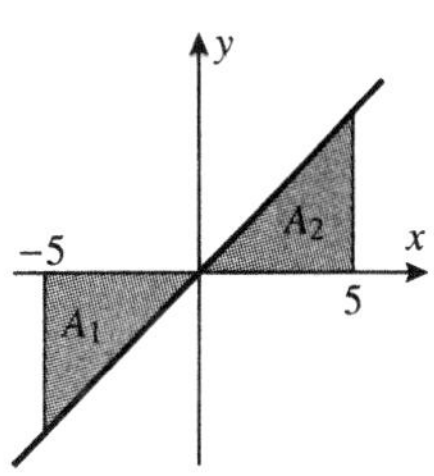

12. **(a)** $A = \frac{1}{2}(1)(2) = 1$

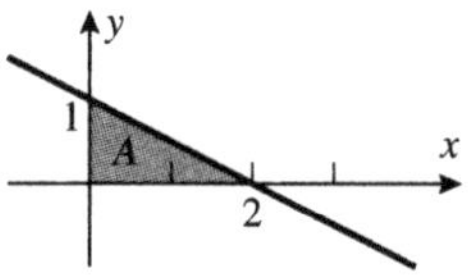

(b) $A = \frac{1}{2}(2)(3/2 + 1/2) = 2$

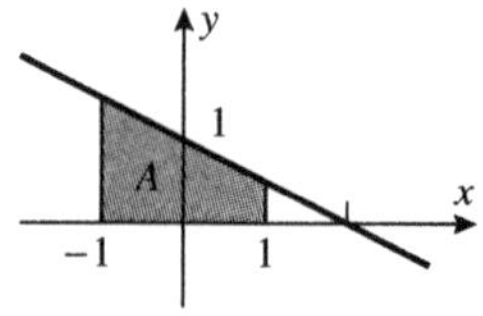

(c) $-A = -\frac{1}{2}(1/2)(1) = -1/4$

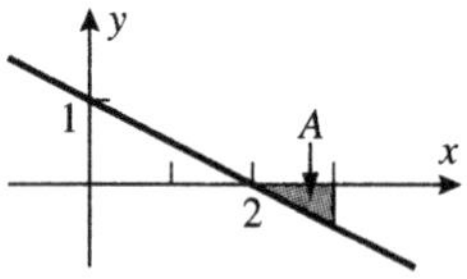

(d) $A_1 - A_2 = 1 - 1/4 = 3/4$

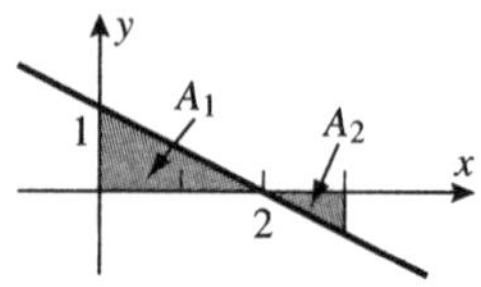

13. **(a)** $A = 2(5) = 10$

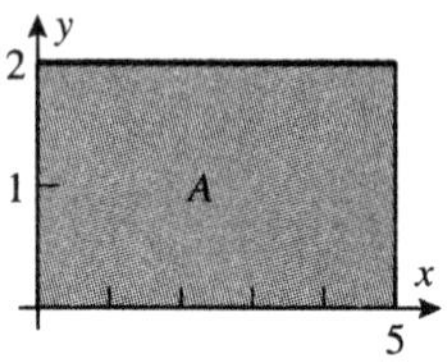

(b) 0; $A_1 = A_2$ by symmetry

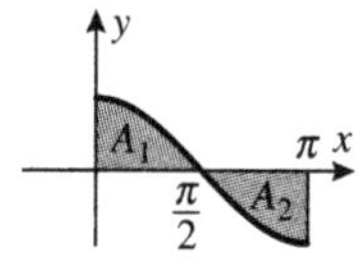

(c) $A_1 + A_2 = \frac{1}{2}(5)(5/2) + \frac{1}{2}(1)(1/2)$
$= 13/2$

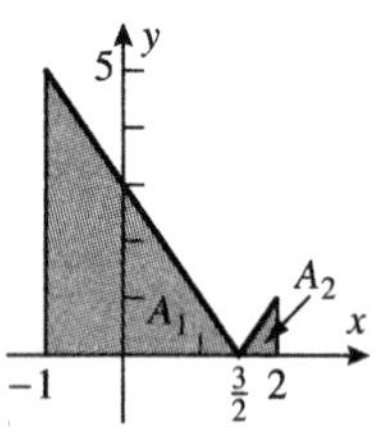

(d) $\frac{1}{2}[\pi(1)^2] = \pi/2$

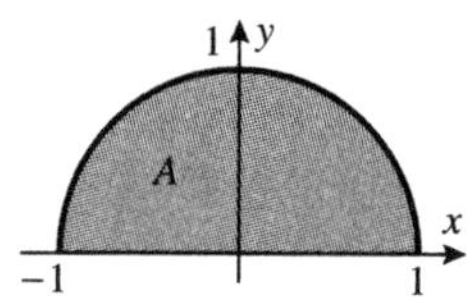

14. **(a)** $A = (6)(5) = 30$

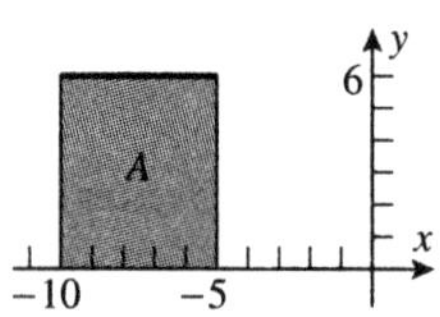

(b) $-A_1 + A_2 = 0$ because
$A_1 = A_2$ by symmetry

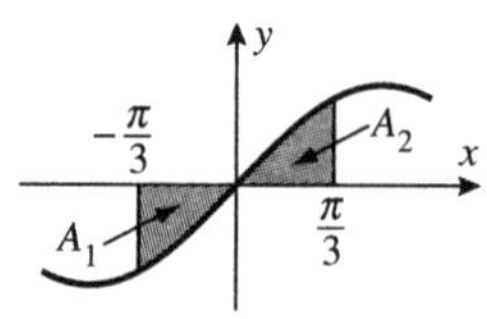

(c) $A_1 + A_2 = \frac{1}{2}(2)(2) + \frac{1}{2}(1)(1) = 5/2$

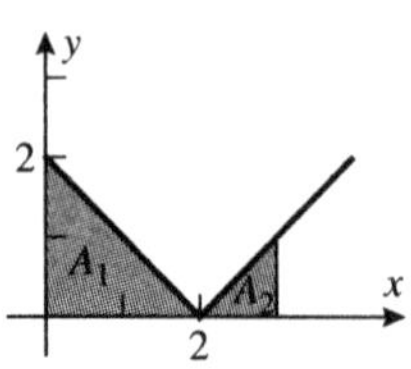

(d) $\frac{1}{4}\pi(2)^2 = \pi$

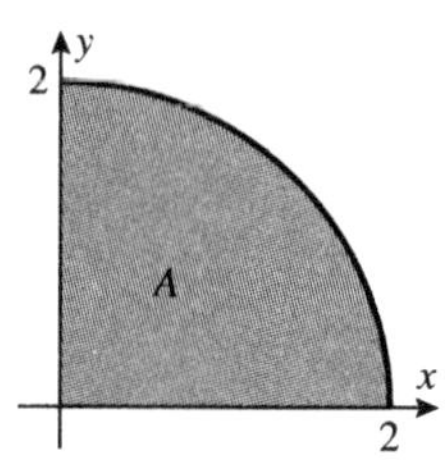

15. (a) 0.8 (b) −2.6 (c) −1.8 (d) −0.3

16. (a) $\int_0^1 f(x)dx = \int_0^1 2xdx = x^2\Big]_0^1 = 1$

(b) $\int_{-1}^1 f(x)dx = \int_{-1}^1 2xdx = x^2\Big]_{-1}^1 = 1^2 - (-1)^2 = 0$

(c) $\int_1^{10} f(x)dx = \int_1^{10} 2dx = 2x\Big]_1^{10} = 18$

(d) $\int_{1/2}^5 f(x)dx = \int_{1/2}^1 2xdx + \int_1^5 2dx = x^2\Big]_{1/2}^1 + 2x\Big]_1^5 = 1^2-(1/2)^2+2{\cdot}5-2{\cdot}1 = 3/4+8 = 35/4$

17. $\int_{-1}^2 f(x)dx + 2\int_{-1}^2 g(x)dx = 5 + 2(-3) = -1$

18. $3\int_1^4 f(x)dx - \int_1^4 g(x)dx = 3(2) - 10 = -4$

19. $\int_1^5 f(x)dx = \int_0^5 f(x)dx - \int_0^1 f(x)dx = 1 - (-2) = 3$

20. $\int_3^{-2} f(x)dx = -\int_{-2}^3 f(x)dx = -\left[\int_{-2}^1 f(x)dx + \int_1^3 f(x)dx\right] = -(2-6) = 4$

21. (a) $\int_0^1 xdx + 2\int_0^1 \sqrt{1-x^2}dx = 1/2 + 2(\pi/4) = (1+\pi)/2$

(b) $4\int_{-1}^3 dx - 5\int_{-1}^3 xdx = 4\cdot 4 - 5(-1/2 + (3\cdot 3)/2) = -4$

22. (a) $\int_{-3}^0 2dx + \int_{-3}^0 \sqrt{9-x^2}dx = 2\cdot 3 + (\pi(3)^2)/4 = 6 + 9\pi/4$

(b) $\int_{-2}^2 dx - 3\int_{-2}^2 |x|dx = 4\cdot 1 - 3(2)(2\cdot 2)/2 = -8$

23. (a) $\sqrt{x} > 0,\ 1 - x < 0$ on $[2,3]$ so the integral is negative

(b) $x^2 > 0,\ 3 - \cos x > 0$ for all x so the integral is positive

24. (a) $x^4 > 0,\ \sqrt{3-x} > 0$ on $[-3,-1]$ so the integral is positive

(b) $x^3 - 9 < 0,\ |x| + 1 > 0$ on $[-2,2]$ so the integral is negative

25. $\int_0^{10} \sqrt{25-(x-5)^2}dx = \pi(5)^2/2 = 25\pi/2$

26. $\int_0^3 \sqrt{9-(x-3)^2}dx = \pi(3)^2/4 = 9\pi/4$

27. $\int_0^1 (3x+1)dx = 5/2$

28. $\int_{-2}^2 \sqrt{4-x^2}dx = \pi(2)^2/2 = 2\pi$

29. **(a)** f is continuous on $[-1,1]$ so f is integrable there by Part (a) of Theorem 5.5.8

(b) $|f(x)| \le 1$ so f is bounded on $[-1,1]$, and f has one point of discontinuity, so by Part (b) of Theorem 5.5.8 f is integrable on $[-1,1]$

(c) f is not bounded on [-1,1] because $\lim_{x\to 0} f(x) = +\infty$, so f is not integrable on [0,1]

(d) $f(x)$ is discontinuous at the point $x=0$ because $\lim_{x\to 0}\sin\dfrac{1}{x}$ does not exist. f is continuous elsewhere. $-1 \le f(x) \le 1$ for x in $[-1,1]$ so f is bounded there. By Part (b), Theorem 5.5.8, f is integrable on $[-1,1]$.

30. Each subinterval of a partition of $[a,b]$ contains both rational and irrational numbers. If all x_k^* are chosen to be rational then

$$\sum_{k=1}^{n} f(x_k^*)\Delta x_k = \sum_{k=1}^{n}(1)\Delta x_k = \sum_{k=1}^{n}\Delta x_k = b-a \text{ so } \lim_{\max \Delta x_k\to 0}\sum_{k=1}^{n} f(x_k^*)\Delta x_k = b-a.$$

If all x_k^* are irrational then $\displaystyle\lim_{\max \Delta x_k\to 0}\sum_{k=1}^{n} f(x_k^*)\Delta x_k = 0$. Thus f is not integrable on $[a,b]$ because the preceding limits are not equal.

31. **(a)** Let $S_n = \displaystyle\sum_{k=1}^{n} f(x_k^*)\Delta x_k$ and $S = \displaystyle\int_a^b f(x)dx$ then $\displaystyle\sum_{k=1}^{n} cf(x_k^*)\Delta x_k = cS_n$ and we want to prove that $\displaystyle\lim_{\max \Delta x_k\to 0} cS_n = cS$. If $c=0$ the result follows immediately, so suppose that $c \ne 0$ then for any $\epsilon > 0$, $|cS_n - cS| = |c||S_n - S| < \epsilon$ if $|S_n - S| < \epsilon/|c|$. But because f is integrable on $[a,b]$, there is a number $\delta > 0$ such that $|S_n - S| < \epsilon/|c|$ whenever $\max \Delta x_k < \delta$ so $|cS_n - cS| < \epsilon$ and hence $\displaystyle\lim_{\max \Delta x_k\to 0} cS_n = cS$.

(b) Let $R_n = \displaystyle\sum_{k=1}^{n} f(x_k^*)\Delta x_k$, $S_n = \displaystyle\sum_{k=1}^{n} g(x_k^*)\Delta x_k$, $T_n = \displaystyle\sum_{k=1}^{n}[f(x_k^*) + g(x_k^*)]\Delta x_k$, $R = \displaystyle\int_a^b f(x)dx$, and $S = \displaystyle\int_a^b g(x)dx$ then $T_n = R_n + S_n$ and we want to prove that $\displaystyle\lim_{\max \Delta x_k\to 0} T_n = R + S$.
$|T_n - (R+S)| = |(R_n - R) + (S_n - S)| \le |R_n - R| + |S_n - S|$
so for any $\epsilon > 0$ $|T_n - (R+S)| < \epsilon$ if $|R_n - R| + |S_n - S| < \epsilon$.
Because f and g are integrable on $[a,b]$, there are numbers δ_1 and δ_2 such that $|R_n - R| < \epsilon/2$ for $\max \Delta x_k < \delta_1$ and $|S_n - S| < \epsilon/2$ for $\max \Delta x_k < \delta_2$.
If $\delta = \min(\delta_1, \delta_2)$ then $|R_n - R| < \epsilon/2$ and $|S_n - S| < \epsilon/2$ for $\max \Delta x_k < \delta$ thus $|R_n - R| + |S_n - S| < \epsilon$ and so $|T_n - (R+S)| < \epsilon$ for $\max \Delta x_k < \delta$ which shows that $\displaystyle\lim_{\max \Delta x_k\to 0} T_n = R + S$.

32. For the smallest, find x_k^* so that $f(x_k^*)$ is minimum on each subinterval: $x_1^* = 1$, $x_2^* = 3/2$, $x_3^* = 3$ so $(2)(1) + (7/4)(2) + (4)(1) = 9.5$. For the largest, find x_k^* so that $f(x_k^*)$ is maximum on each subinterval: $x_1^* = 0$, $x_2^* = 3$, $x_3^* = 4$ so $(4)(1) + (4)(2) + (8)(1) = 20$.

33. $\Delta x_k = \dfrac{4k^2}{n^2} - \dfrac{4(k-1)^2}{n^2} = \dfrac{4}{n^2}(2k-1)$, $x_k^* = \dfrac{4k^2}{n^2}$,

$f(x_k^*) = \dfrac{2k}{n}$, $f(x_k^*)\Delta x_k = \dfrac{8k}{n^3}(2k-1) = \dfrac{8}{n^3}(2k^2 - k)$,

$$\sum_{k=1}^{n} f(x_k^*)\Delta x_k = \frac{8}{n^3}\sum_{k=1}^{n}(2k^2 - k) = \frac{8}{n^3}\left[\frac{1}{3}n(n+1)(2n+1) - \frac{1}{2}n(n+1)\right] = \frac{4}{3}\frac{(n+1)(4n-1)}{n^2},$$

$$\lim_{n\to+\infty}\sum_{k=1}^{n} f(x_k^*)\Delta x_k = \lim_{n\to+\infty}\frac{4}{3}\left(1+\frac{1}{n}\right)\left(4-\frac{1}{n}\right) = \frac{16}{3}.$$

34. For any partition of $[a, b]$ use the right endpoints to form the sum $\sum_{k=1}^{n} f(x_k^*)\Delta x_k$. Since $f(x_k^*) = 0$ for each k, the sum is zero and so is $\int_a^b f(x)\,dx = \lim_{n\to+\infty} \sum_{k=1}^{n} f(x_k^*)\Delta x_k$.

35. With $f(x) = g(x)$ then $f(x) - g(x) = 0$ for $a < x \le b$. By Theorem 5.5.4(b)
$\int_a^b f(x)\,dx = \int_a^b [(f(x) - g(x) + g(x)]dx = \int_a^b [f(x) - g(x)]dx + \int_a^b g(x)dx.$
But the first term on the right hand side is zero (from Exercise 34), so
$\int_a^b f(x)\,dx = \int_a^b g(x)\,dx$

36. Choose any large positive integer N and any partition of $[0, a]$. Then choose x_1^* in the first interval so small that $f(x_1^*)\Delta x_1 > N$. For example choose $x_1^* < \Delta x_1/N$. Then with this partition and choice of x_1^*, $\sum_{k=1}^{n} f(x_k^*)\Delta x_k > f(x_1^*)\Delta x_1 > N$. This shows that the sum is dependent on partition and/or points, so Definition 5.5.1 is not satisfied.

EXERCISE SET 5.6

1. **(a)** $\int_0^2 (2 - x)dx = (2x - x^2/2)\Big]_0^2 = 4 - 4/2 = 2$

(b) $\int_{-1}^1 2dx = 2x\Big]_{-1}^1 = 2(1) - 2(-1) = 4$

(c) $\int_1^3 (x + 1)dx = (x^2/2 + x)\Big]_1^3 = 9/2 + 3 - (1/2 + 1) = 6$

2. **(a)** $\int_0^5 xdx = x^2/2\Big]_0^5 = 25/2$

(b) $\int_3^9 5dx = 5x\Big]_3^9 = 5(9) - 5(3) = 30$

(c) $\int_{-1}^2 (x + 3)dx = (x^2/2 + 3x)\Big]_{-1}^2 = 4/2 + 6 - (1/2 - 3) = 21/2$

3. $\int_2^3 x^3dx = x^4/4\Big]_2^3 = 81/4 - 16/4 = 65/4$

4. $\int_{-1}^1 x^4dx = x^5/5\Big]_{-1}^1 = 1/5 - (-1)/5 = 2/5$

5. $\int_1^9 \sqrt{x}dx = \frac{2}{3}x^{3/2}\Big]_1^9 = \frac{2}{3}(27 - 1) = 52/3$

6. $\int_1^4 x^{-3/5}dx = \frac{5}{2}x^{2/5}\Big]_1^4 = \frac{5}{2}(4^{2/5} - 1)$

7. $\left(\frac{1}{3}x^3 - 2x^2 + 7x\right)\Big]_{-3}^0 = 48$

8. $\left(\frac{1}{2}x^2 + \frac{1}{5}x^5\right)\Big]_{-1}^2 = 81/10$

9. $\int_1^3 x^{-2}dx = -\frac{1}{x}\Big]_1^3 = 2/3$

10. $\int_1^2 x^{-6}dx = -\frac{1}{5x^5}\Big]_1^2 = 31/160$

11. $\displaystyle \frac{4}{5}x^{5/2}\Big]_4^9 = 844/5$

12. $\displaystyle \left(3x^{5/3} + \frac{4}{x}\right)\Big]_1^8 = 179/2$

13. $\displaystyle -\cos\theta\Big]_{-\pi/2}^{\pi/2} = 0$

14. $\displaystyle \tan\theta\Big]_0^{\pi/4} = 1$

15. $\displaystyle \sin x\Big]_{-\pi/4}^{\pi/4} = \sqrt{2}$

16. $\displaystyle \left(\frac{1}{2}x^2 - \sec x\right)\Big]_0^1 = 3/2 - \sec(1)$

17. $\displaystyle \left(6\sqrt{t} - \frac{10}{3}t^{3/2} + \frac{2}{\sqrt{t}}\right)\Big]_1^4 = -55/3$

18. $\displaystyle \left(8\sqrt{y} + \frac{4}{3}y^{3/2} - \frac{2}{3y^{3/2}}\right)\Big]_4^9 = 10819/324$

19. $\displaystyle \left(\frac{1}{2}x^2 - 2\cot x\right)\Big]_{\pi/6}^{\pi/2} = \pi^2/9 + 2\sqrt{3}$

20. $\displaystyle \left(a^{1/2}x - \frac{2}{3}x^{3/2}\right)\Big]_a^{4a} = -\frac{5}{3}a^{3/2}$

21. **(a)** $\displaystyle \int_0^{3/2}(3-2x)dx + \int_{3/2}^2 (2x-3)dx = (3x - x^2)\Big]_0^{3/2} + (x^2 - 3x)\Big]_{3/2}^2 = 9/4 + 1/4 = 5/2$

(b) $\displaystyle \int_0^{\pi/2} \cos x\, dx + \int_{\pi/2}^{3\pi/4}(-\cos x)dx = \sin x\Big]_0^{\pi/2} - \sin x\Big]_{\pi/2}^{3\pi/4} = 2 - \sqrt{2}/2$

22. **(a)** $\displaystyle \int_{-1}^0 \sqrt{2-x}\, dx + \int_0^2 \sqrt{2+x}\, dx = -\frac{2}{3}(2-x)^{3/2}\Big]_{-1}^0 + \frac{2}{3}(2+x)^{3/2}\Big]_0^2$

$$= -\frac{2}{3}(2\sqrt{2} - 3\sqrt{3}) + \frac{2}{3}(8 - 2\sqrt{2}) = \frac{2}{3}(8 - 4\sqrt{2} + 3\sqrt{3})$$

(b) $\displaystyle \int_0^{\pi/6}(1/2 - \sin x)\, dx + \int_{\pi/6}^{\pi/2}(\sin x - 1/2)\, dx$

$$= (x/2 + \cos x)\Big]_0^{\pi/6} - (\cos x + x/2)\Big]_{\pi/6}^{\pi/2}$$

$$= (\pi/12 + \sqrt{3}/2) - 1 - \pi/4 + (\sqrt{3}/2 + \pi/12) = \sqrt{3} - \pi/12 - 1$$

23. **(a)** 17/6

(b) $F(x) = \begin{cases} \dfrac{1}{2}x^2, & x \le 1 \\ \dfrac{1}{3}x^3 + \dfrac{1}{6}, & x > 1 \end{cases}$

24. **(a)** $\displaystyle \int_0^1 \sqrt{x}\, dx + \int_1^4 \frac{1}{x^2}\, dx = \frac{2}{3}x^{3/2}\Big]_0^1 - \frac{1}{x}\Big]_1^4 = 17/12$

(b) $F(x) = \begin{cases} \dfrac{2}{3}x^{3/2}, & x < 1 \\ -\dfrac{1}{x} + \dfrac{5}{3}, & x \ge 1 \end{cases}$

25. 0.665867079; $\displaystyle \int_1^3 \frac{1}{x^2}\, dx = -\frac{1}{x}\Big]_1^3 = 2/3$

26. 1.000257067; $\displaystyle \int_0^{\pi/2} \sin x\, dx = -\cos x\Big]_0^{\pi/2} = 1$

27. 3.106017890; $\int_{-1}^{1} \sec^2 x\,dx = \tan x \Big|_{-1}^{1} = 2\tan 1 \approx 3.114815450$

29. $A = \int_0^3 (x^2+1)dx = \left(\frac{1}{3}x^3 + x\right)\Big]_0^3 = 12$

30. $A = \int_1^2 (-x^2+3x-2)dx = \left(-\frac{1}{3}x^3 + \frac{3}{2}x^2 - 2x\right)\Big]_1^2 = 1/6$

31. $A = \int_0^{2\pi/3} 3\sin x\,dx = -3\cos x\Big]_0^{2\pi/3} = 9/2$

32. $A = -\int_{-2}^{-1} x^3 dx = -\frac{1}{4}x^4\Big]_{-2}^{-1} = 15/4$

33. $A_1 = \int_{-3}^{-2} (x^2-3x-10)dx = \left(\frac{1}{3}x^3 - \frac{3}{2}x^2 - 10x\right)\Big]_{-3}^{-2} = 23/6,$

$A_2 = -\int_{-2}^{5} (x^2-3x-10)dx = 343/6,$

$A_3 = \int_5^8 (x^2-3x-10)dx = 243/6,\ A = A_1 + A_2 + A_3 = 203/2$

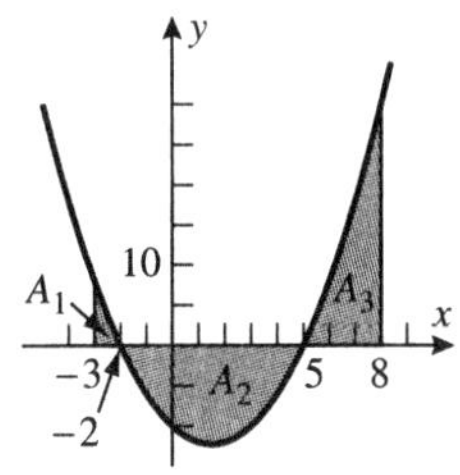

34. **(a)** the area is positive

(b) $\int_{-2}^{5}\left(\frac{1}{100}x^3 - \frac{1}{20}x^2 - \frac{1}{25}x + \frac{1}{5}\right)dx = \left(\frac{1}{400}x^4 - \frac{1}{60}x^3 - \frac{1}{50}x^2 + \frac{1}{5}x\right)\Big]_{-2}^{5} = \frac{343}{1200}$

35. **(a)** the area between the curve and the x-axis breaks into equal parts, one above and one below the x-axis, so the integral is zero

(b) $\int_{-1}^{1} x^3 dx = \frac{1}{4}x^4\Big]_{-1}^{1} = \frac{1}{4}(1^4 - (-1)^4) = 0;$

$\int_{-\pi/2}^{\pi/2} \sin x dx = -\cos x\Big]_{-\pi/2}^{\pi/2} = -\cos(\pi/2) + \cos(-\pi/2) = 0 + 0 = 0$

(c) The area on the left side of the y-axis is equal to the area on the right side, so $\int_{-a}^{a} f(x)dx = 2\int_0^a f(x)dx$

(d) $\int_{-1}^{1} x^2 dx = \frac{1}{3}x^3\Big]_{-1}^{1} = \frac{1}{3}(1^3 - (-1)^3) = \frac{2}{3} = 2\int_0^1 x^2 dx;$

$\int_{-\pi/2}^{\pi/2} \cos x dx = \sin x\Big]_{-\pi/2}^{\pi/2} = \sin(\pi/2) - \sin(-\pi/2) = 1 + 1 = 2 = 2\int_0^{\pi/2} \cos x dx$

36. The numerator is an odd function and the denominator is an even function, so the integrand is an odd function and the integral is zero.

37. **(a)** $x^3 + 1$

(b) $F(x) = \left(\frac{1}{4}t^4 + t\right)\Big]_1^x = \frac{1}{4}x^4 + x - \frac{5}{4};\ F'(x) = x^3 + 1$

38. **(a)** $\cos 2x$

(b) $F(x) = \frac{1}{2}\sin 2t\Big]_{\pi/4}^{x} = \frac{1}{2}\sin 2x - \frac{1}{2},\ F'(x) = \cos 2x$

39. (a) $\sin\sqrt{x}$ (b) $\sqrt{1+\cos^2 x}$

40. (a) $\dfrac{1}{1+\sqrt{x}}$ (b) $\dfrac{1}{1+x+x^2}$

41. $-\dfrac{x}{\cos x}$

42. $|u|$

43. $F'(x) = \sqrt{3x^2+1}$, $F''(x) = \dfrac{3x}{\sqrt{3x^2+1}}$

(a) 0 (b) $\sqrt{13}$ (c) $6/\sqrt{13}$

44. $F'(x) = \dfrac{\cos x}{x^2+3}$, $F''(x) = \dfrac{-(x^2+3)\sin x - 2x\cos x}{(x^2+3)^2}$

(a) 0 (b) 1/3 (c) 0

45. (a) $F'(x) = \dfrac{x-3}{x^2+7} = 0$ when $x = 3$, which is a relative minimum, and hence the absolute minimum, by the first derivative test.

(b) increasing on $[3, +\infty)$, decreasing on $(-\infty, 3]$

(c) $F''(x) = \dfrac{7+6x-x^2}{(x^2+7)^2} = \dfrac{(7-x)(1+x)}{(x^2+7)^2}$; concave up on $(-1, 7)$, concave down on $(-\infty, -1)$ and on $(7, +\infty)$

46.

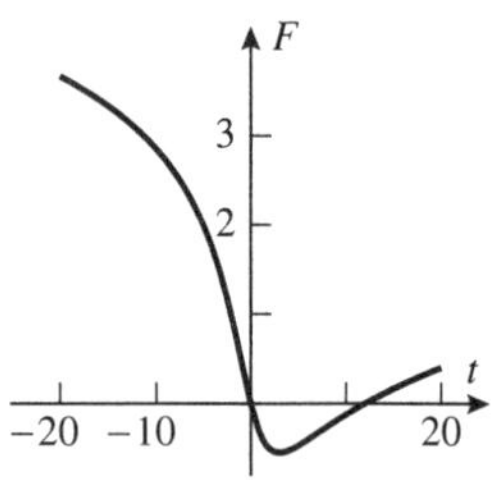

47. (a) $(0, +\infty)$ because f is continuous there and 1 is in $(0, +\infty)$

(b) at $x = 1$ because $F(1) = 0$

48. (a) $(-3, 3)$ because f is continuous there and 1 is in $(-3, 3)$

(b) at $x = 1$ because $F(1) = 0$

49. (a) $f_{\text{ave}} = \dfrac{1}{9}\displaystyle\int_0^9 x^{1/2}dx = 2$; $\sqrt{x^*} = 2$, $x^* = 4$

(b) $f_{\text{ave}} = \dfrac{1}{3}\displaystyle\int_{-1}^{2}(3x^2+2x+1)\,dx = \dfrac{1}{3}(x^3+x^2+x)\Big]_{-1}^{2} = 5$; $3x^{*2}+2x^*+1 = 5$, with solutions $x^* = -(1/3)(1 \pm \sqrt{13})$, but only $x^* = -(1/3)(1-\sqrt{13})$ lies in the interval $[-1, 2]$.

50. (a) $f_{\text{ave}} = \dfrac{1}{2\pi}\displaystyle\int_{-\pi}^{\pi}\sin x\,dx = 0$; $\sin x^* = 0$, $x^* = -\pi, 0, \pi$

(b) $f_{\text{ave}} = \dfrac{1}{2}\displaystyle\int_1^3 \dfrac{1}{x^2}dx = \dfrac{1}{3}$; $\dfrac{1}{(x^*)^2} = \dfrac{1}{3}$, $x^* = \sqrt{3}$

51. $\sqrt{2} \le \sqrt{x^3+2} \le \sqrt{29}$, so $3\sqrt{2} \le \int_0^3 \sqrt{x^3+2}dx \le 3\sqrt{29}$

52. Let $f(x) = x\sin x$, $f(0) = f(1) = 0$, $f'(x) = \sin x + x\cos x = 0$ when $x = -\tan x$, $x \approx 2.0288$, so f has an absolute maximum at $x \approx 2.0288$; $f(2.0288) \approx 1.8197$, so $0 \le x\sin x \le 1.82$ and $0 \le \int_0^\pi x\sin x dx \le 1.82\pi = 5.72$

53. **(a)** $\left[cF(x)\right]_a^b = cF(b) - cF(a) = c[F(b) - F(a)] = c\left[F(x)\right]_a^b$

(b) $$\begin{aligned}\left[F(x)+G(x)\right]_a^b &= [F(b)+G(b)] - [F(a)+G(a)]\\ &= [F(b)-F(a)] + [G(b)-G(a)] = F(x)\Big]_a^b + G(x)\Big]_a^b\end{aligned}$$

(c) $$\begin{aligned}\left[F(x)-G(x)\right]_a^b &= [F(b)-G(b)] - [F(a)-G(a)]\\ &= [F(b)-F(a)] - [G(b)-G(a)] = F(x)\Big]_a^b - G(x)\Big]_a^b\end{aligned}$$

54. Let f be continuous on a closed interval $[a,b]$ and let F be an antiderivative of f on $[a,b]$. By Theorem 4.8.2, $\dfrac{F(b)-F(a)}{b-a} = F'(x^*)$ for some x^* in (a,b). By Theorem 5.6.1, $\int_a^b f(x)\,dx = F(b) - F(a)$, i.e. $\int_a^b f(x)\,dx = F'(x^*)(b-a) = f(x^*)(b-a)$.

EXERCISE SET 5.7

1. **(a)** the increase in height in inches, during the first ten years

(b) the change in the radius in centimeters, during the time interval $t = 1$ to $t = 2$ seconds

(c) the change in the speed of sound in ft/s, during an increase in temperature from $t = 32°$F to $t = 100°$F

(d) the displacement of the particle in cm, during the time interval $t = t_1$ to $t = t_2$ seconds

2. **(a)** $\int_0^1 V(t)dt$ gal

(b) the change $f(x_1) - f(x_2)$ in the values of f over the interval

3. **(a)** $$\begin{aligned}\text{displ} &= s(3) - s(0)\\ &= \int_0^3 v(t)dt = \int_0^2 (1-t)dt + \int_2^3 (t-3)dt = (t - t^2/2)\Big]_0^2 + (t^2/2 - 3t)\Big]_2^3 = -1/2;\\ \text{dist} &= \int_0^3 |v(t)|dt = (t - t^2/2)\Big]_0^1 + (t^2/2 - t)\Big]_1^2 - (t^2/2 - 3t)\Big]_2^3 = 3/2\end{aligned}$$

(b) $$\begin{aligned}\text{displ} &= s(3) - s(0)\\ &= \int_0^3 v(t)dt = \int_0^1 tdt + \int_1^2 dt + \int_2^3 (5-2t)dt = t^2/2\Big]_0^1 + t\Big]_1^2 + (5t - t^2)\Big]_2^3 = 3/2;\\ \text{dist} &= \int_0^1 tdt + \int_1^2 dt + \int_2^{5/2} (5-2t)dt + \int_{5/2}^3 (2t-5)dt\\ &= t^2/2\Big]_0^1 + t\Big]_1^2 + (5t - t^2)\Big]_2^{5/2} + (t^2 - 5t)\Big]_{5/2}^3 = 2\end{aligned}$$

4. 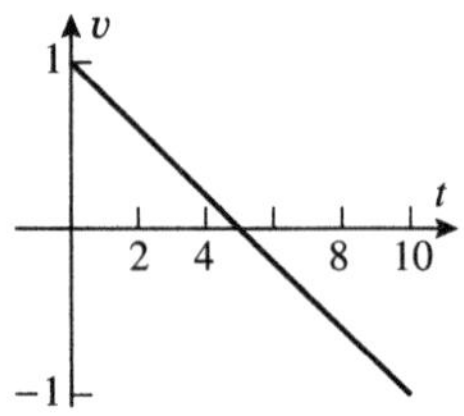

5. **(a)** $v(t) = 20 + \int_0^t a(u)du$; add areas of the small blocks to get

$v(4) \approx 20 + 1.4 + 3.0 + 4.7 + 6.2 = 35.3$ m/s

(b) $v(6) = v(4) + \int_4^6 a(u)du \approx 35.3 + 7.5 + 8.6 = 51.4$ m/s

6. $a > 0$ and therefore (Theorem 5.5.6(a)) $v > 0$, so the particle is always speeding up for $0 < t < 10$

7. **(a)** $s(t) = \int (t^3 - 2t^2 + 1)dt = \frac{1}{4}t^4 - \frac{2}{3}t^3 + t + C$,

$s(0) = \frac{1}{4}(0)^4 - \frac{2}{3}(0)^3 + 0 + C = 1$, $C = 1$, $s(t) = \frac{1}{4}t^4 - \frac{2}{3}t^3 + t + 1$

(b) $v(t) = \int 4\cos 2t\, dt = 2\sin 2t + C_1$, $v(0) = 2\sin 0 + C_1 = -1$, $C_1 = -1$,

$v(t) = 2\sin 2t - 1$, $s(t) = \int (2\sin 2t - 1)dt = -\cos 2t - t + C_2$,

$s(0) = -\cos 0 - 0 + C_2 = -3$, $C_2 = -2$, $s(t) = -\cos 2t - t - 2$

8. **(a)** $s(t) = \int (1 + \sin t)dt = t - \cos t + C$, $s(0) = 0 - \cos 0 + C = -3$, $C = -2$, $s(t) = t - \cos t - 2$

(b) $v(t) = \int (t^2 - 3t + 1)dt = \frac{1}{3}t^3 - \frac{3}{2}t^2 + t + C_1$,

$v(0) = \frac{1}{3}(0)^3 - \frac{3}{2}(0)^2 + 0 + C_1 = 0$, $C_1 = 0$, $v(t) = \frac{1}{3}t^3 - \frac{3}{2}t^2 + t$,

$s(t) = \int \left(\frac{1}{3}t^3 - \frac{3}{2}t^2 + t\right) dt = \frac{1}{12}t^4 - \frac{1}{2}t^3 + \frac{1}{2}t^2 + C_2$,

$s(0) = \frac{1}{12}(0)^4 - \frac{1}{2}(0)^3 + \frac{1}{2}(0)^2 + C_2 = 0$, $C_2 = 0$, $s(t) = \frac{1}{12}t^4 - \frac{1}{2}t^3 + \frac{1}{2}t^2$

9. **(a)** $s(t) = \int (2t - 3)dt = t^2 - 3t + C$, $s(1) = (1)^2 - 3(1) + C = 5$, $C = 7$, $s(t) = t^2 - 3t + 7$

(b) $v(t) = \int \cos t dt = \sin t + C_1$, $v(\pi/2) = 2 = 1 + C_1$, $C_1 = 1$, $v(t) = \sin t + 1$,

$s(t) = \int (\sin t + 1)dt = -\cos t + t + C_2$, $s(\pi/2) = 0 = \pi/2 + C_2$, $C_2 = -\pi/2$,

$s(t) = -\cos t + t - \pi/2$

10. **(a)** $s(t) = \int t^{2/3}dt = \frac{3}{5}t^{5/3} + C$, $s(8) = 0 = \frac{3}{5}32 + C$, $C = -\frac{96}{5}$, $s(t) = \frac{3}{5}t^{5/3} - \frac{96}{5}$

(b) $v(t) = \int \sqrt{t}dt = \frac{2}{3}t^{3/2} + C_1$, $v(4) = 1 = \frac{2}{3}8 + C_1$, $C_1 = -\frac{13}{3}$, $v(t) = \frac{2}{3}t^{3/2} - \frac{13}{3}$,

$s(t)=\int\left(\frac{2}{3}t^{3/2}-\frac{13}{3}\right)dt=\frac{4}{15}t^{5/2}-\frac{13}{3}t+C_2$, $s(4)=-5=\frac{4}{15}32-\frac{13}{3}4+C_2=-\frac{44}{5}+C_2$, $C_2=\frac{19}{5}$, $s(t)=\frac{4}{15}t^{5/2}-\frac{13}{3}t+\frac{19}{5}$

11. **(a)** displacement $=s(\pi/2)-s(0)=\int_0^{\pi/2}\sin t\,dt=-\cos t\Big]_0^{\pi/2}=1$ m

distance $=\int_0^{\pi/2}|\sin t|dt=1$ m

(b) displacement $=s(2\pi)-s(\pi/2)=\int_{\pi/2}^{2\pi}\cos t\,dt=\sin t\Big]_{\pi/2}^{2\pi}=-1$ m

distance $=\int_{\pi/2}^{2\pi}|\cos t|dt=-\int_{\pi/2}^{3\pi/2}\cos t\,dt+\int_{3\pi/2}^{2\pi}\cos t\,dt=3$ m

12. **(a)** displacement $=s(6)-s(0)=\int_0^6(2t-4)dt=(t^2-4t)\Big]_0^6=12$ m

distance $=\int_0^6|2t-4|dt=\int_0^2(4-2t)dt+\int_2^6(2t-4)dt=(4t-t^2)\Big]_0^2+(t^2-4t)\Big]_2^6=20$ m

(b) displacement $=\int_0^5|t-3|dt=\int_0^3-(t-3)dt+\int_3^5(t-3)dt=13/2$ m

distance $=\int_0^5|t-3|dt=13/2$ m

13. **(a)** $v(t)=t^3-3t^2+2t=t(t-1)(t-2)$

displacement $=\int_0^3(t^3-3t^2+2t)dt=9/4$ m

distance $=\int_0^3|v(t)|dt=\int_0^1v(t)dt+\int_1^2-v(t)dt+\int_2^3v(t)dt=11/4$ m

(b) displacement $=\int_0^3(\sqrt{t}-2)dt=2\sqrt{3}-6$ m

distance $=\int_0^3|v(t)|dt=-\int_0^3v(t)dt=6-2\sqrt{3}$ m

14. **(a)** displacement $-\int_1^3(\frac{1}{2}-\frac{1}{t^2})dt=1/3$ m

distance $=\int_1^3|v(t)|dt=-\int_1^{\sqrt{2}}v(t)dt+\int_{\sqrt{2}}^3v(t)dt=10/3-2\sqrt{2}$ m

(b) displacement $=\int_4^9 3t^{-1/2}dt=6$ m

distance $=\int_4^9|v(t)|dt=\int_4^9v(t)dt=6$ m

15. $v(t)=-2t+3$

displacement $=\int_1^4(-2t+3)dt=-6$ m

distance $=\int_1^4|-2t+3|dt=\int_1^{3/2}(-2t+3)dt+\int_{3/2}^4(2t-3)dt=13/2$ m

16. $v(t) = \frac{1}{2}t^2 - 2t$

$$\text{displacement} = \int_1^5 \left(\frac{1}{2}t^2 - 2t\right) dt = -10/3 \text{ m}$$

$$\text{distance} = \int_1^5 \left|\frac{1}{2}t^2 - 2t\right| dt = \int_1^4 -\left(\frac{1}{2}t^2 - 2t\right) dt + \int_4^5 \left(\frac{1}{2}t^2 - 2t\right) dt = 17/3 \text{ m}$$

17. $v(t) = \frac{2}{5}\sqrt{5t+1} + \frac{8}{5}$

$$\text{displacement} = \int_0^3 \left(\frac{2}{5}\sqrt{5t+1} + \frac{8}{5}\right) dt = \frac{4}{75}(5t+1)^{3/2} + \frac{8}{5}t\Big]_0^3 = 204/25 \text{ m}$$

$$\text{distance} = \int_0^3 |v(t)|dt = \int_0^3 v(t)dt = 204/25 \text{ m}$$

18. $v(t) = -\cos t + 2$

$$\text{displacement} = \int_{\pi/4}^{\pi/2} (-\cos t + 2)dt = (\pi + \sqrt{2} - 2)/2 \text{ m}$$

$$\text{distance} = \int_{\pi/4}^{\pi/2} |-\cos t + 2|dt = \int_{\pi/4}^{\pi/2} (-\cos t + 2)dt = (\pi + \sqrt{2} - 2)/2 \text{ m}$$

19. **(a)** $s = \int \sin \frac{1}{2}\pi t\, dt = -\frac{2}{\pi}\cos\frac{1}{2}\pi t + C$

$s = 0$ when $t = 0$ which gives $C = \frac{2}{\pi}$ so $s = -\frac{2}{\pi}\cos\frac{1}{2}\pi t + \frac{2}{\pi}$.

$a = \frac{dv}{dt} = \frac{\pi}{2}\cos\frac{1}{2}\pi t$. When $t = 1 : s = 2/\pi$, $v = 1$, $|v| = 1$, $a = 0$.

(b) $v = -3\int t\, dt = -\frac{3}{2}t^2 + C_1$, $v = 0$ when $t = 0$ which gives $C_1 = 0$ so $v = -\frac{3}{2}t^2$

$s = -\frac{3}{2}\int t^2 dt = -\frac{1}{2}t^3 + C_2$, $s = 1$ when $t = 0$ which gives $C_2 = 1$ so $s = -\frac{1}{2}t^3 + 1$.

When $t = 1 : s = 1/2$, $v = -3/2$, $|v| = 3/2$, $a = -3$.

20. **(a)** negative, because v is decreasing

(b) speeding up when $av > 0$, so $2 < t < 5$; slowing down when $1 < t < 2$

(c) negative, because the area between the graph of $v(t)$ and the t-axis appears to be greater where $v < 0$ compared to where $v > 0$

21. $A = A_1 + A_2 = \int_0^1 (1 - x^2)dx + \int_1^3 (x^2 - 1)dx = 2/3 + 20/3 = 22/3$

22. $A = A_1 + A_2 = \int_0^\pi \sin x dx - \int_\pi^{3\pi/2} \sin x dx = 2 + 1 = 3$

23. $A = A_1 + A_2 = \int_{-1}^0 \left[1 - \sqrt{x+1}\right] dx + \int_0^1 [\sqrt{x+1} - 1]\, dx$

$$= \left(x - \frac{2}{3}(x+1)^{3/2}\right)\Big]_{-1}^0 + \left(\frac{2}{3}(x+1)^{3/2} - x\right)\Big]_0^1 = -\frac{2}{3} + 1 + \frac{4\sqrt{2}}{3} - 1 - \frac{2}{3} = 4\frac{\sqrt{2}-1}{3}$$

24. $A = A_1 + A_2 = \int_{1/2}^{1} \frac{1-x^2}{x^2}\,dx + \int_{1}^{2} \frac{x^2-1}{x^2}\,dx = \left(-\frac{1}{x} - x\right)\Big|_{1/2}^{1} + \left(x + \frac{1}{x}\right)\Big|_{1}^{2}$

$= -2 + 2 + \frac{1}{2} + 2 + \frac{1}{2} - 2 = 1$

25. $s(t) = \frac{20}{3}t^3 - 50t^2 + 50t + s_0$, $s(0) = 0$ gives $s_0 = 0$, so $s(t) = \frac{20}{3}t^3 - 50t^2 + 50t$, $a(t) = 40t - 100$

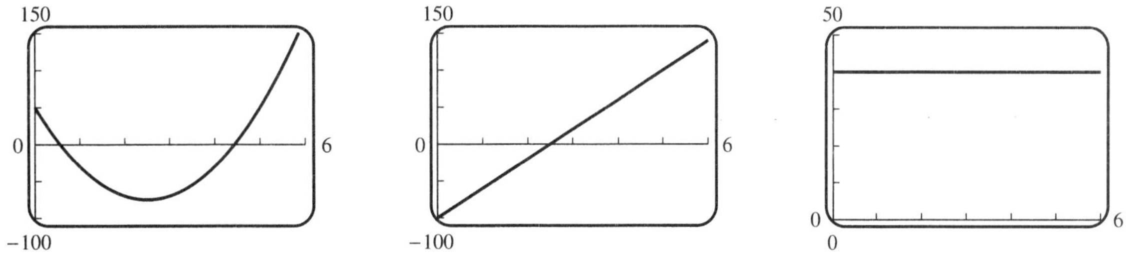

26. $v(t) = 2t^2 - 30t + v_0$, $v(0) = 3 = v_0$, so $v(t) = 2t^2 - 30t + 3$, $s(t) = \frac{2}{3}t^3 - 15t^2 + 3t + s_0$, $s(0) = -5 = s_0$, so $s(t) = \frac{2}{3}t^3 - 15t^2 + 3t - 5$

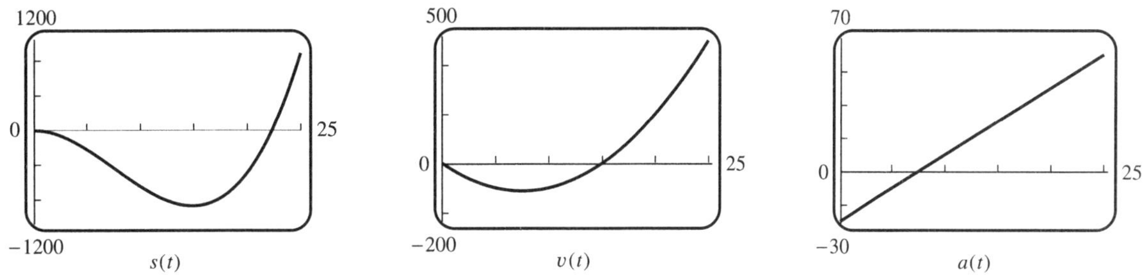

27. **(a)** From the graph the velocity is at first positive, but then turns negative, then positive again. The displacement, which is the cumulative area from $x = 0$ to $x = 5$, starts positive, turns negative, and then turns positive again.

(b) displ $= 5/2 - \sin 5 + 5\cos 5$

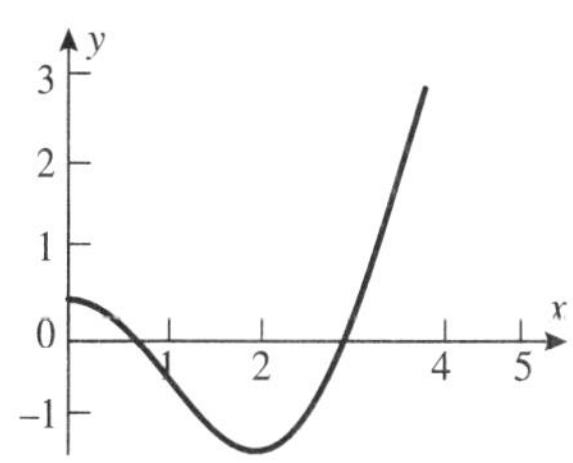

28. **(a)** If $t_0 < 1$ then the area between the velocity curve and the t-axis, between $t = 0$ and $t = t_0$, will always be positive, so the displacement will be positive.

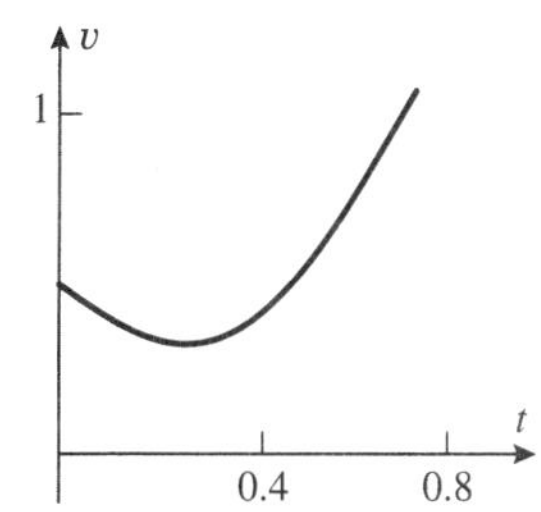

(b) displ $= \frac{\pi^2 + 4}{2\pi^2}$

29. **(a)** $a(t) = \begin{cases} 0, & t < 4 \\ -10, & t > 4 \end{cases}$

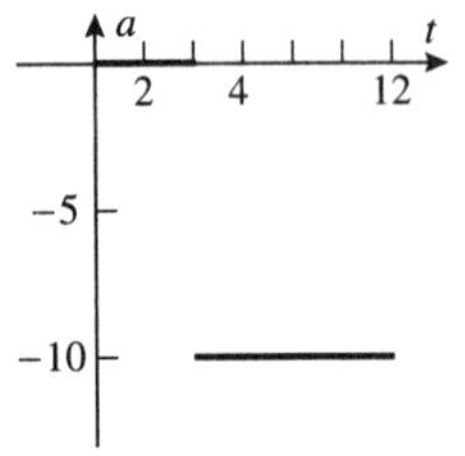

(b) $v(t) = \begin{cases} 25, & t < 4 \\ 65 - 10t, & t > 4 \end{cases}$

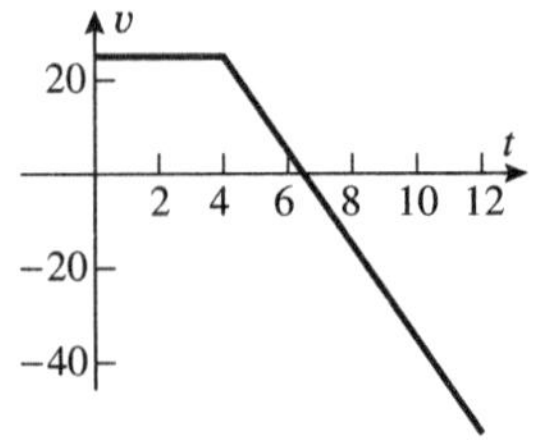

(c) $x(t) = \begin{cases} 25t, & t < 4 \\ 65t - 5t^2 - 80, & t > 4 \end{cases}$, so $x(8) = 120$, $x(12) = -20$

(d) $x(6.5) = 131.25$

30. **(a)** From (9) $t = \dfrac{v - v_0}{a}$; from that and (8)

$s - s_0 = v_0\dfrac{v - v_0}{a} + \dfrac{1}{2}a\dfrac{(v - v_0)^2}{a^2}$; multiply through by a to get

$a(s - s_0) = v_0(v - v_0) + \dfrac{1}{2}(v - v_0)^2 = (v - v_0)\left[v_0 + \dfrac{1}{2}(v - v_0)\right] = \dfrac{1}{2}(v^2 - v_0^2)$. Thus

$a = \dfrac{v^2 - v_0^2}{2(s - s_0)}$.

(b) Put the last result of Part (a) into the first equation of Part (a) to obtain

$t = \dfrac{v - v_0}{a} = (v - v_0)\dfrac{2(s - s_0)}{v^2 - v_0^2} = \dfrac{2(s - s_0)}{v + v_0}$.

(c) From (9) $v_0 = v - at$; use this in (8) to get

$s - s_0 = (v - at)t + \dfrac{1}{2}at^2 = vt - \dfrac{1}{2}at^2$

This expression contains no v_0 terms and so differs from (8).

31. **(a)** $a = -1$ mi/h/s $= -22/15$ ft/s^2 **(b)** $a = 30$ km/h/min $= 1/7200$ km/s^2

32. Take $t = 0$ when deceleration begins, then $a = -10$ so $v = -10t + C_1$, but $v = 88$ when $t = 0$ which gives $C_1 = 88$ thus $v = -10t + 88$, $t \geq 0$

(a) $v = 45$ mi/h $= 66$ ft/s, $66 = -10t + 88$, $t = 2.2$ s

(b) $v = 0$ (the car is stopped) when $t = 8.8$ s

$s = \int v\,dt = \int(-10t + 88)dt = -5t^2 + 88t + C_2$, and taking $s = 0$ when $t = 0$, $C_2 = 0$ so $s = -5t^2 + 88t$. At $t = 8.8$, $s = 387.2$. The car travels 387.2 ft before coming to a stop.

33. $a = a_0$ ft/s^2, $v = a_0t + v_0 = a_0t + 132$ ft/s, $s = a_0t^2/2 + 132t + s_0 = a_0t^2/2 + 132t$ ft; $s = 200$ ft when $v = 88$ ft/s. Solve $88 = a_0t + 132$ and $200 = a_0t^2/2 + 132t$ to get $a_0 = -\dfrac{121}{5}$ when $t = \dfrac{20}{11}$, so $s = -12.1t^2 + 132t$, $v = -\dfrac{121}{5}t + 132$.

(a) $a_0 = -\dfrac{121}{5}$ ft/s^2 **(b)** $v = 55$ mi/h $= \dfrac{242}{3}$ ft/s when $t = \dfrac{70}{33}$ s

(c) $v = 0$ when $t = \dfrac{60}{11}$ s

34. $dv/dt = 3$, $v = 3t + C_1$, but $v = v_0$ when $t = 0$ so $C_1 = v_0$, $v = 3t + v_0$. From $ds/dt = v = 3t + v_0$ we get $s = 3t^2/2 + v_0 t + C_2$ and, with $s = 0$ when $t = 0$, $C_2 = 0$ so $s = 3t^2/2 + v_0 t$. $s = 40$ when $t = 4$ thus $40 = 3(4)^2/2 + v_0(4)$, $v_0 = 4$ m/s

35. Suppose $s = s_0 = 0$, $v = v_0 = 0$ at $t = t_0 = 0$; $s = s_1 = 120$, $v = v_1$ at $t = t_1$; and $s = s_2$, $v = v_2 = 12$ at $t = t_2$. From Exercise 30(a),

$2.6 = a = \dfrac{v_1^2 - v_0^2}{2(s_1 - s_0)}$, $v_1^2 = 2as_1 = 5.2(120) = 624$. Applying the formula again,

$-1.5 = a = \dfrac{v_2^2 - v_1^2}{2(s_2 - s_1)}$, $v_2^2 = v_1^2 - 3(s_2 - s_1)$, so

$$s_2 = s_1 - (v_2^2 - v_1^2)/3 = 120 - (144 - 624)/3 = 280 \text{ m}.$$

36. $a(t) = \begin{cases} 4, & t < 2 \\ 0, & t > 2 \end{cases}$, so, with $v_0 = 0$, $v(t) = \begin{cases} 4t, & t < 2 \\ 8, & t > 2 \end{cases}$ and,

since $s_0 = 0$, $s(t) = \begin{cases} 2t^2, & t < 2 \\ 8t - 8, & t > 2 \end{cases}$ $s = 100$ when $8t - 8 = 100$, $t = 108/8 = 13.5$ s

37. The truck's velocity is $v_T = 50$ and its position is $s_T = 50t + 5000$. The car's acceleration is $a_C = 2$, so $v_C = 2t$, $s_C = t^2$ (initial position and initial velocity of the car are both zero). $s_T = s_C$ when $50t + 5000 = t^2$, $t^2 - 50t - 5000 = (t+50)(t-100) = 0$, $t = 100$ s and $s_C = s_T = t^2 = 10{,}000$ ft.

38. Let $t = 0$ correspond to the time when the leader is 100 m from the finish line; let $s = 0$ correspond to the finish line. Then $v_C = 12$, $s_C = 12t - 115$; $a_L = 0.5$ for $t > 0$, $v_L = 0.5t + 8$, $s_L = 0.25t^2 + 8t - 100$. $s_C = 0$ at $t = 115/12 \approx 9.58$ s, and $s_L = 0$ at $t = -16 + 4\sqrt{41} \approx 9.61$, so the challenger wins.

39. $s = 0$ and $v = 112$ when $t = 0$ so $v(t) = -32t + 112$, $s(t) = -16t^2 + 112t$

(a) $v(3) = 16$ ft/s, $v(5) = -48$ ft/s

(b) $v = 0$ when the projectile is at its maximum height so $-32t + 112 = 0$, $t = 7/2$ s, $s(7/2) = -16(7/2)^2 + 112(7/2) = 196$ ft.

(c) $s = 0$ when it reaches the ground so $-16t^2 + 112t = 0$, $-16t(t - 7) = 0$, $t = 0, 7$ of which $t = 7$ is when it is at ground level on its way down. $v(7) = -112$, $|v| = 112$ ft/s.

40. $s = 112$ when $t = 0$ so $s(t) = -16t^2 + v_0 t + 112$. But $s = 0$ when $t = 2$ thus $-16(2)^2 + v_0(2) + 112 = 0$, $v_0 = -24$ ft/s.

41. **(a)** $s(t) = 0$ when it hits the ground, $s(t) = -16t^2 + 16t = -16t(t - 1) = 0$ when $t = 1$ s.

(b) The projectile moves upward until it gets to its highest point where $v(t) = 0$, $v(t) = -32t + 16 = 0$ when $t = 1/2$ s.

42. **(a)** $s(t) = 0$ when the rock hits the ground, $s(t) = -16t^2 + 555 = 0$ when $t = \sqrt{555}/4$ s

(b) $v(t) = -32t$, $v(\sqrt{555}/4) = -8\sqrt{555}$, the speed at impact is $8\sqrt{555}$ ft/s

43. **(a)** $s(t) = 0$ when the package hits the ground, $s(t) = -16t^2 + 20t + 200 = 0$ when $t = (5 + 5\sqrt{33})/8$ s

(b) $v(t) = -32t + 20$, $v[(5 + 5\sqrt{33})/8] = -20\sqrt{33}$, the speed at impact is $20\sqrt{33}$ ft/s

44. **(a)** $s(t) = 0$ when the stone hits the ground, $s(t) = -16t^2 - 96t + 112 = -16(t^2 + 6t - 7) = -16(t + 7)(t - 1) = 0$ when $t = 1$ s

(b) $v(t) = -32t - 96$, $v(1) = -128$, the speed at impact is 128 ft/s

45. $s(t) = -4.9t^2 + 49t + 150$ and $v(t) = -9.8t + 49$

(a) the projectile reaches its maximum height when $v(t) = 0$, $-9.8t + 49 = 0$, $t = 5$ s

(b) $s(5) = -4.9(5)^2 + 49(5) + 150 = 272.5$ m

(c) the projectile reaches its starting point when $s(t) = 150$, $-4.9t^2 + 49t + 150 = 150$, $-4.9t(t - 10) = 0$, $t = 10$ s

(d) $v(10) = -9.8(10) + 49 = -49$ m/s

(e) $s(t) = 0$ when the projectile hits the ground, $-4.9t^2 + 49t + 150 = 0$ when (use the quadratic formula) $t \approx 12.46$ s

(f) $v(12.46) = -9.8(12.46) + 49 \approx -73.1$, the speed at impact is about 73.1 m/s

46. take $s = 0$ at the water level and let h be the height of the bridge, then $s = h$ and $v = 0$ when $t = 0$ so $s(t) = -16t^2 + h$

(a) $s = 0$ when $t = 4$ thus $-16(4)^2 + h = 0$, $h = 256$ ft

(b) First, find how long it takes for the stone to hit the water (find t for $s = 0$) : $-16t^2 + h = 0$, $t = \sqrt{h}/4$. Next, find how long it takes the sound to travel to the bridge: this time is $h/1080$ because the speed is constant at 1080 ft/s. Finally, use the fact that the total of these two times must be 4 s: $\dfrac{h}{1080} + \dfrac{\sqrt{h}}{4} = 4$, $h + 270\sqrt{h} = 4320$, $h + 270\sqrt{h} - 4320 = 0$, and by the quadratic formula $\sqrt{h} = \dfrac{-270 \pm \sqrt{(270)^2 + 4(4320)}}{2}$, reject the negative value to get $\sqrt{h} \approx 15.15$, $h \approx 229.5$ ft.

47. $g = 9.8/6 = 4.9/3$ m/s^2, so $v = -(4.9/3)t$, $s = -(4.9/6)t^2 + 5$, $s = 0$ when $t = \sqrt{30/4.9}$ and $v = -(4.9/3)\sqrt{30/4.9} \approx -4.04$, so the speed of the module upon landing is 4.04 m/s

48. $s(t) = -\frac{1}{2}gt^2 + v_0t$; $s = 1000$ when $v = 0$, so $0 = v = -gt + v_0$, $t = v_0/g$,
$1000 = s(v_0/g) = -\frac{1}{2}g(v_0/g)^2 + v_0(v_0/g) = \frac{1}{2}v_0^2/g$, so $v_0^2 = 2000g$, $v_0 = \sqrt{2000g}$.
The initial velocity on the Earth would have to be $\sqrt{6}$ times faster than that on the Moon.

49. $f_{\text{ave}} = \dfrac{1}{3-1}\displaystyle\int_1^3 3x\,dx = \frac{3}{4}x^2\Big]_1^3 = 6$

50. $f_{\text{ave}} = \dfrac{1}{2-(-1)}\displaystyle\int_{-1}^2 x^2\,dx = \frac{1}{9}x^3\Big]_{-1}^2 = 1$

51. $f_{\text{ave}} = \dfrac{1}{\pi - 0}\displaystyle\int_0^\pi \sin x\,dx = -\frac{1}{\pi}\cos x\Big]_0^\pi = 2/\pi$

52. $f_{\text{ave}} = \dfrac{1}{\pi - 0}\displaystyle\int_0^\pi \cos x\,dx = \frac{1}{\pi}\sin x\Big]_0^\pi = 0$

53. **(a)** $f_{\text{ave}} = \dfrac{1}{2-0}\displaystyle\int_0^2 x^2 dx = 4/3$

(b) $(x^*)^2 = 4/3, x^* = \pm 2/\sqrt{3}$, but only $2/\sqrt{3}$ is in $[0, 2]$

(c)

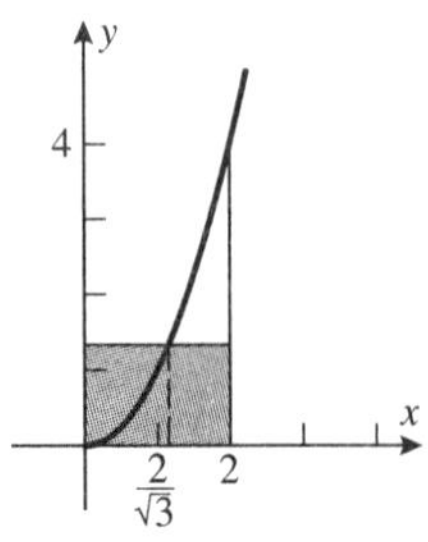

54. (a) $f_{\text{ave}} = \dfrac{1}{4-0}\displaystyle\int_0^4 2x\,dx = 4$ (b) $2x^* = 4,\ x^* = 2$

(c) 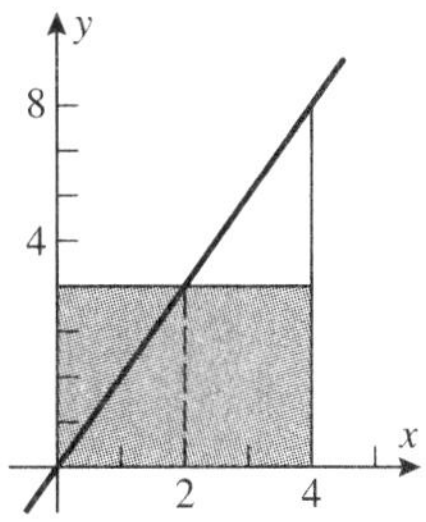

55. (a) $v_{\text{ave}} = \dfrac{1}{4-1}\displaystyle\int_1^4 (3t^3+2)dt = \dfrac{1}{3}\dfrac{789}{4} = \dfrac{263}{4}$

(b) $v_{\text{ave}} = \dfrac{s(4)-s(1)}{4-1} = \dfrac{100-7}{3} = 31$

56. (a) $a_{\text{ave}} = \dfrac{1}{5-0}\displaystyle\int_0^5 (t+1)dt = 7/2$

(b) $a_{\text{ave}} = \dfrac{v(\pi/4)-v(0)}{\pi/4-0} = \dfrac{\sqrt{2}/2-1}{\pi/4} = (2\sqrt{2}-4)/\pi$

57. time to fill tank = (volume of tank)/(rate of filling) = $[\pi(3)^2 5]/(1) = 45\pi$, weight of water in tank at time $t = (62.4)$ (rate of filling)(time) $= 62.4t$,

$\text{weight}_{\text{ave}} = \dfrac{1}{45\pi}\displaystyle\int_0^{45\pi} 62.4t\,dt = 1404\pi$ lb

58. (a) If x is the distance from the cooler end, then the temperature is $T(x) = (15+1.5x)^\circ$ C, and $T_{\text{ave}} = \dfrac{1}{10-0}\displaystyle\int_0^{10} (15+1.5x)dx = 22.5^\circ$ C

(b) By the Mean-Value Theorem for Integrals there exists x^* in $[0, 10]$ such that $f(x^*) = \dfrac{1}{10-0}\displaystyle\int_0^{10} (15+1.5x)dx = 22.5$, $15+1.5x^* = 22.5$, $x^* = 5$

59. (a) amount of water = (rate of flow)(time) = $4t$ gal, total amount = $4(30) = 120$ gal

(b) amount of water $= \displaystyle\int_0^{60} (4+t/10)dt = 420$ gal

(c) amount of water $= \displaystyle\int_0^{120} (10+\sqrt{t})dt = 1200 + 160\sqrt{30} \approx 2076.36$ gal

60. (a) The maximum value of R occurs at 4:30 P.M. when $t = 0$.

(b) $\displaystyle\int_0^{60} 100(1-0.0001t^2)dt = 5280$ cars

61. (a) $\displaystyle\int_a^b [f(x) - f_{\text{ave}}]\,dx = \int_a^b f(x)dx - \int_a^b f_{\text{ave}}dx = \int_a^b f(x)dx - f_{\text{ave}}(b-a) = 0$

because $f_{\text{ave}}(b-a) = \displaystyle\int_a^b f(x)dx$

(b) no, because if $\int_a^b [f(x) - c]dx = 0$ then $\int_a^b f(x)dx - c(b-a) = 0$ so $c = \frac{1}{b-a}\int_a^b f(x)dx = f_{\text{ave}}$ is the only value

EXERCISE SET 5.8

1. **(a)** $\int_1^3 u^7 du$ **(b)** $-\frac{1}{2}\int_7^4 u^{1/2}du$ **(c)** $\frac{1}{\pi}\int_{-\pi}^{\pi} \sin u\, du$ **(d)** $\int_{-3}^{0}(u+5)u^{20}du$

2. **(a)** $\frac{1}{2}\int_{-3}^{7} u^8\, du$ **(b)** $\int_{3/2}^{5/2}\frac{1}{\sqrt{u}}du$

(c) $\int_0^1 u^2 du$ **(d)** $\frac{1}{2}\int_3^4 (u-3)u^{1/2}du$

3. $u = 2x+1$, $\frac{1}{2}\int_1^3 u^4 du = \frac{1}{10}u^5\Big]_1^3 = 121/5$, or $\frac{1}{10}(2x+1)^5\Big]_0^1 = 121/5$

4. $u = 4x-2$, $\frac{1}{4}\int_2^6 u^3 du = \frac{1}{16}u^4\Big]_2^6 = 80$, or $\frac{1}{16}(4x-2)^4\Big]_1^2 = 80$

5. $u = 1-2x$, $-\frac{1}{2}\int_3^1 u^3 du = -\frac{1}{8}u^4\Big]_3^1 = 10$, or $-\frac{1}{8}(1-2x)^4\Big]_{-1}^0 = 10$

6. $u = 4-3x$, $-\frac{1}{3}\int_1^{-2} u^8 du = -\frac{1}{27}u^9\Big]_1^{-2} = 19$, or $-\frac{1}{27}(4-3x)^9\Big]_1^2 = 19$

7. $u = 1+x$, $\int_1^9 (u-1)u^{1/2}du = \int_1^9 (u^{3/2} - u^{1/2})du = \frac{2}{5}u^{5/2} - \frac{2}{3}u^{3/2}\Big]_1^9 = 1192/15$,
or $\frac{2}{5}(1+x)^{5/2} - \frac{2}{3}(1+x)^{3/2}\Big]_0^8 = 1192/15$

8. $u = 4-x$, $\int_9^4 (u-4)u^{1/2}du = \int_9^4 (u^{3/2} - 4u^{1/2})du = \frac{2}{5}u^{5/2} - \frac{8}{3}u^{3/2}\Big]_9^4 = -506/15$
or $\frac{2}{5}(4-x)^{5/2} - \frac{8}{3}(4-x)^{3/2}\Big]_{-5}^0 = -506/15$

9. $u = x/2$, $8\int_0^{\pi/4} \sin u\, du = -8\cos u\Big]_0^{\pi/4} = 8 - 4\sqrt{2}$, or $-8\cos(x/2)\Big]_0^{\pi/2} = 8 - 4\sqrt{2}$

10. $u = 3x$, $\frac{2}{3}\int_0^{\pi/2} \cos u\, du = \frac{2}{3}\sin u\Big]_0^{\pi/2} = 2/3$, or $\frac{2}{3}\sin 3x\Big]_0^{\pi/6} = 2/3$

11. $u = x^2+2$, $\frac{1}{2}\int_6^3 u^{-3}du = -\frac{1}{4u^2}\Big]_6^3 = -1/48$, or $-\frac{1}{4}\frac{1}{(x^2+2)^2}\Big]_{-2}^{-1} = -1/48$

12. $u = \frac{1}{4}x - \frac{1}{4},\ 4\int_{-\pi/4}^{\pi/4} \sec^2 u\, du \ = 4\tan u\Big]_{-\pi/4}^{\pi/4} = 8$, or $4\tan\left(\frac{1}{4}x - \frac{1}{4}\right)\Big]_{1-\pi}^{1+\pi} = 8$

13. $\frac{1}{3}\int_0^5 \sqrt{25-u^2}\, du = \frac{1}{3}\left[\frac{1}{4}\pi(5)^2\right] = \frac{25}{12}\pi$

14. $\frac{1}{2}\int_0^4 \sqrt{16-u^2}\, du = \frac{1}{2}\left[\frac{1}{4}\pi(4)^2\right] = 2\pi$

15. $-\frac{1}{2}\int_1^0 \sqrt{1-u^2}\, du = \frac{1}{2}\int_0^1 \sqrt{1-u^2}\, du = \frac{1}{2}\cdot\frac{1}{4}[\pi(1)^2] = \pi/8$

16. $\int_{-2}^2 \sqrt{4-u^2}\, du = \frac{1}{2}[\pi(2)^2] = 2\pi$

17. $\int_0^1 \sin\pi x dx = -\frac{1}{\pi}\cos\pi x\Big]_0^1 = -\frac{1}{\pi}(-1-1) = 2/\pi$

18. $A = \int_0^{\pi/8} 3\ \cos 2x\, dx = \frac{3}{2}\sin 2x\Big]_0^{\pi/8} = 3\sqrt{2}/4$

19. $\int_3^7 (x+5)^{-2} dx = -(x+5)^{-1}\Big]_3^7 = -\frac{1}{12} + \frac{1}{8} = \frac{1}{24}$

20. $A = \int_0^1 \frac{dx}{(3x+1)^2} = -\frac{1}{3(3x+1)}\Big]_0^1 = \frac{1}{4}$

21. $\frac{1}{2-0}\int_0^2 \frac{x}{(5x^2+1)^2}\, dx = -\frac{1}{2}\frac{1}{10}\frac{1}{5x^2+1}\Bigg|_0^2 = \frac{1}{21}$

22. $f_{\text{ave}} = \frac{1}{1/4-(-1/4)}\int_{-1/4}^{1/4} \sec^2\pi x dx = \frac{2}{\pi}\tan\pi x\Big]_{-1/4}^{1/4} = \frac{4}{\pi}$

23. $\frac{2}{3}(3x+1)^{1/2}\Big]_0^1 = 2/3$

24. $\frac{2}{15}(5x-1)^{3/2}\Big]_1^2 = 38/15$

25. $\frac{2}{3}(x^3+9)^{1/2}\Big]_{-1}^1 = \frac{2}{3}(\sqrt{10} - 2\sqrt{2})$

26. $\frac{1}{10}(t^3+1)^{20}\Big]_{-1}^0 = 1/10$

27. $u = x^2+4x+7,\ \frac{1}{2}\int_{12}^{28} u^{-1/2} du = u^{1/2}\Big]_{12}^{28} = \sqrt{28} - \sqrt{12} = 2(\sqrt{7} - \sqrt{3})$

28. $\int_1^2 \frac{1}{(x-3)^2} dx = -\frac{1}{x-3}\Big]_1^2 = 1/2$

29. $\frac{1}{2}\sin^2 x\Big]_{-3\pi/4}^{\pi/4} = 0$

30. $\frac{2}{3}(\tan x)^{3/2}\Big]_0^{\pi/4} = 2/3$

31. $\frac{5}{2}\sin(x^2)\Big]_0^{\sqrt{\pi}} = 0$

32. $u = \sqrt{x},\ 2\int_\pi^{2\pi} \sin u\, du = -2\cos u\Big]_\pi^{2\pi} = -4$

33. $u = 3\theta$, $\frac{1}{3}\int_{\pi/4}^{\pi/3} \sec^2 u\, du = \frac{1}{3}\tan u\Big]_{\pi/4}^{\pi/3} = (\sqrt{3}-1)/3$

34. $u = \sin 3\theta$, $\frac{1}{3}\int_0^{-1} u^2 du = \frac{1}{9}u^3\Big]_0^{-1} = -1/9$

35. $u = 4-3y$, $y = \frac{1}{3}(4-u)$, $dy = -\frac{1}{3}du$

$$-\frac{1}{27}\int_4^1 \frac{16-8u+u^2}{u^{1/2}}du = \frac{1}{27}\int_1^4 (16u^{-1/2} - 8u^{1/2} + u^{3/2})du$$
$$= \frac{1}{27}\left[32u^{1/2} - \frac{16}{3}u^{3/2} + \frac{2}{5}u^{5/2}\right]_1^4 = 106/405$$

36. $u = 5+x$, $\int_4^9 \frac{u-5}{\sqrt{u}}du = \int_4^9 (u^{1/2} - 5u^{-1/2})du = \frac{2}{3}u^{3/2} - 10u^{1/2}\Big]_4^9 = 8/3$

37. **(b)** $\int_0^{\pi/6} \sin^4 x(1-\sin^2 x)\cos x\, dx = \left(\frac{1}{5}\sin^5 x - \frac{1}{7}\sin^7 x\right)\Big|_0^{\pi/6} = \frac{1}{160} - \frac{1}{896} = \frac{23}{4480}$

38. **(b)** $\int_{-\pi/4}^{\pi/4} \tan^2 x(\sec^2 x - 1)\, dx = \frac{1}{3}\tan^3 x\Big|_{-\pi/4}^{\pi/4} - \int_{-\pi/4}^{\pi/4}(\sec^2 x - 1)\, dx$

$$= \frac{2}{3} + (-\tan x + x)\Big|_{-\pi/4}^{\pi/4} = \frac{2}{3} - 2 + \frac{\pi}{2} = -\frac{4}{3} + \frac{\pi}{2}$$

39. **(a)** $u = 3x+1$, $\frac{1}{3}\int_1^4 f(u)du = 5/3$ **(b)** $u = 3x$, $\frac{1}{3}\int_0^9 f(u)du = 5/3$

(c) $u = x^2$, $1/2\int_4^0 f(u)du = -1/2\int_0^4 f(u)du = -1/2$

40. $u = 1-x$, $\int_0^1 x^m(1-x)^n dx = -\int_1^0 (1-u)^m u^n du = \int_0^1 u^n(1-u)^m du = \int_0^1 x^n(1-x)^m dx$

41. $\sin x = \cos(\pi/2 - x)$,

$$\int_0^{\pi/2} \sin^n x\, dx = \int_0^{\pi/2} \cos^n(\pi/2 - x)dx = -\int_{\pi/2}^0 \cos^n u\, du \quad (u = \pi/2 - x)$$
$$= \int_0^{\pi/2} \cos^n u\, du = \int_0^{\pi/2} \cos^n x\, dx \quad \text{(by replacing } u \text{ by } x\text{)}$$

42. $u = 1-x$, $-\int_1^0 (1-u)u^n du = \int_0^1 (1-u)u^n du = \int_0^1 (u^n - u^{n+1})du = \frac{1}{n+1} - \frac{1}{n+2}$

$$= \frac{1}{(n+1)(n+2)}$$

43. **(a)** $V_{\text{rms}}^2 = \dfrac{1}{1/f - 0}\displaystyle\int_0^{1/f} V_p^2 \sin^2(2\pi ft)dt = \dfrac{1}{2}fV_p^2 \int_0^{1/f}[1 - \cos(4\pi ft)]dt$

$= \dfrac{1}{2}fV_p^2 \left[t - \dfrac{1}{4\pi f}\sin(4\pi ft)\right]\Bigg]_0^{1/f} = \dfrac{1}{2}V_p^2$, so $V_{\text{rms}} = V_p/\sqrt{2}$

(b) $V_p/\sqrt{2} = 120, V_p = 120\sqrt{2} \approx 169.7$ V

44. Let $u = t - x$, then $du = -dx$ and

$\displaystyle\int_0^t f(t-x)g(x)dx = -\int_t^0 f(u)g(t-u)du = \int_0^t f(u)g(t-u)du;$

the result follows by replacing u by x in the last integral.

45. **(a)** $I = -\displaystyle\int_a^0 \frac{f(a-u)}{f(a-u)+f(u)}du = \int_0^a \frac{f(a-u)+f(u)-f(u)}{f(a-u)+f(u)}du$

$= \displaystyle\int_0^a du - \int_0^a \frac{f(u)}{f(a-u)+f(u)}du, I = a - I$ so $2I = a, I = a/2$

(b) $3/2$ **(c)** $\pi/4$

46. $x = \dfrac{1}{u}$, $dx = -\dfrac{1}{u^2}du$, $I = \displaystyle\int_{-1}^1 \frac{1}{1+1/u^2}(-1/u^2)du = -\int_{-1}^1 \frac{1}{u^2+1}du = -I$ so $I = 0$ which is impossible because $\dfrac{1}{1+x^2}$ is positive on $[-1,1]$. The substitution $u = 1/x$ is not valid because u is not continuous for all x in $[-1,1]$.

47. $\displaystyle\int_0^1 \sin\pi x dx = 2/\pi$

49. **(a)** Let $u = -x$ then

$\displaystyle\int_{-a}^a f(x)dx = -\int_a^{-a} f(-u)du = \int_{-a}^a f(-u)du = -\int_{-a}^a f(u)du$

so, replacing u by x in the latter integral,

$\displaystyle\int_{-a}^a f(x)dx = -\int_{-a}^a f(x)dx,\ 2\int_{-a}^a f(x)dx = 0,\ \int_{-a}^a f(x)dx = 0$

The graph of f is symmetric about the origin so $\displaystyle\int_{-a}^0 f(x)dx$ is the negative of $\displaystyle\int_0^a f(x)dx$ thus $\displaystyle\int_{-a}^a f(x)dx = \int_{-a}^0 f(x)\,dx + \int_0^a f(x)dx = 0$

(b) $\displaystyle\int_{-a}^a f(x)dx = \int_{-a}^0 f(x)dx + \int_0^a f(x)dx$, let $u = -x$ in $\displaystyle\int_{-a}^0 f(x)dx$ to get

$\displaystyle\int_{-a}^0 f(x)dx = -\int_a^0 f(-u)du = \int_0^a f(-u)du = \int_0^a f(u)du = \int_0^a f(x)dx$

so $\displaystyle\int_{-a}^a f(x)dx = \int_0^a f(x)dx + \int_0^a f(x)dx = 2\int_0^a f(x)dx$

The graph of $f(x)$ is symmetric about the y-axis so there is as much signed area to the left of the y-axis as there is to the right.

50. **(a)** By Exercise 49(a), $\int_{-1}^{1} x\sqrt{\cos(x^2)}\,dx = 0$

(b) $u = x - \pi/2, du = dx, \sin(u + \pi/2) = \sin u, \cos(u + \pi/2) = -\sin u$

$$\int_0^{\pi} \sin^8 x \cos^5 x\,dx = \int_{-\pi/2}^{\pi/2} \sin^8 u(-\sin^5 u)\,du = -\int_{-\pi/2}^{\pi/2} \sin^{13} u\,du = 0 \text{ by Exercise 49(a).}$$

CHAPTER 5 SUPPLEMENTARY EXERCISES

5. If the acceleration $a =$ const, then $v(t) = at + v_0$, $s(t) = \frac{1}{2}at^2 + v_0 t + s_0$.

6. **(a)** Divide the base into n equal subintervals. Above each subinterval choose the lowest and highest points on the curved top. Draw a rectangle above the subinterval going through the lowest point, and another through the highest point. Add the rectangles that go through the lowest points to obtain a lower estimate of the area; add the rectangles through the highest points to obtain an upper estimate of the area.

(b) $n = 10$: 25.0 cm, 22.4 cm

(c) $n = 20$: 24.4 cm, 23.1 cm

7. **(a)** $\frac{1}{2} + \frac{1}{4} = \frac{3}{4}$ **(b)** $-1 - \frac{1}{2} = -\frac{3}{2}$

(c) $5\left(-1 - \frac{3}{4}\right) = -\frac{35}{4}$ **(d)** -2

(e) not enough information **(f)** not enough information

8. **(a)** $\frac{1}{2} + 2 = \frac{5}{2}$ **(b)** not enough information

(c) not enough information **(d)** $4(2) - 3\frac{1}{2} = \frac{13}{2}$

9. **(a)** $\int_{-1}^{1} dx + \int_{-1}^{1} \sqrt{1 - x^2}\,dx = 2(1) + \pi(1)^2/2 = 2 + \pi/2$

(b) $\frac{1}{3}(x^2 + 1)^{3/2}\Big]_0^3 - \pi(3)^2/4 = \frac{1}{3}(10^{3/2} - 1) - 9\pi/4$

(c) $u = x^2$, $du = 2xdx$; $\frac{1}{2}\int_0^1 \sqrt{1 - u^2}du = \frac{1}{2}\pi(1)^2/4 = \pi/8$

10. $\frac{1}{2}$

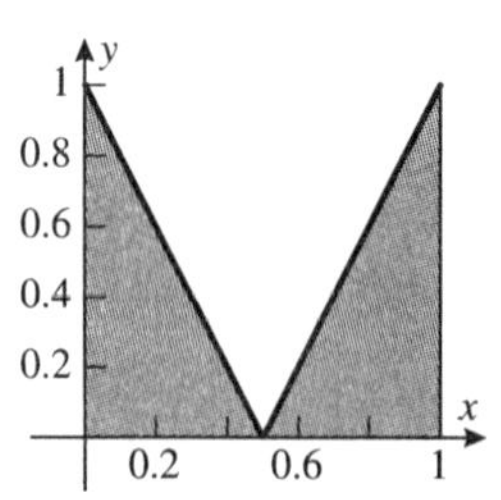

11. The rectangle with vertices $(0,0)$, $(\pi,0)$, $(\pi,1)$ and $(0,1)$ has area π and is much too large; so is the triangle with vertices $(0,0)$, $(\pi,0)$ and $(\pi,1)$ which has area $\pi/2$; $1-\pi$ is negative; so the answer is $35\pi/128$.

12. (a) $\dfrac{1}{n}\displaystyle\sum_{k=1}^{n}\sqrt{\frac{k}{n}}=\sum_{k=1}^{n}f(x_k^*)\Delta x$ where $f(x)=\sqrt{x}$, $x_k^*=k/n$, and $\Delta x=1/n$ for $0\le x\le 1$. Thus

$$\lim_{n\to+\infty}\frac{1}{n}\sum_{k=1}^{n}\sqrt{\frac{k}{n}}=\int_0^1 x^{1/2}dx=\frac{2}{3}$$

(b) $\dfrac{1}{n}\displaystyle\sum_{k=1}^{n}\left(\frac{k}{n}\right)^4=\sum_{k=1}^{n}f(x_k^*)\Delta x$ where $f(x)=x^4$, $x_k^*=k/n$, and $\Delta x=1/n$ for $0\le x\le 1$. Thus

$$\lim_{n\to+\infty}\frac{1}{n}\sum_{k=1}^{n}\left(\frac{k}{n}\right)^4=\int_0^1 x^4dx=\frac{1}{5}$$

13. left endpoints: $x_k^*=1,2,3,4$; $\displaystyle\sum_{k=1}^{4}f(x_k^*)\Delta x=(2+3+2+1)(1)=8$

right endpoints: $x_k^*=2,3,4,5$; $\displaystyle\sum_{k=1}^{4}f(x_k^*)\Delta x=(3+2+1+2)(1)=8$

14. The direction field is clearly an even function, which means that the solution is even, its derivative is odd. Since $\sin x$ is periodic and the direction field is not, that eliminates all but x, the solution of which is the family $y=x^2/2+C$.

15. (a)
$$1\cdot 2+2\cdot 3+\cdots+n(n+1)=\sum_{k=1}^{n}k(k+1)=\sum_{k=1}^{n}k^2+\sum_{k=1}^{n}k$$
$$=\frac{1}{6}n(n+1)(2n+1)+\frac{1}{2}n(n+1)=\frac{1}{3}n(n+1)(n+2)$$

(b)
$$\sum_{k=1}^{n-1}\left(\frac{9}{n}-\frac{k}{n^2}\right)=\frac{9}{n}\sum_{k=1}^{n-1}1-\frac{1}{n^2}\sum_{k=1}^{n-1}k=\frac{9}{n}(n-1)-\frac{1}{n^2}\cdot\frac{1}{2}(n-1)(n)=\frac{17}{2}\left(\frac{n-1}{n}\right);$$
$$\lim_{n\to+\infty}\frac{17}{2}\left(\frac{n-1}{n}\right)=\frac{17}{2}$$

(c)
$$\sum_{i=1}^{3}\left[\sum_{j=1}^{2}i+\sum_{j=1}^{2}j\right]=\sum_{i=1}^{3}\left[2i+\frac{1}{2}(2)(3)\right]=2\sum_{i=1}^{3}i+\sum_{i=1}^{3}3=2\cdot\frac{1}{2}(3)(4)+(3)(3)=21$$

16. (a) $\displaystyle\sum_{k=0}^{14}(k+4)(k+1)$ (b) $\displaystyle\sum_{k=5}^{19}(k-1)(k-4)$

17. For $1\le k\le n$ the k-th L-shaped strip consists of the corner square, a strip above and a strip to the right for a combined area of $1+(k-1)+(k-1)=2k-1$, so the total area is $\displaystyle\sum_{k=1}^{n}(2k-1)=n^2$.

18. $1+3+5+\cdots+(2n-1)=\sum_{k=1}^{n}(2k-1)=2\sum_{k=1}^{n}k-\sum_{k=1}^{n}1=2\cdot\frac{1}{2}n(n+1)-n=n^2$

19. $(3^5-3^4)+(3^6-3^5)+\cdots+(3^{17}-3^{16})=3^{17}-3^4$

20. $\left(1-\frac{1}{2}\right)+\left(\frac{1}{2}-\frac{1}{3}\right)+\cdots+\left(\frac{1}{50}-\frac{1}{51}\right)=\frac{50}{51}$

21. $\left(\frac{1}{2^2}-\frac{1}{1^2}\right)+\left(\frac{1}{3^2}-\frac{1}{2^2}\right)+\cdots+\left(\frac{1}{20^2}-\frac{1}{19^2}\right)=\frac{1}{20^2}-1=-\frac{399}{400}$

22. $(2^2-2)+(2^3-2^2)+\cdots+(2^{101}-2^{100})=2^{101}-2$

23. **(a)**
$$\begin{aligned}\sum_{k=1}^{n}\frac{1}{(2k-1)(2k+1)}&=\frac{1}{2}\sum_{k=1}^{n}\left(\frac{1}{2k-1}-\frac{1}{2k+1}\right)\\&=\frac{1}{2}\left[\left(1-\frac{1}{3}\right)+\left(\frac{1}{3}-\frac{1}{5}\right)+\left(\frac{1}{5}-\frac{1}{7}\right)+\cdots+\left(\frac{1}{2n-1}-\frac{1}{2n+1}\right)\right]\\&=\frac{1}{2}\left[1-\frac{1}{2n+1}\right]=\frac{n}{2n+1}\end{aligned}$$

(b) $\lim_{n\to+\infty}\frac{n}{2n+1}=\frac{1}{2}$

24. **(a)**
$$\begin{aligned}\sum_{k=1}^{n}\frac{1}{k(k+1)}&=\sum_{k=1}^{n}\left(\frac{1}{k}-\frac{1}{k+1}\right)\\&=\left(1-\frac{1}{2}\right)+\left(\frac{1}{2}-\frac{1}{3}\right)+\left(\frac{1}{3}-\frac{1}{4}\right)+\cdots+\left(\frac{1}{n}-\frac{1}{n+1}\right)\\&=1-\frac{1}{n+1}=\frac{n}{n+1}\end{aligned}$$

(b) $\lim_{n\to+\infty}\frac{n}{n+1}=1$

25. $\sum_{i=1}^{n}(x_i-\bar{x})=\sum_{i=1}^{n}x_i-\sum_{i=1}^{n}\bar{x}=\sum_{i=1}^{n}x_i-n\bar{x}$ but $\bar{x}=\frac{1}{n}\sum_{i=1}^{n}x_i$ thus

$\sum_{i=1}^{n}x_i=n\bar{x}$ so $\sum_{i=1}^{n}(x_i-\bar{x})=n\bar{x}-n\bar{x}=0$

26.
$$\begin{aligned}S-rS&=\sum_{k=0}^{n}ar^k-\sum_{k=0}^{n}ar^{k+1}\\&=(a+ar+ar^2+\cdots+ar^n)-(ar+ar^2+ar^3+\cdots+ar^{n+1})\\&=a-ar^{n+1}=a(1-r^{n+1})\end{aligned}$$

so $(1-r)S=a(1-r^{n+1})$, hence $S=a(1-r^{n+1})/(1-r)$

27. (a) $\displaystyle\sum_{k=0}^{19} 3^{k+1} = \sum_{k=0}^{19} 3(3^k) = \frac{3(1-3^{20})}{1-3} = \frac{3}{2}(3^{20}-1)$

(b) $\displaystyle\sum_{k=0}^{25} 2^{k+5} = \sum_{k=0}^{25} 2^5 2^k = \frac{2^5(1-2^{26})}{1-2} = 2^{31} - 2^5$

(c) $\displaystyle\sum_{k=0}^{100} (-1)\left(\frac{-1}{2}\right)^k = \frac{(-1)(1-(-1/2)^{101})}{1-(-1/2)} = -\frac{2}{3}(1+1/2^{101})$

28. (a) $1.999023438, 1.999999046, 2.000000000; 2$

(b) $2.831059456, 2.990486364, 2.999998301; 3$

29. (a) If $u = \sec x$, $du = \sec x \tan x dx$, $\displaystyle\int \sec^2 x \tan x dx = \int u du = u^2/2 + C_1 = (\sec^2 x)/2 + C_1$;

if $u = \tan x$, $du = \sec^2 x dx$, $\displaystyle\int \sec^2 x \tan x dx = \int u du = u^2/2 + C_2 = (\tan^2 x)/2 + C_2$.

(b) They are equal only if $\sec^2 x$ and $\tan^2 x$ differ by a constant, which is true.

30. $\displaystyle\frac{1}{2}\sec^2 x\Big]_0^{\pi/4} = \frac{1}{2}(2-1) = 1/2$ and $\frac{1}{2}\tan^2 x\Big]_0^{\pi/4} = \frac{1}{2}(1-0) = 1/2$

31. $\displaystyle\int \sqrt{1+x^{-2/3}}dx = \int x^{-1/3}\sqrt{x^{2/3}+1}dx$; $u = x^{2/3}+1$, $du = \frac{2}{3}x^{-1/3}dx$

$\displaystyle\frac{3}{2}\int u^{1/2}du = u^{3/2} + C = (x^{2/3}+1)^{3/2} + C$

32. (a) $\displaystyle\int_a^b \sum_{k=1}^{n} f_k(x)dx = \sum_{k=1}^{n}\int_a^b f_k(x)dx$

(b) yes; substitute $c_k f_k(x)$ for $f_k(x)$ in part (a), and then use $\displaystyle\int_a^b c_k f_k(x)dx = c_k\int_a^b f_k(x)dx$ from Theorem 5.5.4

33. (a) $\displaystyle\int_1^x \frac{1}{1+t^2}dt$ (b) $\displaystyle\int_{\tan(\pi/4-2)}^x \frac{1}{1+t^2}dt$

34. (a) $F'(x) = \dfrac{x-3}{x^2+7}$; increasing on $[3,+\infty)$, decreasing on $(-\infty,3]$

(b) $F''(x) = \dfrac{7+6x-x^2}{(x^2+7)^2} = \dfrac{(7-x)(1+x)}{(x^2+7)^2}$; concave up on $(-1,7)$, concave down on $(-\infty,-1)$ and $(7,+\infty)$

(c) $F'(x) = \dfrac{x-3}{x^2+7} = 0$ when $x = 3$, which is a relative minimum, and hence the absolute minimum, by the first derivative test.

(d)

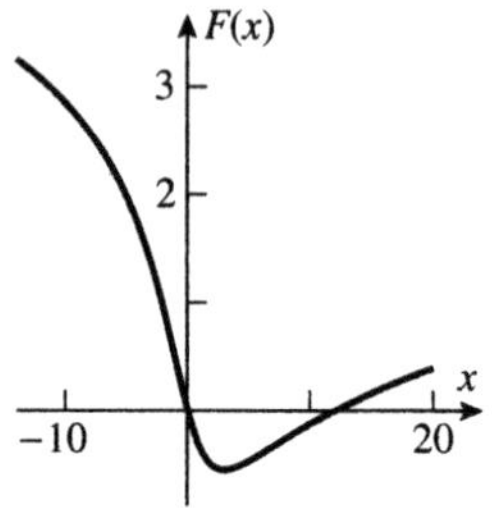

35. $F'(x) = \dfrac{1}{1+x^2} + \dfrac{1}{1+(1/x)^2}(-1/x^2) = 0$ so F is constant on $(0, +\infty)$.

36. $(-3, 3)$ because f is continuous there and 1 is in $(-3, 3)$

37. **(a)** The domain is $(-\infty, +\infty)$; $F(x)$ is 0 if $x = 1$, positive if $x > 1$, and negative if $x < 1$, because the integrand is positive, so the sign of the integral depends on the orientation (forwards or backwards).

(b) The domain is $[-2, 2]$; $F(x)$ is 0 if $x = -1$, positive if $-1 < x \le 2$, and negative if $-2 \le x < -1$; same reasons as in Part (a).

38. The left endpoint of the top boundary is $((b-a)/2, h)$ and the right endpoint of the top boundary is $((b+a)/2, h)$ so

$$f(x) = \begin{cases} 2hx/(b-a), & x < (b-a)/2 \\ h, & (b-a)/2 < x < (b+a)/2 \\ 2h(x-b)/(a-b), & x > (a+b)/2 \end{cases}$$

The area of the trapezoid is given by

$$\int_0^{(b-a)/2} \frac{2hx}{b-a}dx + \int_{(b-a)/2}^{(b+a)/2} h\,dx + \int_{(b+a)/2}^{b} \frac{2h(x-b)}{a-b}dx = (b-a)h/4 + ah + (b-a)h/4 = h(a+b)/2.$$

39. **(a)** no, since the velocity curve is not a straight line

(b) $25 < t < 40$ **(c)** 3.54 ft/s **(d)** 141.5 ft

(e) no since the velocity is positive and the acceleration is never negative

(f) need the position at any one given time (e.g. s_0)

40. $w(t) = \int_0^t \tau/7\,d\tau = t^2/14$, assuming $w_0 = w(0) = 0$; $w_{\text{ave}} = \dfrac{1}{26}\int_{26}^{52} t^2/7\,dt = \dfrac{1}{26}\dfrac{t^3}{21}\Big]_{26}^{52} = 676/3$
Set $676/3 = t^2/14$, $t = \pm\dfrac{26}{3}\sqrt{21}$, so $t \approx 39.716$, so during the 40th week.

41. $u = 5 + 2\sin 3x$, $du = 6\cos 3x dx$; $\displaystyle\int \frac{1}{6\sqrt{u}}du = \frac{1}{3}u^{1/2} + C = \frac{1}{3}\sqrt{5 + 2\sin 3x} + C$

42. $u = 3 + \sqrt{x}$, $du = \dfrac{1}{2\sqrt{x}}dx$; $\displaystyle\int 2\sqrt{u}du = \frac{4}{3}u^{3/2} + C = \frac{4}{3}(3 + \sqrt{x})^{3/2} + C$

43. $u = ax^3 + b$, $du = 3ax^2dx$; $\displaystyle\int \frac{1}{3au^2}du = -\frac{1}{3au} + C = -\frac{1}{3a^2x^3 + 3ab} + C$

44. $u = ax^2$, $du = 2axdx$; $\displaystyle\frac{1}{2a}\int \sec^2 u du = \frac{1}{2a}\tan u + C = \frac{1}{2a}\tan(ax^2) + C$

45. $\displaystyle\left(-\frac{1}{3u^3} - \frac{3}{u} + \frac{1}{4u^4}\right)\Bigg]_{-2}^{-1} = 389/192$

46. $\displaystyle\frac{1}{3\pi}\sin^3 \pi x\Bigg]_0^1 = 0$

47. With $b = 1.618034$, area $= \displaystyle\int_0^b (x + x^2 - x^3)dx = 1.007514$.

48. (a) $f(x) = \frac{1}{3}x^2 \sin 3x - \frac{2}{27}\sin 3x + \frac{2}{9}x\cos 3x - 0.251607$

(b) $f(x) = \sqrt{4 + x^2} + \dfrac{4}{\sqrt{4 + x^2}} - 6$

49. (a) Solve $\frac{1}{4}k^4 - k - k^2 + \frac{7}{4} = 0$ to get $k = 2.073948$.

(b) Solve $-\frac{1}{2}\cos 2k + \frac{1}{3}k^3 + \frac{1}{2} = 3$ to get $k = 1.837992$.

50. $F(x) = \displaystyle\int_{-1}^{x} \frac{t}{\sqrt{2 + t^3}}dt$, $F'(x) = \dfrac{x}{\sqrt{2 + x^3}}$, so F is increasing on $[1, 3]$; $F_{\max} = F(3) \approx 1.152082854$ and $F_{\min} = F(1) \approx -0.07649493141$

51. (a)

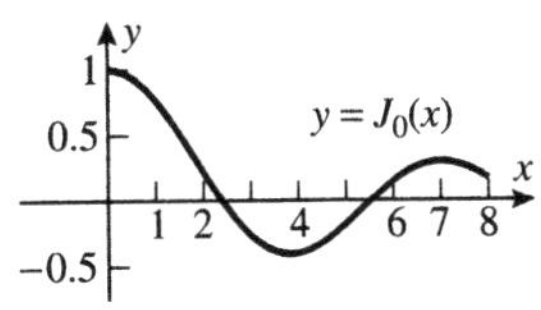

(b) 0.7651976866

(c) $J_0(x) = 0$ if $x = 2.404826$

52. $\displaystyle\lim_{n\to+\infty} \sum_{k=1}^{n} \left[\frac{25(k-1)}{n} - \frac{25(k-1)^2}{n^2}\right]\frac{5}{n} = \frac{125}{6}$

CHAPTER 5 HORIZON MODULE

1. $v_x(0) = 35\cos\alpha$, so from Equation (1), $x(t) = (35\cos\alpha)t$; $v_y(0) = 35\sin\alpha$, so from Equation (2), $y(t) = (35\sin\alpha)t - 4.9t^2$.

2. (a) $v_x(t) = \dfrac{dx(t)}{dt} = 35\cos\alpha$, $v_y(t) = \dfrac{dy(t)}{dt} = 35\sin\alpha - 9.8t$

(b) $v_y(t) = 35\sin\alpha - 9.8t$, $v_y(t) = 0$ when $t = 35\sin\alpha/9.8$;
$y = v_y(0)t - 4.9t^2 = (35\sin\alpha)(35\sin\alpha)/9.8 - 4.9((35\sin\alpha)/9.8)^2 = 62.5\sin^2\alpha$, so
$y_{\max} = 62.5\sin^2\alpha$.

3. $t = x/(35\cos\alpha)$ so $y = (35\sin\alpha)(x/(35\cos\alpha)) - 4.9(x/(35\cos\alpha))^2 = (\tan\alpha)x - \dfrac{0.004}{\cos^2\alpha}x^2$; the trajectory is a parabola because y is a quadratic function of x.

4.

15°	25°	35°	45°	55°	65°	75°	85°
no	yes	no	no	no	yes	no	no

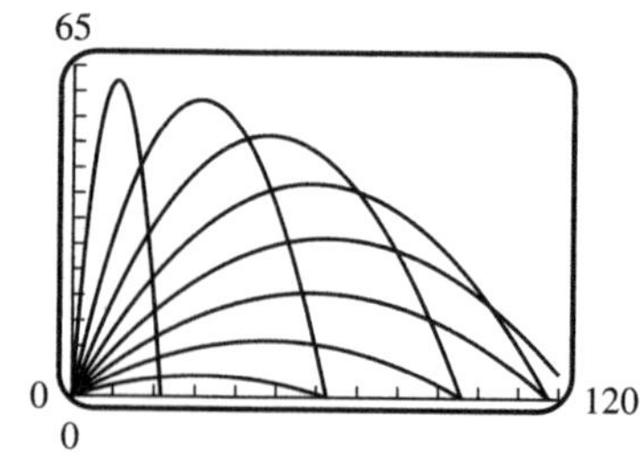

5. $y(t) = (35\sin\alpha \text{ s})t - 4.9t^2 = 0$ when $t = 35\sin\alpha/4.9$, at which time $x = (35\cos\alpha)(35\sin\alpha/4.9) = 125\sin 2\alpha$; this is the maximum value of x, so $R = 125\sin 2\alpha$ m.

6. **(a)** $R = 95$ when $\sin 2\alpha = 95/125 = 0.76$, $\alpha = 0.4316565575, 1.139139769$ rad $\approx 24.73°, 65.27°$.

(b) $y(t) < 50$ is required; but $y(1.139) \approx 51.56$ m, so his height would be 56.56 m.

7. $0.4019 < \alpha < 0.4636$ (radians), or $23.03° < \alpha < 26.57°$

CHAPTER 6

Applications of the Definite Integral in Geometry, Science, and Engineering

EXERCISE SET 6.1

1. $A = \int_{-1}^{2}(x^2+1-x)dx = (x^3/3+x-x^2/2)\Big]_{-1}^{2} = 9/2$

2. $A = \int_{0}^{4}(\sqrt{x}+x/4)dx = (2x^{3/2}/3+x^2/8)\Big]_{0}^{4} = 22/3$

3. $A = \int_{1}^{2}(y-1/y^2)dy = (y^2/2+1/y)\Big]_{1}^{2} = 1$

4. $A = \int_{0}^{2}(2-y^2+y)dy = (2y-y^3/3+y^2/2)\Big]_{0}^{2} = 10/3$

5. **(a)** $A = \int_{0}^{4}(4x-x^2)dx = 32/3$

(b) $A = \int_{0}^{16}(\sqrt{y}-y/4)dy = 32/3$

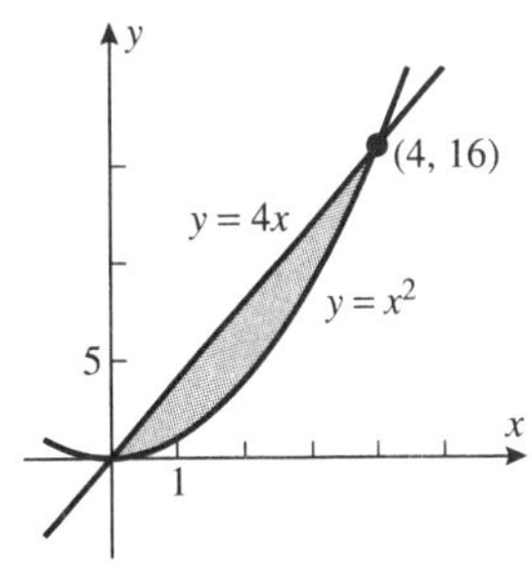

6. Eliminate x to get $y^2 = 4(y+4)/2$, $y^2-2y-8=0$, $(y-4)(y+2)=0$; $y=-2,4$ with corresponding values of $x=1,4$.

(a) $A = \int_{0}^{1}[2\sqrt{x}-(-2\sqrt{x})]dx + \int_{1}^{4}[2\sqrt{x}-(2x-4)]dx$

$= \int_{0}^{1}4\sqrt{x}dx + \int_{1}^{4}(2\sqrt{x}-2x+4)dx = 8/3+19/3 = 9$

(b) $A = \int_{-2}^{4}[(y/2+2)-y^2/4]dy = 9$

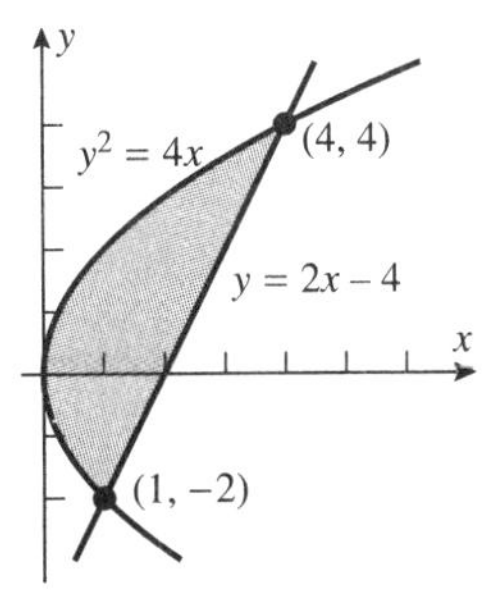

7. $A = \int_{1/4}^{1}(\sqrt{x}-x^2)dx = 49/192$

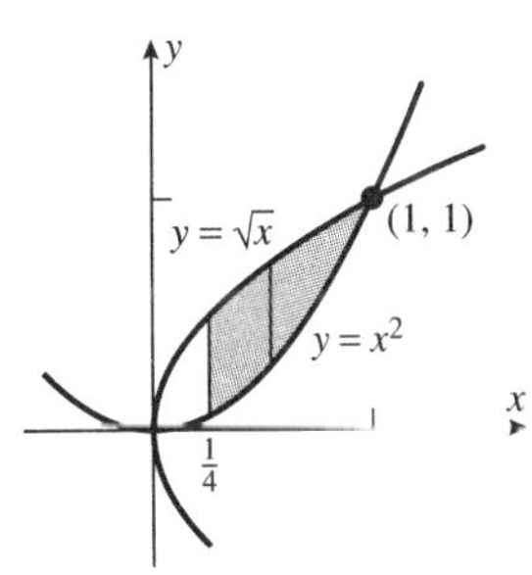

8. $A = \int_0^2 [0 - (x^3 - 4x)]dx$
$= \int_0^2 (4x - x^3)dx = 4$

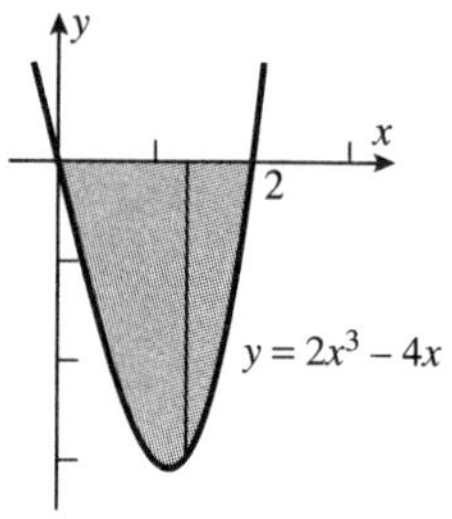

9. $A = \int_{\pi/4}^{\pi/2} (0 - \cos 2x)dx$
$= -\int_{\pi/4}^{\pi/2} \cos 2x\, dx = 1/2$

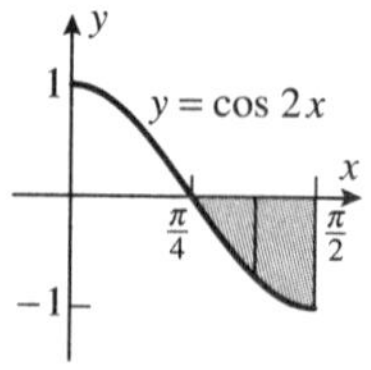

10. Equate $\sec^2 x$ and 2 to get $\sec^2 x = 2$,

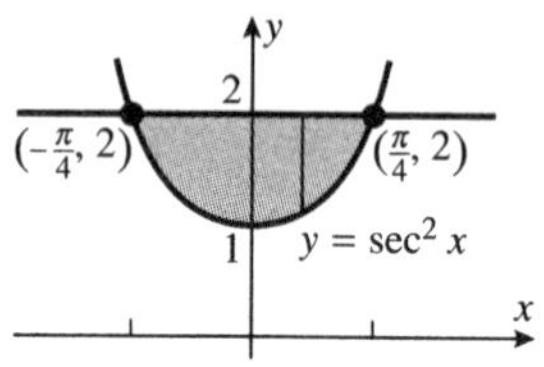

$\sec x = \pm\sqrt{2}$, $x = \pm\pi/4$
$A = \int_{-\pi/4}^{\pi/4} (2 - \sec^2 x)dx = \pi - 2$

11. $A = \int_{\pi/4}^{3\pi/4} \sin y\, dy = \sqrt{2}$

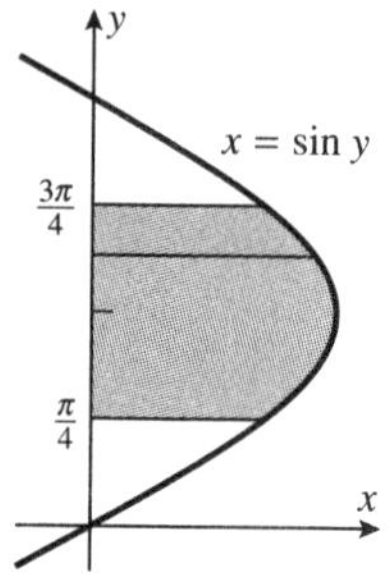

12. $A = \int_{-1}^{2} [(x + 2) - x^2]dx = 9/2$

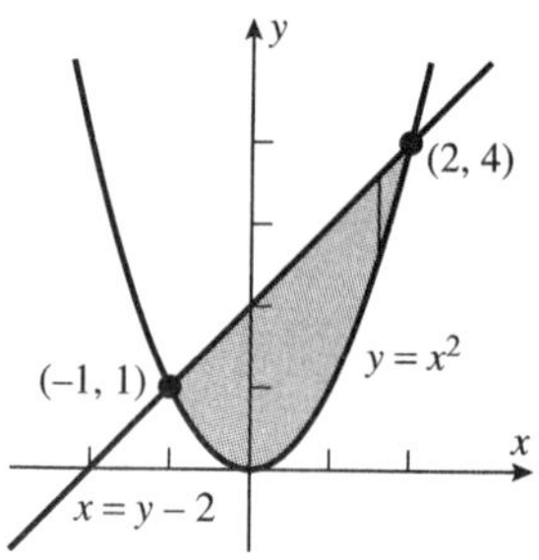

13. $y = 2 + |x - 1| = \begin{cases} 3 - x, & x \le 1 \\ 1 + x, & x \ge 1 \end{cases}$,

$A = \int_{-5}^{1} \left[\left(-\frac{1}{5}x + 7\right) - (3 - x)\right] dx$
$+ \int_{1}^{5} \left[\left(-\frac{1}{5}x + 7\right) - (1 + x)\right] dx$
$= \int_{-5}^{1} \left(\frac{4}{5}x + 4\right) dx + \int_{1}^{5} \left(6 - \frac{6}{5}x\right) dx$
$= 72/5 + 48/5 = 24$

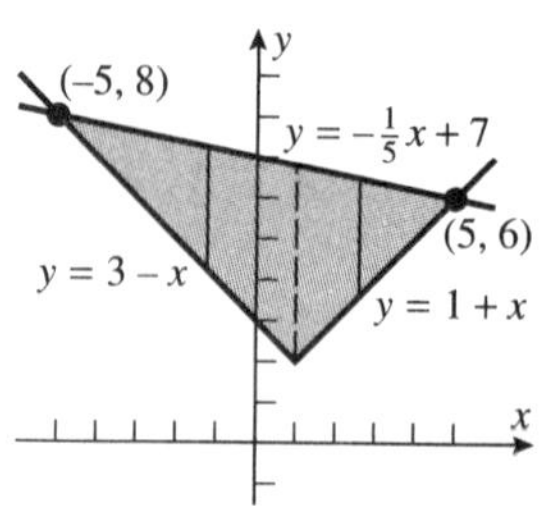

14. $A = \int_0^{2/5} (4x - x)dx$
$+ \int_{2/5}^{1} (-x + 2 - x)dx$
$= \int_0^{2/5} 3x\,dx + \int_{2/5}^{1} (2 - 2x)dx = 3/5$

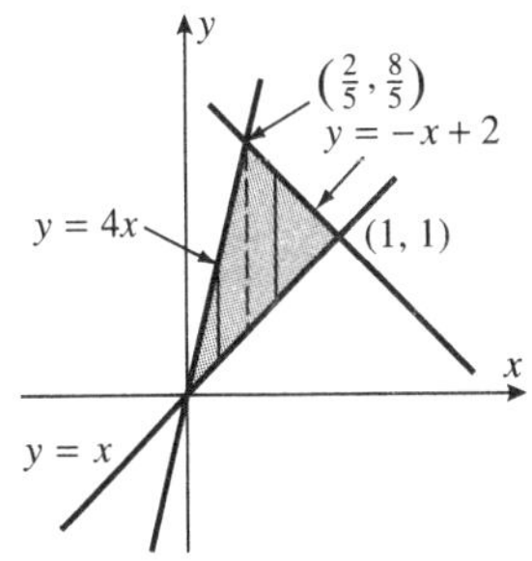

15. $A = \int_0^1 (x^3 - 4x^2 + 3x)dx$
$+ \int_1^3 [-(x^3 - 4x^2 + 3x)]dx$
$= 5/12 + 32/12 = 37/12$

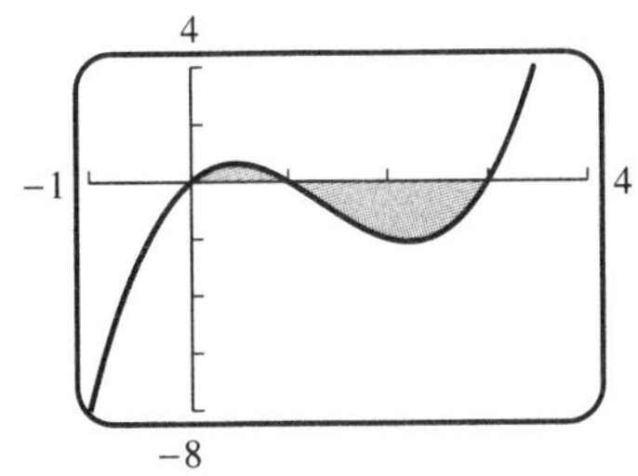

16. Equate $y = x^3 - 2x^2$ and $y = 2x^2 - 3x$ to get $x^3 - 4x^2 + 3x = 0$, $x(x-1)(x-3) = 0; x = 0, 1, 3$ with corresponding values of $y = 0, -1.9$.

$A = \int_0^1 [(x^3 - 2x^2) - (2x^2 - 3x)]dx$
$+ \int_1^3 [(2x^3 - 3x) - (x^3 - 2x^2)]dx$
$= \int_0^1 (x^3 - 4x^2 + 3x)dx + \int_1^3 (-x^3 + 4x^2 - 3x)dx$
$= \frac{5}{12} + \frac{8}{3} = \frac{37}{12}$

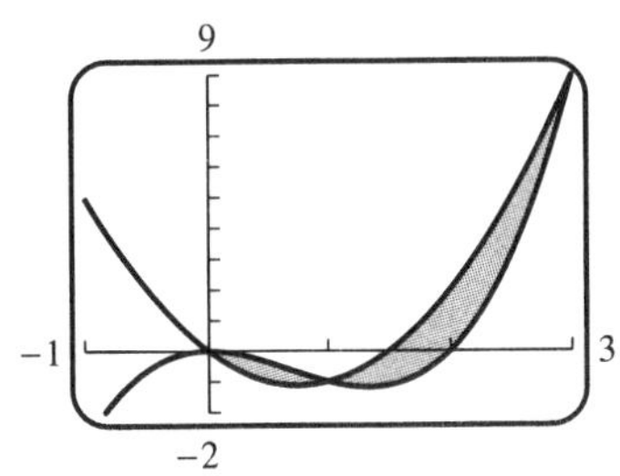

17. From the symmetry of the region

$A = 2\int_{\pi/4}^{5\pi/4} (\sin x - \cos x)dx = 4\sqrt{2}$

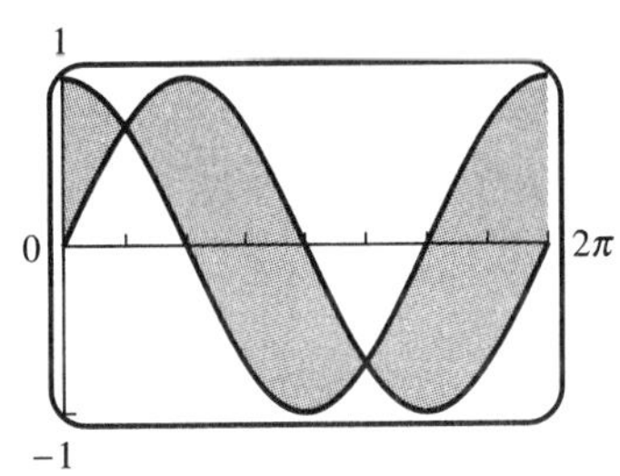

18. The region is symmetric about the origin so

$$A = 2\int_0^2 |x^3 - 4x|dx = 8$$

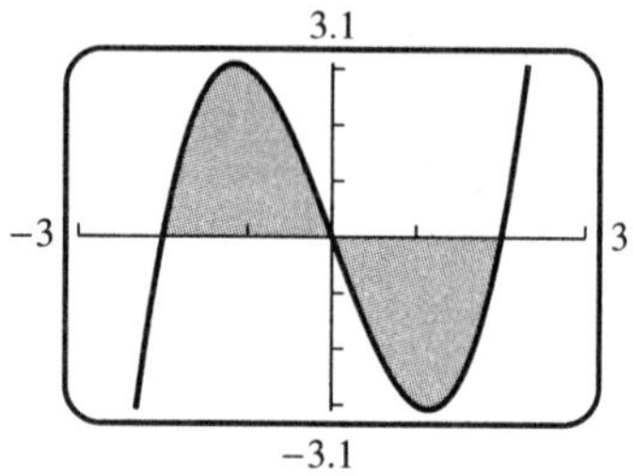

19. $A = \int_{-1}^0 (y^3 - y)dy + \int_0^1 -(y^3 - y)dy$
$= 1/2$

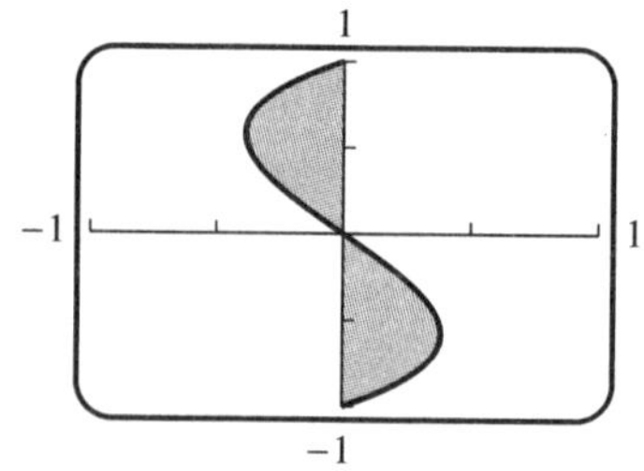

20. $A = \int_0^1 \left[y^3 - 4y^2 + 3y - (y^2 - y)\right] dy$
$+ \int_1^4 \left[y^2 - y - (y^3 - 4y^2 + 3y)\right] dy$
$= 7/12 + 45/4 = 71/6$

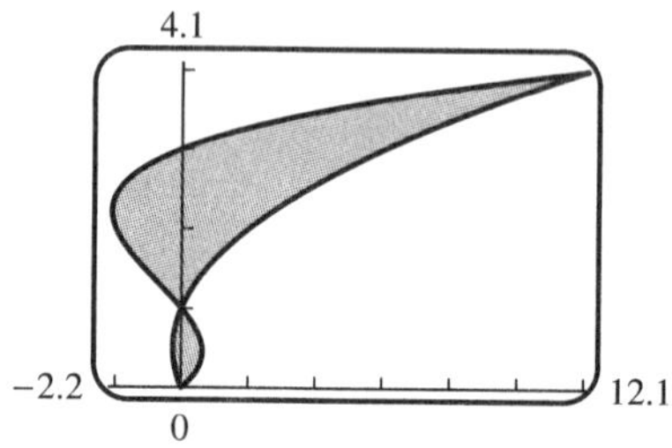

21. Solve $3-2x = x^6+2x^5-3x^4+x^2$ to find the real roots $x = -3, 1$; from a plot it is seen that the line is above the polynomial when $-3 < x < 1$, so $A = \int_{-3}^1 (3-2x-(x^6+2x^5-3x^4+x^2))\, dx = 9152/105$

22. Solve $x^5 - 2x^3 - 3x = x^3$ to find the roots $x = 0, \pm\frac{1}{2}\sqrt{6 + 2\sqrt{21}}$. Thus, by symmetry,

$$A = 2\int_0^{\sqrt{(6+2\sqrt{21})/2}} (x^3 - (x^5 - 2x^3 - 3x))\, dx = \frac{27}{4} + \frac{7}{4}\sqrt{21}$$

23. $\int_0^k 2\sqrt{y}dy = \int_k^9 2\sqrt{y}dy$

$\int_0^k y^{1/2}dy = \int_k^9 y^{1/2}dy$

$\frac{2}{3}k^{3/2} = \frac{2}{3}(27 - k^{3/2})$

$k^{3/2} = 27/2$

$k = (27/2)^{2/3} = 9/\sqrt[3]{4}$

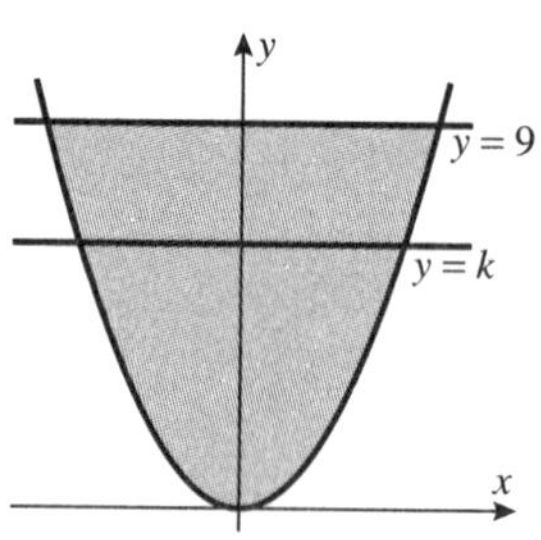

24. $\int_0^k x^2 dx = \int_k^2 x^2 dx$

$\frac{1}{3}k^3 = \frac{1}{3}(8 - k^3)$

$k^3 = 4$

$k = \sqrt[3]{4}$

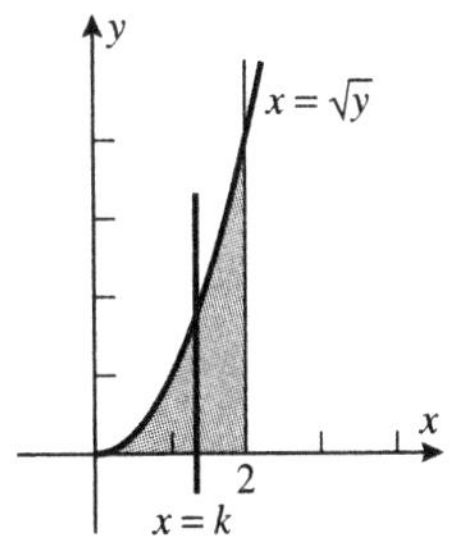

25. **(a)** $A = \int_0^2 (2x - x^2)dx = 4/3$

(b) $y = mx$ intersects $y = 2x - x^2$ where $mx = 2x - x^2, x^2 + (m-2)x = 0, x(x+m-2) = 0$ so $x = 0$ or $x = 2 - m$. The area below the curve and above the line is

$$\int_0^{2-m} (2x - x^2 - mx)dx = \int_0^{2-m} [(2-m)x - x^2]dx = \left[\frac{1}{2}(2-m)x^2 - \frac{1}{3}x^3\right]_0^{2-m} = \frac{1}{6}(2-m)^3$$

so $(2-m)^3/6 = (1/2)(4/3) = 2/3, (2-m)^3 = 4, m = 2 - \sqrt[3]{4}$.

26. The line through $(0,0)$ and $(5\pi/6, 1/2)$ is $y = \frac{3}{5\pi}x$;

$$A = \int_0^{5\pi/6} \left(\sin x - \frac{3}{5\pi}x\right) dx = \frac{\sqrt{3}}{2} - \frac{5}{24}\pi + 1$$

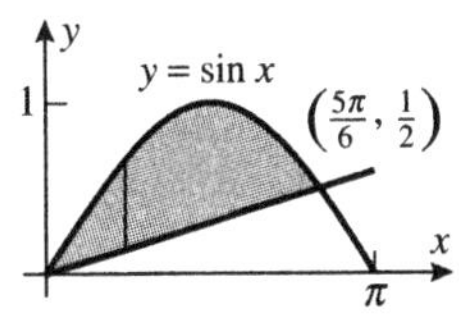

27. **(a)** It gives the area of the region that is between f and g when $f(x) > g(x)$ minus the area of the region between f and g when $f(x) < g(x)$, for $a \le x \le b$.

(b) It gives the area of the region that is between f and g for $a \le x \le b$.

28. **(b)** $\lim_{n\to+\infty} \int_0^1 (x^{1/n} - x)\, dx = \lim_{n\to+\infty} \left[\frac{n}{n+1}x^{(n+1)/n} - \frac{x^2}{2}\right]_0^1 = \lim_{n\to+\infty} \left(\frac{n}{n+1} - \frac{1}{2}\right) = 1/2$

29. The curves intersect at $x = 0$ and, by Newton's Method, at $x \approx 2.595739080 = b$, so $A \approx \int_0^b (\sin x - 0.2x)dx = -\left[\cos x + 0.1x^2\right]_0^b \approx 1.180898334$

30. By Newton's Method, the points of intersection are at $x \approx \pm 0.824132312$, so with $b = 0.824132312$ we have $A \approx 2\int_0^b (\cos x - x^2)dx = 2(\sin x - x^3/3)\Big]_0^b \approx 1.094753609$

31. distance $= \int |v|\, dt$, so

(a) distance $= \int_0^{60} (3t - t^2/20)\, dt = 1800$ ft.

(b) If $T \le 60$ then distance $= \int_0^T (3t - t^2/20)\, dt = \frac{3}{2}T^2 - \frac{1}{60}T^3$ ft.

32. Since $a_1(0) = a_2(0) = 0$, $A = \int_0^T (a_2(t) - a_1(t))\,dt = v_2(T) - v_1(T)$ is the difference in the velocities of the two cars at time T.

33. Solve $x^{1/2} + y^{1/2} = a^{1/2}$ for y to get

$$y = (a^{1/2} - x^{1/2})^2 = a - 2a^{1/2}x^{1/2} + x$$

$$A = \int_0^a (a - 2a^{1/2}x^{1/2} + x)dx = a^2/6$$

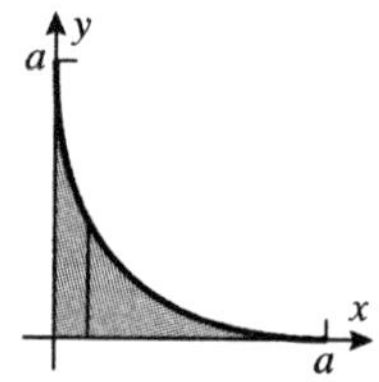

34. Solve for y to get $y = (b/a)\sqrt{a^2 - x^2}$ for the upper half of the ellipse; make use of symmetry to get $A = 4\int_0^a \frac{b}{a}\sqrt{a^2 - x^2}dx = \frac{4b}{a}\int_0^a \sqrt{a^2 - x^2}dx = \frac{4b}{a}\cdot\frac{1}{4}\pi a^2 = \pi ab.$

35. Let A be the area between the curve and the x-axis and A_R the area of the rectangle, then $A = \int_0^b kx^m dx = \frac{k}{m+1}x^{m+1}\Big]_0^b = \frac{kb^{m+1}}{m+1}$, $A_R = b(kb^m) = kb^{m+1}$, so $A/A_R = 1/(m+1)$.

EXERCISE SET 6.2

1. $V = \pi\int_{-1}^3 (3 - x)dx = 8\pi$

2. $V = \pi\int_0^1 [(2 - x^2)^2 - x^2]dx$

$$= \pi\int_0^1 (4 - 5x^2 + x^4)dx$$

$$= 38\pi/15$$

3. $V = \pi\int_0^2 \frac{1}{4}(3 - y)^2 dy = 13\pi/6$

4. $V = \pi\int_{1/2}^2 (4 - 1/y^2)dy = 9\pi/2$

5. $V = \pi\int_0^2 x^4 dx = 32\pi/5$

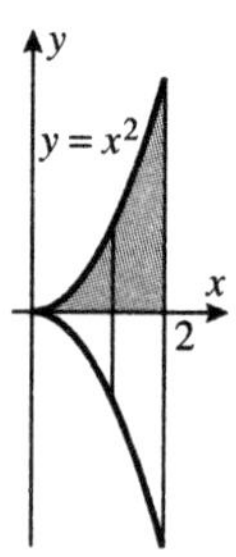

6. $V = \pi\int_{\pi/4}^{\pi/3} \sec^2 x\,dx = \pi(\sqrt{3} - 1)$

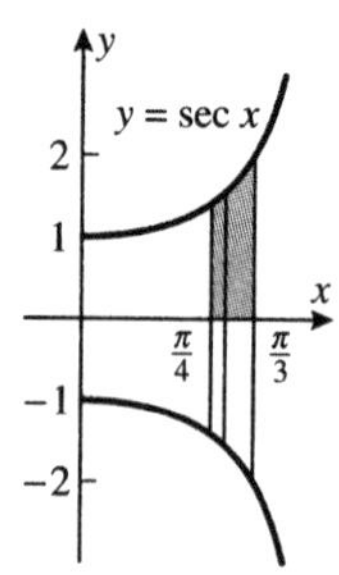

7. $V = \pi \int_{\pi/4}^{\pi/2} \cos x \, dx = (1 - \sqrt{2}/2)\pi$

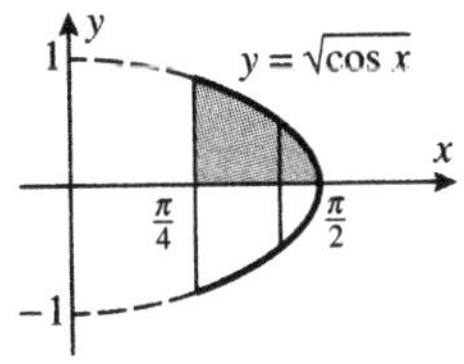

8. $V = \pi \int_0^1 [(x^2)^2 - (x^3)^2] dx$

$= \pi \int_0^1 (x^4 - x^6) dx = 2\pi/35$

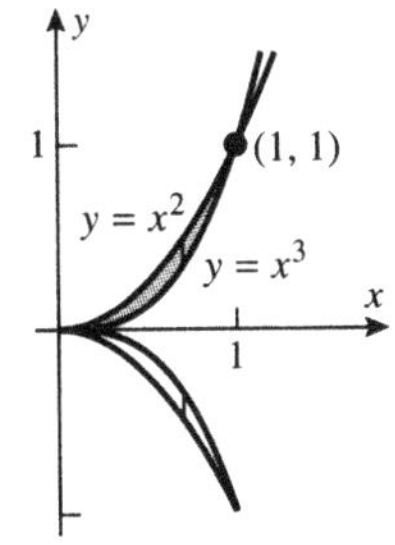

9. $V = \pi \int_{-4}^{4} [(25 - x^2) - 9] dx$

$= 2\pi \int_0^4 (16 - x^2) dx = 256\pi/3$

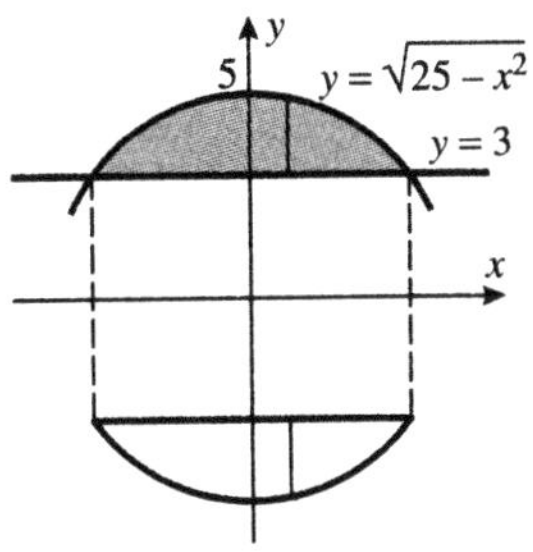

10. $V = \pi \int_{-3}^{3} (9 - x^2)^2 dx$

$= \pi \int_{-3}^{3} (81 - 18x^2 + x^4) dx = 1296\pi/5$

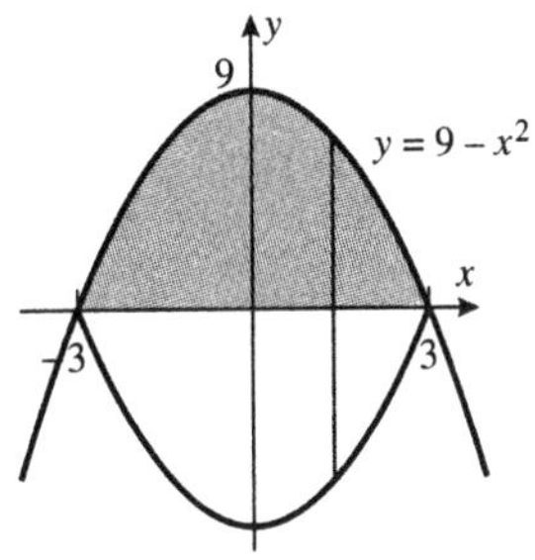

11. $V = \pi \int_0^4 [(4x)^2 - (x^2)^2] dx$

$= \pi \int_0^4 (16x^2 - x^4) dx = 2048\pi/15$

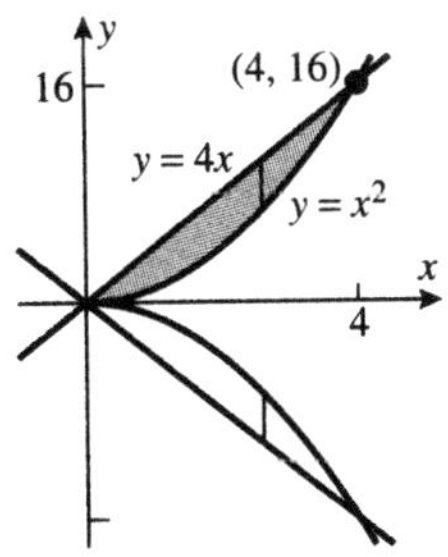

12. $V = \pi \int_0^{\pi/4} (\cos^2 x - \sin^2 x) dx$

$= \pi \int_0^{\pi/4} \cos 2x \, dx = \pi/2$

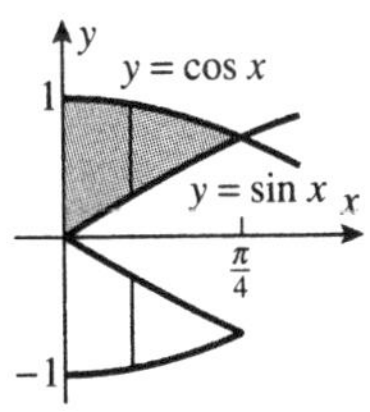

13. $V = \pi \int_0^1 y^{2/3} dy = 3\pi/5$

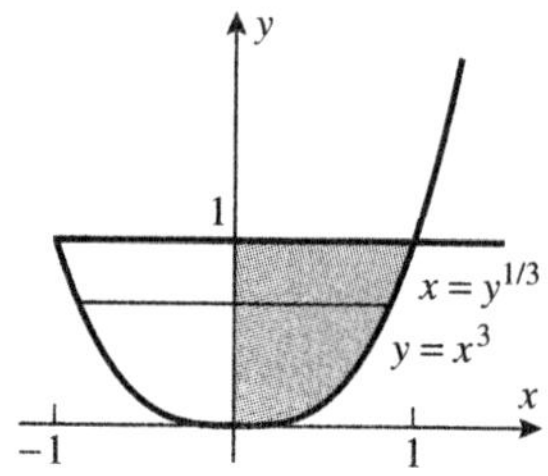

14. $V = \pi \int_{-1}^1 (1 - y^2)^2 dy$

$= \pi \int_{-1}^1 (1 - 2y^2 + y^4) dy = 16\pi/15$

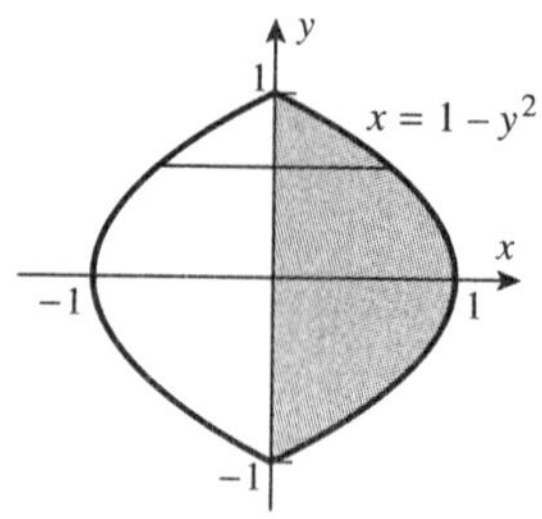

15. $V = \pi \int_{-1}^3 (1 + y) dy = 8\pi$

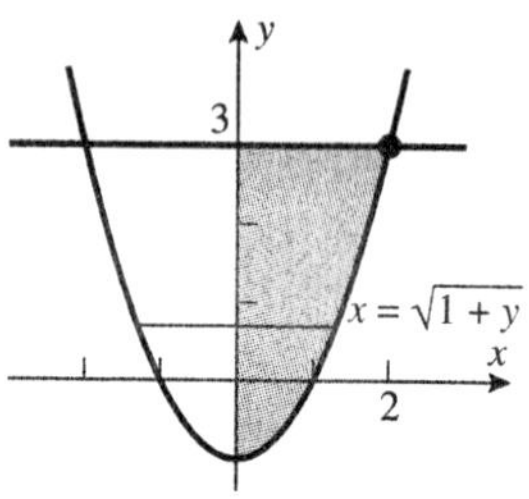

16. $V = \pi \int_0^3 [2^2 - (y + 1)] dy$

$= \pi \int_0^3 (3 - y) dy = 9\pi/2$

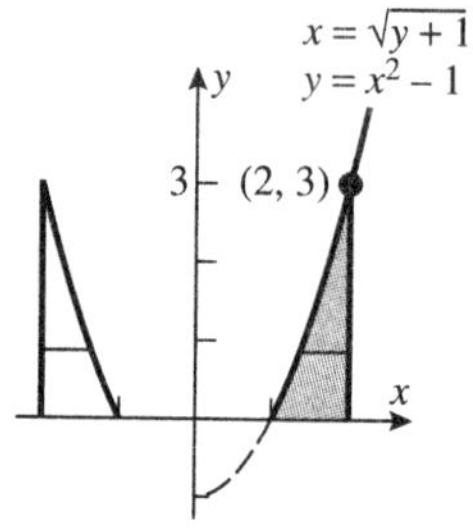

17. $V = \pi \int_{\pi/4}^{3\pi/4} \csc^2 y \, dy = 2\pi$

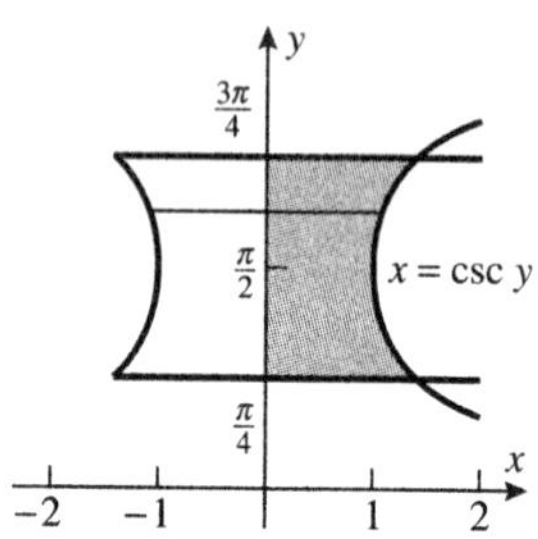

18. $V = \pi \int_0^1 (y - y^4) dy = 3\pi/10$

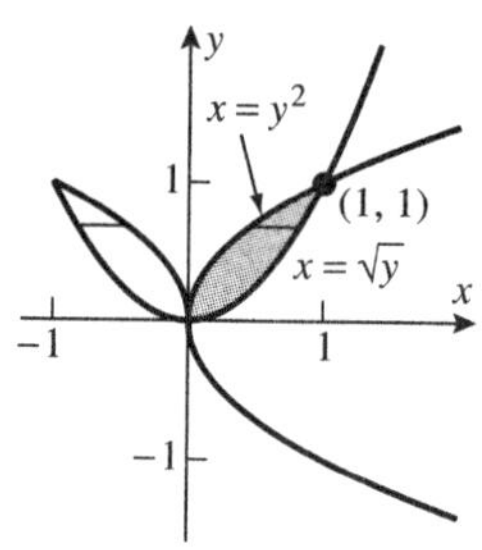

19. $V = \pi \int_{-1}^{2} [(y+2)^2 - y^4]dy = 72\pi/5$

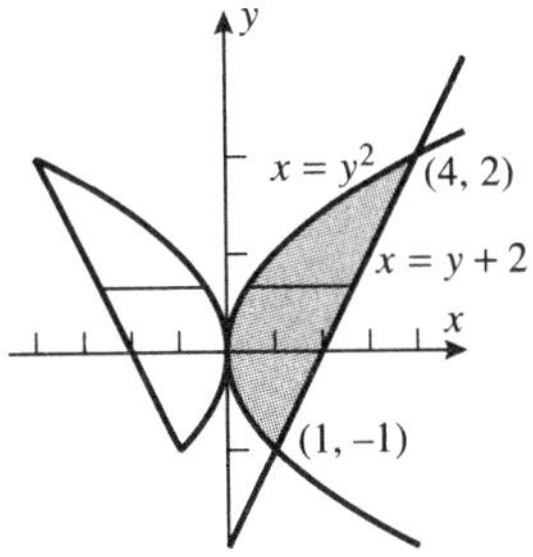

20. $V = \pi \int_{-1}^{1} \left[(2+y^2)^2 - (1-y^2)^2\right] dy$

$= \pi \int_{-1}^{1} (3 + 6y^2)dy = 10\pi$

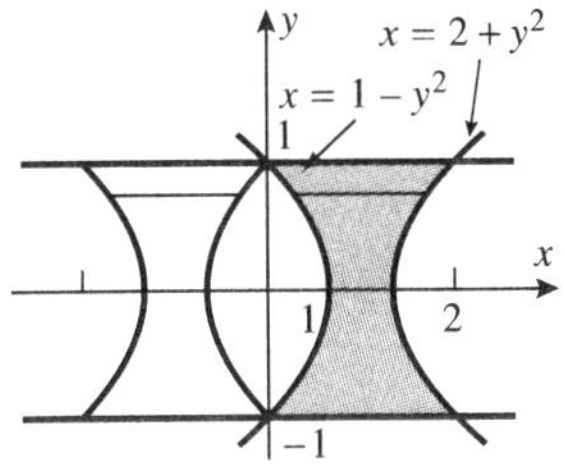

21. $V = \pi \int_{-a}^{a} \frac{b^2}{a^2}(a^2 - x^2)dx = 4\pi ab^2/3$

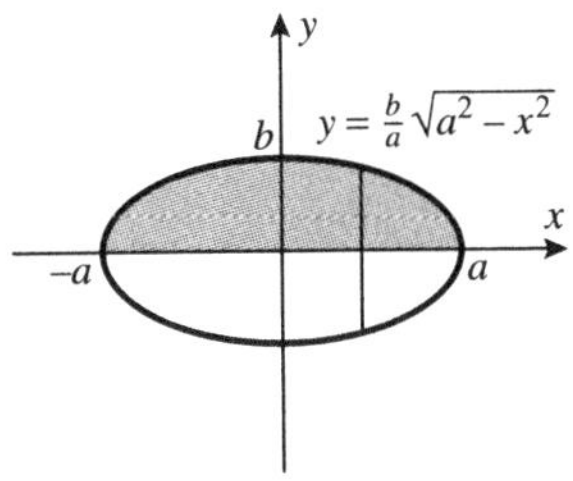

22. $V = \pi \int_{b}^{2} \frac{1}{x^2} dx = \pi(1/b - 1/2);$

$\pi(1/b - 1/2) = 3, b = 2\pi/(\pi + 6)$

23. $V = \pi \int_{-1}^{0} (x+1)dx$

$+ \pi \int_{0}^{1} [(x+1) - 2x]dx$

$= \pi/2 + \pi/2 = \pi$

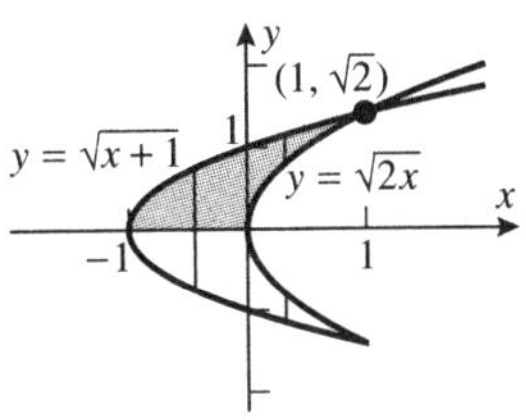

24. $V = \pi \int_{0}^{4} x\,dx + \pi \int_{4}^{6} (6-x)^2 dx$

$= 8\pi + 8\pi/3 = 32\pi/3$

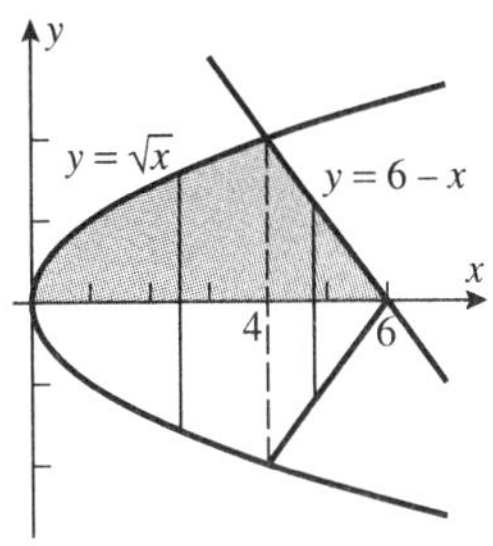

25. $V = \pi \int_{0}^{3} (9 - y^2)^2 dy$

$= \pi \int_{0}^{3} (81 - 18y^2 + y^4)dy$

$= 648\pi/5$

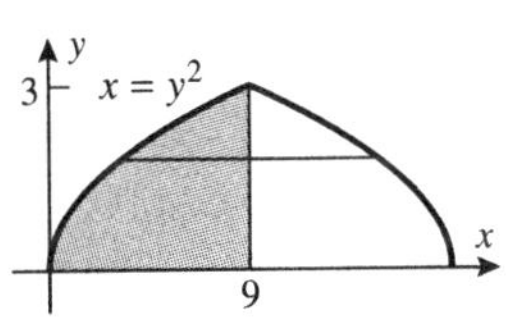

26. $V = \pi \int_{0}^{9} [3^2 - (3 - \sqrt{x})^2]dx$

$= \pi \int_{0}^{9} (6\sqrt{x} - x)dx$

$= 135\pi/2$

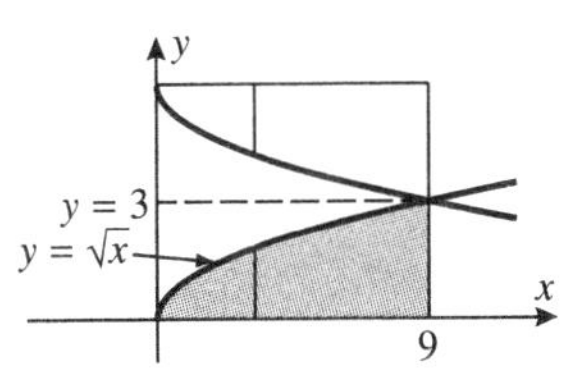

27. $V = \pi \int_0^1 [(\sqrt{x}+1)^2 - (x+1)^2]dx$

$= \pi \int_0^1 (2\sqrt{x} - x - x^2)dx = \pi/2$

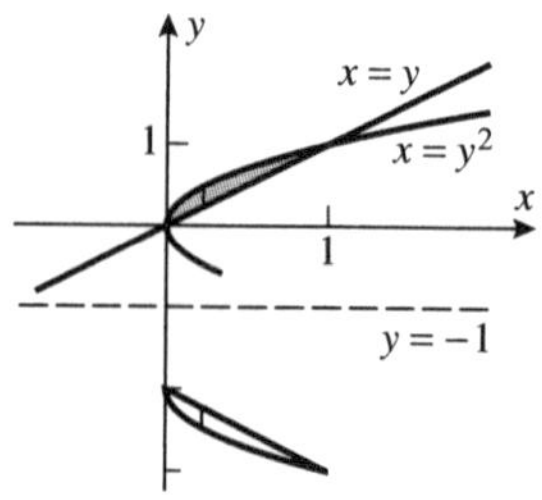

28. $V = \pi \int_0^1 [(y+1)^2 - (y^2+1)^2]dy$

$= \pi \int_0^1 (2y - y^2 - y^4)dy = 7\pi/15$

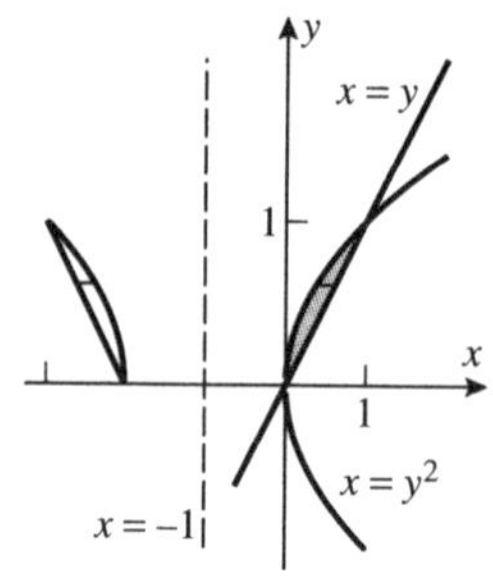

29. $A(x) = \pi(x^2/4)^2 = \pi x^4/16,$

$V = \int_0^{20} (\pi x^4/16)dx = 40,000\pi \text{ ft}^3$

30. $V = \pi \int_0^1 (x - x^4)dx = 3\pi/10$

31. $V = \int_0^1 (x - x^2)^2 dx$

$= \int_0^1 (x^2 - 2x^3 + x^4)dx = 1/30$

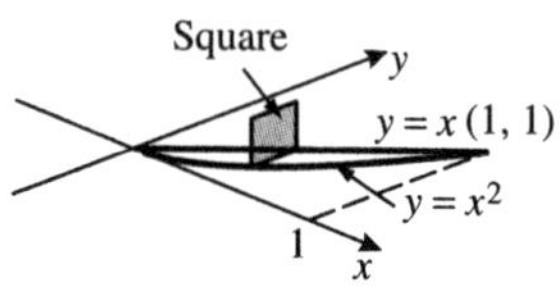

32. $A(x) = \frac{1}{2}\pi\left(\frac{1}{2}\sqrt{x}\right)^2 = \frac{1}{8}\pi x,$

$V = \int_0^4 \frac{1}{8}\pi x \, dx = \pi$

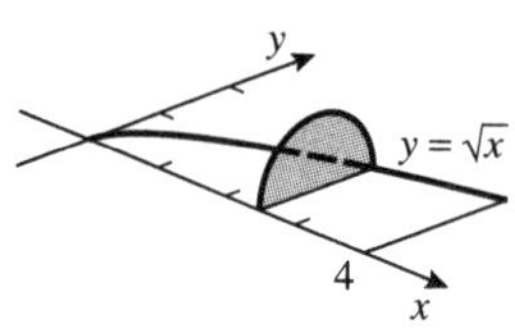

33. On the upper half of the circle, $y = \sqrt{1-x^2}$, so:

(a) $A(x)$ is the area of a semicircle of radius y, so

$A(x) = \pi y^2/2 = \pi(1-x^2)/2;\ V = \frac{\pi}{2}\int_{-1}^{1}(1-x^2)\,dx = \pi\int_0^1 (1-x^2)\,dx = 2\pi/3$

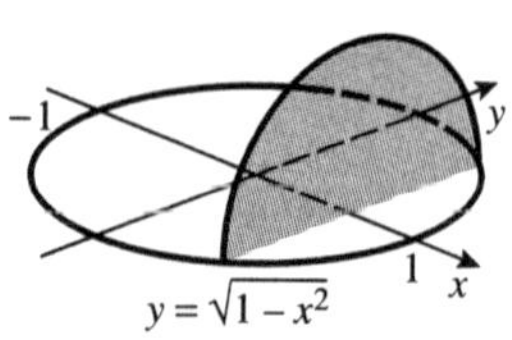

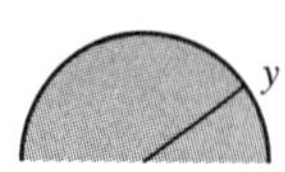

(b) $A(x)$ is the area of a square of side $2y$, so

$$A(x) = 4y^2 = 4(1 - x^2);\ V = 4\int_{-1}^{1}(1 - x^2)\,dx = 8\int_{0}^{1}(1 - x^2)\,dx = 16/3$$

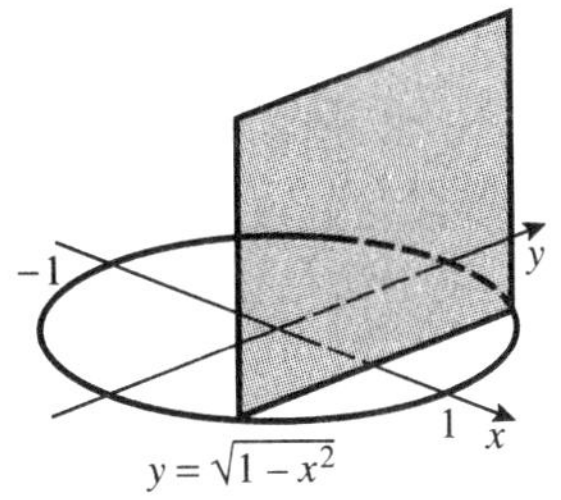

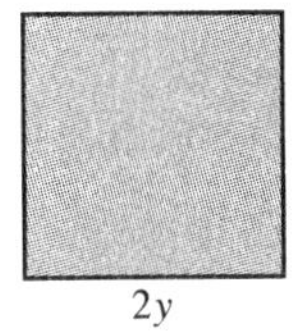

(c) $A(x)$ is the area of an equilateral triangle with sides $2y$, so

$$A(x) = \frac{\sqrt{3}}{4}(2y)^2 = \sqrt{3}y^2 = \sqrt{3}(1 - x^2);$$

$$V = \int_{-1}^{1}\sqrt{3}(1 - x^2)\,dx = 2\sqrt{3}\int_{0}^{1}(1 - x^2)\,dx = 4\sqrt{3}/3$$

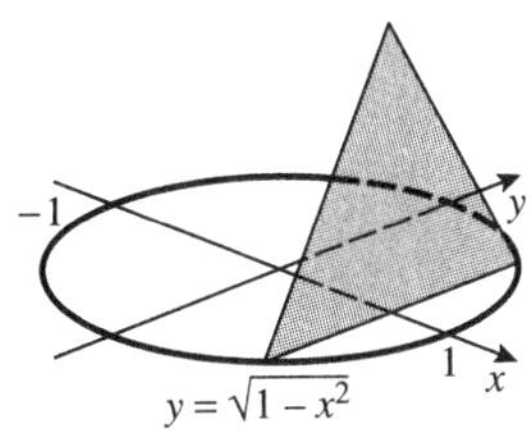

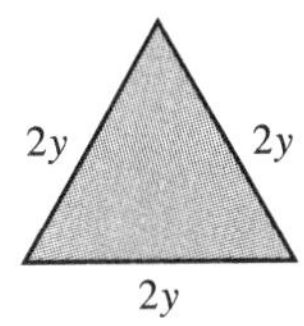

34. By similar triangles, $R/r = y/h$ so $R = ry/h$ and $A(y) = \pi r^2 y^2/h^2$.

$$V = (\pi r^2/h^2)\int_0^h y^2 dy = \pi r^2 h/3$$

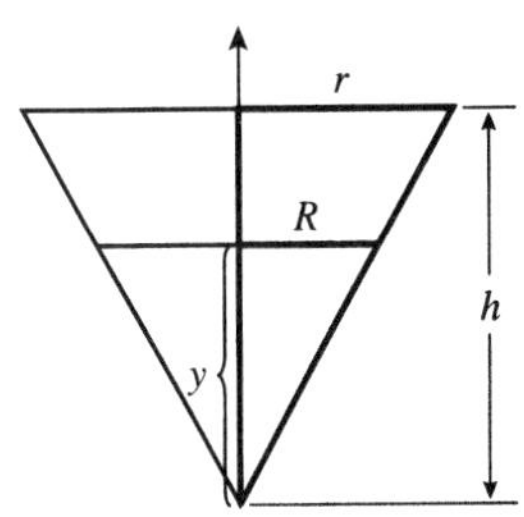

35. The two curves cross at $x = b \approx 1.403288534$, so

$$V = \pi\int_0^b((2x/\pi)^2 - \sin^{16}x)\,dx + \pi\int_b^{\pi/2}(\sin^{16}x - (2x/\pi)^2)\,dx \approx 0.710172176.$$

36. Note that $\pi^2\sin x\cos^3 x = 4x^2$ for $x = \pi/4$. From the graph it is apparent that this is the first positive solution, thus the curves don't cross on $(0, \pi/4)$ and

$$V = \pi\int_0^{\pi/4}[(\pi^2\sin x\cos^3 x)^2 - (4x^2)^2]\,dx = \frac{1}{48}\pi^5 + \frac{17}{2560}\pi^6$$

37. **(a)** $V = \pi \int_{r-h}^{r} (r^2 - y^2)\, dy = \pi(rh^2 - h^3/3) = \frac{1}{3}\pi h^2(3r - h)$

(b) By the Pythagorean Theorem,

$r^2 = (r-h)^2 + \rho^2$, $2hr = h^2 + \rho^2$; from Part (a),

$$V = \frac{\pi h}{3}(3hr - h^2) = \frac{\pi h}{3}\left(\frac{3}{2}(h^2 + \rho^2) - h^2\right)$$
$$= \frac{1}{6}\pi h(h^2 + 3\rho^2).$$

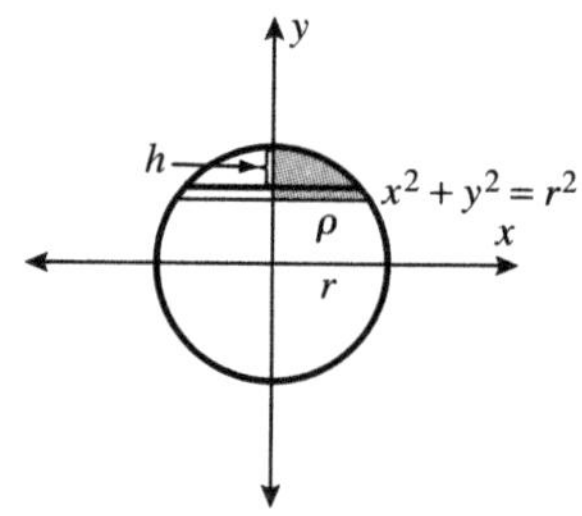

38. Find the volume generated by revolving the shaded region about the y-axis.

$$V = \pi \int_{-10}^{-10+h} (100 - y^2)dy = \frac{\pi}{3}h^2(30 - h)$$

Find dh/dt when $h = 5$ given that $dV/dt = 1/2$.

$$V = \frac{\pi}{3}(30h^2 - h^3),\ \frac{dV}{dt} = \frac{\pi}{3}(60h - 3h^2)\frac{dh}{dt},$$
$$\frac{1}{2} = \frac{\pi}{3}(300 - 75)\frac{dh}{dt},\ \frac{dh}{dt} = 1/(150\pi) \text{ ft/min}$$

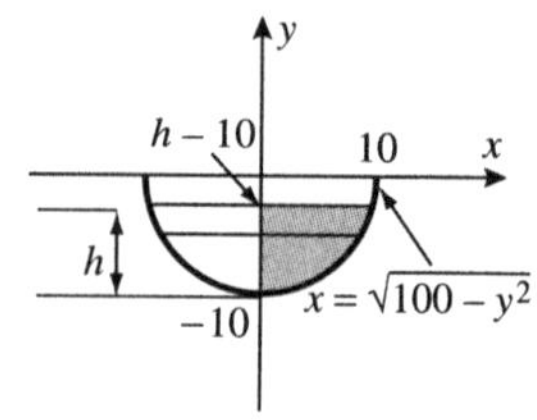

39. **(b)** $\Delta x = \frac{5}{10} = 0.5$; $\{y_0, y_1, \cdots, y_{10}\} = \{0, 2.00, 2.45, 2.45, 2.00, 1.46, 1.26, 1.25, 1.25, 1.25, 1.25\}$;

$$\text{left} = \pi \sum_{i=0}^{9} \left(\frac{y_i}{2}\right)^2 \Delta x \approx 11.157;$$

$$\text{right} = \pi \sum_{i=1}^{10} \left(\frac{y_i}{2}\right)^2 \Delta x \approx 11.771;\ V \approx \text{average} = 11.464 \text{ cm}^3$$

40. If $x = r/2$ then from $y^2 = r^2 - x^2$ we get $y = \pm\sqrt{3}r/2$ as limits of integration; for $-\sqrt{3} \le y \le \sqrt{3}$,

$A(y) = \pi[(r^2 - y^2) - r^2/4] = \pi(3r^2/4 - y^2)$, thus

$$V = \pi \int_{-\sqrt{3}r/2}^{\sqrt{3}r/2} (3r^2/4 - y^2)dy$$
$$= 2\pi \int_{0}^{\sqrt{3}r/2} (3r^2/4 - y^2)dy = \sqrt{3}\pi r^3/2.$$

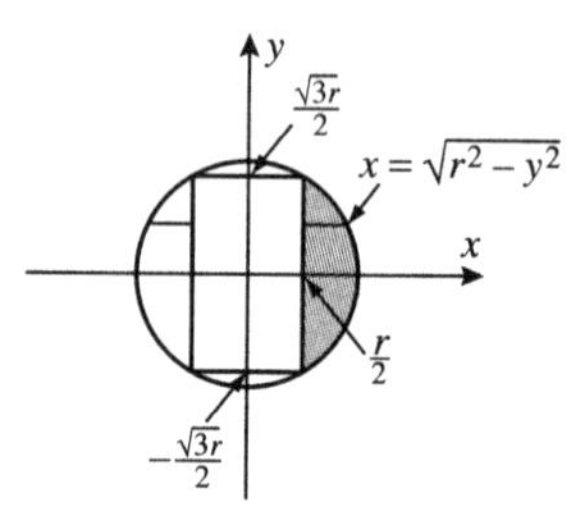

41. **(a)**

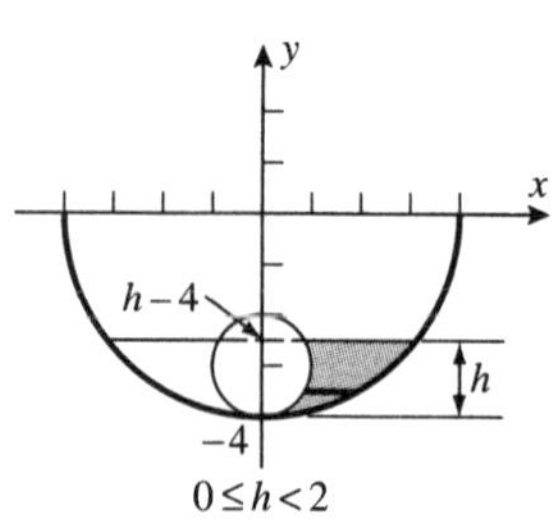

(b)

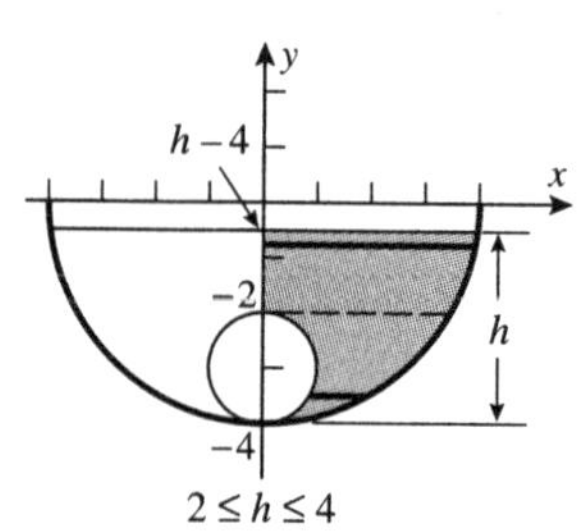

If the cherry is partially submerged then $0 \le h < 2$ as shown in Figure (a); if it is totally submerged then $2 \le h \le 4$ as shown in Figure (b). The radius of the glass is 4 cm and that of the cherry is 1 cm so points on the sections shown in the figures satisfy the equations $x^2 + y^2 = 16$ and $x^2 + (y+3)^2 = 1$. We will find the volumes of the solids that are generated when the shaded regions are revolved about the y-axis.

For $0 \le h < 2$,

$$V = \pi \int_{-4}^{h-4} [(16 - y^2) - (1 - (y+3)^2)]dy = 6\pi \int_{-4}^{h-4} (y+4)dy = 3\pi h^2;$$

for $2 \le h \le 4$,

$$\begin{aligned} V &= \pi \int_{-4}^{-2} [(16 - y^2) - (1 - (y+3)^2)]dy + \pi \int_{-2}^{h-4} (16 - y^2)dy \\ &= 6\pi \int_{-4}^{-2} (y+4)dy + \pi \int_{-2}^{h-4} (16 - y^2)dy = 12\pi + \frac{1}{3}\pi(12h^2 - h^3 - 40) \\ &= \frac{1}{3}\pi(12h^2 - h^3 - 4) \end{aligned}$$

so

$$V = \begin{cases} 3\pi h^2 & \text{if } 0 \le h < 2 \\ \frac{1}{3}\pi(12h^2 - h^3 - 4) & \text{if } 2 \le h \le 4 \end{cases}$$

42. $x = h \pm \sqrt{r^2 - y^2}$,

$$\begin{aligned} V &= \pi \int_{-r}^{r} \left[(h + \sqrt{r^2 - y^2})^2 - (h - \sqrt{r^2 - y^2})^2\right] dy \\ &= 4\pi h \int_{-r}^{r} \sqrt{r^2 - y^2} dy \\ &= 4\pi h \left(\frac{1}{2}\pi r^2\right) = 2\pi^2 r^2 h \end{aligned}$$

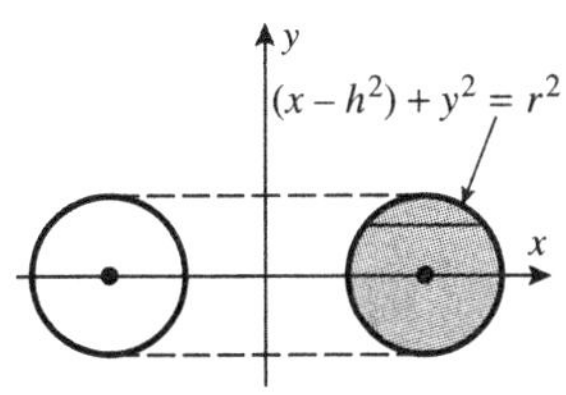

43. $\tan\theta = h/x$ so $h = x\tan\theta$,

$$A(y) = \frac{1}{2}hx = \frac{1}{2}x^2\tan\theta = \frac{1}{2}(r^2 - y^2)\tan\theta$$

because $x^2 = r^2 - y^2$,

$$\begin{aligned} V &= \frac{1}{2}\tan\theta \int_{-r}^{r} (r^2 - y^2)dy \\ &= \tan\theta \int_{0}^{r} (r^2 - y^2)dy = \frac{2}{3}r^3\tan\theta \end{aligned}$$

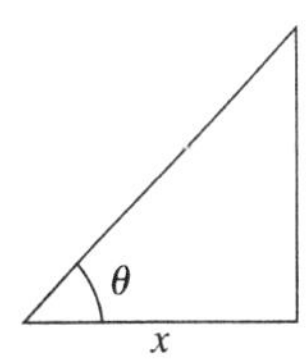

44. $A(x) = (x\tan\theta)(2\sqrt{r^2 - x^2})$
$= 2(\tan\theta)x\sqrt{r^2 - x^2},$
$V = 2\tan\theta\int_0^r x\sqrt{r^2 - x^2}dx$
$= \frac{2}{3}r^3\tan\theta$

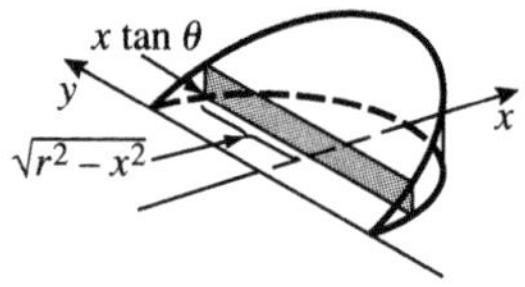

45. Each cross section perpendicular to the y-axis is a square so
$A(y) = x^2 = r^2 - y^2,$
$\frac{1}{8}V = \int_0^r (r^2 - y^2)dy$
$V = 8(2r^3/3) = 16r^3/3$

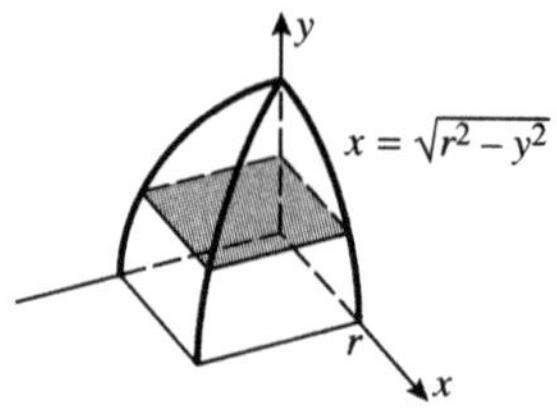

46. The regular cylinder of radius r and height h has the same circular cross sections as do those of the oblique clinder, so by Cavalieri's Principle, they have the same volume: $\pi r^2 h$.

EXERCISE SET 6.3

1. $V = \int_1^2 2\pi x(x^2)dx = 2\pi\int_1^2 x^3 dx = 15\pi/2$

2. $V = \int_0^{\sqrt{2}} 2\pi x(\sqrt{4 - x^2} - x)dx = 2\pi\int_0^{\sqrt{2}}(x\sqrt{4 - x^2} - x^2)dx = \frac{8\pi}{3}(2 - \sqrt{2})$

3. $V = \int_0^1 2\pi y(2y - 2y^2)dy = 4\pi\int_0^1 (y^2 - y^3)dy = \pi/3$

4. $V = \int_0^2 2\pi y[y - (y^2 - 2)]dy = 2\pi\int_0^2 (y^2 - y^3 + 2y)dy = 16\pi/3$

5. $V = \int_0^1 2\pi(x)(x^3)dx$
$= 2\pi\int_0^1 x^4 dx = 2\pi/5$

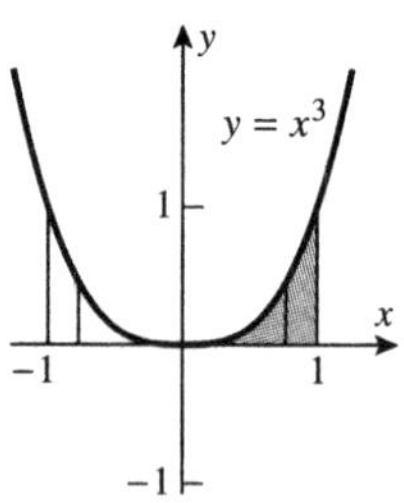

6. $V = \int_4^9 2\pi x(\sqrt{x})dx$
$= 2\pi\int_4^9 x^{3/2}dx = 844\pi/5$

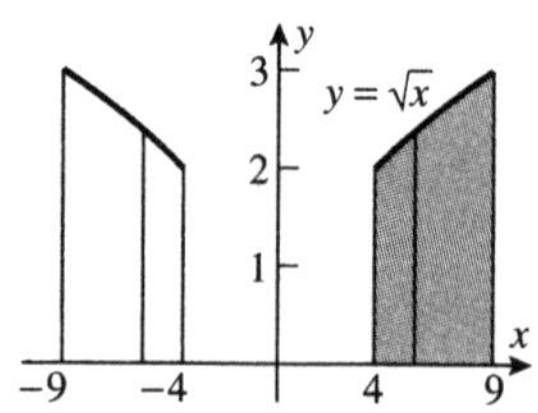

7. $V = \int_1^3 2\pi x(1/x)dx = 2\pi\int_1^3 dx = 4\pi$

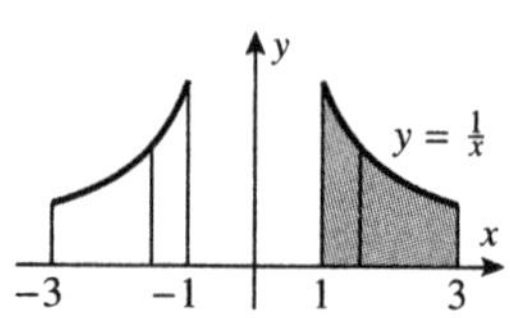

8. $V = \int_0^{\sqrt{\pi}/2} 2\pi x \cos(x^2)dx = \pi/\sqrt{2}$

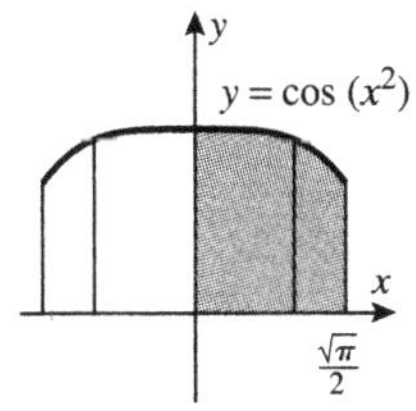

9. $V = \int_1^2 2\pi x[(2x-1) - (-2x+3)]dx$

$= 8\pi \int_1^2 (x^2 - x)dx = 20\pi/3$

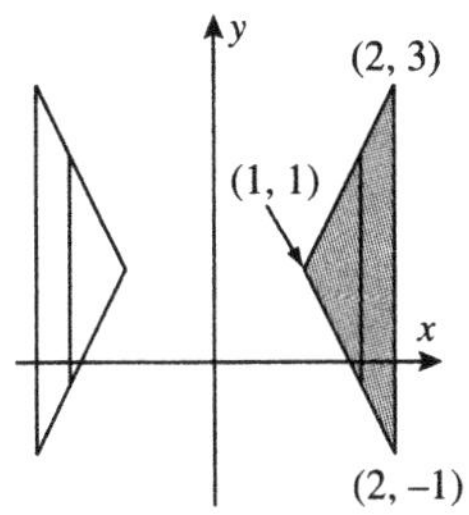

10. $V = \int_0^2 2\pi x(2x - x^2)dx$

$= 2\pi \int_0^2 (2x^2 - x^3)dx = \frac{8}{3}\pi$

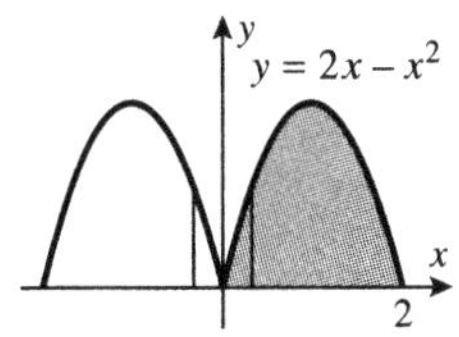

11. $V = \int_0^1 2\pi y^3 dy = \pi/2$

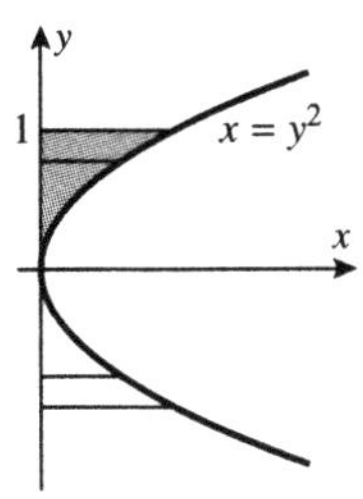

12. $V = \int_2^3 2\pi y(2y)dy = 4\pi \int_2^3 y^2 dy = 76\pi/3$

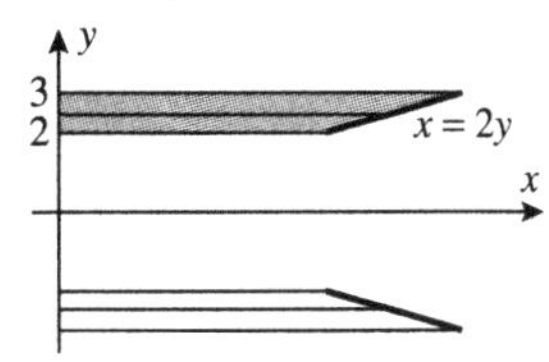

13. $V = \int_0^1 2\pi y(1 - \sqrt{y})dy$

$= 2\pi \int_0^1 (y - y^{3/2})dy = \pi/5$

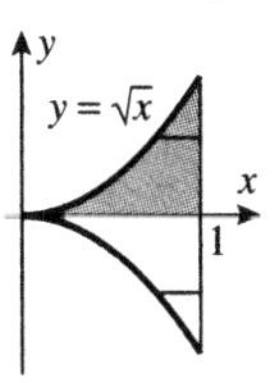

14. $V = \int_1^4 2\pi y(5 - y - 4/y)dy$

$= 2\pi \int_1^4 (5y - y^2 - 4)dy = 9\pi$

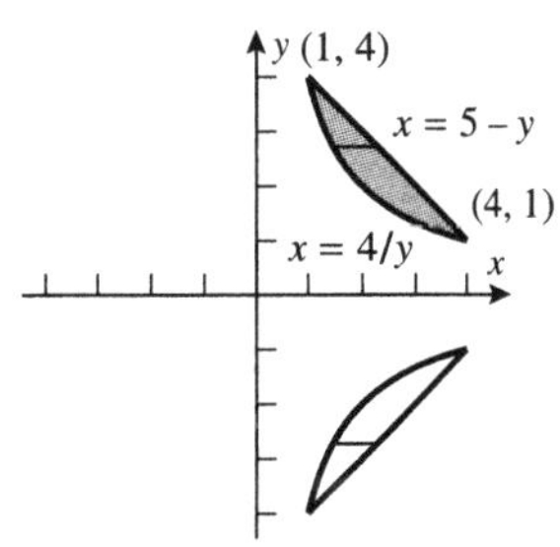

15. $V = 2\pi \int_0^{\pi} x \sin x dx = 2\pi^2$

16. $V = 2\pi \int_0^{\pi/2} x \cos x dx = \pi^2 - 2\pi$

17. (a) $V = \int_0^1 2\pi x(x^3 - 3x^2 + 2x)dx = 7\pi/30$

(b) much easier; the method of slicing would require that x be expressed in terms of y.

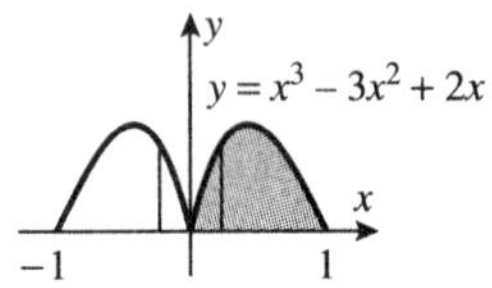

18. $V = \int_1^2 2\pi(x + 1)(1/x^3)dx$

$= 2\pi \int_1^2 (x^{-2} + x^{-3})dx = 7\pi/4$

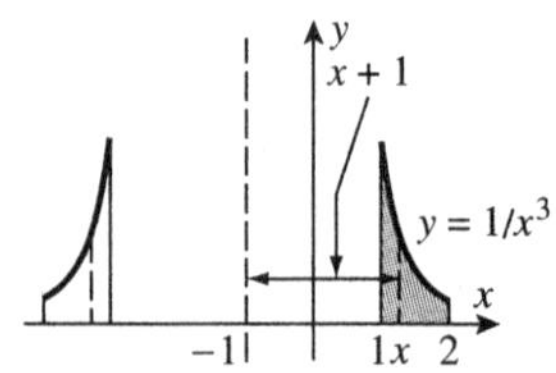

19. $V = \int_0^1 2\pi(1 - y)y^{1/3}dy$

$= 2\pi \int_0^1 (y^{1/3} - y^{4/3})dy = 9\pi/14$

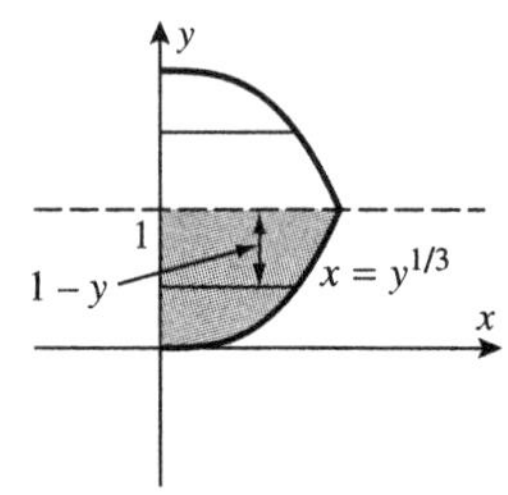

20. (a) $\int_a^b 2\pi x[f(x) - g(x)]dx$ (b) $\int_c^d 2\pi y[f(y) - g(y)]dy$

21. $x = \dfrac{h}{r}(r - y)$ is an equation of the line through $(0, r)$ and $(h, 0)$ so

$V = \int_0^r 2\pi y \left[\dfrac{h}{r}(r - y)\right] dy$

$= \dfrac{2\pi h}{r} \int_0^r (ry - y^2)dy = \pi r^2 h/3$

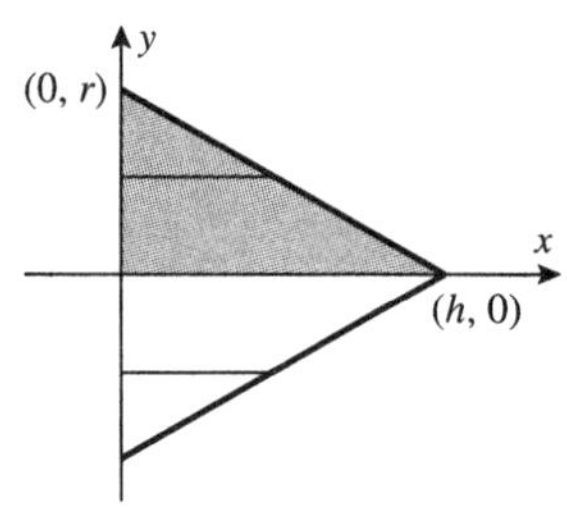

22. $V = \int_0^{k/4} 2\pi(k/2 - x)2\sqrt{kx}dx$

$= 2\pi\sqrt{k} \int_0^{k/4} (kx^{1/2} - 2x^{3/2})dx = 7\pi k^3/60$

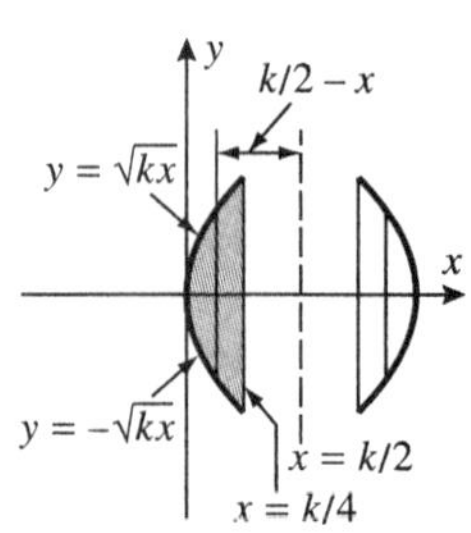

23. $V = \int_0^a 2\pi x(2\sqrt{r^2 - x^2})dx = 4\pi \int_0^a x(r^2 - x^2)^{1/2}dx$

$= -\frac{4\pi}{3}(r^2 - x^2)^{3/2}\Big]_0^a = \frac{4\pi}{3}\left[r^3 - (r^2 - a^2)^{3/2}\right]$

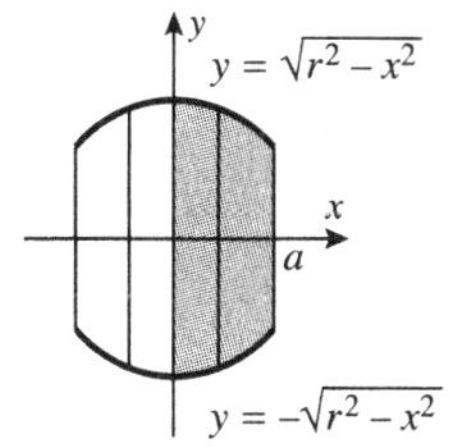

24. $V = \int_{-a}^{a} 2\pi(b - x)(2\sqrt{a^2 - x^2})dx$

$= 4\pi b \int_{-a}^{a} \sqrt{a^2 - x^2}dx - 4\pi \int_{-a}^{a} x\sqrt{a^2 - x^2}dx$

$= 4\pi b \cdot (\text{area of a semicircle of radius } a) - 4\pi(0)$

$= 2\pi^2 a^2 b$

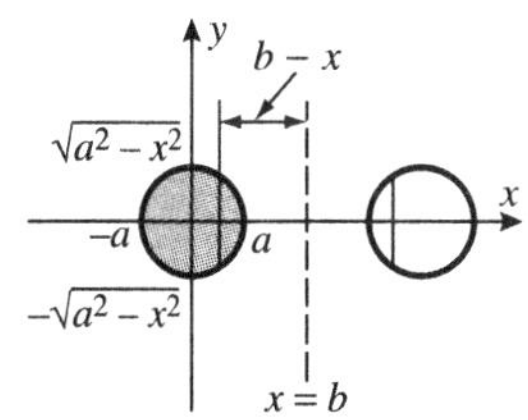

25. $V_x = \pi \int_{1/2}^{b} \frac{1}{x^2}dx = \pi(2 - 1/b)$, $V_y = 2\pi \int_{1/2}^{b} dx = \pi(2b - 1)$;

$V_x = V_y$ if $2 - 1/b = 2b - 1$, $2b^2 - 3b + 1 = 0$, solve to get $b = 1/2$ (reject) or $b = 1$.

EXERCISE SET 6.4

1. (a) $\frac{dy}{dx} = 2$, $L = \int_1^2 \sqrt{1 + 4}dx = \sqrt{5}$

(b) $\frac{dx}{dy} = \frac{1}{2}$, $L = \int_2^4 \sqrt{1 + 1/4}\,dy = 2\sqrt{5}/2 = \sqrt{5}$

2. $\frac{dx}{dt} = 1$, $\frac{dy}{dt} = 5$, $L = \int_0^1 \sqrt{1^2 + 5^2}\,dt = \sqrt{26}$

3. $f'(x) = \frac{9}{2}x^{1/2}$, $1 + [f'(x)]^2 = 1 + \frac{81}{4}x$,

$L = \int_0^1 \sqrt{1 + 81x/4}\,dx = \frac{8}{243}\left(1 + \frac{81}{4}x\right)^{3/2}\Big]_0^1 = (85\sqrt{85} - 8)/243$

4. $g'(y) = y(y^2 + 2)^{1/2}$, $1 + [g'(y)]^2 = 1 + y^2(y^2 + 2) = y^4 + 2y^2 + 1 = (y^2 + 1)^2$,

$L = \int_0^1 \sqrt{(y^2 + 1)^2}dy = \int_0^1 (y^2 + 1)dy = 4/3$

5. $\dfrac{dy}{dx} = \dfrac{2}{3}x^{-1/3},\ 1 + \left(\dfrac{dy}{dx}\right)^2 = 1 + \dfrac{4}{9}x^{-2/3} = \dfrac{9x^{2/3}+4}{9x^{2/3}},$

$$L = \int_1^8 \frac{\sqrt{9x^{2/3}+4}}{3x^{1/3}}dx = \frac{1}{18}\int_{13}^{40} u^{1/2}du,\ u = 9x^{2/3}+4$$

$$= \frac{1}{27}u^{3/2}\Big]_{13}^{40} = \frac{1}{27}(40\sqrt{40} - 13\sqrt{13}) = \frac{1}{27}(80\sqrt{10} - 13\sqrt{13})$$

or (alternate solution)

$$x = y^{3/2},\ \frac{dx}{dy} = \frac{3}{2}y^{1/2},\ 1 + \left(\frac{dx}{dy}\right)^2 = 1 + \frac{9}{4}y = \frac{4+9y}{4},$$

$$L = \frac{1}{2}\int_1^4 \sqrt{4+9y}\,dy = \frac{1}{18}\int_{13}^{40} u^{1/2}du = \frac{1}{27}(80\sqrt{10} - 13\sqrt{13})$$

6. $f'(x) = \dfrac{1}{4}x^3 - x^{-3},\ 1 + [f'(x)]^2 = 1 + \left(\dfrac{1}{16}x^6 - \dfrac{1}{2} + x^{-6}\right) = \dfrac{1}{16}x^6 + \dfrac{1}{2} + x^{-6} = \left(\dfrac{1}{4}x^3 + x^{-3}\right)^2,$

$$L = \int_2^3 \sqrt{\left(\frac{1}{4}x^3 + x^{-3}\right)^2}dx = \int_2^3 \left(\frac{1}{4}x^3 + x^{-3}\right)dx = 595/144$$

7. $x = g(y) = \dfrac{1}{24}y^3 + 2y^{-1},\ g'(y) = \dfrac{1}{8}y^2 - 2y^{-2},$

$$1 + [g'(y)]^2 = 1 + \left(\frac{1}{64}y^4 - \frac{1}{2} + 4y^{-4}\right) = \frac{1}{64}y^4 + \frac{1}{2} + 4y^{-4} = \left(\frac{1}{8}y^2 + 2y^{-2}\right)^2,$$

$$L = \int_2^4 \left(\frac{1}{8}y^2 + 2y^{-2}\right)dy = 17/6$$

8. $g'(y) = \dfrac{1}{2}y^3 - \dfrac{1}{2}y^{-3},\ 1 + [g'(y)]^2 = 1 + \left(\dfrac{1}{4}y^6 - \dfrac{1}{2} + \dfrac{1}{4}y^{-6}\right) = \left(\dfrac{1}{2}y^3 + \dfrac{1}{2}y^{-3}\right)^2,$

$$L = \int_1^4 \left(\frac{1}{2}y^3 + \frac{1}{2}y^{-3}\right)dy = 2055/64$$

9. $(dx/dt)^2 + (dy/dt)^2 = (t^2)^2 + (t)^2 = t^2(t^2+1),\ L = \displaystyle\int_0^1 t(t^2+1)^{1/2}dt = (2\sqrt{2} - 1)/3$

10. $(dx/dt)^2 + (dy/dt)^2 = [2(1+t)]^2 + [3(1+t)^2]^2 = (1+t)^2[4 + 9(1+t)^2],$

$$L = \int_0^1 (1+t)[4 + 9(1+t)^2]^{1/2}dt = (80\sqrt{10} - 13\sqrt{13})/27$$

11. $(dx/dt)^2 + (dy/dt)^2 = (-2\sin 2t)^2 + (2\cos 2t)^2 = 4,\ L = \displaystyle\int_0^{\pi/2} 2\,dt = \pi$

12. $(dx/dt)^2 + (dy/dt)^2 = (-\sin t + \sin t + t\cos t)^2 + (\cos t - \cos t + t\sin t)^2 = t^2,$

$$L = \int_0^{\pi} t\,dt = \pi^2/2$$

13. **(a)** $(dx/d\theta)^2 + (dy/d\theta)^2 = (a(1-\cos\theta))^2 + (a\sin\theta)^2 = a^2(2-2\cos\theta)$, so

$$L = \int_0^{2\pi} \sqrt{(dx/d\theta)^2 + (dy/d\theta)^2}\, d\theta = a\int_0^{2\pi} \sqrt{2(1-\cos\theta)}\, d\theta$$

14. **(a)** Use the interval $0 \le \phi < 2\pi$.

(b) $(dx/d\phi)^2 + (dy/d\phi)^2 = (-3a\cos^2\phi\sin\phi)^2 + (3a\sin^2\phi\cos\phi)^2$

$= 9a^2\cos^2\phi\sin^2\phi(\cos^2\phi + \sin^2\phi) = (9a^2/4)\sin^2 2\phi$, so

$$L = (3a/2)\int_0^{2\pi} |\sin 2\phi|\, d\phi = 6a\int_0^{\pi/2} \sin 2\phi\, d\phi = -3a\cos 2\phi\Big]_0^{\pi/2} = 6a$$

15. **(a)**

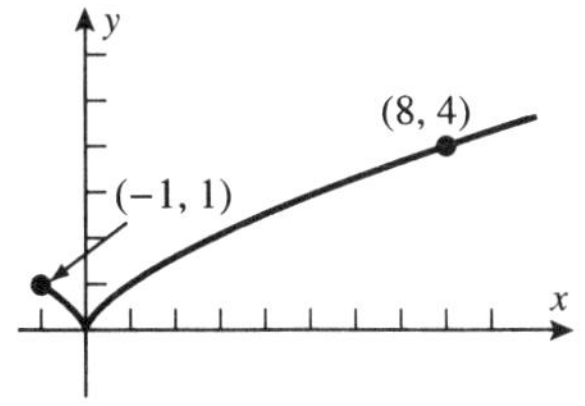

(b) dy/dx does not exist at $x = 0$.

(c) $x = g(y) = y^{3/2}$, $g'(y) = \dfrac{3}{2}y^{1/2}$,

$$L = \int_0^1 \sqrt{1+9y/4}\, dy \quad (\text{portion for } -1 \le x \le 0)$$

$$+ \int_0^4 \sqrt{1+9y/4}\, dy \quad (\text{portion for } 0 \le x \le 8)$$

$$= \frac{8}{27}\left(\frac{13}{8}\sqrt{13} - 1\right) + \frac{8}{27}(10\sqrt{10} - 1) = (13\sqrt{13} + 80\sqrt{10} - 16)/27$$

16. For (4), express the curve $y = f(x)$ in the parametric form $x = t, y = f(t)$ so $dx/dt = 1$ and $dy/dt = f'(t) = f'(x) = dy/dx$. For (5), express $x = g(y)$ as $x = g(t), y = t$ so $dx/dt = g'(t) = g'(y) = dx/dy$ and $dy/dt = 1$.

17. $L = \displaystyle\int_0^2 \sqrt{1+4x^2}\, dx \approx 4.645975301$

18. $L = \displaystyle\int_0^\pi \sqrt{1+\cos^2 y}\, dy \approx 3.820197789$

19. Numerical integration yields: in Exercise 17, $L \approx 4.646783762$; in Exercise 18, $L \approx 3.820197788$.

20. $0 \le m \le f'(x) \le M$, so $m^2 \le [f'(x)]^2 \le M^2$, and $1 + m^2 \le 1 + [f'(x)]^2 \le 1 + M^2$; thus

$\sqrt{1+m^2} \le \sqrt{1+[f'(x)]^2} \le \sqrt{1+M^2}$,

$\displaystyle\int_a^b \sqrt{1+m^2}dx \le \int_a^b \sqrt{1+[f'(x)]^2}dx \le \int_a^b \sqrt{1+M^2}\, dx$, and

$(b-a)\sqrt{1+m^2} \le L \le (b-a)\sqrt{1+M^2}$

21. $f'(x) = \cos x$, $\sqrt{2}/2 \le \cos x \le 1$ for $0 \le x \le \pi/4$ so

$(\pi/4)\sqrt{1+1/2} \le L \le (\pi/4)\sqrt{1+1}$, $\dfrac{\pi}{4}\sqrt{3/2} \le L \le \dfrac{\pi}{4}\sqrt{2}$.

22. $(dx/dt)^2 + (dy/dt)^2 = (-a\sin t)^2 + (b\cos t)^2 = a^2\sin^2 t + b^2\cos^2 t$

$$= a^2(1-\cos^2 t) + b^2\cos^2 t = a^2 - (a^2-b^2)\cos^2 t$$

$$= a^2\left[1 - \frac{a^2-b^2}{a^2}\cos^2 t\right] = a^2[1-k^2\cos^2 t],$$

$$L = \int_0^{2\pi} a\sqrt{1-k^2\cos^2 t}\,dt = 4a\int_0^{\pi/2}\sqrt{1-k^2\cos^2 t}\,dt$$

23. **(a)** $(dx/dt)^2 + (dy/dt)^2 = 4\sin^2 t + \cos^2 t = 4\sin^2 t + (1-\sin^2 t) = 1 + 3\sin^2 t,$

$$L = \int_0^{2\pi}\sqrt{1+3\sin^2 t}\,dt = 4\int_0^{\pi/2}\sqrt{1+3\sin^2 t}\,dt$$

(b) 9.69

(c) distance traveled $= \int_{1.5}^{4.8}\sqrt{1+3\sin^2 t}\,dt \approx 5.16$ cm

24. The distance is $\int_0^{4.6}\sqrt{1+(2.09-0.82x)^2}\,dx \approx 6.65$ m

25. $L = \int_0^{\pi}\sqrt{1+(k\cos x)^2}\,dx$

k	1	2	1.84	1.83	1.832
L	3.8202	5.2704	5.0135	4.9977	5.0008

Experimentation yields the values in the table, which by the Intermediate-Value Theorem show that the true solution k to $L = 5$ lies between $k = 1.83$ and $k = 1.832$, so $k = 1.83$ to two decimal places.

EXERCISE SET 6.5

1. $S = \int_0^1 2\pi(7x)\sqrt{1+49}dx = 70\pi\sqrt{2}\int_0^1 x\,dx = 35\pi\sqrt{2}$

2. $f'(x) = \dfrac{1}{2\sqrt{x}},\ 1+[f'(x)]^2 = 1 + \dfrac{1}{4x}$

$$S = \int_1^4 2\pi\sqrt{x}\sqrt{1+\frac{1}{4x}}dx = 2\pi\int_1^4\sqrt{x+1/4}dx = \pi(17\sqrt{17}-5\sqrt{5})/6$$

3. $f'(x) = -x/\sqrt{4-x^2},\ 1+[f'(x)]^2 = 1 + \dfrac{x^2}{4-x^2} = \dfrac{4}{4-x^2},$

$$S = \int_{-1}^1 2\pi\sqrt{4-x^2}(2/\sqrt{4-x^2})dx = 4\pi\int_{-1}^1 dx = 8\pi$$

4. $y = f(x) = x^3$ for $1 \le x \le 2$, $f'(x) = 3x^2$,

$$S = \int_1^2 2\pi x^3\sqrt{1+9x^4}dx = \frac{\pi}{27}(1+9x^4)^{3/2}\Big]_1^2 = 5\pi(29\sqrt{145}-2\sqrt{10})/27$$

5. $S=\int_0^2 2\pi(9y+1)\sqrt{82}dy=2\pi\sqrt{82}\int_0^2(9y+1)dy=40\pi\sqrt{82}$

6. $g'(y)=3y^2$, $S=\int_0^1 2\pi y^3\sqrt{1+9y^4}dy=\pi(10\sqrt{10}-1)/27$

7. $g'(y)=-y/\sqrt{9-y^2}$, $1+[g'(y)]^2=\dfrac{9}{9-y^2}$, $S=\int_{-2}^2 2\pi\sqrt{9-y^2}\cdot\dfrac{3}{\sqrt{9-y^2}}dy=6\pi\int_{-2}^2 dy=24\pi$

8. $g'(y)=-(1-y)^{-1/2}$, $1+[g'(y)]^2=\dfrac{2-y}{1-y}$,

$$S=\int_{-1}^0 2\pi(2\sqrt{1-y})\frac{\sqrt{2-y}}{\sqrt{1-y}}dy=4\pi\int_{-1}^0\sqrt{2-y}\,dy=8\pi(3\sqrt{3}-2\sqrt{2})/3$$

9. $f'(x)=\dfrac{1}{2}x^{-1/2}-\dfrac{1}{2}x^{1/2}$, $1+[f'(x)]^2=1+\dfrac{1}{4}x^{-1}-\dfrac{1}{2}+\dfrac{1}{4}x=\left(\dfrac{1}{2}x^{-1/2}+\dfrac{1}{2}x^{1/2}\right)^2$,

$$S=\int_1^3 2\pi\left(x^{1/2}-\frac{1}{3}x^{3/2}\right)\left(\frac{1}{2}x^{-1/2}+\frac{1}{2}x^{1/2}\right)dx=\frac{\pi}{3}\int_1^3(3+2x-x^2)dx=16\pi/9$$

10. $f'(x)=x^2-\dfrac{1}{4}x^{-2}$, $1+[f'(x)]^2=1+\left(x^4-\dfrac{1}{2}+\dfrac{1}{16}x^{-4}\right)=\left(x^2+\dfrac{1}{4}x^{-2}\right)^2$,

$$S=\int_1^2 2\pi\left(\frac{1}{3}x^3+\frac{1}{4}x^{-1}\right)\left(x^2+\frac{1}{4}x^{-2}\right)dx=2\pi\int_1^2\left(\frac{1}{3}x^5+\frac{1}{3}x+\frac{1}{16}x^{-3}\right)dx=515\pi/64$$

11. $x=g(y)=\dfrac{1}{4}y^4+\dfrac{1}{8}y^{-2}$, $g'(y)=y^3-\dfrac{1}{4}y^{-3}$,

$$1+[g'(y)]^2=1+\left(y^6-\frac{1}{2}+\frac{1}{16}y^{-6}\right)=\left(y^3+\frac{1}{4}y^{-3}\right)^2,$$

$$S=\int_1^2 2\pi\left(\frac{1}{4}y^4+\frac{1}{8}y^{-2}\right)\left(y^3+\frac{1}{4}y^{-3}\right)dy=\frac{\pi}{16}\int_1^2(8y^7+6y+y^{-5})dy=16{,}911\pi/1024$$

12. $x=g(y)=\sqrt{16-y}$; $g'(y)=-\dfrac{1}{2\sqrt{16-y}}$, $1+[g'(y)]^2=\dfrac{65-4y}{4(16-y)}$,

$$S=\int_0^{15}2\pi\sqrt{16-y}\sqrt{\frac{65-4y}{4(16-y)}}dy=\pi\int_0^{15}\sqrt{65-4y}\,dy=(65\sqrt{65}-5\sqrt{5})\frac{\pi}{6}$$

13. $f'(x)=\cos x$, $1+[f'(x)]^2=1+\cos^2 x$, $S=\int_0^\pi 2\pi\sin x\sqrt{1+\cos^2 x}\,dx=2\pi(\sqrt{2}+\ln(\sqrt{2}+1))$

14. $x=g(y)=\tan y$, $g'(y)=\sec^2 y$, $1+[g'(y)]^2=1+\sec^4 y$;

$$S=\int_0^{\pi/4}2\pi\tan y\sqrt{1+\sec^4 y}\,dy\approx 3.84$$

15. Revolve the line segment joining the points $(0,0)$ and (h,r) about the x-axis. An equation of the line segment is $y=(r/h)x$ for $0\le x\le h$ so

$$S=\int_0^h 2\pi(r/h)x\sqrt{1+r^2/h^2}dx=\frac{2\pi r}{h^2}\sqrt{r^2+h^2}\int_0^h x\,dx=\pi r\sqrt{r^2+h^2}$$

16. $f(x)=\sqrt{r^2-x^2}$, $f'(x)=-x/\sqrt{r^2-x^2}$, $1+[f'(x)]^2=r^2/(r^2-x^2)$,

$$S=\int_{-r}^{r}2\pi\sqrt{r^2-x^2}(r/\sqrt{r^2-x^2})dx=2\pi r\int_{-r}^{r}dx=4\pi r^2$$

17. $g(y)=\sqrt{r^2-y^2}$, $g'(y)=-y/\sqrt{r^2-y^2}$, $1+[g'(y)]^2=r^2/(r^2-y^2)$,

(a) $S=\int_{r-h}^{r}2\pi\sqrt{r^2-y^2}\sqrt{r^2/(r^2-y^2)}\,dy=2\pi r\int_{r-h}^{r}dy=2\pi rh$

(b) From Part (a), the surface area common to two polar caps of height $h_1>h_2$ is $2\pi rh_1-2\pi rh_2=2\pi r(h_1-h_2)$.

18. For (4), express the curve $y=f(x)$ in the parametric form $x=t, y=f(t)$ so $dx/dt=1$ and $dy/dt=f'(t)=f'(x)=dy/dx$. For (5), express $x=g(y)$ as $x=g(t), y=t$ so $dx/dt=g'(t)=g'(y)=dx/dy$ and $dy/dt=1$.

19. $x'=2t, y'=2, (x')^2+(y')^2=4t^2+4$

$$S=2\pi\int_0^4(2t)\sqrt{4t^2+4}dt=8\pi\int_0^4 t\sqrt{t^2+1}dt=\frac{8\pi}{3}(17\sqrt{17}-1)$$

20. $x'=-2\cos t\sin t, y'=5\cos t, (x')^2+(y')^2=4\cos^2 t\sin^2 t+25\cos^2 t$,

$$S=2\pi\int_0^{\pi/2}5\sin t\sqrt{4\cos^2 t\sin^2 t+25\cos^2 t}\,dt=\frac{\pi}{6}(145\sqrt{29}-625)$$

21. $x'=1$, $y'=4t$, $(x')^2+(y')^2=1+16t^2$, $S=2\pi\int_0^1 t\sqrt{1+16t^2}\,dt=\frac{\pi}{24}(17\sqrt{17}-1)$

22. $x'=-2\sin t\cos t, y'=2\sin t\cos t, (x')^2+(y')^2=8\sin^2 t\cos^2 t$

$$S=2\pi\int_0^{\pi/2}\cos^2 t\sqrt{8\sin^2 t\cos^2 t}\,dt=4\sqrt{2}\pi\int_0^{\pi/2}\cos^3 t\sin t\,dt=\sqrt{2}\pi$$

23. $x'=-r\sin t$, $y'=r\cos t$, $(x')^2+(y')^2=r^2$,

$$S=2\pi\int_0^{\pi}r\sin t\sqrt{r^2}\,dt=2\pi r^2\int_0^{\pi}\sin t\,dt=4\pi r^2$$

24. $\dfrac{dx}{d\phi}=a(1-\cos\phi)$, $\dfrac{dy}{d\phi}=a\sin\phi$, $\left(\dfrac{dx}{d\phi}\right)^2+\left(\dfrac{dy}{d\phi}\right)^2=2a^2(1-\cos\phi)$

$$S=2\pi\int_0^{2\pi}a(1-\cos\phi)\sqrt{2a^2(1-\cos\phi)}\,d\phi=2\sqrt{2}\pi a^2\int_0^{2\pi}(1-\cos\phi)^{3/2}d\phi,$$

but $1-\cos\phi=2\sin^2\dfrac{\phi}{2}$ so $(1-\cos\phi)^{3/2}=2\sqrt{2}\sin^3\dfrac{\phi}{2}$ for $0\le\phi\le\pi$ and, taking advantage of the symmetry of the cycloid, $S=16\pi a^2\int_0^{\pi}\sin^3\dfrac{\phi}{2}d\phi=64\pi a^2/3$.

25. **(a)** length of arc of sector = circumference of base of cone,

$\ell\theta=2\pi r, \theta=2\pi r/\ell$; $S=$ area of sector $=\dfrac{1}{2}\ell^2(2\pi r/\ell)=\pi r\ell$

(b) $S = \pi r_2 \ell_2 - \pi r_1 \ell_1 = \pi r_2(\ell_1 + \ell) - \pi r_1 \ell_1 = \pi[(r_2 - r_1)\ell_1 + r_2 \ell]$;
Using similar triangles $\ell_2/r_2 = \ell_1/r_1, r_1\ell_2 = r_2\ell_1, r_1(\ell_1 + \ell) = r_2\ell_1, (r_2 - r_1)\ell_1 = r_1\ell$
so $S = \pi\,(r_1\ell + r_2\ell) = \pi\,(r_1 + r_2)\,\ell$.

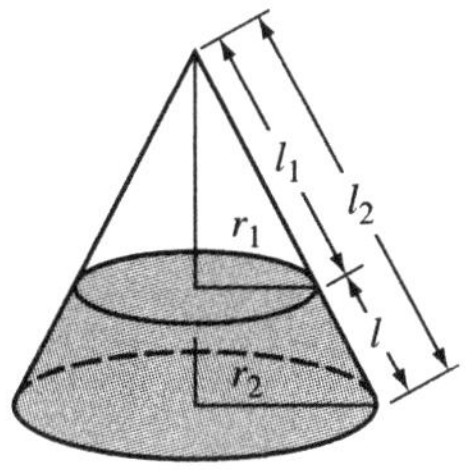

26. $S = \displaystyle\int_a^b 2\pi[f(x) + k]\sqrt{1 + [f'(x)]^2}dx$

27. $2\pi k\sqrt{1 + [f'(x)]^2} \le 2\pi f(x)\sqrt{1 + [f'(x)]^2} \le 2\pi K\sqrt{1 + [f'(x)]^2}$, so

$$\int_a^b 2\pi k\sqrt{1 + [f'(x)]^2}dx \le \int_a^b 2\pi f(x)\sqrt{1 + [f'(x)]^2}dx \le \int_a^b 2\pi K\sqrt{1 + [f'(x)]^2}dx,$$

$$2\pi k\int_a^b \sqrt{1 + [f'(x)]^2}dx \le S \le 2\pi K\int_a^b \sqrt{1 + [f'(x)]^2}dx,\ 2\pi kL \le S \le 2\pi KL$$

28. **(a)** $1 \le \sqrt{1 + [f'(x)]^2}$ so $2\pi f(x) \le 2\pi f(x)\sqrt{1 + [f'(x)]^2}$,

$$\int_a^b 2\pi f(x)dx \le \int_a^b 2\pi f(x)\sqrt{1 + [f'(x)]^2}dx,\ 2\pi\int_a^b f(x)dx \le S, 2\pi A \le S$$

(b) $2\pi A = S$ if $f'(x) = 0$ for all x in $[a, b]$ so $f(x)$ is constant on $[a, b]$.

EXERCISE SET 6.6

1. **(a)** $W = F \cdot d = 30(7) = 210$ ft·lb

(b) $W = \displaystyle\int_1^6 F(x)\,dx = \int_1^6 x^{-2}\,dx = -\frac{1}{x}\Big]_1^6 = 5/6$ ft·lb

2. $W = \displaystyle\int_0^5 F(x)\,dx = \int_0^2 40\,dx - \int_2^5 \frac{40}{3}(x - 5)\,dx = 80 + 60 = 140$ J

3. distance traveled $= \displaystyle\int_0^5 v(t)\,dt = \int_0^5 \frac{4t}{5}\,dt = \frac{2}{5}t^2\Big]_0^5 = 10$ ft. The force is a constant 10 lb, so the work done is $10 \cdot 10 = 100$ ft·lb.

4. **(a)** $F(x) = kx, F(0.05) = 0.05k = 45, k = 900$ N/m

(b) $W = \displaystyle\int_0^{0.03} 900x\,dx = 0.405$ J

(c) $W = \displaystyle\int_{0.05}^{0.10} 900x\,dx = 3.375$ J

5. $F(x) = kx$, $F(0.2) = 0.2k = 100$, $k = 500$ N/m, $W = \int_0^{0.8} 500x dx = 160$ J

6. $F(x) = kx$, $F(1/2) = k/2 = 6$, $k = 12$ N/m, $W = \int_0^2 12x\,dx = 24$ J

7. $W = \int_0^1 kx\,dx = k/2 = 10$, $k = 20$ lb/ft

8. $W = \int_0^6 (9 - x)62.4(25\pi)dx$

$= 1560\pi \int_0^6 (9 - x)dx = 56,160\pi$ ft·lb

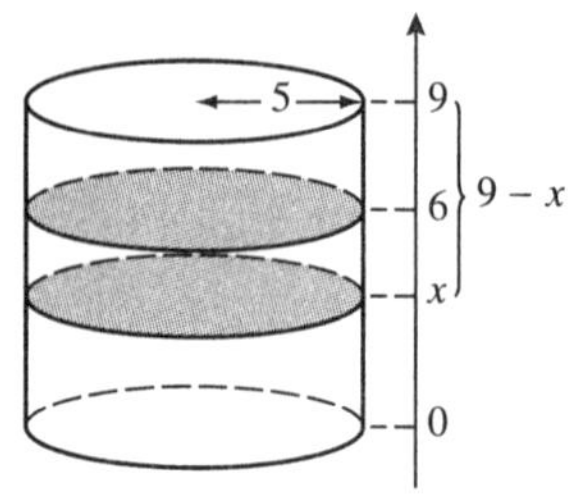

9. $W = \int_0^6 (9 - x)\rho(25\pi)dx = 900\pi\rho$ ft·lb

10. $r/10 = x/15$, $r = 2x/3$,

$W = \int_0^{10} (15 - x)62.4(4\pi x^2/9)dx$

$= \frac{83.2}{3}\pi \int_0^{10} (15x^2 - x^3)dx$

$= 208,000\pi/3$ ft·lb

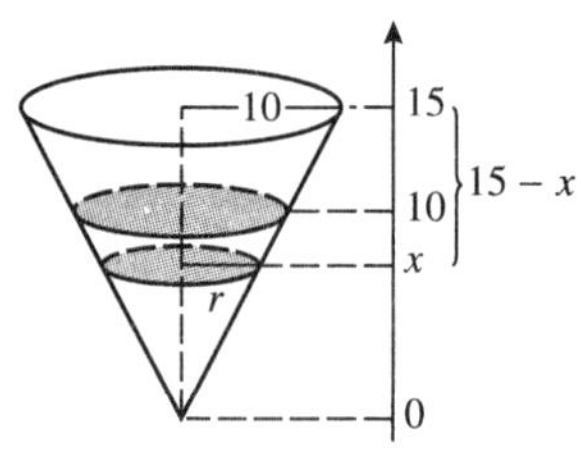

11. $w/4 = x/3, w = 4x/3$,

$W = \int_0^2 (3 - x)(9810)(4x/3)(6)dx$

$= 78480 \int_0^2 (3x - x^2)dx$

$= 261,600$ J

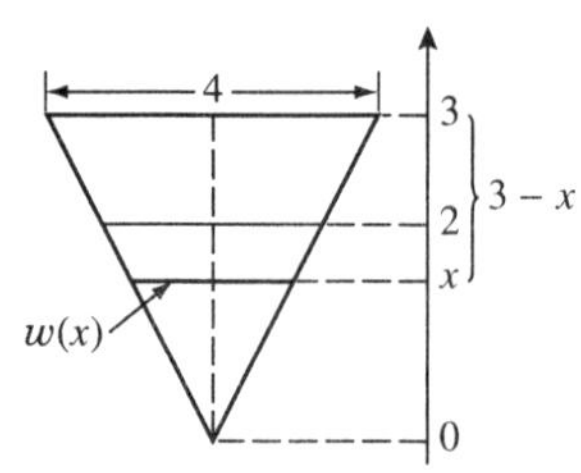

12. $w = 2\sqrt{4 - x^2}$

$W = \int_{-2}^2 (3 - x)(50)(2\sqrt{4 - x^2})(10)dx$

$= 3000 \int_{-2}^2 \sqrt{4 - x^2}dx - 1000 \int_{-2}^2 x\sqrt{4 - x^2}dx$

$= 3000[\pi(2)^2/2] - 0 = 6000\pi$ ft·lb

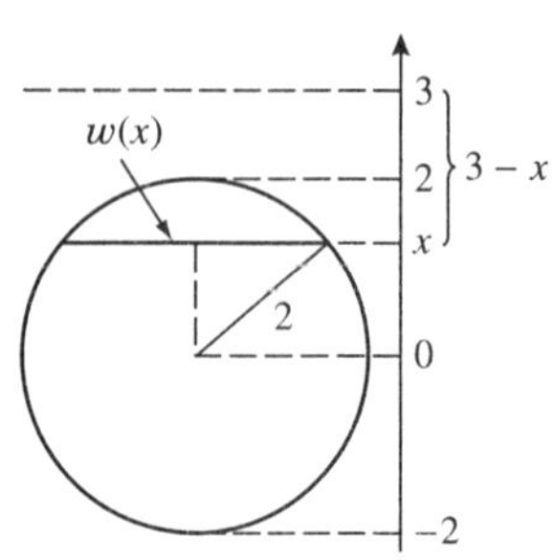

13. (a) $W = \int_0^9 (10 - x)62.4(300)dx$

$= 18{,}720 \int_0^9 (10 - x)dx$

$= 926{,}640$ ft·lb

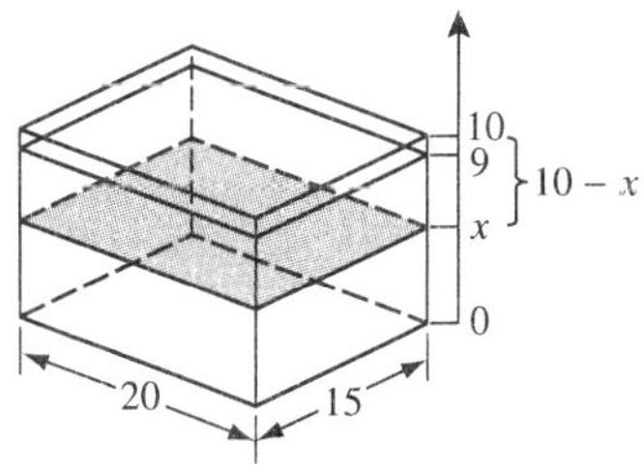

(b) to empty the pool in one hour would require $926{,}640/3600 = 257.4$ ft·lb of work per second so hp of motor $= 257.4/550 = 0.468$

14. $W = \int_0^9 x(62.4)(300)\,dx = 18{,}720 \int_0^9 x\,dx = (81/2)18{,}720 = 758{,}160$ ft·lb

15. $W = \int_0^{100} 15(100 - x)dx$

$= 75{,}000$ ft·lb

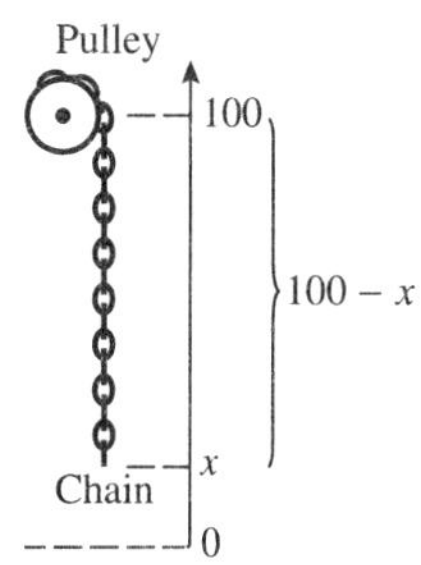

16. The total time of winding the rope is (20 ft)/(2 ft/s) = 10 s. During the time interval from time t to time $t + \Delta t$ the work done is $\Delta W = F(t) \cdot \Delta x$.
The distance $\Delta x = 2\Delta t$, and the force $F(t)$ is given by the weight $w(t)$ of the bucket, rope and water at time t. The bucket and its remaining water together weigh $(3+20) - t/2$ lb, and the rope is $20 - 2t$ ft long and weighs $4(20 - 2t)$ oz or $5 - t/2$ lb. Thus at time t the bucket, water and rope together weigh $w(t) = 23 - t/2 + 5 - t/2 = 28 - t$ lb.
The amount of work done in the time interval from time t to time $t + \Delta t$ is thus $\Delta W = (28 - t)2\Delta t$, and the total work done is

$$W = \lim_{n\to+\infty} \sum (28 - t)2\Delta t = \int_0^{10} (28 - t)2\,dt = 2(28t - t^2/2)\Big|_0^{10} = 460 \text{ ft·lb}.$$

17. When the rocket is x ft above the ground

total weight = weight of rocket + weight of fuel

$= 3 + [40 - 2(x/1000)]$

$= 43 - x/500$ tons,

$W = \int_0^{3000} (43 - x/500)dx = 120{,}000$ ft·tons

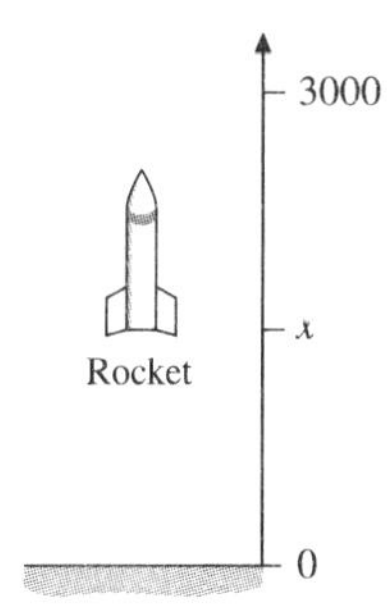

18. Let $F(x)$ be the force needed to hold charge A at position x, then

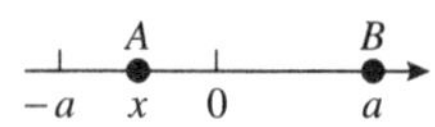

$$F(x) = \frac{c}{(a-x)^2},\ F(-a) = \frac{c}{4a^2} = k,$$

so $c = 4a^2 k$.

$$W = \int_{-a}^{0} 4a^2 k(a-x)^{-2} dx = 2ak \text{ J}$$

19. **(a)** $150 = k/(4000)^2$, $k = 2.4 \times 10^9$, $w(x) = k/x^2 = 2{,}400{,}000{,}000/x^2$ lb

(b) $6000 = k/(4000)^2$, $k = 9.6 \times 10^{10}$, $w(x) = \left(9.6 \times 10^{10}\right)/(x+4000)^2$ lb

(c) $W = \int_{4000}^{5000} 9.6(10^{10})x^{-2}dx = 4{,}800{,}000$ mi·lb $= 2.5344 \times 10^{10}$ ft·lb

20. **(a)** $20 = k/(1080)^2$, $k = 2.3328 \times 10^7$, weight $= w(x+1080) = 2.3328 \cdot 10^7/(x+1080)^2$ lb

(b) $W = \int_0^{10.8} [2.3328 \cdot 10^7/(x+1080)^2]\, dx = 213.86$ mi·lb $= 1{,}129{,}188$ ft·lb

21. $W = F \cdot d = (6.40 \times 10^5)(3.00 \times 10^3) = 1.92 \times 10^9$ J; from the Work-Energy Relationship (5),

$v_f^2 = 2W/m + v_i^2 = 2(1.92 \cdot 10^9)/(4 \cdot 10^5) + 20^2 = 10{,}000$, $v_f = 100$ m/s

22. $W = F \cdot d = (2.00 \times 10^5)(2.00 \times 10^5) = 4 \times 10^{10}$ J; from the Work-Energy Relationship (5), $v_f^2 = 2W/m + v_i^2 = 8 \cdot 10^{10}/(2 \cdot 10^3) + 10^8 \approx 11.832$ m/s.

23. **(a)** The kinetic energy would have decreased by $\frac{1}{2}mv^2 = \frac{1}{2}4 \cdot 10^6(15000)^2 = 4.5 \times 10^{14}$ J

(b) $(4.5 \times 10^{14})/(4.2 \times 10^{15}) \approx 0.107$

(c) $\frac{1000}{13}(0.107) \approx 8.24$ bombs

EXERCISE SET 6.7

1. **(a)** $F = \rho h A = 62.4(5)(100) = 31{,}200$ lb
$P = \rho h = 62.4(5) = 312$ lb/ft^2

(b) $F = \rho h A = 9810(10)(25) = 2{,}452{,}500$ N
$P = \rho h = 9810(10) = 98.1$ kPa

2. **(a)** $F = PA = 6 \cdot 10^5(160) = 9.6 \times 10^7$ N

(b) $F = PA = 100(60) = 6000$ lb

3. $F = \int_0^2 62.4x(4)dx$

$= 249.6 \int_0^2 x\, dx = 499.2$ lb

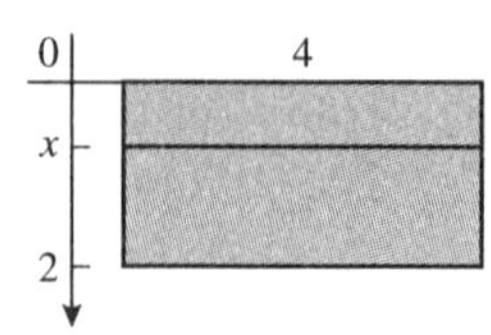

4. $F = \int_1^3 9810x(4)dx$

$= 39{,}240 \int_1^3 x\,dx$

$= 156{,}960\,\text{N}$

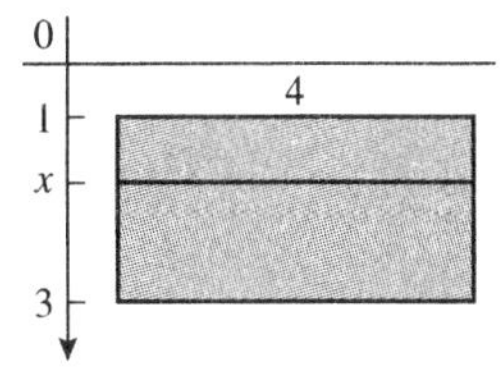

5. $F = \int_0^5 9810x(2\sqrt{25-x^2})dx$

$= 19{,}620 \int_0^5 x(25-x^2)^{1/2}dx$

$= 8.175 \times 10^5\,\text{N}$

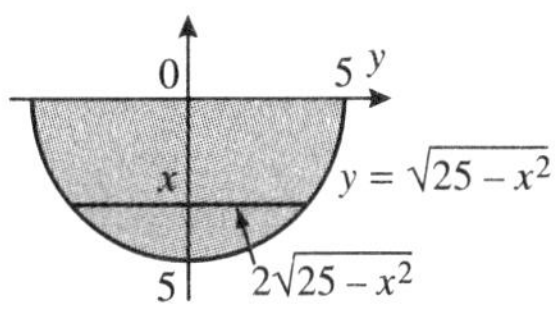

6. By similar triangles

$\frac{w(x)}{4} = \frac{2\sqrt{3}-x}{2\sqrt{3}}$, $w(x) = \frac{2}{\sqrt{3}}(2\sqrt{3}-x)$,

$F = \int_0^{2\sqrt{3}} 62.4x\left[\frac{2}{\sqrt{3}}(2\sqrt{3}-x)\right]dx$

$= \frac{124.8}{\sqrt{3}}\int_0^{2\sqrt{3}}(2\sqrt{3}x - x^2)dx = 499.2\,\text{lb}$

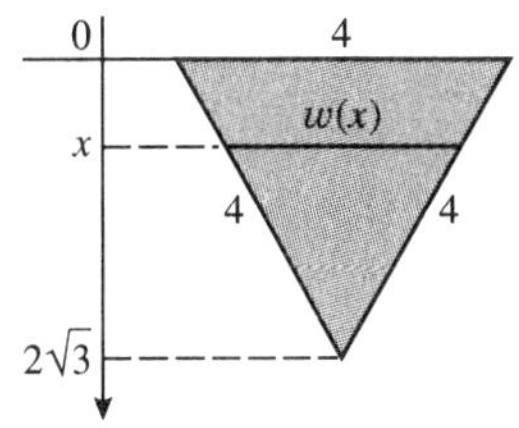

7. By similar triangles

$\frac{w(x)}{6} = \frac{10-x}{8}$

$w(x) = \frac{3}{4}(10-x)$,

$F = \int_2^{10} 9810x\left[\frac{3}{4}(10-x)\right]dx$

$= 7357.5\int_2^{10}(10x - x^2)dx = 1{,}098{,}720\,\text{N}$

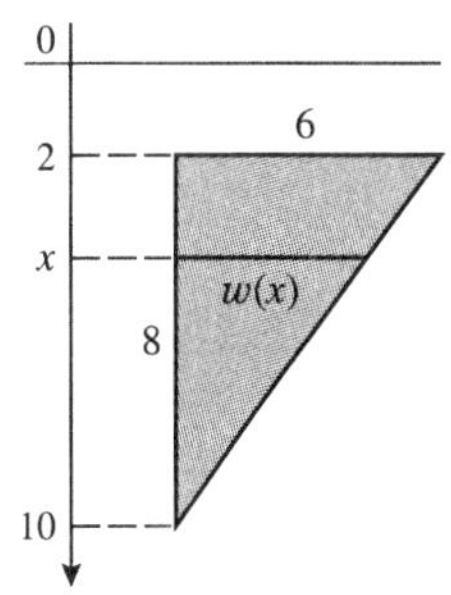

8. $w(x) = 16 + 2u(x)$, but

$\frac{u(x)}{4} = \frac{12-x}{8}$ so $u(x) = \frac{1}{2}(12-x)$,

$w(x) = 16 + (12-x) = 28 - x$,

$F = \int_4^{12} 62.4x(28-x)dx$

$= 62.4\int_4^{12}(28x - x^2)dx = 77{,}209.6$ lb.

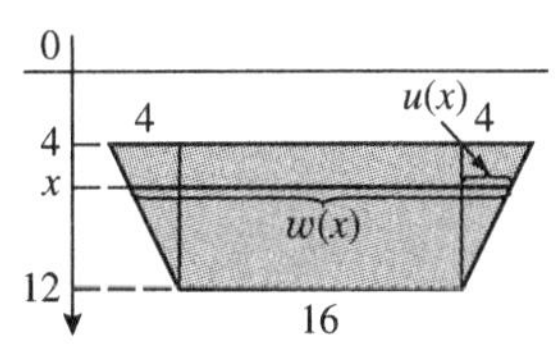

9. Yes: if $\rho_2 = 2\rho_1$ then $F_2 = \int_a^b \rho_2 h(x)w(x)\,dx = \int_a^b 2\rho_1 h(x)w(x)\,dx = 2\int_a^b \rho_1 h(x)w(x)\,dx = 2F_1$.

10. $F = \int_0^2 50x(2\sqrt{4-x^2})dx$

$= 100\int_0^2 x(4-x^2)^{1/2}dx$

$= 800/3\,\text{lb}$

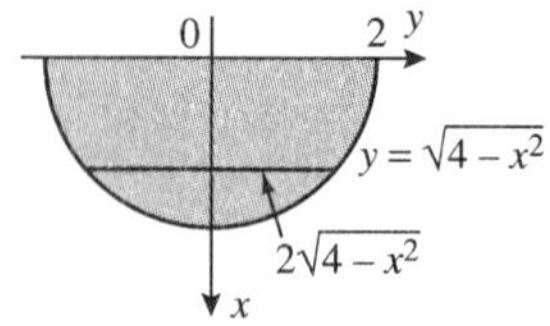

11. Find the forces on the upper and lower halves and add them:

$\dfrac{w_1(x)}{\sqrt{2}a} = \dfrac{x}{\sqrt{2}a/2}$, $w_1(x) = 2x$

$F_1 = \int_0^{\sqrt{2}a/2} \rho x(2x)dx = 2\rho\int_0^{\sqrt{2}a/2} x^2 dx = \sqrt{2}\rho a^3/6$,

$\dfrac{w_2(x)}{\sqrt{2}a} = \dfrac{\sqrt{2}a - x}{\sqrt{2}a/2}$, $w_2(x) = 2(\sqrt{2}a - x)$

$F_2 = \int_{\sqrt{2}a/2}^{\sqrt{2}a} \rho x[2(\sqrt{2}a - x)]dx = 2\rho\int_{\sqrt{2}a/2}^{\sqrt{2}a}(\sqrt{2}ax - x^2)dx = \sqrt{2}\rho a^3/3$,

$F = F_1 + F_2 = \sqrt{2}\rho a^3/6 + \sqrt{2}\rho a^3/3 = \rho a^3/\sqrt{2}$ lb

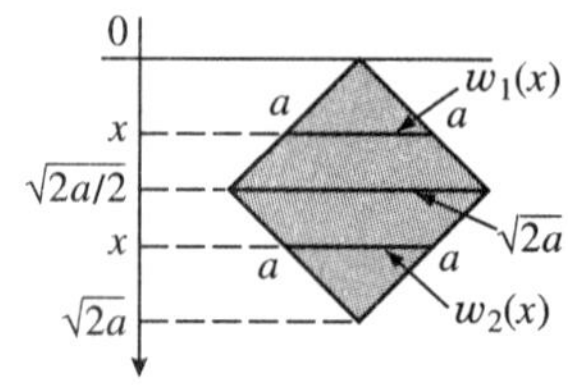

12. If a constant vertical force is applied to a flat plate which is horizontal and the magnitude of the force is F, then, if the plate is tilted so as to form an angle θ with the vertical, the magnitude of the force on the plate decreases to $F\cos\theta$.
Suppose that a flat surface is immersed, at an angle θ with the vertical, in a fluid of weight density ρ, and that the submerged portion of the surface extends from $x = a$ to $x = b$ along an x-axis whose positive diretion is not necessarily down, but is slanted.
Following the derivation of equation (8), we divide the interval $[a, b]$ into n subintervals
$a = x_0 < x_1 < \ldots < x_{n-1} < x_n = b$. Then the magnitude F_k of the force on the plate satisfies the inequalities $\rho h(x_{k-1})A_k\cos\theta \le F_k \le \rho h(x_k)A_k\cos\theta$, or equivalently that
$h(x_{k-1}) \le \dfrac{F_k\sec\theta}{\rho A_k} \le h(x_k)$. Following the argument in the text we arrive at the desired equation
$F = \int_a^b \rho h(x)w(x)\sec\theta\,dx$.

13. $\sqrt{16^2 + 4^2} = \sqrt{272} = 4\sqrt{17}$ is the other dimension of the bottom.

$(h(x) - 4)/4 = x/(4\sqrt{17})$

$h(x) = x/\sqrt{17} + 4$,

$\sec\theta = 4\sqrt{17}/16 = \sqrt{17}/4$

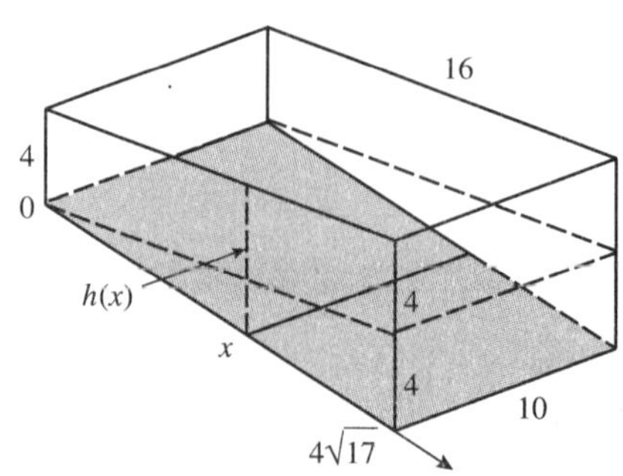

$$F = \int_0^{4\sqrt{17}} 62.4(x/\sqrt{17} + 4)10(\sqrt{17}/4)\,dx$$
$$= 156\sqrt{17}\int_0^{4\sqrt{17}} (x/\sqrt{17} + 4)dx$$
$$= 63{,}648\text{ lb}$$

14. If we lower the water level by y ft then the force F_1 is computed as in Exercise 13, but with $h(x)$ replaced by $h_1(x) = x/\sqrt{17} + 4 - y$, and we obtain

$$F_1 = F - y\int_0^{4\sqrt{17}} 62.4(10)\sqrt{17}/4\,dx = F - 624(17)y = 63{,}648 - 10{,}608y.$$

If $F_1 = F/2$ then $63{,}648/2 = 63{,}648 - 10{,}608y$, $y = 63{,}648/(2\cdot 10{,}608) = 3$, so the water level should be reduced by 3 ft.

15. $h(x) = x\sin 60^\circ = \sqrt{3}x/2$,
$\theta = 30^\circ$, $\sec\theta = 2/\sqrt{3}$,

$$F = \int_0^{100} 9810(\sqrt{3}x/2)(200)(2/\sqrt{3})\,dx$$
$$= 1{,}962{,}000\int_0^{100} x\,dx$$
$$= 9.81\times 10^9\text{ N}$$

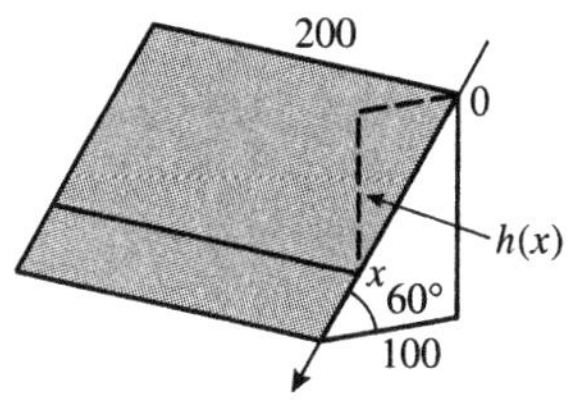

16. $$F = \int_h^{h+2} \rho_0 x(2)dx$$
$$= 2\rho_0\int_h^{h+2} x\,dx$$
$$= 4\rho_0(h+1)$$

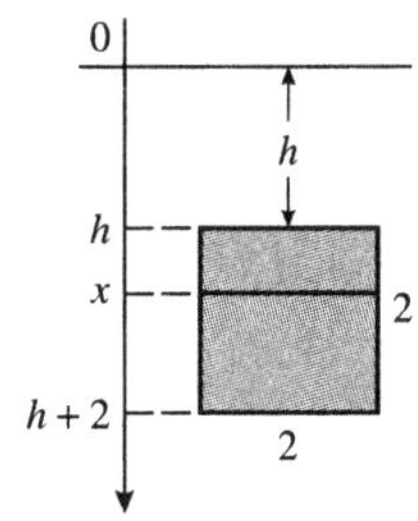

17. (a) From Exercise 16, $F = 4\rho_0(h+1)$ so (assuming that ρ_0 is constant) $dF/dt = 4\rho_0(dh/dt)$ which is a positive constant if dh/dt is a positive constant.

(b) If $dh/dt = 20$ then $dF/dt = 80\rho_0$ lb/min from Part (a).

18. (a) Let h_1 and h_2 be the maximum and minimum depths of the disk D_r. The pressure $P(r)$ on one side of the disk satisfies inequality (5):
$\rho h_1 \le P(r) \le \rho h_2$. But
$\lim_{r\to 0^+} h_1 = \lim_{r\to 0^+} h_2 = h$, and hence
$\rho h = \lim_{r\to 0^+}\rho h_1 \le \lim_{r\to 0^+} P(r) \le \lim_{r\to 0^+}\rho h_2 = \rho h$, so $\lim_{r\to 0^+} P(r) = \rho h$.

(b) The disks D_r in Part (a) have no particular direction (the axes of the disks have arbitrary direction). Thus P, the limiting value of $P(r)$, is independent of direction.

CHAPTER 6 SUPPLEMENTARY EXERCISES

6. **(a)** $A = \int_0^2 (2 + x - x^2)\,dx$

(b) $A = \int_0^2 \sqrt{y}\,dy + \int_2^4 [(\sqrt{y} - (y-2)]\,dy$

(c) $V = \pi \int_0^2 [(2+x)^2 - x^4]\,dx$

(d) $V = 2\pi \int_0^2 y\sqrt{y}\,dy + 2\pi \int_2^4 y[\sqrt{y} - (y-2)]\,dy$

(e) $V = 2\pi \int_0^2 x(2 + x - x^2)\,dx$

(f) $V = \pi \int_0^2 y\,dy + \int_2^4 \pi(y - (y-2)^2)\,dy$

7. **(a)** $A = \int_a^b (f(x) - g(x))\,dx + \int_b^c (g(x) - f(x))\,dx + \int_c^d (f(x) - g(x))\,dx$

(b) $A = \int_{-1}^0 (x^3 - x)\,dx + \int_0^1 (x - x^3)\,dx + \int_1^2 (x^3 - x)\,dx = \frac{1}{4} + \frac{1}{4} + \frac{9}{4} = \frac{11}{4}$

8. **(a)** $S = \int_0^{8/27} 2\pi x\sqrt{1 + x^{-4/3}}\,dx$ **(b)** $S = \int_0^2 2\pi \frac{y^3}{27}\sqrt{1 + y^4/81}\,dy$

(c) $S = \int_0^2 2\pi(y+2)\sqrt{1 + y^4/81}\,dy$

9. By implicit differentiation $\frac{dy}{dx} = -\left(\frac{y}{x}\right)^{1/3}$, so $1 + \left(\frac{dy}{dx}\right)^2 = 1 + \left(\frac{y}{x}\right)^{2/3} = \frac{x^{2/3} + y^{2/3}}{x^{2/3}} = \frac{a^{2/3}}{x^{2/3}}$,

$$L = \int_{-a}^{-a/8} \frac{a^{1/3}}{(-x^{1/3})}dx = -a^{1/3}\int_{-a}^{-a/8} x^{-1/3}dx = 9a/8.$$

10. The base of the dome is a hexagon of side r. An equation of the circle of radius r that lies in a vertical x-y plane and passes through two opposite vertices of the base hexagon is $x^2 + y^2 = r^2$. A horizontal, hexagonal cross section at height y above the base has area $A(y) = \frac{3\sqrt{3}}{2}x^2 = \frac{3\sqrt{3}}{2}(r^2 - y^2)$, hence the volume is $V = \int_0^r \frac{3\sqrt{3}}{2}(r^2 - y^2)\,dy = \sqrt{3}r^3$.

11. Let the sphere have radius R, the hole radius r. By the Pythagorean Theorem, $r^2 + (L/2)^2 = R^2$. Use cylindrical shells to calculate the volume of the solid obtained by rotating about the y-axis the region $r < x < R$, $-\sqrt{R^2 - x^2} < y < \sqrt{R^2 - x^2}$:

$$V = \int_r^R (2\pi x)2\sqrt{R^2 - x^2}\,dx = -\frac{4}{3}\pi(R^2 - x^2)^{3/2}\Big]_r^R = \frac{4}{3}\pi(L/2)^3,$$

so the volume is independent of R.

12. $V = 2\int_0^{L/2} \pi \frac{16R^2}{L^4}(x^2 - L^2/4)^2 = \frac{4\pi}{15}LR^2$

13. (a)

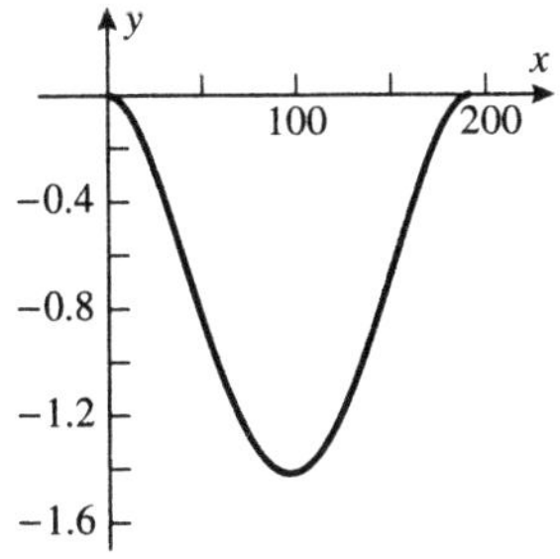

(b) The maximum deflection occurs at $x = 96$ inches (the midpoint of the beam) and is about 1.42 in.

(c) The length of the centerline is $\int_0^{192} \sqrt{1 + (dy/dx)^2}\, dx = 192.026$ in.

14. $y = 0$ at $x = b = 30.585$; distance $= \int_0^b \sqrt{1 + (12.54 - 0.82x)^2}\, dx = 196.306$ yd

15. (a) $F = kx$, $\frac{1}{2} = k\frac{1}{4}$, $k = 2$, $W = \int_0^{1/4} kx\, dx = 1/16$ J

(b) $25 = \int_0^L kx\, dx = kL^2/2$, $L = 5$ m

16. $F = 30x + 2000$, $W = \int_0^{150} (30x + 2000)\, dx = 15 \cdot 150^2 + 2000 \cdot 150 = 637{,}500$ lb·ft

17. (a) $F = \int_0^1 \rho x 3\, dx$ N

(b) By similar triangles $\frac{w(x)}{4} = \frac{x}{2}$, $w(x) = 2x$, so

$F = \int_1^4 \rho(1 + x)2x\, dx$ lb/ft^2.

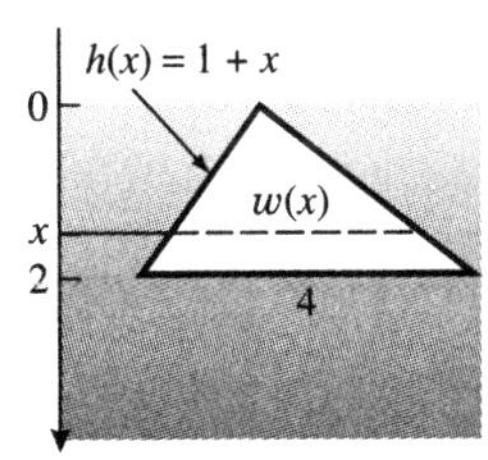

(c) A formula for the parabola is $y = \frac{8}{125}x^2 - 10$, so $F = \int_{-10}^{0} 9810|y|2\sqrt{\frac{125}{8}(y + 10)}\, dy$ N.

18. The x-coordinates of the points of intersection are $a \approx -0.423028$ and $b \approx 1.725171$; the area is $\int_a^b (2\sin x - x^2 + 1)dx \approx 2.542696$.

19. Let (a, k), where $\pi/2 < a < \pi$, be the coordinates of the point of intersection of $y = k$ with $y = \sin x$. Thus $k = \sin a$ and if the shaded areas are equal,

$$\int_0^a (k - \sin x)dx = \int_0^a (\sin a - \sin x)\, dx = a \sin a + \cos a - 1 = 0$$

Solve for a to get $a \approx 2.331122$, so $k = \sin a \approx 0.724611$.

20. The volume is given by $2\pi \int_0^k x \sin x\, dx = 2\pi(\sin k - k \cos k) = 8$; solve for k to get $k = 1.736796$.

CHAPTER 7

Exponential, Logarithmic, and Inverse Trigonometric Functions

EXERCISE SET 7.1

1. **(a)** $f(g(x)) = 4(x/4) = x$, $g(f(x)) = (4x)/4 = x$, f and g are inverse functions
(b) $f(g(x)) = 3(3x - 1) + 1 = 9x - 2 \neq x$ so f and g are not inverse functions
(c) $f(g(x)) = \sqrt[3]{(x^3 + 2) - 2} = x$, $g(f(x)) = (x - 2) + 2 = x$, f and g are inverse functions
(d) $f(g(x)) = (x^{1/4})^4 = x$, $g(f(x)) = (x^4)^{1/4} = |x| \neq x$, f and g are not inverse functions

2. **(a)** They are inverse functions.

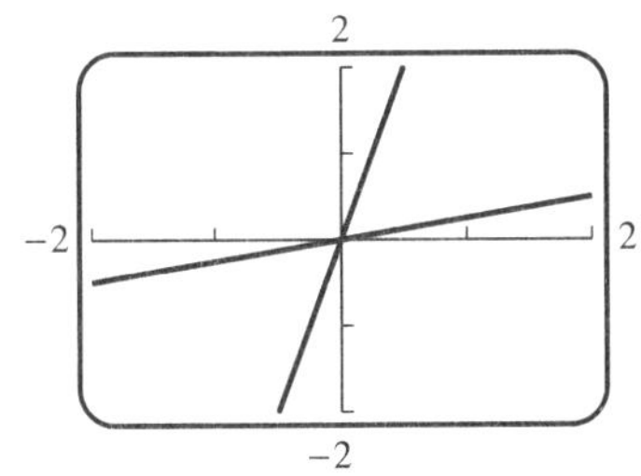

(b) The graphs are not reflections of each other about the line $y = x$.

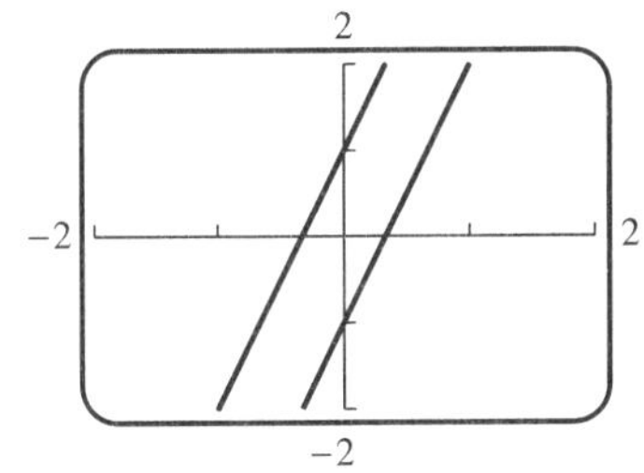

(c) They are inverse functions provided the domain of g is restricted to $[0, +\infty)$

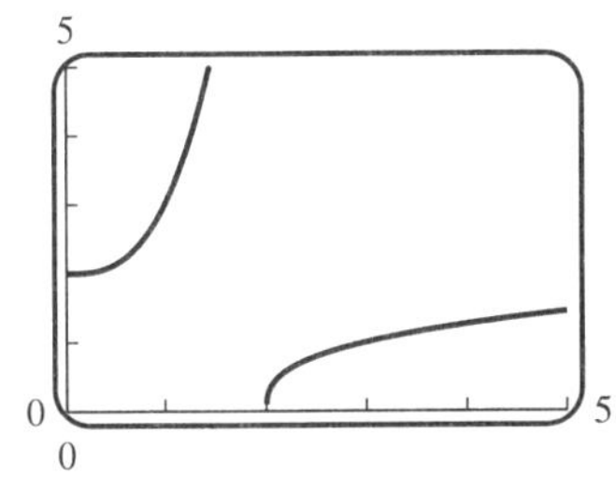

(d) They are inverse functions provided the domain of $f(x)$ is restricted to $[0, +\infty)$

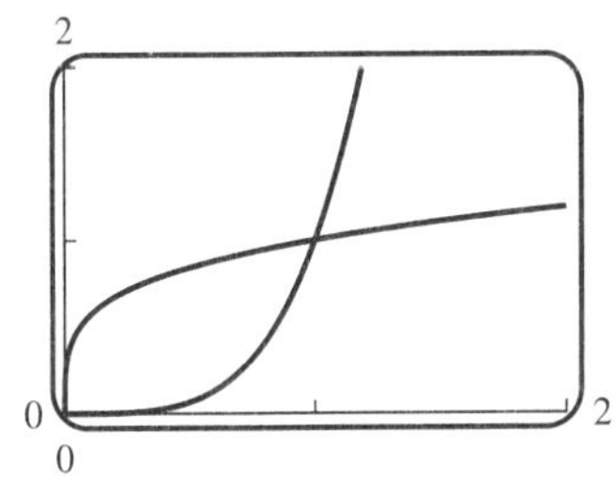

3. **(a)** yes; all outputs (the elements of row two) are distinct
(b) no; $f(1) = f(6)$

4. (a) no; it is easy to conceive of, say, 8 people in line at two different times
   (b) no; perhaps your weight remains constant for more than a year
   (c) yes, since the function is increasing, in the sense that the greater the volume, the greater the weight

5. (a) yes (b) yes (c) no (d) yes (e) no (f) no

6. (a) no, the horizontal line test fails

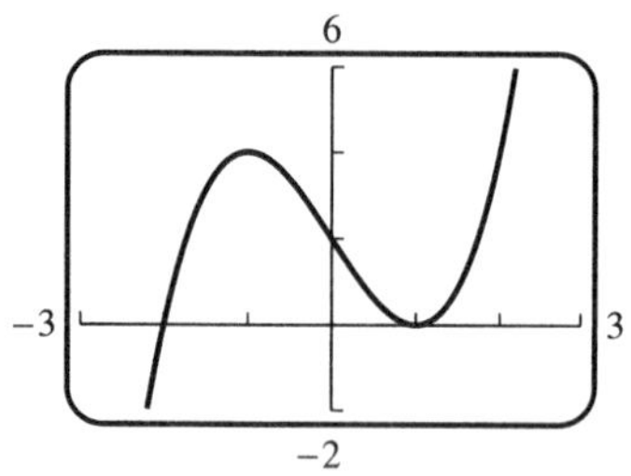

(b) yes, horizontal line test

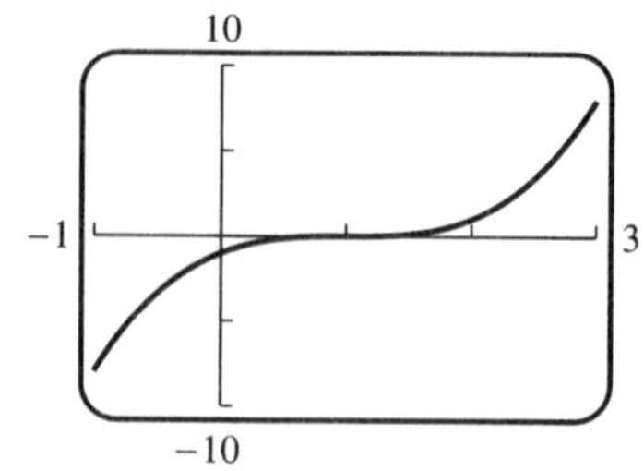

7. (a) no, the horizontal line test fails
   (b) no, the horizontal line test fails
   (c) yes, horizontal line test

8. (d) no, the horizontal line test fails
   (e) no, the horizontal line test fails
   (f) yes, horizontal line test

9. (a) f has an inverse because the graph passes the horizontal line test. To compute $f^{-1}(2)$ start at 2 on the y-axis and go to the curve and then down, so $f^{-1}(2) = 8$; similarly, $f^{-1}(-1) = -1$ and $f^{-1}(0) = 0$.
   (b) domain of f^{-1} is $[-2, 2]$, range is $[-8, 8]$
   (c)

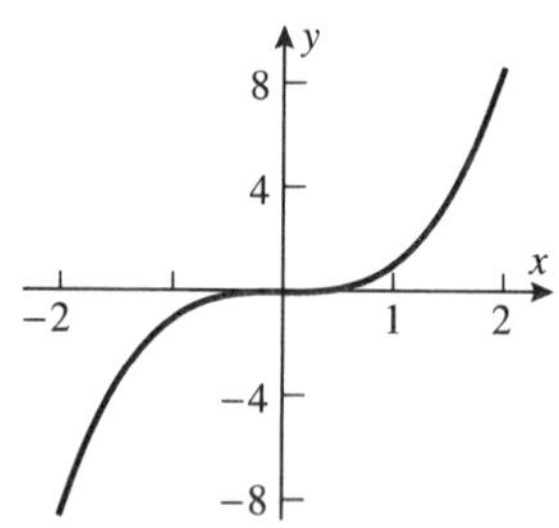

10. (a) the horizontal line test fails
    (b) $-\infty < x \le -1$; $-1 \le x \le 2$; and $2 \le x < 4$.

11. (a) $f'(x) = 2x + 8$; $f' < 0$ on $(-\infty, -4)$ and $f' > 0$ on $(-4, +\infty)$; not one-to-one
    (b) $f'(x) = 10x^4 + 3x^2 + 3 \ge 3 > 0$; $f'(x)$ is positive for all x, so f is one-to-one
    (c) $f'(x) = 2 + \cos x \ge 1 > 0$ for all x, so f is one-to-one

12. (a) $f'(x) = 3x^2 + 6x = x(3x + 6)$ changes sign at $x = -2, 0$, so f is not one-to-one
    (b) $f'(x) = 5x^4 + 24x^2 + 2 \ge 2 > 0$; f' is positive for all x, so f is one-to-one
    (c) $f'(x) = \dfrac{1}{(x+1)^2}$; f is one-to-one because:
    if $x_1 < x_2 < -1$ then $f' > 0$ on $[x_1, x_2]$, so $f(x_1) \ne f(x_2)$
    if $-1 < x_1 < x_2$ then $f' > 0$ on $[x_1, x_2]$, so $f(x_1) \ne f(x_2)$
    if $x_1 < -1 < x_2$ then $f(x_1) > 1 > f(x_2)$ since $f(x) > 1$ on $(-\infty, -1)$ and $f(x) < 1$ on $(-1, +\infty)$

13. $y = f^{-1}(x)$, $x = f(y) = y^5$, $y = x^{1/5} = f^{-1}(x)$

14. $y = f^{-1}(x)$, $x = f(y) = 6y$, $y = \frac{1}{6}x = f^{-1}(x)$

15. $y = f^{-1}(x)$, $x = f(y) = 7y - 6$, $y = \frac{1}{7}(x+6) = f^{-1}(x)$

16. $y = f^{-1}(x)$, $x = f(y) = \frac{y+1}{y-1}$, $xy - x = y + 1$, $(x-1)y = x+1$, $y = \frac{x+1}{x-1} = f^{-1}(x)$

17. $y = f^{-1}(x)$, $x = f(y) = 3y^3 - 5$, $y = \sqrt[3]{(x+5)/3} = f^{-1}(x)$

18. $y = f^{-1}(x)$, $x = f(y) = \sqrt[5]{4y+2}$, $y = \frac{1}{4}(x^5 - 2) = f^{-1}(x)$

19. $y = f^{-1}(x)$, $x = f(y) = \sqrt[3]{2y-1}$, $y = (x^3+1)/2 = f^{-1}(x)$

20. $y = f^{-1}(x)$, $x = f(y) = \frac{5}{y^2+1}$, $y = \sqrt{\frac{5-x}{x}} = f^{-1}(x)$

21. $y = f^{-1}(x)$, $x = f(y) = 3/y^2$, $y = -\sqrt{3/x} = f^{-1}(x)$

22. $y = f^{-1}(x)$, $x = f(y) = \begin{cases} 2y, & y \le 0 \\ y^2, & y > 0 \end{cases}$, $\quad y = f^{-1}(x) = \begin{cases} x/2, & x \le 0 \\ \sqrt{x}, & x > 0 \end{cases}$

23. $y = f^{-1}(x), x = f(y) = \begin{cases} 5/2 - y, & y < 2 \\ 1/y, & y \ge 2 \end{cases}$, $\quad y = f^{-1}(x) = \begin{cases} 5/2 - x, & x > 1/2 \\ 1/x, & 0 < x \le 1/2 \end{cases}$

24. $y = p^{-1}(x)$, $x = p(y) = y^3 - 3y^2 + 3y - 1 = (y-1)^3$, $y = x^{1/3} + 1 = p^{-1}(x)$

25. $y = f^{-1}(x)$, $x = f(y) = (y+2)^4$ for $y \ge 0$, $y = f^{-1}(x) = x^{1/4} - 2$ for $x \ge 16$

26. $y = f^{-1}(x)$, $x = f(y) = \sqrt{y+3}$ for $y \ge -3$, $y = f^{-1}(x) = x^2 - 3$ for $x \ge 0$

27. $y = f^{-1}(x)$, $x = f(y) = -\sqrt{3-2y}$ for $y \le 3/2$, $y = f^{-1}(x) = (3 - x^2)/2$ for $x \le 0$

28. $y = f^{-1}(x)$, $x = f(y) = 3y^2 + 5y - 2$ for $y \ge 0$, $3y^2 + 5y - 2 - x = 0$ for $y \ge 0$,
$y = f^{-1}(x) = (-5 + \sqrt{12x+49})/6$ for $x \ge -2$

29. $y = f^{-1}(x)$, $x = f(y) = y - 5y^2$ for $y \ge 1$, $5y^2 - y + x = 0$ for $y \ge 1$,
$y = f^{-1}(x) = (1 + \sqrt{1-20x})/10$ for $x \le -4$

30. **(a)** $C = \frac{5}{9}(F - 32)$
(b) how many degrees Celsius given the Fahrenheit temperature
(c) $C = -273.15°$ C is equivalent to $F = -459.67°$ F, so the domain is $F \ge -459.67$, the range is $C \ge -273.15$

31. **(a)** $y = f(x) = (6.214 \times 10^{-4})x$ **(b)** $x = f^{-1}(y) = \frac{10^4}{6.214}y$
(c) how many meters in y miles

32. f and f^{-1} are continuous so $f(3) = \lim_{x\to 3} f(x) = 7$; then $f^{-1}(7) = 3$, and
$\lim_{x\to 7} f^{-1}(x) = f^{-1}\left(\lim_{x\to 7} x\right) = f^{-1}(7) - 3$

33. (a) $f(g(x)) = f(\sqrt{x})$
$= (\sqrt{x})^2 = x, x > 1;$
$g(f(x)) = g(x^2)$
$= \sqrt{x^2} = x, x > 1$

(b) 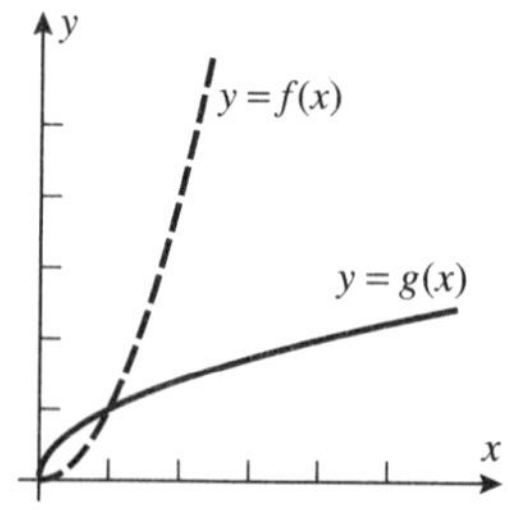

(c) no, because $f(g(x)) = x$ for every x in the domain of g is not satisfied (the domain of g is $x \geq 0$)

34. $y = f^{-1}(x)$, $x = f(y) = ay^2 + by + c$, $ay^2 + by + c - x = 0$, use the quadratic formula to get $y = \dfrac{-b \pm \sqrt{b^2 - 4a(c-x)}}{2a}$;

(a) $f^{-1}(x) = \dfrac{-b + \sqrt{b^2 - 4a(c-x)}}{2a}$ (b) $f^{-1}(x) = \dfrac{-b - \sqrt{b^2 - 4a(c-x)}}{2a}$

35. (a) $f(f(x)) = \dfrac{3 - \dfrac{3-x}{1-x}}{1 - \dfrac{3-x}{1-x}} = \dfrac{3 - 3x - 3 + x}{1 - x - 3 + x} = x$ so $f = f^{-1}$

(b) symmetric about the line $y = x$

36. $y = m(x - x_0)$ is an equation of the line. The graph of the inverse of $f(x) = m(x - x_0)$ will be the reflection of this line about $y = x$. Solve $y = m(x - x_0)$ for x to get $x = y/m + x_0 = f^{-1}(y)$ so $y = f^{-1}(x) = x/m + x_0$.

37. (a) $f(x) = x^3 - 3x^2 + 2x = x(x-1)(x-2)$ so $f(0) = f(1) = f(2) = 0$ thus f is not one-to-one.

(b) $f'(x) = 3x^2 - 6x + 2$, $f'(x) = 0$ when $x = \dfrac{6 \pm \sqrt{36 - 24}}{6} = 1 \pm \sqrt{3}/3$. $f'(x) > 0$ (f is increasing) if $x < 1 - \sqrt{3}/3$, $f'(x) < 0$ (f is decreasing) if $1 - \sqrt{3}/3 < x < 1 + \sqrt{3}/3$, so $f(x)$ takes on values less than $f(1 - \sqrt{3}/3)$ on both sides of $1 - \sqrt{3}/3$ thus $1 - \sqrt{3}/3$ is the largest value of k.

38. (a) $f(x) = x^3(x-2)$ so $f(0) = f(2) = 0$ thus f is not one to one.

(b) $f'(x) = 4x^3 - 6x^2 = 4x^2(x - 3/2)$, $f'(x) = 0$ when $x = 0$ or $3/2$; f is decreasing on $(-\infty, 3/2]$ and increasing on $[3/2, +\infty)$ so $3/2$ is the smallest value of k.

39. if $f^{-1}(x) = 1$, then $x = f(1) = 2(1)^3 + 5(1) + 3 = 10$

40. if $f^{-1}(x) = 2$, then $x = f(2) = (2)^3/[(2)^2 + 1] = 8/5$

41. 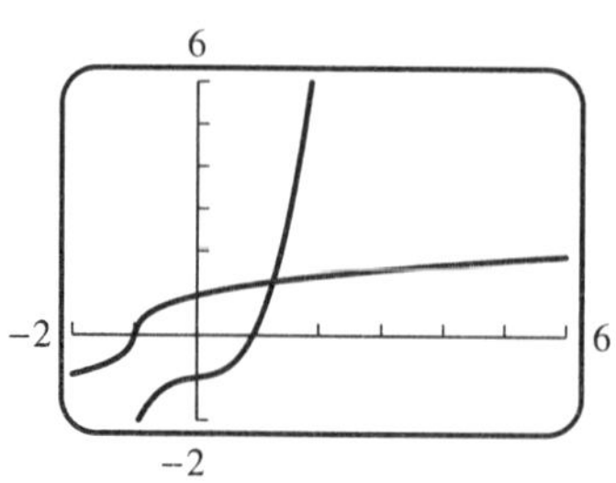

42. 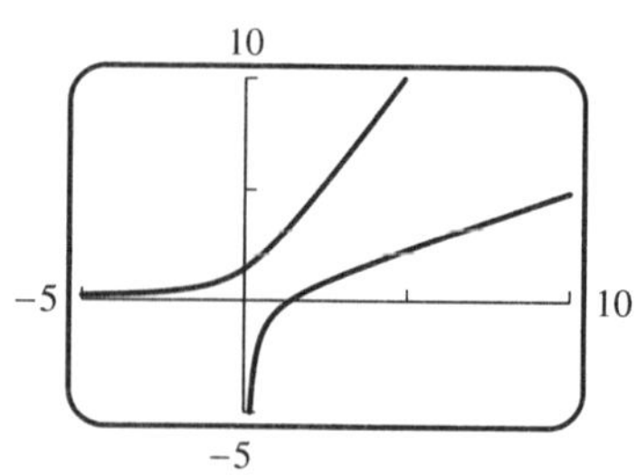

43.

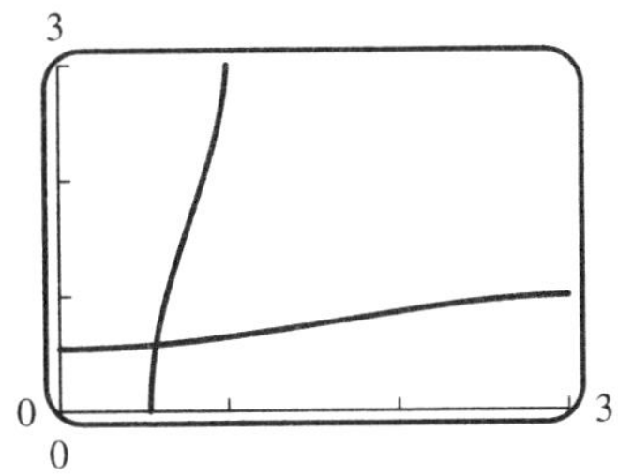

44.

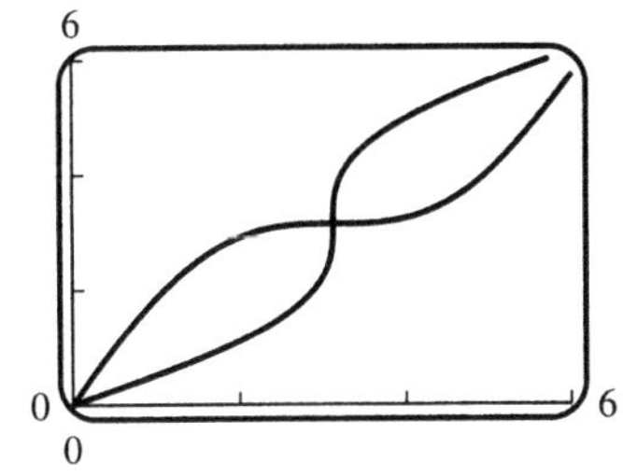

45. $y = f^{-1}(x)$, $x = f(y) = 5y^3 + y - 7$, $\dfrac{dx}{dy} = 15y^2 + 1$, $\dfrac{dy}{dx} = \dfrac{1}{15y^2 + 1}$;

check: $1 = 15y^2\dfrac{dy}{dx} + \dfrac{dy}{dx}$, $\dfrac{dy}{dx} = \dfrac{1}{15y^2 + 1}$

46. $y = f^{-1}(x)$, $x = f(y) = 1/y^2$, $\dfrac{dx}{dy} = -2y^{-3}$, $\dfrac{dy}{dx} = -y^3/2$;

check: $1 = -2y^{-3}\dfrac{dy}{dx}$, $\dfrac{dy}{dx} = -y^3/2$

47. $y = f^{-1}(x)$, $x = f(y) = 2y^5 + y^3 + 1$, $\dfrac{dx}{dy} = 10y^4 + 3y^2$, $\dfrac{dy}{dx} = \dfrac{1}{10y^4 + 3y^2}$;

check: $1 = 10y^4\dfrac{dy}{dx} + 3y^2\dfrac{dy}{dx}$, $\dfrac{dy}{dx} = \dfrac{1}{10y^4 + 3y^2}$

48. $y = f^{-1}(x)$, $x = f(y) = 5y - \sin 2y$, $\dfrac{dx}{dy} = 5 - 2\cos 2y$, $\dfrac{dy}{dx} = \dfrac{1}{5 - 2\cos 2y}$;

check: $1 = (5 - 2\cos 2y)\dfrac{dy}{dx}$, $\dfrac{dy}{dx} = \dfrac{1}{5 - 2\cos 2y}$

49. $f(f(x)) = x$ thus $f = f^{-1}$ so the graph is symmetric about $y = x$.

50. **(a)** Suppose $x_1 \neq x_2$ where x_1 and x_2 are in the domain of g and $g(x_1)$, $g(x_2)$ are in the domain of f then $g(x_1) \neq g(x_2)$ because g is one-to-one so $f(g(x_1)) \neq f(g(x_2))$ because f is one-to-one thus $f \circ g$ is one-to-one because $(f \circ g)(x_1) \neq (f \circ g)(x_2)$ if $x_1 \neq x_2$.

(b) f, g, and $f \circ g$ all have inverses because they are all one-to-one. Let $h = (f \circ g)^{-1}$ then $(f \circ g)(h(x)) = f[g(h(x))] = x$, apply f^{-1} to both sides to get $g(h(x)) = f^{-1}(x)$, then apply g^{-1} to get $h(x) = g^{-1}(f^{-1}(x)) = (g^{-1} \circ f^{-1})(x)$, so $h = g^{-1} \circ f^{-1}$

51.

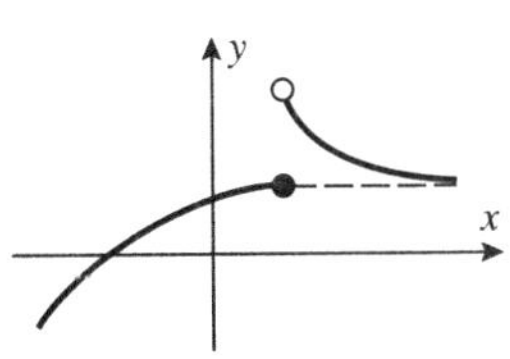

52. Suppose that g and h are both inverses of f then $f(g(x)) = x$, $h[f(g(x))] = h(x)$, but $h[f(g(x))] = g(x)$ because h is an inverse of f so $g(x) = h(x)$.

53. $F'(x) = 2f'(2g(x))g'(x)$ so $F'(3) = 2f'(2g(3))g'(3)$. By inspection $f(1) = 3$, so $g(3) = f^{-1}(3) = 1$ and $g'(3) = (f^{-1})'(3) = 1/f'(f^{-1}(3)) = 1/f'(1) = 1/7$ because $f'(x) = 4x^3 + 3x^2$. Thus $F'(3) = 2f'(2)(1/7) = 2(44)(1/7) = 88/7$.
$F(3) = f(2g(3)), g(3) - f^{-1}(3)$; by inspection $f(1) = 3$, so $g(3) - f^{-1}(3) = 1, F(3) = f(2) = 25$.

EXERCISE SET 7.2

1. **(a)** -4 **(b)** 4 **(c)** $1/4$

2. **(a)** $1/16$ **(b)** 8 **(c)** $1/3$

3. **(a)** 2.9690 **(b)** 0.0341

4. **(a)** 1.8882 **(b)** 0.9381

5. **(a)** $\log_2 16 = \log_2(2^4) = 4$ **(b)** $\log_2\left(\frac{1}{32}\right) = \log_2(2^{-5}) = -5$
(c) $\log_4 4 = 1$ **(d)** $\log_9 3 = \log_9(9^{1/2}) = 1/2$

6. **(a)** $\log_{10}(0.001) = \log_{10}(10^{-3}) = -3$ **(b)** $\log_{10}(10^4) = 4$
(c) $\ln(e^3) = 3$ **(d)** $\ln(\sqrt{e}) = \ln(e^{1/2}) = 1/2$

7. **(a)** 1.3655 **(b)** -0.3011

8. **(a)** -0.5229 **(b)** 1.1447

9. **(a)** $2\ln a + \frac{1}{2}\ln b + \frac{1}{2}\ln c = 2r + s/2 + t/2$ **(b)** $\ln b - 3\ln a - \ln c = s - 3r - t$

10. **(a)** $\frac{1}{3}\ln c - \ln a - \ln b = t/3 - r - s$ **(b)** $\frac{1}{2}(\ln a + 3\ln b - 2\ln c) = r/2 + 3s/2 - t$

11. **(a)** $1 + \log x + \frac{1}{2}\log(x-3)$ **(b)** $2\ln|x| + 3\ln\sin x - \frac{1}{2}\ln(x^2+1)$

12. **(a)** $\frac{1}{3}\log(x+2) - \log\cos 5x$ **(b)** $\frac{1}{2}\ln(x^2+1) - \frac{1}{2}\ln(x^3+5)$

13. $\log\dfrac{2^4(16)}{3} = \log(256/3)$

14. $\log\sqrt{x} - \log(\sin^3 2x) + \log 100 = \log\dfrac{100\sqrt{x}}{\sin^3 2x}$

15. $\ln\dfrac{\sqrt[3]{x}(x+1)^2}{\cos x}$

16. $1 + x = 10^3 = 1000$, $x = 999$

17. $\sqrt{x} = 10^{-1} = 0.1$, $x = 0.01$

18. $x^2 = e^4$, $x = \pm e^2$

19. $1/x = e^{-2}$, $x = e^2$

20. $x = 7$

21. $2x = 8$, $x = 4$

22. $\log_{10} x^3 = 30$, $x^3 = 10^{30}$, $x = 10^{10}$

23. $\log_{10} x = 5$, $x = 10^5$

24. $\ln 4x - \ln x^6 = \ln 2$, $\ln\dfrac{4}{x^5} = \ln 2$, $\dfrac{4}{x^5} = 2$, $x^5 = 2$, $x = \sqrt[5]{2}$

25. $\ln 2x^2 = \ln 3$, $2x^2 = 3$, $x^2 = 3/2$, $x = \sqrt{3/2}$ (we discard $-\sqrt{3/2}$ because it does not satisfy the original equation)

26. $\ln 3^x = \ln 2$, $x\ln 3 = \ln 2$, $x = \dfrac{\ln 2}{\ln 3}$

27. $\ln 5^{-2x} = \ln 3,\ -2x\ln 5 = \ln 3,\ x = -\dfrac{\ln 3}{2\ln 5}$

28. $e^{-2x} = 5/3,\ -2x = \ln(5/3),\ x = -\dfrac{1}{2}\ln(5/3)$

29. $e^{3x} = 7/2,\ 3x = \ln(7/2),\ x = \dfrac{1}{3}\ln(7/2)$

30. $e^x(1-2x) = 0$ so $e^x = 0$ (impossible) or $1 - 2x = 0,\ x = 1/2$

31. $e^{-x}(x+2) = 0$ so $e^{-x} = 0$ (impossible) or $x + 2 = 0,\ x = -2$

32. $e^{2x} - e^x - 6 = (e^x - 3)(e^x + 2) = 0$ so $e^x = -2$ (impossible) or $e^x = 3,\ x = \ln 3$

33. $e^{-2x} - 3e^{-x} + 2 = (e^{-x} - 2)(e^{-x} - 1) = 0$ so $e^{-x} = 2,\ x = -\ln 2$ or $e^{-x} = 1,\ x = 0$

34. (a)

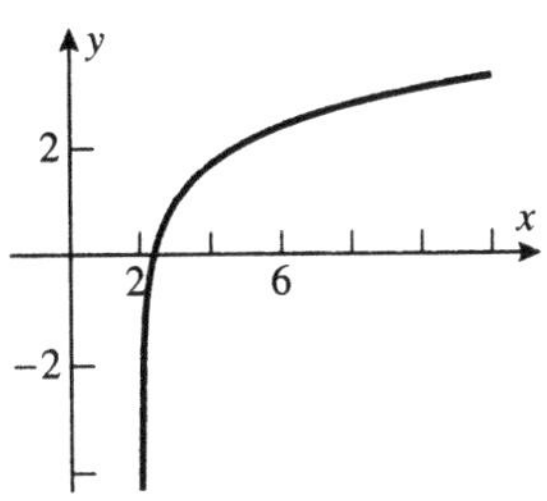

(b)

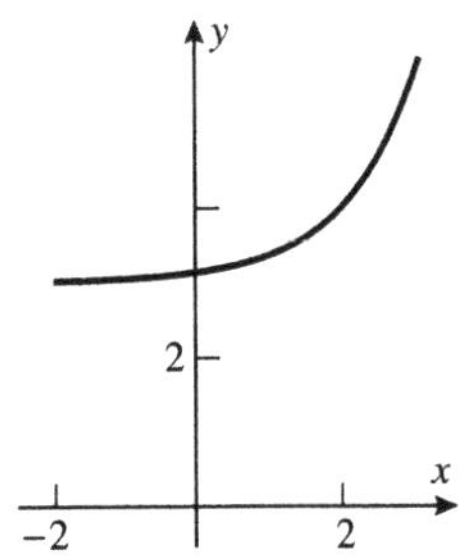

35. (a)

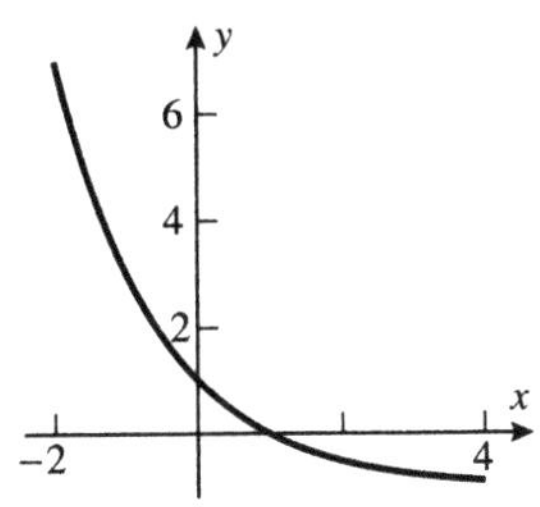

(b)

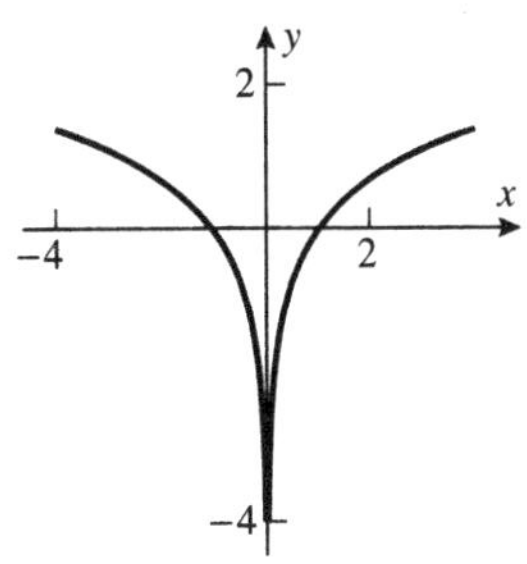

36. (a)

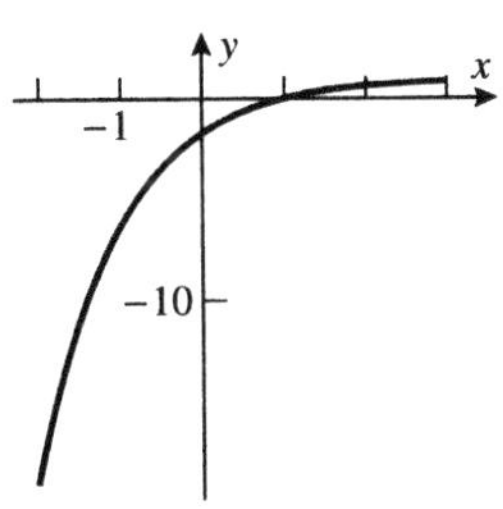

(b)

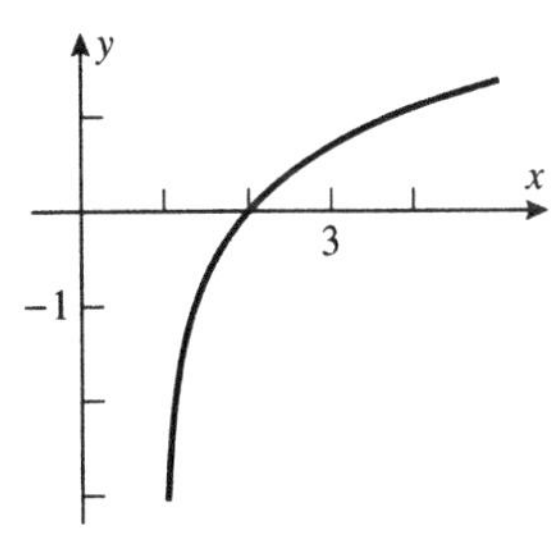

37. $\log_2 7.35 = (\log 7.35)/(\log 2) = (\ln 7.35)/(\ln 2) \approx 2.8777$;
$\log_5 0.6 = (\log 0.6)/(\log 5) = (\ln 0.6)/(\ln 5) \approx -0.3174$

38.

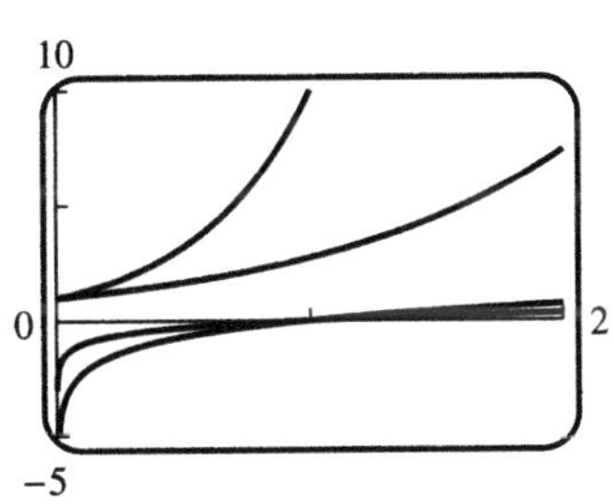

39.

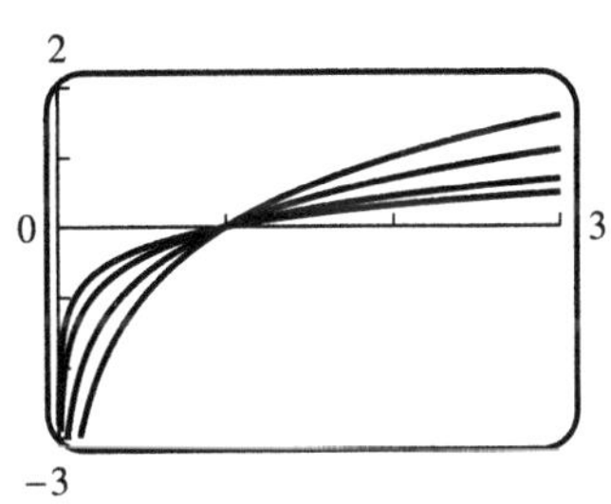

40. (a) Let $X = \log_b x$ and $Y = \log_a x$. Then $b^X = x$ and $a^Y = x$ so $a^Y = b^X$, or $a^{Y/X} = b$, which means $\log_a b = Y/X$. Substituting for Y and X yields $\dfrac{\log_a x}{\log_b x} = \log_a b, \log_b x = \dfrac{\log_a x}{\log_a b}$.

(b) Let $x = a$ to get $\log_b a = (\log_a a)/(\log_a b) = 1/(\log_a b)$ so $(\log_a b)(\log_b a) = 1$.
$(\log_2 81)(\log_3 32) = (\log_2[3^4])(\log_3[2^5]) = (4\log_2 3)(5\log_3 2) = 20(\log_2 3)(\log_3 2) = 20$

41. (a) $x = 3.6541, y = 1.2958$

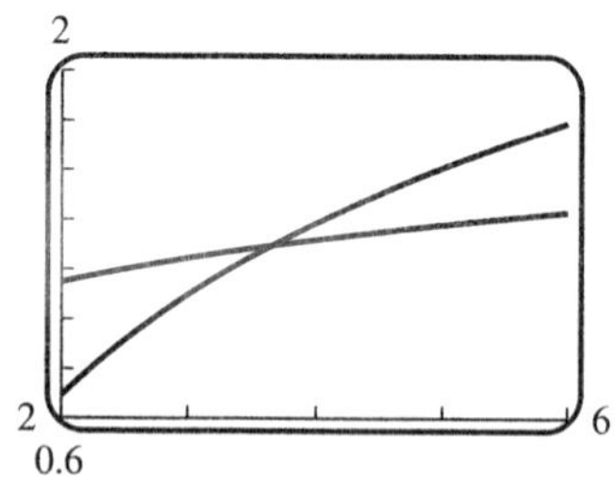

(b) $x \approx 332105.11, y \approx 12.7132$

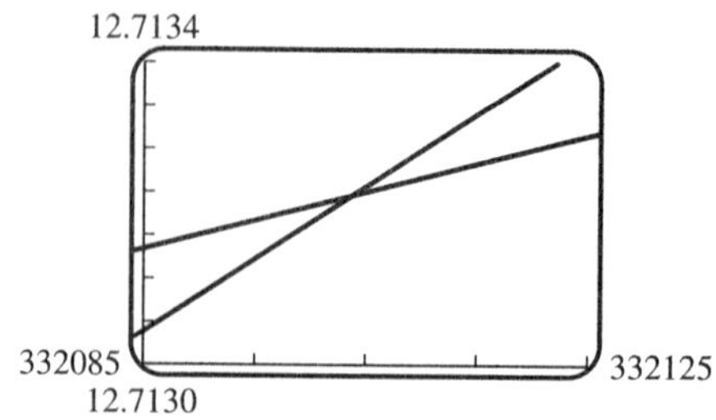

42. Since the units are billions, one trillion is 1,000 units. Solve $1000 = 0.051517(1.1306727)^x$ for x by taking common logarithms, resulting in $3 = \log 0.051517 + x \log 1.1306727$, which yields $x \approx 77.4$, so the debt first reached one trillion dollars around 1977.

43. (a) no, the curve passes through the origin

(b) $y = 2^{x/4}$

(c) $y = 2^{-x}$

(d) $y = (\sqrt{5})^x$

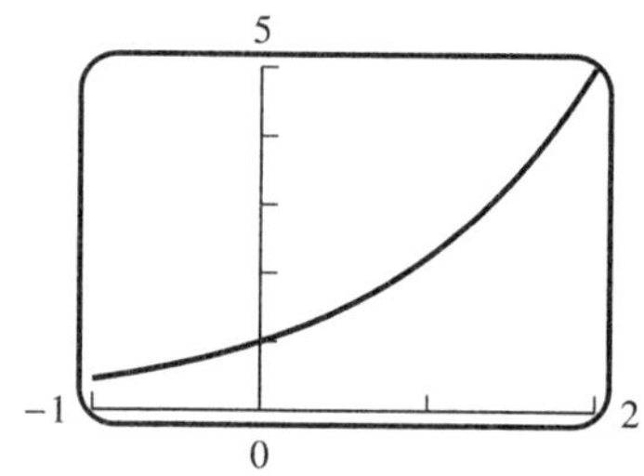

44. (a) As $x \to +\infty$ the function grows very slowly, but it is always increasing and tends to $+\infty$. As $x \to 1^+$ the function tends to $-\infty$.

(b)

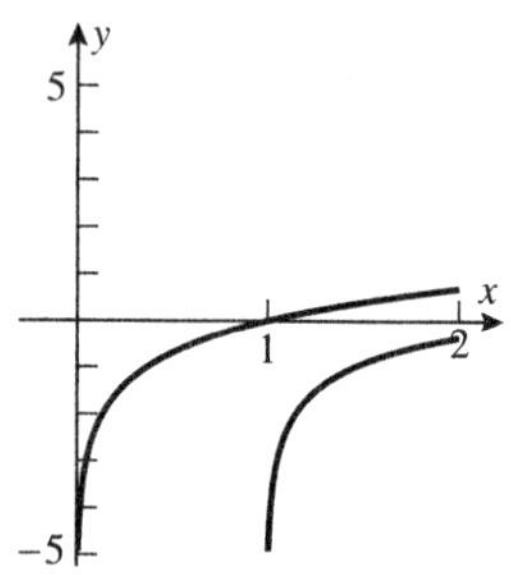

45. $\log(1/2) < 0$ so $3\log(1/2) < 2\log(1/2)$

46. Let $x = \log_b a$ and $y = \log_b c$, so $a = b^x$ and $c = b^y$.
First, $ac = b^x b^y = b^{x+y}$ or equivalently, $\log_b(ac) = x + y = \log_b a + \log_b c$.
Secondly, $a/c = b^x/b^y = b^{x-y}$ or equivalently, $\log_b(a/c) = x - y = \log_b a - \log_b c$.
Next, $a^r = (b^x)^r = b^{rx}$ or equivalently, $\log_b a^r = rx = r\log_b a$.
Finally, $1/c = 1/b^y = b^{-y}$ or equivalently, $\log_b(1/c) = -y = -\log_b c$.

47. $75e^{-t/125} = 15, t = -125\ln(1/5) = 125\ln 5 \approx 201$ days.

48. (a) a If $t = 0$, then $Q = 12$ grams

(b) b $Q = 12e^{-0.055(4)} = 12e^{-0.22} \approx 9.63$ grams

(c) $12e^{-0.055t} = 6, e^{-0.055t} = 0.5, t = -(\ln 0.5)/(0.055) \approx 12.6$ hours

49. (a) 7.4; basic (b) 4.2; acidic (c) 6.4; acidic (d) 5.9; acidic

50. (a) $\log[H^+] = -2.44, [H^+] = 10^{-2.44} \approx 3.6 \times 10^{-3}$ mol/L
(b) $\log[H^+] = -8.06, [H^+] = 10^{-8.06} \approx 8.7 \times 10^{-9}$ mol/L

51. (a) 140 dB; damage (b) 120 dB; damage
(c) 80 dB; no damage (d) 75 dB; no damage

52. Suppose that $I_1 = 3I_2$ and $\beta_1 = 10\log_{10} I_1/I_0$, $\beta_2 = 10\log_{10} I_2/I_0$. Then $I_1/I_0 = 3I_2/I_0$, $\log_{10} I_1/I_0 = \log_{10} 3I_2/I_0 = \log_{10} 3 + \log_{10} I_2/I_0$, $\beta_1 = 10\log_{10} 3 + \beta_2$, $\beta_1 - \beta_2 = 10\log_{10} 3 \approx 4.8$ decibels.

53. Let I_A and I_B be the intensities of the automobile and blender, respectively. Then $\log_{10} I_A/I_0 = 7$ and $\log_{10} I_B/I_0 = 9.3$, $I_A = 10^7 I_0$ and $I_B = 10^{9.3} I_0$, so $I_B/I_A = 10^{2.3} \approx 200$.

54. The decibel level of the nth echo is $120(2/3)^n$;
$120(2/3)^n < 10$ if $(2/3)^n < 1/12, n < \dfrac{\log(1/12)}{\log(2/3)} = \dfrac{\log 12}{\log 1.5} \approx 6.13$ so 6 echoes can be heard.

55. (a) $\log E = 4.4 + 1.5(8.2) = 16.7, E = 10^{16.7} \approx 5 \times 10^{16}$ J

(b) Let M_1 and M_2 be the magnitudes of earthquakes with energies of E and $10E$, respectively. Then $1.5(M_2 - M_1) = \log(10E) - \log E = \log 10 = 1$, $M_2 - M_1 = 1/1.5 = 2/3 \approx 0.67$.

56. Let E_1 and E_2 be the energies of earthquakes with magnitudes M and $M + 1$, respectively. Then $\log E_2 - \log E_1 = \log(E_2/E_1) = 1.5, E_2/E_1 = 10^{1.5} \approx 31.6$.

57. If $t = -2x$, then $x = -t/2$ and $\lim_{x\to 0}(1 - 2x)^{1/x} = \lim_{t\to 0}(1 + t)^{-2/t} = \lim_{t\to 0}[(1 + t)^{1/t}]^{-2} = e^{-2}$.

58. If $t = 3/x$, then $x = 3/t$ and $\lim_{x\to +\infty}(1 + 3/x)^x = \lim_{t\to 0^+}(1 + t)^{3/t} = \lim_{t\to 0^+}[(1 + t)^{1/t}]^3 = e^3$.

EXERCISE SET 7.3

1. $\dfrac{1}{2x}(2) = 1/x$

2. $\dfrac{1}{x^3}(3x^2) = 3/x$

3. $2(\ln x)\left(\dfrac{1}{x}\right) = \dfrac{2\ln x}{x}$

4. $\dfrac{1}{\sin x}(\cos x) = \cot x$

5. $\dfrac{1}{\tan x}(\sec^2 x) = \dfrac{\sec^2 x}{\tan x}$

6. $\dfrac{1}{2 + \sqrt{x}}\left(\dfrac{1}{2\sqrt{x}}\right) = \dfrac{1}{2\sqrt{x}(2 + \sqrt{x})}$

7. $\dfrac{1}{x/(1 + x^2)}\left[\dfrac{(1 + x^2)(1) - x(2x)}{(1 + x^2)^2}\right] = \dfrac{1 - x^2}{x(1 + x^2)}$

8. $\dfrac{1}{\ln x}\left(\dfrac{1}{x}\right) = \dfrac{1}{x\ln x}$

9. $\dfrac{3x^2 - 14x}{x^3 - 7x^2 - 3}$

10. $x^3\left(\frac{1}{x}\right)+(3x^2)\ln x = x^2(1+3\ln x)$

11. $\frac{1}{2}(\ln x)^{-1/2}\left(\frac{1}{x}\right)=\frac{1}{2x\sqrt{\ln x}}$

12. $\frac{\frac{1}{2}2(\ln x)(1/x)}{\sqrt{1+\ln^2 x}}=\frac{\ln x}{x\sqrt{1+\ln^2 x}}$

13. $-\frac{1}{x}\sin(\ln x)$

14. $2\sin(\ln x)\cos(\ln x)\frac{1}{x}=\frac{\sin(2\ln x)}{x}=\frac{\sin(\ln x^2)}{x}$

15. $3x^2\log_2(3-2x)+\frac{-2x^3}{(\ln 2)(3-2x)}$

16. $\left[\log_2(x^2-2x)\right]^3+3x\left[\log_2(x^2-2x)\right]^2\frac{2x-2}{(x^2-2x)\ln 2}$

17. $\frac{2x(1+\log x)-x/(\ln 10)}{(1+\log x)^2}$

18. $1/[x(\ln 10)(1+\log x)^2]$

19. $7e^{7x}$

20. $-10xe^{-5x^2}$

21. $x^3e^x+3x^2e^x=x^2e^x(x+3)$

22. $-\frac{1}{x^2}e^{1/x}$

23. $$\frac{dy}{dx}=\frac{(e^x+e^{-x})(e^x+e^{-x})-(e^x-e^{-x})(e^x-e^{-x})}{(e^x+e^{-x})^2}$$
$$=\frac{(e^{2x}+2+e^{-2x})-(e^{2x}-2+e^{-2x})}{(e^x+e^{-x})^2}=4/(e^x+e^{-x})^2$$

24. $e^x\cos(e^x)$

25. $(x\sec^2 x+\tan x)e^{x\tan x}$

26. $\frac{dy}{dx}=\frac{(\ln x)e^x-e^x(1/x)}{(\ln x)^2}=\frac{e^x(x\ln x-1)}{x(\ln x)^2}$

27. $(1-3e^{3x})e^{(x-e^{3x})}$

28. $\frac{15}{2}x^2(1+5x^3)^{-1/2}\exp(\sqrt{1+5x^3})$

29. $\frac{(x-1)e^{-x}}{1-xe^{-x}}=\frac{x-1}{e^x-x}$

30. $\frac{1}{\cos(e^x)}[-\sin(e^x)]e^x=-e^x\tan(e^x)$

31. $\frac{dy}{dx}+\frac{1}{xy}\left(x\frac{dy}{dx}+y\right)=0,\ \frac{dy}{dx}=-\frac{y}{x(y+1)}$

32. $\frac{dy}{dx}=\frac{1}{x\tan y}\left(x\sec^2 y\frac{dy}{dx}+\tan y\right),\ \frac{dy}{dx}=\frac{\tan y}{x(\tan y-\sec^2 y)}$

33. $\frac{d}{dx}\left[\ln\cos x-\frac{1}{2}\ln(4-3x^2)\right]=-\tan x+\frac{3x}{4-3x^2}$

34. $\frac{d}{dx}\left(\frac{1}{2}[\ln(x-1)-\ln(x+1)]\right)=\frac{1}{2}\left(\frac{1}{x-1}-\frac{1}{x+1}\right)$

35. $\ln|y|=\ln|x|+\frac{1}{3}\ln|1+x^2|,\ \frac{dy}{dx}=x\sqrt[3]{1+x^2}\left[\frac{1}{x}+\frac{2x}{3(1+x^2)}\right]$

36. $\ln|y| = \frac{1}{5}[\ln|x-1| - \ln|x+1|]$, $\frac{dy}{dx} = \frac{1}{5}\sqrt[5]{\frac{x-1}{x+1}}\left[\frac{1}{x-1} - \frac{1}{x+1}\right]$

37. $\ln|y| = \frac{1}{3}\ln|x^2-8| + \frac{1}{2}\ln|x^3+1| - \ln|x^6 - 7x + 5|$

$$\frac{dy}{dx} = \frac{(x^2-8)^{1/3}\sqrt{x^3+1}}{x^6-7x+5}\left[\frac{2x}{3(x^2-8)} + \frac{3x^2}{2(x^3+1)} - \frac{6x^5-7}{x^6-7x+5}\right]$$

38. $\ln|y| = \ln|\sin x| + \ln|\cos x| + 3\ln|\tan x| - \frac{1}{2}\ln|x|$

$$\frac{dy}{dx} = \frac{\sin x\cos x\tan^3 x}{\sqrt{x}}\left[\cot x - \tan x + \frac{3\sec^2 x}{\tan x} - \frac{1}{2x}\right]$$

39. $f'(x) = 2^x\ln 2$; $y = 2^x$, $\ln y = x\ln 2$, $\frac{1}{y}y' = \ln 2$, $y' = y\ln 2 = 2^x\ln 2$

40. $f'(x) = -3^{-x}\ln 3$; $y = 3^{-x}$, $\ln y = -x\ln 3$, $\frac{1}{y}y' = -\ln 3$, $y' = -y\ln 3 = -3^{-x}\ln 3$

41. $f'(x) = \pi^{\sin x}(\ln\pi)\cos x$;

$y = \pi^{\sin x}$, $\ln y = (\sin x)\ln\pi$, $\frac{1}{y}y' = (\ln\pi)\cos x$, $y' = \pi^{\sin x}(\ln\pi)\cos x$

42. $f'(x) = \pi^{x\tan x}(\ln\pi)(x\sec^2 x + \tan x)$;

$y = \pi^{x\tan x}$, $\ln y = (x\tan x)\ln\pi$, $\frac{1}{y}y' = (\ln\pi)(x\sec^2 x + \tan x)$

$y' = \pi^{x\tan x}(\ln\pi)(x\sec^2 x + \tan x)$

43. $\ln y = (\ln x)\ln(x^3 - 2x)$, $\frac{1}{y}\frac{dy}{dx} = \frac{3x^2-2}{x^3-2x}\ln x + \frac{1}{x}\ln(x^3-2x)$,

$$\frac{dy}{dx} = (x^3-2x)^{\ln x}\left[\frac{3x^2-2}{x^3-2x}\ln x + \frac{1}{x}\ln(x^3-2x)\right]$$

44. $\ln y = (\sin x)\ln x$, $\frac{1}{y}\frac{dy}{dx} = \frac{\sin x}{x} + (\cos x)\ln x$, $\frac{dy}{dx} = x^{\sin x}\left[\frac{\sin x}{x} + (\cos x)\ln x\right]$

45. $\ln y = (\tan x)\ln(\ln x)$, $\frac{1}{y}\frac{dy}{dx} = \frac{1}{x\ln x}\tan x + (\sec^2 x)\ln(\ln x)$,

$$\frac{dy}{dx} = (\ln x)^{\tan x}\left[\frac{\tan x}{x\ln x} + (\sec^2 x)\ln(\ln x)\right]$$

46. $\ln y = (\ln x)\ln(x^2+3)$, $\frac{1}{y}\frac{dy}{dx} = \frac{2x}{x^2+3}\ln x + \frac{1}{x}\ln(x^2+3)$,

$$\frac{dy}{dx} = (x^2+3)^{\ln x}\left[\frac{2x}{x^2+3}\ln x + \frac{1}{x}\ln(x^2+3)\right]$$

47. $f'(x) = ex^{e-1}$

48. **(a)** because x^x is not of the form a^x where a is constant

(b) $y = x^x$, $\ln y = x\ln x$, $\dfrac{1}{y}y' = 1 + \ln x$, $y' = x^x(1 + \ln x)$

49. **(a)** $\log_x e = \dfrac{\ln e}{\ln x} = \dfrac{1}{\ln x}$, $\dfrac{d}{dx}[\log_x e] = -\dfrac{1}{x(\ln x)^2}$

(b) $\log_x 2 = \dfrac{\ln 2}{\ln x}$, $\dfrac{d}{dx}[\log_x 2] = -\dfrac{\ln 2}{x(\ln x)^2}$

50. **(a)** e^{x^2} **(b)** $\ln x$

51. **(a)** $f'(x) = ke^{kx}$, $f''(x) = k^2e^{kx}$, $f'''(x) = k^3e^{kx}, \ldots, f^{(n)}(x) = k^ne^{kx}$

(b) $f'(x) = -ke^{-kx}$, $f''(x) = k^2e^{-kx}$, $f'''(x) = -k^3e^{-kx}, \ldots, f^{(n)}(x) = (-1)^nk^ne^{-kx}$

52. $$\frac{dy}{dt} = e^{-\lambda t}(\omega A\cos\omega t - \omega B\sin\omega t) + (-\lambda)e^{-\lambda t}(A\sin\omega t + B\cos\omega t)$$
$$= e^{-\lambda t}[(\omega A - \lambda B)\cos\omega t - (\omega B + \lambda A)\sin\omega t]$$

53. $$f'(x) = \frac{1}{\sqrt{2\pi}\sigma}\exp\left[-\frac{1}{2}\left(\frac{x-\mu}{\sigma}\right)^2\right]\frac{d}{dx}\left[-\frac{1}{2}\left(\frac{x-\mu}{\sigma}\right)^2\right]$$
$$= \frac{1}{\sqrt{2\pi}\sigma}\exp\left[-\frac{1}{2}\left(\frac{x-\mu}{\sigma}\right)^2\right]\left[-\left(\frac{x-\mu}{\sigma}\right)\left(\frac{1}{\sigma}\right)\right]$$
$$= -\frac{1}{\sqrt{2\pi}\sigma^3}(x-\mu)\exp\left[-\frac{1}{2}\left(\frac{x-\mu}{\sigma}\right)^2\right]$$

54. $y = Ae^{kt}$, $dy/dt = kAe^{kt} = k(Ae^{kt}) = ky$

55. $y = Ae^{2x} + Be^{-4x}$, $y' = 2Ae^{2x} - 4Be^{-4x}$, $y'' = 4Ae^{2x} + 16Be^{-4x}$ so
$y'' + 2y' - 8y = (4Ae^{2x} + 16Be^{-4x}) + 2(2Ae^{2x} - 4Be^{-4x}) - 8(Ae^{2x} + Be^{-4x}) = 0$

56. **(a)** $y' = -xe^{-x} + e^{-x} = e^{-x}(1-x)$, $xy' = xe^{-x}(1-x) = y(1-x)$

(b) $y' = -x^2e^{-x^2/2} + e^{-x^2/2} = e^{-x^2/2}(1-x^2)$, $xy' = xe^{-x^2/2}(1-x^2) = y(1-x^2)$

57. **(a)** $f(w) = \ln w$; $f'(1) = \lim\limits_{h\to 0}\dfrac{\ln(1+h) - \ln 1}{h} = \lim\limits_{h\to 0}\dfrac{\ln(1+h)}{h} = \left.\dfrac{1}{w}\right|_{w=1} = 1$

(b) $f(w) = 10^w$; $f'(0) = \lim\limits_{h\to 0}\dfrac{10^h - 1}{h} = \left.\dfrac{d}{dw}(10^w)\right|_{w=0} = \left.10^w\ln 10\right|_{w=0} = \ln 10$

58. **(a)** $f(x) = \ln x$; $f'(e^2) = \lim\limits_{\Delta x\to 0}\dfrac{\ln(e^2 + \Delta x) - 2}{\Delta x} = \left.\dfrac{d}{dx}(\ln x)\right|_{x=e^2} = \left.\dfrac{1}{x}\right|_{x=e^2} = e^{-2}$

(b) $f(w) = 2^w$; $f'(1) = \lim\limits_{w\to 1}\dfrac{2^w - 2}{w - 1} = \left.\dfrac{d}{dw}(2^w)\right|_{w=1} = \left.2^w\ln 2\right|_{w=1} = 2\ln 2$

59. $2\ln x + 3e^x + C$

60. $\displaystyle\int\left[\frac{1}{2}t^{-1} - \sqrt{2}e^t\right]dt = \frac{1}{2}\ln t - \sqrt{2}e^t + C$

61. (a) $\displaystyle\int \frac{1}{u}du = \ln|u| + C = \ln|\ln x| + C$ (b) $\displaystyle -\frac{1}{5}\int e^u\,du = -\frac{1}{5}e^u + C = -\frac{1}{5}e^{-5x} + C$

62. (a) $\displaystyle -\frac{1}{3}\int \frac{1}{u}du = -\frac{1}{3}\ln|u| + C = -\frac{1}{3}\ln|(1+\cos 3\theta)| + C$

(b) $\displaystyle\int \frac{du}{u} = \ln u + C = \ln(1+e^x) + C$

63. $u = 2x$, $du = 2dx$; $\displaystyle\frac{1}{2}\int e^u\,du = \frac{1}{2}e^u + C = \frac{1}{2}e^{2x} + C$

64. $u = 2x$, $du = 2dx$; $\displaystyle\frac{1}{2}\int \frac{1}{u}du = \frac{1}{2}\ln|u| + C = \frac{1}{2}\ln|2x| + C$

65. $u = \sin x$, $du = \cos x\,dx$; $\displaystyle\int e^u\,du = e^u + C = e^{\sin x} + C$

66. $u = x^4$, $du = 4x^3dx$; $\displaystyle\frac{1}{4}\int e^u\,du = \frac{1}{4}e^u + C = \frac{1}{4}e^{x^4} + C$

67. $u = -2x^3$, $du = -6x^2$, $\displaystyle -\frac{1}{6}\int e^u du = -\frac{1}{6}e^u + C = -\frac{1}{6}e^{-2x^3} + C$

68. $u = e^x - e^{-x}$, $du = (e^x + e^{-x})dx$, $\displaystyle\int \frac{1}{u}du = \ln|u| + C = \ln\left|e^x - e^{-x}\right| + C$

69. $\displaystyle\int e^{-x}dx$; $u = -x$, $du = -dx$; $\displaystyle -\int e^u du = -e^u + C = -e^{-x} + C$

70. $\displaystyle\int e^{x/2}dx$; $u = x/2$, $du = dx/2$; $\displaystyle 2\int e^u du = 2e^u + C = 2e^{x/2} + C = 2\sqrt{e^x} + C$

71. $\displaystyle u = \sqrt{y+1}$, $\displaystyle du = \frac{1}{2\sqrt{y+1}}dy$, $\displaystyle 2\int e^u du = 2e^u + C = 2e^{\sqrt{y+1}} + C$

72. $u = \sqrt{y}$, $\displaystyle du = \frac{1}{2\sqrt{y}}dy$, $\displaystyle 2\int \frac{1}{e^u}du = 2\int e^{-u}du = -2e^{-u} + C = -2e^{-\sqrt{y}} + C$

73. $\displaystyle\int\left(1 + \frac{1}{t}\right)dt = t + \ln|t| + C$

74. $e^{2\ln x} = e^{\ln x^2} = x^2$, $x > 0$, so $\displaystyle\int e^{2\ln x}dx = \int x^2dx = \frac{1}{3}x^3 + C$

75. $\ln(e^x) + \ln(e^{-x}) = \ln(e^xe^{-x}) = \ln 1 = 0$ so $\displaystyle\int[\ln(e^x) + \ln(e^{-x})]dx = C$

76. $\displaystyle\int \frac{\cos x}{\sin x}dx$; $u = \sin x$, $du = \cos x dx$; $\displaystyle\int \frac{1}{u}du = \ln|u| + C = \ln|\sin x| + C$

77. $\displaystyle 5e^x\Big]_{\ln 2}^{3} = 5e^3 - 5(2) = 5e^3 - 10$ **78.** $\displaystyle (\ln x)/2\Big]_{1/2}^{1} = (\ln 2)/2$

79. (a) $\displaystyle\frac{1}{2}\int_{-1}^{1} e^u du = \frac{1}{2}e^u\Big|_{-1}^{1} = \frac{1}{2}(e - e^{-1})$ (b) $\displaystyle\int_1^2 u\,du = \frac{3}{2}$

80. $\int_{-6}^{6} \sqrt{36-u^2}du = \pi(6)^2/2 = 18\pi$

81. $u = e^x + 4$, $du = e^x dx$, $u = e^{-\ln 3} + 4 = \frac{1}{3} + 4 = \frac{13}{3}$ when $x = -\ln 3$,

$u = e^{\ln 3} + 4 = 3 + 4 = 7$ when $x = \ln 3$, $\int_{13/3}^{7} \frac{1}{u} du = \ln u \Big]_{13/3}^{7} = \ln(7) - \ln(13/3) = \ln(21/13)$

82. $u = 3 - 4e^x$, $du = -4e^x dx$, $u = -1$ when $x = 0$, $u = -17$ when $x = \ln 5$

$-\frac{1}{4}\int_{-1}^{-17} u\,du = -\frac{1}{8}u^2\Big]_{-1}^{-17} = -36$

83. $\ln(x+e)\Big]_0^e = \ln(2e) - \ln e = \ln 2$

84. $-\frac{1}{2}e^{-x^2}\Big]_1^{\sqrt{2}} = (e^{-1} - e^{-2})/2$

85. $-\frac{1}{3}e^{-3x}\Big]_0^{\ln 2} = -\frac{1}{3}(e^{-3\ln 2} - e^0) = -\frac{1}{3}\left(\frac{1}{8} - 1\right) = 7/24$

86. $\int_{-1}^{0}(1-e^x)dx + \int_0^1 (e^x - 1)dx = (x - e^x)\Big]_{-1}^{0} + (e^x - x)\Big]_0^1 = -1-(-1-e^{-1})+e-1-1 = e+1/e-2$

87. (a)

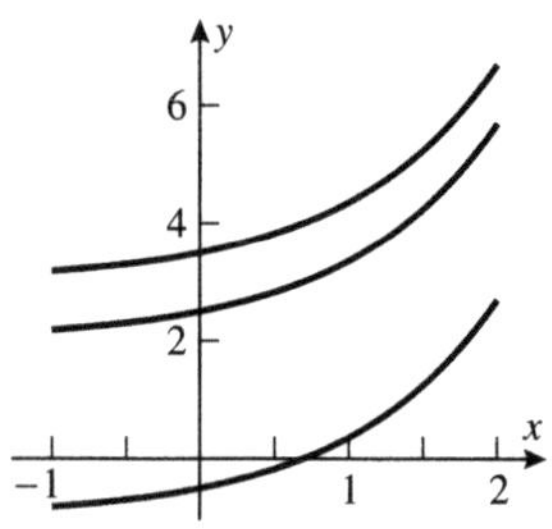

88.

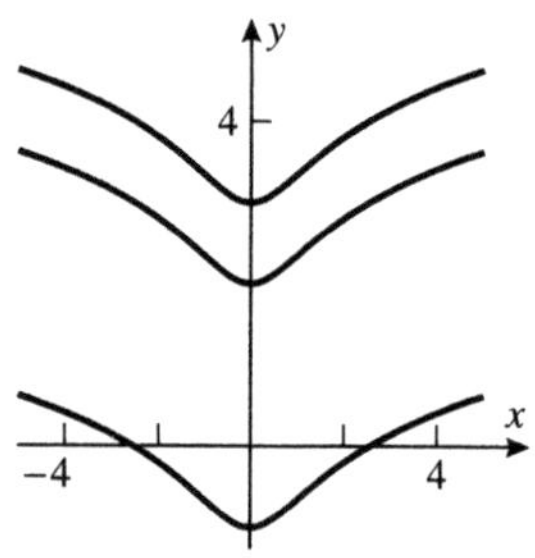

(b) $f(x) = e^x/2 + 1/2$

89. (a) $y(t) = \int 2e^{-t}\,dt = -2e^{-t} + C$, $y(1) = -\frac{2}{e} + C = 3 - \frac{2}{e}, C = 3$; $y(t) = -2e^{-t} + 3$

(b) $y(t) = \int t^{-1}dt = \ln|t| + C$, $y(-1) = C = 5$, $C = 5$; $y(t) = \ln|t| + 5$

EXERCISE SET 7.4

1. (a) critical point $x = 0$; f':
$x = 0$: relative minimum

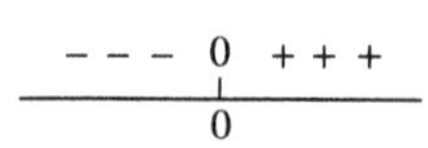

(b) critical point $x = \ln 2$: f':
$x = \ln 2$: relative minimum

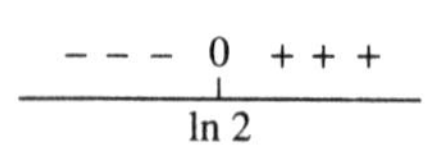

2. (a) critical points $x = -1, 1$: f':
$x = -1$: relative minimum;
$x = 1$: relative maximum

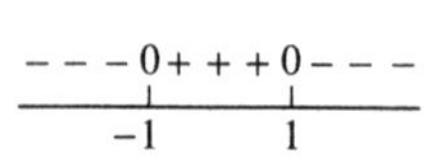

(b) $x = 1$: neither

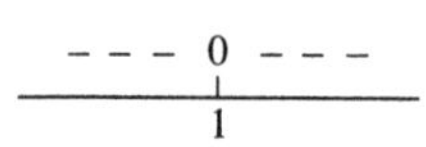

3. $f'(x) = x^2(2x-3)e^{-2x}$, $f'(x) = 0$ for x in $[1,4]$ when $x = 3/2$;
if $x = 1, 3/2, 4$, then $f(x) = e^{-2}, \dfrac{27}{8}e^{-3}, 64e^{-8}$;
critical point at $x = 3/2$; absolute maximum of $\dfrac{27}{8}e^{-3}$ at $x = 3/2$,
absolute minimum of $64e^{-8}$ at $x = 4$

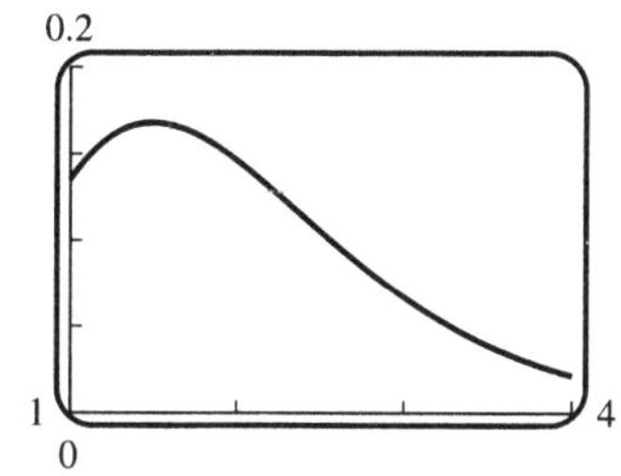

4. $f'(x) = (1 - \ln 2x)/x^2$, $f'(x) = 0$ on $[1, e]$ for $x = e/2$;
if $x = 1, e/2, e$ then $f(x) = \ln 2, 2/e, (\ln 2 + 1)/e$;
absolute minimum of $\dfrac{1+\ln 2}{e}$ at $x = e$,
absolute maximum of $2/e$ at $x = e/2$

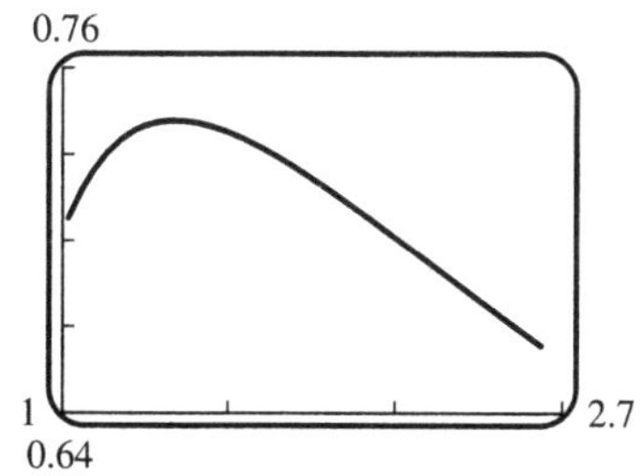

5. **(a)** $\lim_{x\to+\infty} xe^x = +\infty$, $\lim_{x\to-\infty} xe^x = 0$

(b) $y = xe^x$;
$y' = (x+1)e^x$;
$y'' = (x+2)e^x$

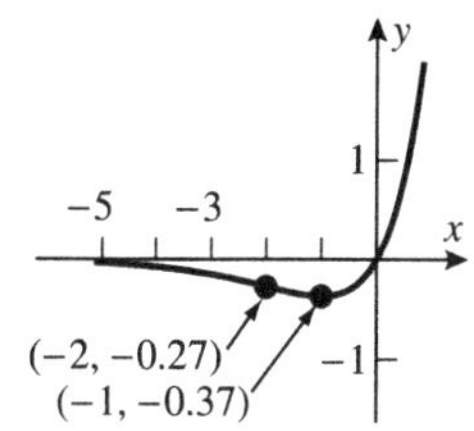

6. **(a)** $\lim_{x\to+\infty} xe^{-2x} = 0$, $\lim_{x\to-\infty} xe^{-2x} = -\infty$

(b) $y = xe^{-2x}$; $y' = -2\left(x - \dfrac{1}{2}\right)e^{-2x}$; $y'' = 4(x-1)e^{-2x}$

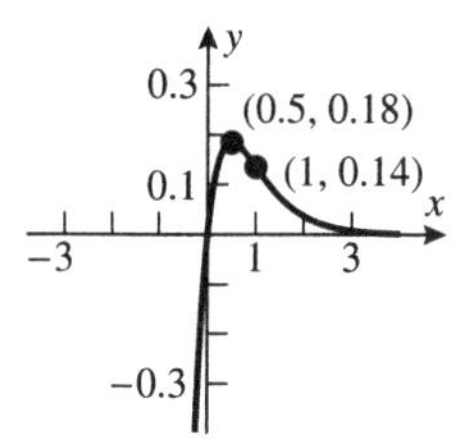

7. **(a)** $\lim_{x\to+\infty} \dfrac{x^2}{e^{2x}} = 0$, $\lim_{x\to-\infty} \dfrac{x^2}{e^{2x}} = +\infty$

(b) $y = x^2/e^{2x} = x^2e^{-2x}$;
$y' = 2x(1-x)e^{-2x}$;
$y'' = 2(2x^2 - 4x + 1)e^{-2x}$;
$y'' = 0$ if $2x^2 - 4x + 1 = 0$, when
$x = \dfrac{4 \pm \sqrt{16-8}}{4} = 1 \pm \sqrt{2}/2 \approx 0.29, 1.71$

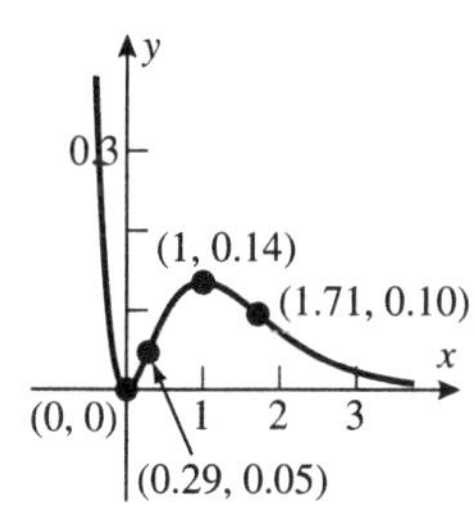

8. **(a)** $\lim_{x\to+\infty} x^2e^{2x} = +\infty$, $\lim_{x\to-\infty} x^2e^{2x} = 0$.

(b) $y = x^2e^{2x}$;
$y' = 2x(x+1)e^{2x}$;
$y'' = 2(2x^2 + 4x + 1)e^{2x}$;
$y'' = 0$ if $2x^2 + 4x + 1 = 0$, when
$x = \dfrac{-4 \pm \sqrt{16-8}}{4} = -1 \pm \sqrt{2}/2 \approx -0.29, -1.71$

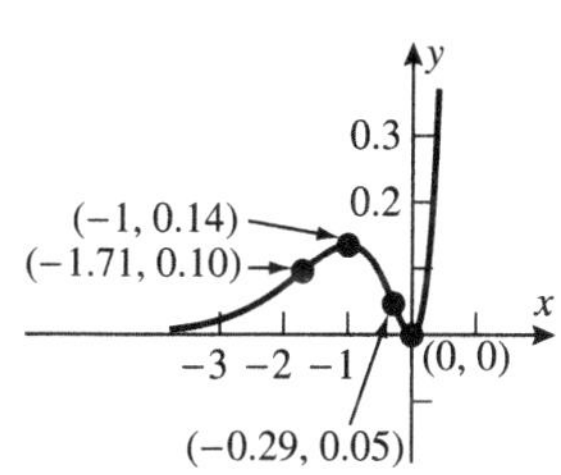

9. (a) $\lim_{x\to+\infty} f(x) = +\infty$, $\lim_{x\to-\infty} f(x) = -\infty$

(b) $y = xe^{x^2}$;
$y' = (1+2x^2)e^{x^2}$;
$y'' = 2x(3+2x^2)e^{x^2}$
no relative extrema, inflection point at $(0,0)$

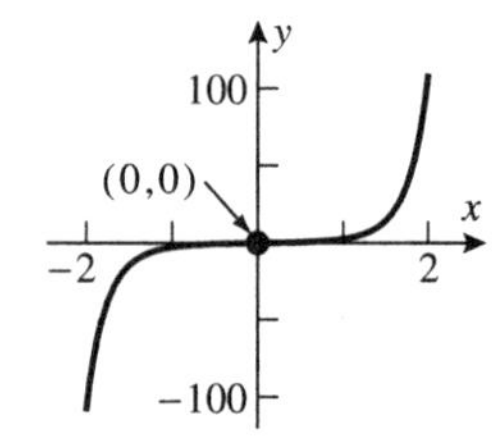

10. (a) $\lim_{x\to\pm\infty} f(x) = 1$

(b) $f'(x) = 2x^{-3}e^{-1/x^2}$ so $f'(x) < 0$ for $x < 0$ and $f'(x) > 0$ for $x > 0$. Set $u = x^2$ and use the given result to find $\lim_{x\to 0} f'(x) = 0$, so (by the First Derivative Test) $f(x)$ has a minimum at $x = 0$. $f''(x) = (-6x^{-4} + 4x^{-6})e^{-1/x^2}$, so $f(x)$ has points of inflection at $x = \pm\sqrt{2/3}$.

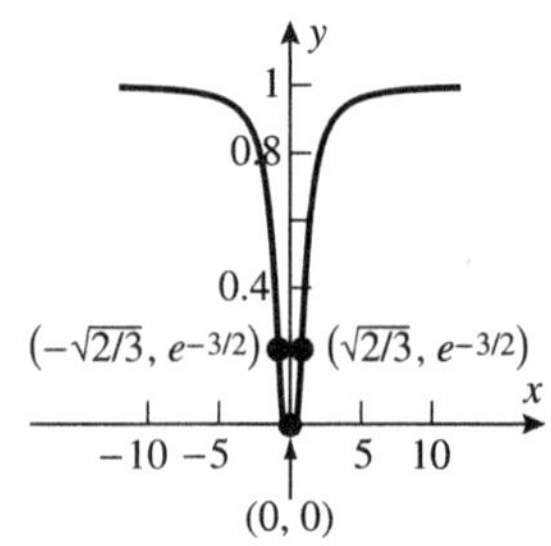

11. $\lim_{x\to+\infty} f(x) = +\infty$, $\lim_{x\to-\infty} f(x) = 0$
$f'(x) = e^x \dfrac{x-1}{x^2}$, $f''(x) = e^x \dfrac{x^2-2x+2}{x^3}$
critical point at $x = 1$;
relative minimum at $x = 1$
no points of inflection
vertical asymptote $x = 0$,
horizontal asymptote $y = 0$ for $x \to -\infty$

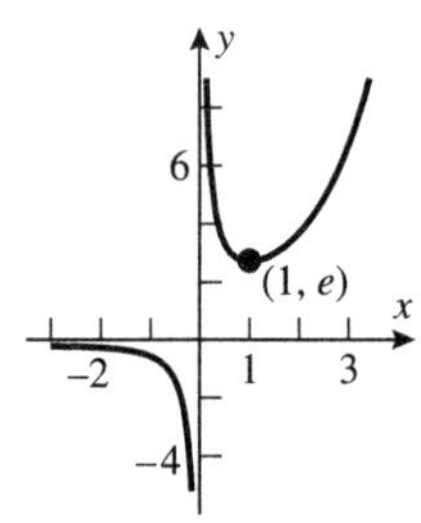

12. $\lim_{x\to+\infty} f(x) = 0$, $\lim_{x\to-\infty} f(x) = -\infty$
$f'(x) = (1-x)e^{-x}$, $f''(x) = (x-2)e^{-x}$
critical point at $x = 1$; relative maximum at $x = 1$
point of inflection at $x = 2$
horizontal asymptote $y = 0$ as $x \to +\infty$

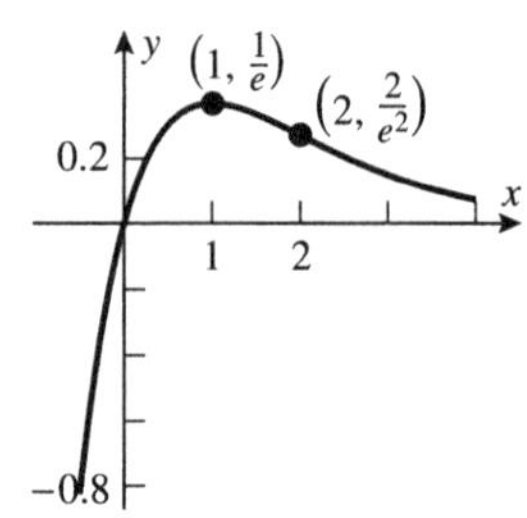

13. $\lim_{x\to+\infty} f(x) = 0$, $\lim_{x\to-\infty} f(x) = +\infty$
$f'(x) = x(2-x)e^{1-x}$, $f''(x) = (x^2-4x+2)e^{1-x}$
critical points at $x = 0, 2$;
relative minimum at $x = 0$,
relative maximum at $x = 2$
points of inflection at $x = 2 \pm \sqrt{2} \approx 0.59, 3.41$
horizontal asymptote $y = 0$ as $x \to +\infty$

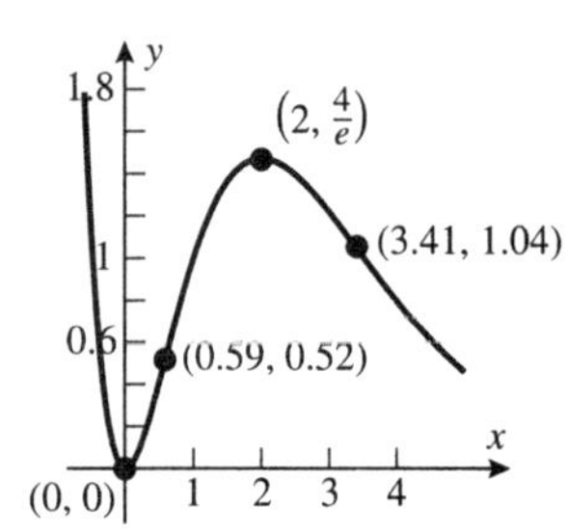

14. $\lim_{x\to+\infty} f(x) = +\infty$, $\lim_{x\to-\infty} f(x) = 0$
$f'(x) = x^2(3+x)e^{x-1}$, $f''(x) = x(x^2+6x+6)e^{x-1}$
critical points at $x = -3, 0$;
relative minimum at $x = -3$
points of inflection at $x = 0, -3 \pm \sqrt{3} \approx 0, -4.7, -1.27$
horizontal asymptote $y = 0$ as $x \to -\infty$

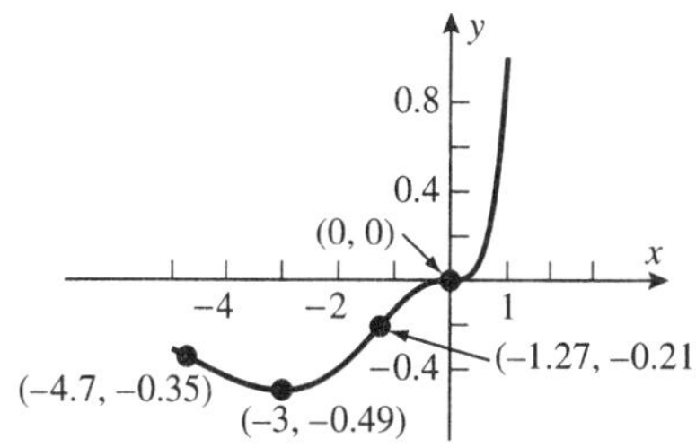

15. (a) $\lim_{x\to0^+} y = \lim_{x\to0^+} x\ln x = \lim_{x\to0^+} \dfrac{\ln x}{1/x} = \lim_{x\to0^+} \dfrac{1/x}{-1/x^2} = 0$;
$\lim_{x\to+\infty} y = +\infty$

(b) $y = x\ln x$,
$y' = 1 + \ln x$,
$y'' = 1/x$,
$y' = 0$ when $x = e^{-1}$

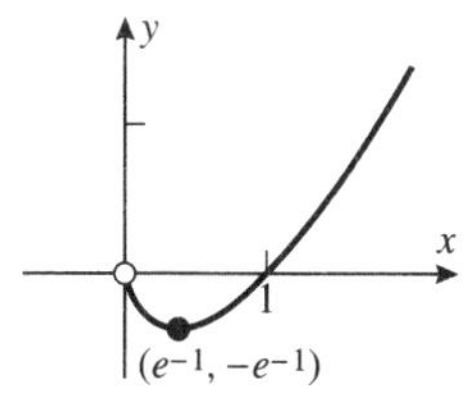

16. (a) $\lim_{x\to0^+} y = \lim_{x\to0^+} \dfrac{\ln x}{1/x^2} = \lim_{x\to0^+} \dfrac{1/x}{-2/x^3} = 0$,
$\lim_{x\to+\infty} y = +\infty$

(b) $y = x^2\ln x, y' = x(1+2\ln x)$,
$y'' = 3 + 2\ln x$,
$y' = 0$ if $x = e^{-1/2}$,
$y'' = 0$ if $x = e^{-3/2}$,
$\lim_{x\to0^+} y' = 0$

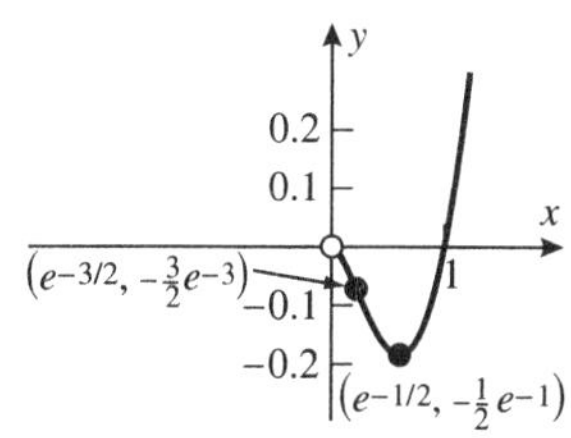

17. (a) $\lim_{x\to0^+} y = \lim_{x\to0^+} \dfrac{\ln x}{x^2} = -\infty$;
$\lim_{x\to+\infty} y = \lim_{x\to+\infty} \dfrac{\ln x}{x^2} = \lim_{x\to+\infty} \dfrac{1/x}{2x} = 0$

(b) $y = \dfrac{\ln x}{x^2}, y' = \dfrac{1-2\ln x}{x^3}$,
$y'' = \dfrac{6\ln x - 5}{x^4}$,
$y' = 0$ if $x = e^{1/2}$,
$y'' = 0$ if $x = e^{5/6}$

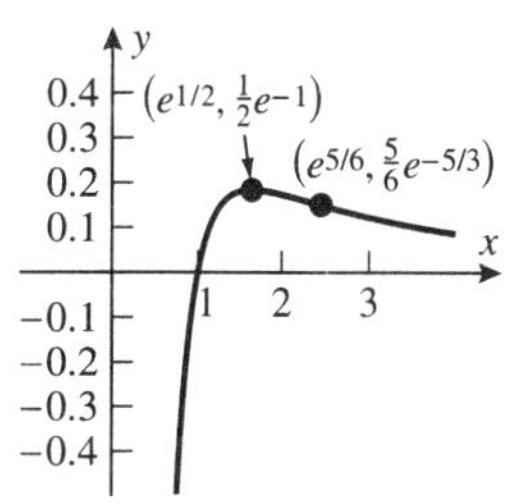

18. (a) Let $u = 1/x$, $\lim_{x\to0^+} (\ln x)/\sqrt{x} = \lim_{u\to+\infty} -\sqrt{u}\ln u = -\infty$ by inspection,
$\lim_{x\to+\infty} (\ln x)/\sqrt{x} = 0$, by the rule given.

(b) $y = \dfrac{\ln x}{\sqrt{x}}$, $y' = \dfrac{2-\ln x}{2x^{3/2}}$
$y'' = \dfrac{-8+3\ln x}{4x^{5/2}}$
$y' = 0$ if $x = e^2$,
$y'' = 0$ if $x = e^{8/3}$

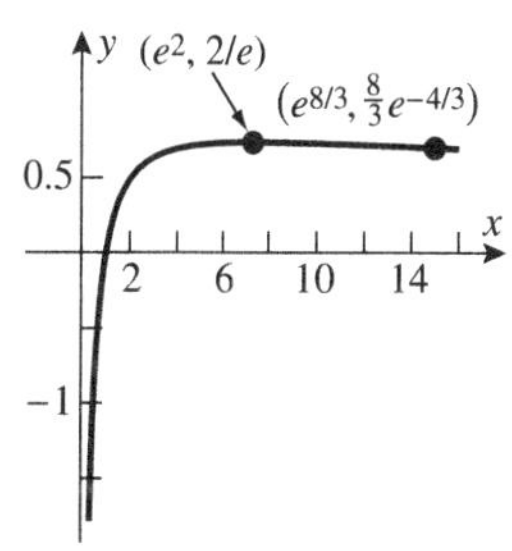

19. **(a)** $\lim_{x\to 0^+} x^2 \ln x = 0$ by the rule given, $\lim_{x\to+\infty} x^2\ln x = +\infty$ by inspection, and $f(x)$ not defined for $x<0$

(b) $y = x^2 \ln 2x$, $y' = 2x\ln 2x + x$
$y'' = 2\ln 2x + 3$
$y' = 0$ if $x = 1/(2\sqrt{e})$,
$y'' = 0$ if $x = 1/(2e^{3/2})$

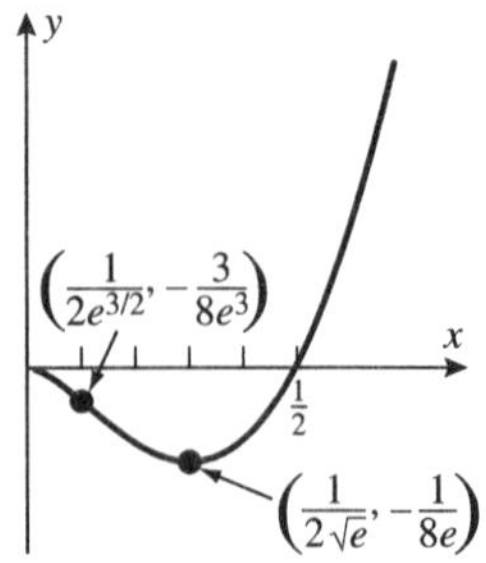

20. **(a)** $\lim_{x\to+\infty} f(x) = +\infty$; $\lim_{x\to 0} f(x) = 0$

(b) $y = \ln(x^2+1)$, $y' = 2x/(x^2+1)$
$y'' = -2\dfrac{x^2-1}{(x^2+1)^2}$
$y' = 0$ if $x = 0$
$y'' = 0$ if $x = \pm 1$

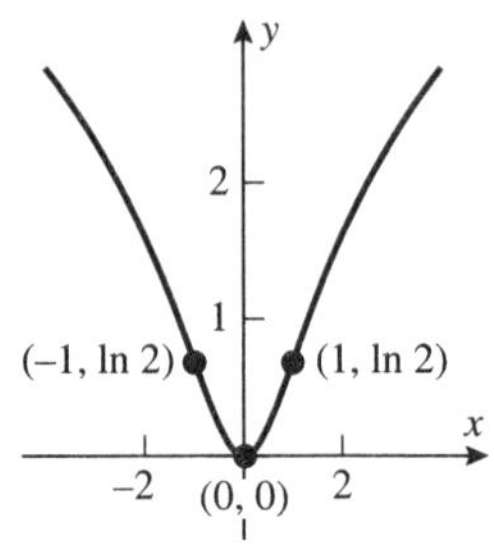

21. **(a)**

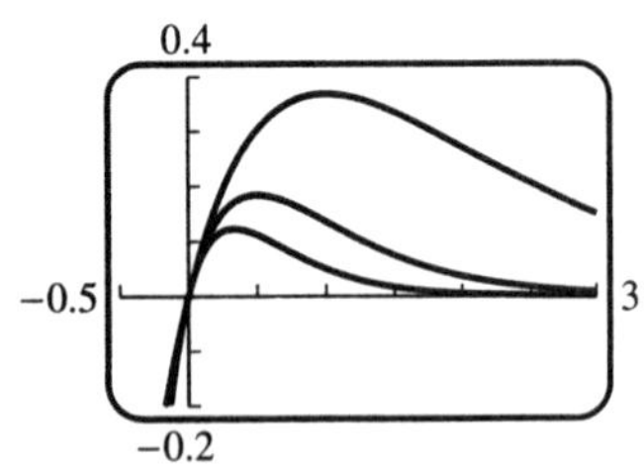

(b) $y' = (1-bx)e^{-bx}$, $y'' = b^2(x-2/b)e^{-bx}$; relative maximum at $x = 1/b$, $y = 1/be$; point of inflection at $x = 2/b$, $y = 2/be^2$. Increasing b moves the relative maximum and the point of inflection to the left and down, i.e. towards the origin.

22. **(a)**

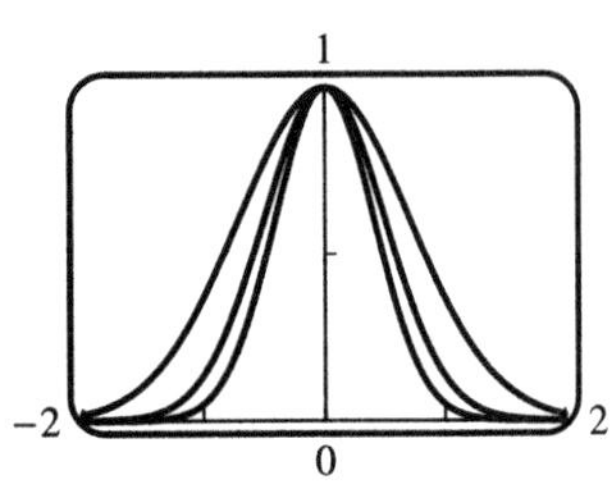

(b) $y' = -2bxe^{-bx^2}$,
$y'' = 2b(-1+2bx^2)e^{-bx^2}$;
relative maximum at $x = 0$, $y = 1$; points of inflection at $x = \pm\sqrt{1/2b}$, $y = 1/\sqrt{e}$. Increasing b moves the points of inflection towards the y-axis; the relative maximum doesn't move.

23. **(a)** The oscillations of $e^x \cos x$ about zero increase as $x \to +\infty$ so the limit does not exist, and $\lim_{x\to-\infty} e^x \cos x = 0$.

(b)

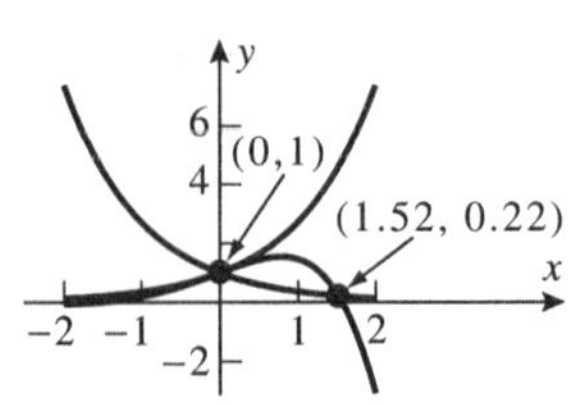

(c) The curve $y = e^{ax}\cos bx$ oscillates between $y = e^{ax}$ and $y = -e^{ax}$. The frequency of oscillation increases when b increases.

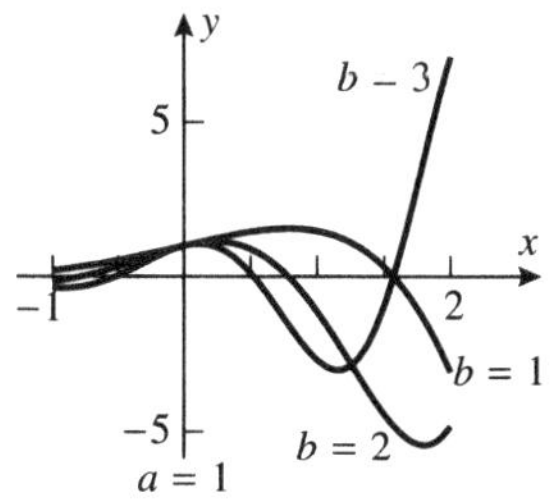

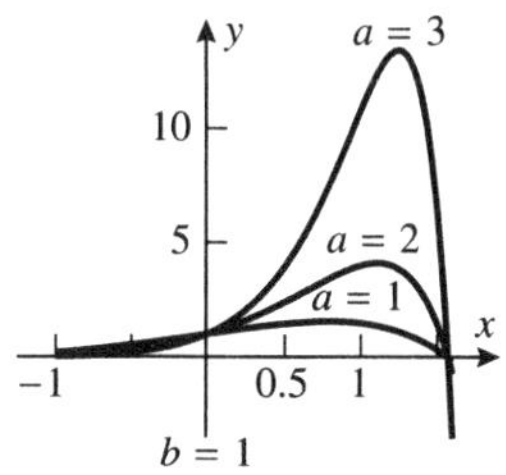

24. Let $P(x_0, y_0)$ be a point on $y = e^{3x}$ then $y_0 = e^{3x_0}$. $dy/dx = 3e^{3x}$ so $m_{\tan} = 3e^{3x_0}$ at P and an equation of the tangent line at P is $y - y_0 = 3e^{3x_0}(x - x_0)$, $y - e^{3x_0} = 3e^{3x_0}(x - x_0)$. If the line passes through the origin then $(0,0)$ must satisfy the equation so $-e^{3x_0} = -3x_0e^{3x_0}$ which gives $x_0 = 1/3$ and thus $y_0 = e$. The point is $(1/3, e)$.

25. (b)

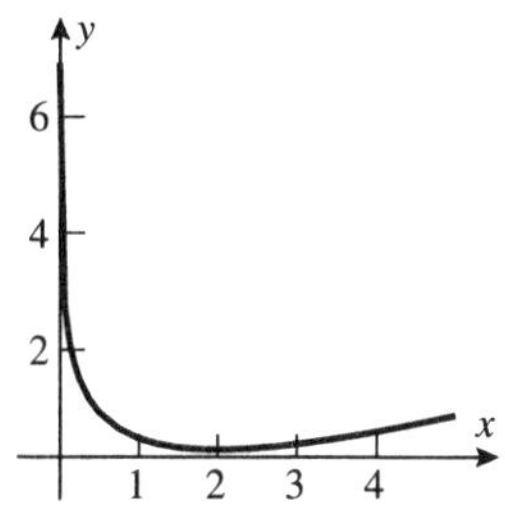

(c) $\dfrac{dy}{dx} = \dfrac{1}{2} - \dfrac{1}{x}$ so $\dfrac{dy}{dx} < 0$ at $x = 1$ and $\dfrac{dy}{dx} > 0$ at $x = e$

(d) The slope is a continuous function which goes from a negative value to a positive value; therefore it must take the value zero in between, by the Intermediate Value Theorem.

(e) $\dfrac{dy}{dx} = 0$ when $x = 2$

26. (a) $\dfrac{dC}{dt} = \dfrac{K}{a-b}\left(ae^{-at} - be^{-bt}\right)$ so $\dfrac{dC}{dt} = 0$ at $t = \dfrac{\ln(a/b)}{a-b}$. This is the only stationary point and $C(0) = 0$, $\lim\limits_{t\to+\infty} C(t) = 0$, $C(t) > 0$ for $0 < t < +\infty$, so it is an absolute maximum.

(b)

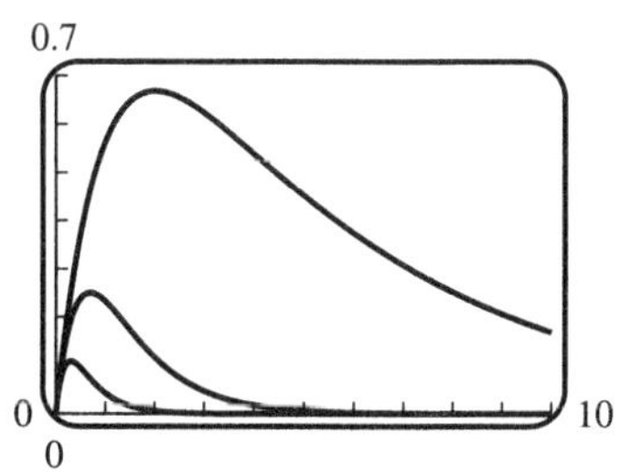

27. (a)

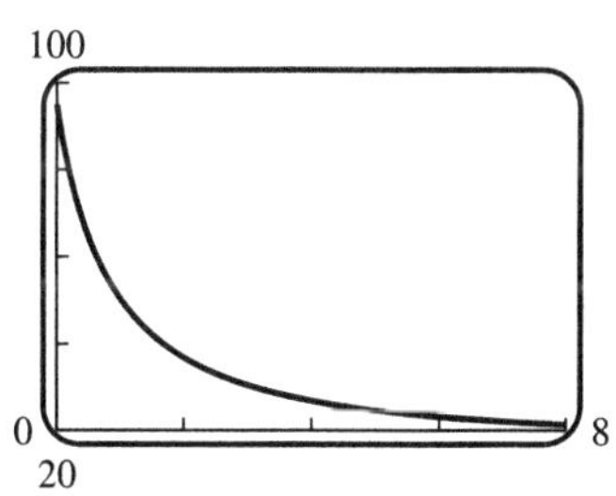

(b) as t tends to $+\infty$, the population tends to 19

$$\lim_{t\to+\infty} P(t) = \lim_{t\to+\infty} \frac{95}{5-4e^{-t/4}} = \frac{95}{5-4\lim\limits_{t\to+\infty} e^{-t/4}} = \frac{95}{5} = 19$$

(c) the rate of population growth tends to zero

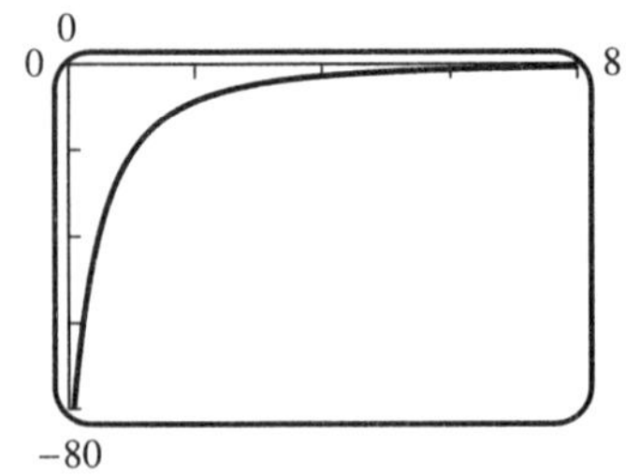

28. **(a)**

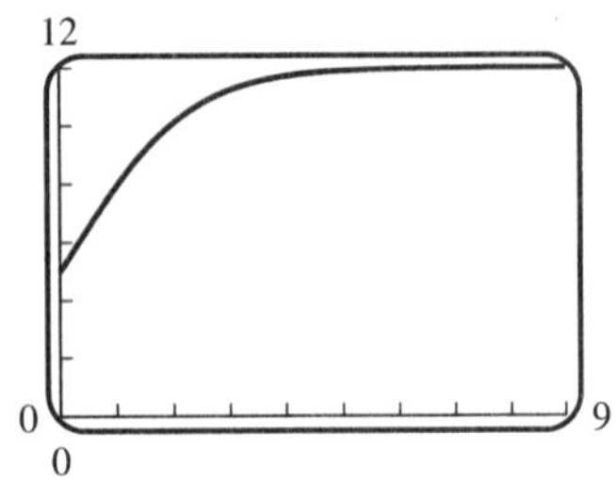

(b) P tends to 12 as t gets large; $\lim\limits_{t\to+\infty} P(t) = \lim\limits_{t\to+\infty} \dfrac{60}{5+7e^{-t}} = \dfrac{60}{5+7\lim\limits_{t\to+\infty} e^{-t}} = \dfrac{60}{5} = 12$

(c) the rate of population growth tends to zero

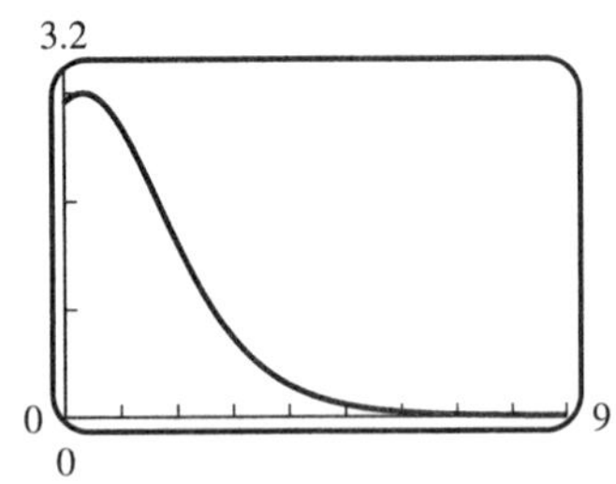

29. $t = 7.67$

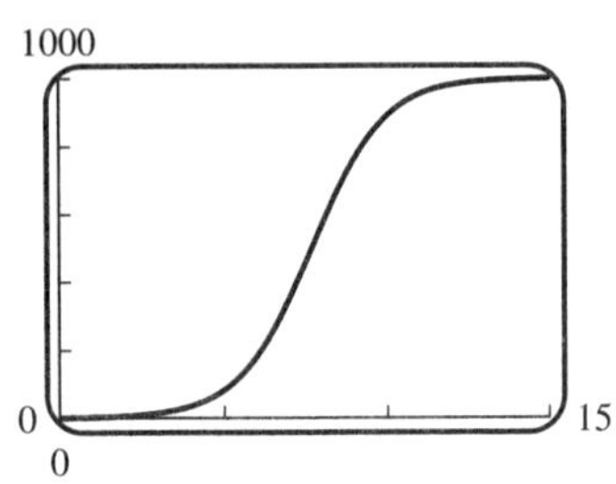

30. **(a)** $\dfrac{dN}{dt} = 250(20-t)e^{-t/20} = 0$ at $t = 20$, $N(0) = 125{,}000$, $N(20) \approx 161{,}788$, and $N(100) \approx 128{,}369$; the absolute maximum is $N = 161{,}788$ at $t = 20$, the absolute minimum is $N = 125{,}000$ at $t = 0$.

(b) The absolute minimum of $\dfrac{dN}{dt}$ occurs when $\dfrac{d^2N}{dt^2} = 12.5(t-40)e^{-t/20} = 0$, $t = 40$.

31. (a) $y'(t) = \dfrac{LAke^{-kt}}{(1+Ae^{-kt})^2}S$, so $y'(0) = \dfrac{LAk}{(1+A)^2}$

(b) The rate of growth increases to its maximum, which occurs when y is halfway between 0 and L, or when $t = \dfrac{1}{k}\ln A$; it then decreases back towards zero.

(c) From (2) one sees that $\dfrac{dy}{dt}$ is maximized when y lies half way between 0 and L, i.e. $y = L/2$. This follows since the right side of (2) is a parabola (with y as independent variable) with y-intercepts $y = 0, L$. The value $y = L/2$ corresponds to $t = \dfrac{1}{k}\ln A$, from (4).

32. Since $0 < y < L$ the right-hand side of (3) can change sign only if the factor $L - 2y$ changes sign, which it does when $y = L/2$. From (1) we have $\dfrac{L}{2} = \dfrac{L}{1+Ae^{-kt}}$, $1 = Ae^{-kt}$, $t = \dfrac{1}{k}\ln A$.

33. $\dfrac{dk}{dT} = k_0 \exp\left[-\dfrac{q(T-T_0)}{2T_0T}\right]\left(-\dfrac{q}{2T^2}\right) = -\dfrac{qk_0}{2T^2}\exp\left[-\dfrac{q(T-T_0)}{2T_0T}\right]$

34. $\beta = 10\log I - 10\log I_0$, $\dfrac{d\beta}{dI} = \dfrac{10}{I\ln 10}$

(a) $\left.\dfrac{d\beta}{dI}\right]_{I=10I_0} = \dfrac{1}{I_0\ln 10}$ db/W/m^2

(b) $\left.\dfrac{d\beta}{dI}\right]_{I=100I_0} = \dfrac{1}{10I_0\ln 10}$ db/W/m^2

(c) $\left.\dfrac{d\beta}{dI}\right]_{I=100I_0} = \dfrac{1}{100I_0\ln 10}$ db/W/m^2

35. Solve $\dfrac{dy}{dt} = 3\dfrac{dx}{dt}$ given $y = x\ln x$. Then $\dfrac{dy}{dt} = \dfrac{dy}{dx}\dfrac{dx}{dt} = (1+\ln x)\dfrac{dx}{dt}$, so $1 + \ln x = 3$, $\ln x = 2$, $x = e^2$.

36. $v(t) = (1-t)e^{-t}$, $a(t) = (t-2)e^{-t}$

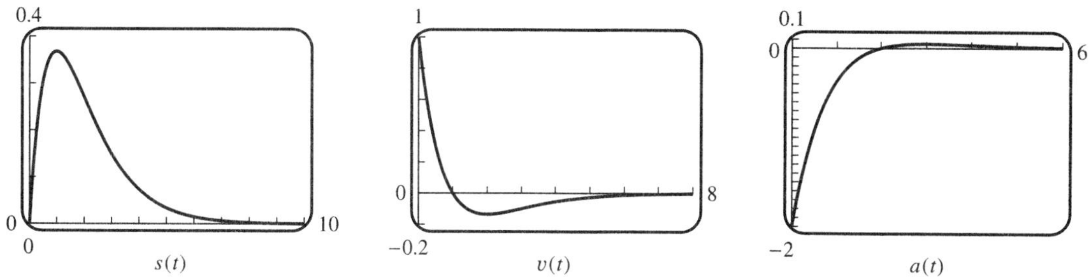

(a) $v = 0$ at $t = 1$

(b) $s = 1/e$ at $t = 1$

(c) a changes sign at $t = 2$, so the particle is speeding up for $1 < t < 2$ and slowing down for $0 < t < 1$ and $2 < t$

37. $\displaystyle\int_1^3 e^x\,dx = e^x\Big]_1^3 = e^3 - e$

38. $\displaystyle\int_1^5 \frac{1}{x}\,dx = \ln x\Big]_1^5 = \ln 5 - \ln 1 = \ln 5$

39. $A = \int_0^{\ln 2} (e^{2x} - e^x)\, dx$

$= \left(\frac{1}{2}e^{2x} - e^x\right)\Big]_0^{\ln 2} = 1/2$

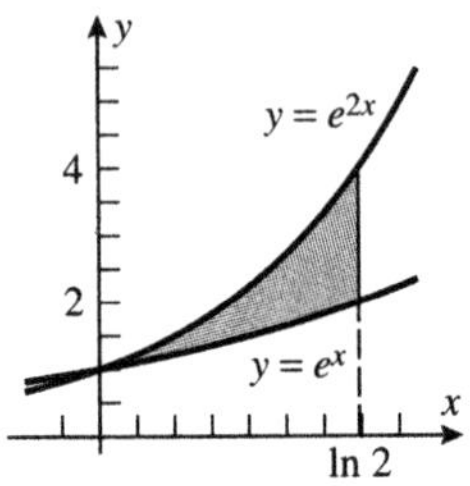

40. $A = \int_1^e \frac{dy}{y} = \ln y\Big]_1^e = 1$

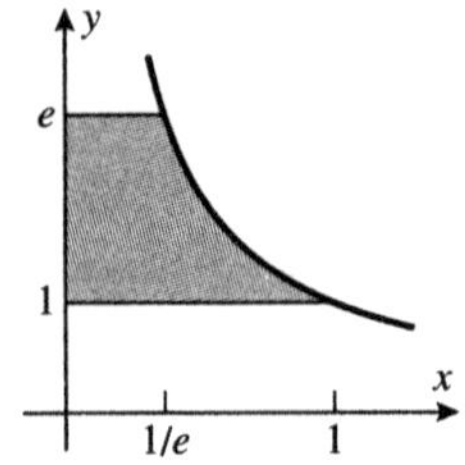

41. $A = A_1 + A_2 = \int_{-1}^0 (1 - e^x)dx + \int_0^1 (e^x - 1)dx = 1/e + e - 2$

42. $A = A_1 + A_2 = \int_{1/2}^1 \frac{1-x}{x}dx + \int_1^2 \frac{x-1}{x}dx = -\left(\frac{1}{2} - \ln 2\right) + (1 - \ln 2) = 1/2$

43. $f_{\text{ave}} = \frac{1}{e-1}\int_1^e \frac{1}{x}dx = \frac{1}{e-1}(\ln e - \ln 1) = \frac{1}{e-1}$

44. $f_{\text{ave}} = \frac{1}{\ln 5 - (-1)}\int_{-1}^{\ln 5} e^x dx = \frac{1}{\ln 5 + 1}(5 - e^{-1}) = \frac{5 - e^{-1}}{1 + \ln 5}$

45. $f_{\text{ave}} = \frac{1}{4-0}\int_0^4 e^{-2x}dx = -\frac{1}{8}e^{-2x}\Big]_0^4 = \frac{1 - e^{-8}}{8}$

46. $\int_0^k e^{2x}dx = 3,\ \frac{1}{2}e^{2x}\Big]_0^k = 3,\ \frac{1}{2}(e^{2k} - 1) = 3,\ e^{2k} = 7,\ k = \frac{1}{2}\ln 7$

47. $y(t) = (802.137)\int e^{1.528t}dt = 524.959e^{1.528t} + C;\ y(0) = 750 = 524.959 + C,\ C = 225.041,$

$y(t) = 524.959e^{1.528t} + 225.041,\ y(12) = 48,233,525,650$

48. $V_{\text{ave}} = \frac{275000}{10-0}\int_0^{10} e^{-0.17t}dt = -161764.7059e^{-0.17t}\Big]_0^{10} = \$132,212.96$

49. $s(t) = \int (25 + 10e^{-0.05t})dt = 25t - 200e^{-0.05t} + C$

(a) $s(10) - s(0) = 250 - 200(e^{-0.5} - 1) = 450 - 200/\sqrt{e} \approx 328.69$ ft

(b) yes; without it the distance would have been 250 ft

50. **(a)** displacement $= \int_0^3 (e^t - 2)dt = e^3 - 7$

distance $= \int_0^3 |v(t)|dt = -\int_0^{\ln 2} v(t)dt + \int_{\ln 2}^3 v(t)dt = e^3 - 9 + 4\ln 2$

(b) displacement $= \int_1^3 \left(\frac{1}{2} - \frac{1}{t}\right) dt = 1 - \ln 3$

distance $= \int_1^3 |v(t)|dt = -\int_1^2 v(t)dt + \int_2^3 v(t)dt = 2\ln 2 - \ln 3$

51. **(a)** From the graph the velocity is positive, so the displacement is always increasing and is therefore positive.

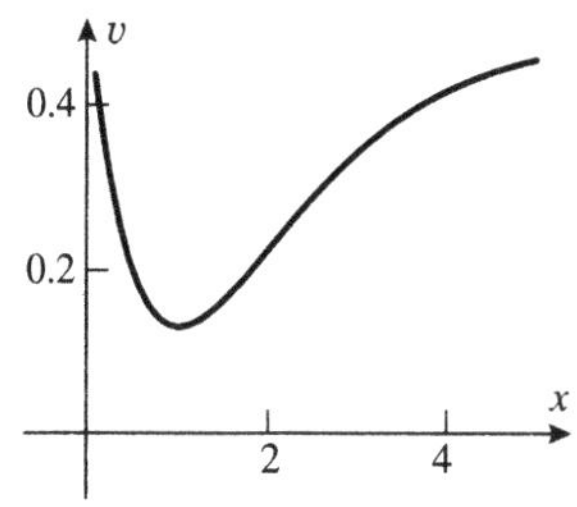

(b) $s(t) = t/2 + (t+1)e^{-t}$

52. **(a)** If $t_0 < 1$ then the area between the velocity curve and the t-axis, between $t = 0$ and $t = t_0$, will always be negative, so the displacement will be negative.

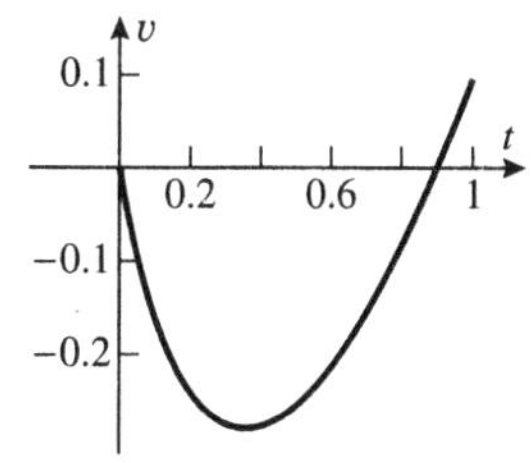

(b) $s(t) = \left(\frac{t^2}{2} - \frac{1}{200}\right)\ln(t+0.1) - \frac{t^2}{4} + \frac{t}{20} - \frac{1}{200}\ln 10$

53. $x = 0$; also set
$f(x) = 1 - e^x \cos x$, $f'(x) = e^x(\sin x - \cos x)$,
$x_{n+1} = x_n - \dfrac{1 - e^x \cos x}{e^x(\sin x - \cos x)}$
$x_1 = 1$, $x_2 = 1.572512605$,
$x_3 = 1.363631415$, $x_7 = x_8 = 1.292695719$

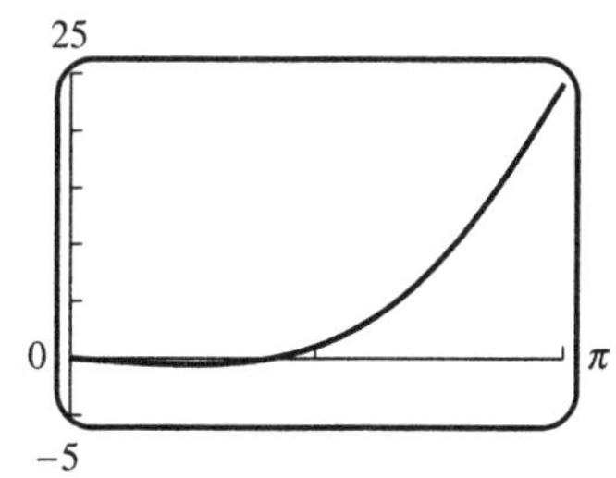

54. The graphs of $y = e^{-x}$ and $y = \ln x$ intersect near $x = 1.3$;
let $f(x) = e^{-x} - \ln x$, $f'(x) = -e^{-x} - 1/x$, $x_1 = 1.3$,
$x_{n+1} = x_n + \dfrac{e^{-x_n} - \ln x_n}{e^{-x_n} + 1/x_n}$, $x_2 = 1.309759929$,
$x_4 = x_5 = 1.309799586$

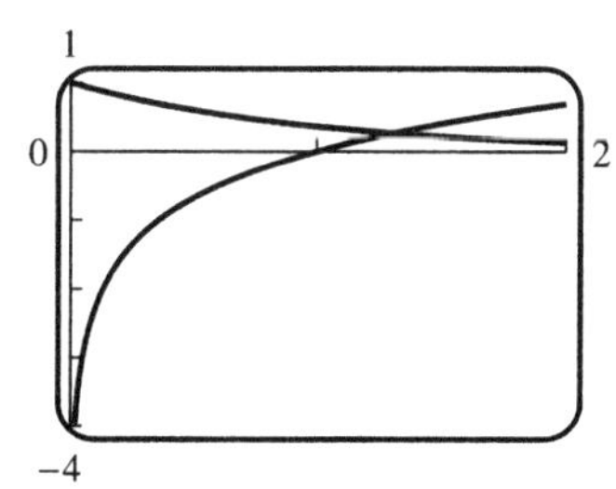

55. A graphing utility shows that there are two inflection points at $x \approx -0.25, 1.25$. These points are the zeros of $f''(x) = (x^4 - 4x^3 + 8x^2 - 4x - 1)\dfrac{e^x}{(x^2+1)^3}$. It is equivalent to find the zeros of $g(x) = x^4 - 4x^3 + 8x^2 - 4x - 1$. One root is $x = 1$ by inspection. Since $g'(x) = 4x^3 - 12x^2 + 16x - 4$,

Newton's Method becomes

$$x_n = x_{n-1} - \frac{x_{n-1}^4 - 4x_{n-1}^3 + 8x_{n-1}^2 - 4x_{n-1} - 1}{4x_{n-1}^3 - 12x_{n-1}^2 + 16x_{n-1} - 4}$$

With $x_0 = -0.25$, $x_1 = -0.18572695, x_2 = -0.179563312, x_3 = -0.179509029,$ $x_4 = x_5 = -0.179509025$. So the points of inflection are at $x \approx -0.18, x = 1$.

56. **(a)** Let $h(x) = e^x - 1 - x$ for $x \geq 0$. Then $h(0) = 0$ and $h'(x) = e^x - 1 \geq 0$ for $x \geq 0$, so $h(x)$ is increasing.

(b) Let $h(x) = e^x - 1 - x - \frac{1}{2}x^2$. Then $h(0) = 0$ and $h'(x) = e^x - 1 - x$. By Part (a), $e^x - 1 - x \geq 0$ for $x \geq 0$, so $h(x)$ is increasing.

(c)

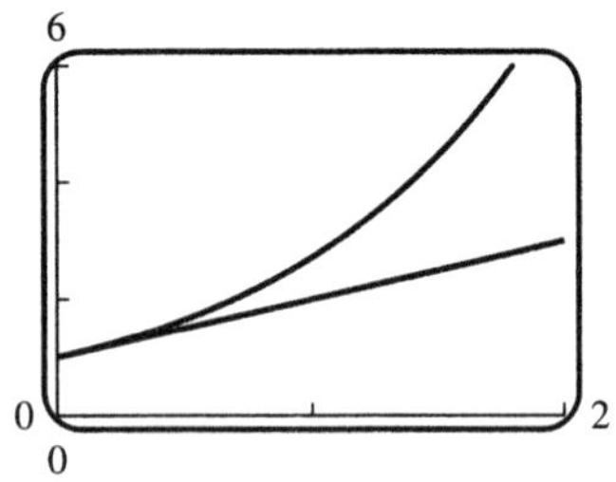

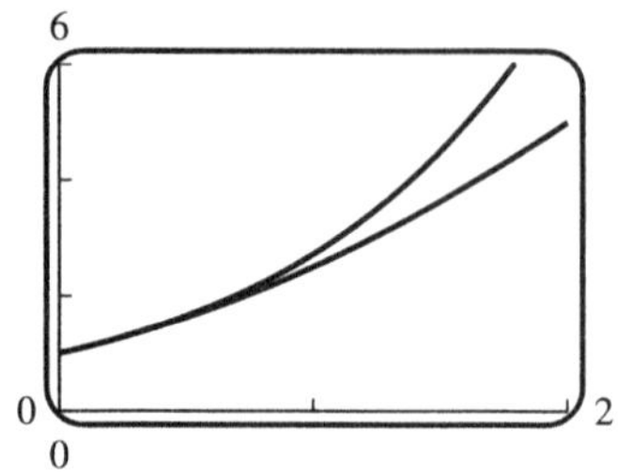

57. $V = \pi \int_0^{\ln 3} e^{2x}\,dx = \frac{\pi}{2}e^{2x}\Big]_0^{\ln 3} = 4\pi$

58. $V = \pi \int_0^1 e^{-4x}\,dx = \frac{\pi}{4}(1 - e^{-4})$

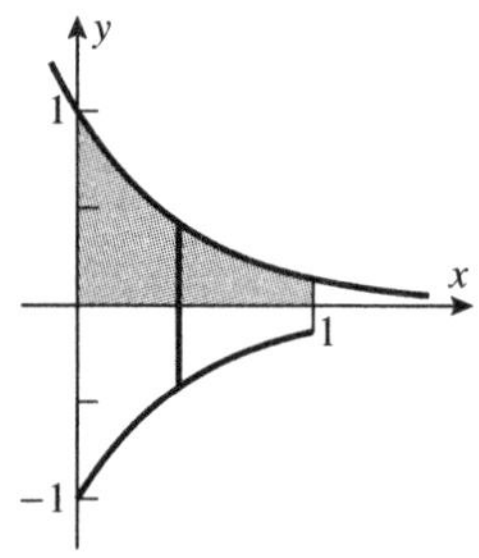

59. $V = 2\pi \int_0^1 \frac{x}{x^2+1}\,dx$

$= \pi \ln(x^2+1)\Big]_0^1 = \pi \ln 2$

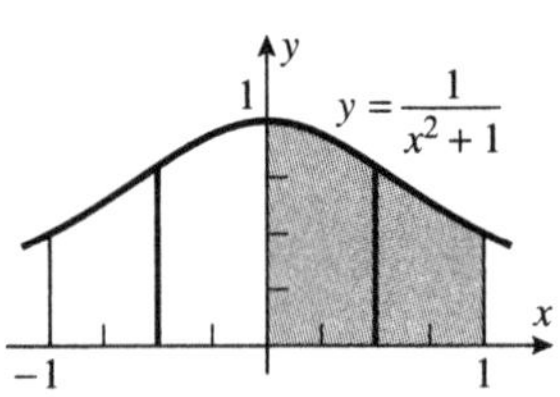

60. $V = \int_1^{\sqrt{3}} 2\pi x e^{x^2}\,dx = \pi e^{x^2}\Big]_1^{\sqrt{3}} = \pi(e^3 - e)$

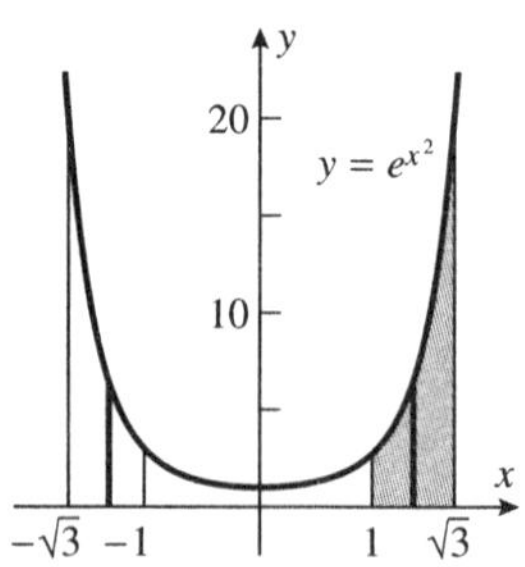

61. $(dx/dt)^2 + (dy/dt)^2 = [e^t(\cos t - \sin t)]^2 + [e^t(\cos t + \sin t)]^2 = 2e^{2t}$,

$L = \int_0^{\pi/2} \sqrt{2}e^t dt = \sqrt{2}(e^{\pi/2} - 1)$

62. $(dx/dt)^2 + (dy/dt)^2 = (2e^t \cos t)^2 + (-2e^t \sin t)^2 = 4e^{2t}$, $L = \int_1^4 2e^t dt = 2(e^4 - e)$

63. $dy/dx = \dfrac{\sec x \tan x}{\sec x} = \tan x$, $\sqrt{1 + (y')^2} = \sqrt{1 + \tan^2 x} = \sec x$ when $0 < x < \pi/4$, so

$$L = \int_0^{\pi/4} \sec x\, dx = \ln(1 + \sqrt{2})$$

64. $dy/dx = \dfrac{\cos x}{\sin x} = \cot x$, $\sqrt{1 + (y')^2} = \sqrt{1 + \cot^2 x} = \csc x$ when $\pi/4 < x < \pi/2$, so

$$L = \int_{\pi/4}^{\pi/2} \csc x\, dx = -\ln(\sqrt{2} - 1) = -\ln\left(\frac{\sqrt{2}-1}{\sqrt{2}+1}(\sqrt{2}+1)\right) = \ln(1 + \sqrt{2})$$

65. $f'(x) = e^x$, $\quad 1 + [f'(x)]^2 = 1 + e^{2x}$, $S = \int_0^1 2\pi e^x \sqrt{1 + e^{2x}}\, dx \approx 22.94$

66. $x = g(y) = \ln y$, $g'(y) = 1/y$, $1 + [g'(y)]^2 = 1 + 1/y^2$; $S = \int_1^e 2\pi \sqrt{1 + 1/y^2} \ln y\, dy \approx 7.05$

67. $x' = e^t(\cos t - \sin t), y' = e^t(\cos t + \sin t), (x')^2 + (y')^2 = 2e^{2t}$

$$S = 2\pi \int_0^{\pi/2} (e^t \sin t)\sqrt{2e^{2t}} dt = 2\sqrt{2}\pi \int_0^{\pi/2} e^{2t} \sin t\, dt$$

$$= 2\sqrt{2}\pi \left[\frac{1}{5} e^{2t}(2\sin t - \cos t)\right]_0^{\pi/2} = \frac{2\sqrt{2}}{5}\pi(2e^{\pi} + 1)$$

EXERCISE SET 7.5

1. (a)

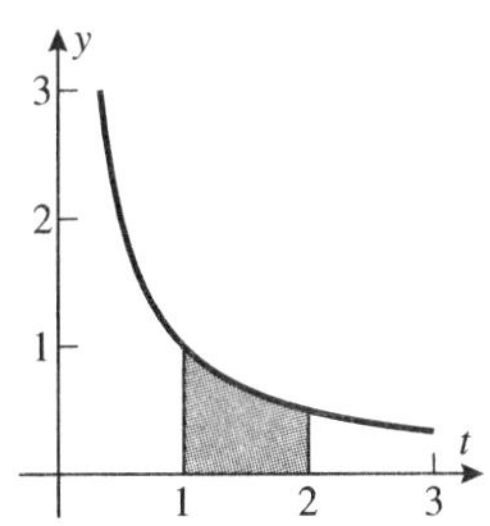

(b)

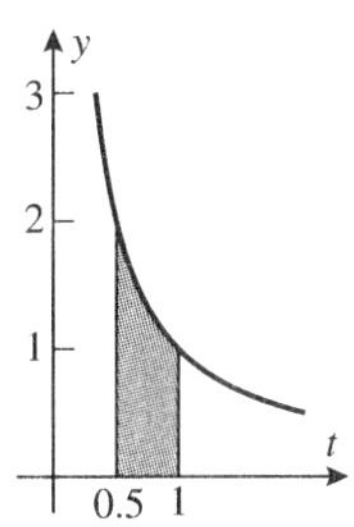

(c)

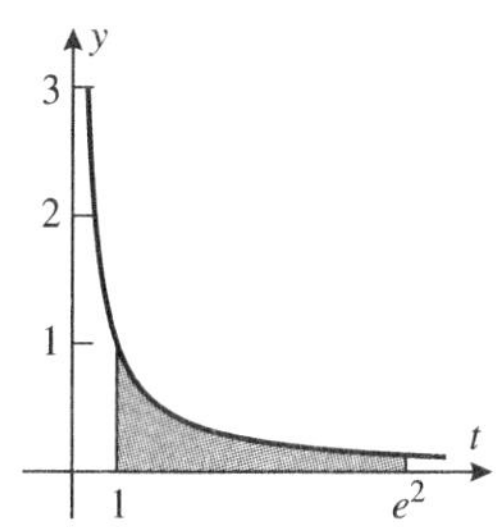

2.

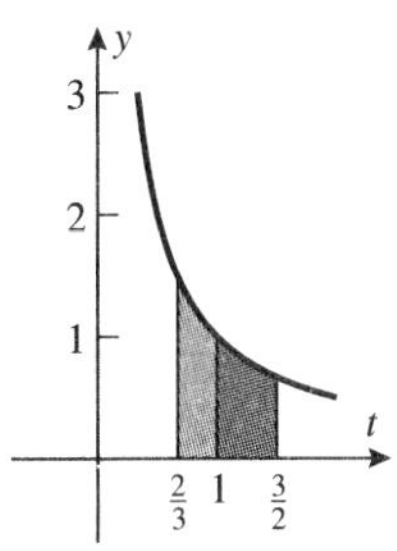

3. (a) $\ln t\Big]_1^{ac} = \ln(ac) = \ln a + \ln c = 7$ **(b)** $\ln t\Big]_1^{1/c} = \ln(1/c) = -5$

(c) $\ln t\Big]_1^{a/c} = \ln(a/c) = 2 - 5 = -3$ **(d)** $\ln t\Big]_1^{a^3} = \ln a^3 = 3\ln a = 6$

4. **(a)** $\ln t\Big]_1^{\sqrt{a}} = \ln a^{1/2} = \frac{1}{2}\ln a = 9/2$ **(b)** $\ln t\Big]_1^{2a} = \ln 2 + 9$

(c) $\ln t\Big]_1^{2/a} = \ln 2 - 9$ **(d)** $\ln t\Big]_2^{a} = 9 - \ln 2$

5. $\ln 5 \approx 1.603210678$; $\ln 5 = 1.609437912$; magnitude of error is < 0.0063

6. $\ln 3 \approx 1.098242635$; $\ln 3 = 1.098612289$; magnitude of error is < 0.0004

7. **(a)** x^{-1}, $x > 0$ **(b)** x^2, $x \neq 0$

(c) $-x^2$, $-\infty < x < +\infty$ **(d)** $-x$, $-\infty < x < +\infty$

(e) x^3, $x > 0$ **(f)** $\ln x + x$, $x > 0$

(g) $x - \sqrt[3]{x}$, $-\infty < x < +\infty$ **(h)** $\frac{e^x}{x}$, $x > 0$

8. **(a)** $f(\ln 3) = e^{-2\ln 3} = e^{\ln(1/9)} = 1/9$

(b) $f(\ln 2) = e^{\ln 2} + 3e^{-\ln 2} = 2 + 3e^{\ln(1/2)} = 2 + 3/2 = 7/2$

9. **(a)** $3^\pi = e^{\pi \ln 3}$ **(b)** $2^{\sqrt{2}} = e^{\sqrt{2}\ln 2}$

10. **(a)** $\pi^{-x} = e^{-x\ln \pi}$ **(b)** $x^{2x} = e^{2x\ln x}$

11. **(a)** $\lim_{x\to+\infty}\left[\left(1+\frac{1}{x}\right)^x\right]^2 = \left[\lim_{x\to+\infty}\left(1+\frac{1}{x}\right)^x\right]^2 = e^2$

(b) $y = 2x$, $\lim_{y\to 0}(1+y)^{2/y} = \lim_{y\to 0}\left[(1+y)^{1/y}\right]^2 = e^2$

12. **(a)** $y = 3x$, $\lim_{y\to+\infty}\left(1+\frac{1}{y}\right)^{y/3} = \lim_{y\to+\infty}\left[\left(1+\frac{1}{y}\right)^y\right]^{1/3} = \left[\lim_{y\to+\infty}\left(1+\frac{1}{y}\right)^y\right]^{1/3} = e^{1/3}$

(b) $\lim_{x\to 0}(1+x)^{1/3x} = \lim_{x\to 0}\left[(1+x)^{1/x}\right]^{1/3} = e^{1/3}$

13. $g'(x) = x^2 - x$ **14.** $g'(x) = 1 - \cos x$

15. **(a)** $\frac{1}{x^3}(3x^2) = \frac{3}{x}$ **(b)** $e^{\ln x}\frac{1}{x} = 1$

16. **(a)** $2x\sqrt{x^2+1}$ **(b)** $-\left(\frac{1}{x^2}\right)\sin\left(\frac{1}{x}\right)$

17. $F'(x) = \frac{\cos x}{x^2+3}$, $F''(x) = \frac{-(x^2+3)\sin x - 2x\cos x}{(x^2+3)^2}$

(a) 0 **(b)** 1/3 **(c)** 0

18. $F'(x) = \sqrt{3x^2+1}$, $F''(x) = \frac{3x}{\sqrt{3x^2+1}}$

(a) 0 **(b)** $\sqrt{13}$ **(c)** $6/\sqrt{13}$

19. **(a)** $\frac{d}{dx}\int_1^{x^2} t\sqrt{1+t}dt = x^2\sqrt{1+x^2}(2x) = 2x^3\sqrt{1+x^2}$

(b) $\int_1^{x^2} t\sqrt{1+t}dt = -\frac{2}{3}(x^2+1)^{3/2} + \frac{2}{5}(x^2+1)^{5/2} - \frac{4\sqrt{2}}{15}$

20. **(a)** $\frac{d}{dx}\int_x^a f(t)dt = -\frac{d}{dx}\int_a^x f(t)dt = -f(x)$

(b) $\frac{d}{dx}\int_{g(x)}^a f(t)dt = -\frac{d}{dx}\int_a^{g(x)} f(t)dt = -f(g(x))g'(x)$

21. **(a)** $-\sin x^2$ **(b)** $-\frac{\tan^2 x}{1+\tan^2 x}\sec^2 x = -\tan^2 x$

22. **(a)** $-(x^2+1)^{40}$ **(b)** $-\cos^3\left(\frac{1}{x}\right)\left(-\frac{1}{x^2}\right) = \frac{\cos^3(1/x)}{x^2}$

23. $-3\frac{3x-1}{9x^2+1} + 2x\frac{x^2-1}{x^4+1}$

24. If f is continuous on an open interval I and $g(x)$, $h(x)$, and a are in I then

$$\int_{h(x)}^{g(x)} f(t)dt = \int_{h(x)}^a f(t)dt + \int_a^{g(x)} f(t)dt = -\int_a^{h(x)} f(t)dt + \int_a^{g(x)} f(t)dt$$

so $\frac{d}{dx}\int_{h(x)}^{g(x)} f(t)dt = -f(h(x))h'(x) + f(g(x))g'(x)$

25. **(a)** $\sin^2(x^3)(3x^2) - \sin^2(x^2)(2x) = 3x^2\sin^2(x^3) - 2x\sin^2(x^2)$

(b) $\frac{1}{1+x}(1) - \frac{1}{1-x}(-1) = \frac{2}{1-x^2}$

26. $F'(x) = \frac{1}{3x}(3) - \frac{1}{x}(1) = 0$ so $F(x)$ is constant on $(0,+\infty)$. $F(1) = \ln 3$ so $F(x) = \ln 3$ for all $x > 0$.

27. from geometry, $\int_0^3 f(t)dt = 0$, $\int_3^5 f(t)dt = 6$, $\int_5^7 f(t)dt = 0$; and $\int_7^{10} f(t)dt$

$$= \int_7^{10} (4t-37)/3dt = -3$$

(a) $F(0) = 0$, $F(3) = 0$, $F(5) = 6$, $F(7) = 6$, $F(10) = 3$

(b) F is increasing where $F' = f$ is positive, so on $[3/2, 6]$ and $[37/4, 10]$, decreasing on $[0, 3/2]$ and $[6, 37/4]$

(c) critical points when $F'(x) = f(x) = 0$, so $x = 3/2, 6, 37/4$; maximum $15/2$ at $x = 6$, minimum $-9/4$ at $x = 3/2$

(d)

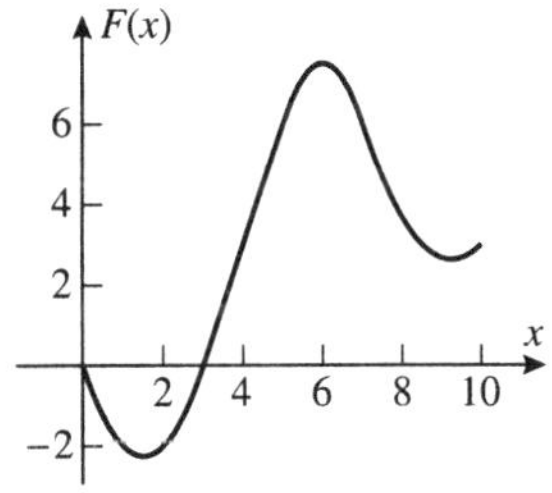

28. $f_{\text{ave}} = \dfrac{1}{10-0}\displaystyle\int_0^{10} f(t)dt = \dfrac{1}{10}F(10) = 0.3$

29. $x < 0: F(x) = \displaystyle\int_{-1}^{x}(-t)dt = -\dfrac{1}{2}t^2\Big]_{-1}^{x} = \dfrac{1}{2}(1-x^2),$

$x \ge 0: F(x) = \displaystyle\int_{-1}^{0}(-t)dt + \int_0^x t\,dt = \dfrac{1}{2} + \dfrac{1}{2}x^2;\ F(x) = \begin{cases} (1-x^2)/2, & x<0 \\ (1+x^2)/2, & x \ge 0 \end{cases}$

30. $0 \le x \le 2: F(x) = \displaystyle\int_0^x t\,dt = \dfrac{1}{2}x^2,$

$x > 2: F(x) = \displaystyle\int_0^2 t\,dt + \int_2^x 2\,dt = 2 + 2(x-2) = 2x - 2;\ F(x) = \begin{cases} x^2/2, & 0 \le x \le 2 \\ 2x-2, & x > 2 \end{cases}$

31. $y(x) = 2 + \displaystyle\int_1^x t^{1/3}dt = 2 + \dfrac{3}{4}t^{4/3}\Big]_1^x = \dfrac{5}{4} + \dfrac{3}{4}x^{4/3}$

32. $y(x) = \displaystyle\int_1^x (t^{1/2} + t^{-1/2})dt = \dfrac{2}{3}x^{3/2} - \dfrac{2}{3} + 2x^{1/2} - 2 = \dfrac{2}{3}x^{3/2} + 2x^{1/2} - \dfrac{8}{3}$

33. $y(x) = 1 + \displaystyle\int_{\pi/4}^{x}(\sec^2 t - \sin t)dt = \tan x + \cos x - \sqrt{2}/2$

34. $y(x) = \displaystyle\int_0^x te^{t^2}dt = \dfrac{1}{2}e^{-x^2} - \dfrac{1}{2}$

35. $P(x) = P_0 + \displaystyle\int_0^x r(t)dt$ individuals

36. $s(T) = s_1 + \displaystyle\int_1^T v(t)dt$

37. II has a minimum at $x = 12$, and I has a zero there, so I could be the derivative of II; on the other hand I has a minimum near $x = 1/3$, but II is not zero there, so II could not be the derivative of I, so I is the graph of $f(x)$ and II is the graph of $\int_0^x f(t)\,dt$.

38. **(b)** $\displaystyle\lim_{k\to 0}\dfrac{1}{k}(x^k - 1) = \dfrac{d}{dt}x^t\Big]_{t=0} = \ln x$

39. **(a)** where $f(t) = 0$; by the First Derivative Test, at $t = 3$

(b) where $f(t) = 0$; by the First Derivative Test, at $t = 1, 5$

(c) at $t = 0, 1$ or 5; from the graph it is evident that it is at $t = 5$

(d) at $t = 0, 3$ or 5; from the graph it is evident that it is at $t = 3$

(e) F is concave up when $F'' = f'$ is positive, i.e. where f is increasing, so on $(0, 1/2)$ and $(2, 4)$; it is concave down on $(1/2, 2)$ and $(4, 5)$

(f)

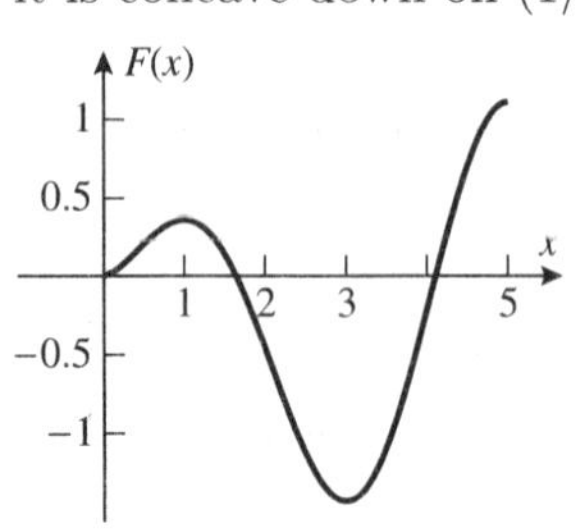

40. (a)

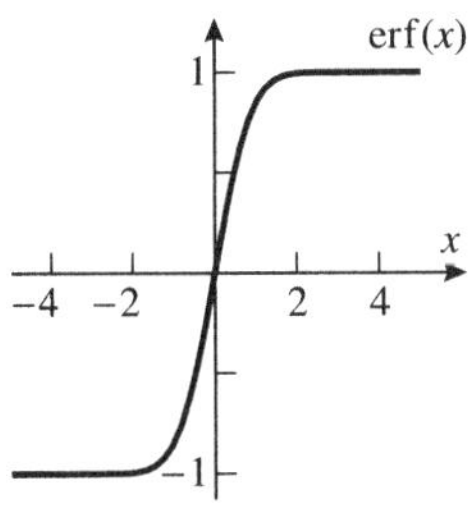

(c) $\text{erf}'(x) > 0$ for all x, so there are no relative extrema

(e) $\text{erf}''(x) = -4xe^{-x^2}/\sqrt{\pi}$ changes sign only at $x = 0$ so that is the only point of inflection

(g) $\lim_{x\to+\infty} \text{erf}(x) = +1$, $\lim_{x\to-\infty} \text{erf}(x) = -1$

41. $C'(x) = \cos(\pi x^2/2)$, $C''(x) = -\pi x \sin(\pi x^2/2)$

(a) $\cos t$ goes from negative to positive at $2k\pi - \pi/2$, and from positive to negative at $t = 2k\pi + \pi/2$, so $C(x)$ has relative minima when $\pi x^2/2 = 2k\pi - \pi/2$, $x = \pm\sqrt{4k-1}$, $k = 1, 2, \ldots$, and $C(x)$ has relative maxima when $\pi x^2/2 = (4k+1)\pi/2$, $x = \pm\sqrt{4k+1}$, $k = 0, 1, \ldots$.

(b) $\sin t$ changes sign at $t = k\pi$, so $C(x)$ has inflection points at $\pi x^2/2 = k\pi$, $x = \pm\sqrt{2k}$, $k = 1, 2, \ldots$; the case $k = 0$ is distinct due to the factor of x in $C''(x)$, but x changes sign at $x = 0$ and $\sin(\pi x^2/2)$ does not, so there is also a point of inflection at $x = 0$

42. Let $F(x) = \int_1^x \ln t\,dt$, $F'(x) = \lim_{h\to 0} \dfrac{F(x+h) - F(x)}{h} = \lim_{h\to 0} \dfrac{1}{h}\int_x^{x+h} \ln t\,dt$; but $F'(x) = \ln x$ so $\lim_{h\to 0} \dfrac{1}{h}\int_x^{x+h} \ln t\,dt = \ln x$

43. Differentiate: $f(x) = 3e^{3x}$, so $2 + \int_a^x f(t)dt = 2 + \int_a^x 3e^{3t}dt = 2 + e^{3t}\Big]_a^x = 2 + e^{3x} - e^{3a} = e^{3x}$ provided $e^{3a} = 2$, $a = (\ln 2)/3$.

44. (a) The area under $1/t$ for $x \le t \le x+1$ is less than the area of the rectangle with altitude $1/x$ and base 1, but greater than the area of the rectangle with altitude $1/(x+1)$ and base 1.

(b) $\int_x^{x+1} \dfrac{1}{t}dt = \ln t\Big]_x^{x+1} = \ln(x+1) - \ln x = \ln(1 + 1/x)$, so $1/(x+1) < \ln(1 + 1/x) < 1/x$ for $x > 0$.

(c) from Part (b), $e^{1/(x+1)} < e^{\ln(1+1/x)} < e^{1/x}$, $e^{1/(x+1)} < 1 + 1/x < e^{1/x}$, $e^{x/(x+1)} < (1 + 1/x)^x < e$; by the Squeezing Theorem, $\lim_{x\to+\infty}(1 + 1/x)^x = e$.

(d) Use the inequality $e^{x/(x+1)} < (1 + 1/x)^x$ to get $e < (1 + 1/x)^{x+1}$ so $(1 + 1/x)^x < e < (1 + 1/x)^{x+1}$.

45. From Exercise 44(d) $\left|e - \left(1 + \dfrac{1}{50}\right)^{50}\right| < y(50)$, and from the graph $y(50) < 0.06$

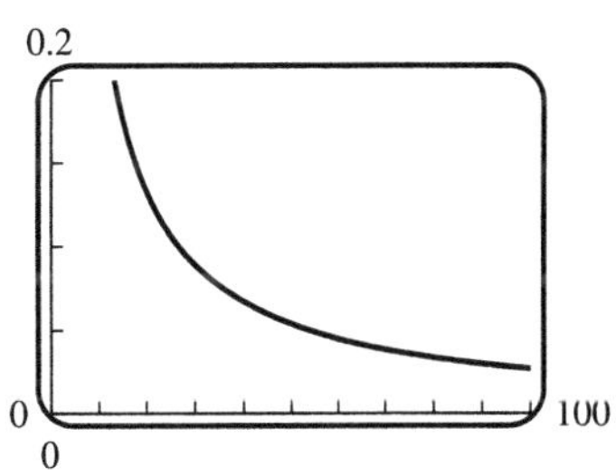

46. $F'(x) = f(x)$, thus $F'(x)$ has a value at each x in I because f is continuous on I so F is continuous on I because a function that is differentiable at a point is also continuous at that point

EXERCISE SET 7.6

1. (a) $-\pi/2$ (b) π (c) $-\pi/4$ (d) 0

2. (a) $\pi/3$ (b) $\pi/3$ (c) $\pi/4$ (d) $2\pi/3$

3. $\theta = -\pi/3$; $\cos\theta = 1/2$, $\tan\theta = -\sqrt{3}$, $\cot\theta = -1/\sqrt{3}$, $\sec\theta = 2$, $\csc\theta = -2/\sqrt{3}$

4. $\theta = \pi/3$; $\sin\theta = \sqrt{3}/2$, $\tan\theta = \sqrt{3}$, $\cot\theta = 1/\sqrt{3}$, $\sec\theta = 2$, $\csc\theta = 2/\sqrt{3}$

5. $\tan\theta = 4/3$, $0 < \theta < \pi/2$; use the triangle shown to get $\sin\theta = 4/5$, $\cos\theta = 3/5$, $\cot\theta = 3/4$, $\sec\theta = 5/3$, $\csc\theta = 5/4$

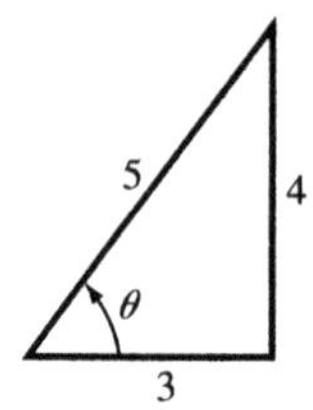

6. $\sec\theta = 2.6, 0 < \theta < \pi/2$; use the triangle shown to get $\sin\theta = 2.4/2.6 = 12/13, \cos\theta = 1/2.6 = 5/13,$ $\tan\theta = 2.4 = 12/5, \cot\theta = 5/12, \csc\theta = 13/12$

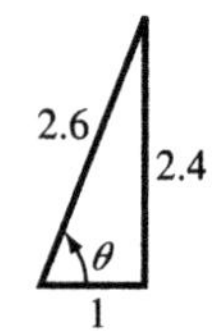

7. (a) $\pi/7$
   (b) $\sin^{-1}(\sin\pi) = \sin^{-1}(\sin 0) = 0$
   (c) $\sin^{-1}(\sin(5\pi/7)) = \sin^{-1}(\sin(2\pi/7)) = 2\pi/7$
   (d) Note that $\pi/2 < 630 - 200\pi < \pi$ so $\sin(630) = \sin(630 - 200\pi) = \sin(\pi - (630 - 200\pi)) = \sin(201\pi - 630)$ where $0 < 201\pi - 630 < \pi/2$; $\sin^{-1}(\sin 630) = \sin^{-1}(\sin(201\pi - 630)) = 201\pi - 630$.

8. (a) $\pi/7$
   (b) π
   (c) $\cos^{-1}(\cos(12\pi/7)) = \cos^{-1}(\cos(2\pi/7)) = 2\pi/7$
   (d) Note that $-\pi/2 < 200 - 64\pi < 0$ so $\cos(200) = \cos(200 - 64\pi) = \cos(64\pi - 200)$ where $0 < 64\pi - 200 < \pi/2$; $\cos^{-1}(\cos 200) = \cos^{-1}(\cos(64\pi - 200)) = 64\pi - 200$.

9. (a) $0 \le x \le \pi$ (b) $-1 \le x \le 1$
   (c) $-\pi/2 < x < \pi/2$ (d) $-\infty < x < +\infty$

10. Let $\theta = \sin^{-1}(-3/4)$ then $\sin\theta = -3/4$, $-\pi/2 < \theta < 0$ and (see figure) $\sec\theta = 4/\sqrt{7}$

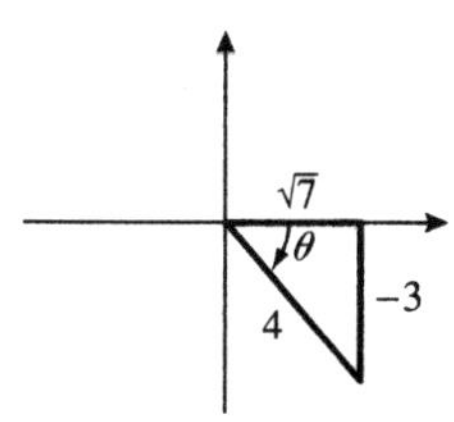

11. Let $\theta = \cos^{-1}(3/5)$, $\sin 2\theta = 2\sin\theta\cos\theta = 2(4/5)(3/5) = 24/25$

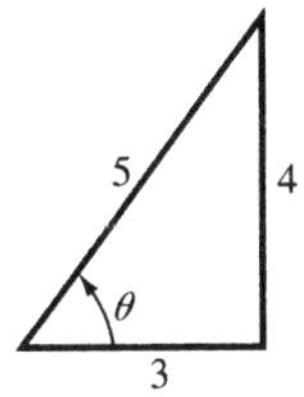

12. **(a)** $\sin(\cos^{-1} x) = \sqrt{1 - x^2}$

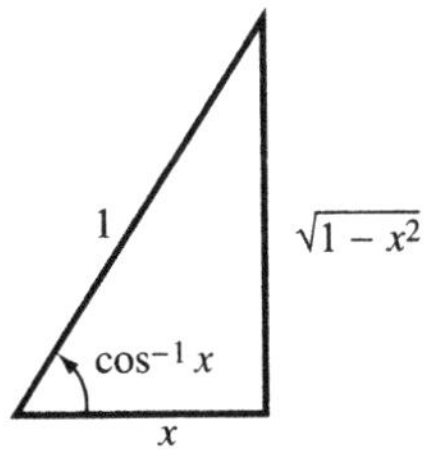

(b) $\tan(\cos^{-1} x) = \dfrac{\sqrt{1 - x^2}}{x}$

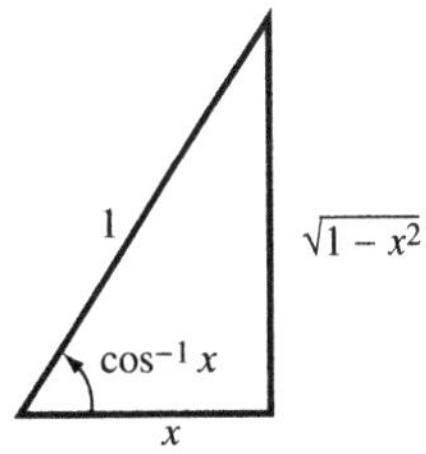

(a) $\csc(\tan^{-1} x) = \dfrac{\sqrt{1 + x^2}}{x}$

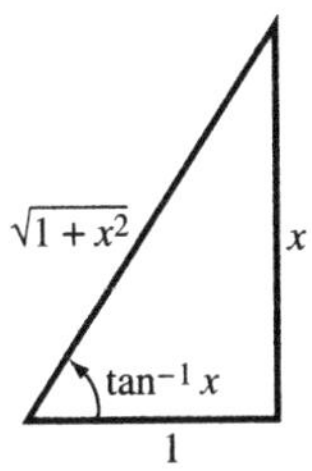

(d) $\sin(\tan^{-1} x) = \dfrac{x}{\sqrt{1 + x^2}}$

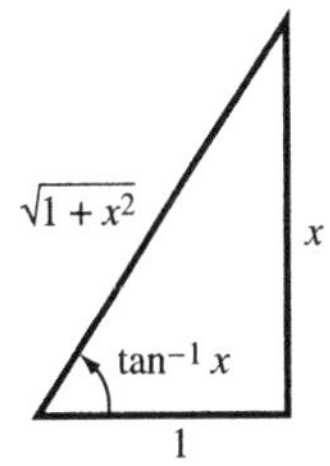

13. **(a)** $\cos(\tan^{-1} x) = \dfrac{1}{\sqrt{1 + x^2}}$

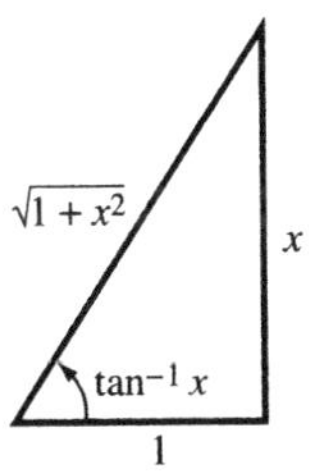

(b) $\tan(\cos^{-1} x) = \dfrac{\sqrt{1 - x^2}}{x}$

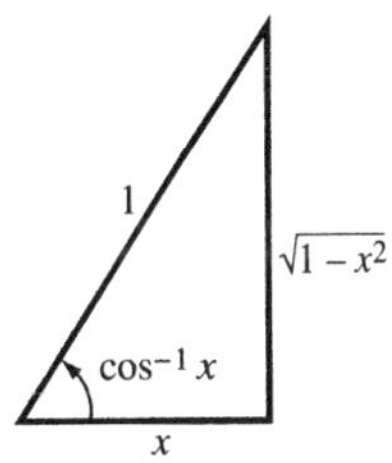

(c) $\sin(\sec^{-1} x) = \dfrac{\sqrt{x^2 - 1}}{x}$

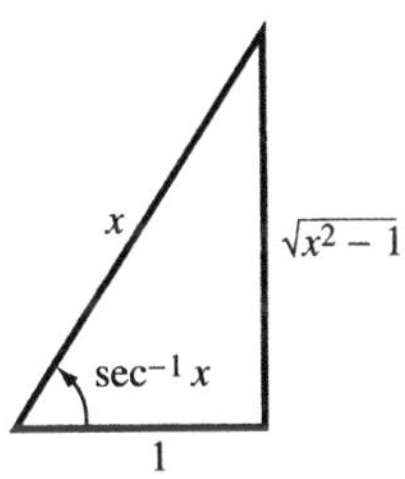

(d) $\cot(\sec^{-1} x) = \dfrac{1}{\sqrt{x^2 - 1}}$

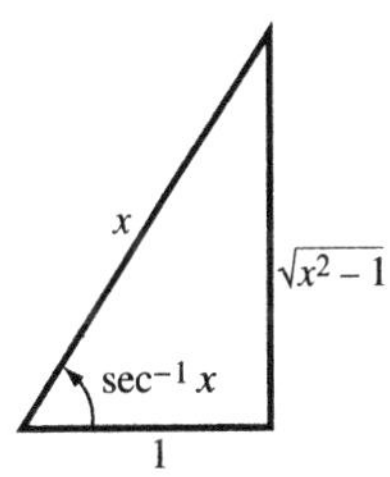

14. **(a)**

x	-1.00	-0.80	-0.6	-0.40	-0.20	0.00	0.20	0.40	0.60	0.80	1.00
$\sin^{-1} x$	-1.57	-0.93	-0.64	-0.41	-0.20	0.00	0.20	0.41	0.64	0.93	1.57
$\cos^{-1} x$	3.14	2.50	2.21	1.98	1.77	1.57	1.37	1.16	0.93	0.64	0.00

(b)

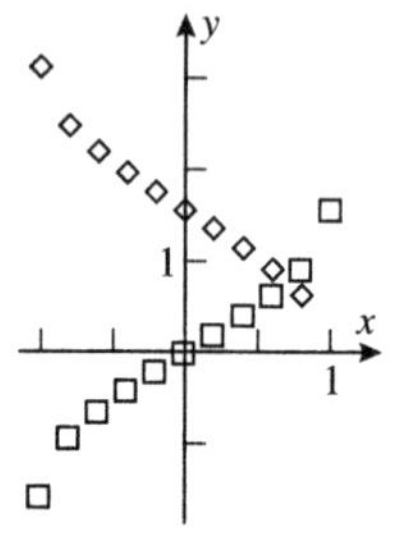

(c)

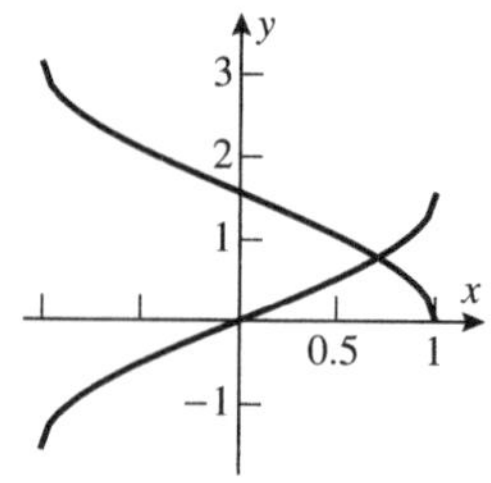

15. **(a)**

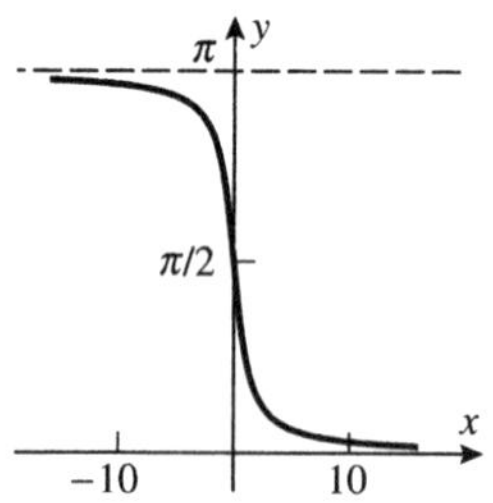

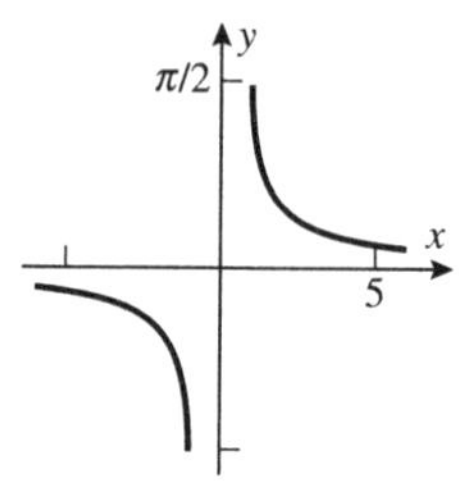

(b) The domain of $\cot^{-1} x$ is $(-\infty, +\infty)$, the range is $(0, \pi)$; the domain of $\csc^{-1} x$ is $(-\infty, -1] \cup [1, +\infty)$, the range is $[-\pi/2, 0) \cup (0, \pi/2]$.

16. **(a)** $y = \cot^{-1} x$; if $x > 0$ then $0 < y < \pi/2$ and $x = \cot y$, $\tan y = 1/x$, $y = \tan^{-1}(1/x)$;
if $x < 0$ then $\pi/2 < y < \pi$ and $x = \cot y = \cot(y - \pi), \tan(y - \pi) = 1/x, y = \pi + \tan^{-1} \dfrac{1}{x}$

(b) $y = \sec^{-1} x$, $x = \sec y$, $\cos y = 1/x$, $y = \cos^{-1}(1/x)$

(c) $y = \csc^{-1} x$, $x = \csc y$, $\sin y = 1/x$, $y = \sin^{-1}(1/x)$

17. **(a)** $55.0°$ **(b)** $33.6°$ **(c)** $25.8°$

18. **(a)** Let $x = f(y) = \cot y, 0 < y < \pi, -\infty < x < +\infty$. Then f is differentiable and one-to-one and $f'(f^{-1}(x)) = \cot(\cot^{-1} x)\cos(\cot^{-1} x) = -x\dfrac{\sqrt{x^2+1}}{x} = -\sqrt{x^2+1} \neq 0$, and

$$\frac{d}{dx}[\cot^{-1} x]\bigg|_{x=0} = \lim_{x\to 0} \frac{1}{f'(f^{-1}(x))} = -\lim_{x\to 0} \sqrt{x^2+1} = -1.$$

(b) If $x \neq 0$ then, from Exercise 16(a),

$\dfrac{d}{dx}\cot^{-1} x = \dfrac{d}{dx}\tan^{-1}\dfrac{1}{x} = -\dfrac{1}{x^2}\dfrac{1}{\sqrt{1+(1/x)^2}} = -\dfrac{1}{\sqrt{x^2+1}}$. For $x = 0$, Part (a) shows the same; thus for $-\infty < x < +\infty$, $\dfrac{d}{dx}[\cot^{-1} x] = -\dfrac{1}{\sqrt{x^2+1}}$.

(c) For $-\infty < u < +\infty$, by the chain rule it follows that $\dfrac{d}{dx}[\cot^{-1} u] = -\dfrac{1}{\sqrt{u^2+1}}\dfrac{du}{dx}$.

19. **(a)** By the chain rule, $\frac{d}{dx}[\csc^{-1} x] = -\frac{1}{x^2}\frac{1}{\sqrt{1-(1/x)^2}} = \frac{-1}{|x|\sqrt{x^2-1}}$

(b) By the chain rule, $\frac{d}{dx}[\csc^{-1} u] = \frac{du}{dx}\frac{d}{du}[\csc^{-1} u] = \frac{-1}{|u|\sqrt{u^2-1}}\frac{du}{dx}$

20. **(a)** $x = \pi - \sin^{-1}(0.37) \approx 2.7626$ rad **(b)** $\theta = 180° + \sin^{-1}(0.61) \approx 217.6°$

21. **(a)** $x = \pi + \cos^{-1}(0.85) \approx 3.6964$ rad **(b)** $\theta = -\cos^{-1}(0.23) \approx -76.7°$

22. **(a)** $x = \tan^{-1}(3.16) - \pi \approx -1.8773$ **(b)** $\theta = 180° - \tan^{-1}(0.45) \approx 155.8°$

23. **(a)** $\frac{1}{\sqrt{1-x^2/9}}(1/3) = 1/\sqrt{9-x^2}$ **(b)** $-2/\sqrt{1-(2x+1)^2}$

24. **(a)** $2x/(1+x^4)$ **(b)** $-\frac{1}{1+x}\left(\frac{1}{2}x^{-1/2}\right) = -\frac{1}{2(1+x)\sqrt{x}}$

25. **(a)** $\frac{1}{|x|^7\sqrt{x^{14}-1}}(7x^6) = \frac{7}{|x|\sqrt{x^{14}-1}}$ **(b)** $-1/\sqrt{e^{2x}-1}$

26. **(a)** $y = 1/\tan x = \cot x$, $dy/dx = -\csc^2 x$

(b) $y = (\tan^{-1} x)^{-1}$, $dy/dx = -(\tan^{-1} x)^{-2}\left(\frac{1}{1+x^2}\right)$

27. **(a)** $\frac{1}{\sqrt{1-1/x^2}}(-1/x^2) = -\frac{1}{|x|\sqrt{x^2-1}}$ **(b)** $\frac{\sin x}{\sqrt{1-\cos^2 x}} = \frac{\sin x}{|\sin x|} = \begin{cases} 1, & \sin x > 0 \\ -1, & \sin x < 0 \end{cases}$

28. **(a)** $-\frac{1}{(\cos^{-1} x)\sqrt{1-x^2}}$ **(b)** $-\frac{1}{2\sqrt{\cot^{-1} x}(1+x^2)}$

29. **(a)** $\frac{e^x}{|x|\sqrt{x^2-1}} + e^x \sec^{-1} x$ **(b)** $\frac{3x^2(\sin^{-1} x)^2}{\sqrt{1-x^2}} + 2x(\sin^{-1} x)^3$

30. **(a)** 0 **(b)** 0

31. $x^3 + x\tan^{-1} y = e^y$, $3x^2 + \frac{x}{1+y^2}y' + \tan^{-1} y = e^y y'$, $y' = \frac{(3x^2+\tan^{-1} y)(1+y^2)}{(1+y^2)e^y - x}$

32. $\sin^{-1}(xy) = \cos^{-1}(x-y)$, $\frac{1}{\sqrt{1-x^2y^2}}(xy'+y) = -\frac{1}{\sqrt{1-(x-y)^2}}(1-y')$,

$y' = \frac{y\sqrt{1-(x-y)^2}+\sqrt{1-x^2y^2}}{\sqrt{1-x^2y^2}-x\sqrt{1-(x-y)^2}}$

33. $\sin^{-1} x\Big]_0^{1/\sqrt{2}} = \sin^{-1}(1/\sqrt{2}) - \sin^{-1} 0 = \pi/4$

34. $u = 2x$, $\frac{1}{2}\int \frac{1}{\sqrt{1-u^2}}du = \frac{1}{2}\sin^{-1}(2x) + C$

35. $\tan^{-1} x\Big]_{-1}^{1} = \tan^{-1} 1 - \tan^{-1}(-1) = \pi/4 - (-\pi/4) = \pi/2$

36. $u = 4x$, $\dfrac{1}{4}\displaystyle\int \frac{1}{1+u^2}du = \frac{1}{4}\tan^{-1}(4x) + C$

37. $\sec^{-1} x\Big]_{\sqrt{2}}^{2} = \sec^{-1} 2 - \sec^{-1}\sqrt{2} = \pi/3 - \pi/4 = \pi/12$

38. $-\sec^{-1} x\Big]_{-\sqrt{2}}^{-2/\sqrt{3}} = -\sec^{-1}(-2/\sqrt{3}) + \sec^{-1}(-\sqrt{2}) = -5\pi/6 + 3\pi/4 = -\pi/12$

39. $u = \tan x$, $\displaystyle\int \frac{1}{\sqrt{1-u^2}}du = \sin^{-1}(\tan x) + C$

40. $u = e^{-x}$, $-\displaystyle\int_{1/2}^{\sqrt{3}/2} \frac{1}{\sqrt{1-u^2}}du = -\sin^{-1} u\Big]_{1/2}^{\sqrt{3}/2} = -\sin^{-1}\frac{\sqrt{3}}{2} + \sin^{-1}\frac{1}{2} = -\frac{\pi}{3} + \frac{\pi}{6} = -\frac{\pi}{6}$

41. $u = e^x$, $\displaystyle\int \frac{1}{1+u^2}du = \tan^{-1}(e^x) + C$

42. $u = t^2$, $\dfrac{1}{2}\displaystyle\int \frac{1}{u^2+1}du = \frac{1}{2}\tan^{-1}(t^2) + C$

43. $u = \sqrt{x}$, $2\displaystyle\int_1^{\sqrt{3}} \frac{1}{u^2+1}du = 2\tan^{-1} u\Big]_1^{\sqrt{3}} = 2(\tan^{-1}\sqrt{3} - \tan^{-1} 1) = 2(\pi/3 - \pi/4) = \pi/6$

44. $u = \cos\theta$, $-\displaystyle\int \frac{1}{u^2+1}du = -\tan^{-1}(\cos\theta) + C$

45. $u = \ln x$, $\displaystyle\int \frac{1}{\sqrt{1-u^2}}du = \sin^{-1}(\ln x) + C$

46. $u = 3x$, $\displaystyle\int \frac{1}{u\sqrt{u^2-1}}du = \sec^{-1}(3x) + C$ if $x > 0$; $-\sec^{-1}(3x) + C$ if $x < 0$

47. $u = a\sin\theta, du = a\cos\theta\, d\theta$; $\displaystyle\int \frac{du}{\sqrt{a^2-u^2}} = a\theta + C = \sin^{-1}\frac{u}{a} + C$

48. If $u > 0$ then $u = a\sec\theta, du = a\sec\theta\tan\theta\, d\theta$, $\displaystyle\int \frac{du}{u\sqrt{u^2-a^2}} = \frac{1}{a}\theta = \frac{1}{a}\sec^{-1}\frac{u}{a} + C$

49. **(a)** $\sin^{-1}(x/3) + C$ **(b)** $(1/\sqrt{5})\tan^{-1}(x/\sqrt{5}) + C$
(c) $(1/\sqrt{\pi})\sec^{-1}(x/\sqrt{\pi}) + C$

50. **(a)** $u = e^x$, $\displaystyle\int \frac{1}{4+u^2}du = \frac{1}{2}\tan^{-1}(e^x/2) + C$

(b) $u = 2x$, $\dfrac{1}{2}\displaystyle\int \frac{1}{\sqrt{9-u^2}}du = \frac{1}{2}\sin^{-1}(2x/3) + C$,

(c) $u = \sqrt{5}y$, $\displaystyle\int \frac{1}{u\sqrt{u^2-3}}du = \frac{1}{\sqrt{3}}\sec^{-1}(\sqrt{5}y/\sqrt{3}) + C$

51. $u = \sqrt{3}x^2$, $\dfrac{1}{2\sqrt{3}}\displaystyle\int_0^{\sqrt{3}} \dfrac{1}{\sqrt{4-u^2}}du = \dfrac{1}{2\sqrt{3}}\sin^{-1}\dfrac{u}{2}\Bigg]_0^{\sqrt{3}} = \dfrac{1}{2\sqrt{3}}\left(\dfrac{\pi}{3}\right) = \dfrac{\pi}{6\sqrt{3}}$

52. $u = \sqrt{x}$, $2\displaystyle\int_1^{\sqrt{2}} \dfrac{1}{\sqrt{4-u^2}}du = 2\sin^{-1}\dfrac{u}{2}\Bigg]_1^{\sqrt{2}} = 2(\pi/4 - \pi/6) = \pi/6$

53. $u = 3x$, $\dfrac{1}{3}\displaystyle\int_0^{2\sqrt{3}} \dfrac{1}{4+u^2}du = \dfrac{1}{6}\tan^{-1}\dfrac{u}{2}\Bigg]_0^{2\sqrt{3}} = \dfrac{1}{6}\dfrac{\pi}{3} = \dfrac{1}{18\pi}$

54. $u = x^2$, $\dfrac{1}{2}\displaystyle\int_1^{3} \dfrac{1}{3+u^2}du = \dfrac{1}{2\sqrt{3}}\tan^{-1}\dfrac{u}{\sqrt{3}}\Bigg]_1^{3} = \dfrac{1}{2\sqrt{3}}(\pi/3 - \pi/6) = \dfrac{\pi}{12\sqrt{3}}$

55. **(a)**

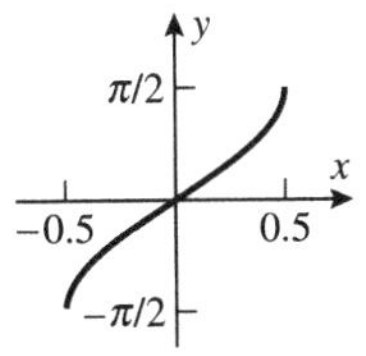

(b)

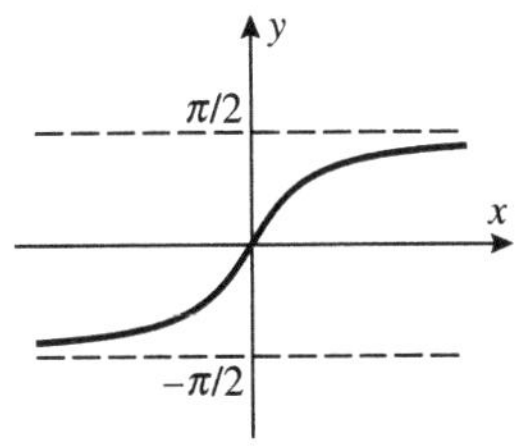

56. **(a)** $\sin^{-1} 0.9 > 1$, so it is not in the domain of $\sin^{-1} x$

(b) $-1 \le \sin^{-1} x \le 1$ is necessary, or $-0.841471 \le x \le 0.841471$

57. **(b)** $\theta = \sin^{-1}\dfrac{R}{R+h} = \sin^{-1}\dfrac{6378}{16,378} \approx 23°$

58. **(a)** If $\gamma = 90°$, then $\sin\gamma = 1$, $\sqrt{1-\sin^2\phi\sin^2\gamma} = \sqrt{1-\sin^2\phi} = \cos\phi$, $D = \tan\phi\tan\lambda = (\tan 23.45°)(\tan 65°) \approx 0.93023374$ so $h \approx 21.1$ hours.

(b) If $\gamma = 270°$, then $\sin\gamma = -1$, $D = -\tan\phi\tan\lambda \approx -0.93023374$ so $h \approx 2.9$ hours.

59. $\sin 2\theta = gR/v^2 = (9.8)(18)/(14)^2 = 0.9$, $2\theta = \sin^{-1}(0.9)$ or $2\theta = 180° - \sin^{-1}(0.9)$ so $\theta = \frac{1}{2}\sin^{-1}(0.9) \approx 32°$ or $\theta = 90° - \frac{1}{2}\sin^{-1}(0.9) \approx 58°$. The ball will have a lower parabolic trajectory for $\theta = 32°$ and hence will result in the shorter time of flight.

60. $4^2 = 2^2 + 3^2 - 2(2)(3)\cos\theta$, $\cos\theta = -1/4$, $\theta = \cos^{-1}(-1/4) \approx 104°$

61. $y = 0$ when $x^2 = 6000v^2/g$, $x = 10v\sqrt{60/g} = 1000\sqrt{30}$ for $v = 400$ and $g = 32$; $\tan\theta = 3000/x = 3/\sqrt{30}$, $\theta = \tan^{-1}(3/\sqrt{30}) \approx 29°$.

62. **(a)** $\theta = \alpha - \beta$, $\cot\alpha = \dfrac{x}{a+b}$ and $\cot\beta = \dfrac{x}{b}$ so

$$\theta = \cot^{-1}\frac{x}{a+b} - \cot^{-1}\left(\frac{x}{b}\right)$$

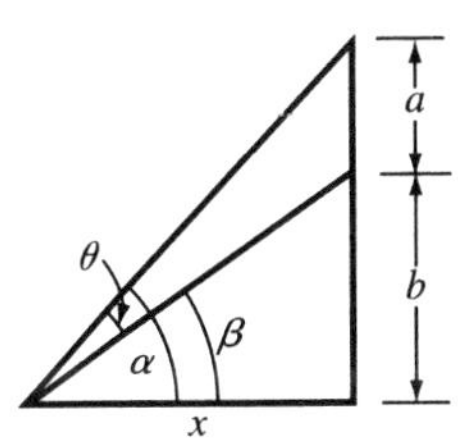

(b) $$\frac{d\theta}{dx} = -\frac{1}{a+b}\left(\frac{1}{1+x^2/(a+b)^2}\right) - \frac{1}{b}\frac{1}{1+(x/b)^2}$$

$$= -\frac{a+b}{(a+b)^2+x^2} - \frac{b}{b^2+x^2}$$

which is negative for all x. Thus θ is a decreasing function of x, and it has no maximum since $\lim\limits_{x\to 0^+} \theta = +\infty$.

63. **(a)** $A = \int_0^{0.8} \frac{1}{\sqrt{1-x^2}} dx = \sin^{-1} x \Big]_0^{0.8} = \sin^{-1}(0.8)$

(b) The calculator was in degree mode instead of radian mode; the correct answer is 0.93.

64. $A = \int_0^{1/6} \frac{1}{\sqrt{1-9x^2}} dx = \frac{1}{3}\int_0^{1/2} \frac{1}{\sqrt{1-u^2}} du = \frac{1}{3}\sin^{-1} u \Big]_0^{1/2} = \pi/18$

65. The area is given by $\int_0^k (1/\sqrt{1-x^2} - x)dx = \sin^{-1} k - k^2/2 = 1$; solve for k to get $k = 0.997301$.

66. $x = \sin y$, $A = \int_0^{\pi/2} \sin y\, dy = -\cos y \Big]_0^{\pi/2} = 1$

67. The curves intersect at $x = a = 0$ and $x = b = 0.838422$ so the area is $\int_a^b (\sin 2x - \sin^{-1} x)dx = 0.174192$.

68. The displacement of the particle during the time interval $[0, T]$ is given by $\int_0^T v(t)dt = 3\tan^{-1} T - 0.25T^2$. The particle is 2 cm from its starting position when $3\tan^{-1} T - 0.25T^2 = 2$ or when $3\tan^{-1} T - 0.25T^2 = -2$; solve for T to get $T = 0.90$, 2.51, and 4.95 sec.

69. $V = \int_{-2}^{2} \pi \frac{1}{4+x^2} dx = \frac{\pi}{2}\tan^{-1}(x/2)\Big]_{-2}^{2} = \pi^2/4$

70. **(a)** $V = 2\pi \int_1^b \frac{x}{1+x^4} dx = \pi \tan^{-1}(x^2)\Big]_1^b = \pi\left[\tan^{-1}(b^2) - \frac{\pi}{4}\right]$

(b) $\lim_{b\to+\infty} V = \pi\left(\frac{\pi}{2} - \frac{\pi}{4}\right) = \frac{1}{4}\pi^2$

71. The area is given by $\int_0^2 1/(1+kx^2)dx = (1/\sqrt{k})\tan^{-1}(2\sqrt{k}) = 0.6$; solve for k to get $k = 5.081435$.

72. **(a)** $\pi \int_0^1 (\sin^{-1} x)^2 dx = 1.468384$.

(b) $2\pi \int_0^{\pi/2} y(1 - \sin y)dy = 1.468384$.

73. $\theta = \pi - (\alpha + \beta)$

$$= \pi - \cot^{-1}(x-2) - \cot^{-1}\frac{5-x}{4},$$

$$\frac{d\theta}{dx} = \frac{1}{1+(x-2)^2} + \frac{-1/4}{1+(5-x)^2/16}$$

$$= -\frac{3(x^2-2x-7)}{[1+(x-2)^2][16+(5-x)^2]}$$

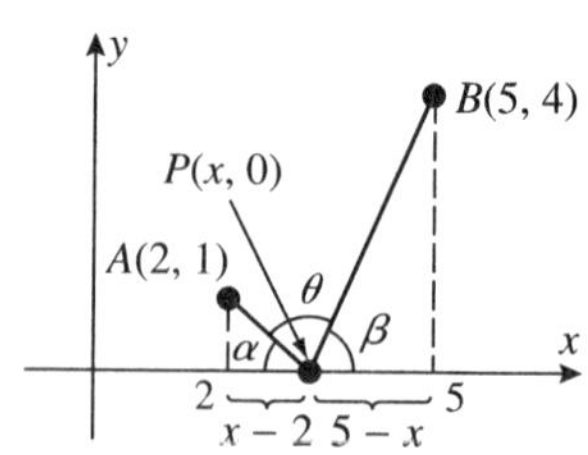

$d\theta/dx = 0$ when $x = \frac{2 \pm \sqrt{4+28}}{2} = 1 \pm 2\sqrt{2}$, only $1 + 2\sqrt{2}$ is in $[2, 5]$; $d\theta/dx > 0$ for x in $[2, 1+2\sqrt{2})$, $d\theta/dx < 0$ for x in $(1+2\sqrt{2}, 5]$, θ is maximum when $x = 1 + 2\sqrt{2}$.

74. $\theta = \alpha - \beta$

$= \cot^{-1}(x/12) - \cot^{-1}(x/2)$

$$\frac{d\theta}{dx} = -\frac{12}{144 + x^2} + \frac{2}{4 + x^2}$$

$$= \frac{10(24 - x^2)}{(144 + x^2)(4 + x^2)}$$

$d\theta/dx = 0$ when $x = \sqrt{24} = 2\sqrt{6}$, by the first derivative test θ is maximum there.

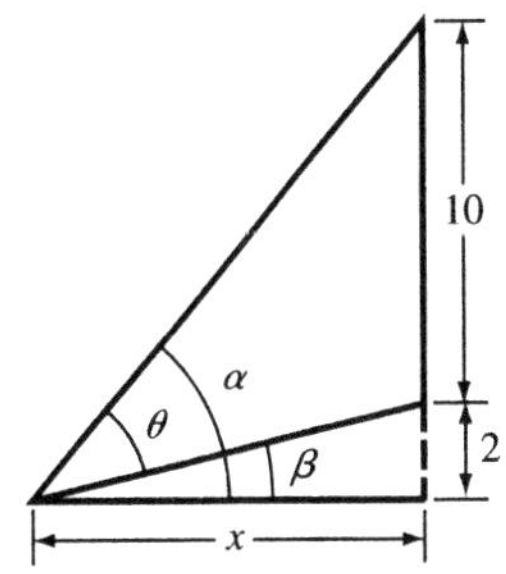

75. By the Mean-Value Theorem on the interval $[0, x]$,

$$\frac{\tan^{-1} x - \tan^{-1} 0}{x - 0} = \frac{\tan^{-1} x}{x} = \frac{1}{1 + c^2} \text{ for } c \text{ in } (0, x), \text{ but}$$

$$\frac{1}{1 + x^2} < \frac{1}{1 + c^2} < 1 \text{ for } c \text{ in } (0, x) \text{ so } \frac{1}{1 + x^2} < \frac{\tan^{-1} x}{x} < 1, \ \frac{x}{1 + x^2} < \tan^{-1} x < x.$$

76. $\dfrac{n}{n^2 + k^2} = \dfrac{1}{1 + k^2/n^2} \dfrac{1}{n}$ so $\displaystyle\sum_{k=1}^{n} \frac{n}{n^2 + k^2} = \sum_{k=1}^{n} f(x_k^*)\Delta x$ where $f(x) = \dfrac{1}{1 + x^2}$, $x_k^* = \dfrac{k}{n}$, and $\Delta x = \dfrac{1}{n}$ for $0 \le x \le 1$. Thus $\displaystyle\lim_{n \to +\infty} \sum_{k=1}^{n} \frac{n}{n^2 + k^2} = \lim_{n \to +\infty} \sum_{k=1}^{n} f(x_k^*)\Delta x = \int_0^1 \frac{1}{1 + x^2} dx = \frac{\pi}{4}$.

77. **(a)** Let $\theta = \sin^{-1}(-x)$ then $\sin\theta = -x$, $-\pi/2 \le \theta \le \pi/2$. But $\sin(-\theta) = -\sin\theta$ and $-\pi/2 \le -\theta \le \pi/2$ so $\sin(-\theta) = -(-x) = x$, $-\theta = \sin^{-1} x$, $\theta = -\sin^{-1} x$.

(b) proof is similar to that in Part (a)

78. **(a)** Let $\theta = \cos^{-1}(-x)$ then $\cos\theta = -x$, $0 \le \theta \le \pi$. But $\cos(\pi - \theta) = -\cos\theta$ and $0 \le \pi - \theta \le \pi$ so $\cos(\pi - \theta) = x$, $\pi - \theta = \cos^{-1} x$, $\theta = \pi - \cos^{-1} x$

(b) Let $\theta = \sec^{-1}(-x)$ for $x \ge 1$; then $\sec\theta = -x$ and $\pi/2 < \theta \le \pi$. So $0 \le \pi - \theta < \pi/2$ and $\pi - \theta = \sec^{-1}\sec(\pi - \theta) = \sec^{-1}(-\sec\theta) = \sec^{-1} x$, or $\sec^{-1}(-x) = \pi - \sec^{-1} x$.

79. **(a)** $\sin^{-1} x = \tan^{-1} \dfrac{x}{\sqrt{1 - x^2}}$ (see figure)

(b) $\sin^{-1} x + \cos^{-1} x = \pi/2$; $\cos^{-1} x = \pi/2 - \sin^{-1} x = \pi/2 - \tan^{-1} \dfrac{x}{\sqrt{1 - x^2}}$

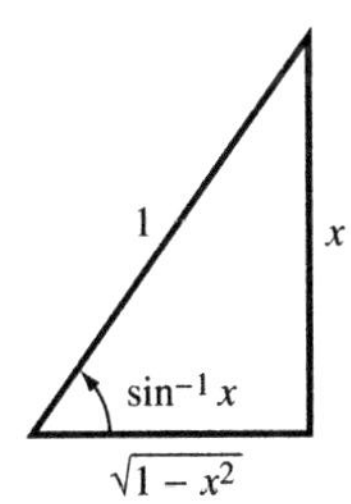

80. $\tan(\alpha + \beta) = \dfrac{\tan\alpha + \tan\beta}{1 - \tan\alpha\tan\beta}$,

$$\tan(\tan^{-1} x + \tan^{-1} y) = \frac{\tan(\tan^{-1} x) + \tan(\tan^{-1} y)}{1 - \tan(\tan^{-1} x)\tan(\tan^{-1} y)} = \frac{x + y}{1 - xy}$$

so $\tan^{-1} x + \tan^{-1} y = \tan^{-1} \dfrac{x + y}{1 - xy}$

81. (a) $\tan^{-1}\frac{1}{2}+\tan^{-1}\frac{1}{3}=\tan^{-1}\frac{1/2+1/3}{1-(1/2)(1/3)}=\tan^{-1}1=\pi/4$

(b) $2\tan^{-1}\frac{1}{3}=\tan^{-1}\frac{1}{3}+\tan^{-1}\frac{1}{3}=\tan^{-1}\frac{1/3+1/3}{1-(1/3)(1/3)}=\tan^{-1}\frac{3}{4}$,

$2\tan^{-1}\frac{1}{3}+\tan^{-1}\frac{1}{7}=\tan^{-1}\frac{3}{4}+\tan^{-1}\frac{1}{7}=\tan^{-1}\frac{3/4+1/7}{1-(3/4)(1/7)}=\tan^{-1}1=\pi/4$

82. $\sin(\sec^{-1}x)=\sin(\cos^{-1}(1/x))=\sqrt{1-\left(\frac{1}{x}\right)^2}=\frac{\sqrt{x^2-1}}{|x|}$

EXERCISE SET 7.7

1. (a) $\lim_{x\to 2}\frac{x^2-4}{x^2+2x-8}=\lim_{x\to 2}\frac{(x-2)(x+2)}{(x+4)(x-2)}=\lim_{x\to 2}\frac{x+2}{x+4}=\frac{2}{3}$

(b) $\lim_{x\to+\infty}\frac{2x-5}{3x+7}=\frac{2-\lim_{x\to+\infty}\frac{5}{x}}{3+\lim_{x\to+\infty}\frac{7}{x}}=\frac{2}{3}$

2. (a) $\frac{\sin x}{\tan x}=\sin x\frac{\cos x}{\sin x}=\cos x$ so $\lim_{x\to 0}\frac{\sin x}{\tan x}=\lim_{x\to 0}\cos x=1$

(b) $\frac{x^2-1}{x^3-1}=\frac{(x-1)(x+1)}{(x-1)(x^2+x+1)}=\frac{x+1}{x^2+x+1}$ so $\lim_{x\to 1}\frac{x^2-1}{x^3-1}=\frac{2}{3}$

3. $\lim_{x\to 1}\frac{1/x}{1}=1$

4. $\lim_{x\to 0}\frac{2\cos 2x}{5\cos 5x}=2/5$

5. $\lim_{x\to 0}\frac{e^x}{\cos x}=1$

6. $\lim_{x\to 3}\frac{1}{6x-13}=1/5$

7. $\lim_{\theta\to 0}\frac{\sec^2\theta}{1}=1$

8. $\lim_{t\to 0}\frac{te^t+e^t}{-e^t}=-1$

9. $\lim_{x\to\pi^+}\frac{\cos x}{1}=-1$

10. $\lim_{x\to 0^+}\frac{\cos x}{2x}=+\infty$

11. $\lim_{x\to+\infty}\frac{1/x}{1}=0$

12. $\lim_{x\to+\infty}\frac{3e^{3x}}{2x}=\lim_{x\to+\infty}\frac{9e^{3x}}{2}=+\infty$

13. $\lim_{x\to 0^+}\frac{-\csc^2 x}{1/x}=\lim_{x\to 0^+}\frac{-x}{\sin^2 x}=\lim_{x\to 0^+}\frac{-1}{2\sin x\cos x}=-\infty$

14. $\lim_{x\to 0^+}\frac{-1/x}{(-1/x^2)e^{1/x}}=\lim_{x\to 0^+}\frac{x}{e^{1/x}}=0$

15. $\lim_{x\to+\infty}\frac{100x^{99}}{e^x}=\lim_{x\to+\infty}\frac{(100)(99)x^{98}}{e^x}=\cdots=\lim_{x\to+\infty}\frac{(100)(99)(98)\cdots(1)}{e^x}=0$

16. $\lim_{x\to 0^+} \dfrac{\cos x/\sin x}{\sec^2 x/\tan x} = \lim_{x\to 0^+} \cos^2 x = 1$

17. $\lim_{x\to 0} \dfrac{2/\sqrt{1-4x^2}}{1} = 2$

18. $\lim_{x\to 0} \dfrac{1 - \dfrac{1}{1+x^2}}{3x^2} = \lim_{x\to 0} \dfrac{1}{3(1+x^2)} = \dfrac{1}{3}$

19. $\lim_{x\to +\infty} xe^{-x} = \lim_{x\to +\infty} \dfrac{x}{e^x} = \lim_{x\to +\infty} \dfrac{1}{e^x} = 0$

20. $\lim_{x\to \pi} (x-\pi)\tan(x/2) = \lim_{x\to \pi} \dfrac{x-\pi}{\cot(x/2)} = \lim_{x\to \pi} \dfrac{1}{-(1/2)\csc^2(x/2)} = -2$

21. $\lim_{x\to +\infty} x\sin(\pi/x) = \lim_{x\to +\infty} \dfrac{\sin(\pi/x)}{1/x} = \lim_{x\to +\infty} \dfrac{(-\pi/x^2)\cos(\pi/x)}{-1/x^2} = \lim_{x\to +\infty} \pi\cos(\pi/x) = \pi$

22. $\lim_{x\to 0^+} \tan x \ln x = \lim_{x\to 0^+} \dfrac{\ln x}{\cot x} = \lim_{x\to 0^+} \dfrac{1/x}{-\csc^2 x} = \lim_{x\to 0^+} \dfrac{-\sin^2 x}{x} = \lim_{x\to 0^+} \dfrac{-2\sin x\cos x}{1} = 0$

23. $\lim_{x\to (\pi/2)^-} \sec 3x \cos 5x = \lim_{x\to (\pi/2)^-} \dfrac{\cos 5x}{\cos 3x} = \lim_{x\to (\pi/2)^-} \dfrac{-5\sin 5x}{-3\sin 3x} = \dfrac{-5(+1)}{(-3)(-1)} = -\dfrac{5}{3}$

24. $\lim_{x\to \pi} (x-\pi)\cot x = \lim_{x\to \pi} \dfrac{x-\pi}{\tan x} = \lim_{x\to \pi} \dfrac{1}{\sec^2 x} = 1$

25. $y = (1-3/x)^x$, $\lim_{x\to +\infty} \ln y = \lim_{x\to +\infty} \dfrac{\ln(1-3/x)}{1/x} = \lim_{x\to +\infty} \dfrac{-3}{1-3/x} = -3$, $\lim_{x\to +\infty} y = e^{-3}$

26. $y = (1+2x)^{-3/x}$, $\lim_{x\to 0} \ln y = \lim_{x\to 0} -\dfrac{3\ln(1+2x)}{x} = \lim_{x\to 0} -\dfrac{6}{1+2x} = -6$, $\lim_{x\to 0} y = e^{-6}$

27. $y = (e^x + x)^{1/x}$, $\lim_{x\to 0} \ln y = \lim_{x\to 0} \dfrac{\ln(e^x + x)}{x} = \lim_{x\to 0} \dfrac{e^x + 1}{e^x + x} = 2$, $\lim_{x\to 0} y = e^2$

28. $y = (1+a/x)^{bx}$, $\lim_{x\to +\infty} \ln y = \lim_{x\to +\infty} \dfrac{b\ln(1+a/x)}{1/x} = \lim_{x\to +\infty} \dfrac{ab}{1+a/x} = ab$, $\lim_{x\to +\infty} y = e^{ab}$

29. $y = (2-x)^{\tan(\pi x/2)}$, $\lim_{x\to 1} \ln y = \lim_{x\to 1} \dfrac{\ln(2-x)}{\cot(\pi x/2)} = \lim_{x\to 1} \dfrac{2\sin^2(\pi x/2)}{\pi(2-x)} = 2/\pi$, $\lim_{x\to 1} y = e^{2/\pi}$

30. $y = [\cos(2/x)]^{x^2}$, $\lim_{x\to +\infty} \ln y = \lim_{x\to +\infty} \dfrac{\ln\cos(2/x)}{1/x^2} = \lim_{x\to +\infty} \dfrac{(-2/x^2)(-\tan(2/x))}{-2/x^3}$

$= \lim_{x\to +\infty} \dfrac{-\tan(2/x)}{1/x} = \lim_{x\to +\infty} \dfrac{(2/x^2)\sec^2(2/x)}{-1/x^2} = -2$, $\lim_{x\to +\infty} y = e^{-2}$

31. $\lim_{x\to 0} \left(\dfrac{1}{\sin x} - \dfrac{1}{x}\right) = \lim_{x\to 0} \dfrac{x - \sin x}{x\sin x} = \lim_{x\to 0} \dfrac{1-\cos x}{x\cos x + \sin x} = \lim_{x\to 0} \dfrac{\sin x}{2\cos x - x\sin x} = 0$

32. $\lim_{x\to 0} \dfrac{1-\cos 3x}{x^2} = \lim_{x\to 0} \dfrac{3\sin 3x}{2x} = \lim_{x\to 0} \dfrac{9}{2}\cos 3x = \dfrac{9}{2}$

33. $\lim_{x\to +\infty} \dfrac{(x^2+x) - x^2}{\sqrt{x^2+x}+x} = \lim_{x\to +\infty} \dfrac{x}{\sqrt{x^2+x}+x} = \lim_{x\to +\infty} \dfrac{1}{\sqrt{1+1/x}+1} = 1/2$

34. $\lim_{x\to 0} \dfrac{e^x - 1 - x}{xe^x - x} = \lim_{x\to 0} \dfrac{e^x - 1}{xe^x + e^x - 1} = \lim_{x\to 0} \dfrac{e^x}{xe^x + 2e^x} = 1/2$

35. $\lim_{x\to+\infty}[x-\ln(x^2+1)] = \lim_{x\to+\infty}[\ln e^x - \ln(x^2+1)] = \lim_{x\to+\infty} \ln\frac{e^x}{x^2+1}$,

$\lim_{x\to+\infty}\frac{e^x}{x^2+1} = \lim_{x\to+\infty}\frac{e^x}{2x} = \lim_{x\to+\infty}\frac{e^x}{2} = +\infty$ so $\lim_{x\to+\infty}[x-\ln(x^2+1)] = +\infty$

36. $\lim_{x\to+\infty}\ln\frac{x}{1+x} = \lim_{x\to+\infty}\ln\frac{1}{1/x+1} = \ln(1) = 0$

38. **(a)** $\lim_{x\to+\infty}\frac{\ln x}{x^n} = \lim_{x\to+\infty}\frac{1/x}{nx^{n-1}} = \lim_{x\to+\infty}\frac{1}{nx^n} = 0$

(b) $\lim_{x\to+\infty}\frac{x^n}{\ln x} = \lim_{x\to+\infty}\frac{nx^{n-1}}{1/x} = \lim_{x\to+\infty} nx^n = +\infty$

39. **(a)** L'Hôpital's Rule does not apply to the problem $\lim_{x\to1}\frac{3x^2-2x+1}{3x^2-2x}$ because it is not a $\frac{0}{0}$ form.

(b) $\lim_{x\to1}\frac{3x^2-2x+1}{3x^2-2x} = 2$

40. $\lim_{x\to1}\frac{4x^3-12x^2+12x-4}{4x^3-9x^2+6x-1} = \lim_{x\to1}\frac{12x^2-24x+12}{12x^2-18x+6} = \lim_{x\to1}\frac{24x-24}{24x-18} = 0$

41. $\lim_{x\to+\infty}\frac{1/(x\ln x)}{1/(2\sqrt{x})} = \lim_{x\to+\infty}\frac{2}{\sqrt{x}\ln x} = 0$

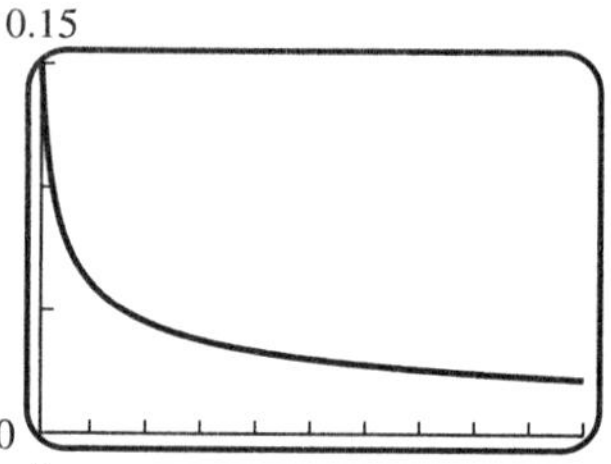

42. $y = x^x$, $\lim_{x\to0^+}\ln y = \lim_{x\to0^+}\frac{\ln x}{1/x} = \lim_{x\to0^+} -x = 0$, $\lim_{x\to0^+} y = 1$

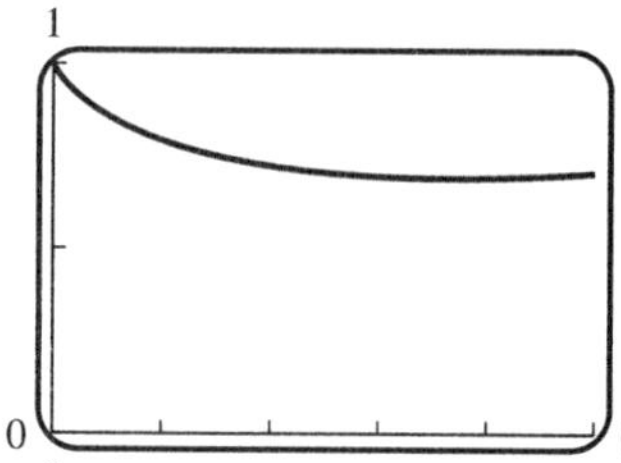

43. $y = (\sin x)^{3/\ln x}$,

$\lim_{x\to0^+}\ln y = \lim_{x\to0^+}\frac{3\ln\sin x}{\ln x} = \lim_{x\to0^+}(3\cos x)\frac{x}{\sin x} = 3$,

$\lim_{x\to0^+} y = e^3$

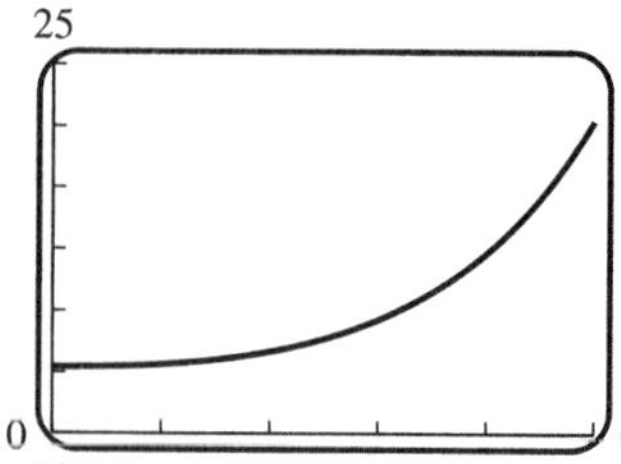

44. $\lim_{x\to\pi/2^-}\frac{4\sec^2 x}{\sec x\tan x} = \lim_{x\to\pi/2^-}\frac{4}{\sin x} = 4$

45. $\ln x - e^x = \ln x - \dfrac{1}{e^{-x}} = \dfrac{e^{-x}\ln x - 1}{e^{-x}};$

$\lim\limits_{x\to+\infty} e^{-x}\ln x = \lim\limits_{x\to+\infty}\dfrac{\ln x}{e^x} = \lim\limits_{x\to+\infty}\dfrac{1/x}{e^x} = 0$ by L'Hôpital's Rule,

so $\lim\limits_{x\to+\infty}[\ln x - e^x] = \lim\limits_{x\to+\infty}\dfrac{e^{-x}\ln x - 1}{e^{-x}} = -\infty$

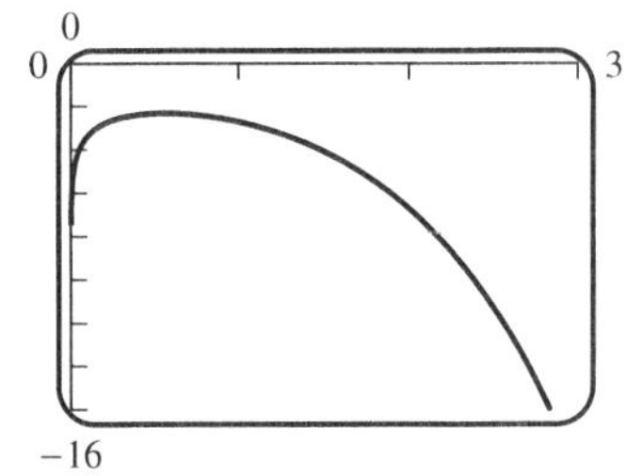

46. $\lim\limits_{x\to+\infty}[\ln e^x - \ln(1+2e^x)] = \lim\limits_{x\to+\infty}\ln\dfrac{e^x}{1+2e^x}$

$= \lim\limits_{x\to+\infty}\ln\dfrac{1}{e^{-x}+2} = \ln\dfrac{1}{2};$

horizontal asymptote $y = -\ln 2$

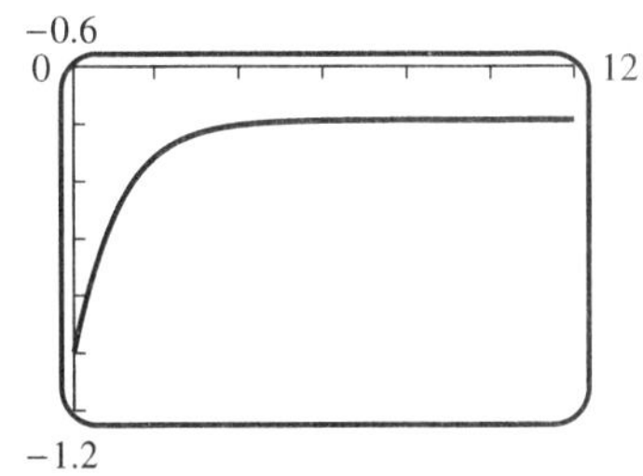

47. $y = (\ln x)^{1/x}$,

$\lim\limits_{x\to+\infty}\ln y = \lim\limits_{x\to+\infty}\dfrac{\ln(\ln x)}{x} = \lim\limits_{x\to+\infty}\dfrac{1}{x\ln x} = 0;$

$\lim\limits_{x\to+\infty} y = 1$, $y = 1$ is the horizontal asymptote

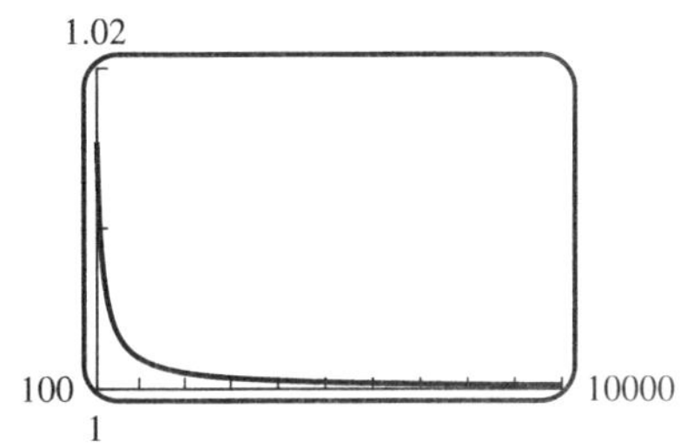

48. $y = \left(\dfrac{x+1}{x+2}\right)^x$, $\lim\limits_{x\to+\infty}\ln y = \lim\limits_{x\to+\infty}\dfrac{\ln\dfrac{x+1}{x+2}}{1/x}$

$= \lim\limits_{x\to+\infty}\dfrac{-x^2}{(x+1)(x+2)} = -1;$

$\lim\limits_{x\to+\infty} y = e^{-1}$ is the horizontal asymptote

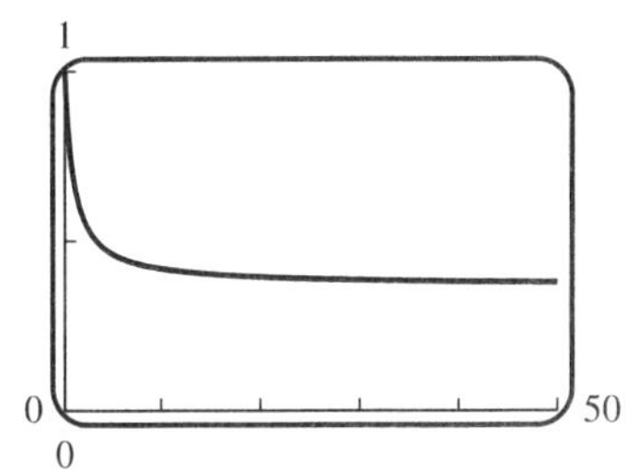

49. **(a)** 0 **(b)** $+\infty$ **(c)** 0 **(d)** $-\infty$ **(e)** $+\infty$ **(f)** $-\infty$

50. **(a)** Type 0^0; $y = x^{(\ln a)/(1+\ln x)}$; $\lim\limits_{x\to0^+}\ln y = \lim\limits_{x\to0^+}\dfrac{(\ln a)\ln x}{1+\ln x} = \lim\limits_{x\to0^+}\dfrac{(\ln a)/x}{1/x} = \lim\limits_{x\to0^+}\ln a = \ln a$,

$\lim\limits_{x\to0^+} y = e^{\ln a} = a$

(b) Type ∞^0; same calculation as Part (a) with $x\to+\infty$

(c) Type 1^∞; $y = (x+1)^{(\ln a)/x}$, $\lim\limits_{x\to0}\ln y = \lim\limits_{x\to0}\dfrac{(\ln a)\ln(x+1)}{x} = \lim\limits_{x\to0}\dfrac{\ln a}{x+1} = \ln a$,

$\lim\limits_{x\to0} y = e^{\ln a} = a$

51. $\lim_{x\to+\infty} \dfrac{1+2\cos 2x}{1}$ does not exist, nor is it $\pm\infty$; $\lim_{x\to+\infty} \dfrac{x+\sin 2x}{x} = \lim_{x\to+\infty}\left(1+\dfrac{\sin 2x}{x}\right) = 1$

52. $\lim_{x\to+\infty} \dfrac{2-\cos x}{3+\cos x}$ does not exist, nor is it $\pm\infty$; $\lim_{x\to+\infty} \dfrac{2x-\sin x}{3x+\sin x} = \lim_{x\to+\infty} \dfrac{2-(\sin x)/x}{3+(\sin x)/x} = \dfrac{2}{3}$

53. $\lim_{x\to+\infty}(2+x\cos 2x+\sin 2x)$ does not exist, nor is it $\pm\infty$; $\lim_{x\to+\infty} \dfrac{x(2+\sin 2x)}{x+1} = \lim_{x\to+\infty} \dfrac{2+\sin 2x}{1+1/x}$, which does not exist because $\sin 2x$ oscillates between -1 and 1 as $x\to+\infty$

54. $\lim_{x\to+\infty}\left(\dfrac{1}{x}+\dfrac{1}{2}\cos x+\dfrac{\sin x}{2x}\right)$ does not exist, nor is it $\pm\infty$;

$$\lim_{x\to+\infty} \frac{x(2+\sin x)}{x^2+1} = \lim_{x\to+\infty} \frac{2+\sin x}{x+1/x} = 0$$

55. $\lim_{R\to 0^+} \dfrac{\frac{Vt}{L}e^{-Rt/L}}{1} = \dfrac{Vt}{L}$

56. **(a)** $\lim_{x\to\pi/2}(\pi/2-x)\tan x = \lim_{x\to\pi/2} \dfrac{\pi/2-x}{\cot x} = \lim_{x\to\pi/2} \dfrac{-1}{-\csc^2 x} = \lim_{x\to\pi/2} \sin^2 x = 1$

(b)
$$\lim_{x\to\pi/2}\left(\frac{1}{\pi/2-x}-\tan x\right) = \lim_{x\to\pi/2}\left(\frac{1}{\pi/2-x}-\frac{\sin x}{\cos x}\right) = \lim_{x\to\pi/2} \frac{\cos x-(\pi/2-x)\sin x}{(\pi/2-x)\cos x}$$
$$= \lim_{x\to\pi/2} \frac{-(\pi/2-x)\cos x}{-(\pi/2-x)\sin x-\cos x}$$
$$= \lim_{x\to\pi/2} \frac{(\pi/2-x)\sin x+\cos x}{-(\pi/2-x)\cos x+2\sin x} = 0$$

(c) $1/(\pi/2-1.57)\approx 1255.765849, \tan 1.57\approx 1255.765592$;
$1/(\pi/2-1.57)-\tan 1.57\approx 0.000265$

57. **(b)** $\lim_{x\to+\infty} x(k^{1/x}-1) = \lim_{t\to 0^+} \dfrac{k^t-1}{t} = \lim_{t\to 0^+} \dfrac{(\ln k)k^t}{1} = \ln k$

(c) $\ln 0.3 = -1.20397$, $1024\left(\sqrt[1024]{0.3}-1\right) = -1.20327$;
$\ln 2 = 0.69315$, $1024\left(\sqrt[1024]{2}-1\right) = 0.69338$

58. **(a)** No; $\sin(1/x)$ oscillates as $x\to 0$.

(b)

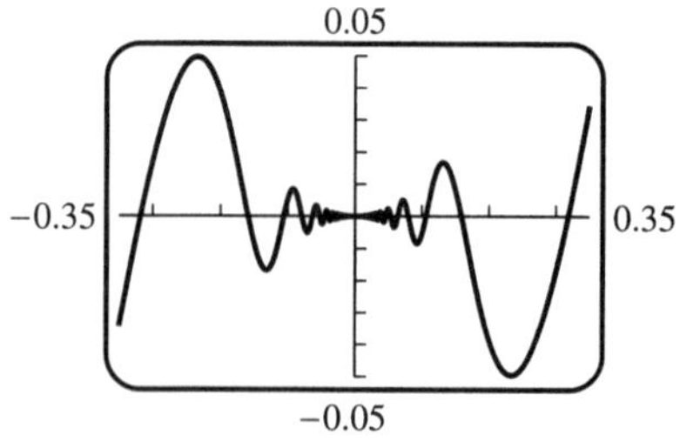

(c) For the limit as $x\to 0^+$ use the Squeezing Theorem together with the inequalities $-x^2 \le x^2\sin(1/x) \le x^2$. For $x\to 0^-$ do the same; thus $\lim_{x\to 0} f(x) = 0$.

59. If $k\ne -1$ then $\lim_{x\to 0}(k+\cos \ell x) = k+1\ne 0$, so $\lim_{x\to 0} \dfrac{k+\cos \ell x}{x^2} = \pm\infty$. Hence $k=-1$, and by the rule

$$\lim_{x\to 0} \frac{-1+\cos \ell x}{x^2} = \lim_{x\to 0} \frac{-\ell\sin \ell x}{2x} = \lim_{x\to 0} \frac{-\ell^2\cos \ell x}{2} = -\frac{\ell^2}{2} = -4 \text{ if } \ell = \pm 2\sqrt{2}.$$

60. (a) Apply the rule to get $\lim_{x\to 0}\frac{-\cos(1/x)+2x\sin(1/x)}{\cos x}$ which does not exist (nor is it $\pm\infty$).

(b) Rewrite as $\lim_{x\to 0}\left[\frac{x}{\sin x}\right][x\sin(1/x)]$, but $\lim_{x\to 0}\frac{x}{\sin x}=\lim_{x\to 0}\frac{1}{\cos x}=1$ and $\lim_{x\to 0}x\sin(1/x)=0$, thus $\lim_{x\to 0}\left[\frac{x}{\sin x}\right][x\sin(1/x)]=(1)(0)=0$

61. $\lim_{x\to 0^+}\frac{\sin(1/x)}{(\sin x)/x}$, $\lim_{x\to 0^+}\frac{\sin x}{x}=1$ but $\lim_{x\to 0^+}\sin(1/x)$ does not exist because $\sin(1/x)$ oscillates between -1 and 1 as $x\to+\infty$, so $\lim_{x\to 0^+}\frac{x\sin(1/x)}{\sin x}$ does not exist.

EXERCISE SET 7.8

1. (a) $\sinh 3\approx 10.0179$
(b) $\cosh(-2)\approx 3.7622$
(c) $\tanh(\ln 4)=15/17\approx 0.8824$
(d) $\sinh^{-1}(-2)\approx -1.4436$
(e) $\cosh^{-1}3\approx 1.7627$
(f) $\tanh^{-1}\frac{3}{4}\approx 0.9730$

2. (a) $\operatorname{csch}(-1)\approx -0.8509$
(b) $\operatorname{sech}(\ln 2)=0.8$
(c) $\coth 1\approx 1.3130$
(d) $\operatorname{sech}^{-1}\frac{1}{2}\approx 1.3170$
(e) $\coth^{-1}3\approx 0.3466$
(f) $\operatorname{csch}^{-1}(-\sqrt{3})\approx -0.5493$

3. (a) $\sinh(\ln 3)=\frac{1}{2}(e^{\ln 3}-e^{-\ln 3})=\frac{1}{2}\left(3-\frac{1}{3}\right)=\frac{4}{3}$

(b) $\cosh(-\ln 2)=\frac{1}{2}(e^{-\ln 2}+e^{\ln 2})=\frac{1}{2}\left(\frac{1}{2}+2\right)=\frac{5}{4}$

(c) $\tanh(2\ln 5)=\frac{e^{2\ln 5}-e^{-2\ln 5}}{e^{2\ln 5}+e^{-2\ln 5}}=\frac{25-1/25}{25+1/25}=\frac{312}{313}$

(d) $\sinh(-3\ln 2)=\frac{1}{2}(e^{-3\ln 2}-e^{3\ln 2})=\frac{1}{2}\left(\frac{1}{8}-8\right)=-\frac{63}{16}$

4. (a) $\frac{1}{2}(e^{\ln x}+e^{-\ln x})=\frac{1}{2}\left(x+\frac{1}{x}\right)=\frac{x^2+1}{2x},\ x>0$

(b) $\frac{1}{2}(e^{\ln x}-e^{-\ln x})=\frac{1}{2}\left(x-\frac{1}{x}\right)=\frac{x^2-1}{2x},\ x>0$

(c) $\frac{e^{2\ln x}-e^{-2\ln x}}{e^{2\ln x}+e^{-2\ln x}}=\frac{x^2-1/x^2}{x^2+1/x^2}=\frac{x^4-1}{x^4+1},\ x>0$

(d) $\frac{1}{2}(e^{-\ln x}+e^{\ln x})=\frac{1}{2}\left(\frac{1}{x}+x\right)=\frac{1+x^2}{2x},\ x>0$

5.

	$\sinh x_0$	$\cosh x_0$	$\tanh x_0$	$\coth x_0$	$\operatorname{sech} x_0$	$\operatorname{csch} x_0$
(a)	2	$\sqrt{5}$	$2/\sqrt{5}$	$\sqrt{5}/2$	$1/\sqrt{5}$	1/2
(b)	3/4	5/4	3/5	5/3	4/5	4/3
(c)	4/3	5/3	4/5	5/4	3/5	3/4

(a) $\cosh^2 x_0 = 1 + \sinh^2 x_0 = 1 + (2)^2 = 5$, $\cosh x_0 = \sqrt{5}$

(b) $\sinh^2 x_0 = \cosh^2 x_0 - 1 = \dfrac{25}{16} - 1 = \dfrac{9}{16}$, $\sinh x_0 = \dfrac{3}{4}$ (because $x_0 > 0$)

(c) $\text{sech}^2 x_0 = 1 - \tanh^2 x_0 = 1 - \left(\dfrac{4}{5}\right)^2 = 1 - \dfrac{16}{25} = \dfrac{9}{25}$, $\text{sech } x_0 = \dfrac{3}{5}$,

$\cosh x_0 = \dfrac{1}{\text{sech } x_0} = \dfrac{5}{3}$, from $\dfrac{\sinh x_0}{\cosh x_0} = \tanh x_0$ we get $\sinh x_0 = \left(\dfrac{5}{3}\right)\left(\dfrac{4}{5}\right) = \dfrac{4}{3}$

6. $\dfrac{d}{dx}\text{csch}x = \dfrac{d}{dx}\dfrac{1}{\sinh x} = -\dfrac{\cosh x}{\sinh^2 x} = -\coth x \,\text{csch } x$ for $x \neq 0$

$\dfrac{d}{dx}\text{sech } x = \dfrac{d}{dx}\dfrac{1}{\cosh x} = -\dfrac{\sinh x}{\cosh^2 x} = -\tanh x \,\text{sech } x$ for all x

$\dfrac{d}{dx}\coth x = \dfrac{d}{dx}\dfrac{\cosh x}{\sinh x} = \dfrac{\sinh^2 x - \cosh^2 x}{\sinh^2 x} = -\text{csch}^2 x$ for $x \neq 0$

7. (a) $y = \sinh^{-1} x$ if and only if $x = \sinh y$; $1 = \dfrac{dy}{dx}\dfrac{dx}{dy} = \dfrac{dy}{dx}\cosh y$; so

$\dfrac{d}{dx}[\sinh^{-1} x] = \dfrac{dy}{dx} = \dfrac{1}{\cosh y} = \dfrac{1}{\sqrt{1+\sinh^2 y}} = \dfrac{1}{\sqrt{1+x^2}}$ for all x.

(b) Let $x \geq 1$. Then $y = \cosh^{-1} x$ if and only if $x = \cosh y$; $1 = \dfrac{dy}{dx}\dfrac{dx}{dy} = \dfrac{dy}{dx}\sinh y$, so

$\dfrac{d}{dx}[\cosh^{-1} x] = \dfrac{dy}{dx} = \dfrac{1}{\sinh y} = \dfrac{1}{\sqrt{\cosh^2 y - 1}} = \dfrac{1}{x^2 - 1}$ for $x \geq 1$.

(c) Let $-1 < x < 1$. Then $y = \tanh^{-1} x$ if and only if $x = \tanh y$; thus
$1 = \dfrac{dy}{dx}\dfrac{dx}{dy} = \dfrac{dy}{dx}\text{sech}^2 y = \dfrac{dy}{dx}(1 - \tanh^2 y) = 1 - x^2$, so $\dfrac{d}{dx}[\tanh^{-1} x] = \dfrac{dy}{dx} = \dfrac{1}{1-x^2}$.

9. $4\cosh(4x - 8)$

10. $4x^3 \sinh(x^4)$

11. $-\dfrac{1}{x}\text{csch}^2(\ln x)$

12. $2\dfrac{\text{sech}^2 2x}{\tanh 2x}$

13. $\dfrac{1}{x^2}\text{csch}(1/x)\coth(1/x)$

14. $-2e^{2x}\text{sech}(e^{2x})\tanh(e^{2x})$

15. $\dfrac{2 + 5\cosh(5x)\sinh(5x)}{\sqrt{4x + \cosh^2(5x)}}$

16. $6\sinh^2(2x)\cosh(2x)$

17. $x^{5/2}\tanh(\sqrt{x})\text{sech}^2(\sqrt{x}) + 3x^2\tanh^2(\sqrt{x})$

18. $-3\cosh(\cos 3x)\sin 3x$

19. $\dfrac{1}{\sqrt{1 + x^2/9}}\left(\dfrac{1}{3}\right) = 1/\sqrt{9 + x^2}$

20. $\dfrac{1}{\sqrt{1 + 1/x^2}}(-1/x^2) = -\dfrac{1}{|x|\sqrt{x^2+1}}$

21. $1/\left[(\cosh^{-1} x)\sqrt{x^2 - 1}\right]$

22. $1/\left[\sqrt{(\sinh^{-1} x)^2 - 1}\,\sqrt{1 + x^2}\right]$

23. $-(\tanh^{-1} x)^{-2}/(1 - x^2)$

24. $2(\coth^{-1} x)/(1 - x^2)$

25. $\dfrac{\sinh x}{\sqrt{\cosh^2 x - 1}} = \dfrac{\sinh x}{|\sinh x|} = \begin{cases} 1, & x > 0 \\ -1, & x < 0 \end{cases}$

26. $(\operatorname{sech}^2 x)/\sqrt{1+\tanh^2 x}$

27. $-\dfrac{e^x}{2x\sqrt{1-x}} + e^x \operatorname{sech}^{-1} x$

28. $10(1+x\operatorname{csch}^{-1}x)^9\left(-\dfrac{x}{|x|\sqrt{1+x^2}} + \operatorname{csch}^{-1}x\right)$

31. $\dfrac{1}{7}\sinh^7 x + C$

32. $\dfrac{1}{2}\sinh(2x-3) + C$

33. $\dfrac{2}{3}(\tanh x)^{3/2} + C$

34. $-\dfrac{1}{3}\coth(3x) + C$

35. $\ln(\cosh x) + C$

36. $-\dfrac{1}{3}\coth^3 x + C$

37. $-\dfrac{1}{3}\operatorname{sech}^3 x\Big]_{\ln 2}^{\ln 3} = 37/375$

38. $\ln(\cosh x)\Big]_0^{\ln 3} = \ln 5 - \ln 3$

39. $u = 3x,\ \dfrac{1}{3}\displaystyle\int \frac{1}{\sqrt{1+u^2}}du = \frac{1}{3}\sinh^{-1} 3x + C$

40. $x = \sqrt{2}u,\ \displaystyle\int \frac{\sqrt{2}}{\sqrt{2u^2-2}}du = \int \frac{1}{\sqrt{u^2-1}}du = \cosh^{-1}(x/\sqrt{2}) + C$

41. $u = e^x,\ \displaystyle\int \frac{1}{u\sqrt{1-u^2}}du = -\operatorname{sech}^{-1}(e^x) + C$

42. $u = \cos\theta,\ -\displaystyle\int \frac{1}{\sqrt{1+u^2}}du = -\sinh^{-1}(\cos\theta) + C$

43. $u = 2x,\ \displaystyle\int \frac{du}{u\sqrt{1+u^2}} = -\operatorname{csch}^{-1}|u| + C = -\operatorname{csch}^{-1}|2x| + C$

44. $x = 5u/3,\ \displaystyle\int \frac{5/3}{\sqrt{25u^2-25}}du = \frac{1}{3}\int \frac{1}{\sqrt{u^2-1}}du = \frac{1}{3}\cosh^{-1}(3x/5) + C$

45. $\tanh^{-1} x\Big]_0^{1/2} = \tanh^{-1}(1/2) - \tanh^{-1}(0) = \dfrac{1}{2}\ln\dfrac{1+1/2}{1-1/2} = \dfrac{1}{2}\ln 3$

46. $\sinh^{-1} t\Big]_0^{\sqrt{3}} = \sinh^{-1}\sqrt{3} - \sinh^{-1} 0 = \ln(\sqrt{3}+2)$

49. $A = \displaystyle\int_0^{\ln 3} \sinh 2x\, dx = \frac{1}{2}\cosh 2x\Big]_0^{\ln 3} = \frac{1}{2}[\cosh(2\ln 3) - 1]$,

but $\cosh(2\ln 3) = \cosh(\ln 9) = \dfrac{1}{2}(e^{\ln 9} + e^{-\ln 9}) = \dfrac{1}{2}(9+1/9) = 41/9$ so $A = \dfrac{1}{2}[41/9 - 1] = 16/9$.

50. $V = \pi\displaystyle\int_0^{\ln 2} \operatorname{sech}^2 x\, dx = \pi\tanh x\Big]_0^{\ln 2} = \pi\tanh(\ln 2) = 3\pi/5$

51. $V = \pi\displaystyle\int_0^5 (\cosh^2 2x - \sinh^2 2x)dx = \pi\int_0^5 dx = 5\pi$

52. $\displaystyle\int_0^1 \cosh ax\,dx = 2, \frac{1}{a}\sinh ax\Big]_0^1 = 2, \frac{1}{a}\sinh a = 2, \sinh a = 2a;$

let $f(a) = \sinh a - 2a$, then $a_{n+1} = a_n - \dfrac{\sinh a_n - 2a_n}{\cosh a_n - 2}$, $a_1 = 2.2, \ldots, a_4 = a_5 = 2.177318985$.

53. $y' = \sinh x$, $1 + (y')^2 = 1 + \sinh^2 x = \cosh^2 x$

$$L = \int_0^{\ln 2} \cosh x\,dx = \sinh x\Big]_0^{\ln 2} = \sinh(\ln 2) = \frac{1}{2}(e^{\ln 2} - e^{-\ln 2}) = \frac{1}{2}\left(2 - \frac{1}{2}\right) = \frac{3}{4}$$

54. $y' = \sinh(x/a)$, $1 + (y')^2 = 1 + \sinh^2(x/a) = \cosh^2(x/a)$

$$L = \int_0^{x_1} \cosh(x/a)dx = a\sinh(x/a)\Big]_0^{x_1} = a\sinh(x_1/a)$$

55. $\sinh(-x) = \dfrac{1}{2}(e^{-x} - e^{x}) = -\dfrac{1}{2}(e^x - e^{-x}) = -\sinh x$

$\cosh(-x) = \dfrac{1}{2}(e^{-x} + e^{x}) = \dfrac{1}{2}(e^x + e^{-x}) = \cosh x$

56. **(a)** $\cosh x + \sinh x = \dfrac{1}{2}(e^x + e^{-x}) + \dfrac{1}{2}(e^x - e^{-x}) = e^x$

(b) $\cosh x - \sinh x = \dfrac{1}{2}(e^x + e^{-x}) - \dfrac{1}{2}(e^x - e^{-x}) = e^{-x}$

(c) $\sinh x\cosh y + \cosh x\sinh y = \dfrac{1}{4}(e^x - e^{-x})(e^y + e^{-y}) + \dfrac{1}{4}(e^x + e^{-x})(e^y - e^{-y})$

$$= \frac{1}{4}[(e^{x+y} - e^{-x+y} + e^{x-y} - e^{-x-y}) + (e^{x+y} + e^{-x+y} - e^{x-y} - e^{-x-y})]$$

$$= \frac{1}{2}[e^{(x+y)} - e^{-(x+y)}] = \sinh(x+y)$$

(d) Let $y = x$ in Part (c).

(e) The proof is similar to Part (c), or: treat x as variable and y as constant, and differentiate the result in Part (c) with respect to x.

(f) Let $y = x$ in Part (e).

(g) Use $\cosh^2 x = 1 + \sinh^2 x$ together with Part (f).

(h) Use $\sinh^2 x = \cosh^2 x - 1$ together with Part (f).

57. **(a)** Divide $\cosh^2 x - \sinh^2 x = 1$ by $\cosh^2 x$.

(b) $\tanh(x+y) = \dfrac{\sinh x\cosh y + \cosh x\sinh y}{\cosh x\cosh y + \sinh x\sinh y} = \dfrac{\dfrac{\sinh x}{\cosh x} + \dfrac{\sinh y}{\cosh y}}{1 + \dfrac{\sinh x\sinh y}{\cosh x\cosh y}} = \dfrac{\tanh x + \tanh y}{1 + \tanh x\tanh y}$

(c) Let $y = x$ in Part (b).

58. **(a)** Let $y = \cosh^{-1} x$; then $x = \cosh y = \dfrac{1}{2}(e^y + e^{-y})$, $e^y - 2x + e^{-y} = 0$, $e^{2y} - 2xe^y + 1 = 0$, $e^y = \dfrac{2x \pm \sqrt{4x^2 - 4}}{2} = x \pm \sqrt{x^2 - 1}$. To determine which sign to take, note that $y \geq 0$ so $e^{-y} \leq e^y$, $x = (e^y + e^{-y})/2 \leq (e^y + e^y)/2 = e^y$, hence $e^y \geq x$ thus $e^y = x + \sqrt{x^2 - 1}$, $y = \cosh^{-1} x = \ln(x + \sqrt{x^2 - 1})$.

(b) Let $y = \tanh^{-1} x$; then $x = \tanh y = \dfrac{e^y - e^{-y}}{e^y + e^{-y}} = \dfrac{e^{2y} - 1}{e^{2y} + 1}$, $xe^{2y} + x = e^{2y} - 1$,

$1 + x = e^{2y}(1 - x)$, $e^{2y} = (1 + x)/(1 - x)$, $2y = \ln \dfrac{1 + x}{1 - x}$, $y = \dfrac{1}{2} \ln \dfrac{1 + x}{1 - x}$.

59. **(a)** $\dfrac{d}{dx}(\cosh^{-1} x) = \dfrac{1 + x/\sqrt{x^2 - 1}}{x + \sqrt{x^2 - 1}} = 1/\sqrt{x^2 - 1}$

(b) $\dfrac{d}{dx}(\tanh^{-1} x) = \dfrac{d}{dx}\left[\dfrac{1}{2}(\ln(1 + x) - \ln(1 - x))\right] = \dfrac{1}{2}\left(\dfrac{1}{1 + x} + \dfrac{1}{1 - x}\right) = 1/(1 - x^2)$

60. Let $y = \operatorname{sech}^{-1} x$ then $x = \operatorname{sech} y = 1/\cosh y$, $\cosh y = 1/x$, $y = \cosh^{-1}(1/x)$; the proofs for the remaining two are similar.

61. If $|u| < 1$ then, by Theorem 8.8.6, $\displaystyle\int \frac{du}{1 - u^2} = \tanh^{-1} u + C$.

For $|u| > 1$, $\displaystyle\int \frac{du}{1 - u^2} = \coth^{-1} u + C = \tanh^{-1}(1/u) + C$.

62. **(a)** $\dfrac{d}{dx}(\operatorname{sech}^{-1}|x|) = \dfrac{d}{dx}(\operatorname{sech}^{-1}\sqrt{x^2}) = -\dfrac{1}{\sqrt{x^2}\sqrt{1 - x^2}}\dfrac{x}{\sqrt{x^2}} = -\dfrac{1}{x\sqrt{1 - x^2}}$

(b) Similar to solution of Part (a)

63. **(a)** $\displaystyle\lim_{x \to +\infty} \sinh x = \lim_{x \to +\infty} \frac{1}{2}(e^x - e^{-x}) = +\infty - 0 = +\infty$

(b) $\displaystyle\lim_{x \to -\infty} \sinh x = \lim_{x \to -\infty} \frac{1}{2}(e^x - e^{-x}) = 0 - \infty = -\infty$

(c) $\displaystyle\lim_{x \to +\infty} \tanh x = \lim_{x \to +\infty} \frac{e^x - e^{-x}}{e^x + e^{-x}} = 1$ **(d)** $\displaystyle\lim_{x \to -\infty} \tanh x = \lim_{x \to -\infty} \frac{e^x - e^{-x}}{e^x + e^{-x}} = -1$

(e) $\displaystyle\lim_{x \to +\infty} \sinh^{-1} x = \lim_{x \to +\infty} \ln(x + \sqrt{x^2 + 1}) = +\infty$

(f) $\displaystyle\lim_{x \to 1^-} \tanh^{-1} x = \lim_{x \to 1^-} \frac{1}{2}[\ln(1 + x) - \ln(1 - x)] = +\infty$

64. **(a)** $\displaystyle\lim_{x \to +\infty} (\cosh^{-1} x - \ln x) = \lim_{x \to +\infty} [\ln(x + \sqrt{x^2 - 1}) - \ln x]$

$$= \lim_{x \to +\infty} \ln \frac{x + \sqrt{x^2 - 1}}{x} = \lim_{x \to +\infty} \ln(1 + \sqrt{1 - 1/x^2}) = \ln 2$$

(b) $\displaystyle\lim_{x \to +\infty} \frac{\cosh x}{e^x} = \lim_{x \to +\infty} \frac{e^x + e^{-x}}{2e^x} = \lim_{x \to +\infty} \frac{1}{2}(1 + e^{-2x}) = 1/2$

65. For $|x| < 1$, $y = \tanh^{-1} x$ is defined and $dy/dx = 1/(1 - x^2) > 0$; $y'' = 2x/(1 - x^2)^2$ changes sign at $x = 0$, so there is a point of inflection there.

66. Let $x = -u/a$, $\displaystyle\int \frac{1}{\sqrt{u^2 - a^2}} du = -\int \frac{a}{a\sqrt{x^2 - 1}} dx = -\cosh^{-1} x + C = -\cosh^{-1}(-u/a) + C$.

$$-\cosh^{-1}(-u/a) = -\ln(-u/a + \sqrt{u^2/a^2 - 1}) = \ln\left[\frac{a}{-u + \sqrt{u^2 - a^2}}\frac{u + \sqrt{u^2 - a^2}}{u + \sqrt{u^2 - a^2}}\right]$$

$$= \ln\left|u + \sqrt{u^2 - a^2}\right| - \ln a = \ln|u + \sqrt{u^2 - a^2}| + C_1$$

so $\displaystyle\int \frac{1}{\sqrt{u^2 - a^2}} du = \ln\left|u + \sqrt{u^2 - a^2}\right| + C_2$.

67. Using $\sinh x + \cosh x = e^x$ (Exercise 56a), $(\sinh x + \cosh x)^n = (e^x)^n = e^{nx} = \sinh nx + \cosh nx$.

68. $\displaystyle\int_{-a}^{a} e^{tx}dx = \frac{1}{t}e^{tx}\Big]_{-a}^{a} = \frac{1}{t}(e^{at} - e^{-at}) = \frac{2\sinh at}{t}$ for $t \neq 0$.

69. **(a)** $y' = \sinh(x/a), 1 + (y')^2 = 1 + \sinh^2(x/a) = \cosh^2(x/a)$

$$L = 2\int_0^b \cosh(x/a)\,dx = 2a\sinh(x/a)\Big]_0^b = 2a\sinh(b/a)$$

(b) The highest point is at $x = b$, the lowest at $x = 0$, so $S = a\cosh(b/a) - a\cosh(0) = a\cosh(b/a) - a$.

70. From Part (a) of Exercise 69, $L = 2a\sinh(b/a)$ so $120 = 2a\sinh(50/a), a\sinh(50/a) = 60$. Let $u = 50/a$, then $a = 50/u$ so $(50/u)\sinh u = 60, \sinh u = 1.2u$. If $f(u) = \sinh u - 1.2u$, then $u_{n+1} = u_n - \dfrac{\sinh u_n - 1.2u_n}{\cosh u_n - 1.2}; u_1 = 1, \ldots, u_5 = u_6 = 1.064868548 \approx 50/a$ so $a \approx 46.95415231$.
From Part (b), $S = a\cosh(b/a) - a \approx 46.95415231[\cosh(1.064868548) - 1] \approx 29.2$ ft.

71. From Part (b) of Exercise 69, $S = a\cosh(b/a) - a$ so $30 = a\cosh(200/a) - a$. Let $u = 200/a$, then $a = 200/u$ so $30 = (200/u)[\cosh u - 1], \cosh u - 1 = 0.15u$. If $f(u) = \cosh u - 0.15u - 1$, then $u_{n+1} = u_n - \dfrac{\cosh u_n - 0.15u_n - 1}{\sinh u_n - 0.15}; u_1 = 0.3, \ldots, u_4 = u_5 = 0.297792782 \approx 200/a$ so $a \approx 671.6079505$. From Part (a), $L = 2a\sinh(b/a) \approx 2(671.6079505)\sinh(0.297792782) \approx 405.9$ ft.

72. **(a)** When the bow of the boat is at the point (x, y) and the person has walked a distance D, then the person is located at the point $(0, D)$, the line segment connecting $(0, D)$ and (x, y) has length a; thus $a^2 = x^2 + (D - y)^2$, $D = y + \sqrt{a^2 - x^2} = a\,\mathrm{sech}^{-1}(x/a)$.

(b) Find D when $a = 15$, $x = 10$: $D = 15\,\mathrm{sech}^{-1}(10/15) = 15\ln\left(\dfrac{1 + \sqrt{5/9}}{2/3}\right) \approx 14.44$ m.

(c) $dy/dx = -\dfrac{a^2}{x\sqrt{a^2 - x^2}} + \dfrac{x}{\sqrt{a^2 - x^2}} = \dfrac{1}{\sqrt{a^2 - x^2}}\left[-\dfrac{a^2}{x} + x\right] = -\dfrac{1}{x}\sqrt{a^2 - x^2}$,

$1 + [y']^2 = 1 + \dfrac{a^2 - x^2}{x^2} = \dfrac{a^2}{x^2}$; with $a = 15$ and $x = 5$, $L = \displaystyle\int_5^{15} \frac{225}{x^2}\,dx = -\frac{225}{x}\Big]_5^{15} = 30$ m.

CHAPTER 7 SUPPLEMENTARY EXERCISES

1. **(a)** $f(g(x)) = x$ for all x in the domain of g, and $g(f(x)) = x$ for all x in the domain of f.
(b) They are reflections of each other through the line $y = x$.
(c) The domain of one is the range of the other and vice versa.
(d) The equation $y = f(x)$ can always be solved for x as a function of y. Functions with no inverses include $y = x^2$, $y = \sin x$.
(e) Yes, g is continuous; this is evident from the statement about the graphs in Part (b) above.
(f) Yes, g must be differentiable (where $f' \neq 0$); this can be inferred from the graphs. Note that if $f' = 0$ at a point then g' cannot exist (infinite slope).

2. **(a)** For $\sin x$, $-\pi/2 \leq x \leq \pi/2$; for $\cos x$, $0 \leq x \leq \pi$; for $\tan x$, $-\pi/2 < x < \pi/2$; for $\sec x$, $0 \leq x < \pi/2$ or $\pi/2 < x \leq \pi$.

(b)

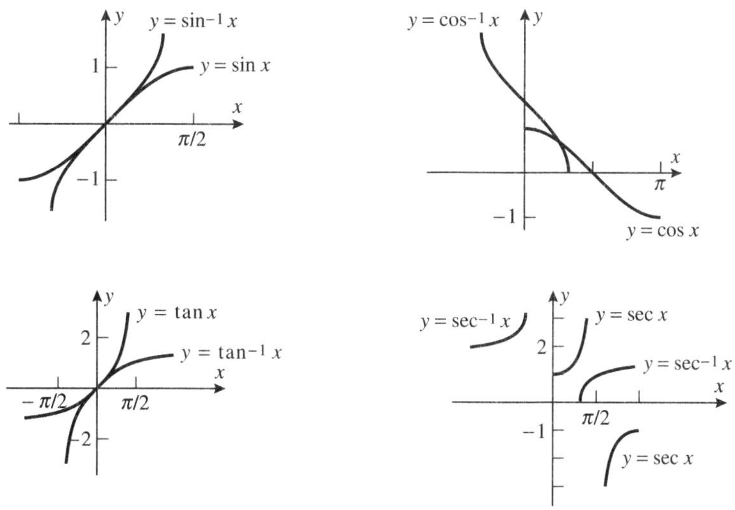

3. **(a)** $x = f(y) = 8y^3 - 1$; $y = f^{-1}(x) = \left(\dfrac{x+1}{8}\right)^{1/3} = \dfrac{1}{2}(x+1)^{1/3}$

(b) $f(x) = (x-1)^2$; f does not have an inverse because f is not one-to-one, for example $f(0) = f(2) = 1$.

(c) $x = f(y) = (e^y)^2 + 1$; $y = f^{-1}(x) = \ln\sqrt{x-1} = \frac{1}{2}\ln(x-1)$

(d) $x = f(y) = \dfrac{y+2}{y-1}$; $y = f^{-1}(x) = \dfrac{x+2}{x-1}$

4. $f'(x) = \dfrac{ad-bc}{(cx+d)^2}$; if $ad - bc = 0$ then the function represents a horizontal line, no inverse. If $ad - bc \neq 0$ then $f'(x) > 0$ or $f'(x) < 0$ so f is invertible. If $x = f(y) = \dfrac{ay+b}{cy+d}$ then $y = f^{-1}(x) = \dfrac{b-xd}{xc-a}$.

5. $3\ln\left(e^{2x}(e^x)^3\right) + 2\exp(\ln 1) = 3\ln e^{2x} + 3\ln(e^x)^3 + 2\cdot 1 = 3(2x) + (3\cdot 3)x + 2 = 15x + 2$

6. Draw equilateral triangles of sides 5, 12, 13, and 3, 4, 5. Then $\sin[\cos^{-1}(4/5)] = 3/5$, $\sin[\cos^{-1}(5/13)] = 12/13$, $\cos[\sin^{-1}(4/5)] = 3/5$, $\cos[\sin^{-1}(5/13)] = 12/13$

(a)
$$\begin{aligned}\cos[\cos^{-1}(4/5) + \sin^{-1}(5/13)] &= \cos(\cos^{-1}(4/5))\cos(\sin^{-1}(5/13)) \\ &\quad - \sin(\cos^{-1}(4/5))\sin(\sin^{-1}(5/13)) \\ &= \frac{4}{5}\frac{12}{13} - \frac{3}{5}\frac{5}{13} = \frac{33}{65}.\end{aligned}$$

(b)
$$\begin{aligned}\sin[\sin^{-1}(4/5) + \cos^{-1}(5/13)] &= \sin(\sin^{-1}(4/5))\cos(\cos^{-1}(5/13)) \\ &\quad + \cos(\sin^{-1}(4/5))\sin(\cos^{-1}(5/13)) \\ &= \frac{4}{5}\frac{5}{13} + \frac{3}{5}\frac{12}{13} = \frac{56}{65}.\end{aligned}$$

7. **(a)**
$$\begin{aligned}\cosh 3x &= \cosh(2x + x) = \cosh 2x\cosh x + \sinh 2x\sinh x \\ &= (2\cosh^2 x - 1)\cosh x + (2\sinh x\cosh x)\sinh x \\ &= 2\cosh^3 x - \cosh x + 2\sinh^2 x\cosh x \\ &= 2\cosh^3 x - \cosh x + 2(\cosh^2 x - 1)\cosh x = 4\cosh^3 x - 3\cosh x\end{aligned}$$

(b) from Theorem 7.8.2 with x replaced by $\dfrac{x}{2}$: $\cosh x = 2\cosh^2\dfrac{x}{2} - 1$,

$2\cosh^2\dfrac{x}{2} = \cosh x + 1$, $\cosh^2\dfrac{x}{2} = \dfrac{1}{2}(\cosh x + 1)$,

$\cosh\dfrac{x}{2} = \sqrt{\dfrac{1}{2}(\cosh x + 1)}$ (because $\cosh\dfrac{x}{2} > 0$)

(c) from Theorem 7.8.2 with x replaced by $\dfrac{x}{2}$: $\cosh x = 2\sinh^2\dfrac{x}{2} + 1$,

$2\sinh^2\dfrac{x}{2} = \cosh x - 1$, $\sinh^2\dfrac{x}{2} = \dfrac{1}{2}(\cosh x - 1)$, $\sinh\dfrac{x}{2} = \pm\sqrt{\dfrac{1}{2}(\cosh x - 1)}$

8. $Y = \ln(Ce^{kt}) = \ln C + \ln e^{kt} = \ln C + kt$, a line with slope k and Y-intercept $\ln C$

9. (a)

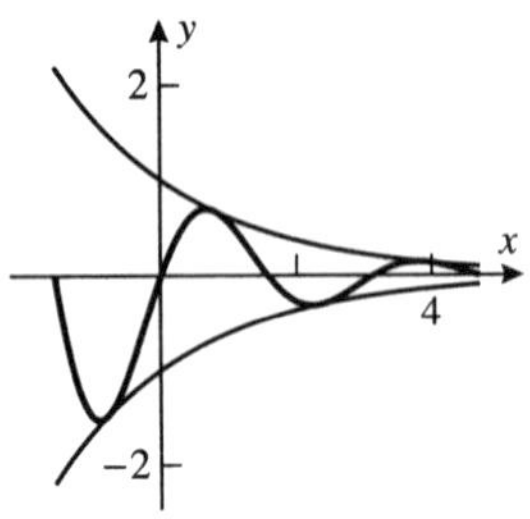

(b) The curve $y = e^{-x/2}\sin 2x$ has x-intercepts at $x = -\pi/2, 0, \pi/2, \pi, 3\pi/2$. It intersects the curve $y = e^{-x/2}$ at $x = \pi/4, 5\pi/4$, and it intersects the curve $y = -e^{-x/2}$ at $x = -\pi/4, 3\pi/4$.

10. (a)

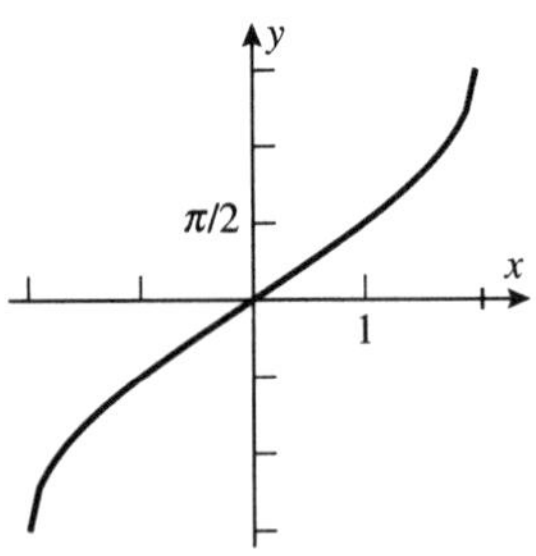

(b)

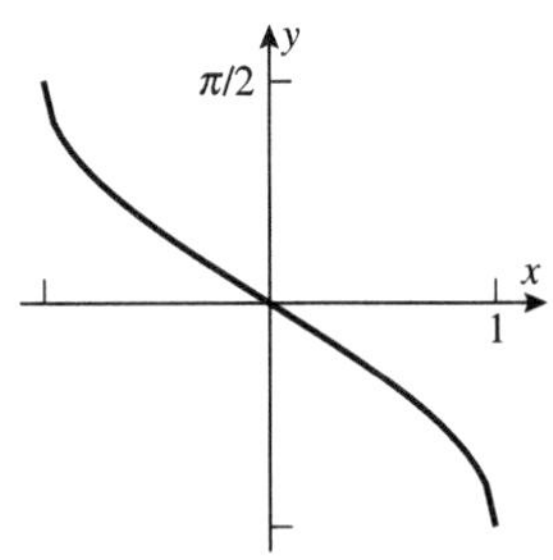

(c)

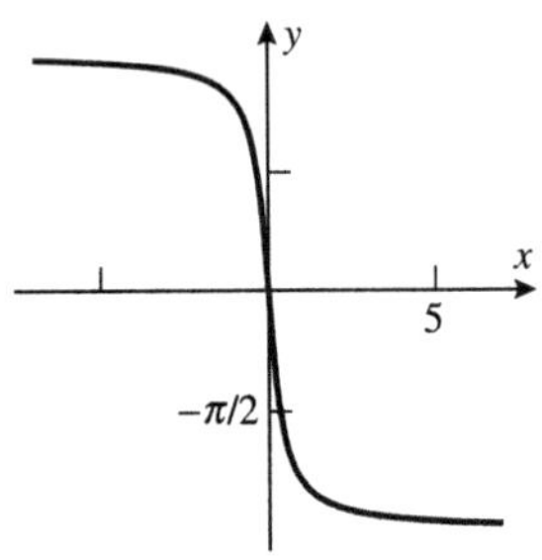

(d)

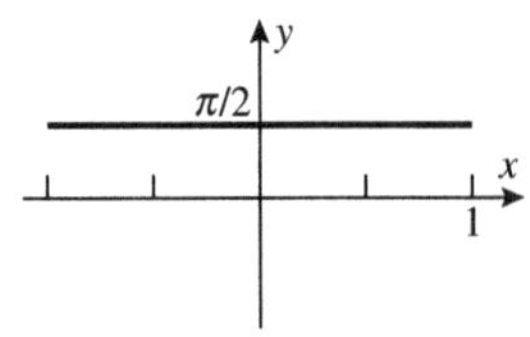

11. Set $a = 68.7672$, $b = 0.0100333$, $c = 693.8597$, $d = 299.2239$.

(a)

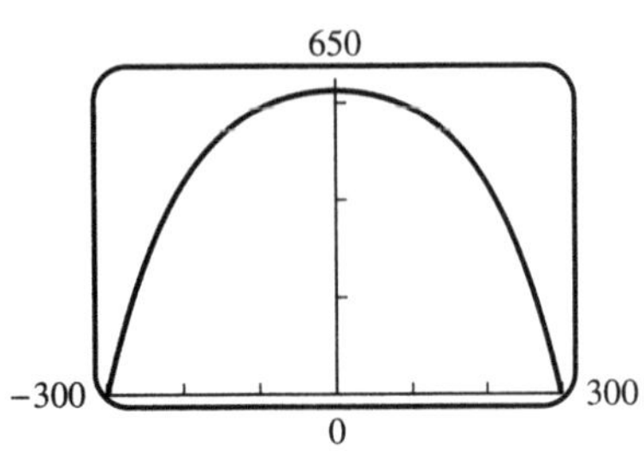

(b) $L = 2\displaystyle\int_0^d \sqrt{1 + a^2b^2\sinh^2 bx}\,dx$

$= 1480.2798$ ft

(c) $x = 283.6249$ ft

(d) $82°$

12. (a) If $x^k = e^x$ then $k \ln x = x$, or $\dfrac{\ln x}{x} = \dfrac{1}{k}$. The steps are reversible.

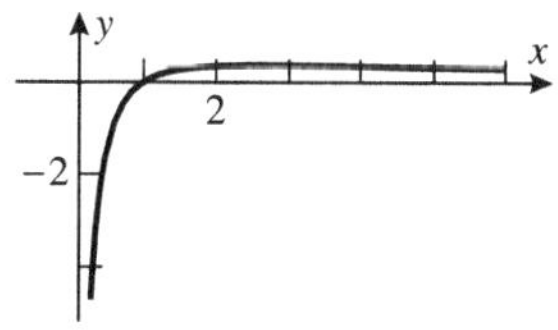

(b) By zooming it is seen that the maximum value of y is approximately 0.368 (actually, $1/e$), so there are two distinct solutions of $x^k = e^x$ whenever $k > 1/0.368 \approx 2.717$.

(c) $x \approx 1.155$

13. (a) The function $\ln x - x^{0.2}$ is negative at $x = 1$ and positive at $x = 4$, so it must be zero in between (IVT).

(b) $x = 3.654$

14. (a)

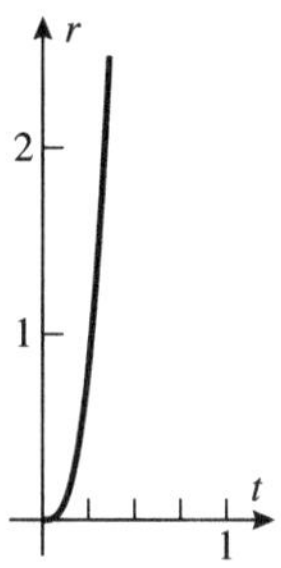

(b) $r = 1$ when $t \approx 0.673080$ s.

(c) $dr/dt = 4.48$ m/s.

15. (a) $y = x^3 + 1$ so $y' = 3x^2$.

(b) $y' = \dfrac{abe^{-x}}{(1 + be^{-x})^2}$

(c) $y = \dfrac{1}{2}\ln x + \dfrac{1}{3}\ln(x+1) - \ln\sin x + \ln\cos x$, so

$$y' = \frac{1}{2x} + \frac{1}{3(x+1)} - \frac{\cos x}{\sin x} - \frac{\sin x}{\cos x} = \frac{5x+3}{6x(x+1)} - \cot x - \tan x.$$

(d) $\ln y = \dfrac{\ln(1+x)}{x}$, $\dfrac{y'}{y} = \dfrac{x/(1+x) - \ln(1+x)}{x^2} = \dfrac{1}{x(1+x)} - \dfrac{\ln(1+x)}{x^2}$,

$$\frac{dy}{dx} = \frac{1}{x}(1+x)^{(1/x)-1} - \frac{(1+x)^{(1/x)}}{x^2}\ln(1+x)$$

(e) $\ln y = e^x \ln x$, $\dfrac{y'}{y} = e^x\left(\dfrac{1}{x} + \ln x\right)$, $\dfrac{dy}{dx} = x^{e^x} e^x\left(\dfrac{1}{x} + \ln x\right) = e^x\left[x^{e^x - 1} + x^{e^x}\ln x\right]$

(f) $y = \sinh^{-1}\dfrac{1}{x^2}$, $\dfrac{dy}{dx} = \dfrac{1}{\sqrt{1 + (1/x^2)^2}}\dfrac{-2}{x^3} = -\dfrac{2}{x\sqrt{x^4+1}}$

16. $y' = ae^{ax}\sin bx + be^{ax}\cos bx$ and $y'' = (a^2 - b^2)e^{ax}\sin bx + 2abe^{ax}\cos bx$, so $y'' - 2ay' + (a^2 + b^2)y$ $= (a^2 - b^2)e^{ax}\sin bx + 2abe^{ax}\cos bx - 2a(ae^{ax}\sin bx + be^{ax}\cos bx) + (a^2 + b^2)e^{ax}\sin bx = 0$.

17. $\sin(\tan^{-1} x) = x/\sqrt{1+x^2}$ and $\cos(\tan^{-1} x) = 1/\sqrt{1+x^2}$, and $y' = \dfrac{1}{1+x^2}$, $y'' = \dfrac{-2x}{(1+x^2)^2}$, hence

$$y'' + 2\sin y\cos^3 y = \frac{-2x}{(1+x^2)^2} + 2\frac{x}{\sqrt{1+x^2}}\frac{1}{(1+x^2)^{3/2}} = 0.$$

18. $y' = a\cosh ax, y'' = a^2\sinh ax = a^2y$

19. Set $y = \log_b x$ and solve $y' = 1$: $y' = \dfrac{1}{x\ln b} = 1$ so $x = \dfrac{1}{\ln b}$. The curves intersect when (x, x) lies on the graph of $y = \log_b x$, so $x = \log_b x$. From Formula (9), Section 7.2, $\log_b x = \dfrac{\ln x}{\ln b}$ from which $\ln x = 1$, $x = e$, $\ln b = 1/e$, $b = e^{1/e} \approx 1.4447$.

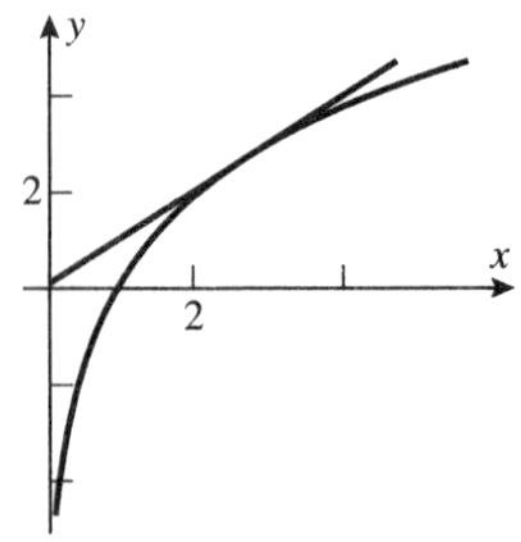

20. **(a)** Find the point of intersection: $f(x) = \sqrt{x} + k = \ln x$. The slopes are equal, so $m_1 = \dfrac{1}{x} = m_2 = \dfrac{1}{2\sqrt{x}}$, $\sqrt{x} = 2$, $x = 4$. Then $\ln 4 = \sqrt{4} + k$, $k = \ln 4 - 2$.

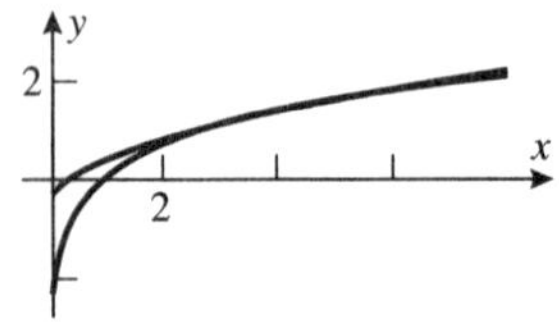

(b) Since the slopes are equal $m_1 = \dfrac{k}{2\sqrt{x}} = m_2 = \dfrac{1}{x}$, so $k\sqrt{x} = 2$. At the point of intersection $k\sqrt{x} = \ln x$, $2 = \ln x$, $x = e^2$, $k = 2/e$.

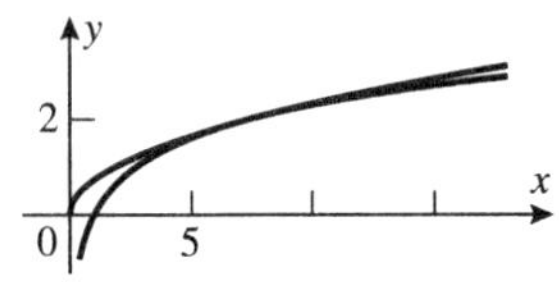

21. $\lim\limits_{x\to 0^+} f(x) = \lim\limits_{x\to+\infty} f(x) = +\infty$ and $f'(x) = \dfrac{e^x(x-2)}{x^3}$, stationary point at $x = 2$. We know $f(x)$ has no maximum and an absolute minimum; by Theorem 4.5.5 $f(x)$ has an absolute minimum at $x = 2$, and $m = e^2/4$.

22. $f'(x) = (1 + \ln x)x^x$, critical point at $x = 1/e$; $\lim\limits_{x\to 0^+} f(x) = \lim\limits_{x\to 0^+} e^{x\ln x} = 1$, $\lim\limits_{x\to+\infty} f(x) = +\infty$; no absolute maximum, absolute minimum $m = e^{-1/e}$ at $x = 1/e$

23. $f_{\text{ave}} = \dfrac{1}{e-1}\displaystyle\int_1^e \frac{1}{x}dx = \frac{1}{e-1}\ln x\Big]_1^e = \frac{1}{e-1}$; $\dfrac{1}{x^*} = \dfrac{1}{e-1}$, $x^* = e - 1$

24. Find t so that $N'(t)$ is maximum. The size of the population is increasing most rapidly when $t = 8.4$ years.

25. $u = \ln x$, $du = (1/x)dx$; $\displaystyle\int_1^2 \frac{1}{u}du = \ln u\Big]_1^2 = \ln 2$

26. $\displaystyle\int_0^1 e^{-x/2}dx = 2(1 - 1/\sqrt{e})$

27. $u = e^{-2x}$, $du = -2e^{-2x}dx$; $-\dfrac{1}{2}\displaystyle\int_1^{1/4}(1 + \cos u)du = \frac{3}{8} + \frac{1}{2}\left(\sin 1 - \sin\frac{1}{4}\right)$

28. Divide $e^x + 3$ into e^{2x} to get $\dfrac{e^{2x}}{e^x + 3} = e^x - \dfrac{3e^x}{e^x + 3}$ so

$$\int \frac{e^{2x}}{e^x + 3}dx = \int e^x dx - 3\int \frac{e^x}{e^x + 3}dx = e^x - 3\ln(e^x + 3) + C$$

29. Since $y = e^x$ and $y = \ln x$ are inverse functions, their graphs are symmetric with respect to the line $y = x$; consequently the areas A_1 and A_3 are equal (see figure). But $A_1 + A_2 = e$, so

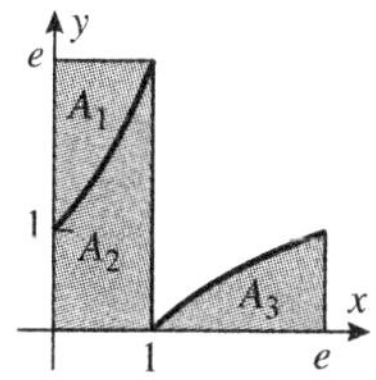

$$\int_1^e \ln x\,dx + \int_0^1 e^x\,dx = A_2 + A_3 = A_2 + A_1 = e$$

30. $\displaystyle\sum_{k=1}^{n} \frac{e^{k/n}}{n} = \sum_{k=1}^{n} f(x_k^*)\Delta x$ where $f(x) = e^x, x_k^* = k/n$, and $\Delta x = 1/n$ for $0 \le x \le 1$. Thus

$$\lim_{n\to+\infty} \sum_{k=1}^{n} \frac{e^{k/n}}{n} = \lim_{n\to+\infty} \sum_{k=1}^{n} f(x_k^*)\Delta x = \int_0^1 e^x\,dx = e - 1.$$

31. Since $f(x) = \dfrac{1}{x}$ is positive and increasing on the interval $[1, 2]$, the left endpoint approximation overestimates the integral of $\dfrac{1}{x}$ and the right endpoint approximation underestimates it.

(a) For $n = 5$ this becomes

$$0.2\left[\frac{1}{1.2} + \frac{1}{1.4} + \frac{1}{1.6} + \frac{1}{1.8} + \frac{1}{2.0}\right] < \int_1^2 \frac{1}{x}dx < 0.2\left[\frac{1}{1.0} + \frac{1}{1.2} + \frac{1}{1.4} + \frac{1}{1.6} + \frac{1}{1.8}\right]$$

(b) For general n the left endpoint approximation to $\displaystyle\int_1^2 \frac{1}{x}dx = \ln 2$ is

$\displaystyle\frac{1}{n}\sum_{k=1}^{n} \frac{1}{1 + (k-1)/n} = \sum_{k=1}^{n} \frac{1}{n + k - 1} = \sum_{k=0}^{n-1} \frac{1}{n + k}$ and the right endpoint approximation is $\displaystyle\sum_{k=1}^{n} \frac{1}{n + k}$. This yields $\displaystyle\sum_{k=1}^{n} \frac{1}{n + k} < \int_1^2 \frac{1}{x}dx < \sum_{k=0}^{n-1} \frac{1}{n + k}$ which is the desired inequality.

(c) By telescoping, the difference is $\dfrac{1}{n} - \dfrac{1}{2n} = \dfrac{1}{2n}$ so $\dfrac{1}{2n} \le 0.1$, $n \ge 5$

(d) $n \ge 1000$

32. **(a)** $x_k^* = 0, 1, 2, 3, 4$

$$\sum_{k=1}^{4} f(x_k^*)\Delta x = \left(e^0 + e^1 + e^2 + e^3 + e^4\right)(1) = (1 - e^5)/(1 - e) = 85.791$$

(b) $x_k^* = 1, 2, 3, 4, 5$

$$\sum_{k=1}^{4} f(x_k^*)\Delta x = \left(e^1 + e^2 + e^3 + e^4 + e^5\right)(1) = e(1 - e^5)/(1 - e) = 233.204$$

(c) $x_k^* = 1/2, 3/2, 5/2, 7/2, 9/2$

$$\sum_{k=1}^{4} f(x_k^*)\Delta x = \left(e^{1/2} + e^{3/2} + e^{5/2} + e^{7/2} + e^{9/2}\right)(1) = e^{1/2}(1 - e^5)/(1 - e) = 141.446$$

33. $0.351220577, 0.420535296, 0.386502483$

34. $1.63379940, 1.805627583, 1.717566087$

35. $f(x) = e^x, [a, b] = [0, 1], \Delta x = \frac{1}{n}; \lim_{n\to+\infty} \sum_{k=1}^{n} f(x_k^*) = \int_0^1 e^x\,dx = e - 1$

36. In the case $+\infty - (-\infty)$ the limit is $+\infty$; in the case $-\infty - (+\infty)$ the limit is $-\infty$, because large positive (negative) quantities are added to large positive (negative) quantities. The cases $+\infty - (+\infty)$ and $-\infty - (-\infty)$ are indeterminate; large numbers of opposite sign are subtracted, and more information about the sizes is needed.

37. **(a)** when the limit takes the form $0/0$ or ∞/∞

(b) Not necessarily; only if $\lim_{x\to a} f(x) = 0$. Consider $g(x) = x$; $\lim_{x\to 0} g(x) = 0$. For $f(x)$ choose $\cos x$, x^2, and $|x|^{1/2}$. Then: $\lim_{x\to 0} \frac{\cos x}{x}$ does not exist, $\lim_{x\to 0} \frac{x^2}{x} = 0$, and $\lim_{x\to 0} \frac{|x|^{1/2}}{x^2} = +\infty$.

38. **(a)** $\lim_{x\to+\infty} (e^x - x^2) = \lim_{x\to+\infty} x^2(e^x/x^2 - 1)$, but $\lim_{x\to+\infty} \frac{e^x}{x^2} = \lim_{x\to+\infty} \frac{e^x}{2x} = \lim_{x\to+\infty} \frac{e^x}{2} = +\infty$

so $\lim_{x\to+\infty} (e^x/x^2 - 1) = +\infty$ and thus $\lim_{x\to+\infty} x^2(e^x/x^2 - 1) = +\infty$

(b) $\lim_{x\to 1} \frac{\ln x}{x^4 - 1} = \lim_{x\to 1} \frac{1/x}{4x^3} = \frac{1}{4}; \lim_{x\to 1} \sqrt{\frac{\ln x}{x^4 - 1}} = \sqrt{\lim_{x\to 1} \frac{\ln x}{x^4 - 1}} = \frac{1}{2}$

(c) $\lim_{x\to 0} a^x \ln a = \ln a$

CHAPTER 8
Principles of Integral Evaluation

EXERCISE SET 8.1

1. $u = 3 - 2x, du = -2dx, \quad -\frac{1}{2}\int u^3\,du = -\frac{1}{8}u^4 + C = -\frac{1}{8}(3-2x)^4 + C$

2. $u = 4 + 9x, du = 9dx, \quad \frac{1}{9}\int u^{1/2}\,du = \frac{2}{3\cdot 9}u^{3/2} + C = \frac{2}{27}(4+9x)^{3/2} + C$

3. $u = x^2, du = 2xdx, \quad \frac{1}{2}\int \sec^2 u\,du = \frac{1}{2}\tan u + C = \frac{1}{2}\tan(x^2) + C$

4. $u = x^2, du = 2xdx, \quad 2\int \tan u\,du = -2\ln|\cos u| + C = -2\ln|\cos(x^2)| + C$

5. $u = 2 + \cos 3x, du = -3\sin 3xdx, \quad -\frac{1}{3}\int \frac{du}{u} = -\frac{1}{3}\ln|u| + C = -\frac{1}{3}\ln(2+\cos 3x) + C$

6. $u = \frac{3x}{2}, du = \frac{3}{2}dx, \quad \frac{2}{3}\int \frac{du}{4+4u^2} = \frac{1}{6}\int \frac{du}{1+u^2} = \frac{1}{6}\tan^{-1}u + C = \frac{1}{6}\tan^{-1}(3x/2) + C$

7. $u = e^x, du = e^x dx, \quad \int \sinh u\,du = \cosh u + C = \cosh e^x + C$

8. $u = \ln x, du = \frac{1}{x}dx, \quad \int \sec u\tan u\,du = \sec u + C = \sec(\ln x) + C$

9. $u = \cot x, du = -\csc^2 xdx, \quad -\int e^u\,du = -e^u + C = -e^{\cot x} + C$

10. $u = x^2, du = 2xdx, \quad \frac{1}{2}\int \frac{du}{\sqrt{1-u^2}} = \frac{1}{2}\sin^{-1}u + C = \frac{1}{2}\sin^{-1}(x^2) + C$

11. $u = \cos 7x, du = -7\sin 7xdx, \quad -\frac{1}{7}\int u^5\,du = -\frac{1}{42}u^6 + C = -\frac{1}{42}\cos^6 7x + C$

12. $u = \sin x, du = \cos x\,dx, \quad \int \frac{du}{u\sqrt{u^2+1}} = -\ln\left|\frac{1+\sqrt{1+u^2}}{u}\right| + C = -\ln\left|\frac{1+\sqrt{1+\sin^2 x}}{\sin x}\right| + C$

13. $u = e^x, du = e^x dx, \quad \int \frac{du}{\sqrt{4+u^2}} = \ln\left(u + \sqrt{u^2+4}\right) + C = \ln\left(e^x + \sqrt{e^{2x}+4}\right) + C$

14. $u = \tan^{-1}x, du = \frac{1}{1+x^2}dx, \quad \int e^u\,du = e^u + C = e^{\tan^{-1}x} + C$

15. $u = \sqrt{x-2}, du = \frac{1}{2\sqrt{x-2}}dx, \quad 2\int e^u\,du = 2e^u + C = 2e^{\sqrt{x-2}} + C$

16. $u = 3x^2 + 2x, du = (6x+2)dx, \quad \frac{1}{2}\int \cot u\,du = \frac{1}{2}\ln|\sin u| + C = \frac{1}{2}\ln\sin|3x^2+2x| + C$

17. $u = \sqrt{x}, du = \frac{1}{2\sqrt{x}}dx, \quad \int 2\cosh u\,du = 2\sinh u + C = 2\sinh\sqrt{x} + C$

18. $u = \ln x, du = \frac{dx}{x}, \quad \int \frac{du}{u} = \ln|u| + C = \ln|\ln x| + C$

19. $u = \sqrt{x}, du = \frac{1}{2\sqrt{x}}\,dx, \quad \int \frac{2\,du}{3^u} = 2\int e^{-u\ln 3}\,du = -\frac{2}{\ln 3}e^{-u\ln 3} + C = -\frac{2}{\ln 3}3^{-\sqrt{x}} + C$

20. $u = \sin\theta, du = \cos\theta d\theta, \quad \int \sec u \tan u\,du = \sec u + C = \sec(\sin\theta) + C$

21. $u = \frac{2}{x}, du = -\frac{2}{x^2}\,dx, \quad -\frac{1}{2}\int \operatorname{csch}^2 u\,du = \frac{1}{2}\coth u + C = \frac{1}{2}\coth\frac{2}{x} + C$

22. $\int \frac{dx}{\sqrt{x^2-3}} = \ln\left|x + \sqrt{x^2-3}\right| + C$

23. $u = e^{-x}, du = -e^{-x}dx, \quad -\int \frac{du}{4-u^2} = -\frac{1}{4}\ln\left|\frac{2+u}{2-u}\right| + C = -\frac{1}{4}\ln\left|\frac{2+e^{-x}}{2-e^{-x}}\right| + C$

24. $u = \ln x, du = \frac{1}{x}dx, \quad \int \cos u\,du = \sin u + C = \sin(\ln x) + C$

25. $u = e^x, du = e^x dx, \quad \int \frac{e^x\,dx}{\sqrt{1-e^{2x}}} = \int \frac{du}{\sqrt{1-u^2}} = \sin^{-1} u + C = \sin^{-1} e^x + C$

26. $u = x^{-1/2}, du = -\frac{1}{2x^{3/2}}\,dx, \quad -\int 2\sinh u\,du = -2\cosh u + C = -2\cosh(x^{-1/2}) + C$

27. $u = x^2, du = 2xdx, \quad \frac{1}{2}\int \frac{du}{\sec u} = \frac{1}{2}\int \cos u\,du = \frac{1}{2}\sin u + C = \frac{1}{2}\sin(x^2) + C$

28. $2u = e^x, 2du = e^x dx, \quad \int \frac{2du}{\sqrt{4-4u^2}} = \sin^{-1} u + C = \sin^{-1}(e^x/2) + C$

29. $4^{-x^2} = e^{-x^2\ln 4}, u = -x^2\ln 4, du = -2x\ln 4\,dx = -x\ln 16\,dx,$

$-\frac{1}{\ln 16}\int e^u\,du = -\frac{1}{\ln 16}e^u + C = -\frac{1}{\ln 16}e^{-x^2\ln 4} + C = -\frac{1}{\ln 16}4^{-x^2} + C$

30. $2^{\pi x} = e^{\pi x\ln 2}, \quad \int 2^{\pi x}\,dx = \frac{1}{\pi\ln 2}e^{\pi x\ln 2} + C = \frac{1}{\pi\ln 2}2^{\pi x} + C$

EXERCISE SET 8.2

1. $u = x,\ dv = e^{-x}dx,\ du = dx,\ v = -e^{-x};\ \int xe^{-x}dx = -xe^{-x} + \int e^{-x}dx = -xe^{-x} - e^{-x} + C$

2. $u = x,\ dv = e^{3x}dx,\ du = dx,\ v = \frac{1}{3}e^{3x};\ \int xe^{3x}dx = \frac{1}{3}xe^{3x} - \frac{1}{3}\int e^{3x}dx = \frac{1}{3}xe^{3x} - \frac{1}{9}e^{3x} + C$

3. $u = x^2,\ dv = e^x dx,\ du = 2x\,dx,\ v = e^x;\ \int x^2e^x dx = x^2e^x - 2\int xe^x dx.$

For $\int xe^x dx$ use $u = x,\ dv = e^x dx,\ du = dx,\ v = e^x$ to get

$\int xe^x dx = xe^x - e^x + C_1$ so $\int x^2e^x dx = x^2e^x - 2xe^x + 2e^x + C$

4. $u = x^2$, $dv = e^{-2x}dx$, $du = 2x\,dx$, $v = -\dfrac{1}{2}e^{-2x}$; $\displaystyle\int x^2e^{-2x}dx = -\frac{1}{2}x^2e^{-2x} + \int xe^{-2x}dx$

For $\displaystyle\int xe^{-2x}dx$ use $u = x$, $dv = e^{-2x}dx$ to get

$$\int xe^{-2x}dx = -\frac{1}{2}xe^{-2x} + \frac{1}{2}\int e^{-2x}dx = -\frac{1}{2}xe^{-2x} - \frac{1}{4}e^{-2x} + C$$

so $\displaystyle\int x^2e^{-2x}dx = -\frac{1}{2}x^2e^{-2x} - \frac{1}{2}xe^{-2x} - \frac{1}{4}e^{-2x} + C$

5. $u = x$, $dv = \sin 2x\,dx$, $du = dx$, $v = -\dfrac{1}{2}\cos 2x$;

$$\int x\sin 2x\,dx = -\frac{1}{2}x\cos 2x + \frac{1}{2}\int \cos 2x\,dx = -\frac{1}{2}x\cos 2x + \frac{1}{4}\sin 2x + C$$

6. $u = x$, $dv = \cos 3x\,dx$, $du = dx$, $v = \dfrac{1}{3}\sin 3x$;

$$\int x\cos 3x\,dx = \frac{1}{3}x\sin 3x - \frac{1}{3}\int \sin 3x\,dx = \frac{1}{3}x\sin 3x + \frac{1}{9}\cos 3x + C$$

7. $u = x^2$, $dv = \cos x\,dx$, $du = 2x\,dx$, $v = \sin x$; $\displaystyle\int x^2\cos x\,dx = x^2\sin x - 2\int x\sin x\,dx$

For $\displaystyle\int x\sin x\,dx$ use $u = x$, $dv = \sin x\,dx$ to get

$\displaystyle\int x\sin x\,dx = -x\cos x + \sin x + C_1$ so $\displaystyle\int x^2\cos x\,dx = x^2\sin x + 2x\cos x - 2\sin x + C$

8. $u = x^2$, $dv = \sin x\,dx$, $du = 2x\,dx$, $v = -\cos x$;

$\displaystyle\int x^2\sin x\,dx = -x^2\cos x + 2\int x\cos x\,dx$; for $\displaystyle\int x\cos x\,dx$ use $u = x$, $dv = \cos x\,dx$ to get

$\displaystyle\int x\cos x\,dx = x\sin x + \cos x + C_1$ so $\displaystyle\int x^2\sin x\,dx = -x^2\cos x + 2x\sin x + 2\cos x + C$

9. $u = \ln x$, $dv = \sqrt{x}\,dx$, $du = \dfrac{1}{x}dx$, $v = \dfrac{2}{3}x^{3/2}$;

$$\int \sqrt{x}\ln x\,dx = \frac{2}{3}x^{3/2}\ln x - \frac{2}{3}\int x^{1/2}dx = \frac{2}{3}x^{3/2}\ln x - \frac{4}{9}x^{3/2} + C$$

10. $u = \ln x$, $dv = x\,dx$, $du = \dfrac{1}{x}dx$, $v = \dfrac{1}{2}x^2$; $\displaystyle\int x\ln x\,dx = \frac{1}{2}x^2\ln x - \frac{1}{2}\int x\,dx = \frac{1}{2}x^2\ln x - \frac{1}{4}x^2 + C$

11. $u = (\ln x)^2$, $dv = dx$, $du = 2\dfrac{\ln x}{x}dx$, $v = x$; $\displaystyle\int (\ln x)^2dx = x(\ln x)^2 - 2\int \ln x\,dx.$

Use $u = \ln x$, $dv = dx$ to get $\displaystyle\int \ln x\,dx = x\ln x - \int dx = x\ln x - x + C_1$ so

$$\int (\ln x)^2dx = x(\ln x)^2 - 2x\ln x + 2x + C$$

12. $u = \ln x$, $dv = \dfrac{1}{\sqrt{x}}dx$, $du = \dfrac{1}{x}dx$, $v = 2\sqrt{x}$; $\displaystyle\int \frac{\ln x}{\sqrt{x}}dx = 2\sqrt{x}\ln x - 2\int \frac{1}{\sqrt{x}}dx = 2\sqrt{x}\ln x - 4\sqrt{x} + C$

13. $u = \ln(2x+3)$, $dv = dx$, $du = \dfrac{2}{2x+3}dx$, $v = x$; $\displaystyle\int \ln(2x+3)dx = x\ln(2x+3) - \int \frac{2x}{2x+3}dx$

but $\displaystyle\int \frac{2x}{2x+3}dx = \int \left(1 - \frac{3}{2x+3}\right)dx = x - \frac{3}{2}\ln(2x+3) + C_1$ so

$$\int \ln(2x+3)dx = x\ln(2x+3) - x + \frac{3}{2}\ln(2x+3) + C$$

14. $u = \ln(x^2+4)$, $dv = dx$, $du = \dfrac{2x}{x^2+4}dx$, $v = x$; $\displaystyle\int \ln(x^2+4)dx = x\ln(x^2+4) - 2\int \frac{x^2}{x^2+4}dx$

but $\displaystyle\int \frac{x^2}{x^2+4}dx = \int \left(1 - \frac{4}{x^2+4}\right)dx = x - 2\tan^{-1}\frac{x}{2} + C_1$ so

$$\int \ln(x^2+4)dx = x\ln(x^2+4) - 2x + 4\tan^{-1}\frac{x}{2} + C$$

15. $u = \sin^{-1}x$, $dv = dx$, $du = 1/\sqrt{1-x^2}dx$, $v = x$;

$$\int \sin^{-1}x\,dx = x\sin^{-1}x - \int x/\sqrt{1-x^2}dx = x\sin^{-1}x + \sqrt{1-x^2} + C$$

16. $u = \cos^{-1}(2x)$, $dv = dx$, $du = -\dfrac{2}{\sqrt{1-4x^2}}dx$, $v = x$;

$$\int \cos^{-1}(2x)dx = x\cos^{-1}(2x) + \int \frac{2x}{\sqrt{1-4x^2}}dx = x\cos^{-1}(2x) - \frac{1}{2}\sqrt{1-4x^2} + C$$

17. $u = \tan^{-1}(2x)$, $dv = dx$, $du = \dfrac{2}{1+4x^2}dx$, $v = x$;

$$\int \tan^{-1}(2x)dx = x\tan^{-1}(2x) - \int \frac{2x}{1+4x^2}dx = x\tan^{-1}(2x) - \frac{1}{4}\ln(1+4x^2) + C$$

18. $u = \tan^{-1}x$, $dv = x\,dx$, $du = \dfrac{1}{1+x^2}dx$, $v = \dfrac{1}{2}x^2$; $\displaystyle\int x\tan^{-1}x\,dx = \frac{1}{2}x^2\tan^{-1}x - \frac{1}{2}\int \frac{x^2}{1+x^2}dx$

but $\displaystyle\int \frac{x^2}{1+x^2}dx = \int \left(1 - \frac{1}{1+x^2}\right)dx = x - \tan^{-1}x + C_1$ so

$$\int x\tan^{-1}x\,dx = \frac{1}{2}x^2\tan^{-1}x - \frac{1}{2}x + \frac{1}{2}\tan^{-1}x + C$$

19. $u = e^x$, $dv = \sin x\,dx$, $du = e^x dx$, $v = -\cos x$; $\displaystyle\int e^x\sin x\,dx = -e^x\cos x + \int e^x\cos x\,dx.$

For $\displaystyle\int e^x\cos x\,dx$ use $u = e^x$, $dv = \cos x\,dx$ to get $\displaystyle\int e^x\cos x = e^x\sin x - \int e^x\sin x\,dx$ so

$$\int e^x\sin x\,dx = -e^x\cos x + e^x\sin x - \int e^x\sin x\,dx,$$

$$2\int e^x\sin x\,dx = e^x(\sin x - \cos x) + C_1,\quad \int e^x\sin x\,dx = \frac{1}{2}e^x(\sin x - \cos x) + C$$

20. $u = e^{2x}$, $dv = \cos 3x\,dx$, $du = 2e^{2x}dx$, $v = \dfrac{1}{3}\sin 3x$;

$\displaystyle\int e^{2x}\cos 3x\,dx = \frac{1}{3}e^{2x}\sin 3x - \frac{2}{3}\int e^{2x}\sin 3x\,dx.$ Use $u = e^{2x}$, $dv = \sin 3x\,dx$ to get

$$\int e^{2x}\sin 3x\,dx = -\frac{1}{3}e^{2x}\cos 3x + \frac{2}{3}\int e^{2x}\cos 3x\,dx \text{ so}$$

$$\int e^{2x}\cos 3x\,dx = \frac{1}{3}e^{2x}\sin 3x + \frac{2}{9}e^{2x}\cos 3x - \frac{4}{9}\int e^{2x}\cos 3x\,dx,$$

$$\frac{13}{9}\int e^{2x}\cos 3x\,dx = \frac{1}{9}e^{2x}(3\sin 3x + 2\cos 3x) + C_1, \int e^{2x}\cos 3x\,dx = \frac{1}{13}e^{2x}(3\sin 3x + 2\cos 3x) + C$$

21. $u = e^{ax}$, $dv = \sin bx\,dx$, $du = ae^{ax}dx$, $v = -\frac{1}{b}\cos bx \quad (b \neq 0)$;

$\int e^{ax}\sin bx\,dx = -\frac{1}{b}e^{ax}\cos bx + \frac{a}{b}\int e^{ax}\cos bx\,dx$. Use $u = e^{ax}$, $dv = \cos bx\,dx$ to get

$$\int e^{ax}\cos bx\,dx = \frac{1}{b}e^{ax}\sin bx - \frac{a}{b}\int e^{ax}\sin bx\,dx \text{ so}$$

$$\int e^{ax}\sin bx\,dx = -\frac{1}{b}e^{ax}\cos bx + \frac{a}{b^2}e^{ax}\sin bx - \frac{a^2}{b^2}\int e^{ax}\sin bx\,dx,$$

$$\int e^{ax}\sin bx\,dx = \frac{e^{ax}}{a^2+b^2}(a\sin bx - b\cos bx) + C$$

22. From Exercise 21 with $a = -3, b = 5, x = \theta$, answer $= \frac{e^{-3\theta}}{\sqrt{34}}(-3\sin 5\theta - 5\cos 5\theta) + C$

23. $u = \sin(\ln x)$, $dv = dx$, $du = \frac{\cos(\ln x)}{x}dx$, $v = x$;

$\int \sin(\ln x)dx = x\sin(\ln x) - \int \cos(\ln x)dx$. Use $u = \cos(\ln x)$, $dv = dx$ to get

$$\int \cos(\ln x)dx = x\cos(\ln x) + \int \sin(\ln x)dx \text{ so}$$

$$\int \sin(\ln x)dx = x\sin(\ln x) - x\cos(\ln x) - \int \sin(\ln x)dx,$$

$$\int \sin(\ln x)dx = \frac{1}{2}x[\sin(\ln x) - \cos(\ln x)] + C$$

24. $u = \cos(\ln x)$, $dv = dx$, $du = -\frac{1}{x}\sin(\ln x)dx$, $v = x$;

$\int \cos(\ln x)dx = x\cos(\ln x) + \int \sin(\ln x)dx$. Use $u = \sin(\ln x)$, $dv = dx$ to get

$$\int \sin(\ln x)dx = x\sin(\ln x) - \int \cos(\ln x)dx \text{ so}$$

$$\int \cos(\ln x)dx = x\cos(\ln x) + x\sin(\ln x) - \int \cos(\ln x)dx,$$

$$\int \cos(\ln x)dx = \frac{1}{2}x[\cos(\ln x) + \sin(\ln x)] + C$$

25. $u = x$, $dv = \sec^2 x\,dx$, $du = dx$, $v = \tan x$;

$$\int x\sec^2 x\,dx = x\tan x - \int \tan x\,dx = x\tan x - \int \frac{\sin x}{\cos x}dx = x\tan x + \ln|\cos x| + C$$

26. $u = x$, $dv = \tan^2 x\,dx = (\sec^2 x - 1)dx$, $du = dx$, $v = \tan x - x$;

$$\int x\tan^2 x\,dx = x\tan x - x^2 - \int(\tan x - x)dx$$

$$= x\tan x - x^2 + \ln|\cos x| + \frac{1}{2}x^2 + C = x\tan x - \frac{1}{2}x^2 + \ln|\cos x| + C$$

27. $u = x^2$, $dv = xe^{x^2}dx$, $du = 2x\,dx$, $v = \frac{1}{2}e^{x^2}$;

$$\int x^3e^{x^2}dx = \frac{1}{2}x^2e^{x^2} - \int xe^{x^2}dx = \frac{1}{2}x^2e^{x^2} - \frac{1}{2}e^{x^2} + C$$

28. $u = xe^x$, $dv = \dfrac{1}{(x+1)^2}dx$, $du = (x+1)e^x\,dx$, $v = -\dfrac{1}{x+1}$;

$$\int \frac{xe^x}{(x+1)^2}dx = -\frac{xe^x}{x+1} + \int e^x dx = -\frac{xe^x}{x+1} + e^x + C = \frac{e^x}{x+1} + C$$

29. $u = x$, $dv = e^{-5x}dx$, $du = dx$, $v = -\frac{1}{5}e^{-5x}$;

$$\int_0^1 xe^{-5x}dx = -\frac{1}{5}xe^{-5x}\Big]_0^1 + \frac{1}{5}\int_0^1 e^{-5x}dx$$

$$= -\frac{1}{5}e^{-5} - \frac{1}{25}e^{-5x}\Big]_0^1 = -\frac{1}{5}e^{-5} - \frac{1}{25}(e^{-5} - 1) = (1 - 6e^{-5})/25$$

30. $u = x$, $dv = e^{2x}dx$, $du = dx$, $v = \frac{1}{2}e^{2x}$;

$$\int_0^2 xe^{2x}dx = \frac{1}{2}xe^{2x}\Big]_0^2 - \frac{1}{2}\int_0^2 e^{2x}dx = e^4 - \frac{1}{4}e^{2x}\Big]_0^2 = e^4 - \frac{1}{4}(e^4 - 1) = (3e^4 + 1)/4$$

31. $u = \ln x$, $dv = x^2dx$, $du = \frac{1}{x}dx$, $v = \frac{1}{3}x^3$;

$$\int_1^e x^2\ln x\,dx = \frac{1}{3}x^3\ln x\Big]_1^e - \frac{1}{3}\int_1^e x^2dx = \frac{1}{3}e^3 - \frac{1}{9}x^3\Big]_1^e = \frac{1}{3}e^3 - \frac{1}{9}(e^3 - 1) = (2e^3 + 1)/9$$

32. $u = \ln x$, $dv = \frac{1}{x^2}dx$, $du = \frac{1}{x}dx$, $v = -\frac{1}{x}$;

$$\int_{\sqrt{e}}^e \frac{\ln x}{x^2}dx = -\frac{1}{x}\ln x\Big]_{\sqrt{e}}^e + \int_{\sqrt{e}}^e \frac{1}{x^2}dx$$

$$= -\frac{1}{e} + \frac{1}{\sqrt{e}}\ln\sqrt{e} - \frac{1}{x}\Big]_{\sqrt{e}}^e = -\frac{1}{e} + \frac{1}{2\sqrt{e}} - \frac{1}{e} + \frac{1}{\sqrt{e}} = \frac{3\sqrt{e} - 4}{2e}$$

33. $u = \ln(x+3)$, $dv = dx$, $du = \dfrac{1}{x+3}dx$, $v = x$;

$$\int_{-2}^{2} \ln(x+3)dx = x\ln(x+3)\Big]_{-2}^{2} - \int_{-2}^{2} \frac{x}{x+3}dx = 2\ln 5 + 2\ln 1 - \int_{-2}^{2}\left[1 - \frac{3}{x+3}\right]dx$$

$$= 2\ln 5 - [x - 3\ln(x+3)]\Big]_{-2}^{2} = 2\ln 5 - (2 - 3\ln 5) + (-2 - 3\ln 1) = 5\ln 5 - 4$$

34. $u = \sin^{-1} x$, $dv = dx$, $du = \dfrac{1}{\sqrt{1-x^2}}dx$, $v = x$;

$$\int_{0}^{1/2} \sin^{-1} x\,dx = x\sin^{-1} x\Big]_{0}^{1/2} - \int_{0}^{1/2} \frac{x}{\sqrt{1-x^2}}dx = \frac{1}{2}\sin^{-1}\frac{1}{2} + \sqrt{1-x^2}\Big]_{0}^{1/2}$$

$$= \frac{1}{2}\left(\frac{\pi}{6}\right) + \sqrt{\frac{3}{4}} - 1 = \frac{\pi}{12} + \frac{\sqrt{3}}{2} - 1$$

35. $u = \sec^{-1}\sqrt{\theta}$, $dv = d\theta$, $du = \dfrac{1}{2\theta\sqrt{\theta-1}}d\theta$, $v = \theta$;

$$\int_{2}^{4} \sec^{-1}\sqrt{\theta}d\theta = \theta\sec^{-1}\sqrt{\theta}\Big]_{2}^{4} - \frac{1}{2}\int_{2}^{4} \frac{1}{\sqrt{\theta-1}}d\theta = 4\sec^{-1} 2 - 2\sec^{-1}\sqrt{2} - \sqrt{\theta-1}\Big]_{2}^{4}$$

$$= 4\left(\frac{\pi}{3}\right) - 2\left(\frac{\pi}{4}\right) - \sqrt{3} + 1 = \frac{5\pi}{6} - \sqrt{3} + 1$$

36. $u = \sec^{-1} x$, $dv = x\,dx$, $du = \dfrac{1}{x\sqrt{x^2-1}}dx$, $v = \dfrac{1}{2}x^2$;

$$\int_{1}^{2} x\sec^{-1} x\,dx = \frac{1}{2}x^2\sec^{-1} x\Big]_{1}^{2} - \frac{1}{2}\int_{1}^{2} \frac{x}{\sqrt{x^2-1}}dx$$

$$= \frac{1}{2}[(4)(\pi/3) - (1)(0)] - \frac{1}{2}\sqrt{x^2-1}\Big]_{1}^{2} = 2\pi/3 - \sqrt{3}/2$$

37. $u = x$, $dv = \sin 4x\,dx$, $du = dx$, $v = -\dfrac{1}{4}\cos 4x$;

$$\int_{0}^{\pi/2} x\sin 4x\,dx = -\frac{1}{4}x\cos 4x\Big]_{0}^{\pi/2} + \frac{1}{4}\int_{0}^{\pi/2} \cos 4x\,dx = -\pi/8 + \frac{1}{16}\sin 4x\Big]_{0}^{\pi/2} = -\pi/8$$

38. $$\int_{0}^{\pi} (x + x\cos x)dx = \frac{1}{2}x^2\Big]_{0}^{\pi} + \int_{0}^{\pi} x\cos x\,dx = \frac{\pi^2}{2} + \int_{0}^{\pi} x\cos x\,dx;$$

$u = x$, $dv = \cos x\,dx$, $du = dx$, $v = \sin x$

$$\int_{0}^{\pi} x\cos x\,dx = x\sin x\Big]_{0}^{\pi} - \int_{0}^{\pi} \sin x\,dx = \cos x\Big]_{0}^{\pi} = -2 \text{ so } \int_{0}^{\pi} (x + x\cos x)dx = \pi^2/2 - 2$$

39. $u = \tan^{-1}\sqrt{x}$, $dv = \sqrt{x}dx$, $du = \dfrac{1}{2\sqrt{x}(1+x)}dx$, $v = \dfrac{2}{3}x^{3/2}$;

$$\int_1^3 \sqrt{x}\tan^{-1}\sqrt{x}dx = \frac{2}{3}x^{3/2}\tan^{-1}\sqrt{x}\bigg]_1^3 - \frac{1}{3}\int_1^3 \frac{x}{1+x}dx$$

$$= \frac{2}{3}x^{3/2}\tan^{-1}\sqrt{x}\bigg]_1^3 - \frac{1}{3}\int_1^3 \left[1 - \frac{1}{1+x}\right] dx$$

$$= \left[\frac{2}{3}x^{3/2}\tan^{-1}\sqrt{x} - \frac{1}{3}x + \frac{1}{3}\ln|1+x|\right]_1^3 = (2\sqrt{3}\pi - \pi/2 - 2 + \ln 2)/3$$

40. $u = \ln(x^2+1)$, $dv = dx$, $du = \dfrac{2x}{x^2+1}dx$, $v = x$;

$$\int_0^2 \ln(x^2+1)dx = x\ln(x^2+1)\bigg]_0^2 - \int_0^2 \frac{2x^2}{x^2+1}dx = 2\ln 5 - 2\int_0^2 \left(1 - \frac{1}{x^2+1}\right) dx$$

$$= 2\ln 5 - 2(x - \tan^{-1}x)\bigg]_0^2 = 2\ln 5 - 4 + 2\tan^{-1}2$$

41. $t = \sqrt{x}$, $t^2 = x$, $dx = 2t\,dt$

(a) $\displaystyle\int e^{\sqrt{x}}dx = 2\int te^t\,dt$; $u = t, dv = e^t dt, du = dt, v = e^t$,

$$\int e^{\sqrt{x}}dx = 2te^t - 2\int e^t\,dt = 2(t-1)e^t + C = 2(\sqrt{x}-1)e^{\sqrt{x}} + C$$

(b) $\displaystyle\int \cos\sqrt{x}\,dx = 2\int t\cos t\,dt$; $u = t, dv = \cos t dt, du = dt, v = \sin t$,

$$\int \cos\sqrt{x}\,dx = 2t\sin t - 2\int \sin t dt = 2t\sin t + 2\cos t + C = 2\sqrt{x}\sin\sqrt{x} + 2\cos\sqrt{x} + C$$

42. Let $q_1(x), q_2(x), q_3(x)$ denote successive antiderivatives of $q(x)$, so that $q_3'(x) = q_2(x), q_2'(x) = q_1(x), q_1'(x) = q(x)$. Let $p(x) = ax^2 + bx + c$.

Repeated Differentiation		Repeated Antidifferentiation
ax^2+bx+c		$q(x)$
	$+$	
$2ax+b$		$q_1(x)$
	$-$	
$2a$		$q_2(x)$
	$+$	
0		$q_3(x)$

Then $\displaystyle\int p(x)q(x)\,dx = (ax^2+bx+c)q_1(x) - (2ax+b)q_2(x) + 2aq_3(x) + C$. Check:

$$\frac{d}{dx}[(ax^2+bx+c)q_1(x) - (2ax+b)q_2(x) + 2aq_3(x)]$$

$$= (2ax+b)q_1(x) + (ax^2+bx+c)q(x) - 2aq_2(x) - (2ax+b)q_1(x) + 2aq_2(x) = p(x)q(x)$$

43.

Repeated Differentiation		Repeated Antidifferentiation
$3x^2 - x + 2$		e^{-x}
	$+$	
$6x - 1$		$-e^{-x}$
	$-$	
6		e^{-x}
	$+$	
0		$-e^{-x}$

$$\int (3x^2 - x + 2)e^{-x} = -(3x^2 - x + 2)e^{-x} - (6x - 1)e^{-x} - 6e^{-x} + C = -e^{-x}[3x^2 + 5x + 7] + C$$

44.

Repeated Differentiation		Repeated Antidifferentiation
$x^2 + x + 1$		$\sin x$
	$+$	
$2x + 1$		$-\cos x$
	$-$	
2		$-\sin x$
	$+$	
0		$\cos x$

$$\begin{aligned}\int (x^2 + x + 1)\sin x\, dx &= -(x^2 + x + 1)\cos x + (2x + 1)\sin x + 2\cos x + C \\ &= -(x^2 + x - 1)\cos x + (2x + 1)\sin x + C\end{aligned}$$

45.

Repeated Differentiation		Repeated Antidifferentiation
$8x^4$		$\cos 2x$
	$+$	
$32x^3$		$\frac{1}{2}\sin 2x$
	$-$	
$96x^2$		$-\frac{1}{4}\cos 2x$
	$+$	
$192x$		$-\frac{1}{8}\sin 2x$
	$-$	
192		$\frac{1}{16}\cos 2x$
	$+$	
0		$\frac{1}{32}\sin 2x$

$$\int 8x^4 \cos 2x\, dx = (4x^4 - 12x^2 + 6)\sin 2x + (8x^3 - 12x)\cos 2x + C$$

46.

Repeated Differentiation		Repeated Antidifferentiation
x^3		$\sqrt{2x+1}$
	$+$	
$3x^2$		$\frac{1}{3}(2x+1)^{3/2}$
	$-$	
$6x$		$\frac{1}{15}(2x+1)^{5/2}$
	$+$	
6		$\frac{1}{105}(2x+1)^{7/2}$
	$-$	
0		$\frac{1}{945}(2x+1)^{9/2}$

$$\int x^3\sqrt{2x+1}\,dx = \frac{1}{3}x^3(2x+1)^{3/2} - \frac{1}{5}x^2(2x+1)^{5/2} + \frac{2}{35}x(2x+1)^{7/2} - \frac{2}{315}(2x+1)^{9/2} + C$$

47. **(a)** $A = \int_1^e \ln x\,dx = (x\ln x - x)\Big]_1^e = 1$

(b) $V = \pi\int_1^e (\ln x)^2 dx = \pi\Big[(x(\ln x)^2 - 2x\ln x + 2x)\Big]_1^e = \pi(e-2)$

48. $A = \int_0^{\pi/2}(x - x\sin x)dx = \frac{1}{2}x^2\Big]_0^{\pi/2} - \int_0^{\pi/2} x\sin x\,dx = \frac{\pi^2}{8} - (-x\cos x + \sin x)\Big]_0^{\pi/2} = \pi^2/8 - 1$

49. $V = 2\pi\int_0^{\pi} x\sin x\,dx = 2\pi(-x\cos x + \sin x)\Big]_0^{\pi} = 2\pi^2$

50. $V = 2\pi\int_0^{\pi/2} x\cos x\,dx = 2\pi(\cos x + x\sin x)\Big]_0^{\pi/2} = \pi(\pi - 2)$

51. distance $= \int_0^5 t^2e^{-t}dt; u = t^2, dv = e^{-t}dt, du = 2tdt, v = -e^{-t},$

distance $= -t^2e^{-t}\Big]_0^5 + 2\int_0^5 te^{-t}dt; u = 2t, dv = e^{-t}dt, du = 2dt, v = -e^{-t},$

distance $= -25e^{-5} - 2te^{-t}\Big]_0^5 + 2\int_0^5 e^{-t}dt = -25e^{-5} - 10e^{-5} - 2e^{-t}\Big]_0^5$

$= -25e^{-5} - 10e^{-5} - 2e^{-5} + 2 = -37e^{-5} + 2$

52. $u = 2t, dv = \sin(k\omega t)dt,\ du = 2dt,\ v = -\dfrac{1}{k\omega}\cos(k\omega t)$; the integrand is an even function of t so

$$\int_{-\pi/\omega}^{\pi/\omega} t\sin(k\omega t)\,dt = 2\int_0^{\pi/\omega} t\sin(k\omega t)\,dt = -\frac{2}{k\omega}t\cos(k\omega t)\Big]_0^{\pi/\omega} + 2\int_0^{\pi/\omega}\frac{1}{k\omega}\cos(k\omega t)\,dt$$

$$= \frac{2\pi(-1)^{k+1}}{k\omega^2} + \frac{2}{k^2\omega^2}\sin(k\omega t)\Big]_0^{\pi/\omega} = \frac{2\pi(-1)^{k+1}}{k\omega^2}$$

53. (a) $\displaystyle\int \sin^3 x\,dx = -\frac{1}{3}\sin^2 x\cos x + \frac{2}{3}\int \sin x\,dx = -\frac{1}{3}\sin^2 x\cos x - \frac{2}{3}\cos x + C$

(b) $\displaystyle\int \sin^4 x\,dx = -\frac{1}{4}\sin^3 x\cos x + \frac{3}{4}\int \sin^2 x\,dx,\ \int \sin^2 x\,dx = -\frac{1}{2}\sin x\cos x + \frac{1}{2}x + C_1$ so

$$\int_0^{\pi/4}\sin^4 x\,dx = \left[-\frac{1}{4}\sin^3 x\cos x - \frac{3}{8}\sin x\cos x + \frac{3}{8}x\right]_0^{\pi/4}$$

$$= -\frac{1}{4}(1/\sqrt{2})^3(1/\sqrt{2}) - \frac{3}{8}(1/\sqrt{2})(1/\sqrt{2}) + 3\pi/32 = 3\pi/32 - 1/4$$

54. (a) $\displaystyle\int \cos^5 x\,dx = \frac{1}{5}\cos^4 x\sin x + \frac{4}{5}\int \cos^3 x\,dx = \frac{1}{5}\cos^4 x\sin x + \frac{4}{5}\left[\frac{1}{3}\cos^2 x\sin x + \frac{2}{3}\sin x\right] + C$

$$= \frac{1}{5}\cos^4 x\sin x + \frac{4}{15}\cos^2 x\sin x + \frac{8}{15}\sin x + C$$

(b) $\displaystyle\int \cos^6 x\,dx = \frac{1}{6}\cos^5 x\sin x + \frac{5}{6}\int \cos^4 x\,dx$

$$= \frac{1}{6}\cos^5 x\sin x + \frac{5}{6}\left[\frac{1}{4}\cos^3 x\sin x + \frac{3}{4}\int \cos^2 x\,dx\right]$$

$$= \frac{1}{6}\cos^5 x\sin x + \frac{5}{24}\cos^3 x\sin x + \frac{5}{8}\left[\frac{1}{2}\cos x\sin x + \frac{1}{2}x\right] + C,$$

$$\left[\frac{1}{6}\cos^5 x\sin x + \frac{5}{24}\cos^3 x\sin x + \frac{5}{16}\cos x\sin x + \frac{5}{16}x\right]_0^{\pi/2} = 5\pi/32$$

55. $u = \sin^{n-1}x,\ dv = \sin x\,dx,\ du = (n-1)\sin^{n-2}x\cos x\,dx,\ v = -\cos x;$

$$\int \sin^n x\,dx = -\sin^{n-1}x\cos x + (n-1)\int \sin^{n-2}x\cos^2 x\,dx$$

$$= -\sin^{n-1}x\cos x + (n-1)\int \sin^{n-2}x\,(1-\sin^2 x)dx$$

$$= -\sin^{n-1}x\cos x + (n-1)\int \sin^{n-2}x\,dx - (n-1)\int \sin^n x\,dx,$$

$$n\int \sin^n x\,dx = -\sin^{n-1}x\cos x + (n-1)\int \sin^{n-2}x\,dx,$$

$$\int \sin^n x\,dx = -\frac{1}{n}\sin^{n-1}x\cos x + \frac{n-1}{n}\int \sin^{n-2}x\,dx$$

56. **(a)** $u = \sec^{n-2} x$, $dv = \sec^2 x\,dx$, $du = (n-2)\sec^{n-2} x \tan x\,dx$, $v = \tan x$;

$$\int \sec^n x\,dx = \sec^{n-2} x \tan x - (n-2)\int \sec^{n-2} x \tan^2 x\,dx$$

$$= \sec^{n-2} x \tan x - (n-2)\int \sec^{n-2} x\,(\sec^2 x - 1)dx$$

$$= \sec^{n-2} x \tan x - (n-2)\int \sec^n x\,dx + (n-2)\int \sec^{n-2} x\,dx,$$

$$(n-1)\int \sec^n x\,dx = \sec^{n-2} x \tan x + (n-2)\int \sec^{n-2} x\,dx,$$

$$\int \sec^n x\,dx = \frac{1}{n-1}\sec^{n-2} x \tan x + \frac{n-2}{n-1}\int \sec^{n-2} x\,dx$$

(b) $$\int \tan^n x\,dx = \int \tan^{n-2} x\,(\sec^2 x - 1)\,dx = \int \tan^{n-1} x \sec^2 x\,dx - \int \tan^{n-2} x\,dx$$

$$= \frac{1}{n-1}\tan^{n-1} x - \int \tan^{n-2} x\,dx$$

(c) $u = x^n$, $dv = e^x dx$, $du = nx^{n-1}dx$, $v = e^x$; $\int x^n e^x dx = x^n e^x - n\int x^{n-1}e^x dx$

57. **(a)** $$\int \tan^4 x\,dx = \frac{1}{3}\tan^3 x - \int \tan^2 x\,dx = \frac{1}{3}\tan^3 x - \tan x + \int dx = \frac{1}{3}\tan^3 x - \tan x + x + C$$

(b) $$\int \sec^4 x\,dx = \frac{1}{3}\sec^2 x \tan x + \frac{2}{3}\int \sec^2 x\,dx = \frac{1}{3}\sec^2 x \tan x + \frac{2}{3}\tan x + C$$

(c) $$\int x^3 e^x dx = x^3 e^x - 3\int x^2 e^x dx = x^3 e^x - 3\left[x^2 e^x - 2\int x e^x dx\right]$$

$$= x^3 e^x - 3x^2 e^x + 6\left[x e^x - \int e^x dx\right] = x^3 e^x - 3x^2 e^x + 6x e^x - 6e^x + C$$

58. **(a)** $u = 3x$,

$$\int x^2 e^{3x} dx = \frac{1}{27}\int u^2 e^u du = \frac{1}{27}\left[u^2 e^u - 2\int u e^u du\right] = \frac{1}{27}u^2 e^u - \frac{2}{27}\left[u e^u - \int e^u du\right]$$

$$= \frac{1}{27}u^2 e^u - \frac{2}{27}u e^u + \frac{2}{27}e^u + C = \frac{1}{3}x^2 e^{3x} - \frac{2}{9}x e^{3x} + \frac{2}{27}e^{3x} + C$$

(b) $u = -\sqrt{x}$,

$$\int_0^1 x e^{-\sqrt{x}} dx = 2\int_0^{-1} u^3 e^u du,$$

$$\int u^3 e^u du = u^3 e^u - 3\int u^2 e^u du = u^3 e^u - 3\left[u^2 e^u - 2\int u e^u du\right]$$

$$= u^3 e^u - 3u^2 e^u + 6\left[u e^u - \int e^u du\right] = u^3 e^u - 3u^2 e^u + 6u e^u - 6e^u + C,$$

$$2\int_0^{-1} u^3 e^u du = 2(u^3 - 3u^2 + 6u - 6)e^u\Big]_0^{-1} = 12 - 32e^{-1}$$

59. $u = x$, $dv = f''(x)dx$, $du = dx$, $v = f'(x)$;

$$\int_{-1}^{1} x\,f''(x)dx = xf'(x)\Big]_{-1}^{1} - \int_{-1}^{1} f'(x)dx$$

$$= f'(1) + f'(-1) - f(x)\Big]_{-1}^{1} = f'(1) + f'(-1) - f(1) + f(-1)$$

60. **(a)** $u = f(x), dv = dx, du = f'(x), v = x;$

$$\int_a^b f(x)\,dx = xf(x)\Big]_a^b - \int_a^b xf'(x)\,dx = bf(b) - af(a) - \int_a^b xf'(x)\,dx$$

(b) Substitute $y = f(x), dy = f'(x)\,dx, x = a$ when $y = f(a)$, $x = b$ when $y = f(b)$,

$$\int_a^b xf'(x)\,dx = \int_{f(a)}^{f(b)} x\,dy = \int_{f(a)}^{f(b)} f^{-1}(y)\,dy$$

(c) From $a = f^{-1}(\alpha)$ and $b = f^{-1}(\beta)$ we get $bf(b) - af(a) = \beta f^{-1}(\beta) - \alpha f^{-1}(\alpha)$; then

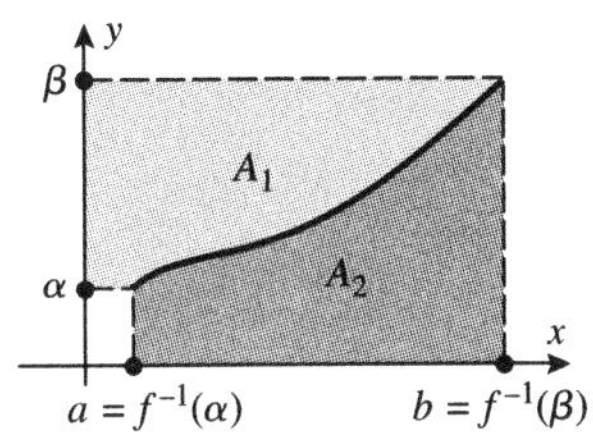

$$\int_\alpha^\beta f^{-1}(x)\,dx = \int_\alpha^\beta f^{-1}(y)\,dy = \int_{f(a)}^{f(b)} f^{-1}(y)\,dy,$$

which, by Part (b), yields

$$\int_\alpha^\beta f^{-1}(x)\,dx = bf(b) - af(a) - \int_a^b f(x)\,dx$$

$$= \beta f^{-1}(\beta) - \alpha f^{-1}(\alpha) - \int_{f^{-1}(\alpha)}^{f^{-1}(\beta)} f(x)\,dx$$

Note from the figure that $A_1 = \int_\alpha^\beta f^{-1}(x)\,dx, A_2 = \int_{f^{-1}(\alpha)}^{f^{-1}(\beta)} f(x)\,dx$, and $A_1 + A_2 = \beta f^{-1}(\beta) - \alpha f^{-1}(\alpha)$, a "picture proof".

61. **(a)** Use Exercise 60(c);

$$\int_0^{1/2} \sin^{-1} x\,dx = \frac{1}{2}\sin^{-1}\left(\frac{1}{2}\right) - 0\cdot\sin^{-1} 0 - \int_{\sin^{-1}(0)}^{\sin^{-1}(1/2)} \sin x\,dx = \frac{1}{2}\sin^{-1}\left(\frac{1}{2}\right) - \int_0^{\pi/6} \sin x\,dx$$

(b) Use Exercise 60(b);

$$\int_e^{e^2} \ln x\,dx = e^2 \ln e^2 - e\ln e - \int_{\ln e}^{\ln e^2} f^{-1}(y)\,dy = 2e^2 - e - \int_1^2 e^y\,dy - 2e^2 - e - \int_1^2 e^x\,dx$$

62. **(a)** $\int u\,dv = uv - \int v\,du = x(\sin x + C_1) + \cos x - C_1 x + C_2 = x\sin x + \cos x + C_2$; the constant C_1 cancels out and hence plays no role in the answer.

(b) $u(v + C_1) - \int (v + C_1)du = uv + C_1 u - \int v\,du - C_1 u = uv - \int v\,du$

63. $u = \ln(x+1), dv = dx, du = \dfrac{dx}{x+1}, v = x+1;$

$$\int \ln(x+1)\,dx = \int u\,dv = uv - \int v\,du = (x+1)\ln(x+1) - \int dx = (x+1)\ln(x+1) - x + C$$

64. $u = \ln(2x+3), dv = dx, du = \dfrac{2dx}{2x+3}, v = x + \dfrac{3}{2};$

$$\int \ln(2x+3)\,dx = \int u\,dv = uv - \int v\,du = (x+\frac{3}{2})\ln(2x+3) - \int dx$$
$$= \frac{1}{2}(2x+3)\ln(2x+3) - x + C$$

65. $u = \tan^{-1} x, dv = x\,dx, du = \dfrac{1}{1+x^2}\,dx, v = \dfrac{1}{2}(x^2+1)$

$$\int x\tan^{-1} x\,dx = \int u\,dv = uv - \int v\,du = \frac{1}{2}(x^2+1)\tan^{-1} x - \frac{1}{2}\int dx$$
$$= \frac{1}{2}(x^2+1)\tan^{-1} x - \frac{1}{2}x + C$$

66. $u = \dfrac{1}{\ln x}, \quad dv = \dfrac{1}{x}\,dx, du = -\dfrac{1}{x(\ln x)^2}\,dx, v = \ln x$

$$\int \frac{1}{x\ln x}\,dx = 1 + \int \frac{1}{x\ln x}\,dx.$$

This seems to imply that $1 = 0$, but recall that both sides represent a function *plus an arbitrary constant*; these two arbitrary constants will take care of the 1.

EXERCISE SET 8.3

1. $u = \cos x, \ -\int u^5 du = -\dfrac{1}{6}\cos^6 x + C$

2. $u = \sin 3x, \ \dfrac{1}{3}\int u^4 du = \dfrac{1}{15}\sin^5 3x + C$

3. $u = \sin ax, \ \dfrac{1}{a}\int u\,du = \dfrac{1}{2a}\sin^2 ax + C, \quad a \neq 0$

4. $\displaystyle\int \cos^2 3x\,dx = \frac{1}{2}\int (1+\cos 6x)dx = \frac{1}{2}x + \frac{1}{12}\sin 6x + C$

5. $\displaystyle\int \sin^2 5\theta\,d\theta = \frac{1}{2}\int (1-\cos 10\theta)d\theta = \frac{1}{2}\theta - \frac{1}{20}\sin 10\theta + C$

6. $\displaystyle\int \cos^3 at\,dt = \int (1-\sin^2 at)\cos at\,dt$

$$= \int \cos at\,dt - \int \sin^2 at\cos at\,dt = \frac{1}{a}\sin at - \frac{1}{3a}\sin^3 at + C \quad (a \neq 0)$$

7. $\displaystyle\int \cos^5 \theta d\theta = \int (1-\sin^2\theta)^2\cos\theta d\theta = \int (1 - 2\sin^2\theta + \sin^4\theta)\cos\theta d\theta$

$$= \sin\theta - \frac{2}{3}\sin^3\theta + \frac{1}{5}\sin^5\theta + C$$

8. $\int \sin^3 x \cos^3 x\, dx = \int \sin^3 x(1-\sin^2 x)\cos x\, dx$

$= \int (\sin^3 x - \sin^5 x)\cos x\, dx = \frac{1}{4}\sin^4 x - \frac{1}{6}\sin^6 x + C$

9. $\int \sin^2 2t \cos^3 2t\, dt = \int \sin^2 2t(1-\sin^2 2t)\cos 2t\, dt = \int (\sin^2 2t - \sin^4 2t)\cos 2t\, dt$

$= \frac{1}{6}\sin^3 2t - \frac{1}{10}\sin^5 2t + C$

10. $\int \sin^3 2x \cos^2 2x\, dx = \int (1-\cos^2 2x)\cos^2 2x \sin 2x\, dx$

$= \int (\cos^2 2x - \cos^4 2x)\sin 2x\, dx = -\frac{1}{6}\cos^3 2x + \frac{1}{10}\cos^5 2x + C$

11. $\int \sin^2 x \cos^2 x\, dx = \frac{1}{4}\int \sin^2 2x\, dx = \frac{1}{8}\int (1-\cos 4x)dx = \frac{1}{8}x - \frac{1}{32}\sin 4x + C$

12. $\int \sin^2 x \cos^4 x\, dx = \frac{1}{8}\int (1-\cos 2x)(1+\cos 2x)^2 dx = \frac{1}{8}\int (1-\cos^2 2x)(1+\cos 2x)dx$

$= \frac{1}{8}\int \sin^2 2x\, dx + \frac{1}{8}\int \sin^2 2x \cos 2x\, dx = \frac{1}{16}\int (1-\cos 4x)dx + \frac{1}{48}\sin^3 2x$

$= \frac{1}{16}x - \frac{1}{64}\sin 4x + \frac{1}{48}\sin^3 2x + C$

13. $\int \sin x \cos 2x\, dx = \frac{1}{2}\int (\sin 3x - \sin x)dx = -\frac{1}{6}\cos 3x + \frac{1}{2}\cos x + C$

14. $\int \sin 3\theta \cos 2\theta d\theta = \frac{1}{2}\int (\sin 5\theta + \sin\theta)d\theta = -\frac{1}{10}\cos 5\theta - \frac{1}{2}\cos\theta + C$

15. $\int \sin x \cos(x/2)dx = \frac{1}{2}\int [\sin(3x/2) + \sin(x/2)]dx = -\frac{1}{3}\cos(3x/2) - \cos(x/2) + C$

16. $u = \cos x,\ -\int u^{1/5}du = -\frac{5}{6}\cos^{6/5} x + C$

17. $\int_0^{\pi/4} \cos^3 x\, dx = \int_0^{\pi/4} (1-\sin^2 x)\cos x\, dx$

$- \left[\sin x - \frac{1}{3}\sin^3 x\right]_0^{\pi/4} = (\sqrt{2}/2) - \frac{1}{3}(\sqrt{2}/2)^3 = 5\sqrt{2}/12$

18. $\int_0^{\pi/2} \sin^2(x/2)\cos^2(x/2)dx = \frac{1}{4}\int_0^{\pi/2} \sin^2 x\, dx = \frac{1}{8}\int_0^{\pi/2} (1-\cos 2x)dx$

$= \frac{1}{8}\left(x - \frac{1}{2}\sin 2x\right)\Big]_0^{\pi/2} = \pi/16$

19. $\int_0^{\pi/3} \sin^4 3x \cos^3 3x\, dx = \int_0^{\pi/3} \sin^4 3x(1-\sin^2 3x)\cos 3x\, dx = \left[\frac{1}{15}\sin^5 3x - \frac{1}{21}\sin^7 3x\right]_0^{\pi/3} = 0$

20. $\int_{-\pi}^{\pi} \cos^2 5\theta\, d\theta = \frac{1}{2}\int_{-\pi}^{\pi}(1+\cos 10\theta)d\theta = \frac{1}{2}\left(\theta + \frac{1}{10}\sin 10\theta\right)\Big]_{-\pi}^{\pi} = \pi$

21. $\int_0^{\pi/6} \sin 2x \cos 4x\, dx = \frac{1}{2}\int_0^{\pi/6}(\sin 6x - \sin 2x)dx = \left[-\frac{1}{12}\cos 6x + \frac{1}{4}\cos 2x\right]_0^{\pi/6}$

$= [(-1/12)(-1) + (1/4)(1/2)] - [-1/12 + 1/4] = 1/24$

22. $\int_0^{2\pi} \sin^2 kx\, dx = \frac{1}{2}\int_0^{2\pi}(1-\cos 2kx)dx = \frac{1}{2}\left(x - \frac{1}{2k}\sin 2kx\right)\Big]_0^{2\pi} = \pi - \frac{1}{4k}\sin 4\pi k \quad (k \neq 0)$

23. $\frac{1}{3}\tan(3x+1) + C$

24. $-\frac{1}{5}\ln|\cos 5x| + C$

25. $u = e^{-2x}, du = -2e^{-2x}\,dx;\ -\frac{1}{2}\int \tan u\, du = \frac{1}{2}\ln|\cos u| + C = \frac{1}{2}\ln|\cos(e^{-2x})| + C$

26. $\frac{1}{3}\ln|\sin 3x| + C$

27. $\frac{1}{2}\ln|\sec 2x + \tan 2x| + C$

28. $u = \sqrt{x}, du = \frac{1}{2\sqrt{x}}dx;\ \int 2\sec u\, du = 2\ln|\sec u + \tan u| + C = 2\ln\left|\sec\sqrt{x} + \tan\sqrt{x}\right| + C$

29. $u = \tan x,\ \int u^2 du = \frac{1}{3}\tan^3 x + C$

30. $\int \tan^5 x(1+\tan^2 x)\sec^2 x\, dx = \int(\tan^5 x + \tan^7 x)\sec^2 x\, dx = \frac{1}{6}\tan^6 x + \frac{1}{8}\tan^8 x + C$

31. $\int \tan^3 4x(1+\tan^2 4x)\sec^2 4x\, dx = \int(\tan^3 4x + \tan^5 4x)\sec^2 4x\, dx = \frac{1}{16}\tan^4 4x + \frac{1}{24}\tan^6 4x + C$

32. $\int \tan^4\theta(1+\tan^2\theta)\sec^2\theta\, d\theta = \frac{1}{5}\tan^5\theta + \frac{1}{7}\tan^7\theta + C$

33. $\int \sec^4 x(\sec^2 x - 1)\sec x \tan x\, dx = \int(\sec^6 x - \sec^4 x)\sec x\tan x\, dx = \frac{1}{7}\sec^7 x - \frac{1}{5}\sec^5 x + C$

34. $\int(\sec^2\theta - 1)^2\sec\theta\tan\theta d\theta = \int(\sec^4\theta - 2\sec^2\theta + 1)\sec\theta\tan\theta d\theta = \frac{1}{5}\sec^5\theta - \frac{2}{3}\sec^3\theta + \sec\theta + C$

35. $\int(\sec^2 x - 1)^2\sec x\, dx = \int(\sec^5 x - 2\sec^3 x + \sec x)dx = \int \sec^5 x\, dx - 2\int\sec^3 x\, dx + \int \sec x\, dx$

$= \frac{1}{4}\sec^3 x\tan x + \frac{3}{4}\int\sec^3 x\, dx - 2\int\sec^3 x\, dx + \ln|\sec x + \tan x|$

$= \frac{1}{4}\sec^3 x\tan x - \frac{5}{4}\left[\frac{1}{2}\sec x\tan x + \frac{1}{2}\ln|\sec x + \tan x|\right] + \ln|\sec x + \tan x| + C$

$= \frac{1}{4}\sec^3 x\tan x - \frac{5}{8}\sec x\tan x + \frac{3}{8}\ln|\sec x + \tan x| + C$

36. $\int [\sec^2(x/2) - 1]\sec^3(x/2)dx = \int [\sec^5(x/2) - \sec^3(x/2)]dx$

$$= 2\left[\int \sec^5 u\,du - \int \sec^3 u\,du\right] \qquad (u = x/2)$$

$$= 2\left[\left(\frac{1}{4}\sec^3 u\tan u + \frac{3}{4}\int \sec^3 u\,du\right) - \int \sec^3 u\,du\right] \qquad \text{(equation (20))}$$

$$= \frac{1}{2}\sec^3 u\tan u - \frac{1}{2}\int \sec^3 u\,du$$

$$= \frac{1}{2}\sec^3 u\tan u - \frac{1}{4}\sec u\tan u - \frac{1}{4}\ln|\sec u + \tan u| + C \qquad \text{(equation (20), (22))}$$

$$= \frac{1}{2}\sec^3\frac{x}{2}\tan\frac{x}{2} - \frac{1}{4}\sec\frac{x}{2}\tan\frac{x}{2} - \frac{1}{4}\ln\left|\sec\frac{x}{2} + \tan\frac{x}{2}\right| + C$$

37. $\int \sec^2 2t(\sec 2t\tan 2t)dt = \frac{1}{6}\sec^3 2t + C$

38. $\int \sec^4 x(\sec x\tan x)dx = \frac{1}{5}\sec^5 x + C$

39. $\int \sec^4 x\,dx = \int (1+\tan^2 x)\sec^2 x\,dx = \int (\sec^2 x + \tan^2 x\sec^2 x)dx = \tan x + \frac{1}{3}\tan^3 x + C$

40. Using equation (20),

$$\int \sec^5 x\,dx = \frac{1}{4}\sec^3 x\tan x + \frac{3}{4}\int \sec^3 x\,dx$$

$$= \frac{1}{4}\sec^3 x\tan x + \frac{3}{8}\sec x\tan x + \frac{3}{8}\ln|\sec x + \tan x| + C$$

41. Use equation (19) to get $\int \tan^4 x\,dx = \frac{1}{3}\tan^3 x - \tan x + x + C$

42. $u = 4x$, use equation (19) to get

$$\frac{1}{4}\int \tan^3 u\,du = \frac{1}{4}\left[\frac{1}{2}\tan^2 u + \ln|\cos u|\right] + C = \frac{1}{8}\tan^2 4x + \frac{1}{4}\ln|\cos 4x| + C$$

43. $\int \sqrt{\tan x}(1+\tan^2 x)\sec^2 x\,dx = \frac{2}{3}\tan^{3/2} x + \frac{2}{7}\tan^{7/2} x + C$

44. $\int \sec^{1/2} x(\sec x\tan x)dx = \frac{2}{3}\sec^{3/2} x + C$

45. $\int_0^{\pi/6} (\sec^2 2x - 1)dx = \left[\frac{1}{2}\tan 2x - x\right]_0^{\pi/6} = \sqrt{3}/2 - \pi/6$

46. $\int_0^{\pi/6} \sec^2\theta(\sec\theta\tan\theta)d\theta = \frac{1}{3}\sec^3\theta\Big]_0^{\pi/6} = (1/3)(2/\sqrt{3})^3 - 1/3 = 8\sqrt{3}/27 - 1/3$

47. $u = x/2$,

$$2\int_0^{\pi/4} \tan^5 u\,du = \left[\frac{1}{2}\tan^4 u - \tan^2 u - 2\ln|\cos u|\right]_0^{\pi/4} = 1/2 - 1 - 2\ln(1/\sqrt{2}) = -1/2 + \ln 2$$

48. $u = \pi x,\ \dfrac{1}{\pi}\displaystyle\int_0^{\pi/4} \sec u \tan u\, du = \dfrac{1}{\pi}\sec u\Big]_0^{\pi/4} = (\sqrt{2}-1)/\pi$

49. $\displaystyle\int (\csc^2 x - 1)\csc^2 x(\csc x \cot x)dx = \int (\csc^4 x - \csc^2 x)(\csc x \cot x)dx = -\frac{1}{5}\csc^5 x + \frac{1}{3}\csc^3 x + C$

50. $\displaystyle\int \frac{\cos^2 3t}{\sin^2 3t}\cdot\frac{1}{\cos 3t}dt = \int \csc 3t \cot 3t\, dt = -\frac{1}{3}\csc 3t + C$

51. $\displaystyle\int (\csc^2 x - 1)\cot x\, dx = \int \csc x(\csc x \cot x)dx - \int \frac{\cos x}{\sin x}dx = -\frac{1}{2}\csc^2 x - \ln|\sin x| + C$

52. $\displaystyle\int (\cot^2 x + 1)\csc^2 x\, dx = -\frac{1}{3}\cot^3 x - \cot x + C$

53. (a) $\displaystyle\int_0^{2\pi} \sin mx \cos nx\, dx = \frac{1}{2}\int_0^{2\pi}[\sin(m+n)x + \sin(m-n)x]dx = \left[-\frac{\cos(m+n)x}{2(m+n)} - \frac{\cos(m-n)x}{2(m-n)}\right]_0^{2\pi}$

but $\cos(m+n)x\Big]_0^{2\pi} = 0, \cos(m-n)x\Big]_0^{2\pi} = 0.$

(b) $\displaystyle\int_0^{2\pi} \cos mx \cos nx\, dx = \frac{1}{2}\int_0^{2\pi}[\cos(m+n)x + \cos(m-n)x]dx;$
since $m \neq n$, evaluate sin at integer multiples of 2π to get 0.

(c) $\displaystyle\int_0^{2\pi} \sin mx \sin nx\, dx = \frac{1}{2}\int_0^{2\pi}[\cos(m-n)x - \cos(m+n)x]\ dx;$
since $m \neq n$, evaluate sin at integer multiples of 2π to get 0.

54. (a) $\displaystyle\int_0^{2\pi} \sin mx \cos mx\, dx = \frac{1}{2}\int_0^{2\pi} \sin 2mx\, dx = -\frac{1}{4m}\cos 2mx\Big|_0^{2\pi} = 0$

(b) $\displaystyle\int_0^{2\pi} \cos^2 mx\, dx = \frac{1}{2}\int_0^{2\pi}(1 + \cos 2mx)\, dx = \frac{1}{2}\left(x + \frac{1}{2m}\sin 2mx\right)\Big|_0^{2\pi} = \pi$

(c) $\displaystyle\int_0^{2\pi} \sin^2 mx\, dx = \frac{1}{2}\int_0^{2\pi}(1 - \cos 2mx)\ dx = \frac{1}{2}\left(x - \frac{1}{2m}\sin 2mx\right)\Big|_0^{2\pi} = \pi$

55. $y' = \tan x,\ 1 + (y')^2 = 1 + \tan^2 x = \sec^2 x,$

$\displaystyle L = \int_0^{\pi/4}\sqrt{\sec^2 x}\, dx = \int_0^{\pi/4}\sec x\, dx = \ln|\sec x + \tan x|\Big]_0^{\pi/4} = \ln(\sqrt{2}+1)$

56. $\displaystyle V = \pi\int_0^{\pi/4}(1 - \tan^2 x)dx = \pi\int_0^{\pi/4}(2 - \sec^2 x)dx = \pi(2x - \tan x)\Big]_0^{\pi/4} = \frac{1}{2}\pi(\pi - 2)$

57. $V = \pi \int_0^{\pi/4} (\cos^2 x - \sin^2 x)dx = \pi \int_0^{\pi/4} \cos 2x\, dx = \frac{1}{2}\pi \sin 2x \Big]_0^{\pi/4} = \pi/2$

58. $V = \pi \int_0^{\pi} \sin^2 x\, dx = \frac{\pi}{2}\int_0^{\pi} (1 - \cos 2x)dx = \frac{\pi}{2}\left(x - \frac{1}{2}\sin 2x\right)\Big]_0^{\pi} = \pi^2/2$

59. With $0 < \alpha < \beta, D = D_\beta - D_\alpha = \frac{L}{2\pi}\int_\alpha^\beta \sec x\, dx = \frac{L}{2\pi}\ln|\sec x + \tan x|\Big]_\alpha^\beta = \frac{L}{2\pi}\ln\left|\frac{\sec\beta + \tan\beta}{\sec\alpha + \tan\alpha}\right|$

60. **(a)** $D = \frac{100}{2\pi}\ln(\sec 25^\circ + \tan 25^\circ) = 7.18$ cm

(b) $D = \frac{100}{2\pi}\ln\left|\frac{\sec 50^\circ + \tan 50^\circ}{\sec 30^\circ + \tan 30^\circ}\right| = 7.34$ cm

61. **(a)**
$$\int \csc x\, dx = \int \sec(\pi/2 - x)dx = -\ln|\sec(\pi/2 - x) + \tan(\pi/2 - x)| + C$$
$$= -\ln|\csc x + \cot x| + C$$

(b) $-\ln|\csc x + \cot x| = \ln\frac{1}{|\csc x + \cot x|} = \ln\frac{|\csc x - \cot x|}{|\csc^2 x - \cot^2 x|} = \ln|\csc x - \cot x|$,

$$-\ln|\csc x + \cot x| = -\ln\left|\frac{1}{\sin x} + \frac{\cos x}{\sin x}\right| = \ln\left|\frac{\sin x}{1 + \cos x}\right|$$
$$= \ln\left|\frac{2\sin(x/2)\cos(x/2)}{2\cos^2(x/2)}\right| = \ln|\tan(x/2)|$$

62.
$$\sin x + \cos x = \sqrt{2}\left[(1/\sqrt{2})\sin x + (1/\sqrt{2})\cos x\right]$$
$$= \sqrt{2}\,[\sin x \cos(\pi/4) + \cos x \sin(\pi/4)] = \sqrt{2}\sin(x + \pi/4),$$
$$\int \frac{dx}{\sin x + \cos x} = \frac{1}{\sqrt{2}}\int \csc(x + \pi/4)dx = -\frac{1}{\sqrt{2}}\ln|\csc(x + \pi/4) + \cot(x + \pi/4)| + C$$
$$= -\frac{1}{\sqrt{2}}\ln\left|\frac{\sqrt{2} + \cos x - \sin x}{\sin x + \cos x}\right| + C$$

63. $a\sin x + b\cos x = \sqrt{a^2 + b^2}\left[\frac{a}{\sqrt{a^2 + b^2}}\sin x + \frac{b}{\sqrt{a^2 + b^2}}\cos x\right] = \sqrt{a^2 + b^2}(\sin x \cos\theta + \cos x \sin\theta)$

where $\cos\theta = a/\sqrt{a^2 + b^2}$ and $\sin\theta = b/\sqrt{a^2 + b^2}$ so $a\sin x + b\cos x = \sqrt{a^2 + b^2}\sin(x + \theta)$

and $\int \frac{dx}{a\sin x + b\cos x} = \frac{1}{\sqrt{a^2 + b^2}}\int \csc(x + \theta)dx = -\frac{1}{\sqrt{a^2 + b^2}}\ln|\csc(x + \theta) + \cot(x + \theta)| + C$

$$= -\frac{1}{\sqrt{a^2 + b^2}}\ln\left|\frac{\sqrt{a^2 + b^2} + a\cos x - b\sin x}{a\sin x + b\cos x}\right| + C$$

64. (a) $\displaystyle\int_0^{\pi/2} \sin^n x\,dx = -\frac{1}{n}\sin^{n-1} x\cos x\Big]_0^{\pi/2} + \frac{n-1}{n}\int_0^{\pi/2}\sin^{n-2} x\,dx = \frac{n-1}{n}\int_0^{\pi/2}\sin^{n-2} x\,dx$

(b) By repeated application of the formula in Part (a)

$$\int_0^{\pi/2}\sin^n x\,dx = \left(\frac{n-1}{n}\right)\left(\frac{n-3}{n-2}\right)\int_0^{\pi/2}\sin^{n-4} x\,dx$$

$$= \begin{cases} \left(\frac{n-1}{n}\right)\left(\frac{n-3}{n-2}\right)\left(\frac{n-5}{n-4}\right)\cdots\left(\frac{1}{2}\right)\int_0^{\pi/2} dx, n \text{ even} \\ \left(\frac{n-1}{n}\right)\left(\frac{n-3}{n-2}\right)\left(\frac{n-5}{n-4}\right)\cdots\left(\frac{2}{3}\right)\int_0^{\pi/2} \sin x\,dx, n \text{ odd} \end{cases}$$

$$= \begin{cases} \dfrac{1\cdot 3\cdot 5\cdots(n-1)}{2\cdot 4\cdot 6\cdots n}\cdot\dfrac{\pi}{2}, n \text{ even} \\ \dfrac{2\cdot 4\cdot 6\cdots(n-1)}{3\cdot 5\cdot 7\cdots n}, n \text{ odd} \end{cases}$$

65. (a) $\displaystyle\int_0^{\pi/2}\sin^3 x\,dx = \frac{2}{3}$ (b) $\displaystyle\int_0^{\pi/2}\sin^4 x\,dx = \frac{1\cdot 3}{2\cdot 4}\cdot\frac{\pi}{2} = 3\pi/16$

(c) $\displaystyle\int_0^{\pi/2}\sin^5 x\,dx = \frac{2\cdot 4}{3\cdot 5} = 8/15$ (d) $\displaystyle\int_0^{\pi/2}\sin^6 x\,dx = \frac{1\cdot 3\cdot 5}{2\cdot 4\cdot 6}\cdot\frac{\pi}{2} = 5\pi/32$

66. Similar to proof in Exercise 64.

EXERCISE SET 8.4

1. $x = 2\sin\theta,\ dx = 2\cos\theta\,d\theta,$

$$4\int\cos^2\theta\,d\theta = 2\int(1+\cos 2\theta)d\theta = 2\theta + \sin 2\theta + C$$

$$= 2\theta + 2\sin\theta\cos\theta + C = 2\sin^{-1}(x/2) + \frac{1}{2}x\sqrt{4-x^2} + C$$

2. $x = \frac{1}{2}\sin\theta,\ dx = \frac{1}{2}\cos\theta\,d\theta,$

$$\frac{1}{2}\int\cos^2\theta\,d\theta = \frac{1}{4}\int(1+\cos 2\theta)d\theta = \frac{1}{4}\theta + \frac{1}{8}\sin 2\theta + C$$

$$= \frac{1}{4}\theta + \frac{1}{4}\sin\theta\cos\theta + C = \frac{1}{4}\sin^{-1}2x + \frac{1}{2}x\sqrt{1-4x^2} + C$$

3. $x = 3\sin\theta,\ dx = 3\cos\theta\,d\theta,$

$$9\int\sin^2\theta\,d\theta = \frac{9}{2}\int(1-\cos 2\theta)d\theta = \frac{9}{2}\theta - \frac{9}{4}\sin 2\theta + C = \frac{9}{2}\theta - \frac{9}{2}\sin\theta\cos\theta + C$$

$$= \frac{9}{2}\sin^{-1}(x/3) - \frac{1}{2}x\sqrt{9-x^2} + C$$

4. $x = 4\sin\theta,\ dx = 4\cos\theta\,d\theta,$

$$\frac{1}{16}\int\frac{1}{\sin^2\theta}d\theta = \frac{1}{16}\int\csc^2\theta\,d\theta = -\frac{1}{16}\cot\theta + C = -\frac{\sqrt{16-x^2}}{16x} + C$$

5. $x = 2\tan\theta$, $dx = 2\sec^2\theta\, d\theta$,

$$\frac{1}{8}\int \frac{1}{\sec^2\theta}d\theta = \frac{1}{8}\int \cos^2\theta\, d\theta = \frac{1}{16}\int (1+\cos 2\theta)d\theta = \frac{1}{16}\theta + \frac{1}{32}\sin 2\theta + C$$

$$= \frac{1}{16}\theta + \frac{1}{16}\sin\theta\cos\theta + C = \frac{1}{16}\tan^{-1}\frac{x}{2} + \frac{x}{8(4+x^2)} + C$$

6. $x = \sqrt{5}\tan\theta$, $dx = \sqrt{5}\sec^2\theta\, d\theta$,

$$5\int \tan^2\theta\sec\theta\, d\theta = 5\int (\sec^3\theta - \sec\theta)d\theta = 5\left(\frac{1}{2}\sec\theta\tan\theta - \frac{1}{2}\ln|\sec\theta + \tan\theta|\right) + C_1$$

$$= \frac{1}{2}x\sqrt{5+x^2} - \frac{5}{2}\ln\frac{\sqrt{5+x^2}+x}{\sqrt{5}} + C_1 = \frac{1}{2}x\sqrt{5+x^2} - \frac{5}{2}\ln(\sqrt{5+x^2}+x) + C$$

7. $x = 3\sec\theta$, $dx = 3\sec\theta\tan\theta\, d\theta$,

$$3\int \tan^2\theta\, d\theta = 3\int (\sec^2\theta - 1)d\theta = 3\tan\theta - 3\theta + C = \sqrt{x^2-9} - 3\sec^{-1}\frac{x}{3} + C$$

8. $x = 4\sec\theta$, $dx = 4\sec\theta\tan\theta\, d\theta$,

$$\frac{1}{16}\int \frac{1}{\sec\theta}d\theta = \frac{1}{16}\int \cos\theta\, d\theta = \frac{1}{16}\sin\theta + C = \frac{\sqrt{x^2-16}}{16x} + C$$

9. $x = \sqrt{2}\sin\theta$, $dx = \sqrt{2}\cos\theta\, d\theta$,

$$2\sqrt{2}\int \sin^3\theta\, d\theta = 2\sqrt{2}\int \left[1 - \cos^2\theta\right]\sin\theta\, d\theta$$

$$= 2\sqrt{2}\left(-\cos\theta + \frac{1}{3}\cos^3\theta\right) + C = -2\sqrt{2-x^2} + \frac{1}{3}(2-x^2)^{3/2} + C$$

10. $x = \sqrt{5}\sin\theta$, $dx = \sqrt{5}\cos\theta\, d\theta$,

$$25\sqrt{5}\int \sin^3\theta\cos^2\theta\, d\theta = 25\sqrt{5}\left(-\frac{1}{3}\cos^3\theta + \frac{1}{5}\cos^5\theta\right) + C = -\frac{5}{3}(5-x^2)^{3/2} + \frac{1}{5}(5-x^2)^{5/2} + C$$

11. $x = \frac{3}{2}\sec\theta$, $dx = \frac{3}{2}\sec\theta\tan\theta\, d\theta$, $\frac{2}{9}\int \frac{1}{\sec\theta}d\theta = \frac{2}{9}\int \cos\theta\, d\theta = \frac{2}{9}\sin\theta + C = \frac{\sqrt{4x^2-9}}{9x} + C$

12. $t = \tan\theta$, $dt = \sec^2\theta\, d\theta$,

$$\int \frac{\sec^3\theta}{\tan\theta}d\theta = \int \frac{\tan^2\theta + 1}{\tan\theta}\sec\theta\, d\theta = \int (\sec\theta\tan\theta + \csc\theta)d\theta$$

$$= \sec\theta + \ln|\csc\theta - \cot\theta| + C = \sqrt{1+t^2} + \ln\frac{\sqrt{1+t^2} - 1}{|t|} + C$$

13. $x = \sin\theta$, $dx = \cos\theta\, d\theta$, $\int \frac{1}{\cos^2\theta}d\theta = \int \sec^2\theta\, d\theta = \tan\theta + C = x/\sqrt{1-x^2} + C$

14. $x = 5\tan\theta$, $dx = 5\sec^2\theta\, d\theta$, $\frac{1}{25}\int \frac{\sec\theta}{\tan^2\theta}d\theta = \frac{1}{25}\int \csc\theta\cot\theta\, d\theta = -\frac{1}{25}\csc\theta + C = -\frac{\sqrt{x^2+25}}{25x} + C$

15. $x = \sec\theta$, $dx = \sec\theta\tan\theta\, d\theta$, $\int \sec\theta\, d\theta = \ln|\sec\theta + \tan\theta| + C = \ln\left|x + \sqrt{x^2-1}\right| + C$

16. $1 + 2x^2 + x^4 = (1 + x^2)^2$, $x = \tan\theta$, $dx = \sec^2\theta\, d\theta$,

$$\int \frac{1}{\sec^2\theta} d\theta = \int \cos^2\theta\, d\theta = \frac{1}{2}\int (1 + \cos 2\theta) d\theta = \frac{1}{2}\theta + \frac{1}{4}\sin 2\theta + C$$
$$= \frac{1}{2}\theta + \frac{1}{2}\sin\theta\cos\theta + C = \frac{1}{2}\tan^{-1} x + \frac{x}{2(1 + x^2)} + C$$

17. $x = \frac{1}{3}\sec\theta$, $dx = \frac{1}{3}\sec\theta\tan\theta\, d\theta$,

$$\frac{1}{3}\int \frac{\sec\theta}{\tan^2\theta} d\theta = \frac{1}{3}\int \csc\theta\cot\theta\, d\theta = -\frac{1}{3}\csc\theta + C = -x/\sqrt{9x^2 - 1} + C$$

18. $x = 5\sec\theta$, $dx = 5\sec\theta\tan\theta\, d\theta$,

$$25\int \sec^3\theta\, d\theta = \frac{25}{2}\sec\theta\tan\theta + \frac{25}{2}\ln|\sec\theta + \tan\theta| + C_1$$
$$= \frac{1}{2}x\sqrt{x^2 - 25} + \frac{25}{2}\ln|x + \sqrt{x^2 - 25}| + C$$

19. $e^x = \sin\theta$, $e^x dx = \cos\theta\, d\theta$,

$$\int \cos^2\theta\, d\theta = \frac{1}{2}\int (1 + \cos 2\theta) d\theta = \frac{1}{2}\theta + \frac{1}{4}\sin 2\theta + C = \frac{1}{2}\sin^{-1}(e^x) + \frac{1}{2}e^x\sqrt{1 - e^{2x}} + C$$

20. $u = \sin\theta$, $\displaystyle\int \frac{1}{\sqrt{2 - u^2}} du = \sin^{-1}\left(\frac{\sin\theta}{\sqrt{2}}\right) + C$

21. $x = 4\sin\theta$, $dx = 4\cos\theta\, d\theta$,

$$1024\int_0^{\pi/2} \sin^3\theta\cos^2\theta\, d\theta = 1024\left[-\frac{1}{3}\cos^3\theta + \frac{1}{5}\cos^5\theta\right]_0^{\pi/2} = 1024(1/3 - 1/5) = 2048/15$$

22. $x = \frac{2}{3}\sin\theta$, $dx = \frac{2}{3}\cos\theta\, d\theta$,

$$\frac{1}{24}\int_0^{\pi/6} \frac{1}{\cos^3\theta} d\theta = \frac{1}{24}\int_0^{\pi/6} \sec^3\theta\, d\theta = \left[\frac{1}{48}\sec\theta\tan\theta + \frac{1}{48}\ln|\sec\theta + \tan\theta|\right]_0^{\pi/6}$$
$$= \frac{1}{48}[(2/\sqrt{3})(1/\sqrt{3}) + \ln|2/\sqrt{3} + 1/\sqrt{3}|] = \frac{1}{48}\left(\frac{2}{3} + \frac{1}{2}\ln 3\right)$$

23. $x = \sec\theta$, $dx = \sec\theta\tan\theta\, d\theta$, $\displaystyle\int_{\pi/4}^{\pi/3} \frac{1}{\sec\theta} d\theta = \int_{\pi/4}^{\pi/3} \cos\theta\, d\theta = \sin\theta\Big]_{\pi/4}^{\pi/3} = (\sqrt{3} - \sqrt{2})/2$

24. $x = \sqrt{2}\sec\theta$, $dx = \sqrt{2}\sec\theta\tan\theta\, d\theta$, $\displaystyle 2\int_0^{\pi/4} \tan^2\theta\, d\theta = \Big[2\tan\theta - 2\theta\Big]_0^{\pi/4} = 2 - \pi/2$

25. $x = \sqrt{3}\tan\theta$, $dx = \sqrt{3}\sec^2\theta\, d\theta$,

$$\frac{1}{9}\int_{\pi/6}^{\pi/3} \frac{\sec\theta}{\tan^4\theta} d\theta = \frac{1}{9}\int_{\pi/6}^{\pi/3} \frac{\cos^3\theta}{\sin^4\theta} d\theta = \frac{1}{9}\int_{\pi/6}^{\pi/3} \frac{1 - \sin^2\theta}{\sin^4\theta}\cos\theta\, d\theta = \frac{1}{9}\int_{1/2}^{\sqrt{3}/2} \frac{1 - u^2}{u^4} du \ (u = \sin\theta)$$
$$= \frac{1}{9}\int_{1/2}^{\sqrt{3}/2} (u^{-4} - u^{-2}) du = \frac{1}{9}\left[-\frac{1}{3u^3} + \frac{1}{u}\right]_{1/2}^{\sqrt{3}/2} = \frac{10\sqrt{3} + 18}{243}$$

26. $x = \sqrt{3}\tan\theta,\ dx = \sqrt{3}\sec^2\theta\, d\theta,$

$$\frac{\sqrt{3}}{3}\int_0^{\pi/3}\frac{\tan^3\theta}{\sec^3\theta}d\theta = \frac{\sqrt{3}}{3}\int_0^{\pi/3}\sin^3\theta\, d\theta = \frac{\sqrt{3}}{3}\int_0^{\pi/3}\left[1-\cos^2\theta\right]\sin\theta\, d\theta$$

$$= \frac{\sqrt{3}}{3}\left[-\cos\theta + \frac{1}{3}\cos^3\theta\right]_0^{\pi/3} = \frac{\sqrt{3}}{3}\left[\left(-\frac{1}{2}+\frac{1}{24}\right)-\left(-1+\frac{1}{3}\right)\right] = 5\sqrt{3}/72$$

27. $u = x^2+4,\ du = 2x\, dx,$

$$\frac{1}{2}\int\frac{1}{u}du = \frac{1}{2}\ln|u| + C = \frac{1}{2}\ln(x^2+4)+C;\ \text{or } x = 2\tan\theta,\ dx = 2\sec^2\theta\, d\theta,$$

$$\int\tan\theta\, d\theta = \ln|\sec\theta| + C_1 = \ln\frac{\sqrt{x^2+4}}{2} + C_1 = \ln(x^2+4)^{1/2} - \ln 2 + C_1$$

$$= \frac{1}{2}\ln(x^2+4) + C \text{ with } C = C_1 - \ln 2$$

28. $x = 2\tan\theta, dx = 2\sec^2\theta\, d\theta,\ \displaystyle\int 2\tan^2\theta\, d\theta = 2\tan\theta - 2\theta + C = x - 2\tan^{-1}\frac{x}{2} + C$; alternatively

$$\int\frac{x^2}{x^2+4}dx = \int dx - 4\int\frac{dx}{x^2+4} = x - 2\tan^{-1}\frac{x}{2} + C$$

29. $y' = \dfrac{1}{x},\ 1+(y')^2 = 1+\dfrac{1}{x^2} = \dfrac{x^2+1}{x^2},$

$$L = \int_1^2\sqrt{\frac{x^2+1}{x^2}}dx = \int_1^2\frac{\sqrt{x^2+1}}{x}dx;\ x = \tan\theta,\ dx = \sec^2\theta\, d\theta,$$

$$L = \int_{\pi/4}^{\tan^{-1}(2)}\frac{\sec^3\theta}{\tan\theta}d\theta = \int_{\pi/4}^{\tan^{-1}(2)}\frac{\tan^2\theta+1}{\tan\theta}\sec\theta\, d\theta = \int_{\pi/4}^{\tan^{-1}(2)}(\sec\theta\tan\theta + \csc\theta)d\theta$$

$$= \left[\sec\theta + \ln|\csc\theta - \cot\theta|\right]_{\pi/4}^{\tan^{-1}(2)} = \sqrt{5} + \ln\left(\frac{\sqrt{5}}{2}-\frac{1}{2}\right) - \left[\sqrt{2}+\ln|\sqrt{2}-1|\right]$$

$$= \sqrt{5}-\sqrt{2}+\ln\frac{2+2\sqrt{2}}{1+\sqrt{5}}$$

30. $y' = 2x,\ 1+(y')^2 = 1+4x^2,$

$$L = \int_0^1\sqrt{1+4x^2}dx;\ x = \frac{1}{2}\tan\theta,\ dx = \frac{1}{2}\sec^2\theta\, d\theta,$$

$$L = \frac{1}{2}\int_0^{\tan^{-1}2}\sec^3\theta\, d\theta = \frac{1}{2}\left(\frac{1}{2}\sec\theta\tan\theta + \frac{1}{2}\ln|\sec\theta+\tan\theta|\right)\Bigg]_0^{\tan^{-1}2}$$

$$= \frac{1}{4}(\sqrt{5})(2) + \frac{1}{4}\ln|\sqrt{5}+2| = \frac{1}{2}\sqrt{5} + \frac{1}{4}\ln(2+\sqrt{5})$$

31. $y' = 2x,\ 1+(y')^2 = 1+4x^2,$

$$S = 2\pi\int_0^1 x^2\sqrt{1+4x^2}dx;\ x = \frac{1}{2}\tan\theta,\ dx = \frac{1}{2}\sec^2\theta\, d\theta,$$

$$S = \frac{\pi}{4}\int_0^{\tan^{-1}2} \tan^2\theta\sec^3\theta\, d\theta = \frac{\pi}{4}\int_0^{\tan^{-1}2}(\sec^2\theta - 1)\sec^3\theta\, d\theta = \frac{\pi}{4}\int_0^{\tan^{-1}2}(\sec^5\theta - \sec^3\theta)d\theta$$

$$= \frac{\pi}{4}\left[\frac{1}{4}\sec^3\theta\tan\theta - \frac{1}{8}\sec\theta\tan\theta - \frac{1}{8}\ln|\sec\theta + \tan\theta|\right]_0^{\tan^{-1}2} = \frac{\pi}{32}[18\sqrt{5} - \ln(2+\sqrt{5})]$$

32. $V = \pi\int_0^1 y^2\sqrt{1-y^2}dy$; $y = \sin\theta$, $dy = \cos\theta\, d\theta$,

$$V = \pi\int_0^{\pi/2}\sin^2\theta\cos^2\theta\, d\theta = \frac{\pi}{4}\int_0^{\pi/2}\sin^2 2\theta\, d\theta = \frac{\pi}{8}\int_0^{\pi/2}(1-\cos 4\theta)d\theta = \frac{\pi}{8}\left(\theta - \frac{1}{4}\sin 4\theta\right)\Big]_0^{\pi/2} = \frac{\pi^2}{16}$$

33. **(a)** $x = 3\sinh u$, $dx = 3\cosh u\, du$, $\int du = u + C = \sinh^{-1}(x/3) + C$

(b) $x = 3\tan\theta$, $dx = 3\sec^2\theta\, d\theta$,

$$\int \sec\theta\, d\theta = \ln|\sec\theta + \tan\theta| + C = \ln\left(\sqrt{x^2+9}/3 + x/3\right) + C$$

but $\sinh^{-1}(x/3) = \ln\left(x/3 + \sqrt{x^2/9 + 1}\right) = \ln\left(x/3 + \sqrt{x^2+9}/3\right)$ so the results agree.

(c) $x = \cosh u$, $dx = \sinh u\, du$,

$$\int \sinh^2 u\, du = \frac{1}{2}\int(\cosh 2u - 1)du = \frac{1}{4}\sinh 2u - \frac{1}{2}u + C$$

$$= \frac{1}{2}\sinh u\cosh u - \frac{1}{2}u + C = \frac{1}{2}x\sqrt{x^2-1} - \frac{1}{2}\cosh^{-1}x + C$$

because $\cosh u = x$, and $\sinh u = \sqrt{\cosh^2 u - 1} = \sqrt{x^2-1}$

34. $A = \frac{4b}{a}\int_0^a \sqrt{a^2-x^2}\, dx$; $x = a\cos\theta, dx = -a\sin\theta\, d\theta$,

$$A = -\frac{4b}{a}\int_{\pi/2}^0 a^2\sin^2\theta\, d\theta = 4ab\int_0^{\pi/2}\sin^2\theta\, d\theta = 2ab\int_0^{\pi/2}(1-\cos 2\theta)\, d\theta = \pi ab$$

35. $\int \frac{1}{(x-2)^2+9}dx = \frac{1}{3}\tan^{-1}\left(\frac{x-2}{3}\right) + C$

36. $\int \frac{1}{\sqrt{1-(x-1)^2}}dx = \sin^{-1}(x-1) + C$

37. $\int \frac{1}{\sqrt{9-(x-1)^2}}dx = \sin^{-1}\left(\frac{x-1}{3}\right) + C$

38. $\int \frac{1}{16(x+1/2)^2+1}dx = \frac{1}{16}\int \frac{1}{(x+1/2)^2+1/16}dx = \frac{1}{4}\tan^{-1}(4x+2) + C$

39. $\int \frac{1}{\sqrt{(x-3)^2+1}}dx = \ln\left(x - 3 + \sqrt{(x-3)^2+1}\right) + C$

40. $\int \frac{x}{(x+3)^2+1}dx$, let $u = x+3$,

$$\int \frac{u-3}{u^2+1}du = \int\left(\frac{u}{u^2+1} - \frac{3}{u^2+1}\right)du = \frac{1}{2}\ln(u^2+1) - 3\tan^{-1}u + C$$
$$= \frac{1}{2}\ln(x^2+6x+10) - 3\tan^{-1}(x+3) + C$$

41. $\int \sqrt{4-(x+1)^2}dx$, let $x+1 = 2\sin\theta$,

$$4\int \cos^2\theta\, d\theta = 2\theta + \sin 2\theta + C = 2\theta + 2\sin\theta\cos\theta + C$$
$$= 2\sin^{-1}\left(\frac{x+1}{2}\right) + \frac{1}{2}(x+1)\sqrt{3-2x-x^2} + C$$

42. $\int \frac{e^x}{\sqrt{(e^x+1/2)^2+3/4}}dx$, let $u = e^x + 1/2$,

$$\int \frac{1}{\sqrt{u^2+3/4}}du = \sinh^{-1}(2u/\sqrt{3}) + C = \sinh^{-1}\left(\frac{2e^x+1}{\sqrt{3}}\right) + C$$

Alternate solution: let $e^x + 1/2 = \frac{\sqrt{3}}{2}\tan\theta$,

$$\int \sec\theta\, d\theta = \ln|\sec\theta + \tan\theta| + C = \ln\left(\frac{2\sqrt{e^{2x}+e^x+1}}{\sqrt{3}} + \frac{2e^x+1}{\sqrt{3}}\right) + C_1$$
$$= \ln(2\sqrt{e^{2x}+e^x+1} + 2e^x + 1) + C$$

43. $$\int \frac{1}{2(x+1)^2+5}dx = \frac{1}{2}\int \frac{1}{(x+1)^2+5/2}dx = \frac{1}{\sqrt{10}}\tan^{-1}\sqrt{2/5}(x+1) + C$$

44. $\int \frac{2x+3}{4(x+1/2)^2+4}dx$, let $u = x+1/2$,

$$\int \frac{2u+2}{4u^2+4}du = \frac{1}{2}\int\left(\frac{u}{u^2+1} + \frac{1}{u^2+1}\right)du = \frac{1}{4}\ln(u^2+1) + \frac{1}{2}\tan^{-1}u + C$$
$$= \frac{1}{4}\ln(x^2+x+5/4) + \frac{1}{2}\tan^{-1}(x+1/2) + C$$

45. $$\int_1^2 \frac{1}{\sqrt{4x-x^2}}dx = \int_1^2 \frac{1}{\sqrt{4-(x-2)^2}}dx = \sin^{-1}\frac{x-2}{2}\Big]_1^2 = \pi/6$$

46. $\int_0^1 \sqrt{4x-x^2}dx = \int_0^1 \sqrt{4-(x-2)^2}dx$, let $x-2 = 2\sin\theta$,

$$4\int_{-\pi/2}^{-\pi/6}\cos^2\theta\, d\theta = \left[2\theta + \sin 2\theta\right]_{-\pi/2}^{-\pi/6} = \frac{2\pi}{3} - \frac{\sqrt{3}}{2}$$

47. $u = \sin^2 x$, $du = 2\sin x\cos x\, dx$;

$$\frac{1}{2}\int\sqrt{1-u^2}\,du = \frac{1}{4}\left[u\sqrt{1-u^2} + \sin^{-1}u\right] + C = \frac{1}{4}\left[\sin^2 x\sqrt{1-\sin^4 x} + \sin^{-1}(\sin^2 x)\right] + C$$

48. $u = x\sin x$, $du = (x\cos x + \sin x)\,dx$;

$$\int \sqrt{1+u^2}\,du = \frac{1}{2}u\sqrt{1+u^2} + \frac{1}{2}\sinh^{-1} u + C = \frac{1}{2}x\sin x\sqrt{1+x^2\sin^2 x} + \frac{1}{2}\sinh^{-1}(x\sin x) + C$$

EXERCISE SET 8.5

1. $\dfrac{A}{(x-2)} + \dfrac{B}{(x+5)}$

2. $\dfrac{5}{x(x-3)(x+3)} = \dfrac{A}{x} + \dfrac{B}{x-3} + \dfrac{C}{x+3}$

3. $\dfrac{2x-3}{x^2(x-1)} = \dfrac{A}{x} + \dfrac{B}{x^2} + \dfrac{C}{x-1}$

4. $\dfrac{A}{x+2} + \dfrac{B}{(x+2)^2} + \dfrac{C}{(x+2)^3}$

5. $\dfrac{A}{x} + \dfrac{B}{x^2} + \dfrac{C}{x^3} + \dfrac{Dx+E}{x^2+1}$

6. $\dfrac{A}{x-1} + \dfrac{Bx+C}{x^2+5}$

7. $\dfrac{Ax+B}{x^2+5} + \dfrac{Cx+D}{(x^2+5)^2}$

8. $\dfrac{A}{x-2} + \dfrac{Bx+C}{x^2+1} + \dfrac{Dx+E}{(x^2+1)^2}$

9. $\dfrac{1}{(x+4)(x-1)} = \dfrac{A}{x+4} + \dfrac{B}{x-1}$; $A = -\dfrac{1}{5}$, $B = \dfrac{1}{5}$ so

$$-\frac{1}{5}\int \frac{1}{x+4}dx + \frac{1}{5}\int \frac{1}{x-1}dx = -\frac{1}{5}\ln|x+4| + \frac{1}{5}\ln|x-1| + C = \frac{1}{5}\ln\left|\frac{x-1}{x+4}\right| + C$$

10. $\dfrac{1}{(x+1)(x+7)} = \dfrac{A}{x+1} + \dfrac{B}{x+7}$; $A = \dfrac{1}{6}$, $B = -\dfrac{1}{6}$ so

$$\frac{1}{6}\int \frac{1}{x+1}dx - \frac{1}{6}\int \frac{1}{x+7}dx = \frac{1}{6}\ln|x+1| - \frac{1}{6}\ln|x+7| + C = \frac{1}{6}\ln\left|\frac{x+1}{x+7}\right| + C$$

11. $\dfrac{11x+17}{(2x-1)(x+4)} = \dfrac{A}{2x-1} + \dfrac{B}{x+4}$; $A = 5$, $B = 3$ so

$$5\int \frac{1}{2x-1}dx + 3\int \frac{1}{x+4}dx = \frac{5}{2}\ln|2x-1| + 3\ln|x+4| + C$$

12. $\dfrac{5x-5}{(x-3)(3x+1)} = \dfrac{A}{x-3} + \dfrac{B}{3x+1}$; $A = 1$, $B = 2$ so

$$\int \frac{1}{x-3}dx + 2\int \frac{1}{3x+1}dx = \ln|x-3| + \frac{2}{3}\ln|3x+1| + C$$

13. $\dfrac{2x^2-9x-9}{x(x+3)(x-3)} = \dfrac{A}{x} + \dfrac{B}{x+3} + \dfrac{C}{x-3}$; $A = 1$, $B = 2$, $C = -1$ so

$$\int \frac{1}{x}\,dx + 2\int \frac{1}{x+3}\,dx - \int \frac{1}{x-3}\,dx = \ln|x| + 2\ln|x+3| - \ln|x-3| + C = \ln\left|\frac{x(x+3)^2}{x-3}\right| + C$$

Note that the symbol C has been recycled; to save space this recycling is usually not mentioned.

14. $\dfrac{1}{x(x+1)(x-1)} = \dfrac{A}{x} + \dfrac{B}{x+1} + \dfrac{C}{x-1}$; $A = -1$, $B = \dfrac{1}{2}$, $C = \dfrac{1}{2}$ so

$$-\int \frac{1}{x}\,dx + \frac{1}{2}\int \frac{1}{x+1}\,dx + \frac{1}{2}\int \frac{1}{x-1}\,dx = -\ln|x| + \frac{1}{2}\ln|x+1| + \frac{1}{2}\ln|x-1| + C$$

$$= \frac{1}{2}\ln\left|\frac{(x+1)(x-1)}{x^2}\right| + C = \frac{1}{2}\ln\frac{|x^2-1|}{x^2} + C$$

15. $\dfrac{x^2+2}{x+2} = x - 2 + \dfrac{6}{x+2}$, $\displaystyle\int\left(x - 2 + \frac{6}{x+2}\right)dx = \frac{1}{2}x^2 - 2x + 6\ \ln|x+2| + C$

16. $\dfrac{x^2-4}{x-1} = x + 1 - \dfrac{3}{x-1}$, $\displaystyle\int\left(x + 1 - \frac{3}{x-1}\right)dx = \frac{1}{2}x^2 + x - 3\ \ln|x-1| + C$

17. $\dfrac{3x^2-10}{x^2-4x+4} = 3 + \dfrac{12x-22}{x^2-4x+4}$, $\dfrac{12x-22}{(x-2)^2} = \dfrac{A}{x-2} + \dfrac{B}{(x-2)^2}$; $A = 12$, $B = 2$ so

$$\int 3dx + 12\int \frac{1}{x-2}dx + 2\int \frac{1}{(x-2)^2}dx = 3x + 12\ln|x-2| - 2/(x-2) + C$$

18. $\dfrac{x^2}{x^2-3x+2} = 1 + \dfrac{3x-2}{x^2-3x+2}$, $\dfrac{3x-2}{(x-1)(x-2)} = \dfrac{A}{x-1} + \dfrac{B}{x-2}$; $A = -1$, $B = 4$ so

$$\int dx - \int \frac{1}{x-1}dx + 4\int \frac{1}{x-2}dx = x - \ln|x-1| + 4\ln|x-2| + C$$

19. $\dfrac{x^5+2x^2+1}{x^3-x} = x^2 + 1 + \dfrac{2x^2+x+1}{x^3-x}$,

$\dfrac{2x^2+x+1}{x(x+1)(x-1)} = \dfrac{A}{x} + \dfrac{B}{x+1} + \dfrac{C}{x-1}$; $A = -1$, $B = 1$, $C = 2$ so

$$\int (x^2+1)dx - \int \frac{1}{x}dx + \int \frac{1}{x+1}dx + 2\int \frac{1}{x-1}dx$$

$$= \frac{1}{3}x^3 + x - \ln|x| + \ln|x+1| + 2\ln|x-1| + C = \frac{1}{3}x^3 + x + \ln\left|\frac{(x+1)(x-1)^2}{x}\right| + C$$

20. $\dfrac{2x^5-x^3-1}{x^3-4x} = 2x^2 + 7 + \dfrac{28x-1}{x^3-4x}$,

$\dfrac{28x-1}{x(x+2)(x-2)} = \dfrac{A}{x} + \dfrac{B}{x+2} + \dfrac{C}{x-2}$; $A = \dfrac{1}{4}$, $B = -\dfrac{57}{8}$, $C = \dfrac{55}{8}$ so

$$\int (2x^2+7)\,dx + \frac{1}{4}\int \frac{1}{x}\,dx - \frac{57}{8}\int \frac{1}{x+2}\,dx + \frac{55}{8}\int \frac{1}{x-2}\,dx$$

$$= \frac{2}{3}x^3 + 7x + \frac{1}{4}\ln|x| - \frac{57}{8}\ln|x+2| + \frac{55}{8}\ln|x-2| + C$$

21. $\dfrac{2x^2+3}{x(x-1)^2} = \dfrac{A}{x} + \dfrac{B}{x-1} + \dfrac{C}{(x-1)^2}$; $A = 3$, $B = -1$, $C = 5$ so

$$3\int \frac{1}{x}\,dx - \int \frac{1}{x-1}\,dx + 5\int \frac{1}{(x-1)^2}dx = 3\ln|x| - \ln|x-1| - 5/(x-1) + C$$

22. $\dfrac{3x^2-x+1}{x^2(x-1)}=\dfrac{A}{x}+\dfrac{B}{x^2}+\dfrac{C}{x-1}$; $A=0,\ B=-1,\ C=3$ so

$$-\int\frac{1}{x^2}dx+3\int\frac{1}{x-1}dx=1/x+3\ln|x-1|+C$$

23. $\dfrac{x^2+x-16}{(x+1)(x-3)^2}=\dfrac{A}{x+1}+\dfrac{B}{x-3}+\dfrac{C}{(x-3)^2}$; $A=-1,\ B=2,\ C=-1$ so

$$-\int\frac{1}{x+1}\,dx+2\int\frac{1}{x-3}\,dx-\int\frac{1}{(x-3)^2}\,dx$$

$$=-\ln|x+1|+2\ln|x-3|+\frac{1}{x-3}+C=\ln\frac{(x-3)^2}{|x+1|}+\frac{1}{x-3}+C$$

24. $\dfrac{2x^2-2x-1}{x^2(x-1)}=\dfrac{A}{x}+\dfrac{B}{x^2}+\dfrac{C}{x-1}$; $A=3,\ B=1,\ C=-1$ so

$$3\int\frac{1}{x}\,dx+\int\frac{1}{x^2}\,dx-\int\frac{1}{x-1}\,dx=3\ln|x|-\frac{1}{x}-\ln|x-1|+C$$

25. $\dfrac{x^2}{(x+2)^3}=\dfrac{A}{x+2}+\dfrac{B}{(x+2)^2}+\dfrac{C}{(x+2)^3}$; $A=1,\ B=-4,\ C=4$ so

$$\int\frac{1}{x+2}\,dx-4\int\frac{1}{(x+2)^2}\,dx+4\int\frac{1}{(x+2)^3}\,dx=\ln|x+2|+\frac{4}{x+2}-\frac{2}{(x+2)^2}+C$$

26. $\dfrac{2x^2+3x+3}{(x+1)^3}=\dfrac{A}{x+1}+\dfrac{B}{(x+1)^2}+\dfrac{C}{(x+1)^3}$; $A=2,\ B=-1,\ C=2$ so

$$2\int\frac{1}{x+1}\,dx-\int\frac{1}{(x+1)^2}\,dx+2\int\frac{1}{(x+1)^3}\,dx=2\ln|x+1|+\frac{1}{x+1}-\frac{1}{(x+1)^2}+C$$

27. $\dfrac{2x^2-1}{(4x-1)(x^2+1)}=\dfrac{A}{4x-1}+\dfrac{Bx+C}{x^2+1}$; $A=-14/17,\ B=12/17,\ C=3/17$ so

$$\int\frac{2x^2-1}{(4x-1)(x^2+1)}dx=-\frac{7}{34}\ln|4x-1|+\frac{6}{17}\ln(x^2+1)+\frac{3}{17}\tan^{-1}x+C$$

28. $\dfrac{1}{x(x^2+1)}=\dfrac{A}{x}+\dfrac{Bx+C}{x^2+1}$; $A=1,\ B=-1,\ C=0$ so

$$\int\frac{1}{x^3+x}\,dx=\ln|x|-\frac{1}{2}\ln(x^2+1)+C=\frac{1}{2}\ln\frac{x^2}{x^2+1}+C$$

29. $\dfrac{x^3+3x^2+x+9}{(x^2+1)(x^2+3)}=\dfrac{Ax+B}{x^2+1}+\dfrac{Cx+D}{x^2+3}$; $A=0,\ B=3,\ C=1,\ D=0$ so

$$\int\frac{x^3+3x^2+x+9}{(x^2+1)(x^2+3)}\,dx=3\tan^{-1}x+\frac{1}{2}\ln(x^2+3)+C$$

30. $\dfrac{x^3+x^2+x+2}{(x^2+1)(x^2+2)}=\dfrac{Ax+B}{x^2+1}+\dfrac{Cx+D}{x^2+2}$; $A=D=0,\ B=C=1$ so

$$\int\frac{x^3+x^2+x+2}{(x^2+1)(x^2+2)}\,dx=\tan^{-1}x+\frac{1}{2}\ln(x^2+2)+C$$

31. $\dfrac{x^3-3x^2+2x-3}{x^2+1} = x-3+\dfrac{x}{x^2+1}$,

$$\int \frac{x^3-3x^2+2x-3}{x^2+1}\,dx = \frac{1}{2}x^2-3x+\frac{1}{2}\ln(x^2+1)+C$$

32. $\dfrac{x^4+6x^3+10x^2+x}{x^2+6x+10} = x^2+\dfrac{x}{x^2+6x+10}$,

$$\int \frac{x}{x^2+6x+10}\,dx = \int \frac{x}{(x+3)^2+1}\,dx = \int \frac{u-3}{u^2+1}du, \quad u=x+3$$

$$= \frac{1}{2}\ln(u^2+1)-3\tan^{-1}u+C_1$$

so $\displaystyle\int \frac{x^4+6x^3+10x^2+x}{x^2+6x+10}\,dx = \frac{1}{3}x^3+\frac{1}{2}\ln(x^2+6x+10)-3\tan^{-1}(x+3)+C$

33. Let $x=\sin\theta$ to get $\displaystyle\int \frac{1}{x^2+4x-5}\,dx$, and $\dfrac{1}{(x+5)(x-1)} = \dfrac{A}{x+5}+\dfrac{B}{x-1}$; $A=-1/6$,

$B=1/6$ so we get $\displaystyle -\frac{1}{6}\int \frac{1}{x+5}\,dx+\frac{1}{6}\int \frac{1}{x-1}\,dx = \frac{1}{6}\ln\left|\frac{x-1}{x+5}\right|+C = \frac{1}{6}\ln\left(\frac{1-\sin\theta}{5+\sin\theta}\right)+C.$

34. Let $x=e^t$; then $\displaystyle\int \frac{e^t}{e^{2t}-4}dt = \int \frac{1}{x^2-4}\,dx$,

$\dfrac{1}{(x+2)(x-2)} = \dfrac{A}{x+2}+\dfrac{B}{x-2}$; $A=-1/4$, $B=1/4$ so

$$-\frac{1}{4}\int \frac{1}{x+2}\,dx+\frac{1}{4}\int \frac{1}{x-2}\,dx = \frac{1}{4}\ln\left|\frac{x-2}{x+2}\right|+C = \frac{1}{4}\ln\left|\frac{e^t-2}{e^t+2}\right|+C.$$

35. $\displaystyle V=\pi\int_0^2 \frac{x^4}{(9-x^2)^2}\,dx,\ \frac{x^4}{x^4-18x^2+81} = 1+\frac{18x^2-81}{x^4-18x^2+81}$,

$$\frac{18x^2-81}{(9-x^2)^2} = \frac{18x^2-81}{(x+3)^2(x-3)^2} = \frac{A}{x+3}+\frac{B}{(x+3)^2}+\frac{C}{x-3}+\frac{D}{(x-3)^2};$$

$A=-\dfrac{9}{4}$, $B=\dfrac{9}{4}$, $C=\dfrac{9}{4}$, $D=\dfrac{9}{4}$ so

$$V=\pi\left[x-\frac{9}{4}\ln|x+3|-\frac{9/4}{x+3}+\frac{9}{4}\ln|x-3|-\frac{9/4}{x-3}\right]_0^2 = \pi\left(\frac{19}{5}-\frac{9}{4}\ln 5\right)$$

36. Let $u=e^x$ to get $\displaystyle\int_{-\ln 5}^{\ln 5}\frac{dx}{1+e^x} = \int_{-\ln 5}^{\ln 5}\frac{e^x\,dx}{e^x(1+e^x)} = \int_{1/5}^{5}\frac{du}{u(1+u)}$,

$$\frac{1}{u(1+u)} = \frac{A}{u}+\frac{B}{1+u};\ A=1,\ B=-1;\ \int_{1/5}^{5}\frac{du}{u(1+u)} = (\ln u-\ln(1+u))\Big]_{1/5}^{5} = \ln 5$$

37. $\dfrac{x^2+1}{(x^2+2x+3)^2} = \dfrac{Ax+B}{x^2+2x+3}+\dfrac{Cx+D}{(x^2+2x+3)^2}$; $A=0$, $B=1$, $C=D=-2$ so

$$\int \frac{x^2+1}{(x^2+2x+3)^2}dx = \int \frac{1}{(x+1)^2+2}dx - \int \frac{2x+2}{(x^2+2x+3)^2}dx$$

$$= \frac{1}{\sqrt{2}}\tan^{-1}\frac{x+1}{\sqrt{2}}+1/(x^2+2x+3)+C$$

38. $\dfrac{x^5+x^4+4x^3+4x^2+4x+4}{(x^2+2)^3}=\dfrac{Ax+B}{x^2+2}+\dfrac{Cx+D}{(x^2+2)^2}+\dfrac{Ex+F}{(x^2+2)^3}$;

$A=B=1$, $C=D=E=F=0$ so

$$\int\frac{x+1}{x^2+2}dx=\frac{1}{2}\ln(x^2+2)+\frac{1}{\sqrt{2}}\tan^{-1}(x/\sqrt{2})+C$$

39. $x^4-3x^3-7x^2+27x-18=(x-1)(x-2)(x-3)(x+3)$,

$\dfrac{1}{(x-1)(x-2)(x-3)(x+3)}=\dfrac{A}{x-1}+\dfrac{B}{x-2}+\dfrac{C}{x-3}+\dfrac{D}{x+3}$;

$A=1/8$, $B=-1/5$, $C=1/12$, $D=-1/120$ so

$$\int\frac{dx}{x^4-3x^3-7x^2+27x-18}=\frac{1}{8}\ln|x-1|-\frac{1}{5}\ln|x-2|+\frac{1}{12}\ln|x-3|-\frac{1}{120}\ln|x+3|+C$$

40. $16x^3-4x^2+4x-1=(4x-1)(4x^2+1)$,

$\dfrac{1}{(4x-1)(4x^2+1)}=\dfrac{A}{4x-1}+\dfrac{Bx+C}{4x^2+1}$; $A=4/5$, $B=-4/5$, $C=-1/5$ so

$$\int\frac{dx}{16x^3-4x^2+4x-1}=\frac{1}{5}\ln|4x-1|-\frac{1}{10}\ln(4x^2+1)-\frac{1}{10}\tan^{-1}(2x)+C$$

41. **(a)**
$$\begin{aligned}x^4+1&=(x^4+2x^2+1)-2x^2=(x^2+1)^2-2x^2\\&=[(x^2+1)+\sqrt{2}x][(x^2+1)-\sqrt{2}x]\\&=(x^2+\sqrt{2}x+1)(x^2-\sqrt{2}x+1);\ a=\sqrt{2},b=-\sqrt{2}\end{aligned}$$

(b) $\dfrac{x}{(x^2+\sqrt{2}x+1)(x^2-\sqrt{2}x+1)}=\dfrac{Ax+B}{x^2+\sqrt{2}x+1}+\dfrac{Cx+D}{x^2-\sqrt{2}x+1}$;

$A=0$, $B=-\dfrac{\sqrt{2}}{4}$, $C=0$, $D=\dfrac{\sqrt{2}}{4}$ so

$$\begin{aligned}\int_0^1\frac{x}{x^4+1}dx&=-\frac{\sqrt{2}}{4}\int_0^1\frac{1}{x^2+\sqrt{2}x+1}dx+\frac{\sqrt{2}}{4}\int_0^1\frac{1}{x^2-\sqrt{2}x+1}dx\\&=-\frac{\sqrt{2}}{4}\int_0^1\frac{1}{(x+\sqrt{2}/2)^2+1/2}dx+\frac{\sqrt{2}}{4}\int_0^1\frac{1}{(x-\sqrt{2}/2)^2+1/2}dx\\&=-\frac{\sqrt{2}}{4}\int_{\sqrt{2}/2}^{1+\sqrt{2}/2}\frac{1}{u^2+1/2}du+\frac{\sqrt{2}}{4}\int_{-\sqrt{2}/2}^{1-\sqrt{2}/2}\frac{1}{u^2+1/2}du\\&=-\frac{1}{2}\tan^{-1}\sqrt{2}u\Big]_{\sqrt{2}/2}^{1+\sqrt{2}/2}+\frac{1}{2}\tan^{-1}\sqrt{2}u\Big]_{-\sqrt{2}/2}^{1-\sqrt{2}/2}\\&=-\frac{1}{2}\tan^{-1}(\sqrt{2}+1)+\frac{1}{2}\left(\frac{\pi}{4}\right)+\frac{1}{2}\tan^{-1}(\sqrt{2}-1)-\frac{1}{2}\left(-\frac{\pi}{4}\right)\\&=\frac{\pi}{4}-\frac{1}{2}[\tan^{-1}(\sqrt{2}+1)-\tan^{-1}(\sqrt{2}-1)]\\&=\frac{\pi}{4}-\frac{1}{2}[\tan^{-1}(1+\sqrt{2})+\tan^{-1}(1-\sqrt{2})]\\&=\frac{\pi}{4}-\frac{1}{2}\tan^{-1}\left[\frac{(1+\sqrt{2})+(1-\sqrt{2})}{1-(1+\sqrt{2})(1-\sqrt{2})}\right]\quad\text{(Exercise 78, Section 7.6)}\\&=\frac{\pi}{4}-\frac{1}{2}\tan^{-1}1=\frac{\pi}{4}-\frac{1}{2}\left(\frac{\pi}{4}\right)=\frac{\pi}{8}\end{aligned}$$

42. $\dfrac{1}{a^2-x^2}=\dfrac{A}{a-x}+\dfrac{B}{a+x}$; $A=\dfrac{1}{2a}$, $B=\dfrac{1}{2a}$ so

$$\frac{1}{2a}\int\left(\frac{1}{a-x}+\frac{1}{a+x}\right)dx=\frac{1}{2a}\left(-\ln|a-x|+\ln|a+x|\right)+C=\frac{1}{2a}\ln\left|\frac{a+x}{a-x}\right|+C$$

EXERCISE SET 8.6

1. Formula (60): $\dfrac{3}{16}\left[4x+\ln|-1+4x|\right]+C$

2. Formula (62): $\dfrac{1}{9}\left[\dfrac{2}{2-3x}+\ln|2-3x|\right]+C$

3. Formula (65): $\dfrac{1}{5}\ln\left|\dfrac{x}{5+2x}\right|+C$

4. Formula (66): $-\dfrac{1}{x}-5\ln\left|\dfrac{1-5x}{x}\right|+C$

5. Formula (102): $\dfrac{1}{5}(x+1)(-3+2x)^{3/2}+C$

6. Formula (105): $\dfrac{2}{3}(-x-4)\sqrt{2-x}+C$

7. Formula (108): $\dfrac{1}{2}\ln\left|\dfrac{\sqrt{4-3x}-2}{\sqrt{4-3x}+2}\right|+C$

8. Formula (108): $\tan^{-1}\dfrac{\sqrt{3x-4}}{2}+C$

9. Formula (69): $\dfrac{1}{2\sqrt{5}}\ln\left|\dfrac{x+\sqrt{5}}{x-\sqrt{5}}\right|+C$

10. Formula (70): $\dfrac{1}{6}\ln\left|\dfrac{x-3}{x+3}\right|+C$

11. Formula (73): $\dfrac{x}{2}\sqrt{x^2-3}-\dfrac{3}{2}\ln\left|x+\sqrt{x^2-3}\right|+C$

12. Formula (93): $-\dfrac{\sqrt{x^2+5}}{x}+\ln(x+\sqrt{x^2+5})+C$

13. Formula (95): $\dfrac{x}{2}\sqrt{x^2+4}-2\ln(x+\sqrt{x^2+4})+C$

14. Formula (90): $-\dfrac{\sqrt{x^2-2}}{2x}+C$

15. Formula (74): $\dfrac{x}{2}\sqrt{9-x^2}+\dfrac{9}{2}\sin^{-1}\dfrac{x}{3}+C$

16. Formula (80): $-\dfrac{\sqrt{4-x^2}}{x}-\sin^{-1}\dfrac{x}{2}+C$

17. Formula (79): $\sqrt{3-x^2}-\sqrt{3}\ln\left|\dfrac{\sqrt{3}+\sqrt{9-x^2}}{x}\right|+C$

18. Formula (117): $-\dfrac{\sqrt{6x-x^2}}{3x}+C$

19. Formula (38): $-\dfrac{1}{10}\sin(5x)+\dfrac{1}{2}\sin x+C$

20. Formula (40): $-\dfrac{1}{14}\cos(7x)+\dfrac{1}{6}\cos(3x)+C$

21. Formula (50): $\dfrac{x^4}{16}\left[4\ln x-1\right]+C$

22. Formula (50): $4\sqrt{x}\left[\dfrac{1}{2}\ln x-1\right]+C$

23. Formula (42): $\dfrac{e^{-2x}}{13}(-2\sin(3x)-3\cos(3x))+C$

24. Formula (43): $\dfrac{e^x}{5}(\cos(2x) + 2\sin(2x)) + C$

25. $u = e^{2x}, du = 2e^{2x}dx$, Formula (62): $\dfrac{1}{2}\displaystyle\int \frac{u\,du}{(4-3u)^2} = \frac{1}{18}\left[\frac{4}{4-3e^{2x}} + \ln\left|4 - 3e^{2x}\right|\right] + C$

26. $u = \sin 2x, du = 2\cos 2x dx$, Formula (116): $\displaystyle\int \frac{du}{2u(3-u)} = \frac{1}{6}\ln\left|\frac{\sin 2x}{3 - \sin 2x}\right| + C$

27. $u = 3\sqrt{x}, du = \dfrac{3}{2\sqrt{x}}dx$, Formula (68): $\dfrac{2}{3}\displaystyle\int \frac{du}{u^2+4} = \frac{1}{3}\tan^{-1}\frac{3\sqrt{x}}{2} + C$

28. $u = \sin 4x, du = 4\cos 4x dx$, Formula (68): $\dfrac{1}{4}\displaystyle\int \frac{du}{9+u^2} = \frac{1}{12}\tan^{-1}\frac{\sin 4x}{3} + C$

29. $u = 3x, du = 3dx$, Formula (76): $\dfrac{1}{3}\displaystyle\int \frac{du}{\sqrt{u^2-4}} = \frac{1}{3}\ln\left|3x + \sqrt{9x^2-4}\right| + C$

30. $u = \sqrt{2}x^2, du = 2\sqrt{2}xdx$, Formula (72):

$$\frac{1}{2\sqrt{2}}\int \sqrt{u^2+3}\,du = \frac{x^2}{4}\sqrt{2x^4+3} + \frac{3}{4\sqrt{2}}\ln\left(\sqrt{2}x^2 + \sqrt{2x^4+3}\right) + C$$

31. $u = 3x^2, du = 6xdx, u^2du = 54x^5dx$, Formula (81):

$$\frac{1}{54}\int \frac{u^2\,du}{\sqrt{5-u^2}} = -\frac{x^2}{36}\sqrt{5-9x^4} + \frac{5}{108}\sin^{-1}\frac{3x^2}{\sqrt{5}} + C$$

32. $u = 2x, du = 2dx$, Formula (83): $2\displaystyle\int \frac{du}{u^2\sqrt{3-u^2}} = -\frac{1}{3x}\sqrt{3-4x^2} + C$

33. $u = \ln x, du = dx/x$, Formula (26): $\displaystyle\int \sin^2 u\,du = \frac{1}{2}\ln x + \frac{1}{4}\sin(2\ln x) + C$

34. $u = e^{-2x}, du = -2e^{-2x}$, Formula (27): $-\dfrac{1}{2}\displaystyle\int \cos^2 u\,du = -\frac{1}{4}e^{-2x} - \frac{1}{8}\sin\left(2e^{-2x}\right) + C$

35. $u = -2x, du = -2dx$, Formula (51): $\dfrac{1}{4}\displaystyle\int ue^u\,du = \frac{1}{4}(-2x-1)e^{-2x} + C$

36. $u = 5x-1, du = 5dx$, Formula (50): $\dfrac{1}{5}\displaystyle\int \ln u\,du = \frac{1}{5}(u\ln u - u) + C = \frac{1}{5}(5x-1)[\ln(5x-1) - 1] + C$

37. $u = \cos 3x, du = -3\sin 3x$, Formula (67): $-\displaystyle\int \frac{du}{u(u+1)^2} = -\frac{1}{3}\left[\frac{1}{1+\cos 3x} + \ln\left|\frac{\cos 3x}{1+\cos 3x}\right|\right] + C$

38. $u = \ln x, du = \dfrac{1}{x}dx$, Formula (105): $\displaystyle\int \frac{u\,du}{\sqrt{4u-1}} = \frac{1}{12}(2\ln x + 1)\sqrt{4\ln x - 1} + C$

39. $u = 4x^2, du = 8xdx$, Formula (70): $\dfrac{1}{8}\displaystyle\int \frac{du}{u^2-1} = \frac{1}{16}\ln\left|\frac{4x^2-1}{4x^2+1}\right| + C$

40. $u = 2e^x, du = 2e^xdx$, Formula (69): $\dfrac{1}{2}\displaystyle\int \frac{du}{3-u^2} = \frac{1}{4\sqrt{3}}\ln\left|\frac{2e^x+\sqrt{3}}{2e^x-\sqrt{3}}\right| + C$

41. $u = 2e^x, du = 2e^x dx$, Formula (74):

$$\frac{1}{2}\int \sqrt{3-u^2}\,du = \frac{1}{4}u\sqrt{3-u^2} + \frac{3}{4}\sin^{-1}(u/\sqrt{3}) + C = \frac{1}{2}e^x\sqrt{3-4e^{2x}} + \frac{3}{4}\sin^{-1}(2e^x/\sqrt{3}) + C$$

42. $u = 3x, du = 3dx$, Formula (80):

$$3\int \frac{\sqrt{4-u^2}du}{u^2} = -3\frac{\sqrt{4-u^2}}{u} - 3\sin^{-1}(u/2) + C = -\frac{\sqrt{4-9x^2}}{x} - 3\sin^{-1}(3x/2) + C$$

43. $u = 3x, du = 3dx$, Formula (112):

$$\frac{1}{3}\int \sqrt{\frac{5}{3}u - u^2}\,du = \frac{1}{6}\left(u - \frac{5}{6}\right)\sqrt{\frac{5}{3}u - u^2} + \frac{25}{216}\sin^{-1}\left(\frac{u-5}{5}\right) + C$$

$$= \frac{18x-5}{36}\sqrt{5x-9x^2} + \frac{25}{216}\sin^{-1}\left(\frac{18x-5}{5}\right) + C$$

44. $u = \sqrt{5}x, du = \sqrt{5}\,dx$, Formula (117):

$$\int \frac{du}{u\sqrt{(u/\sqrt{5}) - u^2}} = -\frac{\sqrt{(u/\sqrt{5}) - u^2}}{u/(2\sqrt{5})} + C = -2\frac{\sqrt{x-5x^2}}{x} + C$$

45. $u = 3x, du = 3dx$, Formula (44):

$$\frac{1}{9}\int u\sin u\,du = \frac{1}{9}(\sin u - u\cos u) + C = \frac{1}{9}(\sin 3x - 3x\cos 3x) + C$$

46. $u = \sqrt{x}, u^2 = x, 2udu = dx$, Formula (45): $2\int u\cos u\,du = 2\cos\sqrt{x} + 2\sqrt{x}\sin\sqrt{x} + C$

47. $u = -\sqrt{x}, u^2 = x, 2udu = dx$, Formula (51): $2\int ue^u du = -2(\sqrt{x}+1)e^{-\sqrt{x}} + C$

48. $u = 2 - 3x^2, du = -6xdx$, Formula (50):

$$-\frac{1}{6}\int \ln u\,du = -\frac{1}{6}(u\ln u - u) + C = -\frac{1}{6}((2-3x^2)\ln(2-3x^2) + \frac{1}{6}(2-3x^2) + C$$

49. $x^2 + 4x - 5 = (x+2)^2 - 9; u = x+2, du = dx$, Formula (70):

$$\int \frac{du}{u^2-9} = \frac{1}{6}\ln\left|\frac{u-3}{u+3}\right| + C = \frac{1}{6}\ln\left|\frac{x-1}{x+5}\right| + C$$

50. $x^2 + 2x - 3 = (x+1)^2 - 4, u = x+1, du = dx$, Formula (77):

$$\int \sqrt{4-u^2}\,du = \frac{1}{2}u\sqrt{4-u^2} + 2\sin^{-1}(u/2) + C$$

$$= \frac{1}{2}(x+1)\sqrt{3-2x-x^2} + 2\sin^{-1}((x+1)/2) + C$$

51. $x^2 - 4x - 5 = (x-2)^2 - 9, u = x-2, du = dx$, Formula (77):

$$\int \frac{u+2}{\sqrt{9-u^2}}du = \int \frac{u\,du}{\sqrt{9-u^2}} + 2\int \frac{du}{\sqrt{9-u^2}} = -\sqrt{9-u^2} + 2\sin^{-1}\frac{u}{3} + C$$

$$= -\sqrt{5+4x-x^2} + 2\sin^{-1}\left(\frac{x-2}{3}\right) + C$$

52. $x^2+6x+13=(x+3)^2+4, u=x+3, du=dx$, Formula (71):

$$\int \frac{(u-3)\,du}{u^2+4}=\frac{1}{2}\ln(u^2+4)-\frac{3}{2}\tan^{-1}(u/2)+C=\frac{1}{2}\ln(x^2+6x+13)-\frac{3}{2}\tan^{-1}((x+3)/2)+C$$

53. $u=\sqrt{x-2},\ x=u^2+2,\ dx=2u\,du;$

$$\int 2u^2(u^2+2)du=2\int(u^4+2u^2)du=\frac{2}{5}u^5+\frac{4}{3}u^3+C=\frac{2}{5}(x-2)^{5/2}+\frac{4}{3}(x-2)^{3/2}+C$$

54. $u=\sqrt{x+1},\ x=u^2-1,\ dx=2u\,du;$

$$2\int(u^2-1)du=\frac{2}{3}u^3-2u+C=\frac{2}{3}(x+1)^{3/2}-2\sqrt{x+1}+C$$

55. $u=\sqrt{x^3+1},\ x^3=u^2-1,\ 3x^2dx=2u\,du;$

$$\frac{2}{3}\int u^2(u^2-1)du=\frac{2}{3}\int(u^4-u^2)du=\frac{2}{15}u^5-\frac{2}{9}u^3+C=\frac{2}{15}(x^3+1)^{5/2}-\frac{2}{9}(x^3+1)^{3/2}+C$$

56. $u=\sqrt{x^3-1},\ x^3=u^2+1,\ 3x^2dx=2u\,du;$

$$\frac{2}{3}\int\frac{1}{u^2+1}du=\frac{2}{3}\tan^{-1}u+C=\frac{2}{3}\tan^{-1}\sqrt{x^3-1}+C$$

57. $u=x^{1/6},\ x=u^6,\ dx=6u^5du;$

$$\int\frac{6u^5}{u^3+u^2}du=6\int\frac{u^3}{u+1}du=6\int\left[u^2-u+1-\frac{1}{u+1}\right]du$$
$$=2x^{1/2}-3x^{1/3}+6x^{1/6}-6\ \ln(x^{1/6}+1)+C$$

58. $u=x^{1/5},\ x=u^5,\ dx=5u^4du;\ \int\frac{5u^4}{u^5-u^3}du=5\int\frac{u}{u^2-1}du=\frac{5}{2}\ln|x^{2/5}-1|+C$

59. $u=x^{1/4},\ x=u^4,\ dx=4u^3du;\ 4\int\frac{1}{u(1-u)}du=4\int\left[\frac{1}{u}+\frac{1}{1-u}\right]du=4\ln\frac{x^{1/4}}{|1-x^{1/4}|}+C$

60. $u=x^{1/3},\ x=u^3,\ dx=3u^2du;\ 3\int\frac{u^4}{u^3+1}du=3\int\left(u-\frac{u}{u^3+1}\right)du,$

$$\frac{u}{u^3+1}=\frac{u}{(u+1)(u^2-u+1)}=\frac{-1/3}{u+1}+\frac{(1/3)u+1/3}{u^2-u+1}\ \text{so}$$

$$3\int\left(u-\frac{u}{u^3+1}\right)du=\int\left(3u+\frac{1}{u+1}-\frac{u+1}{u^2-u+1}\right)du$$
$$=\frac{3}{2}u^2+\ln|u+1|-\frac{1}{2}\ln(u^2-u+1)-\sqrt{3}\tan^{-1}\frac{2u-1}{\sqrt{3}}+C$$
$$=\frac{3}{2}x^{2/3}+\ln|x^{1/3}+1|-\frac{1}{2}\ln(x^{2/3}-x^{1/3}+1)-\sqrt{3}\tan^{-1}\frac{2x^{1/3}-1}{\sqrt{3}}+C$$

61. $u=x^{1/6},\ x=u^6,\ dx=6u^5du;$

$$6\int\frac{u^3}{u-1}du=6\int\left[u^2+u+1+\frac{1}{u-1}\right]du=2x^{1/2}+3x^{1/3}+6x^{1/6}+6\ \ln|x^{1/6}-1|+C$$

62. $u = \sqrt{x}$, $x = u^2$, $dx = 2u\,du$;

$$-2\int \frac{u^2+u}{u-1}du = -2\int\left(u+2+\frac{2}{u-1}\right)du = -x - 4\sqrt{x} - 4\ln|\sqrt{x}-1| + C$$

63. $u = \sqrt{1+x^2}$, $x^2 = u^2-1$, $2x\,dx = 2u\,du$, $x\,dx = u\,du$;

$$\int (u^2-1)du = \frac{1}{3}(1+x^2)^{3/2} - (1+x^2)^{1/2} + C$$

64. $u = (x+3)^{1/5}$, $x = u^5 - 3$, $dx = 5u^4 du$;

$$5\int (u^8 - 3u^3)du = \frac{5}{9}(x+3)^{9/5} - \frac{15}{4}(x+3)^{4/5} + C$$

65. $u = \sqrt{x}$, $x = u^2$, $dx = 2u\,du$, Formula (44): $2\int u\sin u\,du = 2\sin\sqrt{x} - 2\sqrt{x}\cos\sqrt{x} + C$

66. $u = \sqrt{x}$, $x = u^2$, $dx = 2u\,du$, Formula (51): $2\int ue^u du = 2\sqrt{x}e^{\sqrt{x}} - 2e^{\sqrt{x}} + C$

67. $$\int \frac{1}{1+\dfrac{2u}{1+u^2}+\dfrac{1-u^2}{1+u^2}}\frac{2}{1+u^2}du = \int\frac{1}{u+1}du = \ln|\tan(x/2)+1| + C$$

68. $$\int \frac{1}{2+\dfrac{2u}{1+u^2}}\frac{2}{1+u^2}du = \int\frac{1}{u^2+u+1}du$$

$$= \int\frac{1}{(u+1/2)^2+3/4}du = \frac{2}{\sqrt{3}}\tan^{-1}\left(\frac{2\tan(x/2)+1}{\sqrt{3}}\right) + C$$

69. $u = \tan(\theta/2)$, $\displaystyle\int\frac{d\theta}{1-\cos\theta} = \int\frac{1}{u^2}du = -\frac{1}{u} + C = -\cot(\theta/2) + C$

70. $u = \tan(x/2)$,

$$\int\frac{2}{3u^2+8u-3}du = \frac{2}{3}\int\frac{1}{(u+4/3)^2-25/9}du = \frac{2}{3}\int\frac{1}{z^2-25/9}dz \quad (z = u+4/3)$$

$$= \frac{1}{5}\ln\left|\frac{z-5/3}{z+5/3}\right| + C = \frac{1}{5}\ln\left|\frac{\tan(x/2)-1/3}{\tan(x/2)+3}\right| + C$$

71. $u = \tan(x/2)$, $\displaystyle 2\int\frac{1\ \ u^2}{(3u^2+1)(u^2+1)}du$;

$$\frac{1-u^2}{(3u^2+1)(u^2+1)} = \frac{(0)u+2}{3u^2+1} + \frac{(0)u-1}{u^2+1} = \frac{2}{3u^2+1} - \frac{1}{u^2+1} \text{ so}$$

$$\int\frac{\cos x}{2-\cos x}dx = \frac{4}{\sqrt{3}}\tan^{-1}\left[\sqrt{3}\tan(x/2)\right] - x + C$$

72. $u = \tan(x/2)$, $\displaystyle\frac{1}{2}\int\frac{1-u^2}{u}du = \frac{1}{2}\int(1/u - u)du = \frac{1}{2}\ln|\tan(x/2)| - \frac{1}{4}\tan^2(x/2) + C$

73. $\int_2^x \frac{1}{t(4-t)}\,dt = \frac{1}{4}\ln\frac{t}{4-t}\bigg]_2^x$ (Formula (65), $a=4, b=-1$)

$$= \frac{1}{4}\left[\ln\frac{x}{4-x} - \ln 1\right] = \frac{1}{4}\ln\frac{x}{4-x}, \frac{1}{4}\ln\frac{x}{4-x} = 0.5, \ln\frac{x}{4-x} = 2,$$

$\frac{x}{4-x} = e^2$, $x = 4e^2 - e^2x$, $x(1+e^2) = 4e^2$, $x = 4e^2/(1+e^2) \approx 3.523188312$

74. $\int_1^x \frac{1}{t\sqrt{2t-1}}\,dt = 2\tan^{-1}\sqrt{2t-1}\bigg]_1^x$ (Formula (108), $a=-1, b=2$)

$$= 2\left(\tan^{-1}\sqrt{2x-1} - \tan^{-1}1\right) = 2\left(\tan^{-1}\sqrt{2x-1} - \pi/4\right),$$

$2(\tan^{-1}\sqrt{2x-1} - \pi/4) = 1$, $\tan^{-1}\sqrt{2x-1} = 1/2 + \pi/4$, $\sqrt{2x-1} = \tan(1/2 + \pi/4)$,

$x = [1 + \tan^2(1/2 + \pi/4)]/2 \approx 6.307993516$

75. $A = \int_0^4 \sqrt{25-x^2}\,dx = \left(\frac{1}{2}x\sqrt{25-x^2} + \frac{25}{2}\sin^{-1}\frac{x}{5}\right)\bigg]_0^4$ (Formula (74), $a=5$)

$$= 6 + \frac{25}{2}\sin^{-1}\frac{4}{5} \approx 17.59119023$$

76. $A = \int_{2/3}^2 \sqrt{9x^2-4}\,dx$; $u = 3x$,

$A = \frac{1}{3}\int_2^6 \sqrt{u^2-4}\,du = \frac{1}{3}\left(\frac{1}{2}u\sqrt{u^2-4} - 2\ln\left|u + \sqrt{u^2-4}\right|\right)\bigg]_2^6$ (Formula (73), $a^2 = 4$)

$$= \frac{1}{3}\left(3\sqrt{32} - 2\ln(6+\sqrt{32}) + 2\ln 2\right) = 4\sqrt{2} - \frac{2}{3}\ln(3+2\sqrt{2}) \approx 4.481689467$$

77. $A = \int_0^1 \frac{1}{25-16x^2}\,dx$; $u = 4x$,

$A = \frac{1}{4}\int_0^4 \frac{1}{25-u^2}\,du = \frac{1}{40}\ln\left|\frac{u+5}{u-5}\right|\bigg]_0^4 = \frac{1}{40}\ln 9 \approx 0.054930614$ (Formula (69), $a=5$)

78. $A = \int_1^4 \sqrt{x}\ln x\,dx = \frac{4}{9}x^{3/2}\left(\frac{3}{2}\ln x - 1\right)\bigg]_1^4$ (Formula (50), $n = 1/2$)

$$= \frac{4}{9}(12\ln 4 - 7) \approx 4.282458815$$

79. $V = 2\pi\int_0^{\pi/2} x\cos x\,dx = 2\pi(\cos x + x\sin x)\bigg]_0^{\pi/2} = \pi(\pi - 2) \approx 3.586419094$ (Formula (45))

80. $V = 2\pi\int_4^8 x\sqrt{x-4}\,dx = \frac{4\pi}{15}(3x+8)(x-4)^{3/2}\bigg]_4^8$ (Formula (102), $a=-4, b=1$)

$$= \frac{1024}{15}\pi \approx 214.4660585$$

81. $V = 2\pi \int_0^3 xe^{-x}dx;\ u = -x,$

$$V = 2\pi \int_0^{-3} ue^u du - 2\pi e^u(u-1)\Big]_0^{-3} = 2\pi(1 - 4c^{-3}) \approx 5.031899801 \quad \text{(Formula (51))}$$

82. $V = 2\pi \int_1^5 x \ln x\, dx = \frac{\pi}{2}x^2(2\ln x - 1)\Big]_1^5$

$$= \pi(25\ln 5 - 12) \approx 88.70584621 \quad \text{(Formula (50), } n = 1\text{)}$$

83. $L = \int_0^2 \sqrt{1+16x^2}\, dx;\ u = 4x,$

$$L = \frac{1}{4}\int_0^8 \sqrt{1+u^2}\, du = \frac{1}{4}\left(\frac{u}{2}\sqrt{1+u^2} + \frac{1}{2}\ln\left(u + \sqrt{1+u^2}\right)\right)\Bigg]_0^8 \quad \text{(Formula (72), } a^2 = 1\text{)}$$

$$= \sqrt{65} + \frac{1}{8}\ln(8 + \sqrt{65}) \approx 8.409316783$$

84. $L = \int_1^3 \sqrt{1 + 9/x^2}\, dx = \int_1^3 \frac{\sqrt{x^2+9}}{x} dx = \left(\sqrt{x^2+9} - 3\ln\left|\frac{3+\sqrt{x^2+9}}{x}\right|\right)\Bigg]_1^3$

$$= 3\sqrt{2} - \sqrt{10} + 3\ln\frac{3+\sqrt{10}}{1+\sqrt{2}} \approx 3.891581644 \quad \text{(Formula (89), } a = 3\text{)}$$

85. $S = 2\pi \int_0^{\pi} (\sin x)\sqrt{1+\cos^2 x}\, dx;\ u = \cos x,$

$$S = -2\pi\int_1^{-1}\sqrt{1+u^2}\, du = 4\pi\int_0^1 \sqrt{1+u^2}\, du = 4\pi\left(\frac{u}{2}\sqrt{1+u^2} + \frac{1}{2}\ln\left(u+\sqrt{1+u^2}\right)\right)\Bigg]_0^1 a^2 = 1$$

$$= 2\pi\left[\sqrt{2} + \ln(1+\sqrt{2})\right] \approx 14.42359945 \quad \text{(Formula (72))}$$

86. $S = 2\pi\int_1^4 \frac{1}{x}\sqrt{1+1/x^4}\, dx = 2\pi\int_1^4 \frac{\sqrt{x^4+1}}{x^3}\, dx;\ u = x^2,$

$$S = \pi\int_1^{16}\frac{\sqrt{u^2+1}}{u^2}\, du = \pi\left(-\frac{\sqrt{u^2+1}}{u} + \ln\left(u + \sqrt{u^2+1}\right)\right)\Bigg]_1^{16}$$

$$= \pi\left(\sqrt{2} - \frac{\sqrt{257}}{16} + \ln\frac{16+\sqrt{257}}{1+\sqrt{2}}\right) \approx 9.417237485 \text{ (Formula (93), } a^2 = 1\text{)}$$

87. **(a)** $s(t) = 2 + \int_0^t 20\cos^6 u \sin^3 u\, du$

$$= -\frac{20}{9}\sin^2 t\cos^7 t - \frac{40}{63}\cos^7 t + \frac{166}{63}$$

(b)

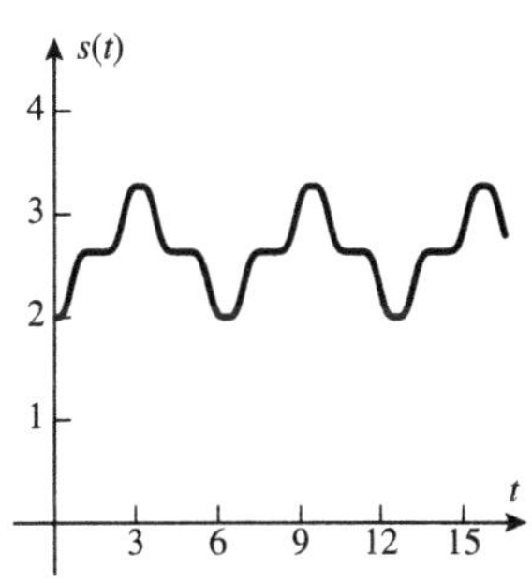

88. **(a)** $v(t) = \int_0^t a(u)\,du = -\frac{1}{10}e^{-t}\cos 2t + \frac{1}{5}e^{-t}\sin 2t + \frac{1}{74}e^{-t}\cos 6t - \frac{3}{37}e^{-t}\sin 6t + \frac{1}{10} - \frac{1}{74}$

$$s(t) = 10 + \int_0^t v(u)\,du$$
$$= -\frac{3}{50}e^{-t}\cos 2t - \frac{2}{25}e^{-t}\sin 2t + \frac{35}{2738}e^{-t}\cos 6t + \frac{6}{1369}e^{-t}\sin 6t + \frac{16}{185}t + \frac{343866}{34225}$$

(b)

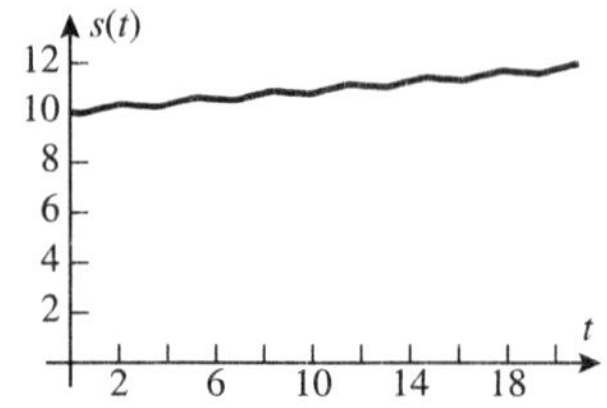

89. **(a)** $\int \sec x\,dx = \int \frac{1}{\cos x}dx = \int \frac{2}{1-u^2}du = \ln\left|\frac{1+u}{1-u}\right| + C = \ln\left|\frac{1+\tan(x/2)}{1-\tan(x/2)}\right| + C$

$$= \ln\left\{\left|\frac{\cos(x/2)+\sin(x/2)}{\cos(x/2)-\sin(x/2)}\right|\left|\frac{\cos(x/2)+\sin(x/2)}{\cos(x/2)+\sin(x/2)}\right|\right\} + C = \ln\left|\frac{1+\sin x}{\cos x}\right| + C$$
$$= \ln|\sec x + \tan x| + C$$

(b) $\tan\left(\frac{\pi}{4}+\frac{x}{2}\right) = \frac{\tan\frac{\pi}{4}+\tan\frac{x}{2}}{1-\tan\frac{\pi}{4}\tan\frac{x}{2}} = \frac{1+\tan\frac{x}{2}}{1-\tan\frac{x}{2}}$

90. $\int \csc x\,dx = \int \frac{1}{\sin x}dx = \int 1/u\,du = \ln|\tan(x/2)| + C$ but

$\ln|\tan(x/2)| = \frac{1}{2}\ln\frac{\sin^2(x/2)}{\cos^2(x/2)} = \frac{1}{2}\ln\frac{(1-\cos x)/2}{(1+\cos x)/2} = \frac{1}{2}\ln\frac{1-\cos x}{1+\cos x}$; also,

$\frac{1-\cos x}{1+\cos x} = \frac{1-\cos^2 x}{(1+\cos x)^2} = \frac{1}{(\csc x + \cot x)^2}$ so $\frac{1}{2}\ln\frac{1-\cos x}{1+\cos x} = -\ln|\csc x + \cot x|$

91. Let $u = \tanh(x/2)$ then $\cosh(x/2) = 1/\operatorname{sech}(x/2) = 1/\sqrt{1-\tanh^2(x/2)} = 1/\sqrt{1-u^2}$,

$\sinh(x/2) = \tanh(x/2)\cosh(x/2) = u/\sqrt{1-u^2}$, so $\sinh x = 2\sinh(x/2)\cosh(x/2) = 2u/(1-u^2)$,

$\cosh x = \cosh^2(x/2) + \sinh^2(x/2) = (1+u^2)/(1-u^2)$, $x = 2\tanh^{-1}u$, $dx = [2/(1-u^2)]du$;

$$\int \frac{dx}{2\cosh x + \sinh x} = \int \frac{1}{u^2+u+1}du = \frac{2}{\sqrt{3}}\tan^{-1}\frac{2u+1}{\sqrt{3}} + C = \frac{2}{\sqrt{3}}\tan^{-1}\frac{2\tanh(x/2)+1}{\sqrt{3}} + C.$$

EXERCISE SET 8.7

1. exact value $= 14/3 \approx 4.666666667$

(a) 4.667600663, $|E_M| \approx 0.000933996$
(b) 4.664795679, $|E_T| \approx 0.001870988$
(c) 4.666651630, $|E_S| \approx 0.000015037$

2. exact value $= 2$

(a) 1.998377048, $|E_M| \approx 0.001622952$
(b) 2.003260982, $|E_T| \approx 0.003260982$
(c) 2.000072698, $|E_S| \approx 0.000072698$

3. exact value $= 2$

(a) 2.008248408, $|E_M| \approx 0.008248408$
(b) 1.983523538, $|E_T| \approx 0.016476462$
(c) 2.000109517, $|E_S| \approx 0.000109517$

4. exact value $= \sin(1) \approx 0.841470985$

(a) 0.841821700, $|E_M| \approx 0.000350715$
(b) 0.840769642, $|E_T| \approx 0.000701343$
(c) 0.841471453, $|E_S| \approx 0.000000468$

5. exact value $= e^{-1} - e^{-3} \approx 0.318092373$

(a) 0.317562837, $|E_M| \approx 0.000529536$
(b) 0.319151975, $|E_T| \approx 0.001059602$
(c) 0.318095187, $|E_S| \approx 0.000002814$

6. exact value $= \frac{1}{2}\ln 5 \approx 0.804718956$

(a) 0.801605339, $|E_M| \approx 0.003113617$
(b) 0.811019505, $|E_T| \approx 0.006300549$
(c) 0.805041497, $|E_S| \approx 0.000322541$

7. $f(x) = \sqrt{x+1}$, $f''(x) = -\frac{1}{4}(x+1)^{-3/2}$, $f^{(4)}(x) = -\frac{15}{16}(x+1)^{-7/2}$; $K_2 = 1/4$, $K_4 = 15/16$

(a) $|E_M| \le \frac{27}{2400}(1/4) = 0.002812500$

(b) $|E_T| \le \frac{27}{1200}(1/4) = 0.005625000$

(c) $|E_S| \le \frac{243}{180 \times 10^4}(15/16) \approx 0.000126563$

8. $f(x) = 1/\sqrt{x}$, $f''(x) = \frac{3}{4}x^{-5/2}$, $f^{(4)}(x) = \frac{105}{16}x^{-9/2}$; $K_2 = 3/4$, $K_4 = 105/16$

(a) $|E_M| \le \frac{27}{2400}(3/4) = 0.008437500$

(b) $|E_T| \le \frac{27}{1200}(3/4) = 0.016875000$

(c) $|E_S| \le \frac{243}{180 \times 10^4}(105/16) \approx 0.000885938$

9. $f(x) = \sin x$, $f''(x) = -\sin x$, $f^{(4)}(x) = \sin x$; $K_2 = K_4 = 1$

(a) $|E_M| \le \frac{\pi^3}{2400}(1) \approx 0.012919282$

(b) $|E_T| \le \frac{\pi^3}{1200}(1) \approx 0.025838564$

(c) $|E_S| \le \frac{\pi^5}{180 \times 10^4}(1) \approx 0.000170011$

10. $f(x) = \cos x$, $f''(x) = -\cos x$, $f^{(4)}(x) = \cos x$; $K_2 = K_4 = 1$

(a) $|E_M| \le \frac{1}{2400}(1) \approx 0.000416667$

(b) $|E_T| \le \frac{1}{1200}(1) \approx 0.000833333$

(c) $|E_S| \le \frac{1}{180 \times 10^4}(1) \approx 0.000000556$

11. $f(x) = e^{-x}$, $f''(x) = f^{(4)}(x) = e^{-x}$; $K_2 = K_4 = e^{-1}$

(a) $|E_M| \le \frac{8}{2400}(e^{-1}) \approx 0.001226265$

(b) $|E_T| \le \frac{8}{1200}(e^{-1}) \approx 0.002452530$

(c) $|E_S| \le \frac{32}{180 \times 10^4}(e^{-1}) \approx 0.000006540$

12. $f(x) = 1/(2x+3)$, $f''(x) = 8(2x+3)^{-3}$, $f^{(4)}(x) = 384(2x+3)^{-5}$; $K_2 = 8$, $K_4 = 384$

(a) $|E_M| \le \frac{8}{2400}(8) \approx 0.026666667$

(b) $|E_T| \le \frac{8}{1200}(8) \approx 0.053333333$

(c) $|E_S| \le \frac{32}{180 \times 10^4}(384) \approx 0.006826667$

13. **(a)** $n > \left[\frac{(27)(1/4)}{(24)(5\times 10^{-4})}\right]^{1/2} \approx 23.7;\ n = 24$ **(b)** $n > \left[\frac{(27)(1/4)}{(12)(5\times 10^{-4})}\right]^{1/2} \approx 33.5;\ n = 34$

(c) $n > \left[\frac{(243)(15/16)}{(180)(5\times 10^{-4})}\right]^{1/4} \approx 7.1;\ n = 8$

14. **(a)** $n > \left[\frac{(27)(3/4)}{(24)(5\times 10^{-4})}\right]^{1/2} \approx 41.1;\ n = 42$ **(b)** $n > \left[\frac{(27)(3/4)}{(12)(5\times 10^{-4})}\right]^{1/2} \approx 58.1;\ n = 59$

(c) $n > \left[\frac{(243)(105/16)}{(180)(5\times 10^{-4})}\right]^{1/4} \approx 11.5;\ n = 12$

15. **(a)** $n > \left[\frac{(\pi^3)(1)}{(24)(10^{-3})}\right]^{1/2} \approx 35.9;\ n = 36$ **(b)** $n > \left[\frac{(\pi^3)(1)}{(12)(10^{-3})}\right]^{1/2} \approx 50.8;\ n = 51$

(c) $n > \left[\frac{(\pi^5)(1)}{(180)(10^{-3})}\right]^{1/4} \approx 6.4;\ n = 8$

16. **(a)** $n > \left[\frac{(1)(1)}{(24)(10^{-3})}\right]^{1/2} \approx 6.5;\ n = 7$ **(b)** $n > \left[\frac{(1)(1)}{(12)(10^{-3})}\right]^{1/2} \approx 9.1;\ n = 10$

(c) $n > \left[\frac{(1)(1)}{(180)(10^{-3})}\right]^{1/4} \approx 1.5;\ n = 2$

17. **(a)** $n > \left[\frac{(8)(e^{-1})}{(24)(10^{-6})}\right]^{1/2} \approx 350.2;\ n = 351$ **(b)** $n > \left[\frac{(8)(e^{-1})}{(12)(10^{-6})}\right]^{1/2} \approx 495.2;\ n = 496$

(c) $n > \left[\frac{(32)(e^{-1})}{(180)(10^{-6})}\right]^{1/4} \approx 15.99;\ n = 16$

18. **(a)** $n > \left[\frac{(8)(8)}{(24)(10^{-6})}\right]^{1/2} \approx 1632.99;\ n = 1633$ **(b)** $n > \left[\frac{(8)(8)}{(12)(10^{-6})}\right]^{1/2} \approx 2309.4;\ n = 2310$

(c) $n > \left[\frac{(32)(384)}{(180)(10^{-6})}\right]^{1/4} \approx 90.9;\ n = 92$

19. $g(X_0) = aX_0^2 + bX_0 + c = 4a + 2b + c = f(X_0) = 1/X_0 = 1/2$; similarly $9a + 3b + c = 1/3, 16a + 4b + c = 1/4$. Three equations in three unknowns, with solution $a = 1/24, b = -3/8, c = 13/12, g(x) = x^2/24 - 3x/8 + 13/12$.

$$\int_0^4 g(x)\,dx = \int \left(\frac{x^2}{24} - \frac{3x}{8} + \frac{13}{12}\right) dx = \frac{25}{36}$$

$$\frac{\Delta x}{3}[f(X_0) + 4f(X_1) + f(X_2)] = \frac{1}{3}\left[\frac{1}{2} + \frac{4}{3} + \frac{1}{4}\right] = \frac{25}{36}$$

20. $f(X_0) = 1 = g(X_0) = c, f(X_1) = 3/4 = g(X_1) = a/36 + b/6 + c$,
$f(X_2) = 1/4 = g(X_2) = a/9 + b/3 + c$,
with solution $a = -9/2, b = -3/4, c = 1, g(x) = -9x^2/2 - 3x/4 + 1$,

$$\int_0^{1/3} g(x)\,dx = 17/72$$

$$\frac{\Delta x}{3}[f(X_0) + 4f(X_1) + f(X_2)] = \frac{1}{18}[1 + 3 + 1/4] = 17/72$$

21. 0.746824948, 0.746824133

22. 1.137631378, 1.137630147

23. 2.129861595, 2.129861293

24. 2.418388347, 2.418399152

25. 0.805376152, 0.804776489

26. 1.536963087, 1.544294774

27. **(a)** 3.142425985, $|E_M| \approx 0.000833331$
(b) 3.139925989, $|E_T| \approx 0.001666665$
(c) 3.141592614, $|E_S| \approx 0.000000040$

28. **(a)** 3.152411433, $|E_M| \approx 0.010818779$
(b) 3.104518326, $|E_T| \approx 0.037074328$
(c) 3.127008159, $|E_S| \approx 0.014584495$

29. $S_{14} = 0.693147984$, $|E_S| \approx 0.000000803 = 8.03 \times 10^{-7}$; the method used in Example 6 results in a value of n which ensures that the magnitude of the error will be less than 10^{-6}, this is not necessarily the *smallest* value of n.

30. **(a)** greater, because the graph of e^{-x^2} is concave up on the interval $(1, 2)$
(b) less, because the graph of e^{-x^2} is concave down on the interval $(0, 0.5)$

31. $f(x) = x\sin x$, $f''(x) = 2\cos x - x\sin x$, $|f''(x)| \le 2|\cos x| + |x|\,|\sin x| \le 2 + 2 = 4$ so $K_2 \le 4$,

$$n > \left[\frac{(8)(4)}{(24)(10^{-4})}\right]^{1/2} \approx 115.5;\ n = 116 \text{ (a smaller } n \text{ might suffice)}$$

32. $f(x) = e^{\cos x}$, $f''(x) = (\sin^2 x)e^{\cos x} - (\cos x)e^{\cos x}$, $|f''(x)| \le e^{\cos x}(\sin^2 x + |\cos x|) \le 2e$ so

$$K_2 \le 2e,\ n > \left[\frac{(1)(2e)}{(24)(10^{-4})}\right]^{1/2} \approx 47.6;\ n = 48 \text{ (a smaller } n \text{ might suffice)}$$

33. $f(x) = \sqrt{x}$, $f''(x) = -\dfrac{1}{4x^{3/2}}$, $\lim\limits_{x\to 0^+} |f''(x)| = +\infty$

34. $f(x) = \sin\sqrt{x}$, $f''(x) = -\dfrac{\sqrt{x}\sin\sqrt{x} + \cos\sqrt{x}}{4x^{3/2}}$, $\lim\limits_{x\to 0^+} |f''(x)| = +\infty$

35. $L = \displaystyle\int_0^{\pi} \sqrt{1 + \cos^2 x}\, dx \approx 3.820187623$

36. $L = \displaystyle\int_1^{3} \sqrt{1 + 1/x^4}\, dx \approx 2.146822803$

37.

t (s)	0	5	10	15	20
v (mi/hr)	0	40	60	73	84
v (ft/s)	0	58.67	88	107.07	123.2

$$\int_0^{20} v\, dt \approx \frac{20}{(3)(4)}[0 + 4(58.67) + 2(88) + 4(107.07) + 123.2] \approx 1604 \text{ ft}$$

38.

t	0	1	2	3	4	5	6	7	8
a	0	0.02	0.08	0.20	0.40	0.60	0.70	0.60	0

$$\int_0^{8} a\, dt \approx \frac{8}{(3)(8)}[0 + 4(0.02) + 2(0.08) + 4(0.20) + 2(0.40) + 4(0.60) + 2(0.70) + 4(0.60) + 0]$$
$$\approx 2.7 \text{ cm/s}$$

39. $\int_0^{180} v\,dt \approx \frac{180}{(3)(6)}[0.00 + 4(0.03) + 2(0.08) + 4(0.16) + 2(0.27) + 4(0.42) + 0.65] = 37.9$ mi

40. $\int_0^{1800} (1/v)dx \approx \frac{1800}{(3)(6)}\left[\frac{1}{3100} + \frac{4}{2908} + \frac{2}{2725} + \frac{4}{2549} + \frac{2}{2379} + \frac{4}{2216} + \frac{1}{2059}\right] \approx 0.71$ s

41. $$V = \int_0^{16} \pi r^2 dy = \pi \int_0^{16} r^2 dy \approx \pi \frac{16}{(3)(4)}[(8.5)^2 + 4(11.5)^2 + 2(13.8)^2 + 4(15.4)^2 + (16.8)^2]$$
$$\approx 9270 \text{ cm}^3 \approx 9.3 \text{ L}$$

42. $A = \int_0^{600} h\,dx \approx \frac{600}{(3)(6)}[0 + 4(7) + 2(16) + 4(24) + 2(25) + 4(16) + 0] = 9000 \text{ ft}^2$,
$V = 75A \approx 75(9000) = 675,000 \text{ ft}^3$

43. $$\int_a^b f(x)\,dx \approx A_1 + A_2 + \cdots + A_n = \frac{b-a}{n}\left[\frac{1}{2}(y_0 + y_1) + \frac{1}{2}(y_1 + y_2) + \cdots + \frac{1}{2}(y_{n-1} + y_n)\right]$$
$$= \frac{b-a}{2n}[y_0 + 2y_1 + 2y_2 + \cdots + 2y_{n-1} + y_n]$$

44. right endpoint, trapezoidal, midpoint, left endpoint

45. **(a)** The maximum value of $|f''(x)|$ is approximately 3.844880.
(b) $n = 18$
(c) 0.904741

46. **(a)** The maximum value of $|f''(x)|$ is approximately 1.467890.
(b) $n = 12$
(c) 1.112062

47. **(a)** The maximum value of $|f^{(4)}(x)|$ is approximately 42.551816.
(b) $n = 8$
(c) 0.904524

48. **(a)** The maximum value of $|f^{(4)}(x)|$ is approximately 7.022710.
(b) $n = 8$
(c) 1.111443

EXERCISE SET 8.8

1. **(a)** improper; infinite discontinuity at $x = 3$
(b) continuous integrand, not improper
(c) improper; infinite discontinuity at $x = 0$
(d) improper; infinite interval of integration
(e) improper; infinite interval of integration and infinite discontinuity at $x = 1$
(f) continuous integrand, not improper

2. **(a)** improper if $p > 0$ **(b)** improper if $1 \le p \le 2$
(c) integrand is continuous for all p, not improper

3. $\lim_{\ell\to+\infty}(-e^{-x})\Big]_0^{\ell} = \lim_{\ell\to+\infty}(-e^{-\ell}+1) = 1$

4. $\lim_{\ell\to+\infty}\frac{1}{2}\ln(1+x^2)\Big]_{-1}^{\ell} = \lim_{\ell\to+\infty}\frac{1}{2}[\ln(1+\ell^2)-\ln 2] = +\infty$, divergent

5. $\lim_{\ell\to+\infty}\ln\frac{x-1}{x+1}\Big]_4^{\ell} = \lim_{\ell\to+\infty}\left(\ln\frac{\ell-1}{\ell+1}-\ln\frac{3}{5}\right) = -\ln\frac{3}{5} = \ln\frac{5}{3}$

6. $\lim_{\ell\to+\infty}-\frac{1}{2}e^{-x^2}\Big]_0^{\ell} = \lim_{\ell\to+\infty}\frac{1}{2}\left(-e^{-\ell^2}+1\right) = 1/2$

7. $\lim_{\ell\to+\infty}-\frac{1}{2\ln^2 x}\Big]_e^{\ell} = \lim_{\ell\to+\infty}\left[-\frac{1}{2\ln^2\ell}+\frac{1}{2}\right] = \frac{1}{2}$

8. $\lim_{\ell\to+\infty}2\sqrt{\ln x}\Big]_2^{\ell} = \lim_{\ell\to+\infty}(2\sqrt{\ln\ell}-2\sqrt{\ln 2}) = +\infty$, divergent

9. $\lim_{\ell\to-\infty}-\frac{1}{4(2x-1)^2}\Big]_{\ell}^{0} = \lim_{\ell\to-\infty}\frac{1}{4}[-1+1/(2\ell-1)^2] = -1/4$

10. $\lim_{\ell\to-\infty}\frac{1}{2}\tan^{-1}\frac{x}{2}\Big]_{\ell}^{2} = \lim_{\ell\to-\infty}\frac{1}{2}\left[\frac{\pi}{4}-\tan^{-1}\frac{\ell}{2}\right] = \frac{1}{2}[\pi/4-(-\pi/2)] = 3\pi/8$

11. $\lim_{\ell\to-\infty}\frac{1}{3}e^{3x}\Big]_{\ell}^{0} = \lim_{\ell\to-\infty}\left[\frac{1}{3}-\frac{1}{3}e^{3\ell}\right] = \frac{1}{3}$

12. $\lim_{\ell\to-\infty}-\frac{1}{2}\ln(3-2e^x)\Big]_{\ell}^{0} = \lim_{\ell\to-\infty}\frac{1}{2}\ln(3-2e^{\ell}) = \frac{1}{2}\ln 3$

13. $\int_{-\infty}^{+\infty}x^3dx$ converges if $\int_{-\infty}^{0}x^3dx$ and $\int_{0}^{+\infty}x^3dx$ both converge; it diverges if either (or both) diverges. $\int_0^{+\infty}x^3dx = \lim_{\ell\to+\infty}\frac{1}{4}x^4\Big]_0^{\ell} = \lim_{\ell\to+\infty}\frac{1}{4}\ell^4 = +\infty$ so $\int_{-\infty}^{+\infty}x^3dx$ is divergent.

14. $\int_0^{+\infty}\frac{x}{\sqrt{x^2+2}}dx = \lim_{\ell\to+\infty}\sqrt{x^2+2}\Big]_0^{\ell} = \lim_{\ell\to+\infty}(\sqrt{\ell^2+2}-\sqrt{2}) = +\infty$

so $\int_{-\infty}^{\infty}\frac{x}{\sqrt{x^2+2}}dx$ is divergent.

15. $\int_0^{+\infty}\frac{x}{(x^2+3)^2}dx = \lim_{\ell\to+\infty}-\frac{1}{2(x^2+3)}\Big]_0^{\ell} = \lim_{\ell\to+\infty}\frac{1}{2}[-1/(\ell^2+3)+1/3] = \frac{1}{6}$,

similarly $\int_{-\infty}^{0}\frac{x}{(x^2+3)^2}dx = -1/6$ so $\int_{-\infty}^{\infty}\frac{x}{(x^2+3)^2}dx = 1/6+(-1/6) = 0$

16. $\int_0^{+\infty} \frac{e^{-t}}{1+e^{-2t}} dt = \lim_{\ell\to+\infty} -\tan^{-1}(e^{-t})\Big]_0^{\ell} = \lim_{\ell\to+\infty}\left[-\tan^{-1}(e^{-\ell}) + \frac{\pi}{4}\right] = \frac{\pi}{4}$,

$\int_{-\infty}^{0} \frac{e^{-t}}{1+e^{-2t}} dt = \lim_{\ell\to-\infty} -\tan^{-1}(e^{-t})\Big]_{\ell}^{0} = \lim_{\ell\to-\infty}\left[-\frac{\pi}{4} + \tan^{-1}(e^{-\ell})\right] = \frac{\pi}{4}$,

$\int_{-\infty}^{+\infty} \frac{e^{-t}}{1+e^{-2t}} dt = \frac{\pi}{4} + \frac{\pi}{4} = \frac{\pi}{2}$

17. $\lim_{\ell\to3^+} -\frac{1}{x-3}\Big]_{\ell}^{4} = \lim_{\ell\to3^+}\left[-1 + \frac{1}{\ell-3}\right] = +\infty$, divergent

18. $\lim_{\ell\to0^+} \frac{3}{2}x^{2/3}\Big]_{\ell}^{8} = \lim_{\ell\to0^+} \frac{3}{2}(4 - \ell^{2/3}) = 6$

19. $\lim_{\ell\to\pi/2^-} -\ln(\cos x)\Big]_0^{\ell} = \lim_{\ell\to\pi/2^-} -\ln(\cos \ell) = +\infty$, divergent

20. $\lim_{\ell\to9^-} -2\sqrt{9-x}\Big]_0^{\ell} = \lim_{\ell\to9^-} 2(-\sqrt{9-\ell}+3) = 6$

21. $\lim_{\ell\to1^-} \sin^{-1} x\Big]_0^{\ell} = \lim_{\ell\to1^-} \sin^{-1}\ell = \pi/2$

22. $\lim_{\ell\to-3^+} -\sqrt{9-x^2}\Big]_{\ell}^{1} = \lim_{\ell\to-3^+}(-\sqrt{8} + \sqrt{9-\ell^2}) = -\sqrt{8}$

23. $\lim_{\ell\to\pi/6^-} -\sqrt{1-2\sin x}\Big]_0^{\ell} = \lim_{\ell\to\pi/6^-}(-\sqrt{1-2\sin\ell}+1) = 1$

24. $\lim_{\ell\to\pi/4^-} -\ln(1-\tan x)\Big]_0^{\ell} = \lim_{\ell\to\pi/4^-} -\ln(1-\tan\ell) = +\infty$, divergent

25. $\int_0^2 \frac{dx}{x-2} = \lim_{\ell\to2^-} \ln|x-2|\Big]_0^{\ell} = \lim_{\ell\to2^-}(\ln|\ell-2| - \ln 2) = -\infty$, divergent

26. $\int_0^2 \frac{dx}{x^2} = \lim_{\ell\to0^+} -1/x\Big]_{\ell}^{2} = \lim_{\ell\to0^+}(-1/2 + 1/\ell) = +\infty$ so $\int_{-2}^{2} \frac{dx}{x^2}$ is divergent

27. $\int_0^8 x^{-1/3} dx = \lim_{\ell\to0^+} \frac{3}{2}x^{2/3}\Big]_{\ell}^{8} = \lim_{\ell\to0^+} \frac{3}{2}(4-\ell^{2/3}) = 6$,

$\int_{-1}^{0} x^{-1/3} dx = \lim_{\ell\to0^-} \frac{3}{2}x^{2/3}\Big]_{-1}^{\ell} = \lim_{\ell\to0^-} \frac{3}{2}(\ell^{2/3} - 1) = -3/2$

so $\int_{-1}^{8} x^{-1/3} dx = 6 + (-3/2) = 9/2$

28. $\int_0^2 \frac{dx}{(x-2)^{2/3}} = \lim_{\ell\to2^-} 3(x-2)^{1/3}\Big]_0^{\ell} = \lim_{\ell\to2^-} 3[(\ell-2)^{1/3} - (-2)^{1/3}] = 3\sqrt[3]{2}$,

similarly $\int_2^4 \frac{dx}{(x-2)^{2/3}} = \lim_{\ell\to2^+} 3(x-2)^{1/3}\Big]_{\ell}^{4} = 3\sqrt[3]{2}$ so $\int_0^4 \frac{dx}{(x-2)^{2/3}} = 6\sqrt[3]{2}$

29. Define $\int_0^{+\infty} \frac{1}{x^2}dx = \int_0^a \frac{1}{x^2}dx + \int_a^{+\infty} \frac{1}{x^2}dx$ where $a > 0$; take $a = 1$ for convenience,

$\int_0^1 \frac{1}{x^2}dx = \lim_{\ell\to 0^+} (-1/x)\Big]_\ell^1 = \lim_{\ell\to 0^+} (1/\ell - 1) = +\infty$ so $\int_0^{+\infty} \frac{1}{x^2}dx$ is divergent.

30. Define $\int_1^{+\infty} \frac{dx}{x\sqrt{x^2-1}} = \int_1^a \frac{dx}{x\sqrt{x^2-1}} + \int_a^{+\infty} \frac{dx}{x\sqrt{x^2-1}}$ where $a > 1$,

take $a = 2$ for convenience to get

$$\int_1^2 \frac{dx}{x\sqrt{x^2-1}} = \lim_{\ell\to 1^+} \sec^{-1} x\Big]_\ell^2 = \lim_{\ell\to 1^+} (\pi/3 - \sec^{-1}\ell) = \pi/3,$$

$$\int_2^{+\infty} \frac{dx}{x\sqrt{x^2-1}} = \lim_{\ell\to+\infty} \sec^{-1} x\Big]_2^\ell = \pi/2 - \pi/3 \text{ so } \int_1^{+\infty} \frac{dx}{x\sqrt{x^2-1}} = \pi/2.$$

31. $\int_0^{+\infty} \frac{e^{-\sqrt{x}}}{\sqrt{x}}dx = 2\int_0^{+\infty} e^{-u}du = 2\lim_{\ell\to+\infty} \left(-e^{-u}\right)\Big]_0^\ell = 2\lim_{\ell\to+\infty} \left(1 - e^{-\ell}\right) = 2$

32. $\int_0^{+\infty} \frac{dx}{\sqrt{x}(x+4)} = 2\int_0^{+\infty} \frac{du}{u^2+4} = 2\lim_{\ell\to+\infty} \frac{1}{2}\tan^{-1}\frac{u}{2}\Big]_0^\ell = \lim_{\ell\to+\infty} \tan^{-1}\frac{\ell}{2} = \frac{\pi}{2}$

33. $\int_0^{+\infty} \frac{e^{-x}}{\sqrt{1-e^{-x}}}dx = \int_0^1 \frac{du}{\sqrt{u}} = \lim_{\ell\to 0^+} 2\sqrt{u}\Big]_\ell^1 = \lim_{\ell\to 0^+} 2(1-\sqrt{\ell}) = 2$

34. $\int_0^{+\infty} \frac{e^{-x}}{\sqrt{1-e^{-2x}}}dx = -\int_1^0 \frac{du}{\sqrt{1-u^2}} = \int_0^1 \frac{du}{\sqrt{1-u^2}} = \lim_{\ell\to 1} \sin^{-1} u\Big]_0^\ell = \lim_{\ell\to 1} \sin^{-1}\ell = \frac{\pi}{2}$

35. $\lim_{\ell\to+\infty} \int_0^\ell e^{-x}\cos x\,dx = \lim_{\ell\to+\infty} \frac{1}{2}e^{-x}(\sin x - \cos x)\Big]_0^\ell = 1/2$

36. $A = \int_0^{+\infty} xe^{-3x}dx = \lim_{\ell\to+\infty} -\frac{1}{9}(3x+1)e^{-3x}\Big]_0^\ell = 1/3$

37. **(a)** 2.726585 **(b)** 2.804364 **(c)** 0.219384 **(d)** 0.504067

39. $1 + \left(\frac{dy}{dx}\right)^2 = 1 + \frac{4 - x^{2/3}}{x^{2/3}} = \frac{4}{x^{2/3}}$; the arc length is $\int_0^8 \frac{2}{x^{1/3}}dx = 3x^{2/3}\Big|_0^8 = 12$

40. $1 + \left(\frac{dy}{dx}\right)^2 = 1 + \frac{16x^2}{9-4x^2} = \frac{9+12x^2}{9-4x^2}$; the arc length is $\int_0^{3/2} \sqrt{\frac{9+12x^2}{9-4x^2}}\,dx \approx 3.633168$

41. $\int \ln x\,dx = x\ln x - x + C$,

$$\int_0^1 \ln x\,dx = \lim_{\ell\to 0^+} \int_\ell^1 \ln x\,dx = \lim_{\ell\to 0^+} (x\ln x - x)\Big]_\ell^1 = \lim_{\ell\to 0^+} (-1 - \ell\ln\ell + \ell),$$

but $\lim_{\ell\to 0^+} \ell\ln\ell = \lim_{\ell\to 0^+} \frac{\ln\ell}{1/\ell} = \lim_{\ell\to 0^+} (-\ell) = 0$ so $\int_0^1 \ln x\,dx = -1$

42. $\displaystyle\int \frac{\ln x}{x^2}dx = -\frac{\ln x}{x} - \frac{1}{x} + C,$

$$\int_1^{+\infty} \frac{\ln x}{x^2}dx = \lim_{\ell\to+\infty}\int_1^{\ell}\frac{\ln x}{x^2}dx = \lim_{\ell\to+\infty}\left(-\frac{\ln x}{x} - \frac{1}{x}\right)\Big]_1^{\ell} = \lim_{\ell\to+\infty}\left(-\frac{\ln \ell}{\ell} - \frac{1}{\ell} + 1\right),$$

but $\displaystyle\lim_{\ell\to+\infty}\frac{\ln\ell}{\ell} = \lim_{\ell\to+\infty}\frac{1}{\ell} = 0$ so $\displaystyle\int_1^{+\infty}\frac{\ln x}{x^2} = 1$

43. $\displaystyle\int_0^{+\infty} e^{-3x}dx = \lim_{\ell\to+\infty}\int_0^{\ell} e^{-3x}dx = \lim_{\ell\to+\infty}\left(-\frac{1}{3}e^{-3x}\right)\Big]_0^{\ell} = \lim_{\ell\to+\infty}\left(-\frac{1}{3}e^{-3\ell} + \frac{1}{9}\right) = \frac{1}{3}$

44. $\displaystyle A = \int_3^{+\infty}\frac{8}{x^2-4}dx = \lim_{\ell\to+\infty} 2\ln\frac{x-2}{x+2}\Big]_3^{\ell} = \lim_{\ell\to+\infty} 2\left[\ln\frac{\ell-2}{\ell+2} - \ln\frac{1}{5}\right] = 2\ln 5$

45. **(a)** $\displaystyle V = \pi\int_0^{+\infty} e^{-2x}dx = -\frac{\pi}{2}\lim_{\ell\to+\infty} e^{-2x}\Big]_0^{\ell} = \pi/2$

(b) $\displaystyle S = 2\pi\int_0^{+\infty} e^{-x}\sqrt{1+e^{-2x}}dx$, let $u = e^{-x}$ to get

$$S = -2\pi\int_1^0\sqrt{1+u^2}du = 2\pi\left[\frac{u}{2}\sqrt{1+u^2} + \frac{1}{2}\ln\left|u+\sqrt{1+u^2}\right|\right]_0^1 = \pi\left[\sqrt{2} + \ln(1+\sqrt{2})\right]$$

47. **(a)** For $x \ge 1, x^2 \ge x, e^{-x^2} \le e^{-x}$

(b) $\displaystyle\int_1^{+\infty} e^{-x}\,dx = \lim_{\ell\to+\infty}\int_1^{\ell} e^{-x}\,dx = \lim_{\ell\to+\infty} -e^{-x}\Big]_1^{\ell} = \lim_{\ell\to+\infty}(e^{-1} - e^{-\ell}) = 1/e$

(c) By Parts (a) and (b) and Exercise 46(b), $\displaystyle\int_1^{+\infty} e^{-x^2}\,dx$ is convergent and is $\le 1/e$.

48. **(a)** If $x \ge 0$ then $e^x \ge 1, \dfrac{1}{2x+1} \le \dfrac{e^x}{2x+1}$

(b) $\displaystyle\lim_{\ell\to+\infty}\int_0^{\ell}\frac{dx}{2x+1} = \lim_{\ell\to+\infty}\frac{1}{2}\ln(2x+1)\Big]_0^{\ell} = +\infty$

(c) By Parts (a) and (b) and Exercise 46(a), $\displaystyle\int_0^{+\infty}\frac{e^x}{2x+1}\,dx$ is divergent.

49. $\displaystyle V = \lim_{\ell\to+\infty}\int_1^{\ell}(\pi/x^2)\,dx = \lim_{\ell\to+\infty} -(\pi/x)\Big]_1^{\ell} = \lim_{\ell\to+\infty}(\pi - \pi/\ell) = \pi$

$\displaystyle A = \lim_{\ell\to+\infty}\int_1^{\ell} 2\pi(1/x)\sqrt{1+1/x^4}\,dx$; use Exercise 46(a) with $f(x) = 2\pi/x, g(x) = (2\pi/x)\sqrt{1+1/x^4}$ and $a = 1$ to see that the area is infinite.

50. **(a)** $\displaystyle 1 \le \frac{\sqrt{x^3+1}}{x}$ for $x \ge 2$, $\displaystyle\int_2^{+\infty} 1dx = +\infty$

(b) $\displaystyle\int_2^{+\infty}\frac{x}{x^5+1}dx \le \int_2^{+\infty}\frac{dx}{x^4} = \lim_{\ell\to+\infty} -\frac{1}{3x^3}\Big]_2^{\ell} = 1/24$

(c) $\displaystyle\int_0^{\infty}\frac{xe^x}{2x+1}dx \ge \int_1^{+\infty}\frac{xe^x}{2x+1} \ge \int_1^{+\infty}\frac{dx}{2x+1} = +\infty$

51. $\displaystyle\int_0^{2x}\sqrt{1+t^3}dt \ge \int_0^{2x} t^{3/2}dt = \frac{2}{5}t^{5/2}\bigg]_0^{2x} = \frac{2}{5}(2x)^{5/2},$

$\displaystyle\lim_{x\to+\infty}\int_0^{2x} t^{3/2}dt = \lim_{x\to+\infty}\frac{2}{5}(2x)^{5/2} = +\infty$ so $\displaystyle\int_0^{+\infty}\sqrt{1+t^3}dt = +\infty$; by L'Hôpital's Rule

$$\lim_{x\to+\infty}\frac{\displaystyle\int_0^{2x}\sqrt{1+t^3}dt}{x^{5/2}} = \lim_{x\to+\infty}\frac{2\sqrt{1+(2x)^3}}{(5/2)x^{3/2}} = \lim_{x\to+\infty}\frac{2\sqrt{1/x^3+8}}{5/2} = 8\sqrt{2}/5$$

52. (b) $u=\sqrt{x}$, $\displaystyle\int_0^{+\infty}\frac{\cos\sqrt{x}}{\sqrt{x}}dx = 2\int_0^{+\infty}\cos u\,du$; $\displaystyle\int_0^{+\infty}\cos u\,du$ diverges by Part (a).

53. Let $x = r\tan\theta$ to get $\displaystyle\int\frac{dx}{(r^2+x^2)^{3/2}} = \frac{1}{r^2}\int\cos\theta\,d\theta = \frac{1}{r^2}\sin\theta + C = \frac{x}{r^2\sqrt{r^2+x^2}} + C$

$$\text{so } u = \frac{2\pi NIr}{k}\lim_{\ell\to+\infty}\frac{x}{r^2\sqrt{r^2+x^2}}\bigg]_a^{\ell} = \frac{2\pi NI}{kr}\lim_{\ell\to+\infty}(\ell/\sqrt{r^2+\ell^2} - a/\sqrt{r^2+a^2})$$
$$= \frac{2\pi NI}{kr}(1 - a/\sqrt{r^2+a^2}).$$

54. Let $a^2 = \dfrac{M}{2RT}$ to get

(a) $\displaystyle\bar{v} = \frac{4}{\sqrt{\pi}}\left(\frac{M}{2RT}\right)^{3/2}\frac{1}{2}\left(\frac{M}{2RT}\right)^{-2} = \frac{2}{\sqrt{\pi}}\sqrt{\frac{2RT}{M}} = \sqrt{\frac{8RT}{\pi M}}$

(b) $\displaystyle v_{\rm rms}^2 = \frac{4}{\sqrt{\pi}}\left(\frac{M}{2RT}\right)^{3/2}\frac{3\sqrt{\pi}}{8}\left(\frac{M}{2RT}\right)^{-5/2} = \frac{3RT}{M}$ so $\displaystyle v_{\rm rms} = \sqrt{\frac{3RT}{M}}$

55. (a) Satellite's weight $= w(x) = k/x^2$ lb when x = distance from center of Earth; $w(4000) = 6000$ so $k = 9.6\times10^{10}$ and $\displaystyle W = \int_{4000}^{4000+\ell} 9.6\times10^{10}x^{-2}dx$ mi·lb.

(b) $\displaystyle\int_{4000}^{+\infty} 9.6\times10^{10}x^{-2}dx = \lim_{\ell\to+\infty} -9.6\times10^{10}/x\bigg]_{4000}^{\ell} = 2.4\times10^7$ mi·lb

56. (a) $\displaystyle\mathcal{L}\{1\} = \int_0^{+\infty}e^{-st}dt = \lim_{\ell\to+\infty} -\frac{1}{s}e^{-st}\bigg]_0^{\ell} = \frac{1}{s}$

(b) $\displaystyle\mathcal{L}\{e^{2t}\} = \int_0^{+\infty}e^{-st}e^{2t}dt = \int_0^{+\infty}e^{-(s-2)t}dt = \lim_{\ell\to+\infty} -\frac{1}{s-2}e^{-(s-2)t}\bigg]_0^{\ell} = \frac{1}{s-2}$

(c) $\displaystyle\mathcal{L}\{\sin t\} = \int_0^{+\infty}e^{-st}\sin t\,dt = \lim_{\ell\to+\infty}\frac{e^{-st}}{s^2+1}(-s\sin t - \cos t)\bigg]_0^{\ell} = \frac{1}{s^2+1}$

(d) $\displaystyle\mathcal{L}\{\cos t\} = \int_0^{+\infty}e^{-st}\cos t\,dt = \lim_{\ell\to+\infty}\frac{e^{-st}}{s^2+1}(-s\cos t + \sin t)\bigg]_0^{\ell} = \frac{s}{s^2+1}$

57. (a) $\displaystyle\mathcal{L}\{f(t)\} = \int_0^{+\infty}te^{-st}\,dt = \lim_{\ell\to+\infty} -(t/s + 1/s^2)e^{-st}\bigg]_0^{\ell} = \frac{1}{s^2}$

(b) $\displaystyle\mathcal{L}\{f(t)\} = \int_0^{+\infty}t^2e^{-st}\,dt = \lim_{\ell\to+\infty} -(t^2/s + 2t/s^2 + 2/s^3)e^{-st}\bigg]_0^{\ell} = \frac{2}{s^3}$

(c) $\displaystyle\mathcal{L}\{f(t)\} = \int_3^{+\infty}e^{-st}dt = \lim_{\ell\to+\infty} -\frac{1}{s}e^{-st}\bigg]_3^{\ell} = \frac{e^{-3s}}{s}$

58.

10	100	1000	10,000
0.8862269	0.8862269	0.8862269	0.8862269

59. **(a)** $u = \sqrt{a}x, du = \sqrt{a}\,dx, 2\int_0^{+\infty} e^{-ax^2}\,dx = \frac{2}{\sqrt{a}}\int_0^{+\infty} e^{-u^2}\,du = \sqrt{\pi/a}$

(b) $x = \sqrt{2}\sigma u, dx = \sqrt{2}\sigma\,du, \frac{2}{\sqrt{2\pi}\sigma}\int_0^{+\infty} e^{-x^2/2\sigma^2}\,dx = \frac{2}{\sqrt{\pi}}\int_0^{+\infty} e^{-u^2}\,du = 1$

60. **(a)** $\int_0^3 e^{-x^2}\,dx \approx 0.8862;\ \sqrt{\pi}/2 \approx 0.8862$

(b) $\int_0^{+\infty} e^{-x^2}\,dx = \int_0^3 e^{-x^2}\,dx + \int_3^{+\infty} e^{-x^2}\,dx$ so $E = \int_3^{+\infty} e^{-x^2}\,dx < \int_3^{+\infty} xe^{-x^2}\,dx = \frac{1}{2}e^{-9} < 7 \times 10^{-5}$

61. **(a)** $\int_0^4 \frac{1}{x^6+1}dx \approx 1.047;\ \pi/3 \approx 1.047$

(b) $\int_0^{+\infty} \frac{1}{x^6+1}dx = \int_0^4 \frac{1}{x^6+1}dx + \int_4^{+\infty} \frac{1}{x^6+1}dx$ so

$$E = \int_4^{+\infty} \frac{1}{x^6+1}dx < \int_4^{+\infty} \frac{1}{x^6}dx = \frac{1}{5(4)^5} < 2 \times 10^{-4}$$

62. If $p = 0$, then $\int_0^{+\infty} (1)dx = \lim_{\ell\to+\infty} x\Big]_0^\ell = +\infty$,

if $p \neq 0$, then $\int_0^{+\infty} e^{px}dx = \lim_{\ell\to+\infty} \frac{1}{p}e^{px}\Big]_0^\ell = \lim_{\ell\to+\infty} \frac{1}{p}(e^{p\ell} - 1) = \begin{cases} -1/p, & p < 0 \\ +\infty, & p > 0 \end{cases}$.

63. If $p = 1$, then $\int_0^1 \frac{dx}{x} = \lim_{\ell\to0^+} \ln x\Big]_\ell^1 = +\infty$;

if $p \neq 1$, then $\int_0^1 \frac{dx}{x^p} = \lim_{\ell\to0^+} \frac{x^{1-p}}{1-p}\Big]_\ell^1 = \lim_{\ell\to0^+} [(1 - \ell^{1-p})/(1-p)] = \begin{cases} 1/(1-p), & p < 1 \\ +\infty, & p > 1 \end{cases}$.

64. $u = \sqrt{1-x}, u^2 = 1 - x, 2u\,du = -dx$;

$$-2\int_1^0 \sqrt{2-u^2}\,du = 2\int_0^1 \sqrt{2-u^2}\,du = \left[u\sqrt{2-u^2} + 2\sin^{-1}(u/\sqrt{2})\right]_0^1 = \sqrt{2} + \pi/2$$

65. $2\int_0^1 \cos(u^2)du \approx 1.809$

66. $-2\int_1^0 \sin(1-u^2)du = 2\int_0^1 \sin(1-u^2)du \approx 1.187$

CHAPTER 8 SUPPLEMENTARY EXERCISES

1. **(a)** integration by parts, $u = x$, $dv = \sin x\,dx$
(b) u-substitution: $u = \sin x$
(c) reduction formula
(d) u-substitution: $u = \tan x$
(e) u-substitution: $u = x^3 + 1$
(f) u-substitution: $u = x + 1$
(g) integration by parts: $dv = dx, u = \tan^{-1} x$
(h) trigonometric substitution: $x = 2\sin\theta$
(i) u-substitution: $u = 4 - x^2$

2. **(a)** $x = 3\tan\theta$ **(b)** $x = 3\sin\theta$ **(c)** $x = \frac{1}{3}\sin\theta$
(d) $x = 3\sec\theta$ **(e)** $x = \sqrt{3}\tan\theta$ **(f)** $x = \frac{1}{9}\tan\theta$

5. **(a)** #40 **(b)** #57 **(c)** #113
(d) #108 **(e)** #52 **(f)** #71

6. **(a)** $u = x^2$, $dv = \dfrac{x}{\sqrt{x^2+1}}dx$, $du = 2x\,dx$, $v = \sqrt{x^2+1}$;

$$\int_0^1 \frac{x^3}{\sqrt{x^2+1}}dx = x^2\sqrt{x^2+1}\Big]_0^1 - 2\int_0^1 x(x^2+1)^{1/2}dx$$
$$= \sqrt{2} - \frac{2}{3}(x^2+1)^{3/2}\Big]_0^1 = \sqrt{2} - \frac{2}{3}[2\sqrt{2}-1] = (2-\sqrt{2})/3$$

(b) $u^2 = x^2+1$, $x^2 = u^2-1$, $2x\,dx = 2u\,du$, $x\,dx = u\,du$;

$$\int_0^1 \frac{x^3}{\sqrt{x^2+1}}dx = \int_0^1 \frac{x^2}{\sqrt{x^2+1}}x\,dx = \int_1^{\sqrt{2}} \frac{u^2-1}{u}u\,du$$
$$= \int_1^{\sqrt{2}}(u^2-1)du = \left(\frac{1}{3}u^3 - u\right)\Big]_1^{\sqrt{2}} = (2-\sqrt{2})/3$$

7. **(a)** $u = 2x$,

$$\int \sin^4 2x\,dx = \frac{1}{2}\int \sin^4 u\,du = \frac{1}{2}\left[-\frac{1}{4}\sin^3 u\cos u + \frac{3}{4}\int \sin^2 u\,du\right]$$
$$= -\frac{1}{8}\sin^3 u\cos u + \frac{3}{8}\left[-\frac{1}{2}\sin u\cos u + \frac{1}{2}\int du\right]$$
$$= -\frac{1}{8}\sin^3 u\cos u - \frac{3}{16}\sin u\cos u + \frac{3}{16}u + C$$
$$= -\frac{1}{8}\sin^3 2x\cos 2x - \frac{3}{16}\sin 2x\cos 2x + \frac{3}{8}x + C$$

(b) $u = x^2$,

$$\int x\cos^5(x^2)dx = \frac{1}{2}\int \cos^5 u\,du = \frac{1}{2}\int (\cos u)(1-\sin^2 u)^2\,du$$
$$= \frac{1}{2}\int \cos u\,du - \int \cos u\sin^2 u\,du + \frac{1}{2}\int \cos u\sin^4 u\,du$$
$$= \frac{1}{2}\sin u - \frac{1}{3}\sin^3 u + \frac{1}{10}\sin^5 u + C$$
$$= \frac{1}{2}\sin(x^2) - \frac{1}{3}\sin^3(x^2) + \frac{1}{10}\sin^5(x^2) + C$$

8. **(a)** With $x = \sec\theta$:

$$\int \frac{1}{x^3-x}dx = \int \cot\theta\,d\theta = \ln|\sin\theta| + C = \ln\frac{\sqrt{x^2-1}}{|x|} + C;\ \text{valid for } |x| > 1.$$

(b) With $x = \sin\theta$:

$$\int \frac{1}{x^3-x}dx = -\int \frac{1}{\sin\theta\cos\theta}d\theta = -\int 2\csc 2\theta\,d\theta$$
$$= -\ln|\csc 2\theta - \cot 2\theta| + C = \ln|\cot\theta| + C = \ln\frac{\sqrt{1-x^2}}{|x|} + C,\quad 0 < |x| < 1.$$

(c) By partial fractions:

$$\int\left(-\frac{1}{x}+\frac{1/2}{x+1}+\frac{1/2}{x-1}\right)dx=-\ln|x|+\frac{1}{2}\ln|x+1|+\frac{1}{2}\ln|x-1|+C$$

$$=\ln\frac{\sqrt{|x^2-1|}}{|x|}+C;\ \text{valid for all } x \text{ except } x=0,\pm 1.$$

9. (a) With $u=\sqrt{x}$:

$$\int\frac{1}{\sqrt{x}\,\sqrt{2-x}}\,dx=2\int\frac{1}{\sqrt{2-u^2}}\,du=2\sin^{-1}(u/\sqrt{2})+C=2\sin^{-1}(\sqrt{x/2})+C;$$

with $u=\sqrt{2-x}$:

$$\int\frac{1}{\sqrt{x}\,\sqrt{2-x}}\,dx=-2\int\frac{1}{\sqrt{2-u^2}}\,du=-2\sin^{-1}(u/\sqrt{2})+C=-2\sin^{-1}(\sqrt{2-x}/\sqrt{2})+C;$$

completing the square:

$$\int\frac{1}{\sqrt{1-(x-1)^2}}\,dx=\sin^{-1}(x-1)+C.$$

(b) In the three results in Part (a) the antiderivatives differ by a constant, in particular $2\sin^{-1}(\sqrt{x/2})=\pi-2\sin^{-1}(\sqrt{2-x}/\sqrt{2})=\pi/2+\sin^{-1}(x-1)$.

10. $A=\displaystyle\int_1^2\frac{3-x}{x^3+x^2}dx,\ \frac{3-x}{x^2(x+1)}=\frac{A}{x}+\frac{B}{x^2}+\frac{C}{x+1};\ A=-4,\ B=3,\ C=4$

$$A=\left[-4\ln|x|-\frac{3}{x}+4\ln|x+1|\right]_1^2$$

$$=(-4\ln 2-\frac{3}{2}+4\ln 3)-(-4\ln 1-3+4\ln 2)=\frac{3}{2}-8\ln 2+4\ln 3=\frac{3}{2}+4\ln\frac{3}{4}$$

11. Solve $y=1/(1+x^2)$ for x to get $x=\sqrt{\dfrac{1-y}{y}}$ and integrate with respect to y to get $A=\displaystyle\int_0^1\sqrt{\frac{1-y}{y}}\,dy$ (see figure)

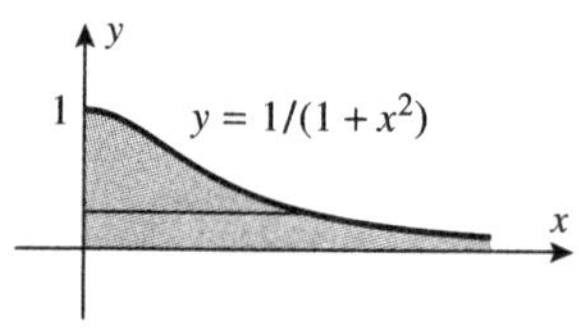

12. $A=\displaystyle\int_e^{+\infty}\frac{\ln x-1}{x^2}\,dx=\lim_{\ell\to+\infty}-\frac{\ln x}{x}\Big]_e^{\ell}=1/e$

13. $V=2\pi\displaystyle\int_0^{+\infty}xe^{-x}dx=2\pi\lim_{\ell\to+\infty}-e^{-x}(x+1)\Big]_0^{\ell}=2\pi\lim_{\ell\to+\infty}\left[1-e^{-\ell}(\ell+1)\right]$

but $\displaystyle\lim_{\ell\to+\infty}e^{-\ell}(\ell+1)=\lim_{\ell\to+\infty}\frac{\ell+1}{e^{\ell}}=\lim_{\ell\to+\infty}\frac{1}{e^{\ell}}=0$ so $V=2\pi$

14. $\displaystyle\int_0^{+\infty}\frac{dx}{x^2+a^2}=\lim_{\ell\to+\infty}\frac{1}{a}\tan^{-1}(x/a)\Big]_0^{\ell}=\lim_{\ell\to+\infty}\frac{1}{a}\tan^{-1}(\ell/a)=\frac{\pi}{2a}=1, a=\pi/2$

15. $u=\cos\theta,\ -\displaystyle\int u^{1/2}du=-\frac{2}{3}\cos^{3/2}\theta+C$

16. Use Endpaper Formula (31) to get $\displaystyle\int\tan^7\theta d\theta=\frac{1}{6}\tan^6\theta-\frac{1}{4}\tan^4\theta+\frac{1}{2}\tan^2\theta+\ln|\cos\theta|+C.$

17. $u = \tan(x^2)$, $\dfrac{1}{2}\displaystyle\int u^2 du = \dfrac{1}{6}\tan^3(x^2) + C$

18. $x = (1/\sqrt{2})\sin\theta$, $dx = (1/\sqrt{2})\cos\theta\, d\theta$,

$$\frac{1}{\sqrt{2}}\int_{-\pi/2}^{\pi/2}\cos^4\theta\, d\theta = \frac{1}{\sqrt{2}}\left\{\frac{1}{4}\cos^3\theta\sin\theta\Big]_{-\pi/2}^{\pi/2} + \frac{3}{4}\int_{-\pi/2}^{\pi/2}\cos^2\theta\, d\theta\right\}$$

$$= \frac{3}{4\sqrt{2}}\left\{\frac{1}{2}\cos\theta\sin\theta\Big]_{-\pi/2}^{\pi/2} + \frac{1}{2}\int_{-\pi/2}^{\pi/2} d\theta\right\} = \frac{3}{4\sqrt{2}}\frac{1}{2}\pi = \frac{3\pi}{8\sqrt{2}}$$

19. $x = \sqrt{3}\tan\theta$, $dx = \sqrt{3}\sec^2\theta\, d\theta$,

$$\frac{1}{3}\int \frac{1}{\sec\theta}d\theta = \frac{1}{3}\int\cos\theta\, d\theta = \frac{1}{3}\sin\theta + C = \frac{x}{3\sqrt{3+x^2}} + C$$

20. $\displaystyle\int \frac{\cos\theta}{(\sin\theta - 3)^2 + 3}d\theta$, let $u = \sin\theta - 3$, $\displaystyle\int\frac{1}{u^2+3}du = \frac{1}{\sqrt{3}}\tan^{-1}[(\sin\theta - 3)/\sqrt{3}] + C$

21. $\displaystyle\int \frac{x+3}{\sqrt{(x+1)^2+1}}dx$, let $u = x + 1$,

$$\int\frac{u+2}{\sqrt{u^2+1}}du = \int\left[u(u^2+1)^{-1/2} + \frac{2}{\sqrt{u^2+1}}\right]du = \sqrt{u^2+1} + 2\sinh^{-1}u + C$$

$$= \sqrt{x^2+2x+2} + 2\sinh^{-1}(x+1) + C$$

<u>Alternate solution</u>: let $x + 1 = \tan\theta$,

$$\int(\tan\theta + 2)\sec\theta\, d\theta = \int\sec\theta\tan\theta\, d\theta + 2\int\sec\theta\, d\theta = \sec\theta + 2\ln|\sec\theta + \tan\theta| + C$$

$$= \sqrt{x^2+2x+2} + 2\ln(\sqrt{x^2+2x+2} + x + 1) + C.$$

22. Let $x = \tan\theta$ to get $\displaystyle\int\frac{1}{x^3 - x^2}dx$.

$$\frac{1}{x^2(x-1)} = \frac{A}{x} + \frac{B}{x^2} + \frac{C}{x-1};\ A = -1,\ B = -1,\ C = 1 \text{ so}$$

$$-\int\frac{1}{x}dx - \int\frac{1}{x^2}dx + \int\frac{1}{x-1}dx = -\ln|x| + \frac{1}{x} + \ln|x-1| + C$$

$$= \frac{1}{x} + \ln\left|\frac{x-1}{x}\right| + C = \cot\theta + \ln\left|\frac{\tan\theta - 1}{\tan\theta}\right| + C = \cot\theta + \ln|1 - \cot\theta| + C$$

23. $\dfrac{1}{(x-1)(x+2)(x-3)} = \dfrac{A}{x-1} + \dfrac{B}{x+2} + \dfrac{C}{x-3}$; $A = -\dfrac{1}{6}$, $B = \dfrac{1}{15}$, $C = \dfrac{1}{10}$ so

$$-\frac{1}{6}\int\frac{1}{x-1}dx + \frac{1}{15}\int\frac{1}{x+2}dx + \frac{1}{10}\int\frac{1}{x-3}dx$$

$$= -\frac{1}{6}\ln|x-1| + \frac{1}{15}\ln|x+2| + \frac{1}{10}\ln|x-3| + C$$

24. $\dfrac{1}{x(x^2+x+1)} = \dfrac{A}{x} + \dfrac{Bx+C}{x^2+x+1}$; $A = 1$, $B = C = -1$ so

$$\int \frac{-x-1}{x^2+x+1}dx = -\int \frac{x+1}{(x+1/2)^2+3/4}dx = -\int \frac{u+1/2}{u^2+3/4}du, \qquad u = x+1/2$$

$$= -\frac{1}{2}\ln(u^2+3/4) - \frac{1}{\sqrt{3}}\tan^{-1}(2u/\sqrt{3}) + C_1$$

so $\displaystyle\int \frac{dx}{x(x^2+x+1)} = \ln|x| - \frac{1}{2}\ln(x^2+x+1) - \frac{1}{\sqrt{3}}\tan^{-1}\frac{2x+1}{\sqrt{3}} + C$

25. $u = \sqrt{x-4}$, $x = u^2+4$, $dx = 2u\,du$,

$$\int_0^2 \frac{2u^2}{u^2+4}du = 2\int_0^2 \left[1 - \frac{4}{u^2+4}\right]du = \left[2u - 4\tan^{-1}(u/2)\right]_0^2 = 4 - \pi$$

26. $u = \sqrt{x}$, $x = u^2$, $dx = 2u\,du$,

$$2\int_0^3 \frac{u^2}{u^2+9}du = 2\int_0^3 \left(1 - \frac{9}{u^2+9}\right)du = \left(2u - 6\tan^{-1}\frac{u}{3}\right)\Big]_0^3 = 6 - \frac{3}{2}\pi$$

27. $u = \sqrt{e^x+1}$, $e^x = u^2-1$, $x = \ln(u^2-1)$, $dx = \dfrac{2u}{u^2-1}du$,

$$\int \frac{2}{u^2-1}du = \int \left[\frac{1}{u-1} - \frac{1}{u+1}\right]du = \ln|u-1| - \ln|u+1| + C = \ln\frac{\sqrt{e^x+1}-1}{\sqrt{e^x+1}+1} + C$$

28. $u = \sqrt{e^x-1}$, $e^x = u^2+1$, $x = \ln(u^2+1)$, $dx = \dfrac{2u}{u^2+1}du$,

$$\int_0^1 \frac{2u^2}{u^2+1}du = 2\int_0^1 \left(1 - \frac{1}{u^2+1}\right)du = (2u - 2\tan^{-1}u)\Big]_0^1 = 2 - \frac{\pi}{2}$$

29. $\displaystyle\lim_{\ell\to+\infty} -\frac{1}{2(x^2+1)}\Big]_a^\ell = \lim_{\ell\to+\infty}\left[-\frac{1}{2(\ell^2+1)} + \frac{1}{2(a^2+1)}\right] = \frac{1}{2(a^2+1)}$

30. $\displaystyle\lim_{\ell\to+\infty} \frac{1}{ab}\tan^{-1}\frac{bx}{a}\Big]_0^\ell = \lim_{\ell\to+\infty}\frac{1}{ab}\tan^{-1}\frac{b\ell}{a} = \frac{\pi}{2ab}$

31. Let $u = x^4$ to get $\displaystyle\frac{1}{4}\int \frac{1}{\sqrt{1-u^2}}du = \frac{1}{4}\sin^{-1}u + C = \frac{1}{4}\sin^{-1}(x^4) + C.$

32. $\displaystyle\int (\cos^{32}x\sin^{30}x - \cos^{30}x\sin^{32}x)dx = \int \cos^{30}x\sin^{30}x(\cos^2x - \sin^2x)dx$

$$= \frac{1}{2^{30}}\int \sin^{30}2x\cos 2x\,dx = \frac{\sin^{31}2x}{31(2^{31})} + C$$

33. $\displaystyle\int \sqrt{x - \sqrt{x^2-4}}dx = \frac{1}{\sqrt{2}}\int(\sqrt{x+2} - \sqrt{x-2})dx = \frac{\sqrt{2}}{3}[(x+2)^{3/2} - (x-2)^{3/2}] + C$

34. $\displaystyle\int \frac{1}{x^{10}(1+x^{-9})}dx = -\frac{1}{9}\int \frac{1}{u}du = -\frac{1}{9}\ln|u| + C = -\frac{1}{9}\ln|1+x^{-9}| + C$

35. **(a)** $(x+4)(x-5)(x^2+1)^2$; $\dfrac{A}{x+4}+\dfrac{B}{x-5}+\dfrac{Cx+D}{x^2+1}+\dfrac{Ex+F}{(x^2+1)^2}$

(b) $-\dfrac{3}{x+4}+\dfrac{2}{x-5}-\dfrac{x-2}{x^2+1}-\dfrac{3}{(x^2+1)^2}$

(c) $-3\ln|x+4|+2\ln|x-5|+2\tan^{-1}x-\dfrac{1}{2}\ln(x^2+1)-\dfrac{3}{2}\left(\dfrac{x}{x^2+1}+\tan^{-1}x\right)+C$

36. **(a)** $\Gamma(1)=\displaystyle\int_0^{+\infty}e^{-t}dt=\lim_{\ell\to+\infty}-e^{-t}\Big]_0^{\ell}=\lim_{\ell\to+\infty}(-e^{-\ell}+1)=1$

(b) $\Gamma(x+1)=\displaystyle\int_0^{+\infty}t^xe^{-t}dt$; let $u=t^x$, $dv=e^{-t}dt$ to get

$$\Gamma(x+1)=-t^xe^{-t}\Big]_0^{+\infty}+x\int_0^{+\infty}t^{x-1}e^{-t}dt=-t^xe^{-t}\Big]_0^{+\infty}+x\Gamma(x)$$

$\displaystyle\lim_{t\to+\infty}t^xe^{-t}=\lim_{t\to+\infty}\frac{t^x}{e^t}=0$ (by multiple applications of L'Hôpital's rule)

so $\Gamma(x+1)=x\Gamma(x)$

(c) $\Gamma(2)=(1)\Gamma(1)=(1)(1)=1$, $\Gamma(3)=2\Gamma(2)=(2)(1)=2$, $\Gamma(4)=3\Gamma(3)=(3)(2)=6$

It appears that $\Gamma(n)=(n-1)!$ if n is a positive integer.

(d) $\Gamma\left(\dfrac{1}{2}\right)=\displaystyle\int_0^{+\infty}t^{-1/2}e^{-t}dt=2\int_0^{+\infty}e^{-u^2}du$ (with $u=\sqrt{t}$) $=2(\sqrt{\pi}/2)=\sqrt{\pi}$

(e) $\Gamma\left(\dfrac{3}{2}\right)=\dfrac{1}{2}\Gamma\left(\dfrac{1}{2}\right)=\dfrac{1}{2}\sqrt{\pi}$, $\Gamma\left(\dfrac{5}{2}\right)=\dfrac{3}{2}\Gamma\left(\dfrac{3}{2}\right)=\dfrac{3}{4}\sqrt{\pi}$

37. **(a)** $t=-\ln x, x=e^{-t}, dx=-e^{-t}dt$,

$$\int_0^1(\ln x)^n dx=-\int_{+\infty}^0(-t)^ne^{-t}dt=(-1)^n\int_0^{+\infty}t^ne^{-t}dt=(-1)^n\Gamma(n+1)$$

(b) $t=x^n, x=t^{1/n}, dx=(1/n)t^{1/n-1}dt$,

$$\int_0^{+\infty}e^{-x^n}dx=(1/n)\int_0^{+\infty}t^{1/n-1}e^{-t}dt=(1/n)\Gamma(1/n)=\Gamma(1/n+1)$$

38. **(a)** $\sqrt{\cos\theta-\cos\theta_0}=\sqrt{2\left[\sin^2(\theta_0/2)-\sin^2(\theta/2)\right]}=\sqrt{2(k^2-k^2\sin^2\phi)}=\sqrt{2k^2\cos^2\phi}$

$=\sqrt{2}\,k\,\cos\phi$; $k\sin\phi=\sin(\theta/2)$ so $k\cos\phi\,d\phi=\dfrac{1}{2}\cos(\theta/2)\,d\theta=\dfrac{1}{2}\sqrt{1-\sin^2(\theta/2)}\,d\theta$

$=\dfrac{1}{2}\sqrt{1-k^2\sin^2\phi}\,d\theta$, thus $d\theta=\dfrac{2k\cos\phi}{\sqrt{1-k^2\sin^2\phi}}\,d\phi$ and hence

$$T=\sqrt{\frac{8L}{g}}\int_0^{\pi/2}\frac{1}{\sqrt{2}k\cos\phi}\cdot\frac{2k\cos\phi}{\sqrt{1-k^2\sin^2\phi}}\,d\phi=4\sqrt{\frac{L}{g}}\int_0^{\pi/2}\frac{1}{\sqrt{1-k^2\sin^2\phi}}\,d\phi$$

(b) If $L=1.5$ ft and $\theta_0=(\pi/180)(20)=\pi/9$, then

$$T=\frac{\sqrt{3}}{2}\int_0^{\pi/2}\frac{d\phi}{\sqrt{1-\sin^2(\pi/18)\sin^2\phi}}\approx 1.37\text{ s.}$$

CHAPTER 8 HORIZON MODULE

1. The depth of the cut equals the terrain elevation minus the track elevation. From Figure 2, the cross sectional area of a cut of depth D meters is $10D + 2 \cdot \frac{1}{2}D^2 = D^2 + 10D$ square meters.

Distance from town A (m)	Terrain elevation (m)	Track elevation (m)	Depth of cut (m)	Cross-sectional area $f(x)$ of cut (m^2)
0	100	100	0	0
2000	105	101	4	56
4000	108	102	6	96
6000	110	103	7	119
8000	104	104	0	0
10,000	106	105	1	11
12,000	120	106	14	336
14,000	122	107	15	375
16,000	124	108	16	416
18,000	128	109	19	551
20,000	130	110	20	600

The total volume of dirt to be excavated, in cubic meters, is $\int_0^{2000} f(x)\,dx$.

By Simpson's Rule, this is approximately

$$\frac{20{,}000 - 0}{3 \cdot 10}[0 + 4 \cdot 56 + 2 \cdot 96 + 4 \cdot 119 + 2 \cdot 0 + 4 \cdot 11 + 2 \cdot 336 + 4 \cdot 375 + 2 \cdot 416 + 4 \cdot 551 + 600]$$

$$= 4{,}496{,}000 \text{ m}^3.$$

Excavation costs \$4 per m^3, so the total cost of the railroad from kA to M is about

$4 \cdot 4{,}496{,}000 = 17{,}984{,}000$ dollars.

2. (a)

Distance from town A (m)	Terrain elevation (m)	Track elevation (m)	Depth of cut (m)	Cross-sectional area $f(x)$ of cut (m^2)
20,000	130	110	20	300
20,100	135	109.8	25.2	887.04
20,200	139	109.6	29.4	1158.36
20,300	142	109.4	32.6	1388.76
20,400	145	109.2	35.8	1639.64
20,500	147	109	38	1824
20,600	148	108.8	39.2	1928.64
20,700	146	108.6	37.4	1772.76
20,800	143	108.4	34.6	1543.16
20,900	139	108.2	30.8	1256.64
21,000	133	108	25	875

The total volume of dirt to be excavated, in cubic meters, is $\int_{20{,}000}^{21{,}000} f(x)\,dx$.

By Simpson's Rule this is approximately

$$\frac{21{,}000 - 20{,}000}{3 \cdot 10}[600 + 4 \cdot 887.04 + 2 \cdot 1158.36 + \ldots + 4 \cdot 1256.64 + 875] = 1{,}417{,}713.33 \text{ m}^3.$$

The total cost of a trench from M to N is about $4 \cdot 1{,}417{,}713.33 \approx 5{,}670{,}853$ dollars.

(b)

Distance from town A (m)	Terrain elevation (m)	Track elevation (m)	Depth of cut (m)	Cross-sectional area $f(x)$ of cut (m^2)
21,000	133	108	25	875
22,000	120	106	14	336
23,000	106	104	2	24
24,000	108	102	6	96
25,000	106	100	6	96
26,000	98	98	0	0
27,000	100	96	4	56
28,000	102	94	8	144
29,000	96	92	4	56
30,000	91	90	1	11
31,000	88	88	0	0

The total volume of dirt to be excavated, in cubic meters, is $\int_{21{,}000}^{31{,}000} f(x)\,dx$. By Simpson's Rule this is approximately

$$\frac{31{,}000 - 21{,}000}{3 \cdot 10}[875 + 4 \cdot 336 + 2 \cdot 24 + \ldots + 4 \cdot 11 + 0] = 1{,}229{,}000 \text{ m}^3.$$

The total cost of the railroad from N to B is about $4 \cdot 1{,}229{,}000 \approx 4{,}916{,}000$ dollars.

3. The total cost if trenches are used everywhere is about $17{,}984{,}000 + 5{,}670{,}853 + 4{,}916{,}000 = 28{,}570{,}853$ dollars.

4. **(a)** The cross-sectional area of a tunnel is $A_T = 80 + \frac{1}{2}\pi 5^2 \approx 119.27$ m^2. The length of the tunnel is 1000 m, so the volume of dirt to be removed is about $1000A_T \approx 1{,}119{,}269.91$ m^3, and the drilling and dirt-piling costs are $8 \cdot 1000A_T \approx 954{,}159$ dollars.

(b) To extend the tunnel from a length of x meters to a length of $x + dx$ meters, we must move a volume of $A_T\,dx$ cubic meters of dirt a distance of about x meters. So the cost of this extension is about $0.06 \times A_T\,dx$ dollars. The cost of moving all of the dirt in the tunnel is therefore

$$\int_0^{1000} 0.06 \times A_T\,dx = 0.06A_T \frac{x^2}{2}\Big]_0^{1000} = 30{,}000A_T \approx 3{,}578{,}097 \text{ dollars.}$$

(c) The total cost of the tunnel is about $954{,}159 + 3{,}578{,}097 \approx 4{,}532{,}257$ dollars.

5. The total cost of the railroad, using a tunnel, is $17{,}894{,}000 + 4{,}532{,}257 + 4{,}916{,}000 + 27{,}432{,}257$ dollars, which is smaller than the cost found in Exercise 3. It will be cheaper to build the railroad if a tunnel is used.

CHAPTER 9

Mathematical Modeling with Differential Equations

EXERCISE SET 9.1

1. $y' = 2x^2 e^{x^3/3} = x^2 y$ and $y(0) = 2$ by inspection.

2. $y' = x^3 - 2\sin x,\ y(0) = 3$ by inspection.

3. **(a)** first order; $\dfrac{dy}{dx} = c;\ (1+x)\dfrac{dy}{dx} = (1+x)c = y$

(b) second order; $y' = c_1 \cos t - c_2 \sin t,\ y'' + y = -c_1 \sin t - c_2 \cos t + (c_1 \sin t + c_2 \cos t) = 0$

4. **(a)** first order; $2\dfrac{dy}{dx} + y = 2\Big(-\dfrac{c}{2}e^{-x/2} + 1\Big) + ce^{-x/2} + x - 3 = x - 1$

(b) second order; $y' = c_1 e^t - c_2 e^{-t},\ y'' - y = c_1 e^t + c_2 e^{-t} - \left(c_1 e^t + c_2 e^{-t}\right) = 0$

5. $\dfrac{1}{y}\dfrac{dy}{dx} = x\dfrac{dy}{dx} + y,\ \dfrac{dy}{dx}(1 - xy) = y^2,\ \dfrac{dy}{dx} = \dfrac{y^2}{1 - xy}$

6. $2x + y^2 + 2xy\dfrac{dy}{dx} = 0$, by inspection.

7. **(a)** IF: $\mu = e^{3\int dx} = e^{3x},\ \dfrac{d}{dx}\left[ye^{3x}\right] = 0, ye^{3x} = C, y = Ce^{-3x}$

separation of variables: $\dfrac{dy}{y} = -3dx, \ln|y| = -3x + C_1, y = \pm e^{-3x}e^{C_1} = Ce^{-3x}$
including $C = 0$ by inspection

(b) IF: $\mu = e^{-2\int dt} = e^{-2t},\ \dfrac{d}{dt}[ye^{-2t}] = 0, ye^{-2t} = C, y = Ce^{2t}$

separation of variables: $\dfrac{dy}{y} = 2dt,\ \ln|y| = 2t + C_1,\ y = \pm e^{C_1}e^{2t} = Ce^{2t}$
including $C = 0$ by inspection

8. **(a)** IF: $\mu = e^{-4\int x\,dx} = e^{-2x^2},\ \dfrac{d}{dx}\left[ye^{-2x^2}\right] = 0,\ y = Ce^{2x^2}$

separation of variables: $\dfrac{dy}{y} = 4x\,dx,\ \ln|y| = 2x^2 + C_1,\ y = \pm e^{C_1}e^{2x^2} = Ce^{2x^2}$
including $C = 0$ by inspection

(b) IF: $\mu = e^{\int dt} = e^t,\ \dfrac{d}{dt}\left[ye^t\right] = 0,\ y = Ce^{-t}$

separation of variables: $\dfrac{dy}{y} = -dt,\ \ln|y| = -t + C_1, y = \pm e^{C_1}e^{-t} = Ce^{-t}$
including $C = 0$ by inspection

9. $\mu = e^{\int 3dx} = e^{3x},\ e^{3x}y = \displaystyle\int e^x\,dx = e^x + C,\ y = e^{-2x} + Ce^{-3x}$

10. $\mu = e^{2\int x\,dx} = e^{x^2}, \dfrac{d}{dx}\left[ye^{x^2}\right] = xe^{x^2}, ye^{x^2} = \dfrac{1}{2}e^{x^2} + C, y = \dfrac{1}{2} + Ce^{-x^2}$

11. $\mu = e^{\int dx} = e^x$, $e^x y = \int e^x \cos(e^x) dx = \sin(e^x) + C$, $y = e^{-x}\sin(e^x) + Ce^{-x}$

12. $\dfrac{dy}{dx} + 2y = \dfrac{1}{2}$, $\mu = e^{\int 2dx} = e^{2x}$, $e^{2x}y = \int \dfrac{1}{2}e^{2x}dx = \dfrac{1}{4}e^{2x} + C$, $y = \dfrac{1}{4} + Ce^{-2x}$

13. $\dfrac{dy}{dx} + \dfrac{x}{x^2+1}y = 0, \mu = e^{\int (x/(x^2+1))dx} = e^{\frac{1}{2}\ln(x^2+1)} = \sqrt{x^2+1}$,
$\dfrac{d}{dx}\left[y\sqrt{x^2+1}\right] = 0,\ y\sqrt{x^2+1} = C,\ y = \dfrac{C}{\sqrt{x^2+1}}$

14. $\dfrac{dy}{dx} + y = \dfrac{1}{1+e^x}$, $\mu = e^{\int dx} = e^x$, $e^x y = \int \dfrac{e^x}{1+e^x}dx = \ln(1+e^x) + C$, $y = e^{-x}\ln(1+e^x) + Ce^{-x}$

15. $\dfrac{1}{y}dy = \dfrac{1}{x}dx$, $\ln|y| = \ln|x| + C_1$, $\ln\left|\dfrac{y}{x}\right| = C_1$, $\dfrac{y}{x} = \pm e^{C_1} = C$, $y = Cx$
including $C = 0$ by inspection

16. $\dfrac{dy}{1+y^2} = x^2\,dx, \tan^{-1}y = \dfrac{1}{3}x^3 + C, y = \tan\left(\dfrac{1}{3}x^3 + C\right)$

17. $\dfrac{dy}{1+y} = -\dfrac{x}{\sqrt{1+x^2}}dx, \ln|1+y| = -\sqrt{1+x^2} + C_1, 1+y = \pm e^{-\sqrt{1+x^2}}e^{C_1} = Ce^{-\sqrt{1+x^2}}$,
$y = Ce^{-\sqrt{1+x^2}} - 1, C \neq 0$

18. $y\,dy = \dfrac{x^3\,dx}{1+x^4}, \dfrac{y^2}{2} = \dfrac{1}{4}\ln(1+x^4) + C_1, 2y^2 = \ln(1+x^4) + C,\ y = \pm\sqrt{[\ln(1+x^4)+C]/2}$

19. $\left(\dfrac{1}{y} + y\right)dy = e^x dx$, $\ln|y| + y^2/2 = e^x + C$; by inspection, $y = 0$ is also a solution

20. $\dfrac{dy}{y} = -x\,dx$, $\ln|y| = -x^2/2 + C_1$, $y = \pm e^{C_1}e^{-x^2/2} = Ce^{-x^2/2}$, including $C = 0$ by inspection

21. $e^y dy = \dfrac{\sin x}{\cos^2 x}dx = \sec x \tan x\,dx$, $e^y = \sec x + C$, $y = \ln(\sec x + C)$

22. $\dfrac{dy}{1+y^2} = (1+x)\,dx, \tan^{-1}y = x + \dfrac{x^2}{2} + C,\ y = \tan(x + x^2/2 + C)$

23. $\dfrac{dy}{y^2-y} = \dfrac{dx}{\sin x}, \int\left[-\dfrac{1}{y} + \dfrac{1}{y-1}\right]dy = \int \csc x\,dx,\ \ln\left|\dfrac{y-1}{y}\right| = \ln|\csc x - \cot x| + C_1$,
$\dfrac{y-1}{y} = \pm e^{C_1}(\csc x - \cot x) = C(\csc x - \cot x),\ y = \dfrac{1}{1 - C(\csc x - \cot x)}, C \neq 0$;
by inspection, $y = 0$ is also a solution, as is $y = 1$.

24. $\dfrac{1}{\tan y}dy = \dfrac{3}{\sec x}dx,\ \dfrac{\cos y}{\sin y}dy = 3\cos x\,dx,\ \ln|\sin y| = 3\sin x + C_1$,
$\sin y = \pm e^{3\sin x + C_1} = \pm e^{C_1}e^{3\sin x} = Ce^{3\sin x}, C \neq 0$,
$y = \sin^{-1}\left(Ce^{3\sin x}\right)$, as is $y = 0$ by inspection

25. $\dfrac{dy}{dx} + \dfrac{1}{x}y = 1,\ \mu = e^{\int (1/x)dx} = e^{\ln x} = x,\ \dfrac{d}{dx}[xy] = x,\ xy = \dfrac{1}{2}x^2 + C,\ y = x/2 + C/x$

(a) $2 = y(1) = \dfrac{1}{2} + C, C = \dfrac{3}{2}, y = x/2 + 3/(2x)$

(b) $2 = y(-1) = -1/2 - C, C = -5/2, y = x/2 - 5/(2x)$

26. $\dfrac{dy}{y} = x\,dx,\ \ln|y| = \dfrac{x^2}{2} + C_1,\ y = \pm e^{C_1}e^{x^2/2} = Ce^{x^2/2}$

(a) $1 = y(0) = C$ so $C = 1,\ y = e^{x^2/2}$

(b) $\dfrac{1}{2} = y(0) = C$, so $y = \dfrac{1}{2}e^{x^2/2}$

27. $\mu = e^{-\int x\,dx} = e^{-x^2/2},\ e^{-x^2/2}y = \displaystyle\int xe^{-x^2/2}dx = -e^{-x^2/2} + C,$

$y = -1 + Ce^{x^2/2},\ 3 = -1 + C,\ C = 4,\ y = -1 + 4e^{x^2/2}$

28. $\mu = e^{\int dt} = e^t,\ e^t y = \displaystyle\int 2e^t\,dt = 2e^t + C,\ y = 2 + Ce^{-t},\ 1 = 2 + C,\ C = -1,\ y = 2 - e^{-t}$

29. $(y + \cos y)\,dy = 4x^2\,dx, \dfrac{y^2}{2} + \sin y = \dfrac{4}{3}x^3 + C, \dfrac{\pi^2}{2} + \sin\pi = \dfrac{4}{3}(1)^3 + C, \dfrac{\pi^2}{2} = \dfrac{4}{3} + C,$

$C = \dfrac{\pi^2}{2} - \dfrac{4}{3}, 3y^2 + 6\sin y = 8x^3 + 3\pi^2 - 8$

30. $\dfrac{dy}{dx} = (x+2)e^y,\ e^{-y}dy = (x+2)dx,\ -e^{-y} = \dfrac{1}{2}x^2 + 2x + C,\ -1 = C,$

$-e^{-y} = \dfrac{1}{2}x^2 + 2x - 1,\ e^{-y} = -\dfrac{1}{2}x^2 - 2x + 1,\ y = -\ln\left(1 - 2x - \dfrac{1}{2}x^2\right)$

31. $2(y-1)\,dy = (2t+1)\,dt, y^2 - 2y = t^2 + t + C, 1 + 2 = C, C = 3, y^2 - 2y = t^2 + t + 3$

32. $y' + \dfrac{\sinh x}{\cosh x}y = \cosh x,\ \mu = e^{\int(\sinh x/\cosh x)dx} = e^{\ln\cosh x} = \cosh x,$

$(\cosh x)y = \displaystyle\int \cosh^2 x\,dx = \int \frac{1}{2}(\cosh 2x + 1)dx = \frac{1}{4}\sinh 2x + \frac{1}{2}x + C = \frac{1}{2}\sinh x\cosh x + \frac{1}{2}x + C,$

$y = \dfrac{1}{2}\sinh x + \dfrac{1}{2}x\,\text{sech}\,x + C\,\text{sech}\,x,\ \dfrac{1}{4} = C,\ y = \dfrac{1}{2}\sinh x + \dfrac{1}{2}x\,\text{sech}\,x + \dfrac{1}{4}\,\text{sech}\,x$

33. **(a)** $\dfrac{dy}{y} = \dfrac{dx}{2x},\ \ln|y| = \dfrac{1}{2}\ln|x| + C_1,$

$|y| = C|x|^{1/2},\ y^2 = Cx;$

by inspection $y = 0$ is also a solution.

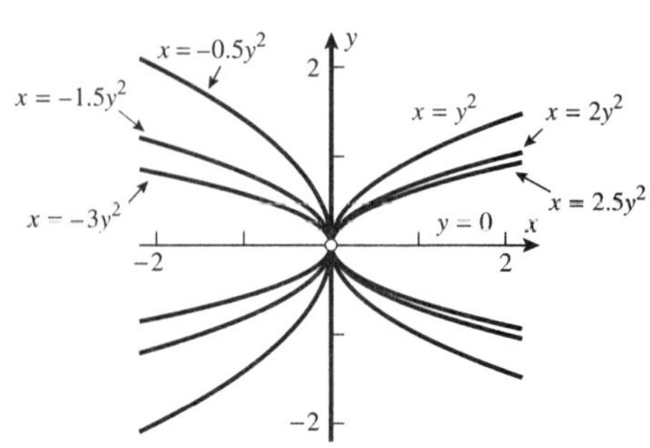

(b) $1 = C(2)^2, C = 1/4, y^2 = x/4$

34. **(a)** $y\,dy = -x\,dx, \dfrac{y^2}{2} = -\dfrac{x^2}{2} + C_1, y = \pm\sqrt{C^2 - x^2}$

(b) $y = \sqrt{25 - x^2}$

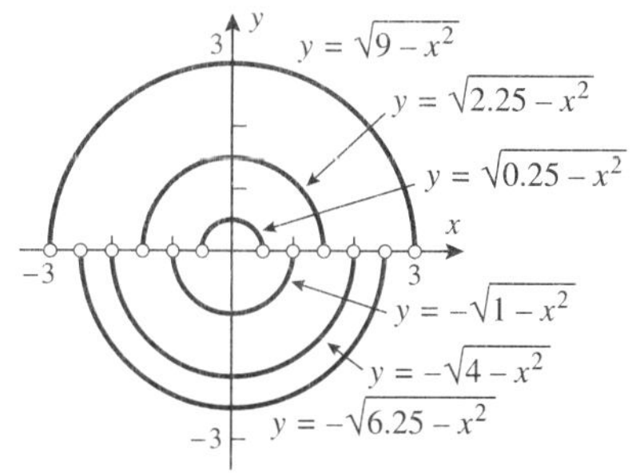

35. $\dfrac{dy}{y} = -\dfrac{x\,dx}{x^2+4},$

$\ln|y| = -\dfrac{1}{2}\ln(x^2+4) + C_1,$

$y = \dfrac{C}{\sqrt{x^2+4}}$

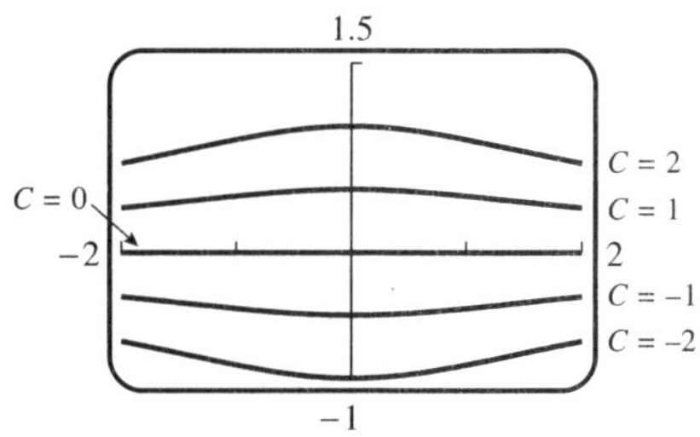

36. $y' + 2y = 3e^t, \mu = e^{2\int dt} = e^{2t},$

$\dfrac{d}{dt}\left[ye^{2t}\right] = 3e^{3t}, ye^{2t} = e^{3t} + C,$

$y = e^t + Ce^{-2t}$

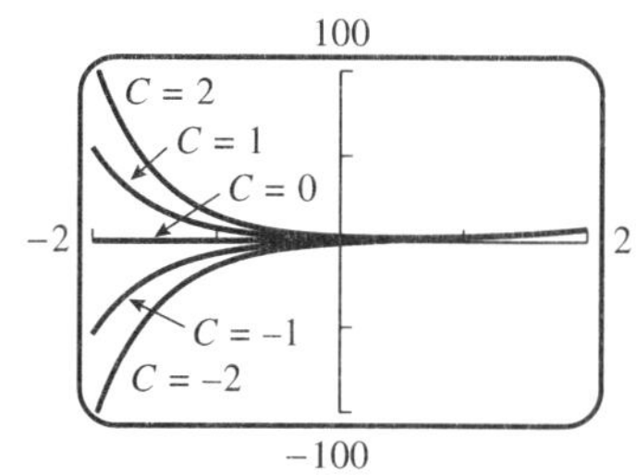

37. $(1 - y^2)\,dy = x^2\,dx,$

$y - \dfrac{y^3}{3} = \dfrac{x^3}{3} + C_1, x^3 + y^3 - 3y = C$

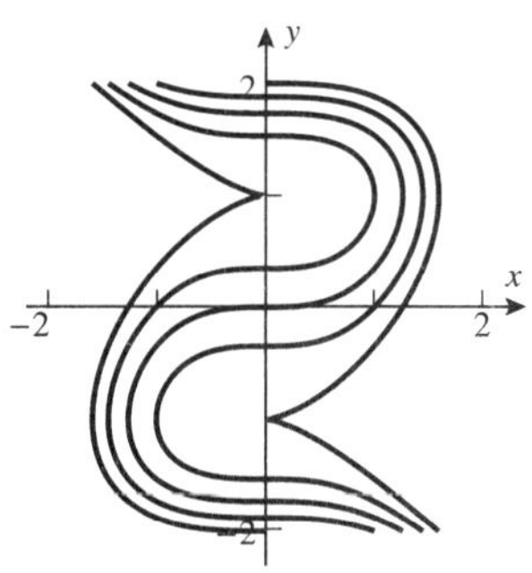

38. $\left(\dfrac{1}{y} + y\right) dy = dx, \ \ln|y| + \dfrac{y^2}{2} = x + C_1,$

$ye^{y^2/2} = \pm e^{C_1}e^x = Ce^x$ including $C = 0$

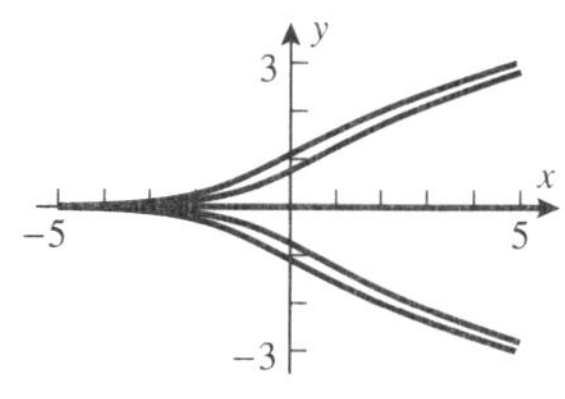

39. Of the solutions $y = \dfrac{1}{2x^2 - C}$, all pass through the point $\left(0, -\dfrac{1}{C}\right)$ and thus never through $(0, 0)$. A solution of the initial value problem with $y(0) = 0$ is (by inspection) $y = 0$. The methods of Example 4 fail because the integrals there become divergent when the point $x = 0$ is included in the integral.

40. If $y_0 \neq 0$ then, proceeding as before, we get $C = 2x^2 - \dfrac{1}{y}, C = 2x_0^2 - \dfrac{1}{y_0}$, and

$y = \dfrac{1}{2x^2 - 2x_0^2 + 1/y_0}$, which is defined for all x provided $2x^2$ is never equal to $2x_0^2 - 1/y_0$; this last condition will be satisfied if and only if $2x_0^2 - 1/y_0 < 0$, or $0 < 2x_0^2 y_0 < 1$. If $y_0 = 0$ then $y = 0$ is, by inspection, also a solution for all real x.

41. $\frac{dy}{dx} = xe^y, e^{-y}\,dy = x\,dx, -e^{-y} = \frac{x^2}{2} + C, x = 2$ when $y = 0$ so $-1 = 2 + C, C = -3, x^2 + 2e^{-y} = 6$

42. $\frac{dy}{dx} = \frac{3x^2}{2y}, 2y\,dy = 3x^2\,dx, y^2 = x^3 + C, 1 = 1 + C, C = 0,$
$y^2 = x^3,\ y = x^{3/2}$ passes through $(1, 1)$.

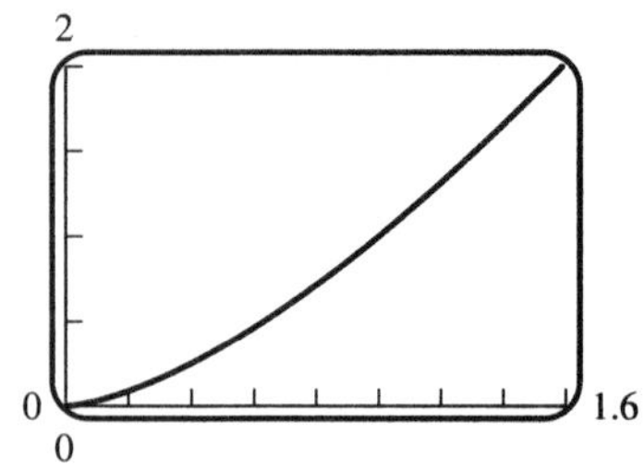

43. $\frac{dy}{dt}$ = rate in − rate out, where y is the amount of salt at time t,

$\frac{dy}{dt} = (4)(2) - \left(\frac{y}{50}\right)(2) = 8 - \frac{1}{25}y,$ so $\frac{dy}{dt} + \frac{1}{25}y = 8$ and $y(0) = 25$.

$\mu = e^{\int (1/25)dt} = e^{t/25},\ e^{t/25}y = \int 8e^{t/25}dt = 200e^{t/25} + C,$

$y = 200 + Ce^{-t/25},\ 25 = 200 + C,\ C = -175,$

(a) $y = 200 - 175e^{-t/25}$ oz **(b)** when $t = 25$, $y = 200 - 175e^{-1} \approx 136$ oz

44. $\frac{dy}{dt} = (5)(10) - \frac{y}{200}(10) = 50 - \frac{1}{20}y,$ so $\frac{dy}{dt} + \frac{1}{20}y = 50$ and $y(0) = 0$.

$\mu = e^{\int \frac{1}{20}dt} = e^{t/20},\ e^{t/20}y = \int 50e^{t/20}dt = 1000e^{t/20} + C,$

$y = 1000 + Ce^{-t/20},\ 0 = 1000 + C,\ C = -1000;$

(a) $y = 1000 - 1000e^{-t/20}$ lb **(b)** when $t = 30$, $y = 1000 - 1000e^{-1.5} \approx 777$ lb

45. The volume V of the (polluted) water is $V(t) = 500 + (20 - 10)t = 500 + 10t$;
if $y(t)$ is the number of pounds of particulate matter in the water,
then $y(0) = 50$, and $\frac{dy}{dt} = 0 - 10\frac{y}{V} = -\frac{1}{50+t}y,\ \frac{dy}{dt} + \frac{1}{50+t}y = 0;\ \mu = e^{\int \frac{dt}{50+t}} = 50 + t;$
$\frac{d}{dt}[(50 + t)y] = 0,\ (50 + t)y = C,\ 2500 = 50y(0) = C,\ y(t) = 2500/(50 + t).$
The tank reaches the point of overflowing when $V = 500 + 10t = 1000$, $t = 50$ min, so
$y = 2500/(50 + 50) = 25$ lb.

46. The volume of the lake (in gallons) is $V = 264\pi r^2 h = 264\pi(15)^2 3 = 178{,}200\pi$ gals. Let $y(t)$ denote the number of pounds of mercury salts at time t, then $\frac{dy}{dt} = 0 - 10^3\frac{y}{V} = -\frac{y}{178.2\pi}$ lb/h and $y_0 = 10^{-5}V = 1.782\pi$ lb; $\frac{dy}{y} = -\frac{dt}{178.2\pi}, \ln y = -\frac{t}{178.2\pi} + C_1,\ y = Ce^{-t/(178.2\pi)}$, and
$C = y(0) = y_0 10^{-5}\ \mathrm{V} = 1.782\pi,\ y = 1.782\pi e^{-t/(178.2\pi)}$ lb of mercury salts.

t	1	2	3	4	5	6	7	8	9	10	11	12
$y(t)$	5.588	5.578	5.568	5.558	5.549	5.539	5.529	5.519	5.509	5.499	5.489	5.480

47. (a) $\dfrac{dv}{dt} + \dfrac{c}{m}v = -g, \mu = e^{(c/m)\int dt} = e^{ct/m}, \dfrac{d}{dt}\left[ve^{ct/m}\right] = -ge^{ct/m}, ve^{ct/m} = -\dfrac{gm}{c}e^{ct/m} + C,$

$v = -\dfrac{gm}{c} + Ce^{-ct/m}$, but $v_0 = v(0) = -\dfrac{gm}{c} + C, C = v_0 + \dfrac{gm}{c}, v = -\dfrac{gm}{c} + \left(v_0 + \dfrac{gm}{c}\right)e^{-ct/m}$

(b) Replace $\dfrac{mg}{c}$ with v_τ and $-ct/m$ with $-gt/v_\tau$ in (23).

(c) From Part (b), $s(t) = C - v_\tau t - (v_0 + v_\tau)\dfrac{v_\tau}{g}e^{-gt/v_\tau}$;

$s_0 = s(0) = C - (v_0 + v_\tau)\dfrac{v_\tau}{g}$, $C = s_0 + (v_0 + v_\tau)\dfrac{v_\tau}{g}$, $s(t) = s_0 - v_\tau t + \dfrac{v_\tau}{g}(v_0 + v_\tau)\left(1 - e^{-gt/v_\tau}\right)$

48. Given $m = 240, g = 32, v_\tau = mg/c$: with a closed parachute $v_\tau = 120$ so $c = 64$, and with an open parachute $v_\tau = 24$, $c = 320$.

(a) Let t denote time elapsed in seconds after the moment of the drop. From Exercise 47(b), while the parachute is closed $v(t) = e^{-gt/v_\tau}(v_0 + v_\tau) - v_\tau = e^{-32t/120}(0 + 120) - 120 = 120\left(e^{-4t/15} - 1\right)$ and thus $v(25) = 120\left(e^{-20/3} - 1\right) \approx -119.85$, so the parachutist is falling at a speed of 119.85 ft/s when the parachute opens. From Exercise 47(c), $s(t) = s_0 - 120t + \dfrac{120}{32}120\left(1 - e^{-4t/15}\right)$, $s(25) = 10000 - 120 \cdot 25 + 450\left(1 - e^{-20/3}\right) \approx 7449.43$ ft.

(b) If t denotes time elapsed after the parachute opens, then, by Exercise 47(c), $s(t) = 7449.43 - 24t + \dfrac{24}{32}(-119.85 + 24)\left(1 - e^{-32t/24}\right) = 0$, with the solution (Newton's Method) $t = 307.4$ s, so the sky diver is in the air for about $25 + 307.4 = 332.4$ s.

49. $\dfrac{dI}{dt} + \dfrac{R}{L}I = \dfrac{V(t)}{L}, \mu = e^{(R/L)\int dt} = e^{Rt/L}, \dfrac{d}{dt}(e^{Rt/L}I) = \dfrac{V(t)}{L}e^{Rt/L},$

$Ie^{Rt/L} = I(0) + \dfrac{1}{L}\displaystyle\int_0^t V(u)e^{Ru/L}du, I(t) = I(0)e^{-Rt/L} + \dfrac{1}{L}e^{-Rt/L}\int_0^t V(u)e^{Ru/L}du.$

(a) $I(t) = \dfrac{1}{4}e^{-5t/2}\displaystyle\int_0^t 12e^{5u/2}du = \dfrac{6}{5}e^{-5t/2}e^{5u/2}\Big]_0^t = \dfrac{6}{5}\left(1 - e^{-5t/2}\right)$ A.

(b) $\displaystyle\lim_{t\to+\infty} I(t) = \dfrac{6}{5}$ A

50. From Exercise 49 and Endpaper Table #42,

$$I(t) = 15e^{-2t} + \frac{1}{3}e^{-2t}\int_0^t 3e^{2u}\sin u\, du = 15e^{-2t} + e^{-2t}\frac{e^{2u}}{5}(2\sin u - \cos u)\Big]_0^t$$
$$= 15e^{-2t} + \frac{1}{5}(2\sin t - \cos t) + \frac{1}{5}e^{-2t}.$$

51. (a) $\dfrac{dv}{dt} = \dfrac{ck}{m_0 - kt} - g, v = -c\ln(m_0 - kt) - gt + C; v = 0$ when $t = 0$ so $0 = -c\ln m_0 + C$,

$C = c\ln m_0, v = c\ln m_0 - c\ln(m_0 - kt) - gt = c\ln\dfrac{m_0}{m_0 - kt} - gt.$

(b) $m_0 - kt = 0.2m_0$ when $t = 100$ so

$v = 2500\ln\dfrac{m_0}{0.2m_0} - 9.8(100) = 2500\ln 5 - 980 \approx 3044\,\text{m/s}.$

52. **(a)** By the chain rule, $\frac{dv}{dt} = \frac{dv}{dx}\frac{dx}{dt} = \frac{dv}{dx}v$ so $m\frac{dv}{dt} = mv\frac{dv}{dx}$.

(b) $\frac{mv}{kv^2 + mg}dv = -dx$, $\frac{m}{2k}\ln(kv^2 + mg) = -x + C$; $v = v_0$ when $x = 0$ so

$C = \frac{m}{2k}\ln(kv_0^2 + mg)$, $\frac{m}{2k}\ln(kv^2 + mg) = -x + \frac{m}{2k}\ln(kv_0^2 + mg)$, $x = \frac{m}{2k}\ln\frac{kv_0^2 + mg}{kv^2 + mg}$.

(c) $x = x_{max}$ when $v = 0$ so

$$x_{max} = \frac{m}{2k}\ln\frac{kv_0^2 + mg}{mg} = \frac{3.56 \times 10^{-3}}{2(7.3 \times 10^{-6})}\ln\frac{(7.3 \times 10^{-6})(988)^2 + (3.56 \times 10^{-3})(9.8)}{(3.56 \times 10^{-3})(9.8)} \approx 1298\,\text{m}$$

53. **(a)** $A(h) = \pi(1)^2 = \pi$, $\pi\frac{dh}{dt} = -0.025\sqrt{h}$, $\frac{\pi}{\sqrt{h}}dh = -0.025dt$, $2\pi\sqrt{h} = -0.025t + C$; $h = 4$ when $t = 0$, so $4\pi = C$, $2\pi\sqrt{h} = -0.025t + 4\pi$, $\sqrt{h} = 2 - \frac{0.025}{2\pi}t$, $h \approx (2 - 0.003979\,t)^2$.

(b) $h = 0$ when $t \approx 2/0.003979 \approx 502.6$ s ≈ 8.4 min.

54. **(a)** $A(h) = 6\left[2\sqrt{4 - (h-2)^2}\right] = 12\sqrt{4h - h^2}$,

$12\sqrt{4h - h^2}\frac{dh}{dt} = -0.025\sqrt{h}$, $12\sqrt{4 - h}\,dh = -0.025dt$,

$-8(4 - h)^{3/2} = -0.025t + C$; $h = 4$ when $t = 0$ so $C = 0$,

$(4 - h)^{3/2} = (0.025/8)t$, $4 - h = (0.025/8)^{2/3}t^{2/3}$,

$h \approx 4 - 0.021375t^{2/3}$ ft

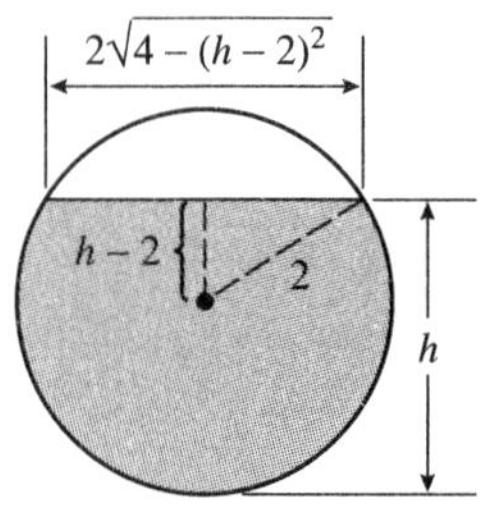

(b) $h = 0$ when $t = \frac{8}{0.025}(4 - 0)^{3/2} = 2560$ s ≈ 42.7 min

55. $\frac{dv}{dt} = -0.04v^2$, $\frac{1}{v^2}dv = -0.04dt$, $-\frac{1}{v} = -0.04t + C$; $v = 50$ when $t = 0$ so $-\frac{1}{50} = C$,

$-\frac{1}{v} = -0.04t - \frac{1}{50}$, $v = \frac{50}{2t+1}$ cm/s. But $v = \frac{dx}{dt}$ so $\frac{dx}{dt} = \frac{50}{2t+1}$, $x = 25\ln(2t+1) + C_1$;

$x = 0$ when $t = 0$ so $C_1 = 0$, $x = 25\ln(2t+1)$ cm.

56. $\frac{dv}{dt} = -0.02\sqrt{v}$, $\frac{1}{\sqrt{v}}dv = -0.02dt$, $2\sqrt{v} = -0.02t + C$; $v = 9$ when $t = 0$ so $6 = C$,

$2\sqrt{v} = -0.02t + 6$, $v = (3 - 0.01t)^2$ cm/s. But $v = \frac{dx}{dt}$ so $\frac{dx}{dt} = (3 - 0.01t)^2$,

$x = -\frac{100}{3}(3 - 0.01t)^3 + C_1$; $x = 0$ when $t = 0$ so $C_1 = 900$, $x = 900 - \frac{100}{3}(3 - 0.01t)^3$ cm.

57. Differentiate to get $\frac{dy}{dx} = -\sin x + e^{-x^2}$, $y(0) = 1$.

58. **(a)** Let $y = \frac{1}{\mu}[H(x) + C]$ where $\mu = e^{P(x)}$, $\frac{dP}{dx} = p(x)$, $\frac{d}{dx}H(x) = \mu q$, and C is an arbitrary constant. Then

$$\frac{dy}{dx} + p(x)y = \frac{1}{\mu}H'(x) - \frac{\mu'}{\mu^2}[H(x) + C] + p(x)y = q - \frac{p}{\mu}[H(x) + C] + p(x)y = q$$

(b) Given the initial value problem, let $C = \mu(x_0)y_0 - H(x_0)$. Then $y = \frac{1}{\mu}[H(x)+C]$ is a solution of the initial value problem with $y(x_0) = y_0$. This shows that the initial value problem has a solution.

To show uniqueness, suppose $u(x)$ also satisfies (5) together with $u(x_0) = y_0$. Following the arguments in the text we arrive at $u(x) = \frac{1}{\mu}[H(x) + C]$ for some constant C. The initial condition requires $C = \mu(x_0)y_0 - H(x_0)$, and thus $u(x)$ is identical with $y(x)$.

59. Suppose that $H(y) = G(x) + C$. Then $\frac{dH}{dy}\frac{dy}{dx} = G'(x)$. But $\frac{dH}{dy} = h(y)$ and $\frac{dG}{dx} = g(x)$, hence $y(x)$ is a solution of (10).

60. (a) $y = x$ and $y = -x$ are both solutions of the given initial value problem.

(b) $\int y\,dy = -\int x\,dx, y^2 = -x^2 + C$; but $y(0) = 0$, so $C = 0$. Thus $y^2 = -x^2$, which is impossible.

61. Suppose $I_1 \subset I$ is an interval with $I_1 \neq I$,and suppose $Y(x)$ is defined on I_1 and is a solution of (5) there. Let x_0 be a point of I_1. Solve the initial value problem on I with initial value $y(x_0) = Y(x_0)$. Then $y(x)$ is an extension of $Y(x)$ to the interval I, and by Exercise 58(b) applied to the interval I_1, it follows that $y(x) = Y(x)$ for x in I_1.

EXERCISE SET 9.2

1.

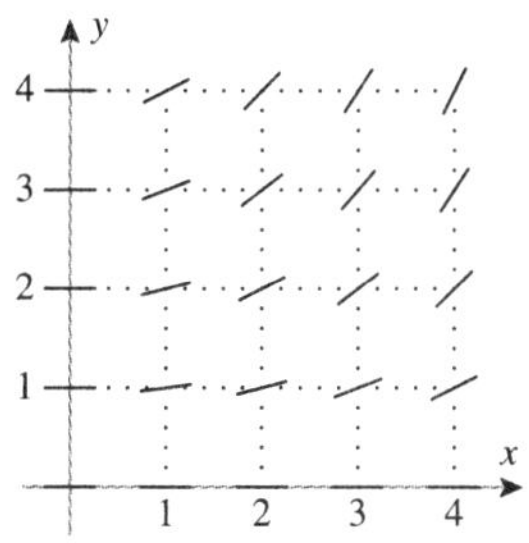

2.

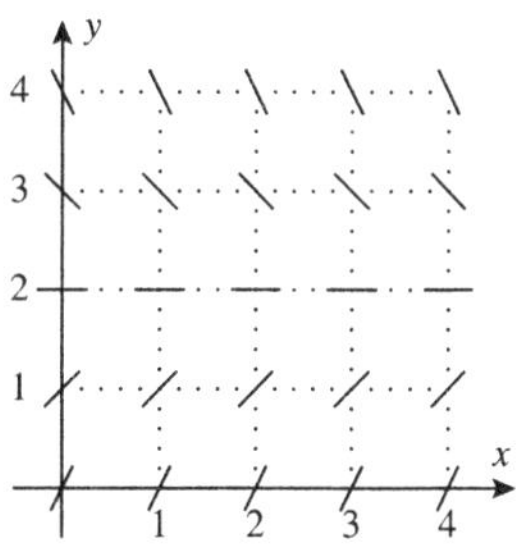

3.

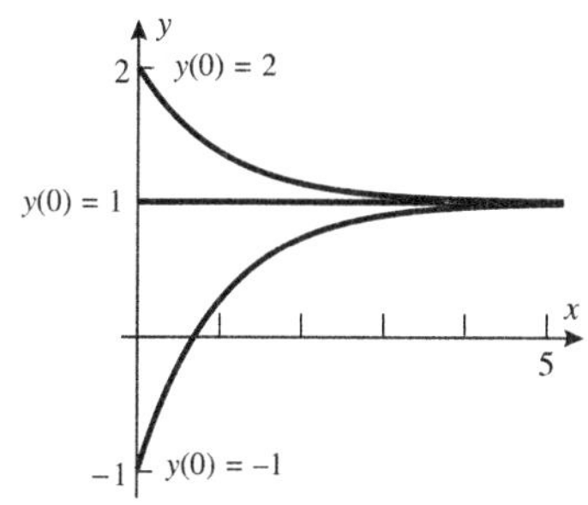

4. $\frac{dy}{dx} + y = 1, \mu = e^{\int dx} = e^x, \frac{d}{dx}[ye^x] = e^x, ye^x = e^x + C, y = 1 + Ce^{-x}$

(a) $-1 = 1 + C, C = -2, y = 1 - 2e^{-x}$

(b) $1 = 1 + C, C = 0, y = 1$

(c) $2 = 1 + C, C = 1, y = 1 + e^{-x}$

5.

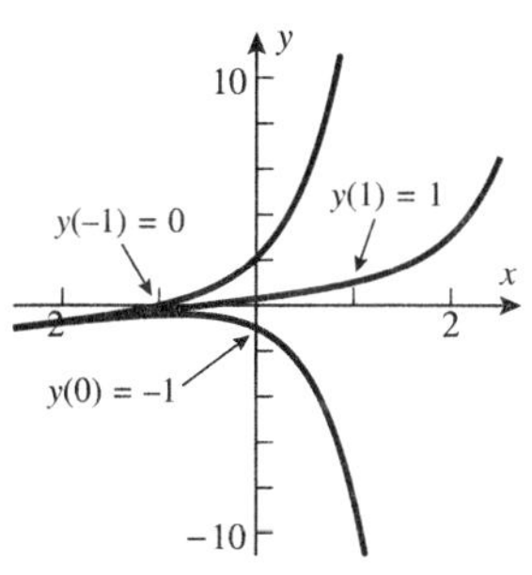

6. $\dfrac{dy}{dx} - 2y = -x, \quad \mu = e^{-2\int dx} = e^{-2x}, \quad \dfrac{d}{dx}\left[ye^{-2x}\right] = -xe^{-2x},$

$ye^{-2x} = \dfrac{1}{4}(2x+1)e^{-2x} + C, \quad y = \dfrac{1}{4}(2x+1) + Ce^{2x}$

(a) $1 = 3/4 + Ce^2,\ C = 1/(4e^2),\ y = \dfrac{1}{4}(2x+1) + \dfrac{1}{4}e^{2x-2}$

(b) $-1 = 1/4 + C, C = -5/4, y = \dfrac{1}{4}(2x+1) - \dfrac{5}{4}e^{2x}$

(c) $0 = -1/4 + Ce^{-2},\ C = e^2/4,\ y = \dfrac{1}{4}(2x+1) + \dfrac{1}{4}e^{2x+2}$

7. $\lim\limits_{x\to+\infty} y = 1$

8. $\lim\limits_{x\to+\infty} y = \begin{cases} +\infty & \text{if } y_0 \geq 1/4 \\ -\infty, & \text{if } y_0 < 1/4 \end{cases}$

9. **(a)** IV, since the slope is positive for $x > 0$ and negative for $x < 0$.
(b) VI, since the slope is positive for $y > 0$ and negative for $y < 0$.
(c) V, since the slope is always positive.
(d) II, since the slope changes sign when crossing the lines $y = \pm 1$.
(e) I, since the slope can be positive or negative in each quadrant but is not periodic.
(f) III, since the slope is periodic in both x and y.

11. **(a)** $y_0 = 1,$
$y_{n+1} = y_n + (x_n + y_n)(0.2) = (x_n + 6y_n)/5$

n	0	1	2	3	4	5
x_n	0	0.2	0.4	0.6	0.8	1.0
y_n	1	1.20	1.48	1.86	2.35	2.98

(b) $y' - y = x,\ \mu = e^{-x},\ \dfrac{d}{dx}\left[ye^{-x}\right] = xe^{-x},$
$ye^{-x} = -(x+1)e^{-x} + C,\ 1 = -1 + C,$
$C = 2,\ y = -(x+1) + 2e^x$

x_n	0	0.2	0.4	0.6	0.8	1.0
$y(x_n)$	1	1.24	1.58	2.04	2.65	3.44
abs. error	0	0.04	0.10	0.19	0.30	0.46
perc. error	0	3	6	9	11	13

(c)

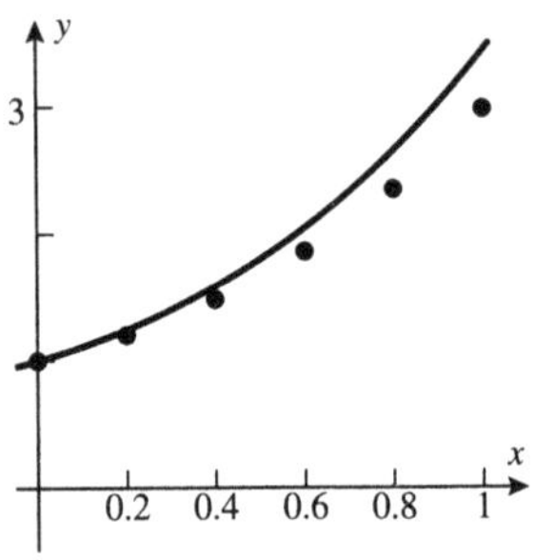

12. $h = 0.1, y_{n+1} = (x_n + 11y_n)/10$

n	0	1	2	3	4	5	6	7	8	9	10
x_n	0	0.1	0.2	0.3	0.4	0.5	0.6	0.7	0.8	0.9	1.0
y_n	1.00	1.10	1.22	1.36	1.53	1.72	1.94	2.20	2.49	2.82	3.19

In Exercise 11, $y(1) \approx 2.98$; in Exercise 12, $y(1) \approx 3.19$; the true solution is $y(1) \approx 3.44$; so the absolute errors are approximately 0.46 and 0.25 respectively.

13. $y_0 = 1, y_{n+1} = y_n + \sqrt{y_n}/2$

n	0	1	2	3	4	5	6	7	8
x_n	0	0.5	1	1.5	2	2.5	3	3.5	4
y_n	1	1.50	2.11	2.84	3.68	4.64	5.72	6.91	8.23

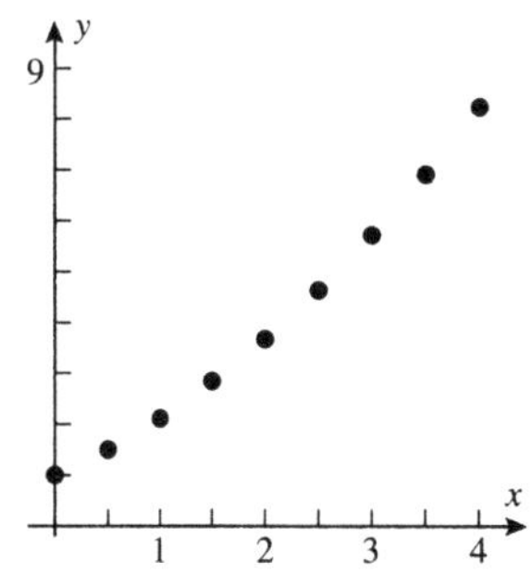

14. $y_0 = 1, y_{n+1} = y_n + (x_n - y_n^2)/4$

n	0	1	2	3	4	5	6	7	8
x_n	0	0.25	0.50	0.75	1.00	1.25	1.50	1.75	2.00
y_n	1	0.75	0.67	0.68	0.75	0.86	0.99	1.12	1.24

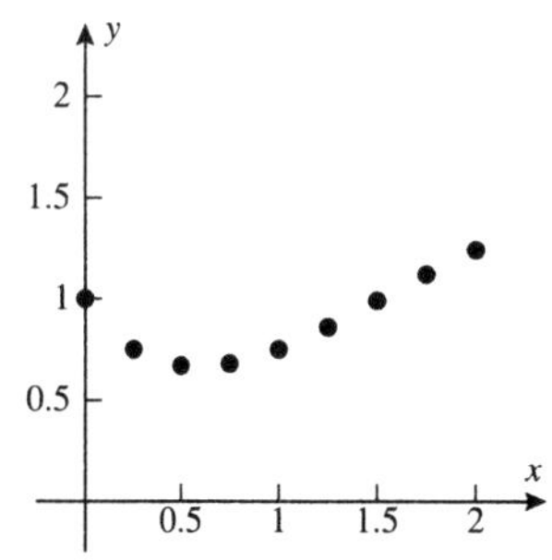

15. $y_0 = 1, y_{n+1} = y_n + \frac{1}{2}\sin y_n$

n	0	1	2	3	4
t_n	0	0.5	1	1.5	2
y_n	1	1.42	1.92	2.39	2.73

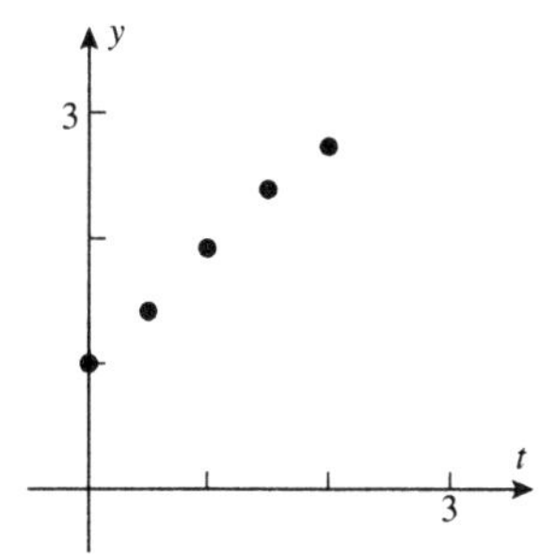

16. $y_0 = 0,\ y_{n+1} = y_n + e^{-y_n}/10$

n	0	1	2	3	4	5	6	7	8	9	10
t_n	0	0.1	0.2	0.3	0.4	0.5	0.6	0.7	0.8	0.9	1.0
y_n	0	0.10	0.19	0.27	0.35	0.42	0.49	0.55	0.60	0.66	0.71

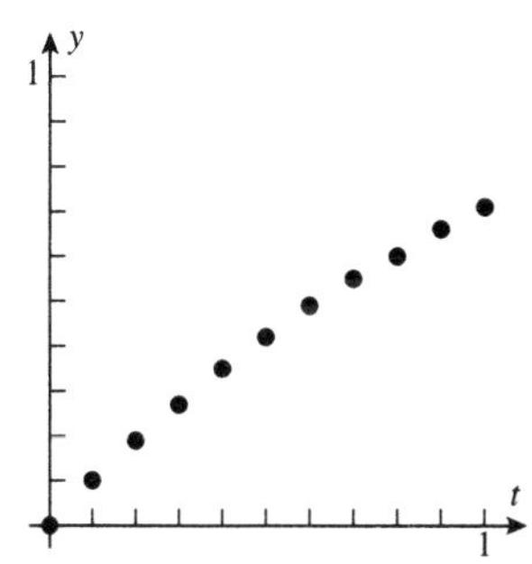

17. $h = 1/5,\ y_0 = 1,\ y_{n+1} = y_n + \frac{1}{5}\cos(2\pi n/5)$

n	0	1	2	3	4	5
t_n	0	0.2	0.4	0.6	0.8	1.0
y_n	1.00	1.06	0.90	0.74	0.80	1.00

18. **(a)** By inspection, $\frac{dy}{dx} = e^{-x^2}$ and $y(0) = 0$.

(b) $y_{n+1} = y_n + e^{-x_n^2}/20 = y_n + e^{-(n/20)^2}/20$ and $y_{20} = 0.7625$. From a CAS, $y(1) = 0.7468$.

19. **(b)** $y\,dy = -x\,dx$, $y^2/2 = -x^2/2 + C_1$, $x^2 + y^2 = C$; if $y(0) = 1$ then $C = 1$ so $y(1/2) = \sqrt{3}/2$.

20. **(a)** $y_0 = 1, y_{n+1} = y_n + (\sqrt{y_n}/2)\Delta x$

$\Delta x = 0.2: \ y_{n+1} = y_n + \sqrt{y_n}/10; \ y_5 \approx 1.5489$

$\Delta x = 0.1: y_{n+1} = y_n + \sqrt{y_n}/20; \ y_{10} \approx 1.5556$

$\Delta x = 0.05: y_{n+1} = y_n + \sqrt{y_n}/40; \ y_{20} \approx 1.5590$

(c) $\dfrac{dy}{\sqrt{y}} = \dfrac{1}{2}dx$, $2\sqrt{y} = x/2 + C$, $2 = C$,

$\sqrt{y} = x/4 + 1, y = (x/4 + 1)^2$,

$y(1) = 25/16 = 1.5625$

EXERCISE SET 9.3

1. **(a)** $\dfrac{dy}{dt} = ky^2$, $y(0) = y_0, k > 0$

(b) $\dfrac{dy}{dt} = -ky^2$, $y(0) = y_0, k > 0$

3. **(a)** $\dfrac{ds}{dt} = \dfrac{1}{2}s$

(b) $\dfrac{d^2s}{dt^2} = 2\dfrac{ds}{dt}$

4. **(a)** $\dfrac{dv}{dt} = -2v^2$

(b) $\dfrac{d^2s}{dt^2} = -2\left(\dfrac{ds}{dt}\right)^2$

5. **(a)** $\dfrac{dy}{dt} = 0.01y$, $y_0 = 10{,}000$

(b) $y = 10{,}000e^{t/100}$

(c) $T = \dfrac{1}{k}\ln 2 = \dfrac{1}{0.01}\ln 2 \approx 69.31$ h

(d) $45{,}000 = 10{,}000e^{t/100}$,

$t = 100 \ln \dfrac{45{,}000}{10{,}000} \approx 150.41$ h

6. $k = \dfrac{1}{T}\ln 2 = \dfrac{1}{20}\ln 2$

(a) $\dfrac{dy}{dt} = ((\ln 2)/20)y$, $y(0) = 1$

(b) $y(t) = e^{t(\ln 2)/20} = 2^{t/20}$

(c) $y(120) = 2^6 = 64$

(d) $1{,}000{,}000 = 2^{t/20}$,

$t = 20\dfrac{\ln 10^6}{\ln 2} \approx 398.63$ min

7. **(a)** $\dfrac{dy}{dt} = -ky$, $y(0) = 5.0 \times 10^7$; $3.83 = T = \dfrac{1}{k}\ln 2$, so $k = \dfrac{\ln 2}{3.83} \approx 0.1810$

(b) $y = 5.0 \times 10^7 e^{-0.181t}$

(c) $y(30) = 5.0 \times 10^7 e^{-0.1810(30)} \approx 219{,}000$

(d) $y(t) = (0.1)y_0 = y_0 e^{-kt}$, $-kt = \ln 0.1$, $t = -\dfrac{\ln 0.1}{0.1810} = 12.72$ days

8. **(a)** $k = \dfrac{1}{T}\ln 2 = \dfrac{1}{140}\ln 2 \approx 0.0050$, so $\dfrac{dy}{dt} = -0.0050y$, $y_0 = 10$.

(b) $y = 10e^{-0.0050t}$

(c) 10 weeks = 70 days so $y = 10e^{-0.35} \approx 7$ mg.

(d) $0.3y_0 = y_0 e^{-kt}$, $t = -\dfrac{\ln 0.3}{0.0050} \approx 240.8$ days

9. $100e^{0.02t} = 5000$, $e^{0.02t} = 50$, $t = \dfrac{1}{0.02}\ln 50 \approx 196$ days

10. $y = 10{,}000e^{kt}$, but $y = 12{,}000$ when $t = 10$ so $10{,}000e^{10k} = 12{,}000$, $k = \dfrac{1}{10}\ln 1.2$. $y = 20{,}000$ when $2 = e^{kt}$, $t = \dfrac{\ln 2}{k} = 10\dfrac{\ln 2}{\ln 1.2} \approx 38$, in the year 2025.

11. $y(t) = y_0 e^{-kt} = 10.0e^{-kt}$, $3.5 = 10.0e^{-k(5)}$, $k = -\dfrac{1}{5}\ln\dfrac{3.5}{10.0} \approx 0.2100$, $T = \dfrac{1}{k}\ln 2 \approx 3.30$ days

12. $y = y_0 e^{-kt}$, $0.6y_0 = y_0 e^{-5k}$, $k = -\dfrac{1}{5}\ln 0.6 \approx 0.10$

(a) $T = \dfrac{\ln 2}{k} \approx 6.9$ yr

(b) $y(t) \approx y_0 e^{-0.10t}$, $\dfrac{y}{y_0} \approx e^{-0.10t}$, so $e^{-0.10t} \times 100$ percent will remain.

13. **(a)** $k = \dfrac{\ln 2}{5} \approx 0.1386; y \approx 2e^{0.1386t}$ **(b)** $y(t) = 5e^{0.015t}$

(c) $y = y_0 e^{kt}$, $1 = y_0 e^k$, $100 = y_0 e^{10k}$. Divide: $100 = e^{9k}$, $k = \dfrac{1}{9}\ln 100 \approx 0.5117$,
$y \approx y_0 e^{0.5117t}$; also $y(1) = 1$, so $y_0 = e^{-0.5117} \approx 0.5995, y \approx 0.5995e^{0.5117t}$.

(d) $k = \dfrac{\ln 2}{T} \approx 0.1386$, $1 = y(1) \approx y_0 e^{0.1386}$, $y_0 \approx e^{-0.1386} \approx 0.8706$, $y \approx 0.8706e^{0.1386t}$

14. **(a)** $k = \dfrac{\ln 2}{T} \approx 0.1386$, $y \approx 10e^{-0.1386t}$ **(b)** $y = 10e^{-0.015t}$

(c) $100 = y_0 e^{-k}$, $1 = y_0 e^{-10k}$. Divide: $e^{9k} = 100$, $k = \dfrac{1}{9}\ln 100 \approx 0.5117$;
$y_0 = e^{10k} \approx e^{5.117} \approx 166.83$, $y = 166.83e^{-0.5117t}$.

(d) $k = \dfrac{\ln 2}{T} \approx 0.1386$, $10 = y(1) \approx y_0 e^{-0.1386}$, $y_0 \approx 10e^{0.1386} \approx 11.4866$, $y \approx 11.4866e^{-0.1386t}$

16. **(a)** None; the half-life is independent of the initial amount.
(b) $kT = \ln 2$, so T is inversely proportional to k.

17. **(a)** $T = \dfrac{\ln 2}{k}$; and $\ln 2 \approx 0.6931$. If k is measured in percent, $k' = 100k$, then $T = \dfrac{\ln 2}{k} \approx \dfrac{69.31}{k'} \approx \dfrac{70}{k'}$.

(b) 70 yr **(c)** 20 yr **(d)** 7%

18. Let $y = y_0 e^{kt}$ with $y = y_1$ when $t = t_1$ and $y = 3y_1$ when $t = t_1 + T$; then $y_0 e^{kt_1} = y_1$ (i) and $y_0 e^{k(t_1+T)} = 3y_1$ (ii). Divide (ii) by (i) to get $e^{kT} = 3$, $T = \dfrac{1}{k}\ln 3$.

19. From (12), $y(t) = y_0 e^{-0.000121t}$. If $0.27 = \dfrac{y(t)}{y_0} = e^{-0.000121t}$ then $t = -\dfrac{\ln 0.27}{0.000121} \approx 10{,}820$ yr, and if $0.30 = \dfrac{y(t)}{y_0}$ then $t = -\dfrac{\ln 0.30}{0.000121} \approx 9950$, or roughly between 9000 B.C. and 8000 B.C.

20. (a) 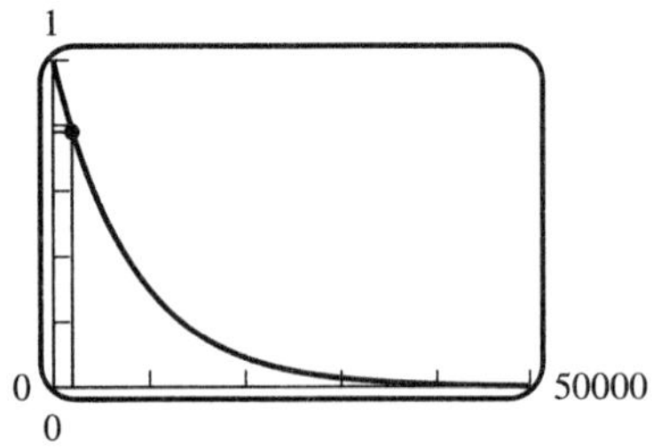

(b) $t = 1988$ yields

$y/y_0 = e^{-0.000121(1988)} \approx 79\%.$

21. $y_0 \approx 2,\ L \approx 8$; since the curve $y = \dfrac{2 \cdot 8}{2 + 6e^{-kt}}$ passes through the point $(2, 4)$, $4 = \dfrac{16}{2 + 6e^{-2k}}$, $6e^{-2k} = 2,\ k = \dfrac{1}{2}\ln 3 \approx 0.5493.$

22. $y_0 \approx 400,\ L \approx 1000$; since the curve $y = \dfrac{400{,}000}{400 + 600e^{-kt}}$ passes through the point $(200, 600)$, $600 = \dfrac{400{,}000}{400 + 600e^{-200k}},\ 600e^{-200k} = \dfrac{800}{3},\ k = \dfrac{1}{200}\ln 2.25 \approx 0.00405.$

23. (a) $y_0 = 5$ (b) $L = 12$ (c) $k = 1$

(d) $L/2 = 6 = \dfrac{60}{5 + 7e^{-t}},\ 5 + 7e^{-t} = 10,\ t = -\ln(5/7) \approx 0.3365$

(e) $\dfrac{dy}{dt} = \dfrac{1}{12}y(12 - y),\ y(0) = 5$

24. (a) $y_0 = 1$ (b) $L = 1000$ (c) $k = 0.9$

(d) $750 = \dfrac{1000}{1 + 999e^{-0.9t}},\ 3(1 + 999e^{-0.9t}) = 4,\ t = \dfrac{1}{0.9}\ln(3 \cdot 999) \approx 8.8949$

(e) $\dfrac{dy}{dt} = \dfrac{0.9}{1000}y(1000 - y),\ y(0) = 1$

25. See (13):

(a) $L = 10$ (b) $k = 10$

(c) $\dfrac{dy}{dt} = 10(1 - 0.1y)y = 25 - (y - 5)^2$ is maximized when $y = 5$.

26. $\dfrac{dy}{dt} = 50y\left(1 - \dfrac{1}{50{,}000}y\right)$; from (13), $k = 50,\ L = 50{,}000$.

(a) $L = 50{,}000$ (b) $k = 50$

(c) $\dfrac{dy}{dt}$ is maximized when $0 = \dfrac{d}{dy}\left(\dfrac{dy}{dt}\right) = 50 - y/500, y = 25{,}000$

27. Assume $y(t)$ students have had the flu t days after semester break. Then $y(0) = 20,\ y(5) = 35$.

(a) $\dfrac{dy}{dt} = ky(L - y) = ky(1000 - y),\ y_0 = 20$

(b) Part (a) has solution $y = \dfrac{20000}{20 + 980e^{-kt}} = \dfrac{1000}{1 + 49e^{-kt}}$;

$35 = \dfrac{1000}{1 + 49e^{-5k}}$, $k = 0.115$, $y \approx \dfrac{1000}{1 + 49e^{-0.115t}}$.

(c)

t	0	1	2	3	4	5	6	7	8	9	10	11	12	13	14
$y(t)$	20	22	25	28	31	35	39	44	49	54	61	67	75	83	93

(d)

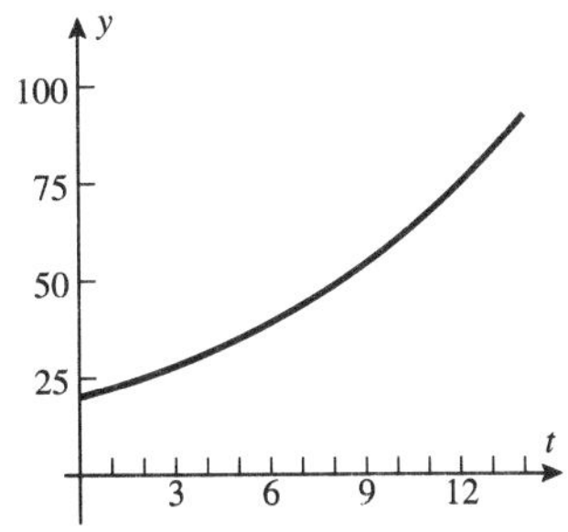

28. **(a)** $\dfrac{dp}{dh} = -kp$, $p(0) = p_0$

(b) $p_0 = 1$, so $p = e^{-kh}$, but $p = 0.83$ when $h = 5000$ thus $e^{-5000k} = 0.83$,

$k = -\dfrac{\ln 0.83}{5000} \approx 0.0000373$, $p \approx e^{-0.0000373h}$ atm.

29. **(a)** $\dfrac{dT}{dt} = -k(T - 21)$, $T(0) = 95$, $\dfrac{dT}{T - 21} = -k\,dt$, $\ln(T - 21) = -kt + C_1$,

$T = 21 + e^{C_1}e^{-kt} = 21 + Ce^{-kt}$, $95 = T(0) = 21 + C, C = 74$, $T = 21 + 74e^{-kt}$

(b) $85 = T(1) = 21 + 74e^{-k}$, $k = -\ln\dfrac{64}{74} = -\ln\dfrac{32}{37}$, $T = 21 + 74e^{t\ln(32/37)} = 21 + 74\left(\dfrac{32}{37}\right)^t$,

$T = 51$ when $\dfrac{30}{74} = \left(\dfrac{32}{37}\right)^t$, $t = \dfrac{\ln(30/74)}{\ln(32/37)} \approx 6.22$ min

30. $\dfrac{dT}{dt} = k(70 - T), T(0) = 40$; $-\ln(70 - T) = kt + C$, $70 - T = e^{-kt}e^{-C}$, $T = 40$ when $t = 0$, so $30 = e^{-C}$, $T = 70 - 30e^{-kt}$; $52 = T(1) = 70 - 30e^{-k}$, $k = -\ln\dfrac{70 - 52}{30} = \ln\dfrac{5}{3} \approx 0.5$, $T \approx 70 - 30e^{-0.5t}$

31. Let T denote the body temperature of McHam's body at time t, the number of hours elapsed after 10:06 P.M.; then $\dfrac{dT}{dt} = -k(T - 72)$, $\dfrac{dT}{T - 72} = -kdt$, $\ln(T - 72) = -kt + C, T = 72 + e^Ce^{-kt}$, $77.9 = 72 + e^C$, $e^C = 5.9$, $T = 72 + 5.9e^{-kt}$, $75.6 = 72 + 5.9e^{-k}$, $k = -\ln\dfrac{3.6}{5.9} \approx 0.4940$, $T = 72 + 5.9e^{-0.4940t}$. McHam's body temperature was last 98.6° when $t = -\dfrac{\ln(26.6/5.9)}{0.4940} \approx -3.05$, so around 3 hours and 3 minutes before 10:06; the death took place at approximately 7:03 P.M., while Moore was on stage.

32. If $T_0 < T_a$ then $\dfrac{dT}{dt} = k(T_a - T)$ where $k > 0$. If $T_0 > T_a$ then $\dfrac{dT}{dt} = -k(T - T_a)$ where $k > 0$; both cases yield $T(t) = T_a + (T_0 - T_a)e^{-kt}$ with $k > 0$.

33. **(a)** $y = y_0 b^t = y_0 e^{t\ln b} = y_0 e^{kt}$ with $k = \ln b > 0$ since $b > 1$.

(b) $y = y_0 b^t = y_0 e^{t\ln b} = y_0 e^{-kt}$ with $k = -\ln b > 0$ since $0 < b < 1$.

(c) $y = 4(2^t) = 4e^{t\ln 2}$

(d) $y = 4(0.5^t) = 4e^{t\ln 0.5} = 4e^{-t\ln 2}$

34. If $y = y_0 e^{kt}$ and $y = y_1 = y_0 e^{kt_1}$ then $y_1/y_0 = e^{kt_1}$, $k = \dfrac{\ln(y_1/y_0)}{t_1}$; if $y = y_0 e^{-kt}$ and $y = y_1 = y_0 e^{-kt_1}$ then $y_1/y_0 = e^{-kt_1}$, $k = -\dfrac{\ln(y_1/y_0)}{t_1}$.

EXERCISE SET 9.4

1. **(a)** $y = e^{2x}$, $y' = 2e^{2x}$, $y'' = 4e^{2x}$; $y'' - y' - 2y = 0$
$y = e^{-x}$, $y' = -e^{-x}$, $y'' = e^{-x}$; $y'' - y' - 2y = 0$.

(b) $y = c_1e^{2x} + c_2e^{-x}$, $y' = 2c_1e^{2x} - c_2e^{-x}$, $y'' = 4c_1e^{2x} + c_2e^{-x}$; $y'' - y' - 2y = 0$

2. **(a)** $y = e^{-2x}$, $y' = -2e^{-2x}$, $y'' = 4e^{-2x}$; $y'' + 4y' + 4y = 0$
$y = xe^{-2x}$, $y' = (1 - 2x)e^{-2x}$, $y'' = (4x - 4)e^{-2x}$; $y'' + 4y' + 4y = 0$.

(b) $y = c_1e^{-2x} + c_2xe^{-2x}$, $y' = -2c_1e^{-2x} + c_2(1 - 2x)e^{-2x}$,
$y'' = 4c_1e^{-2x} + c_2(4x - 4)e^{-2x}$; $y'' + 4y' + 4y = 0$.

3. $m^2 + 3m - 4 = 0$, $(m - 1)(m + 4) = 0$; $m = 1, -4$ so $y = c_1e^x + c_2e^{-4x}$.

4. $m^2 + 6m + 5 = 0$, $(m + 1)(m + 5) = 0$; $m = -1, -5$ so $y = c_1e^{-x} + c_2e^{-5x}$.

5. $m^2 - 2m + 1 = 0$, $(m - 1)^2 = 0$; $m = 1$, so $y = c_1e^x + c_2xe^x$.

6. $m^2 + 6m + 9 = 0$, $(m + 3)^2 = 0$; $m = -3$ so $y = c_1e^{-3x} + c_2xe^{-3x}$.

7. $m^2 + 5 = 0$, $m = \pm\sqrt{5}\,i$ so $y = c_1\cos\sqrt{5}\,x + c_2\sin\sqrt{5}\,x$.

8. $m^2 + 1 = 0$, $m = \pm i$ so $y = c_1\cos x + c_2\sin x$.

9. $m^2 - m = 0$, $m(m - 1) = 0$; $m = 0, 1$ so $y = c_1 + c_2e^x$.

10. $m^2 + 3m = 0$, $m(m + 3) = 0$; $m = 0, -3$ so $y = c_1 + c_2e^{-3x}$.

11. $m^2 + 4m + 4 = 0$, $(m + 2)^2 = 0$; $m = -2$ so $y = c_1e^{-2t} + c_2te^{-2t}$.

12. $m^2 - 10m + 25 = 0$, $(m - 5)^2 = 0$; $m = 5$ so $y = c_1e^{5t} + c_2te^{5t}$.

13. $m^2 - 4m + 13 = 0$, $m = 2 \pm 3i$ so $y = e^{2x}(c_1\cos 3x + c_2\sin 3x)$.

14. $m^2 - 6m + 25 = 0$, $m = 3 \pm 4i$ so $y = e^{3x}(c_1\cos 4x + c_2\sin 4x)$.

15. $8m^2 - 2m - 1 = 0$, $(4m + 1)(2m - 1) = 0$; $m = -1/4, 1/2$ so $y = c_1e^{-x/4} + c_2e^{x/2}$.

16. $9m^2 - 6m + 1 = 0$, $(3m - 1)^2 = 0$; $m = 1/3$ so $y = c_1e^{x/3} + c_2xe^{x/3}$.

17. $m^2 + 2m - 3 = 0$, $(m + 3)(m - 1) = 0$; $m = -3, 1$ so $y = c_1e^{-3x} + c_2e^x$ and $y' = -3c_1e^{-3x} + c_2e^x$.
Solve the system $c_1 + c_2 = 1$, $-3c_1 + c_2 = 5$ to get $c_1 = -1$, $c_2 = 2$ so $y = -e^{-3x} + 2e^x$.

18. $m^2 - 6m - 7 = 0$, $(m+1)(m-7) = 0$; $m = -1, 7$ so $y = c_1e^{-x} + c_2e^{7x}$, $y' = -c_1e^{-x} + 7c_2e^{7x}$. Solve the system $c_1 + c_2 = 5$, $-c_1 + 7c_2 = 3$ to get $c_1 = 4$, $c_2 = 1$ so $y = 4e^{-x} + e^{7x}$.

19. $m^2 - 6m + 9 = 0$, $(m-3)^2 = 0$; $m = 3$ so $y = (c_1 + c_2x)e^{3x}$ and $y' = (3c_1 + c_2 + 3c_2x)e^{3x}$. Solve the system $c_1 = 2$, $3c_1 + c_2 = 1$ to get $c_1 = 2$, $c_2 = -5$ so $y = (2 - 5x)e^{3x}$.

20. $m^2 + 4m + 1 = 0$, $m = -2 \pm \sqrt{3}$ so $y = c_1e^{(-2+\sqrt{3})x} + c_2e^{(-2-\sqrt{3})x}$,
$y' = (-2+\sqrt{3})c_1e^{(-2+\sqrt{3})x} + (-2-\sqrt{3})c_2e^{(-2-\sqrt{3})x}$. Solve the system $c_1 + c_2 = 5$,
$(-2+\sqrt{3})c_1 + (-2-\sqrt{3})c_2 = 4$ to get $c_1 = \frac{5}{2} + \frac{7}{3}\sqrt{3}$, $c_2 = \frac{5}{2} - \frac{7}{3}\sqrt{3}$ so
$y = \left(\frac{5}{2} + \frac{7}{3}\sqrt{3}\right)e^{(-2+\sqrt{3})x} + \left(\frac{5}{2} - \frac{7}{3}\sqrt{3}\right)e^{(-2-\sqrt{3})x}$.

21. $m^2 + 4m + 5 = 0$, $m = -2 \pm i$ so $y = e^{-2x}(c_1\cos x + c_2\sin x)$,
$y' = e^{-2x}[(c_2 - 2c_1)\cos x - (c_1 + 2c_2)\sin x]$. Solve the system $c_1 = -3$, $c_2 - 2c_1 = 0$ to get $c_1 = -3$, $c_2 = -6$ so $y = -e^{-2x}(3\cos x + 6\sin x)$.

22. $m^2 - 6m + 13 = 0$, $m = 3 \pm 2i$ so $y = e^{3x}(c_1\cos 2x + c_2\sin 2x)$,
$y' = e^{3x}[(3c_1 + 2c_2)\cos 2x - (2c_1 - 3c_2)\sin 2x]$. Solve the system $c_1 = -1$, $3c_1 + 2c_2 = 1$ to get $c_1 = -1$, $c_2 = 2$ so $y = e^{3x}(-\cos 2x + 2\sin 2x)$.

23. **(a)** $m = 5, -2$ so $(m-5)(m+2) = 0$, $m^2 - 3m - 10 = 0$; $y'' - 3y' - 10y = 0$.

(b) $m = 4, 4$ so $(m-4)^2 = 0$, $m^2 - 8m + 16 = 0$; $y'' - 8y' + 16y = 0$.

(c) $m = -1 \pm 4i$ so $(m + 1 - 4i)(m + 1 + 4i) = 0$, $m^2 + 2m + 17 = 0$; $y'' + 2y' + 17y = 0$.

24. $c_1e^x + c_2e^{-x}$ is the general solution, but $\cosh x = \frac{1}{2}e^x + \frac{1}{2}e^{-x}$ and $\sinh x = \frac{1}{2}e^x - \frac{1}{2}e^{-x}$ so $\cosh x$ and $\sinh x$ are also solutions.

25. $m^2 + km + k = 0$, $m = \left(-k \pm \sqrt{k^2 - 4k}\right)/2$

(a) $k^2 - 4k > 0$, $k(k-4) > 0$; $k < 0$ or $k > 4$

(b) $k^2 - 4k = 0$; $k = 0, 4$

(c) $k^2 - 4k < 0$, $k(k-4) < 0$; $0 < k < 4$

26. $z = \ln x$; $\dfrac{dy}{dx} = \dfrac{dy}{dz}\dfrac{dz}{dx} = \dfrac{1}{x}\dfrac{dy}{dz}$ and

$$\frac{d^2y}{dx^2} = \frac{d}{dx}\left(\frac{dy}{dx}\right) = \frac{d}{dx}\left(\frac{1}{x}\frac{dy}{dz}\right) = \frac{1}{x}\frac{d^2y}{dz^2}\frac{dz}{dx} - \frac{1}{x^2}\frac{dy}{dz} = \frac{1}{x^2}\frac{d^2y}{dz^2} - \frac{1}{x^2}\frac{dy}{dz},$$

substitute into the original equation to get $\dfrac{d^2y}{dz^2} + (p-1)\dfrac{dy}{dz} + qy = 0$.

27. **(a)** $\dfrac{d^2y}{dz^2} + 2\dfrac{dy}{dz} + 2y = 0$, $m^2 + 2m + 2 = 0$; $m = -1 \pm i$ so
$y = e^{-z}(c_1\cos z + c_2\sin z) = \dfrac{1}{x}[c_1\cos(\ln x) + c_2\sin(\ln x)]$.

(b) $\dfrac{d^2y}{dz^2} - 2\dfrac{dy}{dz} - 2y = 0$, $m^2 - 2m - 2 = 0$; $m = 1 \pm \sqrt{3}$ so
$y = c_1e^{(1+\sqrt{3})z} + c_2e^{(1-\sqrt{3})z} = c_1x^{1+\sqrt{3}} + c_2x^{1-\sqrt{3}}$

28. $m^2 + pm + q = 0$, $m = \frac{1}{2}(-p \pm \sqrt{p^2 - 4q})$. If $0 < q < p^2/4$ then $y = c_1e^{m_1x} + c_2e^{m_2x}$ where $m_1 < 0$ and $m_2 < 0$, if $q = p^2/4$ then $y = c_1e^{-px/2} + c_2xe^{-px/2}$, if $q > p^2/4$ then $y = e^{-px/2}(c_1\cos kx + c_2\sin kx)$ where $k = \frac{1}{2}\sqrt{4q - p^2}$. In all cases $\lim_{x\to+\infty} y(x) = 0$.

29. **(a)** Neither is a constant multiple of the other, since, e.g. if $y_1 = ky_2$ then $e^{m_1x} = ke^{m_2x}$, $e^{(m_1-m_2)x} = k$. But the right hand side is constant, and the left hand side is constant only if $m_1 = m_2$, which is false.

(b) If $y_1 = ky_2$ then $e^{mx} = kxe^{mx}, kx = 1$ which is impossible. If $y_2 = y_1$ then $xe^{mx} = ke^{mx}$, $x = k$ which is impossible.

30. $y_1 = e^{ax}\cos bx$, $y_1' = e^{ax}(a\cos bx - b\sin bx)$, $y_1'' = e^{ax}[(a^2 - b^2)\cos bx - 2ab\sin bx]$ so $y_1'' + py_1' + qy_1 = e^{ax}[(a^2 - b^2 + ap + q)\cos bx - (2ab + bp)\sin bx]$. But $a = -\frac{1}{2}p$ and $b = \frac{1}{2}\sqrt{4q - p^2}$ so $a^2 - b^2 + ap + q = 0$ and $2ab + bp = 0$ thus $y_1'' + py_1' + qy_1 = 0$. Similarly, $y_2 = e^{ax}\sin bx$ is also a solution.
Since $y_1/y_2 = \cot bx$ and $y_2/y_1 = \tan bx$ it is clear that the two solutions are linearly independent.

31. (a) The general solution is $c_1 e^{\mu x} + c_2 e^{mx}$; let $c_1 = 1/(\mu - m)$, $c_2 = -1/(\mu - m)$.

(b) $\lim\limits_{\mu\to m} \dfrac{e^{\mu x} - e^{mx}}{\mu - m} = \lim\limits_{\mu\to m} xe^{\mu x} = xe^{mx}$.

32. (a) If $\lambda = 0$, then $y'' = 0$, $y = c_1 + c_2 x$. Use $y(0) = 0$ and $y(\pi) = 0$ to get $c_1 = c_2 = 0$. If $\lambda < 0$, then let $\lambda = -a^2$ where $a > 0$ so $y'' - a^2 y = 0$, $y = c_1 e^{ax} + c_2 e^{-ax}$. Use $y(0) = 0$ and $y(\pi) = 0$ to get $c_1 = c_2 = 0$.

(b) If $\lambda > 0$, then $m^2 + \lambda = 0$, $m^2 = -\lambda = \lambda i^2$, $m = \pm\sqrt{\lambda}i$, $y = c_1\cos\sqrt{\lambda}\,x + c_2\sin\sqrt{\lambda}\,x$. If $y(0) = 0$ and $y(\pi) = 0$, then $c_1 = 0$ and $c_1\cos\pi\sqrt{\lambda} + c_2\sin\pi\sqrt{\lambda} = 0$ so $c_2\sin\pi\sqrt{\lambda} = 0$. But $c_2\sin\pi\sqrt{\lambda} = 0$ for arbitrary values of c_2 if $\sin\pi\sqrt{\lambda} = 0$, $\pi\sqrt{\lambda} = n\pi$, $\lambda = n^2$ for $n = 1, 2, 3, \ldots$, otherwise $c_2 = 0$.

33. $k/M = 0.25/1 = 0.25$

(a) From (20), $y = 0.3\cos(t/2)$

(b) $T = 2\pi \cdot 2 = 4\pi$ s, $f = 1/T = 1/(4\pi)$ Hz

(c)

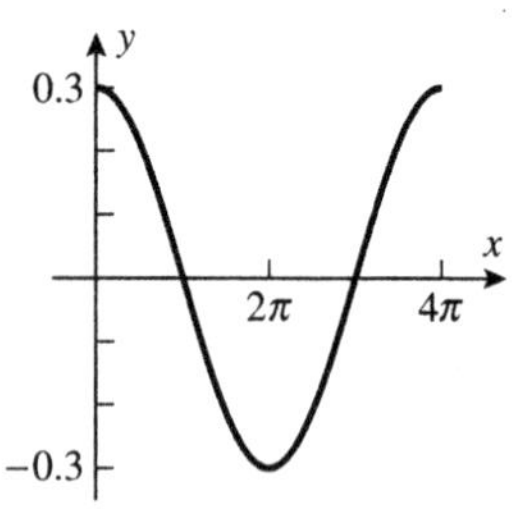

(d) $y = 0$ at the equilibrium position, so $t/2 = \pi/2, t = \pi$ s.

(e) $t/2 = \pi$ at the maximum position below the equlibrium position, so $t = 2\pi$ s.

34. $64 = w = -Mg$, $M = 2$, $k/M = 0.25/2 = 1/8$, $\sqrt{k/M} = 1/(2\sqrt{2})$

(a) From (20), $y = \cos(t/(2\sqrt{2}))$

(b) $T = 2\pi\sqrt{\dfrac{M}{k}} = 2\pi(2\sqrt{2}) = 4\pi\sqrt{2}$ s,
$f = 1/T = 1/(4\pi\sqrt{2})$ Hz

(c)

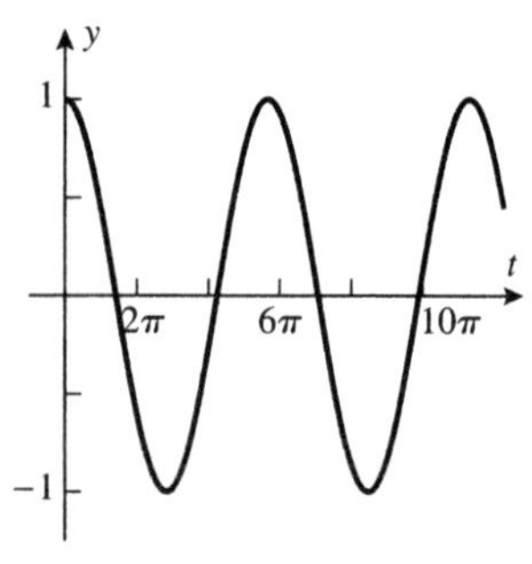

(d) $y = 0$ at the equilibrium position, so $t/(2\sqrt{2}) = \pi/2, t = \pi\sqrt{2}$ s

(e) $t/(2\sqrt{2}) = \pi, t = 2\pi\sqrt{2}$ s

35. $l = 0.05$, $k/M = g/l = 9.8/0.05 = 196\ \text{s}^{-2}$

(a) From (20), $y = -0.12\cos 14t$.

(b) $T = 2\pi\sqrt{M/k} = 2\pi/14 = \pi/7$ s, $f = 7/\pi$ Hz

(c)

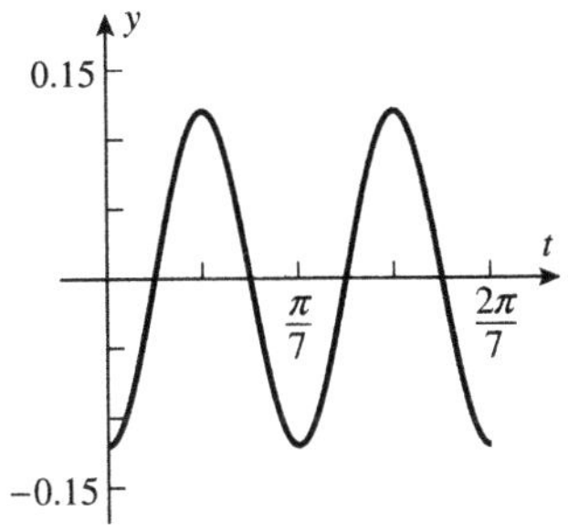

(d) $14t = \pi/2$, $t = \pi/28$ s

(e) $14t = \pi$, $t = \pi/14$ s

36. $l = 0.5$, $k/M = g/l = 32/0.5 = 64$, $\sqrt{k/M} = 8$

(a) From (20), $y = -1.5\cos 8t$.

(b) $T = 2\pi\sqrt{M/k} = 2\pi/8 = \pi/4$ s; $f = 1/T = 4/\pi$ Hz

(c)

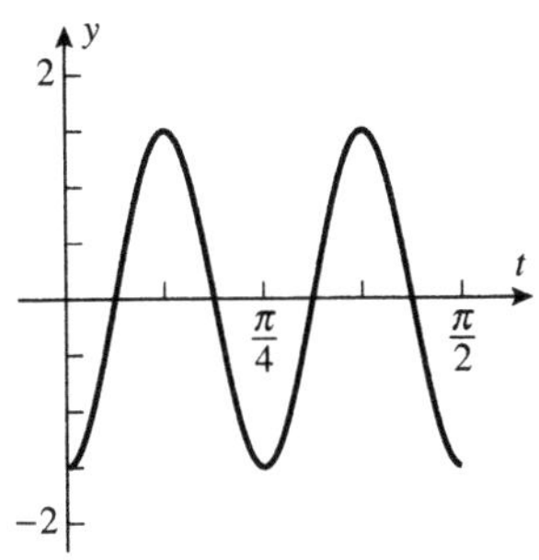

(d) $8t = \pi/2$, $t = \pi/16$ s

(e) $8t = \pi$, $t = \pi/8$ s

37. Assume $y = y_0\cos\sqrt{\dfrac{k}{M}}\,t$, so $v = \dfrac{dy}{dt} = -y_0\sqrt{\dfrac{k}{M}}\sin\sqrt{\dfrac{k}{M}}\,t$

(a) The maximum speed occurs when $\sin\sqrt{\dfrac{k}{M}}\,t = \pm 1$, $\sqrt{\dfrac{k}{M}}\,t = n\pi + \pi/2$, so $\cos\sqrt{\dfrac{k}{M}}\,t = 0$, $y = 0$.

(b) The minimum speed occurs when $\sin\sqrt{\dfrac{k}{M}}\,t = 0$, $\sqrt{\dfrac{k}{M}}\,t = n\pi$, so $\cos\sqrt{\dfrac{k}{M}}\,t = \pm 1$, $y = \pm y_0$.

38. **(a)** $T = 2\pi\sqrt{\dfrac{M}{k}}$, $k = \dfrac{4\pi^2}{T^2}M = \dfrac{4\pi^2}{T^2}\dfrac{w}{g}$, so $k = \dfrac{4\pi^2}{g}\dfrac{w}{9} = \dfrac{4\pi^2}{g}\dfrac{w+4}{25}$, $25w = 9(w+4)$,

$25w = 9w + 36, w = \dfrac{9}{4}, k = \dfrac{4\pi^2}{g}\dfrac{w}{9} = \dfrac{4\pi^2}{32}\dfrac{1}{4} = \dfrac{\pi^2}{32}$

(b) From Part (a), $w = \dfrac{9}{4}$

39. By Hooke's Law, $F(t) = -kx(t)$, since the only force is the restoring force of the spring. Newton's Second Law gives $F(t) = Mx''(t)$, so $Mx''(t) + kx(t) = 0$, $x(0) = x_0, x'(0) = 0$.

40. $0 = v(0) = y'(0) = c_2\sqrt{\dfrac{k}{M}}$, so $c_2 = 0$; $y_0 = y(0) = c_1$, so $y = y_0\cos\sqrt{\dfrac{k}{M}}\,t$.

41. **(a)** $m^2 + 2.4m + 1.44 = 0, (m + 1.2)^2 = 0, m = -1.2,\ y = C_1e^{-6t/5} + C_2te^{-6t/5},$

$C_1 = 1,\ 2 = y'(0) = -\frac{6}{5}C_1 + C_2, C_2 = \frac{16}{5},\ y = e^{-6t/5} + \frac{16}{5}te^{-6t/5}$

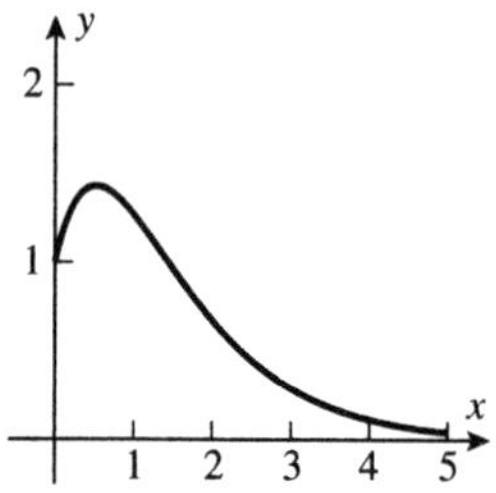

(b) $y'(t) = 0$ when $t = t_1 = 25/48 \approx 0.520833,\ y(t_1) = 1.427364$ cm

(c) $y = \frac{16}{5}\ e^{-6t/5}(t + 5/16) = 0$ only if $t = -5/16$, so $y \neq 0$ for $t \geq 0$.

42. **(a)** $m^2 + 5m + 2 = (m + 5/2)^2 - 17/4 = 0, m = -5/2 \pm \sqrt{17}/2,$

$y = C_1e^{(-5+\sqrt{17})t/2} + C_2e^{(-5-\sqrt{17})t/2},$

$C_1 + C_2 = 1/2,\ -4 = y'(0) = \frac{-5+\sqrt{17}}{2}C_1 + \frac{-5-\sqrt{17}}{2}C_2$

$C_1 = \frac{17 - 11\sqrt{17}}{68}, C_2 = \frac{17 + 11\sqrt{17}}{68}$

$y = \frac{17 - 11\sqrt{17}}{68}e^{(-5+\sqrt{17})t/2} + \frac{17 + 11\sqrt{17}}{68}e^{-(5+\sqrt{17})t/2}$

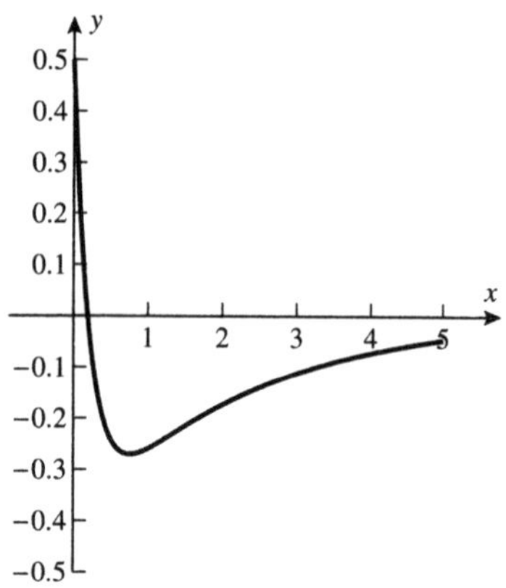

(b) $y'(t) = 0$ when $t = t_1 = 0.759194, y(t_1) = -0.270183$ cm so the maximum distance below the equilibrium position is 0.270183 cm.

(c) $y(t) = 0$ when $t = t_2 = 0.191132,\ y'(t_2) = -1.581022$ cm/sec so the speed is $|y'(t_2)| = 1.581022$ cm/s.

43. **(a)** $m^2 + m + 5 = 0, m = -1/2 \pm (\sqrt{19}/2)i,\ y = e^{-t/2}\left[C_1 \cos(\sqrt{19}t/2) + C_2 \sin(\sqrt{19}t/2)\right],$
$1 = y(0) = C_1, -3.5 = y'(0) = -(1/2)C_1 + (\sqrt{19}/2)C_2,\ C_2 = -6/\sqrt{19},$
$y = e^{-t/2}\cos(\sqrt{19}\,t/2) - (6/\sqrt{19}\,)e^{-t/2}\sin(\sqrt{19}\,t/2)$

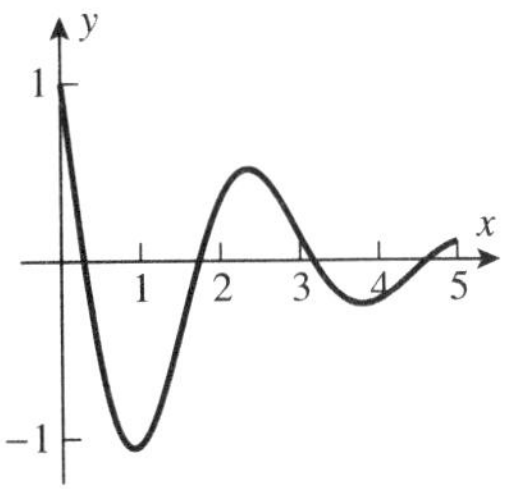

(b) $y'(t) = 0$ for the first time when $t = t_1 = 0.905533$, $y(t_1) = -1.054466$ cm so the maximum distance below the equilibrium position is 1.054466 cm.

(c) $y(t) = 0$ for the first time when $t = t_2 = 0.288274$, $y'(t_2) = -3.210357$ cm/s.

(d) The acceleration is $y''(t)$ so from the differential equation $y'' = -y' - 5y$. But $y = 0$ when the object passes through the equilibrium position, thus $y'' = -y' = 3.210357$ cm/s^2.

44. **(a)** $m^2 + m + 3m = 0, m = -1/2 \pm \sqrt{11}i/2,\ y = e^{-t/2}\left[C_1 \cos(\sqrt{11}t/2) + C_2 \sin(\sqrt{11}t/2)\right],$
$-2 = y(0) = C_1, v_0 = y'(0) = -(1/2)C_1 + (\sqrt{11}/2)C_2,\ C_2 = (v_0 - 1)(2/\sqrt{11}),$
$y(t) = e^{-t/2}\left[-2\cos(\sqrt{11}\,t/2) + (2/\sqrt{11}\,)(v_0 - 1)\sin(\sqrt{11}\,t/2)\right]$
$y'(t) = e^{-t/2}\left[v_0 \cos(\sqrt{11}t/2) + \left[(12 - v_0)/\sqrt{11}\right]\sin(\sqrt{11}t/2)\right]$

(b) We wish to find v_0 such that $y(t) = 1$ but no greater. This implies that $y'(t) = 0$ at that point. So find the largest value of v_0 such that there is a solution of $y'(t) = 0, y(t) = 1$. Note that $y'(t) = 0$ when $\tan\dfrac{\sqrt{11}}{2}t = \dfrac{v_0\sqrt{11}}{v_0 - 12}$. Choose the smallest positive solution t_0 of this equation. Then

$$\sec^2\frac{\sqrt{11}}{2}t_0 = 1 + \tan^2\frac{\sqrt{11}}{2}t_0 = \frac{12[(v_0 - 1)^2 + 11]}{(v_0 - 12)^2}.$$

Assume for now that $v_0 < 12$; if not, we will deal with it later. Then $\tan\dfrac{\sqrt{11}}{2}t_0 < 0$, so

$$\frac{\pi}{2} < \frac{\sqrt{11}}{2}t_0 < \pi, \text{ and } \sec\frac{\sqrt{11}}{2}t_0 = \frac{2\sqrt{3}\sqrt{(v_0 - 1)^2 + 11}}{v_0 - 12}$$

$$\text{and } \cos\frac{\sqrt{11}}{2}t_0 = \frac{v_0 - 12}{\sqrt{3}\sqrt{(v_0 - 1)^2 + 11}},$$

$$\sin\frac{\sqrt{11}}{2}t_0 = \tan\frac{\sqrt{11}}{2}t_0 \cos\frac{\sqrt{11}}{2}t_0 = \frac{v_0\sqrt{11}}{2\sqrt{3}\sqrt{(v_0 - 1)^2 + 11}}, \text{ and}$$

$$y(t_0) = e^{-t_0/2}\left[-2\cos\frac{\sqrt{11}}{2}t_0 + \frac{2(v_0 - 1)}{\sqrt{11}}\sin\frac{\sqrt{11}}{2}t_0\right] = e^{-t_0/2}\frac{\sqrt{(v_0 - 1)^2 + 11}}{\sqrt{3}}.$$

Use various values of $v_0, 0 < v_0 < 12$ to determine the transition point from $y < 1$ to $y > 1$ and then refine the partition on the values of v to arrive at $v \approx 2.44$ cm/s.

(c)

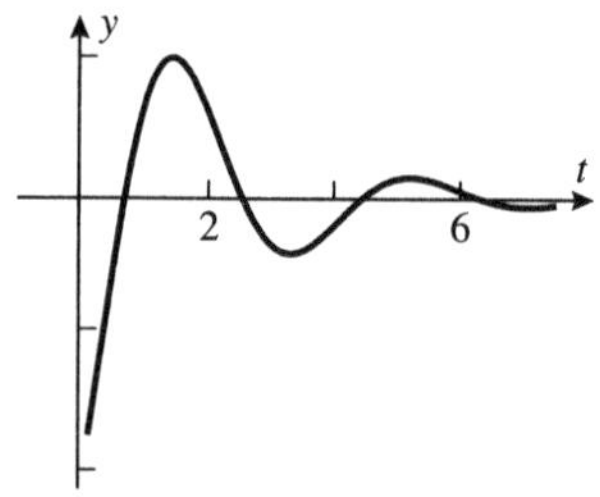

45. (a) $m^2 + 3.5m + 3 = (m + 1.5)(m + 2), y = C_1 e^{-3t/2} + C_2 e^{-2t}$,
$1 = y(0) = C_1 + C_2, v_0 = y'(0) = -(3/2)C_1 - 2C_2,\ C_1 = 4 + 2v_0, C_2 = -3 - 2v_0$,
$y(t) = (4 + 2v_0)e^{-3t/2} - (3 + 2v_0)e^{-2t}$

(b) $v_0 = 2, y(t) = 8e^{-3t/2} - 7e^{-2t},\ v_0 = -1, y(t) = 2e^{-3t/2} - e^{-2t}$,
$v_0 = -4, y(t) = -4e^{-3t/2} + 5e^{-2t}$

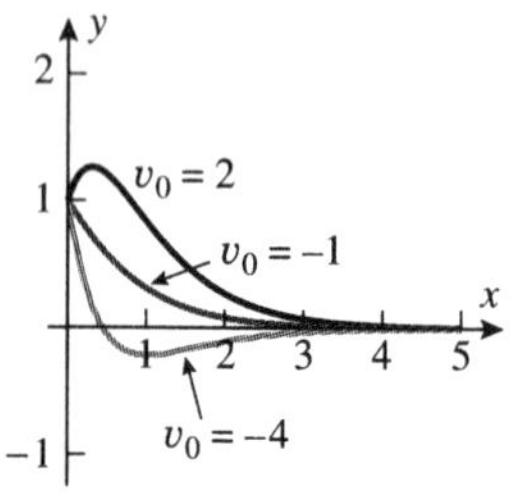

46. $\dfrac{dy}{dt} + p(x)y = c\dfrac{dy_1}{dt} + p(x)(cy_1) = c\left[\dfrac{dy_1}{dt} + p(x)y_1\right] = c \cdot 0 = 0$

CHAPTER 9 SUPPLEMENTARY EXERCISES

4. The differential equation in Part (c) is not separable; the others are.

5. (a) linear (b) linear and separable (c) separable (d) neither

6. IF: $\mu = e^{-2x^2},\ \dfrac{d}{dx}\left[ye^{-2x^2}\right] = xe^{-2x^2},\ ye^{-2x^2} = -\dfrac{1}{4}e^{-2x^2} + C,\ y = -\dfrac{1}{4} + Ce^{2x^2}$

Sep of var: $\dfrac{dy}{4y+1} = x\,dx,\ \dfrac{1}{4}\ln|4y+1| = \dfrac{x^2}{2} + C_1,\ 4y+1 = \pm e^{4C_1}e^{2x^2} = C_2e^{2x^2};\ y = -\dfrac{1}{4} + Ce^{2x^2}$, including $C = 0$

7. The parabola $ky(L - y)$ opens down and has its maximum midway between the y-intercepts, that is, at the point $y = \dfrac{1}{2}(0 + L) = L/2$, where $\dfrac{dy}{dt} = k(L/2)^2 = kL^2/4$.

8. **(a)** If $y = y_0 e^{kt}$, then $y_1 = y_0 e^{kt_1}$, $y_2 = y_0 e^{kt_2}$, divide: $y_2/y_1 = e^{k(t_2-t_1)}$, $k = \dfrac{1}{t_2 - t_1}\ln(y_2/y_1)$,

$T = \dfrac{\ln 2}{k} = \dfrac{(t_2 - t_1)\ln 2}{\ln(y_2/y_1)}$. If $y = y_0 e^{-kt}$, then $y_1 = y_0 e^{-kt_1}$, $y_2 = y_0 e^{-kt_2}$,

$y_2/y_1 = e^{-k(t_2-t_1)}, k = -\dfrac{1}{t_2 - t_1}\ln(y_2/y_1)$, $T = \dfrac{\ln 2}{k} = -\dfrac{(t_2 - t_1)\ln 2}{\ln(y_2/y_1)}$.

In either case, T is positive, so $T = \left|\dfrac{(t_2 - t_1)\ln 2}{\ln(y_2/y_1)}\right|$.

(b) In Part (a) assume $t_2 = t_1 + 1$ and $y_2 = 1.25y_1$. Then $T = \dfrac{\ln 2}{\ln 1.25} \approx 3.1$ h.

9. $\dfrac{dV}{dt} = -kS$; but $V = \dfrac{4\pi}{3}r^3, \dfrac{dV}{dt} = 4\pi r^2\dfrac{dr}{dt}$, $S = 4\pi r^2$, so $dr/dt = -k, r = -kt + C, 4 = C$, $r = -kt + 4, 3 = -k + 4, k = 1, r = 4 - t$ m.

10. Assume the tank contains $y(t)$ oz of salt at time t. Then $y_0 = 0$ and for $0 < t < 15$, $\dfrac{dy}{dt} = 5\cdot 10 - \dfrac{y}{1000}10 = (50 - y/100)$ oz/min, with solution $y = 5000 + Ce^{-t/100}$. But $y(0) = 0$ so $C = -5000$, $y = 5000(1 - e^{-t/100})$ for $0 \le t \le 15$, and $y(15) = 5000(1 - e^{-0.15})$. For $15 < t < 30$, $\dfrac{dy}{dt} = 0 - \dfrac{y}{1000}5$, $y = C_1 e^{-t/200}$, $C_1 e^{-0.075} = y(15) = 5000(1 - e^{-0.15})$, $C_1 = 5000(e^{0.075} - e^{-0.075})$, $y = 5000(e^{0.075} - e^{-0.075})e^{-t/100}, y(30) = 5000(e^{0.075} - e^{-0.075})e^{-0.3} \approx 556.13$ oz.

11. **(a)** Assume the air contains $y(t)$ ft^3 of carbon monoxide at time t. Then $y_0 = 0$ and for $t > 0$, $\dfrac{dy}{dt} = 0.04(0.1) - \dfrac{y}{1200}(0.1) = 1/250 - y/12000$, $\dfrac{d}{dt}\left[ye^{t/12000}\right] = \dfrac{1}{250}e^{t/12000}$, $ye^{t/12000} = 48e^{t/12000} + C$, $y(0) = 0$, $C = -48$; $y = 48(1 - e^{-t/12000})$. Thus the percentage of carbon monoxide is $P = \dfrac{y}{1200}100 = 4(1 - e^{-t/12000})$ percent.

(b) $0.012 = 4(1 - e^{-t/12000})$, $t = 36.05$ min

12. $\dfrac{dy}{y^2+1} = dx$, $\tan^{-1} y = x + C$, $\pi/4 = C$; $y = \tan(x + \pi/4)$

13. $\left(\dfrac{1}{y^5} + \dfrac{1}{y}\right)dy = \dfrac{dx}{x}$, $-\dfrac{1}{4}y^{-4} + \ln|y| = \ln|x| + C$; $-\dfrac{1}{4} = C$, $y^{-4} + 4\ln(x/y) = 1$

14. $\dfrac{dy}{dx} + \dfrac{2}{x}y = 4x$, $\mu = e^{\int(2/x)dx} = x^2$, $\dfrac{d}{dx}\left[yx^2\right] = 4x^3$, $yx^2 = x^4 + C$, $y = x^2 + Cx^{-2}$, $2 = y(1) = 1 + C$, $C = 1$, $y = x^2 + 1/x^2$

15. $\dfrac{dy}{y^2} = 4\sec^2 2x\,dx$, $-\dfrac{1}{y} = 2\tan 2x + C$, $-1 = 2\tan\left(2\dfrac{\pi}{8}\right) + C = 2\tan\dfrac{\pi}{4} + C = 2 + C$, $C = -3$,

$y = \dfrac{1}{3 - 2\tan 2x}$

16. $\dfrac{dy}{y^2 - 5y + 6} = dx$, $\dfrac{dy}{(y-3)(y-2)} = dx$, $\left[\dfrac{1}{y-3} - \dfrac{1}{y-2}\right] dy = dx$, $\ln\left|\dfrac{y-3}{y-2}\right| = x + C_1$,

$\dfrac{y-3}{y-2} = Ce^x$; $y = \ln 2$ if $x = 0$, so $C = \dfrac{\ln 2 - 3}{\ln 2 - 2}$; $y = \dfrac{3 - 2Ce^x}{1 - Ce^x} = \dfrac{3\ln 2 - 6 - (2\ln 2 - 6)e^x}{\ln 2 - 2 - (\ln 2 - 3)e^x}$

17. (a) $\mu = e^{-\int dx} = e^{-x}$, $\dfrac{d}{dx}\left[ye^{-x}\right] = xe^{-x}\sin 3x$,

$$ye^{-x} = \int xe^{-x}\sin 3x\,dx = \left(-\frac{3}{10}x - \frac{3}{50}\right)e^{-x}\cos 3x + \left(-\frac{1}{10}x + \frac{2}{25}\right)e^{-x}\sin 3x + C;$$

$$1 = y(0) = -\frac{3}{50} + C,\ C = \frac{53}{50},\ y = \left(-\frac{3}{10}x - \frac{3}{50}\right)\cos 3x + \left(-\frac{1}{10}x + \frac{2}{25}\right)\sin 3x + \frac{53}{50}e^x$$

(c)

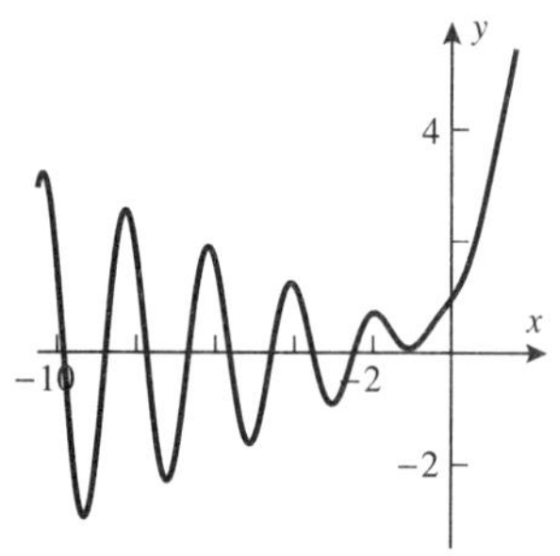

19. (a) Let $T_1 = 5730 - 40 = 5690$, $k_1 = \dfrac{\ln 2}{T_1} \approx 0.00012182$; $T_2 = 5730 + 40 = 5770$, $k_2 \approx 0.00012013$. With $y/y_0 = 0.92, 0.93$, $t_1 = -\dfrac{1}{k_1}\ln\dfrac{y}{y_0} = 684.5, 595.7$; $t_2 = -\dfrac{1}{k_2}\ln(y/y_0) = 694.1, 604.1$; in 1988 the shroud was at most 695 years old, which places its creation in or after the year 1293.

(b) Suppose T is the true half-life of carbon-14 and $T_1 = T(1 + r/100)$ is the false half-life. Then with $k = \dfrac{\ln 2}{T}$, $k_1 = \dfrac{\ln 2}{T_1}$ we have the formulae $y(t) = y_0 e^{-kt}$, $y_1(t) = y_0 e^{-k_1 t}$. At a certain point in time a reading of the carbon-14 is taken resulting in a certain value y, which in the case of the true formula is given by $y = y(t)$ for some t, and in the case of the false formula is given by $y = y_1(t_1)$ for some t_1.

If the true formula is used then the time t since the beginning is given by $t = -\dfrac{1}{k}\ln\dfrac{y}{y_0}$. If the false formula is used we get a false value $t_1 = -\dfrac{1}{k_1}\ln\dfrac{y}{y_0}$; note that in both cases the value y/y_0 is the same. Thus $t_1/t = k/k_1 = T_1/T = 1 + r/100$, so the percentage error in the time to be measured is the same as the percentage error in the half-life.

20. (a) $y_{n+1} - y_n + 0.1(1 + 5t_n - y_n)$, $y_0 = 5$

n	0	1	2	3	4	5	6	7	8	9	10
t_n	1	1.1	1.2	1.3	1.4	1.5	1.6	1.7	1.8	1.9	2
y_n	5.00	5.10	5.24	5.42	5.62	5.86	6.13	6.41	6.72	7.05	7.39

(b) The true solution is $y(t) = 5t - 4 + 4e^{1-t}$, so the percentage errors are given by

t_n	1	1.1	1.2	1.3	1.4	1.5	1.6	1.7	1.8	1.9	2
y_n	5.00	5.10	5.24	5.42	5.62	5.86	6.13	6.41	6.72	7.05	7.39
$y(t_n)$	5.00	5.12	5.27	5.46	5.68	5.93	6.20	6.49	6.80	7.13	7.47
abs. error	0.00	0.02	0.03	0.05	0.06	0.06	0.07	0.07	0.08	0.08	0.08
rel. error (%)	0.00	0.38	0.66	0.87	1.00	1.08	1.12	1.13	1.11	1.07	1.03

21. **(a)** $y = C_1 e^x + C_2 e^{2x}$ **(b)** $y = C_1 e^{x/2} + C_2 x e^{x/2}$

(c) $y = e^{-x/2}\left[C_1 \cos \frac{\sqrt{7}}{2}x + C_2 \sin \frac{\sqrt{7}}{2}x\right]$

22. **(a)** $2y dy = dx, y^2 = x + C$; if $y(0) = 1$ then $C = 1$, $y^2 = x + 1, y = \sqrt{x+1}$; if $y(0) = -1$ then $C = 1, y^2 = x + 1, y = -\sqrt{x+1}$.

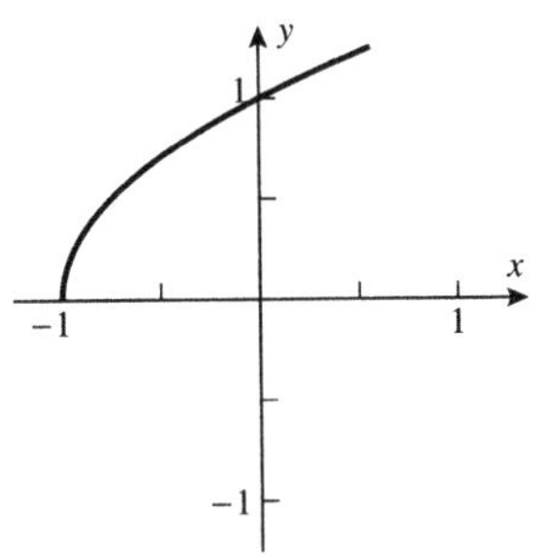

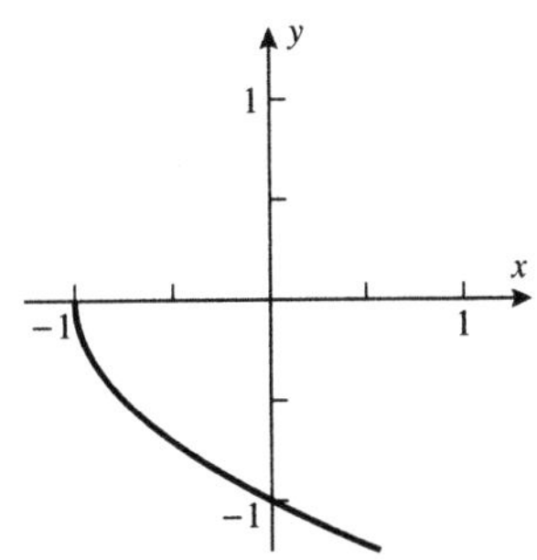

(b) $\frac{dy}{y^2} = -2x\,dx,\ -\frac{1}{y} = -x^2 + C,\ -1 = C,\ y = 1/(x^2+1)$

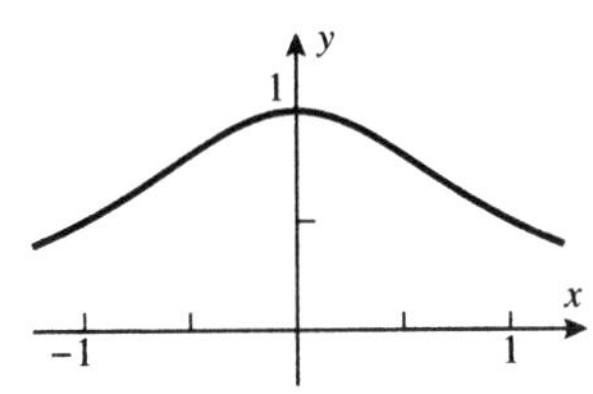

23. **(a)** Use (15) in Section 9.3 with $y_0 = 19, L = 95$: $y(t) = \frac{1805}{19 + 76e^{-kt}}$, $25 = y(1) = \frac{1805}{19 + 76e^{-k}}$, $k \approx 0.3567$; when $0.8L = y(t) = \frac{y_0 L}{19 + 76e^{-kt}}$, $19 + 76e^{-kt} = \frac{5}{4}y_0 = \frac{95}{4}$, $t \approx 7.77$ yr.

(b) From (13), $\frac{dy}{dt} = k\left(1 - \frac{y}{95}\right)y,\ y(0) = 19.$

24. **(a)** $y_0 = y(0) = c_1,\ v_0 = y'(0) = c_2\sqrt{\frac{k}{M}},\ c_2 = \sqrt{\frac{M}{k}}v_0,\ y = y_0 \cos\sqrt{\frac{k}{M}}t + v_0\sqrt{\frac{M}{k}}\sin\sqrt{\frac{k}{M}}t$

(b) $l = 0.5, k/M = g/l = 9.8/0.5 = 19.6,$

$y = -\cos(\sqrt{19.6}\,t) + 0.25\frac{1}{\sqrt{19.6}}\sin(\sqrt{19.6}\,t)$

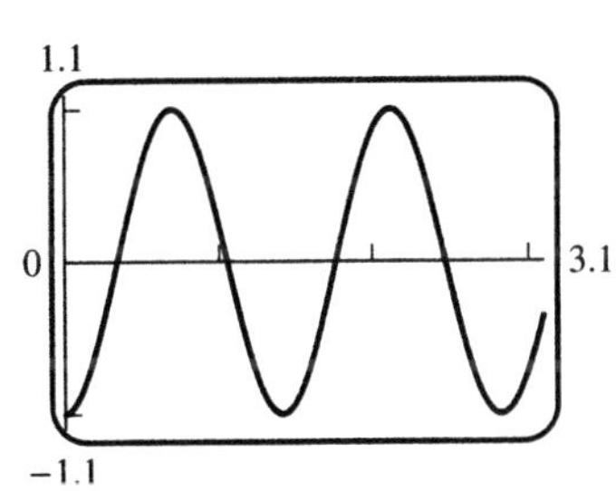

(c) $y = -\cos(\sqrt{19.6}\,t) + 0.25\frac{1}{\sqrt{19.6}}\sin(\sqrt{19.6}\,t)$, so

$$|y\text{max}| = \sqrt{(-1)^2 + \left(\frac{0.25}{\sqrt{19.6}}\right)^2} \approx 1.10016 \text{ m is the maximum displacement.}$$

25. $y = y_0 \cos\sqrt{\frac{k}{M}}\,t,\ T = 2\pi\sqrt{\frac{M}{k}},\ y = y_0\cos\frac{2\pi t}{T}$

(a) $v = y'(t) = -\frac{2\pi}{T}y_0 \sin\frac{2\pi t}{T}$ has maximum magnitude $2\pi|y_0|/T$ and occurs when $2\pi t/T = n\pi + \pi/2,\ y = y_0\cos(n\pi + \pi/2) = 0$.

(b) $a = y''(t) = -\frac{4\pi^2}{T^2}y_0\cos\frac{2\pi t}{T}$ has maximum magnitude $4\pi^2|y_0|/T^2$ and occurs when $2\pi t/T = j\pi,\ y = y_0\cos j\pi = \pm y_0$.

26. (a) In t years the interest will be compounded nt times at an interest rate of r/n each time. The value at the end of 1 interval is $P + (r/n)P = P(1 + r/n)$, at the end of 2 intervals it is $P(1+r/n)+(r/n)P(1+r/n) = P(1+r/n)^2$, and continuing in this fashion the value at the end of nt intervals is $P(1 + r/n)^{nt}$.

(b) Let $x = r/n$, then $n = r/x$ and $\lim_{n\to+\infty} P(1+r/n)^{nt} = \lim_{x\to 0+} P(1+x)^{rt/x} = \lim_{x\to 0+} P[(1+x)^{1/x}]^{rt} = Pe^{rt}$.

(c) The rate of increase is $dA/dt = rPe^{rt} = rA$.

27. (a) $A = 1000e^{(0.08)(5)} = 1000e^{0.4} \approx \$1,491.82$

(b) $Pe^{(0.08)(10)} = 10,000,\ Pe^{0.8} = 10,000,\ P = 10,000e^{-0.8} \approx \$4,493.29$

(c) From (11) of Section 9.3 with $k = r = 0.08$, $T = (\ln 2)/0.08 \approx 8.7$ years.

28. The case $p(x) = 0$ has solutions $y = C_1y_1 + C_2y_2 = C_1x + C_2$. So assume now that $p(x) \neq 0$. The differential equation becomes $\frac{d^2y}{dx^2} + p(x)\frac{dy}{dx} = 0$. Let $Y = \frac{dy}{dx}$ so that the equation becomes $\frac{dY}{dx} + p(x)Y = 0$, which is a first order separable equation in the unknown Y. We get $\frac{dY}{Y} = -p(x)\,dx, \ln|Y| = -\int p(x)\,dx,\ Y = \pm e^{-\int p(x)dx}$.
Let $P(x)$ be a specific antiderivative of $p(x)$; then any solution Y is given by $Y = \pm e^{-P(x)+C_1}$ for some C_1. Thus all solutions are given by $Y(t) = C_2e^{-P(x)}$ including $C_2 = 0$. Consequently $\frac{dy}{dx} = C_2e^{-P(x)},\ y = C_2\int e^{-P(x)}\,dx + C_3$. If we let $y_1(x) = \int e^{-P(x)}\,dx$ and $y_2(x) = 1$ then y_1 and y_2 are both solutions, and they are linearly independent (recall $P(x) \neq 0$) and hence $y(x) = c_1y_1(x) + c_2y_2(x)$.

29. $\frac{d}{dt}\left[\frac{1}{2}k[y(t)]^2 + \frac{1}{2}M(y'(t))^2\right] = ky(t)y'(t) + My'(t)y''(t) = My'(t)[\frac{k}{M}y(t) + y''(t)] = 0$, as required.

CHAPTER 10
Infinite Series

EXERCISE SET 10.1

1. **(a)** $f^{(k)}(x) = (-1)^k e^{-x}$, $f^{(k)}(0) = (-1)^k$; $e^{-x} \approx 1 - x + x^2/2$ (quadratic), $e^{-x} \approx 1 - x$ (linear)

(b) $f'(x) = -\sin x, f''(x) = -\cos x, f(0) = 1, f'(0) = 0, f''(0) = -1,$
$\cos x \approx 1 - x^2/2$ (quadratic), $\cos x \approx 1$ (linear)

(c) $f'(x) = \cos x, f''(x) = -\sin x, f(\pi/2) = 1, f'(\pi/2) = 0, f''(\pi/2) = -1,$
$\sin x \approx 1 - (x - \pi/2)^2/2$ (quadratic), $\sin x \approx 1$ (linear)

(d) $f(1) = 1, f'(1) = 1/2, f''(1) = -1/4;$
$\sqrt{x} = 1 + \frac{1}{2}(x-1) - \frac{1}{8}(x-1)^2$ (quadratic), $\sqrt{x} \approx 1 + \frac{1}{2}(x-1)$ (linear)

2. **(a)** $p_2(x) = 1 + x + x^2/2$, $p_1(x) = 1 + x$

(b) $p_2(x) = 3 + \frac{1}{6}(x-9) - \frac{1}{216}(x-9)^2$, $p_1(x) = 3 + \frac{1}{6}(x-9)$

(c) $p_2(x) = \frac{\pi}{3} + \frac{\sqrt{3}}{6}(x-2) - \frac{7}{72}\sqrt{3}(x-2)^2$, $p_1(x) = \frac{\pi}{3} + \frac{\sqrt{3}}{6}(x-2)$

(d) $p_2(x) = x$, $p_1(x) = x$

3. **(a)** $f'(x) = \frac{1}{2}x^{-1/2}$, $f''(x) = -\frac{1}{4}x^{-3/2}$; $f(1) = 1, f'(1) = \frac{1}{2}, f''(1) = -\frac{1}{4};$
$\sqrt{x} \approx 1 + \frac{1}{2}(x-1) - \frac{1}{8}(x-1)^2$

(b) $x = 1.1, x_0 = 1, \sqrt{1.1} \approx 1 + \frac{1}{2}(0.1) - \frac{1}{8}(0.1)^2 = 1.04875$, calculator value ≈ 1.0488088

4. **(a)** $\cos x \approx 1 - x^2/2$

(b) $2° = \pi/90$ rad, $\cos 2° = \cos(\pi/90) \approx 1 - \dfrac{\pi^2}{2 \cdot 90^2} \approx 0.99939077$, calculator value ≈ 0.99939083

5. $f(x) = \tan x$, $61° = \pi/3 + \pi/180$ rad; $x_0 = \pi/3$, $f'(x) = \sec^2 x$, $f''(x) = 2\sec^2 x \tan x$;
$f(\pi/3) = \sqrt{3}, f'(\pi/3) = 4, f''(x) = 8\sqrt{3}$; $\tan x \approx \sqrt{3} + 4(x - \pi/3) + 4\sqrt{3}(x - \pi/3)^2$,
$\tan 61° = \tan(\pi/3 + \pi/180) \approx \sqrt{3} + 4\pi/180 + 4\sqrt{3}(\pi/180)^2 \approx 1.80397443$,
calculator value ≈ 1.80404776

6. $f(x) = \sqrt{x}$, $x_0 = 36, f'(x) = \frac{1}{2}x^{-1/2}$, $f''(x) = -\frac{1}{4}x^{-3/2}$;
$f(36) = 6, f'(36) = \frac{1}{12}, f''(36) = -\frac{1}{864}$; $\sqrt{x} \approx 6 + \frac{1}{12}(x-36) - \frac{1}{1728}(x-36)^2$;
$\sqrt{36.03} \approx 6 + \dfrac{0.03}{12} - \dfrac{(0.03)^2}{1728} \approx 6.00249947917$, calculator value ≈ 6.00249947938

7. $f^{(k)}(x) = (-1)^k e^{-x}$, $f^{(k)}(0) = (-1)^k; p_0(x) = 1$, $p_1(x) = 1 - x$, $p_2(x) = 1 - x + \frac{1}{2}x^2$,
$p_3(x) = 1 - x + \frac{1}{2}x^2 - \frac{1}{3!}x^3$, $p_4(x) = 1 - x + \frac{1}{2}x^2 - \frac{1}{3!}x^3 + \frac{1}{4!}x^4$; $\displaystyle\sum_{k=0}^{n} \frac{(-1)^k}{k!}x^k$

8. $f^{(k)}(x) = a^k e^{ax}$, $f^{(k)}(0) = a^k$; $p_0(x) = 1$, $p_1(x) = 1 + ax$, $p_2(x) = 1 + ax + \dfrac{a^2}{2}x^2$,

$p_3(x) = 1 + ax + \dfrac{a^2}{2}x^2 + \dfrac{a^3}{3!}x^3$, $p_4(x) = 1 + ax + \dfrac{a^2}{2}x^2 + \dfrac{a^3}{3!}x^3 + \dfrac{a^4}{4!}x^4$; $\displaystyle\sum_{k=0}^{n} \frac{a^k}{k!}x^k$

9. $f^{(k)}(0) = 0$ if k is odd, $f^{(k)}(0)$ is alternately π^k and $-\pi^k$ if k is even; $p_0(x) = 1$, $p_1(x) = 1$,

$p_2(x) = 1 - \dfrac{\pi^2}{2!}x^2$; $p_3(x) = 1 - \dfrac{\pi^2}{2!}x^2$, $p_4(x) = 1 - \dfrac{\pi^2}{2!}x^2 + \dfrac{\pi^4}{4!}x^4$; $\displaystyle\sum_{k=0}^{[\frac{n}{2}]} \frac{(-1)^k \pi^{2k}}{(2k)!}x^{2k}$

NB: The function $[x]$ defined for real x indicates the greatest integer which is $\leq x$.

10. $f^{(k)}(0) = 0$ if k is even, $f^{(k)}(0)$ is alternately π^k and $-\pi^k$ if k is odd; $p_0(x) = 0$, $p_1(x) = \pi x$,

$p_2(x) = \pi x$; $p_3(x) = \pi x - \dfrac{\pi^3}{3!}x^3$, $p_4(x) = \pi x - \dfrac{\pi^3}{3!}x^3$; $\displaystyle\sum_{k=0}^{[\frac{n-1}{2}]} \frac{(-1)^k \pi^{2k+1}}{(2k+1)!}x^{2k+1}$

NB If $n = 0$ then $[\frac{n-1}{2}] = -1$; by definition any sum which runs from $k = 0$ to $k = -1$ is called the 'empty sum' and has value 0.

11. $f^{(0)}(0) = 0$; for $k \geq 1$, $f^{(k)}(x) = \dfrac{(-1)^{k+1}(k-1)!}{(1+x)^k}$, $f^{(k)}(0) = (-1)^{k+1}(k-1)!$; $p_0(x) = 0$,

$p_1(x) = x$, $p_2(x) = x - \dfrac{1}{2}x^2$, $p_3(x) = x - \dfrac{1}{2}x^2 + \dfrac{1}{3}x^3$, $p_4(x) = x - \dfrac{1}{2}x^2 + \dfrac{1}{3}x^3 - \dfrac{1}{4}x^4$; $\displaystyle\sum_{k=1}^{n} \frac{(-1)^{k+1}}{k}x^k$

12. $f^{(k)}(x) = (-1)^k \dfrac{k!}{(1+x)^{k+1}}$; $f^{(k)}(0) = (-1)^k k!$; $p_0(x) = 1$, $p_1(x) = 1 - x$,

$p_2(x) = 1 - x + x^2$, $p_3(x) = 1 - x + x^2 - x^3$, $p_4(x) = 1 - x + x^2 - x^3 + x^4$; $\displaystyle\sum_{k=0}^{n}(-1)^k x^k$

13. $f^{(k)}(0) = 0$ if k is odd, $f^{(k)}(0) = 1$ if k is even; $p_0(x) = 1, p_1(x) = 1$,

$p_2(x) = 1 + x^2/2$, $p_3(x) = 1 + x^2/2$, $p_4(x) = 1 + x^2/2 + x^4/4!$; $\displaystyle\sum_{k=0}^{[\frac{n}{2}]} \frac{1}{(2k)!}x^{2k}$

14. $f^{(k)}(0) = 0$ if k is even, $f^{(k)}(0) = 1$ if k is odd; $p_0(x) = 0$, $p_1(x) = x$, $p_2(x) = x$,

$p_3(x) = x + x^3/3!$, $p_4(x) = x + x^3/3!$; $\displaystyle\sum_{k=0}^{[\frac{n-1}{2}]} \frac{1}{(2k+1)!}x^{2k+1}$

15. $f^{(k)}(x) = \begin{cases} (-1)^{k/2}(x \sin x - k \cos x) & k \text{ even} \\ (-1)^{(k-1)/2}(x \cos x + k \sin x) & k \text{ odd} \end{cases}$, $\quad f^{(k)}(0) = \begin{cases} (-1)^{1+k/2}k & k \text{ even} \\ 0 & k \text{ odd} \end{cases}$

$p_0(x) = 0$, $p_1(x) = 0$, $p_2(x) = x^2, p_3(x) = x^2$, $p_4(x) = x^2 - \dfrac{1}{6}x^4$; $\displaystyle\sum_{k=0}^{[\frac{n}{2}]-1} \frac{(-1)^k}{(2k+1)!}x^{2k+2}$

16. $f^{(k)}(x) = (k+x)e^x$, $f^{(k)}(0) = k$; $p_0(x) = 0$, $p_1(x) = x$, $p_2(x) = x + x^2$,

$p_3(x) = x + x^2 + \dfrac{1}{2}x^3$, $p_4(x) = x + x^2 + \dfrac{1}{2}x^3 + \dfrac{1}{3!}x^4$; $\displaystyle\sum_{k=1}^{n} \frac{1}{(k-1)!}x^k$

17. $f^{(k)}(x_0) = e$; $p_0(x) = e$, $p_1(x) = e + e(x-1)$,

$p_2(x) = e + e(x-1) + \frac{e}{2}(x-1)^2$, $p_3(x) = e + e(x-1) + \frac{e}{2}(x-1)^2 + \frac{e}{3!}(x-1)^3$,

$p_4(x) = e + e(x-1) + \frac{e}{2}(x-1)^2 + \frac{e}{3!}(x-1)^3 + \frac{e}{4!}(x-1)^4$; $\sum_{k=0}^{n} \frac{e}{k!}(x-1)^k$

18. $f^{(k)}(x) = (-1)^k e^{-x}$, $f^{(k)}(\ln 2) = (-1)^k \frac{1}{2}$; $p_0(x) = \frac{1}{2}$, $p_1(x) = \frac{1}{2} - \frac{1}{2}(x - \ln 2)$,

$p_2(x) = \frac{1}{2} - \frac{1}{2}(x-\ln 2) + \frac{1}{2\cdot 2}(x-\ln 2)^2$, $p_3(x) = \frac{1}{2} - \frac{1}{2}(x-\ln 2) + \frac{1}{2\cdot 2}(x-\ln 2)^2 - \frac{1}{2\cdot 3!}(x-\ln 2)^3$,

$p_4(x) = \frac{1}{2} - \frac{1}{2}(x-\ln 2) + \frac{1}{2\cdot 2}(x-\ln 2)^2 - \frac{1}{2\cdot 3!}(x-\ln 2)^3 + \frac{1}{2\cdot 4!}(x-\ln 2)^4$;

$\sum_{k=0}^{n} \frac{(-1)^k}{2\cdot k!}(x-\ln 2)^k$

19. $f^{(k)}(x) = \frac{(-1)^k k!}{x^{k+1}}$, $f^{(k)}(-1) = -k!$; $p_0(x) = -1$; $p_1(x) = -1 - (x+1)$;

$p_2(x) = -1 - (x+1) - (x+1)^2$; $p_3(x) = -1 - (x+1) - (x+1)^2 - (x+1)^3$;

$p_4(x) = -1 - (x+1) - (x+1)^2 - (x+1)^3 - (x+1)^4$; $\sum_{k=0}^{n}(-1)(x+1)^k$

20. $f^{(k)}(x) = \frac{(-1)^k k!}{(x+2)^{k+1}}$, $f^{(k)}(3) = \frac{(-1)^k k!}{5^{k+1}}$; $p_0(x) = \frac{1}{5}$; $p_1(x) = \frac{1}{5} - \frac{1}{25}(x-3)$;

$p_2(x) = \frac{1}{5} - \frac{1}{25}(x-3) + \frac{1}{125}(x-3)^2$; $p_3(x) = \frac{1}{5} - \frac{1}{25}(x-3) + \frac{1}{125}(x-3)^2 - \frac{1}{625}(x-3)^3$;

$p_4(x) = \frac{1}{5} - \frac{1}{25}(x-3) + \frac{1}{125}(x-3)^2 - \frac{1}{625}(x-3)^3 + \frac{1}{3125}(x-3)^4$; $\sum_{k=0}^{n} \frac{(-1)^k}{5^{k+1}}(x-3)^k$

21. $f^{(k)}(1/2) = 0$ if k is odd, $f^{(k)}(1/2)$ is alternately π^k and $-\pi^k$ if k is even;

$p_0(x) = p_1(x) = 1, p_2(x) = p_3(x) = 1 - \frac{\pi^2}{2}(x-1/2)^2$,

$p_4(x) = 1 - \frac{\pi^2}{2}(x-1/2)^2 + \frac{\pi^4}{4!}(x-1/2)^4$; $\sum_{k=0}^{[\frac{n}{2}]} \frac{(-1)^k \pi^{2k}}{(2k)!}(x-1/2)^{2k}$

22. $f^{(k)}(\pi/2) = 0$ if k is even, $f^{(k)}(\pi/2)$ is alternately -1 and 1 if k is odd; $p_0(x) = 0$,

$p_1(x) = -(x-\pi/2)$, $p_2(x) = -(x-\pi/2)$, $p_3(x) = -(x-\pi/2) + \frac{1}{3!}(x-\pi/2)^3$,

$p_4(x) = -(x-\pi/2) + \frac{1}{3!}(x-\pi/2)^3$; $\sum_{k=0}^{[\frac{n-1}{2}]} \frac{(-1)^{k+1}}{(2k+1)!}(x-\pi/2)^{2k+1}$

23. $f(1)=0$, for $k\ge 1$, $f^{(k)}(x)=\dfrac{(-1)^{k-1}(k-1)!}{x^k}$; $f^{(k)}(1)=(-1)^{k-1}(k-1)!$;

$p_0(x)=0,\ p_1(x)=(x-1);\ p_2(x)=(x-1)-\frac{1}{2}(x-1)^2;\ p_3(x)=(x-1)-\frac{1}{2}(x-1)^2+\frac{1}{3}(x-1)^3,$

$p_4(x)=(x-1)-\frac{1}{2}(x-1)^2+\frac{1}{3}(x-1)^3-\frac{1}{4}(x-1)^4;\ \sum_{k=1}^{n}\frac{(-1)^{k-1}}{k}(x-1)^k$

24. $f(e)=1$, for $k\ge 1$, $f^{(k)}(x)=\dfrac{(-1)^{k-1}(k-1)!}{x^k}$; $f^{(k)}(e)=\dfrac{(-1)^{k-1}(k-1)!}{e^k}$;

$p_0(x)=1,\ p_1(x)=1+\frac{1}{e}(x-e);\ p_2(x)=1+\frac{1}{e}(x-e)-\frac{1}{2e^2}(x-e)^2;$

$p_3(x)=1+\frac{1}{e}(x-e)-\frac{1}{2e^2}(x-e)^2+\frac{1}{3e^3}(x-e)^3,$

$p_4(x)=1+\frac{1}{e}(x-e)-\frac{1}{2e^2}(x-e)^2+\frac{1}{3e^3}(x-e)^3-\frac{1}{4e^4}(x-e)^4;\ 1+\sum_{k=1}^{n}\frac{(-1)^{k-1}}{ke^k}(x-e)^k$

25. **(a)** $f(0)=1, f'(0)=2, f''(0)=-2, f'''(0)=6$, the third MacLaurin polynomial for $f(x)$ is $f(x)$.
(b) $f(1)=1, f'(1)=2, f''(1)=-2, f'''(1)=6$, the third Taylor polynomial for $f(x)$ is $f(x)$.

26. **(a)** $f^{(k)}(0)=k!c_k$ for $k\le n$; the nth Maclaurin polynomial for $f(x)$ is $f(x)$.
(b) $f^{(k)}(x_0)=k!c_k$ for $k\le n$; the nth Taylor polynomial about $x=1$ for $f(x)$ is $f(x)$.

27. $f^{(k)}(0)=(-2)^k;\ p_0(x)=1,\ p_1(x)=1-2x,$

$p_2(x)=1-2x+2x^2,\ p_3(x)=1-2x+2x^2-\frac{4}{3}x^3$

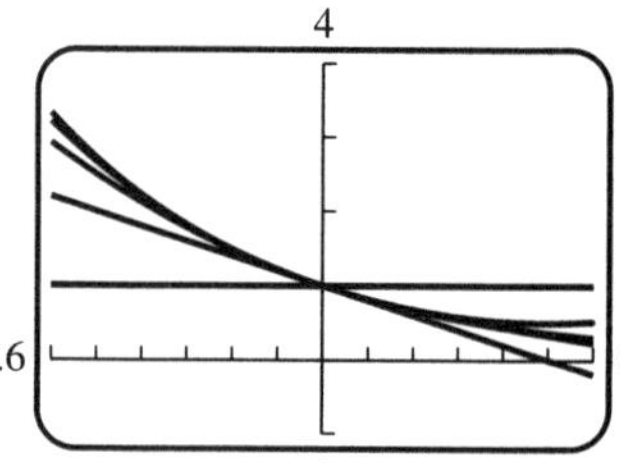

28. $f^{(k)}(\pi/2)=0$ if k is odd, $f^{(k)}(\pi/2)$ is alternately 1 and -1 if k is even; $p_0(x)=1,\ p_2(x)=1-\frac{1}{2}(x-\pi/2)^2,$

$p_4(x)=1-\frac{1}{2}(x-\pi/2)^2+\frac{1}{24}(x-\pi/2)^4,$

$p_6(x)=1-\frac{1}{2}(x-\pi/2)^2+\frac{1}{24}(x-\pi/2)^4-\frac{1}{720}(x-\pi/2)^6$

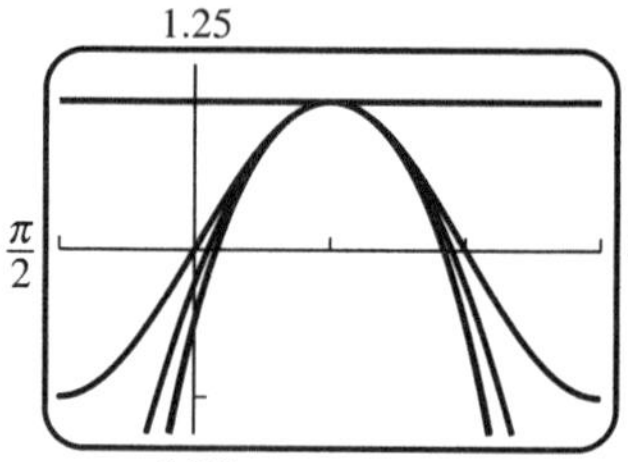

29. $f^{(k)}(\pi)=0$ if k is odd, $f^{(k)}(\pi)$ is alternately -1 and 1 if k is even; $p_0(x)=-1,\ p_2(x)=-1+\frac{1}{2}(x-\pi)^2,$

$p_4(x)=-1+\frac{1}{2}(x-\pi)^2-\frac{1}{24}(x-\pi)^4,$

$p_6(x)=-1+\frac{1}{2}(x-\pi)^2-\frac{1}{24}(x-\pi)^4+\frac{1}{720}(x-\pi)^6$

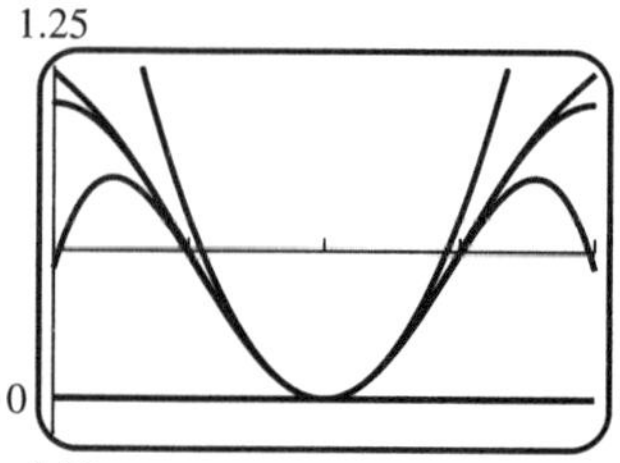

30. $f(0) = 0$; for $k \geq 1$, $f^{(k)}(x) = \dfrac{(-1)^{k-1}(k-1)!}{(x+1)^k}$,

$f^{(k)}(0) = (-1)^{k-1}(k-1)!$; $p_0(x) = 0$, $p_1(x) = x$,

$p_2(x) = x - \dfrac{1}{2}x^2$, $p_3(x) = x - \dfrac{1}{2}x^2 + \dfrac{1}{3}x^3$

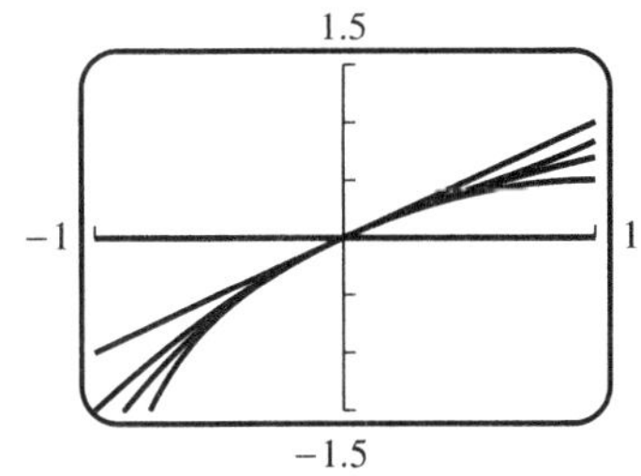

31. $f^{(k)}(x) = e^x$, $|f^{(k)}(x)| \leq e^{1/2} < 2$ on $[0, 1/2]$, let $M = 2$,

$$e^{1/2} = 1 + \frac{1}{2} + \frac{1}{8} + \frac{1}{48} + \frac{1}{24 \cdot 16} + \cdots + \frac{1}{n!2^n} + R_n(1/2);$$

$$|R_n(1/2)| \leq \frac{M}{(n+1)!}(1/2)^{n+1} \leq \frac{2}{(n+1)!}(1/2)^{n+1} \leq 0.00005 \text{ for } n = 5;$$

$$e^{1/2} \approx 1 + \frac{1}{2} + \frac{1}{8} + \frac{1}{48} + \frac{1}{24 \cdot 16} + \frac{1}{120 \cdot 32} \approx 1.64870, \text{ calculator value } 1.64872$$

32. $f(x) = e^x$, $f^{(k)}(x) = e^x$, $|f^{(k)}(x)| \leq 1$ on $[-1, 0]$, $|R_n(x)| \leq \dfrac{1}{(n+1)!}(1)^{n+1} = \dfrac{1}{(n+1)!} < 0.5 \times 10^{-3}$

if $n = 6$, so $e^{-1} \approx 1 - 1 + \dfrac{1}{2!} - \dfrac{1}{3!} + \dfrac{1}{4!} - \dfrac{1}{5!} + \dfrac{1}{6!} \approx 0.3681$, calculator value 0.3679

33. $p(0) = 1$, $p(x)$ has slope -1 at $x = 0$, and $p(x)$ is concave up at $x = 0$, eliminating I, II and III respectively and leaving IV.

34. Let $p_0(x) = 2$, $p_1(x) = p_2(x) = 2 - 3(x-1)$, $p_3(x) = 2 - 3(x-1) + (x-1)^3$.

35. $f^{(k)}(\ln 4) = 15/8$ for k even, $f^{(k)}(\ln 4) = 17/8$ for k odd, which can be written as

$$f^{(k)}(\ln 4) = \frac{16 - (-1)^k}{8}; \quad \sum_{k=0}^{n} \frac{16 - (-1)^k}{8k!}(x - \ln 4)^k$$

36. **(a)** $\cos\alpha \approx 1 - \alpha^2/2$; $x = r - r\cos\alpha = r(1 - \cos\alpha) \approx r\alpha^2/2$

(b) In Figure Ex-36 let $r = 4000$ mi and $\alpha = 1/80$ so that the arc has length $2r\alpha = 100$ mi. Then $x \approx r\alpha^2/2 = \dfrac{4000}{2 \cdot 80^2} = 5/16$ mi.

37. From Exercise 2(a), $p_1(x) = 1 + x$, $p_2(x) = 1 + x + x^2/2$

(a)

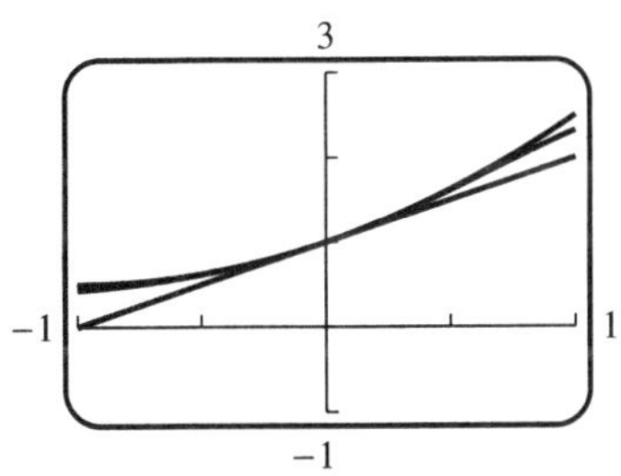

(b)

x	-1.000	-0.750	-0.500	-0.250	0.000	0.250	0.500	0.750	1.000
$f(x)$	0.431	0.506	0.619	0.781	1.000	1.281	1.615	1.977	2.320
$p_1(x)$	0.000	0.250	0.500	0.750	1.000	1.250	1.500	1.750	2.000
$p_2(x)$	0.500	0.531	0.625	0.781	1.000	1.281	1.625	2.031	2.500

(c) $|e^{\sin x} - (1 + x)| < 0.01$ for $-0.14 < x < 0.14$

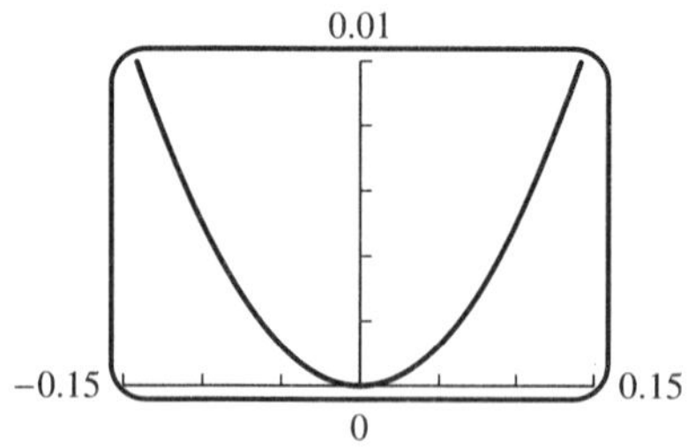

(d) $|e^{\sin x} - (1 + x + x^2/2)| < 0.01$ for $-0.50 < x < 0.50$

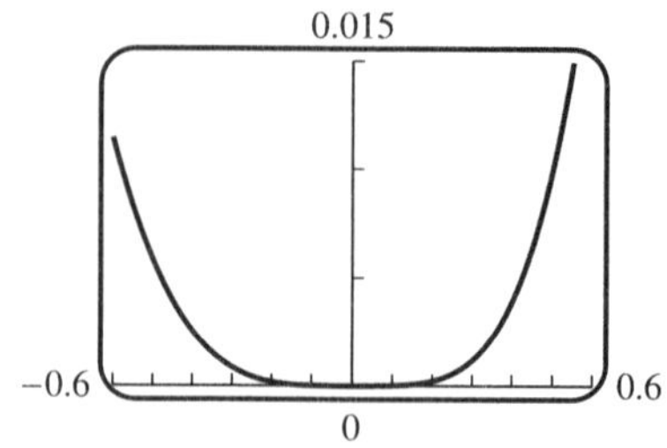

38. (a) $f^{(k)}(x) = e^x \le e^b$,

$$|R_2(x)| \le \frac{e^b b^3}{3!} < 0.0005,$$

$e^b b^3 < 0.003$ if $b \le 0.137$ (by trial and error with a hand calculator), so $[0, 0.137]$.

(b)

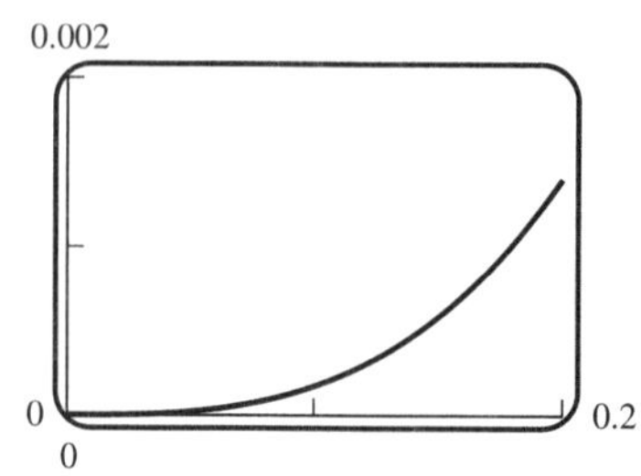

39. (a) $\sin x = x - \dfrac{x^3}{3!} + 0 \cdot x^4 + R_4(x)$,

$$|R_4(x)| \le \frac{|x|^5}{5!} < 0.5 \times 10^{-3} \text{ if } |x|^5 < 0.06,$$

$|x| < (0.06)^{1/5} \approx 0.569, (-0.569, 0.569)$

(b)

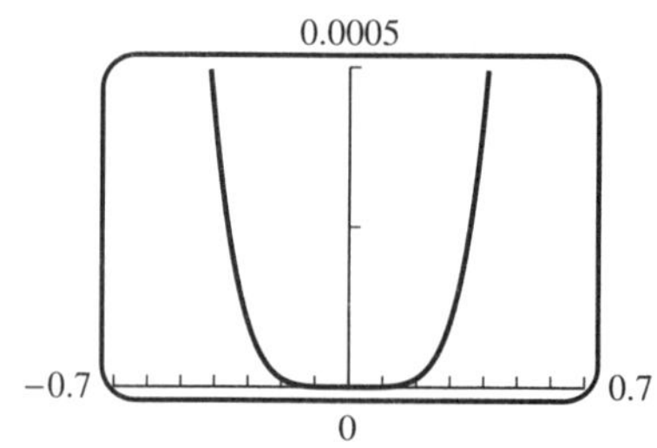

EXERCISE SET 10.2

1. (a) $\dfrac{1}{3^{n-1}}$ (b) $\dfrac{(-1)^{n-1}}{3^{n-1}}$ (c) $\dfrac{2n-1}{2n}$ (d) $\dfrac{n^2}{\pi^{1/(n+1)}}$

2. (a) $(-r)^{n-1}$; $(-r)^n$ (b) $(-1)^{n+1}r^n$; $(-1)^n r^{n+1}$

3. (a) $2, 0, 2, 0$ (b) $1, -1, 1, -1$ (c) $2(1 + (-1)^n)$; $2 + 2\cos n\pi$

4. (a) $(2n)!$ (b) $(2n-1)!$

5. $1/3, 2/4, 3/5, 4/6, 5/7, \ldots$; $\displaystyle\lim_{n\to+\infty} \frac{n}{n+2} = 1$, converges

6. $1/3,\ 4/5,\ 9/7,\ 16/9,\ 25/11, \ldots;\ \lim_{n\to+\infty} \dfrac{n^2}{2n+1} = +\infty$, diverges

7. $2, 2, 2, 2, 2, \ldots;\ \lim_{n\to+\infty} 2 = 2$, converges

8. $\ln 1,\ \ln\dfrac{1}{2},\ \ln\dfrac{1}{3},\ \ln\dfrac{1}{4},\ \ln\dfrac{1}{5}, \ldots;\ \lim_{n\to+\infty} \ln(1/n) = -\infty$, diverges

9. $\dfrac{\ln 1}{1},\ \dfrac{\ln 2}{2},\ \dfrac{\ln 3}{3},\ \dfrac{\ln 4}{4},\ \dfrac{\ln 5}{5}, \ldots;$

$\lim_{n\to+\infty} \dfrac{\ln n}{n} = \lim_{n\to+\infty} \dfrac{1}{n} = 0$ $\left(\text{apply L'Hôpital's Rule to } \dfrac{\ln x}{x}\right)$, converges

10. $\sin\pi,\ 2\sin(\pi/2),\ 3\sin(\pi/3),\ 4\sin(\pi/4),\ 5\sin(\pi/5), \ldots;$

$$\lim_{n\to+\infty} n\sin(\pi/n) = \lim_{n\to+\infty} \frac{\sin(\pi/n)}{1/n} = \lim_{n\to+\infty} \frac{(-\pi/n^2)\cos(\pi/n)}{-1/n^2} = \pi, \text{ converges}$$

11. $0, 2, 0, 2, 0, \ldots;$ diverges

12. $1,\ -1/4,\ 1/9,\ -1/16,\ 1/25, \ldots;\ \lim_{n\to+\infty} \dfrac{(-1)^{n+1}}{n^2} = 0$, converges

13. $-1,\ 16/9,\ -54/28,\ 128/65,\ -250/126, \ldots;$ diverges because odd-numbered terms approach -2, even-numbered terms approach 2.

14. $1/2,\ 2/4,\ 3/8,\ 4/16,\ 5/32, \ldots;\ \lim_{n\to+\infty} \dfrac{n}{2^n} = \lim_{n\to+\infty} \dfrac{1}{2^n \ln 2} = 0$, converges

15. $6/2,\ 12/8,\ 20/18,\ 30/32,\ 42/50, \ldots;\ \lim_{n\to+\infty} \dfrac{1}{2}(1+1/n)(1+2/n) = 1/2$, converges

16. $\pi/4,\ \pi^2/4^2,\ \pi^3/4^3,\ \pi^4/4^4,\ \pi^5/4^5, \ldots;\ \lim_{n\to+\infty} (\pi/4)^n = 0$, converges

17. $\cos(3),\ \cos(3/2),\ \cos(1),\ \cos(3/4),\ \cos(3/5), \ldots;\ \lim_{n\to+\infty} \cos(3/n) = 1$, converges

18. $0,\ -1,\ 0,\ 1,\ 0, \ldots;$ diverges

19. $e^{-1},\ 4e^{-2},\ 9e^{-3},\ 16e^{-4},\ 25e^{-5}, \ldots;\ \lim_{x\to+\infty} x^2e^{-x} = \lim_{x\to+\infty} \dfrac{x^2}{e^x} = 0$, so $\lim_{n\to+\infty} n^2e^{-n} = 0$, converges

20. $1,\ \sqrt{10}-2,\ \sqrt{18}-3,\ \sqrt{28}-4,\ \sqrt{40}-5, \ldots;$

$$\lim_{n\to+\infty} (\sqrt{n^2+3n}-n) = \lim_{n\to+\infty} \frac{3n}{\sqrt{n^2+3n}+n} = \lim_{n\to+\infty} \frac{3}{\sqrt{1+3/n}+1} = \frac{3}{2}, \text{ converges}$$

21. $2,\ (5/3)^2,\ (6/4)^3,\ (7/5)^4,\ (8/6)^5, \ldots;$ let $y = \left[\dfrac{x+3}{x+1}\right]^x$, converges because

$$\lim_{x\to+\infty} \ln y = \lim_{x\to+\infty} \frac{\ln\dfrac{x+3}{x+1}}{1/x} = \lim_{x\to+\infty} \frac{2x^2}{(x+1)(x+3)} = 2, \text{ so } \lim_{n\to+\infty} \left[\frac{n+3}{n+1}\right]^n = e^2$$

22. $-1, 0, (1/3)^3, (2/4)^4, (3/5)^5, \ldots$; let $y = (1-2/x)^x$, converges because

$$\lim_{x\to+\infty} \ln y = \lim_{x\to+\infty} \frac{\ln(1-2/x)}{1/x} = \lim_{x\to+\infty} \frac{-2}{1-2/x} = -2, \lim_{n\to+\infty}(1-2/n)^n = \lim_{x\to+\infty} y = e^{-2}$$

23. $\left\{\dfrac{2n-1}{2n}\right\}_{n=1}^{+\infty}$; $\lim\limits_{n\to+\infty} \dfrac{2n-1}{2n} = 1$, converges

24. $\left\{\dfrac{n-1}{n^2}\right\}_{n=1}^{+\infty}$; $\lim\limits_{n\to+\infty} \dfrac{n-1}{n^2} = 0$, converges

25. $\left\{\dfrac{1}{3^n}\right\}_{n=1}^{+\infty}$; $\lim\limits_{n\to+\infty} \dfrac{1}{3^n} = 0$, converges

26. $\{(-1)^n n\}_{n=1}^{+\infty}$; diverges because odd-numbered terms tend toward $-\infty$, even-numbered terms tend toward $+\infty$.

27. $\left\{\dfrac{1}{n} - \dfrac{1}{n+1}\right\}_{n=1}^{+\infty}$; $\lim\limits_{n\to+\infty} \left(\dfrac{1}{n} - \dfrac{1}{n+1}\right) = 0$, converges

28. $\{3/2^{n-1}\}_{n=1}^{+\infty}$; $\lim\limits_{n\to+\infty} 3/2^{n-1} = 0$, converges

29. $\{\sqrt{n+1} - \sqrt{n+2}\}_{n=1}^{+\infty}$; converges because

$$\lim_{n\to+\infty}(\sqrt{n+1} - \sqrt{n+2}) = \lim_{n\to+\infty} \frac{(n+1)-(n+2)}{\sqrt{n+1}+\sqrt{n+2}} = \lim_{n\to+\infty} \frac{-1}{\sqrt{n+1}+\sqrt{n+2}} = 0$$

30. $\{(-1)^{n+1}/3^{n+4}\}_{n=1}^{+\infty}$; $\lim\limits_{n\to+\infty}(-1)^{n+1}/3^{n+4} = 0$, converges

31. **(a)** $1, 2, 1, 4, 1, 6$ **(b)** $a_n = \begin{cases} n, & n \text{ odd} \\ 1/2^n, & n \text{ even} \end{cases}$ **(c)** $a_n = \begin{cases} 1/n, & n \text{ odd} \\ 1/(n+1), & n \text{ even} \end{cases}$

(d) In Part (a) the sequence diverges, since the even terms diverge to $+\infty$ and the odd terms equal 1; in Part (b) the sequence diverges, since the odd terms diverge to $+\infty$ and the even terms tend to zero; in Part (c) $\lim\limits_{n\to+\infty} a_n = 0$.

32. The even terms are zero, so the odd terms must converge to zero, and this is true if and only if $\lim\limits_{n\to+\infty} b^n = 0$, or $-1 < b < 1$.

33. $\lim\limits_{n\to+\infty} \sqrt[n]{n} = 1$, so $\lim\limits_{n\to+\infty} \sqrt[n]{n^3} = 1^3 = 1$

35. $\lim\limits_{n\to+\infty} x_{n+1} = \dfrac{1}{2} \lim\limits_{n\to+\infty} \left(x_n + \dfrac{a}{x_n}\right)$ or $L = \dfrac{1}{2}\left(L + \dfrac{a}{L}\right)$, $2L^2 - L^2 - a = 0, L = \sqrt{a}$ (we reject $-\sqrt{a}$ because $x_n > 0$, thus $L \geq 0$.

36. **(a)** $a_{n+1} = \sqrt{6 + a_n}$

(b) $\lim\limits_{n\to+\infty} a_{n+1} = \lim\limits_{n\to+\infty} \sqrt{6+a_n}$, $L = \sqrt{6+L}$, $L^2 - L - 6 = 0$, $(L-3)(L+2) = 0$,

$L = -2$ (reject, because the terms in the sequence are positive) or $L = 3$; $\lim\limits_{n\to+\infty} a_n = 3$.

37. **(a)** $1, \frac{1}{4}+\frac{2}{4}, \frac{1}{9}+\frac{2}{9}+\frac{3}{9}, \frac{1}{16}+\frac{2}{16}+\frac{3}{16}+\frac{4}{16} = 1, \frac{3}{4}, \frac{2}{3}, \frac{5}{8}$

(c) $a_n = \frac{1}{n^2}(1+2+\cdots+n) = \frac{1}{n^2}\frac{1}{2}n(n+1) = \frac{1}{2}\frac{n+1}{n}, \lim_{n\to+\infty} a_n = 1/2$

38. **(a)** $1, \frac{1}{8}+\frac{4}{8}, \frac{1}{27}+\frac{4}{27}+\frac{9}{27}, \frac{1}{64}+\frac{4}{64}+\frac{9}{64}+\frac{16}{64} = 1, \frac{5}{8}, \frac{14}{27}, \frac{15}{32}$

(c) $a_n = \frac{1}{n^3}(1^2+2^2+\cdots+n^2) = \frac{1}{n^3}\frac{1}{6}n(n+1)(2n+1) = \frac{1}{6}\frac{(n+1)(2n+1)}{n^2},$

$\lim_{n\to+\infty} a_n = \lim_{n\to+\infty} \frac{1}{6}(1+1/n)(2+1/n) = 1/3$

39. Let $a_n = 0, b_n = \frac{\sin^2 n}{n}, c_n = \frac{1}{n}$; then $a_n \le b_n \le c_n$, $\lim_{n\to+\infty} a_n = \lim_{n\to+\infty} c_n = 0$, so $\lim_{n\to+\infty} b_n = 0$.

40. Let $a_n = 0, b_n = \left(\frac{1+n}{2n}\right)^n, c_n = \left(\frac{3}{4}\right)^n$; then (for $n \ge 2$), $a_n \le b_n \le \left(\frac{n/2+n}{2n}\right)^n = c_n$, $\lim_{n\to+\infty} a_n = \lim_{n\to+\infty} c_n = 0$, so $\lim_{n\to+\infty} b_n = 0$.

41. **(a)** $a_1 = (0.5)^2, a_2 = a_1^2 = (0.5)^4, \ldots, a_n = (0.5)^{2^n}$

(c) $\lim_{n\to+\infty} a_n = \lim_{n\to+\infty} e^{2^n \ln(0.5)} = 0$, since $\ln(0.5) < 0$.

(d) Replace 0.5 in Part (a) with a_0; then the sequence converges for $-1 \le a_0 \le 1$, because if $a_0 = \pm 1$, then $a_n = 1$ for $n \ge 1$; if $a_0 = 0$ then $a_n = 0$ for $n \ge 1$; and if $0 < |a_0| < 1$ then $a_1 = a_0^2 > 0$ and $\lim_{n\to+\infty} a_n = \lim_{n\to+\infty} e^{2^{n-1}\ln a_1} = 0$ since $0 < a_1 < 1$. This same argument proves divergence to $+\infty$ for $|a| > 1$ since then $\ln a_1 > 0$.

42. $f(0.2) = 0.4$, $f(0.4) = 0.8$, $f(0.8) = 0.6$, $f(0.6) = 0.2$ and then the cycle repeats, so the sequence does not converge.

43. **(a)**

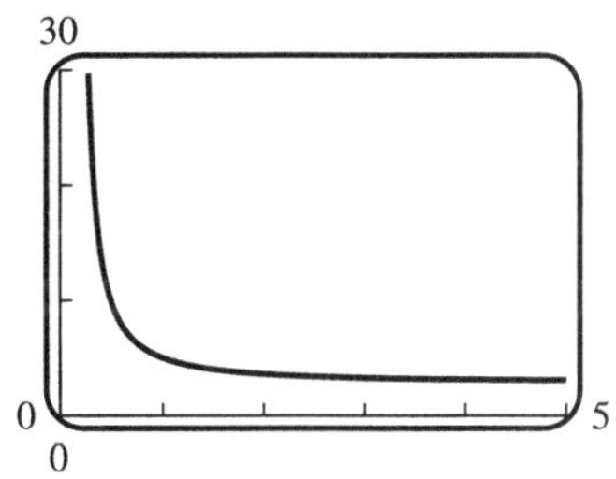

(b) Let $y = (2^x+3^x)^{1/x}$, $\lim_{x\to+\infty} \ln y = \lim_{x\to+\infty} \frac{\ln(2^x+3^x)}{x} = \lim_{x\to+\infty} \frac{2^x \ln 2 + 3^x \ln 3}{2^x+3^x}$

$= \lim_{x\to+\infty} \frac{(2/3)^x \ln 2 + \ln 3}{(2/3)^x + 1} = \ln 3$, so $\lim_{n\to+\infty} (2^n+3^n)^{1/n} = e^{\ln 3} = 3$

Alternate proof: $3 = (3^n)^{1/n} < (2^n+3^n)^{1/n} < (2\cdot 3^n)^{1/n} = 3\cdot 2^{1/n}$. Then apply the Squeezing Theorem.

44. Let $f(x) = 1/(1+x)$, $0 \le x \le 1$. Take $\Delta x_k = 1/n$ and $x_k^* = k/n$ then

$a_n = \sum_{k=1}^{n} \frac{1}{1+(k/n)}(1/n) = \sum_{k=1}^{n} \frac{1}{1+x_k^*}\Delta x_k$ so $\lim_{n\to+\infty} a_n = \int_0^1 \frac{1}{1+x}dx = \ln(1+x)\Big]_0^1 = \ln 2$

45. $a_n = \dfrac{1}{n-1}\displaystyle\int_1^n \dfrac{1}{x}\,dx = \dfrac{\ln n}{n-1}, \lim_{n\to+\infty} a_n = \lim_{n\to+\infty}\dfrac{\ln n}{n-1} = \lim_{n\to+\infty}\dfrac{1}{n} = 0,$

$\left(\text{apply L'Hôpital's Rule to } \dfrac{\ln n}{n-1}\right)$, converges

46. **(a)** If $n \geq 1$, then $a_{n+2} = a_{n+1} + a_n$, so $\dfrac{a_{n+2}}{a_{n+1}} = 1 + \dfrac{a_n}{a_{n+1}}$.

(c) With $L = \lim_{n\to+\infty}(a_{n+2}/a_{n+1}) = \lim_{n\to+\infty}(a_{n+1}/a_n)$, $L = 1 + 1/L$, $L^2 - L - 1 = 0$, $L = (1 \pm \sqrt{5})/2$, so $L = (1+\sqrt{5})/2$ because the limit cannot be negative.

47. $\left|\dfrac{1}{n} - 0\right| = \dfrac{1}{n} < \epsilon$ if $n > 1/\epsilon$

(a) $1/\epsilon = 1/0.5 = 2$, $N = 3$

(b) $1/\epsilon = 1/0.1 = 10$, $N = 11$

(c) $1/\epsilon = 1/0.001 = 1000$, $N = 1001$

48. $\left|\dfrac{n}{n+1} - 1\right| = \dfrac{1}{n+1} < \epsilon$ if $n+1 > 1/\epsilon$, $n > 1/\epsilon - 1$

(a) $1/\epsilon - 1 = 1/0.25 - 1 = 3$, $N = 4$

(b) $1/\epsilon - 1 = 1/0.1 - 1 = 9$, $N = 10$

(c) $1/\epsilon - 1 = 1/0.001 - 1 = 999$, $N = 1000$

49. **(a)** $\left|\dfrac{1}{n} - 0\right| = \dfrac{1}{n} < \epsilon$ if $n > 1/\epsilon$, choose any $N > 1/\epsilon$.

(b) $\left|\dfrac{n}{n+1} - 1\right| = \dfrac{1}{n+1} < \epsilon$ if $n > 1/\epsilon - 1$, choose any $N > 1/\epsilon - 1$.

50. If $|r| < 1$ then $\lim_{n\to+\infty} r^n = 0$; if $r > 1$ then $\lim_{n\to+\infty} r^n = +\infty$, if $r < -1$ then r^n oscillates between positive and negative values that grow in magnitude so $\lim_{n\to+\infty} r^n$ does not exist for $|r| > 1$; if $r = 1$ then $\lim_{n\to+\infty} 1^n = 1$; if $r = -1$ then $(-1)^n$ oscillates between -1 and 1 so $\lim_{n\to+\infty}(-1)^n$ does not exist.

EXERCISE SET 10.3

1. $a_{n+1} - a_n = \dfrac{1}{n+1} - \dfrac{1}{n} = -\dfrac{1}{n(n+1)} < 0$ for $n \geq 1$, so strictly decreasing.

2. $a_{n+1} - a_n = (1 - \dfrac{1}{n+1}) - (1 - \dfrac{1}{n}) = \dfrac{1}{n(n+1)} > 0$ for $n \geq 1$, so strictly increasing.

3. $a_{n+1} - a_n = \dfrac{n+1}{2n+3} - \dfrac{n}{2n+1} = \dfrac{1}{(2n+1)(2n+3)} > 0$ for $n \geq 1$, so strictly increasing.

4. $a_{n+1} - a_n = \dfrac{n+1}{4n+3} - \dfrac{n}{4n-1} = -\dfrac{1}{(4n-1)(4n+3)} < 0$ for $n \geq 1$, so strictly decreasing.

5. $a_{n+1} - a_n = (n + 1 - 2^{n+1}) - (n - 2^n) = 1 - 2^n < 0$ for $n \geq 1$, so strictly decreasing.

6. $a_{n+1} - a_n = [(n+1) - (n+1)^2] - (n - n^2) = -2n < 0$ for $n \geq 1$, so strictly decreasing.

7. $\dfrac{a_{n+1}}{a_n} = \dfrac{(n+1)/(2n+3)}{n/(2n+1)} = \dfrac{(n+1)(2n+1)}{n(2n+3)} = \dfrac{2n^2+3n+1}{2n^2+3n} > 1$ for $n \geq 1$, so strictly increasing.

8. $\dfrac{a_{n+1}}{a_n} = \dfrac{2^{n+1}}{1+2^{n+1}} \cdot \dfrac{1+2^n}{2^n} = \dfrac{2+2^{n+1}}{1+2^{n+1}} = 1 + \dfrac{1}{1+2^{n+1}} > 1$ for $n \geq 1$, so strictly increasing.

9. $\dfrac{a_{n+1}}{a_n} = \dfrac{(n+1)e^{-(n+1)}}{ne^{-n}} = (1+1/n)e^{-1} < 1$ for $n \geq 1$, so strictly decreasing.

10. $\dfrac{a_{n+1}}{a_n} = \dfrac{10^{n+1}}{(2n+2)!} \cdot \dfrac{(2n)!}{10^n} = \dfrac{10}{(2n+2)(2n+1)} < 1$ for $n \geq 1$, so strictly decreasing.

11. $\dfrac{a_{n+1}}{a_n} = \dfrac{(n+1)^{n+1}}{(n+1)!} \cdot \dfrac{n!}{n^n} = \dfrac{(n+1)^n}{n^n} = (1+1/n)^n > 1$ for $n \geq 1$, so strictly increasing.

12. $\dfrac{a_{n+1}}{a_n} = \dfrac{5^{n+1}}{2^{(n+1)^2}} \cdot \dfrac{2^{n^2}}{5^n} = \dfrac{5}{2^{2n+1}} < 1$ for $n \geq 1$, so strictly decreasing.

13. $f(x) = x/(2x+1)$, $f'(x) = 1/(2x+1)^2 > 0$ for $x \geq 1$, so strictly increasing.

14. $f(x) = 3 - 1/x$, $f'(x) = 1/x^2 > 0$ for $x \geq 1$, so strictly increasing.

15. $f(x) = 1/(x + \ln x)$, $f'(x) = -\dfrac{1+1/x}{(x+\ln x)^2} < 0$ for $x \geq 1$, so strictly decreasing.

16. $f(x) = xe^{-2x}$, $f'(x) = (1-2x)e^{-2x} < 0$ for $x \geq 1$, so strictly decreasing.

17. $f(x) = \dfrac{\ln(x+2)}{x+2}$, $f'(x) = \dfrac{1-\ln(x+2)}{(x+2)^2} < 0$ for $x \geq 1$, so strictly decreasing.

18. $f(x) = \tan^{-1} x$, $f'(x) = 1/(1+x^2) > 0$ for $x \geq 1$, so strictly increasing.

19. $f(x) = 2x^2 - 7x$, $f'(x) = 4x - 7 > 0$ for $x \geq 2$, so eventually strictly increasing.

20. $f(x) = x^3 - 4x^2$, $f'(x) = 3x^2 - 8x = x(3x-8) > 0$ for $x \geq 3$, so eventually strictly increasing.

21. $f(x) = \dfrac{x}{x^2+10}$, $f'(x) = \dfrac{10-x^2}{(x^2+10)^2} < 0$ for $x \geq 4$, so eventually strictly decreasing.

22. $f(x) = x + \dfrac{17}{x}$, $f'(x) = \dfrac{x^2-17}{x^2} > 0$ for $x \geq 5$, so eventually strictly increasing.

23. $\dfrac{a_{n+1}}{a_n} = \dfrac{(n+1)!}{3^{n+1}} \cdot \dfrac{3^n}{n!} = \dfrac{n+1}{3} > 1$ for $n \geq 3$, so eventually strictly increasing.

24. $f(x) = x^5 e^{-x}$, $f'(x) = x^4(5-x)e^{-x} < 0$ for $x \geq 6$, so eventually strictly decreasing.

25. **(a)** Yes: a monotone sequence is increasing or decreasing; if it is increasing, then it is increasing and bounded above, so by Theorem 10.3.3 it converges; if decreasing, then use Theorem 10.3.4. The limit lies in the interval $[1, 2]$.

(b) Such a sequence may converge, in which case, by the argument in Part (a), its limit is ≤ 2. But convergence may not happen: for example, the sequence $\{-n\}_{n=1}^{+\infty}$ diverges.

26. (a) $a_{n+1} = \dfrac{|x|^{n+1}}{(n+1)!} = \dfrac{|x|}{n+1}\dfrac{|x|^n}{n!} = \dfrac{|x|}{n+1}a_n$

(b) $a_{n+1}/a_n = |x|/(n+1) < 1$ if $n > |x| - 1$.

(c) From Part (b) the sequence is eventually decreasing, and it is bounded below by 0, so by Theorem 10.3.4 it converges.

(d) If $\lim_{n\to+\infty} a_n = L$ then from Part (a), $L = \dfrac{|x|}{\lim_{n\to+\infty}(n+1)}L = 0$.

(e) $\lim_{n\to+\infty} \dfrac{|x|^n}{n!} = \lim_{n\to+\infty} a_n = 0$

27. (a) $\sqrt{2}, \sqrt{2+\sqrt{2}}, \sqrt{2+\sqrt{2+\sqrt{2}}}$

(b) $a_1 = \sqrt{2} < 2$ so $a_2 = \sqrt{2+a_1} < \sqrt{2+2} = 2$, $a_3 = \sqrt{2+a_2} < \sqrt{2+2} = 2$, and so on indefinitely.

(c) $a_{n+1}^2 - a_n^2 = (2+a_n) - a_n^2 = 2 + a_n - a_n^2 = (2-a_n)(1+a_n)$

(d) $a_n > 0$ and, from Part (b), $a_n < 2$ so $2 - a_n > 0$ and $1 + a_n > 0$ thus, from Part (c), $a_{n+1}^2 - a_n^2 > 0$, $a_{n+1} - a_n > 0$, $a_{n+1} > a_n$; $\{a_n\}$ is a strictly increasing sequence.

(e) The sequence is increasing and has 2 as an upper bound so it must converge to a limit L, $\lim_{n\to+\infty} a_{n+1} = \lim_{n\to+\infty}\sqrt{2+a_n}$, $L = \sqrt{2+L}$, $L^2 - L - 2 = 0$, $(L-2)(L+1) = 0$ thus $\lim_{n\to+\infty} a_n = 2$.

28. (a) If $f(x) = \frac{1}{2}(x + 3/x)$, then $f'(x) = (x^2-3)/(2x^2)$ and $f'(x) = 0$ for $x = \sqrt{3}$; the minimum value of $f(x)$ for $x > 0$ is $f(\sqrt{3}) = \sqrt{3}$. Thus $f(x) \geq \sqrt{3}$ for $x > 0$ and hence $a_n \geq \sqrt{3}$ for $n \geq 2$.

(b) $a_{n+1} - a_n = (3 - a_n^2)/(2a_n) \leq 0$ for $n \geq 2$ since $a_n \geq \sqrt{3}$ for $n \geq 2$; $\{a_n\}$ is eventually decreasing.

(c) $\sqrt{3}$ is a lower bound for a_n so $\{a_n\}$ converges; $\lim_{n\to+\infty} a_{n+1} = \lim_{n\to+\infty}\frac{1}{2}(a_n + 3/a_n)$, $L = \frac{1}{2}(L + 3/L)$, $L^2 - 3 = 0$, $L = \sqrt{3}$.

29. (a) The altitudes of the rectangles are $\ln k$ for $k = 2$ to n, and their bases all have length 1 so the sum of their areas is $\ln 2 + \ln 3 + \cdots + \ln n = \ln(2\cdot 3\cdots n) = \ln n!$. The area under the curve $y = \ln x$ for x in the interval $[1, n]$ is $\int_1^n \ln x\,dx$, and $\int_1^{n+1} \ln x\,dx$ is the area for x in the interval $[1, n+1]$ so, from the figure, $\int_1^n \ln x\,dx < \ln n! < \int_1^{n+1} \ln x\,dx$.

(b) $\int_1^n \ln x\,dx = (x\ln x - x)\Big]_1^n = n\ln n - n + 1$ and $\int_1^{n+1} \ln x\,dx = (n+1)\ln(n+1) - n$ so from Part (a), $n\ln n - n + 1 < \ln n! < (n+1)\ln(n+1) - n$, $e^{n\ln n - n + 1} < n! < e^{(n+1)\ln(n+1) - n}$,

$e^{n\ln n}e^{1-n} < n! < e^{(n+1)\ln(n+1)}e^{-n}$, $\dfrac{n^n}{e^{n-1}} < n! < \dfrac{(n+1)^{n+1}}{e^n}$

(c) From Part (b), $\left[\dfrac{n^n}{e^{n-1}}\right]^{1/n} < \sqrt[n]{n!} < \left[\dfrac{(n+1)^{n+1}}{e^n}\right]^{1/n}$,

$\dfrac{n}{e^{1-1/n}} < \sqrt[n]{n!} < \dfrac{(n+1)^{1+1/n}}{e}$, $\dfrac{1}{e^{1-1/n}} < \dfrac{\sqrt[n]{n!}}{n} < \dfrac{(1+1/n)(n+1)^{1/n}}{e}$,

but $\dfrac{1}{e^{1-1/n}} \to \dfrac{1}{e}$ and $\dfrac{(1+1/n)(n+1)^{1/n}}{e} \to \dfrac{1}{e}$ as $n \to +\infty$ (why?), so $\lim_{n\to+\infty}\dfrac{\sqrt[n]{n!}}{n} = \dfrac{1}{e}$.

30. $n! > \dfrac{n^n}{e^{n-1}}$, $\sqrt[n]{n!} > \dfrac{n}{e^{1-1/n}}$, $\lim\limits_{n\to+\infty} \dfrac{n}{e^{1-1/n}} = +\infty$ so $\lim\limits_{n\to+\infty} \sqrt[n]{n!} = +\infty$.

EXERCISE SET 10.4

1. (a) $s_1 = 2$, $s_2 = 12/5$, $s_3 = \dfrac{62}{25}$, $s_4 = \dfrac{312}{125}$ $s_n = \dfrac{2 - 2(1/5)^n}{1 - 1/5} = \dfrac{5}{2} - \dfrac{5}{2}(1/5)^n$,

$\lim\limits_{n\to+\infty} s_n = \dfrac{5}{2}$, converges

(b) $s_1 = \dfrac{1}{4}$, $s_2 = \dfrac{3}{4}$, $s_3 = \dfrac{7}{4}$, $s_4 = \dfrac{15}{4}$ $s_n = \dfrac{(1/4) - (1/4)2^n}{1 - 2} = -\dfrac{1}{4} + \dfrac{1}{4}(2^n)$,

$\lim\limits_{n\to+\infty} s_n = +\infty$, diverges

(c) $\dfrac{1}{(k+1)(k+2)} = \dfrac{1}{k+1} - \dfrac{1}{k+2}$, $s_1 = \dfrac{1}{6}$, $s_2 = \dfrac{1}{4}$, $s_3 = \dfrac{3}{10}$, $s_4 = \dfrac{1}{3}$;

$s_n = \dfrac{1}{2} - \dfrac{1}{n+2}$, $\lim\limits_{n\to+\infty} s_n = \dfrac{1}{2}$, converges

2. (a) $s_1 = 1/4, s_2 = 5/16, s_3 = 21/64, s_4 = 85/256$

$s_n = \dfrac{1}{4}\left(1 + \dfrac{1}{4} + \cdots + \left(\dfrac{1}{4}\right)^{n-1}\right) = \dfrac{1}{4}\dfrac{1 - (1/4)^n}{1 - 1/4} = \dfrac{1}{3}\left(1 - \left(\dfrac{1}{4}\right)^n\right)$; $\lim\limits_{n\to+\infty} s_n = \dfrac{1}{3}$

(b) $s_1 = 1, s_2 = 5, s_3 = 21, s_4 = 85$; $s_n = \dfrac{4^n - 1}{3}$, diverges

(c) $s_1 = 1/20, s_2 = 1/12, s_3 = 3/28, s_4 = 1/8$;

$s_n = \sum\limits_{k=1}^{n}\left(\dfrac{1}{k+3} - \dfrac{1}{k+4}\right) = \dfrac{1}{4} - \dfrac{1}{n+4}$, $\lim\limits_{n\to+\infty} s_n = 1/4$

3. geometric, $a = 1$, $r = -3/4$, sum $= \dfrac{1}{1 - (-3/4)} = 4/7$

4. geometric, $a = (2/3)^3$, $r = 2/3$, sum $= \dfrac{(2/3)^3}{1 - 2/3} = 8/9$

5. geometric, $a = 7$, $r = -1/6$, sum $= \dfrac{7}{1 + 1/6} = 6$

6. geometric, $r = -3/2$, diverges

7. $s_n = \sum\limits_{k=1}^{n}\left(\dfrac{1}{k+2} - \dfrac{1}{k+3}\right) = \dfrac{1}{3} - \dfrac{1}{n+3}$, $\lim\limits_{n\to+\infty} s_n = 1/3$

8. $s_n = \sum\limits_{k=1}^{n}\left(\dfrac{1}{2^k} - \dfrac{1}{2^{k+1}}\right) = \dfrac{1}{2} - \dfrac{1}{2^{n+1}}$, $\lim\limits_{n\to+\infty} s_n = 1/2$

9. $s_n = \sum\limits_{k=1}^{n}\left(\dfrac{1/3}{3k-1} - \dfrac{1/3}{3k+2}\right) = \dfrac{1}{6} - \dfrac{1/3}{3n+2}$, $\lim\limits_{n\to+\infty} s_n = 1/6$

10. $s_n = \sum_{k=2}^{n+1}\left[\frac{1/2}{k-1} - \frac{1/2}{k+1}\right] = \frac{1}{2}\left[\sum_{k=2}^{n+1}\frac{1}{k-1} - \sum_{k=2}^{n+1}\frac{1}{k+1}\right]$

$= \frac{1}{2}\left[\sum_{k=2}^{n+1}\frac{1}{k-1} - \sum_{k=4}^{n+3}\frac{1}{k-1}\right] = \frac{1}{2}\left[1 + \frac{1}{2} - \frac{1}{n+1} - \frac{1}{n+2}\right]; \lim_{n\to+\infty} s_n = \frac{3}{4}$

11. $\sum_{k=3}^{\infty}\frac{1}{k-2} = \sum_{k=1}^{\infty} 1/k$, the harmonic series, so the series diverges.

12. geometric, $a = (e/\pi)^4$, $r = e/\pi < 1$, sum $= \frac{(e/\pi)^4}{1-e/\pi} = \frac{e^4}{\pi^3(\pi - e)}$

13. $\sum_{k=1}^{\infty}\frac{4^{k+2}}{7^{k-1}} = \sum_{k=1}^{\infty} 64\left(\frac{4}{7}\right)^{k-1}$; geometric, $a = 64$, $r = 4/7$, sum $= \frac{64}{1-4/7} = 448/3$

14. geometric, $a = 125, r = 125/7$, diverges

15. $0.4444\cdots = 0.4 + 0.04 + 0.004 + \cdots = \frac{0.4}{1-0.1} = 4/9$

16. $0.9999\cdots = 0.9 + 0.09 + 0.009 + \cdots = \frac{0.9}{1-0.1} = 1$

17. $5.373737\cdots = 5 + 0.37 + 0.0037 + 0.000037 + \cdots = 5 + \frac{0.37}{1-0.01} = 5 + 37/99 = 532/99$

18. $0.159159159\cdots = 0.159 + 0.000159 + 0.000000159 + \cdots = \frac{0.159}{1-0.001} = 159/999 = 53/333$

19. $0.782178217821\cdots = 0.7821 + 0.00007821 + 0.000000007821 + \cdots = \frac{0.7821}{1-0.0001} = \frac{7821}{9999} = \frac{79}{101}$

20. $0.451141414\cdots = 0.451 + 0.00014 + 0.0000014 + 0.000000014 + \cdots = 0.451 + \frac{0.00014}{1-0.01} = \frac{44663}{99000}$

21. $d = 10 + 2\cdot\frac{3}{4}\cdot 10 + 2\cdot\frac{3}{4}\cdot\frac{3}{4}\cdot 10 + 2\cdot\frac{3}{4}\cdot\frac{3}{4}\cdot\frac{3}{4}\cdot 10 + \cdots$

$= 10 + 20\left(\frac{3}{4}\right) + 20\left(\frac{3}{4}\right)^2 + 20\left(\frac{3}{4}\right)^3 + \cdots = 10 + \frac{20(3/4)}{1-3/4} = 10 + 60 = 70$ meters

22. volume $= 1^3 + \left(\frac{1}{2}\right)^3 + \left(\frac{1}{4}\right)^3 + \cdots + \left(\frac{1}{2^n}\right)^3 + \cdots = 1 + \frac{1}{8} + \left(\frac{1}{8}\right)^2 + \cdots + \left(\frac{1}{8}\right)^n + \cdots$

$= \frac{1}{1-(1/8)} = 8/7$

23. **(a)** $s_n = \ln\frac{1}{2} + \ln\frac{2}{3} + \ln\frac{3}{4} + \cdots + \ln\frac{n}{n+1} = \ln\left(\frac{1}{2}\cdot\frac{2}{3}\cdot\frac{3}{4}\cdots\frac{n}{n+1}\right) = \ln\frac{1}{n+1} = -\ln(n+1)$,

$\lim_{n\to+\infty} s_n = -\infty$, series diverges.

(b) $\ln(1-1/k^2) = \ln\dfrac{k^2-1}{k^2} = \ln\dfrac{(k-1)(k+1)}{k^2} = \ln\dfrac{k-1}{k} + \ln\dfrac{k+1}{k} = \ln\dfrac{k-1}{k} - \ln\dfrac{k}{k+1}$,

$$s_n = \sum_{k=2}^{n+1}\left[\ln\frac{k-1}{k} - \ln\frac{k}{k+1}\right]$$

$$= \left(\ln\frac{1}{2} - \ln\frac{2}{3}\right) + \left(\ln\frac{2}{3} - \ln\frac{3}{4}\right) + \left(\ln\frac{3}{4} - \ln\frac{4}{5}\right) + \cdots + \left(\ln\frac{n}{n+1} - \ln\frac{n+1}{n+2}\right)$$

$$= \ln\frac{1}{2} - \ln\frac{n+1}{n+2}, \ \lim_{n\to+\infty} s_n = \ln\frac{1}{2} = -\ln 2$$

24. **(a)** $\displaystyle\sum_{k=0}^{\infty}(-1)^k x^k = 1 - x + x^2 - x^3 + \cdots = \frac{1}{1-(-x)} = \frac{1}{1+x}$ if $|-x| < 1$, $|x| < 1$, $-1 < x < 1$.

(b) $\displaystyle\sum_{k=0}^{\infty}(x-3)^k = 1 + (x-3) + (x-3)^2 + \cdots = \frac{1}{1-(x-3)} = \frac{1}{4-x}$ if $|x-3| < 1$, $2 < x < 4$.

(c) $\displaystyle\sum_{k=0}^{\infty}(-1)^k x^{2k} = 1 - x^2 + x^4 - x^6 + \cdots = \frac{1}{1-(-x^2)} = \frac{1}{1+x^2}$ if $|-x^2| < 1$, $|x| < 1$, $-1 < x < 1$.

25. **(a)** Geometric series, $a = x$, $r = -x^2$. Converges for $|-x^2| < 1$, $|x| < 1$;
$S = \dfrac{x}{1-(-x^2)} = \dfrac{x}{1+x^2}$.

(b) Geometric series, $a = 1/x^2$, $r = 2/x$. Converges for $|2/x| < 1$, $|x| > 2$;
$S = \dfrac{1/x^2}{1-2/x} = \dfrac{1}{x^2-2x}$.

(c) Geometric series, $a = e^{-x}$, $r = e^{-x}$. Converges for $|e^{-x}| < 1$, $e^{-x} < 1$, $e^x > 1$, $x > 0$;
$S = \dfrac{e^{-x}}{1-e^{-x}} = \dfrac{1}{e^x - 1}$.

26. $\dfrac{\sqrt{k+1}-\sqrt{k}}{\sqrt{k^2+k}} = \dfrac{\sqrt{k+1}-\sqrt{k}}{\sqrt{k}\sqrt{k+1}} = \dfrac{1}{\sqrt{k}} - \dfrac{1}{\sqrt{k+1}}$,

$$s_n = \sum_{k=1}^{n}\left(\frac{1}{\sqrt{k}} - \frac{1}{\sqrt{k+1}}\right) = \left(\frac{1}{\sqrt{1}} - \frac{1}{\sqrt{2}}\right) + \left(\frac{1}{\sqrt{2}} - \frac{1}{\sqrt{3}}\right) + \left(\frac{1}{\sqrt{3}} - \frac{1}{\sqrt{4}}\right)$$

$$+ \cdots + \left(\frac{1}{\sqrt{n}} - \frac{1}{\sqrt{n+1}}\right) = 1 - \frac{1}{\sqrt{n+1}}; \ \lim_{n\to+\infty} s_n = 1$$

27. $s_n = (1 - 1/3) + (1/2 - 1/4) + (1/3 - 1/5) + (1/4 - 1/6) + \cdots + [1/n - 1/(n+2)]$
$= (1 + 1/2 + 1/3 + \cdots + 1/n) - (1/3 + 1/4 + 1/5 + \cdots + 1/(n+2))$
$= 3/2 - 1/(n+1) - 1/(n+2), \ \lim_{n\to+\infty} s_n = 3/2$

28. $\displaystyle s_n = \sum_{k=1}^{n}\frac{1}{k(k+2)} = \sum_{k=1}^{n}\left[\frac{1/2}{k} - \frac{1/2}{k+2}\right] = \frac{1}{2}\left[\sum_{k=1}^{n}\frac{1}{k} - \sum_{k=1}^{n}\frac{1}{k+2}\right]$

$$= \frac{1}{2}\left[\sum_{k=1}^{n}\frac{1}{k} - \sum_{k=3}^{n+2}\frac{1}{k}\right] = \frac{1}{2}\left[1 + \frac{1}{2} - \frac{1}{n+1} - \frac{1}{n+2}\right]; \ \lim_{n\to+\infty} s_n = \frac{3}{4}$$

29. $s_n = \sum_{k=1}^{n} \frac{1}{(2k-1)(2k+1)} = \sum_{k=1}^{n} \left[\frac{1/2}{2k-1} - \frac{1/2}{2k+1}\right] = \frac{1}{2}\left[\sum_{k=1}^{n} \frac{1}{2k-1} - \sum_{k=1}^{n} \frac{1}{2k+1}\right]$

$= \frac{1}{2}\left[\sum_{k=1}^{n} \frac{1}{2k-1} - \sum_{k=2}^{n+1} \frac{1}{2k-1}\right] = \frac{1}{2}\left[1 - \frac{1}{2n+1}\right]; \lim_{n\to+\infty} s_n = \frac{1}{2}$

30. Geometric series, $a = \sin\, x$, $r = -\frac{1}{2}\sin x$. Converges for $|-\frac{1}{2}\sin x| < 1$, $|\sin x| < 2$, so converges for all values of x. $S = \frac{\sin x}{1 + \frac{1}{2}\sin x} = \frac{2\sin x}{2 + \sin x}$.

31. $a_2 = \frac{1}{2}a_1 + \frac{1}{2}$, $a_3 = \frac{1}{2}a_2 + \frac{1}{2} = \frac{1}{2^2}a_1 + \frac{1}{2^2} + \frac{1}{2}$, $a_4 = \frac{1}{2}a_3 + \frac{1}{2} = \frac{1}{2^3}a_1 + \frac{1}{2^3} + \frac{1}{2^2} + \frac{1}{2}$,

$a_5 = \frac{1}{2}a_4 + \frac{1}{2} = \frac{1}{2^4}a_1 + \frac{1}{2^4} + \frac{1}{2^3} + \frac{1}{2^2} + \frac{1}{2}, \ldots, a_n = \frac{1}{2^{n-1}}a_1 + \frac{1}{2^{n-1}} + \frac{1}{2^{n-2}} + \cdots + \frac{1}{2}$,

$\lim_{n\to+\infty} a_n = \lim_{n\to+\infty} \frac{a_1}{2^{n-1}} + \sum_{n=1}^{\infty}\left(\frac{1}{2}\right)^n = 0 + \frac{1/2}{1-1/2} = 1$

32. $0.a_1a_2\cdots a_n9999\cdots = 0.a_1a_2\cdots a_n + 0.9\,(10^{-n}) + 0.09\,(10^{-n}) + \cdots$

$= 0.a_1a_2\cdots a_n + \frac{0.9\,(10^{-n})}{1-0.1} = 0.a_1a_2\cdots a_n + 10^{-n}$

$= 0.a_1a_2\cdots(a_n+1) = 0.a_1a_2\cdots(a_n+1)\,0000\cdots$

33. The series converges to $1/(1-x)$ only if $-1 < x < 1$.

34. $P_0P_1 = a\sin\theta$,
$P_1P_2 = a\sin\theta\cos\theta$,
$P_2P_3 = a\sin\theta\cos^2\theta$,
$P_3P_4 = a\sin\theta\cos^3\theta, \ldots$
(see figure)
Each sum is a geometric series.

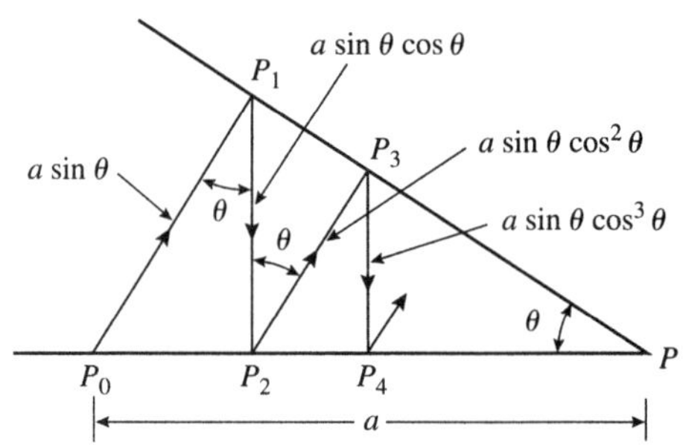

(a) $P_0P_1 + P_1P_2 + P_2P_3 + \cdots = a\sin\theta + a\sin\theta\cos\theta + a\sin\theta\cos^2\theta + \cdots = \frac{a\sin\theta}{1-\cos\theta}$

(b) $P_0P_1 + P_2P_3 + P_4P_5 + \cdots = a\sin\theta + a\sin\theta\cos^2\theta + a\sin\theta\cos^4\theta + \cdots$

$= \frac{a\sin\theta}{1-\cos^2\theta} = \frac{a\sin\theta}{\sin^2\theta} = a\csc\theta$

(c) $P_1P_2 + P_3P_4 + P_5P_6 + \cdots = a\sin\theta\cos\theta + a\sin\theta\cos^3\theta + \cdots$

$= \frac{a\sin\theta\cos\theta}{1-\cos^2\theta} = \frac{a\sin\theta\cos\theta}{\sin^2\theta} = a\cot\theta$

35. By inspection, $\frac{\theta}{2} - \frac{\theta}{4} + \frac{\theta}{8} - \frac{\theta}{16} + \cdots = \frac{\theta/2}{1-(-1/2)} = \theta/3$

36. $A_1 + A_2 + A_3 + \cdots = 1 + 1/2 + 1/4 + \cdots = \dfrac{1}{1-(1/2)} = 2$

37. **(b)** $\dfrac{2^k A}{3^k - 2^k} + \dfrac{2^k B}{3^{k+1} - 2^{k+1}} = \dfrac{2^k\left(3^{k+1} - 2^{k+1}\right)A + 2^k\left(3^k - 2^k\right)B}{\left(3^k - 2^k\right)\left(3^{k+1} - 2^{k+1}\right)}$

$$= \frac{\left(3\cdot 6^k - 2\cdot 2^{2k}\right)A + \left(6^k - 2^{2k}\right)B}{\left(3^k - 2^k\right)\left(3^{k+1} - 2^{k+1}\right)} = \frac{(3A+B)6^k - (2A+B)2^{2k}}{\left(3^k - 2^k\right)\left(3^{k+1} - 2^{k+1}\right)}$$

so $3A + B = 1$ and $2A + B = 0$, $A = 1$ and $B = -2$.

(c) $s_n = \displaystyle\sum_{k=1}^{n}\left[\frac{2^k}{3^k - 2^k} - \frac{2^{k+1}}{3^{k+1} - 2^{k+1}}\right] = \sum_{k=1}^{n}(a_k - a_{k+1})$ where $a_k = \dfrac{2^k}{3^k - 2^k}$.

But $s_n = (a_1 - a_2) + (a_2 - a_3) + (a_3 - a_4) + \cdots + (a_n - a_{n+1})$ which is a telescoping sum,

$s_n = a_1 - a_{n+1} = 2 - \dfrac{2^{n+1}}{3^{n+1} - 2^{n+1}}$, $\displaystyle\lim_{n\to+\infty} s_n = \lim_{n\to+\infty}\left[2 - \frac{(2/3)^{n+1}}{1 - (2/3)^{n+1}}\right] = 2.$

38. **(a)** geometric; 18/5 **(b)** geometric; diverges **(c)** $\displaystyle\sum_{k=1}^{\infty}\frac{1}{2}\left(\frac{1}{2k-1} - \frac{1}{2k+1}\right) = 1/2$

EXERCISE SET 10.5

1. **(a)** $\displaystyle\sum_{k=1}^{\infty}\frac{1}{2^k} = \frac{1/2}{1-1/2} = 1$; $\displaystyle\sum_{k=1}^{\infty}\frac{1}{4^k} = \frac{1/4}{1-1/4} = 1/3$; $\displaystyle\sum_{k=1}^{\infty}\left(\frac{1}{2^k} + \frac{1}{4^k}\right) = 1 + 1/3 = 4/3$

(b) $\displaystyle\sum_{k=1}^{\infty}\frac{1}{5^k} = \frac{1/5}{1-1/5} = 1/4$; $\displaystyle\sum_{k=1}^{\infty}\frac{1}{k(k+1)} = 1$ (Example 5, Section 10 .4);

$$\sum_{k=1}^{\infty}\left[\frac{1}{5^k} - \frac{1}{k(k+1)}\right] = 1/4 - 1 = -3/4$$

2. **(a)** $\displaystyle\sum_{k=2}^{\infty}\frac{1}{k^2-1} = 3/4$ (Exercise 10, Section 10.4); $\displaystyle\sum_{k=2}^{\infty}\frac{7}{10^{k-1}} = \frac{7/10}{1-1/10} = 7/9$;

so $\displaystyle\sum_{k=2}^{\infty}\left[\frac{1}{k^2-1} - \frac{7}{10^{k-1}}\right] = 3/4 - 7/9 = -1/36$

(b) with $a = 9/7, r = 3/7$, geometric, $\displaystyle\sum_{k=1}^{\infty}7^{-k}3^{k+1} = \frac{9/7}{1-(3/7)} = 9/4$;

with $a = 4/5, r = 2/5$, geometric, $\displaystyle\sum_{k=1}^{\infty}\frac{2^{k+1}}{5^k} = \frac{4/5}{1-(2/5)} = 4/3$;

$$\sum_{k=1}^{\infty}\left[7^{-k}3^{k+1} - \frac{2^{k+1}}{5^k}\right] = 9/4 - 4/3 = 11/12$$

3. **(a)** $p=3$, converges **(b)** $p=1/2$, diverges **(c)** $p=1$, diverges **(d)** $p=2/3$, diverges

4. **(a)** $p=4/3$, converges **(b)** $p=1/4$, diverges **(c)** $p=5/3$, converges **(d)** $p=\pi$, converges

5. (a) $\lim_{k\to+\infty} \dfrac{k^2+k+3}{2k^2+1} = \dfrac{1}{2}$; the series diverges. (b) $\lim_{k\to+\infty} \left(1+\dfrac{1}{k}\right)^k = e$; the series diverges.

(c) $\lim_{k\to+\infty} \cos k\pi$ does not exist; the series diverges. (d) $\lim_{k\to+\infty} \dfrac{1}{k!} = 0$; no information

6. (a) $\lim_{k\to+\infty} \dfrac{k}{e^k} = 0$; no information (b) $\lim_{k\to+\infty} \ln k = +\infty$; the series diverges.

(c) $\lim_{k\to+\infty} \dfrac{1}{\sqrt{k}} = 0$; no information (d) $\lim_{k\to+\infty} \dfrac{\sqrt{k}}{\sqrt{k+3}} = 1$; the series diverges.

7. (a) $\displaystyle\int_1^{+\infty} \frac{1}{5x+2} = \lim_{\ell\to+\infty} \frac{1}{5}\ln(5x+2)\Big]_1^\ell = +\infty$, the series diverges by the Integral Test.

(b) $\displaystyle\int_1^{+\infty} \frac{1}{1+9x^2}dx = \lim_{\ell\to+\infty} \frac{1}{3}\tan^{-1}3x\Big]_1^\ell = \frac{1}{3}\left(\pi/2 - \tan^{-1}3\right)$, the series converges by the Integral Test.

8. (a) $\displaystyle\int_1^{+\infty} \frac{x}{1+x^2}dx = \lim_{\ell\to+\infty} \frac{1}{2}\ln(1+x^2)\Big]_1^\ell = +\infty$, the series diverges by the Integral Test.

(b) $\displaystyle\int_1^{+\infty} (4+2x)^{-3/2}dx = \lim_{\ell\to+\infty} -1/\sqrt{4+2x}\Big]_1^\ell = 1/\sqrt{6}$, the series converges by the Integral Test.

9. $\displaystyle\sum_{k=1}^{\infty} \frac{1}{k+6} = \sum_{k=7}^{\infty} \frac{1}{k}$, diverges because the harmonic series diverges.

10. $\displaystyle\sum_{k=1}^{\infty} \frac{3}{5k} = \sum_{k=1}^{\infty} \frac{3}{5}\left(\frac{1}{k}\right)$, diverges because the harmonic series diverges.

11. $\displaystyle\sum_{k=1}^{\infty} \frac{1}{\sqrt{k+5}} = \sum_{k=6}^{\infty} \frac{1}{\sqrt{k}}$, diverges because the p-series with $p = 1/2 \le 1$ diverges.

12. $\lim_{k\to+\infty} \dfrac{1}{e^{1/k}} = 1$, the series diverges because $\lim_{k\to+\infty} u_k = 1 \ne 0$.

13. $\displaystyle\int_1^{+\infty} (2x-1)^{-1/3}dx = \lim_{\ell\to+\infty} \frac{3}{4}(2x-1)^{2/3}\Big]_1^\ell = +\infty$, the series diverges by the Integral Test.

14. $\dfrac{\ln x}{x}$ is decreasing for $x \ge e$, and $\displaystyle\int_3^{+\infty} \frac{\ln x}{x} = \lim_{\ell\to+\infty} \frac{1}{2}(\ln x)^2\Big]_3^\ell = +\infty$, so the series diverges by the Integral Test.

15. $\lim_{k\to+\infty} \dfrac{k}{\ln(k+1)} = \lim_{k\to+\infty} \dfrac{1}{1/(k+1)} = +\infty$, the series diverges because $\lim_{k\to+\infty} u_k \ne 0$.

16. $\displaystyle\int_1^{+\infty} xe^{-x^2}dx = \lim_{\ell\to+\infty} -\frac{1}{2}e^{-x^2}\Big]_1^\ell = e^{-1}/2$, the series converges by the Integral Test.

17. $\lim_{k\to+\infty}(1+1/k)^{-k} = 1/e \neq 0$, the series diverges.

18. $\lim_{k\to+\infty}\dfrac{k^2+1}{k^2+3} = 1 \neq 0$, the series diverges.

19. $\displaystyle\int_1^{+\infty}\frac{\tan^{-1}x}{1+x^2}dx = \lim_{\ell\to+\infty}\frac{1}{2}\left(\tan^{-1}x\right)^2\Big]_1^{\ell} = 3\pi^2/32$, the series converges by the Integral Test, since

$\dfrac{d}{dx}\dfrac{\tan^{-1}x}{1+x^2} = \dfrac{1-2x\tan^{-1}x}{(1+x^2)^2} < 0$ for $x \geq 1$.

20. $\displaystyle\int_1^{+\infty}\frac{1}{\sqrt{x^2+1}}dx = \lim_{\ell\to+\infty}\sinh^{-1}x\Big]_1^{\ell} = +\infty$, the series diverges by the Integral Test.

21. $\lim_{k\to+\infty}k^2\sin^2(1/k) = 1 \neq 0$, the series diverges.

22. $\displaystyle\int_1^{+\infty}x^2e^{-x^3}dx = \lim_{\ell\to+\infty}-\frac{1}{3}e^{-x^3}\Big]_1^{\ell} = e^{-1}/3$,

the series converges by the Integral Test ($x^2e^{-x^3}$ is decreasing for $x \geq 1$).

23. $7\displaystyle\sum_{k=5}^{\infty}k^{-1.01}$, p-series with $p > 1$, converges

24. $\displaystyle\int_1^{+\infty}\text{sech}^2 x\,dx = \lim_{\ell\to+\infty}\tanh x\Big]_1^{\ell} = 1-\tanh(1)$, the series converges by the Integral Test.

25. $\dfrac{1}{x(\ln x)^p}$ is decreasing for $x \geq e^p$, so use the Integral Test with $\displaystyle\int_{e^p}^{+\infty}\frac{dx}{x(\ln x)^p}$ to get

$$\lim_{\ell\to+\infty}\ln(\ln x)\Big]_{e^p}^{\ell} = +\infty \text{ if } p=1, \qquad \lim_{\ell\to+\infty}\frac{(\ln x)^{1-p}}{1-p}\Big]_{e^p}^{\ell} = \begin{cases} +\infty & \text{if } p<1 \\ \dfrac{p^{1-p}}{p-1} & \text{if } p>1 \end{cases}$$

Thus the series converges for $p > 1$.

26. If $p > 0$ set $g(x) = x(\ln x)[\ln(\ln x)]^p$, $g'(x) = (\ln(\ln x))^{p-1}\left[(1+\ln x)\ln(\ln x)+p\right]$, and, for $x > e^e$, $g'(x) > 0$, thus $1/g(x)$ is decreasing for $x > e^e$; use the Integral Test with $\displaystyle\int_{e^e}^{+\infty}\frac{dx}{x(\ln x)[\ln(\ln x)]^p}$ to get

$$\lim_{\ell\to+\infty}\ln[\ln(\ln x)]\Big]_{e^e}^{\ell} = +\infty \text{ if } p=1, \quad \lim_{\ell\to+\infty}\frac{[\ln(\ln x)]^{1-p}}{1-p}\Big]_{e^e}^{\ell} = \begin{cases} +\infty & \text{if } p<1, \\ \dfrac{1}{p-1} & \text{if } p>1 \end{cases}$$

Thus the series converges for $p > 1$ and diverges for $0 < p \leq 1$. If $p \leq 0$ then $\dfrac{[\ln(\ln x)]^p}{x\ln x} \geq \dfrac{1}{x\ln x}$ for $x > e^e$ so the series diverges.

27. **(a)** $3\displaystyle\sum_{k=1}^{\infty}\frac{1}{k^2} - \sum_{k=1}^{\infty}\frac{1}{k^4} = \pi^2/2 - \pi^4/90$ **(b)** $\displaystyle\sum_{k=1}^{\infty}\frac{1}{k^2} - 1 - \frac{1}{2^2} = \pi^2/6 - 5/4$

(c) $\displaystyle\sum_{k=2}^{\infty}\frac{1}{(k-1)^4} = \sum_{k=1}^{\infty}\frac{1}{k^4} = \pi^4/90$

28. **(a)** Suppose $\Sigma(u_k + v_k)$ converges; then so does $\Sigma[(u_k + v_k) - u_k]$, but $\Sigma[(u_k + v_k) - u_k] = \Sigma v_k$, so Σv_k converges which contradicts the assumption that Σv_k diverges. Suppose $\Sigma(u_k - v_k)$ converges; then so does $\Sigma[u_k - (u_k - v_k)] = \Sigma v_k$ which leads to the same contradiction as before.

(b) Let $u_k = 2/k$ and $v_k = 1/k$; then both $\Sigma(u_k + v_k)$ and $\Sigma(u_k - v_k)$ diverge; let $u_k = 1/k$ and $v_k = -1/k$ then $\Sigma(u_k + v_k)$ converges; let $u_k = v_k = 1/k$ then $\Sigma(u_k - v_k)$ converges.

29. **(a)** diverges because $\sum_{k=1}^{\infty}(2/3)^{k-1}$ converges and $\sum_{k=1}^{\infty} 1/k$ diverges.

(b) diverges because $\sum_{k=1}^{\infty} 1/(3k+2)$ diverges and $\sum_{k=1}^{\infty} 1/k^{3/2}$ converges.

(c) converges because both $\sum_{k=2}^{\infty} \frac{1}{k(\ln k)^2}$ (Exercise 25) and $\sum_{k=2}^{\infty} 1/k^2$ converge.

30. **(a)** If $S = \sum_{k=1}^{\infty} u_k$ and $s_n = \sum_{k=1}^{n} u_k$, then $S - s_n = \sum_{k=n+1}^{\infty} u_k$. Interpret u_k, $k = n+1, n+2, \ldots$, as the areas of inscribed or circumscribed rectangles with height u_k and base of length one for the curve $y = f(x)$ to obtain the result.

(b) Add $s_n = \sum_{k=1}^{n} u_k$ to each term in the conclusion of Part (a) to get the desired result:

$$s_n + \int_{n+1}^{+\infty} f(x)\,dx < \sum_{k=1}^{+\infty} u_k < s_n + \int_{n}^{+\infty} f(x)\,dx$$

31. **(a)** In Exercise 30 above let $f(x) = \dfrac{1}{x^2}$. Then $\displaystyle\int_n^{+\infty} f(x)\,dx = -\frac{1}{x}\bigg]_n^{+\infty} = \frac{1}{n}$; use this result and the same result with $n+1$ replacing n to obtain the desired result.

(b) $s_3 = 1 + 1/4 + 1/9 = 49/36$; $58/36 = s_3 + \dfrac{1}{4} < \dfrac{1}{6}\pi^2 < s_3 + \dfrac{1}{3} = 61/36$

(d) $1/11 < \dfrac{1}{6}\pi^2 - s_{10} < 1/10$

33. Apply Exercise 30 in each case:

(a) $f(x) = \dfrac{1}{(2x+1)^2}$, $\displaystyle\int_n^{+\infty} f(x)\,dx = \frac{1}{2(2n+1)}$, so $\displaystyle\frac{1}{46} < \sum_{k=1}^{\infty} \frac{1}{(2k+1)^2} - s_{10} < \frac{1}{42}$

(b) $f(x) = \dfrac{1}{k^2+1}$, $\displaystyle\int_n^{+\infty} f(x)\,dx = \frac{\pi}{2} - \tan^{-1}(n)$, so

$$\pi/2 - \tan^{-1}(11) < \sum_{k=1}^{\infty} \frac{1}{k^2+1} - s_{10} < \pi/2 - \tan^{-1}(10)$$

(c) $f(x) = \dfrac{x}{e^x}$, $\displaystyle\int_n^{+\infty} f(x)\,dx = (n+1)e^{-n}$, so $\displaystyle 12e^{-11} < \sum_{k=1}^{\infty} \frac{k}{e^k} - s_{10} < 11e^{-10}$

34. **(a)** $\displaystyle\int_n^{+\infty} \frac{1}{x^3}\,dx = \frac{1}{2n^2}$; use Exercise 30(b)

(b) $\displaystyle\frac{1}{2n^2} - \frac{1}{2(n+1)^2} < 0.01$ for $n = 5$.

(c) From Part (a) with $n = 5$ obtain $1.200 < S < 1.206$, so $S \approx 1.203$.

35. **(a)** $\displaystyle\int_n^{+\infty} \frac{1}{x^4}\,dx = \frac{1}{3n^3}$; choose n so that $\displaystyle\frac{1}{3n^3} - \frac{1}{3(n+1)^3} < 0.005$, $n = 4$; $S \approx 1.08$

36. **(a)** Let $F(x) = \dfrac{1}{x}$, then $\displaystyle\int_1^n \frac{1}{x}dx = \ln n$ and $\displaystyle\int_1^{n+1} \frac{1}{x}dx = \ln(n+1)$, $u_1 = 1$ so $\ln(n+1) < s_n < 1 + \ln n$.

(b) $\ln(1,000,001) < s_{1,000,000} < 1 + \ln(1,000,000)$, $13 < s_{1,000,000} < 15$

(c) $s_{10^9} < 1 + \ln 10^9 = 1 + 9\ln 10 < 22$

(d) $s_n > \ln(n+1) \geq 100$, $n \geq e^{100} - 1 \approx 2.688 \times 10^{43}$; $n = 2.69 \times 10^{43}$

37. p-series with $p = \ln a$; convergence for $p > 1, a > e$

38. x^2e^{-x} is decreasing and positive for $x > 2$ so the Integral Test applies:
$\displaystyle\int_1^{\infty} x^2e^{-x}\,dx = -(x^2+2x+2)e^{-x}\Big]_1^{\infty} = 5e^{-1}$ so the series converges.

39. **(a)** $f(x) = 1/(x^3+1)$ is decreasing and continuous on the interval $[1, +\infty]$, so the Integral Test applies.

(c)

n	10	20	30	40	50
s_n	0.681980	0.685314	0.685966	0.686199	0.686307

n	60	70	80	90	100
s_n	0.686367	0.686403	0.686426	0.686442	0.686454

(e) Set $\displaystyle g(n) = \int_n^{+\infty} \frac{1}{x^3+1}\,dx = \frac{\sqrt{3}}{6}\pi + \frac{1}{6}\ln\frac{n^3+1}{(n+1)^3} - \frac{\sqrt{3}}{3}\tan^{-1}\left(\frac{2n-1}{\sqrt{3}}\right)$; for $n \geq 13$, $g(n) - g(n+1) \leq 0.0005$; $s_{13} + (g(13) + g(14))/2 \approx 0.6865$, so the sum ≈ 0.6865 to three decimal places.

EXERCISE SET 10.6

1. **(a)** $\displaystyle\frac{1}{5k^2-k} \leq \frac{1}{5k^2-k^2} = \frac{1}{4k^2}, \sum_{k=1}^{\infty} \frac{1}{4k^2}$ converges

(b) $\displaystyle\frac{3}{k-1/4} > \frac{3}{k}, \sum_{k=1}^{\infty} 3/k$ diverges

2. **(a)** $\displaystyle\frac{k+1}{k^2-k} > \frac{k}{k^2} = \frac{1}{k}, \sum_{k=2}^{\infty} 1/k$ diverges

(b) $\displaystyle\frac{2}{k^4+k} < \frac{2}{k^4}, \sum_{k=1}^{\infty} \frac{2}{k^4}$ converges

3. (a) $\dfrac{1}{3^k+5} < \dfrac{1}{3^k}$, $\displaystyle\sum_{k=1}^{\infty} \frac{1}{3^k}$ converges (b) $\dfrac{5\sin^2 k}{k!} < \dfrac{5}{k!}$, $\displaystyle\sum_{k=1}^{\infty} \frac{5}{k!}$ converges

4. (a) $\dfrac{\ln k}{k} > \dfrac{1}{k}$ for $k \geq 3$, $\displaystyle\sum_{k=1}^{\infty} \frac{1}{k}$ diverges

(b) $\dfrac{k}{k^{3/2}-1/2} > \dfrac{k}{k^{3/2}} = \dfrac{1}{\sqrt{k}}$; $\displaystyle\sum_{k=1}^{\infty} \frac{1}{\sqrt{k}}$ diverges

5. compare with the convergent series $\displaystyle\sum_{k=1}^{\infty} 1/k^5$, $\rho = \displaystyle\lim_{k\to+\infty} \frac{4k^7-2k^6+6k^5}{8k^7+k-8} = 1/2$, converges

6. compare with the divergent series $\displaystyle\sum_{k=1}^{\infty} 1/k$, $\rho = \displaystyle\lim_{k\to+\infty} \frac{k}{9k+6} = 1/9$, diverges

7. compare with the convergent series $\displaystyle\sum_{k=1}^{\infty} 5/3^k$, $\rho = \displaystyle\lim_{k\to+\infty} \frac{3^k}{3^k+1} = 1$, converges

8. compare with the divergent series $\displaystyle\sum_{k=1}^{\infty} 1/k$, $\rho = \displaystyle\lim_{k\to+\infty} \frac{k^2(k+3)}{(k+1)(k+2)(k+5)} = 1$, diverges

9. compare with the divergent series $\displaystyle\sum_{k=1}^{\infty} \frac{1}{k^{2/3}}$,

$$\rho = \lim_{k\to+\infty} \frac{k^{2/3}}{(8k^2-3k)^{1/3}} = \lim_{k\to+\infty} \frac{1}{(8-3/k)^{1/3}} = 1/2, \text{ diverges}$$

10. compare with the convergent series $\displaystyle\sum_{k=1}^{\infty} 1/k^{17}$,

$$\rho = \lim_{k\to+\infty} \frac{k^{17}}{(2k+3)^{17}} = \lim_{k\to+\infty} \frac{1}{(2+3/k)^{17}} = 1/2^{17}, \text{ converges}$$

11. $\rho = \displaystyle\lim_{k\to+\infty} \frac{3^{k+1}/(k+1)!}{3^k/k!} = \lim_{k\to+\infty} \frac{3}{k+1} = 0$, the series converges

12. $\rho = \displaystyle\lim_{k\to+\infty} \frac{4^{k+1}/(k+1)^2}{4^k/k^2} = \lim_{k\to+\infty} \frac{4k^2}{(k+1)^2} = 4$, the series diverges

13. $\rho = \displaystyle\lim_{k\to+\infty} \frac{k}{k+1} = 1$, the result is inconclusive

14. $\rho = \displaystyle\lim_{k\to+\infty} \frac{(k+1)(1/2)^{k+1}}{k(1/2)^k} = \lim_{k\to+\infty} \frac{k+1}{2k} = 1/2$, the series converges

15. $\rho = \displaystyle\lim_{k\to+\infty} \frac{(k+1)!/(k+1)^3}{k!/k^3} = \lim_{k\to+\infty} \frac{k^3}{(k+1)^2} = +\infty$, the series diverges

16. $\rho = \displaystyle\lim_{k\to+\infty} \frac{(k+1)/[(k+1)^2+1]}{k/(k^2+1)} = \lim_{k\to+\infty} \frac{(k+1)(k^2+1)}{k(k^2+2k+2)} = 1$, the result is inconclusive.

17. $\rho = \lim_{k\to+\infty} \dfrac{3k+2}{2k-1} = 3/2$, the series diverges

18. $\rho = \lim_{k\to+\infty} k/100 = +\infty$, the series diverges

19. $\rho = \lim_{k\to+\infty} \dfrac{k^{1/k}}{5} = 1/5$, the series converges

20. $\rho = \lim_{k\to+\infty} (1 - e^{-k}) = 1$, the result is inconclusive

21. Ratio Test, $\rho = \lim_{k\to+\infty} 7/(k+1) = 0$, converges

22. Limit Comparison Test, compare with the divergent series $\sum_{k=1}^{\infty} 1/k$

23. Ratio Test, $\rho = \lim_{k\to+\infty} \dfrac{(k+1)^2}{5k^2} = 1/5$, converges

24. Ratio Test, $\rho = \lim_{k\to+\infty} (10/3)(k+1) = +\infty$, diverges

25. Ratio Test, $\rho = \lim_{k\to+\infty} e^{-1}(k+1)^{50}/k^{50} = e^{-1} < 1$, converges

26. Limit Comparison Test, compare with the divergent series $\sum_{k=1}^{\infty} 1/k$

27. Limit Comparison Test, compare with the convergent series $\sum_{k=1}^{\infty} 1/k^{5/2}$, $\rho = \lim_{k\to+\infty} \dfrac{k^3}{k^3+1} = 1$, converges

28. $\dfrac{4}{2+3^k k} < \dfrac{4}{3^k k}$, $\sum_{k=1}^{\infty} \dfrac{4}{3^k k}$ converges (Ratio Test) so $\sum_{k=1}^{\infty} \dfrac{4}{2+k3^k}$ converges by the Comparison Test

29. Limit Comparison Test, compare with the divergent series $\sum_{k=1}^{\infty} 1/k$, $\rho = \lim_{k\to+\infty} \dfrac{k}{\sqrt{k^2+k}} = 1$, diverges

30. $\dfrac{2+(-1)^k}{5^k} \le \dfrac{3}{5^k}$, $\sum_{k=1}^{\infty} 3/5^k$ converges so $\sum_{k=1}^{\infty} \dfrac{2+(-1)^k}{5^k}$ converges

31. Limit Comparison Test, compare with the convergent series $\sum_{k=1}^{\infty} 1/k^{5/2}$,

$\rho = \lim_{k\to+\infty} \dfrac{k^3 + 2k^{5/2}}{k^3+3k^2+3k} = 1$, converges

32. $\dfrac{4+|\cos k|}{k^3} < \dfrac{5}{k^3}$, $\sum_{k=1}^{\infty} 5/k^3$ converges so $\sum_{k=1}^{\infty} \dfrac{4+|\cos k|}{k^3}$ converges

33. Limit Comparison Test, compare with the divergent series $\sum_{k=1}^{\infty} 1/\sqrt{k}$

34. Ratio Test, $\rho = \lim_{k\to+\infty}(1+1/k)^{-k} = 1/e < 1$, converges

35. Ratio Test, $\rho = \lim_{k\to+\infty}\dfrac{\ln(k+1)}{e\ln k} = \lim_{k\to+\infty}\dfrac{k}{e(k+1)} = 1/e < 1$, converges

36. Ratio Test, $\rho = \lim_{k\to+\infty}\dfrac{k+1}{e^{2k+1}} = \lim_{k\to+\infty}\dfrac{1}{2e^{2k+1}} = 0$, converges

37. Ratio Test, $\rho = \lim_{k\to+\infty}\dfrac{k+5}{4(k+1)} = 1/4$, converges

38. Root Test, $\rho = \lim_{k\to+\infty}(\dfrac{k}{k+1})^k = \lim_{k\to+\infty}\dfrac{1}{(1+1/k)^k} = 1/e$, converges

39. diverges because $\lim_{k\to+\infty}\dfrac{1}{4+2^{-k}} = 1/4 \neq 0$

40. $\displaystyle\sum_{k=1}^{\infty}\frac{\sqrt{k}\ln k}{k^3+1} = \sum_{k=2}^{\infty}\frac{\sqrt{k}\ln k}{k^3+1}$ because $\ln 1 = 0$, $\dfrac{\sqrt{k}\ln k}{k^3+1} < \dfrac{k\ln k}{k^3} = \dfrac{\ln k}{k^2}$,

$\displaystyle\int_2^{+\infty}\frac{\ln x}{x^2}dx = \lim_{\ell\to+\infty}\left(-\frac{\ln x}{x}-\frac{1}{x}\right)\Big]_2^{\ell} = \frac{1}{2}(\ln 2+1)$ so $\displaystyle\sum_{k=2}^{\infty}\frac{\ln k}{k^2}$ converges and so does $\displaystyle\sum_{k=1}^{\infty}\frac{\sqrt{k}\ln k}{k^3+1}$.

41. $\dfrac{\tan^{-1}k}{k^2} < \dfrac{\pi/2}{k^2}$, $\displaystyle\sum_{k=1}^{\infty}\frac{\pi/2}{k^2}$ converges so $\displaystyle\sum_{k=1}^{\infty}\frac{\tan^{-1}k}{k^2}$ converges

42. $\dfrac{5^k+k}{k!+3} < \dfrac{5^k+5^k}{k!} = \dfrac{2\left(5^k\right)}{k!}$, $\displaystyle\sum_{k=1}^{\infty}2\left(\frac{5^k}{k!}\right)$ converges (Ratio Test) so $\displaystyle\sum_{k=1}^{\infty}\frac{5^k+k}{k!+3}$ converges

43. Ratio Test, $\rho = \lim_{k\to+\infty}\dfrac{(k+1)^2}{(2k+2)(2k+1)} = 1/4$, converges

44. Ratio Test, $\rho = \lim_{k\to+\infty}\dfrac{2(k+1)^2}{(2k+4)(2k+3)} = 1/2$, converges

45. $u_k = \dfrac{k!}{1\cdot 3\cdot 5\cdots(2k-1)}$, by the Ratio Test $\rho = \lim_{k\to+\infty}\dfrac{k+1}{2k+1} = 1/2$; converges

46. $u_k = \dfrac{1\cdot 3\cdot 5\cdots(2k-1)}{(2k-1)!}$, by the Ratio Test $\rho = \lim_{k\to+\infty}\dfrac{1}{2k} = 0$; converges

47. Root Test: $\rho = \lim_{k\to+\infty}\dfrac{1}{3}(\ln k)^{1/k} = 1/3$, converges

48. Root Test: $\rho = \lim_{k\to+\infty}\dfrac{\pi(k+1)}{k^{1+1/k}} = \lim_{k\to+\infty}\pi\dfrac{k+1}{k} = \pi$, diverges

49. **(b)** $\rho = \lim_{k\to+\infty}\dfrac{\sin(\pi/k)}{\pi/k} = 1$ and $\displaystyle\sum_{k=1}^{\infty}\pi/k$ diverges

50. **(a)** $\cos x \approx 1 - x^2/2, 1-\cos\left(\dfrac{1}{k}\right) \approx \dfrac{1}{2k^2}$ **(b)** $\rho = \lim_{k\to+\infty}\dfrac{1-\cos(1/k)}{1/k^2} = 2$, converges

51. Set $g(x) = \sqrt{x} - \ln x$; $\dfrac{d}{dx}g(x) = \dfrac{1}{2\sqrt{x}} - \dfrac{1}{x} = 0$ when $x = 4$. Since $\lim_{x\to 0+} g(x) = \lim_{x\to+\infty} g(x) = +\infty$ it follows that $g(x)$ has its minimum at $x = 4$, $g(4) = \sqrt{4} - \ln 4 > 0$, and thus $\sqrt{x} - \ln x > 0$ for $x > 0$.

(a) $\dfrac{\ln k}{k^2} < \dfrac{\sqrt{k}}{k^2} = \dfrac{1}{k^{3/2}}$, $\displaystyle\sum_{k=1}^{\infty} \frac{1}{k^{3/2}}$ converges so $\displaystyle\sum_{k=1}^{\infty} \frac{\ln k}{k^2}$ converges.

(b) $\dfrac{1}{(\ln k)^2} > \dfrac{1}{k}$, $\displaystyle\sum_{k=2}^{\infty} \frac{1}{k}$ diverges so $\displaystyle\sum_{k=2}^{\infty} \frac{1}{(\ln k)^2}$ diverges.

52. By the Root Test, $\rho = \lim_{k\to+\infty} \dfrac{\alpha}{(k^{1/k})^{\alpha}} = \dfrac{\alpha}{1^{\alpha}} = \alpha$, the series converges if $\alpha < 1$ and diverges if $\alpha > 1$. If $\alpha = 1$ then the series is $\displaystyle\sum_{k=1}^{\infty} 1/k$ which diverges.

53. (a) If $\sum b_k$ converges, then set $M = \sum b_k$. Then $a_1 + a_2 + \cdots + a_n \le b_1 + b_2 + \cdots + b_n \le M$; apply Theorem 10.5.6 to get convergence of $\sum a_k$.

(b) Assume the contrary, that $\sum b_k$ converges; then use Part (a) of the Theorem to show that $\sum a_k$ converges, a contradiction.

54. (a) If $\lim_{k\to+\infty}(a_k/b_k) = 0$ then for $k \ge K$, $a_k/b_k < 1$, $a_k < b_k$ so $\sum a_k$ converges by the Comparison Test.

(b) If $\lim_{k\to+\infty}(a_k/b_k) = +\infty$ then for $k \ge K$, $a_k/b_k > 1$, $a_k > b_k$ so $\sum a_k$ diverges by the Comparison Test.

EXERCISE SET 10.7

1. $a_{k+1} < a_k$, $\lim_{k\to+\infty} a_k = 0, a_k > 0$

2. $\dfrac{a_{k+1}}{a_k} = \dfrac{k+1}{3k} \le \dfrac{2k}{3k} = \dfrac{2}{3}$ for $k \ge 1$, so $\{a_k\}$ is decreasing and tends to zero.

3. diverges because $\lim_{k\to+\infty} a_k = \lim_{k\to+\infty} \dfrac{k+1}{3k+1} = 1/3 \ne 0$

4. diverges because $\lim_{k\to+\infty} a_k = \lim_{k\to+\infty} \dfrac{k+1}{\sqrt{k}+1} = +\infty \ne 0$

5. $\{e^{-k}\}$ is decreasing and $\lim_{k\to+\infty} e^{-k} = 0$, converges

6. $\left\{\dfrac{\ln k}{k}\right\}$ is decreasing and $\lim_{k\to+\infty} \dfrac{\ln k}{k} = 0$, converges

7. $\rho = \lim_{k\to+\infty} \dfrac{(3/5)^{k+1}}{(3/5)^k} = 3/5$, converges absolutely

8. $\rho = \lim_{k\to+\infty} \dfrac{2}{k+1} = 0$, converges absolutely

9. $\rho = \lim_{k\to+\infty} \frac{3k^2}{(k+1)^2} = 3$, diverges

10. $\rho = \lim_{k\to+\infty} \frac{k+1}{5k} = 1/5$, converges absolutely

11. $\rho = \lim_{k\to+\infty} \frac{(k+1)^3}{ek^3} = 1/e$, converges absolutely

12. $\rho = \lim_{k\to+\infty} \frac{(k+1)^{k+1}k!}{(k+1)!k^k} = \lim_{k\to+\infty} (1+1/k)^k = e$, diverges

13. conditionally convergent, $\sum_{k=1}^{\infty} \frac{(-1)^{k+1}}{3k}$ converges by the Alternating Series Test but $\sum_{k=1}^{\infty} \frac{1}{3k}$ diverges

14. absolutely convergent, $\sum_{k=1}^{\infty} \frac{1}{k^{4/3}}$ converges

15. divergent, $\lim_{k\to+\infty} a_k \neq 0$

16. absolutely convergent, Ratio Test for absolute convergence

17. $\sum_{k=1}^{\infty} \frac{\cos k\pi}{k} = \sum_{k=1}^{\infty} \frac{(-1)^k}{k}$ is conditionally convergent, $\sum_{k=1}^{\infty} \frac{(-1)^k}{k}$ converges by the Alternating Series Test but $\sum_{k=1}^{\infty} 1/k$ diverges.

18. conditionally convergent, $\sum_{k=3}^{\infty} \frac{(-1)^k \ln k}{k}$ converges by the Alternating Series Test but $\sum_{k=3}^{\infty} \frac{\ln k}{k}$ diverges (Limit Comparison Test with $\sum 1/k$).

19. conditionally convergent, $\sum_{k=1}^{\infty} (-1)^{k+1} \frac{k+2}{k(k+3)}$ converges by the Alternating Series Test but $\sum_{k=1}^{\infty} \frac{k+2}{k(k+3)}$ diverges (Limit Comparison Test with $\sum 1/k$)

20. conditionally convergent, $\sum_{k=1}^{\infty} \frac{(-1)^{k+1}k^2}{k^3+1}$ converges by the Alternating Series Test but $\sum_{k=1}^{\infty} \frac{k^2}{k^3+1}$ diverges (Limit Comparison Test with $\sum(1/k)$)

21. $\sum_{k=1}^{\infty} \sin(k\pi/2) = 1+0-1+0+1+0-1+0+\cdots$, divergent ($\lim_{k\to+\infty} \sin(k\pi/2)$ does not exist)

22. absolutely convergent, $\sum_{k=1}^{\infty} \frac{|\sin k|}{k^3}$ converges (compare with $\sum 1/k^3$)

23. conditionally convergent, $\sum_{k=2}^{\infty} \frac{(-1)^k}{k \ln k}$ converges by the Alternating Series Test but $\sum_{k=2}^{\infty} \frac{1}{k \ln k}$ diverges (Integral Test)

24. conditionally convergent, $\sum_{k=1}^{\infty} \frac{(-1)^k}{\sqrt{k(k+1)}}$ converges by the Alternating Series Test but $\sum_{k=1}^{\infty} \frac{1}{\sqrt{k(k+1)}}$ diverges (Limit Comparison Test with $\sum 1/k$)

25. absolutely convergent, $\sum_{k=2}^{\infty} (1/\ln k)^k$ converges by the Root Test

26. conditionally convergent, $\sum_{k=1}^{\infty} \frac{(-1)^{k+1}}{\sqrt{k+1}+\sqrt{k}}$ converges by the Alternating Series Test but $\sum_{k=1}^{\infty} \frac{1}{\sqrt{k+1}+\sqrt{k}}$ diverges (Limit Comparison Test with $\sum 1/\sqrt{k}$)

27. conditionally convergent, let $f(x) = \frac{x^2+1}{x^3+2}$ then $f'(x) = \frac{x(4-3x-x^3)}{(x^3+2)^2} \le 0$ for $x \ge 1$ so $\{a_k\}_{k=2}^{+\infty} = \left\{\frac{k^2+1}{k^3+2}\right\}_{k=2}^{+\infty}$ is decreasing, $\lim_{k\to+\infty} a_k = 0$; the series converges by the Alternating Series Test but $\sum_{k=2}^{\infty} \frac{k^2+1}{k^3+2}$ diverges (Limit Comparison Test with $\sum 1/k$)

28. $\sum_{k=1}^{\infty} \frac{k\cos k\pi}{k^2+1} = \sum_{k=1}^{\infty} \frac{(-1)^k k}{k^2+1}$ is conditionally convergent, $\sum_{k=1}^{\infty} \frac{(-1)^k k}{k^2+1}$ converges by the Alternating Series Test but $\sum_{k=1}^{\infty} \frac{k}{k^2+1}$ diverges

29. absolutely convergent by the Ratio Test, $\rho = \lim_{k\to+\infty} \frac{k+1}{(2k+1)(2k)} = 0$

30. divergent, $\lim_{k\to+\infty} a_k = +\infty$

31. $|\text{error}| < a_8 = 1/8 = 0.125$

32. $|\text{error}| < a_6 = 1/6! < 0.0014$

33. $|\text{error}| < a_{100} = 1/\sqrt{100} = 0.1$

34. $|\text{error}| < a_4 = 1/(5\ln 5) < 0.125$

35. $|\text{error}| < 0.0001$ if $a_{n+1} \le 0.0001$, $1/(n+1) \le 0.0001$, $n+1 \ge 10{,}000$, $n \ge 9{,}999$, $n = 9{,}999$

36. $|\text{error}| < 0.00001$ if $a_{n+1} \le 0.00001$, $1/(n+1)! \le 0.00001$, $(n+1)! \ge 100{,}000$. But $8! = 40{,}320$, $9! = 362{,}880$ so $(n+1)! \ge 100{,}000$ if $n+1 \ge 9$, $n \ge 8$, $n = 8$

37. $|\text{error}| < 0.005$ if $a_{n+1} \le 0.005$, $1/\sqrt{n+1} \le 0.005$, $\sqrt{n+1} \ge 200$, $n+1 \ge 40{,}000$, $n \ge 39{,}999$, $n = 39{,}999$

38. $|\text{error}| < 0.05$ if $a_{n+1} \le 0.05$, $1/[(n+2)\ln(n+2)] \le 0.05$, $(n+2)\ln(n+2) \ge 20$. But $9\ln 9 \approx 19.8$ and $10\ln 10 \approx 23.0$ so $(n+2)\ln(n+2) \ge 20$ if $n+2 \ge 10$, $n \ge 8$, $n = 8$

39. $a_k = \dfrac{3}{2^{k+1}}$, $|\text{error}| < a_{11} = \dfrac{3}{2^{12}} < 0.00074$; $s_{10} \approx 0.4995$; $S = \dfrac{3/4}{1-(-1/2)} = 0.5$

40. $a_k = \left(\dfrac{2}{3}\right)^{k-1}$, $|\text{error}| < a_{11} = \left(\dfrac{2}{3}\right)^{10} < 0.01735$; $s_{10} \approx 0.5896$; $S = \dfrac{1}{1-(-2/3)} = 0.6$

41. $a_k = \dfrac{1}{(2k-1)!}$, $a_{n+1} = \dfrac{1}{(2n+1)!} \le 0.005$, $(2n+1)! \ge 200$, $2n+1 \ge 6$, $n \ge 2.5$; $n = 3$, $s_3 = 1 - 1/6 + 1/120 \approx 0.84$

42. $a_k = \dfrac{1}{(2k-2)!}$, $a_{n+1} = \dfrac{1}{(2n)!} \le 0.005$, $(2n)! \ge 200$, $2n \ge 6$, $n \ge 3$; $n = 3$, $s_3 \approx 0.54$

43. $a_k = \dfrac{1}{k2^k}$, $a_{n+1} = \dfrac{1}{(n+1)2^{n+1}} \le 0.005$, $(n+1)2^{n+1} \ge 200$, $n+1 \ge 6$, $n \ge 5$; $n = 5$, $s_5 \approx 0.41$

44. $a_k = \dfrac{1}{(2k-1)^5 + 4(2k-1)}$, $a_{n+1} = \dfrac{1}{(2n+1)^5 + 4(2n+1)} \le 0.005$,
$(2n+1)^5 + 4(2n+1) \ge 200$, $2n+1 \ge 3$, $n \ge 1$; $n = 1$, $s_1 = 0.20$

45. **(c)** $a_k = \dfrac{1}{2k-1}$, $a_{n+1} = \dfrac{1}{2n+1} \le 10^{-2}$, $2n+1 \ge 100$, $n \ge 49.5$; $n = 50$

46. $\sum(1/k^p)$ converges if $p > 1$ and diverges if $p \le 1$, so $\displaystyle\sum_{k=1}^{\infty}(-1)^k\frac{1}{k^p}$ converges absolutely if $p > 1$, and converges conditionally if $0 < p \le 1$ since it satisfies the Alternating Series Test; it diverges for $p \le 0$ since $\displaystyle\lim_{k\to+\infty} a_k \ne 0$.

47.
$$1+\frac{1}{3^2}+\frac{1}{5^2}+\cdots = \left[1+\frac{1}{2^2}+\frac{1}{3^2}+\cdots\right] - \left[\frac{1}{2^2}+\frac{1}{4^2}+\frac{1}{6^2}+\cdots\right]$$
$$= \frac{\pi^2}{6} - \frac{1}{2^2}\left[1+\frac{1}{2^2}+\frac{1}{3^2}+\cdots\right] = \frac{\pi^2}{6} - \frac{1}{4}\frac{\pi^2}{6} = \frac{\pi^2}{8}$$

48.
$$1+\frac{1}{3^4}+\frac{1}{5^4}+\cdots = \left[1+\frac{1}{2^4}+\frac{1}{3^4}+\cdots\right] - \left[\frac{1}{2^4}+\frac{1}{4^4}+\frac{1}{6^4}+\cdots\right]$$
$$= \frac{\pi^4}{90} - \frac{1}{2^4}\left[1+\frac{1}{2^4}+\frac{1}{3^4}+\cdots\right] = \frac{\pi^4}{90} - \frac{1}{16}\frac{\pi^4}{90} = \frac{\pi^4}{96}$$

49. Every positive integer can be written in exactly one of the three forms $2k-1$ or $4k-2$ or $4k$, so a rearrangement is
$$\left(1-\frac{1}{2}-\frac{1}{4}\right)+\left(\frac{1}{3}-\frac{1}{6}-\frac{1}{8}\right)+\left(\frac{1}{5}-\frac{1}{10}-\frac{1}{12}\right)+\cdots+\left(\frac{1}{2k-1}-\frac{1}{4k-2}-\frac{1}{4k}\right)+\cdots$$
$$=\left(\frac{1}{2}-\frac{1}{4}\right)+\left(\frac{1}{6}-\frac{1}{8}\right)+\left(\frac{1}{10}-\frac{1}{12}\right)+\cdots+\left(\frac{1}{4k-2}-\frac{1}{4k}\right)+\cdots = \frac{1}{2}\ln 2$$

50. (a)

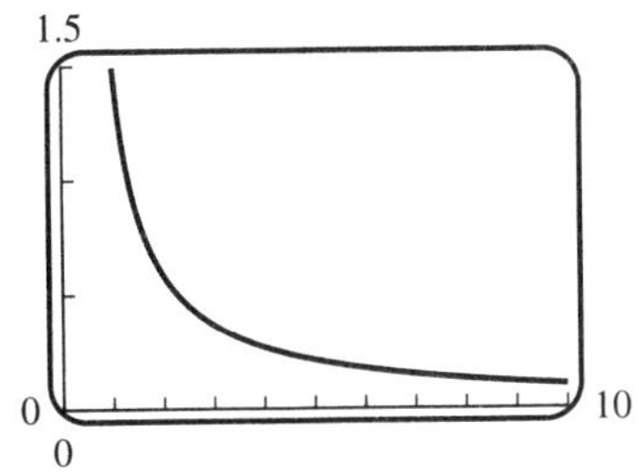

(b) Yes; since $f(x)$ is decreasing for $x \geq 1$ and $\lim_{x\to+\infty} f(x) = 0$, the series satisfies the Alternating Series Test.

51. (a) The distance d from the starting point is

$$d = 180 - \frac{180}{2} + \frac{180}{3} - \cdots - \frac{180}{1000} = 180\left[1 - \frac{1}{2} + \frac{1}{3} - \cdots - \frac{1}{1000}\right].$$

From Theorem 10.7.2, $1 - \frac{1}{2} + \frac{1}{3} - \cdots - \frac{1}{1000}$ differs from $\ln 2$ by less than $1/1001$ so $180(\ln 2 - 1/1001) < d < 180 \ln 2$, $124.58 < d < 124.77$.

(b) The total distance traveled is $s = 180 + \frac{180}{2} + \frac{180}{3} + \cdots + \frac{180}{1000}$, and from inequality (2) in Section 10.5,

$$\int_1^{1001} \frac{180}{x}dx < s < 180 + \int_1^{1000} \frac{180}{x}dx$$

$$180 \ln 1001 < s < 180(1 + \ln 1000)$$

$$1243 < s < 1424$$

52. (a) Suppose $\Sigma|a_k|$ converges, then $\lim_{k\to+\infty} |a_k| = 0$ so $|a_k| < 1$ for $k \geq K$ and thus $|a_k|^2 < |a_k|$, $a_k^2 < |a_k|$ hence Σa_k^2 converges by the Comparison Test.

(b) Let $a_k = \frac{1}{k}$, then $\sum a_k^2$ converges but $\sum a_k$ diverges.

EXERCISE SET 10.8

1. $f^{(k)}(x) = (-1)^k e^{-x}$, $f^{(k)}(0) = (-1)^k$; $\sum_{k=0}^{\infty} \frac{(-1)^k}{k!}x^k$

2. $f^{(k)}(x) = a^k e^{ax}$, $f^{(k)}(0) = a^k$; $\sum_{k=0}^{\infty} \frac{a^k}{k!}x^k$

3. $f^{(k)}(0) = 0$ if k is odd, $f^{(k)}(0)$ is alternately π^k and $-\pi^k$ if k is even; $\sum_{k=0}^{\infty} \frac{(-1)^k \pi^{2k}}{(2k)!}x^{2k}$

4. $f^{(k)}(0) = 0$ if k is even, $f^{(k)}(0)$ is alternately π^k and $-\pi^k$ if k is odd; $\sum_{k=0}^{\infty} \frac{(-1)^k \pi^{2k+1}}{(2k+1)!}x^{2k+1}$

5. $f^{(0)}(0) = 0$; for $k \geq 1$, $f^{(k)}(x) = \frac{(-1)^{k+1}(k-1)!}{(1+x)^k}$, $f^{(k)}(0) = (-1)^{k+1}(k-1)!$; $\sum_{k=1}^{\infty} \frac{(-1)^{k+1}}{k}x^k$

6. $f^{(k)}(x) = (-1)^k \dfrac{k!}{(1+x)^{k+1}}$; $f^{(k)}(0) = (-1)^k k!$; $\sum_{k=0}^{\infty} (-1)^k x^k$

7. $f^{(k)}(0) = 0$ if k is odd, $f^{(k)}(0) = 1$ if k is even; $\sum_{k=0}^{\infty} \dfrac{1}{(2k)!} x^{2k}$

8. $f^{(k)}(0) = 0$ if k is even, $f^{(k)}(0) = 1$ if k is odd; $\sum_{k=0}^{\infty} \dfrac{1}{(2k+1)!} x^{2k+1}$

9. $f^{(k)}(x) = \begin{cases} (-1)^{k/2}(x \sin x - k \cos x) & k \text{ even} \\ (-1)^{(k-1)/2}(x \cos x + k \sin x) & k \text{ odd} \end{cases}$, $f^{(k)}(0) = \begin{cases} (-1)^{1+k/2} k & k \text{ even} \\ 0 & k \text{ odd} \end{cases}$

$\sum_{k=0}^{\infty} \dfrac{(-1)^k}{(2k+1)!} x^{2k+2}$

10. $f^{(k)}(x) = (k+x)e^x$, $f^{(k)}(0) = k$; $\sum_{k=1}^{\infty} \dfrac{1}{(k-1)!} x^k$

11. $f^{(k)}(x_0) = e$; $\sum_{k=0}^{\infty} \dfrac{e}{k!} (x-1)^k$

12. $f^{(k)}(x) = (-1)^k e^{-x}$, $f^{(k)}(\ln 2) = (-1)^k \dfrac{1}{2}$; $\sum_{k=0}^{\infty} \dfrac{(-1)^k}{2 \cdot k!} (x - \ln 2)^k$

13. $f^{(k)}(x) = \dfrac{(-1)^k k!}{x^{k+1}}$, $f^{(k)}(-1) = -k!$; $\sum_{k=0}^{\infty} (-1)(x+1)^k$

14. $f^{(k)}(x) = \dfrac{(-1)^k k!}{(x+2)^{k+1}}$, $f^{(k)}(3) = \dfrac{(-1)^k k!}{5^{k+1}}$; $\sum_{k=0}^{\infty} \dfrac{(-1)^k}{5^{k+1}} (x-3)^k$

15. $f^{(k)}(1/2) = 0$ if k is odd, $f^{(k)}(1/2)$ is alternately π^k and $-\pi^k$ if k is even;

$\sum_{k=0}^{\infty} \dfrac{(-1)^k \pi^{2k}}{(2k)!} (x - 1/2)^{2k}$

16. $f^{(k)}(\pi/2) = 0$ if k is even, $f^{(k)}(\pi/2)$ is alternately -1 and 1 if k is odd; $\sum_{k=0}^{\infty} \dfrac{(-1)^{k+1}}{(2k+1)!} (x - \pi/2)^{2k+1}$

17. $f(1) = 0$, for $k \geq 1$, $f^{(k)}(x) = \dfrac{(-1)^{k-1}(k-1)!}{x^k}$; $f^{(k)}(1) = (-1)^{k-1}(k-1)!$;

$\sum_{k=1}^{\infty} \dfrac{(-1)^{k-1}}{k} (x-1)^k$

18. $f(e) = 1$, for $k \geq 1$, $f^{(k)}(x) = \dfrac{(-1)^{k-1}(k-1)!}{x^k}$; $f^{(k)}(e) = \dfrac{(-1)^{k-1}(k-1)!}{e^k}$;

$1 + \sum_{k=1}^{\infty} \dfrac{(-1)^{k-1}}{k e^k} (x-e)^k$

19. geometric series, $\rho = \lim_{k\to+\infty}\left|\frac{u_{k+1}}{u_k}\right| = |x|$, so the interval of convergence is $-1 < x < 1$, converges there to $\frac{1}{1+x}$ (the series diverges for $x = \pm 1$)

20. geometric series, $\rho = \lim_{k\to+\infty}\left|\frac{u_{k+1}}{u_k}\right| = |x|^2$, so the interval of convergence is $-1 < x < 1$, converges there to $\frac{1}{1-x^2}$ (the series diverges for $x = \pm 1$)

21. geometric series, $\rho = \lim_{k\to+\infty}\left|\frac{u_{k+1}}{u_k}\right| = |x-2|$, so the interval of convergence is $1 < x < 3$, converges there to $\frac{1}{1-(x-2)} = \frac{1}{3-x}$ (the series diverges for $x = 1, 3$)

22. geometric series, $\rho = \lim_{k\to+\infty}\left|\frac{u_{k+1}}{u_k}\right| = |x+3|$, so the interval of convergence is $-4 < x < -2$, converges there to $\frac{1}{1+(x+3)} = \frac{1}{4+x}$ (the series diverges for $x = -4, -2$)

23. **(a)** geometric series, $\rho = \lim_{k\to+\infty}\left|\frac{u_{k+1}}{u_k}\right| = |x/2|$, so the interval of convergence is $-2 < x < 2$, converges there to $\frac{1}{1+x/2} = \frac{2}{2+x}$; (the series diverges for $x = -2, 2$)

(b) $f(0) = 1$; $f(1) = 2/3$

24. **(a)** geometric series, $\rho = \lim_{k\to+\infty}\left|\frac{u_{k+1}}{u_k}\right| = \left|\frac{x-5}{3}\right|$, so the interval of convergence is $2 < x < 8$, converges to $\frac{1}{1+(x-5)/3} = \frac{3}{x-2}$ (the series diverges for $x = 2, 8$)

(b) $f(3) = 3$, $f(6) = 3/4$

25. $\rho = \lim_{k\to+\infty}\frac{k+1}{k+2}|x| = |x|$, the series converges if $|x| < 1$ and diverges if $|x| > 1$. If $x = -1$, $\sum_{k=0}^{\infty}\frac{(-1)^k}{k+1}$ converges by the Alternating Series Test; if $x = 1$, $\sum_{k=0}^{\infty}\frac{1}{k+1}$ diverges. The radius of convergence is 1, the interval of convergence is $[-1, 1)$.

26. $\rho = \lim_{k\to+\infty} 3|x| = 3|x|$, the series converges if $3|x| < 1$ or $|x| < 1/3$ and diverges if $|x| > 1/3$. If $x = -1/3$, $\sum_{k=0}^{\infty}(-1)^k$ diverges, if $x = 1/3$, $\sum_{k=0}^{\infty}(1)$ diverges. The radius of convergence is 1/3, the interval of convergence is $(-1/3, 1/3)$.

27. $\rho = \lim_{k\to+\infty}\frac{|x|}{k+1} = 0$, the radius of convergence is $+\infty$, the interval is $(-\infty, +\infty)$.

28. $\rho = \lim_{k\to+\infty} \frac{k+1}{2}|x| = +\infty$, the radius of convergence is 0, the series converges only if $x = 0$.

29. $\rho = \lim_{k\to+\infty} \frac{5k^2|x|}{(k+1)^2} = 5|x|$, converges if $|x| < 1/5$ and diverges if $|x| > 1/5$. If $x = -1/5$, $\sum_{k=1}^{\infty} \frac{(-1)^k}{k^2}$ converges; if $x = 1/5$, $\sum_{k=1}^{\infty} 1/k^2$ converges. Radius of convergence is 1/5, interval of convergence is $[-1/5, 1/5]$.

30. $\rho = \lim_{k\to+\infty} \frac{\ln k}{\ln(k+1)}|x| = |x|$, the series converges if $|x| < 1$ and diverges if $|x| > 1$. If $x = -1$, $\sum_{k=2}^{\infty} \frac{(-1)^k}{\ln k}$ converges; if $x = 1$, $\sum_{k=2}^{\infty} 1/(\ln k)$ diverges (compare to $\sum(1/k)$). Radius of convergence is 1, interval of convergence is $[-1, 1)$.

31. $\rho = \lim_{k\to+\infty} \frac{k|x|}{k+2} = |x|$, converges if $|x| < 1$, diverges if $|x| > 1$. If $x = -1$, $\sum_{k=1}^{\infty} \frac{(-1)^k}{k(k+1)}$ converges; if $x = 1$, $\sum_{k=1}^{\infty} \frac{1}{k(k+1)}$ converges. Radius of convergence is 1, interval of convergence is $[-1, 1]$.

32. $\rho = \lim_{k\to+\infty} 2\frac{k+1}{k+2}|x| = 2|x|$, converges if $|x| < 1/2$, diverges if $|x| > 1/2$. If $x = -1/2$, $\sum_{k=0}^{\infty} \frac{-1}{2(k+1)}$ diverges; if $x = 1/2$, $\sum_{k=0}^{\infty} \frac{(-1)^k}{2(k+1)}$ converges. Radius of convergence is 1/2, interval of convergence is $(-1/2, 1/2]$.

33. $\rho = \lim_{k\to+\infty} \frac{\sqrt{k}}{\sqrt{k+1}}|x| = |x|$, converges if $|x| < 1$, diverges if $|x| > 1$. If $x = -1$, $\sum_{k=1}^{\infty} \frac{-1}{\sqrt{k}}$ diverges; if $x = 1$, $\sum_{k=1}^{\infty} \frac{(-1)^{k-1}}{\sqrt{k}}$ converges. Radius of convergence is 1, interval of convergence is $(-1, 1]$.

34. $\rho = \lim_{k\to+\infty} \frac{|x|^2}{(2k+2)(2k+1)} = 0$, radius of convergence is $+\infty$, interval of convergence is $(-\infty, +\infty)$.

35. $\rho = \lim_{k\to+\infty} \frac{|x|^2}{(2k+3)(2k+2)} = 0$, radius of convergence is $+\infty$, interval of convergence is $(-\infty, +\infty)$.

36. $\rho = \lim_{k\to+\infty} \frac{k^{3/2}|x|^3}{(k+1)^{3/2}} = |x|^3$, converges if $|x| < 1$, diverges if $|x| > 1$. If $x = -1$, $\sum_{k=0}^{\infty} \frac{1}{k^{3/2}}$ converges; if $x = 1$, $\sum_{k=0}^{\infty} \frac{(-1)^k}{k^{3/2}}$ converges. Radius of convergence is 1, interval of convergence is $[-1, 1]$.

37. $\rho = \lim_{k\to+\infty} \frac{3|x|}{k+1} = 0$, radius of convergence is $+\infty$, interval of convergence is $(-\infty, +\infty)$.

38. $\rho = \lim_{k\to+\infty} \dfrac{k(\ln k)^2|x|}{(k+1)[\ln(k+1)]^2} = |x|$, converges if $|x| < 1$, diverges if $|x| > 1$. If $x = -1$, then, by Exercise 10.5.25, $\sum_{k=2}^{\infty} \dfrac{-1}{k(\ln k)^2}$ converges; if $x = 1$, $\sum_{k=2}^{\infty} \dfrac{(-1)^{k+1}}{k(\ln k)^2}$ converges. Radius of convergence is 1, interval of convergence is $[-1, 1]$.

39. $\rho = \lim_{k\to+\infty} \dfrac{1+k^2}{1+(k+1)^2}|x| = |x|$, converges if $|x| < 1$, diverges if $|x| > 1$. If $x = -1$, $\sum_{k=0}^{\infty} \dfrac{(-1)^k}{1+k^2}$ converges; if $x = 1$, $\sum_{k=0}^{\infty} \dfrac{1}{1+k^2}$ converges. Radius of convergence is 1, interval of convergence is $[-1, 1]$.

40. $\rho = \lim_{k\to+\infty} \dfrac{1}{2}|x-3| = \dfrac{1}{2}|x-3|$, converges if $|x-3| < 2$, diverges if $|x-3| > 2$. If $x = 1$, $\sum_{k=0}^{\infty}(-1)^k$ diverges; if $x = 5$, $\sum_{k=0}^{\infty} 1$ diverges. Radius of convergence is 2, interval of convergence is $(1, 5)$.

41. $\rho = \lim_{k\to+\infty} \dfrac{k|x+1|}{k+1} = |x+1|$, converges if $|x+1| < 1$, diverges if $|x+1| > 1$. If $x = -2$, $\sum_{k=1}^{\infty} \dfrac{-1}{k}$ diverges; if $x = 0$, $\sum_{k=1}^{\infty} \dfrac{(-1)^{k+1}}{k}$ converges. Radius of convergence is 1, interval of convergence is $(-2, 0]$.

42. $\rho = \lim_{k\to+\infty} \dfrac{(k+1)^2}{(k+2)^2}|x-4| = |x-4|$, converges if $|x-4| < 1$, diverges if $|x-4| > 1$. If $x = 3$, $\sum_{k=0}^{\infty} 1/(k+1)^2$ converges; if $x = 5$, $\sum_{k=0}^{\infty}(-1)^k/(k+1)^2$ converges. Radius of convergence is 1, interval of convergence is $[3, 5]$.

43. $\rho = \lim_{k\to+\infty} (3/4)|x+5| = \dfrac{3}{4}|x+5|$, converges if $|x+5| < 4/3$, diverges if $|x+5| > 4/3$. If $x = -19/3$, $\sum_{k=0}^{\infty}(-1)^k$ diverges; if $x = -11/3$, $\sum_{k=0}^{\infty} 1$ diverges. Radius of convergence is $4/3$, interval of convergence is $(-19/3, -11/3)$.

44. $\rho = \lim_{k\to+\infty} \dfrac{(2k+3)(2k+2)k^3}{(k+1)^3}|x-2| = +\infty$, radius of convergence is 0, series converges only at $x = 2$.

45. $\rho = \lim_{k\to+\infty} \dfrac{k^2+4}{(k+1)^2+4}|x+1|^2 = |x+1|^2$, converges if $|x+1| < 1$, diverges if $|x+1| > 1$. If $x = -2$, $\sum_{k=1}^{\infty} \dfrac{(-1)^{3k+1}}{k^2+4}$ converges; if $x = 0$, $\sum_{k=1}^{\infty} \dfrac{(-1)^k}{k^2+4}$ converges. Radius of convergence is 1, interval of convergence is $[-2, 0]$.

46. $\rho = \lim_{k\to+\infty} \dfrac{k\ln(k+1)}{(k+1)\ln k}|x-3| = |x-3|$, converges if $|x-3|<1$, diverges if $|x-3|>1$. If $x=2$, $\displaystyle\sum_{k=1}^{\infty}\frac{(-1)^k\ln k}{k}$ converges; if $x=4$, $\displaystyle\sum_{k=1}^{\infty}\frac{\ln k}{k}$ diverges. Radius of convergence is 1, interval of convergence is $[2,4)$.

47. $\rho = \lim_{k\to+\infty} \dfrac{\pi|x-1|^2}{(2k+3)(2k+2)} = 0$, radius of convergence $+\infty$, interval of convergence $(-\infty,+\infty)$.

48. $\rho = \lim_{k\to+\infty} \dfrac{1}{16}|2x-3| = \dfrac{1}{16}|2x-3|$, converges if $\dfrac{1}{16}|2x-3|<1$ or $|x-3/2|<8$, diverges if $|x-3/2|>8$. If $x=-13/2$, $\displaystyle\sum_{k=0}^{\infty}(-1)^k$ diverges; if $x=19/2$, $\displaystyle\sum_{k=0}^{\infty}1$ diverges. Radius of convergence is 8, interval of convergence is $(-13/2, 19/2)$.

49. $\rho = \lim_{k\to+\infty}\sqrt[k]{|u_k|} = \lim_{k\to+\infty}\dfrac{|x|}{\ln k} = 0$, the series converges absolutely for all x so the interval of convergence is $(-\infty,+\infty)$.

50. $\rho = \lim_{k\to+\infty}\dfrac{2k+1}{(2k)(2k-1)}|x| = 0$ so $R=+\infty$.

51. (a)

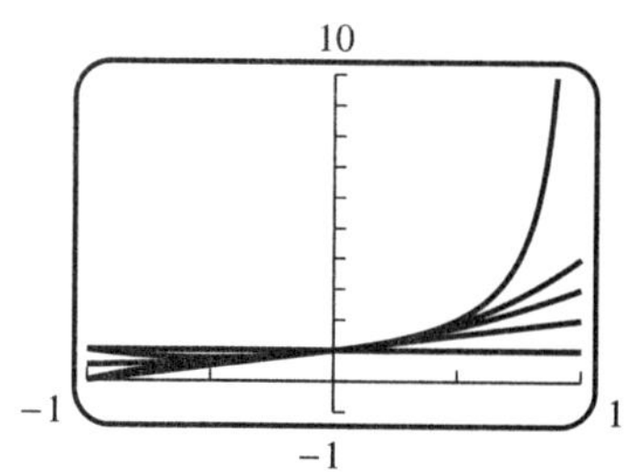

52. Ratio Test: $\rho = \lim_{k\to+\infty}\dfrac{|x|^2}{4(k+1)(k+2)} = 0$, $R=+\infty$

53. By the Ratio Test for absolute convergence,

$$\rho = \lim_{k\to+\infty}\frac{(pk+p)!(k!)^p}{(pk)![(k+1)!]^p}|x| = \lim_{k\to+\infty}\frac{(pk+p)(pk+p-1)(pk+p-2)\cdots(pk+p-[p-1])}{(k+1)^p}|x|$$

$$= \lim_{k\to+\infty} p\left(p-\frac{1}{k+1}\right)\left(p-\frac{2}{k+1}\right)\cdots\left(p-\frac{p-1}{k+1}\right)|x| = p^p|x|,$$

converges if $|x|<1/p^p$, diverges if $|x|>1/p^p$. Radius of convergence is $1/p^p$.

54. By the Ratio Test for absolute convergence,

$$\rho = \lim_{k\to+\infty}\frac{(k+1+p)!k!(k+q)!}{(k+p)!(k+1)!(k+1+q)!}|x| = \lim_{k\to+\infty}\frac{k+1+p}{(k+1)(k+1+q)}|x| = 0,$$

radius of convergence is $+\infty$.

55. (a) By Theorem 10.5.3(b) both series converge or diverge together, so they have the same radius of convergence.

(b) By Theorem 10.5.3(a) the series $\sum(c_k+d_k)(x-x_0)^k$ converges if $|x-x_0|<R$; if $|x-x_0|>R$ then $\sum(c_k+d_k)(x-x_0)^k$ cannot converge, as otherwise $\sum c_k(x-x_0)^k$ would converge by the same Theorem. Hence the radius of convergence of $\sum(c_k+d_k)(x-x_0)^k$ is R.

(c) Let r be the radius of convergence of $\sum(c_k+d_k)(x-x_0)^k$. If $|x-x_0|<\min(R_1,R_2)$ then $\sum c_k(x-x_0)^k$ and $\sum d_k(x-x_0)^k$ converge, so $\sum(c_k+d_k)(x-x_0)^k$ converges. Hence $r\geq\min(R_1,R_2)$ (to see that $r>\min(R_1,R_2)$ is possible consider the case $c_k=-d_k=1$). If in addition $R_1\neq R_2$, and $R_1<|x-x_0|<R_2$ (or $R_2<|x-x_0|<R_1$) then $\sum(c_k+d_k)(x-x_0)^k$ cannot converge, as otherwise all three series would converge. Thus in this case $r=\min(R_1,R_2)$.

56. By the Root Test for absolute convergence,
$\rho=\lim_{k\to+\infty}|c_k|^{1/k}|x|=L|x|, L|x|<1$ if $|x|<1/L$ so the radius of convergence is $1/L$.

57. By assumption $\sum_{k=0}^{\infty}c_kx^k$ converges if $|x|<R$ so $\sum_{k=0}^{\infty}c_kx^{2k}=\sum_{k=0}^{\infty}c_k(x^2)^k$ converges if $|x^2|<R$, $|x|<\sqrt{R}$. Moreover, $\sum_{k=0}^{\infty}c_kx^{2k}=\sum_{k=0}^{\infty}c_k(x^2)^k$ diverges if $|x^2|>R$, $|x|>\sqrt{R}$. Thus $\sum_{k=0}^{\infty}c_kx^{2k}$ has radius of convergence $\sqrt{R}$.

58. The assumption is that $\sum_{k=0}^{\infty}c_kR^k$ is convergent and $\sum_{k=0}^{\infty}c_k(-R)^k$ is divergent. Suppose that $\sum_{k=0}^{\infty}c_kR^k$ is absolutely convergent then $\sum_{k=0}^{\infty}c_k(-R)^k$ is also absolutely convergent and hence convergent because $|c_kR^k|=|c_k(-R)^k|$, which contradicts the assumption that $\sum_{k=0}^{\infty}c_k(-R)^k$ is divergent so $\sum_{k=0}^{\infty}c_kR^k$ must be conditionally convergent.

EXERCISE SET 10.9

1. $\sin 4^\circ=\sin\left(\dfrac{\pi}{45}\right)=\dfrac{\pi}{45}-\dfrac{(\pi/45)^3}{3!}+\dfrac{(\pi/45)^5}{5!}-\cdots$

(a) Method 1: $|R_n(\pi/45)|\leq\dfrac{(\pi/45)^{n+1}}{(n+1)!}<0.000005$ for $n+1=4, n=3$;

$\sin 4^\circ\approx\dfrac{\pi}{45}-\dfrac{(\pi/45)^3}{3!}\approx 0.069756$

(b) Method 2: The first term in the alternating series that is less than 0.000005 is $\dfrac{(\pi/45)^5}{5!}$, so the result is the same as in Part (a).

2. $\cos 3^\circ=\cos\left(\dfrac{\pi}{60}\right)=1-\dfrac{(\pi/60)^2}{2}+\dfrac{(\pi/60)^4}{4!}-\cdots$

(a) Method 1: $|R_n(\pi/60)|\leq\dfrac{(\pi/60)^{n+1}}{(n+1)!}<0.0005$ for $n=2$; $\cos 3^\circ\approx 1-\dfrac{(\pi/60)^2}{2}\approx 0.9986$.

(b) Method 2: The first term in the alternating series that is less than 0.0005 is $\dfrac{(\pi/60)^4}{4!}$, so the result is the same as in Part (a).

3. $|R_n(0.1)| \le \dfrac{(0.1)^{n+1}}{(n+1)!} \le 0.000005$ for $n = 3$; $\cos 0.1 \approx 1 - (0.1)^2/2 = 0.99500$, calculator value $0.995004\ldots$

4. $(0.1)^3/3 < 0.5 \times 10^{-3}$ so $\tan^{-1}(0.1) \approx 0.100$, calculator value ≈ 0.0997

5. Expand about $\pi/2$ to get $\sin x = 1 - \dfrac{1}{2!}(x - \pi/2)^2 + \dfrac{1}{4!}(x - \pi/2)^4 - \cdots$, $85^\circ = 17\pi/36$ radians,

$|R_n(x)| \le \dfrac{|x - \pi/2|^{n+1}}{(n+1)!}$, $|R_n(17\pi/36)| \le \dfrac{|17\pi/36 - \pi/2|^{n+1}}{(n+1)!} = \dfrac{(\pi/36)^{n+1}}{(n+1)!} < 0.5 \times 10^{-4}$

if $n = 3$, $\sin 85^\circ \approx 1 - \dfrac{1}{2}(-\pi/36)^2 \approx 0.99619$, calculator value $0.99619\ldots$

6. $-175^\circ = -\pi + \pi/36$ rad; $x_0 = -\pi, x = -\pi + \pi/36$, $\cos x = -1 + \dfrac{(x+\pi)^2}{2} - \dfrac{(x+\pi)^4}{4!} - \cdots$;

$|R_n| \le \dfrac{(\pi/36)^{n+1}}{(n+1)!} \le 0.00005$ for $n = 3$; $\cos(-\pi + \pi/36) = -1 + \dfrac{(\pi/36)^2}{2} \approx -0.99619$,

calculator value $-0.99619\ldots$

7. $f^{(k)}(x) = \cosh x$ or $\sinh x, |f^{(k)}(x)| \le \cosh x \le \cosh 0.5 = \dfrac{1}{2}\left(e^{0.5} + e^{-0.5}\right) < \dfrac{1}{2}(2+1) = 1.5$

so $|R_n(x)| < \dfrac{1.5(0.5)^{n+1}}{(n+1)!} \le 0.5 \times 10^{-3}$ if $n = 4$, $\sinh 0.5 \approx 0.5 + \dfrac{(0.5)^3}{3!} \approx 0.5208$, calculator

value $0.52109\ldots$

8. $f^{(k)}(x) = \cosh x$ or $\sinh x, |f^{(k)}(x)| \le \cosh x \le \cosh 0.1 = \dfrac{1}{2}\left(e^{0.1} + e^{-0.1}\right) < 1.06$ so

$|R_n(x)| < \dfrac{1.06(0.1)^{n+1}}{(n+1)!} \le 0.5 \times 10^{-3}$ for $n = 2$, $\cosh 0.1 \approx 1 + \dfrac{(0.1)^2}{2!} = 1.005$, calculator value $1.0050\ldots$

9. $f(x) = \sin x$, $f^{(n+1)}(x) = \pm \sin x$ or $\pm \cos x$, $|f^{(n+1)}(x)| \le 1$, $|R_n(x)| \le \dfrac{|x - \pi/4|^{n+1}}{(n+1)!}$,

$\lim\limits_{n\to+\infty} \dfrac{|x - \pi/4|^{n+1}}{(n+1)!} = 0$; by the Squeezing Theorem, $\lim\limits_{n\to+\infty} |R_n(x)| = 0$

so $\lim\limits_{n\to+\infty} R_n(x) = 0$ for all x.

10. $f(x) = e^x$, $f^{(n+1)}(x) = e^x$; if $x > 1$ then $|R_n(x)| \le \dfrac{e^x}{(n+1)!}|x - 1|^{n+1}$; if $x < 1$ then

$|R_n(x)| \le \dfrac{e}{(n+1)!}|x - 1|^{n+1}$. But $\lim\limits_{n\to+\infty} \dfrac{|x - 1|^{n+1}}{(n+1)!} = 0$ so $\lim\limits_{n\to+\infty} R_n(x) = 0$.

11. **(a)** Let $x = 1/9$ in series (13).

(b) $\ln 1.25 \approx 2\left(1/9 + \dfrac{(1/9)^3}{3}\right) = 2(1/9 + 1/3^7) \approx 0.223$, which agrees with the calculator value $0.22314\ldots$ to three decimal places.

12. **(a)** Let $x = 1/2$ in series (13).

(b) $\ln 3 \approx 2\left(1/2 + \dfrac{(1/2)^3}{3}\right) = 2(1/2 + 1/24) = 13/12 \approx 1.083$; the calculator value is 1.099 to three decimal places.

13. **(a)** $(1/2)^9/9 < 0.5 \times 10^{-3}$ and $(1/3)^7/7 < 0.5 \times 10^{-3}$ so

$$\tan^{-1}(1/2) \approx 1/2 - \frac{(1/2)^3}{3} + \frac{(1/2)^5}{5} - \frac{(1/2)^7}{7} \approx 0.4635$$

$$\tan^{-1}(1/3) \approx 1/3 - \frac{(1/3)^3}{3} + \frac{(1/3)^5}{5} \approx 0.3218$$

(b) From Formula (17), $\pi \approx 4(0.4635 + 0.3218) = 3.1412$

(c) Let $a = \tan^{-1}\dfrac{1}{2}, b = \tan^{-1}\dfrac{1}{3}$; then $|a - 0.4635| < 0.0005$ and $|b - 0.3218| < 0.0005$, so $|4(a+b) - 3.1412| \le 4|a - 0.4635| + 4|b - 0.3218| < 0.004$, so two decimal-place accuracy is guaranteed, but not three.

14. $(27+x)^{1/3} = 3(1+x/3^3)^{1/3} = 3\left(1 + \dfrac{1}{3^4}x - \dfrac{1\cdot 2}{3^8 2}x^2 + \dfrac{1\cdot 2\cdot 5}{3^{12}3!}x^3 + \ldots\right)$, alternates after first term,

$\dfrac{3\cdot 2}{3^8 2} < 0.0005$, $\sqrt{28} \approx 3\left(1 + \dfrac{1}{3^4}\right) \approx 3.0370$

15. **(a)** $\cos x = 1 - \dfrac{x^2}{2!} + \dfrac{x^4}{4!} + (0)x^5 + R_5(x)$,

$|R_5(x)| \le \dfrac{|x|^6}{6!} \le \dfrac{(0.2)^6}{6!} < 9 \times 10^{-8}$

(b)

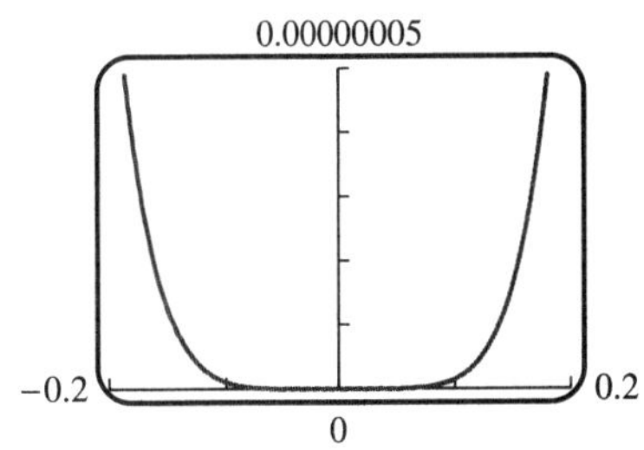

16. **(a)** $f''(x) = -1/(1+x)^2$,

$|f''(x)| < 1/(0.99)^2 \le 1.03$,

$|R_1(x)| \le \dfrac{1.03|x|^2}{2} \le \dfrac{1.03(0.01)^2}{2}$

$\le 5.15 \times 10^{-5}$ for $-0.01 \le x \le 0.01$

(b)

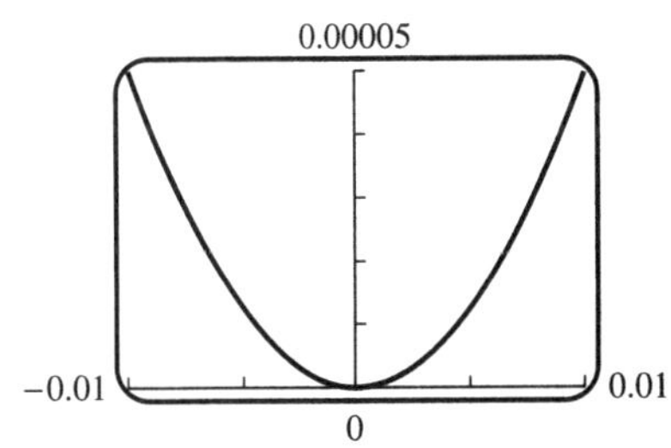

17. **(a)** $(1+x)^{-1} = 1 - x + \dfrac{-1(-2)}{2!}x^2 + \dfrac{-1(-2)(-3)}{3!}x^3 + \cdots + \dfrac{-1(-2)(-3)\cdots(-k)}{k!}x^k + \cdots$

$$= \sum_{k=0}^{\infty}(-1)^k x^k$$

(b) $(1+x)^{1/3} = 1 + (1/3)x + \dfrac{(1/3)(-2/3)}{2!}x^2 + \dfrac{(1/3)(-2/3)(-5/3)}{3!}x^3 + \cdots$

$$+ \frac{(1/3)(-2/3)\cdots(4-3k)/3}{k!}x^k + \cdots = 1 + x/3 + \sum_{k=2}^{\infty}(-1)^{k-1}\frac{2\cdot 5\cdots(3k-4)}{3^k k!}x^k$$

(c) $(1+x)^{-3} = 1 - 3x + \dfrac{(-3)(-4)}{2!}x^2 + \dfrac{(-3)(-4)(-5)}{3!}x^3 + \cdots + \dfrac{(-3)(-4)\cdots(-2-k)}{k!}x^k + \cdots$

$$= \sum_{k=0}^{\infty}(-1)^k\frac{(k+2)!}{2\cdot k!}x^k = \sum_{k=0}^{\infty}(-1)^k\frac{(k+2)(k+1)}{2}x^k$$

18. $(1+x)^m = \binom{m}{0} + \sum_{k=1}^{\infty}\binom{m}{k}x^k = \sum_{k=0}^{\infty}\binom{m}{k}x^k$

19. **(a)** $\dfrac{d}{dx}\ln(1+x) = \dfrac{1}{1+x}, \dfrac{d^k}{dx^k}\ln(1+x) = (-1)^{k-1}\dfrac{(k-1)!}{(1+x)^k}$; similarly $\dfrac{d}{dx}\ln(1-x) = -\dfrac{(k-1)!}{(1-x)^k}$,

so $f^{(n+1)}(x) = n!\left[\dfrac{(-1)^n}{(1+x)^{n+1}} + \dfrac{1}{(1-x)^{n+1}}\right]$.

(b) $\left|f^{(n+1)}(x)\right| \le n!\left|\dfrac{(-1)^n}{(1+x)^{n+1}}\right| + n!\left|\dfrac{1}{(1-x)^{n+1}}\right| = n!\left[\dfrac{1}{(1+x)^{n+1}} + \dfrac{1}{(1-x)^{n+1}}\right]$

(c) If $\left|f^{(n+1)}(x)\right| \le M$ on the interval $[0, 1/3]$ then $|R_n(1/3)| \le \dfrac{M}{(n+1)!}\left(\dfrac{1}{3}\right)^{n+1}$.

(d) If $0 \le x \le 1/3$ then $1+x \ge 1, 1-x \ge 2/3$, $\left|f^{(n+1)}(x)\right| \le M = n!\left[1 + \dfrac{1}{(2/3)^{n+1}}\right]$.

(e) $0.000005 \ge \dfrac{M}{(n+1)!}\left(\dfrac{1}{3}\right)^{n+1} = \dfrac{1}{n+1}\left[\left(\dfrac{1}{3}\right)^{n+1} + \dfrac{(1/3)^{n+1}}{(2/3)^{n+1}}\right] = \dfrac{1}{n+1}\left[\left(\dfrac{1}{3}\right)^{n+1} + \left(\dfrac{1}{2}\right)^{n+1}\right]$

20. Set $x = 1/4$ in Formula (13). Follow the argument of Exercise 19: Parts (a) and (b) remain unchanged; in Part (c) replace (1/3) with (1/4):

$\left|R_n\left(\dfrac{1}{4}\right)\right| \le \dfrac{M}{(n+1)!}\left(\dfrac{1}{4}\right)^{n+1} \le 0.000005$ for x in the interval $[0, 1/4]$. From Part (b), together with $0 \le x \le 1/4, 1+x \ge 1, 1-x \ge 3/4$, follows Part (d): $M = n!\left[1 + \dfrac{1}{(3/4)^{n+1}}\right]$. Part (e) now becomes $0.000005 \ge \dfrac{M}{(n+1)!}\left(\dfrac{1}{4}\right)^{n+1} = \dfrac{1}{n+1}\left[\left(\dfrac{1}{4}\right)^{n+1} + \left(\dfrac{1}{3}\right)^{n+1}\right]$, which is true for $n = 9$.

21. $f(x) = \cos x, f^{(n+1)}(x) = \pm\sin x$ or $\pm\cos x, |f^{(n+1)}(x)| \le 1$, set $M = 1$,

$|R_n(x)| \le \dfrac{1}{(n+1)!}|x-a|^{n+1}, \lim_{n\to+\infty}\dfrac{|x-a|^{n+1}}{(n+1)!} = 0$ so $\lim_{n\to+\infty} R_n(x) = 0$ for all x.

22. $f(x) = \sin x, f^{(n+1)}(x) = \pm\sin x$ or $\pm\cos x, |f^{(n+1)}(x)| \le 1$, follow Exercise 21.

23. **(a)** From Machin's formula and a CAS, $\dfrac{\pi}{4} \approx 0.7853981633974483096156608$, accurate to the 25th decimal place.

(b)

n	s_n
0	0.3183098 78...
1	0.3183098 861837906 067...
2	0.3183098 861837906 7153776 695...
3	0.3183098 861837906 7153776 752674502 34...
$1/\pi$	0.3183098 861837906 7153776 752674502 87...

24. **(a)** $f'(0) = \lim_{h\to 0} \dfrac{f(h) - f(0)}{h} = \lim_{h\to 0} \dfrac{e^{-1/h^2}}{h}$, let $t = 1/h$ then $h = 1/t$ and $\lim_{h\to 0^+} \dfrac{e^{-1/h^2}}{h} = \lim_{t\to+\infty} te^{-t^2} = \lim_{t\to+\infty} \dfrac{t}{e^{t^2}} = \lim_{t\to+\infty} \dfrac{1}{2te^{t^2}} = 0$, similarly $\lim_{h\to 0^-} \dfrac{e^{-1/h^2}}{h} = 0$ so $f'(0) = 0$.

(b) The Maclaurin series is $0 + 0\cdot x + 0\cdot x^2 + \cdots = 0$, but $f(0) = 0$ and $f(x) > 0$ if $x \neq 0$ so the series converges to $f(x)$ only at the point $x = 0$.

EXERCISE SET 10.10

1. **(a)** Replace x with $-x$: $\dfrac{1}{1+x} = 1 - x + x^2 - \cdots + (-1)^k x^k + \cdots$; $R = 1$.

(b) Replace x with x^2 : $\dfrac{1}{1-x^2} = 1 + x^2 + x^4 + \cdots + x^{2k} + \cdots$; $R = 1$.

(c) Replace x with $2x$: $\dfrac{1}{1-2x} = 1 + 2x + 4x^2 + \cdots + 2^k x^k + \cdots$; $R = 1/2$.

(d) $\dfrac{1}{2-x} = \dfrac{1/2}{1-x/2}$; replace x with $x/2$: $\dfrac{1}{2-x} = \dfrac{1}{2} + \dfrac{1}{2^2}x + \dfrac{1}{2^3}x^2 + \cdots + \dfrac{1}{2^{k+1}}x^k + \cdots$; $R = 2$.

2. **(a)** Replace x with $-x$: $\ln(1-x) = -x - x^2/2 - x^3/3 - \cdots - x^k/k - \cdots$; $R = 1$.

(b) Replace x with x^2 : $\ln(1+x^2) = x^2 - x^4/2 + x^6/3 - \cdots + (-1)^{k-1}x^{2k}/k + \cdots$; $R = 1$.

(c) Replace x with $2x$: $\ln(1+2x) = 2x - (2x)^2/2 + (2x)^3/3 - \cdots + (-1)^{k-1}(2x)^k/k + \cdots$; $R = 1/2$.

(d) $\ln(2+x) = \ln 2 + \ln(1+x/2)$; replace x with $x/2$:
$\ln(2+x) = \ln 2 + x/2 - (x/2)^2/2 + (x/2)^3/3 + \cdots + (-1)^{k-1}(x/2)^k/k + \cdots$; $R = 2$.

3. **(a)** From Section 10.9, Example 5(b), $\dfrac{1}{\sqrt{1+x}} = 1 - \dfrac{1}{2}x + \dfrac{1\cdot 3}{2^2\cdot 2!}x^2 - \dfrac{1\cdot 3\cdot 5}{2^3\cdot 3!}x^3 + \cdots$, so

$$(2+x)^{-1/2} = \frac{1}{\sqrt{2}\sqrt{1+x/2}} = \frac{1}{2^{1/2}} - \frac{1}{2^{5/2}}x + \frac{1\cdot 3}{2^{9/2}\cdot 2!}x^2 - \frac{1\cdot 3\cdot 5}{2^{13/2}\cdot 3!}x^3 + \cdots$$

(b) Example 5(a): $\dfrac{1}{(1+x)^2} = 1 - 2x + 3x^2 - 4x^3 + \cdots$, so $\dfrac{1}{(1-x^2)^2} = 1 + 2x^2 + 3x^4 + 4x^6 + \cdots$

4. **(a)** $\dfrac{1}{a-x} = \dfrac{1/a}{1-x/a} = 1/a + x/a^2 + x^2/a^3 + \cdots + x^k/a^{k+1} + \cdots$; $R = |a|$.

(b)
$$1/(a+x)^2 = \frac{1}{a^2}\frac{1}{(1+x/a)^2} = \frac{1}{a^2}\left(1 - 2(x/a) + 3(x/a)^2 - 4(x/a)^3 + \cdots\right)$$
$$= \frac{1}{a^2} - \frac{2}{a^3}x + \frac{3}{a^4}x^2 - \frac{4}{a^5}x^3 + \cdots; \quad R = |a|$$

5. **(a)** $2x - \frac{2^3}{3!}x^3 + \frac{2^5}{5!}x^5 - \frac{2^7}{7!}x^7 + \cdots;\ R = +\infty$

(b) $1 - 2x + 2x^2 - \frac{4}{3}x^3 + \cdots;\ R = +\infty$

(c) $1 + x^2 + \frac{1}{2!}x^4 + \frac{1}{3!}x^6 + \cdots;\ R = +\infty$

(d) $x^2 - \frac{\pi^2}{2}x^4 + \frac{\pi^4}{4!}x^6 - \frac{\pi^6}{6!}x^8 + \cdots;\ R = +\infty$

6. **(a)** $1 - \frac{2^2}{2!}x^2 + \frac{2^4}{4!}x^4 - \frac{2^6}{6!}x^6 + \cdots;\ R = +\infty$

(b) $x^2\left(1 + x + \frac{1}{2!}x^2 + \frac{1}{3!}x^3 + \cdots\right) = x^2 + x^3 + \frac{1}{2!}x^4 + \frac{1}{3!}x^5 + \cdots;\ R = +\infty$

(c) $x\left(1 - x + \frac{1}{2!}x^2 - \frac{1}{3!}x^3 + \cdots\right) = x - x^2 + \frac{1}{2!}x^3 - \frac{1}{3!}x^4 + \cdots;\ R = +\infty$

(d) $x^2 - \frac{1}{3!}x^6 + \frac{1}{5!}x^{10} - \frac{1}{7!}x^{14} + \cdots;\ R = +\infty$

7. **(a)** $x^2\left(1 - 3x + 9x^2 - 27x^3 + \cdots\right) = x^2 - 3x^3 + 9x^4 - 27x^5 + \cdots;\ R = 1/3$

(b) $x\left(2x + \frac{2^3}{3!}x^3 + \frac{2^5}{5!}x^5 + \frac{2^7}{7!}x^7 + \cdots\right) = 2x^2 + \frac{2^3}{3!}x^4 + \frac{2^5}{5!}x^6 + \frac{2^7}{7!}x^8 + \cdots;\ R = +\infty$

(c) Substitute $3/2$ for m and $-x^2$ for x in Equation (18) of Section 10.9, then multiply by x: $x - \frac{3}{2}x^3 + \frac{3}{8}x^5 + \frac{1}{16}x^7 + \cdots;\ R = 1$

8. **(a)** $\frac{x}{x-1} = \frac{-x}{1-x} = -x\left(1 + x + x^2 + x^3 + \cdots\right) = -x - x^2 - x^3 - x^4 - \cdots;\ R = 1.$

(b) $3 + \frac{3}{2!}x^4 + \frac{3}{4!}x^8 + \frac{3}{6!}x^{12} + \cdots;\ R = +\infty$

(c) From Table 10.9.1 with $m = -3$, $(1+x)^{-3} = 1 - 3x + 6x^2 - 10x^3 + \cdots$, so $x(1+2x)^{-3} = x - 6x^2 + 24x^3 - 80x^4 + \cdots;\ R = 1/2$

9. **(a)** $\sin^2 x = \frac{1}{2}(1 - \cos 2x) = \frac{1}{2}\left[1 - \left(1 - \frac{2^2}{2!}x^2 + \frac{2^4}{4!}x^4 - \frac{2^6}{6!}x^6 + \cdots\right)\right]$

$$= x^2 - \frac{2^3}{4!}x^4 + \frac{2^5}{6!}x^6 - \frac{2^7}{8!}x^8 + \cdots$$

(b) $\ln\left[(1+x^3)^{12}\right] = 12\ln(1+x^3) = 12x^3 - 6x^6 + 4x^9 - 3x^{12} + \cdots$

10. **(a)** $\cos^2 x = \frac{1}{2}(1 + \cos 2x) = \frac{1}{2}\left[1 + \left(1 - \frac{2^2}{2!}x^2 + \frac{2^4}{4!}x^4 - \frac{2^6}{6!}x^6 + \cdots\right)\right]$

$$= 1 - x^2 + \frac{2^3}{4!}x^4 - \frac{2^5}{6!}x^6 + \cdots$$

(b) In Equation (13) of Section 10.9 replace x with $-x$: $\ln\left(\frac{1-x}{1+x}\right) = -2\left(x + \frac{1}{3}x^3 + \frac{1}{5}x^5 + \cdots\right)$

11. **(a)** $\dfrac{1}{x} = \dfrac{1}{1-(1-x)} = 1 + (1-x) + (1-x)^2 + \cdots + (1-x)^k + \cdots$
$= 1 - (x-1) + (x-1)^2 - \cdots + (-1)^k(x-1)^k + \cdots$

(b) $(0, 2)$

12. **(a)** $\dfrac{1}{x} = \dfrac{1/x_0}{1 + (x-x_0)/x_0} = 1/x_0 - (x-x_0)/x_0^2 + (x-x_0)^2/x_0^3 - \cdots + (-1)^k(x-x_0)^k/x_0^{k+1} + \cdots$

(b) $(0, 2x_0)$

13. **(a)** $(1 + x + x^2/2 + x^3/3! + x^4/4! + \cdots)(x - x^3/3! + x^5/5! - \cdots) = x + x^2 + x^3/3 - x^5/30 + \cdots$

(b) $(1 + x/2 - x^2/8 + x^3/16 - (5/128)x^4 + \cdots)(x - x^2/2 + x^3/3 - x^4/4 + x^5/5 - \cdots)$
$= x - x^3/24 + x^4/24 - (71/1920)x^5 + \cdots$

14. **(a)** $(1 - x^2 + x^4/2 - x^6/6 + \cdots)\left(1 - \dfrac{1}{2}x^2 + \dfrac{1}{24}x^4 - \dfrac{1}{720}x^6 \cdots\right) = 1 - \dfrac{3}{2}x^2 + \dfrac{25}{24}x^4 - \dfrac{331}{720}x^6 + \cdots$

(b) $\left(1 + \dfrac{4}{3}x^2 + \cdots\right)\left(1 + \dfrac{1}{3}x - \dfrac{1}{9}x^2 + \dfrac{5}{81}x^3 - \cdots\right) = 1 + \dfrac{1}{3}x + \dfrac{11}{9}x^2 + \dfrac{41}{81}x^3 + \cdots$

15. **(a)** $\dfrac{1}{\cos x} = 1 \Big/ \left(1 - \dfrac{1}{2!}x^2 + \dfrac{1}{4!}x^4 - \dfrac{1}{6!}x^6 + \cdots\right) = 1 + \dfrac{1}{2}x^2 + \dfrac{5}{24}x^4 + \dfrac{61}{720}x^6 + \cdots$

(b) $\dfrac{\sin x}{e^x} = \left(x - \dfrac{x^3}{3!} + \dfrac{x^5}{5!} - \cdots\right) \Big/ \left(1 + x + \dfrac{x^2}{2!} + \dfrac{x^3}{3!} + \dfrac{x^4}{4!} + \cdots\right) = x - x^2 + \dfrac{1}{3}x^3 - \dfrac{1}{30}x^5 + \cdots$

16. **(a)** $\dfrac{\tan^{-1} x}{1+x} = \left(x - x^3/3 + x^5/5 - \cdots\right)/(1+x) = x - x^2 + \dfrac{2}{3}x^3 - \dfrac{2}{3}x^4 \cdots$

(b) $\dfrac{\ln(1+x)}{1-x} = \left(x - x^2/2 + x^3/3 - x^4/4 + \cdots\right)/(1-x) = x + \dfrac{1}{2}x^2 + \dfrac{5}{6}x^3 + \dfrac{7}{12}x^4 + \cdots$

17. $e^x = 1 + x + x^2/2 + x^3/3! + \cdots + x^k/k! + \cdots$, $e^{-x} = 1 - x + x^2/2 - x^3/3! + \cdots + (-1)^k x^k/k! + \cdots$;

$\sinh x = \dfrac{1}{2}\left(e^x - e^{-x}\right) = x + x^3/3! + x^5/5! + \cdots + x^{2k+1}/(2k+1)! + \cdots, R = +\infty$

$\cosh x = \dfrac{1}{2}\left(e^x + e^{-x}\right) = 1 + x^2/2 + x^4/4! + \cdots + x^{2k}/(2k)! + \cdots, R = +\infty$

18. $\tanh x = \dfrac{x + x^3/3! + x^5/5! + x^7/7! + \cdots}{1 + x^2/2 + x^4/4! + x^6/6! \cdots} = x - \dfrac{1}{3}x^3 + \dfrac{2}{15}x^5 - \dfrac{17}{315}x^7 + \cdots$

19. $\dfrac{4x-2}{x^2-1} = \dfrac{-1}{1-x} + \dfrac{3}{1+x} = -\left(1 + x + x^2 + x^3 + x^4 + \cdots\right) + 3\left(1 - x + x^2 - x^3 + x^4 + \cdots\right)$
$= 2 - 4x + 2x^2 - 4x^3 + 2x^4 + \cdots$

20. $\dfrac{x^3 + x^2 + 2x - 2}{x^2 - 1} = x + 1 - \dfrac{1}{1-x} + \dfrac{2}{1+x}$
$= x + 1 - \left(1 + x + x^2 + x^3 + x^4 + \cdots\right) + 2\left(1 - x + x^2 - x^3 + x^4 + \cdots\right)$
$= 2 - 2x + x^2 - 3x^3 + x^4 - \cdots$

21. (a) $\dfrac{d}{dx}\left(1 - x^2/2! + x^4/4! - x^6/6! + \cdots\right) = -x + x^3/3! - x^5/5! + \cdots = -\sin x$

(b) $\dfrac{d}{dx}\left(x - x^2/2 + x^3/3 - \cdots\right) = 1 - x + x^2 - \cdots = 1/(1+x)$

22. (a) $\dfrac{d}{dx}\left(x + x^3/3! + x^5/5! + \cdots\right) = 1 + x^2/2! + x^4/4! + \cdots = \cosh x$

(b) $\dfrac{d}{dx}\left(x - x^3/3 + x^5/5 - x^7/7 + \cdots\right) = 1 - x^2 + x^4 - x^6 + \cdots = \dfrac{1}{1+x^2}$

23. (a) $$\begin{aligned}\int\left(1 + x + x^2/2! + \cdots\right)dx &= \left(x + x^2/2! + x^3/3! + \cdots\right) + C_1\\ &= \left(1 + x + x^2/2! + x^3/3! + \cdots\right) + C_1 - 1 = e^x + C\end{aligned}$$

(b) $$\begin{aligned}\int\left(x + x^3/3! + x^5/5! + \cdots\right) &= x^2/2! + x^4/4! + \cdots + C_1\\ &= 1 + x^2/2! + x^4/4! + \cdots + C_1 - 1 = \cosh x + C\end{aligned}$$

24. (a) $$\begin{aligned}\int\left(x - x^3/3! + x^5/5! - \cdots\right)dx &= \left(x^2/2! - x^4/4! + x^6/6! - \cdots\right) + C_1\\ &= -\left(1 - x^2/2! + x^4/4! - x^6/6! + \cdots\right) + C_1 + 1\\ &= -\cos x + C\end{aligned}$$

(b) $\displaystyle\int\left(1 - x + x^2 - \cdots\right)dx = \left(x - x^2/2 + x^3/3 - \cdots\right) + C = \ln(1+x) + C$

(*Note:* $-1 < x < 1$, so $|1+x| = 1+x$)

25. (a) Substitute x^2 for x in the Maclaurin Series for $1/(1-x)$ (Table 10.9.1)

and then multiply by x: $\dfrac{x}{1-x^2} = x\displaystyle\sum_{k=0}^{\infty}(x^2)^k = \sum_{k=0}^{\infty}x^{2k+1}$

(b) $f^{(5)}(0) = 5!c_5 = 5!$, $f^{(6)}(0) = 6!c_6 = 0$ (c) $f^{(n)}(0) = n!c_n = \begin{cases} n! & \text{if } n \text{ odd}\\ 0 & \text{if } n \text{ even}\end{cases}$

26. $x^2\cos 2x = \displaystyle\sum_{k=0}^{\infty}\frac{(-1)^k 2^{2k}}{(2k)!}x^{2k+2}$; $f^{(99)}(0) = 0$ because $c_{99} = 0$.

27. (a) $\displaystyle\lim_{x\to 0}\frac{\sin x}{x} = \lim_{x\to 0}\left(1 - x^2/3! + x^4/5! - \cdots\right) = 1$

(b) $\displaystyle\lim_{x\to 0}\frac{\tan^{-1}x - x}{x^3} = \lim_{x\to 0}\frac{\left(x - x^3/3 + x^5/5 - x^7/7 + \cdots\right) - x}{x^3} = -1/3$

28. (a) $$\begin{aligned}\frac{1-\cos x}{\sin x} &= \frac{1 - \left(1 - x^2/2! + x^4/4! - x^6/6! + \cdots\right)}{x - x^3/3! + x^5/5! - \cdots} = \frac{x^2/2! - x^4/4! + x^6/6! - \cdots}{x - x^3/3! + x^5/5! - \cdots}\\ &= \frac{x/2! - x^3/4! + x^5/6! - \cdots}{1 - x^2/3! + x^4/5! - \cdots}, x \neq 0;\ \lim_{x\to 0}\frac{1-\cos x}{\sin x} = \frac{0}{1} = 0\end{aligned}$$

(b) $$\lim_{x\to 0}\frac{1}{x}\left[\ln\sqrt{1+x}-\sin 2x\right]=\lim_{x\to 0}\frac{1}{x}\left[\frac{1}{2}\ln(1+x)-\sin 2x\right]$$
$$=\lim_{x\to 0}\frac{1}{x}\left[\frac{1}{2}\left(x-\frac{1}{2}x^2+\frac{1}{3}x^3-\cdots\right)-\left(2x-\frac{4}{3}x^3+\frac{4}{15}x^5-\cdots\right)\right]$$
$$=\lim_{x\to 0}\left(-\frac{3}{2}-\frac{1}{4}x+\frac{3}{2}x^2+\cdots\right)=-3/2$$

29. $$\int_0^1 \sin\left(x^2\right)dx=\int_0^1\left(x^2-\frac{1}{3!}x^6+\frac{1}{5!}x^{10}-\frac{1}{7!}x^{14}+\cdots\right)dx$$
$$=\frac{1}{3}x^3-\frac{1}{7\cdot 3!}x^7+\frac{1}{11\cdot 5!}x^{11}-\frac{1}{15\cdot 7!}x^{15}+\cdots\Big]_0^1$$
$$=\frac{1}{3}-\frac{1}{7\cdot 3!}+\frac{1}{11\cdot 5!}-\frac{1}{15\cdot 7!}+\cdots,$$

but $\dfrac{1}{15\cdot 7!}<0.5\times 10^{-3}$ so $\displaystyle\int_0^1 \sin(x^2)dx\approx\frac{1}{3}-\frac{1}{7\cdot 3!}+\frac{1}{11\cdot 5!}\approx 0.3103$

30. $$\int_0^{1/2}\tan^{-1}\left(2x^2\right)dx=\int_0^{1/2}\left(2x^2-\frac{8}{3}x^6+\frac{32}{5}x^{10}-\frac{128}{7}x^{14}+\cdots\right)dx$$
$$=\frac{2}{3}x^3-\frac{8}{21}x^7+\frac{32}{55}x^{11}-\frac{128}{105}x^{15}+\cdots\Big]_0^{1/2}$$
$$=\frac{2}{3}\frac{1}{2^3}-\frac{8}{21}\frac{1}{2^7}+\frac{32}{55}\frac{1}{2^{11}}-\frac{128}{105}\frac{1}{2^{15}}-\cdots,$$

but $\dfrac{32}{55\cdot 2^{11}}<0.5\times 10^{-3}$ so $\displaystyle\int_0^{1/2}\tan^{-1}(2x^2)dx\approx\frac{2}{3\cdot 2^3}-\frac{8}{21\cdot 2^7}\approx 0.0804$

31. $$\int_0^{0.2}\left(1+x^4\right)^{1/3}dx=\int_0^{0.2}\left(1+\frac{1}{3}x^4-\frac{1}{9}x^8+\cdots\right)dx$$
$$=x+\frac{1}{15}x^5-\frac{1}{81}x^9+\cdots\Big]_0^{0.2}=0.2+\frac{1}{15}(0.2)^5-\frac{1}{81}(0.2)^9+\cdots,$$

but $\dfrac{1}{15}(0.2)^5<0.5\times 10^{-3}$ so $\displaystyle\int_0^{0.2}(1+x^4)^{1/3}dx\approx 0.200$

32. $$\int_0^{1/2}(1+x^2)^{-1/4}dx=\int_0^{1/2}\left(1-\frac{1}{4}x^2+\frac{5}{32}x^4-\frac{15}{128}x^6+\cdots\right)dx$$
$$=x-\frac{1}{12}x^3+\frac{1}{32}x^5-\frac{15}{896}x^7+\cdots\Big]_0^{1/2}$$
$$=1/2-\frac{1}{12}(1/2)^3+\frac{1}{32}(1/2)^5-\frac{15}{896}(1/2)^7+\cdots,$$

but $\dfrac{15}{896}(1/2)^7<0.5\times 10^{-3}$ so $\displaystyle\int_0^{1/2}(1+x^2)^{-1/4}dx\approx 1/2-\frac{1}{12}(1/2)^3+\frac{1}{32}(1/2)^5\approx 0.4906$

33. **(a)** $\dfrac{x}{(1-x)^2} = x\dfrac{d}{dx}\left[\dfrac{1}{1-x}\right] = x\dfrac{d}{dx}\left[\sum_{k=0}^{\infty} x^k\right] = x\left[\sum_{k=1}^{\infty} kx^{k-1}\right] = \sum_{k=1}^{\infty} kx^k$

(b) $-\ln(1-x) = \int \dfrac{1}{1-x}dx - C = \int \left[\sum_{k=0}^{\infty} x^k\right] dx - C$

$$= \sum_{k=0}^{\infty} \frac{x^{k+1}}{k+1} - C = \sum_{k=1}^{\infty} \frac{x^k}{k} - C, -\ln(1-0) = 0 \text{ so } C = 0.$$

(c) Replace x with $-x$ in Part (b): $\ln(1+x) = -\sum_{k=1}^{+\infty} \dfrac{(-1)^k}{k}x^k = \sum_{k=1}^{+\infty} \dfrac{(-1)^{k+1}}{k}x^k$

(d) $\sum_{k=1}^{+\infty} \dfrac{(-1)^{k+1}}{k}$ converges by the Alternating Series Test.

(e) By Parts (c) and (d) and the remark, $\sum_{k=1}^{+\infty} \dfrac{(-1)^{k+1}}{k}x^k$ converges to $\ln(1+x)$ for $-1 < x \le 1$.

34. **(a)** In Exercise 33(a), set $x = \dfrac{1}{3}$, $S = \dfrac{1/3}{(1-1/3)^2} = \dfrac{3}{4}$

(b) In Part (b) set $x = 1/4, S = \ln(4/3)$

(c) In Part (e) set $x = 1, S = \ln 2$

35. **(a)** $\sinh^{-1} x = \int (1+x^2)^{-1/2}\, dx - C = \int \left(1 - \dfrac{1}{2}x^2 + \dfrac{3}{8}x^4 - \dfrac{5}{16}x^6 + \cdots\right) dx - C$

$$= \left(x - \frac{1}{6}x^3 + \frac{3}{40}x^5 - \frac{5}{112}x^7 + \cdots\right) - C;\ \sinh^{-1} 0 = 0 \text{ so } C = 0.$$

(b) $(1+x^2)^{-1/2} = 1 + \sum_{k=1}^{\infty} \dfrac{(-1/2)(-3/2)(-5/2)\cdots(-1/2-k+1)}{k!}(x^2)^k$

$$= 1 + \sum_{k=1}^{\infty} (-1)^k \frac{1\cdot 3\cdot 5\cdots(2k-1)}{2^k k!}x^{2k},$$

$$\sinh^{-1} x = x + \sum_{k=1}^{\infty} (-1)^k \frac{1\cdot 3\cdot 5\cdots(2k-1)}{2^k k!(2k+1)}x^{2k+1}$$

(c) $R = 1$

36. **(a)** $\sin^{-1} x = \int (1-x^2)^{-1/2} dx - C = \int \left(1 + \dfrac{1}{2}x^2 + \dfrac{3}{8}x^4 + \dfrac{5}{16}x^6 + \cdots\right) dx - C$

$$= \left(x + \frac{1}{6}x^3 + \frac{3}{40}x^5 + \frac{5}{112}x^7 + \cdots\right) - C, \sin^{-1} 0 = 0 \text{ so } C = 0$$

(b) $\left(1-x^2\right)^{-1/2} = 1 + \sum_{k=1}^{\infty} \frac{(-1/2)(-3/2)(-5/2)\cdots(-1/2-k+1)}{k!}\left(-x^2\right)^k$

$$= 1 + \sum_{k=1}^{\infty} \frac{(-1)^k(1/2)^k(1)(3)(5)\cdots(2k-1)}{k!}(-1)^k x^{2k}$$

$$= 1 + \sum_{k=1}^{\infty} \frac{1\cdot 3\cdot 5\cdots(2k-1)}{2^k k!}x^{2k}$$

$$\sin^{-1} x = x + \sum_{k=1}^{\infty} \frac{1\cdot 3\cdot 5\cdots(2k-1)}{2^k k!(2k+1)}x^{2k+1}$$

(c) $R = 1$

37. **(a)** $y(t) = y_0 \sum_{k=0}^{\infty} \frac{(-1)^k(0.000121)^k t^k}{k!}$

(b) $y(1) \approx y_0(1 - 0.000121t)\Big]_{t=1} = 0.999879y_0$

(c) $y_0 e^{-0.000121} \approx 0.9998790073y_0$

38. **(a)** If $\frac{ct}{m} \approx 0$ then $e^{-ct/m} \approx 1 - \frac{ct}{m}$, and $v(t) \approx \left(1-\frac{ct}{m}\right)\left(v_0+\frac{mg}{c}\right)-\frac{mg}{c} = v_0 - \left(\frac{cv_0}{m}+g\right)t.$

(b) The quadratic approximation is

$$v_0 \approx \left(1-\frac{ct}{m}+\frac{(ct)^2}{2m^2}\right)\left(v_0+\frac{mg}{c}\right)-\frac{mg}{c} = v_0-\left(\frac{cv_0}{m}+g\right)t+\frac{c^2}{2m^2}\left(v_0+\frac{mg}{c}\right)t^2.$$

39. $\theta_0 = 5° = \pi/36$ rad, $k = \sin(\pi/72)$

(a) $T \approx 2\pi\sqrt{\frac{L}{g}} = 2\pi\sqrt{1/9.8} \approx 2.00709$

(b) $T \approx 2\pi\sqrt{\frac{L}{g}}\left(1+\frac{k^2}{4}\right) \approx 2.008044621$

(c) 2.008045644

40. The third order model gives the same result as the second, because there is no term of degree three in (5). By the Wallis sine formula, $\int_0^{\pi/2} \sin^4\phi\, d\phi = \frac{1\cdot 3}{2\cdot 4}\frac{\pi}{2}$, and

$$T \approx 4\sqrt{\frac{L}{g}}\int_0^{\pi/2}\left(1+\frac{1}{2}k^2\sin^2\phi+\frac{1\cdot 3}{2^2 2!}k^4\sin^4\phi\right)d\phi = 4\sqrt{\frac{L}{g}}\left(\frac{\pi}{2}+\frac{k^2}{2}\frac{\pi}{4}+\frac{3k^4}{8}\frac{3\pi}{16}\right)$$

$$= 2\pi\sqrt{\frac{L}{g}}\left(1+\frac{k^2}{4}+\frac{9k^4}{64}\right)$$

41. (a) $F = \dfrac{mgR^2}{(R+h)^2} = \dfrac{mg}{(1+h/R)^2} = mg\left(1 - 2h/R + 3h^2/R^2 - 4h^3/R^3 + \cdots\right)$

(b) If $h = 0$, then the binomial series converges to 1 and $F = mg$.

(c) Sum the series to the linear term, $F \approx mg - 2mgh/R$.

(d) $\dfrac{mg - 2mgh/R}{mg} = 1 - \dfrac{2h}{R} = 1 - \dfrac{2 \cdot 29{,}028}{4000 \cdot 5280} \approx 0.9973$, so about 0.27% less.

42. (a) We can differentiate term-by-term:

$$y' = \sum_{k=1}^{\infty} \frac{(-1)^k x^{2k-1}}{2^{2k-1}k!(k-1)!} = \sum_{k=0}^{\infty} \frac{(-1)^{k+1} x^{2k+1}}{2^{2k+1}(k+1)!k!},\ y'' = \sum_{k=0}^{\infty} \frac{(-1)^{k+1}(2k+1)x^{2k}}{2^{2k+1}(k+1)!k!}, \text{ and}$$

$$xy'' + y' + xy = \sum_{k=0}^{\infty} \frac{(-1)^{k+1}(2k+1)x^{2k+1}}{2^{2k+1}(k+1)!k!} + \sum_{k=0}^{\infty} \frac{(-1)^{k+1} x^{2k+1}}{2^{2k+1}(k+1)!k!} + \sum_{k=0}^{\infty} \frac{(-1)^k x^{2k+1}}{2^{2k}(k!)^2},$$

$$xy'' + y' + xy = \sum_{k=0}^{\infty} \frac{(-1)^{k+1} x^{2k+1}}{2^{2k}(k!)^2}\left[\frac{2k+1}{2(k+1)} + \frac{1}{2(k+1)} - 1\right] = 0$$

(b) $$y' = \sum_{k=0}^{\infty} \frac{(-1)^k (2k+1) x^{2k}}{2^{2k+1}k!(k+1)!},\ y'' = \sum_{k=1}^{\infty} \frac{(-1)^k (2k+1) x^{2k-1}}{2^{2k}(k-1)!(k+1)!}.$$

Since $J_1(x) = \displaystyle\sum_{k=0}^{\infty} \frac{(-1)^k x^{2k+1}}{2^{2k+1}k!(k+1)!}$ and $x^2 J_1(x) = \displaystyle\sum_{k=1}^{\infty} \frac{(-1)^{k-1} x^{2k+1}}{2^{2k-1}(k-1)!k!}$, it follows that

$$x^2y'' + xy' + (x^2-1)y$$

$$= \sum_{k=1}^{\infty} \frac{(-1)^k (2k+1) x^{2k+1}}{2^{2k}(k-1)!(k+1)!} + \sum_{k=0}^{\infty} \frac{(-1)^k (2k+1) x^{2k+1}}{2^{2k+1}(k!)(k+1)!} + \sum_{k=1}^{\infty} \frac{(-1)^{k-1} x^{2k+1}}{2^{2k-1}(k-1)!k!}$$

$$- \sum_{k=0}^{\infty} \frac{(-1)^k x^{2k+1}}{2^{2k+1}k!(k+1)!}$$

$$= \frac{x}{2} - \frac{x}{2} + \sum_{k=1}^{\infty} \frac{(-1)^k x^{2k+1}}{2^{2k-1}(k-1)!k!}\left(\frac{2k+1}{2(k+1)} + \frac{2k+1}{4k(k+1)} - 1 - \frac{1}{4k(k+1)}\right) = 0.$$

(c) From Part (a), $J_0'(x) = \displaystyle\sum_{k=0}^{\infty} \frac{(-1)^{k+1} x^{2k+1}}{2^{2k+1}(k+1)!k!} = -J_1(x)$.

43. Let $f(x) = \displaystyle\sum_{k=0}^{\infty} a_k x^k = \sum_{k=0}^{\infty} b_k x^k$ for $-r < x < r$. Then $a_k = f^{(k)}(0)/k! = b_k$ for all k.

CHAPTER 10 SUPPLEMENTARY EXERCISES

4. (a) $\displaystyle\sum_{k=0}^{\infty} \frac{f^{(k)}(0)}{k!} x^k$ (b) $\displaystyle\sum_{k=0}^{\infty} \frac{f^{(k)}(x_0)}{k!} (x - x_0)^k$

9. (a) always true by Theorem 10.5.2

(b) sometimes false, for example the harmonic series diverges but $\sum(1/k^2)$ converges

(c) sometimes false, for example $f(x) = \sin \pi x, a_k = 0, L = 0$

(d) always true by the comments which follow Example 3(d) of Section 10.2

(e) sometimes false, for example $a_n = \dfrac{1}{2} + (-1)^n \dfrac{1}{4}$

(f) sometimes false, for example $u_k = 1/2$

(g) always false by Theorem 10.5.3

(h) sometimes false, for example $u_k = 1/k, v_k = 2/k$

(i) always true by the Comparison Test

(j) always true by the Comparison Test

(k) sometimes false, for example $\sum(-1)^k/k$

(l) sometimes false, for example $\sum(-1)^k/k$

10. (a) false, $f(x)$ is not differentiable at $x = 0$, Definition 10.8.1

(b) true: $s_n = 1$ if n is odd and $s_{2n} = 1 + 1/(n+1)$; $\lim\limits_{n\to+\infty} s_n = 1$

(c) false, $\lim a_k \neq 0$

11. (a) geometric, $r = 1/5$, converges

(b) $1/(5^k + 1) < 1/5^k$, converges

(c) $\dfrac{9}{\sqrt{k}+1} \geq \dfrac{9}{\sqrt{k}+\sqrt{k}} = \dfrac{9}{2\sqrt{k}}$, $\displaystyle\sum_{k=1}^{\infty} \frac{9}{2\sqrt{k}}$ diverges

12. (a) converges by Alternating Series Test

(b) absolutely convergent, $\displaystyle\sum_{k=1}^{\infty} \left[\frac{k+2}{3k-1}\right]^k$ converges by the Root Test.

(c) $\dfrac{k^{-1/2}}{2+\sin^2 k} > \dfrac{k^{-1}}{2+1} = \dfrac{1}{3k}$, $\displaystyle\sum_{k=1}^{\infty} \frac{1}{3k}$ diverges

13. (a) $\dfrac{1}{k^3+2k+1} < \dfrac{1}{k^3}$, $\displaystyle\sum_{k=1}^{\infty} 1/k^3$ converges, so $\displaystyle\sum_{k=1}^{\infty} \frac{1}{k^3+2k+1}$ converges by the Comparison Test

(b) Limit Comparison Test, compare with the divergent series $\displaystyle\sum_{k=1}^{\infty} \frac{1}{k^{2/5}}$, diverges

(c) $\left|\dfrac{\cos(1/k)}{k^2}\right| < \dfrac{1}{k^2}$, $\displaystyle\sum_{k=1}^{\infty} \frac{1}{k^2}$ converges, so $\displaystyle\sum_{k=1}^{\infty} \frac{\cos(1/k)}{k^2}$ converges absolutely

14. (a) $\displaystyle\sum_{k=1}^{\infty} \frac{\ln k}{k\sqrt{k}} = \sum_{k=2}^{\infty} \frac{\ln k}{k\sqrt{k}}$ because $\ln 1 = 0$,

$$\int_2^{+\infty} \frac{\ln x}{x^{3/2}}\,dx = \lim_{\ell\to+\infty} \left[-\frac{2\ln x}{x^{1/2}} - \frac{4}{x^{1/2}}\right]_2^{\ell} = \sqrt{2}(\ln 2 + 2) \text{ so } \sum_{k=2}^{\infty} \frac{\ln k}{k^{3/2}} \text{ converges}$$

(b) $\dfrac{k^{4/3}}{8k^2+5k+1} \geq \dfrac{k^{4/3}}{8k^2+5k^2+k^2} = \dfrac{1}{14k^{2/3}}$, $\displaystyle\sum_{k=1}^{\infty} \frac{1}{14k^{2/3}}$ diverges

(c) absolutely convergent, $\displaystyle\sum_{k=1}^{\infty} \frac{1}{k^2+1}$ converges (compare with $\sum 1/k^2$)

15. $\sum_{k=0}^{\infty}\frac{1}{5^k}-\sum_{k=0}^{99}\frac{1}{5^k}=\sum_{k=100}^{\infty}\frac{1}{5^k}=\frac{1}{5^{100}}\sum_{k=0}^{\infty}\frac{1}{5^k}=\frac{1}{4\cdot 5^{99}}$

16. no, $\lim_{k\to+\infty}a_k=\frac{1}{2}\neq 0$ (Divergence Test)

17. **(a)** $p_0(x)=1, p_1(x)=1-7x, p_2(x)=1-7x+5x^2, p_3(x)=1-7x+5x^2+4x^3$, $p_4(x)=1-7x+5x^2+4x^3$

(b) If $f(x)$ is a polynomial of degree n and $k\geq n$ then the Maclaurin polynomial of degree k is the polynomial itself; if $k<n$ then it is the truncated polynomial.

18. $\ln(1+x)=x-x^2/2+\cdots$; so $|\ln(1+x)-x|\leq x^2/2$ by Theorem 10.7.2.

19. $\sin x=x-x^3/3!+x^5/5!-x^7/7!+\cdots$ is an alternating series, so $|\sin x-x+x^3/3!-x^5/5!|\leq x^7/7!\leq \pi^7/(4^7 7!)\leq 0.00005$

20. $\int_0^1\frac{1-\cos x}{x}\,dx=\left[\frac{x^2}{2\cdot 2!}-\frac{x^4}{4\cdot 4!}+\frac{x^6}{6\cdot 6!}-\cdots\right]_0^1=\frac{1}{2\cdot 2!}-\frac{1}{4\cdot 4!}+\frac{1}{6\cdot 6!}-\cdots$, and $\frac{1}{6\cdot 6!}<0.0005$,

so $\int_0^1\frac{1-\cos x}{x}\,dx=\frac{1}{2\cdot 2!}-\frac{1}{4\cdot 4!}=0.2396$ to three decimal-place accuracy.

21. **(a)** $\rho=\lim_{k\to+\infty}\left(\frac{2^k}{k!}\right)^{1/k}=\lim_{k\to+\infty}\frac{2}{\sqrt[k]{k!}}=0$, converges

(b) $\rho=\lim_{k\to+\infty}u_k^{1/k}=\lim_{k\to+\infty}\frac{k}{\sqrt[k]{k!}}=e$, diverges

22. **(a)** $1\leq k, 2\leq k, 3\leq k,\ldots,k\leq k$, therefore $1\cdot 2\cdot 3\cdots k\leq k\cdot k\cdot k\cdots k$, or $k!\leq k^k$.

(b) $\sum\frac{1}{k^k}\leq\sum\frac{1}{k!}$, converges

(c) $\lim_{k\to+\infty}\left(\frac{1}{k^k}\right)^{1/k}=\lim_{k\to+\infty}\frac{1}{k}=0$, converges

23. **(a)** $u_{100}=\sum_{k=1}^{100}u_k-\sum_{k=1}^{99}u_k=\left(2-\frac{1}{100}\right)-\left(2-\frac{1}{99}\right)=\frac{1}{9900}$

(b) $u_1=1$; for $k\geq 2, u_k=\left(2-\frac{1}{k}\right)-\left(2-\frac{1}{k-1}\right)=\frac{1}{k(k-1)}$, $\lim_{k\to+\infty}u_k=0$

(c) $\sum_{k=1}^{\infty}u_k=\lim_{n\to+\infty}\sum_{k=1}^{n}u_k=\lim_{n\to+\infty}\left(2-\frac{1}{n}\right)=2$

24. **(a)** $\sum_{k=1}^{\infty}\left(\frac{3}{2^k}-\frac{2}{3^k}\right)=\sum_{k=1}^{\infty}\frac{3}{2^k}-\sum_{k=1}^{\infty}\frac{2}{3^k}=\left(\frac{3}{2}\right)\frac{1}{1-(1/2)}-\left(\frac{2}{3}\right)\frac{1}{1-(1/3)}=2$ (geometric series)

(b) $\sum_{k=1}^{n}[\ln(k+1)-\ln k]=\ln(n+1)$, so $\sum_{k=1}^{\infty}[\ln(k+1)-\ln k]=\lim_{n\to+\infty}\ln(n+1)=+\infty$, diverges

(c) $\lim_{n\to+\infty}\sum_{k=1}^{n}\frac{1}{2}\left(\frac{1}{k}-\frac{1}{k+2}\right)=\lim_{n\to+\infty}\frac{1}{2}\left(1+\frac{1}{2}-\frac{1}{n+1}-\frac{1}{n+2}\right)=\frac{3}{4}$

(d) $\lim_{n\to+\infty}\sum_{k=1}^{n}\left[\tan^{-1}(k+1)-\tan^{-1}k\right]=\lim_{n\to+\infty}\left[\tan^{-1}(n+1)-\tan^{-1}(1)\right]=\frac{\pi}{2}-\frac{\pi}{4}=\frac{\pi}{4}$

25. **(a)** e^2-1 **(b)** $\sin\pi=0$ **(c)** $\cos e$ **(d)** $e^{-\ln 3}=1/3$

26. $a_k=\sqrt{a_{k-1}}=a_{k-1}^{1/2}=a_{k-2}^{1/4}=\cdots=a_1^{1/2^{k-1}}=c^{1/2^k}$

(a) If $c=1/2$ then $\lim_{k\to+\infty}a_k=1$. **(b)** if $c=3/2$ then $\lim_{k\to+\infty}a_k=1$.

27. $e^{-x}=1-x+x^2/2!+\cdots$. Replace x with $-(\frac{x-100}{16})^2/2$ to obtain

$e^{-(\frac{x-100}{16})^2/2}=1-\frac{(x-100)^2}{2\cdot 16^2}+\frac{(x-100)^4}{8\cdot 16^4}+\cdots$, thus

$p\approx\frac{1}{16\sqrt{2\pi}}\int_{100}^{110}\left[1-\frac{(x-100)^2}{2\cdot 16^2}+\frac{(x-100)^4}{8\cdot 16^4}\right]dx\approx 0.23406$ or 23.406%.

28. $f(x)=xe^x=x+x^2+\frac{x^3}{2!}+\frac{x^4}{3!}+\cdots=\sum_{k=0}^{\infty}\frac{x^{k+1}}{k!}$,

$f'(x)=(x+1)e^x=1+2x+\frac{3x^2}{2!}+\frac{4x^3}{3!}+\cdots=\sum_{k=0}^{\infty}\frac{k+1}{k!}x^k$; $\sum_{k=0}^{\infty}\frac{k+1}{k!}=f'(1)=2e$.

29. Let $A=1-\frac{1}{2^2}+\frac{1}{3^2}-\frac{1}{4^2}+\cdots$; since the series all converge absolutely,

$\frac{\pi^2}{6}-A=2\frac{1}{2^2}+2\frac{1}{4^2}+2\frac{1}{6^2}+\cdots=\frac{1}{2}\left(1+\frac{1}{2^2}+\frac{1}{3^2}+\cdots\right)=\frac{1}{2}\frac{\pi^2}{6}$, so $A=\frac{1}{2}\frac{\pi^2}{6}=\frac{\pi^2}{12}$.

30. Compare with $1/k^p$: converges if $p>1$, diverges otherwise.

31. **(a)** $x+\frac{1}{2}x^2+\frac{3}{14}x^3+\frac{3}{35}x^4+\cdots$; $\rho=\lim_{k\to+\infty}\frac{k+1}{3k+1}|x|=\frac{1}{3}|x|$,

converges if $\frac{1}{3}|x|<1$, $|x|<3$ so $R=3$.

(b) $-x^3+\frac{2}{3}x^5-\frac{2}{5}x^7+\frac{8}{35}x^9-\cdots$; $\rho=\lim_{k\to+\infty}\frac{k+1}{2k+1}|x|^2=\frac{1}{2}|x|^2$,

converges if $\frac{1}{2}|x|^2<1$, $|x|^2<2$, $|x|<\sqrt{2}$ so $R=\sqrt{2}$.

32. By the Ratio Test for absolute convergence, $\rho=\lim_{k\to+\infty}\frac{|x-x_0|}{b}=\frac{|x-x_0|}{b}$; converges if $|x-x_0|<b$, diverges if $|x-x_0|>b$. If $x=x_0-b$, $\sum_{k=0}^{\infty}(-1)^k$ diverges; if $x=x_0+b$, $\sum_{k=0}^{\infty}1$ diverges. The interval of convergence is (x_0-b,x_0+b).

33. If $x \geq 0$, then $\cos\sqrt{x} = 1 - \frac{(\sqrt{x})^2}{2!} + \frac{(\sqrt{x})^4}{4!} - \frac{(\sqrt{x})^6}{6!} + \cdots = 1 - \frac{x}{2!} + \frac{x^2}{4!} - \frac{x^3}{6!} + \cdots$; if $x \leq 0$, then $\cosh(\sqrt{-x}) = 1 + \frac{(\sqrt{-x})^2}{2!} + \frac{(\sqrt{-x})^4}{4!} + \frac{(\sqrt{-x})^6}{6!} + \cdots = 1 - \frac{x}{2!} + \frac{x^2}{4!} - \frac{x^3}{6!} + \cdots$.

34. By Exercise 74 of Section 3.5, the derivative of an odd (even) function is even (odd); hence all odd-numbered derivatives of an odd function are even, all even-numbered derivatives of an odd function are odd; a similar statement holds for an even function.

(a) If $f(x)$ is an even function, then $f^{(2k-1)}(x)$ is an odd function, so $f^{(2k-1)}(0) = 0$, and thus the MacLaurin series coefficients $a_{2k-1} = 0, k = 1, 2, \cdots$.

(b) If $f(x)$ is an odd function, then $f^{(2k)}(x)$ is an odd function, so $f^{(2k)}(0) = 0$, and thus the MacLaurin series coefficients $a_{2k} = 0, k = 1, 2, \cdots$.

35. $\left(1 - \frac{v^2}{c^2}\right)^{-1/2} \approx 1 + \frac{v^2}{2c^2}$, so $K = m_0c^2\left[\frac{1}{\sqrt{1 - v^2/c^2}} - 1\right] \approx m_0c^2(v^2/(2c^2)) = m_0v^2/2$

36. **(a)** $\int_n^{+\infty} \frac{1}{x^{3.7}}dx < 0.005$ if $n > 4.93$; let $n = 5$.

(b) $s_n \approx 1.1062$; CAS: 1.10628824

CHAPTER 11
Analytic Geometry in Calculus

EXERCISE SET 11.1

1. 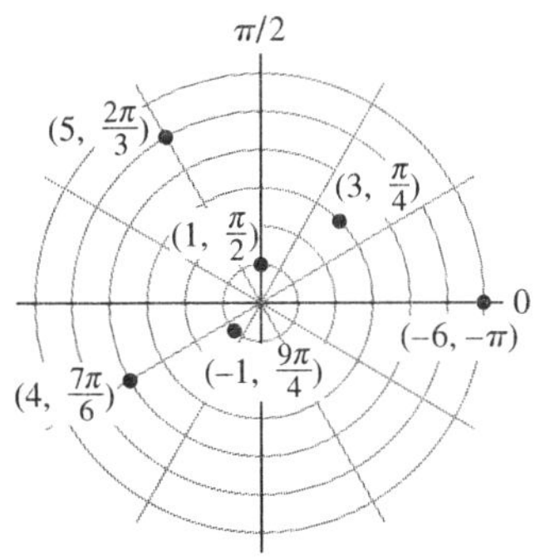

2. 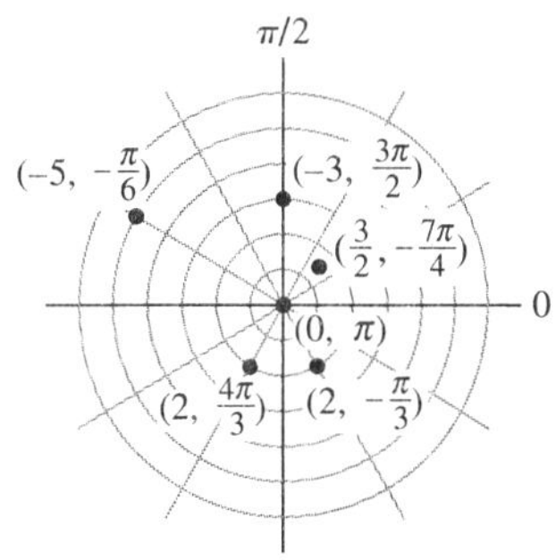

3. **(a)** $(3\sqrt{3}, 3)$ **(b)** $(-7/2, 7\sqrt{3}/2)$ **(c)** $(3\sqrt{3}, 3)$
(d) $(0, 0)$ **(e)** $(-7\sqrt{3}/2, 7/2)$ **(f)** $(-5, 0)$

4. **(a)** $(-4\sqrt{2}, -4\sqrt{2})$ **(b)** $(7\sqrt{2}/2, -7\sqrt{2}/2)$ **(c)** $(4\sqrt{2}, 4\sqrt{2})$
(d) $(5, 0)$ **(e)** $(0, -2)$ **(f)** $(0, 0)$

5. **(a)** both $(5, \pi)$ **(b)** $(4, 11\pi/6), (4, -\pi/6)$ **(c)** $(2, 3\pi/2), (2, -\pi/2)$
(d) $(8\sqrt{2}, 5\pi/4), (8\sqrt{2}, -3\pi/4)$ **(e)** both $(6, 2\pi/3)$ **(f)** both $(\sqrt{2}, \pi/4)$

6. **(a)** $(2, 5\pi/6)$ **(b)** $(-2, 11\pi/6)$ **(c)** $(2, -7\pi/6)$ **(d)** $(-2, -\pi/6)$

7. **(a)** $(5, 0.6435)$ **(b)** $(\sqrt{29}, 5.0929)$ **(c)** $(1.2716, 0.6658)$

8. **(a)** $(5, 2.2143)$ **(b)** $(3.4482, 2.6260)$ **(c)** $(\sqrt{4 + \pi^2/36}, 0.2561)$

9. **(a)** $r^2 = x^2 + y^2 = 4$; circle **(b)** $y = 4$; horizontal line
(c) $r^2 = 3r\cos\theta$, $x^2 + y^2 = 3x$, $(x - 3/2)^2 + y^2 = 9/4$; circle
(d) $3r\cos\theta + 2r\sin\theta = 6$, $3x + 2y = 6$; line

10. **(a)** $r\cos\theta = 5$, $x = 5$; vertical line
(b) $r^2 = 2r\sin\theta$, $x^2 + y^2 = 2y$, $x^2 + (y - 1)^2 = 1$; circle
(c) $r^2 = 4r\cos\theta + 4r\sin\theta$, $x^2 + y^2 = 4x + 4y$, $(x - 2)^2 + (y - 2)^2 = 8$; circle
(d) $r = \dfrac{1}{\cos\theta}\dfrac{\sin\theta}{\cos\theta}$, $r\cos^2\theta = \sin\theta$, $r^2\cos^2\theta = r\sin\theta$, $x^2 = y$; parabola

11. **(a)** $r\cos\theta = 7$ **(b)** $r = 3$
(c) $r^2 - 6r\sin\theta = 0$, $r = 6\sin\theta$
(d) $4(r\cos\theta)(r\sin\theta) = 9$, $4r^2\sin\theta\cos\theta = 9$, $r^2\sin 2\theta = 9/2$

12. **(a)** $r\sin\theta = -3$ **(b)** $r = \sqrt{5}$
(c) $r^2 + 4r\cos\theta = 0$, $r = -4\cos\theta$
(d) $r^4\cos^2\theta = r^2\sin^2\theta$, $r^2 = \tan^2\theta$, $r = \tan\theta$

13.

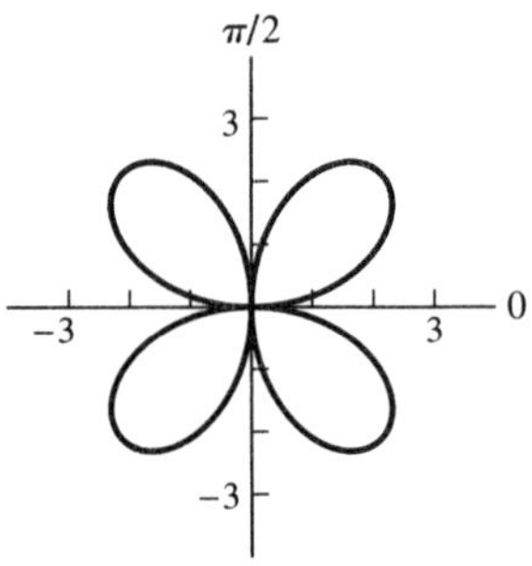

$r = 3\sin 2\theta$

14.

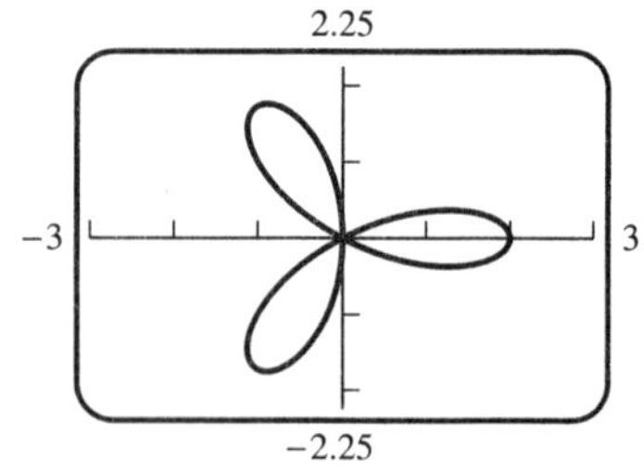

$r = 2\cos 3\theta$

15.

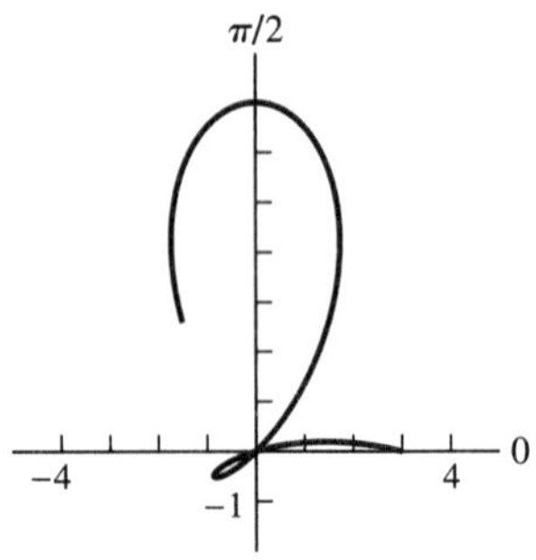

$r = 3 - 4\sin 3\theta$

16.

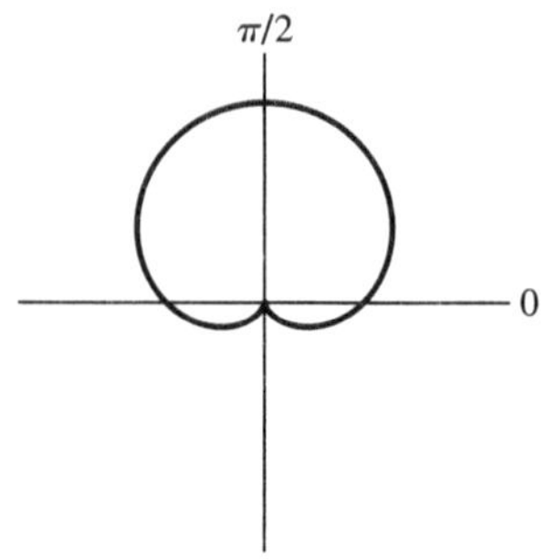

$r = 2 + 2\sin\theta$

17. (a) $r = 5$

(b) $(x-3)^2 + y^2 = 9,\ r = 6\cos\theta$

(c) Example 6, $r = 1 - \cos\theta$

18. (a) From (8-9), $r = a \pm b\sin\theta$ or $r = a \pm b\cos\theta$. The curve is not symmetric about the y-axis, so Theorem 11.2.1(a) eliminates the sine function, thus $r = a \pm b\cos\theta$. The cartesian point $(-3, 0)$ is either the polar point $(3, \pi)$ or $(-3, 0)$, and the cartesian point $(-1, 0)$ is either the polar point $(1, \pi)$ or $(-1, 0)$. A solution is $a = 1, b = -2$; we may take the equation as $r = 1 - 2\cos\theta$.

(b) $x^2 + (y + 3/2)^2 = 9/4, r = -3\sin\theta$

(c) Figure 11.1.18, $a = 1, n = 3, r = \sin 3\theta$

19. (a) Figure 11.1.18, $a = 3, n = 2, r = 3\sin 2\theta$

(b) From (8-9), symmetry about the y-axis and Theorem 11.1.1(b), the equation is of the form $r = a \pm b\sin\theta$. The cartesian points $(3, 0)$ and $(0, 5)$ give $a = 3$ and $5 = a + b$, so $b = 2$ and $r = 3 + 2\sin\theta$.

(c) Example 8, $r^2 = 9\cos 2\theta$

20. (a) Example 6 rotated through $\pi/2$ radian: $a = 3, r = 3 - 3\sin\theta$

(b) Figure 11.1.18, $a = 1, r = \cos 5\theta$

(c) $x^2 + (y - 2)^2 = 4,\ r = 4\sin\theta$

21.
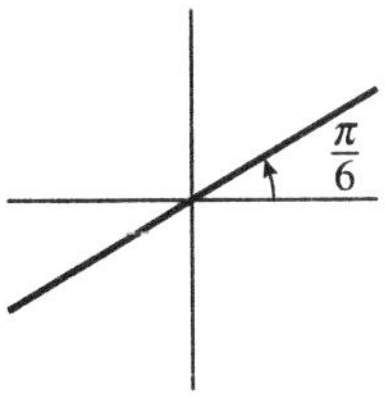

Line

22.
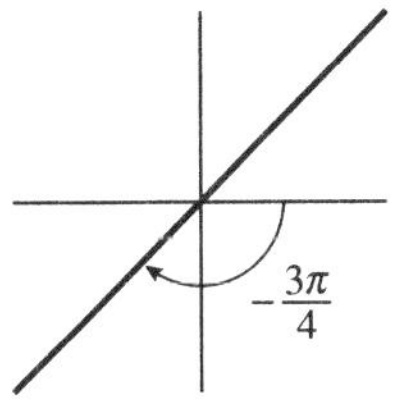

Line

23.
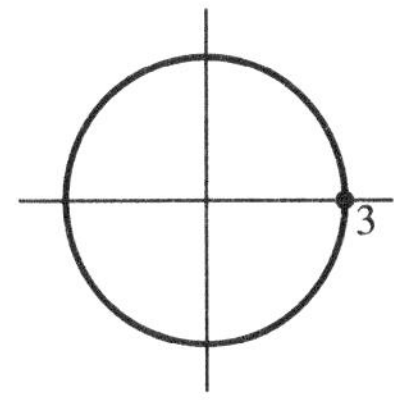

Circle

24.
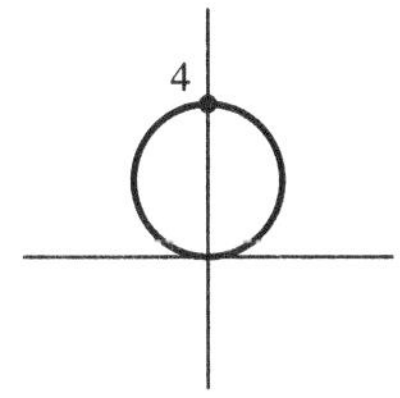

Circle

25.
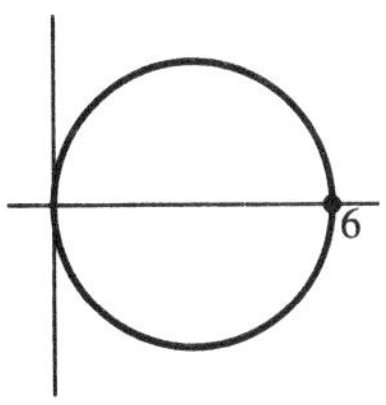

Circle

26.
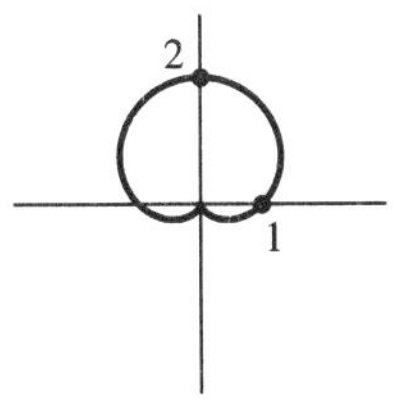

Cardioid

27.
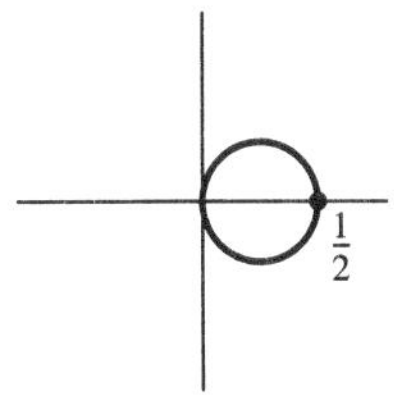

Circle

28.
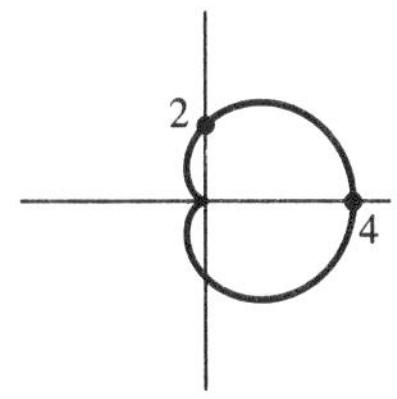

Cardioid

29.
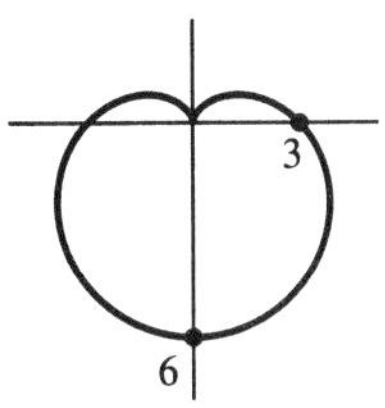

Cardioid

30.
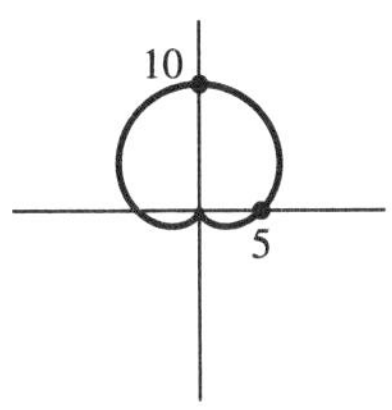

Cardioid

31.
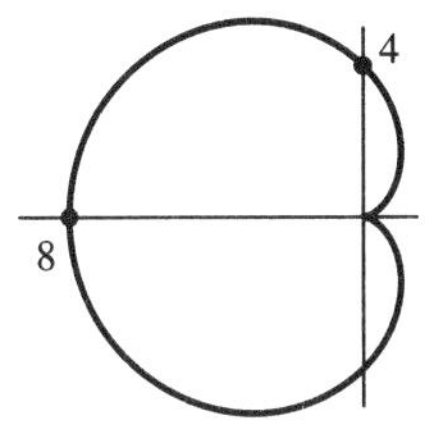

Cardioid

32.
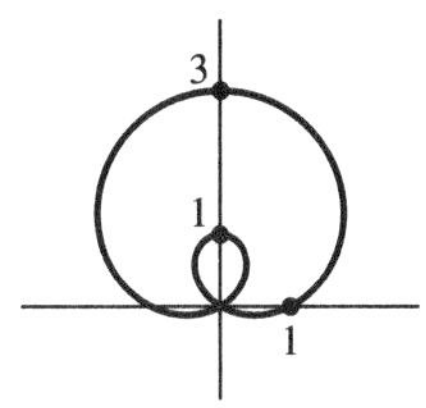

Limaçon

33.
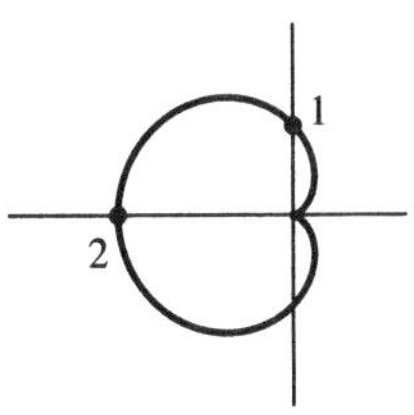

Cardioid

34.
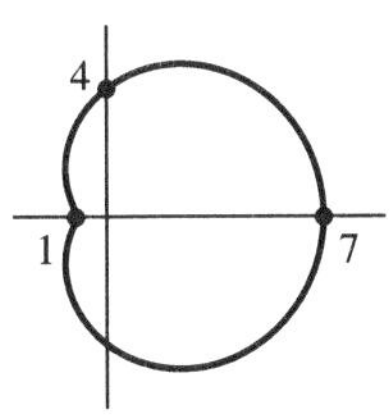

Limaçon

35.
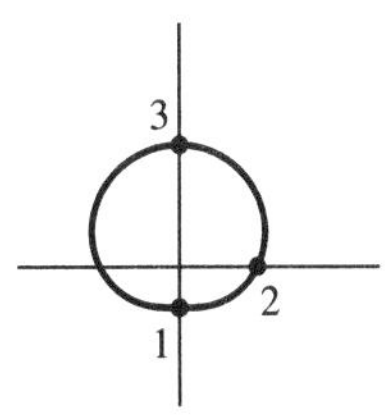

Limaçon

36.
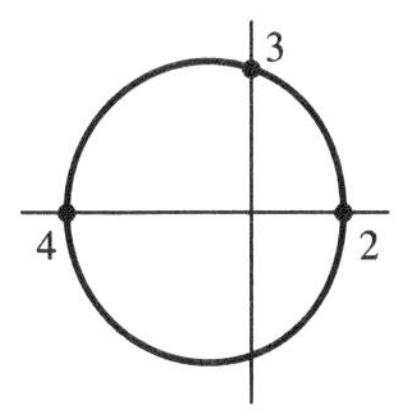

Limaçon

37.
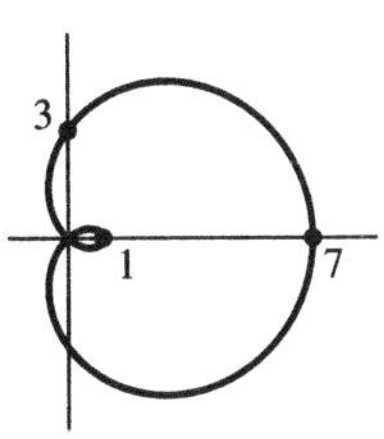

Limaçon

38.
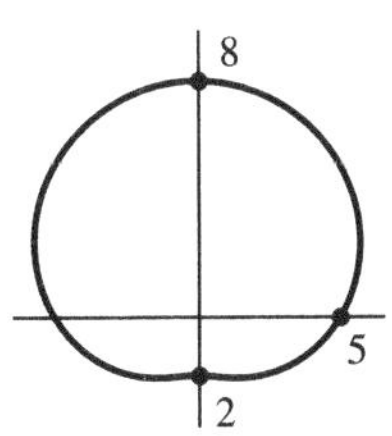

Limaçon

39.
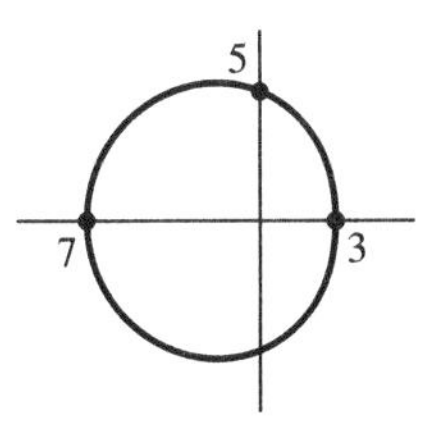

Limaçon

40.
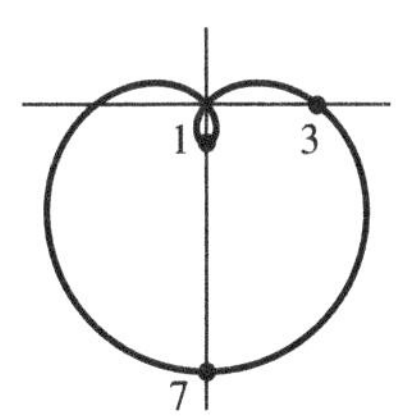

Limaçon

41.
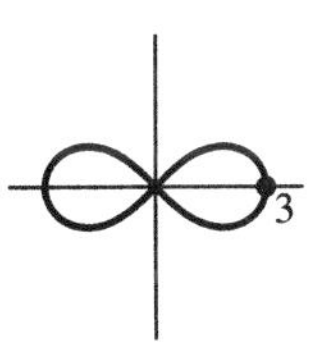

Lemniscate

42.
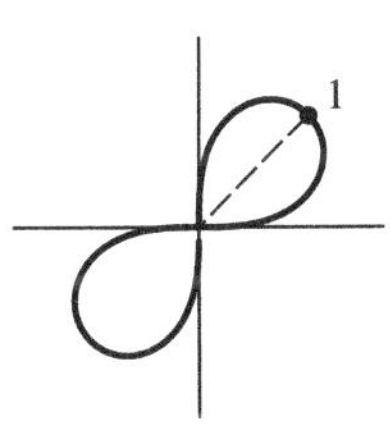

Lemniscate

43.
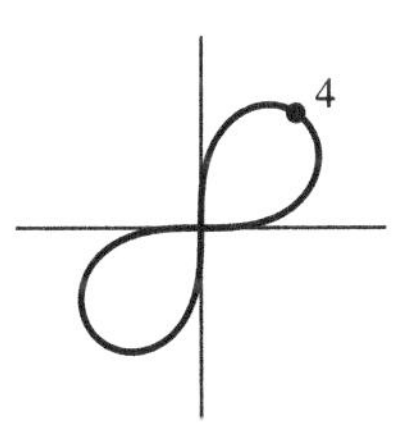

Lemniscate

44.
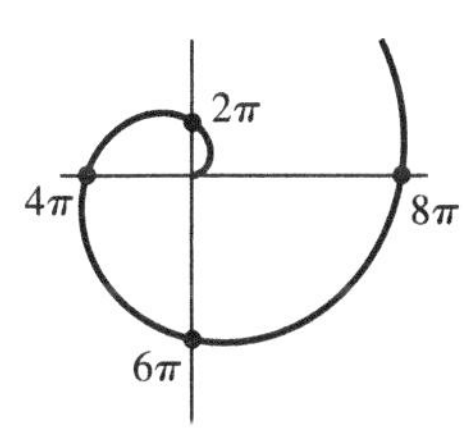

Spiral

45. 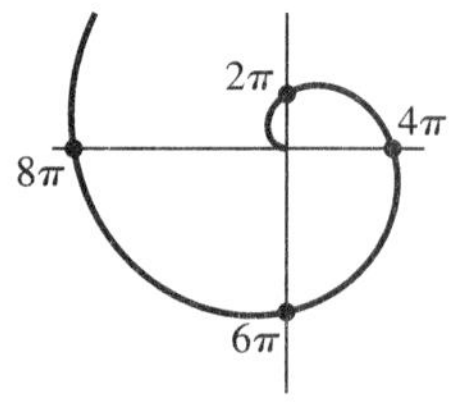

Spiral

46. 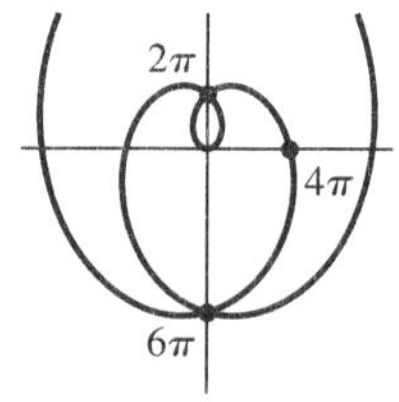

Spiral

47. 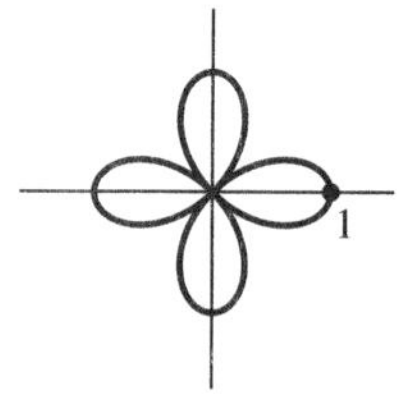

Four-petal rose

48. 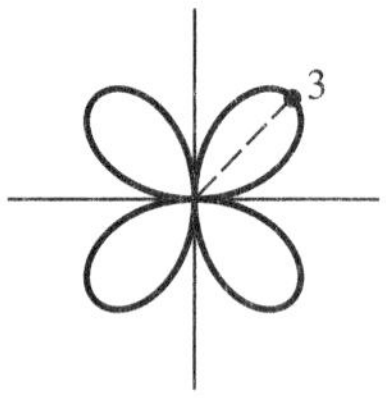

Four-petal rose

49. 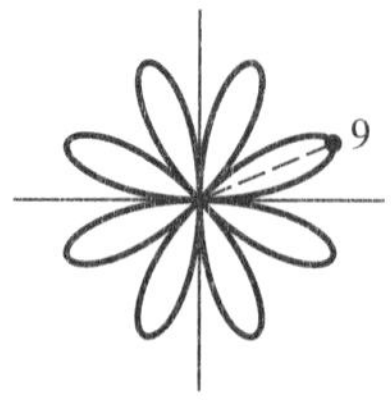

Eight-petal rose

50. 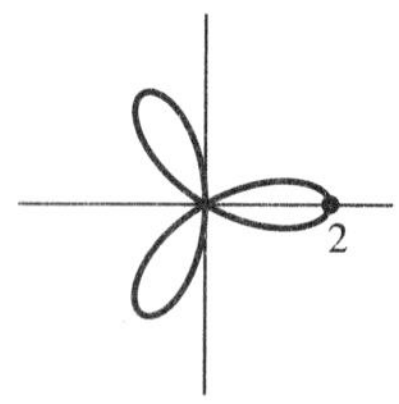

Three-petal rose

51. 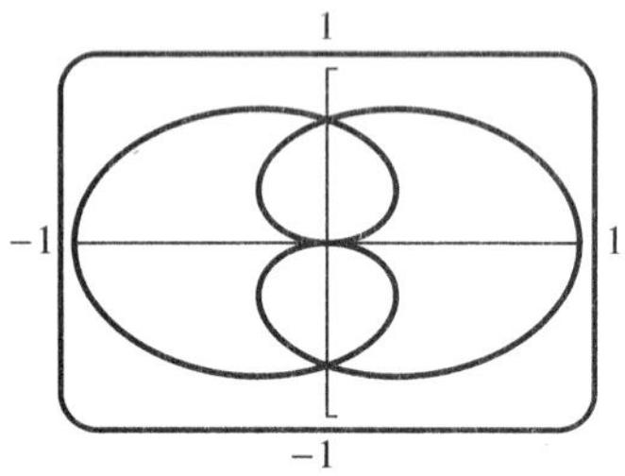

52. 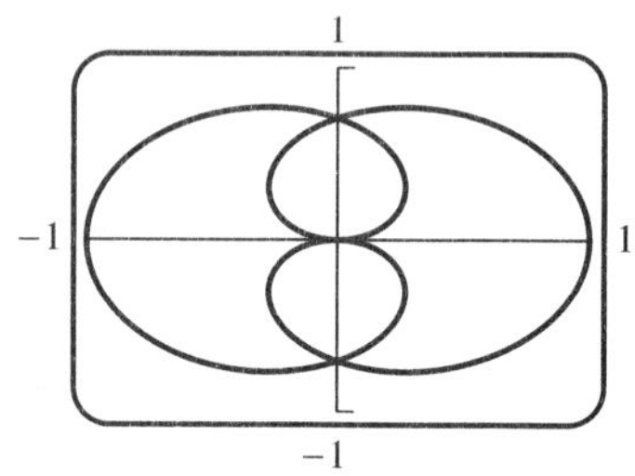

53. 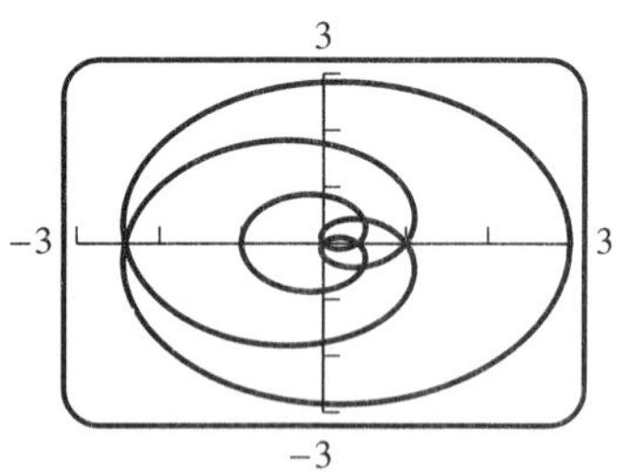

54. 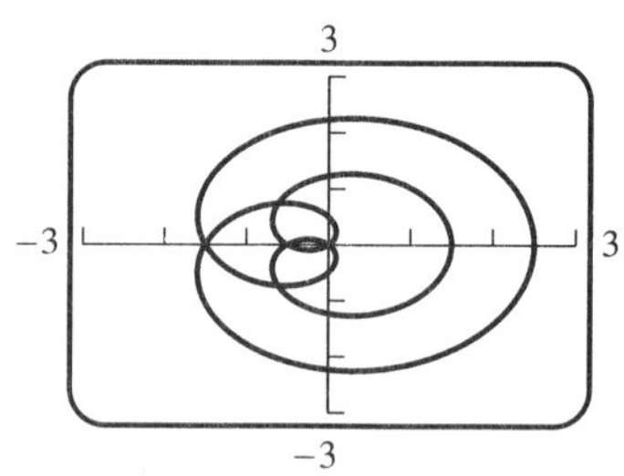

55. 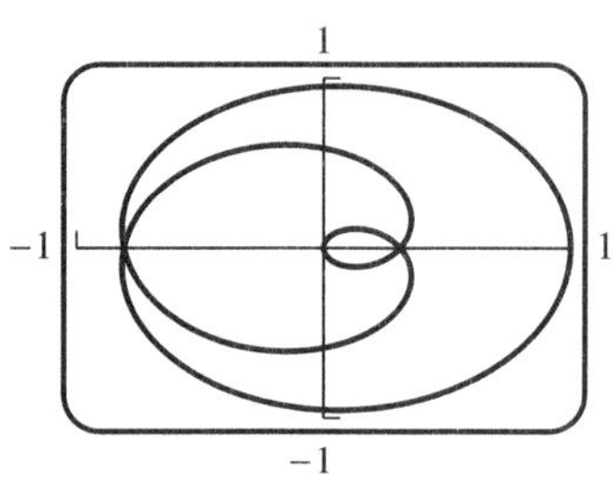

56. $0 \le \theta \le 8\pi$

57. **(a)** $-4\pi < \theta < 4\pi$

58. In I, along the x-axis, $x = r$ grows ever slower with θ. In II $x = r$ grows linearly with θ. Hence I: $r = \sqrt{\theta}$; II: $r = \theta$.

59. **(a)** $r = a/\cos\theta, x = r\cos\theta = a$, a family of vertical lines

(b) $r = b/\sin\theta, y = r\sin\theta = b$, a family of horizontal lines

60. The image of (r_0, θ_0) under a rotation through an angle α is $(r_0, \theta_0 + \alpha)$. Hence $(f(\theta), \theta)$ lies on the original curve if and only if $(f(\theta), \theta + \alpha)$ lies on the rotated curve, i.e. (r, θ) lies on the rotated curve if and only if $r = f(\theta - \alpha)$.

61. (a) $r = 1 + \cos(\theta - \pi/4) = 1 + \dfrac{\sqrt{2}}{2}(\cos\theta + \sin\theta)$

(b) $r = 1 + \cos(\theta - \pi/2) = 1 + \sin\theta$

(c) $r = 1 + \cos(\theta - \pi) = 1 - \cos\theta$

(d) $r = 1 + \cos(\theta - 5\pi/4) = 1 - \dfrac{\sqrt{2}}{2}(\cos\theta + \sin\theta)$

62. $r^2 = 4\cos 2(\theta - \pi/2) = -4\cos 2\theta$

63. Either $r - 1 = 0$ or $\theta - 1 = 0$, so the graph consists of the circle $r = 1$ and the line $\theta = 1$.

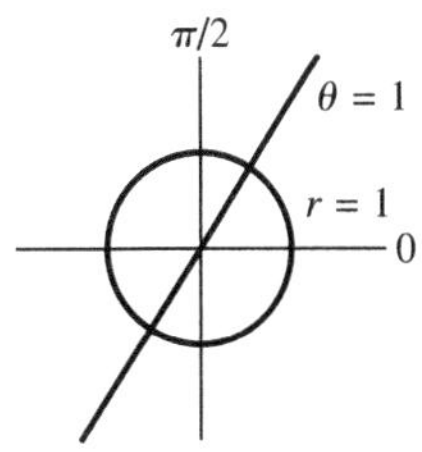

64. (a) $r^2 = Ar\sin\theta + Br\cos\theta$, $x^2 + y^2 = Ay + Bx$, $(x - B/2)^2 + (y - A/2)^2 = (A^2 + B^2)/4$, which is a circle of radius $\dfrac{1}{2}\sqrt{A^2 + B^2}$.

(b) Formula (4) follows by setting $A = 0, B = 2a$, $(x - a)^2 + y^2 = a^2$, the circle of radius a about $(a, 0)$. Formula (5) is derived in a similar fashion.

65. $y = r\sin\theta = (1 + \cos\theta)\sin\theta = \sin\theta + \sin\theta\cos\theta$,
$dy/d\theta = \cos\theta - \sin^2\theta + \cos^2\theta = 2\cos^2\theta + \cos\theta - 1 = (2\cos\theta - 1)(\cos\theta + 1)$;
$dy/d\theta = 0$ if $\cos\theta = 1/2$ or if $\cos\theta = -1$; $\theta = \pi/3$ or π (or $\theta = -\pi/3$, which leads to the minimum point).
If $\theta = \pi/3, \pi$, then $y = 3\sqrt{3}/4, 0$ so the maximum value of y is $3\sqrt{3}/4$ and the polar coordinates of the highest point are $(3/2, \pi/3)$.

66. $x = r\cos\theta = (1 + \cos\theta)\cos\theta = \cos\theta + \cos^2\theta$, $dx/d\theta = -\sin\theta - 2\sin\theta\cos\theta = -\sin\theta(1 + 2\cos\theta)$, $dx/d\theta = 0$ if $\sin\theta = 0$ or if $\cos\theta = -1/2$; $\theta = 0,\ 2\pi/3$, or π. If $\theta = 0,\ 2\pi/3,\ \pi$, then $x = 2, -1/4, 0$ so the minimum value of x is $-1/4$. The leftmost point has polar coordinates $(1/2, 2\pi/3)$.

67. (a) Let (x_1, y_1) and (x_2, y_2) be the rectangular coordinates of the points (r_1, θ_1) and (r_2, θ_2) then

$$d = \sqrt{(x_2 - x_1)^2 + (y_2 - y_1)^2} = \sqrt{(r_2\cos\theta_2 - r_1\cos\theta_1)^2 + (r_2\sin\theta_2 - r_1\sin\theta_1)^2}$$
$$= \sqrt{r_1^2 + r_2^2 - 2r_1r_2(\cos\theta_1\cos\theta_2 + \sin\theta_1\sin\theta_2)} = \sqrt{r_1^2 + r_2^2 - 2r_1r_2\cos(\theta_1 - \theta_2)}.$$

An alternate proof follows directly from the Law of Cosines.

(b) Let P and Q have polar coordinates $(r_1, \theta_1), (r_2, \theta_2)$, respectively, then the perpendicular from OQ to OP has length $h = r_2\sin(\theta_2 - \theta_1)$ and $A = \frac{1}{2}hr_1 = \frac{1}{2}r_1r_2\sin(\theta_2 - \theta_1)$.

(c) From Part (a), $d = \sqrt{9 + 4 - 2\cdot 3\cdot 2\cos(\pi/6 - \pi/3)} = \sqrt{13 - 6\sqrt{3}} \approx 1.615$

(d) $A = \dfrac{1}{2}2\sin(5\pi/6 - \pi/3) = 1$

68. (a) $0 = (r^2 + a^2)^2 - a^4 - 4a^2r^2\cos^2\theta = r^4 + a^4 + 2r^2a^2 - a^4 - 4a^2r^2\cos^2\theta$
$= r^4 + 2r^2a^2 - 4a^2r^2\cos^2\theta$, so $r^2 = 2a^2(2\cos^2\theta - 1) = 2a^2\cos 2\theta$.

(b) The distance from the point (r,θ) to $(a,0)$ is (from Exercise 67(a)) $\sqrt{r^2+a^2-2ra\cos(\theta-0)}=\sqrt{r^2-2ar\cos\theta+a^2}$, and to the point (a,π) is $\sqrt{r^2+a^2-2ra\cos(\theta-\pi)}=\sqrt{r^2+2ar\cos\theta+a^2}$, and their product is $\sqrt{(r^2+a^2)^2-4a^2r^2\cos^2\theta}=\sqrt{r^4+a^4+2a^2r^2(1-2\cos^2\theta)}$
$=\sqrt{4a^4\cos^2 2\theta+a^4+2a^2(2a^2\cos 2\theta)(-\cos 2\theta)}=a^2$

69. $\lim_{\theta\to 0^+} y=\lim_{\theta\to 0^+} r\sin\theta=\lim_{\theta\to 0^+}\frac{\sin\theta}{\theta}=1$, and $\lim_{\theta\to 0^+} x=\lim_{\theta\to 0^+} r\cos\theta=\lim_{\theta\to 0^+}\frac{\cos\theta}{\theta}=+\infty$.

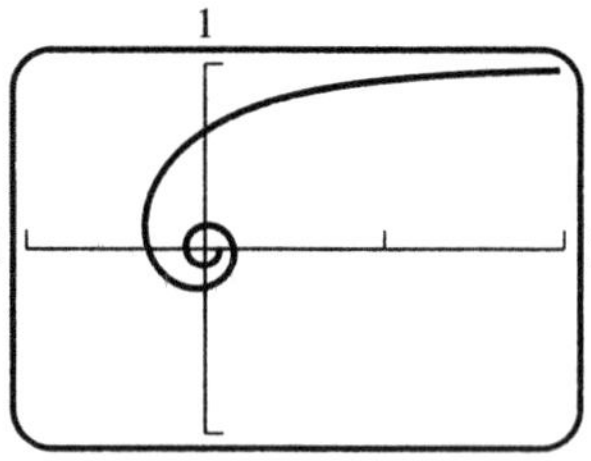

70. $\lim_{\theta\to 0^\pm} y=\lim_{\theta\to 0^\pm} r\sin\theta=\lim_{\theta\to 0^\pm}\frac{\sin\theta}{\theta^2}=\lim_{\theta\to 0^\pm}\frac{\sin\theta}{\theta}\lim_{\theta\to 0^\pm}\frac{1}{\theta}=1\cdot\lim_{\theta\to 0^\pm}\frac{1}{\theta}$, so $\lim_{\theta\to 0^\pm} y$ does not exist.

71. Note that $r\to\pm\infty$ as θ approaches odd multiples of $\pi/2$;
$x=r\cos\theta=4\tan\theta\cos\theta=4\sin\theta$,
$y=r\sin\theta=4\tan\theta\sin\theta$
so $x\to\pm 4$ and $y\to\pm\infty$ as θ approaches odd multiples of $\pi/2$.

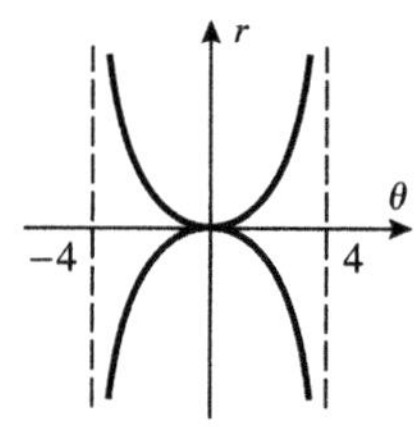

72. $\lim_{\theta\to(\pi/2)^-} x=\lim_{\theta\to(\pi/2)^-} r\cos\theta=\lim_{\theta\to(\pi/2)^-} 2\sin^2\theta=2$, and $\lim_{\theta\to(\pi/2)^-} y=+\infty$,
so $x=2$ is a vertical asymptote.

73. Let $r=a\sin n\theta$ (the proof for $r=a\cos n\theta$ is similar). If θ starts at 0, then θ would have to increase by some positive integer multiple of π radians in order to reach the starting point and begin to retrace the curve. Let (r,θ) be the coordinates of a point P on the curve for $0\le\theta<2\pi$. Now $a\sin n(\theta+2\pi)=a\sin(n\theta+2\pi n)=a\sin n\theta=r$ so P is reached again with coordinates $(r,\theta+2\pi)$ thus the curve is traced out either exactly once or exactly twice for $0\le\theta<2\pi$. If for $0\le\theta<\pi$, $P(r,\theta)$ is reached again with coordinates $(-r,\theta+\pi)$ then the curve is traced out exactly once for $0\le\theta<\pi$, otherwise exactly once for $0\le\theta<2\pi$. But

$$a\sin n(\theta+\pi)=a\sin(n\theta+n\pi)=\begin{cases} a\sin n\theta, & n \text{ even}\\ -a\sin n\theta, & n\text{ odd}\end{cases}$$

so the curve is traced out exactly once for $0\le\theta<2\pi$ if n is even, and exactly once for $0\le\theta<\pi$ if n is odd.

EXERCISE SET 11.2

1. **(a)** $dy/dx=\dfrac{1/2}{2t}=1/(4t)$; $dy/dx\big|_{t=-1}=-1/4$; $dy/dx\big|_{t=1}=1/4$

(b) $x=(2y)^2+1, dx/dy=8y, dy/dx\big|_{y=\pm(1/2)}=\pm 1/4$

2. **(a)** $dy/dx = (4\cos t)/(-3\sin t) = -(4/3)\cot t;\ dy/dx\big|_{t=\pi/4} = -4/3,\ dy/dx\big|_{t=7\pi/4} = 4/3$

(b) $(x/3)^2 + (y/4)^2 = 1,\ 2x/9 + (2y/16)(dy/dx) = 0,\ dy/dx = -16x/9y,$

$dy/dx\big|_{\substack{x=3/\sqrt{2}\\ y=4/\sqrt{2}}} = -4/3; dy/dx\big|_{\substack{x=3/\sqrt{2}\\ y=-4/\sqrt{2}}} = 4/3$

3. $\dfrac{d^2y}{dx^2} = \dfrac{d}{dx}\dfrac{dy}{dx} = \dfrac{d}{dt}\left(\dfrac{dy}{dx}\right)\dfrac{dt}{dx} = -\dfrac{1}{4t^2}(1/2t) = -1/(8t^3)$; positive when $t = -1$, negative when $t = 1$

4. $\dfrac{d^2y}{dx^2} = \dfrac{d}{dt}\left(\dfrac{dy}{dx}\right)\dfrac{dt}{dx} = \dfrac{-(4/3)(-\csc^2 t)}{-3\sin t} = -\dfrac{4}{9}\csc^3 t$; negative at $t = \pi/4$, positive at $t = 7\pi/4$.

5. $dy/dx = \dfrac{2}{1/(2\sqrt{t})} = 4\sqrt{t},\ d^2y/dx^2 = \dfrac{2/\sqrt{t}}{1/(2\sqrt{t})} = 4,\ dy/dx\big|_{t=1} = 4,\ d^2y/dx^2\big|_{t=1} = 4$

6. $dy/dx = \dfrac{t^2}{t} = t,\ d^2y/dx^2 = \dfrac{1}{t},\ dy/dx\big|_{t=2} = 2,\ d^2y/dx^2\big|_{t=2} = 1/2$

7. $dy/dx = \dfrac{\sec^2 t}{\sec t\tan t} = \csc t,\ d^2y/dx^2 = \dfrac{-\csc t\cot t}{\sec t\tan t} = -\cot^3 t,$

$dy/dx\big|_{t=\pi/3} = 2/\sqrt{3},\ d^2y/dx^2\big|_{t=\pi/3} = -1/(3\sqrt{3})$

8. $dy/dx = \dfrac{\sinh t}{\cosh t} = \tanh t,\ \dfrac{d^2y}{dx^2} = \operatorname{sech}^2 t/\cosh t = \operatorname{sech}^3 t,\ dy/dx\big|_{t=0} = 0, d^2y/dx^2\big|_{t=0} = 1$

9. $\dfrac{dy}{dx} = \dfrac{dy/d\theta}{dx/d\theta} = \dfrac{-\cos\theta}{2-\sin\theta};\ \dfrac{d^2y}{dx^2} = \dfrac{d}{d\theta}\left(\dfrac{dy}{dx}\right)/\dfrac{dx}{d\theta} = \dfrac{1}{(2-\sin\theta)^2}\dfrac{1}{2-\sin\theta} = \dfrac{1}{(2-\sin\theta)^3};$

$\dfrac{dy}{dx}\bigg|_{\theta=\pi/3} = \dfrac{-1/2}{2-\sqrt{3}/2} = \dfrac{-1}{4-\sqrt{3}};\ \dfrac{d^2y}{dx^2}\bigg|_{\theta=\pi/3} = \dfrac{1}{(2-\sqrt{3}/2)^3} = \dfrac{8}{(4-\sqrt{3})^3}$

10. $\dfrac{dy}{dx} = \dfrac{3\cos\phi}{-\sin\phi} = -3\cot\phi;\ \dfrac{d^2y}{dx^2} = \dfrac{d}{d\phi}(-3\cot\phi)\dfrac{d\phi}{dx} = -3(-\csc^2\phi)(-\csc\phi) = -3\csc^3\phi;$

$\dfrac{dy}{dx}\bigg|_{\phi=5\pi/6} = 3\sqrt{3};\ \dfrac{d^2y}{dx^2}\bigg|_{\phi=5\pi/6} = -24$

11. **(a)** $dy/dx = \dfrac{-e^{-t}}{e^t} = -e^{-2t}$; for $t = 1$, $dy/dx = -e^{-2}$, $(x, y) = (e, e^{-1})$; $y - e^{-1} = -e^{-2}(x - e)$, $y = -e^{-2}x + 2e^{-1}$

(b) $y = 1/x, dy/dx = -1/x^2, m = -1/e^2, y - e^{-1} = -\dfrac{1}{e^2}(x - e), y = -\dfrac{1}{e^2}x + \dfrac{2}{e}$

12. $dy/dx = \dfrac{16t-2}{2} = 8t - 1$; for $t = 1$, $dy/dx = 7$, $(x, y) = (6, 10)$; $y - 10 = 7(x - 6)$, $y = 7x - 32$

13. $dy/dx = \dfrac{4\cos t}{-2\sin t} = -2\cot t$

(a) $dy/dx = 0$ if $\cot t = 0$, $t = \pi/2 + n\pi$ for $n = 0, \pm 1, \cdots$

(b) $dx/dy = -\dfrac{1}{2}\tan t = 0$ if $\tan t = 0$, $t = n\pi$ for $n = 0, \pm 1, \cdots$

14. $dy/dx = \dfrac{2t+1}{6t^2-30t+24} = \dfrac{2t+1}{6(t-1)(t-4)}$

(a) $dy/dx = 0$ if $t = -1/2$

(b) $dx/dy = \dfrac{6(t-1)(t-4)}{2t+1} = 0$ if $t = 1, 4$

15. $x = y = 0$ when $t = 0, \pi$; $\dfrac{dy}{dx} = \dfrac{2\cos 2t}{\cos t}$; $\left.\dfrac{dy}{dx}\right|_{t=0} = 2$, $\left.\dfrac{dy}{dx}\right|_{t=\pi} = -2$, the equations of the tangent lines are $y = -2x, y = 2x$.

16. $y(t) = 0$ has three solutions, $t = 0, \pm\pi/2$; the last two correspond to the crossing point. For $t = \pm\pi/2$, $m = \dfrac{dy}{dx} = \dfrac{2}{\pm\pi}$; the tangent lines are given by $y = \pm\dfrac{2}{\pi}(x-2)$.

17. If $y = 4$ then $t^2 = 4$, $t = \pm 2$, $x = 0$ for $t = \pm 2$ so $(0, 4)$ is reached when $t = \pm 2$. $dy/dx = 2t/(3t^2-4)$. For $t = 2$, $dy/dx = 1/2$ and for $t = -2$, $dy/dx = -1/2$. The tangent lines are $y = \pm x/2 + 4$.

18. If $x = 3$ then $t^2 - 3t + 5 = 3$, $t^2 - 3t + 2 = 0$, $(t-1)(t-2) = 0$, $t = 1$ or 2. If $t = 1$ or 2 then $y = 1$ so $(3, 1)$ is reached when $t = 1$ or 2. $dy/dx = (3t^2+2t-10)/(2t-3)$. For $t = 1$, $dy/dx = 5$, the tangent line is $y - 1 = 5(x-3)$, $y = 5x - 14$. For $t = 2$, $dy/dx = 6$, the tangent line is $y - 1 = 6(x-3)$, $y = 6x - 17$.

19. (a)

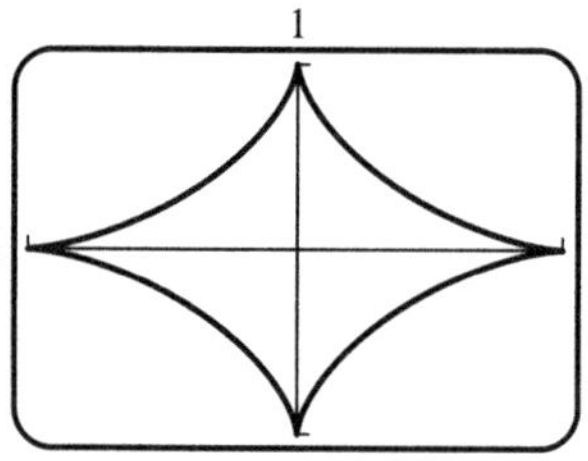

(b) $\dfrac{dx}{dt} = -3\cos^2 t \sin t$ and $\dfrac{dy}{dt} = 3\sin^2 t \cos t$ are both zero when $t = 0, \pi/2, \pi, 3\pi/2, 2\pi$, so singular points occur at these values of t.

20. (a) when $y = 0$

(b) $\dfrac{dx}{dy} = \dfrac{a - a\cos\theta}{a\sin\theta} = 0$ when $\theta = 2n\pi, n = 0, 1, \ldots$ (which is when $y = 0$).

21. Substitute $\theta = \pi/3$, $r = 1$, and $dr/d\theta = -\sqrt{3}$ in equation (7) gives slope $m = 1/\sqrt{3}$.

22. As in Exercise 21, $\theta = \pi/4$, $dr/d\theta = \sqrt{2}/2$, $r = 1 + \sqrt{2}/2$, $m = -1 - \sqrt{2}$

23. As in Exercise 21, $\theta = 2$, $dr/d\theta = -1/4$, $r = 1/2$, $m = \dfrac{\tan 2 - 2}{2\tan 2 + 1}$

24. As in Exercise 21, $\theta = \pi/6$, $dr/d\theta = 4\sqrt{3}a$, $r = 2a$, $m = 3\sqrt{3}/5$

25. As in Exercise 21, $\theta = 3\pi/4$, $dr/d\theta = -3\sqrt{2}/2$, $r = \sqrt{2}/2$, $m = -2$

26. As in Exercise 21, $\theta = \pi$, $dr/d\theta = 3$, $r = 4$, $m = 4/3$

27. $m = \dfrac{dy}{dx} = \dfrac{r\cos\theta + (\sin\theta)(dr/d\theta)}{-r\sin\theta + (\cos\theta)(dr/d\theta)} = \dfrac{\cos\theta + 2\sin\theta\cos\theta}{-\sin\theta + \cos^2\theta - \sin^2\theta}$; if $\theta = 0, \pi/2, \pi$, then $m = 1, 0, -1$.

28. $m = \dfrac{dy}{dx} = \dfrac{\cos\theta(4\sin\theta - 1)}{4\cos^2\theta + \sin\theta - 2}$; if $\theta = 0, \pi/2, \pi$ then $m = -1/2, 0, 1/2$.

29. $dx/d\theta = -a\sin\theta(1 + 2\cos\theta)$, $dy/d\theta = a(2\cos\theta - 1)(\cos\theta + 1)$

(a) horizontal if $dy/d\theta = 0$ and $dx/d\theta \neq 0$. $dy/d\theta = 0$ when $\cos\theta = 1/2$ or $\cos\theta = -1$ so $\theta = \pi/3$, $5\pi/3$, or π; $dx/d\theta \neq 0$ for $\theta = \pi/3$ and $5\pi/3$. For the singular point $\theta = \pi$ we find that $\lim_{\theta\to\pi} dy/dx = 0$. There is a horizontal tangent line at $(3a/2, \pi/3), (0, \pi)$, and $(3a/2, 5\pi/3)$.

(b) vertical if $dy/d\theta \neq 0$ and $dx/d\theta = 0$. $dx/d\theta = 0$ when $\sin\theta = 0$ or $\cos\theta = -1/2$ so $\theta = 0$, π, $2\pi/3$, or $4\pi/3$; $dy/d\theta \neq 0$ for $\theta = 0$, $2\pi/3$, and $4\pi/3$. The singular point $\theta = \pi$ was discussed in Part (a). There is a vertical tangent line at $(2a, 0), (a/2, 2\pi/3)$, and $(a/2, 4\pi/3)$.

30. $dx/d\theta = a(\cos^2\theta - \sin^2\theta) = a\cos 2\theta, dy/d\theta = 2a\sin\theta\cos\theta = a\sin 2\theta$

(a) horizontal if $dy/d\theta = 0$ and $dx/d\theta \neq 0$. $dy/d\theta = 0$ when $\theta = 0, \pi/2, \pi, 3\pi/2$;
$dx/d\theta \neq 0$ for $(0,0), (a, \pi/2), (0, \pi), (-a, 3\pi/2)$; in reality only two distinct points

(b) vertical if $dy/d\theta \neq 0$ and $dx/d\theta = 0$. $dx/d\theta = 0$ when $\theta = \pi/4, 3\pi/4, 5\pi/4, 7\pi/4$; $dy/d\theta \neq 0$ there, so vertical tangent line at $(a/\sqrt{2}, \pi/4), (a/\sqrt{2}, 3\pi/4), (-a/\sqrt{2}, 5\pi/4), (-a/\sqrt{2}, 7\pi/4)$, only two distinct points

31. $dy/d\theta = (d/d\theta)(\sin^2\theta\cos^2\theta) = (\sin 4\theta)/2 = 0$ at $\theta = 0, \pi/4, \pi/2, 3\pi/4, \pi$; at the same points, $dx/d\theta = (d/d\theta)(\sin\theta\cos^3\theta) = \cos^2\theta(4\cos^2\theta - 3)$. Next, $\dfrac{dx}{d\theta} = 0$ at $\theta = \pi/2$, a singular point; and $\theta = 0, \pi$ both give the same point, so there are just three points with a horizontal tangent.

32. $dx/d\theta = 4\sin^2\theta - \sin\theta - 2$, $dy/d\theta = \cos\theta(1 - 4\sin\theta)$. $dy/d\theta = 0$ when $\cos\theta = 0$ or $\sin\theta = 1/4$ so $\theta = \pi/2$, $3\pi/2$, $\sin^{-1}(1/4)$, or $\pi - \sin^{-1}(1/4)$; $dx/d\theta \neq 0$ at these points, so there is a horizontal tangent at each one.

33.

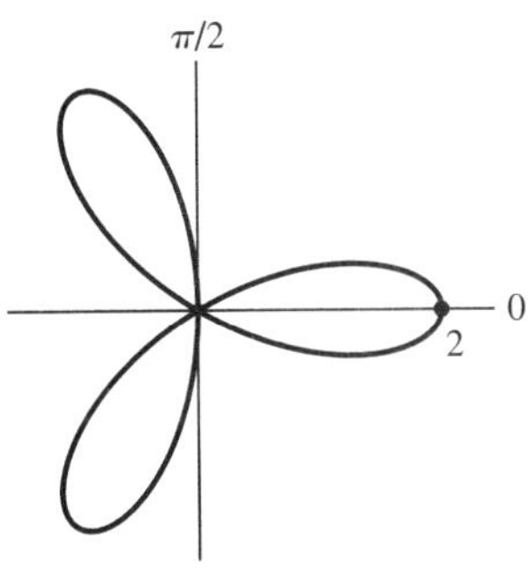

$\theta_0 = \pi/6, \pi/2, 5\pi/6$

34.

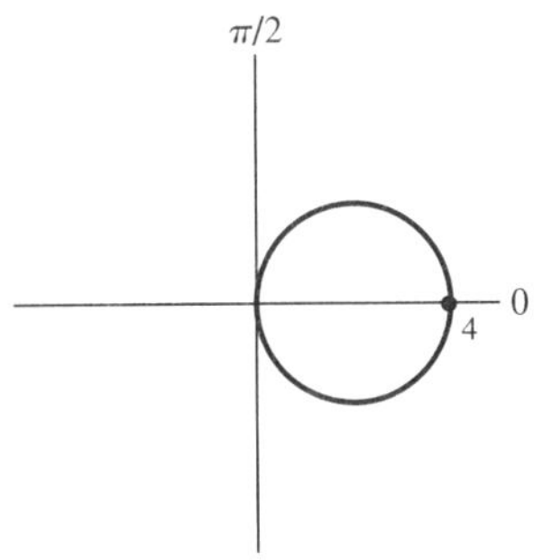

$\theta_0 = \pi/2$

35.

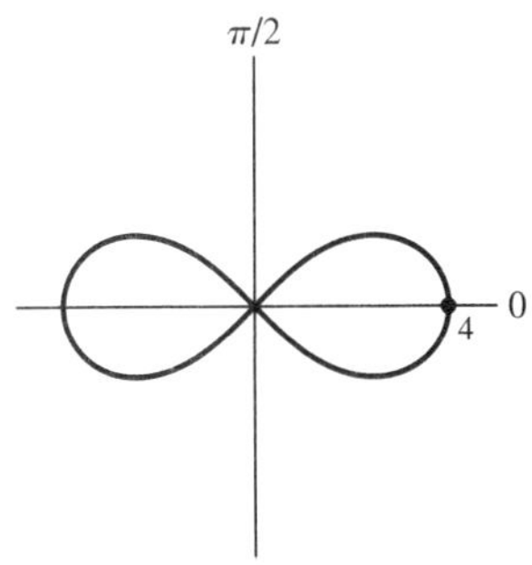

$\theta_0 = \pm\pi/4$

36.

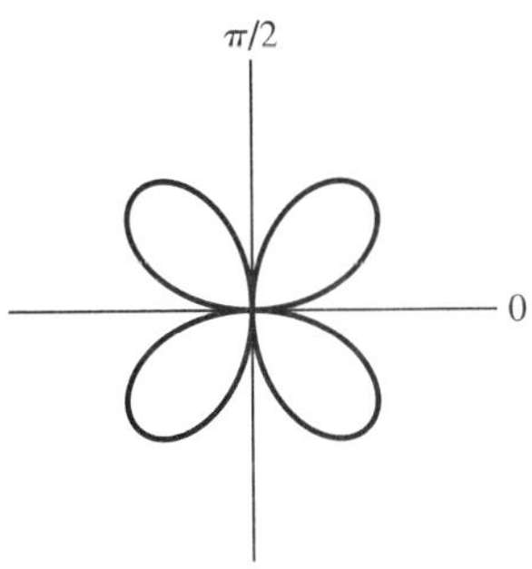

$\theta_0 = 0, \pi/2$

37.

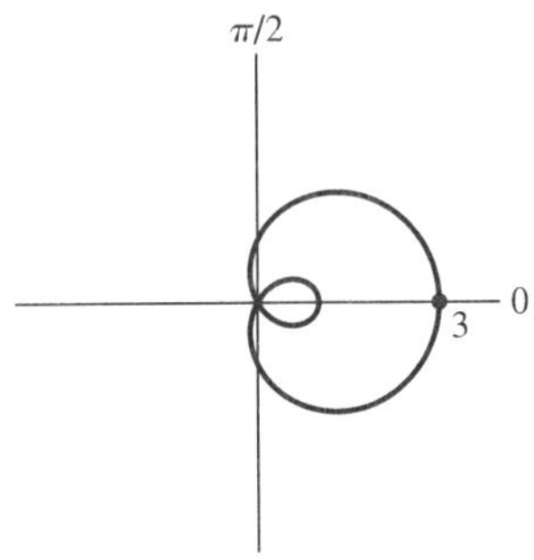

$\theta_0 = 2\pi/3, 4\pi/3$

38.

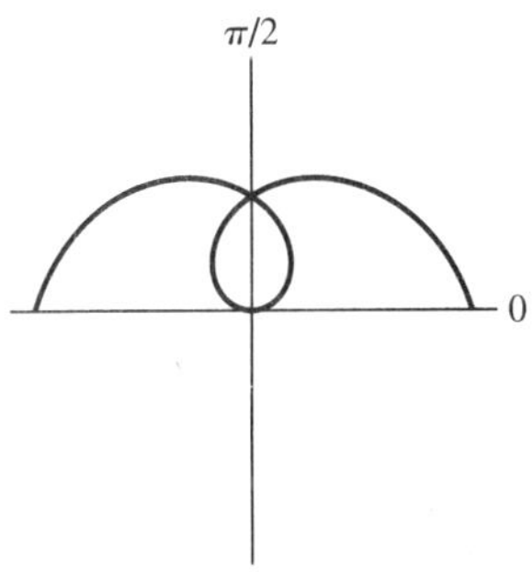

$\theta_0 = 0$

39. $r^2+(dr/d\theta)^2=a^2+0^2=a^2$, $L=\int_0^{2\pi} a\,d\theta=2\pi a$

40. $r^2+(dr/d\theta)^2=(2a\cos\theta)^2+(-2a\sin\theta)^2=4a^2$, $L=\int_{-\pi/2}^{\pi/2} 2a\,d\theta=2\pi a$

41. $r^2+(dr/d\theta)^2=[a(1-\cos\theta)]^2+[a\sin\theta]^2=4a^2\sin^2(\theta/2)$, $L=2\int_0^{\pi} 2a\sin(\theta/2)d\theta=8a$

42. $r^2+(dr/d\theta)^2=[\sin^2(\theta/2)]^2+[\sin(\theta/2)\cos(\theta/2)]^2=\sin^2(\theta/2)$, $L=\int_0^{\pi}\sin(\theta/2)d\theta=2$

43. $r^2+(dr/d\theta)^2=(e^{3\theta})^2+(3e^{3\theta})^2=10e^{6\theta}$, $L=\int_0^2\sqrt{10}e^{3\theta}d\theta=\sqrt{10}(e^6-1)/3$

44. $r^2+(dr/d\theta)^2=[\sin^3(\theta/3)]^2+[\sin^2(\theta/3)\cos(\theta/3)]^2=\sin^4(\theta/3)$,
$L=\int_0^{\pi/2}\sin^2(\theta/3)d\theta=(2\pi-3\sqrt{3})/8$

45. **(a)** From (3), $\dfrac{dy}{dx}=\dfrac{3\sin t}{1-3\cos t}$

(b) At $t=10$, $\dfrac{dy}{dx}=\dfrac{3\sin 10}{1-3\cos 10}\approx-0.46402$, $\theta\approx\tan^{-1}(-0.46402)=-0.4345$

46. **(a)** $\dfrac{dy}{dx}=0$ when $\dfrac{dy}{dt}=2\sin t=0, t=0,\pi,2\pi,3\pi$

(b) $\dfrac{dx}{dt}=0$ when $1-2\cos t=0, \cos t=1/2, t=\pi/3,5\pi/3,7\pi/3$

47. **(a)** $r^2+(dr/d\theta)^2=(\cos n\theta)^2+(-n\sin n\theta)^2=\cos^2 n\theta+n^2\sin^2 n\theta$
$=(1-\sin^2 n\theta)+n^2\sin^2 n\theta=1+(n^2-1)\sin^2 n\theta$,
$L=2\int_0^{\pi/(2n)}\sqrt{1+(n^2-1)\sin^2 n\theta d\theta}$

(b) $L=2\int_0^{\pi/4}\sqrt{1+3\sin^2 2\theta d\theta}\approx 2.42$

(c)

n	2	3	4	5	6	7	8	9	10	11
L	2.42211	2.22748	2.14461	2.10100	2.07501	2.05816	2.04656	2.03821	2.03199	2.02721

n	12	13	14	15	16	17	18	19	20
L	2.02346	2.02046	2.01802	2.01600	2.01431	2.01288	2.01167	2.01062	2.00971

48. **(a)**

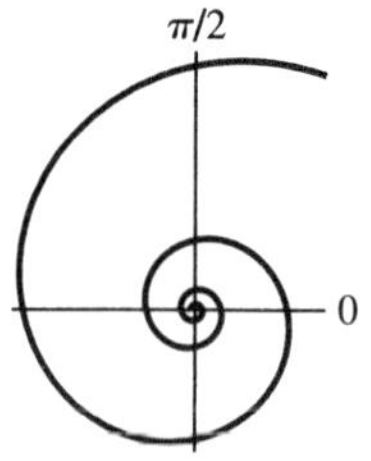

(b) $r^2+(dr/d\theta)^2=(e^{-\theta})^2+(-e^{-\theta})^2=2e^{-2\theta}$, $L=2\int_0^{+\infty}e^{-2\theta}d\theta$

(c) $L=\lim_{\theta_0\to+\infty}2\int_0^{\theta_0}e^{-2\theta}d\theta=\lim_{\theta_0\to+\infty}(1-e^{-2\theta_0})=1$

49. $x' = 2t$, $y' = 2$, $(x')^2 + (y')^2 = 4t^2 + 4$

$$S = 2\pi \int_0^4 (2t)\sqrt{4t^2+4}dt = 8\pi \int_0^4 t\sqrt{t^2+1}dt = \frac{8\pi}{3}(t^2+1)^{3/2}\Big]_0^4 = \frac{8\pi}{3}(17\sqrt{17}-1)$$

50. $x' = e^t(\cos t - \sin t)$, $y' = e^t(\cos t + \sin t)$, $(x')^2 + (y')^2 = 2e^{2t}$

$$S = 2\pi \int_0^{\pi/2} (e^t \sin t)\sqrt{2e^{2t}}dt = 2\sqrt{2}\pi \int_0^{\pi/2} e^{2t}\sin t\, dt$$

$$= 2\sqrt{2}\pi \left[\frac{1}{5}e^{2t}(2\sin t - \cos t)\right]_0^{\pi/2} = \frac{2\sqrt{2}}{5}\pi(2e^{\pi}+1)$$

51. $x' = -2\sin t\cos t$, $y' = 2\sin t\cos t$, $(x')^2 + (y')^2 = 8\sin^2 t\cos^2 t$

$$S = 2\pi \int_0^{\pi/2} \cos^2 t\sqrt{8\sin^2 t\cos^2 t}\, dt = 4\sqrt{2}\pi \int_0^{\pi/2} \cos^3 t\sin t\, dt = -\sqrt{2}\pi\cos^4 t\Big]_0^{\pi/2} = \sqrt{2}\pi$$

52. $x' = 1$, $y' = 4t$, $(x')^2 + (y')^2 = 1 + 16t^2$, $S = 2\pi \int_0^1 t\sqrt{1+16t^2}\, dt = \frac{\pi}{24}(17\sqrt{17}-1)$

53. $x' = -r\sin t$, $y' = r\cos t$, $(x')^2 + (y')^2 = r^2$, $S = 2\pi \int_0^{\pi} r\sin t\sqrt{r^2}\, dt = 2\pi r^2 \int_0^{\pi} \sin t\, dt = 4\pi r^2$

54. $\frac{dx}{d\phi} = a(1-\cos\phi)$, $\frac{dy}{d\phi} = a\sin\phi$, $\left(\frac{dx}{d\phi}\right)^2 + \left(\frac{dy}{d\phi}\right)^2 = 2a^2(1-\cos\phi)$

$$S = 2\pi \int_0^{2\pi} a(1-\cos\phi)\sqrt{2a^2(1-\cos\phi)}\, d\phi = 2\sqrt{2}\pi a^2 \int_0^{2\pi} (1-\cos\phi)^{3/2}d\phi,$$

but $1-\cos\phi = 2\sin^2\frac{\phi}{2}$ so $(1-\cos\phi)^{3/2} = 2\sqrt{2}\sin^3\frac{\phi}{2}$ for $0 \le \phi \le \pi$ and, taking advantage of the symmetry of the cycloid, $S = 16\pi a^2 \int_0^{\pi} \sin^3\frac{\phi}{2}d\phi = 64\pi a^2/3$

55. (a) $\frac{dr}{dt} = 2$ and $\frac{d\theta}{dt} = 1$ so $\frac{dr}{d\theta} = \frac{dr/dt}{d\theta/dt} = \frac{2}{1} = 2$, $r = 2\theta + C$, $r = 10$ when $\theta = 0$ so $10 = C, r = 2\theta + 10$.

(b) $r^2 + (dr/d\theta)^2 = (2\theta+10)^2 + 4$, during the first 5 seconds the rod rotates through an angle of $(1)(5) = 5$ radians so $L = \int_0^5 \sqrt{(2\theta+10)^2+4}d\theta$, let $u = 2\theta + 10$ to get

$$L = \frac{1}{2}\int_{10}^{20} \sqrt{u^2+4}du = \frac{1}{2}\left[\frac{u}{2}\sqrt{u^2+4} + 2\ln|u+\sqrt{u^2+4}|\right]_{10}^{20}$$

$$= \frac{1}{2}\left[10\sqrt{404} - 5\sqrt{104} + 2\ln\frac{20+\sqrt{404}}{10+\sqrt{104}}\right] \approx 75.7 \text{ mm}$$

56. $x = r\cos\theta, y = r\sin\theta$, $\frac{dx}{d\theta} = \frac{dr}{d\theta}\cos\theta - r\sin\theta$, $\frac{dy}{d\theta} = r\cos\theta + \frac{dr}{d\theta}\sin\theta$,

$\left(\frac{dx}{d\theta}\right)^2 + \left(\frac{dy}{d\theta}\right)^2 = r^2 + \left(\frac{dr}{d\theta}\right)^2$, and Formula (6) of Section 8.4 becomes

$$L = \int_{\alpha}^{\beta} \sqrt{r^2 + \left(\frac{dr}{d\theta}\right)^2}\, d\theta$$

EXERCISE SET 11.3

1. (a) $\displaystyle\int_{\pi/2}^{\pi} \frac{1}{2}(1-\cos\theta)^2\,d\theta$ (b) $\displaystyle\int_0^{\pi/2} \frac{1}{2}4\cos^2\theta\,d\theta$ (c) $\displaystyle\int_0^{\pi/2} \frac{1}{2}\sin^2 2\theta\,d\theta$

(d) $\displaystyle\int_0^{2\pi} \frac{1}{2}\theta^2\,d\theta$ (e) $\displaystyle\int_{-\pi/2}^{\pi/2} \frac{1}{2}(1-\sin\theta)^2\,d\theta$ (f) $\displaystyle 2\int_0^{\pi/4} \frac{1}{2}\cos^2 2\theta\,d\theta$

2. (a) $3\pi/8+1$ (b) $\pi/2$ (c) $\pi/8$

(d) $4\pi^3/3$ (e) $3\pi/4$ (f) $\pi/8$

3. (a) $\displaystyle A=\int_0^{2\pi}\frac{1}{2}a^2\,d\theta=\pi a^2$ (b) $\displaystyle A=\int_0^{\pi}\frac{1}{2}4a^2\sin^2\theta\,d\theta=\pi a^2$

(c) $\displaystyle A=\int_{-\pi/2}^{\pi/2}\frac{1}{2}4a^2\cos^2\theta\,d\theta=\pi a^2$

4. (a) $r^2=r\sin\theta+r\cos\theta,\ x^2+y^2-y-x=0,\ \left(x-\frac{1}{2}\right)^2+\left(y-\frac{1}{2}\right)^2=\frac{1}{2}$

(b) $\displaystyle A=\int_{-\pi/4}^{3\pi/4}\frac{1}{2}(\sin\theta+\cos\theta)^2\,d\theta=\pi/2$

5. $\displaystyle A=2\int_0^{\pi}\frac{1}{2}(2+2\cos\theta)^2d\theta=6\pi$ **6.** $\displaystyle A=\int_0^{\pi/2}\frac{1}{2}(1+\sin\theta)^2d\theta=3\pi/8+1$

7. $\displaystyle A=6\int_0^{\pi/6}\frac{1}{2}(16\cos^2 3\theta)d\theta=4\pi$

8. The petal in the first quadrant has area $\displaystyle\int_0^{\pi/2}\frac{1}{2}4\sin^2 2\theta\,d\theta=\pi/2$, so total area $=2\pi$.

9. $\displaystyle A=2\int_{2\pi/3}^{\pi}\frac{1}{2}(1+2\cos\theta)^2d\theta=\pi-3\sqrt{3}/2$ **10.** $\displaystyle A=\int_1^3\frac{2}{\theta^2}d\theta=4/3$

11. area $\displaystyle =A_1-A_2=\int_0^{\pi/2}\frac{1}{2}4\cos^2\theta\,d\theta-\int_0^{\pi/4}\frac{1}{2}\cos 2\theta\,d\theta=\pi/2-\frac{1}{4}$

12. area $\displaystyle =A_1-A_2=\int_0^{\pi}\frac{1}{2}(1+\cos\theta)^2\,d\theta-\int_0^{\pi/2}\frac{1}{2}\cos^2\theta\,d\theta=5\pi/8$

13. The circles intersect when $\cos t=\sqrt{3}\sin t, \tan t=1/\sqrt{3}, t=\pi/6$, so

$$A=A_1+A_2=\int_0^{\pi/6}\frac{1}{2}(4\sqrt{3}\sin t)^2\,dt+\int_{\pi/6}^{\pi/2}\frac{1}{2}(4\cos t)^2\,dt=2\pi-3\sqrt{3}+4\pi/3-\sqrt{3}=10\pi/3-4\sqrt{3}.$$

14. The curves intersect when $1+\cos t=3\cos t, \cos t=1/2, t=\pm\pi/3$, and hence

$$\text{total area}=2\int_0^{\pi/3}\frac{1}{2}(1+\cos t)^2\,dt+2\int_{\pi/3}^{\pi/2}\frac{1}{2}9\cos^2 t\,dt=2(\pi/4+9\sqrt{3}/16+3\pi/8-9\sqrt{3}/16)=5\pi/4.$$

15. $\displaystyle A=2\int_{\pi/6}^{\pi/2}\frac{1}{2}[25\sin^2\theta-(2+\sin\theta)^2]d\theta=8\pi/3+\sqrt{3}$

16. $A = 2\int_0^{\pi} \frac{1}{2}[16 - (2 - 2\cos\theta)^2]d\theta = 10\pi$

17. $A = 2\int_0^{\pi/3} \frac{1}{2}[(2+2\cos\theta)^2 - 9]d\theta = 9\sqrt{3}/2 - \pi$

18. $A = 2\int_0^{\pi/4} \frac{1}{2}(16\sin^2\theta)d\theta = 2\pi - 4$

19. $A = 2\left[\int_0^{2\pi/3} \frac{1}{2}(1/2 + \cos\theta)^2 d\theta - \int_{2\pi/3}^{\pi} \frac{1}{2}(1/2 + \cos\theta)^2 d\theta\right] = (\pi + 3\sqrt{3})/4$

20. $A = 2\int_0^{\pi/3} \frac{1}{2}\left[(2 + 2\cos\theta)^2 - \frac{9}{4}\sec^2\theta\right] d\theta = 2\pi + \frac{9}{4}\sqrt{3}$

21. $A = 2\int_0^{\cos^{-1}(3/5)} \frac{1}{2}(100 - 36\sec^2\theta)d\theta = 100\cos^{-1}(3/5) - 48$

22. $A = 8\int_0^{\pi/8} \frac{1}{2}(4a^2\cos^2 2\theta - 2a^2)d\theta = 2a^2$

23. (a) r is not real for $\pi/4 < \theta < 3\pi/4$ and $5\pi/4 < \theta < 7\pi/4$

(b) $A = 4\int_0^{\pi/4} \frac{1}{2}a^2\cos 2\theta\, d\theta = a^2$

(c) $A = 4\int_0^{\pi/6} \frac{1}{2}\left[4\cos 2\theta - 2\right] d\theta = 2\sqrt{3} - \frac{2\pi}{3}$

24. $A = 2\int_0^{\pi/2} \frac{1}{2}\sin 2\theta\, d\theta = 1$

25. $A = \int_{2\pi}^{4\pi} \frac{1}{2}a^2\theta^2\, d\theta - \int_0^{2\pi} \frac{1}{2}a^2\theta^2\, d\theta = 8\pi^3 a^2$

26. (a) $x = r\cos\theta, y = r\sin\theta,$

$(dx/d\theta)^2 + (dy/d\theta)^2 = (f'(\theta)\cos\theta - f(\theta)\sin\theta)^2 + (f'(\theta)\sin\theta + f(\theta)\cos\theta)^2 = f'(\theta)^2 + f(\theta)^2;$

$S = \int_\alpha^\beta 2\pi f(\theta)\sin\theta\sqrt{f'(\theta)^2 + f(\theta)^2}\, d\theta$ if about $\theta = 0$; similarly for $\theta = \pi/2$

(b) f', g' are continuous and no segment of the curve is traced more than once.

27. $r^2 + \left(\frac{dr}{d\theta}\right)^2 = \cos^2\theta + \sin^2\theta = 1,$

so $S = \int_{-\pi/2}^{\pi/2} 2\pi\cos^2\theta\, d\theta = \pi^2.$

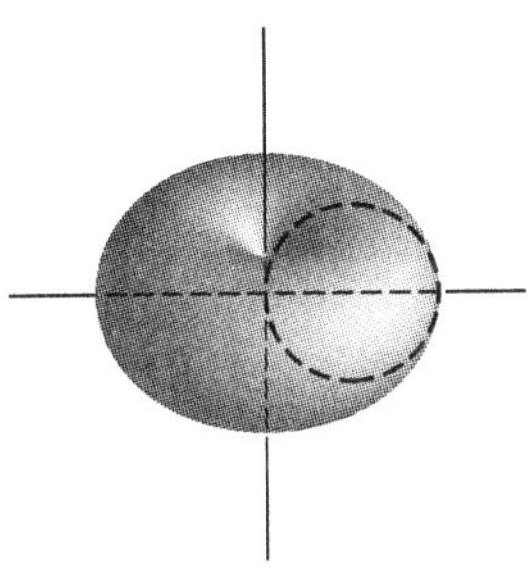

28. $S = \int_0^{\pi/2} 2\pi e^\theta \cos\theta\sqrt{2e^{2\theta}}\, d\theta$

$= 2\sqrt{2}\pi\int_0^{\pi/2} e^{2\theta}\cos\theta\, d\theta = \frac{2\sqrt{2}\pi}{5}(e^\pi - 2)$

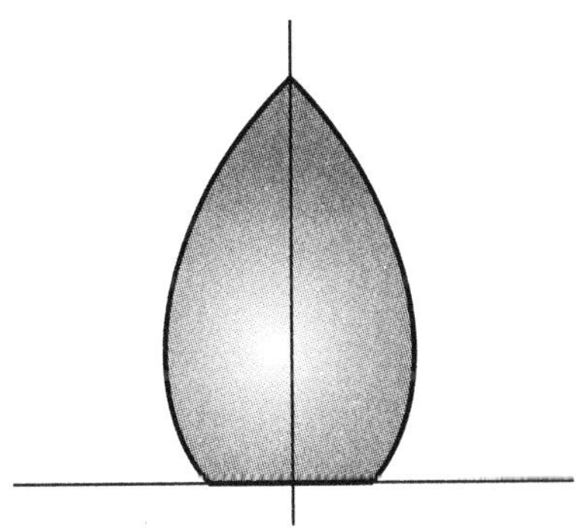

29. $S = \int_0^{\pi} 2\pi(1-\cos\theta)\sin\theta\sqrt{1-2\cos\theta+\cos^2\theta+\sin^2\theta}\,d\theta$

$= 2\sqrt{2}\pi\int_0^{\pi}\sin\theta(1-\cos\theta)^{3/2}\,d\theta = \frac{2}{5}2\sqrt{2}\pi(1-\cos\theta)^{5/2}\Big]_0^{\pi} = 32\pi/5$

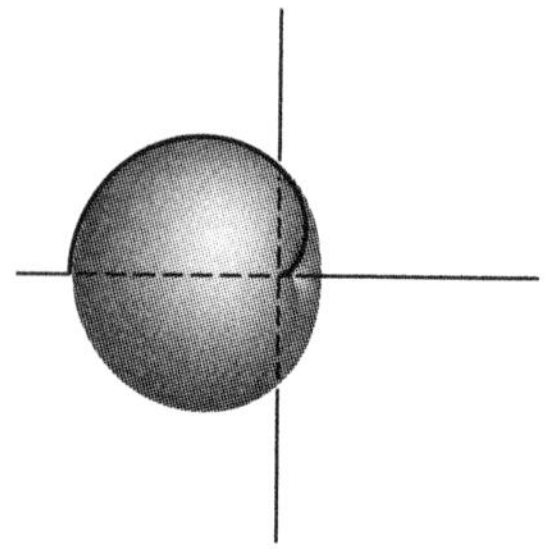

30. $S = \int_0^{\pi} 2\pi a(\sin\theta)a\,d\theta = 4\pi a^2$

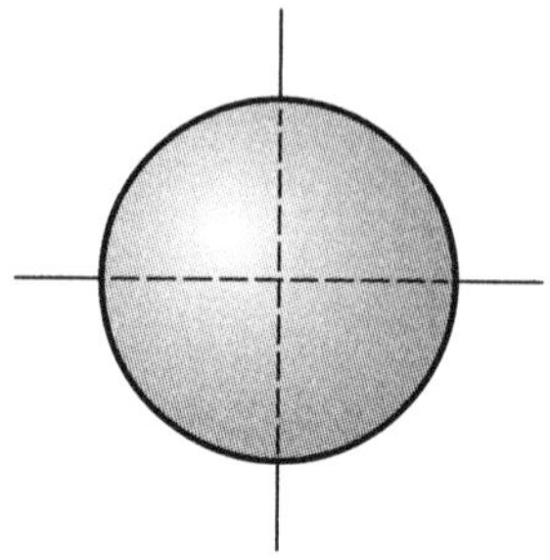

31. **(a)** $r^3\cos^3\theta - 3r^2\cos\theta\sin\theta + r^3\sin^3\theta = 0,\ r = \dfrac{3\cos\theta\sin\theta}{\cos^3\theta+\sin^3\theta}$

32. **(a)** $A = 2\int_0^{\pi/(2n)}\frac{1}{2}a^2\cos^2 n\theta\,d\theta = \dfrac{\pi a^2}{4n}$ **(b)** $A = 2\int_0^{\pi/(2n)}\frac{1}{2}a^2\cos^2 n\theta\,d\theta = \dfrac{\pi a^2}{4n}$

(c) $\dfrac{1}{2n}\times$ total area $= \dfrac{\pi a^2}{4n}$ **(d)** $\dfrac{1}{n}\times$ total area $= \dfrac{\pi a^2}{4n}$

33. If the upper right corner of the square is the point (a,a) then the large circle has equation $r = \sqrt{2}a$ and the small circle has equation $(x-a)^2+y^2 = a^2,\ r = 2a\cos\theta$, so

area of crescent $= 2\int_0^{\pi/4}\frac{1}{2}\left[(2a\cos\theta)^2 - (\sqrt{2}a)^2\right]d\theta = a^2 =$ area of square.

34. $A = \int_0^{2\pi}\frac{1}{2}(\cos 3\theta + 2)^2\,d\theta = 9\pi/2$

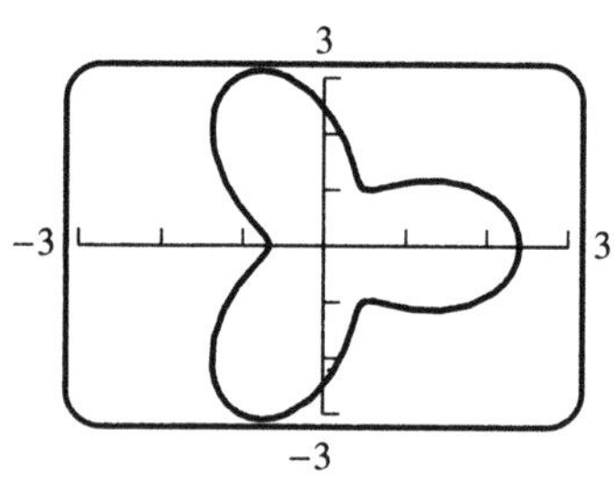

35. $A = \int_0^{\pi/2}\frac{1}{2}4\cos^2\theta\sin^4\theta\,d\theta = \pi/16$

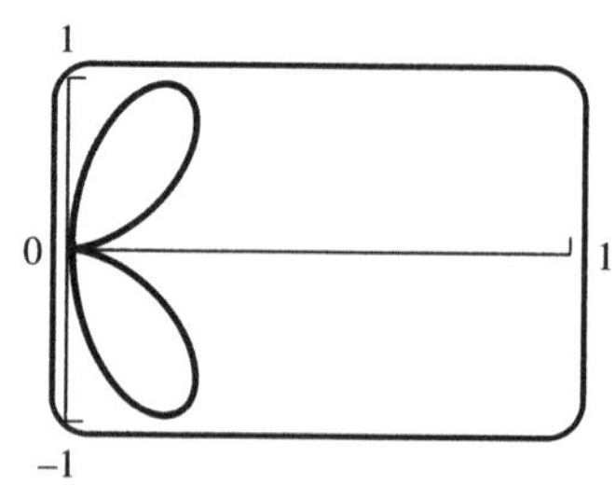

EXERCISE SET 11.4

1. **(a)** $4px = y^2$, point $(1,1), 4p = 1, x = y^2$ **(b)** $-4py = x^2$, point $(3,-3), 12p = 9, -3y = x^2$

(c) $a = 3, b = 2, \dfrac{x^2}{9}+\dfrac{y^2}{4} = 1$ **(d)** $a = 3, b = 2, \dfrac{x^2}{4}+\dfrac{y^2}{9} = 1$

(e) asymptotes: $y = \pm x$, so $a = b$; point $(0, 1)$, so $y^2 - x^2 = 1$

(f) asymptotes: $y = \pm x$, so $b = a$; point $(2, 0)$, so $\dfrac{x^2}{4} - \dfrac{y^2}{4} = 1$

2. **(a)** Part (a), vertex $(0, 0), p = 1/4$; focus $(1/4, 0)$, directrix: $x = -1/4$
Part (b), vertex $(0, 0), p = 3/4$; focus $(0, -3/4)$, directrix: $y = 3/4$

(b) Part (c), $c = \sqrt{a^2 - b^2} = \sqrt{5}$, foci $(\pm\sqrt{5}, 0)$
Part (d), $c = \sqrt{a^2 - b^2} = \sqrt{5}$, foci $(0, \pm\sqrt{5})$

(c) Part (e), $c = \sqrt{a^2 + b^2} = \sqrt{2}$, foci at $(0, \pm\sqrt{2})$; asymptotes: $y^2 - x^2 = 0, y = \pm x$
Part (f), $c = \sqrt{a^2 + b^2} = \sqrt{8} = 2\sqrt{2}$, foci at $(\pm 2\sqrt{2}, 0)$; asymptotes: $\dfrac{x^2}{4} - \dfrac{y^2}{4} = 0, y = \pm x$

3. **(a)**

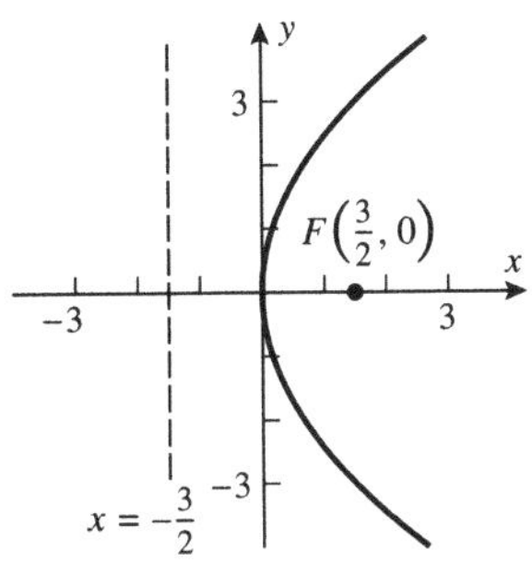

(b)

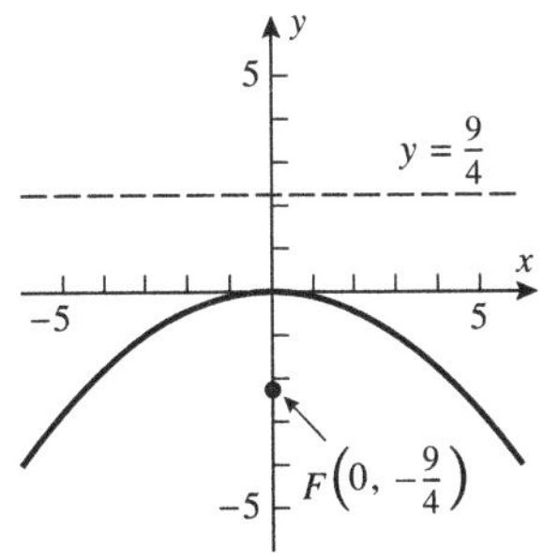

4. **(a)**

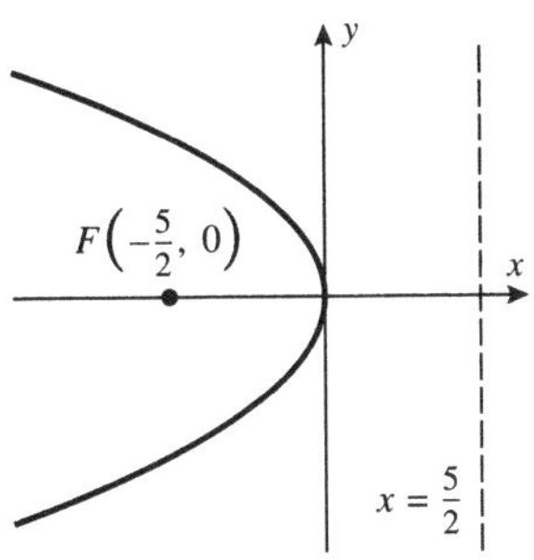

(b)

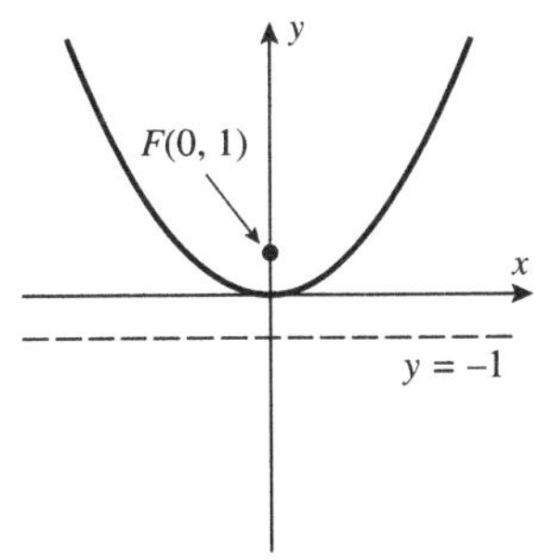

5. **(a)**

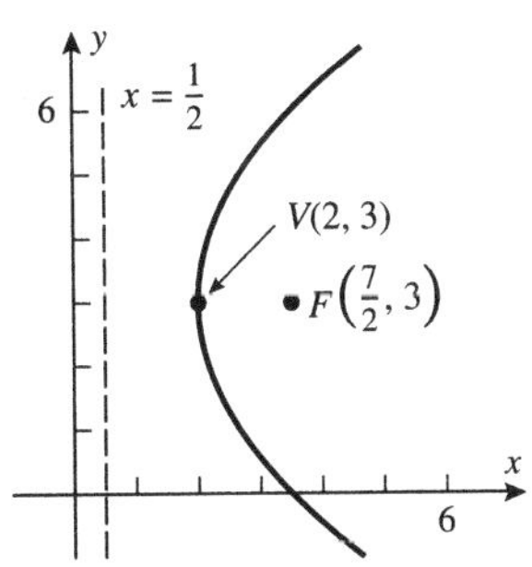

(b)

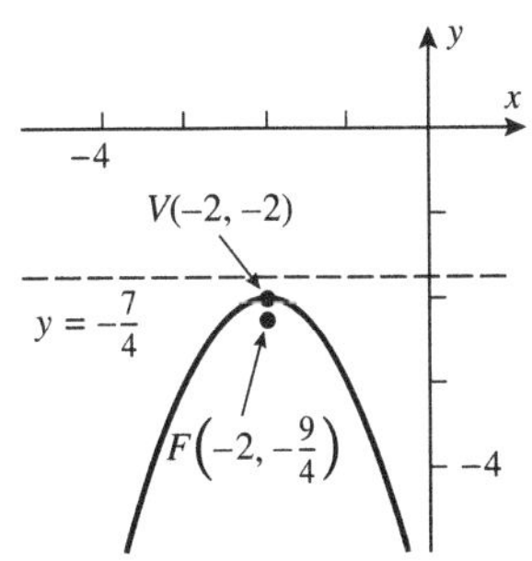

6. **(a)**

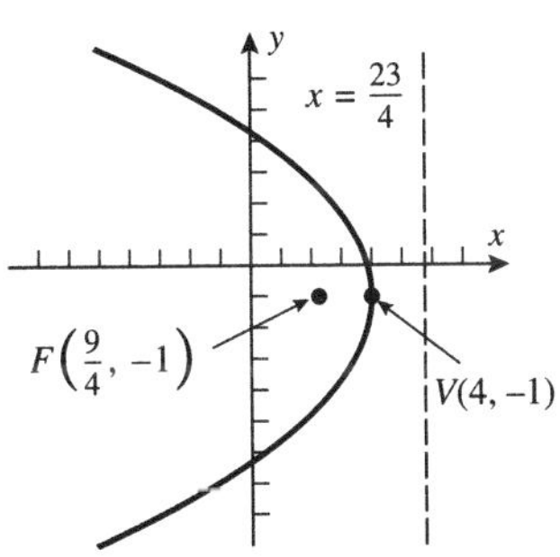

(b)

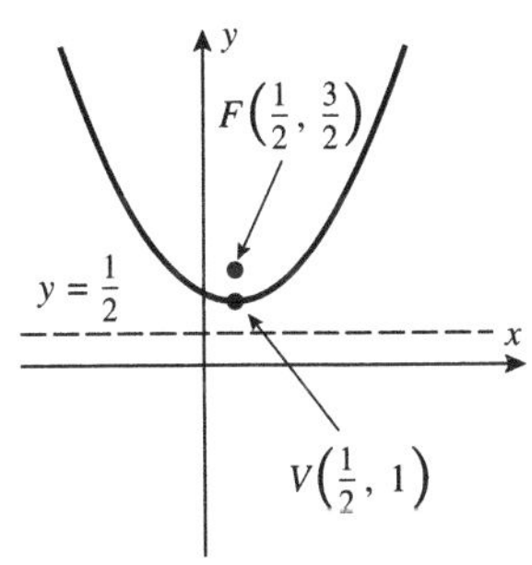

7. (a)

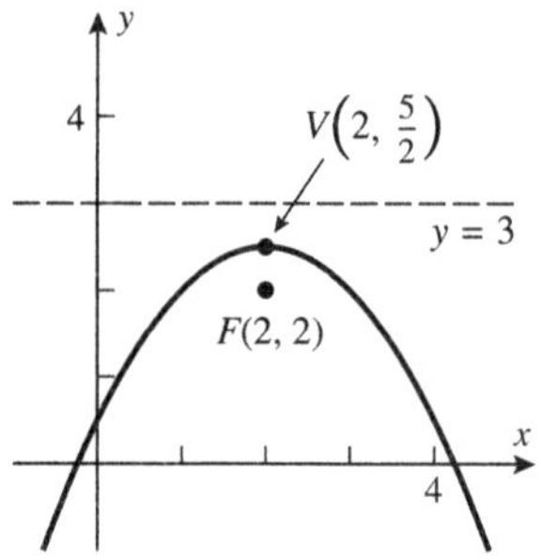

(b)

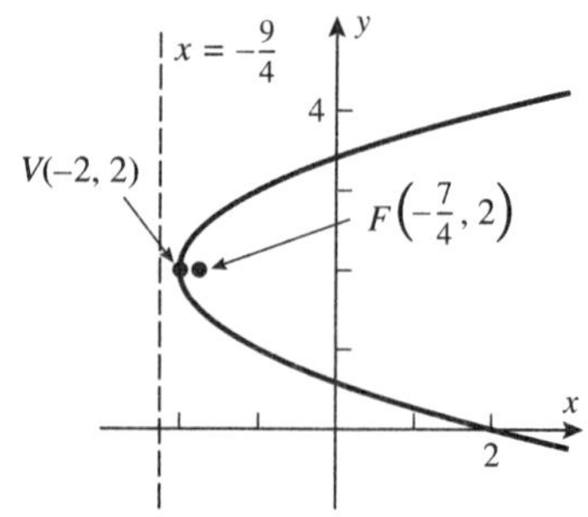

8. (a)

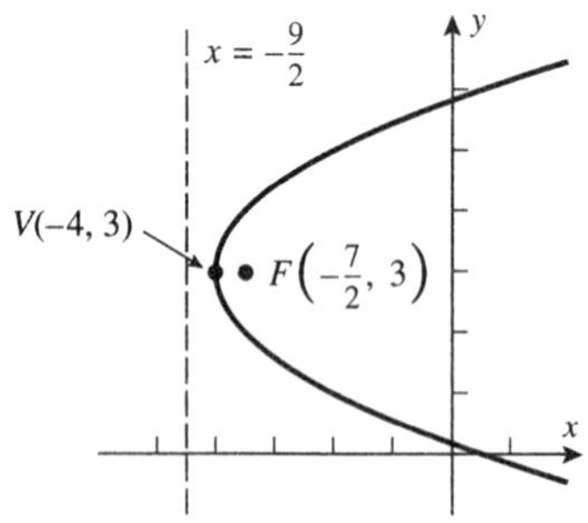

(b)

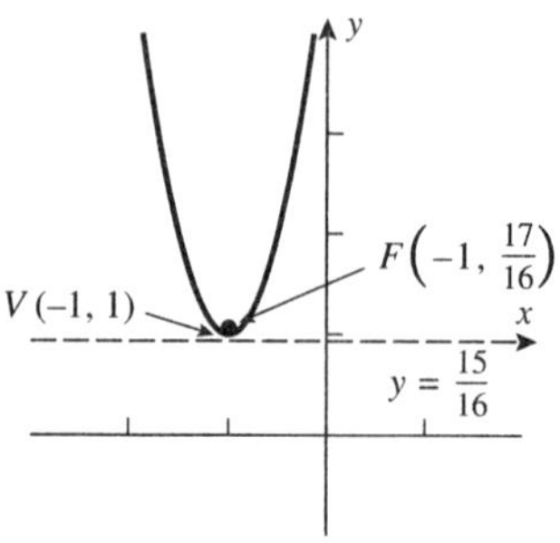

9. (a) $c^2 = 16 - 9 = 7,\ c = \sqrt{7}$

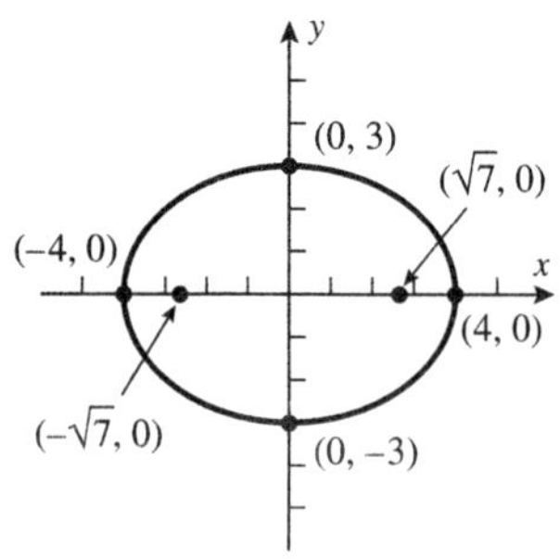

(b) $\dfrac{x^2}{1} + \dfrac{y^2}{9} = 1$

$c^2 = 9 - 1 = 8, c = \sqrt{8}$

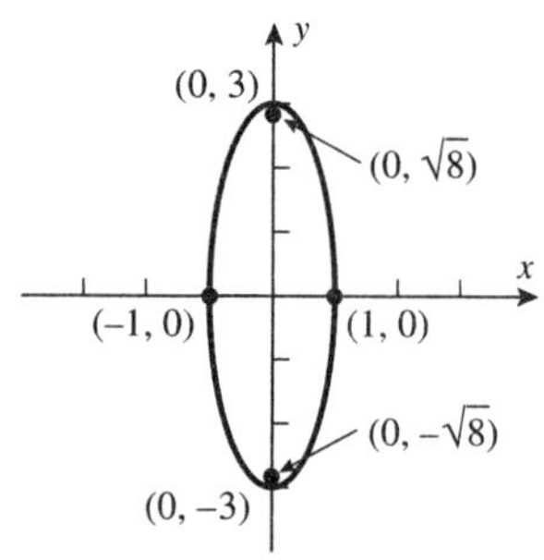

10. (a) $c^2 = 25 - 4 = 21,\ c = \sqrt{21}$

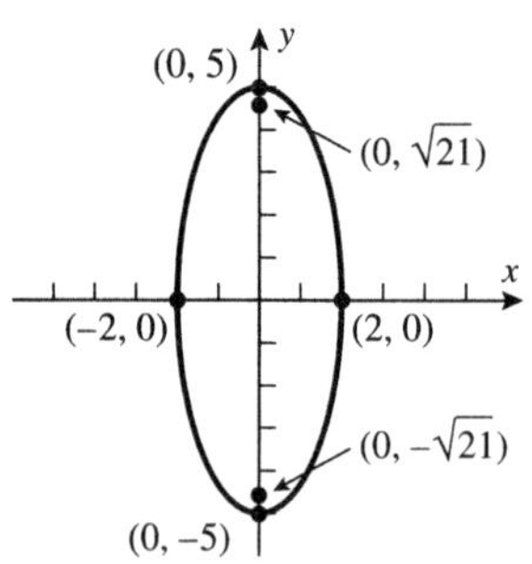

(b) $\dfrac{x^2}{9} + \dfrac{y^2}{4} = 1$

$c^2 = 9 - 4 = 5, c = \sqrt{5}$

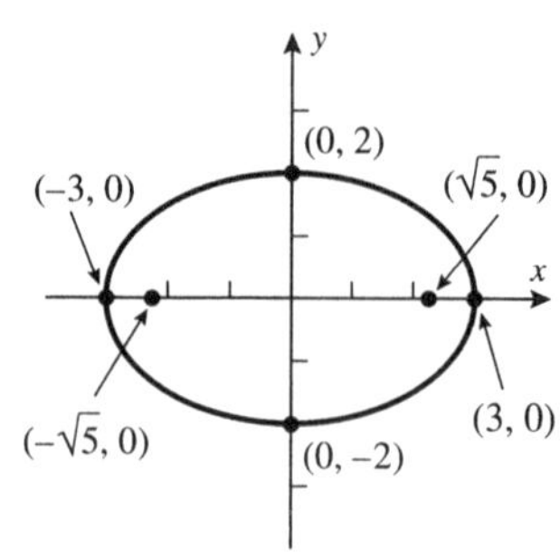

11. (a) $\dfrac{(x-1)^2}{16}+\dfrac{(y-3)^2}{9}=1$

$c^2 = 16 - 9 = 7, c = \sqrt{7}$

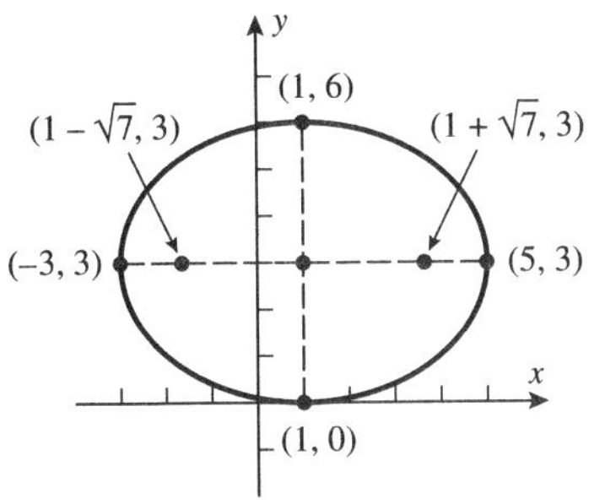

(b) $\dfrac{(x+2)^2}{4}+\dfrac{(y+1)^2}{3}=1$

$c^2 = 4 - 3 = 1, c = 1$

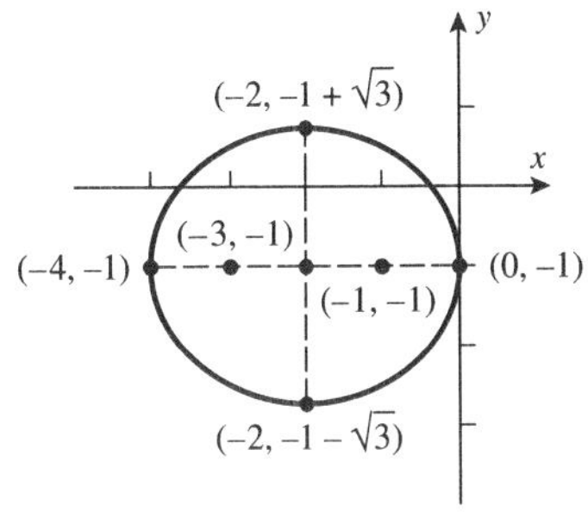

12. (a) $\dfrac{(x+3)^2}{16}+\dfrac{(y-5)^2}{4}=1$

$c^2 = 16 - 4 = 12, c = 2\sqrt{3}$

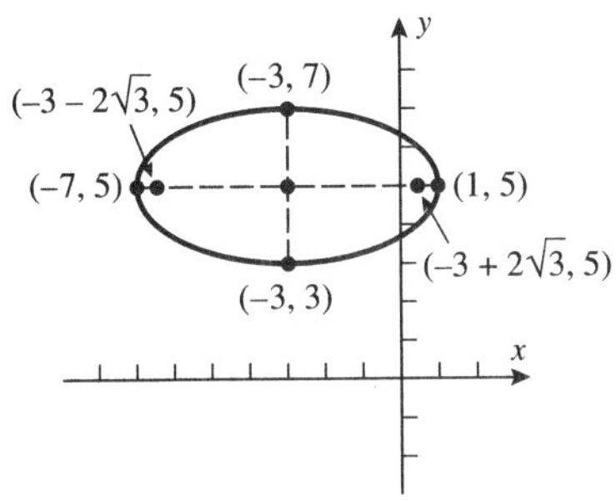

(b) $\dfrac{x^2}{4}+\dfrac{(y+2)^2}{9}=1$

$c^2 = 9 - 4 = 5, c = \sqrt{5}$

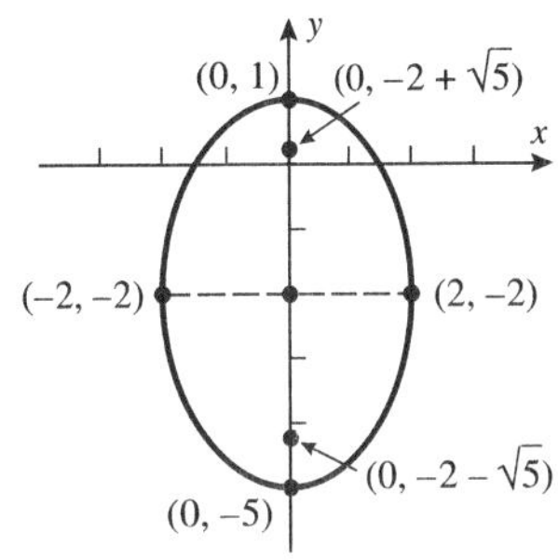

13. (a) $\dfrac{(x+1)^2}{9}+\dfrac{(y-1)^2}{1}=1$

$c^2 = 9 - 1 = 8, c = \sqrt{8}$

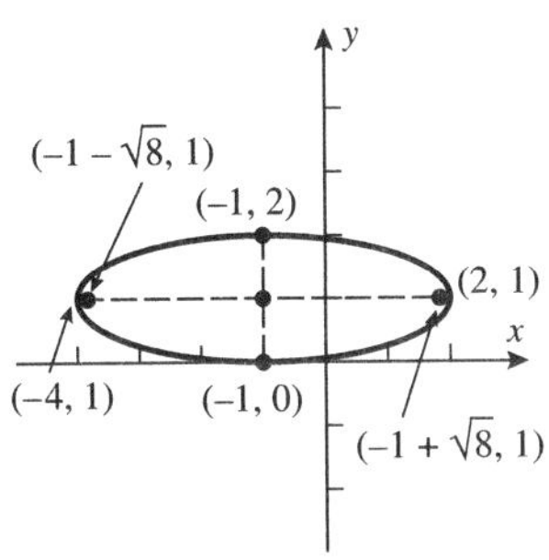

(b) $\dfrac{(x+1)^2}{4}+\dfrac{(y-5)^2}{16}=1$

$c^2 = 16 - 4 = 12, c = 2\sqrt{3}$

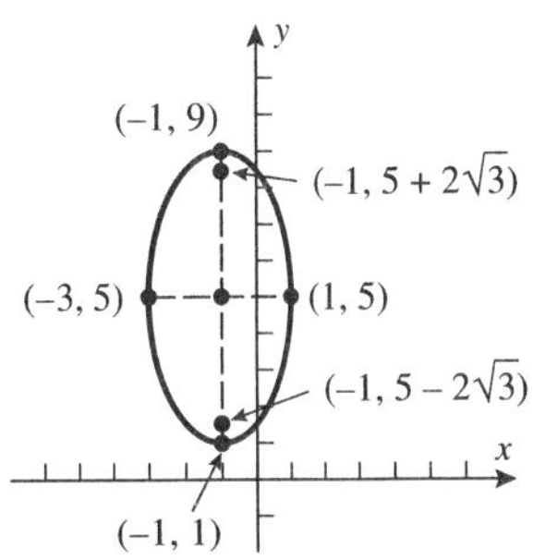

14. (a) $\dfrac{(x+1)^2}{4}+\dfrac{(y-3)^2}{9}=1$

$c^2 = 9 - 4 = 5, c = \sqrt{5}$

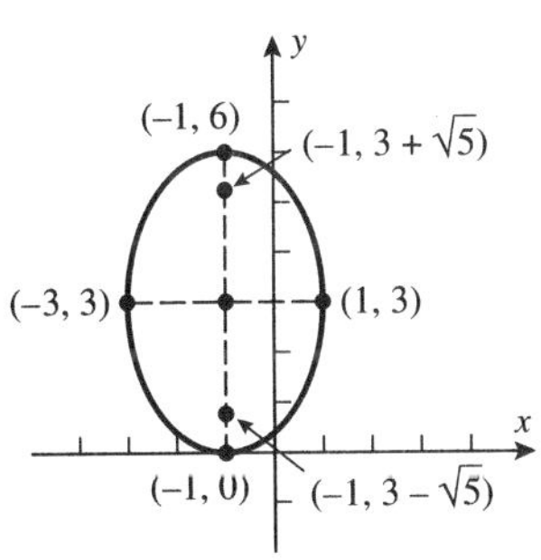

(b) $\dfrac{(x-2)^2}{9}+\dfrac{(y+3)^2}{5}=1$

$c^2 = 9 - 5 = 4, c = 2$

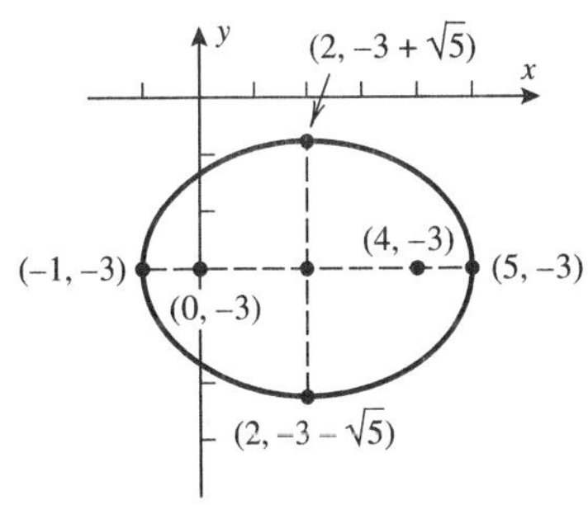

15. **(a)** $c^2 = a^2 + b^2 = 16 + 4 = 20, c = 2\sqrt{5}$

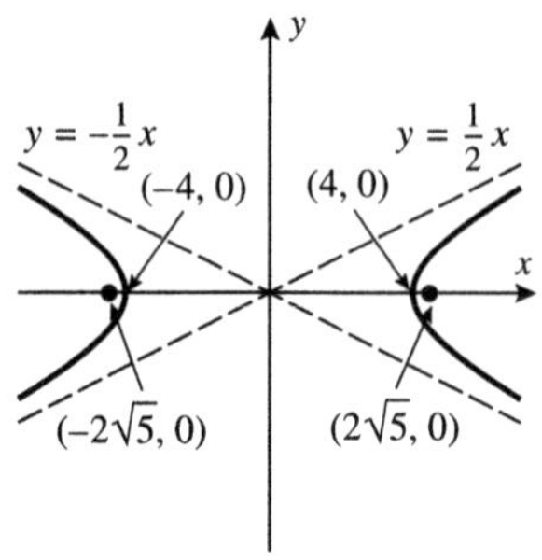

(b) $y^2/4 - x^2/9 = 1$
$c^2 = 4 + 9 = 13, c = \sqrt{13}$

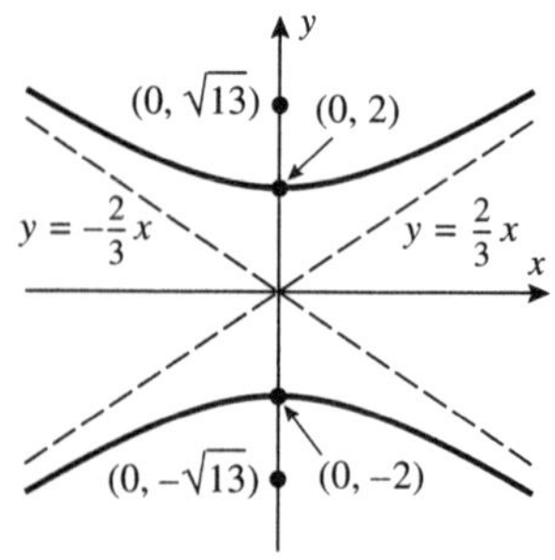

16. **(a)** $c^2 = a^2 + b^2 = 9 + 25 = 34, c = \sqrt{34}$

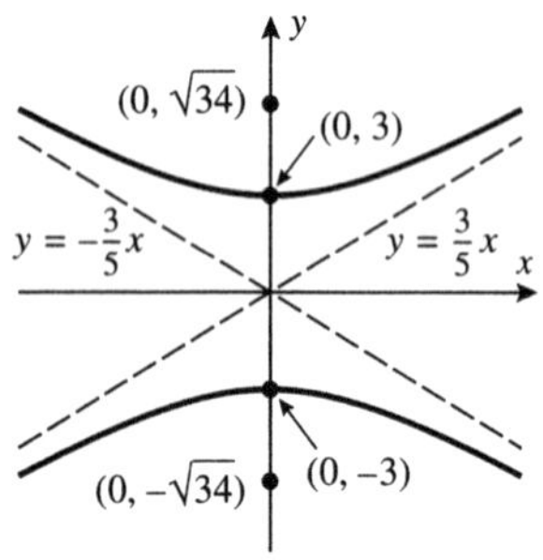

(b) $x^2/25 - y^2/16 = 1$
$c^2 = 25 + 16 = 41, c = \sqrt{41}$

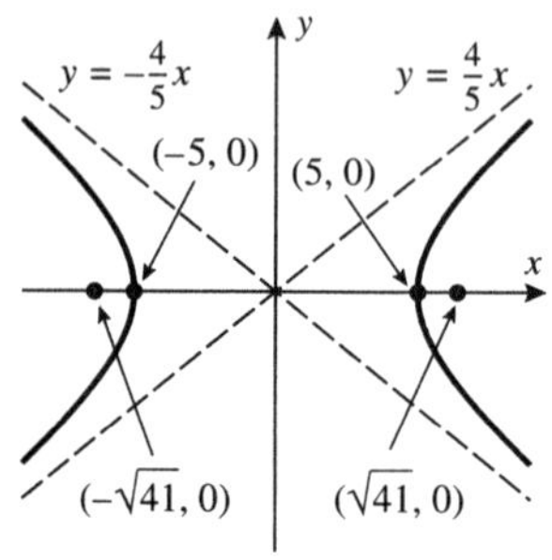

17. **(a)** $c^2 = 9 + 4 = 13, c = \sqrt{13}$

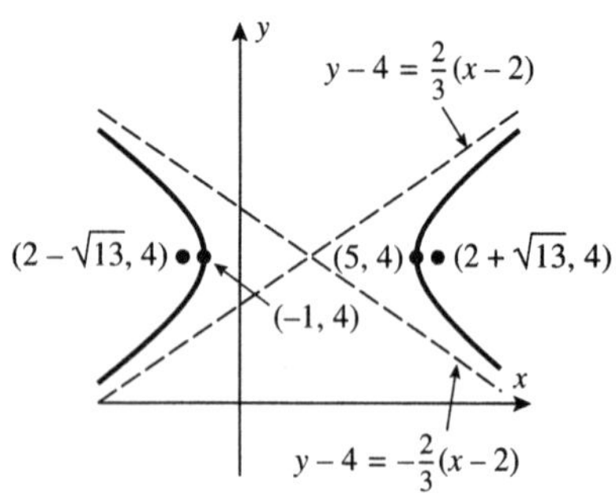

(b) $(y + 3)^2/36 - (x + 2)^2/4 = 1$
$c^2 = 36 + 4 = 40, c = 2\sqrt{10}$

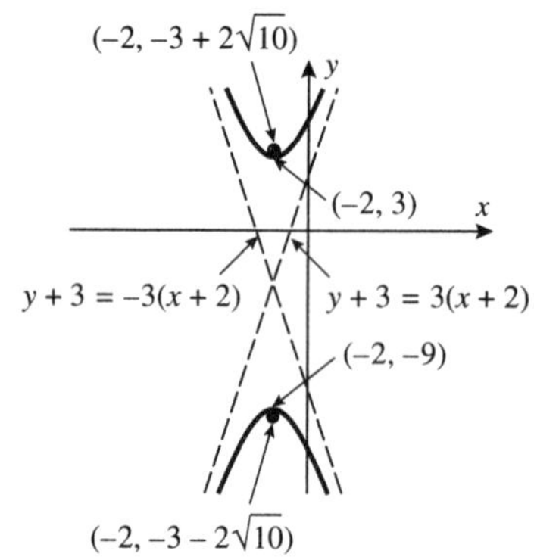

18. **(a)** $c^2 = 3 + 5 = 8, c = 2\sqrt{2}$

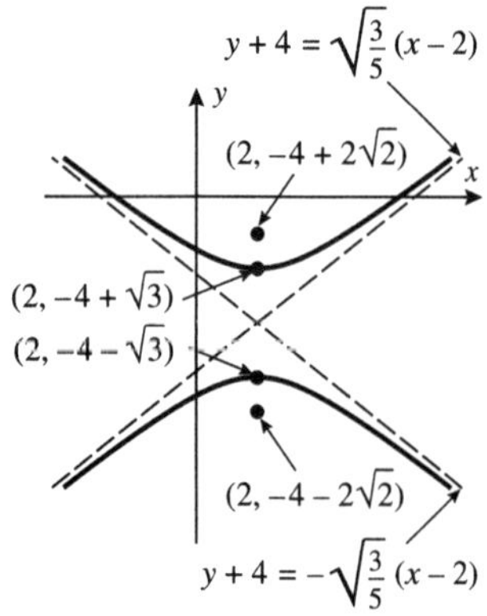

(b) $(x + 1)^2/1 - (y - 3)^2/2 = 1$
$c^2 = 1 + 2 = 3, c = \sqrt{3}$

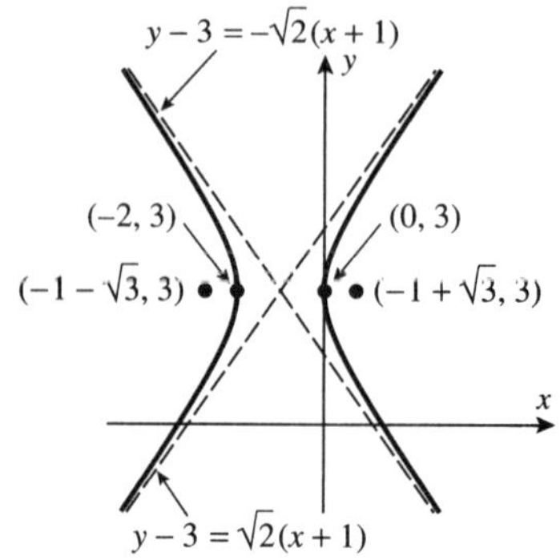

19. **(a)** $(x+1)^2/4-(y-1)^2/1=1$
$c^2=4+1=5, c=\sqrt{5}$

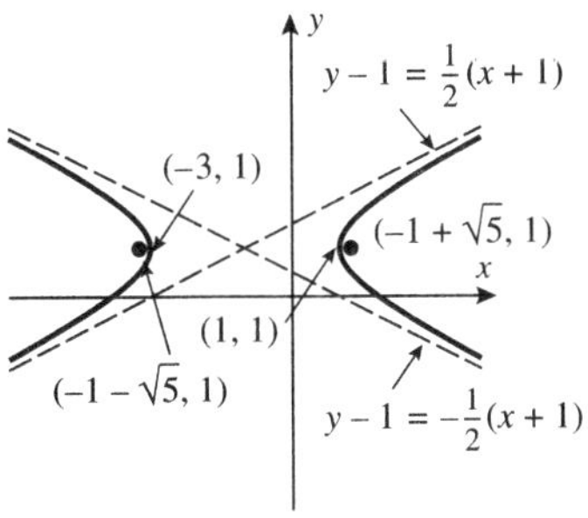

(b) $(x-1)^2/4-(y+3)^2/64=1$
$c^2=4+64=68, c=2\sqrt{17}$

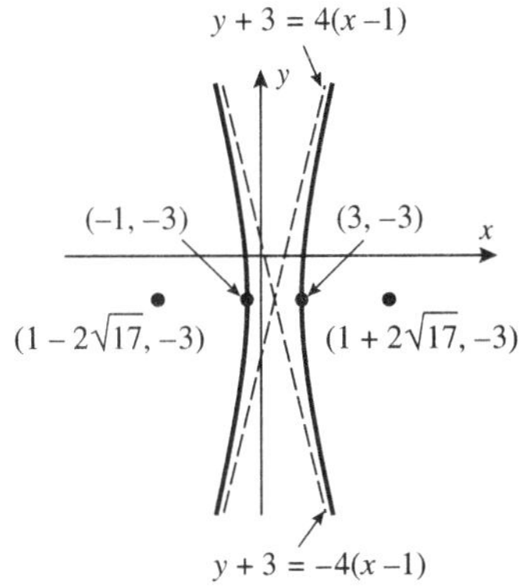

20. **(a)** $(y-3)^2/4-(x+2)^2/9=1$
$c^2=4+9=13, c=\sqrt{13}$

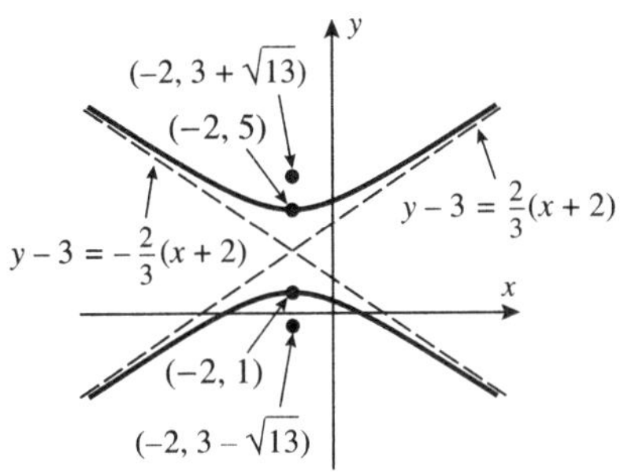

(b) $(y+5)^2/9-(x+2)^2/36=1$
$c^2=9+36=45, c=3\sqrt{5}$

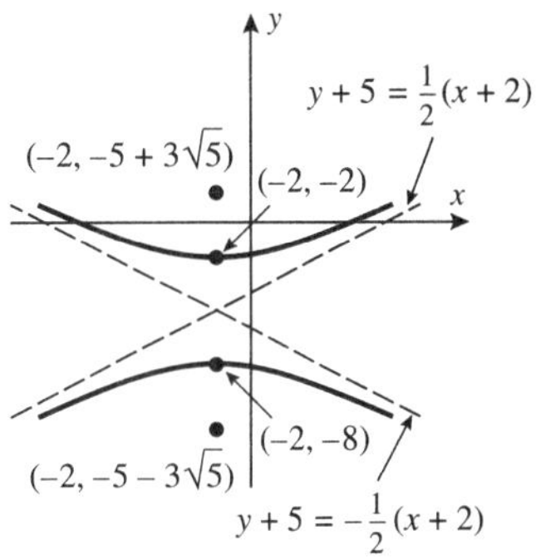

21. **(a)** $y^2=4px,\ p=3,\ y^2=12x$

(b) $y^2=-4px,\ p=7,\ y^2=-28x$

22. **(a)** $x^2=-4py,\ p=4,\ x^2=-16y$

(b) $x^2=-4py,\ p=1/2,\ x^2=-2y$

23. **(a)** $x^2=-4py,\ p=3,\ x^2=-12y$

(b) The vertex is 3 units above the directrix so $p=3$, $(x-1)^2=12(y-1)$.

24. **(a)** $y^2=4px,\ p=6,\ y^2=24x$

(b) The vertex is half way between the focus and directrix so the vertex is at $(2,4)$, the focus is 3 units to the left of the vertex so $p=3$, $(y-4)^2=-12(x-2)$

25. $y^2=a(x-h)$, $4=a(3-h)$ and $9=a(2-h)$, solve simultaneously to get $h=19/5$, $a=-5$ so $y^2=-5(x-19/5)$

26. $(x-5)^2=a(y+3)$, $(9-5)^2=a(5+3)$ so $a=2$, $(x-5)^2=2(y+3)$

27. **(a)** $x^2/9+y^2/4=1$

(b) $a=26/2=13,\ c=5,\ b^2=a^2-c^2=169-25=144;\ x^2/169+y^2/144=1$

28. **(a)** $x^2+y^2/5=1$

(b) $b=8,\ c=6,\ a^2=b^2+c^2=64+36=100;\ x^2/64+y^2/100=1$

29. **(a)** $c = 1$, $a^2 = b^2 + c^2 = 2 + 1 = 3$; $x^2/3 + y^2/2 = 1$

(b) $b^2 = 16 - 12 = 4$; $x^2/16 + y^2/4 = 1$ and $x^2/4 + y^2/16 = 1$

30. **(a)** $c = 3$, $b^2 = a^2 - c^2 = 16 - 9 = 7$; $x^2/16 + y^2/7 = 1$

(b) $a^2 = 9 + 16 = 25$; $x^2/25 + y^2/9 = 1$ and $x^2/9 + y^2/25 = 1$

31. **(a)** $a = 6$, $(2, 3)$ satisfies $x^2/36 + y^2/b^2 = 1$ so $4/36 + 9/b^2 = 1$, $b^2 = 81/8$; $x^2/36 + y^2/(81/8) = 1$

(b) The center is midway between the foci so it is at $(1, 3)$, thus $c = 1$, $b = 1$, $a^2 = 1 + 1 = 2$; $(x - 1)^2 + (y - 3)^2/2 = 1$

32. **(a)** Substitute $(3, 2)$ and $(1, 6)$ into $x^2/A + y^2/B = 1$ to get $9/A + 4/B = 1$ and $1/A + 36/B = 1$ which yields $A = 10$, $B = 40$; $x^2/10 + y^2/40 = 1$

(b) The center is at $(2, -1)$ thus $c = 2$, $a = 3$, $b^2 = 9 - 4 = 5$; $(x - 2)^2/5 + (y + 1)^2/9 = 1$

33. **(a)** $a = 2$, $c = 3$, $b^2 = 9 - 4 = 5$; $x^2/4 - y^2/5 = 1$

(b) $a = 1$, $b/a = 2$, $b = 2$; $x^2 - y^2/4 = 1$

34. **(a)** $a = 3$, $c = 5$, $b^2 = 25 - 9 = 16$; $y^2/9 - x^2/16 = 1$

(b) $a = 3$, $a/b = 1$, $b = 3$; $y^2/9 - x^2/9 = 1$

35. **(a)** vertices along x-axis: $b/a = 3/2$ so $a = 8/3$; $x^2/(64/9) - y^2/16 = 1$
vertices along y-axis: $a/b = 3/2$ so $a = 6$; $y^2/36 - x^2/16 = 1$

(b) $c = 5$, $a/b = 2$ and $a^2 + b^2 = 25$, solve to get $a^2 = 20$, $b^2 = 5$; $y^2/20 - x^2/5 = 1$

36. **(a)** foci along the x-axis: $b/a = 3/4$ and $a^2 + b^2 = 25$, solve to get $a^2 = 16$, $b^2 = 9$; $x^2/16 - y^2/9 = 1$ foci along the y-axis: $a/b = 3/4$ and $a^2 + b^2 = 25$ which results in $y^2/9 - x^2/16 = 1$

(b) $c = 3$, $b/a = 2$ and $a^2 + b^2 = 9$ so $a^2 = 9/5$, $b^2 = 36/5$; $x^2/(9/5) - y^2/(36/5) = 1$

37. **(a)** the center is at $(6, 4)$, $a = 4$, $c = 5$, $b^2 = 25 - 16 = 9$; $(x - 6)^2/16 - (y - 4)^2/9 = 1$

(b) The asymptotes intersect at $(1/2, 2)$ which is the center, $(y - 2)^2/a^2 - (x - 1/2)^2/b^2 = 1$ is the form of the equation because $(0, 0)$ is below both asymptotes, $4/a^2 - (1/4)/b^2 = 1$ and $a/b = 2$ which yields $a^2 = 3$, $b^2 = 3/4$; $(y - 2)^2/3 - (x - 1/2)^2/(3/4) = 1$.

38. **(a)** the center is at $(1, -2)$; $a = 2$, $c = 10$, $b^2 = 100 - 4 = 96$; $(y + 2)^2/4 - (x - 1)^2/96 = 1$

(b) the center is at $(1, -1)$; $2a = 5 - (-3) = 8, a = 4$, $\dfrac{(x-1)^2}{16} - \dfrac{(y+1)^2}{16} = 1$

39. **(a)** $y = ax^2 + b$, $(20, 0)$ and $(10, 12)$ are on the curve so $400a + b = 0$ and $100a + b = 12$. Solve for b to get $b = 16$ ft = height of arch.

(b) $\dfrac{x^2}{a^2} + \dfrac{y^2}{b^2} = 1$, $400 = a^2, a = 20$; $\dfrac{100}{400} + \dfrac{144}{b^2} = 1$,
$b = 8\sqrt{3}$ ft = height of arch.

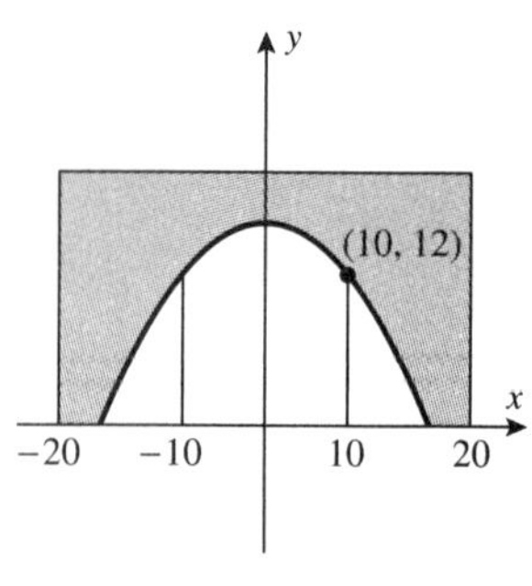

40. **(a)** $(x - b/2)^2 = a(y - h)$, but $(0, 0)$ is on the parabola so $b^2/4 = -ah$, $a = -\dfrac{b^2}{4h}$,

$(x - b/2)^2 = -\dfrac{b^2}{4h}(y - h)$

(b) As in Part (a), $y = -\dfrac{4h}{b^2}(x - b/2)^2 + h$, $A = \displaystyle\int_0^b \left[-\frac{4h}{b^2}(x - b/2)^2 + h\right] dx = \frac{2}{3}bh$

41. We may assume that the vertex is $(0, 0)$ and the parabola opens to the right. Let $P(x_0, y_0)$ be a point on the parabola $y^2 = 4px$, then by the definition of a parabola, PF = distance from P to directrix $x = -p$, so $PF = x_0 + p$ where $x_0 \geq 0$ and PF is a minimum when $x_0 = 0$ (the vertex).

42. Let p = distance (in millions of miles) between the vertex (closest point) and the focus F, then $PD = PF$, $2p + 20 = 40$, $p = 10$ million miles.

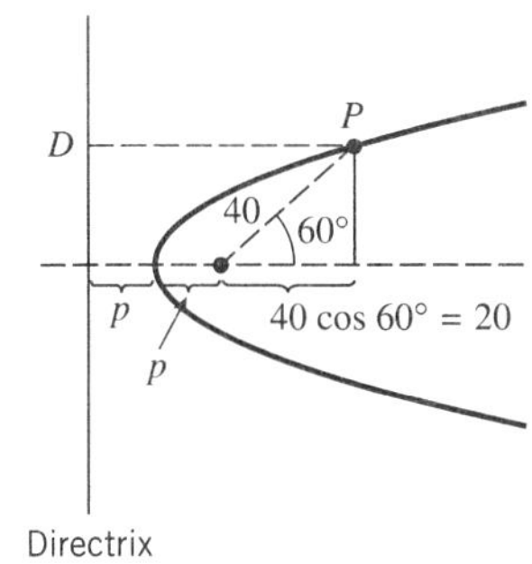

43. Use an xy-coordinate system so that $y^2 = 4px$ is an equation of the parabola, then $(1, 1/2)$ is a point on the curve so $(1/2)^2 = 4p(1)$, $p = 1/16$. The light source should be placed at the focus which is 1/16 ft. from the vertex.

44. **(a)** Substitute $x^2 = y/2$ into $y^2 - 8x^2 = 5$ to get $y^2 - 4y - 5 = 0$; $y = -1, 5$. Use $x^2 = y/2$ to find that there is no solution if $y = -1$ and that $x = \pm\sqrt{5/2}$ if $y = 5$. The curves intersect at $(\sqrt{5/2}, 5)$ and $(-\sqrt{5/2}, 5)$, and thus the area is

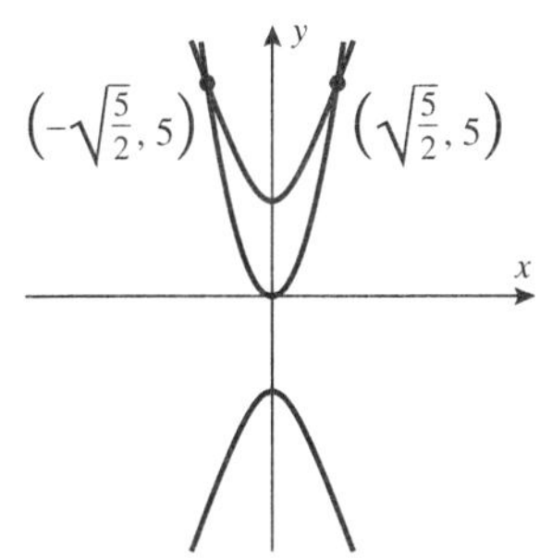

$$A = 2\int_0^{\sqrt{5/2}} (\sqrt{5 + 8x^2} - 2x^2)\, dx$$

$$= \left[x\sqrt{5 + 8x^2} + (5/4)\sqrt{2}\sinh^{-1}(2/5)\sqrt{10}x) - (4/3)x^3\right]_0^{5/2}$$

$$= \frac{5\sqrt{10}}{6} + \frac{5\sqrt{2}}{4}\ln(2 + \sqrt{5})$$

(b) Eliminate x to get $y^2 = 1$, $y = \pm 1$. Use either equation to find that $x = \pm 2$ if $y = 1$ or if $y = -1$. The curves intersect at $(2, 1)$, $(2, -1)$, $(-2, 1)$, and $(-2, -1)$, and thus the area is

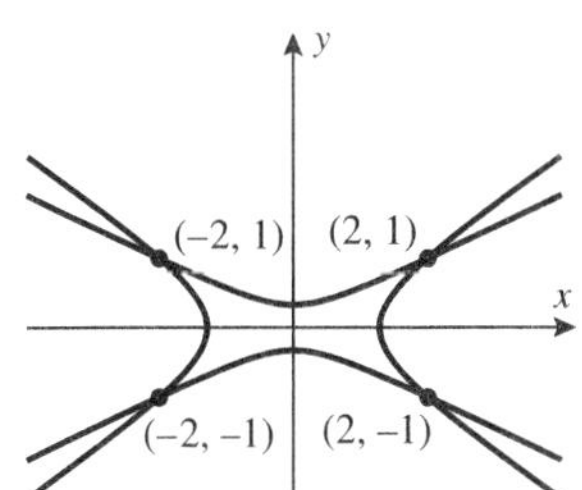

$$A = 4\int_0^{\sqrt{5/3}} \frac{1}{3}\sqrt{1 + 2x^2}\, dx$$

$$+ 4\int_{\sqrt{5/3}}^{2} \left[\frac{1}{3}\sqrt{1 + 2x^2} - \frac{1}{\sqrt{7}}\sqrt{3x^2 - 5}\right] dx$$

$$= \frac{1}{3}\sqrt{2}\ln(2\sqrt{2} + 3) + \frac{10}{21}\sqrt{21}\ln(2\sqrt{3} + \sqrt{7}) - \frac{5}{21}\ln 5$$

(c) Add both equations to get $x^2 = 4$, $x = \pm 2$.
Use either equation to find that $y = \pm\sqrt{3}$ if $x = 2$ or if $x = -2$. The curves intersect at $(2, \sqrt{3})$, $(2, -\sqrt{3})$, $(-2, \sqrt{3})$, $(-2, -\sqrt{3})$ and thus

$$A = 4\int_0^1 \sqrt{7 - x^2}\,dx + 4\int_1^2 \left[\sqrt{7 - x^2} - \sqrt{x^2 - 1}\right] dx$$
$$= 14\sin^{-1}\left(\frac{2}{7}\sqrt{7}\right) + 2\ln(2 + \sqrt{3})$$

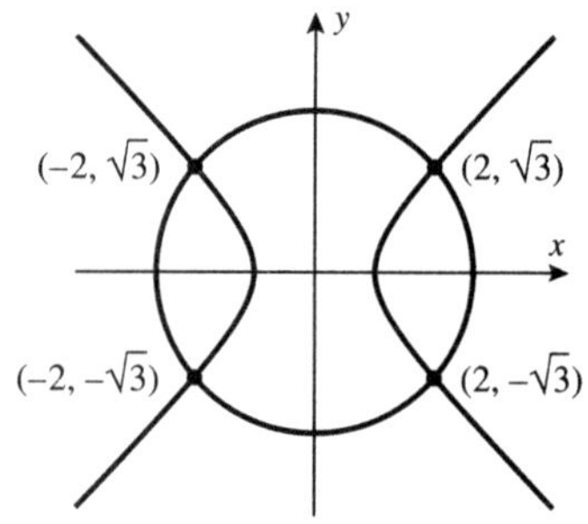

45. (a) $P : (b\cos t, b\sin t)$; $Q : (a\cos t, a\sin t)$; $R : (a\cos t, b\sin t)$

(b) For a circle, t measures the angle between the positive x-axis and the line segment joining the origin to the point. For an ellipse, t measures the angle between the x-axis and OPQ, not OR.

46. (a) For any point (x, y), the equation $y = b\sinh t$ has a unique solution t, $-\infty < t < +\infty$. On the hyperbola,
$$\frac{x^2}{a^2} = 1 + \frac{y^2}{b^2} = 1 + \sinh^2 t$$
$$= \cosh^2 t, \text{ so } x = \pm a\cosh t.$$

(b)

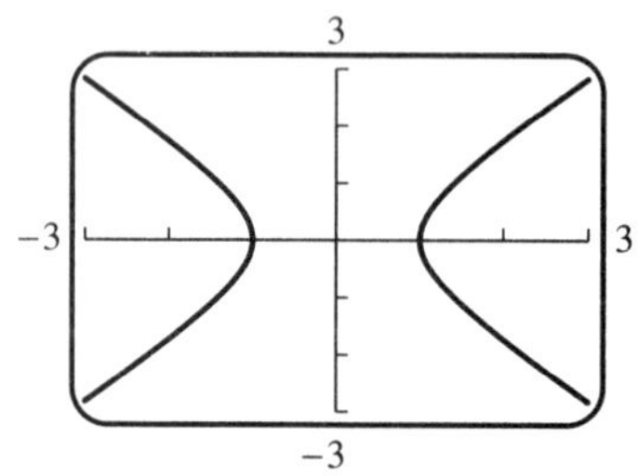

47. (a) For any point (x, y), the equation $y = b\tan t$ has a unique solution t where $-\pi/2 < t < \pi/2$. On the hyperbola, $\dfrac{x^2}{a^2} = 1 + \dfrac{y^2}{b^2} = 1 + \tan^2 t = \sec^2 t$, so $x = \pm a\sec t$.

(b)

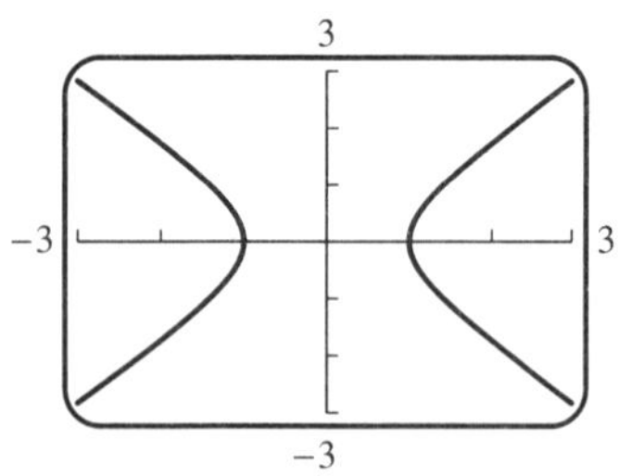

48. By Definition 11.4.1, $(x + 1)^2 + (y - 4)^2 = (y - 1)^2$, $(x + 1)^2 = 6y - 15$, $(x + 1)^2 = 6(y - 5/2)$

49. $(4, 1)$ and $(4, 5)$ are the foci so the center is at $(4, 3)$ thus $c = 2$, $a = 12/2 = 6$, $b^2 = 36 - 4 = 32$; $(x - 4)^2/32 + (y - 3)^2/36 = 1$

50. From the definition of a hyperbola, $\left|\sqrt{(x-1)^2 + (y-1)^2} - \sqrt{x^2 + y^2}\right| = 1$, $\sqrt{(x-1)^2 + (y-1)^2} - \sqrt{x^2 + y^2} = \pm 1$, transpose the second radical to the right hand side of the equation and square and simplify to get $\pm 2\sqrt{x^2 + y^2} = -2x - 2y + 1$, square and simplify again to get $8xy - 4x - 4y + 1 = 0$.

51. Let the ellipse have equation $\dfrac{4}{81}x^2 + \dfrac{y^2}{4} = 1$, then $A(x) = (2y)^2 = 16\left(1 - \dfrac{4x^2}{81}\right)$,
$$V = 2\int_0^{9/2} 16\left(1 - \frac{4x^2}{81}\right) dx = 96$$

52. See Exercise 51, $A(y) = \sqrt{3}x^2 = \sqrt{3}\frac{81}{4}\left(1 - \frac{y^2}{4}\right), V = \sqrt{3}\frac{81}{2}\int_0^2 \left(1 - \frac{y^2}{4}\right) dy = 54\sqrt{3}$

53. Assume $\frac{x^2}{a^2} + \frac{y^2}{b^2} = 1, A = 4\int_0^a b\sqrt{1 - x^2/a^2}\, dx = \pi ab$

54. **(a)** Assume $\frac{x^2}{a^2} + \frac{y^2}{b^2} = 1, V = 2\int_0^a \pi b^2\left(1 - x^2/a^2\right) dx = \frac{4}{3}\pi ab^2$

(b) In Part (a) interchange a and b to obtain the result.

55. Assume $\frac{x^2}{a^2} + \frac{y^2}{b^2} = 1, \frac{dy}{dx} = -\frac{bx}{a\sqrt{a^2 - x^2}}, 1 + \left(\frac{dy}{dx}\right)^2 = \frac{a^4 - (a^2 - b^2)x^2}{a^2(a^2 - x^2)}$,

$$S = 2\int_0^a \frac{2\pi b}{a}\sqrt{1 - x^2/a^2}\sqrt{\frac{a^4 - (a^2 - b^2)x^2}{a^2 - x^2}}\, dx = 2\pi ab\left(\frac{b}{a} + \frac{a}{c}\sin^{-1}\frac{c}{a}\right), c = \sqrt{a^2 - b^2}$$

56. As in Exercise 55, $1 + \left(\frac{dx}{dy}\right)^2 = \frac{b^4 + (a^2 - b^2)y^2}{b^2(b^2 - y^2)}$,

$$S = 2\int_0^b 2\pi a\sqrt{1 - y^2/b^2}\sqrt{\frac{b^4 + (a^2 - b^2)y^2}{b^2(b^2 - y^2)}}\, dy = 2\pi ab\left(\frac{a}{b} + \frac{b}{c}\ln\frac{a + c}{b}\right), c = \sqrt{a^2 - b^2}$$

57. Open the compass to the length of half the major axis, place the point of the compass at an end of the minor axis and draw arcs that cross the major axis to both sides of the center of the ellipse. Place the tacks where the arcs intersect the major axis.

58. Let P denote the pencil tip, and let $R(x, 0)$ be the point below Q and P which lies on the line L. Then $QP + PF$ is the length of the string and $QR = QP + PR$ is the length of the side of the triangle. These two are equal, so $PF = PR$. But this is the definition of a parabola according to Definition 11.4.1.

59. Let P denote the pencil tip, and let k be the difference between the length of the ruler and that of the string. Then $QP + PF_2 + k = QF_1$, and hence $PF_2 + k = PF_1, PF_1 - PF_2 = k$. But this is the definition of a hyperbola according to Definition 11.4.3.

60. In the $x'y'$-plane an equation of the circle is $x'^2 + y'^2 = r^2$ where r is the radius of the cylinder. Let $P(x, y)$ be a point on the curve in the xy-plane, then $x' = x\cos\theta$ and $y' = y$ so $x^2\cos^2\theta + y^2 = r^2$ which is an equation of an ellipse in the xy-plane.

61. $L = 2a = \sqrt{D^2 + p^2D^2} = D\sqrt{1 + p^2}$ (see figure), so $a = \frac{1}{2}D\sqrt{1 + p^2}$, but $b = \frac{1}{2}D$,

$T = c = \sqrt{a^2 - b^2} = \sqrt{\frac{1}{4}D^2(1 + p^2) - \frac{1}{4}D^2} = \frac{1}{2}pD.$

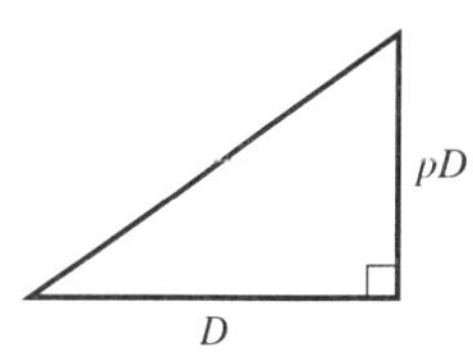

62. $y = \frac{1}{4p}x^2$, $dy/dx = \frac{1}{2p}x$, $dy/dx\Big|_{x=x_0} = \frac{1}{2p}x_0$, the tangent line at (x_0, y_0) has the formula

$y - y_0 = \frac{x_0}{2p}(x - x_0) = \frac{x_0}{2p}x - \frac{x_0^2}{2p}$, but $\frac{x_0^2}{2p} = 2y_0$ because (x_0, y_0) is on the parabola $y = \frac{1}{4p}x^2$.

Thus the tangent line is $y - y_0 = \frac{x_0}{2p}x - 2y_0$, $y = \frac{x_0}{2p}x - y_0$.

63. By implicit differentiation, $\left.\frac{dy}{dx}\right|_{(x_0,y_0)} = -\frac{b^2}{a^2}\frac{x_0}{y_0}$ if $y_0 \neq 0$, the tangent line is

$y - y_0 = -\frac{b^2}{a^2}\frac{x_0}{y_0}(x - x_0)$, $a^2y_0y - a^2y_0^2 = -b^2x_0x + b^2x_0^2$, $b^2x_0x + a^2y_0y = b^2x_0^2 + a^2y_0^2$, but (x_0, y_0) is on the ellipse so $b^2x_0^2 + a^2y_0^2 = a^2b^2$; thus the tangent line is $b^2x_0x + a^2y_0y = a^2b^2$, $x_0x/a^2 + y_0y/b^2 = 1$. If $y_0 = 0$ then $x_0 = \pm a$ and the tangent lines are $x = \pm a$ which also follows from $x_0x/a^2 + y_0y/b^2 = 1$.

64. By implicit differentiation, $\left.\frac{dy}{dx}\right|_{(x_0,y_0)} = \frac{b^2}{a^2}\frac{x_0}{y_0}$ if $y_0 \neq 0$, the tangent line is $y - y_0 = \frac{b^2}{a^2}\frac{x_0}{y_0}(x - x_0)$, $b^2x_0x - a^2y_0y = b^2x_0^2 - a^2y_0^2 = a^2b^2$, $x_0x/a^2 - y_0y/b^2 = 1$. If $y_0 = 0$ then $x_0 = \pm a$ and the tangent lines are $x = \pm a$ which also follow from $x_0x/a^2 - y_0y/b^2 = 1$.

65. Use $\frac{x^2}{a^2} + \frac{y^2}{b^2} = 1$ and $\frac{x^2}{A^2} - \frac{y^2}{B^2} = 1$ as the equations of the ellipse and hyperbola. If (x_0, y_0) is a point of intersection then $\frac{x_0^2}{a^2} + \frac{y_0^2}{b^2} = 1 = \frac{x_0^2}{A^2} - \frac{y_0^2}{B^2}$, so $x_0^2\left(\frac{1}{A^2} - \frac{1}{a^2}\right) = y_0^2\left(\frac{1}{B^2} + \frac{1}{b^2}\right)$ and $a^2A^2y_0^2(b^2 + B^2) = b^2B^2x_0^2(a^2 - A^2)$. Since the conics have the same foci, $a^2 - b^2 = c^2 = A^2 + B^2$, so $a^2 - A^2 = b^2 + B^2$. Hence $a^2A^2y_0^2 = b^2B^2x_0^2$. From Exercises 63 and 64, the slopes of the tangent lines are $-\frac{b^2x_0}{a^2y_0}$ and $\frac{B^2x_0}{A^2y_0}$, whose product is $-\frac{b^2B^2x_0^2}{a^2A^2y_0^2} = -1$. Hence the tangent lines are perpendicular.

66. Use implicit differentiation on $x^2 + 4y^2 = 8$ to get $\left.\frac{dy}{dx}\right|_{(x_0,y_0)} = -\frac{x_0}{4y_0}$ where (x_0, y_0) is the point of tangency, but $-x_0/(4y_0) = -1/2$ because the slope of the line is $-1/2$ so $x_0 = 2y_0$. (x_0, y_0) is on the ellipse so $x_0^2 + 4y_0^2 = 8$ which when solved with $x_0 = 2y_0$ yields the points of tangency $(2, 1)$ and $(-2, -1)$. Substitute these into the equation of the line to get $k = \pm 4$.

67. Let (x_0, y_0) be such a point. The foci are at $(-\sqrt{5}, 0)$ and $(\sqrt{5}, 0)$, the lines are perpendicular if the product of their slopes is -1 so $\frac{y_0}{x_0 + \sqrt{5}} \cdot \frac{y_0}{x_0 - \sqrt{5}} = -1$, $y_0^2 = 5 - x_0^2$ and $4x_0^2 - y_0^2 = 4$. Solve to get $x_0 = \pm 3/\sqrt{5}$, $y_0 = \pm 4/\sqrt{5}$. The coordinates are $(\pm 3/\sqrt{5}, 4/\sqrt{5})$, $(\pm 3/\sqrt{5}, -4/\sqrt{5})$.

68. Let (x_0, y_0) be one of the points; then $\left. dy/dx\right|_{(x_0,y_0)} = 4x_0/y_0$, the tangent line is $y = (4x_0/y_0)x + 4$, but (x_0, y_0) is on both the line and the curve which leads to $4x_0^2 - y_0^2 + 4y_0 = 0$ and $4x_0^2 - y_0^2 = 36$, solve to get $x_0 = \pm 3\sqrt{13}/2$, $y_0 = -9$.

69. Let d_1 and d_2 be the distances of the first and second observers, respectively, from the point of the explosion. Then t = (time for sound to reach the second observer) − (time for sound to reach the first observer) $= d_2/v - d_1/v$ so $d_2 - d_1 = vt$. For constant v and t the difference of distances, d_2 and d_1 is constant so the explosion occurred somewhere on a branch of a hyperbola whose foci are where the observers are. Since $d_2 - d_1 = 2a$, $a = \frac{vt}{2}$, $b^2 = c^2 - \frac{v^2t^2}{4}$, and $\frac{x^2}{v^2t^2/4} - \frac{y^2}{c^2 - (v^2t^2/4)} = 1$.

70. As in Exercise 69, $d_2 - d_1 = 2a = vt = (299{,}792{,}458 \text{ m/s})(10^{-7} \text{ s}) \approx 29.9792$ m.
$a^2 = (vt/2)^2 \approx 449.3762 \text{ m}^2$; $c^2 = (50)^2 = 2500 \text{ m}^2$
$b^2 = c^2 - a^2 = 2050.6238$, $\frac{x^2}{449.3762} - \frac{y^2}{2050.6238} = 1$
But $y = 200$ km $= 200{,}000$ m, so $x \approx 93{,}625.05$ m $= 93.62505$ km. The ship is located at (93.62505,200).

71. **(a)** Use $\dfrac{x^2}{9}+\dfrac{y^2}{4}=1$, $x=\dfrac{3}{2}\sqrt{4-y^2}$,

$$V=\int_{-2}^{-2+h}(2)(3/2)\sqrt{4-y^2}(18)dy=54\int_{-2}^{-2+h}\sqrt{4-y^2}\,dy$$

$$=54\left[\frac{y}{2}\sqrt{4-y^2}+2\sin^{-1}\frac{y}{2}\right]_{-2}^{-2+h}=27\left[4\sin^{-1}\frac{h-2}{2}+(h-2)\sqrt{4h-h^2}+2\pi\right]\ \text{ft}^3$$

(b) When $h=4$ ft, $V_{\text{full}}=108\sin^{-1}1+54\pi=108\pi$ ft^3, so solve for h when $V=(k/4)V_{\text{full}}$, $k=1,2,3$, to get $h=1.19205, 2, 2.80795$ ft or $14.30465, 24, 33.69535$ in.

72. We may assume $A>0$, since if $A<0$ then one can multiply the equation by -1, and if $A=0$ then one can exchange A with C (C cannot be zero simultaneously with A). Then

$$Ax^2+Cy^2+Dx+Ey+F=A\left(x+\frac{D}{2A}\right)^2+C\left(y+\frac{E}{2C}\right)^2+F-\frac{D^2}{4A}-\frac{E^2}{4C}=0.$$

(a) Let $AC>0$. If $F<\dfrac{D^2}{4A}+\dfrac{E^2}{4C}$ the equation represents an ellipse (a circle if $A=C$); if $F=\dfrac{D^2}{4A}+\dfrac{E^2}{4C}$, the point $x=-D/(2A), y=-E/(2C)$; and if $F>\dfrac{D^2}{4A}+\dfrac{E^2}{4C}$ then there is no graph.

(b) If $AC<0$ and $F=\dfrac{D^2}{4A}+\dfrac{E^2}{4C}$, then

$$\left[\sqrt{A}\left(x+\frac{D}{2A}\right)+\sqrt{-C}\left(y+\frac{E}{2C}\right)\right]\left[\sqrt{A}\left(x+\frac{D}{2A}\right)-\sqrt{-C}\left(y+\frac{E}{2C}\right)\right]=0,$$

a pair of lines; otherwise a hyperbola

(c) Assume $C=0$, so $Ax^2+Dx+Ey+F=0$. If $E\neq 0$, parabola; if $E=0$ then $Ax^2+Dx+F=0$. If this polynomial has roots $x=x_1,x_2$ with $x_1\neq x_2$ then a pair of parallel lines; if $x_1=x_2$ then one line; if no roots, then no graph. If $A=0, C\neq 0$ then a similar argument applies.

73. **(a)** $(x-1)^2-5(y+1)^2=5$, hyperbola

(b) $x^2-3(y+1)^2=0, x=\pm\sqrt{3}(y+1)$, two lines

(c) $4(x+2)^2+8(y+1)^2=4$, ellipse

(d) $3(x+2)^2+(y+1)^2=0$, the point $(-2,-1)$ (degenerate case)

(e) $(x+4)^2+2y=2$, parabola

(f) $5(x+4)^2+2y=-14$, parabola

74. distance from the point (x,y) to the focus $(0,p)$ = distance to the directrix $y=-p$, so $x^2+(y-p)^2=(y+p)^2, x^2=4py$

75. distance from the point (x,y) to the focus $(0,-c)$ plus distance to the focus $(0,c)$ = const $=2a$, $\sqrt{x^2+(y+c)^2}+\sqrt{x^2+(y-c)^2}=2a, x^2+(y+c)^2=4a^2+x^2+(y-c)^2-4a\sqrt{x^2+(y-c)^2}$, $\sqrt{x^2+(y-c)^2}=a-\dfrac{c}{a}y$, and since $a^2-c^2=b^2$, $\dfrac{x^2}{b^2}+\dfrac{y^2}{a^2}=1$

76. distance from the point (x,y) to the focus $(-c,0)$ less distance to the focus $(c,0)$ is equal to $2a$, $\sqrt{(x+c)^2+y^2}-\sqrt{(x-c)^2+y^2}=\pm 2a, (x+c)^2+y^2=(x-c)^2+y^2+4a^2\pm 4a\sqrt{(x-c)^2+y^2}$, $\sqrt{(x-c)^2+y^2}=\pm\left(\dfrac{cx}{a}-a\right)$, and, since $c^2-a^2=b^2$, $\dfrac{x^2}{a^2}-\dfrac{y^2}{b^2}=1$

77. Assume the equation of the parabola is $x^2 = 4py$. The tangent line at $P(x_0, y_0)$ (see figure) is given by $(y - y_0)/(x - x_0) = m = x_0/2p$. To find the y-intercept set $x = 0$ and obtain $y = -y_0$. Thus $Q : (0, -y_0)$. The focus is $(0, p) = (0, x_0^2/4y_0)$, the distance from P to the focus is $\sqrt{x_0^2 + (y_0 - p)^2} = \sqrt{4py_0 + (y_0 - p)^2} = \sqrt{(y_0 + p)^2} = y_0 + p$, and the distance from the focus to the y-intercept is $p + y_0$, so triangle FPQ is isosceles, and angles FPQ and FQP are equal.

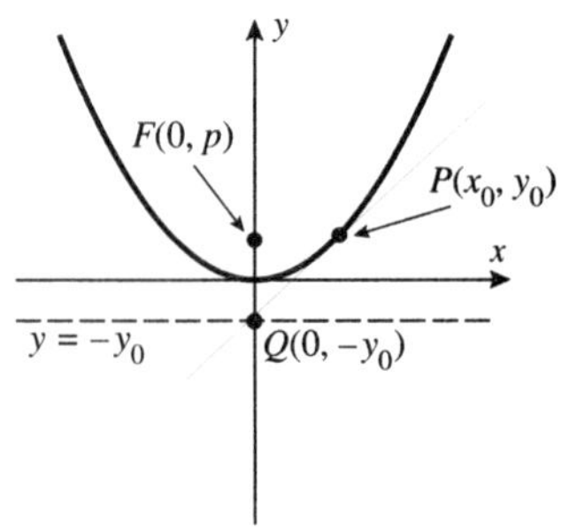

78. **(a)** $\tan\theta = \tan(\phi_2 - \phi_1) = \dfrac{\tan\phi_2 - \tan\phi_1}{1 + \tan\phi_2 \tan\phi_1} = \dfrac{m_2 - m_1}{1 + m_1 m_2}$

(b) By implicit differentiation, $m = dy/dx\Big|_{P(x_0,y_0)} = -\dfrac{b^2}{a^2}\dfrac{x_0}{y_0}$ if $y_0 \neq 0$. Let m_1 and m_2 be the slopes of the lines through P and the foci at $(-c, 0)$ and $(c, 0)$ respectively, then $m_1 = y_0/(x_0 + c)$ and $m_2 = y_0/(x_0 - c)$. For P in the first quadrant,

$$\tan\alpha = \frac{m - m_2}{1 + mm_2} = \frac{-(b^2x_0)/(a^2y_0) - y_0/(x_0 - c)}{1 - (b^2x_0)/[a^2(x_0 - c)]}$$

$$= \frac{-b^2x_0^2 - a^2y_0^2 + b^2cx_0}{[(a^2 - b^2)x_0 - a^2c]\, y_0} = \frac{-a^2b^2 + b^2cx_0}{(c^2x_0 - a^2c)y_0} = \frac{b^2}{cy_0}$$

similarly $\tan(\pi - \beta) = \dfrac{m - m_1}{1 + mm_1} = -\dfrac{b^2}{cy_0} = -\tan\beta$ so $\tan\alpha = \tan\beta$, $\alpha = \beta$. The proof for the case $y_0 = 0$ follows trivially. By symmetry, the result holds for P in the other three quadrants as well.

(c) Let $P(x_0, y_0)$ be in the third quadrant. Suppose $y_0 \neq 0$ and let m = slope of the tangent line at P, m_1 = slope of the line through P and $(-c, 0)$, m_2 = slope of the line through P and $(c, 0)$ then $m = \dfrac{dy}{dx}\Big|_{(x_0,y_0)} = (b^2x_0)/(a^2y_0)$, $m_1 = y_0/(x_0 + c)$, $m_2 = y_0/(x_0 - c)$. Use $\tan\alpha = (m_1 - m)/(1 + m_1m)$ and $\tan\beta = (m - m_2)/(1 + mm_2)$ to get $\tan\alpha = \tan\beta = -b^2/(cy_0)$ so $\alpha = \beta$. If $y_0 = 0$ the result follows trivially and by symmetry the result holds for P in the other three quadrants as well.

EXERCISE SET 11.5

1. **(a)** $\sin\theta = \sqrt{3}/2$, $\cos\theta = 1/2$

$x' = (-2)(1/2) + (6)(\sqrt{3}/2) = -1 + 3\sqrt{3}$, $y' = -(-2)(\sqrt{3}/2) + 6(1/2) = 3 + \sqrt{3}$

(b) $x = \dfrac{1}{2}x' - \dfrac{\sqrt{3}}{2}y' = \dfrac{1}{2}(x' - \sqrt{3}y')$, $y = \dfrac{\sqrt{3}}{2}x' + \dfrac{1}{2}y' = \dfrac{1}{2}(\sqrt{3}x' + y')$

$$\sqrt{3}\left[\frac{1}{2}(x' - \sqrt{3}y')\right]\left[\frac{1}{2}(\sqrt{3}x' + y')\right] + \left[\frac{1}{2}(\sqrt{3}x' + y')\right]^2 = 6$$

$$\frac{\sqrt{3}}{4}(\sqrt{3}x'^2 - 2x'y' - \sqrt{3}y'^2) + \frac{1}{4}(3x'^2 + 2\sqrt{3}x'y' + y'^2) = 6$$

$$\frac{3}{2}x'^2 - \frac{1}{2}y'^2 = 6,\ 3x'^2 - y'^2 = 12$$

(c) 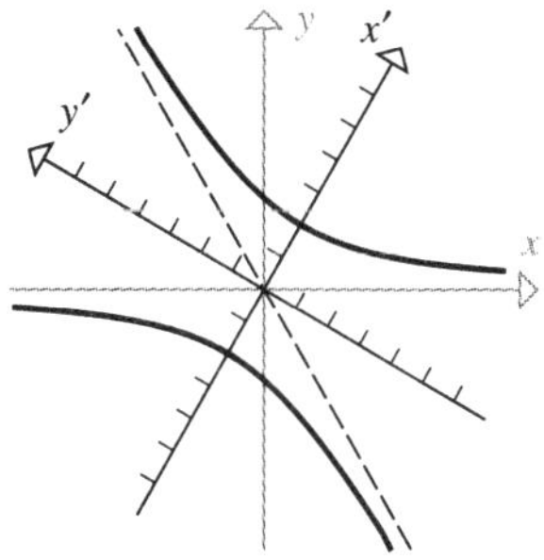

2. (a) $\sin\theta = 1/2$, $\cos\theta = \sqrt{3}/2$
$x' = (1)(\sqrt{3}/2) + (-\sqrt{3})(1/2) = 0$, $y' = -(1)(1/2) + (-\sqrt{3})(\sqrt{3}/2) = -2$

(b) $x = \dfrac{\sqrt{3}}{2}x' - \dfrac{1}{2}y' = \dfrac{1}{2}(\sqrt{3}x' - y')$, $y = \dfrac{1}{2}x' + \dfrac{\sqrt{3}}{2}y' = \dfrac{1}{2}(x' + \sqrt{3}y')$

$$2\left[\frac{1}{2}(\sqrt{3}x' - y')\right]^2 + 2\sqrt{3}\left[\frac{1}{2}(\sqrt{3}x' - y')\right]\left[\frac{1}{2}(x' + \sqrt{3}y')\right] = 3$$

$$\frac{1}{2}(3x'^2 - 2\sqrt{3}x'y' + y'^2) + \frac{\sqrt{3}}{2}(\sqrt{3}x'^2 + 2x'y' - \sqrt{3}y'^2) = 3$$

$3x'^2 - y'^2 = 3$, $x'^2/1 - y'^2/3 = 1$

(c) 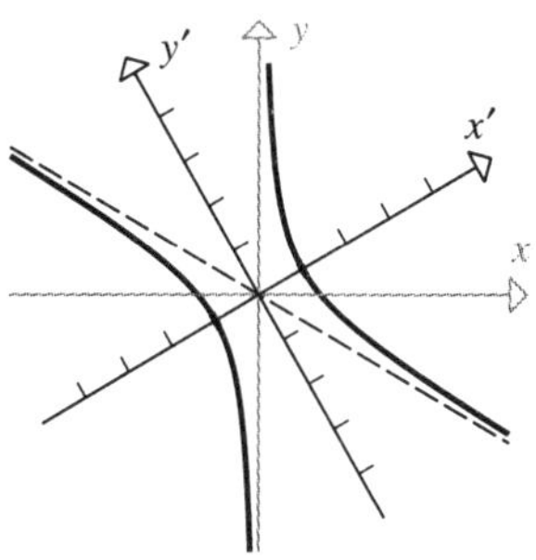

3. $\cot 2\theta = (0-0)/1 = 0$, $2\theta = 90°$, $\theta = 45°$
$x = (\sqrt{2}/2)(x' - y')$, $y = (\sqrt{2}/2)(x' + y')$
$y'^2/18 - x'^2/18 = 1$, hyperbola

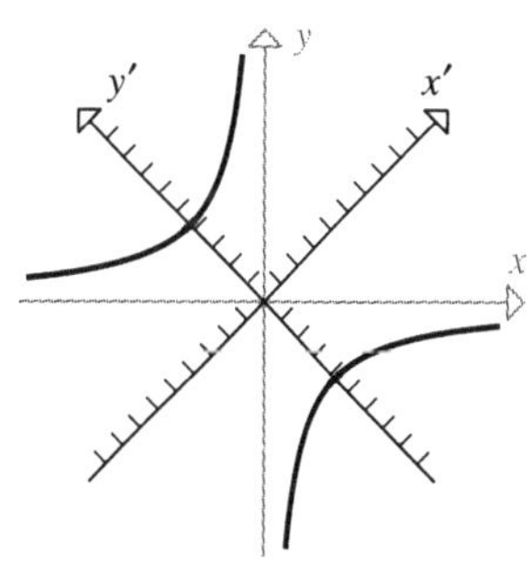

4. $\cot 2\theta = (1-1)/(-1) = 0$, $\theta = 45°$
$x = (\sqrt{2}/2)(x' - y')$, $y = (\sqrt{2}/2)(x' + y')$
$x'^2/4 + y'^2/(4/3) = 1$, ellipse

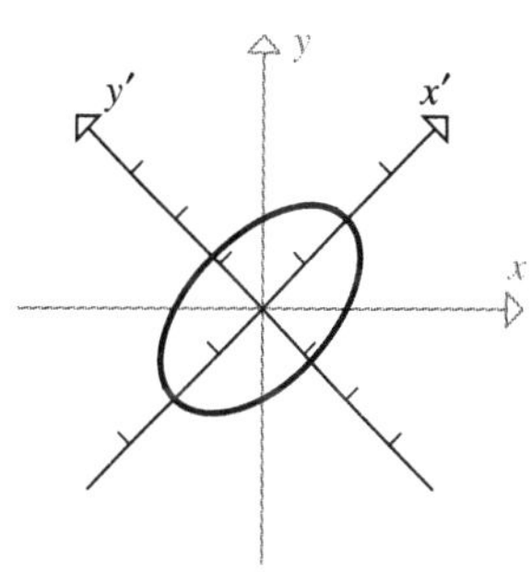

5. $\cot 2\theta = [1 - (-2)]/4 = 3/4$
$\cos 2\theta = 3/5$
$\sin\theta = \sqrt{(1 - 3/5)/2} = 1/\sqrt{5}$
$\cos\theta = \sqrt{(1 + 3/5)/2} = 2/\sqrt{5}$
$x = (1/\sqrt{5})(2x' - y')$
$y = (1/\sqrt{5})(x' + 2y')$
$x'^2/3 - y'^2/2 = 1$, hyperbola

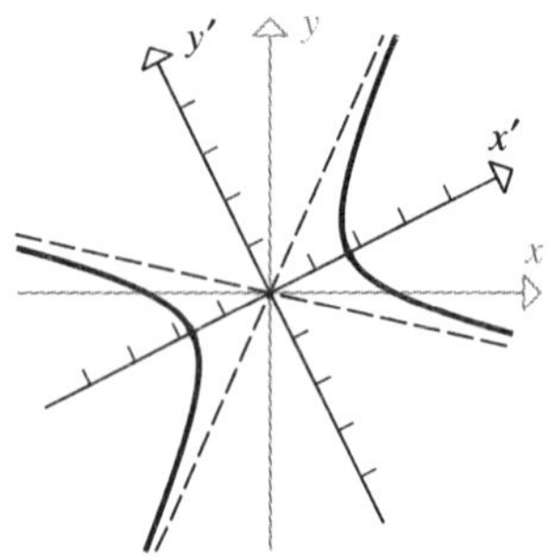

6. $\cot 2\theta = (31 - 21)/(10\sqrt{3}) = 1/\sqrt{3}$,
$2\theta = 60°, \theta = 30°$
$x = (1/2)(\sqrt{3}x' - y')$,
$y = (1/2)(x' + \sqrt{3}y')$
$x'^2/4 + y'^2/9 = 1$, ellipse

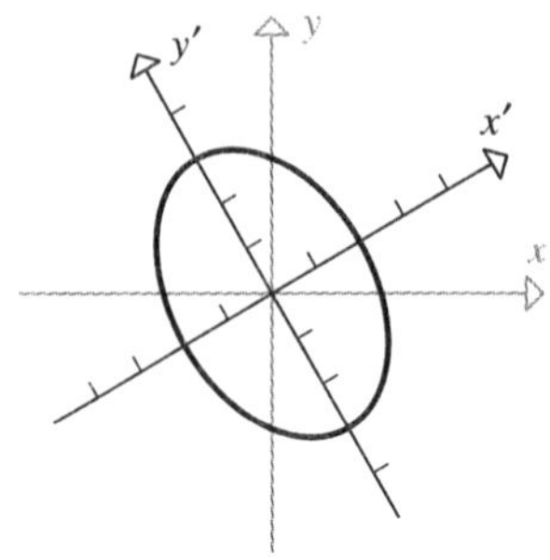

7. $\cot 2\theta = (1 - 3)/(2\sqrt{3}) = -1/\sqrt{3}$,
$2\theta = 120°, \theta = 60°$
$x = (1/2)(x' - \sqrt{3}y')$
$y = (1/2)(\sqrt{3}x' + y')$
$y' = x'^2$, parabola

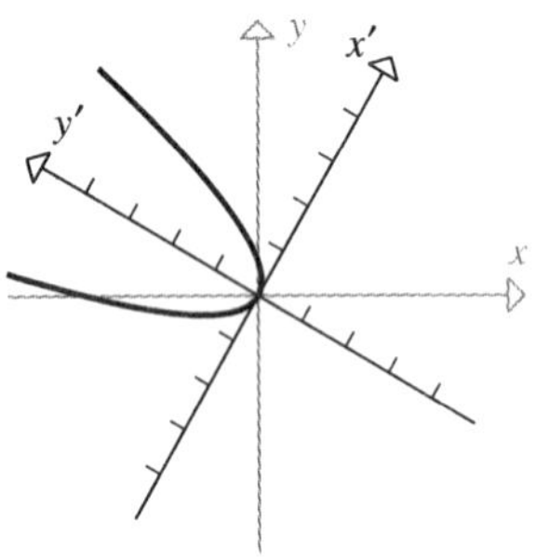

8. $\cot 2\theta = (34 - 41)/(-24) = 7/24$
$\cos 2\theta = 7/25$
$\sin\theta = \sqrt{(1 - 7/25)/2} = 3/5$
$\cos\theta = \sqrt{(1 + 7/25)/2} = 4/5$
$x = (1/5)(4x' - 3y')$,
$y = (1/5)(3x' + 4y')$
$x'^2 + y'^2/(1/2) = 1$, ellipse

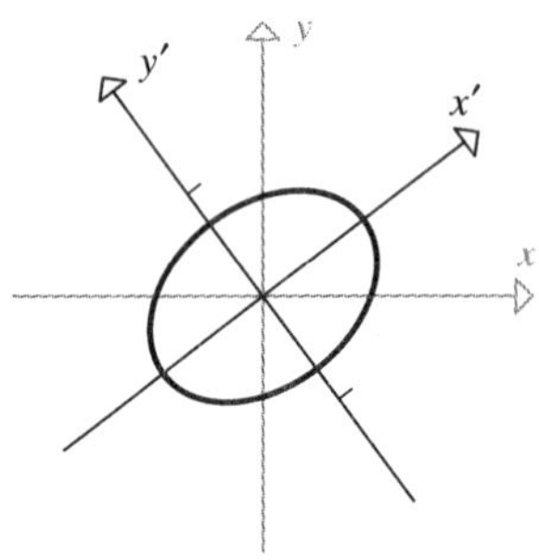

9. $\cot 2\theta = (9 - 16)/(-24) = 7/24$
$\cos 2\theta = 7/25$,
$\sin\theta = 3/5, \qquad \cos\theta = 4/5$
$x = (1/5)(4x' - 3y')$,
$y = (1/5)(3x' + 4y')$
$y'^2 = 4(x' - 1)$, parabola

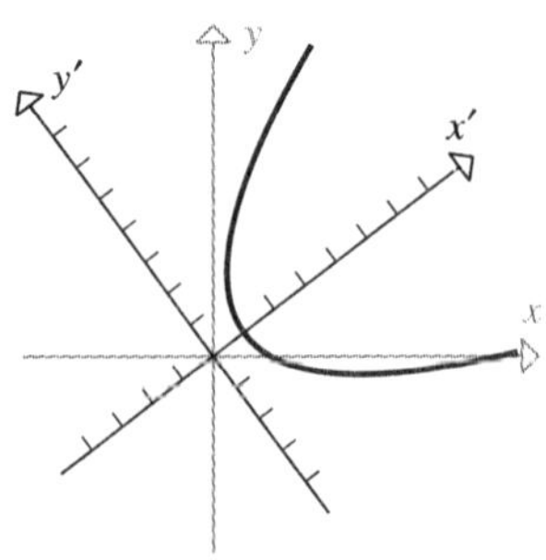

10. $\cot 2\theta = (5-5)/(-6) = 0,$

$\theta = 45°$

$x = (\sqrt{2}/2)(x' - y'),$

$y = (\sqrt{2}/2)(x' + y'),$

$x'^2/8 + (y'+1)^2/2 = 1$, ellipse

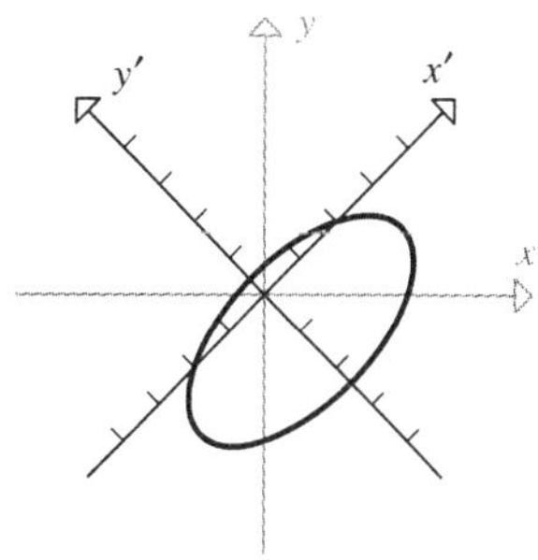

11. $\cot 2\theta = (52-73)/(-72) = 7/24$

$\cos 2\theta = 7/25, \quad \sin\theta = 3/5,$

$\cos\theta = 4/5$

$x = (1/5)(4x' - 3y'),$

$y = (1/5)(3x' + 4y')$

$(x'+1)^2/4 + y'^2 = 1$, ellipse

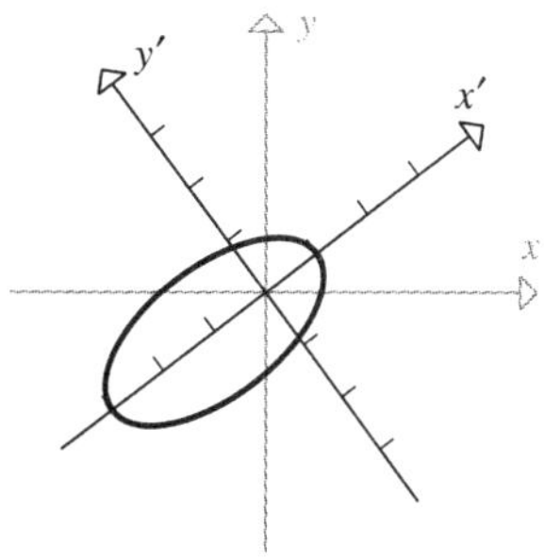

12. $\cot 2\theta = [6-(-1)]/24 = 7/24$

$\cos 2\theta = 7/25, \quad \sin\theta = 3/5,$

$\cos\theta = 4/5$

$x = (1/5)(4x' - 3y'),$

$y = (1/5)(3x' + 4y')$

$(y' - 7/5)^2/3 - (x' + 1/5)^2/2 = 1$, hyperbola

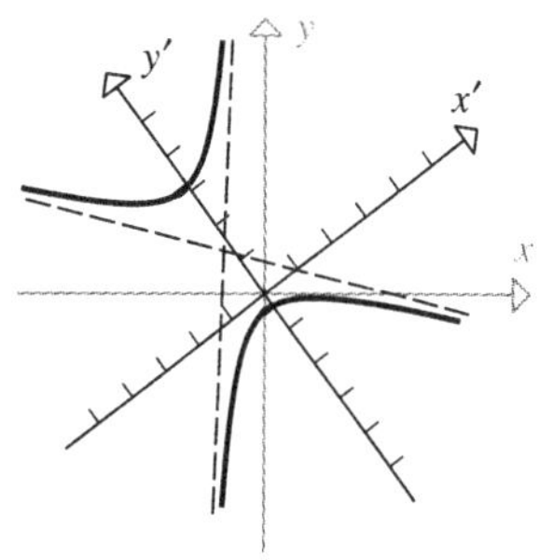

13. Let $x = x'\cos\theta - y'\sin\theta$, $y = x'\sin\theta + y'\cos\theta$ then $x^2 + y^2 = r^2$ becomes $(\sin^2\theta + \cos^2\theta)x'^2 + (\sin^2\theta + \cos^2\theta)y'^2 = r^2$, $x'^2 + y'^2 = r^2$. Under a rotation transformation the center of the circle stays at the origin of both coordinate systems.

14. Multiply the first equation through by $\cos\theta$ and the second by $\sin\theta$ and add to get $x\cos\theta + y\sin\theta = (\cos^2\theta + \sin^2\theta)x' = x'$. Multiply the first by $-\sin\theta$ and the second by $\cos\theta$ and add to get y'.

15. $x' = (\sqrt{2}/2)(x+y)$, $y' = (\sqrt{2}/2)(-x+y)$ which when substituted into $3x'^2 + y'^2 = 6$ yields $x^2 + xy + y^2 = 3$.

16. From (5), $x = \dfrac{1}{2}(\sqrt{3}x' - y')$ and $y = \dfrac{1}{2}(x' + \sqrt{3}y')$ so $y = x^2$ becomes $\dfrac{1}{2}(x' + \sqrt{3}y') = \dfrac{1}{4}(\sqrt{3}x' - y')^2$; simplify to get $3x'^2 - 2\sqrt{3}x'y' + y'^2 - 2x' - 2\sqrt{3}y' = 0$.

17. $\sqrt{x} + \sqrt{y} = 1$, $\sqrt{x} = 1 - \sqrt{y}$, $x = 1 - 2\sqrt{y} + y$, $2\sqrt{y} = 1 - x + y$, $4y = 1 + x^2 + y^2 - 2x + 2y - 2xy$, $x^2 - 2xy + y^2 - 2x - 2y + 1 = 0$. $\cot 2\theta = \dfrac{1-1}{-2} = 0$, $2\theta = \pi/2$, $\theta = \pi/4$. Let $x = x'/\sqrt{2} - y'/\sqrt{2}$, $y = x'/\sqrt{2} + y'/\sqrt{2}$ to get $2y'^2 - 2\sqrt{2}x' + 1 = 0$, which is a parabola. From $\sqrt{x} + \sqrt{y} = 1$ we see that $0 \le x \le 1$ and $0 \le y \le 1$, so the graph is just a portion of a parabola.

18. Let $x = x'\cos\theta - y'\sin\theta$ and $y = x'\sin\theta + y'\cos\theta$ in (7), expand and add all the coefficients of the terms that contain $x'y'$ to get B'.

19. Use (9) to express $B' - 4A'C'$ in terms of A, B, C, and θ, then simplify.

20. Use (9) to express $A' + C'$ in terms of A, B, C, and θ and then simplify.

21. $\cot 2\theta = (A - C)/B = 0$ if $A = C$ so $2\theta = 90°$, $\theta = 45°$.

22. If $F = 0$ then $x^2 + Bxy = 0$, $x(x + By) = 0$ so $x = 0$ or $x + By = 0$ which are lines that intersect at $(0, 0)$. Suppose $F \neq 0$, rotate through an angle θ where $\cot 2\theta = 1/B$ eliminating the cross product term to get $A'x'^2 + C'y'^2 + F' = 0$, and note that $F' = F$ so $F' \neq 0$. From (9),
$A' = \cos^2\theta + B\cos\theta\sin\theta = \cos\theta(\cos\theta + B\sin\theta)$ and
$C' = \sin^2\theta - B\sin\theta\cos\theta = \sin\theta(\sin\theta - B\cos\theta)$ so

$$\begin{aligned} A'C' &= \sin\theta\cos\theta[\sin\theta\cos\theta - B(\cos^2\theta - \sin^2\theta) - B^2\sin\theta\cos\theta] \\ &= \frac{1}{2}\sin 2\theta\left[\frac{1}{2}\sin 2\theta - B\cos 2\theta - \frac{1}{2}B^2\sin 2\theta\right] = \frac{1}{4}\sin^2 2\theta[1 - 2B\cot 2\theta - B^2] \\ &= \frac{1}{4}\sin^2 2\theta[1 - 2B(1/B) - B^2] = -\frac{1}{4}\sin^2 2\theta(1 + B^2) < 0 \end{aligned}$$

thus A' and C' have unlike signs so the graph is a hyperbola.

23. $B^2 - 4AC = (-1)^2 - 4(1)(1) = -3 < 0$; ellipse, point, or no graph. By inspection $(0, \pm\sqrt{2})$ lie on the curve, so it's an ellipse.

24. $B^2 - 4AC = (4)^2 - 4(1)(-2) = 24 > 0$; hyperbola or pair of intersecting lines

25. $B^2 - 4AC = (2\sqrt{3})^2 - 4(1)(3) = 0$; parabola, line, pair of parallel lines, or no graph. By inspection $(-\sqrt{3}, 3), (-\sqrt{3}, -1/3), (0, 0), (-2\sqrt{3}, 0), (0, 2/3)$ lie on the graph; since no three of these points are collinear, it's a parabola.

26. $B^2 - 4AC = (24)^2 - 4(6)(-1) = 600 > 0$; hyperbola or pair of intersecting lines

27. $B^2 - 4AC = (-24)^2 - 4(34)(41) = -5000 < 0$; ellipse, point, or no graph. By inspection $x = \pm 5/\sqrt{34}, y = 0$ satisfy the equation, so it's an ellipse.

28. **(a)** $(x - y)(x + y) = 0$
$y = x$ or $y = -x$
(two intersecting lines)

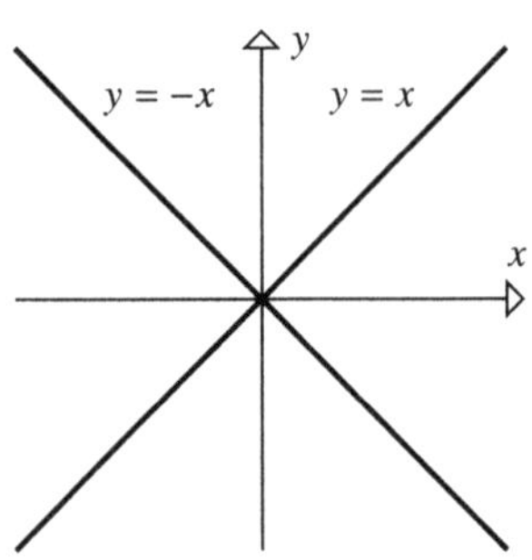

(b) $x^2 + 3y^2 = -7$ which has no real solutions, no graph

(c) $8x^2 + 7y^2 = 0$
$x = 0$ and $y = 0$, (a point)

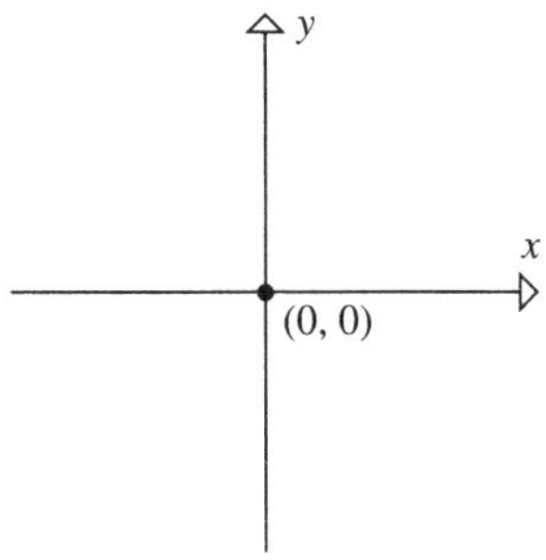

(d) $(x - y)^2 = 0$,
$y = x$ (a line)

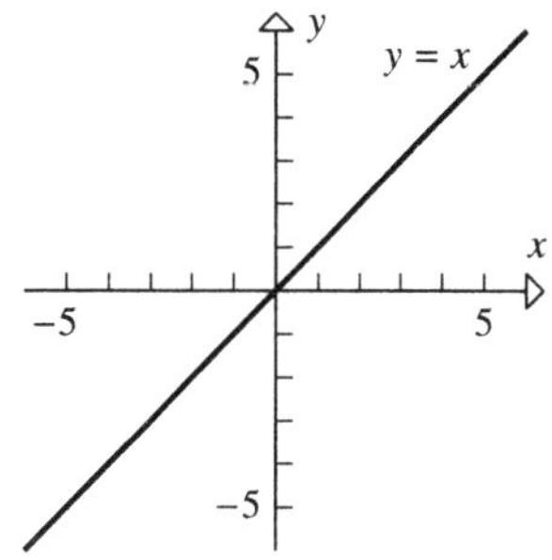

(e) $(3x + 2y)^2 = 36$,
$3x + 2y = 6$ or $3x + 2y = -6$
(a pair of parallel lines)

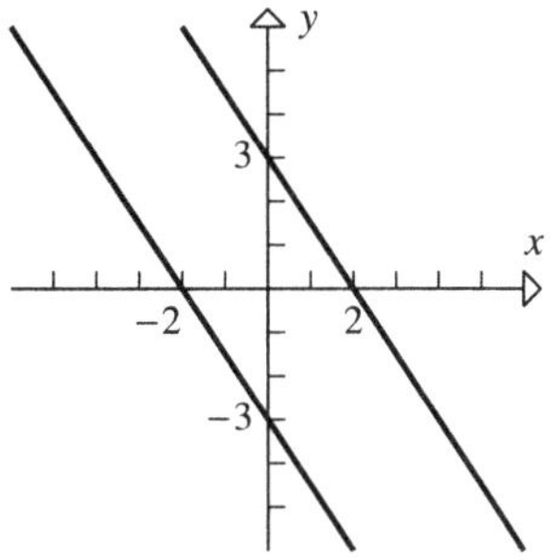

(f) $(x - 1)^2 + (y - 2)^2 = 0$,
the point $(1, 2)$

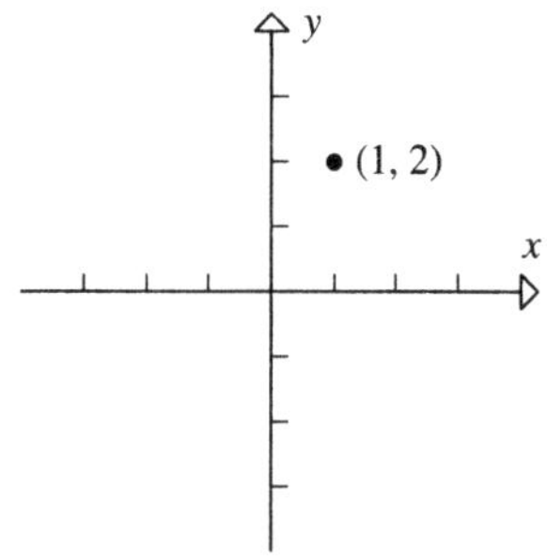

29. Part (b): from (15), $A'C' < 0$ so A' and C' have opposite signs. By multiplying (14) through by -1, if necessary, assume that $A' < 0$ and $C' > 0$ so $(x' - h)^2/C' - (y' - k)^2/|A'| = K$. If $K \neq 0$ then the graph is a hyperbola (divide both sides by K), if $K = 0$ then we get the pair of intersecting lines $(x' - h)/\sqrt{C'} = \pm(y' - k)/\sqrt{|A'|}$.

Part (c): from (15), $A'C' = 0$ so either $A' = 0$ or $C' = 0$ but not both (this would imply that $A = B = C = 0$ which results in (14) being linear). Suppose $A' \neq 0$ and $C' = 0$ then complete the square to get $(x' - h)^2 = -E'y'/A' + K$. If $E' \neq 0$ the graph is a parabola, if $E' = 0$ and $K = 0$ the graph is the line $x' = h$, if $E' = 0$ and $K > 0$ the graph is the pair of parallel lines $x' = h \pm \sqrt{K}$, if $E' = 0$ and $K < 0$ there is no graph.

30. **(a)** $B^2 - 4AC = (1)^2 - 4(1)(2) < 0$ so it is an ellipse (it contains the points $x = 0, y = -1, -1/2$).

(b) $y = -\dfrac{1}{4}x - \dfrac{3}{4} - \dfrac{1}{4}\sqrt{1 + 14x - 7x^2}$ or $y = -\dfrac{1}{4}x - \dfrac{3}{4} + \dfrac{1}{4}\sqrt{1 + 14x - 7x^2}$

31. **(a)** $B^2 - 4AC = (9)^2 - 4(2)(1) > 0$ so the conic is a hyperbola (it contains the points $(2, -1), (2, -3/2)$).

(b) $y = -\dfrac{9}{2}x - \dfrac{1}{2} - \dfrac{1}{2}\sqrt{73x^2 + 42x + 17}$ or $y = -\dfrac{9}{2}x - \dfrac{1}{2} + \dfrac{1}{2}\sqrt{73x^2 + 42x + 17}$

(c)

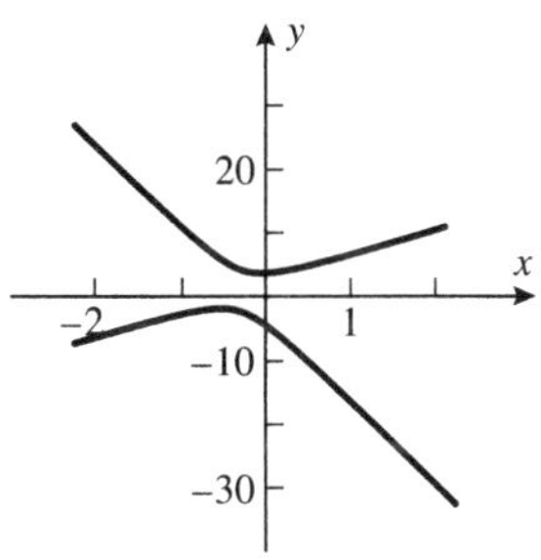

EXERCISE SET 11.6

1. **(a)** $r = \dfrac{3/2}{1 - \cos\theta}, e = 1, d = 3/2$

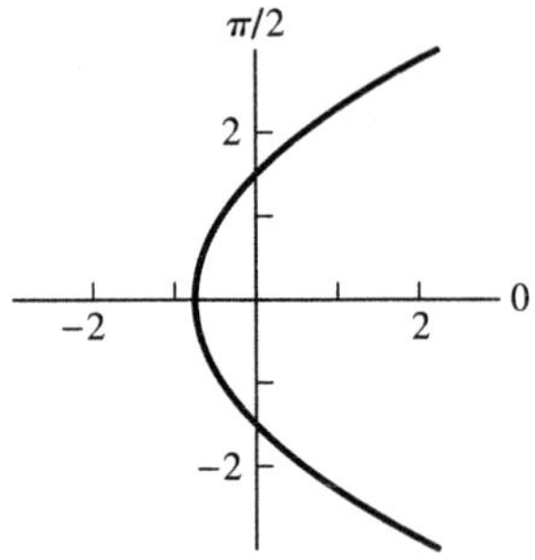

(b) $r = \dfrac{3/2}{1 + \frac{1}{2}\sin\theta}, e = 1/2, d = 3$

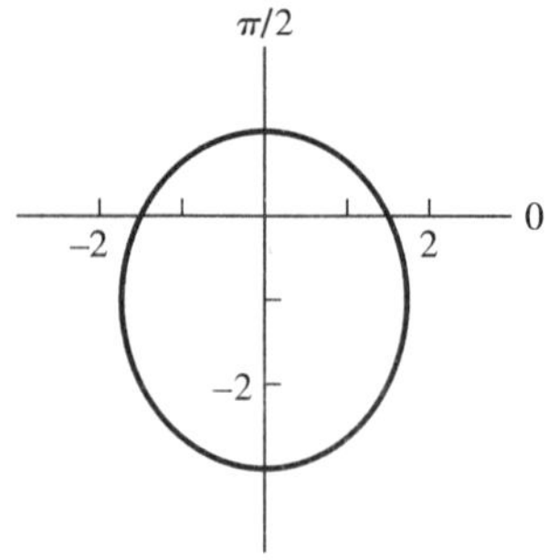

(c) $r = \dfrac{2}{1 + \frac{3}{2}\cos\theta}, e = 3/2, d = 4/3$

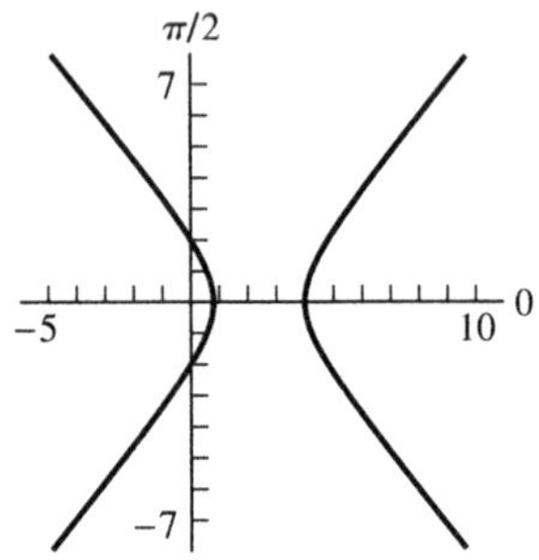

(d) $r = \dfrac{5/3}{1 + \sin\theta}, e = 1, d = 5/3$

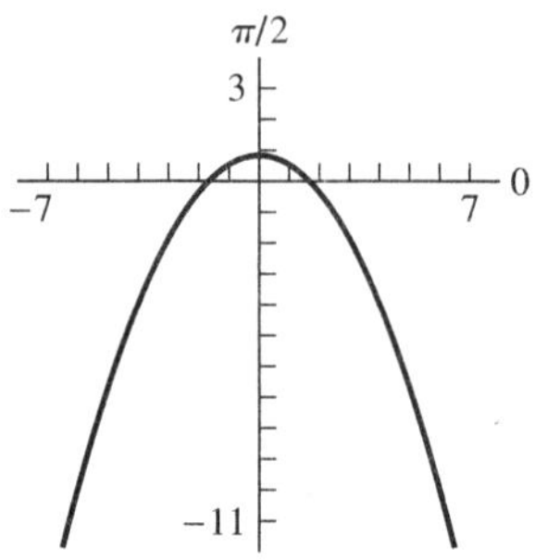

2. **(a)** $r = \dfrac{4/3}{1 - \frac{2}{3}\cos\theta}, e = 2/3, d = 2$

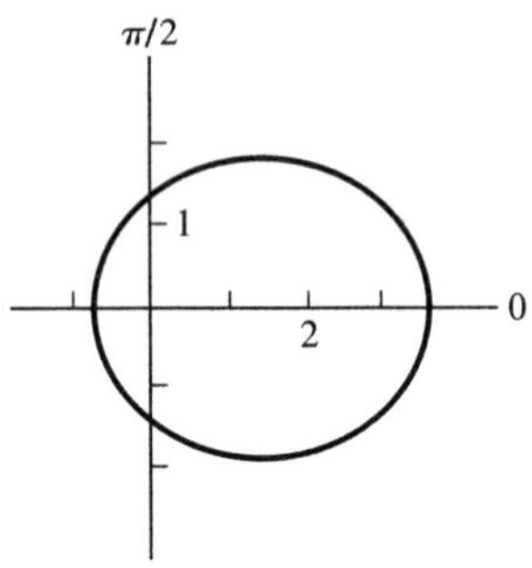

(b) $r = \dfrac{1}{1 - \frac{4}{3}\sin\theta}, e = 4/3, d = 3/4$

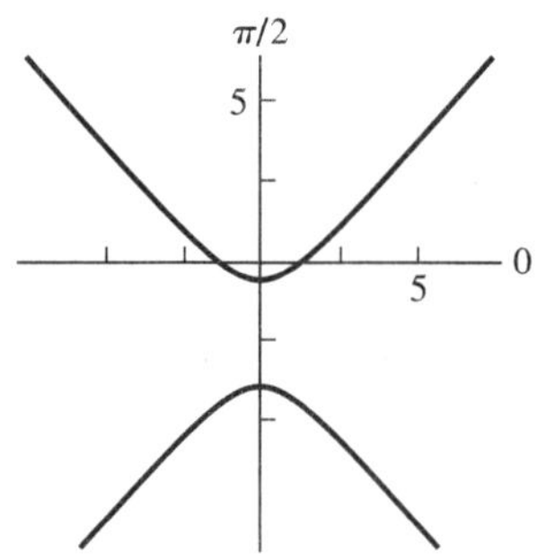

(c) $r = \dfrac{1/3}{1 + \sin\theta}, e = 1, d = 1/3$

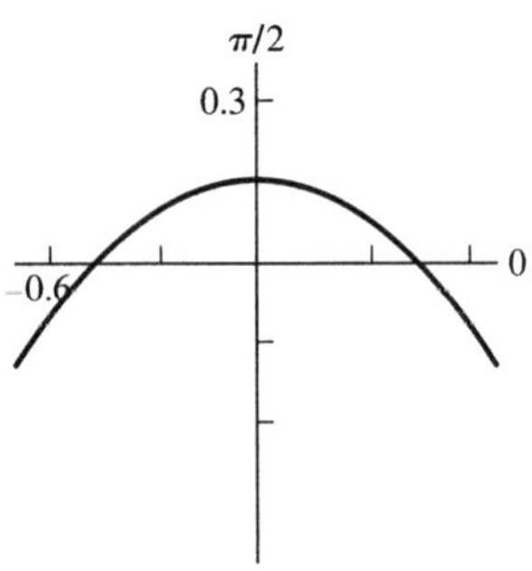

(d) $r = \dfrac{1/2}{1 + 3\sin\theta}, e = 3, d = 1/6$

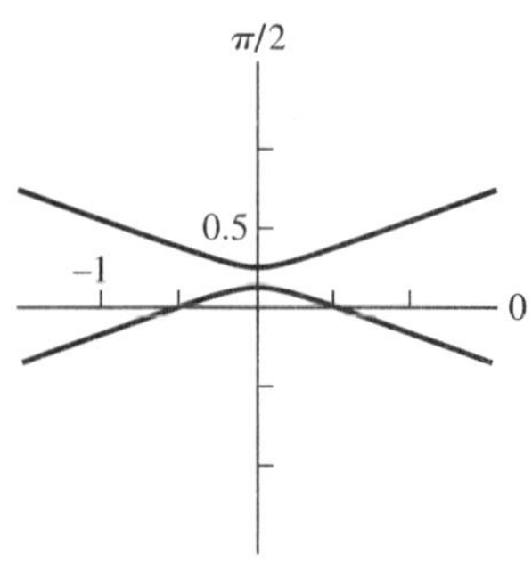

3. **(a)** $e = 1, d = 8$, parabola, opens up

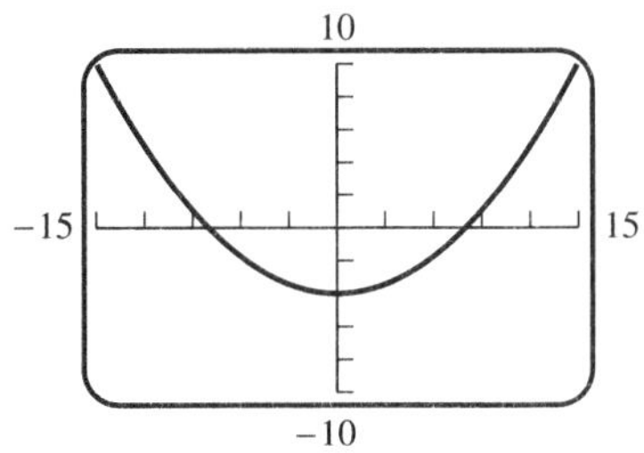

(b) $r = \dfrac{4}{1 + \frac{3}{4}\sin\theta}, e = 3/4, d = 16/3,$
ellipse, directrix 16/3 units above the pole

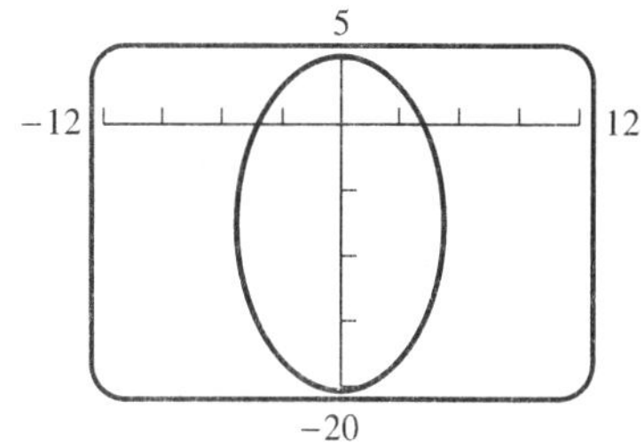

(c) $r = \dfrac{2}{1 - \frac{3}{2}\sin\theta}, e = 3/2, d = 4/3,$
hyperbola, directrix 4/3 units below the pole

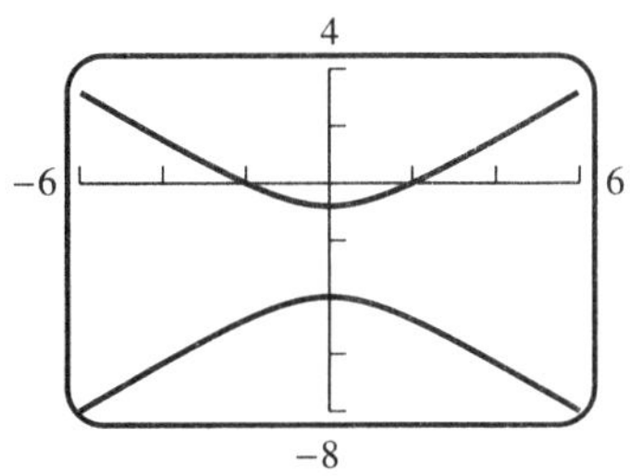

(d) $r = \dfrac{3}{1 + \frac{1}{4}\cos\theta}, e = 1/4, d = 12,$
ellipse, directrix 12 units to the right of the pole

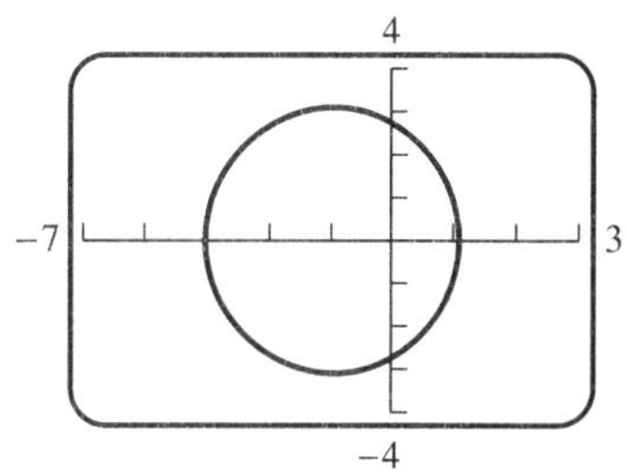

4. **(a)** $e = 1, d = 15$, parabola, opens left

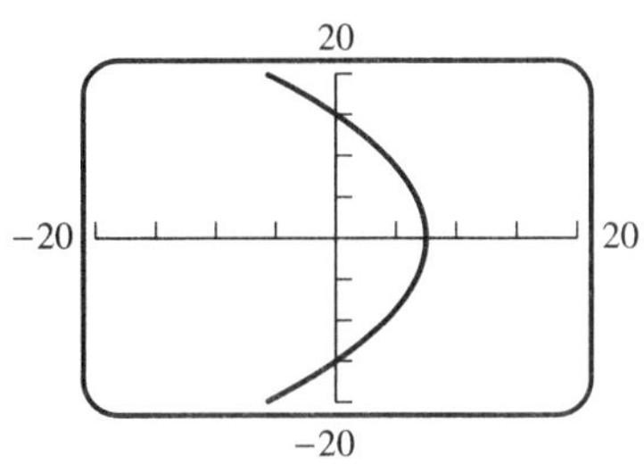

(b) $r = \dfrac{2/3}{1 + \cos\theta}, e = 1,$
$d = 2/3$, parabola, opens left

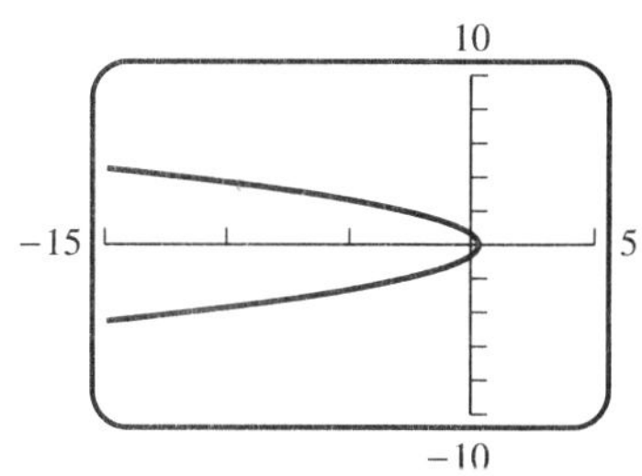

(c) $r = \dfrac{64/7}{1 - \frac{12}{7}\sin\theta}, e = 12/7, d = 16/3,$
hyperbola, directrix 16/3 units below pole

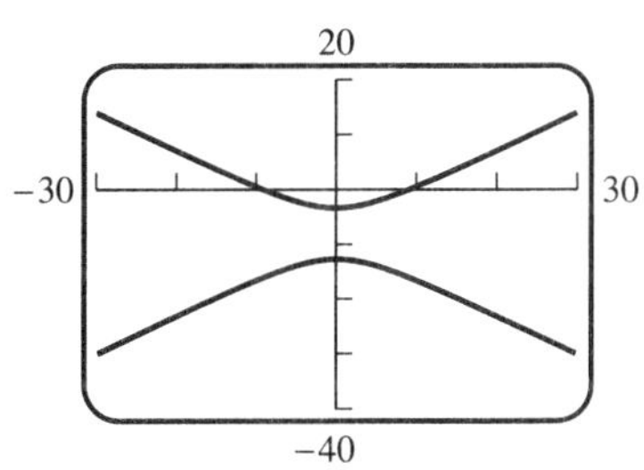

(d) $r = \dfrac{4}{1 - \frac{2}{3}\cos\theta}, e = 2/3, d = 6,$
ellipse, directrix 6 units left of the pole

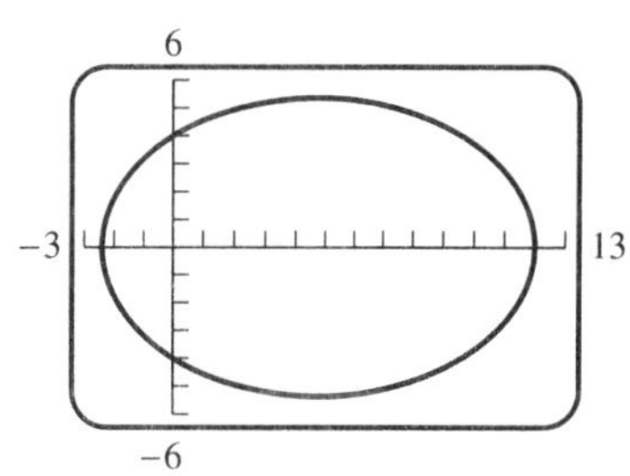

5. **(a)** $d = 1, r = \dfrac{ed}{1 + e\cos\theta} = \dfrac{2/3}{1 + \frac{2}{3}\cos\theta} = \dfrac{2}{3 + 2\cos\theta}$

(b) $e = 1, d = 1, r = \dfrac{ed}{1 - e\cos\theta} = \dfrac{1}{1 - \cos\theta}$

(c) $e = 3/2, d = 1, r = \dfrac{ed}{1 + e\sin\theta} = \dfrac{3/2}{1 + \frac{3}{2}\sin\theta} = \dfrac{3}{2 + 3\sin\theta}$

6. **(a)** $e = 2/3, d = 1, r = \dfrac{ed}{1 - e\sin\theta} = \dfrac{2/3}{1 - \frac{2}{3}\sin\theta} = \dfrac{2}{3 - 2\sin\theta}$

(b) $e = 1, d = 1, r = \dfrac{ed}{1 + e\sin\theta} = \dfrac{1}{1 + \sin\theta}$

(c) $e = 4/3, d = 1, r = \dfrac{ed}{1 - e\cos\theta} = \dfrac{4/3}{1 - \frac{4}{3}\cos\theta} = \dfrac{4}{3 - 4\cos\theta}$

7. **(a)** $r = \dfrac{ed}{1 \pm e\cos\theta}, \theta = 0 : 6 = \dfrac{ed}{1 \pm e}, \theta = \pi : 4 = \dfrac{ed}{1 \mp e}, 6 \pm 6e = 4 \mp 4e, 2 = \mp 10e$, use bottom sign to get $e = 1/5, d = 24$, $r = \dfrac{24/5}{1 - \cos\theta} = \dfrac{24}{5 - 5\cos\theta}$

(b) $e = 1, r = \dfrac{d}{1 - \sin\theta}, 1 = \dfrac{d}{2}, d = 2, r = \dfrac{2}{1 - \sin\theta}$

(c) $r = \dfrac{ed}{1 \pm e\sin\theta}, \theta = \pi/2 : 3 = \dfrac{ed}{1 \pm e}, \theta = 3\pi/2 : -7 = \dfrac{ed}{1 \mp e}, ed = 3 \pm 3e = -7 \pm 7e, 10 = \pm 4e,$

$e = 5/2, d = 21/5, r = \dfrac{21/2}{1 + (5/2)\sin\theta} = \dfrac{21}{2 + 5\sin\theta}$

8. **(a)** $r = \dfrac{ed}{1 \pm e\sin\theta}, 1 = \dfrac{ed}{1 \pm e}, 4 = \dfrac{ed}{1 \mp e}, 1 \pm e = 4 \mp 4e$, upper sign yields $e = 3/5, d = 8/3$,

$r = \dfrac{8/5}{1 + \frac{3}{5}\sin\theta} = \dfrac{8}{5 + 3\sin\theta}$

(b) $e = 1, r = \dfrac{d}{1 - \cos\theta}, 3 = \dfrac{d}{2}, d = 6,\ r = \dfrac{6}{1 - \cos\theta}$

(c) $a = b = 5, e = c/a = \sqrt{50}/5 = \sqrt{2}, r = \dfrac{\sqrt{2}d}{1 + \sqrt{2}\cos\theta}$; $r = 5$ when $\theta = 0$, so $d = 5 + \dfrac{5}{\sqrt{2}}$,

$r = \dfrac{5\sqrt{2} + 5}{1 + \sqrt{2}\cos\theta}$.

9. **(a)** $r = \dfrac{3}{1 + \frac{1}{2}\sin\theta}, e = 1/2, d = 6$, directrix 6 units above pole; if $\theta = \pi/2 : r_0 = 2$; if $\theta = 3\pi/2 : r_1 = 6, a = (r_0 + r_1)/2 = 4, b = \sqrt{r_0 r_1} = 2\sqrt{3}$, center $(0, -2)$ (rectangular coordinates), $\dfrac{x^2}{12} + \dfrac{(y + 2)^2}{16} = 1$

(b) $r = \dfrac{1/2}{1 - \frac{1}{2}\cos\theta}, e = 1/2, d = 1$, directrix 1 unit left of pole; if $\theta = \pi : r_0 = \dfrac{1/2}{3/2} = 1/3$; if $\theta = 0 : r_1 = 1, a = 2/3, b = 1/\sqrt{3}$, center $= (1/3, 0)$ (rectangular coordinates), $\dfrac{9}{4}(x - 1/3)^2 + 3y^2 = 1$

10. (a) $r = \dfrac{6/5}{1 + \frac{2}{5}\cos\theta}, e = 2/5, d = 3$, directrix 3 units right of pole, if $\theta = 0 : r_0 = 6/7$, if $\theta = \pi : r_1 = 2, a = 10/7, b = 2\sqrt{3}/\sqrt{7}$, center $(-4/7, 0)$ (rectangular coordinates), $\dfrac{49}{100}(x + 4/7)^2 + \dfrac{7}{12}y^2 = 1$

(b) $r = \dfrac{2}{1 - \frac{3}{4}\sin\theta}, e = 3/4, d = 8/3$, directrix 8/3 units below pole, if $\theta = 3\pi/2 : r_0 = 8/7$, if $\theta = \pi/2;\ \ r_1 = 8, a = 32/7, b = 8/\sqrt{7}$, center: $(0, 24/7)$ (rectangular coordinates), $\dfrac{7}{64}x^2 + \dfrac{49}{1024}\left(y - \dfrac{24}{7}\right)^2 = 1$

11. (a) $r = \dfrac{2}{1 + 3\sin\theta}, e = 3, d = 2/3$, hyperbola, directrix 2/3 units above pole, if $\theta = \pi/2 :$ $r_0 = 1/2; \theta = 3\pi/2 : r_1 = 1$, center $(0, 3/4), a = 1/4, b = 1/\sqrt{2}, -2x^2 + 16\left(y - \dfrac{3}{4}\right)^2 = 1$

(b) $r = \dfrac{5/3}{1 - \frac{3}{2}\cos\theta}, e = 3/2, d = 10/9$, hyperbola, directrix 10/9 units left of pole, if $\theta = \pi :$ $r_0 = 2/3; \theta = 0 : r_1 = \dfrac{5/3}{1/2} = 10/3$, center $(-2, 0)$, $\ a = 4/3, b = \sqrt{20/9}, \dfrac{9}{16}(x+2)^2 - \dfrac{9}{20}y^2 = 1$

12. (a) $r = \dfrac{4}{1 - 2\sin\theta}, e = 2, d = 2$, hyperbola, directrix 2 units below pole, if $\theta = 3\pi/2 : r_0 = 4/3;$ $\theta = \pi/2 : r_1 = \left|\dfrac{4}{1-2}\right| = 4$, center $(0, -8/3), a = 4/3, b = 4/\sqrt{3},\ \dfrac{9}{16}\left(y + \dfrac{8}{3}\right)^2 - \dfrac{3}{16}x^2 = 1$

(b) $r = \dfrac{15/2}{1 + 4\cos\theta}, e = 4, d = 15/8$, hyperbola, directrix 15/8 units right of pole, if $\theta = 0 :$ $r_0 = 3/2; \theta = \pi : r_1 = \left|-\dfrac{5}{2}\right| = 5/2, a = 1/2, b = \dfrac{\sqrt{15}}{2}$, center $(2, 0)$, $\ 4(x-2)^2 - \dfrac{4}{15}y^2 = 1$

13. (a) $r = \dfrac{\frac{1}{2}d}{1 + \frac{1}{2}\cos\theta} = \dfrac{d}{2 + \cos\theta}$, if $\theta = 0 : r_0 = d/3; \theta = \pi, r_1 = d,$ $8 = a = \dfrac{1}{2}(r_1 + r_0) = \dfrac{2}{3}d, d = 12,\ \ r = \dfrac{12}{2 + \cos\theta}$

(b) $r = \dfrac{\frac{3}{5}d}{1 - \frac{3}{5}\sin\theta} = \dfrac{3d}{5 - 3\sin\theta}$, if $\theta = 3\pi/2 : r_0 = \dfrac{3}{8}d; \theta = \pi/2, r_1 = \dfrac{3}{2}d,$ $4 = a = \dfrac{1}{2}(r_1 + r_0) = \dfrac{15}{16}d, d = \dfrac{64}{15}, r = \dfrac{3(64/15)}{5 - 3\sin\theta} = \dfrac{64}{25 - 15\sin\theta}$

(c) $r = \dfrac{\frac{3}{5}d}{1 - \frac{3}{5}\cos\theta} = \dfrac{3d}{5 - 3\cos\theta}$, if $\theta = \pi : r_0 = \dfrac{3}{8}d; \theta = 0, r_1 = \dfrac{3}{2}d, 4 = b = \dfrac{3}{4}d,$ $d = 16/3,\ \ r = \dfrac{16}{5 - 3\cos\theta}$

(d) $r = \dfrac{\frac{1}{5}d}{1 + \frac{1}{5}\sin\theta} = \dfrac{d}{5 + \sin\theta}$, if $\theta = \pi/2 : r_0 = d/6; \theta = 3\pi/2, r_1 = d/4,$ $5 = c = \dfrac{1}{2}d\left(\dfrac{1}{4} - \dfrac{1}{6}\right) = \dfrac{1}{24}d, d = 120,\ \ r = \dfrac{120}{5 + \sin\theta}$

14. **(a)** $r = \dfrac{\frac{1}{2}d}{1+\frac{1}{2}\sin\theta} = \dfrac{d}{2+\sin\theta}$, if $\theta = \pi/2 : r_0 = d/3; \theta = 3\pi/2 : r_1 = d$,

$10 = a = \dfrac{1}{2}(r_0 + r_1) = \dfrac{2}{3}d, d = 15,\ \ r = \dfrac{15}{2+\sin\theta}$

(b) $r = \dfrac{\frac{1}{5}d}{1-\frac{1}{5}\cos\theta} = \dfrac{d}{5-\cos\theta}$, if $\theta = \pi : r_0 = d/6, \theta = 0 : r_1 = d/4$,

$6 = a = \dfrac{1}{2}(r_1 + r_0) = \dfrac{1}{2}d\left(\dfrac{1}{4}+\dfrac{1}{6}\right) = \dfrac{5}{24}d, d = 144/5,\ \ r = \dfrac{144/5}{5-\cos\theta} = \dfrac{144}{25-5\cos\theta}$

(c) $r = \dfrac{\frac{3}{4}d}{1-\frac{3}{4}\sin\theta} = \dfrac{3d}{4-3\sin\theta}$, if $\theta = 3\pi/2 : r_0 = \dfrac{3}{7}d, \theta = \pi/2 : r_1 = 3d, 4 = b = 3d/\sqrt{7}$,

$d = \dfrac{4}{3}\sqrt{7},\ \ r = \dfrac{4\sqrt{7}}{4-3\sin\theta}$

(d) $r = \dfrac{\frac{4}{5}d}{1+\frac{4}{5}\cos\theta} = \dfrac{4d}{5+4\cos\theta}$, if $\theta = 0 : r_0 = \dfrac{4}{9}d; \theta = \pi : r_1 = 4d$,

$c = 10 = \dfrac{1}{2}(r_1 - r_0) = \dfrac{1}{2}d\left(4 - \dfrac{4}{9}\right) = \dfrac{16}{9}d, d = \dfrac{45}{8},\ \ r = \dfrac{45/2}{5+4\cos\theta} = \dfrac{45}{10+8\cos\theta}$

15. **(a)** $e = c/a = \dfrac{\frac{1}{2}(r_1 - r_0)}{\frac{1}{2}(r_1 + r_0)} = \dfrac{r_1 - r_0}{r_1 + r_0}$

(b) $e = \dfrac{r_1/r_0 - 1}{r_1/r_0 + 1}, e(r_1/r_0 + 1) = r_1/r_0 - 1, \dfrac{r_1}{r_0} = \dfrac{1+e}{1-e}$

16. **(a)** $e = c/a = \dfrac{\frac{1}{2}(r_1 + r_0)}{\frac{1}{2}(r_1 - r_0)} = \dfrac{r_1 + r_0}{r_1 - r_0}$

(b) $e = \dfrac{r_1/r_0 + 1}{r_1/r_0 - 1}, e(r_1/r_0 - 1) = r_1/r_0 + 1, \dfrac{r_1}{r_0} = \dfrac{e+1}{e-1}$

17. **(a)** $T = a^{3/2} = 39.5^{1.5} \approx 248$ yr

(b) $r_0 = a(1-e) = 39.5(1-0.249) = 29.6645$ AU $\approx$ 4,449,675,000 km
$r_1 = a(1+e) = 39.5(1+0.249) = 49.3355$ AU $\approx$ 7,400,325,000 km

(c) $r = \dfrac{a(1-e^2)}{1+e\cos\theta} \approx \dfrac{39.5(1-(0.249)^2)}{1+0.249\cos\theta} \approx \dfrac{37.05}{1+0.249\cos\theta}$ AU

(d)

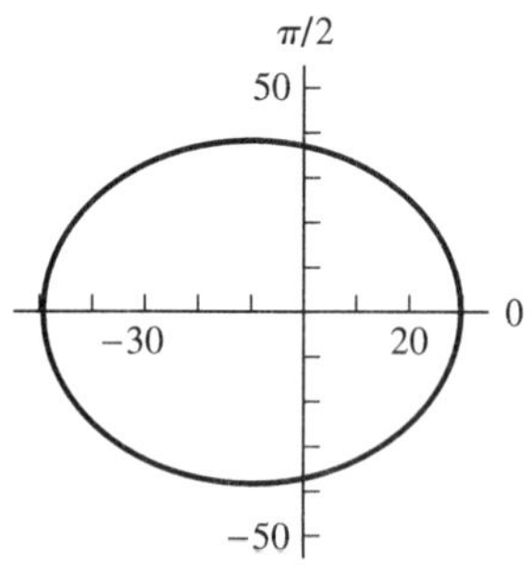

18. **(a)** In yr and AU, $T = a^{3/2}$; in days and km, $\dfrac{T}{365} = \left(\dfrac{a}{150 \times 10^6}\right)^{3/2}$,

so $T = 365 \times 10^{-9}\left(\dfrac{a}{150}\right)^{3/2}$ days.

(b) $T = 365 \times 10^{-9}\left(\dfrac{57.95 \times 10^6}{150}\right)^{3/2} \approx 87.6$ days

(c) $r = \dfrac{55490833.8}{1 + .206\cos\theta}$

From (17) the polar equation of the orbit has the form $r = \dfrac{a(1-e^2)}{1+e\cos\theta} = \dfrac{55490833.8}{1+.206\cos\theta}$ km, or $r = \dfrac{0.3699}{1+0.206\cos\theta}$ AU.

(d)

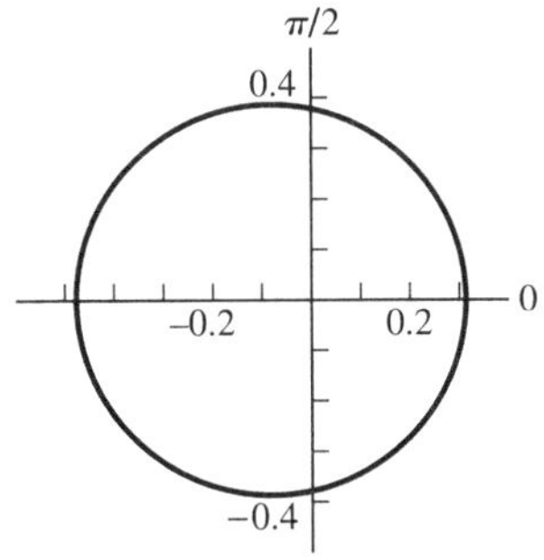

19. **(a)** $a = T^{2/3} = 2380^{2/3} \approx 178.26$ AU

(b) $r_0 = a(1-e) \approx 0.8735$ AU, $r_1 = a(1+e) \approx 355.64$ AU

(c) $r = \dfrac{a(1-e^2)}{1+e\cos\theta} \approx \dfrac{1.74}{1+0.9951\cos\theta}$ AU

(d)

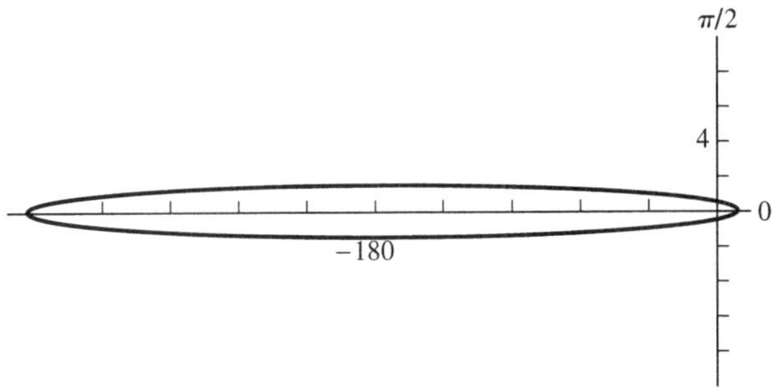

20. **(a)** By Exercise 15(a), $e = \dfrac{r_1 - r_0}{r_1 + r_0} \approx 0.092635$

(b) $a = \frac{1}{2}(r_0 + r_1) = 225{,}400{,}000$ km ≈ 1.503 AU, so $T = a^{3/2} \approx 1.84$ yr

(c) $r = \dfrac{a(1-e^2)}{1+e\cos\theta} \approx \dfrac{223465774.6}{1+0.092635\cos\theta}$ km, or $\approx \dfrac{1.48977}{1+0.092635\cos\theta}$ AU

(d)

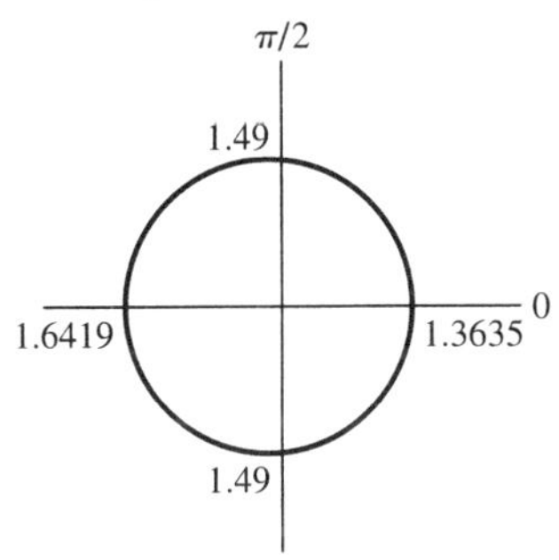

21. $r_0 = a(1-e) \approx 7003$ km, $h_{\min} \approx 7003 - 6440 = 563$ km,
$r_1 = a(1+e) \approx 10{,}726$ km, $h_{\max} \approx 10{,}726 - 6440 = 4286$ km

22. $r_0 = a(1-e) \approx 651{,}736$ km, $h_{\min} \approx 581{,}736$ km; $r_1 = a(1+e) \approx 6{,}378{,}102$ km, $h_{\max} \approx 6{,}308{,}102$ km

23. Since the foci are fixed, c is constant; since $e \to 0$, the distance $\dfrac{a}{e} = \dfrac{c}{e^2} \to +\infty$.

24. **(a)** From Figure 11.4.22, $\dfrac{x^2}{a^2} - \dfrac{y^2}{b^2} = 1$, $\dfrac{x^2}{a^2} - \dfrac{y^2}{c^2 - a^2} = 1$, $\left(1 - \dfrac{c^2}{a^2}\right)x^2 + y^2 = a^2 - c^2$,

$c^2 + x^2 + y^2 = \left(\dfrac{c}{a}x\right) + a^2$, $(x-c)^2 + y^2 = \left(\dfrac{c}{a}x - a\right)^2$,

$\sqrt{(x-c)^2 + y^2} = \dfrac{c}{a}x - a$ for $x > a^2/c$.

(b) From Part (a) and Figure 11.6.1, $PF = \dfrac{c}{a}PD$, $\dfrac{PF}{PD} = c/a$.

CHAPTER 11 SUPPLEMENTARY EXERCISES

2. **(a)** $(\sqrt{2}, 3\pi/4)$ **(b)** $(-\sqrt{2}, 7\pi/4)$ **(c)** $(\sqrt{2}, 3\pi/4)$ **(d)** $(-\sqrt{2}, -\pi/4)$

3. **(a)** circle **(b)** rose **(c)** line **(d)** limaçon
(e) limaçon **(f)** none **(g)** none **(h)** spiral

4. **(a)** $r = \dfrac{1/3}{1 + \frac{1}{3}\cos\theta}$, ellipse, right of pole, distance $= 1$

(b) hyperbola, left of pole, distance $= 1/3$

(c) $r = \dfrac{1/3}{1 + \sin\theta}$, parabola, above pole, distance $= 1/3$

(d) parabola, below pole, distance $= 3$

5. **(a)**

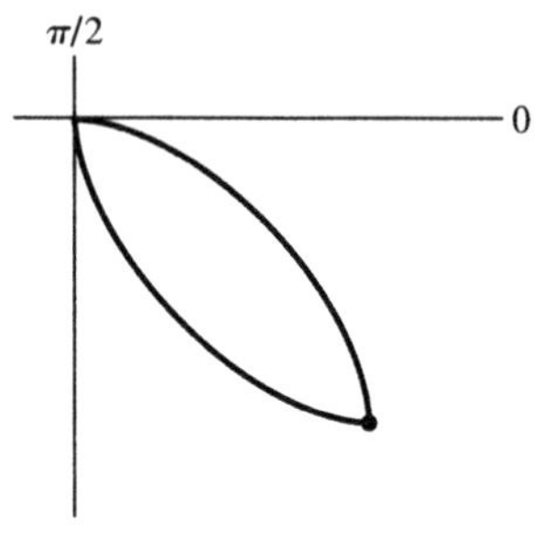

(b)

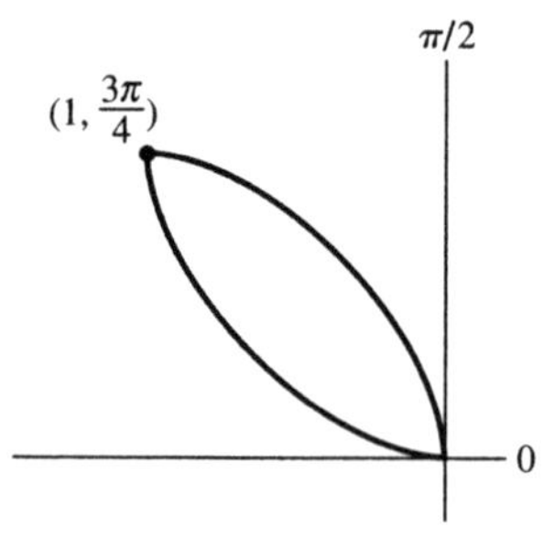

(c)

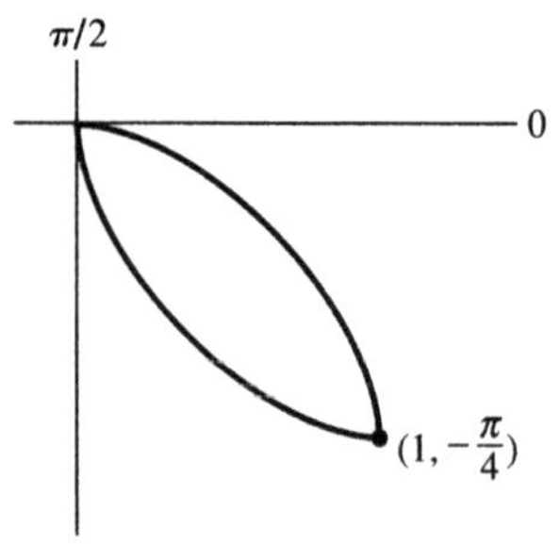

(d)

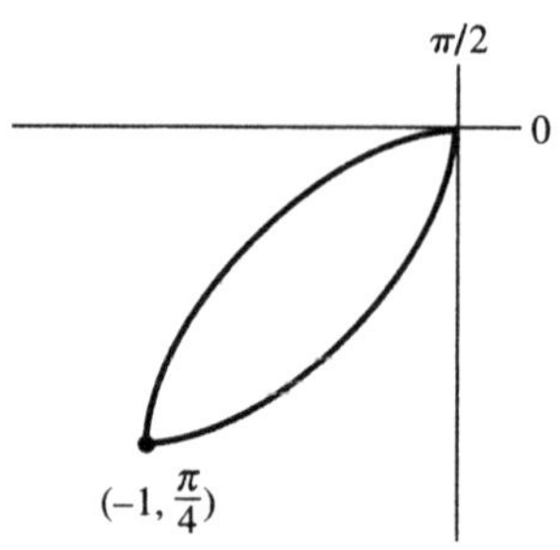

(e)

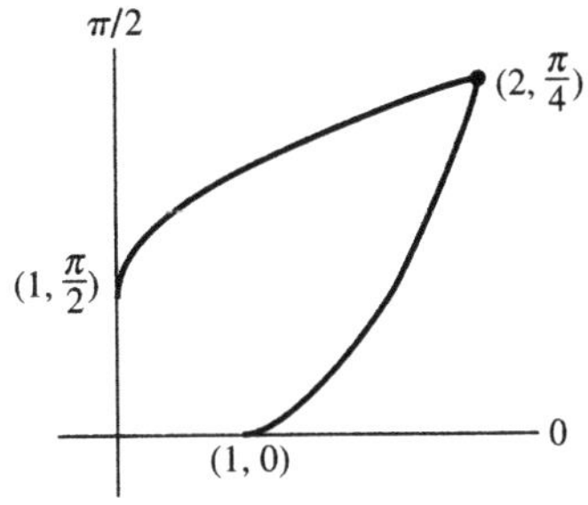

6. Family I: $x^2 + (y - b)^2 = b^2, b < 0$, or $r = 2b\sin\theta$; Family II: $(x - a)^2 + y^2 = a^2, a < 0$, or $r = 2a\cos\theta$

7. **(a)** $r = 2a/(1 + \cos\theta), r + x = 2a, x^2 + y^2 = (2a - x)^2, y^2 = -4ax + 4a^2$, parabola

(b) $r^2(\cos^2\theta - \sin^2\theta) = x^2 - y^2 = a^2$, hyperbola

(c) $r\sin(\theta - \pi/4) = (\sqrt{2}/2)r(\sin\theta - \cos\theta) = 4, y - x = 4\sqrt{2}$, line

(d) $r^2 = 4r\cos\theta + 8r\sin\theta, x^2 + y^2 = 4x + 8y, (x - 2)^2 + (y - 4)^2 = 20$, circle

9. **(a)** $\frac{c}{a} = e = \frac{2}{7}$ and $2b = 6, b = 3, a^2 = b^2 + c^2 = 9 + \frac{4}{49}a^2, \frac{45}{49}a^2 = 9, a = \frac{7}{\sqrt{5}}, \frac{5}{49}x^2 + \frac{1}{9}y^2 = 1$

(b) $x^2 = -4py$, directrix $y = 4$, focus $(-4, 0), 2p = 8, x^2 = -16y$

(c) For the ellipse, $a = 4, b = \sqrt{3}, c^2 = a^2 - b^2 = 16 - 3 = 13$, foci $(\pm\sqrt{13}, 0)$;
for the hyperbola, $c = \sqrt{13}, b/a = 2/3, b = 2a/3, 13 = c^2 = a^2 + b^2 = a^2 + \frac{4}{9}a^2 = \frac{13}{9}a^2$,
$a = 3, b = 2, \ \frac{x^2}{9} - \frac{y^2}{4} = 1$

10. **(a)** $e = 4/5 = c/a, c = 4a/5$, but $a = 5$ so $c = 4, b = 3, \frac{(x+3)^2}{25} + \frac{(y-2)^2}{9} = 1$

(b) directrix $y = 2, p = 2, (x + 2)^2 = -8y$

(c) center $(-1, 5)$, vertices $(-1, 7)$ and $(-1, 3), a = 2, a/b = 8, b = 1/4, \ \frac{(y-5)^2}{4} - 16(x+1)^2 = 1$

11. **(a)**

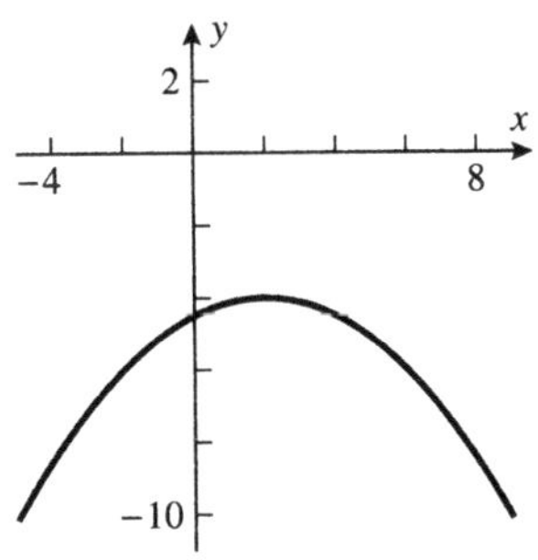

(b)

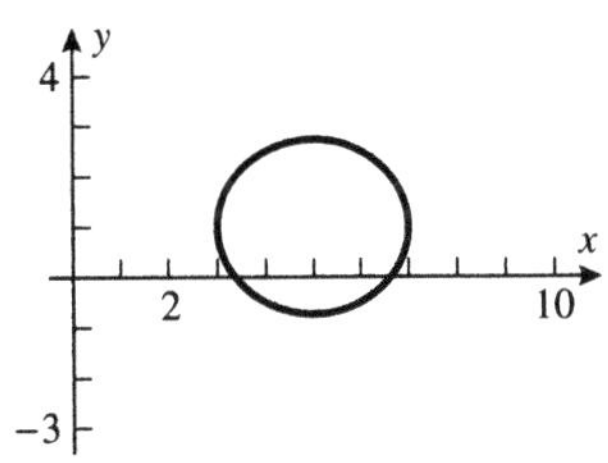

(c)

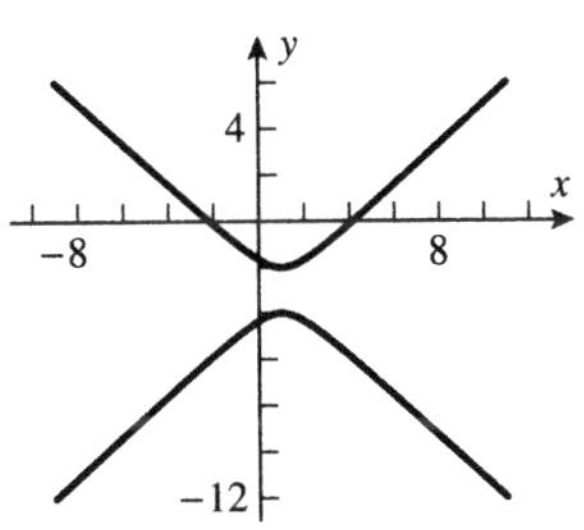

(d)

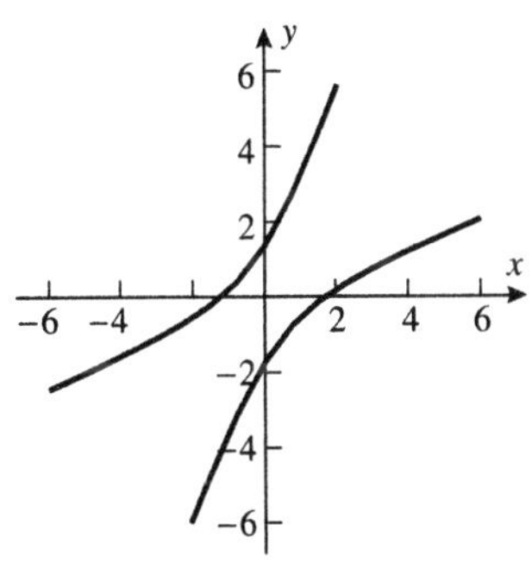

13. **(a)** The equation of the parabola is $y = ax^2$ and it passes through $(2100, 470)$, thus $a = \dfrac{470}{2100^2}$, $y = \dfrac{470}{2100^2}x^2$.

(b) $L = 2\displaystyle\int_0^{2100} \sqrt{1 + \left(2\frac{470}{2100^2}x\right)^2}\, dx$

$$= \frac{x}{220500}\sqrt{48620250000 + 2209x^2} + \frac{220500}{47}\sinh^{-1}\left(\frac{47}{220500}x\right) \approx 4336.3 \text{ ft}$$

14. **(a)** As t runs from 0 to π, the upper portion of the curve is traced out from right to left; as t runs from π to 2π the bottom portion of the curve is traced out from right to left. The loop occurs for $\pi + \sin^{-1}\dfrac{1}{4} < t < 2\pi - \sin^{-1}\dfrac{1}{4}$.

(b) $\displaystyle\lim_{t\to 0^+} x = +\infty$, $\displaystyle\lim_{t\to 0^+} y = 1$; $\displaystyle\lim_{t\to \pi^-} x = -\infty$, $\displaystyle\lim_{t\to \pi^-} y = 1$; $\displaystyle\lim_{t\to \pi^+} x = +\infty$, $\displaystyle\lim_{t\to \pi^+} y = 1$; $\displaystyle\lim_{t\to 2\pi^-} x = -\infty$, $\displaystyle\lim_{t\to 2\pi^-} y = 1$; the horizontal asymptote is $y = 1$.

(c) horizontal tangent line when $dy/dx = 0$, or $dy/dt = 0$, so $\cos t = 0, t = \pi/2, 3\pi/2$; vertical tangent line when $dx/dt = 0$, so $-\csc^2 t - 4\sin t = 0, t = \pi + \sin^{-1}\dfrac{1}{\sqrt[3]{4}}, 2\pi - \sin^{-1}\dfrac{1}{\sqrt[3]{4}}$, $t = 3.823, 5.602$

(d) $r^2 = x^2 + y^2 = (\cot t + 4\cos t)^2 + (1 + 4\sin t)^2 = (4 + \csc t)^2, r = 4 + \csc t$; with $t = \theta$, $f(\theta) = 4 + \csc\theta; m = dy/dx = (f(\theta)\cos\theta + f'(\theta)\sin\theta)/(-f(\theta)\sin\theta + f'(\theta)\cos\theta)$; when $\theta = \pi + \sin^{-1}(1/4), m = \sqrt{15}/15$, when $\theta = 2\pi - \sin^{-1}(1/4), m = -\sqrt{15}/15$, so the tangent lines to the conchoid at the pole have polar equations $\theta = \pm\tan^{-1}\dfrac{1}{\sqrt{15}}$.

15. $$= \int_0^{\pi/6} 4\sin^2\theta\, d\theta + \int_{\pi/6}^{\pi/4} 1\, d\theta = \int_0^{\pi/6} 2(1 - \cos 2\theta)\, d\theta + \frac{\pi}{12} = (2\theta - \sin 2\theta)\Big]_0^{\pi/6} + \frac{\pi}{12}$$
$$= \frac{\pi}{3} - \frac{\sqrt{3}}{2} + \frac{\pi}{12} = \frac{5\pi}{12} - \frac{\sqrt{3}}{2}$$

16. The circle has radius $a/2$ and lies entirely inside the cardioid, so
$$A = \int_0^{2\pi} \frac{1}{2}a^2(1 + \sin\theta)^2\, d\theta - \pi a^2/4 = \frac{3a^2}{2}\pi - \frac{a^2}{4}\pi = \frac{5a^2}{4}\pi$$

17. **(a)** $r = 1/\theta, dr/d\theta = -1/\theta^2, r^2 + (dr/d\theta)^2 = 1/\theta^2 + 1/\theta^4$, $L = \displaystyle\int_{\pi/4}^{\pi/2} \frac{1}{\theta^2}\sqrt{1 + \theta^2}\, d\theta \approx 0.9457$ by Endpaper Table Formula 93.

(b) The integral $\displaystyle\int_1^{+\infty} \frac{1}{\theta^2}\sqrt{1 + \theta^2}\, d\theta$ diverges by the comparison test (with $1/\theta$), and thus the arc length is infinite.

18. **(a)** When the point of departure of the thread from the circle has traversed an angle θ, the amount of thread that has been unwound is equal to the arc length traversed by the point of departure, namely $a\theta$. The point of departure is then located at $(a\cos\theta, a\sin\theta)$, and the tip of the string, located at (x, y), satisfies the equations $x - a\cos\theta = a\theta\sin\theta, y - a\sin\theta = -a\theta\cos\theta$; hence $x = a(\cos\theta + \theta\sin\theta), y = a(\sin\theta - \theta\cos\theta)$.

(b) Assume for simplicity that $a = 1$. Then $dx/d\theta = \theta\cos\theta, dy/d\theta = \theta\sin\theta; dx/d\theta = 0$ has solutions $\theta = 0, \pi/2, 3\pi/2$; and $dy/d\theta = 0$ has solutions $\theta = 0, \pi, 2\pi$. At $\theta = \pi/2, dy/d\theta > 0$, so the direction is North; at $\theta = \pi, dx/d\theta < 0$, so West; at $\theta = 3\pi/2, dy/d\theta < 0$, so South; at $\theta = 2\pi, dx/d\theta > 0$, so East. Finally, $\displaystyle\lim_{\theta\to 0^+} dy/dx = \lim_{\theta\to 0^+} \tan\theta = 0$, so East.

(c)

θ	0	$\pi/2$	π	$3\pi/2$	2π
x	1	$\pi/2$	-1	$-3\pi/2$	1
y	0	1	π	-1	-2π

Note that the parameter θ in these equations does not satisfy equations (1) and (2) of Section 11.1, since it measures the angle of the point of departure and not the angle of the tip of the thread.

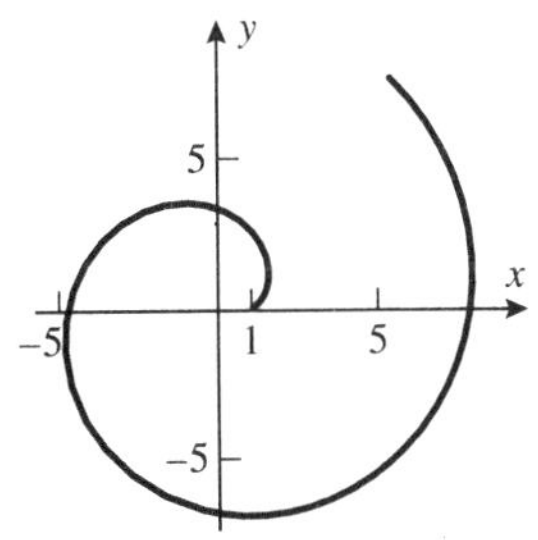

$a = 1$

19. (a) $V = \int_a^{\sqrt{a^2+b^2}} \pi\left(b^2x^2/a^2 - b^2\right)\,dx$
$= \dfrac{\pi b^2}{3a^2}(b^2 - 2a^2)\sqrt{a^2+b^2} + \dfrac{2}{3}ab^2\pi$

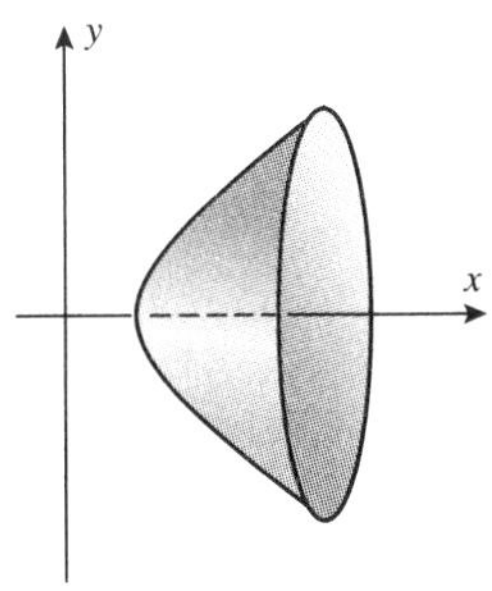

(b) $V = 2\pi \int_a^{\sqrt{a^2+b^2}} x\sqrt{b^2x^2/a^2 - b^2}\,dx = (2b^4/3a)\pi$

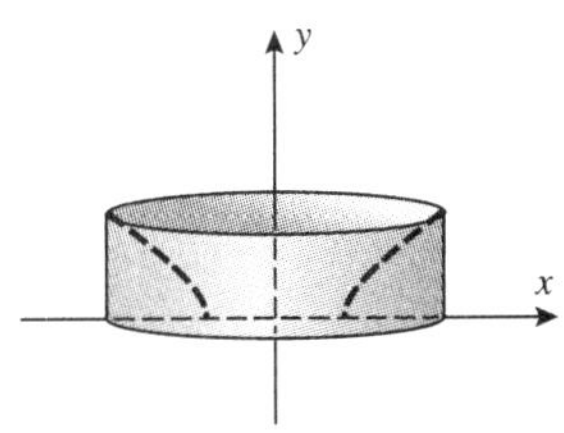

20. (a)

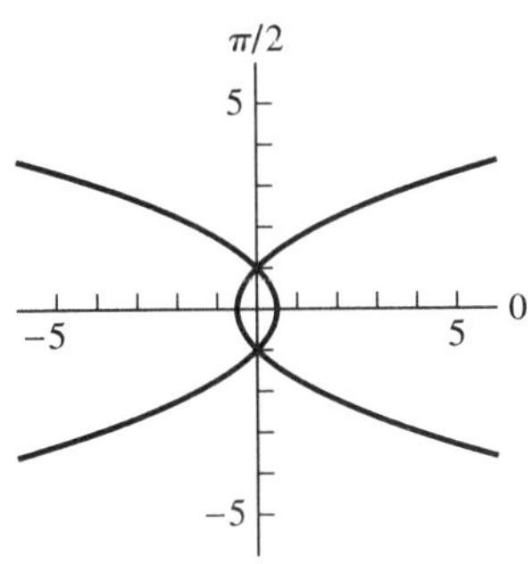

(b) $\theta - \pi/2, 3\pi/2, r = 1$

(c) $dy/dx = \dfrac{r\cos\theta + (dr/d\theta)\sin\theta}{-r\sin\theta + (dr/d\theta)\cos\theta}$; at $\theta = \pi/2, m_1 = (-1)/(-1) = 1, m_2 = 1/(-1) = -1$, $m_1m_2 = -1$; and at $\theta = 3\pi/2, m_1 = -1, m_2 = 1, m_1m_2 = -1$

22. The tips are located at $r=1, \theta=\pi/6, 5\pi/6, 3\pi/2$ and, for example,
$d=\sqrt{1+1-2\cos(5\pi/6-\pi/6)}=\sqrt{2(1-\cos(2\pi/3))}=\sqrt{3}$

23. **(a)** $x=r\cos\theta=\cos\theta+\cos^2\theta, dx/d\theta=-\sin\theta-2\sin\theta\cos\theta=-\sin\theta(1+2\cos\theta)=0$ if $\sin\theta=0$ or $\cos\theta=-1/2$, so $\theta=0,\pi,2\pi/3,4\pi/3$; maximum $x=2$ at $\theta=0$, minimum $x=-1/4$ at $\theta=2\pi/3,4\pi/3$; $\theta=\pi$ is a local maximum for x

(b) $y=r\sin\theta=\sin\theta+\sin\theta\cos\theta, dy/d\theta=2\cos^2\theta+\cos\theta-1=0$ at $\cos\theta=1/2,-1$, so $\theta=\pi/3,5\pi/3,\pi$; maximum $y=3\sqrt{3}/4$ at $\theta=\pi/3$, minimum $y=-3\sqrt{3}/4$ at $\theta=5\pi/3$

24. **(a)** $y=r\sin\theta=(\sin\theta)/\sqrt{\theta}, dy/d\theta=\dfrac{2\theta\cos\theta-\sin\theta}{2\theta^{3/2}}=0$ if $2\theta\cos\theta=\sin\theta, \tan\theta=2\theta$ which only happens once on $(0,\pi]$. Since $\lim\limits_{\theta\to0^+} y=0$ and $y=0$ at $\theta=\pi$, y has a maximum when $\tan\theta=2\theta$.

(b) $\theta\approx 1.16556$

(c) $y_{\max}=(\sin\theta)/\sqrt{\theta}\approx 0.85124$

25. The width is twice the maximum value of y for $0\le\theta\le\pi/4$:
$y=r\sin\theta=\sin\theta\cos2\theta=\sin\theta-2\sin^3\theta, dy/d\theta=\cos\theta-6\sin^2\theta\cos\theta=0$ when $\cos\theta=0$ or $\sin\theta=1/\sqrt{6}, y=1/\sqrt{6}-2/(6\sqrt{6})=\sqrt{6}/9$, so the width of the petal is $2\sqrt{6}/9$.

26. **(a)** $\dfrac{x^2}{225}-\dfrac{y^2}{1521}=1$, so $V=2\displaystyle\int_0^{h/2}225\pi\left(1+\frac{y^2}{1521}\right)dy=\frac{25}{2028}\pi h^3+225\pi h \text{ ft}^3$.

(b) $$S=2\int_0^{h/2}2\pi x\sqrt{1+(dx/dy)^2}\,dy=4\pi\int_0^{h/2}\sqrt{225+y^2\left(\frac{225}{1521}+\left(\frac{225}{1521}\right)^2\right)}\,dy$$

$$=\frac{5\pi h}{338}\sqrt{1028196+194h^2}+\frac{7605\sqrt{194}}{97}\pi\ln\left[\frac{\sqrt{194}h+\sqrt{1028196+194h^2}}{1014}\right]\text{ ft}^2$$

27. **(a)** The end of the inner arm traces out the circle $x_1=\cos t, y_1=\sin t$. Relative to the end of the inner arm, the outer arm traces out the circle $x_2=\cos 2t, y_2=-\sin 2t$. Add to get the motion of the center of the rider cage relative to the center of the inner arm:
$x=\cos t+\cos 2t, y=\sin t-\sin 2t$.

(b) Same as Part (a), except $x_2=\cos 2t, y_2=\sin 2t$, so $x=\cos t+\cos 2t, y=\sin t+\sin 2t$

(c) $$L_1=\int_0^{2\pi}\left[\left(\frac{dx}{dt}\right)^2+\left(\frac{dy}{dt}\right)^2\right]^{1/2}dt=\int_0^{2\pi}\sqrt{5-4\cos 3t}\,dt\approx 13.36489321,$$

$L_2=\displaystyle\int_0^{2\pi}\sqrt{5+4\cos t}\,dt\approx 13.36489322$; L_1 and L_2 appear to be equal, and indeed, with the substitution $u=3t-\pi$ and the periodicity of $\cos u$,

$$L_1=\frac{1}{3}\int_{-\pi}^{5\pi}\sqrt{5-4\cos(u+\pi)}\,du=\int_0^{2\pi}\sqrt{5+4\cos u}\,du=L_2.$$

29. $$C=4\int_0^{\pi/2}\left[\left(\frac{dx}{dt}\right)^2+\left(\frac{dy}{dt}\right)^2\right]^{1/2}dt=4\int_0^{\pi/2}(a^2\sin^2 t+b^2\cos^2 t)^{1/2}dt$$

$$=4\int_0^{\pi/2}(a^2\sin^2 t+(a^2-c^2)\cos^2 t)^{1/2}dt=4a\int_0^{\pi/2}(1-e^2\cos^2 t)^{1/2}dt$$

Set $u=\dfrac{\pi}{2}-t$, $C=4a\displaystyle\int_0^{\pi/2}(1-e^2\sin^2 t)^{1/2}dt$

30. $a = 3, b = 2, c = \sqrt{5},\ C = 4(3)\int_0^{\pi/2} \sqrt{1-(5/9)\cos^2 u}\, du \approx 15.86543959$

31. **(a)** $\dfrac{r_0}{r_1} = \dfrac{59}{61} = \dfrac{1-e}{1+e}, e = \dfrac{1}{60}$

(b) $a = 93 \times 10^6, r_0 = a(1-e) = \dfrac{59}{60}\left(93 \times 10^6\right) = 91{,}450{,}000$ mi

(c) $C = 4 \times 93 \times 10^6 \displaystyle\int_0^{\pi/2} \left[1 - \left(\frac{\cos\theta}{60}\right)^2\right]^{1/2} d\theta \approx 584{,}295{,}652.5$ mi

32. **(a)** $y = y_0 + (v_0 \sin\alpha)\dfrac{x}{v_0\cos\alpha} - \dfrac{g}{2}\left(\dfrac{x}{v_0\cos\alpha}\right)^2 = y_0 + x\tan\alpha - \dfrac{g}{2v_0^2\cos^2\alpha}x^2$

(b) $\dfrac{dy}{dx} = \tan\alpha - \dfrac{g}{v_0^2\cos^2\alpha}x, dy/dx = 0$ at $x = \dfrac{v_0^2}{g}\sin\alpha\cos\alpha,$

$$y = y_0 + \frac{v_0^2}{g}\sin^2\alpha - \frac{g}{2v_0^2\cos^2\alpha}\left(\frac{v_0^2\sin\alpha\cos\alpha}{g}\right)^2 = y_0 + \frac{v_0^2}{2g}\sin^2\alpha$$

33. $\alpha = \pi/4, y_0 = 3, x = v_0 t/\sqrt{2}, y = 3 + v_0 t/\sqrt{2} - 16t^2$

(a) Assume the ball passes through $x = 391, y = 50$, then $391 = v_0 t/\sqrt{2}, 50 = 3 + 391 - 16t^2$, $16t^2 = 344,\ t = \sqrt{21.5},\ v_0 = \sqrt{2}x/t \approx 119.2538820$ ft/s

(b) $\dfrac{dy}{dt} = \dfrac{v_0}{\sqrt{2}} - 32t = 0$ at $t = \dfrac{v_0}{32\sqrt{2}},\ y_{\max} = 3 + \dfrac{v_0}{\sqrt{2}}\dfrac{v_0}{32\sqrt{2}} - 16\dfrac{v_0^2}{2^{11}} = 3 + \dfrac{v_0^2}{128} \approx 114.1053779$ ft

(c) $y = 0$ when $t = \dfrac{-v_0/\sqrt{2} \pm \sqrt{v_0^2/2 + 192}}{-32},\ t \approx -0.035339577$ (discard) and 5.305666365, dist $= 447.4015292$ ft

34. **(a)**

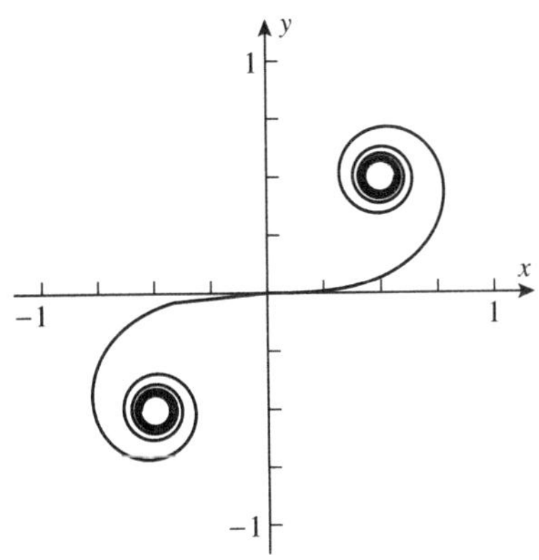

(c) $L = \displaystyle\int_{-1}^{1}\left[\cos^2\left(\frac{\pi t^2}{2}\right) + \sin^2\left(\frac{\pi t^2}{2}\right)\right] dt = 2$

35. $\tan\psi = \tan(\phi - \theta) = \dfrac{\tan\phi - \tan\theta}{1 + \tan\phi\tan\theta} = \dfrac{\dfrac{dy}{dx} - \dfrac{y}{x}}{1 + \dfrac{y}{x}\dfrac{dy}{dx}}$

$$= \frac{\dfrac{r\cos\theta + (dr/d\theta)\sin\theta}{-r\sin\theta + (dr/d\theta)\cos\theta} - \dfrac{\sin\theta}{\cos\theta}}{1 + \left(\dfrac{r\cos\theta + (dr/d\theta)\sin\theta)}{-r\sin\theta + (dr/d\theta)\cos\theta)}\right)\left(\dfrac{\sin\theta}{\cos\theta}\right)} = \frac{r}{dr/d\theta}$$

36. **(a)** From Exercise 35,

$$\tan \psi = \frac{r}{dr/d\theta} = \frac{1-\cos\theta}{\sin\theta} = \tan\frac{\theta}{2},$$

so $\psi = \theta/2$.

(b)

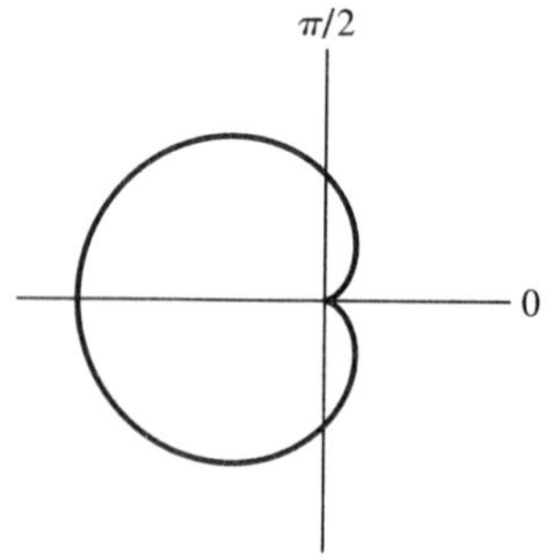

(c) At $\theta = \pi/2, \psi = \theta/2 = \pi/4$. At $\theta = 3\pi/2, \psi = \theta/2 = 3\pi/4$.

37. $\tan\psi = \dfrac{r}{dr/d\theta} = \dfrac{ae^{b\theta}}{abe^{b\theta}} = \dfrac{1}{b}$ is constant, so ψ is constant.

CHAPTER 11 HORIZON MODULE

1. For the Earth, $a_E(1-e_E^2) = 1(1-0.017^2) = 0.999711$, so the polar equation is

$$r = \frac{a_E(1-e_E^2)}{1-e_E\cos\theta} = \frac{0.999711}{1-0.017\cos\theta}.$$

For Rogue 2000, $a_R(1-e_R^2) = 5(1-0.98^2) = 0.198$, so the polar equation is

$$r = \frac{a_R(1-e_R^2)}{1-e_R\cos\theta} = \frac{0.198}{1-0.98\cos\theta}.$$

2.

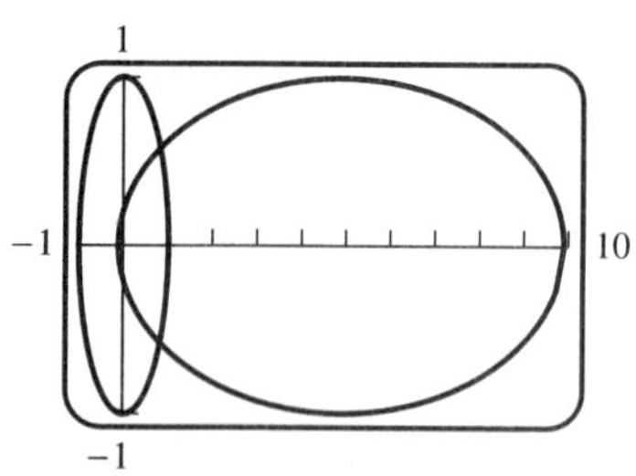

3. At the intersection point A, $\dfrac{k_E}{1-e_E\cos\theta} = \dfrac{k_R}{1-e_R\cos\theta}$, so $k_E - k_Ee_R\cos\theta = k_R - k_Re_E\cos\theta$.

Solving for $\cos\theta$ gives $\cos\theta = \dfrac{k_E - k_R}{k_Ee_R - k_Re_E}$.

4. From Exercise 1, $k_E = 0.999711$ and $k_R = 0.198$, so

$$\cos\theta = \frac{k_E - k_R}{k_Ee_R - k_Re_E} = \frac{0.999711 - 0.198}{0.999711(0.98) - 0.198(0.017)} \approx 0.821130$$

and $\theta = \cos^{-1} 0.821130 \approx 0.607408$ radian.

5. Substituting $\cos\theta \approx 0.821130$ into the polar equation for the Earth gives

$$r \approx \frac{0.999711}{1-0.017(0.821130)} \approx 1.013864,$$

so the polar coordinates of intersection A are approximately $(1.013864, 0.607408)$.

6. By Theorem 11.3.2 the area of the elliptic sector is $\int_{\theta_I}^{\theta_F} \frac{1}{2} r^2\,d\theta$. By Exercise 11.4.53 the area of the entire ellipse is πab, where a is the semimajor axis and b is the semiminor axis. But $b = \sqrt{a^2 - c^2} = \sqrt{a^2 - (ea)^2} = a\sqrt{1 - e^2}$,

so Formula (1) becomes $\dfrac{t}{T} = \dfrac{\int_{\theta_I}^{\theta_F} r^2\,d\theta}{2\pi a^2\sqrt{1-e^2}}$, which implies Formula (2).

7. In Formula (2) substitute $T = 1, \theta_I = 0$, and $\theta_F \approx 0.607408$, and use the polar equation of the Earth's orbit found in Exercise 1:

$$t = \frac{\int_0^{\theta_F} \left(\dfrac{k_E}{1 - e_E\cos\theta}\right)^2 d\theta}{2\pi\sqrt{1 - e_E^2}} \approx \frac{\int_0^{0.607408} \left(\dfrac{0.999711}{1 - 0.017\cos\theta}\right)^2 d\theta}{2\pi\sqrt{0.999711}} \approx 0.099793 \text{ yr.}$$

Note: This calculation can be done either by numerical integration or by using the integration formula

$$\int \frac{d\theta}{(1 - e\cos\theta)^2} = \frac{2\tan^{-1}\left(\sqrt{\dfrac{1+e}{1-e}}\tan\dfrac{\theta}{2}\right)}{(1 - e^2)^{3/2}} + \frac{e\sin\theta}{(1 - e^2)(1 - e\cos\theta)} + C,$$

obtained by using a CAS or by the substitution $u = \tan(\theta/2)$.

8. In Formula (2) we substitute $T = 5\sqrt{5}$ and $\theta_I = 0.45$, and use the polar equation of Rogue 2000's orbit found in Exercise 1:

$$t = \frac{T\int_{\theta_I}^{\theta_F} \left(\dfrac{a_R(1 - e_R^2)}{1 - e_R\cos\theta}\right)^2 d\theta}{2\pi a_R^2\sqrt{1 - e_R^2}} = \frac{5\sqrt{5}\int_{0.45}^{\theta_F} \left(\dfrac{a_R(1 - e_R^2)}{1 - e_R\cos\theta}\right)^2 d\theta}{2\pi a_R^2\sqrt{1 - e_R^2}},$$

so

$$\int_{0.45}^{\theta_F} \left(\frac{a_R(1 - e_R^2)}{1 - e_R\cos\theta}\right)^2 d\theta = \frac{2t\pi a_R^2\sqrt{1 - e_R^2}}{5\sqrt{5}}.$$

9. **(a)** A CAS shows that

$$\int \left(\frac{a_R(1 - e_R^2)}{1 - e_R\cos\theta}\right)^2 d\theta = a_R^2\left(2\sqrt{1 - e_R^2}\tan^{-1}\left(\sqrt{\frac{1+e_R}{1-e_R}}\tan\frac{\theta}{2}\right) + \frac{e_R(1 - e_R^2)\sin\theta}{1 - e_R\cos\theta}\right) + C$$

(b) Evaluating the integral above from $\theta = 0.45$ to $\theta = \theta_F$, setting the result equal to the right side of (3), and simplifying gives

$$\tan^{-1}\left(\sqrt{\frac{1+e_R}{1-e_R}}\tan\frac{\theta}{2}\right) + \frac{e_R\sqrt{1 - e_R^2}\sin\theta}{2(1 - e_R\cos\theta)}\Bigg]_{0.45}^{\theta_F} = \frac{t\pi}{5\sqrt{5}}.$$

Using the known values of e_R and t, and solving numerically, $\theta_F \approx 0.611346$.

10. Substituting $\theta_F \approx 0.611346$ in the equation for Rogue 2000's orbit gives $r \approx 1.002525$ AU. So the polar coordinates of Rogue 2000 when the Earth is at intersection A are about $(1.002525, 0.611346)$.

11. Substituting the values found in Exercises 5 and 10 into the distance formula in Exercise 67 of Section 11.1 gives $d = \sqrt{r_1^2 + r_2^2 - 2r_1r_2\cos(\theta_1 - \theta_2)} \approx 0.012014$ AU $\approx 1.797201 \times 10^6$ km. Since this is less than 4 million kilometers, a notification should be issued. (Incidentally, Rogue 2000's closest approach to the Earth does not occur when the Earth is at A, but about 9 hours earlier, at $t \approx 0.098768$ yr, at which time the distance is about 1.219435 million kilometers.)

CHAPTER 12
Three-Dimensional Space; Vectors

EXERCISE SET 12.1

1. **(a)** $(0,0,0),(3,0,0),(3,5,0),(0,5,0),(0,0,4),(3,0,4),(3,5,4),(0,5,4)$
(b) $(0,1,0),(4,1,0),(4,6,0),(0,6,0),(0,1,-2),(4,1,-2),(4,6,-2),(0,6,-2)$

2. corners: $(2,2,\pm 2)$, $(2,-2,\pm 2)$, $(-2,2,\pm 2)$, $(-2,-2,\pm 2)$

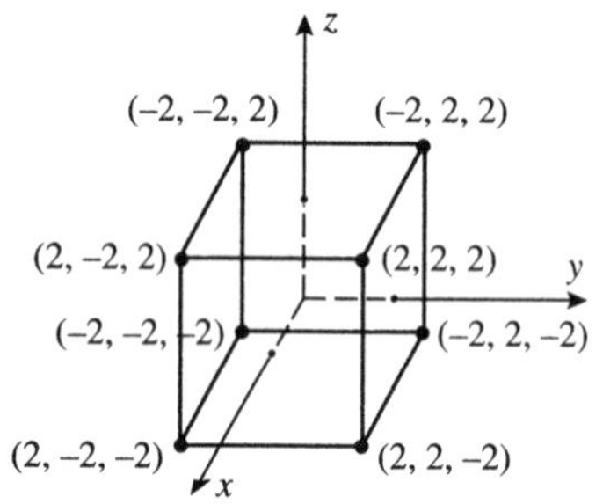

3. corners: $(4,2,-2)$, $(4,2,1)$, $(4,1,1)$, $(4,1,-2)$, $(-6,1,1)$, $(-6,2,1)$, $(-6,2,-2)$, $(-6,1,-2)$

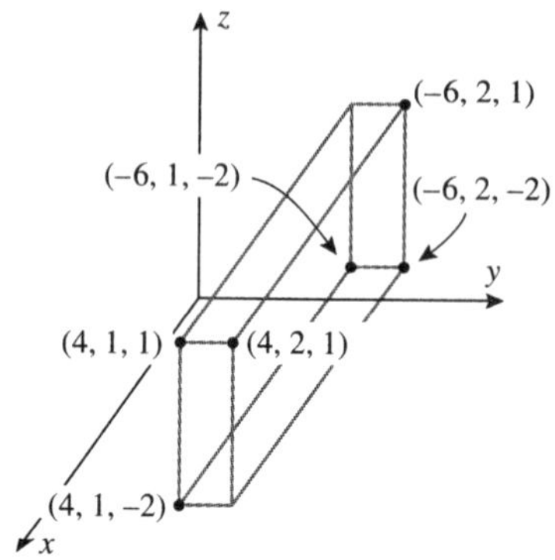

4. **(a)** $(x_2,y_1,z_1),(x_2,y_2,z_1),(x_1,y_2,z_1)(x_1,y_1,z_2),(x_2,y_1,z_2),(x_1,y_2,z_2)$
(b) The midpoint of the diagonal has coordinates which are the coordinates of the midpoints of the edges. The midpoint of the edge (x_1,y_1,z_1) and (x_2,y_1,z_1) is $\left(\frac{1}{2}(x_1+x_2),y_1,z_1\right)$; the midpoint of the edge (x_2,y_1,z_1) and (x_2,y_2,z_1) is $\left(x_2,\frac{1}{2}(y_1+y_2),z_1\right)$; the midpoint of the edge (x_2,y_2,z_1) and (x_2,y_2,z_2)) is $\left(x_2,y_2,\frac{1}{2}(z_1+z_2)\right)$. Thus the coordinates of the midpoint of the diagonal are $\frac{1}{2}(x_1+x_2),\frac{1}{2}(y_1+y_2),\frac{1}{2}(z_1+z_2)$.

5. The diameter is $d=\sqrt{(1-3)^2+(-2-4)^2+(4+12)^2}=\sqrt{296}$, so the radius is $\sqrt{296}/2=\sqrt{74}$. The midpoint $(2,1,-4)$ of the endpoints of the diameter is the center of the sphere.

6. Each side has length $\sqrt{14}$ so the triangle is equilateral.

7. **(a)** The sides have lengths 7, 14, and $7\sqrt{5}$; it is a right triangle because the sides satisfy the Pythagorean theorem, $(7\sqrt{5})^2=7^2+14^2$.
(b) (2,1,6) is the vertex of the 90° angle because it is opposite the longest side (the hypotenuse).
(c) area $=(1/2)(\text{altitude})(\text{base})=(1/2)(7)(14)=49$

8. **(a)** 3 **(b)** 2 **(c)** 5
(d) $\sqrt{(2)^2+(-3)^2}=\sqrt{13}$ **(e)** $\sqrt{(-5)^2+(-3)^2}=\sqrt{34}$ **(f)** $\sqrt{(-5)^2+(2)^2}=\sqrt{29}$

9. **(a)** $(x-1)^2+y^2+(z+1)^2=16$
(b) $r=\sqrt{(-1-0)^2+(3-0)^2+(2-0)^2}=\sqrt{14}$, $(x+1)^2+(y-3)^2+(z-2)^2=14$

(c) $r = \frac{1}{2}\sqrt{(-1-0)^2 + (2-2)^2 + (1-3)^2} = \frac{1}{2}\sqrt{5}$, center $(-1/2, 2, 2)$,
$(x + 1/2)^2 + (y-2)^2 + (z-2)^2 = 5/4$

10. $r = |[\text{distance between } (0{,}0{,}0) \text{ and } (3, -2, 4)] \pm 1| = \sqrt{29} \pm 1$,
$x^2 + y^2 + z^2 = r^2 = \left(\sqrt{29} \pm 1\right)^2 = 30 \pm 2\sqrt{29}$

11. $(x-2)^2 + (y+1)^2 + (z+3)^2 = r^2$,
(a) $r^2 = 3^2 = 9$ (b) $r^2 = 1^2 = 1$ (c) $r^2 = 2^2 = 4$

12. (a) The sides have length 1, so the radius is $\frac{1}{2}$; hence $(x+2)^2 + (y-1)^2 + (z-3)^2 = \frac{1}{4}$

(b) The diagonal has length $\sqrt{1+1+1} = \sqrt{3}$ and is a diameter, so $(x+2)^2 + (y-1)^2 + (z-3)^2 = \frac{3}{4}$.

13. $(x+5)^2 + (y+2)^2 + (z+1)^2 = 49$; sphere, $C(-5, -2, -1)$, $r = 7$

14. $x^2 + (y - 1/2)^2 + z^2 = 1/4$; sphere, $C(0, 1/2, 0)$, $r = 1/2$

15. $(x - 1/2)^2 + (y - 3/4)^2 + (z + 5/4)^2 = 54/16$; sphere, $C(1/2, 3/4, -5/4)$, $r = 3\sqrt{6}/4$

16. $(x+1)^2 + (y-1)^2 + (z+1)^2 = 0$; the point $(-1, 1, -1)$

17. $(x - 3/2)^2 + (y+2)^2 + (z-4)^2 = -11/4$; no graph

18. $(x-1)^2 + (y-3)^2 + (z-4)^2 = 25$; sphere, $C(1, 3, 4)$, $r = 5$

19. (a)

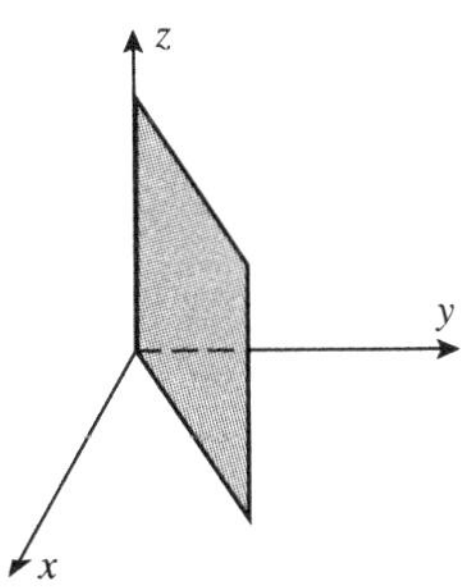

(b)

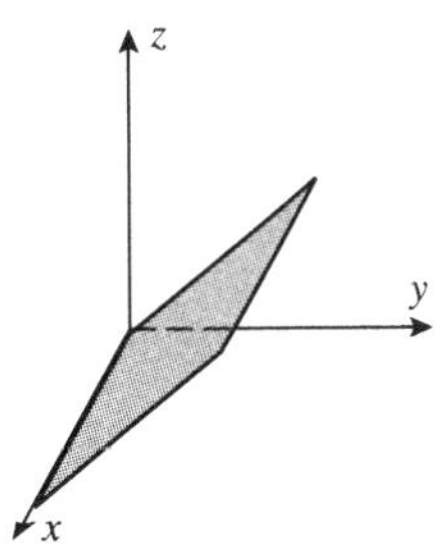

(c)

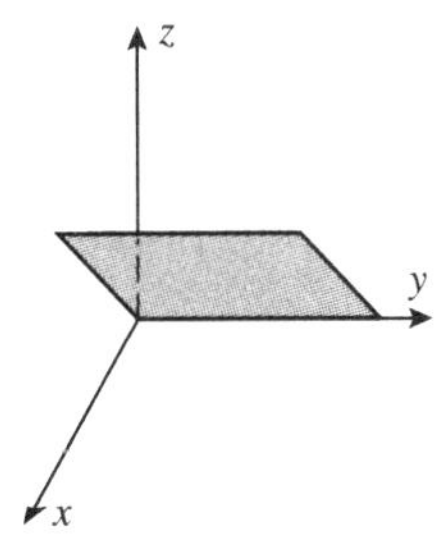

20. (a)

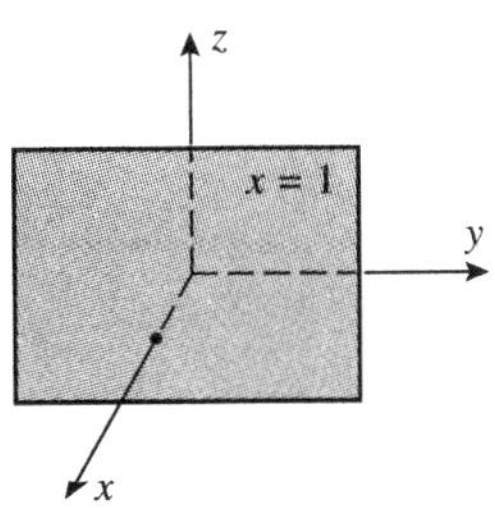

(b)

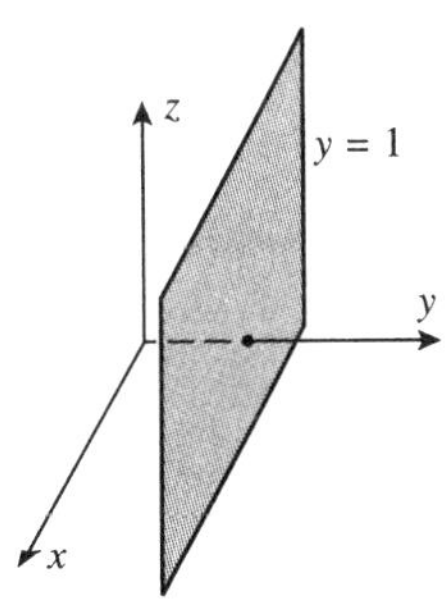

(c)

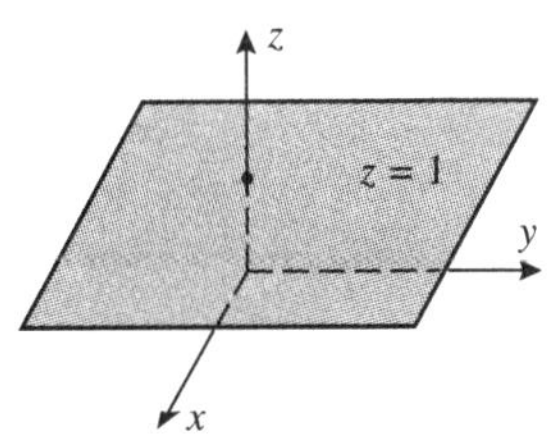

21. (a)

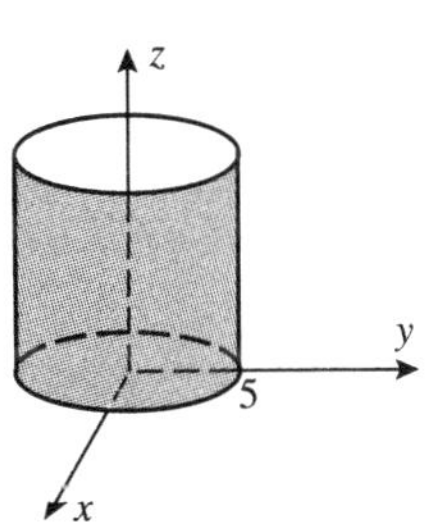

(b)

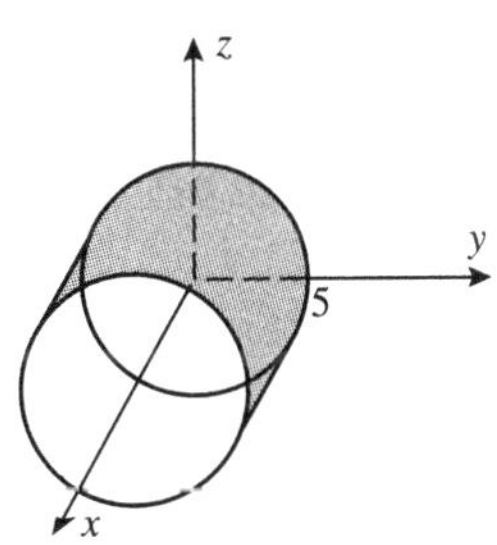

(c)

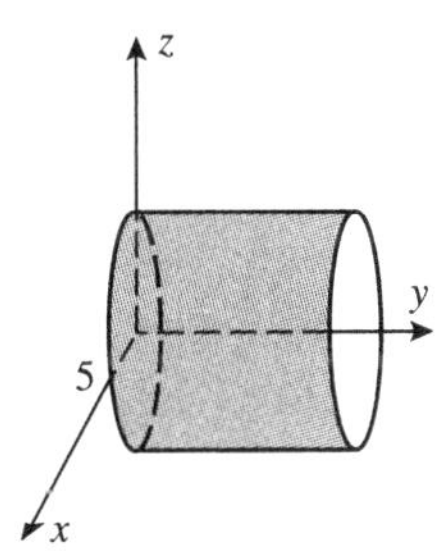

22. **(a)**

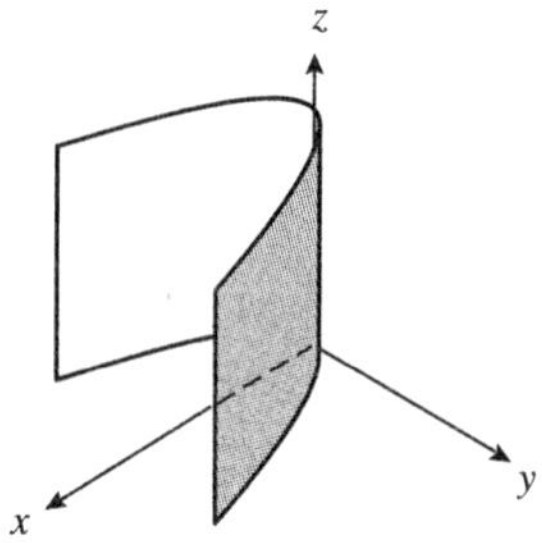

(b)

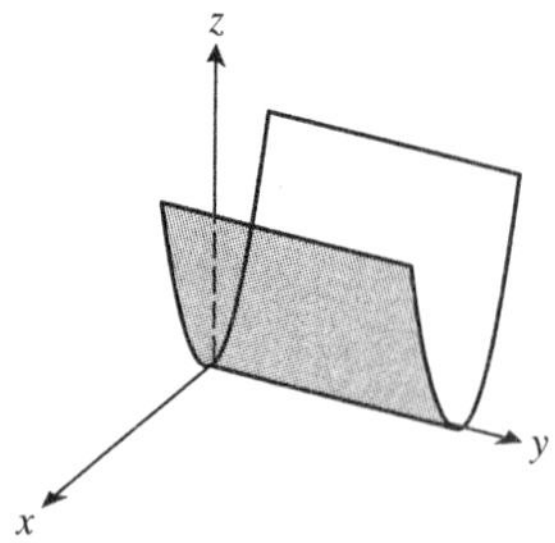

(c)

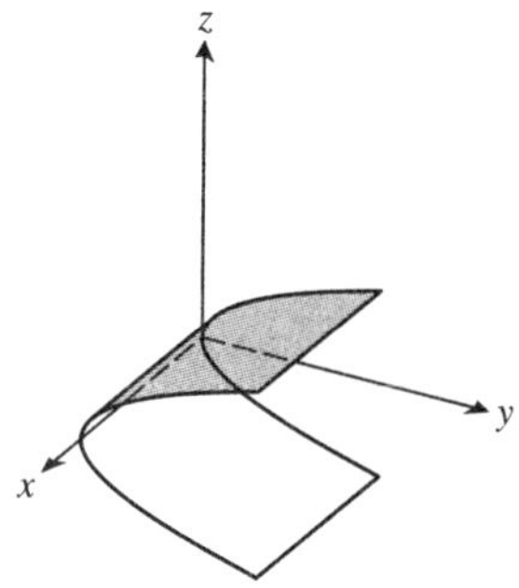

23. **(a)** $-2y + z = 0$ **(b)** $-2x + z = 0$

(c) $(x-1)^2 + (y-1)^2 = 1$ **(d)** $(x-1)^2 + (z-1)^2 = 1$

24. **(a)** $(x-a)^2 + (z-a)^2 = a^2$ **(b)** $(x-a)^2 + (y-a)^2 = a^2$

(c) $(y-a)^2 + (z-a)^2 = a^2$

25.

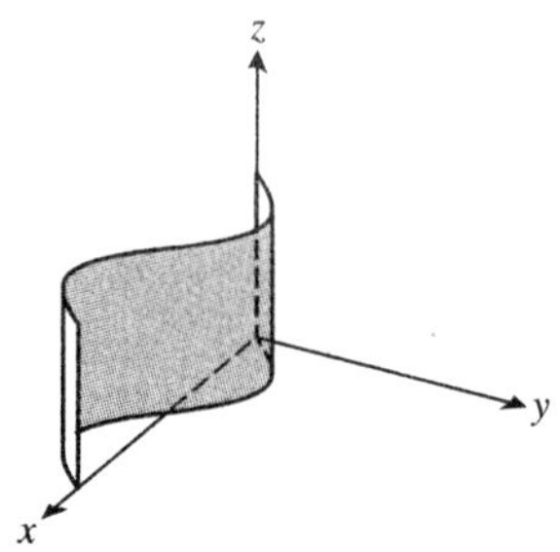

26.

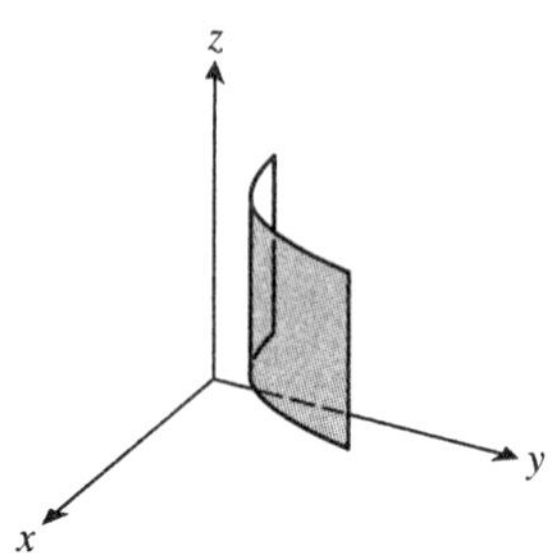

27.

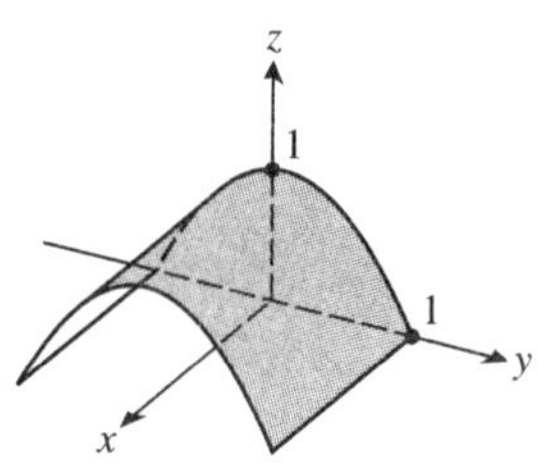

28.

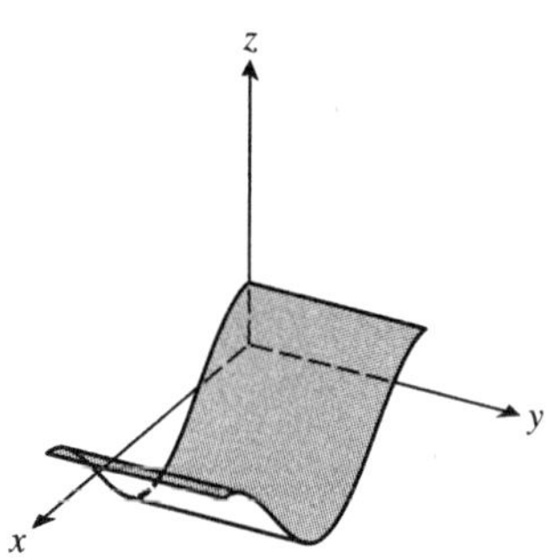

29.

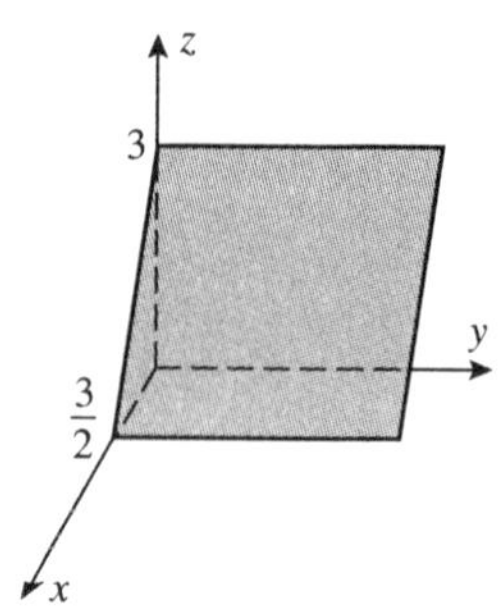

30.

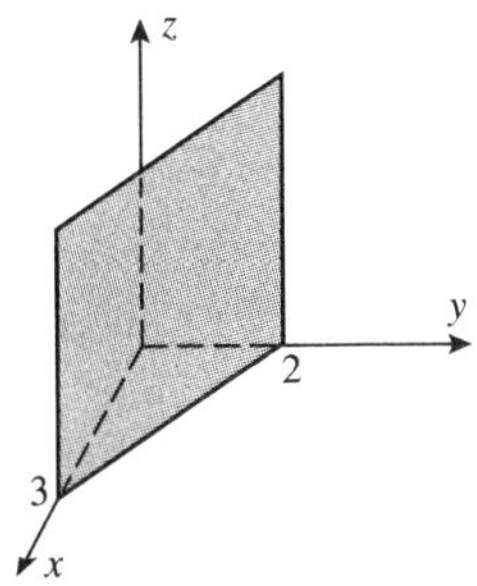

31.

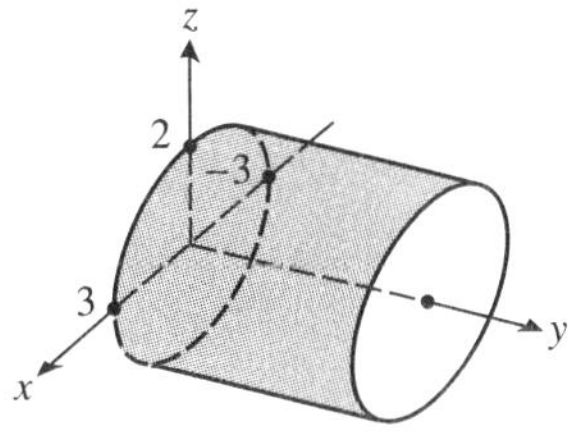

32.

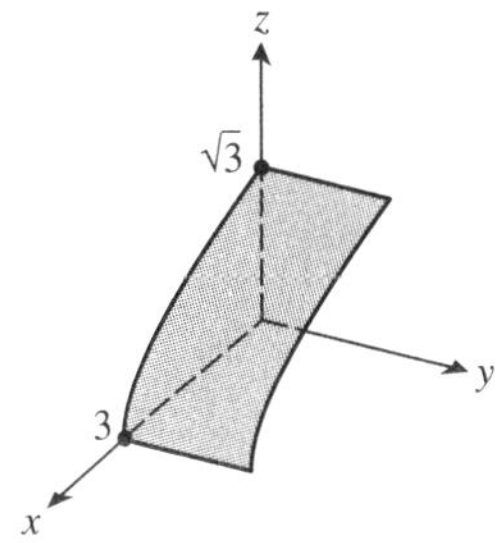

33.

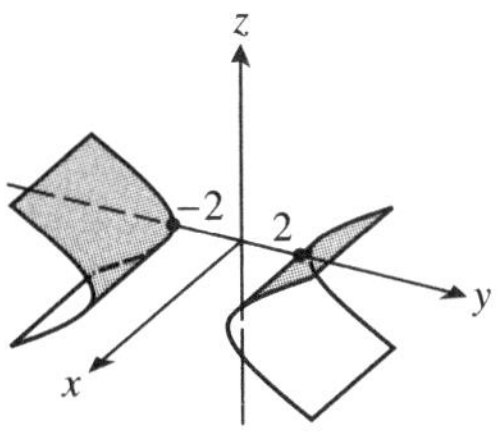

34.

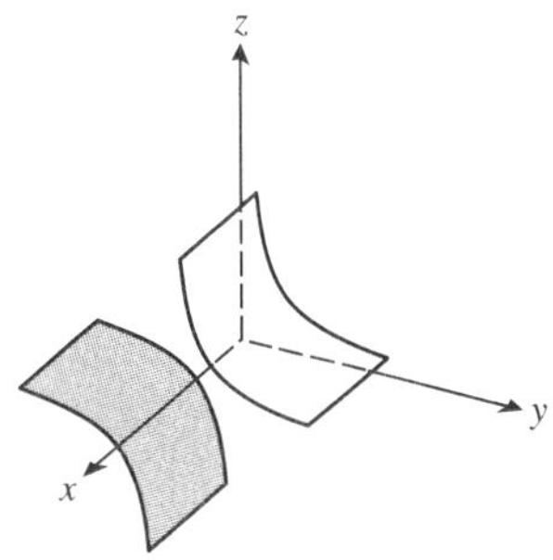

35. (a)

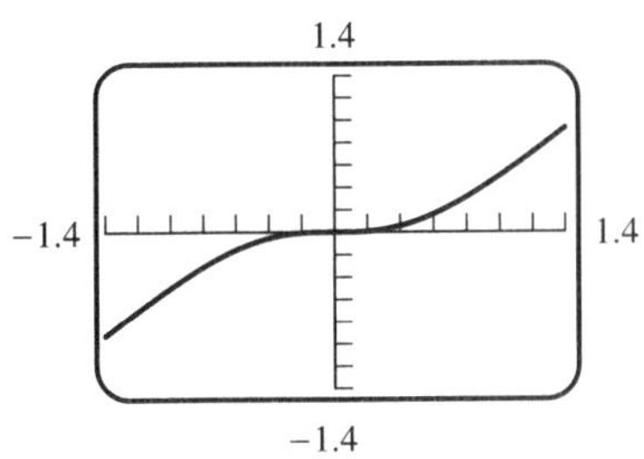

(b)

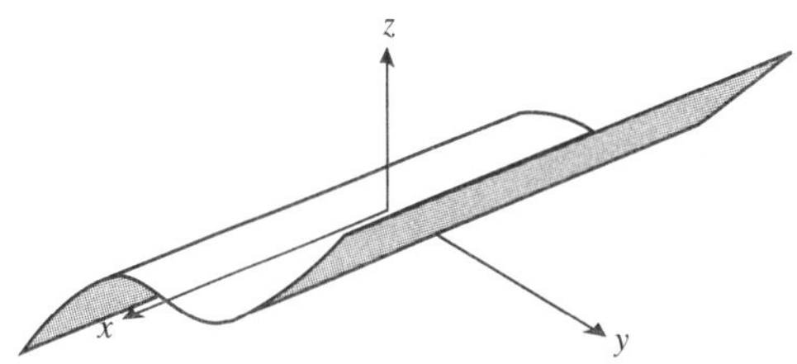

36. (a)

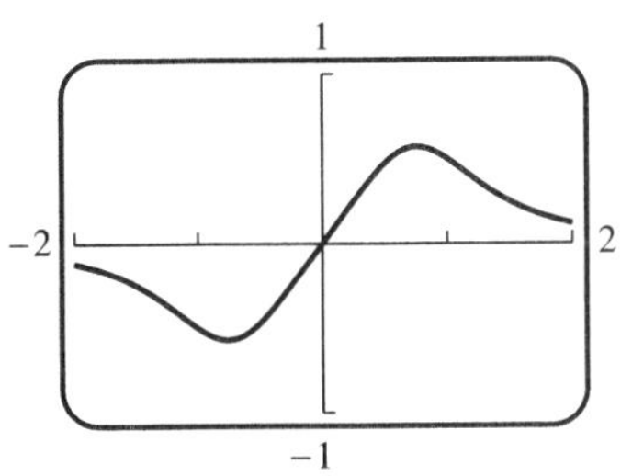

(b)

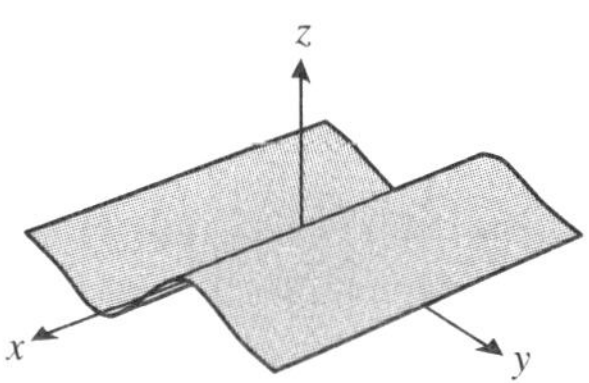

37. Complete the square to get $(x+1)^2 + (y-1)^2 + (z-2)^2 = 9$; center $(-1, 1, 2)$, radius 3. The distance between the origin and the center is $\sqrt{6} < 3$ so the origin is inside the sphere. The largest distance is $3 + \sqrt{6}$, the smallest is $3 - \sqrt{6}$.

38. $(x-1)^2 + y^2 + (z+4)^2 \le 25$; all points on and inside the sphere of radius 5 with center at $(1, 0, -4)$.

39. $(y+3)^2 + (z-2)^2 > 16$; all points outside the circular cylinder $(y+3)^2 + (z-2)^2 = 16$.

40. $\sqrt{(x-1)^2 + (y+2)^2 + z^2} = 2\sqrt{x^2 + (y-1)^2 + (z-1)^2}$, square and simplify to get $3x^2 + 3y^2 + 3z^2 + 2x - 12y - 8z + 3 = 0$, then complete the square to get $(x + 1/3)^2 + (y-2)^2 + (z - 4/3)^2 = 44/9$; center $(-1/3, 2, 4/3)$, radius $2\sqrt{11}/3$.

41. Let r be the radius of a styrofoam sphere. The distance from the origin to the center of the bowling ball is equal to the sum of the distance from the origin to the center of the styrofoam sphere nearest the origin and the distance between the center of this sphere and the center of the bowling ball so $\sqrt{3}R = \sqrt{3}r + r + R$, $(\sqrt{3}+1)r = (\sqrt{3}-1)R$, $r = \dfrac{\sqrt{3}-1}{\sqrt{3}+1}R = (2-\sqrt{3})R$.

42. **(a)** Complete the square to get $(x + G/2)^2 + (y + H/2)^2 + (z + I/2)^2 = K/4$, so the equation represents a sphere when $K > 0$, a point when $K = 0$, and no graph when $K < 0$.

(b) $C(-G/2, -H/2, -I/2)$, $r = \sqrt{K}/2$

43.
$$\begin{aligned}(a\sin\phi\cos\theta)^2 + (a\sin\phi\sin\theta)^2 + (a\cos\phi)^2 &= a^2\sin^2\phi\cos^2\theta + a^2\sin^2\phi\sin^2\theta + a^2\cos^2\phi \\ &= a^2\sin^2\phi(\cos^2\theta + \sin^2\theta) + a^2\cos^2\phi \\ &= a^2\sin^2\phi + a^2\cos^2\phi = a^2(\sin^2\phi + \cos^2\phi) = a^2\end{aligned}$$

EXERCISE SET 12.2

1. **(a–c)**

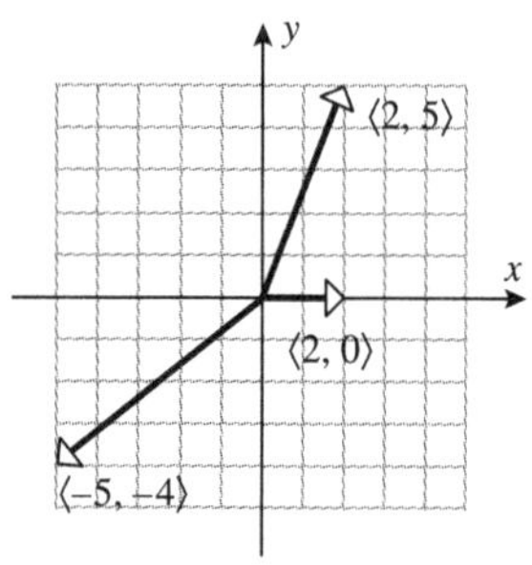

(d–f)

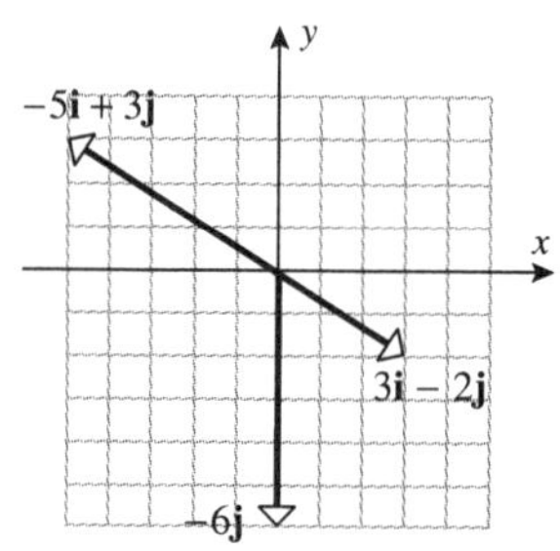

2. **(a–c)**

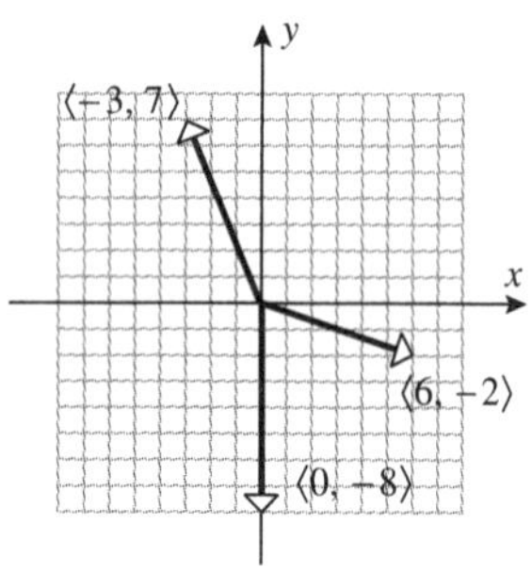

(d–f)

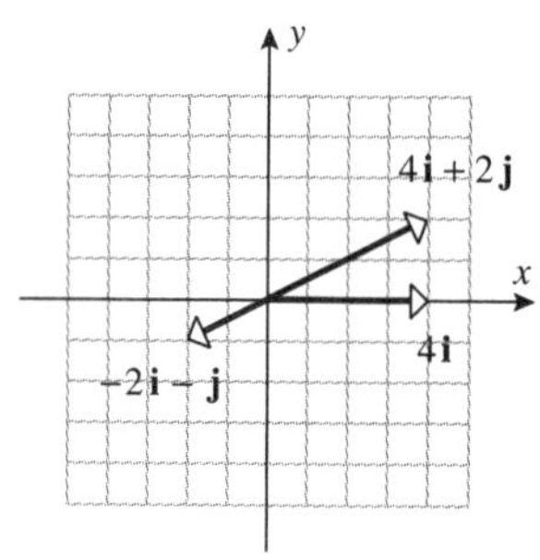

3. **(a–b)**

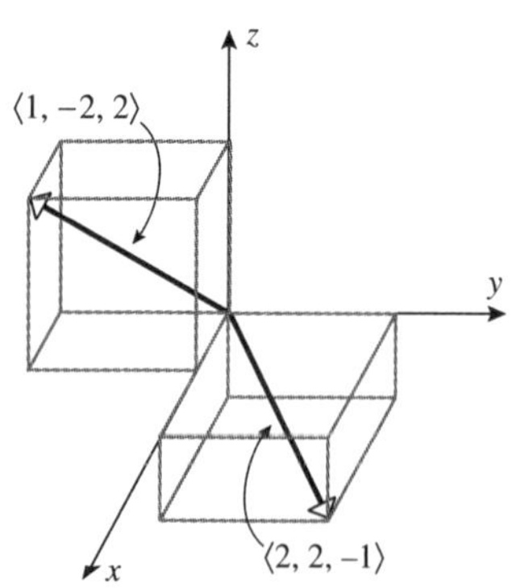

(c–d)

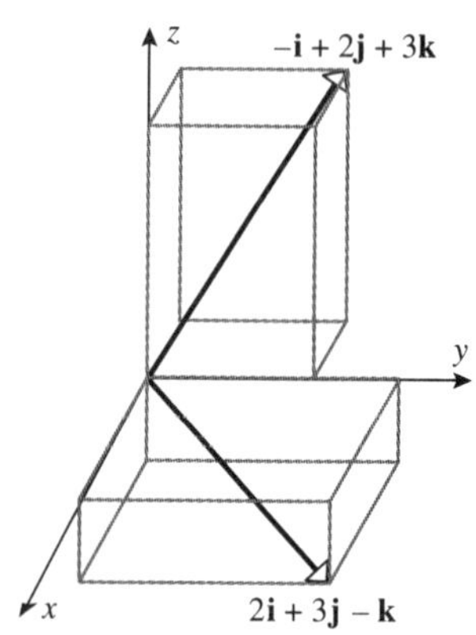

4. (a–b)

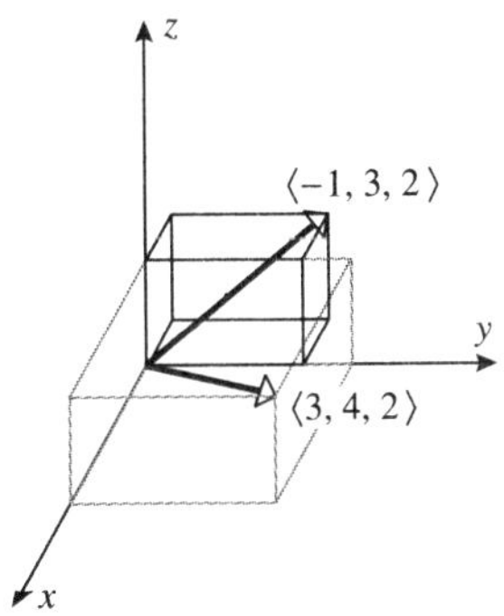

(c–d)

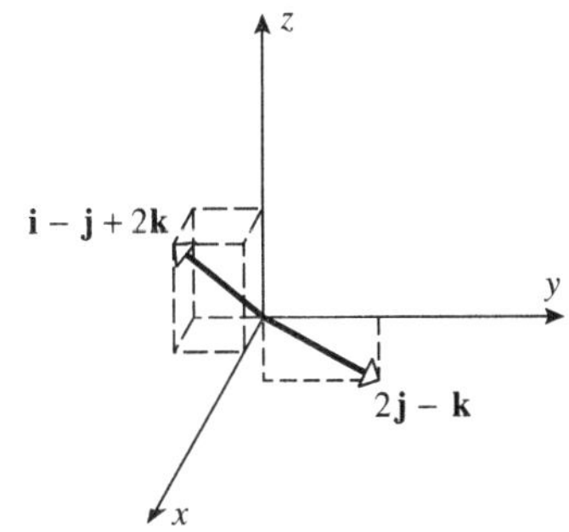

5. (a) $\langle 4-1, 1-5\rangle = \langle 3, -4\rangle$

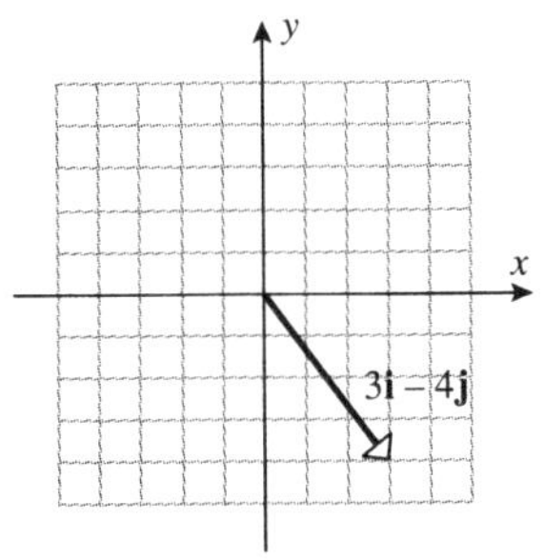

(b) $\langle 0-2, 0-3, 4-0\rangle = \langle -2, -3, 4\rangle$

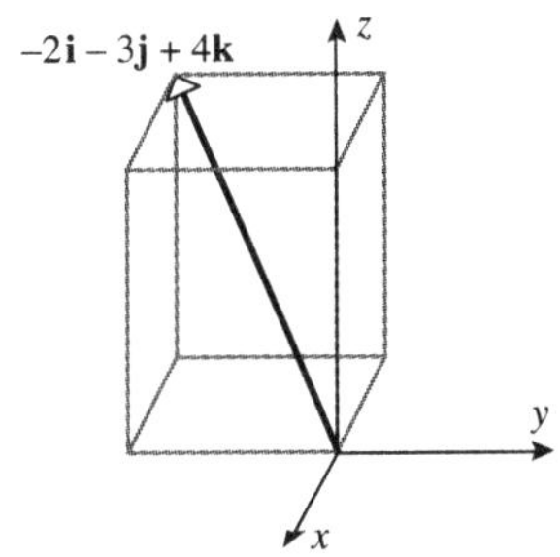

6. (a) $\langle -3-2, 3-3\rangle = \langle -5, 0\rangle$

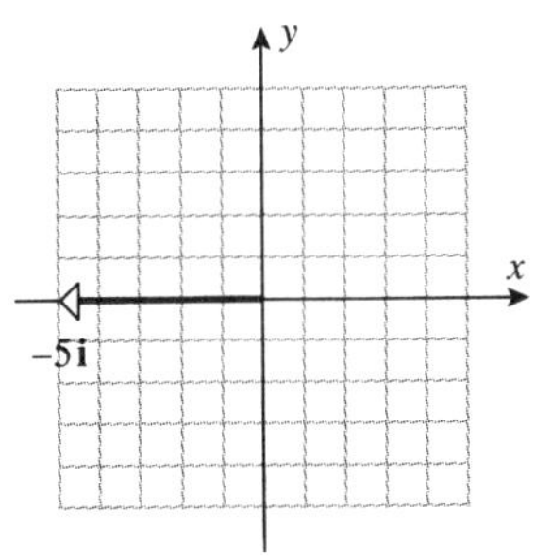

(b) $\langle 0-3, 4-0, 4-4\rangle = \langle -3, 4, 0\rangle$

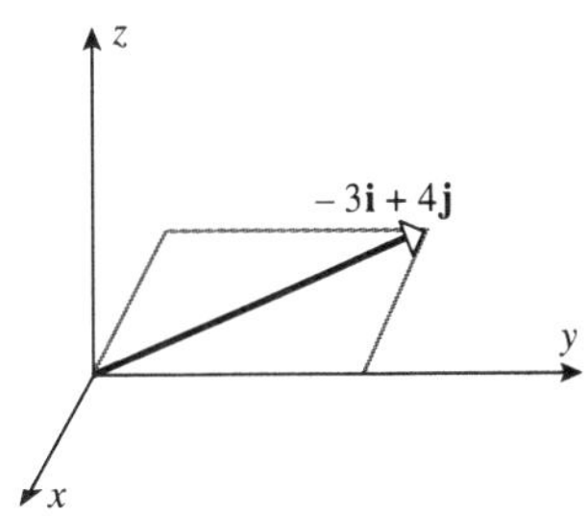

7. (a) $\langle 2-3,\ 8-5\rangle = \langle -1, 3\rangle$
(b) $\langle 0-7,\ 0-(-2)\rangle = \langle -7, 2\rangle$
(c) $\langle -3, 6, 1\rangle$

8. (a) $\langle -4-(-6),\ -1-(-2)\rangle = \langle 2, 1\rangle$
(b) $\langle -1, 6, 1\rangle$
(c) $\langle 5, 0, 0\rangle$

9. (a) Let (x, y) be the terminal point, then $x-1=3$, $x=4$ and $y-(-2)=-2$, $y=-4$. The terminal point is $(4, -4)$.
(b) Let (x, y, z) be the initial point, then $5-x=-3$, $-y=1$, and $-1-z=2$ so $x=8$, $y=-1$, and $z=-3$. The initial point is $(8, -1, -3)$.

10. (a) Let (x, y) be the terminal point, then $x-2=7$, $x=9$ and $y-(-1)=6$, $y=5$. The terminal point is (9,5).
(b) Let (x, y, z) be the terminal point, then $x+2=1$, $y-1=2$, and $z-4=-3$ so $x=-1$, $y=3$, and $z=1$. The terminal point is $(-1, 3, 1)$.

11. (a) $-\mathbf{i}+4\mathbf{j}-2\mathbf{k}$ (b) $18\mathbf{i}+12\mathbf{j}-6\mathbf{k}$ (c) $-\mathbf{i}-5\mathbf{j}-2\mathbf{k}$
(d) $40\mathbf{i}-4\mathbf{j}-4\mathbf{k}$ (e) $-2\mathbf{i}-16\mathbf{j}-18\mathbf{k}$ (f) $-\mathbf{i}+13\mathbf{j}-2\mathbf{k}$

12. **(a)** $\langle 1,-2,0\rangle$ **(b)** $\langle 28,0,-14\rangle+\langle 3,3,9\rangle=\langle 31,3,-5\rangle$
(c) $\langle 3,-1,-5\rangle$ **(d)** $3(\langle 2,-1,3\rangle-\langle 28,0,-14\rangle)=3\langle -26,-1,17\rangle=\langle -78,-3,51\rangle$
(e) $\langle -12,0,6\rangle-\langle 8,8,24\rangle=\langle -20,-8,-18\rangle$
(f) $\langle 8,0,-4\rangle-\langle 3,0,6\rangle=\langle 5,0,-10\rangle$

13. **(a)** $\|\mathbf{v}\|=\sqrt{1+1}=\sqrt{2}$ **(b)** $\|\mathbf{v}\|=\sqrt{1+49}=5\sqrt{2}$
(c) $\|\mathbf{v}\|=\sqrt{21}$ **(d)** $\|\mathbf{v}\|=\sqrt{14}$

14. **(a)** $\|\mathbf{v}\|=\sqrt{9+16}=5$ **(b)** $\|\mathbf{v}\|=\sqrt{2+7}=3$
(c) $\|\mathbf{v}\|=3$ **(d)** $\|\mathbf{v}\|=\sqrt{3}$

15. **(a)** $\|\mathbf{u}+\mathbf{v}\|=\|2\mathbf{i}-2\mathbf{j}+2\mathbf{k}\|=2\sqrt{3}$ **(b)** $\|\mathbf{u}\|+\|\mathbf{v}\|=\sqrt{14}+\sqrt{2}$
(c) $\|-2\mathbf{u}\|+2\|\mathbf{v}\|=2\sqrt{14}+2\sqrt{2}$ **(d)** $\|3\mathbf{u}-5\mathbf{v}+\mathbf{w}\|=\|-12\mathbf{j}+2\mathbf{k}\|=2\sqrt{37}$
(e) $(1/\sqrt{6})\mathbf{i}+(1/\sqrt{6})\mathbf{j}-(2/\sqrt{6})\mathbf{k}$ **(f)** 1

16. If one vector is a positive multiple of the other, say $\mathbf{u}=\alpha\mathbf{v}$ with $\alpha>0$, then $\mathbf{u},\mathbf{v}$ and $\mathbf{u}+\mathbf{v}$ are parallel and $\|\mathbf{u}+\mathbf{v}\|=(1+\alpha)\|\mathbf{v}\|=\|\mathbf{u}\|+\|\mathbf{v}\|$.

17. **(a)** $\|-\mathbf{i}+4\mathbf{j}\|=\sqrt{17}$ so the required vector is $\left(-1/\sqrt{17}\right)\mathbf{i}+\left(4/\sqrt{17}\right)\mathbf{j}$
(b) $\|6\mathbf{i}-4\mathbf{j}+2\mathbf{k}\|=2\sqrt{14}$ so the required vector is $(-3\mathbf{i}+2\mathbf{j}-\mathbf{k})/\sqrt{14}$
(c) $\overrightarrow{AB}=4\mathbf{i}+\mathbf{j}-\mathbf{k}$, $\|\overrightarrow{AB}\|=3\sqrt{2}$ so the required vector is $(4\mathbf{i}+\mathbf{j}-\mathbf{k})/\left(3\sqrt{2}\right)$

18. **(a)** $\|3\mathbf{i}-4\mathbf{j}\|=5$ so the required vector is $-\frac{1}{5}(3\mathbf{i}-4\mathbf{j})=-\frac{3}{5}\mathbf{i}+\frac{4}{5}\mathbf{j}$
(b) $\|2\mathbf{i}-\mathbf{j}-2\mathbf{k}\|=3$ so the required vector is $\frac{2}{3}\mathbf{i}-\frac{1}{3}\mathbf{j}-\frac{2}{3}\mathbf{k}$
(c) $\overrightarrow{AB}=4\mathbf{i}-3\mathbf{j}$, $\|\overrightarrow{AB}\|=5$ so the required vector is $\frac{4}{5}\mathbf{i}-\frac{3}{5}\mathbf{j}$

19. **(a)** $-\frac{1}{2}\mathbf{v}=\langle -3/2,2\rangle$ **(b)** $\|\mathbf{v}\|=\sqrt{85}$, so $\frac{\sqrt{17}}{\sqrt{85}}\mathbf{v}=\frac{1}{\sqrt{5}}\langle 7,0,-6\rangle$ has length $\sqrt{17}$

20. **(a)** $3\mathbf{v}=-6\mathbf{i}+9\mathbf{j}$ **(b)** $-\frac{2}{\|v\|}\mathbf{v}=\frac{6}{\sqrt{26}}\mathbf{i}-\frac{8}{\sqrt{26}}\mathbf{j}-\frac{2}{\sqrt{26}}\mathbf{k}$

21. **(a)** $\mathbf{v}=\|\mathbf{v}\|\langle\cos(\pi/4),\sin(\pi/4)\rangle=\langle 3\sqrt{2}/2,3\sqrt{2}/2\rangle$
(b) $\mathbf{v}=\|\mathbf{v}\|\langle\cos 90^\circ,\sin 90^\circ\rangle=\langle 0,2\rangle$
(c) $\mathbf{v}=\|\mathbf{v}\|\langle\cos 120^\circ,\sin 120^\circ\rangle=\langle -5/2,5\sqrt{3}/2\rangle$
(d) $\mathbf{v}=\|\mathbf{v}\|\langle\cos\pi,\sin\pi\rangle=\langle -1,0\rangle$

22. From (12), $\mathbf{v}=\langle\cos(\pi/6),\sin(\pi/6)\rangle=\langle\sqrt{3}/2,1/2\rangle$ and $\mathbf{w}=\langle\cos(3\pi/4),\sin(3\pi/4)\rangle=\langle -\sqrt{2}/2,\sqrt{2}/2\rangle$, so $\mathbf{v}+\mathbf{w}=((\sqrt{3}-\sqrt{2})/2,(1+\sqrt{2})/2$, $\mathbf{v}-\mathbf{w}=((\sqrt{3}+\sqrt{2})/2,(1-\sqrt{2})/2)$

23. From (12), $\mathbf{v}=\langle\cos 30^\circ,\sin 30^\circ\rangle=\langle\sqrt{3}/2,1/2\rangle$ and $\mathbf{w}=\langle\cos 135^\circ,\sin 135^\circ\rangle=\langle -\sqrt{2}/2,\sqrt{2}/2\rangle$, so $\mathbf{v}+\mathbf{w}=((\sqrt{3}-\sqrt{2})/2,(1+\sqrt{2})/2)$

24. $\mathbf{w}=\langle 1,0\rangle$, and from (12), $\mathbf{v}=\langle\cos 120^\circ,\sin 120^\circ\rangle=\langle -1/2,\sqrt{3}/2\rangle$, so $\mathbf{v}+\mathbf{w}=\langle 1/2,\sqrt{3}/2\rangle$

25. **(a)** The initial point of $\mathbf{u}+\mathbf{v}+\mathbf{w}$ is the origin and the endpoint is $(-2,5)$, so $\mathbf{u}+\mathbf{v}+\mathbf{w}=\langle -2,5\rangle$.

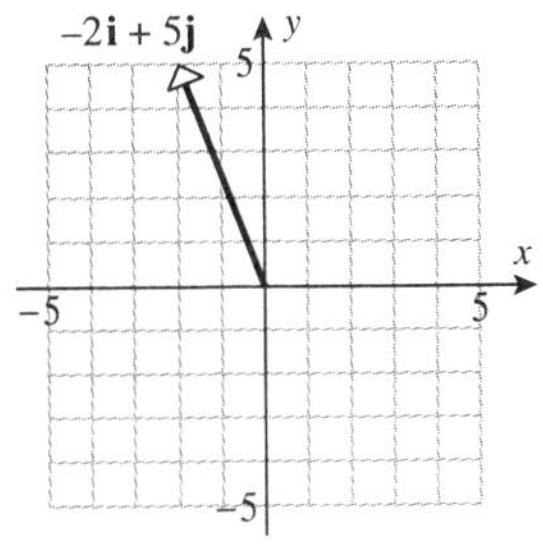

(b) The initial point of $\mathbf{u}+\mathbf{v}+\mathbf{w}$ is $(-5,4)$ and the endpoint is $(-2,-4)$, so $\mathbf{u}+\mathbf{v}+\mathbf{w}=\langle 3,-8\rangle$.

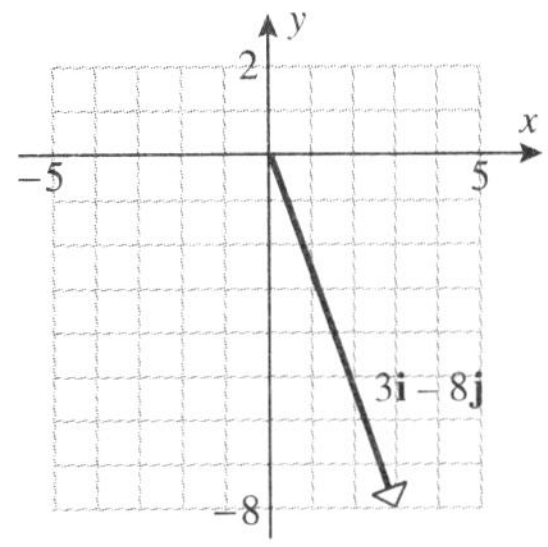

26. **(a)** $\mathbf{v}=\langle -10,2\rangle$ by inspection, so $\mathbf{u}-\mathbf{v}+\mathbf{w}=\mathbf{u}+\mathbf{v}+\mathbf{w}-2\mathbf{v}=$ $\langle -2,5\rangle+\langle 20,-4\rangle=\langle 18,1\rangle$.

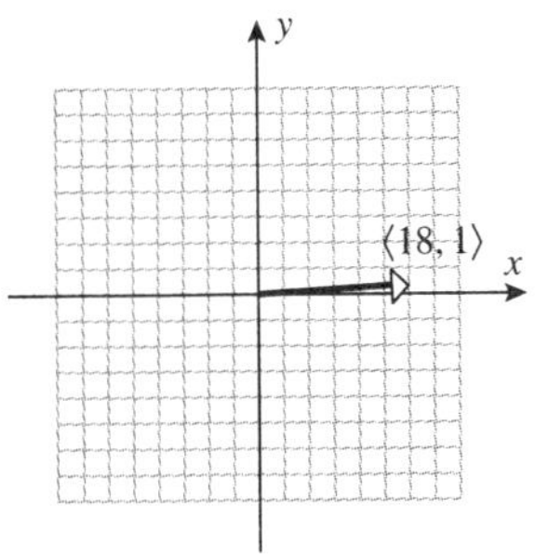

(b) $\mathbf{v}=\langle -3,8\rangle$ by inspection, so $\mathbf{u}-\mathbf{v}+\mathbf{w}=\mathbf{u}+\mathbf{v}+\mathbf{w}-2\mathbf{v}=$ $\langle 3,-8\rangle+\langle 6,-16\rangle=\langle 9,-24\rangle$.

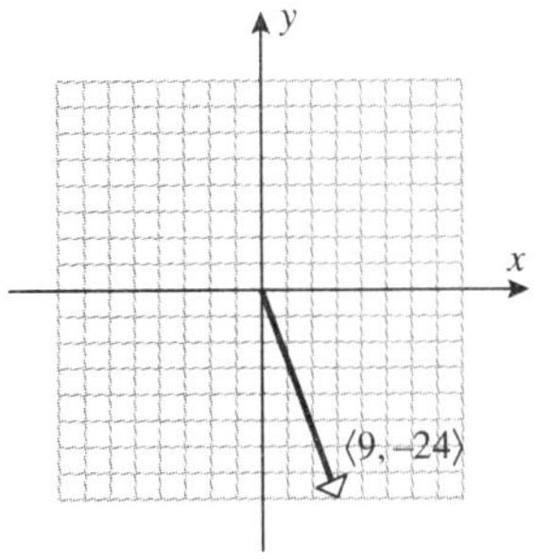

27. $6\mathbf{x}=2\mathbf{u}-\mathbf{v}-\mathbf{w}=\langle -4,6\rangle, \mathbf{x}=\langle -2/3,1\rangle$

28. $\mathbf{u}-2\mathbf{x}=\mathbf{x}-\mathbf{w}+3\mathbf{v}$, $3\mathbf{x}=\mathbf{u}+\mathbf{w}-3\mathbf{v}$, $\mathbf{x}=\dfrac{1}{3}(\mathbf{u}+\mathbf{w}-3\mathbf{v})=\langle 2/3,2/3\rangle$

29. $\mathbf{u}=\dfrac{5}{7}\mathbf{i}+\dfrac{2}{7}\mathbf{j}+\dfrac{1}{7}\mathbf{k}$, $\mathbf{v}=\dfrac{8}{7}\mathbf{i}-\dfrac{1}{7}\mathbf{j}-\dfrac{4}{7}\mathbf{k}$

30. $\mathbf{u}=\langle -5,8\rangle, \mathbf{v}=\langle 7,-11\rangle$

31. $\|(\mathbf{i}+\mathbf{j})+(\mathbf{i}-2\mathbf{j})\|=\|2\mathbf{i}-\mathbf{j}\|=\sqrt{5}$, $\|(\mathbf{i}+\mathbf{j}-(\mathbf{i}-2\mathbf{j})\|=\|3\mathbf{j}\|=3$

32. Let A, B, C be the vertices (0,0), (1,3), (2,4) and D the fourth vertex (x,y). For the parallelogram ABCD, $\overrightarrow{AD}=\overrightarrow{BC}$, $\langle x,y\rangle=\langle 1,1\rangle$ so $x=1$, $y=1$ and D is at (1,1). For the parallelogram ACBD, $\overrightarrow{AD}=\overrightarrow{CB}$, $\langle x,y\rangle=\langle -1,-1\rangle$ so $x=-1$, $y=-1$ and D is at $(-1,-1)$. For the parallelogram $ABDC$, $\overrightarrow{AC}=\overrightarrow{BD}$, $\langle x-1,y-3\rangle=\langle 2,4\rangle$, so $x=3, y=7$ and D is at $(3,7)$.

33. **(a)** $5=\|k\mathbf{v}\|=|k|\|\mathbf{v}\|=\pm 3k$, so k $=\pm 5/3$

(b) $6=\|k\mathbf{v}\|=|k|\|\mathbf{v}\|=2\|\mathbf{v}\|$, so $\|\mathbf{v}\|=3$

34. If $\|k\mathbf{v}\|=0$ then $|k|\|\mathbf{v}\|=0$ so either $k=0$ or $\|\mathbf{v}\|=0$; in the latter case, by (9) or (10), $\mathbf{v}=\mathbf{0}$.

35. **(a)** Choose two points on the line, for example $P_1(0,2)$ and $P_2(1,5)$; then $\overrightarrow{P_1P_2}=\langle 1,3\rangle$ is parallel to the line, $\|\langle 1,3\rangle\|=\sqrt{10}$, so $\langle 1/\sqrt{10},3/\sqrt{10}\rangle$ and $\langle -1/\sqrt{10},-3/\sqrt{10}\rangle$ are unit vectors parallel to the line.

(b) Choose two points on the line, for example $P_1(0,4)$ and $P_2(1,3)$; then $\overrightarrow{P_1P_2} = \langle 1,-1\rangle$ is parallel to the line, $\|\langle 1,-1\rangle\| = \sqrt{2}$ so $\langle 1/\sqrt{2}, -1/\sqrt{2}\rangle$ and $\langle -1/\sqrt{2}, 1/\sqrt{2}\rangle$ are unit vectors parallel to the line.

(c) Pick any line that is perpendicular to the line $y = -5x+1$, for example $y = x/5$; then $P_1(0,0)$ and $P_2(5,1)$ are on the line, so $\overrightarrow{P_1P_2} = \langle 5,1\rangle$ is perpendicular to the line, so $\pm\dfrac{1}{\sqrt{26}}\langle 5,1\rangle$ are unit vectors perpendicular to the line.

36. (a) $\pm\mathbf{k}$ (b) $\pm\mathbf{j}$ (c) $\pm\mathbf{i}$

37. (a) the circle of radius 1 about the origin

(b) the closed disk of radius 1 about the origin

(c) all points outside the closed disk of radius 1 about the origin

38. (a) the circle of radius 1 about the tip of $\mathbf{r}_0$

(b) the closed disk of radius 1 about the tip of $\mathbf{r}_0$

(c) all points outside the closed disk of radius 1 about the tip of $\mathbf{r}_0$

39. (a) the (hollow) sphere of radius 1 about the origin

(b) the closed ball of radius 1 about the origin

(c) all points outside the closed ball of radius 1 about the origin

40. The sum of the distances between (x,y) and the points (x_1,y_1), (x_2,y_2) is the constant k, so the set consists of all points on the ellipse with foci at (x_1,y_1) and (x_2,y_2), and major axis of length k.

41. Since $\phi = \pi/2$, from (14) we get $\|\mathbf{F}_1+\mathbf{F}_2\|^2 = \|\mathbf{F}_1\|^2 + \|\mathbf{F}_2\|^2 = 3600 + 900$,

so $\|\mathbf{F}_1+\mathbf{F}_2\| = 30\sqrt{5}$ lb, and $\sin\alpha = \dfrac{\|\mathbf{F}_2\|}{\|\mathbf{F}_1+\mathbf{F}_2\|}\sin\phi = \dfrac{30}{30\sqrt{5}}, \alpha \approx 26.57°, \theta = \alpha \approx 26.57°$.

42. $\|\mathbf{F}_1+\mathbf{F}_2\|^2 = \|\mathbf{F}_1\|^2 + \|\mathbf{F}_2\|^2 + 2\|\mathbf{F}_1\|\|\mathbf{F}_2\|\cos\phi = 14{,}400 + 10{,}000 + 2(120)(100)\dfrac{1}{2} = 36{,}400$, so

$\|\mathbf{F}_1+\mathbf{F}_2\| = 20\sqrt{91}$ N, $\sin\alpha = \dfrac{\|\mathbf{F}_2\|}{\|\mathbf{F}_1+\mathbf{F}_2\|}\sin\phi = \dfrac{100}{20\sqrt{91}}\sin 60° = \dfrac{5\sqrt{3}}{2\sqrt{91}}, \alpha \approx 27.00°$,

$\theta = \alpha \approx 27.00°$.

43. $\|\mathbf{F}_1+\mathbf{F}_2\|^2 = \|\mathbf{F}_1\|^2 + \|\mathbf{F}_2\|^2 + 2\|\mathbf{F}_1\|\|\mathbf{F}_2\|\cos\phi = 160{,}000 + 160{,}000 - 2(400)(400)\dfrac{\sqrt{3}}{2}$,

so $\|\mathbf{F}_1+\mathbf{F}_2\| \approx 207.06$ N, and $\sin\alpha = \dfrac{\|\mathbf{F}_2\|}{\|\mathbf{F}_1+\mathbf{F}_2\|}\sin\phi \approx \dfrac{400}{207.06}\left(\dfrac{1}{2}\right), \alpha = 75.00°$,

$\theta = \alpha - 30° = 45.00°$.

44. $\|\mathbf{F}_1+\mathbf{F}_2\|^2 = \|\mathbf{F}_1\|^2 + \|\mathbf{F}_2\|^2 + 2\|\mathbf{F}_1\|\|\mathbf{F}_2\|\cos\phi = 16 + 4 + 2(4)(2)\cos 77°$, so

$\|\mathbf{F}_1+\mathbf{F}_2\| \approx 4.86$ lb, and $\sin\alpha = \dfrac{\|\mathbf{F}_2\|}{\|\mathbf{F}_1+\mathbf{F}_2\|}\sin\phi = \dfrac{2}{4.86}\sin 77°, \alpha \approx 23.64°, \theta = \alpha - 27° \approx -3.36°$.

45. Let $\mathbf{F}_1, \mathbf{F}_2, \mathbf{F}_3$ be the forces in the diagram with magnitudes 40, 50, 75 respectively. Then $\mathbf{F}_1+\mathbf{F}_2+\mathbf{F}_3 = (\mathbf{F}_1+\mathbf{F}_2)+\mathbf{F}_3$. Following the examples, $\mathbf{F}_1+\mathbf{F}_2$ has magnitude 45.83 N and makes an angle 79.11° with the positive x-axis. Then $\|(\mathbf{F}_1+\mathbf{F}_2)+\mathbf{F}_3\|^2 \approx 45.83^2 + 75^2 + 2(45.83)(75)\cos 79.11°$, so $\mathbf{F}_1+\mathbf{F}_2+\mathbf{F}_3$ has magnitude ≈ 94.995 N and makes an angle $\theta = \alpha \approx 28.28°$ with the positive x-axis.

46. Let $\mathbf{F}_1, \mathbf{F}_2, \mathbf{F}_3$ be the forces in the diagram with magnitudes 150, 200, 100 respectively. Then $\mathbf{F}_1 + \mathbf{F}_2 + \mathbf{F}_3 = (\mathbf{F}_1 + \mathbf{F}_2) + \mathbf{F}_3$. Following the examples, $\mathbf{F}_1 + \mathbf{F}_2$ has magnitude 279.34 N and makes an angle 91.24° with the positive x-axis. Then
$\|\mathbf{F}_1 + \mathbf{F}_2 + \mathbf{F}_3\|^2 \approx 279.34^2 + 100^2 + 2(279.34)(100)\cos(270 - 91.24)^\circ$, and $\mathbf{F}_1 + \mathbf{F}_2 + \mathbf{F}_3$ has magnitude ≈ 179.37 N and makes an angle 91.94° with the positive x-axis.

47. Let $\mathbf{F}_1, \mathbf{F}_2$ be the forces in the diagram with magnitudes 8, 10 respectively. Then $\|\mathbf{F}_1 + \mathbf{F}_2\|$ has magnitude $\sqrt{8^2 + 10^2 + 2 \cdot 8 \cdot 10 \cos 120^\circ} = 2\sqrt{21} \approx 9.165$ lb, and makes an angle $60^\circ + \sin^{-1} \dfrac{\|\mathbf{F}_1\|}{\|\mathbf{F}_1 + \mathbf{F}_2\|} \sin 120 \approx 109.11^\circ$ with the positive x-axis, so $\mathbf{F}$ has magnitude 9.165 lb and makes an angle -70.89° with the positive x-axis.

48. $\|\mathbf{F}_1 + \mathbf{F}_2\| = \sqrt{120^2 + 150^2 + 2 \cdot 120 \cdot 150 \cos 75^\circ} = 214.98$ N and makes an angle 92.63° with the positive x-axis, and $\|\mathbf{F}_1 + \mathbf{F}_2 + \mathbf{F}_3\| = 232.90$ N and makes an angle 67.23° with the positive x-axis, hence $\mathbf{F}$ has magnitude 232.90 N and makes an angle -112.77° with the positive x-axis.

49. $\mathbf{F}_1 + \mathbf{F}_2 + \mathbf{F} = \mathbf{0}$, where $\mathbf{F}$ has magnitude 250 and makes an angle -90° with the positive x-axis. Thus $\|\mathbf{F}_1 + \mathbf{F}_2\|^2 = \|\mathbf{F}_1\|^2 + \|\mathbf{F}_2\|^2 + 2\|\mathbf{F}_1\|\|\mathbf{F}_2\| \cos 105^\circ = 250^2$ and

$$45^\circ = \alpha = \sin^{-1}\left(\frac{\|\mathbf{F}_2\|}{250} \sin 105^\circ\right), \text{ so } \frac{\sqrt{2}}{2} \approx \frac{\|\mathbf{F}_2\|}{250} 0.9659, \|\mathbf{F}_2\| \approx 183.02 \text{ lb},$$

$\|\mathbf{F}_1\|^2 + 2(183.02)(-0.2588)\|\mathbf{F}_1\| + (183.02)^2 = 62{,}500$, $\|\mathbf{F}_1\| = 224.13$ lb.

50. Similar to Exercise 49, $\|\mathbf{F}_1\| = 100\sqrt{3}$ N, $\|\mathbf{F}_2\| = 100$ N

51. **(a)** $c_1\mathbf{v}_1 + c_2\mathbf{v}_2 = (2c_1 + 4c_2)\,\mathbf{i} + (-c_1 + 2c_2)\,\mathbf{j} = 4\mathbf{j}$, so $2c_1 + 4c_2 = 0$ and $-c_1 + 2c_2 = 4$ which gives $c_1 = -2$, $c_2 = 1$.

(b) $c_1\mathbf{v}_1 + c_2\mathbf{v}_2 = \langle c_1 - 2c_2, -3c_1 + 6c_2\rangle = \langle 3, 5\rangle$, so $c_1 - 2c_2 = 3$ and $-3c_1 + 6c_2 = 5$ which has no solution.

52. **(a)** Equate corresponding components to get the system of equations $c_1 + 3c_2 = -1$, $2c_2 + c_3 = 1$, and $c_1 + c_3 = 5$. Solve to get $c_1 = 2$, $c_2 = -1$, and $c_3 = 3$.

(b) Equate corresponding components to get the system of equations $c_1 + 3c_2 + 4c_3 = 2$, $-c_1 - c_3 = 1$, and $c_2 + c_3 = -1$. From the second and third equations, $c_1 = -1 - c_3$ and $c_2 = -1 - c_3$; substitute these into the first equation to get $-4 = 2$, which is false so the system has no solution.

53. Place $\mathbf{u}$ and $\mathbf{v}$ tip to tail so that $\mathbf{u} + \mathbf{v}$ is the vector from the initial point of $\mathbf{u}$ to the terminal point of $\mathbf{v}$. The shortest distance between two points is along the line joining these points so $\|\mathbf{u} + \mathbf{v}\| \le \|\mathbf{u}\| + \|\mathbf{v}\|$.

54. (a): $\mathbf{u} + \mathbf{v} = (u_1\mathbf{i} + u_2\mathbf{j}) + (v_1\mathbf{i} + v_2\mathbf{j}) = (v_1\mathbf{i} + v_2\mathbf{j}) + (u_1\mathbf{i} + u_2\mathbf{j}) = \mathbf{v} + \mathbf{u}$
(c): $\mathbf{u} + \mathbf{0} = (u_1\mathbf{i} + u_2\mathbf{j}) + 0\mathbf{i} + 0\mathbf{j} = u_1\mathbf{i} + u_2\mathbf{j} = \mathbf{u}$
(e): $k(l\mathbf{u}) = k(l(u_1\mathbf{i} + u_2\mathbf{j})) = k(lu_1\mathbf{i} + lu_2\mathbf{j}) = klu_1\mathbf{i} + klu_2\mathbf{j} = (kl)\mathbf{u}$

55. (d): $\mathbf{u} + (-\mathbf{u}) = (u_1\mathbf{i} + u_2\mathbf{j}) + (-u_1\mathbf{i} - u_2\mathbf{j}) = (u_1 - u_1)\mathbf{i} + (u_1 - u_1)\mathbf{j} = \mathbf{0}$
(g): $(k + l)\mathbf{u} = (k + l)(u_1\mathbf{i} + u_2\mathbf{j}) = ku_1\mathbf{i} + ku_2\mathbf{j} + lu_1\mathbf{i} + lu_2\mathbf{j} = k\mathbf{u} + l\mathbf{u}$
(h): $1\mathbf{u} = 1(u_1\mathbf{i} + u_2\mathbf{j}) = 1u_1\mathbf{i} + 1u_2\mathbf{j} = u_1\mathbf{i} + u_2\mathbf{j} = \mathbf{u}$

56. Draw the triangles with sides formed by the vectors $\mathbf{u}$, $\mathbf{v}$, $\mathbf{u} + \mathbf{v}$ and $k\mathbf{u}$, $k\mathbf{v}$, $k\mathbf{u} + k\mathbf{v}$. By similar triangles, $k(\mathbf{u} + \mathbf{v}) = k\mathbf{u} + k\mathbf{v}$.

57. Let **a**, **b**, **c** be vectors along the sides of the triangle and A,B the midpoints of **a** and **b**, then $\mathbf{u} = \frac{1}{2}\mathbf{a} - \frac{1}{2}\mathbf{b} = \frac{1}{2}(\mathbf{a} - \mathbf{b}) = \frac{1}{2}\mathbf{c}$ so **u** is parallel to **c** and half as long.

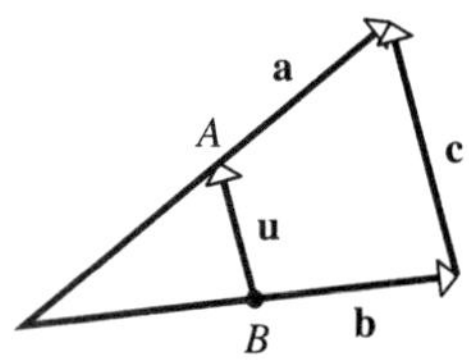

58. Let **a**, **b**, **c**, **d** be vectors along the sides of the quadrilateral and A, B, C, D the corresponding midpoints, then
$\mathbf{u} = \frac{1}{2}\mathbf{b} + \frac{1}{2}\mathbf{c}$ and $\mathbf{v} = \frac{1}{2}\mathbf{d} - \frac{1}{2}\mathbf{a}$ but $\mathbf{d} = \mathbf{a} + \mathbf{b} + \mathbf{c}$ so
$\mathbf{v} = \frac{1}{2}(\mathbf{a} + \mathbf{b} + \mathbf{c}) - \frac{1}{2}\mathbf{a} = \frac{1}{2}\mathbf{b} + \frac{1}{2}\mathbf{c} = \mathbf{u}$ thus ABCD is a parallelogram because sides AD and BC are equal and parallel.

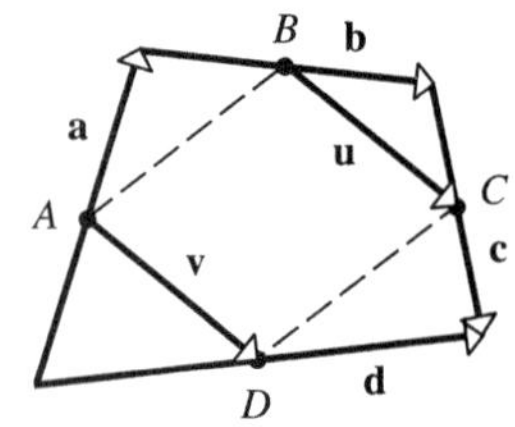

EXERCISE SET 12.3

1. (a) $(1)(6) + (2)(-8) = -10$; $\cos\theta = (-10)/[(\sqrt{5})(10)] = -1/\sqrt{5}$
(b) $(-7)(0) + (-3)(1) = -3$; $\cos\theta = (-3)/[(\sqrt{58})(1)] = -3/\sqrt{58}$
(c) $(1)(8) + (-3)(-2) + (7)(-2) = 0$; $\cos\theta = 0$
(d) $(-3)(4) + (1)(2) + (2)(-5) = -20$; $\cos\theta = (-20)/[(\sqrt{14})(\sqrt{45})] = -20/(3\sqrt{70})$

2. (a) $\mathbf{u}\cdot\mathbf{v} = 1(2)\cos(\pi/6) = \sqrt{3}$ (b) $\mathbf{u}\cdot\mathbf{v} = 2(3)\cos 135^\circ = -3\sqrt{2}$

3. (a) $\mathbf{u}\cdot\mathbf{v} = -34 < 0$, obtuse (b) $\mathbf{u}\cdot\mathbf{v} = 6 > 0$, acute
(c) $\mathbf{u}\cdot\mathbf{v} = -1 < 0$, obtuse (d) $\mathbf{u}\cdot\mathbf{v} = 0$, orthogonal

4. Let the points be P, Q, R in order, then $\overrightarrow{PQ} = \langle 2-(-1), -2-2, 0-3\rangle = \langle 3,-4,-3\rangle$, $\overrightarrow{QR} = \langle 3-2, 1-(-2), -4-0\rangle = \langle 1,3,-4\rangle$, $\overrightarrow{RP} = \langle -1-3, 2-1, 3-(-4)\rangle = \langle -4,1,7\rangle$;
since $\overrightarrow{QP}\cdot\overrightarrow{QR} = -3(1) + 4(3) + 3(-4) = -3 < 0$, $\angle PQR$ is obtuse;
since $\overrightarrow{RP}\cdot\overrightarrow{RQ} = -4(-1) + (-3) + 7(4) = 29 > 0$, $\angle PRQ$ is acute;
since $\overrightarrow{PR}\cdot\overrightarrow{PQ} = 4(3) - 1(-4) - 7(-3) = 37 > 0$, $\angle RPQ$ is acute

5. Since $\mathbf{v}_1\cdot\mathbf{v}_i = \cos\phi_i$, the answers are, in order, $\sqrt{2}/2, 0, -\sqrt{2}/2, -1, -\sqrt{2}/2, 0, \sqrt{2}/2$

6. Proceed as in Exercise 5; $25/2, -25/2, -25, -25/2, 25/2$

7. (a) $\overrightarrow{AB} = \langle 1,3,-2\rangle$, $\overrightarrow{BC} = \langle 4,-2,-1\rangle$, $\overrightarrow{AB}\cdot\overrightarrow{BC} = 0$ so $\overrightarrow{AB}$ and $\overrightarrow{BC}$ are orthogonal; it is a right triangle with the right angle at vertex B.
(b) Let A, B, and C be the vertices $(-1,0)$, $(2,-1)$, and (1,4) with corresponding interior angles α, β, and γ, then

$$\cos\alpha = \frac{\overrightarrow{AB}\cdot\overrightarrow{AC}}{\|\overrightarrow{AB}\|\,\|\overrightarrow{AC}\|} = \frac{\langle 3,-1\rangle\cdot\langle 2,4\rangle}{\sqrt{10}\sqrt{20}} = 1/(5\sqrt{2}),\ \alpha\approx 82^\circ$$

$$\cos\beta = \frac{\overrightarrow{BA}\cdot\overrightarrow{BC}}{\|\overrightarrow{BA}\|\,\|\overrightarrow{BC}\|} = \frac{\langle -3,1\rangle\cdot\langle -1,5\rangle}{\sqrt{10}\sqrt{26}} = 4/\sqrt{65},\ \beta\approx 60^\circ$$

$$\cos\gamma = \frac{\overrightarrow{CA}\cdot\overrightarrow{CB}}{\|\overrightarrow{CA}\|\,\|\overrightarrow{CB}\|} = \frac{\langle -2,-4\rangle\cdot\langle 1,-5\rangle}{\sqrt{20}\sqrt{26}} = 9/\sqrt{130},\ \gamma\approx 38^\circ$$

8. $\overrightarrow{AB}\cdot\overrightarrow{AP} = [2\mathbf{i}+\mathbf{j}+2\mathbf{k}]\cdot[(k-1)\mathbf{i}+(k+1)\mathbf{j}+(k-3)\mathbf{k}]$
$= 2(k-1)+(k+1)+2(k-3) = 5k-7 = 0, k = 7/5.$

9. **(a)** $\mathbf{v}\cdot\mathbf{v}_1 = -ab+ba = 0;\ \ \mathbf{v}\cdot\mathbf{v}_2 = ab+b(-a) = 0$

(b) Let $\mathbf{v}_1 = 2\mathbf{i}+3\mathbf{j},\ \ \mathbf{v}_2 = -2\mathbf{i}-3\mathbf{j};$

take $\mathbf{u}_1 = \dfrac{\mathbf{v}_1}{\|\mathbf{v}_1\|} = \dfrac{2}{\sqrt{13}}\mathbf{i}+\dfrac{3}{\sqrt{13}}\mathbf{j},\ \mathbf{u}_2 = -\mathbf{u}_1.$

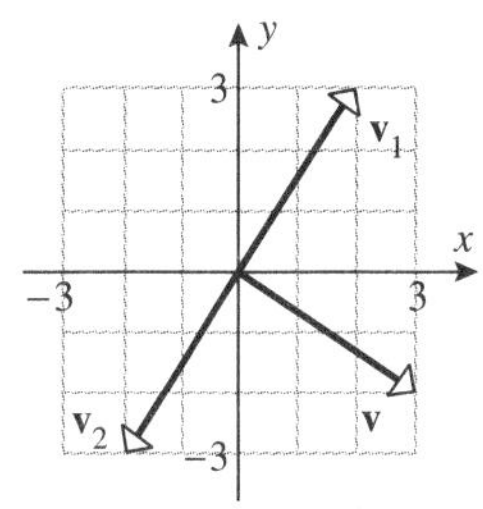

10. By inspection, $3\mathbf{i}-4\mathbf{j}$ is orthogonal to and has the same length as $4\mathbf{i}+3\mathbf{j}$ so $\mathbf{u}_1 = (4\mathbf{i}+3\mathbf{j})+(3\mathbf{i}-4\mathbf{j}) = 7\mathbf{i}-\mathbf{j}$ and $\mathbf{u}_2 = (4\mathbf{i}+3\mathbf{j})+(-1)(3\mathbf{i}-4\mathbf{j}) = \mathbf{i}+7\mathbf{j}$ each make an angle of 45° with $4\mathbf{i}+3\mathbf{j}$; unit vectors in the directions of $\mathbf{u}_1$ and $\mathbf{u}_2$ are $(7\mathbf{i}-\mathbf{j})/\sqrt{50}$ and $(\mathbf{i}+7\mathbf{j})/\sqrt{50}$.

11. **(a)** The dot product of a vector $\mathbf{u}$ and a scalar $\mathbf{v}\cdot\mathbf{w}$ is not defined.

(b) The sum of a scalar $\mathbf{u}\cdot\mathbf{v}$ and a vector $\mathbf{w}$ is not defined.

(c) $\mathbf{u}\cdot\mathbf{v}$ is not a vector.

(d) The dot product of a scalar k and a vector $\mathbf{u}+\mathbf{v}$ is not defined.

12. (b): $\mathbf{u}\cdot(\mathbf{v}+\mathbf{w}) = (6\mathbf{i}-\mathbf{j}+2\mathbf{k})\cdot((2\mathbf{i}+7\mathbf{j}+4\mathbf{k})+(\mathbf{i}+\mathbf{j}-3\mathbf{k})) = (6\mathbf{i}-\mathbf{j}+2\mathbf{k})\cdot(3\mathbf{i}+8\mathbf{j}+\mathbf{k}) = 12;$
$\mathbf{u}\cdot\mathbf{v}+\mathbf{u}\cdot\mathbf{w} = (6\mathbf{i}-\mathbf{j}+2\mathbf{k})\cdot(2\mathbf{i}+7\mathbf{j}+4\mathbf{k})+(6\mathbf{i}-\mathbf{j}+2\mathbf{k})\cdot(\mathbf{i}+\mathbf{j}-3\mathbf{k}) = 13-1 = 12$
(c): $k(\mathbf{u}\cdot\mathbf{v}) = -5(13) = -65;\ (k\mathbf{u})\cdot\mathbf{v} = (-30\mathbf{i}+5\mathbf{j}-10\mathbf{k})\cdot(2\mathbf{i}+7\mathbf{j}+4\mathbf{k}) = -65;$
$\mathbf{u}\cdot(k\mathbf{v}) = (6\mathbf{i}-\mathbf{j}+2\mathbf{k})\cdot(-10\mathbf{i}-35\mathbf{j}-20\mathbf{k}) - -65$

13. **(a)** $\langle 1,2\rangle\cdot(\langle 28,-14\rangle+\langle 6,0\rangle) = \langle 1,2\rangle\cdot\langle 34,-14\rangle = 6$

(b) $\|6\mathbf{w}\| = 6\|\mathbf{w}\| = 36$ **(c)** $24\sqrt{5}$ **(d)** $24\sqrt{5}$

14. false, for example $\mathbf{a} = \langle 1,2\rangle, \mathbf{b} = \langle -1,0\rangle, \mathbf{c} = \langle 5,-3\rangle$

15. **(a)** $\|\mathbf{v}\| = \sqrt{3}$ so $\cos\alpha = \cos\beta = 1/\sqrt{3},\ \cos\gamma = -1/\sqrt{3},\ \alpha = \beta\approx 55^\circ,\ \gamma\approx 125^\circ$

(b) $\|\mathbf{v}\| = 3$ so $\cos\alpha = 2/3,\ \cos\beta = -2/3,\ \cos\gamma = 1/3,\ \alpha\approx 48^\circ,\ \beta\approx 132^\circ,\ \gamma\approx 71^\circ$

16. **(a)** $\|\mathbf{v}\| = 7$ so $\cos\alpha = 3/7,\ \cos\beta = -2/7,\ \cos\gamma = -6/7,\ \alpha\approx 65^\circ,\ \beta\approx 107^\circ,\ \gamma\approx 149^\circ$

(b) $\|\mathbf{v}\| = 5,\ \cos\alpha = 3/5,\ \cos\beta = 0,\ \cos\gamma = -4/5,\ \alpha\approx 53^\circ,\ \beta = 90^\circ,\ \gamma\approx 143^\circ$

17. $\cos^2\alpha + \cos^2\beta + \cos^2\gamma = \dfrac{v_1^2}{\|\mathbf{v}\|^2} + \dfrac{v_2^2}{\|\mathbf{v}\|^2} + \dfrac{v_3^2}{\|\mathbf{v}\|^2} = \left(v_1^2 + v_2^2 + v_3^2\right)/\|\mathbf{v}\|^2 = \|\mathbf{v}\|^2/\|\mathbf{v}\|^2 = 1$

18. Let $\mathbf{v} = \langle x, y, z\rangle$, then $x = \sqrt{x^2+y^2}\cos\theta$, $y = \sqrt{x^2+y^2}\sin\theta$, $\sqrt{x^2+y^2} = \|\mathbf{v}\|\cos\lambda$, and $z = \|\mathbf{v}\|\sin\lambda$, so $x/\|\mathbf{v}\| = \cos\theta\cos\lambda$, $y/\|\mathbf{v}\| = \sin\theta\cos\lambda$, and $z/\|\mathbf{v}\| = \sin\lambda$.

19. $\cos\alpha = \dfrac{\sqrt{3}}{2}\dfrac{1}{2} = \dfrac{\sqrt{3}}{4}$, $\cos\beta = \dfrac{\sqrt{3}}{2}\dfrac{\sqrt{3}}{2} = \dfrac{3}{4}$, $\cos\gamma = \dfrac{1}{2}$; $\alpha \approx 64^\circ, \beta \approx 41^\circ, \gamma = 60^\circ$

20. Let $\mathbf{u}_1 = \|\mathbf{u}_1\|\langle\cos\alpha_1, \cos\beta_1, \cos\gamma_1\rangle$, $\mathbf{u}_2 = \|\mathbf{u}_2\|\langle\cos\alpha_2, \cos\beta_2, \cos\gamma_2\rangle$, $\mathbf{u}_1$ and $\mathbf{u}_2$ are perpendicular if and only if $\mathbf{u}_1 \cdot \mathbf{u}_2 = 0$ so $\|\mathbf{u}_1\|\,\|\mathbf{u}_2\|(\cos\alpha_1\cos\alpha_2 + \cos\beta_1\cos\beta_2 + \cos\gamma_1\cos\gamma_2) = 0$,
$\cos\alpha_1\cos\alpha_2 + \cos\beta_1\cos\beta_2 + \cos\gamma_1\cos\gamma_2 = 0$.

21. (a) $\dfrac{\mathbf{b}}{\|\mathbf{b}\|} = \langle 3/5, 4/5\rangle$, so $\text{proj}_{\mathbf{b}}\mathbf{v} = \langle 6/25, 8/25\rangle$
and $\mathbf{v} - \text{proj}_{\mathbf{b}}\mathbf{v} = \langle 44/25, -33/25\rangle$

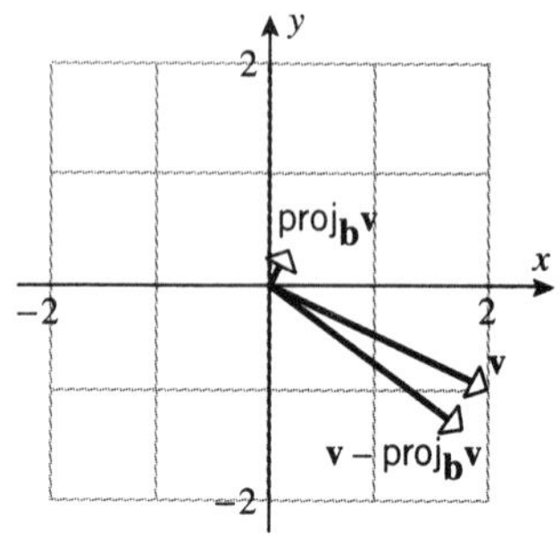

(b) $\dfrac{\mathbf{b}}{\|\mathbf{b}\|} = \langle 1/\sqrt{5}, -2/\sqrt{5}\rangle$, so $\text{proj}_{\mathbf{b}}\mathbf{v} = \langle -6/5, 12/5\rangle$
and $\mathbf{v} - \text{proj}_{\mathbf{b}}\mathbf{v} = \langle 26/5, 13/5\rangle$

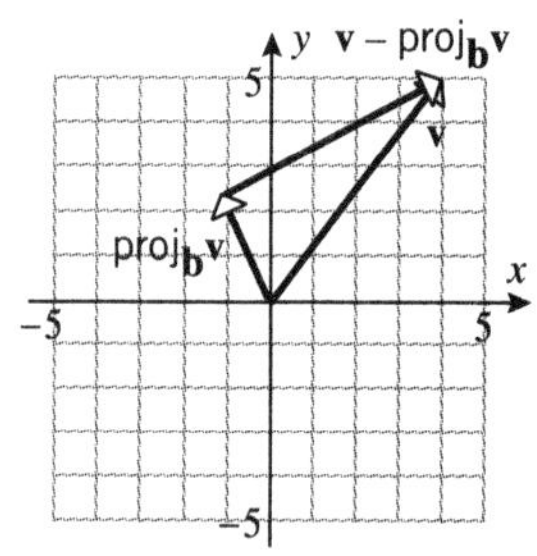

(c) $\dfrac{\mathbf{b}}{\|\mathbf{b}\|} = \langle 2/\sqrt{5}, 1/\sqrt{5}\rangle$, so $\text{proj}_{\mathbf{b}}\mathbf{v} = \langle -16/5, -8/5\rangle$
and $\mathbf{v} - \text{proj}_{\mathbf{b}}\mathbf{v} = \langle 1/5, -2/5\rangle$

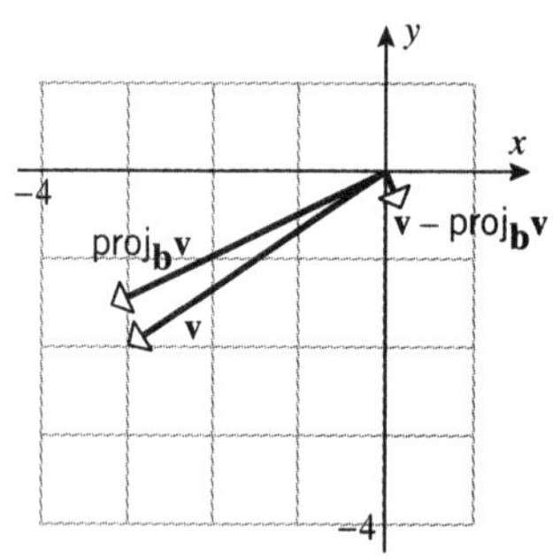

22. (a) $\dfrac{\mathbf{b}}{\|\mathbf{b}\|} = \langle 1/3, 2/3, 2/3\rangle$, so $\text{proj}_{\mathbf{b}}\mathbf{v} = \langle 2/3, 4/3, 4/3\rangle$ and $\mathbf{v} - \text{proj}_{\mathbf{b}}\mathbf{v} = \langle 4/3, -7/3, 5/3\rangle$

(b) $\dfrac{\mathbf{b}}{\|\mathbf{b}\|} = \langle 2/7, 3/7, -6/7\rangle$, so $\text{proj}_{\mathbf{b}}\mathbf{v} = \langle -74/49, -111/49, 222/49\rangle$
and $\mathbf{v} - \text{proj}_{\mathbf{b}}\mathbf{v} = \langle 270/49, 62/49, 121/49\rangle$

23. (a) $\text{proj}_{\mathbf{b}}\mathbf{v} = \langle -1, -1\rangle$, so $\mathbf{v} = \langle -1, -1\rangle + \langle 3, -3\rangle$

(b) $\text{proj}_{\mathbf{b}}\mathbf{v} = \langle 16/5, 0, -8/5\rangle$, so $\mathbf{v} = \langle 16/5, 0, -8/5\rangle + \langle -1/5, 1, -2/5\rangle$

24. **(a)** $\text{proj}_{\mathbf{b}}\mathbf{v} = \langle 1, 1\rangle$, so $\mathbf{v} = \langle 1, 1\rangle + \langle -4, 4\rangle$

(b) $\text{proj}_{\mathbf{b}}\mathbf{v} = \langle 0, -8/5, 4/5\rangle$, so $\mathbf{v} = \langle 0, -8/5, 4/5\rangle + \langle -2, 13/5, 26/5\rangle$

25. $\overrightarrow{AP} = -\mathbf{i} + 3\mathbf{j}$, $\overrightarrow{AB} = 3\mathbf{i} + 4\mathbf{j}$, $\|\text{proj}_{\overrightarrow{AB}}\overrightarrow{AP}\| = |\overrightarrow{AP} \cdot \overrightarrow{AB}|/\|\overrightarrow{AB}\| = 9/5$

$\|\overrightarrow{AP}\| = \sqrt{10}$, $\sqrt{10 - 81/25} = 13/5$

26. $\overrightarrow{AP} = -4\mathbf{i} + 2\mathbf{k}$, $\overrightarrow{AB} = -3\mathbf{i} + 2\mathbf{j} - 4\mathbf{k}$, $\|\text{proj}_{\overrightarrow{AB}}\overrightarrow{AP}\| = |\overrightarrow{AP} \cdot \overrightarrow{AB}|/\|\overrightarrow{AB}\| = 4/\sqrt{29}$.

$\|\overrightarrow{AP}\| = \sqrt{20}$, $\sqrt{20 - 16/29} = \sqrt{564/29}$

27. Let $\mathbf{F}$ be the downward force of gravity on the block, then $\|\mathbf{F}\| = 10(9.8) = 98$ N, and if $\mathbf{F}_1$ and $\mathbf{F}_2$ are the forces parallel to and perpendicular to the ramp, then $\|\mathbf{F}_1\| = \|\mathbf{F}_2\| = 49\sqrt{2}$ N. Thus the block exerts a force of $49\sqrt{2}$ N against the ramp and it requires a force of $49\sqrt{2}$ N to prevent the block from sliding down the ramp.

28. Let x denote the magnitude of the force in the direction of $\mathbf{Q}$. Then the force $\mathbf{F}$ acting on the block is $\mathbf{F} = x\mathbf{i} - 98\mathbf{j}$. Let $\mathbf{u} = -\dfrac{1}{\sqrt{2}}(\mathbf{i} + \mathbf{j})$ and $\mathbf{v} = \dfrac{1}{\sqrt{2}}(\mathbf{i} - \mathbf{j})$ be the unit vectors in the directions along and against the ramp. Then $\mathbf{F}$ decomposes as $\mathbf{F} = -\dfrac{x - 98}{\sqrt{2}}\mathbf{u} + \dfrac{x + 98}{\sqrt{2}}\mathbf{v}$, and thus the block will not slide down the ramp provided $x \geq 98$ N.

29. Three forces act on the block: its weight $-300\mathbf{j}$; the tension in cable A, which has the form $a(-\mathbf{i} + \mathbf{j})$; and the tension in cable B, which has the form $b(\sqrt{3}\mathbf{i} - \mathbf{j})$, where a, b are positive constants. The sum of these forces is zero, which yields $a = 450 + 150\sqrt{3}, b = 150 + 150\sqrt{3}$. Thus the forces along cables A and B are, respectively,
$\|150(3 + \sqrt{3})(\mathbf{i} - \mathbf{j})\| = 450\sqrt{2} + 150\sqrt{6}$ lb, and $\|150(\sqrt{3} + 1)(\sqrt{3}\mathbf{i} - \mathbf{j})\| = 300 + 300\sqrt{3}$ lb.

30. **(a)** Let $\mathbf{T}_A$ and $\mathbf{T}_B$ be the forces exerted on the block by cables A and B. Then $\mathbf{T}_A = a(-10\mathbf{i} + d\mathbf{j})$ and $\mathbf{T}_B = b(20\mathbf{i} + d\mathbf{j})$ for some positive a, b. Since $\mathbf{T}_A + \mathbf{T}_B - 100\mathbf{j} = \mathbf{0}$, we find $a = \dfrac{200}{3d}, b = \dfrac{100}{3d}$, $\mathbf{T}_A = -\dfrac{2000}{3d}\mathbf{i} + \dfrac{200}{3}\mathbf{j}$, and $\mathbf{T}_B = \dfrac{2000}{3d}\mathbf{i} + \dfrac{100}{3}\mathbf{j}$. Thus

$\mathbf{T}_A = \dfrac{200}{3}\sqrt{1 + \dfrac{100}{d^2}}$, $\mathbf{T}_B = \dfrac{100}{3}\sqrt{1 + \dfrac{400}{d^2}}$, and the graphs are:

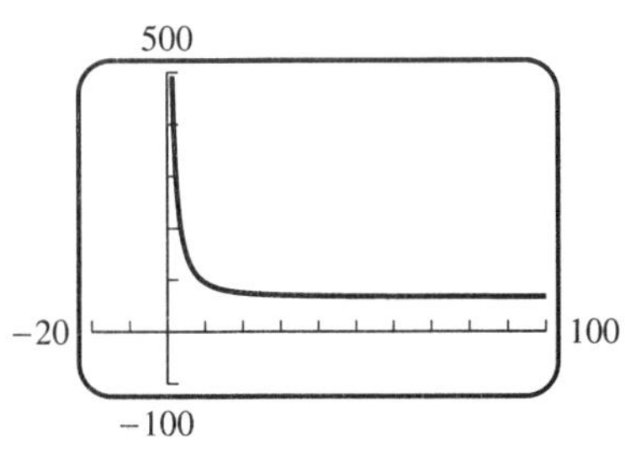

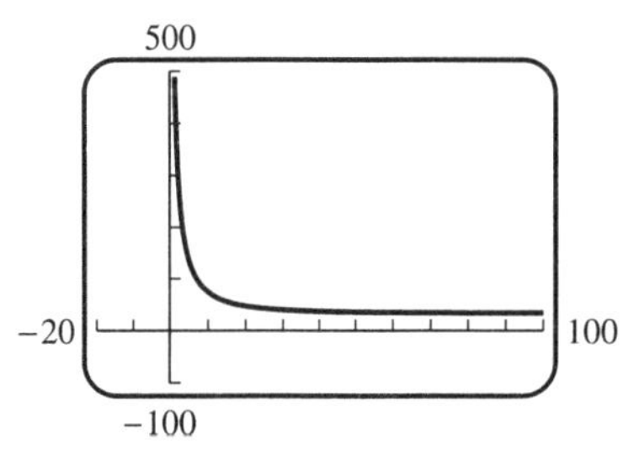

(b) An increase in d will decrease both forces.

(c) The inequality $\|\mathbf{T}_A\| \leq 150$ is equivalent to $d \geq \dfrac{40}{\sqrt{65}}$, and $\|\mathbf{T}_B\| \leq 150$ is equivalent to $d \geq \dfrac{40}{\sqrt{77}}$. Hence we must have $d \geq \dfrac{40}{65}$.

31. Let P and Q be the points (1,3) and (4,7) then $\overrightarrow{PQ} = 3\mathbf{i} + 4\mathbf{j}$ so $W = \mathbf{F} \cdot \overrightarrow{PQ} = -12$ ft·lb.

32. $W = \mathbf{F} \cdot \overrightarrow{PQ} = \|\mathbf{F}\|\,\|\overrightarrow{PQ}\|\cos 45^\circ = (500)(100)\left(\sqrt{2}/2\right) = 25{,}000\sqrt{2}$ N · m $= 25{,}000\sqrt{2}$ J

33. $W = \mathbf{F}\cdot 15\mathbf{i} = 15 \cdot 50\cos 60° = 375$ ft · lb.

34. $W = \mathbf{F}\cdot(15/\sqrt{3})(\mathbf{i}+\mathbf{j}+\mathbf{k}) = -15/\sqrt{3}$ N · m $= -5\sqrt{3}$ J

35. With the cube as shown in the diagram, and a the length of each edge, $\mathbf{d}_1 = a\mathbf{i} + a\mathbf{j} + a\mathbf{k}, \mathbf{d}_2 = a\mathbf{i} + a\mathbf{j} - a\mathbf{k}$, $\cos\theta = (\mathbf{d}_1 \cdot \mathbf{d}_2) / (\|\mathbf{d}_1\|\,\|\mathbf{d}_2\|) = 1/3, \theta \approx 71°$

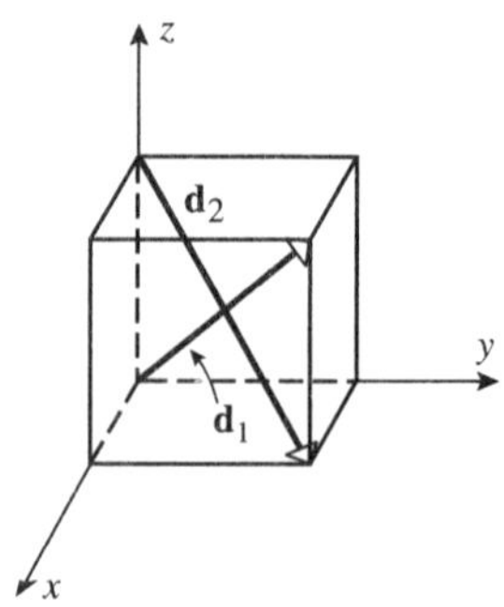

36. Take $\mathbf{i}$, $\mathbf{j}$, and $\mathbf{k}$ along adjacent edges of the box, then $10\mathbf{i} + 15\mathbf{j} + 25\mathbf{k}$ is along a diagonal, and a unit vector in this direction is $\frac{2}{\sqrt{38}}\mathbf{i} + \frac{3}{\sqrt{38}}\mathbf{j} + \frac{5}{\sqrt{38}}\mathbf{k}$. The direction cosines are $\cos\alpha = 2/\sqrt{38}$, $\cos\beta = 3/\sqrt{38}$, and $\cos\gamma = 5/\sqrt{38}$ so $\alpha \approx 71°$, $\beta \approx 61°$, and $\gamma \approx 36°$.

37. $\mathbf{u}+\mathbf{v}$ and $\mathbf{u}-\mathbf{v}$ are vectors along the diagonals,
$(\mathbf{u}+\mathbf{v})\cdot(\mathbf{u}-\mathbf{v}) = \mathbf{u}\cdot\mathbf{u} - \mathbf{u}\cdot\mathbf{v} + \mathbf{v}\cdot\mathbf{u} - \mathbf{v}\cdot\mathbf{v} = \|\mathbf{u}\|^2 - \|\mathbf{v}\|^2$ so $(\mathbf{u}+\mathbf{v})\cdot(\mathbf{u}-\mathbf{v}) = 0$ if and only if $\|\mathbf{u}\| = \|\mathbf{v}\|$.

38. The diagonals have lengths $\|\mathbf{u}+\mathbf{v}\|$ and $\|\mathbf{u}-\mathbf{v}\|$ but
$\|\mathbf{u}+\mathbf{v}\|^2 = (\mathbf{u}+\mathbf{v})\cdot(\mathbf{u}+\mathbf{v}) = \|\mathbf{u}\|^2 + 2\mathbf{u}\cdot\mathbf{v} + \|\mathbf{v}\|^2$, and
$\|\mathbf{u}-\mathbf{v}\|^2 = (\mathbf{u}-\mathbf{v})\cdot(\mathbf{u}-\mathbf{v}) = \|\mathbf{u}\|^2 - 2\mathbf{u}\cdot\mathbf{v} + \|\mathbf{v}\|^2$. If the parallelogram is a rectangle then $\mathbf{u}\cdot\mathbf{v} = 0$ so $\|\mathbf{u}+\mathbf{v}\|^2 = \|\mathbf{u}-\mathbf{v}\|^2$; the diagonals are equal. If the diagonals are equal, then $4\mathbf{u}\cdot\mathbf{v} = 0$, $\mathbf{u}\cdot\mathbf{v} = 0$ so $\mathbf{u}$ is perpendicular to $\mathbf{v}$ and hence the parallelogram is a rectangle.

39. $\|\mathbf{u}+\mathbf{v}\|^2 = (\mathbf{u}+\mathbf{v})\cdot(\mathbf{u}+\mathbf{v}) = \|\mathbf{u}\|^2 + 2\mathbf{u}\cdot\mathbf{v} + \|\mathbf{v}\|^2$ and
$\|\mathbf{u}-\mathbf{v}\|^2 = (\mathbf{u}-\mathbf{v})\cdot(\mathbf{u}-\mathbf{v}) = \|\mathbf{u}\|^2 - 2\mathbf{u}\cdot\mathbf{v} + \|\mathbf{v}\|^2$, add to get
$\|\mathbf{u}+\mathbf{v}\|^2 + \|\mathbf{u}-\mathbf{v}\|^2 = 2\|\mathbf{u}\|^2 + 2\|\mathbf{v}\|^2$

The sum of the squares of the lengths of the diagonals of a parallelogram is equal to twice the sum of the squares of the lengths of the sides.

40. $\|\mathbf{u}+\mathbf{v}\|^2 = (\mathbf{u}+\mathbf{v})\cdot(\mathbf{u}+\mathbf{v}) = \|\mathbf{u}\|^2 + 2\mathbf{u}\cdot\mathbf{v} + \|\mathbf{v}\|^2$ and
$\|\mathbf{u}-\mathbf{v}\|^2 = (\mathbf{u}-\mathbf{v})\cdot(\mathbf{u}-\mathbf{v}) = \|\mathbf{u}\|^2 - 2\mathbf{u}\cdot\mathbf{v} + \|\mathbf{v}\|^2$, subtract to get
$\|\mathbf{u}+\mathbf{v}\|^2 - \|\mathbf{u}-\mathbf{v}\|^2 = 4\mathbf{u}\cdot\mathbf{v}$, the result follows by dividing both sides by 4.

41. $\mathbf{v} = c_1\mathbf{v}_1 + c_2\mathbf{v}_2 + c_3\mathbf{v}_3$ so $\mathbf{v}\cdot\mathbf{v}_i = c_i\mathbf{v}_i\cdot\mathbf{v}_i$ because $\mathbf{v}_i\cdot\mathbf{v}_j = 0$ if $i \neq j$,
thus $\mathbf{v}\cdot\mathbf{v}_i = c_i\|\mathbf{v}_i\|^2, c_i = \mathbf{v}\cdot\mathbf{v}_i/\|\mathbf{v}_i\|^2$ for $i = 1, 2, 3$.

42. $\mathbf{v}_1\cdot\mathbf{v}_2 = \mathbf{v}_1\cdot\mathbf{v}_3 = \mathbf{v}_2\cdot\mathbf{v}_3 = 0$ so they are mutually perpendicular. Let $\mathbf{v} = \mathbf{i} - \mathbf{j} + \mathbf{k}$, then
$c_1 = \frac{\mathbf{v}\cdot\mathbf{v}_1}{\|\mathbf{v}_1\|^2} = \frac{3}{7}$, $c_2 = \frac{\mathbf{v}\cdot\mathbf{v}_2}{\|\mathbf{v}_2\|^2} = -\frac{1}{3}$, and $c_3 = \frac{\mathbf{v}\cdot\mathbf{v}_3}{\|\mathbf{v}_3\|^2} = \frac{1}{21}$.

43. (a) $\mathbf{u} = x\mathbf{i} + (x^2+1)\mathbf{j}, \mathbf{v} = x\mathbf{i} - (x+1)\mathbf{j}, \theta = \cos^{-1}[(\mathbf{u}\cdot\mathbf{v})/(\|\mathbf{u}\|\|\mathbf{v}\|)]$.

Use a CAS to solve $d\theta/dx = 0$ to find that the minimum value of θ occurs when $x \approx -3.136742$ so the minimum angle is about $40°$. NB: Since $\cos^{-1} u$ is a decreasing function of u, it suffices to maximize $(\mathbf{u}\cdot\mathbf{v})/(\|\mathbf{u}\|\|\mathbf{v}\|)$, or, what is easier, its square.

(b) Solve $\mathbf{u}\cdot\mathbf{v} = 0$ for x to get $x \approx -0.682328$.

44. (a) $\mathbf{u} = \cos\theta_1\mathbf{i} \pm \sin\theta_1\mathbf{j}, \mathbf{v} = \pm\sin\theta_2\mathbf{j} + \cos\theta_2\mathbf{k}, \cos\theta = \mathbf{u}\cdot\mathbf{v} = \pm\sin\theta_1\sin\theta_2$

(b) $\cos\theta = \pm\sin^2 45° = \pm 1/2,\ \theta = 60°$

(c) Let $\theta(t) = \cos^{-1}(\sin t \sin 2t)$; solve $\theta'(t) = 0$ for t to find that $\theta_{\max} \approx 140°$ (reject, since θ is acute) when $t \approx 2.186276$ and that $\theta_{\min} \approx 40°$ when $t \approx 0.955317$; for $\theta_{\max}$ check the endpoints $t = 0, \pi/2$ to obtain $\theta_{\max} = \cos^{-1}(0) = \pi/2$.

45. Let $\mathbf{u} = \langle u_1, u_2, u_3\rangle, \mathbf{v} = \langle v_1, v_2, v_3\rangle, \mathbf{w} = \langle w_1, w_2, w_3\rangle$. Then

$$\mathbf{u}\cdot(\mathbf{v}+\mathbf{w}) = \langle u_1(v_1+w_1), u_2(v_2+w_2), u_3(v_3+w_3)\rangle = \langle u_1v_1 + u_1w_1, u_2v_2 + u_2w_2, u_3v_3 + u_3w_3\rangle$$
$$= \langle u_1v_1, u_2v_2, u_3v_3\rangle + \langle u_1w_1, u_2w_2, u_3w_3\rangle = \mathbf{u}\cdot\mathbf{v} + \mathbf{u}\cdot\mathbf{w}$$

$\mathbf{0}\cdot\mathbf{v} = 0\cdot v_1 + 0\cdot v_2 + 0\cdot v_3 = 0$

EXERCISE SET 12.4

1. (a) $\mathbf{i}\times(\mathbf{i}+\mathbf{j}+\mathbf{k}) = \begin{vmatrix} \mathbf{i} & \mathbf{j} & \mathbf{k} \\ 1 & 0 & 0 \\ 1 & 1 & 1 \end{vmatrix} = -\mathbf{j}+\mathbf{k}$

(b) $\mathbf{i}\times(\mathbf{i}+\mathbf{j}+\mathbf{k}) = (\mathbf{i}\times\mathbf{i}) + (\mathbf{i}\times\mathbf{j}) + (\mathbf{i}\times\mathbf{k}) = -\mathbf{j}+\mathbf{k}$

2. (a) $\mathbf{j}\times(\mathbf{i}+\mathbf{j}+\mathbf{k}) = \begin{vmatrix} \mathbf{i} & \mathbf{j} & \mathbf{k} \\ 0 & 1 & 0 \\ 1 & 1 & 1 \end{vmatrix} = \mathbf{i}-\mathbf{k}$

$\mathbf{j}\times(\mathbf{i}+\mathbf{j}+\mathbf{k}) = (\mathbf{j}\times\mathbf{i}) + (\mathbf{j}\times\mathbf{j}) + (\mathbf{j}\times\mathbf{k}) = \mathbf{i}-\mathbf{k}$

(b) $\mathbf{k}\times(\mathbf{i}+\mathbf{j}+\mathbf{k}) = \begin{vmatrix} \mathbf{i} & \mathbf{j} & \mathbf{k} \\ 0 & 0 & 1 \\ 1 & 1 & 1 \end{vmatrix} = -\mathbf{i}+\mathbf{j}$

$\mathbf{k}\times(\mathbf{i}+\mathbf{j}+\mathbf{k}) = (\mathbf{k}\times\mathbf{i}) + (\mathbf{k}\times\mathbf{j}) + (\mathbf{k}\times\mathbf{k}) = \mathbf{j}-\mathbf{i}+\mathbf{0} = -\mathbf{i}+\mathbf{j}$

3. $\langle 7, 10, 9\rangle$ **4.** $-\mathbf{i}-2\mathbf{j}-7\mathbf{k}$ **5.** $\langle -4, -6, -3\rangle$ **6.** $\mathbf{i}+2\mathbf{j}-4\mathbf{k}$

7. (a) $\mathbf{v}\times\mathbf{w} = \langle -23, 7, -1\rangle, \mathbf{u}\times(\mathbf{v}\times\mathbf{w}) = \langle -20, -67, -9\rangle$

(b) $\mathbf{u}\times\mathbf{v} = \langle -10, -14, 2\rangle, (\mathbf{u}\times\mathbf{v})\times\mathbf{w} = \langle -78, 52, -26\rangle$

(c) $(\mathbf{u}\times\mathbf{v})\times(\mathbf{v}\times\mathbf{w}) = \langle -10, -14, 2\rangle\times\langle -23, 7, -1\rangle = \langle 0, -56, -392\rangle$

(d) $(\mathbf{v}\times\mathbf{w})\times(\mathbf{u}\times\mathbf{v}) = \langle 0, 56, 392\rangle$

9. $\mathbf{u}\times\mathbf{v} = (\mathbf{i}+\mathbf{j})\times(\mathbf{i}+\mathbf{j}+\mathbf{k}) = \mathbf{k}-\mathbf{j}-\mathbf{k}+\mathbf{i} = \mathbf{i}-\mathbf{j}$, the direction cosines are $\dfrac{1}{\sqrt{2}}, -\dfrac{1}{\sqrt{2}}, 0$

10. $\mathbf{u}\times\mathbf{v} = 12\mathbf{i}+30\mathbf{j}-6\mathbf{k}$, so $\pm\left(\dfrac{2}{\sqrt{30}}\mathbf{i} + \dfrac{\sqrt{5}}{\sqrt{6}}\mathbf{j} - \dfrac{1}{\sqrt{30}}\mathbf{k}\right)$

11. $\mathbf{n} = \overrightarrow{AB} \times \overrightarrow{AC} = \langle 1,1,-3\rangle \times \langle -1,3,-1\rangle = \langle 8,4,4\rangle$, unit vectors are $\pm\frac{1}{\sqrt{6}}\langle 2,1,1\rangle$

12. A vector parallel to the yz-plane must be perpendicular to $\mathbf{i}$;
$\mathbf{i} \times (\mathbf{3i} - \mathbf{j} + \mathbf{2k}) = -2\mathbf{j} - \mathbf{k}$, $\|-2\mathbf{j} - \mathbf{k}\| = \sqrt{5}$, the unit vectors are $\pm(2\mathbf{j} + \mathbf{k})/\sqrt{5}$.

13. $A = \|\mathbf{u} \times \mathbf{v}\| = \|-7\mathbf{i} - \mathbf{j} + 3\mathbf{k}\| = \sqrt{59}$

14. $A = \|\mathbf{u} \times \mathbf{v}\| = \|-6\mathbf{i} + 4\mathbf{j} + 7\mathbf{k}\| = \sqrt{101}$

15. $A = \frac{1}{2}\|\overrightarrow{PQ} \times \overrightarrow{PR}\| = \frac{1}{2}\|\langle -1,-5,2\rangle \times \langle 2,0,3\rangle\| = \frac{1}{2}\|\langle -15,7,10\rangle\| = \sqrt{374}/2$

16. $A = \frac{1}{2}\|\overrightarrow{PQ} \times \overrightarrow{PR}\| = \frac{1}{2}\|\langle -1,4,8\rangle \times \langle 5,2,12\rangle\| = \frac{1}{2}\|\langle 32,52,-22\rangle\| = 9\sqrt{13}$

17. 80 **18.** 29 **19.** -3 **20.** 1

21. $V = |\mathbf{u} \cdot (\mathbf{v} \times \mathbf{w})| = |-16| = 16$ **22.** $V = |\mathbf{u} \cdot (\mathbf{v} \times \mathbf{w})| = |45| = 45$

23. (a) $\mathbf{u} \cdot (\mathbf{v} \times \mathbf{w}) = 0$, yes (b) $\mathbf{u} \cdot (\mathbf{v} \times \mathbf{w}) = 0$, yes (c) $\mathbf{u} \cdot (\mathbf{v} \times \mathbf{w}) = 245$, no

24. (a) $\mathbf{u} \cdot (\mathbf{w} \times \mathbf{v}) = -\mathbf{u} \cdot (\mathbf{v} \times \mathbf{w}) = -3$ (b) $(\mathbf{v} \times \mathbf{w}) \cdot \mathbf{u} = \mathbf{u} \cdot (\mathbf{v} \times \mathbf{w}) = 3$
(c) $\mathbf{w} \cdot (\mathbf{u} \times \mathbf{v}) = \mathbf{u} \cdot (\mathbf{v} \times \mathbf{w}) = 3$ (d) $\mathbf{v} \cdot (\mathbf{u} \times \mathbf{w}) = \mathbf{u} \cdot (\mathbf{w} \times \mathbf{v}) = -3$
(e) $(\mathbf{u} \times \mathbf{w}) \cdot \mathbf{v} = \mathbf{u} \cdot (\mathbf{w} \times \mathbf{v}) = -3$ (f) $\mathbf{v} \cdot (\mathbf{w} \times \mathbf{w}) = 0$ because $\mathbf{w} \times \mathbf{w} = \mathbf{0}$

25. (a) $V = |\mathbf{u} \cdot (\mathbf{v} \times \mathbf{w})| = |-9| = 9$ (b) $A = \|\mathbf{u} \times \mathbf{w}\| = \|3\mathbf{i} - 8\mathbf{j} + 7\mathbf{k}\| = \sqrt{122}$
(c) $\mathbf{v} \times \mathbf{w} = -3\mathbf{i} - \mathbf{j} + 2\mathbf{k}$ is perpendicular to the plane determined by $\mathbf{v}$ and $\mathbf{w}$; let θ be the angle between $\mathbf{u}$ and $\mathbf{v} \times \mathbf{w}$ then

$$\cos\theta = \frac{\mathbf{u} \cdot (\mathbf{v} \times \mathbf{w})}{\|\mathbf{u}\|\,\|\mathbf{v} \times \mathbf{w}\|} = \frac{-9}{\sqrt{14}\sqrt{14}} = -9/14$$

so the acute angle ϕ that $\mathbf{u}$ makes with the plane determined by $\mathbf{v}$ and $\mathbf{w}$ is $\phi = \theta - \pi/2 = \sin^{-1}(9/14)$.

26. From the diagram,

$$d = \|\mathbf{u}\|\sin\theta = \frac{\|\mathbf{u}\|\|\mathbf{v}\|\sin\theta}{\|\mathbf{v}\|} = \frac{\|\mathbf{u} \times \mathbf{v}\|}{\|\mathbf{v}\|}$$

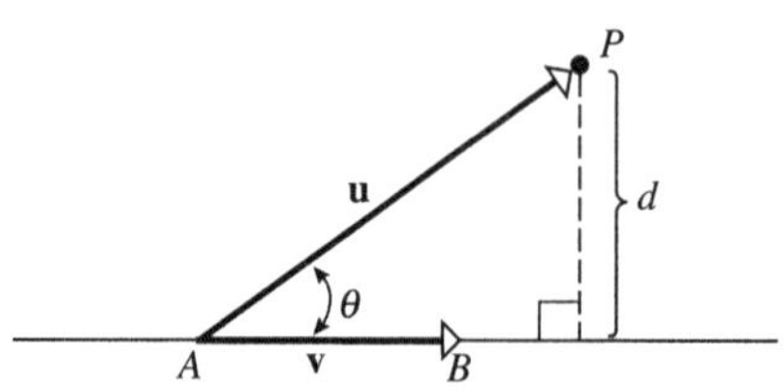

27. (a) $\mathbf{u} = \overrightarrow{AP} = -4\mathbf{i} + 2\mathbf{k}$, $\mathbf{v} = \overrightarrow{AB} = -3\mathbf{i} + 2\mathbf{j} - 4\mathbf{k}$, $\mathbf{u} \times \mathbf{v} = -4\mathbf{i} - 22\mathbf{j} - 8\mathbf{k}$;
distance $= \|\mathbf{u} \times \mathbf{v}\|/\|\mathbf{v}\| = 2\sqrt{141/29}$
(b) $\mathbf{u} = \overrightarrow{AP} = 2\mathbf{i} + 2\mathbf{j}$, $\mathbf{v} = \overrightarrow{AB} = -2\mathbf{i} + \mathbf{j}$, $\mathbf{u} \times \mathbf{v} = 6\mathbf{k}$; distance $= \|\mathbf{u} \times \mathbf{v}\|/\|\mathbf{v}\| = 6/\sqrt{5}$

28. Take $\mathbf{v}$ and $\mathbf{w}$ as sides of the (triangular) base, then area of base $= \frac{1}{2}\|\mathbf{v} \times \mathbf{w}\|$ and height $= \|\text{proj}_{\mathbf{v} \times \mathbf{w}}\mathbf{u}\| = \frac{|\mathbf{u} \cdot (\mathbf{v} \times \mathbf{w})|}{\|\mathbf{v} \times \mathbf{w}\|}$ so $V = \frac{1}{3}$ (area of base) (height) $= \frac{1}{6}|\mathbf{u} \cdot (\mathbf{v} \times \mathbf{w})|$

29. $\overrightarrow{PQ} = \langle 3,-1,-3\rangle, \overrightarrow{PR} = \langle 2,-2,1\rangle, \overrightarrow{PS} = \langle 4,-4,3\rangle$,

$$V = \frac{1}{6}|\overrightarrow{PQ} \cdot (\overrightarrow{PR} \times \overrightarrow{PS})| = \frac{1}{6}|-4| = 2/3$$

30. **(a)** $\cos\theta = \dfrac{\mathbf{u}\cdot\mathbf{v}}{\|\mathbf{u}\|\|\mathbf{v}\|} = -\dfrac{23}{49}$ **(b)** $\sin\theta = \dfrac{\|\mathbf{u}\times\mathbf{v}\|}{\|\mathbf{u}\|\|\mathbf{v}\|} = \dfrac{\|36\mathbf{i} - 24\mathbf{j}\|}{49} = \dfrac{12\sqrt{13}}{49}$

(c) $\dfrac{23^2}{49^2} + \dfrac{144\cdot 13}{49^2} = \dfrac{2401}{49^2} = 1$

31. From Theorems 12.3.3 and 12.4.5a it follows that $\sin\theta = \cos\theta$, so $\theta = \pi/4$.

32. $\|\mathbf{u}\times\mathbf{v}\|^2 = \|\mathbf{u}\|^2\|\mathbf{v}\|^2\sin^2\theta = \|\mathbf{u}\|^2\|\mathbf{v}\|^2(1-\cos^2\theta) = \|\mathbf{u}\|^2\|\mathbf{v}\|^2 - (\mathbf{u}\cdot\mathbf{v})^2$

33. **(a)** $\mathbf{F} = 10\mathbf{j}$ and $\overrightarrow{PQ} = \mathbf{i}+\mathbf{j}+\mathbf{k}$, so the vector moment of $\mathbf{F}$ about P is

$\overrightarrow{PQ}\times\mathbf{F} = \begin{vmatrix} \mathbf{i} & \mathbf{j} & \mathbf{k} \\ 1 & 1 & 1 \\ 0 & 10 & 0 \end{vmatrix} = -10\mathbf{i} + 10\mathbf{k}$, and the scalar moment is $10\sqrt{2}$ lb·ft.

The direction of rotation of the cube about P is counterclockwise looking along $\overrightarrow{PQ}\times\mathbf{F} = -10\mathbf{i} + 10\mathbf{k}$ toward its initial point.

(b) $\mathbf{F} = 10\mathbf{j}$ and $\overrightarrow{PQ} = \mathbf{j}+\mathbf{k}$, so the vector moment of $\mathbf{F}$ about P is

$\overrightarrow{PQ}\times\mathbf{F} = \begin{vmatrix} \mathbf{i} & \mathbf{j} & \mathbf{k} \\ 0 & 1 & 1 \\ 0 & 10 & 0 \end{vmatrix} = -10\mathbf{i}$, and the scalar moment is 10 lb·ft. The direction of rotation of the cube about P is counterclockwise looking along $-10\mathbf{i}$ toward its initial point.

(c) $\mathbf{F} = 10\mathbf{j}$ and $\overrightarrow{PQ} = \mathbf{j}$, so the vector moment of $\mathbf{F}$ about P is

$\overrightarrow{PQ}\times\mathbf{F} = \begin{vmatrix} \mathbf{i} & \mathbf{j} & \mathbf{k} \\ 0 & 1 & 0 \\ 0 & 10 & 0 \end{vmatrix} = \mathbf{0}$, and the scalar moment is 0 lb·ft. Since the force is parallel to the direction of motion, there is no rotation about P.

34. **(a)** $\mathbf{F} = \dfrac{1000}{\sqrt{2}}(-\mathbf{i}+\mathbf{k})$ and $\overrightarrow{PQ} = 2\mathbf{j} - \mathbf{k}$, so the vector moment of $\mathbf{F}$ about P is

$\overrightarrow{PQ}\times\mathbf{F} = 500\sqrt{2}\begin{vmatrix} \mathbf{i} & \mathbf{j} & \mathbf{k} \\ 0 & 2 & -1 \\ -1 & 0 & 1 \end{vmatrix} = 500\sqrt{2}(2\mathbf{i}+\mathbf{j}+2\mathbf{k})$, and the scalar moment is $1500\sqrt{2}$ N·m.

(b) The direction angles of the vector moment of $\mathbf{F}$ about the point P are $\cos^{-1}(2/3) \approx 48^\circ, \cos^{-1}(1/3) \approx 71^\circ$, and $\cos^{-1}(2/3) \approx 48^\circ$.

35. Take the center of the bolt as the origin of the plane. Then $\mathbf{F}$ makes an angle 72° with the positive x-axis, so $\mathbf{F} = 200\cos 72^\circ\mathbf{i} + 200\sin 72^\circ\mathbf{j}$ and $\overrightarrow{PQ} = 0.2\,\mathbf{i} + 0.03\,\mathbf{j}$. The scalar moment is given by

$$\left\| \begin{vmatrix} \mathbf{i} & \mathbf{j} & \mathbf{k} \\ 0.2 & 0.03 & 0 \\ 200\cos 72^\circ & 200\sin 72^\circ & 0 \end{vmatrix} \right\| = \left| 40\frac{1}{4}(\sqrt{5}-1) - 6\frac{1}{4}\sqrt{10+2\sqrt{5}} \right| \approx 36.1882 \text{ N·m}.$$

36. Part (b) : let $\mathbf{u} = \langle u_1, u_2, u_3 \rangle$, $\mathbf{v} = \langle v_1, v_2, v_3 \rangle$, and $\mathbf{w} = \langle w_1, w_2, w_3 \rangle$; show that $\mathbf{u} \times (\mathbf{v} + \mathbf{w})$ and $(\mathbf{u} \times \mathbf{v}) + (\mathbf{u} \times \mathbf{w})$ are the same.

Part (c) :
$$\begin{aligned}(\mathbf{u}+\mathbf{v}) \times \mathbf{w} &= -[\mathbf{w} \times (\mathbf{u}+\mathbf{v})] \text{ from Part (a)}\\ &= -[(\mathbf{w} \times \mathbf{u}) + (\mathbf{w} \times \mathbf{v})] \text{ from Part (b)}\\ &= (\mathbf{u} \times \mathbf{w}) + (\mathbf{v} \times \mathbf{w}) \text{ from Part (a)}\end{aligned}$$

37. Let $\mathbf{u} = \langle u_1, u_2, u_3 \rangle$ and $\mathbf{v} = \langle v_1, v_2, v_3 \rangle$; show that $k(\mathbf{u} \times \mathbf{v})$, $(k\mathbf{u}) \times \mathbf{v}$, and $\mathbf{u} \times (k\mathbf{v})$ are all the same; Part (e) is proved in a similar fashion.

38. Suppose the first two rows are interchanged. Then by definition,

$$\begin{vmatrix} b_1 & b_2 & b_3 \\ a_1 & a_2 & a_3 \\ c_1 & c_2 & c_3 \end{vmatrix} = b_1 \begin{vmatrix} a_2 & a_3 \\ c_2 & c_3 \end{vmatrix} - b_2 \begin{vmatrix} a_1 & a_3 \\ c_1 & c_3 \end{vmatrix} + b_3 \begin{vmatrix} a_1 & a_2 \\ c_1 & c_2 \end{vmatrix}$$
$$= b_1(a_2c_3 - a_3c_2) - b_2(a_1c_3 - a_3c_1) + b_3(a_1c_2 - a_2c_1),$$

which is the negative of the right hand side of (2) after expansion. If two other rows were to be exchanged, a similar proof would hold. Finally, suppose Δ were a determinant with two identical rows. Then the value is unchanged if we interchange those two rows, yet $\Delta = -\Delta$ by Part (b) of Theorem 12.4.1. Hence $\Delta = -\Delta, \Delta = 0$.

39. $-8\mathbf{i} - 8\mathbf{k}, -8\mathbf{i} - 20\mathbf{j} + 2\mathbf{k}$

40. **(a)** From the first formula in Exercise 39 it follows that $\mathbf{u} \times (\mathbf{v} \times \mathbf{w})$ is a linear combination of $\mathbf{v}$ and $\mathbf{w}$ and hence lies in the plane determined by them, and from the second formula it follows that $(\mathbf{u} \times \mathbf{v}) \times \mathbf{w}$ is a linear combination of $\mathbf{u}$ and $\mathbf{v}$ and hence lies in their plane.

(b) $\mathbf{u} \times (\mathbf{v} \times \mathbf{w})$ is orthogonal to $\mathbf{v} \times \mathbf{w}$ and hence lies in the plane of $\mathbf{v}$ and $\mathbf{w}$; similarly for $(\mathbf{u} \times \mathbf{v}) \times \mathbf{w}$.

41. If $\mathbf{a}$, $\mathbf{b}$, $\mathbf{c}$, and $\mathbf{d}$ lie in the same plane then $\mathbf{a} \times \mathbf{b}$ and $\mathbf{c} \times \mathbf{d}$ are parallel so $(\mathbf{a} \times \mathbf{b}) \times (\mathbf{c} \times \mathbf{d}) = \mathbf{0}$

42. Let $\mathbf{u}$ and $\mathbf{v}$ be the vectors from a point on the curve to the points $(2, -1, 0)$ and $(3, 2, 2)$, respectively. Then $\mathbf{u} = (2-x)\mathbf{i} + (-1 - \ln x)\mathbf{j}$ and $\mathbf{v} = (3-x)\mathbf{i} + (2 - \ln x)\mathbf{j} + 2\mathbf{k}$. The area of the triangle is given by $A = (1/2)\|\mathbf{u} \times \mathbf{v}\|$; solve $dA/dx = 0$ for x to get $x = 2.091581$. The minimum area is 1.887850.

43. $\overrightarrow{PQ'} \times \mathbf{F} = \overrightarrow{PQ} \times \mathbf{F} + \overrightarrow{QQ'} \times \mathbf{F} = \overrightarrow{PQ} \times \mathbf{F}$, since $\mathbf{F}$ and $\overrightarrow{QQ'}$ are parallel.

EXERCISE SET 12.5

In many of the Exercises in this section other answers are also possible.

1. **(a)** L_1: $P(1,0), \mathbf{v} = \mathbf{j}, x = 1, y = t$
L_2: $P(0,1), \mathbf{v} = \mathbf{i}, x = t, y = 1$
L_3: $P(0,0), \mathbf{v} = \mathbf{i} + \mathbf{j}, x = t, y = t$

(b) L_1: $P(1,1,0), \mathbf{v} = \mathbf{k}, x = 1, y = 1, z = t$
L_2: $P(0,1,1), \mathbf{v} = \mathbf{i}, x = t, y = 1, z = 1$
L_3: $P(1,0,1), \mathbf{v} = \mathbf{j}, x = 1, y = t, z = 1$
L_4: $P(0,0,0), \mathbf{v} = \mathbf{i} + \mathbf{j} + \mathbf{k}, x = t, y = t, z = t$

2. **(a)** L_1: $x = t, y = 1, 0 \le t \le 1$
L_2: $x = 1, y = t, 0 \le t \le 1$
L_3: $x = t, y = t, 0 \le t \le 1$

(b) L_1: $x = 1, y = 1, z = t, 0 \le t \le 1$
L_2: $x = t, y = 1, z = 1, 0 \le t \le 1$
L_3: $x = 1, y = t, z = 1, 0 \le t \le 1$
L_4: $x = t, y = t, z = t, 0 \le t \le 1$

3. (a) $\overrightarrow{P_1P_2} = \langle 2,3\rangle$ so $x = 3+2t$, $y = -2+3t$ for the line; for the line segment add the condition $0 \le t \le 1$.

(b) $\overrightarrow{P_1P_2} = \langle -3,6,1\rangle$ so $x = 5-3t$, $y = -2+6t$, $z = 1+t$ for the line; for the line segment add the condition $0 \le t \le 1$.

4. (a) $\overrightarrow{P_1P_2} = \langle -3,-5\rangle$ so $x = -3t$, $y = 1-5t$ for the line; for the line segment add the condition $0 \le t \le 1$.

(b) $\overrightarrow{P_1P_2} = \langle 0,0,-3\rangle$ so $x = -1$, $y = 3, z = 5-3t$ for the line; for the line segment add the condition $0 \le t \le 1$.

5. (a) $x = 2+t, y = -3-4t$ (b) $x = t, y = -t, z = 1+t$

6. (a) $x = 3+2t, y = -4+t$ (b) $x = -1-t, y = 3t, z = 2$

7. (a) $\mathbf{r}_0 = 2\mathbf{i} - \mathbf{j}$ so $P(2,-1)$ is on the line, and $\mathbf{v} = 4\mathbf{i} - \mathbf{j}$ is parallel to the line.
(b) At $t = 0, P(-1,2,4)$ is on the line, and $\mathbf{v} = 5\mathbf{i} + 7\mathbf{j} - 8\mathbf{k}$ is parallel to the line.

8. (a) At $t = 0, P(-1,5)$ is on the line, and $\mathbf{v} = 2\mathbf{i} + 3\mathbf{j}$ is parallel to the line.
(b) $\mathbf{r}_0 = \mathbf{i} + \mathbf{j} - 2\mathbf{k}$ so $P(1,1,-2)$ is on the line, and $\mathbf{v} = \mathbf{j}$ is parallel to the line.

9. (a) $\langle x,y\rangle = \langle -3,4\rangle + t\langle 1,5\rangle; \mathbf{r} = -3\mathbf{i} + 4\mathbf{j} + t(\mathbf{i} + 5\mathbf{j})$
(b) $\langle x,y,z\rangle = \langle 2,-3,0\rangle + t\langle -1,5,1\rangle; \mathbf{r} = 2\mathbf{i} - 3\mathbf{j} + t(-\mathbf{i} + 5\mathbf{j} + \mathbf{k})$

10. (a) $\langle x,y\rangle = \langle 0,-2\rangle + t\langle 1,1\rangle; \mathbf{r} = -2\mathbf{j} + t(\mathbf{i} + \mathbf{j})$
(b) $\langle x,y,z\rangle = \langle 1,-7,4\rangle + t\langle 1,3,-5\rangle; \mathbf{r} = \mathbf{i} - 7\mathbf{j} + 4\mathbf{k} + t(\mathbf{i} + 3\mathbf{j} - 5\mathbf{k})$

11. $x = -5+2t$, $y = 2-3t$

12. $x = t$, $y = 3-2t$

13. $2x + 2yy' = 0$, $y' = -x/y = -(3)/(-4) = 3/4$, $\mathbf{v} = 4\mathbf{i} + 3\mathbf{j}$; $x = 3+4t$, $y = -4+3t$

14. $y' = 2x = 2(-2) = -4$, $\mathbf{v} = \mathbf{i} - 4\mathbf{j}$; $x = -2+t$, $y = 4-4t$

15. $x = -1+3t$, $y = 2-4t$, $z = 4+t$

16. $x = 2-t$, $y = -1+2t$, $z = 5+7t$

17. The line is parallel to the vector $\langle 2,-1,2\rangle$ so $x = -2+2t$, $y = -t$, $z = 5+2t$.

18. The line is parallel to the vector $\langle 1,1,0\rangle$ so $x = t$, $y = t$, $z = 0$.

19. (a) $y = 0, 2-t = 0, t = 2, x = 7$ (b) $x = 0, 1+3t = 0, t = -1/3, y = 7/3$

(c) $y = x^2, 2-t = (1+3t)^2, 9t^2 + 7t - 1 = 0, t = \dfrac{-7 \pm \sqrt{85}}{18}, x = \dfrac{-1 \pm \sqrt{85}}{6}, y = \dfrac{43 \mp \sqrt{85}}{18}$

20. $(4t)^2 + (3t)^2 = 25, 25t^2 = 25, t = \pm 1$, the line intersects the circle at $\pm\langle 4,3\rangle$

21. (a) $z = 0$ when $t = 3$ so the point is $(-2,10,0)$
(b) $y = 0$ when $t = -2$ so the point is $(-2,0,-5)$
(c) x is always -2 so the line does not intersect the yz-plane

22. (a) $z = 0$ when $t = 4$ so the point is (7,7,0)
(b) $y = 0$ when $t = -3$ so the point is $(-7,0,7)$
(c) $x = 0$ when $t = 1/2$ so the point is $(0,7/2,7/2)$

23. $(1+t)^2+(3-t)^2=16$, $t^2-2t-3=0$, $(t+1)(t-3)=0$; $t=-1, 3$. The points of intersection are $(0,4,-2)$ and (4,0,6).

24. $2(3t)+3(-1+2t)=6$, $12t=9$; $t=3/4$. The point of intersection is $(5/4, 9/4, 1/2)$.

25. The lines intersect if we can find values of t_1 and t_2 that satisfy the equations $2+t_1=2+t_2$, $2+3t_1=3+4t_2$, and $3+t_1=4+2t_2$. Solutions of the first two of these equations are $t_1=-1$, $t_2=-1$ which also satisfy the third equation so the lines intersect at $(1,-1,2)$.

26. Solve the equations $-1+4t_1=-13+12t_2$, $3+t_1=1+6t_2$, and $1=2+3t_2$. The third equation yields $t_2=-1/3$ which when substituted into the first and second equations gives $t_1=-4$ in both cases; the lines intersect at $(-17,-1,1)$.

27. The lines are parallel, respectively, to the vectors $\langle 7,1,-3\rangle$ and $\langle -1,0,2\rangle$. These vectors are not parallel so the lines are not parallel. The system of equations $1+7t_1=4-t_2$, $3+t_1=6$, and $5-3t_1=7+2t_2$ has no solution so the lines do not intersect.

28. The vectors $\langle 8,-8,10\rangle$ and $\langle 8,-3,1\rangle$ are not parallel so the lines are not parallel. The lines do not intersect because the system of equations $2+8t_1=3+8t_2$, $6-8t_1=5-3t_2$, $10t_1=6+t_2$ has no solution.

29. The lines are parallel, respectively, to the vectors $\mathbf{v}_1=\langle -2,1,-1\rangle$ and $\mathbf{v}_2=\langle -4,2,-2\rangle$; $\mathbf{v}_2=2\mathbf{v}_1$, $\mathbf{v}_1$ and $\mathbf{v}_2$ are parallel so the lines are parallel.

30. The lines are not parallel because the vectors $\langle 3,-2,3\rangle$ and $\langle 9,-6,8\rangle$ are not parallel.

31. $\overrightarrow{P_1P_2}=\langle 3,-7,-7\rangle$, $\overrightarrow{P_2P_3}=\langle -9,-7,-3\rangle$; these vectors are not parallel so the points do not lie on the same line.

32. $\overrightarrow{P_1P_2}=\langle 2,-4,-4\rangle$, $\overrightarrow{P_2P_3}=\langle 1,-2,-2\rangle$; $\overrightarrow{P_1P_2}=2\,\overrightarrow{P_2P_3}$ so the vectors are parallel and the points lie on the same line.

33. If t_2 gives the point $\langle -1+3t_2, 9-6t_2\rangle$ on the second line, then $t_1=4-3t_2$ yields the point $\langle 3-(4-3t_2), 1+2(4-3t_2)\rangle=\langle -1+3t_2, 9-6t_2\rangle$ on the first line, so each point of L_2 is a point of L_1; the converse is shown with $t_2=(4-t_1)/3$.

34. If t_1 gives the point $\langle 1+3t_1, -2+t_1, 2t_1\rangle$ on L_1, then $t_2=(1-t_1)/2$ gives the point $\langle 4-6(1-t_1)/2, -1-2(1-t_1)/2, 2-4(1-t_1)/2\rangle=\langle 1+3t_1, -2+t_1, 2t_1\rangle$ on L_2, so each point of L_1 is a point of L_2; the converse is shown with $t_1=1-2t_2$.

35. The line segment joining the points (1,0) and $(-3,6)$.

36. The line segment joining the points $(-2,1,4)$ and (7,1,1).

37. $A(3,0,1)$ and $B(2,1,3)$ are on the line, and (method of Exercise 25) $\overrightarrow{AP}=-5\mathbf{i}+\mathbf{j}$, $\overrightarrow{AB}=-\mathbf{i}+\mathbf{j}+2\mathbf{k}$, $\|\text{proj}_{\overrightarrow{AB}}\overrightarrow{AP}\|=|\overrightarrow{AP}\cdot\overrightarrow{AB}|/\|\overrightarrow{AB}\|=\sqrt{6}$ and $\|\overrightarrow{AP}\|=\sqrt{26}$, so distance $=\sqrt{26-6}=2\sqrt{5}$. Using the method of Exercise 26, distance $=\dfrac{\|\overrightarrow{AP}\times\overrightarrow{AB}\|}{\|\overrightarrow{AB}\|}=2\sqrt{5}$.

38. $A(2,-1,0)$ and $B(3,-2,3)$ are on the line, and (method of Exercise 25) $\overrightarrow{AP} = -\mathbf{i}+5\mathbf{j}-3\mathbf{k}$, $\overrightarrow{AB} = \mathbf{i}-\mathbf{j}+3\mathbf{k}$, $\|\text{proj}_{\overrightarrow{AB}}\overrightarrow{AP}\| = |\overrightarrow{AP}\cdot\overrightarrow{AB}|/\|\overrightarrow{AB}\| = \dfrac{15}{\sqrt{11}}$ and $\|\overrightarrow{AP}\| = \sqrt{35}$, so distance $= \sqrt{35-225/11} = 4\sqrt{10/11}$. Using the method of Exercise 26, distance $= \dfrac{\|\overrightarrow{AP}\times\overrightarrow{AB}\|}{\|\overrightarrow{AB}\|} = 4\sqrt{10/11}$.

39. The vectors $\mathbf{v}_1 = -\mathbf{i}+2\mathbf{j}+\mathbf{k}$ and $\mathbf{v}_2 = 2\mathbf{i}-4\mathbf{j}-2\mathbf{k}$ are parallel to the lines, $\mathbf{v}_2 = -2\mathbf{v}_1$ so $\mathbf{v}_1$ and $\mathbf{v}_2$ are parallel. Let $t=0$ to get the points $P(2,0,1)$ and $Q(1,3,5)$ on the first and second lines, respectively. Let $\mathbf{u} = \overrightarrow{PQ} = -\mathbf{i}+3\mathbf{j}+4\mathbf{k}$, $\mathbf{v} = \frac{1}{2}\mathbf{v}_2 = \mathbf{i}-2\mathbf{j}-\mathbf{k}$; $\mathbf{u}\times\mathbf{v} = 5\mathbf{i}+3\mathbf{j}-\mathbf{k}$; by the method of Exercise 26 of Section 12.4, distance $= \|\mathbf{u}\times\mathbf{v}\|/\|\mathbf{v}\| = \sqrt{35/6}$.

40. The vectors $\mathbf{v}_1 = 2\mathbf{i}+4\mathbf{j}-6\mathbf{k}$ and $\mathbf{v}_2 = 3\mathbf{i}+6\mathbf{j}-9\mathbf{k}$ are parallel to the lines, $\mathbf{v}_2 = (3/2)\mathbf{v}_1$ so $\mathbf{v}_1$ and $\mathbf{v}_2$ are parallel. Let $t=0$ to get the points $P(0,3,2)$ and $Q(1,0,0)$ on the first and second lines, respectively. Let $\mathbf{u} = \overrightarrow{PQ} = \mathbf{i}-3\mathbf{j}-2\mathbf{k}$, $\mathbf{v} = \frac{1}{2}\mathbf{v}_1 = \mathbf{i}+2\mathbf{j}-3\mathbf{k}$; $\mathbf{u}\times\mathbf{v} = 13\mathbf{i}+\mathbf{j}+5\mathbf{k}$, distance $= \|\mathbf{u}\times\mathbf{v}\|/\|\mathbf{v}\| = \sqrt{195/14}$ (Exer. 26, Section 12.4).

41. **(a)** The line is parallel to the vector $\langle x_1-x_0, y_1-y_0, z_1-z_0\rangle$ so $x = x_0+(x_1-x_0)\,t$, $y = y_0+(y_1-y_0)\,t$, $z = z_0+(z_1-z_0)\,t$

(b) The line is parallel to the vector $\langle a,b,c\rangle$ so $x = x_1+at$, $y = y_1+bt$, $z = z_1+ct$

42. Solve each of the given parametric equations (2) for t to get $t = (x-x_0)/a$, $t = (y-y_0)/b$, $t = (z-z_0)/c$, so (x,y,z) is on the line if and only if $(x-x_0)/a = (y-y_0)/b = (z-z_0)/c$.

43. **(a)** It passes through the point $(1,-3,5)$ and is parallel to $\mathbf{v} = 2\mathbf{i}+4\mathbf{j}+\mathbf{k}$

(b) $\langle x,y,z\rangle = \langle 1+2t, -3+4t, 5+t\rangle$

44. Let the desired point be $P(x_0,y_0,z_0)$, then $\overrightarrow{P_1P} = (2/3)\,\overrightarrow{P_1P_2}$, $\langle x_0-1, y_0-4, z_0+3\rangle = (2/3)\langle 0,1,2\rangle = \langle 0, 2/3, 4/3\rangle$; equate corresponding components to get $x_0 = 1$, $y_0 = 14/3$, $z_0 = -5/3$.

45. **(a)** Let $t=3$ and $t=-2$, respectively, in the equations for L_1 and L_2.

(b) $\mathbf{u} = 2\mathbf{i}-\mathbf{j}-2\mathbf{k}$ and $\mathbf{v} = \mathbf{i}+3\mathbf{j}-\mathbf{k}$ are parallel to L_1 and L_2, $\cos\theta = \mathbf{u}\cdot\mathbf{v}/(\|\mathbf{u}\|\,\|\mathbf{v}\|) = 1/(3\sqrt{11}), \theta \approx 84°$.

(c) $\mathbf{u}\times\mathbf{v} = 7\mathbf{i}+7\mathbf{k}$ is perpendicular to both L_1 and L_2, and hence so is $\mathbf{i}+\mathbf{k}$, thus $x = 7+t$, $y = -1$, $z = -2+t$.

46. **(a)** Let $t = 1/2$ and $t = 1$, respectively, in the equations for L_1 and L_2.

(b) $\mathbf{u} = 4\mathbf{i}-2\mathbf{j}+2\mathbf{k}$ and $\mathbf{v} = \mathbf{i}-\mathbf{j}+4\mathbf{k}$ are parallel to L_1 and L_2, $\cos\theta = \mathbf{u}\cdot\mathbf{v}/(\|\mathbf{u}\|\,\|\mathbf{v}\|) = 14/\sqrt{432}, \theta \approx 48°$.

(c) $\mathbf{u}\times\mathbf{v} = -6\mathbf{i}-14\mathbf{j}-2\mathbf{k}$ is perpendicular to both L_1 and L_2, and hence so is $3\mathbf{i}+7\mathbf{j}+\mathbf{k}$, thus $x = 2+3t$, $y = 7t$, $z = 3+t$.

47. (0,1,2) is on the given line ($t=0$) so $\mathbf{u} = \mathbf{j}-\mathbf{k}$ is a vector from this point to the point (0,2,1), $\mathbf{v} = 2\mathbf{i}-\mathbf{j}+\mathbf{k}$ is parallel to the given line. $\mathbf{u}\times\mathbf{v} = -2\mathbf{j}-2\mathbf{k}$, and hence $\mathbf{w} = \mathbf{j}+\mathbf{k}$, is perpendicular to both lines so $\mathbf{v}\times\mathbf{w} = -2\mathbf{i}-2\mathbf{j}+2\mathbf{k}$, and hence $\mathbf{i}+\mathbf{j}-\mathbf{k}$, is parallel to the line we seek. Thus $x = t$, $y = 2+t$, $z = 1-t$ are parametric equations of the line.

48. $(-2,4,2)$ is on the given line ($t=0$) so $\mathbf{u}=5\mathbf{i}-3\mathbf{j}-4\mathbf{k}$ is a vector from this point to the point $(3,1,-2)$, $\mathbf{v}=2\mathbf{i}+2\mathbf{j}+\mathbf{k}$ is parallel to the given line. $\mathbf{u}\times\mathbf{v}=5\mathbf{i}-13\mathbf{j}+16\mathbf{k}$ is perpendicular to both lines so $\mathbf{v}\times(\mathbf{u}\times\mathbf{v})=45\mathbf{i}-27\mathbf{j}-36\mathbf{k}$, and hence $5\mathbf{i}-3\mathbf{j}-4\mathbf{k}$ is parallel to the line we seek. Thus $x=3+5t$, $y=1-3t$, $z=-2-4t$ are parametric equations of the line.

49. **(a)** When $t=0$ the bugs are at $(4,1,2)$ and (0,1,1) so the distance between them is $\sqrt{4^2+0^2+1^2}=\sqrt{17}$ cm.

(b)

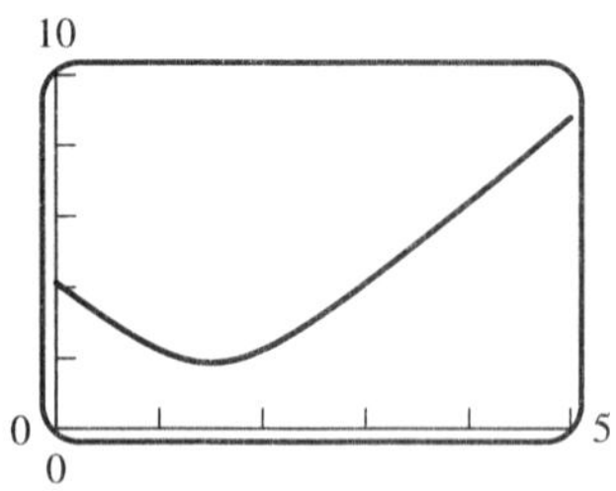

(c) The distance has a minimum value.

(d) Minimize D^2 instead of D (the distance between the bugs).
$D^2=[t-(4-t)]^2+[(1+t)-(1+2t)]^2+[(1+2t)-(2+t)]^2=6t^2-18t+17$,
$d(D^2)/dt=12t-18=0$ when $t=3/2$; the minimum
distance is $\sqrt{6(3/2)^2-18(3/2)+17}=\sqrt{14}/2$ cm.

50. The line intersects the xz-plane when $t=-1$, the xy-plane when $t=3/2$. Along the line, $T=25t^2(1+t)(3-2t)$ for $-1\le t\le 3/2$. Solve $dT/dt=0$ for t to find that the maximum value of T is about 50.96 when $t\approx 1.073590$.

EXERCISE SET 12.6

1. $x=3, y=4, z=5$

2. $x=x_0, y=y_0, z=z_0$

3. $(x-2)+4(y-6)+2(z-1)=0,\ x+4y+2z=28$

4. $-(x+1)+7(y+1)+6(z-2)=0,\ -x+7y+6z=6$

5. $z=0$

6. $2x-3y-4z=0$

7. $\mathbf{n}=\mathbf{i}-\mathbf{j}, x-y=0$

8. $\mathbf{n}=\mathbf{i}+\mathbf{j}, P(1,0,0), (x-1)+y=0, x+y=1$

9. $\mathbf{n}=\mathbf{j}+\mathbf{k}, P(0,1,0), (y-1)+z=0, y+z=1$

10. $\mathbf{n}=\mathbf{j}-\mathbf{k}, y-z=0$

11. $\overrightarrow{P_1P_2}\times\overrightarrow{P_1P_3}=\langle 2,1,2\rangle\times\langle 3,-1,-2\rangle=\langle 0,10,-5\rangle$,for convenience choose $\langle 0,2,-1\rangle$ which is also normal to the plane. Use any of the given points to get $2y-z=1$

12. $\overrightarrow{P_1P_2}\times\overrightarrow{P_1P_3}=\langle -1,-1,-2\rangle\times\langle -4,1,1\rangle=\langle 1,9,-5\rangle, x+9y-5z=16$

13. **(a)** parallel, because $\langle 2,-8,-6\rangle$ and $\langle -1,4,3\rangle$ are parallel

(b) perpendicular, because $\langle 3,-2,1\rangle$ and $\langle 4,5,-2\rangle$ are orthogonal

(c) neither, because $\langle 1,-1,3\rangle$ and $\langle 2,0,1\rangle$ are neither parallel nor orthogonal

14. **(a)** neither, because $\langle 3,-2,1\rangle$ and $\langle 6,-4,3\rangle$ are neither parallel nor orthogonal

(b) parallel, because $\langle 4,-1,-2\rangle$ and $\langle 1,-1/4,-1/2\rangle$ are parallel

(c) perpendicular, because $\langle 1,4,7\rangle$ and $\langle 5,-3,1\rangle$ are orthogonal

15. **(a)** parallel, because $\langle 2,-1,-4\rangle$ and $\langle 3,2,1\rangle$ are orthogonal

(b) neither, because $\langle 1,2,3\rangle$ and $\langle 1,-1,2\rangle$ are neither parallel nor orthogonal

(c) perpendicular, because $\langle 2,1,-1\rangle$ and $\langle 4,2,-2\rangle$ are parallel

16. **(a)** parallel, because $\langle -1,1,-3\rangle$ and $\langle 2,2,0\rangle$ are orthogonal

(b) perpendicular, because $\langle -2,1,-1\rangle$ and $\langle 6,-3,3\rangle$ are parallel

(c) neither, because $\langle 1,-1,1\rangle$ and $\langle 1,1,1\rangle$ are neither parallel nor orthogonal

17. **(a)** $3t-2t+t-5=0$, $t=5/2$ so $x=y=z=5/2$, the point of intersection is $(5/2,5/2,5/2)$

(b) $2(2-t)+(3+t)+t=1$ has no solution so the line and plane do not intersect

18. **(a)** $2(3t)-5t+(-t)+1=0$, $1=0$ has no solution so the line and the plane do not intersect.

(b) $(1+t)-(-1+3t)+4(2+4t)=7$, $t=-3/14$ so $x=1-3/14=11/14$, $y=-1-9/14=-23/14$, $z=2-12/14=8/7$, the point is $(11/14,-23/14,8/7)$

19. $\mathbf{n}_1=\langle 1,0,0\rangle, \mathbf{n}_2=\langle 2,-1,1\rangle, \mathbf{n}_1\cdot\mathbf{n}_2=2$ so

$$\cos\theta=\frac{\mathbf{n}_1\cdot\mathbf{n}_2}{\|\mathbf{n}_1\|\,\|\mathbf{n}_2\|}=\frac{2}{\sqrt{1}\sqrt{6}}=2/\sqrt{6},\theta=\cos^{-1}(2/\sqrt{6})\approx 35^\circ$$

20. $\mathbf{n}_1=\langle 1,2,-2\rangle, \mathbf{n}_2=\langle 6,-3,2\rangle, \mathbf{n}_1\cdot\mathbf{n}_2=-4$ so

$$\cos\theta=\frac{(-\mathbf{n}_1)\cdot\mathbf{n}_2}{\|-\mathbf{n}_1\|\,\|\mathbf{n}_2\|}=\frac{4}{(3)(7)}=4/21,\theta=\cos^{-1}(4/21)\approx 79^\circ$$

(Note: $-\mathbf{n}_1$ is used instead of $\mathbf{n}_1$ to get a value of θ in the range $[0,\pi/2]$)

21. $\langle 4,-2,7\rangle$ is normal to the desired plane and (0,0,0) is a point on it; $4x-2y+7z=0$

22. $\mathbf{v}=\langle 3,2,-1\rangle$ is parallel to the line and $\mathbf{n}=\langle 1,-2,1\rangle$ is normal to the given plane so $\mathbf{v}\times\mathbf{n}=\langle 0,-4,-8\rangle$ is normal to the desired plane. Let $t=0$ in the line to get $(-2,4,3)$ which is also a point on the desired plane, use this point and (for convenience) the normal $\langle 0,1,2\rangle$ to find that $y+2z=10$.

23. Find two points P_1 and P_2 on the line of intersection of the given planes and then find an equation of the plane that contains P_1, P_2, and the given point $P_0(-1,4,2)$. Let (x_0,y_0,z_0) be on the line of intersection of the given planes; then $4x_0-y_0+z_0-2=0$ and $2x_0+y_0-2z_0-3=0$, eliminate y_0 by addition of the equations to get $6x_0-z_0-5=0$; if $x_0=0$ then $z_0=-5$, if $x_0=1$ then $z_0=1$. Substitution of these values of x_0 and z_0 into either of the equations of the planes gives the corresponding values $y_0=-7$ and $y_0=3$ so $P_1(0,-7,-5)$ and $P_2(1,3,1)$ are on the line of intersection of the planes. $\overrightarrow{P_0P_1}\times\overrightarrow{P_0P_2}=\langle 4,-13,21\rangle$ is normal to the desired plane whose equation is $4x-13y+21z=-14$.

24. $\langle 1,2,-1\rangle$ is parallel to the line and hence normal to the plane $x+2y-z=10$

25. $\mathbf{n}_1=\langle 2,1,1\rangle$ and $\mathbf{n}_2=\langle 1,2,1\rangle$ are normals to the given planes, $\mathbf{n}_1\times\mathbf{n}_2=\langle -1,-1,3\rangle$ so $\langle 1,1,-3\rangle$ is normal to the desired plane whose equation is $x+y-3z=6$.

26. $\mathbf{n} = \langle 4, -1, 3\rangle$ is normal to the given plane, $\overrightarrow{P_1P_2} = \langle 3, -1, -1\rangle$ is parallel to the line through the given points, $\mathbf{n} \times \overrightarrow{P_1P_2} = \langle 4, 13, -1\rangle$ is normal to the desired plane whose equation is $4x + 13y - z = 1$.

27. $\mathbf{n}_1 = \langle 2, -1, 1\rangle$ and $\mathbf{n}_2 = \langle 1, 1, -2\rangle$ are normals to the given planes, $\mathbf{n}_1 \times \mathbf{n}_2 = \langle 1, 5, 3\rangle$ is normal to the desired plane whose equation is $x + 5y + 3z = -6$.

28. Let $t = 0$ and $t = 1$ to get the points $P_1(-1, 0, -4)$ and $P_2(0, 1, -2)$ that lie on the line. Denote the given point by P_0, then $\overrightarrow{P_0P_1} \times \overrightarrow{P_0P_2} = \langle 7, -1, -3\rangle$ is normal to the desired plane whose equation is $7x - y - 3z = 5$.

29. The plane is the perpendicular bisector of the line segment that joins $P_1(2, -1, 1)$ and $P_2(3, 1, 5)$. The midpoint of the line segment is $(5/2, 0, 3)$ and $\overrightarrow{P_1P_2} = \langle 1, 2, 4\rangle$ is normal to the plane so an equation is $x + 2y + 4z = 29/2$.

30. $\mathbf{n}_1 = \langle 2, -1, 1\rangle$ and $\mathbf{n}_2 = \langle 0, 1, 1\rangle$ are normals to the given planes, $\mathbf{n}_1 \times \mathbf{n}_2 = \langle -2, -2, 2\rangle$ so $\mathbf{n} = \langle 1, 1, -1\rangle$ is parallel to the line of intersection of the planes. $\mathbf{v} = \langle 3, 1, 2\rangle$ is parallel to the given line, $\mathbf{v} \times \mathbf{n} = \langle -3, 5, 2\rangle$ so $\langle 3, -5, -2\rangle$ is normal to the desired plane. Let $t = 0$ to find the point (0,1,0) that lies on the given line and hence on the desired plane. An equation of the plane is $3x - 5y - 2z = -5$.

31. The line is parallel to the line of intersection of the planes if it is parallel to both planes. Normals to the given planes are $\mathbf{n}_1 = \langle 1, -4, 2\rangle$ and $\mathbf{n}_2 = \langle 2, 3, -1\rangle$ so $\mathbf{n}_1 \times \mathbf{n}_2 = \langle -2, 5, 11\rangle$ is parallel to the line of intersection of the planes and hence parallel to the desired line whose equations are $x = 5 - 2t$, $y = 5t$, $z = -2 + 11t$.

32. Denote the points by A, B, C, and D, respectively. The points lie in the same plane if $\overrightarrow{AB} \times \overrightarrow{AC}$ and $\overrightarrow{AB} \times \overrightarrow{AD}$ are parallel (method 1). $\overrightarrow{AB} \times \overrightarrow{AC} = \langle 0, -10, 5\rangle, \overrightarrow{AB} \times \overrightarrow{AD} = \langle 0, 16, -8\rangle$, these vectors are parallel because $\langle 0, -10, 5\rangle = (-10/16)\langle 0, 16, -8\rangle$. The points lie in the same plane if D lies in the plane determined by A, B, C (method 2), and since $\overrightarrow{AB} \times \overrightarrow{AC} = \langle 0, -10, 5\rangle$, an equation of the plane is $-2y + z + 1 = 0$, $2y - z = 1$ which is satisfied by the coordinates of D.

33. $\mathbf{v} = \langle 0, 1, 1\rangle$ is parallel to the line.

(a) For any t, $6{\cdot}0 + 4t - 4t = 0$, so $(0, t, t)$ is in the plane.

(b) $\mathbf{n} = \langle 5, -3, 3\rangle$ is normal to the plane, $\mathbf{v} \cdot \mathbf{n} = 0$ so the line is parallel to the plane. (0,0,0) is on the line, $(0, 0, 1/3)$ is on the plane. The line is below the plane because (0,0,0) is below $(0, 0, 1/3)$.

(c) $\mathbf{n} = \langle 6, 2, -2\rangle$, $\mathbf{v} \cdot \mathbf{n} = 0$ so the line is parallel to the plane. (0,0,0) is on the line, $(0, 0, -3/2)$ is on the plane. The line is above the plane because (0,0,0) is above $(0, 0, -3/2)$.

34. The intercepts correspond to the points $A(a, 0, 0), B(0, b, 0)$, and $C(0, 0, c)$. $\overrightarrow{AB} \times \overrightarrow{AC} = \langle bc, ac, ab\rangle$ is normal to the plane so $bcx + acy + abz = abc$ or $x/a + y/b + z/c = 1$.

35. $\mathbf{v}_1 = \langle 1, 2, -1\rangle$ and $\mathbf{v}_2 = \langle -1, -2, 1\rangle$ are parallel, respectively, to the given lines and to each other so the lines are parallel. Let $t = 0$ to find the points $P_1(-2, 3, 4)$ and $P_2(3, 4, 0)$ that lie, respectively, on the given lines. $\mathbf{v}_1 \times \overrightarrow{P_1P_2} = \langle -7, -1, -9\rangle$ so $\langle 7, 1, 9\rangle$ is normal to the desired plane whose equation is $7x + y + 9z = 25$.

36. The system $4t_1 - 1 = 12t_2 - 13$, $t_1 + 3 = 6t_2 + 1$, $1 = 3t_2 + 2$ has the solution (Exercise 26, Section 12.5) $t_1 = -4$, $t_2 = -1/3$ so $(-17, -1, 1)$ is the point of intersection. $\mathbf{v}_1 = \langle 4, 1, 0\rangle$ and $\mathbf{v}_2 = \langle 12, 6, 3\rangle$ are (respectively) parallel to the lines, $\mathbf{v}_1 \times \mathbf{v}_2 = \langle 3, -12, 12\rangle$ so $\langle 1, -4, 4\rangle$ is normal to the desired plane whose equation is $x - 4y + 4z = -9$.

37. $\mathbf{n}_1 = \langle -2, 3, 7\rangle$ and $\mathbf{n}_2 = \langle 1, 2, -3\rangle$ are normals to the planes, $\mathbf{n}_1 \times \mathbf{n}_2 = \langle -23, 1, -7\rangle$ is parallel to the line of intersection. Let $z = 0$ in both equations and solve for x and y to get $x = -11/7$, $y = -12/7$ so $(-11/7, -12/7, 0)$ is on the line, a parametrization of which is
$x = -11/7 - 23t$, $y = -12/7 + t$, $z = -7t$.

38. Similar to Exercise 37 with $\mathbf{n}_1 = \langle 3, -5, 2\rangle$, $\mathbf{n}_2 = \langle 0, 0, 1\rangle$, $\mathbf{n}_1 \times \mathbf{n}_2 = \langle -5, -3, 0\rangle$. $z = 0$ so $3x - 5y = 0$, let $x = 0$ then $y = 0$ and (0,0,0) is on the line, a parametrization of which is
$x = -5t$, $y = -3t$, $z = 0$.

39. $D = |2(1) - 2(-2) + (3) - 4|/\sqrt{4+4+1} = 5/3$

40. $D = |3(0) + 6(1) - 2(5) - 5|/\sqrt{9+36+4} = 9/7$

41. (0,0,0) is on the first plane so $D = |6(0) - 3(0) - 3(0) - 5|/\sqrt{36+9+9} = 5/\sqrt{54}$.

42. (0,0,1) is on the first plane so $D = |(0) + (0) + (1) + 1|/\sqrt{1+1+1} = 2/\sqrt{3}$.

43. (1,3,5) and (4,6,7) are on L_1 and L_2, respectively. $\mathbf{v}_1 = \langle 7, 1, -3\rangle$ and $\mathbf{v}_2 = \langle -1, 0, 2\rangle$ are, respectively, parallel to L_1 and L_2, $\mathbf{v}_1 \times \mathbf{v}_2 = \langle 2, -11, 1\rangle$ so the plane $2x - 11y + z + 51 = 0$ contains L_2 and is parallel to L_1, $D = |2(1) - 11(3) + (5) + 51|/\sqrt{4+121+1} = 25/\sqrt{126}$.

44. (3,4,1) and (0,3,0) are on L_1 and L_2, respectively. $\mathbf{v}_1 = \langle -1, 4, 2\rangle$ and $\mathbf{v}_2 = \langle 1, 0, 2\rangle$ are parallel to L_1 and L_2, $\mathbf{v}_1 \times \mathbf{v}_2 = \langle 8, 4, -4\rangle = 4\langle 2, 1, -1\rangle$ so $2x + y - z - 3 = 0$ contains L_2 and is parallel to L_1, $D = |2(3) + (4) - (1) - 3|/\sqrt{4+1+1} = \sqrt{6}$.

45. The distance between $(2, 1, -3)$ and the plane is $|2 - 3(1) + 2(-3) - 4|/\sqrt{1+9+4} = 11/\sqrt{14}$ which is the radius of the sphere; an equation is $(x-2)^2 + (y-1)^2 + (z+3)^2 = 121/14$.

46. The vector $2\mathbf{i} + \mathbf{j} - \mathbf{k}$ is normal to the plane and hence parallel to the line so parametric equations of the line are $x = 3 + 2t$, $y = 1 + t$, $z = -t$. Substitution into the equation of the plane yields $2(3 + 2t) + (1 + t) - (-t) = 0$, $t = -7/6$; the point of intersection is $(2/3, -1/6, 7/6)$.

47. $\mathbf{v} = \langle 1, 2, -1\rangle$ is parallel to the line, $\mathbf{n} = \langle 2, -2, -2\rangle$ is normal to the plane, $\mathbf{v} \cdot \mathbf{n} = 0$ so $\mathbf{v}$ is parallel to the plane because $\mathbf{v}$ and $\mathbf{n}$ are perpendicular. $(-1, 3, 0)$ is on the line so
$D = |2(-1) - 2(3) - 2(0) + 3|/\sqrt{4+4+4} = 5/\sqrt{12}$

48. (a)

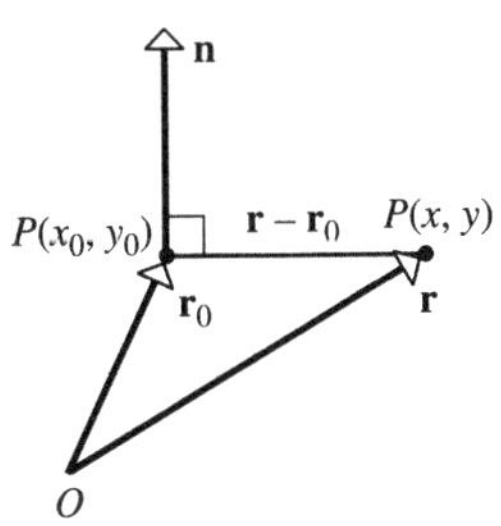

(b) $\mathbf{n} \cdot (\mathbf{r} - \mathbf{r}_0) = a(x - x_0) + b(y - y_0) = 0$

(c) See the proof of Theorem 12.6.1. Since a and b are not both zero, there is at least one point (x_0, y_0) that satisfies $ax + by + d = 0$, so $ax_0 + by_0 + d = 0$. If (x, y) also satisfies $ax + by + d = 0$ then, subtracting, $a(x - x_0) + b(y - y_0) = 0$, which is the equation of a line with $\mathbf{n} = \langle a, b\rangle$ as normal.

(d) Let $Q(x_1, y_1)$ be a point on the line, and position the normal $\mathbf{n} = \langle a, b \rangle$, with length $\sqrt{a^2 + b^2}$, so that its initial point is at Q. The distance is the orthogonal projection of $\overrightarrow{QP_0} = \langle x_0 - x_1, y_0 - y_1 \rangle$ onto $\mathbf{n}$. Then

$$D = \|\text{proj}_{\mathbf{n}} \overrightarrow{QP_0}\| = \left\| \frac{\overrightarrow{QP_0} \cdot \mathbf{n}}{\|\mathbf{n}\|^2} \mathbf{n} \right\| = \frac{|ax_0 + by_0 + d|}{\sqrt{a^2 + b^2}}.$$

49. $D = |2(-3) + (5) - 1|/\sqrt{4+1} = 2/\sqrt{5}$

50. **(a)** If $\langle x_0, y_0, z_0 \rangle$ lies on the second plane, so that $ax_0 + by_0 + cz_0 + d_2 = 0$, then by Theorem 12.6.2, the distance between the planes is $D = \dfrac{|ax_0 + by_0 + cz_0 + d_1|}{\sqrt{a^2 + b^2 + c^2}} = \dfrac{|-d_2 + d_1|}{\sqrt{a^2 + b^2 + c^2}}$

(b) The distance between the planes $-2x + y + z = 0$ and $-2x + y + z + \dfrac{5}{3} = 0$ is $D = \dfrac{|0 - 5/3|}{\sqrt{4+1+1}} = \dfrac{5}{3\sqrt{6}}$.

EXERCISE SET 12.7

1. **(a)** elliptic paraboloid, $a = 2, b = 3$
(b) hyperbolic paraboloid, $a = 1, b = 5$
(c) hyperboloid of one sheet, $a = b = c = 4$
(d) circular cone, $a = b = 1$
(e) elliptic paraboloid, $a = 2, b = 1$
(f) hyperboloid of two sheets, $a = b = c = 1$

2. **(a)** ellipsoid, $a = \sqrt{2}, b = 2, c = \sqrt{3}$
(b) hyperbolic paraboloid, $a = b = 1$
(c) hyperboloid of one sheet, $a = 1, b = 3, c = 1$
(d) hyperboloid of two sheets, $a = 1, b = 2, c = 1$
(e) elliptic paraboloid, $a = \sqrt{2}, b = \sqrt{2}/2$
(f) elliptic cone, $a = 2, b = \sqrt{3}$

3. **(a)** $-z = x^2 + y^2$, circular paraboloid opening down the negative z-axis

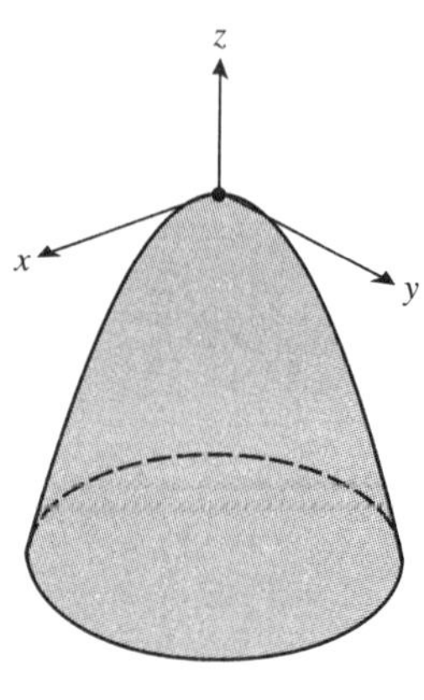

(b) $z = x^2 + y^2$, circular paraboloid, no change
(c) $z = x^2 + y^2$, circular paraboloid, no change

(d) $z = x^2 + y^2$, circular paraboloid, no change

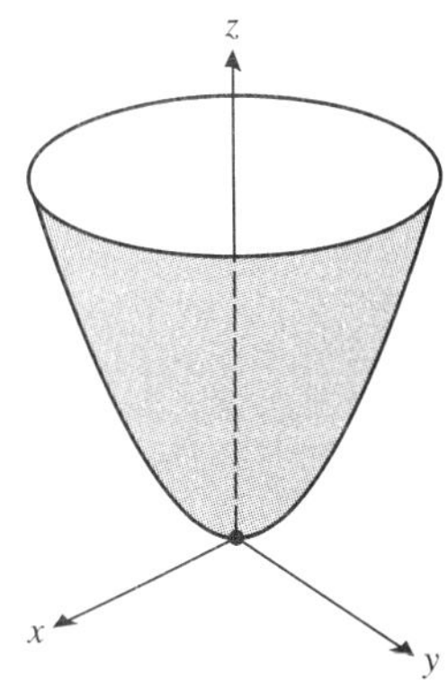

(e) $x = y^2 + z^2$, circular paraboloid opening along the positive x-axis

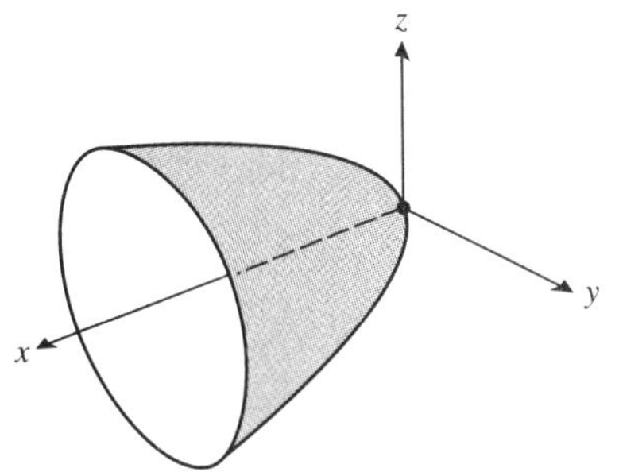

(f) $y = x^2 + z^2$, circular paraboloid opening along the positive y-axis

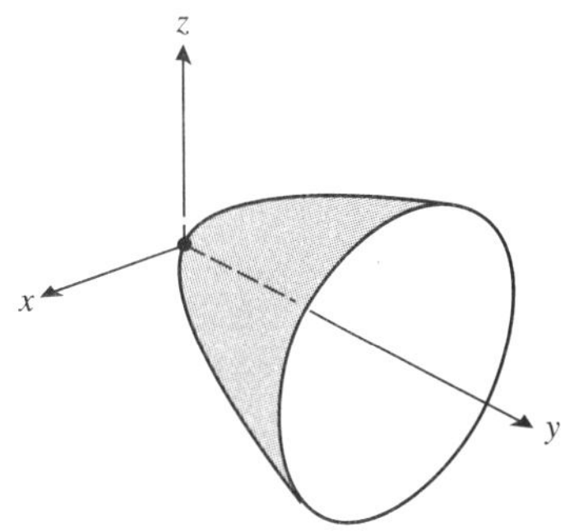

4. **(a)** $x^2 + y^2 - z^2 = 1$, no change
(b) $x^2 + y^2 - z^2 = 1$, no change
(c) $x^2 + y^2 - z^2 = 1$, no change
(d) $x^2 + y^2 - z^2 = 1$, no change

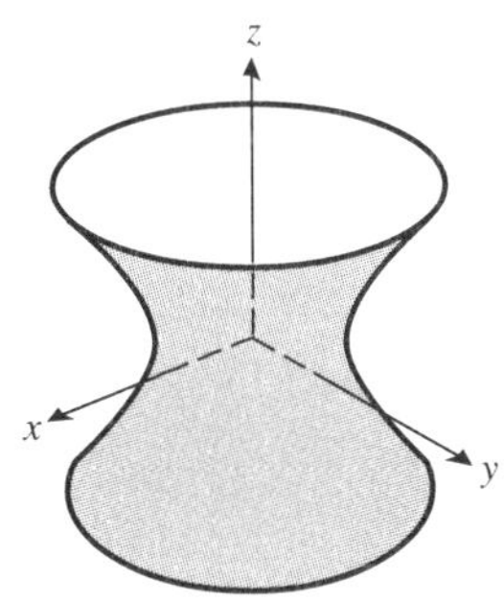

(e) $-x^2 + y^2 + z^2 = 1$, hyperboloid of one sheet with x-axis as axis

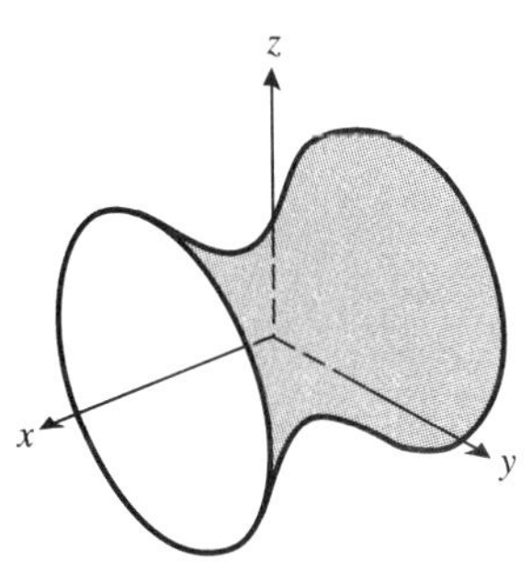

(f) $x^2 \quad y^2 + z^2 = 1$, hyperboloid of one sheet with y-axis as axis

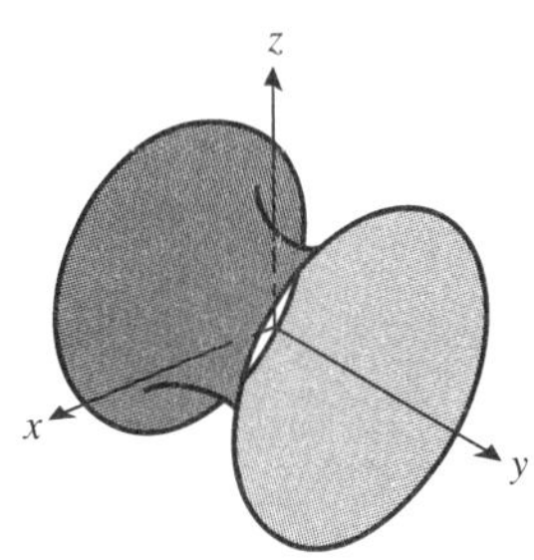

5. **(a)** hyperboloid of one sheet, axis is y-axis
(b) hyperboloid of two sheets separated by yz-plane

(c) elliptic paraboloid opening along the positive x-axis
(d) elliptic cone with x-axis as axis
(e) hyperbolic paraboloid straddling the z-axis
(f) paraboloid opening along the negative y-axis

6. (a) same (b) same (c) same

(d) same (e) $y = \dfrac{x^2}{a^2} - \dfrac{z^2}{c^2}$ (f) $y = \dfrac{x^2}{a^2} + \dfrac{z^2}{c^2}$

7. (a) $x = 0: \dfrac{y^2}{25} + \dfrac{z^2}{4} = 1; y = 0: \dfrac{x^2}{9} + \dfrac{z^2}{4} = 1;$

$z = 0: \dfrac{x^2}{9} + \dfrac{y^2}{25} = 1$

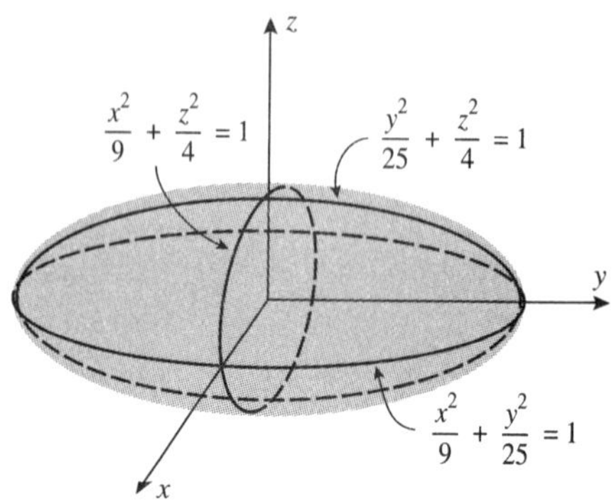

(b) $x = 0: z = 4y^2; y = 0: z = x^2;$
$z = 0: x = y = 0$

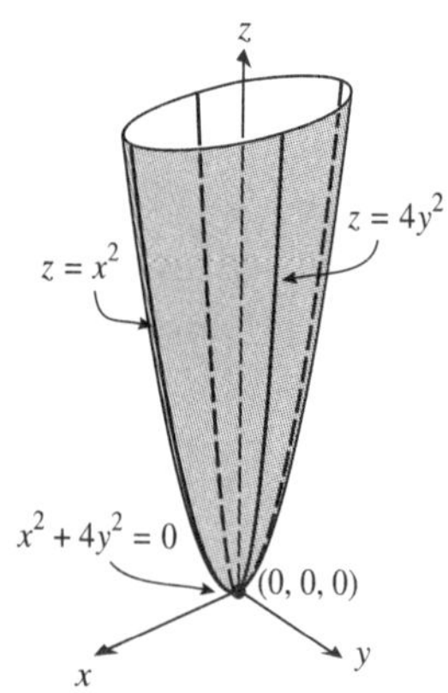

(c) $x = 0: \dfrac{y^2}{16} - \dfrac{z^2}{4} = 1; y = 0: \dfrac{x^2}{9} - \dfrac{z^2}{4} = 1;$

$z = 0: \dfrac{x^2}{9} + \dfrac{y^2}{16} = 1$

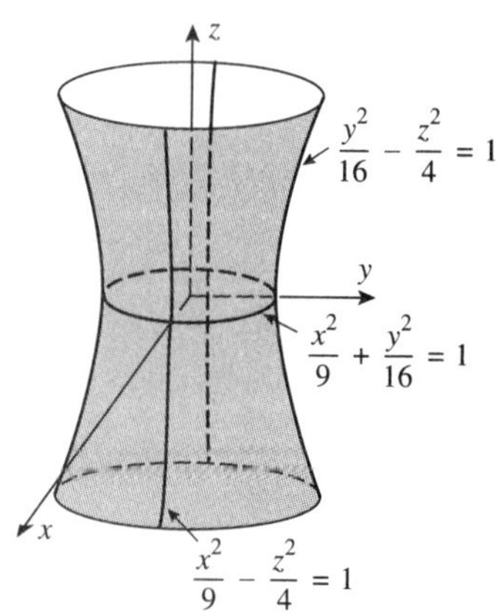

8. **(a)** $x = 0 : y = z = 0; y = 0 : x = 9z^2; z = 0 : x = y^2$

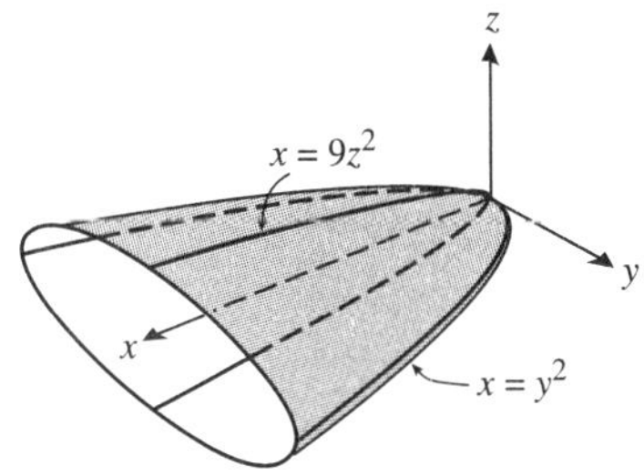

(b) $x = 0 : -y^2 + 4z^2 = 4; y = 0 : x^2 + z^2 = 1;$
$z = 0 : 4x^2 - y^2 = 4$

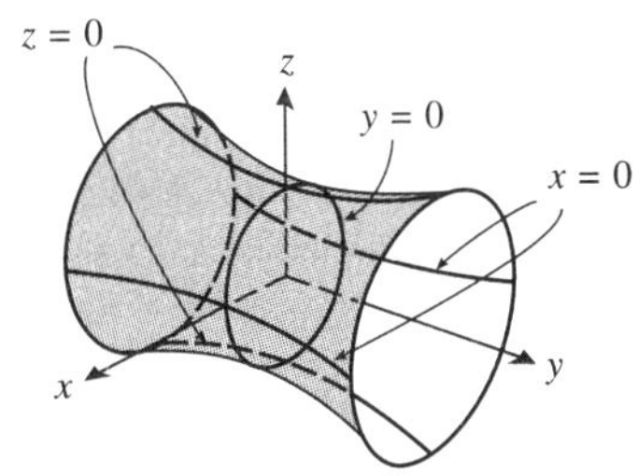

(c) $x = 0 : z = \pm\dfrac{y}{2}; y = 0 : z = \pm x; z = 0 : x = y = 0$

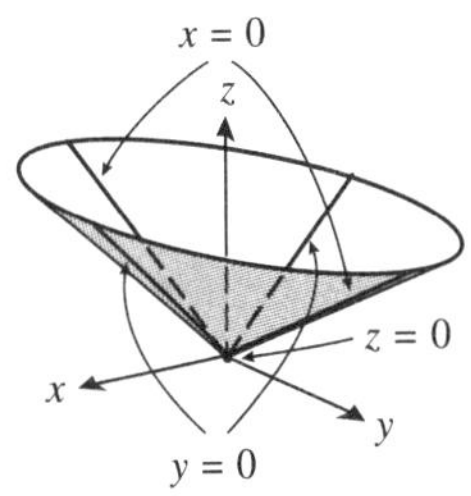

9. **(a)** $4x^2 + z^2 = 3$; ellipse **(b)** $y^2 + z^2 = 3$; circle **(c)** $y^2 + z^2 = 20$; circle
(d) $9x^2 - y^2 = 20$; hyperbola **(e)** $z = 9x^2 + 16$; parabola **(f)** $9x^2 + 4y^2 = 4$; ellipse

10. **(a)** $y^2 - 4z^2 = 27$; hyperbola **(b)** $9x^2 + 4z^2 = 25$; ellipse **(c)** $9z^2 - x^2 = 4$; hyperbola
(d) $x^2 + 4y^2 = 9$; ellipse **(e)** $z = 1 - 4y^2$; parabola **(f)** $x^2 - 4y^2 = 4$; hyperbola

11.

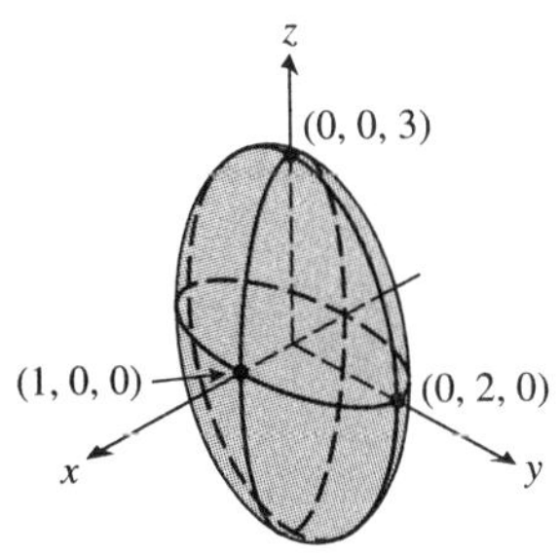

Ellipsoid

12.

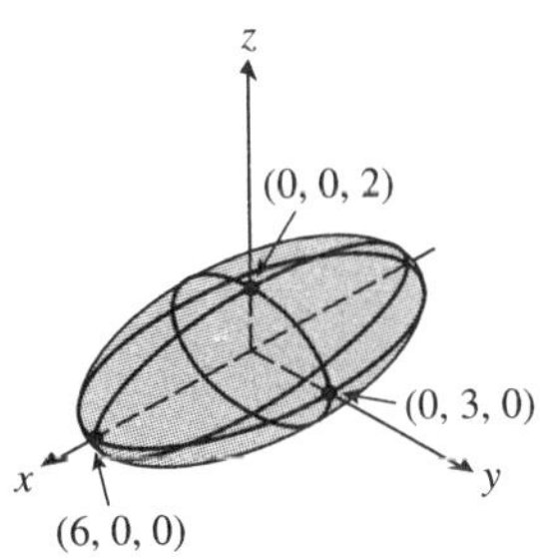

Ellipsoid

13.

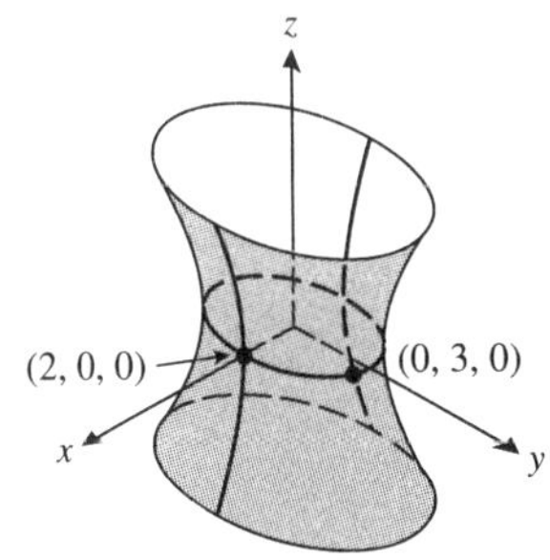

Hyperboloid
of one sheet

14.

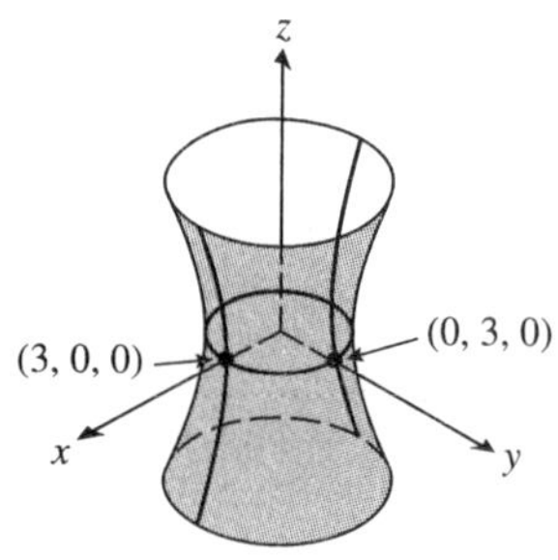

Hyperboloid of one sheet

15.

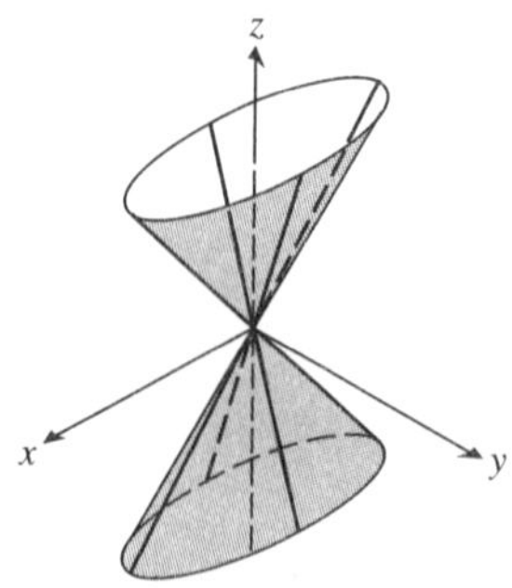

Elliptic cone

16.

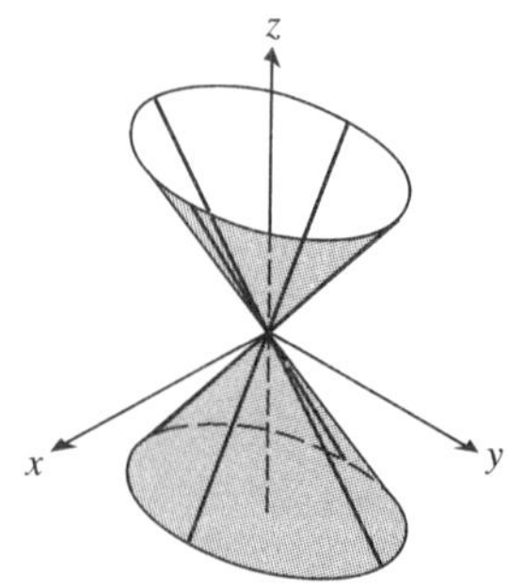

Elliptic cone

17.

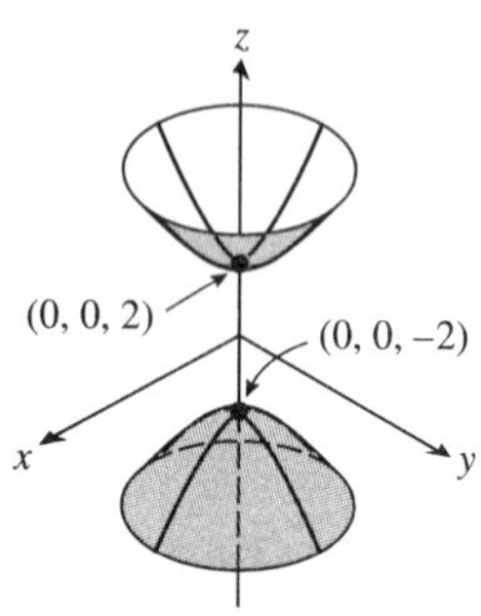

Hyperboloid of two sheets

18.

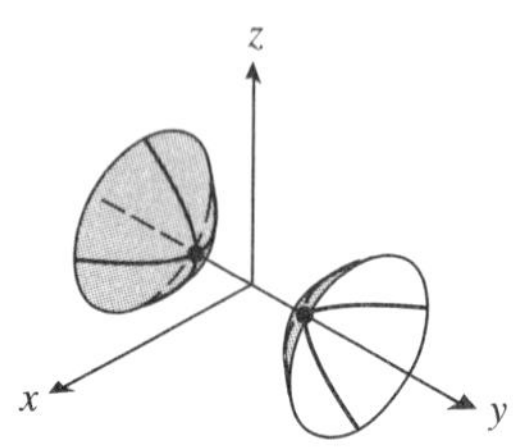

Hyperboloid of two sheets

19.

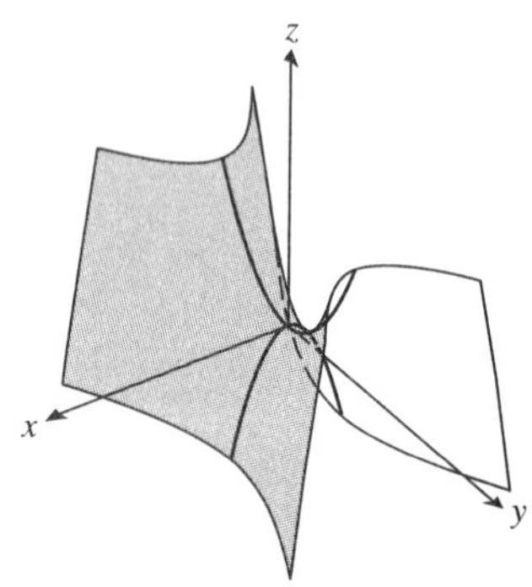

Hyperbolic paraboloid

20.

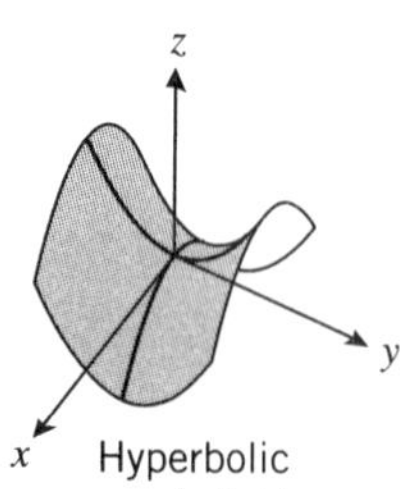

Hyperbolic paraboloid

21.

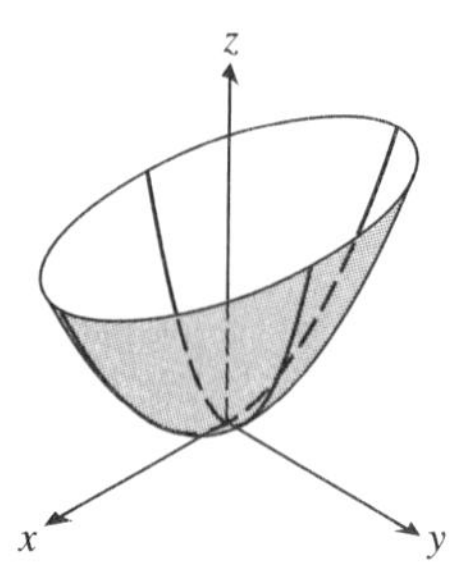

Elliptic paraboloid

22.

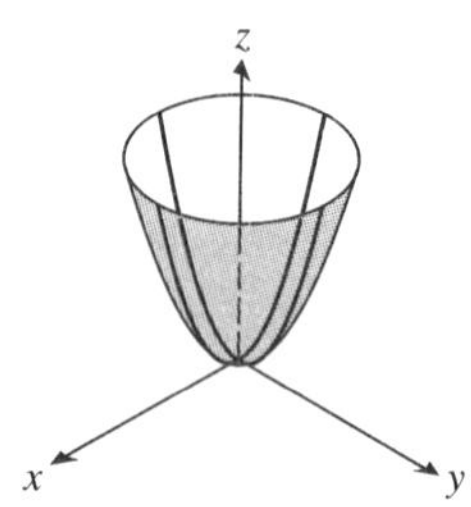

Circular paraboloid

23.

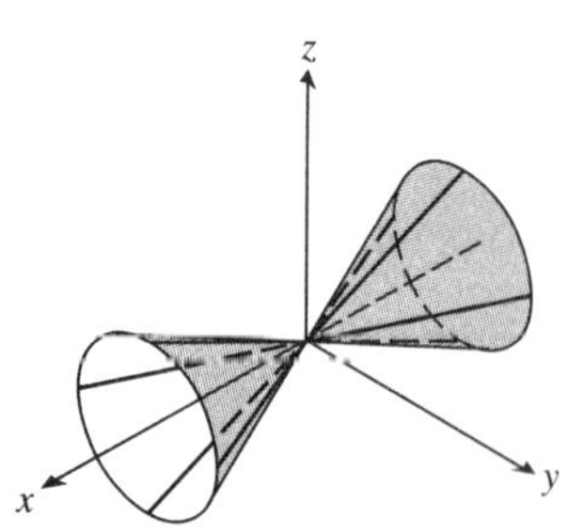

Circular cone

24.

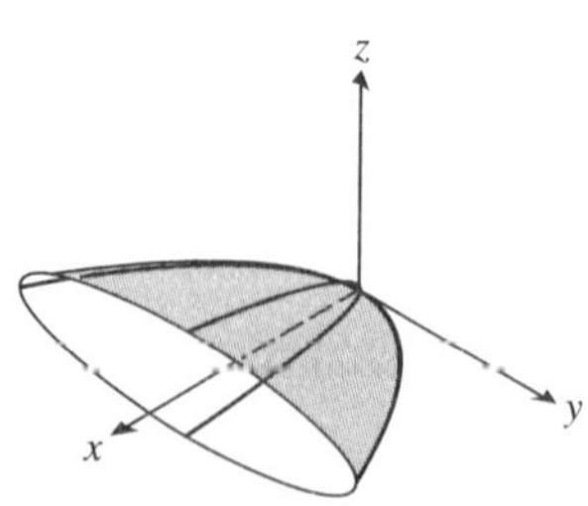

Elliptic paraboloid

25.

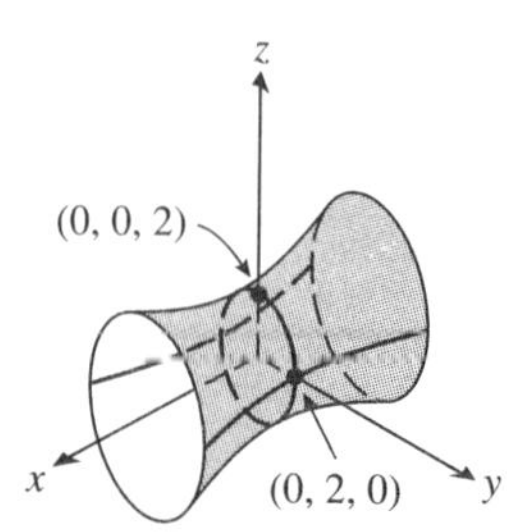

Hyperboloid of one sheet

26.

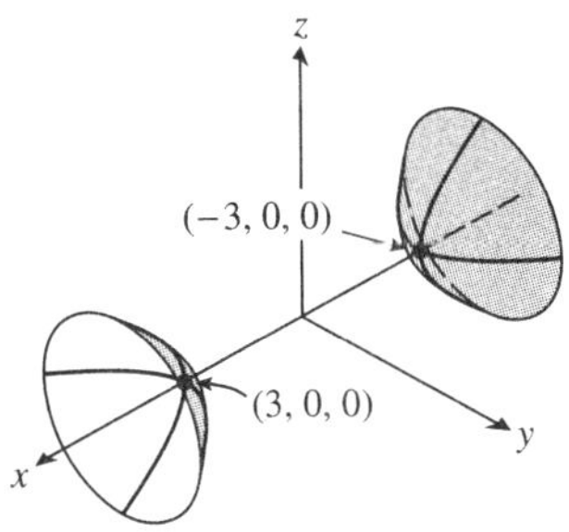

Hyperboloid of two sheets

27.

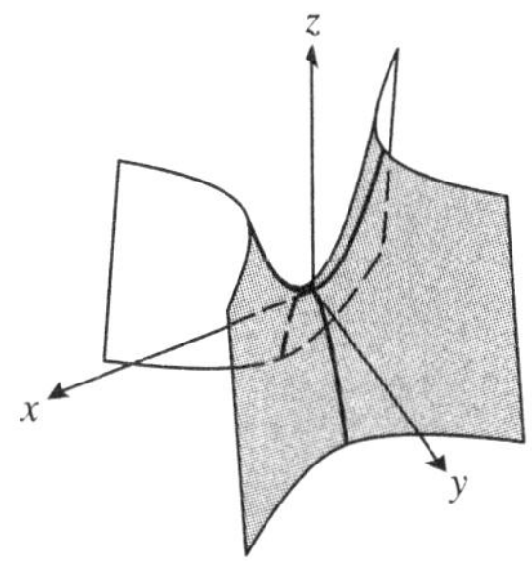

Hyperbolic paraboloid

28.

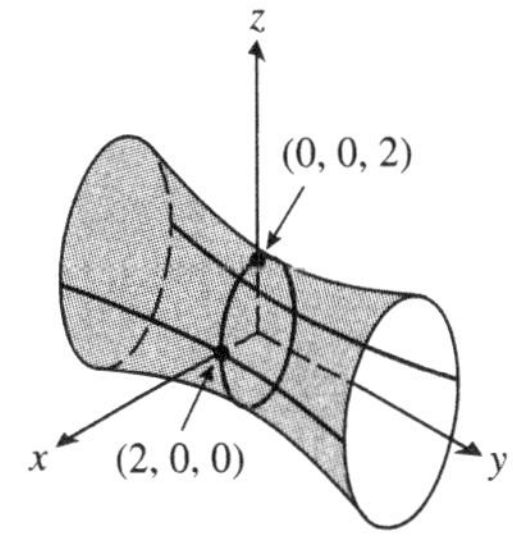

Hyperboloid of one sheet

29.

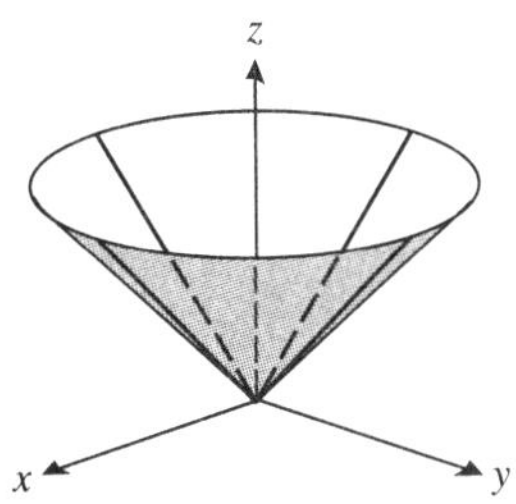

30.

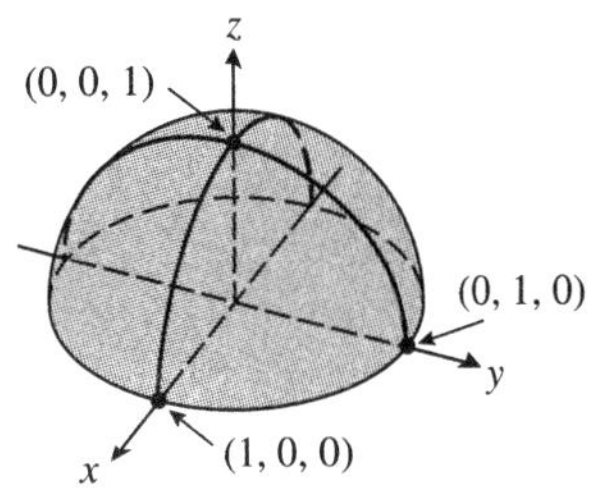

31.

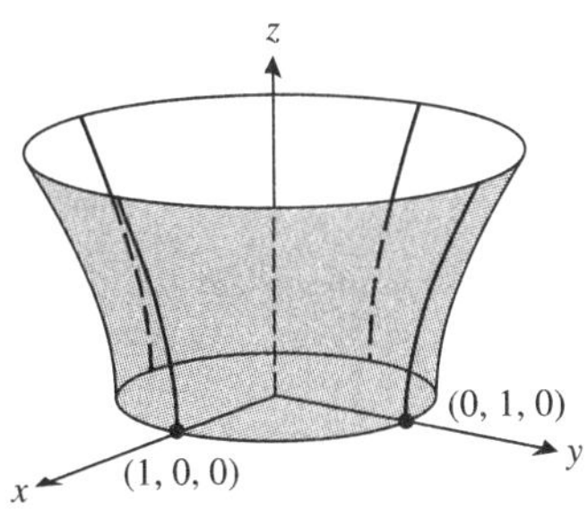

32.

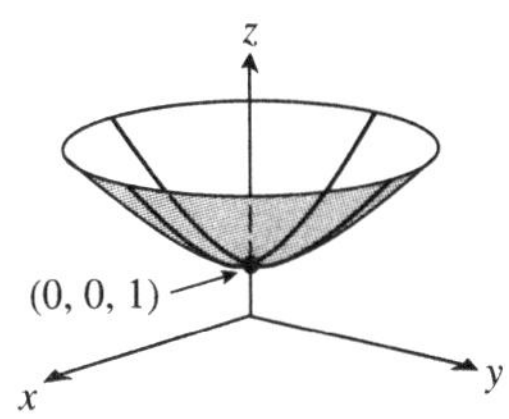

33.

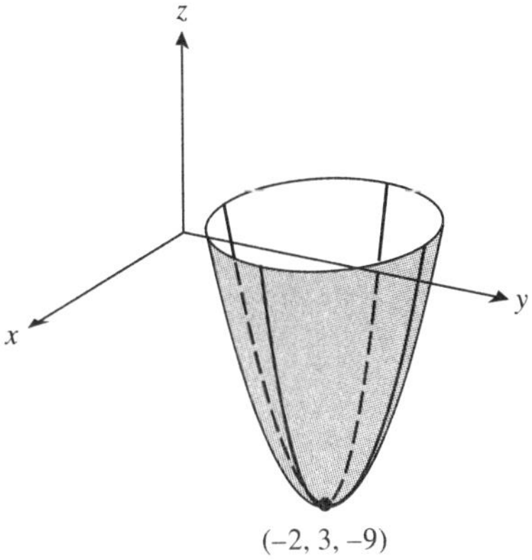

Circular paraboloid

34.

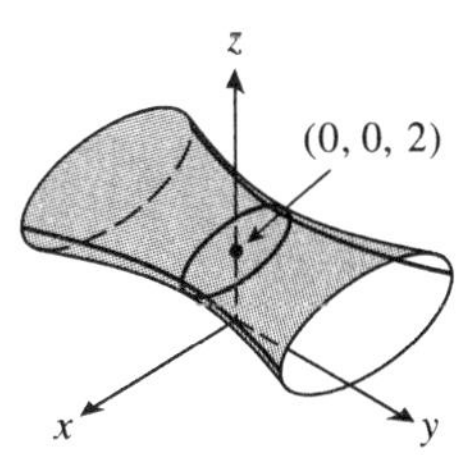

Hyperboloid of one sheet

35.

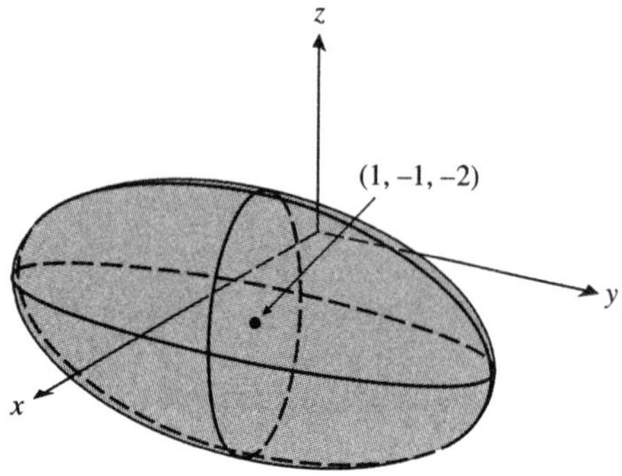

Ellipsoid

36.

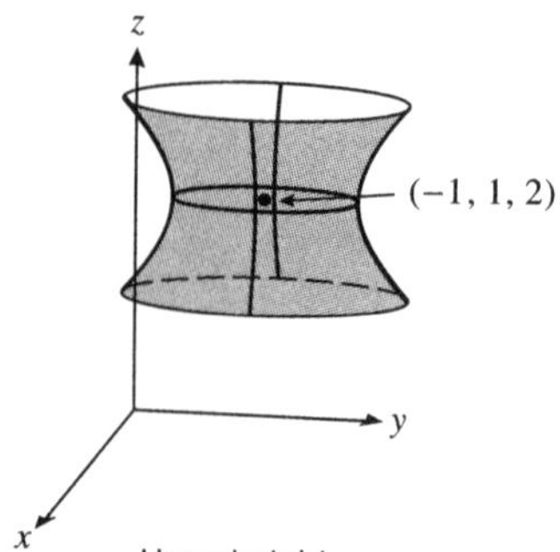

Hyperboloid of one sheet

37. **(a)** $\dfrac{x^2}{9}+\dfrac{y^2}{4}=1$ **(b)** $6, 4$ **(c)** $(\pm\sqrt{5}, 0, \sqrt{2})$
(d) The focal axis is parallel to the x-axis.

38. **(a)** $\dfrac{y^2}{4}+\dfrac{z^2}{2}=1$ **(b)** $4, 2\sqrt{2}$ **(c)** $(3, \pm\sqrt{2}, 0)$
(d) The focal axis is parallel to the y-axis.

39. **(a)** $\dfrac{y^2}{4}-\dfrac{x^2}{4}=1$ **(b)** $(0, \pm 2, 4)$ **(c)** $(0, \pm 2\sqrt{2}, 4)$
(d) The focal axis is parallel to the y-axis.

40. **(a)** $\dfrac{x^2}{4}-\dfrac{y^2}{4}=1$ **(b)** $(\pm 2, 0, -4)$ **(c)** $(\pm 2\sqrt{2}, 0, -4)$
(e) The focal axis is parallel to the x-axis.

41. **(a)** $z+4=y^2$ **(b)** $(2, 0, -4)$ **(c)** $(2, 0, -15/4)$
(d) The focal axis is parallel to the z-axis.

42. **(a)** $z-4=-x^2$ **(b)** $(0, 2, 4)$ **(c)** $(0, 2, 15/4)$
(d) The focal axis is parallel to the z-axis.

43. $x^2+y^2=4-x^2-y^2, x^2+y^2=2$; circle of radius $\sqrt{2}$ in the plane $z=2$, centered at $(0, 0, 2)$

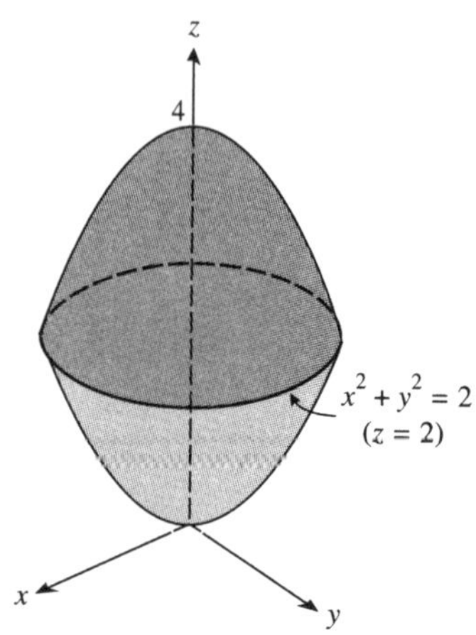

44. $y^2 + z = 4 - 2(y^2 + z), y^2 + z = 4/3$;
parabolas in the planes $x = \pm 2/\sqrt{3}$
which open in direction of the negative z-axis

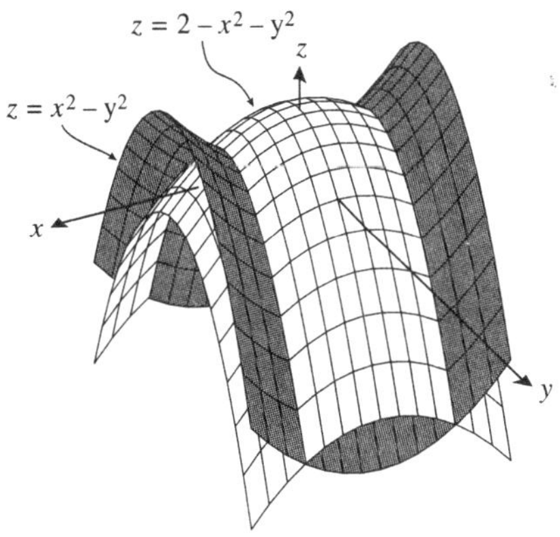

45. $y = 4(x^2 + z^2)$

46. $y^2 = 4(x^2 + z^2)$

47. $|z - (-1)| = \sqrt{x^2 + y^2 + (z-1)^2}$, $z^2 + 2z + 1 = x^2 + y^2 + z^2 - 2z + 1$, $z = (x^2 + y^2)/4$; circular paraboloid

48. $|z + 1| = 2\sqrt{x^2 + y^2 + (z-1)^2}$, $z^2 + 2z + 1 = 4\left(x^2 + y^2 + z^2 - 2z + 1\right)$,
$4x^2 + 4y^2 + 3z^2 - 10z + 3 = 0$, $\dfrac{x^2}{4/3} + \dfrac{y^2}{4/3} + \dfrac{(z - 5/3)^2}{16/9} = 1$; ellipsoid, center at $(0, 0, 5/3)$.

49. If $z = 0$, $\dfrac{x^2}{a^2} + \dfrac{y^2}{a^2} = 1$; if $y = 0$ then $\dfrac{x^2}{a^2} + \dfrac{z^2}{c^2} = 1$; since $c < a$ the major axis has length $2a$, the minor axis length $2c$.

50. $\dfrac{x^2}{a^2} + \dfrac{y^2}{a^2} + \dfrac{z^2}{b^2} = 1$, where $a = 6378.1370, b = 6356.5231$.

51. Each slice perpendicular to the z-axis for $|z| < c$ is an ellipse whose equation is
$\dfrac{x^2}{a^2} + \dfrac{y^2}{b^2} = \dfrac{c^2 - z^2}{c^2}$, or $\dfrac{x^2}{(a^2/c^2)(c^2 - z^2)} + \dfrac{y^2}{(b^2/c^2)(c^2 - z^2)} = 1$, the area of which is
$\pi\left(\dfrac{a}{c}\sqrt{c^2 - z^2}\right)\left(\dfrac{b}{c}\sqrt{c^2 - z^2}\right) = \pi\dfrac{ab}{c^2}\left(c^2 - z^2\right)$ so $V = 2\displaystyle\int_0^c \pi\dfrac{ab}{c^2}\left(c^2 - z^2\right) dz = \dfrac{4}{3}\pi abc$.

EXERCISE SET 12.8

1. **(a)** $(8, \pi/6, -4)$ **(b)** $(5\sqrt{2}, 3\pi/4, 6)$ **(c)** $(2, \pi/2, 0)$ **(d)** $(8, 5\pi/3, 6)$

2. **(a)** $(2, 7\pi/4, 1)$ **(b)** $(1, \pi/2, 1)$ **(c)** $(4\sqrt{2}, 3\pi/4, -7)$ **(d)** $(2\sqrt{2}, 7\pi/4, -2)$

3. **(a)** $(2\sqrt{3}, 2, 3)$ **(b)** $(-4\sqrt{2}, 4\sqrt{2}, -2)$ **(c)** $(5, 0, 4)$ **(d)** $(-7, 0, -9)$

4. **(a)** $(3, -3\sqrt{3}, 7)$ **(b)** $(0, 1, 0)$ **(c)** $(0, 3, 5)$ **(d)** $(0, 4, -1)$

5. **(a)** $(2\sqrt{2}, \pi/3, 3\pi/4)$ **(b)** $(2, 7\pi/4, \pi/4)$ **(c)** $(6, \pi/2, \pi/3)$ **(d)** $(10, 5\pi/6, \pi/2)$

6. **(a)** $(8\sqrt{2}, \pi/4, \pi/6)$ **(b)** $(2\sqrt{2}, 5\pi/3, 3\pi/4)$ **(c)** $(2, 0, \pi/2)$ **(d)** $(4, \pi/6, \pi/6)$

7. **(a)** $(5\sqrt{6}/4, 5\sqrt{2}/4, 5\sqrt{2}/2)$ **(b)** $(7,0,0)$
(c) $(0,0,1)$ **(d)** $(0, -2, 0)$

8. **(a)** $(-\sqrt{2}/4, \sqrt{6}/4, -\sqrt{2}/2)$ **(b)** $(3\sqrt{2}/4, -3\sqrt{2}/4, -3\sqrt{3}/2)$
(c) $(2\sqrt{6}, 2\sqrt{2}, 4\sqrt{2})$ **(d)** $(0, 2\sqrt{3}, 2)$

9. **(a)** $(2\sqrt{3}, \pi/6, \pi/6)$ **(b)** $(\sqrt{2}, \pi/4, 3\pi/4)$
(c) $(2, 3\pi/4, \pi/2)$ **(d)** $(4\sqrt{3}, 1, 2\pi/3)$

10. **(a)** $(4\sqrt{2}, 5\pi/6, \pi/4)$ **(b)** $(2\sqrt{2}, 0, 3\pi/4)$
(c) $(5, \pi/2, \tan^{-1}(4/3))$ **(d)** $(2\sqrt{10}, \pi, \tan^{-1} 3)$

11. **(a)** $(5\sqrt{3}/2, \pi/4, -5/2)$ **(b)** $(0, 7\pi/6, -1)$
(c) $(0, 0, 3)$ **(d)** $(4, \pi/6, 0)$

12. **(a)** $(0, \pi/2, 5)$ **(b)** $(3\sqrt{2}, 0, -3\sqrt{2})$
(c) $(0, 3\pi/4, -\sqrt{2})$ **(d)** $(5/2, 2\pi/3, -5\sqrt{3}/2)$

15.

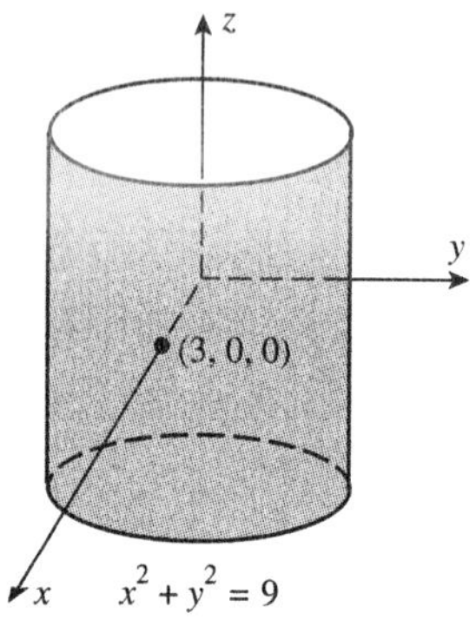

16.

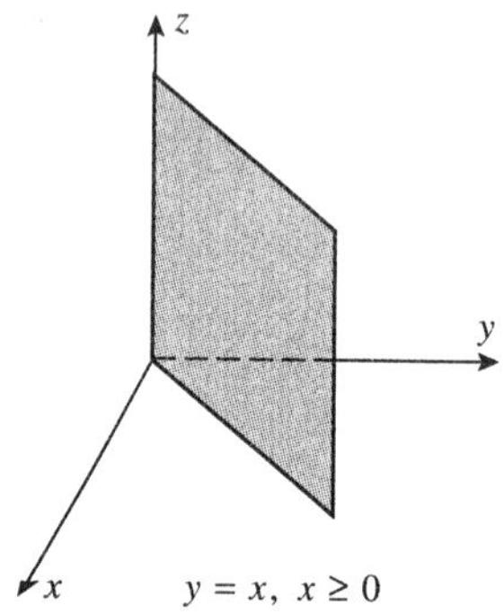

17.

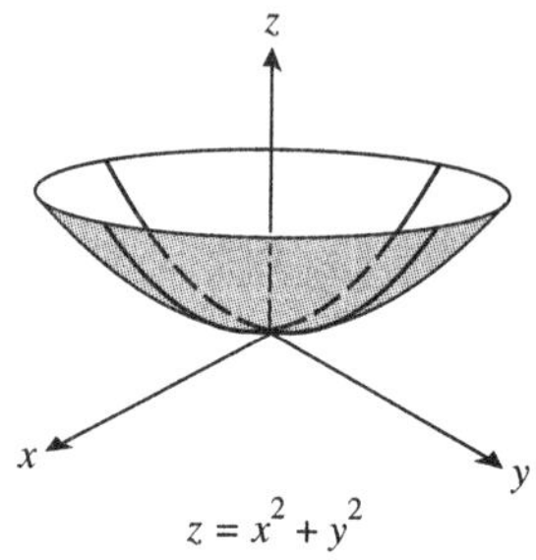

18.

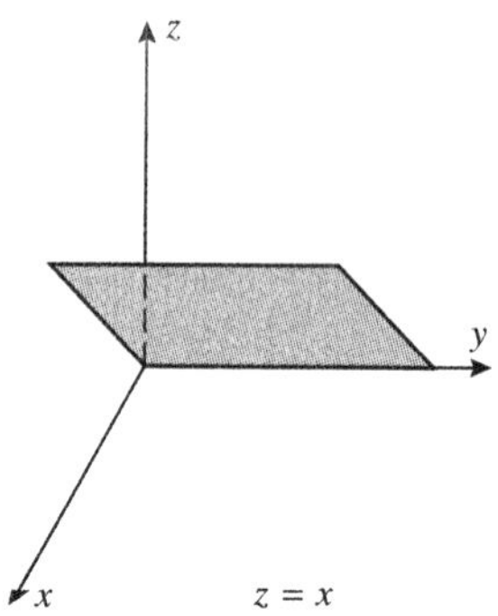

19.

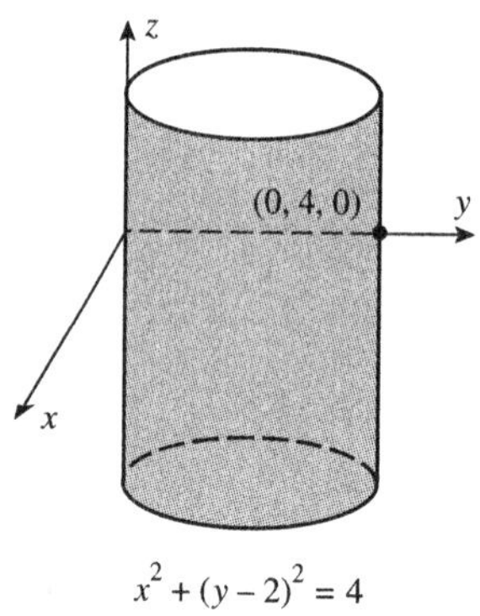

20.

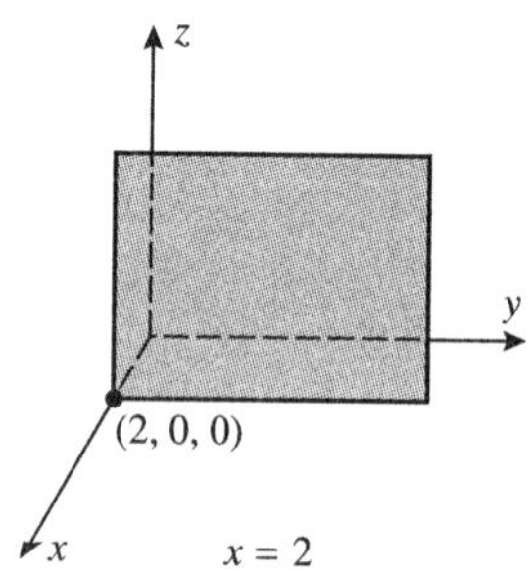

21.

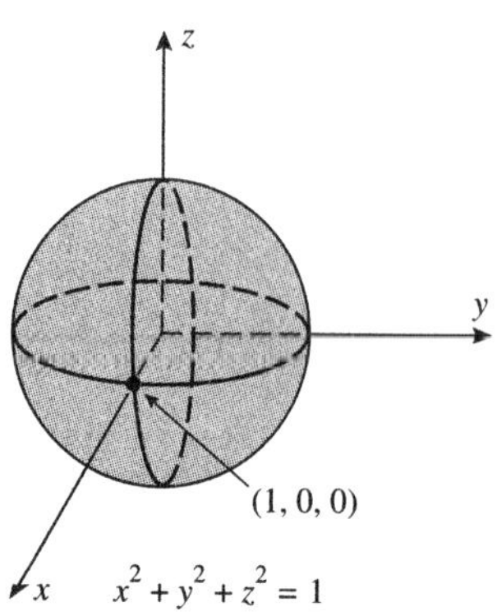

22.

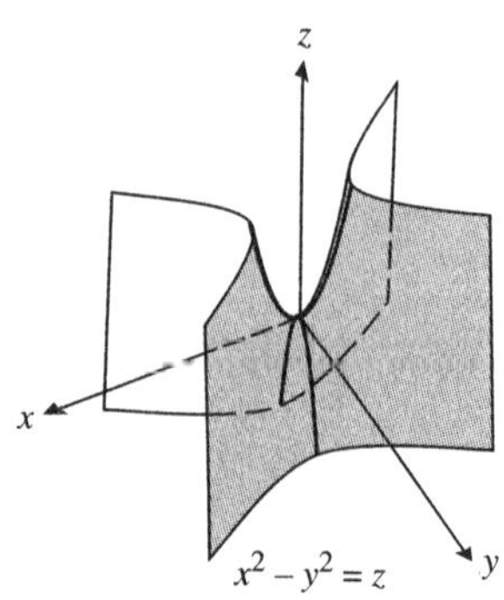

23.

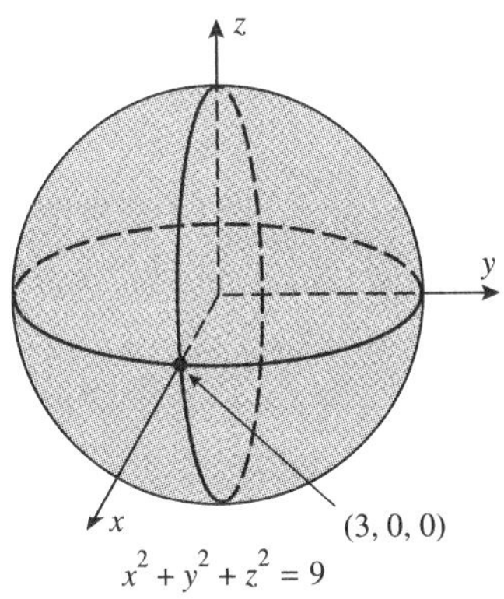

24.

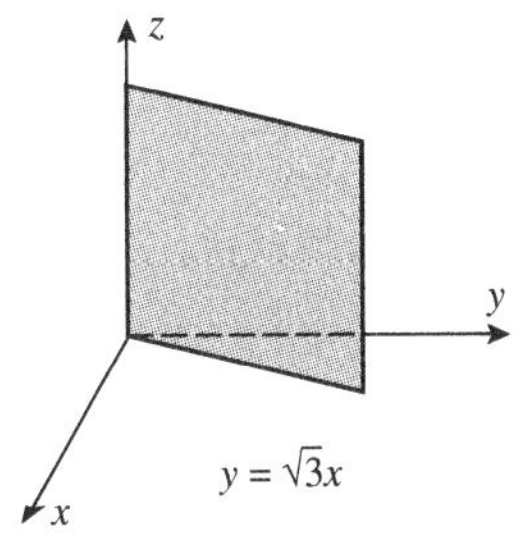

25.

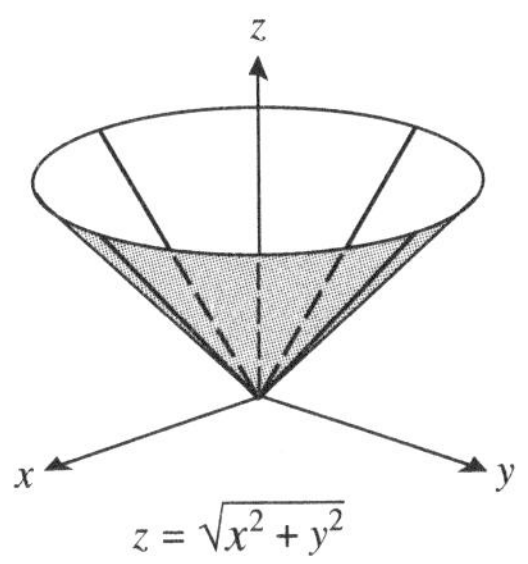

26.

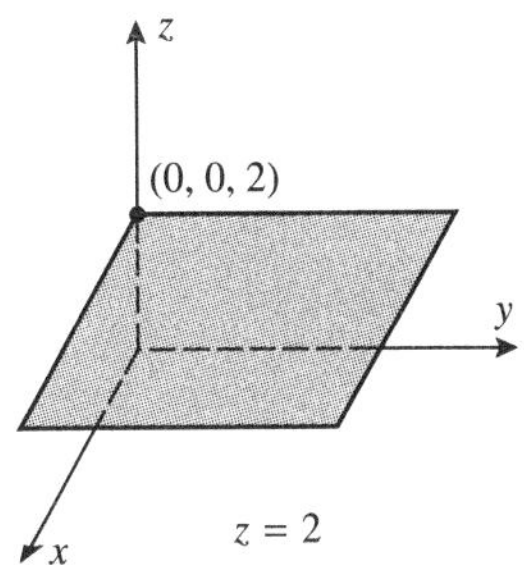

27.

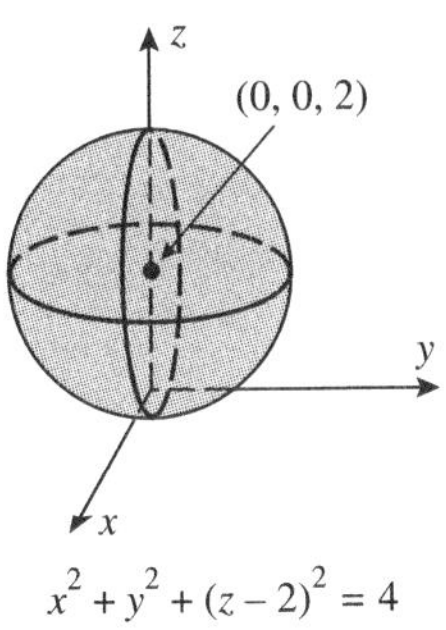

28.

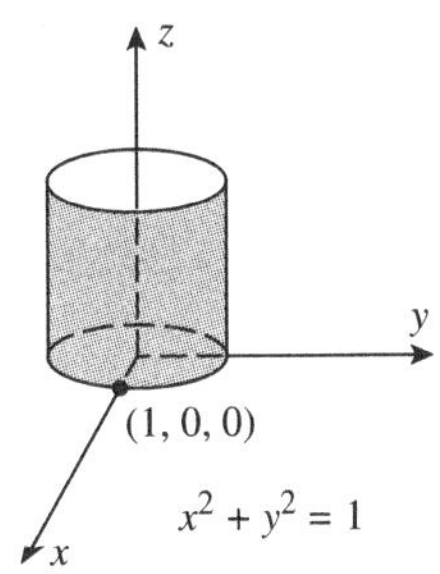

29.

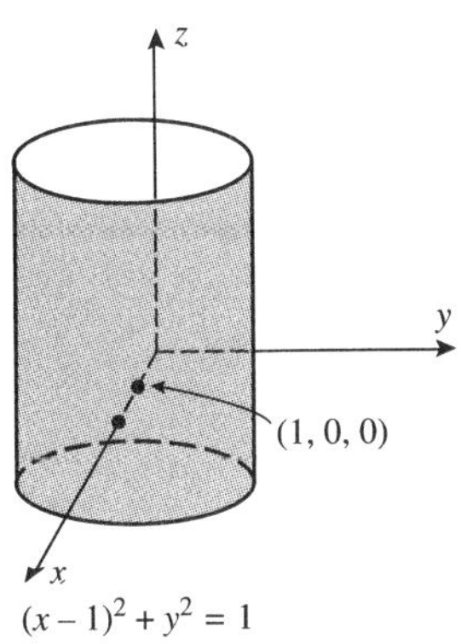

30.

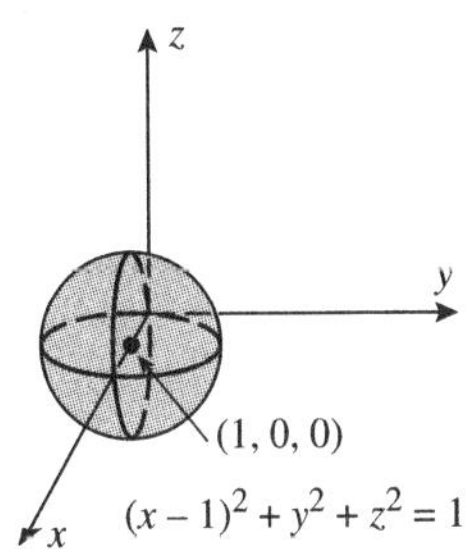

31. (a) $z = 3$ (b) $\rho \cos\phi = 3, \rho = 3 \sec\phi$

32. (a) $r \sin\theta = 2, r = 2 \csc\theta$ (b) $\rho \sin\phi \sin\theta = 2, \rho = 2 \csc\phi \csc\theta$

33. (a) $z = 3r^2$ (b) $\rho \cos\phi = 3\rho^2 \sin^2\phi, \rho = \dfrac{1}{3} \csc\phi \cot\phi$

34. **(a)** $z = \sqrt{3}r$ **(b)** $\rho\cos\phi = \sqrt{3}\rho\sin\phi, \tan\phi = \dfrac{1}{\sqrt{3}}, \phi = \dfrac{\pi}{6}$

35. **(a)** $r = 2$ **(b)** $\rho\sin\phi = 2, \rho = 2\csc\phi$

36. **(a)** $r^2 - 6r\sin\theta = 0,\ r = 6\sin\theta$ **(b)** $\rho\sin\phi = 6\sin\theta, \rho = 6\sin\theta\csc\phi$

37. **(a)** $r^2 + z^2 = 9$ **(b)** $\rho = 3$

38. **(a)** $z^2 = r^2\cos^2\theta - r^2\sin^2\theta = r^2(\cos^2\theta - \sin^2\theta),\ z^2 = r^2\cos 2\theta$
(b) Use the result in Part (a) with $r = \rho\sin\phi$, $z = \rho\cos\phi$ to get $\rho^2\cos^2\phi = \rho^2\sin^2\phi\cos 2\theta$, $\cot^2\phi = \cos 2\theta$

39. **(a)** $2r\cos\theta + 3r\sin\theta + 4z = 1$
(b) $2\rho\sin\phi\cos\theta + 3\rho\sin\phi\sin\theta + 4\rho\cos\phi = 1$

40. **(a)** $r^2 - z^2 = 1$
(b) Use the result of Part (a) with $r = \rho\sin\phi$, $z = \rho\cos\phi$ to get $\rho^2\sin^2\phi - \rho^2\cos^2\phi = 1$, $\rho^2\cos 2\phi = -1$

41. **(a)** $r^2\cos^2\theta = 16 - z^2$
(b) $x^2 = 16 - z^2,\ x^2 + y^2 + z^2 = 16 + y^2,\ \rho^2 = 16 + \rho^2\sin^2\phi\sin^2\theta,\ \rho^2\left(1 - \sin^2\phi\sin^2\theta\right) = 16$

42. **(a)** $r^2 + z^2 = 2z$ **(b)** $\rho^2 = 2\rho\cos\phi,\ \rho = 2\cos\phi$

43. all points on or above the paraboloid $z = x^2 + y^2$, that are also on or below the plane $z = 4$

44. a right circular cylindrical solid of height 3 and radius 1 whose axis is the line $x = 0, y = 1$

45. all points on or between concentric spheres of radii 1 and 3 centered at the origin

46. all points on or above the cone $\phi = \pi/6$, that are also on or below the sphere $\rho = 2$

47. $\theta = \pi/6,\ \phi = \pi/6$, spherical $(4000, \pi/6, \pi/6)$, rectangular $\left(1000\sqrt{3}, 1000, 2000\sqrt{3}\right)$

48. **(a)** $y = r\sin\theta = a\sin\theta$ but $az = a\sin\theta$ so $y = az$, which is a plane that contains the curve of intersection of $z = \sin\theta$ and the circular cylinder $r = a$. From Exercise 60, Section 11.4, the curve of intersection of a plane and a circular cylinder is an ellipse.

(b)

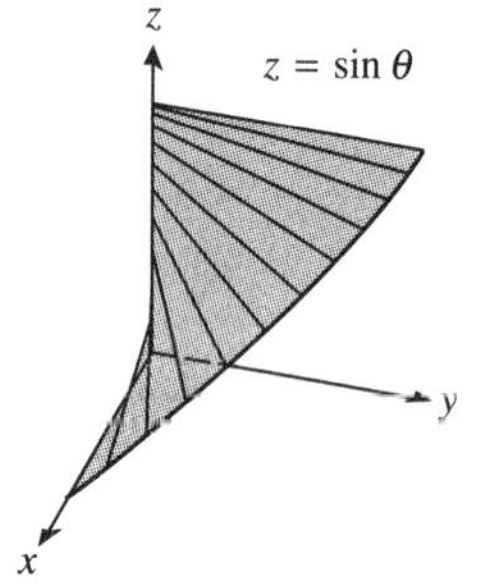

49. **(a)** $(10, \pi/2, 1)$ **(b)** $(0, 10, 1)$ **(c)** $(\sqrt{101}, \pi/2, \tan^{-1} 10)$

50. 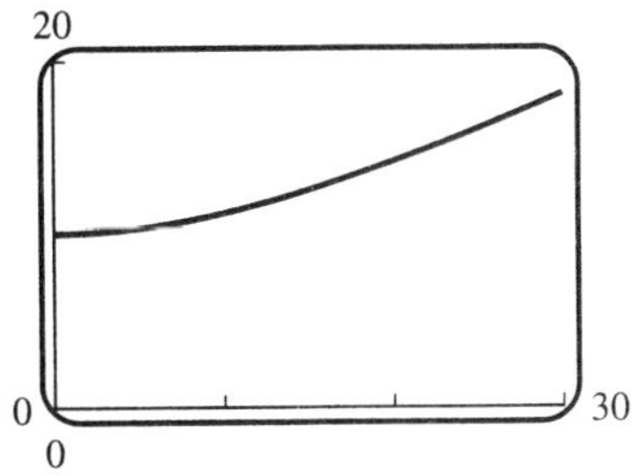

51. Using spherical coordinates: for point A, $\theta_A = 360° - 60° = 300°, \phi_A = 90° - 40° = 50°$; for point B, $\theta_B = 360° - 40° = 320°, \phi_B = 90° - 20° = 70°$. Unit vectors directed from the origin to the points A and B, respectively, are

$$\mathbf{u}_A = \sin 50° \cos 300°\mathbf{i} + \sin 50° \sin 300°\mathbf{j} + \cos 50°\mathbf{k},$$
$$\mathbf{u}_B = \sin 70° \cos 320°\mathbf{i} + \sin 70° \sin 320°\mathbf{j} + \cos 70°\mathbf{k}$$

The angle α between $\mathbf{u}_A$ and $\mathbf{u}_B$ is $\alpha = \cos^{-1}(\mathbf{u}_A \cdot \mathbf{u}_B) \approx 0.459486$ so the shortest distance is $6370\alpha \approx 2927$ km.

CHAPTER 12 SUPPLEMENTARY EXERCISES

2. **(c)** $\mathbf{F} = -\mathbf{i} - \mathbf{j}$
(d) $\|\langle 1, -2, 2\rangle\| = 3$, so $\|\mathbf{r} - \langle 1, -2, 2\rangle\| = 3$, or $(x-1)^2 + (y+2)^2 + (z-2)^2 = 9$

3. **(b)** $x = \cos 120° = -1/2, y = \pm \sin 120° = \pm\sqrt{3}/2$
(d) true: $\|\mathbf{u} \times \mathbf{v}\| = \|\mathbf{u}\|\|\mathbf{v}\||\sin(\theta)| = 1$

4. **(d)** $x + 2y - z = 0$

5. **(b)** $(y, x, z), (x, z, y), (z, y, x)$
(c) the set of points $\{(5, \theta, 1)\}, 0 \le \theta \le 2\pi$
(d) the set of points $\{(\rho, \pi/4, 0)\}, 0 \le \rho < +\infty$

6. $(x+3)^2 + (y-5)^2 + (z+4)^2 = r^2$,
(a) $r^2 = 4^2 = 16$ **(b)** $r^2 = 5^2 = 25$ **(c)** $r^2 = 3^2 = 9$

7. **(a)** $\overrightarrow{AB} = -\mathbf{i} + 2\mathbf{j} + 2\mathbf{k}$, $\overrightarrow{AC} = \mathbf{i} + \mathbf{j} - \mathbf{k}$, $\overrightarrow{AB} \times \overrightarrow{AC} = -4\mathbf{i} + \mathbf{j} - 3\mathbf{k}$, area $= \frac{1}{2}\|\overrightarrow{AB} \times \overrightarrow{AC}\| = \sqrt{26}/2$
(b) area $= \frac{1}{2}h\|\overrightarrow{AB}\| = \frac{3}{2}h = \frac{1}{2}\sqrt{26}$, $h = \sqrt{26}/3$

8. The sphere $x^2 + (y-1)^2 + (z+3)^2 = 16$ has center $Q(0, 1, -3)$ and radius 4, and $\|\overrightarrow{PQ}\| = \sqrt{1^2 + 4^2} = \sqrt{17}$, so minimum distance is $\sqrt{17} - 4$, maximum distance is $\sqrt{17} + 4$.

9. **(a)** $\mathbf{a} \cdot \mathbf{b} = 0$, $4c + 3 = 0$, $c = -3/4$
(b) Use $\mathbf{a} \cdot \mathbf{b} = \|\mathbf{a}\|\ \|\mathbf{b}\| \cos\theta$ to get $4c + 3 = \sqrt{c^2+1}(5)\cos(\pi/4)$, $4c + 3 = 5\sqrt{c^2+1}/\sqrt{2}$ Square both sides and rearrange to get $7c^2 + 48c - 7 = 0$, $(7c-1)(c+7) = 0$ so $c = -7$ (invalid) or $c = 1/7$.
(c) Proceed as in (b) with $\theta = \pi/6$ to get $11c^2 - 96c + 39 = 0$ and use the quadratic formula to get $c = \left(48 \pm 25\sqrt{3}\right)/11$.
(d) $\mathbf{a}$ must be a scalar multiple of $\mathbf{b}$, so $c\mathbf{i} + \mathbf{j} = k(4\mathbf{i} + 3\mathbf{j}), k = 1/3, c = 4/3$.

10. $\overrightarrow{OS} = \overrightarrow{OP} + \overrightarrow{PS} = 3\mathbf{i} + 4\mathbf{j} + \overrightarrow{QR} = 3\mathbf{i} + 4\mathbf{j} + (4\mathbf{i} + \mathbf{j}) = 7\mathbf{i} + 5\mathbf{j}$

11. (a) the plane through the origin which is perpendicular to $\mathbf{r}_0$

(b) the plane through the tip of $\mathbf{r}_0$ which is perpendicular to $\mathbf{r}_0$

12. The normals to the planes are given by $\langle a_1, b_1, c_1\rangle$ and $\langle a_2, b_2, c_2\rangle$, so the condition is $a_1a_2 + b_1b_2 + c_1c_2 = 0$.

13. Since $\overrightarrow{AC} \cdot (\overrightarrow{AB} \times \overrightarrow{AD}) = \overrightarrow{AC} \cdot (\overrightarrow{AB} \times \overrightarrow{CD}) + \overrightarrow{AC} \cdot (\overrightarrow{AB} \times \overrightarrow{AC}) = \mathbf{0} + \mathbf{0} = \mathbf{0}$, the volume of the parallelopiped determined by $\overrightarrow{AB}, \overrightarrow{AC}$, and $\overrightarrow{AD}$ is zero, thus A, B, C, and D are coplanar (lie in the same plane). Since $\overrightarrow{AB} \times \overrightarrow{CD} \neq \mathbf{0}$, the lines are not parallel. Hence they must intersect.

14. The points P lie on the plane determined by A, B and C.

15. (a) false, for example $\mathbf{i} \cdot \mathbf{j} = 0$

(b) false, for example $\mathbf{i} \times \mathbf{i} = \mathbf{0}$

(c) true; $0 = \|\mathbf{u}\| \cdot \|\mathbf{v}\| \cos\theta = \|\mathbf{u}\| \cdot \|\mathbf{v}\| \sin\theta$, so either $\mathbf{u} = \mathbf{0}$ or $\mathbf{v} = \mathbf{0}$ since $\cos\theta = \sin\theta = 0$ is impossible.

16. (a) Replace $\mathbf{u}$ with $\mathbf{a} \times \mathbf{b}$, $\mathbf{v}$ with $\mathbf{c}$, and $\mathbf{w}$ with $\mathbf{d}$ in the first formula of Exercise 39.

(b) From the second formula of Exercise 39,
$(\mathbf{a} \times \mathbf{b}) \times \mathbf{c} + (\mathbf{b} \times \mathbf{c}) \times \mathbf{a} + (\mathbf{c} \times \mathbf{a}) \times \mathbf{b}$
$= (\mathbf{c} \cdot \mathbf{a})\mathbf{b} - (\mathbf{c} \cdot \mathbf{b})\mathbf{a} + (\mathbf{a} \cdot \mathbf{b})\mathbf{c} - (\mathbf{a} \cdot \mathbf{c})\mathbf{b} + (\mathbf{b} \cdot \mathbf{c})\mathbf{a} - (\mathbf{b} \cdot \mathbf{a})\mathbf{c} = \mathbf{0}$

17. $\|\mathbf{u} - \mathbf{v}\|^2 = (\mathbf{u} - \mathbf{v}) \cdot (\mathbf{u} - \mathbf{v}) = \|\mathbf{u}\|^2 + \|\mathbf{v}\|^2 - 2\|\mathbf{u}\|\|\mathbf{v}\| \cos\theta = 2(1 - \cos\theta) = 4\sin^2(\theta/2)$, so $\|\mathbf{u} - \mathbf{v}\| = 2\sin(\theta/2)$

18. $\overrightarrow{AB} = \mathbf{i} - 2\mathbf{j} - 2\mathbf{k}$, $\overrightarrow{AC} = -2\mathbf{i} - \mathbf{j} - 2\mathbf{k}$, $\overrightarrow{AD} = \mathbf{i} + 2\mathbf{j} - 3\mathbf{k}$

(a) From Theorem 12.4.6 and formula (9) of Section 12.4, $\begin{vmatrix} 1 & -2 & -2 \\ -2 & -1 & -2 \\ 1 & 2 & -3 \end{vmatrix} = 29$, so $V = 29$.

(b) The plane containing A, B, and C has normal $\overrightarrow{AB} \times \overrightarrow{AC} = 2\mathbf{i} + 6\mathbf{j} - 5\mathbf{k}$, so the equation of the plane is $2(x - 1) + 6(y + 1) - 5(z - 2) = 0, 2x + 6y - 5z = -14$. From Theorem 12.6.2, $D = \dfrac{|2(2) + 6(1) - 5(-1) + 14|}{\sqrt{65}} = \dfrac{29}{\sqrt{65}}$.

19. (a) $\langle 2, 1, -1\rangle \times \langle 1, 2, 1\rangle = \langle 3, -3, 3\rangle$, so the line is parallel to $\mathbf{i} - \mathbf{j} + \mathbf{k}$. By inspection, $(0, 2, -1)$ lies on both planes, so the line has an equation $\mathbf{r} = 2\mathbf{j} - \mathbf{k} + t(\mathbf{i} - \mathbf{j} + \mathbf{k})$, that is, $x = t, y = 2 - t, z = -1 + t$.

(b) $\cos\theta = \dfrac{\langle 2, 1, -1\rangle \cdot \langle 1, 2, 1\rangle}{\|\langle 2, 1, -1\rangle\|\|\langle 1, 2, 1\rangle\|} = 1/2$, so $\theta = \pi/3$

20. Let $\alpha = 50°, \beta = 70°$, then $\gamma = \cos^{-1}\sqrt{1 - \cos^2\alpha - \cos^2\beta} \approx 47°$.

21. $5\langle \cos 60°, \cos 120°, \cos 135°\rangle = \langle 5/2, -5/2, -5\sqrt{2}/2\rangle$

22. **(a)** Let k be the length of an edge and introduce a coordinate system as shown in the figure, then $\mathbf{d} = \langle k, k, k\rangle, \mathbf{u} = \langle k, k, 0\rangle$, $\cos\theta = \dfrac{\mathbf{d}\cdot\mathbf{u}}{\|\mathbf{d}\|\,\|\mathbf{u}\|} = \dfrac{2k^2}{(k\sqrt{3})\,(k\sqrt{2})} = 2/\sqrt{6}$

so $\theta = \cos^{-1}(2/\sqrt{6}) \approx 35°$

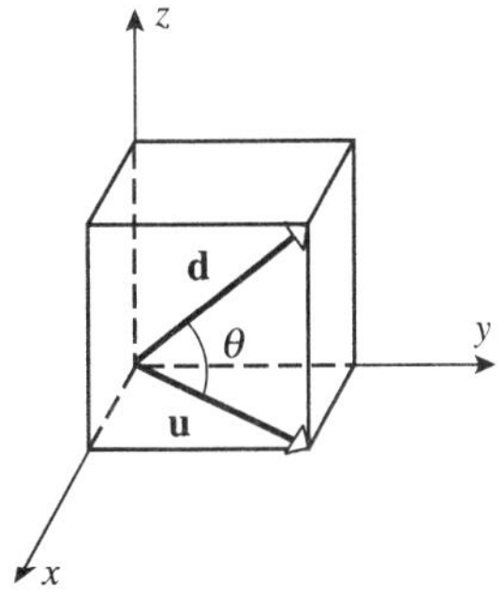

(b) $\mathbf{v} = \langle -k, 0, k\rangle, \cos\theta = \dfrac{\mathbf{d}\cdot\mathbf{v}}{\|\mathbf{d}\|\,\|\mathbf{v}\|} = 0$ so $\theta = \pi/2$ radians.

23. **(a)** $(x-3)^2 + 4(y+1)^2 - (z-2)^2 = 9$, hyperboloid of one sheet

(b) $(x+3)^2 + (y-2)^2 + (z+6)^2 = 49$, sphere

(c) $(x-1)^2 + (y+2)^2 - z^2 = 0$, circular cone

24. **(a)** perpendicular, since $\langle 2, 1, 2\rangle \cdot \langle -1, -2, 2\rangle = 0$

(b) L_1: $\langle x, y, z\rangle = \langle 1+2t, -\dfrac{3}{2}+t, -1+2t\rangle$; L_2: $\langle x, y, z\rangle = \langle 4-t, 3-2t, -4+2t\rangle$

(c) Solve simultaneously $1 + 2t_1 = 4 - t_2, -\dfrac{3}{2} + t_1 = 3 - 2t_2, -1 + 2t_1 = -4 + 2t_2$, solution $t_1 = \dfrac{1}{2}, t_2 = 2, x = 2, y = -1, z = 0$

25. **(a)** $r^2 = z; \rho^2\sin^2\phi = \rho\cos\phi, \rho = \cot\phi\csc\phi$

(b) $r^2(\cos^2\theta - \sin^2\theta) - z^2 = 0, z^2 = r^2\cos 2\theta;$
$\rho^2\sin^2\phi\cos^2\theta - \rho^2\sin^2\phi\sin^2\theta - \rho^2\cos^2\phi = 0, \cos 2\theta = \cot^2\phi$
$\sin^2\phi(\cos^2\theta - \sin^2\theta) - \cos^2\phi = 0$

26. **(a)** $z = r^2\cos^2\theta - r^2\sin^2\theta = x^2 - y^2$ **(b)** $(\rho\sin\phi\cos\theta)(\rho\cos\phi) = 1, xz = 1$

27. **(a)**

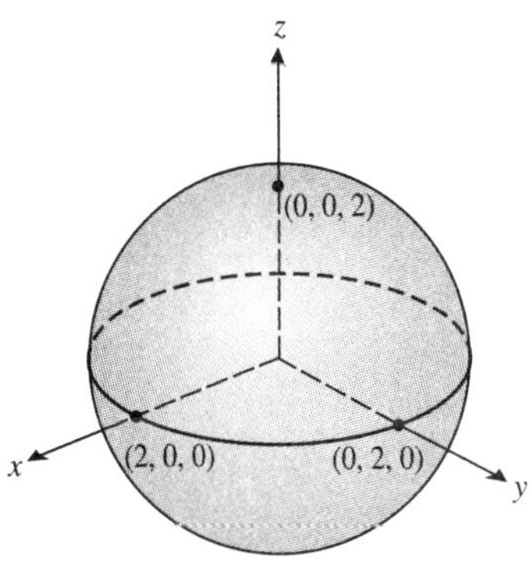

(b)

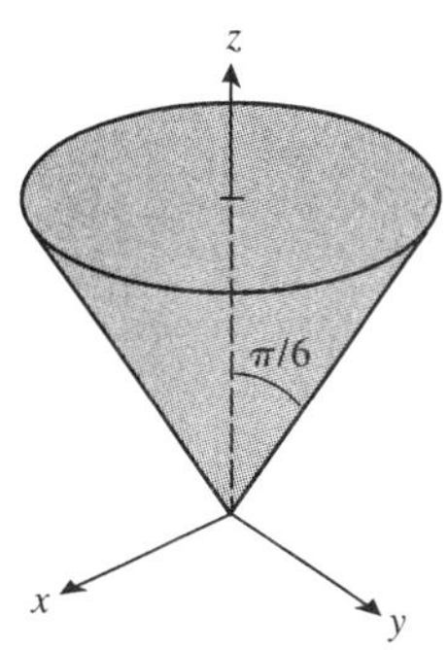

(c)

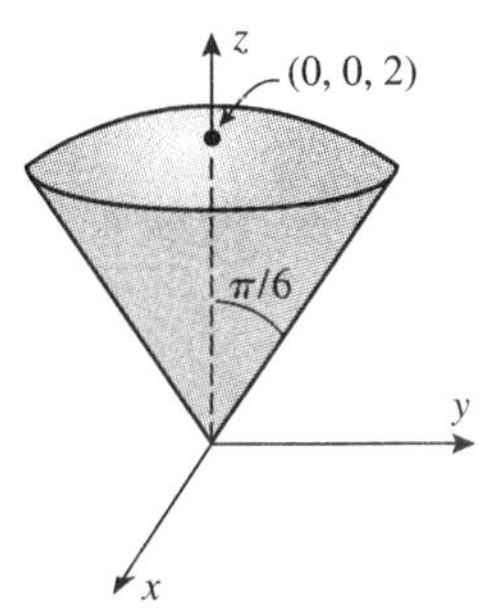

28. (a)

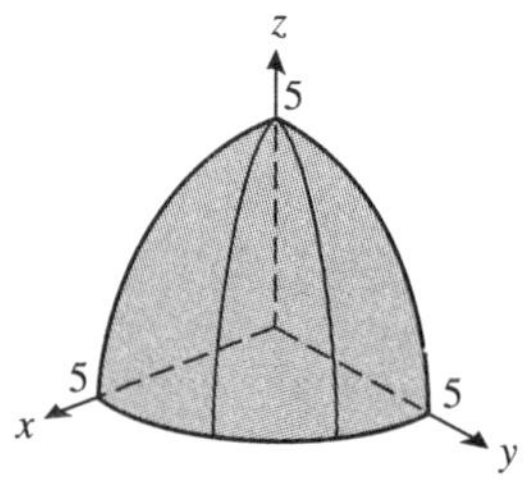

(b)

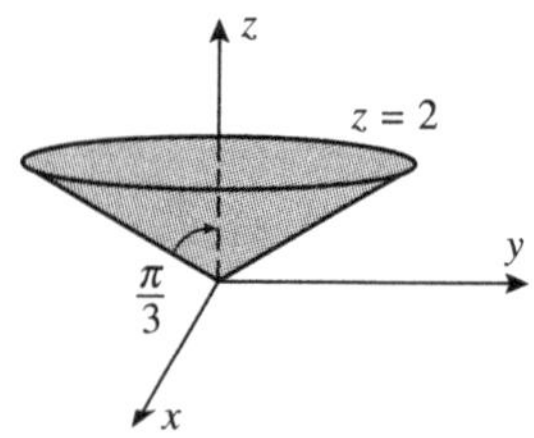

(c)

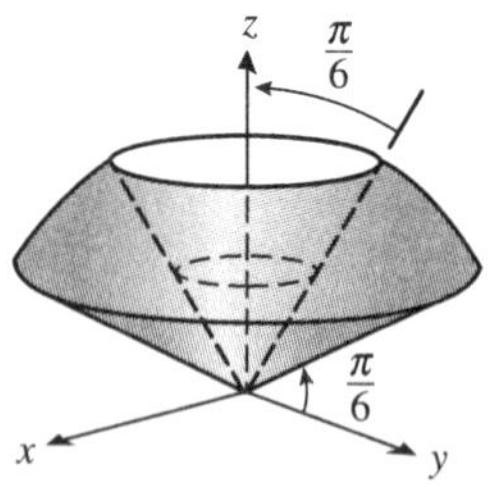

29. (a)

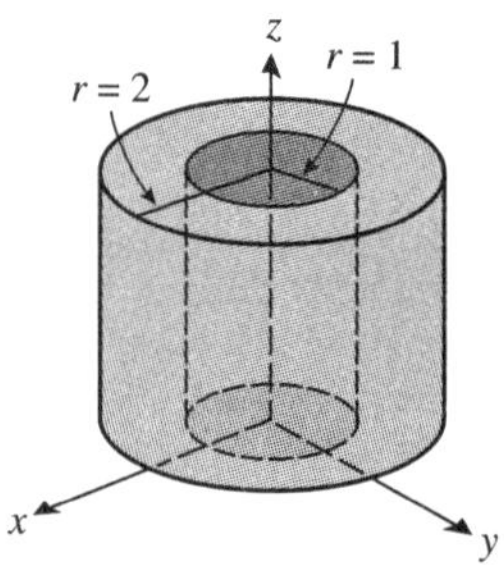

(b)

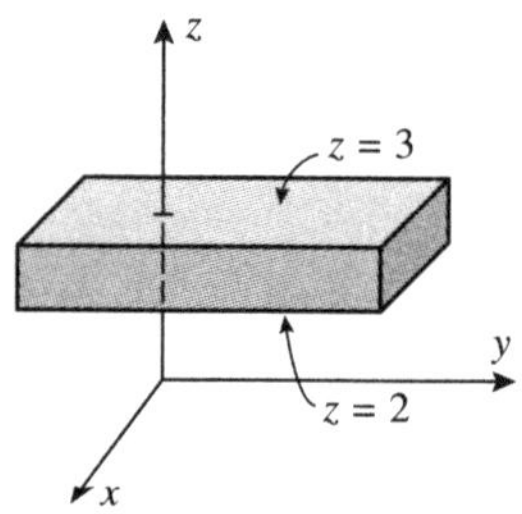

(c)

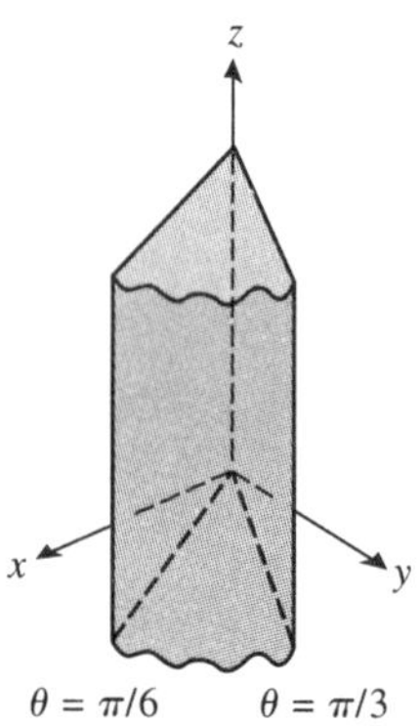

(d)

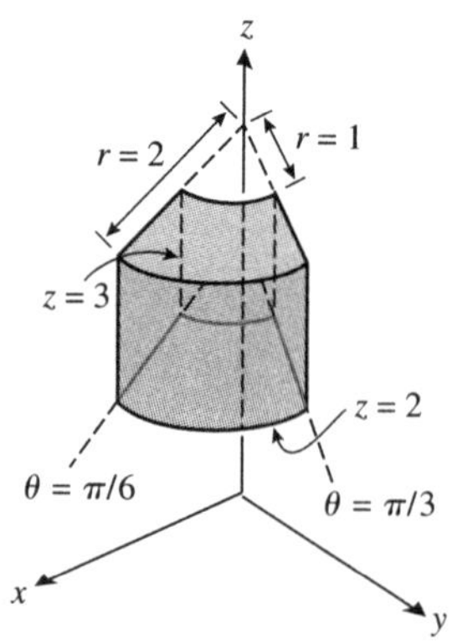

30. (a)

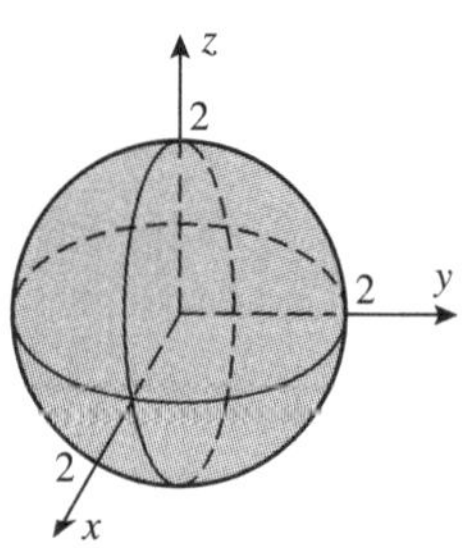

(b)

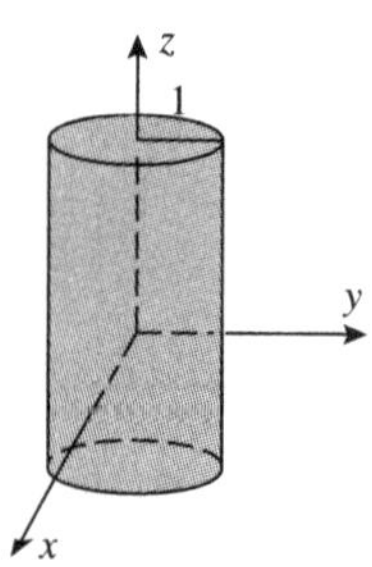

(c)

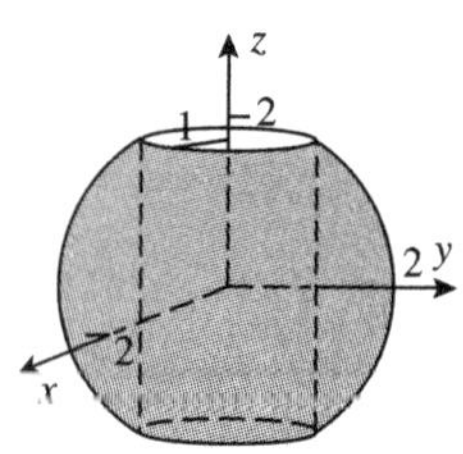

31. **(a)** At $x = c$ the trace of the surface is the circle $y^2 + z^2 = [f(c)]^2$, so the surface is given by $y^2 + z^2 = [f(x)]^2$

(b) $y^2 + z^2 = e^{2x}$ **(c)** $y^2 + z^2 = 4 - \frac{3}{4}x^2$, so let $f(x) = \sqrt{4 - \frac{3}{4}x^2}$

32. **(a)** Permute x and y in Exercise 31a: $x^2 + z^2 = [f(y)]^2$

(b) Permute x and z in Exercise 31a: $x^2 + y^2 = [f(z)]^2$

(c) Permute y and z in Exercise 31a: $y^2 + z^2 = [f(x)]^2$

33.

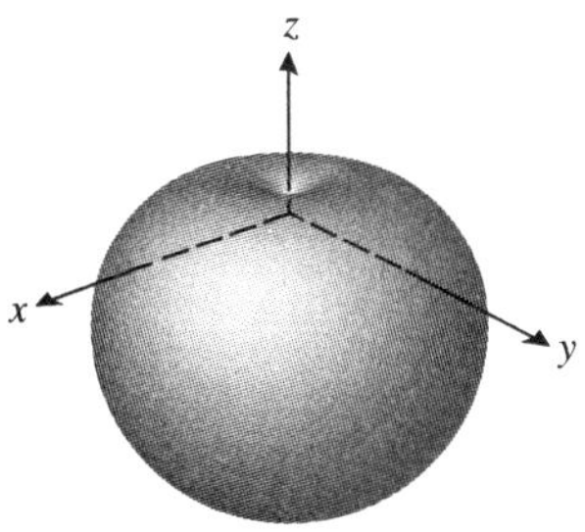

34. $\overrightarrow{PQ} = \langle 1, -1, 6 \rangle$, and $W = \mathbf{F} \cdot \overrightarrow{\mathbf{PQ}} = 13$ lb·ft

35. $\mathbf{F} = \mathbf{F}_1 + \mathbf{F}_2 = 2\mathbf{i} - \mathbf{j} + 3\mathbf{k}, \overrightarrow{PQ} = \mathbf{i} + 4\mathbf{j} - 3\mathbf{k}, W = \mathbf{F} \cdot \overrightarrow{PQ} = -11$ N·m $= -11$ J

36. $\mathbf{F}_1 = 250\cos 38°\mathbf{i} + 250\sin 38°\mathbf{j}, \mathbf{F} = 1000\mathbf{i}, \mathbf{F}_2 = \mathbf{F} - \mathbf{F}_1 = (1000 - 250\cos 38°)\mathbf{i} - 250\sin 38°\mathbf{j}$;

$\|F_2\| = 1000\sqrt{\frac{17}{16} - \frac{1}{2}\cos 38°} \approx 817.62$ N, $\theta = \tan^{-1}\frac{250\sin 38°}{250\cos 38° - 1000} \approx -11°$

37. **(a)** $\mathbf{F} = -6\mathbf{i} + 3\mathbf{j} - 6\mathbf{k}$

(b) $\overrightarrow{OA} = \langle 5, 0, 2 \rangle$, so the vector moment is $\overrightarrow{OA} \times \mathbf{F} = -6\mathbf{i} + 18\mathbf{j} + 15\mathbf{k}$

CHAPTER 13
Vector-Valued Functions

EXERCISE SET 13.1

1. $(-\infty, +\infty)$; $\mathbf{r}(\pi) = -\mathbf{i} - 3\pi\mathbf{j}$

2. $[-1/3, +\infty)$; $\mathbf{r}(1) = \langle 2, 1\rangle$

3. $[2, +\infty)$; $\mathbf{r}(3) = -\mathbf{i} - \ln 3\mathbf{j} + \mathbf{k}$

4. $[-1, 1)$; $\mathbf{r}(0) = \langle 2, 0, 0\rangle$

5. $\mathbf{r} = 3\cos t\mathbf{i} + (t + \sin t)\mathbf{j}$

6. $\mathbf{r} = (t^2 + 1)\mathbf{i} + e^{-2t}\mathbf{j}$

7. $\mathbf{r} = 2t\mathbf{i} + 2\sin 3t\mathbf{j} + 5\cos 3t\mathbf{k}$

8. $\mathbf{r} = t\sin t\mathbf{i} + \ln t\mathbf{j} + \cos^2 t\mathbf{k}$

9. $x = 3t^2$, $y = -2$

10. $x = \sin^2 t$, $y = 1 - \cos 2t$

11. $x = 2t - 1$, $y = -3\sqrt{t}$, $z = \sin 3t$

12. $x = te^{-t}$, $y = 0$, $z = -5t^2$

13. the line in 2-space through the point $(2, 0)$ and parallel to the vector $-3\mathbf{i} - 4\mathbf{j}$

14. the circle of radius 3 in the xy-plane, with center at the origin

15. the line in 3-space through the point $(0, -3, 1)$ and parallel to the vector $2\mathbf{i} + 3\mathbf{k}$

16. the circle of radius 2 in the plane $x = 3$, with center at $(3, 0, 0)$

17. an ellipse in the plane $z = -1$, center at $(0, 0, -1)$, major axis of length 6 parallel to x-axis, minor axis of length 4 parallel to y-axis

18. a parabola in the plane $x = -2$, vertex at $(-2, 0, -1)$, opening upward

19. (a) The line is parallel to the vector $-2\mathbf{i} + 3\mathbf{j}$; the slope is $-3/2$.
    (b) $y = 0$ in the xz-plane so $1 - 2t = 0$, $t = 1/2$ thus $x = 2 + 1/2 = 5/2$ and $z = 3(1/2) = 3/2$; the coordinates are $(5/2, 0, 3/2)$.

20. (a) $x = 3 + 2t = 0$, $t = -3/2$ so $y = 5(-3/2) = -15/2$
    (b) $x = t$, $y = 1 + 2t$, $z = -3t$ so $3(t) - (1 + 2t) - (-3t) = 2$, $t = 3/4$; the point of intersection is $(3/4, 5/2, -9/4)$.

21. (a)

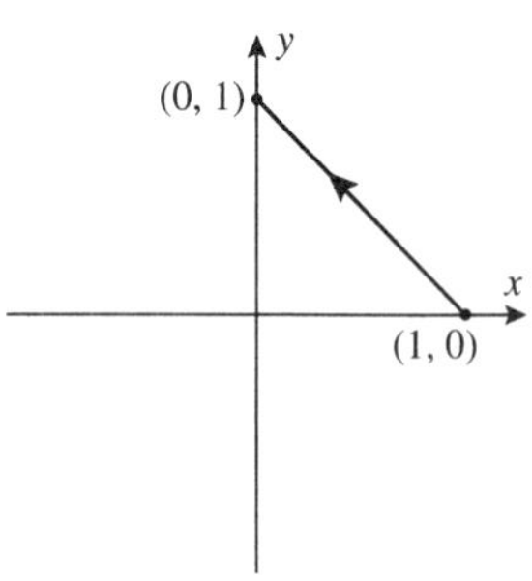

(b)

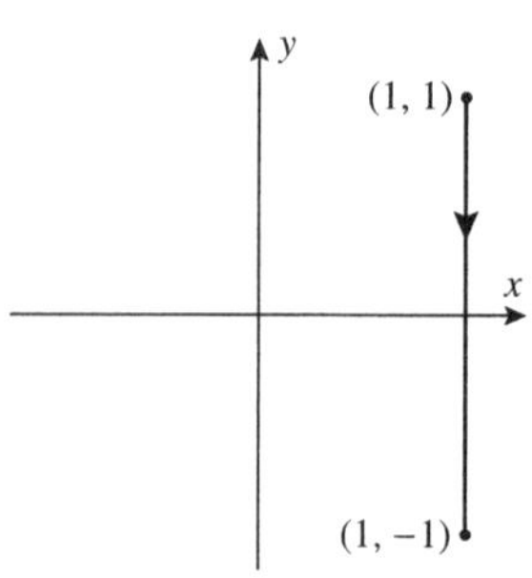

22. (a)

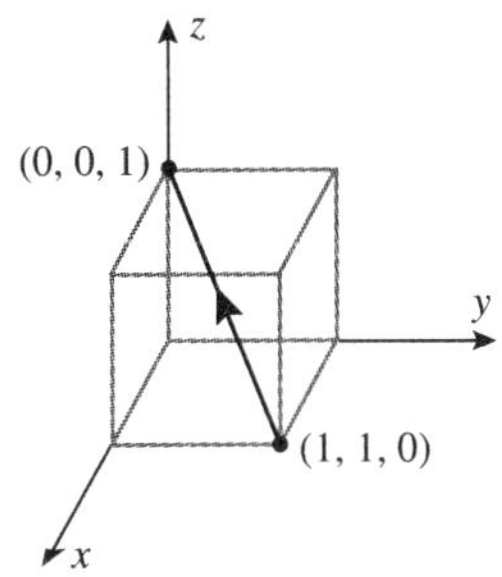

(b)

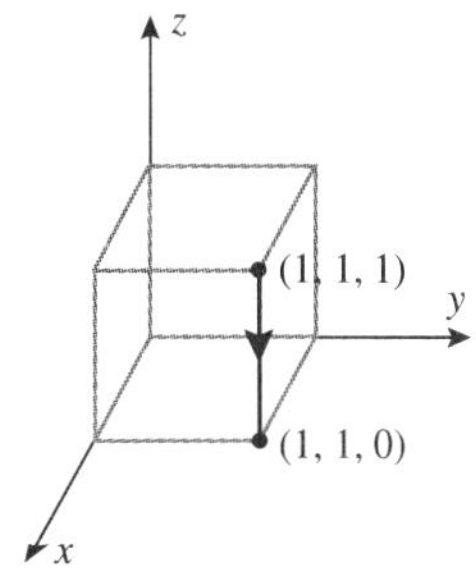

23. $\mathbf{r} = (1-t)(3\mathbf{i}+4\mathbf{j}), 0 \le t \le 1$

24. $\mathbf{r} = (1-t)4\mathbf{k} + t(2\mathbf{i}+3\mathbf{j}), 0 \le t \le 1$

25. $x = 2$

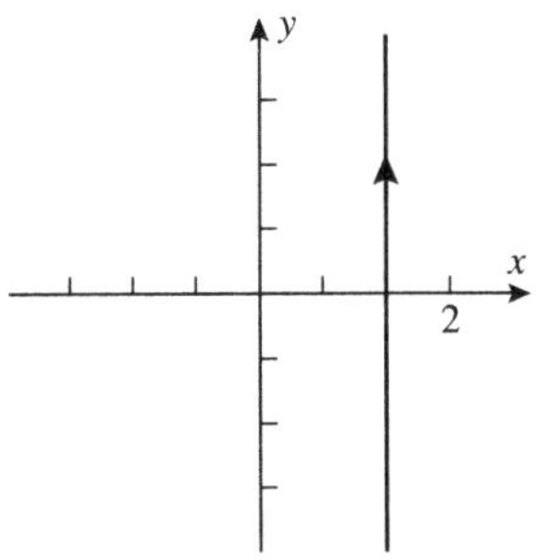

26. $y = 2x + 10$

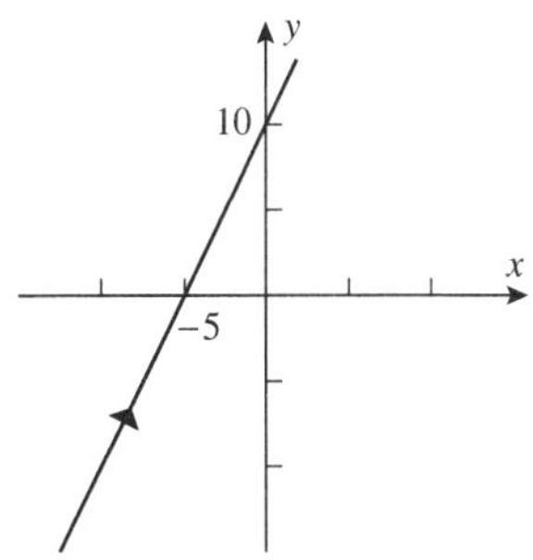

27. $(x-1)^2 + (y-3)^2 = 1$

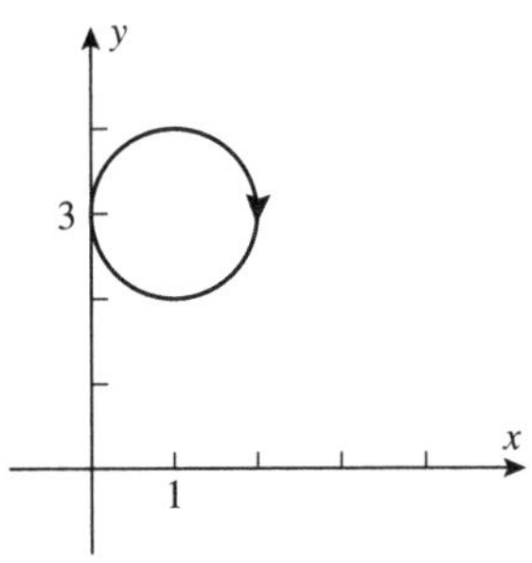

28. $x^2/4 + y^2/25 = 1$

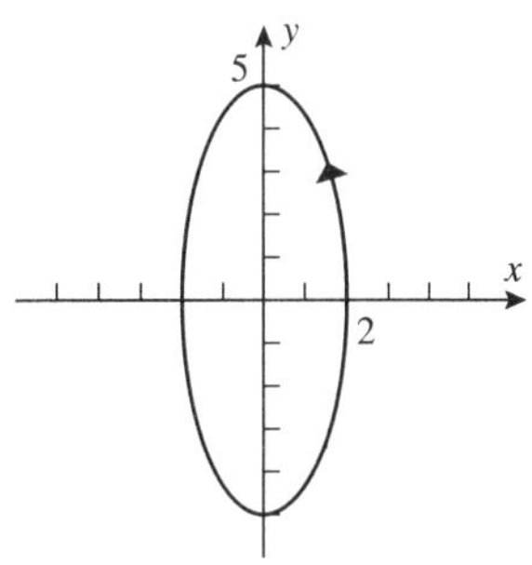

29. $x^2 - y^2 = 1,\ x \ge 1$

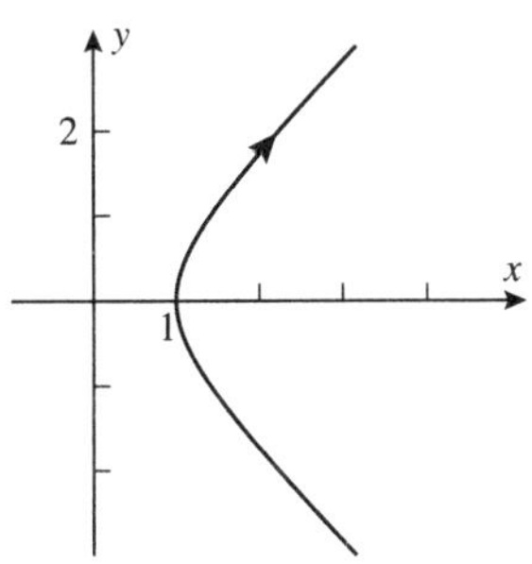

30. $y = 2x^2 + 4,\ x \ge 0$

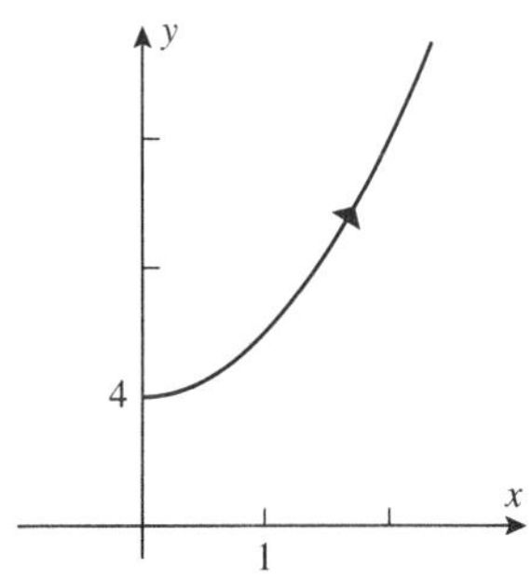

31.

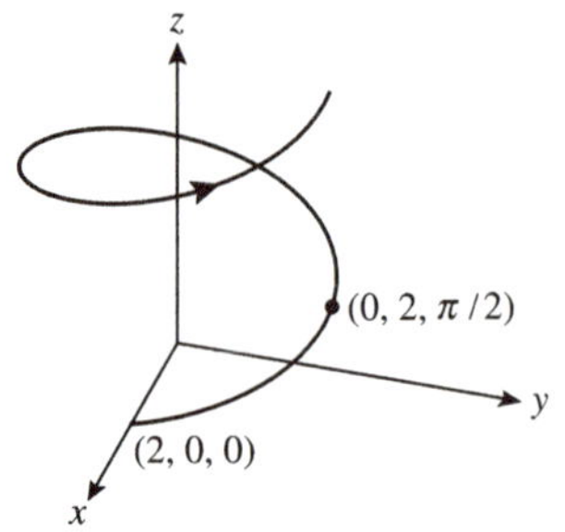

32.

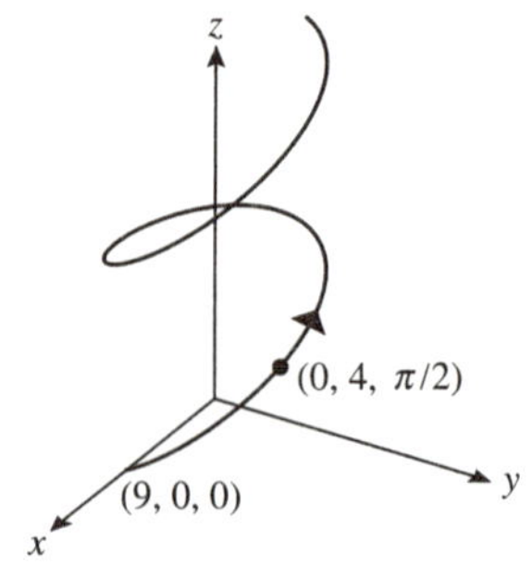

33.

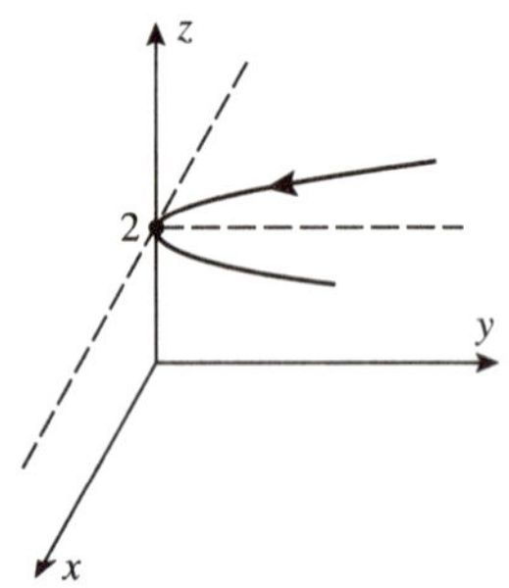

34.

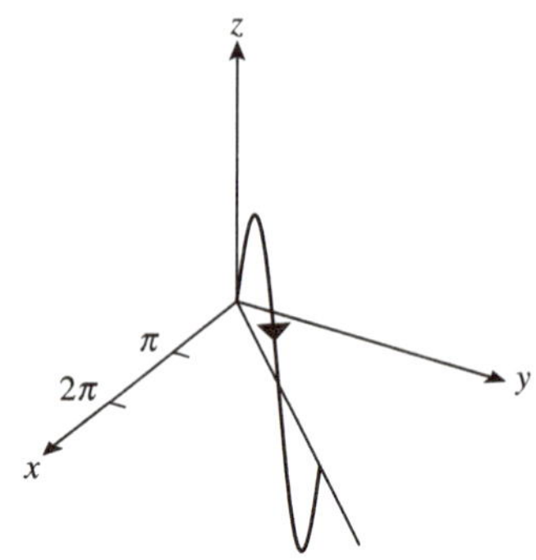

35. $x = t, y = t, z = 2t^2$

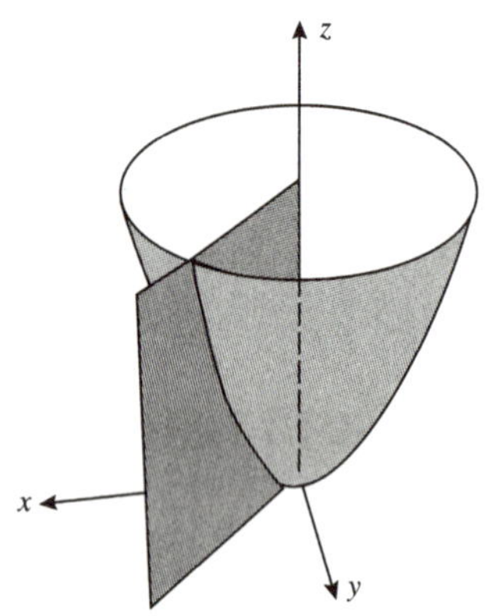

36. $x = t, y = -t, z = \sqrt{2}\sqrt{1 - t^2}$

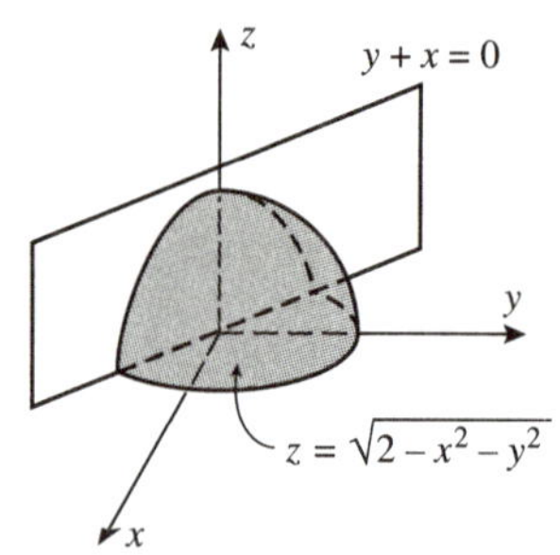

37. $\mathbf{r} = t\mathbf{i} + t^2\mathbf{j} \pm \frac{1}{3}\sqrt{81 - 9t^2 - t^4}\mathbf{k}$

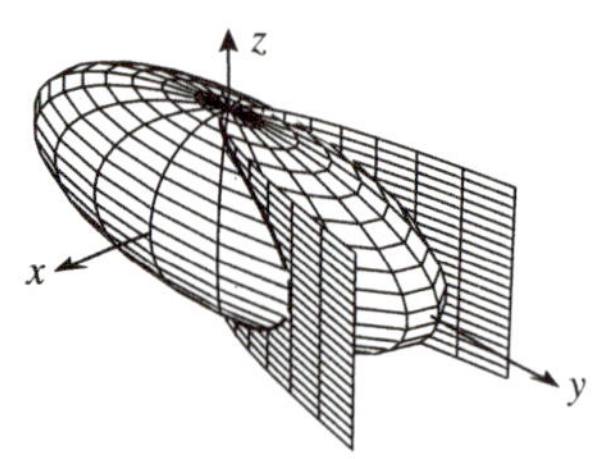

38. $\mathbf{r} = t\mathbf{i} + t\mathbf{j} + (1 - 2t)\mathbf{k}$

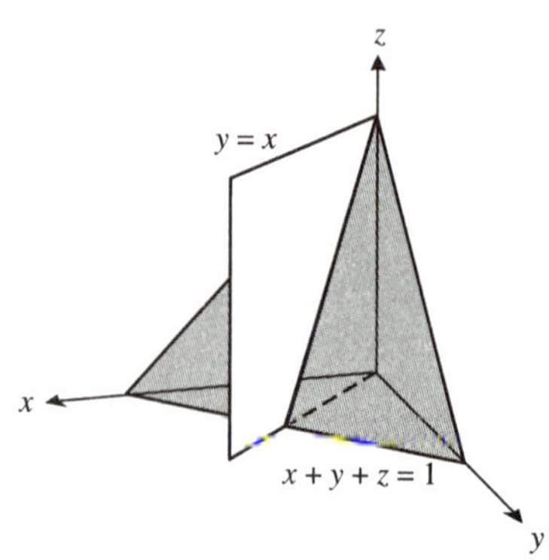

39. $x^2 + y^2 = (t \sin t)^2 + (t \cos t)^2 = t^2(\sin^2 t + \cos^2 t) = t^2 = z$

40. $x - y + z + 1 = t - (1+t)/t + (1-t^2)/t + 1 = [t^2 - (1+t) + (1-t^2) + t]/t = 0$

41. $x = \sin t$, $y = 2\cos t$, $z = \sqrt{3}\sin t$ so $x^2 + y^2 + z^2 = \sin^2 t + 4\cos^2 t + 3\sin^2 t = 4$ and $z = \sqrt{3}x$; it is the curve of intersection of the sphere $x^2 + y^2 + z^2 = 4$ and the plane $z = \sqrt{3}x$, which is a circle with center at $(0,0,0)$ and radius 2.

42. $x = 3\cos t$, $y = 3\sin t$, $z = 3\sin t$ so $x^2 + y^2 = 9\cos^2 t + 9\sin^2 t = 9$ and $z = y$; it is the curve of intersection of the circular cylinder $x^2 + y^2 = 9$ and the plane $z = y$, which is an ellipse with major axis of length $6\sqrt{2}$ and minor axis of length 6.

43. The helix makes one turn as t varies from 0 to 2π so $z = c(2\pi) = 3$, $c = 3/(2\pi)$.

44. $0.2t = 10$, $t = 50$; the helix has made one revolution when $t = 2\pi$ so when $t = 50$ it has made $50/(2\pi) = 25/\pi \approx 7.96$ revolutions.

45. $x^2 + y^2 = t^2\cos^2 t + t^2 \sin^2 t = t^2$, $\sqrt{x^2 + y^2} = t = z$; a conical helix.

46. The curve wraps around an elliptic cylinder with axis along the z-axis; an elliptical helix.

47. **(a)** III, since the curve is a subset of the plane $y = -x$

(b) IV, since only x is periodic in t, and y, z increase without bound

(c) II, since all three components are periodic in t

(d) I, since the projection onto the yz-plane is a circle and the curve increases without bound in the x-direction

49. **(a)** Let $x = 3\cos t$ and $y = 3\sin t$, then $z = 9\cos^2 t$.

(b)

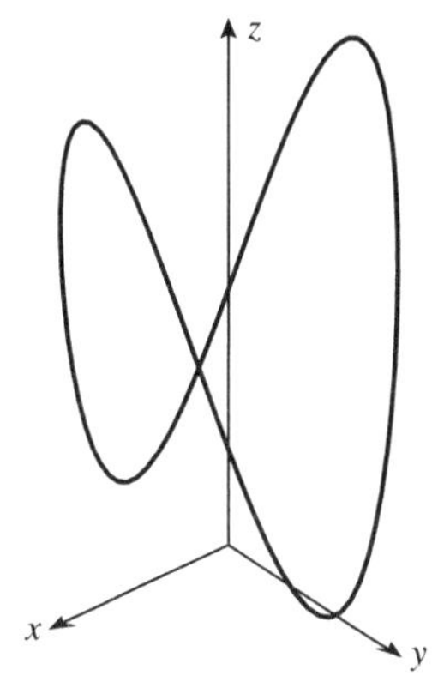

50. The plane is parallel to a line on the surface of the cone and does not go through the vertex so the curve of intersection is a parabola. Eliminate z to get $y + 2 = \sqrt{x^2 + y^2}$, $(y+2)^2 = x^2 + y^2$, $y = x^2/4 - 1$; let $x = t$, then $y = t^2/4 - 1$ and $z = t^2/4 + 1$.

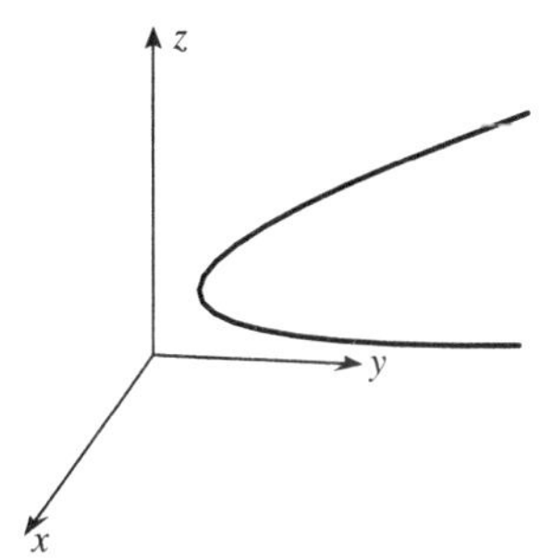

51. **(a)** 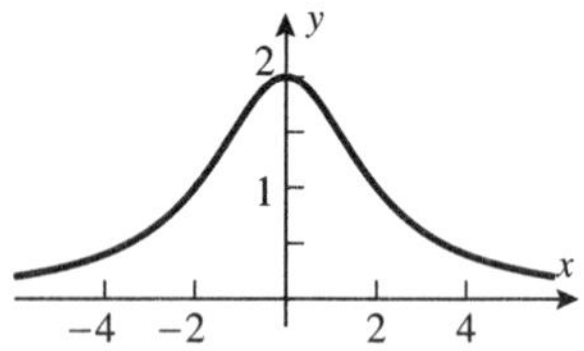

(b) In Part (a) set $x = 2t$; then $y = 2/(1 + (x/2)^2) = 8/(4 + x^2)$

EXERCISE SET 13.2

1. $9\mathbf{i} + 6\mathbf{j}$

2. $\langle\sqrt{2}/2, \sqrt{2}/2\rangle$

3. $\langle 1/3, 0\rangle$

4. $\mathbf{j}$

5. $2\mathbf{i} - 3\mathbf{j} + 4\mathbf{k}$

6. $\langle 3, 1/2, \sin 2\rangle$

7. **(a)** continuous, $\lim_{t\to 0} \mathbf{r}(t) = \mathbf{0} = \mathbf{r}(0)$ **(b)** not continuous, $\lim_{t\to 0} \mathbf{r}(t)$ does not exist

8. **(a)** not continuous, $\lim_{t\to 0} \mathbf{r}(t)$ does not exist. **(b)** continuous, $\lim_{t\to 0} \mathbf{r}(t) = 5\mathbf{i} - \mathbf{j} + \mathbf{k} = \mathbf{r}(0)$

9. 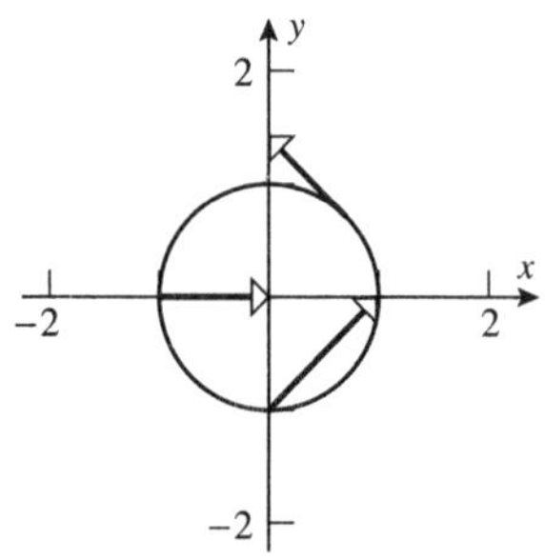

10. 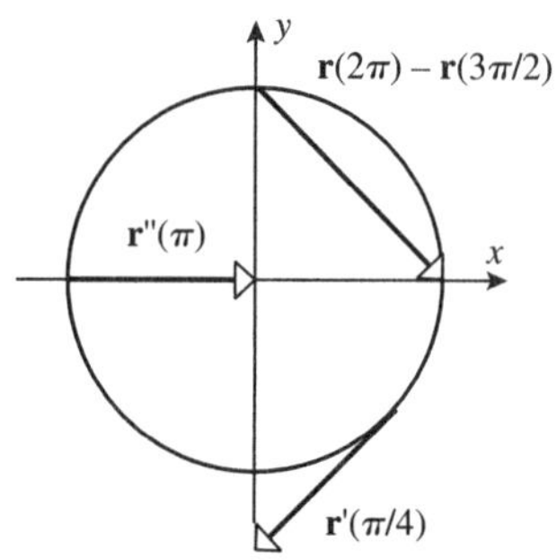

11. $\mathbf{r}'(t) = 5\mathbf{i} + (1 - 2t)\mathbf{j}$

12. $\mathbf{r}'(t) = \sin t\mathbf{j}$

13. $\mathbf{r}'(t) = -\dfrac{1}{t^2}\mathbf{i} + \sec^2 t\mathbf{j} + 2e^{2t}\mathbf{k}$

14. $\mathbf{r}'(t) = \dfrac{1}{1+t^2}\mathbf{i} + (\cos t - t\sin t)\mathbf{j} - \dfrac{1}{2\sqrt{t}}\mathbf{k}$

15. $\mathbf{r}'(t) = \langle 1, 2t\rangle$,
$\mathbf{r}'(2) = \langle 1, 4\rangle$,
$\mathbf{r}(2) = \langle 2, 4\rangle$

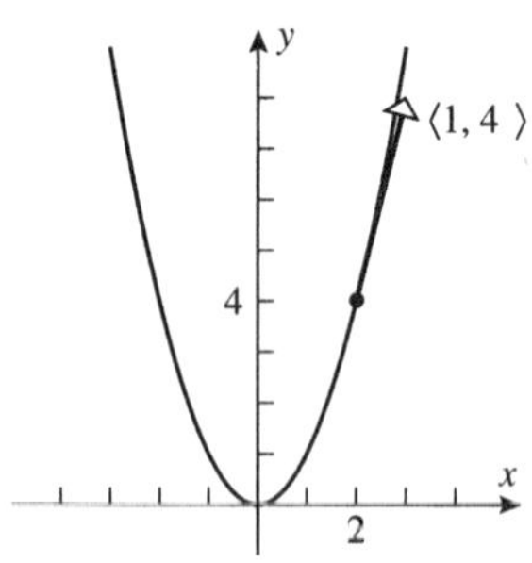

16. $\mathbf{r}'(t) = 3t^2\mathbf{i} + 2t\mathbf{j}$,
$\mathbf{r}'(1) = 3\mathbf{i} + 2\mathbf{j}$
$\mathbf{r}(1) = \mathbf{i} + \mathbf{j}$

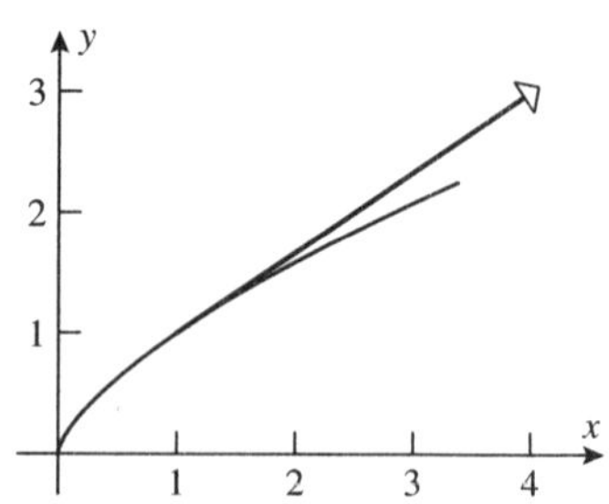

17. $\mathbf{r}'(t) = \sec t \tan t\mathbf{i} + \sec^2 t\mathbf{j}$,
$\mathbf{r}'(0) = \mathbf{j}$
$\mathbf{r}(0) = \mathbf{i}$

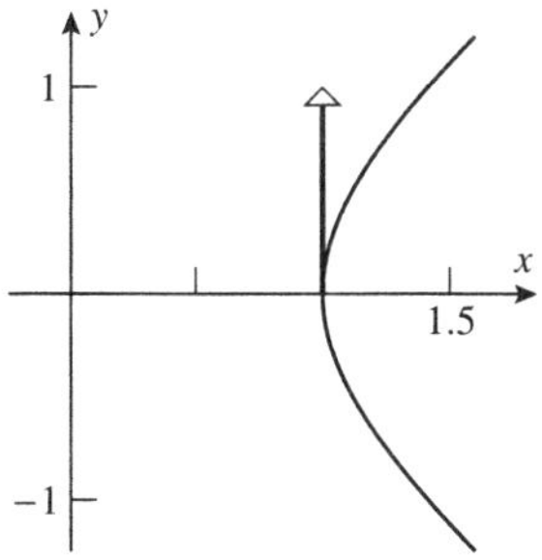

18. $\mathbf{r}'(t) = 2\cos t\mathbf{i} - 3\sin t\mathbf{j}$,
$\mathbf{r}'\left(\frac{\pi}{6}\right) = \sqrt{3}\mathbf{i} - \frac{3}{2}\mathbf{j}$
$\mathbf{r}\left(\frac{\pi}{6}\right) = \mathbf{i} + \frac{3\sqrt{3}}{2}\mathbf{j}$

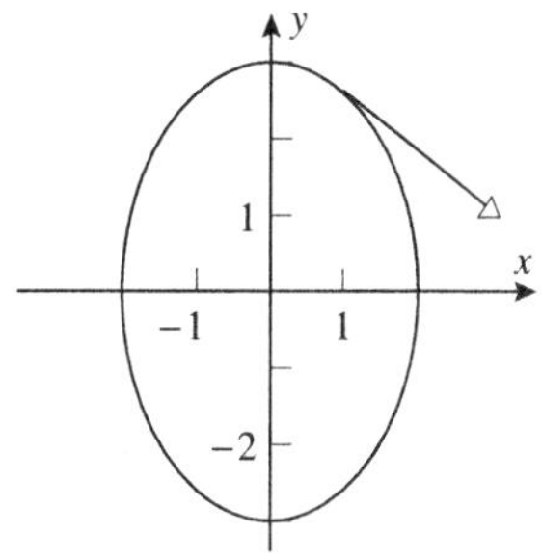

19. $\mathbf{r}'(t) = 2\cos t\mathbf{i} - 2\sin t\mathbf{k}$,
$\mathbf{r}'(\pi/2) = -2\mathbf{k}$,
$\mathbf{r}(\pi/2) = 2\mathbf{i} + \mathbf{j}$

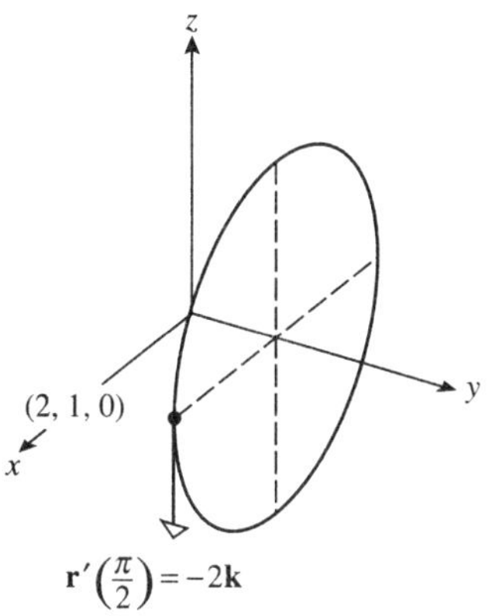

20. $\mathbf{r}'(t) = -\sin t\mathbf{i} + \cos t\mathbf{j} + \mathbf{k}$,
$\mathbf{r}'(\pi/4) = -\frac{1}{\sqrt{2}}\mathbf{i} + \frac{1}{\sqrt{2}}\mathbf{j} + \mathbf{k}$,
$\mathbf{r}(\pi/4) = \frac{1}{\sqrt{2}}\mathbf{i} + \frac{1}{\sqrt{2}}\mathbf{j} + \frac{\pi}{4}\mathbf{k}$

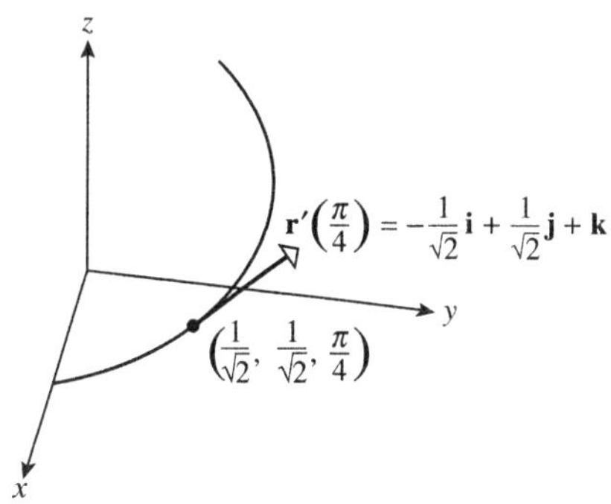

21.

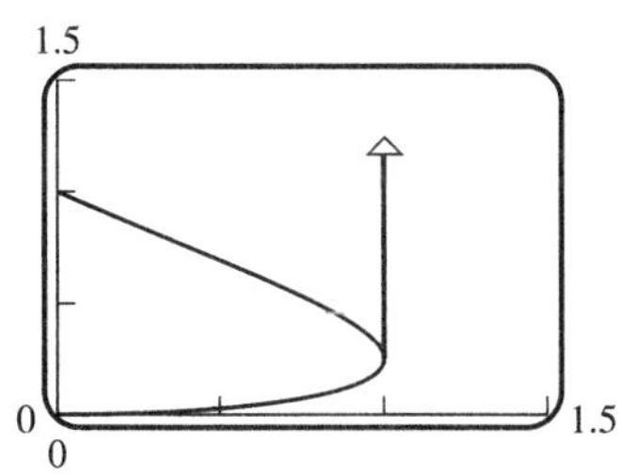

22.

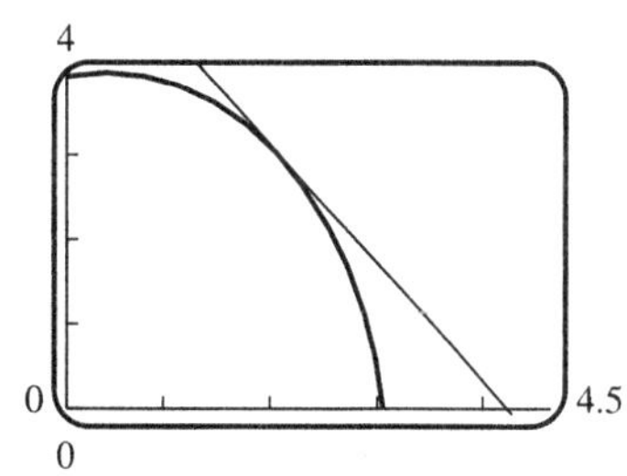

23. $\mathbf{r}'(t) = 2t\mathbf{i} - \frac{1}{t}\mathbf{j}$, $\mathbf{r}'(1) = 2\mathbf{i} - \mathbf{j}$, $\mathbf{r}(1) = \mathbf{i} + 2\mathbf{j}$; $x = 1 + 2t$, $y = 2 - t$

24. $\mathbf{r}'(t) = 2e^{2t}\mathbf{i} + 6\sin 3t\mathbf{j}$, $\mathbf{r}'(0) = 2\mathbf{i}$, $\mathbf{r}(0) = \mathbf{i} - 2\mathbf{j}$; $x = 1 + 2t$, $y = -2$

25. $\mathbf{r}'(t) = -2\pi\sin\pi t\mathbf{i} + 2\pi\cos\pi t\mathbf{j} + 3\mathbf{k}$, $\mathbf{r}'(1/3) = -\sqrt{3}\,\pi\mathbf{i} + \pi\mathbf{j} + 3\mathbf{k}$,
$\mathbf{r}(1/3) = \mathbf{i} + \sqrt{3}\,\mathbf{j} + \mathbf{k}$; $x = 1 - \sqrt{3}\,\pi t$, $y = \sqrt{3} + \pi t$, $z = 1 + 3t$

26. $\mathbf{r}'(t) = \frac{1}{t}\mathbf{i} - e^{-t}\mathbf{j} + 3t^2\mathbf{k}$, $\mathbf{r}'(2) = \frac{1}{2}\mathbf{i} - e^{-2}\mathbf{j} + 12\mathbf{k}$,

$\mathbf{r}(2) = \ln 2\mathbf{i} + e^{-2}\mathbf{j} + 8\mathbf{k}$; $x = \ln 2 + \frac{1}{2}t$, $y = e^{-2} - e^{-2}t$, $z = 8 + 12t$

27. $\mathbf{r}'(t) = 2\mathbf{i} + \frac{3}{2\sqrt{3t+4}}\mathbf{j}$, $t = 0$ at P_0 so $\mathbf{r}'(0) = 2\mathbf{i} + \frac{3}{4}\mathbf{j}$,

$\mathbf{r}(0) = -\mathbf{i} + 2\mathbf{j}$; $\mathbf{r} = (-\mathbf{i} + 2\mathbf{j}) + t\left(2\mathbf{i} + \frac{3}{4}\mathbf{j}\right)$

28. $\mathbf{r}'(t) = -4\sin t\mathbf{i} - 3\mathbf{j}$, $t = \pi/3$ at P_0 so $\mathbf{r}'(\pi/3) = -2\sqrt{3}\mathbf{i} - 3\mathbf{j}$,
$\mathbf{r}(\pi/3) = 2\mathbf{i} - \pi\mathbf{j}$; $\mathbf{r} = (2\mathbf{i} - \pi\mathbf{j}) + t(-2\sqrt{3}\mathbf{i} - 3\mathbf{j})$

29. $\mathbf{r}'(t) = 2t\mathbf{i} + \frac{1}{(t+1)^2}\mathbf{j} - 2t\mathbf{k}$, $t = -2$ at P_0 so $\mathbf{r}'(-2) = -4\mathbf{i} + \mathbf{j} + 4\mathbf{k}$,

$\mathbf{r}(-2) = 4\mathbf{i} + \mathbf{j}$; $\mathbf{r} = (4\mathbf{i} + \mathbf{j}) + t(-4\mathbf{i} + \mathbf{j} + 4\mathbf{k})$

30. $\mathbf{r}'(t) = \cos t\mathbf{i} + \sinh t\mathbf{j} + \frac{1}{1+t^2}\mathbf{k}$, $t = 0$ at P_0 so $\mathbf{r}'(0) = \mathbf{i} + \mathbf{k}$, $\mathbf{r}(0) = \mathbf{j}$; $\mathbf{r} = t\mathbf{i} + \mathbf{j} + t\mathbf{k}$

31. (a) $\lim_{t\to 0}(\mathbf{r}(t) - \mathbf{r}'(t)) = \mathbf{i} - \mathbf{j} + \mathbf{k}$

(b) $\lim_{t\to 0}(\mathbf{r}(t) \times \mathbf{r}'(t)) = \lim_{t\to 0}(-\cos t\mathbf{i} - \sin t\mathbf{j} + k) = -\mathbf{i} + \mathbf{k}$

(c) $\lim_{t\to 0}(\mathbf{r}(t) \cdot \mathbf{r}'(t)) = 0$

32. $\mathbf{r}(t) \cdot (\mathbf{r}'(t) \times \mathbf{r}''(t)) = \begin{vmatrix} t & t^2 & t^3 \\ 1 & 2t & 3t^2 \\ 0 & 2 & 6t \end{vmatrix} = 2t^3$, so $\lim_{t\to 1}\mathbf{r}(t) \cdot (\mathbf{r}'(t) \times \mathbf{r}''(t)) = 2$

33. (a) $\mathbf{r}_1' = 2\mathbf{i} + 6t\mathbf{j} + 3t^2\mathbf{k}$, $\mathbf{r}_2' = 4t^3\mathbf{k}$, $\mathbf{r}_1 \cdot \mathbf{r}_2 = t^7$; $\frac{d}{dt}(\mathbf{r}_1 \cdot \mathbf{r}_2) = 7t^6 = \mathbf{r}_1 \cdot \mathbf{r}_2' + \mathbf{r}_1' \cdot \mathbf{r}_2$

(b) $\mathbf{r}_1 \times \mathbf{r}_2 = 3t^6\mathbf{i} - 2t^5\mathbf{j}$, $\frac{d}{dt}(\mathbf{r}_1 \times \mathbf{r}_2) = 18t^5\mathbf{i} - 10t^4\mathbf{j} = \mathbf{r}_1 \times \mathbf{r}_2' + \mathbf{r}_1' \times \mathbf{r}_2$

34. (a) $\mathbf{r}_1' = -\sin t\mathbf{i} + \cos t\mathbf{j} + \mathbf{k}$, $\mathbf{r}_2' = \mathbf{k}$, $\mathbf{r}_1 \cdot \mathbf{r}_2 = \cos t + t^2$; $\frac{d}{dt}(\mathbf{r}_1 \cdot \mathbf{r}_2) = -\sin t + 2t = \mathbf{r}_1 \cdot \mathbf{r}_2' + \mathbf{r}_1' \cdot \mathbf{r}_2$

(b) $\mathbf{r}_1 \times \mathbf{r}_2 = t\sin t\mathbf{i} + t(1 - \cos t)\mathbf{j} - \sin t\mathbf{k}$,

$\frac{d}{dt}(\mathbf{r}_1 \times \mathbf{r}_2) = (\sin t + t\cos t)\mathbf{i} + (1 + t\sin t - \cos t)\mathbf{j} - \cos t\mathbf{k} = \mathbf{r}_1 \times \mathbf{r}_2' + \mathbf{r}_1' \times \mathbf{r}_2$

35. $3t\mathbf{i} + 2t^2\mathbf{j} + \mathbf{C}$

36. $(\sin t)\mathbf{i} - (\cos t)\mathbf{j} + \mathbf{C}$

37. $(-t\cos t + \sin t)\mathbf{i} + t\mathbf{j} + \mathbf{C}$

38. $\langle (t-1)e^t, t(\ln t - 1)\rangle + \mathbf{C}$

39. $(t^3/3)\mathbf{i} - t^2\mathbf{j} + \ln|t|\mathbf{k} + \mathbf{C}$

40. $\langle -e^{-t}, e^t, t^3\rangle + \mathbf{C}$

41. $\left.\left\langle \frac{1}{3}\sin 3t, \frac{1}{3}\cos 3t \right\rangle\right]_0^{\pi/3} = \langle 0, -2/3\rangle$

42. $\left.\left(\frac{1}{3}t^3\mathbf{i} + \frac{1}{4}t^4\mathbf{j}\right)\right]_0^1 = \frac{1}{3}\mathbf{i} + \frac{1}{4}\mathbf{j}$

43. $\displaystyle\int_0^2 \sqrt{t^2+t^4}\,dt = \int_0^2 t(1+t^2)^{1/2}dt = \frac{1}{3}\left(1+t^2\right)^{3/2}\Big]_0^2 = (5\sqrt{5}-1)/3$

44. $\displaystyle\left\langle -\frac{2}{5}(3-t)^{5/2}, \frac{2}{5}(3+t)^{5/2}, t\right\rangle\Big]_{-3}^{3} = \langle 72\sqrt{6}/5, 72\sqrt{6}/5, 6\rangle$

45. $\displaystyle\left(\frac{2}{3}t^{3/2}\mathbf{i}+2t^{1/2}\mathbf{j}\right)\Big]_1^9 = \frac{52}{3}\mathbf{i}+4\mathbf{j}$

46. $\displaystyle\frac{1}{2}(e^2-1)\mathbf{i}+(1-e^{-1})\mathbf{j}+\frac{1}{2}\mathbf{k}$

47. $\mathbf{y}(t) = \int \mathbf{y}'(t)dt = \frac{1}{3}t^3\mathbf{i}+t^2\mathbf{j}+\mathbf{C}, \mathbf{y}(0) = \mathbf{C} = \mathbf{i}+\mathbf{j}, \mathbf{y}(t) = (\frac{1}{3}t^3+1)\mathbf{i}+(t^2+1)\mathbf{j}$

48. $\mathbf{y}(t) = \int \mathbf{y}'(t)dt = (\sin t)\mathbf{i}-(\cos t)\mathbf{j}+\mathbf{C}$,
$\mathbf{y}(0) = -\mathbf{j}+\mathbf{C} = \mathbf{i}-\mathbf{j}$ so $\mathbf{C} = \mathbf{i}$ and $\mathbf{y}(t) = (1+\sin t)\mathbf{i}-(\cos t)\mathbf{j}$.

49. $\mathbf{y}'(t) = \int \mathbf{y}''(t)dt = t\mathbf{i}+e^t\mathbf{j}+\mathbf{C}_1, \mathbf{y}'(0) = \mathbf{j}+\mathbf{C}_1 = \mathbf{j}$ so $\mathbf{C}_1 = \mathbf{0}$ and $\mathbf{y}'(t) = t\mathbf{i}+e^t\mathbf{j}$.

$\mathbf{y}(t) = \displaystyle\int \mathbf{y}'(t)dt = \frac{1}{2}t^2\mathbf{i}+e^t\mathbf{j}+\mathbf{C}_2$, $\mathbf{y}(0) = \mathbf{j}+\mathbf{C}_2 = 2\mathbf{i}$ so $\mathbf{C}_2 = 2\mathbf{i}-\mathbf{j}$ and

$\mathbf{y}(t) = \displaystyle\left(\frac{1}{2}t^2+2\right)\mathbf{i}+(e^t-1)\mathbf{j}$

50. $\mathbf{y}'(t) = \int \mathbf{y}''(t)dt = 4t^3\mathbf{i}-t^2\mathbf{j}+\mathbf{C}_1$, $\mathbf{y}'(0) = \mathbf{C}_1 = \mathbf{0}$, $\mathbf{y}'(t) = 4t^3\mathbf{i}-t^2\mathbf{j}$

$\mathbf{y}(t) = \displaystyle\int \mathbf{y}'(t)dt = t^4\mathbf{i}-\frac{1}{3}t^3\mathbf{j}+\mathbf{C}_2$, $\mathbf{y}(0) = \mathbf{C}_2 = 2\mathbf{i}-4\mathbf{j}$, $\mathbf{y}(t) = (t^4+2)\mathbf{i}-(\frac{1}{3}t^3+4)\mathbf{j}$

51. $\mathbf{r}'(t) = -4\sin t\mathbf{i}+3\cos t\mathbf{j}$, $\mathbf{r}(t)\cdot\mathbf{r}'(t) = -7\cos t\sin t$, so $\mathbf{r}$ and $\mathbf{r}'$ are perpendicular for $t = 0, \pi/2, \pi, 3\pi/2, 2\pi$. Since

$\|\mathbf{r}(t)\| = \sqrt{16\cos^2 t+9\sin^2 t}$, $\|\mathbf{r}'(t)\| = \sqrt{16\sin^2 t+9\cos^2 t}$,

$\|\mathbf{r}\|\|\mathbf{r}'\| = \sqrt{144+337\sin^2 t\cos^2 t}$, $\theta = \cos^{-1}\left[\dfrac{-7\sin t\cos t}{\sqrt{144+337\sin^2 t\cos^2 t}}\right]$, with the graph

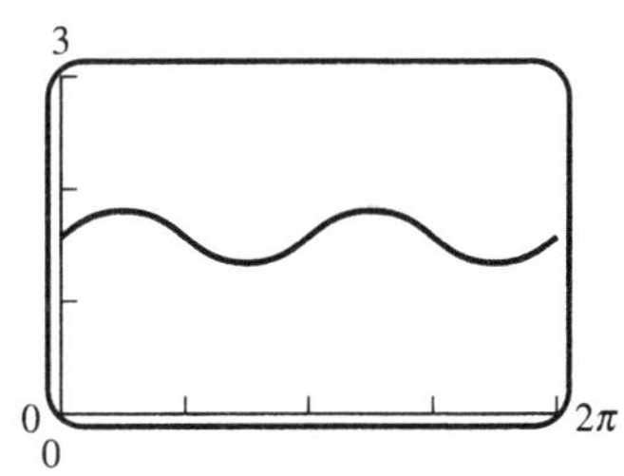

From the graph it appears that θ is bounded away from 0 and π, meaning that $\mathbf{r}$ and $\mathbf{r}'$ are never parallel. We can check this by considering them as vectors in 3-space, and then $\mathbf{r}\times\mathbf{r}' = 12\,\mathbf{k} \neq \mathbf{0}$, so they are never parallel.

52. $\mathbf{r}'(t) = 2t\mathbf{i} + 3t^2\mathbf{j}$, $\mathbf{r}(t) \cdot \mathbf{r}'(t) = 2t^3 + 3t^5 = 0$ only for $t = 0$ since $2 + 3t^2 > 0$.

$\|\mathbf{r}(t)\| = t^2\sqrt{1+t^2}$, $\|\mathbf{r}'(t)\| = t\sqrt{4+9t^2}$, $\theta = \cos^{-1}\left[\dfrac{2+3t^2}{\sqrt{1+t^2}\sqrt{4+9t^2}}\right]$ with the graph

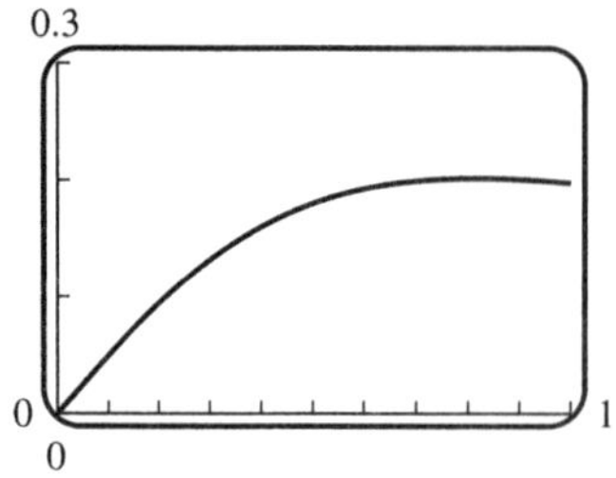

θ appears to be bounded away from π and is zero only for $t = 0$, at which point $\mathbf{r} = \mathbf{r}' = \mathbf{0}$.

53. **(a)** $2t - t^2 - 3t = -2$, $t^2 + t - 2 = 0$, $(t+2)(t-1) = 0$ so $t = -2, 1$. The points of intersection are $(-2, 4, 6)$ and $(1, 1, -3)$.

(b) $\mathbf{r}' = \mathbf{i} + 2t\mathbf{j} - 3\mathbf{k}$; $\mathbf{r}'(-2) = \mathbf{i} - 4\mathbf{j} - 3\mathbf{k}$, $\mathbf{r}'(1) = \mathbf{i} + 2\mathbf{j} - 3\mathbf{k}$, and $\mathbf{n} = 2\mathbf{i} - \mathbf{j} + \mathbf{k}$ is normal to the plane. Let θ be the acute angle, then
for $t = -2$: $\cos\theta = |\mathbf{n} \cdot \mathbf{r}'|/(\|\mathbf{n}\|\ \|\mathbf{r}'\|) = 3/\sqrt{156}$, $\theta \approx 76°$;
for $t = 1$: $\cos\theta = |\mathbf{n} \cdot \mathbf{r}'|/(\|\mathbf{n}\|\ \|\mathbf{r}'\|) = 3/\sqrt{84}$, $\theta \approx 71°$.

54. $\mathbf{r}' = -2e^{-2t}\mathbf{i} - \sin t\mathbf{j} + 3\cos t\mathbf{k}$, $t = 0$ at the point $(1, 1, 0)$ so $\mathbf{r}'(0) = -2\mathbf{i} + 3\mathbf{k}$ and hence the tangent line is $x = 1 - 2t$, $y = 1$, $z = 3t$. But $x = 0$ in the yz-plane so $1 - 2t = 0$, $t = 1/2$. The point of intersection is $(0, 1, 3/2)$.

55. $\mathbf{r}_1(1) = \mathbf{r}_2(2) = \mathbf{i} + \mathbf{j} + 3\mathbf{k}$ so the graphs intersect at P; $\mathbf{r}_1'(t) = 2t\mathbf{i} + \mathbf{j} + 9t^2\mathbf{k}$ and $\mathbf{r}_2'(t) = \mathbf{i} + \frac{1}{2}t\mathbf{j} - \mathbf{k}$ so $\mathbf{r}_1'(1) = 2\mathbf{i} + \mathbf{j} + 9\mathbf{k}$ and $\mathbf{r}_2'(2) = \mathbf{i} + \mathbf{j} - \mathbf{k}$ are tangent to the graphs at P, thus $\cos\theta = \dfrac{\mathbf{r}_1'(1) \cdot \mathbf{r}_2'(2)}{\|\mathbf{r}_1'(1)\|\ \|\mathbf{r}_2'(2)\|} = -\dfrac{6}{\sqrt{86}\sqrt{3}}$, $\theta = \cos^{-1}(6/\sqrt{258}) \approx 68°$.

56. $\mathbf{r}_1(0) = \mathbf{r}_2(-1) = 2\mathbf{i} + \mathbf{j} + 3\mathbf{k}$ so the graphs intersect at P; $\mathbf{r}_1'(t) = -2e^{-t}\mathbf{i} - (\sin t)\mathbf{j} + 2t\mathbf{k}$ and $\mathbf{r}_2'(t) = -\mathbf{i} + 2t\mathbf{j} + 3t^2\mathbf{k}$ so $\mathbf{r}_1'(0) = -2\mathbf{i}$ and $\mathbf{r}_2'(-1) = -\mathbf{i} - 2\mathbf{j} + 3\mathbf{k}$ are tangent to the graphs at P, thus $\cos\theta = \dfrac{\mathbf{r}_1'(0) \cdot \mathbf{r}_2'(-1)}{\|\mathbf{r}_1'(0)\|\ \|\mathbf{r}_2'(-1)\|} = \dfrac{1}{\sqrt{14}}$, $\theta \approx 74°$.

57. $\dfrac{d}{dt}[\mathbf{r}(t) \times \mathbf{r}'(t)] = \mathbf{r}(t) \times \mathbf{r}''(t) + \mathbf{r}'(t) \times \mathbf{r}'(t) = \mathbf{r}(t) \times \mathbf{r}''(t) + \mathbf{0} = \mathbf{r}(t) \times \mathbf{r}''(t)$

58. $\dfrac{d}{dt}[\mathbf{u} \cdot (\mathbf{v} \times \mathbf{w})] = \mathbf{u} \cdot \dfrac{d}{dt}[\mathbf{v} \times \mathbf{w}] + \dfrac{d\mathbf{u}}{dt} \cdot [\mathbf{v} \times \mathbf{w}] = \mathbf{u} \cdot \left(\mathbf{v} \times \dfrac{d\mathbf{w}}{dt} + \dfrac{d\mathbf{v}}{dt} \times \mathbf{w}\right) + \dfrac{d\mathbf{u}}{dt} \cdot [\mathbf{v} \times \mathbf{w}]$

$$= \mathbf{u} \cdot \left[\mathbf{v} \times \frac{d\mathbf{w}}{dt}\right] + \mathbf{u} \cdot \left[\frac{d\mathbf{v}}{dt} \times \mathbf{w}\right] + \frac{d\mathbf{u}}{dt} \cdot [\mathbf{v} \times \mathbf{w}]$$

59. In Exercise 58, write each scalar triple product as a determinant.

60. Let $\mathbf{c} = c_1\mathbf{i} + c_2\mathbf{j}$, $\mathbf{r}(t) = x(t)\mathbf{i} + y(t)\mathbf{j}$, $\mathbf{r}_1(t) = x_1(t)\mathbf{i} + y_1(t)\mathbf{j}$, $\mathbf{r}_2(t) = x_2(t)\mathbf{i} + y_2(t)\mathbf{j}$ and use properties of derivatives.

61. Let $\mathbf{r}_1(t) = x_1(t)\mathbf{i} + y_1(t)\mathbf{j} + z_1(t)\mathbf{k}$ and $\mathbf{r}_2(t) = x_2(t)\mathbf{i} + y_2(t)\mathbf{j} + z_2(t)\mathbf{k}$, in both (6) and (7); show that the left and right members of the equalities are the same.

62. **(a)** $\displaystyle\int k\mathbf{r}(t)\,dt = \int k(x(t)\mathbf{i} + y(t)\mathbf{j} + z(t)\mathbf{k})\,dt$

$$= k\int x(t)\,dt\,\mathbf{i} + k\int y(t)\,dt\,\mathbf{j} + k\int z(t)\,dt\,\mathbf{k} = k\int \mathbf{r}(t)\,dt$$

(b) Similar to Part (a) **(c)** Use Part (a) on Part (b) with $k = -1$

EXERCISE SET 13.3

1. **(a)** The tangent vector reverses direction at the four cusps.

(b) $\mathbf{r}'(t) = -3\cos^2 t\sin t\mathbf{i} + 3\sin^2 t\cos t\mathbf{j} = \mathbf{0}$ when $t = 0, \pi/2, \pi, 3\pi/2, 2\pi$.

2. $\mathbf{r}'(t) = \cos t\mathbf{i} + 2\sin t\cos t\mathbf{j} = \mathbf{0}$ when $t = \pi/2, 3\pi/2$. The tangent vector reverses direction at $(1, 1)$ and $(-1, 1)$.

3. $\mathbf{r}'(t) = 3t^2\mathbf{i} + (6t - 2)\mathbf{j} + 2t\mathbf{k}$; smooth

4. $\mathbf{r}'(t) = -2t\sin(t^2)\mathbf{i} + 2t\cos(t^2)\mathbf{j} - e^{-t}\mathbf{k}$; smooth

5. $\mathbf{r}'(t) = (1 - t)e^{-t}\mathbf{i} + (2t - 2)\mathbf{j} - \pi\sin(\pi t)\mathbf{k}$; not smooth, $\mathbf{r}'(1) = \mathbf{0}$

6. $\mathbf{r}'(t) = \pi\cos(\pi t)\mathbf{i} + (2 - 1/t)\mathbf{j} + (2t - 1)\mathbf{k}$; not smooth, $\mathbf{r}'(1/2) = \mathbf{0}$

7. $(dx/dt)^2 + (dy/dt)^2 + (dz/dt)^2 = (-3\cos^2 t\sin t)^2 + (3\sin^2 t\cos t)^2 + 0^2 = 9\sin^2 t\cos^2 t$,

$\displaystyle L = \int_0^{\pi/2} 3\sin t\cos t\,dt = 3/2$

8. $\displaystyle (dx/dt)^2 + (dy/dt)^2 + (dz/dt)^2 = (-3\sin t)^2 + (3\cos t)^2 + 16 = 25,\ L = \int_0^{\pi} 5dt = 5\pi$

9. $\displaystyle \mathbf{r}'(t) = \langle e^t, -e^{-t}, \sqrt{2}\rangle,\ \|\mathbf{r}'(t)\| = e^t + e^{-t},\ L = \int_0^1 (e^t + e^{-t})dt = e - e^{-1}$

10. $\displaystyle (dx/dt)^2 + (dy/dt)^2 + (dz/dt)^2 = 1/4 + (1 - t)/4 + (1 + t)/4 = 3/4,\ L = \int_{-1}^{1} (\sqrt{3}/2)dt = \sqrt{3}$

11. $\displaystyle \mathbf{r}'(t) = 3t^2\mathbf{i} + \mathbf{j} + \sqrt{6}\,t\mathbf{k},\ \|\mathbf{r}'(t)\| = 3t^2 + 1,\ L = \int_1^3 (3t^2 + 1)dt = 28$

12. $\displaystyle \mathbf{r}'(t) = 3\mathbf{i} - 2\mathbf{j} + \mathbf{k},\ \|\mathbf{r}'(t)\| = \sqrt{14},\ L = \int_3^4 \sqrt{14}\,dt = \sqrt{14}$

13. $\displaystyle \mathbf{r}'(t) = -3\sin t\mathbf{i} + 3\cos t\mathbf{j} + \mathbf{k},\ \|\mathbf{r}'(t)\| = \sqrt{10},\ L = \int_0^{2\pi} \sqrt{10}\,dt = 2\pi\sqrt{10}$

14. $\mathbf{r}'(t) = 2t\mathbf{i} + t\cos t\mathbf{j} + t\sin t\mathbf{k}, \|\mathbf{r}'(t)\| = \sqrt{5}t, L = \int_0^{\pi} \sqrt{5}t\,dt = \pi^2\sqrt{5}/2$

15. $(d\mathbf{r}/dt)(dt/d\tau) = (\mathbf{i} + 2t\mathbf{j})(4) = 4\mathbf{i} + 8t\mathbf{j} = 4\mathbf{i} + 8(4\tau + 1)\mathbf{j};$
$\mathbf{r}(\tau) = (4\tau + 1)\mathbf{i} + (4\tau + 1)^2\,\mathbf{j},\ \mathbf{r}'(\tau) = 4\mathbf{i} + 2(4)(4\tau + 1)\,\mathbf{j}$

16. $(d\mathbf{r}/dt)(dt/d\tau) = \langle -3\sin t, 3\cos t\rangle(\pi) = \langle -3\pi\sin\pi\tau,\ 3\pi\cos\pi\tau\rangle;$
$\mathbf{r}(\tau) = \langle 3\cos\pi\tau, 3\sin\pi\tau\rangle,\ \mathbf{r}'(\tau) = \langle -3\pi\sin\pi\tau, 3\pi\cos\pi\tau\rangle$

17. $(d\mathbf{r}/dt)(dt/d\tau) = (e^t\mathbf{i} - 4e^{-t}\mathbf{j})(2\tau) = 2\tau e^{\tau^2}\mathbf{i} - 8\tau e^{-\tau^2}\mathbf{j};$
$\mathbf{r}(\tau) = e^{\tau^2}\mathbf{i} + 4e^{-\tau^2}\mathbf{j},\ \mathbf{r}'(\tau) = 2\tau e^{\tau^2}\mathbf{i} - 4(2)\tau e^{-\tau^2}\mathbf{j}$

18. $(d\mathbf{r}/dt)(dt/d\tau) = \left(\frac{9}{2}t^{1/2}\mathbf{j} + \mathbf{k}\right)(-1/\tau^2) = -\frac{9}{2\tau^{5/2}}\mathbf{j} - \frac{1}{\tau^2}\mathbf{k};$

$\mathbf{r}(\tau) = \mathbf{i} + 3\tau^{-3/2}\mathbf{j} + \frac{1}{\tau}\mathbf{k},\ \mathbf{r}'(\tau) = -\frac{9}{2}\tau^{-5/2}\mathbf{j} - \frac{1}{\tau^2}\mathbf{k}$

19. **(a)** $\|\mathbf{r}'(t)\| = \sqrt{2}, s = \int_0^t \sqrt{2}\,dt = \sqrt{2}t; \mathbf{r} = \frac{s}{\sqrt{2}}\mathbf{i} + \frac{s}{\sqrt{2}}\mathbf{j}, x = \frac{s}{\sqrt{2}}, y = \frac{s}{\sqrt{2}}$

(b) Similar to Part (a), $x = y = z = \frac{s}{\sqrt{3}}$

20. **(a)** $x = -\frac{s}{\sqrt{2}}, y = -\frac{s}{\sqrt{2}}$

(b) $x = -\frac{s}{\sqrt{3}}, y = -\frac{s}{\sqrt{3}}, z = -\frac{s}{\sqrt{3}}$

21. **(a)** $\mathbf{r}(t) = \langle 1, 3, 4\rangle$ when $t = 0$,

so $s = \int_0^t \sqrt{1 + 4 + 4}\,du = 3t, x = 1 + s/3, y = 3 - 2s/3, z = 4 + 2s/3$

(b) $\mathbf{r}\Big]_{s=25} = \langle 28/3, -41/3, 62/3\rangle$

22. **(a)** $\mathbf{r}(t) = \langle -5, 0, 1\rangle$ when $t = 0$, so $s = \int_0^t \sqrt{9 + 4 + 1}\,du = \sqrt{14}t,$

$x = -5 + 3s/\sqrt{14}, y = 2s/\sqrt{14}, z = 5 + s/\sqrt{14}$

(b) $\mathbf{r}(s)\Big]_{s=10} = \langle -5 + 30/\sqrt{14}, 20/\sqrt{14}, 5 + 10/\sqrt{14}\rangle$

23. $x = 3 + \cos t,\ y = 2 + \sin t,\ (dx/dt)^2 + (dy/dt)^2 = 1,$

$s = \int_0^t du = t$ so $t = s,\ x = 3 + \cos s,\ y = 2 + \sin s$ for $0 \le s \le 2\pi$.

24. $x = \cos^3 t,\ y = \sin^3 t,\ (dx/dt)^2 + (dy/dt)^2 = 9\sin^2 t\cos^2 t,$

$s = \int_0^t 3\sin u\cos u\,du = \frac{3}{2}\sin^2 t$ so $\sin t = (2s/3)^{1/2},\ \cos t = (1 - 2s/3)^{1/2},$

$x = (1 - 2s/3)^{3/2},\ y = (2s/3)^{3/2}$ for $0 \le s \le 3/2$

25. $x = t^3/3$, $y = t^2/2$, $(dx/dt)^2 + (dy/dt)^2 = t^2(t^2+1)$,

$$s = \int_0^t u(u^2+1)^{1/2}du = \frac{1}{3}[(t^2+1)^{3/2} - 1] \text{ so } t = [(3s+1)^{2/3} - 1]^{1/2},$$

$$x = \frac{1}{3}[(3s+1)^{2/3} - 1]^{3/2},\ y = \frac{1}{2}[(3s+1)^{2/3} - 1] \text{ for } s \geq 0$$

26. $x = (1+t)^2$, $y = (1+t)^3$, $(dx/dt)^2 + (dy/dt)^2 = (1+t)^2[4 + 9(1+t)^2]$,

$$s = \int_0^t (1+u)[4+9(1+u)^2]^{1/2}du = \frac{1}{27}([4+9(1+t)^2]^{3/2} - 13\sqrt{13}) \text{ so}$$

$$1 + t = \frac{1}{3}[(27s + 13\sqrt{13})^{2/3} - 4]^{1/2},\ x = \frac{1}{9}[(27s + 13\sqrt{13})^{2/3} - 4],$$

$$y = \frac{1}{27}[(27s + 13\sqrt{13})^{2/3} - 4]^{3/2} \text{ for } 0 \leq s \leq (80\sqrt{10} - 13\sqrt{13})/27$$

27. $x = e^t \cos t$, $y = e^t \sin t$, $(dx/dt)^2 + (dy/dt)^2 = 2e^{2t}$, $s = \int_0^t \sqrt{2}\, e^u\, du = \sqrt{2}(e^t - 1)$ so

$t = \ln(s/\sqrt{2} + 1)$, $x = (s/\sqrt{2} + 1)\cos[\ln(s/\sqrt{2} + 1)]$, $y = (s/\sqrt{2} + 1)\sin[\ln(s/\sqrt{2} + 1)]$
for $0 \leq s \leq \sqrt{2}(e^{\pi/2} - 1)$

28. $x = \sin(e^t)$, $y = \cos(e^t)$, $z = \sqrt{3}e^t$,

$$(dx/dt)^2 + (dy/dt)^2 + (dz/dt)^2 = 4e^{2t},\ s = \int_0^t 2e^u\, du = 2(e^t - 1) \text{ so}$$

$e^t = 1 + s/2$; $x = \sin(1 + s/2)$, $y = \cos(1 + s/2)$, $z = \sqrt{3}(1 + s/2)$ for $s \geq 0$

29. $dx/dt = -a\sin t$, $dy/dt = a\cos t$, $dz/dt = c$,

$$s(t_0) = L = \int_0^{t_0} \sqrt{a^2\sin^2 t + a^2\cos^2 t + c^2}\, dt = \int_0^{t_0} \sqrt{a^2 + c^2}\, dt = t_0\sqrt{a^2 + c^2}$$

30. From Exercise 29, $s(t_0) = t_0\sqrt{a^2 + c^2} = wt_0$, so $s(t) = wt$ and

$$\mathbf{r} = a\cos\frac{s}{w}\mathbf{i} + \sin\frac{s}{w}\mathbf{j} + \frac{bs}{w}\mathbf{k}.$$

31. $x = at - a\sin t$, $y = a - a\cos t$, $(dx/dt)^2 + (dy/dt)^2 = 4a^2\sin^2(t/2)$,

$$s = \int_0^t 2a\sin(u/2)du = 4a[1 - \cos(t/2)] \text{ so } \cos(t/2) = 1 - s/(4a),\ t = 2\cos^{-1}[1 - s/(4a)],$$

$\cos t = 2\cos^2(t/2) - 1 = 2[1 - s/(4a)]^2 - 1$,

$\sin t = 2\sin(t/2)\cos(t/2) = 2(1 - [1 - s/(4a)]^2)^{1/2}(2[1 - s/(4a)]^2 - 1)$,

$x = 2a\cos^{-1}[1 - s/(4a)] - 2a(1 - [1 - s/(4a)]^2)^{1/2}(2[1 - s/(4a)]^2 - 1)$,

$$y = \frac{s(8a - s)}{8a} \text{ for } 0 \leq s \leq 8a$$

32. $\dfrac{dx}{dt} = \cos\theta\dfrac{dr}{dt} - r\sin\theta\dfrac{d\theta}{dt}$, $\dfrac{dy}{dt} = \sin\theta\dfrac{dr}{dt} + r\cos\theta\dfrac{d\theta}{dt}$,

$$\left(\frac{dx}{dt}\right)^2 + \left(\frac{dy}{dt}\right)^2 + \left(\frac{dz}{dt}\right)^2 = \left(\frac{dr}{dt}\right)^2 + r^2\left(\frac{d\theta}{dt}\right)^2 + \left(\frac{dz}{dt}\right)^2$$

33. (a) $(dr/dt)^2 + r^2(d\theta/dt)^2 + (dz/dt)^2 = 9e^{4t}$, $L = \int_0^{\ln 2} 3e^{2t}dt = \frac{3}{2}e^{2t}\Big]_0^{\ln 2} = 9/2$

(b) $(dr/dt)^2 + r^2(d\theta/dt)^2 + (dz/dt)^2 = 5t^2 + t^4 = t^2(5+t^2)$,
$L = \int_1^2 t(5+t^2)^{1/2}dt = 9 - 2\sqrt{6}$

34. $\frac{dx}{dt} = \sin\phi\cos\theta\frac{d\rho}{dt} + \rho\cos\phi\cos\theta\frac{d\phi}{dt} - \rho\sin\phi\sin\theta\frac{d\theta}{dt}$,
$\frac{dy}{dt} = \sin\phi\sin\theta\frac{d\rho}{dt} + \rho\cos\phi\sin\theta\frac{d\phi}{dt} + \rho\sin\phi\cos\theta\frac{d\theta}{dt}$, $\frac{dz}{dt} = \cos\phi\frac{d\rho}{dt} - \rho\sin\phi\frac{d\phi}{dt}$,
$\left(\frac{dx}{dt}\right)^2 + \left(\frac{dy}{dt}\right)^2 + \left(\frac{dz}{dt}\right)^2 = \left(\frac{d\rho}{dt}\right)^2 + \rho^2\sin^2\phi\left(\frac{d\theta}{dt}\right)^2 + \rho^2\left(\frac{d\phi}{dt}\right)^2$

35. (a) $(d\rho/dt)^2 + \rho^2\sin^2\phi(d\theta/dt)^2 + \rho^2(d\phi/dt)^2 = 3e^{-2t}$, $L = \int_0^2 \sqrt{3}e^{-t}dt = \sqrt{3}(1-e^{-2})$

(b) $(d\rho/dt)^2 + \rho^2\sin^2\phi(d\theta/dt)^2 + \rho^2(d\phi/dt)^2 = 5$, $L = \int_1^5 \sqrt{5}dt = 4\sqrt{5}$

36. (a) $\frac{d}{dt}\mathbf{r}(t) = \mathbf{i} + 2t\mathbf{j}$ is never zero, but $\frac{d}{d\tau}\mathbf{r}(\tau^3) = \frac{d}{d\tau}(\tau^3\mathbf{i} + \tau^6\mathbf{j}) = 3\tau^2\mathbf{i} + 6\tau^5\mathbf{j}$ is zero at $\tau = 0$.

(b) $\frac{d\mathbf{r}}{d\tau} = \frac{d\mathbf{r}}{dt}\frac{dt}{d\tau}$, and since $t = \tau^3$, $\frac{dt}{d\tau} = 0$ when $\tau = 0$.

37. (a) $g(\tau) = \pi\tau$ (b) $g(\tau) = \pi(1-\tau)$

38. $t = 1 - \tau$

39. Represent the helix by $x = a\cos t$, $y = a\sin t$, $z = ct$ with $a = 6.25$ and $c = 10/\pi$, so that the radius of the helix is the distance from the axis of the cylinder to the center of the copper cable, and the helix makes one turn in a distance of 20 in. ($t = 2\pi$). From Exercise 29 the length of the helix is $2\pi\sqrt{6.25^2 + (10/\pi)^2} \approx 44$ in.

40. $\mathbf{r}(t) = \cos t\mathbf{i} + \sin t\mathbf{j} + t^{3/2}\mathbf{k}$, $\mathbf{r}'(t) = -\sin t\mathbf{i} + \cos t\mathbf{j} + \frac{3}{2}t^{1/2}\mathbf{k}$

(a) $\|\mathbf{r}'(t)\| = \sqrt{\sin^2 t + \cos^2 t + 9t/4} = \frac{1}{2}\sqrt{4+9t}$

(b) $\frac{ds}{dt} = \frac{1}{2}\sqrt{4+9t}$

(c) $\int_0^2 \frac{1}{2}\sqrt{4+9t}\,dt = \frac{2}{27}(11\sqrt{22} - 4)$

41. $\mathbf{r}'(t) = (1/t)\mathbf{i} + 2\mathbf{j} + 2t\mathbf{k}$

(a) $\|\mathbf{r}'(t)\| = \sqrt{1/t^2 + 4 + 4t^2} = \sqrt{(2t+1/t)^2} = 2t + 1/t$

(b) $\frac{ds}{dt} = 2t + 1/t$

(c) $\int_1^3 (2t + 1/t)dt = 8 + \ln 3$

42. If $\mathbf{r}(t) = x(t)\mathbf{i} + y(t)\mathbf{j} + z(t)\mathbf{k}$ is smooth, then $\|\mathbf{r}'(t)\|$ is continuous and nonzero. Thus the angle between $\mathbf{r}'(t)$ and $\mathbf{i}$, given by $\cos^{-1}(x'(t)/\|\mathbf{r}'(t)\|)$, is a continuous function of t. Similarly, the angles between $\mathbf{r}'(t)$ and the vectors $\mathbf{j}$ and $\mathbf{k}$ are continuous functions of t.

43. Let $\mathbf{r}(t) = x(t)\mathbf{i} + y(t)\mathbf{j}$ and use the chain rule.

EXERCISE SET 13.4

1. (a)

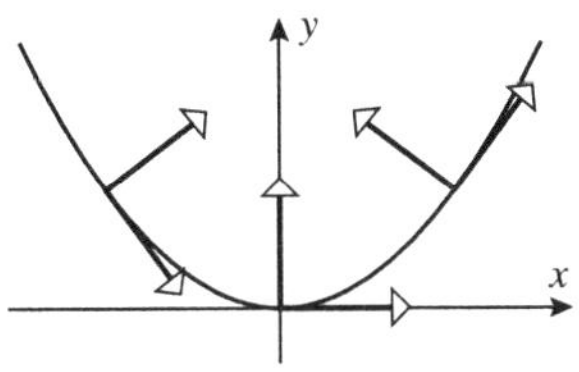

(b)

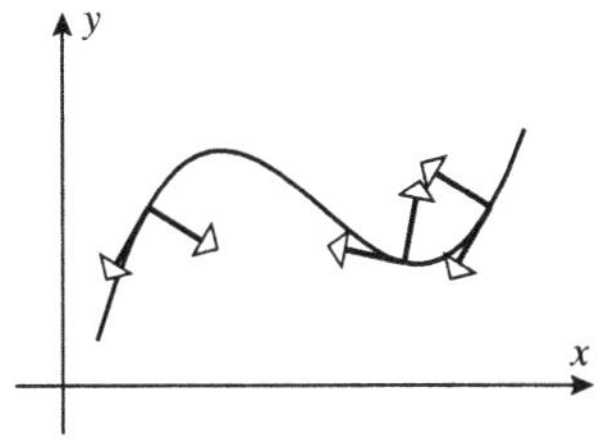

2.

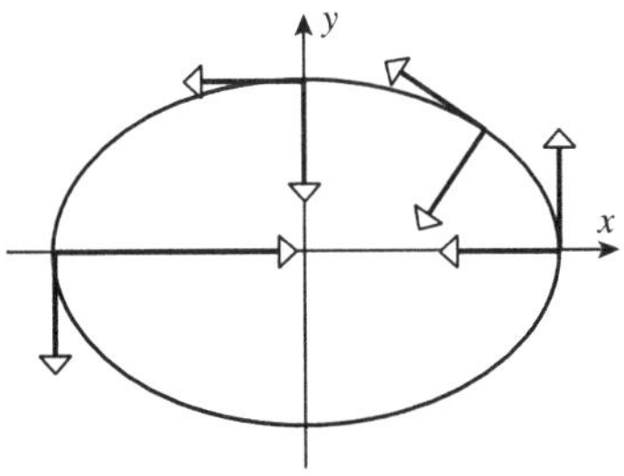

3. $\mathbf{r}'(t) = 2t\mathbf{i} + \mathbf{j}$, $\|\mathbf{r}'(t)\| = \sqrt{4t^2+1}$, $\mathbf{T}(t) = (4t^2+1)^{-1/2}(2t\mathbf{i}+\mathbf{j})$,
$\mathbf{T}'(t) = (4t^2+1)^{-1/2}(2\mathbf{i}) - 4t(4t^2+1)^{-3/2}(2t\mathbf{i}+\mathbf{j})$;
$\mathbf{T}(1) = \frac{2}{\sqrt{5}}\mathbf{i} + \frac{1}{\sqrt{5}}\mathbf{j}$, $\mathbf{T}'(1) = \frac{2}{5\sqrt{5}}(\mathbf{i} - 2\mathbf{j})$, $\mathbf{N}(1) = \frac{1}{\sqrt{5}}\mathbf{i} - \frac{2}{\sqrt{5}}\mathbf{j}$.

4. $\mathbf{r}'(t) = t\mathbf{i} + t^2\mathbf{j}$, $\mathbf{T}(t) = (t^2+t^4)^{-1/2}(t\mathbf{i}+t^2\mathbf{j})$,
$\mathbf{T}'(t) = (t^2+t^4)^{-1/2}(\mathbf{i}+2t\mathbf{j}) - (t+2t^3)(t^2+t^4)^{-3/2}(t\mathbf{i}+t^2\mathbf{j})$;
$\mathbf{T}(1) = \frac{1}{\sqrt{2}}\mathbf{i} + \frac{1}{\sqrt{2}}\mathbf{j}$, $\mathbf{T}'(1) = \frac{1}{2\sqrt{2}}(-\mathbf{i}+\mathbf{j})$, $\mathbf{N}(1) = -\frac{1}{\sqrt{2}}\mathbf{i} + \frac{1}{\sqrt{2}}\mathbf{j}$

5. $\mathbf{r}'(t) = -5\sin t\mathbf{i} + 5\cos t\mathbf{j}$, $\|\mathbf{r}'(t)\| = 5$, $\mathbf{T}(t) = -\sin t\mathbf{i} + \cos t\mathbf{j}$, $\mathbf{T}'(t) = -\cos t\mathbf{i} - \sin t\mathbf{j}$;
$\mathbf{T}(\pi/3) = -\frac{\sqrt{3}}{2}\mathbf{i} + \frac{1}{2}\mathbf{j}$, $\mathbf{T}'(\pi/3) = -\frac{1}{2}\mathbf{i} - \frac{\sqrt{3}}{2}\mathbf{j}$, $\mathbf{N}(\pi/3) = -\frac{1}{2}\mathbf{i} - \frac{\sqrt{3}}{2}\mathbf{j}$

6. $\mathbf{r}'(t) = \frac{1}{t}\mathbf{i} + \mathbf{j}$, $\|\mathbf{r}'(t)\| = \frac{\sqrt{1+t^2}}{t}$, $\mathbf{T}(t) = (1+t^2)^{-1/2}(\mathbf{i}+t\mathbf{j})$,
$\mathbf{T}'(t) = (1+t^2)^{-1/2}(\mathbf{j}) - t(1+t^2)^{-3/2}(\mathbf{i}+t\mathbf{j})$; $\mathbf{T}(e) = \frac{1}{\sqrt{1+e^2}}\mathbf{i} + \frac{e}{\sqrt{1+e^2}}\mathbf{j}$,
$\mathbf{T}'(e) = \frac{1}{(1+e^2)^{3/2}}(-e\mathbf{i}+\mathbf{j})$, $\mathbf{N}(e) = -\frac{e}{\sqrt{1+e^2}}\mathbf{i} + \frac{1}{\sqrt{1+e^2}}\mathbf{j}$

7. $\mathbf{r}'(t) = -4\sin t\mathbf{i} + 4\cos t\mathbf{j} + \mathbf{k}$, $\mathbf{T}(t) = \frac{1}{\sqrt{17}}(-4\sin t\mathbf{i} + 4\cos t\mathbf{j} + \mathbf{k})$,
$\mathbf{T}'(t) = \frac{1}{\sqrt{17}}(-4\cos t\mathbf{i} - 4\sin t\mathbf{j})$, $\mathbf{T}(\pi/2) = -\frac{4}{\sqrt{17}}\mathbf{i} + \frac{1}{\sqrt{17}}\mathbf{k}$
$\mathbf{T}'(\pi/2) = -\frac{4}{\sqrt{17}}\mathbf{j}$, $\mathbf{N}(\pi/2) = -\mathbf{j}$

8. $\mathbf{r}'(t) = \mathbf{i} + t\mathbf{j} + t^2\mathbf{k}$, $\mathbf{T}(t) = (1+t^2+t^4)^{-1/2}(\mathbf{i}+t\mathbf{j}+t^2\mathbf{k})$,
$\mathbf{T}'(t) = (1+t^2+t^4)^{-1/2}(\mathbf{j}+2t\mathbf{k}) - (t+2t^3)(1+t^2+t^4)^{-3/2}(\mathbf{i}+t\mathbf{j}+t^2\mathbf{k})$,
$\mathbf{T}(0) = \mathbf{i}$, $\mathbf{T}'(0) = \mathbf{j} = \mathbf{N}(0)$

9. $\mathbf{r}'(t) = e^t[(\cos t - \sin t)\mathbf{i} + (\cos t + \sin t)\mathbf{j} + \mathbf{k}]$, $\mathbf{T}(t) = \dfrac{1}{\sqrt{3}}[(\cos t - \sin t)\mathbf{i} + (\cos t + \sin t)\mathbf{j} + \mathbf{k}]$,

$\mathbf{T}'(t) = \dfrac{1}{\sqrt{3}}[(-\sin t - \cos t)\mathbf{i} + (-\sin t + \cos t)\mathbf{j}]$,

$\mathbf{T}(0) = \dfrac{1}{\sqrt{3}}\mathbf{i} + \dfrac{1}{\sqrt{3}}\mathbf{j} + \dfrac{1}{\sqrt{3}}\mathbf{k}$, $\mathbf{T}'(0) = \dfrac{1}{\sqrt{3}}(-\mathbf{i} + \mathbf{j})$, $\mathbf{N}(0) = -\dfrac{1}{\sqrt{2}}\mathbf{i} + \dfrac{1}{\sqrt{2}}\mathbf{j}$

10. $\mathbf{r}'(t) = \sinh t\mathbf{i} + \cosh t\mathbf{j} + \mathbf{k}$, $\|\mathbf{r}'(t)\| = \sqrt{\sinh^2 t + \cosh^2 t + 1} = \sqrt{2}\cosh t$,

$\mathbf{T}(t) = \dfrac{1}{\sqrt{2}}(\tanh t\mathbf{i} + \mathbf{j} + \operatorname{sech} t\mathbf{k})$, $\mathbf{T}'(t) = \dfrac{1}{\sqrt{2}}(\operatorname{sech}^2 t\mathbf{i} - \operatorname{sech} t \tanh t\mathbf{k})$, at $t = \ln 2$,

$\tanh(\ln 2) = \dfrac{3}{5}$ and $\operatorname{sech}(\ln 2) = \dfrac{4}{5}$ so $\mathbf{T}(\ln 2) = \dfrac{3}{5\sqrt{2}}\mathbf{i} + \dfrac{1}{\sqrt{2}}\mathbf{j} + \dfrac{4}{5\sqrt{2}}\mathbf{k}$,

$\mathbf{T}'(\ln 2) = \dfrac{4}{25\sqrt{2}}(4\mathbf{i} - 3\mathbf{k})$, $\mathbf{N}(\ln 2) = \dfrac{4}{5}\mathbf{i} - \dfrac{3}{5}\mathbf{k}$

11. From the remark, the line is parametrized by normalizing $\mathbf{v}$, but $\mathbf{T}(t_0) = \mathbf{v}/\|\mathbf{v}\|$, so $\mathbf{r} = \mathbf{r}(t_0) + t\mathbf{v}$ becomes $\mathbf{r} = \mathbf{r}(t_0) + s\mathbf{T}(t_0)$.

12. $\mathbf{r}'(t)\Big]_{t=1} = \langle 1, 2t\rangle\Big]_{t=1} = \langle 1, 2\rangle$, and $\mathbf{T}(1) = \langle \dfrac{1}{\sqrt{5}}, \dfrac{2}{\sqrt{5}}\rangle$, so the tangent line can be parametrized as

$\mathbf{r} = \langle 1, 1\rangle + s\left\langle \dfrac{1}{\sqrt{5}}, \dfrac{2}{\sqrt{5}}\right\rangle$, so $x = 1 + \dfrac{s}{\sqrt{5}}, y = 1 + \dfrac{2s}{\sqrt{5}}$.

13. $\mathbf{r}'(t) = \cos t\mathbf{i} - \sin t\mathbf{j} + t\mathbf{k}$, $\mathbf{r}'(0) = \mathbf{i}$, $\mathbf{r}(0) = \mathbf{j}$, $\mathbf{T}(0) = \mathbf{i}$, so the tangent line has the parametrization $x = s, y = 1$.

14. $\mathbf{r}(1) = \mathbf{i} + \mathbf{j} + \sqrt{8}\mathbf{k}$, $\mathbf{r}'(t) = \mathbf{i} + \mathbf{j} - \dfrac{t}{\sqrt{9 - t^2}}\mathbf{k}$, $\mathbf{r}'(1) = \mathbf{i} + \mathbf{j} - \dfrac{1}{\sqrt{8}}\mathbf{k}$, $\|\mathbf{r}'(1)\| = \dfrac{\sqrt{17}}{\sqrt{8}}$, so the tangent line has parametrizations $\mathbf{r} = \mathbf{i} + \mathbf{j} + \sqrt{8}\mathbf{k} + t\left(\mathbf{i} + \mathbf{j} - \dfrac{1}{\sqrt{8}}\mathbf{k}\right) = \mathbf{i} + \mathbf{j} + \sqrt{8}\mathbf{k} + \dfrac{s\sqrt{8}}{\sqrt{17}}\left(\mathbf{i} + \mathbf{j} - \dfrac{1}{\sqrt{8}}\mathbf{k}\right)$.

15. $\mathbf{T} = \dfrac{3}{5}\cos t\,\mathbf{i} - \dfrac{3}{5}\sin t\,\mathbf{j} + \dfrac{4}{5}\mathbf{k}$, $\mathbf{N} = -\sin t\,\mathbf{i} - \cos t\,\mathbf{j}$, $\mathbf{B} = \mathbf{T} \times \mathbf{N} = \dfrac{4}{5}\cos t\,\mathbf{i} - \dfrac{4}{5}\sin t\,\mathbf{j} - \dfrac{3}{5}\mathbf{k}$. Check:

$\mathbf{r}' = 3\cos t\,\mathbf{i} - 3\sin t\,\mathbf{j} + 4\,\mathbf{k}$, $\mathbf{r}'' = -3\sin t\,\mathbf{i} - 3\cos t\,\mathbf{j}$, $\mathbf{r}' \times \mathbf{r}'' = 12\cos t\,\mathbf{i} - 12\sin t\,\mathbf{j} - 9\,\mathbf{k}$,

$\|\mathbf{r}' \times \mathbf{r}''\| = 15$, $(\mathbf{r}' \times \mathbf{r}'')/\|\mathbf{r}' \times \mathbf{r}''\| = \frac{4}{5}\cos t\,\mathbf{i} - \frac{4}{5}\sin t\,\mathbf{j} - \frac{3}{5}\,\mathbf{k} = \mathbf{B}$.

16. $\mathbf{T}'(t) = \dfrac{1}{\sqrt{2}}[(\cos t + \sin t)\,\mathbf{i} + (-\sin t + \cos t)\,\mathbf{j}], \mathbf{N} = \dfrac{1}{\sqrt{2}}[(-\sin t + \cos t)\,\mathbf{i} - (\cos t + \sin t)\,\mathbf{j}]$,

$\mathbf{B} = \mathbf{T} \times \mathbf{N} = -\mathbf{k}$. Check: $\mathbf{r}' = e^t(\cos t + \sin t)\,\mathbf{i} + e^t(\cos t - \sin t)\,\mathbf{j}$,

$\mathbf{r}'' = 2e^t\cos t\,\mathbf{i} - 2e^t\sin t\,\mathbf{j}$, $\mathbf{r}' \times \mathbf{r}'' = -2e^{2t}\,\mathbf{k}$, $\|\mathbf{r}' \times \mathbf{r}''\| = 2e^{2t}$, $(\mathbf{r}' \times \mathbf{r}'')/\|\mathbf{r}' \times \mathbf{r}''\| = -\mathbf{k} = \mathbf{B}$.

17. $\mathbf{r}'(t) = t\sin t\mathbf{i} + t\cos t\mathbf{j}, \|\mathbf{r}'\| = t, \mathbf{T} = \sin t\mathbf{i} + \cos t\mathbf{j}, \mathbf{N} = \cos t\mathbf{i} - \sin t\mathbf{j}, \mathbf{B} = \mathbf{T} \times \mathbf{N} = -\mathbf{k}$. Check:

$\mathbf{r}' = t\sin t\,\mathbf{i} + t\cos t\,\mathbf{j}$, $\mathbf{r}'' = (\sin t + t\cos t)\,\mathbf{i} + (\cos t - t\sin t)\,\mathbf{j}$, $\mathbf{r}' \times \mathbf{r}'' = -2e^{2t}\,\mathbf{k}$,

$\|\mathbf{r}' \times \mathbf{r}''\| = 2e^{2t}$, $(\mathbf{r}' \times \mathbf{r}'')/\|\mathbf{r}' \times \mathbf{r}''\| = -\mathbf{k} = \mathbf{B}$.

18. $\mathbf{T} = (-a\sin t\,\mathbf{i} + a\cos t\,\mathbf{j} + c\,\mathbf{k})/\sqrt{a^2 + c^2}$, $\mathbf{N} = -\cos t\,\mathbf{i} - \sin t\,\mathbf{j}$,

$\mathbf{B} = \mathbf{T} \times \mathbf{N} = (c\sin t\,\mathbf{i} - c\cos t\,\mathbf{j} + a\,\mathbf{k})/\sqrt{a^2 + c^2}$. Check:

$\mathbf{r}' = -a\sin t\,\mathbf{i} + a\cos t\,\mathbf{j} + c\,\mathbf{k}$, $\mathbf{r}'' = -a\cos t\,\mathbf{i} - a\sin t\,\mathbf{j}$, $\mathbf{r}' \times \mathbf{r}'' = ca\sin t\,\mathbf{i} - ca\cos t\,\mathbf{j} + a^2\,\mathbf{k}$,

$\|\mathbf{r}' \times \mathbf{r}''\| = a\sqrt{a^2 + c^2}$, $(\mathbf{r}' \times \mathbf{r}'')/\|\mathbf{r}' \times \mathbf{r}''\| = \mathbf{B}$.

19. $\mathbf{r}(\pi/4) = \frac{\sqrt{2}}{2}\mathbf{i} + \frac{\sqrt{2}}{2}\mathbf{j} + \mathbf{k}, \mathbf{T} = -\sin t\mathbf{i} + \cos t\mathbf{j} = \frac{\sqrt{2}}{2}(-\mathbf{i} + \mathbf{j}), \mathbf{N} = -(\cos t\mathbf{i} + \sin t\mathbf{j}) = -\frac{\sqrt{2}}{2}(\mathbf{i} + \mathbf{j})$, $\mathbf{B} = \mathbf{k}$; the rectifying, osculating, and normal planes are given (respectively) by $x + y = \sqrt{2}$, $z = 1, -x + y = 0$.

20. $\mathbf{r}(0) = \mathbf{i} + \mathbf{j}, \mathbf{T} = \frac{1}{\sqrt{3}}(\mathbf{i} + \mathbf{j} + \mathbf{k}), \mathbf{N} = \frac{1}{\sqrt{2}}(-\mathbf{j} + \mathbf{k}), \mathbf{B} = \frac{1}{\sqrt{6}}(2\mathbf{i} - \mathbf{j} - \mathbf{k})$; the rectifying, osculating, and normal planes are given (respectively) by $-y + z = -1, 2x - y - z = 1, x + y + z = 2$.

21. **(a)** By formulae (1) and (11), $\mathbf{N}(t) = \mathbf{B}(t) \times \mathbf{T}(t) = \frac{\mathbf{r}'(t) \times \mathbf{r}''(t)}{\|\mathbf{r}'(t) \times \mathbf{r}''(t)\|} \times \frac{\mathbf{r}'(t)}{\|\mathbf{r}'(t)\|}$.

(b) Since $\mathbf{r}'$ is perpendicular to $\mathbf{r}' \times \mathbf{r}''$ it follows from Lagrange's Identity (Exercise 32 of Section 12.4) that $\|(\mathbf{r}'(t) \times \mathbf{r}''(t)) \times \mathbf{r}'(t)\| = \|\mathbf{r}'(t) \times \mathbf{r}''(t)\|\|\mathbf{r}'(t)\|$, and the result follows.

(c) From Exercise 39 of Section 12.4, $(\mathbf{r}'(t) \times \mathbf{r}''(t)) \times \mathbf{r}'(t) = \|\mathbf{r}'(t)\|^2\mathbf{r}''(t) - (\mathbf{r}'(t) \cdot \mathbf{r}''(t))\mathbf{r}'(t) = \mathbf{u}(t)$, so $\mathbf{N}(t) = \mathbf{u}(t)/\|\mathbf{u}(t)\|$

22. **(a)** $\mathbf{r}'(t) = 2t\mathbf{i} + \mathbf{j}, \mathbf{r}'(1) = 2\mathbf{i} + \mathbf{j}, \mathbf{r}''(t) = 2\mathbf{i}, \mathbf{u} = 2\mathbf{i} - 4\mathbf{j}, \mathbf{N} = \frac{1}{\sqrt{5}}\mathbf{i} - \frac{2}{\sqrt{5}}\mathbf{j}$

(b) $\mathbf{r}'(t) = -4\sin t\mathbf{i} + 4\cos t\mathbf{j} + \mathbf{k},\ \mathbf{r}'(\frac{\pi}{2}) = -4\mathbf{i} + \mathbf{k},\ \mathbf{r}''(t) = -4\cos t\mathbf{i} - 4\sin t\mathbf{j}$, $\mathbf{r}''(\frac{\pi}{2}) = -4\mathbf{j}, \mathbf{u} = 17(-4\mathbf{j}), \mathbf{N} = -\mathbf{j}$

23. $\mathbf{r}'(t) = \cos t\mathbf{i} - \sin t\mathbf{j} + \mathbf{k}, \mathbf{r}''(t) = -\sin t\mathbf{i} - \cos t\mathbf{j}, \mathbf{u} = -2(\sin t\mathbf{i} + \cos t\mathbf{j}), \|\mathbf{u}\| = 2, \mathbf{N} = -\sin t\mathbf{i} - \cos t\mathbf{j}$

24. $\mathbf{r}'(t) = \mathbf{i} + 2t\mathbf{j} + 3t^2\mathbf{k}, \mathbf{r}''(t) = 2\mathbf{j} + 6t\mathbf{k}, \mathbf{u}(t) = -(4t + 18t^3)\mathbf{i} + (2 - 18t^4)\mathbf{j} + (6t + 12t^3)\mathbf{k}$,

$$\mathbf{N} = \frac{1}{2\sqrt{81t^8 + 117t^6 + 54t^4 + 13t^2 + 1}}\left(-(4t + 18t^3)\mathbf{i} + (2 - 18t^4)\mathbf{j} + (6t + 12t^3)\mathbf{k}\right)$$

EXERCISE SET 13.5

1. $\kappa \approx \frac{1}{0.5} = 2$

2. $\kappa \approx \frac{1}{4/3} = \frac{3}{4}$

3. $\mathbf{r}'(t) = 2t\mathbf{i} + 3t^2\mathbf{j},\ \mathbf{r}''(t) = 2\mathbf{i} + 6t\mathbf{j},\ \kappa = \|\mathbf{r}'(t) \times \mathbf{r}''(t)\|/\|\mathbf{r}'(t)\|^3 = \frac{6}{t(4 + 9t^2)^{3/2}}$

4. $\mathbf{r}'(t) = -4\sin t\mathbf{i} + \cos t\mathbf{j},\ \mathbf{r}''(t) = -4\cos t\mathbf{i} - \sin t\mathbf{j},\ \kappa = \|\mathbf{r}'(t) \times \mathbf{r}''(t)\|/\|\mathbf{r}'(t)\|^3 = \frac{4}{(16\sin^2 t + \cos^2 t)^{3/2}}$

5. $\mathbf{r}'(t) = 3e^{3t}\mathbf{i} - e^{-t}\mathbf{j},\ \mathbf{r}''(t) = 9e^{3t}\mathbf{i} + e^{-t}\mathbf{j},\ \kappa = \|\mathbf{r}'(t) \times \mathbf{r}''(t)\|/\|\mathbf{r}'(t)\|^3 = \frac{12e^{2t}}{(9e^{6t} + e^{-2t})^{3/2}}$

6. $\mathbf{r}'(t) = -3t^2\mathbf{i} + (1 - 2t)\mathbf{j},\ \mathbf{r}''(t) = -6t\mathbf{i} - 2\mathbf{j},\ \kappa = \|\mathbf{r}'(t) \times \mathbf{r}''(t)\|/\|\mathbf{r}'(t)\|^3 = \frac{6|t^2 - t|}{(9t^4 + 4t^2 - 4t + 1)^{3/2}}$

7. $\mathbf{r}'(t) = -4\sin t\mathbf{i} + 4\cos t\mathbf{j} + \mathbf{k},\ \mathbf{r}''(t) = -4\cos t\mathbf{i} - 4\sin t\mathbf{j}$, $\kappa = \|\mathbf{r}'(t) \times \mathbf{r}''(t)\|/\|\mathbf{r}'(t)\|^3 = 4/17$

8. $\mathbf{r}'(t) = \mathbf{i} + t\mathbf{j} + t^2\mathbf{k},\ \mathbf{r}''(t) = \mathbf{j} + 2t\mathbf{k},\ \kappa = \|\mathbf{r}'(t) \times \mathbf{r}''(t)\|/\|\mathbf{r}'(t)\|^3 = \frac{\sqrt{t^4 + 4t^2 + 1}}{(t^4 + t^2 + 1)^{3/2}}$

9. $\mathbf{r}'(t) = \sinh t\mathbf{i} + \cosh t\mathbf{j} + \mathbf{k}$, $\mathbf{r}''(t) = \cosh t\mathbf{i} + \sinh t\mathbf{j}$, $\kappa = \|\mathbf{r}'(t) \times \mathbf{r}''(t)\|/\|\mathbf{r}'(t)\|^3 = \dfrac{1}{2\cosh^2 t}$

10. $\mathbf{r}'(t) = \mathbf{j} + 2t\mathbf{k}$, $\mathbf{r}''(t) = 2\mathbf{k}$, $\kappa = \|\mathbf{r}'(t) \times \mathbf{r}''(t)\|/\|\mathbf{r}'(t)\|^3 = \dfrac{2}{(4t^2+1)^{3/2}}$

11. $\mathbf{r}'(t) = -3\sin t\mathbf{i} + 4\cos t\mathbf{j} + \mathbf{k}$, $\mathbf{r}''(t) = -3\cos t\mathbf{i} - 4\sin t\mathbf{j}$,
$\mathbf{r}'(\pi/2) = -3\mathbf{i} + \mathbf{k}$, $\mathbf{r}''(\pi/2) = -4\mathbf{j}$; $\kappa = \|4\mathbf{i} + 12\mathbf{k}\|/\| - 3\mathbf{i} + \mathbf{k}\|^3 = 2/5, \rho = 5/2$

12. $\mathbf{r}'(t) = e^t\mathbf{i} - e^{-t}\mathbf{j} + \mathbf{k}$, $\mathbf{r}''(t) = e^t\mathbf{i} + e^{-t}\mathbf{j}$,
$\mathbf{r}'(0) = \mathbf{i} - \mathbf{j} + \mathbf{k}$, $\mathbf{r}''(0) = \mathbf{i} + \mathbf{j}$; $\kappa = \| - \mathbf{i} + \mathbf{j} + 2\mathbf{k}\|/\|\mathbf{i} - \mathbf{j} + \mathbf{k}\|^3 = \sqrt{2}/3, \rho = 3/\sqrt{2}$

13. $\mathbf{r}'(t) = e^t(\cos t - \sin t)\mathbf{i} + e^t(\cos t + \sin t)\mathbf{j} + e^t\mathbf{k}$,
$\mathbf{r}''(t) = -2e^t\sin t\mathbf{i} + 2e^t\cos t\mathbf{j} + e^t\mathbf{k}$, $\mathbf{r}'(0) = \mathbf{i} + \mathbf{j} + \mathbf{k}$,
$\mathbf{r}''(0) = 2\mathbf{j} + \mathbf{k}$; $\kappa = \| - \mathbf{i} - \mathbf{j} + 2\mathbf{k}\|/\|\mathbf{i} + \mathbf{j} + \mathbf{k}\|^3 = \sqrt{2}/3, \rho = 3\sqrt{2}/2$

14. $\mathbf{r}'(t) = \cos t\mathbf{i} - \sin t\mathbf{j} + t\mathbf{k}$, $\mathbf{r}''(t) = -\sin t\mathbf{i} - \cos t\mathbf{j} + \mathbf{k}$,
$\mathbf{r}'(0) = \mathbf{i}$, $\mathbf{r}''(0) = -\mathbf{j} + \mathbf{k}$; $\kappa = \| - \mathbf{j} - \mathbf{k}\|/\|\mathbf{i}\|^3 = \sqrt{2}, \rho = \sqrt{2}/2$

15. $\mathbf{r}'(s) = \dfrac{1}{2}\cos\left(1 + \dfrac{s}{2}\right)\mathbf{i} - \dfrac{1}{2}\sin\left(1 + \dfrac{s}{2}\right)\mathbf{j} + \dfrac{\sqrt{3}}{2}\mathbf{k}$, $\|\mathbf{r}'(s)\| = 1$, so

$\dfrac{d\mathbf{T}}{ds} = -\dfrac{1}{4}\sin\left(1 + \dfrac{s}{2}\right)\mathbf{i} - \dfrac{1}{4}\cos\left(1 + \dfrac{s}{2}\right)\mathbf{j}$, $\kappa = \left\|\dfrac{d\mathbf{T}}{ds}\right\| = \dfrac{1}{4}$

16. $\mathbf{r}'(s) = -\sqrt{\dfrac{3-2s}{3}}\mathbf{i} + \sqrt{\dfrac{2s}{3}}\mathbf{j}$, $\|\mathbf{r}'(s)\| = 1$, so

$\dfrac{d\mathbf{T}}{ds} = \dfrac{1}{\sqrt{9-6s}}\mathbf{i} + \dfrac{1}{\sqrt{6s}}\mathbf{j}, \kappa = \left\|\dfrac{d\mathbf{T}}{ds}\right\| = \sqrt{\dfrac{1}{9-6s} + \dfrac{1}{6s}} = \sqrt{\dfrac{3}{2s(9-6s)}}$

17. **(a)** $\mathbf{r}' = x'\mathbf{i} + y'\mathbf{j}, \mathbf{r}'' = x''\mathbf{i} + y''\mathbf{j}$, $\|\mathbf{r}' \times \mathbf{r}''\| = |x'y'' - x''y'|$, $\kappa = \dfrac{|x'y'' - y'x''|}{(x'^2 + y'^2)^{3/2}}$

(b) Set $x = t$, $y = f(x) = f(t)$, $x' = 1$, $x'' = 0$, $y' = \dfrac{dy}{dx}$, $y'' = \dfrac{d^2y}{dx^2}, \kappa = \dfrac{|d^2y/dx^2|}{(1 + (dy/dx)^2)^{3/2}}$

18. $\dfrac{dy}{dx} = \tan\phi$, $(1 + \tan^2\phi)^{3/2} = (\sec^2\phi)^{3/2} = |\sec\phi|^3$, $\kappa(x) = \dfrac{|y''|}{|\sec\phi|^3} = |y''\cos^3\phi|$

19. $\kappa(x) = \dfrac{|\sin x|}{(1 + \cos^2 x)^{3/2}}$, $\kappa(\pi/2) = 1$

20. $\kappa(x) = \dfrac{2|x|}{(1 + x^4)^{3/2}}$, $\kappa(0) = 0$

21. $\kappa(x) = \dfrac{2|x|^3}{(x^4 + 1)^{3/2}}$, $\kappa(1) = 1/\sqrt{2}$

22. $\kappa(x) = \dfrac{e^{-x}}{(1 + e^{-2x})^{3/2}}$, $\kappa(1) = \dfrac{e^{-1}}{(1 + e^{-2})^{3/2}}$

23. $\kappa(x) = \dfrac{2\sec^2 x|\tan x|}{(1 + \sec^4 x)^{3/2}}$, $\kappa(\pi/4) = 4/(5\sqrt{5})$

24. By implicit differentiation, $dy/dx = 4x/y$, $d^2y/dx^2 = 36/y^3$ so $\kappa = \dfrac{36/|y|^3}{(1 + 16x^2/y^2)^{3/2}}$;
if $(x, y) = (2, 5)$ then $\kappa = \dfrac{36/125}{(1 + 64/25)^{3/2}} = \dfrac{36}{89\sqrt{89}}$

25. $x'(t) = 2t,\ y'(t) = 3t^2,\ x''(t) = 2,\ y''(t) = 6t,$
$x'(1/2) = 1,\ y'(1/2) = 3/4,\ x''(1/2) = 2,\ y''(1/2) = 3;\ \kappa = 96/125$

26. $x'(t) = -4\sin t,\ y'(t) = \cos t,\ x''(t) - -4\cos t,\ y''(t) = -\sin t,$
$x'(\pi/2) = -4,\ y'(\pi/2) = 0,\ x''(\pi/2) = 0,\ y''(\pi/2) = -1;\ \kappa = 1/16$

27. $x'(t) = 3e^{3t},\ y'(t) = -e^{-t},\ x''(t) = 9e^{3t},\ y''(t) = e^{-t},$
$x'(0) = 3,\ y'(0) = -1,\ x''(0) = 9,\ y''(0) = 1;\ \kappa = 6/(5\sqrt{10})$

28. $x'(t) = -3t^2,\ y'(t) = 1 - 2t,\ x''(t) = -6t,\ y''(t) = -2,$
$x'(1) = -3,\ y'(1) = -1,\ x''(1) = -6,\ y''(1) = -2;\ \kappa = 0$

29. $x'(t) = 1, y'(t) = -1/t^2, x''(t) = 0, y''(t) = 2/t^3$
$x'(1) = 1, y'(1) = -1, x''(1) = 0, y''(1) = 2; \kappa = 1/\sqrt{2}$

30. $x'(t) = 4\cos 2t, y'(t) = 3\cos t, x''(t) = -8\sin 2t, y''(t) = -3\sin t,$
$x'(\pi/2) = -4, y'(\pi/2) = 0, x''(\pi/2) = 0, y''(\pi/2) = -3, \kappa = 12/4^{3/2} = 3/2$

31. **(a)** $\kappa(x) = \dfrac{|\cos x|}{(1+\sin^2 x)^{3/2}},$

$$\rho(x) = \frac{(1+\sin^2 x)^{3/2}}{|\cos x|}$$

$\rho(0) = \rho(\pi) = 1.$

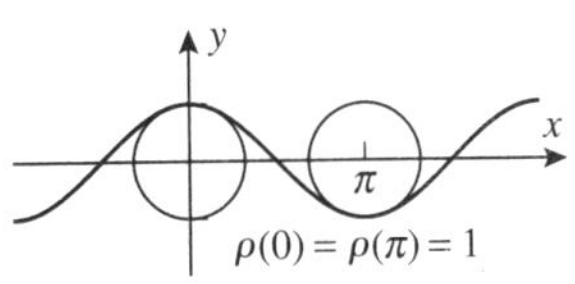

(b) $\kappa(t) = \dfrac{2}{(4\sin^2 t + \cos^2 t)^{3/2}},$

$$\rho(t) = \frac{1}{2}(4\sin^2 t + \cos^2 t)^{3/2},$$

$\rho(0) = 1/2,\ \rho(\pi/2) = 4$

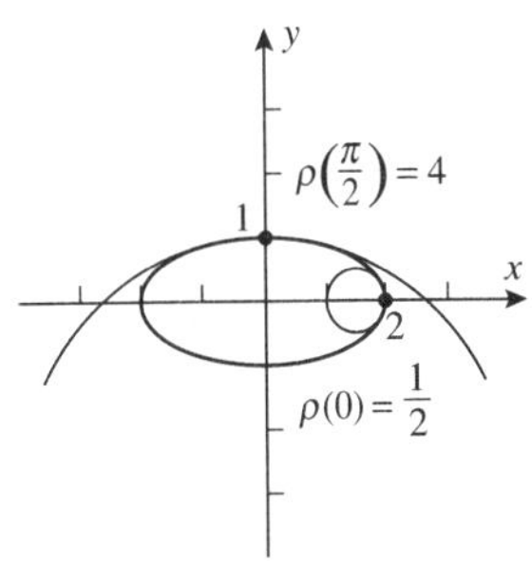

32. $x'(t) = -e^{-t}(\cos t + \sin t),$
$y'(t) = e^{-t}(\cos t - \sin t),$
$x''(t) = 2e^{-t}\sin t,$
$y''(t) = -2e^{-t}\cos t;$
using the formula of Exercise 17(a),
$\kappa = \dfrac{1}{\sqrt{2}}e^t.$

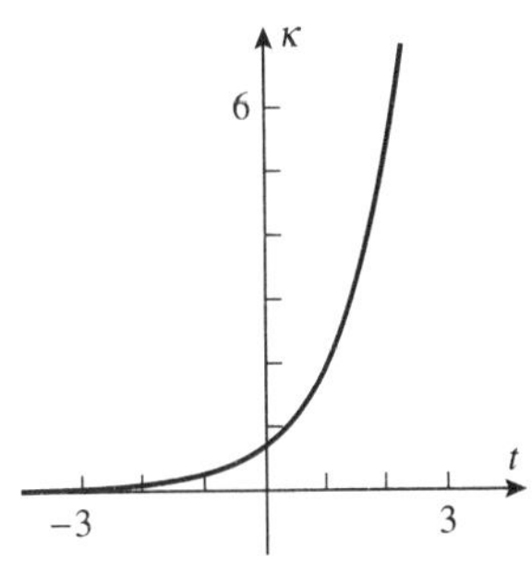

33. **(a)** At $x = 0$ the curvature of I has a large value, yet the value of II there is zero, so II is not the curvature of I; hence I is the curvature of II.

(b) I has points of inflection where the curvature is zero, but II is not zero there, and hence is not the curvature of I; so I is the curvature of II.

34. (a) II takes the value zero at $x = 0$, yet the curvature of I is large there; hence I is the curvature of II.

(b) I has constant zero curvature; II has constant, positive curvature; hence I is the curvature of II.

35. (a)

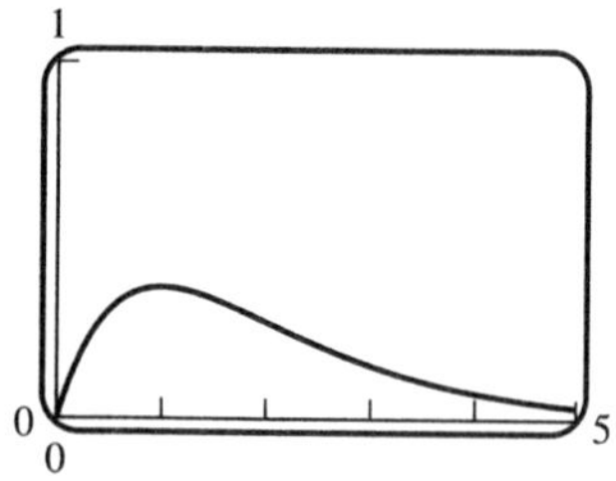

(b)

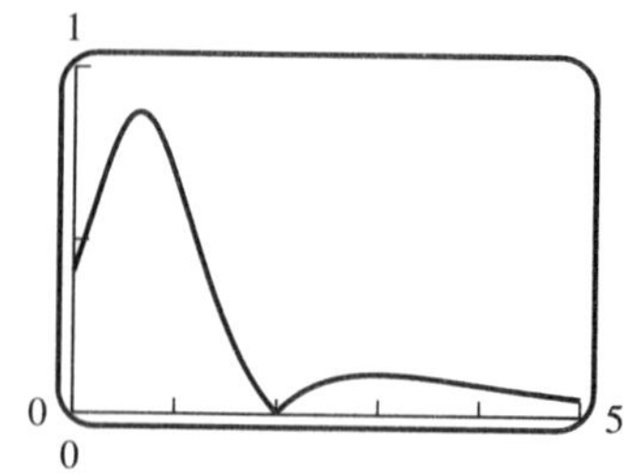

36. (a)

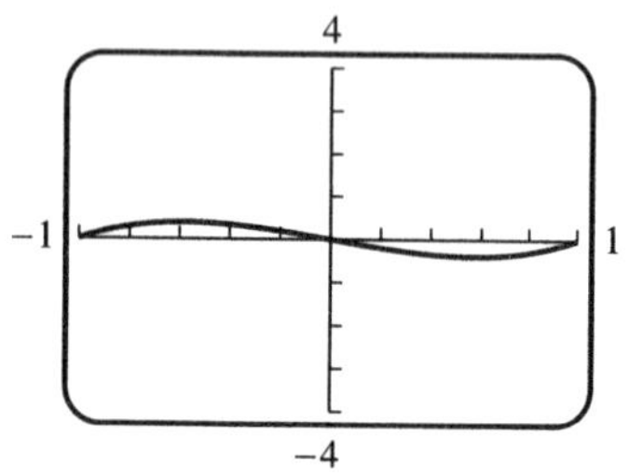

(b)

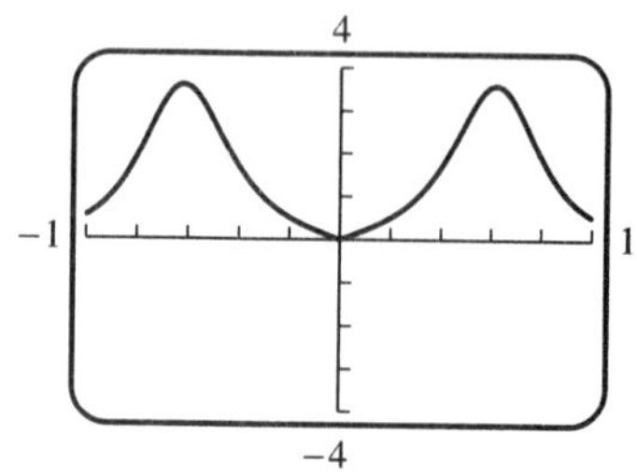

37. (a) $\kappa = \dfrac{|12x^2 - 4|}{\left(1 + (4x^3 - 4x)^2\right)^{3/2}}$

(b)

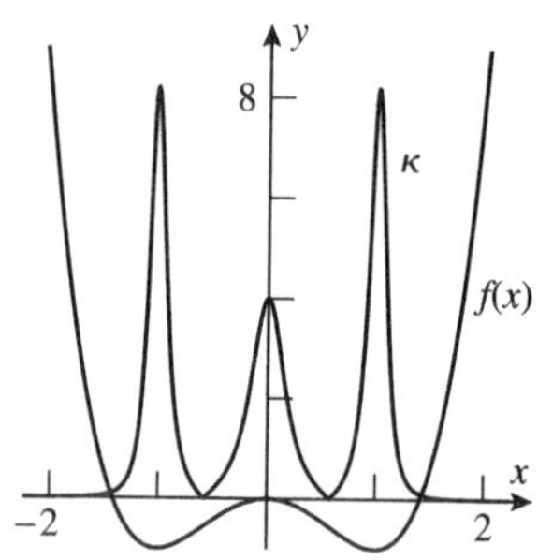

(c) $f'(x) = 4x^3 - 4x = 0$ at $x = 0, \pm 1, f''(x) = 12x^2 - 4$, so extrema at $x = 0, \pm 1$, and $\rho = 1/4$ for $x = 0$ and $\rho = 1/8$ when $x = \pm 1$.

38. (a)

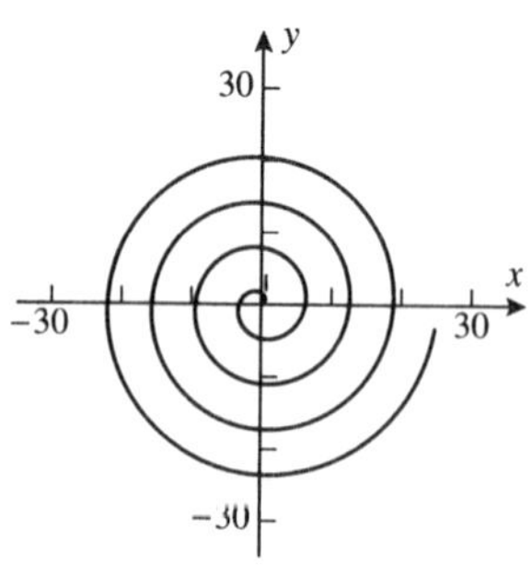

(c) $\kappa(t) = \dfrac{t^2 + 2}{(t^2 + 1)^{3/2}}$

(d) $\lim_{t \to +\infty} \kappa(t) = 0$

39. $\mathbf{r}'(\theta) = \left(-r\sin\theta + \cos\theta\dfrac{dr}{d\theta}\right)\mathbf{i} + \left(r\cos\theta + \sin\theta\dfrac{dr}{d\theta}\right)\mathbf{j}$;

$$\mathbf{r}''(\theta) = \left(-r\cos\theta - 2\sin\theta\frac{dr}{d\theta} + \cos\theta\frac{d^2r}{d\theta^2}\right)\mathbf{i} + \left(-r\sin\theta + 2\cos\theta\frac{dr}{d\theta} + \sin\theta\frac{d^2r}{d\theta^2}\right)\mathbf{j};$$

$$\kappa = \frac{\left|r^2 + 2\left(\dfrac{dr}{d\theta}\right)^2 - r\dfrac{d^2r}{d\theta^2}\right|}{\left[r^2 + \left(\dfrac{dr}{d\theta}\right)^2\right]^{3/2}}.$$

40. Let $r = a$ be the circle, so that $dr/d\theta = 0$, and $\kappa(\theta) = \dfrac{1}{r} = \dfrac{1}{a}$

41. $\kappa(\theta) = \dfrac{3}{2\sqrt{2}(1+\cos\theta)^{1/2}}$, $\kappa(\pi/2) = \dfrac{3}{2\sqrt{2}}$

42. $\kappa(\theta) = \dfrac{1}{\sqrt{5e^{2\theta}}}$, $\kappa(1) = \dfrac{1}{\sqrt{5e^2}}$

43. $\kappa(\theta) = \dfrac{10 + 8\cos^2 3\theta}{(1 + 8\cos^2\theta)^{3/2}}$, $\kappa(0) = \dfrac{2}{3}$

44. $\kappa(\theta) = \dfrac{\theta^2 + 2}{(\theta^2+1)^{3/2}}$, $\kappa(1) = \dfrac{3}{2\sqrt{2}}$

45. The radius of curvature is zero when $\theta = \pi$, so there is a cusp there.

46. $\dfrac{dr}{d\theta} = -\sin\theta$, $\dfrac{d^2r}{d\theta^2} = -\cos\theta$, $\kappa(\theta) = \dfrac{3}{2^{3/2}\sqrt{1+\cos\theta}}$

47. Let $y = t$, then $x = \dfrac{t^2}{4p}$ and $\kappa(t) = \dfrac{1/|2p|}{[t^2/(4p^2)+1]^{3/2}}$;

$t = 0$ when $(x, y) = (0, 0)$ so $\kappa(0) = 1/|2p|$, $\rho = 2|p|$.

48. $\kappa(x) = \dfrac{e^x}{(1+e^{2x})^{3/2}}$, $\kappa'(x) = \dfrac{e^x(1-2e^{2x})}{(1+e^{2x})^{5/2}}$; $\kappa'(x) = 0$ when $e^{2x} = 1/2$, $x = -(\ln 2)/2$. By the first derivative test, $\kappa(-\frac{1}{2}\ln 2)$ is maximum so the point is $(-\frac{1}{2}\ln 2, 1/\sqrt{2})$.

49. Let $x = 3\cos t$, $y = 2\sin t$ for $0 \le t < 2\pi$, $\kappa(t) = \dfrac{6}{(9\sin^2 t + 4\cos^2 t)^{3/2}}$ so

$\rho(t) = \dfrac{1}{6}(9\sin^2 t + 4\cos^2 t)^{3/2} = \dfrac{1}{6}(5\sin^2 t + 4)^{3/2}$ which, by inspection, is minimum when $t = 0$ or π. The radius of curvature is minimum at $(3, 0)$ and $(-3, 0)$.

50. $\kappa(x) = \dfrac{6x}{(1+9x^4)^{3/2}}$ for $x > 0$, $\kappa'(x) = \dfrac{6(1-45x^4)}{(1+9x^4)^{5/2}}$; $\kappa'(x) = 0$ when $x = 45^{-1/4}$ which, by the first derivative test, yields the maximum.

51. $\mathbf{r}'(t) = -\sin t\mathbf{i} + \cos t\mathbf{j} - \sin t\mathbf{k}$, $\mathbf{r}''(t) = -\cos t\mathbf{i} - \sin t\mathbf{j} - \cos t\mathbf{k}$,
$\|\mathbf{r}'(t) \times \mathbf{r}''(t)\| = \|-\mathbf{i} + \mathbf{k}\| = \sqrt{2}$, $\|\mathbf{r}'(t)\| = (1 + \sin^2 t)^{1/2}$; $\kappa(t) = \sqrt{2}/(1+\sin^2 t)^{3/2}$,
$\rho(t) = (1+\sin^2 t)^{3/2}/\sqrt{2}$. The minimum value of ρ is $1/\sqrt{2}$; the maximum value is 2.

52. $\mathbf{r}'(t) = e^t\mathbf{i} - e^{-t}\mathbf{j} + \sqrt{2}\mathbf{k}$, $\mathbf{r}''(t) = e^t\mathbf{i} + e^{-t}\mathbf{j}$;

$\kappa(t) = \dfrac{\sqrt{2}}{e^{2t} + e^{-2t} + 2}$, $\rho(t) = \dfrac{1}{\sqrt{2}}(e^t + e^{-t})^2 = 2\sqrt{2}\cosh^2 t$. The minimum value of ρ is $2\sqrt{2}$.

53. From Exercise 39: $dr/d\theta = ae^{a\theta} = ar$, $d^2r/d\theta^2 = a^2e^{a\theta} = a^2r$; $\kappa = 1/[\sqrt{1+a^2}\,r]$.

54. Use implicit differentiation on $r^2 = a^2\cos 2\theta$ to get $2r\dfrac{dr}{d\theta} = -2a^2\sin 2\theta$, $r\dfrac{dr}{d\theta} = -a^2\sin 2\theta$, and again to get $r\dfrac{d^2r}{d\theta^2} + \left(\dfrac{dr}{d\theta}\right)^2 = -2a^2\cos 2\theta$ so $r\dfrac{d^2r}{d\theta^2} = -\left(\dfrac{dr}{d\theta}\right)^2 - 2a^2\cos 2\theta = -\left(\dfrac{dr}{d\theta}\right)^2 - 2r^2$, thus

$\left|r^2 + 2\left(\dfrac{dr}{d\theta}\right)^2 - r\dfrac{d^2r}{d\theta^2}\right| = 3\left[r^2 + \left(\dfrac{dr}{d\theta}\right)^2\right]$, $\kappa = \dfrac{3}{[r^2 + (dr/d\theta)^2]^{1/2}}$; $\dfrac{dr}{d\theta} = -\dfrac{a^2\sin 2\theta}{r}$ so

$r^2 + \left(\dfrac{dr}{d\theta}\right)^2 = r^2 + \dfrac{a^4\sin^2 2\theta}{r^2} = \dfrac{r^4 + a^4\sin^2 2\theta}{r^2} = \dfrac{a^4\cos^2 2\theta + a^4\sin^2 2\theta}{r^2} = \dfrac{a^4}{r^2}$, hence $\kappa = \dfrac{3r}{a^2}$.

55. (a) $d^2y/dx^2 = 2$, $\kappa(\phi) = |2\cos^3\phi|$

(b) $dy/dx = \tan\phi = 1, \phi = \pi/4$, $\kappa(\pi/4) = |2\cos^3(\pi/4)| = 1/\sqrt{2}$, $\rho = \sqrt{2}$

(c)

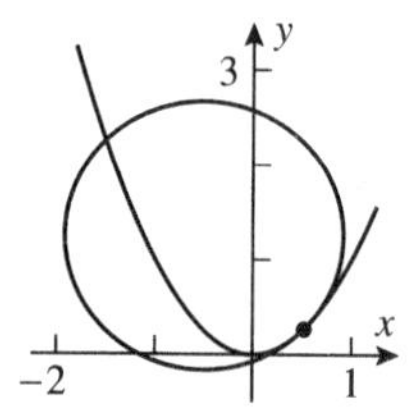

56. (a) $\left(\dfrac{5}{3}, 0\right), \left(0, -\dfrac{5}{2}\right)$ (b) clockwise (c) it is a point, namely the center of the circle

57. $\kappa = 0$ along $y = 0$; along $y = x^2$, $\kappa(x) = 2/(1+4x^2)^{3/2}$, $\kappa(0) = 2$. Along $y = x^3$, $\kappa(x) = 6|x|/(1+9x^4)^{3/2}$, $\kappa(0) = 0$.

58. (a)

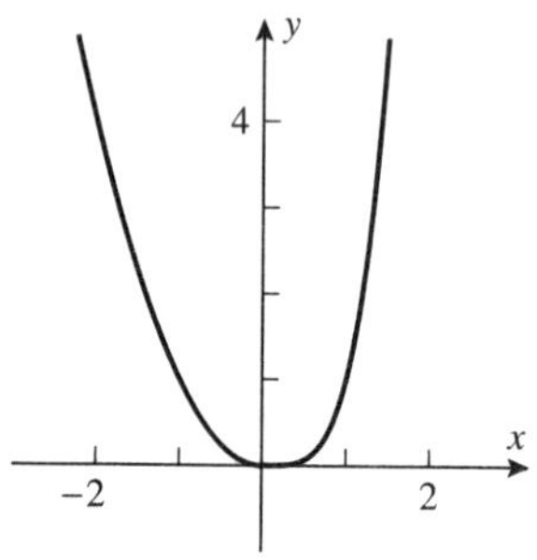

(b) For $y = x^2$, $\kappa(x) = \dfrac{2}{(1+4x^2)^{3/2}}$ so $\kappa(0) = 2$; for $y = x^4$, $\kappa(x) = \dfrac{12x^2}{(1+16x^6)^{3/2}}$ so $\kappa(0) = 0$. κ is not continuous at $x = 0$.

59. $\kappa = 1/r$ along the circle; along $y = ax^2$, $\kappa(x) = 2a/(1+4a^2x^2)^{3/2}$, $\kappa(0) = 2a$ so $2a = 1/r$, $a = 1/(2r)$.

60. $\kappa(x) = \dfrac{|y''|}{(1+y'^2)^{3/2}}$ so the transition will be smooth if the values of y are equal, the values of y' are equal, and the values of y'' are equal at $x = 0$. If $y = e^x$, then $y' = y'' = e^x$; if $y = ax^2 + bx + c$, then $y' = 2ax + b$ and $y'' = 2a$. Equate y, y', and y'' at $x = 0$ to get $c = 1$, $b = 1$, and $a = 1/2$.

61. The result follows from the definitions $\mathbf{N} = \dfrac{\mathbf{T}'(s)}{\|\mathbf{T}'(s)\|}$ and $\kappa = \|\mathbf{T}'(s)\|$.

62. (a) $\mathbf{B} \cdot \dfrac{d\mathbf{B}}{ds} = 0$ because $\|\mathbf{B}(s)\| = 1$ so $\dfrac{d\mathbf{B}}{ds}$ is perpendicular to $\mathbf{B}(s)$.

(b) $\mathbf{B}(s) \cdot \mathbf{T}(s) = 0$, $\mathbf{B}(s) \cdot \dfrac{d\mathbf{T}}{ds} + \dfrac{d\mathbf{B}}{ds} \cdot \mathbf{T}(s) = 0$, but $\dfrac{d\mathbf{T}}{ds} = \kappa\mathbf{N}(s)$ so $\kappa\mathbf{B}(s) \cdot \mathbf{N}(s) + \dfrac{d\mathbf{B}}{ds} \cdot \mathbf{T}(s) = 0$, $\dfrac{d\mathbf{B}}{ds} \cdot \mathbf{T}(s) = 0$ because $\mathbf{B}(s) \cdot \mathbf{N}(s) = 0$; thus $\dfrac{d\mathbf{B}}{ds}$ is perpendicular to $\mathbf{T}(s)$.

(c) $\dfrac{d\mathbf{B}}{ds}$ is perpendicular to both $\mathbf{B}(s)$ and $\mathbf{T}(s)$ but so is $\mathbf{N}(s)$, thus $\dfrac{d\mathbf{B}}{ds}$ is parallel to $\mathbf{N}(s)$ and hence a scalar multiple of $\mathbf{N}(s)$.

(d) If C lies in a plane, then $\mathbf{T}(s)$ and $\mathbf{N}(s)$ also lie in the plane; $\mathbf{B}(s) = \mathbf{T}(s) \times \mathbf{N}(s)$ so $\mathbf{B}(s)$ is always perpendicular to the plane and hence $d\mathbf{B}/ds = \mathbf{0}$, thus $\tau = 0$.

63. $\dfrac{d\mathbf{N}}{ds} = \mathbf{B} \times \dfrac{d\mathbf{T}}{ds} + \dfrac{d\mathbf{B}}{ds} \times \mathbf{T} = \mathbf{B} \times (\kappa\mathbf{N}) + (-\tau\mathbf{N}) \times \mathbf{T} = \kappa\mathbf{B} \times \mathbf{N} - \tau\mathbf{N} \times \mathbf{T}$, but $\mathbf{B} \times \mathbf{N} = -\mathbf{T}$ and $\mathbf{N} \times \mathbf{T} = -\mathbf{B}$ so $\dfrac{d\mathbf{N}}{ds} = -\kappa\mathbf{T} + \tau\mathbf{B}$

64. $\mathbf{r}''(s) = d\mathbf{T}/ds = \kappa\mathbf{N}$ so $\mathbf{r}'''(s) = \kappa d\mathbf{N}/ds + (d\kappa/ds)\mathbf{N}$ but $d\mathbf{N}/ds = -\kappa\mathbf{T} + \tau\mathbf{B}$ so
$\mathbf{r}'''(s) = -\kappa^2\mathbf{T} + (d\kappa/ds)\mathbf{N} + \kappa\tau\mathbf{B}$, $\mathbf{r}'(s) \times \mathbf{r}''(s) = \mathbf{T} \times (\kappa\mathbf{N}) = \kappa\mathbf{T} \times \mathbf{N} = \kappa\mathbf{B}$,
$[\mathbf{r}'(s) \times \mathbf{r}''(s)] \cdot \mathbf{r}'''(s) = -\kappa^3\mathbf{B} \cdot \mathbf{T} + \kappa(d\kappa/ds)\mathbf{B} \cdot \mathbf{N} + \kappa^2\tau\mathbf{B} \cdot \mathbf{B} = \kappa^2\tau$,
$\tau = [\mathbf{r}'(s) \times \mathbf{r}''(s)] \cdot \mathbf{r}'''(s)/\kappa^2 = [\mathbf{r}'(s) \times \mathbf{r}''(s)] \cdot \mathbf{r}'''(s)/\|\mathbf{r}''(s)\|^2$ and
$\mathbf{B} = \mathbf{T} \times \mathbf{N} = [\mathbf{r}'(s) \times \mathbf{r}''(s)]/\|\mathbf{r}''(s)\|$

65. $\mathbf{r} = a\cos(s/w)\mathbf{i} + a\sin(s/w)\mathbf{j} + (cs/w)\mathbf{k}$, $\mathbf{r}' = -(a/w)\sin(s/w)\mathbf{i} + (a/w)\cos(s/w)\mathbf{j} + (c/w)\mathbf{k}$,
$\mathbf{r}'' = -(a/w^2)\cos(s/w)\mathbf{i} - (a/w^2)\sin(s/w)\mathbf{j}$, $\mathbf{r}''' = (a/w^3)\sin(s/w)\mathbf{i} - (a/w^3)\cos(s/w)\mathbf{j}$,
$\mathbf{r}' \times \mathbf{r}'' = (ac/w^3)\sin(s/w)\mathbf{i} - (ac/w^3)\cos(s/w)\mathbf{j} + (a^2/w^3)\mathbf{k}$, $(\mathbf{r}' \times \mathbf{r}'') \cdot \mathbf{r}''' = a^2c/w^6$,
$\|\mathbf{r}''(s)\| = a/w^2$, so $\tau = c/w^2$ and $\mathbf{B} = (c/w)\sin(s/w)\mathbf{i} - (c/w)\cos(s/w)\mathbf{j} + (a/w)\mathbf{k}$

66. (a) $\mathbf{T}' = \dfrac{d\mathbf{T}}{dt} = \dfrac{d\mathbf{T}}{ds}\dfrac{ds}{dt} = (\kappa\mathbf{N})s' = \kappa s'\mathbf{N}$,

$\mathbf{N}' = \dfrac{d\mathbf{N}}{dt} = \dfrac{d\mathbf{N}}{ds}\dfrac{ds}{dt} = (-\kappa\mathbf{T} + \tau\mathbf{B})s' = -\kappa s'\mathbf{T} + \tau s'\mathbf{B}$.

(b) $\|\mathbf{r}'(t)\| = s'$ so $\mathbf{r}'(t) = s'\mathbf{T}$ and $\mathbf{r}''(t) = s''\mathbf{T} + s'\mathbf{T}' = s''\mathbf{T} + s'(\kappa s'\mathbf{N}) = s''\mathbf{T} + \kappa(s')^2\mathbf{N}$.

(c) $\mathbf{r}'''(t) = s''\mathbf{T}' + s'''\mathbf{T} + \kappa(s')^2\mathbf{N}' + [2\kappa s's'' + \kappa'(s')^2]\mathbf{N}$
$= s''(\kappa s'\mathbf{N}) + s'''\mathbf{T} + \kappa(s')^2(\quad\kappa s'\mathbf{T} + \tau s'\mathbf{B}) + [2\kappa s's'' + \kappa'(s')^2]\mathbf{N}$
$= [s''' - \kappa^2(s')^3]\mathbf{T} + [3\kappa s's'' + \kappa'(s')^2]\mathbf{N} + \kappa\tau(s')^3\mathbf{B}$.

(d) $\mathbf{r}'(t) \times \mathbf{r}''(t) = s's''\mathbf{T} \times \mathbf{T} + \kappa(s')^3\mathbf{T} \times \mathbf{N} = \kappa(s')^3\mathbf{B}$, $[\mathbf{r}'(t) \times \mathbf{r}''(t)] \cdot \mathbf{r}'''(t) = \kappa^2\tau(s')^6$ so
$$\tau = \frac{[\mathbf{r}'(t) \times \mathbf{r}''(t)] \cdot \mathbf{r}'''(t)}{\kappa^2(s')^6} = \frac{[\mathbf{r}'(t) \times \mathbf{r}''(t)] \cdot \mathbf{r}'''(t)}{\|\mathbf{r}'(t) \times \mathbf{r}''(t)\|^2}$$

67. $\mathbf{r}' = 2\mathbf{i} + 2t\mathbf{j} + t^2\mathbf{k}$, $\mathbf{r}'' = 2\mathbf{j} + 2t\mathbf{k}$, $\mathbf{r}''' = 2\mathbf{k}$, $\mathbf{r}' \times \mathbf{r}'' = 2t^2\mathbf{i} - 4t\mathbf{j} + 4\mathbf{k}$, $\|\mathbf{r}' \times \mathbf{r}''\| = 2(t^2 + 2)$,
$\tau = 8/[2(t^2 + 2)]^2 = 2/(t^2 + 2)^2$

68. $\mathbf{r}' = -a\sin t\mathbf{i} + a\cos t\mathbf{j} + c\mathbf{k}$, $\mathbf{r}'' = -a\cos t\mathbf{i} - a\sin t\mathbf{j}$, $\mathbf{r}''' = a\sin t\mathbf{i} - a\cos t\mathbf{j}$,
$\mathbf{r}' \times \mathbf{r}'' = ac\sin t\mathbf{i} - ac\cos t\mathbf{j} + a^2\mathbf{k}$, $\|\mathbf{r}' \times \mathbf{r}''\| = \sqrt{a^2(a^2+c^2)}$,
$\tau = a^2c/[a^2(a^2+c^2)] = c/(a^2+c^2)$

69. $\mathbf{r}' = e^t\mathbf{i} - e^{-t}\mathbf{j} + \sqrt{2}\mathbf{k}$, $\mathbf{r}'' = e^t\mathbf{i} + e^{-t}\mathbf{j}$, $\mathbf{r}''' = e^t\mathbf{i} - e^{-t}\mathbf{j}$, $\mathbf{r}' \times \mathbf{r}'' = -\sqrt{2}e^{-t}\mathbf{i} + \sqrt{2}e^t\mathbf{j} + 2\mathbf{k}$,
$\|\mathbf{r}' \times \mathbf{r}''\| = \sqrt{2}(e^t + e^{-t})$, $\tau = (-2\sqrt{2})/[2(e^t+e^{-t})^2] = -\sqrt{2}/(e^t+e^{-t})^2$

70. $\mathbf{r}' = (1-\cos t)\mathbf{i} + \sin t\mathbf{j} + \mathbf{k}$, $\mathbf{r}'' = \sin t\mathbf{i} + \cos t\mathbf{j}$, $\mathbf{r}''' = \cos t\mathbf{i} - \sin t\mathbf{j}$,
$\mathbf{r}' \times \mathbf{r}'' = -\cos t\mathbf{i} + \sin t\mathbf{j} + (\cos t - 1)\mathbf{k}$,
$\|\mathbf{r}' \times \mathbf{r}''\| = \sqrt{\cos^2 t + \sin^2 t + (\cos t - 1)^2} = \sqrt{1 + 4\sin^4(t/2)}$, $\tau = -1/[1 + 4\sin^4(t/2)]$

EXERCISE SET 13.6

1. $\mathbf{v}(t) = -3\sin t\mathbf{i} + 3\cos t\mathbf{j}$
$\mathbf{a}(t) = -3\cos t\mathbf{i} - 3\sin t\mathbf{j}$
$\|\mathbf{v}(t)\| = \sqrt{9\sin^2 t + 9\cos^2 t} = 3$
$\mathbf{r}(\pi/3) = (3/2)\mathbf{i} + (3\sqrt{3}/2)\mathbf{j}$
$\mathbf{v}(\pi/3) = -(3\sqrt{3}/2)\mathbf{i} + (3/2)\mathbf{j}$
$\mathbf{a}(\pi/3) = -(3/2)\mathbf{i} - (3\sqrt{3}/2)\mathbf{j}$

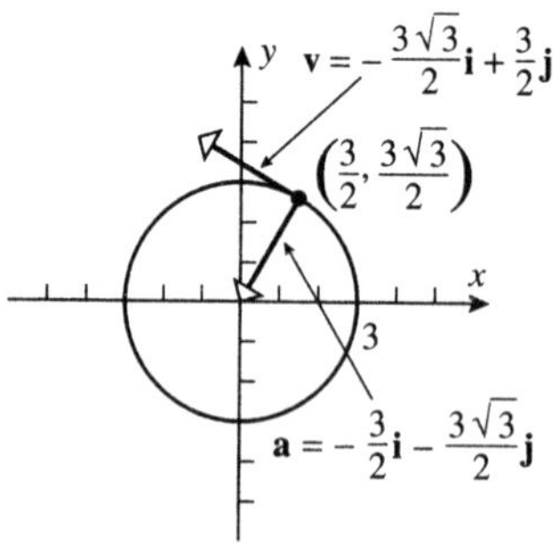

2. $\mathbf{v}(t) = \mathbf{i} + 2t\mathbf{j}$
$\mathbf{a}(t) = 2\mathbf{j}$
$\|\mathbf{v}(t)\| = \sqrt{1 + 4t^2}$
$\mathbf{r}(2) = 2\mathbf{i} + 4\mathbf{j}$
$\mathbf{v}(2) = \mathbf{i} + 4\mathbf{j}$
$\mathbf{a}(2) = 2\mathbf{j}$

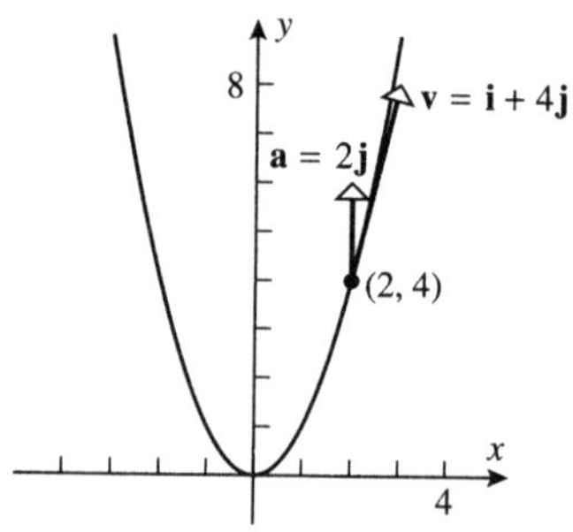

3. $\mathbf{v}(t) = e^t\mathbf{i} - e^{-t}\mathbf{j}$
$\mathbf{a}(t) = e^t\mathbf{i} + e^{-t}\mathbf{j}$
$\|\mathbf{v}(t)\| = \sqrt{e^{2t} + e^{-2t}}$
$\mathbf{r}(0) = \mathbf{i} + \mathbf{j}$
$\mathbf{v}(0) = \mathbf{i} - \mathbf{j}$
$\mathbf{a}(0) = \mathbf{i} + \mathbf{j}$

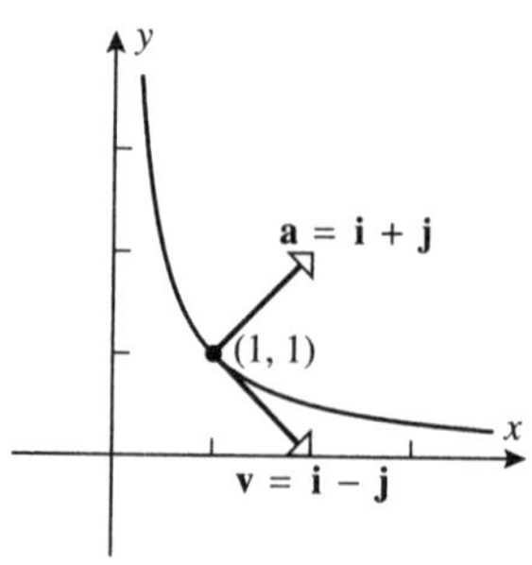

4. $\mathbf{v}(t) = 4\mathbf{i} - \mathbf{j}$

$\mathbf{a}(t) = \mathbf{0}$

$\|\mathbf{v}(t)\| = \sqrt{17}$

$\mathbf{r}(1) = 6\mathbf{i}$

$\mathbf{v}(1) = 4\mathbf{i} - \mathbf{j}$

$\mathbf{a}(1) = \mathbf{0}$

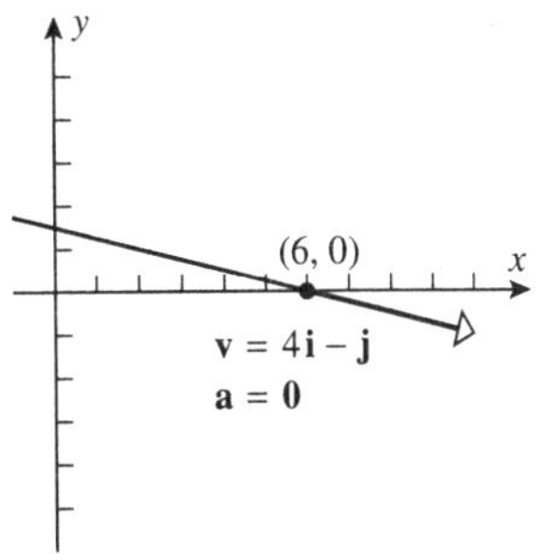

5. $\mathbf{v} = \mathbf{i} + t\mathbf{j} + t^2\mathbf{k}$, $\mathbf{a} = \mathbf{j} + 2t\mathbf{k}$; at $t = 1$, $\mathbf{v} = \mathbf{i} + \mathbf{j} + \mathbf{k}$, $\|\mathbf{v}\| = \sqrt{3}$, $\mathbf{a} = \mathbf{j} + 2\mathbf{k}$

6. $\mathbf{r} = (1 + 3t)\mathbf{i} + (2 - 4t)\mathbf{j} + (7 + t)\mathbf{k}$, $\mathbf{v} = 3\mathbf{i} - 4\mathbf{j} + \mathbf{k}$,
$\mathbf{a} = \mathbf{0}$; at $t = 2$, $\mathbf{v} = 3\mathbf{i} - 4\mathbf{j} + \mathbf{k}$, $\|\mathbf{v}\| = \sqrt{26}$, $\mathbf{a} = \mathbf{0}$

7. $\mathbf{v} = -2\sin t\mathbf{i} + 2\cos t\mathbf{j} + \mathbf{k}$, $\mathbf{a} = -2\cos t\mathbf{i} - 2\sin t\mathbf{j}$;
at $t = \pi/4$, $\mathbf{v} = -\sqrt{2}\mathbf{i} + \sqrt{2}\mathbf{j} + \mathbf{k}$, $\|\mathbf{v}\| = \sqrt{5}$, $\mathbf{a} = -\sqrt{2}\mathbf{i} - \sqrt{2}\mathbf{j}$

8. $\mathbf{v} = e^t(\cos t + \sin t)\mathbf{i} + e^t(\cos t - \sin t)\mathbf{j} + \mathbf{k}$, $\mathbf{a} = 2e^t\cos t\mathbf{i} - 2e^t\sin t\mathbf{j}$; at $t = \pi/2$,
$\mathbf{v} = e^{\pi/2}\mathbf{i} - e^{\pi/2}\mathbf{j} + \mathbf{k}$, $\|\mathbf{v}\| = (1 + 2e^{\pi})^{1/2}$, $\mathbf{a} = -2e^{\pi/2}\mathbf{j}$

9. **(a)** $\mathbf{v} = -a\omega\sin\omega t\mathbf{i} + b\omega\cos\omega t\mathbf{j}$, $\mathbf{a} = -a\omega^2\cos\omega t\mathbf{i} - b\omega^2\sin\omega t\mathbf{j} = -\omega^2\mathbf{r}$

(b) From Part (a), $\|\mathbf{a}\| = \omega^2\|\mathbf{r}\|$

10. **(a)** $\mathbf{v} = 16\pi\cos\pi t\mathbf{i} - 8\pi\sin 2\pi t\mathbf{j}$, $\mathbf{a} = -16\pi^2\sin\pi t\mathbf{i} - 16\pi^2\cos 2\pi t\mathbf{j}$;
at $t = 1$, $\mathbf{v} = -16\pi\mathbf{i}$, $\|\mathbf{v}\| = 16\pi$, $\mathbf{a} = -16\pi^2\mathbf{j}$

(b) $x = 16\sin\pi t, y = 4\cos 2\pi t = 4\cos^2\pi t - 4\sin^2\pi t = 4 - 8\sin^2\pi t,\ y = 4 - x^2/32$

(c) Both $x(t)$ and $y(t)$ are periodic and have period 2, so after 2 s the particle retraces its path.

11. $\mathbf{v} = (6/\sqrt{t})\mathbf{i} + (3/2)t^{1/2}\mathbf{j}$, $\|\mathbf{v}\| = \sqrt{36/t + 9t/4}, d\|\mathbf{v}\|/dt = (-36/t^2 + 9/4)/(2\sqrt{36/t + 9t/4}) = 0$ if $t = 4$ which yields a minimum by the first derivative test. The minimum speed is $3\sqrt{2}$ when $\mathbf{r} = 24\mathbf{i} + 8\mathbf{j}$.

12. $\mathbf{v} = (1 - 2t)\mathbf{i} - 2t\mathbf{j}$, $\|\mathbf{v}\| = \sqrt{(1 - 2t)^2 + 4t^2} = \sqrt{8t^2 - 4t + 1}$,
$\dfrac{d}{dt}\|\mathbf{v}\| = \dfrac{8t - 2}{\sqrt{8t^2 - 4t + 1}} = 0$ if $t = \dfrac{1}{4}$ which yields a minimum by the first derivative test. The minimum speed is $1/\sqrt{2}$ when the particle is at $\mathbf{r} = \dfrac{3}{16}\mathbf{i} - \dfrac{1}{16}\mathbf{j}$.

13. **(a)**

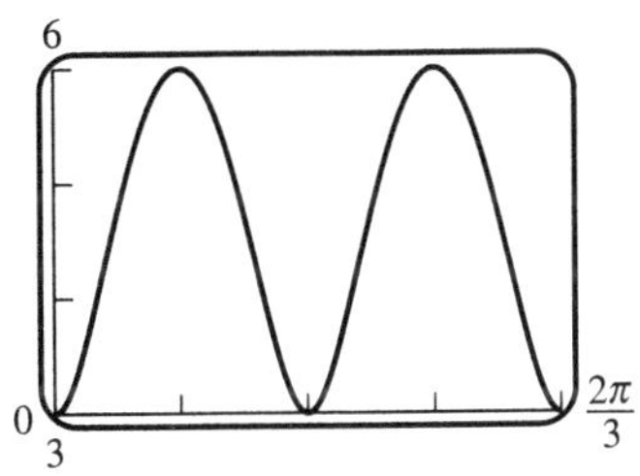

(b) $\mathbf{v} = 3\cos 3t\mathbf{i} + 6\sin 3t\mathbf{j}$, $\|\mathbf{v}\| = \sqrt{9\cos^2 3t + 36\sin^2 3t} = 3\sqrt{1 + 3\sin^2 3t}$; by inspection, maximum speed is 6 and minimum speed is 3

(d) $\dfrac{d}{dt}\|\mathbf{v}\| = \dfrac{27\sin 6t}{2\sqrt{1 + 3\sin^2 3t}} = 0$ when $t = 0, \pi/6, \pi/3, \pi/2, 2\pi/3$; the maximum speed is 6 which occurs first when $\sin 3t = 1, t = \pi/6$.

14. (a)

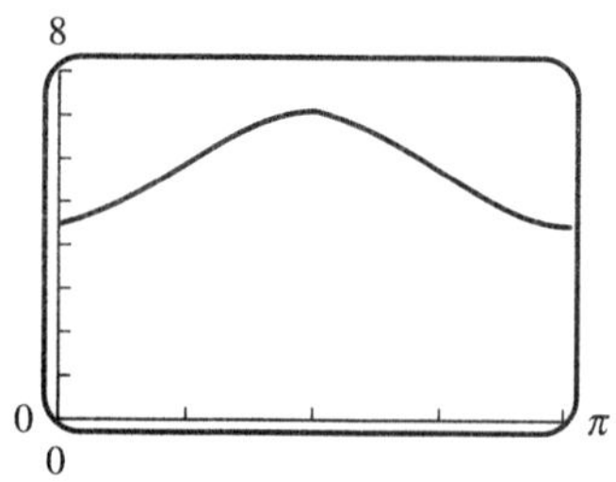

(d) $\mathbf{v} = -6\sin 2t\mathbf{i} + 2\cos 2t\mathbf{j} + 4\mathbf{k}$, $\|\mathbf{v}\| = \sqrt{36\sin^2 2t + 4\cos^2 2t + 16} = 2\sqrt{8\sin^2 t + 5}$; by inspection the maximum speed is $2\sqrt{13}$ when $t = \pi/2$, the minimum speed is $2\sqrt{5}$ when $t = 0$ or π.

15. $\mathbf{v}(t) = -\sin t\mathbf{i} + \cos t\mathbf{j} + \mathbf{C}_1$, $\mathbf{v}(0) = \mathbf{j} + \mathbf{C}_1 = \mathbf{i}$, $\mathbf{C}_1 = \mathbf{i} - \mathbf{j}$, $\mathbf{v}(t) = (1 - \sin t)\mathbf{i} + (\cos t - 1)\mathbf{j}$;
$\mathbf{r}(t) = (t + \cos t)\mathbf{i} + (\sin t - t)\mathbf{j} + \mathbf{C}_2$, $\mathbf{r}(0) = \mathbf{i} + \mathbf{C}_2 = \mathbf{j}$,
$\mathbf{C}_2 = -\mathbf{i} + \mathbf{j}$ so $\mathbf{r}(t) = (t + \cos t - 1)\mathbf{i} + (\sin t - t + 1)\mathbf{j}$

16. $\mathbf{v}(t) = t\mathbf{i} - e^{-t}\mathbf{j} + \mathbf{C}_1$, $\mathbf{v}(0) = -\mathbf{j} + \mathbf{C}_1 = 2\mathbf{i} + \mathbf{j}$; $\mathbf{C}_1 = 2\mathbf{i} + 2\mathbf{j}$ so
$\mathbf{v}(t) = (t + 2)\mathbf{i} + (2 - e^{-t})\mathbf{j}$; $\mathbf{r}(t) = (t^2/2 + 2t)\mathbf{i} + (2t + e^{-t})\mathbf{j} + \mathbf{C}_2$
$\mathbf{r}(0) = \mathbf{j} + \mathbf{C}_2 = \mathbf{i} - \mathbf{j}$, $\mathbf{C}_2 = \mathbf{i} - 2\mathbf{j}$ so $\mathbf{r}(t) = (t^2/2 + 2t + 1)\mathbf{i} + (2t + e^{-t} - 2)\mathbf{j}$

17. $\mathbf{v}(t) = -\cos t\mathbf{i} + \sin t\mathbf{j} + e^t\mathbf{k} + \mathbf{C}_1$, $\mathbf{v}(0) = -\mathbf{i} + \mathbf{k} + \mathbf{C}_1 = \mathbf{k}$ so
$\mathbf{C}_1 = \mathbf{i}$, $\mathbf{v}(t) = (1 - \cos t)\mathbf{i} + \sin t\mathbf{j} + e^t\mathbf{k}$; $\mathbf{r}(t) = (t - \sin t)\mathbf{i} - \cos t\mathbf{j} + e^t\mathbf{k} + \mathbf{C}_2$,
$\mathbf{r}(0) = -\mathbf{j} + \mathbf{k} + \mathbf{C}_2 = -\mathbf{i} + \mathbf{k}$ so $\mathbf{C}_2 = -\mathbf{i} + \mathbf{j}$, $\mathbf{r}(t) = (t - \sin t - 1)\mathbf{i} + (1 - \cos t)\mathbf{j} + e^t\mathbf{k}$.

18. $\mathbf{v}(t) = -\dfrac{1}{t+1}\mathbf{j} + \dfrac{1}{2}e^{-2t}\mathbf{k} + \mathbf{C}_1$, $\mathbf{v}(0) = -\mathbf{j} + \dfrac{1}{2}\mathbf{k} + \mathbf{C}_1 = 3\mathbf{i} - \mathbf{j}$ so

$\mathbf{C}_1 = 3\mathbf{i} - \dfrac{1}{2}\mathbf{k}$, $\mathbf{v}(t) = 3\mathbf{i} - \dfrac{1}{t+1}\mathbf{j} + \left(\dfrac{1}{2}e^{-2t} - \dfrac{1}{2}\right)\mathbf{k}$;

$\mathbf{r}(t) = 3t\mathbf{i} - \ln(t+1)\mathbf{j} - \left(\dfrac{1}{4}e^{-2t} + \dfrac{1}{2}t\right)\mathbf{k} + \mathbf{C}_2$,

$\mathbf{r}(0) = -\dfrac{1}{4}\mathbf{k} + \mathbf{C}_2 = 2\mathbf{k}$ so $\mathbf{C}_2 = \dfrac{9}{4}\mathbf{k}$, $\mathbf{r}(t) = 3t\mathbf{i} - \ln(t+1)\mathbf{j} + \left(\dfrac{9}{4} - \dfrac{1}{4}e^{-2t} - \dfrac{1}{2}t\right)\mathbf{k}$.

19. If $\mathbf{a} = \mathbf{0}$ then $x''(t) = y''(t) = z''(t) = 0$, so $x(t) = x_1 t + x_0, y(t) = y_1 t + y_0, z(t) = z_1 t + z_0$, the motion is along a straight line and has constant speed.

20. (a) If $\|\mathbf{r}\|$ is constant then so is $\|\mathbf{r}\|^2$, but then $x^2 + y^2 = c^2$ (2-space) or $x^2 + y^2 + z^2 = c^2$ (3-space), so the motion is along a circle or a sphere of radius c centered at the origin, and the velocity vector is always perpendicular to the position vector.

(b) If $\|\mathbf{v}\|$ is constant then by the Theorem, $\mathbf{v}(t) \cdot \mathbf{a}(t) = 0$, so the velocity is always perpendicular to the acceleration.

21. $\mathbf{v} = 3t^2\mathbf{i} + 2t\mathbf{j}$, $\mathbf{a} = 6t\mathbf{i} + 2\mathbf{j}$; $\mathbf{v} = 3\mathbf{i} + 2\mathbf{j}$ and $\mathbf{a} = 6\mathbf{i} + 2\mathbf{j}$ when $t = 1$ so $\cos\theta = (\mathbf{v}\cdot\mathbf{a})/(\|\mathbf{v}\|\,\|\mathbf{a}\|) = 11/\sqrt{130}$, $\theta \approx 15°$.

22. $\mathbf{v} = e^t(\cos t - \sin t)\mathbf{i} + e^t(\cos t + \sin t)\mathbf{j}$, $\mathbf{a} = -2e^t \sin t\mathbf{i} + 2e^t \cos t\mathbf{j}$, $\mathbf{v}\cdot\mathbf{a} = 2e^{2t}$, $\|\mathbf{v}\| = \sqrt{2}e^t$, $\|\mathbf{a}\| = 2e^t$, $\cos\theta = (\mathbf{v}\cdot\mathbf{a})/(\|\mathbf{v}\|\,\|\mathbf{a}\|) = 1/\sqrt{2}$, $\theta = 45°$.

23. (a) displacement $= \mathbf{r}_1 - \mathbf{r}_0 = 0.7\mathbf{i} + 2.7\mathbf{j} - 3.4\mathbf{k}$

(b) $\Delta\mathbf{r} = \mathbf{r}_1 - \mathbf{r}_0$, so $\mathbf{r}_0 = \mathbf{r}_1 - \Delta\mathbf{r} = -0.7\mathbf{i} - 2.9\mathbf{j} + 4.8\mathbf{k}$.

24. (a)

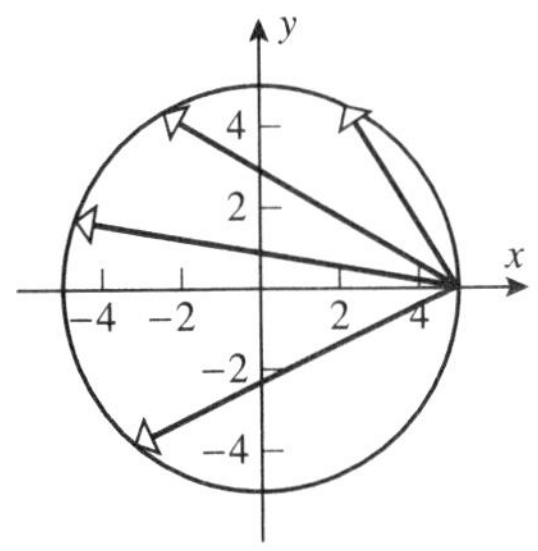

(b) one revolution, or 10π

25. $\Delta\mathbf{r} = \mathbf{r}(3) - \mathbf{r}(1) = 8\mathbf{i} + (26/3)\mathbf{j}$; $\mathbf{v} = 2t\mathbf{i} + t^2\mathbf{j}$, $s = \int_1^3 t\sqrt{4+t^2}\,dt = (13\sqrt{13} - 5\sqrt{5})/3$.

26. $\Delta\mathbf{r} = \mathbf{r}(3\pi/2) - \mathbf{r}(0) = 3\mathbf{i} - 3\mathbf{j}$; $\mathbf{v} = -3\cos t\mathbf{i} - 3\sin t\mathbf{j}$, $s = \int_0^{3\pi/2} 3dt = 9\pi/2$.

27. $\Delta\mathbf{r} = \mathbf{r}(\ln 3) - \mathbf{r}(0) = 2\mathbf{i} - (2/3)\mathbf{j} + \sqrt{2}(\ln 3)\mathbf{k}$; $\mathbf{v} = e^t\mathbf{i} - e^{-t}\mathbf{j} + \sqrt{2}\,\mathbf{k}$, $s = \int_0^{\ln 3}(e^t + e^{-t})dt = 8/3$.

28. $\Delta\mathbf{r} = \mathbf{r}(\pi) - \mathbf{r}(0) = \mathbf{0}$; $\mathbf{v} = -2\sin 2t\mathbf{i} + 2\sin 2t\mathbf{j} - \sin 2t\mathbf{k}$, $\|\mathbf{v}\| = 3|\sin 2t|$, $s = \int_0^{\pi} 3|\sin 2t|dt = 6\int_0^{\pi/2}\sin 2t\,dt = 6$.

29. In both cases, the equation of the path in rectangular coordinates is $x^2 + y^2 = 4$, the particles move counterclockwise around this circle; $\mathbf{v}_1 = -6\sin 3t\mathbf{i} + 6\cos 3t\mathbf{j}$ and $\mathbf{v}_2 = -4t\sin(t^2)\mathbf{i} + 4t\cos(t^2)\mathbf{j}$ so $\|\mathbf{v}_1\| = 6$ and $\|\mathbf{v}_2\| = 4t$.

30. Let $u = 1 - t^3$ in $\mathbf{r}_2$ to get
$\mathbf{r}_1(u) = (3 + 2(1 - t^3))\mathbf{i} + (1 - t^3)\mathbf{j} + (1 - (1 - t^3))\mathbf{k} = (5 - 2t^3)\mathbf{i} + (1 - t^3)\mathbf{j} + t^3\mathbf{k} = \mathbf{r}_2(t)$
so both particles move along the same path; $\mathbf{v}_1 = 2\mathbf{i} + \mathbf{j} - \mathbf{k}$ and $\mathbf{v}_2 = -6t^2\mathbf{i} - 3t^2\mathbf{j} + 3t^2\mathbf{k}$ so $\|\mathbf{v}_1\| = \sqrt{6}$ and $\|\mathbf{v}_2\| = 3\sqrt{6}t^2$.

31. (a) $\mathbf{v} = -e^{-t}\mathbf{i} + e^t\mathbf{j}$, $\mathbf{a} = e^{-t}\mathbf{i} + e^t\mathbf{j}$; when $t = 0$, $\mathbf{v} = -\mathbf{i} + \mathbf{j}$, $\mathbf{a} = \mathbf{i} + \mathbf{j}$, $\|\mathbf{v}\| = \sqrt{2}$, $\mathbf{v}\cdot\mathbf{a} = 0$, $\mathbf{v}\times\mathbf{a} = -2\mathbf{k}$ so $a_T = 0$, $a_N = \sqrt{2}$.

(b) $a_T\mathbf{T} = \mathbf{0}, a_N\mathbf{N} = \mathbf{a} - a_T\mathbf{T} = \mathbf{i} + \mathbf{j}$

(c) $\kappa = 1/\sqrt{2}$

32. (a) $\mathbf{v} = -2t\sin(t^2)\mathbf{i} + 2t\cos(t^2)\mathbf{j}$, $\mathbf{a} = [-4t^2\cos(t^2) - 2\sin(t^2)]\mathbf{i} + [-4t^2\sin(t^2) + 2\cos(t^2)]\mathbf{j}$; when $t = \sqrt{\pi}/2$, $\mathbf{v} = -\sqrt{\pi/2}\mathbf{i} + \sqrt{\pi/2}\mathbf{j}$, $\mathbf{a} = (-\pi/\sqrt{2} - \sqrt{2})\mathbf{i} + (-\pi/\sqrt{2} + \sqrt{2})\mathbf{j}$, $\|\mathbf{v}\| = \sqrt{\pi}$, $\mathbf{v}\cdot\mathbf{a} = 2\sqrt{\pi}$, $\mathbf{v}\times\mathbf{a} = \pi^{3/2}\mathbf{k}$ so $a_T = 2$, $a_N = \pi$

(b) $a_T\mathbf{T} = -\sqrt{2}(\mathbf{i} - \mathbf{j})$, $a_N\mathbf{N} = \mathbf{a} - a_T\mathbf{T} = -(\pi/\sqrt{2})(\mathbf{i} + \mathbf{j})$

(c) $\kappa = 1$

33. **(a)** $\mathbf{v} = (3t^2-2)\mathbf{i} + 2t\mathbf{j}$, $\mathbf{a} = 6t\mathbf{i} + 2\mathbf{j}$; when $t = 1$, $\mathbf{v} = \mathbf{i} + 2\mathbf{j}$, $\mathbf{a} = 6\mathbf{i} + 2\mathbf{j}$, $\|\mathbf{v}\| = \sqrt{5}$, $\mathbf{v} \cdot \mathbf{a} = 10$, $\mathbf{v} \times \mathbf{a} = -10\mathbf{k}$ so $a_T = 2\sqrt{5}$, $a_N = 2\sqrt{5}$

(b) $a_T\mathbf{T} = \dfrac{2\sqrt{5}}{\sqrt{5}}(\mathbf{i} + 2\mathbf{j}) = 2\mathbf{i} + 4\mathbf{j}$, $a_N\mathbf{N} = \mathbf{a} - a_T\mathbf{T} = 4\mathbf{i} - 2\mathbf{j}$

(c) $\kappa = 2/\sqrt{5}$

34. **(a)** $\mathbf{v} = e^t(-\sin t + \cos t)\mathbf{i} + e^t(\cos t + \sin t)\mathbf{j}$, $\mathbf{a} = -2e^t \sin t\mathbf{i} + 2e^t \cos t\mathbf{j}$; when $t = \pi/4$, $\mathbf{v} = \sqrt{2}e^{\pi/4}\mathbf{j}$, $\mathbf{a} = -\sqrt{2}e^{\pi/4}\mathbf{i} + \sqrt{2}e^{\pi/4}\mathbf{j}$, $\|\mathbf{v}\| = \sqrt{2}e^{\pi/4}$, $\mathbf{v} \cdot \mathbf{a} = 2e^{\pi/2}$, $\mathbf{v} \times \mathbf{a} = 2e^{\pi/2}\mathbf{k}$ so $a_T = \sqrt{2}e^{\pi/4}$, $a_N = \sqrt{2}e^{\pi/4}$

(b) $a_T\mathbf{T} = \sqrt{2}e^{\pi/4}\mathbf{j}$, $a_N\mathbf{N} = \mathbf{a} - a_T\mathbf{T} = -\sqrt{2}e^{\pi/4}\mathbf{i}$

(c) $\kappa = \dfrac{1}{\sqrt{2}e^{\pi/4}}$

35. **(a)** $\mathbf{v} = (-1/t^2)\mathbf{i} + 2t\mathbf{j} + 3t^2\mathbf{k}$, $\mathbf{a} = (2/t^3)\,\mathbf{i} + 2\mathbf{j} + 6t\mathbf{k}$; when $t = 1$, $\mathbf{v} = -\mathbf{i} + 2\mathbf{j} + 3\mathbf{k}$, $\mathbf{a} = 2\mathbf{i} + 2\mathbf{j} + 6\mathbf{k}$, $\|\mathbf{v}\| = \sqrt{14}$, $\mathbf{v} \cdot \mathbf{a} = 20$, $\mathbf{v} \times \mathbf{a} = 6\mathbf{i} + 12\mathbf{j} - 6\mathbf{k}$ so $a_T = 20/\sqrt{14}$, $a_N = 6\sqrt{3}/\sqrt{7}$

(b) $a_T\mathbf{T} = -\dfrac{10}{7}\mathbf{i} + \dfrac{20}{7}\mathbf{j} + \dfrac{30}{7}\mathbf{k}$, $a_N\mathbf{N} = \mathbf{a} - a_T\mathbf{T} = \dfrac{24}{7}\mathbf{i} - \dfrac{6}{7}\mathbf{j} + \dfrac{12}{7}\mathbf{k}$

(c) $\kappa = \dfrac{6\sqrt{6}}{14^{3/2}} = \left(\dfrac{3}{7}\right)^{3/2}$

36. **(a)** $\mathbf{v} = e^t\mathbf{i} - 2e^{-2t}\mathbf{j} + \mathbf{k}$, $\mathbf{a} = e^t\mathbf{i} + 4e^{-2t}\mathbf{j}$; when $t = 0$, $\mathbf{v} = \mathbf{i} - 2\mathbf{j} + \mathbf{k}$, $\mathbf{a} = \mathbf{i} + 4\mathbf{j}$, $\|\mathbf{v}\| = \sqrt{6}$, $\mathbf{v} \cdot \mathbf{a} = -7$, $\mathbf{v} \times \mathbf{a} = -4\mathbf{i} + \mathbf{j} + 6\mathbf{k}$ so $a_T = -7/\sqrt{6}$, $a_N = \sqrt{53/6}$

(b) $a_T\mathbf{T} = -\dfrac{7}{6}(\mathbf{i} - 2\mathbf{j} + \mathbf{k}), a_N\mathbf{N} = \mathbf{a} - a_T\mathbf{T} = \dfrac{13}{6}\mathbf{i} + \dfrac{19}{3}\mathbf{j} + \dfrac{7}{6}\mathbf{k}$

(c) $\kappa = \dfrac{\sqrt{53}}{6\sqrt{6}}$

37. **(a)** $\mathbf{v} = 3\cos t\mathbf{i} - 2\sin t\mathbf{j} - 2\cos 2t\mathbf{k}$, $\mathbf{a} = -3\sin t\mathbf{i} - 2\cos t\mathbf{j} + 4\sin 2t\mathbf{k}$; when $t = \pi/2$, $\mathbf{v} = -2\mathbf{j} + 2\mathbf{k}$, $\mathbf{a} = -3\mathbf{i}$, $\|\mathbf{v}\| = 2\sqrt{2}$, $\mathbf{v} \cdot \mathbf{a} = 0$, $\mathbf{v} \times \mathbf{a} = -6\mathbf{j} - 6\mathbf{k}$ so $a_T = 0$, $a_N = 3$

(b) $a_T\mathbf{T} = \mathbf{0}, a_N\mathbf{N} = \mathbf{a} = -3\mathbf{i}$

(c) $\kappa = \dfrac{3}{8}$

38. **(a)** $\mathbf{v} = 3t^2\mathbf{j} - (16/t)\mathbf{k}$, $\mathbf{a} = 6t\mathbf{j} + (16/t^2)\mathbf{k}$; when $t = 1$, $\mathbf{v} = 3\mathbf{j} - 16\mathbf{k}$, $\mathbf{a} = 6\mathbf{j} + 16\mathbf{k}$, $\|\mathbf{v}\| = \sqrt{265}$, $\mathbf{v} \cdot \mathbf{a} = -238$, $\mathbf{v} \times \mathbf{a} = 144\mathbf{i}$ so $a_T = -238/\sqrt{265}$, $a_N = 144/\sqrt{265}$

(b) $a_T\mathbf{T} = -\dfrac{714}{265}\mathbf{j} + \dfrac{3808}{265}\mathbf{k}$, $a_N\mathbf{N} = \mathbf{a} - a_T\mathbf{T} = \dfrac{2304}{265}\mathbf{j} + \dfrac{432}{265}\mathbf{k}$

(c) $\kappa = \dfrac{144}{265^{3/2}}$

39. $\|\mathbf{v}\| = 4$, $\mathbf{v} \cdot \mathbf{a} = -12$, $\mathbf{v} \times \mathbf{a} = 8\mathbf{k}$ so $a_T = -3$, $a_N = 2$, $\mathbf{T} = -\mathbf{j}$, $\mathbf{N} = (\mathbf{a} - a_T\mathbf{T})/a_N = \mathbf{i}$

40. $\|\mathbf{v}\| = \sqrt{5}$, $\mathbf{v} \cdot \mathbf{a} = 3$, $\mathbf{v} \times \mathbf{a} = -6\mathbf{k}$ so $a_T = 3/\sqrt{5}$, $a_N = 6/\sqrt{5}$, $\mathbf{T} = (1/\sqrt{5})(\mathbf{i} + 2\mathbf{j})$, $\mathbf{N} = (\mathbf{a} - a_T\mathbf{T})/a_N = (1/\sqrt{5})(2\mathbf{i} - \mathbf{j})$

41. $\|\mathbf{v}\| = 3$, $\mathbf{v} \cdot \mathbf{a} = 4$, $\mathbf{v} \times \mathbf{a} = 4\mathbf{i} - 3\mathbf{j} - 2\mathbf{k}$ so $a_T = 4/3$, $a_N = \sqrt{29}/3$, $\mathbf{T} = (1/3)(2\mathbf{i} + 2\mathbf{j} + \mathbf{k})$, $\mathbf{N} = (\mathbf{a} - a_T\mathbf{T})/a_N = (\mathbf{i} - 8\mathbf{j} + 14\mathbf{k})/(3\sqrt{29})$

42. $\|\mathbf{v}\| = 5$, $\mathbf{v} \cdot \mathbf{a} = -5$, $\mathbf{v} \times \mathbf{a} = -4\mathbf{i} - 10\mathbf{j} - 3\mathbf{k}$ so $a_T = -1$, $a_N = \sqrt{5}$, $\mathbf{T} = (1/5)(3\mathbf{i} - 4\mathbf{k})$, $\mathbf{N} = (\mathbf{a} - a_T\mathbf{T})/a_N = (8\mathbf{i} - 5\mathbf{j} + 6\mathbf{k})/(5\sqrt{5})$

43. $a_T = \dfrac{d^2s}{dt^2} = \dfrac{d}{dt}\sqrt{3t^2+4} = 3t/\sqrt{3t^2+4}$ so when $t = 2$, $a_T = 3/2$.

44. $a_T = \dfrac{d^2s}{dt^2} = \dfrac{d}{dt}\sqrt{t^2+e^{-3t}} = (2t - 3e^{-3t})/[2\sqrt{t^2+e^{-3t}}]$ so when $t = 0$, $a_T = -3/2$.

45. $a_T = \dfrac{d^2s}{dt^2} = \dfrac{d}{dt}\sqrt{(4t-1)^2+\cos^2\pi t} = [4(4t-1) - \pi\cos\pi t\sin\pi t]/\sqrt{(4t-1)^2+\cos^2\pi t}$ so when $t = 1/4$, $a_T = -\pi/\sqrt{2}$.

46. $a_T = \dfrac{d^2s}{dt^2} = \dfrac{d}{dt}\sqrt{t^4+5t^2+3} = (2t^3+5t)/\sqrt{t^4+5t^2+3}$ so when $t = 1$, $a_T = 7/3$.

47. $a_N = \kappa(ds/dt)^2 = (1/\rho)(ds/dt)^2 = (1/1)(2.9 \times 10^5)^2 = 8.41 \times 10^{10}$ km/s^2

48. $\mathbf{a} = (d^2s/dt^2)\mathbf{T} + \kappa(ds/dt)^2\mathbf{N}$ where $\kappa = \dfrac{|d^2y/dx^2|}{[1+(dy/dx)^2]^{3/2}}$. If $d^2y/dx^2 = 0$, then $\kappa = 0$ and $\mathbf{a} = (d^2s/dt^2)\mathbf{T}$ so $\mathbf{a}$ is tangent to the curve.

49. $a_N = \kappa(ds/dt)^2 = [2/(1+4x^2)^{3/2}](3)^2 = 18/(1+4x^2)^{3/2}$

50. $y = e^x$, $a_N = \kappa(ds/dt)^2 = [e^x/(1+e^{2x})^{3/2}](2)^2 = 4e^x/(1+e^{2x})^{3/2}$

51. $\mathbf{a} = a_T\mathbf{T} + a_N\mathbf{N}$; by the Pythagorean Theorem $a_N = \sqrt{\|\mathbf{a}\|^2 - a_T^2} = \sqrt{9-9} = 0$

52. As in Exercise 51, $\|\mathbf{a}\|^2 = a_T^2 + a_N^2, 81 = 9 + a_N^2, a_N = \sqrt{72} = 6\sqrt{2}$.

53. Let $c = ds/dt, a_N = \kappa\left(\dfrac{ds}{dt}\right)^2, a_N = \dfrac{1}{1000}c^2$, so $c^2 = 1000a_N$, $c \le 10\sqrt{10}\sqrt{1.5} \approx 38.73$ m/s.

54. 10 km/h is the same as $\dfrac{100}{36}$ m/s, so $\|\mathbf{F}\| = 500\dfrac{1}{15}\left(\dfrac{100}{36}\right)^2 \approx 257.20$ N.

55. **(a)** $v_0 = 320$, $\alpha = 60°$, $s_0 = 0$ so $x = 160t$, $y = 160\sqrt{3}t - 16t^2$.

(b) $dy/dt = 160\sqrt{3} - 32t$, $dy/dt = 0$ when $t = 5\sqrt{3}$ so $y_{\max} = 160\sqrt{3}(5\sqrt{3}) - 16(5\sqrt{3})^2 = 1200$ ft.

(c) $y = 16t(10\sqrt{3} - t)$, $y = 0$ when $t = 0$ or $10\sqrt{3}$ so $x_{\max} = 160(10\sqrt{3}) = 1600\sqrt{3}$ ft.

(d) $\mathbf{v}(t) = 160\mathbf{i} + (160\sqrt{3} - 32t)\mathbf{j}$, $\mathbf{v}(10\sqrt{3}) = 160(\mathbf{i} - \sqrt{3}\mathbf{j})$, $\|\mathbf{v}(10\sqrt{3})\| = 320$ ft/s.

56. **(a)** $v_0 = 980$, $\alpha = 45°$, $s_0 = 0$ so $x = 490\sqrt{2}\,t$, $y = 490\sqrt{2}t - 4.9t^2$

(b) $dy/dt = 490\sqrt{2} - 9.8t$, $dy/dt = 0$ when $t = 50\sqrt{2}$ so $y_{\max} = 490\sqrt{2}(50\sqrt{2}) - 4.9(50\sqrt{2})^2 = 24,500$ m.

(c) $y = 4.9t(100\sqrt{2} - t)$, $y = 0$ when $t = 0$ or $100\sqrt{2}$ so $x_{\max} = 490\sqrt{2}(100\sqrt{2}) = 98,000$ m.

(d) $\mathbf{v}(t) = 490\sqrt{2}\,\mathbf{i} + (490\sqrt{2} - 9.8t)\mathbf{j}$, $\mathbf{v}(100\sqrt{2}) = 490\sqrt{2}(\mathbf{i} - \mathbf{j})$, $\|\mathbf{v}(100\sqrt{2})\| = 980$ m/s.

57. $v_0 = 80$, $\alpha = -60°$, $s_0 = 168$ so $x = 40t$, $y = 168 - 40\sqrt{3}\,t - 16t^2$; $y = 0$ when $t = -7\sqrt{3}/2$ (invalid) or $t = \sqrt{3}$ so $x(\sqrt{3}) = 40\sqrt{3}$ ft.

58. $v_0 = 80$, $\alpha = 0°$, $s_0 = 168$ so $x = 80t$, $y = 168 - 16t^2$; $y = 0$ when $t = -\sqrt{42}/2$ (invalid) or $t = \sqrt{42}/2$ so $x(\sqrt{42}/2) = 40\sqrt{42}$ ft.

59. $\alpha = 30°$, $s_0 = 0$ so $x = \sqrt{3}v_0 t/2$, $y = v_0 t/2 - 16t^2$; $dy/dt = v_0/2 - 32t$, $dy/dt = 0$ when $t = v_0/64$ so $y_{\max} = v_0^2/256 = 2500$, $v_0 = 800$ ft/s.

60. $\alpha = 45°$, $s_0 = 0$ so $x = \sqrt{2}\,v_0 t/2$, $y = \sqrt{2}v_0 t/2 - 4.9t^2$; $y = 0$ when $t = 0$ or $\sqrt{2}v_0/9.8$ so $x_{\max} = v_0^2/9.8 = 24,500$, $v_0 = 490$ m/s.

61. $v_0 = 800$, $s_0 = 0$ so $x = (800\cos\alpha)t$, $y = (800\sin\alpha)t - 16t^2 = 16t(50\sin\alpha - t)$; $y = 0$ when $t = 0$ or $50\sin\alpha$ so $x_{\max} = 40,000\sin\alpha\cos\alpha = 20,000\sin 2\alpha = 10,000$, $2\alpha = 30°$ or $150°$, $\alpha = 15°$ or $75°$.

62. **(a)** $v_0 = 5$, $\alpha = 0°$, $s_0 = 4$ so $x = 5t$, $y = 4 - 16t^2$; $y = 0$ when $t = -1/2$ (invalid) or $1/2$ so it takes the ball $1/2$ s to hit the floor.

(b) $\mathbf{v}(t) = 5\mathbf{i} - 32t\mathbf{j}$, $\mathbf{v}(1/2) = 5\mathbf{i} - 16\mathbf{j}$, $\|\mathbf{v}(1/2)\| = \sqrt{281}$ so the ball hits the floor with a speed of $\sqrt{281}$ ft/s.

(c) $v_0 = 0$, $\alpha = -90°$, $s_0 = 4$ so $x = 0$, $y = 4 - 16t^2$; $y = 0$ when $t = 1/2$ so both balls would hit the ground at the same instant.

63. **(a)** Let $\mathbf{r}(t) = x(t)\mathbf{i} + y(t)\mathbf{j}$ with $\mathbf{j}$ pointing up. Then $\mathbf{a} = -32\mathbf{j} = x''(t)\mathbf{i} + y''(t)\mathbf{j}$, so $x(t) = At + B, y(t) = -16t^2 + Ct + D$. Next, $x(0) = 0, y(0) = 4$ so $x(t) = At, y(t) = -16t^2 + Ct + 4$; $y'(0)/x'(0) = \tan 60° = \sqrt{3}$, so $C = \sqrt{3}A$; and $40 = v_0 = \sqrt{x'(0)^2 + y'(0)^2} = \sqrt{A^2 + 3A^2}, A = 20$, thus $\mathbf{r}(t) = 20t\,\mathbf{i} + (-16t^2 + 20\sqrt{3}t + 4)\,\mathbf{j}$. When $x = 15$, $t = \dfrac{3}{4}$, and $y = 4 + 20\sqrt{3}\dfrac{3}{4} - 16\left(\dfrac{3}{4}\right)^2 \approx 20.98$ ft, so the water clears the corner point A with 0.98 ft to spare.

(b) $y = 20$ when $-16t^2 + 20\sqrt{3}t - 16 = 0, t = 0.668$ (reject) or $1.497, x(1.497) \approx 29.942$ ft, so the water hits the roof.

(c) about $29.942 - 15 = 14.942$ ft

64. $x = (v_0/2)t, y = 4 + (v_0\sqrt{3}/2)t - 16t^2$, solve $x = 15, y = 20$ simultaneously for v_0 and t, $v_0/2 = 15/t, t^2 = \dfrac{15}{16}\sqrt{3} - 1, t \approx 0.7898, v_0 \approx 30/0.7898 \approx 37.98$ ft/s.

65. **(a)** $x = (35\sqrt{2}/2)t$, $y = (35\sqrt{2}/2)t - 4.9t^2$, from Exercise 17a in Section 13.5

$$\kappa = \frac{|x'y'' - x''y'|}{[(x')^2 + (y')^2]^{3/2}},\ \kappa(0) = \frac{9.8}{35^2\sqrt{2}} = 0.004\sqrt{2} \approx 0.00565685; \rho = 1/\kappa \approx 176.78 \text{ m}$$

(b) $y'(t) = 0$ when $t = \dfrac{25}{14}\sqrt{2}, y = \dfrac{125}{4}$ m

66. **(a)** $\mathbf{a} = a_T\mathbf{T} + a_N\mathbf{N}, a_T = \dfrac{d^2s}{dt^2} = -7.5 \text{ ft/s}^2,\ a_N = \kappa\left(\dfrac{ds}{dt}\right)^2 = \dfrac{1}{\rho}(132)^2 = \dfrac{132^2}{3000} \text{ ft/s}^2,$

$$\|\mathbf{a}\| = \sqrt{a_T^2 + a_N^2} = \sqrt{(7.5)^2 + \left(\frac{132^2}{3000}\right)^2} \approx 9.49 \text{ ft/s}^2$$

(b) $\cos\theta = \dfrac{\mathbf{a}\cdot\mathbf{T}}{\|\mathbf{a}\|\|\mathbf{T}\|} = \dfrac{a_T}{\|\mathbf{a}\|} \approx -\dfrac{7.5}{9.49} \approx -0.79, \theta \approx 2.48 \text{ radians} \approx 142°$

67. $s_0 = 0$ so $x = (v_0\cos\alpha)t,\ y = (v_0\sin\alpha)t - gt^2/2$

(a) $dy/dt = v_0\sin\alpha - gt$ so $dy/dt = 0$ when $t = (v_0\sin\alpha)/g,\ y_{\max} = (v_0\sin\alpha)^2/(2g)$

(b) $y = 0$ when $t = 0$ or $(2v_0\sin\alpha)/g$, so $x = R = (2v_0^2\sin\alpha\cos\alpha)/g = (v_0^2\sin 2\alpha)/g$ when $t = (2v_0\sin\alpha)/g$; R is maximum when $2\alpha = 90°$, $\alpha = 45°$, and the maximum value of R is v_0^2/g.

68. The range is $(v_0^2\sin 2\alpha)/g$ and the maximum range is v_0^2/g so $(v_0^2\sin 2\alpha)/g = (3/4)v_0^2/g$, $\sin 2\alpha = 3/4$, $\alpha = (1/2)\sin^{-1}(3/4) \approx 24.3°$ or $\alpha = (1/2)[180° - \sin^{-1}(3/4)] \approx 65.7°$.

69. $v_0 = 80,\ \alpha = 30°,\ s_0 = 5$ so $x = 40\sqrt{3}t,\ y = 5 + 40t - 16t^2$

(a) $y = 0$ when $t = (-40 \pm \sqrt{(40)^2 - 4(-16)(5)})/(-32) = (5 \pm \sqrt{30})/4$, reject $(5 - \sqrt{30})/4$ to get $t = (5 + \sqrt{30})/4 \approx 2.62$ s.

(b) $x \approx 40\sqrt{3}(2.62) \approx 181.5\,\text{ft}$.

70. **(a)** $v_0 = v,\ s_0 = h$ so $x = (v\cos\alpha)t,\ y = h + (v\sin\alpha)t - \dfrac{1}{2}gt^2$. If $x = R$, then $(v\cos\alpha)t = R$, $t = \dfrac{R}{v\cos\alpha}$ but $y = 0$ for this value of t so $h + (v\sin\alpha)[R/(v\cos\alpha)] - \dfrac{1}{2}g[R/(v\cos\alpha)]^2 = 0$, $h + (\tan\alpha)R - g(\sec^2\alpha)R^2/(2v^2) = 0,\ g(\sec^2\alpha)R^2 - 2v^2(\tan\alpha)R - 2v^2h = 0.$

(b) $2g\sec^2\alpha\tan\alpha R^2 + 2g\sec^2\alpha R\dfrac{dR}{d\alpha} - 2v^2\sec^2\alpha R - 2v^2\tan\alpha\dfrac{dR}{d\alpha} = 0$; if $\dfrac{dR}{d\alpha} = 0$ and $\alpha = \alpha_0$ when $R = R_0$, then $2g\sec^2\alpha_0\tan\alpha_0 R_0^2 - 2v^2\sec^2\alpha_0 R_0 = 0,\ g\tan\alpha_0 R_0 - v^2 = 0$, $\tan\alpha_0 = v^2/(gR_0)$.

(c) If $\alpha = \alpha_0$ and $R = R_0$, then from Part (a) $g(\sec^2\alpha_0)R_0^2 - 2v^2(\tan\alpha_0)R_0 - 2v^2h = 0$, but from Part (b) $\tan\alpha_0 = v^2/(gR_0)$ so $\sec^2\alpha_0 = 1 + \tan^2\alpha_0 = 1 + v^4/(gR_0)^2$ thus $g[1 + v^4/(gR_0)^2]R_0^2 - 2v^2[v^2/(gR_0)]R_0 - 2v^2h = 0,\ gR_0^2 - v^4/g - 2v^2h = 0$, $R_0^2 = v^2(v^2 + 2gh)/g^2,\ R_0 = (v/g)\sqrt{v^2 + 2gh}$ and $\tan\alpha_0 = v^2/(v\sqrt{v^2 + 2gh}) = v/\sqrt{v^2 + 2gh},\ \alpha_0 = \tan^{-1}(v/\sqrt{v^2 + 2gh})$.

71. **(a)** $v_0(\cos\alpha)(2.9) = 259\cos 23°$ so $v_0\cos\alpha \approx 82.21061,\ v_0(\sin\alpha)(2.9) - 16(2.9)^2 = -259\sin 23°$ so $v_0\sin\alpha \approx 11.50367$; divide $v_0\sin\alpha$ by $v_0\cos\alpha$ to get $\tan\alpha \approx 0.139929$, thus $\alpha \approx 8°$ and $v_0 \approx 82.21061/\cos 8° \approx 83$ ft/s.

(b) From Part (a), $x \approx 82.21061t$ and $y \approx 11.50367t - 16t^2$ for $0 \le t \le 2.9$; the distance traveled is $\displaystyle\int_0^{2.9} \sqrt{(dx/dt)^2 + (dy/dt)^2}\,dt \approx 268.76$ ft.

EXERCISE SET 13.7

1. The results follow from formulae (1) and (7) of Section 11.6.

2. **(a)** $(r_{\max} - r_{\min})/(r_{\max} + r_{\min}) = 2ae/(2a) = e$

(b) $r_{\max}/r_{\min} = (1+e)/(1-e)$, and the result follows.

3. **(a)** From (15) and (6), at $t = 0$,
$\mathbf{C} = \mathbf{v}_0 \times \mathbf{b}_0 - GM\mathbf{u} = v_0\mathbf{j} \times r_0v_0\mathbf{k} - GM\mathbf{u} = r_0v_0^2\mathbf{i} - GM\mathbf{i} = (r_0v_0^2 - GM)\mathbf{i}$

(b) From (22), $r_0v_0^2 - GM = GMe$, so from (7) and (17), $\mathbf{v} \times \mathbf{b} = GM(\cos\theta\mathbf{i} + \sin\theta\mathbf{j}) + GMe\mathbf{i}$, and the result follows.

(c) From (10) it follows that $\mathbf{b}$ is perpendicular to $\mathbf{v}$, and the result follows.

(d) From Part (c) and (10), $\|\mathbf{v} \times \mathbf{b}\| = \|\mathbf{v}\|\|\mathbf{b}\| = vr_0v_0$. From Part (b),
$\|\mathbf{v} \times \mathbf{b}\| = GM\sqrt{(e+\cos\theta)^2 + \sin^2\theta} = GM\sqrt{e^2 + 2e\cos\theta + 1}$. By (10) and Part (c), $\|\mathbf{v} \times \mathbf{b}\| = \|\mathbf{v}\|\|\mathbf{b}\| = v(r_0v_0)$ thus $v = \dfrac{GM}{r_0v_0}\sqrt{e^2 + 2e\cos\theta + 1}$. From (22), $r_0v_0^2/(GM) = 1 + e$, $GM/(r_0v_0) = v_0/(1+e)$ so $v = \dfrac{v_0}{1+e}\sqrt{e^2 + 2e\cos\theta + 1}$.

4. At the end of the minor axis, $\cos\theta = -c/a = -e$ so
$v = \dfrac{v_0}{1+e}\sqrt{e^2 + 2e(-e) + 1} = \dfrac{v_0}{1+e}\sqrt{1 - e^2} = v_0\sqrt{\dfrac{1-e}{1+e}}$.

5. $v_{\max}$ occurs when $\theta = 0$ so $v_{\max} = v_0$; $v_{\min}$ occurs when $\theta = \pi$ so
$v_{\min} = \dfrac{v_0}{1+e}\sqrt{e^2 - 2e + 1} = v_{\max}\dfrac{1-e}{1+e}$, thus $v_{\max} = v_{\min}\dfrac{1+e}{1-e}$.

6. If the orbit is a circle then $e = 0$ so from Part (d) of Exercise 3, $v = v_0$ at all points on the orbit. Use (22) with $e = 0$ to get $v_0 = \sqrt{GM/r_0}$ so $v = \sqrt{GM/r_0}$.

7. $r_0 = 6440 + 200 = 6640$ km so $v = \sqrt{3.99 \times 10^5/6640} \approx 7.75$ km/s.

8. From Example 1, the orbit is 22,250 mi above the Earth, thus $v \approx \sqrt{\dfrac{1.24 \times 10^{12}}{26{,}250}} \approx 6873$ mi/h.

9. From (23) with $r_0 = 6440 + 300 = 6740$ km, $v_{\text{esc}} = \sqrt{\dfrac{2(3.99) \times 10^5}{6740}} \approx 10.88$ km/s.

10. From (29), $T = \dfrac{2\pi}{\sqrt{GM}}a^{3/2}$. But $T = 1$ yr $= 365 \cdot 24 \cdot 3600$ s, thus $M = \dfrac{4\pi^2a^3}{GT^2} \approx 1.99 \times 10^{30}$ kg.

11. **(a)** At perigee, $r = r_{\min} = a(1-e) = 238{,}900\ (1 - 0.055) \approx 225{,}760$ mi; at apogee,
$r = r_{\max} = a(1+e) = 238{,}900(1 + 0.055) \approx 252{,}040$ mi. Subtract the sum of the radius of the Moon and the radius of the Earth to get
minimum distance $= 225{,}760 - 5080 = 220{,}680$ mi,
and maximum distance $= 252{,}040 - 5080 = 246{,}960$ mi.

(b) $T = 2\pi\sqrt{a^3/(GM)} = 2\pi\sqrt{(238{,}900)^3/(1.24 \times 10^{12})} \approx 659$ hr ≈ 27.5 days.

12. **(a)** $r_{\text{min}} = 6440 + 649 = 7{,}089$ km, $r_{\text{max}} = 6440 + 4{,}340 = 10{,}780$ km so $a = (r_{\text{min}} + r_{\text{max}})/2 = 8934.5$ km.

(b) $e = (10{,}780 - 7{,}089)/(10{,}780 + 7{,}089) \approx 0.207$.

(c) $T = 2\pi\sqrt{a^3/(GM)} = 2\pi\sqrt{(8934.5)^3/(3.99 \times 10^5)} \approx 8400 \text{ s} \approx 140 \text{ min}$

13. **(a)** $r_0 = 4000 + 180 = 4180$ mi, $v = \sqrt{\dfrac{GM}{r_0}} = \sqrt{1.24 \times 10^{12}/4180} \approx 17{,}224$ mi/h

(b) $r_0 = 4180$ mi, $v_0 = \sqrt{\dfrac{GM}{r_0}} + 600$; $e = \dfrac{r_0 v_0^2}{GM} - 1 = 1200\sqrt{\dfrac{r_0}{GM}} + (600)^2 \dfrac{r_0}{GM} \approx 0.071$; $r_{\text{max}} = 4180(1 + 0.071)/(1 - 0.071) \approx 4819$ mi; the apogee altitude is $4819 - 4000 = 819$ mi.

14. By equation (20), $r = \dfrac{k}{1 + e\cos\theta}$, where $k > 0$. By assumption, r is minimal when $\theta = 0$, hence $e \geq 0$.

CHAPTER 13 SUPPLEMENTARY EXERCISES

2. **(a)** the line through the tips of $\mathbf{r}_0$ and $\mathbf{r}_1$

(b) the line segment connecting the tips of $\mathbf{r}_0$ and $\mathbf{r}_1$

(c) the line through the tip of $\mathbf{r}_0$ which is parallel to $\mathbf{r}'(t_0)$

4. **(a)** speed **(b)** distance traveled **(c)** distance of the particle from the origin

7. **(a)** $\mathbf{r}(t) = \displaystyle\int_0^t \cos\left(\frac{\pi u^2}{2}\right) du\,\mathbf{i} + \int_0^t \sin\left(\frac{\pi u^2}{2}\right) du\,\mathbf{j}$;

$$\left\|\frac{d\mathbf{r}}{dt}\right\|^2 = x'(t)^2 + y'(t)^2 = \cos^2\left(\frac{\pi t^2}{2}\right) + \sin^2\left(\frac{\pi t^2}{2}\right) = 1 \text{ and } \mathbf{r}(0) = \mathbf{0}$$

(b) $\mathbf{r}'(s) = \cos\left(\dfrac{\pi s^2}{2}\right)\mathbf{i} + \sin\left(\dfrac{\pi s^2}{2}\right)\mathbf{j}, \mathbf{r}''(s) = -\pi s\sin\left(\dfrac{\pi s^2}{2}\right)\mathbf{i} + \pi s\cos\left(\dfrac{\pi s^2}{2}\right)\mathbf{j}$, $\kappa = \|\mathbf{r}''(s)\| = \pi|s|$

(c) $\kappa(s) \to +\infty$, so the spiral winds ever tighter.

8. **(a)** The tangent vector to the curve is always tangent to the sphere.

(b) $\|\mathbf{v}\| = \text{const}$, so $\mathbf{v} \cdot \mathbf{a} = 0$; the acceleration vector is always perpendicular to the velocity vector

(c) $\|\mathbf{r}(t)\|^2 = \left(1 - \dfrac{1}{4}\cos^2 t\right)(\cos^2 t + \sin^2 t) + \dfrac{1}{4}\cos^2 t = 1$

9. **(a)** $\|\mathbf{r}(t)\| = 1$, so by Theorem 13.2.9, $\mathbf{r}'(t)$ is always perpendicular to the vector $\mathbf{r}(t)$. Then $\mathbf{v}(t) = R\omega(-\sin\omega t\mathbf{i} + \cos\omega t\mathbf{j}), v = \|\mathbf{v}(t)\| = R\omega$

(b) $\mathbf{a} = -R\omega^2(\cos\omega t\mathbf{i} + \sin\omega t\mathbf{j}), a = \|\mathbf{a}\| = R\omega^2$, and $\mathbf{a} = -\omega^2\mathbf{r}$ is directed toward the origin.

(c) The smallest value of t for which $\mathbf{r}(t) = \mathbf{r}(0)$ satisfies $\omega t = 2\pi$, so $T = t = \dfrac{2\pi}{\omega}$.

10. (a) $F = \|\mathbf{F}\| = m\|\mathbf{a}\| = mR\omega^2 = mR\dfrac{v^2}{R^2} = \dfrac{mv^2}{R}$

(b) $R = 6440 + 3200 = 9640$ km, $6.43 = v = R\omega = 9640\omega$, $\omega = \dfrac{6.43}{9640} \approx 0.000667$,

$a = R\omega^2 = v\omega = \dfrac{6.43^2}{9640} \approx 0.00429$ km/s^2

$\mathbf{a} = -a(\cos\omega t\mathbf{i} + \sin\omega t\mathbf{j}) \approx -0.00429[\cos(0.000667t)\mathbf{i} + \sin(0.000667t)\mathbf{j}]$

(c) $F = ma \approx 70(0.00429)$ kg $\cdot$ km/s$^2 \approx 0.30030$ kN $= 300.30$ N

11. (a) Let $\mathbf{r} = x\mathbf{i} + y\mathbf{j} + z\mathbf{k}$, then $x^2 + z^2 = t^2(\sin^2 \pi t + \cos^2 \pi t) = t^2 = y^2$

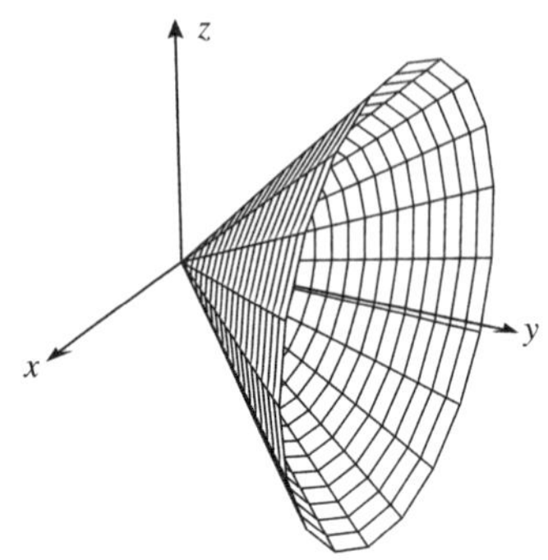

(b) Let $x = t$, then $y = t^2, z = \pm\sqrt{4 - t^2/3 - t^4/6}$

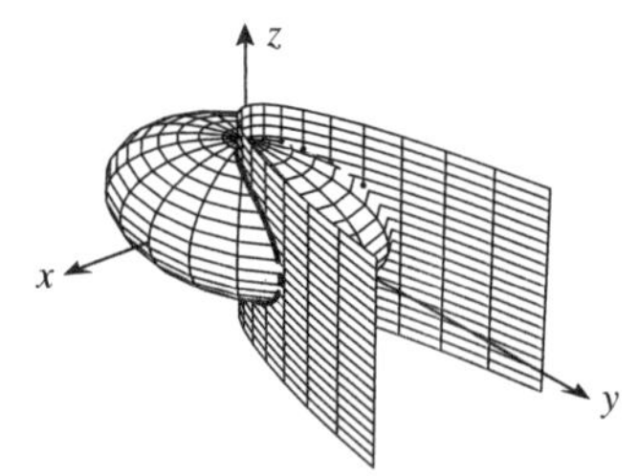

12.

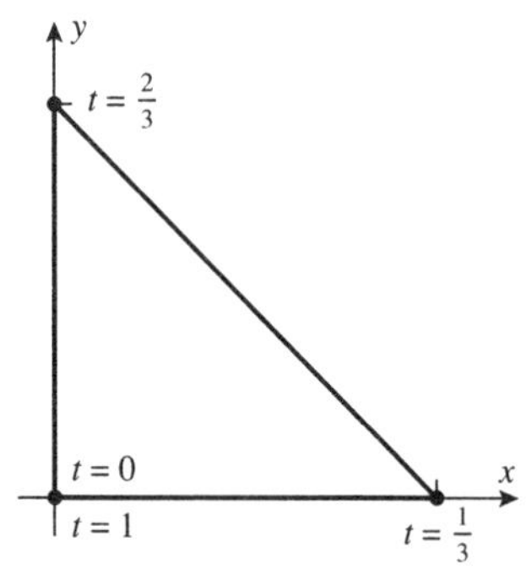

13. (a) $\|\mathbf{e}_r(t)\|^2 = \cos^2\theta + \sin^2\theta = 1$, so $\mathbf{e}_r(t)$ is a unit vector; $\mathbf{r}(t) = r(t)\mathbf{e}(t)$, so they have the same direction if $r(t) > 0$, opposite if $r(t) < 0$. $\mathbf{e}_\theta(t)$ is perpendicular to $\mathbf{e}_r(t)$ since $\mathbf{e}_r(t) \cdot \mathbf{e}_\theta(t) = 0$, and it will result from a counterclockwise rotation of $\mathbf{e}_r(t)$ provided $\mathbf{e}(t) \times \mathbf{e}_\theta(t) = \mathbf{k}$, which is true.

(b) $\dfrac{d}{dt}\mathbf{e}_r(t) = \dfrac{d\theta}{dt}(-\sin\theta\mathbf{i} + \cos\theta\mathbf{j}) = \dfrac{d\theta}{dt}\mathbf{e}_\theta(t)$ and $\dfrac{d}{dt}\mathbf{e}_\theta(t) = -\dfrac{d\theta}{dt}(\cos\theta\mathbf{i} + \sin\theta\mathbf{j}) = -\dfrac{d\theta}{dt}\mathbf{e}_r(t)$, so

$$\mathbf{v}(t) = \frac{d}{dt}\mathbf{r}(t) = \frac{d}{dt}(r(t)\mathbf{e}_r(t)) = r'(t)\mathbf{e}_r(t) + r(t)\frac{d\theta}{dt}\mathbf{e}_\theta(t)$$

(c) From Part (b), $\mathbf{a} = \dfrac{d}{dt}\mathbf{v}(t)$

$$= r''(t)\mathbf{e}_r(t) + r'(t)\frac{d\theta}{dt}\mathbf{e}_\theta(t) + r'(t)\frac{d\theta}{dt}\mathbf{e}_\theta(t) + r(t)\frac{d^2\theta}{dt^2}\mathbf{e}_\theta(t) - r(t)\left(\frac{d\theta}{dt}\right)^2\mathbf{e}_r(t)$$

$$= \left[\frac{d^2r}{dt^2} - r\left(\frac{d\theta}{dt}\right)^2\right]\mathbf{e}_r(t) + \left[r\frac{d^2\theta}{dt^2} + 2\frac{dr}{dt}\frac{d\theta}{dt}\right]\mathbf{e}_\theta(t)$$

14. The height $y(t)$ of the rocket satisfies $\tan\theta = y/b, y = b\tan\theta, v = \dfrac{dy}{dt} = \dfrac{dy}{d\theta}\dfrac{d\theta}{dt} = b\sec^2\theta\dfrac{d\theta}{dt}$.

15. $\mathbf{r} = \mathbf{r}_0 + t\overrightarrow{PQ} = (t-1)\mathbf{i} + (4-2t)\mathbf{j} + (3+2t)\mathbf{k}; \left\|\dfrac{d\mathbf{r}}{dt}\right\| = 3, \mathbf{r}(s) = \dfrac{s-3}{3}\mathbf{i} + \dfrac{12-2s}{3}\mathbf{j} + \dfrac{9+2s}{3}\mathbf{k}$

16. By equation (26) of Section 13.6, $\mathbf{r}(t) = (60\cos\alpha)t\mathbf{i} + ((60\sin\alpha)t - 16t^2 + 4)\mathbf{j}$, and the maximum height of the baseball occurs when $y'(t) = 0$, $60\sin\alpha = 32t$, $t = \dfrac{15}{8}\sin\alpha$, so the ball clears the ceiling if $y_{\max} = (60\sin\alpha)\dfrac{15}{8}\sin\alpha - 16\dfrac{15^2}{8^2}\sin^2\alpha + 4 \le 25$, $\dfrac{15^2\sin^2\alpha}{4} \le 21, \sin^2\alpha \le \dfrac{28}{75}$. The ball hits the wall when $x = 60, t = \sec\alpha$, and $y(\sec\alpha) = 60\sin\alpha\sec\alpha - 16\sec^2\alpha + 4$. Maximize the height $h(\alpha) = y(\sec\alpha) = 60\tan\alpha - 16\sec^2\alpha + 4$, subject to the constraint $\sin^2\alpha \le \dfrac{28}{75}$. Then $h'(\alpha) = 60\sec^2\alpha - 32\sec^2\alpha\tan\alpha = 0, \tan\alpha = \dfrac{60}{32} = \dfrac{15}{8}$, so $\sin\alpha = \dfrac{15}{\sqrt{8^2+15^2}} = \dfrac{15}{17}$, but for this value of α the constraint is not satisfied (the ball hits the ceiling). Hence the maximum value of h occurs at one of the endpoints of the α-interval on which the ball clears the ceiling, i.e. $\left[0, \sin^{-1}\sqrt{28/75}\right]$. Since $h'(0) = 60$, it follows that h is increasing throughout the interval, since $h' > 0$ inside the interval. Thus $h_{\max}$ occurs when $\sin^2\alpha = \dfrac{28}{75}$, $h_{\max} = 60\tan\alpha - 16\sec^2\alpha + 4 = 60\dfrac{\sqrt{28}}{\sqrt{47}} - 16\dfrac{75}{47} + 4 = \dfrac{120\sqrt{329} - 1012}{47} \approx 24.78$ ft. Note: the possibility that the baseball keeps climbing until it hits the wall can be rejected as follows: if so, then $y'(t) = 0$ after the ball hits the wall, i.e. $t = \dfrac{15}{8}\sin\alpha$ occurs after $t = \sec\alpha$, hence $\dfrac{15}{8}\sin\alpha \ge \sec\alpha$, $15\sin\alpha\cos\alpha \ge 8$, $15\sin 2\alpha \ge 16$, impossible.

17. $\mathbf{r}'(1) = 3\mathbf{i} + 10\mathbf{j} + 10\mathbf{k}$, so if $\mathbf{r}'(t) = 3t^2\mathbf{i} + 10\mathbf{j} + 10t\mathbf{k}$ is perpendicular to $\mathbf{r}'(1)$, then
$9t^2 + 100 + 100t = 0, t = -10, -10/9$,
so $\mathbf{r} = -1000\mathbf{i} - 100\mathbf{j} + 500\mathbf{k}, -(1000/729)\mathbf{i} - (100/9)\mathbf{j} + (500/81)\mathbf{k}$.

18. Let $\mathbf{r}(t) = x(t)\mathbf{i} + y(t)\mathbf{j}$, then $\dfrac{dx}{dt} = x(t), \dfrac{dy}{dt} = y(t), x(0) = x_0, y(0) = y_0$, so
$x(t) = x_0e^t, y(t) = y_0e^t$, $\mathbf{r}(t) = e^t\mathbf{r}_0$. If $\mathbf{r}(t)$ is a vector in 3 space then an analogous solution holds.

19. **(a)** $\dfrac{d\mathbf{v}}{dt} = 2t^2\mathbf{i} + \mathbf{j} + \cos 2t\mathbf{k}, \mathbf{v}_0 = \mathbf{i} + 2\mathbf{j} - \mathbf{k}$, so $x'(t) = \dfrac{2}{3}t^3 + 1, y'(t) = t + 2, z'(t) = \dfrac{1}{2}\sin 2t - 1,$

$x(t) = \dfrac{1}{6}t^4 + t, y(t) = \dfrac{1}{2}t^2 + 2t, z(t) = -\dfrac{1}{4}\cos 2t - t + \dfrac{1}{4}$, since $\mathbf{r}(0) = \mathbf{0}$. Hence

$$\mathbf{r}(t) = \left(\frac{1}{6}t^4 + t\right)\mathbf{i} + \left(\frac{1}{2}t^2 + 2t\right)\mathbf{j} - \left(\frac{1}{4}\cos 2t + t - \frac{1}{4}\right)\mathbf{k}$$

(b) $\left.\dfrac{ds}{dt}\right]_{t=1} = \|\mathbf{r}'(t)\|\Big]_{t=1} \sqrt{(5/3)^2 + 9 + (1 - (\sin 2)/2)^2} \approx 3.475$

20. $\|\mathbf{v}\|^2 = \mathbf{v}(t) \cdot \mathbf{v}(t),\ 2\|\mathbf{v}\|\dfrac{d}{dt}\|\mathbf{v}\| = 2\mathbf{v} \cdot \mathbf{a},\ \dfrac{d}{dt}(\|\mathbf{v}\|) = \dfrac{1}{\|\mathbf{v}\|}(\mathbf{v} \cdot \mathbf{a})$

CHAPTER 14
Partial Derivatives

EXERCISE SET 14.1

1. **(a)** $f(2,1)=(2)^2(1)+1=5$ **(b)** $f(1,2)=(1)^2(2)+1=3$
(c) $f(0,0)=(0)^2(0)+1=1$ **(d)** $f(1,-3)=(1)^2(-3)+1=-2$
(e) $f(3a,a)=(3a)^2(a)+1=9a^3+1$ **(f)** $f(ab,\, a-b)=(ab)^2(a-b)+1=a^3b^2-a^2b^3+1$

2. **(a)** $2t$ **(b)** $2x$ **(c)** $2y^2+2y$

3. **(a)** $f(x+y,x-y)=(x+y)(x-y)+3=x^2-y^2+3$
(b) $f\left(xy,3x^2y^3\right)=(xy)\left(3x^2y^3\right)+3=3x^3y^4+3$

4. **(a)** $(x/y)\sin(x/y)$ **(b)** $xy\sin(xy)$ **(c)** $(x-y)\sin(x-y)$

5. $F(g(x),h(y))=F\left(x^3,3y+1\right)=x^3e^{x^3(3y+1)}$

6. $g(u(x,y),v(x,y))=g\left(x^2y^3,\pi xy\right)=\pi xy\sin\left[\left(x^2y^3\right)^2(\pi xy)\right]=\pi xy\sin\left(\pi x^5y^7\right)$

7. **(a)** t^2+3t^{10} **(b)** 0 **(c)** 3076

8. $\sqrt{t}e^{-3\ln\left(t^2+1\right)}=\dfrac{\sqrt{t}}{\left(t^2+1\right)^3}$

9. **(a)** At $T=25$ there is a drop in temperature of 12 degrees when v changes from 5 to 10, thus $WCI\approx(2/5)(-12)+22=22-24/5=17.2°$ F.
(b) At $v=5$ there is an increase in temperature of 5 degrees as T changes from 25 to 30 degrees, thus $WCI\approx(3/5)5+22=25°$ F.

10. **(a)** $T\approx(4/5)(-7)+22=22-5.6=16.4°$ F
(b) $T\approx(2/5)6+16=16+2.4=18.4°$ F

11. **(a)** The depression is $20-16=4$, so the relative humidity is 66%.
(b) The relative humidity $\approx 77-(1/2)7=73.5\%$.
(c) The relative humidity $\approx 59+(2/5)4=60.6\%$.

12. **(a)** 4° C
(b) The relative humidity $\approx 62-(1/4)9=59.75\%$.
(c) The relative humidity $\approx 77+(1/5)(79-77)=77.4\%$.

13. **(a)** 19 **(b)** -9 **(c)** 3
(d) a^6+3 **(e)** $-t^8+3$ **(f)** $(a+b)(a-b)^2b^3+3$

14. **(a)** $x^2(x+y)(x-y)+(x+y)=x^2\left(x^2-y^2\right)+(x+y)=x^4-x^2y^2+x+y$
(b) $(xz)(xy)(y/x)+xy=xy^2z+xy$

15. $F\left(x^2,y+1,z^2\right)=(y+1)e^{x^2(y+1)z^2}$ **16.** $g\left(x^2z^3,\pi xyz,xy/z\right)=(xy/z)\sin\left(\pi x^3yz^4\right)$

17. **(a)** $f(\sqrt{5},2,\pi,-3\pi)=80\sqrt{\pi}$ **(b)** $f(1,1,\ldots,1)=\sum_{k=1}^{n}k=n(n+1)/2$

18. **(a)** $f(-2, 2, 0, \pi/4) = 1$
(b) $f(1, 2, \ldots, n) = n(n+1)(2n+1)/6$, see Theorem 2(b), Section 5.4

19.

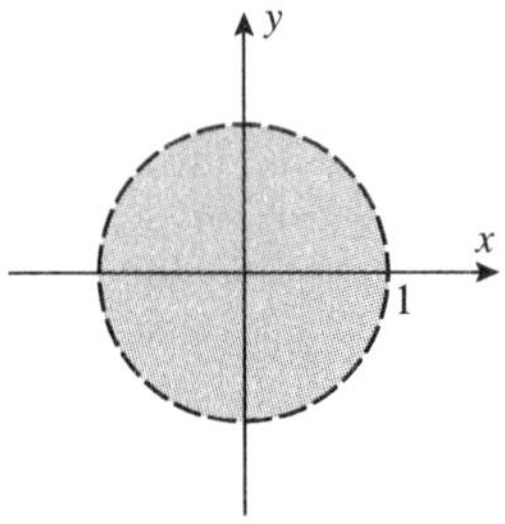

20.

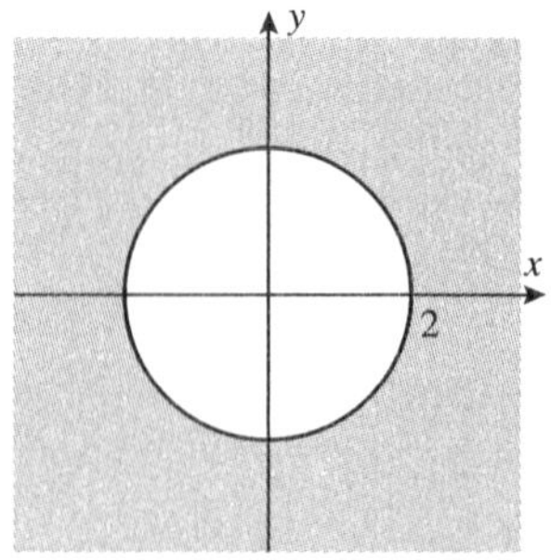

21.

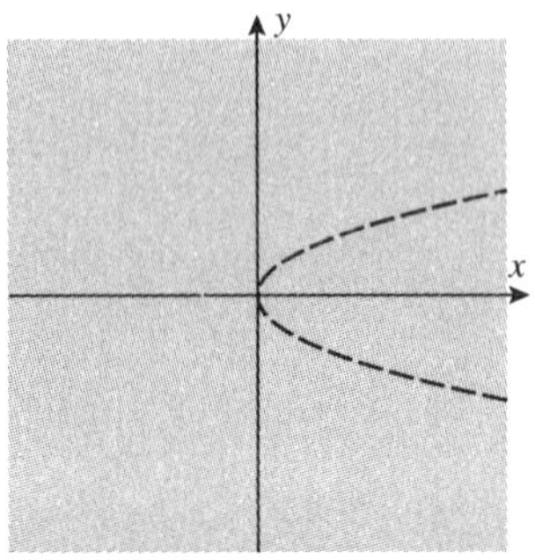

22.

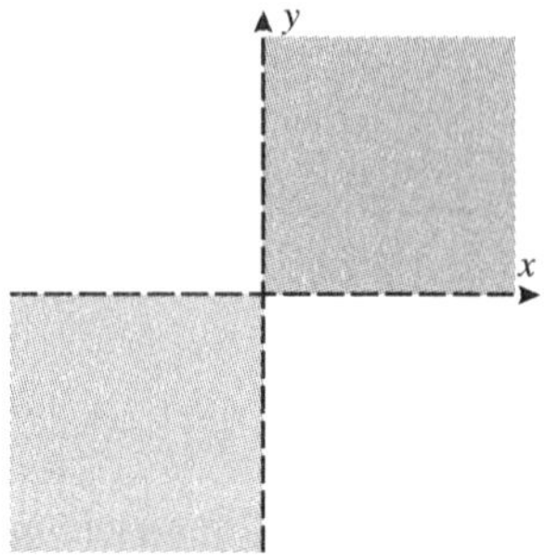

23. **(a)** all points above or on the line $y = -2$
(b) all points on or within the sphere $x^2 + y^2 + z^2 = 25$
(c) all points in 3-space

24. **(a)** all points on or between the vertical lines $x = \pm 2$.
(b) all points above the line $y = 2x$
(c) all points not on the plane $x + y + z = 0$

25.

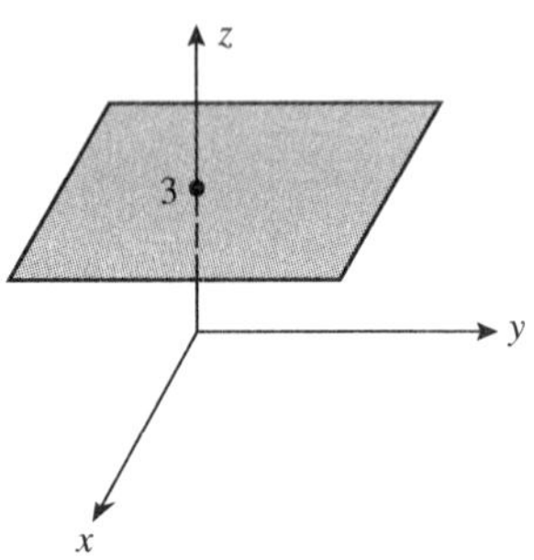

26.

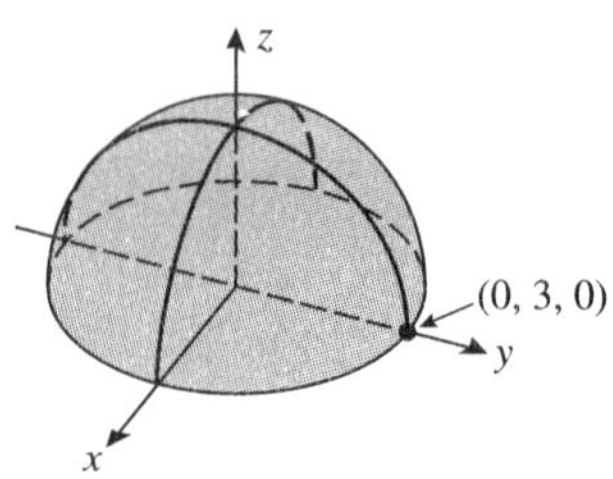

27.

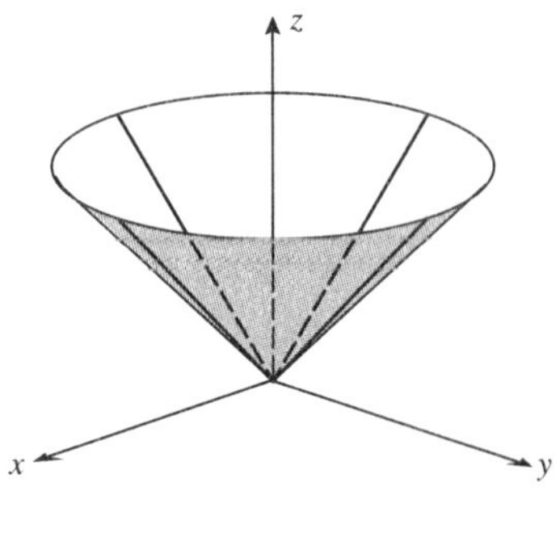

28.

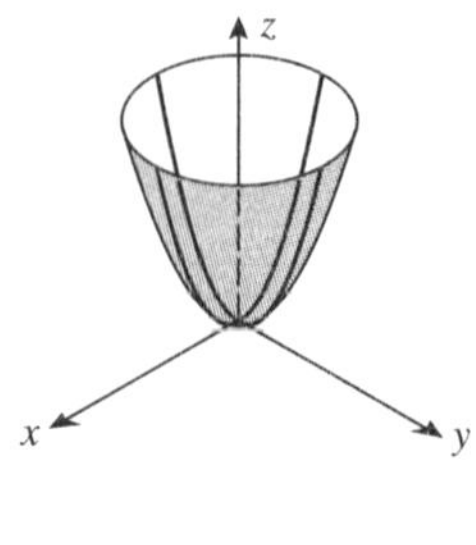

29.

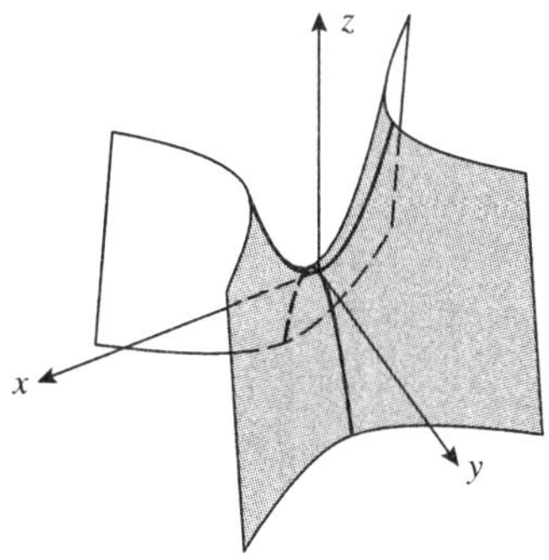

30.

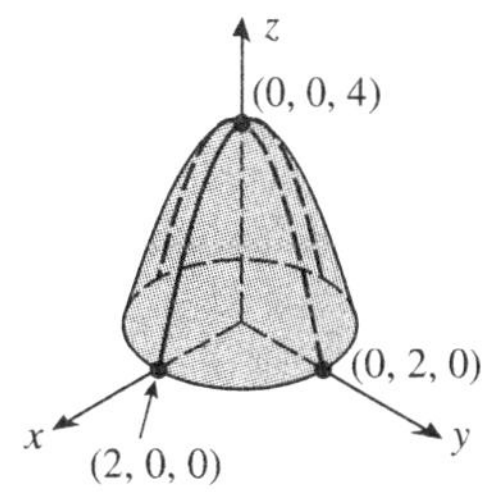

31.

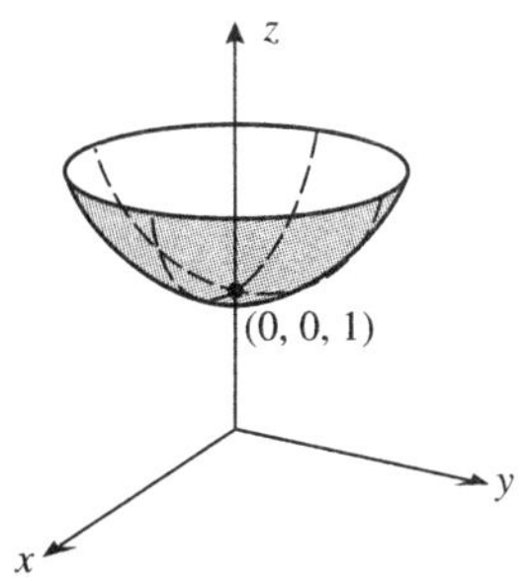

32.

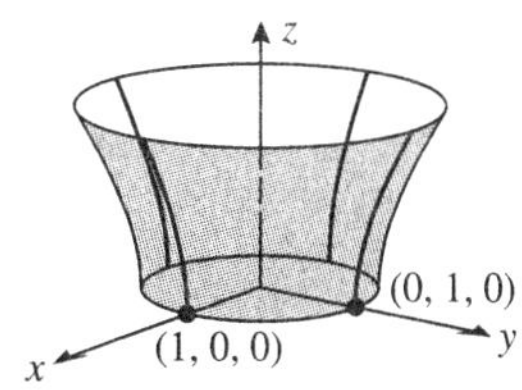

33.

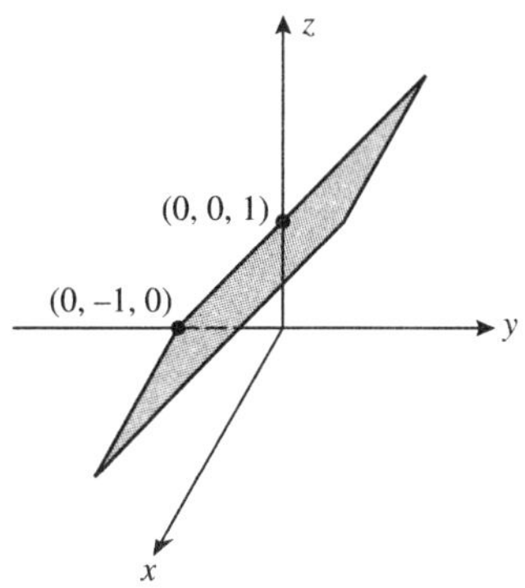

34.

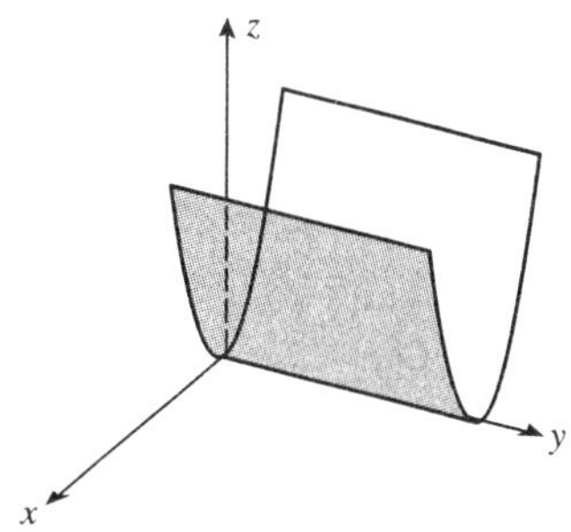

35. **(a)** $f(x, y) = 1 - x^2 - y^2$, because $f = c$ is a circle of radius $\sqrt{1 - c}$ (provided $c \leq 1$), and the radii in (a) decrease as c increases.

(b) $f(x, y) = \sqrt{x^2 + y^2}$ because $f = c$ is a circle of radius c, and the radii increase uniformly.

(c) $f(x, y) = x^2 + y^2$ because $f = c$ is a circle of radius $\sqrt{c}$ and the radii in the plot grow like the square root function.

36. **(a)** III, because the surface has 9 peaks along the edges, three peaks to each edge

(b) IV, because the center is relatively flat and the deep valley in the first quadrant points in the direction of the positive x-axis

(c) I, because the deep valley in the first quadrant points in the direction of the positive y-axis

(d) II, because the surface has four peaks

37. **(a)** A **(b)** B **(c)** increase

(d) decrease **(e)** increase **(f)** decrease

38. (a) Medicine Hat, since the contour lines are closer together near Medicine Hat than they are near Chicago.

(b) The change in atmospheric pressure is about $\Delta p \approx 999 - 1010 = -11$, so the average rate of change is $\Delta p/1400 \approx -0.0079$.

39.

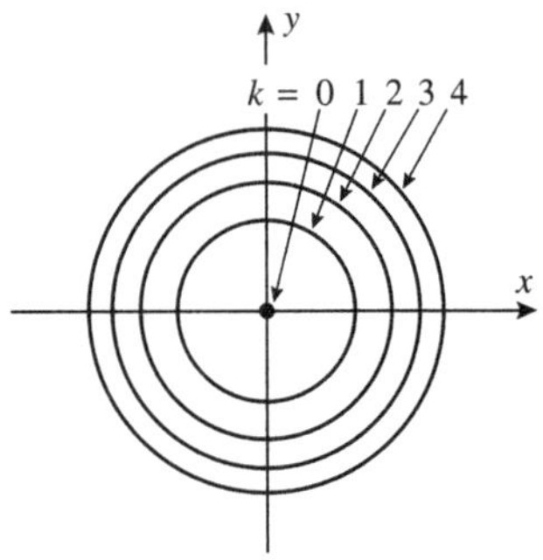

40.

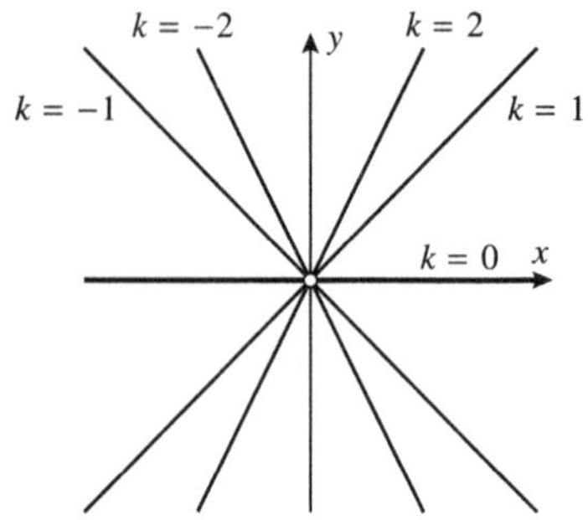

41.

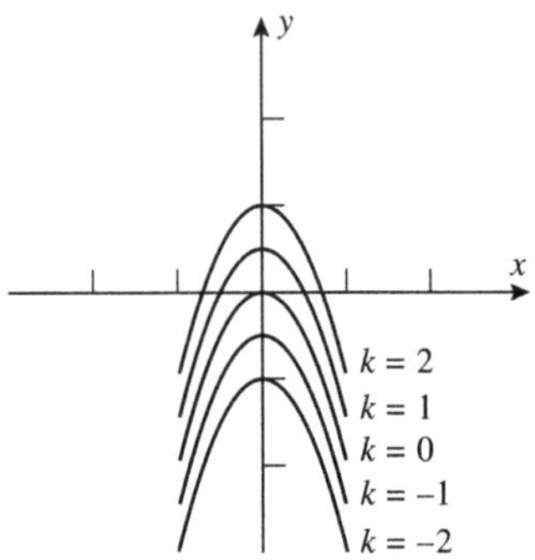

42.

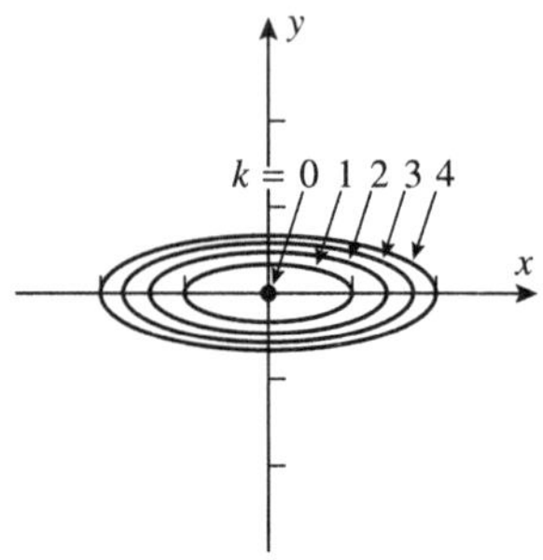

43.

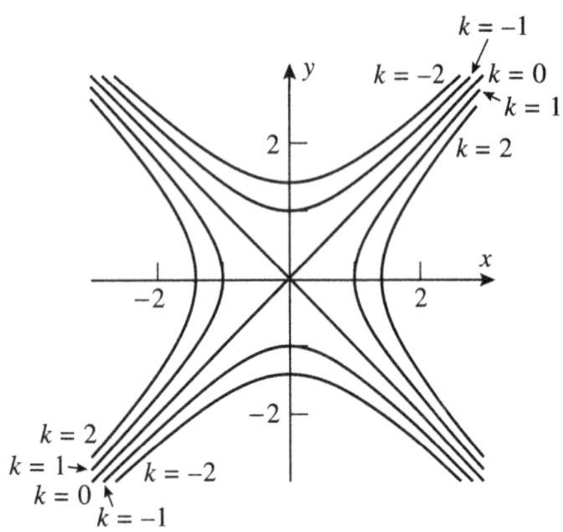

44.

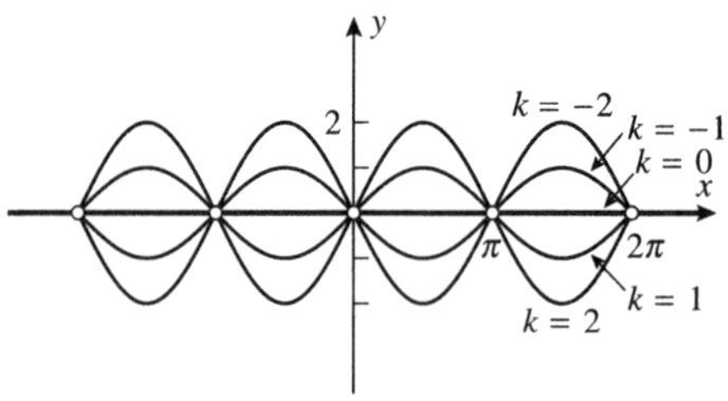

45.

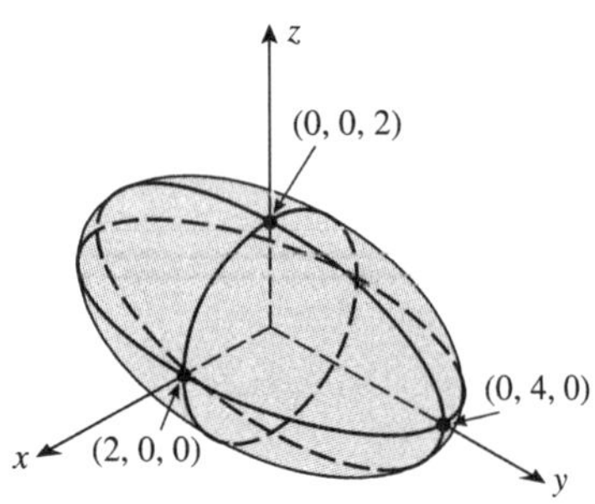

46.

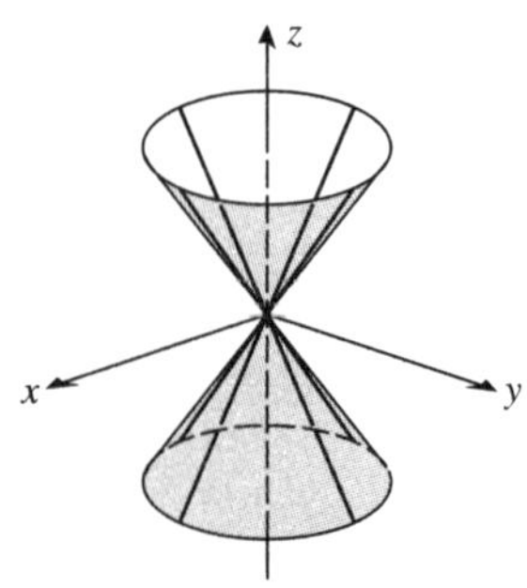

47.

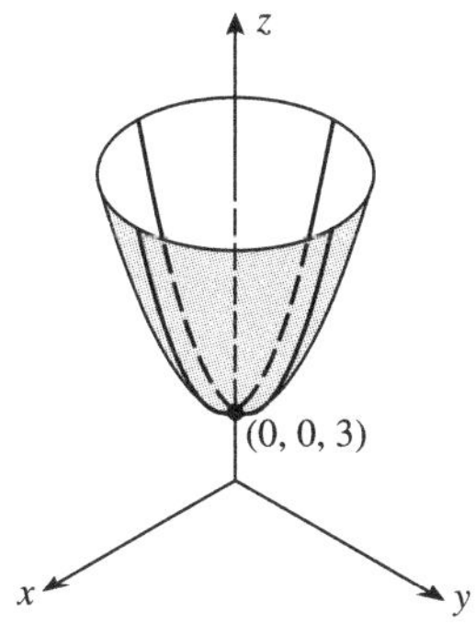

48.

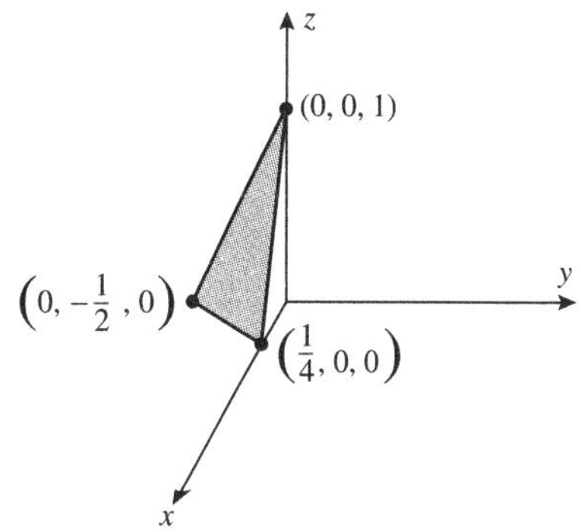

49. concentric spheres, common center at (2,0,0)

50. parallel planes, common normal $3\mathbf{i} - \mathbf{j} + 2\mathbf{k}$

51. concentric cylinders, common axis the y-axis

52. circular paraboloids, common axis the z-axis, all the same shape but with different vertices along z-axis.

53. **(a)** $f(-1, 1) = 0$; $x^2 - 2x^3 + 3xy = 0$ **(b)** $f(0, 0) = 0$; $x^2 - 2x^3 + 3xy = 0$
(c) $f(2, -1) = -18$; $x^2 - 2x^3 + 3xy = -18$

54. **(a)** $f(\ln 2, 1) = 2$; $ye^x = 2$ **(b)** $f(0, 3) = 3$; $ye^x = 3$
(c) $f(1, -2) = -2e$; $ye^x = -2e$

55. **(a)** $f(1, -2, 0) = 5$; $x^2 + y^2 - z = 5$ **(b)** $f(1, 0, 3) = -2$; $x^2 + y^2 - z = -2$
(c) $f(0, 0, 0) = 0$; $x^2 + y^2 - z = 0$

56. **(a)** $f(1, 0, 2) = 3$; $xyz + 3 = 3$, $xyz = 0$ **(b)** $f(-2, 4, 1) = -5$; $xyz + 3 = -5$, $xyz = -8$
(c) $f(0, 0, 0) = 3$; $xyz = 0$

57. **(a)**

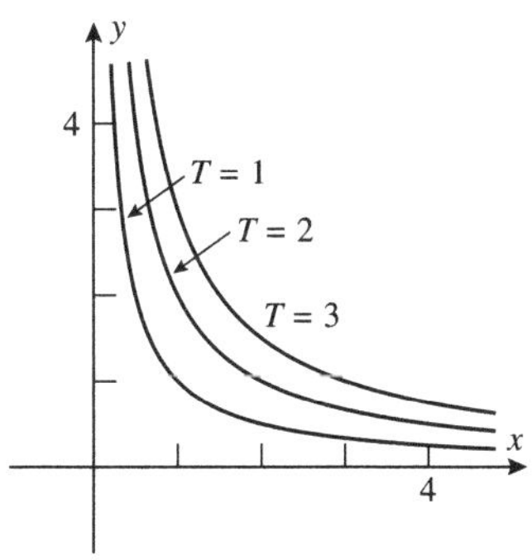

(b) At $(1, 4)$ the temperature is $T(1, 4) = 4$ so the temperature will remain constant along the path $xy = 4$.

58. $V = \dfrac{8}{\sqrt{16 + x^2 + y^2}}$

$x^2 + y^2 = \dfrac{64}{V^2} - 16$

the equipotential curves are circles.

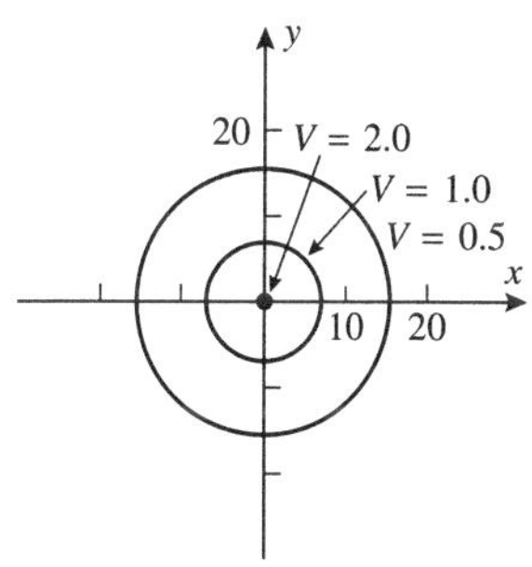

59. **(a)**

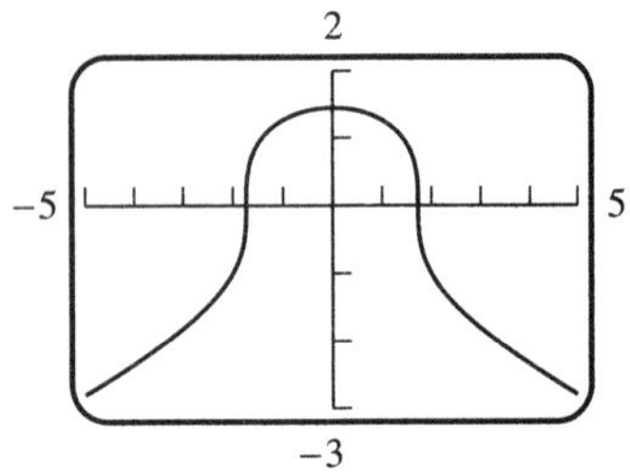

(b)

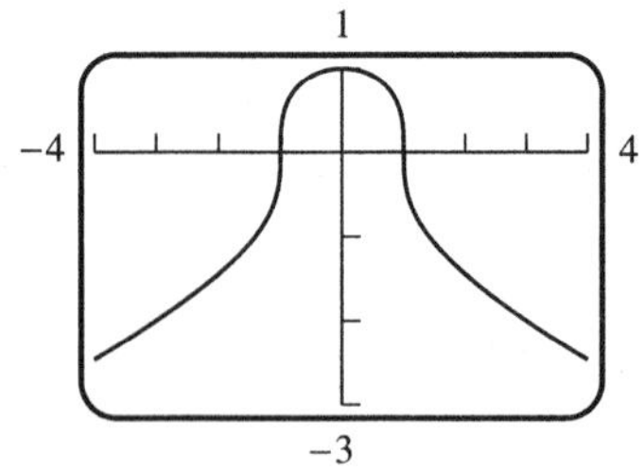

60. **(a)**

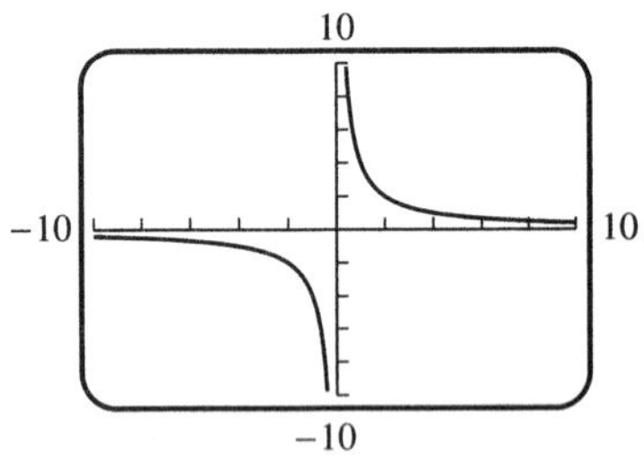

(b)

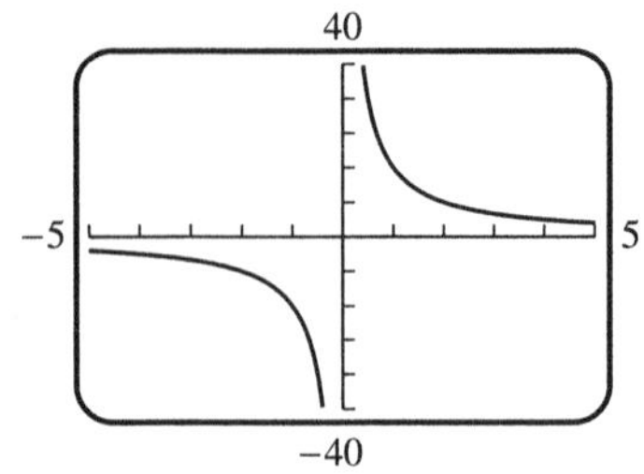

61. **(a)**

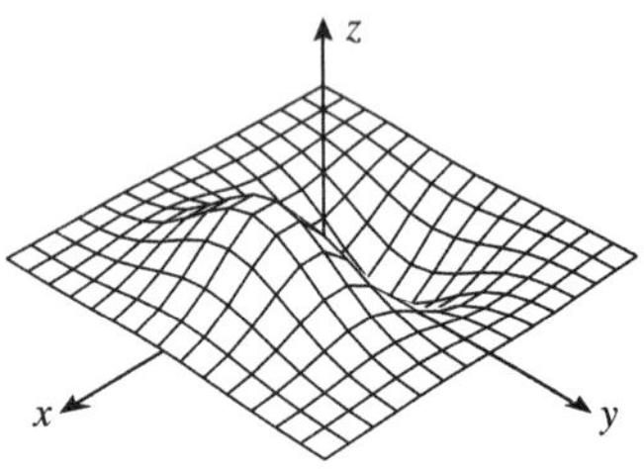

(b)

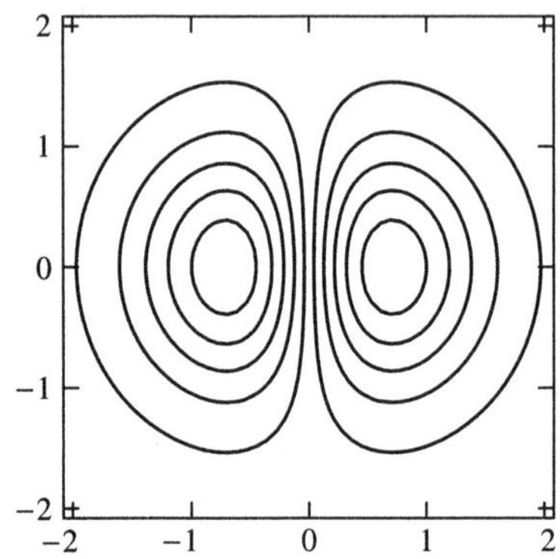

62. **(a)**

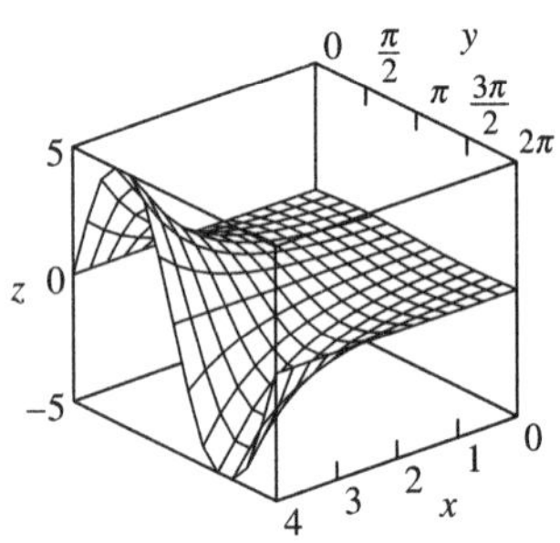

(b)

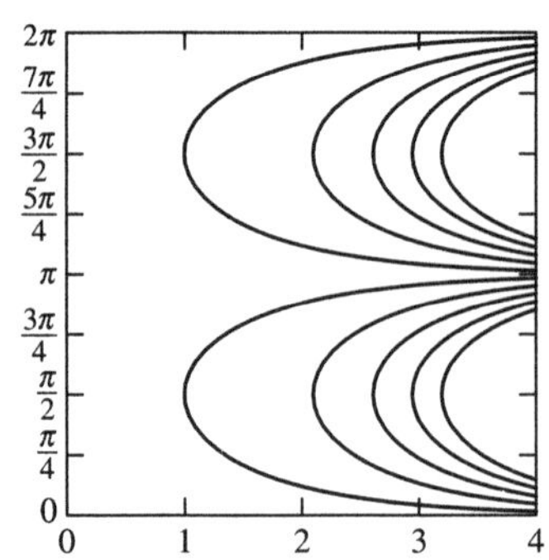

63. **(a)** The graph of g is the graph of f shifted one unit in the positive x-direction.

(b) The graph of g is the graph of f shifted one unit up the z-axis.

(c) The graph of g is the graph of f shifted one unit down the y-axis and then inverted with respect to the plane $z = 0$.

64. **(a)**

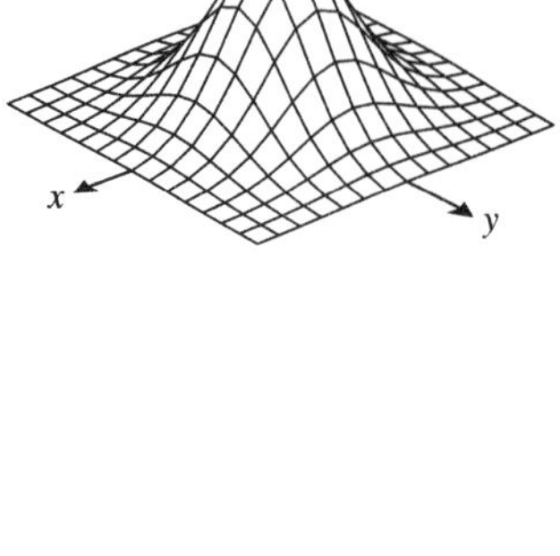

(b) If a is positive and increasing then the graph of g is more pointed, and in the limit as $a \to +\infty$ the graph approaches a 'spike' on the z-axis of height 1. As a decreases to zero the graph of g gets flatter until it finally approaches the plane $z = 1$.

EXERCISE SET 14.2

1. 35 **2.** $\pi^2/2$ **3.** -8 **4.** e^{-7} **5.** 0 **6.** 0

7. (a) Along $x = 0$ $\lim\limits_{(x,y)\to(0,0)} \dfrac{3}{x^2+2y^2} = \lim\limits_{y\to 0} \dfrac{3}{2y^2}$ does not exist.

(b) Along $x = 0$, $\lim\limits_{(x,y)\to(0,0)} \dfrac{x+y}{x+y^2} = \lim\limits_{y\to 0} \dfrac{1}{y}$ does not exist.

8. (a) Along $y = 0$: $\lim\limits_{x\to 0} \dfrac{x}{x^2} = \lim\limits_{x\to 0} \dfrac{1}{x}$ does not exist because $\left|\dfrac{1}{x}\right| \to +\infty$ as $x \to 0$ so the original limit does not exist.

(b) Along $y = 0$: $\lim\limits_{x\to 0} \dfrac{1}{x}$ does not exist, so the original limit does not exist.

9. Let $z = x^2 + y^2$, then $\lim\limits_{(x,y)\to(0,0)} \dfrac{\sin\left(x^2+y^2\right)}{x^2+y^2} = \lim\limits_{z\to 0^+} \dfrac{\sin z}{z} = 1$

10. Let $z = x^2 + y^2$, then $\lim\limits_{(x,y)\to(0,0)} \dfrac{1-\cos\left(x^2+y^2\right)}{x^2+y^2} = \lim\limits_{z\to 0^+} \dfrac{1-\cos z}{z} = \lim\limits_{z\to 0^+} \dfrac{\sin z}{1} = 0$

11. Let $z = x^2 + y^2$, then $\lim\limits_{(x,y)\to(0,0)} e^{-1/\left(x^2+y^2\right)} = \lim\limits_{z\to 0^+} e^{-1/z} = 0$

12. With $z = x^2 + y^2$, $\lim\limits_{z\to+\infty} \dfrac{1}{\sqrt{z}} e^{-1/\sqrt{z}}$; let $w = \dfrac{1}{\sqrt{z}}$, $\lim\limits_{w\to+\infty} \dfrac{w}{e^w} = 0$

13. $\lim\limits_{(x,y)\to(0,0)} \dfrac{\left(x^2+y^2\right)\left(x^2-y^2\right)}{x^2+y^2} = \lim\limits_{(x,y)\to(0,0)} \left(x^2-y^2\right) = 0$

14. $\lim\limits_{(x,y)\to(0,0)} \dfrac{\left(x^2+4y^2\right)\left(x^2-4y^2\right)}{x^2+4y^2} = \lim\limits_{(x,y)\to(0,0)} \left(x^2-4y^2\right) = 0$

15. along $y = 0$: $\lim\limits_{x\to 0} \dfrac{0}{3x^2} = \lim\limits_{x\to 0} 0 = 0$; along $y = x$: $\lim\limits_{x\to 0} \dfrac{x^2}{5x^2} = \lim\limits_{x\to 0} 1/5 = 1/5$ so the limit does not exist.

16. Let $z = x^2 + y^2$, then $\lim\limits_{(x,y)\to(0,0)} \dfrac{1-x^2-y^2}{x^2+y^2} = \lim\limits_{z\to 0^+} \dfrac{1-z}{z} = +\infty$ so the limit does not exist.

17. 8/3 **18.** $\ln 5$

19. Let $t = \sqrt{x^2+y^2+z^2}$, then $\lim\limits_{(x,y,z)\to(0,0,0)} \dfrac{\sin\left(x^2+y^2+z^2\right)}{\sqrt{x^2+y^2+z^2}} = \lim\limits_{t\to 0^+} \dfrac{\sin\left(t^2\right)}{t} = 0$

20. With $t = \sqrt{x^2+y^2+z^2}$, $\lim_{t\to 0^+} \frac{\sin t}{t^2} = \lim_{t\to 0^+} \frac{\cos t}{2t} = +\infty$ so the limit does not exist.

21. $y\ln(x^2+y^2) = r\sin\theta \ln r^2 = 2r(\ln r)\sin\theta$, so $\lim_{(x,y)\to(0,0)} y\ln(x^2+y^2) = \lim_{r\to 0^+} 2r(\ln r)\sin\theta = 0$

22. $\frac{x^2y^2}{\sqrt{x^2+y^2}} = \frac{(r^2\cos^2\theta)(r^2\sin^2\theta)}{r} = r^3\cos^2\theta\sin^2\theta$, so $\lim_{(x,y)\to(0,0)} \frac{x^2y^2}{\sqrt{x^2+y^2}} = 0$

23. $\frac{e^{\sqrt{x^2+y^2+z^2}}}{\sqrt{x^2+y^2+z^2}} = \frac{e^\rho}{\rho}$, so $\lim_{(x,y,z)\to(0,0,0)} \frac{e^{\sqrt{x^2+y^2+z^2}}}{\sqrt{x^2+y^2+z^2}} = \lim_{\rho\to 0^+} \frac{e^\rho}{\rho}$ does not exist.

24. $\lim_{(x,y,z)\to(0,0,0)} \tan^{-1}\left[\frac{1}{x^2+y^2+z^2}\right] = \lim_{\rho\to 0^+} \tan^{-1}\frac{1}{\rho^2} = \frac{\pi}{2}$

25. **(a)** No, since there seem to be points near $(0,0)$ with $z = 0$ and other points near $(0,0)$ with $z \approx 1/2$.

(b) $\lim_{x\to 0} \frac{mx^3}{x^4+m^2x^2} = \lim_{x\to 0} \frac{mx}{x^2+m^2} = 0$ **(c)** $\lim_{x\to 0} \frac{x^4}{2x^4} = \lim_{x\to 0} 1/2 = 1/2$

(d) A limit must be unique if it exists, so $f(x,y)$ cannot have a limit as $(x,y)\to(0,0)$.

26. **(a)** Along $y = mx$: $\lim_{x\to 0} \frac{mx^4}{2x^6+m^2x^2} = \lim_{x\to 0} \frac{mx^2}{2x^4+m^2} = 0$;

along $y = kx^2$: $\lim_{x\to 0} \frac{kx^5}{2x^6+k^2x^4} = \lim_{x\to 0} \frac{kx}{2x^2+k^2} = 0$.

(b) $\lim_{x\to 0} \frac{x^6}{2x^6+x^6} = \lim_{x\to 0} \frac{1}{3} = \frac{1}{3} \neq 0$

27. **(a)** $\lim_{t\to 0} \frac{abct^3}{a^2t^2+b^4t^4+c^4t^4} = \lim_{t\to 0} \frac{abct}{a^2+b^4t^2+c^4t^2} = 0$

(b) $\lim_{t\to 0} \frac{t^4}{t^4+t^4+t^4} = \lim_{t\to 0} 1/3 = 1/3$

28. $\pi/2$ because $\frac{x^2+1}{x^2+(y-1)^2} \to +\infty$ as $(x,y)\to(0,1)$

29. $-\pi/2$ because $\frac{x^2-1}{x^2+(y-1)^2} \to -\infty$ as $(x,y)\to(0,1)$

30. with $z = x^2+y^2$, $\lim_{z\to 0^+} \frac{\sin z}{z} = 1 = f(0,0)$

31. No, because $\lim_{(x,y)\to(0,0)} \frac{x^2}{x^2+y^2}$ does not exist.

Along $x = 0$: $\lim_{y\to 0}(0/y^2) = \lim_{y\to 0} 0 = 0$; along $y = 0$: $\lim_{x\to 0}(x^2/x^2) = \lim_{x\to 0} 1 = 1$.

32. Using polar coordinates with $r > 0$, $xy = r^2\sin\theta\cos\theta$ and $x^2+y^2 = r^2$ so

$|xy\ln(x^2+y^2)| = |r^2\sin\theta\cos\theta\ln r^2| \le |2r^2\ln r|$, but $\lim_{r\to 0^+} 2r^2\ln r = 0$ thus

$\lim_{(x,y)\to(0,0)} xy\ln(x^2+y^2) = 0$; $f(x,y)$ will be continuous at (0,0) if we define $f(0,0) = 0$.

33.

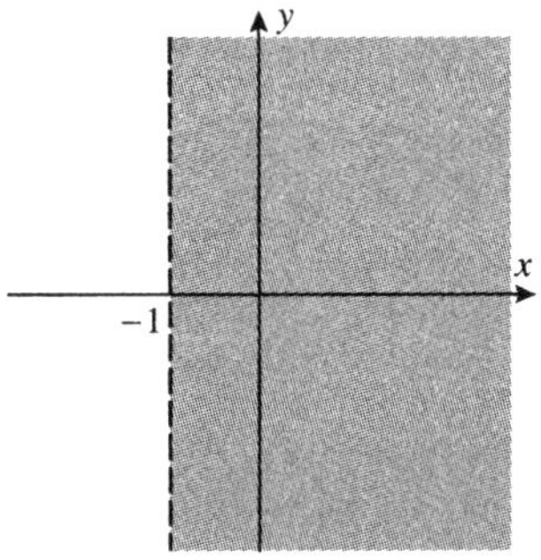

34.

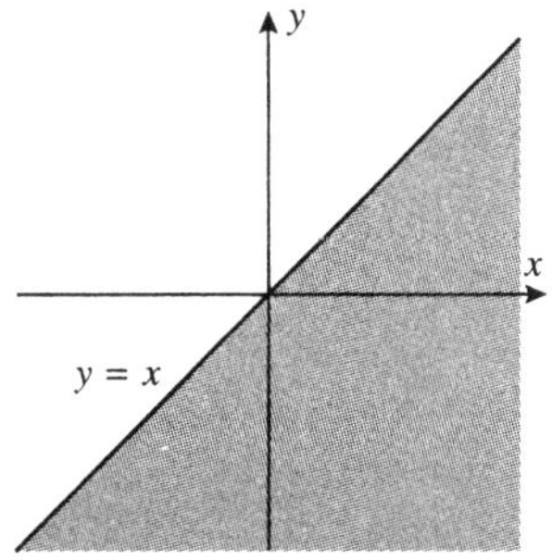

35.

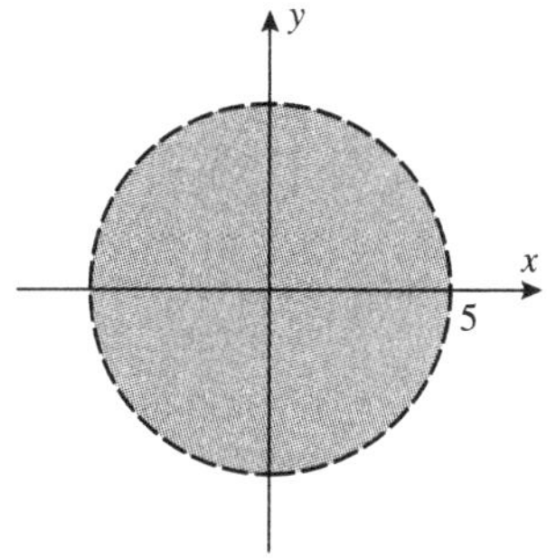

36.

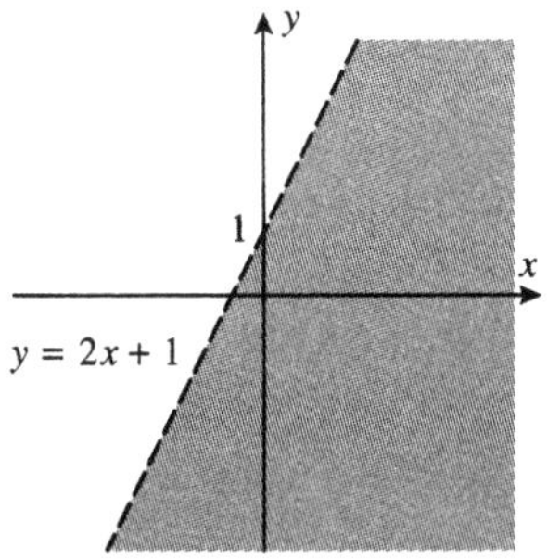

37.

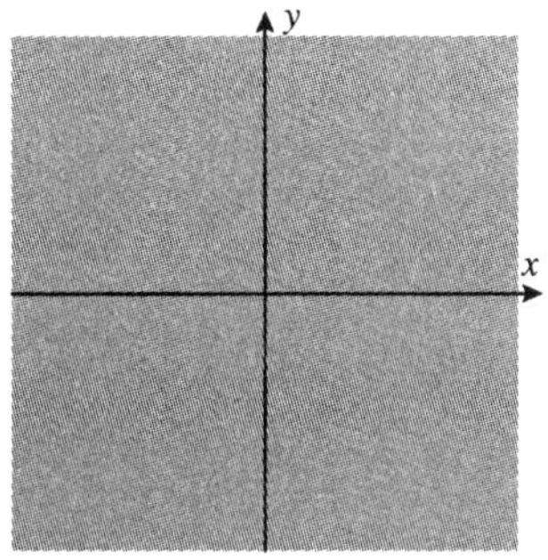

38.

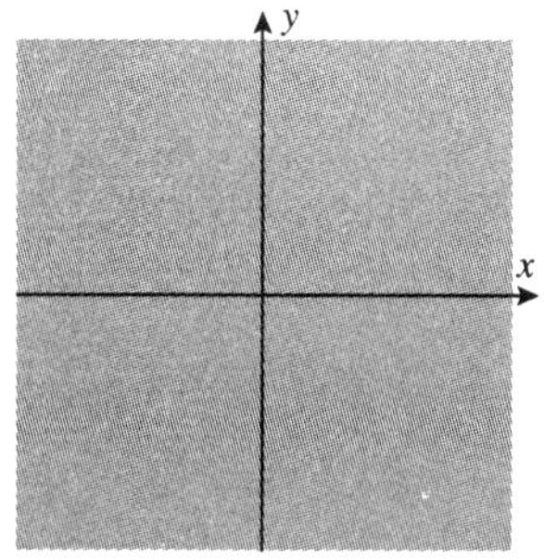

39.

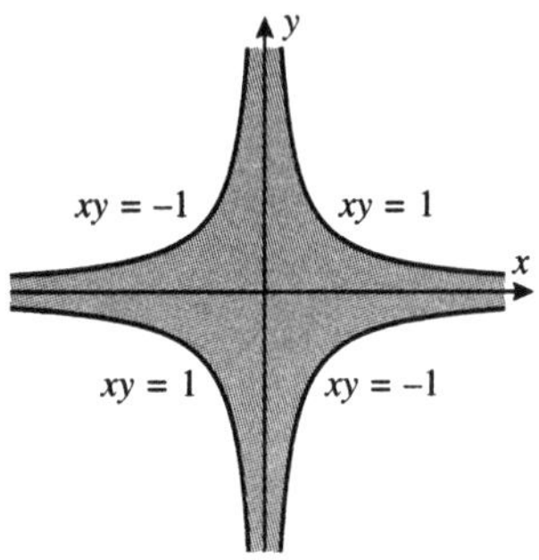

40.

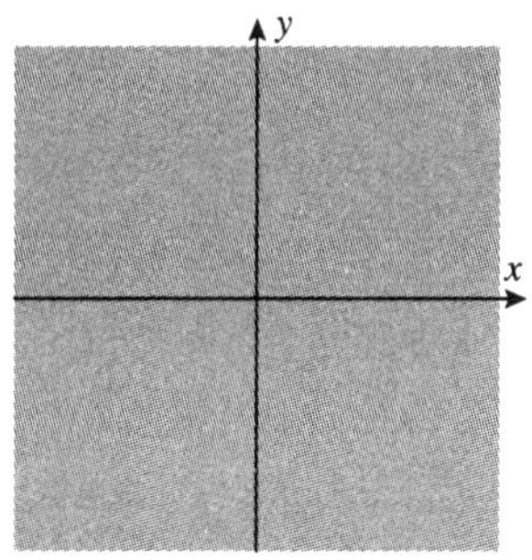

41. all of 3-space

42. all points inside the sphere with radius 2 and center at the origin

43. all points not on the cylinder $x^2 + z^2 = 1$

44. all of 3-space

EXERCISE SET 14.3

1. **(a)** $9x^2y^2$ **(b)** $6x^3y$ **(c)** $9y^2$ **(d)** $9x^2$
(e) $6y$ **(f)** $6x^3$ **(g)** 36 **(h)** 12

2. **(a)** $2e^{2x}\sin y$ **(b)** $e^{2x}\cos y$ **(c)** $2\sin y$ **(d)** 0
(e) $\cos y$ **(f)** e^{2x} **(g)** 0 **(h)** 4

3. **(a)** $\dfrac{\partial z}{\partial x} = \dfrac{3}{2\sqrt{3x+2y}}$; slope $= \dfrac{3}{8}$ **(b)** $\dfrac{\partial z}{\partial y} = \dfrac{1}{\sqrt{3x+2y}}$; slope $= \dfrac{1}{4}$

4. **(a)** $\dfrac{\partial z}{\partial x} = e^{-y}$; slope $= 1$ **(b)** $\dfrac{\partial z}{\partial y} = -xe^{-y} + 5$; slope $= 2$

5. **(a)** $\dfrac{\partial z}{\partial x} = -4\cos(y^2 - 4x)$; rate of change $= -4\cos 7$

(b) $\dfrac{\partial z}{\partial y} = 2y\cos(y^2 - 4x)$; rate of change $= 2\cos 7$

6. **(a)** $\dfrac{\partial z}{\partial x} = -\dfrac{1}{(x+y)^2}$; rate of change $= -\dfrac{1}{4}$ **(b)** $\dfrac{\partial z}{\partial y} = -\dfrac{1}{(x+y)^2}$; rate of change $= -\dfrac{1}{4}$

7. $\partial z/\partial x =$ slope of line parallel to xz-plane $= -4$; $\partial z/\partial y =$ slope of line parallel to yz-plane $= 1/2$

8. Moving to the right from (x_0, y_0) decreases $f(x,y)$, so $f_x < 0$; moving up increases f, so $f_y > 0$.

9. **(a)** The right-hand estimate is $\partial r/\partial v \approx (222 - 197)/(85 - 80) = 5$; the left-hand estimate is $\partial r/\partial v \approx (197 - 173)/(80 - 75) = 4.8$; the average is $\partial r/\partial v \approx 4.9$.

(b) The right-hand estimate is $\partial r/\partial \theta \approx (200 - 197)/(45 - 40) = 0.6$; the left-hand estimate is $\partial r/\partial \theta \approx (197 - 188)/(40 - 35) = 1.8$; the average is $\partial r/\partial \theta \approx 1.2$.

10. **(a)** The right-hand estimate is $\partial r/\partial v \approx (253 - 226)/(90 - 85) = 5.4$; the left-hand estimate is $(226 - 200)/(85 - 80) = 5.2$; the average is $\partial r/\partial v \approx 5.3$.

(b) The right-hand estimate is $\partial r/\partial \theta \approx (222 - 226)/(50 - 45) = -0.8$; the left-hand estimate is $(226 - 222)/(45 - 40) = 0.8$; the average is $\partial r/\partial v \approx 0$.

11. $\partial z/\partial x = 8xy^3e^{x^2y^3}$, $\partial z/\partial y = 12x^2y^2e^{x^2y^3}$

12. $\partial z/\partial x = -5x^4y^4\sin(x^5y^4)$, $\partial z/\partial y = -4x^5y^3\sin(x^5y^4)$

13. $\partial z/\partial x = x^3/(y^{3/5} + x) + 3x^2\ln(1 + xy^{-3/5})$, $\partial z/\partial y = -(3/5)x^4/(y^{8/5} + xy)$

14. $\partial z/\partial x = ye^{xy}\sin(4y^2)$, $\partial z/\partial y = 8ye^{xy}\cos(4y^2) + xe^{xy}\sin(4y^2)$

15. $\dfrac{\partial z}{\partial x} = -\dfrac{y(x^2 - y^2)}{(x^2 + y^2)^2}$, $\dfrac{\partial z}{\partial y} = \dfrac{x(x^2 - y^2)}{(x^2 + y^2)^2}$ **16.** $\dfrac{\partial z}{\partial x} = \dfrac{xy^3(3x + 4y)}{2(x+y)^{3/2}}$, $\dfrac{\partial z}{\partial y} = \dfrac{x^2y^2(6x + 5y)}{2(x+y)^{3/2}}$

17. $f_x(x,y) = (3/2)x^2y\left(5x^2 - 7\right)\left(3x^5y - 7x^3y\right)^{-1/2}$

$f_y(x,y) = (1/2)x^3\left(3x^2 - 7\right)\left(3x^5y - 7x^3y\right)^{-1/2}$

18. $f_x(x,y) = -2y/(x-y)^2$, $f_y(x,y) = 2x/(x-y)^2$

19. $f_x(x,y) = \dfrac{y^{-1/2}}{y^2 + x^2}$, $f_y(x,y) = -\dfrac{xy^{-3/2}}{y^2 + x^2} - \dfrac{3}{2}y^{-5/2}\tan^{-1}(x/y)$

20. $f_x(x,y) = 3x^2e^{-y} + (1/2)x^{-1/2}y^3\sec\sqrt{x}\tan\sqrt{x}$, $f_y(x,y) = -x^3e^{-y} + 3y^2\sec\sqrt{x}$

21. $f_x(x,y) = -(4/3)y^2\sec^2 x\left(y^2\tan x\right)^{-7/3}$, $f_y(x,y) = -(8/3)y\tan x\left(y^2\tan x\right)^{-7/3}$

22. $f_x(x,y) = 2y^2\cosh\sqrt{x}\sinh\left(xy^2\right)\cosh\left(xy^2\right) + \dfrac{1}{2}x^{-1/2}\sinh\sqrt{x}\sinh^2\left(xy^2\right)$

$f_y(x,y) = 4xy\cosh\sqrt{x}\sinh\left(xy^2\right)\cosh\left(xy^2\right)$

23. $f_x(x,y) = -2x,\ f_x(3,1) = -6;\ f_y(x,y) = -21y^2,\ f_y(3,1) = -21$

24. $\partial f/\partial x = x^2y^2e^{xy} + 2xye^{xy},\ \partial f/\partial x\,\big|_{(1,1)} = 3e;\ \partial f/\partial y = x^3ye^{xy} + x^2e^{xy},\ \partial f/\partial y\,\big|_{(1,1)} = 2e$

25. $\partial z/\partial x = x(x^2+4y^2)^{-1/2},\ \partial z/\partial x\,\big|_{(1,2)} = 1/\sqrt{17}\ ;\ \partial z/\partial y = 4y(x^2+4y^2)^{-1/2},\ \partial z/\partial y\,\big|_{(1,2)} = 8/\sqrt{17}$

26. $\partial w/\partial x = -x^2y\sin xy + 2x\cos xy,\ \dfrac{\partial w}{\partial x}(1/2,\pi) = -\pi/4;\ \partial w/\partial y = -x^3\sin xy,\ \dfrac{\partial w}{\partial y}(1/2,\pi) = -1/8$

27. (a) $2xy^4z^3 + y$ (b) $4x^2y^3z^3 + x$ (c) $3x^2y^4z^2 + 2z$
(d) $2y^4z^3 + y$ (e) $32z^3 + 1$ (f) 438

28. (a) $2xy\cos z$ (b) $x^2\cos z$ (c) $-x^2y\sin z$
(d) $4y\cos z$ (e) $4\cos z$ (f) 0

29. $f_x = 2z/x,\ f_y = z/y,\ f_z = \ln(x^2y\cos z) - z\tan z$

30. $f_x = y^{-5/2}z\sec(xz/y)\tan(xz/y),\ f_y = -xy^{-7/2}z\sec(xz/y)\tan(xz/y) - (3/2)y^{-5/2}\sec(xz/y),$
$f_z = xy^{-5/2}\sec(xz/y)\tan(xz/y)$

31. $f_x = -y^2z^3/\left(1 + x^2y^4z^6\right),\ f_y = -2xyz^3/\left(1 + x^2y^4z^6\right),\ f_z = -3xy^2z^2/\left(1 + x^2y^4z^6\right)$

32. $f_x = 4xyz\cosh\sqrt{z}\sinh\left(x^2yz\right)\cosh\left(x^2yz\right),\ f_y = 2x^2z\cosh\sqrt{z}\sinh\left(x^2yz\right)\cosh\left(x^2yz\right),$
$f_z = 2x^2y\cosh\sqrt{z}\sinh\left(x^2yz\right)\cosh\left(x^2yz\right) + (1/2)z^{-1/2}\sinh\sqrt{z}\sinh^2\left(x^2yz\right)$

33. $\partial w/\partial x = yze^z\cos xz,\ \partial w/\partial y = e^z\sin xz,\ \partial w/\partial z = ye^z(\sin xz + x\cos xz)$

34. $\partial w/\partial x = 2x/\left(y^2 + z^2\right),\ \partial w/\partial y = -2y\left(x^2 + z^2\right)/\left(y^2 + z^2\right)^2,\ \partial w/\partial z = 2z\left(y^2 - x^2\right)/\left(y^2 + z^2\right)^2$

35. $\partial w/\partial x = x/\sqrt{x^2 + y^2 + z^2},\ \partial w/\partial y = y/\sqrt{x^2 + y^2 + z^2},\ \partial w/\partial z = z/\sqrt{x^2 + y^2 + z^2}$

36. $\partial w/\partial x = 2y^3e^{2x+3z},\ \partial w/\partial y = 3y^2e^{2x+3z},\ \partial w/\partial z = 3y^3e^{2x+3z}$

37. (a) e (b) $2e$ (c) e

38. (a) $2/\sqrt{7}$ (b) $4/\sqrt{7}$ (c) $1/\sqrt{7}$

39. (a)

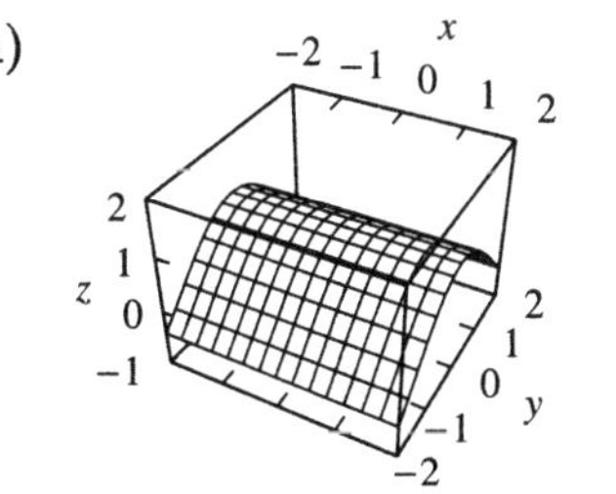

(b)

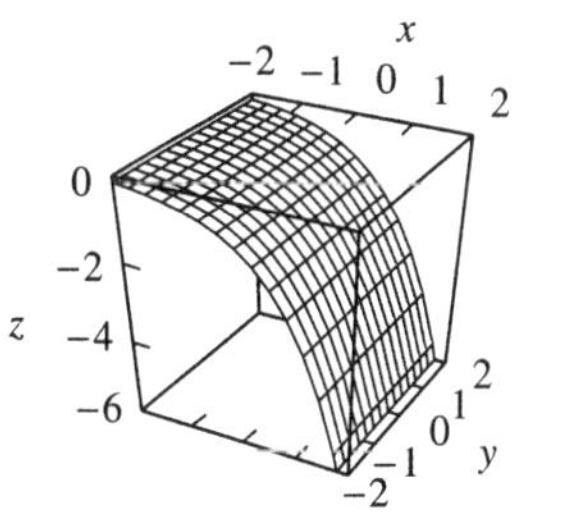

40.

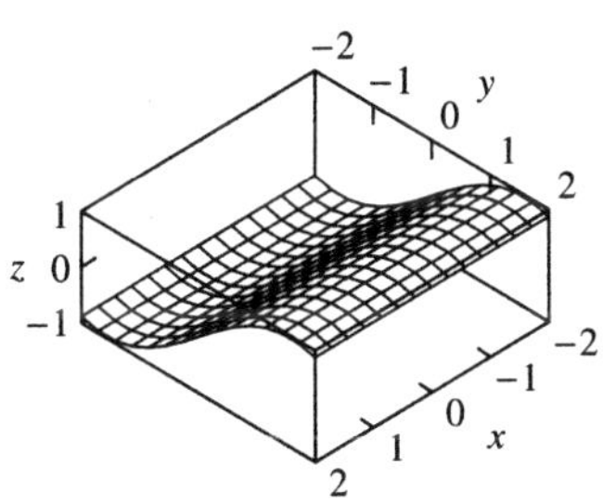

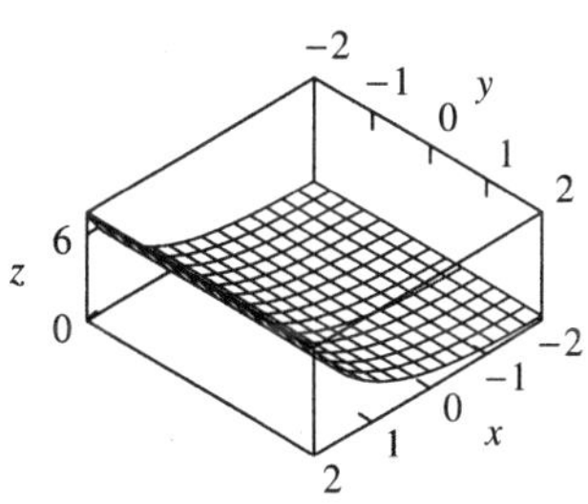

41. $\partial z/\partial x = 2x + 6y(\partial y/\partial x) = 2x$, $\partial z/\partial x]_{(2,1)} = 4$

42. $\partial z/\partial y = 6y$, $\partial z/\partial y\,|_{(2,1)} = 6$

43. $\partial z/\partial x = -x\left(29 - x^2 - y^2\right)^{-1/2}$, $\partial z/\partial x]_{(4,3)} = -2$

44. (a) $\partial z/\partial y = 8y$, $\partial z/\partial y]_{(-1,1)} = 8$ (b) $\partial z/\partial x = 2x$, $\partial z/\partial x]_{(-1,1)} = -2$

45. (a) $\partial V/\partial r = 2\pi rh$ (b) $\partial V/\partial h = \pi r^2$
(c) $\partial V/\partial r]_{r=6,\ h=4} = 48\pi$ (d) $\partial V/\partial h]_{r=8,\ h=10} = 64\pi$

46. (a) $\partial V/\partial s = \dfrac{\pi s d^2}{6\sqrt{4s^2 - d^2}}$ (b) $\partial V/\partial d = \dfrac{\pi d(8s^2 - 3d^2)}{24\sqrt{4s^2 - d^2}}$
(c) $\partial V/\partial s]_{s=10,\ d=16} = 320\pi/9$ (d) $\partial V/\partial d]_{s=10,\ d=16} = 16\pi/9$

47. (a) $P = 10T/V$, $\partial P/\partial T = 10/V$, $\partial P/\partial T]_{T=80,\ V=50} = 1/5\ \text{lb}/(\text{in}^2\text{K})$
(b) $V = 10T/P, \partial V/\partial P = -10T/P^2$, if $V = 50$ and $T = 80$ then
$P = 10(80)/(50) = 16$, $\partial V/\partial P]_{T=80,\ P=16} = -25/8(\text{in}^5/\text{lb})$

48. (a) $\partial T/\partial x = 3x^2 + 1$, $\partial T/\partial x]_{(1,2)} = 4$ (b) $\partial T/\partial y = 4y$, $\partial T/\partial y]_{(1,2)} = 8$

49. (a) $V = lwh, \partial V/\partial l = wh = 6$
(b) $\partial V/\partial w = lh = 15$
(c) $\partial V/\partial h = lw = 10$

50. (a) $\partial A/\partial a = (1/2)b\sin\theta = (1/2)(10)\left(\sqrt{3}/2\right) = 5\sqrt{3}/2$
(b) $\partial A/\partial \theta = (1/2)ab\cos\theta = (1/2)(5)(10)(1/2) = 25/2$
(c) $b = (2A\csc\theta)/a$, $\partial b/\partial a = -(2A\csc\theta)/a^2 = -b/a = -2$

51. $\partial V/\partial r = \dfrac{2}{3}\pi rh = \dfrac{2}{r}(\dfrac{1}{3}\pi r^2 h) = 2V/r$

52. (a) $\partial z/\partial y = x^2$, $\partial z/\partial y]_{(1,3)} = 1$, $\mathbf{j} + \mathbf{k}$ is parallel to the tangent line so $x = 1$, $y = 3 + t$, $z = 3 + t$
(b) $\partial z/\partial x = 2xy$, $\partial z/\partial x]_{(1,3)} = 6$, $\mathbf{i} + 6\mathbf{k}$ is parallel to the tangent line so $x = 1 + t$, $y = 3$, $z = 3 + 6t$

53. (a) $2x - 2z(\partial z/\partial x) = 0$, $\partial z/\partial x = x/z = \pm 3/(2\sqrt{6}) = \pm\sqrt{6}/4$
(b) $z = \pm\sqrt{x^2 + y^2 - 1}$, $\partial z/\partial x = \pm x/\sqrt{x^2 + y^2 - 1} = \pm\sqrt{6}/4$

54. (a) $2y - 2z(\partial z/\partial y) = 0$, $\partial z/\partial y = y/z = \pm 4/(2\sqrt{6}) = \pm\sqrt{6}/3$
(b) $z = \pm\sqrt{x^2 + y^2 - 1}$, $\partial z/\partial y = \pm y/\sqrt{x^2 + y^2 - 1} = \pm\sqrt{6}/3$

55. $\dfrac{3}{2}\left(x^2 + y^2 + z^2\right)^{1/2}\left(2x + 2z\dfrac{\partial z}{\partial x}\right) = 0$, $\partial z/\partial x = -x/z$; similarly, $\partial z/\partial y = -y/z$

56. $\dfrac{4x - 3z^2(\partial z/\partial x)}{2x^2 + y - z^3} = 1$, $\dfrac{\partial z}{\partial x} = \dfrac{4x - 2x^2 - y + z^3}{3z^2}$; $\dfrac{1 - 3z^2(\partial z/\partial y)}{2x^2 + y - z^3} = 0$, $\dfrac{\partial z}{\partial y} = \dfrac{1}{3z^2}$

57. $2x + z\left(xy\dfrac{\partial z}{\partial x} + yz\right)\cos xyz + \dfrac{\partial z}{\partial x}\sin xyz = 0,\ \dfrac{\partial z}{\partial x} = -\dfrac{2x + yz^2\cos xyz}{xyz\cos xyz + \sin xyz};$

$z\left(xy\dfrac{\partial z}{\partial y} + xz\right)\cos xyz + \dfrac{\partial z}{\partial y}\sin xyz = 0,\ \dfrac{\partial z}{\partial y} = -\dfrac{xz^2\cos xyz}{xyz\cos xyz + \sin xyz}$

58. $e^{xy}(\cosh z)\dfrac{\partial z}{\partial x} + ye^{xy}\sinh z - z^2 - 2xz\dfrac{\partial z}{\partial x} = 0,\ \dfrac{\partial z}{\partial x} = \dfrac{z^2 - ye^{xy}\sinh z}{e^{xy}\cosh z - 2xz};$

$e^{xy}(\cosh z)\dfrac{\partial z}{\partial y} + xe^{xy}\sinh z - 2xz\dfrac{\partial z}{\partial y} = 0,\ \dfrac{\partial z}{\partial y} = -\dfrac{xe^{xy}\sinh z}{e^{xy}\cosh z - 2xz}$

59. $(3/2)\left(x^2 + y^2 + z^2 + w^2\right)^{1/2}\left(2x + 2w\dfrac{\partial w}{\partial x}\right) = 0$, $\partial w/\partial x = -x/w$; similarly, $\partial w/\partial y = -y/w$ and $\partial w/\partial z = -z/w$

60. $\partial w/\partial x = -4x/3$, $\partial w/\partial y = -1/3$, $\partial w/\partial z = (2x^2 + y - z^3 + 3z^2 + 3w)/3$

61. $\dfrac{\partial w}{\partial x} = -\dfrac{yzw\cos xyz}{2w + \sin xyz},\ \dfrac{\partial w}{\partial y} = -\dfrac{xzw\cos xyz}{2w + \sin xyz},\ \dfrac{\partial w}{\partial z} = -\dfrac{xyw\cos xyz}{2w + \sin xyz}$

62. $\dfrac{\partial w}{\partial x} = \dfrac{ye^{xy}\sinh w}{z^2 - e^{xy}\cosh w},\ \dfrac{\partial w}{\partial y} = \dfrac{xe^{xy}\sinh w}{z^2 - e^{xy}\cosh w},\ \dfrac{\partial w}{\partial z} = \dfrac{2zw}{e^{xy}\cosh w - z^2}$

63. $f_x = e^{x^2}$, $f_y = -e^{y^2}$

64. $f_x = ye^{x^2y^2}$, $f_y = xe^{x^2y^2}$

65. (a) $-\dfrac{1}{4x^{3/2}}\cos y$ (b) $-\sqrt{x}\cos y$ (c) $-\dfrac{\sin y}{2\sqrt{x}}$ (d) $-\dfrac{\sin y}{2\sqrt{x}}$

66. (a) $8 + 84x^2y^5$ (b) $140x^4y^3$ (c) $140x^3y^4$ (d) $140x^3y^4$

67. $f_x = 8x - 8y^4$, $f_y = -32xy^3 + 35y^4$, $f_{xy} = f_{yx} = -32y^3$

68. $f_x = x/\sqrt{x^2 + y^2}$, $f_y = y/\sqrt{x^2 + y^2}$, $f_{xy} = f_{yx} = -xy(x^2 + y^2)^{-3/2}$

69. $f_x = e^x\cos y$, $f_y = -e^x\sin y$, $f_{xy} = f_{yx} = -e^x\sin y$

70. $f_x = e^{x-y^2}$, $f_y = -2ye^{x-y^2}$, $f_{xy} = f_{yx} = -2ye^{x-y^2}$

71. $f_x = 4/(4x - 5y)$, $f_y = -5/(4x - 5y)$, $f_{xy} = f_{yx} = 20/(4x - 5y)^2$

72. $f_x = 2x/(x^2 + y^2), f_y = 2y/(x^2 + y^2), f_{xy} = -4xy/(x^2 + y^2)^2$

73. $f_x = 2y/(x + y)^2, f_y = -2x/(x + y)^2, f_{xy} = f_{yx} = 2(x - y)/(x + y)^3$

74. $f_x = 4xy^2/(x^2 + y^2)^2$, $f_y = -4x^2y/(x^2 + y^2)^2$, $f_{xy} = f_{yx} = 8xy(x^2 - y^2)/(x^2 + y^2)^3$

75. III is a plane, and its partial derivatives are constants, so III cannot be $f(x, y)$. If I is the graph of $z = f(x, y)$ then (by inspection) f_y is constant as y varies, but neither II nor III is constant as y varies. Hence $z = f(x, y)$ has II as its graph, and as II seems to be an odd function of x and an even function of y, f_x has I as its graph and f_y has III as its graph.

76. The slope at P in the positive x-direction is negative, the slope in the positive y-direction is negative, thus $\partial z/\partial x < 0, \partial z/\partial y < 0$; the curve through P which is parallel to the x-axis is concave down, so $\partial^2 z/\partial x^2 < 0$; the curve parallel to the y-axis is concave down, so $\partial^2 z/\partial y^2 < 0$.

77. (a) $\dfrac{\partial^3 f}{\partial x^3}$ (b) $\dfrac{\partial^3 f}{\partial y^2 \partial x}$ (c) $\dfrac{\partial^4 f}{\partial x^2 \partial y^2}$ (d) $\dfrac{\partial^4 f}{\partial y^3 \partial x}$

78. (a) f_{xyy} (b) f_{xxxx} (c) f_{xxyy} (d) f_{yyyxx}

79. (a) $30xy^4 - 4$ (b) $60x^2y^3$ (c) $60x^3y^2$

80. (a) $120(2x-y)^2$ (b) $-240(2x-y)^2$ (c) $480(2x-y)$

81. (a) $f_{xyy}(0,1) = -30$ (b) $f_{xxx}(0,1) = -125$ (c) $f_{yyxx}(0,1) = 150$

82. (a) $\dfrac{\partial^3 w}{\partial y^2 \partial x} = -e^y \sin\, x,\ \left.\dfrac{\partial^3 w}{\partial y^2 \partial x}\right|_{(\pi/4,0)} = -1/\sqrt{2}$

(b) $\dfrac{\partial^3 w}{\partial x^2 \partial y} = -e^y \cos\, x,\ \left.\dfrac{\partial^3 w}{\partial x^2 \partial y}\right|_{(\pi/4,0)} = -1/\sqrt{2}$

83. (a) $f_{xy} = 15x^2y^4z^7 + 2y$ (b) $f_{yz} = 35x^3y^4z^6 + 3y^2$
(c) $f_{xz} = 21x^2y^5z^6$ (d) $f_{zz} = 42x^3y^5z^5$
(e) $f_{zyy} = 140x^3y^3z^6 + 6y$ (f) $f_{xxy} = 30xy^4z^7$
(g) $f_{zyx} = 105x^2y^4z^6$ (h) $f_{xxyz} = 210xy^4z^6$

84. (a) $160(4x-3y+2z)^3$ (b) $-1440(4x-3y+2z)^2$ (c) $-5760(4x-3y+2z)$

85. (a) $f_x = 2x+2y,\ f_{xx} = 2,\ f_y = -2y+2x,\ f_{yy} = -2;\ f_{xx} + f_{yy} = 2 - 2 = 0$

(b) $z_x = e^x \sin y - e^y \sin x,\ z_{xx} = e^x \sin y - e^y \cos x,\ z_y = e^x \cos y + e^y \cos x,$
$z_{yy} = -e^x \sin y + e^y \cos x;\ z_{xx} + z_{yy} = e^x \sin y - e^y \cos x - e^x \sin y + e^y \cos x = 0$

(c) $z_x = \dfrac{2x}{x^2+y^2} - 2\dfrac{y}{x^2}\dfrac{1}{1+(y/x)^2} = \dfrac{2x-2y}{x^2+y^2},\ z_{xx} = -2\dfrac{x^2-y^2-2xy}{(x^2+y^2)^2},$

$z_y = \dfrac{2y}{x^2+y^2} + 2\dfrac{1}{x}\dfrac{1}{1+(y/x)^2} = \dfrac{2y+2x}{x^2+y^2},\ z_{yy} = -2\dfrac{y^2-x^2+2xy}{(x^2+y^2)^2};$

$z_{xx} + z_{yy} = -2\dfrac{x^2-y^2-2xy}{(x^2+y^2)^2} - 2\dfrac{y^2-x^2+2xy}{(x^2+y^2)^2} = 0$

86. (a) $z_t = -e^{-t}\sin(x/c),\ z_x = (1/c)e^{-t}\cos(x/c),\ z_{xx} = -(1/c^2)e^{-t}\sin(x/c);$
$z_t - c^2 z_{xx} = -e^{-t}\sin(x/c) - c^2(-(1/c^2)e^{-t}\sin(x/c)) = 0$

(b) $z_t = -e^{-t}\cos(x/c),\ z_x = -(1/c)e^{-t}\sin(x/c),\ z_{xx} = -(1/c^2)e^{-t}\cos(x/c);$
$z_t - c^2 z_{xx} = -e^{-t}\cos(x/c) - c^2(-(1/c^2)e^{-t}\cos(x/c)) = 0$

87. $u_x = \omega \sin c\omega t \cos \omega x,\ u_{xx} = -\omega^2 \sin c\omega t \sin \omega x,\ u_t = c\omega \cos c\omega t \sin \omega x,\ u_{tt} = -c^2\omega^2 \sin c\omega t \sin \omega x;$
$u_{xx} - \dfrac{1}{c^2}u_{tt} = -\omega^2 \sin c\omega t \sin \omega x - \dfrac{1}{c^2}(-c^2)\omega^2 \sin c\omega t \sin \omega x = 0$

88. (a) $\partial u/\partial x = \partial v/\partial y = 2x,\ \partial u/\partial y = -\partial v/\partial x = -2y$

(b) $\partial u/\partial x = \partial v/\partial y = e^x \cos y,\ \partial u/\partial y = -\partial v/\partial x = -e^x \sin y$

(c) $\partial u/\partial x = \partial v/\partial y = 2x/(x^2+y^2),\ \partial u/\partial y = -\partial v/\partial x = 2y/(x^2+y^2)$

89. $\partial u/\partial x = \partial v/\partial y$ and $\partial u/\partial y = -\partial v/\partial x$ so $\partial^2 u/\partial x^2 = \partial^2 v/\partial x\partial y$, and $\partial^2 u/\partial y^2 = -\partial^2 v/\partial y\partial x$, $\partial^2 u/\partial x^2 + \partial^2 u/\partial y^2 = \partial^2 v/\partial x\partial y - \partial^2 v/\partial y\partial x$, if $\partial^2 v/\partial x\partial y = \partial^2 v/\partial y\partial x$ then $\partial^2 u/\partial x^2 + \partial^2 u/\partial y^2 = 0$; thus u satisfies Laplace's equation. The proof that v satisfies Laplace's equation is similar. Adding Laplace's equations for u and v gives Laplaces' equation for $u + v$.

90. $\partial^2 R/\partial R_1^2 = -2R_2^2/(R_1+R_2)^3$, $\partial^2 R/\partial R_2^2 = -2R_1^2/(R_1+R_2)^3$,

$$\left(\partial^2 R/\partial R_1^2\right)\left(\partial^2 R/\partial R_2^2\right) = 4R_1^2R_2^2/\left(R_1+R_2\right)^6 = \left[4/\left(R_1+R_2\right)^4\right]\left[R_1R_2/\left(R_1+R_2\right)\right]^2$$
$$= 4R^2/\left(R_1+R_2\right)^4$$

91. $\partial f/\partial v = 8vw^3x^4y^5$, $\partial f/\partial w = 12v^2w^2x^4y^5$, $\partial f/\partial x = 16v^2w^3x^3y^5$, $\partial f/\partial y = 20v^2w^3x^4y^4$

92. $\partial w/\partial r = \cos st + ue^u \cos ur$, $\partial w/\partial s = -rt\sin st$,
$\partial w/\partial t = -rs\sin st$, $\partial w/\partial u = re^u\cos ur + e^u \sin ur$

93. $\partial f/\partial v_1 = 2v_1/\left(v_3^2+v_4^2\right)$, $\partial f/\partial v_2 = -2v_2/\left(v_3^2+v_4^2\right)$, $\partial f/\partial v_3 = -2v_3\left(v_1^2-v_2^2\right)/\left(v_3^2+v_4^2\right)^2$,
$\partial f/\partial v_4 = -2v_4\left(v_1^2-v_2^2\right)/\left(v_3^2+v_4^2\right)^2$

94. $\dfrac{\partial V}{\partial x} = 2xe^{2x-y} + e^{2x-y}$, $\dfrac{\partial V}{\partial y} = -xe^{2x-y} + w$, $\dfrac{\partial V}{\partial z} = w^2e^{zw}$, $\dfrac{\partial V}{\partial w} = wze^{zw} + e^{zw} + y$

95. (a) 0 (b) 0 (c) 0 (d) 0
(e) $2(1+yw)e^{yw}\sin z\cos z$ (f) $2xw(2+yw)e^{yw}\sin z\cos z$

96. 128, −512, 32, 64/3

97. $\partial w/\partial x_i = -i\sin(x_1 + 2x_2 + \ldots + nx_n)$

98. $\partial w/\partial x_i = \dfrac{1}{n}\left(\displaystyle\sum_{k=1}^{n} x_k\right)^{(1/n)-1}$

99. (a) xy-plane, $f_x = 12x^2y + 6xy$, $f_y = 4x^3 + 3x^2$, $f_{xy} = f_{yx} = 12x^2 + 6x$
(b) $y \neq 0, f_x = 3x^2/y$, $f_y = -x^3/y^2$, $f_{xy} = f_{yx} = -3x^2/y^2$

100. (a) $x^2 + y^2 > 1$, (the exterior of the circle of radius 1 about the origin);
$f_x = x/\sqrt{x^2+y^2-1}$, $f_y = y/\sqrt{x^2+y^2-1}$, $f_{xy} = f_{yx} = -xy\left(x^2+y^2-1\right)^{-3/2}$
(b) xy-plane, $f_x = 2x\cos(x^2+y^3)$, $f_y = 3y^2\cos(x^2+y^3)$, $f_{xy} = f_{yx} = -6xy^2\sin\left(x^2+y^3\right)$

101. $f_x(2,-1) = \displaystyle\lim_{x\to 2}\frac{f(x,-1)-f(2,-1)}{x-2} = \lim_{x\to 2}\frac{2x^2+3x+1-15}{x-2} = \lim_{x\to 2}(2x+7) = 11$ and

$$f_y(2,-1) = \lim_{y\to -1}\frac{f(2,y)-f(2,-1)}{y+1} = \lim_{y\to -1}\frac{8-6y+y^2-15}{y+1} = \lim_{y\to -1} y-7 = -8$$

102. $f_x(x,y) = \dfrac{2}{3}(x^2+y^2)^{-1/3}(2x) = \dfrac{4x}{3(x^2+y^2)^{1/3}}$, $(x,y) \neq (0,0)$;

$$f_x(0,0) = \frac{d}{dx}[f(x,0)]\bigg]_{x=0} = \frac{d}{dx}[x^{4/3}]\bigg]_{x=0} = \frac{4}{3}x^{1/3}\bigg]_{x=0} = 0.$$

103. **(a)** $f_y(0,0) = \frac{d}{dy}[f(0,y)]\Big]_{y=0} = \frac{d}{dy}[y]\Big]_{y=0} = 1$

(b) If $(x,y) \neq (0,0)$, then $f_y(x,y) = \frac{1}{3}(x^3+y^3)^{-2/3}(3y^2) = \frac{y^2}{(x^3+y^3)^{2/3}}$; $f_y(x,y)$ does not exist when $y \neq 0$ and $y = -x$

EXERCISE SET 14.4

1. **(a)** Let $f(x,y) = e^x \sin y$; $f(0,0) = 0$, $f_x(0,0) = 0$, $f_y(0,0) = 1$, so $e^x \sin y \approx y$

(b) Let $f(x,y) = \frac{2x+1}{y+1}$; $f(0,0) = 1$, $f_x(0,0) = 2$, $f_y(0,0) = -1$, so $\frac{2x+1}{y+1} \approx 1 + 2x - y$

2. $f(1,1) = 1$, $f_x(x,y) = \alpha x^{\alpha-1}y^{\beta}$, $f_x(1,1) = \alpha$, $f_y(x,y) = \beta x^{\alpha}y^{\beta-1}$, $f_y(1,1) = \beta$, so $x^{\alpha}y^{\beta} \approx 1 + \alpha(x-1) + \beta(y-1)$

3. **(a)** Let $f(x,y,z) = xyz + 2$, then $f_x = f_y = f_z = 1$ at $x = y = z = 1$, and $L(x,y,z) = f(1,1,1) + f_x(x-1) + f_y(y-1) + f_z(z-1) = 3 + x - 1 + y - 1 + z - 1 = x + y + z$

(b) Let $f(x,y,z) = \frac{4x}{y+z}$, then $f_x = 2$, $f_y = -1$, $f_z = -1$ at $x = y = z = 1$, and

$$\begin{aligned} L(x,y,z) &= f(1,1,1) + f_x(x-1) + f_y(y-1) + f_z(z-1) \\ &= 2 + 2(x-1) - (y-1) - (z-1) = 2x - y - z + 2 \end{aligned}$$

4. Let $f(x,y,z) = x^{\alpha}y^{\beta}z^{\gamma}$, then $f_x = \alpha$, $f_y = \beta$, $f_z = \gamma$ at $x = y = z = 1$, and $f(x,y,z) \approx f(1,1,1) + f_x(x-1) + f_y(y-1) + f_z(z-1) = 1 + \alpha(x-1) + \beta(y-1) + \gamma(z-1)$

5. $f(x,y) \approx f(3,4) + f_x(x-3) + f_y(y-4) = 5 + 2(x-3) - (y-4)$ and $f(3.01, 3.98) \approx 5 + 2(0.01) - (-0.02) = 5.04$

6. $f(x,y) \approx f(-1,2) + f_x(x+1) + f_y(y-2) = 2 + (x+1) + 3(y-2)$ and $f(-0.99, 2.02) \approx 2 + 0.01 + 3(0.02) = 2.07$

7. $L(x,y) = f(1,1) + f_x(1,1)(x-1) + f_y(1,1)(y-1)$ and $L(1.1, 0.9) = 3.15 = 3 + 2(0.1) + f_y(1,1)(-0.1)$ so $f_y(1,1) = -0.05/(-0.1) = 0.5$

8. $L(x,y) = 3 + f_x(0,-1)x - 2(y+1)$, $3.3 = 3 + f_x(0,-1)(0.1) - 2(-0.1)$, so $f_x(0,-1) = 0.1/0.1 = 1$

9. $L(x,y,z) = f(1,2,3) + (x-1) + 2(y-2) + 3(z-3)$, $f(1.01, 2.02, 3.03) \approx 4 + 0.01 + 2(0.02) + 3(0.03) = 4.14$

10. $L(x,y,z) = f(2,1,-2) - (x-2) + (y-1) - 2(z+2)$, $f(1.98, 0.99, -1.97) \approx 0.02 - 0.01 - 2(0.03) = -0.05$

11. $x - y + 2z - 2 = L(x,y,z) = f(3,2,1) + f_x(3,2,1)(x-3) + f_y(3,2,1)(y-2) + f_z(3,2,1)(z-1)$, so $f_x(3,2,1) = 1$, $f_y(3,2,1) = -1$, $f_z(3,2,1) = 2$ and $f(3,2,1) = L(3,2,1) = 1$

12. $L(x,y,z) = x + 2y + 3z + 4 = (x-0) + 2(y+1) + 3(z+2) - 4$, $f(0,-1,-2) = -4$, $f_x(0,-1,-2) = 1$, $f_y(0,-1,-2) = 2$, $f_z(0,-1,-2) = 3$

13. $L(x, y) = f(x_0, y_0) + f_x(x_0, y_0)(x - x_0) + f_y(x_0, y_0)(y - y_0)$,
$2y - 2x - 2 = x_0^2 + y_0^2 + 2x_0(x - x_0) + 2y_0(y - y_0)$, from which it follows that $x_0 = -1, y_0 = 1$.

14. $f(x, y) = x^2y$, so $f_x(x_0, y_0) = 2x_0y_0$, $f_y(x_0, y_0) = x_0^2$, and
$L(x, y) = f(x_0, y_0) + 2x_0y_0(x - x_0) + x_0^2(y - y_0)$. But $L(x, y) = 8 - 4x + 4y$, hence
$-4 = 2x_0y_0$, $4 = x_0^2$ and $8 = f(x_0, y_0) - 2x_0^2y_0 - x_0^2y_0 = -2x_0^2y_0$. Thus either $x_0 = -2, y_0 = 1$ from which it follows that $8 = -8$, a contradiction, or $x_0 = 2, y_0 = -1$, which is a solution since then $8 = -2x_0^2y_0 = 8$ is true.

15. $L(x, y, z) = f(x_0, y_0, z_0) + f_x(x_0, y_0, z_0)(x - x_0) + f_y(x_0, y_0, z_0)(y - y_0) + f_z(x_0, y_0, z_0)(z - z_0)$,
$y + 2z - 1 = x_0y_0 + z_0^2 + y_0(x - x_0) + x_0(y - y_0) + 2z_0(z - z_0)$, so that $x_0 = 1, y_0 = 0, z_0 = 1$.

16. $L(x, y, z) = f(x_0, y_0, z_0) + f_x(x_0, y_0, z_0)(x - x_0) + f_y(x_0, y_0, z_0)(y - y_0) + f_z(x_0, y_0, z_0)(z - z_0)$.
Then $x - y - z - 2 = x_0y_0z_0 + y_0z_0(x - x_0) + x_0z_0(y - y_0) + x_0y_0(z - z_0)$, hence
$y_0z_0 = 1, x_0z_0 = -1, x_0y_0 = -1$, and $-2 = x_0y_0z_0 - 3x_0y_0z_0$, or $x_0y_0z_0 = 1$. Since now $x_0 = -y_0 = -z_0$, we must have $|x_0| = |y_0| = |z_0| = 1$ or else $|x_0y_0z_0| \neq 1$, impossible. Thus $x_0 = 1, y_0 = z_0 = -1$ (note that $(-1, 1, 1)$ is not a solution).

17. **(a)** $f(P) = 1/5, f_x(P) = -x/(x^2 + y^2)^{-3/2}\Big]_{(x,y)=(4,3)} = -4/125$,

$f_y(P) = -y/(x^2 + y^2)^{-3/2}\Big]_{(x,y)=(4,3)} = -3/125, L(x, y) = \frac{1}{5} - \frac{4}{125}(x - 4) - \frac{3}{125}(y - 3)$

(b) $L(Q) - f(Q) = \frac{1}{5} - \frac{4}{125}(-0.08) - \frac{3}{125}(0.01) - 0.2023342382 \approx -0.0000142382$,
$|PQ| = \sqrt{0.08^2 + 0.01^2} \approx 0.0008062257748, |L(Q) - f(Q)|/|PQ| \approx 0.000176603$

18. **(a)** $f(P) = 1, f_x(P) = 0.5, f_y(P) = 0.3, L(x, y) = 1 + 0.5(x - 1) + 0.3(y - 1)$
(b) $L(Q) - f(Q) = 1 + 0.5(0.05) + 0.3(-0.03) - 1.05^{0.5}0.97^{0.3} \approx 0.00063$,
$|PQ| = \sqrt{0.05^2 + 0.03^2} \approx 0.05831, |L(Q) - f(Q)|/|PQ| \approx 0.0107$

19. **(a)** $f(P) = 0, f_x(P) = 0, f_y(P) = 0, L(x, y) = 0$
(b) $L(Q) - f(Q) = -0.003\sin(0.004) \approx -0.000012, |PQ| = \sqrt{0.003^2 + 0.004^2} = 0.005$,
$|L(Q) - f(Q)|/|PQ| \approx 0.0024$

20. **(a)** $f(P) = \ln 2, f_x(P) = 1, f_y(P) = 1/2, L(x, y) = \ln 2 + (x - 1) + \frac{1}{2}(y - 2)$
(b) $L(Q) - f(Q) = \ln 2 + 0.01 + (1/2)(0.02) - \ln 2.0402 \approx 0.0000993383$,
$|PQ| = \sqrt{0.01^2 + 0.02^2} \approx 0.02236067978$, $|L(Q) - f(Q)|/|PQ| \approx 0.0044425$

21. **(a)** $f(P) = 6, f_x(P) = 6, f_y(P) = 3, f_z(P) = 2, L(x, y) = 6 + 6(x - 1) + 3(y - 2) + 2(z - 3)$
(b) $L(Q) - f(Q) = 6 + 6(0.001) + 3(0.002) + 2(0.003) - 6.018018006 = -.000018006$,
$|PQ| = \sqrt{0.001^2 + 0.002^2 + 0.003^2} \approx .0003741657387$; $|L(Q) - f(Q)|/|PQ| \approx -0.000481$

22. **(a)** $f(P) = 0, f_x(P) = 1/2, f_y(P) = 1/2, f_z(P) = 0, L(x, y) = \frac{1}{2}(x + 1) + \frac{1}{2}(y - 1)$
(b) $L(Q) - f(Q) = 0, |L(Q) - f(Q)|/|PQ| = 0$

23. **(a)** $f(P) = e, f_x(P) = e, f_y(P) = -e, f_z(P) = -e, L(x, y) = e + e(x - 1) - e(y + 1) - e(z + 1)$
(b) $L(Q) - f(Q) = e - 0.01e + 0.01e - 0.01e - 0.99e^{0.9999} = 0.99(e - e^{0.9999})$,
$|PQ| = \sqrt{0.01^2 + 0.01^2 + 0.01^2} \approx 0.01732, |L(Q) - f(Q)|/|PQ| \approx 0.01554$

24. **(a)** $f(P) = 0, f_x(P) = 1, f_y(P) = -1, f_z(P) = 1,\ L(x, y, z) = (x - 2) - (y - 1) + (z + 1)$
(b) $L(Q) - f(Q) = 0.02 + 0.03 - 0.01 - \ln 1.0403 \approx 0.00049086691$,
$|PQ| = \sqrt{0.02^2 + 0.03^2 + 0.01^2} \approx 0.03742,\ |L(Q) - f(Q)|/|PQ| \approx 0.01312$

25. $dz = 7dx - 2dy$ **26.** $dz = ye^{xy}dx + xe^{xy}dy$ **27.** $dz = 3x^2y^2dx + 2x^3ydy$

28. $dz = (10xy^5 - 2)dx + (25x^2y^4 + 4)dy$

29. $dz = \left[y/\left(1 + x^2y^2\right)\right] dx + \left[x/\left(1 + x^2y^2\right)\right] dy$

30. $dz = 2\sec^2(x - 3y)\tan(x - 3y)dx - 6\sec^2(x - 3y)\tan(x - 3y)dy$

31. $dw = 8dx - 3dy + 4dz$ **32.** $dw = yze^{xyz}dx + xze^{xyz}dy + xye^{xyz}dz$

33. $dw = 3x^2y^2zdx + 2x^3yzdy + x^3y^2dz$

34. $dw = \left(8xy^3z^7 - 3y\right)dx + \left(12x^2y^2z^7 - 3x\right)dy + \left(28x^2y^3z^6 + 1\right)dz$

35. $dw = \dfrac{yz}{1 + x^2y^2z^2}dx + \dfrac{xz}{1 + x^2y^2z^2}dy + \dfrac{xy}{1 + x^2y^2z^2}dz$

36. $dw = \dfrac{1}{2\sqrt{x}}dx + \dfrac{1}{2\sqrt{y}}dy + \dfrac{1}{2\sqrt{z}}dz$

37. $df = (2x + 2y - 4)dx + 2xdy;\ x = 1,\ y = 2,\ dx = 0.01,\ dy = 0.04$ so
$df = 0.10$ and $\Delta f = 0.1009$

38. $df = (1/3)x^{-2/3}y^{1/2}dx + (1/2)x^{1/3}y^{-1/2}dy;\ x = 8,\ y = 9,\ dx = -0.22,\ dy = 0.03$ so $df = -0.045$ and $\Delta f \approx -0.045613$

39. $df = -x^{-2}dx - y^{-2}dy;\ x = -1,\ y = -2,\ dx = -0.02,\ dy = -0.04$ so
$df = 0.03$ and $\Delta f \approx 0.029412$

40. $df = \dfrac{y}{2(1 + xy)}dx + \dfrac{x}{2(1 + xy)}dy;\ x = 0,\ y = 2,\ dx = -0.09,\ dy = -0.02$ so
$df = -0.09$ and $\Delta f \approx -0.098129$

41. $df = 2y^2z^3dx + 4xyz^3dy + 6xy^2z^2dz, x = 1, y = -1, z = 2, dx = -0.01, dy = -0.02, dz = 0.02$ so $df = 0.96$ and $\Delta f \approx 0.97929$

42. $df = \dfrac{yz(y + z)}{(x + y + z)^2}dx + \dfrac{xz(x + z)}{(x + y + z)^2}dy + \dfrac{xy(x + y)}{(x + y + z)^2}dz, x = -1, y = -2, z = 4, dx = -0.04,$
$dy = 0.02, dz = -0.03$ so $df = 0.58$ and $\Delta f \approx 0.60529$

43. Label the four smaller rectangles A, B, C, D starting with the lower left and going clockwise. Then the increase in the area of the rectangle is represented by B, C and D; and the portions B and D represent the approximation of the increase in area given by the total differential.

44. $V + \Delta V = (\pi/3)4.05^2(19.95) \approx 109.0766250\pi, V = 320\pi/3, \Delta V \approx 2.40996\pi;$
$dV = (2/3)\pi rhdr + (1/3)\pi r^2 dh;\ r = 4,\ h = 20,\ dr = 0.05,\ dh = -0.05$ so $dV = 2.4\pi$, and $\Delta V/dV \approx 1.00415$.

45. $A = xy,\ dA = ydx + xdy,\ dA/A = dx/x + dy/y,\ |dx/x| \le 0.03$ and $|dy/y| \le 0.05$,
$|dA/A| \le |dx/x| + |dy/y| \le 0.08 = 8\%$

46. $V = (1/3)\pi r^2 h,\ dV = (2/3)\pi rhdr + (1/3)\pi r^2 dh,\ dV/V = 2(dr/r) + dh/h,\ |dr/r| \le 0.01$ and $|dh/h| \le 0.04,\ |dV/V| \le 2|dr/r| + |dh/h| \le 0.06 = 6\%$.

47. $z = \sqrt{x^2+y^2},\ dz = \dfrac{x}{\sqrt{x^2+y^2}}dx + \dfrac{y}{\sqrt{x^2+y^2}}dy,$

$$\frac{dz}{z} = \frac{x}{x^2+y^2}dx + \frac{y}{x^2+y^2}dy = \frac{x^2}{x^2+y^2}\left(\frac{dx}{x}\right) + \frac{y^2}{x^2+y^2}\left(\frac{dy}{y}\right),$$

$\left|\dfrac{dz}{z}\right| \le \dfrac{x^2}{x^2+y^2}\left|\dfrac{dx}{x}\right| + \dfrac{y^2}{x^2+y^2}\left|\dfrac{dy}{y}\right|$, if $\left|\dfrac{dx}{x}\right| \le r/100$ and $\left|\dfrac{dy}{y}\right| \le r/100$ then

$\left|\dfrac{dz}{z}\right| \le \dfrac{x^2}{x^2+y^2}(r/100) + \dfrac{y^2}{x^2+y^2}(r/100) = \dfrac{r}{100}$ so the percentage error in z is at most about $r\%$.

48. **(a)** $z = \sqrt{x^2+y^2},\ dz = x\left(x^2+y^2\right)^{-1/2}dx + y\left(x^2+y^2\right)^{-1/2}dy,$
$|dz| \le x\left(x^2+y^2\right)^{-1/2}|dx| + y\left(x^2+y^2\right)^{-1/2}|dy|$; if $x = 3,\ y = 4,\ |dx| \le 0.05$, and $|dy| \le 0.05$ then $|dz| \le (3/5)(0.05) + (4/5)(0.05) = 0.07$ cm

(b) $A = (1/2)xy,\ dA = (1/2)ydx + (1/2)xdy,$
$|dA| \le (1/2)y|dx| + (1/2)x|dy| \le 2(0.05) + (3/2)(0.05) = 0.175\ \text{cm}^2$.

49. $dT = \dfrac{\pi}{g\sqrt{L/g}}dL - \dfrac{\pi L}{g^2\sqrt{L/g}}dg,\ \dfrac{dT}{T} = \dfrac{1}{2}\dfrac{dL}{L} - \dfrac{1}{2}\dfrac{dg}{g};\ |dL/L| \le 0.005$ and $|dg/g| \le 0.001$ so
$|dT/T| \le (1/2)(0.005) + (1/2)(0.001) = 0.003 = 0.3\%$

50. $dP = (k/V)dT - (kT/V^2)dV,\ dP/P = dT/T - dV/V$; if $dT/T = 0.03$ and $dV/V = 0.05$ then $dP/P = -0.02$ so there is about a 2% decrease in pressure.

51. **(a)** $\left|\dfrac{d(xy)}{xy}\right| - \left|\dfrac{y\,dx + x\,dy}{xy}\right| = \left|\dfrac{dx}{x} + \dfrac{dy}{y}\right| \le \left|\dfrac{dx}{x}\right| + \left|\dfrac{dy}{y}\right| \le \dfrac{r}{100} + \dfrac{s}{100};\ (r+s)\%$

(b) $\left|\dfrac{d(x/y)}{x/y}\right| = \left|\dfrac{y\,dx - x\,dy}{xy}\right| = \left|\dfrac{dx}{x} - \dfrac{dy}{y}\right| \le \left|\dfrac{dx}{x}\right| + \left|\dfrac{dy}{y}\right| \le \dfrac{r}{100} + \dfrac{s}{100};\ (r+s)\%$

(c) $\left|\dfrac{d(x^2y^3)}{x^2y^3}\right| = \left|\dfrac{2xy^3\,dx + 3x^2y^2\,dy}{x^2y^3}\right| = \left|2\dfrac{dx}{x} + 3\dfrac{dy}{y}\right| \le 2\left|\dfrac{dx}{x}\right| + 3\left|\dfrac{dy}{y}\right|$
$\le 2\dfrac{r}{100} + 3\dfrac{s}{100};\ (2r+3s)\%$

(d) $\left|\dfrac{d(x^3y^{1/2})}{x^3y^{1/2}}\right| = \left|\dfrac{3x^2y^{1/2}\,dx + (1/2)x^3y^{-1/2}\,dy}{x^3y^{1/2}}\right| = \left|3\dfrac{dx}{x} + \dfrac{1}{2}\dfrac{dy}{y}\right| \le 3\left|\dfrac{dx}{x}\right| + \dfrac{1}{2}\left|\dfrac{dy}{y}\right|$
$\le 3\dfrac{r}{100} + \dfrac{1}{2}\dfrac{s}{100};\ (3r + \tfrac{1}{2}s)\%$

52. $R = 1/(1/R_1 + 1/R_2 + 1/R_3)$, $\partial R/\partial R_1 = \dfrac{1}{R_1^2(1/R_1 + 1/R_2 + 1/R_3)^2} = R^2/R_1^2$, similarly $\partial R/\partial R_2 = R^2/R_2^2$ and $\partial R/\partial R_3 = R^2/R_3^2$ so $\dfrac{dR}{R} = (R/R_1)\dfrac{dR_1}{R_1} + (R/R_2)\dfrac{dR_2}{R_2} + (R/R_3)\dfrac{dR_3}{R_3}$,

$$\begin{aligned}\left|\frac{dR}{R}\right| &\le (R/R_1)\left|\frac{dR_1}{R_1}\right| + (R/R_2)\left|\frac{dR_2}{R_2}\right| + (R/R_3)\left|\frac{dR_3}{R_3}\right| \\ &\le (R/R_1)(0.10) + (R/R_2)(0.10) + (R/R_3)(0.10) \\ &= R(1/R_1 + 1/R_2 + 1/R_3)(0.10) = (1)(0.10) = 0.10 = 10\%\end{aligned}$$

53. $dA = \dfrac{1}{2}b\sin\theta da + \dfrac{1}{2}a\sin\theta db + \dfrac{1}{2}ab\cos\theta d\theta$,

$$\begin{aligned}|dA| &\le \frac{1}{2}b\sin\theta|da| + \frac{1}{2}a\sin\theta|db| + \frac{1}{2}ab\cos\theta|d\theta| \\ &\le \frac{1}{2}(50)(1/2)(1/2) + \frac{1}{2}(40)(1/2)(1/4) + \frac{1}{2}(40)(50)\left(\sqrt{3}/2\right)(\pi/90) \\ &= 35/4 + 50\pi\sqrt{3}/9 \approx 39 \text{ ft}^2\end{aligned}$$

54. $V = \ell wh$, $dV = whd\ell + \ell hdw + \ell wdh$, $|dV/V| \le |d\ell/\ell| + |dw/w| + |dh/h| \le 3(r/100) = 3r\%$

55. If $f(x,y) = f(x_0,y_0)$ for all (x,y) then $L(x,y) = f(x_0,y_0)$ since the first partial derivatives of f are zero. Thus the error E is zero and f is differentiable. The proof for three variables is analogous.

56. Let $f(x,y) = ax + by + c$. Then $L(x,y) = f(x_0,y_0) + f_x(x_0,y_0)(x - x_0) + f_y(x_0,y_0)(y - y_0) = ax_0 + by_0 + c + a(x - x_0) + b(y - y_0) = ax + by + c$, so $L = f$ and thus E is zero. For three variables the proof is analogous.

57. $f_x = 2x\sin y$, $f_y = x^2\cos y$ are both continuous everywhere, so f is differentiable everywhere.

58. $f_x = y\sin z$, $f_y = x\sin z$, $f_z = xy\cos z$ are all continuous everywhere, so f is differentiable everywhere.

59. $f_x = 2x$, $f_y = 2y$, $f_z = 2z$ so $L(x,y,z) = 0$, $E = f - L = x^2 + y^2 + z^2$, and

$$\lim_{(x,y,z)\to(0,0,0)} \frac{E(x,y,z)}{\sqrt{x^2+y^2+z^2}} = \lim_{(x,y,z)\to(0,0,0)} \sqrt{x^2+y^2+z^2} = 0, \text{ so } f \text{ is differentiable at } (0,0,0).$$

60. $f_x = 2xr(x^2+y^2+z^2)^{r-1}$, $f_y = 2yr(x^2+y^2+z^2)^{r-1}$, $f_z = 2zr(x^2+y^2+z^2)^{r-1}$, so the partials of f exist only if $r \ge 1$. If so then $L(x,y,z) = 0$, $E(x,y,z) = f(x,y,z)$ and

$\dfrac{E(x,y,z)}{\sqrt{x^2+y^2+z^2}} = (x^2+y^2+z^2)^{r-1/2}$, so f is differentiable at $(0,0,0)$ if and only if $r > 1/2$.

61. Let $\epsilon > 0$. Then $\displaystyle\lim_{x\to x_0}\frac{f(x)}{g(x)} = 0$ if and only if there exists $\delta > 0$ such that $\left|\dfrac{f(x)}{g(x)}\right| < \epsilon$ whenever $|x - x_0| < \delta$. But this condition is equivalent to $\left|\dfrac{f(x)}{|g(x)|}\right| < \epsilon$, and thus the two limits both exist or neither exists.

62. f is continuous at (x_0,y_0) if and only if $\displaystyle\lim_{(x,y)\to(x_0,y_0)} f(x,y) = f(x_0,y_0)$. Since the limit of M is clearly $f(x_0,y_0)$, the limit of f will be $f(x_0,y_0)$ if and only if $\displaystyle\lim_{(x,y)\to(x_0,y_0)} E(x,y) = 0$.

63. If f is differentiable at (x_0,y_0) then $L(x,y)$ exists and is a linear function and thus differentiable, and thus the difference $E = f - L$ is also differentiable.

64. That f is differentiable means that $\lim_{(x,y)\to(x_0,y_0)} \dfrac{E_f(x,y)}{\sqrt{(x-x_0)^2+(y-y_0)^2}} = 0$, where $E_f(x,y) = f(x,y) - L_f(x,y)$; here $L_f(x,y)$ is the linear approximation to f at (x_0, y_0). Let f_x and f_y denote $f_x(x_0,y_0), f_y(x_0,y_0)$ respectively. Then $g(x,y,z) = z - f(x,y)$,

$$L_f(x,y) = f(x_0,y_0) + f_x(x-x_0) + f_y(y-y_0),$$

$$\begin{aligned} L_g(x,y,z) &= g(x_0,y_0,z_0) + g_x(x-x_0) + g_y(y-y_0) + g_z(z-z_0) \ , \\ &= 0 - f_x(x-x_0) - f_y(y-y_0) + (z-z_0) \end{aligned}$$

and

$$\begin{aligned} E_g(x,y,z) &= g(x,y,z) - L_g(x,y,z) = (z - f(x,y)) + f_x(x-x_0) + f_y(y-y_0) - (z-z_0) \\ &= f(x_0,y_0) + f_x(x_0,y_0)(x-x_0) + f_y(x_0,y_0)(y-y_0) - f(x,y) = -E_f(x,y) \end{aligned}$$

Thus $\dfrac{|E_g(x,y,z)|}{\sqrt{(x-x_0)^2+(y-y_0)^2+(z-z_0)^2}} \le \dfrac{|E_f(x,y)|}{\sqrt{(x-x_0)^2+(y-y_0)^2}}$

so $\lim_{(x,y,z)\to(x_0,y_0,z_0)} \dfrac{E_g(x,y,z)}{\sqrt{(x-x_0)^2+(y-y_0)^2+(z-z_0)^2}} = 0$

and g is differentiable at (x_0, y_0, z_0).

65. Let $x > 0$. Then $\dfrac{f(x,y) - f(x,0)}{y - 0}$ can be $-1/y$ or 0 depending on whether $y > 0$ or $y < 0$. Thus the partial derivative $f_y(x,0)$ cannot exist. A similar argument works for $f_x(0,y)$ if $y > 0$.

66. The condition $\lim_{(x,y)\to(x_0,y_0)} \dfrac{E(x,y)}{\sqrt{(x-x_0)^2+(y-y_0)^2}} = 0$ is equivalent to $\lim_{(x,y)\to(x_0,y_0)} \epsilon(x,y) = 0$ which is equivalent to ϵ being continuous at (x_0, y_0) with $\epsilon(0,0) = 0$. Since ϵ is continuous, f is differentiable.

EXERCISE SET 14.5

1. $42t^{13}$

2. $\dfrac{2(3+t^{-1/3})}{3(2t+t^{2/3})}$

3. $3t^{-2}\sin(1/t)$

4. $\dfrac{1-2t^4-8t^4\ln t}{2t\sqrt{1+\ln t - 2t^4\ln t}}$

5. $-\dfrac{10}{3}t^{7/3}e^{1-t^{10/3}}$

6. $(1+t)e^t\cosh(te^t/2)\sinh(te^t/2)$

7. $165t^{32}$

8. $\dfrac{3-(4/3)t^{-1/3}-24t^{-7}}{3t-2t^{2/3}+4t^{-6}}$

9. $-2t\cos\left(t^2\right)$

10. $\dfrac{1-512t^5-2560t^5\ln t}{2t\sqrt{1+\ln t-512t^5\ln t}}$

11. 3264

12. 0

13. $\partial z/\partial u = 24u^2v^2 - 16uv^3 - 2v + 3$, $\partial z/\partial v = 16u^3v - 24u^2v^2 - 2u - 3$

14. $\partial z/\partial u = 2u/v^2 - u^2v\sec^2(u/v) - 2uv^2\tan(u/v)$
$\partial z/\partial v = -2u^2/v^3 + u^3\sec^2(u/v) - 2u^2v\tan(u/v)$

15. $\partial z/\partial u = -\dfrac{2\sin u}{3\sin v}$, $\partial z/\partial v = -\dfrac{2\cos u\cos v}{3\sin^2 v}$

16. $\partial z/\partial u = 3 + 3v/u - 4u$, $\partial z/\partial v = 2 + 3\ln u + 2\ln v$

17. $\partial z/\partial u = e^u$, $\partial z/\partial v = 0$

18. $\partial z/\partial u = -\sin(u-v)\sin\left(u^2+v^2\right) + 2u\cos(u-v)\cos\left(u^2+v^2\right)$
$\partial z/\partial v = \sin(u-v)\sin\left(u^2+v^2\right) + 2v\cos(u-v)\cos\left(u^2+v^2\right)$

19. $\partial T/\partial r = 3r^2\sin\theta\cos^2\theta - 4r^3\sin^3\theta\cos\theta$
$\partial T/\partial\theta = -2r^3\sin^2\theta\cos\theta + r^4\sin^4\theta + r^3\cos^3\theta - 3r^4\sin^2\theta\cos^2\theta$

20. $dR/d\phi = 5e^{5\phi}$

21. $\partial t/\partial x = \left(x^2+y^2\right)/\left(4x^2y^3\right)$, $\partial t/\partial y = \left(y^2-3x^2\right)/\left(4xy^4\right)$

22. $\partial w/\partial u = \dfrac{2v^2\left[u^2v^2-(u-2v)^2\right]}{\left[u^2v^2+(u-2v)^2\right]^2}$, $\partial w/\partial v = \dfrac{u^2\left[(u-2v)^2-u^2v^2\right]}{\left[u^2v^2+(u-2v)^2\right]^2}$

23. $\partial z/\partial r = (dz/dx)(\partial x/\partial r) = 2r\cos^2\theta/\left(r^2\cos^2\theta+1\right)$,
$\partial z/\partial\theta = (dz/dx)(\partial x/\partial\theta) = -2r^2\sin\theta\cos\theta/\left(r^2\cos^2\theta+1\right)$

24. $\partial u/\partial x = (\partial u/\partial r)(dr/dx) + (\partial u/\partial t)(\partial t/\partial x)$
$= \left(s^2\ln t\right)(2x) + \left(rs^2/t\right)\left(y^3\right) = x(4y+1)^2\left(1+2\ln xy^3\right)$
$\partial u/\partial y = (\partial u/\partial s)(ds/dy) + (\partial u/\partial t)(\partial t/\partial y)$
$= (2rs\ln t)(4) + \left(rs^2/t\right)\left(3xy^2\right) = 8x^2(4y+1)\ln xy^3 + 3x^2(4y+1)^2/y$

25. $\partial w/\partial\rho = 2\rho\left(4\sin^2\phi+\cos^2\phi\right)$, $\partial w/\partial\phi = 6\rho^2\sin\phi\cos\phi$, $\partial w/\partial\theta = 0$

26. $\dfrac{dw}{dx} = \dfrac{\partial w}{\partial x} + \dfrac{\partial w}{\partial y}\dfrac{dy}{dx} + \dfrac{\partial w}{\partial z}\dfrac{dz}{dx} = 3y^2z^3 + (6xyz^3)(6x) + 9xy^2z^2\dfrac{1}{2\sqrt{x-1}}$
$= 3(3x^2+2)^2(x-1)^{3/2} + 36x^2(3x^2+2)(x-1)^{3/2} + \dfrac{9}{2}x(3x^2+2)^2\sqrt{x-1}$
$= \dfrac{3}{2}(3x^2+2)(39x^3-30x^2+10x-4)\sqrt{x-1}$

27. $-\pi$

28. $351/2$, -168

29. $\sqrt{3}e^{\sqrt{3}}$, $\left(2-4\sqrt{3}\right)e^{\sqrt{3}}$

30. 1161

31. $F(x,y) = x^2y^3 + \cos y$, $\dfrac{dy}{dx} = -\dfrac{2xy^3}{3x^2y^2-\sin y}$

32. $F(x,y) = x^3 - 3xy^2 + y^3 - 5$, $\dfrac{dy}{dx} = -\dfrac{3x^2-3y^2}{-6xy+3y^2} = \dfrac{x^2-y^2}{2xy-y^2}$

33. $F(x,y) = e^{xy} + ye^y - 1$, $\dfrac{dy}{dx} = -\dfrac{ye^{xy}}{xe^{xy}+ye^y+e^y}$

34. $F(x,y) = x - (xy)^{1/2} + 3y - 4$, $\dfrac{dy}{dx} = -\dfrac{1-(1/2)(xy)^{-1/2}y}{-(1/2)(xy)^{-1/2}x+3} = \dfrac{2\sqrt{xy}-y}{x-6\sqrt{xy}}$

35. $\dfrac{\partial F}{\partial x} + \dfrac{\partial F}{\partial z}\dfrac{\partial z}{\partial x} = 0$ so $\dfrac{\partial z}{\partial x} = -\dfrac{\partial F/\partial x}{\partial F/\partial z}$.

36. $\dfrac{\partial F}{\partial y} + \dfrac{\partial F}{\partial z}\dfrac{\partial z}{\partial y} = 0$ so $\dfrac{\partial z}{\partial y} = -\dfrac{\partial F/\partial y}{\partial F/\partial z}$.

37. $\dfrac{\partial z}{\partial x} = \dfrac{2x+yz}{6yz-xy}$, $\dfrac{\partial z}{\partial y} = \dfrac{xz-3z^2}{6yz-xy}$

38. $\ln(1+z) + xy^2 + z - 1 = 0$; $\dfrac{\partial z}{\partial x} = -\dfrac{y^2(1+z)}{2+z}$, $\dfrac{\partial z}{\partial y} = -\dfrac{2xy(1+z)}{2+z}$

39. $ye^x - 5\sin 3z - 3z = 0$; $\dfrac{\partial z}{\partial x} = -\dfrac{ye^x}{-15\cos 3z - 3} = \dfrac{ye^x}{15\cos 3z + 3}$, $\dfrac{\partial z}{\partial y} = \dfrac{e^x}{15\cos 3z + 3}$

40. $\dfrac{\partial z}{\partial x} = -\dfrac{ze^{yz}\cos xz - ye^{xy}\cos yz}{ye^{xy}\sin yz + xe^{yz}\cos xz + ye^{yz}\sin xz}$, $\dfrac{\partial z}{\partial y} = -\dfrac{ze^{xy}\sin yz - xe^{xy}\cos yz + ze^{yz}\sin xz}{ye^{xy}\sin yz + xe^{yz}\cos xz + ye^{yz}\sin xz}$

41. $D = \left(x^2+y^2\right)^{1/2}$ where x and y are the distances of cars A and B, respectively, from the intersection and D is the distance between them.
$dD/dt = \left[x/\left(x^2+y^2\right)^{1/2}\right](dx/dt) + \left[y/\left(x^2+y^2\right)^{1/2}\right](dy/dt)$, $dx/dt = -25$ and $dy/dt = -30$ when $x = 0.3$ and $y = 0.4$ so $dD/dt = (0.3/0.5)(-25) + (0.4/0.5)(-30) = -39$ mph.

42. $T = (1/10)PV$, $dT/dt = (V/10)(dP/dt) + (P/10)(dV/dt)$, $dV/dt = 4$ and $dP/dt = -1$ when $V = 200$ and $P = 5$ so $dT/dt = (20)(-1) + (1/2)(4) = -18$ K/s.

43. $A = \dfrac{1}{2}ab\sin\theta$ but $\theta = \pi/6$ when $a = 4$ and $b = 3$ so $A = \dfrac{1}{2}(4)(3)\sin(\pi/6) = 3$.

Solve $\dfrac{1}{2}ab\sin\theta = 3$ for θ to get $\theta = \sin^{-1}\left(\dfrac{6}{ab}\right)$, $0 \le \theta \le \pi/2$.

$$\frac{d\theta}{dt} = \frac{\partial\theta}{\partial a}\frac{da}{dt} + \frac{\partial\theta}{\partial b}\frac{db}{dt} = \frac{1}{\sqrt{1-\dfrac{36}{a^2b^2}}}\left(-\frac{6}{a^2b}\right)\frac{da}{dt} + \frac{1}{\sqrt{1-\dfrac{36}{a^2b^2}}}\left(-\frac{6}{ab^2}\right)\frac{db}{dt}$$

$$= -\frac{6}{\sqrt{a^2b^2-36}}\left(\frac{1}{a}\frac{da}{dt} + \frac{1}{b}\frac{db}{dt}\right), \quad \frac{da}{dt} = 1 \text{ and } \frac{db}{dt} = 1$$

when $a = 4$ and $b = 3$ so $\dfrac{d\theta}{dt} = -\dfrac{6}{\sqrt{144-36}}\left(\dfrac{1}{4}+\dfrac{1}{3}\right) = -\dfrac{7}{12\sqrt{3}} = -\dfrac{7}{36}\sqrt{3}$ radians/s

44. From the law of cosines, $c = \sqrt{a^2+b^2-2ab\cos\theta}$ where c is the length of the third side.

$\theta = \pi/3$ so $c = \sqrt{a^2+b^2-ab}$,

$$\frac{dc}{dt} = \frac{\partial c}{\partial a}\frac{da}{dt} + \frac{\partial c}{\partial b}\frac{db}{dt} = \frac{1}{2}(a^2+b^2-ab)^{-1/2}(2a-b)\frac{da}{dt} + \frac{1}{2}\left(a^2+b^2-ab\right)^{-1/2}(2b-a)\frac{db}{dt}$$

$$= \frac{1}{2\sqrt{a^2+b^2-ab}}\left[(2a-b)\frac{da}{dt} + (2b-a)\frac{db}{dt}\right], \quad \frac{da}{dt} = 2 \text{ and } \frac{db}{dt} = 1 \text{ when } a = 5 \text{ and } b = 10$$

so $\dfrac{dc}{dt} = \dfrac{1}{2\sqrt{75}}[(0)(2) + (15)(1)] = \sqrt{3}/2$ cm/s. The third side is increasing.

45. $V = (\pi/4)D^2h$ where D is the diameter and h is the height, both measured in inches, $dV/dt = (\pi/2)Dh(dD/dt) + (\pi/4)D^2(dh/dt)$, $dD/dt = 3$ and $dh/dt = 24$ when $D = 30$ and $h = 240$, so $dV/dt = (\pi/2)(30)(240)(3) + (\pi/4)(30)^2(24) = 16{,}200\pi$ in^3/year.

46. $\dfrac{dT}{dt} = \dfrac{\partial T}{\partial x}\dfrac{dx}{dt} + \dfrac{\partial T}{\partial y}\dfrac{dy}{dt} = \dfrac{y^2}{x}\dfrac{dx}{dt} + 2y\ln x\dfrac{dy}{dt}$, $dx/dt = 1$ and $dy/dt = -4$ at (3,2) so
$dT/dt = (4/3)(1) + (4\ln 3)(-4) = 4/3 - 16\ln 3°$ C/s.

47. (a) $V = \ell wh, \dfrac{dV}{dt} = \dfrac{\partial V}{\partial \ell}\dfrac{d\ell}{dt} + \dfrac{\partial V}{\partial w}\dfrac{dw}{dt} + \dfrac{\partial V}{\partial h}\dfrac{dh}{dt} = wh\dfrac{d\ell}{dt} + \ell h\dfrac{dw}{dt} + \ell w\dfrac{dh}{dt}$

$= (3)(6)(1) + (2)(6)(2) + (2)(3)(3) = 60 \text{ in}^3/\text{s}$

(b) $D = \sqrt{\ell^2 + w^2 + h^2}$; $dD/dt = (\ell/D)d\ell/dt + (w/D)dw/dt + (h/D)dh/dt$

$= (2/7)(1) + (3/7)(2) + (6/7)(3) = 26/7 \text{ in/s}$

48. $S = 2(lw + wh + lh), \dfrac{dS}{dt} = \dfrac{\partial S}{\partial w}\dfrac{dw}{dt} + \dfrac{\partial S}{\partial l}\dfrac{dl}{dt} + \dfrac{\partial S}{\partial h}\dfrac{dh}{dt}$

$= 2(l + h)\dfrac{dw}{dt} + 2(w + h)\dfrac{dl}{dt} + 2(w + l)\dfrac{dh}{dt} = 80 \text{ in}^2/\text{s}$

49. (a) $f(tx, ty) = 3t^2x^2 + t^2y^2 = t^2 f(x, y)$; $n = 2$

(b) $f(tx, ty) = \sqrt{t^2x^2 + t^2y^2} = tf(x, y)$; $n = 1$

(c) $f(tx, ty) = t^3x^2y - 2t^3y^3 = t^3 f(x, y)$; $n = 3$

(d) $f(tx, ty) = 5/\left(t^2x^2 + 2t^2y^2\right)^2 = t^{-4} f(x, y)$; $n = -4$

50. (a) If $f(u, v) = t^n f(x, y)$, then $\dfrac{\partial f}{\partial u}\dfrac{du}{dt} + \dfrac{\partial f}{\partial v}\dfrac{dv}{dt} = nt^{n-1}f(x, y)$, $x\dfrac{\partial f}{\partial u} + y\dfrac{\partial f}{\partial v} = nt^{n-1}f(x, y)$;

let $t = 1$ to get $x\dfrac{\partial f}{\partial x} + y\dfrac{\partial f}{\partial y} = nf(x, y)$.

(b) If $f(x, y) = 3x^2 + y^2$ then $xf_x + yf_y = 6x^2 + 2y^2 = 2f(x, y)$;
If $f(x, y) = \sqrt{x^2 + y^2}$ then $xf_x + yf_y = x^2/\sqrt{x^2 + y^2} + y^2/\sqrt{x^2 + y^2} = \sqrt{x^2 + y^2} = f(x, y)$;
If $f(x, y) = x^2y - 2y^3$ then $xf_x + yf_y = 3x^2y - 6y^3 = 3f(x, y)$;
If $f(x, y) = \dfrac{5}{(x^2 + 2y^2)^2}$ then $xf_x + yf_y = x\dfrac{5(-2)2x}{(x^2 + 2y^2)^3} + y\dfrac{5(-2)4y}{(x^2 + 2y^2)^3} = -4f(x, y)$

51. (a) $\dfrac{\partial z}{\partial x} = \dfrac{dz}{du}\dfrac{\partial u}{\partial x}, \dfrac{\partial z}{\partial y} = \dfrac{dz}{du}\dfrac{\partial u}{\partial y}$

(b) $\dfrac{\partial^2 z}{\partial x^2} = \dfrac{dz}{du}\dfrac{\partial^2 u}{\partial x^2} + \dfrac{\partial}{\partial x}\left(\dfrac{dz}{du}\right)\dfrac{\partial u}{\partial x} = \dfrac{dz}{du}\dfrac{\partial^2 u}{\partial x^2} + \dfrac{d^2 z}{du^2}\left(\dfrac{\partial u}{\partial x}\right)^2$;

$\dfrac{\partial^2 z}{\partial y \partial x} = \dfrac{dz}{du}\dfrac{\partial^2 u}{\partial y \partial x} + \dfrac{\partial}{\partial y}\left(\dfrac{dz}{du}\right)\dfrac{\partial u}{\partial x} = \dfrac{dz}{du}\dfrac{\partial^2 u}{\partial y \partial x} + \dfrac{d^2 z}{du^2}\dfrac{\partial u}{\partial x}\dfrac{\partial u}{\partial y}$

$\dfrac{\partial^2 z}{\partial y^2} = \dfrac{dz}{du}\dfrac{\partial^2 u}{\partial y^2} + \dfrac{\partial}{\partial y}\left(\dfrac{dz}{du}\right)\dfrac{\partial u}{\partial y} = \dfrac{dz}{du}\dfrac{\partial^2 u}{\partial y^2} + \dfrac{d^2 z}{du^2}\left(\dfrac{\partial u}{\partial y}\right)^2$

52. (a) $z = f(u)$, $u = x^2 - y^2$; $\partial z/\partial x = (dz/du)(\partial u/\partial x) = 2x dz/du$
$\partial z/\partial y = (dz/du)(\partial u/\partial y) = -2y dz/du$, $y\partial z/\partial x + x\partial z/\partial y = 2xy dz/du - 2xy dz/du = 0$

(b) $z = f(u)$, $u = xy$; $\dfrac{\partial z}{\partial x} = \dfrac{dz}{du}\dfrac{\partial u}{\partial x} = y\dfrac{dz}{du}, \dfrac{\partial z}{\partial y} = \dfrac{dz}{du}\dfrac{\partial u}{\partial y} = x\dfrac{dz}{du}$,

$x\dfrac{\partial z}{\partial x} - y\dfrac{\partial z}{\partial y} = xy\dfrac{dz}{du} - xy\dfrac{dz}{du} = 0.$

(c) $yz_x + xz_y = y(2x\cos(x^2 - y^2)) - x(2y\cos(x^2 - y^2)) = 0$

(d) $xz_x - yz_y = xye^{xy} - yxe^{xy} = 0$

53. Let $z = f(u)$ where $u = x + 2y$; then $\partial z/\partial x = (dz/du)(\partial u/\partial x) = dz/du$,
$\partial z/\partial y = (dz/du)(\partial u/\partial y) = 2dz/du$ so $2\partial z/\partial x - \partial z/\partial y = 2dz/du - 2dz/du = 0$

54. Let $z = f(u)$ where $u = x^2 + y^2$; then $\partial z/\partial x = (dz/du)(\partial u/\partial x) = 2x\ dz/du$,
$\partial z/\partial y = (dz/du)(\partial u/\partial y) = 2ydz/du$ so $y\ \partial z/\partial x - x\partial z/\partial y = 2xydz/du - 2xydz/du = 0$

55. $\dfrac{\partial w}{\partial x} = \dfrac{dw}{du}\dfrac{\partial u}{\partial x} = \dfrac{dw}{du}, \dfrac{\partial w}{\partial y} = \dfrac{dw}{du}\dfrac{\partial u}{\partial y} = 2\dfrac{dw}{du}, \dfrac{\partial w}{\partial z} = \dfrac{dw}{du}\dfrac{\partial u}{\partial z} = 3\dfrac{dw}{du}$, so $\dfrac{\partial w}{\partial x} + \dfrac{\partial w}{\partial y} + \dfrac{\partial w}{\partial z} = 6\dfrac{dw}{du}$

56. $\partial w/\partial x = (dw/d\rho)(\partial \rho/\partial x) = (x/\rho)dw/d\rho$, similarly $\partial w/\partial y = (y/\rho)dw/d\rho$ and
$\partial w/\partial z = (z/\rho)dw/d\rho$ so $(\partial w/\partial x)^2 + (\partial w/\partial y)^2 + (\partial w/\partial z)^2 = (dw/d\rho)^2$

57. $z = f(u, v)$ where $u = x - y$ and $v = y - x$,
$\dfrac{\partial z}{\partial x} = \dfrac{\partial z}{\partial u}\dfrac{\partial u}{\partial x} + \dfrac{\partial z}{\partial v}\dfrac{\partial v}{\partial x} = \dfrac{\partial z}{\partial u} - \dfrac{\partial z}{\partial v}$ and $\dfrac{\partial z}{\partial y} = \dfrac{\partial z}{\partial u}\dfrac{\partial u}{\partial y} + \dfrac{\partial z}{\partial v}\dfrac{\partial v}{\partial y} = -\dfrac{\partial z}{\partial u} + \dfrac{\partial z}{\partial v}$ so $\dfrac{\partial z}{\partial x} + \dfrac{\partial z}{\partial y} = 0$

58. Let $w = f(r, s, t)$ where $r = x - y$, $s = y - z$, $t = z - x$;
$\partial w/\partial x = (\partial w/\partial r)(\partial r/\partial x) + (\partial w/\partial t)(\partial t/\partial x) = \partial w/\partial r - \partial w/\partial t$, similarly
$\partial w/\partial y = -\partial w/\partial r + \partial w/\partial s$ and $\partial w/\partial z = -\partial w/\partial s + \partial w/\partial t$ so $\partial w/\partial x + \partial w/\partial y + \partial w/\partial z = 0$

59. **(a)** $1 = -r\sin\theta\dfrac{\partial \theta}{\partial x} + \cos\theta\dfrac{\partial r}{\partial x}$ and $0 = r\cos\theta\dfrac{\partial \theta}{\partial x} + \sin\theta\dfrac{\partial r}{\partial x}$; solve for $\partial r/\partial x$ and $\partial\theta/\partial x$.

(b) $0 = -r\sin\theta\dfrac{\partial \theta}{\partial y} + \cos\theta\dfrac{\partial r}{\partial y}$ and $1 = r\cos\theta\dfrac{\partial \theta}{\partial y} + \sin\theta\dfrac{\partial r}{\partial y}$; solve for $\partial r/\partial y$ and $\partial\theta/\partial y$.

(c) $\dfrac{\partial z}{\partial x} = \dfrac{\partial z}{\partial r}\dfrac{\partial r}{\partial x} + \dfrac{\partial z}{\partial \theta}\dfrac{\partial \theta}{\partial x} = \dfrac{\partial z}{\partial r}\cos\theta - \dfrac{1}{r}\dfrac{\partial z}{\partial \theta}\sin\theta.$

$\dfrac{\partial z}{\partial y} = \dfrac{\partial z}{\partial r}\dfrac{\partial r}{\partial y} + \dfrac{\partial z}{\partial \theta}\dfrac{\partial \theta}{\partial y} = \dfrac{\partial z}{\partial r}\sin\theta + \dfrac{1}{r}\dfrac{\partial z}{\partial \theta}\cos\theta.$

(d) Square and add the results of Parts (a) and (b).

(e) From Part (c),

$$\begin{aligned}
\frac{\partial^2 z}{\partial x^2} &= \frac{\partial}{\partial r}\left(\frac{\partial z}{\partial r}\cos\theta - \frac{1}{r}\frac{\partial z}{\partial \theta}\sin\theta\right)\frac{\partial r}{\partial x} + \frac{\partial}{\partial \theta}\left(\frac{\partial z}{\partial r}\cos\theta - \frac{1}{r}\frac{\partial z}{\partial \theta}\sin\theta\right)\frac{\partial \theta}{\partial x}\\
&= \left(\frac{\partial^2 z}{\partial r^2}\cos\theta + \frac{1}{r^2}\frac{\partial z}{\partial \theta}\sin\theta - \frac{1}{r}\frac{\partial^2 z}{\partial r\partial\theta}\sin\theta\right)\cos\theta\\
&\quad + \left(\frac{\partial^2 z}{\partial\theta\partial r}\cos\theta - \frac{\partial z}{\partial r}\sin\theta - \frac{1}{r}\frac{\partial^2 z}{\partial \theta^2}\sin\theta - \frac{1}{r}\frac{\partial z}{\partial \theta}\cos\theta\right)\left(-\frac{\sin\theta}{r}\right)\\
&= \frac{\partial^2 z}{\partial r^2}\cos^2\theta + \frac{2}{r^2}\frac{\partial z}{\partial \theta}\sin\theta\cos\theta - \frac{2}{r}\frac{\partial^2 z}{\partial\theta\partial r}\sin\theta\cos\theta + \frac{1}{r^2}\frac{\partial^2 z}{\partial \theta^2}\sin^2\theta + \frac{1}{r}\frac{\partial z}{\partial r}\sin^2\theta.
\end{aligned}$$

Similarly, from Part (c),

$$\frac{\partial^2 z}{\partial y^2} = \frac{\partial^2 z}{\partial r^2}\sin^2\theta - \frac{2}{r^2}\frac{\partial z}{\partial \theta}\sin\theta\cos\theta + \frac{2}{r}\frac{\partial^2 z}{\partial\theta\partial r}\sin\theta\cos\theta + \frac{1}{r^2}\frac{\partial^2 z}{\partial \theta^2}\cos^2\theta + \frac{1}{r}\frac{\partial z}{\partial r}\cos^2\theta.$$

Add to get $\dfrac{\partial^2 z}{\partial x^2} + \dfrac{\partial^2 z}{\partial y^2} = \dfrac{\partial^2 z}{\partial r^2} + \dfrac{1}{r^2}\dfrac{\partial^2 z}{\partial \theta^2} + \dfrac{1}{r}\dfrac{\partial z}{\partial r}.$

60. $z_x = \dfrac{-2y}{x^2+y^2}, z_{xx} = \dfrac{4xy}{(x^2+y^2)^2}, z_y = \dfrac{2x}{x^2+y^2}, z_{yy} = -\dfrac{4xy}{(x^2+y^2)^2}, z_{xx}+z_{yy}=0;$

$z = \tan^{-1}\dfrac{2r^2\cos\theta\sin\theta}{r^2(\cos^2\theta-\sin^2\theta)} = \tan^{-1}\tan 2\theta = 2\theta + k\pi$ for some fixed k; $z_r = 0, z_{\theta\theta} = 0$

61. **(a)** By the chain rule, $\dfrac{\partial u}{\partial r} = \dfrac{\partial u}{\partial x}\cos\theta + \dfrac{\partial u}{\partial y}\sin\theta$ and $\dfrac{\partial v}{\partial \theta} = -\dfrac{\partial v}{\partial x} r\sin\theta + \dfrac{\partial v}{\partial y} r\cos\theta$, use the Cauchy-Riemann conditions $\dfrac{\partial u}{\partial x} = \dfrac{\partial v}{\partial y}$ and $\dfrac{\partial u}{\partial y} = -\dfrac{\partial v}{\partial x}$ in the equation for $\dfrac{\partial u}{\partial r}$ to get $\dfrac{\partial u}{\partial r} = \dfrac{\partial v}{\partial y}\cos\theta - \dfrac{\partial v}{\partial x}\sin\theta$ and compare to $\dfrac{\partial v}{\partial \theta}$ to see that $\dfrac{\partial u}{\partial r} = \dfrac{1}{r}\dfrac{\partial v}{\partial \theta}$. The result $\dfrac{\partial v}{\partial r} = -\dfrac{1}{r}\dfrac{\partial u}{\partial \theta}$ can be obtained by considering $\dfrac{\partial v}{\partial r}$ and $\dfrac{\partial u}{\partial \theta}$.

(b) $u_x = \dfrac{2x}{x^2+y^2}, \ v_y = 2\dfrac{1}{x}\dfrac{1}{1+(y/x)^2} = \dfrac{2x}{x^2+y^2} = u_x;$

$u_y = \dfrac{2y}{x^2+y^2}, \ v_x = -2\dfrac{y}{x^2}\dfrac{1}{1+(y/x)^2} = -\dfrac{2y}{x^2+y^2} = -u_y;$

$u = \ln r^2, v = 2\theta, u_r = 2/r, v_\theta = 2$, so $u_r = \dfrac{1}{r}v_\theta, \ u_\theta = 0, v_r = 0$, so $v_r = -\dfrac{1}{r}u_\theta$

62. **(a)** $u_x = f'(x+ct), \ u_{xx} = f''(x+ct), \ u_t = cf'(x+ct), \ u_{tt} = c^2 f''(x+ct); u_{tt} = c^2 u_{xx}$

(b) Substitute g for f and $-c$ for c in Part (a).

(c) Since the sum of derivatives equals the derivative of the sum, the result follows from Parts (a) and (b).

(d) $\sin t \sin x = \dfrac{1}{2}(-\cos(x+t) + \cos(x-t))$

63. $\partial w/\partial\rho = (\sin\phi\cos\theta)\partial w/\partial x + (\sin\phi\sin\theta)\partial w/\partial y + (\cos\phi)\,\partial w/\partial z$
$\partial w/\partial\phi = (\rho\cos\phi\cos\theta)\partial w/\partial x + (\rho\cos\phi\sin\theta)\partial w/\partial y - (\rho\sin\phi)\partial w/\partial z$
$\partial w/\partial\theta = -(\rho\sin\phi\sin\theta)\partial w/\partial x + (\rho\sin\phi\cos\theta)\partial w/\partial y$

64. **(a)** $\dfrac{\partial w}{\partial x} = \dfrac{\partial f}{\partial x} + \dfrac{\partial f}{\partial z}\dfrac{\partial z}{\partial x}$ **(b)** $\dfrac{\partial w}{\partial y} = \dfrac{\partial f}{\partial y} + \dfrac{\partial f}{\partial z}\dfrac{\partial z}{\partial y}$

65. $w_r = e^r/\left(e^r+e^s+e^t+e^u\right), \ w_{rs} = -e^r e^s/\left(e^r+e^s+e^t+e^u\right)^2,$
$w_{rst} = 2e^r e^s e^t/\left(e^r+e^s+e^t+e^u\right)^3,$
$w_{rstu} = -6e^r e^s e^t e^u/\left(e^r+e^s+e^t+e^u\right)^4 = -6e^{r+s+t+u}/e^{4w} = -6e^{r+s+t+u-4w}$

66. $\partial w/\partial y_1 = a_1\partial w/\partial x_1 + a_2\partial w/\partial x_2 + a_3\partial w/\partial x_3,$
$\partial w/\partial y_2 = b_1\partial w/\partial x_1 + b_2\partial w/\partial x_2 + b_3\partial w/\partial x_3$

67. **(a)** $dw/dt = \displaystyle\sum_{i=1}^{4}(\partial w/\partial x_i)\,(dx_i/dt)$

(b) $\partial w/\partial v_j = \displaystyle\sum_{i=1}^{4}(\partial w/\partial x_i)\,(\partial x_i/\partial v_j)$ for $j = 1, 2, 3$

68. Let $u = x_1^2 + x_2^2 + ... + x_n^2$; then $w = u^k$, $\partial w/\partial x_i = ku^{k-1}(2x_i) = 2k\ x_i u^{k-1}$,
$\partial^2 w/\partial x_i^2 = 2k(k-1)x_i u^{k-2}(2x_i) + 2ku^{k-1} = 4k(k-1)x_i^2 u^{k-2} + 2ku^{k-1}$ for $i = 1, 2, \ldots, n$

so $\displaystyle\sum_{i=1}^{n} \partial^2 w/\partial x_i^2 = 4k(k-1)u^{k-2}\sum_{i=1}^{n} x_i^2 + 2kn\,u^{k-1}$

$$= 4k(k-1)u^{k-2}u + 2kn\,u^{k-1} = 2ku^{k-1}[2(k-1) + n]$$

which is 0 if $k = 0$ or if $2(k-1) + n = 0$, $k = 1 - n/2$.

69. $dF/dx = (\partial F/\partial u)(du/dx) + (\partial F/\partial v)(dv/dx)$
$= f(u)g'(x) - f(v)h'(x) = f(g(x))g'(x) - f(h(x))h'(x)$

70. Represent the line segment C that joins A and B by $x = x_0 + (x_1 - x_0)t$, $y = y_0 + (y_1 - y_0)t$ for $0 \le t \le 1$. Let $F(t) = f(x_0 + (x_1 - x_0)t, y_0 + (y_1 - y_0)t)$ for $0 \le t \le 1$; then $f(x_1, y_1) - f(x_0, y_0) = F(1) - F(0)$. Apply the Mean Value Theorem to $F(t)$ on the interval [0,1] to get $[F(1) - F(0)]/(1-0) = F'(t^*)$, $F(1) - F(0) = F'(t^*)$ for some t^* in (0,1) so $f(x_1, y_1) - f(x_0, y_0) = F'(t^*)$. By the chain rule, $F'(t) = f_x(x,y)(dx/dt) + f_y(x,y)(dy/dt) = f_x(x,y)(x_1 - x_0) + f_y(x,y)(y_1 - y_0)$. Let (x^*, y^*) be the point on C for $t = t^*$ then $f(x_1, y_1) - f(x_0, y_0) = F'(t^*) = f_x(x^*, y^*)(x_1 - x_0) + f_y(x^*, y^*)(y_1 - y_0)$.

71. Let (a, b) be any point in the region, if (x, y) is in the region then by the result of Exercise 70 $f(x,y) - f(a,b) = f_x(x^*, y^*)(x-a) + f_y(x^*, y^*)(y-b)$ where (x^*, y^*) is on the line segment joining (a, b) and (x, y). If $f_x(x,y) = f_y(x,y) = 0$ throughout the region then $f(x,y) - f(a,b) = (0)(x-a) + (0)(y-b) = 0$, $f(x,y) = f(a,b)$ so $f(x,y)$ is constant on the region.

EXERCISE SET 14.6

1. $\nabla f(x,y) = (3y/2)(1+xy)^{1/2}\mathbf{i} + (3x/2)(1+xy)^{1/2}\mathbf{j}$, $\nabla f(3,1) = 3\mathbf{i} + 9\mathbf{j}$,
$D_{\mathbf{u}}f = \nabla f \cdot \mathbf{u} = 12/\sqrt{2} = 6\sqrt{2}$

2. $\nabla f(x,y) = 2ye^{2xy}\mathbf{i} + 2xe^{2xy}\mathbf{j}$, $\nabla f(4,0) = 8\mathbf{j}$, $D_{\mathbf{u}}f = \nabla f \cdot \mathbf{u} = 32/5$

3. $\nabla f(x,y) = \left[2x/\left(1+x^2+y\right)\right]\mathbf{i} + \left[1/\left(1+x^2+y\right)\right]\mathbf{j}$, $\nabla f(0,0) = \mathbf{j}$, $D_{\mathbf{u}}f = -3/\sqrt{10}$

4. $\nabla f(x,y) = -\left[(c+d)y/(x-y)^2\right]\mathbf{i} + \left[(c+d)x/(x-y)^2\right]\mathbf{j}$,
$\nabla f(3,4) = -4(c+d)\mathbf{i} + 3(c+d)\mathbf{j}$, $D_{\mathbf{u}}f = -(7/5)(c+d)$

5. $\nabla f(x,y,z) = 20x^4y^2z^3\mathbf{i} + 8x^5yz^3\mathbf{j} + 12x^5y^2z^2\mathbf{k}$, $\nabla f(2,-1,1) = 320\mathbf{i} - 256\mathbf{j} + 384\mathbf{k}$, $D_{\mathbf{u}}f = -320$

6. $\nabla f(x,y,z) = yze^{xz}\mathbf{i} + e^{xz}\mathbf{j} + (xye^{xz} + 2z)\mathbf{k}$, $\nabla f(0,2,3) = 6\mathbf{i} + \mathbf{j} + 6\mathbf{k}$, $D_{\mathbf{u}}f = 45/7$

7. $\nabla f(x,y,z) = \dfrac{2x}{x^2+2y^2+3z^2}\mathbf{i} + \dfrac{4y}{x^2+2y^2+3z^2}\mathbf{j} + \dfrac{6z}{x^2+2y^2+3z^2}\mathbf{k}$,
$\nabla f(-1,2,4) = (-2/57)\mathbf{i} + (8/57)\mathbf{j} + (24/57)\mathbf{k}$, $D_{\mathbf{u}}f = -314/741$

8. $\nabla f(x,y,z) = yz\cos xyz\mathbf{i} + xz\cos xyz\mathbf{j} + xy\cos xyz\mathbf{k}$,
$\nabla f(1/2, 1/3, \pi) = (\pi\sqrt{3}/6)\mathbf{i} + (\pi\sqrt{3}/4)\mathbf{j} + (\sqrt{3}/12)\mathbf{k}$, $D_{\mathbf{u}}f = (1-\pi)/12$

9. $\nabla f(x,y) = 12x^2y^2\mathbf{i} + 8x^3y\mathbf{j}$, $\nabla f(2,1) = 48\mathbf{i} + 64\mathbf{j}$, $\mathbf{u} = (4/5)\mathbf{i} - (3/5)\mathbf{j}$, $D_{\mathbf{u}}f = \nabla f \cdot \mathbf{u} = 0$

10. $\nabla f(x,y) = (2x-3y)\mathbf{i} + \left(-3x+12y^2\right)\mathbf{j}$, $\nabla f(-2,0) = -4\mathbf{i}+6\mathbf{j}$, $\mathbf{u} = (\mathbf{i}+2\mathbf{j})/\sqrt{5}$, $D_{\mathbf{u}}f = 8/\sqrt{5}$

11. $\nabla f(x,y) = \left(y^2/x\right)\mathbf{i} + 2y\ln x\mathbf{j}$, $\nabla f(1,4) = 16\mathbf{i}$, $\mathbf{u} = (-\mathbf{i}+\mathbf{j})/\sqrt{2}$, $D_{\mathbf{u}}f = -8\sqrt{2}$

12. $\nabla f(x,y) = e^x\cos y\mathbf{i} - e^x\sin y\mathbf{j}$, $\nabla f(0,\pi/4) = (\mathbf{i}-\mathbf{j})/\sqrt{2}$, $\mathbf{u} = (5\mathbf{i}-2\mathbf{j})/\sqrt{29}$, $D_{\mathbf{u}}f = 7/\sqrt{58}$

13. $\nabla f(x,y) = -\left[y/\left(x^2+y^2\right)\right]\mathbf{i} + \left[x/\left(x^2+y^2\right)\right]\mathbf{j}$,
$\nabla f(-2,2) = -(\mathbf{i}+\mathbf{j})/4$, $\mathbf{u} = -(\mathbf{i}+\mathbf{j})/\sqrt{2}$, $D_{\mathbf{u}}f = \sqrt{2}/4$

14. $\nabla f(x,y) = (e^y - ye^x)\mathbf{i} + (xe^y - e^x)\mathbf{j}$, $\nabla f(0,0) = \mathbf{i}-\mathbf{j}$, $\mathbf{u} = (5\mathbf{i}-2\mathbf{j})/\sqrt{29}$, $D_{\mathbf{u}}f = 7/\sqrt{29}$

15. $\nabla f(x,y,z) = \left(3x^2z - 2xy\right)\mathbf{i} - x^2\mathbf{j} + \left(x^3+2z\right)\mathbf{k}$, $\nabla f(2,-1,1) = 16\mathbf{i}-4\mathbf{j}+10\mathbf{k}$,
$\mathbf{u} = (3\mathbf{i}-\mathbf{j}+2\mathbf{k})/\sqrt{14}$, $D_{\mathbf{u}}f = 72/\sqrt{14}$

16. $\nabla f(x,y,z) = -x\left(x^2+z^2\right)^{-1/2}\mathbf{i} + \mathbf{j} - z\left(x^2+z^2\right)^{-1/2}\mathbf{k}$, $\nabla f(-3,1,4) = (3/5)\mathbf{i}+\mathbf{j}-(4/5)\mathbf{k}$,
$\mathbf{u} = (2\mathbf{i}-2\mathbf{j}-\mathbf{k})/3$, $D_{\mathbf{u}}f = 0$

17. $\nabla f(x,y,z) = -\dfrac{1}{z+y}\mathbf{i} - \dfrac{z-x}{(z+y)^2}\mathbf{j} + \dfrac{y+x}{(z+y)^2}\mathbf{k}$, $\nabla f(1,0,-3) = (1/3)\mathbf{i} + (4/9)\mathbf{j} + (1/9)\mathbf{k}$,
$\mathbf{u} = (-6\mathbf{i}+3\mathbf{j}-2\mathbf{k})/7$, $D_{\mathbf{u}}f = -8/63$

18. $\nabla f(x,y,z) = e^{x+y+3z}(\mathbf{i}+\mathbf{j}+3\mathbf{k})$, $\nabla f(-2,2,-1) = e^{-3}(\mathbf{i}+\mathbf{j}+3\mathbf{k})$, $\mathbf{u} = (20\mathbf{i}-4\mathbf{j}+5\mathbf{k})/21$,
$D_{\mathbf{u}}f = (31/21)e^{-3}$

19. $\nabla f(x,y) = (y/2)(xy)^{-1/2}\mathbf{i} + (x/2)(xy)^{-1/2}\mathbf{j}$, $\nabla f(1,4) = \mathbf{i} + (1/4)\mathbf{j}$,
$\mathbf{u} = \cos\theta\mathbf{i} + \sin\theta\mathbf{j} = (1/2)\mathbf{i} + \left(\sqrt{3}/2\right)\mathbf{j}$, $D_{\mathbf{u}}f = 1/2 + \sqrt{3}/8$

20. $\nabla f(x,y) = [2y/(x+y)^2]\mathbf{i} - [2x/(x+y)^2]\mathbf{j}$, $\nabla f(-1,-2) = -(4/9)\mathbf{i} + (2/9)\mathbf{j}$, $\mathbf{u} = \mathbf{j}$, $D_{\mathbf{u}}f = 2/9$

21. $\nabla f(x,y) = 2\sec^2(2x+y)\mathbf{i} + \sec^2(2x+y)\mathbf{j}$, $\nabla f(\pi/6,\pi/3) = 8\mathbf{i}+4\mathbf{j}$, $\mathbf{u} = (\mathbf{i}-\mathbf{j})/\sqrt{2}$, $D_{\mathbf{u}}f = 2\sqrt{2}$

22. $\nabla f(x,y) = \cosh x\cosh y\mathbf{i} + \sinh x\sinh y\mathbf{j}$, $\nabla f(0,0) = \mathbf{i}$, $\mathbf{u} = -\mathbf{i}$, $D_{\mathbf{u}}f = -1$

23. $\nabla f(x,y) = y(x+y)^{-2}\mathbf{i} - x(x+y)^{-2}\mathbf{j}$, $\nabla f(1,0) = -\mathbf{j}$, $\overrightarrow{PQ} = -2\mathbf{i}-\mathbf{j}$, $\mathbf{u} = (-2\mathbf{i}-\mathbf{j})/\sqrt{5}$,
$D_{\mathbf{u}}f = 1/\sqrt{5}$

24. $\nabla f(x,y) = -e^{-x}\sec y\mathbf{i} + e^{-x}\sec y\tan y\mathbf{j}$,
$\nabla f(0,\pi/4) = \sqrt{2}(-\mathbf{i}+\mathbf{j})$, $\overrightarrow{PO} = -(\pi/4)\mathbf{j}$, $\mathbf{u} = -\mathbf{j}$, $D_{\mathbf{u}}f = -\sqrt{2}$

25. $\nabla f(x,y) = \dfrac{ye^y}{2\sqrt{xy}}\mathbf{i} + \left(\sqrt{xy}e^y + \dfrac{xe^y}{2\sqrt{xy}}\right)\mathbf{j}$, $\nabla f(1,1) = (e/2)(\mathbf{i}+3\mathbf{j})$, $\mathbf{u} = -\mathbf{j}$, $D_{\mathbf{u}}f = -3e/2$

26. $\nabla f(x,y) = -y(x+y)^{-2}\mathbf{i} + x(x+y)^{-2}\mathbf{j}$, $\nabla f(2,3) = (-3\mathbf{i}+2\mathbf{j})/25$, if $D_{\mathbf{u}}f = 0$ then $\mathbf{u}$ and ∇f are orthogonal, by inspection $2\mathbf{i}+3\mathbf{j}$ is orthogonal to $\nabla f(2,3)$ so $\mathbf{u} = \pm(2\mathbf{i}+3\mathbf{j})/\sqrt{13}$.

27. $\nabla f(2,1,-1) = -\mathbf{i} + \mathbf{j} - \mathbf{k}$. $\overrightarrow{PQ} = -3\mathbf{i} + \mathbf{j} + \mathbf{k}$, $\mathbf{u} = (-3\mathbf{i} + \mathbf{j} + \mathbf{k})/\sqrt{11}$, $D_{\mathbf{u}}f = 3/\sqrt{11}$

28. $\nabla f(-1,-2,1) = 13\mathbf{i} + 5\mathbf{j} - 20\mathbf{k}$, $\mathbf{u} = -\mathbf{k}$, $D_{\mathbf{u}}f = 20$

29. Solve the system $(3/5)f_x(1,2) - (4/5)f_y(1,2) = -5$, $(4/5)f_x(1,2) + (3/5)f_y(1,2) = 10$ for

(a) $f_x(1,2) = 5$ **(b)** $f_y(1,2) = 10$

(c) $\nabla f(1,2) = 5\mathbf{i} + 10\mathbf{j}$, $\mathbf{u} = (-\mathbf{i} - 2\mathbf{j})/\sqrt{5}$, $D_{\mathbf{u}}f = -5\sqrt{5}$.

30. $\nabla f(-5,1) = -3\mathbf{i} + 2\mathbf{j}$, $\overrightarrow{PQ} = \mathbf{i} + 2\mathbf{j}$, $\mathbf{u} = (\mathbf{i} + 2\mathbf{j})/\sqrt{5}$, $D_{\mathbf{u}}f = 1/\sqrt{5}$

31. f increases the most in the direction of III.

32. The contour lines are closer at P, so the function is increasing more rapidly there, hence ∇f is larger at P.

33. $\nabla z = 4\mathbf{i} - 8\mathbf{j}$

34. $\nabla z = -4e^{-3y}\sin 4x\mathbf{i} - 3e^{-3y}\cos 4x\mathbf{j}$

35. $\nabla w = \dfrac{x}{x^2+y^2+z^2}\mathbf{i} + \dfrac{y}{x^2+y^2+z^2}\mathbf{j} + \dfrac{z}{x^2+y^2+z^2}\mathbf{k}$

36. $\nabla w = e^{-5x}\sec(x^2yz)\left[(2xyz\tan(x^2yz) - 5)\,\mathbf{i} + x^2z\tan(x^2yz)\,\mathbf{j} + x^2y\tan(x^2yz)\,\mathbf{k}\right]$

37. $\nabla f(x,y) = 3(2x+y)\left(x^2+xy\right)^2\mathbf{i} + 3x\left(x^2+xy\right)^2\mathbf{j}$, $\nabla f(-1,-1) = -36\mathbf{i} - 12\mathbf{j}$

38. $\nabla f(x,y) = -x\left(x^2+y^2\right)^{-3/2}\mathbf{i} - y\left(x^2+y^2\right)^{-3/2}\mathbf{j}$, $\nabla f(3,4) = -(3/125)\mathbf{i} - (4/125)\mathbf{j}$

39. $\nabla f(x,y,z) = [y/(x+y+z)]\mathbf{i} + [y/(x+y+z) + \ln(x+y+z)]\mathbf{j} + [y/(x+y+z)]\mathbf{k}$,
$\nabla f(-3,4,0) = 4\mathbf{i} + 4\mathbf{j} + 4\mathbf{k}$

40. $\nabla f(x,y,z) = 3y^2z\tan^2 x\sec^2 x\,\mathbf{i} + 2yz\tan^3 x\,\mathbf{j} + y^2\tan^3 x\,\mathbf{k}$, $\nabla f(\pi/4,-3) = 54\,\mathbf{i} - 6\,\mathbf{j} + 9\,\mathbf{k}$

41. $f(1,2) = 3$,
level curve $4x - 2y + 3 = 3$,
$2x - y = 0$;
$\nabla f(x,y) = 4\mathbf{i} - 2\mathbf{j}$
$\nabla f(1,2) = 4\mathbf{i} - 2\mathbf{j}$

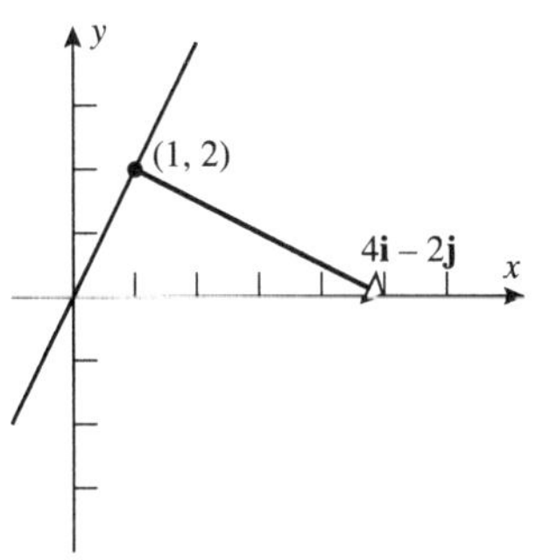

42. $f(-2,2) = 1/2$,
level curve $y/x^2 = 1/2$,
$y = x^2/2$ for $x \neq 0$.
$\nabla f(x,y) = -\left(2y/x^3\right)\mathbf{i} + \left(1/x^2\right)\mathbf{j}$
$\nabla f(-2,2) = (1/2)\mathbf{i} + (1/4)\mathbf{j}$

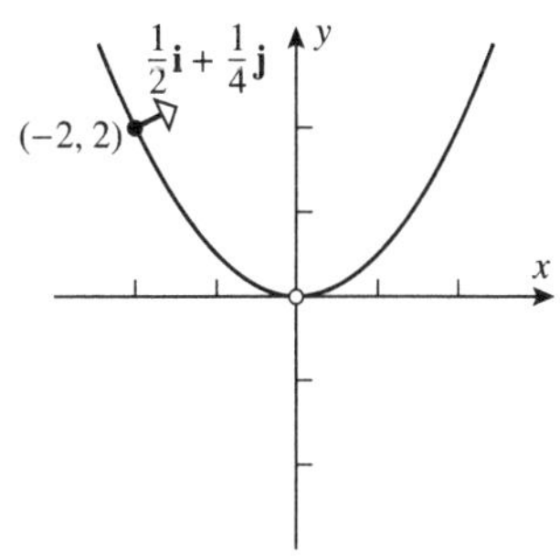

43. $f(-2,0)=4$,
level curve $x^2+4y^2=4$,
$x^2/4+y^2=1$.
$\nabla f(x,y)=2x\mathbf{i}+8y\mathbf{j}$
$\nabla f(-2,0)=-4\mathbf{i}$

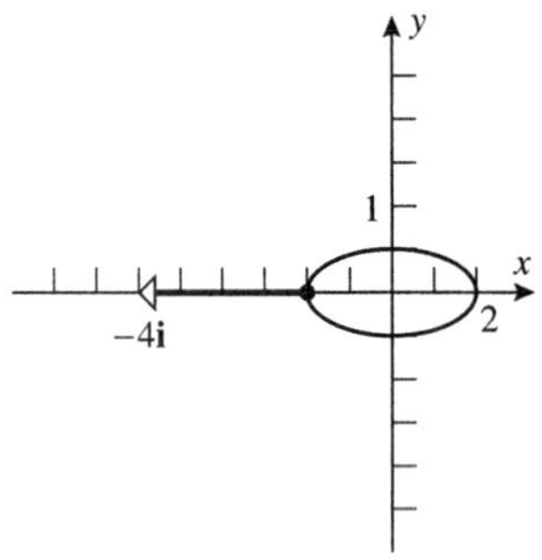

44. $f(2,-1)=3$,
level curve $x^2-y^2=3$.
$\nabla f(x,y)=2x\mathbf{i}-2y\mathbf{j}$
$\nabla f(2,-1)=4\mathbf{i}+2\mathbf{j}$

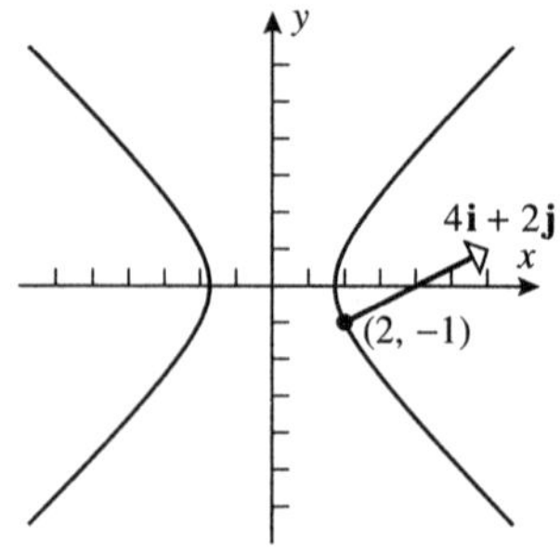

45. $\nabla f(x,y)=8xy\mathbf{i}+4x^2\mathbf{j}$, $\nabla f(1,-2)=-16\mathbf{i}+4\mathbf{j}$ is normal to the level curve through P so $\mathbf{u}=\pm(-4\mathbf{i}+\mathbf{j})/\sqrt{17}$.

46. $\nabla f(x,y)=(6xy-y)\mathbf{i}+\left(3x^2-x\right)\mathbf{j}$, $\nabla f(2,-3)=-33\mathbf{i}+10\mathbf{j}$ is normal to the level curve through P so $\mathbf{u}=\pm(-33\mathbf{i}+10\mathbf{j})/\sqrt{1189}$.

47. $\nabla f(x,y)=12x^2y^2\mathbf{i}+8x^3y\mathbf{j}$, $\nabla f(-1,1)=12\mathbf{i}-8\mathbf{j}$, $\mathbf{u}=(3\mathbf{i}-2\mathbf{j})/\sqrt{13}$, $\|\nabla f(-1,1)\|=4\sqrt{13}$

48. $\nabla f(x,y)=3\mathbf{i}-(1/y)\mathbf{j}$, $\nabla f(2,4)=3\mathbf{i}-(1/4)\mathbf{j}$, $\mathbf{u}=(12\mathbf{i}-\mathbf{j})/\sqrt{145}$, $\|\nabla f(2,4)\|=\sqrt{145}/4$

49. $\nabla f(x,y)=x\left(x^2+y^2\right)^{-1/2}\mathbf{i}+y\left(x^2+y^2\right)^{-1/2}\mathbf{j}$,
$\nabla f(4,-3)=(4\mathbf{i}-3\mathbf{j})/5$, $\mathbf{u}=(4\mathbf{i}-3\mathbf{j})/5$, $\|\nabla f(4,-3)\|=1$

50. $\nabla f(x,y)=y(x+y)^{-2}\mathbf{i}-x(x+y)^{-2}\mathbf{j}$, $\nabla f(0,2)=(1/2)\mathbf{i}$, $\mathbf{u}=\mathbf{i}$, $\|\nabla f(0,2)\|=1/2$

51. $\nabla f(1,1,-1)=3\mathbf{i}-3\mathbf{j}$, $\mathbf{u}=(\mathbf{i}-\mathbf{j})/\sqrt{2}$, $\|\nabla f(1,1,-1)\|=3\sqrt{2}$

52. $\nabla f(0,-3,0)=(\mathbf{i}-3\mathbf{j}+4\mathbf{k})/6$, $\mathbf{u}=(\mathbf{i}-3\mathbf{j}+4\mathbf{k})/\sqrt{26}$, $\|\nabla f(0,-3,0)\|=\sqrt{26}/6$

53. $\nabla f(1,2,-2)=(-\mathbf{i}+\mathbf{j})/2$, $\mathbf{u}=(-\mathbf{i}+\mathbf{j})/\sqrt{2}$, $\|\nabla f(1,2,-2)\|=1/\sqrt{2}$

54. $\nabla f(4,2,2)=(\mathbf{i}-\mathbf{j}-\mathbf{k})/8$, $\mathbf{u}=(\mathbf{i}-\mathbf{j}-\mathbf{k})/\sqrt{3}$, $\|\nabla f(4,2,2)\|=\sqrt{3}/8$

55. $\nabla f(x,y)=-2x\mathbf{i}-2y\mathbf{j}$, $\nabla f(-1,-3)=2\mathbf{i}+6\mathbf{j}$, $\mathbf{u}=-(\mathbf{i}+3\mathbf{j})/\sqrt{10}$, $-\|\nabla f(-1,-3)\|=-2\sqrt{10}$

56. $\nabla f(x,y)=ye^{xy}\mathbf{i}+xe^{xy}\mathbf{j}$; $\nabla f(2,3)=e^6(3\mathbf{i}+2\mathbf{j})$, $\mathbf{u}=-(3\mathbf{i}+2\mathbf{j})/\sqrt{13}$, $-\|\nabla f(2,3)\|=-\sqrt{13}e^6$

57. $\nabla f(x,y)=-3\sin(3x-y)\mathbf{i}+\sin(3x-y)\mathbf{j}$,
$\nabla f(\pi/6,\pi/4)=(-3\mathbf{i}+\mathbf{j})/\sqrt{2}$, $\mathbf{u}=(3\mathbf{i}-\mathbf{j})/\sqrt{10}$, $-\|\nabla f(\pi/6,\pi/4)\|=-\sqrt{5}$

58. $\nabla f(x,y)=\dfrac{y}{(x+y)^2}\sqrt{\dfrac{x+y}{x-y}}\mathbf{i}-\dfrac{x}{(x+y)^2}\sqrt{\dfrac{x+y}{x-y}}\mathbf{j}$, $\nabla f(3,1)=(\sqrt{2}/16)(\mathbf{i}-3\mathbf{j})$,
$\mathbf{u}=-(\mathbf{i}-3\mathbf{j})/\sqrt{10}$, $-\|\nabla f(3,1)\|=-\sqrt{5}/8$

59. $\nabla f(5,7,6)=-\mathbf{i}+11\mathbf{j}-12\mathbf{k}$, $\mathbf{u}=(\mathbf{i}-11\mathbf{j}+12\mathbf{k})/\sqrt{266}$, $-\|\nabla f(5,7,6)\|=-\sqrt{266}$

60. $\nabla f(0,1,\pi/4) = 2\sqrt{2}(\mathbf{i}-\mathbf{k})$, $\mathbf{u} = -(\mathbf{i}-\mathbf{k})/\sqrt{2}$, $-\|\nabla f(0,1,\pi/4)\| = -4$

61. $\nabla f(4,-5) = 2\mathbf{i}-\mathbf{j}$, $\mathbf{u} = (5\mathbf{i}+2\mathbf{j})/\sqrt{29}$, $D_{\mathbf{u}}f = 8/\sqrt{29}$

62. Let $\mathbf{u} = u_1\mathbf{i} + u_2\mathbf{j}$ where $u_1^2 + u_2^2 = 1$, but $D_{\mathbf{u}}f = \nabla f \cdot \mathbf{u} = u_1 - 2u_2 = -2$ so $u_1 = 2u_2 - 2$, $(2u_2-2)^2 + u_2^2 = 1$, $5u_2^2 - 8u_2 + 3 = 0$, $u_2 = 1$ or $u_2 = 3/5$ thus $u_1 = 0$ or $u_1 = -4/5$; $\mathbf{u} = \mathbf{j}$ or $\mathbf{u} = -\frac{4}{5}\mathbf{i} + \frac{3}{5}\mathbf{j}$.

63. (a) At $(1,2)$ the steepest ascent seems to be in the direction $\mathbf{i}+\mathbf{j}$ and the slope in that direction seems to be $0.5/(\sqrt{2}/2) = 1/\sqrt{2}$, so $\nabla f \approx \frac{1}{2}\mathbf{i} + \frac{1}{2}\mathbf{j}$, which has the required direction and magnitude.

(b) The direction of $-\nabla f(4,4)$ appears to be $-\mathbf{i}-\mathbf{j}$ and its magnitude appears to be $1/0.8 = 5/4$.

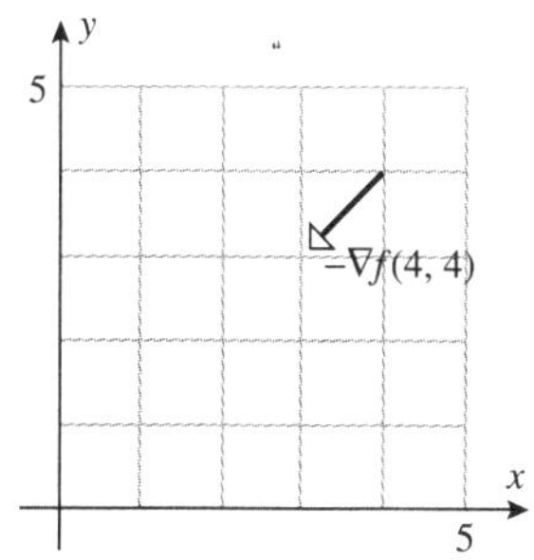

64. (a)

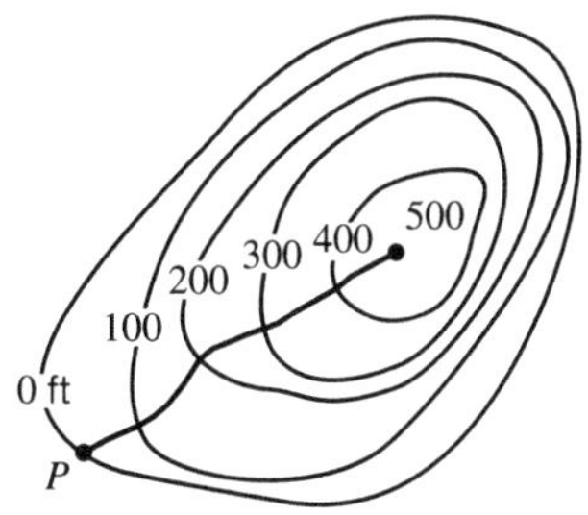

Depart from each contour line in a direction orthogonal to that contour line, as an approximation to the optimal path.

(b)

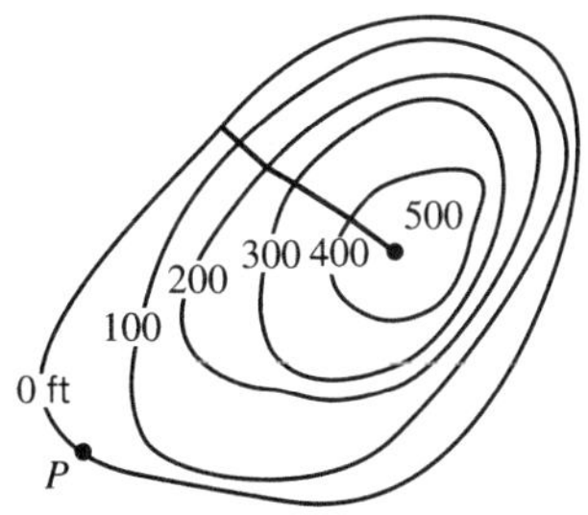

At the top there is no contour line, so head for the nearest contour line. From then on depart from each contour line in a direction orthogonal to that contour line, as in Part (a).

65. $\nabla z = 6x\mathbf{i} - 2y\mathbf{j}$, $\|\nabla z\| = \sqrt{36x^2+4y^2} = 6$ if $36x^2+4y^2 = 36$; all points on the ellipse $9x^2+y^2 = 9$.

66. $\nabla z = 3\mathbf{i} + 2y\mathbf{j}$, $\|\nabla z\| = \sqrt{9+4y^2}$, so $\nabla\|\nabla z\| = \dfrac{4y}{\sqrt{9+4y^2}}\mathbf{j}$, and $\nabla\|\nabla z\|\Big|_{(x,y)=(5,2)} = \dfrac{8}{5}\mathbf{j}$

67. $\mathbf{r} = t\mathbf{i} - t^2\mathbf{j}$, $d\mathbf{r}/dt = \mathbf{i} - 2t\mathbf{j} = \mathbf{i} - 4\mathbf{j}$ at the point $(2,-4)$, $\mathbf{u} = (\mathbf{i}-4\mathbf{j})/\sqrt{17}$;
$\nabla z = 2x\mathbf{i} + 2y\mathbf{j} = 4\mathbf{i} - 8\mathbf{j}$ at $(2,-4)$, hence $dz/ds = D_{\mathbf{u}}z = \nabla z \cdot \mathbf{u} = 36/\sqrt{17}$.

68. **(a)** $\nabla T(x,y) = \dfrac{y\left(1-x^2+y^2\right)}{\left(1+x^2+y^2\right)^2}\mathbf{i} + \dfrac{x\left(1+x^2-y^2\right)}{\left(1+x^2+y^2\right)^2}\mathbf{j}$, $\nabla T(1,1) = (\mathbf{i}+\mathbf{j})/9$, $\mathbf{u} = (2\mathbf{i}-\mathbf{j})/\sqrt{5}$, $D_{\mathbf{u}}T = 1/\left(9\sqrt{5}\right)$

(b) $\mathbf{u} = -(\mathbf{i}+\mathbf{j})/\sqrt{2}$, opposite to $\nabla T(1,1)$

69. **(a)** $\nabla V(x,y) = -2e^{-2x}\cos 2y\mathbf{i} - 2e^{-2x}\sin 2y\mathbf{j}$, $\mathbf{E} = -\nabla V(\pi/4, 0) = 2e^{-\pi/2}\mathbf{i}$

(b) $V(x,y)$ decreases most rapidly in the direction of $-\nabla V(x,y)$ which is $\mathbf{E}$.

70. $\nabla z = -0.04x\mathbf{i} - 0.08y\mathbf{j}$, if $x = -20$ and $y = 5$ then $\nabla z = 0.8\mathbf{i} - 0.4\mathbf{j}$.

(a) $\mathbf{u} = -\mathbf{i}$ points due west, $D_{\mathbf{u}}z = -0.8$, the climber will descend because z is decreasing.

(b) $\mathbf{u} = (\mathbf{i}+\mathbf{j})/\sqrt{2}$ points northeast, $D_{\mathbf{u}}z = 0.2\sqrt{2}$, the climber will ascend at the rate of $0.2\sqrt{2}$ m per m of travel in the $xy-$plane.

(c) The climber will travel a level path in a direction perpendicular to $\nabla z = 0.8\mathbf{i} - 0.4\mathbf{j}$, by inspection $\pm(\mathbf{i}+2\mathbf{j})/\sqrt{5}$ are unit vectors in these directions; $(\mathbf{i}+2\mathbf{j})/\sqrt{5}$ makes an angle of $\tan^{-1}(1/2) \approx 27°$ with the positive y-axis so $-(\mathbf{i}+2\mathbf{j})/\sqrt{5}$ makes the same angle with the negative y-axis. The compass direction should be N 27° E or S 27° W.

71. Let $\mathbf{u}$ be the unit vector in the direction of $\mathbf{a}$, then $D_{\mathbf{u}}f(3,-2,1) = \nabla f(3,-2,1)\cdot\mathbf{u} = \|\nabla f(3,-2,1)\|\cos\theta = 5\cos\theta = -5$, $\cos\theta = -1$, $\theta = \pi$ so $\nabla f(3,-2,1)$ is oppositely directed to $\mathbf{u}$; $\nabla f(3,-2,1) = -5\mathbf{u} = -10/3\mathbf{i} + 5/3\mathbf{j} + 10/3\mathbf{k}$.

72. **(a)** $\nabla T(1,1,1) = (\mathbf{i}+\mathbf{j}+\mathbf{k})/8$, $\mathbf{u} = -(\mathbf{i}+\mathbf{j}+\mathbf{k})/\sqrt{3}$, $D_{\mathbf{u}}T = -\sqrt{3}/8$

(b) $(\mathbf{i}+\mathbf{j}+\mathbf{k})/\sqrt{3}$ **(c)** $\sqrt{3}/8$

73. **(a)** $\nabla r = \dfrac{x}{\sqrt{x^2+y^2}}\mathbf{i} + \dfrac{y}{\sqrt{x^2+y^2}}\mathbf{j} = \mathbf{r}/r$

(b) $\nabla f(r) = \dfrac{\partial f(r)}{\partial x}\mathbf{i} + \dfrac{\partial f(r)}{\partial y}\mathbf{j} = f'(r)\dfrac{\partial r}{\partial x}\mathbf{i} + f'(r)\dfrac{\partial r}{\partial y}\mathbf{j} = f'(r)\nabla r$

74. **(a)** $\nabla\left(re^{-3r}\right) = \dfrac{(1-3r)}{r}e^{-3r}\mathbf{r}$

(b) $3r^2\mathbf{r} = \dfrac{f'(r)}{r}\mathbf{r}$ so $f'(r) = 3r^3$, $f(r) = \dfrac{3}{4}r^4 + C$, $f(2) = 12 + C = 1$, $C = -11$; $f(r) = \dfrac{3}{4}r^4 - 11$

75. $\mathbf{u}_r = \cos\theta\mathbf{i} + \sin\theta\mathbf{j}$, $\mathbf{u}_\theta = -\sin\theta\mathbf{i} + \cos\theta\mathbf{j}$,

$$\nabla z = \frac{\partial z}{\partial x}\mathbf{i} + \frac{\partial z}{\partial y}\mathbf{j} = \left(\frac{\partial z}{\partial r}\cos\theta - \frac{1}{r}\frac{\partial z}{\partial\theta}\sin\theta\right)\mathbf{i} + \left(\frac{\partial z}{\partial r}\sin\theta + \frac{1}{r}\frac{\partial z}{\partial\theta}\cos\theta\right)\mathbf{j}$$

$$= \frac{\partial z}{\partial r}(\cos\theta\mathbf{i} + \sin\theta\mathbf{j}) + \frac{1}{r}\frac{\partial z}{\partial\theta}(-\sin\theta\mathbf{i} + \cos\theta\mathbf{j}) = \frac{\partial z}{\partial r}\mathbf{u}_r + \frac{1}{r}\frac{\partial z}{\partial\theta}\mathbf{u}_\theta$$

76. **(a)** $\nabla(f+g) - (f_x+g_x)\mathbf{i} + (f_y+g_y)\mathbf{j} = (f_x\mathbf{i} + f_y\mathbf{j}) + (g_x\mathbf{i} + g_y\mathbf{j}) = \nabla f + \nabla g$

(b) $\nabla(cf) = (cf_x)\mathbf{i} + (cf_y)\mathbf{j} = c\left(f_x\mathbf{i} + f_y\mathbf{j}\right) = c\nabla f$

(c) $\nabla(fg) = (fg_x + gf_x)\mathbf{i} + (fg_y + gf_y)\mathbf{j} = f\left(g_x\mathbf{i} + g_y\mathbf{j}\right) + g\left(f_x\mathbf{i} + f_y\mathbf{j}\right) = f\nabla g + g\nabla f$

(d) $\nabla(f/g) = \dfrac{gf_x - fg_x}{g^2}\mathbf{i} + \dfrac{gf_y - fg_y}{g^2}\mathbf{j} = \dfrac{g\,(f_x\mathbf{i} + f_y\mathbf{j}) - f\,(g_x\mathbf{i} + g_y\mathbf{j})}{g^2} = \dfrac{g\nabla f - f\nabla g}{g^2}$

(e) $\nabla(f^n) = \left(nf^{n-1}f_x\right)\mathbf{i} + \left(nf^{n-1}f_y\right)\mathbf{j} = nf^{n-1}\left(f_x\mathbf{i} + f_y\mathbf{j}\right) = nf^{n-1}\nabla f$

77. $\mathbf{r}'(t) = \mathbf{v}(t) = k(x,y)\nabla\mathbf{T} = -8k(x,y)x\mathbf{i} - 2k(x,y)y\mathbf{j}$; $\dfrac{dx}{dt} = -8kx$, $\dfrac{dy}{dt} = -2ky$. Divide and solve to get $y^4 = 256x$; one parametrization is $x(t) = e^{-8t}$, $y(t) = 4e^{-2t}$.

78. $\mathbf{r}'(t) = \mathbf{v}(t) = k\nabla\mathbf{T} = -2k(x,y)x\mathbf{i} - 4k(x,y)y\mathbf{j}$. Divide and solve to get $y = \dfrac{3}{25}x^2$; one parametrization is $x(t) = 5e^{-2t}$, $y(t) = 3e^{-4t}$.

79.

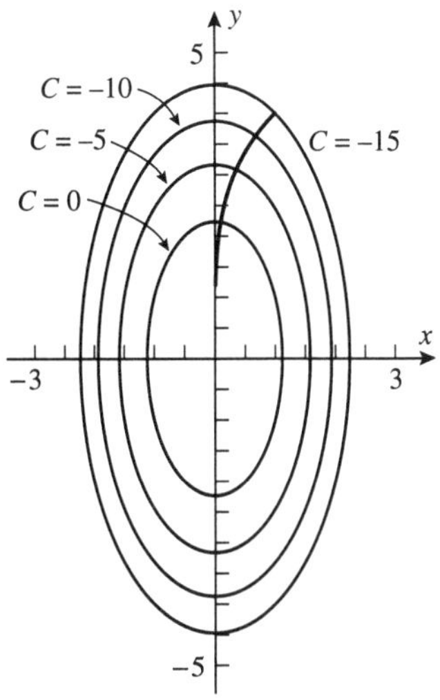

80.

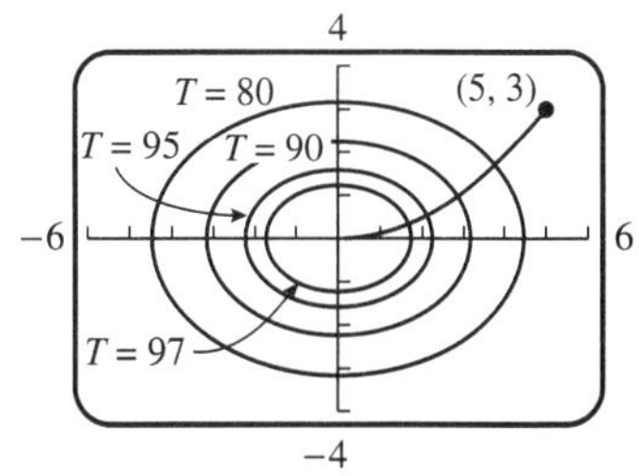

81. (a)

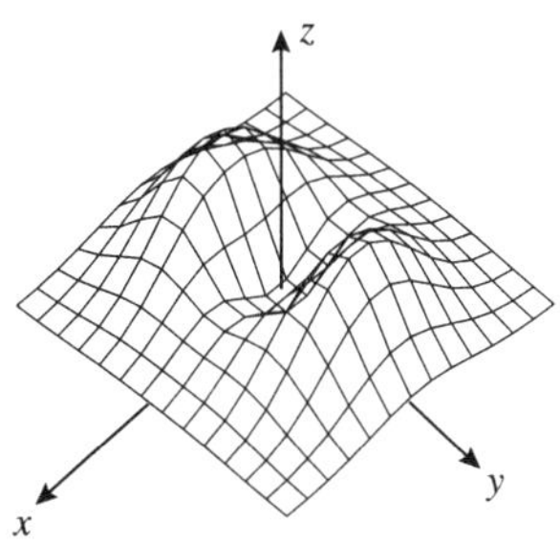

(c) $\nabla f = [2x - 2x(x^2+3y^2)]e^{-(x^2+y^2)}\mathbf{i} + [6y - 2y(x^2+3y^2)]e^{-(x^2+y^2)}\mathbf{j}$

(d) $\nabla f = \mathbf{0}$ if $x = y = 0$ or $x = 0, y = \pm 1$ or $x = \pm 1, y = 0$.

82. $dz/dt = (\partial z/\partial x)(dx/dt) + (\partial z/\partial y)(dy/dt)$
$= (\partial z/\partial x\mathbf{i} + \partial z/\partial y\mathbf{j}) \cdot (dx/dt\mathbf{i} + dy/dt\mathbf{j}) = \nabla z \cdot \mathbf{r}'(t)$

83. $\nabla f(x,y) = f_x(x,y)\mathbf{i} + f_y(x,y)\mathbf{j}$, if $\nabla f(x,y) = 0$ throughout the region then $f_x(x,y) = f_y(x,y) = 0$ throughout the region, the result follows from Exercise 71, Section 14.5.

84. Let $\mathbf{u}_1$ and $\mathbf{u}_2$ be nonparallel unit vectors for which the directional derivative is zero. Let $\mathbf{u}$ be any other unit vector, then $\mathbf{u} = c_1\mathbf{u}_1 + c_2\mathbf{u}_2$ for some choice of scalars c_1 and c_2,

$D_{\mathbf{u}}f(x,y) = \nabla f(x,y) \cdot \mathbf{u} = c_1\nabla f(x,y) \cdot \mathbf{u}_1 + c_2\nabla f(x,y) \cdot \mathbf{u}_2$
$= c_1D_{\mathbf{u}_1}f(x,y) + c_2D_{\mathbf{u}_2}f(x,y) = 0.$

85. $\nabla f(u,v,w) = \frac{\partial f}{\partial x}\mathbf{i} + \frac{\partial f}{\partial y}\mathbf{j} + \frac{\partial f}{\partial z}\mathbf{k}$

$$= \left(\frac{\partial f}{\partial u}\frac{\partial u}{\partial x} + \frac{\partial f}{\partial v}\frac{\partial v}{\partial x} + \frac{\partial f}{\partial w}\frac{\partial w}{\partial x}\right)\mathbf{i} + \left(\frac{\partial f}{\partial u}\frac{\partial u}{\partial y} + \frac{\partial f}{\partial v}\frac{\partial v}{\partial y} + \frac{\partial f}{\partial w}\frac{\partial w}{\partial y}\right)\mathbf{j}$$

$$+ \left(\frac{\partial f}{\partial u}\frac{\partial u}{\partial z} + \frac{\partial f}{\partial v}\frac{\partial v}{\partial z} + \frac{\partial f}{\partial w}\frac{\partial w}{\partial z}\right)\mathbf{k} = \frac{\partial f}{\partial u}\nabla u + \frac{\partial f}{\partial v}\nabla v + \frac{\partial f}{\partial w}\nabla w$$

86. **(a)** The distance between $(x_0 + su_1, y_0 + su_2)$ and (x_0, y_0) is $|s|\sqrt{u_1^2 + u_2^2} = |s|$, so the condition $\lim_{s \to 0} \frac{E(s)}{|s|} = 0$ is exactly the condition of Definition 14.4.1, with the local linear approximation of f given by $L(s) = f(x_0, y_0) + f_x(x_0, y_0)su_1 + f_y(x_0, y_0)su_2$, which in turn says that $g'(0) = f_x(x_0, y_0) + f_y(x_0, y_0)$.

(b) The function $E(s)$ of Part (a) has the same values as the function $E(x, y)$ when $x = x_0 + su_1, y = y_0 + su_2$, and the distance between (x, y) and (x_0, y_0) is $|s|$, so the limit in Part (a) is equivalent to the limit (5) of Definition 14.4.2.

(c) Let $f(x, y)$ be differentiable at (x_0, y_0) and let $\mathbf{u} = u_1\mathbf{i} + u_2\mathbf{j}$ be a unit vector. Then by Parts (a) and (b) the directional derivative $D_{\mathbf{u}}\frac{d}{ds}[f(x_0 + su_1, y_0 + su_2)]_{s=0}$ exists and is given by $f_x(x_0, y_0)u_1 + f_y(x_0, y_0)u_2$.

87. **(a)** $\frac{d}{ds}f(x_0+su_1, y_0+su_2)$ at $s = 0$ is by definition equal to $\lim_{s \to 0} \frac{f(x_0 + su_1, y_0 + su_2) - f(x_0, y_0)}{s}$, and from Exercise 86(a) this value is equal to $f_x(x_0, y_0)u_1 + f_y(x_0, y_0)u_2$.

(b) For any number $\epsilon > 0$ a number $\delta > 0$ exists such that whenever $0 < |s| < \delta$ then $\left|\frac{f(x_0 + su_1, y_0 + su_2) - f(x_0, y_0) - f_x(x_0, y_0)su_1 - f_y(x_0, y_0)su_2}{s}\right| < \epsilon$.

(c) For any number $\epsilon > 0$ there exists a number $\delta > 0$ such that $\frac{|E(x, y)|}{\sqrt{(x - x_0)^2 + (y - y_0)^2}} < \epsilon$ whenever $0 < \sqrt{(x - x_0)^2 + (y - y_0)^2} < \delta$.

(d) For any number $\epsilon > 0$ there exists a number $\delta > 0$ such that $\left|\frac{f(x_0 + su_1, y_0 + su_2) - f(x_0, y_0) - f_x(x_0, y_0)su_1 - f_y(x_0, y_0)su_2|}{s}\right| < \epsilon$ when $0 < |s| < \delta$.

(e) Since f is differentiable at (x_0, y_0), by Part (c) the Equation (5) of Definition 14.2.1 holds. By Part (d), for any $\epsilon > 0$ there exists $\delta > 0$ such that $\left|\frac{f(x_0 + su_1, y_0 + su_2) - f(x_0, y_0) - f_x(x_0, y_0)su_1 - f_y(x_0, y_0)su_2}{s}\right| < \epsilon$ when $0 < |s| < \delta$.

By Part (a) it follows that the limit in Part (a) holds, and thus that $\frac{d}{ds}f(x_0 + su_1, y_0 + su_2)\Big]_{s=0} = f_x(x_0, y_0)u_1 + f_y(x_0, y_0)u_2$, which proves Equation (4) of Theorem 14.6.3.

EXERCISE SET 14.7

1. At P, $\partial z/\partial x = 48$ and $\partial z/\partial y = -14$, tangent plane $48x - 14y - z = 64$, normal line $x = 1 + 48t$, $y = -2 - 14t$, $z = 12 - t$.

2. At P, $\partial z/\partial x = 14$ and $\partial z/\partial y = -2$, tangent plane $14x - 2y - z = 16$, normal line $x = 2 + 14t$, $y = 4 - 2t$, $z = 4 - t$.

3. At P, $\partial z/\partial x = 1$ and $\partial z/\partial y = -1$, tangent plane $x - y - z = 0$, normal line $x = 1 + t$, $y = -t$, $z = 1 - t$.

4. At P, $\partial z/\partial x = -1$ and $\partial z/\partial y = 0$, tangent plane $x + z = -1$, normal line $x = -1 - t$, $y = 0$, $z = -t$.

5. At P, $\partial z/\partial x = 0$ and $\partial z/\partial y = 3$, tangent plane $3y - z = -1$, normal line $x = \pi/6$, $y = 3t$, $z = 1 - t$.

6. At P, $\partial z/\partial x = 1/4$ and $\partial z/\partial y = 1/6$, tangent plane $3x + 2y - 12z = -30$, normal line $x = 4 + t/4$, $y = 9 + t/6$, $z = 5 - t$.

7. By implicit differentiation $\partial z/\partial x = -x/z$, $\partial z/\partial y = -y/z$ so at P, $\partial z/\partial x = 3/4$ and $\partial z/\partial y = 0$, tangent plane $3x - 4z = -25$, normal line $x = -3 + 3t/4$, $y = 0$, $z = 4 - t$.

8. By implicit differentiation $\partial z/\partial x = (xy)/(4z)$, $\partial z/\partial y = x^2/(8z)$ so at P, $\partial z/\partial x = 3/8$ and $\partial z/\partial y = -9/16$, tangent plane $6x - 9y - 16z = 5$, normal line $x = -3 + 3t/8$, $y = 1 - 9t/16$, $z = -2 - t$.

9. The tangent plane is horizontal if the normal $\partial z/\partial x\mathbf{i} + \partial z/\partial y\mathbf{j} - \mathbf{k}$ is parallel to $\mathbf{k}$ which occurs when $\partial z/\partial x = \partial z/\partial y = 0$.

   (a) $\partial z/\partial x = 3x^2y^2$, $\partial z/\partial y = 2x^3y$; $3x^2y^2 = 0$ and $2x^3y = 0$ for all (x, y) on the x-axis or y-axis, and $z = 0$ for these points, the tangent plane is horizontal at all points on the x-axis or y-axis.

   (b) $\partial z/\partial x = 2x - y - 2$, $\partial z/\partial y = -x + 2y + 4$; solve the system $2x - y - 2 = 0$, $-x + 2y + 4 = 0$, to get $x = 0$, $y = -2$. $z = -4$ at $(0, -2)$, the tangent plane is horizontal at $(0, -2, -4)$.

10. $\partial z/\partial x = 6x$, $\partial z/\partial y = -2y$, so $6x_0\mathbf{i} - 2y_0\mathbf{j} - \mathbf{k}$ is normal to the surface at a point (x_0, y_0, z_0) on the surface. $6\mathbf{i} + 4\mathbf{j} - \mathbf{k}$ is normal to the given plane. The tangent plane and the given plane are parallel if their normals are parallel so $6x_0 = 6$, $x_0 = 1$ and $-2y_0 = 4$, $y_0 = -2$. $z = -1$ at $(1, -2)$, the point on the surface is $(1, -2, -1)$.

11. $\partial z/\partial x = -6x$, $\partial z/\partial y = -4y$ so $-6x_0\mathbf{i} - 4y_0\mathbf{j} - \mathbf{k}$ is normal to the surface at a point (x_0, y_0, z_0) on the surface. This normal must be parallel to the given line and hence to the vector $-3\mathbf{i} + 8\mathbf{j} - \mathbf{k}$ which is parallel to the line so $-6x_0 = -3$, $x_0 = 1/2$ and $-4y_0 = 8$, $y_0 = -2$. $z = -3/4$ at $(1/2, -2)$. The point on the surface is $(1/2, -2, -3/4)$.

12. (3,4,5) is a point of intersection because it satisfies both equations. Both surfaces have $(3/5)\mathbf{i} + (4/5)\mathbf{j} - \mathbf{k}$ as a normal so they have a common tangent plane at (3,4,5).

13. (a) $2t + 7 = (-1 + t)^2 + (2 + t)^2$, $t^2 = 1$, $t = \pm 1$ so the points of intersection are $(-2, 1, 5)$ and $(0, 3, 9)$.

   (b) $\partial z/\partial x = 2x$, $\partial z/\partial y = 2y$ so at $(-2, 1, 5)$ the vector $\mathbf{n} = -4\mathbf{i} + 2\mathbf{j} - \mathbf{k}$ is normal to the surface. $\mathbf{v} = \mathbf{i} + \mathbf{j} + 2\mathbf{k}$ is parallel to the line; $\mathbf{n} \cdot \mathbf{v} = -4$ so the cosine of the acute angle is $[\mathbf{n} \cdot (-\mathbf{v})]/(\|\mathbf{n}\|\ \| - \mathbf{v}\|) = 4/\left(\sqrt{21}\sqrt{6}\right) = 4/\left(3\sqrt{14}\right)$. Similarly, at (0,3,9) the vector $\mathbf{n} = 6\mathbf{j} - \mathbf{k}$ is normal to the surface, $\mathbf{n} \cdot \mathbf{v} = 4$ so the cosine of the acute angle is $4/\left(\sqrt{37}\sqrt{6}\right) = 4/\sqrt{222}$.

14. $z = xf(u)$ where $u = x/y$, $\partial z/\partial x = xf'(u)\partial u/\partial x + f(u) = (x/y)f'(u) + f(u) = uf'(u) + f(u)$, $\partial z/\partial y = xf'(u)\partial u/\partial y = -(x^2/y^2)f'(u) = -u^2 f'(u)$. If (x_0, y_0, z_0) is on the surface then, with $u_0 = x_0/y_0$, $[u_0 f'(u_0) + f(u_0)]\,\mathbf{i} - u_0^2 f'(u_0)\,\mathbf{j} - \mathbf{k}$ is normal to the surface so the tangent plane is $[u_0 f'(u_0) + f(u_0)]\,x - u_0^2 f'(u_0)y - z = [u_0 f'(u_0) + f(u_0)]x_0 - u_0^2 f'(u_0)y_0 - z_0$

$$= \left[\frac{x_0}{y_0} f'(u_0) + f(u_0)\right] x_0 - \frac{x_0^2}{y_0^2} f'(u_0)\, y_0 - z_0$$
$$= x_0 f(u_0) - z_0 = 0$$

so all tangent planes pass through the origin.

15. **(a)** $f(x, y, z) = x^2 + y^2 + 4z^2, \nabla f = 2x\mathbf{i} + 2y\mathbf{j} + 8z\mathbf{k}, \nabla f(2, 2, 1) = 4\mathbf{i} + 4\mathbf{j} + 8\mathbf{k}$, $\mathbf{n} = \mathbf{i} + \mathbf{j} + 2\mathbf{k},\ x + y + 2z = 6$

(b) $\mathbf{r}(t) = 2\mathbf{i} + 2\mathbf{j} + \mathbf{k} + t(\mathbf{i} + \mathbf{j} + 2\mathbf{k}), x(t) = 2 + t, y(t) = 2 + t, z(t) = 1 + 2t$

(c) $\cos\theta = \dfrac{\mathbf{n} \cdot \mathbf{k}}{\|\mathbf{n}\|} = \dfrac{\sqrt{2}}{\sqrt{3}}, \theta \approx 35.26^\circ$

16. **(a)** $f(x, y, z) = xz - yz^3 + yz^2$, $\mathbf{n} = \nabla f(2, -1, 1) = \mathbf{i} + 3\mathbf{k}$; tangent plane $x + 3z = 5$

(b) normal line $x = 2 + t,\ y = -1,\ z = 1 + 3t$

(c) $\cos\theta = \dfrac{\mathbf{n} \cdot \mathbf{k}}{\|\mathbf{n}\|} = \dfrac{3}{\sqrt{10}}, \theta \approx 18.43^\circ$

17. Set $f(x, y) = z + x - z^4(y - 1)$, then $f(x, y, z) = 0, \mathbf{n} = \pm\nabla f(3, 5, 1) = \pm(\mathbf{i} - \mathbf{j} - 19\mathbf{k})$, unit vectors $\pm\dfrac{1}{\sqrt{363}}(\mathbf{i} - \mathbf{j} - 19\mathbf{k})$

18. $f(x, y, z) = \sin xz - 4\cos yz$, $\nabla f(\pi, \pi, 1) = -\mathbf{i} - \pi\mathbf{k}$; unit vectors $\pm\dfrac{1}{\sqrt{1 + \pi^2}}(\mathbf{i} + \pi\mathbf{k})$

19. $f(x, y, z) = x^2 + y^2 + z^2$, if (x_0, y_0, z_0) is on the sphere then $\nabla f(x_0, y_0, z_0) = 2(x_0\mathbf{i} + y_0\mathbf{j} + z_0\mathbf{k})$ is normal to the sphere at (x_0, y_0, z_0), the normal line is $x = x_0 + x_0 t,\ y = y_0 + y_0 t,\ z = z_0 + z_0 t$ which passes through the origin when $t = -1$.

20. $f(x, y, z) = 2x^2 + 3y^2 + 4z^2$, if (x_0, y_0, z_0) is on the ellipsoid then $\nabla f(x_0, y_0, z_0) = 2(2x_0\mathbf{i} + 3y_0\mathbf{j} + 4z_0\mathbf{k})$ is normal there and hence so is $\mathbf{n}_1 = 2x_0\mathbf{i} + 3y_0\mathbf{j} + 4z_0\mathbf{k}$; $\mathbf{n}_1$ must be parallel to $\mathbf{n}_2 = \mathbf{i} - 2\mathbf{j} + 3\mathbf{k}$ which is normal to the given plane so $\mathbf{n}_1 = c\mathbf{n}_2$ for some constant c. Equate corresponding components to get $x_0 = c/2$, $y_0 = -2c/3$, and $z_0 = 3c/4$; substitute into the equation of the ellipsoid yields $2\left(c^2/4\right) + 3\left(4c^2/9\right) + 4\left(9c^2/16\right) = 9$, $c^2 = 108/49$, $c = \pm 6\sqrt{3}/7$. The points on the ellipsoid are $\left(3\sqrt{3}/7, -4\sqrt{3}/7, 9\sqrt{3}/14\right)$ and $\left(-3\sqrt{3}/7, 4\sqrt{3}/7, -9\sqrt{3}/14\right)$.

21. $f(x, y, z) = x^2 + y^2 - z^2$, if (x_0, y_0, z_0) is on the surface then $\nabla f(x_0, y_0, z_0) = 2(x_0\mathbf{i} + y_0\mathbf{j} - z_0\mathbf{k})$ is normal there and hence so is $\mathbf{n}_1 = x_0\mathbf{i} + y_0\mathbf{j} - z_0\mathbf{k}$; $\mathbf{n}_1$ must be parallel to $\overrightarrow{PQ} = 3\mathbf{i} + 2\mathbf{j} - 2\mathbf{k}$ so $\mathbf{n}_1 = c\,\overrightarrow{PQ}$ for some constant c. Equate components to get $x_0 = 3c$, $y_0 = 2c$ and $z_0 = 2c$ which when substituted into the equation of the surface yields $9c^2 + 4c^2 - 4c^2 = 1$, $c^2 = 1/9$, $c = \pm 1/3$ so the points are $(1, 2/3, 2/3)$ and $(-1, -2/3, -2/3)$.

22. $f_1(x, y, z) = 2x^2 + 3y^2 + z^2$, $f_2(x, y, z) = x^2 + y^2 + z^2 - 6x - 8y - 8z + 24$, $\mathbf{n}_1 = \nabla f_1(1, 1, 2) = 4\mathbf{i} + 6\mathbf{j} + 4\mathbf{k}$, $\mathbf{n}_2 = \nabla f_2(1, 1, 2) = -4\mathbf{i} - 6\mathbf{j} - 4\mathbf{k}$, $\mathbf{n}_1 = -\mathbf{n}_2$ so $\mathbf{n}_1$ and $\mathbf{n}_2$ are parallel.

23. $\mathbf{n}_1 = 2\mathbf{i} - 2\mathbf{j} - \mathbf{k}, \mathbf{n}_2 = 2\mathbf{i} - 8\mathbf{j} + 4\mathbf{k}, \mathbf{n}_1 \times \mathbf{n}_2 = -16\mathbf{i} - 10\mathbf{j} - 12\mathbf{k}$ is tangent to the line, so $x(t) = 1 + 8t, y(t) = -1 + 5t, z(t) = 2 + 6t$

24. $f(x, y, z) = \sqrt{x^2 + y^2} - z, \mathbf{n}_1 = \nabla f(4, 3, 5) = \frac{4}{5}\mathbf{i} + \frac{3}{5}\mathbf{j} - \mathbf{k}, \mathbf{n}_2 = \mathbf{i} + 2\mathbf{j} + 2\mathbf{k}, \mathbf{n}_1 \times \mathbf{n}_2 = (16\mathbf{i} - 13\mathbf{j} + 5\mathbf{k})/5$ is tangent to the line, $x(t) = 4 + 16t, y(t) = 3 - 13t, z(t) = 5 + 5t$

25. $f(x, y, z) = x^2 + z^2 - 25,\ g(x, y, z) = y^2 + z^2 - 25, \mathbf{n}_1 = \nabla f(3, -3, 4) = 6\mathbf{i} + 8\mathbf{k}$, $\mathbf{n}_2 = \nabla g(3, -3, 4) = -6\mathbf{j} + 8\mathbf{k}, \mathbf{n}_1 \times \mathbf{n}_2 = 48\mathbf{i} - 48\mathbf{j} - 36\mathbf{k}$ is tangent to the line, $x(t) = 3 + 4t, y(t) = -3 - 4t, z(t) = 4 - 3t$

26. (a) $f(x, y, z) = z - 8 + x^2 + y^2, g(x, y, z) = 4x + 2y - z, \mathbf{n}_1 = 4\mathbf{j} + \mathbf{k}, \mathbf{n}_2 = 4\mathbf{i} + 2\mathbf{j} - \mathbf{k}$, $\mathbf{n}_1 \times \mathbf{n}_2 = -6\mathbf{i} + 4\mathbf{j} - 16\mathbf{k}$ is tangent to the line, $x(t) = 3t, y(t) = 2 - 2t, z(t) = 4 + 8t$

27. Use implicit differentiation to get $\partial z/\partial x = -c^2x/\left(a^2z\right),\ \partial z/\partial y = -c^2y/\left(b^2z\right)$. At (x_0, y_0, z_0), $z_0 \neq 0$, a normal to the surface is $-\left[c^2x_0/\left(a^2z_0\right)\right]\mathbf{i} - \left[c^2y_0/\left(b^2z_0\right)\right]\mathbf{j} - \mathbf{k}$ so the tangent plane is $-\frac{c^2x_0}{a^2z_0}x - \frac{c^2y_0}{b^2z_0}y - z = -\frac{c^2x_0^2}{a^2z_0} - \frac{c^2y_0^2}{b^2z_0} - z_0,\ \frac{x_0x}{a^2} + \frac{y_0y}{b^2} + \frac{z_0z}{c^2} = \frac{x_0^2}{a^2} + \frac{y_0^2}{b^2} + \frac{z_0^2}{c^2} = 1$

28. $\partial z/\partial x = 2x/a^2,\ \partial z/\partial y = 2y/b^2$. At (x_0, y_0, z_0) the vector $\left(2x_0/a^2\right)\mathbf{i} + \left(2y_0/b^2\right)\mathbf{j} - \mathbf{k}$ is normal to the surface so the tangent plane is $\left(2x_0/a^2\right)x + \left(2y_0/b^2\right)y - z = 2x_0^2/a^2 + 2y_0^2/b^2 - z_0$, but $z_0 = x_0^2/a^2 + y_0^2/b^2$ so $\left(2x_0/a^2\right)x + \left(2y_0/b^2\right)y - z = 2z_0 - z_0 = z_0,\ 2x_0x/a^2 + 2y_0y/b^2 = z + z_0$

29. $\mathbf{n}_1 = f_x(x_0, y_0)\mathbf{i} + f_y(x_0, y_0)\mathbf{j} - \mathbf{k}$ and $\mathbf{n}_2 = g_x(x_0, y_0)\mathbf{i} + g_y(x_0, y_0)\mathbf{j} - \mathbf{k}$ are normal, respectively, to $z = f(x, y)$ and $z = g(x, y)$ at P; $\mathbf{n}_1$ and $\mathbf{n}_2$ are perpendicular if and only if $\mathbf{n}_1 \cdot \mathbf{n}_2 = 0$,
$f_x(x_0, y_0)\, g_x(x_0, y_0) + f_y(x_0, y_0)\, g_y(x_0, y_0) + 1 = 0$,
$f_x(x_0, y_0)\, g_x(x_0, y_0) + f_y(x_0, y_0)\, g_y(x_0, y_0) = -1$.

30. $\mathbf{n}_1 = f_x\mathbf{i} + f_y\mathbf{j} - \mathbf{k} = \frac{x_0}{\sqrt{x_0^2 + y_0^2}}\mathbf{i} + \frac{y_0}{\sqrt{x_0^2 + y_0^2}}\mathbf{j} - \mathbf{k}$; similarly $\mathbf{n}_2 = -\frac{x_0}{\sqrt{x_0^2 + y_0^2}}\mathbf{i} - \frac{y_0}{\sqrt{x_0^2 + y_0^2}}\mathbf{j} - \mathbf{k}$; since a normal to the sphere is $\mathbf{N} = x_0\mathbf{i} + y_0\mathbf{j} + z_0\mathbf{k}$, and $\mathbf{n}_1 \cdot \mathbf{N} = \sqrt{x_0^2 + y_0^2} - z_0 = 0$, $\mathbf{n}_2 \cdot \mathbf{N} = -\sqrt{x_0^2 + y_0^2} - z_0 = 0$, the result follows.

31. $\nabla f = f_x\mathbf{i} + f_y\mathbf{j} + f_z\mathbf{k}$ and $\nabla g = g_x\mathbf{i} + g_y\mathbf{j} + g_z\mathbf{k}$ evaluated at (x_0, y_0, z_0) are normal, respectively, to the surfaces $f(x, y, z) = 0$ and $g(x, y, z) = 0$ at (x_0, y_0, z_0). The surfaces are orthogonal at (x_0, y_0, z_0) if and only if $\nabla f \cdot \nabla g = 0$ so $f_xg_x + f_yg_y + f_zg_z = 0$.

32. $f(x, y, z) = x^2 + y^2 + z^2 - a^2 = 0, g(x, y, z) = z^2 - x^2 - y^2 = 0$,
$f_xg_x + f_yg_y + f_zg_z = -4x^2 - 4y^2 + 4z^2 = 4g(x, y, z) = 0$

33. $z = \frac{k}{xy}$; at a point $\left(a, b, \frac{k}{ab}\right)$ on the surface, $\left\langle -\frac{k}{a^2b}, -\frac{k}{ab^2}, -1\right\rangle$ and hence $\left\langle bk, ak, a^2b^2\right\rangle$ is normal to the surface so the tangent plane is $bkx + aky + a^2b^2z = 3abk$. The plane cuts the x, y, and z-axes at the points 3a, 3b, and $\frac{3k}{ab}$, respectively, so the volume of the tetrahedron that is formed is $V = \frac{1}{3}\left(\frac{3k}{ab}\right)\left[\frac{1}{2}(3a)(3b)\right] = \frac{9}{2}k$, which does not depend on a and b.

EXERCISE SET 14.8

1. (a) minimum at $(2, -1)$, no maxima (b) maximum at $(0, 0)$, no minima
   (c) no maxima or minima

2. (a) maximum at $(-1, 5)$, no minima (b) no maxima or minima
   (c) no maxima or minima

3. $f(x, y) = (x-3)^2 + (y+2)^2$, minimum at $(3, -2)$, no maxima

4. $f(x, y) = -(x+1)^2 - 2(y-1)^2 + 4$, maximum at $(-1, 1)$, no minima

5. $f_x = 6x + 2y = 0$, $f_y = 2x + 2y = 0$; critical point (0,0); $D = 8 > 0$ and $f_{xx} = 6 > 0$ at (0,0), relative minimum.

6. $f_x = 3x^2 - 3y = 0$, $f_y = -3x - 3y^2 = 0$; critical points (0,0) and $(-1, 1)$; $D = -9 < 0$ at (0,0), saddle point; $D = 27 > 0$ and $f_{xx} = -6 < 0$ at $(-1, 1)$, relative maximum.

7. $f_x = 2x - 2xy = 0$, $f_y = 4y - x^2 = 0$; critical points (0,0) and $(\pm 2, 1)$; $D = 8 > 0$ and $f_{xx} = 2 > 0$ at (0,0), relative minimum; $D = -16 < 0$ at $(\pm 2, 1)$, saddle points.

8. $f_x = 3x^2 - 3 = 0$, $f_y = 3y^2 - 3 = 0$; critical points $(-1, \pm 1)$ and $(1, \pm 1)$; $D = -36 < 0$ at $(-1, 1)$ and $(1, -1)$, saddle points; $D = 36 > 0$ and $f_{xx} = 6 > 0$ at (1,1), relative minimum; $D = 36 > 0$ and $f_{xx} = -36 < 0$ at $(-1, -1)$, relative maximum.

9. $f_x = y + 2 = 0$, $f_y = 2y + x + 3 = 0$; critical point $(1, -2)$; $D = -1 < 0$ at $(1, -2)$, saddle point.

10. $f_x = 2x + y - 2 = 0$, $f_y = x - 2 = 0$; critical point $(2, -2)$; $D = -1 < 0$ at $(2, -2)$, saddle point.

11. $f_x = 2x + y - 3 = 0$, $f_y = x + 2y = 0$; critical point $(2, -1)$; $D = 3 > 0$ and $f_{xx} = 2 > 0$ at $(2, -1)$, relative minimum.

12. $f_x = y - 3x^2 = 0$, $f_y = x - 2y = 0$; critical points (0,0) and $(1/6, 1/12)$; $D = -1 < 0$ at (0,0), saddle point; $D = 1 > 0$ and $f_{xx} = -1 < 0$ at $(1/6, 1/12)$, relative maximum.

13. $f_x = 2x - 2/\left(x^2 y\right) = 0$, $f_y = 2y - 2/\left(xy^2\right) = 0$; critical points $(-1, -1)$ and $(1, 1)$; $D = 32 > 0$ and $f_{xx} = 6 > 0$ at $(-1, -1)$ and $(1, 1)$, relative minima.

14. $f_x = e^y = 0$ is impossible, no critical points.

15. $f_x = 2x = 0$, $f_y = 1 - e^y = 0$; critical point $(0, 0)$; $D = -2 < 0$ at $(0, 0)$, saddle point.

16. $f_x = y - 2/x^2 = 0$, $f_y = x - 4/y^2 = 0$; critical point (1,2); $D = 3 > 0$ and $f_{xx} = 4 > 0$ at $(1, 2)$, relative minimum.

17. $f_x = e^x \sin y = 0$, $f_y = e^x \cos y = 0$, $\sin y = \cos y = 0$ is impossible, no critical points.

18. $f_x = y \cos x = 0$, $f_y = \sin x = 0$; $\sin x = 0$ if $x = n\pi$ for $n = 0, \pm 1, \pm 2, \ldots$ and $\cos x \neq 0$ for these values of x so $y = 0$; critical points $(n\pi, 0)$ for $n = 0, \pm 1, \pm 2, \ldots$; $D = -1 < 0$ at $(n\pi, 0)$, saddle points.

19. $f_x = -2(x+1)e^{-(x^2+y^2+2x)} = 0$, $f_y = -2ye^{-(x^2+y^2+2x)} = 0$; critical point $(-1, 0)$; $D = 4e^2 > 0$ and $f_{xx} = -2e < 0$ at $(-1, 0)$, relative maximum.

20. $f_x = y - a^3/x^2 = 0$, $f_y = x - b^3/y^2 = 0$; critical point $(a^2/b, b^2/a)$; if $ab > 0$ then $D = 3 > 0$ and $f_{xx} = 2b^3/a^3 > 0$ at $(a^2/b, b^2/a)$, relative minimum; if $ab < 0$ then $D = 3 > 0$ and $f_{xx} = 2b^3/a^3 < 0$ at $(a^2/b, b^2/a)$, relative maximum.

21. 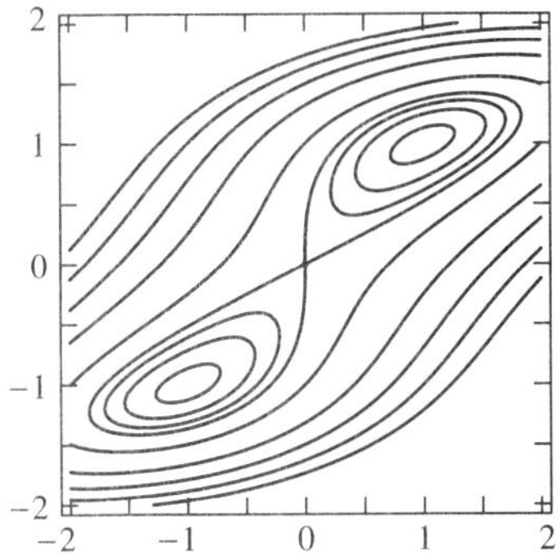

$\nabla f = (4x - 4y)\mathbf{i} - (4x - 4y^3)\mathbf{j} = \mathbf{0}$ when $x = y, x = y^3$, so $x = y = 0$ or $x = y = \pm 1$. At $(0,0), D = -16$, a saddle point; at $(1,1)$ and $(-1,-1), D = 32 > 0, f_{xx} = 4$, a relative minimum.

22. 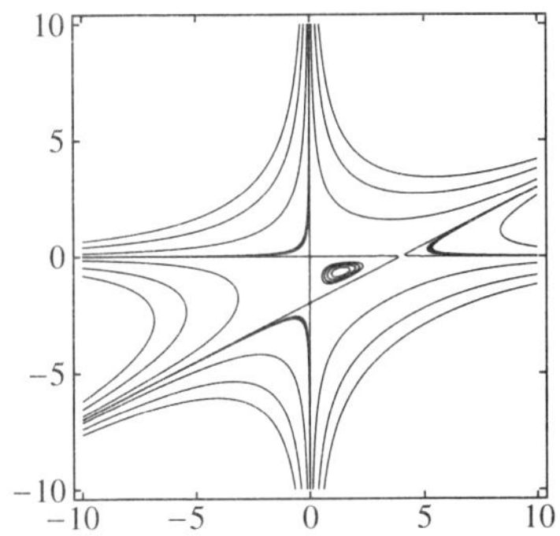

$\nabla f = (2y^2 - 2xy + 4y)\mathbf{i} + (4xy - x^2 + 4x)\mathbf{j} = \mathbf{0}$ when $2y^2 - 2xy + 4y = 0, 4xy - x^2 + 4x = 0$, with solutions $(0,0), (0,-2), (4,0), (4/3,-2/3)$. At $(0,0)$, $D = -16$, a saddle point. At $(0,-2)$, $D = -16$, a saddle point. At $(4,0), D = -16$, a saddle point. At $(4/3,-2/3), D = 16/3$, $f_{xx} = 4/3 > 0$, a relative minimum.

23. (a) critical point (0,0); $D = 0$

(b) $f(0,0) = 0$, $x^4 + y^4 \geq 0$ so $f(x,y) \geq f(0,0)$, relative minimum.

24. (a) critical point (0,0); $D = 0$

(b) The trace of the surface on the plane $x = 0$ has equation $z = -y^4$, which has a maximum at $(0,0,0)$; the trace of the surface on the plane $y = 0$ has equation $z = x^4$, which has a minimum at $(0,0,0)$.

25. (a) $f_x = 3e^y - 3x^2 = 3\left(e^y - x^2\right) = 0$, $f_y = 3xe^y - 3e^{3y} = 3e^y\left(x - e^{2y}\right) = 0$, $e^y = x^2$ and $e^{2y} = x$, $x^4 = x$, $x\left(x^3 - 1\right) = 0$ so $x = 0, 1$; critical point $(1,0)$; $D = 27 > 0$ and $f_{xx} = -6 < 0$ at $(1,0)$, relative maximum.

(b) $\lim_{x \to -\infty} f(x,0) = \lim_{x \to -\infty} \left(3x - x^3 - 1\right) = +\infty$ so no absolute maximum.

26. $f_x = 8xe^y - 8x^3 = 8x(e^y - x^2) = 0$, $f_y = 4x^2e^y - 4e^{4y} = 4e^y(x^2 - e^{3y}) = 0$, $x^2 = e^y$ and $x^2 = e^{3y}$, $e^{3y} = e^y$, $e^{2y} = 1$, so $y = 0$ and $x = \pm 1$; critical points (1,0) and $(-1,0)$. $D = 128 > 0$ and $f_{xx} = -16 < 0$ at both points so a relative maximum occurs at each one.

27. $f_x = y - 1 = 0$, $f_y = x - 3 = 0$; critical point (3,1).

$$\begin{aligned} \text{Along } y = 0: &\quad u(x) = -x; \text{ no critical points,} \\ \text{along } x = 0: &\quad v(y) = -3y; \text{ no critical points,} \\ \text{along } y = -\frac{4}{5}x + 4: &\quad w(x) = -\frac{4}{5}x^2 + \frac{27}{5}x - 12; \text{ critical point } (27/8, 13/10). \end{aligned}$$

(x, y)	$(3, 1)$	$(0, 0)$	$(5, 0)$	$(0, 4)$	$(27/8, 13/10)$
$f(x, y)$	-3	0	-5	-12	$-231/80$

Absolute maximum value is 0, absolute minimum value is -12.

28. $f_x = y - 2 = 0$, $f_y = x = 0$; critical point (0,2), but (0,2) is not in the interior of R.

$$\begin{aligned} \text{Along } y = 0: &\quad u(x) = -2x; \text{ no critical points,} \\ \text{along } x = 0: &\quad v(y) = 0; \text{ no critical points,} \\ \text{along } y = 4 - x: &\quad w(x) = 2x - x^2; \text{ critical point } (1, 3). \end{aligned}$$

(x, y)	$(0, 0)$	$(0, 4)$	$(4, 0)$	$(1, 3)$
$f(x, y)$	0	0	-8	1

Absolute maximum value is 1, absolute minimum value is -8.

29. $f_x = 2x - 2 = 0$, $f_y = -6y + 6 = 0$; critical point (1,1).

$$\begin{aligned} \text{Along } y = 0: &\quad u_1(x) = x^2 - 2x; \text{ critical point } (1, 0), \\ \text{along } y = 2: &\quad u_2(x) = x^2 - 2x; \text{ critical point } (1, 2) \\ \text{along } x = 0: &\quad v_1(y) = -3y^2 + 6y; \text{ critical point } (0, 1), \\ \text{along } x = 2: &\quad v_2(y) = -3y^2 + 6y; \text{ critical point } (2, 1) \end{aligned}$$

(x, y)	$(1, 1)$	$(1, 0)$	$(1, 2)$	$(0, 1)$	$(2, 1)$	$(0, 0)$	$(0, 2)$	$(2, 0)$	$(2, 2)$
$f(x, y)$	2	-1	-1	3	3	0	0	0	0

Absolute maximum value is 3, absolute minimum value is -1.

30. $f_x = e^y - 2x = 0$, $f_y = xe^y - e^y = e^y(x - 1) = 0$; critical point $(1, \ln 2)$.

$$\begin{aligned} \text{Along } y = 0: &\quad u_1(x) = x - x^2 - 1; \text{ critical point } (1/2, 0), \\ \text{along } y = 1: &\quad u_2(x) = ex - x^2 - e; \text{ critical point } (e/2, 1), \\ \text{along } x = 0: &\quad v_1(y) = -e^y; \text{ no critical points,} \\ \text{along } x = 2: &\quad v_2(y) = e^y - 4; \text{ no critical points.} \end{aligned}$$

(x, y)	$(0, 0)$	$(0, 1)$	$(2, 1)$	$(2, 0)$	$(1, \ln 2)$	$(1/2, 0)$	$(e/2, 1)$
$f(x, y)$	-1	$-e$	$e - 4$	-3	-1	$-3/4$	$e(e - 4)/4 \approx -0.87$

Absolute maximum value is $-3/4$, absolute minimum value is -3.

31. $f_x = 2x - 1 = 0$, $f_y = 4y = 0$; critical point $(1/2, 0)$.
Along $x^2 + y^2 = 4 : y^2 = 4 - x^2$, $u(x) = 8 - x - x^2$ for $-2 \le x \le 2$; critical points $(-1/2, \pm\sqrt{15}/2)$.

(x, y)	$(1/2, 0)$	$(-1/2, \sqrt{15}/2)$	$(-1/2, -\sqrt{15}/2)$	$(-2, 0)$	$(2, 0)$
$f(x, y)$	$-1/4$	$33/4$	$33/4$	6	2

Absolute maximum value is $33/4$, absolute minimum value is $-1/4$.

32. $f_x = y^2 = 0$, $f_y = 2xy = 0$; no critical points in the interior of R.

Along $y = 0$: $u(x) = 0$; no critical points,
along $x = 0$: $v(y) = 0$; no critical points
along $x^2 + y^2 = 1$: $w(x) = x - x^3$ for $0 \le x \le 1$; critical point $\left(1/\sqrt{3}, \sqrt{2/3}\right)$.

(x, y)	$(0,0)$	$(0,1)$	$(1,0)$	$\left(1/\sqrt{3}, \sqrt{2/3}\right)$
$f(x, y)$	0	0	0	$2\sqrt{3}/9$

Absolute maximum value is $\frac{2}{9}\sqrt{3}$, absolute minimum value is 0.

33. Maximize $P = xyz$ subject to $x + y + z = 48$, $x > 0$, $y > 0$, $z > 0$. $z = 48 - x - y$ so $P = xy(48 - x - y) = 48xy - x^2y - xy^2$, $P_x = 48y - 2xy - y^2 = 0$, $P_y = 48x - x^2 - 2xy = 0$. But $x \ne 0$ and $y \ne 0$ so $48 - 2x - y = 0$ and $48 - x - 2y = 0$; critical point (16,16). $P_{xx}P_{yy} - P_{xy}^2 > 0$ and $P_{xx} < 0$ at (16,16), relative maximum. $z = 16$ when $x = y = 16$, the product is maximum for the numbers 16,16,16.

34. Minimize $S = x^2 + y^2 + z^2$ subject to $x + y + z = 27$, $x > 0$, $y > 0$, $z > 0$. $z = 27 - x - y$ so $S = x^2 + y^2 + (27 - x - y)^2$, $S_x = 4x + 2y - 54 = 0$, $S_y = 2x + 4y - 54 = 0$; critical point (9,9); $S_{xx}S_{yy} - S_{xy}^2 = 12 > 0$ and $S_{xx} = 4 > 0$ at (9,9), relative minimum. $z = 9$ when $x = y = 9$, the sum of the squares is minimum for the numbers 9,9,9.

35. Maximize $w = xy^2z^2$ subject to $x + y + z = 5$, $x > 0$, $y > 0$, $z > 0$. $x = 5 - y - z$ so $w = (5 - y - z)y^2z^2 = 5y^2z^2 - y^3z^2 - y^2z^3$, $w_y = 10yz^2 - 3y^2z^2 - 2yz^3 = yz^2(10 - 3y - 2z) = 0$, $w_z = 10y^2z - 2y^3z - 3y^2z^2 = y^2z(10 - 2y - 3z) = 0$, $10 - 3y - 2z = 0$ and $10 - 2y - 3z = 0$; critical point when $y = z = 2$; $w_{yy}w_{zz} - w_{yz}^2 = 320 > 0$ and $w_{yy} = -24 < 0$ when $y = z = 2$, relative maximum. $x = 1$ when $y = z = 2$, xy^2z^2 is maximum at (1,2,2).

36. Minimize $w = D^2 = x^2 + y^2 + z^2$ subject to $x^2 - yz = 5$. $x^2 = 5 + yz$ so $w = 5 + yz + y^2 + z^2$, $w_y = z + 2y = 0$, $w_z = y + 2z = 0$; critical point when $y = z = 0$; $w_{yy}w_{zz} - w_{yz}^2 = 3 > 0$ and $w_{yy} = 2 > 0$ when $y = z = 0$, relative minimum. $x^2 = 5$, $x = \pm\sqrt{5}$ when $y = z = 0$. The points $\left(\pm\sqrt{5}, 0, 0\right)$ are closest to the origin.

37. The diagonal of the box must equal the diameter of the sphere, thus we maximize $V = xyz$ or, for convenience, $w = V^2 = x^2y^2z^2$ subject to $x^2 + y^2 + z^2 = 4a^2$, $x > 0$, $y > 0$, $z > 0$; $z^2 = 4a^2 - x^2 - y^2$ hence $w = 4a^2x^2y^2 - x^4y^2 - x^2y^4$, $w_x = 2xy^2(4a^2 - 2x^2 - y^2) = 0$, $w_y = 2x^2y\left(4a^2 - x^2 - 2y^2\right) = 0$, $4a^2 - 2x^2 - y^2 = 0$ and $4a^2 - x^2 - 2y^2 = 0$; critical point $\left(2a/\sqrt{3}, 2a/\sqrt{3}\right)$;
$w_{xx}w_{yy} - w_{xy}^2 = \frac{4096}{27}a^8 > 0$ and $w_{xx} = -\frac{128}{9}a^4 < 0$ at $\left(2a/\sqrt{3}, 2a/\sqrt{3}\right)$, relative maximum.
$z = 2a/\sqrt{3}$ when $x = y = 2a/\sqrt{3}$, the dimensions of the box of maximum volume are $2a/\sqrt{3}, 2a/\sqrt{3}, 2a/\sqrt{3}$.

38. Maximize $V = xyz$ subject to $x + y + z = 1$, $x > 0$, $y > 0$, $z > 0$. $z = 1 - x - y$ so $V = xy - x^2y - xy^2$, $V_x = y(1 - 2x - y) = 0$, $V_y = x(1 - x - 2y) = 0$, $1 - 2x - y = 0$ and $1 - x - 2y = 0$; critical point $(1/3, 1/3)$; $V_{xx}V_{yy} - V_{xy}^2 = 1/3 > 0$ and $V_{xx} = -2/3 < 0$ at $(1/3, 1/3)$, relative maximum. The maximum volume is $V = (1/3)(1/3)(1/3) = 1/27$.

39. Let x, y, and z be, respectively, the length, width, and height of the box. Minimize $C = 10(2xy) + 5(2xz + 2yz) = 10(2xy + xz + yz)$ subject to $xyz = 16$. $z = 16/(xy)$ so $C = 20(xy + 8/y + 8/x)$, $C_x = 20(y - 8/x^2) = 0$, $C_y = 20(x - 8/y^2) = 0$;
critical point (2,2); $C_{xx}C_{yy} - C_{xy}^2 = 1200 > 0$
and $C_{xx} = 40 > 0$ at (2,2), relative minimum. $z = 4$ when $x = y = 2$. The cost of materials is minimum if the length and width are 2 ft and the height is 4 ft.

40. Maximize the profit $P = 500(y-x)(x-40) + [45{,}000 + 500(x-2y)](y-60)$
$= 500(-x^2 - 2y^2 + 2xy - 20x + 170y - 5400)$.
$P_x = 1000(-x+y-10) = 0$, $P_y = 1000(-2y+x+85) = 0$; critical point (65,75);
$P_{xx}P_{yy} - P_{xy}^2 = 1{,}000{,}000 > 0$ and $P_{xx} = -1000 < 0$ at (65,75), relative maximum. The profit will be maximum when $x = 65$ and $y = 75$.

41. **(a)** $x = 0 : f(0,y) = -3y^2$, minimum -3, maximum 0;
$x = 1$, $f(1,y) = 4 - 3y^2 + 2y$, $\dfrac{\partial f}{\partial y}(1,y) = -6y + 2 = 0$ at $y = 1/3$, minimum 3, maximum 13/3;
$y = 0$, $f(x,0) = 4x^2$, minimum 0, maximum 4;
$y = 1$, $f(x,1) = 4x^2 + 2x - 3$, $\dfrac{\partial f}{\partial x}(x,1) = 8x + 2 \neq 0$ for $0 < x < 1$, minimum -3, maximum 3

(b) $f(x,x) = 3x^2$, minimum 0, maximum 3; $f(x,1-x) = -x^2+8x-3$, $\dfrac{d}{dx}f(x,1-x) = -2x+8 \neq 0$ for $0 < x < 1$, maximum 4, minimum -3

(c) $f_x(x,y) = 8x + 2y = 0$, $f_y(x,y) = -6y + 2x = 0$, solution is $(0,0)$, which is not an interior point of the square, so check the sides: minimum -3, maximum 13/3.

42. Maximize $A = ab\sin\alpha$ subject to $2a + 2b = \ell$, $a > 0$, $b > 0$, $0 < \alpha < \pi$. $b = (\ell - 2a)/2$ so $A = (1/2)(\ell a - 2a^2)\sin\alpha$, $A_a = (1/2)(\ell - 4a)\sin\alpha$, $A_\alpha = (a/2)(\ell - 2a)\cos\alpha$; $\sin\alpha \neq 0$ so from $A_a = 0$ we get $a = \ell/4$ and then from $A_\alpha = 0$ we get $\cos\alpha = 0$, $\alpha = \pi/2$. $A_{aa}A_{\alpha\alpha} - A_{a\alpha}^2 = \ell^2/8 > 0$ and $A_{aa} = -2 < 0$ when $a = \ell/4$ and $\alpha = \pi/2$, the area is maximum.

43. Minimize $S = xy + 2xz + 2yz$ subject to $xyz = V$, $x > 0$, $y > 0$, $z > 0$ where x, y, and z are, respectively, the length, width, and height of the box. $z = V/(xy)$ so $S = xy + 2V/y + 2V/x$, $S_x = y - 2V/x^2 = 0$, $S_y = x - 2V/y^2 = 0$; critical point $(\sqrt[3]{2V}, \sqrt[3]{2V})$; $S_{xx}S_{yy} - S_{xy}^2 = 3 > 0$ and $S_{xx} = 2 > 0$ at this point so there is a relative minimum there. The length and width are each $\sqrt[3]{2V}$, the height is $z = \sqrt[3]{2V}/2$.

44. The altitude of the trapezoid is $x\sin\phi$ and the lengths of the lower and upper bases are, respectively, $27 - 2x$ and $27 - 2x + 2x\cos\phi$ so we want to maximize
$A = (1/2)(x\sin\phi)[(27-2x) + (27 - 2x + 2x\cos\phi)] = 27x\sin\phi - 2x^2\sin\phi + x^2\sin\phi\cos\phi$.
$A_x = \sin\phi(27 - 4x + 2x\cos\phi)$,
$A_\phi = x(27\cos\phi - 2x\cos\phi - x\sin^2\phi + x\cos^2\phi) = x(27\cos\phi - 2x\cos\phi + 2x\cos^2\phi - x)$.
$\sin\phi \neq 0$ so from $A_x = 0$ we get $\cos\phi = (4x - 27)/(2x)$, $x \neq 0$ so from $A_\phi = 0$ we get $(27 - 2x + 2x\cos\phi)\cos\phi - x = 0$ which, for $\cos\phi = (4x-27)/(2x)$, yields $4x - 27 - x = 0$, $x = 9$. If $x = 9$ then $\cos\phi = 1/2$, $\phi = \pi/3$. The critical point occurs when $x = 9$ and $\phi = \pi/3$; $A_{xx}A_{\phi\phi} - A_{x\phi}^2 = 729/2 > 0$ and $A_{xx} = -3\sqrt{3}/2 < 0$ there, the area is maximum when $x = 9$ and $\phi = \pi/3$.

45. **(a)** $\dfrac{\partial g}{\partial m} = \displaystyle\sum_{i=1}^{n} 2(mx_i + b - y_i)x_i = 2\left(m\sum_{i=1}^{n} x_i^2 + b\sum_{i=1}^{n} x_i - \sum_{i=1}^{n} x_i y_i\right) = 0$ if

$$\left(\sum_{i=1}^{n} x_i^2\right)m + \left(\sum_{i=1}^{n} x_i\right)b = \sum_{i=1}^{n} x_i y_i,$$

$$\frac{\partial g}{\partial b} = \sum_{i=1}^{n} 2(mx_i + b - y_i) = 2\left(m\sum_{i=1}^{n} x_i + bn - \sum_{i=1}^{n} y_i\right) = 0 \text{ if } \left(\sum_{i=1}^{n} x_i\right)m + nb = \sum_{i=1}^{n} y_i$$

(b) $\displaystyle\sum_{i=1}^{n}(x_i-\bar{x})^2=\sum_{i=1}^{n}(x_i^2-2\bar{x}x_i+\bar{x}^2)=\sum_{i=1}^{n}x_i^2-2\bar{x}\sum_{i=1}^{n}x_i+n\bar{x}^2$

$$=\sum_{i=1}^{n}x_i^2-\frac{2}{n}\left(\sum_{i=1}^{n}x_i\right)^2+\frac{1}{n}\left(\sum_{i=1}^{n}x_i\right)^2$$

$$=\sum_{i=1}^{n}x_i^2-\frac{1}{n}\left(\sum_{i=1}^{n}x_i\right)^2\ge 0 \text{ so } n\sum_{i=1}^{n}x_i^2-\left(\sum_{i=1}^{n}x_i\right)^2\ge 0$$

This is an equality if and only if $\displaystyle\sum_{i=1}^{n}(x_i-\bar{x})^2=0$, which means $x_i=\bar{x}$ for each i.

(c) The system of equations $Am+Bb=C, Dm+Eb=F$ in the unknowns m and b has a unique solution provided $AE\ne BD$, and if so the solution is $m=\dfrac{CE-BF}{AE-BD}, b=\dfrac{F-Dm}{E}$, which after the appropriate substitution yields the desired result.

46. (a) $\displaystyle g_{mm}=2\sum_{i=1}^{n}x_i^2,\ g_{bb}=2n,\ g_{mb}=2\sum_{i=1}^{n}x_i,$

$$D=g_{mm}g_{bb}-g_{mb}^2=4\left[n\sum_{i=1}^{n}x_i^2-\left(\sum_{i=1}^{n}x_i\right)^2\right]>0 \text{ and } g_{mm}>0$$

(b) $g(m,b)$ is of the second-degree in m and b so the graph of $z=g(m,b)$ is a quadric surface.

(c) The function $z=g(m,b)$, as a function of m and b, has only one critical point, found in Exercise 47, and tends to $+\infty$ as either $|m|$ or $|b|$ tends to infinity, since g_{mm} and g_{bb} are both positive. Thus the only critical point must be a minimum.

47. $\displaystyle n=3,\sum_{i=1}^{3}x_i=3,\sum_{i=1}^{3}y_i=7,\sum_{i=1}^{3}x_iy_i=13,\sum_{i=1}^{3}x_i^2=11, y=\frac{3}{4}x+\frac{19}{12}$

48. $\displaystyle n=4,\sum_{i=1}^{4}x_i=7,\sum_{i=1}^{4}y_i=4,\sum_{i=1}^{4}x_i^2=21,\sum_{i=1}^{4}x_iy_i=-2, y=-\frac{36}{35}x+\frac{14}{5}$

49. $\displaystyle\sum_{i=1}^{4}x_i=10,\sum_{i=1}^{4}y_i=8.2,\sum_{i=1}^{4}x_i^2=30,\sum_{i=1}^{4}x_iy_i=23, n=4;\ m=0.5,\ b=0.8,\ y=0.5x+0.8.$

50. $\displaystyle\sum_{i=1}^{5}x_i=15,\sum_{i=1}^{5}y_i=15.1,\sum_{i=1}^{5}x_i^2=55,\sum_{i=1}^{5}x_iy_i=39.8, n=5; m=-0.55, b=4.67, y=4.67-0.55x$

51. (a) $y=\dfrac{8843}{140}+\dfrac{57}{200}t\approx 63.1643+0.285t$

(b)

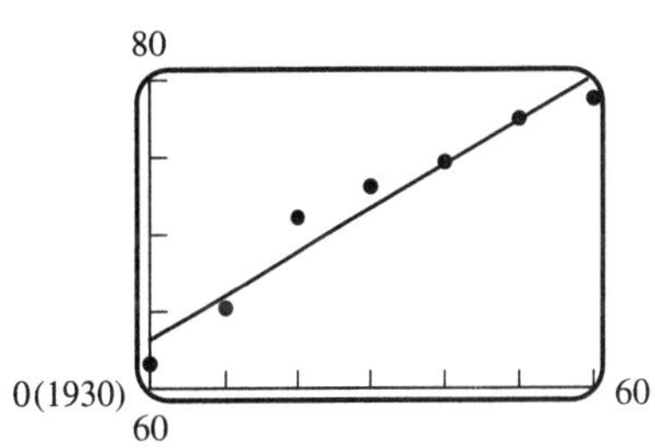

(c) $y=\dfrac{2909}{35}\approx 83.1143$

52. (a) $y \approx 119.84 - 1.13x$

(b) 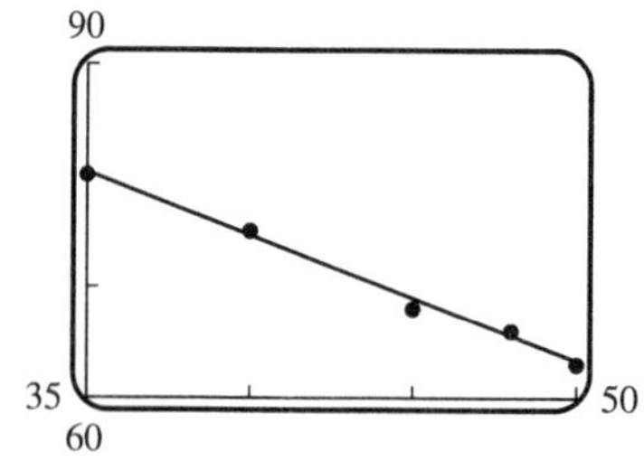

(c) about 52 units

53. (a) $P = \dfrac{2798}{21} + \dfrac{171}{350}T \approx 133.2381 + 0.4886T$

(b) 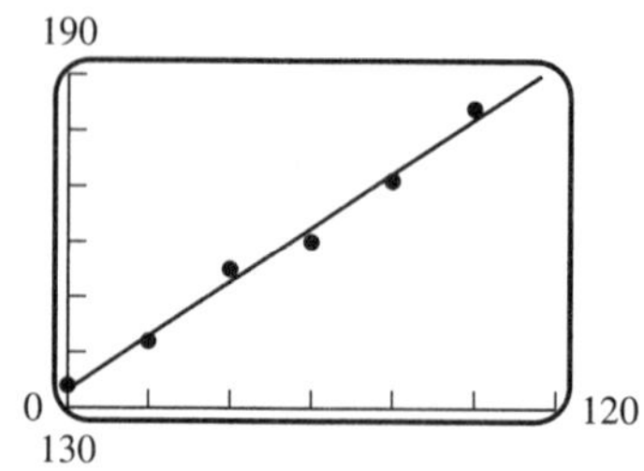

(c) $T \approx -\dfrac{139{,}900}{513} \approx -272.7096°$ C

54. (a) for example, $z = y$

(b) For example, on $0 \le x \le 1, 0 \le y \le 1$ let $z = \begin{cases} y & \text{if } 0 < x < 1, 0 < y < 1 \\ 1/2 & \text{if } x = 0, 1 \text{ or } y = 0, 1 \end{cases}$

55. $f(x_0, y_0) \ge f(x, y)$ for all (x, y) inside a circle centered at (x_0, y_0) by virtue of Definition 14.8.1. If r is the radius of the circle, then in particular $f(x_0, y_0) \ge f(x, y_0)$ for all x satisfying $|x - x_0| < r$ so $f(x, y_0)$ has a relative maximum at x_0. The proof is similar for the function $f(x_0, y)$.

EXERCISE SET 14.9

1. (a) $xy = 4$ is tangent to the line, so the maximum value of f is 4.
(b) $xy = 2$ intersects the curve and so gives a smaller value of f.
(c) Maximize $f(x, y) = xy$ subject to the constraint $g(x, y) = x + y - 4 = 0, \nabla f = \lambda \nabla g$, $y\mathbf{i} + x\mathbf{j} = \lambda(\mathbf{i} + \mathbf{j})$, so solve the equations $y = \lambda, x = \lambda$ with solution $x = y = \lambda$, but $x + y = 4$, so $x = y = 2$, and the maximum value of f is $f = xy = 4$.

2. (a) $x^2 + y^2 = 25$ is tangent to the line at $(3, 4)$, so the minimum value of f is 25.
(b) A larger value of f yields a circle of a larger radius, and hence intersects the line.
(c) Minimize $f(x, y) = x^2 + y^2$ subject to the constraint $g(x, y) = 3x + 4y - 25 = 0, \nabla f = \lambda \nabla g$, $2x\mathbf{i} + 2y\mathbf{j} = 3\lambda\mathbf{i} + 4\lambda\mathbf{j}$, so solve $2x = 3\lambda, 2y = 4\lambda$ and $3x + 4y - 25 = 0$; solution is $x = 3, y = 4$, minimum $= 25$.

3. (a) 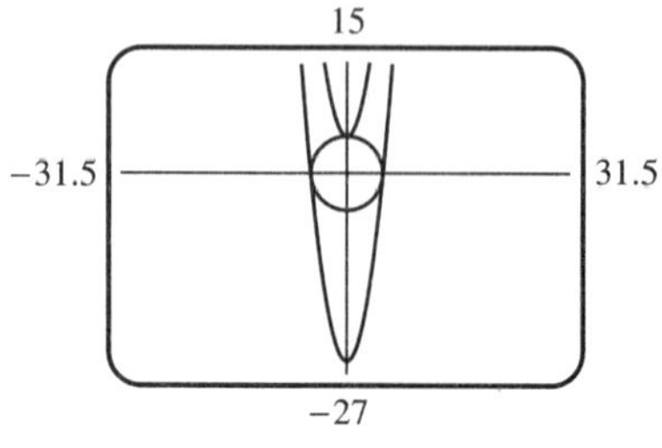

(b) one extremum at $(0, 5)$ and one at approximately $(\pm 5, 0)$, so minimum value -5, maximum value ≈ 25

(c) Find the minimum and maximum values of $f(x,y) = x^2 - y$ subject to the constraint $g(x,y) = x^2 + y^2 - 25 = 0, \nabla f = \lambda \nabla g, 2x\mathbf{i} - \mathbf{j} = 2\lambda x\mathbf{i} + 2\lambda y\mathbf{j}$, so solve $2x = 2\lambda x, -1 = 2\lambda y, x^2 + y^2 - 25 = 0$. If $x = 0$ then $y = \pm 5, f = \mp 5$, and if $x \neq 0$ then $\lambda = 1, y = -1/2, x^2 - 25 - 1/4 = 99/4, f = 99/4 + 1/2 - 101/4$, so the maximum value of f is $101/4$ at $(\pm 3\sqrt{11}/2, -1/2)$ and the minimum value of f is -5 at $(0,5)$.

4. **(a)**

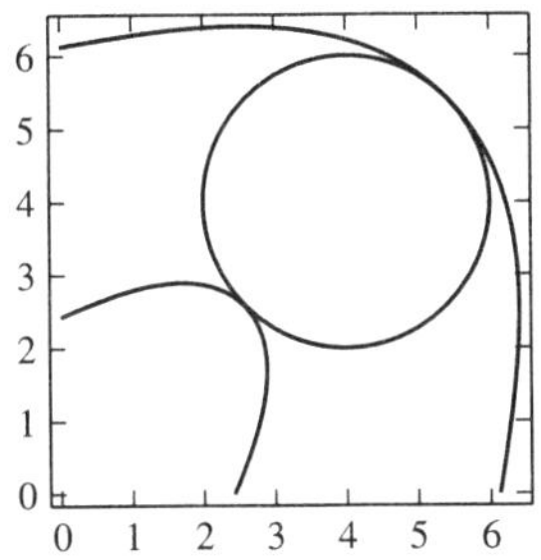

(b) $f \approx 15$

(d) Set $f(x,y) = x^3 + y^3 - 3xy$, $g(x,y) = (x-4)^2 + (y-4)^2 - 4$; minimize f subject to the constraint $g = 0: \nabla f = \lambda g, (3x^2 - 3y)\mathbf{i} + (3y^2 - 3x)\mathbf{j} = 2\lambda(x-4)\mathbf{i} + 2\lambda(y-4)\mathbf{j}$, so solve (use a CAS) $3x^2 - 3y = 2\lambda(x-4), 3y^2 - 3x = 2\lambda(y-4)$ and $(x-4)^2 + (y-4)^2 - 4 = 0$; minimum value $f = 14.52$ at $(2.5858, 2.5858)$

5. $y = 8x\lambda$, $x = 16y\lambda$; $y/(8x) = x/(16y)$, $x^2 = 2y^2$ so $4\left(2y^2\right) + 8y^2 = 16$, $y^2 = 1$, $y = \pm 1$. Test $\left(\pm\sqrt{2}, -1\right)$ and $\left(\pm\sqrt{2}, 1\right)$. $f\left(-\sqrt{2}, -1\right) = f\left(\sqrt{2}, 1\right) = \sqrt{2}$, $f\left(-\sqrt{2}, 1\right) = f\left(\sqrt{2}, -1\right) = -\sqrt{2}$. Maximum $\sqrt{2}$ at $\left(-\sqrt{2}, -1\right)$ and $\left(\sqrt{2}, 1\right)$, minimum $-\sqrt{2}$ at $\left(-\sqrt{2}, 1\right)$ and $\left(\sqrt{2}, -1\right)$.

6. $2x = 2x\lambda, -2y = 2y\lambda, x^2 + y^2 = 25$. If $x \neq 0$ then $\lambda = 1$ and $y = 0$ so $x^2 + 0^2 = 25$, $x = \pm 5$. If $x = 0$ then $0^2 + y^2 = 25$, $y = \pm 5$. Test $(\pm 5, 0)$ and $(0, \pm 5)$: $f(\pm 5, 0) = 25$, $f(0, \pm 5) = -25$, maximum 25 at $(\pm 5, 0)$, minimum -25 at $(0, \pm 5)$.

7. $12x^2 = 4x\lambda$, $2y = 2y\lambda$. If $y \neq 0$ then $\lambda = 1$ and $12x^2 = 4x$, $12x(x - 1/3) = 0$, $x = 0$ or $x = 1/3$ so from $2x^2 + y^2 = 1$ we find that $y = \pm 1$ when $x = 0$, $y = \pm\sqrt{7}/3$ when $x = 1/3$. If $y = 0$ then $2x^2 + (0)^2 = 1$, $x = \pm 1/\sqrt{2}$. Test $(0, \pm 1)$, $\left(1/3, \pm\sqrt{7}/3\right)$, and $\left(\pm 1/\sqrt{2}, 0\right)$. $f(0, \pm 1) = 1$, $f\left(1/3, \pm\sqrt{7}/3\right) = 25/27$, $f\left(1/\sqrt{2}, 0\right) = \sqrt{2}$, $f\left(-1/\sqrt{2}, 0\right) = -\sqrt{2}$. Maximum $\sqrt{2}$ at $\left(1/\sqrt{2}, 0\right)$, minimum $-\sqrt{2}$ at $\left(-1/\sqrt{2}, 0\right)$.

8. $1 = 2x\lambda$, $-3 = 6y\lambda$; $1/(2x) = -1/(2y)$, $y = -x$ so $x^2 + 3(-x)^2 = 16$, $x = \pm 2$. Test $(-2, 2)$ and $(2, -2)$. $f(-2, 2) = -9$, $f(2, -2) = 7$. Maximum 7 at $(2, -2)$, minimum -9 at $(-2, 2)$.

9. $2 = 2x\lambda, 1 = 2y\lambda, -2 = 2z\lambda$; $1/x = 1/(2y) = -1/z$ thus $x = 2y$, $z = -2y$ so $(2y)^2 + y^2 + (-2y)^2 = 4$, $y^2 = 4/9$, $y = \pm 2/3$. Test $(-4/3, -2/3, 4/3)$ and $(4/3, 2/3, -4/3)$. $f(-4/3, -2/3, 4/3) = -6$, $f(4/3, 2/3, -4/3) = 6$. Maximum 6 at $(4/3, 2/3, -4/3)$, minimum -6 at $(-4/3, -2/3, 4/3)$.

10. $3 = 4x\lambda, 6 = 8y\lambda, 2 = 2z\lambda$; $3/(4x) = 3/(4y) = 1/z$ thus $y = x$, $z = 4x/3$, so $2x^2 + 4x^2 + (4x/3)^2 = 70$, $x^2 = 9$, $x = \pm 3$. Test $(-3, -3, -4)$ and (3,3,4). $f(-3, -3, -4) = -35$, $f(3, 3, 4) = 35$. Maximum 35 at (3, 3, 4), minimum -35 at $(-3, -3, -4)$.

11. $yz = 2x\lambda$, $xz = 2y\lambda$, $xy = 2z\lambda$; $yz/(2x) = xz/(2y) = xy/(2z)$ thus $y^2 = x^2$, $z^2 = x^2$ so $x^2 + x^2 + x^2 = 1$, $x = \pm 1/\sqrt{3}$. Test the eight possibilities with $x = \pm 1/\sqrt{3}$, $y = \pm 1/\sqrt{3}$, and $z = \pm 1/\sqrt{3}$ to find the maximum is $1/\left(3\sqrt{3}\right)$ at $\left(1/\sqrt{3}, 1/\sqrt{3}, 1/\sqrt{3}\right)$, $\left(1/\sqrt{3}, -1/\sqrt{3}, -1/\sqrt{3}\right)$, $\left(-1/\sqrt{3}, 1/\sqrt{3}, -1/\sqrt{3}\right)$, and $\left(-1/\sqrt{3}, -1/\sqrt{3}, 1/\sqrt{3}\right)$; the minimum is $-1/\left(3\sqrt{3}\right)$ at $\left(1/\sqrt{3}, 1/\sqrt{3}, -1/\sqrt{3}\right)$, $\left(1/\sqrt{3}, -1/\sqrt{3}, 1/\sqrt{3}\right)$, $\left(-1/\sqrt{3}, 1/\sqrt{3}, 1/\sqrt{3}\right)$, and $\left(-1/\sqrt{3}, -1/\sqrt{3}, -1/\sqrt{3}\right)$.

12. $4x^3 = 2\lambda x, 4y^3 = 2\lambda y, 4z^3 = 2\lambda z$; if x (or y or z) $\neq 0$ then $\lambda = 2x^2$ (or $2y^2$ or $2z^2$).
Assume for the moment that $|x| \leq |y| \leq |z|$. Then:
Case I: $x, y, z \neq 0$ so $\lambda = 2x^2 = 2y^2 = 2z^2, x = \pm y = \pm z$, $3x^2 = 1, x = \pm 1/\sqrt{3}$,
$f(x, y, z) = 3/9 = 1/3$
Case II: $x = 0, y, z \neq 0$; then $y = \pm z, 2y^2 = 1, y = \pm z = \pm 1/\sqrt{2}, f(x, y, z) = 2/4 = 1/2$
Case III: $x = y = 0, z \neq 0$; then $z^2 = 1, z = \pm 1, f(x, y, z) = 1$
Thus f has a maximum value of 1 at $(0, 0, \pm 1), (0, \pm 1, 0)$, and $(\pm 1, 0, 0)$ and a minimum value of $1/3$ at $(\pm 1/\sqrt{3}, \pm 1/\sqrt{3}, \pm 1/\sqrt{3})$.

13. $f(x, y) = x^2 + y^2$; $2x = 2\lambda$, $2y = -4\lambda$; $y = -2x$ so $2x - 4(-2x) = 3$, $x = 3/10$. The point is $(3/10, -3/5)$.

14. $f(x, y) = (x-4)^2 + (y-2)^2$, $g(x, y) = y - 2x - 3$; $2(x-4) = -2\lambda$, $2(y-2) = \lambda$; $x - 4 = -2(y-2)$, $x = -2y + 8$ so $y = 2(-2y+8) + 3$, $y = 19/5$. The point is $(2/5, 19/5)$.

15. $f(x, y, z) = x^2 + y^2 + z^2$; $2x = \lambda$, $2y = 2\lambda$, $2z = \lambda$; $y = 2x$, $z = x$ so $x + 2(2x) + x = 1$, $x = 1/6$. The point is $(1/6, 1/3, 1/6)$.

16. $f(x, y, z) = (x-1)^2 + (y+1)^2 + (z-1)^2$; $2(x-1) = 4\lambda$, $2(y+1) = 3\lambda$, $2(z-1) = \lambda$; $x = 4z - 3$, $y = 3z - 4$ so $4(4z-3) + 3(3z-4) + z = 2$, $z = 1$. The point is $(1, -1, 1)$.

17. $f(x, y) = (x-1)^2 + (y-2)^2$; $2(x-1) = 2x\lambda$, $2(y-2) = 2y\lambda$; $(x-1)/x = (y-2)/y$, $y = 2x$ so $x^2 + (2x)^2 = 45$, $x = \pm 3$. $f(-3, -6) = 80$ and $f(3, 6) = 20$ so (3,6) is closest and $(-3, -6)$ is farthest.

18. $f(x, y, z) = x^2 + y^2 + z^2$; $2x = y\lambda$, $2y = x\lambda$, $2z = -2z\lambda$. If $z \neq 0$ then $\lambda = -1$ so $2x = -y$ and $2y = -x$, $x = y = 0$; substitute into $xy - z^2 = 1$ to get $z^2 = -1$ which has no real solution. If $z = 0$ then $xy - (0)^2 = 1$, $y = 1/x$, and also (from $2x = y\lambda$ and $2y = x\lambda$), $2x/y = 2y/x$, $y^2 = x^2$ so $(1/x)^2 = x^2$, $x^4 = 1$, $x = \pm 1$. Test (1,1,0) and $(-1, -1, 0)$ to see that they are both closest to the origin.

19. $f(x, y, z) = x + y + z$, $x^2 + y^2 + z^2 = 25$ where x, y, and z are the components of the vector; $1 = 2x\lambda$, $1 = 2y\lambda$, $1 = 2z\lambda$; $1/(2x) = 1/(2y) = 1/(2z)$; $y = x$, $z = x$ so $x^2 + x^2 + x^2 = 25$, $x = \pm 5/\sqrt{3}$. $f\left(-5/\sqrt{3}, -5/\sqrt{3}, -5/\sqrt{3}\right) = -5\sqrt{3}$ and $f\left(5/\sqrt{3}, 5/\sqrt{3}, 5/\sqrt{3}\right) = 5\sqrt{3}$ so the vector is $5(\mathbf{i} + \mathbf{j} + \mathbf{k})/\sqrt{3}$.

20. $x^2 + y^2 = 25$ is the constraint; solve $8x - 4y = 2x\lambda$, $-4x + 2y = 2y\lambda$. If $x = 0$ then $y = 0$ and conversely; but $x^2 + y^2 = 25$, so x and y are nonzero. Thus $\lambda = (4x - 2y)/x = (-2x + y)/y$, so $0 = 2x^2 + 3xy - 2y^2 = (2x - y)(x + 2y)$, hence $y = 2x$ or $x = -2y$. If $y = 2x$ then $x^2 + (2x)^2 = 25$, $x = \pm\sqrt{5}$. If $x = -2y$ then $(-2y^2) + y^2 = 25$, $y = \pm\sqrt{5}$. $T\left(-\sqrt{5}, -2\sqrt{5}\right) = T\left(\sqrt{5}, 2\sqrt{5}\right) = 0$ and $T\left(2\sqrt{5}, -\sqrt{5}\right) = T\left(-2\sqrt{5}, \sqrt{5}\right) = 125$. The highest temperature is 125 and the lowest is 0.

21. Minimize $f = x^2 + y^2 + z^2$ subject to $g(x, y, z) = x + y + z - 27 = 0$. $\nabla f = \lambda \nabla g$, $2x\mathbf{i} + 2y\mathbf{j} + 2z\mathbf{k} = \lambda\mathbf{i} + \lambda\mathbf{j} + \lambda\mathbf{k}$, solution $x = y = z = 9$, minimum value 243

22. Maximize $f(x, y, z) = xy^2z^2$ subject to $g(x, y, z) = x + y + z - 5 = 0, \nabla f = \lambda \nabla g = \lambda(\mathbf{i} + \mathbf{j} + \mathbf{k})$, $\lambda = y^2z^2 = 2xyz^2 = 2xy^2z$, $\lambda = 0$ is impossible, hence $x, y, z \neq 0$, and $z = y = 2x, 5x - 5 = 0$, $x = 1, y = z = 2$, maximum value 16 at $(1, 2, 2)$

23. Minimize $f = x^2 + y^2 + z^2$ subject to $x^2 - yz = 5, \nabla f = \lambda \nabla g, 2x = 2x\lambda, 2y = -z\lambda, 2z = -y\lambda$. If $\lambda \neq \pm 2$, then $y = z = 0, x = \pm\sqrt{5}, f = 5$; if $\lambda = \pm 2$ then $x = 0$, and since $-yz = 5$, $y = -z = \pm\sqrt{5}, f = 10$, thus the minimum value is 5 at $(\pm\sqrt{5}, 0, 0)$.

24. The diagonal of the box must equal the diameter of the sphere so maximize $V = xyz$ or, for convenience, maximize $f = V^2 = x^2y^2z^2$ subject to $g(x,y,z) = x^2 + y^2 + z^2 - 4a^2 = 0, \nabla f = \lambda\nabla g$, $2xy^2z^2 = 2\lambda x, 2x^2yz^2 = 2\lambda y, 2x^2y^2z = 2\lambda z$. Since $V \neq 0$ it follows that $x, y, z \neq 0$, hence $x = \pm y = \pm z, 3x^2 = 4a^2, x = \pm 2a/\sqrt{3}$, maximum volume $8a^3/(3\sqrt{3})$.

25. Let x, y, and z be, respectively, the length, width, and height of the box. Minimize $f(x,y,z) = 10(2xy) + 5(2xz + 2yz) = 10(2xy + xz + yz)$ subject to $g(x,y,z) = xyz - 16 = 0$, $\nabla f = \lambda\nabla g, 20y + 10z = \lambda yz, 20x + 10z = \lambda xz, 10x + 10y = \lambda xy$. Since $V = xyz = 16, x, y, z \neq 0$, thus $\lambda z = 20 + 10(z/y) = 20 + 10(z/x)$, so $x = y$. From this and $10x + 10y = \lambda xy$ it follows that $20 = \lambda x$, so $10z = 20x, z = 2x = 2y, V = 2x^3 = 16$ and thus $x = y = 2$ ft, $z = 4$ ft, $f(2,2,4) = 240$ cents.

26. **(a)** If $g(x,y) = x = 0$ then $8x + 2y = \lambda, -6y + 2x = 0$; but $x = 0$, so $y = \lambda = 0$,

$f(0,0) = 0$ maximum, $f(0,1) = -3$, minimum.

If $g(x,y) = x - 1 = 0$ then $8x + 2y = \lambda, -6y + 2x = 0$; but $x = 1$, so $y = 1/3$,

$f(1,1/3) = 13/3$ maximum, $f(1,0) = 4$, $f(1,1) = 3$ minimum.

If $g(x,y) = y = 0$ then $8x + 2y = 0, -6y + 2x = \lambda$; but $y = 0$ so $x = \lambda = 0$,

$f(0,0) = 0$ minimum, $f(1,0) = 4$, maximum.

If $g(x,y) = y - 1 = 0$ then $8x + 2y = 0, -6y + 2x = \lambda$; but $y = 1$ so $x = -1/4$, no solution,

$f(0,1) = -3$ minimum, $f(1,1) = 3$ maximum.

(b) If $g(x,y) = x - y = 0$ then $8x + 2y = \lambda, -6y + 2x = -\lambda$; but $x = y$ so solution $x = y = \lambda = 0, f(0,0) = 0$ minimum, $f(1,1) = 3$ maximum. If $g(x,y) = 1 - x - y = 0$ then $8x + 2y = -1, -6y + 2x = -1$; but $x + y = 1$ so solution is $x = -2/13, y = 3/2$ which is not on diagonal, $f(0,1) = -3$ minimum, $f(1,0) = 4$ maximum.

27. Maximize $A(a,b,\alpha) = ab\sin\alpha$ subject to $g(a,b,\alpha) = 2a + 2b - \ell = 0, \nabla_{(a,b,\alpha)}f = \lambda\nabla_{(a,b,\alpha)}g$, $b\sin\alpha = 2\lambda, a\sin\alpha = 2\lambda, ab\cos\alpha = 0$ with solution $a = b\ (= \ell/4), \alpha = \pi/2$ maximum value if parallelogram is a square.

28. Minimize $f(x,y,z) = xy + 2xz + 2yz$ subject to $g(x,y,z) = xyz - V = 0, \nabla f = \lambda\nabla g$, $y + 2z = \lambda yz, x + 2z = \lambda xz, 2x + 2y = \lambda xy; \lambda = 0$ leads to $x = y = z = 0$, impossible, so solve for $\lambda = 1/z + 2/x = 1/z + 2/y = 2/y + 2/x$, so $x = y = 2z, x^3 = 2V$, minimum value $3(2V)^{2/3}$

29. **(a)** Maximize $f(\alpha,\beta,\gamma) = \cos\alpha\cos\beta\cos\gamma$ subject to $g(\alpha,\beta,\gamma) = \alpha + \beta + \gamma - \pi = 0$, $\nabla f = \lambda\nabla g, -\sin\alpha\cos\beta\cos\gamma = \lambda, -\cos\alpha\sin\beta\cos\gamma = \lambda, -\cos\alpha\cos\beta\sin\gamma = \lambda$ with solution $\alpha = \beta = \gamma = \pi/3$, maximum value 1/8

(b) for example, $f(\alpha,\beta) = \cos\alpha\cos\beta\cos(\pi - \alpha - \beta)$

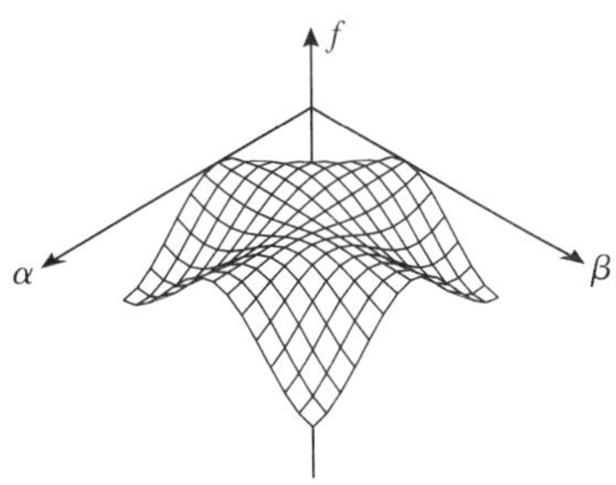

30. Find maxima and minima $z = x^2 + 4y^2$ subject to the constraint $g(x, y) = x^2 + y^2 - 1 = 0$, $\nabla z = \lambda \nabla g, 2x\mathbf{i} + 8y\mathbf{j} = 2\lambda x\mathbf{i} + 2\lambda y\mathbf{j}$, solve $2x = 2\lambda x, 8y = 2\lambda y$. If $y \neq 0$ then $\lambda = 4, x = 0, y^2 = 1$ and $z = x^2 + 4y^2 = 4$. If $y = 0$ then $x^2 = 1$ and $z = 1$, so the maximum height is obtained for $(x, y) = (0, \pm 1), z = 4$ and the minimum height is $z = 1$ at $(\pm 1, 0)$.

CHAPTER 14 SUPPLEMENTARY EXERCISES

1. (a) They approximate the profit per unit of any additional sales of the standard or high-resolution monitors, respectively.

   (b) The rates of change with respect to the two directions x and y, and with respect to time.

3. $z = \sqrt{x^2 + y^2} = c$ implies $x^2 + y^2 = c^2$, which is the equation of a circle; $x^2 + y^2 = c$ is also the equation of a circle (for $c > 0$).

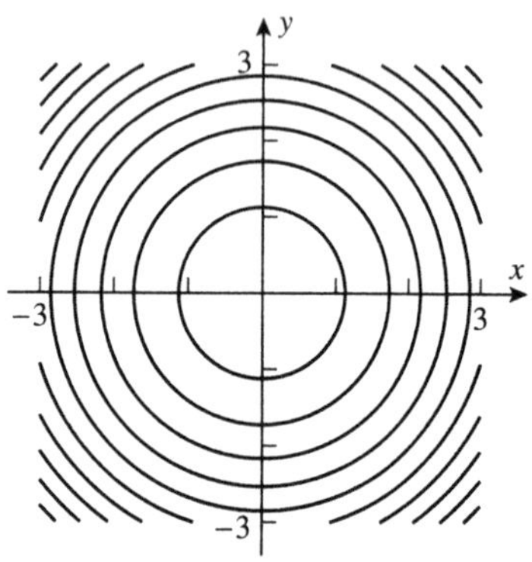

$z = x^2 + y^2$

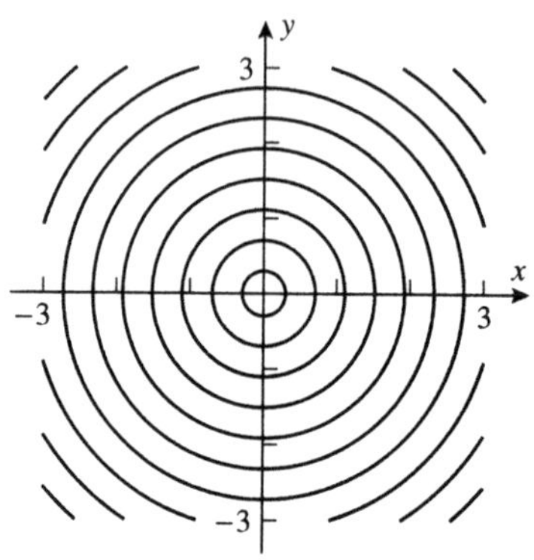

$z = \sqrt{x^2 + y^2}$

5. (b) $f(x, y, z) = z - x^2 - y^2$

7. (a) $f(\ln y, e^x) = e^{\ln y} \ln e^x = xy$ (b) $e^{r+s} \ln(rs)$

8. (a)

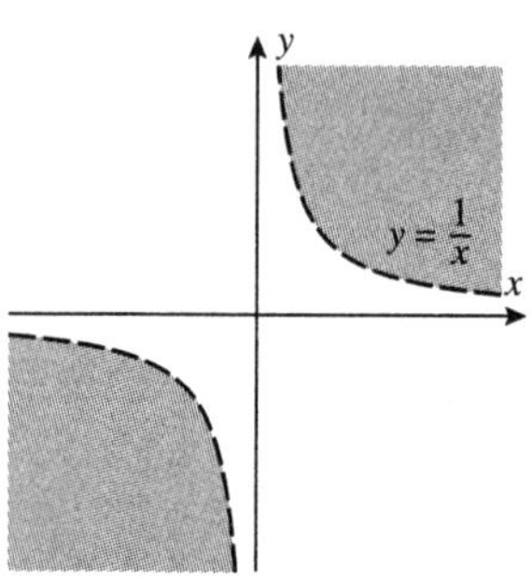

(b)

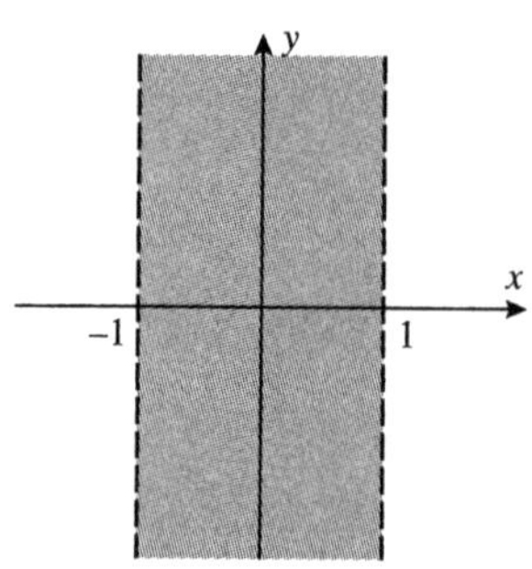

9. $w_x = 2x \sec^2(x^2 + y^2) + \sqrt{y}$, $w_{xy} = 8xy \sec^2(x^2 + y^2) \tan(x^2 + y^2) + \frac{1}{2} y^{-1/2}$,

$w_y = 2y \sec^2(x^2 + y^2) + \frac{1}{2} xy^{-1/2}$, $w_{yx} = 8xy \sec^2(x^2 + y^2) \tan(x^2 + y^2) + \frac{1}{2} y^{-1/2}$

10. $\partial w/\partial x = \dfrac{1}{x - y} - \sin(x + y), \partial^2 w/\partial x^2 = -\dfrac{1}{(x - y)^2} - \cos(x + y)$,

$\partial w/\partial y = -\dfrac{1}{x - y} - \sin(x + y)$, $\partial^2 w/\partial y^2 = -\dfrac{1}{(x - y)^2} - \cos(x + y) = \partial^2 w/\partial x^2$

11. $F_x = -6xz, F_{xx} = -6z, F_y = -6yz, F_{yy} = -6z, F_z = 6z^2 - 3x^2 - 3y^2,$
$F_{zz} = 12z, F_{xx} + F_{yy} + F_{zz} = -6z - 6z + 12z = 0$

12. $f_x = yz + 2x, f_{xy} = z, f_{xyz} = 1, f_{xyzx} = 0;\ f_z = xy - (1/z), f_{zx} = y, f_{zxx} = 0, f_{zxxy} = 0$

13. (a) $P = \dfrac{10T}{V}$,
$$\frac{dP}{dt} = \frac{\partial P}{\partial T}\frac{dT}{dt} + \frac{\partial P}{\partial V}\frac{dV}{dt} = \frac{10}{V}\cdot 3 - \frac{10T}{V^2}\cdot 0 = \frac{30}{V} = \frac{30}{2.5} = 12\ \text{N}/(\text{m}^2\text{min}) = 12\ \text{Pa/min}$$
(b) $$\frac{dP}{dt} = \frac{\partial P}{\partial T}\frac{dT}{dt} + \frac{\partial P}{\partial V}\frac{dV}{dt} = \frac{10}{V}\cdot 0 - \frac{10T}{V^2}\cdot(-3) = \frac{30T}{V^2} = \frac{30\cdot 50}{(2.5)^2} = 240\ \text{Pa/min}$$

14. (a) $z = 1 - y^2$, slope $= \dfrac{\partial z}{\partial y} = -2y = 4$ (b) $z = 1 - 4x^2, \dfrac{\partial z}{\partial x} = -8x = -8$

15. $x^4 - x + y - x^3y = (x^3 - 1)(x - y)$, limit $= -1$, not defined on the line $y = x$ so not continuous at $(0, 0)$

16. $\dfrac{x^4 - y^4}{x^2 + y^2} = x^2 - y^2$, limit $= \lim\limits_{(x,y)\to(0,0)} (x^2 - y^2) = 0$, continuous

17. Use the unit vectors $\mathbf{u} = \langle \frac{1}{\sqrt{2}}, \frac{1}{\sqrt{2}}\rangle, \mathbf{v} = \langle 0, -1\rangle, \mathbf{w} = \langle -\frac{1}{\sqrt{5}}, -\frac{2}{\sqrt{5}}\rangle = -\frac{\sqrt{2}}{\sqrt{5}}\mathbf{u} + \frac{1}{\sqrt{5}}\mathbf{v}$, so that
$$D_{\mathbf{w}}f = -\frac{\sqrt{2}}{\sqrt{5}}D\mathbf{u}f + \frac{1}{\sqrt{5}}D\mathbf{v}f = -\frac{\sqrt{2}}{\sqrt{5}}2\sqrt{2} + \frac{1}{\sqrt{5}}(-3) = -\frac{7}{\sqrt{5}}$$

18. (a) $\mathbf{n} = z_x\mathbf{i} + z_y\mathbf{j} - \mathbf{k} = 8\mathbf{i} + 8\mathbf{j} - \mathbf{k}$, tangent plane $8x + 8y - z = 4 + 8\ln 2$, normal line $x(t) = 1 + 8t, y(t) = \ln 2 + 8t, z(t) = 4 - t$

(b) $\mathbf{n} = 3\mathbf{i} + 10\mathbf{j} - 14\mathbf{k}$, tangent plane $3x + 10y - 14z = 30$, normal line $x(t) = 2 + 3t, y(t) = 1 + 10t, z(t) = -1 - 14t$

19. The origin is not such a point, so assume that the normal line at $(x_0, y_0, z_0) \neq (0, 0, 0)$ passes through the origin, then $\mathbf{n} = z_x\mathbf{i} + z_y\mathbf{j} - \mathbf{k} = -y_0\mathbf{i} - x_0\mathbf{j} - \mathbf{k}$; the line passes through the origin and is normal to the surface if it has the form $\mathbf{r}(t) = -y_0t\mathbf{i} - x_0t\mathbf{j} - t\mathbf{k}$ and $(x_0, y_0, z_0) = (x_0, y_0, 2 - x_0y_0)$ lies on the line if $-y_0t = x_0, -x_0t = y_0, -t = 2 - x_0y_0$, with solutions $x_0 = y_0 = -1$, $x_0 = y_0 = 1, x_0 = y_0 = 0$; thus the points are $(0, 0, 2), (1, 1, 1), (-1, -1, 1)$.

20. $\mathbf{n} = \frac{2}{3}x_0^{-1/3}\mathbf{i} + \frac{2}{3}y_0^{-1/3}\mathbf{j} + \frac{2}{3}z_0^{-1/3}\mathbf{k}$, tangent plane $x_0^{-1/3}x + y_0^{-1/3}y + z_0^{-1/3}z = x_0^{2/3} + y_0^{2/3} + z_0^{2/3} = 1$; intercepts are $x = x_0^{1/3}, y = y_0^{1/3}, z = z_0^{1/3}$, sum of squares of intercepts is $x_0^{2/3} + y_0^{2/3} + z_0^{2/3} = 1$.

21. A tangent to the line is $6\mathbf{i} + 4\mathbf{j} + \mathbf{k}$, a normal to the surface is $\mathbf{n} = 18x\mathbf{i} + 8y\mathbf{j} - \mathbf{k}$, so solve $18x = 6k, 8y = 4k, -1 = k;\ k = -1, x = -1/3, y = -1/2, z = 2$

22. $\Delta w = (1.1)^2(-0.1) - 2(1.1)(-0.1) + (-0.1)^2(1.1) - 0 = 0.11$,
$dw = (2xy - 2y + y^2)dx + (x^2 - 2x + 2yx)dy = -(-0.1) = 0.1$

23. $dV = \frac{2}{3}xh\,dx + \frac{1}{3}x^2dh = \frac{2}{3}2(-0.1) + \frac{1}{3}(0.2) = -0.06667\ \text{m}^3; \Delta V = -0.07267\ \text{m}^3$

24. $\nabla f = (2x + 3y - 6)\mathbf{i} + (3x + 6y + 3)\mathbf{j} = \mathbf{0}$ if $2x + 3y = 6, x + 2y = -1, x = 15, y = -8, D = 3 > 0$, $f_{xx} = 2 > 0$, so f has a relative minimum at $(15, -8)$.

25. $\nabla f = (2xy - 6x)\mathbf{i} + (x^2 - 12y)\mathbf{j} = \mathbf{0}$ if $2xy - 6x = 0, x^2 - 12y = 0$; if $x = 0$ then $y = 0$, and if $x \neq 0$ then $y = 3, x = \pm 6$, thus the gradient vanishes at $(0,0), (-6,3), (6,3)$; $f_{xx} = 0$ at all three points, $f_{yy} = -12 < 0, D = -4x^2$, so $(\pm 6, 3)$ are saddle points, and near the origin we write $f(x,y) = (y-3)x^2 - 6y^2$; since $y - 3 < 0$ when $|y| < 3$, f has a local maximum by inspection.

26. $\nabla f = (3x^2 - 3y)\mathbf{i} - (3x - y)\mathbf{j} = \mathbf{0}$ if $y = x^2, 3x = y$, so $x = y = 0$ or $x = 3, y = 9$; at $x = y = 0, D = -9$, saddle point; at $x = 3, y = 9, D = 9, f_{xx} = 18 > 0$, relative minimum

27. (a)

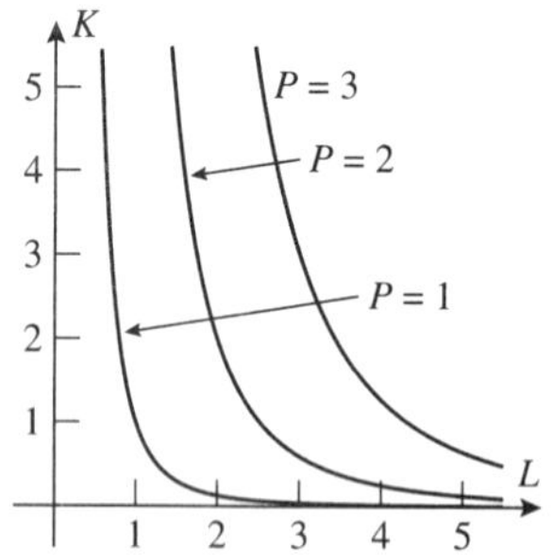

(b)

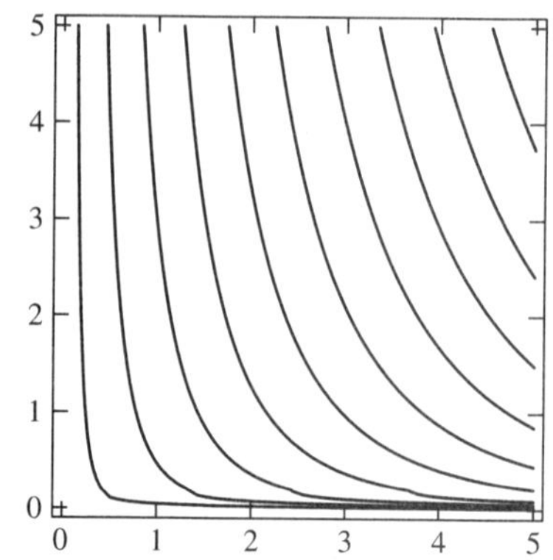

28. (a) $\partial P/\partial L = c\alpha L^{\alpha-1}K^{\beta}, \partial P/\partial K = c\beta L^{\alpha}K^{\beta-1}$

(b) the rates of change of output with respect to labor and capital equipment, respectively

(c) $K(\partial P/\partial K) + L(\partial P/\partial L) = c\beta L^{\alpha}K^{\beta} + c\alpha L^{\alpha}K^{\beta} = (\alpha + \beta)P = P$

29. (a) Maximize $P = 1000L^{0.6}(200{,}000 - L)^{0.4}$ subject to $50L + 100K = 200{,}000$ or $L = 2K = 4000$.
$600\left(\dfrac{K}{L}\right)^{0.4} = \lambda,\ 400\left(\dfrac{L}{K}\right)^{0.6} = 2\lambda,\ L + 2K = 4000$; so $\dfrac{2}{3}\left(\dfrac{L}{K}\right) = 2$, thus $L = 3K$, $L = 2400, K = 800, P(2400, 800) = 1000 \cdot 2400^{0.6} \cdot 800^{0.4} = 1000 \cdot 3^{0.6} \cdot 800 = 800{,}000 \cdot 3^{0.6} \approx$ \$1,546,545.64

(b) The value of labor is $50L = 120{,}000$ and the value of capital is $100K = 80{,}000$.

30. (a) $y^2 = 8 - 4x^2$, find extrema of $f(x) = x^2(8 - 4x^2) = -4x^4 + 8x^2$ defined for $-\sqrt{2} \le x \le \sqrt{2}$. Then $f'(x) = -16x^3 + 16x = 0$ when $x = 0, \pm 1, f''(x) = -48x^2 + 16$, so f has a relative maximum at $x = \pm 1, y = \pm 2$ and a relative minimum at $x = 0, y = \pm 2\sqrt{2}$. At the endpoints $x = \pm\sqrt{2}, y = 0$ we obtain the minimum $f = 0$ again.

(b) $f(x,y) = x^2y^2, g(x,y) = 4x^2 + y^2 - 8 = 0, \nabla f = 2xy^2\mathbf{i} + 2x^2y\mathbf{j} = \lambda\nabla g = 8\lambda x\mathbf{i} + 2\lambda y\mathbf{j}$, so solve $2xy^2 = \lambda 8x, 2x^2y = \lambda 2y$. If $x = 0$ then $y = \pm 2\sqrt{2}$, and if $y = 0$ then $x = \pm\sqrt{2}$. In either case f has a relative and absolute minimum. Assume $x, y \neq 0$, then $y^2 = 4\lambda, x^2 = \lambda$, use $g = 0$ to obtain $x^2 = 1, x = \pm 1, y = \pm 2$, and $f = 4$ is a relative and absolute maximum at $(\pm 1, \pm 2)$.

31. Let the first octant corner of the box be (x,y,z), so that $(x/a)^2 + (y/b)^2 + (z/c)^2 = 1$. Maximize $V = 8xyz$ subject to $g(x,y,z) = (x/a)^2 + (y/b)^2 + (z/c)^2 = 1$, solve $\nabla V = \lambda\nabla g$, or $8(yz\mathbf{i} + xz\mathbf{j} + xy\mathbf{k}) = (2\lambda x/a^2)\mathbf{i} + (2\lambda y/b^2)\mathbf{j} + (2\lambda z/c^2)\mathbf{k}, 8a^2yz = 2\lambda x, 8b^2xz = 2\lambda y, 8c^2xy = 2\lambda z$. For the maximum volume, $x, y, z \neq 0$; divide the first equation by the second to obtain $a^2y^2 = b^2x^2$; the first by the third to obtain $a^2z^2 = c^2x^2$, and finally $b^2z^2 = c^2y^2$. From $g = 1$ get
$3(x/a)^2 = 1, x = \pm a/\sqrt{3}$, and then $y = \pm b/\sqrt{3}, z = \pm c/\sqrt{3}$. The dimensions of the box are $\dfrac{2a}{\sqrt{3}} \times \dfrac{2b}{\sqrt{3}} \times \dfrac{2c}{\sqrt{3}}$, and the maximum volume is $8abc/(3\sqrt{3})$.

32. (a) $\frac{dy}{dx} = -\frac{6x - 5y + y\sec^2 xy}{-5x + x\sec^2 xy}$. (b) $\frac{dy}{dx} = -\frac{\ln y + \cos(x-y)}{x/y - \cos(x-y)}$

33. $\frac{dy}{dx} = -\frac{f_x}{f_y}$, $\frac{d^2y}{dx^2} = -\frac{f_y(d/dx)f_x - f_x(d/dx)f_y}{f_y^2} = -\frac{f_y(f_{xx} + f_{xy}(dy/dx)) - f_x(f_{xy} + f_{yy}(dy/dx))}{f_y^2}$

$$= -\frac{f_y(f_{xx} + f_{xy}(-f_x/f_y)) - f_x(f_{xy} + f_{yy}(-f_x/f_y))}{f_y^2} = \frac{-f_y^2 f_{xx} + 2f_xf_yf_{xy} - f_x^2 f_{yy}}{f_y^3}$$

34. Denote the currents I_1, I_2, I_3 by x, y, z respectively. Then minimize $F(x,y,z) = x^2R_1 + y^2R_2 + z^2R_3$ subject to $g(x,y,z) = x + y + z - I = 0$, so solve $\nabla F = \lambda\nabla g, 2xR_1\mathbf{i} + 2yR_2\mathbf{j} + 2zR_3\mathbf{k} = \lambda(\mathbf{i}+\mathbf{j}+\mathbf{k})$, $\lambda = 2xR_1 = 2yR_2 = 2zR_3$, so the minimum value of F occurs when $I_1 : I_2 : I_3 = \frac{1}{R_1} : \frac{1}{R_2} : \frac{1}{R_3}$.

35. Solve $(t-1)^2/4 + 16e^{-2t} + (2-\sqrt{t})^2 = 1$ for t to get $t = 1.833223, 2.839844$; the particle strikes the surface at the points $P_1(0.83322, 0.639589, 0.646034)$, $P_2(1.83984, 0.233739, 0.314816)$. The velocity vectors are given by $\mathbf{v} = \frac{dx}{dt}\mathbf{i} + \frac{dy}{dt}\mathbf{j} + \frac{dz}{dt}\mathbf{k} = \mathbf{i} - 4e^{-t}\mathbf{j} - 1/(2\sqrt{t})\mathbf{k}$, and a normal to the surface is $\mathbf{n} = \nabla(x^2/4 + y^2 + z^2) = x/2\mathbf{i} + 2y\mathbf{j} + 2z\mathbf{k}$. At the points P_i these are
$\mathbf{v}_1 = \mathbf{i} - 0.639589\mathbf{j} - 0.369286\mathbf{k}$, $\mathbf{v}_2 = \mathbf{i} - 0.233739\mathbf{j} + 0.296704\mathbf{k}$;
$\mathbf{n}_1 = 0.41661\mathbf{i} + 1.27918\mathbf{j} + 1.29207\mathbf{k}$ and $\mathbf{n}_2 = 0.91992\mathbf{i} + 0.46748\mathbf{j} + 0.62963\mathbf{k}$ so
$\cos^{-1}[(\mathbf{v}_i \cdot \mathbf{n}_i)/(\|\mathbf{v}_i\|\,\|\mathbf{n}_i\|)] = 112.3°, 61.1°$; the acute angles are $67.7°, 61.1°$.

36. (a) $F'(x) = \int_0^1 e^y\cos(xe^y)dy = \frac{\sin(ex) - \sin x}{x}$

(b) Use a CAS to get $x = \frac{\pi}{e+1}$ so the maximum value of $F(x)$ is

$$F\left(\frac{\pi}{e+1}\right) = \int_0^1 \sin\left(\frac{\pi}{e+1}e^y\right) dy \approx 0.909026.$$

37. Let x, y, z be the lengths of the sides opposite angles α, β, γ, located at A, B, C respectively. Then $x^2 = y^2 + z^2 - 2yz\cos\alpha$ and $x^2 = 100 + 400 - 2(10)(20)/2 = 300$, $x = 10\sqrt{3}$ and

$$2x\frac{dx}{dt} = 2y\frac{dy}{dt} + 2z\frac{dz}{dt} - 2\left(y\frac{dz}{dt}\cos\alpha + z\frac{dy}{dt}\cos\alpha - yz(\sin\alpha)\frac{d\alpha}{dt}\right)$$

$$= 2(10)(4) + 2(20)(2) - 2\left(10(2)\frac{1}{2} + 20(4)\frac{1}{2} - 10(20)\frac{\sqrt{3}}{2}\frac{\pi}{60}\right) = 60 + \frac{10\pi}{\sqrt{3}}$$

so $\frac{dx}{dt} = \sqrt{3} + \frac{\pi}{6}$, the length of BC is increasing.

38. (a) $\frac{d}{dt}\left(\frac{\partial z}{\partial x}\right) = \frac{\partial}{\partial x}\left(\frac{\partial z}{\partial x}\right)\frac{dx}{dt} + \frac{\partial}{\partial y}\left(\frac{\partial z}{\partial x}\right)\frac{dy}{dt} = \frac{\partial^2 z}{\partial x^2}\frac{dx}{dt} + \frac{\partial^2 z}{\partial y\partial x}\frac{dy}{dt}$ by the Chain Rule, and

$$\frac{d}{dt}\left(\frac{\partial z}{\partial y}\right) = \frac{\partial}{\partial x}\left(\frac{\partial z}{\partial y}\right)\frac{dx}{dt} + \frac{\partial}{\partial y}\left(\frac{\partial z}{\partial y}\right)\frac{dy}{dt} = \frac{\partial^2 z}{\partial x\partial y}\frac{dx}{dt} + \frac{\partial^2 z}{\partial y^2}\frac{dy}{dt}$$

(b) $\frac{dz}{dt} = \frac{\partial z}{\partial x}\frac{dx}{dt} + \frac{\partial z}{\partial y}\frac{dy}{dt}$,

$$\frac{d^2z}{dt^2} = \frac{dx}{dt}\left(\frac{\partial^2 z}{\partial x^2}\frac{dx}{dt} + \frac{\partial^2 z}{\partial y\partial x}\frac{dy}{dt}\right) + \frac{\partial z}{\partial x}\frac{d^2x}{dt^2} + \frac{dy}{dt}\left(\frac{\partial^2 z}{\partial x\partial y}\frac{dx}{dt} + \frac{\partial^2 z}{\partial y^2}\frac{dy}{dt}\right) + \frac{\partial z}{\partial y}\frac{d^2y}{dt^2}$$

CHAPTER 15
Multiple Integrals

EXERCISE SET 15.1

1. $\int_0^1 \int_0^2 (x+3)dy\,dx = \int_0^1 (2x+6)dx = 7$

2. $\int_1^3 \int_{-1}^1 (2x-4y)dy\,dx = \int_1^3 4x\,dx = 16$

3. $\int_2^4 \int_0^1 x^2 y\,dx\,dy = \int_2^4 \frac{1}{3}y\,dy = 2$

4. $\int_{-2}^0 \int_{-1}^2 (x^2+y^2)dx\,dy = \int_{-2}^0 (3+3y^2)dy = 14$

5. $\int_0^{\ln 3} \int_0^{\ln 2} e^{x+y}dy\,dx = \int_0^{\ln 3} e^x\,dx = 2$

6. $\int_0^2 \int_0^1 y\sin x\,dy\,dx = \int_0^2 \frac{1}{2}\sin x\,dx = (1-\cos 2)/2$

7. $\int_{-1}^0 \int_2^5 dx\,dy = \int_{-1}^0 3\,dy = 3$

8. $\int_4^6 \int_{-3}^7 dy\,dx = \int_4^6 10dx = 20$

9. $\int_0^1 \int_0^1 \frac{x}{(xy+1)^2}dy\,dx = \int_0^1 \left(1-\frac{1}{x+1}\right)dx = 1-\ln 2$

10. $\int_{\pi/2}^{\pi} \int_1^2 x\cos xy\,dy\,dx = \int_{\pi/2}^{\pi} (\sin 2x - \sin x)dx = -2$

11. $\int_0^{\ln 2} \int_0^1 xy\,e^{y^2x}dy\,dx = \int_0^{\ln 2} \frac{1}{2}(e^x-1)dx = (1-\ln 2)/2$

12. $\int_3^4 \int_1^2 \frac{1}{(x+y)^2}dy\,dx = \int_3^4 \left(\frac{1}{x+1}-\frac{1}{x+2}\right)dx = \ln(25/24)$

13. $\int_{-1}^1 \int_{-2}^2 4xy^3dy\,dx = \int_{-1}^1 0\,dx = 0$

14. $\int_0^1 \int_0^1 \frac{xy}{\sqrt{x^2+y^2+1}}dy\,dx = \int_0^1 [x(x^2+2)^{1/2} - x(x^2+1)^{1/2}]dx = (3\sqrt{3}-4\sqrt{2}+1)/3$

15. $\int_0^1 \int_2^3 x\sqrt{1-x^2}\,dy\,dx = \int_0^1 x(1-x^2)^{1/2}dx = 1/3$

16. $\int_0^{\pi/2} \int_0^{\pi/3} (x\sin y - y\sin x)dy\,dx = \int_0^{\pi/2} \left(\frac{x}{2}-\frac{\pi^2}{18}\sin x\right)dx = \pi^2/144$

17. (a) $x_k^* = k/2-1/4, k=1,2,3,4; y_l^* = l/2-1/4, l=1,2,3,4,$

$$\iint_R f(x,y)\,dxdy \approx \sum_{k=1}^4\sum_{l=1}^4 f(x_k^*,y_l^*)\Delta A_{kl} = \sum_{k=1}^4\sum_{l=1}^4[(k/2-1/4)^2+(l/2-1/4)](1/2)^2 = 37/4$$

(b) $\int_0^2 \int_0^2 (x^2+y)\,dxdy = 28/3$; the error is $|37/4-28/3| = 1/12$

18. **(a)** $x_k^* = k/2 - 1/4, k = 1, 2, 3, 4; y_l^* = l/2 - 1/4, l = 1, 2, 3, 4,$

$$\int\int_R f(x, y)\,dxdy \approx \sum_{k=1}^{4}\sum_{l=1}^{4} f(x_k^*, y_l^*)\Delta A_{kl} = \sum_{k=1}^{4}\sum_{l=1}^{4}[(k/2 - 1/4) - 2(l/2 - 1/4)](1/2)^2 = -4$$

(b) $\int_0^2\int_0^2 (x - 2y)\,dxdy = -4$; the error is zero

19. $V = \int_3^5\int_1^2 (2x + y)dy\,dx = \int_3^5 (2x + 3/2)dx = 19$

20. $V = \int_1^3\int_0^2 (3x^3 + 3x^2y)dy\,dx = \int_1^3 (6x^3 + 6x^2)dx = 172$

21. $V = \int_0^2\int_0^3 x^2 dy\,dx = \int_0^2 3x^2 dx = 8$

22. $V = \int_0^3\int_0^4 5(1 - x/3)dy\,dx = \int_0^3 5(4 - 4x/3)dx = 30$

23. **(a)**

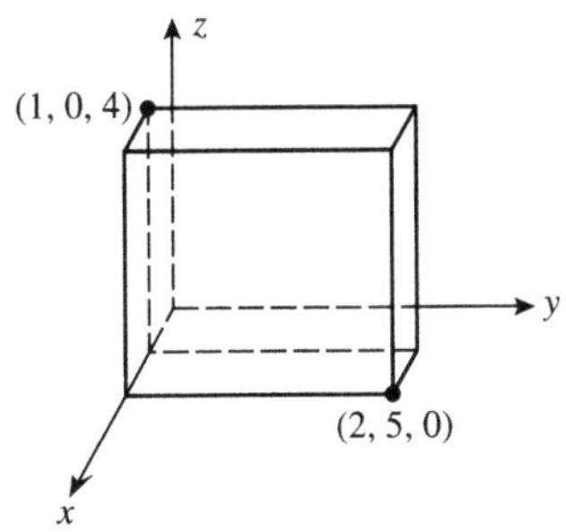

(b)

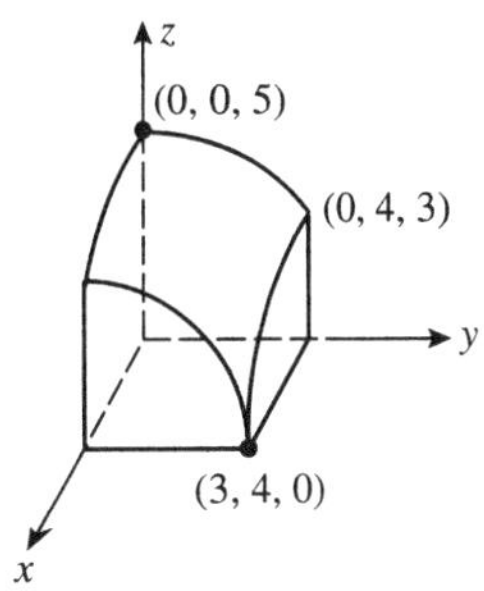

24. **(a)**

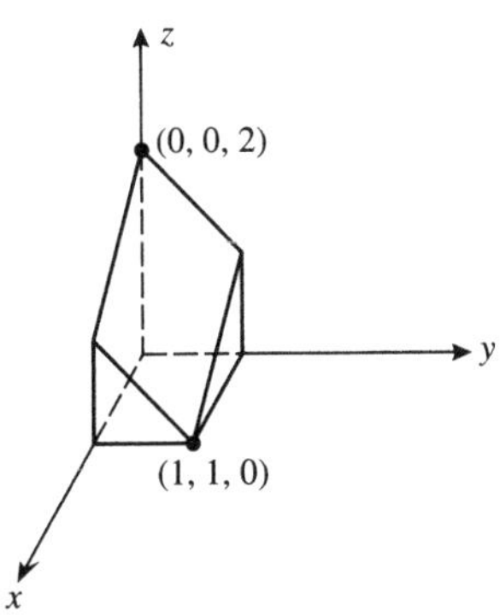

(b)

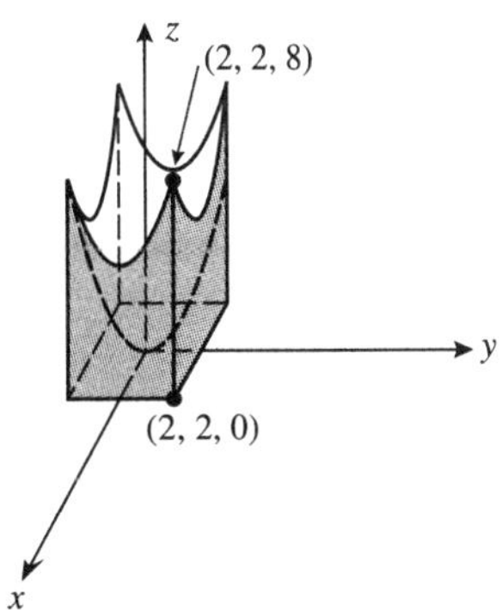

25. $\int_0^{1/2}\int_0^{\pi} x\cos(xy)\cos^2\pi x\,dy\,dx = \int_0^{1/2} \cos^2\pi x\sin(xy)\Big]_0^{\pi} dx$

$$= \int_0^{1/2} \cos^2\pi x \sin\pi x\,dx = -\frac{1}{3\pi}\cos^3\pi x\Big]_0^{1/2} = \frac{1}{3\pi}$$

26. (a)

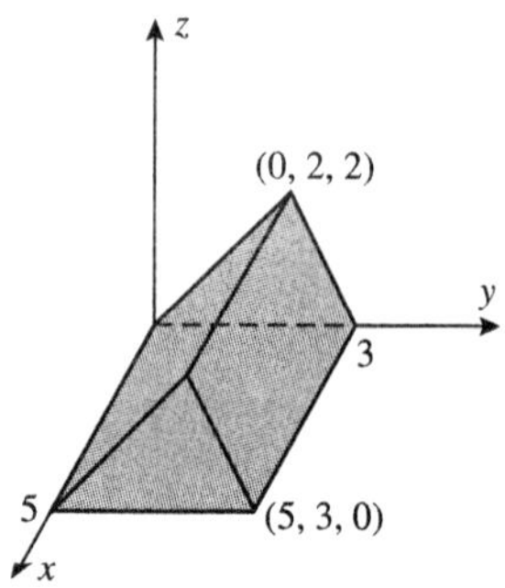

(b) $$V = \int_0^5 \int_0^2 y\,dy\,dx + \int_0^5 \int_2^3 (-2y+6)\,dy\,dx$$
$$= 10 + 5 = 15$$

27. $$f_{\text{ave}} = \frac{2}{\pi}\int_0^{\pi/2}\int_0^1 y\sin xy\,dx\,dy = \frac{2}{\pi}\int_0^{\pi/2}\left(-\cos xy\Big]_{x=0}^{x=1}\right)dy = \frac{2}{\pi}\int_0^{\pi/2}(1-\cos y)\,dy = 1-\frac{2}{\pi}$$

28. $$\text{average} = \frac{1}{3}\int_0^3\int_0^1 x(x^2+y)^{1/2}dx\,dy = \int_0^3 \frac{1}{9}[(1+y)^{3/2}-y^{3/2}]dy = 2(31-9\sqrt{3})/45$$

29. $$T_{\text{ave}} = \frac{1}{2}\int_0^1\int_0^2 \left(10-8x^2-2y^2\right)dy\,dx = \frac{1}{2}\int_0^1\left(\frac{44}{3}-16x^2\right)dx = \left(\frac{14}{3}\right)^\circ$$

30. $$f_{\text{ave}} = \frac{1}{A(R)}\int_a^b\int_c^d k\,dy\,dx = \frac{1}{A(R)}(b-a)(d-c)k = k$$

31. 1.381737122

32. 2.230985141

33. $$\iint_R f(x,y)dA = \int_a^b\left[\int_c^d g(x)h(y)dy\right]dx = \int_a^b g(x)\left[\int_c^d h(y)dy\right]dx$$
$$= \left[\int_a^b g(x)dx\right]\left[\int_c^d h(y)dy\right]$$

34. The integral of $\tan x$ (an odd function) over the interval $[-1, 1]$ is zero.

35. The first integral equals 1/2, the second equals $-1/2$. No, because the integrand is not continuous.

EXERCISE SET 15.2

1. $$\int_0^1\int_{x^2}^x xy^2dy\,dx = \int_0^1 \frac{1}{3}(x^4-x^7)dx = 1/40$$

2. $$\int_1^{3/2}\int_y^{3-y} y\,dx\,dy = \int_1^{3/2}(3y-2y^2)dy = 7/24$$

3. $$\int_0^3\int_0^{\sqrt{9-y^2}} y\,dx\,dy = \int_0^3 y\sqrt{9-y^2}\,dy = 9$$

4. $$\int_{1/4}^1\int_{x^2}^x \sqrt{x/y}\,dy\,dx = \int_{1/4}^1\int_{x^2}^x x^{1/2}y^{-1/2}dy\,dx = \int_{1/4}^1 2(x-x^{3/2})dx = 13/80$$

5. $\displaystyle\int_{\sqrt{\pi}}^{\sqrt{2\pi}}\int_0^{x^3}\sin(y/x)dy\,dx = \int_{\sqrt{\pi}}^{\sqrt{2\pi}}[-x\cos(x^2)+x]dx = \pi/2$

6. $\displaystyle\int_{-1}^{1}\int_{-x^2}^{x^2}(x^2-y)dy\,dx = \int_{-1}^{1}2x^4dx = 4/5$

7. $\displaystyle\int_{\pi/2}^{\pi}\int_0^{x^2}\frac{1}{x}\cos(y/x)dy\,dx = \int_{\pi/2}^{\pi}\sin x\,dx = 1$

8. $\displaystyle\int_0^1\int_0^x e^{x^2}dy\,dx = \int_0^1 xe^{x^2}dx = (e-1)/2$

9. $\displaystyle\int_0^1\int_0^x y\sqrt{x^2-y^2}\,dy\,dx = \int_0^1\frac{1}{3}x^3dx = 1/12$

10. $\displaystyle\int_1^2\int_0^{y^2}e^{x/y^2}dx\,dy = \int_1^2(e-1)y^2dy = 7(e-1)/3$

11. (a) $\displaystyle\int_0^2\int_0^{x^2}xy\,dy\,dx = \int_0^2\frac{1}{2}x^5dx = \frac{16}{3}$

(b) $\displaystyle\int_1^3\int_{-(y-5)/2}^{(y+7)/2}xy\,dx\,dy = \int_1^3(3y^2+3y)dy = 38$

12. (a) $\displaystyle\int_0^1\int_{x^2}^{\sqrt{x}}(x+y)dy\,dx = \int_0^1(x^{3/2}+x/2-x^3-x^4/2)dx = 3/10$

(b) $\displaystyle\int_{-1}^1\int_{-\sqrt{1-x^2}}^{\sqrt{1-x^2}}x\,dy\,dx + \int_{-1}^1\int_{-\sqrt{1-x^2}}^{\sqrt{1-x^2}}y\,dy\,dx = \int_{-1}^1 2x\sqrt{1-x^2}\,dx + 0 = 0$

13. (a) $\displaystyle\int_4^8\int_{16/x}^{x}x^2dy\,dx = \int_4^8(x^3-16x)dx = 576$

(b) $\displaystyle\int_2^4\int_{16/y}^{8}x^2\,dxdy + \int_4^8\int_y^8 x^2\,dx\,dy = \int_4^8\left[\frac{512}{3}-\frac{4096}{3y^3}\right]dy + \int_4^8\frac{512-y^3}{3}dy$

$\displaystyle = \frac{640}{3}+\frac{1088}{3} = 576$

14. (a) $\displaystyle\int_1^2\int_0^y xy^2dx\,dy = \int_1^2\frac{1}{2}y^4dy = 31/10$

(b) $\displaystyle\int_0^1\int_1^2 xy^2\,dydx + \int_1^2\int_x^2 xy^2\,dydx = \int_0^1 7x/3\,dx + \int_1^2\frac{8x-x^4}{3}dx = 7/6+29/15 = 31/10$

15. (a) $\displaystyle\int_{-1}^1\int_{-\sqrt{1-x^2}}^{\sqrt{1-x^2}}(3x-2y)dy\,dx = \int_{-1}^1 6x\sqrt{1-x^2}\,dx = 0$

(b) $\displaystyle\int_1^1\int_{-\sqrt{1-y^2}}^{\sqrt{1-y^2}}(3x-2y)\,dxdy = \int_{-1}^1 -4y\sqrt{1-y^2}\,dy = 0$

16. (a) $\displaystyle\int_0^5\int_{5-x}^{\sqrt{25-x^2}}y\,dy\,dx = \int_0^5(5x-x^2)dx = 125/6$

(b) $\displaystyle\int_0^5\int_{5-y}^{\sqrt{25-y^2}}y\,dxdy = \int_0^5 y\left(\sqrt{25-y^2}-5+y\right)dy = 125/6$

17. $\displaystyle\int_0^4\int_0^{\sqrt{y}}x(1+y^2)^{-1/2}dx\,dy = \int_0^4\frac{1}{2}y(1+y^2)^{-1/2}dy = (\sqrt{17}-1)/2$

18. $\displaystyle\int_0^{\pi}\int_0^{x} x\cos y\,dy\,dx = \int_0^{\pi} x\sin x\,dx = \pi$

19. $\displaystyle\int_0^{2}\int_{y^2}^{6-y} xy\,dx\,dy = \int_0^{2} \frac{1}{2}(36y - 12y^2 + y^3 - y^5)dy = 50/3$

20. $\displaystyle\int_0^{\pi/4}\int_{\sin y}^{1/\sqrt{2}} x\,dx\,dy = \int_0^{\pi/4} \frac{1}{4}\cos 2y\,dy = 1/8$

21. $\displaystyle\int_0^{1}\int_{x^3}^{x} (x-1)dy\;dx = \int_0^{1} (-x^4 + x^3 + x^2 - x)dx = -7/60$

22. $\displaystyle\int_0^{1/\sqrt{2}}\int_x^{2x} x^2 dy\,dx + \int_{1/\sqrt{2}}^{1}\int_x^{1/x} x^2 dy\,dx = \int_0^{1/\sqrt{2}} x^3 dx + \int_{1/\sqrt{2}}^{1} (x - x^3)dx = 1/8$

23. **(a)**

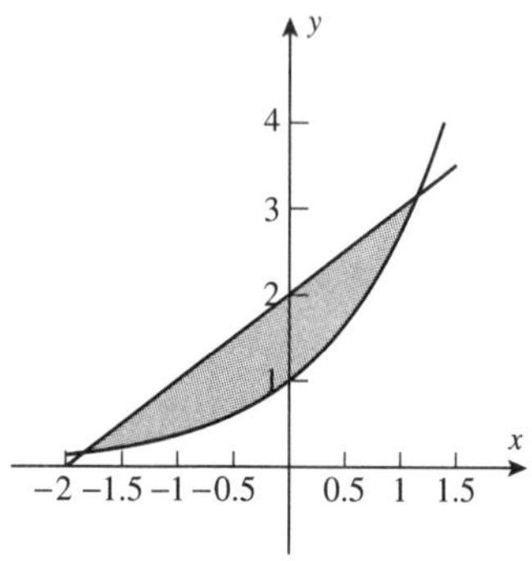

(b) $x = (-1.8414, 0.1586), (1.1462, 3.1462)$

(c) $\displaystyle\iint_R x\,dA \approx \int_{-1.8414}^{1.1462}\int_{e^x}^{x+2} x\,dydx = \int_{-1.8414}^{1.1462} x(x + 2 - e^x)\,dx \approx -0.4044$

(d) $\displaystyle\iint_R x\,dA \approx \int_{0.1586}^{3.1462}\int_{y-2}^{\ln y} x\,dxdy = \int_{0.1586}^{3.1462} \left[\frac{\ln^2 y}{2} - \frac{(y-2)^2}{2}\right] dy \approx -0.4044$

24. **(a)**

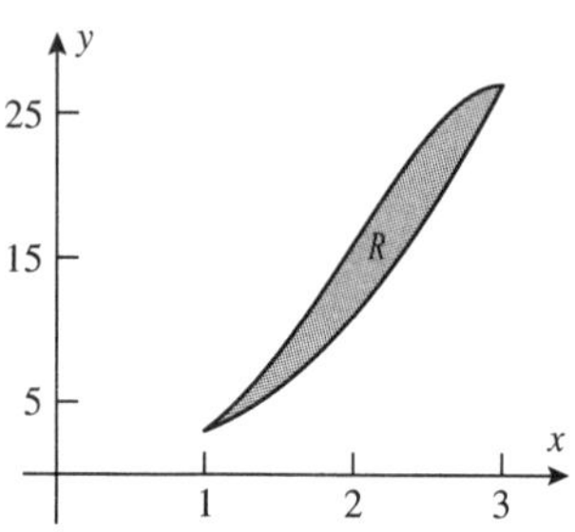

(b) $(1, 3), (3, 27)$

(c) $\displaystyle\int_1^{3}\int_{3-4x+4x^2}^{4x^3-x^4} x\,dy\,dx = \int_1^{3} x[(4x^3 - x^4) - (3 - 4x + 4x^2)]\,dx = \frac{224}{15}$

25. $\displaystyle A = \int_0^{\pi/4}\int_{\sin x}^{\cos x} dy\,dx = \int_0^{\pi/4} (\cos x - \sin x)dx = \sqrt{2} - 1$

26. $\displaystyle A = \int_{-4}^{1}\int_{3y-4}^{-y^2} dx\,dy = \int_{-4}^{1} (-y^2 - 3y + 4)dy = 125/6$

27. $A = \int_{-3}^{3}\int_{1-y^2/9}^{9-y^2} dx\, dy = \int_{-3}^{3} 8(1-y^2/9)dy = 32$

28. $A = \int_0^1 \int_{\sinh x}^{\cosh x} dy\, dx = \int_0^1 (\cosh x - \sinh x)dx = 1 - e^{-1}$

29. $\int_0^4 \int_0^{6-3x/2} (3 - 3x/4 - y/2)\, dy\, dx = \int_0^4 [(3-3x/4)(6-3x/2) - (6-3x/2)^2/4]\, dx = 12$

30. $\int_0^2 \int_0^{\sqrt{4-x^2}} \sqrt{4-x^2}\, dy\, dx = \int_0^2 (4-x^2)\, dx = 16/3$

31. $V = \int_{-3}^{3}\int_{-\sqrt{9-x^2}}^{\sqrt{9-x^2}} (3-x)dy\, dx = \int_{-3}^{3} (6\sqrt{9-x^2} - 2x\sqrt{9-x^2})dx = 27\pi$

32. $V = \int_0^1 \int_{x^2}^{x} (x^2+3y^2)dy\, dx = \int_0^1 (2x^3 - x^4 - x^6)dx = 11/70$

33. $V = \int_0^3 \int_0^2 (9x^2+y^2)dy\, dx = \int_0^3 (18x^2 + 8/3)dx = 170$

34. $V = \int_{-1}^{1}\int_{y^2}^{1} (1-x)dx\, dy = \int_{-1}^{1} (1/2 - y^2 + y^4/2)dy = 8/15$

35. $V = \int_{-3/2}^{3/2}\int_{-\sqrt{9-4x^2}}^{\sqrt{9-4x^2}} (y+3)dy\, dx = \int_{-3/2}^{3/2} 6\sqrt{9-4x^2}\, dx = 27\pi/2$

36. $V = \int_0^3 \int_{y^2/3}^{3} (9-x^2)dx\, dy = \int_0^3 (18 - 3y^2 + y^6/81)dy = 216/7$

37. $V = 8\int_0^5 \int_0^{\sqrt{25-x^2}} \sqrt{25-x^2}dy\, dx = 8\int_0^5 (25-x^2)dx = 2000/3$

38. $V = 2\int_0^2 \int_0^{\sqrt{1-(y-1)^2}} (x^2+y^2)dx\, dy = 2\int_0^2 \left(\frac{1}{3}[1-(y-1)^2]^{3/2} + y^2[1-(y-1)^2]^{1/2}\right) dy,$

let $y - 1 = \sin\theta$ to get $V = 2\int_{-\pi/2}^{\pi/2} \left[\frac{1}{3}\cos^3\theta + (1+\sin\theta)^2\cos\theta\right]\cos\theta\, d\theta$ which eventually yields $V = 3\pi/2$

39. $V = 4\int_0^1 \int_0^{\sqrt{1-x^2}} (1-x^2-y^2)dy\, dx = \frac{8}{3}\int_0^1 (1-x^2)^{3/2}dx = \pi/2$

40. $V = \int_0^2 \int_0^{\sqrt{4-x^2}} (x^2+y^2)dy\, dx = \int_0^2 \left[x^2\sqrt{4-x^2} + \frac{1}{3}(4-x^2)^{3/2}\right] dx = 2\pi$

41. $\int_0^{\sqrt{2}} \int_{y^2}^{2} f(x,y)dx\, dy$ **42.** $\int_0^8 \int_0^{x/2} f(x,y)dy\, dx$ **43.** $\int_1^{e^2} \int_{\ln x}^{2} f(x,y)dy\, dx$

44. $\int_0^1 \int_{e^y}^e f(x,y)dx\,dy$ **45.** $\int_0^{\pi/2} \int_0^{\sin x} f(x,y)dy\,dx$ **46.** $\int_0^1 \int_{x^2}^{\sqrt{x}} f(x,y)dy\,dx$

47. $\int_0^4 \int_0^{y/4} e^{-y^2}dx\,dy = \int_0^4 \frac{1}{4}ye^{-y^2}dy = (1-e^{-16})/8$

48. $\int_0^1 \int_0^{2x} \cos(x^2)dy\,dx = \int_0^1 2x\cos(x^2)dx = \sin 1$

49. $\int_0^2 \int_0^{x^2} e^{x^3}dy\,dx = \int_0^2 x^2e^{x^3}dx = (e^8-1)/3$

50. $\int_0^{\ln 3} \int_{e^y}^3 x\,dx\,dy = \frac{1}{2}\int_0^{\ln 3}(9-e^{2y})dy = \frac{1}{2}(9\ln 3 - 4)$

51. $\int_0^2 \int_0^{y^2} \sin(y^3)dx\,dy = \int_0^2 y^2\sin(y^3)dy = (1-\cos 8)/3$

52. $\int_0^1 \int_{e^x}^e x\,dy\,dx = \int_0^1 (ex - xe^x)dx = e/2 - 1$

53. (a) $\int_0^4 \int_{\sqrt{x}}^2 \sin \pi y^3\,dy\,dx$; the inner integral is non-elementary.

$$\int_0^2 \int_0^{y^2} \sin\left(\pi y^3\right)\,dx\,dy = \int_0^2 y^2\sin\left(\pi y^3\right)\,dy = -\frac{1}{3\pi}\cos\left(\pi y^3\right)\Big]_0^2 = 0$$

(b) $\int_0^1 \int_{\sin^{-1}y}^{\pi/2} \sec^2(\cos x)dx\,dy$; the inner integral is non-elementary.

$$\int_0^{\pi/2} \int_0^{\sin x} \sec^2(\cos x)dy\,dx = \int_0^{\pi/2} \sec^2(\cos x)\sin x\,dx = \tan 1$$

54. $V = 4\int_0^2 \int_0^{\sqrt{4-x^2}} (x^2+y^2)\,dy\,dx = 4\int_0^2 \left(x^2\sqrt{4-x^2} + \frac{1}{3}(4-x^2)^{3/2}\right)\,dx \qquad (x = 2\sin\theta)$

$$= \int_0^{\pi/2} \left(\frac{64}{3} + \frac{64}{3}\sin^2\theta - \frac{128}{3}\sin^4\theta\right)\,d\theta = \frac{64}{3}\frac{\pi}{2} + \frac{64}{3}\frac{\pi}{4} - \frac{128}{3}\frac{\pi}{2}\frac{1\cdot 3}{2\cdot 4} = 8\pi$$

55. The region is symmetric with respect to the y-axis, and the integrand is an odd function of x, hence the answer is zero.

56. This is the volume in the first octant under the surface $z = \sqrt{1-x^2-y^2}$, so 1/8 of the volume of the sphere of radius 1, thus $\frac{\pi}{6}$.

57. Area of triangle is 1/2, so $\bar{f} = 2\int_0^1 \int_x^1 \frac{1}{1+x^2}dy\,dx = 2\int_0^1 \left[\frac{1}{1+x^2} - \frac{x}{1+x^2}\right]dx = \frac{\pi}{2} - \ln 2$

58. Area $= \int_0^2 (3x - x^2 - x)\,dx = 4/3$, so

$$\bar{f} = \frac{3}{4}\int_0^2 \int_x^{3x-x^2} (x^2 - xy)dy\,dx = \frac{3}{4}\int_0^2 (-2x^3 + 2x^4 - x^5/2)dx = -\frac{3}{4}\frac{8}{15} = -\frac{2}{5}$$

59. $T_{\text{ave}} = \dfrac{1}{A(R)}\displaystyle\iint_R (5xy + x^2)\,dA$. The diamond has corners $(\pm 2, 0), (0, \pm 4)$ and thus has area $A(R) = 4\frac{1}{2}2(4) = 16\text{m}^2$. Since $5xy$ is an odd function of x (as well as y), $\displaystyle\iint_R 5xy\,dA = 0$. Since x^2 is an even function of both x and y,

$$T_{\text{ave}} = \frac{4}{16}\iint_{\substack{R\\x,y>0}} x^2\,dA = \frac{1}{4}\int_0^2\int_0^{4-2x} x^2\,dydx = \frac{1}{4}\int_0^2 (4-2x)x^2\,dx = \frac{1}{4}\left(\frac{4}{3}x^3 - \frac{1}{2}x^4\right)\bigg]_0^2 = \frac{2}{3}{}^\circ\text{C}$$

60. The area of the lens is $\pi R^2 = 4\pi$ and the average thickness T_{ave} is

$$T_{\text{ave}} = \frac{4}{4\pi}\int_0^2\int_0^{\sqrt{4-x^2}} \left(1 - (x^2+y^2)/4\right)\,dydx = \frac{1}{\pi}\int_0^2 \frac{1}{6}(4-x^2)^{3/2}dx \qquad (x = 2\cos\theta)$$

$$= \frac{8}{3\pi}\int_0^{\pi} \sin^4\theta\,d\theta = \frac{8}{3\pi}\frac{1\cdot 3}{2\cdot 4}\frac{\pi}{2} = \frac{1}{2}\text{ in}$$

61. $y = \sin x$ and $y = x/2$ intersect at $x = 0$ and $x = a = 1.895494$, so

$$V = \int_0^a\int_{x/2}^{\sin x} \sqrt{1+x+y}\,dy\,dx = 0.676089$$

EXERCISE SET 15.3

1. $\displaystyle\int_0^{\pi/2}\int_0^{\sin\theta} r\cos\theta dr\,d\theta = \int_0^{\pi/2}\frac{1}{2}\sin^2\theta\cos\theta\,d\theta = 1/6$

2. $\displaystyle\int_0^{\pi}\int_0^{1+\cos\theta} r\,dr\,d\theta = \int_0^{\pi}\frac{1}{2}(1+\cos\theta)^2 d\theta = 3\pi/4$

3. $\displaystyle\int_0^{\pi/2}\int_0^{a\sin\theta} r^2 dr\,d\theta = \int_0^{\pi/2}\frac{a^3}{3}\sin^3\theta\,d\theta = \frac{2}{9}a^3$

4. $\displaystyle\int_0^{\pi/6}\int_0^{\cos 3\theta} r\,dr\,d\theta = \int_0^{\pi/6}\frac{1}{2}\cos^2 3\theta\,d\theta = \pi/24$

5. $\displaystyle\int_0^{\pi}\int_0^{1-\sin\theta} r^2\cos\theta\,dr\,d\theta = \int_0^{\pi}\frac{1}{3}(1-\sin\theta)^3\cos\theta\,d\theta = 0$

6. $\displaystyle\int_0^{\pi/2}\int_0^{\cos\theta} r^3 dr\,d\theta = \int_0^{\pi/2}\frac{1}{4}\cos^4\theta\,d\theta = 3\pi/64$

7. $\displaystyle A = \int_0^{2\pi}\int_0^{1-\cos\theta} r\,dr\,d\theta = \int_0^{2\pi}\frac{1}{2}(1-\cos\theta)^2 d\theta = 3\pi/2$

8. $\displaystyle A = 4\int_0^{\pi/2}\int_0^{\sin 2\theta} r\,dr\,d\theta = 2\int_0^{\pi/2}\sin^2 2\theta\,d\theta = \pi/2$

9. $\displaystyle A = \int_{\pi/4}^{\pi/2}\int_{\sin 2\theta}^{1} r\,dr\,d\theta = \int_{\pi/4}^{\pi/2}\frac{1}{2}(1-\sin^2 2\theta)d\theta = \pi/16$

10. $A = 2\int_0^{\pi/3}\int_{\sec\theta}^{2} r\,dr\,d\theta = \int_0^{\pi/3}(4-\sec^2\theta)d\theta = 4\pi/3-\sqrt{3}$

11. $A = 2\int_{\pi/6}^{\pi/2}\int_{2}^{4\sin\theta} r\,dr\,d\theta = \int_{\pi/6}^{\pi/2}(16\sin^2\theta-4)d\theta = 4\pi/3+2\sqrt{3}$

12. $A = 2\int_{\pi/2}^{\pi}\int_{1+\cos\theta}^{1} r\,dr\,d\theta = \int_{\pi/2}^{\pi}(-2\cos\theta-\cos^2\theta)d\theta = 2-\pi/4$

13. $V = 8\int_0^{\pi/2}\int_1^3 r\sqrt{9-r^2}\,dr\,d\theta = \frac{128}{3}\sqrt{2}\int_0^{\pi/2}d\theta = \frac{64}{3}\sqrt{2}\pi$

14. $V = 2\int_0^{\pi/2}\int_0^{2\sin\theta} r^2 dr\,d\theta = \frac{16}{3}\int_0^{\pi/2}\sin^3\theta\,d\theta = 32/9$

15. $V = 2\int_0^{\pi/2}\int_0^{\cos\theta}(1-r^2)r\,dr\,d\theta = \frac{1}{2}\int_0^{\pi/2}(2\cos^2\theta-\cos^4\theta)d\theta = 5\pi/32$

16. $V = 4\int_0^{\pi/2}\int_1^3 dr\,d\theta = 8\int_0^{\pi/2}d\theta = 4\pi$

17. $V = \int_0^{\pi/2}\int_0^{3\sin\theta} r^2\sin\theta\,dr\,d\theta = 9\int_0^{\pi/2}\sin^4\theta\,d\theta = 27\pi/16$

18. $V = 4\int_0^{\pi/2}\int_{2\cos\theta}^{2} r\sqrt{4-r^2}\,dr\,d\theta + \int_{\pi/2}^{\pi}\int_0^2 r\sqrt{4-r^2}\,dr\,d\theta$

$$= \frac{32}{3}\int_0^{\pi/2}\sin^3\theta\,d\theta + \frac{32}{3}\int_{\pi/2}^{\pi}d\theta = \frac{64}{9}+\frac{16}{3}\pi$$

19. $\int_0^{2\pi}\int_0^1 e^{-r^2}r\,dr\,d\theta = \frac{1}{2}(1-e^{-1})\int_0^{2\pi}d\theta = (1-e^{-1})\pi$

20. $\int_0^{\pi/2}\int_0^3 r\sqrt{9-r^2}\,dr\,d\theta = 9\int_0^{\pi/2}d\theta = 9\pi/2$

21. $\int_0^{\pi/4}\int_0^2 \frac{1}{1+r^2}r\,dr\,d\theta = \frac{1}{2}\ln 5\int_0^{\pi/4}d\theta = \frac{\pi}{8}\ln 5$

22. $\int_{\pi/4}^{\pi/2}\int_0^{2\cos\theta} 2r^2\sin\theta\,dr\,d\theta = \frac{16}{3}\int_{\pi/4}^{\pi/2}\cos^3\theta\sin\theta\,d\theta = 1/3$

23. $\int_0^{\pi/2}\int_0^1 r^3 dr\,d\theta = \frac{1}{4}\int_0^{\pi/2}d\theta = \pi/8$

24. $\int_0^{2\pi}\int_0^2 e^{-r^2}r\,dr\,d\theta = \frac{1}{2}(1-e^{-4})\int_0^{2\pi}d\theta = (1-e^{-4})\pi$

25. $\int_0^{\pi/2}\int_0^{2\cos\theta} r^2 dr\,d\theta = \frac{8}{3}\int_0^{\pi/2}\cos^3\theta\,d\theta = 16/9$

26. $\displaystyle\int_0^{\pi/2}\int_0^1 \cos(r^2)r\,dr\,d\theta = \frac{1}{2}\sin 1\int_0^{\pi/2} d\theta = \frac{\pi}{4}\sin 1$

27. $\displaystyle\int_0^{\pi/2}\int_0^a \frac{r}{(1+r^2)^{3/2}}dr\,d\theta = \frac{\pi}{2}\left(1 - 1/\sqrt{1+a^2}\right)$

28. $\displaystyle\int_0^{\pi/4}\int_0^{\sec\theta\tan\theta} r^2 dr\,d\theta = \frac{1}{3}\int_0^{\pi/4} \sec^3\theta\tan^3\theta\,d\theta = 2(\sqrt{2}+1)/45$

29. $\displaystyle\int_0^{\pi/4}\int_0^2 \frac{r}{\sqrt{1+r^2}}dr\,d\theta = \frac{\pi}{4}(\sqrt{5}-1)$

30. $\displaystyle\int_{\tan^{-1}(3/4)}^{\pi/2}\int_{3\csc\theta}^5 r\,dr\,d\theta = \frac{1}{2}\int_{\tan^{-1}(3/4)}^{\pi/2} (25 - 9\csc^2\theta)d\theta$

$$= \frac{25}{2}\left[\frac{\pi}{2} - \tan^{-1}(3/4)\right] - 6 = \frac{25}{2}\tan^{-1}(4/3) - 6$$

31. $\displaystyle V = \int_0^{2\pi}\int_0^a hr\,dr\,d\theta = \int_0^{2\pi} h\frac{a^2}{2}\,d\theta = \pi a^2 h$

32. (a) $\displaystyle V = 8\int_0^{\pi/2}\int_0^a \frac{c}{a}(a^2-r^2)^{1/2}\,r\,dr\,d\theta = -\frac{4c}{3a}\pi(a^2-r^2)^{3/2}\Big]_0^a = \frac{4}{3}\pi a^2 c$

(b) $V \approx \frac{4}{3}\pi(6378.1370)^2 6356.5231 \approx 1{,}083{,}168{,}200{,}000\text{ km}^3$

33. $\displaystyle V = 2\int_0^{\pi/2}\int_0^{a\sin\theta} \frac{c}{a}(a^2-r^2)^{1/2}r\,dr\,d\theta = \frac{2}{3}a^2c\int_0^{\pi/2}(1-\cos^3\theta)d\theta = (3\pi-4)a^2c/9$

34. $\displaystyle A = 4\int_0^{\pi/4}\int_0^{a\sqrt{2\cos 2\theta}} r\,dr\,d\theta = 4a^2\int_0^{\pi/4}\cos 2\theta\,d\theta = 2a^2$

35. $\displaystyle A = \int_{\pi/6}^{\pi/4}\int_{\sqrt{8\cos 2\theta}}^{4\sin\theta} r\,dr\,d\theta + \int_{\pi/4}^{\pi/2}\int_0^{4\sin\theta} r\,dr\,d\theta$

$$= \int_{\pi/6}^{\pi/4}(8\sin^2\theta - 4\cos 2\theta)d\theta + \int_{\pi/4}^{\pi/2} 8\sin^2\theta\,d\theta = 4\pi/3 + 2\sqrt{3} - 2$$

36. $\displaystyle A = \int_0^{\phi}\int_0^{2a\sin\theta} r\,dr\,d\theta = 2a^2\int_0^{\phi}\sin^2\theta\,d\theta = a^2\phi - \frac{1}{2}a^2\sin 2\phi$

37. (a) $\displaystyle I^2 = \left[\int_0^{+\infty} e^{-x^2}dx\right]\left[\int_0^{+\infty} e^{-y^2}dy\right] = \int_0^{+\infty}\left[\int_0^{+\infty} e^{-x^2}dx\right]e^{-y^2}dy$

$$= \int_0^{+\infty}\int_0^{+\infty} e^{-x^2}e^{-y^2}dx\,dy = \int_0^{+\infty}\int_0^{+\infty} e^{-(x^2+y^2)}dx\,dy$$

(b) $\displaystyle I^2 = \int_0^{\pi/2}\int_0^{+\infty} e^{-r^2}r\,dr\,d\theta = \frac{1}{2}\int_0^{\pi/2} d\theta = \pi/4$ (c) $I = \sqrt{\pi}/2$

38. (a) 1.173108605 (b) $\displaystyle\int_0^{\pi}\int_0^1 re^{-r^4}\,dr\,d\theta = \pi\int_0^1 re^{-r^4}\,dr \approx 1.173108605$

39. $V = \int_0^{2\pi} \int_0^R D(r) r\, dr\, d\theta = \int_0^{2\pi} \int_0^R k e^{-r} r\, dr\, d\theta = -2\pi k(1+r)e^{-r}\Big]_0^R = 2\pi k[1 - (R+1)e^{-R}]$

40. $\int_{\tan^{-1}(1/3)}^{\tan^{-1}(2)} \int_0^2 r^3 \cos^2\theta\, dr\, d\theta = 4\int_{\tan^{-1}(1/3)}^{\tan^{-1}(2)} \cos^2\theta\, d\theta = \frac{1}{5} + 2[\tan^{-1}(2) - \tan^{-1}(1/3)] = \frac{1}{5} + \frac{\pi}{2}$

EXERCISE SET 15.4

1. **(a)**

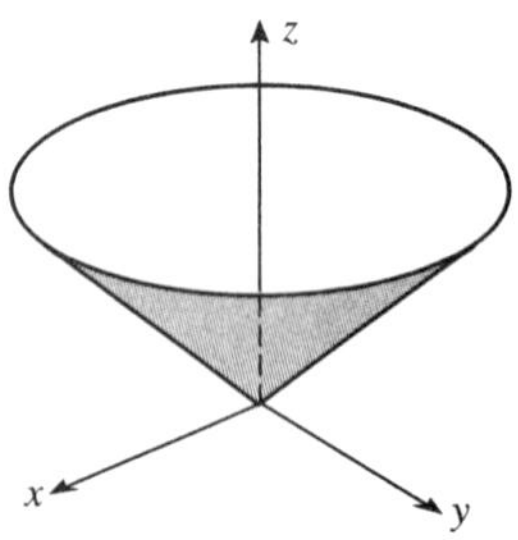

(b)

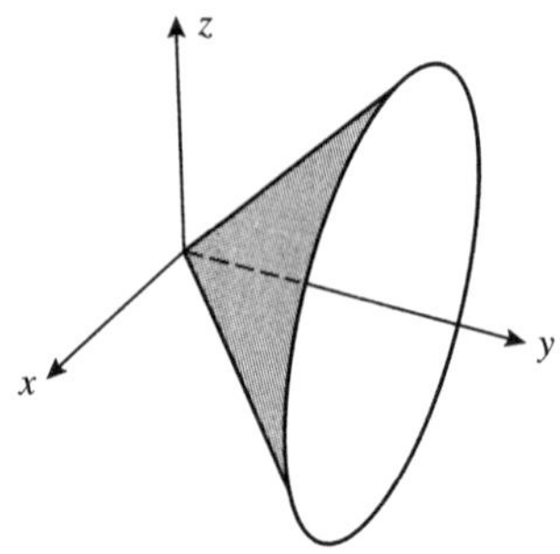

(c)

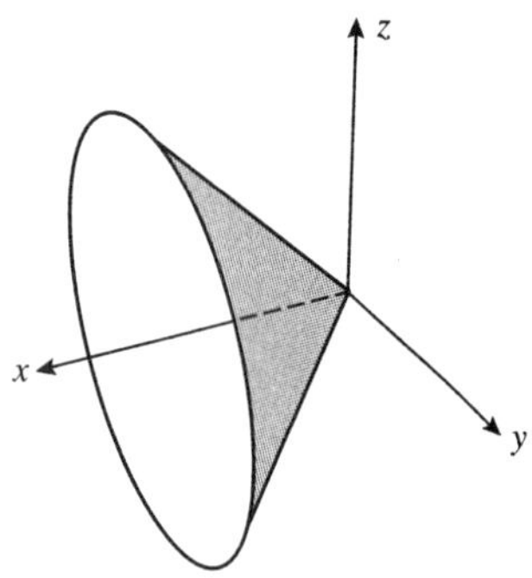

2. **(a)**

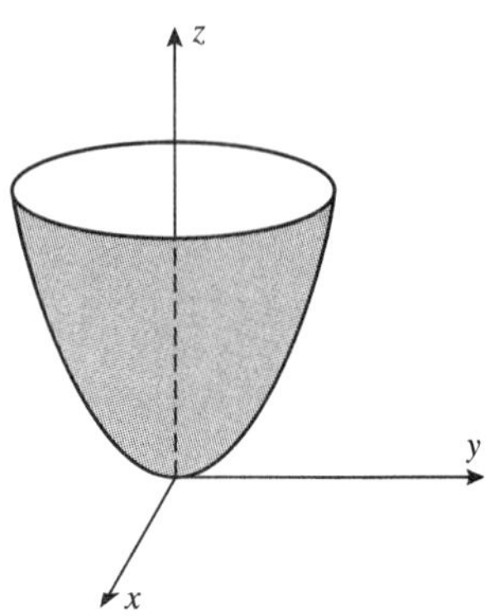

(b)

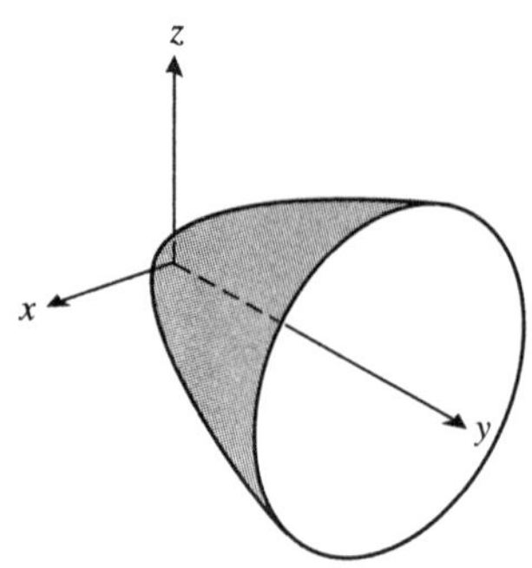

(c)

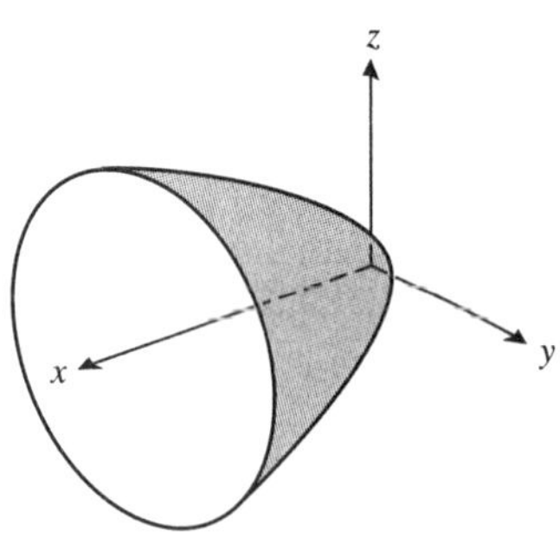

3. (a) $x = u, y = v, z = \frac{5}{2} + \frac{3}{2}u - 2v$ (b) $x = u, y = v, z = u^2$

4. (a) $x = u, y = v, z = \frac{v}{1+u^2}$ (b) $x = u, y = v, z = \frac{1}{3}v^2 - \frac{5}{3}$

5. (a) $x = 5\cos u, y = 5\sin u, z = v; 0 \le u \le 2\pi, 0 \le v \le 1$

(b) $x = 2\cos u, y = v, z = 2\sin u; 0 \le u \le 2\pi, 1 \le v \le 3$

6. (a) $x = u, y = 1 - u, z = v; -1 \le v \le 1$ (b) $x = u, y = 5 + 2v, z = v; 0 \le u \le 3$

7. $x = u, y = \sin u \cos v, z = \sin u \sin v$

8. $x = u, y = e^u \cos v, z = e^u \sin v$

9. $x = r\cos\theta, y = r\sin\theta, z = \frac{1}{1+r^2}$

10. $x = r\cos\theta, y = r\sin\theta, z = e^{-r^2}$

11. $x = r\cos\theta, y = r\sin\theta, z = 2r^2\cos\theta\sin\theta$

12. $x = r\cos\theta, y = r\sin\theta, z = r^2(\cos^2\theta - \sin^2\theta)$

13. $x = r\cos\theta, y = r\sin\theta, z = \sqrt{9 - r^2}; r \le \sqrt{5}$

14. $x = r\cos\theta, y = r\sin\theta, z = r; r \le 3$

15. $x = \frac{1}{2}\rho\cos\theta, y = \frac{1}{2}\rho\sin\theta, z = \frac{\sqrt{3}}{2}\rho$

16. $x = 3\cos\theta, y = 3\sin\theta, z = 3\cot\phi$

17. $z = x - 2y$; a plane

18. $y = x^2 + z^2, 0 \le y \le 4$; part of a circular paraboloid

19. $(x/3)^2 + (y/2)^2 = 1; 2 \le z \le 4$; part of an elliptic cylinder

20. $z = x^2 + y^2; 0 \le z \le 4$; part of a circular paraboloid

21. $(x/3)^2 + (y/4)^2 = z^2; 0 \le z \le 1$; part of an elliptic cone

22. $x^2 + (y/2)^2 + (z/3)^2 = 1$; an ellipsoid

23. (a) $x = r\cos\theta, y = r\sin\theta, z = r, 0 \le r \le 2; x = u, y = v, z = \sqrt{u^2 + v^2}; 0 \le u^2 + v^2 \le 4$

24. (a) I: $x = r\cos\theta, y = r\sin\theta, z = r^2, 0 \le r \le \sqrt{2}$; II: $x = u, y = v, z = u^2 + v^2; u^2 + v^2 \le 2$

25. (a) $0 \le u \le 3, 0 \le v \le \pi$ (b) $0 \le u \le 4, -\pi/2 \le v \le \pi/2$

26. (a) $0 \le u \le 6, -\pi \le v \le 0$ (b) $0 \le u \le 5, \pi/2 \le v \le 3\pi/2$

27. (a) $0 \le \phi \le \pi/2, 0 \le \theta \le 2\pi$ (b) $0 \le \phi \le \pi, 0 \le \theta \le \pi$

28. (a) $\pi/2 \le \phi \le \pi, 0 \le \theta \le 2\pi$ (b) $0 \le \theta \le \pi/2, 0 \le \phi \le \pi/2$

29. $u = 1, v = 2, \mathbf{r}_u \times \mathbf{r}_v = -2\mathbf{i} - 4\mathbf{j} + \mathbf{k}; 2x + 4y - z = 5$

30. $u = 1, v = 2, \mathbf{r}_u \times \mathbf{r}_v = -4\mathbf{i} - 2\mathbf{j} + 8\mathbf{k}; 2x + y - 4z = -6$

31. $u = 0, v = 1, \mathbf{r}_u \times \mathbf{r}_v = 6\mathbf{k}; z = 0$

32. $\mathbf{r}_u \times \mathbf{r}_v = 2\mathbf{i} - \mathbf{j} - 3\mathbf{k}; 2x - y - 3z = -4$

33. $\mathbf{r}_u \times \mathbf{r}_v = (\sqrt{2}/2)\mathbf{i} - (\sqrt{2}/2)\mathbf{j} + (1/2)\mathbf{k}; x - y + \dfrac{\sqrt{2}}{2}z = \dfrac{\pi\sqrt{2}}{8}$

34. $\mathbf{r}_u \times \mathbf{r}_v = 2\mathbf{i} - \ln 2\mathbf{k}; 2x - (\ln 2)z = 0$

35. $z = \sqrt{9-y^2},\ z_x = 0,\ z_y = -y/\sqrt{9-y^2},\ z_x^2 + z_y^2 + 1 = 9/(9-y^2),$

$$S = \int_0^2 \int_{-3}^3 \frac{3}{\sqrt{9-y^2}}\,dy\,dx = \int_0^2 3\pi\,dx = 6\pi$$

36. $z = 8 - 2x - 2y,\ z_x^2 + z_y^2 + 1 = 4 + 4 + 1 = 9,\ S = \displaystyle\int_0^4 \int_0^{4-x} 3\,dy\,dx = \int_0^4 3(4-x)dx = 24$

37. $z^2 = 4x^2 + 4y^2,\ 2zz_x = 8x$ so $z_x = 4x/z$, similarly $z_y = 4y/z$ thus

$$z_x^2 + z_y^2 + 1 = (16x^2 + 16y^2)/z^2 + 1 = 5,\ S = \int_0^1 \int_{x^2}^x \sqrt{5}\,dy\,dx = \sqrt{5}\int_0^1 (x - x^2)dx = \sqrt{5}/6$$

38. $z^2 = x^2 + y^2,\ z_x = x/z,\ z_y = y/z,\ z_x^2 + z_y^2 + 1 = (z^2 + y^2)/z^2 + 1 = 2,$

$$S = \iint_R \sqrt{2}\,dA = 2\int_0^{\pi/2} \int_0^{2\cos\theta} \sqrt{2}\,r\,dr\,d\theta = 4\sqrt{2}\int_0^{\pi/2} \cos^2\theta\,d\theta = \sqrt{2}\pi$$

39. $z_x = -2x,\ z_y = -2y,\ z_x^2 + z_y^2 + 1 = 4x^2 + 4y^2 + 1,$

$$S = \iint_R \sqrt{4x^2 + 4y^2 + 1}\,dA = \int_0^{2\pi} \int_0^1 r\sqrt{4r^2 + 1}\,dr\,d\theta$$

$$= \frac{1}{12}(5\sqrt{5} - 1)\int_0^{2\pi} d\theta = (5\sqrt{5} - 1)\pi/6$$

40. $z_x = 2,\ z_y = 2y,\ z_x^2 + z_y^2 + 1 = 5 + 4y^2,$

$$S = \int_0^1 \int_0^y \sqrt{5 + 4y^2}\,dx\,dy = \int_0^1 y\sqrt{5 + 4y^2}\,dy = (27 - 5\sqrt{5})/12$$

41. $\partial\mathbf{r}/\partial u = \cos v\mathbf{i} + \sin v\mathbf{j} + 2u\mathbf{k},\ \partial\mathbf{r}/\partial v = -u\sin v\mathbf{i} + u\cos v\mathbf{j},$

$$\|\partial\mathbf{r}/\partial u \times \partial\mathbf{r}/\partial v\| = u\sqrt{4u^2 + 1};\ S = \int_0^{2\pi} \int_1^2 u\sqrt{4u^2 + 1}\,du\ dv = (17\sqrt{17} - 5\sqrt{5})\pi/6$$

42. $\partial\mathbf{r}/\partial u = \cos v\mathbf{i} + \sin v\mathbf{j} + \mathbf{k},\ \partial\mathbf{r}/\partial v = -u\sin v\mathbf{i} + u\cos v\mathbf{j},$

$$\|\partial\mathbf{r}/\partial u \times \partial\mathbf{r}/\partial v\| = \sqrt{2}u;\ S = \int_0^{\pi/2} \int_0^{2v} \sqrt{2}\ u\ du\ dv = \frac{\sqrt{2}}{12}\pi^3$$

43. $z_x = y,\ z_y = x,\ z_x^2 + z_y^2 + 1 = x^2 + y^2 + 1,$

$$S = \iint_R \sqrt{x^2 + y^2 + 1}\,dA = \int_0^{\pi/6} \int_0^3 r\sqrt{r^2 + 1}\,dr\,d\theta = \frac{1}{3}(10\sqrt{10} - 1)\int_0^{\pi/6} d\theta = (10\sqrt{10} - 1)\pi/18$$

44. $z_x = x$, $z_y = y$, $z_x^2 + z_y^2 + 1 = x^2 + y^2 + 1$,

$$S = \iint_R \sqrt{x^2+y^2+1}\, dA = \int_0^{2\pi}\int_0^{\sqrt{8}} r\sqrt{r^2+1}\, dr\, d\theta = \frac{26}{3}\int_0^{2\pi} d\theta = 52\pi/3$$

45. On the sphere, $z_x = -x/z$ and $z_y = -y/z$ so $z_x^2 + z_y^2 + 1 = (x^2+y^2+z^2)/z^2 = 16/(16 - x^2 - y^2)$; the planes $z = 1$ and $z = 2$ intersect the sphere along the circles $x^2 + y^2 = 15$ and $x^2 + y^2 = 12$;

$$S = \iint_R \frac{4}{\sqrt{16 - x^2 - y^2}} dA = \int_0^{2\pi}\int_{\sqrt{12}}^{\sqrt{15}} \frac{4r}{\sqrt{16-r^2}}\, dr\, d\theta = 4\int_0^{2\pi} d\theta = 8\pi$$

46. On the sphere, $z_x = -x/z$ and $z_y = -y/z$ so $z_x^2 + z_y^2 + 1 = (x^2+y^2+z^2)/z^2 = 8/(8 - x^2 - y^2)$; the cone cuts the sphere in the circle $x^2 + y^2 = 4$;

$$S = \int_0^{2\pi}\int_0^2 \frac{2\sqrt{2}r}{\sqrt{8-r^2}} dr\, d\theta = (8 - 4\sqrt{2})\int_0^{2\pi} d\theta = 8(2-\sqrt{2})\pi$$

47. $\mathbf{r}(u,v) = a\cos u \sin v\mathbf{i} + a\sin u\sin v\mathbf{j} + a\cos v\mathbf{k}$, $\|\mathbf{r}_u \times \mathbf{r}_v\| = a^2\sin v$,

$$S = \int_0^{\pi}\int_0^{2\pi} a^2 \sin v\, du\, dv = 2\pi a^2 \int_0^{\pi} \sin v\, dv = 4\pi a^2$$

48. $\mathbf{r} = r\cos u\mathbf{i} + r\sin u\mathbf{j} + v\mathbf{k}$, $\|\mathbf{r}_u \times \mathbf{r}_v\| = r$; $S = \displaystyle\int_0^h\int_0^{2\pi} r\, du\, dv = 2\pi r h$

49. $z_x = \dfrac{h}{a}\dfrac{x}{\sqrt{x^2+y^2}}$, $z_y = \dfrac{h}{a}\dfrac{y}{\sqrt{x^2+y^2}}$, $z_x^2 + z_y^2 + 1 = \dfrac{h^2x^2 + h^2y^2}{a^2(x^2+y^2)} + 1 = (a^2+h^2)/a^2$,

$$S = \int_0^{2\pi}\int_0^a \frac{\sqrt{a^2+h^2}}{a} r\, dr\, d\theta = \frac{1}{2}a\sqrt{a^2+h^2}\int_0^{2\pi} d\theta = \pi a\sqrt{a^2+h^2}$$

50. Revolving a point $(a_0, 0, b_0)$ of the xz-plane around the z-axis generates a circle, an equation of which is $\mathbf{r} = a_0\cos u\mathbf{i} + a_0 \sin u\mathbf{j} + b_0\mathbf{k}, 0 \le u \le 2\pi$. A point on the circle $(x-a)^2 + z^2 = b^2$ which generates the torus can be written $\mathbf{r} = (a + b\cos v)\mathbf{i} + b\sin v\mathbf{k}, 0 \le v \le 2\pi$. Set $a_0 = a + b\cos v$ and $b_0 = a + b\sin v$ and use the first result: any point on the torus can thus be written in the form $\mathbf{r} = (a + b\cos v)\cos u\mathbf{i} + (a + b\cos v)\sin u\mathbf{j} + b\sin v\mathbf{k}$, which yields the result.

51. $\partial\mathbf{r}/\partial u = -(a + b\cos v)\sin u\mathbf{i} + (a + b\cos v)\cos u\mathbf{j}$,
$\partial\mathbf{r}/\partial v = -b\sin v\cos u\mathbf{i} - b\sin v\sin u\mathbf{j} + b\cos v\mathbf{k}$, $\|\partial\mathbf{r}/\partial u \times \partial\mathbf{r}/\partial v\| = b(a + b\cos v)$;

$$S = \int_0^{2\pi}\int_0^{2\pi} b(a + b\cos v)du\ dv = 4\pi^2 ab$$

52. $\|\mathbf{r}_u \times \mathbf{r}_v\| = \sqrt{u^2+1}$; $S = \displaystyle\int_0^{4\pi}\int_0^5 \sqrt{u^2+1}\, du\, dv = 4\pi\int_0^5 \sqrt{u^2+1}\, du = 174.7199011$

53. $z = -1$ when $v \approx 0.27955$, $z = 1$ when $v \approx 2.86204$, $\|\mathbf{r}_u \times \mathbf{r}_v\| = |\cos v|$;

$$S = \int_0^{2\pi}\int_{0.27955}^{2.86204} |\cos v|\, dv\, du \approx 9.099$$

54. **(a)** $x = v\cos u, y = v\sin u, z = f(v)$, for example **(b)** $x = v\cos u, y = v\sin u, z = 1/v^2$

(c)

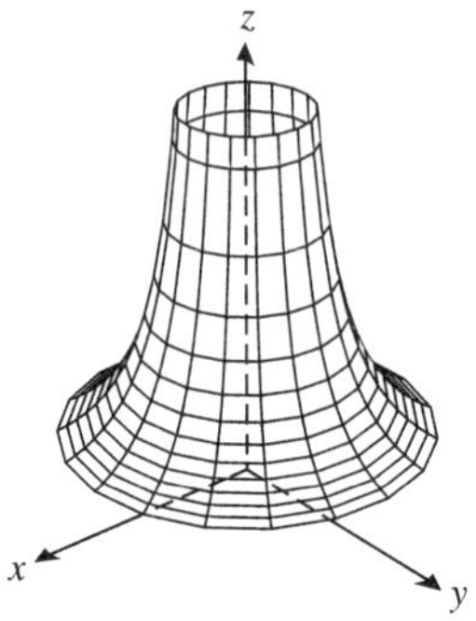

55. $(x/a)^2 + (y/b)^2 + (z/c)^2 = \cos^2 v(\cos^2 u + \sin^2 u) + \sin^2 v = 1$, ellipsoid

56. $(x/a)^2 + (y/b)^2 - (z/c)^2 = \cos^2 u\cosh^2 v + \sin^2 u\cosh^2 v - \sinh^2 v = 1$, hyperboloid of one sheet

57. $(x/a)^2 + (y/b)^2 - (z/c)^2 = \sinh^2 v + \cosh^2 v(\sinh^2 u - \cosh^2 u) = -1$, hyperboloid of two sheets

EXERCISE SET 15.5

1. $\displaystyle\int_{-1}^{1}\int_0^2\int_0^1 (x^2+y^2+z^2)dx\,dy\,dz = \int_{-1}^{1}\int_0^2 (1/3+y^2+z^2)dy\,dz = \int_{-1}^{1}(10/3+2z^2)dz = 8$

2. $\displaystyle\int_{1/3}^{1/2}\int_0^{\pi}\int_0^1 zx\sin xy\,dz\,dy\,dx = \int_{1/3}^{1/2}\int_0^{\pi}\frac{1}{2}x\sin xy\,dy\,dx = \int_{1/3}^{1/2}\frac{1}{2}(1-\cos\pi x)dx = \frac{1}{12}+\frac{\sqrt{3}-2}{4\pi}$

3. $\displaystyle\int_0^2\int_{-1}^{y^2}\int_{-1}^{z} yz\,dx\,dz\,dy = \int_0^2\int_{-1}^{y^2}(yz^2+yz)dz\,dy = \int_0^2\left(\frac{1}{3}y^7+\frac{1}{2}y^5-\frac{1}{6}y\right)dy = \frac{47}{3}$

4. $\displaystyle\int_0^{\pi/4}\int_0^1\int_0^{x^2} x\cos y\,dz\,dx\,dy = \int_0^{\pi/4}\int_0^1 x^3\cos y\,dx\,dy = \int_0^{\pi/4}\frac{1}{4}\cos y\,dy = \sqrt{2}/8$

5. $\displaystyle\int_0^3\int_0^{\sqrt{9-z^2}}\int_0^x xy\,dy\,dx\,dz = \int_0^3\int_0^{\sqrt{9-z^2}}\frac{1}{2}x^3dx\,dz = \int_0^3\frac{1}{8}(81-18z^2+z^4)dz = 81/5$

6. $\displaystyle\int_1^3\int_x^{x^2}\int_0^{\ln z} xe^y\,dy\,dz\,dx = \int_1^3\int_x^{x^2}(xz-x)dz\,dx = \int_1^3\left(\frac{1}{2}x^5-\frac{3}{2}x^3+x^2\right)dx = 118/3$

7. $\displaystyle\int_0^2\int_0^{\sqrt{4-x^2}}\int_{-5+x^2+y^2}^{3-x^2-y^2} x\,dz\,dy\,dx = \int_0^2\int_0^{\sqrt{4-x^2}}[2x(4-x^2)-2xy^2]dy\,dx$

$$= \int_0^2\frac{4}{3}x(4-x^2)^{3/2}dx = 128/15$$

8. $\displaystyle\int_1^2\int_z^2\int_0^{\sqrt{3}y}\frac{y}{x^2+y^2}dx\,dy\,dz = \int_1^2\int_z^2\frac{\pi}{3}dy\,dz = \int_1^2\frac{\pi}{3}(2-z)dz = \pi/6$

9. $\int_0^{\pi}\int_0^1\int_0^{\pi/6} xy\sin yz\,dz\,dy\,dx = \int_0^{\pi}\int_0^1 x[1-\cos(\pi y/6)]dy\,dx = \int_0^{\pi}(1-3/\pi)x\,dx = \pi(\pi-3)/2$

10. $\int_{-1}^1\int_0^{1-x^2}\int_0^y y\,dz\,dy\,dx = \int_{-1}^1\int_0^{1-x^2} y^2dy\,dx = \int_{-1}^1 \frac{1}{3}(1-x^2)^3dx = 32/105$

11. $\int_0^{\sqrt{2}}\int_0^x\int_0^{2-x^2} xyz\,dz\,dy\,dx = \int_0^{\sqrt{2}}\int_0^x \frac{1}{2}xy(2-x^2)^2dy\,dx = \int_0^{\sqrt{2}}\frac{1}{4}x^3(2-x^2)^2dx = 1/6$

12. $\int_{\pi/6}^{\pi/2}\int_y^{\pi/2}\int_0^{xy}\cos(z/y)dz\,dx\,dy = \int_{\pi/6}^{\pi/2}\int_y^{\pi/2} y\sin x\,dx\,dy = \int_{\pi/6}^{\pi/2} y\cos y\,dy = (5\pi-6\sqrt{3})/12$

13. $\int_0^3\int_1^2\int_{-2}^1 \frac{\sqrt{x+z^2}}{y}\,dzdydx \approx 9.425$

14. $8\int_0^1\int_0^{\sqrt{1-x^2}}\int_0^{\sqrt{1-x^2-y^2}} e^{-x^2-y^2-z^2}\,dz\,dy\,dx \approx 2.381$

15. $V = \int_0^4\int_0^{(4-x)/2}\int_0^{(12-3x-6y)/4} dz\,dy\,dx = \int_0^4\int_0^{(4-x)/2}\frac{1}{4}(12-3x-6y)dy\,dx$

$= \int_0^4 \frac{3}{16}(4-x)^2dx = 4$

16. $V = \int_0^1\int_0^{1-x}\int_0^{\sqrt{y}} dz\,dy\,dx = \int_0^1\int_0^{1-x}\sqrt{y}\,dy\,dx = \int_0^1\frac{2}{3}(1-x)^{3/2}dx = 4/15$

17. $V = 2\int_0^2\int_{x^2}^4\int_0^{4-y} dz\,dy\,dx = 2\int_0^2\int_{x^2}^4(4-y)dy\,dx = 2\int_0^2\left(8-4x^2+\frac{1}{2}x^4\right)dx = 256/15$

18. $V = \int_0^1\int_0^y\int_0^{\sqrt{1-y^2}} dz\,dx\,dy = \int_0^1\int_0^y\sqrt{1-y^2}\,dx\,dy = \int_0^1 y\sqrt{1-y^2}\,dy = 1/3$

19. The projection of the curve of intersection onto the xy-plane is $x^2+y^2=1$,

$V = 4\int_0^1\int_0^{\sqrt{1-x^2}}\int_{4x^2+y^2}^{4-3y^2} dz\,dy\,dx$

20. The projection of the curve of intersection onto the xy-plane is $2x^2+y^2=4$,

$V = 4\int_0^{\sqrt{2}}\int_0^{\sqrt{4-2x^2}}\int_{3x^2+y^2}^{8-x^2-y^2} dz\,dy\,dx$

21. $V = 2\int_{-3}^3\int_0^{\sqrt{9-x^2}/3}\int_0^{x+3} dz\,dy\,dx$

22. $V = 8\int_0^1\int_0^{\sqrt{1-x^2}}\int_0^{\sqrt{1-x^2}} dz\,dy\,dx$

23. (a)

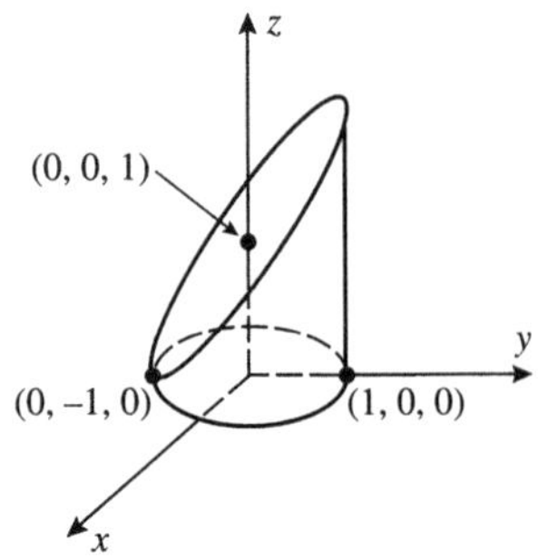

(b)

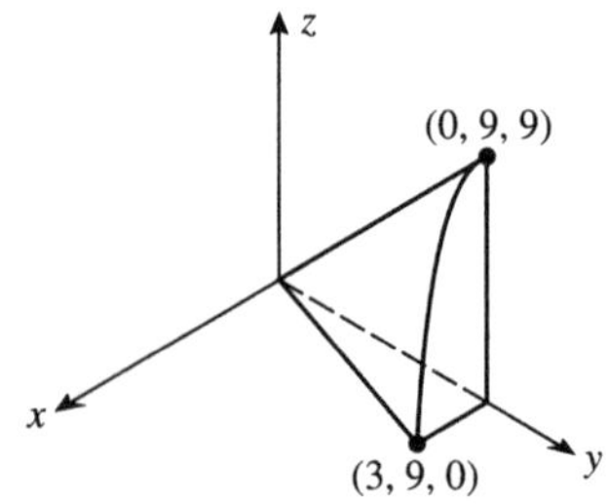

(c)

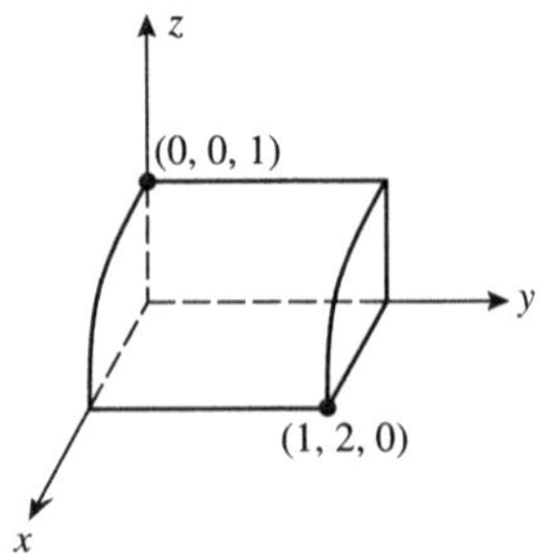

24. (a)

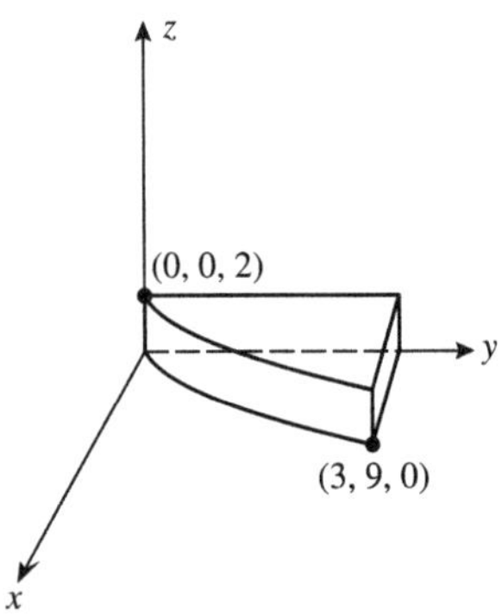

(b)

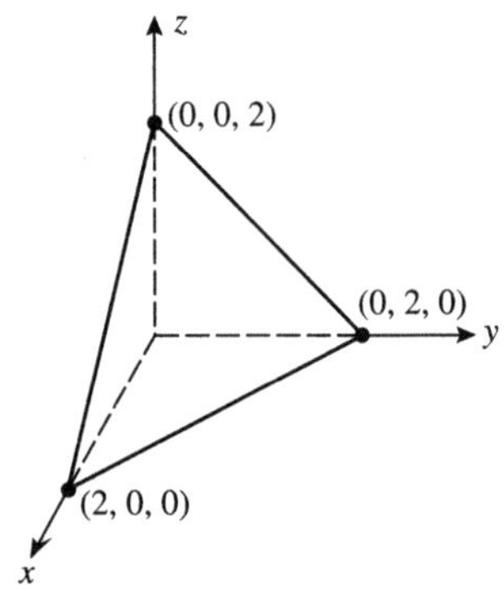

(c)

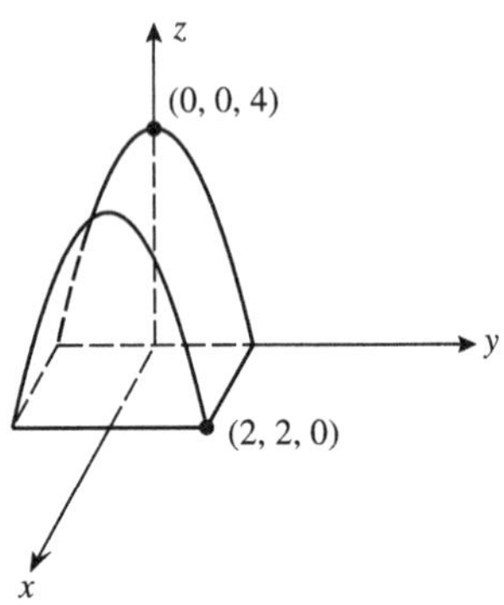

25. $V = \int_0^1 \int_0^{1-x} \int_0^{1-x-y} dz\,dy\,dx = 1/6,\ f_{\text{ave}} = 6 \int_0^1 \int_0^{1-x} \int_0^{1-x-y} (x+y+z)\,dz\,dy\,dx = \dfrac{3}{4}$

26. The integrand is an odd function of each of x, y, and z, so the answer is zero.

27. The volume $V = \dfrac{3\pi}{\sqrt{2}}$, and thus

$$r_{\text{ave}} = \frac{\sqrt{2}}{3\pi} \iiint\limits_G \sqrt{x^2+y^2+z^2}\,dV = \frac{\sqrt{2}}{3\pi} \int_{-1/\sqrt{2}}^{1/\sqrt{2}} \int_{-\sqrt{1-2x^2}}^{\sqrt{1-2x^2}} \int_{5x^2+5y^2}^{6-7x^2-y^2} \sqrt{x^2+y^2+z^2}\,dz dy dx \approx 3.291$$

28. $V = 1, d_{\text{ave}} = \dfrac{1}{V}\displaystyle\int_0^1\int_0^1\int_0^1 \sqrt{(x-z)^2+(y-z)^2+z^2}\,dxdydz \approx 0.771$

29. (a) $\displaystyle\int_0^a\int_0^{b(1-x/a)}\int_0^{c(1-x/a-y/b)} dz\,dy\,dx,\ \int_0^b\int_0^{a(1-y/b)}\int_0^{c(1-x/a-y/b)} dz\,dx\,dy,$

$\displaystyle\int_0^c\int_0^{a(1-z/c)}\int_0^{b(1-x/a-z/c)} dy\,dx\,dz,\ \int_0^a\int_0^{c(1-x/a)}\int_0^{b(1-x/a-z/c)} dy\,dz\,dx,$

$\displaystyle\int_0^c\int_0^{b(1-z/c)}\int_0^{a(1-y/b-z/c)} dx\,dy\,dz,\ \int_0^b\int_0^{c(1-y/b)}\int_0^{a(1-y/b-z/c)} dx\,dz\,dy$

(b) Use the first integral in Part (a) to get

$$\int_0^a\int_0^{b(1-x/a)} c\left(1-\frac{x}{a}-\frac{y}{b}\right) dy\,dx = \int_0^a \frac{1}{2}bc\left(1-\frac{x}{a}\right)^2 dx = \frac{1}{6}abc$$

30. $V = 8\displaystyle\int_0^a\int_0^{b\sqrt{1-x^2/a^2}}\int_0^{c\sqrt{1-x^2/a^2-y^2/b^2}} dz\,dy\,dx$

31. (a) $\displaystyle\int_0^2\int_0^{\sqrt{4-x^2}}\int_0^5 f(x,y,z)\,dzdydx$

(b) $\displaystyle\int_0^9\int_0^{3-\sqrt{x}}\int_y^{3-\sqrt{x}} f(x,y,z)\,dzdydx$

(c) $\displaystyle\int_0^2\int_0^{4-x^2}\int_y^{8-y} f(x,y,z)\,dzdydx$

32. (a) $\displaystyle\int_0^3\int_0^{\sqrt{9-x^2}}\int_0^{\sqrt{9-x^2-y^2}} f(x,y,z)dz\,dy\,dx$

(b) $\displaystyle\int_0^4\int_0^{x/2}\int_0^2 f(x,y,z)dz\,dy\,dx$

(c) $\displaystyle\int_0^2\int_0^{4-x^2}\int_{x^2}^{4-y} f(x,y,z)dz\,dy\,dx$

33. (a) At any point outside the closed sphere $\{x^2+y^2+z^2 \le 1\}$ the integrand is negative, so to maximize the integral it suffices to include all points inside the sphere; hence the maximum value is taken on the region $G = \{x^2+y^2+z^2 \le 1\}$.

(b) 4.934802202

(c) $\displaystyle\int_0^{2\pi}\int_0^{\pi}\int_0^1 (1-\rho^2)\rho\,d\rho\,d\phi\,d\theta = \frac{\pi^2}{2}$

34.
$$\int_a^b\int_c^d\int_k^\ell f(x)g(y)h(z)dz\,dy\,dx = \int_a^b\int_c^d f(x)g(y)\left[\int_k^\ell h(z)dz\right]dy\,dx$$
$$= \left[\int_a^b f(x)\left[\int_c^d g(y)dy\right]dx\right]\left[\int_k^\ell h(z)dz\right]$$
$$= \left[\int_a^b f(x)dx\right]\left[\int_c^d g(y)dy\right]\left[\int_k^\ell h(z)dz\right]$$

35. (a) $\displaystyle\left[\int_{-1}^1 x\,dx\right]\left[\int_0^1 y^2dy\right]\left[\int_0^{\pi/2}\sin z\,dz\right] = (0)(1/3)(1) = 0$

(b) $\displaystyle\left[\int_0^1 e^{2x}dx\right]\left[\int_0^{\ln 3} e^y dy\right]\left[\int_0^{\ln 2} e^{-z}dz\right] = [(e^2-1)/2](2)(1/2) = (e^2-1)/2$

EXERCISE SET 15.6

1. Let a be the unknown coordinate of the fulcrum; then the total moment about the fulcrum is $5(0-a)+10(5-a)+20(10-a)=0$ for equilibrium, so $250-35a=0$, $a=50/7$. The fulcrum should be placed $50/7$ units to the right of m_1.

2. At equilibrium, $10(0-4)+3(2-4)+4(3-4)+m(6-4)=0, m=25$

3. $A=1,\ \overline{x}=\int_0^1\int_0^1 x\,dy\,dx=\frac{1}{2},\ \overline{y}=\int_0^1\int_0^1 y\,dy\,dx=\frac{1}{2}$

4. $A=2, \overline{x}=\frac{1}{2}\iint\limits_G x\,dy\,dx$, and the region of integration is symmetric with respect to the x-axes and the integrand is an odd function of x, so $\overline{x}=0$. Likewise, $\overline{y}=0$.

5. $A=1/2,\ \iint\limits_R x\,dA=\int_0^1\int_0^x x\,dy\,dx=1/3,\ \iint\limits_R y\,dA=\int_0^1\int_0^x y\,dy\,dx=1/6;$

centroid $(2/3,1/3)$

6. $A=\int_0^1\int_0^{x^2}dy\,dx=1/3,\ \iint\limits_R x\,dA=\int_0^1\int_0^{x^2}x\,dy\,dx=1/4,$

$\iint\limits_R y\,dA=\int_0^1\int_0^{x^2}y\,dy\,dx=1/10$; centroid $(3/4,3/10)$

7. $A=\int_0^1\int_x^{2-x^2}dy\,dx=7/6,\ \iint\limits_R x\,dA=\int_0^1\int_x^{2-x^2}x\,dy\,dx=5/12,$

$\iint\limits_R y\,dA=\int_0^1\int_x^{2-x^2}y\,dy\,dx=19/15$; centroid $(5/14,38/35)$

8. $A=\frac{\pi}{4},\ \iint\limits_R x\,dA=\int_0^1\int_0^{\sqrt{1-x^2}}x\,dy\,dx=\frac{1}{3},\ \overline{x}=\frac{4}{3\pi},\ \overline{y}=\frac{4}{3\pi}$ by symmetry

9. $\overline{x}=0$ from the symmetry of the region,

$A=\frac{1}{2}\pi(b^2-a^2),\ \iint\limits_R y\,dA=\int_0^\pi\int_a^b r^2\sin\theta\,dr\,d\theta=\frac{2}{3}(b^3-a^3)$; centroid $\overline{x}=0,\ \overline{y}=\frac{4(b^3-a^3)}{3\pi(b^2-a^2)}$.

10. $\overline{y}=0$ from the symmetry of the region, $A=\pi a^2/2$,

$\iint\limits_R x\,dA=\int_{-\pi/2}^{\pi/2}\int_0^a r^2\cos\theta\,dr\,d\theta=2a^3/3$; centroid $\left(\frac{4a}{3\pi},0\right)$

11. $M=\iint\limits_R \delta(x,y)dA=\int_0^1\int_0^1|x+y-1|\,dxdy$

$$=\int_0^1\left[\int_0^{1-x}(1-x-y)\,dy+\int_{1-x}^1(x+y-1)\,dy\right]dx=\frac{1}{3}$$

$$\overline{x} = 3\int_0^1\int_0^1 x\delta(x,y)\,dy\,dx = 3\int_0^1\left[\int_0^{1-x} x(1-x-y)\,dy + \int_{1-x}^1 x(x+y-1)\,dy\right]dx = \frac{1}{2}$$

By symmetry, $\overline{y} = \dfrac{1}{2}$ as well; center of gravity $(1/2, 1/2)$

12. $\overline{x} = \dfrac{1}{M}\displaystyle\iint_G x\delta(x,y)\,dA$, and the integrand is an odd function of x while the region is symmetric with respect to the y-axis, thus $\overline{x} = 0$; likewise $\overline{y} = 0$.

13. $M = \displaystyle\int_0^1\int_0^{\sqrt{x}}(x+y)dy\,dx = 13/20$, $M_x = \displaystyle\int_0^1\int_0^{\sqrt{x}}(x+y)y\,dy\,dx = 3/10$,

$M_y = \displaystyle\int_0^1\int_0^{\sqrt{x}}(x+y)x\,dy\,dx = 19/42$, $\overline{x} = M_y/M = 190/273$, $\overline{y} = M_x/M = 6/13$;

the mass is $13/20$ and the center of gravity is at $(190/273, 6/13)$.

14. $M = \displaystyle\int_0^\pi\int_0^{\sin x} y\,dy\,dx = \pi/4$, $\overline{x} = \pi/2$ from the symmetry of the density and the region,

$M_x = \displaystyle\int_0^\pi\int_0^{\sin x} y^2 dy\,dx = 4/9$, $\overline{y} = M_x/M = \dfrac{16}{9\pi}$; mass $\pi/4$, center of gravity $\left(\dfrac{\pi}{2}, \dfrac{16}{9\pi}\right)$.

15. $M = \displaystyle\int_0^{\pi/2}\int_0^a r^3\sin\theta\cos\theta\,dr\,d\theta = a^4/8$, $\overline{x} = \overline{y}$ from the symmetry of the density and the region, $M_y = \displaystyle\int_0^{\pi/2}\int_0^a r^4\sin\theta\cos^2\theta\,dr\,d\theta = a^5/15$, $\overline{x} = 8a/15$; mass $a^4/8$, center of gravity $(8a/15, 8a/15)$.

16. $M = \displaystyle\int_0^\pi\int_0^1 r^3 dr\,d\theta = \pi/4$, $\overline{x} = 0$ from the symmetry of density and region,

$M_x = \displaystyle\int_0^\pi\int_0^1 r^4\sin\theta\,dr\,d\theta = 2/5$, $\overline{y} = \dfrac{8}{5\pi}$; mass $\pi/4$, center of gravity $\left(0, \dfrac{8}{5\pi}\right)$.

17. $V = 1, \overline{x} = \displaystyle\int_0^1\int_0^1\int_0^1 x\,dz\,dy\,dx = \dfrac{1}{2}$, similarly $\overline{y} = \overline{z} = \dfrac{1}{2}$; centroid $\left(\dfrac{1}{2}, \dfrac{1}{2}, \dfrac{1}{2}\right)$

18. symmetry, $\displaystyle\iiint_G z\,dz\,dy\,dx = \int_0^2\int_0^{2\pi}\int_0^1 rz\,dr\,d\theta\,dz = 2\pi$, centroid $= (0,0,1)$

19. $\overline{x} = \overline{y} = \overline{z}$ from the symmetry of the region, $V = 1/6$,

$\overline{x} = \dfrac{1}{V}\displaystyle\int_0^1\int_0^{1-x}\int_0^{1-x-y} x\,dz\,dy\,dx = (6)(1/24) = 1/4$; centroid $(1/4, 1/4, 1/4)$

20. The solid is described by $-1 \le y \le 1, 0 \le z \le 1-y^2, 0 \le x \le 1-z$;

$V = \displaystyle\int_{-1}^1\int_0^{1-y^2}\int_0^{1-z} dx\,dz\,dy = \dfrac{4}{5}, \overline{x} = \dfrac{1}{V}\int_{-1}^1\int_0^{1-y^2}\int_0^{1-z} x\,dx\,dz\,dy = \dfrac{5}{14}, \overline{y} = 0$ by symmetry,

$\overline{z} = \dfrac{1}{V}\displaystyle\int_{-1}^1\int_0^{1-y^2}\int_0^{1-z} z\,dx\,dz\,dy = \dfrac{2}{7}$; the centroid is $\left(\dfrac{5}{14}, 0, \dfrac{2}{7}\right)$.

21. $\overline{x} = 1/2$ and $\overline{y} = 0$ from the symmetry of the region,

$$V = \int_0^1 \int_{-1}^1 \int_{y^2}^1 dz\,dy\,dx = 4/3,\ \overline{z} = \frac{1}{V}\iiint\limits_G z\,dV = (3/4)(4/5) = 3/5;\ \text{centroid } (1/2, 0, 3/5)$$

22. $\overline{x} = \overline{y}$ from the symmetry of the region,

$$V = \int_0^2 \int_0^2 \int_0^{xy} dz\,dy\,dx = 4,\ \overline{x} = \frac{1}{V}\iiint\limits_G x\,dV = (1/4)(16/3) = 4/3,$$

$$\overline{z} = \frac{1}{V}\iiint\limits_G z\,dV = (1/4)(32/9) = 8/9;\ \text{centroid } (4/3, 4/3, 8/9)$$

23. $\overline{x} = \overline{y} = \overline{z}$ from the symmetry of the region, $V = \pi a^3/6$,

$$\overline{x} = \frac{1}{V}\int_0^a \int_0^{\sqrt{a^2-x^2}} \int_0^{\sqrt{a^2-x^2-y^2}} x\,dz\,dy\,dx = \frac{1}{V}\int_0^a \int_0^{\sqrt{a^2-x^2}} x\sqrt{a^2-x^2-y^2}\,dy\,dx$$

$$= \frac{1}{V}\int_0^{\pi/2} \int_0^a r^2\sqrt{a^2-r^2}\cos\theta\,dr\,d\theta = \frac{6}{\pi a^3}(\pi a^4/16) = 3a/8;\ \text{centroid } (3a/8, 3a/8, 3a/8)$$

24. $\overline{x} = \overline{y} = 0$ from the symmetry of the region, $V = 2\pi a^3/3$

$$\overline{z} = \frac{1}{V}\int_{-a}^a \int_{-\sqrt{a^2-x^2}}^{\sqrt{a^2-x^2}} \int_0^{\sqrt{a^2-x^2-y^2}} z\,dz\,dy\,dx = \frac{1}{V}\int_{-a}^a \int_{-\sqrt{a^2-x^2}}^{\sqrt{a^2-x^2}} \frac{1}{2}(a^2-x^2-y^2)dy\,dx$$

$$= \frac{1}{V}\int_0^{2\pi} \int_0^a \frac{1}{2}(a^2-r^2)r\,dr\,d\theta = \frac{3}{2\pi a^3}(\pi a^4/4) = 3a/8;\ \text{centroid } (0, 0, 3a/8)$$

25. $M = \displaystyle\int_0^a \int_0^a \int_0^a (a-x)dz\,dy\,dx = a^4/2$, $\overline{y} = \overline{z} = a/2$ from the symmetry of density and region, $\overline{x} = \dfrac{1}{M}\displaystyle\int_0^a \int_0^a \int_0^a x(a-x)dz\,dy\,dx = (2/a^4)(a^5/6) = a/3$;
mass $a^4/2$, center of gravity $(a/3, a/2, a/2)$

26. $M = \displaystyle\int_{-a}^a \int_{-\sqrt{a^2-x^2}}^{\sqrt{a^2-x^2}} \int_0^h (h-z)dz\,dy\,dx = \frac{1}{2}\pi a^2h^2$, $\overline{x} = \overline{y} = 0$ from the symmetry of density and region, $\overline{z} = \dfrac{1}{M}\displaystyle\iiint\limits_G z(h-z)dV = \frac{2}{\pi a^2h^2}(\pi a^2h^3/6) = h/3$;

mass $\pi a^2h^2/2$, center of gravity $(0, 0, h/3)$

27. $M = \displaystyle\int_{-1}^1 \int_0^1 \int_0^{1-y^2} yz\,dz\,dy\,dx = 1/6$, $\overline{x} = 0$ by the symmetry of density and region,

$$\overline{y} = \frac{1}{M}\iiint\limits_G y^2z\,dV = (6)(8/105) = 16/35,\ \overline{z} = \frac{1}{M}\iiint\limits_G yz^2dV = (6)(1/12) = 1/2;$$

mass $1/6$, center of gravity $(0, 16/35, 1/2)$

28. $M = \int_0^3 \int_0^{9-x^2} \int_0^1 xz\,dz\,dy\,dx = 81/8$, $\bar{x} = \frac{1}{M}\iiint\limits_G x^2 z\,dV = (8/81)(81/5) = 8/5$,

$\bar{y} = \frac{1}{M}\iiint\limits_G xyz\,dV = (8/81)(243/8) = 3$, $\bar{z} = \frac{1}{M}\iiint\limits_G xz^2 dV = (8/81)(27/4) = 2/3$;

mass 81/8, center of gravity $(8/5, 3, 2/3)$

29. **(a)** $M = \int_0^1 \int_0^1 k(x^2+y^2)dy\,dx = 2k/3$, $\bar{x} = \bar{y}$ from the symmetry of density and region,

$\bar{x} = \frac{1}{M}\iint\limits_R kx(x^2+y^2)dA = \frac{3}{2k}(5k/12) = 5/8$; center of gravity $(5/8, 5/8)$

(b) $\bar{y} = 1/2$ from the symmetry of density and region,

$M = \int_0^1 \int_0^1 kx\,dy\,dx = k/2$, $\bar{x} = \frac{1}{M}\iint\limits_R kx^2 dA = (2/k)(k/3) = 2/3$,

center of gravity $(2/3, 1/2)$

30. **(a)** $\bar{x} = \bar{y} = \bar{z}$ from the symmetry of density and region,

$M = \int_0^1 \int_0^1 \int_0^1 k(x^2+y^2+z^2)dz\,dy\,dx = k$,

$\bar{x} = \frac{1}{M}\iiint\limits_G kx(x^2+y^2+z^2)dV = (1/k)(7k/12) = 7/12$; center of gravity $(7/12, 7/12, 7/12)$

(b) $\bar{x} = \bar{y} = \bar{z}$ from the symmetry of density and region,

$M = \int_0^1 \int_0^1 \int_0^1 k(x+y+z)dz\,dy\,dx = 3k/2$,

$\bar{x} = \frac{1}{M}\iiint\limits_G kx(x+y+z)dV = \frac{2}{3k}(5k/6) = 5/9$; center of gravity $(5/9, 5/9, 5/9)$

31. $V = \iiint\limits_G dV = \int_0^\pi \int_0^{\sin x} \int_0^{1/(1+x^2+y^2)} dz\,dy\,dx = 0.666633$,

$\bar{x} = \frac{1}{V}\iiint\limits_G xdV = 1.177406$, $\bar{y} = \frac{1}{V}\iiint\limits_G ydV = 0.353554$, $\bar{z} = \frac{1}{V}\iiint\limits_G zdV = 0.231557$

32. **(b)** Use polar coordinates for x and y to get

$$V = \iiint\limits_G dV = \int_0^{2\pi} \int_0^a \int_0^{1/(1+r^2)} r\,dz\,dr\,d\theta = \pi\ln(1+a^2),$$

$$\bar{z} = \frac{1}{V}\iiint\limits_G zdV = \frac{a^2}{2(1+a^2)\ln(1+a^2)}$$

Thus $\lim_{a\to 0^+} \bar{z} = \frac{1}{2}$; $\lim_{a\to+\infty} \bar{z} = 0$.

$\lim_{a\to 0^+} \bar{z} = \frac{1}{2}$; $\lim_{a\to+\infty} \bar{z} = 0$

(c) Solve $\bar{z} = 1/4$ for a to obtain $a \approx 1.980291$.

33. Let $x = r\cos\theta$, $y = r\sin\theta$, and $dA = r\,dr\,d\theta$ in formulas (11) and (12).

34. $\overline{x} = 0$ from the symmetry of the region, $A = \int_0^{2\pi}\int_0^{a(1+\sin\theta)} r\,dr\,d\theta = 3\pi a^2/2$,

$$\overline{y} = \frac{1}{A}\int_0^{2\pi}\int_0^{a(1+\sin\theta)} r^2\sin\theta\,dr\,d\theta = \frac{2}{3\pi a^2}(5\pi a^3/4) = 5a/6;\text{ centroid } (0, 5a/6)$$

35. $\overline{x} = \overline{y}$ from the symmetry of the region, $A = \int_0^{\pi/2}\int_0^{\sin 2\theta} r\,dr\,d\theta = \pi/8$,

$$\overline{x} = \frac{1}{A}\int_0^{\pi/2}\int_0^{\sin 2\theta} r^2\cos\theta\,dr\,d\theta = (8/\pi)(16/105) = \frac{128}{105\pi};\text{ centroid } \left(\frac{128}{105\pi}, \frac{128}{105\pi}\right)$$

36. $\overline{x} = 3/2$ and $\overline{y} = 1$ from the symmetry of the region,

$$\iint_R x\,dA = \overline{x}A = (3/2)(6) = 9,\ \iint_R y\,dA = \overline{y}A = (1)(6) = 6$$

37. $\overline{x} = 0$ from the symmetry of the region, $\pi a^2/2$ is the area of the semicircle, $2\pi\overline{y}$ is the distance traveled by the centroid to generate the sphere so $4\pi a^3/3 = (\pi a^2/2)(2\pi\overline{y})$, $\overline{y} = 4a/(3\pi)$

38. (a) $V = \left[\frac{1}{2}\pi a^2\right]\left[2\pi\left(a + \frac{4a}{3\pi}\right)\right] = \frac{1}{3}\pi(3\pi + 4)a^3$

(b) the distance between the centroid and the line is $\frac{\sqrt{2}}{2}\left(a + \frac{4a}{3\pi}\right)$ so

$$V = \left[\frac{1}{2}\pi a^2\right]\left[2\pi\frac{\sqrt{2}}{2}\left(a + \frac{4a}{3\pi}\right)\right] = \frac{1}{6}\sqrt{2}\pi(3\pi + 4)a^3$$

39. $\overline{x} = k$ so $V = (\pi ab)(2\pi k) = 2\pi^2 abk$

40. $\overline{y} = 4$ from the symmetry of the region,

$$A = \int_{-2}^{2}\int_{x^2}^{8-x^2} dy\,dx = 64/3 \text{ so } V = (64/3)[2\pi(4)] = 512\pi/3$$

41. The region generates a cone of volume $\frac{1}{3}\pi ab^2$ when it is revolved about the x-axis, the area of the region is $\frac{1}{2}ab$ so $\frac{1}{3}\pi ab^2 = \left(\frac{1}{2}ab\right)(2\pi\overline{y})$, $\overline{y} = b/3$. A cone of volume $\frac{1}{3}\pi a^2 b$ is generated when the region is revolved about the y-axis so $\frac{1}{3}\pi a^2 b = \left(\frac{1}{2}ab\right)(2\pi\overline{x})$, $\overline{x} = a/3$. The centroid is $(a/3, b/3)$.

42. $I_x = \int_0^a\int_0^b y^2\delta\,dy\,dx = \frac{1}{3}\delta ab^3$, $I_y = \int_0^a\int_0^b x^2\delta\,dy\,dx = \frac{1}{3}\delta a^3 b$,

$$I_z = \int_0^a\int_0^b (x^2 + y^2)\delta\,dy\,dx = \frac{1}{3}\delta ab(a^2 + b^2)$$

43. $I_x = \int_0^{2\pi}\int_0^a r^3\sin^2\theta\,\delta\,dr\,d\theta = \delta\pi a^4/4$; $I_y = \int_0^{2\pi}\int_0^a r^3\cos^2\theta\,\delta\,dr\,d\theta = \delta\pi a^4/4 = I_x$;

$$I_z = I_x + I_y = \delta\pi a^4/2$$

EXERCISE SET 15.7

1. $\displaystyle\int_0^{2\pi}\int_0^1\int_0^{\sqrt{1-r^2}} zr\,dz\,dr\,d\theta = \int_0^{2\pi}\int_0^1 \frac{1}{2}(1-r^2)r\,dr\,d\theta = \int_0^{2\pi}\frac{1}{8}d\theta = \pi/4$

2. $\displaystyle\int_0^{\pi/2}\int_0^{\cos\theta}\int_0^{r^2} r\sin\theta\,dz\,dr\,d\theta = \int_0^{\pi/2}\int_0^{\cos\theta} r^3\sin\theta\,dr\,d\theta = \int_0^{\pi/2}\frac{1}{4}\cos^4\theta\sin\theta\,d\theta = 1/20$

3. $\displaystyle\int_0^{\pi/2}\int_0^{\pi/2}\int_0^1 \rho^3\sin\phi\cos\phi\,d\rho\,d\phi\,d\theta = \int_0^{\pi/2}\int_0^{\pi/2}\frac{1}{4}\sin\phi\cos\phi\,d\phi\,d\theta = \int_0^{\pi/2}\frac{1}{8}d\theta = \pi/16$

4. $\displaystyle\int_0^{2\pi}\int_0^{\pi/4}\int_0^{a\sec\phi} \rho^2\sin\phi\,d\rho\,d\phi\,d\theta = \int_0^{2\pi}\int_0^{\pi/4}\frac{1}{3}a^3\sec^3\phi\sin\phi\,d\phi\,d\theta = \int_0^{2\pi}\frac{1}{6}a^3 d\theta = \pi a^3/3$

5. $\displaystyle V = \int_0^{2\pi}\int_0^3\int_{r^2}^9 r\,dz\,dr\,d\theta = \int_0^{2\pi}\int_0^3 r(9-r^2)dr\,d\theta = \int_0^{2\pi}\frac{81}{4}d\theta = 81\pi/2$

6. $$V = 2\int_0^{2\pi}\int_0^2\int_0^{\sqrt{9-r^2}} r\,dz\,dr\,d\theta = 2\int_0^{2\pi}\int_0^2 r\sqrt{9-r^2}dr\,d\theta$$
$$= \frac{2}{3}(27-5\sqrt{5})\int_0^{2\pi} d\theta = 4(27-5\sqrt{5})\pi/3$$

7. $r^2+z^2=20$ intersects $z=r^2$ in a circle of radius 2; the volume consists of two portions, one inside the cylinder $r=\sqrt{20}$ and one outside that cylinder:

$$V = \int_0^{2\pi}\int_0^2\int_{-\sqrt{20-r^2}}^{r^2} r\,dz\,dr\,d\theta + \int_0^{2\pi}\int_2^{\sqrt{20}}\int_{-\sqrt{20-r^2}}^{\sqrt{20-r^2}} r\,dz\,dr\,d\theta$$
$$= \int_0^{2\pi}\int_0^2 r\left(r^2+\sqrt{20-r^2}\right)dr\,d\theta + \int_0^{2\pi}\int_2^{\sqrt{20}} 2r\sqrt{20-r^2}\,dr\,d\theta$$
$$= \frac{4}{3}(10\sqrt{5}-13)\int_0^{2\pi} d\theta + \frac{128}{3}\int_0^{2\pi} d\theta = \frac{152}{3}\pi + \frac{80}{3}\pi\sqrt{5}$$

8. $z=hr/a$ intersects $z=h$ in a circle of radius a,

$$V = \int_0^{2\pi}\int_0^a\int_{hr/a}^h r\,dz\,dr\,d\theta = \int_0^{2\pi}\int_0^a \frac{h}{a}(ar-r^2)dr\,d\theta = \int_0^{2\pi}\frac{1}{6}a^2h\,d\theta = \pi a^2h/3$$

9. $\displaystyle V = \int_0^{2\pi}\int_0^{\pi/3}\int_0^4 \rho^2\sin\phi\,d\rho\,d\phi\,d\theta = \int_0^{2\pi}\int_0^{\pi/3}\frac{64}{3}\sin\phi\,d\phi\,d\theta = \frac{32}{3}\int_0^{2\pi} d\theta = 64\pi/3$

10. $\displaystyle V = \int_0^{2\pi}\int_0^{\pi/4}\int_1^2 \rho^2\sin\phi\,d\rho\,d\phi\,d\theta = \int_0^{2\pi}\int_0^{\pi/4}\frac{7}{3}\sin\phi\,d\phi\,d\theta = \frac{7}{6}(2-\sqrt{2})\int_0^{2\pi} d\theta = 7(2-\sqrt{2})\pi/3$

11. In spherical coordinates the sphere and the plane $z=a$ are $\rho=2a$ and $\rho=a\sec\phi$, respectively. They intersect at $\phi=\pi/3$,

$$V = \int_0^{2\pi}\int_0^{\pi/3}\int_0^{a\sec\phi} \rho^2\sin\phi\,d\rho\,d\phi\,d\theta + \int_0^{2\pi}\int_{\pi/3}^{\pi/2}\int_0^{2a} \rho^2\sin\phi\,d\rho\,d\phi\,d\theta$$
$$= \int_0^{2\pi}\int_0^{\pi/3}\frac{1}{3}a^3\sec^3\phi\sin\phi\,d\phi\,d\theta + \int_0^{2\pi}\int_{\pi/3}^{\pi/2}\frac{8}{3}a^3\sin\phi\,d\phi\,d\theta$$
$$= \frac{1}{2}a^3\int_0^{2\pi} d\theta + \frac{4}{3}a^3\int_0^{2\pi} d\theta = 11\pi a^3/3$$

12. $V = \int_0^{2\pi}\int_{\pi/4}^{\pi/2}\int_0^3 \rho^2 \sin\phi \, d\rho\, d\phi\, d\theta = \int_0^{2\pi}\int_{\pi/4}^{\pi/2} 9\sin\phi\, d\phi\, d\theta = \frac{9\sqrt{2}}{2}\int_0^{2\pi} d\theta = 9\sqrt{2}\pi$

13. $$\int_0^{\pi/2}\int_0^a\int_0^{a^2-r^2} r^3\cos^2\theta\, dz\, dr\, d\theta = \int_0^{\pi/2}\int_0^a (a^2r^3 - r^5)\cos^2\theta\, dr\, d\theta$$
$$= \frac{1}{12}a^6\int_0^{\pi/2}\cos^2\theta\, d\theta = \pi a^6/48$$

14. $\int_0^{\pi}\int_0^{\pi/2}\int_0^1 e^{-\rho^3}\rho^2\sin\phi\, d\rho\, d\phi\, d\theta = \frac{1}{3}(1-e^{-1})\int_0^{\pi}\int_0^{\pi/2}\sin\phi\, d\phi\, d\theta = (1-e^{-1})\pi/3$

15. $\int_0^{\pi/2}\int_0^{\pi/4}\int_0^{\sqrt{8}} \rho^4\cos^2\phi\sin\phi\, d\rho\, d\phi\, d\theta = 32(2\sqrt{2}-1)\pi/15$

16. $\int_0^{2\pi}\int_0^{\pi}\int_0^3 \rho^3\sin\phi\, d\rho\, d\phi\, d\theta = 81\pi$

17. (a) $$\int_{\pi/3}^{\pi/2}\int_1^4\int_{-2}^2 \frac{r\tan^3\theta}{\sqrt{1+z^2}}\, dz\, dr\, d\theta = \left(\int_{\pi/6}^{\pi/3}\tan^3\theta\, d\theta\right)\left(\int_1^4 r\, dr\right)\left(\int_{-2}^2 \frac{1}{\sqrt{1+z^2}}\, dz\right)$$
$$= \left(\frac{4}{3} - \frac{1}{2}\ln 3\right)\frac{15}{2}\left(-2\ln(\sqrt{5}-2)\right) = \frac{5}{2}(-8+3\ln 3)\ln(\sqrt{5}-2)$$

(b) $\int_{\pi/3}^{\pi/2}\int_1^4\int_{-2}^2 \frac{y\tan^3 z}{\sqrt{1+x^2}}\, dx\, dy\, dz$; the region is a rectangular solid with sides $\pi/6$, 3, 4.

18. $\int_0^{\pi/2}\int_0^{\pi/4}\frac{1}{18}\cos^{37}\theta\cos\phi\, d\phi\, d\theta = \frac{\sqrt{2}}{36}\int_0^{\pi/2}\cos^{37}\theta\, d\theta = \frac{4{,}294{,}967{,}296}{755{,}505{,}013{,}725}\sqrt{2} \approx 0.008040$

19. (a) $V = 2\int_0^{2\pi}\int_0^a\int_0^{\sqrt{a^2-r^2}} r\, dz\, dr\, d\theta = 4\pi a^3/3$

(b) $V = \int_0^{2\pi}\int_0^{\pi}\int_0^a \rho^2\sin\phi\, d\rho\, d\phi\, d\theta = 4\pi a^3/3$

20. (a) $$\int_0^2\int_0^{\sqrt{4-x^2}}\int_0^{\sqrt{4-x^2-y^2}} xyz\, dz\, dy\, dx$$
$$= \int_0^2\int_0^{\sqrt{4-x^2}}\frac{1}{2}xy(4-x^2-y^2)dy\, dx = \frac{1}{8}\int_0^2 x(4-x^2)^2 dx = 4/3$$

(b) $$\int_0^{\pi/2}\int_0^2\int_0^{\sqrt{4-r^2}} r^3 z\sin\theta\cos\theta\, dz\, dr\, d\theta$$
$$= \int_0^{\pi/2}\int_0^2 \frac{1}{2}(4r^3 - r^5)\sin\theta\cos\theta\, dr\, d\theta = \frac{8}{3}\int_0^{\pi/2}\sin\theta\cos\theta\, d\theta = 4/3$$

(c) $$\int_0^{\pi/2}\int_0^{\pi/2}\int_0^2 \rho^5\sin^3\phi\cos\phi\sin\theta\cos\theta\, d\rho\, d\phi\, d\theta$$
$$= \int_0^{\pi/2}\int_0^{\pi/2}\frac{32}{3}\sin^3\phi\cos\phi\sin\theta\cos\theta\, d\phi\, d\theta = \frac{8}{3}\int_0^{\pi/2}\sin\theta\cos\theta\, d\theta = 4/3$$

21. $M = \int_0^{2\pi}\int_0^3\int_r^3 (3-z)r\,dz\,dr\,d\theta = \int_0^{2\pi}\int_0^3 \frac{1}{2}r(3-r)^2 dr\,d\theta = \frac{27}{8}\int_0^{2\pi} d\theta = 27\pi/4$

22. $M = \int_0^{2\pi}\int_0^a\int_0^h k\,zr\,dz\,dr\,d\theta = \int_0^{2\pi}\int_0^a \frac{1}{2}kh^2 r\,dr\,d\theta = \frac{1}{4}ka^2h^2\int_0^{2\pi} d\theta = \pi ka^2h^2/2$

23. $M = \int_0^{2\pi}\int_0^{\pi}\int_0^a k\rho^3 \sin\phi\,d\rho\,d\phi\,d\theta = \int_0^{2\pi}\int_0^{\pi} \frac{1}{4}ka^4 \sin\phi\,d\phi\,d\theta = \frac{1}{2}ka^4\int_0^{2\pi} d\theta = \pi ka^4$

24. $M = \int_0^{2\pi}\int_0^{\pi}\int_1^2 \rho\sin\phi\,d\rho\,d\phi\,d\theta = \int_0^{2\pi}\int_0^{\pi} \frac{3}{2}\sin\phi\,d\phi\,d\theta = 3\int_0^{2\pi} d\theta = 6\pi$

25. $\bar{x} = \bar{y} = 0$ from the symmetry of the region,

$$V = \int_0^{2\pi}\int_0^1\int_{r^2}^{\sqrt{2-r^2}} r\,dz\,dr\,d\theta = \int_0^{2\pi}\int_0^1 (r\sqrt{2-r^2} - r^3)dr\,d\theta = (8\sqrt{2}-7)\pi/6,$$

$$\bar{z} = \frac{1}{V}\int_0^{2\pi}\int_0^1\int_{r^2}^{\sqrt{2-r^2}} zr\,dz\,dr\,d\theta = \frac{6}{(8\sqrt{2}-7)\pi}(7\pi/12) = 7/(16\sqrt{2}-14);$$

centroid $\left(0, 0, \dfrac{7}{16\sqrt{2}-14}\right)$

26. $\bar{x} = \bar{y} = 0$ from the symmetry of the region, $V = 8\pi/3$,

$$\bar{z} = \frac{1}{V}\int_0^{2\pi}\int_0^2\int_r^2 zr\,dz\,dr\,d\theta = \frac{3}{8\pi}(4\pi) = 3/2; \text{ centroid } (0,0,3/2)$$

27. $\bar{x} = \bar{y} = \bar{z}$ from the symmetry of the region, $V = \pi a^3/6$,

$$\bar{z} = \frac{1}{V}\int_0^{\pi/2}\int_0^{\pi/2}\int_0^a \rho^3\cos\phi\sin\phi\,d\rho\,d\phi\,d\theta = \frac{6}{\pi a^3}(\pi a^4/16) = 3a/8;$$

centroid $(3a/8, 3a/8, 3a/8)$

28. $\bar{x} = \bar{y} = 0$ from the symmetry of the region, $V = \int_0^{2\pi}\int_0^{\pi/3}\int_0^4 \rho^2\sin\phi\,d\rho\,d\phi\,d\theta = 64\pi/3,$

$$\bar{z} = \frac{1}{V}\int_0^{2\pi}\int_0^{\pi/3}\int_0^4 \rho^3\cos\phi\sin\phi\,d\rho\,d\phi\,d\theta = \frac{3}{64\pi}(48\pi) = 9/4; \text{ centroid } (0,0,9/4)$$

29. $\bar{y} = 0$ from the symmetry of the region, $V = 2\int_0^{\pi/2}\int_0^{2\cos\theta}\int_0^{r^2} r\,dz\,dr\,d\theta = 3\pi/2,$

$$\bar{x} = \frac{2}{V}\int_0^{\pi/2}\int_0^{2\cos\theta}\int_0^{r^2} r^2\cos\theta\,dz\,dr\,d\theta = \frac{4}{3\pi}(\pi) = 4/3,$$

$$\bar{z} = \frac{2}{V}\int_0^{\pi/2}\int_0^{2\cos\theta}\int_0^{r^2} rz\,dz\,dr\,d\theta = \frac{4}{3\pi}(5\pi/6) = 10/9; \text{ centroid } (4/3, 0, 10/9)$$

30. $M = \int_0^{\pi/2}\int_0^{2\cos\theta}\int_0^{4-r^2} zr\,dz\,dr\,d\theta = \int_0^{\pi/2}\int_0^{2\cos\theta} \frac{1}{2}r(4-r^2)^2 dr\,d\theta$

$$= \frac{16}{3}\int_0^{\pi/2}(1-\sin^6\theta)d\theta = (16/3)(11\pi/32) = 11\pi/6$$

31. $V = \int_0^{\pi/2}\int_{\pi/6}^{\pi/3}\int_0^2 \rho^2\sin\phi\,d\rho\,d\phi\,d\theta = \int_0^{\pi/2}\int_{\pi/6}^{\pi/3}\frac{8}{3}\sin\phi\,d\phi\,d\theta = \frac{4}{3}(\sqrt{3}-1)\int_0^{\pi/2}d\theta$
$= 2(\sqrt{3}-1)\pi/3$

32. $M = \int_0^{2\pi}\int_0^{\pi/4}\int_0^1 \rho^3\sin\phi\,d\rho\,d\phi\,d\theta = \int_0^{2\pi}\int_0^{\pi/4}\frac{1}{4}\sin\phi\,d\phi\,d\theta = \frac{1}{8}(2-\sqrt{2})\int_0^{2\pi}d\theta = (2-\sqrt{2})\pi/4$

33. $\bar{x} = \bar{y} = 0$ from the symmetry of density and region,
$M = \int_0^{2\pi}\int_0^1\int_0^{1-r^2}(r^2+z^2)r\,dz\,dr\,d\theta = \pi/4,$
$\bar{z} = \frac{1}{M}\int_0^{2\pi}\int_0^1\int_0^{1-r^2} z(r^2+z^2)r\,dz\,dr\,d\theta = (4/\pi)(11\pi/120) = 11/30$; center of gravity $(0,0,11/30)$

34. $\bar{x} = \bar{y} = 0$ from the symmetry of density and region, $M = \int_0^{2\pi}\int_0^1\int_0^r zr\,dz\,dr\,d\theta = \pi/4,$
$\bar{z} = \frac{1}{M}\int_0^{2\pi}\int_0^1\int_0^r z^2r\,dz\,dr\,d\theta = (4/\pi)(2\pi/15) = 8/15$; center of gravity $(0,0,8/15)$

35. $\bar{x} = \bar{y} = 0$ from the symmetry of density and region,
$M = \int_0^{2\pi}\int_0^{\pi/2}\int_0^a k\rho^3\sin\phi\,d\rho\,d\phi\,d\theta = \pi ka^4/2,$
$\bar{z} = \frac{1}{M}\int_0^{2\pi}\int_0^{\pi/2}\int_0^a k\rho^4\sin\phi\cos\phi\,d\rho\,d\phi\,d\theta = \frac{2}{\pi ka^4}(\pi ka^5/5) = 2a/5$; center of gravity $(0,0,2a/5)$

36. $\bar{x} = \bar{z} = 0$ from the symmetry of the region, $V = 54\pi/3 - 16\pi/3 = 38\pi/3,$
$\bar{y} = \frac{1}{V}\int_0^{\pi}\int_0^{\pi}\int_2^3 \rho^3\sin^2\phi\sin\theta\,d\rho\,d\phi\,d\theta = \frac{1}{V}\int_0^{\pi}\int_0^{\pi}\frac{65}{4}\sin^2\phi\sin\theta\,d\phi\,d\theta$
$= \frac{1}{V}\int_0^{\pi}\frac{65\pi}{8}\sin\theta\,d\theta = \frac{3}{38\pi}(65\pi/4) = 195/152$; centroid $(0,195/152,0)$

37. $M = \int_0^{2\pi}\int_0^{\pi}\int_0^R \delta_0 e^{-(\rho/R)^3}\rho^2\sin\phi\,d\rho\,d\phi\,d\theta = \int_0^{2\pi}\int_0^{\pi}\frac{1}{3}(1-e^{-1})R^3\delta_0\sin\phi\,d\phi\,d\theta$
$= \frac{4}{3}\pi(1-e^{-1})\delta_0 R^3$

38. **(a)** The sphere and cone intersect in a circle of radius $\rho_0\sin\phi_0$,
$$V = \int_{\theta_1}^{\theta_2}\int_0^{\rho_0\sin\phi_0}\int_{r\cot\phi_0}^{\sqrt{\rho_0^2-r^2}} r\,dz\,dr\,d\theta = \int_{\theta_1}^{\theta_2}\int_0^{\rho_0\sin\phi_0}\left(r\sqrt{\rho_0^2-r^2} - r^2\cot\phi_0\right)dr\,d\theta$$
$$= \int_{\theta_1}^{\theta_2}\frac{1}{3}\rho_0^3(1-\cos^3\phi_0-\sin^3\phi_0\cot\phi_0)d\theta = \frac{1}{3}\rho_0^3(1-\cos^3\phi_0-\sin^2\phi_0\cos\phi_0)(\theta_2-\theta_1)$$
$$= \frac{1}{3}\rho_0^3(1-\cos\phi_0)(\theta_2-\theta_1).$$

(b) From Part (a), the volume of the solid bounded by $\theta=\theta_1$, $\theta=\theta_2$, $\phi=\phi_1$, $\phi=\phi_2$, and $\rho=\rho_0$ is $\frac{1}{3}\rho_0^3(1-\cos\phi_2)(\theta_2-\theta_1)-\frac{1}{3}\rho_0^3(1-\cos\phi_1)(\theta_2-\theta_1)=\frac{1}{3}\rho_0^3(\cos\phi_1-\cos\phi_2)(\theta_2-\theta_1)$ so the volume of the spherical wedge between $\rho=\rho_1$ and $\rho=\rho_2$ is

$$\Delta V=\frac{1}{3}\rho_2^3(\cos\phi_1-\cos\phi_2)(\theta_2-\theta_1)-\frac{1}{3}\rho_1^3(\cos\phi_1-\cos\phi_2)(\theta_2-\theta_1)$$
$$=\frac{1}{3}(\rho_2^3-\rho_1^3)(\cos\phi_1-\cos\phi_2)(\theta_2-\theta_1)$$

(c) $\frac{d}{d\phi}\cos\phi=-\sin\phi$ so from the Mean-Value Theorem $\cos\phi_2-\cos\phi_1=-(\phi_2-\phi_1)\sin\phi^*$ where ϕ^* is between ϕ_1 and ϕ_2. Similarly $\frac{d}{d\rho}\rho^3=3\rho^2$ so $\rho_2^3-\rho_1^3=3\rho^{*2}(\rho_2-\rho_1)$ where ρ^* is between ρ_1 and ρ_2. Thus $\cos\phi_1-\cos\phi_2=\sin\phi^*\Delta\phi$ and $\rho_2^3-\rho_1^3=3\rho^{*2}\Delta\rho$ so $\Delta V=\rho^{*2}\sin\phi^*\Delta\rho\Delta\phi\Delta\theta$.

39. $I_z=\int_0^{2\pi}\int_0^a\int_0^h r^2\delta\, r\,dz\,dr\,d\theta=\delta\int_0^{2\pi}\int_0^a\int_0^h r^3dz\,dr\,d\theta=\frac{1}{2}\delta\pi a^4h$

40. $I_y=\int_0^{2\pi}\int_0^a\int_0^h (r^2\cos^2\theta+z^2)\delta r\,dz\,dr\,d\theta=\delta\int_0^{2\pi}\int_0^a(hr^3\cos^2\theta+\frac{1}{3}h^3r)dr\,d\theta$

$$=\delta\int_0^{2\pi}\left(\frac{1}{4}a^4h\cos^2\theta+\frac{1}{6}a^2h^3\right)d\theta=\delta\left(\frac{\pi}{4}a^4h+\frac{\pi}{3}a^2h^3\right)$$

41. $I_z=\int_0^{2\pi}\int_{a_1}^{a_2}\int_0^h r^2\delta\, r\,dz\,dr\,d\theta=\delta\int_0^{2\pi}\int_{a_1}^{a_2}\int_0^h r^3dz\,dr\,d\theta=\frac{1}{2}\delta\pi h(a_2^4-a_1^4)$

42. $I_z=\int_0^{2\pi}\int_0^{\pi}\int_0^a(\rho^2\sin^2\phi)\delta\,\rho^2\sin\phi\,d\rho\,d\phi\,d\theta=\delta\int_0^{2\pi}\int_0^{\pi}\int_0^a\rho^4\sin^3\phi\,d\rho\,d\phi\,d\theta=\frac{8}{15}\delta\pi a^5$

EXERCISE SET 15.8

1. $\frac{\partial(x,y)}{\partial(u,v)}=\begin{vmatrix}1 & 4\\ 3 & -5\end{vmatrix}=-17$

2. $\frac{\partial(x,y)}{\partial(u,v)}=\begin{vmatrix}1 & 4v\\ 4u & -1\end{vmatrix}=-1-16uv$

3. $\frac{\partial(x,y)}{\partial(u,v)}=\begin{vmatrix}\cos u & -\sin v\\ \sin u & \cos v\end{vmatrix}=\cos u\cos v+\sin u\sin v=\cos(u-v)$

4. $\frac{\partial(x,y)}{\partial(u,v)}=\begin{vmatrix}\frac{2(v^2-u^2)}{(u^2+v^2)^2} & -\frac{4uv}{(u^2+v^2)^2}\\ \frac{4uv}{(u^2+v^2)^2} & \frac{2(v^2-u^2)}{(u^2+v^2)^2}\end{vmatrix}=4/(u^2+v^2)^2$

5. $x=\frac{2}{9}u+\frac{5}{9}v,\ y=-\frac{1}{9}u+\frac{2}{9}v;\ \frac{\partial(x,y)}{\partial(u,v)}=\begin{vmatrix}2/9 & 5/9\\ -1/9 & 2/9\end{vmatrix}=\frac{1}{9}$

6. $x=\ln u,\ y=uv;\ \frac{\partial(x,y)}{\partial(u,v)}=\begin{vmatrix}1/u & 0\\ v & u\end{vmatrix}=1$

7. $x = \sqrt{u+v}/\sqrt{2}$, $y = \sqrt{v-u}/\sqrt{2}$; $\dfrac{\partial(x,y)}{\partial(u,v)} = \begin{vmatrix} \dfrac{1}{2\sqrt{2}\sqrt{u+v}} & \dfrac{1}{2\sqrt{2}\sqrt{u+v}} \\ -\dfrac{1}{2\sqrt{2}\sqrt{v-u}} & \dfrac{1}{2\sqrt{2}\sqrt{v-u}} \end{vmatrix} = \dfrac{1}{4\sqrt{v^2-u^2}}$

8. $x = u^{3/2}/v^{1/2}$, $y = v^{1/2}/u^{1/2}$; $\dfrac{\partial(x,y)}{\partial(u,v)} = \begin{vmatrix} \dfrac{3u^{1/2}}{2v^{1/2}} & -\dfrac{u^{3/2}}{2v^{3/2}} \\ -\dfrac{v^{1/2}}{2u^{3/2}} & \dfrac{1}{2u^{1/2}v^{1/2}} \end{vmatrix} = \dfrac{1}{2v}$

9. $\dfrac{\partial(x,y,z)}{\partial(u,v,w)} = \begin{vmatrix} 3 & 1 & 0 \\ 1 & 0 & -2 \\ 0 & 1 & 1 \end{vmatrix} = 5$

10. $\dfrac{\partial(x,y,z)}{\partial(u,v,w)} = \begin{vmatrix} 1-v & -u & 0 \\ v-vw & u-uw & -uv \\ vw & uw & uv \end{vmatrix} = u^2v$

11. $y = v, x = u/y = u/v, z = w - x = w - u/v$; $\dfrac{\partial(x,y,z)}{\partial(u,v,w)} = \begin{vmatrix} 1/v & -u/v^2 & 0 \\ 0 & 1 & 0 \\ -1/v & u/v^2 & 1 \end{vmatrix} = 1/v$

12. $x = (v+w)/2, y = (u-w)/2, z = (u-v)/2$, $\dfrac{\partial(x,y,z)}{\partial(u,v,w)} = \begin{vmatrix} 0 & 1/2 & 1/2 \\ 1/2 & 0 & -1/2 \\ 1/2 & -1/2 & 0 \end{vmatrix} = -\dfrac{1}{4}$

13.

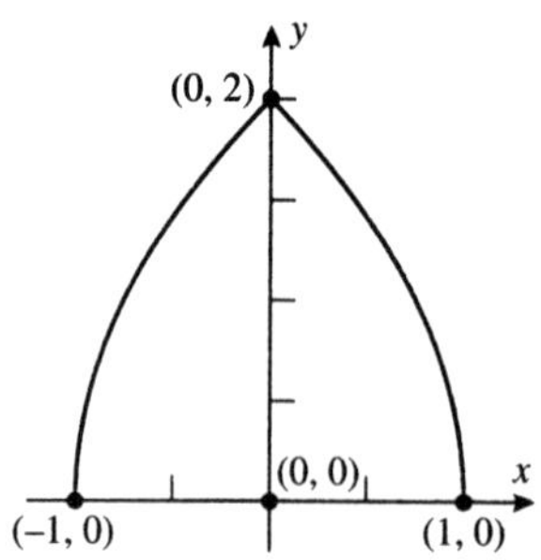

14.

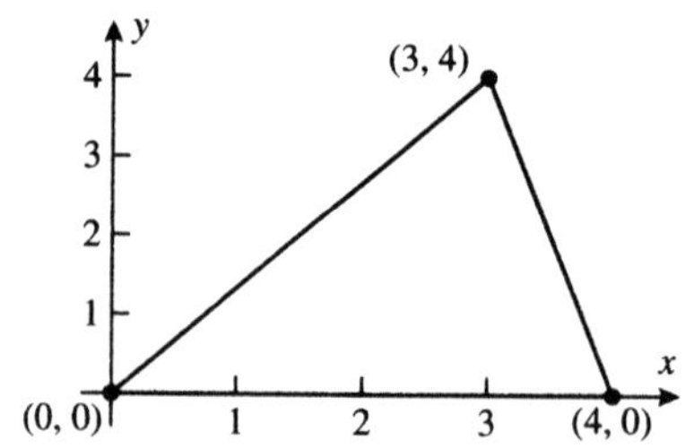

15.

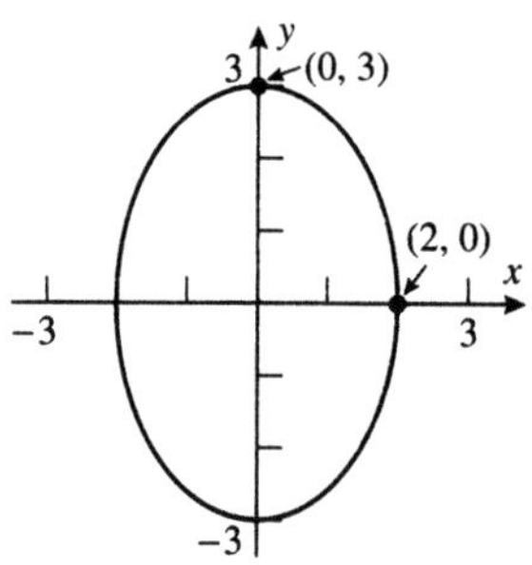

16.

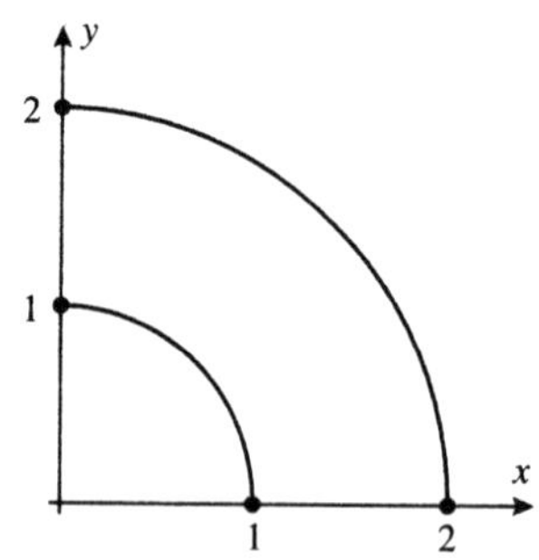

17. $x = \dfrac{1}{5}u + \dfrac{2}{5}v$, $y = -\dfrac{2}{5}u + \dfrac{1}{5}v$, $\dfrac{\partial(x,y)}{\partial(u,v)} = \dfrac{1}{5}$; $\dfrac{1}{5}\iint\limits_S \dfrac{u}{v} dA_{uv} = \dfrac{1}{5}\int_1^3 \int_1^4 \dfrac{u}{v}\, du\, dv = \dfrac{3}{2}\ln 3$

18. $x = \dfrac{1}{2}u + \dfrac{1}{2}v,\ y = \dfrac{1}{2}u - \dfrac{1}{2}v,\ \dfrac{\partial(x,y)}{\partial(u,v)} = -\dfrac{1}{2};\ \dfrac{1}{2}\displaystyle\iint_S ve^{uv}dA_{uv} = \dfrac{1}{2}\int_1^4\int_0^1 ve^{uv}du\,dv = \dfrac{1}{2}(e^4 - e - 3)$

19. $x = u + v,\ y = u - v,\ \dfrac{\partial(x,y)}{\partial(u,v)} = -2$; the boundary curves of the region S in the uv-plane are $v = 0, v = u$, and $u = 1$ so $2\displaystyle\iint_S \sin u\cos v\,dA_{uv} = 2\int_0^1\int_0^u \sin u\cos v\ dv\ du = 1 - \dfrac{1}{2}\sin 2$

20. $x = \sqrt{v/u},\ y = \sqrt{uv}$ so, from Example 3, $\dfrac{\partial(x,y)}{\partial(u,v)} = -\dfrac{1}{2u}$; the boundary curves of the region S in the uv-plane are $u = 1, u = 3, v = 1$, and $v = 4$ so $\displaystyle\iint_S uv^2\left(\frac{1}{2u}\right)dA_{uv} = \frac{1}{2}\int_1^4\int_1^3 v^2du\ dv = 21$

21. $x = 3u, y = 4v,\ \dfrac{\partial(x,y)}{\partial(u,v)} = 12$; S is the region in the uv-plane enclosed by the circle $u^2 + v^2 = 1$. Use polar coordinates to obtain $\displaystyle\iint_S 12\sqrt{u^2+v^2}(12)\,dA_{uv} = 144\int_0^{2\pi}\int_0^1 r^2dr\ d\theta = 96\pi$

22. $x = 2u, y = v,\ \dfrac{\partial(x,y)}{\partial(u,v)} = 2$; S is the region in the uv-plane enclosed by the circle $u^2 + v^2 = 1$. Use polar coordinates to obtain $\displaystyle\iint_S e^{-(4u^2+4v^2)}(2)\,dA_{uv} = 2\int_0^{2\pi}\int_0^1 re^{-4r^2}dr\ d\theta = (1 - e^{-4})\pi/2$

23. Let S be the region in the uv-plane bounded by $u^2 + v^2 = 1$, so $u = 2x, v = 3y$,

$x = u/2, y = v/3,\ \dfrac{\partial(x,y)}{\partial(u,v)} = \begin{vmatrix} 1/2 & 0 \\ 0 & 1/3 \end{vmatrix} = 1/6$, use polar coordinates to get

$$\frac{1}{6}\iint_S \sin(u^2+v^2)du\,dv = \frac{1}{6}\int_0^{\pi/2}\int_0^1 r\sin r^2\,dr\,d\theta = \frac{\pi}{24}(-\cos r^2)\Big]_0^1 = \frac{\pi}{24}(1 - \cos 1)$$

24. $u = x/a, v = y/b, x = au, y = bv;\ \dfrac{\partial(x,y)}{\partial(u,v)} = ab;\ A = ab\displaystyle\int_0^{2\pi}\int_0^1 r\,dr\,d\theta = \pi ab$

25. $x = u/3, y = v/2, z = w,\ \dfrac{\partial(x,y,z)}{\partial(u,v,w)} = 1/6$; S is the region in uvw-space enclosed by the sphere $u^2 + v^2 + w^2 = 36$ so

$$\iiint_S \frac{u^2}{9}\frac{1}{6}dV_{uvw} = \frac{1}{54}\int_0^{2\pi}\int_0^{\pi}\int_0^6 (\rho\sin\phi\cos\theta)^2\rho^2\sin\phi\ d\rho\ d\phi\ d\theta$$

$$= \frac{1}{54}\int_0^{2\pi}\int_0^{\pi}\int_0^6 \rho^4\sin^3\phi\cos^2\theta d\rho\ d\phi\ d\theta = \frac{192}{5}\pi$$

26. Let G_1 be the region $u^2+v^2+w^2 \le 1$, with $x=au, y=bv, z=cw, \dfrac{\partial(x,y,z)}{\partial(u,v,w)} = abc$; then use spherical coordinates in uvw-space:

$$I_x = \iiint\limits_G (y^2+z^2)dx\,dy\,dz = abc\iiint\limits_{G_1}(b^2v^2+c^2w^2)\,du\,dv\,dw$$

$$= \int_0^{2\pi}\int_0^{\pi}\int_0^1 abc(b^2\sin^2\phi\sin^2\theta + c^2\cos^2\phi)\rho^4\sin\phi\,d\rho\,d\phi\,d\theta$$

$$= \int_0^{2\pi}\frac{abc}{15}(4b^2\sin^2\theta+2c^2)d\theta = \frac{4}{15}\pi abc(b^2+c^2)$$

27. $u=\theta=\cot^{-1}(x/y), v=r=\sqrt{x^2+y^2}$

28. $u=r=\sqrt{x^2+y^2}, v=(\theta+\pi/2)/\pi=(1/\pi)\tan^{-1}(y/x)+1/2$

29. $u=\frac{3}{7}x-\frac{2}{7}y, v=-\frac{1}{7}x+\frac{3}{7}y$

30. $u=-x+\frac{4}{3}y, v=y$

31. Let $u=y-4x, v=y+4x$, then $x=\frac{1}{8}(v-u), y=\frac{1}{2}(v+u)$ so $\dfrac{\partial(x,y)}{\partial(u,v)}=-\dfrac{1}{8}$;

$$\frac{1}{8}\iint\limits_S \frac{u}{v}dA_{uv} = \frac{1}{8}\int_2^5\int_0^2\frac{u}{v}du\,dv = \frac{1}{4}\ln\frac{5}{2}$$

32. Let $u=y+x, v=y-x$, then $x=\frac{1}{2}(u-v), y=\frac{1}{2}(u+v)$ so $\dfrac{\partial(x,y)}{\partial(u,v)}=\dfrac{1}{2}$;

$$-\frac{1}{2}\iint\limits_S uv\,dA_{uv} = -\frac{1}{2}\int_0^2\int_0^1 uv\,du\,dv = -\frac{1}{2}$$

33. Let $u=x-y, v=x+y$, then $x=\frac{1}{2}(v+u), y=\frac{1}{2}(v-u)$ so $\dfrac{\partial(x,y)}{\partial(u,v)}=\dfrac{1}{2}$; the boundary curves of the region S in the uv-plane are $u=0, v=u$, and $v=\pi/4$; thus

$$\frac{1}{2}\iint\limits_S \frac{\sin u}{\cos v}dA_{uv} = \frac{1}{2}\int_0^{\pi/4}\int_0^v\frac{\sin u}{\cos v}du\,dv = \frac{1}{2}[\ln(\sqrt{2}+1)-\pi/4]$$

34. Let $u=y-x, v=y+x$, then $x=\frac{1}{2}(v-u), y=\frac{1}{2}(u+v)$ so $\dfrac{\partial(x,y)}{\partial(u,v)}=-\dfrac{1}{2}$; the boundary curves of the region S in the uv-plane are $v=-u, v=u, v=1$, and $v=4$; thus

$$\frac{1}{2}\iint\limits_S e^{u/v}dA_{uv} = \frac{1}{2}\int_1^4\int_{-v}^v e^{u/v}du\,dv = \frac{15}{4}(e-e^{-1})$$

35. Let $u=y/x, v=x/y^2$, then $x=1/(u^2v), y=1/(uv)$ so $\dfrac{\partial(x,y)}{\partial(u,v)}=\dfrac{1}{u^4v^3}$;

$$\iint\limits_S \frac{1}{u^4v^3}dA_{uv} = \int_1^4\int_1^2\frac{1}{u^4v^3}du\,dv = 35/256$$

36. Let $x = 3u, y = 2v, \dfrac{\partial(x,y)}{\partial(u,v)} = 6$; S is the region in the uv-plane enclosed by the circle $u^2 + v^2 = 1$

so $\displaystyle\iint_R (9 - x - y)dA = \iint_S 6(9 - 3u - 2v)dA_{uv} = 6\int_0^{2\pi}\int_0^1 (9 - 3r\cos\theta - 2r\sin\theta)r\,dr\,d\theta = 54\pi$

37. $x = u, y = w/u, z = v + w/u, \dfrac{\partial(x,y,z)}{\partial(u,v,w)} = -\dfrac{1}{u}$;

$$\iiint_S \frac{v^2 w}{u} dV_{uvw} = \int_2^4 \int_0^1 \int_1^3 \frac{v^2 w}{u} du\,dv\,dw = 2\ln 3$$

38. $u = xy, v = yz, w = xz, 1 \le u \le 2, 1 \le v \le 3, 1 \le w \le 4,$

$x = \sqrt{uw/v}, y = \sqrt{uv/w}, z = \sqrt{vw/u}, \dfrac{\partial(x,y,z)}{\partial(u,v,w)} = \dfrac{1}{2\sqrt{uvw}}$

$$V = \iiint_G dV = \int_1^2 \int_1^3 \int_1^4 \frac{1}{2\sqrt{uvw}} dw\,dv\,du = 4(\sqrt{2} - 1)(\sqrt{3} - 1)$$

39. **(b)** If $x = x(u,v), y = y(u,v)$ where $u = u(x,y), v = v(x,y)$, then by the chain rule

$$\frac{\partial x}{\partial u}\frac{\partial u}{\partial x} + \frac{\partial x}{\partial v}\frac{\partial v}{\partial x} = \frac{\partial x}{\partial x} = 1, \quad \frac{\partial x}{\partial u}\frac{\partial u}{\partial y} + \frac{\partial x}{\partial v}\frac{\partial v}{\partial y} = \frac{\partial x}{\partial y} = 0$$

$$\frac{\partial y}{\partial u}\frac{\partial u}{\partial x} + \frac{\partial y}{\partial v}\frac{\partial v}{\partial x} = \frac{\partial y}{\partial x} = 0, \quad \frac{\partial y}{\partial u}\frac{\partial u}{\partial y} + \frac{\partial y}{\partial v}\frac{\partial v}{\partial y} = \frac{\partial y}{\partial y} = 1$$

40. **(a)** $\dfrac{\partial(x,y)}{\partial(u,v)} = \begin{vmatrix} 1-v & -u \\ v & u \end{vmatrix} = u; \quad u = x + y, v = \dfrac{y}{x+y},$

$\dfrac{\partial(u,v)}{\partial(x,y)} = \begin{vmatrix} 1 & 1 \\ -y/(x+y)^2 & x/(x+y)^2 \end{vmatrix} = \dfrac{x}{(x+y)^2} + \dfrac{y}{(x+y)^2} = \dfrac{1}{x+y} = \dfrac{1}{u};$

$\dfrac{\partial(u,v)}{\partial(x,y)}\dfrac{\partial(x,y)}{\partial(u,v)} = 1$

(b) $\dfrac{\partial(x,y)}{\partial(u,v)} = \begin{vmatrix} v & u \\ 0 & 2v \end{vmatrix} = 2v^2; \quad u = x/\sqrt{y}, v = \sqrt{y},$

$\dfrac{\partial(u,v)}{\partial(x,y)} = \begin{vmatrix} 1/\sqrt{y} & -x/(2y^{3/2}) \\ 0 & 1/(2\sqrt{y}) \end{vmatrix} = \dfrac{1}{2y} = \dfrac{1}{2v^2}; \dfrac{\partial(u,v)}{\partial(x,y)}\dfrac{\partial(x,y)}{\partial(u,v)} = 1$

(c) $\dfrac{\partial(x,y)}{\partial(u,v)} = \begin{vmatrix} u & v \\ u & -v \end{vmatrix} = -2uv; \quad u = \sqrt{x+y}, v = \sqrt{x-y},$

$\dfrac{\partial(u,v)}{\partial(x,y)} = \begin{vmatrix} 1/(2\sqrt{x+y}) & 1/(2\sqrt{x+y}) \\ 1/(2\sqrt{x-y}) & -1/(2\sqrt{x-y}) \end{vmatrix} = -\dfrac{1}{2\sqrt{x^2-y^2}} = -\dfrac{1}{2uv}; \dfrac{\partial(u,v)}{\partial(x,y)}\dfrac{\partial(x,y)}{\partial(u,v)} = 1$

41. $\dfrac{\partial(u,v)}{\partial(x,y)} = 3xy^4 = 3v$ so $\dfrac{\partial(x,y)}{\partial(u,v)} = \dfrac{1}{3v}; \dfrac{1}{3}\displaystyle\iint_S \frac{\sin u}{v} dA_{uv} = \frac{1}{3}\int_1^2 \int_\pi^{2\pi} \frac{\sin u}{v} du\,dv = -\frac{2}{3}\ln 2$

42. $\dfrac{\partial(u,v)}{\partial(x,y)} = 8xy$ so $\dfrac{\partial(x,y)}{\partial(u,v)} = \dfrac{1}{8xy}$; $xy\left|\dfrac{\partial(x,y)}{\partial(u,v)}\right| = xy\left(\dfrac{1}{8xy}\right) = \dfrac{1}{8}$ so

$$\frac{1}{8}\iint_S dA_{uv} = \frac{1}{8}\int_9^{16}\int_1^4 du\,dv = 21/8$$

43. $\dfrac{\partial(u,v)}{\partial(x,y)} = -2(x^2+y^2)$ so $\dfrac{\partial(x,y)}{\partial(u,v)} = -\dfrac{1}{2(x^2+y^2)}$;

$$(x^4-y^4)e^{xy}\left|\frac{\partial(x,y)}{\partial(u,v)}\right| = \frac{x^4-y^4}{2(x^2+y^2)}e^{xy} = \frac{1}{2}(x^2-y^2)e^{xy} = \frac{1}{2}ve^u \text{ so}$$

$$\frac{1}{2}\iint_S ve^u dA_{uv} = \frac{1}{2}\int_3^4\int_1^3 ve^u du\,dv = \frac{7}{4}(e^3-e)$$

44. Set $u = x+y+2z, v = x-2y+z, w = 4x+y+z$, then $\dfrac{\partial(u,v,w)}{\partial(x,y,z)} = \begin{vmatrix} 1 & 1 & 2 \\ 1 & -2 & 1 \\ 4 & 1 & 1 \end{vmatrix} = 18$, and

$$V = \iiint_R dx\,dy\,dz = \int_{-6}^{6}\int_{-2}^{2}\int_{-3}^{3}\frac{\partial(x,y,z)}{\partial(u,v,w)}du\,dv\,dw = 6(4)(12)\frac{1}{18} = 16$$

45. **(a)** Let $u = x+y, v = y$, then the triangle R with vertices $(0,0),(1,0)$ and $(0,1)$ becomes the triangle in the uv-plane with vertices $(0,0),(1,0),(1,1)$, and

$$\iint_R f(x+y)dA = \int_0^1\int_0^u f(u)\frac{\partial(x,y)}{\partial(u,v)}dv\,du = \int_0^1 uf(u)\,du$$

(b) $\displaystyle\int_0^1 ue^u\,du = (u-1)e^u\Big]_0^1 = 1$

46. **(a)** $\dfrac{\partial(x,y,z)}{\partial(r,\theta,z)} = \begin{vmatrix} \cos\theta & -r\sin\theta & 0 \\ \sin\theta & r\cos\theta & 0 \\ 0 & 0 & 1 \end{vmatrix} = r$, $\left|\dfrac{\partial(x,y,z)}{\partial(r,\theta,z)}\right| = r$

(b) $\dfrac{\partial(x,y,z)}{\partial(\rho,\phi,\theta)} = \begin{vmatrix} \sin\phi\cos\theta & \rho\cos\phi\cos\theta & -\rho\sin\phi\sin\theta \\ \sin\phi\sin\theta & \rho\cos\phi\sin\theta & \rho\sin\phi\cos\theta \\ \cos\phi & -\rho\sin\phi & 0 \end{vmatrix} = \rho^2\sin\phi$; $\left|\dfrac{\partial(x,y,z)}{\partial(\rho,\phi,\theta)}\right| = \rho^2\sin\phi$

CHAPTER 15 SUPPLEMENTARY EXERCISES

3. **(a)** $\displaystyle\iint_R dA$ **(b)** $\displaystyle\iiint_G dV$ **(c)** $\displaystyle\iint_R \sqrt{1+\left(\frac{\partial z}{\partial x}\right)^2+\left(\frac{\partial z}{\partial y}\right)^2}\,dA$

4. **(a)** $x = a\sin\phi\cos\theta, y = a\sin\phi\sin\theta, z = a\cos\phi, 0\le\theta\le 2\pi, 0\le\phi\le\pi$
(b) $x = a\cos\theta, y = a\sin\theta, z = z, 0\le\theta\le 2\pi, 0\le z\le h$

7. $\displaystyle\int_0^1\int_{1-\sqrt{1-y^2}}^{1+\sqrt{1-y^2}} f(x,y)\,dx\,dy$

8. $\displaystyle\int_0^2\int_x^{2x} f(x,y)\,dy\,dx + \int_2^3\int_x^{6-x} f(x,y)\,dy\,dx$

9. **(a)** $(1,2)=(b,d),(2,1)=(a,c)$, so $a=2, b=1, c=1, d=2$

(b) $\displaystyle\iint_R dA=\int_0^1\int_0^1 \frac{\partial(x,y)}{\partial(u,v)}du\,dv=\int_0^1\int_0^1 3du\,dv=3$

10. If $0<x,y<\pi$ then $0<\sin\sqrt{xy}\le 1$, with equality only on the hyperbola $xy=\pi^2/4$, so

$$0=\int_0^{\pi}\int_0^{\pi} 0\,dy\,dx<\int_0^{\pi}\int_0^{\pi}\sin\sqrt{xy}\,dy\,dx<\int_0^{\pi}\int_0^{\pi}1\,dy\,dx=\pi^2$$

11. $\displaystyle\int_{1/2}^{1} 2x\cos(\pi x^2)\,dx=\frac{1}{\pi}\sin(\pi x^2)\Big]_{1/2}^{1}=-1/(\sqrt{2}\pi)$

12. $\displaystyle\int_0^2 \frac{x^2}{2}e^{y^3}\Big]_{x=-y}^{x=2y}dy=\frac{3}{2}\int_0^2 y^2e^{y^3}\,dy=\frac{1}{2}e^{y^3}\Big]_0^2=\frac{1}{2}\left(e^8-1\right)$

13. $\displaystyle\int_0^1\int_{2y}^{2} e^x e^y\,dx\,dy$

14. $\displaystyle\int_0^{\pi}\int_0^{x}\frac{\sin x}{x}\,dy\,dx$

15.

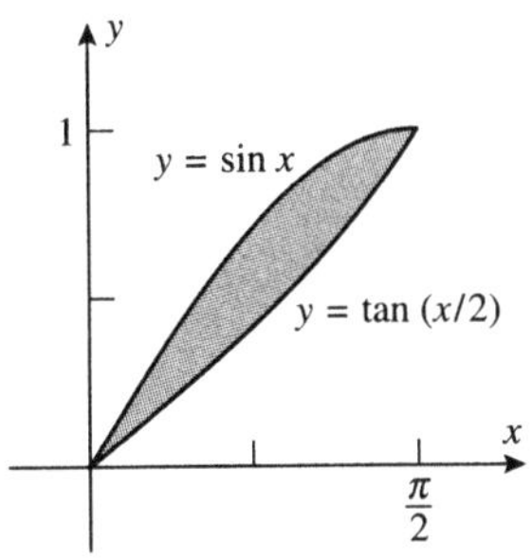

16.

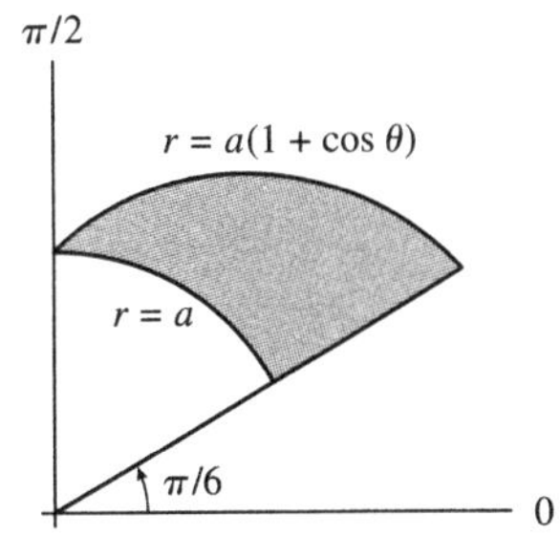

17. $\displaystyle 2\int_0^8\int_0^{y^{1/3}} x^2\sin y^2dx\,dy=\frac{2}{3}\int_0^8 y\sin y^2\,dy=-\frac{1}{3}\cos y^2\Big]_0^8=\frac{1}{3}(1-\cos 64)\approx 0.20271$

18. $\displaystyle\int_0^{\pi/2}\int_0^2 (4-r^2)r\,dr\,d\theta=2\pi$

19. $\displaystyle\sin 2\theta=2\sin\theta\cos\theta=\frac{2xy}{x^2+y^2}$, and $r=2a\sin\theta$ is the circle $x^2+(y-a)^2=a^2$, so

$$\int_0^a\int_{a-\sqrt{a^2-x^2}}^{a+\sqrt{a^2-x^2}}\frac{2xy}{x^2+y^2}dy\,dx=\int_0^a x\left[\ln\left(a+\sqrt{a^2-x^2}\right)-\ln\left(a-\sqrt{a^2-x^2}\right)\right]dx=a^2$$

20. $\displaystyle\int_{\pi/4}^{\pi/2}\int_0^2 4r^2(\cos\theta\sin\theta)\,r\,dr\,d\theta=-4\cos 2\theta\Big]_{\pi/4}^{\pi/2}=4$

21. $\displaystyle\int_0^{2\pi}\int_0^2\int_{r^4}^{16} r^2\cos^2\theta\,r\,dz\,dr\,d\theta=\int_0^{2\pi}\cos^2\theta\,d\theta\int_0^2 r^3(16-r^4)\,dr=32\pi$

22. $\displaystyle\int_0^{\pi/2}\int_0^{\pi/2}\int_0^1\frac{1}{1+\rho^2}\rho^2\sin\phi\,d\rho\,d\phi\,d\theta=\left(1-\frac{\pi}{4}\right)\frac{\pi}{2}\int_0^{\pi/2}\sin\phi\,d\phi$

$$=\left(1-\frac{\pi}{4}\right)\frac{\pi}{2}(-\cos\phi)\Big]_0^{\pi/2}=\left(1-\frac{\pi}{4}\right)\frac{\pi}{2}$$

23. (a) $\displaystyle\int_0^{2\pi}\int_0^{\pi/3}\int_0^a (\rho^2\sin^2\phi)\rho^2\sin\phi\,d\rho\,d\phi\,d\theta = \int_0^{2\pi}\int_0^{\pi/3}\int_0^a \rho^4\sin^3\phi\,d\rho\,d\phi\,d\theta$

(b) $\displaystyle\int_0^{2\pi}\int_0^{\sqrt{3}a/2}\int_{r/\sqrt{3}}^{\sqrt{a^2-r^2}} r^2\,dz\,r dr\,d\theta = \int_0^{2\pi}\int_0^{\sqrt{3}a/2}\int_{r/\sqrt{3}}^{\sqrt{a^2-r^2}} r^3\,dz\,dr\,d\theta$

(c) $\displaystyle\int_{-\sqrt{3}a/2}^{\sqrt{3}a/2}\int_{-\sqrt{(3a^2/4)-x^2}}^{\sqrt{(3a^2/4)-x^2}}\int_{\sqrt{x^2+y^2}/\sqrt{3}}^{\sqrt{a^2-x^2-y^2}} (x^2+y^2)\,dz\,dy\,dx$

24. (a) $\displaystyle\int_0^4\int_{-\sqrt{4x-x^2}}^{\sqrt{4x-x^2}}\int_{x^2+y^2}^{4x} dz\,dy\,dx$

(b) $\displaystyle\int_{-\pi/2}^{\pi/2}\int_0^{4\cos\theta}\int_{r^2}^{4r\cos\theta} r\,dz\,dr\,d\theta$

25. $\displaystyle\int_0^2\int_{(y/2)^{1/3}}^{2-y/2} dx\,dy = \int_0^2\left(2-\frac{y}{2}-\left(\frac{y}{2}\right)^{1/3}\right)dy = \left(2y-\frac{y^2}{4}-\frac{3}{2}\left(\frac{y}{2}\right)^{4/3}\right)\Big]_0^2 = \frac{3}{2}$

26. $\displaystyle A = 6\int_0^{\pi/6}\int_0^{\cos 3\theta} r\,dr\,d\theta = 3\int_0^{\pi/6}\cos^2 3\theta = \pi/4$

27. $\displaystyle V = \int_0^{2\pi}\int_0^{a/\sqrt{3}}\int_{\sqrt{3}r}^{a} r\,dz\,dr\,d\theta = 2\pi\int_0^{a/\sqrt{3}} r(a-\sqrt{3}r)\,dr = \frac{\pi a^3}{9}$

28. The intersection of the two surfaces projects onto the yz-plane as $2y^2+z^2=1$, so

$$V = 4\int_0^{1/\sqrt{2}}\int_0^{\sqrt{1-2y^2}}\int_{y^2+z^2}^{1-y^2} dx\,dz\,dy$$

$$= 4\int_0^{1/\sqrt{2}}\int_0^{\sqrt{1-2y^2}}(1-2y^2-z^2)\,dz\,dy = 4\int_0^{1/\sqrt{2}}\frac{2}{3}(1-2y^2)^{3/2}\,dy = \frac{\sqrt{2}\pi}{4}$$

29. $\|\mathbf{r}_u\times\mathbf{r}_v\| = \sqrt{2u^2+2v^2+4}$,

$$S = \iint_{u^2+v^2\le 4}\sqrt{2u^2+2v^2+4}\,dA = \int_0^{2\pi}\int_0^2\sqrt{2}\sqrt{r^2+2}\,r\,dr\,d\theta = \frac{8\pi}{3}(3\sqrt{3}-1)$$

30. $\displaystyle\|\mathbf{r}_u\times\mathbf{r}_v\| = \sqrt{1+u^2},\ S = \int_0^2\int_0^{3u}\sqrt{1+u^2}dv\,du = \int_0^2 3u\sqrt{1+u^2}du = 5^{3/2}-1$

31. $(\mathbf{r}_u\times\mathbf{r}_v)\Big]_{\substack{u=1\\v=2}} = \langle -2,-4,1\rangle$, tangent plane $2x+4y-z=5$

32. $u=-3, v=0$, $(\mathbf{r}_u\times\mathbf{r}_v)\Big]_{\substack{u=-3\\v=0}} = \langle -18,0,-3\rangle$, tangent plane $6x+z=-9$

33. $\displaystyle A = \int_{-4}^4\int_{y^2/4}^{2+y^2/8} dx\,dy = \int_{-4}^4\left(2-\frac{y^2}{8}\right)dy = \frac{32}{3}$; $\bar{y}=0$ by symmetry;

$\displaystyle\int_{-4}^4\int_{y^2/4}^{2+y^2/8} x\,dx\,dy = \int_{-4}^4\left(2+\frac{1}{4}y^2-\frac{3}{128}y^4\right)dy = \frac{256}{15},\ \bar{x} = \frac{3}{32}\frac{256}{15} = \frac{8}{5}$; centroid $\left(\frac{8}{5},0\right)$

34. $A = \pi ab/2, \bar{x} = 0$ by symmetry,

$$\int_{-a}^{a}\int_{0}^{b\sqrt{1-x^2/a^2}} y\,dy\,dx = \frac{1}{2}\int_{-a}^{a} b^2(1-x^2/a^2)dx = 2ab^2/3, \text{ centroid } \left(0, \frac{4b}{3\pi}\right)$$

35. $V = \dfrac{1}{3}\pi a^2 h, \bar{x} = \bar{y} = 0$ by symmetry,

$$\int_0^{2\pi}\int_0^a\int_0^{h-rh/a} rz\,dz\,dr\,d\theta = \pi\int_0^a rh^2\left(1-\frac{r}{a}\right)^2 dr = \pi a^2h^2/12, \text{ centroid } (0,0,h/4)$$

36. $V = \displaystyle\int_{-2}^{2}\int_{x^2}^{4}\int_{0}^{4-y} dz\,dy\,dx = \int_{-2}^{2}\int_{x^2}^{4}(4-y)dy\,dx = \int_{-2}^{2}\left(8-4x^2+\frac{1}{2}x^4\right)dx = \frac{256}{15},$

$$\int_{-2}^{2}\int_{x^2}^{4}\int_{0}^{4-y} y\,dz\,dy\,dx = \int_{-2}^{2}\int_{x^2}^{4}(4y-y^2)\,dy\,dx = \int_{-2}^{2}\left(\frac{1}{3}x^6 - 2x^4 + \frac{32}{3}\right)dx = \frac{1024}{35}$$

$$\int_{-2}^{2}\int_{x^2}^{4}\int_{0}^{4-y} z\,dz\,dy\,dx = \int_{-2}^{2}\int_{x^2}^{4}\frac{1}{2}(4-y)^2dy\,dx = \int_{-2}^{2}\left(-\frac{x^6}{6}+2x^4-8x^2+\frac{32}{3}\right)dx = \frac{2048}{105}$$

$\bar{x} = 0$ by symmetry, centroid $\left(0, \dfrac{12}{7}, \dfrac{8}{7}\right)$

37. The two quarter-circles with center at the origin and of radius A and $\sqrt{2}A$ lie inside and outside of the square with corners $(0,0), (A,0), (A,A), (0,A)$, so the following inequalities hold:

$$\int_0^{\pi/2}\int_0^A \frac{1}{(1+r^2)^2} r\,dr\,d\theta \le \int_0^A\int_0^A \frac{1}{(1+x^2+y^2)^2}dx\,dy \le \int_0^{\pi/2}\int_0^{\sqrt{2}A}\frac{1}{(1+r^2)^2}r\,dr\,d\theta$$

The integral on the left can be evaluated as $\dfrac{\pi A^2}{4(1+A^2)}$ and the integral on the right equals $\dfrac{2\pi A^2}{4(1+2A^2)}$. Since both of these quantities tend to $\dfrac{\pi}{4}$ as $A \to +\infty$, it follows by sandwiching that

$$\int_0^{+\infty}\int_0^{+\infty}\frac{1}{(1+x^2+y^2)^2}dx\,dy = \frac{\pi}{4}.$$

38. The centroid of the circle which generates the tube travels a distance

$$s = \int_0^{4\pi}\sqrt{\sin^2 t + \cos^2 t + 1/16}\,dt = \sqrt{17}\pi, \text{ so } V = \pi(1/2)^2\sqrt{17}\pi = \sqrt{17}\pi^2/4.$$

39. **(a)** Let S_1 be the set of points (x,y,z) which satisfy the equation $x^{2/3}+y^{2/3}+z^{2/3} = a^{2/3}$, and let S_2 be the set of points (x,y,z) where $x = a(\sin\phi\cos\theta)^3, y = a(\sin\phi\sin\theta)^3, z = a\cos^3\phi$, $0 \le \phi \le \pi, 0 \le \theta < 2\pi$.
If (x,y,z) is a point of S_2 then

$$x^{2/3}+y^{2/3}+z^{2/3} = a^{2/3}[(\sin\phi\cos\theta)^3 + (\sin\phi\sin\theta)^3 + \cos^3\phi] = a^{2/3}$$

so (x,y,z) belongs to S_1.
If (x,y,z) is a point of S_1 then $x^{2/3}+y^{2/3}+z^{2/3} = a^{2/3}$. Let
$x_1 = x^{1/3}, y_1 = y^{1/3}, z_1 = z^{1/3}, a_1 = a^{1/3}$. Then $x_1^2+y_1^2+z_1^2 = a_1^2$, so in spherical coordinates $x_1 = a_1\sin\phi\cos\theta, y_1 = a_1\sin\phi\sin\theta, z_1 = a_1\cos\phi$, with

$$\theta = \tan^{-1}\left(\frac{y_1}{x_1}\right) = \tan^{-1}\left(\frac{y}{x}\right)^{1/3}, \phi = \cos^{-1}\frac{z_1}{a_1} = \cos^{-1}\left(\frac{z}{a}\right)^{1/3}. \text{ Then}$$

$x = x_1^3 = a_1^3(\sin\phi\cos\theta)^3 = a(\sin\phi\cos\theta)^3$, similarly $y = a(\sin\phi\sin\theta)^3, z = a\cos\phi$ so (x,y,z) belongs to S_2. Thus $S_1 = S_2$

(b) Let $a = 1$ and $\mathbf{r} = (\cos\theta\sin\phi)^3\mathbf{i} + (\sin\theta\sin\phi)^3\mathbf{j} + \cos^3\phi\mathbf{k}$, then

$$S = 8\int_0^{\pi/2}\int_0^{\pi/2} \|\mathbf{r}_\theta \times \mathbf{r}_\phi\| d\phi\, d\theta$$

$$= 72\int_0^{\pi/2}\int_0^{\pi/2} \sin\theta\cos\theta\sin^4\phi\cos\phi\sqrt{\cos^2\phi + \sin^2\phi\sin^2\theta\cos^2\theta}\, d\theta\, d\phi \approx 4.4506$$

(c) $$\frac{\partial(x,y,z)}{\partial(\rho,\theta,\phi)} = \left\| \begin{matrix} \sin^3\phi\cos^3\theta & 3\rho\sin^2\phi\cos\phi\cos^3\theta & -3\rho\sin^3\phi\cos^2\theta\sin\theta \\ \sin^3\phi\sin^3\theta & 3\rho\sin^2\phi\cos\phi\sin^3\theta & 3\rho\sin^3\phi\sin^2\theta\cos\theta \\ \cos^3\phi & -3\rho\cos^2\phi\sin\phi & 0 \end{matrix} \right\|$$

$$= 9\rho^2\cos^2\theta\sin^2\theta\cos^2\phi\sin^5\phi,$$

$$V = 9\int_0^{2\pi}\int_0^{\pi}\int_0^{a} \rho^2\cos^2\theta\sin^2\theta\cos^2\phi\sin^5\phi\, d\rho\, d\phi\, d\theta = \frac{4}{35}\pi a^3$$

40. $$V = \frac{4}{3}\pi a^3, \bar{d} = \frac{3}{4\pi a^3}\iiint\limits_{\rho\le a} \rho dV = \frac{3}{4\pi a^3}\int_0^{\pi}\int_0^{2\pi}\int_0^{a} \rho^3\sin\phi\, d\rho\, d\theta\, d\phi = \frac{3}{4\pi a^3}2\pi(2)\frac{a^4}{4} = \frac{3}{4}a$$

41. **(a)** $(x/a)^2 + (y/b)^2 + (z/c)^2 = \sin^2\phi\cos^2\theta + \sin^2\phi\sin^2\theta + \cos^2\phi = \sin^2\phi + \cos^2\phi = 1$, an ellipsoid

(b) $\mathbf{r}(\phi,\theta) = \langle 2\sin\phi\cos\theta, 3\sin\phi\sin\theta, 4\cos\phi\rangle$; $\mathbf{r}_\phi \times \mathbf{r}_\theta = 2\langle 6\sin^2\phi\cos\theta, 4\sin^2\phi\sin\theta, 3\cos\phi\sin\phi\rangle$,

$\|\mathbf{r}_\phi \times \mathbf{r}_\theta\| = 2\sqrt{16\sin^4\phi + 20\sin^4\phi\cos^2\theta + 9\sin^2\phi\cos^2\phi}$,

$$S = \int_0^{2\pi}\int_0^{\pi} 2\sqrt{16\sin^4\phi + 20\sin^4\phi\cos^2\theta + 9\sin^2\phi\cos^2\phi}\, d\phi\, d\theta \approx 111.5457699$$

CHAPTER 16
Topics in Vector Calculus

EXERCISE SET 16.1

1. **(a)** III because the vector field is independent of y and the direction is that of the negative x-axis for negative x, and positive for positive

(b) IV, because the y-component is constant, and the x-component varies priodically with x

2. **(a)** I, since the vector field is constant

(b) II, since the vector field points away from the origin

3. **(a)** true **(b)** true **(c)** true

4. **(a)** false, the lengths are equal to 1 **(b)** false, the y-component is then zero

(c) false, the x-component is then zero

5.

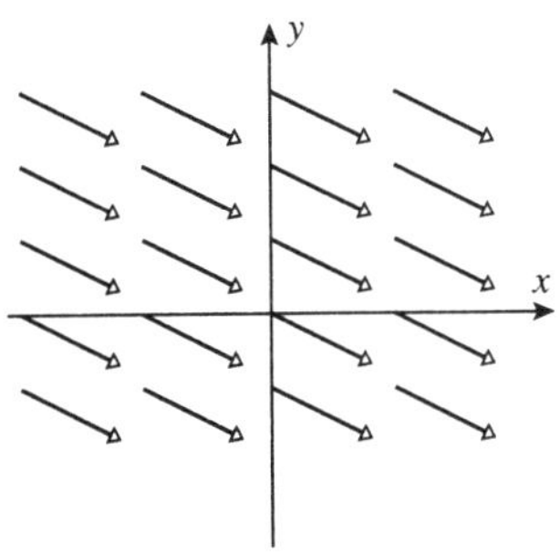

6.

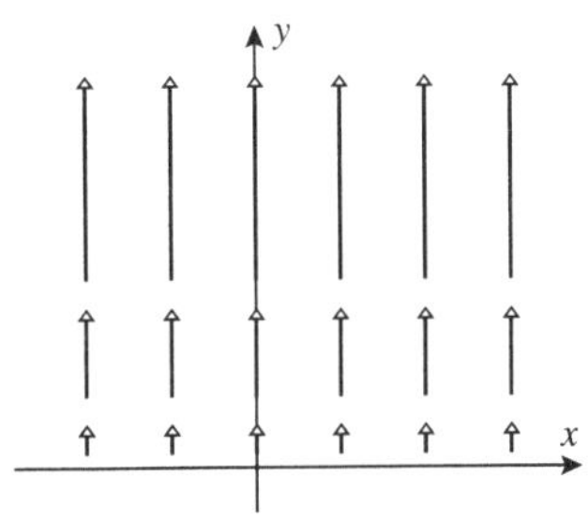

7.

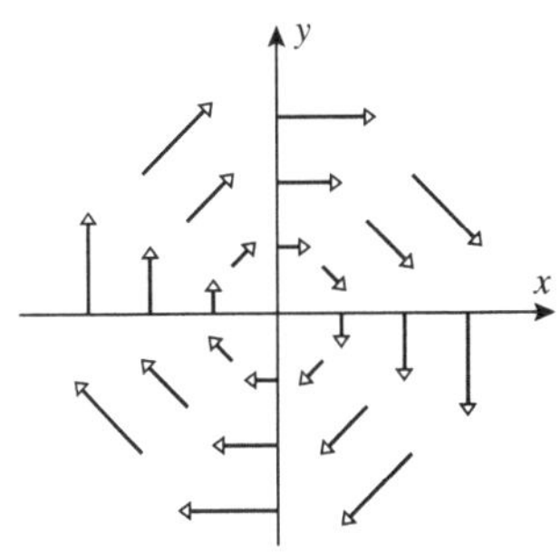

8.

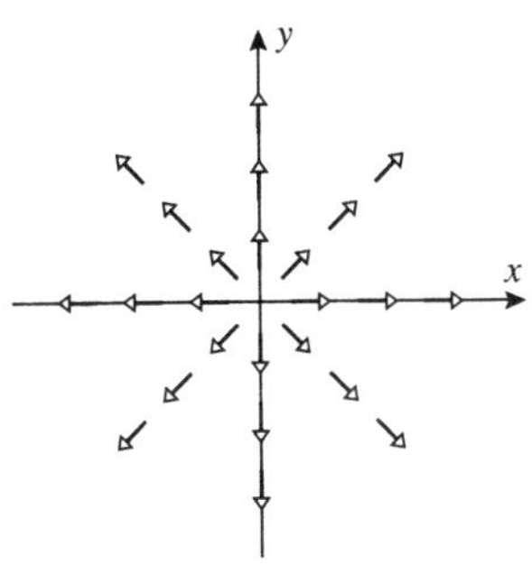

9.

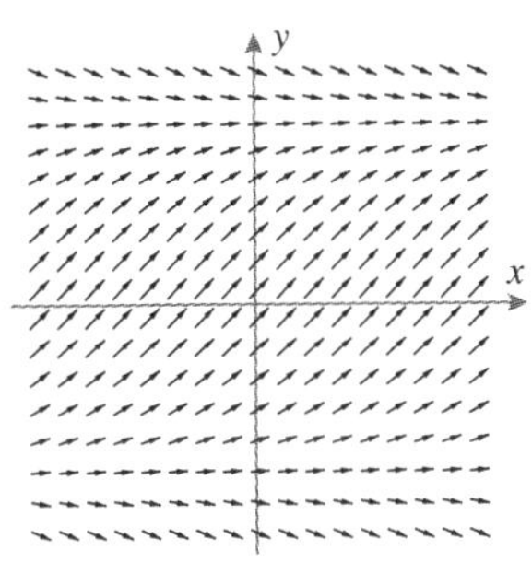

10.

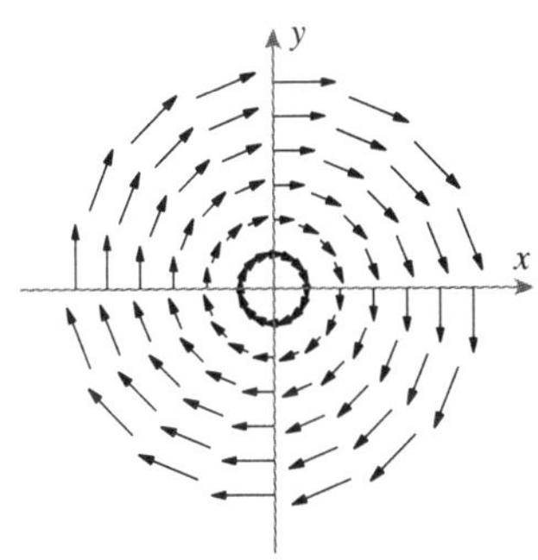

11. **(a)** $\nabla\phi = \phi_x\mathbf{i} + \phi_y\mathbf{j} = \dfrac{y}{1+x^2y^2}\mathbf{i} + \dfrac{x}{1+x^2y^2}\mathbf{j} = \mathbf{F}$, so $\mathbf{F}$ is conservative for all x, y

(b) $\nabla\phi = \phi_x\mathbf{i} + \phi_y\mathbf{j} = 2x\mathbf{i} - 6y\mathbf{j} + 8z\mathbf{k} = \mathbf{F}$ so $\mathbf{F}$ is conservative for all x, y

12. **(a)** $\nabla\phi = \phi_x\mathbf{i} + \phi_y\mathbf{j} = (6xy - y^3)\mathbf{i} + (4y + 3x^2 - 3xy^2)\mathbf{j} = \mathbf{F}$, so $\mathbf{F}$ is conservative for all x, y

(b) $\nabla\phi = \phi_x\mathbf{i} + \phi_y\mathbf{j} + \phi_z\mathbf{k} = (\sin z + y\cos x)\mathbf{i} + (\sin x + z\cos y)\mathbf{j} + (x\cos z + \sin y)\mathbf{k} = \mathbf{F}$, so $\mathbf{F}$ is conservative for all x, y

13. div $\mathbf{F} = 2x + y$, curl $\mathbf{F} = z\mathbf{i}$

14. div $\mathbf{F} = z^3 + 8y^3x^2 + 10zy$, curl $\mathbf{F} = 5z^2\mathbf{i} + 3xz^2\mathbf{j} + 4xy^4\mathbf{k}$

15. div $\mathbf{F} = 0$, curl $\mathbf{F} = (40x^2z^4 - 12xy^3)\mathbf{i} + (14y^3z + 3y^4)\mathbf{j} - (16xz^5 + 21y^2z^2)\mathbf{k}$

16. div $\mathbf{F} = ye^{xy} + \sin y + 2\sin z\cos z$, curl $\mathbf{F} = -xe^{xy}\mathbf{k}$

17. div $\mathbf{F} = \dfrac{2}{\sqrt{x^2+y^2+z^2}}$, curl $\mathbf{F} = \mathbf{0}$

18. div $\mathbf{F} = \dfrac{1}{x} + xze^{xyz} + \dfrac{x}{x^2+z^2}$, curl $\mathbf{F} = -xye^{xyz}\mathbf{i} + \dfrac{z}{x^2+z^2}\mathbf{j} + yze^{xyz}\mathbf{k}$

19. $\nabla\cdot(\mathbf{F}\times\mathbf{G}) = \nabla\cdot(-(z+4y^2)\mathbf{i} + (4xy+2xz)\mathbf{j} + (2xy-x)\mathbf{k}) = 4x$

20. $\nabla\cdot(\mathbf{F}\times\mathbf{G}) = \nabla\cdot((x^2yz^2 - x^2y^2)\mathbf{i} - xy^2z^2\mathbf{j} + xy^2z\mathbf{k}) = -xy^2$

21. $\nabla\cdot(\nabla\times\mathbf{F}) = \nabla\cdot(-\sin(x-y)\mathbf{k}) = 0$

22. $\nabla\cdot(\nabla\times\mathbf{F}) = \nabla\cdot(-ze^{yz}\mathbf{i} + xe^{xz}\mathbf{j} + 3e^y\mathbf{k}) = 0$

23. $\nabla\times(\nabla\times\mathbf{F}) = \nabla\times(xz\mathbf{i} - yz\mathbf{j} + y\mathbf{k}) = (1+y)\mathbf{i} + x\mathbf{j}$

24. $\nabla\times(\nabla\times\mathbf{F}) = \nabla\times((x+3y)\mathbf{i} - y\mathbf{j} - 2xy\mathbf{k}) = -2x\mathbf{i} + 2y\mathbf{j} - 3\mathbf{k}$

27. Let $\mathbf{F} = f\mathbf{i} + g\mathbf{j} + h\mathbf{k}$; div $(k\mathbf{F}) = k\dfrac{\partial f}{\partial x} + k\dfrac{\partial g}{\partial y} + k\dfrac{\partial h}{\partial z} = k$ div $\mathbf{F}$

28. Let $\mathbf{F} = f\mathbf{i} + g\mathbf{j} + h\mathbf{k}$; curl $(k\mathbf{F}) = k\left(\dfrac{\partial h}{\partial y} - \dfrac{\partial g}{\partial z}\right)\mathbf{i} + k\left(\dfrac{\partial f}{\partial z} - \dfrac{\partial h}{\partial x}\right)\mathbf{j} + k\left(\dfrac{\partial g}{\partial x} - \dfrac{\partial f}{\partial y}\right)\mathbf{k} = k$ curl $\mathbf{F}$

29. Let $\mathbf{F} = f(x,y,z)\mathbf{i} + g(x,y,z)\mathbf{j} + h(x,y,z)\mathbf{k}$ and $\mathbf{G} = P(x,y,z)\mathbf{i} + Q(x,y,z)\mathbf{j} + R(x,y,z)\mathbf{k}$, then

$$\begin{aligned} \text{div } (\mathbf{F}+\mathbf{G}) &= \left(\frac{\partial f}{\partial x} + \frac{\partial P}{\partial x}\right) + \left(\frac{\partial g}{\partial y} + \frac{\partial Q}{\partial y}\right) + \left(\frac{\partial h}{\partial z} + \frac{\partial R}{\partial z}\right) \\ &= \left(\frac{\partial f}{\partial x} + \frac{\partial g}{\partial y} + \frac{\partial h}{\partial z}\right) + \left(\frac{\partial P}{\partial x} + \frac{\partial Q}{\partial y} + \frac{\partial R}{\partial z}\right) = \text{div } \mathbf{F} + \text{div } \mathbf{G} \end{aligned}$$

30. Let $\mathbf{F} = f(x,y,z)\mathbf{i} + g(x,y,z)\mathbf{j} + h(x,y,z)\mathbf{k}$ and $\mathbf{G} = P(x,y,z)\mathbf{i} + Q(x,y,z)\mathbf{j} + R(x,y,z)\mathbf{k}$, then

$$\begin{aligned} \text{curl } (\mathbf{F}+\mathbf{G}) = &\left[\frac{\partial}{\partial y}(h+R) - \frac{\partial}{\partial z}(g+Q)\right]\mathbf{i} + \left[\frac{\partial}{\partial z}(f+P) - \frac{\partial}{\partial x}(h+R)\right]\mathbf{j} \\ &+ \left[\frac{\partial}{\partial x}(g+Q) - \frac{\partial}{\partial y}(f+P)\right]\mathbf{k}; \end{aligned}$$

expand and rearrange terms to get curl $\mathbf{F}$ + curl $\mathbf{G}$.

31. Let $\mathbf{F} = f\mathbf{i} + g\mathbf{j} + h\mathbf{k}$;

$$\begin{aligned} \text{div } (\phi\mathbf{F}) &= \left(\phi\frac{\partial f}{\partial x} + \frac{\partial \phi}{\partial x}f\right) + \left(\phi\frac{\partial g}{\partial y} + \frac{\partial \phi}{\partial y}g\right) + \left(\phi\frac{\partial h}{\partial z} + \frac{\partial \phi}{\partial z}h\right) \\ &= \phi\left(\frac{\partial f}{\partial x} + \frac{\partial g}{\partial y} + \frac{\partial h}{\partial z}\right) + \left(\frac{\partial \phi}{\partial x}f + \frac{\partial \phi}{\partial y}g + \frac{\partial \phi}{\partial z}h\right) \\ &= \phi \text{ div } \mathbf{F} + \nabla\phi\cdot\mathbf{F} \end{aligned}$$

32. Let $\mathbf{F} = f\mathbf{i} + g\mathbf{j} + h\mathbf{k}$;

curl $(\phi\mathbf{F}) = \left[\frac{\partial}{\partial y}(\phi h) - \frac{\partial}{\partial z}(\phi g)\right]\mathbf{i} + \left[\frac{\partial}{\partial z}(\phi f) - \frac{\partial}{\partial x}(\phi h)\right]\mathbf{j} + \left[\frac{\partial}{\partial x}(\phi g) - \frac{\partial}{\partial y}(\phi f)\right]\mathbf{k}$; use the product rule to expand each of the partial derivatives, rearrange to get ϕ curl $\mathbf{F} + \nabla\phi \times \mathbf{F}$

33. Let $\mathbf{F} = f\mathbf{i} + g\mathbf{j} + h\mathbf{k}$;

$$\text{div(curl }\mathbf{F}) = \frac{\partial}{\partial x}\left(\frac{\partial h}{\partial y} - \frac{\partial g}{\partial z}\right) + \frac{\partial}{\partial y}\left(\frac{\partial f}{\partial z} - \frac{\partial h}{\partial x}\right) + \frac{\partial}{\partial z}\left(\frac{\partial g}{\partial x} - \frac{\partial f}{\partial y}\right)$$
$$= \frac{\partial^2 h}{\partial x\partial y} - \frac{\partial^2 g}{\partial x\partial z} + \frac{\partial^2 f}{\partial y\partial z} - \frac{\partial^2 h}{\partial y\partial x} + \frac{\partial^2 g}{\partial z\partial x} - \frac{\partial^2 f}{\partial z\partial y} = 0,$$

assuming equality of mixed second partial derivatives

34. curl $(\nabla\phi) = \left(\frac{\partial^2\phi}{\partial y\partial z} - \frac{\partial^2\phi}{\partial z\partial y}\right)\mathbf{i} + \left(\frac{\partial^2\phi}{\partial z\partial x} - \frac{\partial^2\phi}{\partial x\partial z}\right)\mathbf{j} + \left(\frac{\partial^2\phi}{\partial x\partial y} - \frac{\partial^2\phi}{\partial y\partial x}\right)\mathbf{k} = \mathbf{0}$, assuming equality of mixed second partial derivatives

35. $\nabla\cdot(k\mathbf{F}) = k\nabla\cdot\mathbf{F}$, $\nabla\cdot(\mathbf{F}+\mathbf{G}) = \nabla\cdot\mathbf{F} + \nabla\cdot\mathbf{G}$, $\nabla\cdot(\phi\mathbf{F}) = \phi\nabla\cdot\mathbf{F} + \nabla\phi\cdot\mathbf{F}$, $\nabla\cdot(\nabla\times\mathbf{F}) = 0$

36. $\nabla\times(k\mathbf{F}) = k\nabla\times\mathbf{F}$, $\nabla\times(\mathbf{F}+\mathbf{G}) = \nabla\times\mathbf{F} + \nabla\times\mathbf{G}$, $\nabla\times(\phi\mathbf{F}) = \phi\nabla\times\mathbf{F} + \nabla\phi\times\mathbf{F}$, $\nabla\times(\nabla\phi) = \mathbf{0}$

37. (a) curl $\mathbf{r} = 0\mathbf{i} + 0\mathbf{j} + 0\mathbf{k} = \mathbf{0}$

(b) $\nabla\|\mathbf{r}\| = \nabla\sqrt{x^2+y^2+z^2} = \frac{x}{\sqrt{x^2+y^2+z^2}}\mathbf{i} + \frac{y}{\sqrt{x^2+y^2+z^2}}\mathbf{j} + \frac{z}{\sqrt{x^2+y^2+z^2}}\mathbf{k} = \frac{\mathbf{r}}{\|\mathbf{r}\|}$

38. (a) div $\mathbf{r} = 1 + 1 + 1 = 3$

(b) $\nabla\frac{1}{\|\mathbf{r}\|} = \nabla(x^2+y^2+z^2)^{-1/2} = -\frac{x\mathbf{i}+y\mathbf{j}+z\mathbf{k}}{(x^2+y^2+z^2)^{3/2}} = -\frac{\mathbf{r}}{\|\mathbf{r}\|^3}$

39. (a) $\nabla f(r) = f'(r)\frac{\partial r}{\partial x}\mathbf{i} + f'(r)\frac{\partial r}{\partial y}\mathbf{j} + f'(r)\frac{\partial r}{\partial z}\mathbf{k} = f'(r)\nabla r = \frac{f'(r)}{r}\mathbf{r}$

(b) $\text{div}[f(r)\mathbf{r}] = f(r)\text{div }\mathbf{r} + \nabla f(r)\cdot\mathbf{r} = 3f(r) + \frac{f'(r)}{r}\mathbf{r}\cdot\mathbf{r} = 3f(r) + rf'(r)$

40. (a) $\text{curl}[f(r)\mathbf{r}] = f(r)\text{curl }\mathbf{r} + \nabla f(r)\times\mathbf{r} = f(r)\mathbf{0} + \frac{f'(r)}{r}\mathbf{r}\times\mathbf{r} = \mathbf{0}+\mathbf{0} = \mathbf{0}$

(b) $$\nabla^2 f(r) = \text{div}[\nabla f(r)] = \text{div}\left[\frac{f'(r)}{r}\mathbf{r}\right] = \frac{f'(r)}{r}\text{div }\mathbf{r} + \nabla\frac{f'(r)}{r}\cdot\mathbf{r}$$
$$= 3\frac{f'(r)}{r} + \frac{rf''(r)-f'(r)}{r^3}\mathbf{r}\cdot\mathbf{r} = 2\frac{f'(r)}{r} + f''(r)$$

41. $f(r) = 1/r^3, f'(r) = -3/r^4$, $\text{div}(\mathbf{r}/r^3) = 3(1/r^3) + r(-3/r^4) = 0$

42. Multiply $3f(r) + rf'(r) = 0$ through by r^2 to obtain $3r^2f(r) + r^3f'(r) = 0$, $d[r^3f(r)]/dr = 0$, $r^3f(r) = C, f(r) = C/r^3$, so $\mathbf{F} = C\mathbf{r}/r^3$ (an inverse-square field).

43. (a) At the point (x, y) the slope of the line along which the vector $-y\mathbf{i} + x\mathbf{j}$ lies is $-x/y$; the slope of the tangent line to C at (x, y) is dy/dx, so $dy/dx = -x/y$.

(b) $ydy = -xdx$, $y^2/2 = -x^2/2 + K_1$, $x^2 + y^2 = K$

44. $dy/dx = x, y = x^2/2 + K$

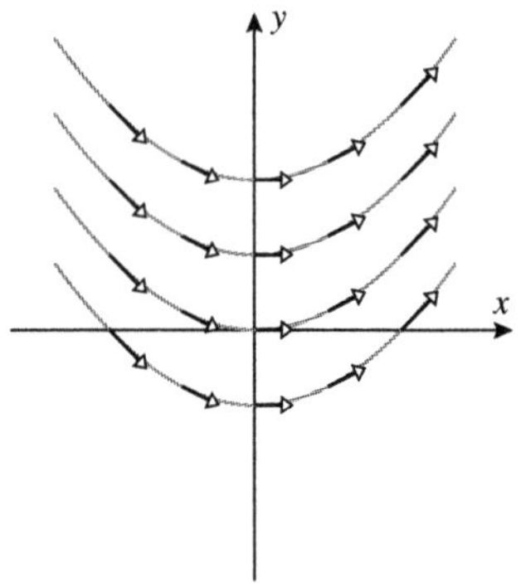

45. $dy/dx = 1/x, y = \ln x + K$

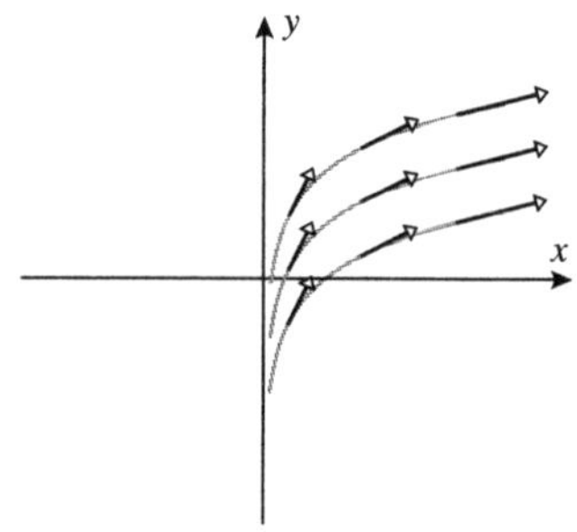

46. $dy/dx = -y/x,\ (1/y)dy = (-1/x)dx,\ \ln y = -\ln x + K_1,$
$y = e^{K_1}e^{-\ln x} = K/x$

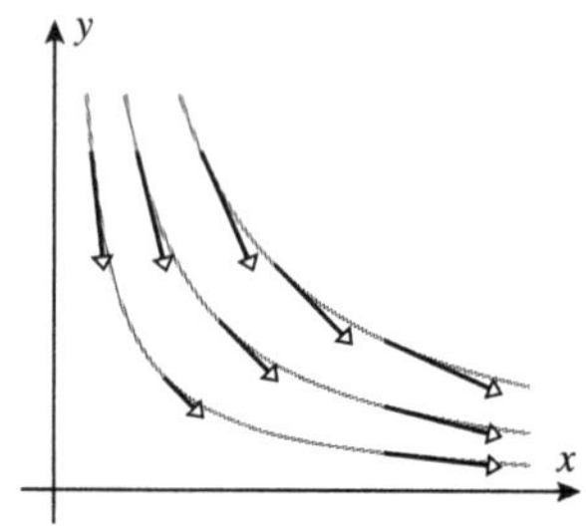

EXERCISE SET 16.2

1. **(a)** $\int_0^1 dy = 1$ because $s = y$ is arclength measured from $(0, 0)$

(b) 0, because $\sin xy = 0$ along C

2. **(a)** $\int_C ds =$ length of line segment $= 2$ **(b)** 0, because x is constant and $dx = 0$

3. **(a)** $ds = \sqrt{\left(\frac{dx}{dt}\right)^2 + \left(\frac{dy}{dt}\right)^2}\,dt$, so $\int_0^1 (2t - 3t^2)\sqrt{4 + 36t^2}\,dt = -\frac{11}{108}\sqrt{10} - \frac{1}{36}\ln(\sqrt{10} - 3) - \frac{4}{27}$

(b) $\int_0^1 (2t - 3t^2)2\,dt = 0$ **(c)** $\int_0^1 (2t - 3t^2)6t\,dt = -\frac{1}{2}$

4. **(a)** $\int_0^1 t(3t^2)(6t^3)^2\sqrt{1 + 36t^2 + 324t^4}\,dt = \frac{864}{5}$ **(b)** $\int_0^1 t(3t^2)(6t^3)^2\,dt = \frac{54}{5}$

(c) $\int_0^1 t(3t^2)(6t^3)^2 6t\,dt = \frac{648}{11}$ **(d)** $\int_0^1 t(3t^2)(6t^3)^2 18t^2\,dt = 162$

5. **(a)** $C : x = t,\ y = t,\ 0 \le t \le 1;\ \int_0^1 6t\,dt = 3$

(b) $C : x = t,\ y = t^2,\ 0 \le t \le 1;\ \int_0^1 (3t + 6t^2 - 2t^3)dt = 3$

(c) $C: x = t,\ y = \sin(\pi t/2),\ 0 \le t \le 1;$
$$\int_0^1 [3t + 2\sin(\pi t/2) + \pi t\cos(\pi t/2) - (\pi/2)\sin(\pi t/2)\cos(\pi t/2)]dt = 3$$

(d) $C: x = t^3,\ y = t,\ 0 \le t \le 1;\ \int_0^1 (9t^5 + 8t^3 - t)dt = 3$

6. (a) $C: x = t, y = t, z = t, 0 \le t \le 1;\ \int_0^1 (t + t - t)\,dt = \frac{1}{2}$

(b) $C: x = t, y = t^2, z = t^3, 0 \le t \le 1;\ \int_0^1 (t^2 + t^3(2t) - t(3t^2))\,dt = -\frac{1}{60}$

(c) $C: x = \cos\pi t, y = \sin\pi t, z = t, 0 \le t \le 1;\ \int_0^1 (-\pi\sin^2\pi t + \pi t\cos\pi t - \cos\pi t)\,dt = -\frac{\pi}{2} - \frac{2}{\pi}$

7. $\int_0^3 \frac{\sqrt{1+t}}{1+t}\,dt = \int_0^3 (1+t)^{-1/2}\,dt = 2$

8. $\sqrt{5}\int_0^1 \frac{1+2t}{1+t^2}\,dt = \sqrt{5}(\pi/4 + \ln 2)$

9. $\int_0^1 3(t^2)(t^2)(2t^3/3)(1+2t^2)\,dt = 2\int_0^1 t^7(1+2t^2)\,dt = 13/20$

10. $\frac{\sqrt{5}}{4}\int_0^{2\pi} e^{-t}\,dt = \sqrt{5}(1 - e^{-2\pi})/4$

11. $\int_0^{\pi/4} (8\cos^2 t - 16\sin^2 t - 20\sin t\cos t)dt = 1 - \pi$

12. $\int_{-1}^1 \left(\frac{2}{3}t - \frac{2}{3}t^{5/3} + t^{2/3}\right)dt = 6/5$

13. $C: x = (3-t)^2/3,\ y = 3 - t,\ 0 \le t \le 3;\ \int_0^3 \frac{1}{3}(3-t)^2dt = 3$

14. $C: x = t^{2/3},\ y = t,\ -1 \le t \le 1;\ \int_{-1}^1 \left(\frac{2}{3}t^{2/3} - \frac{2}{3}t^{1/3} + t^{7/3}\right)dt = 4/5$

15. $C: x = \cos t,\ y = \sin t,\ 0 \le t \le \pi/2;\ \int_0^{\pi/2} (-\sin t - \cos^2 t)dt = -1 - \pi/4$

16. $C: x = 3 - t,\ y = 4 - 3t,\ 0 \le t \le 1;\ \int_0^1 (-37 + 41t - 9t^2)dt = -39/2$

17. $\int_0^1 (-3)e^{3t}dt = 1 - e^3$

18. $\int_0^{\pi/2} (\sin^2 t\cos t - \sin^2 t\cos t + t^4(2t))\,dt = \frac{\pi^6}{192}$

19. (a) $\int_0^{\ln 2} \left(e^{3t} + e^{-3t}\right)\sqrt{e^{2t} + e^{-2t}}\,dt$

$$= \frac{63}{64}\sqrt{17} + \frac{1}{4}\ln(4 + \sqrt{17}) - \frac{1}{8}\ln\frac{\sqrt{17}+1}{\sqrt{17}-1} - \frac{1}{4}\ln(\sqrt{2}+1) + \frac{1}{8}\ln\frac{\sqrt{2}+1}{\sqrt{2}-1}$$

(b) $\int_0^{\pi/2} [\sin t\cos t\,dt - \sin^2 t\,dt] = \frac{1}{2} - \frac{\pi}{4}$

20. (a) $\displaystyle\int_0^{\pi/2} \cos^{21} t \sin^9 t \sqrt{(-3\cos^2 t \sin t)^2 + (3\sin^2 t \cos t)^2}\, dt$

$$= 3\int_0^{\pi/2} \cos^{22} t \sin^{10} t\, dt = \frac{61{,}047}{4{,}294{,}967{,}296}\pi$$

(b) $\displaystyle\int_1^e \left(t^5 \ln t + 7t^2(2t) + t^4(\ln t)\frac{1}{t}\right) dt = \frac{5}{36}e^6 + \frac{59}{16}e^4 - \frac{491}{144}$

21. (a) $C_1 : (0,0)$ to $(1,0); x = t, y = 0, 0 \le t \le 1$
$C_2 : (1,0)$ to $(0,1); x = 1 - t, y = t, 0 \le t \le 1$
$C_3 : (0,1)$ to $(0,0); x = 0, y = 1 - t, 0 \le t \le 1$

$$\int_0^1 (0)dt + \int_0^1 (-1)dt + \int_0^1 (0)dt = -1$$

(b) $C_1 : (0,0)$ to $(1,0); x = t, y = 0, 0 \le t \le 1$
$C_2 : (1,0)$ to $(1,1); x = 1, y = t, 0 \le t \le 1$
$C_3 : (1,1)$ to $(0,1); x = 1 - t, y = 1, 0 \le t \le 1$
$C_4 : (0,1)$ to $(0,0); x = 0, y = 1 - t, 0 \le t \le 1$

$$\int_0^1 (0)dt + \int_0^1 (-1)dt + \int_0^1 (-1)dt + \int_0^1 (0)dt = -2$$

22. (a) $C_1 : (0,0)$ to $(1,1); x = t, y = t, 0 \le t \le 1$
$C_2 : (1,1)$ to $(2,0); x = 1 + t, y = 1 - t, 0 \le t \le 1$
$C_3 : (2,0)$ to $(0,0); x = 2 - 2t, y = 0, 0 \le t \le 1$

$$\int_0^1 (0)dt + \int_0^1 2dt + \int_0^1 (0)dt = 2$$

(b) $C_1 : (-5,0)$ to $(5,0); x = t, y = 0, -5 \le t \le 5$
$C_2 : x = 5\cos t, y = 5\sin t, 0 \le t \le \pi$

$$\int_{-5}^5 (0)dt + \int_0^\pi (-25)dt = -25\pi$$

23. $C_1 : x = t, y = z = 0, 0 \le t \le 1, \displaystyle\int_0^1 0\, dt = 0$; $C_2 : x = 1, y = t, z = 0, 0 \le t \le 1, \displaystyle\int_0^1 (-t)\, dt = -\frac{1}{2}$

$C_3 : x = 1, y = 1, z = t, 0 \le t \le 1, \displaystyle\int_0^1 3\, dt = 3$; $\displaystyle\int_C x^2 z\, dx - yx^2\, dy + 3\, dz = 0 - \frac{1}{2} + 3 = \frac{5}{2}$

24. $C_1 : (0,0,0)$ to $(1,1,0); x = t, y = t, z = 0, 0 \le t \le 1$
$C_2 : (1,1,0)$ to $(1,1,1); x = 1, y = 1, z = t, 0 \le t \le 1$
$C_3 : (1,1,1)$ to $(0,0,0); x = 1 - t, y = 1 - t, z = 1 - t, 0 \le t \le 1$

$$\int_0^1 (-t^3)dt + \int_0^1 3\, dt + \int_0^1 -3dt = -1/4$$

25. $\displaystyle\int_0^\pi (0)dt = 0$

26. $\displaystyle\int_0^1 (e^{2t} - 4e^{-t})dt = e^2/2 + 4e^{-1} - 9/2$

27. $\displaystyle\int_0^1 e^{-t}dt = 1 - e^{-1}$

28. $\displaystyle\int_0^{\pi/2} (7\sin^2 t \cos t + 3\sin t \cos t)dt = 23/6$

29. Represent the circular arc by $x = 3\cos t, y = 3\sin t, 0 \le t \le \pi/2$.
$\int_C x\sqrt{y}ds = 9\sqrt{3}\int_0^{\pi/2} \sqrt{\sin\ t}\cos t\ dt = 6\sqrt{3}$

30. $\delta(x,y) = k\sqrt{x^2+y^2}$ where k is the constant of proportionality,
$\int_C k\sqrt{x^2+y^2}ds = k\int_0^1 e^t(\sqrt{2}e^t)\ dt = \sqrt{2}k\int_0^1 e^{2t}\ dt = (e^2-1)k/\sqrt{2}$

31. $\int_C \frac{kx}{1+y^2}ds = 15k\int_0^{\pi/2} \frac{\cos t}{1+9\sin^2 t}\ dt = 5k\tan^{-1} 3$

32. $\delta(x,y,z) = kz$ where k is the constant of proportionality,
$\int_C kzds = \int_1^4 k(4\sqrt{t})(2+1/t)\ dt = 136k/3$

33. $C : x = t^2,\ y = t,\ 0 \le t \le 1;\ W = \int_0^1 3t^4 dt = 3/5$

34. $W = \int_1^3 (t^2+1-1/t^3+1/t)dt = 92/9 + \ln 3$

35. $W = \int_0^1 (t^3+5t^6)dt = 27/28$

36. $C_1 : (0,0,0)$ to $(1,3,1)$; $x = t,\ y = 3t,\ z = t,\ 0 \le t \le 1$
$C_2 : (1,3,1)$ to $(2,-1,4)$; $x = 1+t,\ y = 3-4t,\ z = 1+3t,\ 0 \le t \le 1$
$W = \int_0^1 (4t+8t^2)dt + \int_0^1 (-11-17t-11t^2)dt = -37/2$

37. Since $\mathbf{F}$ and $\mathbf{r}$ are parallel, $\mathbf{F}\cdot\mathbf{r} = \|\mathbf{F}\|\|\mathbf{r}\|$, and since $\mathbf{F}$ is constant,
$\int \mathbf{F}\cdot d\mathbf{r} = \int_C d(\mathbf{F}\cdot\mathbf{r}) = \int_C d(\|\mathbf{F}\|\|\mathbf{r}\|) = \sqrt{2}\int_{-4}^4 \sqrt{2}dt = 16$

38. $\int_C \mathbf{F}\cdot\mathbf{r} = 0$, since $\mathbf{F}$ is perpendicular to the curve

39. $C : x = 4\cos t, y = 4\sin t, 0 \le t \le \pi/2$
$\int_0^{\pi/2} \left(-\frac{1}{4}\sin t + \cos t\right)\ dt = 3/4$

40. $C_1 : (0,3)$ to $(6,3); x = 6t, y = 3, 0 \le t \le 1$
$C_2 : (6,3)$ to $(6,0); x = 6, y = 3-3t, 0 \le t \le 1$
$\int_0^1 \frac{6}{36t^2+9}dt + \int_0^1 \frac{12}{36+9(1-t)^2}dt = \frac{1}{3}\tan^{-1} 2 - \frac{2}{3}\tan^{-1}(1/2)$

41. Represent the parabola by $x = t, y = t^2, 0 \le t \le 2$.
$\int_C 3xds = \int_0^2 3t\sqrt{1+4t^2}\ dt = (17\sqrt{17}-1)/4$

42. Represent the semicircle by $x = 2\cos t, y = 2\sin t, 0 \le t \le \pi$.
$\int_C x^2yds = \int_0^\pi 16\cos^2 t\sin t\ dt = 32/3$

43. **(a)** $2\pi rh = 2\pi(1)2 = 4\pi$ **(b)** $S = \int_C z(t)\,dt$

(c) $C: x = \cos t, y = \sin t, 0 \le t \le 2\pi; S = \int_0^{2\pi} (2 + (1/2)\sin 3t)\,dt = 4\pi$

44. $C: x = a\cos t, y = -a\sin t, 0 \le t \le 2\pi,$

$$\int_C \frac{x\,dy - y\,dx}{x^2 + y^2} = \int_0^{2\pi} \frac{-a^2\cos^2 t - a^2\sin^2 t}{a^2}\,dt = \int_0^{2\pi} dt = 2\pi$$

45. $W = \int_C \mathbf{F}\cdot d\mathbf{r} = \int_0^1 (\lambda t^2(1-t), t - \lambda t(1-t))\cdot(1, \lambda - 2\lambda t)\,dt = -\lambda/12$, $W = 1$ when $\lambda = -12$

46. The force exerted by the farmer is $\mathbf{F} = \left(150 + 20 - \frac{1}{10}z\right)\mathbf{k} = \left(170 - \frac{3}{4\pi}t\right)\mathbf{k}$, so $\mathbf{F}\cdot d\mathbf{r} = \left(170 - \frac{1}{10}z\right)dz$, and $W = \int_0^{60}\left(170 - \frac{1}{10}z\right)dz = 10{,}020$. Note that the functions $x(z), y(z)$ are irrelevant.

EXERCISE SET 16.3

1. $\partial x/\partial y = 0 = \partial y/\partial x$, conservative so $\partial\phi/\partial x = x$ and $\partial\phi/\partial y = y$, $\phi = x^2/2 + k(y)$, $k'(y) = y$, $k(y) = y^2/2 + K$, $\phi = x^2/2 + y^2/2 + K$

2. $\partial(3y^2)/\partial y = 6y = \partial(6xy)/\partial x$, conservative so $\partial\phi/\partial x = 3y^2$ and $\partial\phi/\partial y = 6xy$, $\phi = 3xy^2 + k(y)$, $6xy + k'(y) = 6xy$, $k'(y) = 0$, $k(y) = K$, $\phi = 3xy^2 + K$

3. $\partial(x^2y)/\partial y = x^2$ and $\partial(5xy^2)/\partial x = 5y^2$, not conservative

4. $\partial(e^x\cos y)/\partial y = -e^x\sin y = \partial(-e^x\sin y)/\partial x$, conservative so $\partial\phi/\partial x = e^x\cos y$ and $\partial\phi/\partial y = -e^x\sin y$, $\phi = e^x\cos y + k(y)$, $-e^x\sin y + k'(y) = -e^x\sin y$, $k'(y) = 0$, $k(y) = K$, $\phi = e^x\cos y + K$

5. $\partial(\cos y + y\cos x)/\partial y = -\sin y + \cos x = \partial(\sin x - x\sin y)/\partial x$, conservative so $\partial\phi/\partial x = \cos y + y\cos x$ and $\partial\phi/\partial y = \sin x - x\sin y$, $\phi = x\cos y + y\sin x + k(y)$, $-x\sin y + \sin x + k'(y) = \sin x - x\sin y$, $k'(y) = 0$, $k(y) = K$, $\phi = x\cos y + y\sin x + K$

6. $\partial(x\ln y)/\partial y = x/y$ and $\partial(y\ln x)/\partial x = y/x$, not conservative

7. **(a)** $\partial(y^2)/\partial y = 2y = \partial(2xy)/\partial x$, independent of path

(b) $C: x = -1 + 2t$, $y = 2 + t$, $0 \le t \le 1$; $\int_0^1 (4 + 14t + 6t^2)dt = 13$

(c) $\partial\phi/\partial x = y^2$ and $\partial\phi/\partial y = 2xy$, $\phi = xy^2 + k(y)$, $2xy + k'(y) = 2xy$, $k'(y) = 0$, $k(y) = K$, $\phi = xy^2 + K$. Let $K = 0$ to get $\phi(1,3) - \phi(-1,2) = 9 - (-4) = 13$

8. **(a)** $\partial(y\sin x)/\partial y = \sin x = \partial(-\cos x)/\partial x$, independent of path

(b) $C_1: x = \pi t$, $y = 1 - 2t$, $0 \le t \le 1$; $\int_0^1 (\pi\sin\pi t - 2\pi t\sin\pi t + 2\cos\pi t)dt = 0$

(c) $\partial\phi/\partial x = y\sin x$ and $\partial\phi/\partial y = -\cos x$, $\phi = -y\cos x + k(y)$, $-\cos x + k'(y) = -\cos x$, $k'(y) = 0$, $k(y) = K$, $\phi = -y\cos x + K$. Let $K = 0$ to get $\phi(\pi,-1) - \phi(0,1) = (-1) - (-1) = 0$

9. $\partial(3y)/\partial y = 3 = \partial(3x)/\partial x$, $\phi = 3xy$, $\phi(4,0) - \phi(1,2) = -6$

10. $\partial(e^x \sin y)/\partial y = e^x \cos y = \partial(e^x \cos y)/\partial x$, $\phi = e^x \sin y$, $\phi(1,\pi/2) - \phi(0,0) = e$

11. $\partial(2xe^y)/\partial y = 2xe^y = \partial(x^2e^y)/\partial x$, $\phi = x^2e^y$, $\phi(3,2) - \phi(0,0) = 9e^2$

12. $\partial(3x - y + 1)/\partial y = -1 = \partial[-(x + 4y + 2)]/\partial x$,
$\phi = 3x^2/2 - xy + x - 2y^2 - 2y$, $\phi(0,1) - \phi(-1,2) = 11/2$

13. $\partial(2xy^3)/\partial y = 6xy^2 = \partial(3x^2y^2)/\partial x$, $\phi = x^2y^3$, $\phi(-1,0) - \phi(2,-2) = 32$

14. $\partial(e^x \ln y - e^y/x)/\partial y = e^x/y - e^y/x = \partial(e^x/y - e^y \ln x)/\partial x$,
$\phi = e^x \ln y - e^y \ln x$, $\phi(3,3) - \phi(1,1) = 0$

15. $\phi = x^2y^2/2$, $W = \phi(0,0) - \phi(1,1) = -1/2$

16. $\phi = x^2y^3, W = \phi(4,1) - \phi(-3,0) = 16$

17. $\phi = e^{xy}$, $W = \phi(2,0) - \phi(-1,1) = 1 - e^{-1}$

18. $\phi = e^{-y}\sin x, W = \phi(-\pi/2,0) - \phi(\pi/2,1) = -1 - 1/e$

19. $\partial(e^y + ye^x)/\partial y = e^y + e^x = \partial(xe^y + e^x)/\partial x$ so $\mathbf{F}$ is conservative, $\phi(x,y) = xe^y + ye^x$ so
$\int_C \mathbf{F} \cdot d\mathbf{r} = \phi(0, \ln 2) - \phi(1,0) = \ln 2 - 1$

20. $\partial(2xy)/\partial y = 2x = \partial(x^2 + \cos y)/\partial x$ so $\mathbf{F}$ is conservative, $\phi(x,y) = x^2y + \sin y$ so
$\int_C \mathbf{F} \cdot dr = \phi(\pi,\pi/2) - \phi(0,0) = \pi^3/2 + 1$

21.
$$\mathbf{F} \cdot dr = [(e^y + ye^x)\mathbf{i} + (xe^y + e^x)\mathbf{j}] \cdot [(\pi/2)\cos(\pi t/2)\mathbf{i} + (1/t)\mathbf{j}]dt$$
$$= \left(\frac{\pi}{2}\cos(\pi t/2)(e^y + ye^x) + (xe^y + e^x)/t\right) dt,$$
so $\displaystyle\int_C \mathbf{F} \cdot d\mathbf{r} = \int_1^2 \left(\frac{\pi}{2}\cos(\pi t/2)\left(t + (\ln t)e^{\sin(\pi t/2)}\right) + \left(\sin(\pi t/2) + \frac{1}{t}e^{\sin(\pi t/2)}\right)\right) dt = \ln 2 - 1$

22. $\mathbf{F} \cdot d\mathbf{r} = \left(2t^2\cos(t/3) + [t^2 + \cos(t\cos(t/3))](\cos(t/3) - (t/3)\sin(t/3))\right) dt$, so
$$\int_C \mathbf{F} \cdot d\mathbf{r} = \int_0^\pi \left(2t^2\cos(t/3) + [t^2 + \cos(t\cos(t/3))](\cos(t/3) - (t/3)\sin(t/3))\right) dt = 1 + \pi^3/2$$

23. No; a closed loop can be found whose tangent everywhere makes an angle $< \pi$ with the vector field, so the line integral $\int_C \mathbf{F} \cdot d\mathbf{r} > 0$, and by Theorem 16.3.2 the vector field is not conservative.

24. The vector field is constant, say $\mathbf{F} = a\mathbf{i} + b\mathbf{j}$, so let $\phi(x,y) = ax + by$ and $\mathbf{F}$ is conservative.

25. If $\mathbf{F}$ is conservative, then $\mathbf{F} = \nabla\phi = \dfrac{\partial\phi}{\partial x}\mathbf{i} + \dfrac{\partial\phi}{\partial y}\mathbf{j} + \dfrac{\partial\phi}{\partial z}\mathbf{k}$ and hence $f = \dfrac{\partial\phi}{\partial x}, g = \dfrac{\partial\phi}{\partial y}$, and $h = \dfrac{\partial\phi}{\partial z}$.

Thus $\dfrac{\partial f}{\partial y} = \dfrac{\partial^2\phi}{\partial y\partial x}$ and $\dfrac{\partial g}{\partial x} = \dfrac{\partial^2\phi}{\partial x\partial y}, \dfrac{\partial f}{\partial z} = \dfrac{\partial^2\phi}{\partial z\partial x}$ and $\dfrac{\partial h}{\partial x} = \dfrac{\partial^2\phi}{\partial x\partial z}, \dfrac{\partial g}{\partial z} = \dfrac{\partial^2\phi}{\partial z\partial y}$ and $\dfrac{\partial h}{\partial y} = \dfrac{\partial^2\phi}{\partial y\partial z}$.
The result follows from the equality of mixed second partial derivatives.

26. Let $f(x, y, z) = yz, g(x, y, z) = xz, h(x, y, z) = yx^2$, then $\partial f/\partial z = y, \partial h/\partial x = 2xy \neq \partial f/\partial z$, thus by Exercise 25, $\mathbf{F} = f\mathbf{i}+g\mathbf{j}+h\mathbf{k}$ is not conservative, and by Theorem 16.3.2, $\int_C yz\,dx+xz\,dy+yx^2\,dz$ is not independent of the path.

27. $\dfrac{\partial}{\partial y}(h(x)[x\sin y + y\cos y]) = h(x)[x\cos y - y\sin y + \cos y]$

$\dfrac{\partial}{\partial x}(h(x)[x\cos y - y\sin y]) = h(x)\cos y + h'(x)[x\cos y - y\sin y]$,

equate these two partial derivatives to get $(x\cos y - y\sin y)(h'(x) - h(x)) = 0$ which holds for all x and y if $h'(x) = h(x)$, $h(x) = Ce^x$ where C is an arbitrary constant.

28. **(a)** $\dfrac{\partial}{\partial y}\dfrac{cx}{(x^2+y^2)^{3/2}} = -\dfrac{3cxy}{(x^2+y^2)^{-5/2}} = \dfrac{\partial}{\partial x}\dfrac{cy}{(x^2+y^2)^{3/2}}$ when $(x, y) \neq (0, 0)$,

so by Theorem 16.3.3, $\mathbf{F}$ is conservative. Set $\partial\phi/\partial x = cx/(x^2+y^2)^{-3/2}$,

then $\phi(x, y) = -c(x^2+y^2)^{-1/2} + k(y), \partial\phi/\partial y = cy/(x^2+y^2)^{-3/2} + k'(y)$, so $k'(y) = 0$.

Thus $\phi(x, y) = -\dfrac{c}{(x^2+y^2)^{1/2}}$ is a potential function.

(b) curl $\mathbf{F} = \mathbf{0}$ is similar to Part (a), so $\mathbf{F}$ is conservative. Let

$\phi(x, y, z) = \displaystyle\int \frac{cx}{(x^2+y^2+z^2)^{3/2}}\,dx = -c(x^2+y^2+z^2)^{-1/2} + k(y, z)$. As in Part (a),

$\partial k/\partial y = \partial k/\partial z = 0$, so $\phi(x, y, z) = -c/(x^2+y^2+z^2)^{1/2}$ is a potential function for $\mathbf{F}$.

29. **(a)** See Exercise 28, $c = 1$; $W = \displaystyle\int_P^Q \mathbf{F}\cdot d\mathbf{r} = \phi(3, 2, 1) - \phi(1, 1, 2) = -\frac{1}{\sqrt{14}} + \frac{1}{\sqrt{6}}$

(b) C begins at $P(1, 1, 2)$ and ends at $Q(3, 2, 1)$ so the answer is again $W = -\dfrac{1}{\sqrt{14}} + \dfrac{1}{\sqrt{6}}$.

(c) The circle is not specified, but cannot pass through $(0, 0, 0)$, so Φ is continuous and differentiable on the circle. Start at any point P on the circle and return to P, so the work is $\Phi(P) - \Phi(P) = 0$.
C begins at, say, $(3, 0)$ and ends at the same point so $W = 0$.

30. **(a)** $\mathbf{F}\cdot d\mathbf{r} = \left(y\dfrac{dx}{dt} - x\dfrac{dy}{dt}\right)dt$ for points on the circle $x^2 + y^2 = 1$, so

$C_1 : x = \cos t, y = \sin t, 0 \le t \le \pi, \displaystyle\int_{C_1} \mathbf{F}\cdot d\mathbf{r} = \int_0^\pi (-\sin^2 t - \cos^2 t)\,dt = -\pi$

$C_2 : x = \cos t, y = -\sin t, 0 \le t \le \pi, \displaystyle\int_{C_2} \mathbf{F}\cdot d\mathbf{r} = \int_0^\pi (\sin^2 t + \cos^2 t)\,dt = \pi$

(b) $\dfrac{\partial f}{\partial y} = \dfrac{x^2 - y^2}{(x^2+y^2)^2}, \dfrac{\partial g}{\partial x} = -\dfrac{y^2 - x^2}{(x^2+y^2)^2} = \dfrac{\partial f}{\partial y}$

(c) The circle about the origin of radius 1, which is formed by traversing C_1 and then traversing C_2 in the reverse direction, does not lie in an open simply connected region inside which $\mathbf{F}$ is continuous, since $\mathbf{F}$ is not defined at the origin, nor can it be defined there in such a way as to make the resulting function continuous there.

31. If C is composed of smooth curves $C_1, C_2, \ldots, C_n$ and curve C_i extends from (x_{i-1}, y_{i-1}) to (x_i, y_i) then $\displaystyle\int_C \mathbf{F}\cdot d\mathbf{r} = \sum_{i=1}^{n}\int_{C_i}\mathbf{F}\cdot d\mathbf{r} = \sum_{i=1}^{n}[\phi(x_i, y_i) - \phi(x_{i-1}, y_{i-1})] = \phi(x_n, y_n) - \phi(x_0, y_0)$
where (x_0, y_0) and (x_n, y_n) are the endpoints of C.

32. $\displaystyle\int_{C_1}\mathbf{F}\cdot d\mathbf{r} + \int_{-C_2}\mathbf{F}\cdot d\mathbf{r} = 0$, but $\displaystyle\int_{-C_2}\mathbf{F}\cdot d\mathbf{r} = -\int_{C_2}\mathbf{F}\cdot d\mathbf{r}$ so $\displaystyle\int_{C_1}\mathbf{F}\cdot d\mathbf{r} = \int_{C_2}\mathbf{F}\cdot d\mathbf{r}$, thus $\displaystyle\int_C \mathbf{F}\cdot d\mathbf{r}$ is independent of path.

33. Let C_1 be an arbitrary piecewise smooth curve from (a, b) to a point (x, y_1) in the disk, and C_2 the vertical line segment from (x, y_1) to (x, y). Then

$$\phi(x,y) = \int_{C_1}\mathbf{F}\cdot d\mathbf{r} + \int_{C_2}\mathbf{F}\cdot d\mathbf{r} = \int_{(a,b)}^{(x,y_1)}\mathbf{F}\cdot d\mathbf{r} + \int_{C_2}\mathbf{F}\cdot d\mathbf{r}.$$

The first term does not depend on y;

hence $\displaystyle\frac{\partial\phi}{\partial y} = \frac{\partial}{\partial y}\int_{C_2}\mathbf{F}\cdot d\mathbf{r} = \frac{\partial}{\partial y}\int_{C_2} f(x,y)dx + g(x,y)dy.$

However, the line integral with respect to x is zero along C_2, so $\displaystyle\frac{\partial\phi}{\partial y} = \frac{\partial}{\partial y}\int_{C_2} g(x,y)\,dy.$

Express C_2 as $x = x, y = t$ where t varies from y_1 to y, then $\displaystyle\frac{\partial\phi}{\partial y} = \frac{\partial}{\partial y}\int_{y_1}^{y} g(x,t)\,dt = g(x,y).$

EXERCISE SET 16.4

1. $\displaystyle\iint_R (2x - 2y)dA = \int_0^1\int_0^1 (2x - 2y)dy\,dx = 0$; for the line integral, on $x = 0, y^2\,dx = 0, x^2\,dy = 0$; on $y = 0, y^2\,dx = x^2\,dy = 0$; on $x = 1, y^2\,dx + x^2\,dy = dy$; and on $y = 1, y^2\,dx + x^2\,dy = dx$,
hence $\displaystyle\oint_C y^2\,dx + x^2\,dy = \int_0^1 dy + \int_1^0 dx = 1 - 1 = 0$

2. $\displaystyle\iint_R (1 - 1)dA = 0$; for the line integral let $x = \cos t, y = \sin t$,

$$\oint_C y\,dx + x\,dy = \int_0^{2\pi}(-\sin^2 t + \cos^2 t)dt = 0$$

3. $\displaystyle\int_{-2}^{4}\int_1^2 (2y - 3x)dy\,dx = 0$

4. $\displaystyle\int_0^{2\pi}\int_0^3 (1 + 2r\sin\theta)r\,dr\,d\theta = 9\pi$

5. $\displaystyle\int_0^{\pi/2}\int_0^{\pi/2}(-y\cos x + x\sin y)dy\,dx = 0$

6. $\displaystyle\iint_R (\sec^2 x - \tan^2 x)dA = \iint_R dA = \pi$

7. $\displaystyle\iint_R [1 - (-1)]dA = 2\iint_R dA = 8\pi$

8. $\displaystyle\int_0^1\int_{x^2}^{x}(2x - 2y)dy\,dx = 1/30$

9. $\displaystyle\iint_R \left(-\frac{y}{1+y} - \frac{1}{1+y}\right) dA = -\iint_R dA = -4$

10. $\displaystyle\int_0^{\pi/2}\int_0^4 (-r^2) r\,dr\,d\theta = -32\pi$

11. $\displaystyle\iint_R \left(-\frac{y^2}{1+y^2} - \frac{1}{1+y^2}\right) dA = -\iint_R dA = -1$

12. $\displaystyle\iint_R (\cos x \cos y - \cos x \cos y) dA = 0$

13. $\displaystyle\int_0^1 \int_{x^2}^{\sqrt{x}} (y^2 - x^2) dy\,dx = 0$

14. **(a)** $\displaystyle\int_0^2 \int_{x^2}^{2x} (-6x + 2y) dy\,dx = -56/15$ **(b)** $\displaystyle\int_0^2 \int_{x^2}^{2x} 6y\,dy\,dx = 64/5$

15. **(a)** $C: x = \cos t, y = \sin t, 0 \le t \le 2\pi;$

$$\oint_C = \int_0^{2\pi} \left(e^{\sin t}(-\sin t) + \sin t \cos t e^{\cos t}\right) dt \approx -3.550999378;$$

$$\iint_R \left[\frac{\partial}{\partial x}(ye^x) - \frac{\partial}{\partial y}e^y\right] dA = \iint_R [ye^x - e^y]\,dA$$

$$= \int_0^{2\pi}\int_0^1 \left[r \sin\theta e^{r\cos\theta} - e^{r\sin\theta}\right] r\,dr\,d\theta \approx -3.550999378$$

(b) $C_1: x = t, y = t^2, 0 \le t \le 1; \displaystyle\int_{C_1} [e^y\,dx + ye^x\,dy] = \int_0^1 \left[e^{t^2} + 2t^3 e^t\right] dt \approx 2.589524432$

$C_2: x = t^2, y = t, 0 \le t \le 1; \displaystyle\int_{C_2} [e^y\,dx + ye^x\,dy] = \int_0^1 \left[2te^t + te^{t^2}\right] dt = \frac{e+3}{2} \approx 2.859140914$

$\displaystyle\int_{C_1} - \int_{C_2} \approx -0.269616482; \iint_R = \int_0^1 \int_{x^2}^{\sqrt{x}} [ye^x - e^y]\,dy\,dx \approx -0.269616482$

16. **(a)** $\displaystyle\oint_C x\,dy = \int_0^{2\pi} ab\cos^2 t\,dt = \pi ab$ **(b)** $\displaystyle\oint_C -y\,dx = \int_0^{2\pi} ab\sin^2 t\,dt = \pi ab$

17. $\displaystyle A = \frac{1}{2}\oint_C -y\,dx + x\,dy = \frac{1}{2}\int_0^{2\pi} (3a^2 \sin^4\phi\cos^2\phi + 3a^2\cos^4\phi\sin^2\phi)d\phi$

$$= \frac{3}{2}a^2 \int_0^{2\pi} \sin^2\phi\cos^2\phi\,d\phi = \frac{3}{8}a^2\int_0^{2\pi}\sin^2 2\phi\,d\phi = 3\pi a^2/8$$

18. $C_1: (0,0)$ to $(a,0)$; $x = at, \quad y = 0, \quad 0 \le t \le 1$
$C_2: (a,0)$ to $(0,b)$; $x = a - at, y = bt, \quad 0 \le t \le 1$
$C_3: (0,b)$ to $(0,0)$; $x = 0, \quad y = b - bt, 0 \le t \le 1$

$$A = \oint_C x\,dy = \int_0^1 (0)dt + \int_0^1 ab(1-t)dt + \int_0^1 (0)dt = \frac{1}{2}ab$$

19. $C_1 : (0,0)$ to $(a,0)$; $x = at,\ y = 0,\ 0 \le t \le 1$
$C_2 : (a,0)$ to $(a\cos t_0, b\sin t_0)$; $x = a\cos t,\ y = b\sin t,\ 0 \le t \le t_0$
$C_3 : (a\cos t_0, b\sin t_0)$ to $(0,0)$; $x = -a(\cos t_0)t,\ y = -b(\sin t_0)t,\ -1 \le t \le 0$

$$A = \frac{1}{2}\oint_C -y\,dx + x\,dy = \frac{1}{2}\int_0^1 (0)\,dt + \frac{1}{2}\int_0^{t_0} ab\,dt + \frac{1}{2}\int_{-1}^0 (0)\,dt = \frac{1}{2}ab\,t_0$$

20. $C_1 : (0,0)$ to $(a,0)$; $x = at,\ y = 0,\ 0 \le t \le 1$
$C_2 : (a,0)$ to $(a\cosh t_0, b\sinh t_0)$; $x = a\cosh t,\ y = b\sinh t,\ 0 \le t \le t_0$
$C_3 : (a\cosh t_0, b\sinh t_0)$ to $(0,0)$; $x = -a(\cosh t_0)t,\ y = -b(\sinh t_0)t,\ -1 \le t \le 0$

$$A = \frac{1}{2}\oint_C -y\,dx + x\,dy = \frac{1}{2}\int_0^1 (0)\,dt + \frac{1}{2}\int_0^{t_0} ab\,dt + \frac{1}{2}\int_{-1}^0 (0)\,dt = \frac{1}{2}ab\,t_0$$

21. $W = \iint\limits_R y\,dA = \int_0^\pi \int_0^5 r^2 \sin\theta\,dr\,d\theta = 250/3$

22. We cannot apply Green's Theorem on the region enclosed by the closed curve C, since $\mathbf{F}$ does not have first order partial derivatives at the origin. However, the curve C_{x_0}, consisting of $y = x_0^3/4, x_0 \le x \le 2; x = 2, x_0^3/4 \le y \le 2$; and $y = x^3/4, x_0 \le x \le 2$ encloses a region R_{x_0} in which Green's Theorem does hold, and

$$W = \oint\limits_C \mathbf{F}\cdot d\mathbf{r} = \lim_{x_0\to 0^+} \oint\limits_{C_{x_0}} \mathbf{F}\cdot d\mathbf{r} = \lim_{x_0\to 0^+} \iint\limits_{R_{x_0}} \nabla\cdot\mathbf{F}\,dA$$

$$= \lim_{x_0\to 0^+} \int_{x_0}^2 \int_{x_0^3/4}^{x^3/4} \left(\frac{1}{2}x^{-1/2} - \frac{1}{2}y^{-1/2}\right) dy\,dx$$

$$= \lim_{x_0\to 0^+} \left(-\frac{18}{35}\sqrt{2} - \frac{\sqrt{2}}{4}x_0^3 + x_0^{3/2} + \frac{3}{14}x_0^{7/2} - \frac{3}{10}x_0^{5/2}\right) = -\frac{18}{35}\sqrt{2}$$

23. $\oint_C y\,dx - x\,dy = \iint\limits_R (-2)dA = -2\int_0^{2\pi}\int_0^{a(1+\cos\theta)} r\,dr\,d\theta = -3\pi a^2$

24. $\bar{x} = \frac{1}{A}\iint\limits_R x\,dA$, but $\oint_C \frac{1}{2}x^2 dy = \iint\limits_R x\,dA$ from Green's Theorem so

$\bar{x} = \frac{1}{A}\oint_C \frac{1}{2}x^2 dy = \frac{1}{2A}\oint_C x^2 dy$. Similarly, $\bar{y} = -\frac{1}{2A}\oint_C y^2 dx$.

25. $A = \int_0^1 \int_{x^3}^x dy\,dx = \frac{1}{4}$; $C_1 : x = t, y = t^3, 0 \le t \le 1$, $\int_{C_1} x^2\,dy = \int_0^1 t^2(3t^2)\,dt = \frac{3}{5}$

$C_2 : x = t, y = t, 0 \le t \le 1$; $\int_{C_2} x^2\,dy = \int_0^1 t^2\,dt = \frac{1}{3}$, $\oint_C x^2\,dy = \int_{C_1} - \int_{C_2} = \frac{3}{5} - \frac{1}{3} = \frac{4}{15}, \bar{x} = \frac{8}{15}$

$\int_C y^2\,dx = \int_0^1 t^6\,dt - \int_0^1 t^2\,dt = \frac{1}{7} - \frac{1}{3} = -\frac{4}{21}, \bar{y} = \frac{8}{21}$, centroid $\left(\frac{8}{15}, \frac{8}{21}\right)$

26. $A = \dfrac{a^2}{2}; C_1 : x = t, y = 0, 0 \le t \le a, C_2 : x = a - t, y = t, 0 \le t \le a; C_3 : x = 0, y = a - t, 0 \le t \le a;$

$$\int_{C_1} x^2\,dy = 0, \int_{C_2} x^2\,dy = \int_0^a (a-t)^2\,dt = \frac{a^3}{3}, \int_{C_3} x^2\,dy = 0, \oint_C x^2\,dy = \int_{C_1} + \int_{C_2} + \int_{C_3} = \frac{a^3}{3},\ \bar{x} = \frac{a}{3};$$

$$\int_C y^2\,dx = 0 - \int_0^a t^2\,dt + 0 = -\frac{a^3}{3},\ \bar{y} = \frac{a}{3},\ \text{centroid } \left(\frac{a}{3}, \frac{a}{3}\right)$$

27. $\bar{x} = 0$ from the symmetry of the region,

$C_1 : (a, 0)$ to $(-a, 0)$ along $y = \sqrt{a^2 - x^2}$; $x = a\cos t,\ y = a\sin t,\ 0 \le t \le \pi$

$C_2 : (-a, 0)$ to $(a, 0)$; $x = t,\ y = 0,\ -a \le t \le a$

$$A = \pi a^2/2, \quad \bar{y} = -\frac{1}{2A}\left[\int_0^\pi -a^3\sin^3 t\,dt + \int_{-a}^a (0)dt\right]$$

$$= -\frac{1}{\pi a^2}\left(-\frac{4a^3}{3}\right) = \frac{4a}{3\pi};\ \text{centroid } \left(0, \frac{4a}{3\pi}\right)$$

28. $A = \dfrac{ab}{2}; C_1 : x = t, y = 0,\ 0 \le t \le a, C_2 : x = a, y = t,\ 0 \le t \le b;$

$C_3 : x = a - at, y = b - bt,\ 0 \le t \le 1;$

$$\int_{C_1} x^2\,dy = 0, \int_{C_2} x^2\,dy = \int_0^b a^2\,dt = ba^2, \int_{C_3} x^2\,dy = \int_0^1 a^2(1-t)^2(-b)\,dt = -\frac{ba^2}{3},$$

$$\oint_C x^2\,dy = \int_{C_1} + \int_{C_2} + \int_{C_3} = \frac{2ba^2}{3},\ \bar{x} = \frac{2a}{3};$$

$$\int_C y^2\,dx = 0 + 0 - \int_0^1 ab^2(1-t)^2\,dt = -\frac{ab^2}{3},\ \bar{y} = \frac{b}{3},\ \text{centroid } \left(\frac{2a}{3}, \frac{b}{3}\right)$$

29. From Green's Theorem, the given integral equals $\displaystyle\iint_R (1 - x^2 - y^2)dA$ where R is the region enclosed by C. The value of this integral is maximum if the integration extends over the largest region for which the integrand $1 - x^2 - y^2$ is nonnegative so we want $1 - x^2 - y^2 \ge 0,\ x^2 + y^2 \le 1$. The largest region is that bounded by the circle $x^2 + y^2 = 1$ which is the desired curve C.

30. **(a)** $C : x = a + (c - a)t,\ y = b + (d - b)t,\ 0 \le t \le 1$

$$\int_C -y\,dx + x\,dy = \int_0^1 (ad - bc)dt = ad - bc$$

(b) Let C_1, C_2, and C_3 be the line segments from (x_1, y_1) to (x_2, y_2), (x_2, y_2) to (x_3, y_3), and (x_3, y_3) to (x_1, y_1), then if C is the entire boundary consisting of C_1, C_2, and C_3

$$A = \frac{1}{2}\int_C -y\,dx + x\,dy = \frac{1}{2}\sum_{i=1}^{3}\int_{C_i} -y\,dx + x\,dy$$

$$= \frac{1}{2}[(x_1y_2 - x_2y_1) + (x_2y_3 - x_3y_2) + (x_3y_1 - x_1y_3)]$$

(c) $A = \dfrac{1}{2}[(x_1y_2 - x_2y_1) + (x_2y_3 - x_3y_2) + \cdots + (x_ny_1 - x_1y_n)]$

(d) $A = \dfrac{1}{2}[(0 - 0) + (6 + 8) + (0 + 2) + (0 - 0)] = 8$

31. $\int_C \mathbf{F} \cdot d\mathbf{r} = \int_C (x^2 + y)\,dx + (4x - \cos y)\,dy = 3 \iint_R dA = 3(25 - 2) = 69$

32. $\int_C \mathbf{F} \cdot d\mathbf{r} = \int_C (e^{-x} + 3y)\,dx + x\,dy = -2 \iint_R dA = -2[\pi(4)^2 - \pi(2)^2] = -24\pi$

EXERCISE SET 16.5

1. R is the annular region between $x^2 + y^2 = 1$ and $x^2 + y^2 = 4$;

$$\iint_\sigma z^2 dS = \iint_R (x^2 + y^2)\sqrt{\frac{x^2}{x^2 + y^2} + \frac{y^2}{x^2 + y^2} + 1}\, dA$$
$$= \sqrt{2} \iint_R (x^2 + y^2) dA = \sqrt{2} \int_0^{2\pi} \int_1^2 r^3 dr\, d\theta = \frac{15}{2}\pi\sqrt{2}.$$

2. $z = 1 - x - y$, R is the triangular region enclosed by $x + y = 1$, $x = 0$ and $y = 0$;

$$\iint_\sigma xy\, dS = \iint_R xy\sqrt{3}\, dA = \sqrt{3} \int_0^1 \int_0^{1-x} xy\, dy\, dx = \frac{\sqrt{3}}{24}.$$

3. Let $\mathbf{r}(u, v) = \cos u\mathbf{i} + v\mathbf{j} + \sin u\mathbf{k}, 0 \le u \le \pi, 0 \le v \le 1$. Then $\mathbf{r}_u = -\sin u\mathbf{i} + \cos u\mathbf{k}, \mathbf{r}_v = \mathbf{j}$,

$\mathbf{r}_u \times \mathbf{r}_v = -\cos u\mathbf{i} - \sin u\mathbf{k}, \|\mathbf{r}_u \times \mathbf{r}_v\| = 1, \iint_\sigma x^2 y\, dS = \int_0^1 \int_0^\pi v \cos^2 u\, du\, dv = \pi/4$

4. $z = \sqrt{4 - x^2 - y^2}$, R is the circular region enclosed by $x^2 + y^2 = 3$;

$$\iint_\sigma (x^2 + y^2) z\, dS = \iint_R (x^2 + y^2)\sqrt{4 - x^2 - y^2}\sqrt{\frac{x^2}{4 - x^2 - y^2} + \frac{y^2}{4 - x^2 - y^2} + 1}\, dA$$
$$= \iint_R 2(x^2 + y^2) dA = 2 \int_0^{2\pi} \int_0^{\sqrt{3}} r^3 dr\, d\theta = 9\pi.$$

5. If we use the projection of σ onto the xz-plane then $y = 1 - x$ and R is the rectangular region in the xz-plane enclosed by $x = 0$, $x = 1$, $z = 0$ and $z = 1$;

$$\iint_\sigma (x - y - z) dS = \iint_R (2x - 1 - z)\sqrt{2} dA = \sqrt{2} \int_0^1 \int_0^1 (2x - 1 - z) dz\, dx = -\sqrt{2}/2$$

6. R is the triangular region enclosed by $2x + 3y = 6$, $x = 0$, and $y = 0$;

$$\iint_\sigma (x + y) dS = \iint_R (x + y)\sqrt{14}\, dA = \sqrt{14} \int_0^3 \int_0^{(6-2x)/3} (x + y) dy\, dx = 5\sqrt{14}.$$

7. There are six surfaces, parametrized by projecting onto planes:
$\sigma_1 : z = 0;\ 0 \le x \le 1,\ 0 \le y \le 1$ (onto xy-plane), $\sigma_2 : x = 0;\ 0 \le y \le 1,\ 0 \le z \le 1$ (onto yz-plane), $\sigma_3 : y = 0;\ 0 \le x \le 1,\ 0 \le z \le 1$ (onto xz-plane), $\sigma_4 : z = 1;\ 0 \le x \le 1,\ 0 \le y \le 1$ (onto xy-plane), $\sigma_5 : x = 1;\ 0 \le y \le 1,\ 0 \le z \le 1$ (onto yz-plane), $\sigma_6 : y = 1;\ 0 \le x \le 1,\ 0 \le z \le 1$ (onto xz-plane).

By symmetry the integrals over σ_1, σ_2 and σ_3 are equal, as are those over σ_4, σ_5 and σ_6, and

$$\iint_{\sigma_1}(x+y+z)dS = \int_0^1\int_0^1 (x+y)dx\,dy = 1; \quad \iint_{\sigma_4}(x+y+z)dS = \int_0^1\int_0^1 (x+y+1)dx\,dy = 2,$$

thus, $\displaystyle\iint_{\sigma}(x+y+z)dS = 3\cdot 1 + 3\cdot 2 = 9.$

8. Let $\mathbf{r}(\phi,\theta) = \sin\phi\cos\theta\mathbf{i} + \sin\phi\sin\theta\mathbf{j} + \cos\phi\mathbf{k}, 0 \le \theta \le 2\pi, 0 \le \phi \le \pi/2; \|\mathbf{r}_\phi \times \mathbf{r}_\theta\| = \sin\phi,$

$$\iint_{\sigma}(1+\cos\phi)\,dS = \int_0^{2\pi}\int_0^{\pi/2}(1+\cos\phi)\sin\phi\,d\phi\,d\theta$$

$$= 2\pi\int_0^{\pi/2}(1+\cos\phi)\sin\phi\,d\phi = 3\pi$$

9. R is the circular region enclosed by $x^2+y^2=1$;

$$\iint_{\sigma}\sqrt{x^2+y^2+z^2}\,dS = \iint_R \sqrt{2(x^2+y^2)}\sqrt{\frac{x^2}{x^2+y^2}+\frac{y^2}{x^2+y^2}+1}\,dA$$

$$= \lim_{r_0\to 0^+} 2\iint_{R'}\sqrt{x^2+y^2}\,dA$$

where R' is the annular region enclosed by $x^2+y^2=1$ and $x^2+y^2=r_0^2$ with r_0 slightly larger than 0 because $\sqrt{\dfrac{x^2}{x^2+y^2}+\dfrac{y^2}{x^2+y^2}+1}$ is not defined for $x^2+y^2=0$, so

$$\iint_{\sigma}\sqrt{x^2+y^2+z^2}\,dS = \lim_{r_0\to 0^+} 2\int_0^{2\pi}\int_{r_0}^1 r^2dr\,d\theta = \lim_{r_0\to 0^+}\frac{4\pi}{3}(1-r_0^3) = \frac{4\pi}{3}.$$

10. Let $\mathbf{r}(\phi,\theta) = a\sin\phi\cos\theta\,\mathbf{i} + a\sin\phi\sin\theta\,\mathbf{j} + a\cos\phi\,\mathbf{k}$,
$0 \le \theta \le 2\pi, 0 \le \phi \le \pi; \|\mathbf{r}_\phi \times \mathbf{r}_\theta\| = a^2\sin\phi,\ x^2+y^2 = a^2\sin^2\phi$

$$\iint_{\sigma} f(x,y,z) = \int_0^{2\pi}\int_0^{\pi} a^4\sin^3\phi\,d\phi\,d\theta = \frac{8}{3}\pi a^4$$

11. (a) $\displaystyle\frac{\sqrt{29}}{16}\int_0^6\int_0^{(12-2x)/3} xy(12-2x-3y)dy\,dx$

(b) $\displaystyle\frac{\sqrt{29}}{4}\int_0^3\int_0^{(12-4z)/3} yz(12-3y-4z)dy\,dz$

(c) $\displaystyle\frac{\sqrt{29}}{9}\int_0^3\int_0^{6-2z} xz(12-2x-4z)dx\,dz$

12. (a) $\displaystyle a\int_0^a\int_0^{\sqrt{a^2-x^2}} x\,dy\,dx$ (b) $\displaystyle a\int_0^a\int_0^{\sqrt{a^2-z^2}} z\,dy\,dz$

(c) $\displaystyle a\int_0^a\int_0^{\sqrt{a^2-z^2}} \frac{xz}{\sqrt{a^2-x^2-z^2}}dx\,dz$

13. $18\sqrt{29}/5$ **14.** $a^4/3$

15. $\int_0^4\int_1^2 y^3 z\sqrt{4y^2+1}\,dy\,dz;\ \frac{1}{2}\int_0^4\int_1^4 xz\sqrt{1+4x}\,dx\,dz$

16. $a\int_0^9\int_{a/\sqrt{5}}^{a/\sqrt{2}} \frac{x^2y}{\sqrt{a^2-y^2}}\,dy\,dx,\ a\int_{a/\sqrt{2}}^{2a/\sqrt{5}}\int_0^9 x^2\,dx\,dz$ **17.** $391\sqrt{17}/15-5\sqrt{5}/3$

18. The region $R: 3x^2+2y^2=5$ is symmetric in y. The integrand is

$x^2yz\,dS = x^2y(5-3x^2-2y^2)\sqrt{1+36x^2+16y^2}\,dy\,dx$, which is odd in y, hence $\iiint\limits_{\sigma} x^2yz\,dS = 0$.

19. $z=\sqrt{4-x^2},\ \frac{\partial z}{\partial x} = -\frac{x}{\sqrt{4-x^2}},\ \frac{\partial z}{\partial y}=0;$

$$\iint\limits_{\sigma} \delta_0 dS = \delta_0\iint\limits_{R}\sqrt{\frac{x^2}{4-x^2}+1}\,dA = 2\delta_0\int_0^4\int_0^1 \frac{1}{\sqrt{4-x^2}}dx\,dy = \frac{4}{3}\pi\delta_0.$$

20. $z=\frac{1}{2}(x^2+y^2)$, R is the circular region enclosed by $x^2+y^2=8$;

$$\iint\limits_{\sigma}\delta_0 dS = \delta_0\iint\limits_{R}\sqrt{x^2+y^2+1}\,dA = \delta_0\int_0^{2\pi}\int_0^{\sqrt{8}}\sqrt{r^2+1}\,r\,dr\,d\theta = \frac{52}{3}\pi\delta_0.$$

21. $z=4-y^2$, R is the rectangular region enclosed by $x=0$, $x=3$, $y=0$ and $y=3$;

$$\iint\limits_{\sigma} y\,dS = \iint\limits_{R} y\sqrt{4y^2+1}\,dA = \int_0^3\int_0^3 y\sqrt{4y^2+1}\,dy\,dx = \frac{1}{4}(37\sqrt{37}-1).$$

22. R is the annular region enclosed by $x^2+y^2=1$ and $x^2+y^2=16$;

$$\iint\limits_{\sigma} x^2z\,dS = \iint\limits_{R} x^2\sqrt{x^2+y^2}\sqrt{\frac{x^2}{x^2+y^2}+\frac{y^2}{x^2+y^2}+1}\,dA$$
$$=\sqrt{2}\iint\limits_{R} x^2\sqrt{x^2+y^2}\,dA = \sqrt{2}\int_0^{2\pi}\int_1^4 r^4\cos^2\theta\,dr\,d\theta = \frac{1023\sqrt{2}}{5}\pi.$$

23. $M=\iint\limits_{\sigma}\delta(x,y,z)dS = \iint\limits_{\sigma}\delta_0 dS = \delta_0\iint\limits_{\sigma} dS = \delta_0 S$

24. $\delta(x,y,z)=|z|$; use $z=\sqrt{a^2-x^2-y^2}$, let R be the circular region enclosed by $x^2+y^2=a^2$, and σ the hemisphere above R. By the symmetry of both the surface and the density function with respect to the xy-plane we have

$$M = 2\iint\limits_{\sigma} z\,dS = 2\iint\limits_{R}\sqrt{a^2-x^2-y^2}\sqrt{\frac{x^2}{a^2-x^2-y^2}+\frac{y^2}{a^2-x^2-y^2}+1}\,dA = \lim_{r_0\to a^-} 2a\iint\limits_{R_{r_0}} dA$$

where R_{r_0} is the circular region with radius r_0 that is slightly less than a. But $\iint\limits_{R_{r_0}} dA$ is simply the area of the circle with radius r_0 so $M=\lim_{r_0\to a^-} 2a(\pi r_0^2) = 2\pi a^3$.

25. By symmetry $\bar{x} = \bar{y} = 0$.

$$\iint_\sigma dS = \iint_R \sqrt{x^2+y^2+1}\, dA = \int_0^{2\pi}\int_0^{\sqrt{8}} \sqrt{r^2+1}\, r\, dr\, d\theta = \frac{52\pi}{3},$$

$$\iint_\sigma z\, dS = \iint_R z\sqrt{x^2+y^2+1}\, dA = \frac{1}{2}\iint_R (x^2+y^2)\sqrt{x^2+y^2+1}\, dA$$

$$= \frac{1}{2}\int_0^{2\pi}\int_0^{\sqrt{8}} r^3\sqrt{r^2+1}\, dr\, d\theta = \frac{596\pi}{15}$$

so $\bar{z} = \dfrac{596\pi/15}{52\pi/3} = \dfrac{149}{65}$. The centroid is $(\bar{x}, \bar{y}, \bar{z}) = (0, 0, 149/65)$.

26. By symmetry $\bar{x} = \bar{y} = 0$.

$$\iint_\sigma dS = \iint_R \frac{2}{\sqrt{4-x^2-y^2}} dA = 2\int_0^{2\pi}\int_0^{\sqrt{3}} \frac{r}{\sqrt{4-r^2}} dr\, d\theta = 4\pi,$$

$$\iint_\sigma z\, dS = \iint_R 2\, dA = (2)(\text{area of circle of radius } \sqrt{3}) = 6\pi$$

so $\bar{z} = \dfrac{6\pi}{4\pi} = \dfrac{3}{2}$. The centroid is $(\bar{x}, \bar{y}, \bar{z}) = (0, 0, 3/2)$.

27. $\partial\mathbf{r}/\partial u = \cos v\mathbf{i} + \sin v\mathbf{j} + 3\mathbf{k}, \partial\mathbf{r}/\partial v = -u\sin v\mathbf{i} + u\cos v\mathbf{j}$, $\|\partial\mathbf{r}/\partial u \times \partial\mathbf{r}/\partial v\| = \sqrt{10}u$;

$$3\sqrt{10}\iint_R u^4 \sin v \cos v\, dA = 3\sqrt{10}\int_0^{\pi/2}\int_1^2 u^4 \sin v \cos v\, du\, dv = 93/\sqrt{10}$$

28. $\partial\mathbf{r}/\partial u = \mathbf{j}, \partial\mathbf{r}/\partial v = -2\sin v\mathbf{i} + 2\cos v\mathbf{k}$, $\|\partial\mathbf{r}/\partial u \times \partial\mathbf{r}/\partial v\| = 2$;

$$8\iint_R \frac{1}{u} dA = 8\int_0^{2\pi}\int_1^3 \frac{1}{u} du\, dv = 16\pi \ln 3$$

29. $\partial\mathbf{r}/\partial u = \cos v\mathbf{i} + \sin v\mathbf{j} + 2u\mathbf{k}, \partial\mathbf{r}/\partial v = -u\sin v\mathbf{i} + u\cos v\mathbf{j}$, $\|\partial\mathbf{r}/\partial u \times \partial\mathbf{r}/\partial v\| = u\sqrt{4u^2+1}$;

$$\iint_R u\, dA = \int_0^{\pi}\int_0^{\sin v} u\, du\, dv = \pi/4$$

30. $\partial\mathbf{r}/\partial u = 2\cos u\cos v\mathbf{i} + 2\cos u \sin v\mathbf{j} - 2\sin u\mathbf{k}$, $\partial\mathbf{r}/\partial v = -2\sin u\sin v\mathbf{i} + 2\sin u\cos v\mathbf{j}$;
$\|\partial\mathbf{r}/\partial u \times \partial\mathbf{r}/\partial v\| = 4\sin u$;

$$4\iint_R e^{-2\cos u}\sin u\, dA = 4\int_0^{2\pi}\int_0^{\pi/2} e^{-2\cos u}\sin u\, du\, dv = 4\pi(1-e^{-2})$$

31. $\partial z/\partial x = -2xe^{-x^2-y^2}, \partial z/\partial y = -2ye^{-x^2-y^2}$,
$(\partial z/\partial x)^2 + (\partial z/\partial y)^2 + 1 = 4(x^2+y^2)e^{-2(x^2+y^2)} + 1$; use polar coordinates to get

$$M = \int_0^{2\pi}\int_0^3 r^2\sqrt{4r^2e^{-2r^2}+1}\, dr\, d\theta \approx 57.895751$$

32. **(b)** $A = \displaystyle\iint_\sigma dS = \int_0^{2\pi}\int_{-1}^{1} \frac{1}{2}\sqrt{40u\cos(v/2) + u^2 + 4u^2\cos^2(v/2) + 100}du\, dv \approx 62.93768644$;

$\bar{x} \approx 0.01663836266$; $\bar{y} = \bar{z} = 0$ by symmetry

EXERCISE SET 16.6

1. **(a)** zero **(b)** zero **(c)** positive
(d) negative **(e)** zero **(f)** zero

2. **(a)** positive **(b)** zero **(c)** zero
(d) zero **(e)** negative **(f)** zero

3. **(a)** positive **(b)** zero **(c)** positive
(d) zero **(e)** positive **(f)** zero

4. 0; the flux is zero on the faces $y = 0, 1$ and $z = 0, 1$; it is 1 on $x = 1$ and -1 on $x = 0$

5. **(a)** $\mathbf{n} = -\cos v\mathbf{i} - \sin v\mathbf{j}$ **(b)** inward, by inspection

6. **(a)** $-r\cos\theta\mathbf{i} - r\sin\theta\mathbf{j} + r\mathbf{k}$ **(b)** inward, by inspection

7. $\mathbf{n} = -z_x\mathbf{i} - z_y\mathbf{j} + \mathbf{k}$, $\displaystyle\iint_R \mathbf{F}\cdot\mathbf{n}\,dS = \iint_R (2x^2 + 2y^2 + 2(1 - x^2 - y^2))\,dS = \int_0^{2\pi}\int_0^1 2r\,dr\,d\theta = 2\pi$

8. With $z = 1 - x - y$, R is the triangular region enclosed by $x + y = 1$, $x = 0$ and $y = 0$; use upward normals to get

$$\iint_\sigma \mathbf{F}\cdot\mathbf{n}\,dS = 2\iint_R (x + y + z)dA = 2\iint_R dA = (2)(\text{area of } R) = 1.$$

9. R is the annular region enclosed by $x^2 + y^2 = 1$ and $x^2 + y^2 = 4$;

$$\iint_\sigma \mathbf{F}\cdot\mathbf{n}\,dS = \iint_R \left(-\frac{x^2}{\sqrt{x^2+y^2}} - \frac{y^2}{\sqrt{x^2+y^2}} + 2z\right)dA$$
$$= \iint_R \sqrt{x^2+y^2}\,dA = \int_0^{2\pi}\int_1^2 r^2 dr\,d\theta = \frac{14\pi}{3}.$$

10. R is the circular region enclosed by $x^2 + y^2 = 4$;

$$\iint_\sigma \mathbf{F}\cdot\mathbf{n}\,dS = \iint_R (2y^2 - 1)dA = \int_0^{2\pi}\int_0^2 (2r^2\sin^2\theta - 1)r\,dr\,d\theta = 4\pi.$$

11. R is the circular region enclosed by $x^2 + y^2 - y = 0$; $\displaystyle\iint_\sigma \mathbf{F}\cdot\mathbf{n}\,dS = \iint_R (-x)dA = 0$ since the region R is symmetric across the y-axis.

12. With $z = \frac{1}{2}(6 - 6x - 3y)$, R is the triangular region enclosed by $2x + y = 2$, $x = 0$, and $y = 0$;

$$\iint_\sigma \mathbf{F}\cdot\mathbf{n}\,dS = \iint_R \left(3x^2 + \frac{3}{2}yx + zx\right)dA = 3\iint_R x\,dA = 3\int_0^1\int_0^{2-2x} x\,dy\,dx = 1.$$

13. $\partial\mathbf{r}/\partial u = \cos v\mathbf{i} + \sin v\mathbf{j} - 2u\mathbf{k}$, $\partial\mathbf{r}/\partial v = -u\sin v\mathbf{i} + u\cos v\mathbf{j}$,
$\partial\mathbf{r}/\partial u \times \partial\mathbf{r}/\partial v = 2u^2\cos v\mathbf{i} + 2u^2\sin v\mathbf{j} + u\mathbf{k}$;

$$\iint_R (2u^3 + u)\,dA = \int_0^{2\pi}\int_1^2 (2u^3 + u)du\,dv = 18\pi$$

14. $\partial\mathbf{r}/\partial u = \mathbf{k}, \partial\mathbf{r}/\partial v = -2\sin v\mathbf{i} + \cos v\mathbf{j}, \partial\mathbf{r}/\partial u \times \partial\mathbf{r}/\partial v = -\cos v\mathbf{i} - 2\sin v\mathbf{j};$

$$\iint_R (2\sin^2 v - e^{-\sin v}\cos v)\, dA = \int_0^{2\pi}\int_0^5 (2\sin^2 v - e^{-\sin v}\cos v) du\, dv = 10\pi$$

15. $\partial\mathbf{r}/\partial u = \cos v\mathbf{i} + \sin v\mathbf{j} + 2\mathbf{k}, \partial\mathbf{r}/\partial v = -u\sin v\mathbf{i} + u\cos v\mathbf{j},$
$\partial\mathbf{r}/\partial u \times \partial\mathbf{r}/\partial v = -2u\cos v\mathbf{i} - 2u\sin v\mathbf{j} + u\mathbf{k};$

$$\iint_R u^2\, dA = \int_0^{\pi}\int_0^{\sin v} u^2 du\, dv = 4/9$$

16. $\partial\mathbf{r}/\partial u = 2\cos u\cos v\mathbf{i} + 2\cos u\sin v\mathbf{j} - 2\sin u\mathbf{k}, \ \partial\mathbf{r}/\partial v = -2\sin u\sin v\mathbf{i} + 2\sin u\cos v\mathbf{j};$
$\partial\mathbf{r}/\partial u \times \partial\mathbf{r}/\partial v = 4\sin^2 u\cos v\mathbf{i} + 4\sin^2 u\sin v\mathbf{j} + 4\sin u\cos u\mathbf{k};$

$$\iint_R 8\sin u\, dA = 8\int_0^{2\pi}\int_0^{\pi/3} \sin u\, du\, dv = 8\pi$$

17. In each part, divide σ into the six surfaces
$\sigma_1 : x = -1$ with $|y| \le 1$, $|z| \le 1$, and $\mathbf{n} = -\mathbf{i}$, $\sigma_2 : x = 1$ with $|y| \le 1$, $|z| \le 1$, and $\mathbf{n} = \mathbf{i}$,
$\sigma_3 : y = -1$ with $|x| \le 1$, $|z| \le 1$, and $\mathbf{n} = -\mathbf{j}$, $\sigma_4 : y = 1$ with $|x| \le 1$, $|z| \le 1$, and $\mathbf{n} = \mathbf{j}$,
$\sigma_5 : z = -1$ with $|x| \le 1$, $|y| \le 1$, and $\mathbf{n} = -\mathbf{k}$, $\sigma_6 : z = 1$ with $|x| \le 1$, $|y| \le 1$, and $\mathbf{n} = \mathbf{k}$,

(a) $\displaystyle\iint_{\sigma_1} \mathbf{F}\cdot\mathbf{n}\, dS = \iint_{\sigma_1} dS = 4, \iint_{\sigma_2} \mathbf{F}\cdot\mathbf{n}\, dS = \iint_{\sigma_2} dS = 4$, and $\displaystyle\iint_{\sigma_i} \mathbf{F}\cdot\mathbf{n}\, dS = 0$ for

$i = 3, 4, 5, 6$ so $\displaystyle\iint_{\sigma} \mathbf{F}\cdot\mathbf{n}\, dS = 4 + 4 + 0 + 0 + 0 + 0 = 8.$

(b) $\displaystyle\iint_{\sigma_1} \mathbf{F}\cdot\mathbf{n}\, dS = \iint_{\sigma_1} dS = 4$, similarly $\displaystyle\iint_{\sigma_i} \mathbf{F}\cdot\mathbf{n}\, dS = 4$ for $i = 2, 3, 4, 5, 6$ so

$\displaystyle\iint_{\sigma} \mathbf{F}\cdot\mathbf{n}\, dS = 4 + 4 + 4 + 4 + 4 + 4 = 24.$

(c) $\displaystyle\iint_{\sigma_1} \mathbf{F}\cdot\mathbf{n}\, dS = -\iint_{\sigma_1} dS = -4, \iint_{\sigma_2} \mathbf{F}\cdot\mathbf{n}\, dS = 4$, similarly $\displaystyle\iint_{\sigma_i} \mathbf{F}\cdot\mathbf{n}\, dS = -4$ for $i = 3, 5$

and $\displaystyle\iint_{\sigma_i} \mathbf{F}\cdot\mathbf{n}\, dS = 4$ for $i = 4, 6$ so $\displaystyle\iint_{\sigma} \mathbf{F}\cdot\mathbf{n}\, dS = -4 + 4 - 4 + 4 - 4 + 4 = 0.$

18. Decompose σ into a top σ_1 (the disk) and a bottom σ_2 (the portion of the paraboloid). Then

$$\mathbf{n}_1 = \mathbf{k}, \iint_{\sigma_1} \mathbf{F}\cdot\mathbf{n}_1\, dS = -\iint_{\sigma_1} y\, dS = -\int_0^{2\pi}\int_0^1 r^2\sin\theta\, dr\, d\theta = 0,$$

$$\mathbf{n}_2 = (2x\mathbf{i} + 2y\mathbf{j} - \mathbf{k})/\sqrt{1 + 4x^2 + 4y^2}, \iint_{\sigma_2} \mathbf{F}\cdot\mathbf{n}_2\, dS = \iint_{\sigma_2} \frac{y(2x^2 + 2y^2 + 1)}{\sqrt{1 + 4x^2 + 4y^2}}\, dS = 0,$$

because the surface σ_2 is symmetric with respect to the xy-plane and the integrand is an odd function of y. Thus the flux is 0.

19. R is the circular region enclosed by $x^2+y^2=1$; $x=r\cos\theta, y=r\sin\theta, z=r$,

$\mathbf{n}=\cos\theta\mathbf{i}+\sin\theta\mathbf{j}-\mathbf{k}$;

$$\iint\limits_{\sigma}\mathbf{F}\cdot\mathbf{n}\,dS=\iint\limits_{R}(\cos\theta+\sin\theta-1)\,dA=\int_0^{2\pi}\int_0^1(\cos\theta+\sin\theta-1)\,r\,dr\,d\theta=-\pi.$$

20. Let $\mathbf{r}=\cos v\mathbf{i}+u\mathbf{j}+\sin v\mathbf{k}, -2\le u\le 1, 0\le v\le 2\pi; \mathbf{r}_u\times\mathbf{r}_v=\cos v\mathbf{i}+\sin v\mathbf{k}$,

$$\iint\limits_{\sigma}\mathbf{F}\cdot\mathbf{n}\,dS=\iint\limits_{R}(\cos^2 v+\sin^2 v)\,dA=\text{area of } R=3\cdot 2\pi=6\pi$$

21. **(a)** $\mathbf{n}=\dfrac{1}{\sqrt{3}}[\mathbf{i}+\mathbf{j}+\mathbf{k}]$,

$$V=\int_{\sigma}\mathbf{F}\cdot\mathbf{n}\,dS=\int_0^1\int_0^{1-x}(2x-3y+1-x-y)\,dy\,dx=0\text{ m}^3/\text{s}$$

(b) $m=0\cdot 806=0$ kg/s

22. **(a)** Let $x=3\sin\phi\cos\theta, y=3\sin\phi\sin\theta, z=3\cos\phi,\ \mathbf{n}=\sin\phi\cos\theta\mathbf{i}+\sin\phi\sin\theta\,\mathbf{j}+\cos\phi\,\mathbf{k}$, so

$$V=\int_{\sigma}\mathbf{F}\cdot\mathbf{n}\,dS=\iint\limits_{A}9\sin\phi\,(-3\sin^2\phi\sin\theta\cos\theta+3\sin\phi\cos\phi\sin\theta+9\sin\phi\cos\phi\cos\theta)\,dA$$

$$=\int_0^{2\pi}\int_0^3 3\sin\phi\cos\theta(-\sin\phi\sin\theta+4\cos\phi)\,r\,dr\,d\theta=0\text{ m}^3$$

(b) $\dfrac{dm}{dt}=0\cdot 1060=0$ kg/s

23. **(a)** $G(x,y,z)=x-g(y,z)$, $\nabla G=\mathbf{i}-\dfrac{\partial g}{\partial y}\mathbf{j}-\dfrac{\partial g}{\partial z}\mathbf{k}$, apply Theorem 16.6.3:

$$\iint\limits_{\sigma}\mathbf{F}\cdot\mathbf{n}dS=\iint\limits_{R}\mathbf{F}\cdot\left(\mathbf{i}-\frac{\partial x}{\partial y}\mathbf{j}-\frac{\partial x}{\partial z}\mathbf{k}\right)dA, \text{ if } \sigma \text{ is oriented by front normals, and}$$

$$\iint\limits_{\sigma}\mathbf{F}\cdot\mathbf{n}dS=\iint\limits_{R}\mathbf{F}\cdot\left(-\mathbf{i}+\frac{\partial x}{\partial y}\mathbf{j}+\frac{\partial x}{\partial z}\mathbf{k}\right)dA, \text{ if } \sigma \text{ is oriented by back normals,}$$

where R is the projection of σ onto the yz-plane.

(b) R is the semicircular region in the yz-plane enclosed by $z=\sqrt{1-y^2}$ and $z=0$;

$$\iint\limits_{\sigma}\mathbf{F}\cdot\mathbf{n}\,dS=\iint\limits_{R}(-y-2yz+16z)dA=\int_{-1}^1\int_0^{\sqrt{1-y^2}}(-y-2yz+16z)dz\,dy=\frac{32}{3}.$$

24. **(a)** $G(x,y,z)=y-g(x,z)$, $\nabla G=-\dfrac{\partial g}{\partial x}\mathbf{i}+\mathbf{j}-\dfrac{\partial g}{\partial z}\mathbf{k}$, apply Theorem 16.6.3:

$$\iint\limits_{R}\mathbf{F}\cdot\left(\frac{\partial y}{\partial x}\mathbf{i}-\mathbf{j}+\frac{\partial y}{\partial z}\mathbf{k}\right)dA, \ \sigma \text{ oriented by left normals,}$$

and $\displaystyle\iint\limits_{R}\mathbf{F}\cdot\left(-\frac{\partial y}{\partial x}\mathbf{i}+\mathbf{j}-\frac{\partial y}{\partial z}\mathbf{k}\right)dA$, σ oriented by right normals,

where R is the projection of σ onto the xz-plane.

(b) R is the semicircular region in the xz-plane enclosed by $z = \sqrt{1-x^2}$ and $z = 0$;

$$\iint_\sigma \mathbf{F}\cdot\mathbf{n}\,dS = \iint_R (-2x^2 + (x^2+z^2) - 2z^2)dA = -\int_{-1}^{1}\int_0^{\sqrt{1-x^2}} (x^2+z^2)dz\,dx = -\frac{\pi}{4}.$$

25. (a) On the sphere, $\|\mathbf{r}\| = a$ so $\mathbf{F} = a^k\mathbf{r}$ and $\mathbf{F}\cdot\mathbf{n} = a^k\mathbf{r}\cdot(\mathbf{r}/a) = a^{k-1}\|\mathbf{r}\|^2 = a^{k-1}a^2 = a^{k+1}$,

hence $\displaystyle\iint_\sigma \mathbf{F}\cdot\mathbf{n}\,dS = a^{k+1}\iint_\sigma dS = a^{k+1}(4\pi a^2) = 4\pi a^{k+3}$.

(b) If $k = -3$, then $\displaystyle\iint_\sigma \mathbf{F}\cdot\mathbf{n}\,dS = 4\pi$.

26. Let $\mathbf{r} = \sin u\cos v\mathbf{i} + \sin u\sin v\mathbf{j} + \cos u\mathbf{k}$, $\mathbf{r}_u\times\mathbf{r}_v = \sin^2 u\cos v\mathbf{i} + \sin^2 u\sin v\mathbf{j} + \sin u\cos u\mathbf{k}$,

$$\mathbf{F}\cdot(\mathbf{r}_u\times\mathbf{r}_v) = a^2\sin^3 u\cos^2 v + \frac{1}{a}\sin^3 u\sin^2 v + a\sin u\cos^3 u,$$

$$\begin{aligned}\iint_\sigma \mathbf{F}\cdot\mathbf{n}\,dS &= \int_0^{2\pi}\int_0^{\pi}\left(a^2\sin^3 u\cos^2 v + \frac{1}{a}\sin^3 u\sin^2 v + a\sin u\cos^3 u\right)du\,dv\\ &= \frac{4}{3a}\int_0^{\pi}(a^3\cos^2 v + \sin^2 v)\,dv\\ &= \frac{4\pi}{3}\left(a^2 + \frac{1}{a}\right) = 10 \text{ if } a \approx -1.722730, 0.459525, 1.263205\end{aligned}$$

EXERCISE SET 16.7

1. $\sigma_1 : x = 0, \mathbf{F}\cdot\mathbf{n} = -x = 0, \displaystyle\iint_{\sigma_1}(0)dA = 0$ $\qquad \sigma_2 : x = 1, \mathbf{F}\cdot\mathbf{n} = x = 1, \displaystyle\iint_{\sigma_2}(1)dA = 1$

$\sigma_3 : y = 0, \mathbf{F}\cdot\mathbf{n} = -y = 0, \displaystyle\iint_{\sigma_3}(0)dA = 0$ $\qquad \sigma_4 : y = 1, \mathbf{F}\cdot\mathbf{n} = y = 1, \displaystyle\iint_{\sigma_4}(1)dA = 1$

$\sigma_5 : z = 0, \mathbf{F}\cdot\mathbf{n} = -z = 0, \displaystyle\iint_{\sigma_5}(0)dA = 0$ $\qquad \sigma_6 : z = 1, \mathbf{F}\cdot\mathbf{n} = z = 1, \displaystyle\iint_{\sigma_6}(1)dA = 1$

$$\iint_\sigma \mathbf{F}\cdot\mathbf{n} = 3;\ \iiint_G \operatorname{div}\mathbf{F}dV = \iiint_G 3dV = 3$$

2. For any point $\mathbf{r} = x\mathbf{i} + y\mathbf{j} + z\mathbf{k}$ on σ let $\mathbf{n} = x\mathbf{i} + y\mathbf{j} + z\mathbf{k}$; then $\mathbf{F}\cdot\mathbf{n} = x^2 + y^2 + z^2 = 1$, so

$$\iint_\sigma \mathbf{F}\cdot\mathbf{n}\,dS = \iint_\sigma dS = 4\pi;\text{ also }\iiint_G \operatorname{div}\mathbf{F}dV = \iiint_G 3dV = 3(4\pi/3) = 4\pi$$

3. $\sigma_1 : z = 1, \mathbf{n} = \mathbf{k}, \mathbf{F}\cdot\mathbf{n} = z^2 = 1, \displaystyle\iint_{\sigma_1}(1)dS = \pi$,

$\sigma_2 : \mathbf{n} = 2x\mathbf{i} + 2y\mathbf{j} - \mathbf{k}, \mathbf{F}\cdot\mathbf{n} = 4x^2 - 4x^2y^2 - x^4 - 3y^4$,

$$\iint_{\sigma_2} \mathbf{F}\cdot\mathbf{n}\,dS = \int_0^{2\pi}\int_0^1 \left[4r^2\cos^2\theta - 4r^4\cos^2\theta\sin^2\theta - r^4\cos^4\theta - 3r^4\sin^4\theta\right] r\,dr\,d\theta = \frac{\pi}{3};$$

$$\iint_{\sigma} = \frac{4\pi}{3}$$

$$\iiint_G \text{div } \mathbf{F} dV = \iiint_G (2+z)dV = \int_0^{2\pi}\int_0^1\int_{r^2}^1 (2+z)dz\,r\,dr\,d\theta = 4\pi/3$$

4. $\sigma_1 : x=0, \mathbf{F}\cdot\mathbf{n} = -xy = 0, \iint_{\sigma_1}(0)dA = 0$ $\quad\sigma_2 : x=2, \mathbf{F}\cdot\mathbf{n} = xy = 2y, \iint_{\sigma_2}(2y)dA = 8$

$\sigma_3 : y=0, \mathbf{F}\cdot\mathbf{n} = -yz = 0, \iint_{\sigma_3}(0)dA = 0$ $\quad\sigma_4 : y=2, \mathbf{F}\cdot\mathbf{n} = yz = 2z, \iint_{\sigma_4}(2z)dA = 8$

$\sigma_5 : z=0, \mathbf{F}\cdot\mathbf{n} = -xz = 0, \iint_{\sigma_5}(0)dA = 0$ $\quad\sigma_6 : z=2, \mathbf{F}\cdot\mathbf{n} = xz = 2x, \iint_{\sigma_6}(2x)dA = 8$

$\iint_{\sigma} \mathbf{F}\cdot\mathbf{n} = 24$; also $\iiint_G \text{div } \mathbf{F}dV = \iiint_G (y+z+x)dV = 24$

5. G is the rectangular solid; $\iiint_G \text{div } \mathbf{F}\,dV = \int_0^2\int_0^1\int_0^3 (2x-1)\,dx\,dy\,dz = 12.$

6. G is the spherical solid enclosed by σ; $\iiint_G \text{div } \mathbf{F}\,dV = \iiint_G 0\,dV = 0\iiint_G dV = 0.$

7. G is the cylindrical solid;

$$\iiint_G \text{div } \mathbf{F}\,dV = 3\iiint_G dV = (3)(\text{volume of cylinder}) = (3)[\pi a^2(1)] = 3\pi a^2.$$

8. G is the solid bounded by $z = 1 - x^2 - y^2$ and the xy-plane;

$$\iiint_G \text{div } \mathbf{F}\,dV = 3\iiint_G dV = 3\int_0^{2\pi}\int_0^1\int_0^{1-r^2} r\,dz\,dr\,d\theta = \frac{3\pi}{2}.$$

9. G is the cylindrical solid;

$$\iiint_G \text{div } \mathbf{F}\,dV = 3\iiint_G (x^2+y^2+z^2)dV = 3\int_0^{2\pi}\int_0^2\int_0^3 (r^2+z^2)r\,dz\,dr\,d\theta = 180\pi.$$

10. G is the tetrahedron; $\iiint_G \text{div } \mathbf{F}\,dV = \iiint_G x\,dV = \int_0^1\int_0^{1-x}\int_0^{1-x-y} x\,dz\,dy\,dx = \frac{1}{24}.$

11. G is the hemispherical solid bounded by $z = \sqrt{4-x^2-y^2}$ and the xy-plane;

$$\iiint_G \text{div } \mathbf{F}\,dV = 3\iiint_G (x^2+y^2+z^2)dV = 3\int_0^{2\pi}\int_0^{\pi/2}\int_0^2 \rho^4\sin\phi\,d\rho\,d\phi\,d\theta = \frac{192\pi}{5}.$$

12. G is the hemispherical solid;

$$\iiint\limits_G \text{div}\,\mathbf{F}\,dV = 5\iiint\limits_G z\,dV = 5\int_0^{2\pi}\int_0^{\pi/2}\int_0^a \rho^3\sin\phi\cos\phi\,d\rho\,d\phi\,d\theta = \frac{5\pi a^4}{4}.$$

13. G is the conical solid;

$$\iiint\limits_G \text{div}\,\mathbf{F}\,dV = 2\iiint\limits_G (x+y+z)dV = 2\int_0^{2\pi}\int_0^1\int_r^1 (r\cos\theta + r\sin\theta + z)r\,dz\,dr\,d\theta = \frac{\pi}{2}.$$

14. G is the solid bounded by $z = 2x$ and $z = x^2 + y^2$;

$$\iiint\limits_G \text{div}\,\mathbf{F}\,dV = \iiint\limits_G dV = 2\int_0^{\pi/2}\int_0^{2\cos\theta}\int_{r^2}^{2r\cos\theta} r\,dz\,dr\,d\theta = \frac{\pi}{2}.$$

15. G is the solid bounded by $z = 4 - x^2$, $y + z = 5$, and the coordinate planes;

$$\iiint\limits_G \text{div}\,\mathbf{F}\,dV = 4\iiint\limits_G x^2 dV = 4\int_{-2}^{2}\int_0^{4-x^2}\int_0^{5-z} x^2 dy\,dz\,dx = \frac{4608}{35}.$$

16. $$\iint\limits_\sigma \mathbf{F}\cdot\mathbf{n}\,dS = \iiint\limits_G \text{div}\,\mathbf{F}\,dV = \iiint\limits_G 0\,dV = 0;$$

since the vector field is constant, the same amount enters as leaves.

17. $$\iint\limits_\sigma \mathbf{r}\cdot\mathbf{n}\,dS = \iiint\limits_G \text{div}\,\mathbf{r}\,dV = 3\iiint\limits_G dV = 3\text{vol}(G)$$

18. $$\iint\limits_\sigma \mathbf{F}\cdot\mathbf{n}\,dS = 3[\pi(3^2)(5)] = 135\pi$$

19. $$\iint\limits_\sigma \text{curl}\,\mathbf{F}\cdot\mathbf{n}\,dS = \iiint\limits_G \text{div}(\text{curl}\,\mathbf{F})dV = \iiint\limits_G (0)dV = 0$$

20. $$\iint\limits_\sigma \nabla f\cdot\mathbf{n}\,dS = \iiint\limits_G \text{div}\,(\nabla f)dV = \iiint\limits_G \nabla^2 f dV$$

21. $$\iint\limits_\sigma (f\nabla g)\cdot\mathbf{n} = \iiint\limits_G \text{div}\,(f\nabla g)dV = \iiint\limits_G (f\nabla^2 g + \nabla f\cdot\nabla g)dV$$ by Exercise 31, Section 16.1.

22. $$\iint\limits_\sigma (f\nabla g)\cdot\mathbf{n}\,dS = \iiint\limits_G (f\nabla^2 g + \nabla f\cdot\nabla g)dV$$ by Exercise 21;

$$\iint\limits_\sigma (g\nabla f)\cdot\mathbf{n}\,dS = \iiint\limits_G (g\nabla^2 f + \nabla g\cdot\nabla f)dV$$ by interchanging f and g;

subtract to obtain the result.

23. Since $\mathbf{v}$ is constant, $\nabla \cdot \mathbf{v} = \mathbf{0}$. Let $\mathbf{F} = f\mathbf{v}$; then $\text{div}\mathbf{F} = (\nabla f)\mathbf{v}$ and by the Divergence Theorem

$$\iint_\sigma f\mathbf{v}\cdot\mathbf{n}\,dS = \iint_\sigma \mathbf{F}\cdot\mathbf{n}\,dS = \iiint_G \text{div}\mathbf{F}\,dV = \iiint_G (\nabla f)\cdot\mathbf{v}\,dV$$

24. Let $\mathbf{r} = u\mathbf{i} + v\mathbf{j} + w\mathbf{k}$ so that, for $\mathbf{r} \neq \mathbf{0}$,

$$\mathbf{F}(x,y,z) = \mathbf{r}/||\mathbf{r}||^k = \frac{u}{(u^2+v^2+w^2)^{k/2}}\mathbf{i} + \frac{v}{(u^2+v^2+w^2)^{k/2}}\mathbf{j} + \frac{w}{(u^2+v^2+w^2)^{k/2}}\mathbf{k}$$

$\dfrac{\partial \mathbf{F}_1}{\partial u} = \dfrac{u^2+v^2+w^2-ku^2}{(u^2+v^2+w^2)^{(k/2)+1}}$; similarly for $\partial\mathbf{F}_2/\partial v, \partial\mathbf{F}_3/\partial w$, so that

$$\text{div }\mathbf{F} = \frac{3(u^2+v^2+w^2) - k(u^2+v^2+w^2)}{(u^2+v^2+w^2)^{(k/2)+1}} = 0 \text{ if and only if } k = 3.$$

25. (a) The flux through any cylinder whose axis is the z-axis is positive by inspection; by the Divergence Theorem, this says that the divergence cannot be negative at the origin, else the flux through a small enough cylinder would also be negative (impossible), hence the divergence at the origin must be ≥ 0.

(b) Similar to Part (a), ≤ 0.

26. (a) $\mathbf{F} = x\mathbf{i} + y\mathbf{j} + z\mathbf{k}$, div $\mathbf{F} = 3$ (b) $\mathbf{F} = -x\mathbf{i} - y\mathbf{j} - z\mathbf{k}$, div $\mathbf{F} = -3$

27. div $\mathbf{F} = 0$; no sources or sinks.

28. div $\mathbf{F} = y - x$; sources where $y > x$, sinks where $y < x$.

29. div $\mathbf{F} = 3x^2 + 3y^2 + 3z^2$; sources at all points except the origin, no sinks.

30. div $\mathbf{F} = 3(x^2+y^2+z^2-1)$; sources outside the sphere $x^2+y^2+z^2 = 1$, sinks inside the sphere $x^2+y^2+z^2 = 1$.

31. Let σ_1 be the portion of the paraboloid $z = 1 - x^2 - y^2$ for $z \geq 0$, and σ_2 the portion of the plane $z = 0$ for $x^2 + y^2 \leq 1$. Then

$$\begin{aligned}\iint_{\sigma_1} \mathbf{F}\cdot\mathbf{n}\,dS &= \iint_R \mathbf{F}\cdot(2x\mathbf{i} + 2y\mathbf{j} + \mathbf{k})\,dA \\ &= \int_{-1}^{1}\int_{-\sqrt{1-x^2}}^{\sqrt{1-x^2}} (2x[x^2y - (1-x^2-y^2)^2] + 2y(y^3-x) + (2x+2-3x^2-3y^2))\,dy\,dx \\ &= 3\pi/4;\end{aligned}$$

$z = 0$ and $\mathbf{n} = -\mathbf{k}$ on σ_2 so $\mathbf{F}\cdot\mathbf{n} = 1 - 2x$, $\displaystyle\iint_{\sigma_2}\mathbf{F}\cdot\mathbf{n}\,dS = \iint_{\sigma_2}(1-2x)dS = \pi$. Thus

$\displaystyle\iint_\sigma \mathbf{F}\cdot\mathbf{n}\,dS = 3\pi/4 + \pi = 7\pi/4$. But div $\mathbf{F} = 2xy + 3y^2 + 3$ so

$$\iiint_G \text{div }\mathbf{F}\,dV = \int_{-1}^{1}\int_{-\sqrt{1-x^2}}^{\sqrt{1-x^2}}\int_0^{1-x^2-y^2} (2xy + 3y^2 + 3)\,dz\,dy\,dx = 7\pi/4.$$

EXERCISE SET 16.8

1. (a) The flow is independent of z and has no component in the direction of $\mathbf{k}$, and so by inspection the only nonzero component of the curl is in the direction of $\mathbf{k}$. However both sides of (9) are zero, as the flow is orthogonal to the curve C_a. Thus the curl is zero.

   (b) Since the flow appears to be tangential to the curve C_a, it seems that the right hand side of (9) is nonzero, and thus the curl is nonzero, and points in the positive z-direction.

2. (a) The only nonzero vector component of the vector field is in the direction of $\mathbf{i}$, and it increases with y and is independent of x. Thus the curl of F is nonzero, and points in the positive z-direction. Alternatively, let $\mathbf{F} = f\mathbf{i}$, and let C be the circle of radius ϵ with positive orientation. Then $\mathbf{T} = -\sin\theta\,\mathbf{i} + \cos\theta\,\mathbf{j}$, and

$$\oint_C \mathbf{F}\cdot\mathbf{T}\,ds = -\epsilon\int_0^{2\pi} f(\epsilon,\theta)\sin\theta\,d\theta = -\epsilon\int_0^{\pi} f(\epsilon,\theta)\sin\theta\,d\theta - \epsilon\int_{-\pi}^{0} f(\epsilon,\theta)\sin\theta\,d\theta$$

$$= -\epsilon\int_0^{\pi} (f(\epsilon,\theta) - f(-\epsilon,\theta))\sin\theta\,d\theta < 0$$

   because from the picture $f(\epsilon,\theta) > f(\epsilon,-\theta)$ for $0 < \theta < \pi$. Thus, from (9), the curl is nonzero and points in the negative z-direction.

   (b) By inspection the vector field is constant, and thus its curl is zero.

3. If σ is oriented with upward normals then C consists of three parts parametrized as $C_1 : \mathbf{r}(t) = (1-t)\mathbf{i} + t\mathbf{j}$ for $0 \le t \le 1$, $C_2 : \mathbf{r}(t) = (1-t)\mathbf{j} + t\mathbf{k}$ for $0 \le t \le 1$, $C_3 : \mathbf{r}(t) = t\mathbf{i} + (1-t)\mathbf{k}$ for $0 \le t \le 1$.

$$\int_{C_1} \mathbf{F}\cdot d\mathbf{r} = \int_{C_2} \mathbf{F}\cdot d\mathbf{r} = \int_{C_3} \mathbf{F}\cdot d\mathbf{r} = \int_0^1 (3t-1)dt = \frac{1}{2} \text{ so}$$

$\oint_C \mathbf{F}\cdot d\mathbf{r} = \frac{1}{2} + \frac{1}{2} + \frac{1}{2} = \frac{3}{2}$. curl $\mathbf{F} = \mathbf{i} + \mathbf{j} + \mathbf{k}$, $z = 1 - x - y$, R is the triangular region in the xy-plane enclosed by $x + y = 1$, $x = 0$, and $y = 0$;

$$\iint_\sigma (\text{curl }\mathbf{F})\cdot\mathbf{n}\,dS = 3\iint_R dA = (3)(\text{area of } R) = (3)\left[\frac{1}{2}(1)(1)\right] = \frac{3}{2}.$$

4. If σ is oriented with upward normals then C can be parametrized as $\mathbf{r}(t) = \cos t\mathbf{i} + \sin t\mathbf{j} + \mathbf{k}$ for $0 \le t \le 2\pi$.

$$\oint_C \mathbf{F}\cdot d\mathbf{r} = \int_0^{2\pi} (\sin^2 t\cos t - \cos^2 t\sin t)dt = 0;$$

curl $\mathbf{F} = \mathbf{0}$ so $\displaystyle\iint_\sigma (\text{curl }\mathbf{F})\cdot\mathbf{n}\,dS = \iint_\sigma 0\,dS = 0.$

5. If σ is oriented with upward normals then C can be parametrized as $\mathbf{r}(t) = a\cos t\mathbf{i} + a\sin t\mathbf{j}$ for $0 \le t \le 2\pi$.

$$\oint_C \mathbf{F}\cdot d\mathbf{r} = \int_0^{2\pi} 0\,dt = 0; \text{ curl } \mathbf{F} = \mathbf{0} \text{ so } \iint_\sigma (\text{curl }\mathbf{F})\cdot\mathbf{n}\,dS = \iint_\sigma 0\,dS = 0.$$

6. If σ is oriented with upward normals then C can be parametrized as $\mathbf{r}(t) = 3\cos t\mathbf{i} + 3\sin t\mathbf{j}$ for $0 \le t \le 2\pi$.

$$\oint_C \mathbf{F}\cdot d\mathbf{r} = \int_0^{2\pi} (9\sin^2 t + 9\cos^2 t)dt = 9\int_0^{2\pi} dt = 18\pi.$$

curl $\mathbf{F} = -2\mathbf{i} + 2\mathbf{j} + 2\mathbf{k}$, R is the circular region in the xy-plane enclosed by $x^2 + y^2 = 9$;

$$\iint_\sigma (\text{curl }\mathbf{F})\cdot\mathbf{n}\,dS = \iint_R (-4x + 4y + 2)dA = \int_0^{2\pi}\int_0^3 (-4r\cos\theta + 4r\sin\theta + 2)r\,dr\,d\theta = 18\pi.$$

7. Take σ as the part of the plane $z = 0$ for $x^2 + y^2 \le 1$ with $\mathbf{n} = \mathbf{k}$; curl $\mathbf{F} = -3y^2\mathbf{i} + 2z\mathbf{j} + 2\mathbf{k}$,

$$\iint_\sigma (\text{curl }\mathbf{F})\cdot\mathbf{n}\,dS = 2\iint_\sigma dS = (2)(\text{area of circle}) = (2)[\pi(1)^2] = 2\pi.$$

8. curl $\mathbf{F} = x\mathbf{i} + (x - y)\mathbf{j} + 6xy^2\mathbf{k}$;

$$\iint_\sigma (\text{curl }\mathbf{F})\cdot\mathbf{n}\,dS = \iint_R (x - y - 6xy^2)dA = \int_0^1\int_0^3 (x - y - 6xy^2)dy\,dx = -30.$$

9. C is the boundary of R and curl $\mathbf{F} = 2\mathbf{i} + 3\mathbf{j} + 4\mathbf{k}$, so

$$\oint \mathbf{F}\cdot\mathbf{r} = \iint_R \text{curl }\mathbf{F}\cdot\mathbf{n}\,dS = \iint_R 4\,dA = 4(\text{area of } R) = 16\pi$$

10. curl $\mathbf{F} = -4\mathbf{i} - 6\mathbf{j} + 6y\mathbf{k}$, $z = y/2$ oriented with upward normals, R is the triangular region in the xy-plane enclosed by $x + y = 2$, $x = 0$, and $y = 0$;

$$\iint_\sigma (\text{curl }\mathbf{F})\cdot\mathbf{n}\,dS = \iint_R (3 + 6y)dA = \int_0^2\int_0^{2-x} (3 + 6y)dy\,dx = 14.$$

11. curl $\mathbf{F} = x\mathbf{k}$, take σ as part of the plane $z = y$ oriented with upward normals, R is the circular region in the xy-plane enclosed by $x^2 + y^2 - y = 0$;

$$\iint_\sigma (\text{curl }\mathbf{F})\cdot\mathbf{n}\,dS = \iint_R x\,dA = \int_0^\pi\int_0^{\sin\theta} r^2\cos\theta\,dr\,d\theta = 0.$$

12. curl $\mathbf{F} = -y\mathbf{i} - z\mathbf{j} - x\mathbf{k}$, $z = 1 - x - y$ oriented with upward normals, R is the triangular region in the xy-plane enclosed by $x + y = 1$, $x = 0$ and $y = 0$;

$$\iint_\sigma (\text{curl }\mathbf{F})\cdot\mathbf{n}\,dS = \iint_R (-y - z - x)dA = -\iint_R dA = -\frac{1}{2}(1)(1) = -\frac{1}{2}.$$

13. curl $\mathbf{F} = \mathbf{i} + \mathbf{j} + \mathbf{k}$, take σ as the part of the plane $z = 0$ with $x^2 + y^2 \le a^2$ and $\mathbf{n} = \mathbf{k}$;

$$\iint_\sigma (\text{curl }\mathbf{F})\cdot\mathbf{n}\,dS = \iint_\sigma dS = \text{ area of circle } = \pi a^2.$$

14. curl $\mathbf{F} = \mathbf{i} + \mathbf{j} + \mathbf{k}$, take σ as the part of the plane $z = 1/\sqrt{2}$ with $x^2 + y^2 \le 1/2$ and $\mathbf{n} = \mathbf{k}$.

$$\iint_\sigma (\text{curl }\mathbf{F})\cdot\mathbf{n}\,dS = \iint_\sigma dS = \text{ area of circle } = \frac{\pi}{2}.$$

15. (a) Take σ as the part of the plane $2x + y + 2z = 2$ in the first octant, oriented with downward normals; curl $\mathbf{F} = -x\mathbf{i} + (y-1)\mathbf{j} - \mathbf{k}$,

$$\oint_C \mathbf{F}\cdot\mathbf{T}\,ds = \iint_\sigma (\text{curl }\mathbf{F})\cdot\mathbf{n}\,dS$$
$$= \iint_R \left(x - \frac{1}{2}y + \frac{3}{2}\right) dA = \int_0^1 \int_0^{2-2x} \left(x - \frac{1}{2}y + \frac{3}{2}\right) dy\,dx = \frac{3}{2}.$$

(b) At the origin curl $\mathbf{F} = -\mathbf{j} - \mathbf{k}$ and with $\mathbf{n} = \mathbf{k}$, curl $\mathbf{F}(0,0,0)\cdot\mathbf{n} = (-\mathbf{j}-\mathbf{k})\cdot\mathbf{k} = -1$.

(c) The rotation of $\mathbf{F}$ has its maximum value at the origin about the unit vector in the same direction as curl $\mathbf{F}(0,0,0)$ so $\mathbf{n} = -\frac{1}{\sqrt{2}}\mathbf{j} - \frac{1}{\sqrt{2}}\mathbf{k}$.

16. (a) Using the hint, the orientation of the curve C with respect to the surface σ_1 is the opposite of the orientation of C with respect to the surface σ_2.
Thus in the expressions

$$\iint_{\sigma_1} (\text{curl }\mathbf{F})\cdot\mathbf{n}\,dS = \int_C \mathbf{F}\cdot\mathbf{T}\,dS \text{ and } \iint_{\sigma_2} (\text{curl }\mathbf{F})\cdot\mathbf{n}\,dS = \oint_C \mathbf{F}\cdot\mathbf{T}\,dS,$$

the two line integrals have oppositely oriented tangents $\mathbf{T}$. Hence

$$\iint_\sigma (\text{curl }\mathbf{F})\cdot\mathbf{n}\,dS = \iint_{\sigma_1} (\text{curl }\mathbf{F})\cdot\mathbf{n}\,dS + \iint_{\sigma_2} (\text{curl }\mathbf{F})\cdot\mathbf{n}\,dS = \mathbf{0}.$$

(b) The flux of the curl field through the boundary of a solid is zero.

17. Since $\oint_C \mathbf{E}\cdot\mathbf{r}d\mathbf{r} = \iint_\sigma \text{curl }\mathbf{E}\cdot\mathbf{n}\,dS$, it follows that $\iint_\sigma \text{curl }\mathbf{E}\cdot\mathbf{n}dS = -\iint_\sigma \frac{\partial\mathbf{B}}{\partial t}\cdot\mathbf{n}dS$. This relationship holds for any surface σ, hence curl $\mathbf{E} = -\frac{\partial\mathbf{B}}{\partial t}$.

18. Parametrize C by $x = \cos t, y = \sin t, 0 \le t \le 2\pi$. But $\mathbf{F} = x^2y\mathbf{i} + (y^3 - x)\mathbf{j} + (2x-1)\mathbf{k}$ along C so $\oint_C \mathbf{F}\cdot d\mathbf{r} = -5\pi/4$. Since curl $\mathbf{F} = (-2z-2)\mathbf{j} + (-1-x^2)\mathbf{k}$,

$$\iint_\sigma (\text{curl }\mathbf{F})\cdot\mathbf{n}\,dS = \iint_R (\text{curl }\mathbf{F})\cdot(2x\mathbf{i} + 2y\mathbf{j} + \mathbf{k})\,dA$$
$$= \int_{-1}^1 \int_{-\sqrt{1-x^2}}^{\sqrt{1-x^2}} [2y(2x^2 + 2y^2 - 4) - 1 - x^2]\,dy\,dx = -5\pi/4$$

CHAPTER 16 SUPPLEMENTARY EXERCISES

2. (b) $\frac{c}{\|\mathbf{r}-\mathbf{r}_0\|^3}(\mathbf{r}-\mathbf{r}_0)$ (c) $c\frac{(x-x_0)\mathbf{i} + (y-y_0)\mathbf{j} + (z-z_0)\mathbf{k}}{\sqrt{(x-x_0)^2 + (y-y_0)^2 + (z-z_0)^2}}$

3. (a) $\int_a^b \left[f(x(t),y(t))\frac{dx}{dt} + g(x(t),y(t))\frac{dy}{dt}\right] dt$

(b) $\displaystyle\int_a^b f(x(t), y(t))\sqrt{x'(t)^2 + y'(t)^2}\,dt$

4. (a) $M = \displaystyle\int_C \delta(x,y,z)\,ds$ (b) $L = \displaystyle\int_C ds$ (c) $S = \displaystyle\iint_\sigma dS$

(d) $A = \displaystyle\oint_C x\,dy = -\oint_C y\,dx = \frac{1}{2}\oint_C -y\,dx + x\,dy$

11. $\displaystyle\iint_\sigma f(x,y,z)dS = \iint_R f(x(u,v), y(u,v), z(u,v))\|\mathbf{r}_u \times \mathbf{r}_v\|\,du\,dv$

13. Let O be the origin, P the point with polar coordinates $\theta = \alpha, r = f(\alpha)$, and Q the point with polar coordinates $\theta = \beta, r = f(\beta)$. Let

$C_1 : O$ to P; $x = t\cos\alpha$, $y = t\sin\alpha$, $0 \le t \le f(\alpha)$, $-y\dfrac{dx}{dt} + x\dfrac{dy}{dt} = 0$

$C_2 : P$ to Q; $x = f(t)\cos t$, $y = f(t)\sin t$, $\alpha \le \theta \le \beta, -y\dfrac{dx}{dt} + x\dfrac{dy}{dt} = f(t)^2$

$C_3 : Q$ to O; $x = -t\cos\beta$, $y = -t\sin\beta$, $-f(\beta) \le t \le 0, -y\dfrac{dx}{dt} + x\dfrac{dy}{dt} = 0$

$A = \dfrac{1}{2}\displaystyle\oint_C -y\,dx + x\,dy = \frac{1}{2}\int_\alpha^\beta f(t)^2\,dt$; set $t = \theta$ and $r = f(\theta) = f(t)$, $A = \dfrac{1}{2}\displaystyle\int_\alpha^\beta r^2\,d\theta.$

14. (a) $\mathbf{F}(x,y,z) = \dfrac{qQ(x\mathbf{i} + y\mathbf{j} + z\mathbf{k})}{4\pi\epsilon_0(x^2+y^2+z^2)^{3/2}}$

(b) $\mathbf{F} = \nabla\phi$, where $\phi = -\dfrac{qQ}{4\pi\epsilon_0(x^2+y^2+z^2)^{1/2}}$, so $W = \phi(3,1,5) - \phi(3,0,0) = \dfrac{qQ}{4\pi\epsilon_0}\left(\dfrac{1}{3} - \dfrac{1}{\sqrt{35}}\right)$.

$C : x = 3, y = t, z = 5t, 0 \le t \le 1$; $\mathbf{F}\cdot d\mathbf{r} = \dfrac{qQ[0 + t + 25t]\,dt}{4\pi\epsilon_0(9 + t^2 + 25t^2)^{3/2}}$

$W = \displaystyle\int_0^1 \frac{26qQt\,dt}{4\pi\epsilon_0(26t^2+9)^{3/2}} = \frac{qQ}{4\pi\epsilon_0}\left(\frac{1}{\sqrt{35}} - \frac{1}{3}\right)$

15. (a) Assume the mass M is located at the origin and the mass m at (x,y,z), then

$\mathbf{r} = x\mathbf{i} + y\mathbf{j} + z\mathbf{k}, \mathbf{F}(x,y,z) = -\dfrac{GmM}{(x^2+y^2+z^2)^{3/2}}\mathbf{r},$

$$W = -\int_{t_1}^{t_2} \frac{GmM}{(x^2+y^2+z^2)^{3/2}}\left(x\frac{dx}{dt} + y\frac{dy}{dt} + z\frac{dz}{dt}\right)dt$$
$$= GmM(x^2+y^2+z^2)^{-1/2}\Big]_{t_1}^{t_2} = GmM\left(\frac{1}{r_2} - \frac{1}{r_1}\right)$$

(b) $W = 3.99 \times 10^5 \times 10^3\left[\dfrac{1}{7170} - \dfrac{1}{6970}\right] \approx -1596.801594\ \text{km}^2\text{kg/s}^2 \approx -1.597 \times 10^9$ J

16. div $\mathbf{F} = \dfrac{y^2 - x^2}{(x^2+y^2)^2} + \dfrac{x^2 - y^2}{(x^2+y^2)^2} + \dfrac{1}{(x^2+y^2)} = \dfrac{1}{x^2+y^2}$, the level surface of div $\mathbf{F} = 1$ is the cylinder about the z-axis of radius 1.

17. $\bar{x} = 0$ by symmetry; by Exercise 16, $\bar{y} = -\dfrac{1}{2A}\displaystyle\int_C y^2\,dx$; $C_1: y = 0, -a \le x \le a, y^2\,dx = 0$;

$C_2: x = a\cos\theta, y = a\sin\theta, 0 \le \theta \le \pi$, so

$$\bar{y} = -\frac{1}{2(\pi a^2/2)}\int_0^\pi a^2\sin^2\theta(-a\sin\theta)\,d\theta = \frac{4a}{3\pi}$$

18. $\bar{y} = \bar{x}$ by symmetry; by Exercise 16, $\bar{x} = \dfrac{1}{2A}\displaystyle\int_C x^2\,dy$; $C_1: y = 0, 0 \le x \le a, x^2\,dy = 0$;

$C_2: x = a\cos\theta, y = a\sin\theta, 0 \le \theta \le \pi/2$; $C_3: x = 0, x^2\,dy = 0$;

$$\bar{x} = \frac{1}{2(\pi a^2/4)}\int_0^{\pi/2} a^2(\cos^2\theta)a\cos\theta\,d\theta = \frac{4a}{3\pi}$$

19. $\bar{y} = 0$ by symmetry; $\bar{x} = \dfrac{1}{2A}\displaystyle\int_C x^2\,dy$; $A = \alpha a^2$; $C_1: x = t\cos\alpha, y = -t\sin\alpha, 0 \le t \le a$;

$C_2: x = a\cos\theta, y = a\sin\theta, -\alpha \le \theta \le \alpha$;

$C_3: x = t\cos\alpha, y = t\sin\alpha, 0 \le t \le a$ (reverse orientation);

$$2A\bar{x} = -\int_0^a t^2\cos^2\alpha\sin\alpha\,dt + \int_{-\alpha}^{\alpha} a^3\cos^3\theta\,d\theta - \int_0^a t^2\cos^2\alpha\sin\alpha\,dt,$$

$$= -\frac{2a^3}{3}\cos^2\alpha\sin\alpha + 2a^3\int_0^\alpha \cos^3\theta\,d\theta = -\frac{2a^3}{3}\cos^2\alpha\sin\alpha + 2a^3\left[\sin\alpha - \frac{1}{3}\sin^3\alpha\right]$$

$$= \frac{4}{3}a^3\sin\alpha;\ \text{since} A = \alpha a^2, \bar{x} = \frac{2a}{3}\frac{\sin\alpha}{\alpha}$$

20. $A = \displaystyle\int_0^a \left(b - \frac{b}{a^2}x^2\right)dx = \frac{2ab}{3}$, $C_1: x = t, y = bt^2/a^2, 0 \le t \le a$;

$C_2: x = a - t, y = b, 0 \le t \le a, x^2\,dy = 0$; $C_3: x = 0, y = b - t, 0 \le t \le b, x^2\,dy = y^2\,dx = 0$;

$$2A\bar{x} = \int_0^a t^2(2bt/a^2)\,dt = \frac{a^2b}{2},\ \bar{x} = \frac{3a}{8};$$

$$2A\bar{y} = -\int_0^a (bt^2/a^2)^2\,dt + \int_0^a b^2\,dt = -\frac{ab^2}{5} + ab^2 = \frac{4ab^2}{5}, \bar{y} = \frac{3b}{5}$$

21. (a) $\displaystyle\int_C f(x)\,dx + g(y)\,dy = \iint_R \left(\frac{\partial}{\partial x}g(y) - \frac{\partial}{\partial y}f(x)\right)dA = 0$

(b) $W = \displaystyle\int_C \mathbf{F}\cdot d\mathbf{r} = \int_C f(x)\,dx + g(y)\,dy = 0$, so the work done by the vector field around any simple closed curve is zero. The field is conservative.

22. (a) Let $\mathbf{r} = d\cos\theta\mathbf{i} + d\sin\theta\mathbf{j} + z\mathbf{k}$ in cylindrical coordinates, so

$$\frac{d\mathbf{r}}{dt} = \frac{d\mathbf{r}}{d\theta}\frac{d\theta}{dt} = \omega(-d\sin\theta\mathbf{i} + d\cos\theta\mathbf{j}), \mathbf{v} = \frac{d\mathbf{r}}{dt} = \omega\mathbf{k}\times\mathbf{r} = \boldsymbol{\omega}\times\mathbf{r}.$$

(b) From Part (a), $\mathbf{v} = \omega d(-\sin\theta\mathbf{i} + \cos\theta\mathbf{j}) = -\omega y\mathbf{i} + \omega x\mathbf{j}$

(c) From Part (b), curl $\mathbf{v} = 2\omega\mathbf{k} = 2\boldsymbol{\omega}$

(d) No; from Exercise 34 in Section 16.1, if ϕ were a potential function for $\mathbf{v}$, then curl $(\nabla\phi) =$ curl $\mathbf{v} = \mathbf{0}$, contradicting Part (c) above.

23. Yes; by imagining a normal vector sliding around the surface it is evident that the surface has two sides.

24. $D_{\mathbf{n}}\phi = \mathbf{n}\cdot\nabla\phi$, so $\displaystyle\iint_\sigma D_{\mathbf{n}}\phi\,dS = \iint_\sigma \mathbf{n}\cdot\nabla\phi\,dS = \iiint_G \nabla\cdot(\nabla\phi)\,dV$

$$= \iiint_G \left[\frac{\partial^2\phi}{\partial x^2} + \frac{\partial^2\phi}{\partial y^2} + \frac{\partial^2\phi}{\partial z^2}\right] dV$$

25. By Exercise 24, $\displaystyle\iint_\sigma D_{\mathbf{n}}f\,dS = -\iiint_G [f_{xx} + f_{yy} + f_{zz}]\,dV = -6\iiint_G dV = -6\text{vol}(G) = -8\pi$

26. **(a)** $f_y - g_x = e^{xy} + xye^{xy} - e^{xy} - xye^{xy} = 0$ so the vector field is conservative.

(b) $\phi_x = ye^{xy} - 1, \phi = e^{xy} - x + k(x), \phi_y = xe^{xy}$, let $k(x) = 0; \phi(x,y) = e^{xy} - x$

(c) $\displaystyle W = \int_C \mathbf{F}\cdot d\mathbf{r} = \phi(x(8\pi), y(8\pi)) - \phi(x(0), y(0)) = \phi(8\pi, 0) - \phi(0,0) = -8\pi$

27. **(a)** If $h(x)\mathbf{F}$ is conservative, then $\dfrac{\partial}{\partial y}(yh(x)) = \dfrac{\partial}{\partial x}(-2xh(x))$, or $h(x) = -2h(x) - 2xh'(x)$ which has the general solution $x^3h(x)^2 = C_1, h(x) = Cx^{-3/2}$, so $C\dfrac{y}{x^{3/2}}\mathbf{i} - C\dfrac{2}{x^{1/2}}\mathbf{j}$ is conservative, with potential function $\phi = -2Cy/\sqrt{x}$.

(b) If $g(y)\mathbf{F}(x,y)$ is conservative then $\dfrac{\partial}{\partial y}(yg(y)) = \dfrac{\partial}{\partial x}(-2xg(y))$, or $g(y) + yg'(y) = -2g(y)$, with general solution $g(y) = C/y^3$, so $\mathbf{F} = C\dfrac{1}{y^2}\mathbf{i} - C\dfrac{2x}{y^3}\mathbf{j}$ is conservative, with potential function Cx/y^2.

28. A computation of curl $\mathbf{F}$ shows that curl $\mathbf{F} = \mathbf{0}$ if and only if the three given equations hold. Moreover the equations hold if $\mathbf{F}$ is conservative, so it remains to show that $\mathbf{F}$ is conservative if curl $\mathbf{F} = \mathbf{0}$. Let C by any simple closed curve in the region. Since the region is simply connected, there is a piecewise smooth, oriented surface σ in the region with boundary C. By Stokes' Theorem,
$\displaystyle\oint_C \mathbf{F}\cdot d\mathbf{r} = \iint_\sigma (\text{curl }\mathbf{F})\cdot\mathbf{n}\,dS = \iint_\sigma 0\,dS = 0.$

By the 3-space analog of Theorem 16.3.2, $\mathbf{F}$ is conservative.

29. **(a)** conservative, $\phi(x,y,z) = xz^2 - e^{-y}$ **(b)** not conservative, $f_y \neq g_x$

30. **(a)** conservative, $\phi(x,y,z) = -\cos x + yz$ **(b)** not conservative, $f_z \neq h_x$

CHAPTER 16 HORIZON MODULE

1. (a) If $\mathbf{r} = x\mathbf{i} + y\mathbf{j}$ denotes the position vector, then $\mathbf{F}_1 \cdot \mathbf{r} = 0$ by inspection, so the velocity field is tangent to the circle. The relationship $\mathbf{F}_1 \times \mathbf{r} = -\dfrac{k}{2\pi}\mathbf{k}$ indicates that $\mathbf{r}, \mathbf{F}_1, \mathbf{k}$ is a right-handed system, so the flow is counterclockwise. The polar form $\mathbf{F}_1 = -\dfrac{k}{2\pi r}(\sin\theta\mathbf{i} - \cos\theta\mathbf{j})$ shows that the speed is the constant $\dfrac{k}{2\pi r}$ on a circle of radius r; and it also shows that the speed is proportional to $\dfrac{1}{r}$ with constant of proportionality $k/(2\pi)$.

   (b) Since $\|\mathbf{F}_1\| = \dfrac{k}{2\pi r}$, when $r = 1$ we get $k = 2\pi\|\mathbf{F}_1\|$

2. 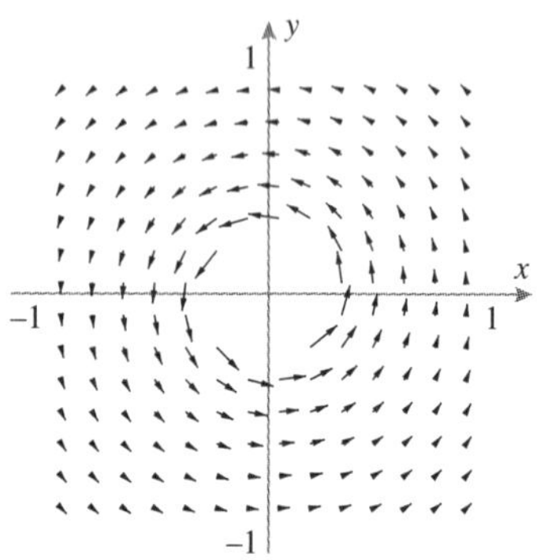

3. (a) $\mathbf{F}_2 = -\dfrac{q}{2\pi\|\mathbf{r}\|^2}\mathbf{r}$ so $\mathbf{F}_2$ is directed toward the origin, and $\|\mathbf{F}_2\| = \dfrac{q}{2\pi r}$ is constant for constant r, and the speed is inversely proportional to the distance from the origin (constant of proportionality $\dfrac{q}{2\pi}$). Since the velocity vector is directed toward the origin, the fluid flows towards the origin, which must therefore be a sink.

   (b) From Part (a) when $r = 1$, $q = 2\pi\|\mathbf{F}_2\|$.

4. 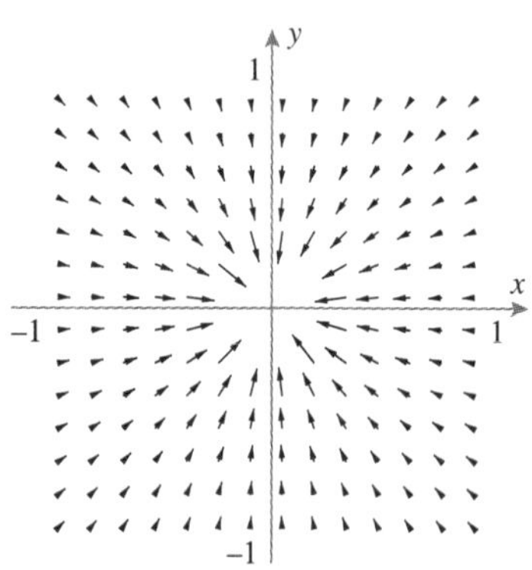

5. (b) The magnitudes of the field vectors increase, and their directions become more tangent to circles about the origin.

   (c) The magnitudes of the field vectors increase, and their directions tend more towards the origin.

6. **(a)** The inward component is $\mathbf{F}_2$, so at $r = 20, 15 = \|\mathbf{F}_2\| = \dfrac{q}{2\pi(20)}$, so $q = 600\pi$; the tangential component is $\mathbf{F}_1$, so at $r = 20, 45 = \|\mathbf{F}_1\| = \dfrac{k}{2\pi(20)}$, so $k = 1800\pi$.

(b) $\mathbf{F} = -\dfrac{1}{x^2+y^2}[(300x + 900y)\mathbf{i} + (300y - 900x)\mathbf{j}]$

(c) $\|\mathbf{F}\| = \dfrac{300\sqrt{10}}{r} \le 5$ km/hr if $r \ge 60\sqrt{10} \approx 189.7$ km.

7. $\mathbf{F} = -\dfrac{1}{2\pi r}[(q\cos\theta + k\sin\theta)\mathbf{i} + (q\sin\theta - k\cos\theta)\mathbf{j}] = -\dfrac{q}{2\pi r}\mathbf{u}_r + \dfrac{k}{2\pi r}\mathbf{u}_\theta = -\dfrac{1}{2\pi r}(q\mathbf{u}_r - k\mathbf{u}_\theta)$

8. $\mathbf{F} \cdot \nabla\psi = -\dfrac{1}{2\pi r}(q\mathbf{u}_r - k\mathbf{u}_\theta) \cdot \left(\dfrac{k}{r}\mathbf{u}_r + \dfrac{1}{r}q\mathbf{u}_\theta\right) = -\dfrac{1}{2\pi r}\left(q\dfrac{k}{r} - k\dfrac{1}{r}q\right) = 0$, since $\mathbf{u}_r$ and $\mathbf{u}_\theta$ are orthogonal unit vectors.

9. From the hypotheses of Exercise 8, $\psi = k\ln r + G(\theta)$, $\dfrac{\partial}{\partial\theta}\psi = G'(\theta) = q$;
let $G = q\theta, \psi = k\ln r + q\theta$

10. The streamline $\psi = c$ becomes $k\ln r + q\theta = c, \ln r = -q\theta/k + c/k$,
$r = e^{-q\theta/k}e^{c/k} = \kappa e^{-q\theta/k}$, where $\kappa > 0$.

11.

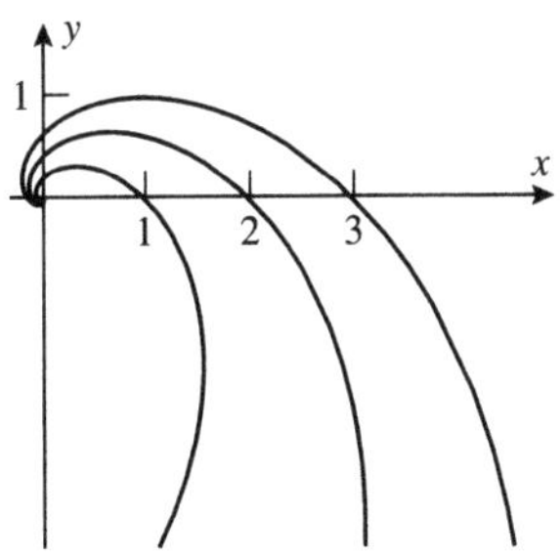

12. $q = 600\pi, k = 1800\pi, r = \kappa e^{-\theta/3}$; at $r = 20, \theta = \pi/4, \kappa = re^{\theta/3} = 20e^{\pi/12} \approx 25.985$; the desired streamline has the polar equation $r = 25.985e^{-\theta/3}$.

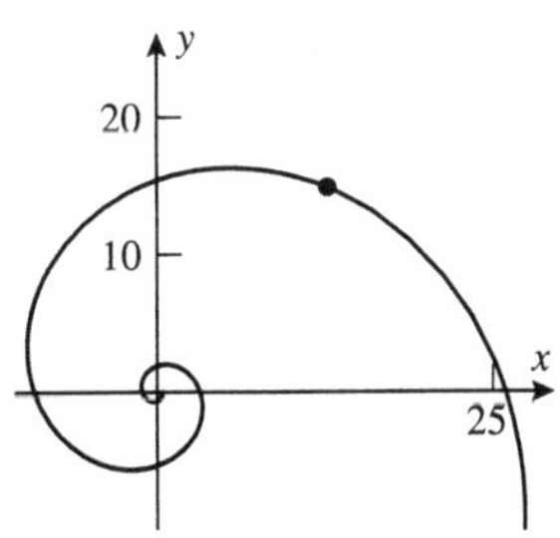

APPENDIX A
Real Numbers, Intervals, and Inequalities

EXERCISE SET A

1. (a) rational (b) integer, rational (c) integer, rational
(d) rational (e) integer, rational (f) irrational
(g) rational (h) 467 integer, rational

2. (a) irrational (b) rational (c) rational (d) rational

3. (a) $x = 0.123123123\ldots$, $1000x = 123 + x$, $x = 123/999 = 41/333$
(b) $x = 12.7777\ldots$, $10(x-12) = 7 + (x-12)$, $9x = 115$, $x = 115/9$
(c) $x = 38.07818181\ldots$, $100x = 3807.81818181\ldots$, $99x = 100x - x = 3769.74$,
$x = \dfrac{3769.74}{99} = \dfrac{376974}{9900} = \dfrac{20943}{550}$
(d) $\dfrac{4296}{10000} = \dfrac{537}{1250}$

4. $x = 0.99999\ldots$, $10x = 9 + x$, $9x = 9$, $x = 1$

5. (a) If r is the radius, then $D = 2r$ so $\left(\dfrac{8}{9}D\right)^2 = \left(\dfrac{16}{9}r\right)^2 = \dfrac{256}{81}r^2$. The area of a circle of radius r is πr^2 so $256/81$ was the approximation used for π.
(b) $22/7 \approx 3.1429$ is better than $256/81 \approx 3.1605$.

6. (a) $\dfrac{223}{71} < \dfrac{333}{106} < \dfrac{63}{25}\left(\dfrac{17+15\sqrt{5}}{7+15\sqrt{5}}\right) < \dfrac{355}{113} < \dfrac{22}{7}$
(b) Ramanujan's (c) Athoniszoon's (d) Ramanujan's

7.

Line	2	3	4	5	6	7
Blocks	3, 4	1, 2	3, 4	2, 4, 5	1, 2	3, 4

8.

Line	1	2	3	4	5
Blocks	all blocks	none	2, 4	2	2, 3

9. (a) always correct (add -3 to both sides of $a \le b$)
(b) not always correct (correct only if $a = b$)
(c) not always correct (correct only if $a = b$)
(d) always correct (multiply both sides of $a \le b$ by 6)
(e) not always correct (correct only if $a \ge 0$ or $a = b$)
(f) always correct (multiply both sides of $a \le b$ by the nonnegative quantity a^2)

10. (a) always correct
(b) not always correct (for example let $a = b = 0$, $c = 1$, $d = 2$)
(c) not always correct (for example let $a = 1$, $b = 2$, $c = d = 0$)

11. (a) all values because $a = a$ is always valid (b) none

12. $a = b$, because if $a \ne b$ then $a < b$ and $b < a$ are contradictory

13. **(a)** yes, because $a \leq b$ means $a < b$ or $a = b$, thus $a < b$ certainly means $a \leq b$
(b) no, because $a < b$ is false if $a = b$ is true

14. **(a)** $x^2 - 5x = 0$, $x(x-5) = 0$ so $x = 0$ or $x = 5$
(b) $-1, 0, 1, 2$ are the only integers that satisfy $-2 < x < 3$

15. **(a)** $\{x : x \text{ is a positive odd integer}\}$ **(b)** $\{x : x \text{ is an even integer}\}$
(c) $\{x : x \text{ is irrational}\}$ **(d)** $\{x : x \text{ is an integer and } 7 \leq x \leq 10\}$

16. **(a)** not equal to A because 0 is not in A **(b)** equal to A
(c) equal to A because $(x-3)(x^2 - 3x + 2) = 0$, $(x-3)(x-2)(x-1) = 0$ so $x = 1$, 2, or 3

17. **(a)** false, there are points inside the triangle that are not inside the circle
(b) true, all points inside the triangle are also inside the square
(c) true **(d)** false **(e)** true
(f) true, a is inside the circle **(g)** true

18. **(a)** $\emptyset$, $\{a_1\}$, $\{a_2\}$, $\{a_3\}$, $\{a_1, a_2\}$, $\{a_1, a_3\}$, $\{a_2, a_3\}$, $\{a_1, a_2, a_3\}$ **(b)** $\emptyset$

19. **(a)** 

(b) 

(c) 

(d) 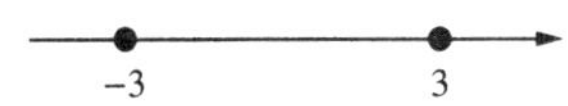

(e) 

(f) 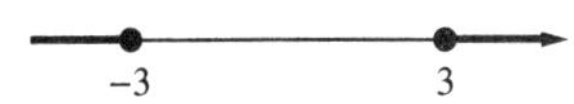

20. **(a)** 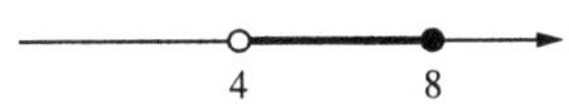

(b) 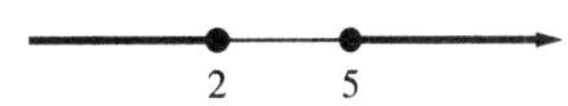

(b) 

(d) none

21. **(a)** $[-2, 2]$ **(b)** $(-\infty, -2) \cup (2, +\infty)$

22. **(a)** 

(b) 

(c) 

(d) 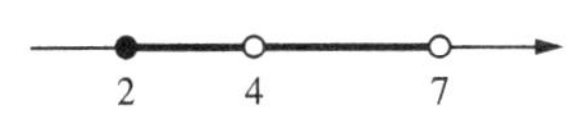

(e) 

(f) 

(g)
(h) 

23. $3x < 10$; $(-\infty, 10/3)$
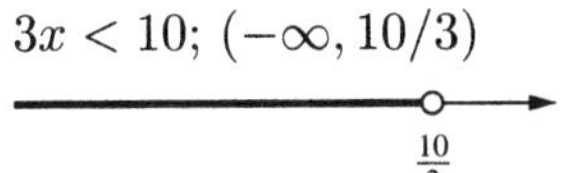

24. $\frac{1}{5}x \geq 8$; $[40, +\infty)$
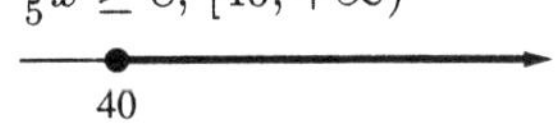

25. $2x \leq -11$; $(-\infty, -11/2]$
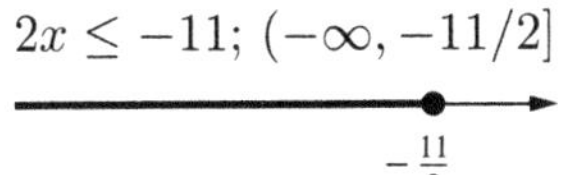

26. $9x < -10$; $(-\infty, -10/9)$
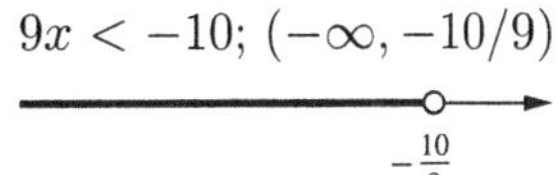

27. $2x \leq 1$ and $2x > -3$; $(-3/2, 1/2]$
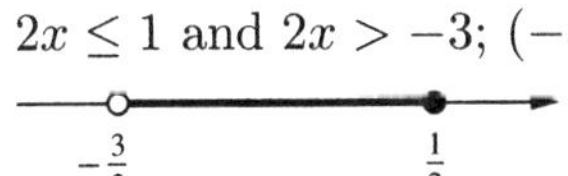

28. $8x \geq 5$ and $8x \leq 14$; $[\frac{5}{8}, \frac{7}{4}]$
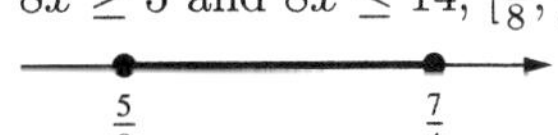

29. $\dfrac{x}{x-3} - 4 < 0,\ \dfrac{12-3x}{x-3} < 0,\ \dfrac{4-x}{x-3} < 0;$
$(-\infty, 3) \cup (4, +\infty)$

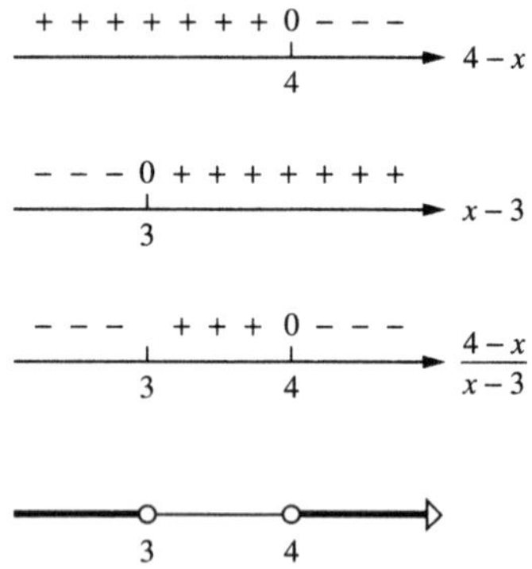

30. $\dfrac{x}{8-x} + 2 = \dfrac{16-x}{8-x} \geq 0;$
$(-\infty, 8) \cup [16, +\infty)$

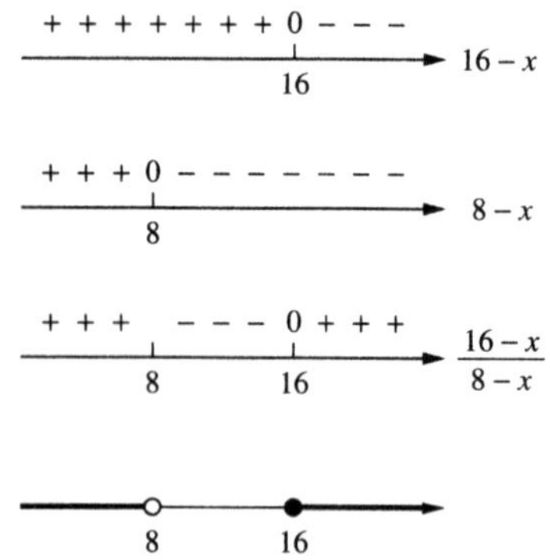

31. $\dfrac{3x+1}{x-2} - 1 = \dfrac{2x+3}{x-2} < 0,\ \dfrac{x+3/2}{x-2} < 0;$
$(-\frac{3}{2}, 2)$

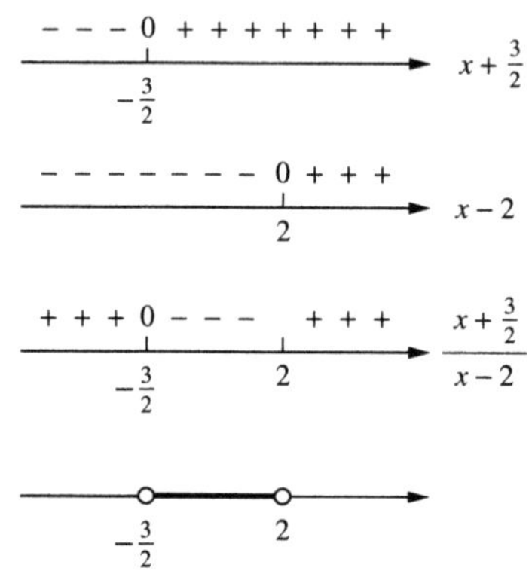

32. $\dfrac{x/2-3}{4+x} - 1 > 0,\ \dfrac{x-6}{4+x} - 2 > 0,\ \dfrac{x+14}{x+4} < 0;$
$(-14, -4)$

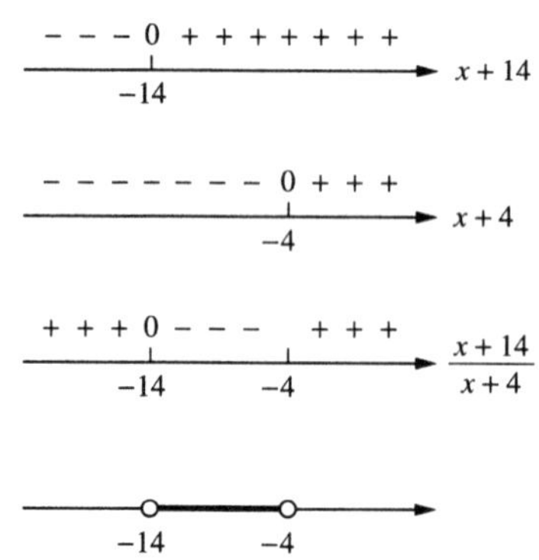

33. $\dfrac{4}{2-x} - 1 = \dfrac{x+2}{2-x} \leq 0;\ (-\infty, -2] \cup (2, +\infty)$

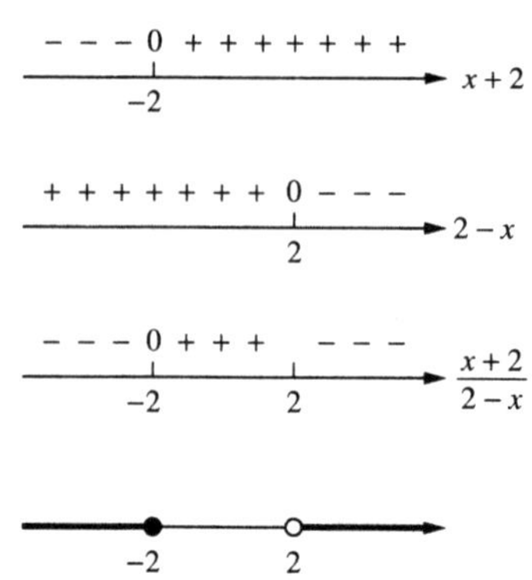

34. $\dfrac{3}{x-5} - 2 = \dfrac{13-2x}{x-5} \le 0,\ \dfrac{13/2 - x}{x-5} \le 0;$
$(-\infty, 5) \cup [\frac{13}{2}, +\infty)$

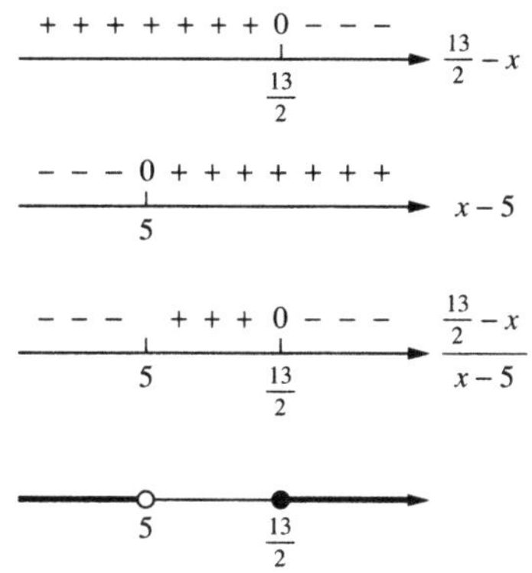

35. $x^2 - 9 = (x+3)(x-3) > 0;$
$(-\infty, -3) \cup (3, +\infty)$

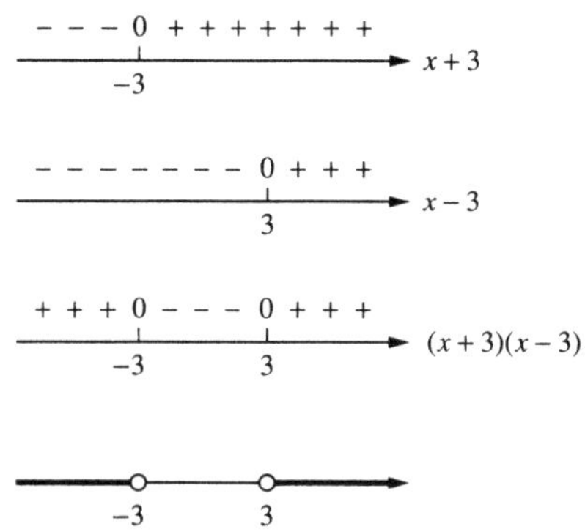

36. $x^2 - 5 = (x-\sqrt{5})(x+\sqrt{5}) \le 0;\ [-\sqrt{5}, \sqrt{5}]$

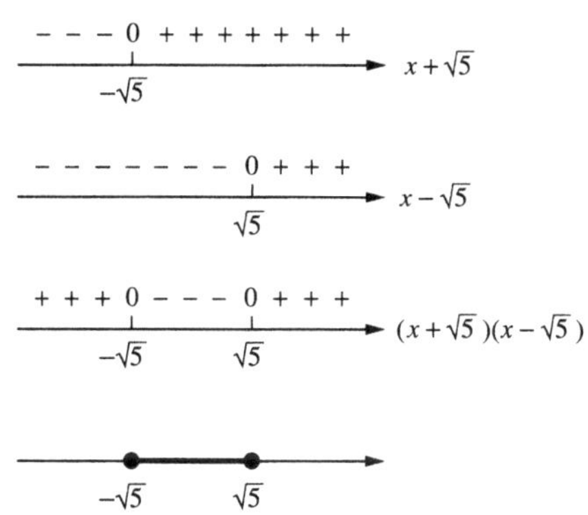

37. $(x-4)(x+2) > 0;\ (-\infty, -2) \cup (4, +\infty)$

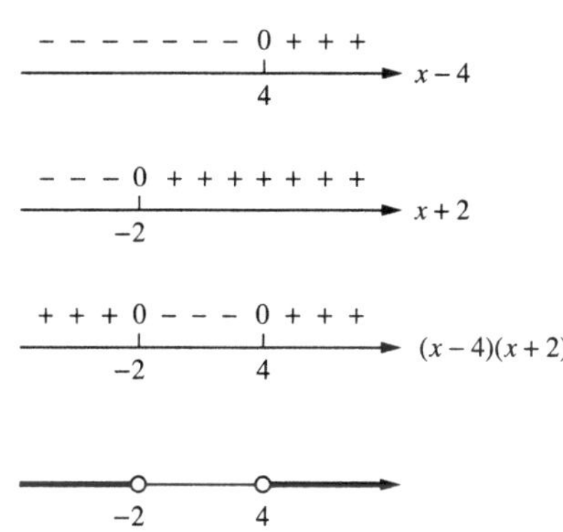

38. $(x-3)(x+4) < 0$; $(-4, 3)$

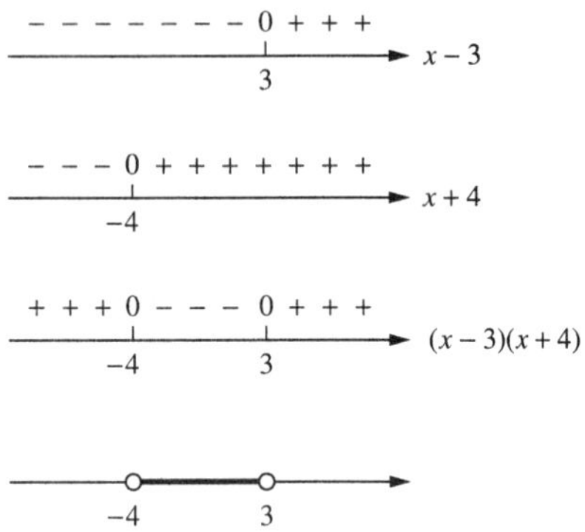

39. $(x-4)(x-5) \leq 0$; $[4, 5]$

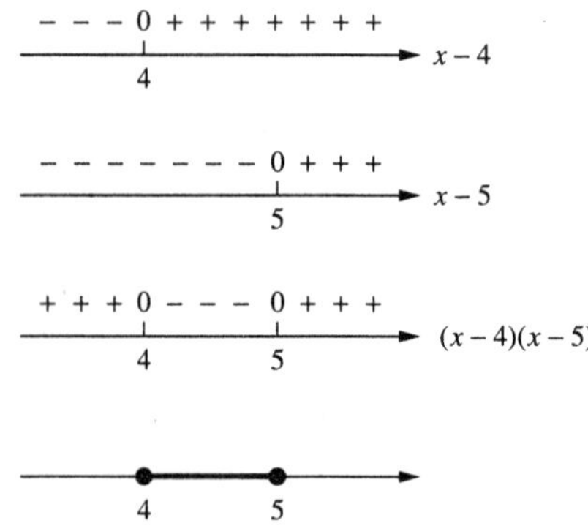

40. $(x-2)(x-1) \geq 0$;
$(-\infty, 1] \cup [2, +\infty)$

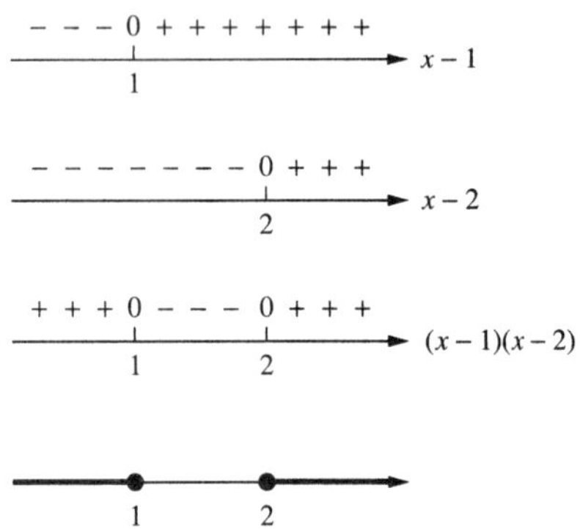

41. $\dfrac{3}{x-4} - \dfrac{2}{x} = \dfrac{x+8}{x(x-4)} > 0$;
$(-8, 0) \cup (4, +\infty)$

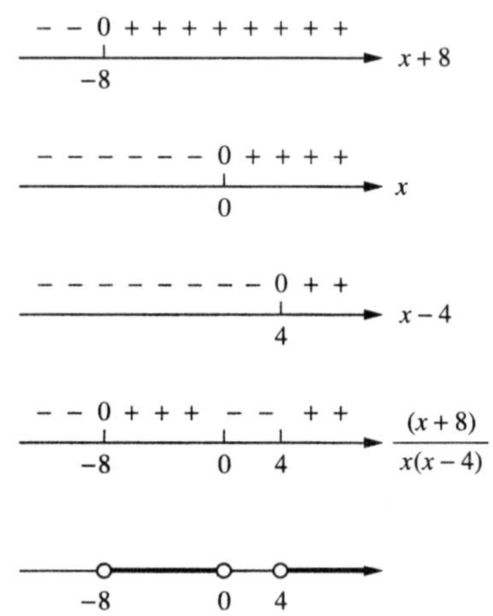

42. $\dfrac{1}{x+1} - \dfrac{3}{x-2} = \dfrac{-2x-5}{(x+1)(x-2)} \geq 0,$
$\dfrac{x+5/2}{(x+1)(x-2)} \leq 0;$
$(-\infty, -\frac{5}{2}] \cup (-1, 2)$

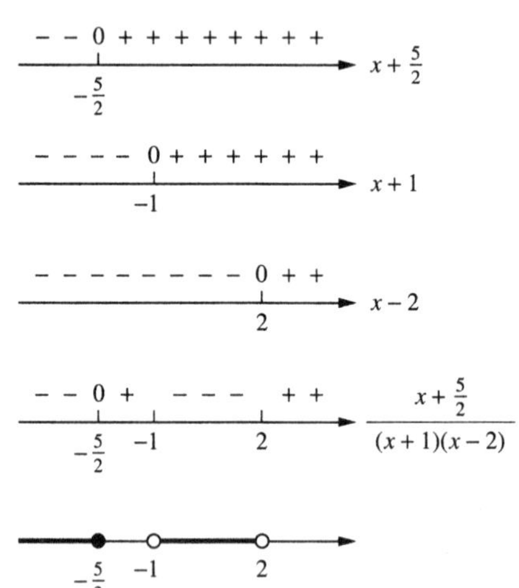

43. By trial-and-error we find that $x = 2$ is a root of the equation $x^3 - x^2 - x - 2 = 0$ so $x - 2$ is a factor of $x^3 - x^2 - x - 2$. By long division we find that $x^2 + x + 1$ is another factor so $x^3 - x^2 - x - 2 = (x-2)(x^2 + x + 1)$. The linear factors of $x^2 + x + 1$ can be determined by first finding the roots of $x^2 + x + 1 = 0$ by the quadratic formula. These roots are complex numbers so $x^2 + x + 1 \neq 0$ for all real x; thus $x^2 + x + 1$ must be always positive or always negative. Since $x^2 + x + 1$ is positive when $x = 0$, it follows that $x^2 + x + 1 > 0$ for all real x. Hence $x^3 - x^2 - x - 2 > 0$, $(x-2)(x^2+x+1) > 0$, $x - 2 > 0$, $x > 2$, so $S = (2, +\infty)$.

44. By trial-and-error we find that $x = 1$ is a root of the equation $x^3 - 3x + 2 = 0$ so $x - 1$ is a factor of $x^3 - 3x + 2$. By long division we find that $x^2 + x - 2$ is another factor so $x^3 - 3x + 2 = (x-1)(x^2+x-2) = (x-1)(x-1)(x+2) = (x-1)^2(x+2)$. Therefore we want to solve $(x-1)^2(x+2) \le 0$. Now if $x \neq 1$, then $(x-1)^2 > 0$ and so $x + 2 \le 0$, $x \le -2$. By inspection, $x = 1$ is also a solution so $S = (-\infty, -2] \cup \{1\}$.

45. $\sqrt{x^2 + x - 6}$ is real if $x^2 + x - 6 \ge 0$. Factor to get $(x+3)(x-2) \ge 0$ which has as its solution $x \le -3$ or $x \ge 2$.

46. $\dfrac{x+2}{x-1} \ge 0$; $(-\infty, -2] \cup (1, +\infty)$

47. $25 \le \dfrac{5}{9}(F - 32) \le 40$, $45 \le F - 32 \le 72$, $77 \le F \le 104$

48. **(a)** $n = 2k$, $n^2 = 4k^2 = 2(2k^2)$ where $2k^2$ is an integer.
(b) $n = 2k + 1$, $n^2 = 4k^2 + 4k + 1 = 2(2k^2 + 2k) + 1$ where $2k^2 + 2k$ is an integer.

49. **(a)** Assume m and n are rational, then $m = \dfrac{p}{q}$ and $n = \dfrac{r}{s}$ where p, q, r, and s are integers so $m + n = \dfrac{p}{q} + \dfrac{r}{s} = \dfrac{ps + rq}{qs}$ which is rational because $ps + rq$ and qs are integers.

(b) (proof by contradiction) Assume m is rational and n is irrational, then $m = \dfrac{p}{q}$ where p and q are integers. Suppose that $m + n$ is rational, then $m + n = \dfrac{r}{s}$ where r and s are integers so $n = \dfrac{r}{s} - m = \dfrac{r}{s} - \dfrac{p}{q} = \dfrac{rq - ps}{sq}$. But $rq - ps$ and sq are integers, so n is rational which contradicts the assumption that n is irrational.

50. **(a)** Assume m and n are rational, then $m = \dfrac{p}{q}$ and $n = \dfrac{r}{s}$ where p, q, r, and s are integers so $mn = \dfrac{p}{q} \cdot \dfrac{r}{s} = \dfrac{pr}{qs}$ which is rational because pr and qs are integers.

(b) (proof by contradiction) Assume m is rational and nonzero and that n is irrational, then $m = \dfrac{p}{q}$ where p and q are integers and $p \neq 0$. Suppose that mn is rational, then $mn = \dfrac{r}{s}$ where r and s are integers so $n = \dfrac{r/s}{m} = \dfrac{r/s}{p/q} = \dfrac{rq}{ps}$. But rq and ps are integers, so n is rational which contradicts the assumption that n is irrational.

51. $a = \sqrt{2}, b = \sqrt{3}, c - \sqrt{6}, d = -\sqrt{2}$ are irrational, and $a + d = 0$, a rational; $a + a = 2\sqrt{2}$, an irrational; $ad = -2$, a rational; and $ab = c$, an irrational.

52. **(a)** irrational (Exercise 49(b)) **(b)** irrational (Exercise 50(b))
(c) rational by inspection; Exercise 51 gives no information
(d) $\sqrt{\pi}$ must be irrational, for if it were rational, then so would be $\pi = \left(\sqrt{\pi}\right)^2$ by Exercise 50(a); but π is known to be irrational.

53. The average of a and b is $\frac{1}{2}(a + b)$; if a and b are rational then so is the average, by Exercise 49(a) and Exercise 50(a). On the other hand if $a = b = \sqrt{2}$ then the average of a and b is irrational, but the average of a and $-b$ is rational.

54. If $10^x = 3$, then $x > 0$ because $10^x \leq 1$ for $x \leq 0$. If $10^{p/q} = 3$ with p, q integers, then $10^p = 3^q$. Following Exercise 48, if $n = 2k$ is even, then $n^2, n^3, n^4, \ldots$ are even; and if $n = 2k + 1$ then $n^2, n^3, n^4, \ldots$ are odd. Since $10^p = 3^q$, the left side is even and the right side is odd, a contradiction.

55. $8x^3 - 4x^2 - 2x + 1$ can be factored by grouping terms: $(8x^3 - 4x^2) - (2x - 1) = 4x^2(2x - 1) - (2x - 1) = (2x - 1)(4x^2 - 1) = (2x - 1)^2(2x + 1)$. The problem, then, is to solve $(2x - 1)^2(2x + 1) < 0$. By inspection, $x = 1/2$ is not a solution. If $x \neq 1/2$, then $(2x - 1)^2 > 0$ and it follows that $2x + 1 < 0$, $2x < -1$, $x < -1/2$, so $S = (-\infty, -1/2)$.

56. Rewrite the inequality as $12x^3 - 20x^2 + 11x - 2 \geq 0$. If a polynomial in x with integer coefficients has a rational zero $\dfrac{p}{q}$, a fraction in lowest terms, then p must be a factor of the constant term and q must be a factor of the coefficient of the highest power of x. By trial-and-error we find that $x = 1/2$ is a zero, thus $(x - 1/2)$ is a factor so

$$\begin{aligned} 12x^3 - 20x^2 + 11x - 2 &= (x - 1/2)\left(12x^2 - 14x + 4\right) \\ &= 2(x - 1/2)\left(6x^2 - 7x + 2\right) \\ &= 2(x - 1/2)(2x - 1)(3x - 2) = (2x - 1)^2(3x - 2). \end{aligned}$$

Now to solve $(2x - 1)^2(3x - 2) \geq 0$ we first note that $x = 1/2$ is a solution. If $x \neq 1/2$ then $(2x - 1)^2 > 0$ and $3x - 2 \geq 0$, $3x \geq 2$, $x \geq 2/3$ so $S = [2/3, +\infty) \cup \{1/2\}$.

57. If $a < b$, then $ac < bc$ because c is positive; if $c < d$, then $bc < bd$ because b is positive, so $ac < bd$ (Theorem A.1(a)). (Note that the result is still true if one of a, b, c, d is allowed to be negative, that is $a < 0$ or $c < 0$.)

58. no, since the decimal representation is not repeating (the string of zeros does not have constant length)

APPENDIX B
Absolute Value

EXERCISE SET B

1. (a) 7 (b) $\sqrt{2}$ (c) k^2 (d) k^2

2. $\sqrt{(x-6)^2} = x-6$ if $x \geq 6$, $\sqrt{(x-6)^2} = -(x-6) = -x+6$ if $x < 6$

3. $|x-3| = |3-x| = 3-x$ if $3-x \geq 0$, which is true if $x \leq 3$

4. $|x+2| = x+2$ if $x+2 \geq 0$ so $x \geq -2$.

5. All real values of x because $x^2+9 > 0$.

6. $|x^2+5x| = x^2+5x$ if $x^2+5x \geq 0$ so $x(x+5) \geq 0$ which is true for $x \leq -5$ or $x \geq 0$.

7. $|3x^2+2x| = |x(3x+2)| = |x||3x+2|$. If $|x||3x+2| = x|3x+2|$, then $|x||3x+2| - x|3x+2| = 0$, $(|x|-x)|3x+2| = 0$, so either $|x|-x = 0$ or $|3x+2| = 0$. If $|x|-x=0$, then $|x| = x$, which is true for $x \geq 0$. If $|3x+2| = 0$, then $x = -2/3$. The statement is true for $x \geq 0$ or $x = -2/3$.

8. $|6-2x| = |2(3-x)| = |2||3-x| = 2|x-3|$ for all real values of x.

9. $\sqrt{(x+5)^2} = |x+5| = x+5$ if $x+5 \geq 0$, which is true if $x \geq -5$.

10. $\sqrt{(3x-2)^2} = |3x-2| = |2-3x| = 2-3x$ if $2-3x \geq 0$ so $x \leq 2/3$.

13. (a) $|7-9| = |-2| = 2$
(b) $|3-2| = |1| = 1$
(c) $|6-(-8)| = |14| = 14$
(d) $|-3-\sqrt{2}| = |-(3+\sqrt{2})| = 3+\sqrt{2}$
(e) $|-4-(-11)| = |7| = 7$
(f) $|-5-0| = |-5| = 5$

14. $\sqrt{a^4} = \sqrt{(a^2)^2} = |a^2|$, but $|a^2| = a^2$ because $a^2 \geq 0$ so it is valid for all values of a.

15. (a) B is 6 units to the left of A; $b = a-6 = -3-6 = -9$.
(b) B is 9 units to the right of A; $b = a+9 = -2+9 = 7$.
(c) B is 7 units from A; either $b = a+7 = 5+7 = 12$ or $b = a-7 = 5-7 = -2$. Since it is given that $b > 0$, it follows that $b = 12$.

16. In each case we solve for e in terms of f:
(a) $e = f-4$; e is to the left of f.
(b) $e = f+4$; e is to the right of f.
(c) $e = f+6$; e is to the right of f.
(d) $e = f-7$; e is to the left of f.

17. $|6x-2| = 7$

Case 1: $6x-2=7$, $6x=9$, $x=3/2$

Case 2: $6x-2=-7$, $6x=-5$, $x=-5/6$

18. $|3+2x| = 11$

Case 1: $3+2x=11$, $2x=8$, $x=4$

Case 2: $3+2x=-11$, $2x=-14$, $x=-7$

19. $|6x-7| = |3+2x|$

Case 1: $6x-7=3+2x$, $4x=10$, $x=5/2$

Case 2: $6x-7=-(3+2x)$, $8x=4$, $x=1/2$

20. $|4x+5| = |8x-3|$

Case 1: $4x+5=8x-3$, $-4x=-8$, $x=2$

Case 2: $4x+5=-(8x-3)$, $12x=-2$, $x=-1/6$

21. $|9x| - 11 = x$

Case 1:
$$9x - 11 = x$$
$$8x = 11$$
$$x = 11/8$$

Case 2:
$$-9x - 11 = x$$
$$-10x = 11$$
$$x = -11/10$$

22. $2x - 7 = |x + 1|$

Case 1:
$$2x - 7 = x + 1$$
$$x = 8$$

Case 2:
$$2x - 7 = -(x + 1)$$
$$3x = 6$$
$x = 2$; not a solution because x must also satisfy $x < -1$

23. $\left|\dfrac{x+5}{2-x}\right| = 6$

Case 1:
$$\frac{x+5}{2-x} = 6$$
$$x + 5 = 12 - 6x$$
$$7x = 7$$
$$x = 1$$

Case 2:
$$\frac{x+5}{2-x} = -6$$
$$x + 5 = -12 + 6x$$
$$-5x = -17$$
$$x = 17/5$$

24. $\left|\dfrac{x-3}{x+4}\right| = 5$

Case 1:
$$\frac{x-3}{x+4} = 5$$
$$x - 3 = 5x + 20$$
$$-4x = 23$$
$$x = -23/4$$

Case 2:
$$\frac{x-3}{x+4} = -5$$
$$x - 3 = -5x - 20$$
$$6x = -17$$
$$x = -17/6$$

25.
$$|x + 6| < 3$$
$$-3 < x + 6 < 3$$
$$-9 < x < -3$$
$$S = (-9, -3)$$

26.
$$|7 - x| \le 5$$
$$-5 \le 7 - x \le 5$$
$$-12 \le -x \le -2$$
$$12 \ge x \ge 2$$
$$S = [2, 12]$$

27.
$$|2x - 3| \le 6$$
$$-6 \le 2x - 3 \le 6$$
$$-3 \le 2x \le 9$$
$$-3/2 \le x \le 9/2$$
$$S = [-3/2, 9/2]$$

28.
$$|3x + 1| < 4$$
$$-4 < 3x + 1 < 4$$
$$-5 < 3x < 3$$
$$-5/3 < x < 1$$
$$S = (-5/3, 1)$$

29. $|x + 2| > 1$

Case 1:
$$x + 2 > 1$$
$$x > -1$$

Case 2:
$$x + 2 < -1$$
$$x < -3$$

$$S = (-\infty, -3) \cup (-1, +\infty)$$

30. $\left|\dfrac{1}{2}x - 1\right| \ge 2$

Case 1:
$$\frac{1}{2}x - 1 \ge 2$$
$$\frac{1}{2}x \ge 3$$
$$x \ge 6$$

Case 2:
$$\frac{1}{2}x - 1 \le -2$$
$$\frac{1}{2}x \le -1$$
$$x \le -2$$

$$S = (-\infty, -2] \cup [6, +\infty)$$

31. $|5 - 2x| \ge 4$

Case 1:
$$5 - 2x \ge 4$$
$$-2x \ge -1$$
$$x \le 1/2$$

Case 2:
$$5 - 2x \le -4$$
$$-2x \le -9$$
$$x \ge 9/2$$

$$S = (-\infty, 1/2] \cup [9/2, +\infty)$$

32. $|7x + 1| > 3$

Case 1:
$$7x + 1 > 3$$
$$7x > 2$$
$$x > 2/7$$

Case 2:
$$7x + 1 < -3$$
$$7x < -4$$
$$x < -4/7$$

$$S = (-\infty, -4/7) \cup (2/7, +\infty)$$

33. $\dfrac{1}{|x-1|} < 2, x \neq 1$

$|x-1| > 1/2$

Case 1: $x - 1 > 1/2$, $x > 3/2$

Case 2: $x - 1 < -1/2$, $x < 1/2$

$S = (-\infty, 1/2) \cup (3/2, +\infty)$

34. $\dfrac{1}{|3x+1|} \geq 5, x \neq -1/3$

$|3x+1| \leq 1/5$

$-1/5 \leq 3x + 1 \leq 1/5$

$-6/5 \leq 3x \leq -4/5$

$-2/5 \leq x \leq -4/15$

$S = [-2/5, -1/3) \cup (-1/3, -4/15]$

35. $\dfrac{3}{|2x-1|} \geq 4, x \neq 1/2$

$\dfrac{|2x-1|}{3} \leq \dfrac{1}{4}$

$|2x-1| \leq 3/4$

$-3/4 \leq 2x - 1 \leq 3/4$

$1/4 \leq 2x \leq 7/4$

$1/8 \leq x \leq 7/8$

$S = [1/8, 1/2) \cup (1/2, 7/8]$

36. $\dfrac{2}{|x+3|} < 1, x \neq -3$

$\dfrac{|x+3|}{2} > 1$

$|x+3| > 2$

Case 1: $x + 3 > 2$, $x > -1$

Case 2: $x + 3 < -2$, $x < -5$

$S = (-\infty, -5) \cup (-1, +\infty)$

37. $\sqrt{(x^2 - 5x + 6)^2} = x^2 - 5x + 6$ if $x^2 - 5x + 6 \geq 0$ or, equivalently, if $(x-2)(x-3) \geq 0$; $x \in (-\infty, 2] \cup [3, +\infty)$.

38. If $x \geq 2$ then $3 \leq x - 2 \leq 7$ so $5 \leq x \leq 9$; if $x < 2$ then $3 \leq 2 - x \leq 7$ so $-5 \leq x \leq -1$. $S = [-5, -1] \cup [5, 9]$.

39. If $u = |x-3|$ then $u^2 - 4u = 12$, $u^2 - 4u - 12 = 0$, $(u-6)(u+2) = 0$, so $u = 6$ or $u = -2$. If $u = 6$ then $|x-3| = 6$, so $x = 9$ or $x = -3$. If $u = -2$ then $|x-3| = -2$ which is impossible. The solutions are -3 and 9.

41.
$$\begin{aligned} |a-b| &= |a + (-b)| \\ &\leq |a| + |-b| \text{ (triangle inequality)} \\ &= |a| + |b|. \end{aligned}$$

42.
$$\begin{aligned} a &= (a-b) + b \\ |a| &= |(a-b) + b| \\ |a| &\leq |a-b| + |b| \text{ (triangle inequality)} \\ |a| - |b| &\leq |a-b|. \end{aligned}$$

43. From Exercise 42
(i) $|a| - |b| \leq |a-b|$; but $|b| - |a| \leq |b-a| = |a-b|$, so (ii) $|a| - |b| \geq -|a-b|$.
Combining (i) and (ii): $-|a-b| \leq |a| - |b| \leq |a-b|$, so $||a| - |b|| \leq |a-b|$.

APPENDIX C
Coordinate Planes and Lines

EXERCISE SET C

1.

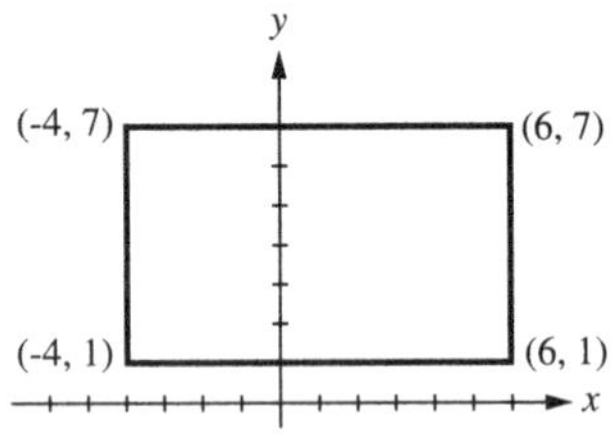

2. area $= \frac{1}{2}bh = \frac{1}{2}(5-(-3))(1) = 4$

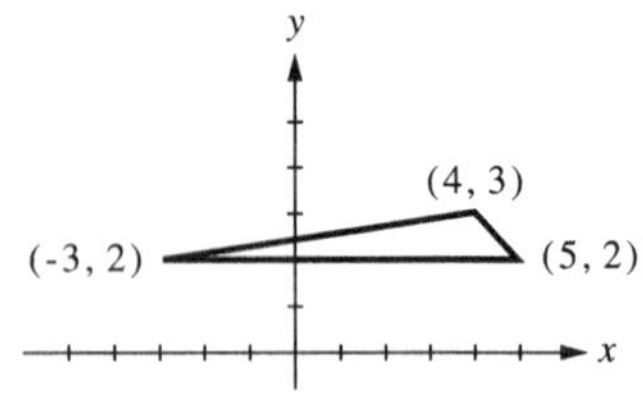

3. **(a)** $x = 2$

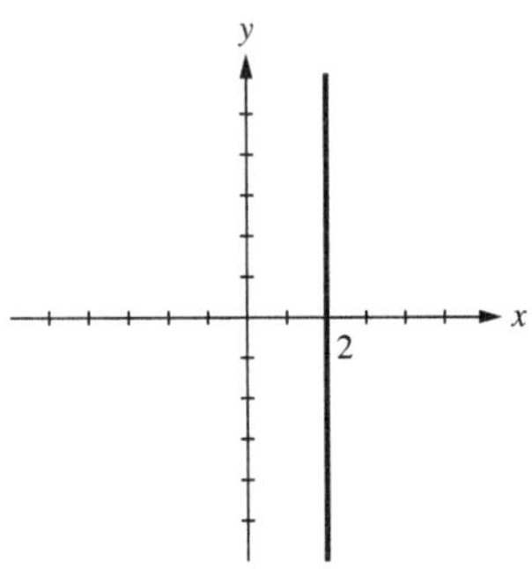

(b) $y = -3$

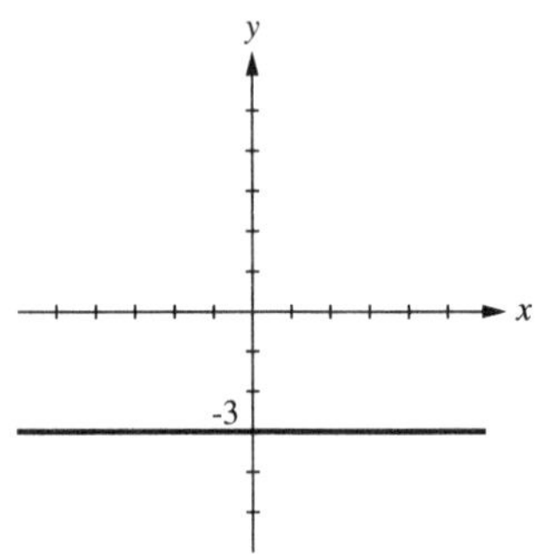

(c) $x \geq 0$

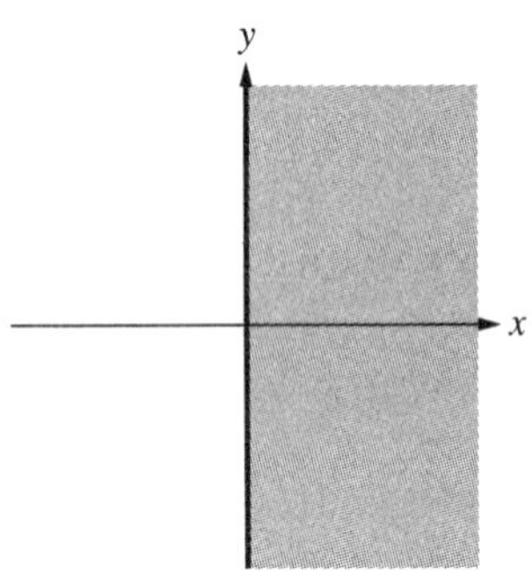

(d) $y = x$

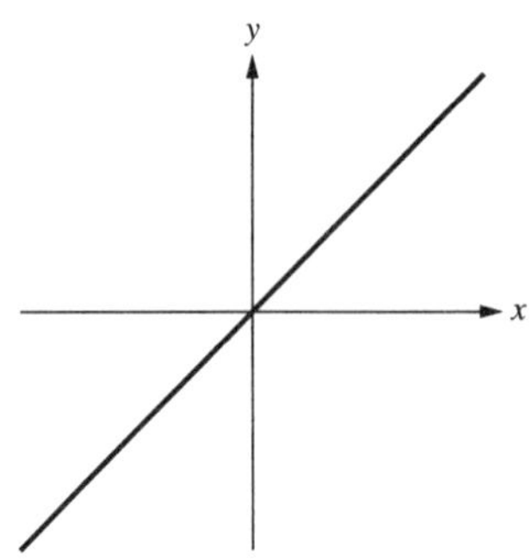

(e) $y \geq x$

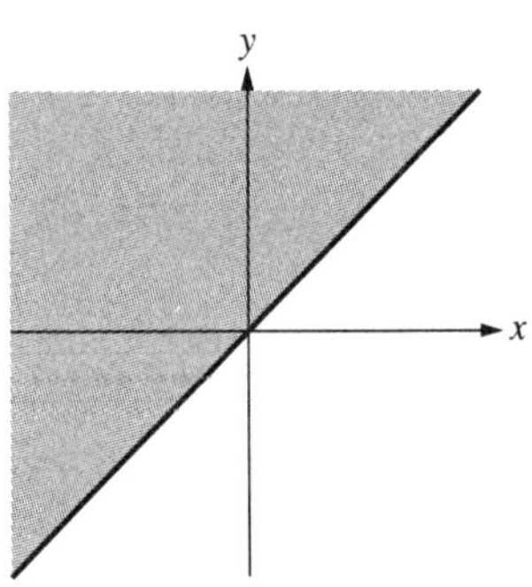

(f) $|x| \geq 1$

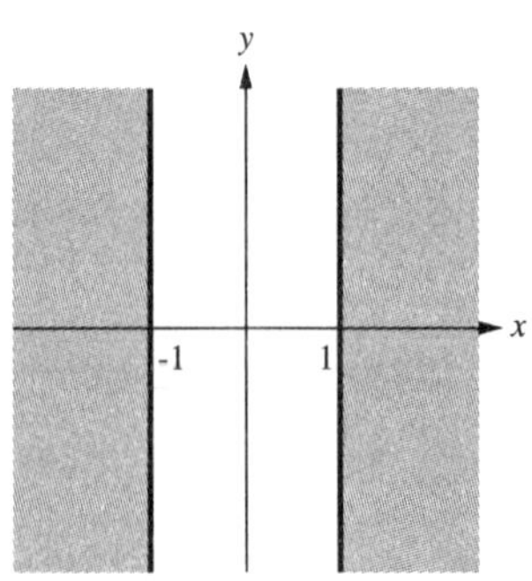

4. (a) $x = 0$

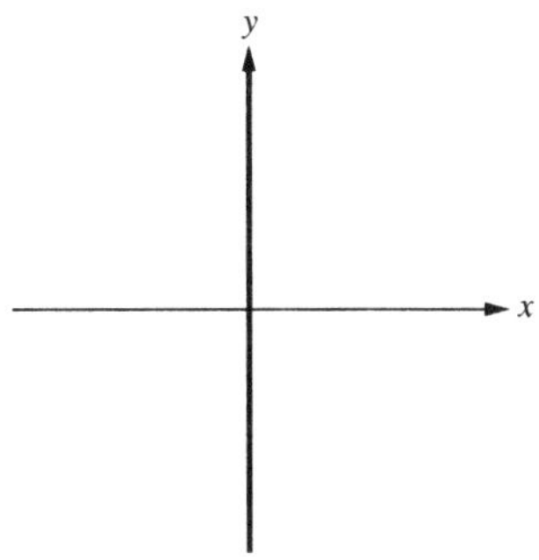

(b) $y = 0$

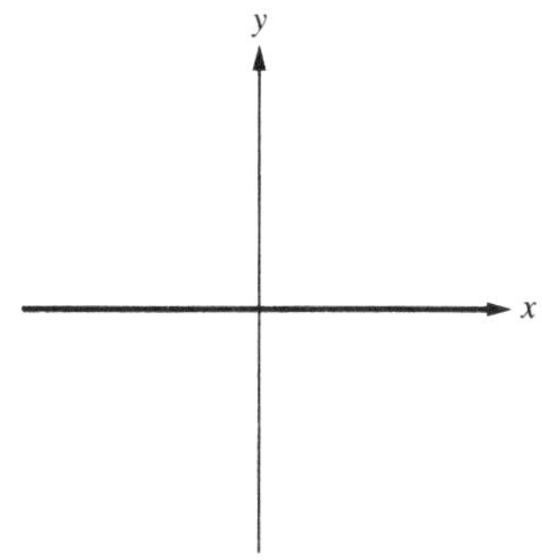

(c) $y < 0$

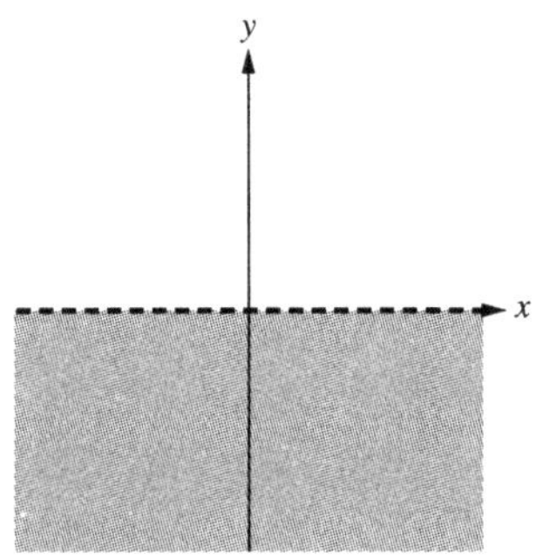

(d) $x \geq 1$ and $y \leq 2$

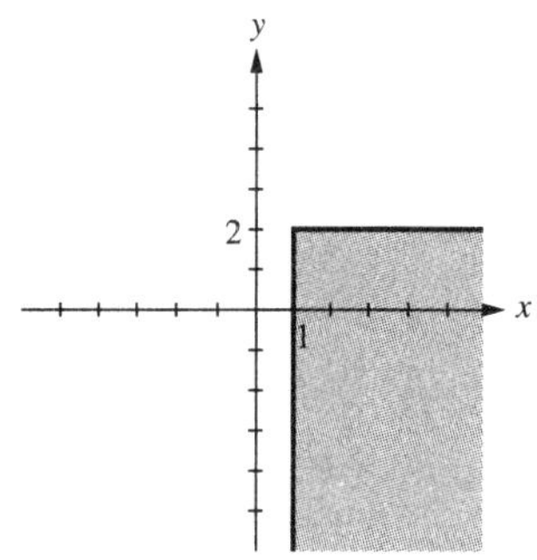

(e) $x = 3$

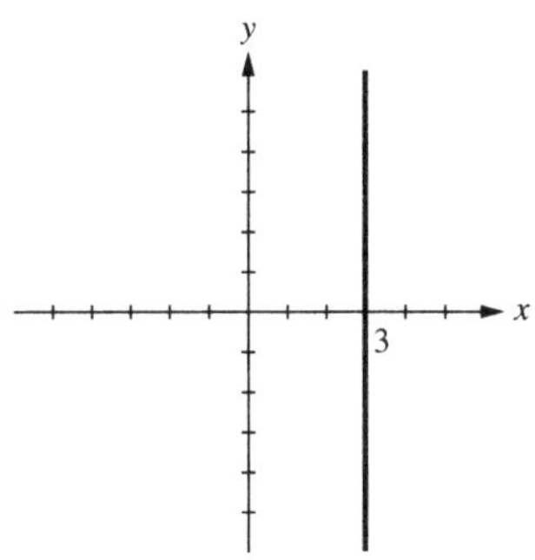

(f) $|x| = 5$

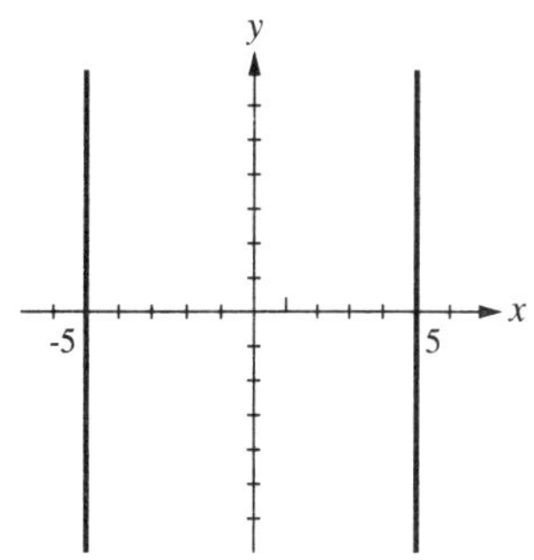

5. $y = 4 - x^2$

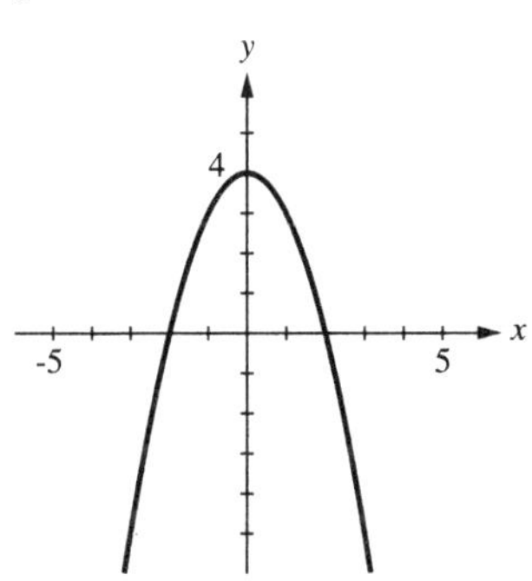

6. $y = 1 + x^2$

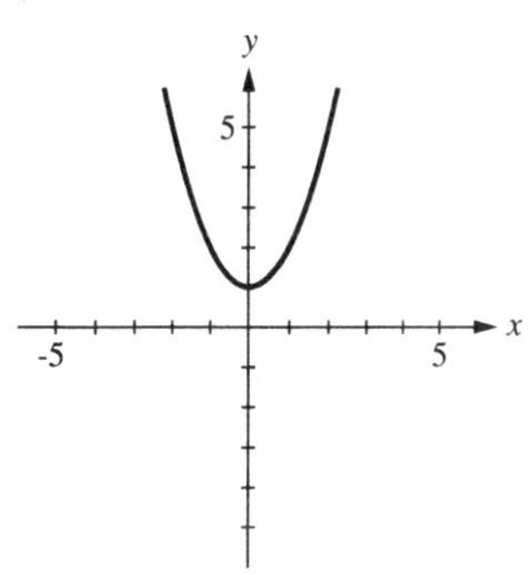

7. $y = \sqrt{x-4}$

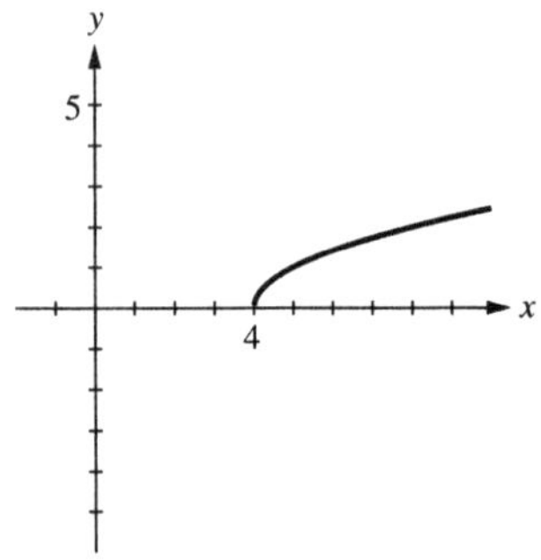

8. $y = -\sqrt{x+1}$

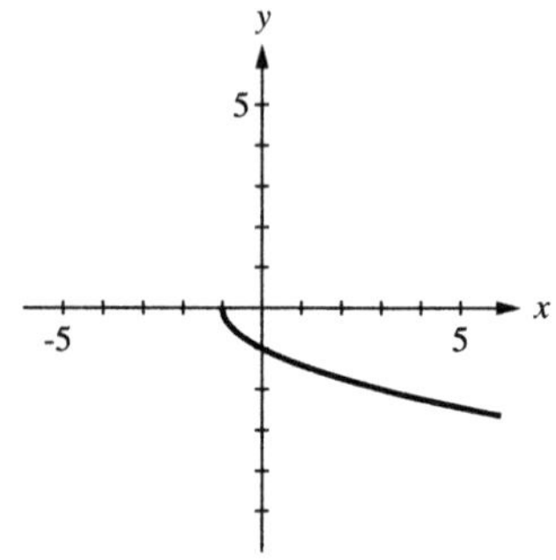

9. $x^2 - x + y = 0$

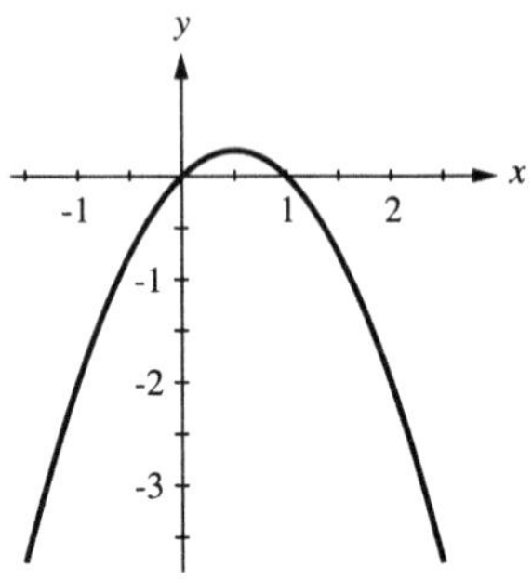

10. $x = y^3 - y^2$

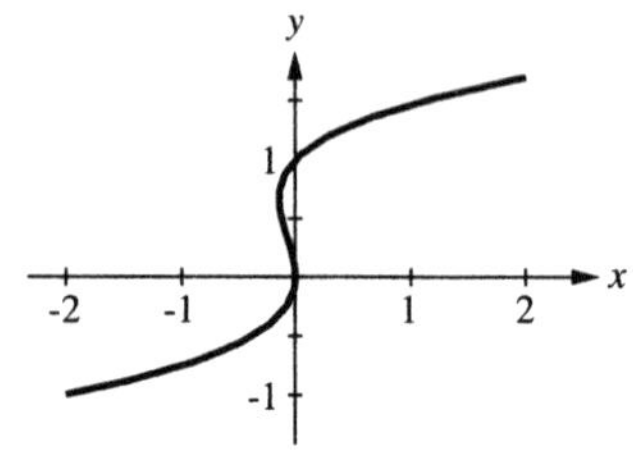

11. $x^2 y = 2$

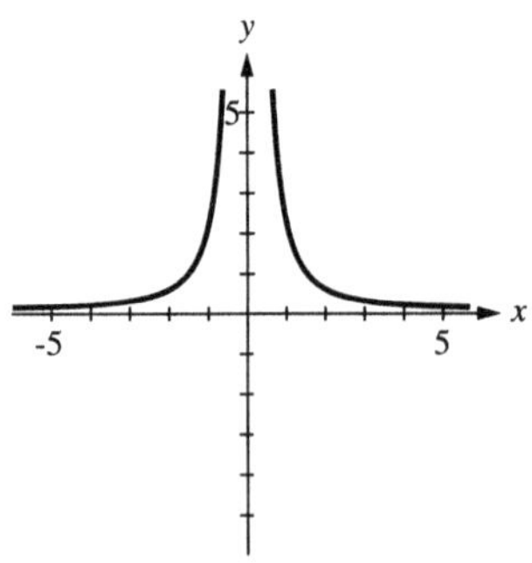

12. $xy = -1$

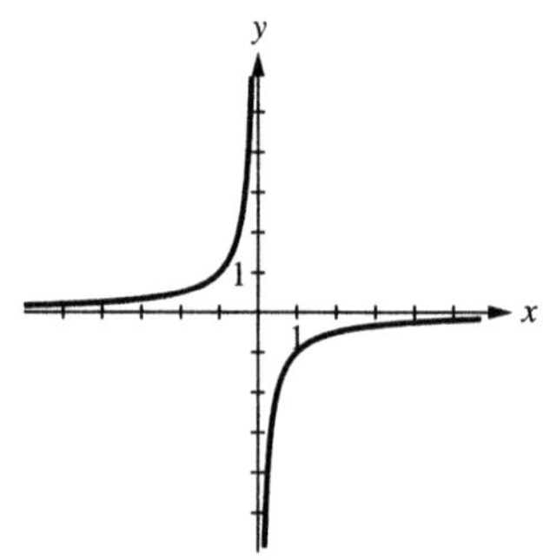

13. **(a)** $m = \dfrac{4-2}{3-(-1)} = \dfrac{1}{2}$ **(b)** $m = \dfrac{1-3}{7-5} = -1$

(c) $m = \dfrac{\sqrt{2}-\sqrt{2}}{-3-4} = 0$ **(d)** $m = \dfrac{12-(-6)}{-2-(-2)} = \dfrac{18}{0}$, not defined

14. $m_1 = \dfrac{5-2}{6-(-1)} = \dfrac{3}{7}$, $m_2 = \dfrac{7-2}{2-(-1)} = \dfrac{5}{3}$, $m_3 = \dfrac{7-5}{2-6} = -\dfrac{1}{2}$

15. **(a)** The line through $(1,1)$ and $(-2,-5)$ has slope $m_1 = \dfrac{-5-1}{-2-1} = 2$, the line through $(1,1)$ and $(0,-1)$ has slope $m_2 = \dfrac{-1-1}{0-1} = 2$. The given points lie on a line because $m_1 = m_2$.

(b) The line through $(-2,4)$ and $(0,2)$ has slope $m_1 = \dfrac{2-4}{0+2} = -1$, the line through $(-2,4)$ and $(1,5)$ has slope $m_2 = \dfrac{5-4}{1+2} = \dfrac{1}{3}$. The given points do not lie on a line because $m_1 \neq m_2$.

16. 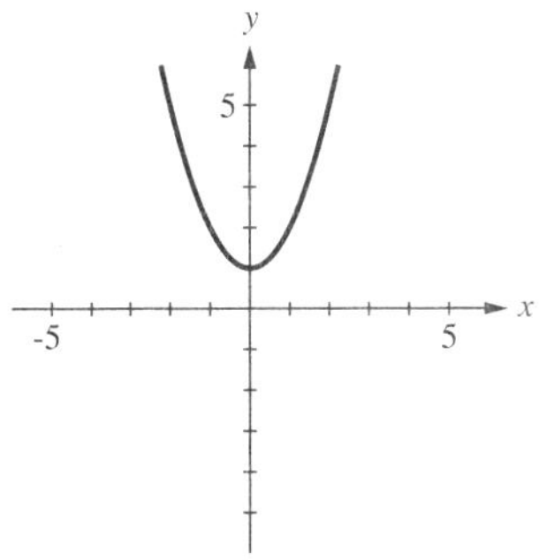

17. 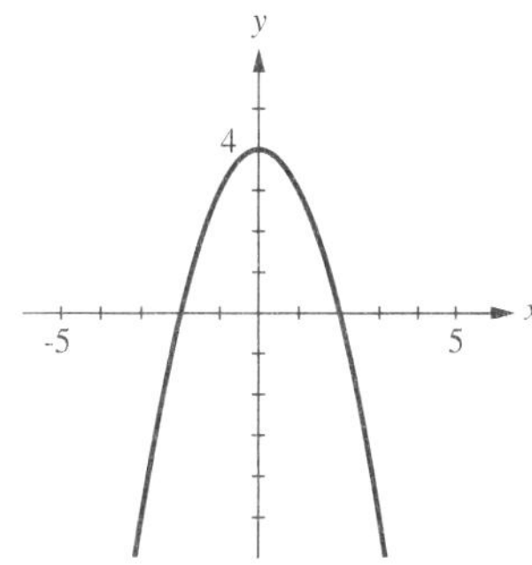

18. The triangle is equiangular because it is equilateral. The angles of inclination of the sides are $0°$, $60°$, and $120°$ (see figure), thus the slopes of its sides are $\tan 0° = 0$, $\tan 60° = \sqrt{3}$, and $\tan 120° = -\sqrt{3}$.

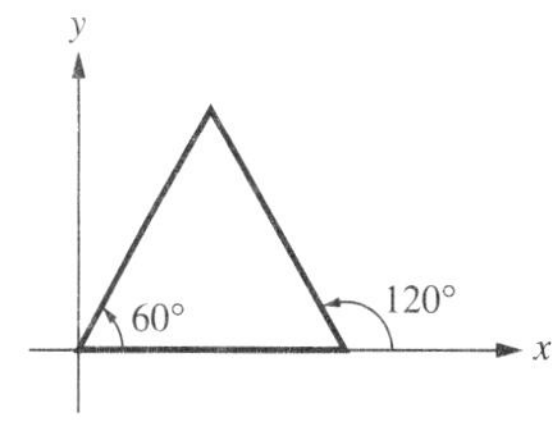

19. III < II < IV < I

20. III < IV < I < II

21. Use the points $(1, 2)$ and (x, y) to calculate the slope: $(y-2)/(x-1) = 3$
(a) if $x = 5$, then $(y-2)/(5-1) = 3$, $y - 2 = 12$, $y = 14$
(b) if $y = -2$, then $(-2-2)/(x-1) = 3$, $x - 1 = -4/3$, $x = -1/3$

22. Use $(7, 5)$ and (x, y) to calculate the slope: $(y-5)/(x-7) = -2$
(a) if $x = 9$, then $(y-5)/(9-7) = -2$, $y - 5 = -4$, $y = 1$
(b) if $y = 12$, then $(12-5)/(x-7) = -2$, $x - 7 = -7/2$, $x = 7/2$

23. Using $(3, k)$ and $(-2, 4)$ to calculate the slope, we find $\dfrac{k-4}{3-(-2)} = 5$, $k - 4 = 25$, $k = 29$.

24. The slope obtained by using the points $(1, 5)$ and $(k, 4)$ must be the same as that obtained from the points $(1, 5)$ and $(2, -3)$ so $\dfrac{4-5}{k-1} = \dfrac{-3-5}{2-1}$, $-\dfrac{1}{k-1} = -8$, $k - 1 = 1/8$, $k = 9/8$.

25. $\dfrac{0-2}{x-1} = -\dfrac{0-5}{x-4}$, $-2x + 8 = 5x - 5$, $7x = 13$, $x = 13/7$

26. Use $(0, 0)$ and (x, y) to get $\dfrac{y-0}{x-0} = \dfrac{1}{2}$, $y = \dfrac{1}{2}x$. Use $(7, 5)$ and (x, y) to get $\dfrac{y-5}{x-7} = 2$, $y - 5 = 2(x-7)$, $y = 2x - 9$. Solve the system of equations $y = \dfrac{1}{2}x$ and $y = 2x - 9$ to get $x = 6$, $y = 3$.

27. Show that opposite sides are parallel by showing that they have the same slope: using $(3, -1)$ and $(6, 4)$, $m_1 = 5/3$; using $(6, 4)$ and $(-3, 2)$, $m_2 = 2/9$; using $(-3, 2)$ and $(-6, -3)$, $m_3 = 5/3$; using $(-6, -3)$ and $(3, -1)$, $m_4 = 2/9$. Opposite sides are parallel because $m_1 = m_3$ and $m_2 = m_4$.

28. The line through $(3, 1)$ and $(6, 3)$ has slope $m_1 = 2/3$, the line through $(3, 1)$ and $(2, 9)$ has slope $m_2 = -8$, the line through $(6, 3)$ and $(2, 9)$ has slope $m_3 = -3/2$. Because $m_1 m_3 = -1$, the corresponding lines are perpendicular so the given points are vertices of a right triangle.

29. **(a)**

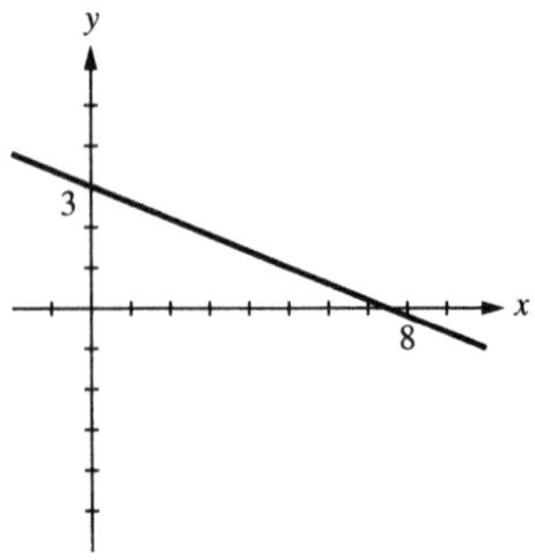

(b)

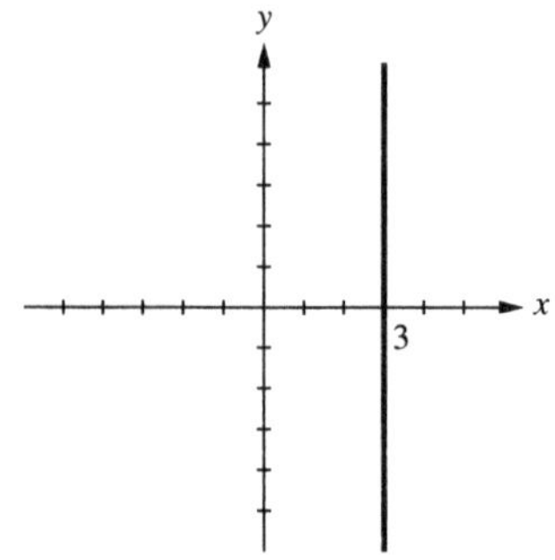

(c)

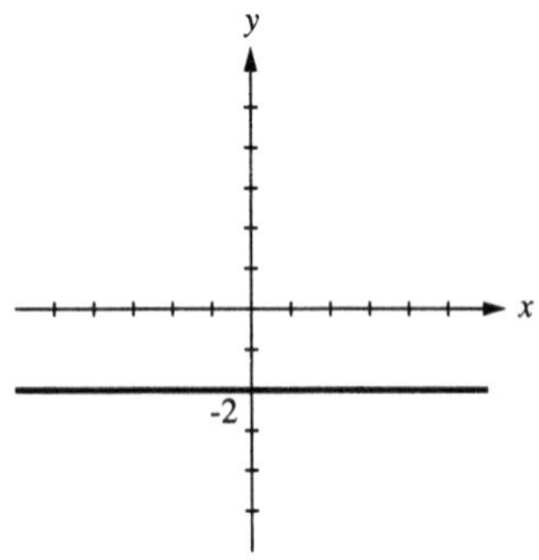

(d)

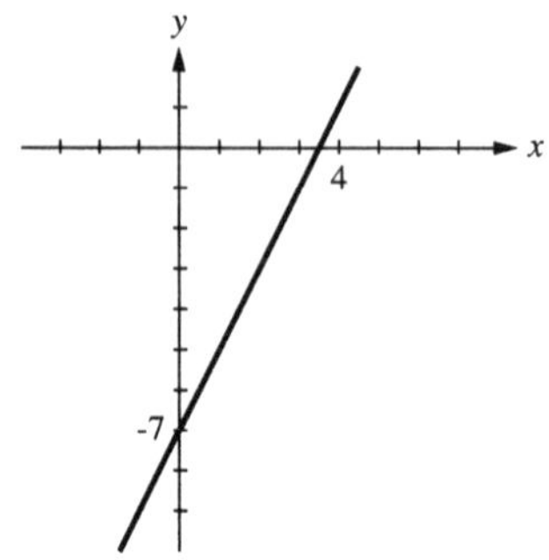

30. **(a)**

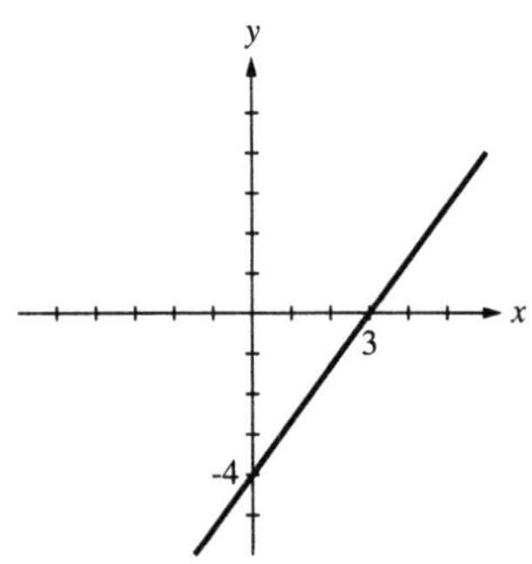

(b)

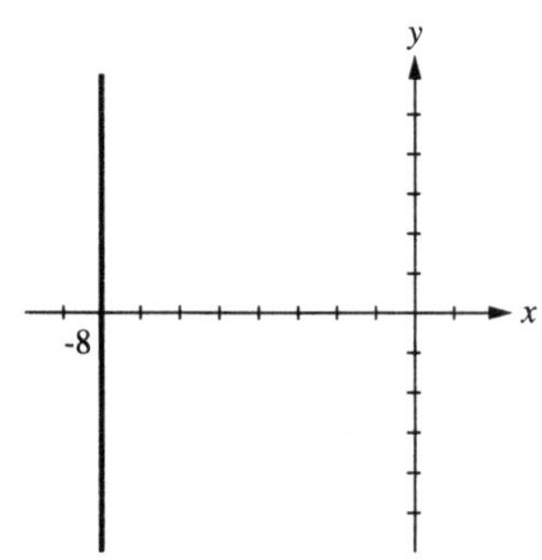

(c)

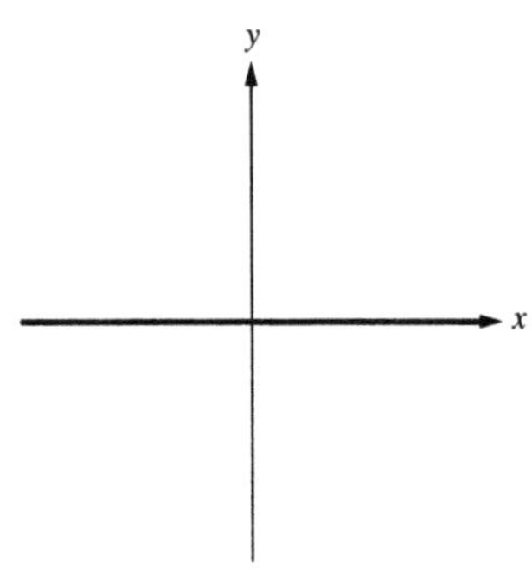

(d)

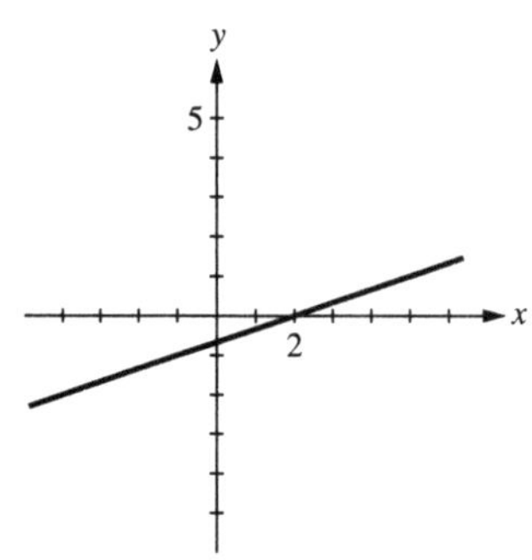

31. **(a)**

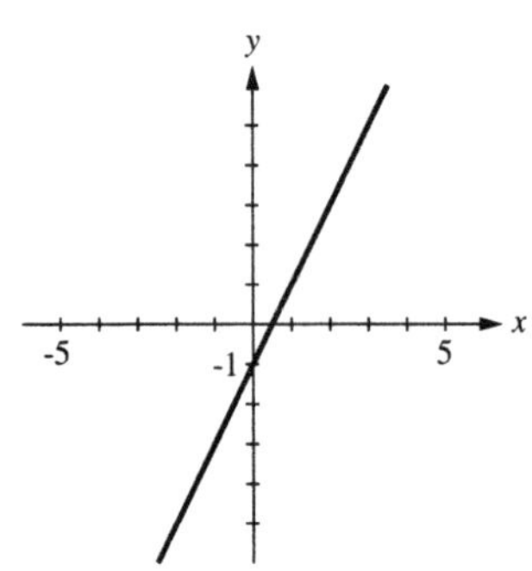

(b)

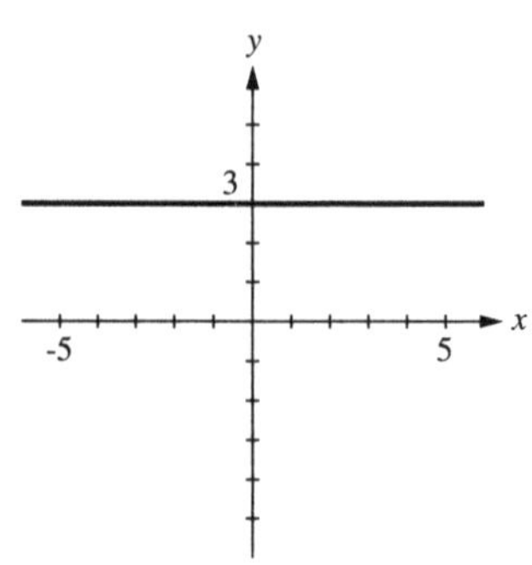

(c)

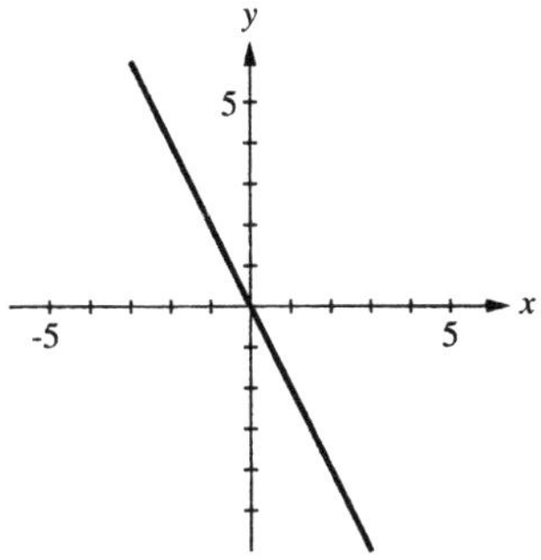

32. (a)

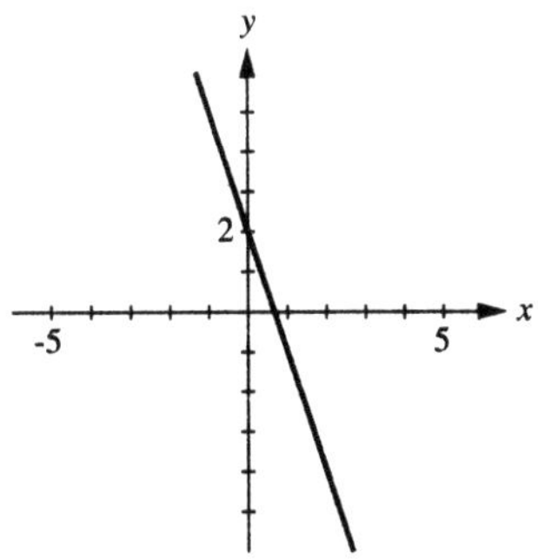

(b)

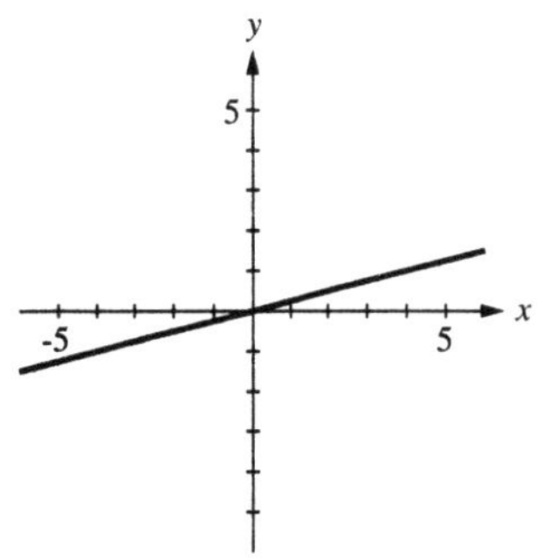

(c)

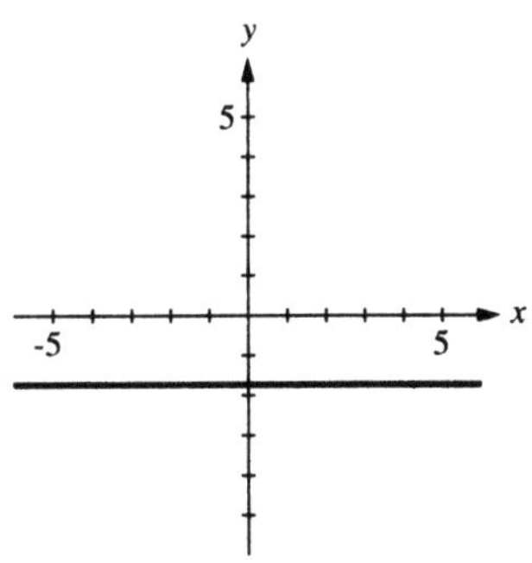

33. (a) $m = 3$, $b = 2$

(b) $m = -\dfrac{1}{4}$, $b = 3$

(c) $y = -\dfrac{3}{5}x + \dfrac{8}{5}$ so $m = -\dfrac{3}{5}$, $b = \dfrac{8}{5}$

(d) $m = 0$, $b = 1$

(c) $y = -\dfrac{b}{a}x + b$ so $m = -\dfrac{b}{a}$, y-intercept b

34. (a) $m = -4$, $b = 2$

(b) $y = \dfrac{1}{3}x - \dfrac{2}{3}$ so $m = \dfrac{1}{3}$, $b = -\dfrac{2}{3}$

(c) $y = -\dfrac{3}{2}x + 3$ so $m = -\dfrac{3}{2}$, $b = 3$

(d) $y = 3$ so $m = 0$, $b = 3$

(e) $y = -\dfrac{a_0}{a_1}x$ so $m = -\dfrac{a_0}{a_1}$, $b = 0$

35. (a) $m = (0 - (-3))/(2 - 0)) = 3/2$ so $y = 3x/2 - 3$

(b) $m = (-3 - 0)/(4 - 0) = -3/4$ so $y = -3x/4$

36. (a) $m = (0 - 2)/(2 - 0)) = -1$ so $y = -x + 2$

(b) $m = (2 - 0)/(3 - 0) = 2/3$ so $y = 2x/3$

37. $y = -2x + 4$

38. $y = 5x - 3$

39. The slope m of the line must equal the slope of $y = 4x-2$, thus $m = 4$ so the equation is $y = 4x+7$.

40. The slope of the line $3x + 2y = 5$ is $-3/2$ so the line through $(-1, 2)$ with this slope is $y - 2 = -\frac{3}{2}(x+1)$; $y = -\frac{3}{2}x + \frac{1}{2}$.

41. The slope m of the line must be the negative reciprocal of the slope of $y = 5x+9$, thus $m = -1/5$ and the equation is $y = -x/5 + 6$.

42. The slope of the line $x - 4y = 7$ is $1/4$ so a line perpendicular to it must have a slope of -4; $y + 4 = -4(x-3)$; $y = -4x + 8$.

43. $y - 4 = \dfrac{-7-4}{1-2}(x-2) = 11(x-2)$, $y = 11x - 18$.

44. $y - 6 = \dfrac{1-6}{-2-(-3)}(x-(-3))$, $y - 6 = -5(x+3)$, $y = -5x - 9$.

45. The line passes through $(0, 2)$ and $(-4, 0)$, thus $m = \dfrac{0-2}{-4-0} = \dfrac{1}{2}$ so $y = \dfrac{1}{2}x + 2$.

46. The line passes through $(0, b)$ and $(a, 0)$, thus $m = \dfrac{0-b}{a-0} = -\dfrac{b}{a}$, so the equation is $y = -\dfrac{b}{a}x + b$.

47. $y = 1$

48. $y = -8$

49. (a) $m_1 = 4$, $m_2 = 4$; parallel because $m_1 = m_2$
(b) $m_1 = 2$, $m_2 = -1/2$; perpendicular because $m_1 m_2 = -1$
(c) $m_1 = 5/3$, $m_2 = 5/3$; parallel because $m_1 = m_2$
(d) If $A \neq 0$ and $B \neq 0$, then $m_1 = -A/B$, $m_2 = B/A$ and the lines are perpendicular because $m_1 m_2 = -1$. If either A or B (but not both) is zero, then the lines are perpendicular because one is horizontal and the other is vertical.
(e) $m_1 = 4$, $m_2 = 1/4$; neither

50. (a) $m_1 = -5$, $m_2 = -5$; parallel because $m_1 = m_2$
(b) $m_1 = 2$, $m_2 = -1/2$; perpendicular because $m_1 m_2 = -1$.
(c) $m_1 = -4/5$, $m_2 = 5/4$; perpendicular because $m_1 m_2 = -1$.
(d) If $B \neq 0$, then $m_1 = m_2 = -A/B$ and the lines are parallel because $m_1 = m_2$. If $B = 0$ (and $A \neq 0$), then the lines are parallel because they are both perpendicular to the x-axis.
(e) $m_1 = 1/2$, $m_2 = 2$; neither

51. $y = (-3/k)x + 4/k$, $k \neq 0$
(a) $-3/k = 2$, $k = -3/2$
(b) $4/k - 5$, $k = 4/5$
(c) $3(-2) + k(4) = 4$, $k = 5/2$
(d) The slope of $2x - 5y = 1$ is $2/5$ so $-3/k = 2/5$, $k = -15/2$.
(e) The slope of $4x + 3y = 2$ is $-4/3$ so the slope of a line perpendicular to it is $3/4$; $-3/k = 3/4$, $k = -4$.

52. $y^2 = 3x$: the union of the graphs of $y = \sqrt{3x}$ and $y = -\sqrt{3x}$

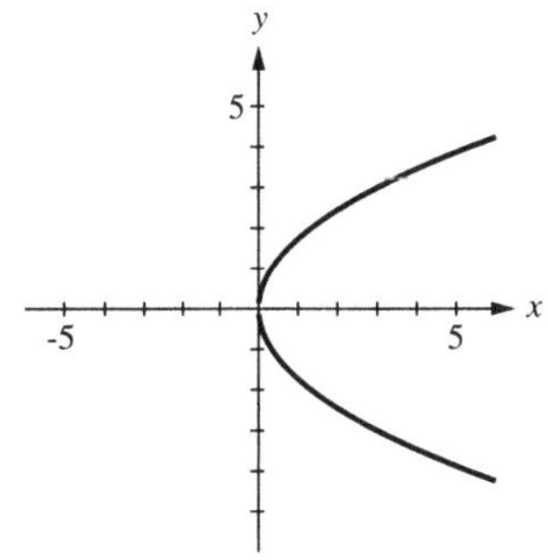

53. $(x-y)(x+y) = 0$: the union of the graphs of $x - y = 0$ and $x + y = 0$

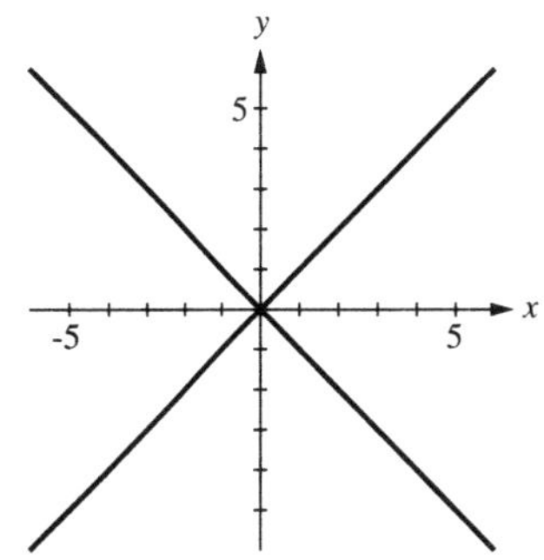

54. $F = \frac{9}{5}C + 32$

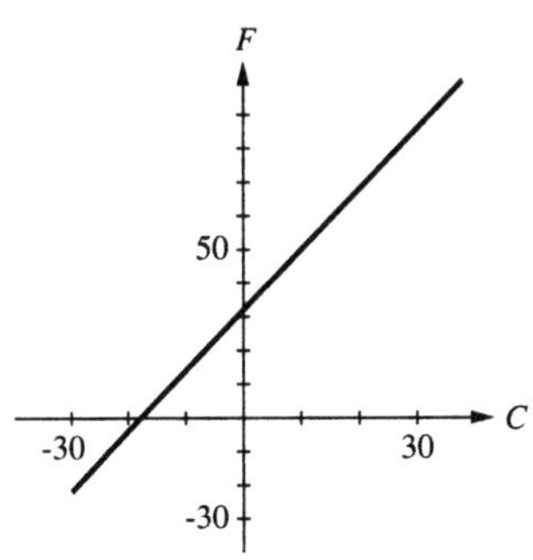

55. $u = 3v^2$

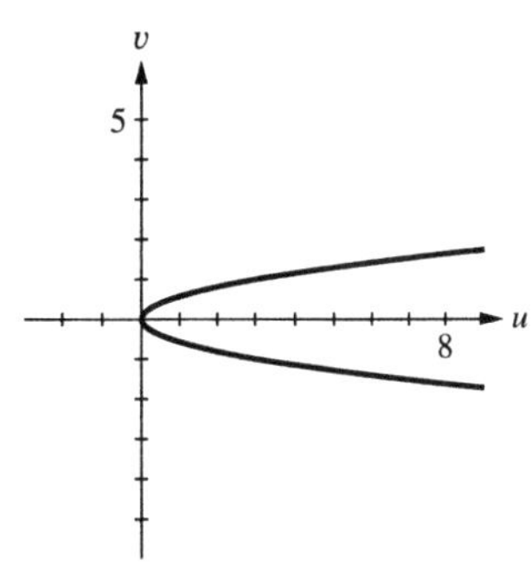

56. $Y = 4X + 5$

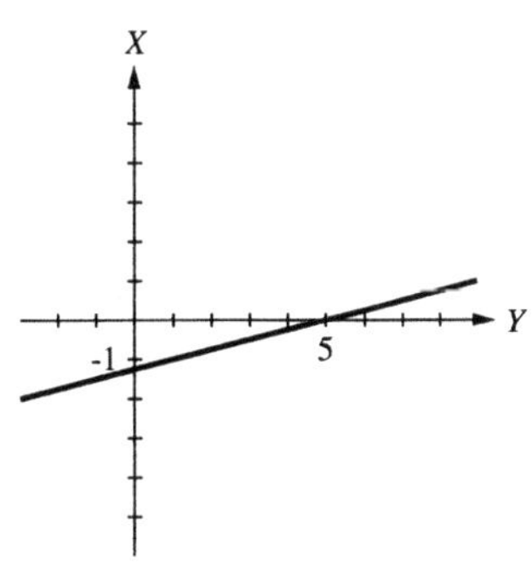

57. Solve $x = 5t + 2$ for t to get $t = \frac{1}{5}x - \frac{2}{5}$, so $y = \left(\frac{1}{5}x - \frac{2}{5}\right) - 3 = \frac{1}{5}x - \frac{17}{5}$, which is a line.

58. Solve $x = 1 + 3t^2$ for t^2 to get $t^2 = \frac{1}{3}x - \frac{1}{3}$, so $y = 2 - \left(\frac{1}{3}x - \frac{1}{3}\right) = -\frac{1}{3}x + \frac{7}{3}$, which is a line; $1 + 3t^2 \geq 1$ for all t so $x \geq 1$.

59. An equation of the line through $(1, 4)$ and $(2, 1)$ is $y = -3x + 7$. It crosses the y-axis at $y = 7$, and the x-axis at $x = 7/3$, so the area of the triangle is $\frac{1}{2}(7)(7/3) = 49/6$.

60. $(2x - 3y)(2x + 3y) = 0$, so
$2x - 3y = 0$, $y = \frac{2}{3}x$ or $2x + 3y = 0$,
$y = -\frac{2}{3}x$. The graph consists of the lines $y = \pm\frac{2}{3}x$.

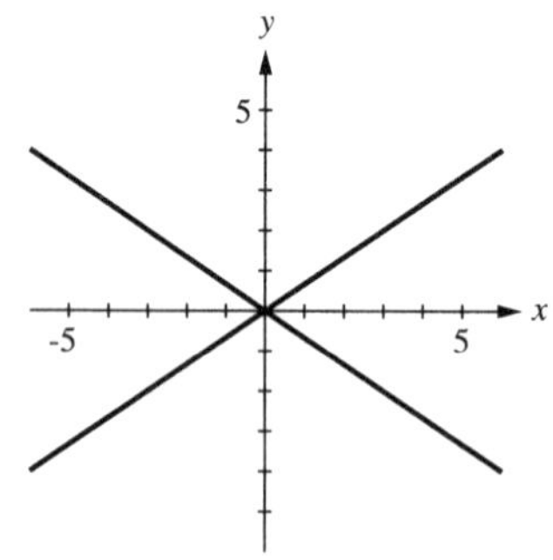

61. **(a)** yes **(b)** yes **(c)** no **(d)** yes
(e) yes **(f)** yes **(g)** no

APPENDIX D
Distances, Circles, and Quadratic Equations

EXERCISE SET D

1. in the proof of Theorem D.1

2. (a) $d = \sqrt{(-1-2)^2 + (1-5)^2} = \sqrt{9+16} = \sqrt{25} = 5$

(b) $\left(\dfrac{2+(-1)}{2}, \dfrac{5+1}{2}\right) = (1/2, 3)$

3. (a) $d = \sqrt{(1-7)^2 + (9-1)^2} = \sqrt{36+64} = \sqrt{100} = 10$

(b) $\left(\dfrac{7+1}{2}, \dfrac{1+9}{2}\right) = (4, 5)$

4. (a) $d = \sqrt{(-3-2)^2 + (6-0)^2} = \sqrt{25+36} = \sqrt{61}$

(b) $\left(\dfrac{2+(-3)}{2}, \dfrac{0+6}{2}\right) = (-1/2, 3)$

5. (a) $d = \sqrt{[-7-(-2)]^2 + [-4-(-6)]^2} = \sqrt{25+4} = \sqrt{29}$

(b) $\left(\dfrac{-2+(-7)}{2}, \dfrac{-6+(-4)}{2}\right) = (-9/2, -5)$

6. Let $A(1,1)$, $B(-2,-8)$, and $C(4,10)$ be the given points (see diagram). A, B, and C lie on a straight line if and only if $d_1 + d_2 = d_3$, where d_1, d_2, and d_3 are the lengths of the line segments AB, AC, and BC. But

$d_1 = \sqrt{(-2-1)^2 + (-8-1)^2} = 3\sqrt{10}$,

$d_2 = \sqrt{(4-1)^2 + (10-1)^2} = 3\sqrt{10}$,

$d_3 = \sqrt{(4+2)^2 + (10+8)^2} = 6\sqrt{10}$; because $d_1 + d_2 = d_3$,

it follows that A, B, and C lie on a straight line.

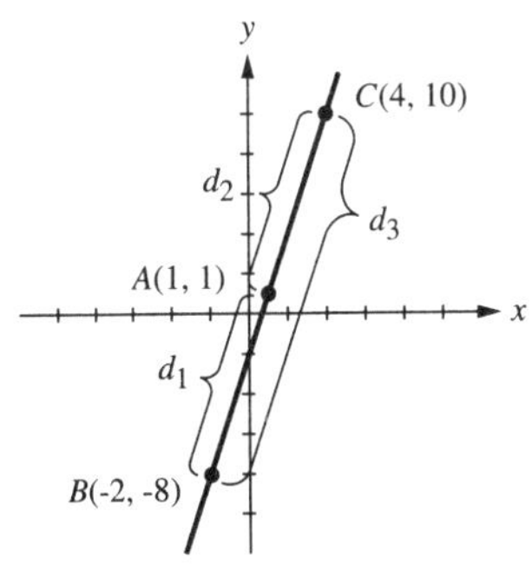

7. Let $A(5,-2)$, $B(6,5)$, and $C(2,2)$ be the given vertices and a, b, and c the lengths of the sides opposite these vertices; then

$a = \sqrt{(2-6)^2 + (2-5)^2} = \sqrt{25} = 5$ and $b = \sqrt{(2-5)^2 + (2+2)^2} = \sqrt{25} = 5$.

Triangle ABC is isosceles because it has two equal sides ($a = b$).

8. A triangle is a right triangle if and only if the square of the longest side is equal to the sum of the squares of the other two sides (Pythagorean theorem). With $A(1,3)$, $B(4,2)$, and $C(-2,-6)$ as vertices and s_1, s_2, and s_3 the lengths of the sides opposite these vertices we find that $s_1^2 = (-2-4)^2 + (-6-2)^2 = 100$, $s_2^2 = (-2-1)^2 + (-6-3)^2 = 90$, $s_3^2 = (4-1)^2 + (2-3)^2 = 10$, and that $s_1^2 = s_2^2 + s_3^2$, so ABC is a right triangle. The right angle occurs at the vertex $A(1,3)$.

9. $P_1(0,-2)$, $P_2(-4,8)$, and $P_3(3,1)$ all lie on a circle whose center is $C(-2,3)$ if the points P_1, P_2 and P_3 are equidistant from C. Denoting the distances between P_1, P_2, P_3 and C by d_1, d_2 and d_3 we find that $d_1 = \sqrt{(0+2)^2 + (-2-3)^2} = \sqrt{29}$, $d_2 = \sqrt{(-4+2)^2 + (8-3)^2} = \sqrt{29}$, and $d_3 = \sqrt{(3+2)^2 + (1-3)^2} = \sqrt{29}$, so P_1, P_2 and P_3 lie on a circle whose center is $C(-2,3)$ because $d_1 = d_2 = d_3$.

10. The distance between $(t, 2t-6)$ and $(0,4)$ is
$\sqrt{(t-0)^2+(2t-6-4)^2} = \sqrt{t^2+(2t-10)^2} = \sqrt{5t^2-40t+100}$;
the distance between $(t, 2t-6)$ and $(8,0)$ is $\sqrt{(t-8)^2+(2t-6)^2} = \sqrt{5t^2-40t+100}$,
so $(t, 2t-6)$ is equidistant from $(0,4)$ and $(8,0)$.

11. If $(2, k)$ is equidistant from (3,7) and (9,1), then
$\sqrt{(2-3)^2+(k-7)^2} = \sqrt{(2-9)^2+(k-1)^2}$, $1+(k-7)^2 = 49+(k-1)^2$,
$1+k^2-14k+49 = 49+k^2-2k+1$, $-12k=0$, $k=0$.

12. $(x-3)/2 = 4$ and $(y+2)/2 = -5$ so $x = 11$ and $y = -12$.

13. The slope of the line segment joining $(2,8)$ and $(-4,6)$ is $\dfrac{6-8}{-4-2} = \dfrac{1}{3}$ so the slope of the perpendicular bisector is -3. The midpoint of the line segment is $(-1,7)$ so an equation of the bisector is $y-7 = -3(x+1)$; $y = -3x+4$.

14. The slope of the line segment joining $(5,-1)$ and $(4,8)$ is $\dfrac{8-(-1)}{4-5} = -9$ so the slope of the perpendicular bisector is $\dfrac{1}{9}$. The midpoint of the line segment is $(9/2, 7/2)$ so an equation of the bisector is
$y - \dfrac{7}{2} = \dfrac{1}{9}\left(x - \dfrac{9}{2}\right)$; $y = \dfrac{1}{9}x + 3$.

15. Method (see figure): Find an equation of the perpendicular bisector of the line segment joining $A(3,3)$ and $B(7,-3)$. All points on this perpendicular bisector are equidistant from A and B, thus find where it intersects the given line.

The midpoint of AB is $(5,0)$, the slope of AB is $-3/2$ thus the slope of the perpendicular bisector is $2/3$ so an equation is

$$y - 0 = \frac{2}{3}(x-5)$$
$$3y = 2x - 10$$
$$2x - 3y - 10 = 0.$$

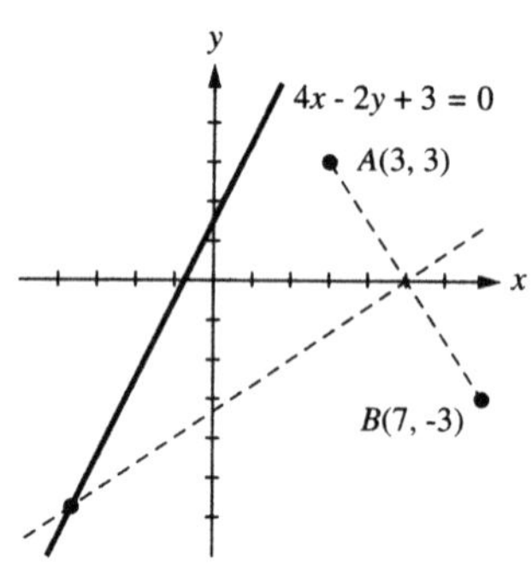

The solution of the system

$$\begin{cases} 4x - 2y + 3 = 0 \\ 2x - 3y - 10 = 0 \end{cases}$$

gives the point $(-29/8, -23/4)$.

16. **(a)** $y = 4$ is a horizontal line, so the vertical distance is $|4-(-2)| = |6| = 6$.
(b) $x = -1$ is a vertical line, so the horizontal distance is $|-1-3| = |-4| = 4$.

17. Method (see figure): write an equation of the line that goes through the given point and that is perpendicular to the given line; find the point P where this line intersects the given line; find the distance between P and the given point.

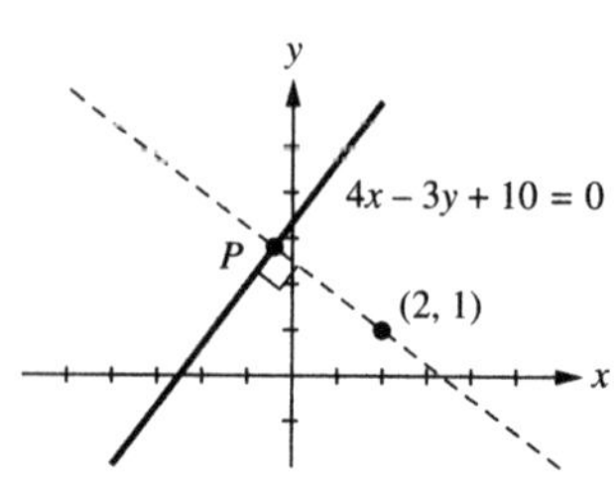

The slope of the given line is $4/3$, so the slope of a line perpendicular to it is $-3/4$.

The line through $(2,1)$ having a slope of $-3/4$ is $y-1=-\frac{3}{4}(x-2)$ or, after simplification, $3x+4y=10$ which when solved simultaneously with $4x-3y+10=0$ yields $(-2/5,14/5)$ as the point of intersection. The distance d between $(-2/5,14/5)$ and $(2,1)$ is $d=\sqrt{(2+2/5)^2+(1-14/5)^2}=3$.

18. (See the solution to Exercise 17 for a description of the method.) The slope of the line $5x+12y-36=0$ is $-5/12$. The line through $(8,4)$ and perpendicular to the given line is $y-4=\frac{12}{5}(x-8)$ or, after simplification, $12x-5y=76$. The point of intersection of this line with the given line is found to be $\left(\frac{84}{13},\frac{4}{13}\right)$ and the distance between it and $(8,4)$ is 4.

19. If $B=0$, then the line $Ax+C=0$ is vertical and $x=-C/A$ for each point on the line. The line through (x_0,y_0) and perpendicular to the given line is horizontal and intersects the given line at the point $(-C/A,\ y_0)$. The distance d between $(-C/A,\ y_0)$ and (x_0,y_0) is

$$d=\sqrt{(x_0+C/A)^2+(y_0-y_0)^2}=\sqrt{\frac{(Ax_0+C)^2}{A^2}}=\frac{|Ax_0+C|}{\sqrt{A^2}}$$

which is the value of $\dfrac{|Ax_0+By_0+C|}{\sqrt{A^2+B^2}}$ for $B=0$.

If $B\neq 0$, then the slope of the given line is $-A/B$ and the line through (x_0,y_0) and perpendicular to the given line is

$y-y_0=\dfrac{B}{A}(x-x_0)$, $Ay-Ay_0=Bx-Bx_0$, $Bx-Ay=Bx_0-Ay_0$.

The point of intersection of this line and the given line is obtained by solving

$Ax+By=-C$ and $Bx-Ay=Bx_0-Ay_0$.

Multiply the first equation through by A and the second by B and add the results to get

$(A^2+B^2)x=B^2x_0-ABy_0-AC$ so $x=\dfrac{B^2x_0-ABy_0-AC}{A^2+B^2}$

Similarly, by multiplying by B and $-A$, we get $y=\dfrac{-ABx_0+A^2y_0-BC}{A^2+B^2}$.

The square of the distance d between (x,y) and (x_0,y_0) is

$$\begin{aligned}
d^2&=\left[x_0-\frac{B^2x_0-ABy_0-AC}{A^2+B^2}\right]^2+\left[y_0-\frac{-ABx_0+A^2y_0-BC}{A^2+B^2}\right]^2\\
&=\frac{(A^2x_0+ABy_0+AC)^2}{(A^2+B^2)^2}+\frac{(ABx_0+B^2y_0+BC)^2}{(A^2+B^2)^2}\\
&=\frac{A^2(Ax_0+By_0+C)^2+B^2(Ax_0+By_0+C)^2}{(A^2+B^2)^2}\\
&=\frac{(Ax_0+By_0+C)^2(A^2+B^2)}{(A^2+B^2)^2}=\frac{(Ax_0+By_0+C)^2}{A^2+B^2}
\end{aligned}$$

so $d=\dfrac{|Ax_0+By_0+C|}{\sqrt{A^2+B^2}}$.

20. $d=\dfrac{|4(2)-3(1)+10|}{\sqrt{4^2+(-3)^2}}=\dfrac{|15|}{\sqrt{25}}=\dfrac{15}{5}=3$.

21. $d=\dfrac{|5(8)+12(4)-36|}{\sqrt{5^2+12^2}}=\dfrac{|52|}{\sqrt{169}}=\dfrac{52}{13}=4$.

22. Method (see figure): Let $A(0,a)$, $B(b,0)$, and $C(c,0)$ be the given vertices; find equations for the perpendicular bisectors L_1, L_2, and L_3 and show that they all intersect at the same point.

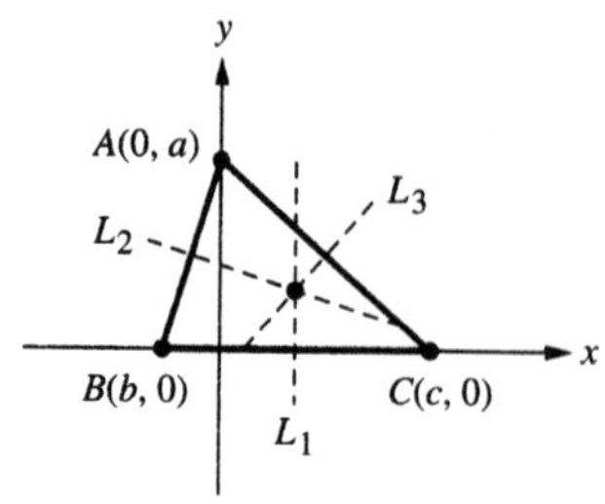

line L_1: The midpoint of BC is $\left(\dfrac{b+c}{2}, 0\right)$ and since L_1 is vertical, an equation for L_1 is $x = \dfrac{b+c}{2}$;

line L_2: The midpoint of AB is $\left(\dfrac{b}{2}, \dfrac{a}{2}\right)$; the slope of AB is $-\dfrac{a}{b}$ (if $b \neq 0$) so the slope of L_2 is $\dfrac{b}{a}$ (even if $b = 0$) and an equation of L_2 is $y - \dfrac{a}{2} = \dfrac{b}{a}\left(x - \dfrac{b}{2}\right)$;

line L_3: The midpoint of AC is $\left(\dfrac{c}{2}, \dfrac{a}{2}\right)$; the slope of AC is $-\dfrac{a}{c}$ (if $c \neq 0$) so the slope of L_3 is $\dfrac{c}{a}$ (even if $c = 0$) and an equation of L_3 is $y - \dfrac{a}{2} = \dfrac{c}{a}\left(x - \dfrac{c}{2}\right)$.

For the point of intersection of L_1 and L_2, solve $x = \dfrac{b+c}{2}$ and $y - \dfrac{a}{2} = \dfrac{b}{a}\left(x - \dfrac{b}{2}\right)$.

The point is found to be $\left(\dfrac{b+c}{2}, \dfrac{a^2+bc}{2a}\right)$. The point of intersection of L_1 and L_3 is obtained by solving the system $x = \dfrac{b+c}{2}$ and $y - \dfrac{a}{2} = \dfrac{c}{a}\left(x - \dfrac{c}{2}\right)$, its solution yields the point $\left(\dfrac{b+c}{2}, \dfrac{a^2+bc}{2a}\right)$.

So L_1, L_2, and L_3 all intersect at the same point.

23. **(a)** center (0,0), radius 5 **(b)** center (1,4), radius 4 **(c)** center $(-1,-3)$, radius $\sqrt{5}$ **(d)** center $(0,-2)$, radius 1

24. **(a)** center $(0,0)$, radius 3 **(b)** center $(3,5)$, radius 6 **(c)** center $(-4,-1)$, radius $\sqrt{8}$ **(d)** center $(-1,0)$, radius 1

25. $(x-3)^2 + (y-(-2))^2 = 4^2$, $(x-3)^2 + (y+2)^2 = 16$

26. $(x-1)^2 + (y-0)^2 = (\sqrt{8}/2)^2$, $(x-1)^2 + y^2 = 2$

27. $r = 8$ because the circle is tangent to the x-axis, so $(x+4)^2 + (y-8)^2 = 64$.

28. $r = 5$ because the circle is tangent to the y-axis, so $(x-5)^2 + (y-8)^2 = 25$.

29. $(0,0)$ is on the circle, so $r = \sqrt{(-3-0)^2 + (-4-0)^2} = 5$; $(x+3)^2 + (y+4)^2 = 25$.

30. $r = \sqrt{(4-1)^2 + (-5-3)^2} = \sqrt{73}$; $(x-4)^2 + (y+5)^2 = 73$.

31. The center is the midpoint of the line segment joining $(2,0)$ and $(0,2)$ so the center is at $(1,1)$. The radius is $r = \sqrt{(2-1)^2 + (0-1)^2} = \sqrt{2}$, so $(x-1)^2 + (y-1)^2 = 2$.

32. The center is the midpoint of the line segment joining $(6,1)$ and $(-2,3)$, so the center is at $(2,2)$. The radius is $r = \sqrt{(6-2)^2 + (1-2)^2} = \sqrt{17}$, so $(x-2)^2 + (y-2)^2 = 17$.

33. $(x^2 - 2x) + (y^2 - 4y) = 11$, $(x^2 - 2x + 1) + (y^2 - 4y + 4) = 11 + 1 + 4$, $(x-1)^2 + (y-2)^2 = 16$; center (1,2) and radius 4

34. $(x^2+8x)+y^2=-8$, $(x^2+8x+16)+y^2=-8+16$, $(x+4)^2+y^2=8$; center $(-4,0)$ and radius $2\sqrt{2}$

35. $2(x^2+2x)+2(y^2-2y)=0$, $2(x^2+2x+1)+2(y^2-2y+1)=2+2$, $(x+1)^2+(y-1)^2=2$; center $(-1,1)$ and radius $\sqrt{2}$

36. $6(x^2-x)+6(y^2+y)=3$, $6(x^2-x+1/4)+6(y^2+y+1/4)=3+6/4+6/4$, $(x-1/2)^2+(y+1/2)^2=1$; center $(1/2,-1/2)$ and radius 1

37. $(x^2+2x)+(y^2+2y)=-2$, $(x^2+2x+1)+(y^2+2y+1)=-2+1+1$, $(x+1)^2+(y+1)^2=0$; the point $(-1,-1)$

38. $(x^2-4x)+(y^2-6y)=-13$, $(x^2-4x+4)+(y^2-6y+9)=-13+4+9$, $(x-2)^2+(y-3)^2=0$; the point $(2,3)$

39. $x^2+y^2=1/9$; center $(0,0)$ and radius $1/3$

40. $x^2+y^2=4$; center (0,0) and radius 2

41. $x^2+(y^2+10y)=-26$, $x^2+(y^2+10y+25)=-26+25$, $x^2+(y+5)^2=-1$; no graph

42. $(x^2-10x)+(y^2-2y)=-29$, $(x^2-10x+25)+(y^2-2y+1)=-29+25+1$, $(x-5)^2+(y-1)^2=-3$; no graph

43. $16\left(x^2+\frac{5}{2}x\right)+16(y^2+y)=7$, $16\left(x^2+\frac{5}{2}x+\frac{25}{16}\right)+16\left(y^2+y+\frac{1}{4}\right)=7+25+4$,
$(x+5/4)^2+(y+1/2)^2=9/4$; center $(-5/4,-1/2)$ and radius $3/2$

44. $4(x^2-4x)+4(y^2-6y)=9$, $4(x^2-4x+4)+4(y^2-6y+9)=9+16+36$, $(x-2)^2+(y-3)^2=61/4$; center $(2,3)$ and radius $\sqrt{61}/2$

45. (a) $y^2=16-x^2$, so $y=\pm\sqrt{16-x^2}$. The bottom half is $y=-\sqrt{16-x^2}$.
(b) Complete the square in y to get $(y-2)^2=3-2x-x^2$, so $y-2=\pm\sqrt{3-2x-x^2}$, or $y=2\pm\sqrt{3-2x-x^2}$. The top half is $y=2+\sqrt{3-2x-x^2}$.

46. (a) $x^2=9-y^2$ so $x=\pm\sqrt{9-y^2}$. The right half is $x=\sqrt{9-y^2}$.
(b) Complete the square in x to get $(x-2)^2=1-y^2$ so $x-2=\pm\sqrt{1-y^2}$, $x=2\pm\sqrt{1-y^2}$. The left half is $x=2-\sqrt{1-y^2}$.

47. (a)

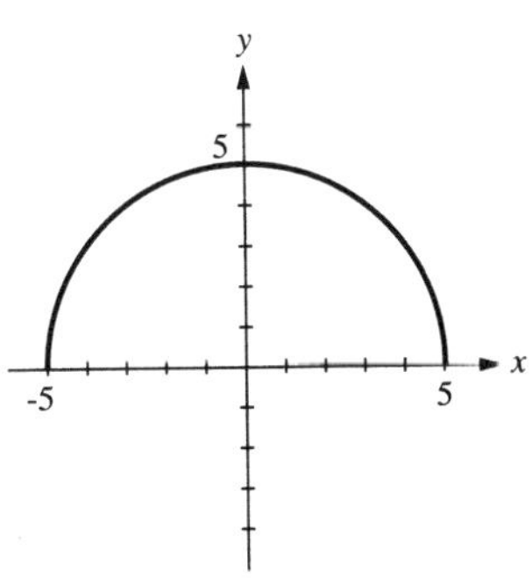

(b)
$$\begin{aligned} y &= \sqrt{5+4x-x^2} \\ &= \sqrt{5-(x^2-4x)} \\ &= \sqrt{5+4-(x^2-4x+4)} \\ &= \sqrt{9-(x-2)^2} \end{aligned}$$

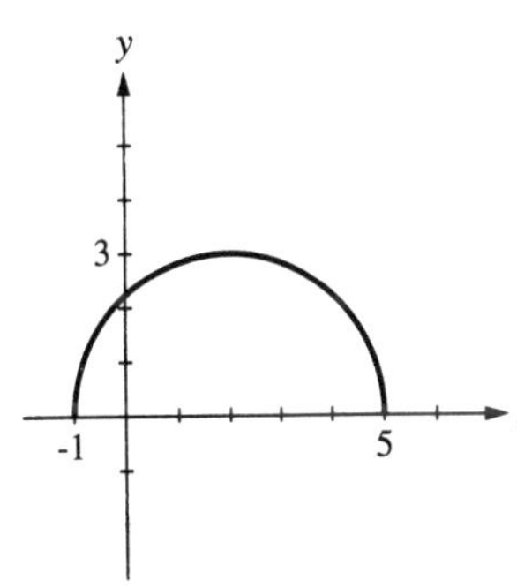

48. (a)

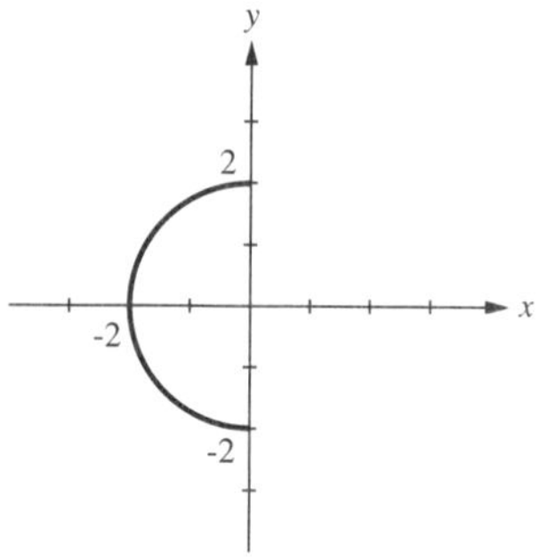

(b)

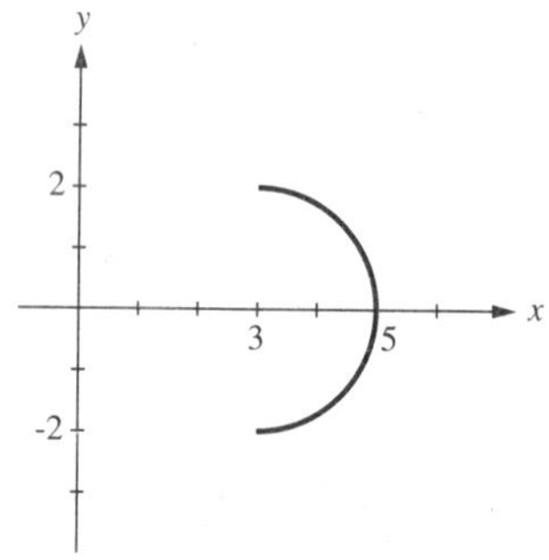

49. The tangent line is perpendicular to the radius at the point. The slope of the radius is $4/3$, so the slope of the perpendicular is $-3/4$. An equation of the tangent line is $y - 4 = -\frac{3}{4}(x-3)$, or $y = -\frac{3}{4}x + \frac{25}{4}$.

50. (a) $(x+1)^2 + y^2 = 10$, center at $C(-1,0)$. The slope of CP is $-1/3$ so the slope of the tangent is 3; $y+1 = 3(x-2)$, $y = 3x-7$.

(b) $(x-3)^2 + (y+2)^2 = 26$, center at $C(3,-2)$. The slope of CP is 5 so the slope of the tangent is $-\frac{1}{5}$; $y - 3 = -\frac{1}{5}(x-4)$, $y = -\frac{1}{5}x + \frac{19}{5}$.

51. (a) The center of the circle is at (0,0) and its radius is $\sqrt{20} = 2\sqrt{5}$. The distance between P and the center is $\sqrt{(-1)^2 + (2)^2} = \sqrt{5}$ which is less than $2\sqrt{5}$, so P is inside the circle.

(b) Draw the diameter of the circle that passes through P, then the shorter segment of the diameter is the shortest line that can be drawn from P to the circle, and the longer segment is the longest line that can be drawn from P to the circle (can you prove it?). Thus, the smallest distance is $2\sqrt{5} - \sqrt{5} = \sqrt{5}$, and the largest is $2\sqrt{5} + \sqrt{5} = 3\sqrt{5}$.

52. (a) $x^2 + (y-1)^2 = 5$, center at $C(0,1)$ and radius $\sqrt{5}$. The distance between P and C is $3\sqrt{5}/2$ so P is outside the circle.

(b) The smallest distance is $\frac{3}{2}\sqrt{5} - \sqrt{5} = \frac{1}{2}\sqrt{5}$, the largest distance is $\frac{3}{2}\sqrt{5} + \sqrt{5} = \frac{5}{2}\sqrt{5}$.

53. Let (a,b) be the coordinates of T (or T'). The radius from $(0,0)$ to T (or T') will be perpendicular to L (or L') so, using slopes, $b/a = -(a-3)/b$, $a^2+b^2 = 3a$. But (a,b) is on the circle so $a^2+b^2 = 1$, thus $3a = 1$, $a = 1/3$. Let $a = 1/3$ in $a^2+b^2 = 1$ to get $b^2 = 8/9$, $b = \pm\sqrt{8}/3$. The coordinates of T and T' are $(1/3, \sqrt{8}/3)$ and $(1/3, -\sqrt{8}/3)$.

54. (a) $\sqrt{(x-2)^2 + (y-0)^2} = \sqrt{2}\sqrt{(x-0)^2 + (y-1)^2}$; square both sides and expand to get $x^2 - 4x + 4 + y^2 = 2(x^2 + y^2 - 2y + 1)$, $x^2 + y^2 + 4x - 4y - 2 = 0$, which is a circle.

(b) $(x^2 + 4x) + (y^2 - 4y) = 2$, $(x^2 + 4x + 4) + (y^2 - 4y + 4) = 2 + 4 + 4$, $(x+2)^2 + (y-2)^2 = 10$; center $(-2,2)$, radius $\sqrt{10}$.

55. (a) $[(x-4)^2 + (y-1)^2] + [(x-2)^2 + (y+5)^2] = 45$
$x^2 - 8x + 16 + y^2 - 2y + 1 + x^2 - 4x + 4 + y^2 + 10y + 25 = 45$
$2x^2 + 2y^2 - 12x + 8y + 1 = 0$, which is a circle.

(b) $2(x^2 - 6x) + 2(y^2 + 4y) = -1$, $2(x^2 - 6x + 9) + 2(y^2 + 4y + 4) = -1 + 18 + 8$, $(x-3)^2 + (y+2)^2 = 25/2$; center $(3,-2)$, radius $5/\sqrt{2}$.

56. If $x^2 - y^2 = 0$, then $y^2 = x^2$ so $y = x$ or $y = -x$. The graph of $x^2 - y^2 = 0$ consists of the graphs of the two lines $y = \pm x$. The graph of $(x-c)^2 + y^2 = 1$ is a circle of radius 1 with center at $(c,0)$.

Examine the figure to see that the system cannot have just one solution, and has 0 solutions if $|c| > \sqrt{2}$, 2 solutions if $|c| = \sqrt{2}$, 3 solutions if $|c| = 1$, and 4 solutions if $|c| < \sqrt{2}$, $|c| \neq 1$.

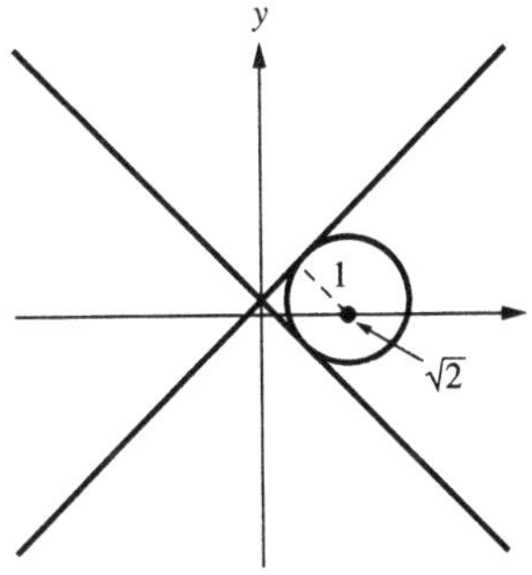

2 solutions

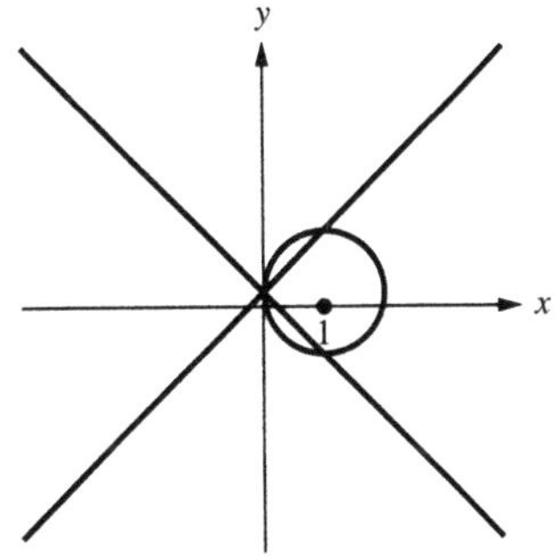

3 solutions

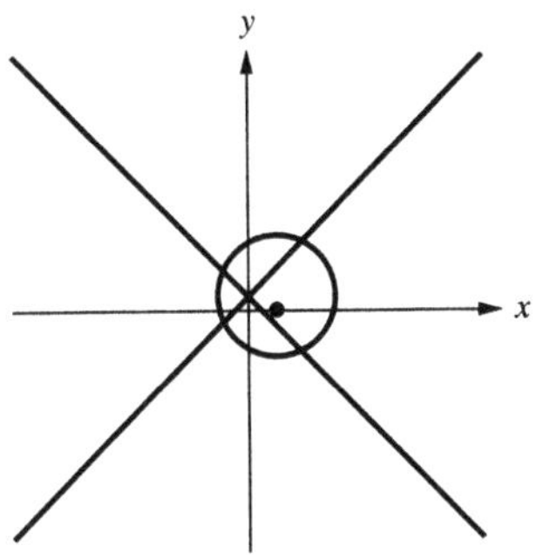

4 solutions

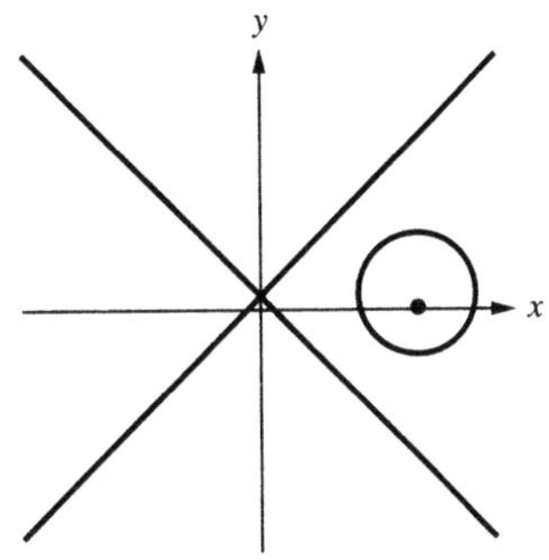

0 solutions

57. $y = x^2 + 2$

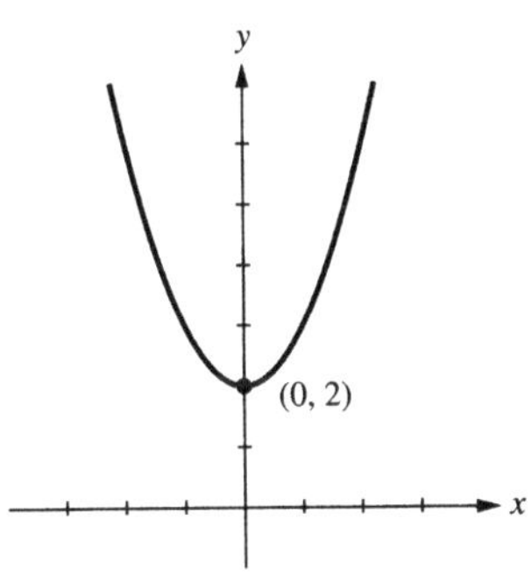

58. $y = x^2 - 3$

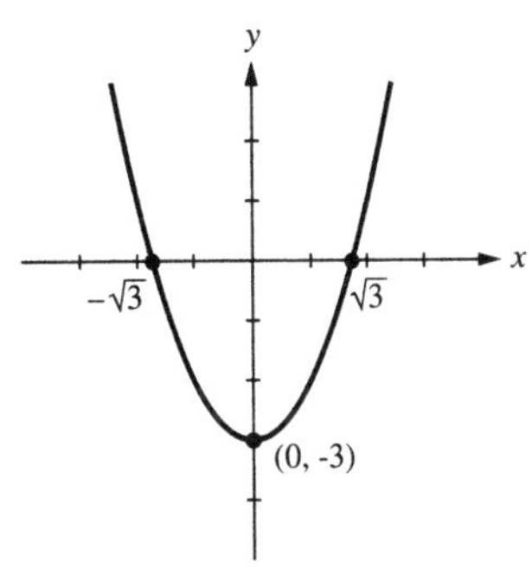

59. $y = x^2 + 2x - 3$

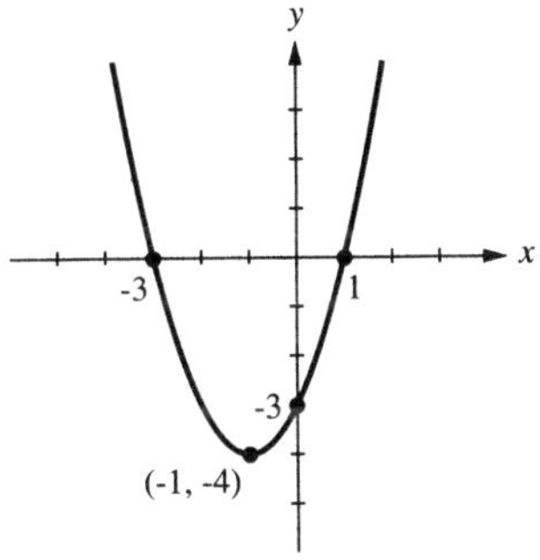

60. $y = x^2 - 3x - 4$

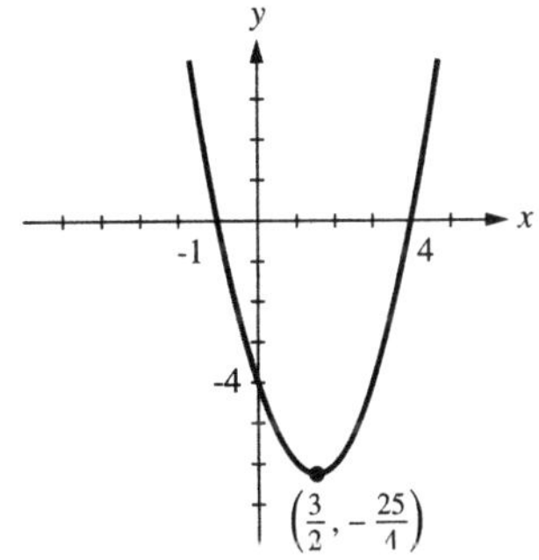

61. $y = -x^2 + 4x + 5$

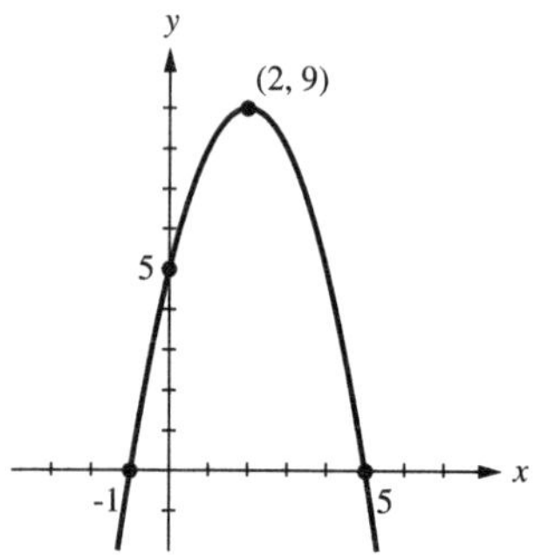

62. $y = -x^2 + x$

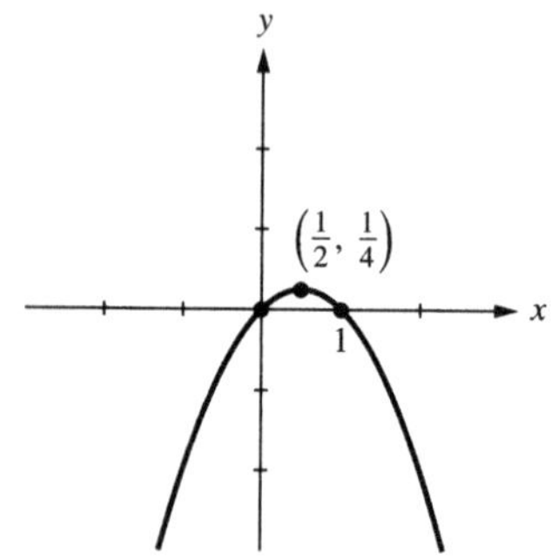

63. $y = (x-2)^2$

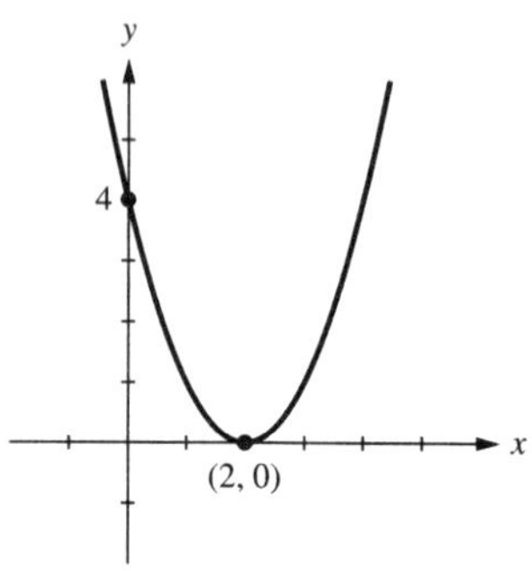

64. $y = (3+x)^2$

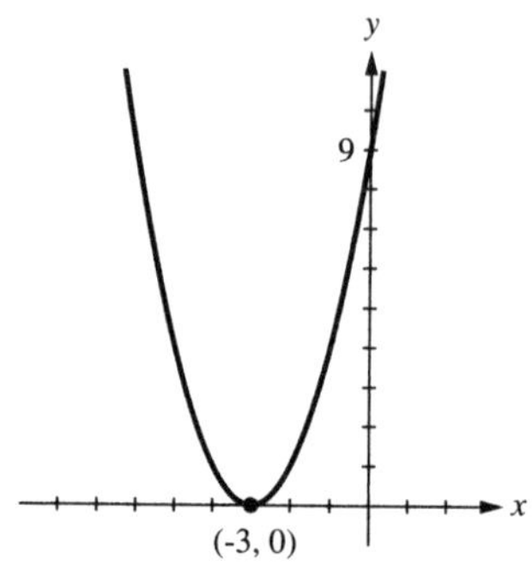

65. $x^2 - 2x + y = 0$

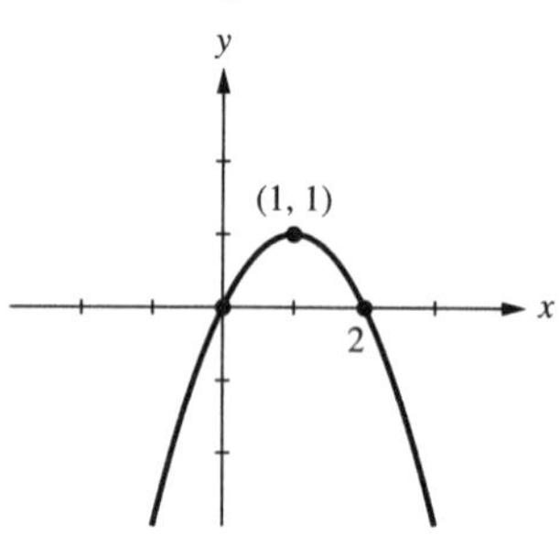

66. $x^2 + 8x + 8y = 0$

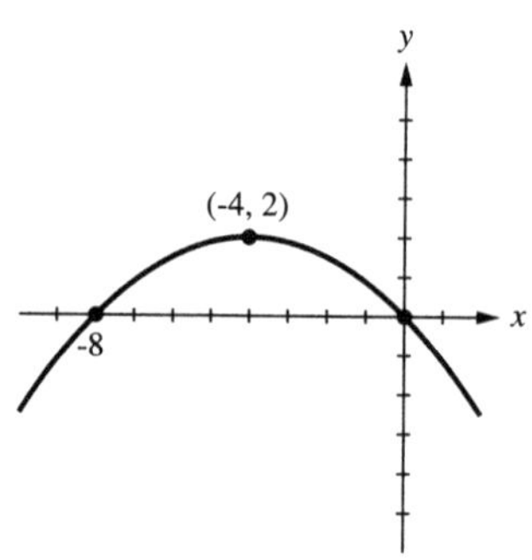

67. $y = 3x^2 - 2x + 1$

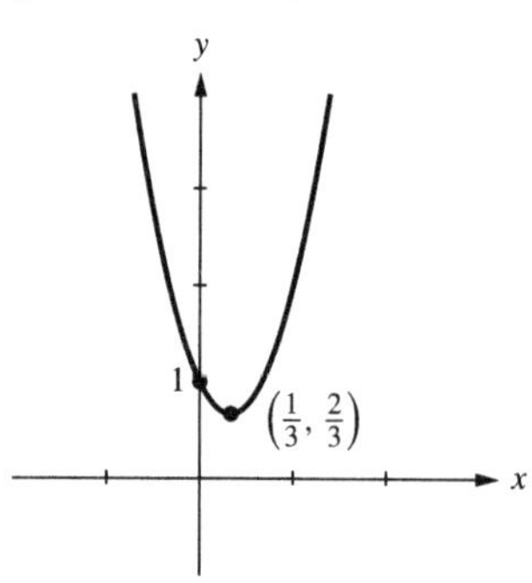

68. $y = x^2 + x + 2$

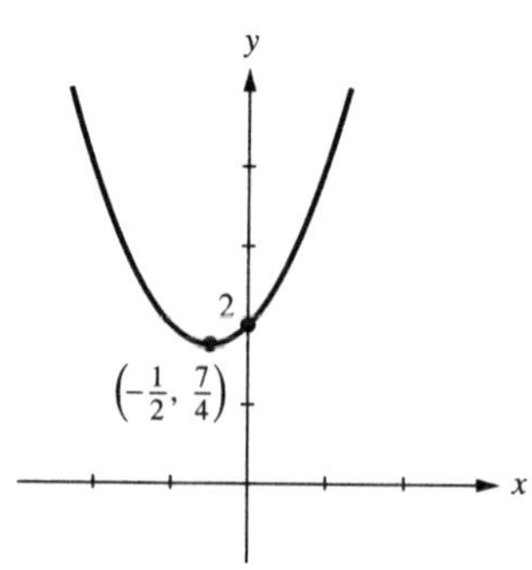

69. $x = -y^2 + 2y + 2$

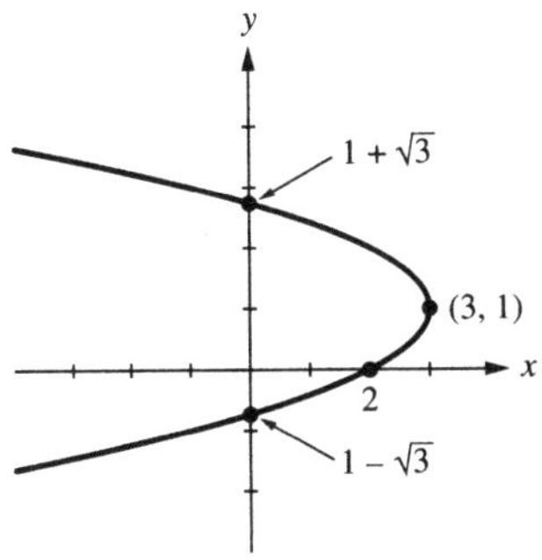

70. $x = y^2 - 4y + 5$

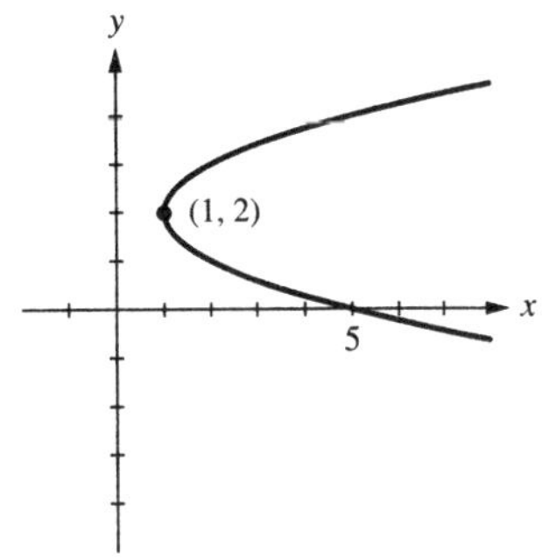

71. **(a)** $x^2 = 3 - y$, $x = \pm\sqrt{3-y}$. The right half is $x = \sqrt{3-y}$.
(b) Complete the square in x to get $(x-1)^2 = y+1$, $x = 1 \pm \sqrt{y+1}$. The left half is $x = 1 - \sqrt{y+1}$.

72. **(a)** $y^2 = x + 5$, $y = \pm\sqrt{x+5}$. The upper half is $y = \sqrt{x+5}$.
(b) Complete the square in y to get $(y - 1/2)^2 = x + 9/4$, $y - 1/2 = \pm\sqrt{x + 9/4}$, $y = 1/2 \pm \sqrt{x + 9/4}$. The lower half is $y = 1/2 - \sqrt{x + 9/4}$.

73. **(a)**

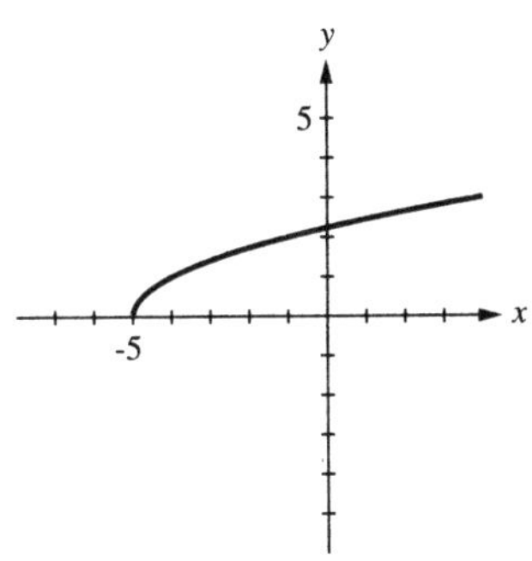

(b)

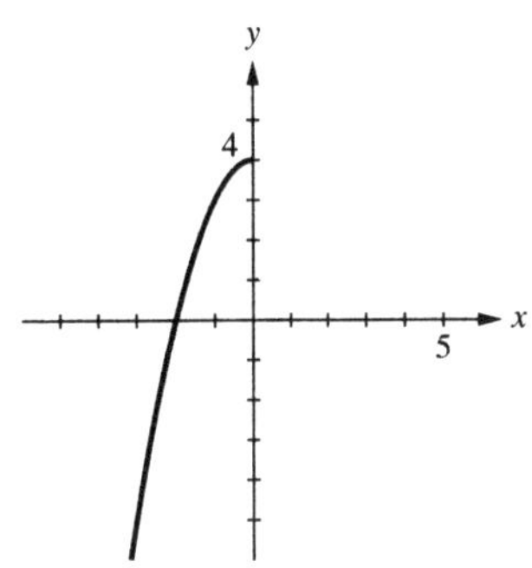

74. **(a)**

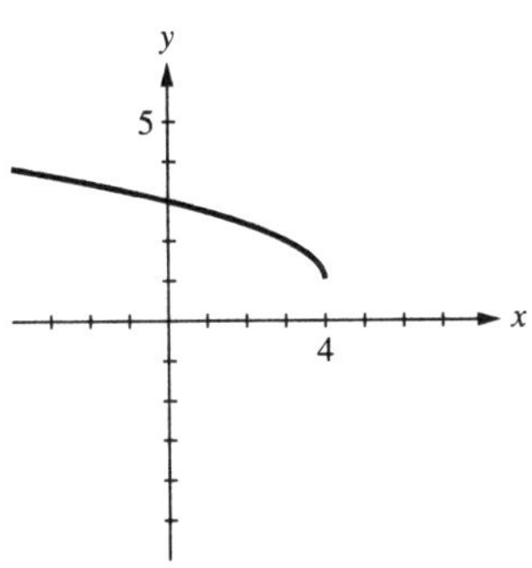

(b)

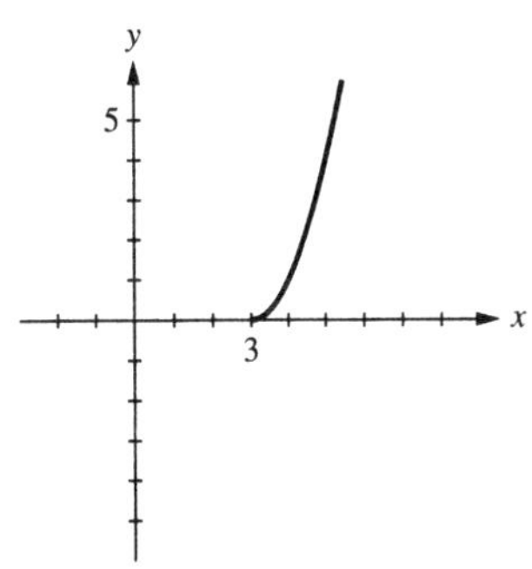

75. **(a)**

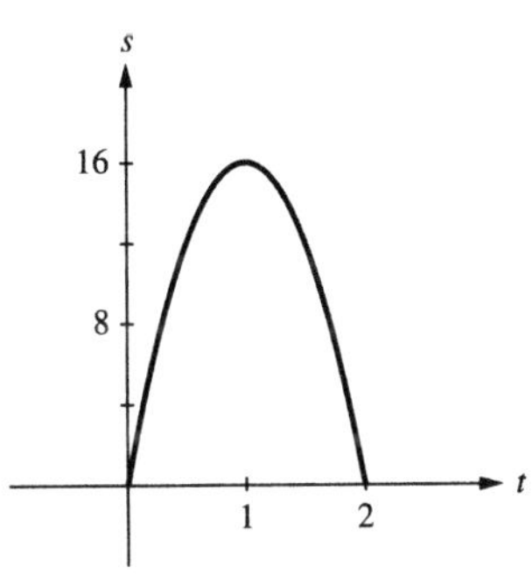

(b) The ball will be at its highest point when $t = 1$ sec; it will rise 16 ft.

76. (a) $2x + y = 500$, $y = 500 - 2x$. (b) $A = xy = x(500 - 2x) = 500x - 2x^2$.
(c) The graph of A versus x is a parabola with its vertex (high point) at $x = -b/(2a) = -500/(-4) = 125$, so the maximum value of A is $A = 500(125) - 2(125)^2 = 31{,}250$ ft^2.

77. (a) $(3)(2x) + (2)(2y) = 600$, $6x + 4y = 600$, $y = 150 - 3x/2$
(b) $A = xy = x(150 - 3x/2) = 150x - 3x^2/2$
(c) The graph of A versus x is a parabola with its vertex (high point) at $x = -b/(2a) = -150/(-3) = 50$, so the maximum value of A is $A = 150(50) - 3(50)^2/2 = 3{,}750$ ft^2.

78. (a) $y = ax^2 + bx + c = a\left(x^2 + \frac{b}{a}x\right) + c$

$$= a\left(x^2 + \frac{b}{a}x + \frac{b^2}{4a^2}\right) + c - \frac{b^2}{4a} = a\left(x + \frac{b}{2a}\right)^2 + \left(c - \frac{b^2}{4a}\right)$$

(b) If $a < 0$ then y is always less than $c - \frac{b^2}{4a}$ except when $x = -\frac{b}{2a}$, so the graph has its high point there. If $a > 0$ then y is always greater than $c - \frac{b^2}{4a}$ except when $x = -\frac{b}{2a}$, so the graph has its low point there.

79. (a) The parabola $y = 2x^2 + 5x - 1$ opens upward and has x-intercepts of $x = (-5 \pm \sqrt{33})/4$, so $2x^2 + 5x - 1 < 0$ if $(-5 - \sqrt{33})/4 < x < (-5 + \sqrt{33})/4$.
(b) The parabola $y = x^2 - 2x + 3$ opens upward and has no x-intercepts, so $x^2 - 2x + 3 > 0$ if $-\infty < x < +\infty$.

80. (a) The parabola $y = x^2 + x - 1$ opens upward and has x-intercepts of $x = (-1 \pm \sqrt{5})/2$, so $x^2 + x - 1 > 0$ if $x < (-1 - \sqrt{5})/2$ or $x > (-1 + \sqrt{5})/2$.
(b) The parabola $y = x^2 - 4x + 6$ opens upward and has no x-intercepts, so $x^2 - 4x + 6 < 0$ has no solution.

81. (a) The t-coordinate of the vertex is $t = -40/[(2)(-16)] = 5/4$, so the maximum height is $s = 5 + 40(5/4) - 16(5/4)^2 = 30$ ft.
(b) $s = 5 + 40t - 16t^2 = 0$ if $t \approx 2.6$ s
(c) $s = 5 + 40t - 16t^2 > 12$ if $16t^2 - 40t + 7 < 0$, which is true if $(5 - 3\sqrt{2})/4 < t < (5 + 3\sqrt{2})/4$. The length of this interval is $(5 + 3\sqrt{2})/4 - (5 - 3\sqrt{2})/4 = 3\sqrt{2}/2 \approx 2.1$ s.

82. $x + 3 - x^2 > 0$, $x^2 - x - 3 < 0$, $(1 - \sqrt{13})/2 < x < (1 + \sqrt{13})/2$

APPENDIX E
Trigonometry Review

EXERCISE SET E

1. (a) $5\pi/12$ (b) $13\pi/6$ (c) $\pi/9$ (d) $23\pi/30$

2. (a) $7\pi/3$ (b) $\pi/12$ (c) $5\pi/4$ (d) $11\pi/12$

3. (a) $12°$ (b) $(270/\pi)°$ (c) $288°$ (d) $540°$

4. (a) $18°$ (b) $(360/\pi)°$ (c) $72°$ (d) $210°$

5.

	$\sin\theta$	$\cos\theta$	$\tan\theta$	$\csc\theta$	$\sec\theta$	$\cot\theta$
(a)	$\sqrt{21}/5$	$2/5$	$\sqrt{21}/2$	$5/\sqrt{21}$	$5/2$	$2/\sqrt{21}$
(b)	$3/4$	$\sqrt{7}/4$	$3/\sqrt{7}$	$4/3$	$4/\sqrt{7}$	$\sqrt{7}/3$
(c)	$3/\sqrt{10}$	$1/\sqrt{10}$	3	$\sqrt{10}/3$	$\sqrt{10}$	$1/3$

6.

	$\sin\theta$	$\cos\theta$	$\tan\theta$	$\csc\theta$	$\sec\theta$	$\cot\theta$
(a)	$1/\sqrt{2}$	$1/\sqrt{2}$	1	$\sqrt{2}$	$\sqrt{2}$	1
(b)	$3/5$	$4/5$	$3/4$	$5/3$	$5/4$	$4/3$
(c)	$1/4$	$\sqrt{15}/4$	$1/\sqrt{15}$	4	$4/\sqrt{15}$	$\sqrt{15}$

7. $\sin\theta = 3/\sqrt{10}$, $\cos\theta = 1/\sqrt{10}$

8. $\sin\theta = \sqrt{5}/3$, $\tan\theta = \sqrt{5}/2$

9. $\tan\theta = \sqrt{21}/2$, $\csc\theta = 5/\sqrt{21}$

10. $\cot\theta = \sqrt{15}$, $\sec\theta = 4/\sqrt{15}$

11. Let x be the length of the side adjacent to θ, then $\cos\theta = x/6 = 0.3$, $x = 1.8$.

12. Let x be the length of the hypotenuse, then $\sin\theta = 2.4/x = 0.8$, $x = 2.4/0.8 = 3$.

13.

	θ	$\sin\theta$	$\cos\theta$	$\tan\theta$	$\csc\theta$	$\sec\theta$	$\cot\theta$
(a)	$225°$	$-1/\sqrt{2}$	$-1/\sqrt{2}$	1	$-\sqrt{2}$	$-\sqrt{2}$	1
(b)	$-210°$	$1/2$	$-\sqrt{3}/2$	$-1/\sqrt{3}$	2	$-2/\sqrt{3}$	$-\sqrt{3}$
(c)	$5\pi/3$	$\sqrt{3}/2$	$1/2$	$-\sqrt{3}$	$-2/\sqrt{3}$	2	$-1/\sqrt{3}$
(d)	$-3\pi/2$	1	0	—	1	—	0

14.

	θ	$\sin\theta$	$\cos\theta$	$\tan\theta$	$\csc\theta$	$\sec\theta$	$\cot\theta$
(a)	$330°$	$-1/2$	$\sqrt{3}/2$	$-1/\sqrt{3}$	-2	$2/\sqrt{3}$	$-\sqrt{3}$
(b)	$-120°$	$-\sqrt{3}/2$	$-1/2$	$\sqrt{3}$	$-2/\sqrt{3}$	-2	$1/\sqrt{3}$
(c)	$9\pi/4$	$1/\sqrt{2}$	$1/\sqrt{2}$	1	$\sqrt{2}$	$\sqrt{2}$	1
(d)	-3π	0	-1	0	—	-1	—

15.

	$\sin\theta$	$\cos\theta$	$\tan\theta$	$\csc\theta$	$\sec\theta$	$\cot\theta$
(a)	$4/5$	$3/5$	$4/3$	$5/4$	$5/3$	$3/4$
(b)	$-4/5$	$3/5$	$-4/3$	$-5/4$	$5/3$	$-3/4$
(c)	$1/2$	$-\sqrt{3}/2$	$-1/\sqrt{3}$	2	$-2\sqrt{3}$	$-\sqrt{3}$
(d)	$-1/2$	$\sqrt{3}/2$	$-1/\sqrt{3}$	-2	$2/\sqrt{3}$	$-\sqrt{3}$
(e)	$1/\sqrt{2}$	$1/\sqrt{2}$	1	$\sqrt{2}$	$\sqrt{2}$	1
(f)	$1/\sqrt{2}$	$-1/\sqrt{2}$	-1	$\sqrt{2}$	$-\sqrt{2}$	-1

16.

	$\sin\theta$	$\cos\theta$	$\tan\theta$	$\csc\theta$	$\sec\theta$	$\cot\theta$
(a)	$1/4$	$\sqrt{15}/4$	$1/\sqrt{15}$	4	$4/\sqrt{15}$	$\sqrt{15}$
(b)	$1/4$	$-\sqrt{15}/4$	$-1/\sqrt{15}$	4	$-4/\sqrt{15}$	$-\sqrt{15}$
(c)	$3/\sqrt{10}$	$1/\sqrt{10}$	3	$\sqrt{10}/3$	$\sqrt{10}$	$1/3$
(d)	$-3/\sqrt{10}$	$-1/\sqrt{10}$	3	$-\sqrt{10}/3$	$-\sqrt{10}$	$1/3$
(e)	$\sqrt{21}/5$	$-2/5$	$-\sqrt{21}/2$	$5/\sqrt{21}$	$-5/2$	$-2/\sqrt{21}$
(f)	$-\sqrt{21}/5$	$-2/5$	$\sqrt{21}/2$	$-5/\sqrt{21}$	$-5/2$	$2/\sqrt{21}$

17. (a) $x = 3\sin 25° \approx 1.2679$ (b) $x = 3/\tan(2\pi/9) \approx 3.5753$

18. (a) $x = 2/\sin 20° \approx 5.8476$ (b) $x = 3/\cos(3\pi/11) \approx 4.5811$

19.

	$\sin\theta$	$\cos\theta$	$\tan\theta$	$\csc\theta$	$\sec\theta$	$\cot\theta$
(a)	$a/3$	$\sqrt{9-a^2}/3$	$a/\sqrt{9-a^2}$	$3/a$	$3/\sqrt{9-a^2}$	$\sqrt{9-a^2}/a$
(b)	$a/\sqrt{a^2+25}$	$5/\sqrt{a^2+25}$	$a/5$	$\sqrt{a^2+25}/a$	$\sqrt{a^2+25}/5$	$5/a$
(c)	$\sqrt{a^2-1}/a$	$1/a$	$\sqrt{a^2-1}$	$a/\sqrt{a^2-1}$	a	$1/\sqrt{a^2-1}$

20. (a) $\theta = 3\pi/4 \pm 2n\pi$ and $\theta = 5\pi/4 \pm 2n\pi$, $n = 0, 1, 2, \ldots$
(b) $\theta = 5\pi/4 \pm 2n\pi$ and $\theta = 7\pi/4 \pm 2n\pi$, $n = 0, 1, 2, \ldots$

21. (a) $\theta = 3\pi/4 \pm n\pi$, $n = 0, 1, 2, \ldots$
(b) $\theta = \pi/3 \pm 2n\pi$ and $\theta = 5\pi/3 \pm 2n\pi$, $n = 0, 1, 2, \ldots$

22. (a) $\theta = 7\pi/6 \pm 2n\pi$ and $\theta = 11\pi/6 \pm 2n\pi$, $n = 0, 1, 2, \ldots$
(b) $\theta = \pi/3 \pm n\pi$, $n = 0, 1, 2, \ldots$

23. (a) $\theta = \pi/6 \pm n\pi$, $n = 0, 1, 2, \ldots$
(b) $\theta = 4\pi/3 \pm 2n\pi$ and $\theta = 5\pi/3 \pm 2n\pi$, $n = 0, 1, 2, \ldots$

24. (a) $\theta = 3\pi/2 \pm 2n\pi$, $n = 0, 1, 2, \ldots$ (b) $\theta = \pi \pm 2n\pi$, $n = 0, 1, 2, \ldots$

25. (a) $\theta = 3\pi/4 \pm n\pi$, $n = 0, 1, 2, \ldots$ (b) $\theta = \pi/6 \pm n\pi$, $n = 0, 1, 2, \ldots$

26. (a) $\theta = 2\pi/3 \pm 2n\pi$ and $\theta = 4\pi/3 \pm 2n\pi$, $n = 0, 1, 2, \ldots$
(b) $\theta = 7\pi/6 \pm 2n\pi$ and $\theta = 11\pi/6 \pm 2n\pi$, $n = 0, 1, 2, \ldots$

27. (a) $\theta = \pi/3 \pm 2n\pi$ and $\theta = 2\pi/3 \pm 2n\pi$, $n = 0, 1, 2, \ldots$
(b) $\theta = \pi/6 \pm 2n\pi$ and $\theta = 11\pi/6 \pm 2n\pi$, $n = 0, 1, 2, \ldots$

28. $\sin\theta = -3/5$, $\cos\theta = -4/5$, $\tan\theta = 3/4$, $\csc\theta = -5/3$, $\sec\theta = -5/4$, $\cot\theta = 4/3$

29. $\sin\theta = 2/5$, $\cos\theta = -\sqrt{21}/5$, $\tan\theta = -2/\sqrt{21}$, $\csc\theta = 5/2$, $\sec\theta = -5/\sqrt{21}$, $\cot\theta = -\sqrt{21}/2$

30. (a) $\theta = \pi/2 \pm 2n\pi$, $n = 0, 1, 2, \ldots$ (b) $\theta = \pm 2n\pi$, $n = 0, 1, 2, \ldots$
(c) $\theta = \pi/4 \pm n\pi$, $n = 0, 1, 2, \ldots$ (d) $\theta = \pi/2 \pm 2n\pi$, $n = 0, 1, 2, \ldots$
(e) $\theta = \pm 2n\pi$, $n = 0, 1, 2, \ldots$ (f) $\theta = \pi/4 \pm n\pi$, $n = 0, 1, 2, \ldots$

31. (a) $\theta = \pm n\pi$, $n = 0, 1, 2, \ldots$ (b) $\theta = \pi/2 \pm n\pi$, $n = 0, 1, 2, \ldots$
(c) $\theta = \pm n\pi$, $n = 0, 1, 2, \ldots$ (d) $\theta = \pm n\pi$, $n = 0, 1, 2, \ldots$
(e) $\theta = \pi/2 \pm n\pi$, $n = 0, 1, 2, \ldots$ (f) $\theta = \pm n\pi$, $n = 0, 1, 2, \ldots$

32. Construct a right triangle with one angle equal to 17°, measure the lengths of the sides and hypotenuse and use formula (6) for $\sin\theta$ and $\cos\theta$ to approximate $\sin 17°$ and $\cos 17°$.

33. (a) $s = r\theta = 4(\pi/6) = 2\pi/3$ cm (b) $s = r\theta = 4(5\pi/6) = 10\pi/3$ cm

34. $r = s/\theta = 7/(\pi/3) = 21/\pi$

35. $\theta = s/r = 2/5$

36. $\theta = s/r$ so $A = \frac{1}{2}r^2\theta = \frac{1}{2}r^2(s/r) = \frac{1}{2}rs$

37. (a) $2\pi r = R(2\pi - \theta)$, $r = \dfrac{2\pi - \theta}{2\pi}R$

(b) $h = \sqrt{R^2 - r^2} = \sqrt{R^2 - (2\pi - \theta)^2 R^2/(4\pi^2)} = \dfrac{\sqrt{4\pi\theta - \theta^2}}{2\pi}R$

38. The circumference of the circular base is $2\pi r$. When cut and flattened, the cone becomes a circular sector of radius L. If θ is the central angle that subtends the arc of length $2\pi r$, then $\theta = (2\pi r)/L$ so the area S of the sector is $S = (1/2)L^2(2\pi r/L) = \pi r L$ which is the lateral surface area of the cone.

39. Let h be the altitude as shown in the figure, then
$h = 3\sin 60° = 3\sqrt{3}/2$ so $A = \frac{1}{2}(3\sqrt{3}/2)(7) = 21\sqrt{3}/4$.

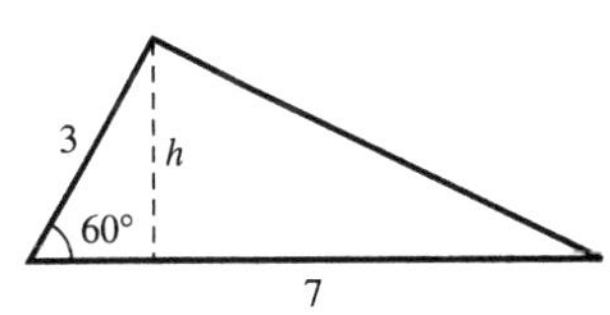

40. Draw the perpendicular from vertex C as shown in the figure, then
$h = 9\sin 30° = 9/2$, $a = h/\sin 45° = 9\sqrt{2}/2$,
$c_1 = 9\cos 30° = 9\sqrt{3}/2$, $c_2 = a\cos 45° = 9/2$,
$c_1 + c_2 = 9(\sqrt{3} + 1)/2$, angle $C = 180° - (30° + 45°) = 105°$

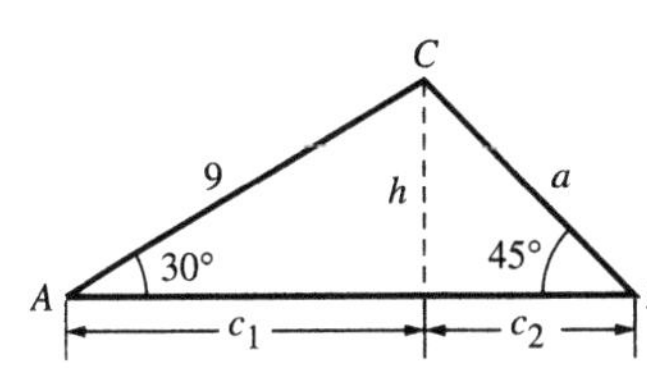

41. Let x be the distance above the ground, then $x = 10\sin 67° \approx 9.2$ ft.

42. Let x be the height of the building, then $x = 120\tan 76° \approx 481$ ft.

43. From the figure, $h = x - y$ but $x = d\tan\beta$, $y = d\tan\alpha$ so $h = d(\tan\beta - \tan\alpha)$.

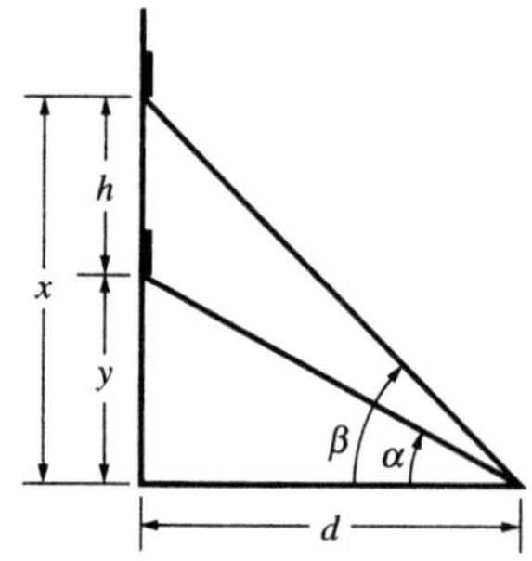

44. From the figure, $d = x - y$ but $x = h\cot\alpha$, $y = h\cot\beta$ so $d = h(\cot\alpha - \cot\beta)$, $h = \dfrac{d}{\cot\alpha - \cot\beta}$.

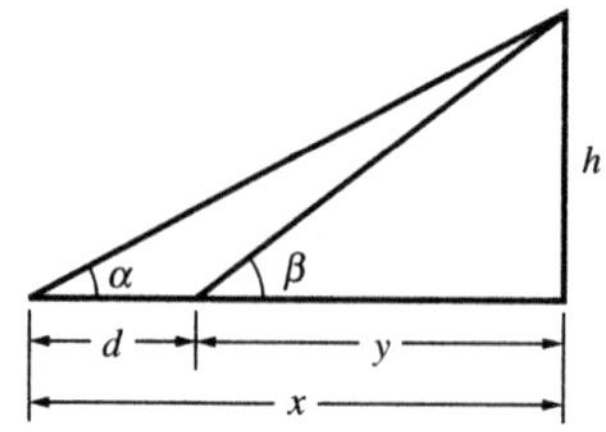

45. (a) $\sin 2\theta = 2\sin\theta\cos\theta = 2(\sqrt{5}/3)(2/3) = 4\sqrt{5}/9$
(b) $\cos 2\theta = 2\cos^2\theta - 1 = 2(2/3)^2 - 1 = -1/9$

46. (a) $\sin(\alpha - \beta) = \sin\alpha\cos\beta - \cos\alpha\sin\beta = (3/5)(1/\sqrt{5}) - (4/5)(2/\sqrt{5}) = -1/\sqrt{5}$
(b) $\cos(\alpha + \beta) = \cos\alpha\cos\beta - \sin\alpha\sin\beta = (4/5)(1/\sqrt{5}) - (3/5)(2/\sqrt{5}) = -2/(5\sqrt{5})$

47. $\sin 3\theta = \sin(2\theta + \theta) = \sin 2\theta\cos\theta + \cos 2\theta\sin\theta = (2\sin\theta\cos\theta)\cos\theta + (\cos^2\theta - \sin^2\theta)\sin\theta$
$= 2\sin\theta\cos^2\theta + \sin\theta\cos^2\theta - \sin^3\theta = 3\sin\theta\cos^2\theta - \sin^3\theta$; similarly, $\cos 3\theta = \cos^3\theta - 3\sin^2\theta\cos\theta$

48. $\dfrac{\cos\theta\sec\theta}{1 + \tan^2\theta} = \dfrac{\cos\theta\sec\theta}{\sec^2\theta} = \dfrac{\cos\theta}{\sec\theta} = \dfrac{\cos\theta}{(1/\cos\theta)} = \cos^2\theta$

49. $\dfrac{\cos\theta\tan\theta + \sin\theta}{\tan\theta} = \dfrac{\cos\theta(\sin\theta/\cos\theta) + \sin\theta}{\sin\theta/\cos\theta} = 2\cos\theta$

50. $2\csc 2\theta = \dfrac{2}{\sin 2\theta} = \dfrac{2}{2\sin\theta\cos\theta} = \left(\dfrac{1}{\sin\theta}\right)\left(\dfrac{1}{\cos\theta}\right) = \csc\theta\sec\theta$

51. $\tan\theta + \cot\theta = \dfrac{\sin\theta}{\cos\theta} + \dfrac{\cos\theta}{\sin\theta} = \dfrac{\sin^2\theta + \cos^2\theta}{\sin\theta\cos\theta} = \dfrac{1}{\sin\theta\cos\theta} = \dfrac{2}{2\sin\theta\cos\theta} = \dfrac{2}{\sin 2\theta} = 2\csc 2\theta$

52. $\dfrac{\sin 2\theta}{\sin\theta} - \dfrac{\cos 2\theta}{\cos\theta} = \dfrac{\sin 2\theta\cos\theta - \cos 2\theta\sin\theta}{\sin\theta\cos\theta} = \dfrac{\sin\theta}{\sin\theta\cos\theta} = \sec\theta$

53. $\dfrac{\sin\theta + \cos 2\theta - 1}{\cos\theta - \sin 2\theta} = \dfrac{\sin\theta + (1 - 2\sin^2\theta) - 1}{\cos\theta - 2\sin\theta\cos\theta} = \dfrac{\sin\theta(1 - 2\sin\theta)}{\cos\theta(1 - 2\sin\theta)} = \tan\theta$

54. Using (47), $2\sin 2\theta\cos\theta = 2(1/2)(\sin\theta + \sin 3\theta) = \sin\theta + \sin 3\theta$

55. Using (47), $2\cos 2\theta\sin\theta = 2(1/2)[\sin(-\theta) + \sin 3\theta] = \sin 3\theta - \sin\theta$

56. $\tan(\theta/2) = \dfrac{\sin(\theta/2)}{\cos(\theta/2)} = \dfrac{2\sin^2(\theta/2)}{2\sin(\theta/2)\cos(\theta/2)} = \dfrac{1 - \cos\theta}{\sin\theta}$

57. $\tan(\theta/2) = \dfrac{\sin(\theta/2)}{\cos(\theta/2)} = \dfrac{2\sin(\theta/2)\cos(\theta/2)}{2\cos^2(\theta/2)} = \dfrac{\sin\theta}{1+\cos\theta}$

58. From (52), $\cos(\pi/3+\theta)+\cos(\pi/3-\theta) = 2\cos(\pi/3)\cos\theta = 2(1/2)\cos\theta = \cos\theta$

59. From the figures, area $= \dfrac{1}{2}hc$ but $h = b\sin A$
so area $= \dfrac{1}{2}bc\sin A$. The formulas
area $= \dfrac{1}{2}ac\sin B$ and area $= \dfrac{1}{2}ab\sin C$
follow by drawing altitudes from vertices B and C, respectively.

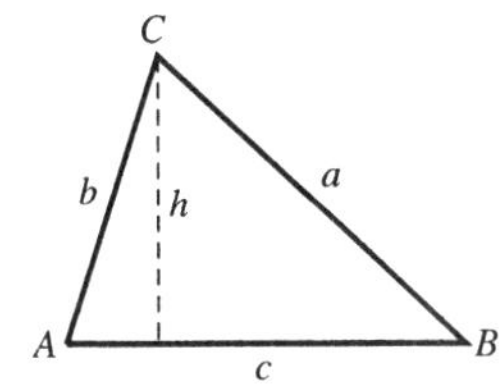

60. From right triangles ADC and BDC,
$h_1 = b\sin A = a\sin B$ so $a/\sin A = b/\sin B$.
From right triangles AEB and CEB,
$h_2 = c\sin A = a\sin C$ so $a/\sin A = c/\sin C$
thus $a/\sin A = b/\sin B = c/\sin C$.

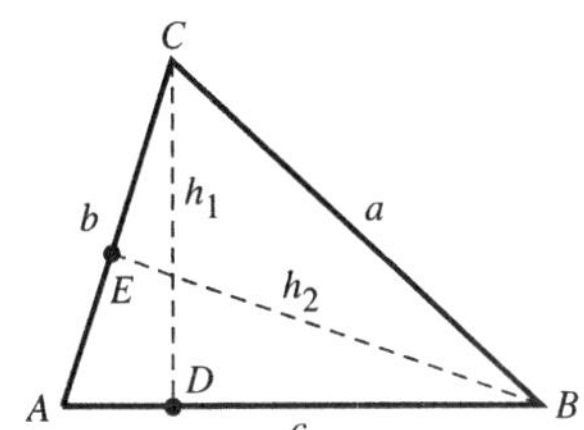

61. **(a)** $\sin(\pi/2+\theta) = \sin(\pi/2)\cos\theta + \cos(\pi/2)\sin\theta = (1)\cos\theta + (0)\sin\theta = \cos\theta$
(b) $\cos(\pi/2+\theta) = \cos(\pi/2)\cos\theta - \sin(\pi/2)\sin\theta = (0)\cos\theta - (1)\sin\theta = -\sin\theta$
(c) $\sin(3\pi/2-\theta) = \sin(3\pi/2)\cos\theta - \cos(3\pi/2)\sin\theta = (-1)\cos\theta - (0)\sin\theta = -\cos\theta$
(d) $\cos(3\pi/2+\theta) = \cos(3\pi/2)\cos\theta - \sin(3\pi/2)\sin\theta = (0)\cos\theta - (-1)\sin\theta = \sin\theta$

62. $\tan(\alpha+\beta) = \dfrac{\sin(\alpha+\beta)}{\cos(\alpha+\beta)} = \dfrac{\sin\alpha\cos\beta+\cos\alpha\sin\beta}{\cos\alpha\cos\beta-\sin\alpha\sin\beta}$, divide numerator and denominator by $\cos\alpha\cos\beta$ and use $\tan\alpha = \dfrac{\sin\alpha}{\cos\alpha}$ and $\tan\beta = \dfrac{\sin\beta}{\cos\beta}$ to get (38);

$\tan(\alpha-\beta) = \tan(\alpha+(-\beta)) = \dfrac{\tan\alpha+\tan(-\beta)}{1-\tan\alpha\tan(-\beta)} = \dfrac{\tan\alpha-\tan\beta}{1+\tan\alpha\tan\beta}$ because

$\tan(-\beta) = -\tan\beta$.

63. **(a)** Add (34) and (36) to get $\sin(\alpha-\beta)+\sin(\alpha+\beta) = 2\sin\alpha\cos\beta$ so
$\sin\alpha\cos\beta = (1/2)[\sin(\alpha-\beta)+\sin(\alpha+\beta)]$.
(b) Subtract (35) from (37). **(c)** Add (35) and (37).

64. **(a)** From (47), $\sin\dfrac{A+B}{2}\cos\dfrac{A-B}{2} = \dfrac{1}{2}(\sin B+\sin A)$ so

$\sin A+\sin B = 2\sin\dfrac{A+B}{2}\cos\dfrac{A-B}{2}$.

(b) Use (49) **(c)** Use (48)

65. $\sin\alpha+\sin(-\beta) = 2\sin\dfrac{\alpha-\beta}{2}\cos\dfrac{\alpha+\beta}{2}$, but $\sin(-\beta) = -\sin\beta$ so

$\sin\alpha - \sin\beta = 2\cos\dfrac{\alpha+\beta}{2}\sin\dfrac{\alpha-\beta}{2}$.

66. **(a)** From (34), $C\sin(\alpha+\phi) = C\sin\alpha\cos\phi + C\cos\alpha\sin\phi$ so $C\cos\phi = 3$ and $C\sin\phi = 5$, square and add to get $C^2(\cos^2\phi+\sin^2\phi) = 9+25$, $C^2 = 34$. If $C = \sqrt{34}$ then $\cos\phi = 3/\sqrt{34}$ and $\sin\phi = 5/\sqrt{34}$ so ϕ is the first-quadrant angle for which $\tan\phi = 5/3$.
$3\sin\alpha + 5\cos\alpha = \sqrt{34}\sin(\alpha+\phi)$.

(b) Follow the procedure of part (a) to get $C\cos\phi = A$ and $C\sin\phi = B$, $C = \sqrt{A^2+B^2}$, $\tan\phi = B/A$ where the quadrant in which ϕ lies is determined by the signs of A and B because $\cos\phi = A/C$ and $\sin\phi = B/C$, so $A\sin\alpha + B\cos\alpha = \sqrt{A^2+B^2}\sin(\alpha+\phi)$.

67. Consider the triangle having a, b, and d as sides. The angle formed by sides a and b is $\pi - \theta$ so from the law of cosines, $d^2 = a^2+b^2-2ab\cos(\pi-\theta) = a^2+b^2+2ab\cos\theta$, $d = \sqrt{a^2+b^2+2ab\cos\theta}$.

APPENDIX F
Solving Polynomial Equations

EXERCISE SET F

1. **(a)** $q(x) = x^2 + 4x + 2, r(x) = -11x + 6$
(b) $q(x) = 2x^2 + 4, r(x) = 9$
(c) $q(x) = x^3 - x^2 + 2x - 2, r(x) = 2x + 1$

2. **(a)** $q(x) = 2x^2 - x + 2, r(x) = 5x + 5$
(b) $q(x) = x^3 + 3x^2 - x + 2, r(x) = 3x - 1$
(c) $q(x) = 5x^3 - 5, r(x) = 4x^2 + 10$

3. **(a)** $q(x) = 3x^2 + 6x + 8, r(x) = 15$
(b) $q(x) = x^3 - 5x^2 + 20x - 100, r(x) = 504$
(c) $q(x) = x^4 + x^3 + x^2 + x + 1, r(x) = 0$

4. **(a)** $q(x) = 2x^2 + x - 1, r(x) = 0$
(b) $q(x) = 2x^3 - 5x^2 + 3x - 39, r(x) = 147$
(c) $q(x) = x^6 + x^5 + x^4 + x^3 + x^2 + x + 1, r(x) = 2$

5.

x	0	1	-3	7
$p(x)$	-4	-3	101	5001

6.

x	1	-1	3	-3	7	-7	21	-21
$p(x)$	-24	-12	12	0	420	-168	10416	-7812

7. **(a)** $q(x) = x^2 + 6x + 13, r = 20$ **(b)** $q(x) = x^2 + 3x - 2, r = -4$

8. **(a)** $q(x) = x^4 - x^3 + x^2 - x + 1, r = -2$ **(b)** $q(x) = x^4 + x^3 + x^2 + x + 1, r = 0$

9. Assume $r = a/b$ a and b integers with $a > 0$:
(a) b divides 1, $b = \pm 1$; a divides 24, $a = 1, 2, 3, 4, 6, 8, 12, 24$;
the possible candidates are $\{\pm 1, \pm 2, \pm 3, \pm 4, \pm 6, \pm 8, \pm 12, \pm 24\}$
(b) b divides 3 so $b = \pm 1, \pm 3$; a divides -10 so $a = 1, 2, 5, 10$;
the possible candidates are $\{\pm 1, \pm 2, \pm 5, \pm 10, \pm 1/3, \pm 2/3, \pm 5/3, \pm 10/3\}$
(c) b divides 1 so $b = \pm 1$; a divides 17 so $a = 1, 17$;
the possible candidates are $\{\pm 1, \pm 17\}$

10. An integer zero c divides -21, so $c = \pm 1, \pm 3, \pm 7, \pm 21$ are the only possibilities; substitution of these candidates shows that the integer zeros are $-7, -1, 3$

11. $(x + 1)(x - 1)(x - 2)$

12. $(x + 2)(3x + 1)(x - 2)$

13. $(x + 3)^3(x + 1)$

14. $2x^4 + x^3 - 19x^2 + 9$

15. $(x + 3)(x + 2)(x + 1)^2(x - 3)$

17. -3 is the only real root.

18. $x = -3/2, 2 \pm \sqrt{3}$ are the real roots.

19. $x = -2, -2/3, -1 \pm \sqrt{3}$ are the real roots.

20. $-2, -1, 1/2, 3$

21. $-2, 2, 3$ are the only real roots.

23. If $x-1$ is a factor then $p(1)=0$, so $k^2-7k+10=0$, $k^2-7k+10=(k-2)(k-5)$, so $k=2,5$.

24. $(-3)^7=-2187$, so -3 is a root and thus by Theorem F.4, $x+3$ is a factor of x^7+2187.

25. If the side of the cube is x then $x^2(x-3)=196$; the only real root of this equation is $x=7$ cm.

26. (a) Try to solve $\frac{a}{b} > \left(\frac{a}{b}\right)^3 + 1$. The polynomial $p(x) = x^3 - x + 1$ has only one real root $c \approx -1.325$, and $p(0) = 1$ so $p(x) > 0$ for all $x > c$; hence there is no positive rational solution of $\frac{a}{b} > \left(\frac{a}{b}\right)^3 + 1$.

(b) From part (a), any real $x < c$ is a solution.

27. Use the Factor Theorem with x as the variable and y as the constant c.

(a) For any positive integer n the polynomial $x^n - y^n$ has $x = y$ as a root.
(b) For any positive even integer n the polynomial $x^n - y^n$ has $x = -y$ as a root.
(c) For any positive odd integer n the polynomial $x^n + y^n$ has $x = -y$ as a root.

Notes

Notes

Notes

Notes

Notes